Wild Mammals of North America

JOSEPH A. CHAPMAN is professor and head of the Appalachian Environmental Laboratory, Center for Environmental and Estuarine Studies, of the University of Maryland. He is one of the authors of *Wild Animals of North America* (1979) and senior editor of the three-volume *Worldwide Furbearer Conference Proceedings* (1981). GEORGE A. FELDHAMER is assistant professor at the Appalachian Environmental Laboratory, Center for Environmental and Estuarine Studies, of the University of Maryland. Drs. Chapman and Feldhamer are contributors to the volume *Evolution of Domesticated Mammals* (1982).

FRANCES P. YOUNGER is illustrator at the Chesapeake Biological Laboratory of the Center for Environmental and Estuarine Studies, of the University of Maryland.

This book was brought to publication with the generous assistance of the Appalachian Environmental Laboratory, Center for Environmental and Estuarine Studies, of the University of Maryland. Additional financial support was provided through a grant from Conoco, Inc.

WILD MAMMALS
OF NORTH AMERICA

Biology, Management, and Economics

EDITED BY JOSEPH A. CHAPMAN, PH.D.
AND GEORGE A. FELDHAMER, PH.D.

THE JOHNS HOPKINS UNIVERSITY PRESS
Baltimore and London

The Johns Hopkins University Press, Baltimore, Maryland 21218
The Johns Hopkins Press Ltd., London

Library of Congress Cataloging in Publication Data
Main entry under title:

Wild mammals of North America.

  Bibliography
  Includes index.
  1. Mammals—North America.   2. Wildlife management—North America.
I. Chapman, Joseph A.   II. Feldhamer, George A.
QL715.W56   639.9'79097   81-8209
ISBN 0-8018-2353-6   AACR2

Figure credits:
Alfred L. Gardner: 1.3, 1.6, 1.7. Terry L. Yates: 2.4, 2.6. Laurie Wilkins: 3.9. Jerry Focht: 5.9A. Leonard Lee Rue III: 5.9B, 5.9C, 6.3, 6.5, 7.2, 15.3. Vagn Flyger: 11.4, 11.5. Walter E. Howard: 13.5, 13.7. William Julian: 15.4. Robert G. Linscombe: 15.5., 15.6. Murray L. Johnson: 16.7. Wendell E. Dodge: 17.2, 17.4, 17.6–17.8. Frank Todd, Hubbs Sea World Research Institute, San Diego: 18.1. Marine Studios, Marineland, Fla.: 18.5. J. G. Mead, U.S. National Museum: 18.6. Edward D. Asper: 18.7, 18.8. Robert Pitman: 18.9. Stephen Leatherwood: 18.10. Donald Lusby, Sea Library: 18.11. U.S. Navy: 18.12. G. C. Pike: 19.2. Hideo Omura, Whales Research Institute, Tokyo: 19.4–19.6, 19.8. W. A. Watkins: 19.7. Michael R. Pelton: 24.7. John J. Craighead and John A. Mitchell: 25.2, 25.9, 25.10. Jack W. Lentfer: 26.3, 26.4. Marjorie A. Stricklund, Carman W. Douglas, Milan Novak, and Nadine P. Hunziger: 28.2, 28.3, 28.5. Frederick G. Lindzey: 33.2, 33.4. Ed Park: 36.2. Karl W. Kenyon: 37.3. Chet M. McCord: 39.5. Keith Ronald, Jane Selley, and Pamela Healey: 40.4–40.31. Miami Seaquarium: 41.2. Daniel K. Odell: 41.4–41.6. John A. Bissonette: 42.3, 42.5. George A. Feldhamer: 45.6. Alaska Department of Fish and Game: 46.5, 46.7. V. Van Ballenberghe: 46.4. L. J. Miller: 47.3–47.6. Lynn Kitchen: 48.2, 48.4. Victor L. Coggins: 50.2. Rolf Johnson: 52.4, 52.5, 52.6. Ronald Wigal: 53.4. James Goldberry: 53.6. James M. Sweeney and John R. Sweeney: 56.1, 56.2.

Skull drawings by Wilma Martin. All other art and skull photographs by Frances P. Younger. Michael J. Reber provided technical assistance and advice in photographing skulls.

Contents

6491

Foreword

There is increasing awareness throughout the world of the need to recognize and understand all factors in the management and conservation of wildlife. Knowledge of a species's biology and ecology is essential if that species is to be managed in any way (and this includes "preservation" as well as "harvesting" or "exploitation"). But we must also consider the economic and cultural pressures that society brings to bear on a species or group of species. This is an extremely important consideration in the so-called emerging countries if wildlife is to survive, and it is also important in societies where modern wildlife management is currently practiced, as in North America.

A text is needed that would address all the above factors for important wildlife species. Such important species or groups of species often require special management attention, so a reference manual collecting pertinent information under one cover would be a valuable asset. *Wild Mammals of North America* represents such a contribution. No other single book can be used as a basic reference for most of the economically important mammals. Up to now, the information brought together in this work could be found only by the most arduous search of many books, periodicals, and journals.

The editors have intended this book to be up to date and comprehensive. The depth of coverage is extensive; the chapter authors not only review the current state of knowledge but also present much new and original information as well. The editors have chosen the species and groups that, in their opinion, biologists and naturalists are most likely to be involved with.

This book truly qualifies as a major reference text and represents an extraordinary achievement by the editors and authors. I expect it to become one of the most important references for both practicing biologists and their teaching counterparts. It also serves as a general reference, valuable to workers in many fields of biological sciences and to those people who simply are interested in wildlife.

MAURICE HORNOCKER

Preface

Wild Mammals of North America was developed to fulfill our needs in a course taught at the Appalachian Environmental Laboratory entitled Mammalian Population Analysis and Management. Initially, we attempted to compile the material presented in this text as a supplemental class project several years ago. However, it quickly became apparent that the sheer magnitude of the literature, and the wealth of new information being published each year, precluded such an approach. Later we discussed the need for a text containing the material with colleagues working in state, federal, and university positions. These discussions overwhelmingly affirmed the need for such a book.

Following these discussions, we approached noted authorities on particular species, such as Marc Bekoff, Jack Kaufmann, Frank Miller, Jim Peek, and Walt Howard, about contributing chapters for the book. Their favorable responses gave early impetus to continue with the project. Upon this instigation, we developed the format and a list of species that should be covered.

The format and subject matter were selected on the basis of our perceived need; they represent a gradually evolved compromise between an ideal and what was feasible given the constraints of preparation time and volume length. Thus, the book is slanted toward ecology and management of species, yet includes pertinent life history information. Authors have brought together up-to-date information dealing with the biology, status, and management of 57 of the economically important mammalian species or species complexes in North America. The general format for each chapter contains distribution, description, physiology, reproduction, ecology, food habits, habitat, behavior, mortality, age determination, management, economic status, current research and management needs, and an extensive citation of literature. This format was devised as an attempt to provide a great deal of information in an easily readable fashion. For certain species, their distribution in Mexico is discussed to provide needed continuity. Geographic limitations had to be placed on coverage, however, and we felt that "north of Mexico" was the most logical unit, especially considering the relative lack of information on many species that occur solely in Mexico.

The contributing authors represent diverse backgrounds and affiliations, but all are recognized authorities on a particular species or group of mammals. The chapters include a synthesis of literature previously available only from numerous scattered scientific journals and unpublished works, and in many cases original information published for the first time. The book offers this information to students, professional management personnel, scientists, and interested lay persons in an easy to read format arranged systematically according to order, family, and genus.

For the most part, previously published distribution records for most species are historic. We have attempted to include the most up-to-date distributions of native and introduced species covered, based on literature and on questionnaires sent to all the U.S. states and the Canadian provinces.

For certain species, it is apparent that very little is known, while for others a vast amount of literature is available. Active management efforts are often lacking because the information on which to base management is simply not known. We hope this volume may provide the impetus or direction for much needed research.

The common and scientific names of this book generally follow J. K. Jones, D. C. Carter, and H. H. Genoway's (1975) *Revised Checklist of North American Mammals North of Mexico* (Occas. Pap. Mus. Texas Tech. Univ. 28:1–14). However, when inconsistencies arose, the preferences of the chapter authors were honored. The revised edition of E. R. Hall's *Mammals of North America* (New York: John Wiley & Sons, 1981) was published while our work was in the galley stage of preparation. We attempted to incorporate changes in nomenclature wherever possible, but in some cases this was prohibitive. Therefore, certain chapters use material from the earlier edition of *Mammals of North America* if there were no changes. Other chapters cite both the original and revised works.

We hope that this book will be a significant contribution to the field of wildlife ecology and that it will be a valuable text and reference work for many years to come. It is a tribute to a relatively new and vitally important field of science, and to the 87 contributors whose efforts have culminated in this work. Because of the sheer volume of this text, we had to rely heavily on the various contributors to help us keep errors to a minimum. However, we accept fully our responsibility as editors for any omissions, inconsistencies, typographical errors, and other similar problems inherent in a book of this size. We welcome constructive criticism and comments on this volume and suggestions for improvement.

Acknowledgments

Obviously, this book would not have been possible without the aid and support of a great many people. For long hours of manuscript processing and typing, we are deeply grateful to the office staff of the Appalachian Environmental Laboratory: secretaries Evelyn Kirk, Mabel Lancaster, and Mona Llewellyn; typists Debbie Lancaster and Kathryn Twigg; and business manager Ann Allen. For devoting appreciable amounts of time to the critical review of various chapters, we are indebted to the AEL faculty.

Thanks are due Dr. Peter E. Wagner, former director, Ian Morris, director, and Dennis L. Taylor, associate director for academic affairs, of the Center for Environmental and Estuarine Studies, University of Maryland, for continuing encouragement. Professor Paul Winn, associate director for administration, UMCEES, made available support for this project. We give special thanks to Wendy Harris, of The Johns Hopkins University Press, the manuscript editor, for laboring untiringly and for keeping the pressure on us to insure the completion of the book. We also acknowledge the gracious support of Anders Richter, Jack Goellner, James Johnston, Allen Carter, and Barbara Lamb Kraft, of The Johns Hopkins University Press. Dr. Ted Bookhout, of The Ohio State University, reviewed an early chapter and gave suggestions for strengthening the development of later chapters.

We are thankful for the cooperation of the wildlife agencies of the various states and provinces for assistance in the preparation of range maps. We are also appreciative of the wildlife and mammalogical journals for the use of a number of figures that are presented in the various chapters.

We are indebted to the Smithsonian Institution and the Canadian Museum of Natural History for furnishing specimens used in the illustrations. In particular, we acknowledge the technical advice of Robert Fisher and Don Wilson, of the U.S. Fish and Wildlife Service, Bird and Mammal Laboratories of the National Museum. We also thank David Gubernick, of Duke University, for providing cranial materials for the skull drawings of the horse.

For the artwork that so richly enhances the book, credit goes to Frances Younger, scientific illustrator for UMCEES. Her diligence and constant eye for detail are obvious in the graphs and drawings. Wilma Martin, a free-lance artist, prepared the skull drawings. Finally, we express our gratitude to the chapter authors, whose diligence and perseverance made everything possible.

I

Marsupialia, Insectivora, Chiroptera, and Xenarthra

1

Virginia Opossum

Didelphis virginiana

Alfred L. Gardner

NOMENCLATURE

COMMON NAMES. Virginia opossum, opossum, possum

SCIENTIFIC NAME. *Didelphis virginiana*

SUBSPECIES. *D. v. virginiana, D. v. pigra, D. v. californica,* and *D. v. yucatanensis.*

Commonly (Reynolds 1953), but erroneously, considered a "living fossil" inhabiting North America since Cretaceous times, *Didelphis virginiana* is the only native member of the family Didelphidae found north of Mexico. The earliest known remains likely representing the Virginia opossum date from the Sangamon Interglacial Stage of the Pleistocene (Hibbard et al. 1965). Post-Wisconsin remains are widespread in the United States and Mexico. The earliest fossil record for *Didelphis* is from Pliocene deposits in South America.

Marsupials were a conspicuous part of the Cretaceous and early Tertiary fauna of North America and persisted there until the Miocene (for summary, see Clemens 1977). Clemens (1968) did not regard *Didelphis* as an archetypal remnant of the Tertiary fauna. Instead, on the basis of the derived nature of several morphological characters, particularly dental, he considered *Didelphis* a relatively late evolutionary product of a South American marsupial radiation. Chromosomal data (Reig et al. 1977) support this point of view.

Gardner (1973) presented geologic, karyologic, behavioral, and distributional evidence supporting his hypothesis that the Virginia opossum evolved from *Didelphis marsupialis* relatively late in the Pleistocene and has only recently (in terms of geologic time) become widely distributed in North and Middle America. Remains from archeological sites (Guilday 1958) and evidence in numerous other reports (Grinnell et al. 1937, and Hamilton 1958, among others) demonstrate that the Virginia opossum has moved northward within historic times and, aided by human activities, continues to expand its range.

The taxonomy of the Virginia opossum has had an intricately complicated history. Traditionally, the name *virginiana* has been associated with the popula-

FIGURE 1.1. Distribution of the Virginia opossum (*Didelphis virginiana*).

tions in the United States and Canada. Much of the recent literature has followed Hershkovitz (1951), who treated *D. virginiana* as a subspecies of *D. marsupialis.* As Gardner (1973) demonstrated, the Virginia opossum is distinct from its widespread Neotropical progenitor; the two are sympatric from northeastern Mexico to northwestern Costa Rica. Gardner (1973) reviewed the nomenclature of the genus in North and Middle America, and his taxonomic scheme is followed here.

DISTRIBUTION

The Virginia opossum occurs from southern Ontario and British Columbia, Canada, through much of the United States and Mexico to northwestern Costa Rica (figure 1.1). When Europeans began to settle North America, this opossum ranged as far north as northern Ohio and northern West Virginia (Guilday 1958). Since then the species has gradually moved northward (figure 1.2). Much of this advance has been encour-

3

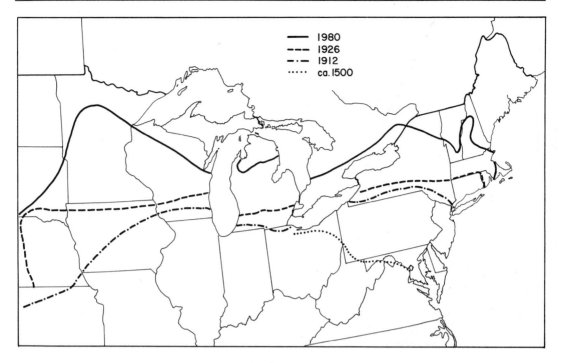

FIGURE 1.2. Changes in the northern distributional limits of the Virginia opossum (*Didelphis virginiana*) in eastern North America. Information on distribution ca. 1500 based on Guilday 1958; information on 1912 based on Cory 1912; information on 1926 based on Ashbrook and Arnold 1927; and information on 1980 drawn from references in the text.

aged by people's activities, which include introductions. All United States and Canadian populations west of the Great Plains and the Rocky Mountains are the result of transplants from the eastern United States. The Virginia opossum has probably moved southward in Central America during historic time, but documentary evidence is lacking.

Reports of distributional records are numerous: Arizona (Hock 1952; Hoffmeister and Goodpaster 1954), Massachusetts and New Hampshire (Kennard 1925), Minnesota (Hazard 1963), New Mexico (Sands 1960), New York (Coleman 1929; Stoner 1939; Severinghaus 1975; Manuel 1977), Ontario (Smith 1935; Peterson and Downing 1956), Oregon (Jewett and Dobyns 1929), South Dakota (Findley 1956), Texas (Bowers and Judd 1969), Vermont (Kirk 1921; Osgood 1938; Davis 1938), Washington (Scheffer 1943), Wisconsin (Hollister 1908; Long and Copes 1968), and Wyoming (Brown 1965; Long 1965). Summarizations of the natural and human-caused spread of the species are found in state, regional, and faunal reports (Miller 1899; Grinnell et al. 1937; Hamilton 1958; Jones 1964; Packard and Judd 1968; Armstrong 1972; Godin 1977). Virginia opossums are indicated for 42 states in the United States and for Ontario and British Columbia, Canada, on the basis of survey questionnaires returned by colleges, universities, and state and provincial wildlife management agencies (Deems and Pursley 1978). New Mexico did not list the species, but Arizona and Maine did. Populations in Arizona and New Mexico may have perished. I have been unable to find any record of specimens from Maine, although Godin (1977) included the southwestern corner of the state within the range of New England populations.

The Virginia opossum can be found in almost any habitat from sea level to elevations over 3,000 m. Northern and elevational limits appear to be controlled by climate and the availability of den sites and winter food.

DESCRIPTION

Size and Weight. Newborn young approximate 14 millimeters in snout–rump length and 0.13 gram in weight. By weaning age, the average head and body (snout–rump) length has increased to near 200 mm and weight to near 160 g. Weights and measurements of post–weaning-age opossums vary greatly. Hartman (1928) listed an extremely small (650 g) but sexually mature female (had pouch young) that measured 345 mm in head and body length. He considered 400 mm and 1,300 g to approximate the size of an average reproductively mature female.

Adult males are larger than adult females (table 1.1), a size distinction that begins at sexual maturity (Gardner 1973). Sexual dimorphism in certain dimensions (such as length of canine) has a primary genetic basis, but most size differences between sexes probably result secondarily from reproductive activity. Opossums grow throughout life (Washburn 1946; Lowrance 1949); however, the energy demands on females rearing young result in smaller size and lighter weight (table 1.2).

TABLE 1.1. External measurements (mm) of *Didelphis virginiana*

Location	Sex	N	Total Length X̄	(Range)	Tail X̄	(Range)	Hind Foot X̄	(Range)	Ear X̄	(Range)	Reference
Pennsylvania	♂	10	779	(698–883)	291	(221–356)	69	(63–75)	54	(44–60)	Blumenthal and Kirkland 1976
	♀	18	753	(666–828)	289	(216–319)	64	(60–68)	55	(49–60)	
Ohio	♀	5	729	(600–817)	298	(255–355)	63	(55–70)			Preble 1942
Arkansas	♂	11	694		280		58		49		Sealander 1979
	♀	7	661		274		55		49		
Georgia	♂	6	560	(513–692)	254	(224–313)	53	(48–63)			Golley 1962[a]
	♀	5	583	(352–808)	261	(222–321)	57	(49–71)			
Texas	♂♂		782		324		66				Davis 1974
	♀♀		710		320		63				
Louisiana	♂	22	821	(751–940)	310	(223–380)	62	(50–85)	51	(40–60)	Lowery 1974
	♀	22	732	(613–900)	302	(260–329)	60	(38–75)	50	(42–58)	
Sinaloa-Sonora	♂	8	805	(740–870)	376	(346–425)	65	(57–70)	53	(48–60)	this report[b]
	♀	8	765	(680–860)	389	(325–470)	58	(50–63)	50	(46–53)	
Oaxaca	♂	22	827	(675–947)	374	(305–446)	65	(50–80)			this report[b]
	♀	18	778	(687–890)	358	(295–412)	62	(52–70)			
Nicaragua	♂	16	824	(730–910)	377	(310–445)	63	(48–73)	52	(40–60)	this report[b]
	♀	20	760	(678–838)	352	(255–409)	60	(50–74)	48	(34–59)	

SOURCE: Measurements of samples from the United States are from the literature. Mexican (Sonora, Sinaloa, and Oaxaca) and Central American (Nicaragua) samples are part of the Latin American series reported on by Gardner (1973).
[a]Either not all adults as stated or errors in these measurements.
[b]Adults (age classes 5 and 6).

Reported external measurements do not reveal any trend in size among United States opossums, except for tail length (shorter in more northern populations). On the average, Latin American populations have longer tails (see table 1.1), but tend to weigh less than United States animals. Cranial measurements of Latin American *D. virginiana* were reported by Gardner (1973).

Few reports include weights and few of these distinguish sexes and ages. Peterson (1966) said Virginia opossums weigh up to 14 pounds (6.35 kilograms), and Jackson (1961) gave 13.5 pounds (6.12 kg) as attainable by excessively fat, old individuals. Because the average female weighs less (see table 1.2), these weights probably represent males. Audubon's and Bachman's (1851) weight of 12 pounds (5.4 kg) for a

TABLE 1.2. Weights (kg) of male and female *Didelphis virginiana*

Location	Males N	X̄	(Range)	Females N	X̄	(Range)	Reference
Michigan	40	3.6	(2.8–4.6)	12	2.4	(2.0–3.2)	Brocke 1970[a]
Michigan	66	2.2	(0.8–3.3)	36	1.6	(0.7–3.1)	Brocke 1970[b]
New York	83	2.8		60	1.9		Hamilton 1958[a]
New York	10	4.1	(3.4–5.0)	9	2.7	(2.5–3.2)	Hamilton 1958[c]
Pennsylvania	10	3.4	(2.2–4.0)	18	2.4	(1.8–3.0)	Blumenthal and Kirkland 1976
Iowa	5	3.1		10	1.8		Wiseman and Hendrickson 1950
Illinois	9	3.8	(1.7–5.9)	7	2.4	(1.8–3.1)	Pippitt 1976
Indiana	1	5.4					Lindsay 1960[c]
Kansas				18[d]	1.4	(0.8–2.0)	Fitch and Sandidge 1953
Kansas	3	2.9	(2.6–3.4)	4	2.6	(2.2–3.2)	Pippitt 1976
Georgia				5	1.1	(0.9–3.7)	Golley 1962
Louisiana	105	2.0	(0.9–3.0)	74	1.8	(0.3–1.6)	Edmunds et al. 1978
Louisiana	22		(2.7–4.7)	22		(1.2–2.2)	Lowery 1974
Nicaragua	20	1.6	(0.9–2.3)	21	1.2	(0.5–2.0)	USNM[e]

NOTE: Locations arranged north to south.
[a]Adults
[b]Juveniles
[c]Heaviest of each sex
[d]Females with pouch young taken in March
[e]From specimens in National Museum of Natural History

female seems excessive. Females at little more than half that weight are unusually large. The heaviest wild-caught male reported (from Illinois) weighed 5.9 kg (13 pounds) (Pippitt 1976). Brocke's (1970) mention of a 7.9 kg (15 pounds, 14 ounces) specimen in Michigan did not include information on sex or whether it was wild caught at that weight or was an obese captive. Comparisons of weights in table 1.2 show a general trend toward lighter animals in southern populations.

Weight is not as closely correlated with size in Virginia opossums as it is in other American marsupials (Eisenberg and Wilson 1981). This opossum is the only didelphid capable of putting on large stores of body fat. Body weights are influenced by season; animals are heavier in fall and early winter than at other seasons. The total lipid fraction of body weight can measure as high as 31 percent in winter-taken animals (Brocke 1970) and should be even higher in the fall. In northern latitudes, overwintering opossums can survive a 40 to 45 percent loss of body weight (Fitch and Sandidge 1953; Brocke 1970); 30 to 40 percent of this loss is fat, the remaining 60 to 70 percent is due to a combination of carbohydrate and protein catabolism.

External Characters. *"An Opassom hath a head like a Swine, & a taile like a Rat, and is of the Bignes of a Cat. Under her belly shee hath a bagge, wherein shee lodgeth, carrieth, and sucketh her young"* (Smith 1612:14). One of the earliest published, Captain John Smith's description of animals from Virginia needs but few details to be complete.

The whitish to pale gray conical head tapers to a pointed snout; the cheek invariably is white, bordered above by a grayish to dusky black eye stripe and eye ring, and behind by darker colored sides of the head and neck. The darker color of the dorsum often extends forward over the crown in a narrow V-shaped wedge to between the eyes; the naked, leathery ears are black with white or flesh-colored tips. The lower legs are black, and the feet are black with white toes; the first toe of the hind foot is large, opposable, and lacking a claw or nail. The long, scaly, and scantily haired tail is black at base but otherwise flesh colored to dirty white. The body fur is long and dense, the hairs white basally, dark brown to black terminally, and interspersed with long white guard hair (gray phase) or with broadly black-tipped guard hair (black phase).

The foregoing description best fits *D. v. virginiana*, which in northern parts of its range may have an entirely white face, white forefeet, and little or no black pigmentation at the base of the tail. Opossums that have overwintered often have lost the tips of the ears and end of the tail from frostbite.

Virginia opossums from gulf coastal states (*D. v. pigra*) are darker, often with all black ears and hind feet. Opossums from southern Texas through Nicaragua (*D. v. californica* and *D. v. yucatanensis*) are even darker and have all black ears and feet. In these populations, the white cheek is conspicuous against the darker face. They also have proportionally longer tails (see table 1.1), of which half or more of the naked portion is black.

Females have an external, fur-lined abdominal pouch (marsupium) enclosing the teats, which are most commonly arranged in a 12-teat horseshoe arc with an additional teat in the center. Teat number may vary from 9 to 17 (Hamilton 1958). Males have permanently descended testes in a pendulous scrotum anterior to a bifurcated penis. A urinogenital sinus adjoins the rectum in both sexes; a common sphincter muscle controls both openings. The chest area of males is often stained yellow to orange from skin glands over the manubrium. Sweat glands are restricted to tail and plantar surfaces of feet (Fortney 1973). Friction ridges are present on plantar surfaces of feet; the tail is prehensile.

Cranial and Skeletal Characteristics. The braincase is small, the postorbital width of the skull is less than interorbital width, and the sagittal crest is well developed. The nasals are long, narrow anteriorly, wide posteriorly (figure 1.3), and often in contact with the lacrimal, which recedes from the outer margin of the jugal before terminating. The orbital extension of the palatine is usually broad, and there are two pairs of palatine vacuities. The angular process of the dentary is inflected medially. Adults have 50 teeth, for which the dental formula is 5/4, 1/1, 3/3, 4/4; the third (last) premolar has a deciduous (milk) precursor. Epipubic bones are present in both sexes; the male lacks a baculum.

Color Phases. Gray is the most common color pattern in the Virginia opossum. The black phase is uncommon to rare north of Georgia and the gulf coastal states (including Texas) but is common in the southeastern United States and southward through Mexico to Costa Rica. Judging from museum specimens, black predominates in some Mexican populations (Gardner 1973). The main difference between black and gray phases is the color of the guard hair, which is all white in the latter and terminally one-third or more black in the former. The iris is black in both. Black-phase opossums often have a few all-white guard hairs scattered through the pelage.

Cinnamon-colored animals have been reported from the United States, but are unrecorded from Latin American populations. This rare variant differs from gray and black animals in that it is cinnamon brown wherever the latter two are normally pigmented black (including ears, irides, and base of tail) and in having cinnamon-colored guard hair. Overall, the fur is softer, perhaps because of the shorter guard hair. Hartman (1922) suggested that two kinds of cinnamon mutant occur, one with and the other lacking guard hair.

Two white mutants are known. True albinos with pink (pigmentless) ears, lips, eyelids, irides, feet, and tail are known throughout the range of the species. The other mutant, called albinotic by Hartman (1952:42), has white fur but normally pigmented skin (i.e., black ears, eyelids, and lips). Albinotic animals, although extremely rare, have been reported from several populations.

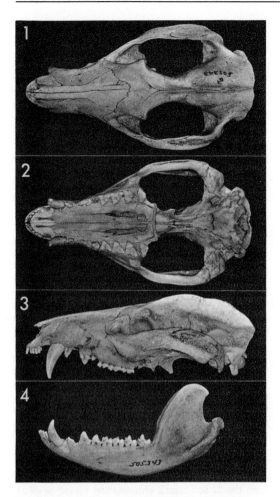

FIGURE 1.3. Skull of the Virginia opossum (*Didelphis virginiana*) (USNM 505343): 1, dorsal view of cranium; 2, ventral view of cranium; 3, lateral view of cranium; 4, lateral view of mandible.

The genetic basis for these color variants and mutants is incompletely known. The allele for black guard hair may be recessive and expressed only in the homozygous condition. This would explain finding black- and gray-phase animals in the same litter when one or both parents are gray. The breeding experiments necessary to verify this hypothesis have not been done.

Anatomy. The most distinctive features of marsupials are in the anatomy of the reproductive tract. Females have two uteri, each connected to the median vagina by lateral vaginae, which receive sperm. A temporary median birth canal forms at parturition to permit direct passage of neonates through the median vagina to the urogenital sinus. The female reproductive tract was first described and illustrated by Tyson (1698). The male was also described by Tyson (1704). These were again described by Hill and Fraser (1925) for the female and by Chase (1939) for the male.

Studies of the skeletal system include those by Washburn (1946) and Nesslinger (1956) on the de-

velopment of ossification centers, by Cheng (1955) on the pectoral girdle, by Elftman (1929) on the pelvic girdle, and by Lowrance (1949, 1957) on skeletal variability and growth. There are 7 cervical, 13 thoracic, 6 lumbar, and 2 sacral vertebrae. Caudal vertebrae vary from 26 to 29. The coracoid becomes fused to the scapula by attainment of reproductive maturity. Certain epiphyses of the cranial and appendicular skeleton never unite; therefore, some parts of the opossum skeleton grow throughout life.

Extensive recent summaries on the anatomy of marsupials (including *D. virginiana*) are found in chapters by Barbour, Parker, and Crompton et al. in Stonehouse and Gilmore (1977). Other recent works include Ellsworth's (1975*a*, 1975*b*) on musculature and Johnson's (1977) on the central nervous system. Bryant (1977) presented a synthesis of the literature on the lymphatic system, the immunohematopoietic complex, and the development of immune mechanisms. Little is known of the structure and function of the endocrine system aside from Kingsbury's (1940) work on the thymus, McDonald's (1977) review of adrenocortical function, and Hearn's (1977) review of pituitary function. The structure and function of the testes and associated reproductive glands were summarized by Setchell (1977). The morphology of the penis and spermatozoa was covered by Biggers (1966). Selenka (1887) was the first to report paired (copulatory) spermatozoa, which he found in the lateral vaginae of a recently bred female. This peculiar configuration (didelphid type) results from pairing of spermatozoa in the epididymides (Biggers and Creed 1962; Biggers and DeLamater 1965). Paired spermatozoa move in a straight line, but if separated, each half swims in a circle.

PHYSIOLOGY

Morrison and Petajan (1962) and Petajan and Morrison (1962) showed that pouch young begin to develop thermoregulatory ability by day 55 and that by day 94 young opossums are able to maintain deep body temperatures at air temperatures as low as 5° C for up to two hours. Dills (1972) and Dills and Manganiello (1973), using surgically implanted transmitters, demonstrated average circadian temperature fluctuations of about 3° C correlated with the activity state of the animal. Higgenbotham and Koon (1955) reported body temperatures of 35.0° C; Morrison and Petajan (1962), 35.2° C; and McManus (1969), 35.5° C. The general conclusion from these studies and those by Nardone et al. (1955), Brocke (1970), and Pippitt (1976) is that body temperatures fluctuate daily and seasonally. Telemetered body temperatures of free-ranging Virginia opossums in northern Illinois ranged from 32.2 to 37.9° C in winter and 35.0 to 37.0° C in summer (Pippitt 1976). The range of body temperature recorded under different circumstances varied according to activity, body size, nutritional state, ambient temperature, and manipulative methods used by the investigators. Pippitt (1976) found that body temperatures of laboratory-held opossums averaged 35.75° C (32.2–

38.0). Large (>2.5 kg) winter-acclimated animals maintained body temperatures at ambient temperatures as low as −5° C. One large Virginia opossum gradually raised its body temperature from about 35.7° C to near 36.3° C when exposed to an ambient temperature of −5° C over a 120-minute period. Under the same conditions, the body temperature of a small (< 2.3 kg) opossum decreased from about 35.9° C to near 34.7° C. Body temperatures of Pippitt's (1976) small (< 2.5 kg) winter-acclimated animals were higher than those of larger opossums over the same range of ambient temperatures, except at −5° C when body temperatures were significantly lower. Small opossums had difficulty regulating body temperatures when ambient temperatures were below 0° C, whereas large animals could tolerate air temperatures as low as −20° C for short periods. Vasoconstriction, piloerection, shivering, and behavioral avoidance of low temperatures are the most important thermoregulatory responses to cold.

McNab (1978) found thermal conductance in *D. virginiana* to be much lower than in Panamanian *D. marsupialis,* although basal metabolic rates were about the same. Thermal conductance levels for McNab's opossums from Florida were close to those reported by Pippitt (1976) for winter-acclimated opossums of comparable weights from northern Illinois. Pippitt also showed that opossums can lower conductance, probably by reducing blood flow to peripheral parts of the body, as the ambient temperature declines. He found, however, that conductance increases at air temperatures equal to or less than 0° C. McNab (1978) reported basal metabolic rates about double those reported by Brocke (1970) and Lustick and Lustick (1972). Pippitt (1976), whose values also exceeded those of Brocke, pointed out that some opossums have lower basal rates in winter than in summer. Brocke's measurements, as low as 0.15 ml O_2/g/hour in a 4.7 kg male, were made on sleeping animals. Pippitt's free-ranging summer animals had a mean oxygen consumption rate of 0.73 ml O_2/g/hour. Mean rates for winter animals varied from 0.511 to 0.702, depending on body weight.

Oppossums experience heat stress at ambient temperatures exceeding the thermal neutral zone (30° C). Typical responses are panting and spreading saliva over the body. Eccrine sweat glands are located only on plantar surfaces and the skin of the tail (Fortney 1973). Because Virginia opossums are nocturnal, they are unlikely to encounter stressfully high ambient temperatures. Water balance was studied by Plakke (1970) and Plakke and Pfeiffer (1965).

Pippitt (1976) subdermally implanted two-channel radio transmitters to monitor heart rate and body temperature in laboratory-held and free-ranging opossums. Heart rate in free-ranging opossums averaged 154.5 (109–220) beats per minute in small summer animals and 120.4 (75–310) to 149.8 (89–214), depending on weight class, in winter animals. Francq (1970) measured heart rate in alert and death-feigning "young adult" opossums. QRS- and T-wave deflection were the same in both activity states. Heart rate averaged 2.76 (2.2–3.6) beats per second in alert and 2.97 (1.7–4.6) beats per second in death-feigning animals. Wilber (1955) recorded 200 beats per minute for

an opossum under light anesthesia. The higher average rates in laboratory tests (Wilber 1955; Francq 1970) probably resulted from the agitated state of the subjects.

Studies on gas-transporting capacity, tolerance to high and low oxygen levels, and wound healing were summarized by McManus (1974). The faculty for wound healing as evidenced by the high frequency of healed broken bones was commented on by Black (1935).

Tamar (1961) found that opossums have low taste sensitivity. Although they are able to perceive color (Friedman 1967) and can discriminate black and white objects under test conditions (James 1960), opossums respond more quickly to audio than to visual cues when pursuing small moving prey (Langley 1979).

GENETICS

Schneider (1977) summarized the knowledge of the karyology and cytogenetics of *D. virginiana.* His sections on cell cycles and DNA, RNA, and protein synthesis are excellent reviews, but the section on chromosomal evolution omitted the studies by Gardner (1973) and Reig et al. (1977).

Although the chromosomes of the Virginia opossum have been studied since Jordan (1911) reported an erroneous diploid number of 17, Shaver (1962) was the first to describe the karyotype correctly. The diploid number is 22, the fundamental number (number of autosomal arms) 32. Autosomes consist of three pairs of large subtelocentric, three pairs of medium-sized subtelocentric, and four pairs of medium-sized acrocentric chromosomes. The X chromosome is a smaller, medium-sized submetacentric, the Y a small acrocentric.

The *D. virginiana* karyotype differs markedly from the 22 acrocentric chromosome karyotype of *D. marsupialis,* its probable progenitor. The chromosomal differences are complex and unique among American marsupials because they are not explainable by whole-arm translocations. Gardner (1973) hypothesized the attainment of reproductive isolation from *D. marsupialis* through a series of rapidly fixed chromosomal rearrangements acquired by an isolate of the parental stock.

REPRODUCTION

Breeding Season. The breeding season, defined as the period beginning with earliest conception and ending with weaning the latest litter, generally begins in January and extends to November. Most published accounts document the onset of reproductive activity, but few offer adequate information on later and terminal phases. Abundant evidence confirms two distinct breeding periods: a short, early period of high breeding activity covering six to seven weeks, followed by a longer, less intense period beginning two to four weeks later and covering at least two months.

The earliest record for probable conception is mid-December (Edmunds et al. 1978) in northern Louisiana. Hartman (1928) also found evidence of ovulation and conception in December, but recorded an ovulation peak during the third and fourth weeks of the year (after 1 January) in the vicinity of Austin,

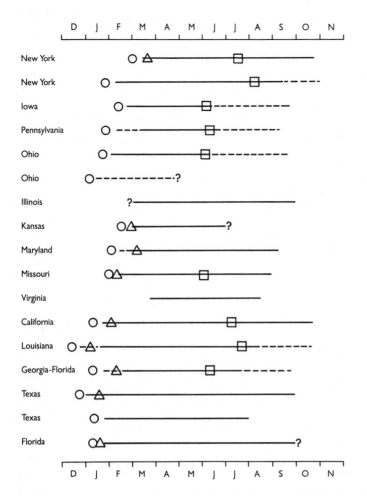

FIGURE 1.4. Breeding season of the Virginia opossum (*Didelphis virginiana*). Circles indicate the time of earliest conception; triangles, the time of birth peaks during the first breeding period; squares, the time of latest recorded birth; a solid line, the time when females had pouch young; and a broken line, the time females probably had pouch young. Numbers correspond to the following references, arranged by state and latitude from north to south: 1, VanDruff 1971 (43° N); 2, Hamilton 1958 (42° 30' N); 3, Wiseman and Hendrickson 1950 (40° 50' N); 4, Blumenthal and Kirkland 1976 (40° N); 5, Petrides 1949 (40° N); 6, Grote and Dalby 1974 (40° N); 7, Sanderson 1961 (40° N); 8, Fitch and Sandidge 1953 (39° 08' N); 9, Llewellyn and Dale 1964 (39° 06' N); 10, Reynolds 1945 (38° 58' N); 11, Stout and Sonenshine 1974 (37° 28' N); 12, Reynolds 1952 (ca. 36° N); 13, Edmunds et al. 1978 (32° 30' N); 14, McKeever 1958 (30° 40' N); 15, Hartman 1928 (30° 15' N); 16, Lay 1942 (30° N); 17, Burns and Burns 1957 (29° 40' N).

Texas. Jurgelski and Porter (1974) found that about 30 percent of North Carolinan females showed signs of estrus from 20 December to 8 January. About half of these cycles were abortive. Most females entered estrus during the three weeks between 8 and 31 January. Early breeding dates were summarized by Grote and Dalby (1973). Accurate fixing of latest conception is difficult from the literature. Hamilton's (1958) data indicate successful breeding as late as August (figure 1.4).

The shift from December–January onset of breeding in the southern United States to January–February at northern latitudes is shown in figure 1.4. The breeding season is unknown for Latin American populations, but may be inferred from figure 1.5.

Estrous Cycle. Hartman (1921*a*, 1923*b*) and Reynolds

(1952) determined the onset of estrus from vaginal smears. Reynolds (1952) and Jurgelski and Porter (1974) found palpation of the mammary area (Hartman 1921*a*) unreliable for determining estrus. The vaginal-smear method is lucidly described and illustrated by Jurgelski and Porter (1974).

Reynolds (1952) concluded that estrus lasts about 36 hours and fertile matings occur only during the first 12. Spontaneous ovulation occurs during estrus. Ova are fertilized in the fallopian tubes before the addition of a coat of albumen and a thin shell membrane on their 24-hour trip to the uterus.

Reynolds (1952) found variability of 22 to 38 days (average 29.5) in the length of estrous cycles in California Virginia opossums. Jurgelski and Porter (1974) reported variation of 17 to 38 days (average 25.5). Reynolds (1952) noted that the time of estrus of sister litter mates did not vary more than 5 days, suggesting genetic variation in response to season. The Virginia opossum is polyestrous and five or six (Reynolds 1952) or seven (Jurgelski and Porter 1974) cycles are possible during a breeding season in the absence of pregnancy. Reynolds (1952) found that older (second year) opossums have fewer (three to four) cycles, but that the length of the estrous cycle is not influenced by age. Females reenter estrus within 2 to 8 days following loss of young. Estrous females breed only once in each cycle. A temporary vaginal plug, which is shed within 36 hours, forms after copulation.

Opossums examined by Reynolds (1952) and Jurgelski and Porter (1974) were all in anestrum by the third week of July. Some opossums in New York (Hamilton 1958) and Nicaragua (Biggers 1966), however, were still fertile in August. Males have spermatozoa in either the testes, epididymides, or both throughout the year but testis weight tends to be lower from September to December (in Pennsylvania), coincident with the anestrous period of females (Biggers 1966).

Gestation. Gestation is about 12 days and 18 hours long (Reynolds 1952, Burns and Burns 1957). Young are born in embryologic stage 34 (McGrady 1938) during the first half of day 13, or in stage 35 during the second half of day 13. Stage 34 embryos are recognizable by their blocklike, elongate heads and fully developed oral shields. The head is rounded or ovoid, and the oral shield has disappeared by stage 35, only

12 hours later. McGrady (1938) did the definitive study on the embryology of the Virginia opossum and reviewed the earlier literature on intrauterine development.

Birth. A few minutes before giving birth, the female usually shows preparturient behavior: restlessness, ears folded against head, and tail drawn forward between the legs. She then assumes a humped sitting position with the tail forward, and licks the vulva as the young are expelled. All young are born within 12 minutes (Reynolds 1952). Most neonates emerge already free of fetal membranes and swinging their forelimbs in a swimming motion. They grasp the hairs of the belly with their forefeet and climb upward to the pouch. The distance to be traveled is only 40 to 50 mm because of the posture of the mother. Upward orientation is maintained because the newborn can only use the forelimbs to crawl; therefore, the weight of the body controls direction. Once the pouch is reached each neonate attaches to a nipple or dies. Hartman (1952) mentioned finding 21 young in the pouch. Twelve of these were attached to nipples; the remaining 9 perished.

Litter Size. The size of a litter is determined by the number of young reaching the pouch and the number of functional nipples. Litter sizes of 15 and 16 (Barton 1823; Hamilton 1958) and 17 (Bailey 1923) are known. Because 13 is the normal number of teats, pouch young can only exceed this number when more nipples are available. Reynolds (1952) believed that only 7 to 10 of the normal number of 13 teats were functional in most of his females from central Califor-

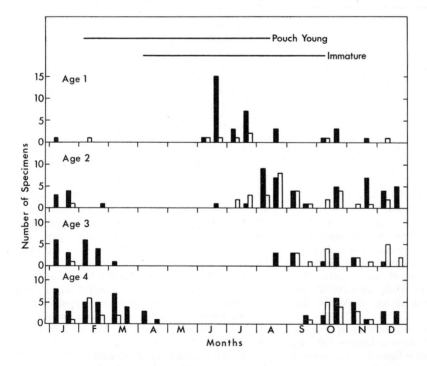

FIGURE 1.5. Seasonal distribution of age classes of the Virginia opossum (*Didelphis virginiana*) in Latin America. Mexican specimens are represented by a solid bar, Central American specimens by an open bar.

TABLE 1.3. Litter sizes of *Didelphis virginiana* in the United States

| Location | N | Litter Sizes | | | Reference |
		Mean	Mode	Range	
New York	156	9.0		1–15	VanDruff 1971
New York	346	8.7	9(22%)	2–15	Hamilton 1958
Pennsylvania	20	9.2		1–14	Blumenthal and Kirkland 1976
Iowa	7	9.0		6–12	Wiseman and Hendrickson 1950
Missouri	42	8.9	8(33%)	5–13	Reynolds 1945
Nebraska	23	8.6			Reynolds 1945
Ohio	5	9.6			Petrides 1949
Illinois	85	7.9			Holmes and Sanderson 1965
Maryland	57	7.7[a]			Llewellyn and Dale 1964
Kansas	28	7.5	7(25%)	1–12	Fitch and Sandidge 1953
Virginia	3	7.3		6–9	Stout and Sonenshine 1974
California	44	7.2	7(25%)	4–11	Reynolds 1952
California	6	10.0		8–12	Grinnell et al. 1937
Georgia-Florida	143	7.1		1–11	McKeever 1958
Louisiana	68	6.8		1–10	Edmunds et al. 1978
Texas	65	6.8			Lay 1942
Florida	50	6.3	7(44%)	3–9	Burns and Burns 1957

NOTE: Mode followed by percentage of litters (in parentheses). Sample size (N) is number of litters. Locations arranged from north to south.

[a]Mean of 8.5 in 26 litters of young $\leqslant$ 40 mm; 6.6 in 23 litters of young $\geqslant$ 60 mm.

nia. He observed births of 4, 13, 15, and 25 young (total of 57), but only 34 (60 percent) reached the pouch, where some failed to find nipples.

There is a general trend toward larger litters in northern populations (table 1.3). Burns and Burns (1957) found the smallest mean litter size (6.3) in Florida. Forty-four percent of the litters they examined contained fewer than 7, the modal number. I found no clear trend toward a higher frequency of smaller litters indicated in other reports. Most investigators did not report modal numbers of pouch young.

Although of general interest, litters of single young (table 1.3) are of little biological importance because these are not usually carried through to weaning age. The stimulus of at least 2 suckling young is normally required to maintain lactation (Reynolds 1952). Two (3 litters) was the lowest number of pouch young among the 346 litters Hamilton (1958) examined from New York. Litter sizes of Virginia opossums from Mexico, based on label information from museum specimens, range from 1 to 14 (mean = 7.6, mode = 6, N = 15).

Llewellyn and Dale (1964) found a mean of 8.5 in 26 litters of young measuring 40 mm or less (up to 25 days of age) and a mean of 6.6 in 23 litters of young measuring 60 mm or more (40 days of age or older). These data suggest progressive mortality of pouch young as litters mature. Fitch and Sandidge (1953) found progressively fewer pouch young being carried by some females when recaptured with the same litter, and reported losses of 23 percent. Sanderson (1961) calculated pouch young mortality as 12 percent. Hamilton (1958) found no evidence of mortality among pouch young and believed that if such losses occurred they were trivial. Enders (1966) said that the critical period when losses occur begins at the time hair

appears. This is about the time when pouch young begin to release their hold on the nipple; therefore, losses may increase because of their newly acquired relative freedom.

Burns and Burns (1957) found no correlation between weight of the mother and the number of young in the pouch. Reynolds (1952) found evidence that older females had decreased fertility, were less able to rear young, and showed a tendency to return to anestrum earlier than young opossums. These older females were probably in their second breeding season.

Number of Litters per Season. Although Audubon and Bachman (1851), Hartman (1928), and McManus (1974) suggested that three litters might occasionally be produced during a single year by some females, all evidence indicates that the Virginia opossum produces a maximum of two litters each breeding season. Tyndale-Biscoe and Mackenzie (1976) found two litters per year in Colombian *D. marsupialis*, except in Valle del Cauca, where there were birth peaks in January, May, and August. They interpreted the third birth peak as evidence of the production of three litters a year. Biggers (1966), who did not distinguish between *D. marsupialis and D. virginiana* in his sample from Nicaragua, also had three peaks of reproductive activity (January, May, and August).

To determine if more than two litters a year were produced by Latin American Virginia opossums, I examined the data on Mexican and Middle American *D. virginiana* gathered during my systematic review of the genus (Gardner 1973). All records for animals that had not yet acquired a full set of teeth, including 60 specimens from Nicaragua, were segregated by age class, and the date of capture (from specimen labels) was plotted by half-month intervals (see figure 1.5). The younger age classes are poorly represented in the sample, and may reflect lower trap susceptibility of younger animals and bias by the preparator. Figure 1.5 shows evidence of births in August or early September, but shows no third peak that would indicate intensive breeding activity in August.

Biggers's (1966) data for Nicaraguan opossums may reflect breeding season differences between sympatric *D. virginiana* and *D. marsupialis*. The western region of Nicaragua has a seasonal climate with a well-marked dry season, whereas the eastern zone is much more humid and the seasonality of rainfall is less extreme. These ecological differences favor *D. virginiana* in the drier Pacific region and undoubtedly influence reproductive success in both species.

The simplest explanation for the three breeding peaks found by Biggers (1966) and Tyndale-Biscoe and Mackenzie (1976), as well as the evidence in figure 1.5 for births in August or early September, is as follows. The first peak (January) represents the start of normal breeding activity synchronized by the October through December anestrous period. January and early February is when most females give birth to their first litter. The second peak (May) represents the return to estrus at the time of weaning the first litter and includes some contribution by females whose earlier breeding efforts were unsuccessful. Females giving birth to first or second litters in May or June are not likely to wean them in time to breed again in August. The third peak (August) is caused by sexually precocial (6- to 7-month-old) females born in January and early February. Females who were previously unable to rear young successfully and females weaning litters born in March or April could also contribute to the August peak. An examination of the timing of the breeding peaks (Biggers 1966; Tyndale-Biscoe and Mackenzie 1976) shows a 4- to 4.5-month interval between the first and second, but a shorter 2.5- to 3-month span between the second and third that argues against any female successfully rearing three litters a year. An analysis of the ages and the conditions of the marsupium in the cohort breeding in August is necessary to test this hypothesis.

Although females can produce two litters a year, Hamilton (1958) showed that not all do. Based on the presence of pouch young, he found barren females in each month of the breeding season. The frequency of barren females varied from a low of 20 percent in April to a high of 46 percent in August. Jackson (1961) said that one litter per year was the rule in Wisconsin and that two were the exception. Data usually are insufficient to estimate the percentage of females that rear two litters a year. In central New York, VanDruff (1971) found that 21 out of 40 (52 percent) recaptured females had produced a second litter. Lay (1942) believed that two litters a year was normal in eastern Texas; however, he had positive evidence from only three females.

The numbers of litters born per female each season is unknown. Although two litters can be raised to weaning age, a female may actually give birth to three or more during a season. Females that have single young and others who happen to lose their litters,

either at birth or after the young are established in the pouch, return to estrus and breed again. Reynolds (1952) fed an inadequate diet to three females carrying pouch young and discovered that each subsequently lost its litter and was back in estrus 3 to 15 days after the young had died. Reynolds's experimental result may mimic natural losses of litters while a female is unable to find adequate food when temporarily incapacitated by injury or during severe drought or harsh spring weather.

It was once believed that a lag of several weeks was required after the young were weaned before estrus recurred because lactation was presumed to inhibit the recurrence of estrus and time was presumably required to allow the enlarged nipples to regress to a size a newborn opossum could take into its mouth (Hartman 1923*a*, 1923*b*; Reynolds 1945). Reynolds (1952) later showed that females may breed while still producing milk and that nipples rapidly regress within a few days, but instead of receding to their original size, each develops a small papilla at the tip to which newborn young attach. These papillae were illustrated by Hamilton (1958:37).

Sexual Maturity. A 6-month-old (186 days) female Reynolds (1952) reared in captivity gave birth to and successfully reared six young. She and a litter mate, which was in estrus at the same time, had been weaned only 90 days earlier. Reynolds, however, claimed that females born in the earliest litter recorded in Central California (29 January) would not have sexually matured by the end of the mating period (early July) of the same season. The 660-g, 345 mm in snout–rump length (S–R) female with 10.5 mm in crown–rump length (C–R) embryos on 11 February recorded by Hartman (1928) could not have been younger than 6.5 to 7 months of age if the latest breeding in Texas is in July. The youngest reproductively mature female cited by Reynolds (1952), which bred at 173 days of age when she measured only 366 mm (S–R), represents the minimum breeding age (5.5 months) for the Virginia opossum.

Males studied by Reynolds (1952) reached sexual maturity (had sperm in the epididymides) by 8.5 months of age. One inseminated a female when he was 247 days old. Biggers (1966), using Reynolds's data, stated that sexual maturity in the male is reached at approximately 240 days (8 months) of age.

Sex Ratios. Opossum sex ratios in United States populations are summarized in table 1.4. With the exception of the ratio given by Petrides (1949), sex ratios among pouch young are at parity. Immature and adult opossums, however, deviate from a 1:1 sex ratio. The percentage of males varies from 20 (VanDruff 1971) and 34 (Blumenthal and Kirkland 1976) to 62 (Llewellyn and Dale 1964). Free-ranging males are more susceptible than females to capture because of their greater cruising range and the higher probability that males will be active during the winter because their larger bulk imparts greater tolerance to colder temperatures.

Under certain circumstances, these traits may contribute to disproportionately high mortality among

TABLE 1.4. Sex ratios of *Didelphis virginiana* in the United States

	Pouch Young		Immature			Adult			Immature and Adult			
Location	N	%	S	N	%	S	N	%	S	N	%	Reference
Wisconsin									R,T,H	161	53	Knudsen and Hale 1970
Michigan			T	153	61	T	48	50	T	201	59	Stuewer 1943
New York						T,R,C	434	27				VanDruff 1971
New York	811	52	T	149	56	T	146	20	T	295	38	VanDruff 1971
New York	911	52							N	846	53	Hamilton 1958
Pennsylvania	183	51							T	62	34[a]	Blumenthal and Kirkland 1976
Missouri									P	2,185	57	Bennitt and Nagel 1937
Missouri									T,D	116	55	Reynolds 1945
Missouri									P	1,076	58	Reynolds 1945
Ohio									P	330	54	Petrides 1949
Ohio									T	48	42	Petrides 1949
Illinois	363	49	T	337	56	T	180	42	T	519	51	Holmes and Sanderson 1965[b]
Maryland			T	114	62	T	106	56	T	220	54	Llewellyn and Dale 1964
Kansas									P	426	56	Sandidge 1953
Kansas									T	62	44	Sandidge 1953
Kansas	16	50							T	56	50	Sandidge 1953
Virginia			T	42	52	T	29	52	T	71	52	Stout and Sonenshine 1974
Louisiana			T	30	60	T	179	59	T	209	59	Edmunds et al. 1978
Texas									T	117	56	Lay 1942

NOTE: Percentages are of males in sample (N). Specimens (S) examined were pelts (P) or whole animals acquired by trapping (T), shooting (H), or predator control (C), caught with aid of dogs (D), found dead on road (R), or not segregated on basis of source (N). Locations arranged approximately north to south.

[a]Percentage males (38.9) reported was an error; correct value was 33.9 (Kirkland personal communication).

[b]Includes data from Sanderson 1961.

males. When local configurations of roads and barriers to dispersal, for example, expose the wider-ranging males to mortality from highway traffic, a lower proportion of males to females in the population would result. Highway mortality may have caused the lower proportion of males reported by VanDruff (1971) and Blumenthal and Kirkland (1976), particularly if that mortality factor was differentially biased. That more males than females are active during the winter was shown by Blumenthal and Kirkland (1976), who reported greater numbers of males in their winter sample, yet the percentage of males for all seasons combined was 34 (table 1.4). Petrides (1949) had a higher winter (December and January) proportion of males (54 percent), based largely on pelts, but had a lower ratio (42 percent), based on year-round live-trapping results. Whatever the cause of higher mortality among males in central New York (VanDruff 1971), the decline occurred among adults because the relative numbers of immature males compares well with results from other studies (table 1.4).

All that is presently known about the sexual composition of populations is that opossums start life about equally divided sexually and that after weaning age males usually predominate.

DEVELOPMENT

The timing of developmental events is reasonably fixed and consistent in most mammals. Opossums are an exception. They exhibit a frustrating degree of varia-

tion in almost every phase of growth and development. Several reports have contributed to the understanding of the sequence of these events in the Virginia opossum. Among the more important are: Hartman's (1928) review of the breeding season and rate of intrauterine and postnatal development; McGrady's (1938) embryological treatise, which includes information on postnatal development; Petrides's (1949) report on sex and age determination; and Reynolds's (1952) study on reproduction. Hartman's review summarized and refined some of his earlier works on the same subject. Petrides relied heavily on the reports by Hartman and McGrady, and erroneously attributed some of McGrady's information on tooth eruption to Hartman. Reynolds's report included useful information on many aspects of development and life history. Although some of the information in these four reports is contradictory, I have extensively used them and a few others summarized by McManus (1974) to construct the outline presented here.

Growth of Young. The tiny, naked, and altricial newborn opossum approximates 14 mm in crown-rump (C–R) length and 0.13 g in weight. Each neonate attaches to a nipple, where it will remain for approximately 60 days. The end of the nipple swells in the mouth, and from the third day through the first few weeks of life the young must be removed carefully to avoid tearing the lips. Young so removed after the third day are difficult to reattach until shortly before the age when jaws open. The nipple gradually lengthens (from about 1 mm to approximately 35 mm) and acts as a tether permitting the attached young freedom of movement in the pouch.

Marked variation in size, weight, and developmental stage occurs among litter mates. Petrides (1949) constructed a growth table based on lengths and weights from Hartman (1928), Moore and Bodian (1940), and Reynolds (cited in Petrides 1949, original not seen). Reynolds (1952) also constructed a growth table for young opossums 1 to 95 days of age. His table does not include weight. The data of Petrides (1949) and Reynolds (1952), at 5-day intervals, are presented in table 1.5.

Age estimation based on size alone becomes more difficult as the young grow older. Reynolds (1952) included a growth curve to show the magnitude of individual variation among pouch young of known age; Hunsaker (1977) presented a similar growth curve. Based on averages of the S–R lengths of litter mates, the accuracy limits of age estimates range from 2 to 3 days for pouch young under 20 days of age to approximately two weeks for animals at weaning age (95 to 100 days old).

The S–R length (essentially the same as snout-anus length and length of head and body) measurement is difficult to apply to attached young. The measurement can always be expected to become more variable as size increases because it is almost impossible to align the spinal column in the same way every time the animal is measured. Tyndale-Biscoe and Mackenzie (1976) used the length of the head as a measure of size.

This technique has not been applied to the Virginia opossum. It could prove a more accurate estimator of age than S–R length.

Development of Young. Although a newborn opossum is embryonic in appearance, it has functional but relatively undeveloped circulatory, respiratory, and digestive systems; a functional mesonephros; a partly functional muscular system; and a wholly cartilagenous skeleton. The ear and eye are nonfunctional and the central nervous system is incomplete. Forelimbs are developed and the fingers, which bear deciduous claws, are capable of gripping hairs and climbing, but the hind limbs and tail are rudimentary. The circular mouth grasps a nipple, which in three days enlarges within the mouth, anchoring the young to the mother while it suckles for another 55 to 60 days. The sequence of developmental stages follows:

Days 1 to 3. If removed, young can reattach themselves to nipples.

Days 11 to 15. Follicles for rostral vibrissae appear.

Day 16. Vibrissae and juvenile body hair appear on nose.

Day 17. Sex organs visible on all young. Some young may be sexed as early as day 11 (McGrady 1938).

Day 20. Cheek vibrissae begin to appear; hind limbs can be used to shift position, but hind toes cannot be flexed.

Days 20 to 25. Pinnae become free.

Days 28 to 34. Pigmentation appears on lateral margins of scrotum. Females still show no pigmentation.

Days 33 to 41. Pigmentation appears on base of tail in both sexes.

Days 34 to 35. Downy body hair is visible under a hand lens; vibrissae begin to lengthen.

Day 37. Young cry when handled; mouth is still sealed.

Day 40. Pigmentation appears on upper neck and shoulders.

Day 43. Young able to flex toes on hind feet.

Day 48. Pigmentation spread over entire dorsum, sparse dark hair emerges on neck and shoulders. This is the earliest age when young begin releasing the nipple; mouth still sealed.

Day 50. Hair short and sparse. This is the youngest age when the eyes and mouth begin to open. Young now are approximately the size of house mice.

Days 55 to 68. Mouth opens.

Days 58 to 72. Eyes open.

Day 60. Body completely covered with short dark hair; the first tooth erupts (deciduous third premolar); nipples (now about 25 mm long) permit young to lie outside of the pouch while suckling.

Day 70. Young conspicuously furry; juvenile hair (sometimes called juvenile guard hair) about 15 mm long and beginning to shed; dark underfur about 10 mm long. Young commence leaving the mother and crawling short distances before returning to pouch. Gray and black phases cannot be distinguished until the juvenile guard hair is shed.

TABLE 1.5. Growth of pouch young *Didelphis virginiana* from the United States

	Petrides (1949)		Reynolds (1952)	
Day	S–R	Weight	$\bar{X}$ S–R	Range
1	13	0.13	13.8	13.0–15.4
5	17	0.4	18.8	17.7–19.6
10	24	0.9	24.5	22.5–27.0
15	30	1.3	30.4	27.2–32.0
20	33	1.7	37.1	33.0–40.0
25	40	2.4	41.7	37.0–44.6
30	45	3.9	48.4	42.6–52.7
35	53	5.4	52.8	50.0–56.0
40	60	7.0	60.7	54.4–65.2
45			68.4	58.4–72.1
50	75	13.0	71.9	64.0–80.8
55			79.6	70.3–93.8
60	100	25.0	91.6	81.2–100.8
65			106.0	89.0–117.3
70	125	45.0	115.8	101.8–123.0
75			128.0	110.0–140.9
80	150	80.0	148.0	134.1–165.0
85			161.1	147.0–183.0
90			185.1	165.0–200.0
95			195.2	171.6–217.7
100	180	125.0		

NOTE: Snout–rump (S–R) measurement given by Petrides is distance between snout and anus. Measurements in millimeters, weights in grams, age in days.

Days 70 to 85. Young usually leave den in mother's pouch and if left behind they begin crying. The female normally makes a clicking sound when the young cry and the young respond by going to her. If the female does not make the sound, some or all of the young may remain in the den.

Day 75. Second premolar is present; young capable of feigning death.

Days 75 to 85. Last four upper incisors, the canines, and the first premolar erupt.

Day 77. Descent of testes is completed.

Days 81 to 82. Young are commonly left in the den while female forages. All young cry if separated from the mother before day 83.

Days 83 to 91. Young do not cry when left in den.

Day 87. Weaning begins; young show strong interest in solid food, but are still suckling. Petrides (1949) and Hartman (1928) said that young are weaned by day 80. This remark may have been based on finding females without their litter (see days 81 and 82), which does not mean that the young have been weaned. Some or all young are still left in the den when the mother forages. If young leave the den with the female, they travel on her back or run beside her.

Days 87 to 92. Young produce clicking noise.

Day 95. First incisor, first lower molar, and sometimes the first upper molar present.

Days 96 to 106. All young are weaned. Young show excited interest in solid food after day 96, at which time they are nursed infrequently. The young are largely independent of the mother after day 96, but will run to her in response to clicks as late as day 108. They now may leave the den alone and may sleep elsewhere. By weaning age the young have developed their full range of hearing (about eight octaves). The most acute hearing is at the upper end of the scale, and the young respond more readily to sounds made by rustling leaves and grass than to the human voice (McGrady 1938).

SEX DETERMINATION

Pouch Young. Sex can be determined from the age of 11 (McGrady 1938) to 15 (Reynolds 1952) days, when the rudimentary pouch and scrotum first become visible. These features were illustrated by McGrady (1938:200) and Petrides (1949:367).

Free-Living Opossums, and Carcasses. The presence of a prominent scrotum in males and a pouch in females is clear evidence of the sex. Carcasses can be sexed by noting the presence or absence of a penis, which is often withdrawn into the urogenital sinus.

Skins. Most opossum skins encountered will be cased pelts collected during the trapping season (late fall and winter). Evidence of a scrotum or a pouch will indicate the sex. Some skins, however, can be difficult to sex because the identifying features have been torn away or discarded during preparation. Bare patches of black-pigmented skin near the ventral edge of the pelt indicate a male even if the scrotum has been removed. The abdominal skin is usually thinner in females and the

outline of a pouch can often be seen from the flesh side. If the abdominal skin has been torn away, nipples can sometimes be found along the margin of the tear, verifying the skin as female. Subadult females and those that did not rear litters during the previous breeding season have small, easily overlooked pouches lined with unstained fur. Conspicuous teats within a patch of sparse, tan to orange brown hair indicate an adult female that has reared young. Winter skins of adult females collected after January have white-haired pouches because the stained pelage has been replaced by new hair. Pouches of adults are conspicuously larger than those of subadults. Petrides (1949:368) illustrated skin and pouch characteristics.

Skulls. The sex of adult (age classes 4 and 5) Virginia opossums can be determined by the size of the canines, which are consistently longer and heavier in the male. To verify this, I measured the anterior–posterior length of the crown at the level of the palate, and the height of the crown (often incorrectly called the length of the tooth) from the level of the palate, on 64 *D. virginiana* selected at random from specimens in the collections of the National Museum of Natural History (USNM) representing populations from several localities in the United States. Although many of the canines are well worn and a few have broken tips, there is little overlap in the measurements. The sex of each specimen always was identified correctly when both dimensions of the canine were used for individuals in the same age class. The mean and range of the measurements (in mm) for the length and height of the crown, respectively, are: age class 4, males 6.6 (6.3–6.7), 13.5 (10.7–14.8), and females 5.1 (4.6–5.6), 10.3 (9.4–11.3); age class 5, males 7.7 (6.4–9.8), 16.3 (13.7–20.5), and females 5.5 (4.5–6.4), 11.2 (9.4–12.8) (see table 1.6 for an explanation of age classes).

Age Determination. Pouch young, immature or juvenile, and adult are the age categories used in most reports. The term *pouch young* is self-explanatory. Present subadult categories are ambiguous, and thus far in this chapter I have simply used the term *immature* when referring to available information on postpouch animals that have been called immature, juvenile, or subadult in the literature (see table 1.4). I prefer to use the term *immature* for animals that still lack an erupted first upper molar (M^1). This age class consists of weaning-age opossums. All other age classes (1 to 6) are based on the molar eruption sequence and the replacement of the deciduous last premolar. Some age class 1 opossums may be still of weaning age. These age classes and their corresponding probable ages in months are summarized in table 1.6.

Reasonably accurate criteria for estimating the ages of pouch young have been published (see table 1.5), but those for determining the ages of older opossums are less exact. Petrides (1949) used a growth curve based on weight and total length of known-age animals to arrive at age estimates for tooth-eruption stages. Similar categories were developed by Lowrance (1949), VanDruff (1971), Gardner (1973), and Tyndale-Biscoe and Mackenzie (1976).

TABLE 1.6. Age classes and approximate ages in months for *Didelphis* spp. based on eruption sequence of molars and replacement of deciduous (d) third (last) premolar

Premolar (third)	Molars 1	Molars 2	Molars 3	Molars 4	Age Class — Tyndale-Biscoe and Mackenzie[a] (1976)	Age Class — Gardner[b] (1973)	Age Class — Lowrance[c] (1949)	Age in Months — Gilmore[a] (1943)	Age in Months — Petrides[c] (1949)	Age in Months — VanDruff[c] (1971)	Age in Months — This[c,d] report
d3/d3	0/(1)	0/0	0/0	0/0	1[e]	immature			80 days +		<4
d3/d3	(1)/1	0/0	0/0	0/0							
d3/d3	1/1	0/(2)	0/0	0/0	2	1			4		4
d3/d3	1/1	(2)/2	0/(3)	0/0				juvenile 6–8		4–6	5
d3/(3)	1/1	2/2	(3)/3	0/(4)	3	2	1		5–8.5	5–7	5–6
(3)/3	1/1	2/2	3/3	0/4	4	3	2	subadult 8–10	7–11	7–8	6–7
3/3	1/1	2/2	3/3	(4)/4		4	3			9–10	7–9
3/3	1/1	2/2	3/3	4/4	5	5	4	adult 10+	10+	10+	10+
Wear on M$^{1-2}$					6	6					
Wear on all molars					7						

NOTE: Parentheses around tooth numbers indicate erupting teeth.
[a] *D. marsupialis* and *D. albiventris*
[b] *D. marsupialis* and *D. virginiana*
[c] *D. virginiana*
[d] Approximate ages based in part on data in **figure** 1.5
[e] See Tyndale-Biscoe and Mackenzie (1976:252, fig. 2)

The dental criteria for the age classes in table 1.6 and illustrated in figure 1.6 are the tooth eruption and replacement sequence in maxillary and mandibular toothrows. The topmost toothrows in figure 1.6 (immature age class) show the upper molariform deciduous premolar. The remaining toothrows (age classes 1 to 6) are aligned on the third premolar. The example for age class 3 lacks the crown of the upper deciduous premolar on the side shown here (probably lost during the skull-cleaning process) but retains it in the left toothrow (not shown). Age class 5 opossums are over 10 months old. The wear facets on the molars of the age class 6 example are typical for older adult opossums. The crowns of the first and second upper premolars in this specimen have been worn away, leaving the separate peglike double roots of each tooth. The minimum age when the last upper molar shows wear has not been determined.

The series of photographs used by Tyndale-Biscoe and Mackenzie (1976) to illustrate the sequence of tooth eruption, replacement, and wear are misleading if their figure was the basis for their system of age classes. Their "dental class 1" toothrow shows no upper molars; therefore, it corresponds to my "immature" age class. Their dental classes 2, 3, and 4 correspond to my age classes 1, 2, and 3, in the same sequence. Other inconsistencies are mainly differences in interpretation.

Petrides (1949) examined the usefulness of the sagittal crest, radial epiphyses, and epipubic bones as guides for estimating age in opossums. He believed that the length of epipubic bones had aging potential because these bones were shorter in specimens still lacking the upper last molar (M^4) than in animals with complete dentition. Lowrance (1949) considered the length of the epipubic bones to be the most variable linear dimension she examined.

The age when the coracoid fuses with the scapula (fused by 7 to 8 months of age), and the ilium with the pubis and ischium (fused by 10 months of age), is variable (Lowrance 1949) and of limited value when aging skeletons.

Some age class 3 and all age class 4 and older females are sexually mature. Males younger than age class 4 (average age is eight months) are not reproductively mature. Age categories based on tooth eruption and replacement sequence need to be calibrated against dental stages of known-age, free-ranging opossums. Samples from throughout the range of the species should be contrasted to determine what effect sex, climate, birth period, and breeding history may have had on dental-stage ages.

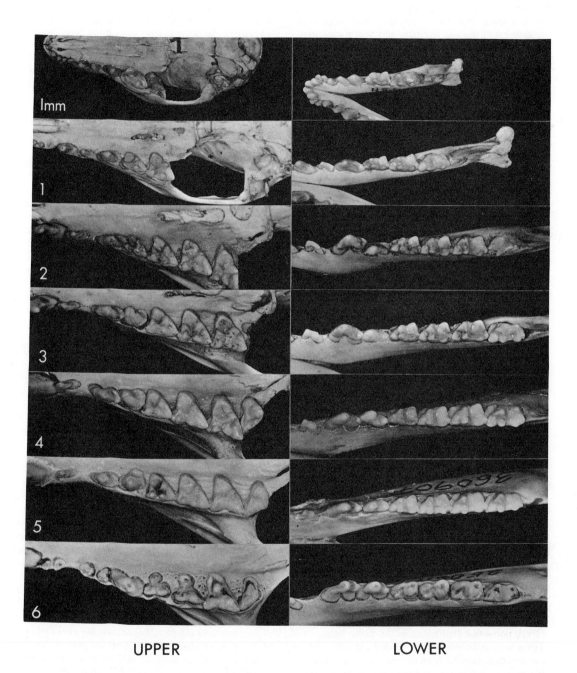

UPPER   **LOWER**

FIGURE 1.6. Right upper and lower toothrows of the different age classes of the Virginia opossum (*Didelphis virginiana*). Toothrows are aligned on the third premolar. Age classes are arranged in sequence from immature at the top to age class 6 at the bottom.

ECOLOGY

To understand the ecology of the Virginia opossum, one must know what habitats are used, how individuals are distributed in these habitats, and how many opossums these habitats can support. This requires basic information on diet, behavior, longevity, litter size, developmental rate, age at sexual maturity, breeding season, functional morphology, and physical and physiological characteristics of opossums. With added information on habitat preference, dispersal pattern, home range, population composition, interspecific interactions, and mortality factors gained from the study of populations, one can estimate density, reproductive rate, and population turnover. Only then will one have an appreciation of the ecological role of this marsupial. A full estimation also requires an understanding of the evolutionary, environmental, and historical factors that have shaped its distribution and life history.

Habitat Preference. Opossums prefer deciduous woodlands in association with streams, but all habitats within their range of ecological tolerances are used (Lay 1942; Reynolds 1945; Llewellyn and Dale 1964; Stout and Sonenshine 1974). The broad ecological tolerances of the Virginia opossum contrast markedly with those of its closest relative, the common opossum (*D. marsupialis*) of Latin America. In the northern extremes of its range, the common opossum is restricted to the humid tropical and subtropical habitats of eastern and southern Mexico, generally below 1,000 m. The Virginia opossum, however, ranges from sea level to over 3,000 m throughout most of Mexico in a variety of humid to arid habitats.

In the United States, opossums have been found in marshlands (VanDruff 1971) and in an array of forested, grassland, agricultural, and suburban habitats (Yeager 1936; Sandidge 1953; Fitch and Sandidge 1953; Verts 1963; Wood and Odum 1964). Lowest numbers are found in residential, agricultural, and grassland habitats, in that order (Verts 1963; Llewellyn and Dale 1964; Blumenthal and Kirkland 1976). Cumberland Valley, Pennsylvania, opossums showed marked preference for agricultural edges (Blumenthal and Kirkland 1976), but the mosaic of land-use patterns in this valley suggests that the animals were responding to habitat diversity and edge effect. Yeager (1936) believed that small farms improved habitat conditions. Packard and Judd (1968) and Bowers and Judd (1969) noted that distribution in the Llano Estacado of Texas was associated with wooded canyons and escarpments that extend from the essentially treeless high plains eastward to the mesquite plains, where opossums are common. Distribution within these habitats is influenced by the availability of dens and water and by the locations of seasonally abundant foods, especially fruits and berries.

Daytime dens include holes, cracks, and crevices in and under trees, stumps, hollow logs, hay stacks, vine tangles, rock outcrops, road culverts, attics and foundations of buildings, and piles of rock, brush, and debris (Yeager 1936; Hamilton 1958; Holmes and Sand-

erson 1965; Fitch and Shirer 1970; Shirer and Fitch 1970). The use of crow and squirrel tree nests (Yeager 1936; Reynold 1945) suggests that opossums may also modify the arboreal nests of the eastern wood rat (*Neotoma floridana*). Opossums have been found in holes dug by turtles (Rand and Host 1942) and commonly use burrows made by skunks, armadillos (*Dasypus novemcinctus*), and woodchucks. Enlargement of these dens by the red fox probably benefits larger opossums in the northern parts of their range.

Winter dens in the central and northern United States are ground burrows, which afford maximum protection in cold weather. Opossums on poorly drained soils must have an adequate supply of dens at or above ground level. Lay (1942) found that ground dens were flooded during the rainy season in east Texas.

Llewellyn and Dale (1964) found higher concentrations of opossums in woodlands near water than in dry upland areas. Lay's (1942) animals temporarily abandoned certain areas and moved several hundred yards to seek water during droughts. The greatest distance from water recorded by Reynolds (1945) was 228.6 m (750 ft). Actual distances may have been greater because Reynolds was using dogs to capture opossums, which ran toward the nearest pond or stream when pursued. Sandidge (1953) said the greatest distance between any den and a source of drinking water was approximately 366 m (1,200 ft). Captives require considerable amounts of water to avoid dessication, and accessibility of surface water may be critical to suitable opossum habitat.

Home Range. Opossums are not territorial and do not maintain separate home ranges. Because they are generally solitary wanderers that rarely remain in any one area for long periods of time, home ranges (activity ranges) are difficult to define.

Lay (1942) based his home-range estimates on 29 (of 117 marked) opossums that were recaptured 3 or more times. One of these 29 was taken 48 times at 11 trap stations, another was taken 30 times at 13 stations, and a third opossum was taken 20 times at 7 stations. The average of the greatest distances traveled was 445 m (1,460 ft). Based on a mean travel radius of 223 m (730 ft), he calculated a circular home range of 15.5 ha (38.4 a). Lay concluded that the home range of an average opossum fell between 4.7 and 17.5 ha (11.5 to 38.4 a).

Wiseman and Hendrickson (1950) said that recaptures of three opossums marked the year before indicated a yearly mobility of 402 m (0.25 mi), and that recaptures of four others were all within 804 m (0.5 mi) of the tagging sites.

Fitch and Sandidge (1953) estimated the usual home range as near 20.2 ha (50 a) or a little less. They found no discernible difference in distances traveled by males and females or by adults and young of the year. Thirteen of their marked opossums were captured five or more times, for an average distance between trapping stations of 596 m (1,954 ft), indicating a home range of nearly 28.4 ha (70 a). This, they believed, was too large and reflected the greater time intervals between captures during which the opossums had shifted their activity into new areas.

Verts (1963) caught 13 opossums two or more times in a study in northwestern Illinois where about 98 percent of the area was under cultivation. The mean distance between captures was 512.4 m (1,680 ft), and the mean of the longest distances traveled by each opossum was 830.8 m (2,724 ft). He used these data to calculate home ranges of between 54.3 and 82.2 ha (134 to 203 a) and surmised that opossums living in intensively cultivated regions had to travel farther as they foraged than those living in wooded areas elsewhere. Verts calculated the home ranges as circular, but stated that home ranges determined on the basis of elliptical shape might be more accurate. Therefore, he used the method of Stumpf and Mohr (1962) on his own data and on those of Fitch and Sandidge (1953) from Kansas, and found that home ranges in Illinois were about 2.9 times longer than broad, and in Kansas, about 2.7 times longer than wide. This method, applied to Lay's (1942) data on Texas opossums, yielded home ranges about 2.7 times longer than broad (Stumpf and Mohr 1962). Using elliptical home ranges as a model, Verts derived opossum home ranges of about 12.5 ha (31 a) in Texas, about 13.4 ha (33 a) in Kansas, and about 38.9 ha (96 a) in intensively cultivated Carroll County, Illinois.

Llewellyn and Dale (1964) did not calculate home ranges in their report but did say that ranges tended to be long and narrow rather than circular. They reported a male that had been caught nine times in eight months within an area about 3.2 km (2 mi) long and 0.6 km (0.4 mi) wide.

Fitch and Shirer (1970), using radio telemetry to study movements in Kansan opossums, said that most of the home ranges tended toward a circular shape. Their opossums tended to distribute activity in different directions from whichever den was in use at the time, instead of moving toward a particular foraging area. When intensity of activity relative to distance from the den was examined, they found that 51.1 percent of the activity was concentrated within the central one-tenth (12 ha) of the hypothetical circular home range. Fitch and Shirer (1970:178) also commented on what they called the "compound nature of the home range," and they suggested that the home range may consist of many overlapping segments, each having a den as its focal point. They also pointed out that the total area increased with time and suggested that a home range be considered as either a lifetime range, a range from one den, or the area used during some arbitrary time period (such as a month). They calculated an average home range of approximately 120 ha on the basis of a circular configuration. Opossums changed dens often; the number of days a den was in use averaged 2.2 days (range: 1 to 26 days), based on 220 den shifts. The den-to-den distance averaged 301 m and was not significantly different for either sex.

Shirer and Fitch (1970) reported 223 den shifts averaging 302 m (range: 9 to 903 m) between successive dens. Seventy-five percent (168) of the 265 den-use

periods were for only one day. They recorded the distances traveled from the den on each night's foray as varying from 9.1 m (30 ft) to 617.6 m (2,025 ft). A radius of 617.6 m from a single den yielded a potential home range of 119.9 ha. For opossums trailed to several dens, the potential home range, again based on a circle, increased to an average of more than 243 ha.

VanDruff (1971) found that some of the opossums he studied on the Montezuma National Wildlife Refuge in central New York were more sedentary than others. One radio-tracked female yearling remained within the same 4.9-ha (12-a) area for over four months. Activity ranges varied in size from approximately 4 ha for a juvenile female to over 40.5 ha for a large adult male. No opossum had a circular home range, although the ranges of juveniles and subadults were more circular in shape than those of adults. VanDruff learned from telemetry bearings that, as the activity range increased in size, the area became linear in shape. The activity ranges of six radio-tracked adults averaged about 2.7 times longer than broad, similar to the dimensions determined by Verts (1963). The Montezuma Refuge opossums showed a preference for ecotonal or edge habitats. Fitch and Sandidge (1953) noted the same preference in their Kansas opossums, which tended to forage along the edges of fields and along creeks and gullies in nonforested areas, and along streams and rock ledges in woodlands.

Pippitt's (1976) radio-tracked opossums in northern Illinois selected den sites near areas of ripening fruit during late summer and fall. Pippitt found seasonal differences in the amount of time spent in a den and the rate at which den shifts were made. During winter, opossums weighing less than 4 kg tended to stay at the same den if the air temperature was below freezing. Larger opossums shifted dens even when the air temperature was in the −11° to −20° C range.

Opossums show little inclination to follow trails and tend to take circuitous and erratic routes as they forage. Wiseman and Hendrickson (1950) trailed five opossums in light snow over the entire course of a night's travel. The total distance traveled each night per opossum ranged from approximately 0.8 km (0.5 mi) to 3.2 km (2 mi). Brocke (1970) followed an opossum that had traveled 2.5 km in 15 cm of snow during a single night. The straight-line distance from any point on this trail to the den never exceeded 400 m.

Density. Lay (1942) estimated an average of one opossum to 1.6 ha (4 a) in poorly drained coastal pine-hardwood forest in eastern Texas. The less productive sandy coastal-prairie habitat had an estimated density of one per 5.9 ha based on the excavation of all dens found on 1.6-ha (4-a) sample plots.

Wiseman and Hendrickson (1950) determined opossum density as about one per 43.2 ha (6 per 640 a) in mixed pasture, woodland, and agricultural habitat in Iowa.

Fitch and Sandidge (1953) roughly estimated the density of the fall population in Kansas as one per 8.1 ha (20 a) and said the population was reduced to about half that by late spring.

Sanderson (1961) and Holmes and Sanderson (1965) estimated population densities in an isolated 1.1-ha (2.8-a) patch of mature oak-hickory forest and in old-field habitats surrounded by cultivated farmland in Illinois. Densities were calculated by two methods: (1) the number of young marked while in the pouch and later trapped as independent animals, and (2) the ratio of marked to unmarked opossums retrapped the year following initial capture. The first method assumed no mortality of juveniles. Both methods used 1 July as an arbitrary date and were based on the average litter size determined for each year, and the assumptions that sex ratios were even and that movements into and out of the area were negligible. Using the first method, they estimated one opossum per ha (259 opossums per mi²) in 1958, one per 0.6 ha (419 per mi²) in 1959, one per 1.9 ha (138 per mi.²) in 1960, and one per ha (261 per mi²) in 1961. Using the second method, they estimated one opossum per 0.4 ha (634 per mi²) in 1957, one per 1.1 ha (232 per mi²) in 1958, and one per 0.8 ha (338 per mi²) in 1959. The numbers actually live trapped per 259.2 ha (1 mi²) varied from 11.8 to 16.8 percent of the population sizes estimated for one year.

Verts (1963) estimated density as one per 25.9 ha (64 a) in a region of intensively cultivated farmland in northwestern Illinois.

VanDruff (1971) found average densities of adult opossums to be one per 7 ha (5.8 adults per 100 a) in waterfowl nesting habitat in the Montezuma National Wildlife Refuge. Fall populations, which included juveniles, may have been as high as one per 0.9 to 1.0 ha (250 to 300 per mi²).

Stout and Sonenshine (1974) reported an average density of one opossum per 20.2 ha in an area of heterogeneous habitat near Richmond, Virginia. Minimum numbers known to be alive were lowest in the first quarter of the year and highest in the second or third quarter. Estimates ranged from as low as zero in the spring quarter of 1965 to a high of approximately one opossum per 5.6 ha in the third quarter of 1967. Their data, although covering only six years, showed peak densities in 1964 and 1967, an interval of four years. Apgar (1934) noted periodicity in the numbers of live Virginia opossums presented to the Philadelphia Zoological Garden over a period of 26 years. On the average, noticeable increases in accessions occurred about every six years, indicating high levels in the wild population at those times.

Hunsaker (1977) reviewed and compared the results presented in most of the reports outlined above.

Longevity and Population Composition. Maximum longevity in captivity is in excess of 4 years and 5 months (Crandall 1964). Longevity in the wild is unknown. The life span determined by Lay (1942) from trapping results was little more than 15 months. The adult male mentioned by Reynolds (1945:375) was at least 2 years old when killed, assuming that it was about a year old when originally tagged. Petrides (1949) estimated the average longevity to be 1.3 years and the probable turnover period as 4.8 years. This

estimate was considerably less than the 7+ years estimated by Hartman (1923a) based on captives. Fitch and Sandidge (1953) listed a female marked as a pouch young on 6 May 1950 that was recaptured on 28 February 1952, indicating an age of 22+ months. They reported that 70.8 percent of the opossums trapped over a 3-year period were yearlings. The breeding population of females consisted mainly of opossums born the preceding spring, and they estimated that 95 percent of the 1949-50 breeding population was replaced by the time of the following breeding season. Sanderson (1961) trapped no opossum older than 15 months after their first capture, and concluded that few live beyond the summer following their birth. Reynolds (1952) and Jurgelski and Porter (1974) believed that females in their second reproductive season were less fecund. Only 3 percent of the opossums reported on by Llewellyn and Dale (1964) were recorded for longer than a year. The longest record was for a male estimated to be about 3 years old when last trapped. One female was estimated as 27 months old; another about 24 months old when last caught. Their results indicated that Petrides's (1949) 1.3 years as the average life expectancy and 4.8 years as the turnover period were too high. VanDruff (1971) calculated average ecological longevity as 1.1 years and population turnover as 3.5 years. Young of the year made up 87.7 percent of the late fall population; the average annual winter carry-over rate for marked opossums was 9.2 percent.

The lowest numbers of opossums are found from January through April (the first period of the breeding season) when the female population consists entirely of reproductively adult animals (not counting pouch young). Assuming that births occurred from January through July in the preceding year, the cohort of females reproducing for the first time would be from 6 to 12 months old in January. Females entering their second breeding season would be from 20 to 26 months old in January. The reproductively adult segment of the male population in January would consist of 8- to 12-month-old animals breeding for the first time and males over 20 months of age that may or may not have bred the previous year. Relative ages of the breeding animals, and the onset of breeding activity, shift a month or more depending on climate and latitude (see figure 1.4).

Independent young of the year should first appear in the population from late April through May and peak population levels including all age classes can be expected in the fall after the latest litters of the season are weaned. Latest-born young in northern extremes of the range may be at a decided disadvantage because of the lower winter-survival potential of smaller animals. Brocke (1970) found that all animals weighing less than 1 kg were absent from his Michigan population after mid-January.

Studies relying on recapturing marked animals are based on the premise that all animals are equally susceptible to traps, and that all animals are lifetime residents on the study area. Ample evidence proves the opposite as far as the Virginia opossum is concerned.

Some animals become habituated to traps, but the majority are caught only once. Wiseman and Hendrickson (1950) noted that some opossums passed about a meter from freshly baited traps without showing apparent interest. Opossums often shift dens and their home range (activity area) every few days. Therefore, long periods between recaptures may mean only that the animal was absent from the study area during that interval.

The difficulty in accurately aging postpouch opossums is the major reason why the age class composition of populations is so poorly known. Perhaps another reason is that investigators have mistakenly attributed a long life span to the Virginia opossum. Persuasive evidence indicates that opossums are not long-lived animals. Although adults in miserable physical condition (lacked parts of ears, tail, and limbs; had worn, broken, and carious teeth) studied by VanDruff (1971) appeared to be several years old, some had been tagged as juveniles the year before. Jurgelski and Porter (1974) considered the effective reproductive life as one year for females. They distinguished females in their second reproductive year by the worn and broken condition of the canines and by the wider skin flaps on the pouch (Petrides 1949). According to Hunsaker (1977) the longevity and reproductive activity of opossums is short compared with similar-sized mammals. He said that they were mature at 9 months and reproductively fit for at least 18 months, but probably not beyond three years. The effective breeding life of males is unknown. Most of the available life history information simply does not permit the determination of the age composition of adults.

Dispersal Patterns. Reynolds (1945) surmised that most opossums are nomadic, on the basis of recovering only 5 of 68 marked animals, the record of an adult male that moved 11.2 km in nine months, and the rapid repopulation of an area cleared of opossums by hunters.

Fitch and Sandidge (1953) found that young males, after becoming independent, tend to wander more widely than young females and tend to settle in areas removed from the mother's home range. All recaptured females marked as pouch young had moved, on the average, a little less than 365 m.

The longest single move recorded by Holmes and Sanderson (1965) was 1,738.5 m (5,700 ft) for an adult male. Ten captures for 8 juvenile females and 33 captures for 20 juvenile males on successive nights yielded average distances of 88 m (290 ft) and 162 m (530 ft), respectively. The 29 captures of 9 adult females on successive nights averaged 279 m (914 ft) apart; 5 captures of 4 adult males averaged 653 m (2,140 ft) apart. Some juvenile females stayed in the general area where they were born; three generations of females were caught in the same area over a period of 25 months.

Fitch and Shirer (1970) said that immatures wandered less widely than adults but occasionally make dispersive movements into new areas. Based on radiotelemetric observations, they found that young opossums progressively ranged farther as they matured

and that adult males moved more widely than adult females.

The longest straightline distance recorded by VanDruff (1971) was 8,050 m during a 15-month period by a male originally marked as a juvenile. An adult female moved 4,930 m in 10 days in June, a second was carrying 9 pouch young when she moved 3,560 m in 7 days, and a third carrying 10 pouch young traveled 3,460 m in 4 days. The longest single move recorded was that of an adult male who moved 3.2 km (2 mi) from one den to another in one night. Most opossums did not travel as extensively. Juveniles moved less than adults and some remained in an area for a few weeks before moving on. Males moved farther than females. The longest and most rapid moves occurred in late spring and early summer. Understanding the dispersal pattern of newly independent young was difficult, partly because the location where they last left the mother could not be determined.

Interspecific Interactions. DENS. Because opossums do not dig burrows, they must rely on ground dens dug by other animals. Opossums have been found sharing dens with an armadillo (Lay 1942), a woodchuck (Sandidge 1953), and a raccoon (*Procyon lotor*) (Stuewer 1943). In addition to the last two mammals, Pippitt (1976) found opossums sharing dens with cottontail rabbits. A complex burrow system in Michigan was simultaneously occupied by an opossum, a woodchuck, a raccoon, and a striped skunk (Brocke 1970). Shrews, mice, weasels, snakes, and other small vertebrates as well as a host of insects and other invertebrates may cohabit parts of the den used by opossums. Some dens are used communally. During the year, the same burrow may be used at different times by red fox, raccoon, woodchuck, opossum, striped skunk, cottontail rabbit, and gopher turtle (*Gopherus polyphemus*).

Shirer and Fitch (1970) found no apparent competition for dens by raccoons, striped skunks, and opossums. Striped skunks dug the ground dens used by the others. Brocke (1970) and Pippitt (1976) suggested that clearing a forest area resulted in higher numbers of woodchucks, striped skunks, and cottontail rabbits, thereby increasing the number of underground dens, which seem to be critical for the winter survival of opossums.

PREDATION. Opossums prey on a number of small vertebrates and invertebrates. They are known to kill cottontail rabbits and chickens, and take eggs and nestlings of ground-nesting birds including waterfowl. This marsupial is one of the few mammals that regularly preys on shrews and moles. Food habits studies, however, demonstrate that most of the other mammals and birds consumed are scavenged as carrion.

Distributional Limits. Minimally adequate habitat for opossums must include accessible water and a sufficient number of dens. Food usually is not a problem where these two requirements are met, except in regions where winter climate hampers foraging activity.

Tyndale-Biscoe (1973) used the $-7°$ C January

isotherm to predict the eventual northern distributional limit. Brocke (1970) suggested that the southern edge of the pine-hemlock ecotone approximates the northern distributional limit for the Virginia opossum in eastern North America. Brocke believed that the lack of foraging success imposed by a combination of the number of days with subfreezing temperatures and those with accumulated snow depths exceeding 28 cm was critical. He found that the northern distributional limit in Michigan coincided with a winter severity level of 70 days of enforced inactivity. Brocke's hypothesis explains the periodic reductions of populations along the northern periphery of the range following severe winters and suggests that the general warming trend over the past century has had as much to do with the northern expansion of the distribution of the Virginia opossum as have human activities in providing den sites and winter food.

FOOD HABITS

Dietary studies confirm omnivory in the Virginia opossum. Well known to scavenge carrion and garbage, and prone to use the most abundant foods available, opossums also eat quantities of grass and other green vegetation, seem to prefer maggots gleaned from rotting flesh over the putrid flesh itself, dig and consume mushrooms from beneath snow, kill and eat poisonous snakes with impunity, cannibalize their less fortunate brethren (thereby aiding survival during harsh northern winters), and eat great quantities of earthworms, insects, and other invertebrates. Yet food habits studies show that opossums do not live up to their reputation as rapacious raiders of chicken coops or as serious predators of rabbits, raccoons, pheasants, waterfowl, and other game.

Several studies have been based on stomach analysis alone. Others dealt with scats or various combinations of scat, stomach, and intestinal contents. The most extensive analysis was by Hamilton (1958), who examined 461 stomachs. His study demonstrated the need to segregate results on a monthly basis to understand food habits, food availability, and dietary requirements. A few studies grouped results by season, but most did not, except by circumstance, as when samples were collected during only part of the year.

Hamilton (1951, 1958) and others have pointed out the omissions possible when only feces are analyzed in food habits studies. Determinations of foods easily identified in the stomach may not be possible from gut contents or from feces, where green vegetation and many soft-bodied invertebrates may go unrecognized.

Food items should be reported by frequency and volume to avoid the biases inherent in each method. Frequency of occurrence may overemphasize a food item when it actually contributes little to the total caloric intake. Volumetric analyses may overrate food items present in large quantity in only one or a few individuals out of the total number examined.

Some reports are contradictory. For example, Worth (1975:517) retrieved "opossum fecal droppings consisting entirely of persimmon seeds held together by . . . undigested pulp" and, after finding that the seeds germinated, commented on the persimmon (*Diospyros virginiana*)-disseminating ability of the Virginia opossum. Reynolds (1945:373), however, said that 19 field-collected scats with remains of persimmons did not contain seeds. Examination of 63 scats recovered from 11 captive opossums fed exclusively on persimmons for a week revealed only 1 with a seed.

Several accounts of food habits mentioned paper, cellophane, leaves, and other trash and detritus found in the feces or intestinal tracts. Some accounts also noted egg shells, isolated feathers, and insect exoskeletons, much of the latter eaten during winter (Hamilton 1958). When gleaning for food, opossums ingest a variety of minute items. Although a few of these items may be consumed accidentally, perhaps most are leaves that held clusters of insect eggs, or are blood- or grease-soiled dirt, leaves, or trash. Except when fighting a trap, opossums probably do not eat seemingly extraneous items unless they contain something desired as food.

Thus far, food habits studies have not explored the possible effect of sex, age, or reproductive status on the diet of the Virginia opossum. The diets of Latin American populations are unknown.

Audubon and Bachman (1851) identified the foods of opossums as corn, chestnuts, acorns (*Quercus* sp.) and other nuts, small tubers, young briar (*Smilax* sp.) shoots, blackberries (*Rubus* sp.), wild cherries (*Prunus* sp.), persimmons and other fruits, mice, cotton rats (*Sigmodon* sp.), bird eggs, nestling birds, and broods of young rabbits (*Sylvilagus* sp.). Sperry (1933) reported an opossum in North Carolina that had eaten a red bat (*Lasiurus borealis*). Additional information on the foods consumed by opossums has been reported for specific states.

Michigan. Dearborn (1932) gave volumetric percentages of foods consumed by 40 opossums as insects 30.4, birds 24.7, and mammals 23.2. Two-thirds of the mammalian remains consisted of opossum. Other animal food items of lesser importance were crayfish (*Cambarus* sp.), snakes, bird eggs, and frogs.

Stuewer (1943, an analysis of 15 stomachs and 9 scats) found that opossums had eaten rabbits, short-tailed shrews (*Blarina brevicauda*), mice, a red squirrel (*Tamiasciurus hudsonicus*), chicken, pheasant (*Phasianus colchicus*), frogs, caterpillers, and small slugs. Plant food items included corn, buckwheat, berries, wild grape (*Vitis* sp.), and several kinds of seed.

Taube (1947) examined the stomach contents of 6 opossums collected in September and 49 collected in November and December 1941 in his evaluation of the predatory status of Michigan populations. He added data on 78 stomachs collected in the fall of 1933 and 8 collected in the spring of 1934. Eastern cottontail (*Sylvilagus floridanus*) made up the greatest volume (31 percent) and occurred in 19 percent of the stomachs. The autumn 1933 sample contained rabbit in 40 percent (12 percent by volume) of the stomachs. Other mammals in order of their importance in the diet were opossum (9 percent by volume, in 4 stomachs), vole (*Microtus* sp., in 6 percent of the stomachs), eastern mole (*Scalopus aquaticus*), shrews, fox squirrels (*Sciurus niger*), and striped skunk (*Mephitis mephitis*). Other animal foods were pheasant (1 stomach), chicken (4 stomachs), miscellaneous birds, snakes, amphibians (frogs, toads, and salamanders), insects, snails and slugs, and earthworms. Of the autumn 1941 opossums, insects were in about 90 percent of the stomachs. Volumes of grasshoppers were 32 percent in September, but only 4 percent in November and December, although present in 88 percent of the samples. Earthworms occurred in 27 percent of the stomachs and made up 8 percent of the food consumed. Plant materials included fruit (about 11 percent of volume, in over 50 percent of the samples) and grass (5 percent of contents, but in 69 percent of the stomachs). Rabbits and pheasants were eaten in the fall, probably as carrion.

Brocke (1970) examined 20 stomachs collected in January, February, and March. He found remains of striped skunk, muskrat (*Ondatra zibethicus*), cottontails, white-tailed deer (*Odocoileus virginianus*), opossums, shrews, mice, a garter snake (*Thamnophis* sp.), a leopard frog (*Rana pipiens*), insects, earthworms, grass and plant fibers, and unidentified mammals. Remains of mammals, much of which was carrion, occurred in 85 percent of the stomachs; plant material in 35 percent.

Texas. Lay (1942) estimated the volumetric percentages of the following food items recovered from 16 opossums taken in September in eastern Texas: insects and worms 45, fruit 11.8, green leaves 11.0, leaf and log litter 10.6, mammals 7.0, acorns 4.7, birds 4.3, crayfish 3.3, snails 0.8, and trace amounts of cellophane and grass seeds.

Wood (1954) analyzed the contents of 23 scats and 25 digestive tracts of opossums collected in oak woodlands of east Texas and recovered 39 food items. Thirty-six of these were in digestive tracts, 26 more than were recorded from scats, indicating that digestive tracts (particularly stomachs) contain more identifiable food items than scats. The volumetric percentages were: plants (fruit and green vegetation) 44.8, insects 25.0, mammals 14.9, amphibians and reptiles 7.4, birds 3.8, unidentified remains 3.2, and a trace of noninsect invertebrates. Cottontails were the most common mammal identified (in 12 percent of the digestive tracts and 4.3 percent of the scats). Other mammals identified were opossums, hispid cotton rats (*Sigmodon hispidus*), and white-footed mice (*Peromyscus leucopus*). Insects, second in importance as food, occurred in 88.0 percent of the digestive tracts and 69.5 percent of the scats. Copperheads (*Agkistrodon contortix*) were the most common reptile in the diet (5.7 percent by volume). Acorns (31.4 percent by volume) were the most important plant food. Other identified plant items were persimmons, pears, blackberries (*Rubus* sp.), French mulberry (*Morus* sp.), grass, watermelon, grape, and hackberry (*Celtis occidentalis*). Although poorly represented in the diet,

birds included the Yellow-Bellied Sapsucker (*Sphyrapicus varius*), Mourning Dove (*Zenaidura macroura*), Spotted Towhee (*Pipilo erythrophthalmus*), and chicken.

Missouri. Reynolds (1945) identified 69 food items in 259 scats of opossums from central Missouri collected between September and May. He listed the percentage frequencies of the following food categories as: insects 87.6, fruits 50.6, noninsect invertebrates 32.4, mammals 28.2, reptiles 18.9, grains and seeds 12.7, and birds and bird eggs 8.9. Reynolds also recorded 52 food items in the stomach contents of 68 opossums collected during the six months from December to May. Of these, he tabulated the following categories by percentage volume: insects 34.2, mammals 32.3, reptiles 10.0, grain and seeds 7.3, fruits 6.8, birds and bird eggs 4.9, noninsect invertebrates 4.5. With the exception of crickets (Gryllidae), squash bugs (Coreidae), and stink bugs (Pentatomidae), more fall and spring scats contained insects than did winter scats. Fruits generally were more commonly eaten during the autumn. Land snail frequencies were 39 and 26 percent for fall and spring scats, respectively. Cottontail rabbits, moles, and fox squirrels were commoner in winter scats than in fall or spring scats. Reptiles were most numerous in the spring, and grains and seeds were most frequently encountered in the fall scats, except corn, which was present in 29 percent of the winter sample. Chicken or eggs, although present in only 12 out of 259 scats, were more common in the winter. Cottontail rabbits were the most important food by volume (12.9 percent) in the stomachs. Next in order of importance by volume were carabid beetles (8.7), scarabaeid beetles (7.5), corn (7.3), eastern mole (6.4), stink bugs (5.3), opossums (4.9), squash bugs (4.8), and short-horned grasshoppers (Locustidae) (4.4).

Pennsylvania. Gifford and Whitebread (1951) found insects, earthworms, mice, an anuran, rabbit hair, snake scales, feathers, and pokeweed (*Phytolacca americana*) seeds in four opossum stomachs from south central Pennsylvania.

Grimm and Roberts (1950) noted the percentage occurrence of the following foods in 18 fall and winter stomachs provided by trappers from southeastern Pennsylvania: insects and spiders 83.3, fruits 44.4, cottontail rabbits 27.7, birds other than poultry 22.2, mice 22.2, carrion 22.2, shrews and moles 11.1, poultry 11.1, woodchuck (*Marmota monax*) 11.1, frogs and toads 5.1, and earthworms 5.1. They believed that most if not all of the rabbit, poultry, and woodchuck was carrion. Some of these items were used by trappers for bait.

Blumenthal and Kirkland (1976) examined the stomach contents of 62 opossums from the Cumberland Valley. The frequency and volumetric percentages, respectively, of the 15 food categories they listed were: mammals 98, 26; grasses and grains 90, 13; insects 85, 9; leaves 78, 8; fruits and seeds 76, 12; sand and stones 72, 7; stems 44, 4; earthworms 43, 8; mollusks 39, 4; birds 30, 4; egg shells 8, < 1;

arachnids 8, < 1; fish 6, < 1; trash 6, < 1; and amphibians 1, < 1. Mammals, grasses and grains, and insects were found in 75 percent of the stomachs. Mammals, grasses and grains, and fruits and seeds made up 51 percent of the total food volume, and analysis by season showed the importance of these foods throughout the year. Seasonal shifts in some foods probably reflected availability (e.g., earthworms in the spring, and birds and insects in the summer). Weed seeds were most commonly found in the winter samples, and fruits and berries were most common in the summer and fall. Mammals were the most important food at all seasons and included the Virginia opossum, short-tailed shrew, white-footed mouse, house mouse (*Mus musculus*), and meadow vole (*Microtus pennsylvanicus*).

Iowa. Wiseman and Hendrickson (1950) examined 81 samples of fecal material recovered at traps and 6 scats found in the field in southeastern Iowa during winter (January to March) and spring (April to June). They listed the percentage frequencies of the following food categories as: insects 87, grain and plant items 66, fruits 31, crayfish 24, mammals 22, millipedes 13, birds and bird eggs 8, and reptiles 1. In order of importance, the most common winter foods were corn, ground beetles (Carabidae), grasshoppers, miscellaneous seeds, ground cherry (*Physalis* sp.), unidentified insects, and mammals (rabbit). The most frequent spring foods in order of importance were ground beetles, crayfish, mulberry (*Morus* sp.), corn, mammals (voles), millipedes, and miscellaneous seeds. Chicken feathers, recovered from the scats of one opossum, were believed to have been eaten with chicken feed.

New York. Hamilton (1951, 1958) presented the most detailed analysis of food habits published to date. The large number of stomachs (180) whose contents were reported in 1951 was increased to 461 by the time of his 1958 report, which included more than 118 identified food items. Percentages of the frequency and volume, respectively, of the 16 major food categories he listed are: insects 42.9, 7.9; mammals 42.1, 22.6; green vegetation 38.6, 8.1; fruits 33.4, 14.1; earthworms 23.6, 10.3; amphibians 23.2, 9.3; birds 18.9, 7.2; mollusks 15.6, 3.0; reptiles 10.8, 5.6; carrion 7.8, 6.0; grain and mast 3.5, 1.9; millipedes 2.2, trace; centipedes 1.3, trace; crustaceans 0.3, 0.9; fungi 0.3, trace; and undetermined 4.6, 3.0. Nearly three-fourths (72.3 percent) of the volume consumed was mammals, fruit, earthworms, amphibians, green vegetation, and insects. The frequencies of these six major food categories varied from 23.0 to 42.7 percent. The frequency of mammals varied seasonally from a high of 77.8 percent in December to a low of 25.0 percent in August; of insects, from 77.1 in August to 10.3 in March; of fruits, from 42.3 in October to 5.6 in April; of amphibians, from 50.0 in August to zero in February; of earthworms, from 51.7 in March to 9.1 in February; and of green vegetation, from 58.6 in March to 31.3 in August.

Principal insects included grasshoppers, beetles, and lepidopterous larvae. The most frequently encountered mammal was the meadow vole (in 17.6 percent

of the stomachs); next was the short-tailed shrew (8.0 percent). Green grasses, clover, and other ground vegetation were a natural and important part of the diet and were found in 178 animals. Grapes were the main fruit eaten; much of the fruit was dried, and fallen berries and drupes were recovered long after their time of ripening. Earthworms were present in 20 percent or more of the stomachs in January, March, April, October, November, and December. Toads (*Bufo* sp.) were found in 10 percent of the stomachs. Most of the birds consumed were probably carrion, except for nestlings, which were probably taken from their nests. Both slugs and snails were eaten. The snapping turtle (*Chelydra serpentina*) and painted turtle (*Chrysemys* sp.) that had been eaten were recently hatched. Carrion was probably underestimated, and many of the animal foods may have been gleaned from roadsides or were unrecovered hunter kills. Grains and acorns, found in 3.7 percent of the stomachs, were not as important as other studies have indicated. Hamilton did not report the frequency of trap debris, dirt, stones, dead leaves, wax paper, and aluminum foil because these were probably not ingested purposely.

According to VanDruff's (1971) data on predation on waterfowl nests in the Montezuma Wildlife Refuge in central New York, opossums were responsible for 15 (24 percent) of 63 destroyed duck nests over a three-year period. Yet 437 (48 percent) of 916 animals killed during nine years (1961 to 1969) of the toxic-egg predator reduction program at the Montezuma Refuge were opossums.

Maryland. Llewellyn and Uhler (1952) examined 37 stomachs and 66 scats of opossums from the Patuxent Wildlife Research Center in southern Maryland. A variety of insects, vertebrates, and plant items as well as snails and millipedes were recovered. Plant foods included persimmons, wild grapes, apples, pokeberry (*Phytolacca americana*), briar, corn, beechnuts (*Fagus grandifolia*), nightshade (*Solanum nigrum*), cherries, dewberries (*Rubus* sp.), and blackberries. The last two made up about 20 percent of the foods consumed in June and July, whereas the first three were among the most important fall foods. Animal food matter (about one-third insects) was 86 percent of the estimated volume.

Kansas. Sandidge (1953) based his analysis of opossum food habits on field observations and the examination of 62 digestive tracts collected from September to March. The percentages of volume, weight (dried), and frequency, respectively, of the major food categories were: insects 42.2, 42.6, 86.7; mammals 41.4, 39.4, 33.3; fruits 8.6, 10.3, 13.3; and birds 3.1, 2.8, 21.7.

The most common insects were carabid beetles and short-horned grasshoppers, the only insects found in winter samples. Other insects present were metallic wood borers (Buprestidae), lady bird beetles (Coccinellidae), horned passalus (Passalidae), lamellicorn beetles (Scarabaeidae), carrion beetles (Silphidae), stink bugs, assassin bugs (Reduviidae), and crickets. Cottontail rabbits were the most important mammalian food. Other mammals identified were the white-footed mouse, muskrat, prairie vole (*Microtus ochrogaster*), eastern mole, and Virginia opossum. Flesh-eating beetles (Silphidae) accompanied rabbit, muskrat, and opossum in 12 of 19 occurrences, thereby indicating carrion.

Fruits occurred in eight digestive tracts but were greatest (pears) in volume and weight in late fall and winter. Fruits and seeds included winter wheat, goosefoot (*Chenopodium* sp.), wild grape, apple, and pear.

Chicken (feathers, bone, and egg shells) was in 9 of the 13 digestive tracts containing birds, but was likely gleaned from garbage or eaten as carrion. Other birds (one each recorded) were the Yellow-shafted Flicker (*Colaptes auratus*), Cardinal (*Richmondina cardinalis*), meadowlark (*Sturnella* sp.), and starling (*Sturnus vulgaris*).

The five-lined skink (*Eumeces fasciatus*), De-Kay's snake (*Storeria dekayi*), worm snake (*Carphophis amoenus*), and leopard frog were the only herpetozoa identified. Centipedes made up 1.6 percent of the total weight and 1.6 percent of the volume, and occurred in 38.3 percent of the samples. Land snails (*Triodopsis albolabris*) and crayfish (*Orconetes* sp.) were identified in one and two stomachs, respectively.

Fitch and Sandidge (1953) examined scats collected on the University of Kansas Natural History Reservation from August to November 1951 and in January, September, October, and November 1952. Wild fruits (principally wild grape and hackberry) made up the bulk of the diet. Blackberry was the important summer food. Crayfish was the most common animal item found. Other foods were wild plum (*Prunus americanus*), wild crabapple (*Pyrus ioensis*), cherry, corn, insects (beetles, cicadas, grasshoppers, yellow jackets—*Vespula* sp.), Blue Jay (*Cyanocitta cristata*, one feather), fox squirrel, eastern cottontail, copperheads, and a snail. A large male opossum was surprised as it killed and had begun to eat a 150-g young rabbit.

Illinois. Stieglitz and Klimstra (1962) reported on the contents of 131 opossum digestive tracts collected from 1958 through 1960 during the six-month period from August to February in southern Illinois. Of the 75 animal and 66 plant items recorded, only 24 animal and 11 plant food items contributed 0.5 percent or more by volume to the diet.

Animal food items made up 76.2 percent of the total volume, plant foods 23.8 percent. Mammals were 48.7 percent of the total volume. Opossums filled four stomachs and contributed 16.3 percent of the volume. Cottontail rabbit was second, with 14.7 percent of the volume, and occurred in 15.3 percent of the digestive tracts. Other mammals included the prairie vole, deer mouse (*Peromyscus maniculatus*), gray fox (*Urocyon cinereoargenteus*), striped skunk, short-tailed shrew, eastern mole, and woodchuck. Other vertebrates included chicken (7.1 percent of volume), grackle (*Quiscalus quiscala*), meadowlark, domestic pigeon, towhee, junco (*Junco* sp.), cardinal, Carolina wren

(*Thryotherus ludovicianus*), blue racer (*Coluber constrictor*), turtle, frogs, and toads. Insects appeared in 93.1 percent of the tracts but contributed only 6.3 percent of the volume. The most important insects were scarabaeid beetles (larvae), short-horned grasshoppers, and lepidopterous larvae. Fly larvae appeared in several tracts but not always in association with carrion. Earthworms appeared in only five digestive tracts.

Plant food items were dominated by persimmon (8.1 percent of total volume), which ranked third among all foods. Other plant items were corn, grasses, pokeberry, wild grape, and other fruits.

The 10 top-ranking foods by volume were opossum, cottontail rabbit, persimmon, chicken, prairie vole, pokeberry, grackle, gray fox, frogs, and scarabaeid larvae. The ranking by percentage frequency was: grasses 82.4, short-horned grasshopper 54.2, opossum 52.7, ground beetles 38.9, snails 31.3, pokeberry 25.2, nightshade 25.2, stinkbugs 22.9, persimmon 21.4, and cottontail rabbit 15.3. Most of the seasonal variation was seen in the decrease in plant materials from fall through winter and the increase in mammalian food consumed during the same period.

Wisconsin. Knudsen and Hale (1970) examined the contents of 151 opossum stomachs represented by at least 5 from each month of the year. They found 65 different identifiable food items, which they grouped in the following major categories (values are percentages of occurrence and total volume, respectively): mammals 25, 41; birds 12, 24; invertebrates 19, 10; reptiles, amphibians, and fish 10, 12; plants 7, 6; and garbage, trash, and litter 27, 6.

Cottontail rabbits and several species of mice were the important mammalian foods. Other mammals were shrews, moles, grey squirrel (*Sciurus carolinensis*), fox squirrel, muskrat, rats (*Rattus* sp.), dogs, house cats, and swine. Most of the animals were taken as carrion, as confirmed by the decayed condition of the flesh and the presence of maggots in the stomachs. Next in importance were birds, of which chicken was most abundant. Other birds represented were ducks, bobwhite quail (*Colinus virginianus*), pheasant, domestic pigeon, screech owl (*Otus asio*), crow (*Corvus brachyrhynchos*), and several species of songbirds. Game birds were in 10 stomachs, but in small amounts.

Invertebrates included earthworms, crayfish, snails, and several kinds of insect. Earthworms, although absent from the winter samples, were the most frequent food in the other three seasons, and in the annual total, where they made up 8 percent of the volume.

The reptiles, amphibians, and fish included small frogs, toads, garter snakes (*Thamnophis* sp.), and a number of fish believed to have been discarded by fishermen. Plant items were fruits, grains, vegetables, and mast. Apples were the important food in this group.

Mammals and insects were the important winter foods. Most of the garbage was noted during the winter, indicating the raiding of refuse and garbage by opossums when foods are scarce. Earthworms, insects, songbirds, frogs, snakes, and chickens were the most commonly eaten foods in the summer. There was no indication that opossums were a serious threat to game birds.

Indiana. Whitaker et al. (1977) identified 71 categories of food from the contents of 83 stomachs. The volumetric percentages of the following major food categories were: mammals 22.2, birds 21.3, other vertebrates 9.0, insects 11.7, other invertebrates 13.6, vegetation 19.0, and garbage 3.1. Most of the vertebrate remains, which included shrews, voles, opossums, chipmunks (*Tamias striatus*), cottontail rabbits, squirrels, deer mice, house mice, bobwhite quail, robins (*Turdus migratorius*), chicken, flickers, toads, turtle, and salamander, were probably acquired as carrion. Carabid beetles, crickets, grasshoppers, and lepidopterous larvae were the most important insects in the diet. Earthworms, at 10.9 percent of the volume and in 34.9 percent of the stomachs, were exceeded in importance only by unidentified birds and unidentified plants. Plant food items were led in importance by unidentified plants; other plant items included grass, wheat, corn and other grain and seeds, and apples and other fruit.

BEHAVIOR

Activity. Opossums are nocturnal except in winter, when they are occasionally seen in the daytime during warm spells in otherwise bitter cold weather (Fitch and Shirer 1970; Brocke 1970; Pippitt 1976). McManus (1971) found seasonal shifts in general activity; activity was greatest in spring and summer. He also noted seasonal variation in the relative frequency of certain activity states. Greatest locomotor activity was noted in the summer and least in the fall. Ingestive activity was greatest in the fall; nest-building activity was greater in the fall and winter than at other times of the year. Hunsaker and Shupe (1977) discussed other aspects of opossum activity patterns.

Death Feigning. Feigning death, although not unique to opossums, is so clearly associated with the animal that the term *playing possum* has become idiomatic. Opossums assume a highly stereotyped catatonic state when faced with inescapable threat situations, as when attacked by dogs or when approached if caught in a steel trap. Tactile stimuli are usually required to produce the response. The opossum falls on its side and lies still with the body slightly flexed. The corners of the mouth are drawn back from the slightly opened jaws, from which drools copius saliva. The animal often defecates and discharges a greenish, foul smelling substance from its anal glands. The eyes remain open and the ears may twitch at sharp sounds, otherwise there is little or no response to movements or prodding. Recovery may be immediate, but appears to be slow as the animal assesses the situation. Added stimulation during this time causes the animal to return to its death-feigning state. Habituated captives often do not feign death.

Behavioral and electrocardiographic features of death feigning were examined by Francq (1969, 1970). He found no significant differences between alert and death-feigning states in heart rate or height of T-wave deflection, QRS-wave configuration, and QT interval.

Locomotion. The Virginia opossum has terrestrial, arboreal, and aquatic locomotor patterns. Limb posture and the characteristic quadrupedal type of terrestrial locomotion were studied cineradiographically by Jenkins (1977). Although not agile or rapid climbers, arboreal locomotion is aided by the opposable hallux, prehensile tail, and palmar and plantar friction ridges. Opossums climb trees to forage, escape danger, and rest during the evening, and to seek tree holes and arboreal leaf nests as daytime retreats. Opossums are slow, strong swimmers and can swim underwater (Moore 1955; McManus 1970). Wilber and Weidenbacher (1961) found that a "wetted" female swam 430 minutes in a 22° C water bath, indicating the ability to swim across moderately large expanses of water. Locomotor speed and behavior were reviewed by Hunsaker and Shupe (1977).

Grooming. The hind foot is extensively used in grooming the back of the head, ears, and sides and upper parts of the body. The hair is combed by the hind toes and claws, which are licked clean after each bout. Opossums lick the forefeet and then rub them over the face and snout in a catlike manner. The abdominal and genital areas are groomed by licking. Females extensively lick the pouch, especially if young are present. Males lick the penis after sexual activity. Additional grooming behavior was extensively covered by Hunsaker and Shupe (1977).

Nest Building. Nest-building materials are passed under the body and packed into the coiled tail, in which they are carried to the nest (Pray 1921; Smith 1941). Hopkins (1977) noted this behavior in 88- to 97-day-old captives.

Reynolds (1945) found that opossums enlarged abandoned crow and squirrel nests by adding grass, leaves, and corn husks. He also used the presence or absence of nest-building materials as a criterion for identifying ground dens used by opossums. Pippitt (1976) said that fall and winter dens were plugged with leaves. The plug in fall dens was loose, but the leaves and other materials were so tightly packed in the entrances of winter dens that he was unable to probe the next cavity to sample CO_2 levels.

Vocalizations. The Virginia opossum makes four distinct vocal sounds (clicking, hissing, growling, and a screech) and all may be used in aggressive interactions. Clicking sounds, uttered by young and adults, are used by males in mating behavior, during aggressive encounters between adults, and by females in the presence of young (McManus 1970). These clicks, described by Hartman (1923a:354) as having a "metallic ring," were discussed by Hunsaker and Shupe (1977), who believed that the sound was made by the upper lips. Although some clicklike sounds may be produced by the lips or other parts of the mouth, I believe that the rapid, metallic-sounding clicks reported most often are made by clashing the canines together. Adults, especially males, have flattened areas worn on the anterior faces of the upper canines. These are evident in the lateral view of the rostrum in figure 1.7, even though this skull is of a young adult (age class 4). Hunsaker and Shupe (1977) have described the other opossum vocalizations and the contexts in which they are used.

Social Behavior. Opossums are basically solitary and most interactions with other animals are avoided except when mating, caring for young, and personal defense are involved.

COPULATION AND SEXUAL BEHAVIOR. Copulation has been described by several observers and perhaps most recently in detail by McManus (1967) and Hunsaker and Shupe (1977). The male is tolerated only briefly by the female. While in pursuit of a female, the male clicks continuously. When she is receptive the male mounts by placing his forefeet on her shoulders, grasping her neck in his teeth and shifting his whole weight onto her, and clasping each of her hind legs with one of his feet. The curious aspect of this behavior is the almost universal observation that the mating pair then tumble over onto their right sides. In the instances when the pair remains upright or falls to their left, subsequent examination failed to find sperm in the female's genital tract (Reynolds 1952). After copulation, which lasts approximately 20 minutes, all additional attempts to mount the female are rebuffed. King (1960) noted an example of Davian behavior (Dickerman 1960) when he observed a male attempting to copulate with a dead female lying in the middle of a highway.

Pippitt (1976) suggested that males may produce a sex-attractant pheromone, which assists an estrous female in locating a potential mate. He reached this conclusion after noting that three radio-tagged females moved relatively long distances during unfavorable weather to dens occupied by males where mating evidently took place.

MATERNAL BEHAVIOR. Although the female supplies milk and provides warmth and protection, her only active contribution to the welfare of her offspring from birth until the time they begin to leave the pouch is cleaning the pouch and contained young. The young travel to the pouch, attach to nipples, and suckle milk without assistance. Each young remains at the nipple during most of this time, anchored by the swollen nipple tip held in the mouth with the aid of a series of small projections on the lips, tongue, and palate. After the young are able to release the nipple and move by themselves, the female responds with the clicking vocalization, which serves to orient the young to her. Weaning may be largely a passive activity by the female. The young probably wean themselves as they acquire more and more interest in solid food and less interest in suckling (Reynolds 1952). The normal lactation period of about 100 days was lengthened by 1.5 months when Reynolds (1952) substituted 60-day-old young for the 94-day-old young of one of his captive

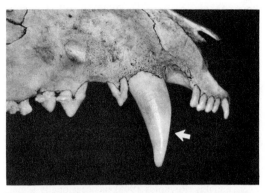

FIGURE 1.7. Right side of the rostrum of a male Virginia opossum (*Didelphis virginiana*) (USNM 505343), showing the worn, flattened anterior face of the upper canine.

females. Another female, after having been penned for a week without her young and with a drying pouch, adopted and successfully nursed another's litter (Wiseman and Hendrickson 1950).

AGONISTIC BEHAVIOR. The Virginia opossum either avoids or shows aggressive displays toward another individual. Males usually are aggressive toward other males but rarely toward females. Females are tolerant of each other unless one is in estrus, which elicits aggression by nonestrous females (Reynolds 1952). Females aggressively repulse males unless sexually receptive, and females have killed males much larger than themselves when penned together in small enclosures. Reynolds (1952) described a fighting dance or dominance display by aggressive males and females. The shuffle-walk dance is silent but is preceded and followed by clicking.

Opened-mouthed threat behavior and aggressive growling and screeching are common in male encounters. Males stand face to face, the head and shoulders of each weaving from side to side as they hold the mouth open and flatten the ears and vibrissae against the head. Each proceeds to feint and lunge at the other. This bluffing routine either is followed by the turning away and retreat of the submissive male or accelerates into savage biting, snapping, and slashing until one gives ground or is killed. McManus (1974) said that fighting may result in death feigning by the weaker individual. At least among captive opossums, death feigning may not be an effective strategy to avoid being killed because of the opossum's propensity for cannibalistic behavior, even when other food is available. Active fighting and cannibalism occur among weaned littermates if they are housed together; young have been known to kill their mother, and vice versa (Raven 1929).

When wild opossums are caught by hand, they growl, defecate, and emit a foul-smelling greenish substance from the pair of anal glands. Although the Virginia opossum is more docile and easier to handle than the common opossum (*D. marsupialis*) of Latin America, it still bites.

SCENT MARKING. Reynolds (1952) described marking by males. The opossum alternately licks and rubs the sides of his head against an object. Although Reynolds noted this behavior throughout the year, marking reached its height at the time of breeding and was elicited by the odor of other males and by the odor of females in estrus. Other males repond to marked objects by licking and rubbing them and by performing the fighting dance. Females also show interest in the marks and may engage in a similar, but less vigorous dance. The marked sites probably serve to advertise the presence of a male and may attract estrous females.

COHABITATION. Opossums usually den alone; however, females (Reynolds 1945) or occasionally a male and female may share the same den (Lay 1942; Fitch and Shirer 1970; Pippitt 1976). Recently weaned littermates may share occupancy for a few days or weeks.

MORTALITY

Several authors have commented on the poor or weakened appearance of many opossums. Opossums in regions subjected to severe winter weather are often missing half or more of the ears and tail due to frostbite and subsequent sloughing of the frozen tissue (Smiley 1938; Stuewer 1943; Wiseman and Hendrickson 1950; Hamilton 1958). Opossums often have excessive numbers of cuts, scratches, ripped ears, lost toes, broken teeth, and broken bones as well as internal parasites and ectoparasites (Black 1935; Lay 1942; Fitch and Sandidge 1953). Fitch and Sandidge (1953) commented that the toe-clip wounds of marked animals were slow to heal, but did not say that marking contributed to mortality. The study of populations some-

times causes death, usually through exposure to cold (Fitch and Sandidge 1953) or heat (Llewellyn and Dale 1964) while the opossum is in a trap.

Opossums are preyed on by dogs, coyotes, foxes, raccoons, bobcats, raptors, and large snakes. Of these, dogs and horned owls are probably the most serious predators. Rand and Host (1942) mentioned a skull found in an eagle's nest in Florida. Some opossums, especially juveniles, may be preyed on by larger adults. Nevertheless, it is likely that losses through natural predation are of little consequence.

Hunting and trapping activities take a high toll (figure 1.8), and there is a tradition among farmers to kill all opossums encountered because of their reputation as destroyers of poultry. Although human activities have generally stimulated the expansion of opossum populations, the rapid urbanization process and concomitant construction of roads and highways during the past half-century have both reduced available habitat and, without doubt, encouraged the most serious cause of mortality, motor vehicle traffic.

Interest in the Virginia opossum as an experimental animal and as a reservoir of zoonoses infecting humans and domestic animals has resulted in a large number of reports on opossum diseases and parasites. Most of these reports have been summarized by Barr (1963) and Potkay (1970, 1977). Those reports not reviewed by 1977 are mainly taxonomic (e.g., Joseph 1974; Pence 1973; Pence and Little 1972; Prestwood 1976), descriptive (Long et al. 1975), or experimental (Stone et al. 1972).

Diseases. VIRAL DISEASES. Viral diseases found in the Virginia opossum include rabies and a number of arboviruses. Experimental infection with rabies virus is difficult, and young opossums are more susceptible

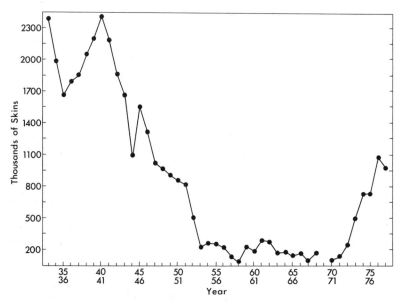

FIGURE 1.8. Reported annual fur harvest of the Virginia opossum (*Didelphis virginiana*) in the United States, from the 1933/34 season to the 1976/77 season. From Deems and Pursley 1978; D. Pursley unpublished data; and records of the annual fur catch in the United States from the files of the National Fish and Wildlife Laboratory, U.S. Fish and Wildlife Service, Washington, D.C.

than adults. The natural incidence of rabies is low and opossums are not considered to be important reservoirs of the disease. Twelve other viral diseases found or tested in opossums (Potkay 1977) include yellow fever, herpes virus (pseudorabies), and a series of encephalitis-causing viruses. Generally refractory to most of these, infection by pseudorabies, vesicular stomatitis, and B virus proves fatal to opossums.

BACTERIAL DISEASES. Mycobacterial (tuberculosis), *Borrelia* (relapsing fever), *Pasteurella* (tularemia), and *Salmonella* (enteritis, typhoid fever) infections are known in opossums. Captives with bacterial endocarditis harbored streptococcal bacteria. Other diseases in opossums include *Bordetella, Stapholococcus, Proteus, Aerobactor, Pseudomonas,* and leptospiral infections. Opossums are considered an important natural reservoir of leptospirosis in wildlife and humans.

RICKETTSIAL DISEASES. Opossums are natural hosts for spotted fever and murine endemic typhus rickettsiae and often show no clinical symptoms of these diseases.

PROTOZOAN DISEASES. Virginia opossums are known to harbor *Toxoplasma, Sarcocystis, Besnoitia, Coccidia, Trichomonas,* and *Trypanosoma.* Animals carrying *Trypanosoma cruzi* (Chagas's disease) have been found in Texas, Georgia, Florida, and Alabama.

MYCOTIC DISEASES. Ringworm and histoplasmosis infect opossums.

NUTRITIONAL DISEASES. Captive opossums often suffer from cage paralysis (usually rickets or osteomalacia) if they receive an inadequate diet.

OTHER DISEASES. A number of neoplasms have been reported in the Virginia opossum. The urogenital, respiratory, and reticuloendothelial systems are those most commonly afflicted by tumors. Fibrous osteodystrophy of the jaws was described by Long et al. (1975). Lesions have been noted in all parts of the digestive, urogenital, respiratory, cardiovascular, musculoskeletal, hematopoietic, central nervous, and endocrine systems. Many of these lesions are presumed to be caused by parasites. The occurrence of these conditions was summarized in tabular form by Potkay (1977).

Response to Snake Venoms. The remarkable resistance of the Virginia opossum to envenomation by poisonous snakes was reported on by Kilmon (1976) and Werner and Vick (1977). Kilmon stressed 15 anesthetized opossums by subjecting them to bites by the eastern diamondback rattlesnake (*Crotalus adamanteus*), timber rattlesnake (*Crotalus horridus*), cottonmouth moccasin (*Agkistrodon piscivorus*), Russell's viper (*Vipera russelli*), and common Asiatic cobra (*Naja naja*). None of the opossums showed any tissue reaction other than the fang punctures. There was a drop in arterial blood pressure of 5 mm Hg (from a norm of 140/105 to 135/100 mm Hg) in another anesthetized, 4.1-kg opossum immediately following the bite of a cottonmouth moccasin. The heart rate increased from 160 to 180 per minute, but the respira-

tion rate was unchanged (12 per minute). Blood pressure recovered after 10 minutes, and by 30 minutes the heart rate had returned to normal limits. A smaller (3.6-kg) opossum received a dose of cottonmouth moccasin venom in the caudal vein equivalent to five times the lethal dosage for a 15-kg dog. A drop in blood pressure and a rise in heart rate were the immediate responses; recovery was complete within 30 minutes. The opossum was then given almost double the first dosage, with similar results. No organ damage was found 24 hours later, when this animal was killed and autopsied.

Werner and Vick (1977) stressed opossums by actual snake bite or by intravenous and intramuscular injection of 4 to 60 times the dosage normally lethal to susceptible mammals. The venoms used were from pit vipers (Crotalidae), cobras and coral snakes (Elapidae), puff adder (Viperidae), and sea snake (Hydrophidae). All opossums, with the exception of those given crotalid venoms, died within 24 hours. Those given crotalid venoms in high dosages responded with small changes in blood pressure and heart rate, but all recovered in 15 to 30 minutes. None died or showed signs of local hemorrhaging, swelling, or tissue damage. Apparently, only adults were used in these experiments. Although young opossums may not be as immune to snake venom as the adults, the general conclusion is that opossums are unaffected by the bites of pit vipers, but are sensitive to venoms of other poisonous snakes. Therefore, opossums can safely prey on copperheads, rattlesnakes, and water moccasins.

Ectoparasites. Opossums are host to a number of ticks, mites, bugs (Hemiptera), lice, fleas, and pentastomids. Many of these ectoparasites are vectors of diseases such as spotted fever, typhus, and Chagas's disease.

Although Hamilton (1958) said that opossums were relatively free of ectoparasites, Hock (1952) considered them to be one of the most heavily parasitized mammals in North America. Morlan (1952) examined 349 opossums from Georgia and recorded two species of lice, nine of fleas, nine of mites, five of ticks, and two of chiggers. The cat flea (*Ctenocephalides felis*) and dog tick (*Dermacentor variabilis*) infested the greatest number of individuals. The dog tick was also found on opossums in Texas (Lay 1942), Kansas (Fitch and Sandidge 1953), Missouri (Reynolds 1945), and Indiana (Whitaker et al. 1977). Other records were summarized by Potkay (1977). Blumenthal and Kirkland (1976) noted seasonal differences in ectoparasite loads. Highest infestations were on winter animals, a finding that they interpreted as the result of cold season denning and cohabitation with other mammals. Whitaker et al. (1977), noting that some of the ectoparasites found are considered specific on other hosts, suggested that opossums acquired these parasites because of their carrion-eating habit. Experimental transfer of sarcoptic mange mites (*Sarcoptes scabiei*) to opossums from red foxes (*Vulpes vulpes*) failed to produce infestations (Stone et al. 1972).

Endoparasites. At least 30 nematodes, 26 trematodes, 7 cestodes, and 7 acanthocephalans are known to inhabit tissues and organs of opossums (Pence and Little 1972, Potkay 1977). Roundworms (*Physaloptera turgida*), common nematode parasites of the digestive tract, were more abundant in females carrying pouch young than in females of the same age class without pouch young. Pinworms (*Cruzia americana*), on the other hand, showed no differences in infection loads, even when hosts were compared by habitat, age class, sex, and reproductive condition (Blumenthal and Kirkland 1976).

ECONOMIC STATUS, RESEARCH, AND MANAGEMENT

Fur. Hamilton (1958) pointed out the numerically high rank opossums have held in the fur trade in the United States. The numbers of opossum pelts reported during the 44-year period between 1933 and 1978 are graphed in figure 1.8. The high numbers of opossum pelts taken in the 1933/34 and 1940/41 seasons (2.40 million and 2.42 million, respectively) represent only part of the annual harvest during those years. Ignoring the unknown numbers killed for food, sport, and because they are considered predators, only 15 of the 31 states then known to have harvestable populations of opossums reported the number of pelts taken in the 1933/34 season. The numbers of skins reported for the 1940/41 season were based on information from only 21 states. Some states, such as Texas and New York, indicated that their actual harvest may have been underestimated by 50 percent or more. The steady decline from 1940 through 1953, and the continued low numbers reported through 1971, reflected the low prices paid and the general unpopularity of long-haired furs during that time. The average price paid for opossum pelts between 1934 and 1971 varied from approximately 30¢ to 58¢ per skin. The average price per skin jumped from 85¢ for furs taken in the 1970/71 season to $2.50 the next season and $3.00 in the 1973/74 season. The average price from 1975 to 1977 was $2.50 (Deems and Pursley 1978).

For comparison, the numbers of opossum pelts and the average price paid per skin in Louisiana (Lowery 1974) between 1913 and 1972 range from a high of 518,295 at $1.10 apiece in the 1928/29 season to a low of 3,009 at 30¢ apiece in the 1967/68 trapping season. The lowest average price paid was 10¢ in the 1938/39, 1940/41, and 1958/59 seasons.

According to the North American fur harvest statistics for the 1976/77 season, 31 of the 39 states assumed to have harvestable populations of opossums reported a combined harvest of 1,064,725 skins with a total estimated value of $2,661,813. The average price per skin ($2.50) was the lowest paid for any North American furbearer, except for the striped skunk ($2.25) and weasels ($1.00). Nevertheless, opossums ranked second in numbers of pelts reported by 6 states (California, Georgia, Mississippi, Missouri, Oklahoma, and Texas), third in 10 (Illinois, Indiana, Iowa, Kansas, Kentucky, New Jersey, North Carolina,

Ohio, Virginia, and West Virginia), and fourth in 2 (Maryland and Nebraska). The total reported harvest of opossums in the 1977/78 season (32 of the 39 states with opossum populations reporting) was 958,818 skins (Pursley personal communication).

In terms of the numbers of skins reported country wide in the United States from the 1970/71 season through the 1976/77 season (the latest for which complete data are available), the 10 most important furbearers were muskrat, raccoon, nutria, opossum, red fox, mink, beaver, coyote, gray fox, and striped skunk. During these seven trapping seasons, opossums ranked sixth in two and fourth in five. Opossum fur is dyed and plucked to simulate more expensive furs or is used in its natural state; most commonly as trim.

Food. Opossums are eaten in many regions of the United States and, until recently, were often available in markets (Bailey 1923). Although eating opossum long has been a tradition in the South, where "possum and taters" is an esteemed dish (Lay 1942; Hartman 1952; Lowery 1974), the meat seems to lack appeal in other areas, particularly in the Northeast and in the Lake States. Hamilton (1958), for example, found the meat greasy and lacking flavor.

Many recipes for the preparation of opossum recommend scalding in hot lye water and scraping off the hair. The obvious purpose is to retain as much fat as possible. The Virginia opossum is the only member of the genus *Didelphis* that accumulates heavy layers of fat, and also is the only didelphid commonly eaten by humans. Many accounts of travels in South America have mentioned the unpalatability of the common opossum, *D. marsupialis* (Hartman 1952:150). Yucatecan Indians say that dark-colored specimens tasted bad, but that paler ones were delectable (Hatt 1938). It is likely that the paler opossums were *D. virginiana* (see description of cheek color) and the others *D. marsupialis*.

Anyone who has spent several weeks or more in field camps, where meat is usually lean and hard to come by, often appreciates the flavor of fat meat. Once out of the field and enjoying a richer diet, fat meat loses its appeal and may taste greasy instead of delicious. Now that larger segments of an increasingly urban society enjoy a richer and more varied diet, opossum meat is not as popular as it once was. This may be particularly true where opossums are hunted for sport instead of for meat and where peoples' tastes are offended at the thought of eating any furbearer, especially one as odd appearing as an opossum.

Sport. Night hunting for opossums, with or without dogs, is a common activity in some parts of the country. But, as pointed out by Allen (1940), Stuewer (1943), Taube (1947), and Hamilton (1958), many sportsmen prefer to consider opossums as an impediment to good raccoon hunting and a detriment to more highly valued wildlife than as a worthy object of pursuit.

Predation. Opossums have enjoyed a long, if not entirely justified, reputation as rapacious destroyers of poultry (Hartman 1952:57). True, they kill chickens and prey on young cottontail rabbits as well as on the eggs and broods of ground-nesting birds. Although these depredations on poultry and game may alarm farmers and sportsmen, the opossum's impact is negligible. Analyses of food habits show the Virginia opossum to be an omnivorous gleaner and scavenger of a wide variety of foods. Most of the mammals and birds eaten (including poultry) are encountered as carrion. Opossums also eat large numbers of voles and mice as well as poisonous snakes.

Those farmers, sportsmen, and game biologists who irrationally consider the animal a serious pest or predator may find it difficult to accept the fact that opossums have comparatively little impact on economically important wildlife. Anthropomorphic generalizing such as "I don't like him either. He is a sluggish, smelly, disreputable critter without a semblance of character or self respect" (Allen 1940:4) can hardly be construed as an objective evaluation of the attributes of the animal.

Research. Opossums have long fascinated biologists interested in the marsupial way of life. Much of the initial stimulus to do research on the Virginia opossum was the belief that the animal was a "living fossil" that retained all of its Cretaceous qualities and characteristics. The major obstacle to using opossums as research animals was the lack of efficient and effective methods for their husbandry. The reports by Jurgelski (1974), Jurgelski et al. (1974), and Jurgelski and Porter (1974) indicate that most of the problems have been solved. Therefore, opossums are expected to assume increasing importance as laboratory animals, particularly in developmental and biomedical research.

Despite great interest, as indicated by the many published reports on the biology and ecology of opossums, relatively little is known about their behavior, dispersal patterns, and population structure. Some unanswered questions, suggesting promising areas of research, are:

1. How does diet, climate, and population density influence the breeding season and reproductive potential?
2. What is the reproductive life of a free-ranging opossum?
3. What is the age structure of populations and how does it vary seasonally?
4. What are the exact causes of mortality?
5. How do mortality factors affect sex and age-class composition of populations?
6. Do the food habits of newly weaned and juvenile age classes differ from those of subadults and adults?
7. What is the role played by opossums in suburban and urban wildlife populations?

One obstacle to the field study of opossums has been the lack of uniformly applicable methods for aging individuals. Methods and the developmental criteria used for determining age are outlined and discussed in the sections on "Development" and "Age Determination." Other more effective methods might be developed with additional study and a little imagination. Measuring the length of the head (Tyndale-Biscoe and Mackenzie 1976) instead of snout–rump length on pouch young to monitor growth rate has not been done on North American populations.

Other problems concern monitoring activity and movements. Live-trapping has not been nearly as effective as radio telemetry in determining the activity ranges of opossums. Radio telemetry was used by Holmes and Sanderson (1965), Fitch and Shirer (1970), Shirer and Fitch (1970), VanDruff (1971), and Pippitt (1976). Pippitt used two-channel implanted radios to monitor physiological functions in free-living animals.

Of the several marking techniques available, those most commonly used have been toe clipping and the affixing of different kinds of ear tags (Lay 1942; Fitch and Sandidge 1953; Sanderson 1961). Another method using a tag encircling the Achilles tendon was described by Cook (1943). Radioisotope labels (Wolff and Holleman 1978) have never been tried on opossums.

Several studies of food habits have included descriptions of the methods used (Hamilton 1958). A more detailed description of processing procedures for food habits studies was given by Korschgen (1971).

Management. The Virginia opossum has limited food and sport value. Its value as a furbearer results from the large numbers of pelts taken and the apparent ability of populations to recover from heavy hunting and trapping pressure. Much of the concern on the part of game management agencies is due to prejudice against opossums by farmers and sportsmen. All serious studies, however, indicate that opossums have little impact on poultry and wildlife. Instead, opossums probably serve a positive role by reducing some noxious species of wildlife and suppressing some wildlife diseases by removing dead animals.

In areas where predation on ground-nesting upland game and waterfowl is a potential problem, the most effective remedy may be to reduce the availability of den sites by controlling woodchuck and skunk populations. Reduction of den sites also would reduce depredations caused by other predators that rely on ground dens.

MISCONCEPTIONS

The Virginia opossum has played a prominent role in the folklore of the southern United States, and is the subject of numerous songs, rhymes, fables, and misconceptions (Hartman 1921*b*, 1952; Lowery 1974). Some of the following more common misconceptions still find their way into the literature.

1. *False:* The Virginia opossum is a living fossil, a holdover in North America from the age of dinosaurs.
   True: All marsupials originally inhabiting North America became extinct by the mid-Tertiary. The first record for the Virginia opossum is late Pleistocene.

2. *False:* The opossum is stupid.

   True: Opossums are inhibited animals, especially under lighted conditions, but are by no means stupid. Results from some learning and discrimination tests rank opossums above dogs and more or less on a par with pigs in intelligence (Hunsaker and Shupe 1977).

3. *False:* Male opossums copulate with the female's nose.

   True: The forked penis permits deposition of spermatozoa into the lateral vaginae of the female, not in the nostrils.

4. *False:* Young opossums are not born as in other mammals, but are formed at the ends of the nipples like buds of a plant.

   True: Young are born as in other mammals, although in a relatively undeveloped condition.

5. *False:* Young are blown through the mother's nostrils into the pouch.

   True: Same as 4 above.

6. *False:* The mother pumps milk into the young.

   True: The young suckle without aid from the mother (Enders 1966).

7. *False:* The lips and tongue fuse to the nipple.

   True: The pouch young become firmly attached to the nipple, but no fusion occurs. The false observation is based on the bloody, sometimes torn mouth that results when young are not carefully removed. To avoid damage to the young, researchers have suggested gently twisting to "unscrew" them from the nipple.

8. *False:* A female rears three litters a year.

   True: There is no evidence that the Virginia opossum successfully rears more than two litters a year. Females may give birth to three or more litters during a season if earlier litters are lost.

LITERATURE CITED

Allen, D. L. 1940. Nobody loves the 'possum. Michigan Conserv. (March):4, 10.

Apgar, C. S., Jr. 1934. Analysis of life records of Didelphis virginiana. Pages 51–55 *in* H. Fox. Report of the Laboratory and Museum of Comparative Pathology of the Zoological Society of Philadelphia [for 1933]. Philadelphia. 55pp.

Armstrong, D. M. 1972. Distribution of mammals in Colorado. Univ. Kansas Mus. Nat. Hist. Monogr. 3. 415pp.

Ashbrook, F. G., and Arnold, B. M. 1927. Fur-bearing animals of the United States: the opossum. Fur J. 1:28–29.

Audubon, J. J., and Bachman, J. 1851. The quadrupeds of North America. Vol. 2. V. G. Audubon, New York. 334pp.

Bailey, V. 1923. Mammals of the District of Columbia. Proc Bio. Soc. Washington 36:103–138.

Barr, T. R. B. 1963. Infectious diseases in the opossum, a review. J. Wildl. Manage. 27:53–71.

Barton, B. S. 1823. Facts, observations, and conjectures relative to the generation of the opossum of North America. Ann. Philos., new ser. 6:349–354.

Bennitt, R., and Nagel, W. O. 1937. A survey of the resident game and furbearers of Missouri. Univ. Missouri Studies 12. 215pp.

Biggers, J. D. 1966. Reproduction in male marsupials. Pages 251–280 *in* I. W. Rowlands, ed. Comparative biology of reproduction in mammals. Zool. Soc. London Symp. 15. 559pp.

Biggers, J. D., and Creed, R. F. S. 1962. Conjugate spermatozoa of the North American opossum. Nature 196:1112–1113.

Biggers, J. D., and DeLamater, E. D. 1965. Marsupial spermatozoa: pairing in the epididymis of the American forms. Nature 208:402–404.

Black, J. D. 1935. Vitality of the Virginia opossum as exhibited in the skeleton. J. Mammal. 16:223.

Blumenthal, E. M., and Kirkland, G. L., Jr. 1976. The biology of the opossum, *Didelphis virginiana*, in southcentral Pennsylvania. Proc. Pennsylvania Acad. Sci. 50:81–85.

Bowers, J. H., and Judd, F. W. 1969. Notes on the distribution of *Didelphis marsupialis* and *Citellus spilosoma* in western Texas. Texas J. Sci. 20:277.

Brocke, R. H. 1970. The winter ecology and bioenergetics of the opossum, Didelphis marsupialis, as distributional factors in Michigan. Ph.D. Thesis. Michigan State Univ., East Lansing. 215pp.

Brown, L. N. 1965. Status of opossum, *Didelphis marsupialis*, in Wyoming. Southwestern Nat. 10:142–143.

Bryant, B. J. 1977. The development of the lymphatic and immunohematopoietic systems. Pages 349–386 *in* D. Hunsaker, II, ed. The biology of marsupials. Academic Press, New York. 537pp.

Burns, R. K., and Burns, L. M. 1957. Observations on the breeding of the American opossum in Florida. Rev. Suisse Zool. 64:595–605.

Chase, E. B. 1939. The reproductive system of the male opossum *Didelphis virginiana* Kerr and its experimental modification. J. Morphol. 65:215–239.

Cheng, C. 1955. The development of the shoulder region of the opossum *Didelphis virginiana* Kerr with special reference to the musculature. J. Morphol. 97:415–472.

Clemens, W. A. 1968. Origin and early evolution of marsupials. Evolution 22:1–18.

————. 1977. Phylogeny of the marsupials. Pages 51–68 *in* B. Stonehouse and D. Gilmore, eds. The biology of marsupials. Univ. Park Press, Baltimore. 486pp.

Coleman, R. H. 1929. Opossum in the lower Hudson Valley, New York. J. Mammal. 10:250.

Cook, A. H. 1943. A technique for marking mammals. J. Mammal. 24:45–47.

Cory, C. B. 1912. The mammals of Illinois and Wisconsin. Field Mus. Nat. Hist., Zool. Ser. 11:1–502.

Crandall, L. S. 1964. The management of wild mammals in captivity. Univ. Chicago Press, Chicago. 769pp.

Davis, G. W. 1938. Virginia opossum in Vermont. J. Mammal. 19:499.

Davis, W. B. 1974. The mammals of Texas. Texas Parks Wildl. Dept., Austin, Bull. 41:1–294.

Dearborn, N. 1932. Foods of some predatory fur-bearing animals in Michigan. Univ. Michigan School For. Conserv. Bull. 1:1–52.

Deems, E. F., Jr., and Pursley, D. 1978. North American furbearers. Int. Assoc. Fish Wildl. Agencies, 171pp.

Dickerman, R. W. 1960. "Davian behavior complex" in ground squirrels. J. Mammal. 41:403.

Dills, G. 1972. Telemetered thermal responses of a specimen of the Virginia opossum. J. Alabama Acad. Sci. 43:55–62.

Dills, G. C., and Manganiello, T. 1973. Diel temperature fluctuations of the Virginia opossum (*Didelphis virginiana*). J. Mammal. 54:763–765.

Edmunds, R. M.; Goertz, J. W.; and Linscombe, G. 1978. Age ratios, weights, and reproduction of the Virginia opossum in north Louisiana. J. Mammal. 59:884–885.

Eisenberg, J. F., and Wilson, D. E. 1981. Relative brain size and demographic strategies in didelphid marsupials. Am. Nat. 118:110–126.

Elftman, H. O. 1929. Functional adaptations of the pelvis in marsupials. Bull. Am. Mus. Nat. Hist. 58:189–232.

Ellsworth, A. F. 1975a. Atlas of the North American opossum, *Didelphis*. R. Krieger Publ. Co., Huntington, N.Y. 160pp.

———. 1975b. Reassessment of muscle homologues and nomenclature in conservative amniotes. R. Krieger Publ. Co., Huntington, N.Y. 84pp.

Enders, R. K. 1966. Attachment, nursing and survival of young in some didelphids. Pages 195–204 in I. W. Rowlands, ed. Comparative biology of reproduction in mammals. Zool. Soc. London Symp. 15. 559pp.

Findley, J. S. 1956. Mammals of Clay County, South Dakota. Univ. South Dakota Publ. Bio. 1:1–45.

Fitch, H. S., and Sandidge, L. L. 1953. Ecology of the opossum on a natural area in northeastern Kansas. Univ. Kansas Publ., Mus. Nat. Hist. 7:305–338.

Fitch, H. S., and Shirer, H. W. 1970. A radiotelemetric study of spatial relationships in the opossum. Am. Midl. Nat. 48:170–186.

Fortney, J. A. 1973. Cytology of eccrine sweat glands in the opossum. Am. J. Anat. 136:205–219.

Francq, E. N. 1969. Behavioral aspects of feigned death in the opossum *Didelphis marsupialis*. Am. Midl. Nat. 81:556–568.

———. 1970. Electrocardiograms of the opossum, *Didelphis marsupialis*, during feigned death. J. Mammal. 51:395.

Friedman, H. 1967. Colour vision in the Virginia opossum. Nature 213:835–836.

Gardner, A. L. 1973. The systematics of the genus Didelphis (Marsupialia: Didelphidae) in North and Middle America. Spec. Publ., Mus. Texas Tech Univ. 4:1–81.

Gifford, C. L., and Whitebread, R. 1951. Mammal survey of south central Pennsylvania. Pennsylvania Game Comm., Harrisburg. 75pp.

Gilmore, R. M. 1943. Mammalogy in an epidemiological study of jungle yellow fever in Brazil. J. Mammal. 24:144–162.

Godin, A. J. 1977. Wild mammals of New England. Johns Hopkins Univ. Press, Baltimore. 304pp.

Golley, F. B. 1962. Mammals of Georgia. Univ. Georgia Press, Athens. 218pp.

Grimm, W. C., and Roberts, H. A. 1950. Mammal survey of southwestern Pennsylvania. Pennsylvania Game Comm., Harrisburg. 99pp.

Grinnell, J.; Dixon, J. S.; and Linsdale, J. M. 1937. Fur-bearing mammals of California. Vol. 1. Univ. California Press, Berkeley. 375pp.

Grote, J. C., and Dalby, P. L. 1973. An early litter for the opossum (*Didelphis marsupialis*) in Ohio. Ohio J. Sci. 73:240–241.

Guilday, J. E. 1958. The prehistoric distribution of the opossum. J. Mammal. 39:39–43.

Hamilton, W. J., Jr. 1951. The food of the opossum in New York state. J. Wildl. Manage. 15:258–264.

———. 1958. Life history and economic relations of the opossum (*Didelphis marsupialis virginiana*) in New York state. Cornell Univ. Agric. Exp. Stn. Mem. 354:1–48.

Hartman, C. G. 1921a. Dioestrous changes in the mammary gland of the opossum and the diagnosis of pregnancy. Am. J. Physiol. 55:308–309.

———. 1921b. Traditional belief concerning the generation of the opossum (*Didelphis virginiana* L.). J. Am. Folklore 34:321–323.

———. 1922. A brown mutation in the opossum (Didelphis virginiana) with remarks upon the gray and the black phases in this species. J. Mammal. 3:146–149.

———. 1923a. Breeding habits, development, and birth of the opossum. Smithsonian Rep. (for 1921). Pp. 347–363.

———. 1923b. The oestrous cycle in the opossum. Am. J. Anat. 32:353–421.

———. 1928. The breeding season of the opossum (Didelphis virginiana) and the rate of intrauterine and postnatal development. J. Morphol. 46:143–215.

———. 1952. Possums. Univ. Texas Press, Austin. 174 pp.

Hatt, R. T. 1938. Notes concerning mammals collected in Yucatan. J. Mammal. 19:333–337.

Hazard, E. B. 1963. Records of the opossum in northern Minnesota. J. Mammal. 44:118.

Hearn, J. P. 1977. Pituitary function in marsupial reproduction. Pages 337–344 in B. Stonehouse and D. Gilmore, eds. The biology of marsupials. Univ. Park Press, Baltimore. 486pp.

Hershkovitz, P. 1951. Mammals from British Honduras, Mexico, Jamaica and Haiti. Fieldiana-Zool. 31:547–569.

Hibbard, C. W.; Ray, C. E.; Savage, D. E.; Taylor, D. W.; and Guilday, J. E. 1965. Quaternary mammals of North America. Pages 509–525 in H. E. Wright, Jr., and D. G. Frey, eds. The Quaternary of the United States. Princeton Univ. Press, Princeton, N.J. 922pp.

Higgenbotham, A. C., and Koon, W. E. 1955. Temperature regulation in the Virginia opossum. Am. J. Physiol. 181:69–71.

Hill, J. P., and Fraser, E. A. 1925. Some observations on the female urogenital organs of Didelphyidae. Proc. Zool. Soc. London (1925), pp.189–219.

Hock, R. J. 1952. The opossum in Arizona. J. Mammal. 33:464–470.

Hoffmeister, D. F., and Goodpaster, W. W. 1954. The mammals of the Huachuca Mountains, southeastern Arizona. Illinois Bio. Monogr. 24:1–152.

Hollister, N. 1908. Notes on Wisconsin mammals. Bull. Wisconsin Nat. Hist. Soc. 4:137–142.

Holmes, A. C. V., and Sanderson, G. C. 1965. Populations and movements of opossums in east-central Illinois. J. Wildl. Manage. 29:287–295.

Hopkins, D. 1977. Nest-building behavior in the immature Virginia opossum (*Didelphis virginiana*). Mammalia 41:361–362.

Hunsaker, D., II. 1977. Ecology of New World marsupials. Pages 95–156 in D. Hunsaker II, ed. The biology of marsupials. Academic Press, New York. 537pp.

Hunsaker, D., II, and Shupe, D. 1977. Behavior of New World marsupials. Pages 279–348 in D. Hunsaker II, ed. The biology of marsupials. Academic Press, New York. 537pp.

Jackson, H. H. T. 1961. Mammals of Wisconsin. Univ. Wisconsin Press, Madison. 504pp.

James, W. T. 1960. A study of visual discrimination in the opossum. J. Genet. Psychol. 97:127–130.

Jenkins, F. A., Jr. 1971. Limb posture and locomotion in the Virginia opossum (*Didelphis marsupialis*) and in other non-cursorial mammals. J. Zool. London 165:303–315.

Jewett, S. G., and Dobyns, H. W. 1929. The Virginia opossum in Oregon. J. Mammal. 10:351.

Johnson, J. I., Jr. 1977. Central nervous system of marsupials. Pages 157–278 in D. Hunsaker II, ed. The biology of marsupials. Academic Press, New York. 537pp.

Jones, J. K., Jr. 1964. Distribution and taxonomy of mammals of Nebraska. Univ. Kansas Publ., Mus. Nat. Hist. 16: 1–356.

Jordan, H. E. 1911. The spermatogenesis of the opossum (Didelphis virginiana) with special reference to the accessory chromosome and chondriosomes. Arch. Zellforschung 7:41–86.

Joseph, T. 1974. *Eimeria indianensis* sp. n. and an *Isospora* sp. from the opossum Didelphis virginiana (Kerr). J. Protozool. 21:12–15.

Jurgelski, W., Jr. 1974. The opossum (*Didelphis virginiana* Kerr) as a biomedical model. I. Research perspective, husbandry, and laboratory techniques. Lab. Anim. Sci. 24:376–403.

Jurgelski, W., Jr.; Forsythe, W.; Dahl, D.; Thomas, L. D.; Moore, J. A.; Kotin, P.; Falk, H. L.; and Vogel, F. S. 1974. The opossum (*Didelphis virginiana* Kerr) as a biomedical model. II. Breeding the opossum in captivity: facility design. Lab. Anim. Sci. 24:404–411.

Jurgelski, W., Jr., and Porter, M. E. 1974. The opossum (*Didelphis virginiana* Kerr) as a biomedical model. III. Breeding the opossum in captivity: methods. Lab. Anim. Sci. 24:412–425.

Kennard, F. H. 1925. The Virginia opossum in Massachusetts and New Hampshire. J. Mammal. 6:196.

Kilmon, J. A., Sr. 1976. High tolerance to snake venom by the Virginia opossum, *Didelphis virginiana*. Toxicon 14:337–340.

King, O. M. 1960. A note on opossum behavior. J. Mammal. 42:397.

Kingsbury, B. F. 1940. The development of the pharyngeal derivatives of the opossum (*Didelphis virginiana*) with special reference to the thymus. Am. J. Anat. 67:393–435.

Kirk, G. L. 1921. Opossum in Vermont. J. Mammal. 2:109.

Knudsen, G. J., and Hale, J. B. 1970. Food habits of opossums in southern Wisconsin. Wisconsin Dept. Nat. Resour. Rep. 61:1–11.

Korschgen, L. J. 1971. Procedures for food-habits analyses. Pages 233–250 in R. H. Giles, Jr., ed. Wildlife management techniques, 3rd ed. Wildl. Soc., Washington, D.C. 633pp.

Langley, W. M. 1979. Preference of the striped skunk and opossum for auditory over visual prey stimuli. Carnivore 2:31–38.

Lay, D. W. 1942. Ecology of the opossum in eastern Texas. J. Mammal. 23:147–159.

Lindsay, D. M. 1960. Mammals of Ripley and Jefferson counties, Indiana. J. Mammal. 41:253–262.

Llewellyn, L. M., and Dale, F. H. 1964. Notes on the ecology of the opossum in Maryland. J. Mammal. 45:113–122.

Llewellyn, L. M., and Uhler, F. M. 1952. The food of fur animals of the Patuxent Research Refuge, Maryland. Am. Midl. Nat. 48:193–203.

Long, C. A. 1965. The mammals of Wyoming. Univ. Kansas Publ., Mus. Nat. Hist. 14:493–758.

Long, C. A., and Copes, F. A. 1968. Note on the rate of dispersion of the opossum in Wisconsin. Am. Midl. Nat. 80:283–284.

Long, G. G.; Stookey, J. L.; Terrell, T. G.; and Whitney, G. D. 1975. Fibrous osteodystrophy in an opossum. J. Wildl. Manage. 11:221–223.

Lowery, G. H., Jr. 1974. The mammals of Louisiana and its adjacent waters. Louisiana State Univ. Press, Baton Rouge. 565pp.

Lowrance, E. W. 1949. Variability and growth of the opossum skeleton. J. Morphol. 85:569–593.

———. 1957. Correlations of certain ponderal and linear skeletal measurements with skull weight and skull length in the opossum. Anat. Rec. 128:69–76.

Lustick, S., and Lustick, D. D. 1972. Energetics in the opossum, *Didelphis marsupialis virginiana*. Comp. Biochem. Physiol. 43:643–647.

McDonald, I. R. 1977. Adrenocortical function in marsupials. Pages 345–378 in B. Stonehouse and D. Gilmore, eds. The biology of marsupials. Univ. Park Press, Baltimore. 486pp.

McGrady, E., Jr. 1938. The embryology of the opossum. Am. Anat. Mem. 16. 233pp.

McKeever, S. 1958. Reproduction in the opossum in southwestern Georgia and northwestern Florida. J. Wildl. Manage. 22:303.

McManus, J. J. 1967. Observations on sexual behavior of the opossum, *Didelphis marsupialis*. J. Mammal. 48:486–487.

———. 1969. Temperature regulation in the opossum, *Didelphis marsupialis virginiana*. J. Mammal. 50:550–558.

———. 1970. Behavior of captive opossums, *Didelphis marsupialis virginiana*. Am. Midl. Nat. 84:144–169.

———. 1971. Activity of captive *Didelphis marsupialis*. J. Mammal. 52:846–848.

———. 1974. Didelphis virginiana. Mammal. Species 40:1–6.

McNab, B. K. 1978. The comparative energetics of Neotropical marsupials. J. Comp. Physiol., B 125:115–128.

Manuel, B. J. 1977. Occurrence of the opossum on the Tug Hill Plateau. New York Fish Game J. 24:98.

Miller, G. S. 1899. Preliminary list of New York mammals. Bull. New York State Mus. 6:271–390.

Mohr, C. O. 1937. Illinois trappers averages reveal coon and possum distribution. Illinois Conserv. 2:3–4,8.

Moore, C. R., and Bodian, D. 1940. Opossum pouch young as experimental material. Anat. Rec. 76:319–327.

Moore, J. C. 1955. Opossum takes refuge under water. J. Mammal. 36:559–561.

Morlan, H. B. 1952. Host relationships and seasonal abundance of some southwest Georgia ectoparasites. Am. Midl. Nat. 48:74–93.

Morrison, P. R., and Petajan, J. H. 1962. The development of temperature regulation in the opossum, *Didelphis marsupialis virginiana*. Physiol. Zool. 35:52–65.

Nardone, R. M.; Wilber, C. G.; and Musacchia, X. J. 1955. Electocardiogram of the opossum during exposure to cold. Am. J. Physiol. 181:352–356.

Nesslinger, C. L. 1956. Ossification centers and skeletal development in the postnatal Virginia opossum. J. Mammal. 37:382–394.

Osgood, F. L. 1938. The mammals of Vermont. J. Mammal. 19:435–441.

Packard, R. L., and Judd, F. W. 1968. Comments on some mammals from western Texas. J. Mammal. 49:535–538.

Pence, D. B. 1973. Notes on two species of hypopial nymphs of the genus *Marsupialichus* (Arcarina:Glycyphagidae) from mammals in Louisiana. J. Med. Ent., Honolulu 10:329–332.

Pence, D. B., and Little, M. D. 1972. *Anatrichosoma buccalis* sp. n. (Nematoda:Trichosomoididae) from the buccal mucosa of the common opossum, *Didelphis marsupialis* L. J. Parasitol. 58:767–773.

Petajan, J. H., and Morrison, P. R. 1962. Physical and physiological factors modifying the development of temperature regulation in the opossum. J. Exp. Zool. 149:45–57.

Peterson, R. L. 1966. The mammals of eastern Canada. Oxford Univ. Press, Toronto. 465pp.

Peterson, R. L., and Downing, S. C. 1956. Distribution records of the opossum in Ontario. J. Mammal. 37:431–435.

Petrides, G. A. 1949. Sex and age determination in the opossum. J. Mammal. 30:364–378.

Pippitt, D. D. 1976. A radiotelemetric study of the winter energetics of the opossum *Didelphis virginiana* Kerr. Ph.D. Thesis. Univ. Kansas, Lawrence. 84pp.

Plakke, R. K. 1970. Urea, electrolyte and total solution excretion following water deprivation in the opossum (*Didelphis marsupialis virginiana*). Comp. Biochem. Physiol. 34:325–332.

Plakke, R. K., and Pfeiffer, E. W. 1965. Influence of plasma urea on urine concentration in the opossum (*Didelphis marsupialis virginiana*). Nature 207:866–867.

Potkay, S. 1970. Diseases of the opossum (*Didelphis marsupialis*): a review. Lab. Anim. Care 20:502–511.

———. 1977. Diseases of marsupials. Pages 415–506 *in* D. Hunsaker, II, ed. The biology of marsupials. Academic Press, New York. 537pp.

Pray, L. 1921. Opossum carries leaves with its tail. J. Mammal. 2:109–110.

Preble, N. A. 1942. Notes on the mammals of Morrow County, Ohio. J. Mammal. 23:82–86.

Prestwood, A. 1976. *Didelphostrongylus hayesi* gen. et sp. n. (Melastrongyloidea:Filaroididae) from the opossum, *Didelphis marsupialis*. J. Parasitol. 62:272–275.

Rand, A. L., and Host, P. 1942. Results of the Archbold Expeditions No. 45. Mammal notes from Highland County, Florida. Bull. Am. Mus. Nat. Hist. 80:1–21.

Raven, H. C. 1929. A case of matricide in the opossum. J. Mammal. 10:168.

Reig, O. A.; Gardner, A. L.; Bianchi, N. O.; and Patton, J. L. 1977. The chromosomes of the Didelphidae (Marsupialia) and their evolutionary significance. Bio. J. Linn. Soc. London 4:191–216.

Reynolds, H. C. 1945. Some aspects of the life history and ecology of the opossum in central Missouri. J. Mammal. 26:361–379.

———. 1952. Studies on reproduction in the opossum (*Didelphis virginiana*). Univ. California Publ. Zool. 52:223–284.

———. 1953. The opossum. Sci. Am. 188:88–94.

Sanderson, G. C. 1961. Estimating opossum populations by marking young. J. Wildl. Manage. 25:20–27.

Sandidge, L. L. 1953. Food and dens of the opossum (*Didelphis virginiana*) in northeastern Kansas. Trans. Kansas Acad. Sci. 56:97–106.

Sands, J. L. 1960. The opossum in New Mexico. J. Mammal. 41:393.

Scheffer, J. B. 1943. The opossum settles in Washington state. Murrelet 24:27–28.

Schneider, L. K. 1977. Marsupial chromosomes, cell cycles, and cytogenetics. Pages 51–94 *in* D. Hunsaker II, ed. The biology of marsupials. Academic Press, New York. 537pp.

Sealander, J. A. 1979. A guide to Arkansas mammals. River Road Press, Conway, Ark. 313pp.

Selenka, E. 1887. Studien über Entwicklungsgeschichte. Vol. 4, Das Opossum (*Didelphys virginiana*). C. W. Kreidel Verlag, Wiesbaden. pp. 101–172, pls. 25–30.

Setchell, P. P. 1977. Reproduction in male marsupials. Pages 411–458 *in* B. Stonehouse and D. Gilmore, eds. The biology of marsupials. Univ. Park Press, Baltimore. 486pp.

Severinghaus, C. W. 1975. Occurrence of the opossum in the central Adirondacks. New York Fish Game J. 22:80.

Shaver, E. L. 1962. The chromosomes of the opossum, Didelphis virginiana. Can. J. Genet. Cytol. 4:62–68.

Shirer, H. W., and Fitch, H. S. 1970. Comparison from radiotracking of movements and denning habits of the raccoon, striped skunk, and opossum in northeastern Kansas. J. Mammal. 51:491–503.

Smiley, D., Jr. 1938. An opossum in New York State feels the effects of winter. J. Mammal. 19:499.

Smith, J. 1612. A map of Virginia. J. Barnes, Oxford. 110pp.

Smith, J. H. 1935. The opossum in Kent County, Ontario. Can. Field Nat. 49:109.

Smith, L. 1941. An observation on the nest-building behavior of the opossum. J. Mammal. 22:201–202.

Sperry, C. C. 1933. Opossum and skunk eat bats. J. Mammal. 14:152–153.

Stieglitz, W. O., and Klimstra, W. D. 1962. Dietary pattern of the Virginia opossum, *Didelphis marsupialis virginianus* Kerr, late summer-winter, southern Illinois. Trans. Illinois Acad. Sci. 55:198–208.

Stone, W. B., Jr.; Parks, E.; Weber, B. L.; and Parks, F. V. 1972. Experimental transfer of sarcoptic mange from red foxes and wild canids to captive wildlife and domestic animals. New York Fish Game J. 19:1–11.

Stonehouse, B., and Gilmore, D., eds. 1977. The biology of marsupials. Univ. Park Press, Baltimore. 486pp.

Stoner, D. 1939. Remarks on abundance and range of the opossum. J. Mammal. 20:250–251.

Stout, J., and Sonenshine, D. E. 1974. Ecology of an opossum population in Virginia, 1963–69. Acta Theriol. 19:235–245.

Stuewer, F. W. 1943. Raccoons: their habits and management in Michigan. Ecol. Monogr. 13:203–258.

Stumpf, W. A., and Mohr, C. O. 1962. Linearity of home ranges of California mice and other animals. J. Wildl. Manage. 26:149–154.

Tamar, H. 1961. Taste reception in the opossum and the bat. Physiol. Zool. 34:86–91.

Taube, C. M. 1947. Food habits of Michigan opossums. J. Wildl. Manage. 11:97–103.

Tyndale-Biscoe, C. H. 1973. Life of marsupials. Am. Elsevier Publ. Co., New York. 254pp.

Tyndale-Biscoe, C. H., and Mackenzie, R. B. 1976. Reproduction in *Didelphis marsupialis* and *D. albiventris* in Colombia. J. Mammal. 57:249–265.

Tyson, E. 1698. Anatomy of an opossum Didelphys. Philos. Trans. R. Soc. London 20:105–164.

———. 1704. Carigueya seu marsupiale anerinanum: or, The anatomy of an opossum. Philos. Trans. R. Soc. London 24:1565–1575.

VanDruff, L. W. 1971. The ecology of the raccoon and opossum, with emphasis on their role as waterfowl nest predators. Ph.D. Thesis. Cornell Univ., Ithaca, N.Y. 140pp.

Verts, B. J. 1963. Movements and populations of opossums in a cultivated area. J. Wildl. Manage. 27:127–129.

Washburn, S. L. 1946. The sequence of epiphysial union in the opossum. Anat. Rec. 95:353–363.

Werner, R. M., and Vick, J. A. 1977. Resistance of the opossum (*Didelphis virginiana*) to envenomation by snakes of the family Crotalidae. Toxicon 15:29–33.

Whitaker, J. O., Jr.; Jones, G. S.; and Goff, R. J. 1977. Ectoparasites and food habits of the opossum, Didelphis Virginiana, in Indiana. Proc. Indiana Acad. Sci. 86:501–507.

Wilber, C. G. 1955. Electrocardiographic studies on the opossum. J. Mammal. 36:284–286.

Wilber, C. G., and Weidenbacher, G. H. 1961. Swimming capacity of some wild mammals. J. Mammal. 42:428–429.

Wiseman, G. L., and Hendrickson, G. O. 1950. Notes on the

life history of the opossum in southeast Iowa. J. Mammal. 31:331–337.

Wolff, J. O., and Holleman, D. F. 1978. Use of radioisotope labels to establish genetic relationships in free-ranging small mammals. J. Mammal. 59:859–860.

Wood, J. E. 1954. Food habits of furbearers of the upland post oak region in Texas. J. Mammal. 35:406–415.

Wood, J. E., and Odum, E. P. 1964. A nine-year history of furbearer populations on the AEC Savannah River Plant area. J. Mammal. 45:540–551.

Worth, C. B. 1975. Virginia opossums (*Didelphis virginiana*) as disseminators of the commom persimmon (*Diospyros virginiana*). J. Mammal. 56:517.

Yeager, L. H. 1936. Winter daytime dens of opossums. J. Mammal. 17:410–411.

ALFRED L. GARDNER, U.S. Fish and Wildlife Service, National Museum of Natural History, Washington, D.C. 20560.

2

Moles

Talpidae

Terry L. Yates
Richard J. Pedersen

NOMENCLATURE

ORDER. Insectivora
FAMILY. Talpidae

Seven species are currently recognized in North America: *Scalopus aquaticus*, the eastern mole; *Parascalops breweri*, the hairy-tailed mole; *Condylura cristata*, the star-nosed mole; *Scapanus latimanus*, the broad-footed mole; *Scapanus townsendii*, the townsend's mole; *Scapanus orarius*, the coast mole; and *Neürotrichus gibbsii*, the shrew mole.

Moles of the family Talpidae are diverse and widespread in the United States, ranging from the tiny shrew mole of the Pacific Northwest to the bizarre, semiaquatic, star-nosed mole of New England. The meandering surface runways made by these animals in their constant search for food are familiar sights to many Americans, and yet few people have more than vague ideas concerning the nature of the creatures that inhabit them. Relatively speaking, little is known scientifically of these animals as well. Moles are probably the least understood major component of the North American mammalian fauna.

DISTRIBUTION

Modern moles of the family Talpidae are distributed over much of the temperate region and are also found in parts of Southeast Asia (Yates 1978). Five genera are represented in North America: *Scalopus, Condylura,* and *Parascalops,* which occur to the east of the Rocky Mountains, and *Scapanus* and *Neürotrichus,* which are restricted to areas west of the Rocky Mountains (Hall and Kelson 1959).

The eastern mole, *Scalopus aquaticus,* has the widest range of any North American mole, occurring throughout much of the eastern United States where soils are favorable (figure 2.1). It ranges from northern Tamaulipus, Mexico, northward to southeastern South Dakota, Minnesota, and Michigan, eastward to Massachusetts and much of southern New England, and then south to the southernmost tip of Florida (Yates 1978). Lowery (1974) reported that the eastern mole

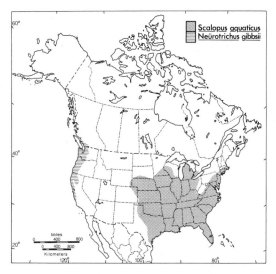

FIGURE 2.1. Distribution of the eastern mole (*Scalopus aquaticus*) and the shrew mole (*Neürotrichus gibbsii*).

occurs throughout the upland portions of Louisiana, but is not common in coastal situations. Two relict populations have been reported, one from northern Coahuila (Baker 1951) and another from Presidio County, Texas (Allen 1891).

The star-nosed mole, *Condylura cristata,* and the hairy-tailed mole, *Parascalops breweri,* occur in the northeastern United States and southeastern Canada, where they are sympatric in many areas (figures 2.2 and 2.3). The range of the former is slightly more extensive than that of the latter. In the northern part of its range, the star-nosed mole is found from extreme eastern Manitoba and Minnesota to as far northeast as Labrador and Nova Scotia. The species ranges southeastward through much of Wisconsin, northern Indiana, and Ohio, along the Atlantic coast as far south as southeastern Georgia, and in the Appalachian Mountains to eastern Tennessee and western North Carolina (Petersen and Yates 1980). The hairy-tailed mole ranges from southern Ontario across southern Quebec, possibly into New Brunswick (Peterson 1966), then south to central Ohio and Connecticut and

FIGURE 2.2. Distribution of the star-nosed mole (*Condylura cristata*) and the coast mole (*Scapanus orarius*).

FIGURE 2.3. Distribution of the hairy-tailed mole (*Parascalops breweri*), the broad-footed mole (*Scapanus latimanus*), and Townsend's mole (*S. townsendii*).

along the Appalachian Mountains to western North Carolina (Hallett 1978).

The remaining four species of North American mole are restricted to areas west of the Rocky Mountains. The broad-footed mole, *Scapanus latimanus*, is found from southern Oregon south along the Pacific coastal regions to the San Pedro Martir Mountains in Baja California (see figure 2.3). Its range also extends throughout much of northern California east into western Nevada, then south in California east of the Central Valley (Yates 1978). The other two species of *Scapanus* have more northerly distributions and occur sympatrically throughout much of their range. The coast mole occurs from southwestern

British Columbia through the western portions of Washington and Oregon to coastal northwestern California (see figure 2.2). It is also known to occur in parts of eastern Washington and Oregon and in one area of extreme west-central Idaho (Yates 1978). Townsend's mole is more restricted in distribution (see figure 2.3), being found from extreme northwestern California along the coastal regions of Oregon and Washington to extreme southwestern British Columbia (Yates 1978). *Scapanus townsendii* is a lowland animal, with the exception of one population that occurs in subalpine meadows in the Olympic Mountains of Washington State (Johnson and Yates 1980).

The remaining species, *Neürotrichus gibbsii*, is restricted to western regions of North America from Santa Cruz County, California, north through western Oregon and Washington to southern British Columbia (see figure 2.1). A small population of shrew moles is also known from Destruction Island, Jefferson County, Washington (Dalquest 1948).

DESCRIPTION

Many of the unique morphological features of North American moles can be attributed to their being highly specialized for a fossorial, secretive existence (Jackson 1915; Slonaker 1920; Campbell 1939; Dalquest 1948; Reed 1954; Lowery 1974). Though rather heterogenous as a family, all species of talpid in North America possess numerous characteristics that distinguish them from their closest North American relatives, the shrews of the family Soricidae.

Externally they present a streamlined body well structured for life underground. Unlike shrews, there is no ear pinnae and the external appendages are short and remain close to the body (figure 2.4), all features that reduce drag when burrowing. The eyes are minute and probably useless except for light detection (Slonaker 1902). In *Scalopus* the eyes are completely covered with skin. In most species the forepaws are broader than they are long and the terminal phalanges of the forefeet in most genera bifurcate, providing additional support for the well-developed claws used in digging. The zygomata and auditory bullae, which are absent in shrews, are always present in moles.

North American moles are most often confused with another group of fossorial mammals, the pocket gophers (Geomyidae). Although the two differ in many respects, their different anatomical adaptations for burrowing are perhaps the most striking. Pocket gophers (and most other fossorial mammals) dig with the forepaws held beneath the body much as a dog would do. Talpids, on the other hand, have developed a system of lateral-stroke burrowing that is unique among Recent mammals. This has been accompanied by a drastic alteration of the pectoral girdle. The pelvic girdle by comparison is relatively narrow and unmodified. One of the most striking features of the pectoral girdle in moles is the presence of a humeroclavicular joint. The clavicle is a small cubical bone that articulates directly with the humerus instead of the scapula. The scapula itself is an extremely elongated bone: the

FIGURE 2.4. Fossorial adaptation of the eastern mole (*Scalopus aquaticus*).

length is as much as four times the greatest width of the blade (Gaughran 1954). The humerus is a massive rectangular bone very unlike the humerus of other mammalian groups. The humerus provides a large surface area for the attachment of the well-developed musculature used in digging. The unusual pectoral girdle and associated muscles tend to obscure the short neck region and cause the forepaws to be rotated parallel to the body (see figure 2.4). The forepaws themselves are broad and covered with a thick layer of skin. Many species have the surface area of the forepaws increased by the presence of a semicircular sesamoid bone, which supports an additional flap of skin.

Although all North American species possess the same general adaptations described above, two genera, *Neürotrichus* and *Condylura,* have undergone various specializations related to life styles other than purely fossorial. The star-nosed mole, *Condylura cristata,* is a semiaquatic species characterized by a ring of 22 fleshy appendages around the nose. The exact function of these nasal rays is not known, but they appear to have a tactile function. They contain highly sensitive tactile organs called Eimer organs (Eimer 1871), and they function as mechanoreceptors similar to Pacinian corpuscles (Van Vleck 1965). The tail of *Condylura* is relatively longer than that of the more fossorial species. Its length averages approximately that of the body. Though not webbed, the hind feet are larger than in other genera, a modification useful in swimming.

The shrew mole, *Neürotrichus gibbsii,* is a semifossorial species that often moves about above ground (Dalquest and Orcutt 1942). The pectoral girdle and forefeet are specialized for digging but are not as well developed as in other genera. The forefeet are longer than they are wide, a characteristic that causes shrew moles to be mistaken for shrews on many occasions.

The overall size of moles is highly variable. Adult weights range from 10 g for *Neürotrichus gibbsii* (Dalquest and Orcutt 1942) to 141.7 g for *Scapanus townsendii* (Pedersen 1963). *Scapanus latimanus* and

Scalopus aquaticus exhibit the most extensive amount of geographic variation in size. Total length in *S. latimanus* females ranges from 135 mm in specimens from Baja California to 193 mm in specimens from northern California (Yates 1978). Males average slightly larger. A similar variation occurs from north to south in *S. aquaticus.*

Individual size measurements such as those presented above can be misleading if sex is not taken into account. All species of fossorial talpid in North America exhibit significant sexual dimorphism, with males averaging larger in all species. The typical secondary sexual variation found in moles is illustrated in table 2.1 for two samples of *Scalopus aquaticus* from Texas. It is interesting to note that the two species of mole that are not strictly fossorial, *Condylura* and *Neürotrichus,* do not show significant size differences due to sex.

Skull and Dentition. The skulls and dentition of all genera, although differing in detail, exhibit the same basic plan (figure 2.5). The cranium is dorsoventrally compressed (less so in *Condylura*) and the rostrum is long and narrow in most species. Sutures in most species ossify early, making age determination using suture closure unreliable. Ossification of the maxillary bones, however, occurs only in adults, leaving the roots of the upper molars exposed in immature specimens.

The first upper incisor is flattened in all species and lacks the elongated crown characteristic of shrew incisors. The number of teeth varies between genera from 36 in *Neürotrichus* and *Scalopus* to 44 in the other three genera. The overall dental formula is 3/2–3, 1/0–1, 2–4/2–4, 3/3 = 36–44. Despite the variation in dental formula, all talpids have premolars that are sharply differentiated from the molars and the upper molars are always dilambdodont.

The zygomatic arch is complete in all species but lacks jugal bones. The auditory bullae, although present, may be complete or incomplete.

Pelage. The most striking characteristic of the pelage of moles is its soft, velvetlike appearance. The velvet texture is more pronounced in fossorial genera, where the hairs are nearly equal in length and no distinct underfur is present (Jackson 1915). In *Neürotrichus* and *Condylura* some of the hairs are longer and coarser than others, producing a less silky appearance. Jackson (1915) reported that the basal pelage consisted of a series of transverse vermiculations (most pronounced in *Scalopus* and *Scapanus,* least in *Neürotrichus*), which were due to structural as well as chromatic differences. Each hair consisted of alternating normally pigmented cylindrical sections 1 to 2 mm long with finer flat sections 0.2 to 0.5 mm long that were unpigmented. The flat sections act as hinges, allowing the hair to bend forward or backward with little friction. This condition allows moles to move forward or backward in their tunnels with ease.

Molt lines, appearing as a sharp line of demarcation between old and new pelage, are often seen in all species of North American mole. The actual time of

TABLE 2.1. Results of analysis of variance between males and females for two samples of *Scalopus aquaticus*

| | Conroe, Texas | | | | | | Rockport, Texas | | | | | |
| | Males | | Females | | | | Males | | Females | | | |
Variate	N	Mean	N	Mean	CV	F	N	Mean	N	Mean	CV	F
Total length	19	151.8	12	140.2	3.9	30.7[a]	8	140.0	10	131.0	4.1	11.8[a]
Tail length	19	24.4	12	23.2	12.4	1.1	8	24.2	10	21.6	6.9	12.5[a]
Hind foot length	19	19.5	12	18.5	5.5	7.1[a]	10	17.6	10	16.3	5.9	7.8[a]
Width of forepaw	19	16.3	12	14.9	7.8	9.8[a]	10	14.8	12	13.4	6.1	13.3[a]
Length of forepaw	19	21.1	12	19.6	4.1	25.8[a]	10	18.4	12	17.9	5.0	1.2
Greatest length of skull	19	32.8	12	31.7	2.1	20.2[a]	10	31.0	12	30.3	2.1	6.3[b]
Basilar length	19	27.3	12	26.3	2.4	16.3[a]	10	25.9	12	24.8	2.3	17.3[a]
Mastoidal breadth	19	17.3	12	16.8	2.3	10.3[a]	10	16.6	12	16.2	1.7	10.7[a]
Interorbital breadth	19	7.1	12	7.0	3.4	0.8	10	6.8	12	6.6	3.3	2.0
Length of maxillary toothrow	19	10.1	12	9.7	2.7	16.5[a]	10	10.2	12	9.8	3.2	9.0[a]
Length of palate	19	14.2	12	13.7	2.8	11.7[a]	10	14.1	12	13.5	3.9	6.2[b]
Width across M2-M2	19	8.9	12	8.6	3.5	11.3	10	9.6	12	8.7	4.1	3.9[c]
Width across canines	19	3.8	12	3.7	3.7	3.6[c]	10	3.8	12	3.6	6.8	3.9[c]
Depth of skull	19	9.8	12	9.4	2.0	22.4[a]	10	9.2	12	9.2	4.1	0.1

SOURCE: Yates and Schmidly 1977
[a]Probability level of 0.01
[b]Probability level of 0.05
[c]Probability level of 0.10

molt appears to be related to climatic factors and is highly variable between genera (Jackson 1915). In most genera there is a molt twice a year roughly corresponding to spring and autumn.

Brown, yellow, orange, and olivaceous tints occur on the snout, chin, wrist, and other parts of the body of moles. This has led many authors to characterize, partially or completely, various subspecific and specific forms using these chromatic variations (Yates and Schmidly 1975). Jackson (1915), Eadie (1954), and Yates and Schmidly (1975) found such conditions common in *Scalopus, Condylura,* and *Parascalops.* With the exception of occasional white spots and lines, which Jackson (1915) referred to as "partial albinism," it appears that the activity of skin glands, not genetic variation in pigmentation, is responsible for these occurrences. Eadie (1954) regarded these chromatic variations (more pronounced in males, especially during the breeding season) as temporary stains produced by subdoriferous and perineal glands.

PHYSIOLOGY

Respiration. The basic physiology of talpids, especially with regard to thermoregulation and metabolism, is poorly known. The few studies that have been conducted in this area are inconclusive. Pearson (1947) reported an oxygen consumption rate for an adult star-nosed mole as 4.2–4.5 cc/g/h. The only other published account, by Wiegert (1961), averaged 1.84–2.37 cc/g/h. The animals involved in the latter study were juveniles, leading Wiegert (1961) to suggest a low activity rate or a decrease in basic net rate of metabolism if the juveniles had only partially functional homeothermy.

Pedersen (1963) reported no unusual physiologi-

cal abnormality in blood or urine analyses performed on Townsend's moles from Tillamook County, Oregon. The results of these analyses are presented in tables 2.2 and 2.3, respectively. Albumin was present in the urine in amounts of 3+ and the red blood count and hematocrit levels were only slightly higher than levels recorded for humans.

Senses. Considerably more data are available regarding the various senses of moles. The eyes of all moles are greatly reduced in size and are apparently useless except for the possible ability to detect light (Slonaker 1902; Godfrey and Crowcroft 1960). Although Dalquest and Orcutt (1942) found that shrew moles did not respond to bright light flashed 2.5 cm from their eyes, others felt that these animals can at least distinguish light from dark (Arlton 1936; Godfrey and Crowcroft 1960). The eye itself is a small black sphere that does not occupy a socket but lies far forward on the maxillary region of the skull. Ocular muscles are present (at least in *Scalopus*) but oculomotor nerves are lacking; the optic nerve is present but attenuated (Gaughran 1954). The optic nerve cells sometimes form a layer several cells thick, or they may be crowded into a mass (Arlton 1936).

Hearing appears to be reasonably well developed in most genera (Slonaker 1920; Hamilton 1931; Arlton 1936; Gaughran 1954). With the exception of loss of the pinna, no atrophy of the ear has occurred. In contrast, the structures of the middle and inner ear are normal and relatively large (Arlton 1936). The acutal degree of acuteness among the various genera is not known. Dalquest and Orcutt (1942) found that *Neürotrichus* did not respond to the sound of the ordinary human voice, but high-pitched sounds in the range of 8,000 to 30,000 vibrations per second produced an contrast, tactile ability is well developed in all species. The tactile function of the nasal rays of *Condylura* has already been discussed. This species makes great efforts to keep the nose clean by frequent washings in available water. Structures similar to those found in the nasal rays of *Condylura* have been reported for the European mole, *Talpa europaea* (Godfrey and Crowcroft 1960), but not for other North American species. All genera of North American mole possess numerous tactile hairs on the snout, on the dorsal surface of the head, along the borders of the forepaws, and possibly along the tail (Hamilton 1931; Eadie 1939; Dalquest and Orcutt 1942). Such a highly developed tactile sense must assuredly provide a substantial adaptive advantage to a blind fossorial mammal by providing a wide array of information regarding orientation to the surroundings, the presence of food items, and potential danger.

An annual swelling of the tail occurs during the winter and spring months in *Condylura* (Hamilton 1931). The function of this swelling is not known but it appears to be caused by deposition of fat in the tail. Eadie and Hamilton (1956) noted the great majority of males had swollen tails prior to and during the breeding season; most did not after the breeding season. This led these authors to postulate that the function of this fatty tissue in male *C. cristata* may be to act as a temporary reservoir of energy during the breeding season. Females, however, also exibit periodic enlargement of the tail during winter and spring (Hamilton 1931).

GENETICS

Lack of information regarding the genetics of moles may be partially attributed to the highly fossorial nature of many species. Such a secretive existence has made it particularly difficult to acquire live specimens for genetic studies. Although several authors (Meylan 1968; Gropp 1969; Brown and Waterbury 1971; Lynch 1971; Yates and Schmidly 1975; Tsuchiya 1979) have reported on the karyology of various species of talpid, our knowledge of the karyotypes of moles is limited for the most part to fundamental and diploid numbers of a few specimens with only implied homology of elements (table 2.4). The only published C- and G-banded karyotype of a mole was by Yates et al. (1976). Electrophoretic and immunological distance information are also lacking for the Talpidae.

Cytological investigations of many fossorial rodents (such as *Geomys, Thomomys, Spalax,* and *Ctenomys*) have revealed considerable chromosomal variation and polymorphism. Moles, however, appear to be karyotypically conservative (Yates and Schmidly 1975). No intraspecific karyotypic difference has been found in any North American species. All North American species except *Neürotrichus* have a diploid number (2N) of 34. *Neürotrichus* has a diploid number of 38. It is interesting that many species of mole from Europe and Japan also have 34 chromosomes. The fundamental number (FN) in North American species ranges from 62 in *Parascalops* to 72 in *Neürotrichus*. (See table 2.4.)

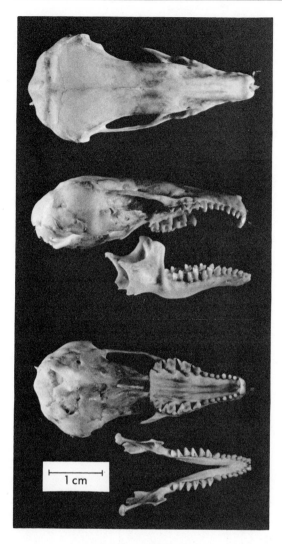

FIGURE 2.5. Skull of the hairy-tailed mole (*Scapanus townsendii*). From top to bottom: lateral view of cranium, lateral view of mandible, dorsal view of cranium, ventral view of cranium, dorsal view of mandible.

immediate response. Pederson (1963) found that Townsend's moles did not respond to normal household noises but would respond to tapping on the side of their cage. Arlton (1936), however, noted that captive *Scalopus* did respond to household noises. How much of these latter responses is actually due to acoustical acuity and how much is due to tactile ability are yet to be determined.

The sense of smell apparently is poorly developed despite the fact that the nasal passages are longer than in most other mammals. Although the sense of smell may play some role in food recognition, it does not appear to be of much use in locating it.

Numerous authors have reported no response until food was placed within a few centimeters of the nose and in most cases acutal contact was required (Hamilton 1931; Arlton 1936; Dalquest and Orcutt 1942). In

TABLE 2.2. Hematology of *Scapanus townsendii* from Oregon

Specimen Number	Hemoglobin[a]	WBC	Hematocrit[b]	RBC[c]	White Blood Cells			Lets
					PMN[d]	SL[e]	Mono[f]	
3	21.63	8,150	60					
60	18.20	5,600	49	7.03	39	54	7	normal[g]
76	17.60	7,900	48	6.23	25	74	1	normal[g]
87	16.80	6,400	45	5.86	50	50	0	normal[g]
88	17.60	4,200	48	6.01	70	30	0	normal[g]
89	15.00	5,050	35	5.48	56	40	3	normal[g]
93	11.60	2,750	40					

[a]Grams/100 cubic centimeters of blood
[b]Packed cell volume in percent
[c]Red blood cells in millions/cubic millimeter of blood
[d]Polymorphonuclearneutrophil/100 cells
[e]Small lymphs/100 cells
[f]Monocyte/100 cells
[g]150,000 to 300,000/cubic millimeter

TABLE 2.3. Urinalysis of *Scapanus townsendii* from Oregon

Specimen Number	Reaction	Sugar[a]	Albumin[b]	Miscellaneous
22	8.0	negative	3	
40	8.0	negative	4	
42	6.5	negative	3	
56	6.5	negative	4	
62	8.0	negative	4	
97	6.5	negative	3	some triple phosphate
137	7.0	1 plus	3	and uric acid crystals
138	6.5	trace	3	
140	7.0	trace	3	
199	7.0	2 plus	4	
200	6.0	trace	3	

[a]Robert's test
[b]Clinitest

TABLE 2.4. Summary of numbers and morphology of chromosomes for selected moles

Taxon	Males	Females	Biarmed	Acrocentric	2N	FN
Scalopus aquaticus (Yates and Schmidly 1975)	22	1	32	0	34	64
Parascalops breweri (Gropp 1969)	2	0	30	2	34	62
Scapanus latimanus (Lynch 1971)	1	1	32	0	34	64
Neurotrichus gibbsii (Brown and Waterbury 1971)	0	1	38	0	38	72
Condylura cristata (Meylan 1968)					34	64
Talpa europaea (Gropp 1969)	10	9	32	0	34	64
Urotrichus talpoides (Tsuchiya 1979)					34	

SOURCE: Modified from Yates and Schmidly 1975

The sex chromosomes in all species examined are the expected XX in females and XY in males. The Y chromosome is a minute dot and appears to be similar in all species of talpid. Yates et al. (1976) reported a secondary constriction corresponding to the nucleolar organizing region on a pair of autosomes in *Scalopus*. Similar constrictions have been reported for *Para-scalops* and *Talpa* (Gropp 1969). The presence of a heterochromatic polymorphism has been found in *Scalopus* (Yates et al. 1976), along with large amounts of heterochromatin (figure 2.6).

Reasons for the uniformity of mole karyotypes are difficult to explain. The possibility remains that the population dynamics of moles are very different from those of other fossorial mammals. Another possiblity is that genetic factors that affect the symmetry of the karyotype are involved. From an evolutionary standpoint talpids are a much older group than pocket gophers and may represent a group whose karyotype has been canalized (Bickham and Baker 1979).

REPRODUCTION

Breeding. Breeding in moles occurs only once a year, usually during late winter or early spring. The peak of the breeding season varies geographically and may differ by several months in a species with an extensive geographic range. Conaway (1959) reported the peak of the breeding season for *Scalopus* in Wisconsin as the last week in March and the first week in April. *Parascalops* from New Hampshire have a similar breeding season (Eadie 1939). Davis (1942) and Lowery (1974) reported that the breeding season of the eastern mole in Texas and Louisiana began in early February. Yates and Schmidly (1977) believed it began as early as January in those regions. Moore (1939) stated that Townsend's moles bred between early February and early March, whereas *Condylura* males are reported to be reproductively active from January through June (Eadie and Hamilton 1956; Hamilton 1931). *Neürotrichus* apparently has a more extensive breeding season, ranging from late February until August (Dalquest and Orcutt 1942), and breeding may occur throughout the year (Dalquest 1948).

Both sexes of *Condylura* and *Parascalops* have been reported to breed when less than a year old (Eadie 1939; Eadie and Hamilton 1956). Conaway (1959) found no evidence that female *Scalopus* breed during the year they are born. He found no females one year old or older, however, to have passed through a breeding season without breeding.

Gestation and Litter Size. The exact gestation period is not known for any North American mole. It is generally assumed to be from four to six weeks. Most genera have two to five young (Eadie 1939; Dalquest 1948; Scheffer 1949; Conaway 1959), although *Condylura* may have up to seven (Davis and Peck 1970); one *Parascalops* was taken with eight embryos (Richmond and Rosland 1949).

Reproductive Cycle. The reproductive cycles of American moles, though differing in details, appear to be quite similar. Few species have been examined in depth, however. In males, the testes and associated glands become greatly enlarged prior to mating. Eadie (1948*a*, 1948*b*) described the male reproductive anatomy and formation of copulatory plugs in *Condylura* and *Parascalops*. He found that the reproductive tract and associated glands may constitute as much as 14 percent of the total body weight in a breeding male. A unique system to assure mating success has been found in these two genera (Eadie 1948*a*) and may operate in other moles as well. Corpora amylacea are secreted by the paried prostate glands and a substance secreted from the Cowper's gland causes it to coagulate. This causes the formation of a copulatory plug in

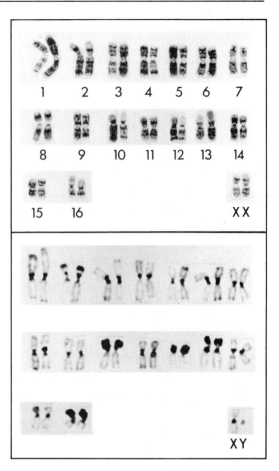

FIGURE 2.6. G-banded (top) and C-banded (bottom) karyotype of the eastern mole (*Scalopus aquaticus*).

the female, which retains the seminal mass, thus assuring mating success. The vagina remains sealed in the female of all species until follicles with antra appear in the ovary. This condition, coupled with the lack of a scrotum in males, makes accurate sex determination of nonbreeding moles difficult without dissection.

The factors that trigger the complex reproductive cycles in these blind subterranean mammals are not known. Neither temperature nor relative day length appears to be involved. Conaway (1959) suggested that internal rhythms or other environmental factors may be of primary importance in regulating the cycles. Pévet et al. (1976) suggested that, at least in the European mole, the pineal gland may play an important role in regulating the reproductive cycle.

ECOLOGY

Habitat. Moles occur in a variety of habitats from subalpine meadows in the case of some populations of *Scapanus townsendii* (Johnson and Yates 1980) to lowland swamps in the case of *Condylura cristata*. The three highly fossorial genera tend to prefer drier soils than do *Neürotrichus* and *Condylura*, but all moles prefer soils where burrowing is easy and are usually absent from heavy clay and stony or gravelly soils (Jackson 1915). The degree of rockiness appears to be unimportant to *Parascalops,* but soils with high clay or moisture content are avoided (Eadie 1939). In addition to soil type, the soil condition, moisture, and availablity of food items appear to be the most important factors affecting the presence of moles (Arlton 1936). Moles tend to be absent altogether from arid lands (Silver and Moore 1941). Soil types otherwise suitable for habitation that are too wet or too dry are often avoided by the highly fossorial species, such as *Scalopus* (Davis 1942; Glass 1943).

Tunnels and Burrows. All species construct two basic types of tunnel: deep, more permanent tunnels, and shallow surface runways. Differences in the extent and nature of these tunnels occur between most species. *Neürotrichus* most commonly constructs shallow burrows beneath decaying leaf litter that form complex interconnecting networks (Dalquest and Orcutt 1942). Although these burrows are little more than shallow troughs, shrew moles do construct some burrows similar to those of other mole species. These tunnels are less frequent and seldom found more than 30 cm below the surface. They are less extensive than those of other species and have open entrances (Dalquest and Orcutt 1942).

The tunnels of star-nosed moles are similar to those of the more fossorial species but are typically constructed near marshy areas or streams (Hamilton 1931; Rust 1966). Characteristic "molehills" are formed during the construction of deep tunnels.

Parascalops and *Scalopus* construct both deep and surface tunnels similar to those of *Condylura,* although loamy, well-drained soils are preferred. Deep tunnels range from 25 to 45 cm below the surface in *Parascalops* and from 15 to 60 cm deep in *Scalopus* (Hisaw 1923*b*). Many of the surface tunnels constructed during foraging are used only once; others are used frequently and may be in use for many years. Wright (1945) reported that some *Parascalops* tunnels may be in use for up to eight years. Breaks in these tunnels are usually repaired when first encountered. *Scalopus* repairs all breaks and will not tolerate openings in the burrow system. Eadie (1939) reported that artificial breaks are repaired by *Parascalops,* but some openings in the tunnel, possibly used by shrews and mice, were not repaired even though the tunnels were being used by moles.

The types of tunnel constructed by the three species of *Scapanus* are similar to those of *Scalopus.* Both surface and deep tunnels are constructed, although *S. townsendii* constructs two types of deep tunnel. The most extensive permanent system is an interconnected intricate system 15-20 cm deep. Less common tunnels are sometimes formed under fence rows, building foundations, and roadbeds. These tunnels may be 1 to 3 meters below the surface. They are frequently used by more than one individual and serve as major travel routes between areas. Deep runs are used much more frequently by this species than surface runs. The result is a more intricate network of deep

tunnels than is common in most species, and a large number of "molehills" in areas where the tunnels occur. Pederson (1963) reported mounds of this species numbering up to 805 per hectare. Silver and Moore (1941) reported that one Townsend's mole constructed 302 mounds in 77 days.

S. orarius and *S. townsendii* occur sympatrically throughout much of their range but appear to prefer different microenvironments. The coast mole seems to prefer deeper burrows and better drained soils and is less colonial than Townsend's mole (Dalquest 1948). This species is less common in agricultural lands and is often found in dense thickets and deciduous woods, which the larger Townsend's mole appears to avoid.

The methods used to construct burrows appear to be basically the same in all species. Hisaw (1923*b*) provided a detailed account of this process for the eastern mole. He found that moles do not "swim" through the soil as was commonly believed but actually dig burrows primarily with the powerful forepaws. In constructing the familiar surface runs, a lateral stroke type of digging is employed that involves one forefoot at a time. The body is rotated 45 degrees to the right if the left foot is involved. The forepaws are brought together several times to position the claws and the left forepaw is thrust upward rapidly. At the same time the opposite foot, which is braced against the burrow, is extended to create more force. The dirt is thus forced upward, forming the surface ridge. The nose is not used in loosening dirt but serves a tactile function in directing the forepaws. The procedure for forming deeper runways is basically the same except that the loose dirt is either brought to the surface to form mounds or deposited in abandoned surface tunnels.

Most talpids construct nest and rest chambers. The form and structure vary with the species, but the functional aspects of these chambers—rearing young and resting—remain rather uniform. *Scapanus townsendii* appears to be the most advanced nest builder. Nest mounds of this species are easily located during March by the above-ground appearance of a single mound 51-76 cm in diameter or larger, or an aggregation of small mounds (Kuhn et al. 1966). A single nest mound was found in Tillamook County, Oregon, measuring 142 cm × 71 cm × 36 cm (Pedersen 1963). A nest chamber 15-20 cm below ground level is most common, but chambers have been located 76-127 cm deep. The nest chamber averages 30-36 cm in diameter, with as many as 11 exists. Frequently, one exit drops vertically immediately below the chamber (Kuhn et al. 1966). The nest is composed of an inner layer of dry grass and an outer layer of green grass that is periodically replaced during the time young occupy the nest (Pedersen 1963).

Harvey (1976) located nests of *Scalopus aquaticus* 15-25 cm beneath the surface composed of coarse grass and/or leaves. The nests were 18-22 cm long, and 10-12 cm wide, and were enlargements of deep runs. He found that some moles used two to seven nests.

Nests of *Parascalops* and *Condylura* are constructed in a similar fashion in both shallow and deep tunnels and are composed of well-packed leaves and grass (Hallett 1978). Nests for the young are larger than those used for resting, but are similar in construction. Nest sites of *Condylura* are by necessity carefully chosen to be above any high water flooding (Hamilton 1931; Davis and Peck 1970).

Neürotrichus gibbsii is apparently the only mole that may nest above ground. Dalquest and Orcutt (1942) found that the species easily climbed shrubby vegetation. One nest composed of damp willow leaves was found 0.6 m above ground in an old alder stump. Racey (1929) noted that the shrew mole frequently enlarged a portion of the tunnel system to form a small sleeping chamber with a vent hole. No vegetation was found in these structures.

Home Range and Social Structure. The population demography of talpids is poorly known. The only comprehensive study of home range in a North American species was for the eastern mole (Harvey 1976), and only 12 specimens were involved in that study. Data from that study suggest that moles may range over larger areas than other fossorial mammals. The average home range of the eastern moles in Harvey's (1976) study exceeded that of many rodents (Yates and Schmidly 1977). The home range of a male eastern mole averaged almost 23 times as large as that of a male Plains pocket gopher (*Geomys bursarius*), 42 times as large as a male Botta's pocket gopher (*Thomomys bottae*), and 5 times as large as a male Ord's kangaroo rat (*Dipodomys ordii*) (table 2.5). Male moles have considerably larger home ranges than females (Harvey 1976).

The effects of home range size on deme size and gene flow are not clear but may be misleading. Moles are solitary for the most part, except during the breeding season, and the home ranges discussed above are essentially linear in nature. Above-ground dispersal is rare in most fossorial genera and usually involves juveniles exclusively (Giger 1965). Flooding may also account for some apparent dispersal. However, Giger

TABLE 2.5. Home range estimates (in square meters) of *Scalopus aquaticus* and five species of rodents

Species	Home Range	
	Male	Female
Scalopus aquaticus (Harvey 1976)	10640.0	2748.5
Dipodomys ordii (Garner 1973)	1951.7	2230.5
Dipodomys elator (Roberts and Packard 1973)	791.1	791.1
Reithrodontomys fulvescens (Packard 1968)	1859.0	2333.7
Geomys bursarius (Wilks 1963)	468.4	144.9
Thomomys bottae (Howard and Childs 1959)	250.9	128.8

SOURCE: Yates and Schmidly 1977

(1965) reported that a tagged *S. townsendii* moved 459 m in returning to its former home range after displacement by flood water. *Condylura* and *Neürotrichus* may disperse above ground more commonly, but data on home range are not available for these species.

It is doubtful that small bodies of water such as rivers and streams present significant barriers to dispersal, because most moles are good swimmers (Hamilton 1931; Schmidt 1931; Arlton 1936; Foote 1941; Dalquest and Orcutt 1942). Yates and Schmidly (1975) felt that the heavy clay soils associated with certain river systems, instead of the rivers themselves, form the real barriers to dispersal. The only species known to take to water freely is the star-nosed mole. This species has been reported swimming under the ice in winter and is frequently caught in muskrat and minnow traps set in streams and lakes (Hamilton 1931).

Hamilton (1931) suggested that *Condylura* is gregarious and perhaps colonial, although these local aggregations of animals may relate more to food supply than to social structuring. The same reasoning may explain the local concentrations of shrew moles, which Dalquest and Orcutt (1942) described as traveling in "loose bands."

Parascalops, Scapanus, and *Scalopus* appear to be solitary except during the breeding season. Giger (1973) found that movements of *S. townsendii* were confined to isolated tunnel systems or portions of larger systems. Eadie (1939) reported similar findings for *Parascalops.* Harvey (1976) found that the ranges of individual *Scalopus* overlapped, but multiple captures from a single tunnel system are rare. Eadie (1939) captured numerous males from female burrow systems during the breeding season. In late summer, after the young have left the nests, both sexes appear to associate freely for a short period of time.

FOOD HABITS

The food habits of moles have received more attention than many aspects of their biology. The highly differentiated dental complement of 44 teeth coupled with rapid digestion, short intestines, and a diet consisting mostly of invertebrates establish moles as insectivores. The diet is highly variable among species, but in general earthworms, insects, and other invertebrates compose the majority of the diet (table 2.6). Vegetation comprises a portion of the diet in most species; however, sufficient data have accumulated (Scheffer 1917; Wight 1928; Moore 1933; Pedersen 1963) to suggest that it is not accidentally ingested, as has been proposed by some investigators (Scheffer 1910; West 1910; Whitaker and Schemeltz 1974).

The eastern mole has a voracious appetite and consumes large percentages of its weight in food daily (Hisaw 1923a; Christian 1950). Hisaw (1923a) found that captive eastern moles preferred earthworms and white grubs, insect larvae, adult insects, and vegetable matter, in that order of preference. Whitaker and Schemeltz (1974) reported similar findings (see table 2.6). In captivity, *Scalopus* has been shown to

TABLE 2.6. Food habits of five species of North American mole, expressed as percentages

Species	Earthworms	Vegetation	Insects	Seeds	Beef	Misc.
Scapanus townsendii						
Wight (1928)	85.6		14.4			
Moore (1933)	76.1	15.9	7.4			
Pedersen (1963)	72.0	28.0				
Whitaker et al.						
(1979)	54.9	9.4	13.9	6.8		
Scapanus orarius						
Moore (1933)	69.9	1.2	28.7			
Whitaker et al.						
(1979)	56.2	2.3	30.0	2.0		
Scalopus aquaticus						
West (1910)	31.0	13.0	52.0			
Hisaw (1923*b*)	33.8	7.9	18.4	12.2	27.3	
Whitaker and						
Schemeltz (1974)	26.8	10.9	47.8	4.5		1.4
Neurotrichus gibbsii						
Dalquest and						
Orcutt (1942)	42.0		12.0			8.0
Whitaker et al.						
(1979)	48.5	0.6	43.3			
Parascalops breweri						
Eadie (1939)	34.0	2.0	47.0			17.0

eat anything from ground beef (Hisaw 1923*a*) to mice and small birds (Christian 1950). Yates and Schmidly (1977) found that this species did well on Alpo dog food. The average consumption rate in captivity ranges from 32.1 percent (Hisaw 1923*a*) to 55.0 percent (Christian 1950) of body weight daily. Hisaw (1923*a*) was able to show that this rate increased to 66.6 percent when an eastern mole was not fed for 24 hours.

Earthworms appear commonly to represent the most important food item in the diet of most species of mole. Whitaker et al. (1979) found this true for *S. townsendii, S. orarius,* and *N. gibbsii* in Oregon. Rust (1966) found that the stomach contents of eight star-nosed moles from Wisconsin contained almost 84 percent earthworms, whereas Hamilton (1931) found *Condylura* feeding almost exclusively on aquatic annelids and insects. The difference between the two analyses may be related to where the animals were caught. Hamilton's moles were taken near large bodies of water, while Rust's specimens came from areas with only small ponds as sources of aquatic food.

Eadie (1939) examined the stomach contents of 100 *Parascalops* and found that 47 percent of the diet consisted of insects and 34 percent of earthworms. Brooks (1923) reported that *Parascalops* in West Virginia often destroyed the nests of ground-dwelling wasps and fed heavily on the larvae and pupae. Brooks (1908) maintained captive *Parascalops* on bird eggs and meat but found that they starved when fed only vegetable matter. In contrast to Hisaw's (1923*a*) conclusions, Fay (1954) found that captive hairy-tailed moles weighing 41, 49, and 45 grams ate an average of 158.9, 132.4, and 116.0 grams of earthworms, respectively, per day. Dalquest and Orcutt (1942) found shrew moles capable of eating 1.4 times their body weight in 12 hours.

Townsend's moles appear to be limited to a narrow range of food items (table 5.5), primarily earthworms and insects (Wight 1928). Pedersen (1963) examined the stomachs of 106 males and 76 females from Tillamook County, Oregon, and found that vegetation, in the form of fleshy-fibrous grass roots, was a year-long component; 81 percent of the stomachs contained vegetation ranging from 200 to 1800 mg. A "root ball" was found in 18 percent of the stomachs. This appeared as a dark brown mass of compacted roots, averaging 10 × 5 mm in size and shaped like a kidney bean. Earthworms were found in 79 percent of all stomachs and ranged in amount from 200 to 4,000 mg. Insects were found only in trace amounts in six stomachs, and one stomach contained 100 percent mollusks. Variation in the diet of this species may be related to locality, time of year, and occurrence of locally abundant food items.

MORTALITY

Predation. Because of their fossorial habits, moles have few major predators except humans. They are, however, taken on occasion by a wide variety of organisms, especially when forced from their burrow systems or during dispersal periods. Avian predators, especially owls, appear to be the most significant predators of North American moles (Hamilton 1931; Von Bloeker 1937; Dalquest and Orcutt 1942; Parmalee 1954; Choate 1971; Giger 1973). Shrew moles appear to be easy prey for most predators, ranging from snakes to raccoons (*Procyon lotor*), dogs, and cats (Dalquest and Orcutt 1942). Star-nosed moles also spend a lot of time above ground and are fairly easy prey for most predators. In addition to avian predators, Hamilton (1931) resported skunks feeding on

Condylura, and Krull (1969) even found an eastern chipmunk (*Tamias striatus*) feeding on a star-nosed mole. Carnivores such as foxes and coyotes (*Canis latrans*) also take moles on occasion. Hamilton et al. (1937) and Eadie (1939) reported red foxes (*Vulpes vulpes*) feeding on hairy-tailed moles. In addition, snakes (Saylor 1938), bullfrogs (Heller 1927), and opossums (*Didelphis virginiana*) (Sperry 1933) have been reported feeding on *Parascalops.*

Parasites. Few comprehensive works on the parasites of moles are available in the literature. Ectoparasites have received the most attention, and a summary is provided in table 2.7. Yates et al. (1979) provided the only account concerning the ectoparasites of all North American species of mole. They found that host specificity was not pronounced among the species of ectoparasite examined, and suggested that most are acquired secondarily by moles from other rodents and insectivores that inhabit mole tunnels.

Fleas and mites are the most common ectoparasites found on moles. Thirteen species of flea and 25 species of mite have been identified from North American moles (see table 2.7). Of these, 15 species have been found only on east coast moles, 16 are restricted to west coast hosts, and 7 have been found on both west coast and east coast species. *Androlaelaps fahrenholzi* has been reported from all species of mole in North America, and *Haemogamasus liponyssoides* has been found on all species except *C. cristata.* It should be noted, however, that few *Condylura* have been examined for parasites.

Other ectoparasites reported from moles include two species of beetle in the genus *Leptinus* (Jameson 1950; Whitaker and Schemeltz 1974; Yates et al. 1979), and the louse, *Haematopinoides squamosus* (Whitaker and Schemeltz 1974; Yates et al. 1979). Densities on most hosts have been reported as light, although Yates et al. (1979) found 477 specimens of the chigger *Neotrombicula brennani* on one *S. townsendii.* Eadie (1939) reported up to 24 fleas and 20 mites per host specimen of *P. breweri.*

Little is known regarding the endoparasites of moles. Eadie (1939) found Acanthocephalid worms of the genus *Moniliformis* infecting the ileum of *Parascalops.* He also reported roundworms in the stomach of 15 percent of the *Parascalops* examined. Nematodes were reported as common in *Condylura* (Hamilton 1931). Ascarids have been reported from the stomach wall and walls of the intestines, and cestodes are occasionally found (Hamilton, 1931). Dalquest and Orcutt (1942) found roundworms in the large intestine of *Neürotrichus*, and also a single large tapeworm of the genus *Hymenoplepis.*

ECONOMIC STATUS AND MANAGEMENT

The economic status of moles has been debated for years with little consensus as to whether they are harmful or beneficial. Because of the diverse life styles and wide range of habitats occupied by moles it seems that any discussion of economic status in North American moles must, by necessity, center on individual genera. Of the five American genera, *Condylura, Neürotrichus,* and *Parascalops* must be considered either economically neutral or beneficial. All three tend to be rather localized in distribution and usually occur in areas not suitable for cultivation. The occasional damage by *Condylura* and *Parascalops* to lawns, flower beds, and golf courses is more than offset by their destruction of harmful insects and tilling of the soil.

The extent of damage caused by *Scapanus* and *Scalopus* is much greater and more difficult to assess. Moles of these genera have been associated with considerable economic loss by farmers, commercial bulb growers, produce growers, and home gardeners. Major damage by Townsend's moles is incurred as a result of one to four "molehills" discharged per day as a mole extends its tunnel system in search of food. Economic loss to dairy farmers in Tillamook County, Oregon, was estimated to be $100,000 per year as a result of forage loss, increased equipment breakdown, and poor silage quality (Wick 1962). Forage production in pastures can be reduced by 10–50 percent because of being covered by molehills. During harvest operations dirt and rocks in the molehills cause equipment breakdown and equipment stoppage, and the resultant mix of forage and dirt affects silage quality.

Commercial bulb growers, produce growers, and home gardeners frequently experience losses as a direct result of damage to bulbs and roots and the indirect effect of underground portions of plants being dessicated. Moles frequently extend tunnel systems under row crops in search of earthworms. This problem is especially pronounced in areas where *Scalopus* is common. During the process of tunneling, some bulbs and roots are injested as food, but more frequently damage to the plant results from removal of dirt around the bulb or root, leading to dessication. Lawns and golf courses are also subject to mole damage. In addition, molehills provide a medium for "weed" seed germination.

Mole damage is often confused with damage done by gophers (*Thomomys* sp.), meadow mice (*Microtus* sp.), and other small mammals. The utilization of mole tunnels by other mammals is common, especially in the Pacific Northwest (Yates et al. 1979). Although most people can tell the difference between moles, gophers, voles, and mice "in hand," their telltale sign is frequently misinterpreted. Molehills can easily be distinguished from gopher mounds by the characteristic shape and position of the burrow exit. Molehills are volcanolike in shape. The dirt is pushed from the burrow exit to form a symmetrical cone with no visible sign of an exit point. Excavation of the molehill reveals an exit point located in the center of the cone at ground level. A gopher expels dirt from its tunnel system by pushing the dirt from the burrow and depositing a mound in a "fan-shaped" configuration with the burrow entrance located at the epicenter of the mound. The gopher burrow entrance is a visible "plug" of dirt, whereas the mole burrow is unmarked by a visible dirt plug.

Meadow mice do not deposit dirt mounds, but

TABLE 2.7. Ectoparasites of North American moles

Parasite	Reference[a]	S. townsendii	S. latimanus	S. orarius	P. breweri	S. aquaticus	N. gibbsii	C. cristata
Coleoptera (beetles)								
Leptinus americana	1, 3				X			
Leptinus testaceus	4				X			
Siphonaptera (fleas)								
Ctenopthalmus pseudagyrtes	1, 3, 4				X	X		
Nearctopsylla genalis	1					X		
Nearctopsylla hygini	3				X			
Nearctopsylla jordani	4	X		X				
Stenoponia americana	1					X		
Doratopsylla blarinae	3				X	X		
Histrichopsylla tahavuana	3				X			
Megabothris acerbus	3				X			
Peromyscopsylla hesperomys	3				X			
Corypsylla jordani	4		X				X	
Corypsylla ornata	4	X	X	X		X	X	
Catallagia decipiens	4			X				
Epitedia scapani	4						X	
Anoplura (sucking lice)								
Haematopinoides squamosus	1, 4				X	X		
Acari (mites)								
Androlaelaps casalis	2			X				
Androlaelaps fahrenholzi	1, 2, 4	X	X	X	X	X	X	X
Duhaematopinus abnormis	3				X			
Eadiae brevihamata	2						X	
Eulaelaps stabularis	1, 2	X		X		X		
Euryparasitus sp.	2						X	
Euschoengastia trigenuala	1					X		
Glycypnagus hypudaei	2	X						
Haemogamasus ambulens	2	X		X				
H. harperi	1					X		
H. keegani	2, 4	X	X				X	
H. liponyssoides	1, 2, 3, 4	X	X	X	X	X	X	
H. reidi	2, 4	X		X	X			
Hirstionyssus blarinae	1, 3, 4	X		X	X	X		
H. obsoletus	2	X		X				
H. utahensis	2						X	
Ixodes angustus	4	X	X	X			X	
Labidophorus talpia	3				X			
Laelaps kochi	2						X	
Neotrombicula brennani	4	X						
Ornithonyssus bacoti	1					X		
Protomyobia brevisetosa	2	X						
Pygmephorus horridus	2			X				
P. sp.	1, 2	X				X		
Scalopacarus obesus	1					X		
Kenoryctes latiporus	1					X		

[a]References
1. Whitaker and Schmeltz (1974)
2. Whitaker et al. (1979)
3. Hallett (1978)
4. Yates et al. (1979)

their presence can be identified by small burrow openings, 1–3 cm in diameter, interconnected by a series of surface runways. Careful examination of the runways will reveal accumulations of fecal droppings at several points in the interconnected surface runway system.

Inspection of damaged plants will help in determining which animal is responsible. Direct mole damage to plant material is confined to underground parts of the plant. Damaged bulbs appear to have been shredded without visible incisor tooth marks. Gophers, because of their incisiform tooth structure and habits of "barking" or "clipping" plants, often cause considerable damage. Incisiform tooth marks and angular stem clippings are characteristic of gopher damage. Meadow mice will occasionally girdle small woody plants and sometimes the roots of fruit trees. However, their incisiform tooth marks are different in appearance, being very narrow striations.

Control Methods. Control methods vary with the animal causing the damage, making it imperative to identify the target species. There are and have been a myriad of control methods applied to moles. The most common method employed in noncommercial situations is trapping. It is practical and positive for the home owner and small properties, but time consuming and uneconomical for large commercial acreages. Trapping requires some location of the tunnel systems by probing, excavation of the trap hole, and trap placement. Traps commonly employed are "cinch" or "squeeze" traps, spear traps, and choker-loop traps. All require aligning the trap with the borrow system, placement of a dirt plug for the trigger to rest upon, and filling the trap hole with loose dirt to exclude light. The mole is trapped when it attempts to clear the runway system, discharging the trap. Live-trapping is possible for genera other than *Scalopus* and *Scapanus* by sinking a one-gallon container below the mole tunnel and covering the top with a board, or by constructing a special trap for that purpose (Yates and Schmidly 1975). But in most cases these methods are too time consuming.

Chemical control agents, including sulfate, strychnine, sodium monofluoroacetate (1080), and phosphorus compounds, were formerly the most common chemicals used commercially to control moles. Recent federal and state restrictions upon bait manufacturers and commercial applicators have diminished the use and manufacture of these chemicals. Thallium sulfate, 1080, and phosphorus compounds are no longer available to the general public or commercial bait manufacturers. Strychnine baits are available in some states for both commercial and public use. However, application of these agents is often restricted by a rigid licensing system.

Common baits used to control moles are usually treated cereal grains or composite mixtures of cereal grains, binders, and other ingredients particular to the individual bait producer. A common control agent found in commercial baits today involves an anticoagulant, a chemical that inhibits normal platelet function in the blood, resulting in hemorrhaging and death.

Application of chemical control agents to moles usually involves deposition of the bait material into the runway system or excavating the tunnel and bait placement. The most efficient and productive method is bait placement in the tunnel system using a probe. The degree of success in mole control is usually dependent upon proper bait placement, intensity of effort, and bait acceptance. Often other occupants of the mole tunnels are killed in greater numbers than the moles.

There are other methods of mole control frequently encountered, usually derived from folklore, home remedies, or other unproven methods. One of the most common home remedies used involves a noise-making device placed in the yard to scare moles from the premises. This method involves placing empty soft drink bottles at an angle with the bottom in the mole tunnel, necks sticking out. Supposedly, the wind blowing in the bottles make a piping sound that causes the mole to desert the runway. A second home remedy involves placing in the tunnel system an offensive odoriferous material or injurious mechanical device that causes physical impairment to the mole. These include broken glass, razor blades, thorns, napthalene flakes, moth balls and exhaust fumes. Unfortunately, these remedies usually prove more hazardous to the applicator than to the moles. Results are usually nonevident.

Whether conflicts with humans are counterbalanced by the beneficial aspects of moles probably varies with each situation. In the vast majority of the cases, moles will probably be judged to be more beneficial than harmful. In cases where the reverse is true, control is difficult at best.

Although moleskins are of little value today, trade in them was once a viable business in both Europe and the United States. From the 17th through 19th centuries, moleskins were used for caps, purses, tobacco pouches, and trimmings for garments. The demand for moleskins was such in the 19th century in Germany that applications were made for state protection of the mole to prevent its extinction (Godfrey and Crowcroft 1960).

Today, moleskins are no longer used in the U.S. clothing industry, although millions were once imported from Europe. Godfrey and Crowcroft (1960) reported that in 1959 approximately one million skins were still trapped in Britain. It is possible that a commercial source for the large Townsend's mole would enhance and complement control efforts.

CURRENT RESEARCH AND MANAGEMENT NEEDS

Research is needed in virtually all aspects of the basic biology of North American moles. Moles compose a major component of the North American mammalian fauna and are often of considerable economic importance, yet published information on this important group is fragmented and incomplete. The lack of information concerning talpid biology may be a major reason why control and management have been essentially ineffective.

One major area where research is needed is population demography. Detailed studies of basic life histories, social interactions, deme size, emigration and immigration, or population size are not available for talpids. Likewise, basic physiological adaptations to the fossorial niche are poorly understood. Data in all of these areas of talpid biology will be required before effective control and management measures can be developed.

Finally, fossorial mammals are distinct and similar in many features of their biology, but in some ways moles seem to differ from those patterns of variability found in other fossorial mammals, such as pocket gophers. Numerous studies need to be conducted that document the extent to which moles fit the patterns of variability found in other fossorial mammals. These kinds of data are critical in delineating those factors that, from an evolutionary standpoint, ultimately result in the patterns of variation observed in fossorial species.

LITERATURE CITED

Allen, J. A. 1891. Allen on mammals from Texas and Mexico. Bull. Am. Mus. Nat. Hist. 3:221.

Arlton, A. V. 1936. An ecological study of the mole. J. Mammal. 17:349-371.

Baker, R. H. 1951. Two new moles (genus *Scalopus*) from Mexico and Texas. Univ. Kans. Mus. Nat. Hist. 5(2):17-24.

Bickham, J. W., and Baker, R. J. 1979. Canalization model of chromosomal evolution. Bull. Carnegie Mus. Nat. Hist. 13:70-84.

Brooks, F. E. 1908. Notes on the habits of mice, moles and shrews. Bull. West Virginia Univ. Agric. Exp. Stn. 113:87-133.

―――. 1923. Moles destroy wasps' nests. J. Mammal. 4:183.

Brown, R. M., and Waterbury, A. M. 1971. Karyotype of a female shrew-mole *Neürotrichus gibbsii gibbsii.* Mammal. Chrom. News. 12:45.

Campbell, B. 1939. The shoulder anatomy of the moles. A study of phylogeny and adaptation. Am. Anat. 64:1-39.

Choate, J. R. 1971. Notes on geographic distribution and habitats of mammals eaten by owls in southern New England. Trans. Kansas Acad. Sci. 74:212-216.

Christian, J. J. 1950. Behavior of the mole (*Scalopus*) and the shrew (*Blarina*). J. Mammal. 31:281-287.

Conaway, C. H. 1959. The reproductive cycle of the Eastern mole. J. Mammal. 40:180-194.

Dalquest, W. W. 1948. Mammals of Washington. Univ. Kansas Publ., Mus. Nat. Hist. 2:1-444.

Dalquest, W. W., and Orcutt, D. R. 1942. The biology of the least shrew-mole, *Neürotrichus gibbsii minor.* Am. Midl. Nat. 27:387-401.

Davis, D. E., and Peck, F. 1970. Litter size of the star-nosed mole (*Condylura cristata*). J. Mammal. 51:156.

Davis, W. B. 1942. The moles of Texas. Am. Midl. Nat. 27:380-386.

Eadie, W. R. 1939. A contribution to the biology of *Parascalops breweri.* J. Mammal. 20:150-173.

―――. 1948a. Corpora amylacea in the prostatic secretion and experiments on the formation of a copulatory plug in some insectivores. Anat. Rec. 102:259-272.

―――. 1948b. The male accessory reproductive glands of *Condylura* with notes on a unique prostatic secretion. Anat. Rec. 101:59-79.

―――. 1954. Skin gland activity and pelage descriptions in moles. J. Mammal. 35:186-196.

Eadie, W. R., and Hamilton, W. J., Jr. 1956. Notes on reproduction in the star-nosed mole. J. Mammal. 37:223-231.

Eimer, T. 1871. Die Schnauze des Maulwurfes als Tastwerkzeug. Arch. Mikr. Anat. 7:181-191.

Fay, F. H. 1954. Quantitative experiments on food consumption of *Parascalops breweri.* J. Mammal. 35:107-109.

Foote, L. E. 1941. A swimming hairy-tailed mole. J. Mammal. 22:452.

Garner, H. W. 1973. Population dynamics, reproduction, and activities of the kangaroo rat, *Dipodomys ordii,* in west Texas. Grad. Studies, Texas Tech. Univ. 7:1-28.

Gaughran, G. R. L. 1954. A comparative study of the osteology and myology of the cranial and cervical regions of the shrew, *Blarina brevicauda,* and the mole, *Scalopus aquaticus.* Univ. Michigan Mus. Zool. Misc. Pub. 80:1-82.

Giger, R. D. 1965. Surface activity of moles as indicated by remains in barn owl pellets. Murrelet 46:32-36.

―――. 1973. Movements and homing in Townsend's mole near Tillamook, Oregon, J. Mammal. 54:648-659.

Glass, B. P. 1943. Factors governing the distribution of *Scalopus aquaticus.* Manuscript located at: W. B. Davis private library, Dept. Wildl. and Fisheries Sci., Texas A&M Univ.

Godfrey, G., and Crowcroft, P. 1960. The life of the mole (*Talpa europaea* Linnaeus). Latimer, Trend, and Co., Ltd., London. 152pp.

Gropp, A. M. 1969. Cytologic mechanisms of karyotype evolution in insectivores. Pages 247-266 in K. Bernirscke, ed. Comparative mammalian cytogenetics. Springer-Verlag, New York. 473pp.

Hall, E. R., and Kelson, K. R. 1959. The mammals of North America. Ronald Press, New York. Vol. 1. 546pp.

Hallett, J. G. 1978. *Parascalops breweri* (Bachman, 1842). Mammal. Species 98:1-4.

Hamilton, W. J., Jr. 1931. Habits of the star-nosed mole, *Condylura cristata.* J. Mammal. 12:345-355.

Hamilton, W. J., Jr.; Howley, N. W.; and MacGregor, A. E. 1937. Late summer and early fall flood foods of the red fox in central Massachusetts. J. Mammal. 18:366-367.

Harvey, M. J. 1976. Home range, movements, and diel activity of the eastern mole, *Scalopus aquaticus.* Am. Midl. Nat. 95:436-445.

Heller, J. A. 1927. Brewer's mole as food of the bullfrog. Copeia 165:116.

Hisaw, F. L. 1923a. Feeding habits of moles, J. Mammal. 4:9-20.

―――. 1923b. Observations on the burrowing habits of moles. J. Mammal. 4:79-88.

Howard, W. E., and Childs, H. E., Jr. 1959. Ecology of pocket gophers with emphasis on *Thomomys bottae mewa.* Hilgardia 29:277-358.

Jackson, H. H. T. 1915. A review of the American moles. North Am. Fauna 38. 100pp.

Jameson, E. W., Jr. 1950. The external parasites of the short-tailed shrew, *Blarina brevicauda* (Say). J. Mammal. 31:138-145.

Johnson, M. L., and Yates, T. L. 1980. A new Townsend's mole (Insectivore: Talpidae) from the state of Washington. Texas Tech. Univ. Mus. Occas. Pap. 63:1-6.

Krull, J. N. 1969. Observation of *Tamias striatus* feeding upon *Condylura cristata*. Trans. Illinois State Acad. Sci. 62:221.

Kuhn, L. W.; Wick, W. Q.; and Pedersen, R. J. 1966. Breeding nests of Townsend's mole in Oregon. J. Mammal. 47:239-249.

Lowery, G. H. 1974. The mammals of Louisiana and its adjacent waters. Louisiana State Univ. Press, Baton Rouge. 565pp.

Lynch, J. R. 1971. The chromosomes of the California mole (*Scapanus latimanus*). Mammal Chrom. News 12:83-84.

Meylan, A. 1968. Formules chromosomiques de quelques petits mammifères nord-américains. Rev. Suisse Zool. 75:691-696.

Moore, A. W. 1933. Food habits of Townsend and coast moles. J. Mammal. 14:36-40.

———. 1939. Notes on the Townsend mole. J. Mammal. 20:499-501.

Packard, R. L. 1968. An ecological study of the fulvous harvest mouse in eastern Texas. Am. Midl. Nat. 79:68-88.

Parmalee, P. W. 1954. Food of the great horned owl and barn owl in East Texas. Auk 71:469-470.

Pearson, O. P. 1947. The rate of metabolism of some small mammals. Ecology 28:127-145.

Pedersen, R. J. 1963. The life history and ecology of Townsend's mole *Scapanus townsendii* (Bachman) in Tillamook, County, Oregon. M. S. Thesis. Oregon State Univ., Corvallis. 66pp.

Petersen, K. E., and Yates, T. L. 1980. *Condylura cristata*. Mammal. Species 129:1-4.

Peterson, R. L. 1966. The mammals of eastern Canada. Oxford Univ. Press, Toronto. 497pp.

Pévet, P.; Juillard, M. T.; Smith, A. R.; and Kappers, J. 1976. The pineal gland of the mole (*Talpa europaea* L.). Part 3: A fluorescence histochemical study. Cell Tissue Res. 165:297-306.

Racey, K. 1929. Observations on *Neurotrichus gibbsii gibbsii*. Murrelet 10:61-62.

Reed, C. A. 1954. The origin of a familiar character: a study in the evolutionary anatomy of moles. Anat. Rec. 118. 343pp. Abstr.

Richmond, N. D., and Rosland [sic], H. R. 1949. Mammal survey of northwestern Pennsylvania. Pennsylvania Game Comm. and U.S. Fish and Wildl. Serv., Harrisburg. 67pp.

Roberts, J. D., and Packard, R. L. 1973. Comments on movements, home range, and ecology of the Texas kangaroo rat, *Dipodomys elater* Merriam. J. Mammal. 54:957-962.

Rust, C. C. 1966. Notes on the star-nosed mole (*Condylura cristata*). J. Mammal. 47:538.

Saylor, L. W. 1938. Hairy-tailed mole in Virginia. J. Mammal. 19:247.

Scheffer, T. H. 1910. The common mole. Kansas State Agric. Coll. Exp. Stn., Bull. 168. 36pp.

———. 1917. The common mole of eastern United States. U.S.D.A. Farmer's Bull. 583:12.

———. 1949. Ecological comparisons of three genera of moles. Trans. Kansas Acad. Sci. 52:30-37.

Schmidt, F. J. W. 1931. Mammals of western Clark County. J. Mammal. 12:99-117.

Silver, J., and Moore, A. W. 1941. Mole control. U.S.D.I. Conserv. Bull. 16. U.S. Gov. Printing Office, Washington, D.C. 17pp.

Slonaker, J. R. 1902. The eye of the common mole, *Scalopus aquaticus machrinus*. J. Comp. Neurol. 12:335-366.

———. 1920. Some morphological changes for adaptation in the mole. J. Morphol. 34:335-363.

Sperry, C. C. 1933. Opossum and skunk eat bats. J. Mammal. 14:152-153.

Tsuchiya, K. 1979. A contribution to the chromosome study in Japanese mammals. Proc. Japan Acad. 55B:191-195.

Van Vleck, D. B. 1965. The anatomy of the nasal rays on *Condylura cristata*. J. Mammal. 46:248-253.

Von Bloeker, C. 1937. Mammal remains from detritus of raptorial birds in California. J. Mammal. 18:360-361.

West, J. A. 1910. A study of the food of moles in Illinois. Bull. Illinois Lab. Nat. Hist. 9:14-22.

Whitaker, J. O., Jr.; Maser, C.; and Pedersen, R. J. 1979. Food and ectoparasitic mites of Oregon moles. Northwest Sci. 53:268-273.

Whitaker, J. O., Jr., and Schmeltz, L. L. 1974. Food and external parasites of the eastern mole, *Scalopus aquaticus,* from Indiana. Proc. Indiana Acad. Sci. 1973.

Wick, W. Q. 1962. Control moles for $1 an acre. Oregon's Agric. Prog. 9:10-11.

Wiegert, R. G. 1961. Nest construction and oxygen consumption of *Condylura*. J. Mammal. 42:528-529.

Wight, H. M. 1928. Food habits of Townsend's mole, *Scapanus townsendii* (Bachman). J. Mammal. 9:19-23.

Wilks, B. J. 1963. Some aspects of the ecology and population dynamics of the pocket gopher (*Geomys bursarius*) in southern Texas. Texas J. Sci. 15:241-283.

Wright, P. L. 1945. *Parascalops* tunnel in use after eight years. J. Mammal. 26:438-439.

Yates, T. L. 1978. Systematics and evolution of North American moles (Insectivora: Talpidae). Ph.D. Dissertation. Texas Tech. Univ. 304pp.

Yates, T. L.; Pence, D. B.; and Launchbaugh, G. K. 1979. Ectoparasites from seven species of North American moles (Insectivora: Talpidae). J. Med. Entomol. 16:166-168.

Yates, T. L., and Schmidly, D. J. 1975. Karyotype of the eastern mole (*Scalopus aquaticus*), with comments on the karyology of the family Talpidae. J. Mammal. 56:902-905.

———. 1977. Systematics of *Scalopus aquaticus* (Linnaeus) in Texas and adjacent states. Texas Tech. Univ. Mus. Occas. Pap. 45:1-36.

Yates, T. L.; Stock, A. D.; and Schmidly, D. J. 1976. Chromosome banding patterns and the nucleolar organizer region of the eastern mole (*Scalopus aquaticus*). Experientia 32:1276-1277.

TERRY L. YATES, Department of Biology and Museum of Southwestern Biology, University of New Mexico, Albuquerque, New Mexico 87131.

RICHARD J. PEDERSEN, U.S. Forest Service, Shasta-Trinity National Forest, Redding, California 96001.

3

Bats

Vespertilionidae and Molossidae   Stephen R. Humphrey

NOMENCLATURE

COMMON NAME. Little brown bat
SCIENTIFIC NAME. *Myotis lucifugus*
SUBSPECIES. *M. l. alascensis, M. l. carissima, M. l. lucifugus,* and *M. l. occultus.*

COMMON NAME. Gray bat
SCIENTIFIC NAME. *Myotis grisescens*
This is a monotypic species, with no subspecies named.

COMMON NAME. Indiana bat
SCIENTIFIC NAME. *Myotis sodalis*
This is a monotypic species.

COMMON NAME. Big brown bat
SCIENTIFIC NAME. *Eptesicus fuscus*
SUBSPECIES. *E. f. bahamensis, E. f. bernardinus, E. f. dutertreus, E. f. fuscus, E. f. miradorensis, E. f. osceola, E. f. pallidus,* and *E. f. peninsulae.*
Of the ten recognized subspecies, eight occur in North America north of Mexico. However, taxonomic revision of the species is in progress.

COMMON NAME. Red bat
SCIENTIFIC NAME. *Lasiurus borealis*
SUBSPECIES. *L. b. borealis, L. b. ornatus,* and *L. b. teliotis.*
Three of the six subspecies occur in North America north of Mexico.

COMMON NAME. Free-tailed bat, Brazilian free-tailed bat, Mexican free-tailed bat, guano bat.
SCIENTIFIC NAME. *Tadarida brasiliensis*
SUBSPECIES. *T. b. bahamensis, T. b. cynocephala,* and *T. b. mexicana.*
These subspecies occur in North America north of Mexico.

DISTRIBUTION

The mobility of bats results in frequent extralimital records of individuals disoriented during migration or wandering away from their normal range. To the extent possible, the distribution maps (figures 3.1–3.6) exclude such records to show areas of normal residence.

The little brown bat occurs in cool temperate and boreal zones across North America. The species is absent from the Great Plains and warm temperate zones, except at sites with permanent water in the Southwest (figure 3.1). Winter and summer ranges are similar in the east, but in the west the winter range is unknown (Barbour and Davis 1969). Errant individuals from the Southeast, Kansas, and Texas have been excluded from the map.

The gray bat lives in cave regions of the Midwest and Southeast (figure 3.2). Summer and winter ranges are similar but not identical, because different caves are occupied seasonally. About 95 percent of all gray bats are thought to hibernate in only nine caves (Tuttle 1979).

The Indiana bat inhabits the cool temperate zone of the eastern United States. Its summer range is shown in figure 3.3. In winter, the range is much smaller, as the species retreats to caves. A total of 53 winter populations is reported (Humphrey 1978), though a few more have now been found. About 87 percent of the Indiana bats known to be alive winter in only seven caves (Humphrey 1978). Excluded from the map are two records from the Florida panhandle.

The big brown bat occurs widely over southern Canada, the United States, and western and central Mexico (figure 3.4). This species is abundant in the Midwest but is rare in the southeastern states, central Texas, and the northernmost portions of its range (Barbour and Davis 1969; Humphrey 1975; Schowalter and Gunson 1979). Its summer and winter ranges are very similar.

The red bat ranges from southern Canada south through Central America (figure 3.5) but is absent from much of Florida and the arid mountain and high plateau country of the western United States and central Mexico. Winter range is roughly south of San Francisco and the Ohio River (Barbour and Davis 1969). Individuals also have been reported at sea off Nova Scotia and on Bermuda (Hall and Kelson 1959). This species is consistently abundant in the eastern

United States but rare in the West (Barbour and Davis 1969; Humphrey 1975; LaVal and LaVal 1979).

The free-tailed bat occurs from the southern United States to South America (figure 3.6). Extralimital records have been excluded from British Columbia, Ohio, Illinois, Nebraska, and most of Kansas. Winter range is similar to summer range in the southeastern and far western states, but populations are absent from Texas and Oklahoma west to Nevada and southernmost California during winter (Cockrum 1969).

DESCRIPTION

Species of bat are relatively difficult to differentiate without experience or comparison with museum specimens. For detailed descriptions, a workable key, and numerous photographs, refer to Barbour and Davis (1969).

The little brown bat has banded, glossy brown fur, because of the yellow tips of the hairs. The forearm measures 34–41 mm. The feet have sparse, long hairs that extend beyond the ends of the claws. The calcar is straight. The facial skin is pigmented dark brown.

The gray bat has unbanded fur on its back, uniformly gray (or sometimes dull orange) from base to tip. The forearm is 40–46 mm long. The feet are relatively large, and the wing skin is attached at the ankle rather than at the base of the toe.

The Indiana bat has banded fur with a surficial

FIGURE 3.1. Distribution of the little brown bat (*Myotis lucifugus*).

FIGURE 3.2. Distribution of the gray bat (*Myotis grisescens*).

FIGURE 3.3. Distribution of the Indiana bat (*Myotis sodalis*).

FIGURE 3.4. Distribution of the big brown bat (*Eptesicus fuscus*).

FIGURE 3.5. Distribution of the red bat (*Lasiurus borealis*).

FIGURE 3.6. Distribution of the free-tailed bat (*Tadarida brasiliensis*).

chestnut color, except at each shoulder, where the hairs lack chestnut tips and appear dark brown. The forearm measures 35–41 mm. The feet are small and contrast with those of the little brown bat by lacking hairs that extend beyond the claws. The calcar has an outward bow or keel. The facial skin is light brown or slightly pink.

The big brown bat has glossy, dark brown fur, with yellow or reddish tips. The forearm is 42–51 mm long. The calcar has a keel on it. (Figure 3.7.)

The red bat ranges in color from bright orange in the north to pale yellow orange in the south. Sexual dichromatism results from the gray-tipped hairs of females. The forearm is 35–45 mm long, the ears are short and round, and the interfemoral skin is heavily furred.

The free-tailed bat has short fur that is uniformly brown or gray in color. The forearm is 36–46 mm long, and the wings are very long and narrow. Half of the tail extends beyond the interfemoral skin. The ears attach to the forehead between the eyes. Stiff hairs as long as the foot extend from the toes.

PHYSIOLOGY

Thermoregulation in Summer. Though active bats have a normal body temperature of about 37° C (Lyman 1970), temperate zone bats are heterothermic during reproduction (Henshaw 1970). Female little brown bats stop regulating their body temperatures for the last 18 days of pregnancy and during lactation; the infants also are poor thermoregulators (Studier and O'Farrell 1972). Cessation of thermoregulation late in gestation is associated with a very rapid increase in energy use for embryo growth. The energy demand is much higher for lactation than for late pregnancy (Studier et al. 1973). Consequently, ambient roost temperature has a

direct effect on the growth rate of embryos (Racey 1973) and young (Tuttle 1975). Interaction of roost temperature and food supply is suggested by Racey's observation that the temperature effect was minimal with unlimited food but strong when food was withheld.

North American bats can tolerate ambient temperatures several degrees higher than most species of small mammal. Hot environments are tolerated by accepting a heat load of 1–2.5° C above ambient temperature and undertaking evaporative cooling by using metabolic water to pant, salivate, and lick exposed skin (Licht and Leitner 1967). However, a temperature of 43.5° C was lethal within one or two hours.

Hibernation. In autumn many species of temperate zone bat accumulate a large supply of body fat, enter a roost with a cool and stable microclimate, and go into torpor. Body temperature drops to conform with the temperature of the major heat sink (the cave wall) (McNab 1974), and heart rate slows to 10–80 beats per minute (Davis and Reite 1967), compared with a norm of about 600 beats per minute in an active bat (Lyman 1970). Metabolism proceeds at a slow rate during hibernation. The value of 0.004 ml $O_2/g^{-1}/hr^{-1}$, artificially produced at a laboratory temperature of $-5°$ C, may be the lowest metabolic rate recorded in a mammal (Henshaw 1970). Temperatures favoring the most frugal use of fat reserves by the little brown bat have been reported as 2° C (Hock 1951) and 5° C (W. Davis 1970). Deep hibernation of bats can last without arousal for two to three months (Folk 1940). The entire period of hibernation in high latitudes may last more than eight months, with brief arousals at irregular intervals. By the end of winter, almost no fat remains (Henshaw 1970), and protein may be catabolized as an energy source (Dodgen and Blood 1956). If ambient temperatures approach freezing, bats can respond

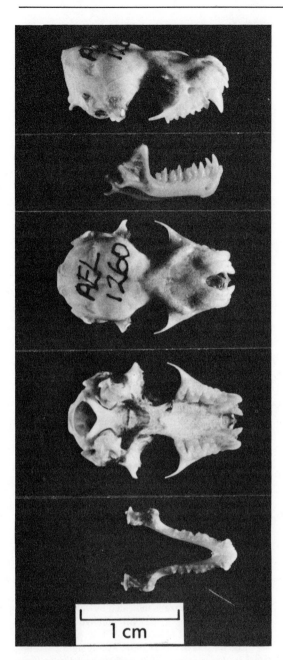

FIGURE 3.7. Skull of the big brown bat (*Eptesicus fuscus*). From top to bottom: lateral view of cranium, lateral view of mandible, dorsal view of cranium, ventral view of cranium, dorsal view of mandible.

the furred tail membrane over the ventral body surface like a blanket. During warm winter days, on the other hand, red bats do not arouse until ambient temperature rises to about 20° C. They do not become active at temperatures below which flying insects are unavailable as food (Davis and Reite 1967).

Arousal from hibernation to a metabolic state supporting sustained flight can occur in as little as seven minutes, depending on the species and ambient conditions. Heat is generated in several ways during arousal. The heart rate increases greatly and rises as high as 700–800 beats per minute (Lyman 1970). Bodies of brown fat, especially a large one between the scapulae, are warmer than the heart during arousal (Smalley and Dryer 1963), indicating their contribution to heat production. Specializations for heat production in brown fat include a high concentration of large mitochondria and a good vascular network (Henshaw 1970). Metabolic activity in the brown fat appears to be under the control of secretions of the adrenal gland (Hayward and Ball 1966). Heat is also generated by shivering of the skeletal muscles, at least in some species (Henshaw and Folk 1966).

REPRODUCTION

Anatomy. Little is remarkable about the reproductive anatomy of male bats. Males in reproductively active condition are readily identified by their descended testes and enlarged epididymides.

Female bats have a bicornuate uterus. In *Myotis* that have only one young a year, the left uterine horn is too small to accommodate an embryo, although both ovaries are functional (Guthrie and Jeffers 1938; Wimsatt 1944*a*). In *Tadarida,* only the right ovary and uterine horn are functional, even though the uterus is only slightly asymmetrical (Wimsatt 1979; Jerrett 1979). All North American bats have two pectoral mammae. Lactating females are easily distinguished by their enlarged mammary glands.

Reproductive Physiology. The reproductive physiology of bat species that hibernate involves a number of specializations. Spermatogenesis takes place in summer and early autumn, and the epididymides fill with sperm in early autumn, remaining functional through the winter and emptying by spring (Guthrie 1933; Miller 1939). The male accessory organs reach maximum development in autumn, remain in secretory distension during mating and hibernation, and quickly regress with the end of hibernation (Gustafson 1979). Females remain in breeding condition from autumn through the end of hibernation, and large, mature follicles survive through the winter. Ovulation is delayed until the bats emerge from hibernation in spring (Guthrie and Jeffers 1938; Wimsatt 1944*a*). These follicles are structurally specialized and associated with large amounts of glycogen (Wimsatt and Parks 1966). Spermatozoa are stored in the uterus and remain fertile through the winter (Wimsatt 1944*b*). A consistent feature of sperm

either by arousing and moving to a warmer site in the cave or remaining in hibernation but increasing their metabolic rate (Davis and Reite 1967). Several species can withstand brief periods of below-freezing body temperatures without ill effects. Red bats respond to subfreezing temperatures by raising the rate of respiration, assuming a spherical body shape, and stretching storage is that spermatozoan heads are oriented toward the uterine epithelium and are in contact with epithelial microvilli (Racey 1979). Here the spermatozoa are metabolically active and have access to glycogen in the epithelium and to a variety of sugars in body fluids. As in other mammals, luteinizing hormone regulates ovulation, and large amounts of this hormone are present only near the end of the hibernation period (Oxberry 1979).

In nonhibernating free-tailed bats, spermatogenesis occurs from September through winter, with spermatozoa available from 20 January to 18 April (Sherman 1937). Thus, the period of sperm storage in the epididymides is much briefer than in hibernating species. Likewise, female free-tails lack the physiological specializations of reproduction noted above for hibernating species. Sherman (1937) noted that vessicular follicles appear in the ovary shortly after birth (by late June) and remain until ovulation in the next spring.

Breeding Season. The prolonged retention of spermatozoa in male epididymides and stored secretions in male accessory glands, and the long survival of spermatozoa in female uteri, make possible a prolonged breeding season for hibernating species. Hence, breeding occurs in autumn, winter, and probably early spring for the little brown, gray, Indiana, and big brown bats. Most matings appear to occur during autumn, after arrival at the caves to be used for hibernation, but supplemental copulation occurs during winter (Guthrie 1933; Miller 1939; Mumford 1958; Hall 1962; Humphrey and Cope 1976). Red bats also mate in autumn (Barbour and Davis 1969). Presumably their brief periods of foraging activity during winter do not stimulate ovulation, so that fertilization is delayed until spring. The free-tailed bat is unlike the above species in that it breeds in spring (Sherman 1937; Davis et al. 1962; Cockrum 1969).

Gestation Period. The length of gestation was estimated at 50–60 days for the little brown bat (Wimsatt 1945), and a similar period probably applies to other species of hibernating bat in North America. The gestation period for red bats was judged to be 80–90 days (Jackson 1961). Gestation in free-tailed bats was estimated at 77–84 days (Sherman 1937) and "a few days in excess of 90 days" (Davis et al. 1962).

Reproductive Rate. Reproductive rate is a problematic measurement for bats because of the possibility that nonreproductive females may be inaccessible if not occupying the nursery roost. For example, little brown bat populations in nonnursery shelters range from 0–59 percent female during June (Schowalter et al. 1979). Females that do not join nurseries may be individuals that have not achieved sexual maturity, and their numbers are difficult to measure because of the obscurity of their roosts.

However, the reproductive rate of females in nurseries is very high in all bat species studied. For the little brown bat, the reproductive rate averaged 98.1 percent (N=420) in Indiana and 97.4 percent (N=901) in other study areas (Humphrey and Cope 1976). However, in central Alberta, near the northern edge of the species' range, only 91 percent (N=183) of the females in nurseries were pregnant (Schowalter et al. 1979). For the Indiana bat, at least 23 of 25 females (92 percent) reproduced (Humphrey et al. 1977). The reproductive rate of gray bats is less than that for other *Myotis* because of the one-year delay in sexual maturity of females, but thereafter it should be similar. Nonnursery summer populations of gray bats are occupied by adult males and yearlings of both sexes (Tuttle 1976a). Two distinctive features of big brown bat reproductive rate are unexplained by existing data. The rate is reduced by half in western populations, where the litter size is one instead of two. What environmental difference would select for this adaptation is unknown. Among eastern populations, the litter size varies among colonies and years (Brenner 1968; Mills et al. 1975), for unknown reasons. The pregnancy rate was recorded as 92 percent (N=14), by Schowalter and Gunson (1979). However, only 50 (Schowalter and Gunson 1979) to 75 percent (Christian 1956) of yearling females are parous. Nonparous adult female red bats have not been noted in published literature. However, red bats have larger litters than most other North American bats. Davis et al. (1962) found 96.4 percent of 1,710 free-tailed bats to be pregnant.

Litter Size. For the little brown, gray, Indiana, and free-tailed bats, litter size is typically one (Humphrey and Cope 1976; Guthrie 1933; Sherman 1937). Exceptions are rare. Two embryos occurred twice in 312 pregnancies of the little brown bat (Schowalter et al. 1979) but not at all in 1,649 pregnant free-tails (Davis et al. 1962).

Litter size of the big brown bat varies regionally. West of the Great Plains, typically one is born, whereas in the East two are the norm (Christian 1956). Christian's data from Maryland showed 94 percent with two young and 6 percent with one (N=17). In eastern Kansas, Phillips (1966) reported one young as typical, but Kunz (1974a) found two in 10 of 11 pregnancies in central Kansas. In central Alberta 15 of 115 (13 percent) females had two embryos and the rest had one (Schowalter and Gunson 1979). Environmental control of litter size is indicated by variation among colonies and years in eastern populations (Brenner 1968; Mills et al. 1975).

Red bats have large litters, averaging 3.24 young (N=87) (Jones et al. 1967; Barbour and Davis 1969; LaVal and LaVal 1979). High natality is adaptive in view of the high preweaning mortality rate of this species (Kunz 1971).

Breeding Synchrony. All species of North American bat bear their young in a two to five-week period in late spring to early summer. Parturition tends to occur later in higher latitudes, but much variation in each area results from differences among roost temperatures, which affect the growth rate of embryos.

Breeding Age. Most North American bats produce their first young at the age of one year, including the little brown, Indiana, and free-tailed bats (Guthrie

1933; Davis et al. 1962). No data are available on this point for the red bat.

Exceptions from this pattern are important because delaying the age at first breeding markedly slows population growth rate. The gray bat is the most specialized bat in North America in this regard. Yearling females have only small follicles in their ovaries during spring migration and some females in nurseries are not pregnant (Guthrie 1933). Males do not produce spermatozoa until their second summer (Miller 1939). Gray bats produce their first young when two years old (Tuttle 1979). Hence, a year is added to the life span required for a gray bat to replace herself and her mate, compared with species that mature earlier. Partial exceptions occur also for the little brown and big brown bats. Though nonparous female big brown bats were not noted in Kansas (Kunz 1974a), only 75 percent of the yearlings in Maryland reproduced (Christian 1956), and only 50 percent in Alberta did (Schowalter and Gunson 1979). Such latitudinal variation suggests that the length of the growing season may affect age at sexual maturity. Data on female little brown bats in nonnursery roosts and on pregnancy rate in nurseries (see "Reproductive Rates") suggest a lesser latitudinal effect on this species as well. Male little brown bats typically do not produce spermatozoa until their second summer (Guthrie 1933; Miller 1939), although some young of the year have been found mating in autumn (Fenton 1969).

ECOLOGY

The adaptations of migration and hibernation appear fully responsive to the extreme but predictable seasonal variation in the food supply of insectivorous North American bats. Merriam's (1894) observation of the importance of the growing season as a limiting factor for vertebrates applies directly to temperate zone bats, for the interaction of physiological requirements with the opportunities provided by summer and winter roosts appears to be crucial (Humphrey 1975).

Numerous lines of evidence suggest that little brown bats have reduced natality and increased mortality (in both the first year and adulthood) at high latitudes. Hence the length of the growing season affects the demographic performance of the species. In New York and Vermont, many young enter hibernation with little stored fat and may not survive the winter (Davis and Hitchcock 1965). Farther south, in Indiana and Kentucky, this phenomenon is not prominent (Humphrey and Cope 1976). At low latitudes no females occupied nonnursery shelters and most of the females in nurseries bore young, as discussed above. By contrast, 13 percent of the occupants of shelters in New York were nonreproducing females, and about half of the one-year-old females did not reproduce (Davis and Hitchcock 1965). Natality was slowed even further in Alberta (Schowalter et al. 1979). Females there comprised 59 percent of the population in nonnatal summer shelters. The pregnancy rate in nurseries was 91 percent, and a very high proportion of nonparous individuals were one year old. Pregnant females in the

yearling age class were low in weight, carried small fetuses, and had a high rate of embryo resorption. These disadvantages in survival and reproduction of young little brown bats in the north may be compensated for by their high rate of survival into old age (Griffin and Hitchcock 1965; Keen and Hitchcock 1980). The longevity record of 30 years in New England/Ontario (Keen and Hitchcock 1980), compared with 14 years in Indiana/Kentucky (Humphrey and Cope 1976), supports the long-standing hypothesis—reexamined inconclusively by Herreid (1964)—that longevity depends partly on metabolism. No reduction in fertility has been noted in old females.

Gray bats are affected by a series of limiting factors during the first year of life. For nonflying young, weight gain depends on roost temperature, and both cluster size and the heat-trapping configuration of the roost are crucial variables (Tuttle 1975). For newly recruited juveniles, growth and mortality rates are dependent on the commuting distance between roosts and foraging areas (Tuttle 1976b). The amount of weight lost during migration depends on the distance between summer and winter caves (Tuttle 1976a). Thus, large gray bat populations are promoted by high nursery cave temperatures, nearby foraging habitat, and proximity of a hibernation cave to the nursery cave (Tuttle 1976a).

Trees used as nursery roosts by Indiana bats were fairly effective at intercepting solar heat, but the resulting microclimate is relatively cool and far more variable than that of cave domes or house attics. During a summer of normal weather, the bats reared their young and left the nursery before severe autumn weather began, with sufficient time remaining to migrate, store fat, and mate before entering hibernation. However, in a summer with unusually cool weather, despite use of alternative roosts, maturation was delayed. Young were recruited over two weeks late and the last migrants departed three weeks late. The cool summer weather may have lowered survival, for some bats were subjected to freezing weather at the nursery, and little time remained for prehibernation fat storage (Humphrey et al. 1977).

Foraging Habitat. Bats forage in a great variety of habitats. They require a suitable supply of the preferred kind of food and an affordable commuting distance from the dayroost. The little brown bat forages primarily over water throughout its range (Barbour and Davis 1969; Findley 1969). However, it also forages in woods and near scattered trees on its way to water bodies and occasionally in areas where water is distant.

The gray bat forages over water, with little use of other habitats. In Tennessee, gray bats use lakes and rivers, with extensive reliance on reservoirs (Tuttle 1976b). Where only streams are available in eastern Missouri, gray bats forage over swift rivers and secluded streams with well-developed riparian vegetation, with brief flights over the adjacent floodplain (LaVal et al. 1977). Tuttle (1976b) has established an inverse relation of the body weight of newly flying young gray bats and the distance to the nearest over-water feeding areas, concluding that excessive commuting distance exacts an energetic cost that slows the weight gain of young during weaning and prior to autumn migration.

Foraging habitat of Indiana bats from nursery colonies occurs near the foliage of riparian and floodplain trees along small streams (Humphrey et al. 1977; J. B. Cope and A. R. Richter personal communication). However, adult males in an area devoid of breeding females foraged in dense forest, mostly on hillsides and ridges, whereas the nearby riparian habitat was occupied by reproducing gray bats, a larger species (LaVal et al. 1977). Pregnant female Indiana bats confined their feeding to the strip of riparian habitat, but foraging range trebled by use of adjacent floodplains after young were recruited (Humphrey et al. 1977). Belwood (1979) concluded that the tunnellike closed canopy of riparian forest is a key habitat feature because it supports enough insects to fuel pregnancy and lactation in unusually cool or late summers during which reproduction otherwise would fail.

The big brown bat is very general in its foraging habitat, feeding over pastures, ponds, creeks, rivers, mountainsides, suburban vegetation, and city streets. Hence, this species is more weakly tied to the air space over aquatic habitats than the three myotine species, above.

Similarly, the red bat forages over a variety of aquatic and terrestrial habitats (LaVal and LaVal 1979). This species differs from the foregoing ones by feeding in the open space high above streams, ponds, pastures, and forests. This high feeding occurs at altitudes of about 100–200 m and is most pronounced early in the evening, shortly after sunset. Red bats are distinctive in routinely foraging in the late afternoon on warm winter days, when temperatures exceed about 10° C (Barbour and Davis 1969; LaVal and LaVal 1979).

The free-tailed bat forages at low to high altitudes over ponds, streams, cattle watering tanks, freshwater marshes, prairies, deserts, and agricultural crops. Davis et al. (1962) thought that free-tails commuted to foraging areas as far as 97 km away from dayroosts. Humphrey (1972) watched nightly dispersal of free-tails gliding at an altitude of about 700 m above the ground, traveling at least 14 km away from the roost. Observations with radar showed flights of distances up to 25 km at altitudes up to 3100 m (Williams et al. 1973). These authors saw from helicopters that foraging was confined to altitudes of 0–200 m.

Roosting Habitat. Roosts are essential features of bat habitat, and several types serve different functions. Nightroosts provide places to rest between feeding bouts and may be arenas for important social behaviors. Dayroosts are of four types, as follows: (1) nursery roosts, which provide protection from predators and have a microclimate satisfactory for gestation, lactation, and development of the young; (2) summer male roosts, which are used as resting sites for animals that are segregated from nurseries; (3) transient roosts, which are used temporarily in spring and autumn, and for some species are major sites for copulation; and (4) winter roosts, which in regions of mild climate may be simply daytime resting sites. Where winter is more

severe, winter roosts must have stable microclimatic features that permit long-term hibernation.

Information on the roosts of the the little brown bat has been summarized by Humphrey and Cope (1976). Nursery roosts are in buildings, usually in the attics of houses or churches or under the roofs of barns. Only five nursery populations have been found in other kinds of structures. For example, Schowalter et al. (1979) found 195 nurseries in buildings and 1 in a tree. Typical nursery roosts are hot, dark, poorly ventilated, and have several small access holes. The bats commonly undergo behavioral thermoregulation, moving to relatively warm or cool spots within the attics depending on the thermal effects of daily weather. Summer male roosts for this species have been found in a great variety of structures. Winter roosts are caves and mines in which the bats hibernate. These sites have cool, stable temperatures, high humidity, and very slow air currents. Mean temperature of hibernation sites in Missouri is 8.1° C (Myers 1964), and sites with similar temperatures are selected elsewhere at comparable latitudes. However, in New Jersey and Ontario, cooler sites between 1° and 3° C are used (Hitchcock 1949; McManus 1974). Many winter roosts also serve as transient roosts in spring and autumn. In autumn, congregation at these sites (referred to as "swarming") of bats from a region's summer roosts facilitates mate finding and outbreeding.

In contrast, the gray bat uses caves for all roosting year round, and only a few cases of nurseries in other structures have been recorded. Because of the constraints of cave microclimates, only a few sites are suitable, and the distribution of gray bats appears to be roost limited (Tuttle 1979). In accordance with Dwyer's (1971) general prediction, few caves in the northeastern United States are warm enough for rearing young and few in the Southeast are cold enough for successful hibernation (Tuttle 1976a). Nursery populations succeed because the bats use caves containing structurally exceptional heat traps or because the bats actively warm their roosts with the metabolic heat from a large number of clustered individuals (Tuttle 1976a). Tuttle (1975) demonstrated that young gray bats reared at an average ambient temperature of 13.9° C reached flight age nine days later than young reared at 16.4° C; weight gain was faster in even warmer sites. In Missouri, Myers (1964) recorded temperatures of summer roosts averaging 17.5° C and those of winter roosts averaging 8.7° C. Winter roosts in Kentucky were at 10–11° C (Hall 1962).

The Indiana bat uses trees for nurseries and hibernates in caves and mines. Small nursery colonies have been found in a hollow limb and under loose bark of a dead tree. Documentation of microclimate in the latter (Humphrey et al. 1977) showed the nursery tree to be moderately effective at trapping solar heat. Nonetheless, reliance on trees for roosts exposes these animals to variation in the weather, with the growth rate of young dependent on the relative warmth of the growing season. A relatively cool year resulted in a dangerous delay of maturity until autumn. Small groups of adult males spend the summer around hibernacular caves (Hall 1962; Myers 1964). Winter habitat consists of caves and mines that have cool and stable temperatures all winter long, with rock temperature optimally 4–8° C (Hall 1962; Myers 1964; Humphrey 1978). As with the gray bat, only a few caves have proper microclimatic characteristics for Indiana bat winter habitat.

The big brown bat roosts in a wide array of natural and man-made structures in both summer and winter (Barbour and Davis 1969). Nursery roosts include attics, barns, bridges, house shutters, tree hollows, hollow cactus, and rock crevices. The microclimate of nursery roosts has not been carefully studied. Building sites are typically much better ventilated than those used by little brown bats, and Schowalter and Gunson (1979) found them to be cooler as well. Big brown bats move down into the walls of buildings to avoid excessive heat on some days (Davis et al. 1968). Solitary males in summer occur in buildings, bridges, rock crevices, and trees. In winter, big brown bats hibernate in caves, mines, storm sewers, rock crevices, and the walls of heated buildings. In the northern extreme of the species' range, the preferred cave temperature is 5.6° C and small clusters are typical (Beer and Richards 1956). However, in most of the winter range big brown bats typically occur singly, deep in rock crevices at colder sites just inside the entrances of caves and mines, or else aggregate in buildings. Individuals arouse and move from subterranean sites frequently during winter, rarely staying in one place for more than two weeks (Mills et al. 1975).

Red bats roost in trees during summer, singly or as a family of mother and young. They hang from a leaf petiole or twig in a spot protected from direct sunlight, rain, wind, and predators but warmed by a southern exposure (McClure 1942; Constantine 1966). Winter roosts have never been found but presumably are located in trees also.

The free-tailed bat forms nursery roosts in caves in the central part of its range, migrating to Mexico and extreme southern United States for the winter (Davis et al. 1962; Cockrum 1969). Free-tails that breed in Mexico migrate to low elevations for the winter (Nelson 1930). However, in California and Oregon (Benson 1947; P. Leitner personal communication), and in the Southeast (LaVal 1973), free-tails locate summer roosts in buildings and remain torpid there or move to other buildings during winter. In Florida most free-tails are absent from nursery building in winter (Sherman 1937). Summer male roosts are formed in caves or buildings. Nursery roosts in caves were found to range seasonally from 8° to 43° C (Twente 1956), but rock temperature when young were being reared was typically 38° C (Cagle 1950). In building nurseries, air temperature next to free-tails rose above 38° C (LaVal 1973) or 42° C (Licht and Leitner 1967) before the bats moved to cooler sites. LaVal (1973) recorded freetails roosting in a hollow tree during autumn.

FOOD HABITS

Most species of North American bat eat insects, although a few in the southwestern United States eat nectar, pollen, and fruit. Insectivorous bats have dental and jaw adaptations for specializing on hard or soft insects, and some species have both kinds of structure (Freeman 1979; Belwood 1979). Consequently, two foraging guilds exist, one with beetles as the dietary staple and the other dominated by soft insects—typically moths in arid regions (Black 1974) and often flies (including mosquitoes) in mesic areas. Use of soft insects other than moths in mesic areas often links bat foraging to aquatic habitat. Some bat species occupy both guilds simultaneously or switch from one to the other seasonally or regionally.

The little brown bat eats primarily soft insects. Its diet usually is dominated by flies (e.g., Whitaker et al. 1977) or sometimes by moths (Whitaker 1972). After pregnancy, the proportion of flies is reduced, and a larger portion of the diet is contributed by mayflies (Ephemeroptera) (Buchler 1976), moths (Lepidoptera), and caddisflies (Trichoptera) (Belwood and Fenton 1976), or by beetles (Coleoptera) and mayflies (Anthony and Kunz 1977). Although adult females feed selectively among the available prey, adult males and juveniles do not (Belwood and Fenton 1976; Anthony and Kunz 1977). These authors attributed the selectivity of adult females to the relatively high energy demands of lactation (Studier et al. 1973). In contrast to the usual pattern, in northern Ontario where food supplies may be scarce, adults of both sexes appeared to be nonselective (Belwood and Fenton 1976).

Studies of gray bat food habits have been conducted but not yet published. Tuttle (1976b) noted remains of mayflies under a gray bat roost.

The Indiana bat eats mostly soft insects. During pregnancy, flies (Diptera), moths, and caddisflies make up most of the diet, but moths are preferred during and after lactation (Belwood 1979). Many of the kinds of fly important in early summer reproduce in aquatic habitats (Belwood 1979).

Beetles are the main food of big brown bats (Hamilton 1933; Phillips 1966; Ross 1967; Whitaker 1972; Black 1974), regardless of region, age, sex, or reproductive condition (Belwood 1979).

The food habits of the red bat are poorly documented. A variety of hard and soft insects have been noted in their diet (Hamilton 1943; Jackson 1961; Ross 1967). Based on jaw morphology, Belwood (1979) predicted a soft diet.

The diet of free-tailed bats is dominated by moths and also includes beetles, ants (Hymenoptera), and other kinds of insect (Ross 1967).

Feeding rate during the initial, early evening foraging period of little brown bats was 7–8 insects per minute, or 140 insects in 20 minutes (Gould 1955; Anthony and Kunz 1977). Pregnant females ate an average of 2.5 g of insects nightly (2.72 kJ/g of prefeeding body weight), lactating females ate 3.7 g (4.23 kJ/g), and juveniles ate 1.8 g (2.47 kJ/g) (Anthony and Kunz 1977).

Understanding the significance of food habits to bat ecology will require a community approach (Belwood 1979) and research on the nutritional value of different insect taxa. Some preliminary data on food quality compiled by Rasweiler (1977) show wet weight percentages for protein content as 27.1 percent for beetles and 10.1 percent for flying ants.

BEHAVIOR

Site Attachment and Homing. Site attachment is the degree to which animals depend on a particular site while carrying out life support functions (Fisler 1969). Many bat species are strongly attached to nursery or hibernation sites, an association reflected in the ability of many species to return to a home roost when displaced some distance away. Site attachment has made feasible the study of long-term survivorship. Careful documentation of site attachment in a few species has been done by measuring return rates to nursery or winter roosts in successive years. Little brown bats return year after year to their natal roosts to bear young. The few adult males present in nurseries also show strong site attachment (Humphrey and Cope 1976). A similar pattern occurs for the gray bat (Tuttle 1976b), except that in some cases attachment was to a summer home range including several roost sites used by the colony. Return rates were high for adults and young females, but young males showed no summer site attachment. Gray bats return to the same winter roosts year after year to hibernate. Female big brown bats also show high attachment to their natal nurseries (Mills et al. 1975). By contrast, free-tailed bats exhibit a lack of attachment to nursery sites, rearing their young in different caves in successive years (Davis et al. 1962).

Homing performance has been demonstrated in many species of bat. Davis (1966) showed that homing usually involves movements within the familiar area encompassed by daily and annual movements. The low frequency of successful homing movements from unfamiliar release points does not clarify whether random movement or orientation ability is involved. He suggested that echolocation may be important in short-distance homing but that long-distance homing requires the use of vision. Homing success of the little brown bat appears to be independent of familiar areas, direction, or sex, and decreases exponentially with distance (Leffler et al. 1979). The Indiana bat also can home from beyond its familiar area, with success enhanced from the direction of normal migration (Hassell and Harvey 1965).

Migration. Migratory behavior differs among bat species with different adaptations to food scarcity in the temperate zone winter. The little brown bat, gray bat, and Indiana bat migrate from summer roosts to hibernation sites in caves and mines. Maximum recorded distances are 805 km for the little brown bat (Fenton 1969), 437 km for the gray bat (Tuttle 1976b), and 483 km for the Indiana bat (Barbour and Davis 1969). These movements are not necessarily latitudinal. Migration disperses the individuals concentrated at the hibernation site to summer range, in any direction (Griffin 1970; Tuttle 1976b). The big brown bat is relatively sedentary, typically undergoing local migration or none at all (Barbour and Davis 1969). The long-distance record for migration in this species is 290 km (Mills et al. 1975). Red bats are thought to have a lengthy southward migration in autumn, roosting in forests during winter and alternately foraging on warm afternoons and staying in torpor during cold periods (Barbour and Davis 1969). Some evidence suggests different movement patterns for the two sexes. However, the pattern of red bat migration has not been clearly documented, and its magnitude has been questioned by LaVal and LaVal (1979). Free-tailed bats in the southwestern states are long-distance migrants, moving to southern Texas or Mexico, with a maximum recorded movement of 1,231 km (Cockrum 1969). Most free-tailed bats migrate far enough to avoid the lack of a food supply caused by temperate zone winters. Often, large groups of free-tailed bats move in a single flock from one cave to another on their way south (Constantine 1967).

Mating. Because of the nocturnal, aerial activity of bats, their mating behavior is observed infrequently. As far as is known, all North American bats have a promiscuous mating system. Species that undergo winter-long hibernation mate primarily in autumn, with a few matings during winter involving individuals temporarily aroused from hibernation (Hall 1962). Associated with autumn mating is "swarming" behavior, a phenomenon in which large numbers of bats fly in and out of cave entrances from dusk to dawn, while relatively few roost in the cave during the day. This behavior appears to be an adaptation that enables individuals that are widely dispersed during summer to come together to breed, facilitating mate finding and outbreeding (Humphrey and Cope 1976; Cope and Humphrey 1977) as well as familiarizing inexperienced young with the location of places where they can survive their first winter (Davis and Hitchcock 1965; Fenton 1969). However, O'Farrell and Studier (1973) reported copulation of little brown bats in the nursery roost prior to migration. Red bats frequently have been reported to attempt copulation in flight (Barbour and Davis 1969), but little more is known of their mating behavior. Free-tailed bats breed in late February and early March in northwestern Mexico, before migrating northward. At this time, females are unusually docile when handled by humans (Cockrum 1969).

Mother-Young Relationships. When young bats are small, they are carried by their mother, with their milk teeth clasping a nipple and their bodies held next to the mother's flank by her folded wing. Here the young presumably nurse at will and share the mother's body heat by direct contact. Young seldom are carried on the mother's nocturnal feeding flight. However, they may be carried, clinging to a nipple, when the mothers move young to a different roost site (R. Davis 1970). At night, when mothers leave the roost to forage, the young form an active cluster that may help retard the loss of metabolic heat. The mothers return several times during the night to nurse the young, with the result that the lactation period is characterized by three pulses of foraging activity per night instead of the usual two.

In most species of bat, mothers nurse only their own young. A strong mother-young bond appears to exist, and big brown, little brown, and Indiana bats are known to retrieve their young if fallen from the roost

(Davis et al. 1968; O'Farrell and Studier 1973; Humphrey et al. 1977). In at least one species, unique vocal signatures develop in the isolation call of individual young, and this may enable individual recognition of young by their mothers; olfactory cues also appear to be important (Brown 1976). Conversely, young *M. lucifugus* can recognize the calls of their own mothers (Turner et al. 1972). By contrast, female free-tailed bats will suckle any young and appear to serve as a communal milk herd to the millions of young in a large colony (Davis et al. 1962). Hence, the mother-young bond may be nonexistent in this species.

Tandem foraging flights of paired Indiana bats at recruitment time suggest that young may develop their foraging behavior by following their mother (Humphrey et al. 1977). During the weaning period, it may be typical for young to feed on both milk and insects (Kunz 1974*b*).

Orientation Behavior. Insectivorous bats detect their prey by echolocation—perception of targets and background objects from the echoes of emitted sounds. As bats fly about in a foraging area, they emit ultrasonic echolocation sounds at the rate of about 10 per second. When they pursue and attempt to capture an insect, the rate increases until the signals run together as an apparent buzz (when the sounds are electronically transformed to the human audible range but are heard in real time). Insects are captured in the mouth, in the interfemoral skin held in a cupped shape, or in the wing skin (Webster 1963, 1967). The process of detection, pursuit, and capture typically takes one second (Simmons et al. 1978).

The free-tailed bat uses a short, constant-frequency signal as it searches for insects in the obstacle-free air in which it typically feeds (Simmons et al. 1978). Echoes from these signals indicate the presence or absence of a target, but little else (Simmons et al. 1978). When free-tails pursue an insect, they shift to frequency-modulated signals—downward sweeps of sound. The broader bandwidth results in echoes containing more information about target features and position (Simmons et al. 1978). In contrast, most North American bats feed near vegetation and routinely use a short, constant-frequency signal immediately followed by a frequency-modulated signal with multiple harmonics during foraging. The constant-frequency component is deleted during pursuit of prey, and the repetition rate of the remaining frequency-modulated signal is increased (Simmons et al. 1978). This pattern applies to the little brown, gray, Indiana, and big brown bats. Frequency-modulated signals are very effective for sensing target direction and range (to within 1 cm) but are not useful at measuring target velocity (Simmons et al. 1975). Bats using these signals can discriminate between flat targets of different size or shape (Simmons and Vernon 1971) and can recognize differences in target texture (Webster and Brazier 1965; Simmons et al. 1974).

Detection of large landscape and background features by echolocation is possible at an extreme range of perhaps 100 m, but effective range may be much less if the atmosphere is humid, the signal is weak, or the surface texture is complex (Griffin 1970; Suthers 1970). Several experiments comparing the orientation performance of blinded and untreated bats indicate dependence on vision for long-distance orientation (Griffin 1970). The visual acuity of bats is fairly good (Suthers 1966, 1970).

Bats have well-developed scent glands of several types, some with odors distinctive even to humans. Olfaction appears to play a minor role in food detection and selection by insectivorous bats (Suthers 1970) but may be very important in social communication (Brown 1976; Bradbury 1977). Little is known of the function of scents in orientation.

MORTALITY

Prenatal Mortality Rates. In the little brown bat, the number of secondary growing follicles in winter ranged from 53 to 305 and averaged 157, counting both ovaries. However, only one tertiary follicle is present, rupturing at the end of hibernation (Guthrie and Jeffers 1938). Maturing of only one follicle was confirmed by Wimsatt (1944*a*). Consequently, only one zygote implants in the uterus. Twin fetuses are rare, and only one twin birth has been reported (Humphrey and Cope 1976). The gray bat and Indiana bat also develop only one ovum, although a single case of two was noted in an Indiana bat (Guthrie 1933). From these observations, Wimsatt (1945) concluded that the number of young in *Myotis* is determined by some factor operating in the ovary to limit the number of ova shed. Because about 98 percent of the female little brown bats in nurseries reproduce each year (Humphrey and Cope 1976), little further prenatal mortality appears to occur in this species. The same appears to be true for other bat species that typically bear only one young.

In the eastern subspecies of the big brown bat a larger number of tertiary follicles develop for release in ovulation. The number of mature follicles in one sample ranged from 2 to 7 and averaged 4.2 (Wimsatt 1944*a*). In another the range was 2 to 4, with an average of 3.4 (Christian 1956). Wimsatt (1945) showed that most of these were shed and succeeded in implanting in the uterine wall, with embryos resorbed as the uterus became overcrowded during gestation. Early in pregnancy, the number of embryos in the uterus ranged from 2 to 5 and averaged 3.0, decreasing to 1 to 2 and an average 1.9 in late pregnancy (Kunz 1974*a*).

Data on the red bat (Kunz 1971) suggest an estimate of 8.6 percent loss of late-term embryos. However, this figure also includes some postnatal mortality.

Mortality Rates from Birth to Weaning. Data summarized by Foster et al. (1978) showed preweaning mortality to be 8 percent or less (including some prenatal mortality) in the Indiana bat, 10.0 percent in the big brown bat, 25.7 percent or more (excluding some prenatal mortality) in the red bat, and 1.3 percent in the free-tailed bat. These preweaning mortality rates were generally related to litter size, and the differential success of parental care was attributed to the levels of environmental stress encountered.

Juvenile and Adult Mortality Rates. The typical mammalian survival curve (Deevey 1947; Caughley 1966) exhibits high mortality of juveniles between weaning and adulthood, and then either (1) a lower but steadily increasing mortality rate, or (2) a two-stage pattern, with low, relatively constant mortality during much of adulthood, followed by sharply increasing mortality as physiological senescence takes place. Bats conform to one or the other of these patterns with one significant difference.

Juvenile mortality appears to be high in all bat species studied, but values may be confounded by a negative response to disturbance associated with the capture-recapture technique employed. Estimates of juvenile mortality (from weaning to the age of 1 year) are between 52.8 and 79.6 percent for female little brown bats (Humphrey and Cope 1976) and between 68.1 and 89.5 percent for female big brown bats (Mills et al. 1975). Adult mortality for these species differs from normal mammalian patterns by remaining constant throughout life. Using winter capture-recapture data, the annual mortality rate for adult female little brown bats is 14.3 percent in Indiana (Humphrey and Cope 1976) and 24.5 percent in Ontario (Keen and Hitchcock 1980). For adult female big brown bats it is 18.3 percent (Goehring 1972). Thus, these species appear to escape the final phase of accelerating mortality associated with aging. These low mortality rates lead to impressive longevity, with the record held by a little brown bat recaptured after 30 years of life (Keen and Hitchcock 1980). However, mean life span is much shorter. Mean life remaining after weaning has been estimated at between 1.17 and 2.15 years for female little brown bats (Humphrey and Cope 1976). From Christian's (1956) data on big brown bats, Davis (1959) calculated mean life span at birth as 1.2 years and mean life remaining at weaning as 2.3 years. Mortality of the Indiana bat is more typically mammalian, with constant mortality of 24.1 percent for adult females up to 6 years after marking, 34 percent up to 10 years, and 95.9 percent thereafter (Humphrey and Cope 1977).

The information on mortality rates given above is summarized in figure 3.8. The differences in demographic performances of the three species reflect adaptations to distinctive environmental opportunities and problems. Humphrey (1975) pointed out a tendency for bat species with large nursery colonies to have low mortality rates. Roosts offering protection from mortality agents and metabolic advantages of a stable microclimate at physiologically optimal levels should contribute to high survival rates. Density-dependent selective pressures could lead to the reduced biotic potential apparent for the little brown bat and Indiana bat. Nurseries of the big brown bat are less well protected from predators and microclimatic extremes, and the exposure of winter roosts and frequent winter movements may contribute to reduced survival of inexperienced young of the year. For this species, a higher biotic potential can accommodate higher mortality rates early in life. How the little and big brown bats escape the increased mortality expected in old age is unknown.

Mortality Factors. The mortality rates discussed above can be viewed as the baseline demographic performance of healthy populations. Markedly higher rates occur when human activities impact bat populations. The combination of natural and human-related mortality factors has caused substantial regional or range-wide reduction in the numbers of several North American species of bat. The little brown bat population in Indiana and northern Kentucky declined from about 16,000 to 3,200 (80 percent) in a decade (Humphrey and Cope 1976). The total population of gray bats has declined from about 5,000,000 to 1,199,000 (76 percent) (Tuttle 1979). The known number of living Indiana bats has declined from about 640,400 to 460,000 (28 percent) in the 15 years prior to 1975 (Humphrey 1978).

Bats are subject to a wide array of mortality agents (Gillette and Kimbrough 1970). Many of these are accidents that happen when individuals are flying or seeking shelter. The effect of these events is slow attrition of the population over time, compensated for by annual pulses of reproduction. Migration may result in many accidental deaths. Little brown and gray bats are frequently found in unnatural circumstances (often dead) in autumn and spring, and juvenile gray bats are 10 times more likely to be found than adults (Tuttle and Stevenson 1977). Most cases of predation have been recorded at roost entrances, involving aerial attacks by hawks and owls or scavenging of ill or injured bats on the ground by skunks, ring-tailed cats, and other Carnivora.

Catastrophic mortality factors that cause major losses are rare, unless caused by people, but they can have a pronounced impact because of their magnitude. Many little brown bats died when a migratory group encountered severe weather (Zimmerman 1937). Winter floods of caves have caused extirpation of little brown and Indiana bat populations (Hall 1962; DeBlase et al. 1965); about 300,000 Indiana bats died in one case. Unusually severe freezing weather also has killed many hibernating Indiana bats (Humphrey 1978). Cave flooding was given as a problem for gray bats (Tuttle 1979), although no recent cases have been documented. Additionally, collapse of a cave entrance caused gray bats to abandon a nursery cave (Tuttle 1979). Mass die-offs of free-tailed bats in caves at the time of autumn migration occur after groups of bats end exhausting migratory flights by roosting in relatively cool shelters (Davis et al. 1962; Constantine 1967); the causal mechanism was unknown but was not rabies or DDT poisoning (Constantine 1967). Davis and Hitchcock (1965) found that many young little brown bats entered hibernation without storing large amounts of fat, and they postulated that these fail to survive the winter. A. R. Richter (personal communication) noted the same phenomenon in Indiana bats along with evidence of their death. When quantified,

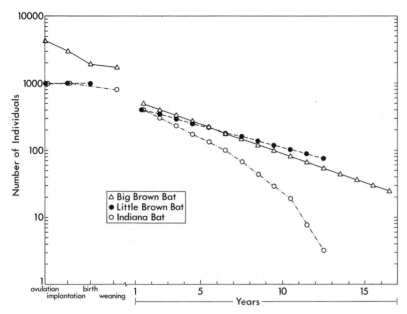

FIGURE 3.8. Survival rates of females of three species of bat. The relative placement of the curves is based on the median litter size (two for the big brown bat, one for the others), to show mortality relative to annual reproduction. Some interpretive judgments were made in selecting data for adult survival and in setting the initial values for the postweaning number.

wintertime death of underweight juveniles could account for a high proportion of the first-year mortality shown (see figure 3.7).

PARASITES. In a world-wide survey of bat endoparasites, Ubelaker (1970) reported 13 genera of protozoan, 42 genera of trematode, 24 genera of nematode (occurring in bats as adults), 10 genera of cestode, and 1 genus of Acanthocephala. Among North American bat endoparasites, only *Trypanosoma cruzi* is of potential significance in public health. This blood flagellate causes Chagas's disease in tropical America. Though *T. cruzi* is found in numerous mammal species in the southwestern United States (Wood 1962), its presence in migratory free-tailed bats and *Myotis velifer* (Cadena 1967) could be important in distributing the trypanosome in the warm temperate zone. That Chagas's disease has not become a problem in the United States may be attributable to exclusion of the assasin bug vectors (Reduviidae) from human dwellings by window screens. Constantine (1970) discussed the possibility that the bat parasite is biologically distinct from *T. cruzi*. No other endoparasite of North American bats is thought to pose a human health problem.

Ectoparasites of North American bats are of no known public health importance, even though a diversity of mites, ticks, fleas, true bugs, and flies are specialized as bat parasites. Bats occupying buildings (particularly little brown bats) commonly are thought to harbor human bed bugs (*Cimex lectularius*). This false impression stems from the frequent presence of bat bugs (*C. adjunctus*) in bat roosts; this ectoparasite is rarely a human pest.

VIRAL, FUNGAL, BACTERIAL, AND RICKETTSIAL DISEASES. Rabies is by far the most significant public health problem associated with bats. Biological and medical aspects of bat rabies have been reviewed in depth by Constantine (1970) and summarized by Trimarchi (1978), and the following is based on their accounts. Rabies is a disease of the central nervous system caused by an RNA virus that normally is transmitted by a bite by an infected mammal. Investigation of the possibility of aerosol rabies transmission in free-tailed bat caves was inconclusive. Long survival of bats capable of aggressive transmission is known to be possible only for a Neotropical vampire bat (*Desmodus rotundus*). In temperate zone bats, subclinical infections can incubate for over a year. Once aggressive and abnormal symptoms develop with virus being shed in the saliva, these bats can transmit the disease for 3–10 days before they die. Rabies infection rates of clinically asymptomatic bats in the United States usually are a fraction of 1 percent. Rates vary seasonally, being low early in the warm season but rising as high as 2 or 3 percent in autumn, when young and adult bats are dispersing and migrating. Rates also vary among species, generally being highest in red bats and other lasiurine species. One rabies epizootic has been documented in big brown bats, with a 10 percent infection rate (N=99). Pre-1970 literature reports of massive die-offs that were speculatively attributed to disease were not accompanied by virology testing.

Whether bats serve as reservoirs of rabies for other wild animals is not clear, although dead end infections of people and livestock obviously occur. Some evidence suggests that bat rabies contributes to rabies cycles in Carnivora. Annual migration to the Neotropics and roosting with vampire bats by free-tailed bats from the United States may be significant in the epidemiology of bat rabies.

Only 10 human deaths from bat-transmitted rabies have been recorded in the United States (Trimarchi 1978). Although most were from bites, 2 resulted from exposure to urine, mites, and possibly saliva in the air under free-tailed bat nursery roosts. One person who contracted rabies from a bat survived after receiving intensive treatment of the symptoms.

A long list of arboviruses known from bats was compiled by Constantine (1970). Among those found in North America are Venezuelan equine encephalitis virus in Mexico and Florida, Rio Bravo virus in the Southwest, St. Louis encephalitis virus in Texas, and Montana *Myotis* leukoencephalitis virus. Several uncharacterized viruses have been found in bats in the United States.

A number of pathogenic fungi are associated with bat roosts (Constantine 1970). Most are not truly disease organisms but are saprophytic occupants of enriched soil. The most important in North America is *Histoplasma capsulatum*, which is widespread in soil of the southern United States and the Mississippi and Ohio river valleys. Histoplasmosis often infects a large portion of a human population in asymptomatic form or as a minor respiratory illness. However, if a large number of spores are inhaled, serious disease can result, and systemic cases can be fatal. High-grade infections have resulted from inhaling dust of chicken houses, bird roosts, and bat roosts. The fungus has been found in tissues of little brown, gray, Indiana, and free-tailed bats, among others.

Little is known about bacterial diseases of bats, but they seem to be of little public health importance. Rickettsial diseases are unknown in North American bats (Constantine 1970).

AGE DETERMINATION

The age of nursing young can be determined by the predictable pattern of tooth eruption (Fenton 1970; Foster et al. 1978). Juvenile bats can be distinguished from adults by the finger joints (figure 3.9), which remain cartilaginous, translucent, and smoothly tapered in outline for some time after weaning (Barbour and Davis 1969). The joints of adults are opaque and distinctly knobby in outline. Because of the rapid development of most bats, distinguishing juveniles from adults usually is no longer consistently possible by the time autumn migration occurs. However, at northern latitudes where the growing season is short, the two age classes remain distinguishable longer, possibly through winter (Davis and Hitchcock 1965; Fenton 1970; Schowalter et al. 1979). The delayed sexual maturity of gray bats makes possible the designation of juveniles during winter and of yearlings by their unfaded pelage and nonreproductive condition during the next summer (M. D. Tuttle personal communication).

Judging the condition of the nipples also allows separation of subadults from females that have reared young. Nipples of the former are slightly pink, small, and densely surrounded by fur, whereas those of the latter are darkly pigmented, prominent, and encircled by an area of sparse fur.

The most effective technique for determining age is to tag the upper arms of juveniles with number 2 aluminum-lipped bands or size XCL celluloid rings for later recapture. Tooth wear is not useful for determining the age of bats that eat soft insects, like the little brown bat (Humphrey and Cope 1976), but it serves as a usable age estimator for species that consistently eat beetles. Christian (1956) distinguished age groups of big brown bats by tooth wear and noted the presence of annular rings in the dentine of canines. Mills et al. (1975) found a consistent relation of tooth wear to the age of banded big brown bats. Schowalter and Gunson (1979) used dental annuli of this species to compare age structure among populations.

ECONOMIC STATUS

Bats have both positive and negative economic values, although neither has been measured. On the negative side are the costs of controlling bats as house pests and dealing with suspected cases of bat-transmitted rabies. These costs appear to be relatively minor in temperate North America, contrasting strikingly with the costs of disease and blood loss in tropical livestock caused by vampire bats. On the positive side are the benefits of bats' insectivory, although our economic system places no value on free services performed in natural ecosystems. A figure of 30 million kg of insects consumed annually by free-tailed bats in the United States is estimated by considering an original population of 100 million individuals (Barbour and Davis 1969; Mohr 1972), 120 days' residence during the warm seasons (Davis et al. 1962), and the 2.5-g nightly consumption of insects by little brown bats, a smaller species (Anthony and Kunz 1977). Similar calculations indicate that documented declines of bats have resulted in the cessation of considerable insect consumption: 32 kg nightly by the 12,800 little brown bats lost from Indiana, 2,260 kg nightly by the 905,400 gray bats lost range wide, and 450 kg nightly by the 180,500 Indiana bats lost throughout their distribution. Some of the prey taken by bats are pest species, including mosquitoes eaten by little brown bats and *Myotis austroriparius* (Buchler 1976; Anthony and Kunz 1977; Zinn 1977) and alfalfa weevils (*Hypera postica*) eaten by big brown bats (Belwood 1979). The total economic cost of losing the services of bats as insect predators has not been measured.

MANAGEMENT

Pest Control. Bats living in buildings are considered pests by some because of noises, odors, droppings, and the possibility of disease. Others have chosen to coexist with bats in their attics and have done so for decades without problems that they considered important. Educating property owners about the rarity of bat-borne disease may reduce the number of requests

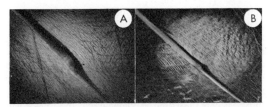

FIGURE 3.9. A, finger joints of a juvenile bat, showing smoothly tapered outlines and translucent, cartilaginous epiphyses (backlit). B, finger joints of an adult bat, showing knobby outlines and solid bone at the joint.

for pest control. If pest control is selected, the available methods include rendering the roost uninhabitable, killing the bats, or closing the entrances used by the bats at times when the bats are not present.

Several methods have been used to make bat roosts uninhabitable, with varying degrees of success. Placing moth-repellent chemicals in the roosting spots can be effective. A method based on bats' high-frequency hearing is to use an aquarium pump to blow a high-frequency dog whistle. Illuminating the roosting area with floodlights has been shown to be effective in some cases (Laidlow and Fenton 1971). Another method is to coat the roost surface with grease or some other sticky substance. The advantages of these methods are their simplicity and low cost. Failure or only temporary effectiveness of these methods would be disadvantages.

Any number of direct methods of killing bats are available, including shotgun, tennis racket, and capture of the animals in the roost or at roost entrances. However, the method preferred by exterminator companies is to spray the bats and the roost with pesticides. The advantage claimed is high effectiveness, though documentation is lacking. Disadvantages include high cost, undocumented risks of burdening a human residence with persistent pesticides, and increased exposure to moribund or convulsing bats (Kunz et al. 1977). These authors reported a case in which use of pesticides resulted in frequent occurrence of dying bats on building floors or in the adjacent yard for at least two years after application.

Methods for closing entrances that bats use to gain access to the roost vary depending on building structure and condition (see Cope 1959). If only a few cracks or holes are present, they may be plugged with oakum (petroleum-soaked rope) or other materials. If the problem involves structurally unsound eaves or roof, renovation is the most effective solution. Advantages of this approach are effectiveness and permanence. Disadvantages are the effort and high cost involved.

A significant though noneconomic advantage of control methods that exclude bats from their roost rather than killing them is that the animals have the opportunity to roost elsewhere, continuing their function as insectivores. Evidence that excluded bats successfully relocate themselves, however, is lacking except in cases where an alternate roost was already in use.

Bat Behavior toward Humans. Bats react aggressively to attempts to handle them. If unable to fly away, they usually emit an irritation buzz and bite whatever touches them. However, healthy bats avoid humans whenever possible. Reports of unprovoked attacks by bats usually result from a disoriented bat's landing on the nearest available object. Nonetheless, an unprovoked approach followed by a bite should be handled as a potential rabies transmission case. To remove a bat that has wandered into the house, the simplest method is to open a window or door to the outside, close doors leading to other rooms, and let the bat fly out. If a bat must be captured it should be picked up with a net, a can with a lid, or a leather work glove. Children should be taught to not attempt to pick up bats or other mammals encountered in unusual circumstances.

Bites by bats should be treated by washing the wound, procuring the bat for rabies testing, and seeking medical attention (Trimarchi 1978). Researchers expecting regular encounters with bats should be treated with preexposure rabies vaccine.

Bat species differ in their response to nursery disturbance by humans. Some individuals of the most tolerant species disperse to alternate roosts for a few days before reoccupying the nursery. The gray bat is one of the least tolerant species (Mohr 1972). Nursery disturbance has been a major cause of population decline and extirpation, apparently uncompensated by reestablishment elsewhere (Tuttle 1979). Likewise, the big brown bat is relatively intolerant of disturbance in the roost (Beer 1955; Hitchcock 1965; Phillips 1966; Mills et al. 1975).

Bats hibernating in caves during winter usually respond to the passing of humans by arousal from torpor, flight within the roost, and eventual reclustering and reentry into torpor. These events may take up to several hours and are fueled by stored fat. Thus, frequent human visits to a winter roost can deplete the energy supply of all population members and lower their chances of surviving the winter.

Conservation. In contrast to natural catastrophes, large-scale bat mortality caused by humans is a frequent occurrence. Reports of killing with stones, torches, and other means are given for hibernating little brown, gray, and Indiana bats (Humphrey and Cope 1976; Humphrey 1978; Tuttle 1979). Extermination of nursery colonies in buildings is a major cause of population loss for the little brown bat, big brown bat, and free-tailed bat in those parts of its range where it normally occupies buildings. Many colonies are excluded from roosts in buildings by reroofing, repairing eaves, or sealing cracks used for entry by the bats. When little brown bats are excluded from their nursery roosts, dispersal movements occur, but marked animals fail to appear in their traditional sites for hibernation; presumably most excluded animals die. Extermination and exclusion resulted in the loss of at least 52 percent of the little brown bats in 23 colonies monitored for about a decade (Humphrey and Cope 1976). Disturbance of gray bat colonies in caves has been a major cause of decline. This involves several phenomena: direct killing, intolerant reaction to nursery disturbance, and possibly mortality resulting from fat depletion by excessive winter visits by caving enthusiasts and researchers. After abandoning one cave, gray bats rarely move to an unoccupied cave, nor do nearby colonies increase in size. A number of gray bat caves also have been submerged by impoundments (Tuttle 1979). Loss of about 60,000 Indiana bats was attributed to frequent disturbance of one hibernating population by humans. About half of the total decline in Indiana bat numbers resulted from habitat alteration, in which structures built at cave entrances interfered with cave thermodynamics and made roosts too warm for bat survival. As with other species, most Indiana bats from impacted populations fail to appear at nearby caves occupied by cogeners, and presumably most die (Humphrey 1978). Whether loss of summer foraging habitat—riparian forest along midwestern streams—has contributed to the decline of Indiana bats is unknown (Humphrey et al. 1977).

Insecticides in the food web have been confirmed as causing population declines in insectivorous bats. Brains of juvenile gray bats found dead in unusual numbers below nursery roosts contained lethal concentrations of dieldrin (Clark et al. 1978). The source was thought to be aldrin sprayed for cutworm control on Missouri cornfields adjacent to gray bat foraging habitat. Clark et al. expressed concern over the persistence of dieldrin residues in the soil (Korschgen 1971) and over potential effects of other insecticides being substituted for aldrin. Geluso et al. (1976), seeking to explain a population decline and die-off of nursing young free-tailed bats, found high concentrations of DDE in brain tissue after animals were starved and heavily exercised as if migrating. In these animals, death of young was attributed to the DDE, whereas adult death was attributed to starvation (Clark and Kroll 1977). The apparent pathway in both species involved receipt of residues through the mother's milk and concentration in fatty tissues, including the brain.

Tuttle (1979) has raised the question of whether ecosystem stresses become mortality factors by impacting the food web of gray bats. Suggested impacts were deforestation and chemical pollution or siltation of waterways, which may reduce forest and aquatic insects on which gray bats feed.

Population declines have placed several species of bat in danger of extinction. Currently protected under the Endangered Species Act are the gray, Indiana, and Hawaiian hoary (*Lasiurus cinereus semotus*) bats (U.S. Fish and Wildlife Service 1978), along with two relictual eastern subspecies of the western big-eared bat (*Plecotus townsendii virginianus* and *P. t. ingens*) from the Appalachian and Ozark mountains (U.S. Fish and Wildlife Service 1979). All are classified as endangered. Although sizable declines also have occurred in other species, these bats are unique in their intolerance of roost disturbance and/or appear to have a density-dependent effect in maintaining an optimal nursery microclimate (Constantine 1967; Humphrey 1975; Tuttle 1975, 1979). The big-eared and gray bats are highly sensitive to disturbance, and roosting re-

quirements are restricted for Indiana bats in winter and gray bats in summer and winter. Loss of foraging habitat may be a big problem for Indiana and gray bats but has not been measured. Finally, all these species are reproductively specialized for slow population growth, with the gray bat being the most stable of all in the absence of human-induced mortality. In addition to those species classified as endangered, the free-tailed bat resembles the gray bat in having enormous populations that may gain a metabolic advantage from body heat warming the nursery roosts. Such a density-dependent effect would reinforce any worsening extrinsic effect. Insecticide poisoning is an apparent problem to the migratory populations of this species that feed over intensively farmed river valleys in the United States in summer and in Mexico in winter (Cockrum 1970; Geluso et al. 1976).

Management recommendations have been made for cave roosting habitat of the Indiana and gray bats (Humphrey 1978; Tuttle 1979). These emphasize acquisition and protection of the most important roosts. Cave protection involves restricted visitor access when bats are present, and avoiding disruption of normal cave microclimate that is caused by artificially reducing the air flow at entrances. In recognition of the deleterious impact of research activities, restrictions are recommended that reduce disturbance to a minimum. These involve limiting visits in hibernacula to one per winter, avoiding major banding activities where movement data already exist, and avoiding visits to gray bat nursery roosts by sampling only at cave entrances or foraging areas. Details of cave microclimate dynamics and acceptable methods of gating caves also have been published (Tuttle 1977; Tuttle and Stevenson 1978). A gate designed for Indiana bats is shown in figure 3.10. In addition, riparian forests that provide summer habitat for the Indiana bat should be protected (Humphrey et al. 1977; Belwood 1979). Both species can regain much of their former abundance if treated with respect.

CURRENT RESEARCH AND MANAGEMENT NEEDS

Many gaps exist in our knowledge of the basic biology of bats. Major areas include the role of olfaction in orientation, the role of metabolism in determining longevity, the nutritional quality of foods eaten by bats, and bat community ecology. A number of applied problems dictate research needs. Nursery populations of the free-tailed bat should be surveyed to determine if the publicized population declines are individual cases or range wide. Food web pathways for insecticide concentration in bat tissues need to be identified for gray and free-tailed bats, including both summer and winter range for the latter. The persistence of pesticides in use and potential threats of new chemicals should be evaluated (Clark et al. 1978). Alternative methods of controlling crop pests without also destroying natural predators need continued study.

The rate and impact of clearing riparian forest on population levels of Indiana and gray bats should be measured and incorporated into management planning

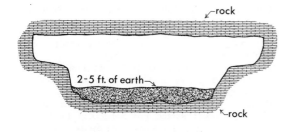

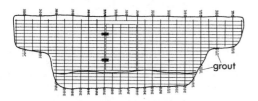

ELEVATION OF CAVE GATE

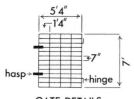

GATE DETAILS

FIGURE 3.10. Gate for a cave entrance, designed for Indiana bats (*Myotis sodalis*). The gate should open into the cave. Steel should conform to ASTM designation A36 structural steel, welding should conform to Structural Welding Code AWS D1.1-72, and grout should be one part cement to two parts sand. U bolts should be tack-welded or battered after tightening. Since installation of this gate, the hibernating population has remained stable, whereas many unprotected populations have declined. A different gate design is required for some other species of bat; this gate is not suitable for gray bats (*Myotis grisescens*). Redrawn from an engineering drawing courtesy of the St. Louis district, U.S. Army Corps of Engineers.

for these species. Research needs to be done on the effectiveness of inexpensive methods to make bat roosts in buildings uninhabitable. More information is needed on the fate of bats that are displaced from their roosts.

The management recommendations cited above are being formalized in recovery plans as part of the federal planning process for endangered species management. Implementation of thorough plans should enable the recovery of the affected species to normal population levels. Responsive management of riparian forest may be the most difficult achievement because of its widespread conversion to other land uses.

OTHER BATS

The United States and Canada are inhabited by 39 species of bat (Findley and Jones 1967; Barbour and Davis 1969, 1970). The 6 species presented above are the most abundant and widespread and are of significance because of their role as pests, insect predators, and carriers of disease, or because of their endangered

status. Included in the remainder are 4 species of a Neotropical family not mentioned above, the Phyllostomatidae. For information on species not covered here, the reader is referred to Barbour and Davis (1969).

LITERATURE CITED

Anthony, E. L. P., and Kunz, T. H. 1977. Feeding strategies of the little brown bat, *Myotis lucifugus*, in southern New Hampshire. Ecology 58:775-786.

Barbour, R. W., and Davis, W. H. 1969. Bats of America. Univ. Kentucky Press, Lexington. 286pp.

———. 1970. The status of *Myotis occultus*. J. Mammal. 51:150-151.

Beer, J. R. 1955. Survival and movements of banded big brown bats. J. Mammal. 36:242-248.

Beer, J. R., and Richards, A. G. 1956. Hibernation of the big brown bat. J. Mammal. 37:31-41.

Belwood, J. J. 1979. Feeding ecology of an Indiana bat community with emphasis on the endangered Indiana bat, *Myotis sodalis*. M.S. Thesis. Univ. Florida, Gainesville. 103pp.

Belwood, J. J., and Fenton, M. B. 1976. Variation in the diet of *Myotis lucifugus* (Chiroptera: Vespertilionidae). Can. J. Zool. 54:1674-1678.

Benson, S. B. 1947. Comments on migration and hibernation in *Tadarida mexicana*. J. Mammal. 28:407-408.

Black, H. L. 1974. A north temperate bat community: structure and prey populations. J. Mammal. 55:138-157.

Bradbury, J. W. 1977. Social organization and communication. Pages 1-72 *in* W. A. Wimsatt, ed. Biology of bats. Vol. 3. Academic Press, New York. 651pp.

Brenner, F. J. 1968. A three-year study of two breeding colonies of the big brown bat, *Eptesicus fuscus*. J. Mammal. 49:775-778.

Brown, P. 1976. Vocal communication in the pallid bat, *Antrozous pallidus*. Z. Tierpsychol. 41:34-54.

Buchler, E. R. 1976. Prey selection by *Myotis lucifugus* (Chiroptera: Vespertilionidae). Am. Nat. 110:619-678.

Cadena, A. A. 1967. Studies on flagellates of the bats *Tadarida brasiliensis* and *Myotis velifer* and of their ectoparasites in Oklahoma. M.S. Thesis. Oklahoma State Univ., Stillwater. 33pp.

Cagle, F. R. 1950. A Texas colony of bats, *Tadarida mexicana*. J. Mammal. 31:400-402.

Caughley, G. 1966. Mortality patterns in mammals. Ecology 47:906-918.

Christian, J. J. 1956. The natural history of a summer aggregation of the big brown bat, *Eptesicus fuscus fuscus*. Am. Midl. Nat. 55:66-95.

Clark, D. R., Jr., and Kroll, J. C. 1977. Effects of DDE on experimentally poisoned free-tailed bats (*Tadarida brasiliensis*): lethal brain concentrations. J. Toxicol. Environ. Health 3:893-901.

Clark, D. R., Jr.; LaVal, R. K.; and Swineford, D. M. 1978. Dieldrin-induced mortality in an endangered species, the gray bat (*Myotis grisescens*). Science 199:1357-1359.

Cockrum, E. L. 1969. Migration in the guano bat, *Tadarida brasiliensis*. Univ. Kansas Mus. Nat. Hist. Misc. Publ. 51:303-336.

———. 1970. Insecticides and guano bats. Ecology 51:761-762.

Constantine, D. G. 1966. Ecological observations on lasiurine bats in Iowa. J. Mammal. 47:34-41.

———. 1967. Activity patterns of the Mexican free-tailed bat. Univ. New Mexico Publ. Bio. 7. 79pp.

———. 1970. Bats in relation to the health, welfare, and economy of man. Pages 319-449 *in* W. A. Wimsatt, ed.

Biology of bats. Vol. 2. Academic Press, New York. 447pp.

Cope, J. B. 1959. Build bats out. Pest Control 27:28–29.

Cope, J. B., and Humphrey, S. R. 1977. Spring and autumn swarming behavior in the Indiana bat, *Myotis sodalis*. J. Mammal. 58:93–95.

Davis, D. E. 1959. Manual for analysis of rodent populations. Located at: Johns Hopkins Univ. 54pp. Mimeogr.

Davis, J. 1966. Homing performance and homing ability in bats. Ecol. Monogr. 36:201–237.

———. 1970. Carrying of young by flying female North American bats. Am. Midl. Nat. 83:186–196.

Davis, R. B.; Herreid C. F., II; and Short, H. L. 1962. Mexican free-tailed bats in Texas. Ecol. Monogr. 32:311–346.

Davis, W. H. 1970. Hibernation: ecology and physiological ecology. Pages 265–300 *in* W. A. Wimsatt, ed. Biology of bats. Vol. 1. Academic Press, New York. 406pp.

Davis, W. H.; Barbour, R. W.; and Hassell, M. D. 1968. Colonial behavior of *Eptesicus fuscus*. J. Mammal. 49:44–50.

Davis, W. H., and Hitchcock, H. B. 1965. Biology and migration of the bat, *Myotis lucifugus*, in New England. J. Mammal. 46:296–313.

Davis, W. H., and Reite, O. B. 1967. Responses of bats from temperate regions to changes in ambient temperature. Bio. Bull. 132:320–328.

DeBlase, A. F.; Humphrey, S. R.; and Drury, K. S. 1965. Cave flooding and mortality in bats in Wind Cave, Kentucky. J. Mammal 46:96.

Deevey, E. S., Jr. 1947. Life tables for natural populations of animals. Q. Rev. Bio. 22:283–314.

Dodgen, C. L., and Blood, F. R. 1956. Energy sources in the bat. Am. J. Physiol. 187:151–154.

Dwyer, P. D. 1971. Temperature regulation and cave-dwelling in bats: an evolutionary perspective. Mammalia 35:424–455.

Fenton, M. B. 1969. Summer activity of *Myotis lucifugus* (Chiroptera: Vespertilionidae) at hibernacula in Ontario and Quebec. Can. J. Zool. 47:597–602.

———. 1970. The deciduous dentition and its replacement in *Myotis lucifugus* (Chiroptera: Vespertilionidae). Can. J. Zool. 48:817–820.

Findley, J. S. 1969. Biogeography of southwestern boreal and desert mammals. Univ. Kansas Mus. Nat. Hist. Misc. Publ. 51:113–128.

Findley, J. S., and Jones, C. 1967. Taxonomic relationships of bats of the species *Myotis fortidens, M. lucifugus,* and *M. occultus*. J. Mammal 48:429–444.

Fisler, G. F. 1969. Mammalian organizational systems. Los Angeles Co. Mus., Contrib. Sci. 167:1–32.

Folk, G. E., Jr. 1940. Shift of population among hibernating bats. J. Mammal 21:306–315.

Foster, G. W.; Humphrey, S. R.; and Humphrey, P. P. 1978. Survival rate of young southeastern brown bats, *Myotis austroriparius*, in Florida. J. Mammal. 59:299–304.

Freeman, P. W. 1979. Specialized insectivory: beetle-eating and moth-eating molossid bats. J. Mammal. 60:467–479.

Gillette, D. D., and Kimbrough, J. D. 1970. Chiropteran mortality. Pages 262–283 *in* B. H. Slaughter and D. W. Walton, eds. About bats. Southern Methodist Univ., Dallas. 339pp.

Geluso, K. N.; Altenbach, J. S.; and Wilson, D. E. 1976. Bat mortality: pesticide poisoning and migratory stress. Science 194:184–186.

Goehring, H. H. 1972. Twenty-year study of *Eptesicus fuscus* in Minnesota. J. Mammal. 53:201–207.

Gould, E. 1955. The feeding efficiency of insectivorous bats. J. Mammal. 36:399–407.

Griffin, D. R. 1970. Migrations and homing of bats. Pages 233–264 *in* W. A. Wimsatt, ed. Biology of bats. Vol. 1. Academic Press, New York. 406pp.

Griffin, D. R., and Hitchcock, H. B. 1965. Probable 24-year longevity records of *Myotis lucifugus*. J. Mammal. 46:332.

Gustafson, A. W. 1979. Male reproductive patterns in hibernating bats. J. Reprod. Fert. Symp. Rep. 14:317–331.

Guthrie, M. J. 1933. The reproductive cycles of some cave bats. J. Mammal. 14:199–216.

Guthrie, M. J., and Jeffers, K. R. 1938. Growth of follicles in the ovaries of the bat *Myotis lucifugus lucifugus*. Anat. Rec. 71:477–496.

Hall, E. R., and Kelson, K. R. 1959. The mammals of North America. Ronald Press, New York. 1083pp.

Hall, J. S. 1962. A life history and taxonomic study of the Indiana bat, *Myotis sodalis*. Sci. Publ. Reading Public Mus. and Art Gallery 12:1–68.

Hamilton, W. J., Jr. 1933. The insect food of the big brown bat. J. Mammal. 14:155–156.

———. 1943. The mammals of eastern United States. Comstock Publ. Associates, Ithaca, N.Y. 432pp.

Hassell, M. D., and Harvey, M. J. 1965. Differential homing in *Myotis sodalis*. Am. Midl. Nat. 74:501–503.

Hayward, J. S., and Ball, E. G. 1966. Quantitative aspects of brown adipose tissue thermogenesis during arousal from hibernation. Bio. Bull. 131:94–103.

Henshaw, R. E. 1970. Thermoregulation in bats. Pages 188–232 and suppl. *in* B. H. Slanghter and D. W. Walton, eds. About bats. Southern Methodist Univ., Dallas. 339pp.

Henshaw, R. E., and Folk, G. E., Jr. 1966. Relation of thermoregulation to seasonally changing microclimate in two species of bats (*Myotis lucifugus* and *Myotis sodalis*). Physiol. Zool. 39:223–236.

Herreid, C. F., II. 1964. Bat longevity and metabolic rate. Exp. Geront. 1:1–9.

Hitchcock, H. B. 1949. Hibernation of bats in southeastern Ontario and adjacent Quebec. Can. Field Nat. 63:47–59.

———. 1965. Twenty-three years of bat-banding in Ontario and Quebec. Can. Field Nat. 79:4–14.

Hock, R. J. 1951. The metabolic rates and body temperatures of bats. Bio. Bull. 101:289–299.

Humphrey, S. R. 1972. Adaptations of refuging free-tailed bats. Bat Res. News 13:19, 21–26.

———. 1975. Nursery roosts and community diversity of Nearctic bats. J. Mammal. 56:321–346.

———. 1978. Status, winter habitat, and management of the endangered Indiana bat, *Myotis sodalis*. Florida Sci. 41:65–76.

Humphrey, S. R., and Cope, J. B. 1976. Population ecology of the little brown bat, *Myotis lucifugus*, in Indiana and north-central Kentucky. Am. Soc. Mamm., Spec. Publ. 4:1–81.

——— 1977. Survival rates of the endangered Indiana bat, *Myotis sodalis*. J. Mammal. 58:32–36.

Humphrey, S. R.; Richter, A. R.; and Cope, J. B. 1977. Summer habitat and ecology of the endangered Indiana bat, *Myotis sodalis*. J. Mammal. 58:334–346.

Jackson, H. H. T. 1961. The mammals of Wisconsin. Univ. Wisconsin Press, Madison. 504pp.

Jerrett, D. P. 1979. Female reproductive patterns in nonhibernating bats. J. Reprod. Fert. Symp. Rep. 14:369–378.

Jones, J. K., Jr.; Fleharty, E. D.; and Dunnigan, P. B. 1967. The distributional status of bats in Kansas. Univ. Kansas Mus. Nat. Hist. Misc. Publ. 46:1–33.

Keen, R., and Hitchcock, H. B. 1980. Survival and longevity of the little brown bat (*Myotis lucifugus*) in southeastern Ontario. J. Mammal. 61:1-7.

Korschgen, L. J. 1971. Disappearance and persistence of aldrin after five annual applications. J. Wildl. Manage. 35:494-500.

Kunz, T. H. 1971. Reproduction of some vespertilionid bats in central Iowa. Am. Midl. Nat. 86:477-486.

———. 1974*a*. Reproduction, growth, and mortality of the vespertilionid bat, *Eptesicus fuscus,* in Kansas. J. Mammal. 55:1-13.

———. 1974*b*. Feeding ecology of a temperate insectivorous bat (*Myotis velifer*). Ecology 55:693-711.

Kunz, T. H.; Anthony, E. L. P.; and Rumage, W. T., III. 1977. Mortality of little brown bats following multiple pesticide applications. J. Wildl. Manage. 41:476-483.

Laidlow, G. W. J., and Fenton, M. B. 1971. Control of nursery colony populations of bats by artificial light. J. Wildl. Manage. 35:843-846.

LaVal, R. K. 1973. Observations on the biology of *Tadarida brasiliensis cynocephala* in southeastern Louisiana. Am. Midl. Nat. 89:112-120.

LaVal, R. K.; Clawson, R. L.; Laval, M. L.; and Caire, W. 1977. Foraging behavior and nocturnal activity patterns of Missouri bats, with emphasis on the endangered species *Myotis grisescens* and *Myotis sodalis*. J. Mammal. 58:592-599.

LaVal, R. K., and LaVal, M. K. 1979. Notes on reproduction, behavior, and abundance of the red bat, *Lasiurus borealis*. J. Mammal. 60:209-212.

Leffler, J. W.; Leffler, L. T.; and Hall, J. S. 1979. Effects of familiar area on the homing ability of the little brown bat, *Myotis lucifugus*. J. Mammal. 60:201-204.

Licht, P., and Leitner, P. 1967. Physiological responses to high environmental temperatures in three species of microchiropteran bats. Comp. Biochem. Physiol. 22:371-387.

Lyman, C. P. 1970. Thermoregulation and metabolism in bats. Pages 301-330 in W. A. Wimsatt, ed. Biology of bats. Vol. 1. Academic Press, New York. 406pp.

McClure, H. E. 1942. Summer activities of bats (genus *Lasiurus*) in Iowa. J. Mammal. 23:430-434.

McManus, J. J. 1974. Activity and thermal preference of the little brown bat, *Myotis lucifugus,* during hibernation. J. Mammal. 55:844-846.

McNab, B. K. 1974. The behavior of temperate cave bats in a subtropical environment. Ecology 55:943-958.

Merriam, C. H. 1894. Laws of temperature control of the geographic distribution of terrestrial animals and plants. Nat. Geogr. 6:229-238, pl. 12-14.

Miller, R. E. 1939. The reproductive cycle in male bats of the species *Myotis lucifugus* and *Myotis grisescens*. J. Morph. 64:267-295.

Mills, R. S.; Barrett, G. W.; and Farrell, M. P. 1975. Population dynamics of the big brown bat (*Eptesicus fuscus*) in southwestern Ohio. J. Mammal. 56:591-604.

Mohr, C. E. 1972. The status of threatened species of cave-dwelling bats. Bull. Nat. Speleol. Soc. 34:33-47.

Mumford, R. E. 1958. Population turnover in wintering bats in Indiana. J. Mammal. 39:253-261.

Myers, R. F. 1964. Ecology of three species of myotine bats in the Ozark Plateau. Ph.D. Dissertation. Univ. Missouri, Columbia. 210pp.

Nelson, E. W. 1930. Wild animals of North America. Nat. Geogr. Soc., Washington, D.C. 254pp.

O'Farrell, M. J., and Studier, E. H. 1973. Reproduction, growth, and development in *Myotis thysanodes* and *M. lucifugus* (Chiroptera: Vespertilionidae). Ecology 54:18-30.

Oxberry, B. A. 1979. Female reproductive patterns in hibernating bats. J. Reprod. Fert. Symp. Rep. 14:359-367.

Phillips, G. L. 1966. Ecology of the big brown bat (Chiroptera: Vespertilionidae) in northeastern Kansas. Am. Midl. Nat. 75:168-198.

Racey, P. A. 1973. Environmental factors affecting the length of gestation in heterothermic bats. J. Reprod. Fert., Suppl. 19:175-189.

———. 1979. The prolonged storage and survival of spermatozoa in Chiroptera. J. Reprod. Fert. Symp. Rep. 14:391-402.

Rasweiler, J. J., IV. 1977. The care and management of bats as laboratory animals. Pages 519-617 in W. A. Wimsatt, ed. Biology of bats. Vol. 3. Academic Press, New York. 651pp.

Ross, A. 1967. Ecological aspects of the food habits of insectivorous bats. Proc. Western Found. Vert. Zool. 1:205-264.

Schowalter, D. B., and Gunson, J. R. 1979. Reproductive biology of the big brown bat (*Eptesicus fuscus*) in Alberta. Can. Field Nat. 93:48-54.

Schowalter, D. B.; Gunson, J. R.; and Harder, L. D. 1979. Life history characteristics of little brown bats (*Myotis lucifugus*) in Alberta. Can. Field Nat. 93:243-251.

Sherman, H. B. 1937. Breeding habits of the free-tailed bat. J. Mammal. 18:176-187.

Simmons, J. A.; Fenton, M. B.; and O'Farrell, M. J. 1978. Echolocation and pursuit of prey by bats. Science 203:16-21.

Simmons, J. A.; Howell, D. J.; and Suga, N. 1975. Information content of bat sonar echoes. Am. Sci. 63:204-215.

Simmons, J. A.; Lavender, W. A.; Lavender, B. A.; Doroshow, C. A.; Kiefer, S. W.; Livingston, R.; Scallet, A. C.; and Crowley, D. E. 1974. Target structure and echo spectral discrimination by echolocating bats. Science 186:1130-1132.

Simmons, J. A., and Vernon, J. A. 1971. Echolcation: discrimination of targets by the bat, *Eptesicus fuscus*. J. Exp. Zool. 176:315-328.

Smalley, R. L., and Dryer, R. L. 1963. Brown fat: thermogenic effect during arousal from hibernation in the bat. Science 140:1333-1334.

Studier, E. H.; Lysengen, V. L.; and O'Farrell, M. J. 1973. Biology of *Myotis thysanodes* and *M. lucifugus* (Chiroptera: Vespertilionidae). Part 2: Bioenergetics of pregnancy and lactation. Comp. Biochem. Physiol. 44A:467-471.

Studier, E. H., and O'Farrell, M. J. 1972. Biology of *Myotis thysanodes* and *M. lucifugus* (Chiroptera: Vespertilionidae). Part 1: Thermoregulation. Comp. Biochem. Physiol. 41A:567-595.

Suthers, R. A. 1966. Optomotor responses by echolocating bats. Science 152:1102-1104.

———. 1970. Vision, olfaction, and taste. Pages 265-309 in W. A. Wimsatt, ed. Biology of bats. Vol. 2. Academic Press, New York. 477pp.

Trimarchi, C. W. 1978. Rabies in insectivorous temperate-zone bats. Bat Res. News 19:7-12.

Turner, D.; Shaugnessy, A.; and Gould, E. 1972. Individual recognition between mother and infant bats (*Myotis*). Pages 365-371 in S. R. Galler, K. Schmidt-Koenig, G. Jacobs, and R. Belleville, eds. Animal orientation and navigation, a symposium. N.A.S.A., Washington, D.C.

Tuttle, M. D. 1975. Population ecology of the gray bat. (*Myotis grisescens*): factors influencing early growth and development. Univ. Kansas Mus. Nat. Hist., Occas. Pap. 36:1-24.

———. 1976*a*. Population ecology of the gray bat (*Myotis grisescens*): philopatry, timing and patterns of movement, weight loss during migration, and seasonal adaptive strategies. Univ. Kansas Mus. Nat. Hist. Occas. Pap. 54:1–38.

———. 1976*b*. Population ecology of the gray bat (*Myotis grisescens*): factors influencing growth and survival of newly volant young. Ecology 57:587–595.

———. 1977. Gating as a means of protecting cave-dwelling bats. Pages 77–82 *in* T. Aley and D. Rhodes, eds. National cave management symposium proceedings, 1976. Speleobooks, Albuquerque, N. Mex.

———. 1979. Status, causes of decline, and management of endangered gray bats. J. Wildl. Manage. 43:1–17.

Tuttle, M. D., and Stevenson, D. E. 1977. An analysis of migration as a mortality factor in the gray bat based on public recoveries of banded bats. Am. Midl. Nat. 97:235–240.

———. 1978. Variation in the cave environment and its biological implications. Pages 108–121 *in* R. Zuber, J. Chester, S. Gilbert, and D. Rhoades, eds. National cave management symposium proceedings, 1977. Adobe Press, Albuquerque, N. Mex.

Twente, J. W., Jr. 1956. Ecological observations on a colony of *Tadarida mexicana*. J. Mammal. 37:42–47.

Ubelaker, J. E. 1970. Some observations on ecto- and endoparasites of Chiroptera. Pages 247–261 *in* B. H. Slaughter and D. W. Walton, eds. About bats. Southern Methodist Univ., Dallas. 339pp.

U.S. Fish and Wildlife Service. 1978. List of endangered and threatened wildlife and plants. Fed. Register 43(238):58030–58048.

———. 1979. Two bats protected as endangered. Endangered Species Tech. Bull. 4(12):10.

Webster, F. A. 1963. Active energy radiating systems: the bat and ultrasonic principles II; acoustical control of airborne interceptions by bats. Proc. Internat. Congr. Technol. and Blindness 1:49–135.

———. 1967. Interception performance of echolocating bats in the presence of interference. Pages 673–713 *in* R. G. Busnel, ed. Les systèmes sonars animaux. Lab. Physiol. Accoust., Paris.

Webster, F. A., and Brazier, O. G. 1965. Experimental studies on target detection, evaluation and interception by echolocating bats. Aerospace Med. Res. Labs. Tech. Rep. 65–172. 135pp.

Whitaker, J. O., Jr. 1972. Food habits of bats from Indiana. Can. J. Zool. 50:877–883.

Whitaker, J. O., Jr.; Maser, C.; and Keller, L. E. 1977. Food habits of bats of western Oregon. Northwest Sci. 51:46–55.

Williams, T. C.; Ireland, L. C.; and Williams, J. M. 1973. High altitude flights of the free-tailed bat, *Tadarida brasiliensis,* observed with radar. J. Mammal. 54:807–821.

Wimsatt, W. A. 1944*a*. Growth of the ovarian follicle and ovulation in *Myotis lucifigus lucifigus.* Am. J. Anat. 74:129–173.

———. 1944*b*. Further studies on the survival of spermatozoa in the female reproductive tract of the bat. Anat. Rec. 88:193–204.

———. 1945. Notes on breeding behavior, pregnancy, and parturition in some vespertilionid bats of the eastern United States. J. Mammal. 26:23–33.

———. 1979. Reproductive asymmetry and unilateral pregnancy in Chiroptera. J. Reprod. Fert. Symp. Rep. 14:345–357.

Wimsatt, W. A., and Parks, H. F. 1966. Ultrastructure of the surviving follicle of hibernation and of the ovum-follicle cell relationship in the vespertilionid bat *Myotis lucifugus.* Pages 419–454 *in* Comparative biology of reproduction in mammals. Symp. Zool. Soc. London 15. Academic Press, New York.

Wood, S. F. 1962. Blood parasites of mammals of the Californian Sierra Nevada foothills, with special reference to *Trypanosoma cruzi* Chagas and *Hepatozoan leptosoma* sp.n. Bull. Southern California Acad. Sci. 61:161–176.

Zimmerman, F. R. 1937. Migration of little brown bats. J. Mammal. 18:363.

Zinn, T. L. 1977. Community ecology of Florida bats with emphasis on *Myotis austroriparius.* M.S. Thesis. Univ. Florida, Gainesville. 87pp.

STEPHEN R. HUMPHREY, Florida State Museum, University of Florida, Gainesville, Florida 32611.

4

Armadillo

Dasypus novemcinctus   Gary J. Galbreath

NOMENCLATURE

COMMON NAME. Common long-nosed armadillo, nine-banded armadillo
SCIENTIFIC NAME. *Dasypus novemcinctus*
SUBSPECIES. *D. n. mexicanus*

The subspecific systematics of the species *Dasypus novemcinctus* are in need of reanalysis. Presently, all North American populations of this species are included within the taxon *D. n. mexicanus*, and all statements about "armadillos" in this chapter refer to only this subspecies.

D. novemcinctus is the only extant North American representative of the mammalian order Xenarthra. The phylogenetic split between an ancestor of this order and an ancestor of most or all other extant placental mammals apparently occurred in the Cretaceous Period. *Dasypus novemcinctus* is distantly related to other North American mammals.

DISTRIBUTION

Around the beginning of this century, *D. n. mexicanus* within the United States was found mainly in semiarid regions of Texas south of latitude 33°N (Bailey 1905). Since then the range has expanded considerably to the north and east. Armadillos have been introduced to some areas, and in Alabama and peninsular Florida demes are known to have resulted from such introductions. By 1972 armadillo demes of the western range segment were found as far north as central Oklahoma and Arkansas and as far east as southern Alabama and the Florida panhandle (Humphrey 1974). Since then, there has been further northward expansion, and the introduction-derived population of peninsular Florida and southern Georgia has merged with the western population. Figure 4.1 displays the approximate distribution of the armadillo in North America.

The possibility cannot be excluded that the north and east spread of the western armadillo population was importantly related to human environmental modification. However, Slaughter (1961) proposed plausibly that severe winter weather, perhaps "just one

FIGURE 4.1. Distribution of the armadillo (*Dasypus novemcinctus*).

severe winter every decade or so," held this species to the south until the mid-19th century.

DESCRIPTION

A typical mature adult armadillo of either sex is about 0.7 meters in length, and weighs 4 kilograms when in peak condition. Armadillos reach 6.3 kg in weight (G. E. Gause, personal communication). Sexual dimorphism in body size is slight or absent; adult females may average somewhat larger than adult males.

The body is largely covered by a rather high vaulted bony carapace. The long tail and the top of the head are also armored. The skull is flattened dorsoventrally and tapers to a long, narrow rostrum. At the end of the snout is a nosepad that includes the external nasal orifices. The dark ears are large relative to the skull.

The legs are short. There are normally five toes on each hind foot and four on each front foot, all bearing strong claws. The two middle claws of each front foot, and the middle claw of each rear foot, are the largest.

The adult carapace is hard but somewhat flexible. Its surface is a patchwork of dark and ivory coloration. The carapace sides are substantially lighter than the dorsum. Proximally, the ventral surface of the tail is ivory over much of its area, but the dorsal surface of the tail has pronounced dark bands. Distally, the tail becomes dark. The underside of the body is generally light colored over most of its surface, but in some individuals the underside is partially bright yellow.

The bony "armor" is composed of small scales. Each scale consists of a dermal bony plate, an epidermal keratinous covering, and the connective tissue linking the two. Each scale increases in size during the animal's growth. The epidermal scutes, reptilian in appearance, wear away rather than being shed.

The top of the head carries a bony plate, which on many animals has a flexure posteriorly. There are two large shields on the body: a continuous scapular shield and a pelvic shield. The latter is partly "bandlike" on its anterior sides but is otherwise continuous. Between the two shields is a series of side-to-side bands, usually eight in number, composed largely of the posterior portions of scales. Partial bands sometimes occur, fused to full bands. The eight bands are normally separated from each other, and from the shields, by unscaled, loosely infolded epidermis, greatly increasing body flexibility. This skin, and that found between the head plate and the scapular shield, is much less tough than that on the ventral body surface. The tail has a scalar covering that does not connect with the pelvic shield. Over the greater part of tail length, the scales form a series of "rings," most of which completely encircle the tail. Flexibility of the ringed portion of the tail is due to flexibility between rings. Ring diameter, like tail diameter, tends to decrease in size distally. The more distal scales do not form clear rings.

The ears are hairless and are densely reticulated with scutes. The face, part of the neck, and the undersides of head and body are studded with hard nuclei, each of which is associated with a group of hairs. The feet and exposed parts of the legs possess many scutes. Hairs arise from the posterior edges of the backbands, with an average of about four such hairs per scale. The medial portions of the legs and those ventral areas protected by the carapace edges often give rise to long coarse hair, which varies in its profusion among individuals.

Females normally possess four mammae, two thoracic and two abdominal.

The armadillo possesses a pair of highly developed anal glands. These produce the musty odor generally noticeable when handling a captured animal. At such times the glands may be partly everted. The anal glands contain secretory products of both apocrine and sebaceous glands, and also sloughed epidermal cells. Haynes and Enders (1961) suggested that the glandular secretion may differ from its usual composition during the preovulation and ovulation periods of the female.

Skull and Dentition. The top of the adult skull is tightly attached to the headplate by connective tissue. The zygomatic arch is complete. The turbinals are single rolled. The premaxillae are small and toothless. A lacrimal foramen is present. The mandible is slim and elongate, with ascending rami extending obliquely and a V-shaped symphysis (figure 4.2). *Dasypus novemcinctus* lacks ossified bullae.

A common adult dental formula of *D. n. mexicanus* is 8/8, but the range in total tooth number is at least 28 to 32. The adult teeth are small, subcylindrical, and without enamel, and project only slightly above the gums. Each tooth has a single root with an open pulp cavity, and the teeth are presumably ever growing. At birth there are 7 teeth on each side of each jaw. These deciduous teeth possess an enamel cover-

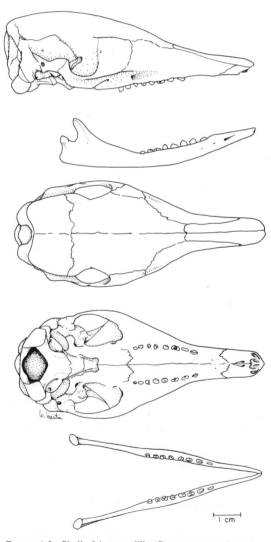

FIGURE 4.2. Skull of the armadillo (*Dasypus novemcinctus*). From top to bottom: lateral view of cranium, lateral view of mandible, dorsal view of cranium, ventral view of cranium, dorsal view of mandible.

ing. An eighth tooth, lacking enamel, appears late and, unlike the others, is not replaced. In adults, the 8 maxillary teeth on a side articulate with the dentary teeth such that the second most anterior lower tooth articulates with the most anterior upper tooth.

Skeletal System. There are typically 11 thoracic and 3 lumbar vertebrae. The scapula possesses a second spine, and the acromion and coracoid processes are strongly developed. A clavicle is present. The femur possesses a prominent third trochanter. The tibia and fibula are fused proximally and distally and give rise to powerful shank muscles.

Extra articulations exist between lumbar and thoracic vertebrae. The axis and cervical vertebrae 3, 4, and 5 are fused in mature animals. There is a great degree of fusion of vertebrae with the ilia and ischia.

The carapace is braced by prominent dorsolateral processes from the ilia and ischia, by elongate metapophyses from the lumbar vertebrae, and by the modified tips of neural spines of all the thoracic and lumbar vertebrae. The strongly developed panniculus carnosus muscles and broad ligaments from expanded dorsolateral flanges of the ilia attach to the carapace. No direct, bony contact is made between the skeleton and carapace. The homologues of the costal cartilages of many mammals are ossified and the ribs are broad; these characteristics, together with heavy intercostal muscles, result in a rigid rib case.

Soft Anatomy. The tongue of the armadillo is long and vermiform. The submaxillary glands are well developed. The postrenal vena cavae are paired, an unusual condition for a mammal (Engelmann 1978). The pyloric portion of the stomach is heavily muscled and presumably triturates food.

PHYSIOLOGY

Scholander et al. (1943) found that an armadillo was capable, under experimental conditions, of developing a considerable oxygen debt during minutes of "violent muscular work carried on during suspended breathing." The nostrils are often pressed against the soil while digging for food items, and breathing is probably difficult or impossible at such times. An armadillo exposed to three different concentrations of CO_2 in air showed relatively slight changes in body temperature, total ventilation, and oxygen consumption, suggesting adaptation to the presumably elevated CO_2 concentrations of burrows (Kay 1977).

Weiss and Wislocki (1956) reported seasonal variation in red blood cell production in backband dermal plates. During autumn, red blood cell production decreased in the marrow cavities of the plates. By midwinter, production had virtually ceased and the marrow cavities were greatly reduced. In March, osteoclastic activity was found to have begun and red blood cell production to have increased.

Armadillo thermoregulation was studied in the laboratory by B. K. McNab (personal communication). He found a rectal temperature of about 34.5°C

during animals' daytime quiescence, at air temperatures within a minimal thermoneutral ambient temperature range of 28° to 38°C. A zone of ambient temperatures (e.g., 24°C) apparently existed within which, at least temporarily, the body temperature was allowed to decrease somewhat. But at yet lower experimental ambient temperatures, body temperature rose. McNab also found that adjustment to a sudden sharp fall in ambient temperature, in terms of metabolic adjustment and of regaining a "normal" body temperature, was a slow process that could last as long as several days. For a mammal of its body mass, the armadillo has a total minimal thermal conductance about double that expected, and a very low basal rate of metabolism.

In cold ambient temperatures, thermoregulatory mechanisms utilized by armadillos include shivering, changing posture to a ballike configuration, and probably vasoconstriction (Johansen 1961). Utilization of a nest helps in keeping body temperature up while minimizing metabolic expense. Specialized vascular arrangements exist that presumably facilitate conservation of body heat through countercurrent heat exchange in the extremities. In high ambient temperatures, thermoregulatory mechanisms include panting, vasodilation, and postural change such that the ventral surface is maximally exposed (Johansen 1961).

GENETICS

The chromosome number of the armadillo is $2n = 64$. Satellited chromosomes are present on one arm of a small metacentric pair. The X chromosome is submetacentric; the Y chromosome is a much smaller, apparently acrocentric, element (Benirschke et al. 1969).

REPRODUCTION AND GROWTH

The reproductive tract of the adult female armadillo consists of paired ovaries and oviducts; a simplex, kite-shaped uterus with the "point" at the fundic end; a vagina; a long urogenital sinus; and a large external clitoris through which the urogenital sinus opens to the outside. The posterior, "cervical," narrow end of the uterus does not project into the vagina.

The adult male armadillo lacks a scrotum. The testes are located internally, at the entrance to the inguinal canal. The large penis possesses a thick body as far down as the end of the prominences of the corpora cavernosa, and then becomes much slimmer. During copulation this slimmer tip probably inserts into the vagina (Enders and Buchanan 1959). Observations of mating in captivity and comparison of the shape of the penis with that of the female's urogenital sinus and vagina both suggest that copulation normally occurs with the male mounting from the rear (contra Newman 1913).

Certain species of the genus *Dasypus* (including *D. novemcinctus*) are unique among the Vertebrata in that they are known to reproduce typically by monozygotic polyembryony. Normal embryonic development in *D. n. mexicanus* is initiated by the fertilization of a single ovum. An inner cell mass blastocyst often remains at the fundic tip of the uterus for a lengthy period of time before implanting. Implantation is restricted to the very small endometrial depression at the fundic tip of the uterus (Buchanan 1967; Galbreath in press). Formation of the four embryos results from a lateral and dorsoventral budding of the inner cell mass. Thickened ectoderm of the four buds becomes four embryonic shields. Four embryos within a single amniotic vesicle are produced, all four being identical with respect to chromosomal genetic complement. Each embryo eventually develops its own amniotic cavity and its own nutritive connection with the maternal mucosa.

The chorioallantoic placenta is of the villous hemochorial type. Initially, the trophoblast penetrates the sinusoids of the uterine endometrium, and development of the träger chords and vascular villi occurs within the sinusoids. After the mature placental pattern is established, the villi increase in number and increasingly distend the sinusoids. The endothelium of the sinusoids is not destroyed; no disruption of endothelium occurs except at the initial invasion site. The original pattern of endometrial circulation is essentially maintained throughout pregnancy (Enders et al. 1958). At parturition the endometrium is torn loose at the sinusoids.

The young are born in an advanced state of development and greatly resemble adults in external body proportions. Neonates lack much of the adult pigmentation and have relatively pliable carapaces. Storrs (1967) reported size and weight data for neonates; an average weight for either sex was about 85 grams. Infants in the wild are probably fully weaned by three months of age.

The number of viable monozygotic embryos within a uterus can vary from two to five; Buchanan (1957) reported a case of six normal embryos plus an additional amniotic sac. One or more embryos may become necrotic in utero, thus reducing litter size at birth. But litter size is usually four.

A typical mature female ovulates no more than once per year. Hamlett (1935) reported that armadillos "in Texas" ovulated primarily during July. Utilizing animals collected over about a 10-year period from an area north of Houston, Enders (1966) found a bimodal distribution in time of ovulation. Most multiparous females ovulated during the summer, but there was another distinct peak of ovulation in the autumn, apparently mainly involving nulliparous females. Implantation occurred largely in the October through December period, apparently with a peak during late November and early December. Parturition in Texas and Florida occurs mainly in March and April.

Most yearling and two-year-old armadillos are distinctly smaller than most mature adults. Both sexes attain maturity (full growth) between 3.0 and 4.0 years of age (Galbreath unpublished data). It is unlikely that many females ovulate prior to 2.0 years of age. Male pubertal changes apparently are usually complete by 2.0

years of age (Gause personal communication). Mature adult males normally produce sperm year round (Storrs et al. 1977; Gause personal communication; McCusker in press).

The potential life span of *D. n. mexicanus* is unknown. One female repeatedly captured by this author at Archbold Biological Station had attained (at time of last capture) an age of at least seven years, and probably eight or more.

ECOLOGY

Population Ecology. It is probable that the armadillo is presently at or near the northern limits of its potential distribution under present climatic conditions in many western areas. McNab (personal communication) estimated that a 5-kg, adult, nonpregnant armadillo, storing 14 percent of its body weight as fat, could survive starvation for about 10 days at a nest temperature of 0°C. This species has a greatly reduced capability of procuring food when the ground is frozen. Young of the year probably tend to be especially sensitive to cold-weather–induced starvation. The high minimal thermal conductance of this species exacerbates the problem of maintaining a body temperature compatible with survival. Local populations in Texas and Louisiana have apparently been severely reduced at times by cold periods (Kalmbach 1943; Fitch et al. 1952).

The armadillo is rarely or never found in truly arid regions. Humphrey (1974) indicated that about 38 centimeters annual precipitation was the lowest rainfall found in areas where local armadillo populations were extant in the United States. He noted both a range contraction for the armadillo in western Texas and an overall trend of decreasing precipitation in the relevant area. The availability of insect food and/or water is probably critical in determining this western boundary.

Instances of apparent armadillo local population decline correlated with drought have been noted by Kalmbach (1943), Taber (1945), and Clark (1951). Taber (1945) found that armadillos in areas with heavily clay-impregnated blackland soils tended to burrow in stream banks and to forage in alluvial bottomlands. These seasonally sun-hardened soils can be very difficult for armadillos to probe.

Slaughter (1961) may have provided an important clue to understanding both geographical range limitation and population density variation in *D. n. mexicanus.* He stated that "by far the majority of insects suitable for armadillos require rapidly rotting wood or deep moist soil." Taber (1945) noted that where a decaying log had added deep humus to the soil, the ground was "honey-combed with probes so close together as to lose all individual shape." Layne and Glover (1977) found that foraging activity was more localized and probings more frequent in areas with either damp soil with grass cover or deep litter. It was "more dispersed with less frequent probings in open, drier situations." The amount of rotting wood and the extent of deep moist soil in a region perhaps jointly constitute an important, geographically widespread determinant of both a local population's ability to persist and its potential maximum density.

I know of no published, reliable density estimates for armadillo populations. The density of mature armadillos, in an area of sand pine scrub, slash pine/turkey oak forest, and developed grassland and orchard, in Florida, was much less than one animal per hectare (Galbreath unpublished data).

Home Range. An armadillo often remains in the same general area over a substantial period of time, so the concept of "home range" is biologically useful. Home ranges of armadillos were investigated at Archbold Biological Station in Florida (Galbreath unpublished data). Home range sizes of two mature armadillos, a male and a female, were relatively thoroughly studied. Mapped home range sizes were 10.8 and 7.6 ha, respectively. The male's total home range, particularly, was probably even larger than that mapped. Each of these animals had the greatest part of its home range in slash pine/turkey oak forest, although both home ranges extended into developed grassland and orchard, and the female's extended into sand pine scrub habitat. The available data for other mature animals of this population (Layne and Glover 1977; Galbreath unpublished data) are compatible with the hypothesis that these home range sizes were not atypical for mature adults.

Developed areas at this site appeared to represent preferred foraging habitat. Moreover, most of the terrain of the two major developed areas was divided, separately among mature adults of each sex, into "territories." Each territory consisted of a zone that was largely exclusive of the presence of other same-sex, mature individuals. Part of the home range of each territory "owner" consisted of natural habitat. Home ranges (and territories) of mature individuals overlapped extensively intersexually, even within developed areas, and the home ranges of younger animals overlapped those of mature adults of both sexes.

At a Mississippi site containing both forested and developed regions, the home ranges were investigated for a particular set of armadillos that often appeared in open, grassy areas (J. Jacobs personal communication). Average minimal home range sizes were 3.3 ha for males and 3.4 ha for females. Mature animals in this sample probably possessed significantly smaller home ranges than did mature armadillos at the Florida site, and the home ranges of these Mississippi animals overlapped extensively, within and between sexes.

In both Florida and Mississippi there were indications of competition for access to "open" areas (Galbreath unpublished data; Jacobs personal communication). Some armadillos are probably essentially restricted to forest habitat by intrasexual competition. In both study areas, a minority of individuals appeared to be "transients," lacking true home ranges or in the process of home range delimitation.

Armadillos generally seem to be well oriented within their home ranges. Olfaction probably is important in recognition of familiar places. There is no strong evidence that distinct odor trails are maintained. Feces are sometimes buried, both in captivity and under natural conditions.

Dens. Armadillos usually sleep in constructed nests. A nest may be located on the ground, in a limestone cavity, or in an underground den. Underground dens range from several centimeters to more than a meter in depth below surface, and approximately from 0.7 to 4.6 m in length (Taber 1945; Clark 1951). There may be one or more branch tunnels that terminate without a nest chamber. An underground den may have one or more entrance tunnels, but usually only one is in regular use. Entrance tunnels often begin to turn at a moderate angle not far from the entrance. Tunnels utilized by adults are typically around 18 cm in diameter. A den usually contains a single, enlarged nest chamber.

An entrance may be either clogged with debris or open. The entrance is often located at the base of a large root, a runner, a fallen tree, or the roots at the base of a tree or shrub. In regularly flooded regions, dens tend to be located in patches of land higher than usual in elevation. Near a lake, Taber (1945) found dens grouped together with as many as twelve entrances on a knoll. In most habitats, dens are much less clumped. An adult usually possesses several dens.

Some relatively short burrows lacking nest chambers may function mainly as escape burrows. Taber suggested that such burrows may serve as food traps, and Clark (1951) recorded the presence of masses of camel-crickets (*Ceuthephilus* sp.) in burrows.

Activity. Mature armadillos are usually largely nocturnal in hot weather. During cold weather increased diurnal activity is often prevalent. Pregnant and yearling armadillos tend to begin foraging especially early in the day during the winter.

Layne and Glover (1978) utilized periodic armadillo track censuses on a smoothed strip along a firelane as a measure of activity. Censuses were conducted from early January through most of April, and from very late May through the end of August. In all seasons, peak activity fell within the 20° to 25°C temperature range, suggesting that behavioral thermoregulation was involved. As measured, much more daily activity occurred in summer than in winter or spring. This result may represent increased movement, and/or more erratic movement related to mating, during the summer.

FOOD HABITS

The armadillo is primarily a generalist predator of invertebrates that live in soil, litter, and rotten wood. Beetle larvae and adults, particularly those of scarabeids, often constitute the most important single food category. A variety of other invertebrate groups is also utilized as food, including (but not limited to): ants and ant pupae, lepidopteran larvae and pupae, termites, millipedes, centipedes, roaches, crickets, grasshoppers, phasmatids, bugs, fly larvae and pupae, arachnids, earthworms, snails, and slugs (Kalmbach 1943; Fitch et al. 1952).

Vertebrates are eaten to a lesser extent, particularly reptiles and amphibians. Species of reptile recorded from armadillo stomachs include three skink species (*Eumeces fasciatus, Lygosoma laterale,* and *Neoseps reynoldsi*), the green anole (*Anolis carolinensis*), the six-lined racerunner (*Cnemidophorus sexlineatus*), the worm lizard (*Rhineura floridana*), the Florida scrub lizard (*Sceloporous woodi*), two blackheaded snake species (*Tantilla gracilis* and *T. relicta*), one blind snake species (*Leptotyphlops dulcis*), and the rough earth snake (*Haldea striatula*) (Kalmbach 1943; Fitch et al. 1952; Layne 1976). Amphibians recorded from armadillo stomachs include the dwarf salamander (*Manculus quadridigitatus*), the amphiuma (*Amphiuma means*), the oak toad (*Bufo quercicus*), at least one species of narrow-mouthed toad (*Gastrophryne carolinensis*), anurans of the genera *Scaphiopus, Rana,* and *Pseudacris,* and salamanders of the genera *Ambystoma* and *Notophthalmus* (Kalmbach 1943; Fitch et al. 1952; Galbreath unpublished data).

Consumption of reptiles occurs principally during cold months of the year, probably both because many reptiles are relatively torpid at low temperatures and because invertebrate food availability is sometimes severely reduced during cold weather. Fitch et al. (1952) suggested that consumption of grasshoppers by armadillos may increase in cold weather due to their torpor.

Eggs of lizards, snakes, turtles, and birds are eaten by armadillos but represent a probably trivial aspect of the diet. On one occasion an armadillo was observed feeding on three very young cottontail rabbits (*Sylvilagus* sp.) (Kalmbach 1943).

Armadillos sometimes consume moderate amounts of plant foods. In Florida, contents of already-fallen citrus fruits are sometimes ingested. Armadillo stomachs have been found to contain remains of white bay fruits (*Magnolia virginiana*), muscadine grapes (*Muscadinia rotundifolia*), blueberries (*Vaccinium* sp.), dewberries (*Rubus trivialis*), mulberry fruits (*Morus* sp.), French mulberry fruits (*Callicarpa* sp.), plum fruits (*Prunus* sp.), juniper berries (*Juniperus* sp.), holly fruits (*Ilex* sp.), longleaf pine seeds (*Pinus palustris*), and mushrooms and other fungi (Kalmbach 1943; Fitch et al. 1952). Hamilton (1946) found that black persimmons (*Diospyros texana*) constituted more than 80 percent of the stomach contents of armadillos collected at one Texas site in mid-August.

An armadillo stomach often contains a large amount of soil. Ingestion of soil, as well as such plant debris as pine needles, bits of rotten wood, twigs, and grass, may often be unavoidable due to the feeding techniques of the armadillo.

BEHAVIOR

General Ethology. My impression is that the eyesight of the armadillo is virtually useless for detecting images or movement at even moderate distances. Eyesight may be better at a distance of several centimeters. Sounds such as a human voice or automobile idling noise typically produce no response from armadillos. But the sound of crackling leaves, or automobile ignition noise, will often alert an armadillo.

Most armadillo vocalizations audible to humans appear to be mere byproducts of breathing, panting, and/or sniffing. The low-frequency, low-amplitude "chuck" sounds sometimes made by members of male-female pairs may be of communicatory significance (Christensen and Waring 1980).

Armadillos possess a reflex mechanism that often causes an animal to "buck" violently in reaction to a sound or touch. A buck may, however, be immediately followed by a return to peaceful foraging. When frightened, an armadillo may "freeze" initially, and may then run, move away more slowly, or resume foraging. The goal of an armadillo being closely pursued is usually to reach a burrow or other protected place.

Armadillos excavate burrows by loosening soil with the nose and forefeet, pushing it beneath the abdomen, and thrusting it backward using the hind feet (Taber 1945). Nest material (largely plant-derived debris) is initially gathered with inward raking movements of the front feet. The armadillo transports it to the den entrance primarily by hopping backward bipedally while clutching the material against its underside and rear legs.

An armadillo will sometimes stand partly or wholly upright, braced by the tail. This posture appears to be associated with sniffing of the air and with attempts to reach potential food. Forward gaits include a bounding run, a shuffling walk, and a piglike trot.

The bony armor of this species often allows an individual to move rapidly through thick or thorny brush with relative ease. The armor perhaps also makes it more difficult for some predators to achieve a firm grasp. Once in a burrow, an armadillo may arch its body such that the carapace presses against the tunnel wall, making the animal difficult to extract.

Intraspecific Behavior. Armadillos are largely solitary. Infants share nest(s) with their mother for some time after birth. Cohesive foraging groups of young littermates occur, with or without the mother. But by nine months of age, and probably much earlier, armadillos typically become solitary. Apart from females with their young, sets of young littermates, and heterosexual pairs, armadillos usually singly occupy nests.

A male armadillo may follow and harass a female that is not reproductively receptive. Such a female sometimes flees, and a lengthy chase may result. Behavioral evidence suggests that a pheromone is involved in the attraction of males to receptive females (Galbreath unpublished data). Some armadillos exhibit heterosexual pair bonding of up to several months' duration (Jacobs personal communication).

Aggressive intraspecific incidents occasionally occur. One or both of the armadillos involved will chase, leap at, and/or kick the other. Typically, one animal is the aggressor, and the other attempts to escape. Significant injury is probably not a common result of such fights. A female armadillo with a nest containing her young nearby may be extremely aggressive toward a mammalian intruder.

Foraging and Feeding Behavior. Daily foraging activity generally begins near the den. The foraging path of an individual may appear random when only a short segment is considered. Over a long segment, a trend in a particular direction is sometimes evident, though the animal may often be headed at an angle to this direction. The nose tends to be held just above the ground during foraging activity. In areas with much litter, a distinct furrow is often plowed.

Olfaction is important in locating food items well beneath the soil surface. The front claws are used to dig to such an item. The nose is pressed against the soil above the food item while the digging occurs, and thus sinks deeper beneath the soil surface. The eyes may become covered when the armadillo is probing for food or is litter plowing. Probe holes tend to be steep and conical. Taber (1945) measured probe holes between 4 and about 17 cm in depth. When a termite colony or ant hill is being attacked, the armadillo's head and shoulders may become hidden by debris (Kalmbach 1943). The front claws are used to tear at rotten wood. After attacking an ant hill an armadillo will sometimes roll violently about, presumably to remove ants.

Small insects such as ants, termites, and the smaller beetles, and many larger but soft-bodied animals, are typically swallowed whole. Larger hard-shelled insects, salamanders, anurans, lizards, and many plant foods are usually chewed (Kalmbach 1943).

MORTALITY

Armadillos are notable for their lack of external parasites. This may be partially due to the armor and the tough ventral body skin. Parasitic organisms recorded from wild armadillos in North America include the flea *Echinophaga gallinacea*, the fluke *Brachylaemus virginianus*, at least one tapeworm species, at least two species of acanthocephalan, one species of oxyuroid nematode, two species of spiruroid nematode, one species of strongyloid nematode, the protozoan *Trypanosoma cruzi*, and one bacterium (probably *Mycobacterium leprae*) (Hightower et al. 1953; Talmage and Buchanan 1954; E. E. Storrs personal communication).

The infective organism of Chagas's disease, *T. cruzi*, was reported from one Texas armadillo (Talmage and Buchanan 1954). Walsh et al. (1975) reported a leprosylike disease infecting wild armadillos in southern Louisiana. The infective organism has been identified as either the human leprosy bacterium (*M. leprae*) or some very similar mycobacterium. Although infected armadillos have been reported largely from Louisiana, an infected Texas animal and another from Mississippi have been found. In Louisiana, Walsh et al. (1977) reported that the percentage of infectivity ranged from 4 to 29.6 percent for small samples from

11 sites. This disease was not found in any of at least 200 Florida specimens (Storrs personal communication).

The armadillo is one of the most common "roadkilled" mammals in Florida; automobiles may be a major cause of mortality for some local populations. Human hunting, including some large-scale, organized hunts, has caused considerable mortality in the southern United States. Dogs are sometimes used in hunting armadillos, and undoubtedly kill many armadillos on their own. Dog attack appears to be a major cause of the stump-tailed condition common among armadillos in certain areas.

Game wardens and conservation officers in a survey listed the dog (36 positive answers), the coyote (*Canis latrans*) (11), the black bear (*Ursus americanus*) (1), and the bobcat (*Felis rufus*) (1) as known predators of armadillos (Humphrey 1974). Kalmbach (1943) noted that examination of 569 coyote stomachs collected in Texas failed to reveal identifiable armadillo remains. He noted that a cougar (*Felis concolor*) had fed on an armadillo. A. Carr (personal communication) watched a bobcat stalk and kill an armadillo in Florida.

ECONOMIC STATUS AND MANAGEMENT

The presence of the armadillo in the United States has had both positive and negative effects. Armadillos consume certain destructive arthropods; some scarabeid beetles, for instance, are destructive to cultivated crops and/or grasslands. The scarabeid *Euetheola rugiceps*, destructive to cane, corn, and rice crops, may at times have been locally controlled in Texas by armadillos (cf. Kalmbach 1943; Fitch et al. 1952).

Armadillo meat was said by Kalmbach (1943) to be an important food for rural poor in Texas. Armadillos have been hunted for sport and food throughout much of their range in the United States. A curio industry based on armadillo skins once was of considerable local economic significance in Texas (Kalmbach 1943). Fitch et al. (1952) reported that this industry "reached a peak prior to the first World War" and was "still thriving on a reduced scale."

Layne (1976) noted that in Florida the armadillo may be useful as a food source for various larger mammalian carnivores and avian carrion feeders whose original food resources may have been diminished by human disturbance and associated environmental changes.

The recently increased use of *Dasypus novemcinctus* as a biomedical research animal may well result in its most significant effect on man. Appropriate diet in captivity was discussed by Meritt (1970, 1973) and Divers (1978). Divers also discussed aspects of husbandry, disease, and clinical pathology.

On the negative side, peanut, corn, and cantaloupe crops have apparently been damaged by armadillos (cf. Kalmbach 1943; Fitch et al. 1952). Horses and cattle are said to suffer harm due to stepping in armadillo burrows (Lowery 1974). Layne

(1976) noted that armadillo burrows can be "a nuisance along road shoulders and may weaken dikes or other earthen structures."

There is concern that armadillo predation may significantly affect populations of rare endemic Florida reptiles. At Archbold Biological Station, in the south-central part of the Florida peninsula, remains of the Florida scrub lizard, the sand skink (*Neoseps reynoldsi*), the worm lizard, and the Florida crowned snake (*Tantilla relicta*) have been found in armadillo stomachs (Layne 1976).

Apart from possible effects on populations of native species through predation, armadillos probably importantly affect natural ecosystems in other ways (e.g., by influencing litter-layer turnover). Florida armadillos sometimes reside in burrows originally excavated by the endangered gopher tortoise (*Gopherus polyphemus*). When a wild-caught armadillo and a gopher tortoise were together in a crate, the armadillo attacked the tortoise in a way that seemed "designed" to attempt to overturn the tortoise. One wonders whether armadillos sometimes evict such tortoises from burrows and/or overturn them.

In some areas where management is oriented toward maintaining a truly natural ecosystem, the armadillo exemplifies a philosophical problem. Should a naturally invading species be allowed to affect an ecosystem without hindrance by man, or should the original absence of the species be considered grounds for its decimation? Extirpation by man of armadillos from most wild areas would be virtually impossible, but controlled shooting could conceivably maintain low population levels in some areas.

On most land managed for hunting purposes, the armadillo can best be regarded as a huntable species that does not require active management. In areas where leprosy is nonexistent, armadillos can provide meat for hunters.

CURRENT RESEARCH NEEDS

Much remains to be learned concerning armadillo behavior, physiology, and evolutionary and population ecology. Future studies of population density variation and its causes, factors limiting geographical range, demography, individual growth, and thermoregulation under natural conditions would be useful. Evidence suggests that intrasexual territoriality may exist in some populations; further studies of exclusive areas and territoriality are needed. Mating behavior and pair bonding are largely unstudied. Methods of aging armadillos are very poorly developed. Research could profitably be conducted on the senses of hearing, sight, and olfaction, and on methods by which armadillos orient within home ranges. Studies of intersibling behavior might be important to the testing of sociobiological theory (Alexander 1974), due to the identical chromosomal genetic complements of littermates.

From the management viewpoint, studies could be conducted of the possible effects of armadillos on populations of rare types of reptile, on litter ecosystems, and on various crops and insect pests.

The current and potential use of *D. novemcinctus* in biomedical research suggests that further studies of armadillo health and maintenance in captivity might be useful. A particular problem concerns the captive breeding of this animal, which so far has met with little success, though wild-caught pregnant females often give birth successfully.

LITERATURE CITED

Alexander, R. D. 1974. The evolution of social behavior. Annu. Rev. Ecol. Syst. 5:325-383.

Benirschke, K.; Low, R. J.; and Ferm, V. H. 1969. Cytogenetic studies of some armadillos. Pages 330-345 in K. Benirschke, ed. Comparative mammalian cytogenetics. Springer Verlag, New York.

Buchanan, G. D. 1957. Variation in litter size of nine-banded armadillos. J. Mammal. 38:529.

———. 1967. The presence of two conceptuses in the uterus of a nine-banded armadillo. J. Reprod. Fert. 13:329-331.

Christensen, C. G., and Waring, G. H. 1980. The "chuck" sound of the nine-banded armadillo (*Dasypus novemcinctus*). J. Mammal. 61:737-738.

Clark, W. K. 1951. Ecological life history of the armadillo in the eastern Edwards Plateau region. Am. Midl. Nat. 46:337-358.

Divers, B. J. 1978. Edentates. Pages 439-448 in M. E. Fowler, ed. Zoo and wild animal medicine. W. B. Saunders Co., Philadelphia.

Enders, A. C. 1966. The reproductive cycle of the nine-banded armadillo (*Dasypus novemcinctus*). Pages 295-310 in I. W. Rowlands, ed. Comparative biology of reproduction in mammals. Academic Press, London.

Enders, A. C., and Buchanan, G. D. 1959. The reproductive tract of the female nine-banded armadillo. Texas Rep. Bio. Med. 17:323-340.

Enders, A. C.; Buchanan, G. D.; and Talmage, R. V. 1958. Histological and histochemical observations on the armadillo uterus during the delayed and post-implantation periods. Anat. Rec. 130:639-651.

Engelmann, G. F. 1978. The logic of phylogenetic analysis and the phylogeny of the Xenarthra (Mammalia). Ph.D. Thesis. Columbia Univ. 329pp.

Fitch, H. S.; Goodrum, P.; and Newman, C. 1952. The armadillo in the southeastern United States. J. Mammal. 33:21-37.

Galbreath, G. J. The evolution of monozygotic polyembryony in *Dasypus*. In G. G. Montgomery, ed. The evolution and ecology of sloths, anteaters, and armadillos (Mammalia, Xenarthra = Edentata). Smithsonian Inst. Press, Washington, D.C. In press.

Hamilton, W. J., Jr. 1946. The black persimmon as a summer food of the Texas armadillo. J. Mammal. 27:175.

Hamlett, G. W. D. 1935. Delayed implantation and discontinuous development in mammals. Q. Rev. Bio. 10:432-447.

Haynes, J. F., and Enders, A. C. 1961. The composition of the anal glands of *Dasypus novemcinctus*. Am. J. Anat. 108:295-301.

Hightower, B. G.; Lehmann, V. W.; and Eads, R. B. 1953. Ectoparasites from mammals and birds on a quail preserve. J. Mammal. 34:268-271.

Humphrey, S. R. 1974. Zoogeography of the nine-banded armadillo (*Dasypus novemcinctus*) in the United States. BioScience 24:457-462.

Johansen, K. 1961. Temperature regulation in the nine-banded armadillo (*Dasypus novemcinctus mexicanus*). Physiol. Zool. 34:126–144.

Kalmbach, E. R. 1943. The armadillo: its relation to agriculture and game. Game, Fish, and Oyster Comm., Austin, Texas. 61pp.

Kay, F. R. 1977. Ventilatory and metabolic responses of an armadillo (*Dasypus novemcinctus*) to elevated CO_2 concentrations. Abstr. of Tech. Pap., 57th Annu. Meet. Am. Soc. Mammal., East Lansing, Mich.

Layne, J. N. 1976. The armadillo, one of Florida's oddest animals. Florida Nat. 49:8–12.

Layne, J. N., ánd Glover, D. 1977. Home range of the armadillo in Florida. J. Mammal. 58:411–413.

———. 1978. Activity cycles of the nine-banded armadillo (*Dasypus novemcinctus*) in southern Florida. Abstr. of Pap., 2nd Int. Theriol. Congr., Brno, Czechoslavakia.

Lowery, G. H., Jr. 1974. The mammals of Louisiana and its adjacent waters. Louisiana State Univ. Press, Baton Rouge. 565pp.

McCusker, J. S. Testicular cycles of the nine-banded armadillo (*Dasypus novemcinctus*) from Texas. *In* G. G. Montgomery, ed. The evolution and ecology of sloths, anteaters, and armadillos (Mammalia, Xenarthra = Edentata). Smithsonian Inst. Press, Washington, D.C. In press.

Meritt, D. A., Jr. 1970. Edentate diets currently in use at Lincoln Park Zoo, Chicago. Int. Zoo Yearb. 10:136–138.

———. 1973. Edentate diets. Part 1, Armadillos. Lab. Anim. Sci. 23:540–542.

Newman, H. H. 1913. The natural history of the nine-banded armadillo of Texas. Am. Nat. 47:513–539.

Scholander, P. F.; Irving, L.; and Grinnell, S. W. 1943. Respiration of the armadillo with possible implication as to its burrowing. J. Cell. Comp. Physiol. 21:53–63.

Slaughter, B. H. 1961. The significance of *Dasypus bellus* (Simpson) in Pleistocene local faunas. Texas J. Sci. 13:311–315.

Storrs, E. E. 1967. Individuality in monozygotic quadruplets of the armadillo, *Dasypus novemcinctus*. Ph.D. Thesis. Univ. Texas. 141pp.

Storrs, E. E.; D'Addamio, G. H.; and Roussel, J. D. 1977. Seasonal variation in semen quality in feral and colony adapted nine-banded armadillos, *Dasypus novemcinctus*, Linn. Abstr. of Tech. Pap., 57th Annu. Meet., Am. Soc. Mammal., East Lansing, Mich.

Taber, F. W. 1945. Contribution on the life history and ecology of the nine-banded armadillo. J. Mammal. 26:211–226.

Talmage, R. V., and Buchanan, G. D. 1954. The armadillo (*Dasypus novemcinctus*): a review of its natural history, ecology, anatomy and reproductive physiology. Rice Inst. Pam. 41 (2):1–135.

Walls, G. L. 1942. The vertebrate eye, and its adaptive radiation. Cranbrook Press, Bloomfield Hills, Mich. 785pp.

Walsh, G. P.; Storrs, E. E.; Burchfield, H. P.; Cottrell, E. H.; Vidrine, M. F.; and Binford, C. H. 1975. Leprosy-like disease occurring naturally in armadillos. J. Reticuloendoth. Soc. 18:347–351.

Walsh, G. P.; Storrs, E. E.; Meyers, W.; and Binford, C. H. 1977. Naturally-acquired leprosy-like disease in the nine-banded armadillo (*Dasypus novemcinctus*): recent epizootiologic findings. J. Reticuloendoth. Soc. 22:363–367.

Weiss, L. P., and Wislocki, G. B. 1956. Seasonal variations in hematopoiesis in the dermal bones of the nine-banded armadillo. Anat. Rec. 126:143–164.

GARY J. GALBREATH, Assistant Curator of Mammals, Department of Zoology, Field Museum of Natural History, Roosevelt Road at Lake Shore Drive, Chicago, Illinois 60605.

II

Lagomorpha

5

Cottontails

Sylvilagus floridanus and Allies

Joseph A. Chapman
J. Gregory Hockman
William R. Edwards

NOMENCLATURE

COMMON NAMES. Eastern cottontail, Florida cottontail
SCIENTIFIC NAME. *Sylvilagus floridanus*
SUBSPECIES NORTH OF MEXICO. *S. f. similis, S. f. mearnsii, S. f. llanensis, S. f. alacer, S. f. mallurus, S. f. hitchensi, S. f. floridanus, S. f. ammophilus, S. f. cognatus, S. f. robustus, S. f. chapmani, S. f. holzneri, S. f. hesperius,* and *S. f. paulsoni.*

COMMON NAMES. Desert cottontail, Audubon's cottontail
SCIENTIFIC NAME. *Sylvilagus audubonii*
SUBSPECIES NORTH OF MEXICO. *S. a. audubonii, S. a. vallicola, S. a. sanctidiegi, S. a. arizonae, S. a. baileyi, S. a. cedrophilus, S. a. neomexicanus, S. a. minor,* and *S. a. parvulus.*

COMMON NAME. Brush Rabbit
SCIENTIFIC NAME. *Sylvilagus bachmani*
SUBSPECIES NORTH OF MEXICO. *S. b. ubericolor, S. b. tehamae, S. b. macrorhinus, S. b. riparius, S. b. mariposae, S. b. bachmani, S. b. virgulti,* and *S. b. cinerascens.*

COMMON NAMES. Nuttall's cottontail, mountain cottontail
SCIENTIFIC NAME. *Sylvilagus nuttallii*
SUBSPECIES. *S. n. nuttallii, S. n. grangeri,* and *S. n. pinetis.*

COMMON NAMES. Swamp rabbit, canecutter
SCIENTIFIC NAME. *Sylvilagus aquaticus*
SUBSPECIES. *S. a. aquaticus* and *S. a. littoralis.*

COMMON NAME. Marsh rabbit
SCIENTIFIC NAME. *Sylvilagus palustris*
SUBSPECIES. *S. p. paludicola* and *S. p. palustris*

COMMON NAME. New England cottontail
SCIENTIFIC NAME. *Sylvilagus transitionalis*
SUBSPECIES. There are currently no recognized subspecies of *S. transitionalis.*

COMMON NAME. Pygmy rabbit
SCIENTIFIC NAME. *Sylvilagus idahoensis.*
Some biologists feel the pygmy rabbit should be placed in the monotypic genus *Brachylagus.*
SUBSPECIES. There are currently no recognized subspecies of *S. idahoensis.*

DISTRIBUTION

Cottontails are widely distributed throughout the United States and extreme southern Canada (figures 5.1–5.6). The most widely distributed of the cottontails is *S. floridanus.* It inhabits diverse areas over broad geographic provinces from southern Canada through the United States and into Mexico and beyond. This rabbit occurs sympatrically with six species of *Sylvilagus* and six species of *Lepus.* The range of no other rabbit overlaps that of so many other leporids (Chapman et al. 1980). The eastern cottontail has been widely transplanted, and for this reason subspecific designations are somewhat meaningless, particularly in eastern North America (Chapman and Morgan 1973). This species is also expanding its range northward at a fairly rapid pace, particularly in New England. Populations of eastern cottontails have been introduced and established in Washington (Dalquest 1941) and Oregon (Graf 1955) (figure 5.1).

The Audubon's cottontail is found throughout the arid Southwest and the high deserts into northern Montana (figure 5.2). The mountain cottontail is found in the intermountain region, and its range broadly overlaps that of the Audubon's cottontail in Montana, Wyoming, Utah, and Colorado (figure 5.3). The range of both these species is stable, although the eastern cottontail appears to be displacing *S. nuttallii* in southwestern North Dakota (Genoways and Jones 1972).

The brush rabbit is confined to the Pacific coast of North America south of the Columbia River (figure 5.4). The range of the species has not changed in recent years. The pygmy rabbit is found mainly in the great basin and is associated with big sagebrush (*Artemisia tridentata*). There is an isolated population of pygmy rabbits in southeastern Washington for which the status is poorly known.

Both the swamp and marsh rabbits are confined to the southeastern United States (figure 5.5). The range of the marsh rabbit has apparently changed little from historic times. However, the swamp rabbit's range has begun to diminish southward, apparently due to drainage and habitat alteration. Potential swamp rabbit habitat in Missouri has decreased from 850,000 hectares in 1870 to fewer than 40,000 ha in 1973, primarily due to the conversion of lowland hardwood forests to row crops (Korte and Fredrickson 1977).

The current distribution of the New England cottontail is a mosaic pattern typical of refugional relicts (figure 5.6). Chapman and Stauffer (in press) believe

that *S. transitionalis* has had a range similar to that in figure 5.6 for the past several decades. Udvardy (1969) defines refugional relicts as "populations that live on an ecological refuge, i.e., on an isolated habitat that has preserved the environmental conditions which were widespread in the past, before the refuge became isolated from the main ecological unit (ecosystem, faunation, community, etc.)." Apparently, *S. transitionalis* retreated southward during the cooling climate and vegetative changes associated with the Pleistocene glaciation. When the warming trend began and vegetative changes reversed, *S. transitionalis* in its extended range was vulnerable to competition from the more ubiquitous *S. floridanus*. *S. transitionalis* was gradually restricted to mountain balds and higher ele-

FIGURE 5.1. Distribution of the eastern cottontail (*Sylvilagus floridanus*).

FIGURE 5.2. Distribution of the desert cottontail (*Sylvilagus audubonii*).

FIGURE 5.3. Distribution of the Nuttall's cottontail (*Sylvilagus nuttallii*).

FIGURE 5.4. Distribution of the brush rabbit (*Sylvilagus bachmani*) and the pygmy rabbit (*S. idahoensis*).

FIGURE 5.5. Distribution of the swamp rabbit (*Sylvilagus aquaticus*) and the marsh rabbit (*S. palustris*).

FIGURE 5.6. Distribution of the New England cottontail (*Sylvilagus transitionalis*).

vations that simulated more northern environments. Because of the limited distribution of *S. transitionalis* and the apparent rapid shrinkage of suitable habitat in New England, Chapman and Stauffer (in press) recommended that this rabbit be listed in a category of "special concern." Several states have suggested that this rabbit should be listed as threatened. However, because of the similarity in appearance between *S. transitionalis* and *S. floridanus*, listing the former as threatened or endangered could cause severe management problems because the latter is the most important game mammal in the eastern United States.

DESCRIPTION

Cottontails are true rabbits; the young are altricial and born naked in a nest. They are members of the family Leporidae, which includes three North American genera: *Sylvilagus*, the cottontails; *Lepus*, the hares and jackrabbits; and *Romerolagus*, the monotypic volcano rabbit found only on volcanic slopes in central Mexico. The *Sylvilagus* are all restricted to the new world.

Most cottontails are in the subgenus *Sylvilagus*, but the pygmy rabbit, *S. idahoensis*, is in the monotypic subgenus *Brachylagus*. Several investigators have proposed putting the pygmy rabbit in its own genus (*Brachylagus*).

Cottontails vary in size from the small *S. idahoensis* to the largest member of the genus *S. aquaticus*. Ranges of body measurements for the genus are as follows: total length 250-538 mm, tail length 18-73 mm, hind foot length 65-110 mm, ear length from notch (dry) 36-74 mm. All cottontails have relatively large ears and feet. The skull of *Sylvilagus* is typically rabbitlike, with a highly fenestrated maxillary bone. The animals possess a supraorbital process of the frontal bone, a straight cutting edge on the upper incisors, and a second set of "pegged" teeth directly behind the upper incisors. The dental formula is 2/1,

0/0, 3/2, 3/3 = 28. The presence of the interparietal bone distinguishes the genus *Sylvilagus* from the genus *Lepus* (figures 5.7 and 5.8). Sexual dimorphism occurs in most species, with females being about 1 to 10 percent larger than males (Orr 1940; Chapman 1971*a*; Chapman and Morgan 1973). Below is a brief description of each of the eight species covered in this chapter.

Sylvilagus floridanus. The eastern cottontail is considered a large rabbit. The pelage is dense and long, brown to gray in color with a white underside of the body and tail (figure 5.9A). The posterior extension of the supraorbital process of the frontal is transversely thickened. This cottontail has the widest distribution of any of the *Sylvilagus*. Diagnostic characteristics vary according to locality, but it is generally easy to distinguish this cottontail from other sympatric species. On the basis of pelage, it may be difficult to distinguish *S. floridanus* from *S. transitionalis* in New England and south along the Appalachians. The eastern cottontail is lighter in color and may possess a white spot on the forehead. The New England cottontail rarely possesses a white spot on the forehead and often has a black spot between the ears. However, these two species are readily distinguished by cranial characteristics (see figures 5.8A and 5.8B). Detailed discussions of body measurements and cranial characteristics may be found in Nelson (1909), Hall (1951), and Chapman et al. (1980).

Sylvilagus audubonii. The Audubon's cottontail is considered large for the genus *Sylvilagus* (total length 370 to 400 mm). The ears are quite long (55 to 67 mm) and sparsely haired on the inner surfaces, and the feet are also sparsely haired. The upper body and large tail are gray in color and the undersurfaces are white (see Figure 5.9B). The rostrum is long (about 31 mm) and the skull has substantially upturned supraorbital processes. An identifying characteristic is the broad postorbital extension of the supraorbital process (see fig-

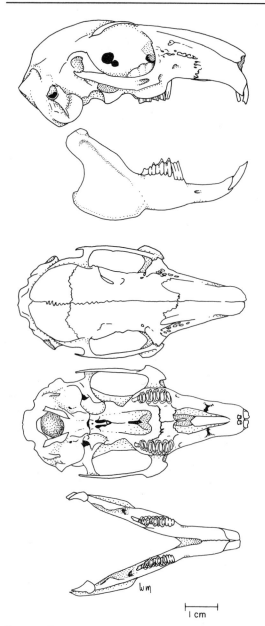

FIGURE 5.7. Skull of the eastern cottontail (*Sylvilagus floridanus*). From top to bottom: lateral view of cranium, lateral view of mandible, dorsal view of cranium, ventral view of cranium, dorsal view of mandible.

ure 5.8C). Detailed descriptions are given in Orr (1940), Hall (1951), and Chapman and Willner (1978).

Sylvilagus bachmani. The brush rabbit is a medium to small cottontail (total length 303 to 369 mm). The color varies from dark brown to gray brown above to white underneath. The hind feet are small (71 to 86 mm) and the legs are short. The feet are not covered with long, dense hair. The ears (45 to 63 mm) and tail (10 to 30 mm) are small. The rostrum is short and the supraorbital processes are small and well separated

from the cranium; the preorbital notch is prominent (figure 5.8D). Detailed descriptions are in Orr (1940), Hall (1951), Chapman (1971*a*), and Chapman (1974).

Sylvilagus nuttallii. The mountain cottontail is medium to large in size (total length 338 to 390 mm). The feet are covered with long, dense hair and the legs are long. The ears are short and rounded at the tip. This rabbit is grayish above and white below, with a large (30 to 54 mm) grizzled tail. The rostrum is long and the supraorbital processes are small and have very pointed anterior projections, with long postorbital processes (see figure 5.8E). Descriptions are given in Orr (1940), Hall (1951), and Chapman (1975*a*).

Sylvilagus aquaticus. The swamp rabbit is the largest member of the genus (total length 452 to 552 mm). The dorsal surface of this rabbit is black to rusty brown in color and the ventral surface is white. The width of the rostrum over the anterior maxillary teeth is greater than the interorbital width. The anterior and posterior projections of the supraorbital process are tightly fused to the skull or lacking (see figure 5.8F). Descriptions are in Hall (1951), Golley (1962), Lowery (1974), and Chapman and Feldhamer (1981).

Sylvilagus palustris. The marsh rabbit is a medium-sized cottontail (total length 425 to 440 mm) (see figure 5.9C). The dorsal surface of this rabbit is blackish brown to reddish brown in color, while the belly is brownish or gray, but never white. The ears, tail, and feet are all very small for the size of the rabbit. The dark color of the underside of the tail distinguishes this rabbit from all other sympatric cottontails. The anterior and posterior extensions of the supraorbital processes are attached to the skull along the full length (Hall 1951) (see figure 5.8G). Descriptions are in Hall (1951) and Chapman and Willner (1981).

Sylvilagus transitionalis. The New England cottontail is medium to large in size (total length 386 to 430 mm). The dorsal surface is dark brown to buff in color overlain with a blackwash that gives a penciled effect. The anterior edges of the ears are covered with black hair. There is a distinct black spot between the ears (see figure 5.9D). To the untrained observer, these pelage characteristics may be more apparent than real.

The supraorbital process is short or missing and the postorbital process is long and slender, rarely touching the skull. The sutures between the frontals and nasals are irregular or jagged in outline (see figure 5.8B). Descriptions are found in Hall (1951), Chapman and Paradiso (1972), Chapman and Morgan (1973), Chapman (1975*b*), and Chapman and Stauffer (in press).

Sylvilagus idahoensis. The pygmy rabbit is the smallest cottontail (total length 250 to 290 mm). The hind legs are short and the hind feet are broad and heavily haired. The ears are short, rounded, and densely haired. The dorsal surface is gray, the tail is small, and the underside is buff rather than white. The rostrum is short and pointed and the anterior and posterior supraorbital processes are long and pointed. The skull appears immature (see figure 5.8H). Descriptions are in Hall (1951) and Green and Flinders (1980).

PHYSIOLOGY

Physiological Cycles. Four morphological indicators have been used to assess the physiological responses of cottontails to environmental factors. These include the adrenal index, the spleen index, the body fat index, and the condition index (Bailey 1968b; Chapman et al. 1977; Bittner and Chapman in press).

ADRENAL AND SPLENIC INDEXES. The adrenal indexes of *S. floridanus* were high in the spring during the peak of the breeding season (Chapman et al. 1977; Bittner and Chapman in press). Male adrenal indexes were consistently higher than those of females. Adrenal hypertrophy has been related to a variety of factors, including increasing and high population densities, breeding activity, varying environmental factors, disease, and increased competition for food (Christian 1963).

Chapman et al. (1977) found that the adrenal indexes in a population of *S. floridanus* from three counties in western Maryland were high in the winter during the period of the most inclement weather. However, the adrenal indexes of *S. transitionalis* from the same region were highest in April to June, during the breeding season. They concluded that *S. transitionalis* was better adapted to the cold boreal environment than *S. floridanus*.

Bittner and Chapman (in press) found that mean adrenal indexes were high in the spring, which was the height of the breeding season on St. Clements Island, Maryland. Male adrenal indexes were also consistently higher than those of females. Cottontails from western Maryland were found to exhibit this same condition (Chapman et al. 1977). It has been shown that social hierarchy is stronger in males than in females (Marsden and Holler 1964; Chapman et al. 1977), with the "pressure to breed" probably the underlying cause of higher male adrenal indexes.

However, adrenal hypertrophy has been related to factors other than breeding pressure. Selye (1973) found that adrenal hypertrophy resulted when organisms were subjected to a variety of stressors. Specific factors causing stress in animals included: increasing and high population densities, varying environmental factors, disease, and increased competition for food (Christian 1963). High density and decreased food availability were evident on St. Clements Island, yet Bittner and Chapman (in press) reported that high stress as indicated by adrenal hypertrophy was not apparent.

Physiological responses have been used to determine whether stress factors were acting upon animal populations. Most of this work has centered around the adrenal gland and its associated hormones (Selye 1956; Christian 1963; Myers 1967; Chapman et al. 1977).

The spleen has also been used to describe responses to stress, but the results vary among species (Selye 1956; Conaway and Wight 1962; Christian 1963; Willner et al. 1979). The work of Bittner and Chapman (in press) tended to contradict some of the studies on the adrenal gland, since high stress as evidenced by adrenal hypertrophy was not apparent. In fact, adrenal indexes for St. Clements Island were significantly lower than those reported by Chapman et al. (1977) for eastern cottontail populations in western Maryland subjected to much lower densities. This suggests that adrenal weight alone is not sufficient evidence of stress when comparing Lagomorph populations from two different locations.

BODY FAT INDEXES. Body fat indexes have been used to assess the general condition of *S. floridanus* (Lord 1963; Chapman et al. 1977; Bittner and Chapman in press) and *S. transitionalis* (Chapman et al. 1977). In western Maryland, both *S. floridanus* and *S. transitionalis* experienced a gradual variation in the body fat index except for two instances: the eastern cottontail sharply increased its body fat index between summer and fall, and the New England cottontail decreased its body fat index between winter and spring (Chapman et al. 1977). However, both species had a relatively high body fat index in the winter. Lord (1963) found that *S. floridanus* in Illinois had more body fat in the winter, as did Bittner and Chapman (in press) for the St. Clements Island population.

CONDITION INDEXES. Bailey (1968b) used length-weight ratios to assess the condition of *S. floridanus*. He believed that winter weight loss was normal for eastern cottontails in northern latitudes. Loss of weight (condition) in eastern cottontails has been reported by several investigators (Allen 1939; Elder and Sowls 1942; Haugen 1942; Chapman et al. 1977). Even with supplemental feeding, Lord and Casteel (1960) were unable to prevent winter weight loss.

Chapman et al. (1977) found that *S. transitionalis*, in contrast to *S. floridanus*, was in its best condition in winter. The difference in the condition indexes of these two species indicated that *S. transitionalis* was better adapted to the colder temperatures characteristic of northern climates. *S. transitionalis* responded totally differently from any of the northern latitude *S. floridanus* populations studied.

INDEX INTERRELATIONSHIPS. Male and female eastern cottontails showed some correlations between condition indexes on St. Clements Island (Bittner and Chapman in press). The adrenal and splenic indexes reached a peak in the spring when the condition index and the body fat index were relatively low. As expected, the body fat index was highest when the condition index was highest.

Chapman et al. (1977) found that *S. floridanus* exhibited a consecutive peaking of physiological indexes, i.e., peak body fat in the fall, peak adrenal index in winter, and peak condition in spring. The seasonal pattern for attaining these peaks was the same for all

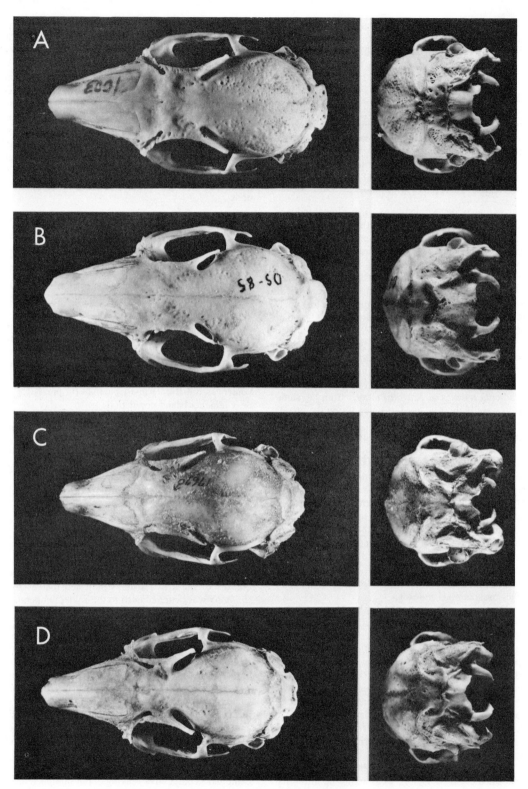

FIGURE 5.8. Dorsal and occipital views of North American *Sylvilagus:* A, *S. floridanus;* B, *S. transitionalis;* C, *S. audubonii;* D, *S. bachmani;* E, *S. nuttallii;* F, *S. aquaticus;* G, *S. palustris;* H, *S. idahoensis.*

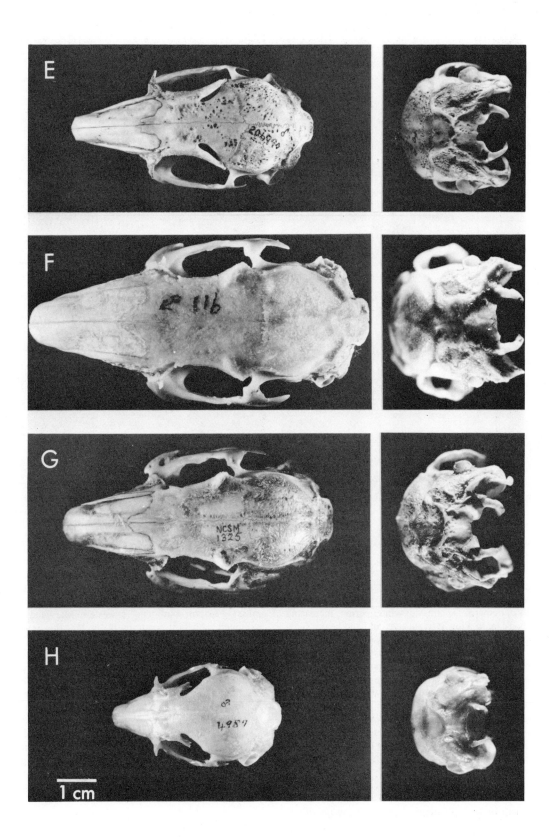

FIGURE 5.9. Four North American species of *Sylvilagus:* A, *S, floridanus;* B, *S. audubonii;* C, *S. palustris;* D, *S. transitionalis.*

indexes, but occurred one season out of phase. Therefore, it appeared that the eastern cottontail physiologically anticipated the stressful winter period, increasing its body fat to allow it to survive cold and heavy snow better and to enter the breeding season in late winter in good condition. However, *S. transitionalis* was more arbitrary and immediate in its responses to stress, the primary stress apparently being reproduction.

This tendency toward immediate reaction to stress as seen by the correlation between the body fat index and the inverse of the adrenal index indicated that *S. transitionalis* was slightly unstable with respect to change; it responded well to small stresses by rapidly metabolizing fat, but not favorably to larger stress factors, presumably because it had metabolized large amounts of body fat, reducing its condition index.

We are dealing here with *S. floridanus* at the northern limits of its range, where *S. transitionalis* is not present. Somewhat different physiological mechanisms would be logical, as food habits and digestive capabilities are probably associated with distribution as a function of evolution.

Cottontails on St. Clements Island bred while in their best physical condition (Bittner and Chapman in press). Eastern cottontails in western Maryland also exhibited their best condition in the breeding season, with this occurring in the period of the most moderate weather (Chapman et al. 1977).

On St. Clements Island, mean adrenal indexes were highest in the spring, which was the height of the breeding season. Male adrenal indexes were consistently higher than those of females. Eastern cottontails from western Maryland were found to exhibit this same condition (Chapman et al. 1977). The social hierarchy of eastern cottontails is stronger in males than in females and may be the underlying cause of higher male adrenal indexes. Chapman et al. (1977) found that reproduction was a high stress factor in both *S. floridanus* and *S. transitionalis.*

CONDITION INDEXES AND ENVIRONMENTAL FACTORS. In general, in western Maryland and nearby West Virginia, environmental conditions appeared favorable during spring and summer for *S. floridanus* and during fall and winter for *S. transitionalis* (Chapman et al. 1977). This was manifested in a peaking of the condition index in the spring for *S. floridanus* and a peaking of the body fat index in the fall for *S. transitionalis*. This was supported by the low adrenal index of *S. transitionalis* in the fall and winter. Myers (1967) reported that *Oryctolagus cuniculus* had its highest adrenal weight in winter and a trough in summer, similar to *S. floridanus*. Therefore, the physiological responses of *S. transitionalis* to seasonal change were different from either *S. floridanus* or *O. cuniculus*.

According to Chapman et al. (1977), measurement of physiological indexes indicated the following general conclusions: (1) *S. transitionalis* was better adapted to colder weather, (2) *S. floridanus* was more eurythermal than *S. transitionalis*, and (3) reproduction was to some degree inherently stressful, as both species were in the best condition just prior to the breeding season.

Flux (1964:485), working on introduced hares, stated that "fat deposits are probably for maintaining breeding condition rather than acting merely as a winter food store." The observations of Chapman et al. (1977) on *S. floridanus* and *S. transitionalis* support that theory. However, Lord (1963) believed that *S. floridanus* in Illinois increased fat reserves as a physiological response to advancing winter.

Tomich (1962) found that adrenal glands of the male California ground squirrel increased during and after the breeding season. He felt this indicated postbreeding recovery from fighting and crowding. For male ground squirrels, summer was the time of greatest stress, but not so for *S. floridanus*. Chapman et al. (1977) felt that crowding, if to the point where it would cause an increase in adrenal size (Christian and Davis 1955), would most likely occur in July–September, but it did not occur in either *S. floridanus* or *S. transitionalis*. Thus, male/male interaction as described by Marsden and Holler (1964) and cold temperatures appeared to be probable causes for increased male adrenal weight in winter in western Maryland and nearby West Virginia. Tomich (1962) found a postbreeding stressor in the California ground squirrel, and a similar phenomenon may occur for *S. transitionalis*. Herrick (1965) found that pregnancy was the highest stress factor identified in female jackrabbits. Tomich (1962) found that pregnancy was a severe stressor for female ground squirrels.

The onset of reproduction of both *S. floridanus* and *S. transitionalis* males was correlated with the ending of adverse weather (Chapman et al. 1977). Males of both species were in breeding condition well before the first warm days of spring. Increased day length induces reproductive behavior, complete libido, and spermatogenesis in *S. transitionalis* (Bissonnette and Csech 1939). In western Maryland and nearby West Virginia, male breeding activity was observed well in advance of the breeding season of females (Chapman et al. 1977). Early male reproductive activity preceding the actual onset of breeding allows cottontail populations to take advantage of fluctuations in the onset of breeding from one year to another. There is minimum delay in breeding after good weather fi-

nally arrives. This is particularly critical in late springs and toward the northern limits of occupied range.

Thermoregulation. As has already been noted, *S. floridanus* is more eurythermal than *S. transitionalis* (Chapman et al. 1977). However, few detailed studies on thermoregulation in *Sylvilagus* have been conducted.

Hinds (1973) studied thermoregulation in *S. audubonii*. He reported that the body temperature was 38.3° C at ambient temperatures below 30° C, and was the same regardless of season. The body temperature equaled the ambient temperature at 41.9° C. Further, *S. audubonii* has a relatively high lethal temperature and evaporative cooling capacity, which allows these leporids to survive the hot conditions of the open desert for short periods of time. According to Hinds (1973:708): "*Sylvilagus audubonii* survives in the desert by taking advantage of every possibility to minimize the heat load and water expenditure. A relatively high evaporative cooling capacity and high lethal body temperature of 44.8° C provides a safety factor for desert cottontails if avoidance is not possible."

Metabolism. Younger *S. floridanus* have a tendency to choose foods that contain more digestable energy and protein. This assists the younger rabbits in satisfying their energy needs. Rose (1974:476) found that "total assimilation and respiration are greater for larger rabbits but that the rates of assimilation and respiration per gram of body weight decrease with increasing body size." Even though *S. floridanus* consumes a relatively small percentage of the vegetation available, digestible energy may be limiting in the winter and early spring (Rose 1973).

S. audubonii showed a definite seasonal change in the basal metabolic rate (Hinds 1973). However, *S. audubonii* did not show a reduction in metabolic rate relative to other mammals. The basal metabolic rate was 0.651 milliliters of oxygen per gram-hour during the summer, and was 18 percent lower than the basal metabolic rate of 0.790 ml of O_2 per gr-hr during the winter. High rates of metabolism may be associated with demands for thermoregulation in winter and reproduction in spring and early summer. High basal metabolism rates would be a disadvantage in bad weather.

Organ Weights. In *S. floridanus*, the left adrenal gland is heavier than the right (Bailey and Schroeder 1967), and a similar relationship has been reported for *S. bachmani* (Chapman 1971*a*). The left kidney of female *S. bachmani* was significantly heavier than that of males. The right kidney of male *S. bachmani* was heavier than the left. Chapman (1971*a*) also weighed the liver, heart, and spleen but found no significant difference in the weight of these organs between the sexes.

Hormonal Studies. Male *S. floridanus* pituitary glands contain more follicle-stimulating hormone (FSH) than those of females. Pituitaries from cottontails less than one year old contain significantly less FSH than those of older rabbits. Since the pituitary secretes very little gonadotropin until several weeks of age, age variation is to be expected. Cottontail pituitaries collected during the breeding season (March to August) contain higher quantities of FSH than those collected in other months of the year (Stevens 1962).

During the breeding season, the luteinizing hormone (LH) levels are apparently higher than at other times. Thyroid gland activity also seems to become progressively higher from January to September. Thyroid gland activity seems to be correlated with seasonal temperatures (Stevens 1962). Numerous studies on a variety of mammals have emphasized interactions that may occur among hormones from the pituitary, thyroid, adrenal cortex, and ovary to modify the responses of the female reproductive system. However, ovarian responses are not just the result of the independent action of the gonadotropins (Stevens 1962).

GENETICS

Chromosome numbers for some North American species of *Sylvilagus* are summarized in table 5.1. For those species karyotyped the diploid chromosome numbers vary from 42 to 52, with most *Sylvilagus* species apparently being 2N=42. Recent karyotyping indicates that there was considerable error in some earlier work.

There have been electrophoretic studies of the serum proteins of several of the *Sylvilagus* (Johnson and Wicks, 1964; Johnson 1968; Chapman and Morgan 1973; Morgan and Chapman in press). An extensive study of the serum proteins of *S. floridanus*, *S. audubonii*, and *S. transitionalis* was conducted by Chapman and Morgan (1973) and Morgan and Chapman (in press). *S. audubonii* normally had 20 serum proteins. The species was considered polymorphic and possessed a polymorphic transferrin. This was not unexpected, since many desert species exhibit varying degrees of polymorphism. On the other extreme, *S. transitionalis* had 18 serum proteins and showed no polymorphism. The patterns of the serum proteins were very consistent among individuals for each species.

Four populations of *S. floridanus* were examined by Chapman and Morgan (1973). Three were considered native: *S. f. alacer* (Kansas), *S. f. mearnsi* (Missouri), and *S. f. chapmani* (Texas). The other population was considered to be of intergrade origin, including *S. f. mallurus* and many other subspecies of *S. floridanus*. In *S. f. alacer*, 23 serum proteins were observed, as well as a polymorphic transferrin. A pre-

transferrin polymorphism was also reported. Twenty-two serum proteins were observed in *S. f. mearnsi* with a pretransferrin polymorphism that was presumably the same one observed in *S. f. alacer*. There were 20 serum proteins in *S. f. chapmani*, but the pattern was distinct from those of *S. f. alacer* and *S. f. mearnsi*.

The serum proteins of the intergrade *S. floridanus* from Maryland are quite variable (Chapman and Morgan 1973). The increased variation in the serum protein patterns of the Maryland *S. floridanus* is attributed to the massive introductions of *S. floridanus* and other *Sylvilagus* species into Maryland from 1920 to 1950. In the Maryland *S. floridanus*, 18 serum proteins were usually found. Ten polymorphic regions were observed, including two pretransferrin, the transferrin, and a posttransferrin. In addition, eight other pretransferrin and two posttransferrin patterns were observed. Three "hybrid" patterns were observed, suggesting that some Maryland *S. floridanus* may have crossbred with another species that had been introduced into the area.

Morgan and Chapman (in press) examined the serum proteins of eastern cottontails introduced from western Maryland to St. Clements Island. They found that the island population exhibited previously unobserved serum protein patterns and a remarkable degree of polymorphism.

REPRODUCTION

Anatomy. The reproductive system of the male *Sylvilagus* is similar to that of the domestic rabbit. Males possess a scrotal sac that is entirely covered with hair and is visible only during the breeding season. The paired testes of adults of some species are large (over 10 grams), and are scrotal during the breeding season. The penis is cylindrical and normally withdrawn into a sheath. The seminal vesicle, vesicular gland, prostate, paraprostates, and bulbourethral glands are similar in gross morphology and histology to that of other lagomorphs. In *S. transitionalis* and *S. floridanus* histological comparisons of the testes revealed no difference between the two species, with the tunica albuginea varying from 37 to 62.5 microns in thickness in both species (Chapman et al. 1977).

The female reproductive system of *Sylvilagus* also is comparable to that of the domestic rabbit. *Sylvilagus* possesses corpora lutea of varying sizes, sometimes reaching 2.48 mm in diameter compared to about 5 mm in the domestic rabbit. Chapman et al.

TABLE 5.1. Summary of chromosome numbers in *Sylvilagus* (as reported in the literature)

| Species | Number of Chromosome Pairs | | | | Source |
	Metacentric	Submetacentric	Acrocentric	Total	
S. floridanus	6	11	3	42	Holden and Eabry 1970
S. audubonii	5	11	4	42	Worthington and Sutton 1966
S. nuttallii	5	11	4	42	Worthington and Sutton 1966
S. bachmani	4	13	6	48	Worthington 1970
S. transitionalis	5	17	3	52	Holden and Eabry 1970

(1977) gave detailed descriptions of both the gross morphology and the histology of the ovaries of *S. transitionalis* and *S. floridanus*. The uterus is duplex, each side having its own distinct cervical canal. The size and condition of the uterus are dependent on the reproductive condition of the animal. Nonpregnant, nonparous females tend to have a smoother, unstriated, less convoluted and less muscular uterus than that of pregnant or parous females. After parturition, the uterus shrinks rapidly in size. The vagina is a single tube leading from the cervix. The clitoris is near the external opening. The color of the vulva is variable, ranging from pinkish white to dark purple. Females possess four to five pairs of mammae; pectoral, thoracic and abdominal mammae are present. These are supplied by two strips of mammary tissue lying on each side of the midline of the abdomen.

Physiology of Reproduction. Ecke (1955) gives an excellent account of the physiology of the reproductive cycle of female *S. floridanus,* and the following discussion is drawn largely from his work.

Female *S. floridanus* are in anestrus (quiescent period) during the winter months. At this time, follicular growth on the ovaries is greatly suppressed. The follicles may develop slightly but maturation does not occur. As the breeding season approaches, an external stimulus (probably increasing day length in combination with temperature) stimulates the pituitary gland to begin secreting follicle-stimulating hormone (FSH) into the blood stream, which in turn reaches the ovaries. FSH acts as a somatic nutrient and stimulates the growth of follicles and the development of ova. The ova develop to a submature stage, at which time the rabbit is in heat. Heat is maintained until copulation occurs (Ecke 1955).

The stimulation of copulation results in the pituitary's secreting luteinizing hormone (LH), which then results in rapid growth of the follicle and ovulation. Ovulation occurs 10 hours after copulation. Ova are fertilized in the fallopian tube. On about the fourth day, the fertilized ova enter the uterus. On or about the seventh day, they are implanted. The blastocyst is at the one- to five-millimeter stage when implanted (Ecke 1955).

The cottontail may remain in preestrus for long periods of time if copulation does not occur. While in preestrus, submature follicles are present and old follicles are either replaced or remain (Ecke 1955).

Cottontails do not experience a true estrous cycle because they are induced ovulators. However, occasionally ovulation is induced by one doe attempting to copulate with another or by copulation with an infertile male. This may result in pseudopregnancy. At this time, follicular growth is retarded and corpora lutea are formed (Ecke 1955).

Once ovulation has occurred, corpora lutea form where the follicles ruptured and are present through the entire pregnancy. The corpora lutea secrete progesterone, which prevents the formation of mature ova. Toward the end of the pregnancy, the placenta begins to secrete progesterone and the corpora lutea decrease

in size. Reduction in corpora lutea signifies reduction in secretion of progesterone. With reduction in ovarian progesterone follicles begin to mature. At parturition fully mature follicles are present (except lost litters). Breeding occurs immediately after parturition. Sex pheromones are secreted prior to parturition, attracting males. These pheromones are probably by-products of ovarian estrogen from maturing follicles. Thus, cottontails usually copulate again immediately following parturition (Ecke 1955). It is this phenomena that results in breeding synchrony in cottontail populations.

The hormones that stimulate follicular and fetal development are also involved in mammary function. Domestic rabbits that become pregnant immediately following parturition may not be able to supply nutrients to suckle young and support fetal growth. This can result in total litter resorption between the 8th and 15th days (Ecke 1955). Total litter resorption is rare in *Sylvilagus*. Cottontails begin to eat green plants on about the 8th day and are weaned by the 15th day.

Total litter resorption occurred in only 1.7 percent of the litters of *S. floridanus* in western Maryland (Chapman et al. 1977). Only two total litter resorptions were found in a large sample of *S. palustris* examined in southern Florida (Holler and Conaway 1979). No total litter resorptions were reported for *S. bachmani* in Oregon (Chapman and Harman 1972) or *S. transitionalis* in West Virginia (Chapman et al. 1977). Ecke (1955) believed total litter resorption was rare because wild cottontails have relatively small litters and their nursing period is shorter than that of the domestic rabbit.

Breeding Season. In *S. floridanus,* initial reproductive activity occurs later at higher elevations and at higher latitudes (Conaway et al. 1974; Chapman et al. 1977). The onset of breeding begins as early as the first week of January in Alabama (Barkalow 1962) to as late as the last week in March in southern Wisconsin (Rongstad 1966). The breeding season lasts from mid-March to mid-September in Connecticut (Dalke 1942), from February through September in New York (Schierbaum 1967), and from late February through August in western Maryland (Chapman et al. 1977). In the southern states, breeding seasons are of longer duration, as in Georgia, which has a nine-month season (Pelton and Provost 1972), and in southern Texas, with a year-round breeding season (Bothma and Teer 1977). An introduced population of *S. floridanus* in western Oregon began to breed in late January and ceased breeding by early September (Trethewey and Verts 1971).

The onset of breeding varies between different populations and within the same population from year to year (Conaway and Wight 1962; Hill 1966; Chapman et al. 1977). Hill (1966) suggested that temperature, rather than diet, is the primary factor controlling the onset (date) of breeding each year. Many researchers have correlated severe weather with delays in the onset of the breeding season (Hamilton 1940; Wight and Conaway 1961; Conaway and Wight 1962).

Ecke (1955) concluded that the limits of the breeding season are closely related to the availability of succulent vegetation. To a degree, the onset of breeding anticipates the availability of succulent green foods 28 days later. Changes in photoperiod are an important factor in regulating cottontail breeding seasons (Bissonnette and Csech 1939), perhaps more so in more northern latitudes. The major environmental factors controlling breeding activity in southern Texas are temperature and rainfall. Rainfall affects the amount of succulent vegetation available (Bothma and Teer 1977). The onset of male reproductive activity for both *S. floridanus* and *S. transitionalis* was closely correlated with temperature (Chapman et al. 1977).

In *S. audubonii* the breeding season is also variable. In California it begins in December and ends in June (Orr 1940), and in Arizona it begins in January and ends in September (Sowls 1957; Stout 1970). Ingles (1941) reported on a California population that bred year around. The breeding season in Texas did not begin until late February or early March (Chapman and Morgan 1974). In *S. nuttallii* in northeastern California, the breeding season began about April and ended in July (Orr 1940). In Oregon, the breeding season lasted from February to July (Powers and Verts 1971).

The breeding season of *S. bachmani* appears to be about the same total length in both the northern and southern parts of the range. The breeding season in Oregon lasted from February through August (Chapman and Harman 1972), and in California from December through about June (Mossman 1955). The breeding season of *S. transitionalis* lasts from March to September in Maryland and West Virginia (Chapman et al. 1977).

Apparently in some regions of Texas, the swamp rabbit breeds year around. The breeding season is longest in the south-central United States and becomes shorter with increasing latitude (Hunt 1959). In Louisiana, breeding has been reported in all months except October (Svihla 1929). In Missouri, the breeding season lasts from February through June (Sorensen et al. 1968). In southern Florida the marsh rabbit breeds year around (Holler and Conaway 1979), while in northern Florida an anestrous period occurs from October to March (Blair 1936).

The breeding season of *S. idahoensis* is very short. In Idaho it lasts from March through May (Wilde et al. 1976) and in Utah, February through March (Janson 1946). Orr (1940) believed that the breeding season of *S. idahoensis* was limited to the spring.

Gestation Period. The mean gestation period of the eastern cottontail is about 28 days (range 25 to 35 days) (Dice 1929; Dalke 1942; Bruna 1952; Evans 1962; Marsden and Conaway 1963). The gestation period of *S. bachmani* is 27 ± 3 days (Chapman and Harman 1972), of *S. nuttallii* 29 ± 1 day (Cowan and Guiquet 1956), of *S. transitionalis* 28 days (Dalke 1942), and of *S. aquaticus* 37 ± 3 days (Hunt 1959; Holler et al. 1963; Sorensen et al. 1968).

Breeding Synchrony. *Sylvilagus*, like other lagomorphs, is an induced ovulator, ovulation occurring only after copulation or other suitable stimulus has occurred. If pregnancy does not follow ovulation, the cottontail may exhibit pseudopregnancy, which may last about one-half the normal gestation period (Conaway and Wight 1962). The total loss of a litter through abortion or resorption also may produce effects similar to pseudopregnancy and may be accompanied by lactation.

The cottontails exhibit a relatively well synchronized breeding season, and conception usually follows almost immediately after parturition of the previous litter (Casteel 1967; Johnson 1973). The breeding season can be divided into conception periods on the basis of this breeding synchrony (Evans 1962; Conaway et al. 1963; Chapman and Harman 1972; Pelton and Provost 1972; Chapman et al. 1977).

Synchrony apparently begins to break down after late June or early July (Johnson 1973; Trethewey and Verts 1971; Pelton and Provost 1972). Breakdown in synchrony is associated with reduced attention of preparturient females and the failure of females to breed for several days or not at all postpartum (Johnson 1973). This suggests the failure of ovarian development during the final week of pregnancy, thus the probable reduced pituitary function. Timing with respect to the summer solstice suggests that photoperiod may be involved.

Litter Size. As with many other mammals, there is an inverse relationship between litter size and latitude in some *Sylvilagus* that is compensated for by an increase in the length of the breeding season at lower latitudes (Lord 1960). Conaway et al. (1974) gave a review of latitude and litter size in *S. floridanus*. The size of first litters varies from 2.95 to 5.10 over the range of this species (table 5.2) and the mean litter size varies from 3.06 to 5.60 (table 5.3).

There is considerable variation in the mean litter size among the species of *Sylvilagus* (tables 5.2, 5.3, and 5.4). The more fecund species is clearly *S. floridanus*; however, the genus as a whole produces relatively large litters.

In most species, the number of young per litter varies with the time of the year. Usually the first and last litters of the year are smaller (tables 5.2 and 5.3). Litters gradually increase in size and peak in midseason and then decline toward the end of the breeding season (Chapman et al. 1977).

Variation in litter size has also been associated with age of the rabbit. In *S. floridanus*, adults appear to produce larger litters (Chapman et al. 1977); conversely, in *S. aquaticus*, younger females produce larger litters (Sorensen et al. 1968). Big does tend to have more young than small does of the same genotype.

The mean litter sizes from areas of high-fertility soils have been shown to be significantly larger than those from areas of low-fertility soils. Exceedingly large litter sizes have been reported by Rongstad (1966) and Barkalow (1961) (a litter of 9 young), Lemke (1957) (a litter of 10), and Kirkpatrick (1960) (12 young in one litter).

TABLE 5.2. Regional comparisons of the mean size of first litters of *Sylvilagus floridanus*

Location	Mean Size of Litter	Sample Size	Year	Source
Western Maryland	4.50	20	1971–72	Chapman et al. 1977
St. Clements Island, Maryland	3.00	1	1976–77	Bittner and Chapman in press
Iowa	4.80	63	1958–61	Kline 1962
Northern Iowa	4.90	43	1958–61	Kline 1962
Southern Iowa	4.60	17	1958–61	Kline 1962
Missouri	3.30	17	1938–40	Schwartz 1942
Northern Missouri	4.10	55	1962	Evans et al. 1965
Southern Missouri	3.60	38	1962	Evans et al. 1965
North Dakota	5.00	55	1964–65	Conaway et al. 1974
United States, (30–35° N)	3.40	50	1964–65	Conaway et al. 1974
United States, (35–40° N)	4.20	158	1964–65	Conaway et al. 1974
United States, (40–45° N)	5.10	36	1964–65	Conaway et al. 1974
Lower coastal plains, Alabama	2.95	80	1959–67	Hill 1972
Piedmont plateau, Alabama	3.00	16	1959–67	Hill 1972
Upper coastal plains, Alabama	3.13	30	1959–67	Hill 1972
Tennessee Valley, Alabama	3.72	103	1959–67	Hill 1972
Black belt, Alabama	3.65	40	1959–67	Hill 1972
Southwestern Texas	4.00	3	1973	Chapman and Morgan 1974
Western Oregon	3.87	16	1969	Trethewey and Verts 1971

Ovulations, Embryo Locations, and Resorptions. The mean ovulation rate for *S. floridanus* in western Maryland was 5.75, while the mean litter size was 5.01. Thus, 12.9 percent of the eggs ovulated either failed to implant or were resorbed (Chapman et al. 1977). For *S. palustris* a preimplantation loss of 8 percent was reported by Holler and Conaway (1979), for *S. aquaticus* 2 percent (Hill 1967).

Conaway and Wight (1962) found that luteinized follicles occurred in only about 1 percent of the ovaries of Missouri cottontails. They found no polyovular follicles past primary stages. Chapman et al. (1977) reported that luteinized follicles occurred in less than 1 percent of the *S. floridanus*, and none were found in *S. transitionalis* they examined. The polyovular condition was found in one *S. floridanus* and four *S. transitionalis*. Six percent of the visible fetuses of *S. floridanus* examined in western Maryland were being resorbed and there was considerable variation in the resorption rate by month. In fact, the resorption rate was inversely related to litter size. Total litter resorption was not common. Trethewey and Verts (1971) reported that 28.3 percent of the introduced female *S. floridanus* in Oregon contained resorbing fetuses.

In *S. transitionalis*, 8 percent of the visible fetuses were resorbed (Chapman et al. 1977). Sowls (1957) reported finding only 1 case of embryo resorption in 56 female *S. audubonii* he examined in Arizona. Similarly, Holler and Conaway (1979) reported that resorption of visible fetuses of *S. palustris*

TABLE 5.3. Regional comparisons of mean annual litter sizes of *Sylvilagus floridanus*

Location	Mean Size of Litter	Sample Size	Year	Source
New York	4.50	28		Hamilton 1940
Pennsylvania	5.42	26	1939	Beule 1940
Michigan	5.10	11	1935–37	Allen 1939
Wisconsin[a]	4.95	20	1961–63	Rongstad 1966
Western Maryland	5.02	65	1971–72	Chapman et al. 1977
St. Clements Island, Maryland	3.57	21	1976–77	Bittner and Chapman in press
Maryland[a]	4.80	35	1955	Sheffer 1957
Western Oregon	5.10	106	1969	Trethewey and Verts 1971
Illinois	5.60	31	1947–48	Ecke 1955
Illinois	5.31	469	1957–59	Lord 1961
Missouri	4.40	42	1968	Schwartz 1942
Virginia	4.70	21	1939–41	Llewellyn and Handley 1945
Alabama	3.47	611		Hill 1972
Georgia (Coastal Plain)	3.18	108	1966–67	Pelton and Jenkins 1971
Georgia (Piedmont)	3.11	85	1966–67	Pelton and Jenkins 1971
Georgia (Mountain)	3.06	16	1966–67	Pelton and Jenkins 1971
South Texas	3.30	279	1965–68	Bothman and Teer 1977

[a]Studies conducted in pens.

TABLE 5.4. Mean size of litters of *Sylvilagus* as reported in the literature

Species	Location	Mean Litter Size	Sample Size	Year	Source
S. transitionalis	West Virginia	3.22 (F)	9	1971–75	Chapman et al. 1977
S. transitionalis	West Virginia	4.00 (S)	7	1971–75	Chapman et al. 1977
S. transitionalis	West Virginia	3.56 (C)	16	1971–75	Chapman et al. 1977
S. transitionalis	Connecticut	5.20 (C)	19	1936	Dalke 1937
S. audubonii	Texas	2.60 (F)	10	1973	Chapman and Morgan 1974
S. audubonii	Arizona	2.90 (C)	56	1951–55	Sowls 1957
S. audubonii	Arizona	3.30 (C)	10	1967–68	Stout 1970
S. audubonii	California	3.33 (C)	119	1966–67	Graves and Asserson 1967
S. audubonii	California	3.60 (C)	19	1939–41	Fitch 1947
S. audubonii	California	3.30 (C)	19		Orr 1940
S. bachmanii	Oregon	2.86 (C)	15	1968–69	Chapman and Harman 1972
S. bachmanii	California	4.00 (C)	14	1950–51	Mossman 1955
S. bachmanii	California	3.50 (C)	11		Orr 1940
S. aquaticus	Alabama	2.89 (F)	95	1960–67	Hill 1967
S. aquaticus	Alabama	3.17 (S)	17	1960–67	Hill 1967
S. aquaticus	Missouri	2.80 (C)	14	1956–57	Toll et al. 1960
S. aquaticus	Texas	2.83 (C)	28		Hunt 1959
S. nuttallii	Oregon	4.00 (F)	5	1969	Powers and Verts 1971
S. nuttallii	Oregon	4.30 (C)	31	1969	Powers and Verts 1971
S. palustris	Florida	4.00 (C)	3		Blair 1936
S. palustris	southern Florida	2.82 (C)	121	1968–69	Holler and Conaway 1979
S. idahoensis	Utah	5.90 (C)	14		Janson 1946

NOTE: F = first litters; S = subsequent litters; C = combined litters

accounted for only 3 percent of the ovulated ova. They also found 2 cases of complete litter resorption. Hill (1967) found that 2 percent of the visible fetuses were resorbed in *S. aquaticus* from Alabama. Conversely, Stout (1970) reported embryos being resorbed in 40 percent of the female *S. audubonii* he examined in Arizona.

Breeding Age. Juvenile breeding is well documented in *S. floridanus*. Juvenile females accounted for 3.9 percent of the pregnancies in western Maryland (Chapman et al. 1977). Conversely, 52.4 percent of the juvenile females were sexually active in western Oregon (Trethewey and Verts 1971). In both Maryland and Oregon, the juvenile females had a smaller litter size than older females. Casteel and Edwards (1964) reported two instances of multiparous juvenile female cottontails.

About 18 percent of the juvenile female *S. transitionalis* bred in West Virginia (Chapman et al. 1977), while only 6.7 percent of the juvenile female *S. nuttallii* bred in eastern Oregon (Powers and Verts 1971). In South Florida, 22 percent of the juvenile female marsh rabbits six to nine months of age were parous (Holler and Conaway 1979). Sowls (1957) believed the juvenile female breeding in *S. audubonii* was an important factor in offsetting the low litter size he found for the species in Arizona.

Juvenile male reproductive activity is reported for *Sylvilagus* but is considered insignificant in terms of populations because of the polygamous nature of the genus. Chapman et al. (1977) reported that a few juvenile male *S. floridanus* were sexually active from July to September in western Maryland. They also reported that all *S. transitionalis* males were potentially

reproductively active during the breeding season following the one in which they were born, but not before.

Johnson (1973) observed apparent reduced libido in adult male cottontails in July and speculated on the basis of their behavior that early-born males may participate in late-season breeding.

Reproductive Rate. The number of litters per year in *Sylvilagus* varies among species as well as on a latitudinal gradient within the same species. However, the number of litters produced per year can be misleading, since many of the species have small litter sizes. In terms of the number of young produced, *S. floridanus* is clearly the most fecund member of the genus. Trethewey and Verts (1971) reported that introduced *S. floridanus* in western Oregon produced 39 young in 8 litters, while Conaway et al. (1963) in Missouri reported 35 young in 7 litters.

For other members of the genus, total productivity varies considerably. *S. nuttallii* produced 22 young in 5 litters (Powers and Verts 1971), *S. transitionalis* produced about 23 young in about 6 litters (Chapman et al. 1977), *S. bachmanii* produced about 15 young in about 6 litters (Chapman and Harman 1972), and *S. palustris* produced about 14–19 young in a year-round breeding season (Holler and Conaway 1979).

Nests and Newborn Cottontails. Female cottontails build elaborate nests in which they give birth to their altricial young. Nests of *S. floridanus* are slanting holes in the ground with average measurements of: length 18.03 cm, width 12.57 cm, and depth 11.94 cm (Friley 1955). In Texas, five nests averaged 12.5 cm long, 10.4 cm wide, and 9.1 cm deep (Bothma and Teer 1977). Casteel (1966) found the average measurements from 21 nests to be: 10.16 cm deep, 12.07

cm wide, and 14.61 cm long. Bothma and Teer (1977) reported no relationship between the size of the nest and the size of the litter. Friley (1955) found that nest holes in southern Michigan farming areas contained an outer lining of grass or herbaceous stems covering all sides and a heavy inner layer of belly or side fur from the female. Illinois nests were lined first with leaves, then with an inner lining of fur plucked from the female (Ecke 1955). Casteel (1966) determined that residual vegetative cover was the preferred nest material, especially grass stems when available. Females pulled fur from almost every part of the body except the abdomen.

The nest of *S. aquaticus* is built on top of the ground and constructed of stalks of dead weeds pulled around an inner lining of fur (Goodpaster and Hoffmeister 1952). The nest has a side entrance and is 4–7 cm deep, 15 cm wide, and 18 cm high (Lowe 1958; Holler et al. 1963). The nest of *S. audubonii* is a pear-shaped excavation in the ground about 150 to 250 mm deep. It is lined with grass and rabbit fur (Ingles 1941). *S. idahoensis* is the only member of the genus that is believed to dig burrows, but no one has yet verified the use of burrows as nesting sites or for rearing young.

Ecke (1955) gave the following description and average measurements for neonates of *S. floridanus*: weight 35 to 45 g, total length 90 to 110 mm, hind foot 21 to 23 mm. Young at birth are covered with fine hair, eyes are tightly closed, and legs are developed enough for them to crawl into their nests. Eyes opened on the 4th or 5th day and young were able to leave the nest between 14 and 16 days after birth. Kentucky nestlings opened their eyes at 7 to 8 days and were able to move out of the nest at 14 days (Bruna 1952). Bothma and Teer (1977) found that nestling rabbits in southeastern Texas opened their eyes between 6 and 7 days of age and moved away from the nest at 12 days. The newborn young of all of the *Sylvilagus* are similar in appearance.

Edwards (1963) reported a nest containing 15 young *S. floridanus*. Hendrickson (1943) reported nests containing 11 and 12 young *S. floridanus* and concluded on the basis of size that there may be occasional common usage of nests by two or more females.

ECOLOGY

Sex Ratios. Some adult and embryonic sex ratios of *Sylvilagus* are given in table 5.5. Differences in the proportions of females among the 7 samples of embryonic young were not significant. However, differences in the proportions of females among the 14 samples of adult cottontails were significant (χ^2_{13} = 23.2; $P <0.05$).

Chapman et al. (1977) suggested that in the Maryland population of *S. floridanus* there is selective mortality against males. However, a phenomenon of selective mortality against males does not now appear typical for the genus. Further, the phenomenon of synchronous breeding and the lack of breeding colonies or harems by cottontails suggest the need for balanced sex

TABLE 5.5. Some adult and embryonic sex ratios for *Sylvilagus*

Species	Location	Sex Ratio (Males/100 Females) Ratio	Number	Sample Size	Year	Source
		Adults				
S. floridanus	Wisconsin	108/100	207/191	398		Elder and Sowls 1942
S. floridanus	western Maryland	83/100	201/243	444	1971–73	Chapman et al. 1977
S. floridanus	St. Clements Island, Maryland	90/100	98/109	207	1976–77	Bittner and Chapman in press
S. floridanus	western Oregon	97/100	239/247	486	1968–69	Trethewey and Verts 1971
S. floridanus	Michigan	103/100	194/189	383	1935–37	Allen 1939
S. audubonii	Arizona	110/100	213/194	407	1951–55	Sowls 1957
S. audubonii	Texas	80/100	8/10	18	1973	Chapman and Morgan 1974
S. transitionalis	western Maryland/West Virginia	58/100	32/56	88	1971–74	Chapman et al. 1977
S. nuttallii	central Oregon	84/100	121/145	266	1968–69	Powers and Verts 1971
S. aquaticus	Alabama	87/100	204/234	438	1960–67	Hill 1967
S. aquaticus	Texas	74/100	64/88	152	1954–55	Hunt 1959
S. aquaticus	Missouri	127/100	107/84	191	1957–59	Holten and Toll 1960
S. bachmani	western Oregon	82/100	46/56	102	1967–69	Chapman and Harman 1972
S. bachmani	California	128/100	46/36	82	1950–51	Mossman 1955
Total		$\bar{X}$ 94.6/100 Σ1,780/1,882				
		embryonic				
S. floridanus	Western Maryland	133/100	117/88	205	1971–73	Chapman et al. 1977
S. floridanus	St. Clements Island, Maryland	65/100	11/17	28	1976–77	Bittner and Chapman in press
S. floridanus	Western Oregon	116/100	93/80	173	1968–69	Trethewey and Verts 1971
S. transitionalis	Western Maryland/West Virginia	50/100	2/4	6	1971–74	Chapman et al. 1977
S. nuttallii	Central Oregon	84/100	35/41	76	1968–69	Powers and Verts 1971
S. bachmani	Western Oregon	85/100	12/14	26	1967–69	Chapman and Harman 1972
S. bachmani	California	92/100	13/14	27	1950–51	Mossman 1955
Total		$\bar{X}$ = 110/100 Σ = 283/258				

ratios or those actually favoring males. Edwards (1962c) found no significant difference in the rates of harvest between males and females in mid-November for cottontails live-trapped in October and early November. For now, any conclusions as to accelerated mortality for a particular sex appear premature, but possible differences warrant further consideration.

Age Structure. Edwards (1962a, 1962c) made an extensive analysis of the age structure of cottontails (*S. floridanus*) in Ohio from lenses of 8,147 animals taken by hunters during 15–18 November of 1958, 1960, and 1961 (see the section "Age Determination"). The proportions of young in the Ohio samples were 0.832, 0.831, and 0.831 for the years 1959, 1960, and 1961, respectively. The consistency of the annual samples suggested that the average age ratio might represent a normal age ratio. In Michigan, the proportion of juveniles in a 10-year collection was 0.840 (Edwards 1964). Thus, the available data suggest that for *S. floridanus* taken by hunters in mid-November in the midwest there is a normal age structure.

Age ratios of local populations of cottontails often vary spatially and temporally (Edwards 1962a, 1962c). Edwards (1962c) reported mid-November juvenile:adult ratios of *S. floridanus* in Ohio that ranged from 1.4:1 to 10.9:1. Therefore, net productivity was estimated to range from about 3:1 to 20:1 young per adult female in mid-November, with normal net productivity about 10 young per adult female.

Seasonal aspects of natality and/or juvenile mortality are variable in space and time. In effect, the relative contribution of spring and summer litters to fall cottontail populations is variable among areas and among years for local areas.

High juvenile:adult ratios are considered indicative of high net productivity. In general, cottontail populations having high juvenile:adult ratios evidence high variances and low mean weights of juvenile lenses (Edwards 1962c). Thus, high age ratios typically reflect proportionally more young from late litters. This probably also reflects good survival of young from early litters, which in turn contribute as breeders to the late litters. Low age ratios are typically associated with relatively low numbers of young from late litters and from early litters.

Relative production and/or survival of the yearling class (the previous year's juvenile class) has a major impact on the age structure of the total adult class. Where numbers of yearlings are high, mean weights and variances of adult lenses tend to be low, and vice versa. In general, low mean weights and variances of adult lenses are usually associated with high survival and increasing abundance of cottontails.

Age characteristics of cottontail populations suggest short-term responses to variable environments. Population density is adjusted by changes in carrying capacity through variations in net productivity, with survival as a major element in determining net productivity (Edwards 1962c).

Abundance and Density. Estimates of the abundance of cottontails are usually derived from live-trapping data or from observed ratios of color-marked individuals. Such estimates are subject to several inherent sources of bias.

Study areas are usually defined arbitrarily on the basis of ownership, cover type, traps and time available, etc. Areas are typically unbounded, animals are free to come and go, and the population to be censused is not discrete but part of a larger population dispersed over a larger, contiguous unit of occupied range. In fall and winter, cottontail populations often contain a dispersing cohort, some of which over a trapping period will move onto and others away from the trapped area. Numerous animals having centers of activity off of, but near, the area will spend some time on the area and in so doing become subject to trapping, marking, and observation. Thus, numbers of cottontails using an area during trapping can be considerably greater than those "on" the area at any given time.

Conversely, some population cohorts, or individuals, under certain circumstances may not be trapable or observable and thus not counted in the census. Population estimates must be interpreted with caution. They relate to individuals available for capture and marking. It is usually not appropriate to divide number estimates by area to obtain density estimates. Regardless, it is easier to think in terms of density.

Local populations of cottontails may occasionally reach densities of 20 per hectare, but are normally considerably lower. The fall density of a confined population of *S. floridanus* on the 210-ha Urbana (Ohio) Wildlife Area in 1964 was 14.9 per ha (Leite 1965). The peak density of *S. floridanus* for St. Clements Island, Maryland, was 10.2 per ha (Bittner and Chapman in press). Based on recovery of tagged cottontails taken by hunters on the 2,400-ha Delaware (Ohio) Wildlife Area in 1959, 1960 and 1961, the mean fall density of cottontails was 3.1 per ha (Edwards unpublished data).

In the canebrakes of southwestern Indiana, fall densities of *S. aquaticus* were reported at about 0.4 per ha (Terrel 1972). Flinders and Hansen (1973) estimated December densities of *S. audubonii* at 0.02 per ha in the shortgrass prairie region of northeastern Colorado. McKay and Verts (1978) reported wide fluctuations of 0.06 to 2.5/ha from monthly estimates of population densities of *S. nuttallii* in shrub-juniper scrubland of central Oregon. Most of the initial declines in density during autumn and early winter appeared to be associated with periods of low ambient temperatures and their effects on survival, principally on those in the youngest cohort.

Edwards et al. (in press) concluded that in Illinois over the past 23 years there has been a consistent long-term decline that has exceeded 70 percent for the statewide cottontail population and 95 percent over extensive areas of intensive grain farming. This decline has been closely associated with massive changes in agricultural land use (table 5.6).

In general, highest densities of cottontails often occur in island (or pen) situations where emigration is restricted and predation is controlled. In contiguous range, high densities of cottontails are typically as-

TABLE 5.6. Linear correlation of land use with numbers of cottontails observed on roadside censuses conducted in July and August in Ford County, Illinois, from 1956 through 1978

Sibley Land Use Parameter (Acreages)	Correlation Coefficient[a]
Corn	−0.746
Soybeans	−0.648
Small grains (wheat and oats)	0.771
Hay, harvested	0.922
Hay, unharvested	0.921
Hay, pastured	0.611
Permanent pasture	0.773
''All'' rowcrops	−0.855
''All'' hay	0.966
''All'' pasture	0.668

SOURCE: Edwards et al. in press
[a] r = 0.505, p = 0.01.

sociated with an abundance of well-distributed escape cover. The effects of extreme weather are often important factors in cottontail mortality and thus density. These observations support a hypothesis that dispersal and mortality are principal mechanisms of population control for cottontails.

Although the human population of North America has been rising at a relatively rapid rate, numbers of rabbit hunters have been declining. Since 1956 the number of hunters and the number of farms in Illinois have both declined by 33 percent (Edwards et al. in press). In fact, the decline in cottontail hunters has been closely correlated with many aspects of agricultural land use in Illinois. It is probable that the decline in hunters reflected both fewer farms on which to hunt and fewer cottontails on those farms. The decline in harvest (58 percent) was almost twice that of the decline of hunters (33 percent) in Illinois from 1956 through 1977 (Edwards et al. in press). Thus, even though hunter numbers have declined in Illinois over the past 20 years, hunting pressure appears to have increased relative to the dwindling habitat and current levels of cottontail abundance.

The close association of changes of cottontail abundance in Illinois with the changing pattern of agriculture suggests the possibility that cottontail numbers may reflect major national economic cycles and growth in gross national product—in this case the agricultural product. The hypothesis is that cottontails decrease with economic growth and prosperity and increase during periods of major economic recession.

The economic history of the United States has been one of prolonged growth and increased productivity with occasional periods of depression and reduced productivity (U.S. Bureau of Economic Analysis 1973). There were significant depressions in the 1890s and the 1930s and a short-lived recession in the late 1940s and early 1950s. This recession is significant in that federal programs such as the Soil Bank, Conservation Reserve, and Cropland Adjustment Program resulted in holding significant acreages out of crop production and in grassland-type cover into the early 1960s.

The early settlers apparently found cottontails scarce, but the animals apparently flourished and extended their range northward during the period of crude pioneer agriculture (Anderson 1940; Jacobson et al. 1978). Cottontail densities were low in Indiana in 1840 but high in the early 1900s (Leopold 1931). Gerstell (1937) provided a picture of low cottontail abundance in Pennsylvania in the 1920s and early 1930s related to clean farming. High densities of cottontails occurred in Ohio during 1938-40 and far lower densities in the late 1940s. These high densities were associated with the high incidence of abandoned and fallow farmland resulting from the Depression of the 1930s.

Bailey (1968a) presented a compilation of published and previously unpublished data on cottontail abundance for the years 1928-65. In general, cottontails were scarce in the late 1920s, relatively abundant in the early and late 1930s and early 1940s, scarce in 1947-48 but abundant in some areas in 1949-50, and scarce in the early 1950s but extensively abundant in the late 1950s followed by a general decline after 1959. Sadler (in press) concluded that over the past 15-18 years there have been continuing declines in cottontail abundance in midwestern states.

While the data at hand are not conclusive, they certainly tend to support the concept of a long-term relationship between cottontail abundance and major economic cycles, which necessarily reflect fluctuations in agricultural productivity and land use intensity. We can expect to see a continuing pattern of fluctuating cottontail abundance. We can expect weather-related, local, and short-term population increases, but the basic pattern of long-term declines in cottontail populations will continue unless there is a major agricultural recession that leads to less intensive land use. The world need for food, the U.S. national need for export commodities, and the current energy-related move to grain alcohol as a fuel appear to preclude any trend to less intensive agriculture.

It is important that losses of grassland habitats be viewed in much the same way as ecologists and conservationists view the loss and destruction of wetland and forest habitats. Conservationists and wildlife managers should strive to develop and preserve grassland habitat wherever possible on public and private lands. The true plight of grassland animals such as the cottontail is only beginning to be appreciated.

Periodicity (Cycles). Numerous workers have reported cyclic tendencies in North American lagomorphs (Leopold 1931; McCabe 1943; Grange 1949; Wight 1959; Keith 1963; Keith and Windberg 1978). Bailey (1968a) advanced the hypothesis of an 8-9-year regional periodicity in cottontail abundance. Kenneth Sadler (in press) suggested highs at 10-year intervals ending in ''6.'' Although the data suggest the possibility of cyclic phenomena, any periodicity over the last 20-25 years has been of far less significance than changing patterns of land use in determining cottontail abundance. Cyclic periodicity in lagomorphs today is probably confined to relatively simple predator/prey systems characteristic of more stable northern and western ecosystems.

Weather. The relationships of cottontail populations and weather have been considered by numerous workers (Allen 1939; Grinnell 1939; Johnson and Hendrickson 1958; Wight and Conaway 1961). However, relationships are typically nebulous and difficult to substantiate. Havera (1973) observed that the weather parameter most consistently related to cottontail harvest in Illinois was total snowfall of the previous winter, especially for February and March.

Havera (1973) and Applegate and Trout (1976) believed that snowfall and rainfall, respectively, primarily affect cottontail populations by influencing production and survival of the first litter in early spring and subsequently the potential for early-born young to contribute both as individuals and as breeders to the fall populations.

The lack of strong evidence for effects of weather on cottontails is not surprising. Weather parameters are arbitrarily defined from climatological data. Relatively few years of cottontail data are available for analysis. In recent years cottontail populations have been responding to major changes in land use. Cottontails are potentially vulnerable to weather throughout the year. No single monthly weather parameter may consistently dominate. It is probable that any significant departure of seasonal weather from the basic climate of the region could have adverse effects on local cottontail populations. Critical weather parameters may change from year to year.

Home Range. The home ranges used by individual cottontails are undoubtedly variable in size. Factors affecting the size of the home range include the species of cottontail; the type, arrangement, and stability of the habitat; the age and sex of the individual; the season of the year and weather patterns; the density of the population; and intraspecific and interspecific competition.

During the season when vegetation is lush and food abundant, and again in winter when weather is severe, daily ranges are relatively small. In fall and late winter cottontails range over larger areas. Janes (1959) concluded that cottontails had home ranges of about 2 ha. Chapman and Trethewey (1972) found that juvenile males were more mobile than adult males and that males moved longer distances than females.

As occupants of successional and disturbed environments it is important that cottontails possess behavioral mechanisms that allow a sizable segment of the population to recognize and remain in relatively secure situations as breeders. It is also important that the population contain a second cohort of potential breeders that disperses to colonize newly favorable habitats or to recolonize understocked habitats. Two types of dispersal are recognized: innate and forced (density or environment induced) (Sakai et al. 1958; Howard 1960). Generally, dispersers are younger animals that have not bred (Grinnell 1922; Howard 1960; Gibb 1977).

During fall and winter, populations of *S. floridanus* contain two cohorts: one having fixed home ranges and the other without such affinities. The resident cohort represents cottontails that have bred, while the dispersing element is composed of younger individuals that have not yet bred. The change from disperser to resident is believed largely related to the onset of reproduction. Early-born young that breed in their natal summer apparently develop strong home range affinities. Young that do not breed are thought not to develop such affinities until the following breeding season in February or March. The probable role of dispersal in the ecology of the cottontail was developed in detail by Edwards et al. (in press).

In general, cottontails do not maintain territories. Home ranges of the different cottontail age and sex classes overlap broadly during much of the year, particularly in late fall and winter, when they tend to concentrate in areas offering the best combination of food and escape cover. However, Trent and Rongstad (1974) found little or no overlap in the ranges of adult females during the breeding season.

Old rabbits probably have larger ranges than young rabbits. Also, during the growing season, ranges of cottontails are relatively small. As cover becomes reduced in fall and early winter, territories appear to increase in size but focus on some element of dense, often thorny, escape cover. Maximum home range size appears to coincide with the end of winter and the onset of breeding. Males in particular may enlarge their range at that time.

Paths, trails, and even roads are important elements of cottontail home ranges. The animals spend considerable time during the twilight and dark hours on such travel lanes. Open areas are used extensively at night, while dense, heavy cover is used more during the day.

It is probable that cottontails use woody cover considerably more in winter than in summer. This is particularly true where dense herbaceous plants provide adequate escape cover in summer. Whether seasonal shifts in habitat are sufficient to regard summer ranges as distinct from winter ranges for some individuals is not known, although it is certainly possible.

Chapman (1971b) calculated home ranges of *S. bachmani* from radiotelemetry. He found that males had larger ranges than females, and juvenile males had the largest range of any sex or age group (table 5.7). Brush rabbits, on the average, had very small home ranges; they restricted their activities to dense brushy cover and rarely occupied bramble clumps smaller than about 450 m². The intensive use of brambles by *S. bachmani* is shown in figure 5.10.

Dixon et al. (in press) compared home range size and shape in *S. floridanus* and *S. bachmani* using activity isopleths (Dixon and Chapman 1980). They found that the difference in home range size between *S. floridanus* and *S. bachmani* appeared to be due to differences in habitat. *S. bachmani* home ranges were characterized by core areas of activity associated with clumps of *Rubus* sp. with little activity outside of the clumps, while home ranges of native *S. floridanus*

TABLE 5.7. Means and ranges of one standard diameter, number of *Sylvilagus bachmani,* and number of radio positions for each sex and age group of brush rabbits on the E. E. Wilson Game Management Area, near Corvallis, Oregon

		Standard Diameter (meters)		Number of Rabbits	Number of Radio Positions	Number of Radio Positions per Rabbit	
Age	Sex	Mean ± SD	Range			Minimum	Maximum
Adult	M	37.3 ± 14.0	19.3 to 72.3	12	345	12	47
Juvenile	M	46.0 ± 23.6	10.9 to 93.4	9	305	13	51
Mean	M	41.6 ± 17.9	10.9 to 93.4	21	650	12	51
Adult	F	33.1 ± 16.8	9.3 to 71.2	24	662	8	78
Juvenile	F	30.2 ± 11.4	12.6 to 48.2	14	356	7	43
Mean	F	32.1 ± 14.9	9.3 to 71.2	38	1018	7	78

SOURCE: Chapman 1971*b*

were larger and lacked the well-defined core activity areas (figure 5.11).

Moreover, introduced *S. floridanus,* which were studied using mark-recapture techniques on the same study area as the *S. bachmani,* had larger home ranges. The *S. floridanus* had home ranges including, but not restricted to, the same bramble clumps utilized by the *S. bachmani* (Chapman and Trethewey 1972). Thus, the introduced *S. floridanus* was able to utilize a more diverse habitat than native *S. bachmani* in Oregon.

Home range size of *S. aquaticus* in Indiana was estimated to range between 1.2 and 12.6 ha (Terrel 1972). In California a home range size of 6.1 ha was reported for male *S. auduboni* and a range size as small as 0.4 ha for females (Ingles 1941). However, Fitch (1947) reported that male and female *S. auduboni* had home ranges of about 3–4 ha. A fall home range of only 0.2–0.7 ha was reported for *S. transitionalis* in Connecticut (Dalke 1937).

It is probable that, in part, the estimated sizes of home ranges have reflected the methods and usually brief time periods used in collecting the data and thus tend to be minimum.

FOOD HABITS

The food habits of cottontails have been studied extensively and have been found to vary greatly, depending on the species of the rabbit and the locality and availability of palatable plants. The food habits of cottontails are essentially cosmopolitan. Below is a brief discussion of the major food items of the eight species of cottontail discussed in this chapter.

Eastern cottontails in New York feed upon a wide variety of plant species. Herbaceous species were chosen during the growing season and woody species were chosen during the dormant season. The most important woody food plants were: apple (*Malus pumila*), staghorn sumac (*Rhus typhina*), red maple (*Acer rubrum*), blackberry (*Rubus alleghreniensis*), and red raspberry (*R. strigosus*). The most important herbaceous food plants were: Kentucky bluegrass (*Poa pratensis*), Canada bluegrass (*P. compressa*), timothy (*Phleum pratense*), quack grass (*Agropyron repens*), orchard grass (*Dactylis glomerata*), red clover

(*Trifolium pratense*), and wild carrot (*Daucus carota*). Staghorn sumac was chosen over smooth sumac (Smith 1950).

Herbaceous plants, except when snow covered, were used almost entirely as a food source for Ohio *S. floridanus.* Bluegrass was the most abundantly taken food all seasons, along with orchard grass, timothy, and nodding wild rye (*Elymus canadensis*). Red clover and Korean lespedeza (*Lespedeza stipulacoa*) were also seasonally utilized (Dusi 1952). Young cottontails preferred succulent weedy forbs such as dandelion (*Taraxacum officinale*) and prickly lettuce (*Lactuca scariola*), with Rugel's plantain (*Plantage rugelii*), curly dock (*Rumex crispus*), ragweed (*Ambrosia* spp.), and red clover also highly preferred (Bailey and Siglin 1966). Klimstra and Corder (1957) reported that *S. floridanus* in southern Illinois relied almost entirely on herbaceous plants as a source of food. Kentucky bluegrass and other perennial grasses were utilized extensively during all seasons. Agricultural crops, primarily corn, soybeans, and wheat, constituted a major portion of the diet during the entire year. Dewberry and blackberry were utilized the greatest during the fall and winter months. Horse nettle (*Solanum carolinse*), a weed species in intertilled crops or recently fallowed land, was also used extensively. It was concluded that soybeans were probably the most desired of any food available until such time as the plants reached maturity.

Dalke and Sime (1941) conducted extensive food habits research on cottontails in Connecticut; however, no distinction was made between *S. floridanus* and *S. transitionalis.* They observed two pronounced feeding periods: the first was three to four hours after sunrise and the second was from sunset to one hour after. Feeding habits changed with changes in seasons and the associated changes in dominant plant species. Spring diets consisted of herbaceous plants, mainly clover, timothy, and alfalfa, which were preferred through October. The fall period, November and December, was a transition period from herbaceous to woody plant materials. Winter diets consisted mainly of: tree species, including gray birch (*Betula populifolia*), red maple, apple, aspen (*Populus tremuloides*), choke cherry (*Prunus virginiana*), and wild black cherry (*P. serotina*); shrubs and vines; blackberry; dewberry (*Rubus villosus*); willow (*Salix* sp.); black alder (*Ilex berticillata*); male-berry (*Lyonia ligustrina*); and highbush blueberry (*Vaccinium corymbosum*).

In the autumn and winter *S. transitionalis* eats most of the common herbaceous and woody plants (Dalke 1937); in the summer, clovers and grasses accounted for 56 percent of the diet of this cottontail. Pringle (1960) believed that *S. transitionalis* and *S. floridanus* had similar food preferences during the summer, but found that *S. transitionalis* had a much more restricted diet during the winter. However, Dalke and Sime (1941) believed that the food habits of the two species were practically identical year around. In comparative feeding trials, Nottage (1972) found that *S. floridanus* is better adapted to a wide variety of diets than *S. transitionalis*.

The seasonal availability of edible plants appears to be the most important influence on the diet of *S. audubonii* (Fitch 1947). This conclusion supported the work of Orr (1940), who found that *S. audubonii* feed seasonally on grasses, sedges, rushes, willows, oaks, blackberries, wild roses, and California mugwort (*Artemisia vulgaris*).

Desert cottontails have been shown to have great dietary adaptability in relation to changing moisture regimes. Apparently, *S. audubonii* can survive drought periods on a diet consisting mainly of dry grasses and forbs by using cactus and forbs with high moisture content as a source of water (Turkowski 1975).

According to Orr (1940), in the spring and summer, *S. nuttallii* selects grasses over all other potential food items. In eastern Lassen County, California, sagebrush (*Artemisia* sp.) is believed to be the most important food item on a year-round basis. Similarly, the primary diet of *S. idahoensis* is sagebrush (Orr 1940).

Grasses are the most important food items of *S. bachmani* throughout its range (Orr 1940); however, when available, clover (*Trifolium involucratun*) was the preferred food. Brush rabbits also feed on the stems and berries of woody plants such as blackberry (*Rubus* sp.).

S. aquaticus feeds on plants in proportion to their abundance (Toll et al. 1960). Sedges and grasses appear to be an important item in the diet throughout their range (Terrel 1972). *S. palustris* feeds extensively on the leaves and twigs of woody plants (Blair 1936), although a major portion of its diet may consist of both terrestrial and aquatic herbaceous plants. The marsh rabbit will also dig and feed on the rhizomes and bulbs of the several plants (Hamilton 1963).

Coprophagy has been reported for most of the species of *Sylvilagus*. Two types of pellet are excreted: hard, brown fecal pellets and soft, green food pellets. About 60 percent of the total fecal excretion of hard

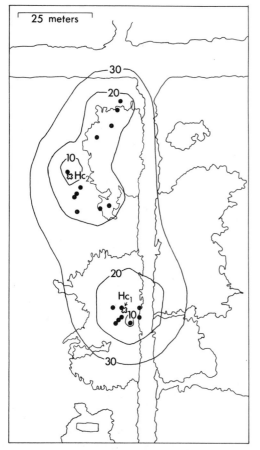

FIGURE 5.10. Home range of an adult female brush rabbit (*Sylvilagus bachmani*) on the E. E. Wilson Game Management Area, near Corvallis, Oregon. Solid dots indicate locations determined by radio telemetry. A home range area of 0.33 ha is defined by the 30-m isopleth. Areas of greater activity are defined by the 10- and 20-m isopleths. The crosses marked Hc_1 and Hc_2 indicate centers of activity. Most of the activity is within or closely associated with the bramble clumps. See Dixon and Chapman 1980 for details.

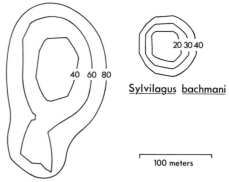

Sylvilagus bachmani

Sylvilagus floridanus

FIGURE 5.11. Size comparison of the winter home range of an eastern cottontail (*Sylvilagus floridanus*) from Grant County, Wisconsin, with the winter home range of a brush rabbit (*S. bachmani*) from Benton County, Oregon, calculated using activity isopleths. See Dixon and Chapman 1980 and Dixon et al. in press for details.

pellets is nutrients (Hamilton 1955; Bailey 1969). Soft pellets are eaten directly from the anus before they touch the ground, usually two or three pellets at a time (Kirkpatrick 1956). Soft pellets are produced in the cecum and provide vitamin B supplementation. In *S. aquaticus,* coprophagy occurs during the daylight hours when the animal is resting, rather than at night, when it is feeding (Toll et al. 1960).

Numerous papers on cottontail food habits stress the importance of grasses. Many such papers are based on the occurrence of undigested tissues in fecal pellets. While the potential significance of the occurrence of grass in fecal pellets of cottontails should not be underestimated, it is very possible that the importance of highly digestible succulent weedy forbs, particularly of young plants, has not been fully appreciated. Although the cottontail has definite food preferences, a wide variety of plants are acceptable and adequate provided base nutritional requirements are met.

The food habits of young rabbits may be particularly critical. The obvious preference of young cottontails for succulent green weedy forbs, particularly such genera as *Plantago, Taraxacum, Lactuca, Sonchus, Erechtites, Cichorium, Rumex,* and *Brassica,* seems particularly significant. It is probable that these foods provide an abundant source of amino acids and primary and trace minerals. Although speculative, the availability of the sulfur amino acids may be particularly important (Hanson and Jones 1976).

HABITAT

Cottontails are widely distributed across North America. No single definitive vegetative community is habitat for the genus *Sylvilagus;* the habitat requirements of cottontails are met in numerous diverse locations. Cottontails are found in a wide variety of disturbed, successional, and transitional habitats often characterized by weedy forbs and bunch-type perennial grasses with an abundance of well-distributed escape sites, often dense, thorny, low-growing, woody perennials.

Cottontails are "r" selected, that is, selected for high rates of reproduction with density regulated primarily through mortality (survival) and dispersal; thus, predation and escape are essential considerations in understanding their habitat relations. Cottontails have evolved survival mechanisms suited to a wide range of communities and climates. The requirements of the different species must necessarily be somewhat different. But because they are cottontails, there are undoubtedly many basic similarities of habitat structure and community wherever they may be found. The crux of the similarity of the habitat requirements for the different species of *Sylvilagus* must certainly lie in their status as "r"-selected herbivores.

In Connecticut, brushpiles were preferred as shelter and resting cover for *S. floridanus* and *S. transitionalis.* When brushpiles were absent, herbaceous and shrubby vegetation was used as hiding and resting places. Brushpiles in cutover woodlands provide winter shelter and an adequate supply of winter food in

the form of stump sprouts and exposed shrubby and herbaceous vegetation (Dalke 1942). In southern Michigan farming areas, *S. floridanus* preferred forms located in native herbaceous vegetation. The basic material of construction was grass, with good herbaceous protection on all sides. In the spring, forms were constructed of brush and briars, and in the winter, forms of herbaceous vegetation were used (Friley 1955).

According to Bruna (1952:41), the "utility of cover probably is determined by what cover is present at a certain time, and its position in relation to the daily and seasonal requirements of a rabbit, rather than any individual preference." Broomsedge and woods were used by *S. floridanus* in Kentucky throughout all seasons.

Smith (1950:23) concluded that the habitat requirements for cottontails in the Hudson Valley of New York should provide the following: "areas permanently in grassland; hedgerows; areas of low, dense brush, such as cutover woodland and overgrown fields; and dens for escape cover." Old field situations were also reported as highly preferred habitat by Friley (1955), Heard (1962), and Nugent (1968). Woodchuck (*Marmota* sp.) holes provide good dens for escape cover and shelter. Interspersion of fields and briar thickets, along with creation of edge by breaking up large, continuous parcels of monotypic habitat, has proven valuable in cottontail habitat improvement.

The habitat preferences of both *S. transitionalis* and *S. floridanus* have been studied in New England (Linkkila 1971; Jackson 1973; Johnston 1972). There were no obvious differences in the habitat of the two species. Further, Johnston (1972:38) stated, "Neither the distribution of *S. floridanus* nor *S. transitionalis* show any relationship to any major forest region or to any land use pattern." Jackson (1973) reported a drastic decline in the *S. transitionalis* population in New England and attributed the decline to changes in habitat. He believed that cover, which had previously been abundant, had grown up to the point that it was no longer suitable for *S. transitionalis.*

In Connecticut the mixed cottontail population of *S. transitionalis* and *S. floridanus* of 1960 changed to one solely of *S. floridanus* (Linkkila 1971). The change was apparently too rapid to attribute to natural succession. However, while the causitive factors were unknown, they obviously resulted in the decline of *S. transitionalis* populations.

The habitat of *S. transitionalis* in the mid-Atlantic region is confined to the higher elevations of the Appalachian region (Llewellyn and Handley 1945; Chapman and Paradiso 1972; Chapman and Morgan 1973; Blymyer 1976; Chapman et al. 1977). The species is found on six- to seven-year-old clear cuts in the higher elevations of Virginia (Blymyer 1976) and old overgrown farmsteads and pockets of heath-conifer habitat in western Maryland and West Virginia (Chapman and Morgan 1973). All of the habitats of *S. transitionalis* appear to have dense cover and conifers associated with them.

Recent information on the habitat preferences of *S. transitionalis* in the southern Appalachians is sketchy. Specimens were collected in Alabama and Georgia in areas that contained conifers and scrubby vegetation, notably *Kalmia* and *Vaccinium* (Chapman and Stauffer in press). A specimen from Kentucky was collected at 1,189 m in "an ivy patch" composed largely of *Kalmia* and *Vaccinium* (Barbour 1951).

According to Chapman and Stauffer (in press), in the mid- and southern Appalachians, *S. transitionalis* inhabits the highlands. It reaches its greatest densities in 5- to 10-year-old clear cuts and older scrubby vegetation. They reported that in all the locations where they found *S. transitionalis,* conifers and one or more ericaceous shrubs were present (i.e., *Kalmia, Vaccinium,* or *Rhododendron*).

S. audubonii, S. nuttallii, and *S. idahoensis* inhabit arid areas. According to Orr (1940), *S. audubonii* is found primarily in the lower Sonoran Life Zone and may show a preference for plants that are often associated with stream bottoms, such as willows. *S. nuttallii* is often associated with rocky ravines and sagebrush-covered hills (Dice 1926). This species also shows a preference for willows along river bottoms. According to Orr (1940), in Nevada *S. nuttallii* prefers higher, rockier, sagebrush-covered areas than *S. audubonii.* The latter occupies adjacent desert valleys. *S. idahoensis* has only been found associated with sagebrush. According to Orr (1940), the sage must also be particularly high and grow in dense clumps. While *S. idahoensis* is probably not always limited to tall sagebrush, most field observations and more recent studies have shown the preference of this rabbit for tall sage.

S. bachmani is an inhabitant of brushy areas, thus its common name, "brush rabbit." In the Willamette Valley, Oregon, *S. bachmani* inhabited distinct bramble clumps (*Rubus* sp.) (Chapman 1971*b*). In California, *S. bachmani* is associated with the Upper Sonoran chaparral belt (Orr 1940). This species also inhabits dense brush along the margins of redwood and Douglas fir forests. In California it is associated with buckbrush (*Ceanothus* sp.), willows (*Salix* sp.), California rose bay (*Rhododendron californicum*), chaparral broom (*Baccharis pilularis*), wild rose (*Rosa californica*), poison oak (*Rhus diversiloba*), scrub oak (*Quercus dumosa*), snowberry (*Symphoricarpus albus*), yellow pine (*Pinus ponderosa*), and manzanita (*Arctostaphylos* sp.) (Orr 1940).

Neither *S. aquaticus* nor *S. palustris* is found far from water. They are associated with floodplains, bottomlands, tributaries and estuaries of larger rivers or streams, and swamps and marshes (Svihla 1929; Blair 1936; Cockrum 1949; Lowe 1958; Hunt 1959; Terrel 1972). In Indiana, *S. aquaticus* was closely confined to canebrakes (*Arundinaria gigantia*) (Harrison and Hickie 1931). Holler and Conaway (1979) found *S. palustris* abundant in sugarcane plantations in southern Florida.

Terrel (1972) found that *S. aquaticus* limits its range to an area no farther than 2 km from a major body of water and prefers a system of small sloughs, low ridges, and grassy marshes. The swamp rabbit also utilizes grain fields as a source of cover and food dur-

ing periods of flooding. Nineteen southern species of flora dominate the vegetation occurring in swamp rabbit habitat: *Populus heterophylla, Celtis laevigata, Liquidambar syraciflua, Carya pecan, Gleditsia aquatica, Adelia acuminata, Fraxinus profunia, Quercus falcata, Q. lyrata, Taxodium distichum, Ilex decidua, Vitis cinera, Arundinaris gigantea, Bignonia capreolata, Dioscorea hirticaudis, Cassia marilandica, Cocculus carolinus, Dassiflora lutea,* and *Viola missouriensis.* During periods of flooding, Conaway et al. (1960) found that high water forced the swamp rabbits out of their normal habitat onto higher ground and more crowded conditions, but as soon as the water receded, the rabbits usually returned to their previously occupied areas.

The fringes of swamp rabbit habitat sometimes overlap *S. floridanus* habitat, with a belt of cohabitation of 64–128 m. Throughout the area of sympatry *S. aquaticus* inhabits wooded swamps, while *S. floridanus* is restricted to grassy rights of way; both species utilize the edge (Toll et al. 1960). Terrel (1972) also found habitat overlap between these two species. *S. aquaticus* were nearly six times as plentiful in their most desired habitat of selectively logged tracts, twice as common in mature forests, and equally as abundant in old field growth, but only in one case (during flood conditions) was *S. floridanus* observed more than 100 m inside a forest.

Optimum habitat of *S. floridanus* in the Midwest is old, weedy, moderately grazed, unimproved, native grassland pasture containing numerous dense clumps of thorny shrubs and small trees. The herbaceous vegetation is dominated by grasses and their associates. Weedy forbs include such genera as *Lespedeza, Aster, Lactuca, Sanchus, Cichorium, Trifolium, Plantago, Taraxacum, Amoranthus, Chenopodium, Ambrosia,* and *Solidago.* The thorny shrubs and trees would include such genera as *Rosa, Rubus, Crataegus, Gleditsia, Zanthoxylum, Robina, Prunus, Malus,* and *Ribes.*

Over much of their range cottontails have become much reduced in numbers over the past 20 to 50 years, or even longer, as previously noted. It is naturally assumed that sustained, long-term declines in animal abundance relate to sustained losses of grassland habitats. When the rural landscapes of the Midwest are viewed in time perspective, we see that we have lost the interspersion of the prairie grasslands, groves, wooded streams, and river bottom forests. We have lost the substitute grasslands of alfalfa, clover, brome, grass, and timothy, and the old weedy and brushy pastures have been plowed or "improved"—or are overgrazed. Crops such as oats and clover are no longer commonly grown. There is little for a cottontail to eat and almost no place to hide. In short, we have already lost much, and will continue to lose more cottontail habitat. It is a matter of increased human density, technology, land use, and economics.

Nest Cover. Cottontails undoubtedly nest in a wide variety of cover types. Hendrickson (1940) reported *S. f. mearnsii* to nest in cover from bare soil to forested land. Most nests of *S. floridanus* undoubtedly occur in grassland. Early nests particularly appear to be placed in dense grassy sods under 15 cm in height. In suburban areas in the Midwest, April and May litters are commonly reported in lawns. Such nests are typically in quite open, exposed situations. Summer litters are rarely reported in lawns and it is presumed that some other, less exposed cover is used for summer nests. Hendrickson (1940) reported summer nests in mixed hayfields in cover more than 20 cm tall.

The fact that cottontail nests are dug into the ground makes them vulnerable to flooding, and nestling cottontails are frequently lost to floods. Well-sodded lawns, in general, have an excellent capacity to absorb rainfall and resist flooding. The placement of early nests in short, dense sods appears to be an adaptation that helps reduce losses to flooding. The relatively brief period that young are in the nest and the high reproductive potential of cottontails are further adaptations that help offset losses to factors such as flooding and nest destruction.

BEHAVIOR

There have been numerous observational studies conducted concerning the behavior of *Sylvilagus*. These studies include work done on *S. floridanus* by Janes (1959), Lord (1964), Marsden and Holler (1964), Casteel (1966), and others. In addition, behavioral observations have been made on *S. audubonii* (Cushing 1939; Orr 1940; Ingles 1941), *S. nuttallii* (Orr 1940), *S. bachmani* (Orr 1940; Chapman and Verts 1969; Zoloth 1969), *S. transitionalis* (Bissonnette and Csech 1939; Dalke 1942; Nugent 1968; Olmstead 1970), *S. idahoensis* (Orr 1940), *S. aquaticus* (Marsden and Holler 1964; Holler and Sorensen 1969; Sorensen 1972), and *S. palustris* (Blair 1936; Hamilton 1963). However, the most complete study was that of Marsden and Holler (1964); they classified and described the basic behavior patterns of *S. floridanus* and *S. aquaticus,* described their social organization, and discussed the relationship between behavior patterns and the reproductive cycle.

Two principal categories of social behavior in *S. floridanus* and *S. aquaticus* were found: (1) basic postures, movements, and vocalization; and (2) adult social interactions. Under the first category, male behavior consisted of the alert posture, submissive posture, approach, dash, rush, mounting, marking, scratching, and paw raking. Female behaviors were the threat, boxing, charge, jump, and presentation.

Adult social interactions were further categorized as reproductive interactions and dominant-subordinant interactions. Marsden and Holler (1964) drew five major conclusions as to the social structure of the eastern cottontail and the swamp rabbit. The basis upon which social structure rested in both species was the dominance hierarchy, particularly among the males. Their conclusions were: (1) the male hierarchy prevents fighting between males while the females are in estrus—the effectiveness of the male hierarchy was demonstrated by the fact that fighting rarely occurred; (2) frequency of male dominance displays was directly correlated with the social status of the rabbit—nearly all male aggression in the population was by the dominant individuals; (3) challenge to the social position of an individual male was greatest from the individual immediately below it in social status; (4) there was a direct correlation between a male swamp rabbit's mounts, its proximity to females, and its social status (this relationship was not apparent in *S. floridanus*); and (5) the number of male-female interactions a male participates in directly correlated with the social status of the male—the majority of the copulations were by the dominant male. Dominant male *S. floridanus* were more far-ranging and frequent explorers than subordinant males (Janes 1959; Marsden and Holler 1964). Territoriality in *S. aquaticus* is restricted to defense of an area surrounding the female (Marsden and Holler 1964).

MORTALITY

Edwards et al. (in press) advanced the thesis that for the cottontail (*S. floridanus*) in the Midwest, population regulation is primarily survival dependent. This view stems in part from the lack of evidence for significant density-dependent regulation of reproduction. To support that contention they noted the following parameters pertaining to cottontails:

1. Synchronous breeding (Conaway and Wight 1962; Stevens 1962; Casteel 1966; Kirkpatrick and Baldwin 1974; Chapman et al. 1977).
2. Onset of breeding dependent on weather (Wight and Conaway 1961; Chapman et al. 1977).
3. Postpartum breeding (Conaway and Wight 1962; Casteel 1966).
4. "Full" participation (Wight and Conaway 1962*b*; Stevens 1962; Chapman et al. 1977; McKay and Verts 1978).
5. Consistency of litter size (Lord 1960; Wight and Conaway 1962*b*; Stevens 1962; Chapman et al. 1977).
6. Subadult breeding (Sowls 1957; Lord 1961; Stevens 1962; Casteel and Edwards 1964; Kibbe and Kirkpatrick 1971; Chapman et al. 1977; Bothma and Teer 1977).
7. No apparent curtailment of breeding in response to density per se (Sheffer 1957; Kirkpatrick and Baldwin 1974).
8. Minimum of energy invested in rearing young (short—28 days—gestation, rapid development, free living at 10–14 days, nursing only 1–2 times daily).

These points were developed extensively by Edwards et al. (in press).

Numerous workers have attempted with little success to find some single (universal) explanation for population regulation (Chitty 1957, 1971; Watson and Moss 1970; Krebs 1971; Krebs and Myers 1974). Our experience with cottontails leads us to Gibb's (1977) conclusion that there is "no need to look for one single limiting factor common to all situations.'' In a broad sense, regulation of cottontail abundance can be seen as a somewhat hierarchical system of interrelated multiple controls the principal components of which include:

1. Limits of tolerance to environmental factors, principally their extreme fluctuation;
2. Innate dispersal that is independent of density;
3. Denial of access to available essential resources by competition with other organisms of the same and different species, thus facilitating density-dependent dispersal;
4. Innate density-dependent regulation of natality (this mechanism is either not present or not effectively operative in the cottontail);
5. Vulnerability to predation;
6. Susceptibility to disease and infection;
7. Consequences of exhaustion of available resources, principally nutritional inadequacies.

The level of control is seen as becoming increasingly severe as additional components of the system become more effective and as density increases. Starvation and stress-related symptoms are seen as physiological consequences, not innate regulatory mechanisms, as suggested by Chitty (1960) and Christian and Davis (1964). Population regulation in the cottontail is a function primarily of density-dependent survival, with dispersion contributing an element of self-regulation to the control process. However, it should be pointed out that Conaway and Wight (1962) and more recently Bittner and Chapman (in press) believed that high density was a factor in reducing the reproductive rate of *S. floridanus*.

Workers are increasingly recognizing the role of predation in the control of herbivore abundance (Pearson 1971; Maclean et al. 1974; Gibb 1977). Keith and Windberg (1978) felt that winter food shortage was the factor responsible for initiating the decline in the hare cycle, with predation prolonging that decline. On the basis of extensive theoretical modeling, Holling (1966) perceived thresholds of behavior and concluded that predation is discontinuous, with search time important in determining when predation will effectively cease. Predation is very probably the major factor in the actual death of cottontails. Predation is seen as a primary direct cause of death in regulation of cottontail abundance.

Many canids, felids, mustelids, and raptors prey on cottontails, as do certain snakes (table 5.8).

Ectoparasites. The cottontails are hosts of a wide variety of ectoparasites. Those that have received the most attention are the ticks, a vector for Rocky Mountain spotted fever. Many of the ticks that cottontails host are readily transmissible to humans and domestic pets and thus are of concern from the public health standpoint. Among the more common ectoparasites of the genus *Sylvilagus* are the ticks of the family *Ixodidae*, the fleas *Pulicidae* and *Leptopsyllidae*, and warbles, *Cuterebridae*. A summary of the most com-

TABLE 5.8. Some predators of the genus *Sylvilagus*

Cottontail Species	Predator	Source
S. floridanus	mammal raccoon (*Procyon lotor*) ring-tailed cat (*Bassariscus astutus*) marten (*Martes americana*) fisher (*M. pennanti*) weasel (*Mustela* sp.) red fox (*Vulpes vulpes*) gray fox (*Urocyon cinereoargenteus*) coyote (*Canis latrans*) bobcat (*Felis lynx*) feral cat (*Felis domesticus*)	Martin, et al. 1961
	birds Red-tailed Hawk (*Buteo jamaicensis*) Red-Shouldered Hawk (*B. lineatus*) Rough-Legged Hawk (*B. lagopus*) Cooper's Hawk (*Accipiter cooperi*) Goshawk (*A. gentilis*) Golden Eagle (*Anguila chrysaetos*) Marsh Hawk (*Circus cyaneus*) Crow (*Corvus brachyrhynchos*)	McAtee 1935; Hager 1957; Gates 1972; Trent and Rongstad 1974
S. audubonii	mammal coyote gray fox badger (*Taxidea taxus*) bobcat raccoon skunk (*Mephitis mephitis*) mink (*Mustela vison*) kit fox (*Vulpes macrotis*)	Orr 1940; Ingles 1941; Fitch 1947

TABLE 5.8—*Continued*

Cottontail Species	Predator	Source
	birds	
	Red-Tailed Hawk	
	Cooper's Hawk	
	Swainson's Hawk (*B. swainsoni*)	
	Golden Eagle	
	reptiles	Grinnell et al. 1930
	rattlesnake (*Crotalus oreganus notalus*)	
S. bachmani	mammal	Dixon 1925; Orr 1940
	bobcat	
	coyote	
	gray fox	
	long-tailed weasel (*M. frenata*)	
	bird	Bryant 1918; Foster 1927; Hall 1927; Sumner
	Red-tailed hawk	1929; Orr 1940
	Cooper's hawk	
	Barn owl (*Tyto alba*)	
	Great horned owl (*B. virginianus*)	
	Marsh hawk	
	Swainson's hawk	
	Golden eagle	
	Horned owl	
	Barn owl	
	reptile	
	rattlesnake (*C. confluenti*)	
	gopher snake (*Pituophis catenifer*)	
S. nuttallii	mammal	Orr 1940
	bobcat	
	coyote	
	bird	Borell and Ellis 1934; Hall 1946
	Great Horned Owl	
	Long-Eared Owl (*Asio wilsonianus*)	
	Marsh Hawk	
	reptile	Orr 1940
	rattlesnake	
	gopher snake	
S. aquaticus	mammal	Svihla 1929; Lowe 1958
	domestic dog	
	reptile	
	American alligator (*Alligator mississip-*	
	piensis)	
S. palustris	mammal	Blair 1936
	bobcat	
	bird	
	Great Horned Owl	
	Barred Owl (*Strix varia*)	
	Barn Owl	
	Marsh Hawk	
	Red-tailed Hawk	
	Bald Eagle (*Haliaeetus lencocephalus*)	
	reptile	
	water moccasin (*Agkistrodon pisciuorus*)	
	diamond back rattlesnake (*C. ademanteus*)	
S. idahoensis	mammal	Wilde 1978; Green and Flinders 1980;
	weasel	Gashwiler et al. 1960
	coyote	
	red fox	
	bobcat	
	bird	
	owls (*Bubo* sp.)	
	Marsh Hawk	

mon ectoparasites of the *Sylvilagus* is given in table 5.9.

Endoparasites. The cottontails are hosts to a wide variety of endoparasites. The endoparasites of the genus *Sylvilagus* are listed in table 5.10.

Viral, Bacterial, and Rickettsial Diseases. The cottontails are known reservoirs of tularemia (*Francisella tularensis*) and Rocky Mountain spotted fever (*Ric-*

kettsia rickettsi). In Illinois, *S. floridanus* is the source of more than 90 percent of the human cases of tularemia (Yeatter and Thompson 1952). The main vector of tularemia is the rabbit tick (*Haemaphysalis leporis-palustris*), but other ticks and fleas may also carry the disease. Humans contract tularemia by coming in direct contact with the flesh and blood of infected rabbits, or by eating infected rabbits that are not properly cooked. Tularemia occurs widely throughout

TABLE 5.9. Some ectoparasites of the genus *Sylvilagus*

Cottontail Species	Parasite	Source
S. floridanus	tick	Morgan and Waller 1940; Stannard and Pietsch
	Ambylyomma americanum	1958; Heard 1962; Jacobson and Kirkpat-
	Haemaphysalis leporis-palustris	rick 1974; Cooney and Burgdorfer 1974
	Dermacentor variabilis	
	Ixodes dentatus	
	I. cookei	
	flea	
	Ceratophyllus multispinosus	
	Cediopsylla simplex	
	Odontopsyllus multispinosus	
	Hoplopsyllus affinis	
	Ctenocephalides canis	
	bot	
	Cuterebra buccata	
	C. cuniculi	
	C. horripilum	
	chigger	
	Euschongustia peromysci	
	Trombicula alfreddugesi	
	T. whartoni	
S. transitionalis	tick	Harman and Chapman 1977
	H. leporis-palustris	
	I. dentatus	
	flea	
	C. simplex	
	C. canis	
	O. multispinosus	
	bot	
	Cuterebra sp.	
S. audubonii	tick	Hall 1921; Ingles 1941; Fitch 1947
	Spilopsyllus sp.	
	C. felis	
	bot	
	Cuterebra sp.	
S. bachmani	flea	Fox 1926
	H. powersii	
	H. minutus	
S. aquaticus	tick	Ward 1934; Harkema 1938
	H. leporis- palustris	
S. palustris	tick	Caldwell 1966
	Haemaphysalis ?	
	bot	
	Cuterebra sp.	
S. idahoensis	tick	Wilde 1978; Janson 1946; Davis 1939; Bacon
	D. parumapertus	et al. 1959
	D. andersoni	
	H. leporis-palustris	
	mite	
	Ornithorodors sp.	
	flea	
	C. inaequalis	
	O. dentatus	
	Orchopeas sexdentatus	

TABLE 5.10. Some endoparasites of the genus *Sylvilagus*

Cottontail Species	Parasite	Source

S. floridanus	cestode
	  Ctenotaenia ctenoids
	  C. variabilis
	  Moscouoyia pectinata
	  Multiceps serialis
	  Taenia pisiformis
	  Cittotaenia perplexa
	nematode
	  Dermatoxys veligera
	  Nematodirus triangularis
	  N. leporis
	  Obeliscoides caniculi
	  Passalurus ambiguous
	  P. nonanulatus
	  Physaloptera sp.
	  Trichostrongylus affinis
	  T. calcaratus
	  Dirofilaria scapiceps
	  Trichuris leporis
	  T. sylvilagi
	  Baylisascaris procyonis
	  Ascaris columnaris
	trematode
	  Hasstilesia tricolor
		Morgan and Waller 1940; Erickson 1947; Heard 1962; Novelsky and Dyer 1970; Jacobson and Kirkpatrick 1974; Jacobson et al. 1974; Nettles et al. 1975

S. transitionalis	cestode
	  Cittotaenia variabilis
	  Cysticersuc pisiformis
	nematode
	  Obeliscoides cunniculi
	  P. ambiguus
	  Dirofilaria scapiceps
	coccidia
	  Eimeria sp.
		Eabry 1968; Dalke 1937, 1942; Erickson 1947; Rankin 1946

S. audubonii	cestode
	  C. variabilis
	  Raillietina retractilis
	  T. pisiformis
	nematode
	  D. veligera
	  N. leporis
	  O. cunniculi
	  P. ambiguus
	protozoan
	  Trichomonas
	  Chilomastix
	coccidia
	  Eimeria sp.
		Hall 1916; Herman and Jankieurcz 1943; Stiles 1896; Erickson 1947

S. nuttallii	cestode
	  C. pectinata
	  C. perplexa
	  C. variabilis
	  R. retractilis
	  T. pisiformis
	nematode
	  D. veligera
	  N. neomexicanus
	  Protostrongylus pulmonalis
	  T. colubriformis
	coccidia
	  Eimeria sp.
		Erickson 1947; Hall 1908; Dikmans 1937; Scott 1943; Honess 1935, 1939

TABLE 5.10—*Continued*

Cottontail Species	Parasite	Source
S. bachmani	cestode	Chapman 1974
	M. pectinata-americana	
	T. pisiformis	
	nematode	
	P. ambiguous	
S. aquaticus	cestode	Price 1928; Chandler 1929; Ward 1934; Smith
	C. ctenoides	1940; Erickson 1947; Lumsden and Zischke
	C. variabilis	1962
	M. serialis	
	R. stilesiella	
	nematode	
	Graphidium strigosum	
	N. leporis	
	O. cuniculi	
	P. ambiguous	
	T. calcaratus	
	T. leporis	
	trematode	
	H. texensis	
	H. tricolor	
S. idahoensis	nematode	Janson 1946
	Nematodirus sp.	

North America and the number of cases in various regions fluctuates year by year and season by season (Yeatter and Thompson 1952). However, tularemia is generally most prevalent in the spring and fall. The disease is always fatal to the rabbit.

When infected cottontails are encountered in the field, they may appear somewhat sluggish. Internally the disease can be recognized by a peppering of tiny white spots on the liver and spleen. If a rabbit with this condition is found, it should be burned and one's hands washed and disinfected. Cuts or abrasions should be treated with iodine. Cottontails may not exhibit any signs of tularemia in early stages (Yeatter and Thompson 1952).

The ectoparasites of the cottontail rabbits are known carriers of Rocky Mountain spotted fever and many other rickettsial diseases of humans (Philip 1946; Parker et al. 1952; Shirai et al. 1961; Burgdorfer et al. 1966; Sonenshine et al. 1966; Burgdorfer 1969; Cooney and Burgdorfer 1974). Rocky Mountain spotted fever is generally thought of as a disease of western North America; however, the disease is widely distributed. So far as is known, the primary natural means of infection in humans is through the bite of a tick, in the case of the cottontail, mainly *H. leporis-palustris, I. dentatus* (Parker et al. 1952), and *Amblyomma americanum* (Philip 1946). Other diseases in which the cottontails have been implicated include *Staphylococcus aureus* (McCoy and Steenbergen 1969) and eastern encephalitis (Hayes et al. 1964).

AGE DETERMINATION

A number of techniques have been used to determine the age of a cottontail. Those most often used include judging (1) prenatal growth, (2) epiphyseal cartilage closure, (3) body measurements, (4) bone growth, and (5) eye lens weight.

Prenatal Development. The age of prenatal *S. floridanus* can be estimated to within one day of age from five days to term based on nose to tail length, hind foot length, and weight. Rongstad (1969) developed a scale based on known-age embryos.

Epiphyseal Cartilage Closure. Thomsen and Mortensen (1946) were the first to study the epiphyseal groove as an age criterion. They found that young of the year could be differentiated from adults by the presence of the epiphyseal cartilage. In newborn cottontails, a thick plate of epiphyseal cartilage is found between the diaphysis, or shaft, and the epipysis, or head, of long bones (figure 5.12). The cartilage area is a growth center where proliferation of cartilage and its replacement by bony tissue results in elongation of bone. As a rabbit grows older, the cartilage plate gradually decreases in thickness due to a regression of cell division and eventually is replaced entirely by

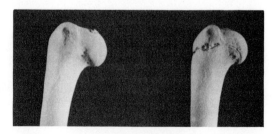

FIGURE 5.12. The proximal humerus of the eastern cottontail (*Sylvilagus floridanus*): adult (left) and juvenile (right). Note the epiphyseal groove present on the humerus of the juvenile and the smooth, fully ossified humerus of the adult.

bone during a rabbit's first winter. Bones of adult cottontails have no epiphyseal cartilage (Hale 1949).

Bones may be examined by x-rays or by external features of bones detached from a carcass to determine if epiphyseal cartilage disappears in the humerus of young cottontails at the age of approximately nine months. Wight and Conaway (1962*a*) compared the x-ray technique with the eye lens technique and conclude that the eye lens weight was the better aging criterion in cottontails. Bothma et al. (1972) concluded that the epiphyseal groove was a useful tool for judging ages up to nine months.

Body Measurements. Hill (1971) studied several body measurements that could be used for age determination in *S. floridanus*. Based on 151 known-age rabbits, he concluded that tarsus length provided the best estimate of age in young cottontails.

The length of the hind foot can be used as a criterion of age (Beule and Studholme 1942). However, Petrides (1951) believed that the length of the hind foot was not useful beyond the age of 3.5 months. Body weight is not useful in determining the age of a cottontail (Bothma et al. 1972).

Bone Growth. Cottontails may be placed into five separate age categories based on the growth of the skull (Hoffmeister and Zimmerman 1967). In rabbits less than 21 days old the teeth are not fully erupted; in rabbits 21–92 days old there is no fusion anywhere along the exoccipital-supraoccipital suture; in rabbits 93–105 days old the suture is visible but slightly fused; in rabbits 106–170 days old the suture is fused but visible; and in rabbits older than 170 days the suture is not visible. The lower jaw was evaluated as an age criterion by Bothma et al. (1972). Based on their data, it does not appear that the lower jaw would be a particularly valuable age criterion for cottontails.

Periosteal Zonations, or Underbone Growth. Sullins et al. (1976) found that stained lines in the periostal zone of *S. floridanus* and *S. nuttallii* mandibles are useful for separation of age classes among the adult cohort. Although the periosteal zonation technique has not been thoroughly tested, it appears that in combination with the lens weight technique this technique may provide a means by which age structures of cottontail populations can be established more precisely.

Eye Lens Weight. The weight of the dried eye lens is clearly the most useful measurement for aging cottontails. Eye lens weight as an indicator of age was first suggested by Lord (1959). The method enables the estimation of age past the period of epiphyseal closure. He correlated the relationship between age (x) and the dry weight of the lens (y) for approximating the ages of cottontails in days. An algebraic equation (Edwards 1967) derived from the lens growth model of Dudzinski and Mykytowycz (1961) can be used to estimate the ages of cottontails in days from dry lens weights. Edwards (1967) gave the details of the procedure for removing and fixing the lenses and a lens weight-age table for *S. floridanus* scaled to 371 days. Epiphyseal growth is compared to lens weight in Pelton (1969).

In practice, the lens technique is generally satisfactory for cottontails up to 9–10 months of age. For populations sampled through November and early December it is possible to distinguish young of the year from "adults." Older year classes do not differentiate satisfactorily.

It is best to take weights of both lenses and use the weight of the heavier. Lenses that appear atypical in color or shape, or those with evident loss of tissue, should be discarded. Tabulation of lens weights by 5-mg classes and coding to log form facilitates statistical analysis.

It is very important to standardize the period for collecting lenses if differences among years or areas are to be evaluated. Eberhardt et al. (1963) and Edwards (1962*a*) demonstrated that juveniles are more vulnerable to hunting, and presumably other causes of mortality, than adults. If hunter-killed rabbits provide a source of lenses, the opening one to three days of the hunting season are often a desirable sampling period, provided that the hunting season opens on about the same date each year at each area.

ECONOMIC STATUS

The cottontail is the principal game animal in the United States. Extremely high reproductive rates enable it to withstand heavy hunting pressures. The pelt is used for clothing and the meat is considered a wild food delicacy. Cottontails are also in demand by beagle clubs across the United States, as they provide excellent sport and challenge for trailing hounds and the people who train and breed them.

Cottontails occasionally, when found in high concentrations, may become depredators of succulent garden crops and nursery and orchard seedlings. Mature orchards are also subject to debarking in the winter. Overall, the cottontail's recreational and economic benefits from sport hunters greatly outweigh the minor damage done to crops, nurseries, and orchards.

MANAGEMENT

Habitat Management. Cottontails utilize a wide range of disturbed, successional, and transitional habitats. The essential ingredients of good cottontail habitat appear to be an abundance of well-distributed escape cover interspersed within a grassland-type community with an abundance of weedy forbs. An idealized old pasture situation has already been described.

The primary emphasis in habitat management for cottontails in almost every instance will be on increasing the number of cottontails that will survive on the managed area. Madson (1959) considered the brushpile the handiest device in rabbit management. It is instant cover. There is no quicker way to increase cottontails in most situations than by building brushpiles.

Brushpiles for cottontails should be at least 4–6 m in diameter and 1–2 m high. They should be placed near the edges of woodlots, in weedy fencerows, along streams and waterways, at the edges of pastures and hay fields, and in other places where weedy forbs and grasses already provide food and limited cover. Brushpiles should be distributed at distances of 50–100 m wherever practical over the managed area.

Although brushpiles are a quick and simple solution to habitat management for cottontails, they are only a temporary, not a permanent solution. Most brushpiles will only last three to five years; primary use may only be one to two years. If brushpiles are to be a primary element in a program of habitat management, one-third to one-quarter of them should be replaced on an annual basis. In the long run, brushpiles should not be considered as a substitute for more permanent, living vegetation.

Succession must be accounted for in any sound program of habitat management. If left unmanaged, today's pasture becomes tomorrow's woodland. The shrub becomes a tree. The habitat of cottontails and bobwhites will become the range of tree squirrels and wild turkeys. Management for cottontails must proceed on a continuing, typically annual, basis.

Several techniques may be used to control succession, usually in combination. These include perscribed burning, sharecropping, and grazing. These techniques have the advantage of being relatively cost effective, as they require a minimum of labor and equipment. Ellis et al. (1969) discuss the theory and economics of using sharecropping and prescribed burning in upland wildlife management on public lands. Bulldozing and herbicides have limited application in cottontail management.

Management typically proceeds on the basis of a long-range land use plan that depicts the existing topography, soils, woody vegetation, watercourses, roadways, crop fields, fencerows, pastures, and other significant cover and habitat types. Fields and tracts are designated for principal management type—woodland, fencerow, pasture, rotation cropland, or prescribed burning.

Tracts in woodland or those not suited to the other designations usually are maintained as woodland or allowed to succeed to woodland. If existing pastures have suitable fences they should usually continue to be pastured, but only moderately, to assure adequate stands of grasses and weedy forbs. Vegetative pattern, structure, and species composition induced by controlled grazing can provide highly desirable habitat for cottontail—interspersed food, escape cover, and travel lanes. It may be desirable to prescribe burning for such pastures in late winter at three- to five-year intervals.

Because of problems relating to topography and soil erosion it is often desirable to manage formerly cropped fields as permanent grasslands. This usually requires reseeding. Reseeding should include a variety of native warm-season prairie grasses and legumes. In the Midwest these might include *Andropogon scoparius, A. gerardi, Panicum vergatum, Sorghastrum nutans,* and the legumes *Lespedeza* spp., *Trifolium repens, T. hybridun,* and *T. pratense.* It is usually desirable to add small amounts of cool-season forage grasses to the mixture, such as *Phleum pratense, Agrostis alba,* and *Dactylis glomerata.*

Bromis innermis is a good grass for seeding on strip cover such as roadsides, ditch banks, and fencerows, but in fields mature sods tend to be too thick and dense for much rabbit use. Tall, dense, clump grasses that stand well in winter under snow, such as *Panicum virgatum,* provide good winter escape cover and reduce the need for woody escape cover.

Reseeding of grasslands, particularly where warm-season grasses are to be established, is best done on fallowed ground in late spring using oats as a nurse crop. It is usually necessary to mow in late July and September for weed control for one to two years after seeding. Mowing should *always* be with a rotary-type mower as high as possible (25+ cm).

Permanent grasslands may be managed by prescribed burning every two to four years, haying in late June, or light grazing. Abused soils should be moderately fertilized to preclude undesirable species such as *Agrostis hyemalis, Aristida* spp., *Andropogon virginicus,* and *Diodea teres,* or their local equivalents. High-fertility sites tend to be dominated initially by coarse weeds such as *Ambrosia artemissifolia* and later by *Eupatorium* spp. It is almost always necessary to control weeds in new grassland seedings. A knowledge of the relationships between soil fertility and secondary plant succession for local soils is important in habitat management.

Prescribed burning is the single most utilitarian tool available to the wildlife manager. In planning to use prescribed burning, one should consider natural barriers as a means of limiting the extent of the fire—roads, streams, ditches, or plowed fields, for example. Plowing or disking of fire lanes is usually not practical and often leads to erosion. Prescribed burning often enhances the quality woodlands for wildlife and rarely results in serious damage to trees. Relatively large tracts can be burned cheaply and effectively for wildlife. If damage is feared one simply does not burn. Damage caused by trying to contain fires is almost invariably worse than that of the fire.

Many management programs include the planting of trees and particularly shrubs for wildlife food and cover. An almost endless roster of plants could be compiled. Many of these are vulnerable to fire. In planting woody vegetation for cottontails one should try to select those plants that are more fire tolerant and select planting sites that naturally exclude fire or where fire protection can be given.

In cottontail management the natural thorny shrubs should be favored. Where planting is to be done, shrubs should be selected that maintain a low, dense, clump-type growth form. Numerous ornamentals have highly desirable form and growth patterns. Select what is available and does well in the local area. Although considered an undesirable exotic woodland weed by foresters, *Lonicera japonica* provides excellent escape cover and winter food for cottontails and certainly should not be discouraged in areas where it naturally occurs.

Rosa multiflora has been extensively planted and is an excellent cover plant for cottontails. Unfortunately, it easily escapes and colonizes untilled land. It has attained virtual pest plant status in many areas.

After 15–20 years many public areas become virtually unhuntable. Multiflora usually sprouts or recolonizes quickly after prescribed burning. Fortunately or unfortunately, multiflora is a plant that often grows on undisturbed wild land whether planted or not.

In cottontail management on public lands, fields suited to periodic cropping are often best managed by sharecropping under a four- to six-year rotation. An effective rotation in the Midwest is corn-oats-meadow. Under the terms of a contract agreement a local farm operator plants corn the first year and takes his share (usually ⅔ or ¾), and then leaves the remainder standing. The following March or April he seeds some combination of forage grasses and legumes with oats in the harvested part of the field and broadcast seeds legumes over the unharvested, still-standing corn. In July he combines the oats (keeps the total crop) and in late August mows the new meadow seeding for weed control. The field lies unmanaged for two to four years and is then replanted to corn. Seeded forage grasses include *Phleum pratense, Agrostis alba, Trifolium hybridum, T. pratense, Medicago sativa, Melilotus* spp., and *Lespedeza* spp. With several fields it is possible to have considerable diversity and interspersion of cover types.

Occasionally rough weedy forbs such as *Solidago* and *Eupatorium* dominate a field to the virtual exclusion of other plants. Such fields usually do not burn effectively, and even if they do, the same undesirable plants exclude the more desirable, and use by cottontails is minimal. When this happens the best solution is usually to try to burn, disk thoroughly, and reseed with desired prairie or forage grasses and legumes.

In the discussion of model cottontail habitat the probable ecological significance of cattle trails was suggested. In the absence of grazing, about the only way to provide a system of trails is by periodic mowing. The hard evidence for providing a system of mowed trails has not been documented. There is no question that such trails are extensively used by cottontails, particularly where grassy vegetation is rank and comprised primarily of the sod grasses such as *Bromus* and *Poa*. Trails facilitate access to different habitat types. They may not necessarily be needed, but they should be considered in cottontail management.

As the preceding discussion implies, habitat diversity and interspersion are key elements in cottontail management. A maxim of management is to allow no large block of uniform cover types.

Opportunities for management of cottontails are considerably less on private lands than on public area—except that there are far more private lands, at least in the east, midwest, and southern states. The basic problems of cottontail management on private lands are not unlike those on public lands: the need to favor an abundance, diversity, and interspersion of grasses, weedy forbs, and escape cover. The primary difficulty is that cover must be limited to such places as roadsides, waterways, ditch and stream banks, field corners, old building sites, fencerows, and around pond sites.

Once we could say that good wildlife management was compatible with good farm management.

This may still be true, but not in terms of prescribed modern farming practices, which favor monotypic or two-crop agriculture, larger fields and farms, weed control, improved ranges, channelization, and all the bigger, cleaner, faster, better practices prescribed today. While opportunity to manage cottontails on private lands still exists, in reality such opportunity is rapidly being lost to increasingly intensive agriculture.

A special management consideration is that of lands managed, usually by sports clubs, for field trials for beagle and basset hounds. Such areas require a high and sustained population of cottontails.

One expensive but generally successful solution to the problem of managing field trial areas for cottontails has been to enclose the fields with fencing to preclude dispersal and exclude ground predators. Fences are usually of 2.5-cm mesh galvanized after weaving, 16-gauge wire, approximately 2 m high with 10–15 cm buried underground. Charged electric wires are placed about 50 cm above the ground and also 10–15 cm down from the top on the outside of the fence.

Enclosures are usually in the range of 25 to 50 ha in size. Numerous brushpiles are provided for escape cover, food plots are usually seeded, and feeders and salt blocks are provided in the enclosures.

A typical problem in managing enclosures is too many, rather than too few, cottontails. It is usually necessary to begin to trap and remove rabbits from enclosures during late summer and fall. If this is not done, overpopulation will develop, which can virtually denude the pen of vegetation.

In programming habitat management it is always desirable to consider costs in relation to anticipated results and realistic objectives. Under a zero management program cottontail abundance will, in time, decline to a very low level as succession advances on unmanaged lands or as people continue to impact habitat on agricultural lands. The objective of management is to sustain cottontails at some desirable level of abundance. This poses the question of what level of abundance can realistically be expected. No definitive answer to this question is possible.

It appears likely that on managed areas of 500 ha or larger where forest does not greatly exceed 25 percent area, fall densities of cottontails approaching 2–3 per ha are a realistic goal for management. On areas of about 50 ha, sustained densities of 3–5 ha appear attainable. Harvest of 40 to 50 percent should also be sustainable—if accomplished by mid-December. The costs of management should relate to these levels of abundance and harvest. When costs are viewed in the light of expected densities and harvests it is obvious that management must rely on such comparatively cheap tools as prescribed burning, grazing, and sharecropping.

Stocking Programs. Because *S. floridanus* is a highly prized game mammal, considerable effort and money has been spent to stock cottontails. States that formerly had massive stocking programs include Maryland, New Jersey, and Pennsylvania (Dice 1927; Chapman and Morgan 1973; Chapman and Fuller

1975). In general, stocking programs have been carried on without regard for disease, impact on native species, or proper evaluation to determine the survival of the introduced cottontails. Dice (1927) was the first to warn of the dangers associated with the massive introduction of alien cottontails and it is safe to say that his concerns have been more or less borne out by more recent work (McDowell 1955; McDonough 1960; Chapman and Morgan 1973) but largely ignored by sportsmen.

The situation in the mid-Atlantic region has been summarized by Chapman and Fuller (1975). Originally, the southeastern subspecies of the eastern cottontail, *S. f. mallurus,* was an abundant inhabitant of the glades and river bottoms, while *S. transitionalis* occurred in the woodlands of the higher elevations (Doutt et al. 1966).

The situation began to change in the 1920s when a decline in cottontail populations was suspected. As a remedial measure, game departments, hunt clubs, and private individuals in several states of the region inaugurated a program of massive cottontail importations. As early as the turn of the century, *S. floridanus* of several subspecies had been introduced into Kansas, Missouri, and Texas (Chapman and Morgan 1973). Large-scale introductions began during the 1920s and continued well into the 1950s. After that the magnitude of the importations diminished, but even today some rabbits are being introduced by private individuals.

In the 1930s Pennsylvania was importing and releasing 50,000 cottontails annually. Gerstell (1937) pointed out the folly of such introductions in terms of their failure to support the annual harvest, the general failure of the phenotype released in Pennsylvania, and the logistical problems of releasing 50,000 rabbits in an area of 45,000 square miles.

Thanks largely to the pioneering work of Gerstell, during the 1940s many persons engaged in the importation programs became concerned about the policy of introducing alien rabbits. They reasoned that perhaps the introduced animals were bringing new diseases and parasites to the area and that, in any event, the animals might not survive. Because of this, the importation programs slowed. Recent research demonstrates that the introduction programs have apparently produced a series of fundamental changes in the character and distribution of the eastern cottontail in the mid-Atlantic region (Chapman and Morgan 1973).

Chapman and Fuller (1975:57–58) concluded "1. The mid-Atlantic region's native subspecies of eastern cottontail, *S. f. mallurus,* no longer exists in pure form within the region; interbreeding with introduced rabbits has obliterated the genotype or form, a fact which can be demonstrated genetically and morphologically"; "2. Intergrade eastern cottontails are the inheritors of a genetic vigor which renders them highly efficient colonizers in both stable and changing environments"; "3. The native New England cottontail has not interbred freely with the introduced eastern cottontails and has not adapted to changing environmental conditions"; and "4. Intergrade eastern cottontails are now displacing New England cottontails in the mountainous woodlands of the mid-Atlantic region."

Changes in the cottontail populations in the mid-Atlantic region are the direct result of importation programs and habitat alteration. Furthermore, introduced *S. floridanus* have infused the native population with new genetic material that has apparently enabled it to adapt better to a changing environment and to compete more successfully with other native species.

Harvest. The administration of an annual hunting season—formulating and enforcing game laws and regulations—is a principal component in the management of cottontails by state conservation agencies.

Measured in terms of numbers of hunters, hours of recreation, or meat in the pan, there can be little doubt that the cottontail is America's number one game animal. Preno and Labisky (1971) estimated that 70 percent of licensed Illinois hunters hunted cottontails. Edwards (1962b) estimated that in 1959 over 90 percent of the upland game hunters in Ohio hunted cottontails. In the states bordering the Mississippi, to the east, and south of pine forests of the lake states and New England, the cottontail is still king, but of a crumbling empire.

A check of seasons and bag limits reveals little consistency in hunting regulations among states. In general, game laws for cottontails tend to be most restrictive in the states most heavily populated and most intensively farmed. Even today in the open range lands of the west there are typically no laws restricting the hunting seasons or bag limits for cottontails.

In the Midwest and the mid-Atlantic seaboard states, where regulations are the most restrictive, seasons tend to open in mid-November and close in January. Elsewhere they tend to open in September or October and close in February or March. In the restrictive states, bag limits are typically four to five per day.

In the Midwest, mid-November openings are usually favored by farm groups as coinciding with the end of harvest. Yeatter and Thompson (1952) recommended late seasons to minimize contact with diseased rabbits and the transmission of tularemia.

Over the past 70–80 years hunting regulations have tended to become increasingly more restrictive as cottontail numbers have continued to decline.

The prevailing view has been pretty much that if we take care of the habitat, the cottontail will take care of itself (interpretation: habitat and weather, not hunting, are the principal determinants of cottontail abundance). Unfortunately, we have not taken very good care of the habitat. In the 1950s and 1960s most wildlife biologists felt that cottontails could not be overhunted and favored more liberal hunting laws. Today some are beginning to question the dogma of 30 years ago.

Hunting seasons for cottontails could begin at the end of the breeding season in September, when numbers are at or near their peak in the annual cycle. The ideal season would end sufficiently early to assure an adequate broodstock, perhaps in late December with the onset of truly cold weather in the northern range, or in January or even February further south. This ignores tradition, and tradition is as important in hunting cottontails as in hunting anything else. The traditional cottontail hunter has a rural heritage in which one does not hunt until the crops are harvested, typically in December or January with a touch of snow on the ground.

Gerstell (1937) reported that the annual harvest of cottontails per hunter in Pennsylvania declined from 8.3 for 1916–20 to 4.5 for 1931–36 even though the bag limit dropped from 10 to 4 or 5 and the season from 39 to 24–30 days. In another early study Allen (1938) estimated cottontail density of 1.21/ha and a harvest of 0.82/ha on a 200-ha farm in Kalamazoo County, Michigan. He estimated that 55 percent of the fall population was harvested. Kline and Hendrickson (1954) estimated mortality of cottontails to have been 86 precent over the period from 1 September 1952 to 1 January 1953 on a study area in Decatur County, Iowa.

In Ohio hunting pressure and harvest occur early in the season, usually over 40 percent on the opening day (Edwards unpublished data). On the basis of aerial census in intensively farmed northwest Ohio, Edwards (unpublished data) concluded an average of about 21 days of hunting per square mile over the first 15 days of the 1952 upland game season. On the basis of coordinated bag checks and questionnaire surveys, Edwards also found that reported hunter success on questionnaires was roughly twice that obtained on bag checks. Reported hunting pressure was about twice that observed on the aerial census. These findings indicate very sizable (roughly 3x) biases in questionnaire surveys.

Probably the best data on cottontail harvest and hunting pressure are available from hunter check stations on public hunting areas. The Delaware Wildlife Area, Delaware County, Ohio, is a 2,400-ha flood control reservoir owned and operated by the U.S. Army Corps of Engineers but managed by the Ohio Department of Natural Resources for multiple use recreation and open each fall to public hunting. During the 1950s the Delaware Area was subjected to heavy hunting pressure due in large measure to a program of propagation and release of pheasants. Most Ohio hunters take cottontails while pheasant hunting, and vice versa. It is not practical to attempt to distinguish different types of Ohio hunters.

The high levels of hunting pressure and harvest at Delaware in the 1950s are readily apparent (table 5.11). During the years 1956–59 hunting pressure averaged 5.02 hunters per ha and harvest 1.50 cottontails per ha. Recovery of cottontails live-trapped prior to the 1959, 1960, and 1961 seasons averaged 44 percent. Population estimates based on live-trapping show declines of 65–70 percent from shortly before to shortly after the hunting season and are considered to reflect primarily hunting-related mortality. The numbers of cottontails harvested demonstrate a strong capacity for the cottontail to sustain heavy hunting pressure. However, in the 1960s and 1970s the cottontail population on the Delaware Area declined markedly, as did the statewide cottontail population in Ohio (E. A. Leite personal communication).

Possible harvest-related declines in cottontail numbers are indicated by data collected at hunter check stations at several Ohio public hunting areas. Leite (1965) and Bachant (1972) report what is certainly a classic example of intensive harvest of cottontails on a public hunting area, the 209-ha Urbana Wildlife Area, Champaign County, Ohio. The Urbana area was a former state game farm fenced to exclude predators and preclude dispersal of game. It was opened to public hunting in 1964, after 33 years of operation as a game farm. Game farm records show an average of 786 cottontails (3.8 per ha) removed each winter over the preceding 22 years. No ground dens were available to serve as escape cover.

During the 1964 hunting season (which extended from 16 November 1964 through 30 January 1965), a total of 2,657 hunters hunted 5,041 hours on the Urbana area and reported 2,300 cottontails bagged. This is equivalent to 12.7 hunter trips, 24.1 hours of hunting, and 11.0 cottontails known harvested per hectare. On the opening day, 304 hunters took 833 cottontails, or 2.74 per hunter. The opening day harvest was 3.98 per ha and 36.2 percent of the season total.

Live-trapping done in October and early November 1964 at Urbana resulted in the tagging of 636 cottontails; 476 of these were among the cottontails taken by the hunters. That represents a known

TABLE 5.11. Summary of cottontail harvest and hunting pressure on the Delaware, Ohio, Wildlife Area

Item	Year								
	1951	1952	1953	1954	1955	1956	1957	1958	1959
Hunters									
opening day	1,032	2,300	2,507	3,626	3,051	2,716	3,110	3,087	2,768
#/ha	0.43	0.96	1.04	1.51	1.27	1.13	1.30	1.29	1.15
season	5,258	8,320	9,381	11,340	11,345	16,229	15,143	15,211	16,288
#/ha	1.29	3.47	3.91	4.73	4.73	6.76	6.30	6.34	6.79
Cottontails harvested									
1st 5,000 hunters	2,848	2,075	2,696	2,210	1,769	1,415	2,131	1,231	1,863
reported[a]	2,909	3,003	3,943	3,461	2,761	3,352	3,715	2,711	3,369
adjusted[b]	2,997	3,161	4,183	3,722	2,949	3,684	4,437	3,128	4,046
adjusted #/ha	1.25	1.32	1.74	1.55	1.23	1.54	1.85	1.30	1.69

SOURCE: Edwards unpublished data

[a]Actually reported to and observed at checking stations (minimum).

[b]Adjusted for nonreporting hunters (an estimate; may be high).

harvest of 74.8 percent. Based on tag ratios, the 1964 fall population at Urbana was estimated at 3,121 cottontails, or 14.93 per ha.

Live-trapping conducted following the hunting season and prior to the breeding season in 1965 indicated a population of 67 cottontails surviving on the Urbana area, or 0.32 per ha. Thus, fall and winter mortality of the prehunt population was estimated at 97.9 percent. Unaccounted mortality, including crippling and other hunting losses plus what might be termed *natural* mortality, amounted to 23 percent of the fall population. In all probability, hunting-related losses were a major element in the unaccounted mortality.

Regardless, the estimated 67 survivors could not be expected to sustain the 1964 harvest in 1965. Live-trapping in the fall of 1965 indicated a prehunt population of only 154 cottontails (or 0.74 per ha) at Urbana.

Hunting on the Urbana area in 1965 showed 748 hunters (3.6 per ha), 1,488 hours (7.1 per ha), and only 69 cottontails bagged (0.33 per ha)—a dramatic change from the results of the previous season. The prebreeding population at Urbana was estimated at about 60 in 1966.

The heavy hunting pressure imposed on the cottontail population on the Urbana Wildlife Area in 1964 is strongly indicated as a probable cause in the reduction of the cottontail population on that area. Lack of suitable escape cover may have compounded the problem.

Several things are apparent. First, given predator control and inhibition of dispersal, high-density cottontail populations develop. Also, under those conditions of sustained high density an annual average of almost four cottontails per hectare was harvested over a period of 22 years with no adverse consequences.

The concept that cottontail populations can support unlimited hunting pressure appears untenable. It appears that on heavily hunted areas managers should consider limiting harvest to about 50 percent as a desirable goal. Edwards's (1962c) data for the Delaware Wildlife Area suggest that the population was sustaining a known harvest of 44 percent. In the long run, however, it is the number of cottontails that survive, not the percentage harvested, that is most critical to the subsequent fall population. Limiting harvest to 50 percent is merely a suggested goal to help assure an adequate abundance of potential breeders relative to the habitat available. Under normal conditions five of six cottontails alive in mid-November will die during the next 12 months. Experience indicates that approximately three of the six can be taken without endangering the potential abundance of cottontails a year later. Second, and what is often not appreciated, is that on the average, under normal conditions we can expect in the fall only five young for each surviving adult. Viewed in this way it becomes obvious that management must stress survival of adults.

Depredations. In general, depredations by *Sylvilagus* are minor compared to their importance as a game species. Nonetheless, at certain times they may damage agricultural crops and tree plantings (Orr 1940; Hayne 1950; Blair and Langlinais 1960). A number of exceptable repellents may be used, and some of these are discussed by Hayne (1950) and Hayne and Cardinell (1958). The habitat may also be manipulated to prevent depredations, and Orr (1940) and Kundaeli and Reynolds (1972) discussed some of these methods.

Transect Indexes. Numerous methods of estimating abundance have been developed for cottontails, particularly *S. floridanus*. Most of these are based on individuals counted along fixed transects, usually roadsides, according to standardized procedures. Behavior and observability of cottontails are highly variable depending on time of day, time of year, and weather. Procedures should recognize cottontail activity periods. The best periods for cottontail roadside census are late March and early April, late June, and November.

The fall and spring counts should begin one hour after sunset on nights with normal or above-normal temperature and zero to light wind. The summer census should begin 30 minutes after sunrise. Again, cold and windy days should be avoided. Fall and spring routes may be 100 km long, whereas June routes should be no longer than 50 km.

Transect counts provide indexes to relative abundance. They do not give density estimates. The best procedure is to use transect counts to compare changes in abundance over a period of years for the same study area. Comparison of roadside counts of rabbits in physiographically different regions is not warranted. Similarly, differences in counts between spring and summer, or between summer and fall, have no basis for comparison because of seasonal differences in cover and cottontail behavior.

Day-to-day variability in transect counts associated with weather patterns is a major problem. Numerous replicate counts of individual routes or large numbers of routes and strict standardization of procedures are necessary to obtain useful data. At least 1,000 km of transect coverage per area per year are usually desirable even under the best of standardized procedures. Rural mail carrier censuses are effective in developing large sample sizes with minimum special effort. However, cooperation may be difficult to obtain.

Live-Trapping. Live-trapping is a relatively efficient means of obtaining an estimate of the relative abundance of cottontails on areas 25–100 ha in size. The suggested procedure is to distribute unbaited wooden live-traps at a rate of about four to five per ha along intersecting mowed grid lines. Random placement of traps is undesirable, as it usually results in low capture success. It is preferable to place traps along open existing travel lanes or along lanes mowed specifically for rabbits to use as travel lanes.

Rabbits travel along open, well-defined routes and traps must be set in the open in plain sight along those routes to assure high rates of capture. Rabbits seemingly enter traps out of curiosity. It is better to have traps in place two to three days prior to trapping.

Probability of capture is neither uniform nor random among individuals and it is variable from day to day. Estimators that assume uniform and random probability of capture should not be used. Otis et al. (1978) and Burnham and Overton (1979) discussed some problems of estimation. Preliminary applications of their improved "jackknife estimator" to field data have been promising, although the calculations are tedious unless done by computer.

Harvest Estimates. Estimates of statewide rabbit harvest are normally derived from questionnaires mailed to a sample of licensed hunters. There are several problems and biases associated with questionnaire surveys.

A major difficulty in mail surveys is obtaining the list of sample names. Three procedures have been used. One procedure is to have enforcement officers and biologists obtain names and addresses from hunters contacted in the field. This procedure has the advantage of having a harvest data base from the original field contact to contrast with success rates reported by the same individuals on the questionnaire survey. The sample may, however, not be representative of the true hunter population.

A second procedure for obtaining a sample is to obtain a working list of the current year hunters from records held by license agents. Random sampling procedures are possible provided names are collected late in the hunting season. This results in late contact and memory bias is increased.

A third procedure is to compile mailing lists from the hunters who bought licenses the previous year. This allows an early preliminary contact but eliminates those hunters from consideration who hunt for the first time or who did not buy a license the previous year.

Biases in harvest estimates derive from several causes. Some hunters fail to respond even after repeated contact. Usually two to three repeat mailings are necessary to obtain 70-90 percent compliance. Late respondents report lower success, suggesting that nonrespondents have below-normal success. Projecting harvest on reported success leads to overestimation of the actual take of game.

Memory is perhaps the major source of bias. Many hunters simply do not remember how often they hunted or what they killed. Reports show disproportionate multiples of 5 and 10 in daily bag limits. Success reported on questionnaires is consistently higher than that from bag checks. Some of the difference may be subconscious; some may be bragging. For whatever reason, the result is a positive bias.

The value of questionnaire surveys lies in the ease of attaining large masses of data that in a general way demonstrate trends and patterns of success and hunting pressure over the season, from year to year, and in various areas of a state. Such information is basic to harvest management for cottontails.

Hunter bag checks provide another means to obtain data on the harvest of cottontails. Where such censuses are used, conservation officers and biologists routinely record data such as number of hunters in a party, hours hunted, and rabbits taken during the routine law enforcement contacts with hunters in the field. Bag check data may be collected on key days or weeks, or over entire hunting seasons. Such data provide useful indexes to cottontail abundance and hunting success for comparisons on time or area bases.

Hunter check stations are operated on some private, leased, and public hunting areas. They can provide excellent data on harvest, success, and hunting pressure. However, voluntary cooperation is not effective, as many hunters will not comply and the resulting data are incomplete. To assure compliance, licenses are best retained at checking stations while hunters are afield, periodic field checks should be made by enforcement officers, and vehicle access and parking should be restricted during hunting hours.

CURRENT RESEARCH NEEDS

"After surveying eight important rabbit states, I am convinced that the characteristics of rabbit populations, and the factors determining their abundance and scarcity, are more difficult to decipher, and are receiving less thought and study from sportsmen and naturalists, than is the case in any other species of small game" (Leopold 1931:89). Strange as it may seem, even after 50 years relatively little progress has been made on research directed to an understanding of natural regulation. Emigration and dispersal are important aspects of cottontail ecology that warrant serious attention.

The basis of cottontail management is habitat management. New approaches to wildlife habitat analysis are developing rapidly, based on computer modeling. The key element is vegetative structure. Discrimination analyses are being used to identify important structural characteristics. These techniques can and should be directed to an understanding of the complex habitat relations of cottontails. This will not be a brief, simple, or cheap form of research.

The cottontail utilizes primarily successional vegetation and occurs through a range of early and midsuccessional series. In managing cottontails it is necessary to devise ways periodically to disrupt or inhibit natural succession, as progression to woodland is often rapid, particularly east of the Mississippi. There is much yet to learn about the various potential types of disturbance as they relate to successional processes in different ecological situations.

Soil fertility is one principal factor in determining plant species composition and rates of change in secondary succession (Edwards 1975). Fertility management has considerable significance to cottontail management and warrants research attention, particularly on areas where soils tend to be infertile or eroded.

One of the keys to good cottontail research is continuity of involvement. This does not mean that the same study should continue forever (although certain basic inventories should be conducted annually). What is important is that individuals with an interest in cottontails and an understanding of community ecology be given sustained support to maintain active programs of research related to the biology and ecology of cottontails in relation to their environments.

LITERATURE CITED

Allen, D. L. 1938. Ecological studies on the vertebrate fauna of a 500-acre farm in Kalamazoo County, Michigan. Ecol. Monogr. 8:347–436.

———. 1939. Michigan cottontails in winter. J. Wildl. Manage. 3:307–322.

Anderson, R. M. 1940. The spread of cottontail rabbits in Canada. Can. Field Nat. 54:70–72.

Applegate, J. E., and Trout, J. R. 1976. Weather and the harvest of cottontails in New Jersey. J. Wildl. Manage. 40:658–662.

Bachant, J. P. 1972. Relation of habitat structure to cottontail rabbit production, survival and harvest rates. Final Job Rep., Ohio F.A.P.R. Proj. W-103-R, Job V-d. 82pp.

Bacon, M.; Drake, C. H.; and Miller, N. G. 1959. Ticks (Acarina: Ixodoidca) on rabbits and rodents of eastern and central Washington. J. Parasitol. 45:281–286.

Bailey, J. A. 1968a. Regionwide fluctuations in the abundance of cottontails. Trans. North Am. Wildl. and Nat. Resour. Conf. 33:265–277.

———. 1968b. A weight-length relationship for evaluating physical condition of cottontails. J. Wildl. Manage. 32:835–841.

———. 1969. Quantity of soft pellets produced by caged cottontails. J. Wildl. Manage. 33:421.

Bailey, J. A., and Schroeder, R. E. 1967. Weights of left and right adrenal glands in cottontails. J. Mammal. 48:475.

Bailey, J. A., and Siglin, R. J. 1966. Some food preferences of young cottontails. J. Mammal. 47:129–130.

Barbour, R. W. 1951. The mammals of Big Black Mountain, Harlan County, Kentucky. J. Mammal. 32:100–110.

Barkalow, F. S. 1962. Latitude related to reproduction in the cottontail rabbit. J. Wildl. Manage. 26:32–37.

Barkalow, F. S., Jr. 1961. A large cottontail litter. J. Mammal. 42:254.

Beule, J. D. 1940. Cottontail nesting-study in Pennsylvania. Trans. North Am. Wildl. Conf. 5:320–328.

Beule, J. D., and Studholme, A. T. 1942. Cottontail rabbit nests and nestlings. J. Wildl. Manage. 6:133–140.

Bissonnette, T. H., and Csech, A. G. 1939. Modified sexual periodicity in cottontail rabbits. Bio. Bull. 17:364–367.

Bittner, S. L., and Chapman, J. A. In press. Reproductive and physiological cycles in an island population of *Sylvilagus floridanus*. Proc. World Lagomorph Conf., Univ. Guelph. Ontario.

Blair, R. M., and Langlinais, M. J. 1960. Nutria and swamp rabbits damage baldcypress plantings. J. For. (May), pp. 388–389.

Blair, W. F. 1936. The Florida marsh rabbit. J. Mammal. 17:197–207.

Blymyer, M. J. 1976. A new elevation record for the New England cottontail (*Sylvilagus transitionalis*) in Virginia. Ches. Sci. 17:197–207.

Borell, A. E., and Ellis, R. 1934. Mammals of the Ruby Mountains region of north-eastern Nevada. J. Mammal. 15:12–44.

Bothma, J. du P., and Teer, J. G. 1977. Reproduction and productivity in South Texas cottontail rabbits. Mammalia 41:253–281.

Bothma, J. du P.; Teer, J. G.; and Gates, C. E. 1972. Growth and age determination of the cottontail in South Texas. J. Wildl. Manage. 36:1209–1210.

Bruna, J. F. 1952. Kentucky rabbit investigations. Fed. Aid Proj. 26-R. Frankfort, Kentucky. 83pp.

Bryant, H. C. 1918. Evidence of the food of hawks and owls in California. Condor 20:126–127.

Burgdorfer, W. 1969. Ecology of tick vectors of American spotted fever. Bull. World Health Organ. 40:375–381.

Burgdorfer, W.; Friedhoff, K. T.; and Lancaster, J. L. 1966. Natural history of tick-borne spotted fever in the USA. Bull. World Health Organ. 35:149–153.

Burnham, K. P., and Overton, W. S. 1979. Robust estimation of population size when capture probabilities vary among animals. Ecology 60:927–936.

Caldwell, L. D. 1966. Marsh rabbit development and ectoparasites. J. Mammal. 47:527–528.

Casteel, D. A. 1966. Nest building, parturition, and copulation in the cottontail rabbit. Am. Midl. Nat. 75:160–167.

———. 1967. Timing of ovulation and implantation in the cottontail rabbit. J. Wildl. Manage. 31:194–197.

Casteel, D. A., and Edwards, W. R. 1964. Two instances of multiparous juvenile cottontails. J. Wildl. Manage. 28:858–859.

Chandler, A. C. 1929. A new species of trematode worm belonging to the genus *Hasstilesia* from rabbits in Texas. Proc. U.S. Natl. Mus. 75:1–5.

Chapman, J. A. 1971a. Organ weights and sexual dimorphism of the brush rabbit. J. Mammal. 52:453–455.

———. 1971b. Orientation and homing of the brush rabbit (*Sylvilagus bachmani*). J. Mammal. 52:686–699.

———. 1974. *Sylvilagus bachmani*. Mammal. Species 34:1–4.

———. 1975a. *Sylvilagus nuttallii*. Mammal. Species 56:1–3.

———. 1975b. *Sylvilagus transitionalis*. Mammal. Species 55:1–4.

Chapman, J. A., and Feldhamer, G. A. 1981. *Sylvilagus aquaticus*. Mammal. Species 151:1–4.

Chapman, J. A., and Fuller, K. B. 1975. Our changing cottontails. Atlantic Nat. 30:54–59.

Chapman, J. A., and Harman, A. L. 1972. The breeding biology of a brush rabbit population. J. Wildl. Manage. 36:816–823.

Chapman, J. A.; Harman, A. L.; and Samuel, D. E. 1977. Reproductive and physiological cycles in the cottontail complex in western Maryland and nearby West Virginia. Wildl. Monogr. 56:1–73.

Chapman, J. A.; Hockman, J. G.; Ojeda, M. M. 1980. *Sylvilagus floridanus*. Mammal. Species. 136:1–8.

Chapman, J. A., and Morgan, R. P., II. 1973. Systematic status of the cottontail complex in western Maryland and nearby West Virginia. Wildl. Monogr. 36:1–54.

———. 1974. Onset of the breeding season and size of first litters in two species of cottontails from southwestern Texas. Southwestern Nat. 19:277–280.

Chapman, J. A., and Paradiso, J. L. 1972. First records of the New England cottontail (*Sylvilagus transitionalis*) from Maryland. Ches. Sci. 13:17–18.

Chapman, J. A., and Stauffer, J. R., Jr. In press. The status and distribution of the New England cottontail, *Sylvilagus transitionalis* (Bangs). Proc. World Lagomorph Conf., Guelph, Ontario.

Chapman, J. A., and Trethewey, D. E. C. 1972. Movements within a population of introduced eastern cottontail rabbits. J. Wildl. Manage. 36:155–158.

Chapman, J. A., and Verts, B. J. 1969. Interspecific aggressive behavior in rabbits. Murrelet 50:17–18.

Chapman, J. A., and Willner, G. R. 1978. *Sylvilagus audubonii*. Mammal. Species 106:1–4.

———. 1981. *Sylvilagus palustris*. Mammal. Species 153:1–3.

Chitty, D. 1957. Self-regulation of numbers through changes in viability. *In* Population studies: animal ecology and demography. Cold Spring Harbor Symp. Quant. Bio. 22:277–280.

———. 1960. Population processes in the vole and their relevance to general theory. Can. J. Zool. 38:99–113.

———. 1971. The natural selection of self-regulatory behavior in animal populations. Pages 136-170 *in* I. A. McLaren, ed. Natural regulation of animal populations. Atherton Press, New York. 195pp.

Christian, J. J. 1963. Endocrine adaptive mechanisms and the physiologic regulation of population growth. Pages 189-353 *in* M. U. Mayer and R. G. VanGelder, eds. Physiological mammalogy. Vol. 1. Academic Press, New York. 381pp.

Christian, J. J., and Davis, D. E. 1955. Reduction of adrenal weight in rodents by reducing population size. Trans. North Am. Wildl. Conf. 20:177-189.

———. 1964. Endocrines, behavior, and population. Science 146:1550-1560.

Cockrum, E. L. 1949. Range extension of the swamp rabbit in Illinois. J. Mammal. 30:427-429.

Conaway, C. H.; Baskett, T. S.; and Toll, J. E. 1960. Embryo resorption in the swamp rabbit. J. Wildl. Manage. 24:197-202.

Conaway, C. H.; Sadler, K. C.; and Hazelwood, D. H. 1974. Geographic variation in litter size and onset of breeding in cottontails. J. Wildl. Manage. 38:473-481.

Conaway, C. H., and Wight, H. M. 1962. Onset of reproductive season and first pregnancy of the season in cottontails. J. Wildl. Manage. 26:278-290.

Conaway, C. H.; Wight, H. M.; and Sadler, K. C. 1963. Annual production by a cottontail population. J. Wildl. Manage. 27:171-175.

Cooney, J. C., and Burgdorfer, W. 1974. Zoonotic potential (Rocky Mountain spotted fever and tularemia) in the Tennessee Valley region. Part 1: Ecologic studies of ticks infesting mammals in land between the lakes. Am. J. Trop. Med. Hyg. 23:99-108.

Cowan, I. M., and Guiquet, C. J. 1956. The mammals of British Columbia. Handb. British Columbia Prov. Mus. 11:1-413.

Cushing, J. E. 1939. The relation of some observations upon predation to theories of protective coloration. Condor 41:100-111.

Dalke, P. D. 1937. A preliminary report of the New England cottontail studies. Trans. North Am. Wildl. Conf. 2:542-548.

———. 1942. The cottontail rabbits in Connecticut. Bull. Connecticut Geol. Nat. Hist. Surv. 65:1-97.

Dalke, P. D., and Sime, P. R. 1941. Food habits of the eastern and New England cottontails. J. Wildl. Manage. 5:216-228.

Dalquest, W. W. 1941. Distribution of cottontail rabbits in Washington State. J. Wildl. Manage. 5:408-411.

Davis, W. B. 1939. The recent mammals of Idaho. Caxton Printers, Ltd., Cadwell, Idaho. 400pp.

Dice, L. R. 1926. Notes on Pacific Coast rabbits and pikas. Occas. Pap. Mus. Zool. Univ. Michigan 166:1-28.

———. 1927. The transfer of game and furbearing mammals from state to state with special reference to the cottontail rabbit. J. Mammal. 8:90-96.

———. 1929. An attempt to breed cottontail rabbits in captivity. J. Mammal. 10:225-229.

Dikmans, G. 1937. A note on the members of the nematode genus *Trichostrongylus* occurring in rodents and lagomorphs, with descriptions of two new species. J. Washington Acad. Sci. 27:203-209.

Dixon, J. 1925. Food predilections of predatory and furbearing mammals. J. Mammal. 6:34-46.

Dixon, K. R., and Chapman, J. A. 1980. Harmonic mean measure of animal activity areas. Ecology 61:1040-1044.

Dixon, K. R.; Chapman, J. A.; Rongstad, O. J.; and Orhelein, K. M. In press. A comparison of home range size in

Sylvilagus floridanus and *S. bachmani*. Proc. World Lagomorph Conf., Guelph, Ontario.

Doutt, J. K.; Hepperstall, C. A.; and Guilday, J. E. 1966. Mammals of Pennsylvania. Pennsylvania Game Comm., Harrisburg, and Carnegie Mus., Carnegie Inst., Pittsburgh, Pa. 273pp.

Dudzinski, M. L., and Mykytowycz, R. 1961. The eye lens as an indicator of the wild rabbit in Australia. CSIRO Wildl. Res. 6:156-159.

Dusi, J. L. 1952. The food habits of several populations of cottontail rabbits in Ohio. J. Wildl. Manage. 16:180-186.

Eabry, H. S. 1968. An ecological study of *Sylvilagus transitionalis* and *S. floridanus* of northeastern Connecticut. Agric. Exp. Stn., Univ. Connecticut, Storrs, Conn. 27pp.

Eberhardt, L. L.; Peterle, T. J.; and Schofield, R. 1963. Problems in a rabbit population study. Wildl. Monogr. 10:1-51.

Ecke, D. H. 1955. The reproductive cycle of the Mearns cottontail in Illinois. Am. Midl. Nat. 53:294-311.

Edwards, W. R. 1962*a*. Age structure of Ohio cottontail populations from weights of lenses. J. Wildl. Manage. 26:125-132.

———. 1962*b*. Farm game hunter questionnaire survey, 1959. Game Res. in Ohio 1:13-15.

———. 1962*c*. A three year age structure study of Ohio cottontail populations from weights of eye lenses. M.S. Thesis. Ohio State Univ., Columbus. 100pp.

———. 1963. Fifteen cottontails in a nest. J. Mammal. 44(3):416-417.

———. 1964. Evidence for a normal age composition for cottontails. J. Wildl. Manage. 28:738-742.

———. 1967. Tables for estimating ages and birth dates of cottontail rabbits. Illinois Nat. Hist. Surv. Bio. Notes 59:1-4.

———. 1975. Soil fertility and competition in first-year secondary succession after cropping. Ph.D. Thesis. Univ. Illinois, Urbana-Champaign. 133pp.

Edwards, W. R.; Havera, S. P.; Labisky, R. F.; Ellis, J. A.; and Warner, R. E. In press. The abundance of cottontails (*Sylvilagus floridanus*) in relation to agricultural landuse in Illinois (U.S.A.), 1956-1978, with comments on mechanisms of regulation. Proc. World Lagomorph Conf., Guelph, Ontario.

Elder, W. H., and Sowls, L. K. 1942. Body weight and sex ratio of cottontail rabbits. J. Wildl. Manage. 6:203-207.

Ellis, J. A.; Thomas, K. P.; and Edwards, W. R. 1969. Responses of bobwhites to management in Illinois. J. Wildl. Manage. 33:749-762.

Erickson, A. B. 1947. Helminth parasites of rabbits of the genus *Sylvilagus*. J. Wildl. Manage. 11:255-263.

Evans, R. D. 1962. Breeding characteristics of southeastern Missouri cottontails. Proc. Ann. Conf. Southeastern Assoc. Game Fish Commissioners 16:140-142.

Evans, R. D.; Sadler, K. C.; Conaway, C. H.; and Baskett, T. S. 1965. Regional comparisons of cottontail reproduction in Missouri. Am. Midl. Nat. 74:176-184.

Fitch, H. S. 1947. Ecology of a cottontail rabbit (*Sylvilagus audubonii*) population in central California. California Fish Game 33:159-184.

Flinders, J. T., and Hansen, R. M. 1973. Abundance and dispersion of leporids within a shortgrass ecosystem. J. Mammal. 54:287-291.

Flux, J. E. C. 1964. Hare reproduction in New Zealand. N.Z. J. Agric. 109:483-486.

Foster, G. L. 1927. A note on dietary habits of the barn owl. Condor 29:246.

Fox, D. 1926. Some new Siphonaptera from California. Pan-Pacific Entomol. 2:182–187.

Friley, C. E. 1955. A study of cottontail habitat preferences on a southern Michigan farming area. Fed. Aid to Wildl. Restoration Proj. W-48-R. Michigan. 20pp.

Gashwiler, J. S.; Robinette, W. L.; and Morris, O. W. 1960. Foods of bobcats in Utah and Eastern Nevada. Wildl. Manage. 24:226–229.

Gates, J. M. 1972. Red-tailed hawk populations and ecology in east-central Wisconsin. Wilson Bull. 84:421–433.

Genoways, H. H., and Jones, J. K., Jr. 1972. Mammals from southwestern North Dakota. Occas. Pap. Mus., Texas Tech. Univ. 6:1–36.

Gerstell, R. 1937. The management of the cottontail rabbit in Pennsylvania. Pennsylvania Game News 7:6,7,27; 8:15–20; 8:8–12; 8:12–15,26.

Gibb, J. A. 1977. Factors affecting population density in the wild rabbit, Oryctolagus cuniculus (L.), and their relevance to small mammals. Pages 33–46 in B. Stonehouse and C. Perrins, eds. Evolutionary ecology. Univ. Park Press, Baltimore. 310pp.

Golley, F. B. 1962. Mammals of Georgia. Univ. Georgia Press, Athens. 218pp.

Goodpaster, W. W., and Hoffmeister, D. F. 1952. Notes on the mammals of western Tennessee. J. Mammal. 33:362–371.

Graf, W. 1955. Cottontail rabbit introductions and distribution in western Oregon. J. Wildl. Manage. 19:184–188.

Grange, W. B. 1949. The way to game abundance with an explanation of game cycles. Chas. Scribner's Sons, New York. 365pp.

Graves, W. C., and Asserson, W. C., III. 1967. Cottontail rabbit investigations. California Dept. Fish Game, Job Completion Rep. P-R Proj. W-47-R-15. 12pp.

Green, J. S., and Flinders, J. T. 1980. Brachylagus idahoensis. Mammal. Species 125:1–4.

Grinnell, J. 1922. The role of the "accidental." Auk 39:373–380.

———. 1939. Effects of a wet year on mammalian populations. J. Mammal. 20:62–64.

Grinnell, J.; Dixon, J.; and Linsdale, J. M. 1930. Vertebrate natural history of a section of northern California through the Lassen Peak region. Univ. California Publ. Zool. 35:1–594.

Hagar, D. C. 1957. Nesting populations of red-tailed hawks and horned owls in central New York state. Wilson Bull. 69:263–272.

Hale, J. B. 1949. Aging cottontail rabbits by bone growth. J. Wildl. Manage. 13:216–225.

Hall, E. R. 1927. The barn owl in its relation to the rodent population at Berkeley California. Condor 29:274–275.

———. 1946. Mammals of Nevada. Univ. California Press, Berkeley and Los Angeles. 710pp.

———. 1951. A synopsis of the North American lagomorpha. Univ. Kansas Publ., Mus. Nat. Hist. 5:119–202.

Hall, M. C. 1908. A new rabbit cestode, Cittotaenia mosaics. Proc. U.S. Natl. Mus. 34:691–699.

———. 1916. Nematode parasites of mammals of the orders Rodentia, Lagomorpha, and Hyracoidea. Proc. U.S. Natl. Mus. 50:1–258.

———. 1921. Cuterebra larvae. Some cats with a list of those recorded on other hosts. J. Am. Vet. Med. Assoc. 59:480–484.

Hamilton, W. J. 1940. Breeding habits of the cottontail rabbit in New York State. J. Mammal. 21:8–11.

———. 1955. Coprophagy in the swamp rabbit. J. Mammal. 36:303–304.

———. 1963. The mammals of eastern United States. Hafner Publ. Co., New York. 380pp.

Hanson, H. C., and Jones, R. L. 1976. The biogeochemistry of blue, snow and Ross' geese. Southern Illinois Univ. Press, Carbondale. 281pp. Illinois Nat. Hist. Surv. Spec. Publ. No. 1.

Harkema, R. 1938. The parasites of some North Carolina rodents. Ecol. Monogr. 6:151–232.

Harman, D. M., and Chapman, J. A. 1977. A seasonal study of the ectoparasites of Sylvilagus transitionalis. Proc. Pennsylvania Acad. Sci. 51:40–42.

Harrison, T., and Hickie, P. F. 1931. Indiana's swamp rabbit. J. Mammal. 12:319–320.

Haugen, A. O. 1942. Life history studies of the cottontail rabbit in southwestern Michigan. Am. Midl. Nat. 28:204–244.

Havera, S. P. 1973. The relationship of Illinois weather and agriculture to the eastern cottontail rabbit. Illinois State Water Surv. Tech. Rep. 4. 92pp. Mimeogr.

Hayes, R. O.; Daniels, J. B.; Maxfield, H. T.; and Wheeler, R. E. 1964. Field and laboratory studies of eastern encephalitis in warm- and cold-blooded vertebrates. Am. J. Trop. Med. Hyg. 13:595–606.

Hayne, D. W. 1950. A further test of cottontail repellents for garden use. Michigan Agric. Exp. Stn. Q. Bull. 32:373–377.

Hayne, D. W., and Cardinell, H. A. 1958. New materials as cottontail repellents. Michigan Agric. Exp. Stn. Q. Bull. 41:88–98.

Heard, L. P. 1962. Job completion report, rabbit studies (cottontail). Pittman-Robertson Proj. W-48-R-8, Mississippi. 25pp.

Hendrickson, G. O. 1940. Nesting cover used by Mearns cottontails. Trans. North Am. Wildl. Conf. 5:328–331.

———. 1943. Mearns cottontail investigations in Iowa. Ames Forester 21:59–73.

Herman, C. M., and Jankieurcz, H. A. 1943. Parasites of cottontail rabbits on the San Joaquin Experimental range, California. J. Wildl. Manage. 7:395–400.

Herrick, E. H. 1965. Endocrine studies. In The black-tailed jackrabbit in Kansas. Kansas State Univ. Agric. Exp. Stn. Tech. Bull. 140:73–75.

Hill, E. P. 1966. Some effects of weather on cottontail reproduction in Alabama. Proc. Ann. Conf. Southeastern Game Fish Commissioners 19:48–57.

———. 1967. Notes on the life history of the swamp rabbit in Alabama. Proc. Southeastern Assoc. Game and Fish Commissioners 21:117–123.

———. 1971. An evaluation of several body measurements for determining age in live juvenile cottontails. Proc. Ann. Conf. Southeastern Assoc. Game Fish Commissioners 25:269–281.

———. 1972. The cottontail rabbit in Alabama. Auburn Univ., Agric. Exp. Stn. Bull. 440. 103pp.

Hinds, D. S. 1973. Acclimation of thermoregulation in the desert cottontail, Sylvilagus audubonii. J. Mammal. 54:708–728.

Hoffmeister, D. F., and Zimmerman, E. G. 1967. Growth of the skull in the cottontail (Sylvilagus floridanus) and its application to age-determination. Am. Midl. Nat. 78:198–206.

Holden, H. E., and Eabry, H. S. 1970. Chromosomes of Sylvilagus floridanus and S. transitionalis. J. Mammal. 51:166–168.

Holler, N. R.; Baskett, T. S.; and Rogers, J. P. 1963. Reproduction in confined swamp rabbits. J. Wildl. Manage. 27:179–183.

Holler, N. R., and Conaway, C. H. 1979. Reproduction of

the marsh rabbit (*Sylvilagus palustris*) in south Florida. J. Mammal. 60:769–777.

Holler, N. R., and Sorensen, M. F. 1969. Changes in behavior of a male swamp rabbit. J. Mammal. 50:832–833.

Holling, C. S. 1966. The functional response of invertebrate predators to prey density. Mem. Entomol. Soc. Can. 48:1–86.

Holten, J. W., and Toll, J. E. 1960. Winter weights of juvenile and adult swamp rabbits in southeastern Missouri. J. Wildl. Manage. 24:229–230.

Honess, R. F. 1935. Studies on the tapeworms of the Black Hills cottontail rabbit, *Sylvilagus nuttallii grangeri* (Allen), with special reference to the life history of *Cittotaenia variabilis* Stiles. Univ. Wyoming Publ. Sci. 2:1–10.

———. 1939. The coccidia infesting the cottontail rabbit, *Sylvilagus nuttallii grangeri* (Allen), with descriptions of two new species. Parasitology 31:281–284.

Howard, W. E. 1960. Innate and environmental dispersal of individual vertebrates. Am. Midl. Nat. 63:152–161.

Hunt, T. P. 1959. Breeding habits of the swamp rabbit with notes on its life history. J. Mammal. 40:82–96.

Ingles, L. G. 1941. Natural history observations on the Audubon cottontail. J. Mammal. 22:227–250.

Jackson, S. N. 1973. Distribution of cottontail rabbits *Sylvilagus* spp. in northern New England. Agric. Exp. Stn., Univ. Connecticut. 48pp.

Jacobson, H. A., and Kirkpatrick, R. L. 1974. Effects of parasitism on selected physiological measurements of the cottontail rabbit. J. Wildl. Dis. 10:384–391.

Jacobson, H. A.; Kirkpatrick, R. L.; and Holliman, R. B. 1974. Emaciation and enteritis of cottontail rabbits infected with *Hasstilesia tricolor* and observations on a fluke to fluke attachment phenomenon. J. Wildl. Dis. 10:111–114.

Jacobson, H. A.; Kirkpatrick, R. L.; and McGinnes, B. S. 1978. Disease and physiologic characteristics of two cottontail populations in Virginia. Wildl. Monogr. 60:1–53.

Janes, D. W. 1959. Home range and movements of the eastern cottontail in Kansas. Univ. Kansas Publ., Mus. Nat. Hist. 10:553–572.

Janson, R. G. 1946. A survey of the native rabbits of Utah with reference to their classification, distribution, life histories and ecology. Master's Thesis. Utah State Univ., Logan. 103pp.

Johnson, A. M., and Hendrickson, G. O. 1958. Effects of weather conditions on the winter activity of Mearns cottontail. Proc. Iowa Acad. Sci. 65:554–558.

Johnson, L. W. 1973. A model for the synchronous breeding of the cottontail. M.S. Thesis. Univ. Illinois, Urbana. 50pp.

Johnson, M. L. 1968. Application of blood protein electrophoretic studies to problems in mammalian taxonomy. Syst. Zool. 17:23–30.

Johnson, M. L., and Wicks, M. J. 1964. Serum-protein electrophoresis in mammals: significance in the higher taxonomic categories. Pages 681–694 *in* C. A. Leone, ed. Taxonomic biochemistry and serology. Ronald Press Co., New York. 728pp.

Johnston, J. E. 1972. Identification and distribution of cottontail rabbits in southern New England. Agric. Exp. Stn. Univ. Connecticut, Storrs, Conn. 70pp.

Keith, L. B. 1963. Wildlife's ten-year cycle. Univ. Wisconsin Press, Madison. 201pp.

Keith, L. B., and Windberg, L. A. 1978. A demographic analysis of the snowshoe hare cycle. Wildl. Monogr. 58:1–70.

Kibbe, D. P., and Kirkpatrick, D. L. 1971. Systematic evaluation of late summer breeding in juvenile cottontails, *Sylvilagus floridanus*. J. Mammal. 52:465–467.

Kirkpatrick, C. M. 1956. Coprophagy in the cottontail. J. Mammal. 37:300.

———. 1960. Unusual cottontail litter. J. Mammal. 41:119–120.

Kirkpatrick, R. L., and Baldwin, D. M. 1974. Population density and reproduction in penned cottontail rabbits. J. Wildl. Manage. 38:482–487.

Klimstra, W. D., and Corder, E. L. 1957. Food of the cottontail in southern Illinois. Trans. Illinois State Acad. Sci. 50:247–256.

Kline, P. D. 1962. Vernal breeding of cottontails in Iowa. J. Iowa Acad. Sci. 69:244–252.

Kline, P. D., and Hendrickson, G. O. 1954. Autumnal decimation of Mearns cottontail, Decatur Co., Iowa, 1952. Proc. Iowa Acad. Sci. 61:524–527.

Korte, P. A., and Fredrickson, L. H. 1977. Swamp rabbit distribution in Missouri. Trans. Missouri Acad. Sci. 10 & 11:72–77.

Krebs, C. J. 1971. Genetic and behavioral studies on fluctuating vole populations. Pages 243–256 *in* P. J. Den Boer and G. R. Gradwell, eds. Proc. Adv. Study Inst. Dynamics of Numbers in Populations. Wageningen, Pudoc, Oosterbeek, Netherlands. 611pp.

Krebs, C. J., and Myers, J. H. 1974. Population cycles in small mammals. Adv. Ecol. Res. 8:267–399.

Kundaeli, J. N., and Reynolds, H. G. 1972. Desert cottontail use of natural and modified pinyon-juniper woodlands. J. Range Manage. 25:116–118.

Leite, E. A. 1965. Relation of habitat structure to cottontail rabbit production, survival and harvest rates. Job Prog. Rep., Ohio F.A.P.R. Proj. W-103-R-8, Job 12. 17pp.

Lemke, G. W. 1957. An unusually late pregnancy in a Wisconsin cottontail. J. Mammal. 38:275.

Leopold, A. 1931. Report on a game survey of the north central states. Sporting Arms and Ammunition Mfrs. Inst., Madison, Wis. 199pp.

Linkkila, T. E. 1971. Influence of habitat upon changes within the interspecific Connecticut cottontail populations. Agric. Exp. Stn., Univ. Connecticut, Storrs, Conn. 21pp.

Llewellyn, L. M., and Handley, C. O. 1945. The cottontail rabbits of Virginia. J. Mammal. 26:379–390.

Lord, R. D. 1959. The lens as an indication of age in cottontail rabbits. J. Wildl. Manage. 23:358–360.

———. 1960. Litter size and latitude in North American mammals. Am. Midl. Nat. 64:488–499.

———. 1961. Magnitudes of reproduction in cottontail rabbits. J. Wildl. Manage. 25:28–33.

———. 1963. The cottontail rabbit in Illinois. Tech. Bull. Illinois Dept. Conserv. 3:1–94.

———. 1964. Seasonal changes in the activity of penned cottontail rabbits. Anim. Behav. 12:38–41.

Lord, R. D., and Casteel, D. A. 1960. Importance of food to cottontail winter mortality. Trans. North Am. Wildl. Nat. Res. Conf. 25:267–274.

Lowe, C. E. 1958. Ecology of the swamp rabbit in Georgia. J. Mammal. 39:116–127.

Lowery, G. H., Jr. 1974. The mammals of Louisiana and its adjacent waters. Louisiana State Univ. Press, Baton Rouge. 565pp.

Lumsden, R. D., and Zischke, J. A. 1962. Seven trematodes from small mammals in Louisiana. Tulane Stud. Zool. 9:87–100.

McAtee, W. L. 1935. Food habits of common hawks. U.S.D.A. Circ. 370:1–36.

McCabe, R. A. 1943. Population trends in Wisconsin cottontails. J. Mammal. 24:18–22.

McCoy, R. H., and Steenbergen, F. 1969. Staphylococcus epizootic in western Oregon cottontails. Bull. Wildl. Dis. Assoc. 5:11.

McDonough, J. L. 1960. The cottontail in Massachusetts. Bull. Massachusetts Div. Fish Game. 22pp.

McDowell, R. D. 1955. Restocking with "native" cottontails. J. Wildl. Manage. 19:61–65.

McKay, D. O., and Verts, B. J. 1978. Estimates of some attributes of a population of Nuttall's cottontails. J. Wildl. Manage. 42:159–168.

Maclean, S. F., Jr.; Fitzgerald, B. M., and Pitelka, F. A. 1974. Population cycles in arctic lemmings: winter reproduction and predation by weasels. Arct. Alpine Res. 6:1–12.

Madson, J. 1959. The cottontail rabbit. Olin Mathieson Chemical Corp., East Alton, Ill. 56pp.

Marsden, H. M., and Conaway, C. H. 1963. Behavior and the reproductive cycle in the cottontail. J. Wildl. Manage. 27:161–170.

Marsden, H. M., and Holler, N. R. 1964. Social behavior in confined populations of the cottontail and swamp rabbit. Wildl. Monogr. 13:1–39.

Martin A. C.; Zim, H. S.; and Nelson, A. L. 1961. American wildlife and plants: a guide to wildlife food habits. Dover Publ., Inc., New York. 500pp.

Morgan, B. B. and Waller, E. F. 1940. A survey of the parasites of the Iowa cottontail (Sylvilagus floridanus mearnsi). J. Wildl. Manage. 4:21–26.

Morgan, R. P., II, and Chapman, J. A. In press. The serum proteins of the Sylvilagus complex. Proc. World Lagomorph Conf., Guelph, Ontario.

Mossman, A. S. 1955. Reproduction of the brush rabbit in California. J. Wildl. Manage. 19:177–184.

Myers, K. 1967. Morphological changes in the adrenal glands of wild rabbits. Nature 213:147–150.

Nelson, E. W. 1909. The rabbits of North America. North Am. Fauna 29:1–314.

Nettles, V. F.; Davidson, W. R.; Fisk, S. F.; and Jacobson, H. A. 1975. An epizootic of cerebrospinal nematodiasis in cottontail rabbits. J. Am. Vet. Med. Assoc. 167:600–602.

Nottage, E. J. 1972. Comparative feeding trials of Sylvilagus floridanus and Sylvilagus transitionalis. Agric. Exp. Stn., Univ. Connecticut, Storrs, Conn. 39pp.

Novelsky, M. A., and Dyer, W. G. 1970. Helminths of the eastern cottontail rabbit, Sylvilagus floridanus, from North Dakota. Am. Midl. Nat. 84:267–269.

Nugent, R. F. 1968. Utilization of fall and winter habitat by the cottontail rabbits of northwestern Connecticut. Agric. Exp. Stn., Univ. Connecticut, Storrs, Conn. 34pp.

Olmstead, D. L. 1970. Behavioral comparisons of two species of cottontails (Sylvilagus floridanus) and (Sylvilagus transitionalis). Trans. Northeastern Sect. Wildl. Soc. 27:115–126.

Orr, R. T. 1940. The rabbits of California. Occas. Pap. California Acad. Sci. 19:1–227.

Otis, D. L.; Burnham, K. P.; White, G. C.; and Anderson, D. R. 1978. Statistical inference from capture data on closed animal populations. Wildl. Monogr. 61:1–135.

Parker, E. R.; Bell, J. F.; Chalgren, W. S.; Thrailkill, F. B.; and McKee, M. T. 1952. The recovery of strains of rocky mountain spotted fever and tularemia from ticks of the eastern United States. J. Infect. Dis. 91:231–237.

Pearson, O. P. 1971. Additional measurements of the impact of carnivores on California voles (Microtus californicus). J. Mammal. 52:41–49.

Pelton, M. R. 1969. The relationship between epiphyseal groove closure and age of the cottontail rabbit (Sylvilagus floridanus). J. Mammal. 50:624–625.

Pelton, M. R., and Jenkins, J. H. 1971. Productivity of Georgia cottontails. Proc. Ann. Conf. Southeastern Assoc. Game Fish Comm. 25:261–268.

Pelton, M. R., and Provost, E. E. 1972. Onset of breeding and breeding synchrony by Georgia cottontails. J. Wildl. Manage. 36:544–549.

Petrides, G. A. 1951. The determination of sex and age ratios in the cottontail rabbit. Am. Midl. Nat. 46:312–336.

Philip, C. B. 1946. Rickottsial diseases in man. Symp. Med. Sci., Am. Assoc. Adv. Sci., Boston, Mass. 97–112.

Powers, R. A., and Verts, B. J. 1971. Reproduction in the mountain cottontail rabbit in Oregon. J. Wildl. Manage. 35:605–613.

Preno, W. L., and Labisky, R. F. 1971. Abundance and harvest of doves, pheasants, bobwhites, squirrels, and cottontails in Illinois, 1956–69. Illinois Dept. Conserv. Tech. Bull. 4. 76pp.

Price, E. W. 1928. List of helminth parasites occurring in Texas. J. Parasitol. 14:200–201.

Pringle, L. P. 1960. A study of the biology and ecology of the New England cottontail (Sylvilagus transitionalis) in Massachusetts. M.S. Thesis. Univ. Mass., Amherst. 57pp.

Rankin, J. S., Jr. 1946. Helminth parasites of birds and mammals in western Massachusetts. Am. Midl. Nat. 35:756–768.

Rongstad, O. J. 1966. Biology of penned cottontail rabbits. J. Wildl. Manage. 30:312–319.

———. 1969. Gross prenatal development of cottontail rabbits. J. Wildl. Manage. 33:164–168.

Rose, G. B. 1973. Energy metabolism of adult cottontail rabbits, Sylvilagus floridanus, in simulated field conditions, Am. Midl. Nat. 89:473–478.

———. 1974. Energy dynamics of immature cottontail rabbits. Am. Midl. Nat. 91:473–477.

Sadler, K. C. In press. Thirty-one years of rabbit population data in Missouri. Proc. World Lagomorph Conf., Guelph, Ontario.

Sakai, K.; Narise, T.; Hiraizumi, Y.; and Iyama, S. 1958. Studies on competition in plants and animals. Part 9: Experimental studies on migration in Drosophilia melanogaster. Evolution 12:93–101.

Schierbaum, D. 1967. Job completion report, evaluation of cottontail rabbit productivity. Pittman-Robertson Proj. W-84-R-12. Albany, N. Y. 21pp.

Schwartz, C. W. 1942. Breeding season of the cottontail in central Missouri. J. Mammal. 23:1–16.

Scott, J. W. 1943. A new lungworm from the Leporidae Protostrongylus sylvilagii, n. sp. Univ. Wyoming Publ. 10:57–71.

Selye, H. 1956. The stress of life. McGraw Hill Co., Inc., New York. 285pp.

———. 1973. The evolution of the stress concept. Am. Sci. 61:692–699.

Sheffer, D. E. 1957. Cottontail rabbit propagation in small breeding pens. J. Wildl. Manage. 21:90.

Shirai, A.; Boxeman, F. M.; Perri, S.; Humphries, J. W.; and Fuller, H. S. 1961. Ecology of Rocky Mountain spotted fever. Part 1: Rickettsia rickettsi recovered from a cottontail rabbit in Virginia. Exp. Bio. Med. 107:211–214.

Smith, C. C. 1940. Notes on the food and parasites of the rabbits of a lowland area in Oklahoma. J. Wildl. Manage. 4:429–431.

Smith, R. H. 1950. Cottontail rabbit investigations. Final Rep., Pittman-Robertson Proj. 1-R. New York. 84pp.

Sonenshine, D. E.; Atwood, E. L.; and Lamb, J. T. 1966.

The ecology of ticks transmitting Rocky Mountain spotted fever in a study area in Virginia. Ann. Entom. Soc. Am. 59:1234–1262.

Sorensen, M. F. 1972. Parental behavior in swamp rabbits. J. Mammal. 53:840–849.

Sorensen, M. F.; Rogers, J. P.; and Baskett, T. S. 1968. Reproduction and development in confined swamp rabbits. J. Wildl. Manage. 32:520–531.

Sowls, L. K. 1957. Reproduction in the Audubon cottontail in Arizona. J. Mammal. 38:234–243.

Stannard, L. J., and Pietsch, L. R. 1958. Ectoparasites of the cottontail rabbit in Lee County, northern Illinois, Illinois Nat. Hist. Surv., Bio. Note 38:1–18.

Stevens, V. C. 1962. Regional variation in productivity and reproduction physiology of the cottontail rabbit in Ohio. Trans. North Am. Wildl. Nat. Res. Conf. 27:243–253.

Stiles, C. W. 1896. A revision of the adult tapeworms of hares and rabbits. Proc. U.S. Natl. Mus. 19:145–235.

Stout, G. G. 1970. The breeding biology of the desert cottontail in the Phoenix region, Arizona. J. Wildl. Manage. 34:47–51.

Sullins, G. L.; McKay, D. O.; and Verts, B. J. 1976. Estimating ages of cottontails by periosteal zonations. Northwest Sci. 50:17–22.

Sumner, E. L., Jr. 1929. Comparative studies in the growth of young raptors. Condor 31:85–111.

Svihla, R. D. 1929. Habits of *Sylvilagus aquaticus littoralis*. J. Mammal. 10:315–319.

Terrel, T. L. 1972. The swamp rabbit (*Sylvilagus aquaticus*) in Indiana. Am. Midl. Nat. 87:283–295.

Thomsen, H. P., and Mortensen, O. A. 1946. Bone growth as an age criterion in the cottontail rabbit. J. Wildl. Manage. 10:171–174.

Toll, J. E.; Baskett, T. S.; and Conaway, C. H. 1960. Home range, reproduction, and foods of the swamp rabbit in Missouri. Am. Midl. Nat. 63:398–412.

Tomich, P. Q. 1962. The annual cycle of the California ground squirrel *Citellus beecheyi*. Univ. California Publ. Zool. 65:213–282.

Trent, T. T., and Rongstad, O. S. 1974. Home range and survival of cottontail rabbits in southwestern Wisconsin. J. Wildl. Manage. 38:459–472.

Trethewey, D. E. C., and Verts, B. J. 1971. Reproduction in eastern cottontail rabbits in western Oregon. Am. Midl. Nat. 86:463–476.

Turkowski, F. J. 1975. Dietary adaptability of the desert cottontail. J. Wildl. Manage. 39:748–756.

Udvardy, M. D. F. 1969. Dynamic zoogeography. Van Nostrand Reinhold Co., New York. 445pp.

Ward, J. W. 1934. A study of some parasites of rabbits of central Oklahoma. Proc. Oklahoma Acad. Sci. 14:31–32.

Watson, A., and Moss, R. 1970. Dominance, spacing behaviour and aggression in relation to population limitation in vertebrates. Pages 167–218 *in* A. Watson, ed. Animal populations in relation to their food resources. Blackwell Scientific Publications, Oxford. 477pp.

Wight, H. M. 1959. Eleven years of rabbit population data in Missouri. J. Wildl. Manage. 23:34–39.

Wight, H. M., and Conaway, C. H. 1961. Weather influences on the onset of breeding in Missouri cottontails. J. Wildl. Manage. 25:87–89.

———. 1962*a*. A comparison of methods for determining age of cottontails. J. Wildl. Manage. 26:160–163.

———. 1962*b*. Determination of pregnancy rates of cottontail rabbits. J. Wildl. Manage. 26:93–95.

Wilde, D. B. 1978. A population analysis of the pygmy rabbit (*Sylvilagus idahoensis*) on the INEL site. Ph.D. Thesis. Idaho State Univ., Pocatoello. 172pp.

Wilde, D. B.; Fisher, J. S.; and Keller, B. L. 1976. A demographic analysis of the pygmy rabbit, *Sylvilagus idahoensis*. Pages 88–105 *in* O. D. Markam, ed. 1975 progress report Idaho National Engineering Laboratory Site Radioecology-Ecology Programs. U.S. Energy Res. Devel. Admin. Idaho Falls. 205pp.

Willner, G. R.; Chapman, J. A.; and Pursley, D. 1979. Reproduction, physiological responses, food habits, and abundance of nutria on Maryland marshes. Wildl. Monogr. 65:1–43.

Worthington, D. A. 1970. The karyotype of the brush rabbit, *Sylvilagus bachmani*. Mammal. Chromosome Newsl. 2:21.

Worthington, D. H., and Sutton, D. A. 1966. Chromosome numbers and karyotypes of three species of Leporidae. Mammal. Chromosome Newsl. 8:282–283.

Yeatter, R. E., and Thompson, D. H. 1952. Tularemia, weather, and rabbit populations. Illinois Nat. Hist. Surv. Bull. 25:351–382.

Zoloth, S. R. 1969. Observations of the population of brush rabbits on Ano Nuevo Island, California. Wassman J. Bio. 27:149–161.

JOSEPH A. CHAPMAN, Appalachian Environmental Laboratory, Center for Environmental and Estuarine Studies, University of Maryland, Frostburg State College Campus, Frostburg, Maryland 21532.

J. GREGORY HOCKMAN, Washington Department of Game, Longview, Washington 98632.

WILLIAM R. EDWARDS, Illinois Natural History Survey, Urbana, Illinois 61801.

6

Jackrabbits

Lepus californicus and Allies

John P. Dunn
Joseph A. Chapman
Rex E. Marsh

NOMENCLATURE

COMMON NAMES. Black-tailed jackrabbit, California jackrabbit.
SCIENTIFIC NAME. *Lepus californicus*
SUBSPECIES NORTH OF MEXICO. *L. c. bennettii, L. c. californicus, L. c. deserticola, L. c. eremicus, L. c. melanotis, L. c. merriami, L. c. richardsonii, L. c. texianus,* and *L. c. wallawalla.*

COMMON NAMES. White-tailed jackrabbit, prairie hare.
SCIENTIFIC NAME. *Lepus townsendii*
SUBSPECIES. *L. t. campanius,* and *L. t. townsendii.*

COMMON NAMES. Antelope jackrabbit, white-sided jackrabbit, Allen's jackrabbit.
SCIENTIFIC NAME. *Lepus alleni*
SUBSPECIES NORTH OF MEXICO. *L. a. alleni*

COMMON NAMES. White-sided jackrabbit, Gaillard's jackrabbit, snow sides.
SCIENTIFIC NAME. *Lepus callotis*
SUBSPECIES NORTH OF MEXICO. *L. c. gaillardi*

DISTRIBUTION

The black-tailed jackrabbit is the most common jackrabbit in the western United States. Its range extends from the Pacific coast to western Missouri and Arkansas, and from the prairie and grassland regions of South Dakota southward to Texas. In the west the black-tail ranges from Washington and Idaho in the north into northern Mexico in the south, inhabiting much of the arid Southwest, including the desert habitat (figure 6.1).

The black-tailed jackrabbit has been successfully introduced into a number of eastern states. DeVoes et al. (1956) reported that *L. californicus* was introduced to Nantucket and Martha's Vineyard, Massachusetts. The black-tailed jackrabbit is also well established in southern Florida, where it utilizes pasture and sand prairie habitats (Layne 1965). Clapp et al. (1976) and Chapman and Sandt (1977) have reported that the black-tailed jackrabbit was successfully introduced on

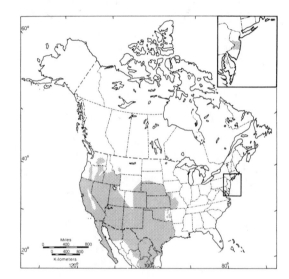

FIGURE 6.1. Distribution of the black-tailed jackrabbit (*Lepus californicus*).

the eastern shore of Virginia and Maryland. *L. californicus* has also been introduced into New Jersey.

The white-tailed jackrabbit's range extends from the prairies of the midwestern states and southern Canada westward through the sagebrush to the high mountain slopes of the Rockies, Cascades, and Sierras (figure 6.2). It may range as far south as the northern borders of New Mexico and Arizona.

During the early part of the 1900s *L. townsendii* was reported to be extending its range as far east as Wisconsin, Iowa, and Missouri (Leopold 1945). However, in recent years these populations have been declining. The white-tailed jackrabbit is now considered extirpated from Kansas (Hall 1955) and southern Nebraska (Jones 1964), and is considered rare in Missouri (Watkins and Novak 1973). Brown (1947) believed the cultivation of land that was formerly open prairie may have caused *L. townsendii* to disappear in Kansas. Kline (1963) and DeVoes (1964) have suggested that the removal of the original forest cover may be respon-

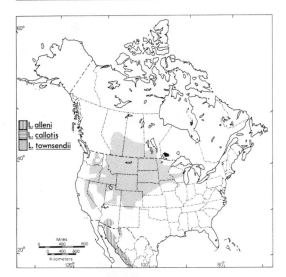

FIGURE 6.2. Distribution of the antelope jackrabbit (*Lepus alleni*), the white-sided jackrabbit (*L. callotis*), and the white-tailed jackrabbit (*L. townsendii*).

sible for the extension of the range of this species into Iowa and Wisconsin.

DESCRIPTION

Jackrabbits are technically hares, since the young are born fully haired, and not rabbits, whose young are born blind, naked, and in a nest. Jackrabbits, along with other hares and rabbits, belong to the family Leporidae, of the order Lagomorpha, which was once a suborder of Rodentia. The large ears and feet, the presence of a supraorbital process of the frontal bone, and a straight cutting edge on the upper incisor distinguish it as belonging to the family Leporidae. The absence of the interparietal bone distinguishes the genus *Lepus* from the genus *Sylvilagus* (Glass 1973). The dental formula for all North American jackrabbits is 2/1, 0/0, 3/2, 3/3 = 28.

The black-tailed jackrabbit (*L. californicus*) can be distinguished from other hares by its large tail with a black middorsal stripe extending onto the back (figure 6.3 left). The upper body parts are gray to blackish. The black-tailed jackrabbit can be distinguished from the white-tailed jackrabbit (*L. townsendii*) by the black-edged ears and the less pronounced area of white on the sides of the body (Hall and Kelson 1959). The hind foot of *L. californicus* averages less than 135 mm in length, while that of *L. townsendii* averages 148 mm (Orr 1940).

These two species of jackrabbit also can be distinguished on the basis of cranial characters. *L. californicus* has a relatively long and slender skull, with the supraorbital process being narrower than in *L. townsendii* and extending slightly above the frontal plane (figure 6.4). *L. townsendii* possesses a shorter, more arched skull, with the supraorbital process broader than in *L. californicus* and noticeably elevated

above the frontal plane (Orr 1940). The pattern of infolding on the first upper incisor of *L. townsendii* can be used to distinguish it from the skull of *L. californicus* east of the Rocky Mountains. *L. californicus* possesses a bifurcation of the tooth enamel, while *L. townsendii* has only a simple groove on the anterior surface of the tooth (Hall and Kelson 1959).

In Colorado the mean body weight of black-tailed jackrabbits was 2.54 kg (Flinders and Hansen 1972). In Arizona, Vorhies and Taylor (1933) found the weight of 23 adult males to average 2.3 kg. Females weigh more than males (Tiemeier 1965; Griffing 1974). Haskell and Reynolds (1947) reported that the black-tailed jackrabbit attains its maximum body weight at 32 weeks of age (see figure 7.6). Body measurements for *L. californicus* are as follows: total length 465–630 mm, tail 50–112 mm, hind foot 112–145 mm, and ear from notch 99–131 mm (Hall and Kelson 1959).

The white-tailed jackrabbit is the only jackrabbit that exhibits two annual molts (figure 6.3 right), although other members of Leporidae have a second annual molt. The summer pelage is grayish brown on the upper parts with the tail all white or with a buffy mid-dorsal stripe. The winter pelage is paler than that in the summer, with hares in the northern part of the range turning completely white. Winter guard hairs are completely white (Hall and Kelson 1959; Orr 1940).

The white winter coat is usually acquired in the geographical part of the range where snow persists throughout the winter. After the spring molt the white coat is replaced with a darker summer pelage. Where snow cover is nonexistent or sparse, summer and winter coats may vary little. A polytypic white pelage occurs in those regions where snow cover is variable in length (Hansen and Bear 1962).

Similarity between the snowshoe hare (*L. americanus*) in its winter pelage and the white-tailed jackrabbit can lead to misidentification in the field. Positive identification is aided by the tricolored hairs of the snowshoe as opposed to the completely white hairs of the jackrabbit (Burt and Grossenheider 1959).

Body measurements of adult white-tailed jackrabbits are as follows: total length 565–655 mm, tail 66–112 mm, hind foot length 145–172 mm, and ear length from notch 96–113 mm (Hall and Kelson 1959). The white-tailed jackrabbit is somewhat larger than the black-tailed jackrabbit, with weights of adults of the former ranging from 2.5 kg to 3.4 kg in California (Orr 1940). Sexual dimorphism has been reported in *L. townsendii*, with females being larger in both total length and weight (Orr 1940; Bear and Hansen 1966).

PHYSIOLOGY

Metabolism and Thermoregulation. The basal metabolic rate of adult black-tailed jackrabbits was estimated to be 54.11 kcal/kg body wt/24 hr, and a similar value of 53.14 kcal/kg body wt/24 hr was reported for the white-tailed jackrabbit (Flinders and Hansen 1972). In the Mojave Desert, where extremes in temperature occur, Shoemaker et al. (1976) estimated an

FIGURE 6.3.   A black-tailed jackrabbit (*Lepus californicus*) (left) and a white-tailed jackrabbit (*L. townsendii*) (right). The black tail of *L. californicus* readily distinguishes it from *L. townsendii*. *L. townsendii* may turn white during the winter, while *L. californicus* does not.

annual energy expenditure of 55,200 kcal/kg/yr for *L. californicus*. Nagy et al. (1976) suggested that jackrabbits were capable of regulating water losses from evaporation through physiological mechanisms. *L. californicus* has been shown to exhibit a diurnal shift in body temperature. An adult jackrabbit at rest produces a body temperature of 37–38° C (Shoemaker et al. 1976). In order to maintain such body temperatures, jackrabbits would normally require substantial amounts of water for evaporative cooling (Costa et al. 1976). However, they avoid this water demand by increasing body temperature to 41° C during the heat of the day, thereby storing heat that otherwise would have to be dissipated (Shoemaker et al. 1976). Jackrabbits also possess some ability to concentrate urine and thereby reduce additional water loss. This may be accomplished through the precipitation of calcium salts in the urine. Jackrabbits can also reduce water loss by excreting dry feces. Fecal water loss was estimated at no more than 15 percent of the total water loss (Nagy et al. 1976). Another physiological adaption for heat loss is the increased flow of blood to the ears, which results in greater convective and radiative heat loss (Schmidt-Nielsen et al. 1965).

Reese and Haines (1978) examined the effects of dehydration on jackrabbits. They found that severe water restriction, which may occur seasonally, resulted in a decreased metabolic rate and a reduction in evaporative water loss. The jackrabbit must then shift its thermoregulation to other nonevaporative mechanisms.

Behavioral mechanisms may aid in the dissipation of heat. Jackrabbits actively seek shade and may adjust their posture to maximize the surface/volume ratio for increasing convective heat loss (Schmidt-Nielsen et al. 1965).

Acclimatization to winter conditions is accomplished by mechanisms that reduce heat loss. Jackrabbits can increase the insulative quality of the fur and thereby increase the total insulation of the body. There is also a slight increase in metabolism during the winter months (Hinds 1977).

Adrenal Weights. Adrenal weights and morphology have been used to determine the endocrine response of the black-tailed jackrabbit (Herrick 1965). Pregnant females exhibited heavier adrenal weights than nonpregnant females, suggesting that pregnancy was a

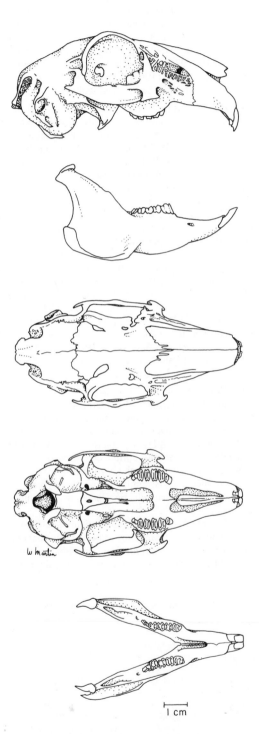

FIGURE 6.4. Skull of the black-tailed jackrabbit (*Lepus californicus*). From top to bottom: lateral view of cranium, lateral view of mandible, dorsal view of cranium, ventral view of cranium, dorsal view of mandible.

major stress factor. Comparison of biotic and environmental factors with adrenal weight indicated that temperature, age, and body weight also were important factors affecting adrenal weight (Herrick 1965).

GENETICS

Worthington and Sutton (1966) determined the diploid chromosome number of *L. californicus* to be 48. This species has two pairs of metacentric autosomes, 14 pairs of submetacentric autosomes, and 7 pairs of acrocentric autosomes. The X sex chromosome is a large submetacentric, and the Y is a "dot" chromosome.

In a study conducted in North Dakota, Jalal et al. (1967) reported that the white-tailed jackrabbit's diploid chromosome number was 48, with 23 pairs of autosomes and 1 pair of sex chromosomes. The chromosome length was 1.5 to 6.5 microns in the male. Hansen and Bear (1962) indicated that the winter pelage of *L. townsendii* may be genetically determined. They postulated that incomplete dominance of two or more alleles may be responsible for the polytypic winter pelage of this species.

REPRODUCTION

Anatomy. The reproductive system of male black-tailed and white-tailed jackrabbits is very similar to that of the domestic rabbit. Jackrabbits possess a scrotal sac that is entirely covered with hair and may be visible only during the breeding season. Scrotums of older individuals usually have a dark blue color that may serve as an indicator of age. The paired testes of adults are large (over 8 cc), and are scrotal during the breeding season. The penis is cylindrical and is normally withdrawn into the sheath. The seminal vesicle, vesicular gland, prostate, paraprostates, and bulbourethral glands are similar in gross morphology and histology to that of other lagomorphs (Lechleitner 1959*a*).

The female reproductive system also is comparable to that of the domestic rabbit. Jackrabbits possess larger corpora lutea than the domestic rabbit, sometimes reaching 12 mm in diameter, compared to about 5 mm in the latter. The uterus is duplex, each side having its own distinct cervical canal. The size and condition of the uterus is dependent on the reproductive condition of the animal. Nonpregnant, nonparous females tend to have a smoother, unstriated, less convoluted and less muscular uterus than that of pregnant or parous females. After parturition, the uterus shrinks rapidly in size. The vagina is a single tube leading from the cervix. The clitoris is near the external opening. The color of the vulva is variable, ranging from almost white to a dark purple. Females usually possess six mammae, two thoracic and four abdominal. These are supplied by two strips of mammary tissue extending down the midline of the abdomen (Lechleitner 1959*a*; James and Seabloom 1969*a*).

Breeding Season. The length of the breeding season in black-tailed jackrabbits is variable, depending on latitude and various environmental factors. In the northern part of its range in Idaho, French et al. (1965) reported a 128-day breeding season, approximately from February through May. In California, the breeding season occurs from late January through August, with occasional breeding possible during any month of the year (Haskell and Reynolds 1947; Lechleitner 1959a; Orr 1940). In Utah, Gross et al. (1974) reported a short breeding season occurring January through July. Over 75 percent of the adult females were pregnant during the period of February through April. Bronson and Tiemeier (1958a) reported a seven-month breeding season of 220 days, extending from January through August, in Kansas. Sperm production in males began in December, and all adult males contained sperm from December until late August. Sixty-five to 90 percent of the adult females were pregnant during the period of March through August (Tiemeier 1965).

In the southwestern United States, breeding began as early as January and ended by late August or September (Griffing and Davis 1976). In Arizona, Vorhies and Taylor (1933) reported jackrabbits breeding 10 months of the year. Two peaks of breeding, which corresponded to the annual cycle of rainfall and the greening of vegetation, have been reported in California (Lechleitner 1959a), Arizona (Vorhies and Taylor 1933), and New Mexico (Davis et al. 1975). Comparison of the lengths of the breeding seasons from California, Kansas, Utah, and Arizona indicate a shorter breeding season for those areas located at higher latitudes with more severe winter climates (French et al. 1965).

The length of the breeding season of L. townsendii appears to be relatively constant throughout its range. In Iowa and North Dakota it averages 148 days in length and extends from late February through mid-July (Kline 1963; James and Seabloom 1969a). In southern Colorado, ovarian size and weight along with uterine width reached a maximum during June (Bear and Hansen 1966).

Testes of white-tailed jackrabbits from North Dakota and Colorado were scrotal from late December until late July. They then ascended into the abdominal cavity during the remainder of the year. Sperm production peaked between March and June and then rapidly

decreased during July (Bear and Hansen 1966; James and Seabloom 1969a).

Gestation Period. Haskell and Reynolds (1947) reported a gestation period ranging from 41 to 47 days, with a mean gestation period of 43 days for black-tailed jackrabbits in California. This differed slightly from the mean gestation period of 40 days reported by Gross et al. (1974) for a black-tailed jackrabbit population in Utah. Kline (1963) estimated a gestation period of 42 days for white-tailed jackrabbits in Iowa.

Reproductive Rate. The number of litters conceived by black-tailed jackrabbits range from two litters per year in Idaho, where there is a 6-month breeding season (Feldhamer 1979), to seven litters per year in Arizona, where a 10-month breeding season occurs (Vorhies and Taylor 1933). The average annual production of young per female ranges from 8 in Arizona (Vorhies and Taylor 1933) to 16.7 for a population in Utah (Gross et al. 1974). The average annual production throughout the range of the black-tailed jackrabbit is about 14 young per female (table 6.1).

In the white-tailed jackrabbit there is also some variation on the total number of litters produced. In North Dakota the species averaged 3.29 litters per year (James and Seabloom 1969a). Kline (1963) reported that females in Iowa may produce three or four litters per year, while Bear and Hansen (1966) reported that L. townsendii in southern Colorado had only one litter per year. James and Seabloom (1969a) estimated the total annual production of the white-tail in North Dakota to be 14.7 young per female.

Breeding Synchrony. The jackrabbit, like other lagomorphs, is an induced ovulator; ovulation occurs only after copulation or other suitable stimulus has occurred. If pregnancy does not follow ovulation, the jackrabbit may exhibit pseudopregnancy, which may last about one-half the normal gestation period. The total loss of a litter through abortion or resorption also may produce effects similar to pseudopregnancy and may be accompanied by lactation (Lechleitner 1959a).

The black-tailed jackrabbit exhibits a relatively well synchronized breeding season, with conception possible immediately following parturition of the previous litter. The breeding season can be divided into

TABLE 6.1. Calculated reproductive rate per female *L. californicus* for several locales based on length of breeding season and mean litter size

Locality	(A) Breeding-Season Length	(B) Potential Number of Litters (A/43)	(C) Unweighted Mean Litter Size	Estimated Reproductive Output/Female (B X C)	Reference
Southern Idaho	128 days	3.0	3.35	10.1	French et al. (1965)
Northern Utah	190 days	4.4	3.8	16.7	Gross et al. (1974)
Southwestern Kansas	240 days	5.6	2.8	15.7	Tiemeier (1965)
Northern California	242 days	5.6	2.3	12.9	Lechleitner (1959a)
Southern Arizona	300 days	7.0	2.24	15.7	Vorhies and Taylor (1933)

SOURCE: Gross et al. 1974

four or five conception periods on the basis of this breeding synchrony (Gross et al. 1974).

Synchronous breeding also occurs in the white-tailed jackrabbit. James and Seabloom (1969a) reported a synchronous breeding pattern in North Dakota, where four well-defined peaks of breeding occurred. They suggested that postpartum breeding, as occurs in the black-tailed jackrabbit, may be the determining factor in synchronization of breeding.

Litter Size. In the black-tailed jackrabbit, there is a direct relationship between litter size and latitude, and there is an increase in the length of the breeding season at lower latitudes (French et al. 1965).

In the northern part of the range, litter sizes have been reported at 4.9 in Idaho (Feldhamer 1979) and 3.8 in Utah (Gross et al. 1974). Tiemeier (1965) reported a litter size of 2.8 in southwest Kansas. In the more southerly portions of the range, average litter sizes have been reported at 2.3 in California (Lechleitner 1959a), 2.2 in Arizona (Vorhies and Taylor 1933), and 1.9 in New Mexico (Griffing and Davis 1976).

There is some change in the number of young per litter associated with the yearly breeding cycle. Litter sizes are small at the onset of breeding. They gradually increase and peak in midseason and then decline toward the end of the breeding season (Lechleitner 1959a; Tiemeier 1965; Gross et al. 1974).

Average litter size throughout the range of the white-tailed jackrabbit is not consistent. Lord (1960) found no significant correlation between litter size and latitude. In Colorado the species averages 5 young per litter (Bear and Hansen 1966), while an average litter from North Dakota is 4.6 young (James and Seabloom 1969a). Kline (1963) found that *L. townsendii* in Iowa averaged 3.6 young per litter.

Variation of litter sizes from successive litter periods was found in North Dakota white-tailed jackrabbits. The greatest number of young per litter occurred during the second litter period. This was due to the variation in ova production and failure of the ova to implant during the first, third, and fourth litter periods (James and Seabloom 1969a). The effects of weather and resource availability on litter size have not yet been investigated.

Breeding Age. Young male black-tailed jackrabbits are first capable of producing spermatozoa between five and seven months of age, when they weigh approximately 1900 grams (Lechleitner 1959a; Bronson and Tiemeier 1958a). However, it is doubtful that males can breed at this age, since their testes do not contain a sufficiently high level of sperm. High sperm levels are not generally attained until the animal reaches a weight close to 2100 grams, between seven and eight months of age (Lechleitner 1959a).

Breeding by juvenile females is negligible in most populations (Haskell and Reynolds 1947; Bronson and Tiemeier 1958a). However, breeding by juvenile females has been reported in a jackrabbit population in Utah. Those juvenile females that were born during the first conception period in the spring were capable of breeding in their first year. Twenty-seven percent of

these young-of-the-year females had bred (Gross et al. 1974).

White-tailed jackrabbits normally breed in the spring following the one in which they were born (Kline 1963; Bear and Hansen 1966; James and Seabloom 1969a). Some juvenile males may reach sexual maturity by late summer but it is doubtful that these juveniles breed to any extent until the following year. No evidence of breeding in juvenile females was reported (James and Seabloom 1969a).

Nests and Newborn Hares. Female jackrabbits build forms in which they give birth to their precocial young (Tiemeier 1965). These forms are merely shallow excavations or depressions in the soil no more than a few cm deep, usually under some sort of protective cover. Tiemeier (1965) observed that litters may be dropped in depressions with little or no form preparation.

Newborn jackrabbits are very precocial. They are born fully furred and with their eyes open and are able to move about readily at birth. The young leave the nest shortly after birth, usually within 24 hours of parturition. Tiemeier's (1965) observations suggested that litters may stay together for a week or more after leaving the nest.

Neonates of the black-tailed jackrabbit have the characteristic black-tipped ears and black tail. Their pelage is darker than that of adults and is gradually replaced by a lighter coat. Often a small patch may be present on the forehead, but it disappears by three months of age. The adult pelage is attained between six and nine months of age (Haskell and Reynolds 1947).

Measurements of newborn black-tailed jackrabbits have been recorded as follows: total length 168 mm, hind foot 43 mm, and a weight of 110 grams. The ear and hind foot reach maximum size at 15 weeks, total length at 28 weeks, and full weight at 32 weeks of age (Haskell and Reynolds 1947).

It appears that neonatal *L. townsendii* grow at the same rate as neonatal *L. californicus*. Average measurements from one-day-old white-tailed jackrabbits in Colorado were: hind foot 58.3 mm, ear from notch 40.7 mm, and weight 105.3 grams (Bear and Hansen 1966).

ECOLOGY

Sex Ratio. The sex ratio in the black-tailed jackrabbit is constant from one region to another. A 1:1 sex ratio for both adults and juveniles has been reported for California (Lechleitner 1959a), New Mexico (Griffing and Davis 1976), Kansas (Tiemeier 1965), and Utah (Gross et al. 1974). Fetal sex ratios are also 1:1 (Lechleitner 1959a; Tiemeier 1965; Gross et al. 1974; Feldhamer, 1979). A sex ratio for the white-tailed jackrabbit has not been reported, but presumably would be similar to that of the black-tailed jackrabbit.

Home Range. Home range size of the jackrabbit is affected by the pattern of food, cover, and water in the surrounding habitat. Lechleitner (1958b) reported that the black-tailed jackrabbit in California had a well-defined home range usually less than 20.2 hectares.

Home ranges of adult females were larger than those of males. French et al. (1965) reported a home range of less than 16.2 ha in Idaho. Tiemeier (1965) also determined the home range of adults in Kansas to be 16.2 ha; the home ranges of adults were significantly greater than those of juveniles. Home ranges of adult females were larger than those of males (Lechleitner 1958b). The greatest movements of jackrabbits occurred at night rather than in the morning or evening.

The home range of the white-tailed jackrabbit averages 89.4 ha and is significantly larger than that of the black-tailed jackrabbit. Males possess a greater home range than females, with the greatest amount of movement being nocturnal (Donoho 1972).

Fluctuations in Population Density. Gross et al. (1974) reported on a pattern of population fluctuation for the black-tailed jackrabbit in the Curlew Valley of northern Utah. Three population cycles were observed during the period 1951–70, and jackrabbit densities ranged from 0.1 per ha to as high as 1.0 per ha. Although there was some variation in the amount of synchrony, density estimates were constant over a broad region of northern Utah and southern Idaho (Gross et al. 1974). Evans et al. (1970) comment on population fluctuations in which peak densities usually occur every 6 to 10 years, and Clark (1975) reported that they varied from 5 to 10 years in California.

Changes in jackrabbit population densities have been reported by many investigators. Bronson and Tiemeier (1959) observed that black-tailed jackrabbits were not cyclic in Kansas, but periods of drought concentrated jackrabbits in cultivated fields. Changes in the amount of food available were believed to be responsible for locally heavy concentrations where densities reached as high as 34.6 hares per ha. During normal precipitation, densities were 1 jackrabbit for every 1.5 to 2.5 ha (Bronson and Tiemeier 1959). Similarly, changes in density in Arizona were related to the process of grazing by cattle. Overgrazing in combination with drought created more favorable habitat by creating more open, weedy areas, which replaced the natural climax grasses (Taylor et al. 1935). Hayden (1966b) found that maximum densities of jackrabbits in the Mojave Desert occurred around sources of water during the winter. These seasonal densities then decreased by 85 percent from spring to summer.

Density estimates of black-tailed jackrabbits in the arid Southwest ranged from 0.2 per ha in Nevada (Hayden 1966b) through 0.9 per ha in the Great Salt Lake Desert of Utah (Woodbury 1955) to 1.2 per ha in Arizona (Vorhies and Taylor 1933). In the more temperate regions, densities ranged from 3.0 per ha for California (Lechleitner 1958b) to as high as 34.6 per ha for the agricultural areas of Kansas (Bronson and Tiemeier 1959).

Flinders and Hansen (1973) reported a density of about 0.01 per ha for the white-tailed jackrabbit on a short-grass prairie in southeastern Colorado. In Colorado the white-tailed jackrabbit did not exhibit as marked a fluctuation in density as the black-tailed jackrabbit in the same area (Donoho 1972).

FOOD HABITS

The food habits of the black-tailed jackrabbit have been studied extensively by many biologists and have been found to be highly variable depending on the locality and availability of palatable plants. Some of the more common food items are reported in table 6.2.

Black-tailed jackrabbits prefer green succulent vegetation when it is available. Depending on the habitat, their diet consists mainly of grasses and forbs during the summer and changes to shrubs during the winter when snow or frost makes succulents unavailable (Hansen and Flinders 1969). Grasses and sedges are just as important as forbs and shrubs in the yearly diet of *L. californicus*. According to Flinders and Hansen (1972), grasses and sedges made up 49 percent of the overall diet of the species in northern Colorado. In the sagebrush (*Artemisia tridentata*) and bitterbrush (*Purshia tridentata*) communities of Washington, forbs provide the mainstay of the jackrabbit's diet. In both communities forbs comprised 75 percent of the diet, with grasses and shrubs providing the remainder. Yarrow (*Achillea millefolium*) was the most common food item, comprising over 25 percent of the diet. Needle-and-thread grass (*Stipa comata*) was the preferred plant in the sagebrush habitat, while yarrow was preferred in the bitterbrush habitat (Uresk 1978).

On the salt desert ranges of Utah, Currie and Goodwin (1966) found that jackrabbits ate grasses, forbs, and shrubs in early spring. Grasses were preferred throughout spring and summer, and shrubs during late fall and winter. In that study, the more common grass species utilized were Indian ricegrass (*Oryzopsis hymenoides*) and squirreltail (*Sitanion hystrix*). The more prominent shrub species were winterfat (*Eurotia lanata*), shadscale (*Atriplex confertifolia*), saltsage (*A. nuttallii*), and sagebrush.

In northeastern Colorado, Flinders and Hansen (1972) observed that seven plant species accounted for 64 percent of the dry plants eaten by black-tailed jackrabbits. In order of importance, these species were western wheatgrass (*Agropyron smithii*), alfalfa (*Medicago sativa*), summer cypress (*Kochia scoparia*), winter wheat (*Triticum aestivum*), crested wheatgrass (*Agropyron cristatum*), rubber rabbitbrush (*Chrysothamnus nauseosus*), and sun sedge (*Carex heliophila*). Sparks (1968) also noted the importance of western wheatgrass in the diet of jackrabbits located on the sandhill ranges of Utah. Jackrabbits were also noted to have fed heavily on cultivated crops (winter wheat, alfalfa, and crested wheatgrass) during the winter months (Flinders and Hansen 1972).

Black-tailed jackrabbits in the Gray Lodge Waterfowl Management Area of California ate almost all of the species of grass and forb weed found there, but the herbaceous weeds were not a preferred item in the diet (Lechleitner 1958a). Their preferred foods were various cereal crops, especially young barley

TABLE 6.2. Major foods of *Lepus californicus*

Locality	Vegetative Species	
	Woody	Herbaceous
Utah	shadscale (*Atriplex confertifolia*) greasewood (*Sarcobatus vermiculatis*)	globemallow (*Sphaeralcea coccinia*) winterfat (*Eurotia lanata*) Indian ricegrass (*Oryzopsis hymenoides*)
Colorado	sagebrush (*Artemisia* sp.)	larkspur (*Delphinium* sp.) spiderwort (*Tradescantia* sp.) blue grama (*Bouteloua gracilis*) yucca (*Yucca glauca*) alfalfa (*Medicago sativa*) western wheatgrass (*Agropyron smithii*)
California Butte Co.		barley (*Hordeum* sp.) mallow plants (*Malvaceae*) curly dock (*Rumex crispus*)
Northeastern Lassen Co.	hopsage (*Grayia spinosa*) four-winged saltbush (*Atriplex canescens*) horsebrush (*Tetradymia* sp.) sagebrush	
Kansas	snakeweed (*Gutierrezia sarothrae*)	yucca (*Yucca glauca*) prickly pear (*Opuntia* sp.) sand dropseed (*Sporobolus cryptandrus*) pigweed (*Amaranthus retroflexus*) Russian thistle (*Salsola kali*) buffalo grass (*Buchloe dactyloides*) blue grama winter wheat (*Triticum aestivum*) buffalo burr (*Solanum rostratum*) lily (*Lilium* sp.) sedge (*Carix* sp.)
Arizona	mesquite (*Prosopis juliflora*)	cactus (*Opuntia* sp.) grasses (*Aristida, Bouteloua, Eragrostis, Trichloris, Panicum, Sporobolus, Echinochloa*)
Nevada	creosote (*Larrea* sp.) white sage (*Artemisia* sp.)	needle and thread (*Stipa comata*) downy brome (*Bromus tectorum*)

SOURCE: Hansen and Flinders 1969

plants (*Hordeum* sp.). All species of weed were eaten by black-tailed jackrabbits, but they were not a preferred item in the diet (Lechleitner 1958*a*). In October, Orr (1940) observed jackrabbits in the Great Basin section of California feeding mainly on sagebrush.

In Arizona, Vorhies and Taylor (1933) determined the food habits of black-tailed jackrabbits from analysis of stomach contents. Mesquite (*Prosopis juliflora*) made up 54 percent of the annual diet, grasses 24 percent, and cacti 3.3 percent. The amount of food available was shown to be related to the cycle of rainy and dry seasons in Arizona.

Examination of stomach contents of *L. californicus* from New Mexico also indicated a preference for mesquite leaves and seedpods. In addition, soaptree yucca (*Yucca glauca*), snakeweed (*Gutierrezia sarothrae*), croton (*Croton* sp.), and spurge (*Euphorbia*

sp.) were present in the stomachs analyzed (Griffing and Davis 1976). Hayden (1966*a*) also determined the food habits of black-tailed jackrabbits in Nevada from analysis of stomach contents. Nineteen species of perennial shrub, six species of annual, and three species of grass were identified in the diet. The annual diet of these jackrabbits could be divided into two major periods: in late summer through winter the diet was dominated by creosotebush (*Larrea* sp.) and winterfat, and in spring through early summer the diet was dominated by Indian ricegrass (*Oryzopsis* sp.), brome (*Bromus* sp.), and needle-and-thread grass (Hayden 1966*a*).

White-tailed jackrabbits feed mainly on plants that are in the prereproductive or early reproductive stages of development (Flinders and Hansen 1972). These plants usually have the greatest nutritive value and contain relatively high proportions of moisture and crude protein.

In the Colorado study seven plant species comprised over 67 percent of the diet (table 6.3). These species, in order of importance, were western wheatgrass, winter wheat, summer cypress, vetches (*Oxytropis* sp. and *Astragalus* sp.), sun sedge, rubber rabbitbrush, and crested wheatgrass (Flinders and Hansen 1972).

In southern Colorado the diet of the white-tailed jackrabbit during the spring is made up of 87 percent shrubs, with Parry's rabbitbrush (*C. parryi*) constituting 70 percent of the diet and fringed sage (*A. frigida*) 15 percent. Grasses and forbs comprised the remaining material (Bear and Hansen 1966). In the short-grass prairie region of northeastern Colorado, western wheatgrass and winter wheat comprised most of the diet during the early spring. During this season, winter wheat is green and succulent and is highly palatable to jackrabbits (Flinders and Hansen 1972).

The summer diet consists of 70 percent forbs, 19 percent grasses, and 7 percent shrubs (Bear and Hansen 1966). The four most common species, in order of importance, were clover (*Trifolium* sp.), dandelion (*Taraxacum officinale*), dryland sedge (*C. obtusata*), and Indian paintbrush (*Castilleja integra*). In the short-grass prairie region, western wheatgrass was the most important food item during the summer, comprising 21 percent of the diet (Flinders and Hansen 1972).

The autumn diet consists mostly of grasses and forbs. Bear and Hansen (1966) reported that grasses made up 43 percent of the fall diet, forbs 34 percent, and shrubs 14 percent. Important plant species included dryland sedge, goosefoot (*Chenopodium* sp.), fringed sage, and winterfat. Western wheatgrass is an important fall food item, constituting as much as 46 percent of the diet (Flinders and Hansen 1972).

The winter diet of Colorado white-tailed jackrabbits is composed of 76 percent shrubs, with rabbitbrush making up 72 percent and clover 12 percent (Bear and Hansen 1966). Cultivated crops may also constitute an important part of the winter diet. White-tailed jackrabbits use much winter wheat and crested wheatgrass during the winter months (Flinders and Hansen 1972; Pittman 1977). They have also been observed feeding on ears of corn that had been left in fields (Findley 1956).

Jackrabbits were believed to be utilizing mineral licks in Arizona (Vorhies and Taylor 1933). They dug or bit portions of earth, which were then ingested.

The feeding times of jackrabbits are affected by various environmental factors, such as weather, time of the year, and light (Orr 1940; Lechleitner 1958*a*). Feeding usually occurs during the early morning and evening hours, and throughout the night. Black-tailed jackrabbits also have been observed feeding during daylight hours of cloudy days, while feeding activity may be delayed several hours during a full moon. Calm, dry evenings are the most favorable for jackrabbit activity. Wind, falling temperatures, and precipitation were factors that significantly altered this feeding behavior (Tiemeier 1965; Lechleitner 1958*a*).

TABLE 6.3. Foods of *Lepus townsendii* in Colorado

Grasses
Crested wheatgrass (*Agropyron cristatum*)
Western wheatgrass (*A. smithii*)
Wheatgrass (*A. trachycaulum*)
Dogtown grass (*Aristida longiseta*)
Oats (*Avena sativa*)
Blue grama (*Bouteloua gracilis*)
Brome (*Bromus anomalus*)
Smooth brome (*B. inermis*)
Downy brome (*B. tectorum*)
Sedges (*Carex* sp.)
Hairgrass (*Deschampsia caespitosa*)
Love grass (*Eragrostis* sp.)
Arizona fescue (*Festuca arizonica*)
Foxtail barley (*Hordeum jubatum*)
Junegrass (*Koeleria cristata*)
Ringgrass (*Muhlenbergia torreyi*)
Bluegrass (*Poa* sp.)
Squirreltail (*Sitanion hystrix*)
Sand dropseed (*Sporobolus cryptandrus*)
Needle and thread (*Stipa comata*)
Letterman needlegrass (*S. lettermani*)
Wheat (*Triticum aestivum*)

Forbs
Aster (*coloradoensis*)
Vetch (*Astragalus* sp)
Saltbush (*Atriplex canescens*)
Indian paintbrush (*Castilleja integra*)
Golden aster (*Chrysopsis villosa*)
Goosefoot (*Chenopodium* sp.)
Corydalis (*Cordydalis aurea*)
Daisy (*Erigeron* sp.)
Gilia (*Gilia aggregata*)
Gaura (*Gaura coccinea*)
Sunflower (*Helianthus annuus*)
Summer cypress (*Kochia* sp.)
Blazing star (*Liatris punctata*)
Lupine (*Lupinus* sp.)
Alfalfa (*Medicago sativa*)
Bluebells (*Mertensia lanceolata*)
Four o'clocks (*Mirabilis linearis*)
Musineon (*Musineon divaricatum*)
Prickly pear cactus (*Opuntia polyacantha*)
Loco weed (*Oxytropis* sp.)
Beard in tongue (*Penstemon* sp.)
Plaintain (*Plantago purshii*)
Cinquefoil (*Potentilla* sp.)
Psoralea (*Psoralea tenluiflora*)
Rabbitbrush (*Chrysothamnus* sp.)
Russian thistle (*Salsola kali*)
Groundsel (*Senecio* sp.)
Scarlet false mallow (*Sphaeralcea coccinia*)
Dandelion (*Taraxacum officinale*)
False lupine (*Thermopsis divaricarpa*)
Clover (*Trifolium* sp.)
Verbenia (*Verbena bracteata*)
Yucca (*Yucca glauca*)

Shrubs
Fringed sage (*Artemisia frigida*)
Winterfat (*Eurotia lanata*)

SOURCE: Bear and Hansen 1966; Flinders and Hansen 1972

The amount and type of food present are the most important factors governing selection of a feeding site (Orr 1940). Black-tailed jackrabbits prefer to feed in areas that are inconspicuous yet enable them to detect danger from a moderate distance. Most black-tailed jackrabbits prefer to feed in the open, sometimes using hollows or shallow depressions. During the time of the year when woody plants make up a larger portion of the diet, jackrabbits utilize the margins of brush tracts adjacent to open areas (Orr 1940).

Coprophagy is common in many lagomorphs, and it has been reported in both the black-tailed and white-tailed jackrabbits (Lechleitner 1957; Bear and Hansen 1966). Black-tailed jackrabbits were observed to take feces directly from the anus by placing the head between the hind legs while in a squatting position (Lechleitner 1957). The greatest amount of reingestion occurred in the morning (0800–1000 hours), when jackrabbits were least active. Fecal pellets were also found in the digestive tract of preweaned young. These pellets were probably taken from the mother, and it is believed that they are the source of intestinal bacteria for the young jackrabbit. Hansen and Flinders (1969) found that fecal pellets are high in protein content and contain large amounts of B vitamins produced by intestinal bacteria.

HABITAT

The black-tailed jackrabbit occupies many diverse habitats. It is primarily an animal of the arid regions of the western United States, where it is found in association with short-grass areas. Orr (1940) reported that in California the jackrabbit could be found anywhere from below sea level to as high as 3,810 m. In the Mojave Desert of California, jackrabbits utilized the sagebrush-creosote regions. They also inhabit many cultivated agricultural areas of California as well as rangeland. In southeastern New Mexico, black-tailed jackrabbits occupied a typical desert shrub community consisting of mesquite, snakeweed, and soaptree yucca (Davis et al. 1975). Black-tails are also important herbivores in the northern desert shrub region of northwestern Utah. The dominant shrub species there were juniper (*Juniperus osteosperma*) and big sagebrush (Gross et al. 1974). The black-tailed jackrabbit has also adapted well to many agricultural situations in the dryer western states.

White-tailed jackrabbits are associated with the plains flora of the northern and midwestern prairie and may also occur in the open areas of the west. Orr (1940) reported that they prefer open flats and rye grass fields in California, while in Wisconsin and Iowa the best range was located in the larger expanses of cropland and pasture with scattered brushy fencerows (Dubke 1973; Schwartz 1973). In Colorado, *L. townsendii* prefer grassland habitat, utilizing both upland and lowland areas (Bear and Hansen 1966). Braun and Streeter (1968) found that the species occurred in the alpine zone above the timberline, where they inhabited tundralike vegetation.

The immigration of black-tailed jackrabbits into areas formerly occupied by the white-tailed jackrabbits has been well documented (Brown 1947; Carter 1939; Couch 1927). The change in habitat brought about by the cultivation of the prairie has been suggested as a major factor in this expansion (Brown 1947; Carter 1939). Flinders and Hansen (1975) reported that when both species came in contact on the short-grass prairie, *L. townsendii* tended to select more sparsely vegetated upland habitats. The black-tailed jackrabbit is more efficient than the white-tailed jackrabbit in utilizing a feeding site. Thus, the black-tailed jackrabbit may be able to displace the white-tailed jackrabbit due to its greater adaptability and efficiency of utilizing different habitats (Hansen and Flinders 1969).

BEHAVIOR

Mating Behavior. Both black-tailed and white-tailed jackrabbits have a complex mating behavior involving long chases, jumping, and frequent fighting between males and females. Lechleitner (1958a) observed that sexually excited males traveled along trails with their nose to the ground. Such males covered large areas usually corresponding to the size of the animal's home range. When a female was encountered, two types of reaction occurred. If the male approached in a hesitant manner, the female would jump at the male, striking with the forefeet until the male retreated. If the male approached in a persistent manner, either the female or the male would jump in the air, while the other would run underneath. During such jumping activity, urine emission was frequently observed. White-tailed jackrabbits usually exhibit a more pronounced jumping behavior than black-tailed jackrabbits (Blackburn 1968; Lechleitner 1958a; Pontrelli 1968).

The jumping behavior was frequently followed by the sexual chase, during which the male would chase the female while running in a rapid zigzag fashion. During the chase, the male would often try to mount the female. Usually this response was met by aggressive action on the part of the female. The purpose of sexual chase probably is to stimulate the female, after which she is receptive to mounting attempts by the male. Usually only two jackrabbits are involved in the chase; however, three or more have been observed in the chase (Lechleitner 1958a; Pontrelli 1968).

Copulation occurs immediately following the chase (Lechleitner 1958a). The female lowers her head and ears and elevates her hindquarters as the male mounts. Copulation lasts only a few seconds and is accompanied by rapid vibrations of the male's hind quarters. After ejaculation, the pair separate and any further attempts to mount are rebuffed by the female.

Aggressive Behavior. It is difficult to separate the sexual behavior patterns of the jackrabbit from those of an aggressive nature. Most aggressive responses take the form of head butting, biting, jumping, running in circles around another animal, or avoidance reactions (Tiemeier 1965).

Lechleitner (1958a) observed that males engaged in the sexual chase sometimes rebuffed other males in the immediate area. Such encounters were short in duration, consisting of a charge by one animal and a rapid retreat by the other.

Females also exhibit a high degree of antagonism during the breeding season. They charged other jackrabbits that came within 5 or 10 meters of their position. This charging behavior resulted in the dispersion of females during the breeding season (Lechleitner 1958a).

Often, aggressive behavior ended very suddenly. If an individual was rebuffed by a male or female during the sexual chase, it would suddenly begin a substitute or displacement behavior. Displacement activities began immediately after the previous behavior was interrupted. The most common displacement activity observed was that the jackrabbit would immediately start to feed. Another consisted of digging in the ground with the forefeet and then rolling in the dirt ("dusting") (Lechleitner 1958a).

Escape Behavior. Jackrabbits apparently depend on hearing more than sight to detect danger. At the first sign of danger, the ears are raised to a vertical position, while shifting and turning until the source of the disturbance is detected. Two types of escape response were noted in *L. californicus*. If the source of danger was distant and the animal was in dense vegetation, it tried to sneak away from the intruder. The jackrabbit would run with the body close to the ground and the ears lowered (Lechleitner 1958a). Jackrabbits may also respond by "freezing" if they are surprised at close distances or in their forms. The head is pressed to the ground, the ears lowered, and the hind quarters slightly elevated in this freezing position. If approached still closer, the black-tailed jackrabbit would respond by rapidly running from the intruder. With its ears laid down on the back, such running consisted of long leaps with an occasional leap higher and longer than the rest. In a full run jackrabbits can reach speeds up to 56 km per hour, and cover two or three meters with each bound (Orr 1940; Lechleitner 1958a).

When closely pursued or cornered, both species of jackrabbit will enter water and readily swim. Only the front feet are used during swimming (Orr 1940; Lechleitner 1958a). Black-tailed jackrabbits forced to swim due to flooding or fright were observed to be poor swimmers and drowned after a few minutes (W. E. Howard personal communication).

Use of Forms and Trails. The use of forms has been reported for both species of jackrabbit. Jackrabbits frequently enter forms or shallow depressions during resting periods. However, in the Mojave Desert, Costa et al. (1976) reported that black-tailed jackrabbits entered short burrows that were dug either by the jackrabbit itself or by tortoises (*Gopherus agassizi*) and then enlarged by the jackrabbit. This rare behavior was noted only on hot summer days when temperatures sometimes exceeded 42°C. The forms of white-tailed jackrabbits are usually excavated to a depth of 10 cm, with the dirt piled at the front of the form. They were usually found under stunted hemlocks (*Tsuga*

canadensis), junipers (*Juniperus* sp.), lodgepole pine (*Pinus contorta*), or sagebrush. White-tailed jackrabbits may use snow tunnels as resting sites (Orr 1940; Bear and Hansen 1966).

In most habitats jackrabbits, if numerous, will leave visible well-defined trails of compacted soil or trampled vegetation (Crouch 1973; Orr 1940). These are used in moving about the habitat and in traveling to and from distant food sources. When flying in light aircraft, it is not uncommon to observe dozens of trails leading from arid rangeland to an alfalfa field. These are used so consistently by jackrabbits that individuals conducting reductional jackrabbit control will use these trails and trail intersections as appropriate places for baiting jackrabbits (Marsh and Salmon in press).

Miscellaneous Behavior. Vocalizations have been reported in the black-tailed jackrabbit (Lechleitner 1958a). When injured or handled, a high-pitched, piercing scream was noted. A low growl or grunt was heard from females acting aggressively toward males.

Jackrabbits spend part of their resting time in body maintenance activities. Black-tailed jackrabbits groom themselves by licking the body with the tongue. Those areas inaccessible to the tongue are washed by licking the feet and then using the feet to moisten the remaining body surfaces (Lechleitner 1958a). Tiemeier (1965) often observed jackrabbits dusting in shallow depressions dug in the soil. Such behavior probably brings relief from ticks and other ectoparasites and keeps the fur from becoming oily.

MORTALITY

Prenatal Mortality Rates. Prenatal mortality is that which occurs either before or after implantation of the ova. Lechleitner (1959a) reported a preimplantation mortality rate of 6.7 percent for all ova shed, and a postimplantation mortality rate of 6.2 percent for black-tailed jackrabbits in California. In Kansas, Tiemeier (1965) found 9.4 percent preimplantation mortality and a resorption rate of 5.1 percent for implanted embryos, for a total intrauterine mortality rate of 14.5 percent. Gross et al. (1974) estimated preimplantation and postimplantation mortality rates to be 8.0 and 3.0 percent, respectively. Feldhamer (1979) calculated an intrauterine mortality rate of 46.3 percent (16.3 percent preimplantation, 30.0 percent postimplantation loss) for the black-tailed jackrabbit in Idaho.

In the white-tailed jackrabbit, James and Seabloom (1969a) found that preimplantation losses affected 6.7 to 28.7 percent of all ova shed, with a mean of 16.7 percent for the years 1964 and 1965. Resorption of embryos affected 4.6 percent of all ova shed and 19 percent of all litters. Mean prenatal loss from implantation failure and resorption was estimated at 21 percent of all ova shed. Total prenatal mortality from all causes was 28 percent.

Juvenile Mortality Rates. Estimates of juvenile mortality from birth to 12 months of age show little variation in the black-tailed jackrabbit. In Kansas, Tiemeier (1965) reported that juvenile losses ranged from 35 to 67 percent, with a mean of 63 percent, for a six-year period. The highest juvenile mortality occurred during October, November, and December. Gross et al. (1974) reported that juvenile mortality rates in Utah ranged from 24 to 71 percent, with a mean of 59 percent. By the fall, more than half of the juveniles had been removed from the population. A first-year mortality rate of 91 percent was estimated by Feldhamer (1979) for the black-tailed jackrabbit in Idaho.

Adult Mortality Rates. The adult mortality rate of the black-tailed jackrabbit was 57 percent for the period of March to October in Utah (Gross et al. 1974). Mortality estimates ranging from 9 to 87 percent were obtained for an eight-year period. The October to March mean mortality rates were estimated to be 56 percent, with values ranging from 34 to 68 percent. There appeared to be no significant difference in mortality rates between the two periods.

Mortality Factors. Jackrabbits are subjected to predation from a wide variety of mammalian, reptilian, and avian predators. Many avian predators have been observed feeding on jackrabbits or having jackrabbit remains present in pellets or fecal matter. These species include the golden eagle (*Aquila chrysaetus*), rough-legged hawk (*Buteo lagopus*), Swainson's hawk (*B. swainsoni*), ferruginous hawk (*B. regalis*), red-tailed hawk (*B. jamaicensis*), marsh hawk (*Circus cyaneus*), and great horned owl (*Bubo virginianus*) (Orr 1940; Tiemeier 1965; Wagner and Stoddart 1972). Of the mammalian predators of the jackrabbit, the coyote (*Canis latrans*) is clearly the most important. Coyote predation is the major mortality factor operating on jackrabbit populations in northern Utah (Stoddart 1970). Clark (1972) estimated that, in terms of volume, jackrabbits made up three-fourths of the diet of coyotes in northern Utah. Stoddart (1970), using radio telemetry to determine mortality rates, reported that 64 percent of transmitter-equipped jackrabbits fell victim to coyote predation. The red fox (*Vulpes vulpes*) and the gray fox (*Urocyon cinereoargentens*) are known to take jackrabbits (Errington 1935; Hamilton 1935; Scott 1955). The bobcat (*Lynx rufus*) is another predator, although of lesser importance than the coyote (Wagner and Stoddart 1972). Gashwiler et al. (1960) determined from stomach analysis of bobcats that rabbits made up 45.2 percent of their diet in Utah and Nevada. House cats are known to prey on young black-tailed rabbits. In one jackrabbit study in the Sacramento Valley of California, dogs (*Canis familaris*) were believed to be the only predator that killed substantial numbers of jackrabbits (Lechleitner 1958b). The badger (*Taxidea taxus*) preys on jackrabbits in Kansas (Tiemeier 1965). Studies of food habits of the cougar (*Felis concolor*) in Utah and Nevada have shown that jackrabbits rank fourth among prey items of the cougar (Robinette et al. 1959). Raccoons (*Procyon lotor*) and striped skunks (*Mephitis mephitis*) probably are capable of capturing

newborn or young jackrabbits. Reptilian predators include garter snakes (*Thamnophis sirtalis*) and gopher snakes (*Pituophis catenifer*), as well as rattlesnakes (*Crotalus* sp.) (Lechleitner 1958b; Vorhies and Taylor 1933). In spite of all the predators, Evans et al. (1970) were of the opinion that predation played a very limited role in regulating jackrabbit numbers to tolerable levels in agricultural situations.

People are an important mortality factor relative to the jackrabbit. Where jackrabbits cause crop damage, people often take remedial action such as poisoning, shooting, or trapping, thereby locally contributing substantially to mortality. The jackrabbit is recognized as a game or huntable nongame animal in many states, and jackrabbit hunting is a fairly popular sport in some regions. Hunters may take substantial numbers annually. Tiemeier (1965) reports as many as 1,800 black-tailed jackrabbits being removed by hunters during the winter from a 1,920-acre farm in Kansas.

Parasites. Many species of ectoparasite are known to occur on the jackrabbit. Some of the more common ectoparasites of the black-tailed jackrabbit are the rabbit fleas *Hoplopsyllus foxi* and *H. glacialis*, the common rabbit tick, *Haemaphysalis leporis-palustris*, and the Rocky Mountain tick, *Dermacentor* sp. Only one genus of louse, *Haemodipus* sp., has been found to occur on *L. californicus*. Warbles or bot fly larvae, *Cuterebra* sp., are abundant parasites of the black-tailed jackrabbit (Phillip et al. 1955; Lechleitner 1959b; Hansen et al. 1965). Up to 50 percent of the population in Arizona was infected with this parasite (Vorhies and Taylor 1933). Multiple infestations on one animal are not uncommon. Ectoparasites of the white-tailed jackrabbit include the fleas *Cediopsylla inaequalis*, *Hoplopsyllus affinis*, and *Pulex irritans;* and the tick *Dermacentor andersoni* (Voth and James 1965).

The jackrabbit is a host for a wide variety of endoparasites. The endoparasites of the black-tailed and white-tailed jackrabbit are listed in table 6.4.

Viral, Bacterial, and Rickettsial Diseases. Black-tailed jackrabbits have been implicated as reservoirs for such diseases as tularemia, equine encephalitis, brucellosis, Q fever, and Rocky Mountain spotted fever. Jackrabbits may occur in large numbers and their proximity to humans and domestic animals makes them an important threat (McMahon 1965). The occurrence of disease in the white-tailed jackrabbit is poorly known; diseases no doubt exists in this species and are probably similar to those reported for other leporids.

Occurrences of tularemia in jackrabbits are known to be sporadic, with ticks serving as vectors (Lechleitner 1959b). Phillip et al. (1955) were able to isolate the tularemia bacterium (*Pasteurella tularensis*) from ticks taken from jackrabbits. Jackrabbits injected with cultures of *P. tularensis* died within six days. Jackrabbits acquiring the disease in the wild die quickly, explaining the lack of antibodies observed for *P. tularensis* (Lechleitner 1959b).

TABLE 6.4 Endoparasites of *Lepus californicus* and *Lepus townsendii*

Investigator	Location	Species
	L. californicus	
Hansen et al. (1965)	Kansas	*Raillietina loeweni*
		Passalurus nonanulatus
		Micipsella brevicauda
		Nematodirus sp.
		Multiceps sp.
		Physaloptera sp.
		Taenia pisiformis
		Dermatoxys veligera
		Nematodirus leporis
		Dirofilaria scapiceps
Lechleitner (1959*b*)	California	*Raillietina retractulus*
		Cittotaenia pectinata
		Biogastranema affinis
		Taenia sp.
		Multiceps sp.
		Fasciola hepatica
		Eimeria sp.
	L. townsendii	
Voth and James (1965)	North Dakota	*Eimeria* sp.
		Cittotaenia pectinata
		Multiceps sp.
		Raillietina loeweni
		Taenia pisiformis

Rocky Mountain spotted fever (*Rickettsia rickettsii*) has been shown to infect *L. californicus* (Lechleitner 1959*b*; Phillip et al. 1955). The animals acquire the disease from infected ticks (McMahon 1965).

The presence of antibodies against Q fever (*Coxiella burnetti*) has been reported in jackrabbits in California. Also, jackrabbits gave positive titers for western equine encephalitis. Brucellosis (*Brucella abortus*) has been reported in the jackrabbit in Nevada. However, it is not known whether the jackrabbit can serve as a reservoir for brucellosis in cattle herds (Phillip et al. 1955; Lechleitner 1959*b*).

AGE DETERMINATION

Several methods have been used to determine age in the jackrabbit, including the closure of the epiphysis of the proximal end of the humerus. Lechleitner (1959*a*) was able to distinguish three age classes for black-tailed jackrabbits: class 1 (2 to 9 months old) had a distinct epiphyseal groove, class 2 (10 to 12 months old) had an epiphyseal groove that was almost closed, and class 3 (greater than one year old) had no evidence of an epiphyseal groove present. Similarly, James and Seabloom (1969*b*) found that the white-tailed jackrabbit could be separated into three age classes very similar to those of *L. californicus*. The change from age class 1 to age class 2 occurred at 6 to 7 months of age. The change from age class 2 to age class 3 occurred when the animal was 13 to 14 months of age.

The weight of the dried eye lens has been used to determine age in both black-tailed and white-tailed jackrabbits (Tiemeier and Plenert 1964; Connolly et al. 1969; Gross et al. 1974; James and Seabloom 1969*b*). The lens weight technique is superior to other methods, because it estimates the month in which the animal was born. However, lens weights may vary throughout the range of the species, possibly because of nutritional differences.

Juvenile males can be separated from adult males by the size and morphology of the penis. In black-tailed jackrabbits less than 9 months of age the penis cannot be everted, while in older males it can. Adult penis length is from 3.5 to 5.0 cm. Males one year or older have warty protuberances on the penis that readily identify them as older adults (Lechleitner 1959*a*). In both species of jackrabbit, individuals older than 10 months tend to have a dark blue scrotal sac that is characteristic of sexually mature adults (Lechleitner 1959*a*; James and Seabloom 1969*b*).

Young-of-the-year females can be classified as juveniles by the condition of the reproductive tract. The presence of longitudinal striations in the uterus can be used to separate parous adults from juvenile non-parous females. Also, the ovary of a juvenile *L. californicus* female consists mostly of interstitial tissue, while the adult ovary contains large follicles and corpora lutea (Lechleitner 1959*a*).

ECONOMIC STATUS

Economically, jackrabbits are both desirable and undesirable, depending on the situation and mankind's relationship to the animals. On a positive side, they provide sport hunting and coursing. Although not considered as desirable to eat as cottontails, they sometimes serve as subsistence food, particularly for some families in the lower socioeconomic groups. Classified as a game or huntable nongame species, both black-tailed and white-tailed jackrabbits are popular among hunters in some regions. Year-round open hunting seasons with no bag limits exist in some states, and in others bag limits and seasons are quite liberal, providing recreation for many. Jackrabbits also serve as prey, providing falconers with a sporting challenge. Over much of its distribution the rabbit's place in the ecological community makes it biologically interesting and, if not too numerous, esthetically desirable.

Although market hunting of jackrabbits for human consumption does not exist today, in the 1800s and early 1900s it occurred on a relatively large scale, with jackrabbits being sent to the larger cities throughout the United States. Residents of San Francisco were thought to consume the most rabbits, with 100 dozen to 150 dozen rabbits sold each day in the winter of 1894–95 at 50¢ to $1 per dozen (Palmer 1896).

Black-tailed jackrabbit pelts, although not highly valued as fur, were sold for felt making in the early days of market hunting. However, the superior fur of other species of leporid, especially those from colder climates that had dense winter pelts, was economically more important in those days. Today, jackrabbits have a minor role in the fur industry. Bear and Hansen (1966) reported that in Colorado, when dense populations of white-tailed jackrabbits existed, hunters marketed unskinned rabbits, whose pelts were used to make felt and carcasses used for mink food. In some years as many as 65,000 jackrabbits were bought at one Colorado collection point.

On the negative side, the jackrabbit is considered an important pest in some regions and situations, especially when its numbers are high. In the western United States, jackrabbits cause a reported $3.2 million of damage annually to agriculture (Hegdal 1966). On pastures and rangeland they compete for forage with grazing livestock. At times they cause extensive damage to a wide variety of orchard trees and other crops (Bronson and Tiemeier 1958*b*; Marsh and Salmon in press). Plantings of forest trees in certain regions may also be damaged from their feeding, but this is relatively minor in scope compared with the damage inflicted on tree seedlings by snowshoe hares (*Lepus americanus*).

On western rangelands jackrabbits can be significant competitors of domestic animals. Vorhies and Taylor (1933) estimated that 148 black-tails ate as much forage as one cow. Similarly, Currie and Goodwin (1966) reported that 6 black-tails could destroy as much forage as one sheep could eat in a day. Bear and Hansen (1966) estimated that 15 white-tailed jackrabbits ate as much forage as one sheep consumed. In Colorado competition between jackrabbits and cattle seemed to be greatest in early spring (Hansen and Flinders 1969).

The most important economic effect of the jackrabbit is its role in crop depredation. It feeds on a wide variety of crops and competes directly with efforts to grow food and fiber. Jackrabbits damage crops including trees, grain, alfalfa, cotton, and a wide variety of vegetables. In Kansas, Bronson and Tiemeier (1958*b*) reported heavy damage to wheat, sorghum, and alfalfa from jackrabbits. Crop damage is often most severe in or is restricted to fields adjacent to land that hares use for resting (Marsh and Salmon in press). In the Southwest, Vorhies and Taylor (1933) reported damage to alfalfa, orchard, cotton, and other crops. Droughts often contribute to excessive crop damage by drawing rabbits from rangeland or uncultivated fields to cultivated crops, especially those under irrigation (Bickler and Shoemaker 1975).

Jackrabbits do frequent some open forest habitats, but have little direct adverse effect on forest trees except during forest regeneration. Crouch (1973) reported jackrabbit damage to ponderosa pine (*Pinus ponderosa*) seedlings in south-central Oregon. Rabbits clipped the main stems on 43 percent of the 1,080 tree seedlings planted in a burnt area that, before a wildfire, was a pine-bitterbrush-needlegrass plant community. Read (1971) reported similar damage to ponderosa pine seedlings on a tree plantation in Nebraska. However, damage from jackrabbits in reforestation efforts is not widespread.

The significance of rabbit diseases transmissible to people is of some concern (Hull 1963; McMahon 1965). Hunters should always take into consideration the possibility of contracting tularemia (a debilitating but rarely fatal disease) from rabbits when handling carcasses of jackrabbits and other rabbits. In North America 70 percent of tularemia cases in humans are attributed to contact with hares and leporids. Eastern cottontails (*Sylvilagus floridanus*) are the direct source of over 55 percent of all cases of the disease in humans (McDowell et al. 1964). Although ticks and deer flies may serve as vectors for tularemia, this route of transmission is less important than direct contact with an infected animal while skinning or dressing it (McDowell et al. 1964). Rubber gloves should be used to avoid infection through the skin, especially through abrasions or broken skin on hands. Well-cooked rabbit meat presents no tularemia hazard. With the use of a few common-sense precautions, the danger of contracting tularemia is remote.

Cases of bubonic plague in humans, though more often associated with ground squirrels (*Spermophilus* spp.), have been linked to rabbits. Some rabbit species, including the black-tailed jackrabbit, have been found to be infected with plague, although human cases have been associated primarily with the hunting and cleaning of cottontails (Graves et al. 1978). Fleas often serve as vectors in transmitting plague from animals to people, particularly when ground squirrels are involved.

The transmission of Rocky Mountain spotted fever from jackrabbits to humans is only a remote possibility because the rabbit ticks do not ordinarily feed on humans (Lechleitner 1959*b*).

Jackrabbits potentially may transmit certain livestock diseases such as Q fever and brucellosis, but little is known of the actual threat to livestock production (McMahon 1965).

MANAGEMENT

Management of jackrabbits in the United States has been directed primarily toward decreasing their populations because of their crop depredation activities, a contrast to the common objective of increasing the numbers of the more desirable lagomorph game species. Management of jackrabbits as a game species, for the most part, is the direct result of state game regulations that govern hunting seasons, possession of bag limits, and methods of taking jackrabbits. Traditional methods of attempting to improve the habitat or available food have not come into play with jackrabbits because the species, when in suitable habitats, is usually in adequate numbers to satisfy hunters. Private, state, and federal land set aside as military reservations, game preserves, wildlife refuges, parks, and other recreational areas, depending on habitat and management policies, often serve well as sanctuaries for jackrabbits. Jackrabbits are adaptable enough to coexist very well with people. As could be expected, in states where jackrabbits are viewed as a serious and widely distributed pest to agriculture, little emphasis is placed on their management as a desirable animal.

The construction of numerous airports has unintentionally and substantially benefited the black-tailed jackrabbit. Many airports in the western United States would support relatively high numbers if rabbits were not controlled (Rohe et al. 1963; Marsh and Salmon in press). These airport populations have also created some perplexing management problems because high numbers of rabbits tend to attract both avian and mammalian predators. While jackrabbits are occasionally run over on runways, this creates little hazard, even for light aircraft. However, dogs or other large mammalian predators chasing rabbits on runways do constitute a hazard, particularly for light aircraft. Avian predators may be sucked into jet engines, causing malfunctions, or may penetrate windshields or damage other parts of planes. In western airports, black-tailed jackrabbits are commonly controlled by fencing, shooting, poisoning, or other means.

Early western farmers quickly became aware of jackrabbits and, while the animals often provided fresh meat, they were more of a liability as a pest to crops than an asset as food. Before farmers could grow crops successfully in many areas, it was essential to have some economical way to prevent crop losses from rabbits. Jackrabbits were controlled in the late 1800s usually with large rabbit drives (Palmer 1896). This period in American history was a time of agricultural expansion in many areas of Midwest and West, and black-tailed jackrabbit populations over wide areas grew to devastating numbers. The introduction of cultivated agriculture and, in some areas, irrigation into prime jackrabbit habitat apparently triggered an enormous upward surge in rabbit populations (Marsh and Salmon in press). In some regions, these populations reached levels where rabbits ate entire crops (Palmer 1896).

In a typical rabbit drive, the majority of the farm population of a district would turn out and surround a territory, often several kilometers in expanse, driving rabbits toward a central corral bordered by wide wing-type fences. After being concentrated in such an enclosure, rabbits were clubbed to death by the hundreds or thousands. The largest rabbit drive on record in California was held about 24 km southwest of Fresno on 12 March 1892. Approximately 8,000 people were present, and an estimated 20,000 to 30,000 jackrabbits were killed (Palmer 1896).

In many regions, early rabbit drives were often social events as much as biological ones. At a time when social activities in rural areas were less diversified than today, rabbit drives were like cattle round-ups or threshing time—an opportunity for camaraderie and neighborly cooperation. Palmer (1896) wrote that rabbit drives often meant a gala day and were a popular way of celebrating some special event or making a day a local holiday. Schools were closed, and women and children often joined to help in the drive or at least to watch the spectacle.

These drives undoubtedly saved many crops, but they were not always highly effective. Success was measured often only by the number of rabbits killed, not by taking into consideration how many were still left to cause damage and repopulate an area (Marsh and Salmon in press). The large rabbit drive as the principal control method reached its peak roughly between 1887 and 1892 (Palmer 1896) and then gradually gave way to hunting and the use of rabbit-proof fencing, protective tree trunk guards, chemical repellents, and poison baits.

Organized group hunts, also popular, outlived rabbit drives. This method of hunting requires a group of relatively good marksmen who systematically transect an area walking parallel to each other, shooting animals that jump and run ahead. Covering blocks of land in this fashion has often reduced rabbit numbers substantially. The technique is still used today, although on a very limited basis.

Individual hunting, as opposed to group hunting used more extensively in past years, remains a very good control method if rabbit numbers are not too great. Shooting is most effective from late afternoon to dusk, when rabbits are feeding and therefore more visible. Early morning hunting is also effective but less so.

The control of rabbits by rabbit drives and large organized hunts has essentially been abandoned. Today damage is prevented largely with the use of fences, tree trunk protectors, chemical repellents, and population reduction. When population reduction is required to prevent damage, it most often involves either poisoning or shooting and only rarely trapping. Sport hunting, if well managed, may offer recreation in addition to providing some relief from crop depredations.

Jackrabbits, as well as other species, are excluded

from fields and gardens with the use of rabbit-proof fence of woven wire or poultry netting of a mesh not larger than 3.8 centimeters, 91 cm to 121 cm high, with the bottom turned 15 cm outward and buried at least 15 cm in the ground. Jackrabbits and cottontails ordinarily will not jump a 60-cm fence, although jackrabbits can easily clear such a height when pursued by humans or dogs. Fence heights of at least 91 cm are normally recommended. Tight-fitting gates with sills or other means of preventing rabbits from digging beneath are essential, and gates must be kept closed except when actually in use.

Rabbit fences are expensive to build and ordinarily are impractical and uneconomical unless the crop is of high value and extensive damage is likely. For example, in a study of the protection of large acreages of alfalfa in California, fencing was found to be economical only when rabbit damage was severe (Bickler and Shoemaker 1975). Today, rabbit fences are most likely to be constructed around specialty plantings such as experimental plots, nurseries, vegetable and flower seed crops, and other high-value crops. Fences are also used around noncommercial plantings such as home vegetable and flower gardens.

Over the years, farmers have used protectors to guard trunks of individual young orchard trees against rabbits (Garlough et al. 1942). Homemade and commercial wraparound trunk protectors have been made from a variety of materials including plastic, cardboard, paper, and aluminum. These are valuable aides in preventing rabbit damage as well as sun scalding and freezing damage to sensitive trees.

Cylinders of poultry netting make excellent protectors for trunks of young orchard trees. Cylinders formed from 30-to-45-cm sections of 60-to-91-cm-wide poultry netting (2.5-cm mesh) are used commonly to protect trees. Cylinders should be staked away from trees so that rabbits cannot press them against trees and gnaw between the wire mesh (Storer 1958). Wrapping the base of a haystack with 91-cm-high poultry netting also provides good protection from rabbits.

Various repellents have been used to discourage rabbits from feeding on plants and gnawing on bark and twigs of trees and vines. Chemical repellents may provide relief from rabbit damage and thus are valuable when other approaches are not practical. Applications must often be repeated, however, to protect new growth or renew repellency lost through rain, snow, or sprinkler irrigation. Rabbit repellents are somewhat unpredictable in effectiveness. They may offer some protection as long as rabbits have acceptable alternative untreated food, but when food is scarce, repellents may have little value.

The most effective repellents for jackrabbits include products made with one of the following chemicals: TMTD (tetramethyl thiuram disulfide), TNBA (trinitrobenzene-anilene), and ZAC (zinc dimethyl-dithiocarbamate cyclohexylamine) (Evans et al. 1970). Repellents containing other active ingredients are also available. Control methods used for black-tailed jack-

rabbits will also be applicable for white-tailed jackrabbits where they are pests.

When jackrabbits become a serious pest problem affecting large acreages of crops, poison baits, where safe to use, may be the most effective and practical means of reducing the damage. Strychnine-poisoned bait has been the most extensively used poison over the years (Johnson 1964; Evans et al. 1970). However, following the development of the safer anticoagulant rodenticides, these more expensive poisons have come into use for jackrabbit control (Clark 1975; Johnson 1978; Marsh and Salmon in press). Zinc phosphide-poisoned bait has been used with success, although it has not come into common use (Evans et al. 1970). In many localities, it is not safe to use poisons of any type. In other situations that meet control criteria, properly conducted poisoning of jackrabbits presents minimal hazard to nontarget species.

"Catch" crops or buffer crops planted around borders of crops to reduce rabbit damage have been tried with only isolated successes (Lewis 1946; Evans et al. 1970). The buffer crop method of damage control is seldom practical or economically sound. Other control methods or materials that have been explored experimentally include numerous kinds of poison bait, toxic foliar spray, chemosterilant, and chemical repellent. Few of them seem to promise much for the immediate future. Controlling rabbits biologically with disease has been considered but ruled out because of potentially undesirable biological consequences (Evans et al. 1970).

Jackrabbits living on rangelands prefer the more open areas lacking dense, tall vegetation, and livestock grazing often helps provide favorable jackrabbit habitat (Vorhies and Taylor 1933; Arnold 1942; Bronson and Tiemeier 1958b). It is not clear, however, whether rabbits prefer one type of site over another because of openness, the type and amount of food available, or a combination of factors (Vorhies and Taylor 1933). A variety of factors such as shallow or poor soil and low rainfall may also influence the plant community and contribute to sparse and low vegetation. Evans et al. (1970) pointed out the inconsistencies in the influence of grazing intensity on jackrabbits. Numerous instances of moderate to dense populations of jackrabbits on lightly grazed or ungrazed land lends little credence to the belief that overgrazing is the major contributor to high jackrabbit populations. As for open conditions, jackrabbits are self-serving; even in moderate numbers they are often capable of keeping certain types of vegetation low in localized areas in the absence of livestock grazing.

There is no evidence that lowering livestock stocking rate on rangeland in hopes of increasing vegetative cover will reduce jackrabbit numbers enough to compensate for the loss in livestock production. However, where ranges are fully stocked and jackrabbit numbers become high, consideration must be given to the possibility that livestock, in combination with a dense population of rabbits, will cause overutilization of the range (Taylor et al. 1935). In such situations if the rabbit population does not crash from natural causes within a couple of years, then good range management practices would dictate that either the rabbit population be reduced artificially or the stocking rate lowered to prevent permanent range deterioration.

The present state of the art in rabbit control offers few solutions to the problem of jackrabbit competition with livestock on rangeland. Control measures such as shooting or poisoning often are not cost effective for much of the rangeland in the western United States. One exception to this is where seeding efforts are under way to improve range conditions (Wetherbee 1967). Although expensive, rabbit control by shooting or poisoning may be the only way to protect newly seeded ranges until they become established.

It has been mentioned earlier that *L. californicus* has been introduced for hunting and other sporting purposes into a number of states, including Massachusetts, Maryland, Virginia, and Florida. The history of rabbit releases in the United States and other countries suggests that such introductions can have lasting undesirable ecological consequences. Rabbits may not only compete with existing fauna in an area but have a profound and unfavorable influence on native vegetation. Equally important, introduced species may become pests, competing with efforts to grow crops and establish forests (Howard 1964, 1965). Proposed introduction of any rabbit or rodent species into new areas should be viewed very critically. Some states have laws or regulations prohibiting such introductions. Air shipments of black-tailed jackrabbits from Idaho to Italy for release for sport hunting are an example of nonnative rabbits continuing to be introduced into new regions, constituting a management concern.

CURRENT RESEARCH AND MANAGEMENT NEEDS

More information is needed on the population genetics and adaptability of the jackrabbit to various habitats. Since the jackrabbit does occupy a wide range of habitats, such information may prove valuable in assessing the importance of habitat manipulation practices and in explaining the recent changes in jackrabbit distributions.

Since the jackrabbit can become a serious pest in many agricultural areas, methods of controlling population numbers are needed. More research is needed on chemical control of jackrabbits in agricultural areas. Some states recognize the jackrabbit as a pest species only, and, therefore, put little emphasis on its importance as a game animal.

Information on the ecology of the white-tailed jackrabbit is essential. Such aspects as physiology, genetics, mortality, parasites, and behavior need further study. Management programs for the white-tailed jackrabbit have been almost nonexistent. If the jackrabbit is to be considered a game species, more information is needed on how to maximize the population to satisfy hunter interest.

Ecological studies are also needed in areas where

the white-tail and black-tail are sympatric. The effects of interspecific competition and habitat utilization of the two species need to be studied in detail. This is particularly important in areas where *L. californicus* is extending its range into habitat formerly occupied by *L. townsendii*. The specific habitat requirements of each species need to be determined to evaluate accurately any changes taking place in jackrabbit distribution.

OTHER JACKRABBITS

Antelope Jackrabbit. The antelope jackrabbit, *Lepus alleni* Mearns, is also referred to as the white-sided jackrabbit. Three subspecies occur in North America: *L. a. alleni*, *L. a. palitans*, and *L. a. tiburonensis*. Its range extends from the arid portions of southern Arizona southward along the coastal plain of the Gulf of California to the Mexican state of Nayarit (see figure 6.2).

The antelope jackrabbit is the largest of the North American lagomorphs, with adults averaging 3.6 kg in weight. Measurements from adults in Arizona were: total length 553–670 mm; tail length 48–76 mm; hind foot length 138–173; and ear from notch 138–173 mm (Vorhies and Taylor 1933). This jackrabbit can be easily identified by its large white-edged ears, 13 to 17 cm in length, and the white rump patch that is characteristic of the species (see figure 6.5). It also lacks the black-tipped ears characteristic of *L. californicus*. The top and sides of the head are buff colored, while the tail is white with a black middorsal stripe extending onto the rump. The shoulders and flanks are a uniform gray color (Hall and Kelson 1959).

This jackrabbit is found exclusively in the lower Sonoran life zone, where it occupies elevations ranging from sea level to 1,200 meters. In Arizona, the preferred habitat consists of grassy slopes with moderate elevations. It also inhabits the cactus belt where species of mesquite, grass, and catclaw (*Acacia greggi*) predominate. Smaller populations also may in-

FIGURE 6.5. The antelope jackrabbit (*Lepus alleni*). Note the pale color, in contrast with the darker dorsal surface, and the long ears and hind legs.

habit the creosote bush desert and valley bottoms (Vorhies and Taylor 1933).

There appears to be little difference between the breeding season of the antelope jackrabbit and that of the black-tailed jackrabbit where they are sympatric. Vorhies and Taylor (1933) reported a 10-month breeding season for both species in Arizona. Mating occurred every month of the year except November and December. The antelope jackrabbit averages 1.93 young per litter, with seven litters possible per year. Vorhies and Taylor (1933) suggested that *L. alleni*, unlike *L. californicus*, does not place the young in a form but scatters them through the surrounding habitat immediately after birth. Like other jackrabbits, this species is fully furred at birth, with the eyes open and neonates able to move about readily. The young possess the white-edged ears characteristic of the species, but lack the white rump patch that is associated with adults. Four individuals from the same litter averaged 121.4 g in weight. Average measurements of five-day-old jackrabbits were: total length 150 mm; tail length 8.2 mm; hindfoot 42.5 mm, and ear from notch 38.5 mm (Vorhies and Taylor 1933). Fetal sex ratios have not been determined, but those of adults did not differ significantly from a 1:1 ratio.

L. alleni does not exhibit any marked population fluctuation. Densities ranged from 0.2 hares per ha in the semidesert habitat to 0.5 hares per ha in the mesa habitat type (Vorhies and Taylor 1933).

The antelope jackrabbit tends to be gregarious during all times of the year. Groups ranging from 6 to over 100 individuals have been observed together. Daily movements are usually comprised of moving to and from feeding areas during the night. Daily movements of 1 to 3 km are not uncommon. The home range of adults is several kilometers in diameter (Vorhies and Taylor 1933).

The escape behavior exhibited by *L. alleni* is the most spectacular of the jackrabbits. When this species runs away from a potential predator it conspicuously flashes its white rump patch toward the predator while running in a zigzag fashion. These white "ruptive" marks on the rump are flashed toward the predator by a set of muscles under the skin. The skin of the hindquarters is pulled up over the back while the hairs are everted exposing the large white patch. The purpose of the rump flash is to confuse the predator and to obliterate the outline of the jackrabbit (Vorhies and Taylor 1933).

Analysis of the stomach contents of *L. alleni* has shown that 36 percent of the diet is composed of mesquite, 45 percent grass, and 7.8 percent cactus, with various forbs and weeds comprising the remainder of the diet (Vorhies and Taylor 1933). The food habits of the antelope jackrabbit are related to the alternating dry and rainy seasons. Following the winter rains the grasses produce new leaves that are highly palatable to jackrabbits. This is reflected in an increase in the occurrence of grass in the diet, which may comprise as much as 80 percent during this period. Mesquite becomes the predominant food item during the dry season when perennial grasses become dry and less palatable.

Species of grass consumed by this jackrabbit include: three awn (*Aristida* sp.), grama (*Bouteloua* sp.), love grass (*Eragrostis* sp.), spike grass (*Trichloris crinita*), panic grass (*Panicum* sp.), sandbur (*Cenchrus* sp.), drop seed (*Sporobolus* sp.), red top (*Agrostis* sp.), barnyard grass (*Echinochloa* sp.), and finger grass (*Chloris* sp.). Cactus is consumed in the drier seasons of the year. The less spiny species are favored, particularly *Opuntia* and *Echinocactus* (Vorhies and Taylor 1933).

The antelope jackrabbit is undoubtedly subjected to the same predators and many of the same parasites and diseases as those reported for the black-tailed jackrabbit. Tularemia is the only disease of any significance reported to affect Arizona jackrabbits (Vorhies and Taylor 1933).

This species may cause extensive damage to crops and rangeland. This is particularly true during periods of drought when other food sources are unavailable (Vorhies and Taylor 1933). As a pest the same damage prevention or control measures that are applied to the black-tailed jackrabbit can be applied to the antelope jackrabbit.

In Arizona, the antelope jackrabbit, although hunted, is considered a nongame mammal with no closed season or bag limit. It is not considered rare, and is frequently hunted along with black-tailed jackrabbits.

More current comprehensive research is needed on the antelope jackrabbit. Information on reproduction, food habits, genetics, and distribution are clearly needed.

White-Sided Jackrabbit. The white-sided jackrabbit was first described as *Lepus gaillardi* by Mearns (1896) from a specimen collected near Whitewater, Chihuahua, Mexico. It has since been reclassified as a subspecies of *L. callotis* by Anderson and Gaunt (1962). Two subspecies of the white-sided jackrabbit are now recognized, *L. c. callotis* and *L. c. gaillardi*.

The range of *L. callotis* extends northward from the Mexican state of Oaxaca along the Sierra Madre to Chihuahua and eastern Sonora (see figure 6.5). In the United States *L. callotis* is considered rare and is known to occur only in the Animas and Playas valleys of southwestern New Mexico (Anderson and Gaunt 1962; Bogan and Jones 1975; Bednarz 1977). Populations of *L. callotis* and *L. alleni* are completely allopatric, although the range of *L. californicus* does overlap the northern distribution of both species. Anderson and Gaunt (1962) have suggested that *L. callotis* probably evolved from an isolated population of *L. californicus* in Mexico. Part of the newly formed species *L. callotis* was then isolated on the western coastal plain, where it evolved into *L. alleni*.

The distinctive characters of *L. callotis* are its pale buffy to blackish hue, brown to blackish nape, white sides and underparts, and grey rump with black on the upper part of the tail (Anderson and Gaunt 1962). Where it occurs with *L. californicus* the two can be separated by examining overall skull characteristics. The skull of *L. callotis* has a higher nasal aperature, a more inclined parietal, smaller auditory bullae, more prominent supraorbital processes, smaller auditory meatuses, and a less constricted basioccipital (Anderson and Gaunt 1962). Average measurements from four adults in New Mexico were: total length 543.2 mm, hind foot 125.8 mm, ear length 137.7 mm, and tail length 81.0 mm. The average weight was 2.7 kg (Bednarz 1977).

Information on reproduction in the white-sided jackrabbit is limited. Black-tailed jackrabbits in southeastern New Mexico are reported to have an eight-month breeding season extending from January to late August (Griffing and Davis 1976). A similar breeding season probably occurs for *L. callotis* in the same area. Bednarz (1977) estimated from a limited number of pregnant females that the absolute minimum breeding season would be 18 weeks, extending from mid-April to mid-August. The average number of young per litter, from a sample of 10 females, was 2.2 (Anderson 1972; Bogan and Jones 1975; Bednarz 1977).

In New Mexico the white-sided jackrabbit inhabits level short-grass habitats where the dominant species of grass are blue grama (*Bouteloua gracilis*), black grama (*B. eriopoda*), buffalo grass (*Buchloe dactyloides*), and wolftail (*Lycurus phleoides*). Bednarz (1977) found a positive correlation between the density of *L. callotis* and grass composition. This jackrabbit preferred habitats composed of 65 percent or more grasses, 25 percent or less forbs, and less than 1 percent shrubs. In this area of New Mexico, the black-tailed and white-sided jackrabbits occupy distinct habitats, with the black-tailed jackrabbit occupying those grasslands with a higher percentage of forbs and shrubs. Densities were estimated at one white-sided jackrabbit per 31.6 ha. Black-tailed jackrabbit densities increased in response to overgrazing as the number of shrub and forb species increased. White-sided jackrabbits preferred perennial grasses that are associated with good range conditions (Bednarz 1977). Competition was minimal between the two species and probably occurred only in areas of marginal habitat.

The white-sided jackrabbit has several unique behavioral patterns when compared to other species of jackrabbit. This species is almost totally nocturnal, as compared to the black-tailed and white-tailed jackrabbit, which are both crepuscular and nocturnal. Bednarz (1977) reported most activity for *L. callotis* occurring between 2200 and 0500 hours.

The escape behavior of the white-sided jackrabbit is similar to that of *L. alleni*. When flushed from its form, *L. callotis* alternately flashes its white side patches while running away from the intruder. Another escape behavior observed was that of leaping straight up into the air while extending the hind legs and flashing the white sides. This behavior was noted when the jackrabbit was startled or alarmed by a predator (Bednarz 1977).

The most conspicuous trait of the white-sided jackrabbit is its tendency to occur in pairs (Anderson 1972; Bogan and Jones 1975; Conway 1976; Bednarz 1977). These pairs consist of a male and female that exhibit a strong affinity toward maintaining an appar-

ent pair bond. It is not known whether this pair bond occurs throughout the year, but it is most prominent during the breeding season (Bednarz 1977). With the exception of the male-female pair bond, the breeding behavior of this species is similar to that reported for the black-tailed jackrabbit (Lechleitner 1958a; Pontrelli 1968). Once the pair bond is established the male takes on the aggressive duties of defending the pair from intruding males. The function of pair bonding may be to keep the sexes together during the breeding season, since densities are very low for this species (Bednarz 1977).

The white-sided jackrabbit also has been observed to construct and utilize forms. These forms were slightly larger than those reported for *L. californicus*, averaging 37 cm in length, 18.3 cm in width, and 6.3 cm in depth. Dense stands of tabosa grass (*Hilaria mutica*) usually surround the form, which is located in clumps of this grass (Bogan and Jones 1975; Bednarz 1977). White-sided jackrabbits may occupy underground shelters, although such behavior is rare. Bednarz (1977) reported flushing *L. callotis* from an abandoned kit fox (*Vulpes macrotis*) den.

Three types of vocalization occur in *L. callotis*. One, an alarm or fear reaction, is characterized by a high-pitched scream. The second occurred when an intruding male approached a pair of jackrabbits. The male of the pair produced harsh grunts until the intruder left or was chased away. The last vocalization occurred during the sexual chase and consisted of a trilling grunt. It was not determined which member of the pair produced this sound (Bednarz 1977).

The feeding habits of *L. callotis* were observed by Bednarz (1977). The diet consisted of over 99 percent grass. The only nongrass item that was found to any degree was the sedge nutgrass (*Cyperus rotundus*). Plant species observed being consumed were: buffalo grass, tabossa grass, fiddleneck (*Amsinckia* sp.), wolftail, blue grama, vine mesquite (*Panicum obtusum*), ring muhly (*Muhlenbergia torreyi*), woolly Indian wheat (*Plantago purshi*), and Wright buckwheat (*Eriogonom wrightii*). Bogan and Jones (1975) also reported that a grazed thistle flower (*Cirsium* sp.) occurred in a form of *L. callotis*.

No information is available on mortality rates for *L. callotis* or on instances of predation. Many predators of the black-tailed jackrabbit occur throughout the range of *L. callotis* and probably feed upon it to some degree. Five species of pathogenic bacterium have been isolated from *L. callotis*. Those include four respiratory pathogens: *Pneumococcus* sp., *Pseudomonas pseudomallei*, *Klebsiella ozanae*, and a Moraxella-like organism. A potential pathogen of the mesentaries and lymph nodes, *Yersinia pseudotuberculosis*, was also isolated. An unidentified coccidiosis infection was reported (Bednarz 1977). The tick *Demacentor paramapertus* and the flea *Pulex simulans* were ectoparasites found in New Mexico. No serious disease or epidemic has been reported occurring in *L. callotis* populations (Bednarz 1977).

The economic effects of *L. callotis* populations are minimal in the United States. While competition between *L. callotis* and cattle does exist, it is negligible, since *L. callotis* occurs in densities only one-twentieth of those attained by *L. californicus* (Bednarz 1977). In addition, *L. callotis* is found only on good range and where competition with cattle would not be severe.

Management of the white-sided jackrabbit has consisted of trying to maintain the desired habitat. Range management that encourages grasslands with a high percentage of perennial grasses and a low percentage of shrubs and forbs is being conducted (Bednarz 1977). Habitat preservation of such areas is crucial to the survival of the species. In New Mexico *L. callotis* is a protected species, with no open hunting season.

Many aspects of the biology and ecology of the white-sided jackrabbit need to be studied. Little is known concerning the physiology, genetics, and reproductive biology of this species. Current research needs include the mortality factors of this species, as well as its food habits and habitat requirements. *L. callotis* is a rare jackrabbit, and management should be centered around maintaining existing populations and preventing habitat deterioration.

LITERATURE CITED

Anderson, S. 1972. Mammals of Chihuahua: taxonomy and distribution. Am. Mus. Nat. Hist. Bull. 148:149–410.

Anderson, S., and Gaunt, A. S. 1962. A classification of the white-sided jackrabbits of Mexico. Am. Mus. Novit. 2088:1–16.

Arnold, J. F. 1942. Forage consumption and preferences of experimentally fed Arizona and antelope jack rabbits. Univ. Arizona Agric. Exp. Stn. Tech. Bull. 98. 86pp.

Bartel, M. H., and Hansen, M. F. 1964. *Raillietina loeweni* sp. from the hare in Kansas, with notes on *Raillietina* of North American mammals. J. Parasitol. 50:448–453.

Bear, G. D., and Hansen, R. M. 1966. Food habits, growth, and reproduction of white-tailed jackrabbits in southern Colorado. Colorado State Univ. Agric. Exp. Stn. Tech. Bull. 90. 39pp.

Bednarz, J. 1977. The white-sided jackrabbit in New Mexico: distribution, numbers, and biology in the grasslands of Hidalgo County. Res. rep. of the New Mexico Dept. of Game and Fish, Endangered Species Program, Santa Fe. 33pp. Manuscript.

Bickler, P. E., and Shoemaker, V. H. 1975. Alfalfa damage by jackrabbits in southern California deserts. California Agric. 29(7):10–12.

Blackburn, D. F. 1968. Courtship behavior among white-tailed and black-tailed jackrabbits. Great Basin Nat. 33(3):203.

Bogan, M. A., and Jones, C. 1975. Observations on *Lepus callotis* in New Mexico. Proc. Bio. Soc. Washington, D.C. 88(5):45–50.

Braun, C. E., and Streeter, R. G. 1968. Observations on the occurrence of the white-tailed jackrabbit in the Alpine zone. J. Mammal. 49 (1):160–161.

Bronson, F. H., and Tiemeier, O. W. 1958a. Reproduction and age distribution of black-tailed jackrabbits in Kansas. J. Wildl. Manage. 22(4):409–414.

———. 1958b. Notes on crop damage by jackrabbits. Trans. Kansas Acad. Sci. 49:455–456.

———. 1959. The relationship of precipitation and black-tailed jackrabbit populations in Kansas. Ecology 40:194–198.

Brown, L. 1947. Why has the white-tailed jackrabbit (*Lepus townsendii campanius*) become scarce in Kansas? Trans. Kansas Acad. Sci. 49:455–456.

Burt, W. H., and Grossenheider, R. P. 1959. A field guide to the mammals. Houghton Mifflon Co., Boston, Mass. 200pp.

Carter, F. L. 1939. A study in jackrabbit shifts in range in western Kansas. Trans. Kansas Acad. Sci. 42:431–435.

Chapman, J. A.; and Sandt, J. L. 1977. The black-tailed jackrabbit, *Lepus californicus*, in Maryland. Ches. Sci. 18(3):319.

Clapp, R. B.; Weske, J. S.; and Clapp, T. C. 1976. Establishment of the black-tailed jackrabbit on the Virginia eastern shore. J. Mammal. 57(1):180–181.

Clark, D. O. 1975. Vertebrate pest control handbook. Calif. Dept. of Food and Agric., Sacramento, Calif. 300pp.

Clark, F. W. 1972. Influence of jackrabbit density on coyote population change. J. Wildl. Manage. 36(2):343–356.

Connolly, G. E.; Dudzinski, M. L.; and Longhurst, W. M. 1969. The eye lens as an indicator of age in the black-tailed jackrabbit. J. Wildl. Manage. 33(1):159–164.

Conway, M. C. 1976. A rare hare. New Mexico Wildl. 21(2):21–23.

Costa, R.; Nagy, K. A.; and Shoemaker, V. H. 1976. Observations of behavior on black-tailed jackrabbits in the Mojave Desert. J. Mammal. 57(2) 399–402.

Couch, L. K. 1927. Migrations of the Washington black-tailed jackrabbit. J. Mammal. 8:313–314.

Crouch, G. L. 1973. Jackrabbits injure ponderosa pine seedlings. Tree Planters' Notes 24(3):15–17.

Currie, P. O., and Goodwin, D. L. 1966. Consumption of forage by black-tailed jackrabbits on salt desert ranges of Utah. J. Wildl. Manage. 30(2):304–311.

Davis, C. A.; Medlin, J. A.; and Griffing, J. P. 1975. Abundance of black-tailed jackrabbits, desert cottontail rabbits, and coyotes in southeastern New Mexico. New Mexico State Univ. Agric. Exp. Stn. Res. Rep. 293.

DeVoes, A. 1964. Range changes of mammals in the Great Lakes region. Am. Midl. Nat. 71:210–231.

DeVoes, A.; Manville, R. H.; and Van Gelder, R. G. 1956. Introduced mammals and their influence on native biota. Zoologica 41(4):172.

Donoho, H. S. 1972. Dispersion and dispersal of white-tailed and black-tailed jackrabbits, Pawnee National Grasslands. M.S. Thesis. Colorado State Univ. 83pp.

Dubke, R. T. 1973. The white-tailed jackrabbit in Wisconsin. Wisconsin Dept. Nat. Res. Final Rep. P.R. Study 106, Project W-141-R-8. 24pp.

Errington, P. L. 1935. Food habits of mid-west fox. J. Mammal. 16:192–200.

Evans, J.; Hegdal, P. L.; and Griffith, R. E., Jr. 1970. Methods of controlling jackrabbits. Pages 109–116. *in* R. H. Dana, ed. Proc. 4th Vertebr. Pest Conf., 3–5 March 1970, West Sacramento, Calif.

Feldhamer, G. A. 1979. Age, sex ratios and reproductive potential in black-tailed jackrabbits. Mammalia 43:473–478.

Findley, J. S. 1956. Mammals of Clay County, South Dakota. Univ. South Dakota Publ. in Bio. 1:45.

Flinders, J. T., and Hansen, R. M. 1972. Diets and habitats of jackrabbits in northeastern Colorado. Colorado State Univ. Range Sci. Dept. Sci. Ser. 12.

———. 1973. Abundance and dispersion of leporids within a shortgrass ecosystem. J. Mammal. 54(1):287–291.

———. 1975. Spring population responses of cottontails and

jackrabbits to cattle grazing shortgrass prairie. J. Range Manage. 28(4):290–293.

French, N. R.; McBride, R.; and Detmer, J. 1965. Fertility and population density of the black-tailed jackrabbit. J. Wildl. Manage. 29(1):14–26.

Garlough, F. E.; Welch, J. F.; and Spencer, H. J. 1942. Rabbits in relation to crops. U.S. Fish and Wildl. Serv. Conserv. Bull. 11. 20pp.

Gashwiler, J. S.; Robinette, W. L.; and Morris, O. W. 1960. Foods of bobcats in Utah and eastern Nevada. J. Wildl. Manage. 24(2):226–228.

Glass, B. P. 1973. A key to the skulls of North American mammals. Oklahoma State Univ. Press, Stillwater. 59pp.

Graves, G. N.; Bennett, W. C.; Wheelers, J. R.; Miller, B. E.; and Forcum, D. L. 1978. Sylvatic plague studies in southeast New Mexico. Part 2, Relationship of the desert cottontail and its fleas. J. Med. Entomol. 14(5):511–522.

Griffing, J. P. 1974. Body measurements of black-tailed jackrabbits of southeastern New Mexico with implications of Allen's rule. J. Mammal. 55(3):674–678.

Griffing, J. P., and C. A. Davis. 1976. Black-tailed jackrabbits in southeastern New Mexico: population structure, reproduction, feeding and use of forms. New Mexico State Univ. Exp. Stn. Res. Rep. 318pp.

Gross, J. E.; Stoddart, L. C.; and Wagner, F. H. 1974. Demographic analysis of a northern Utah jackrabbit population. Wildl. Monogr. 40:1–68.

Hall, E. R. 1955. Handbook of mammals of Kansas. Univ. Kansas Mus. Nat. Hist. Misc. Publ. 7. 303pp.

Hall, E. R., and Kelson, K. R. 1959. The mammals of North America. Vol. 1. Ronald Press Co., New York. 546 pp.

Hamilton, W. J., Jr. 1935. Notes on food of red foxes in New York and New England. J. Mammal. 16:16–21.

Hansen, M. F.; Bartel, M. H.; Lyon, E. T.; and El-Rawi, B. M. 1965. Helminth and arthropod parasites. Pages 41–61 *in* The black-tailed jackrabbit in Kansas. Kansas State Univ. Agric. and Appl. Sci., Agric. Exp. Stn. Tech. Bull. 140.

Hansen, R. M., and Bear, G. D. 1962. Winter coats of white-tailed jackrabbits in southwestern Colorado. J. Mammal. 44(3):420–421.

Hansen, R. M., and Flinders, J. T. 1969. Food habits of North American hares. Range Sci. Dep. Sci. Ser. 1. Colorado State Univ., Fort Collins. 17pp.

Haskell, H. S., and Reynolds, H. G. 1947. Growth, developmental food requirements and breeding of the California jackrabbit. J. Mammal. 28(2):129–136.

Hayden, P. 1966a. Food habits of the black-tailed jackrabbit in southern Nevada. J. Mammal. 47(1):42–46.

———. 1966b. Seasonal occurrence of jackrabbits on Jackass Flat, Nevada. J. Wildl. Manage. 30(4):835–838.

Hegdal, P. L. 1966. Jackrabbit damage in the western United States. Supp. Rep. F-42.2. Denver Wildl. Res. Center, Jackrabbit Res. Stn., Idaho.

Herrick, E. H. 1965. Endocrine studies. Pages 73–75 *in* The black-tailed jackrabbit in Kansas. Kansas State Univ. Agric. and Appl. Sci., Agric. Exp. Stn. Tech. Bull. 140.

Hildreth, A. C., and Brown, G. B. 1955. Repellents to protect trees and shrubs from damage by rabbits. U.S.D.A. Tech. Bull. 1134. 31pp.

Hinds, D. S. 1977. Acclimatization of thermoregulation in desert inhabiting jackrabbits (*Lepus alleni* and *Lepus californicus*). Ecology 58(2):246–264.

Howard, W. E. 1964. Introduced browsing mammals and habitat stability in New Zealand. J. Wildl. Manage. 28(3):421–429.

———. 1965. Interaction of behavior, ecology, and genetics of introduced mammals. Pages 461–484 *in* H. G. Baker and G. L. Stebbins, eds. The genetics of colonizing species. Academic Press, New York. 588pp.

Hull, T. G. 1963. The role of different animals and birds in diseases transmitted to man. Pages 876–924 *in* T. G. Hull, ed. Diseases transmitted from animals to man. 5th ed. Charles C. Thomas, Springfield, Ill. 967pp.

Jalal, S. M.; James, T. R.; and Seabloom, R. W. 1967. Karyotype of the white-tailed jackrabbit. Proc. North Dakota Acad. Sci. 21:92–98.

James, T. R., and Seabloom, R. W. 1969a. Reproductive biology of the white-tailed jackrabbit in North Dakota. J. Wildl. Manage. 33(3):558–568.

———. 1969b. Aspects of growth in the white-tailed jackrabbit. Proc. North Dakota Acad. Sci. 22:7–14.

Johnson, J. C. 1978. Anticoagulant baiting for jackrabbit control. Pages 152–153. W. E. Howard, ed. Proc. 8th Vertebr. Pest Conf., 7–9 March 1978, Sacramento, Calif.

Johnson, W. V. 1964. Rabbit control. Pages 90–96 *in* Proc. 2nd Vertebr. Pest Control Conf., 4 and 5 March 1964, Anaheim, Calif.

Jones, J. K. 1964. Distribution and taxonomy of mammals of Nebraska. Univ. of Kansas Mus Nat. Hist. Publ. 16:1–356.

Kline, P. D. 1963. Notes on the biology of the jackrabbit in Iowa. Proc. Iowa Acad. Sci. 70:196–204.

Layne, J. N. 1965. Occurrence of black-tailed jackrabbits in Florida. J. Mammal. 46(3):502.

Lechleitner, R. R. 1957. Reingestion in the black-tailed jackrabbit. J. Mammal. 38(4):481–485.

———. 1958a. Certain aspects of behavior of the black-tailed jackrabbit. Am. Midl. Nat. 60(1):145–155.

———. 1958b. Movements, density, and mortality in a black-tailed jackrabbit population. J. Wildl. Manage. 22(4):371–384.

———. 1959a. Sex ratio, age classes, and reproduction of the black-tailed jackrabbit. J. Mammal. 40(1):63–81.

———. 1959b. Some parasites and infectious diseases in a black-tailed jackrabbit population in the Sacramento Valley of California. California Fish and Game 45(2):83–91.

Leopold, A. 1945. The distribution of Wisconsin hares. Trans. Wisconsin Acad. Sci. 37:1–14.

Lewis, J. H. 1946. Planting practice to reduce crop damage by jackrabbits. J. Wildl. Manage. 10(3):277.

Lord, R. D. 1960. Litter size and latitude in North American mammals. Am. Midl. Nat. 64(2):488–499.

Marsh, R. E., and Salmon, T. P. The control of jack rabbits in California agriculture. In press.

McDowell, J. W.; Scott, H. G.; Stojanovich, C. J.; and Weinburgh, H. B. 1964. Tularemia. U.S. Dept. of H.E.W., Public Health Service, Atlanta, Ga. 81pp.

McMahon, K. J. 1965. Bacterial and rickettsial diseases. Pages 65–71 *in* The black-tailed jackrabbit in Kansas. Kansas State Univ. Agric. and Appl. Sci. Agric. Exp. Stn. Tech. Bull. 140.

Mearns, E. A. 1896. Preliminary description of a new subgenus and six new species and subspecies of hares, from the Mexican border of the United States. Proc. U.S. Natl. Mus. 18:551–565.

Nagy, K. A.; Shoemaker, V. H.; and Costa, W. R. 1976. Water, electrolyte, and nitrogen budgets in jackrabbits (*Lepus californicus*) in the Mojave Desert. Physiol. Zool. 49(3):351–362.

Orr, R. T. 1940. The rabbits of California. California Acad. Sci. 19. San Francisco, Calif. 227pp.

Palmer, T. S. 1896. The jackrabbits of the United States. U.S.D.A. Bull. 8. 81pp.

Phillip, C. B.; Bell, J. F.; and Larson, C. L. 1955. Evidence of infectious diseases and parasites in a peak population of black-tailed jackrabbits in Nevada. J. Wildl. Manage. 19(2):225–233.

Pittman, V. J. 1977. Winter wheat seedling preference by grazing rodents. Can. J. Plant Sci. 57(3):1009–1012.

Pontrelli, M. J. 1968. Mating behavior of the black-tailed jackrabbit. J. Mammal. 49(4):785–786.

Read, R. A. 1971. Browsing preference by jackrabbits in a ponderosa pine provenance plantation. U.S.D.A. For. Serv. Res. Note RM-186, Rocky Mt. Forest and Range Exp. Stn. 4pp.

Reese, J. B., and Haines, H. 1978. Effects of dehydration on metabolic rate and fluid distribution in the jackrabbit, *Lepus californicus*. Physiol. Zool. 51(2):155–165.

Robinette, W. L.; Gashwiler, J. S.; and Morris, O. W. 1959. Food habits of the cougar in Utah and Nevada. J. Wildl. Manage. 23(3):261–272.

Rohe, D. L.; Dallas, N.; and Estes, L. G. 1963. Control of jack rabbits at a California municipal airport. California Vector Views 10(12):73–75.

Schmidt-Nielsen, K.; Dawson, T. J.; Hammel, H. T.; Hinds, D.; and Jackson, D. C. 1965. The jackrabbit: a study in its desert survival. Hvalradets Skrifter 48:125–142.

Schwartz, C. 1973. The cottontail and the white-tailed jackrabbit in Iowa, 1963–1972. Iowa Wildl. Res. Bull. 6. 23pp.

Scott, T. G. 1955. Dietary patterns of red and gray foxes. Ecology 36(2):366–367.

Shoemaker, V. H.; Nagy, K. A.; and Costa, W. R. 1976. Energy utilization and temperature regulation by jackrabbits (*Lepus californicus*) in the Mojave Desert. Physiol. Zool. 49(3):364–375.

Sparks, D. R. 1968. Diets of black-tailed jackrabbits on the Sandhill Rangeland of Colorado. J. Range Manage. 21:203–208.

Stoddart, L. C. 1970. A telemetric method for detecting jackrabbit mortality. J. Wildl. Manage. 34(3):501–507.

Storer, T. I. 1958. Controlling field rodents in California. Univ. California Exp. Stn. Ext. Ser. Circ. 434 (Rev.) 50pp.

Taylor, W. P.; Vorhies, C. T.; and Lister, P. B. 1935. The relation of jackrabbits to grazing in southern Arizona. J. For. 33:490–498.

Tiemeier, O. W., and Plenert, M. L. 1964. A comparison of three methods for determining the age of black-tailed jackrabbits. J. Mammal. 45(3):409–416.

———. 1965. Bionomics. Pages 5–37 *in* The black-tailed jackrabbit in Kansas. Kansas State Univ. Agric. and Appl. Sci., Agric. Exp. Stn. Tech. Bull. 140.

Uresk, D. W. 1978. Diets of the black-tailed hare in steppe vegetation. J. Range Manage. 31(6):439–442.

Vorhies, C. J., and Taylor, W. P. 1933. The life histories and ecology of the jackrabbits *Lepus alleni* and *Lepus californicus* in relation to grazing in Arizona. Univ. Arizona Agric. Exp. Stn. Tech. Bull. 49:1–117.

Voth, D. R., and James, T. R. 1965. Parasites of the white-tailed jackrabbit in southwestern North Dakota. Proc. North Dakota Acad. Sci. 19:15–18.

Wagner, F. H., and Stoddart, L. C. 1972. Influence of coyote predation on black-tailed jackrabbit populations in Utah. J. Wildl. Manage. 36(2):329–342.

Watkins, L. C., and Novak, R. M. 1973. The white-tailed jackrabbit in Missouri. Southwestern Nat. 18(3): 341–357.

West, R. R.; Bartel, M. H.; and Plenert, M. L. 1961. Use of forms by black-tailed jackrabbits in southwestern Kansas. Trans. Kansas Acad. Sci. 64:344–348.

Wetherbee, F. A. 1967. A method of controlling jack rabbits on a range rehabilitation project in California. Pages 111–117 *in* Proc. 3rd Vertebr. Pest Conf. 7–9 March 1967, San Francisco, Calif.

Woodbury, A. M. 1955. Ecology of the Great Salt Lake Desert. Ecology 36:353–356.

Worthington, D. H., and Sutton, D. A. 1966. Chromosome numbers and analysis in three species of Leporidae. Mammal. Chromosomes Newsl. 22:194–195.

JOHN P. DUNN, Washington Department of Game, Olympia, Washington 98504.

JOSEPH A. CHAPMAN, Appalachian Environmental Laboratory, Center for Environmental and Estuarine Studies, University of Maryland, Frostburg State College Campus, Frostburg, Maryland 21532.

REX E. MARSH, Wildlife and Fisheries Biology, University of California, Davis, California 95616.

7

Snowshoe Hare and Allies

Lepus americanus and Allies

Steven L. Bittner
Orrin J. Rongstad

NOMENCLATURE

COMMON NAMES. Snowshoe hare, varying hare, snowshoe rabbit
SCIENTIFIC NAME. *Lepus americanus*
SUBSPECIES. *L. a. americanus, L. a. bairdii, L. a. cascadensis, L. a. columbiensis, L. a. dalli, L. a. klamathensis, L. a. oregonus, L. a. pallidus, L. a. phaeonotus, L. a. pineus, L. a. seclusus, L. a. struthopus, L. a. tahoensis, L. a. virginianus,* and *L. a. washingtonii.*

The snowshoe hare was originally identified by Erxleben in 1777, from specimens taken around Hudson Bay, Canada. In 1933 and 1934, Orr identified two subspecies as *Lepus washingtonii tahoensis* and *Lepus bairdii oregonus,* but they were reclassified to *Lepus americanus tahoensis* and *Lepus americanus oregonus,* respectively, by Dalquest in 1942 (Hall and Kelson 1959).

Other Hares. The snowshoe hare belongs to the family Leporidae, which includes, besides other native rabbits described in Chapters 5 and 6, the Alaskan hare (*Lepus othus*) and the arctic hare (*L. arcticus*). There are two subspecies of the Alaskan hare, both associated with coastal Alaska, from near the Yukon Territory border to the Aleutian Islands. There are eight subspecies of the arctic hare, which are widely scattered throughout northern Canada and Greenland. The European hare (*L. capensis*) has been introduced into New England and Ontario. The Canadian population has thrived while the U.S. population has remained disjunct. The first introductions of this hare occurred in 1888 from eastern Europe and were the subspecies *L. c. hybridus* (Dean and De Vos 1965).

DISTRIBUTION

The snowshoe hare is common to most of Canada and the northeastern United States. Its domain extends across the Great Lakes region and south through the Allegheny Mountain Range into North Carolina and Tennessee. In western North America, the snowshoe

FIGURE 7.1. Distribution of the snowshoe hare (*Lepus americanus*).

hare ranges down the Rocky Mountains into Utah and New Mexico. It also ranges south into western Nevada and along the Pacific coast into California (figure 7.1).

DESCRIPTION

The snowshoe hare is a member of the order Lagomorpha and the family Leporidae. Although it is rabbitlike in appearance, it is not a true rabbit: there are cranial differences; and rabbits construct nests and their neonates are altricial, whereas the snowshoe and other hares do not construct nests and their young are precocial. The large ears and large hind feet are characteristic of the hare genus. The body measurements of the snowshoe hare are as follows: total length 363–520 mm, tail length 25–55 mm, hind foot length 112–150 mm, and ear length from the notch 62–70 mm (Hall and Kelson 1959). The weight of the snowshoe hare ranges from 1,000 to 2,300 g (Doutt et al. 1967) but averages about 1,300 g. Grange (1932*a*) found Wisconsin snowshoes to range from 1,400 to 1,600 g, while Adams (1959) reported that in Montana they ranged from 1,000 to 1,100 g.

The general appearance of the snowshoe hare in summer is rusty brown, with the nose dark brown to black and the nostrils edged with pure white. The snowshoe's chin is white and the belly is white to grayish (Grange 1932*b*). The tail is white on top and grayish underneath, and the ears are black-tipped (Doutt et al. 1967; Grance 1932*b*).

The winter appearance of the snowshoe hare is pure white except for black-tipped ears (Doutt et al. 1967; Grange 1932*b*) (figure 7.2). Two subspecies of the snowshoe, *L. a. oregonus* and *L. a. washingtonii*, often remain in their summer pelage throughout the winter months (Booth 1961; Orr 1934). The winter pelage of the snowshoe hare appears white, but the hair is actually tricolored. The outermost zone is white, the middle zone is tawny brown, and the innermost zone ranges from dark gray to black (Grange 1932*b*). This tricoloring positively distinguishes the snowshoe from the arctic hare, *Lepus arcticus*, and the white-tailed jackrabbit, *Lepus townsendii*, whose winter pelage is completely white (Burt and Grossenheider 1959).

The dental formula for the snowshoe hare is 2/1 0/0 3/2 3/3 = 28. The skull has a basilar length of less than 67 mm. A second set of upper incisors ("peg teeth") is located immediately behind the first set. The cutting edge of the first upper incisors is straight (figure 7.3).

PHYSIOLOGY

Molts. The snowshoe hare goes through two annual molts, an autumnal change of pelage and a vernal change. The autumnal molt begins in October and lasts until all hares are white, generally in December. The

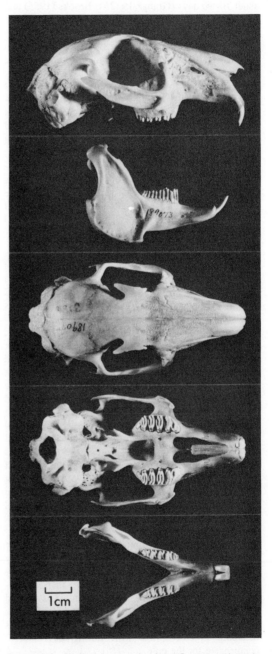

FIGURE 7.3. Skull of the snowshoe hare (*Lepus americanus*). From top to bottom: lateral view of cranium, lateral view of mandible, dorsal view of cranium, ventral view of cranium, dorsal view of mandible.

FIGURE 7.2. The snowshoe hare (*Lepus americanus*) in winter pelage.

sequence of molting is as follows: the tips of ears, wrists, and portions of the feet; the feet and lower legs, together with the lower part of the rump and the tail area; the upper legs, sides, and upper part of the rump; the belly and the bases of the ears; the sides of the face, the chest, shoulders, hips, and lower back; the remainder of the head and upper back, the medias dorsal shoulder area, and crown. This molt is complete in about 70–90 days (Grange 1932*b*). Brooks (1955) and Aldous (1937) in Minnesota report that snowshoes are completely white by December.

The vernal change of pelage generally begins in March, with all snowshoes returning to summer pelage by May. Aldous (1937) found that snowshoes in Minnesota begin their vernal pelage change in March and by April are mostly brown with some white. This vernal molt was complete by May. Grange (1932*a*) has described the molting process as follows: "a general shedding of the pure white guard hairs and a breaking off of the white tips of the winter under hairs; the ears, nose and head area; the upper dorsal region; the shoulders and areas about the forelegs; the hips, upper legs and the chest; the lower legs, rump and probably the belly; the hind feet and tail area." Like the autumnal change, the vernal change is complete within 70–90 days.

Metabolism. Not only does the snowshoe hare's winter fur act as a camouflage agent, but it also serves as a good insulator against the cold. Hart et al. (1965) determined that the insulative capacity of the fur is 27 percent greater in winter than in summer. This was consistent with their observations on the heat conductance of snowshoe hares caught during the summer, which was 35 percent greater than that of hares caught during the winter. Feist and Rosenmann (1975) related these insulative and conductive capacities to certain metabolic adjustments, which were reflected in the snowshoe's greater maximum metabolic capacity and its altered sympathoadrenal response to cold in the winter. More epinephrine was produced by the adrenal medulla cells, and more norepinephrine was produced at the sympathetic nerve endings. These adjustments help reduce shivering in hares during the winter (Feist and Rosenmann 1975). Hart et al. (1965), found that hares shiver more in the cold during the summer than during the winter. The snowshoe hare's lowest critical temperature is 10° C in the summer, as opposed to −5° C in the winter.

Irving et al. (1957) found that the basic metabolic rate of snowshoes was 0.68 ml O₂/gram of body weight/hour. Hart et al. (1965) found that the snowshoe's oxygen uptake is lower in winter than in summer. Furthermore, its mean caloric intake averaged 311 calories per day during the winter and 372 calories per day during the summer. Thus, the snowshoe's metabolic processes slow down during the winter to compensate for the cold.

Body Weight. Rowan and Keith (1959) reported that the average monthly weight of snowshoe hares in Alberta fluctuated (figure 7.4). An increase in weight occurs in the fall months, culminating in December.

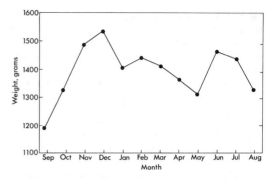

FIGURE 7.4. Mean monthly weights of the snowshoe hare (*Lepus americanus*) from Alberta, Canada. Adapted from Rowan and Keith 1959.

After a weight loss in January, there is a slight increase in weight in February, followed by a gradual but steady decrease during March, April, and May. The weight of the snowshoe rises sharply in June but falls off markedly in August. The females are larger than the males, as is the case with all members of the family Leporidae. Rowan and Keith (1959) found the greatest differences in weight between sexes in November, when females weighed an average of 1554 g and males 1429 g, and from January through April, when the average for females was 1424 g and for males was 1390 g. From May through August, the averages were 1416 g for females and 1419 g for males. These fluctuations were related to food supply, nutritional levels, weather, the reproductive cycle, and possibly endocrine processes.

Cary and Keith (1979) reported that hare mean weights varied as much as 200 g between the high and low in the population cycle. Lowest weights occurred just after the peak in the cycle and peak weights occurred just before the population low. Since snowshoe hare weights vary within seasons and between years, meaningful comparisons of hare weights from different areas must take these factors into account.

REPRODUCTION

Anatomy. The female snowshoe hare has a duplex uterus. It consists of a single vagina, two cervixes, and two uteri. The adult female has ribbonlike uteri 4–6 mm wide with longitudinal striations; juvenile uteri are less than 3.5 mm wide and threadlike. The weight of both ovaries in pregnant females ranges from 0.4 to 3.3 g (Newson 1964). Female snowshoe hares have eight mammae. The male snowshoe hare has paired testes and epididymides. The testes descend annually into the scrotum and are located anterior to the penis (DeBlase and Martin 1974). The adult penis is pointed and easily everted beyond its sheath (Keith et al. 1968).

Breeding Season. The initiation of the breeding season depends on several factors. Severaid (1945) correlated sexual activity with the increasing amount of daily illumination. Meslow and Keith (1971) reported that latitude is related to the onset of the breeding season. The farther north the species, the later its breeding will begin.

The breeding season of the snowshoe hare begins with the change in activity of the male testes. This process starts around February, depending on the area of the range and environmental factors. This has been noted in Ontario (Newson 1964), Maine (Severaid 1945), and Montana (Adams 1959). Bookhout (1965*a*) reported that in Michigan testes weight increased in January, while O'Farrell (1965) stated that in Alaska testes descend in snowshoe hares during March. Maximum testes size is reached at various times in different regions. Newson (1964) found that snowshoe testes reach maximum weight in May on Manitoulin Island, Ontario, while Dodds (1965) reported that maximum weight is reached in June in Newfoundland.

Males testes weight decreases as early as June in Michigan (Bookhout 1965*a*), with most male snowshoes becoming sexually inactive by July (Keith et al. 1966; Newson 1964; Dodds 1965). Some males remain active into August, however, as evidenced by litters born in September (Bookhout 1965*a*). The mean date of testes regression in Alberta varied from 3 July to 11 August in different years (Keith and Windberg 1978). The latest regressions were during the increasing phase in the population cycle and the earliest testes regression as during the decreasing phase. Meslow and Keith (1968) hypothesized that mating stopped when testes regressed to about one-half the breeding season maximum size.

Female snowshoes come into estrus in March in Newfoundland (Dodds 1965). This seems to be the most common period, as found by Meslow and Keith (1968) in Alberta and by Severaid (1945) in Maine. However, Bookhout (1965*a*) reported females coming into first estrus in early April in Michigan, as did Dolbeer and Clark (1975) in Colorado. First litters are usually dropped in mid-April (Dodds 1965; Green and Evans 1940*c*; Adams 1959), with some first litters being born in May (O'Farrell 1965; Dolbeer and Clark 1975).

Cary and Keith (1979) found that the mean date of first conception in Alberta varied between 25 March and 15 April in different years. These changes were highly correlated with spring weather and phase in the population cycle, with earliest litters being born during the increase phase in the population cycle and latest first litters born during the decrease phase.

The breeding season of the snowshoe hare generally lasts through mid-August (Dodds 1965; Newson 1964; Severaid 1945), but the birth of the last litter has been reported as early as July in Alaska (O'Farrell 1965) and as late as September in Michigan (Bookhout 1965*a*).

Pregnancy Rate. Green and Evans (1940*c*) found 25 percent of the female hares in Minnesota pregnant during the period of 25 March to 10 April. Eighty-one percent of the females were reported to be pregnant from 10 April to 30 April. Newson (1964) stated that, throughout the study, 75 percent of the females on Manitoulin Island in Ontario were pregnant, while in Maine, Severaid (1945) reported that all females had at least one litter, with 87.5 percent of the females having two or more litters and 56.2 percent of the females having three or more litters per year. It should be noted, however, that Severaid's data reflect the use of artificial propagation procedures.

Cary and Keith (1979) analyzed pregnancy rates from Alberta hares for each of the four litter groups during the summer and for 16 years of data. They found that pregnancy rates declined significantly by the third and fourth litter groups. There were significant between-year differences in all but the second litter group. Pregnancy rates for the first litter group ranged from 78 to 100 percent and 82 to 100 percent for the second litter group. Pregnancy rates for both the third and fourth litters varied with the cycle, with fourth litters going from near 0 about a year after the population high to about 70 percent just after the low.

Gestation Period. The gestation period of the snowshoe hare has been found to range from 34 days (Meslow and Keith 1971) to 40 days (Severaid 1945). The most widely reported gestation period is 37 days, as noted by Keith et al. (1966) for Alberta, by Dell and Schierbaum (1974) for New York, and by Newson (1964) for Ontario. Dolbeer and Clark (1975) found a 38-day gestation period for snowshoes in Colorado and Utah. Severaid found a mean gestation period of 36 days during one year of his study and 38 days during the next year. It is possible that gestation period may vary between years.

Numbers of Litters. The snowshoe hare is a seasonally polyestrous species. The breeding season begins in mid-March and is characterized by immediate postpartum breeding (Keith et al. 1966), which makes it possible for females to have 4 litters per year. The average number of litters per female is greater in the central portion of the snowshoe's range, while females residing on the periphery of the range have fewer litters per year (Keith et al. 1966). In Newfoundland, Dodds (1965) found that females averaged 2.9 to 3.5 litters per year. In Alberta, Rowan and Keith (1956) reported an average of 2.7 litters per year, while Meslow and Keith (1968) observed an annual average of 3.3 litters. Cary and Keith (1979) reported that the number of litters per year varied from almost 3 just after the population high to almost 4 just after the population low. In Colorado, Dolbeer and Clark (1975) found that female snowshoe hares averaged a little over 2 litters per year, while in Michigan, Bookhout (1965*a*) described females as averaging 2.4 litters per year. Both areas are on the periphery of the snowshoe's range, so Dolbeer and Clark's and Bookhout's data support the findings of Keith et al. (1966) reported in table 7.1.

Litter Size. Litter size of snowshoe hares depends on several factors. Snowshoes from the southern section of the range have smaller litters than those from the more northerly areas (Keith et al. 1966; Bookhout 1965*a*). Meslow and Keith (1971) made the interesting observation that the deeper the snow in the preceding winter, the larger the litters the following spring. They

TABLE 7.1. Reproductive characteristics of *Lepus americanus* in the wild in different sections of its range

Region and Species	Approximate Latitude	Years	Average Testis Weight			Dates of Conception First Litter Group		
			April	May	June	Average	Earliest	Latest
Wisconsin *phaeonotus*	45	1965		6.6		4/7		
Michigan *phaeonotus*	46	1959–61	5.2	6.1	4.5	4/1–15	4/3–10	7/6–30
Minnesota *phaeonotus*	46	1931–35						
	46	1932–38				4/4–18		
Montana *bairdii*	48	1953–54				3/24–4/7	3/13	
Ontario *americanus*	46	1959–61	4.9	5.9	4.9	4/11–17	3/30–4/3	7/13–26
	49	1933–35						
Newfoundland *struthopus*	49	1954–62				4/ 4		8/3–4
Manitoba *phaeonotus*	50	1923–35				3/24–4/7	3/15	7/17
	50	1943				4/4	3/24	
Manitoba *americanus*	51	1933						
Alberta *americanus*	54	1961–65	7.7		7.7	3/31–4/14	3/27–4/3	6/24–7/26
	56	1949–52 1956–57					4/1	7/15
Alaska *macfarlani*	60–66	1937						
	65	1955–56	4.5	6.2	6.2	4/26–30	4/18	6/30
	65	1958–61				4/8–22	4/10	6/29

SOURCE: Keith et al. 1966.

related this to the fact that deeper snow enables the hares to acquire more nutritious food, i.e., the snowshoes reach a higher stratum in the trees and eat younger and more nutritious food. Thus, their ovulation rate increases and litter size increases.

Minnesota hares average 2.4 young per litter (Aldous 1937), while hares in Newfoundland average 3.9 young per litter (Dodds 1965). Dodds (1965) reported that each female averaged 11.3 to 13.7 young per year. This is higher than the 6.5 young per female found by Bookhout in Michigan (1965a). Thus, the fecundity of the snowshoe hare is greater in the northern reaches of its range than in the south (see table 7.1). Keith (1979) stated that reproduction in cyclic snowshoe hare populations varies greatly between years. Total young produced per adult female ranged from 7.5 to 17.8 for different years of the Alberta study.

Severaid (1945), Dodds (1965), Keith et al. (1971), and Dolbeer and Clark (1975) all noted that first litters are the smallest. The average size of first litters in Colorado is 3.0, with subsequent litters averaging 4.75 young (Dolbeer and Clark 1975). Accord-

ing to Severaid (1945), first litters in Maine average 2.6 young, second litters 3.3 young, third litters 3.4 young, and fourth litters 3.5 young. Described on a monthly basis, the average litter size in Maine increases from 2.5 young per litter in April to 3.4 young per litter in July.

Keith (1979) developed a hypothesis to explain the changes in litter size with latitude. The essential components of this hypothesis are: (1) the size of the first litter is determined by the gonadotropin level at the time of conception; (2) the gonadotropin level is regulated by day length; (3) at any given latitude, day length and gonadotropin level will be greater when first litter conception is later; (4) after the vernal equinox (21 March), day length on any given date will increase with latitude; (5) the time of first conception is influenced by spring weather; (6) the increase in litter size between the first and later litters parallels an increase in gonadotropin level; (7) the size of later litters is highly correlated with the size of first litters but the relationship between day length and first-litter size does not persist for later litters; and (8) gonadotropin levels at the time of later litter conceptions are largely

Average Litter Size					Percentage Pregnancy May–July Indicated by		Litters/ Adult Female per Year		Average Young/ Adult Female per Year
First Litters		Later Litters							
Corpora Lutea	Embryos	Corpora Lutea	Embryos	Maximum	Embryos	Corpora Lutea and Embryos	Average	Maximum	
			2.20	4					
			2.83	5	60		2.42	4	6.5
	2.04		2.79	7				4	
	2.40		3.17	7	59		2.35	3	6.8
				5	77		2.94	3	8.2
2.38		3.48		6		88		4	6.3
	2.25		3.39	6					
	2.77		4.23	8	73	93	3.20	4	12.2
	2.52		4.21	8				5	
	3.40								
	4.18								
3.20	2.71	4.94	4.49	9	70		3.15	4	12.8
	2.88			7	69		2.75	5	10.5
				7					
	3.72			7	64		1.79	3	7.2
					57		1.68	3	7.8

determined by the timing of the first conception and not by the amount of day length increases thereafter. Therefore, litters are larger where days are longer due to latitude or where conceptions are later due to spring weather, which may be influenced by latitude or altitude.

Keith and Windberg (1978) found that litter size was also strongly correlated with the condition of the animals the prior winter. When there was no overwinter weight loss, ovulation rates for litters 2–4 were over 6, and when overwinter weight loss was 10–12 percent, ovulation rates were just over 4. Keith et al. (1966) also pointed out that there may be genetically fixed differences in litter sizes between regions.

Prenatal Growth. The prenatal growth of the snowshoe hare has been throughly studied by Bookhout (1964), who described the gestation period as follows: digits are not present before 15 days of gestation, while nails do not appear until after the 19th day; fat is stored after 21 days of gestation, and the vibrissae appear after the 23rd day. From the 12th to the 18th day of gestation, the average daily increase in length is 2.7

mm and the average daily weight increase is 0.24 g, while between the 30th and 34th day of gestation, the average daily increase in length is 4.8 mm and the average daily weight increase is 7.0 g. Newson (1964) also constructed a growth curve for snowshoe embryos that was based on length measurements between 7 and 14 days and weights between 13 and 35 days. She reported that 14-day-old fetuses weighed 0.3 g on the average, while 35-day-old fetuses weighed more than 60.5 g. Embryos are visible after 6–8 days of gestation (Adams 1959; Dodds 1965; Green and Evans 1940c; Keith et al. 1966).

Newborn Hares. Neonatal snowshoe hares are precocial, fully furred, and open eyed (Aldous 1937). Their pelage is dull gray with black-tipped ears and a grayish belly (Grange 1932b). Newborn hares in New York weigh an average of 81.7 g, and range from 64 g to 96 g (Dell and Schierbaum 1974). Their length ranges from 97 mm to 113.4 mm, with an average of 105.3 mm. Adams (1959) found that newborn snowshoes in Montana weighed about 40 g. Grange (1932a) reported that young snowshoes gain an average of 9 g per day and attain adult weight in 90 to 150 days (Aldous 1937; O'Farrell 1965).

Severaid (1942) gave a growth-weight curve of penned hares from birth to 182 days; these animals weighed an average of 67 g at birth and 123, 218, 329, 486, 606, 710, 784, 887 g for weekly intervals to 8 weeks of age and were 1510 g by the 23rd week of age. Keith et al. (1968) also gave growth curves for penned hares and presented evidence that wild hares grew slower than penned animals. Keith and Windberg (1978) found that growth rates of young hares varied greatly between years and were negatively correlated with the mean weight loss of adults during the previous winter to spring.

Rongstad and Tester (1971) found that young hares of a litter separate within a few days of birth and spend their days in separate hiding places. The young get together and the female returns one time each day, the young suckle for 5–10 minutes, and then the female and young separate again until the next day. The female may spend her day in a form as much as 250 meters from her young. Weaning occurs between 25 and 28 days (Severaid 1942), except for the last litter born during the summer, which may nurse for two months or longer (Severaid 1942; Rongstad and Tester 1971).

Breeding Age. Snowshoe hares normally breed in the spring following the one in which they are born. Juvenile breeding is uncommon. During the 16 years of the Alberta study, nine instances of juvenile breeding were observed (Keith and Meslow 1967; Vaughan and Keith 1980). These occurred only in first-litter females, and during the years immediately after the cyclic low. An average of 15 percent of first-litter females had litters during these years.

Sex Ratio. A 1:1 adult sex ratio has been reported in Ontario (Newson and DeVos 1964) and in Colorado and Utah (Dolbeer and Clark 1975). Dodds (1965) found 1.1 males per female in Newfoundland, and Aldous (1937) reported 1.2 males per female in Minnesota. Juvenile sex ratios are 1.4 males per female in Montana (Adams 1959), 1.2 males per female in Newfoundland (Dodds 1965), and 1:1 in Alberta (Meslow and Keith 1968) and Ontario (Newson and DeVos 1964). Fetal sex ratios are 1:1 in Ontario (Newson and DeVos 1964).

ECOLOGY

Home Range. The home range of the snowshoe hare is dependent upon vegetative cover and population density. Bider (1961) found that vegetative structures play an important role in the size of the home range. In Montana, Adams (1959) found that home ranges are smaller in brushy wood than in open woods. In Colorado and Utah, the average home range of both sexes is 8.1 hectares (Dolbeer and Clark 1975). O'Farrell (1965) found no difference in home range (5.9 ha) between sexes in Alaska. According to Adams (1959) in Montana, the home range of males is 10.1 ha, that of females is 7.5 ha, and that of juveniles is 5.5 ha. In

Minnesota monthly home ranges of two radio-tagged adult females in summer were between 8 and 13 ha. Fall and winter home ranges of two juveniles were slightly smaller (Rongstad and Tester 1971).

Population Cycles. The snowshoe hare exhibits a 10-year population cycle. Green and Evans (1940a) and Chitty and Nicholson (1943) reported that hare populations in Minnesota peaked in 1933, declined in 1938, and peaked again in 1942. During peak years, populations may reach considerable densities. Seton (1928) observed 38.6 hares per ha, while Criddle (1938) reported 11.6 per ha. In 1956, Keith and Waring noted that hares reached a maximum density of 23.2 hares per ha. Likewise, when snowshoe populations are low, their densities may drop as low as 0.12 per ha (Green and Evans 1940a). This great variation in snowshoe population gives rise to the cyclic phenomenon. Keith and Windberg (1978) reported density extremes of 0.13 to 0.26 per ha during low populations and 5.9 to 11.8 hares per ha during the high for their Alberta study areas during a 16-year study.

There have been many hypotheses as to the causes of the snowshoe hare's 10-year cycle. Keith (1963) reviewed the evidence that existed for the cycle and all existing hypotheses to explain it. He then started 16 years of intensive study on the population dynamics of the snowshoe hare. In 1974, Keith (1974) described a conceptual model to explain the cycle. In later research, experiments were conducted to test various aspects of this conceptual model (Pease et al. 1979; Windberg and Keith 1976a, 1976b; Vaughan and Keith 1981).

Keith's (1974) conceptual model explains the cause of the 10-year cycle as an interaction between hares and their food supply, followed by a hare-predator interaction. As hare densities increase, the biomass of available food is less than that needed by the population. Excessive plant utilization by hares reduces subsequent annual increments of new growth. The winter food shortage increases mortality rates and decreases the following year's reproduction (small litters, shorter breeding season, lower pregnancy rates). Predator populations that build up have little effect on the high hare populations, but as the hare numbers decrease, rates of predation become significant. Predation carries the hare population below its food-dictated potential and extends the period of the decline.

The declining base population eventually leads to a drop in predator numbers as a result of egress, nonbreeding, or losses of young. Survival of hares then rises sharply and the birth rate is at its maximum due to an abundance of food.

Keith (1974) hypothesized that the synchrony of hare population fluctuations between regions is caused by periodic continent-wide weather patterns and the mobility of predators.

FOOD HABITS

The summer diet of the snowshoe hare consists mainly of green succulent plants, usually clovers (*Trifolium* sp.), grasses (Gramineae), sedges (Cyperaceae), ferns (*Polypodiaceae*), and forbes (Aldous 1936). The snowshoe feeds on these herbaceous plants until the first frost, when woody vegetation replaces green plants as the major portion of the diet.

In Newfoundland, Dodds (1960) reported that snowshoes consumed 31 species of vegetation in the winter. The most common species encountered were paper birch (*Betula papyrifera*), spruce (*Picea* sp.), and pines (*Pinus* sp.). Bider (1961) found that, in summer, snowshoes in Newfoundland fed primarily on white clover, grasses, sedges, ferns, marsh marigold (*Caltha palustris*), and dandelion (*Taraxacum* sp.). DeVos (1964) reported that snowshoes on Manitoulin Island, Ontario, preferred white pine (*Pinus strobus*) and red pine (*P. resinosa*) during the winter months, but also fed on quaking aspen (*Populus tremuloides*), white birch (*Betula alba*), balsam poplar (*Populus balsamifera*), and juneberry (*Amelanchier* sp.). He stated that clipping of woody vegetation is more prominent than stripping of the bark, except when population densities are high.

In the Great Lakes region, woody plants comprise the major portion of the winter diets of snowshoe hares. Young poplars, willows (*Salix* sp.), and birches are preferred foods (Hansen and Flinders 1969). In Wisconsin, snowshoes eat jewelweed (*Impatiens biflora*) and dandelion during the summer, while their winter diet consists mainly of white pine, tamarack (*Larix* sp.), black spruce (*Picea mariana*), balsam fir (*Abies balsamae*), white cedar (*Thuja occidentalis*), willows, birches, and blackberries (*Rubus* sp.) (Grange 1932a).

In West Virginia, snowshoe hares prefer southern highbush cranberry (*Vaccinium erythrocarpum*), but also feed on rosebay rhododendron (*Rhododendron lapponicum*), red spruce (*Picea rubens*), and hemlock (*Tsuga canadensis*) (Brooks 1955).

Radwan and Campbell (1968) reported that snowshoe hares in Washington prefer the spotted cat's-ear (*Hypochoeris radicata*), especially the open flowers. The flowers contain more sugar than the leaves or flower buds. The major food items of snowshoe hares are summarized in table 7.2.

Snowshoe hares have been known to feed on carrion. Brooks (1955) witnessed two hares feeding on a deer carcass in West Virginia. He felt this to be a common practice during the winter. Captive snow-

TABLE 7.2. Major foods of *Lepus americanus*

Locality	Forage Species	
	Woody	Herbaceous
Alaska	willows	
	paper birch	
	alder	
	spruce	
Canada	white pine	clover
	red pine	grasses
	aspen	forbes
	alder	
	hazelnut	
	willow	
Montana	Douglas fir	
Newfoundland	paper birch	white clover
	speckled alder	grasses
	balsam fir	sedges
	spruce	ferns
	aspen	dandelion
	high bush cranberry	
	mountain maple	
	gooseberry	
	raspberry	
	larch	
Lake states	willow	clover
	aspen	pussytoes
	birches	legumes
	jackpine	dandelion
	white pine	jewelweed
	tamarack	
Virginia	southern high bush cranberry	poverty grass
	red spruce	
	larch	
	rhododendron	
Washington	catsear	

SOURCE: Hansen and Flinders 1969

shoes, when fed dry pellets, require 0.1 to 0.25 liters of water per day. In winter, snowshoes in the wild obtain water by eating snow (Hansen and Flinders 1969; Dodds 1960). In winter, snowshoe hares consume about 300 g of choice mixed browse per day (Bookhout 1965; Pease et al. 1979), most browse having a maximum diameter of 3 to 4 mm.

Pease et al. (1979) concluded that what snowshoe hares eat over winter is a function of availability and this varies through the boreal forest and its ecotones. However, some species are highly unpalatable and little utilized even during periods of food scarcity. In Rochester, Alberta, the unpalatable species were members of the family Caprifoliaceae, especially *Lonicera, Symphoricarpos,* and *Virburnum.* Black spruce was also little used.

Basically nocturnal animals, snowshoe hares feed most actively in the early morning hours (Brooks 1955; Hansen and Flinders 1969). Bider (1961) and DeVos (1964) found that hares can browse to a height of 45 cm. In a deep snow, they may reach as high as 2 m (Dodds 1960).

Snowshoe hares are coprophagous (Bookhout 1959; Hansen and Flinders 1976), and produce two types of pellet: a soft, amorphous pellet coated with a thin layer of mucus, and a hard, spherical pellet. The soft pellets, produced during the daylight hours, are picked by the snowshoe hare directly from the anus and reingested. These soft pellets are high in protein and contain large quantities of certain B vitamins, so more energy is obtained from the vegetative matter by passing it through the body a second time.

HABITAT

The snowshoe hare occupies a variety of habitats, but generally prefers dense, second-growth-type forests (Brooks 1955; Cowan et al. unpublished data). In Wisconsin, Grange (1932a) described aspen areas and coniferous swamps as good habitats for snowshoes, but hardwood forests as a relatively poor habitat. In Minnesota, Green and Evans (1940a) found that lowlands and alder swamps comprised excellent snowshoe habitat. Small stands of evergreens are also widely preferred. In some areas of Michigan, the snowshoe shows a marked preference for habitats dominated by subclimax forest types, especially transition zones between lowlands and uplands (Bookhout 1965c).

Richmond and Chien (1976) found snowshoe hares in Connecticut in areas where seedling-sapling seral stages predominated. These snowshoes foraged at the forest's edge and in small clearings, but seldom in open fields. In West Virginia, Brooks (1955) found that hares persisted in second-growth forest composed of birch, beech (*Fagus grandifolia*), and maple (*Acer saccharum*), although the heaviest populations were located in young stands of spruce in the sapling and pole stages. Snowshoes in the central Rocky Mountain region heavily utilize areas where mixed spruce, fir, and lodgepole pine (*Pinus contorta*) forests are located (Dolbeer and Clark 1975). Conroy et al. (1979) reported that in the northern lower penninsula of Michi-

gan high densities of hares will not be found farther than 200–400 m from cedar-fir cover and high densities are less likely to be found in areas with a solid canopy than in areas with high habitat interspersion.

Keith (1963) points out that during population lows snowshoe hares are restricted to islands of favorable habitat and that during high populations they disperse into less favorable habitats. Dense brushy cover is preferred. Keith (1966) concluded that this shift of habitats was mainly caused by movements rather than differential survival or reproduction. Keith and Surrendi (1971) found that hares abandoned severely burned areas but that these areas were reoccupied within 15 months when the vegetation resprouted.

Grange (1965) felt that fire was very important to hares. He stated that snowshoes inhabit all stages of northern forest except those that are primarily hardwoods (maples and oaks). Great abundances or population explosions (which comprise the classical snowshoe hare cycle) are limited to very early succession forest stages not long after the occurrence of fire.

In the absence of fire, forests mature and hare numbers tend to be low. At these times there are only small isolated populations associated with bog or stream edges and places that may support patches of willow, alder, hazel, and other low-growing woody vegetation.

BEHAVIOR

Mating Behavior. The mating behavior of the snowshoe hare is unique. Forcum (1966) described it as follows: "the male snowshoe approached the female, sniffed her and jumped into the air. After landing, the male urinated on the female and left. The male reapproached the female and the female jumped into the air twice, after which the male left. The male returned, jumped into the air and urinated on the female. Both snowshoes then went into the bushes where more jumping occurred." Hares commonly urinated on their partners during premating activities (Bookhout 1965a). One hare runs beneath the other, which is in the air, with the airborne hare urinating on the other. These accounts vary somewhat, but urination is a consistent part of the snowshoe's mating behavior.

Aggressive Behavior. Little aggressive behavior is expressed by snowshoe hares. Grange (1932a) found that males in pens are quarrelsome only during the breeding season, when they usually fight with their teeth. In late pregnancy, penned females are very aggressive toward males, and often will not let males into their cage. Before and after the breeding season, however, snowshoe hares have been seen feeding and playing together, with aggression having little significance in these processes.

Windberg and Keith (1976a) found that about 20 percent of the hares trapped during the breeding season in years of high population had scars (tears in the skin). This was probably related to fighting associated with breeding. Hares with scars were relatively scarce during the winter, but there was a significantly greater amount of scarring on dispersing hares over resident hares during the population peak (Windberg and Keith 1976a). They suspected that the higher incidence of scarring may have indicated low social rank and increased wounding in intraspecific encounters.

Escape Behavior. Escape behavior of the snowshoe hare has not been studied to a great extent. Johnson (1925) observed that snowshoes will resort to swimming. In Montana, Adams (1959) noted that they will use burrows when being pursued, but will not dig their own. Juvenile snowshoes may remain motionless and try to hide to avoid danger, but adults normally rely on flight to escape danger. This is apparently related to the physical development of the animal, as older animals exhibit greater muscle development and can run faster than juveniles (Adams 1959).

Vocalization. Several authors have reported vocalizations by snowshoe hares. Trapp and Trapp (1965) noted that, upon removal from traps, snowshoes make a clicking noise, apparently due to nervousness. Forcum (1966), in describing their mating behavior, found that some males utter a birdlike sound and doglike whines. Snowshoes have been reported to make a grunting noise, usually in association with displeasure, and to squeal in a high-pitched tone, mainly to express terror (Grange 1932a).

Miscellaneous Behavior. Runways are often used by snowshoe hares, especially when snow is present (Richmond and Chien 1976). Forms, or daytime resting places, are usually located in dense cover and sometimes consist of small depressions (Grange 1932a; O'Farrell 1965). While in these forms, snowshoes tuck their front feet back under their rumps and lie with their chests and bellies on their legs. This puts them in a position to leap and start running to avoid danger (Adams 1959). Snowshoes occasionally use dusting places, small depressions of loose, powdery soil, to alleviate tick infestations (Grange 1932a; Aldous 1937).

MORTALITY

Prenatal Mortality Rates. Newson (1964) investigated prenatal mortality in snowshoe hares. Six percent of the ova are lost before implantation. Death also occurs when embryos are lost shortly after implantation and when fetuses do not survive to full term. Dolbeer and Clark (1975) have reported a total prenatal mortality rate of 8.6 percent in Colorado and 9.4 percent in Utah. Cary and Keith (1979) found 7.9 percent of ova shed were lost prior to implantation and 2.8 percent of implanted embryos were resorbed, and overall prenatal mortality of 10.5 percent.

Juvenile Mortality Rates. Juvenile mortality rates of snowshoe hares have been reported in Alberta (Meslow and Keith 1968; Keith and Windberg 1978) and in Colorado and Utah (Dolbeer and Clark 1975). Little variation has been reported in these rates, but the survival rate is slightly better for juveniles in the Rocky Mountain region. Juvenile mortality ranges from 76 percent to 97 percent in Alberta, while mortality rates in the Rocky Mountain region were 66 percent to 100 percent. The mortality rate for juveniles was highest in open areas.

Keith and Windberg (1978) reported that mortality from birth to winter was 89 to 94 percent during years when population decline was greatest and 42 to 76 percent during years of population growth. These changes are mainly a result of differences in mortality after the first 1½ months of life. There was also a suggestion that third and fourth litters have a lower survival rate than early litters.

Adult Mortality Rates. Adult mortality varies from region to region. In the Rocky Mountain region, Dolbeer and Clark (1975) found the annual rate to be 55 percent. Throughout an eight-year study in Minnesota, Green and Evans (1940b) found an annual adult mortality rate of 70 percent, with the greatest number of deaths occurring in winter. Keith and Windberg (1978) gave adult mortality rates of 89 percent during years of declining and low populations, an average of 64 percent during years of increasing populations, and a low of 47 percent in the peak year.

Mortality Factors. Predation is a major threat to the life of the snowshoe hare. Mammalian and avian predators can reduce a hare population. Mammalian predators include the short-tailed shrew (*Blarina brevicauda*) (Rongstad 1965), mink (*Mustela vison*) (Adams 1959), lynx (*Lynx canadensis*), red fox (*Vulpes vulpes*), coyote (*Canis latrans*), black bear (*Ursus americanus*), bobcat (*Lynx rufus*), dogs (*Canis familiarus*) (O'Farrell 1965), and domestic cats (*Felis cattus*) (Doucet 1973). Avian predators include the great horned owl (*Bubo virginianus*) (Adams 1959; Morse 1939; O'Farrell 1965), goshawk (*Accipiter gentilis*), barred owl (*Strix varia*), and ravens (*Corvus corax*) (Morse 1939).

Keith et al. (1977) reported lynx, coyote, great horned owl, and goshawk as the major winter predators on snowshoe hares in Alberta. In some winters these predators killed over 40 percent of the hares, which was 70 percent of the total winter mortality. The red-tailed hawk, rough-legged hawk, snowy owl, and long-tailed weasel were also listed as predators in Alberta.

Brand et al. (1975) in a telemetry study in Alberta found that predation was responsible for 21 of 26 (81 percent) reported deaths. Lynx, coyote, weasel, horned owl, and goshawk were the identified predators.

Hunting by humans also is a major mortality factor for the snowshoe hare. Bookhout (1956b) reported that 200,000–400,000 snowshoes are killed annually by hunters in Michigan. Dodds and Thurber (1965) found that the hunters' harvest is related to the snowshoe population level: the higher the population, the greater the harvest.

PARASITES. Many species of parasite have been associated with the snowshoe hare. Among them are the following ectoparasites: the rabbit tick (*Haemophysalis leporispalustris*), which affects the nose, mouth, and ear region (Dodds and Mackiewicz 1961; Philip 1938); the Rocky Mountain wood tick (*Dermacentor andersoni*); rabbit fleas (*Hoplopsyllus glacialis lynx, Ceratophyllus garei,* and *Spelopsyllus cuniculi*); and the black fly (*Simulium venustum*) (Dodds and Mackiewicz 1961; Green et al. 1939; Burgdorfer et al. 1961).

Endoparasites are very common to the snowshoe hare. Trematodes, nematodes, cestodes, and many larval stages of parasite have been identified. Extensive studies have been conducted by Erickson (1944) in Minnesota and by Bookhout (1971) in Michigan. Of those parasites reported by Bookhout, the stomach worm (*Obeliscoides cuniculi*) occurred in 95.9 percent of the snowshoes. Nematodes (*Trichostrongylus* sp.) were reported in 85.9 percent of the hares in Michigan. Other species identified by Bookhout include the cestode larvae *Taenia pisiformis*; filarial worms (*Dirofilaria scapiceps*), which are found in the subcutaneous facis around the joints; whipworms (*Trichuris leporis*); pinworms (*Passalurus ambiguus*); larvae of *Multiceps* sp.; several nematodes of the *Nematodirus* species; and lungworms (*Protostrongylus boughtoni*). Erickson (1944) also identified these species and reported several others, including several tapeworm cysts (*Cittotaenia pectinata americana, C. variavilis,* and those of the *Hymenolepis* and *Multiceps* species). Other authors have reported additional endoparasites of snowshoe hares (table 7.3).

Erickson (1944) stated that hares parasitized by more than three species generally have a lowered resistance to disease and eventually die sooner. Coprophagy leads to increased accumulation of helminths and other parasite ova, which can result in excessive parasite loads. Some parasites may produce toxins.

DISEASES AND VIRUSES. Seven viruses have been reported in the snowshoe hare in Alberta (Hoff et al. 1970): eastern equine encephalomyelitis, silverwater, California encephalitis, western equine encephalomyelitis, powasson, buttonwillow, and encephalomyocarclitis. California encephalitis and silverwater are most persistently active in the snowshoe.

Tularemia, *Pasteurella tularensis,* has been detected in the snowshoe hare by Green et al. (1939) in snowshoes in Minnesota and by Hoff et al. (1970) in Alberta. Tularemia does not occur frequently in hares. Other diseases include *Pasteurella pseudotuberculosis, Salmonella* sp. (Green et al. 1939), and listerosis, *Listeria moncytogenes* (Dodds and Mackiewicz 1961).

Shock disease has been described in detail by Green and Larson (1938a, 1938b). It is characterized by the sudden onset of convulsions and usually results in rapid death. There are three syndromes of shock disease: (1) convulsions characterized by the onset of sudden running movements of the legs, extension of the hind legs, retraction of the head and neck, and the absence of foaming at the mouth, with eyes usually fixed; (2) a lethargic and comatose state; and (3) characteristics of both of the above syndromes. These convulsive seizures are hypoglycemic in nature, and death is due to abnormally low blood sugar levels. Liver glycogen levels are drastically reduced in snowshoe hares afflicted by shock disease. The normal liver glycogen level is 5.5 percent; hares that die of shock disease have a liver glycogen level of 0.02 to 0.18 percent. The average weight of hares that die of shock disease in Minnesota is 1,177 g, as compared to their average live weight, 1,402 g (Green et al. 1939).

Diagnosis of shock disease is relatively simple. If equipment is not available to test blood glucose levels, the disease can be recognized by necropsy. The liver is dark and hypotrophic, with the capsule of the liver wrinkled and somewhat loose. The pleural and peritoneal cavities contain fluid. The adrenal gland becomes inflamed and the spleen loses all blood and appears brown (Green and Larson 1938b).

Another disease, called "trap sickness," was thought by some workers to be the same as shock disease because of the similarity in terminal symptoms, particularly the convulsions and hypoglycemia. Chitty (1959) felt that shock disease was a physiological consequence of the stress of confinement, and Iverson (1968) noted that animals with trap sickness have gastric ulcers and do not have the liver atrophy and the peritoneal odor described for animals with shock disease.

AGE DETERMINATION

Several methods have proven reliable for determining the age of snowshoe hares. Adams (1959) and Keith et al. (1968) found that adult snowshoes during the summer and fall can be distinguished from juveniles by the amount of white hair on their feet. They reported adults have more white guard hairs on their feet than juveniles, until juveniles go through their first molt prior to winter. After this molt, juveniles cannot be distinguished from adults by this method.

Closure of the epiphysis of the humerus has been used by Newson and DeVos (1964) and Dodds (1965) to determine juveniles from adults. As in many other species of mammal, the weight of the eye lens increases with age, as originally described by Lord (1959).

Specific techniques for distinguishing juveniles from adults within the same sex have been described by several authors. Keith et al. (1968) found that the adult male's penis is pointed and easily everted beyond its sheath, while that of the juvenile male is blunt, smaller, and barely eversible. In the breeding season, the adult male's penis is red (Aldous 1937). During the summer of its birth, the juvenile male's penis is pale.

Adult and juvenile females can be distinguished by several methods. The adult female's uteri are ribbonlike, usually 4–6 mm wide with longitudinal striations (Newson 1964). The juvenile female's uteri are threadlike, less than 3.5 mm in width. The teats of adult females are longer than 3 mm and usually are palpable. Juvenile female teats are less than 3 mm long and are not palpable (Adams 1959; Keith et al. 1968).

TABLE 7.3. Endoparasites of *Lepus americanus*

Investigator and Year	Location	Endoparasite
Boughton 1932	Canada	*Taenia pisiformis*
		Multiceps serialis
		Strongyloides papillosus
		Trichuris leporis
		Nematodirus triangularis
		Trichostrongylus sp.
		Synthetocaulus leporis
Manweiler 1938	Minnesota	*Filaria scapiceps*
Philip 1938	Alaska	*Taenia pisiformis*
		Cittotaenia pectinata americana
		Passalurus nonannulatus
Green et al. 1939	Minnesota	*Cittotaenia pectinata americana*
		Taenia serialis
		Taenia pisiformis
		Obeliscoides cuniculi
		Passalurus nonannulatus
		Trichuris leporis
		Nematodirus triangularis
		Synthetocaulus cuniculi
Goble and Dougherty 1943		*Protostrongylus boughtoni*
Highby 1943	Minnesota	*Dirofilaria scapiceps*
Erickson 1944	Minnesota	*Trichostrongylus calcartus*
		Taenia affinis
		Nematodirus triangularis
		Passalurus nonannulatus
		Taenia pisiformis
		Euparyphium melis
		Cittotaenia pectinata americana
		Cittotaenia variabilis
		Hymenolepis sp.
		Taenia sp.
		Multiceps sp.
		Nematodirus leporis
		Dermatoxys viligera
		Trichuris leporis
		Fannia scalaris
		Aphiochaeta xanthena
		Sarcophaga sp.
Penner 1954	Connecticut	*Dirofilaria scapiceps*
Dodds and Mackiewicz 1961	Newfoundland	*Dicrocoelium dendriticum*
		Mosgovoyia pectinata
		Taenia pisiformis
		Hydatigera taeniaeformis
		Multiceps sp.
		Obeliscoides cuniculi
		Trichostrongylus axei
Bookhout 1971	Michigan	*Taenia pisiformis*
		Cladotaenia circi
		Dirofilaria scapiceps
		Obeliscoides cuniculi
		Trichostrongylus sp.
		Trichuris leporis
		Passalurus ambiguus
		Nematodirus sp.
		Protostrongylus boughtoni
		Multiceps sp.

ECONOMIC STATUS

The economic status of the snowshoe hare depends on the area in question and the attitudes of the people involved. Forest plantations have been severely damaged by snowshoe hares. In the Great Lakes region, nearly 100 percent of the white pines (*Pinus strobus*) in a plantation were damaged by hares (Kitridge 1929). Seventy-eight percent of the white spruce was damaged, as was 53 percent of the red pines. In 1959, Rudolf found in Minnesota that hares had damaged 43 percent of the trees on one plantation. Damage results from nipping of terminal and lateral buds and girdling of trees (Hansen and Flinders 1969).

Competition between deer and snowshoe hares has been studied extensively. In Michigan, Bookhout (1965*b*) found that deer remove 6.8 percent of the total available browse, while snowshoes remove 1.4 percent. These figures are in contrast to those of Krefting and Stoeckeler (1935), who found that hares cause more damage in areas where both species occur.

A beneficial side effect of the snowshoe hare's feeding habits is its ability to thin young forest stands. Roe and Stoeckler (1950) believed that hares can save foresters money by thinning young stands of jackpines in the Great Lakes region. Snowshoes are also beneficial in thinning stands of pine, spruce, and aspen (Cox 1938).

Snowshoe hares are a valuable game species in some areas of the eastern United States and Canada (Webb 1937). As stated previously, between 200,000 and 400,000 snowshoe hares are killed by hunters each year in Michigan. Expenditures to pursue this species are beneficial to communities within the snowshoe's range. Snowshoe hares will probably become increasingly important as a game species as the number of hunters increases and as habitat and wildlife populations decrease near large human populations.

Another positive economic consideration of the snowshoe hare is the large numbers of mammalian predators that build up on the large snowshoe hare populations. Keith (1963) concluded after analyzing fur buyers' records that lynx, coyote, colored fox, fisher, marten, and mink populations fluctuate with snowshoe hare populations in at least some parts of their range. The value of the furs from these animals is astronomical. The fur of the snowshoe hare is not durable and is of little commercial value.

The value of the snowshoe hare as a protein producer for human consumption may someday be important. The American Indian has always utilized it as a food source. The snowshoe hare was introduced into Newfoundland in the 1870s and is currently snared and sold for meat in markets. With snowshoe densities reaching as high as 30 per ha and with an average weight of 1400 g, there could be a standing crop of 42 kg per ha during peak years of the cycle. Wisely managed, this could become a valuable protein source.

An indirect economic consideration of the snowshoe hare is also related to predators, both mammalian and avian. As numbers decline, the predators that have built up must shift to other prey and may widely disperse. Influxes into the United States of goshawks, horned owls, and other avian predators occur at these times. Even lynx show up far south of their normal range during these periods. The impact of these predators on other species and ecosystems is not known but may be great. Keith (1974) concluded that the synchrony between the hare and ruffed grouse (*Bonasa umbellus*) fluctuations was due to predators. He also believed that some of the fluctuations of small-game populations (pinnated grouse [*Tympanuchus cupido*], bobwhite quail [*Colinus virginianus*], cottontail rabbits [*Sylvilagus floridanus*], etc.) south of the snowshoe hare range may be due to emigrating predators. Since these are valuable game species, undoubtedly there are economic effects.

MANAGEMENT

Since the economic benefits of snowshoe hares are both positive and negative, management can be divided into programs to increase hares and those to decrease hares or to decrease their detrimental effects. Habitat management is the ultimate answer to both of these problems; however, this is not feasible on extensive areas.

Damage Control. Past management has centered largely on control of snowshoe hare damage to tree plantations (Herbert 1945; Basser 1955). Aldous and Aldous (1944) suggested several methods to control damage by snowshoes. Hunting can be efficient in small areas, but has no value on large tracts of land. Other suggested methods include snaring, drives, poison, and trapping. Keith (1972) pointed out that high rates of ingress during cyclic peaks negate any effects of direct killing. They removed 700 hares from 10 ha in one year. Szukiel (1973) used repellents to prevent hares from girdling trees in Europe. Although this method worked to some extent, it did not completely deter hares from eating plants treated with these repellents.

Keith (1972) suggested management techniques to help reduce hare damage to forest plantations in Alberta. (1) Planting should be done mainly during the predictable period of relatively low hare populations. (2) Seedling palatability should be investigated. Only species or genetic strains that are least palatable should be used. Keith recommended eliminating pine from the Alberta planting program, since trees up to 5 cm in diameter are commonly girdled and killed. (3) A hare repellent may have to be applied to get the plantings through the period of the first high in the hare cycle. By the second peak following planting, the trees should be safe from hares.

In protecting small areas such as gardens or the edges of agricultural fields, fences could be used. It may also be possible to clear all cover species from the area to be protected, since snowshoes do not forage very far from protective cover.

Habitat Management. Since snowshoe hares are not always a popular game species, literature on snowshoe hare habitat management for the purpose of maximizing snowshoe numbers is scarce.

Adams (1959) suggested that habitat management should involve thinning of dense areas. Light thinning should be done in young pole-size stands where a dense canopy has formed. The establishment of dense cover intermingled with lightly covered areas provides an optimum habitat for the snowshoe. Moreover, Brocke (1974) stated that open areas are used for travel between feeding areas, and not as prime food sources. Continuous cover for travel with dense resting and feeding cover is an essential ingredient of the snowshoe's habitat. If browse is available within the travel, resting, and feeding cover regions, snowshoes will utilize this area maximally (Brocke 1974).

Conroy et al. (1979) found that high densities of hares would not be found in solidly canopied areas in the lower peninsula of Michigan. They recommended that management of hares in less forested areas should emphasize the interspersion of upland and lowland habitats. Cuttings for hare management should be small and situated so that lowland cover is not farther than 200 to 400 m from all parts of the cutting.

Grange (1965) discussed the importance of fire to hare populations and habitat. Controlled burns could undoubtedly be used in habitat management for hares.

Artificial Propagation and Stocking. Restocking has been used extensively in the past, but was of no value in areas with resident snowshoe hare populations. The state of Maine conducted a research project (Severaid 1942) to develop methods of artificial propagation of snowshoes to make stocking programs more dependable. In areas where the habitat is suitable but hares are not present, stocking with wild hares is possible (Grange 1949).

Predator Control. Since predators killed over 40 percent of hares during some winters (Keith et al. 1977), predator control could probably be used to increase hare population on small areas if an efficient control method could be found. This game management tool is becoming almost impossible to use, since avian predators now have federal protection and mammalian predators in many states are being made game species with designated seasons for harvest. Dense cover intermingled with good food species is probably the only solution to heavy predation.

Laws and Regulations. Snowshoe hare densities in most of the range are determined by habitat, food supplies and predators. Hunting laws such as bag limits and season lengths probably have little effect on the overall population. Laws to restrict the kill may be more important on the fringes of the snowshoe range or during cyclic lows, when populations are restricted to pockets of favorable habitat.

CURRENT RESEARCH AND MANAGEMENT NEEDS

Much future work will probably be done in testing various aspects of Keith's (1974, 1979) hypothesis on the mechanisms of wildlife 10-year cycle. Areas that need more research are the effects of the predators on other species within the snowshoe hare range and especially the effects of these predators when they emigrate.

Keith (1979) felt that research was needed to determine what limited the geographic distribution of the snowshoe hare.

More work is needed on the management of the habitat in order to maximize snowshoe hare returns to the hunter. With better habitat management, it may be possible to increase the snowshoe hare's importance as a game species. Not only would this provide more hunting opportunities, but it would also increase monetary output for this sport.

OTHER HARES

Alaskan Hare. The Alaskan hare, *Lepus othus,* is a close relative of the mountain hare, *L. timidus,* of Eurasia, and they may be the same species (Hall and Kelson 1959). The Alaskan hare is one of North America's least known lagomorphs because of its remote and restricted range (figure 7.5). Recent studies have included reproduction and growth (Anderson and Lent 1977), while past studies were observational in nature (Walkinshaw 1947).

The two subspecies of *L. othus* are associated with distinct habitat types, and they tend to be composed of a complex of disjunct populations. The habitat of *L. o. othus* is primarily tundra or alluvial plain, while that of *L. o. poadromus* is primarily coastal lowland areas on the Aleutian Island chain.

Alaskan hares are brown above and light below in the spring and summer and begin turning white in mid-September. By late November they are completely white except for black-tipped ears (Walkinshaw 1947). They exhibit the fastest growth rate of any of the four

FIGURE 7.5. Distribution of the European hare (*Lepus capensis*), the arctic hare (*L. arcticus*), and the Alaskan hare (*L. othus*).

hares studied, including *L. townsendi, L. californicus,* and *L. americanus* (figure 7.6). This rapid growth is believed to be an adaptive response to the short arctic growing season (Anderson and Lent 1977).

Alaskan hares produce a single litter per year and the litter size varies from 3 to 8, with a mean of 6.4. Conception occurs between 28 May and 14 June. Thus, the gestation period appears to be about 40–50 days. Both preimplantation and postimplantation losses have been investigated (Anderson and Lent 1977).

Arctic Hare. The arctic hare (*Lepus arcticus*) is widely distributed in the tundra regions of northern Canada and Labrador (see figure 7.5). This species is found primarily beyond the tree line and may have the most northerly distribution of any Lagomorph. It is highly adapted to its cold and barren habitat (Wang et al. 1973). Nelson (1909) stated that throughout most of its range, the arctic hare spends the summer north of the tree limit, but in winter it may penetrate a hundred miles or more into the northern border of timber.

Bergerud (1967) described the reduction in distribution of the arctic hare in Newfoundland and hypothesized that this was caused by increased lynx predation after snowshoe hares were introduced to the island. He further suggested that the southern edge of the arctic hare's range in North America may depend in part on the availability of escape habitat and the abundance of lynx, which is dependent on hare numbers. Keith (1979) stated that there probably is only one species of arctic hare and that *L. arcticus, L. othus,* and *L. timidus* are all the same. *Lepus timidus* is found in both the tundra and the boreal forest of Europe and Asia. If these species are the same, Bergerud may be right and the arctic hare's range would extend southward farther into the boreal forest if it were not for lynx.

Arctic hares may weigh 5 kg and are generally light in color even during the summer months, being all white with black-tipped ears in the winter. Parker (1977) reported that adult female weights averaged 4.5 kg in summer and 3.9 kg in winter; male weights averaged 4.1 and 4.0 kg.

Arctic hares have highly modified incisors, which they use to feed on the small snow-covered arctic plants; these incisors are diagnostic of the "arctic" hares.

The arctic hare gives birth to a single litter in June or July. The mean litter size is six or seven, with a range of two to seven (Manniche 1910; Soper 1928; Manning and Macpherson 1958; Parker 1977; Pedersen 1966).

Both Alaskan and arctic hares are an important mammalian resource in the north, being a source of food and clothing for native peoples and a potentially valuable recreational resource.

European Hare. The European hare (*Lepus capensis*) has been introduced into many parts of the world including North America, and has attained popularity as a game animal in a few isolated places (see figure 7.3). However, these introduced hares are generally considered pests, damaging forest plantings, orchards, and watersheds.

The European hare is fairly large for the genus, weighing up to 4 kg. It is rusty brown to grey in color. In New Zealand the main part of the breeding season lasts about eight months. The gestation period is 42 days. The litter size is small, about 2 to 3, and the average female in New Zealand produces 9 to 10 young per year (Flux 1967). The home range of the species in New York is apparently quite large (Eabry 1970).

The introduced European hare in North America has not reached the population densities that the species has in other areas. Canadian populations have shown greater densities than those in the United States (Dean and DeVos 1967). The reason for this is not known; however, Jezierski (1968), after studying European hare introductions in Europe, believed that competition with resident hares and increased mortality of introduced individuals affected population densities, as did suitability of habitat.

This chapter is contribution no. 1056-AEL, Center for Environmental and Estuarine Studies, University of Maryland.

LITERATURE CITED

Adams, L. 1959. An analysis of a population of snowshoe hares in northwestern Montana. Ecol. Monogr. 29(2):141–170.

Aldous, C. M. 1936. Food habits of *Lepus americanus phaeonotus.* J. Mammal. 17:175–76.

———. 1937. Notes on the life history of the snowshoe hare. J. Mammal. 18:46–57.

Aldous, C. M., and Aldous, S. E. 1944. The snowshoe hare, a serious enemy of forest plantations. J. For. 42:88–94.

Anderson, H. L., and Lent, P. C. 1977. Reproduction and growth of the Tundra hare (*Lepus othus*). J. Mammal. 58(1):53–57.

Bear, G. D., and Hansen, R. M. 1966. Food habits, growth and reproduction of white-tailed jack rabbits in southern Colorado. Colorado State Univ. Agric. Exp. Stn. Tech. Bull. 90:1–59.

Bergerud, A. T. 1967. The distribution and abundance of arctic hares in Newfoundland. Can. Field Nat. 81(4):242–248.

Besser, J. 1955. Field and enclosure studies with experimental repellents for the protection of trees and shrubs from damage by rabbits and deer. Spec. Rep. U.S. Fish Wildl. Res. Lab., Denver. 19pp.

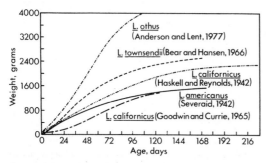

FIGURE 7.6. Weight gain of several species of juvenile hare. Adapted from Anderson and Lent 1977.

Bider, J. R. 1961. An ecological study of the hare *Lepus americanus*. Can. J. Zool. 39(1):81–103.

Bookhout, T. A. 1959. Reingestion by the snowshoe rabbit. J. Mammal. 40(2):250.

———. 1964. Prenatal development of snowshoe hares. J. Wildl. Manage. 28(2):338–345.

———. 1965a. Breeding biology of snowshoe hares in Michigan's Upper Peninsula. J. Wildl. Manage. 29(2):296–303.

———. 1965b. Feeding coactions between snowshoe hares and white-tailed deer in northern Michigan. North Am. Wildl. Nat. Res. Conf. 30:321–335.

———. 1965c. The snowshoe hare in Upper Michigan: its biology and feeding coactions with white-tailed deer. Michigan Dept. Conserv. Res. and Dev. Rep. 438. 191pp.

———. 1971. Helminth parasites in snowshoe hares from northern Michigan. J. Wildl. Dis. 7(Oct.):246–248.

Booth, E. S. 1961. How to know mammals. William C. Brown Co., Dubuque, Iowa. 203pp.

Boughton, R. V. 1932. The influence of helminth parasitism on the abundance of the snowshoe rabbit in western Canada. Can. J. Res. 7:524–547.

Brand, C. J.; Vowles, R. H.; and Keith, L. B. 1975. Snowshoe hare mortality monitored by telemetry. J. Wildl. Manage. 39(4):741–747.

Brocke, R. H. 1975. Preliminary guidelines for snowshoe hare habitat management in the Adirondacks. Trans. Northeast Sect. Wildl. Soc., Fish Wildl. Conf. 32:46–66.

Brooks, M. 1955. An isolated population of the Virginia varying hare. J. Wildl. Manage. 19(1):54–61.

Burgdorfer, W.; Newhouse, V. F.; and Thomas, L. A. 1961. Isolation of California encephalitis virus from the blood of the snowshoe hare (*Lepus americanus*) in western Montana. Am. J. Hyg. 73:344–349.

Burt, W. H., and Grossenheider, R. P. 1959. A field guide to the mammals. Houghton Mifflin Co., Boston. 200pp.

Cary, J. R., and Keith, L. B. 1979. Reproductive change in the 10-year cycle of snowshoe hares. Can. J. Zool. 57(2):375–390.

Chitty, D., and Nicholson, M. 1943. The snowshoe rabbit enquiry, 1940–41. Can. Field Nat. 57(4, 5):64–68.

Cowan, A. B.; Spaulding, W. M.; and Bookhout, T. A. Parasites of ruffed grouse (*Bonasas umbellus*) and the snowshoe hare (*Lepus americanus*) in the Upper Peninsula of Michigan. 12pp. Manuscript.

Cox, W. T. 1938. Snowshoe hare useful in thinning forest stands. J. For. 36:1107–1109.

Criddle, S. 1938. A study of the snowshoe hare. Can. Field Nat. 52:31–40.

Dean, P. B., and DeVos, A. 1965. The spread and present status of the European hare, *Lepus europaeus hybridus* (Desmarest), in North America. Can. Field Nat. 79(1):38–48.

DeBlase, A. F., and Martin, R. E. 1974. A manual of Mammalogy. William C. Brown Co., Dubuque, Iowa. 329pp.

Dell, J., and Schierbaum, D. L. 1974. Prenatal development of snowshoe hares. New York Fish Game J. 21(2):89–104.

DeVos, A. 1964. Food utilization of snowshoe hares on Manitoulin Island, Ontario. J. For 62(4):238–244.

Dodds, D. G. 1960. Food competition and range relationships of moose and snowshoe hare in Newfoundland. J. Wildl. Manage. 24(1):52–60.

———. 1965. Reproduction and productivity of snowshoe hares in Newfoundland. J. Wildl. Manage. 29(2):303–315.

Dodds, D. G., and Thurber, H. G. 1965. Snowshoe hare (*Lepus americanus struthopus*) harvests on Long Island, Nova Scotia. Can. Field Nat. 79(2):130–133.

Dolbeer, R. A., and Clark, W. R. 1975. Population ecology of snowshoe hares in the Central Rocky Mountains. J. Wildl. Manage. 39(3):535–549.

Doucet, G. J. 1973. House cat as predator of snowshoe hare. J. Wildl. Manage. 37(4):591.

Doutt, J. K.; Heppenstall, C. A.; and Guilday, J. E. 1967. Mammals of Pennsylvania. Pennsylvania Game Comm., Harrisburg. 238pp.

Eabry, H. S. 1970. A feasibility study to investigate and evaluate the possible future directions of European hare management in New York. Federal Aid P-R Proj. W-84-R-17. Mimeogr.

Erickson, A. B. 1944. Helminth infections in relation to population fluctuations in snowshoe hares. J. Wildl. Manage. 8(2):134–153.

Feist, D. D., and Rosenmann, M. 1975. Seasonal sympathoadrenal and metabolic responses to cold in the Alaskan snowshoe hare (*Lepus americanus macfarlani*). Comp. Biochem. Physiol. 51A:449–455.

Flux, J. E. C. 1967. Reproduction and body weights of the hare *Lepus europaeus* Pallus, in New Zealand. New Zealand J. Sci. 10(2):357–401.

Forcum, D. L. 1966. Postpartum behavior and vocalizations of snowshoe hares. J. Mammal. 47(3):543.

Gobel, F. C., and Dougherty, E. C. 1943. Notes on the lungworms (Genus *Protostrongylus*) of varying hares (*Lepus americanus*) in eastern North America. J. Parasitol. 29:397–404.

Goodwin, K. L., and Currie, O. P. 1965. Growth and development of black-tailed jack rabbits. J. Mammal. 46(1):96–98.

Grange, W. B. 1932a. Observations on the snowshoe hare, *Lepus americanus phaeonotus*, Allen. J. Mammal. 13(1):1–19.

———. 1932b. The pelages and color changes of the snowshoe hare, *Lepus americanus phaeonotus*, Allen. J. Mammal. 12(2):99–116.

———. 1949. The way to game abundance. Scribner's & Sons, New York. 365pp.

———. 1965. Fire and tree growth relationships to snowshoe rabbits. Pages 111–125 *in* Proc. 4th Annu. Tall Timbers Fire Ecol. Conf.

Green, R. G., and Evans, C. A. 1940a. Studies on a population cycle of snowshoe hares on the Lake Alexander area. Part 1, Gross annual census, 1932–1939. J. Wildl. Manage. 4(2):220–238.

———. 1940b. Studies on the population cycle of snowshoe hares on the Lake Alexander area. Part 2, Mortality according to age groups and seasons. J. Wildl. Manage. 4(3):267–278.

———. 1940c. Studies on the population cycle of snowshoe hares on the Lake Alexander area. Part 3, Effects of reproduction and mortality of young hares on the cycle. J. Wildl. Manage. 4(4):247–258.

Green, R. G., and Larson, C. L. 1938a. A description of shock disease in the snowshoe hare. Am. J. Hyg. 28:190–212.

———. 1938b. Shock disease and the snowshoe hare cycle. Science 87(2257):298–299.

Green, R. G.; Larson, C. L.; and Bell, J. F. 1939. Shock diseases as the cause of the periodic decimation of the snowshoe hare. Am. J. Hyg. 30(3):83–102.

Hall, E. R., and Kelson, K. R. 1959. The mammals of North America. Vol. 1. Ronald Press Co., New York. 546pp.

Hansen, R. M., and Flinders, J. T. 1969. Food habits of

North American hares. Colorado State Univ. Range Sci. Dept. Sci. Ser. 31. 18pp.

Hart, J. S.; Pohl, H.; and Tener, J. S. 1965. Seasonal acclimization in varying hare (*Lepus americanus*). Can. J. Zool. 43(5):731–744.

Haskell, H. S., and Reynolds, H. G. 1947. Growth, developmental food requirements, and breeding activity of the California jack rabbit. J. Mammal. 28(1):129–136.

Herbert, P. A. 1945. Observations on the incompatibility of wood and game protection. Trans. North Am. Wildl. Conf. 10:119–125.

Highby, P. R. 1943. Vectors, transmission, development and incidence of *Dirofilaria scapiceps* (Leidy, 1886) (Nematoda) from the snowshoe hare in Minnesota. J. Parasitol. 29:253–259.

Hoff, G. L.; Yuill, T. M.; Iverson, J. O.; and Hanson, R. P. 1970. Selected microbial agents in snowshoe hares and other vertebrates in Alberta. J. Wildl. Dis. 6(4):472–478.

Irving, L.; Krog, J.; Krog, H.; and Monson, M. 1957. Metabolism of varying hare in winter. J. Mammal. 38(4):527–529.

Jezierski, W. 1968. Some ecological aspects of introduction of the European hare. Acta Theriol. 13(1):1–30.

Johnson, C. E. 1925. The jack and snowshoe rabbits as swimmers. J. Mammal. 6:245–249.

Keith, L. B. 1963. Wildlife's ten-year cycle. Univ. of Wisconsin Press, Madison. 201pp.

———. 1966. Habitat vacancy during a snowshoe hare decline. J. Wildl. Manage. 30(4):828–832.

———. 1972. Snowshoe hare populations and forest regeneration in northern Alberta. Rep. prepared for Alberta For. Serv., Edmonton. 41pp.

———. 1974. Some features of population dynamics in mammals. Proc. Int. Cong. Game Bio. 11:17–58.

———. 1979. Population dynamics of hares. Proc. World Lagomorph Conf., 13–17 August 1979.

Keith, L. B., and Meslow, E. C. 1967. Juvenile breeding in the snowshoe hare. J. Mammal. 48(2):327.

Keith, L. B., and Surrendi, D. C. 1971. Effects of fire on a snowshoe hare population. J. Wildl. Manage. 35(1):16–26.

Keith, L. B., and Waring, J. D. 1956. Evidence of orientation and homing in snowshoe hares. Can. J. Zool. 34(4):579–581.

Keith, L. B., and Windberg, L. A. 1978. A demographic analysis of the snowshoe hare cycle. Wildl. Monogr. 58. 70pp.

Keith, L. B.; Meslow, E. C.; and Rongstad, O. J. 1968. Techniques for snowshoe hare population studies. J. Wildl. Manage. 32(4):501–512.

Keith, L. B.; Rongstad, O. J.; and Meslow, E. C. 1966. Regional differences in reproductive traits of the snowshoe hare. Can. J. Zool. 44(5):953–961.

Keith, L. B.; Todd, A. W.; Brand, C. J., Adamcik, R. S.; and Rusch, D. H. 1977. An analysis of predation during a cyclic fluctuation of snowshoe hares. Proc. Int. Cong. Game Bio. 13:151–175.

Kitridge, J., Jr. 1929. Forest planting in the Lake States. U.S.D.A. Bull. 1497:1–87.

Krefting, L. W., and Stoeckeler, J. H. 1953. Effect of simulated snowshoe hare and deer damage on planted conifers in the Lake States. J. Wildl. Manage. 17(4):487–494.

Lord, R. D., Jr. 1959. The lens as an indicator of age in cottontail rabbits. J. Wildl. Manage. 23(3):358–360.

Manniche, A. L. V. 1910. The terrestrial mammals and birds of northeast Greenland. Medd. om Gronl. 45:1–200.

Manning, T. H., and Macpherson, A. H. 1958. The mammals of Banks Island. Arctic Inst. North Am. Tech. Pap. 2:1–74.

Manweiler, J. 1938. Parasites of the snowshoe hare. J. Mammal. 19:379.

Meslow, E. C., and Keith, L. B. 1968. Demographic parameters of a snowshoe hare population. J. Wildl. Manage. 32(4):812–834.

———. 1971. A correlation of weather versus snowshoe hare population parameters. J. Wildl. Manage. 35(1):1–15.

Morse, M. 1939. A local study of predation upon hares and grouse during the cyclic decimation. J. Wildl. Manage. 3(3):203–211.

Nelson, E. W. 1909. The rabbits of North America. North Am. Fauna 29. 314pp.

Newson, J. 1964. Reproduction and prenatal mortality of snowshoe hares on Manitoulin Island, Ontario. Can. J. Zool. 42(6):987–1005.

Newson, R., and DeVos, A. 1964. Population structure and body weights of snowshoe hares on Manitoulin Island, Ontario. Can. J. Zool. 42(6):975–986.

O'Farrell, T. P. 1965. Home range and ecology of snowshoe hares in interior Alaska. J. Mammal. 46(3):406–418.

Orr, R. T. 1933. A new race of snowshoe rabbit from California. J. Mammal. 14:54–56.

———. 1934. Description of a new snowshoe rabbit from eastern Oregon, with notes on its life history. J. Mammal. 15:152–154.

Parker, G. R. 1977. Morphology, reproduction, diet, and behavior of the Arctic hare (*Lepus arcticus monstrabilis*) on Axel Heiberg Island, Northwest Territories. Can. Field Nat. 91(1):8–18.

Pease, J. L.; Vowles, R. H.; and Keith, L. B. 1979. Interaction of snowshoe hares and woody vegetation. J. Wildl. Manage. 43(1):43–60.

Penner, L. R. 1954. *Dirofilaria scapiceps* from snowshoe hare in Connecticut. J. Mammal. 35(2):458–459.

Philip, C. B. 1938. A parasitological reconnaissance in Alaska with particular reference to varying hares. J. Parasitol. 24:483–488.

Radwan, M. A., and Campbell, D. L. 1968. Snowshoe hare preference for spotted catsear flowers in western Washington. J. Wildl. Manage. 32(1):104–108.

Renolds, J. K., and Stinson, R. H. 1959. Reproduction in the European hare in southern Ontario. Can. J. Zool. 37 (5):627–631.

Richmond, M. E., and Chien, C.-Y. 1976. Status of the snowshoe hare on the Connecticut Hill Wildlife Management Area. New York Fish Game J. 23(1):1–12.

Roe, E. I., and Stoeckler, J. H. 1950. Thinning overdense jackpine seedling in the Lake States. J. For. 48:861–865.

Rongstad, O. J. 1965. Short-tailed shrew attacks young snowshoe hare. J. Mammal. 46(2):328–329.

Rongstad, O. J., and Tester, J. R. 1971. Behavior and maternal relations of young snowshoe hares. J. Wildl. Manage. 35(2):338–346.

Rowan, W., and Keith, L. B. 1956. Reproductive potential and sex ratios of snowshoe hares in northern Alberta. Can. J. Zool. 34(4):273–281.

———. 1959. Monthly weights of snowshoe hares from north-central Alberta. J. Mammal. 40(2):221–226.

Rudolf, P. O. 1950. Forest plantations in the Lake States. U.S.D.A. Tech. Bull. 1010:1–171.

Seton, E. T. 1928. Lives of Game Animals. Doubleday, Doron & Co., Inc., Garden City, N.Y. 4:700–735.

Severaid, J. H. 1942. The snowshoe hare, its life history and artificial propagation. Maine Dept. Inland Fish Game. 95pp.

———. 1945. Breeding potential and artificial propagation of the snowshoe hare. J. Wildl. Manage. 9(4):290–295.

Soper, J. D. 1928. A faunal investigation of southern Baffin Island. Natl. Mus. Can. Bull. 53:1–143.

Szukiel, E. 1973. The effect of repellents on the food preferences of hares. Acta Theriol. 18(26):481–488.

Trapp, G. R., and Trapp, C. 1965. Another vocal sound made by snowshoe hares. J. Mammal. 46(4):705.

Vaughan, M. R., and Keith, L. B. 1980. Breeding by juvenile snowshoe hares. J. Wildl. Manage. 44(4):948–951.

———. 1981. Demographic response of experimental snowshoe hare populations to overwinter food shortage. J. Wildl. Manage. 45(2):354–380.

Walkinshaw, L. H. 1947. Notes on the Arctic hare. J. Mammal. 28(4):353–375.

Wang, L. C. H.; Jones, D. L.; MacArthur, R. A.; and Fuller, W. A. 1973. Adaptation to cold: energy metabolism in an atypical lagomorph, the arctic hare (*Lepus arcticus*). Can. J. Zool. 51(8):841–846.

Webb, W. L. 1937. Notes on the sex ratio of the snowshoe rabbit. J. Mammal. 18:343–347.

Windberg, L. A., and Keith, L. B. 1976a. Snowshoe hare population response to artificial high densities. J. Mammal. 57(3):523–553.

———. 1976b. Experimental analyses of dispersal in snowshoe hare populations. Can. J. Zool. 54(12):2061–2081.

STEVEN L. BITTNER, Appalachian Environmental Laboratory, Center for Environmental and Estuarine Studies, University of Maryland, Frostburg State College Campus, Frostburg, Maryland 21532. Current address: Indian Springs Wildlife Management Area, Maryland Wildlife Administration, Route 1, Box 118, Big Pool, Maryland 21711.

ORRIN J. RONGSTAD, Department of Wildlife Ecology, University of Wisconsin, Madison, Wisconsin 53706.

III

Rodentia

8

Mountain Beaver

Aplodontia rufa

George A. Feldhamer
James A. Rochelle

NOMENCLATURE

COMMON NAMES. Mountain beaver, boomer, sewellel, whistler, chehalis, mountain rat
SCIENTIFIC NAME. *Aplodontia rufa*
SUBSPECIES. *A. r. rufa, A. r. californica, A. r. pacifica, A. r. phaea, A. r. rainieri, A. r. nigra,* and *A. r. humboldtiana.*

The mountain beaver is the only living species of the family Aplodontidae. Because of its zygomasseteric structure, it is generally regarded as the most primitive member of the order Rodentia. Its geologic range extends from the late Eocene to Recent (McLaughlin 1967).

DISTRIBUTION

The present distribution of the mountain beaver is confined exclusively to western North America, and extends from extreme southern British Columbia south to central California and east to the Cascade Mountains and the Sierra Nevada (figure 8.1). Although it may range to elevations of 2,200 m, the mountain beaver is more commonly found in humid, densely vegetated areas at lower elevations (Walker et al. 1975). Limits to distribution are associated with rainfall and edaphic conditions that promote succulent vegetation and relatively high humidity within burrows (Voth 1968).

DESCRIPTION

The general body conformation is compact, thickset, and cylindrical. Total length of mature individuals ranges from about 300 to 500 mm, which includes a stumpy, fully furred tail 10 to 25 mm in length. Forelimbs and hind limbs are pentadactyl and of approximately equal length. The pollex is opposable and prehensile. Mean adult weights vary from 0.8 to 1.8 kg (Hall and Kelson 1959; Lovejoy and Black 1974).

Pelage. In adults, the dark blackish brown coloration of the fur is similar in both sexes. It is coarse, dense, and short with numerous interspersed guard hairs. The guard hairs on the dorsum and limbs appear blackish,

FIGURE 8.1. Distribution of the mountain beaver (*Aplodontia rufa*).

whereas those on the ventral and lateral surfaces have a lighter appearance. Juvenile animals have fine, gray fur with numerous white-tipped hairs (Taylor 1918). Yearlings may retain some of their typical juvenile pelage, but generally have attained adult pelage (Lovejoy and Black 1974). All mountain beavers have a small white spot at the base of the ear. Albino and melanistic animals, although apparently rare, have been described (Godin 1964).

During all seasons, females have a nearly circular area of brownish black hair about 15 mm in diameter that surrounds each of the three pairs of mammae. These patches of hair contrast with the generally lighter coloration of the ventral surface. They are not apparent in males.

Mountain beavers undergo a single annual molt, which generally begins in July or August and continues for two to three months. Pelage replacement begins on the anterior dorsal surface and posterior lateral surfaces. The ventral area is the last to complete the molt (Taylor 1918).

167

Skull and Dentition. The skull of the adult mountain beaver is unusually broad and flattened in appearance. Characteristic features include flask-shaped auditory bullae, a palate that extends posteriorly to the third upper molar, and a mandible with a greatly inflected angular process and a relatively high coronoid process (figure 8.2). In his revision of the genus Aplodontia, Taylor (1918) noted "considerable" variation in the size of adult skulls from the same locality.

The cheek teeth are modified hypsodont, rootless (ever growing), and prismatic. The anteriormost upper premolar (P3) is small and peglike. The remaining cheek teeth are set obliquely and decrease in size from the second premolar (P4) to the last molar. With the exception of P3, each cheek tooth has a unique, spinelike projection; these projections occur on the labial side of the upper cheek teeth and the lingual side of the lower cheek teeth (figure 2). The dental formula is 1/1, 0/0, 2/1, 3/3 = 22.

PHYSIOLOGY

Body Weight and Temperature. The mean body weight of male mountain beavers in Oregon was significantly greater than that of females (Lovejoy and Black 1974). Adult body weights of both sexes vary seasonally. The lowest generally occur during the reproductive season (January through March) and the spring, whereas the highest body weights occur in the summer, when forage is most abundant (Lovejoy 1972).

Johnson (1971) found that the mean body temperature of free-ranging animals was 38° C, and reported hyperthermia at ambient temperatures above 29° C. A lethal body temperature of 42° C resulted from two hours of exposure to ambient temperatures between 32° and 35° C. The apparent inability of the mountain beaver physiologically to thermoregulate to minimize heat stress was suggested as a factor contributing to the restriction of its geographic distribution (Johnson 1971).

The morphology of the cutaneus maximus muscle of several species of rodent, including the mountain beaver, was described by Woods and Howland (1977). In the primitive condition exemplified by *A. rufa,* the origin of this five-parted muscle is on the humerus bone, while the insertion is on the trunk (that is, skin). Secondary insertion also occurs on one or more caudal vertebrae.

Renal. Because it is the most primitive living rodent, the renal physiology of the mountain beaver has been thoroughly investigated. The structure of the kidney itself is very primitive. Pfeiffer (1968) noted that no extension of the renal pelvis penetrated between the outer zone of the medulla and the cortex. Seventy percent of the nephrons lay entirely in the cortex. Long-looped nephrons apparently are minimal (Schmidt-Neilsen and Pfeiffer 1970; Pfeiffer 1968), and those present do not have a thin segment (Pfeiffer et al. 1960). Thus, the ability to concentrate urine is limited and the mountain beaver is one of the least efficient

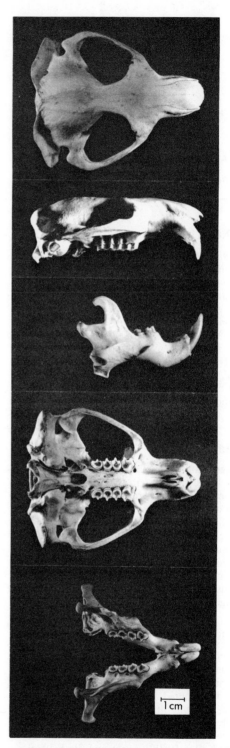

FIGURE 8.2. Skull of the mountain beaver (*Aplodontia rufa*). From top to bottom: lateral view of cranium, lateral view of mandible, dorsal view of cranium, ventral view of cranium, dorsal view of mandible.

mammals in its ability to conserve water (Dicker and Eggleton 1964). Although contradictory observations have been made (Fisler 1965), apparently the mountain beaver must consume considerable amounts of water, possibly in the form of succulent vegetation, for its well-being. This apparent requirement also may act to limit its geographic distribution to regions of abundant rainfall and vegetation.

Neural. Merzenich et al. (1973), in an examination of brain stem auditory nuclei of 106 mammalian species, found that the mountain beaver had "a very large and unique cochlear nuclear complex" four to seven times larger than those of the other 17 species of rodents investigated. They suggested that this system is specialized to detect slow changes in air pressure, which they believed would be of value to a tunnel-dwelling species. No similar specialization of this complex was found in other fossorial species, however, including pocket gophers (Geomyidae) or moles (Talpidae) (Merzenich et al. 1973). Although members of these families all inhabit tunnels, it may be noted that gophers and moles plug all surface openings whereas the mountain beaver does not. Also, only the mountain beaver is active aboveground for extended periods. These factors may be associated with this particular neural specialization (see the section "Activity").

GENETICS

McMillin and Sutton (1972) prepared metaphase chromosome karyograms of five *A. r. californica* and four *A. r. phaea*. Karyotypes were identical for both subspecies. The diploid number (2 N) was 46 chromosomes; 6 pairs were metacentric and 16 pairs were submetacentric. The Y chromosome of males was also submetacentric. Although the species is primitive in relation to other living rodents, these investigators suggested that because their 2 N number was not large, mountain beavers are "advanced within (their) own lineage."

The α and β subunits of the enzyme lactate dehydrogenase (LDH) were investigated electrophoretically for 34 species of rodents, including mountain beaver, by Baur and Pattie (1968). Considering blood samples, mountain beavers possessed both subunits, as did 11 species of hystricomorph rodents tested. Many myomorph rodents, apparently, genetically suppress the expression of the β subunit.

REPRODUCTION

Anatomy. The ovary of the mountain beaver is elliptical, and in anestrous females no follicular development is seen macroscopically. The ovaries of females in estrus, however, have follicles up to 3 mm in diameter, and the corpora lutea are reddish in color (Pfeiffer 1958). Externally, parous females may be distinguished from nulliparous females by their long, pendantlike nipples. The uteri of parous animals are characterized by their compressed, ribbonlike, more vascularized appearance and are much larger than the threadlike uteri of nulliparous individuals. After parturition, placental scars 1 cm in diameter are visible. These gradually regress until early fall, when they "become obscured by gradual hypertrophy and hyperplasia of the uterus as estrous approaches" (Pfeiffer 1958).

The male reproductive tract was described by Pfeiffer (1956), and closely resembles that of certain sciurids. During the breeding season, the testes may be semiscrotal. They are abdominal during the remainder of the year. The testes and accessory sex organs fluctuate in size throughout the year, and attain their maximum size during the breeding season. For the remainder of the year, seminiferous tubules are much reduced in size and relatively few spermatogonia and primary spermatocytes are present.

Breeding. In Oregon, the testes of mountain beavers descend in late December or early January, and males remain in breeding condition through late March. Females are monestrous, and come into estrus about mid-February (Lovejoy 1972; Lovejoy and Black 1974). They apparently remain in breeding condition for a relatively short duration. Only animals two years old and older form corpora lutea of pregnancy (McLaughlin 1967). Although estrus may occur in yearling females, they apparently do not conceive.

Parturition. Parturition generally occurs from late March to early April, after a gestation period of 28 to 30 days. Cramblet and Ridenhour (1956) observed the process in a captive mountain beaver. For 1.5 hours before parturition, the animal periodically gnashed its teeth while pressing its forefeet against the genital region. Immediately before expulsion, it sat on its haunches and lowered its head between its legs. The first fetus was a caudal presentation that it assisted by pressing its forefoot around the vulva and licking this area. Two fetuses followed, one a cephalic presentation, and the actual birth process was completed in about 33 minutes.

Newborn. One litter per year is produced. It may range in size from two to six; two or three are most common. The neonates are altricial; vibrissae and pinnae are seen, although the latter are not extended, and the tail is apparent. Nails develop by 5 days of age, and fine, downy hairs by 1 week. Incisors developed by about 1 month of age (Hooven 1977). The eyes are closed during the first 10 days. The average body length of three neonates described by Cramblet and Ridenhour (1956) was 87.7 mm and body weight was 27.0 g. The weights of 2-day-old young born in captivity in Oregon were 18 and 22 g. Mean weight was 110 g at 1 month of age and 347 g at 2 months. Young mountain beavers grow rapidly. They attain approximately 40 percent of the mean adult body weight during April and May when adult females are lactating. The young are weaned at 2 months of age and emerge from the burrow within the next 2 weeks. By 4 months of age, they reach close to 70 percent of the mean adult body weight. Growth rate eventually declines, and yearling animals attain about 88 percent of the mean adult weight (Lovejoy 1972; Lovejoy and Black 1974).

ECOLOGY

Mountain beavers inhabit densely vegetated areas of high annual precipitation, and commonly are found in the initial seral stages after clearcutting of forest areas. They are not gregarious, and concentrations are a function of suitable habitat rather than sociability of the species. Lovejoy (1972) estimated that population density varied from 6.7 to 8.7 per hectare (ha) on a logged area in the Coast Range of Oregon. Voth (1968) estimated 41 animals per ha in a similar area. He felt this probably was an overestimation, however, despite the habitat's being "among the best available." Hooven (1977) felt that population densities on areas with new growth may increase from less than 1 per ha to 15–20 per ha.

During a two-year period, Lovejoy and Black (1974, 1979) trapped 1.6:1.0 adult males to females. This may have been the result of trapping bias, however, as the sex ratio of 72 juveniles from three successive breeding seasons was 1:1. Extensive trapping over a five-year period in the Pacific Northwest resulted in an adult sex ratio of 60 percent males:40 percent females, however (Borrecco and Anderson 1980).

Activity. Mountain beavers do not hibernate, and, although primarily nocturnal, they may be active at any hour of the day. Ingles (1959) reported that they had six or seven periods of activity every 24 hours during the summer. These varied in duration up to 2.8 hours. Also, 50 to 60 percent more activity occurred at night, at least during the summer. Increased activity may be stimulated by one or a combination of environmental factors, such as changes in light intensity, temperature, relative humidity, or air movement. Increased nocturnal activity probably did not result from an attempt to minimize potential predation, as the two most common predators in the area, coyotes (*Canis latrans*) and great horned owls (*Bubo virginianus*), were themselves most active at night (Ingles 1969). Burrowing activity is most prevalent during the summer. Aboveground activity essentially ceases during late fall and winter.

Home Range and Movements. Martin (1971) found that the home ranges of adult mountain beavers varied in size from 0.03 to 0.2 ha, with no apparent difference in mean ranges of males and females. Other home range values reported are comparably restricted and within the expected range of variation caused by differences in population density, habitat, and other factors. Lovejoy (1972) found that adult males had an average home range of 0.3 ha, adult females about 0.2 ha, and juveniles about 0.1 ha. Although individual home ranges may overlap, mountain beavers defend their nest sites and associated burrows.

Dispersing subadults apparently make extensive linear movements. In an effort to establish permanent home ranges, they follow existing burrows as well as traveling aboveground. Martin (1971) reported that after subadults establish nest sites, their movements are comparable to those of adults. Mountain beavers apparently are faithful to their site selection. Two males remained in the vicinity of their nests for at least 31 and 44 months, respectively, and over 90 percent of the telemetry-determined positions recorded by Martin (1971) occurred within 24 m of the nest site.

Burrow System. Mountain beavers burrow in moderately firm soil where drainage is adequate. They dig by scooping with the forefeet and pushing the soil underneath the body. The incisors also may be used to loosen packed soil and stones. They push the soil toward an opening with the head and shoulders, and ultimately expel it from the burrow by scraping with the hind feet (Voth 1968). The diameter of a tunnel may range between 10 and 20 cm (Camp 1918), depending on the size of the animal and the texture of the soil. Tunnels are seldom more than 120 cm below the surface. Deep tunnels lead to or connect chambers; shallow tunnels, above 25 cm below the surface, lead to burrow openings. An individual network of tunnel systems may be 100 m in diameter, with an opening every 6 or 7 m.

The microclimate within burrow systems is cool and stable. The maximum annual fluctuation in burrow temperature is only about 45 percent of the fluctuation in ambient temperature, and Johnson (1971) found that weekly temperature variation never exceeded 4° C within the burrows.

Within the burrow system, Voth (1968) described five types of chamber: nest, feeding, refuse, fecal pellet, and earth ball. The nest chamber, often 50 to 60 cm in diameter and 36 cm high, usually is located at sites where drainage is good, and, together with the feeding chamber, forms a hub from which other tunnels and chambers radiate (Martin 1971). Nests are roughly circular in shape. Those of adults contain as much as a bushel of vegetative material, whereas those of subadults are usually smaller. Where available, salal (*Gaultheria shallon*) often makes up the soft, dry central portion of the nest. Sword fern (*Polystichum munitum*), bracken fern (*Pteridium aquilinum*), and other readily available vegetative material are common components of the coarser, sometimes moist outer shell (Johnson and Martin 1969; Martin 1971).

Feeding chambers are adjacent to the nest chamber and are often as large or larger in size. Besides feeding, mountain beavers use these areas to store large caches of both wilted and recently cut vegetation. Decayed plant material from the feeding areas is placed in the refuse chambers. These are sometimes only blind tunnels, occasionally no larger than 12 cm in diameter, that open from feeding chambers.

Fecal pellet chambers also may open from feeding chambers, or pellets may be deposited in a portion of a feeding chamber. Voth (1968) reported that the mean diameter of pellet chambers was 18 cm. He found as many as six pellet chambers associated with a single burrow system, each containing fecal pellets in a different state of decomposition.

The earth ball chambers are storage areas for stones or compacted dirt retained by the animals. Voth (1968) felt that these ''balls,'' which averaged 8 cm in diameter and 200 g in weight, were used to plug entrances to nest and feeding chambers. Because the diet of mountain beavers consists mostly of succulent vegetation, the earth balls also may serve as abrasives on which they trim their incisors.

FOOD HABITS

The mountain beaver has a functional cecum that, together with the stomach, provides ample room for large volumes of ingested vegetation. The small intestine averages 1.3 m in length and the large intestine 0.9 m. In free-ranging individuals, the entire digestive tract may account for 25 to 50 percent of the body weight (Voth 1968). Nonetheless, because of the relatively low energy content of its food, at least 75 percent of the activity time of this species is spent gathering and ingesting food (Ingles 1959). Thus, mountain beavers harvest substantial quantities of vegetation, much of which decays before it is eaten. Voth (1968) felt that 2.5 times as much vegetation was cut by this species as was actually eaten. However, a portion of this may be used as nesting material.

The mountain beaver is strictly herbivorous. Sword fern was the main component of the diet of males during winter and spring in the population studied by Voth (1968). Lactating females incorporated conifers and high-protein grasses in the spring. During the summer and fall, bracken fern was the preferred food item for both sexes. The diet of juveniles after weaning is similar to that of adults. The general categories of vegetation in the diets of adult males, females, and juveniles are given in table 8.1.

Although pteridophytes comprise the preferred food species throughout the year, during winter in areas of heavy snowfall, bark and twigs are readily taken. Shrubby vegetation also may be used during the summer months. Crouch (1968) found heavy clipping by mountain beavers on vine maple (*Acer circinatum*),

red huckleberry (*Vaccinium parvifolium*), and red alder (*Alnus rubra*). Five other species of hardwood received moderate to light utlization. These included big-leaf maple (*Acer macrophyllum*), cascara (*Rhamnus purshiana*), hazel (*Corylus cornuta*), ocean spray (*Holodiscus discolor*), and willow (*Salix* sp.). The remaining species of shrub that were common on the study area showed no evidence of use. These included rose (*Rosa gymnocarpa*), cherry (*Prunus emarginata*), currant (*Ribes sanguineum*), and red-stem ceanothus (*Ceanothus sanguineus*). In his review of the earlier literature, Godin (1964:32) listed 26 additional plant species, including cherry, reportedly eaten by mountain beavers, although he did not cite the original literature.

Free-ranging mountain beavers drink water when it is available. However, the species apparently is easily maintained in captivity on a diet of succulent vegetation in the absence of free-flowing water (Fisler 1965).

BEHAVIOR

Foraging and Feeding. The burrows of mountain beavers often open directly into suitable vegetation and the animals move only short distances while foraging. Although they may consume vegetation while aboveground, forage is usually carried or dragged into the burrow, where it is eaten or stored (Martin 1971). The animals may climb small trees and shrubs while foraging, clipping branches as they ascend. They may climb to a height of 7 m. Descent is made head first (Ingles 1960), or by simply releasing the grip and falling to the ground.

While feeding, the animal may sit on its haunches with the hind legs extended in front. Fisler (1965) observed this posture in a captive mountain beaver and noted that food was manipulated mainly with the forefeet, only rarely with the hind feet.

The phenomenon of ''haymaking'' has been noted by numerous early observers (Godin 1964:17). Mountain beavers stack fresh vegetation near burrow

TABLE 8.1. General categories of food items in the diet of mountain beavers as determined from counts of epidermal fragments from fecal pellets

Vegetation Category	Males and Nonpregnant Females (N = 12)	Females	
		Lactating (N = 3)	Juveniles (N = 4)
Pteridophytes (ferns)	84.0	37.7	90.7
Conifers	3.4	33.9	0.0
Grasses	2.5	18.4	4.6
Forbs	1.9	4.8	2.6
Hardwoods	5.4	1.3	1.3
Mosses	1.0	3.5	0.9
Shrubs	1.1	0.0	0.0

SOURCE: Adapted from Voth 1968

[a]Numbers given are percentages of total for each age and sex group.

entrances. These piles may be as deep as 60 cm. Voth (1968) suggested possible explanations for "haymaking" behavior. He felt that it may be related to improved succulence or nutritional quality of the vegetation. Also, it may involve the psychological well-being of the animal by reducing the number of times the nest and feeding chambers are opened to bring in vegetation.

The mountain beaver reingests certain fecal pellets. As it defecates, the animal takes each pellet in its mouth and tosses it onto the fecal pile. Occasionally, discarded pellets are retrieved and chewed or eaten (Ingles 1961). Such reingestion (coprophagy) has been documented in several other species of mammal.

Other Behaviors. Although the hind feet are not webbed, mountain beavers enter pools or streams without hesitation and are excellent swimmers. Walker et al. (1975) stated that the animals will swim through flooded tunnels.

Grooming apparently is minimal. Fisler (1965) reported that a captive mountain beaver groomed infrequently, using its teeth and front claws, for periods of two minutes or less. He felt that the species was hampered in this behavior by the inflexibility of the neck and forelegs.

Although generally silent, several types of vocalization have been attributed to mountain beavers. These include growls, whistles, whines, "coughs," and squeals.

MORTALITY

Apparently, little is known concerning mortality rates of mountain beavers, although numerous predators have been reported. Lovejoy (1972) estimated that their longevity in the wild was five to six years.

Approximately 51 percent of the mountain beavers in a 7-ha area were killed by a controlled broadcast burn that left little residual slash. On a similar 9-ha area that had 1.8 ha of slash and brush left unburned, only 20 percent of the resident population was killed (Motubu 1978).

Predators. Predators cited by previous authors include most of the carnivores—although documentation is often lacking and there is no determination of the effect of predation on population densities of mountain beavers. Taylor (1918) believed that skunks (*Mephitis mephitis, Spilogale putorius*), mink (*Mustela vison*), gray fox (*Urocyon cinereoargentius*), raccoon (*Procyon lotor*), badger (*Taxidea taxus*), bobcat (*Felis rufus*), and fisher (*Martes pennanti*) preyed on mountain beavers. Other investigators have included cougar (*Felis concolor*) and long-tailed weasels (*Mustela frenata*), which may prey on young mountain beavers (Godin 1964). As noted previously, coyotes and great horned owls also prey on mountain beavers. Sweeney (1978) observed the remains of mountain beaver in bobcat scats from western Washington. The skeletal remains of 55 individual mountain beavers were found in a golden eagle (*Aquila chrysaetos*) nest in western Washington (A. M. Bruce personal communication).

Parasites. Eighteen species of ectoparasites and one species of endoparasite from mountain beavers have been described (Godin 1964). Ectoparasites included one species of mammal-nest beetle (Coleoptera), nine species of fleas (Siphonaptera), six species of mites (Acarina), and two species of ticks (Acarina). The only endoparasite reported was the tapeworm *Taenia tenuicollis*. It would be unusual if future research did not identify additional helminth parasites of mountain beavers.

The larval stage of a moth, *Amydria effrentella*, was found to be a common inhabitant of mountain beaver nests (Johnson and Martin 1969). Voth (1968) stated that generally the nests of males were heavily parasitized but those of females were not. However, of five nests investigated, belonging to both males and females, J. E. Borrecco (personal communication) found that each contained "many" ectoparasites.

AGE DETERMINATION

As noted previously, pelage characteristics may be used to differentiate juveniles and adults. Pfeiffer (1958) estimated the age of immature mountain beavers on the basis of body weight, using growth curves of known-age juveniles as a standard. However, Lovejoy (1972) found this method impractical for mature animals because of the extreme variation within age classes.

In males, the length of the baculum may be used to determine age. Pfeiffer (1956) found that the mean length of the baculum in juveniles was about 11 mm, whereas it was 21 mm in yearlings and about 30 mm in adults. He did not indicate the amount of variation within age classes, however. For both males and females, the degree of closure of the epiphyseal femoral suture also is considered a "fairly reliable" method of age determination in this species (Pfeiffer 1958). He also noted that the third upper molar is rooted, and used the degree of wear on m3 as a supplemental determinant of age.

ECONOMIC STATUS AND MANAGEMENT

The mountain beaver is considered neither a game animal nor a furbearer. Its meat is not sought after and even prime pelts sell for only about 20 cents (Ingles 1965). The economic importance of the species involves the extensive damage it does to conifer seedlings and saplings and the resulting losses incurred by the forest products industry.

Interference with forest management, and particularly with reforestation efforts, by mountain beavers has been documented for some time (Couch 1925; Scheffer 1929, 1952; Staebler et al. 1954). Following a recent survey of forest landowners and managers in Washington, Oregon, and northern California, Borrecco et al. (1979) reported that damage by mountain beavers was occurring on about 111,000 ha of forest land. About 75 percent of this damage occurred in Douglas fir (*Pseudotsuga menziesii*) forest types; western hemlock (*Tsuga heterophylla*) and several other conifer species were also damaged.

Mountain beaver damage to conifers takes several forms, depending on tree size. Burrowing activity may occasionally uproot or bury seedlings (Voth 1968) but feeding activities are the primary cause of injury (Hooven 1977; Black et al. 1979). Three primary categories of mountain beaver-caused injury to conifers were described by Lawrence et al. (1961): (1) stem clippings or cutting of seedlings, (2) branch cutting, and (3) basal girdling (removal of bark). Mortality, suppression of growth, and deformity of trees result from injury. Although damage may occur to trees from less than 1 to over 20 years of age, the most common (Borrecco et al. 1979) and most serious form of damage is clipping of small seedlings. This damage, which occurs from immediately after planting to up to 4 years postplanting, often results in seedling mortality and subsequent understocking or failure of plantations. Clipping or basal girdling of larger seedlings does not often result in mortality but causes deformities and growth losses. Borrecco and Anderson (1980) displayed the importance of seedling size to severity and extent of damage. From a 6,000-tree sample taken in 24 randomly selected plantations in western Washington, they observed that, 2 years after planting, of those trees sustaining damage from mountain beavers, 2-0 seedlings had sustained 53 percent mortality and 2-1 seedlings had sustained 36 percent mortality. (Seedlings designated 2-0 were grown in nursery seed beds for 2 years at high densities. Seedlings designated 2-1 were also in seed beds at high densities for 2 years and then transplanted and grown for another year at wider spacing prior to outplanting.) Mortality to 2-1 seedlings occurred only in the first year after planting, while 2-0 seedlings were killed over the 2-year period. The impact on height growth of surviving damaged seedlings was greater on 2-0 than on 2-1 seedlings (figure 8.3), and when damage occurred in the first year rather than the second year after planting (figure 8.4). The relatively greater impact of mountain beaver damage on height growth of Douglas fir seedlings compared to clipping by snowshoe hare (*Lepus americanus*) and browsing by deer (*Odocoileus* spp.) is also shown in figure 8.3.

The economic impact of mountain beaver and other types of wildlife damage to forest crops have been difficult to determine, primarily because of the lack of information on long-term growth effects of damage occurring to trees at juvenile growth stages. Lawrence (1959) estimated wildlife-caused losses on Weyerhaeuser Company lands in the Pacific Northwest at about $900,000 per year; a substantial portion of this total was due to mountain beaver. More recently, Weyerhaeuser Company estimated annual mountain beaver-caused losses at over $1 million on company lands. Black et al. (1979) presented information on long-term growth effects of wildlife damage to conifer plantations, and Brodie et al. (1979) estimated the economic impacts of these damage levels, but did not separate out the effect of mountain beavers in particular.

Control. Because mountain beavers are considered pests from the economic standpoint, management gen-

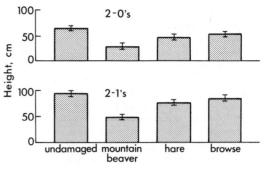

FIGURE 8.3. Mean heights and standard errors of damaged and undamaged 2-0 and 2-1 seedlings after two years in the field.

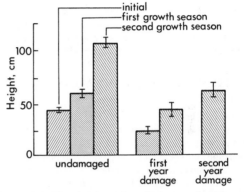

FIGURE 8.4. Mean heights and standard errors of undamaged Douglas fir 2-1 nursery stock compared to stock damaged in the first and second years after planting.

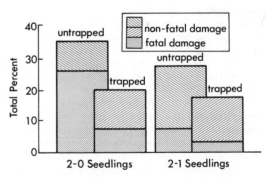

FIGURE 8.5. Levels of mountain beaver (*Aplodontia rufa*)-caused mortality and nonfatal damage on 2-0 and 2-1 nursery stock in trapped and untrapped areas two years after planting. Stippling indicates mortality; cross-hatching indicates nonfatal damage.

erally has involved the application of various methods of damage control. At the present time there are no toxicants registered for mountain beaver control. The repellent Thiram, which successfully prevents clipping by snowshoe hares, provides little or no protection from mountain beaver clipping. Trapping with Conibear no. 110 traps is the most common method used to reduce populations. It effectively reduces damage (figure 8.5). Trained crews place traps in main runways of burrow systems at rates of 20 or more traps per ha depending on mountain beaver density. Trapping costs average $75 to $100 per ha under good conditions, that is, moderate mountain beaver densities, moderate topography, and relatively slash-free sites (Borrecco and Anderson 1980). The high labor requirement for trapping, which competes for manpower with other forestry activities, and the associated high costs are the major disadvantages of trapping. Recently, tubes of polypropylene plastic netting placed around individual seedlings have been shown to reduce damage caused by several wildlife species (Campbell and Evans 1975). In studies evaluating these tubes, Borrecco and Anderson (1980) observed mountain beaver damage levels exceeding 44 percent for untubed trees compared to 3 percent for tubed trees. Cost is the major factor limiting widespread use of plastic netting. Recent Weyerhaeuser Company experience indicates costs of 30 to 40 cents per tree, or $375 or more per ha, depending on plantation stocking levels. Research efforts to reduce these costs are currently under way.

CURRENT RESEARCH AND MANAGEMENT NEEDS

Improved methods of controlling mountain beaver damage to conifer plantations are needed. Approaches that emphasize indirect control, rather than population reduction, will be better accepted by the public in general and have greater potential for authorization by regulatory agencies. Although chemical repellents have been examined with little success in the past, availability of a nursery-applied material to provide first-year protection in the field would be of major value. Novel approaches, such as the use of pheromones to modify behavior, might ultimately provide effective damage control techniques but require major basic research commitments. Unique physiological characteristics of mountain beavers, such as their inability to concentrate urine and thus the requirement for highly succulent forage, might be exploited as to possibilities for chemical interference with controlling mechanisms. An improved understanding of the digestive processes and nutritional requirements of the mountain beaver might provide insights into new control approaches. Although reproductive inhibitors have proven relatively unsuccessful in most rodent control applications, the single litter and relatively small litter size of mountain beavers suggest a possible control opportunity. An improved understanding of population dynamics, and particularly natural mortality factors, is needed. Voth (1968) suggested modification of plant communities to exclude preferred plant species through the introduction of insects or the use of herbicides. These are possible applied research topics. Presently, the application of integrated approaches, including vegetative manipulation through site preparation, selection of appropriately sized planting stock, attention to timing of trapping in relation to plantation establishment, and careful and regular surveys of plantation areas prior to and after planting, should continue to provide acceptable levels of damage control.

An early draft of this chapter was reviewed by H. C. Black, Department of Forest Science, Oregon State University. This is contribution no. 1001-AEL, Center for Environmental and Estuarine Studies, University of Maryland.

LITERATURE CITED

Bauer, E. W., and Pattie, D. L. 1968. Lactate dehydrogenase genes in rodents. Nature 218:341–343.

Black, H. C.; Dimock, E. J.; Evans, J. A.; and Rochelle, J. A. 1979. Animal damage to coniferous plantations in Oregon and Washington. Part 1: A survey, 1963–1975. Oregon State Univ., For. Res. Lab., Corvallis. Res. Bull. 25. 44pp.

Borrecco, J. E. 1976. "Vexar" tubing as a means to protect seedlings from wildlife damage. Weyerhaeuser Co., For. Res. Tech. Rep. n.p.

Borrecco, J. E.; Anderson, H. W.; Black, H. C.; Evans, J.; Guenther, K. S.; Lindsey, G. D.; Matthews, R. P.; and Moore, T. K. 1979. Survey of mountain beaver damage to forests in the Pacific Northwest, 1977. Dept. Nat. Res., Olympia, Wash. DNR Note 26. 16pp.

Borrecco, J. E., and Anderson, R. J. 1980. Mountain beaver problems in the forests of California, Oregon and Washington. Pages 135–142 in 9th Vertebr. Pest Conf., March 1980, Fresno, Calif.

Brodie, D.; Black, H. C.; Dimock, E. J., II; Evans, J; Kao, C.; and Rochelle, J. A. 1979. Animal damage to coniferous plantations in Oregon and Washington. Part 2: An economic evaluation, 1979. Oregon State Univ., For. Res. Lab., Corvallis. Res. Bull. 26. 22pp.

Camp, D. L. 1918. Excavations of burrows of the rodent *Aplodontia,* with observations on habitat of the animal. Univ. Cal. Publ. Zool. 17:517–536.

Campbell, D. L., and Evans, J. 1975. "Vexar" seedling protectors to reduce wildlife damage to Douglas fir. U.S.D.I. Wildl. Leafl. 508. 11pp.

Cramblet, H. M., and Ridenhour, R. L. 1956. Parturition in *Aplodonita.* J. Mammal. 37:87–90.

Couch, L. K. 1925. Rodent damage to young forests. Murrelet 6:39.

Crouch, G. L. 1968. Clipping of woody plants by mountain beaver. J. Mammal. 49:151–152.

Dicker, S. E., and Eggleton, M. G. 1964. Renal function in the primitive mammal, *Aplodontia rufa,* with some observations on squirrels. J. Physiol. 170:186–194.

Fisler, G. F. 1965. A captive mountain beaver. J. Mammal. 46:707–709.

Godin, A. J. 1964. A review of the literature on the mountain beaver. U.S. Fish Wildl. Serv. Spec. Sci. Rep. Wildl. 78. 52pp.

Hall, E. R., and Kelson, K. R. 1959. The mammals of North America. Ronald Press Co., New York. 2 vols.

Hooven, E. F. 1977. The mountain beaver in Oregon: its life history and control. Oregon State Univ., For. Res. Lab., Corvallis. Res. Pap. 30. 20pp.

Ingles, L. G. 1959. A quantitative study of mountain beaver activity. Am. Midl. Nat. 61:419–423.

———. 1960. Tree climbing beavers. J. Mammal. 41:120–121.

———. 1961. Reingestion in the mountain beaver. J. Mammal. 42:411–412.

———. 1965. Mammals of the Pacific states. Stanford Univ. Press, Stanford, Calif. 506pp.

Johnson, N. E., and Martin, P. 1969. *Amydria effrentella* from nests of mountain beaver, *Aplodontia rufa.* Ann. Entomol. Soc. Am. 62:396–399.

Johnson, S. R. 1971. The thermal regulation, microclimate

and distribution of the mountain beaver, *Aplodontia rufa pacifica* Merriam. Ph.D. Thesis. Oregon State Univ., Corvallis. 164pp.

Lawrence, W. H. 1959. Wildlife-damage control problems on Pacific Northwest tree farms. Trans. North Am. Wildl. Conf. 23:146-152.

Lawrence, W. H.; Kverno, N. B.; and Hartwell, H. D. 1961. Guide to wildlife feeding injuries on conifers in the Pacific northwest. Western For. Conserv. Assoc., Portland, Oreg. 44pp.

Lovejoy, B. P. 1972. A capture-recapture analysis of a mountain beaver population in western Oregon. Ph.D. Thesis. Oregon State Univ., Corvallis. 105pp.

Lovejoy, B. P., and Black, H. C. 1974. Growth and weight of the mountain beaver, *Aplodontia rufa pacifica*. J. Mammal. 55:364-369.

Lovejoy, B. P.; Black, H. C.; and Hooven, E. F. 1978. Reproduction, growth and development of the mountain beaver (*Aplodontia rufa pacifica*). Northwest Sci. 52:323-328.

McLaughlin, C. A. 1967. Aplodontoid, Sciuroid, Geomyoid, Castoroid and Anomaluroid rodents. Pages 210-225 *in* S. Anderson and J. K. Jones, Jr., eds. Recent mammals of the world. Ronald Press Co., New York. 453pp.

McMillin, J. H., and Sutton, D. A. 1972. Additional information on chromosomes of *Aplodontia rufa* (Sciuridae). Southwestern Nat. 17:307-308.

Martin, P. 1971. Movements and activities of the mountain beaver (*Aplodontia rufa*). J. Mammal. 52:717-723.

Merzenich, M. M.; Kitzes, L.; and Aitkin, L. 1973. Anatomical and physiological evidence for auditory specialization in the mountain beaver (*Aplodontia rufa*). Brain Res. 58:331-344.

Motubu, D. A. 1978. Effects of controlled slash burning on the mountain beaver (*Aplodontia rufa rufa*). Northwest Sci. 52:92-99.

Pfeiffer, E. W. 1956. The male reproductive tract of a primitive rodent, *Aplodontia rufa*. Anat. Rec. 124:629-635.

————. 1958. The reproductive cycle of the female mountain beaver. J. Mammal. 39:223-235.

————. 1968. Comparative anatomical observations of the mammalian renal pelvis and medulla. J. Anat. 102:321-331.

Pfeiffer, E. W.; Nungesser, N. C.; Iverson, D. A.; and Wallerius, J. F. 1960. The renal anatomy of the primitive rodent, *Aplodontia rufa*, and a consideration of its functional significance. Anat. Rec. 137:227-235.

Scheffer, T. H. 1929. Mountain beavers in the Pacific Northwest. their habits, economic status, and control. U.S.D.A. Farmer's Bull. 1958. 18pp.

Schmidt-Nielson, B., and Pfeiffer, E. W. 1970. Urea and urinary concentrating ability in the mountain beaver *Aplodontia rufa*. Am. J. Physiol. 218:1370-1375.

Staebler, G. R.; Lauterbach, P. G.; and Moore, A. W. 1954. Effect of animal damage on young coniferous plantations in southwest Washington. J. For. 52:730-733.

Sweeney, S. J. 1978. Diet, reproduction and population structure of the bobcat (*Lynx rufus fasciatus*) in Western Washington. M.S. Thesis. Univ. Washington, Seattle. 61pp.

Taylor, W. P. 1918. Revision of the rodent genus *Aplodontia*. Univ. Cal. Publ. Zool. 17:435-504.

Voth, E. H. 1968. Food habits of the Pacific mountain beaver, *Aplodontia rufa pacifica* Merriam. Ph.D. Thesis. Oregon State Univ. Corvallis. 263pp.

Walker, E. P., et al. 1975. Mammals of the world. 3rd ed. rev. J. L. Paradiso. Johns Hopkins Press, Baltimore. 2 vols.

GEORGE A. FELDHAMER, Appalachian Environmental Laboratory, Center for Environmental and Estuarine Studies, University of Maryland, Frostburg State College Campus, Frostburg, Maryland 21532.

JAMES A. ROCHELLE, Weyerhaeuser Company, Western Forest Research Center, Centralia, Washington 98531.

9

Marmots

Marmota monax and Allies

David S. Lee
John B. Funderburg

NOMENCLATURE

COMMON NAMES. *M. monax* group: woodchuck, chuck, groundhog, whistle pig, siffleur (French Canadian); *M. flaviventris* group: yellow-bellied marmot, woodchuck, rockchuck, groundhog marmot; and *M. caligata* group (including *M. broweri*, *M. olympus*, and *M. vancouverensis*): hoary marmot, Alaskan marmot, Olympic Mountain marmot, Vancouver marmot, groundhog, whistler, siffleur, whistling pig, whistling marmot, "badger."

SCIENTIFIC NAMES AND SUBSPECIES. *Marmota monax: M. m. bunkeri, M. m. canadensis, M. m. ignava, M. m. johnsoni, M. m. monax, M. m. ochracea, M. m. petrensis, M. m. preblorum, M. M. rufescens; Marmota flaviventris: M. f. avara, M. f. dacota, M. f. engelhardti, M. f. flaviventris, M. f. fortirostris, M. f. luteola, M. f. nosophora, M. f. notioros, M. f. obscura, M. f. parvula, M. f. sierrae; Marmota caligata: M. c. caligata, M. c. cascadensis, M. c. nivaria, M. c. okanagana, M. c. oxytona, M. c. raceyi, M. c. sheldoni, M. c. vigilis; Marmota broweri; Marmota olympus;* and *Marmota vancouverensis.*

FIGURE 9.1. Distribution of the woodchuck (*Marmota monax*). Adapted from Hall and Kelson 1959.

Howell (1915) noted that the American marmots are composed of three distinct groups (outlined above): *monax*, *flaviventris*, and *caligata*. He noted that *caligata* was closest to the Eurasian forms. Rausch (1953) regarded *M. olympus*, as well as some races of the hoary marmot, to be only subspecifically distinct from the Old World marmot, *M. marmota*. The specific status of *M. olympus* and *M. vancouverensis* has been questioned by several authors. Rausch and Rausch (1965) gave cytogenetic evidence for the elevation of *M. caligata broweri* to full species status.

Hoffman et al. (1979) examined the relationships of *Marmota*, emphasizing cranial morphology. They concluded that *M. camtschatica* of northeastern Eurasia and *M. broweri*, *M. caligata*, *M. olympus*, and *M. vancouverensis* of northwestern North America comprise a related group with an amphiberingian distribution. Multivariate analysis revealed two distinct subgroups, one including *M. camtschatica* and *M. broweri* and the other comprising the remaining North American species of the *caligata* group. *Marmota*

olympus may represent a relict survivor of the Nearctic marmot that gave rise to these groups.

Other Marmots. In addition to the six North American marmots discussed here, there are eight recent Eurasian species that occur in western Europe and northern Asia. Of these the Alpine marmot of the European Alps, *M. marmota*, the type species for the genus, is the best known. Other species include the Steppe (or Siberian) marmot, *M. bobak;* the Himalayan marmot, *M. himalayana;* the gray marmot, *M. baibacina;* the Mongolian bobak, *M. sibirica;* the red marmot, *M. caudata;* Menzbier's (or Talas) marmot, *M. menzbieri;* and the black-capped marmot, *M. camtschatica*. All subsequent discussions will deal exclusively with the North American forms, unless noted.

DISTRIBUTION

M. monax, the widest ranging North American marmot, is distributed from eastern Alaska across the southern half of Canada to Great Slave Lake and the southern shores of Hudson Bay and Quebec. (figure 9.1). An isolated race, *M. m. ignava,* occurs in eastern Labrador. From Canada the range extends southward in the Rocky Mountains to northern Idaho and east of the Great Plains south to Arkansas and central Alabama. On the Atlantic slope the species was apparently absent from the Delmarva Peninsula and the Coastal Plain and Piedmont south of the Roanoke River, Virginia, until recently. Robinson and Lee (1980) documented its recent range expansion onto the Delmarva Peninsula and the Piedmont Plateau and Coastal Plain of North Carolina. Lowery (1974) discussed a record from northern Louisiana, and Scott (1937) noted a northward extension of the woodchuck into Iowa along wooded streams.

M. *flaviventris* is recorded from the interior valleys of southern British Columbia south in the Great Basin to Nevada and Utah, and in the Cascade-Sierra system from Mount Hood, Oregon, to the vicinity of Owens Lake, California. In the Rocky Mountains it occurs from the Canadian border south to the Pecos River Mountains, New Mexico. A disjunct race, *M. f. dacota,* is known from the Black Hills area of Wyoming and South Dakota (figure 9.2).

The hoary marmot occurs at high elevations near the timber line on talus slopes and alpine meadows from Idaho and Washington north to Alaska.

The Olympic marmot is confined to the summits of the Olympic Mountains in Washington above the timber line. Its total range is about 1760 km². The Vancouver marmot is endemic to portions of Vancouver Island, British Columbia. The Alaskan marmot is confined to the north slope of Alaska (figure 9.3).

DESCRIPTION

North American marmots are large, heavy-bodied rodents attaining weights of 2.3 to 11 kg and total lengths of 418 to 820 mm (table 9.1). In the *M. flaviventris* and *M. caligata* groups, adult males are somewhat larger than females. Warren (1926) reported two possible dwarf yellow-bellied marmots. Weights for different age groups vary seasonally. The feet have five claw-bearing digits, although the thumb is rudimentary. The claws are thick and slightly curved; those on the front feet are heavier. The palms are naked with five pads, and the soles are naked with six pads. The

FIGURE 9.2. Distribution of the yellow-bellied marmot (*Marmota flaviventris*). Adapted from Hall and Kelson 1959.

FIGURE 9.3. Distribution of the hoary marmot (*Marmota caligata*), Alaska marmot (*M. broweri*), Vancouver marmot (*M. vancouverensis*), and Olympic marmot (*M. olympus*).

TABLE 9.1. Standard measurements and range of adult weight in North American marmots

Animal	Measurement			
	Total Length	Tail Length	Hind Foot Length	Weight
Woodchuck	418–665 mm	100–155 mm	66–88 mm	2.3–5.4 kg
Yellow-bellied marmot	470–700	130–220	70–92	2.3–4.5
Hoary marmot	620–820	170–250	90–113	3.6–9.1
Olympic marmot	680–785	195–252	94–112	4.1–11
Vancouver marmot	670–750	180–237	91–110	
Alaskan marmot	553–600	150–162	86–88	3.1–3.6

SOURCE: Compiled from various sources.

head is short and broad; all legs are short and thick set, and the densely haired, slightly flattened tail is one-fifth to one-third of the total length. The ears are short, broad, rounded, and well haired. The eyes are circular and small. *Marmota monax* has four pairs of mammae, whereas all other species have five pairs. A cheek pouch is present but is rudimentary and lacks retractor muscles (Hall and Kelson 1959; Howell 1915).

Pelage. Fur color varies considerably among different species. The upper parts are gray, brown, or yellowish. The underparts, although paler, normally differ little in color from the upper parts, but in *M. flaviventris* the belly is yellow. Some species have white markings, and Lambert (1939) reported "white" woodchucks. The occurrence of melanism in *Marmota* was summarized by Howell (1915) and Fryxell (1929). Fur texture, which apparently is influenced by climate, varies from thick to thin, with different degrees of coarseness. There appears to be only a single, summer molt, but no uniform pattern to molting. In woodchucks, the annual molt in yearlings and adults occurred from late May into September (Davis 1964). In young of the year, molting was protracted throughout the population, beginning in early July and completed by 1 October. For individuals, molting was completed in about 3½ weeks. The Olympic marmot changes from brown in the spring to yellow in August, but this is apparently due to bleaching. This is unique to marmots and was described by Walker (1964) and in detail by Barash (1973b). Overall species—specific coloration for North American marmots is available in Burr and Grossenheider (1952), and Hoffmann et al. (1979) described the pelage of adults and juveniles of *M. broweri* and *M. caligata* and compared them with the Eurasian *M. camtschatica*.

Skull and Dentition. Photographs and cranial measurements of series of North American species and subspecies are available in Howell (1915). Condylobasal length is normally 75 to 100 mm. Rostrum and cranium are subequal, with the interorbital region being much wider than the postorbital region (figure 9.4). On all marmots the second upper premolar (p4) is as large or larger than the first molar (m1). The cheek teeth are highly crowned and metaloph is complete on each upper molar. The dental formula is i 1/1, c 0/0, pm 2/1, m 3/3 = 22. Hoffmann et al. (1979) provided comparative cranial dimensions for 24 characters for *M. camtschatica, M. broweri, M. caligata, M. olympus,* and *M. vancouverensis* and discussed cranial sexual dimorphism and geographic and interspecific variation. Their findings led to some of their thoughts on relationships (previously discussed under "Nomenclature").

PHYSIOLOGY

Body Weight and Hibernation. Because of the metabolic changes caused by hibernation, variations in body weight and fat deposition have received considerable investigation (Bailey 1965; Bailey and Davis 1965; Davis 1967a, 1967b; Snyder et al. 1961). Mar-

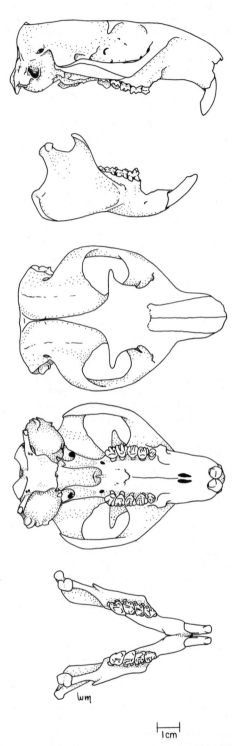

FIGURE 9.4. Skull of the woodchuck (*Marmota monax*). From top to bottom: lateral view of cranium, lateral view of mandible, dorsal view of cranium, ventral view of cranium, dorsal view of mandible.

mots are true hibernators that do not store food for winter but subsist entirely on fat reserves stored in their bodies. In some cases considerable effort has been given to documenting seasonal weight variations for different sexes and age classes. Grizzell (1955) and Hamilton (1934) reported weight losses of 30.4 and 33.3 percent for animals hibernating in Maryland and New York, respectively. The rate at which stored fat is used dictates the duration of survival during the months of normal hibernation. Bailey and Davis (1965) found hibernation to be about seven times more efficient in conserving energy than the aroused state. Laboratory experiments by Davis (1967b) showed that woodchucks maintained at 6° C became torpid when deprived of food but ones that were fed did not. Laboratory animals were retained in hibernation for up to eight months, but in nature hibernation normally lasts only three to four months. In the southeastern portion of its range *M. monax* may exhibit extremely brief hibernation periods (Robinson and Lee 1980). Emergence from burrows in Pennsylvania occurred between 29 January and 8 February in eight years of observation, and the date had no relation to local weather. Woodchucks exposed to very low temperatures ($-20°$ C) aroused promptly.

From a total of 1,700 woodchucks examined from south-central Pennsylvania, Snyder et al. (1961) found that these animals attain progressively greater weights during the first three years of life. They gain and lose weight periodically, gaining for about six months then losing for the next six. Young of the year had a mean increase in weight of 16 to 20 g per day from April through September. Yearlings gained only 14.7 g per day during this period, and adults 11.5 to 13.0 g per day. Weight loss during hibernation varied from 20 to 37 percent. Weight loss immediately after emergence from hibernation was much greater. Fat was used as a food source during the period of food scarcity that followed hibernation, and also was an important energy source for females during gestation and lactation. Anderson et al. (1976) reported that few female yellow-bellied marmots gained weight prior to weaning young, but afterward deposited fat at the same rate regardless of reproductive history.

Other. Many of the classic physiological studies of marmots were conducted on European species and will not be discussed here. Benedict and Lee (1939) gave a thorough presentation of the physiology of marmot hibernation, and most aspects have been considered by more recent authors. Rasmussen (1915, 1916, 1917) discussed blood volume and oxygen and carbon dioxide content (corpuscles, hemoglobin, and specific gravity) of the blood of hibernating and active woodchucks. He observed that blood volume varied from 4.4 to 8.3 percent of the body weight, and that the percentage of blood was highest in midsummer, decreasing nearly 30 percent before maximum fat accumulation and onset of dormancy. He also noted that, although there was little change in the number of red corpuscles, the amount of hemoglobin and specific gravity in arterial blood and the number of circulating

white corpuscles decreased about 50 percent in torpid animals.

Additional studies of marmot blood addressed specific and seasonal variations in biochemical properties, immunology, and effects of altitude. Hoffmann et al. (1979) examined three blood serum proteins and found that only albumin showed no difference in mobility among the four species studied and appeared to be monomorphic in all *Marmota*. The effects of hibernation on the chemical composition of blood plasma, as well as brain tissue and cerebrospinal and intraocular fluids, were examined by Bito and Roberts (1974). They found that plasma [K+] was similar in hibernating and active woodchucks, but was significantly higher than in the blood of resting or anesthetized animals. From this they concluded that active transport of some solutes across the blood-brain and blood-ocular barriers continues during hibernation. No major differences in immunology in hibernating and nonhibernating woodchucks were found, although the researchers noted that fibrinogen was greatly decreased or absent in hibernating individuals. Wenberg et al. (1973) believed that this may be the reason clotting of blood is inhibited during hibernation. Winders (1974) suggested that marmots should be considered for further laboratory investigation of oxygen conductance under conditions of hypoxic stress. He found that yellow-bellied marmots acclimated at 840 feet and then exposed to a simulated 12,000-foot altitude showed the following increases: hemoglobin 12.6 percent; hematocrit 10.8 percent; 2,3–diphosphyglycerate 66,8, and p_{50} 18.9 percent.

Miller and South (1974) studied thermoregulation as it responded to spinal cord cooling in marmots and observed that differential thermoregulatory gains of the cord may be of importance in temperature regulation during phases of the hibernation cycle. Seasonal variations in some hormones of woodchucks led Wenberg and Holland (1973) to conclude that epinephrine, norepinephrine, and vanilmandelic acid excretion rates were lowest in the early phases of hibernation, with increases occurring just prior to arousal. These authors assumed that levels of these hormones assist the animals in preparation for arousal. Smith and Hern (1979) examined the structure of the apocrine–sebaceous anal scent gland of the woodchuck and found the apocrine and sebaceous components emptied into a common duct. Light and electron microscope studies suggested that apocrine cells from different types of gland have distinctive morphologies that reflect differences in function and product. Christian (1966) followed seasonal changes in adrenal glands in woodchucks and found that body length was the base against which to measure change in weight of these glands because total body weight varied so much due to seasonal fat accumulations.

GENETICS

Karyotypes of the North American marmots are as follows: *Marmota monax* 2N=38, FN=66?; *Marmota flaviventris* 2N=42, FN=66; *Marmota caligata*

2N=42, FN=66; *Marmota broweri* 2N=36, FN=66; *Marmota olympus* 2N=40, FN=66; and *Marmota vancouverensis* 2N=42, FN=66.

Hoffman et al. (1979) used the above chromosomal evidence to interpret relations of the western North American and eastern Eurasian marmots. They concluded that *M. olympus* represents a relict of a Nearctic species from which *M. camtschatica* ancestors were derived, and this ancestral population may have also given rise to *M. caligata* (Hoffman and Nadler 1968). Fewer chromosomal changes are required to remodel the karyotype of *camtschatica* to that of *broweri* than to drive *broweri* from *caligata*. Thus, it seems that *broweri* was derived from essentially Eurasian stock. *Marmota vancouverensis,* on the other hand, is an isolated member of the *caligata* group.

REPRODUCTION

Exact dates of parturition are difficult to assess because the activity occurs in a burrow. Periods of mating, gestation, parturition, appearance of young from the burrow, and litter size are not similar for all species. Differences are a result of latitude and altitude (i.e., growing season), but data are somewhat fragmentary. Barash (1973*b*) graphically illustrated the breeding period, time of birth, dispersal of young, and growing season for three species of marmot. Parental behavior of hoary marmots was discussed by Barash (1974*b*). Nee (1969) and Svendsen (1974) discussed aspects of reproduction in yellow-bellied marmots. Grizzell (1955) surveyed the known literature on *M. monax* reproduction. What follows is from these works unless otherwise noted.

Anatomy. Externally determining the sex of adult, nonlactating marmots may be confusing. The penis of the male is about twice as far from the anus as is the vulva of the female. As a result, the normal length of the perineum of the male (25 mm) is about twice that of the female (12 mm). This varies with the size of the animal. Mossman et al. (1932) described the male anatomy of *M. monax* based on 10 specimens. They noted that the genital system is of a considerably smaller proportional size than in the spermophile, and that the testes are abdominal throughout most of the year.

Breeding Age. Marmots normally do not breed until at least two years of age. Only *M. monax* breed as yearlings, when 10 to 25 percent may reproduce. The female Olympic marmot apparently does not breed until its third year. Few positive records of breeding by yearling males or females are available for any species. When yearling *M. monax* breed in their first season, they do so at a later date than adults. Davis (1964) observed that some yearling males had pigmented testes, as do adults, in March and April. By May and June all yearlings' testes were pigmented. Mossman et al. (1932) described male reproductive tracts in Sciuridae, and this is consistent throughout the *Marmota*.

Anderson et al. (1976) found that in adult female yellow-bellied marmots production of young is limited by spring food and hibernacula resources. Females of this species may regularly produce young only every other year.

Mating. Hoyt and Hoyt (1950) provided the only account of copulation. Captive yearling *M. monax* showed interest in each other when awakening from hibernation. The two romped and played each day between 21 and 29 March. On the morning of the 29th they "boxed" for a few seconds and then the male mounted the female. The female assumed a position of nose tucked under chest, eyes closed, and tail up. The male gripped her fur with his teeth about halfway up her back. Copulation continued for about eight minutes. They separated for a few minutes, groomed, and then the male mounted again for about five minutes. After a second separation of several minutes the male again mounted for three minutes. After this they lost interest in each other. Courtship behavior and sequence of the Olympic marmot was described by Barash (1973*b*), and that of the yellow-bellied marmot by Armitage (1965).

Male and female *M. monax* often consort together in the breeding season and may remain together for some time. The breeding period in Maryland is from late February through March, and in central Missouri it begins in mid-February (Twichell 1939). The reproductive cycle of *M. monax* was discussed by Snyder and Christian (1960). Schoonmaker (1938) noted that as parturition nears the female drives the male from her den. Copulation in the yellow-bellied marmot occurs in the two weeks following arousal from hibernation (Armitage 1965; Nee 1969).

Gestation. Hamilton (1934) reported a gestation period of about four weeks in *M. monax* and Grizzell (1955) observed gestation periods of 32 and 33 days for two captive-bred females of this species. Hoyt and Hoyt (1950) reported a 31-day, 10-hour gestation period for a captive woodchuck. Gestation requires 30 days in the yellow-bellied marmot (Armitage 1962). Marmots are monestrous, having one estrous cycle per year, and produce no more than one litter per year. Female Olympic marmots breed every other year and the young remain with the mother for two seasons.

Parturition. Grizzell (1955) caught pregnant female *M. monax* until one or two days before parturition. They would then disappear for about a week. Captives were belligerent for several days after parturition and refused food. By the third or fourth day they began to feed, and they were eating heartily after seven days. Births occurred during a short calendar period: 1 to 21 April (Davis 1964), and 5 April to 10 May (Grizzell 1955). Hoyt and Hoyt (1950) observed that a captive female had her first young sometime between 1900 and 2200 hours and the remaining four throughout the night. Adult *M. flaviventris* with embryos are known from 8 April to 4 May, and a gravid *M. caligata* was found on 27 May (Howell 1915). Young yellow-bellied marmots are born during the first week of June (Downhower and Armitage 1971). The actual birth process apparently has not been described for any species.

Litter Size, Neonates, and Young. Marmots normally produce 4 to 6 young, with extremes of 3 (*M. flaviventris*) and 9 (*M. monax*) recorded. For 29 groundhog females from Maryland, Grizzell (1955) found an average of 4.6 young. Hamilton (1934) reported an average of 4.07 young per female in New York, based on embryo counts and examination of placental scars soon after parturition. *Marmota flaviventris* has 5.6 (N=?) young per litter and *M. caligata* has 5 (N=1), again based on embryo counts (Howell 1915). The average weight of 11 captive-born woodchucks, less than 24 hours after birth, was 27.2 g. The young are born blind, helpless, toothless, and almost naked. The skin, as is typical of young rodents, has a pink red tinge. The forelegs are better developed than the hindlegs. At one week of age, three young weighed from 45 to 62 g. A slight slit opens in the eyes at this age (Grizzell 1955). Hamilton (1934) provided an excellent account of development from birth to seven weeks. Age characteristics for older animals were described by Davis (1964). He found that in Pennsylvania young first emerge from their burrows around 15 May, when they weigh from 300 to 450 g. They gain approximately 19 grams per day from June through September (Snyder et al. 1961). By July the young of *M. monax* begin to disperse, normally to occupy abandoned dens. Davis found that young can be separated from older individuals by size, characters of pelage, shape of incisors and head, and color of testes (see "Age Determination").

Downhower and Armitage (1971) reported that yellow-bellied marmots remain in their burrows 20 to 30 days *post partum* and are first seen above ground in the first week of July. The earlier young marmots are weaned, and the more they weigh at hibernation, the more likely they are to survive their first winter (Anderson et al. 1975). The socioecology of female reproductive strategies was also discussed by Anderson et al. (1976), who concluded that the number of offspring a female weaned was significantly associated with the estimated number that could be produced based on food resources. Delaying pregnancy until optimum forage is available fails as a counterstrategy because young have insufficient time to accumulate fat prior to hibernation.

Barash (1973b) studied the time of appearance of "infant" Olympic marmots and their social interactions with the rest of the colony. Young were first seen above ground during the last ten days in July, and they were probably born in late June. Initially they remained above ground only briefly. After four days they generally remained out for less than 10 minutes at a time and stayed within a meter or so of their burrow. Within two weeks they were remaining outside the burrow nearly as long as the adults. Since they began foraging immediately after emergence, Barash assumed that hunger as a result of weaning may be the prime cause of initial emergence. All colony members were extremely tolerant of the infants, who frequently crawled over the larger animals and chewed their fur.

Woodchucks disperse the same year they are born, yellow-bellied marmots the first year after birth, and Olympic marmots the third year (Barash 1973b). Dispersal factors were discussed by several authors and a good summary provided by Barash (1973b, 1974a). The delayed dispersal is believed to be a progressive modification in response to a shortened growing season. Shirer and Downhower (1969) followed radiocollared young yellow-bellied marmots.

ECOLOGY

The woodchuck primarily inhabits dry soils in open woodlands, thickets, rocky slopes, fields, and clearings. It often persists along road and utility rights of way and on levies and dams where the natural terrain does not provide adequate open habitat. Hoary and yellow-bellied marmots usually occur in areas with rock outcrops or slides having nearby meadows or other subalpine areas with ample vegetation. Often they are present at high elevations (3300 m or more). Olympic marmots inhabit subalpine to alpine, rain-shadowed meadows and talus slopes just above and below the timber line (1500 to 1750 m) (Barash 1973b). Bee and Hall (1956) described the habitat of the Alaskan marmot as the precipitous sides of canyons and valleys in the Brooks Range. They noted that it preferred the base of active talus where boulders are large and have accumulated to a depth sufficient to provide subsurface protection.

Activity Cycles. Most activity of marmots is diurnal, but Hamilton (1934) suggested that shortly after emergence from hibernation *M. monax* may also be active at night. Grizzell (1955) documented only a few cases of nocturnal movement. Woodchucks, and probably all marmots, spend a large part of their lives either resting or sleeping. Even during peak activity seasons, individuals may spend only an hour or two above ground. Activity during the first week after the emergence of spring is highly erratic, perhaps directly influenced by weather. Grizzell (1955) observed that typical midspring activity consisted of sunning in the morning, with maximum activity times gradually changed to morning and evening, reverting back to midday in the fall. Woodchucks living in woodlots are not as restricted by temperatures and often feed throughout the day.

July and August emergence of yellow-bellied marmots from the burrow was roughly correlated with the time sunlight first reached the colony area, and the major period of morning emergence extended over an hour. Animals appeared at about the same time on both cloudy and sunny days (Armitage 1962, 1965). Many animals defecated immediately after emerging. The first 15 to 30 minutes after emergence were spent grooming and sunning. Animals gradually dispersed to feeding areas by 0800 hours; feeding lasted about two hours. The remainder of the day was spent in other activities or in the burrow, with occasional forays to feeding areas. From 1000 to 1600 hours there was a marked reduction in overall activity. From 1600 hours until sunset a second activity peak, involving mainly feeding, was observed. All animals normally entered their burrows 15 to 30 minutes after sunset. Armitage (1962, 1965) graphically illustrated and compared daily activity patterns for May, July, and August.

Seasonal shifts in activity essentially involve hibernating-nonhibernating periods. In Maryland, the last woodchucks entered hibernation in the second week in November, and most adults were hibernating by 1 November (Grizzell 1955). Although midwinter emergence for several marmot species has been documented (Grizzell 1955; Couch 1932; Rutter 1930; Hall 1930; and others), winter activity is generally atypical. Robinson and Lee (1980) concluded that the long season of activity in parts of the Southeast may account for the groundhog's recent range expansion and success. They noted activity in all months of the year in east-central North Carolina. Normally the older and fatter animals go into hibernation first, followed by yearlings and juveniles. Older and larger individuals usually appear first in the spring, too (Grizzell 1955). Armitage (1962) reported that some yellow-bellied marmots were in hibernation by the first and second weeks of August, but that parturient females and young had not hibernated prior to the last week in August. Hoary marmots begin hibernation in late September or early October in British Columbia, and in mid-September in Alaska (Howell 1915).

Home Range and Movements. Svendsen (1974) studied the effects of behavioral and environmental factors on spatial distribution and population dynamics of yellow-bellied marmots. Although *M. flaviventris* is considered colonial, this is not its only option. Seventy-five percent were colonial, 16 percent live at satellite sites, and 8 percent were transients. Factors influencing spatial distribution were individual behavioral profiles and on size and composition of the habitat. Dispersal in yellow-bellied marmots occurred primarily among yearlings. All colonial adult males and 41 percent of adult females were recruited from outside the colony (Armitage and Downhower 1974). The size and shape of the home range depend on the terrain and the location of feeding areas in relation to burrows. Home ranges of females exhibit three patterns: separate, slight overlap, and major overlap.

Olympic marmots were not territorial, but had a general colony range beyond which various members of the colony very rarely strayed. The only exception involved interactions between the resident male and satellite males (Barash 1963b).

Rall (1945), working with *Marmota barbacia* in Russia, remarked that in the steppe-plain habitats burrows were not numerous and were distributed evenly. Here the animals may make long daily journeys of up to 400 m. Colonies in ribbon/ravine and focus/hill habitats have more burrows concentrated in small areas, and maximum daily movement is not over 100 m. Grizzell (1955) noted a similar pattern for *M. monax*. Woodchucks exhibit little territoriality except in the immediate area of the home den (Hamilton 1934; Cahalane 1947; Grizzell 1955). Seasonal variations in home range of woodchucks were studied by Smith (1972), who showed that woodchucks may migrate from winter to summer den sites with corresponding shifts in home range.

Burrows. Burrows of most North American species have been described or partly described. Normally only one adult occupies a burrow, but Alpine yellow-bellied marmots may at some sites commonly occupy the same hibernacula and summer residence burrows.

The most detailed descriptions are for woodchuck burrows. These are easy to recognize, although they may at times be confused with those of the fox, skunk, and badger. A *Marmota monax* den normally has one to five entrances and is nearly always in well-drained soil. A mound of fresh soil at the entrance and numerous trails radiating to and from feeding areas are characteristic. The entrance is often plugged. Since the animals clean out their burrows several times a week, fresh soil is constantly being deposited at the main entrance. When foxes appropriate woodchuck dens, there is an absence of fresh soil at the entrance and animal remains can usually be found around the dens. Unlike woodchucks, badgers do not frequently deposit fresh soil at the entrance and their den entrances are wider.

Merriam (1886) described two different entrance types: (1) the plunge hole dropping abruptly downward, and (2) the main entrance (about a 150-mm opening) descending at a moderate angle. The main entrance is characterized by fresh soil. The plunge hole is excavated from within by burrowing up from one of the main galleries, leaving little soil on the outside. These entrances to the burrow are considerably smaller than the main entrance and provide a well-concealed escape route.

There are two types of den, for winter and for summer. Winter dens normally are in wooded or brush habitats on sloping ground. They usually have only one entrance and serve hibernation, but occasionally are used throughout the year. Summer dens are most often in open areas and have more entrances.

Den design varies considerably. Dens may be simple, short burrows terminating in a nest chamber, or complex ones with numerous entrances, a maze of burrows, blind chambers, and multiple nest chambers. The more complex dens evidently are older ones and have perhaps been enlarged by several succeeding generations of animals. At the bottom of the main entrance there usually is a turnaround chamber, but from there dens assume a variety of diverse forms. The nest chamber is typically constructed on a higher plane than the deepest part of the system, and usually contains dried and shredded leaf material (Grizzell 1955). The deepest den investigated by Grizzell was only slightly over 1.5 m in depth.

Yellow-bellied marmots burrow beneath rocks,

and use them for lookout posts and for sunning (Svendsen 1974). Armitage (1962) recognized two types of burrow, home burrows and auxiliary burrows. The animals normally raise their young and spend the night in home burrows and retreat to them when danger threatens. Home burrows always have at least three entrances. Auxiliary burrows commonly have only one or two openings and are places of refuge for animals unable to return to the home burrow. Home and auxiliary burrows are connected by trail systems. Barash (1973*b*) noted the same two burrow types for the Olypmic marmot and discussed similar aspects of design and use. Auxiliary burrows were quite shallow (three burrows were between 120 and 180 cm deep). Beltz and Booth (1952) briefly described a partly excavated home burrow of an Olympic marmot. Sleeping chambers were lined with sedges and other meadow plants, and distinct latrine chambers were found.

FOOD HABITS

Marmots are herbivorous, but except for documentation of damage to crops their food habits have received little study, most of it done in conjunction with economically oriented biological surveys conducted in the early part of this century. Howell (1915) summarized the principal foods of the woodchuck as clover, alfalfa, and grasses. Marmots do considerable damage to these crops by consuming them and trampling down much that is not eaten. Damage to other cultivated crops and orchards is seldom serious. Howell (1915) reported damage to young corn plants and occasional feeding on leaves of pumpkins, squash, and beans and noted that a specimen collected in a sassafras tree (*Sassafras albidum*) had its stomach packed with leaves from the tree. Marmots probably climb trees (see ''Behavior'') to feed. Gianini (1925) found remains of large numbers of ''June bugs'' in the scats from numerous woodchucks and in the stomach of one individual in New York. The size of these insects and the numbers consumed led Gianini to believe that they were not ingested accidentally. One woodchuck was reported eating aquatic plants obtained by climbing out on fallen trees along a lake shore in Ontario. (Fraser 1979). It was assumed to be attracted to sodium-rich plants. Weeks and Kirkpatric (1978) previously reported sodium-drive phenology in woodchucks. Armitage et al. (1979) described cannibalism in yellow-bellied marmots. Barash (1973*b*) gave evidence that Olympic marmots are cannibalistic and on two occasions also saw marmots carrying dead chipmunks (*Eutamias townsendi*). Water requirements of marmots normally are satisfied by dew and succulent vegetation, but on occasion the animals drink water. Olympic marmots were observed to drink daily when standing water was present and frequently ate snow. Usually, however, such sources were lacking, and animals obtained water from food (Barash 1973*b*).

Armitage (1962) noted that in his study colony of yellow-bellied marmots foraging animals typically crawled through beds of clover, creating a large number of winding and crisscrossing ''crawlways.''

Animals usually raised their heads while chewing, and this behavior occurred between 0.8 and 5.1 times per minute. Food resources for yellow-bellied marmots were also considered by Armitage (1979). Kilgore and Armitage (1978) concluded that, even though these marmots consumed only 2.0 to 6.4 percent of available net primary production, they may be food limited if they forage selectively. These authors illustrated that yellow-bellied marmots were selective as to the species of plant and the parts of the plant that they consumed. All of four rejected plant foods contained alkaloids, and perhaps population sizes of marmot colonies are at least in part limited by chemical defenses of local plants. They recommended that plant-herbivore interactions be studied in several communities because of the variation in plant species composition.

Grizzell (1955) provided a summary of published food habits of woodchucks and some specific information on economics of crop damage (lacking from most earlier reports). He noted that at the Patuxent Research Refuge, Maryland, the animals ate leaves, twigs, and fruit from apple trees, peas, beans, carrots, cabbage, celery, pumpkin, squash, potato vines, young plants and ears of corn, vetch, stonecrop, sedges, blackberry, serviceberry, roses, sunflowers, and dandelions. A common complaint had to do with garden depredations. One or two animals can literally ruin a small garden almost overnight. About $25 in damage was caused by one or two animals in a small garden in less than a week (1955 figures). Three groundhog dens in a cornfield were watched to observe the extent of crop damage. Little damage was noted until a heavy rain knocked over a large number of corn stalks, making the ears readily available. Grizzell also examined 32 stomachs from individuals collected in the spring and summer, and discussed the diversity of 24 different foot items. Red clover, white clover, grass, chickweed, and alfalfa were most frequently found.

Barash (1973*b*) commented on the food habits of the Olympic marmot: ''Upon emerging in late spring, their diet consisted largely of grasses and roots but within a few weeks they ate lupine, avalanche and glacier lilies (*Erythronium grandiflorum*), magenta paintbrush (*Castilleja oreopola*), American harebells, mountain buckwheat, etc. When the meadow flowers became particularly abundant, marmots selectively ate the flower heads. Sapling alpine firs and western white pine (*Pinus monticola*) were occasionally gnawed, perhaps more for their abrasive nature than their nutritive value.''

BEHAVIOR

Social Organization. An excellent review of the possible evolution of marmot societies was provided by Barash (1974*a*). He contrasted the solitary and aggressive woodchuck with the highly social Olympic marmot and with the yellow-bellied marmot, which is intermediate in its social organization. The degree of social behavior is inversely co-related with the length of the growing season. The Olympic marmot, living in areas with growing seasons of 40 to 70 days (contrasted to 150 days for woodchucks in Pennsylvania), shows protracted development and dependence of young that in part affect their social organization. Other North American marmot species fall in between these behavioral extremes.

Barash (1973*b*) studied the composition of 15 colonies of Olympic marmots and found a typical colony to consist of one adult male, two adult females, occasionally some two year olds, yearlings from the previous year's litter, and young of the year. Colonies often have satellite males that avoid the resident adult males. Barash found that individuals lacked territories and home ranges and frequently moved in a loosely organized feeding group of up to eight animals. Some colonies were so isolated that there was seldom if ever any interaction between colony members. Others were nearly contiguous and interactions between occupants occurred almost daily. Only encounters between dominant males resulted in fighting.

The hoary marmot is perhaps most similar in social organization to the Olympic marmot. Barash (1974*b*) examined the social behavior (patterns of burrow use, greetings, play, aggressive chasing) of four colonies of hoary marmots in Montana. A colony was composed of a dominant male with a few adult females (three years or older), two year olds, yearlings, and pups. As with the Olympic marmot, this closely integrated colony organization is believed by Barash to have developed with its reproductive patterns characterized by late maturation and dispersal.

Social organization appears to be least stereotyped in the yellow-bellied marmot but the yellow-bellied marmot is still clearly intermediate in degree of organization between the Olympic marmot and the woodchuck. Barash (1973*a*) believed that the variation in social organization was primarily caused by length of growing season (i.e., altitude), but Armitage (1977) considered population density, age-sex structure of the population, individual behavior characteristics, time of year, the way colony space is shared, and the number of years colony residents have coexisted to be of major importance. Lack of any correlation between colonies in patterns of social behavior suggests that there is no single factor integrating the behavior-population systems. Each colony is essentially a separate behavioral system.

Svendsen (1974) observed that, depending on an individual's behavior profile and the size and composition of habitat, animals may live as solitary individuals. Of the adult-yearling group he studied, 75 percent lived as members of colonies, 16 percent lived at satellite sites, and 1 percent were transients. Aggressive females, for example, tended to occupy small harems at even larger sites or to live as solitary individuals regardless of the size of the site, while social females occurred primarily in the larger sites in harems. Armitage and Downhower (1974) found that, compared to satellites, colonial individuals retain larger residencies and reproduce at a higher rate. Although population trends between studied colonies were dissimilar, they were relatively stable compared to trends of satellites.

Downhower and Armitage (1971) reviewed the evolution of polygamy using *M. flaviventris* as a model. They concluded that while females made their greatest contribution to the next generation when they were monogamous, males are maximally fit when they maintain a harem of two or three females. The social structure of all colonial marmots apparently evolved from these factors, and perhaps in the yellow-bellied marmot has not yet stabilized.

Social organization of other North American species is minimal or remains unstudied (*M. broweri*), although Henderson and Gilbert (1978) reported uniform spacing of the burrows of field-dwelling woodchucks, which they attributed to agonistic behavior of juveniles. It should be pointed out that the social behavior of the woodchuck has not been extensively studied because the aggressive nature of the animals is so extreme that the resulting spacing of adults would provide observers with few opportunities to observe interactions. Merriam (1971) telemetrically recorded 223 interburrow movements of woodchucks and suggested that the low frequency of such movements was a result of agonistic behavior.

Burrow Construction and Maintenance. Armitage (1962) noted that digging and carrying of grasses for nest building by yellow-bellied marmots was infrequent. Only four new burrows were added to his seven-hectare study area in four years. Several animals participated in the digging, which was done with the forelegs, with excavated soil thrown between the hindlegs. Soil accumulated at the entrance of the burrow was pushed away by the animals lying down and shoving loose dirt with its chest and forelegs. Yellow-bellied marmots lie down to gather grass, which they tear loose with the teeth and repeatedly position in the mouth with forelegs until the mouth is full. Normally only one trip was made to the burrow when animals were carrying grass. Most grass gathering by adults was apparently in preparation for hibernation.

Grizzell (1955) noted that succeeding generations of groundhogs use and enlarge established dens. Other marmot species probably do this as well. Goodwin (1935) marked woodchucks as they entered their dens, and after carefully plugging all exits he dug to the end every open tunnel, without finding the animals. He concluded that they were able to dig ahead of him, although other authors implied that the animals sealed themselves inside chambers. Excrement is seldom encountered around the den entrances of marmots (Hamilton 1934), but usually is deposited in a dry chamber in the burrow or buried in the mound at the entrance hole. (Burrow descriptions are provided under "Ecology.")

Communication. Marmots use visual and auditory senses for the main contacts between members of a group, with some olfactory and tactile cues. Although the woodchuck is often referred to as the "whistle pig," this name is somewhat misleading, since the shrill whistle is much more likely to be heard from western marmots than from eastern species. All species whistle either to warn individuals of danger or in response to other excitement. The sound originates in the vocal cords (Muller-Using 1955; Munch 1958). Waring (1966) and Barash (1973b) described auditory communication in yellow-bellied and Olympic marmots. Six different whistles of the yellow-bellied marmot have been reported, constructed around a "primary whistle motif" of 4000 H_z and a duration of 0.037 seconds. Four to five vocalizations were recorded for the Olympic marmot. Its primary call is similar to that of the hoary marmot. Hoffmann et al. (1979) provide audiospectrograms and the only information on *M. broweri* vocalizations, and *M. vancouverensis* calls have not been recorded. It is stated that it rarely vocalizes (Swarth 1911). The whistle of *M. monax* is a short, sharp, piercing note, followed in the eastern species by a series of hisslike noises emitted while the animal trembles all over (Grizzell 1955). The note is preceded by a quick movement of the head. Lloyd (1972) reviewed the descriptive literature of the two-part whistle of *M. monax,* which is a short, sharp whistle followed by a chuckling sort of warble (Seton 1928; and others). Lloyd also provided a detailed description of the sound and analyzed it with audiospectrograms. The vocalization is complex, with the initial shriek intense and slurred upward from 2.7 to 4.8 cps, after a progressive series of alternate two-tone pulses ending in a 3.8-to-6.0-sec warble of decreasing intensity.

Grizzell (1955) observed a female warn her litter by whistling when she heard the barking of dogs; they then scurried to the den. Young learn to whistle about the time they begin to disperse from the natal den, but are not capable of making as loud a sound as adults. Lloyd (1972) reported that a captive individual did not whistle until two years of age.

The basic vocalizations of yellow-bellied marmots served to alert other members of the colony, who normally responded by sitting up (Armitage 1962). The alarm call, although the same basic note, is higher and sharper. On hearing this call neighboring marmots reacted by running to a burrow. No particular animal acted as a sentinel. Marmot whistles are generally regarded as alert or alarm calls. German workers studying *M. marmota,* however, interpreted the call as an acoustical marker (Bopp 1958; Mitteil 1955; Munch 1958). Muller-Using (1955) emphasized the warning nature of the call but suggested that it may function incidentally as a territorial proclamation. There is little evidence for territoriality in North American marmots, however, and whistles have been heard in a variety of circumstances (Trump 1943; Hamilton 1934; Merriam 1886; Lloyd 1972).

Alarm calls are given by marmots when they are highly nervous, startled, or hear or see potential predators. Lloyd (1972) was able to stimulate his captive female woodchuck to vocalize by entering the room in which she was caged and then leaving abruptly, or by walking past the case when it was outdoors. Zimmerman (1955) reported on the cry of a marmot at the approach of a human. Armitage (1965) discussed various animal species that elicited alarm calls from

yellow-bellied marmots. He also noted shrieklike sounds given by young at play, and by submissive female adults when approached by dominant males. Growls were made by animals in burrows.

Olfactory communication has been little studied. Grizzell (1955) illustrated and discussed the three nipplelike glands located just inside the anus of the woodchuck. Smith and Hearn (1979) described the glands' structure and gave evidence of a single cell type producing both apocrine and melocrine secretions. The apocrine-sebaceous anal glands are everted, and a musky odor disseminated, when the animal is excited. The odor probably does not serve as a predator repellent. During the breeding season the odor may aid in mate location, or at least in identifying occupied burrows. Hamilton (1934) noted that characteristic odor just prior to woodchucks' emerging from dens, and Grizzell (1955) was able to identify occupied woodchuck burrows by the scent.

Another olfactoral communication reported and illustrated by Armitage (1962) for *M. flaviventris* was the "greeting." Two individuals directly approached each other with tails arched, and appeared to sniff at each other's cheeks. Usually there was little if any contact. Most encounters observed were between young or a young and an adult. Similar behavioral patterns were described for *M. marmota* (Muller-Using 1955; Munch 1958) and *M. olympus* (Barash 1973b).

Locomotion. Although there is a surprising amount of information on the less typical movements of North American marmots, such as tree climbing (Bowdish 1922; Hickman 1922; Medsger 1922; Stoner 1922; Robb 1926; Johnson 1926; and others) and swimming (Cram 1923; Johnson 1923; Phillips 1923; Twidale 1936), there is little information on terrestrial locomotion. It should be noted that during two years of intensive field investigation, Grizzell (1955) only twice observed woodchucks in trees and only once saw an animal swimming. Marmots are capable of slow walks, rapid, bounding runs, and gallops where the hind feet are set before the forelegs (Munch 1958; Muller-Using 1954; Armitage 1962).

Marmot tracks closely resemble those of the raccoon (*Procyon lotor*), but can be distinguished by the alternate four-toed tracks of the front feet and five-toed tracks of the hind feet. Raccoons have five toes on each foot.

Armitage (1962) noted the behavioral significance of tail positions during locomotion of yellow-bellied marmots. Normally the tail was held in a half-moon position and waved from side to side. When reacting to an alarm signal the tail was held down or pointed posteriorly. Submissive animals always held their tails low.

Sunning. Armitage (1962) noted that yellow-bellied marmots sunned themselves early in the morning when the air was cool. At these times animals lay with their bodies orientated broadside to the sun. During warmer periods of the day these marmots positioned themselves lengthwise toward the sun, and during the hottest periods sunning was restricted and most animals remained in the shade. Woodchucks often sun themselves on warm days near winter burrow entrances in early spring and late fall and probably on cool summer mornings as well.

Posture. All species of marmot frequently sit in an upright position while feeding or when responding to an alarm call or other stimulus. This apparently allows them to see distant objects over surrounding vegetation. Albert and Panuska (1979) discussed postural variations in hibernating woodchucks.

Grooming. Yellow-bellied marmots groom at any time of the day, but the greatest activity observed was shortly after morning emergence from burrows and prior to feeding. Grooming of dominant individuals by submissive marmots was recorded five times, always after agonistic behavior (Armitage 1962).

Fear and Defense. Although marmots normally flee from superior adversaries, they can be very fierce when cornered (Glover 1943; Grizzell 1955). At such times they chatter with their incisors. Holt (1924) reported an encounter between a woodchuck and a pack of dogs in which the dogs were defeated, and Waring (1965) saw an adult yellow-bellied marmot chase a marten (*Martes americana*), a known marmot predator, from its colony.

Play. Play activity in yellow-bellied marmots is frequent only among young, although adult females may play with young of their own litters (Armitage 1962). Barash (1973b) had difficulty distinguishing between social "playing" and "fighting" in *M. olympus,* and combined the two as "play fights," the nature of the activity shifting as the age of the participants increased.

Visual Signals. The upraised tail and tail flicking may serve an important communicatory function in marmots. Muller-Using (1956) described this behavior in *M. marmota,* attributing it to "insecurity." It was considered an intension movement in *M. flaviventris* by Armitage (1962), and this may apply to *M. olympus* as well. Barash (1973b) correlated this behavior with age, activity, and distance from burrow for Olympic marmots.

Greetings. The highly social Olympic marmot averages about one greeting per active individual per hour. Greetings in the yellow-bellied marmot average about one-tenth this number, and among asocial woodchucks the behavior is not even described (Barash 1974a). Greeting behavior is similar to that noted for other sciurids (King 1955; Munch 1958), and is believed to involve olfactory as well as tactile communication, probably related to individual recognition. The greetings—their frequency, duration, and period of occurrence—were described by Armitage (1962) for the yellow-bellied marmot and by Barash 1973a) for the Olympic marmot.

Dominance and Agonistic Behavior. Gordon (1936) reviewed territorial behavior and social dominance in the Sciuridae. Barash (1973b) described agnoistic behavior for the Olympic marmot and generally outlined the basic agonistic profiles of three species in his discussion of the evolution of marmot societies (1974a). He found the solitary species (*M. monax*) most aggressive, the yellow-bellied marmot moderately aggressive, and the highly colonial Olympic marmot least agonistic. Armitage (1965) described vernal variation in agonistic behavior between age classes of yellow-bellied marmots and noted maximum interactions between yearlings and adults in the second and third weeks after the emergence of spring. Armigate (1962, 1974, 1977) demonstrated the complexity and intricacies of agonistic behavior and suggested that each colony may have its own behavioral profile. Armitage and Downhower (1970) described "interment behavior" in yellow-bellied marmots, where dominant animals plug the burrow entrances of other colony members with soil and rocks.

Barash (1974b) discussed the agonistic aspect of social behavior for the hoary marmot. He found that aggressive chases were most abundant in June and decreased steadily through summer. Adult males were the most frequent chasers, while two year olds and yearlings were the most frequent chasees. Barash also reported on the difficulty of separating play from agonistic bouts in this species.

Aspects of agonistic behavior of the woodchuck were discussed by Brown (1948), Gordon (1936), and Grizzell (1955). Bronson (1964) provided the only detailed report of agonistic behavior in the woodchuck. He stated that during the postreproductive period its behavior could essentially be described as ingestive and agonistic. Solitary animals established dominance-subordination relationships with those animals whose home ranges overlapped or met their own, and there was no evidence of territorial defense. The chief behavioral characteristic was avoidance by subordinate animals of their dominant neighbors. Average aggressive interaction rate was about one per animal per day, being most pronounced early in the season and decreasing progressively until hibernation.

MORTALITY

The greatest recorded age for *Marmota monax* is of an individual that lived for almost 10 years in the London Zoo. Linduska (1947) reported wild individuals up to four years of age at the Rose Lake Wildlife Experiment Station, Michigan. Hamilton (1934) believed 4 or 5 years to be the natural life span, and Grizzell (1955), based on limited recapture studies, suggested that 5 or 6 years was typical.

Human activities often create areas where woodchuck populations quickly expand. Robinson and Lee (1980) noted the use of road embankments and river dikes as den sites and avenues of dispersal. In such areas highway mortality and periodic flooding tend to keep populations from expanding. Mortality from automobiles is increased by the favorable foraging habitats created by road shoulders but seems relatively low even in areas of high densities. This is certainly a result of the animal's visual abilities, constant alertness, and lack of nocturnal foraging. In regions of agricultural activities, particularly crops of alfalfa, soybeans, and clover, and dairy and stock-raising areas, groundhogs often become a major pest. Shooting, traps, and gas are successfully used to control populations (see "Economic Status and Management" section). This is less a problem in western North America.

Flooding. Flooding of dens may be a serious problem during hibernation and when the young are helpless. Currier (1949) reported that 7 out of 24 dens were abandoned after two severe floods. Yeager and Anderson (1944) reported on a November flood in central Illinois after which 17 dead woodchucks were found; the first was discovered about two weeks after the peak of the flood. The animals were driven from their dens by the rising water, were unable to find suitable cover, and died primarily from exposure. Flooding is not a major mortality factor in western marmots because of the elevated rocky terrain and sites of den selection.

Fire. Fire has a minimal impact on marmot populations. Leedy (1949) observed woodchucks surviving a hot fire that burned a 40-ha area. About one-third of the dens had been reopened and some of the animals traveled distances of up to 75 m to obtain food. Furthermore, fire would tend to stimulate herbaceous plant growth, eliminate much dense, woody vegetation, and perhaps provide a long-range beneficial effect.

Hibernation. Marmots have an adaptive strategy of rapid growth and weight gain, and high tissue-growth efficiencies, and they extend the use of plant-growing season for their young by reproducing immediately after hibernation. Nevertheless, young weaned late or during years with a short growing season often do not have enough time to gain weight, and die during hibernation. This has been documented only for *M. flaviventris* (Armitage et al. 1975), but it probably applies to other marmot species as well. Mortality during hibernation is probably a factor for adults too, but this has not been investigated to any degree.

Predation. Specific observations of predation on marmots are few. Powell (1972) reported on predation in some western marmots. Barash (1973*b*) reported seeing cougars (*Felis concolor*) and coyotes (*Canis latrans*) feed on Olympic marmots. Coyotes, bobcats (*Felis rufus*), fishers (*Martes pennanti*), bears (*Ursus* sp.), red-tailed hawks (*Buteo jamaicensis*), and other species elicit alarm calls. Grizzly bears (*Ursus arctos*) are reportedly a major predator on the hoary marmot (Murie 1961). Golden eagles (*Aquila chaysaetos*) are probably a major predator on many of the western marmots. Knight and Erickson (1978) found that yellow-bellied marmots represented 71 percent of the total biomass of food remains from two golden eagle nests in north-central Washington. Grizzell (1955) considered dogs and foxes to be the two most serious predators of woodchucks. Black bears (*Ursus americanus*) dug woodchucks out of their burrows in early fall when food was scarce (Beule 1940).

Badgers (*Taxidea taxus*), snakes, hawks, and owls all on occasion eat woodchucks. Most predation is on young animals, since adults are able to defend themselves against most natural predators.

Agonistic Behavior. Armitage (1965) provided circumstantial evidence that a highly aggressive marmot can affect reproduction and/or litter survival of subordinate females. Grizzell (1955) reported loss of litters by *M. monax* under social stress.

Parasites and Diseases. Grizzell (1955) noted that woodchucks were relatively free of external and internal parasites. He found the roundworm *Obeliscoides cuniculi* in only 2 of 40+ stomachs examined. This parasite was previously reported from woodchucks (Twichell 1939; Rausch and Tiner 1946; Wallace 1942; Hamilton 1930). A woodchuck infested with the parasitic mite *Atricholaelaps glasgowi* was described and illustrated by Grizzell (1955). He also reported an apparently fatal infestation of the mite *Haemolaelaps glasgowi* from an animal found in Michigan. Ticks, fleas, and parasitic dipteran larvae have also been reported but not specifically identified. Infestations are generally mild. Woodchucks at times may act as reservoirs for rodent-borne diseases such as tularemia and sylvatic plague. Marmots may be infested with ticks that carry spotted fever (Eadie 1954). Over 200 spotted fever ticks (*Dermacentor venustus*) have been taken from a single yellow-bellied marmot. Pertinent parasite records for the amphiberingian marmots were summarized by Rausch and Rausch (1971). The flea *Oropsylla silantiewi* is known from *M. broweri* and other Palearctic (European) species, and *M. caligata* harbors a Nearctic form, *Thrassis pristinus*. The cestode *Diandrya composita* occurs in all Nearctic marmots except *M. monax*. *Catenotaenia reggiae* is a cestode shared by *M. caligata* and *M. broweri*. Hoffmann et al. (1979) used the parasite fauna of the amphiberingian marmots as additional evidence of specific relationships.

It is of interest that for years the idea prevailed that the central Asian marmots were responsible for the spread of plague. The last great outbreak of pneumonic plague in Manchuria was in the winter of 1910–11, and was reported to have started among tarbagan (marmot) hunters. Efforts were made during the epidemic to locate the disease among the native marmots. Only one infected individual was found, but individuals captured and inoculated were found to be susceptible to plague infestation (Howell 1915; Wu 1923).

AGE DETERMINATION

Davis (1964) found that young-of-the-year, yearling, and adult *M. monax* could be distinguished by size, characteristics of pelage, shape and coloration of incisors, eye lens weight, and shape of head. By late summer the difference in size between young woodchucks is less clearly defined. Yearlings and adults molt from late May well into September, while young of the year do not begin the molt until July. Therefore, their pelage remains short and fine further into the season than that of older animals. The incisors in young of the year are narrow, long, and pointed and lack the brown stain characteristic of incisors of older animals. The head of young woodchucks is small and the muzzle is narrow and pointed. During early spring yearlings can be distinguished from adults by these same criteria. Additionally, from February through April yearling males do not yet have pigmented testes. Although Davis found no definite criteria to determine chronological age after the first year, he did demonstrate a linear increase in the weight of the dried eye lens of known-age animals.

Four age classes of the yellow-bellied marmot can be recognized by size and weight: juvenile, yearling, two year old, and three year old or older (Armitage et al. 1976). Barash (1973*b*) provided mortality rates for five age classes and survivorship curves of *M. olympus*. His five- to six-year maximum age was biased by his inability to determine the age of free-ranging animals past their third year.

ECONOMIC STATUS AND MANAGEMENT

Economic Status. Most problems concerning crop depredation or burrowing activities are localized and often are the work of one or two individuals. Only *M. monax* and *M. flaviventris* occur in areas where crops are grown. In regions where there are extensive areas of "mowed" crops (oats, wheat, etc.), damage can be extensive. The animals consume large amounts of these crops and other grasses. Grizzell (1955) estimated that adult woodchucks ate 0.5 kg of food per day. While making trails they trample additional plants, which cannot then be harvested. They occasionally are a serious problem to home gardens, and in some areas damage fruit trees. Lee saw woodchucks in Maryland climb apple trees and eat apples.

Marmots burrow into earth dams, canal banks, and irrigation ditches, sometimes causing breaks. Their soil mounds may damage the cutting blades of mowers, necessitating high-level cutting and consequent harvest loss in hay and alfalfa fields (Eadie 1954). Their burrows are a potential hazard to some domestic animals. Marmots may act as reservoirs for diseases such as tularemia and sylvatic plague, and harbor the tick vectors of spotted fever (Eadie 1954).

The value of marmot burrows as refuges for rabbits and other small animals has been mentioned by many authors (Phillips 1939; Gerstell 1939; Schoonmaker 1936; Chapman 1938; Tubbs 1936; and others). A good summary of den use by other species was provided by Grizzell (1955). It is apparent that in many areas this use enables large populations of certain species to persist that otherwise would lack adequate den sites. Schmeltz and Whitaker (1977) set live and snap traps in the entrances of 94 woodchuck burrows in Indiana. In addition to 35 woodchucks, the traps yielded 20 opossums (*Didelphis virginiana*), 2 masked shrews (*Sorex cinereus*), 10 short-tailed shrews (*Blarina brevicauda*), 19 cottontails (*Sylvilagus floridanus*), 104 white-footed mice (*Peromyscus*

leucopus), 2 meadow voles (*Microtus pennsylvanicus*), 32 house mice (*Mus musculus*), 2 meadow jumping mice (*Zapus hudsonius*), 8 raccoons (*Procyon lotor*), 1 red fox (*Vulpes vulpes*), and 1 gray fox (*Urocyon cinereoargenteus*). Marmots also provide local, small game hunting opportunities (Gilpin 1946; Eadie 1954).

Marmots are seldom used for food or fur in North America, except by native Indians, even though their flesh is palatable and the fur of the smaller European species is extensively used. The fur of *M. monax* is of poor quality, but the western species possess handsome pelts. Sometimes skins are dyed and sold as mink or sable. Additionally, marmots are good experimental animals for physiological and behavioral studies, and are easily maintained in captivity.

Management. Marmots sometimes are kept under control in western states by dropping poison into or near their dens, but fumigation (with carbon disulphide, carbon bisulfide, calcium cyanide, methyl bromide, gas cartridges, or exhaust from gasoline engines) is the control method in eastern states (de Vos and Merrill 1957; U.S.D.I. 1943; Silver 1928; Eadie 1954; Henderson and Craig 1932). Eadie (1954), Marsh and Howard (1977), Dudderar (1977), and others provided instructions and listed precautions for poisoning and gassing methods. Den gassing is the most practical method of controlling large numbers of marmots. Only active dens with fresh digging should be treated. Spring is the best time of year for control measures, since the animals are active and rearing young are still in the den. In addition, spring vegetation is not so dense as to hide den openings, and this time of year the active dens are also less likely to shelter other wildlife. Fumigants are not effective against hibernating woodchucks. All burrow openings should be closed with sod or soil and packed tightly after the gas has been released. Damp soil provides the most effective seal. Fumigants should not be used if there is danger of fumes escaping into occupied buildings. Carbon bisulfide and methyl bromide are phytotoxic and should not be used beneath or near valued plants. The burrows should be checked a week after fumigation and any opened holes should be treated again. Gas cartridges were effective in temporarily eliminating woodchucks from a 100-acre study area, but the area was recolonized the same year (de Vos and Merrill 1957).

Baiting may be more effective for animals that are numerous and live in rock outcrops or other areas where fumigation is difficult. Apples, carrots, and sweet potatoes treated with strychnine have been used as bait, but great care must be exercised in the use of such baits because of potential hazards to nontarget animals.

Marmots can also be controlled by shooting and trapping. Shooting is effective, and a patient marksman can greatly reduce a local marmot population in a few days. Vacant burrows, however, are quickly occupied by individuals from adjoining areas. Numbers 1½ or 2 steel traps, and probably the larger

Conibear traps, will hold marmots. Unbaited traps should be set in the burrow entrance, concealed with a light cover of dry grass, and secured firmly to prevent their being dragged into the burrows (Trump and Hendrickson 1943; Grizzell 1955). With effort marmots can be captured alive with wire mesh traps, but they do not readily enter solid box traps constructed of wood or metal (Trump and Hendrickson 1943). Traps normally are placed directly in runways and baited with cut vegetables. Leg-hold traps are more effective, especially when more than one or two individuals need to be removed. Marmots may be protected in some areas, and it is advisable to consult with local fish and game authorities before control methods are undertaken.

In past years bounties have provided successful control of marmots (Silver 1928; Howell 1915), but we are not aware of any major control attempt using bounties during the last 50 years.

Davis (1962) and Davis et al. (1964) presented data on the potential harvest of woodchucks and its effects on birth, mortality, and movement rates. These workers concluded that in response to removal of woodchucks (1,040 individuals from 600 acres in four years) there was decreased loss by emigration, increased birth rates, and increase in survival of the young. The populations of the harvested area and a control area were numerically indistinguishable at the end of the study. Maximum harvest could be maintained from a population of 100 individuals with a birth rate of 1.6, a probability of survival of adults of 0.57, and a probability of survival of the young of 0.27 if the harvest replaced all natural mortality.

CURRENT RESEARCH AND MANAGEMENT NEEDS

Early workers devoted most of their efforts to studying various aspects of marmot hibernation and to economic considerations. Taxonomic relationships are still in need of study, and many new taxonomic tools have become available since Howell's (1915) revision of the genus. The recent survey by Hoffman et al. (1979) of the relationships of amphiberingian marmots is a good case in point.

The biology of some marmot species has received little attention; other species have become classic models for behavioral and other disciplines of research.

As in the past, marmots are best managed on a local basis as needs arise, since they seldom cause problems over large parts of their ranges. The relatively low reproductive effort, especially of the western forms, seldom allows marmots to be included in the pest category, as many rodents are.

The authors thank the following personnel of the North Carolina State Museum of Natural History: Mary Kay Clark and Steven P. Platania for assistance in locating pertinent literature, Kathleen Wade and Cynthia Wilkerson for typing various drafts of this chapter, and John E. Cooper for kindly reviewing the text.

LITERATURE CITED

Albert, T. F., and Panuska, J. A. 1979. Postural variation in hibernating woodchucks (*Marmota monax*). Am. Midl. Nat. 101(1):223–225.

Anderson, D. C.,; Armitage, K. B.; and Hoffman, R. S. 1976. Socioecology of marmots: female reproductive strategies. Ecology 57:552–560.

Armitage, K. B. 1962. Social behavior of a colony of the yellow-bellied marmot. Anim. Behav. 10:319–331.

———. 1965. Vernal behavior of the yellow-bellied marmot. Anim. Behav. 13:59–61.

———. 1974. Male behavior and territoriality in the yellow-bellied marmot. J. Zool. London 172:233–265.

———. 1977. Social variety in the yellow-bellied marmot: a population-behavioral system. Anim. Behav. 25(3):585–593.

———. 1979. Food selectivity of yellow-bellied marmots. J. Mammal. 60(3):628–629.

Armitage, K. B., and Downhower, J. F. 1970. Interment behavior in the yellow-bellied marmot (*Marmota flaviventris*). J. Mammal. 51(1):177–178.

———. 1974. Demography of yellow-bellied marmot populations. Ecology 55:1233–1245.

Armitage, K. B.; Downhower, J. F.; and Svendsen, G. E. 1976. Seasonal changes in weights of marmots. Am. Midl. Nat. 96:36–51.

Armitage, K. B.; Johns, D.; and Anderson, D. C. 1979. Cannibalism among yellow-bellied marmots. J. Mammal. 60(1):205–207.

Bailey, E. D. 1965. The influence of social interaction and season on weight change in woodchucks. J. Mammal. 46:438–445.

Bailey, E. D., and Davis, D. E. 1965. The utilization of body fat during hibernation of woodchucks. Can. J. Zool. 43:701–707.

Barash, D. P. 1973a. Social variety in the yellow-bellied marmot (*Marmota flaviventris*). Anim. Behav. 21:579–584.

———. 1973b. The social biology of the Olympic marmot. Anim. Behav. Monogr. 6:171–245.

———. 1974a. The evolution of marmot studies: a general theory. Science 185:415–420.

———. 174b. The social behavior of the hoary marmot (*Marmota caligata*). Anim. Behav. 22:256–261.

Bee, J. W., and Hall, E. R. 1956. Mammals of Northern Alaska. Univ. Kans. Mus. Nat. Hist. Misc. Publ. 8. 309pp.

Beltz, A., and Booth, E. S. 1952. Notes on the burrowing and food habits of the Olympic marmot. J. Mammal. 33(4):495–496.

Beule, J. D. 1940. The ecological relationships between the woodchuck and the cottontail: their life history and management. M.S. Thesis. Penn. State Coll.

Bito, L. Z., and Roberts, J. C. 1974. The effects of hibernation on the chemical composition of cerebrospinal and intraocular fluids, blood plasma and brain tissue of the woodchuck, *Marmota monax*. Comp. Biochem. Physiol. A Comp. Physiol. 47(1):173–193.

Bopp, P. 1958. Fluchtdistanzen und territoriales verhalten beim murmeltier. Mitt. Naturforsch. Ges. Schaffhausen 26:1–6.

Bowdish, B. S. 1922. Tree-climbing woodchucks. J. Mammal. 3(4):259.

Bronson, F. H. 1964. Agonistic behavior in woodchucks. Anim. Behav. 12:470–478.

Brown, C. P. 1948. Woodchucks observed while fighting. J. Mammal. 29(1):70.

Burt, W. H., and Grossenheider, R. P. 1952 A field guide to

the mammals. Houghton Mifflin Co., Boston. 284pp.

Cahalane, V. H. 1947. Mammals of North America. Macmillan Co., New York. 682pp.

Chapman, F. B. 1938. Mr. Woodchuck, master conservationist. Ohio Conserv. Bull. 26–27.

Christian, J. J. 1962. Seasonal changes in the adrenal glands in woodchucks (*Marmota monax*). Endocrinology 71:431–447.

Cram, W. E. 1923. Another swimming woodchuck. J. Mammal. 4(4):256.

Couch, L. K. 1932. Late record of the yellow-bellied marmot. Murrelet 13(1):25.

Currier, W. W. 1949. The effect of den flooding on woodchucks. J. Mammal. 30:429.

Davis, D. E. 1962. The potential harvest of woodchucks. J. Wildl. Manage. 26(2):144–149.

———. 1964. Evaluation of characters for determining age of woodchucks. J. Wildl. Manage. 28(1):9–15.

———. 1967a. The annual rhythm of fat deposition in woodchucks (*Marmota monax*). Physiol. Zool. 40:391–402.

———. 1967b. The role of environmental factors in hibernation of woodchucks (*Marmota monax*). Ecology 48:683–689.

Davis, D. E.; Christian, J. J.; and Bronson, F. 1964. Effects of exploitation on birth, mortality, and movement rates in a woodchuck population. J. Wildl. Manage. 28(1):1–9.

de Vos, A., and Merrill, H. A. 1957. Results of a woodchuck control experiment. J. Wildl. Manage. 21(4):454–456.

de Vos, A.; Merrill, H. A.; and Armitage, K. B. 1971. The yellow-bellied marmot and the evolution of polygamy. Am. Nat. 105:355–370.

Dudderar, G. 1977. Controlling vertebrate damage/ woodchucks. Michigan State Univ. Ext. Bull. E-866. [1p.]

Eadie, W. R. 1954. Animal control in field, farm, and forest. Macmillan Co., New York. 257pp.

Fraser, D. 1979. Aquatic feeding by a woodchuck. Can. Field Nat. 93(3):309–310.

Fryxell, F. M. 1928. Melanism among the marmots of the Teton Range, Wyoming. J. Mammal. 9(4):336–337.

Gerstell, R. 1939. The value of groundhog holes as winter retreats for rabbits. Pennsylvania Game News 10(6):6–9.

Gianini, C. A. 1925. Tree-climbing and insect-eating woodchucks. J. Mammal. 6(4):281–282.

Gilpin, J. J. 1946. Hunters stalk woodchuck. Kentucky Wildl. 1(3):27–28.

Glover, F. A. 1943. A furious woodchuck. J. Mammal. 24(3):402.

Goodwin, G. G. 1935. The mammals of Connecticut. State Geol. Nat. Hist. Surv. Bull. 53:97–100.

Gordon, K. 1936. Territorial behavior and social dominance among Sciuridae. J. Mammal. 17(2):171–172.

Grizzell, R. A., Jr. 1955. A study of the southern woodchuck, *Marmota monax monax*. Am. Midl. Nat. 53:257–293.

Hall, E. R. 1930. Groundhog active in winter. Can. Field Nat. 44:198.

Hall, E. R., and Kelson, K. R. 1958. The mammals of North America. 2 vols. Ronald Press, New York. 1083pp.

Hamilton, W. J., Jr. 1930. Notes on the mammals of Breathitt County, Kentucky. J. Mammal. 11:310.

———. 1934. The life history of the rufescent woodchuck, *Marmota monax rufescens*. Ann. Carnegie Mus. 23:85–178.

Henderson, J., and Craig, E. L. 1932. Economic mammalogy. Charles C. Thomas, Springfield, Ill. 397pp.

Henderson, J.; Craig, E. L.; and Gilbert, F. F. 1978. Distribution and density of woodchuck burrow systems in relation to land-use practices. Can. Field Nat. 92(2):128–136.

Hickman, C. P. 1922. Woodchuck climbs trees. J. Mammal. 3(4):260–261.

Hoffman, R. S., and Nadler, C. F. 1968. Chromosomes and systematics of some North American species of the genus *Marmota*. Experentia 24:740–742.

Hoffman, R. S.; Koeppl, J. W.; and Nadler, C. F. 1979. The relationships of the Amphiberingian marmots (Mammalia: Sciuridae). Univ. Kans. Mus. Nat. Hist. Occas. Pap. 83. 56pp.

Holt, E. G. 1924. Woodchuck in Alabama. J. Mammal. 5:67–68.

Howell, A. H. 1915. Revision of the American marmots. North Am. Fauna 37:1–80.

Hoyt, S. Y., and Hoyt, S. F. 1950. Gestation period of the woodchuck, *Marmota monax*. J. Mammal. 31(4):454.

Johnson, A. M. 1926. Tree-climbing woodchucks again. J. Mammal. 7(2):132–133.

Johnson, C. E. 1923. Aquatic habits of the woodchuck. J. Mammal. 4(2):105–107.

Jones, G. W. 1937. The ear muscles of the woodchuck. J. Mammal. 18(4):517.

Kilgore, D. L., Jr., and Armitage, K. B. 1978. Energetics of yellow-bellied marmot populations. Ecology 59:78–88.

King, J. A. 1955. Social behavior, social organization, and population dynamics in a black-tailed prairie dog town in the Black Hills of South Dakota. Contr. Lab. Vent. Bio. Univ. Michigan 67.

Knight, R. L., and Erickson, A. W. 1978. Marmots as a food source of golden eagles along the Columbia river. Murrelet 59(1):28–30.

Lambert, H. 1939. White woodchucks. Nature 32(3):134.

Leedy, D. L. 1949. Woodchucks survive brush fire and remain in the area. J. Mammal. 30(1):73.

Linduska, J. P. 1947. Longevity of some Michigan farm game mammals. J. Mammal. 28:126–129.

Lloyd, J. E. Vocalization in *Marmota monax*. J. Mammal. 52(1):214–216.

Lowery, G. H. The mammals of Louisiana and its adjacent waters. Louisiana State Univ. Press, Baton Rouge. 565pp.

Marsh, R. E., and Howard, W. E. 1977. Vertebrate pest control manual. Part 2. Pest control 45(9):22–31.

Medsger, O. P. 1922. The tree-climbing habits of woodchucks. J. Mammal. 3(4):261–262.

Merriam, C. H. 1886. The mammals of the Adirondack Region. Henry Holt & Co., New York. 316pp.

Merriam, H. G. 1971. Woodchuck burrow distribution and related movement patterns. J. Mammal. 52:732–746.

Miller, V. M., and South, F. E. 1974. Termogensis in response to spinal cord cooling in the marmot. Fed. Proc. 33(3): part 1, p. 423.

Mitteil, S. 1955. Der schrei des murmeltiers als akustische territoriumsmarkierung. P. Bopp. Saugetierk. Mitt. 3(1):28.

Mossman, H. W.; Lawlay, J. W.,; and Bradley, J. A. 1932. The male reproductive tract of the Sciuridae. Am. J. Anat. 51(1):89–155.

Muller-Using, D. 1954. Beitrage zur Oekologie der *Marmota marmota marmota*. Z. Saeugetierk. 19:166–167.

———. 1955. Vom 'pfeifen' des murmeltieres. Z. Jagdwiss. 1:32–33.

———. 1956. Zum verhalten des murmeltieres. Z. Tierpsychol. 13:135–142.

Munch, H. 1958. Zur okologie und psychologie von *Marmota m. marmota*. Z. Saeugetierk. 23:129–138.

Murie, A. 1961. A naturalist in Alaska. Devin-Adair.

Nee, J. A. 1969. Reproduction in a population of yellow-bellied marmots (*Marmota flaviventris*). J. Mammal. 50:756-765.

Phillips, J. C. 1923. A swimming woodchuck. J. Mammal. 4(4):256.

Phillips, J. M. 1939. Save the woodchuck. Pennsylvania Game News 10(2):12.

Powell, R. A. 1972. Predation on marmots. Rep. to Rocky Mountain Bio. Lab. 9pp.

Rall, Y. M. 1945. The dynamic density of rodents and some methods of studying it. Byelleten Moskovokogo Obshchestva Ispitatcle Prorodi Sect. Zool. 50:62-64.

Rasmussen, A. T. 1915. The oxygen and carbon dioxide content of the blood during hibernation in the woodchuck (*Marmota monax*). Am. J. Physiol. 39(1):20-30.

———. 1916. A further study of the blood gases during hibernation in the woodchuck (*Marmota monax*). Am. J. Physiol. 41(4):464-482.

———. 1917. The volume of the blood during hibernation and other periods of the year in the woodchuck (*Marmota monax*). Am. J. Physiol. 44(2):132-148.

Rausch, R. 1953. On the status of some arctic mammals. Arctic 6:91-148.

Rausch, R., and Rausch, V. R. 1965. Cytogenetic evidence of the specific distinction of an Alaskan marmot, *Marmota broweri* (Hall and Gillmore). Chromosoma 16:618-623.

———. 1971. The sometic chromosomes of some North American marmots.

Rausch, R., and Tiner, J. D. 1946. *Obeliscoides cuniculi* from the woodchuck in Ohio and Michigan. J. Mammal. 27(2):177-178.

Robb, W. H. 1926. Another tree-climbing woodchuck. J. Mammal. 7(2):133.

Robinson, S. S., and Lee, D. S. 1980. Recent range expansion of the woodchuck, *Marmota monax*, on the Southeast. Brimleyana 3:43-48.

Rutter, R. J. 1930. Winter rambling of a woodchuck. Can. Field Nat. 44(9):213.

Schmeltz, L. L., and Whitaker, J. O., Jr. 1977. Use of woodchuck burrows by woodchucks and other mammals. Trans. Kentucky Acad. Sci. 38(1-2):79-82.

Schoonmaker, W. J. 1936. The value of woodchucks. Nature 27(5):302-303.

———. 1938. The woodchuck: lord of the clover field. Bull. New York Zool. Soc. 41(1):3-12.

Scott, T. G. 1937. Mammals of Iowa. J. Sci. 12:67-68.

Seton, E. T. 1928. Lives of game animals. Doubleday, Doran & Co., Garden City, N.Y. 949pp.

Shirer, W. H., and Downhower, J. F. 1969. Radio tracking of dispersing yellow-bellied marmots. Kansas Acad. Sci. Trans. 71(4):463-479.

Silver, J. 1928. Woodchuck control in the eastern states. U.S. Dept. Agric. Leafl. 21. 6pp.; Fur J. 2(5):26, 48-50.

Smith, J. D., and Hearn, G. W. 1979. Ultrastructure of the apocrine-sebaceous anal scent gland of the woodchuck, *Marmota monax*: evidence for apocrine and melocrine secretion by a single cell type. Anat. Rec. 193(2):269-292.

Smith, M. C. 1972. Seasonal variation in home ranges of woodchucks. M.S. Thesis. Univ. Guelph, Ontario, 67pp.

Smith, M. C., and Christian, J. J. 1960. Reproductive cycles and litter size in the woodchuck. Ecology 4:647-655.

Smith, M. C.; Davis, D. E.; and Christian, J. J. 1961. Seasonal changes in the weights of woodchucks. J. Mammal. 42:297-312.

Stoner, D. 1922. Another tree-climbing woodchuck. J. Mammal. 3(4):260.

Svendsen, G. E. 1974. Behavioral and environmental factors in the spatial distribution and population dynamics of a yellow-bellied marmot population. Ecology 55:760-771.

Swarth, H. S. 1911. Two new species of marmots from Northwestern America. Univ. California Publ. Zool. 7:201-204.

Trump, R. F. 1943. Ways of the woodchuck. Nat. Hist. 52(5):221-225.

Trump, R. F., and Hendrickson, G. O. 1943. Methods for trapping and tagging woodchucks. J. Wildl. Manage. 7(4):420-421.

Tubbs, F. F. 1936. Woodchucks may aid the rabbit supply. Mich. Conserv. 5(12):4.

Twichell, A. R. 1939. Notes on the southern woodchuck in Missouri. J. Mammal. 20:71-74.

Twidale, J. K. 1936. A swimming woodchuck. Field London 168(4362):272.

U.S.D.I. 1943. Control of woodchucks. U.S. Fish Wildl. Serv. Wildl. Leafl. 237. 4pp.

Walker, E. P. 1964. Mammals of the world. Johns Hopkins Press, Baltimore.

Wallace, F. C. 1942. The stomach worm, *Obeloscoides cunicali*, in the woodchuck. J. Wildl. Manage. 6:92.

Waring, G. H. 1966. Sounds and communications of the yellow-bellied marmot. Anim. Behav. 14:177-183.

———. 1965. Behavior of a marmot toward a martin. J. Mammal. 46(4):681.

Warren, E. R. 1926. Dwarf marmots. J. Mammal. 7(4):332-333.

———. 1935. Notes on the voice of marmots. J. Mammal. 16(2):152-153.

Weeks, H. P., and Kirkpatrick, C. M. 1978. Salt preferences and sodium drive phenology in fox squirrels and woodchucks. J. Mammal. 59:531-542.

Wenberg, G. M., and Holland, J. C. 1973. The circannual variations of some of the hormones of the woodchuck, *Marmota monax*. Comp. Biochem. Physiol. A Comp. Physiol. 46(3):523-535.

Wenberg, G. M.; Holland, J. C.; and Sewell, J. 1973. Some aspects of the hematology and immunology of the hibernating and nonhibernating woodchuck, *Marmota monax*. Comp. Biochem. Physiol. A Comp. Physiol. 46(3):513-521.

Winders, R. L. 1974. Hematogenous adjustments to simulated high altitude in *Marmota flaviventris*. Fed. Proc. 33(3):part 1, p. 308.

Wu, L. T. 1923. The Tarbagan or Siberian marmot: researches into their relation with bubonic plague in man. China J. Sci. Arts 1:39-50.

Yeager, L. E., and Anderson, H. G. 1944. Some effects of flooding and waterfowl concentrations on mammals of a refuge area in central Illinois. Am. Midl. Nat. 31:166-167.

Zimmerman, K. 1955. Zum murmeltierschrei bei annaeherung eines menschen. Saeugetierk. Mitt. 3(3):125.

DAVID S. LEE, North Carolina State Museum of Natural History, P.O. Box 27647, Raleigh, North Carolina 27611.

JOHN B. FUNDERBURG, North Carolina State Museum of Natural History, P.O. Box 27647, Raleigh, North Carolina 27611.

10

Ground Squirrels

Spermophilus beecheyi and Allies

P. Quentin Tomich

NOMENCLATURE

COMMON NAMES. Ground squirrel, gopher, picket pin, spermophile, antelope squirrel, prairie squirrel, rock squirrel

SCIENTIFIC NAMES. *Ammospermophilus harrisii, A. leucurus, A. interpres, A. nelsoni, Spermophilus townsendii, S. washingtoni, S. brunneus, S. richardsonii, S. armatus, S. beldingi, S. columbianus, S. undulatus, S. tridecemlineatus, S. mexicanus, S. spilosoma, S. franklinii, S. variegatus, S. beecheyi, S. mohavensis, S. tereticaudus, S. lateralis,* and *S. saturatus*

SUBSPECIES. The 22 species are further classified into a formidable array of 93 geographic forms enumerated, but not listed, in table 10.1.

The genera *Ammospermophilus* and *Spermophilus* collectively encompass the "ground squirrels," because of their common terrestrial and burrowing habits, and I follow this convention. Howell (1938) designated the genus *Citellus* Oken, 1816, to include all of the terrestrial squirrels, under several subgenera. Simpson (1945) accepted *Citellus* for the entire assemblage and equated it with *Spermophilus* Cuvier, 1825. Bryant (1945) concluded that full generic status should be accorded *Ammospermophilus* and separated it from *Citellus*, providing two genera of terrestrial squirrel. This revision is widely accepted today.

Hershkovitz (1949) declared that Oken's 1816 list, including *Citellus*, was non-Linnaean and, therefore, unavailable under the international rules of nomenclature. Hence, *Citellus* has dropped out of usage in favor of *Spermophilus*. The latter name is the source of the vernacular "spermophile" applied in the 19th-century literature. Indeed, these are seed-loving rodents and from this propensity derives much of their conflict with the economic interests of humans.

Hall and Kelson (1959) and Hall (1981) provided the most comprehensive review of ground squirrels treated in this chapter, and arranged *Spermophilus* into six subgenera. The genera *Ammospermophilus* (no subgenera) and *Spermophilus* embrace 22 species, 8 of which are monotypic; 14 are ranked into 2 to 13 subspecies, for a total of 93 (see table 10.1). Six additional American species are found in Baja California and Mexico.

MacClintock (1970) effectively describes the various ground squirrels, provides drawings of several species, and presents data on behavior, habitat, and ecological relations.

DISTRIBUTION AND PHYLOGENY

Ground squirrels demonstrate a particularly broad variety of adaptations to various conditions of climate, topography, altitude, and latitude. The species of *Ammospermophilus* range mainly in the southwestern United States and Mexico, especially in arid and mountainous regions. *Spermophilus* sp. are widely distributed. The circumpolar *S. undulatus*, the arctic ground squirrel, is found in North America from St. Lawrence Island and the Aleutians to a remarkable 71° N latitude in the vicinity of Point Barrow, Alaska, south to 56° N latitude in British Columbia and east to Hudson's Bay.

The most wide ranging is *S. tridecemlineatus*, the 13-lined ground squirrel, from Texas to Michigan and eastern Ohio. Ranges of those species of ground squirrel found in the United States and Canada are given in figures 10.1–10.6. They are keyed for reference in table 10.1, which summarizes size characteristics.

The Great Basin pluvial lakes region of the Pleistocene appears to have been the metropolis of differentiation in ground squirrels. The largest and once extensive Lake Bonneville survives principally as Great Salt Lake in Utah (Thornbury 1965). Nineteen species of ground squirrel are found within 500 miles of its present shores. Excluded are only the arctic *S. undulatus* and 8 Mexican species, some of which may represent a more southerly nucleus of evolutionary change in the Mexican highland sector. Fifteen of the 22 species north of Mexico occur in the four Great Basin states and California. Nine are found in Utah, with 8 each in Idaho, Oregon, and Nevada. California supports 9, 7 of which are shared by one or more of the above states. Durrant and Hansen (1954) discussed sympatry, allopatry, and phylogeny in 5 species of the subgenus *Spermophilus* in the Great Basin. They concluded that *S. townsendii* is the most divergent, with *S. richardsonii* and *S. armatus* next; *S. columbianus* and *S. beldingi* are more conservative and closer to the ancestral type.

Chromosome karyotyping tends to confirm previous systematic examination of ground squirrels based on gross morphological characteristics. Nadler and Hughes (1966) examined *Spermophilus mexicanus, S. tridecemlineatus,* and *S. spilosoma* (subgenus *Ictidomys*) and concluded that all were correctly classified. Nadler (1966) did similar work on seven of the eight species of the subgenus *Spermophilus* and produced similar results, suggesting that two species groups are present. *S. townsendii, S. washingtoni,* and possibly *S. brunneus,* are segregated from the other five species.

Starch-gel electrophoresis carries differentiation to the genic level and has extended systematics to much finer details. Cothran et al. (1977), in a study of 34 natural populations of species in *Ictidomys,* showed that *Spermophilus tridecemlineatus* and *S. mexicanus* are not widely diverged as species. These and other studies lead to an improved understanding of the ground squirrel complex and better population management practices.

DESCRIPTION

Ground squirrels are typically short legged and may have rounded ears. The tail is often short and terete, though it may be long and brushlike. The eyes are fairly large and set high in the head. Ventral pelage coloration is usually light. Dorsal coloration is generally a grizzled yellowish brown or gray, and buffy whitish spots may be present. The dorsal pelage of *S. tridecemlineatus* has alternating dark and light stripes with inner rows of yellowish white spots. There are four to seven pairs of mammae. The postorbital processes of ground squirrels are well developed. Unlike tree squirrels (*Sciurus* sp.), ground squirrels have zygomatic arches that expand posteriorly (figure 10.7). Cheek teeth are high crowned and the dental formula is 1/1, 0/0, 1-2/1, 3/3 = 20-22.

ANNUAL CYCLE

Ground squirrels respond strictly to the progression of the seasons and maintain a cycle of activity that is repeated annually. A basic endogenous control of this rhythm may be observed in the laboratory. However, endogenous control results in shortening of the cycle when not reset to external cues of photoperiod and temperature as found in the accustomed natural environment. Blake (1972) reviewed this facet in some detail and found that in members of the genus *Spermophilus* the annual cycle is composed of successive periods of fattening, torpor, and reproduction. This cycle is apparently reset upon emergence from a variable degree of winter dormancy (Michener 1979). The timing of torpor differs and usually is called estivation when the animals go below ground in summer, as in some southern populations, and hibernation when tor-

por begins with cold weather in northern and high-altitude populations. Other aspects of adaptation to climatic conditions are the amounts of fat stored to carry through periods of torpor, food consumption, and the degree of reduction in metabolic functions (Davis 1976; Walker et al. 1979).

Ammospermophilus departs from this cycle and may not have periods of torpidity comparable to estivation and hibernation. *A. harrisii* is active the year around in southern Arizona (Neal 1965*a*), and Hawbecker (1958) found no evidence of dormancy in *A. leucurus* in the western San Joaquin Valley of California. The animals generally are active in the mild winter climates of their range, and remain so in the heat of the summer. Torpidity is induced in *A. leucurus* by cold stress but usually the animals have no physiological means of arousal, lose weight rapidly, and the mortality rate is high (Kramm 1972).

Smith (1973) induced torpor in *Spermophilus spilosoma* and concluded that this southern species is capable of true hibernation. There is some dispute as to which populations of the species normally hibernate. Under controlled conditions the test animals displayed a "typical pattern" of hibernation; that is, periods of about 5 days of activity before again entering hibernation. The golden-mantled ground squirrel (*S. lateralis*), a temperate climate forest species, held at 4 to 6° C aroused every 5 or 6 days near the beginning and end of hibernation, but individual periods of dormancy lasted about 14 days in midwinter (Jameson 1964). In central California Linsdale (1946) observed adult male *S. beecheyi* entering estivation as early as late May. Females followed when maternal duties were completed and fat stores accumulated, from June to August. Young of the year begin to hibernate when they approach adult size and attain stores of fat. In some colonies, however, a few are active through the winter, except on the coldest days. Food stored in the burrows is eaten during periods of arousal, although body fat is highly important as an energy source during seasons of reduced activity. *S. mohavensis,* a true desert species of southern California, remains underground from August to February. However, in the laboratory it demonstrated alternate periods of dormancy and wakefulness for 3 to 5 days at a time (Bartholomew and Hudson 1960).

Loehr and Risser (1977) observed that *S. beldingi* was active only 4 to 5 months of the year in mountain meadows at 1860 meters elevation in the Sierra Nevada where the snow cover is heavy. Alcorn (1940) found *S. townsendii* active for only 4 to 4.5 months in similar severe winter climates of Nevada. He noted that if the summer food supply is not sufficient to fatten the squirrels in 120 to 135 days they may stay out longer or emerge for a time in autumn.

Blake (1972) described two simultaneous cycles of rhythm of the golden-mantled ground squirrel, a hibernal cycle and the reproductive cycle. Overlap results in five physiological stages, and this pattern is also observed in *S. tridecemlineatus.* These cyclic changes in the reproductive system, endocrine glands, body temperature, and water requirements are affirmed

Table 10.1. Summary of American ground squirrels north of Mexico

Scientific Name	Common Name	Number of Subspecies	General Range	Figure	Characteristics of Size (in millimeters)			
					Length	Tail	Ear	Skull
Ammospermophilus								
A. harrisii	Harris's antelope squirrel	2	Arizona, N. Mexico	1	220–250	74–94	38–42	38.2–41.2
A. leucurus	white-tailed antelope squirrel	9	Great Basin to Baja California	1	194–239	54–87	35–43	
A. interpres	Texas antelope squirrel	0	S. New Mexico, W. Texas, and Mexican plateau	1	220–235	68–84	36–40	
A. nelsoni	Nelson's antelope squirrel	0	W. Central California	1	218–240	63–79	40–43	
Spermophilus subgenus *Spermophilus*								
S. townsendii	Townsend's ground squirrel	7	Great Basin region	2	167–271	32–72	29–38	32.4–43.3
S. washingtoni	Washington's ground squirrel	0	E. Washington and N. Oregon	3	185–245	32–65	30–38	35.0–41.4
S. brunneus	Idaho ground squirrel	0	SW. Idaho	3	214–252	46–61	33–37	38.7–40.8
S. richardsonii	Richardson's ground squirrel	4	Great Basin, Canadian and U.S. prairies	3	243–337	65–100	39.5–49	42.0–48.6
S. armatus	Uinta ground squirrel	0	Great Basin states	5	280–303	63–81	42–45.5	46.3–48.5
S. beldingi	Belding's ground squirrel	3	Great Basin and Sierra Nevada	5	254–300	55–76	40–47	41.3–46.3
S. columbianus	Columbian ground squirrel	2	Rocky Mountains of S. Canada and N. United States	4	327–410	80–116	48–58	49.5–57.0

Species	Common name		Geographic range					
S. undulatus	arctic ground squirrel	8	arctic and subarctic Canada	2	332–495	77–153	50–68	50.7–65.8
subgenus *Ictidomys*								
S. tridecemlineatus	13-lined ground squirrel	10	Canadian prairies to Ohio; Great Plains to Gulf of Mexico		170–297	60–132	27–41	34.0–45.8
S. mexicanus	Mexican ground squirrel	2	S. Texas to central Mexico	2	280–380	110–166	38–51	41.0–52.5
S. spilosoma	spotted ground squirrel	13	central U.S. plains to central Mexico	6	185–253	55–92	28–38	34.1–42.7
subgenus *Poliocitellus*								
S. franklinii	Franklin's ground squirrel	0	S. Canadian Plains to central United States	5	381–397	136–153	53–57	52.1–55.1
subgenus *Otospermophilus*								
S. variegatus	rock squirrel	8	SW. United States to S. Mexico and Baja California	4	430–525	172–252	53–65	56.0–67.7
S. beecheyi	California ground squirrel	8	S. Washington to Baja California	4	357–500	145–200	50–64	51.6–62.4
subgenus *Xerospermophilus*								
S. mohavensis	Mohave ground squirrel	0	small region of California desert	5	210–230	57–72	32–38	38.1–40.0
S. tereticaudus	round-tailed ground squirrel	4	S. Nevada to central Mexico	2	204–266	60–107	32–40	34.9–39.3
subgenus *Callospermophilus*								
S. lateralis	golden-mantled ground squirrel	13	disjunct in S. Canadian Rockies to Arizona; California to Colorado	6	230–308	63–118	35–46	39.6–45.6
S. saturatus	cascade golden-mantled ground squirrel	0	S. British Columbia and Washington	6	286–315	92–118	43–49	44.0–48.3

SOURCE: Hall and Kelson (1959), Hall (1981).

FIGURE 10.1. Distribution of the white-tailed antelope squirrel (*Ammospermophilus leucurus*), Nelson's antelope squirrel (*A. nelsoni*), Harris's antelope squirrel (*A. harrisii*), and Texas antelope squirrel (*A. interpres*).

FIGURE 10.2. Distribution of the arctic ground squirrel (*Spermophilus undulatus*), 13-lined ground squirrel (*S. tridecemlineatus*), Townsend's ground squirrel (*S. townsendii*), and round-tailed ground squirrel (*S. tereticaudus*).

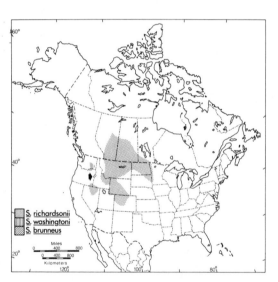

FIGURE 10.3. Distribution of the Richardson's ground squirrel (*Spermophilus richardsonii*), Washington ground squirrel (*S. washingtoni*), and Idaho ground squirrel (*S. brunneus*).

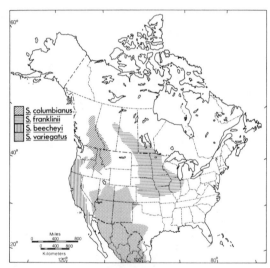

FIGURE 10.4. Distribution of the Columbian ground squirrel (*Spermophilus columbianus*), Franklin's ground squirrel (*S. franklinii*), California ground squirrel (*S. beecheyi*), and rock squirrel (*S. variegatus*).

by data from several other species. In *S. beecheyi* changes are especially remarkable in the testes and accessory glands (Tomich 1962). The testes begin to enlarge in November and reach the maximum combined size of 10 to 12 grams, or about 1.3 percent of body weight, in January (figure 10.8). Accessory glands reach maximum size in March to May, when the testes are rapidly exhausted during the breeding season. All males are producing motile sperm in March and April, and in some individuals as early as January. However, only when the females have reached a simi-

lar state of readiness can effective mating occur. Extreme dates of sexual activity span a period of 81 to 87 days and the pattern of birth dates is bimodal, from April to July in central California (Tomich 1962).

The annual cycle of ground squirrels has been described for most species. The pattern is similar, varying within a species in response to local weather, climate, and availability of foodstuffs as widely as it does among several different species over the entire ranges of *Ammospermophilus* and *Spermophilus*. Rausch (1951) found *S. undulatus* abundant in the

FIGURE 10.5. Distribution of the Belding's ground squirrel (*Spermophilus beldingi*), Uinta ground squirrel (*S. armatus*), Mohave ground squirrel (*S. mohavensis*), and spotted ground squirrel (*S. spilosoma*).

FIGURE 10.6. Distribution of the golden-mantled ground squirrel (*Spermophilus lateralis*), Cascade golden-mantled ground squirrel (*S. saturatus*), and Mexican ground squirrel (*S. mexicanus*).

Brooks Range of central Alaska (68° N latitude) and common to at least 1067 m altitude. Snow cover in this region is light. Hibernation began usually in late September or early October, and squirrels emerged very early in spring, about 15 March to 15 April. The intermittent spring snows and cloudy days kept the squirrels below ground for much of the time, but by late April they were active regularly on bright, sunny days. This species is active for about six months of the year, considerably longer than some more southern species, apparently in response to the light snow cover.

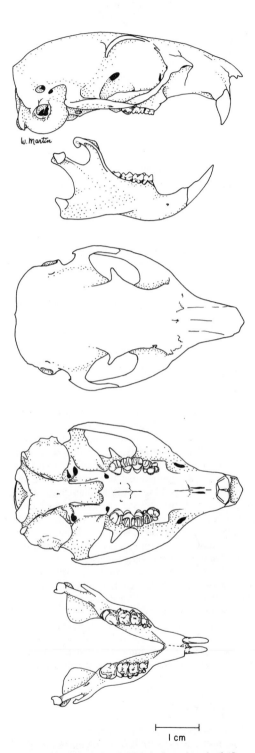

FIGURE 10.7. Skull of the California ground squirrel (*Spermophilus beecheyi*). From top to bottom: lateral view of cranium, lateral view of mandible, dorsal view of cranium, ventral view of cranium, dorsal view of mandible.

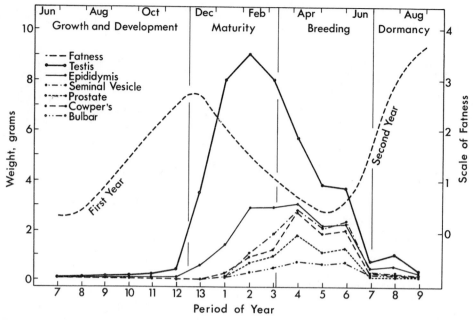

FIGURE 10.8. Annual cycle of growth, reproductive apparatus, torpidity, and fatness in the male California ground squirrel (*Spermophilus beecheyi*). First-year animals leave the dens in June, then reach maturity and are heavy with fat by November. Periods of hibernation are brief, or there may be none. Second-year animals store fat earlier and in larger amounts, then begin estivation by June or July, remaining underground until January or February. Females follow a similar pattern but may begin estivation somewhat later. For convenience, the year was divided into 13 four-week periods. The heavy solid line represents degree of fatness. Adapted from Tomich 1962.

The onset of breeding may be controlled by seasonal temperatures. Michener (1973) found that *S. richardsonii*, after three years of normally cold spring seasons, advanced the breeding season about 10 days to coincide with warmer weather in the fourth year. Tomich (1962) reported that in two successive years, seasonal and site differences in median dates of breeding fluctuated between 9 and 16 days in populations of *S. beecheyi* 20 km apart. This was correlated with lateness or earliness of the season as reflected in phenology of agricultural crops keyed to temperature and rainfall regimes of flat valley lands.

The round-tailed ground squirrel (*S. tereticaudus*), a desert-adapted hibernating species (Neal 1965*a*), responds to variations in winter rainfall, as reflected in the production of spring vegetation, for the timing of its breeding season (Reynolds and Turkowski 1972). Low rainfall delays the breeding season and each additional 12 mm of precipitation in January and February advance the onset of breeding about nine days.

REPRODUCTION

Normal gestation in ground squirrels is known to vary by species from 24 to 30 days, but in many forms exact determinations have not been made (table 10.2). Litter size generally averages 6 to 8, consistent with the high capacity for reproduction in ground squirrels. Shaw (1925) found an average of only 4.6 in five litters of *S.*

columbianus; maximum litter size is 12 to 13 in other species. Several workers have noted geographic variation in the number of young produced by *S. beecheyi*. Chapman and Lind (1973) summarized these data and described an inverse relationship between latitude and mean litter size. Where the freeze-free period is about 325 days in southern California (34° N latitude), litter size averaged 8.4. In central Oregon, where the freeze-free period is 195 days (44° 30′ N latitude), litter size averaged 5.5. They interpreted the larger litter sizes as an adaptation to more southern latitudes where *S. beecheyi* may be active the year around and is susceptible to a higher annual predation.

Growth and Development. Shaw (1918, 1925) was perhaps the first to describe postnatal growth and development of ground squirrels. He studied the Colum-

TABLE 10.2. Examples of gestation periods in ground squirrels

Species	Number of Days	Authority
Ammospermophilus harrisii	at least 29	Neal 1965*b*
Spermophilus townsendii	about 24	Alcorn 1940
S. richardsonii	28	Michener 1973
S. beldingi	28	Sauer 1976
S. columbianus	24	Shaw 1925
S. beecheyi	30	Tomich 1962
S. lateralis	27 to 28	Cameron 1967

bian ground squirrel in a "citellery," or enclosure built over dens of wild squirrels. It also included smaller pens where pregnant females were caged and provided with nest boxes where broods could be examined readily. For its time it was certainly an advanced approach to research. Clark (1970) added the Richardson ground squirrel to a growing list of species for which extensive data were available: Harris's (Neal 1965*b*) and golden-mantled ground squirrels (McKeever 1964; Clark and Skryja 1969). Data are now reported also for the Belding (Morton and Tung 1971) and Franklin species (Turner et al. 1976).

Newborn ground squirrels are typical altricial rodents: the bare skin is reddish in color, eyes and ears are closed, no teeth are erupted, and locomotion is restricted to aimless writhings. Development and growth are rapid, so head and body hair is generally present by 4 to 14 days, incisors erupt at 7 to 21 days, eyes open in 20 to 37 days, and the fully active pups are weaned to independence at 22 to 48 days. Morton and Tung (1971) established a growth rate constant for each of eight species (table 10.3). This factor was 0.043 for the southern nonhibernating spotted ground squirrel; it graded upward to 0.089 for the faster-growing arctic ground squirrel of much higher latitudes. Clark (1970) had described a cline in chronology of physical characteristics and behaviorisms, citing earlier maturity in the hibernators. A general pattern emerges in which the long-period hibernators are weaned at 22 to 35 days and the strictly nonhibernating species at about 48 days. The accelerated growth of the hibernators appears to be an adaptation to a brief season of abundant foods in which the young must develop survival skills and store fat quickly.

ECOLOGY

Populations and Dispersal. Ground squirrels are highly adaptable mammals, within the range of environmental conditions that permit them to live and perhaps to thrive. Given particularly severe ecological stresses, a local or regional population may be extirpated. When conditions are optimal, ground squirrel populations often increase to larger than usual densities or occupy habitats that were formerly vacant. Gener-

ally there is a long-term equilibrium and species survive the interplay of genetic make-up and external factors in their surroundings.

Mankind has played an important part in promoting the abundance of ground squirrels. Disturbance of prairie lands for the planting of cereal crops in mid-America tended to improve the environment of these animals. Similar increases were noted in the western states with the introduction of intensive grazing and the conversion of sagebrush (*Artemisia* sp.) lands to irrigated cropland (White 1972). This process is reversible, as demonstrated by studies in coastal rangelands of California (Linsdale 1946). When farming was discontinued at a site, including streamside grain fields and pasture, a population decline occurred for *S. beecheyi* after three years. In six years the squirrels had abandoned the land. Yet, on adjacent lands where farming was uninterrupted, population densities were stable. In another study, on intensively farmed flat lands of the Sacramento Valley, *S. beecheyi* responded to the rotation of crops (Tomich 1962). Colonies were established usually along fence lines and drainage ditches, with seasonal use of the fields. The squirrels had choices of crops at either side of a fence, and weedy plants in the fence rows and other waste areas. Cultivated barley was the choice food and there was evidence that the vitality of colonies was affected by rotation farming. Barley or oats supported thriving colonies; fields of beans, tomatoes or sugar beets supported smaller or dwindling colonies.

Evans (1951) reported that *S. tridecemlineatus* extended its range appreciably to the north in Michigan and Wisconsin, and eastward in Ohio, since 1900. Apparently, this was the result of clearing more land, yet the species was not considered a serious pest. *S. townsendii*, formerly the most conspicuous mammal of uncultivated lands over much of its range, suffered dramatic reductions in its numbers, beginning in 1936 in Utah and Washington and shortly thereafter in Nevada. Populations had not recovered in numbers and distribution and remained restricted over much of the natural range of the species 18 years later (Hansen 1956). In this instance, sylvatic plague, caused by the bacterium *Yersinia pestis*, was suspected as the major causative agent of population suppression, rather than climatic conditions. A colony of *S. richardsonii* in

TABLE 10.3. Growth rates in 8 species of *Spermophilus*

Species	Life Style	N	Growth Rate Constant (K)	Reference
S. spilosoma	nonhibernator	6	0.043	Blair (1942)
S. harrisii	nonhibernator	13	0.049	Neal (1965)
S. beecheyi	occasional hibernator	13–34	0.052	Tomich (1962)
S. tridecemlineatus	hibernator	12	0.053	Bridgewater (1966)
S. tereticaudus	hibernator	26	0.057	Neal (1965)
S. beldingi	hibernator	31–33	0.066	Morton and Tung 1971
S. richardsoni	hibernator	6–28	0.071	Clark (1970)
S. undulatus	hibernator	9	0.089	Mayer and Roche (1954)

SOURCE: Morton and Tung 1971

North Dakota was flooded for eight days by overflow from an adjacent river, yet the animals survived and began to emerge from hibernation a week after the waters subsided (Quanstrom 1966). The dense black clay soils of the river bottom were frozen at the time of the flooding and protected the nest chambers. The owner of the property stated that the land had been flooded several times in the previous 20 years, but that the squirrels had always survived.

S. lateralis occupies sagebrush lands and forested regions, but not dense virgin forest. Tevis (1956) observed that 4 years after logging of virgin timber 4 km from the nearest squirrel population, the logged area was colonized. He attributed this invasion to dispersal of young squirrels from the parent colony. Young animals normally die in the dense forest, but succeeded in this instance in extending the local range into a newly created suitable habitat. Hansen (1962) recorded a southward extension of range for *S. richardsonii* in Colorado after 1930, at the rate of 2.4 km per year for a period of 28 years. These populations were reoccupying range used before historic times and may have entered Colorado from the north as few as 115 years earlier. Changes in the land because of human occupation were not a factor.

Most ground squirrel species are generally aggressive and tenacious in the expansion of range or in recovery from ecological reverses and control campaigns. However, one innocuous desert species, the Mohave ground squirrel, *Spermophilus mohavensis*, may be in peril for its existence. In a 1972 survey (Hoyt 1973) this squirrel was recorded from only four localities in its limited range in Los Angeles and San Bernardino counties of California. Consequently, it is on the U.S. Fish and Wildlife Service List of Rare and Endangered Species. The nearby Greater Los Angeles metropolitan region and the spread of the human presence into formerly serene desert areas may well be factors in the depletion of this squirrel.

Disease Vectors. According to Barnes (1978), *S. beecheyi* in California is a major host of plague and has been a dominant source of this disease in humans. A closely related species, *S. variegatus*, of the southern Rocky Mountains, is associated with human infections of plague in that region. Both species harbor the flea *Diamanus montanus*, which is an efficient vector of plague. However, the rate of plague cases in California has not risen with the increase in human population. This may be attributable to control measures applied to *S. beecheyi*, a species usually not tolerated near human habitations; *S. variegatus* is a benign species relative to agricultural crops and is associated with rocky lands and stream washes. However, its small clustered populations are often near habitations where they are unmolested and provide a high rate of exposure to humans when there is an epizootic of plague.

Home Range. Ground squirrels tend to have moderate to small home ranges. The requirements for food, shelter, and life's processes usually occur within an area that may change according to daily or seasonal needs. Hawbecker (1958) trapped and released *Ammosper-mophilus nelsoni* for a period of six years and found that 22 of 29 adult males never moved outside a colony area of 4.5 hectares. Ranges of the females and young were similar to those of males. The longest linear movement was 1290 m for a male and about 915 m for a female. Mean home range estimates for another desert species, *Spermophilus tereticaudus*, were near 0.30 ha (Drabek 1973). *S. columbianus* was generally restricted in home range (Evans 1951). Of 11 taken, 8 were undetected outside an area of about 0.65 ha in two seasons. The greatest measured distance of travel was 128 m.

The larger-bodied *S. beecheyi* is also quite sedentary and usually restricted to small home ranges often only 0.12 to 0.25 ha (Evans and Holdenried 1943); 52 percent had lineal ranges of 137 m or less. Adult males had the smallest ranges. Some squirrels had measurably smaller home ranges as adults than as juveniles. Only six moved 1100 m or more, and the maximum recorded distance was about 1200 m. The ranges of adult males seldom overlapped. De Vos and Ray (1959) observed a single male *S. undulatus* for a period of four weeks at an arctic camp and noted that it used a remarkably large range of about 4.0 ha in normal activity.

Colonization. Ground squirrels often have been caged as pets. They also have been transported to new locations. Hall (1940) reported a dozen *S. beecheyi* transported as pets from near Clearlake, California, to the San Francisco peninsula, about 145 km to the south. There, the squirrels escaped and survived to interbreed in the next season with a local population of a different subspecies in that region. An anomolous, disjunct population of *S. richardsonii* was found by Hansen (1962) at a prison farm in Colorado. He suspected that this colony arose from stock released at that site. Burris and McKnight (1973) found that *S. undulatus* was successfully introduced to the Aleutian Islands of Unalaska about 1896 and Kavalga in 1920. The probable intention of these introductions was to provide a prey species for foxes. A remarkable incident of accidental transport of live squirrels from California to Hawaii was detected when one or two *S. beecheyi* escaped from a van of personal effects opened on the island of Kauai in Novemeber 1979. One animal was trapped in the vicinity and there was no further evidence of squirrels. The squirrels had eaten chicken feed in the van during the sea voyage of more than 4000 km. The van had been closed up at an industrial site at Costa Mesa, California, where ground squirrels were numerous, and was opened some 44 days later in Hawaii (Thomas Telfer personal communication). MacClintock (1970) cited additional examples of transport and establishment of ground squirrels within the continental United States.

FOOD HABITS

In the pursuit of food, ground squirrels often conflict with the interests of humans. In studies of *Ammosper-mophilus nelsoni* in the San Joaquin Valley of California, Hawbecker (1947) found that green parts and seeds of filaree (*Erodium cicutarium*) and red brome grass (*Bromus rubens*) were the main food plants. Insects, principally grasshoppers, are a staple in the diet when available. *Pogonomyrmex* ants were sometimes almost the only insect visibly available, as during the hot dry season, and these were noted in the diet at that time. Thus, these squirrels are opportunistic about available foods. *A. leucurus* is more carnivorous, and in southern Nevada Bradley (1968) found six genera of rodent in 43 percent of stomachs examined. Various lizards made up about half of the identified vertebrate remains. However, *A. leucurus* does consume green plants and seasonally these may be the most important food materials.

Spermophilus columbianus of northwestern Montana consumes a great variety of foods, chiefly vegetable (Manville 1959). Early in summer leafy vegetation is taken. Later, fruits and seeds are consumed in quantity. Stems and heads of clover, oats, rye, and barley and garden vegetables are eaten in croplands. Animal matter is taken only in small amounts, including grasshoppers, cicadas, beetles, and caterpillars. Mayer (1953) concluded that *S. undulatus* near Point Barrow, Alaska, "will eat almost anything," but that it is primarily a vegetarian. *S. mexicanus* in Texas ate leaves and beans of mesquite (*Prosopis* sp.) in quantity, and most of the herbaceous plants in the vicinity of their burrows. Johnson grass (*Sorghum halepense*) and filaree (*Erodium*) were also among favored foods (Edwards 1946). The squirrels were observed chasing flies and other insects, and it was concluded that insects encountered in the burrows also were eaten.

Linsdale (1946) listed more than 60 species of plant eaten by *S. beecheyi* at one site in the coastal mountains of California. Seeds as large as acorns were the most frequently used part of the plant. They also seasonally consume bulk green forage. The species is omnivorous, eating insects, especially the most numerous grasshopper, *Melanoplus devastator*. The principal foods of *S. lateralis* in coniferous forests of northwestern California were fungi and leafy herbs. In good years, conifer seeds made up 40 percent of the diet, but they were less than 5 percent of the annual diet. As in other ground squirrels, foods are consumed in a seasonal pattern related to availability (McKeever 1964).

Predation by ground squirrels on other vertebrates may interfere with management and research projects for wildlife species, or may be a direct economic loss. Linsdale (1946) noted that *S. beecheyi* attacked small birds trapped in a bird-banding program. Where there was access to a poultry pen, the squirrels readily adapted their diet to chicken feed and also ate eggs. Olson (1950) described predation by *S. beecheyi* on egg pods of the grunion (*Leuresthes tenuis*), a beach-spawning game fish of southern California. A colony of ground squirrels living in an embankment above the beach dug eggs 5 to 8 cm deep in the sand.

Ground squirrels prey on game bird nests, eggs, and young (Sowls 1948). Based on observation of captive *S. franklinii* fed eggs, Sowls (1948) described the destruction of duck nests. A squirrel grasps a duck egg under its body, thrusts it forward against the incisor teeth by using its hind legs, and thus breaks a small hole in the shell. The hole is enlarged by biting off fragments of the shell. Sowls could distinguish among predation by a crow, skunk, or ground squirrel. Eggs of the canvas-back duck (*Athya valisineria*) were hard shelled and immune to attack. In one study, 12 percent of 96 nests of other ducks were destroyed. Although *S. franklinii* inhabited lake shores and marsh areas it did not swim to reach nests on islets or in emergent vegetation.

Bridgewater and Penny (1966) described the systematic predation by *S. tridecemlineatus* on a nest of the cottontail rabbit (*Sylvilagus floridanus*). At least three young, each about 60 grams in weight, were taken from the nest over a period of days and killed and partly eaten at a site some distance away. Young shrikes (*Lanius ludovicianus*) that fell from the nest and lizards (*Cnemidophorus sexlineatus*) are also eaten by *S. tridecemlineatus*.

TRAPPING AND HANDLING METHODS

Trapping ground squirrels has been successful for control in special situations, but generally it is impractical for large operations. However, ground squirrels can be live-trapped when animals are needed for research or other purposes. A trap about 15 x 15 x 60 cm is large enough for the largest species. Baits of grain, nuts, or fruits are generally successful in attracting squirrels to traps. Prebaiting with open traps tends to improve the catch per unit effort. Treadles, trip wires, bait hooks, and other release mechanisms are incorporated into various models of cage trap. An unbaited number 0 steel jump trap lightly covered with soil, or a noose of fine wire or other filament placed about the entrance to a den, may also be used. Considerable study has been done to improve methods of trapping and handling various species of ground squirrel.

Horn and Fitch (1946) devised a wire mesh trap after experience with box traps, steel traps, and nooses produced less than satisfactory results for their purposes. They found *S. beecheyi* highly cautious and reluctant to enter box traps. Steel traps and nooses resulted in unacceptable injury. The new trap was 15 × 15 × 45 cm, designed from 12-mm hardware cloth except for a sheet metal door hinged at the top and 25 mm mesh at the far end to accentuate the impression of a tunnel. The trigger was a hanging wire rod, which released the drop door upon light contact. A notched stick is successful as a replacement for the original wire rod.

Prychodko (1952) developed a trap made from spiraled steel wire, which can be inserted into a burrow. It is closed after entry by a wire gate that prevents the squirrel from backing out. Disturbance by wind, shyness toward surface traps, and death from exposure required an improved technique for capture of *S. richardsonii*. Shemanchuk and Bergen (1968) devised a trap from a cylinder of sheet metal. The trap is affixed with a screened jar lid at the upper end and a metal drop door at the lower end. The trap was of a proper size to insert into a burrow and satisfactory for capture of squirrels.

Melchior and Iwen (1965) used wire mesh traps of two sizes for *S. undulatus* and fashioned a nylon sling for handling and marking this rather large but docile species. Their report included a review of pelage dye and toe clip–marking procedures. The authors found that deliberate soil disturbance in placing the traps was an attractant to squirrels, which formerly avoided carefully placed traps where no soil was shifted. Balph (1968) studied the behavior of free-living *S. armatus* toward traps and concluded that this species regarded eating of bait as a reward, and capture as punishment. Interplay of these two factors determined whether individual squirrels would be recaptured. Squirrels associated the trap site with capture and were taken more easily when traps were moved to a new nearby location. He concluded that this species is attracted to the novelty of cage traps and that traps should be conspicuous rather than concealed.

CAPTIVITY AND BREEDING

Maintaining ground squirrels in captivity for study and controlled breeding experiments has been a requirement of investigations of various species. Shaw (1925) was perhaps the first to produce offspring from captive *S. columbianus*. In his elaborate complex of cages and yards, he segregated males from females in adjacent pens until sexual interest was intense, permitted the animals to intermingle for a few hours, and then again segregated them. In one series of experiments, four females were successfully bred and a gestation period of 24 days determined. Marsh and Howard (1968) caged *S. beecheyi* in the spring, summer, or fall, then mated pairs, or a male and two females, in the fall or late winter. Cages were 120 × 240 × 60 cm and each of 20 cages was equipped with a two-compartment wooden nest box 30 × 30 × 53 cm. They were placed on the ground outdoors, subject to normal daylight and weather conditions. Laboratory ration, greens, and water were provided. In the second, more successful year of the tests, 5 of 31 mated females produced litters conceived in captivity. Barr and Musacchia (1968) established a colony of five *S. tridecemlineatus* in an earth-filled metal cylinder 1.2 m high and 2.1 m in diameter and screened at the top. The outdoor container was banked with earth for insulation. The squirrels readily dug burrows when caged in October and two surviving females produced young the following spring. It appears that cages or pens of moderate size are needed for successful breeding.

Pregnant females captured in the wild and confined to small cages will bear young and care for them. Tomich (1962) kept *S. beecheyi* in batteries of wire cages each 32 × 35 × 60 cm and equipped with a wooden nest box. He found that females bore their young without incident and usually reared them through infancy and weaning. A laboratory ration, rolled barley, water, and carrots or potatoes were pro-

vided. Similar success was attained by Zimny (1965) with *S. tridecemlineatus* in complementary studies of growth and development.

ECONOMIC STATUS

Ground squirrel meat was a substantial item in the diet of early Americans, and the skins were sometimes used to make clothing. Alcorn (1940) reported that the Piute ground squirrel (*S. townsendii* spp.) in Nevada got its name from the Piute Indians, who used it for food over countless ages. In the 1930s, tribesmen captured squirrels by pouring buckets of water into the shallow burrow systems, forcing the squirrels to the surface, where they were caught by hand. Some were transported alive for long distances to be sold or traded. Jacobsen (1918) noted that California natives ate *S. beecheyi* and that this species was brought to the San Antonio Mission along with other foods as a gift in 1773. A commercial market for ground squirrels was established in San Francisco in the 1870s among Chinese residents of the city. Tons of the animals were shipped to the San Francisco and Oakland markets until 1908, when bubonic plague was proved to be carried by the California ground squirrel and its fleas. In 1877, 50 pairs of gloves were made from squirrel skin and test marketed, but the project failed as a commercial venture.

The Pit River Indians of northeastern California regularly used fat *S. beldingi*. Modern subsistence hunters and gardeners throughout the west and Mexico may still eat various ground squirrels that are available. Eskimos in the vicinity of Point Barrow, Alaska, used the arctic ground squirrel for food and for fur for special items of clothing (Mayer 1953). Early records of museum specimens labeled "Point Barrow" were postulated as having been acquired by collectors from natives who brought the animals to the village for sale or barter from the nearest populations of squirrels, about 45 km inland. Rausch (1951) remarked that when caribou (*Rangifer tarandus*) were scarce, the Nunamiut tribe of Anaktuvuk Pass sometimes killed the arctic ground squirrel for food. Earlier custom required that a period of time elapse after squirrel had been cooked before a pot could be used for other meat.

Bounty payments may be classed as an economic use of ground squirrels. This expensive system of "control" has been used in the past but the principal benefit derived was the cash payment to individuals who brought in the squirrels. Jacobsen (1918) discussed some the shortcomings of bounty systems. A California law providing bounty payments for a number of years was finally repealed in 1877. However, many county governments paid bounties thereafter to pursue destruction of ground squirrels by unspecified means. Authorities often underestimated rodent populations in their districts, and even at a nominal bounty of two to five cents per squirrel, county treasuries were heavily strained when hundreds of thousands of claims were made. Abuses included turning in scalps in one county and tails of the same animals in another, when these parts were so specified as evidence. Also, tails were removed from live squirrels, which were allowed to run free to assure a large new stock the following year. A few clever operators were discovered to make several "tails" from one animal by wrapping squirrel fur on sticks to produce a fair semblance of a tail in stubby-tailed species such as *S. beldingi*.

However, bounty work was most often done by farmers sincerely desiring riddance of squirrels. The chief failure of bounty systems was that when ground squirrels were reduced to only moderate numbers the incentive to pursue them waned, and no long-term control was attained. Sheppard (1979) noted that a bounty was raised against *S. richardsonii* in 1899 in the prairie provinces of Canada, at three cents per tail, but the project was dropped a few years later because it was ineffective. Bounty systems have fallen out of style, and one would be ill advised in this age to institute such a program.

Economic costs of ground squirrels are high. Jacobsen (1918) estimated a $30 million annual loss to California agriculture in that era. Whereas such figures may be frightening, and possibly inflated, they do convey some notion of the seriousness of ground squirrel predation on fruit, nut, and grain crops. Shaw (1920) worked with *S. columbianus* in eastern Washington, where it is a particular pest of winter wheat. In test plots a single male adult reduced yield by 33 percent. A female with a brood of young destroyed or ate 23 kilograms of wheat in a 130-day season. Average destruction was $1.76 per test squirrel. Forage losses in relation to comparative gain in weight by beef cattle on pasture were documented by Howard et al. (1959). In field experiments where *S. beecheyi* was controlled, greater winter gains of 0.34 to 0.47 kg per head per day were measured in comparison to pastures where no squirrel reduction was applied. More elaborate tests of squirrel energy budgets at the same site produced more encouraging results. Diets of cattle and squirrels were generally not similar, in particular on a seasonal basis related to the abundance of various food plants. Squirrels did not seriously limit the availability of the preferred foods of cattle (Schitoskey and Woodmansee 1978).

Soil erosion and damage to irrigation ditches and dams has been attributed to ground squirrels. In northern California, *S. beecheyi* burrows were precursors of gullies. They enlarged first by subsurface flow until they caved in, and then by subsequent overland flow. Grazing encouraged the persistence of squirrels, but only severe overuse by cattle resulted in serious gully formation (Longhurst 1957). Ground squirrels contributed to the failure of at least three earth structures of irrigation systems in Colorado, Nevada, and California in the period 1975–1980. These events were associated with unusually high waters and lack of attention to existing rodent burrows. Damage to farmlands by flooding was estimated at more than one million dollars (Rex Marsh personal communication). With increasing impoundment and transport of water in the western states more attention should be paid to damage of earthen structures by ground squirrels.

MANAGEMENT

Historical Aspects. Early notice of ground squirrels was taken by American agriculturists, as exemplified by Kennicott (1857:73). He stated, "the prairie squirrels are fitted to inhabit the grassy plains which cover most of the western part of our Union, their food being prairie plants with their roots and seeds." In reference to *S. tridecemlineatus* and *S. franklinii* he wrote, "The two species have rendered themselves obnoxious to our prairie farmers" (p. 74). However, he advised caution in control of the *S. franklinii* because of its habits of killing voles (*Microtus* sp.) and insects, it was largely a problem only on newly broken ground, "it disappears before the plow"; and was far less numerous. Control consisted of clubbing, trapping, shooting, and drowning the squirrels in their burrows; these were apparently effective methods among corn farmers and rural gardeners (see also Bailey 1893).

A more advanced application of pioneer ground squirrel control was given by the California State Commission of Horticulture (1918). This comprehensive report dealt with the life histories, habits, and control of ground squirrels in California and nearby western states. The project was a result of combined studies by state, federal, and university agencies during World War 1 to protect and promote agricultural production. Grinnell and Dixon (1918) brought out several points as a result of their studies. Of the species of ground squirrel found in California, only *S. beecheyi* and *S. beldingi* were of major significance. Estimated ratios of importance as agricultural pests were *beecheyi* 17, *beldingi* 5, and all others 1 for the two species and for all other species combined. Damage was heavy not only on cultivated grain, forage, vegetable, fruit, and nut crops, but also on rangelands supporting livestock. This situation applies to the diversity of ground squirrel species today; ground squirrels of lesser economic importance occupy nonagriculture land and generally have a low density.

Jacobsen (1918) produced a classic history of ground squirrel control, citing the earliest description of the California ground squirrel in the vicinity of Monterey. The first recorded campaign for control of *S. beecheyi* was at Santa Barbara Mission, in 1808. Thousands were killed in nine days of May in that year. The 1918 report established a number of precepts of ground squirrel control, paraphrased as follows: (1) differential acceptance of baits was noted, with some species or populations preferring easily available natural foods; (2) the total volume of a burrow system was so large (up to 2.0 m³) that larger than anticipated volumes of fumigants were needed for adequate control; (3) configuration of burrow systems was such that heavy gases would not travel over abrupt rises in the tunnels; (4) squirrels increase when natural predators are reduced in numbers, when bounteous foods are provided on croplands, and with heavy grazing of rangelands, which favors low-growing seed plants. That the campaign against ground squirrels during World War 1 had serious backing is evidenced by the governor's proclamation of "Ground Squirrel Week," during which school children and other citizens were urged to participate in squirrel eradication (figure 10.9).

Some 44 years later, Jacobsen (1962) reviewed further the American pest problem as keynote speaker before the First Vertebrate Pest Control Conference in Sacramento, California. He cited 1901 as the beginning of statewide interest in control programs (in Kansas) as a cooperative venture with the then Bureau of Biological Survey of the U.S. Department of Agriculture. South Dakota followed in 1916. State laws were enacted to authorize agricultural agencies to do pest control work and to conduct educational campaigns. The research capabilities of state colleges and universities became a part of the effort. Infectious diseases of people, such as bubonic plague and Rocky Mountain spotted fever, prompted the U.S. Public Health Service to enter the field. Problems of forest management related to ground squirrels brought the U.S. Forest Service into the programs as early as 1913. By 1917 the county-state-federal system of rodent suppression was well organized.

Some early efforts were directed toward a search for infectious agents that would be selectively effective against rodents. These were unsuccessful, and programs turned to intensified efforts to find an effective poison bait or lethal gas. By about 1910, a strychnine formula was being employed successfully. Operational guidelines such as prebaiting, safeguards in handling and use of hazardous materials, drawing up cooperative agreements, adherence to conservation principles, legal responsibilities, open communication with farm groups, and establishment of agency responsibilities were all a part of suppression programs.

Within the above framework modern control and management of ground squirrel populations has evolved. The usual pattern is that county agricultural commissions and state departments of agriculture have responsibility for private or state lands. The U.S. Fish and Wildlife Service (a successor to the old Bureau of Biological Survey) deals with problems on federal lands, besides usually being the lead agency in research functions.

Current Status. Modern regulation of pest control dates from the passage of the National Environmental Policy Act of 1969 (NEPA), which became law on 1 January 1970. This law places strict guidelines on the development, labeling, registration, and use of toxicants for wildlife species, including ground squirrels, and is administered under the Environmental Protection Agency (EPA). Snow (1974) reviewed these regulations in relation to the Environmental Impact Statement (EIS). The EIS is required and binding on all federal, state, and private agencies for projects that have a significant impact on the environment, and rodent control is among these projects.

Lee (1970) stated that early federal regulation

PROCLAMATION.

EXECUTIVE DEPARTMENT,
STATE OF CALIFORNIA.

The State Commissioner of Horticulture has instituted a campaign to destroy the ground squirrel throughout the State. It is said that these rodents do an annual damage to the amount of $30,000,000, a great part of this damage consisting in the destruction of foodstuffs, and in these times special efforts should be made to prevent such loss.

In connection with the campaign the State Commissioner of Horticulture has personally offered prizes to those schools of all classes which make the best records in killing squirrels. I heartily endorse the plan, and I hope that the efforts of those in charge of the campaign will be crowned with success, and in connection therewith I do hereby set aside the week of April 29th to May 4th as Ground Squirrel Week, and trust that during that time the school children and all other persons will do their utmost to relieve the country of the ground squirrel pest.

WILLIAM D. STEPHENS,
Governor.

Dated: Sacramento, April 8, 1918.

FIGURE 10.9. A page from the monthly bulletin of the California State Commission of Horticulture, demonstrating serious interest in the problem of ground squirrel damage to food crops. From Grinnell and Dixon 1918.

dates from 1910, and that after 1946 registration was required for any toxicant classed as an economic poison. He also outlined the lengthy routine that is required currently in laboratory and field studies for registration and labeling of a product for a general or particular use. Kverno (1970) recommended standardization of testing procedures for comparative purposes. A plea was made in the same year for support of an anticipated vastly expanded research effort for compliance with the registration requirements, emphasizing that exacting data were a necessity (Merrill 1970). A further burden of "efficacy testing," which includes such criteria as taste, color, particle size, and effects on the animal to be controlled at various temperatures, was outlined by Ochs (1972). A response by the pest control industry indicated considerable progress in compliance with NEPA standards (Beck 1974).

The net result of NEPA has been a reduction of interest by private enterprise in the development of new toxicants because of the high cost—generally millions of dollars—of bringing a product onto the market. As an alternative, much private and government effort has been expended for registration of the standard toxicants that have had a long history of use in ground squirrel control. An additional complication is the requirement for consultation with the Fish and Wildlife Service under section 7 of the Endangered Species Act, when federal funding is involved in squirrel control campaigns.

Dana (1962) outlined precepts, up to that period, for control of *S. beecheyi* and *S. beldingi*, the two most important California species. These precepts, along with supporting data from later papers, provide a model for squirrel management. It is beyond the scope of this chapter to provide details on regulation and recommendations for use of toxicants under the specifics of NEPA regulation. Following is a summary of methods and materials being used, or that have been considered for use in controlling ground squirrels in North America.

STRYCHNINE ALKALOID. Strychnine-coated grain has been especially effective because of the seed-gathering habit of most squirrels in the drier seasons. Poison is absorbed through the mucous membrane lining the cheek pouches. As early as 1909 it was observed that strychnine was absorbed more rapidly in the cheek pouches than in the stomach. Squirrels often are killed before returning to their dens. O'Brien (1978) concluded that strychnine was effective against *S. beldingi* and *S. townsendii* in Nevada alfalfa fields when applied on chopped cabbage baits. Record (1978) used it against *S. richardsonii* on Montana grazing lands, with grain baits.

THALLIUM SULFATE. Thallium sulfate (Tl_2SO_4) was introduced from Europe in 1926 and used as a supplement to strychnine as a stomach poison, usually on barley or hulled oats. It is slow-acting and in small doses may require several days for lethal results, but a large dose will kill quickly. In an early example of concern about abuses of toxic materials in the environment, Linsdale (1931) pointed out that at least 30 unintended species of wildlife were susceptible to direct or secondary thallium poisoning. Although advances in baiting techniques prevented some losses, because it is a persistent heavy metal, thallium is unlikely to be registered under NEPA, and is currently banned from use as a pesticide.

ZINC PHOSPHIDE. This poison (Zn_3P_2) was developed for ground squirrel control in the early 1940s. Some problems of adhesion to grain baits were overcome after early failures. It is usually most effective when freshly mixed, applied sparingly, and used in rotation with other baits. Zinc phosphide acts with the release of phosphine gas when moistened by body fluids after ingestion and is rapid acting; residual products are inert.

Seldom is there accidental poisoning of other wildlife species because of the rapid degradation of Zn_3P_2 in the field and the reluctance of birds to take new foods immediately. Hood (1972) reported on registration and use of this substance for various rodents, including ground squirrels, under research performed by the U.S. Fish and Wildlife Service. O'Brien (1978) found it reasonably effective on Nevada squirrel populations. Aversion is acquired when sublethal doses are taken, and prebaiting is generally recommended. Schilling (1976) reported on nine years' experience with aerial application of zinc phosphide baits in Fresno County, California. A complete program was developed, beginning in January, for the start of flying in mid-May.

COMPOUND 1080. Sodium fluoroacetate ($NaC_2H_2FO_2$, or 1080) is an acute toxicant for which there is no known antidote. It is the most effective poison used on squirrels. Thorough treatment every two or three years usually controls a population. Best results are obtained during a four- to six-week period in the breeding season. This compound was first used as a rodenticide in 1946 and was widely acclaimed thereafter. Secondary poisoning is a hazard to carnivores that may eat dead squirrels; carnivores are extremely susceptible to compound 1080. Ground squirrels may avoid eating treated baits in preference to clean baits. In some applications, mixing only 30 percent treated bait with clean bait improved results over 100 percent treated grain applications. Protection of birds is achieved by crimping grain baits and dyeing them green, red, or yellow to render them less attractive to birds. With careful application, accidental mortality is minimized. Compound 1080 is still used under present regulations; however, it is banned from use on federal lands, and along with zinc phosphide its use is under question by the EPA Rebuttable Presumption against Registration (RPAR) process, which is an added safeguard to assure minimal hazards in pesticide use. O'Brien (1978) found 1080 the most effective toxicant he tested, and Record (1978) felt that it could be used against *S. columbianus* if the proper state permits could be obtained.

Marsh (1967) described advances in aerial application of compound 1080 on California rangelands. Fixed-wing aircraft are satisfactory for this work and are less expensive than the more maneuverable helicopters. Pilots must be trained to recognize concentrations of squirrel colonies. An experienced pilot can treat 400 to 800 ha per hour when aided by radio communication with ground personnel. Aerial application against *S. beldingi* in northeastern California was adapted to chopped cabbage bait (Sauer 1976) beginning in 1965, with 1080 as the toxicant. Chopped cabbage was a standard bait by 1972, when it was first applied from the air at a great saving of hand labor. Air application is especially effective on remote or inaccessible pockets of ground squirrels overlooked in hand application of baits, but which are potential sources of reinfestation on adjoining lands. Because *S. beldingi* is partial to green baits, in earlier years dating back to 1915, dandelion (*Taraxacum officinale*) was hand gathered and used as bait. In one instance about 2 ha of dandelion were planted expressly for harvest as green bait.

FUMIGANTS. Carbon disulfide (CS_2) was used in the United States as early as 1876 for ground squirrel control. The programs were patterned after its use on sewer rats in France. The usual method was to dip balls of burlap into the liquid CS_2, force them into the burrow, and close the burrow with tamped earth. An alternative method was to allow the gas to permeate the burrow for a time and then ignite it with a torch. Caution was exercised to prevent fires in grassy areas when this method was used.

The U.S. Public Health Service devised a pump to vaporize CS_2. Subsequent modifications perfected this equipment, allowing it rapidly to force a high concentration of the gas throughout the burrow system. The CS_2 method is best used when soil is moist and burrow systems are closed. But some squirrels have the ability to escape the gas, or may have plugs in place to deter predators and thus avoid the gas.

Methyl bromide (CH_3Br) is effective in ground squirrel control. It is also effective in the control of associated insects such as fleas, which may be of public health significance. It can be used in wet or dry soils and is not flammable. It has these advantages over the carbon disulfide method, but is more expensive.

Automobile exhaust piped into a ground squirrel den is an effective control agent because of its content of carbon monoxide (CO). It has been used most often in orchards to clean up small residual populations. Hydrogen cyanide (HCN), calcium cyanide [$Ca(CN_2)$], sulfur dioxide (SO_2), and gasoline and other petrols are among other agents tried. None has come into general use; however, several gas cartridge preparations are commercially available. At the present time fumigants are extremely limited in operational squirrel control because their use is labor intensive, and in general they have been superceded by less expensive baiting techniques.

ANTIFERTILITY AGENTS. Chemical substances such as synthetic estrogens that could upset hormone balances and block ovulation have been considered (Balser 1964), because ground squirrels breed only once a year. Alsager (1972) applied the idea of chemosterilants to *S. richardsonii* in Alberta. He used diethylstilbesterol and mestranol, with inconclusive results. In more elaborate tests with mestranol, Goulet and Sadlier (1974), after two years of treatment of female *S. richardsonii*, found that the population was reduced. Changes in behavior and physiology were noted as factors in population reduction. The authors cautiously asserted that mestranol could be an effective way of reducing ground squirrel populations.

REPELLENTS. In coniferous forest management, Kverno (1964) found that chemical repellents such as endrin were not especially effective against ground squirrels. Baiting programs adapted to particular local conditions were recommended as an alternative for protecting seed and young plantings.

Electromagnetic devices have been touted as a new weapon against pests, including ground squirrels, and several models have been offered for sale. Manufacturers say such equipment "takes the electromagnetic field that already exists and simply puts a pattern in it." The EPA is unconvinced and has ordered them off the market as ineffective (Smith 1979). It appears that development of ground squirrel repellents is not significantly advanced at the present time.

PHEROMONES. Chemical olfactory signals (pheromones) are important in the social behavior of ground squirrels. Anal, dorsal, and cheek glands apparently are present in all ground squirrels. The glands probably are important in producing signals between squirrels relative to social organization, for identification, and as warnings to individuals and possible predators. Salmon (1978) speculated that synthetic pheromones may have attractant properties that could improve the efficiency of trap and bait stations. Perhaps some could be used as repellents to reduce the efficiency of reproduction in ground squirrels. Research on pheromones is in its early development, but it appears to have some potential for squirrel control.

LITERATURE CITED

Alcorn, J. R. 1940. Life history and notes on the Piute ground squirrel. J. Mammal. 21:160–170.

Alsager, D. E. 1972. Experimental population suppression of Richardson's ground squirrels (*Spermophilus richardsonii*) in Alberta. Pages 93–100 in Proc. 5th Vertebr. Pest Conf., Fresno, Calif.

Bailey, V. 1893. The prairie ground squirrels, or spermophiles, of the Mississippi Valley. U.S. Dept. Agric. Div. Ornithol. Mammal. Bull. 4. 69pp.

Balph, D. F. 1968. Behavioral responses of unconfined Uinta ground squirrels to trapping. J. Wildl. Manage. 32:778–794.

Balser, D. S. 1964. Antifertility agents in vertebrate pest control. Pages 133–137 in Proc. 2nd Vertebr. Pest Control Conf., Anaheim, Calif.

Barnes, A. M. 1978. Rodent population control for public health and safety. Pages 158–160 in Proc. 8th Vertebr. Pest Conf., Sacramento, Calif.

Barr, R. E., and Musacchia, X. J. 1968. Breeding among captive *Citellus tridecemlineatus*. J. Mammal. 49:343–344.

Bartholomew, G. A., and Hudson, J. W. 1960. Aestivation in the Mohave ground squirrel *Citellus mohavensis*. Bull. Mus. Comp. Zool. 124:193–208.

Beck, J. R. 1974. An overview of ASTM's activities in establishing standards for vertebrate pest control methods. Pages 55–57 in Proc. 6th Vertebr. Pest Conf., Anaheim, Calif.

Blair, W. F. 1942. Rate of development of young spotted ground squirrels. J. Mammal. 23:342–343.

Blake, B. H. 1972. The annual cycle and fat storage in two populations of golden-mantled ground squirrels. J. Mammal. 53:157–167.

Bradley, W. G. 1968. Food habits of the antelope ground squirrel in southern Nevada. J. Mammal. 49:14–21.

Bridgewater, D. D. 1966. Laboratory breeding, early growth, development and behavior of *Citellus tridecemlineatus* (Rodentia). Southwestern Nat. 11:325–337.

Bridgewater, D. D., and Penny, D. F. 1966. Predation by *Citellus tridecemlineatus* on other vertebrates. J. Mammal. 47:345–346.

Bryant, M. D. 1945. Phylogeny of the Nearctic Sciuridae. Am. Midl. Nat. 33:257–390.

Burris, O. E., and McKnight, D. E. 1973. Game transplants in Alaska. Alaska Dept. Fish Game Tech. Bull. 4. 57pp.

California State Commission of Horticulture. 1918. California ground squirrels. Monthly Bull. 7:595–807.

Cameron, D. M., Jr. 1967. Gestation period of the golden-mantled ground squirrel (*Citellus lateralis*). J. Mammal. 48:492–493.

Chapman, J. A., and Lind, G. S. 1973. Latitude and litter size of the California ground squirrel, *Spermophilus beecheyi*. Bull. Southern California Acad. Sci. 72:101–105.

Clark, D. O. 1978. Control of ground squirrels in California using anticoagulant treated baits. Pages 98–103 *in* Proc. 8th Vertebr. Pest. Conf., Sacramento, Calif.

Clark, T. W. 1970. Early growth, development, and behavior of the Richardson's ground squirrel (*Spermophilus richardsoni elegans*). Am. Midl. Nat. 83:197–212.

Clark, T. W., and Skryja, D. D. 1969. Postnatal development and growth of the golden-mantled ground squirrel, *Spermophilus lateralis lateralis*. J. Mammal. 50:627–629.

Cothran, E. G.; Zimmerman, E. G.; and Nadler, C. F. 1977. Genic differentiation and evolution in the ground squirrel subgenus *Ictidomys* (genus *Spermophilus*). J. Mammal. 58:610–622.

Dana, R. H. 1962. Ground squirrel control in California. Pages 126–142 *in* Proc. Vertebr. Pest Control Conf., Sacramento, Calif.

Davis, D. E. 1976. Hibernation and circannual rhythm of food consumption in marmots and ground squirrels. Q. Rev. Bio. 51:477–514.

de Vos, A., and Ray, E. M. 1959. Home range and activity pattern of an arctic ground squirrel. J. Mammal. 40:610–611.

Drabek, C. M. 1973. Home range and daily activity of the round-tailed ground squirrel, *Spermophilus tereticaudus neglectus*. Am. Midl. Nat. 89:287–293.

Durrant, S. D., and Hansen, R. M. 1954. Distribution pattern and phylogeny of some western ground squirrels. Syst. Zool. 3:82–85.

Edwards, R. L. 1946. Some notes on the life history of the striped ground squirrel (*Citellus tridecemlineatus*) in an abandoned field in southeastern Michigan. J. Mammal. 32:437–449.

Evans, F. C., and Holdenried, R. 1943. A population study of the Beechey ground squirrel in central California. J. Mammal. 24:231–260.

Goulet, L. A., and Sadlier, R. M. F. S. 1974. The effects of a chemosterilant (Mestranol) on population and behavior in the Richardson's ground squirrel (*Spermophilus richardsonii*) in Alberta. Pages 90–100 *in* Proc. 6th Vertebr. Pest Conf., Anaheim, Calif.

Grinnell, J., and Dixon, J. 1918. Natural history of the ground squirrels of California. Pages 597–709 *in* California State Comm. Hort. Monthly Bull. 7. 212pp.

Hall, E. R. 1940. Transplantation of the Douglas ground squirrel. California Fish Game 26:77.

———. 1981. The mammals of North America. 2nd ed. 2 vols. John Wiley & Sons, New York.

Hall, E. R., and Kelson, K. R. 1959. The mammals of North America. 2 vols. Ronald Press Co., New York.

Hansen, R. M. 1956. Decline in Townsend ground squirrels in Utah. J. Mammal. 37:123–124.

———. 1962. Dispersal of Richardson ground squirrel in Colorado. Am. Midl. Nat. 68:58–66.

Hawbecker, A. C. 1947. Food and moisture requirements of the Nelson ground squirrel. J. Mammal. 28:115–125.

———. 1958. Survival and home range of the Nelson antelope ground squirrel. J. Mammal. 39:207–215.

Herskovitz, P. 1949. Status of the names credited to Oken, 1816. J. Mammal. 30:289–301.

Hood, G. A. 1972. Zinc phosphide: a new look at an old rodenticide for field rodents. Pages 85–92 *in* Proc. 5th Vertebr. Pest Conf., Fresno, Calif.

Horn, E. E., and Fitch, H. S. 1946. Trapping the California ground squirrel. J. Mammal. 27:220–224.

Howard, W. E.; Wagnon, K. A., and Bently, J. R. 1959. Competition between ground squirrels and cattle for range forage. J. Range Manage. 12:110–115.

Howell, A. H. 1938. Revision of the North American ground squirrels. North Am. Fauna 56. 256pp.

Hoyt, D. F. 1973. Mohave ground squirrel survey, 1972. California Dept. Fish Game, Spec. Wild. Invest. W-54-R, Prog. Rep. Job 11-5-5. 10pp.

Jacobsen, W. C. 1918. A history of ground squirrel control in California. Pages 721–761 *in* California State Hort. Comm. Monthly Bull. 7. 212pp.

———. 1962. The pest animal problem. Pages 8–17 *in* Proc. Vertebr. Pest Control Conf., Sacramento, Calif.

Jameson, E. W., Jr. 1964. Patterns of hibernation of captive *Citellus lateralis* and *Eutamias speciosus*. J. Mammal. 45:455–460.

Kennicott, R. 1857. The quadrupeds of Illinois injurious and beneficial to the farmer. Pages 52–110 *in* Pat. Off. Rep. Part 1, Agric. Rep. for 1856.

Kramm, K. R. 1972. Body temperature regulation and torpor in the Antelope ground squirrel, *Ammospermophilus leucurus*. J. Mammal. 53:609–611.

Kverno, N. B. 1964. Forest animal damage control. Pages 81–89 *in* Proc. 2nd Vertebr. Pest Control Conf., Anaheim, Calif.

———. 1970. Standardization of procedures for developing vertebrate control agents. Pages 138–139 *in* Proc. 4th Vertebr. Pest Conf., West Sacramento, Calif.

Lee, J. O., Jr. 1970. Outlook for rodenticides and avicides registration. Pages 5–8 *in* Proc. 4th Vertebr. Pest Conf., West Sacramento, Calif.

Linsdale, J. M. 1931. Facts concerning use of thallium in California to poison rodents: its destructiveness to game birds and other valuable wildlife. Condor 33:92–106.

———. 1946. The California ground squirrel. Univ. California Press, Berkeley. 475pp.

Loehr, K. A., and Risser, A. C., Jr. 1977. Daily and seasonal activity patterns of the Belding ground squirrel in the Sierra Nevada. J. Mammal. 58:445–448.

Longhurst, W. M. 1957. A history of squirrel burrow formation in relation to grazing. J. Range Manage. 10:182–184.

MacClintock, D. 1970. Squirrels of North America. Van Nostrand, New York. 184pp.

McKeever, S. 1964. The biology of the golden-mantled ground squirrel, *Citellus lateralis*. Ecol. Monogr. 34:383–401.

Manville, R. H. 1959. The Columbian ground squirrel in northwestern Montana. J. Mammal. 40:26–45.

Marsh, R. E. 1967. Aircraft as a means of baiting ground squirrels. Pages 2–6 *in* Proc. 3rd. Vertebr. Pest Conf., San Francisco, Calif.

Marsh, R. E., and Howard, W. E. 1968. Breeding ground squirrels, *Spermophilus beecheyi*, in captivity. J. Mammal. 49:781–783.

———. 1979. Stalking the *Spermophilus beecheyi*. Western Fruit Grower 99:W2-3, W7.

Mayer, W. V. 1953. A preliminary study of the Barrow ground squirrel, *Citellus parryi barrowensis*. J. Mammal. 34:334–345.

Mayer, W. V., and Roche, E. T. 1954. Developmental patterns of the barrow ground squirrel, *Spermophilus undulatus barrowensis*. Growth 18:53–69.

Melchior, H. R., and Iwen, F. A. 1965. Trapping, restraining, and marking arctic ground squirrels for behavioral observations. J. Wildl. Manage. 29:671–678.

Merrill, H. A. 1970. The outlook for vertebrate pest control. Pages 144–145 *in* Proc. 4th Vertebr. Pest Conf., West Sacramento, Calif.

Michener, G. R. 1973. Climatic conditions and breeding in Richardson's ground squirrel. J. Mammal. 54:499–503.

———. 1979. The circannual cycle of Richardson's ground squirrels in southern Alberta. J. Mammal. 60:760–768.

Morton, M. L., and Tung, H. L. 1971. Growth and development in the Belding ground squirrel (*Spermophilus beldingi*). J. Mammal. 52:611–616.

Nadler, C. F. 1966. Chromosomes and systematics of American ground squirrels of the subgenus *Spermophilus*. J. Mammal. 47:579–596.

Nadler, C. F., and Hughes, C. E. 1966. Chromosomes and taxonomy of the ground squirrel subgenus *Ictidomys*. J. Mammal. 47:46–53.

Navarrete, F. S. 1978. Rodents as agricultural pests in Mexico: National Rodent Campaign. Pages 118–119 in Proc. 8th Vertebr. Pest Conf., Sacramento, Calif.

Neal, B. J. 1965a. Reproductive habits of round-tailed and Harris antelope ground squirrels. J. Mammal. 46:200–206.

———. 1965b. Growth and development of the round-tailed and Harris antelope ground squirrels. Am. Midl. Nat. 73:479–489.

O'Brien, J. 1978. Chopped cabbage baits for ground squirrel control in Nevada. Pages 25–27 in Proc. 8th Vertebr. Pest Conf., Sacramento, Calif.

Ochs, P. 1972. Efficacy testing of vertebrate pest control agents. Pages 138–142 in Proc. 5th Vertebr. Pest Conf., Fresno, Calif.

Olson, A. C., Jr. 1950. Ground squirrels and horned larks as predators upon grunion eggs. California Fish Game 36:323–327.

Prychodko, W. 1952. A live trap for ground squirrels. J. Mammal. 33:497.

Quanstrom, W. R. 1966. Flood tolerance in Richardson's ground squirrel. J. Mammal. 47:323.

Rausch, R. L. 1951. Notes on the Nunamiut Eskimo and mammals of the Anaktuvik Pass Region, Brooks Range, Alaska. Arctic 4:147–195.

Record, C. R. 1978. Ground squirrel and prairie dog control in Montana. Pages 93–97 in Proc. 8th Vertebr. Pest Conf., Sacramento, Calif.

Reynolds, H. G., and Turkowski, F. 1972. Reproductive variations in the round-tailed ground squirrel as related to winter rainfall. J. Mammal. 53:893–898.

Salmon, T. P. 1978. Pheromones: their potential for ground squirrel control. Pages 112–114 in Proc. 8th Vertebr. Pest Conf., Sacramento, Calif.

Sauer, W. C. 1976. Control of the Oregon ground squirrel (*Spermophilus beldingi oregonus*). Pages 99–109 in Proc. 7th Vertebr. Pest Conf., Monterey, Calif.

Schilling, C. 1976. Operational aspects of successful ground squirrel control by aerial application of grain bait. Pages 110–115 in Proc. 7th Vertebr. Pest Conf., Monterey, Calif.

Schitoskey, F., Jr., and Woodmansee, S. R. 1978. Energy requirements and diet of the California ground squirrel. J. Wildl. Manage. 42:373–382.

Shaw, W. T. 1918. The Columbian ground squirrel. Pages 710–720 in California State Comm. Hort. Monthly Bull. 7. 212pp.

———. 1920. The cost of a squirrel and squirrel control. State Coll. Washington Agric. Exp. Stn. Popular Bull. 118. 19pp.

———. 1925. Breeding and development of the Columbian ground squirrel. J. Mammal. 6:106–113.

Shemanchuk, J. A., and Bergen, H. J. 1968. The Gen trap, a simple, humane trap for Richardson's ground squirrels, *Citellus richardsonii* (Sabine). J. Mammal. 49:553–555.

Sheppard, D. H. 1979. Richardson's ground squirrel: Hinterland who's who. Can. Wildl. Serv. Minister of Supply and Serv. Cat. CW 69–4/64. 4pp.

Simpson, G. G. 1945. The principles of classification and a classification of mammals. Am. Mus. Nat. Hist. Bull. 85:1–350.

Smith, J. R. 1973. Hibernation in the spotted ground squirrel, *Spermophilus spilosoma annectens*. J. Mammal. 54:499.

Smith, R. J. 1979. Rodent repellers attract EPA strictures. Science 204:484–486.

Snow, G. F. 1974. Environmental impact statements. Pages 34–36 in Proc. 6th Vertebr. Pest Conf., Anaheim, Calif.

Sowls, L. K. 1948. The Franklin ground squirrel, *Citellus franklinii* (Sabine), and its relationship to nesting ducks. J. Mammal. 29:113–137.

Svihla, A. 1939. Breeding habits of Townsend's ground squirrel. Murrelet 20:6–10.

Tevis, L. P., Jr. 1956. Invasion of a logged area by golden-mantled ground squirrels. J. Mammal. 37:291–292.

Thornbury, W. D. 1965. Regional geomorphology of the United States. Wiley, New York. 609pp.

Tomich, P. Q. 1962. The annual cycle of the California ground squirrel *Citellus beecheyi*. Univ. California Publ. Zool. 65:213–282.

Turner, B.; Iverson, S. L.; and Severson, K. L. 1976. Postnatal growth and development of captive Franklin's ground squirrels (*Spermophilus franklinii*). Am. Midl. Nat. 95:93–102.

Walker, J. M.; Garber, A.; Berger, R. J.; and Heller, H. C. 1979. Sleep and estivation (shallow torpor): continuous processes of energy conservation. Science 204:1098–1100.

White, L. 1972. The Oregon ground squirrel in northeastern California: its adaptation to a changing agricultural environment. Pages 82–84 in Proc. 5th Vertebr. Pest Conf., Fresno, Calif.

Zimny, M. L. 1965. Thirteen-lined ground squirrels born in captivity. J. Mammal. 46:521–522.

P. QUENTIN TOMICH, Research Unit, State of Hawaii, Department of Health, Honokaa, Hawaii 96727.

11

Fox and Gray Squirrels

Sciurus niger, S. carolinensis, and Allies

Vagn Flyger
J. Edward Gates

NOMENCLATURE

COMMON NAMES. Fox squirrel, stump-eared squirrel, cat squirrel
SCIENTIFIC NAME. *Sciurus niger*
SUBSPECIES. *S. n. avicennia, S. n. bachmani, S. n. cinereus, S. n. limitis, S. n. ludovicianus, S. n. niger, S. n. rufiventer, S. n. shermani, S. n. subauratus,* and *S. n. vulpinus.*

The fox squirrel was described originally in 1758 by Linnaeus in *Systema naturae,* 10th edition, from a black specimen from Georgia or the Carolinas. In the same edition, *S. cinereus* was also described. It was later given subspecific status under *S. niger* (Barkalow 1954). Ten subspecies are currently recognized.

COMMON NAMES. Gray squirrel, cat squirrel, migratory squirrel
SCIENTIFIC NAME. *Sciurus carolinensis*
SUBSPECIES. *S. c. carolinensis, S. c. extimus, S. c. fuliginosus, S. c. hypophaeus,* and *S. c. pennsylvanicus.*

The gray squirrel was originally described by Gmelin in 1788 (*Caroli a Linné systema naturae,* 13th edition). Five subspecies are presently recognized.

DISTRIBUTION

The range of the fox squirrel (figure 11.1) extends slightly farther west than that of the gray squirrel (figure 11.2) but not as far northward. Fox squirrels have been successfully introduced in numerous western cities; expanding introduced populations exist in California, Oregon, and Washington. Introduced populations of gray squirrels also occur in California, Oregon, Washington, Montana, and British Columbia. Well-established introduced populations occur in Great Britain and South Africa, where they are considered pests, and a colony exists in Melbourne, Australia.

DESCRIPTION

Fox and gray squirrels are members of the order Rodentia and family Sciuridae. Both are similar in behavior and body conformation, but the fox squirrel is larger than most other tree squirrels and possesses a larger, more fluffy tail. The rounded ears of the fox squirrel are distinctively shorter and thicker than those of the gray squirrel. Neither fox nor gray squirrels develop tufts on the top edges of their ears. Sexes show no differences in size or color.

The body measurements of adult fox squirrels are as follows: total length 454–698 millimeters (mm), tail length 200–330 mm, hind foot length 51–82 mm (Hall and Kelson 1959). Weights range from 696 to 1,233 grams (g) (Allen 1943; Longley 1963) and average about 800 g. Thoma and Marshall (1960) found that Minnesota fox squirrels range from 507 to 1,000 g, while Dozier and Hall (1944) reported variations of 737 to 1,361 g in Maryland.

Body measurements of adult gray squirrels are: total length 383–525 mm, tail length 150–243 mm, hind foot length 53–76 mm. Weights range from 338 to 750 g (Barkalow and Shorten 1973). The largest gray squirrels occur in the northern portion of their range, while the smallest occupy the southern tip of Florida. Male gray squirrels in West Virginia average 523 g, while females average 518 g—essentially equal in weight (Uhlig 1955). Redmond (1953) found that adults averaged 460 g in Mississippi, while adult squirrels in Illinois averaged 535 g (Brown and Yeager 1945).

Pelage. The coat color of fox squirrels varies greatly, both locally and regionally, from all black in Florida to silver gray with a white belly in Maryland or grizzled rusty brown above with orange ventral surface in Michigan. Extreme color variations often occur among fox squirrels within a single woodlot. The most widespread subspecies, *S. n. rufiventer,* is commonly a reddish black to light grizzled color above and bright to dull rufous or light gray to dirty white below. The tail is mixed black and tawny rufous; the soles of the feet are black (Baumgartner 1943). Melanistic fox squirrels

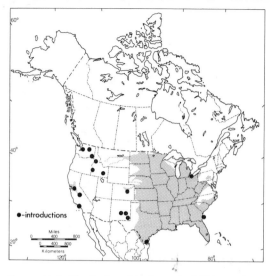

FIGURE 11.1. Distribution of the fox squirrel (*Sciurus niger*) in North America.

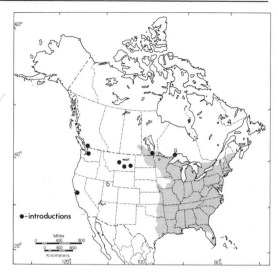

FIGURE 11.2. Distribution of the gray squirrel (*Sciurus carolinensis*) in North America.

frequently occur, especially in southern portions of their range.

The coat color of gray squirrels gives the impression of grayness, but, upon close inspection, it generally has a salt-and-pepper appearance due to the alternating bands of brown, white, and black on the hairs. The middorsal portion is usually slightly more brownish, while the forehead is slightly darker. Brown patches appear on the cheeks. Usually brown or cinnamon patches are present on the flanks behind the forelegs, and sometimes the upper surface of the paws is brown. The soles of the feet are fleshy gray. The ears are light brown or tan colored. Frequently, in the north, there is a patch of white hair behind the ears. This patch is especially prominent in winter. Chin, throat, and belly are white. Tail hairs are banded with brown and black, with white tips. Fox squirrel tail hairs are not white tipped.

Gray squirrel coat colors are also variable. Some individuals have a distinct reddish cast, often with dark belly fur. These may be mistakenly identified as hybrids of fox and gray squirrels. The ventral and dorsal fur of gray squirrels tends to be darker in the southern portions of their range than in the north. Although the sexes are alike, juveniles have less brown and appear more woolly than adults. Melanistic squirrels, varying from pure black to sooty gray, are abundant in some portions of the northern part of the range. White squirrels, including albinos, occasionally are found, especially in urban areas. Olney, Illinois, is noted for its white squirrel colony.

Skull and Dentition. The basilar length of the skull of the fox squirrel is about 66 mm and that of the gray squirrel 63 mm (figure 11.3). Both species have well-developed postorbital processes. Fox squirrel bones have a pink tint and fluoresce bright red under long-

wave ultraviolet light (Flyger and Levin 1977). They can be distinguished from all other normal adult mammalian species by this means.

The dental formula of the fox squirrel is 1/1, 0/0, 1/1, 3/3 = 20; whereas that of the gray squirrel is 1/1, 0/0, 2/1, 3/3 = 22. The third upper premolar of the gray squirrel is peglike and is sometimes absent. The cheek teeth are rooted. Structure is brachyodont and bunodont.

PHYSIOLOGY

Molts. The molt of fox and gray squirrels is almost identical. Both species molt twice yearly, the summer coat not differing much from that of winter. The ears become distinctly furred in winter in the northern portions of their range. The spring molt begins in March on the head and progresses posteriorly. The fall molt begins in mid-September on the flanks and moves anteriorly. The tail molt occurs once yearly, usually in late July and August. Nursing females do not begin molting until their young are weaned. Young animals start molting their juvenile coat when approximately three months old and may molt into either the winter or summer pelt.

Metabolism. Fox squirrels are unique among mammals in the accumulation of uroporphyrin in teeth, bones, and soft tissues. This produces a pink coloration of bones and other tissues that fluoresces under long-wave ultraviolet light. The condition is due to a deficiency of uroporphyrinogen 3 cosynthetase. A similar condition, congenital erythropoietic porphyria, a rare disease among other mammals, including humans, is characterized by skin lesions, extreme photosensitivity, and hemolytic problems. Fox squirrels show none of these symptoms.

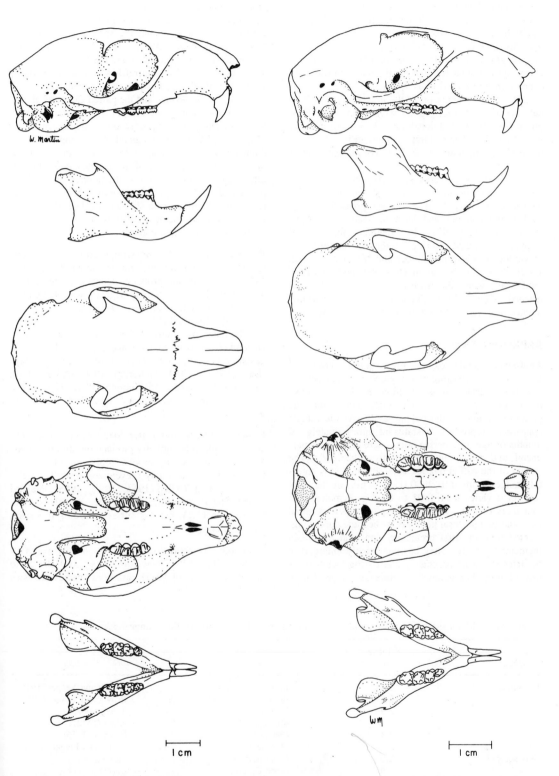

FIGURE 11.3. Skulls of the fox squirrel (*Sciurus niger*) (left) and the gray squirrel (*S. carolinensis*) (right). From top to bottom: lateral view of cranium, lateral view of mandible, dorsal view of cranium, ventral view of cranium, dorsal view of mandible.

Fox and gray squirrels have yellow-pigmented lenses. This pigment is water soluble, with a maximum absorption spectrum at 370 nanometers (nm) (Cooper and Robson 1966). Histologically, the retina of gray squirrels appears to consist only of cones, but electroretinograms demonstrate that the retina has two receptor systems that function as though both rods and cones were present (Gouras 1964).

Oxygen consumption in the gray squirrel is 0.84 cubic centimeters (cc)/g body weight per hour. Heat production is 9.61 kilocalories (kcal)/100 g body weight per day. The respiratory quotient is 0.785, and the body temperature is 38.7° C (Bolls and Perfect 1972).

Captive fox and gray squirrels consume food at high rates in early autumn and at reduced rates in midwinter. Body weights of adults are greatest in September and October and least in March and April (Short and Duke 1971). In Michigan, fox squirrel weights increased during summer, attaining a maximum of 800 g in October and November (Allen 1942). A decline begins in December and continues until May, when the lowest weights of 740 g are reached (Allen 1942). There is no significant difference in weight between the sexes for either species.

REPRODUCTION

Anatomy. Female fox and gray squirrels possess a duplex uterus consisting of a single vagina, one cervix, and two uterine horns. Nulliparous females have threadlike uteri 1.5 mm in diameter coiled into the posterior portion of the pelvic cavity. As the female approaches reproductive status at approximately 10 months of age, the uterine horns increase in diameter, uncoil, and stretch out anteriorly from the cervix. The weight of the paired ovaries in adult female fox squirrels ranges from 0.01 to 0.05 g, while in the gray squirrel it varies from 0.01 to 0.03 g. Females have four pairs of teats: one inguinal pair, one abdominal pair, and two pectoral pairs. Teats become darkly pigmented with the first pregnancy and tend to remain pigmented for life. They become greatly enlarged in the later stages of pregnancy and during lactation. A tiny (2 mm) but distinct os clitoridis is present in female fox and gray squirrels (also in western gray [*S. griseus*] and tassel-eared [*S. aberti*] squirrels).

The testes of male fox and gray squirrels descend around 10 months of age; the scrotal sack becomes heavily pigmented, almost black, at 12 months of age. The testes remain descended and are located posterior to the penis. The fox squirrel has a 10-mm baculum, while that of adult gray squirrels is 8 mm long.

The single prostate and the paired Cowpers glands undergo spectacular seasonal changes. Mature male fox squirrels in breeding condition (mid-December to May–June) have Cowpers glands measuring 26 mm in diameter; whereas in August the same glands measure only 8 mm. The prostate of this species varies from 6 mm to 24 mm in length during the same seasons. Cowpers glands in gray squirrels grow from 6 mm in diameter during September to 24 mm by late December. During this time, the prostate grows from 4 to 20 mm. Testes weights remain constant.

Breeding Season. The breeding seasons of fox squirrels and gray squirrels are similar (Soots 1964; Smith 1967). The initiation probably depends on decreasing day length. Breeding activity begins toward the end of December at the time of the winter solstice, regardless of latitude (Moore 1957).

Female fox squirrels come into estrus in mid-December or early January and again in June. The gestation period of both species is about 44 to 45 days (Baumgartner 1940; Goodrum 1940; Allen 1942; Brown and Yeager 1945; Shorten 1954). The earliest litters appear in late January; most births occur in mid-March and July (Allen 1943).

Pregnancy Rate and Litter Size. Adult female fox and gray squirrels normally produce two litters a year. Yearling females may produce only one litter. Poor food conditions also may prevent some adult females from breeding (Allen 1943; Uhlig 1955; Packard 1956). The average litter size of tree squirrels is about three (table 11.1), but this varies according to season and food conditions.

Neonates. Newborn fox and gray squirrels weigh 14 to 18 g (Uhlig 1955). Newborn fox squirrels are 50 to 60 mm long (Allen 1942), and both species are born

TABLE 11.1. Mean litter size of fox and gray squirrels in different parts of their range

Species	$\bar{X}$ Litter Size	Location	Source
Fox squirrel	3.4	Illinois (farmland)	Brown and Yeager (1945)
	2.5	Illinois (woodland)	Brown and Yeager (1945)
	2.9	Missouri	Terrill (1941)
	2.9	Michigan	Allen (1942)
	2.8	Kansas	Packard (1956)
	2.2	Maryland	Lustig and Flyger (1975)
Gray squirrel	2.9	Great Britain	Shorten (1954)
	2.9	North Carolina	Barkalow and Shorten (1973)
	2.7	Illinois	Brown and Yeager (1945)

hairless, blind, and with their ears closed. Skin color is pink. Distinct vibrissae are present on the nose and chin. Claws are well developed. Neither upper nor lower incisors have erupted.

Development of Young. Tree squirrels develop slowly compared to other rodents. Eyes open at 4 to 5 weeks, and ears open at about 6 weeks of age. Lower incisors erupt at 3 weeks, upper incisors at 5 weeks. Weaning occurs at 8 or 9 weeks of age, but young may not be self-supporting until 12 weeks or older (Allen 1942). Hair begins to appear on the dorsal surface at 8 to 10 days of age in fox squirrels (Brown and Yeager 1945) and at about 14 days of age in gray squirrels (Uhlig 1955).

Breeding Age. Female fox and gray squirrels attain sexual maturity at an age of 10 to 11 months. Females usually produce their first litters when one year old. A female born in February will usually produce her first litter the following February; however, she may not produce a second litter that year (Allen 1943; Uhlig 1955).

Sex Ratios. Sex ratios vary considerably, depending upon how the data were collected, that is, by trapping or shooting. A sex ratio of 99 males/100 females occurred in a total of 1,054 adult fox squirrels trapped (Allen 1943; Longley 1963). A sex ratio of 128 males/100 females was observed from 20,764 adult fox squirrels shot by hunters. A total of 141 nestling gray squirrels from various locations had a sex ratio of 135 males/100 females (Allen 1943; Brown and Yeager 1945; Packard 1956).

ECOLOGY

Home Range. Size of home range is influenced by several variables including population density, sex and age composition, food availability, and habitat. Smaller home ranges often are correlated with higher population densities. Availability of food within the habitat will also influence home range size. Larger ranges are observed in poorer habitats. Thus, comparisons between studies are difficult to make and only broad generalities are possible.

Allen (1943) stated that the average annual home range of a fox squirrel is at least 4.0 hectares (ha) but may exceed 16.2 ha. Males also tended to have larger home ranges than females. In Oklahoma, Chesemore (1975) found that male home ranges averaged 2.87 ha, while females had an average home range of 0.44 ha. Baumgartner (1938) observed fox squirrels in Ohio traveling daily between woodlots 1.2 kilometers (km) apart. McCloskey (1975) found an average home range to be 2.2 ha in Iowa, while fox squirrel home ranges averaged 16.4 ha in Ohio (Donohoe and Beal 1972). In the study by Donohoe and Beal (1972), home ranges were measured by trapping in widely varying habitats, and the small number of captures plus the attraction of bait in the traps undoubtedly influenced the results. The use of radio telemetry should provide better data in the future.

The home range of gray squirrels is generally smaller than that of fox squirrels. Flyger (1960) reported home ranges of the former to be 0.8 ha for males and 0.5 ha for females in Maryland. Cordes and Barkalow (1973) found the average home range to be 0.7 ha, with males again having a slightly larger range than females. Hougart (1975), following radio-collared squirrels, found it to be 1.8 ha. Male gray squirrels in an urban environment in British Columbia ranged over 20.2 ha, while females used 2.0 to 6.1 ha (Robinson and Cowan 1954). Doebel and McGinnes (1974) measured the mean home range of both sexes as 0.5 ha.

Seasonal Movements. Fox and gray squirrels are noticeably more active in September and early October than at any other time of year. During this season, they are burying acorns and nuts for the winter. Some dispersal occurs at this time, mainly by subadults from the spring litters.

Fox and gray squirrels frequently move to new locations, especially in the fall. Allen (1943) found 1 marked fox squirrel that moved 64.4 km and 11 others that each moved over 16.1 km. The dispersal period, frequently referred to as the "fall reshuffle," occurs between late August and mid-December in Michigan (Allen 1943), between 10 August and 15 September in Ohio (Baumgartner 1943b), and before August in Illinois (Jordan 1971). Fox squirrels in Illinois deserted one area after another as food availability shifted. The extent of such shifts depends on the habitat types and the diversity of food species (Brown and Yeager 1945).

Mass Movements. "Migrations" involving vast hoards of gray squirrels were described in historical accounts of the 18th and 19th centuries of eastern North America. Some of the early mass movements were estimated by Seton (1920) to involve more than one billion individuals moving en masse in one direction. Observers described great masses of squirrels drowning in attempts to cross rivers, devastation of crops by the hoards, and a scarcity of squirrels for several years following such a migration. Many of these accounts were possibly exaggerated.

Mass movements on a reduced scale are still reported on rare occasions. September dispersals now sometimes involve enough gray squirrels to receive comments in the news media. Such events usually are reported as increased numbers of dead squirrels on the highways, but in 1968 enough squirrels were involved to generate reports of migrations in eastern New York, Pennsylvania, North Carolina, and eastern Tennessee (Flyger 1969).

Fox squirrels rarely have been described as participating in fall mass movements or emigrations. Schorger (1949) reported three occurrences from Wisconsin: October 1938, September 1944, and fall 1946. Terrill (1941) recorded similar mass movements from Missouri in the fall of 1933, the spring and fall of 1934, and the spring of 1935. He stated that these movements were never as massive as gray squirrel emigrations.

Population Fluctuations. Fox and gray squirrel densities fluctuate irregularly. Uhlig (1956) reported that gray squirrel densities over a six-year period in four state forests in West Virginia varied from 1.1 to 2.5, 0.6 to 2.1, 0.5 to 1.1, and 0.8 to 3.5 per ha. Fluctuations over a three-year period of 2 to 10 squirrels per ha occurred in a suburban Maryland habitat (Flyger 1959). Nixon and McClain (1969) reported a density of 3.2 squirrels (fox and gray) per ha in Ohio. A severe spring frost reduced the available food to such an extent that by fall squirrel density had decreased to about 0.5 per ha. The effect of either a good or a poor acorn crop is reflected in squirrel abundance the following year (Burns et al. 1954).

Baumgartner (1940) believed that irregular 4.5- to 6-year cycles occurred in Ohio fox squirrels. However, Terrill (1941) found no indication of fox squirrel cycles in Missouri, nor did Allen (1943) find any indication of cycles in Michigan. In 1966, a severe spring frost in Ohio was followed by a decline in fox squirrel density that fall due to either increased dispersal or increased mortality (Nixon and McClain 1969). Breeding also virtually ceased from May 1966 to August 1967.

Kline (1965) believed that Iowa fox squirrels are not entirely dependent on mast, and that winter and early spring weather conditions are more important in affecting densities of fox squirrels. Extended periods of deep snow, subnormal temperatures, and high winds may reduce fox squirrel vitality and curtail or delay production of young. D. L. Allen (1943) and J. M. Allen (1952) felt that fall mast crops affected the overwintering populations of squirrels. The annual harvest of fox and gray squirrels fluctuated in the same ratio, thus indicating that the factors affecting population levels work equally on both species (Bakken 1952).

Habitat. The range of fox squirrels extends several hundred kilometers farther west than that of gray squirrels, because the former species occupies the narrow gallery forests along rivers and streams extending into the prairie regions. Shelterbelt plantings of various tree species have provided additional habitat in Kansas and Oklahoma.

Fox squirrels are most abundant in the transition belt between prairie and oak (*Quercus* sp.) woodland in the central portion of Oklahoma and Iowa. The opening of eastern woodlands for grazing and crops created additional habitat. Fox squirrels occupy upland hardwood forests, dense timber along streams and rivers, open pecan (*Carya pecan*) orchards, or wherever a few trees occur. They are even found in the most densely forested bottomlands of rivers (Chesemore 1975). Primary fox squirrel habitat in Wisconsin is oak-hickory woodlots with oak, swamp hardwoods, and mixed upland hardwoods as secondary habitats.

Gray squirrels seem to predominate in more mature timber. As the percentage of woodland increases, the ratio of gray to fox squirrels increases. When 10 percent of the land is wooded, both species seem to be equally abundant. If 70 percent or more of an area is wooded, fox squirrels are absent (Besnady 1957).

The relationship of fox squirrel abundance to the proportion of total forested land was demonstrated in Charles County, Missouri, where 10,026 ha of land were permitted to develop into a forest with very little pasturing of livestock. Fox squirrels declined and gray squirrels increased. In another portion of the same county, with small isolated woodlots and some pasturing by livestock, fox squirrels increased while gray squirrels almost vanished (Terrill 1941).

The same situation occurred in Indiana, in Michigan, and in other parts of the midwest. Much of this area was almost entirely forested when the first settlers arrived. Gray squirrels were abundant and fox squirrels were rare. The development of agriculture at the expense of the forests destroyed most of the gray squirrel habitat. The resulting small woodlots and fencerows became excellent fox squirrel habitat and this species prospered (Allen 1943). Recently, as some areas have reverted to mature woodlands, gray squirrels have again become the dominant species. Fox squirrels are also found in many cities and towns of the midwest (Brown and Yeager 1945).

Only fox squirrels occur in the prairie region of Illinois. This land is almost entirely agricultural. The small amount of woodland is mostly oak-hickory with a few scattered black walnuts (*Juglans nigra*). Most of these woodlots are grazed. In other portions of the state fox and gray squirrels occur together, both in bottoms and on the bluffs along the rivers; however, grays are more restricted than fox squirrels.

In Ohio, fox squirrels occur primarily on glaciated soils, and gray squirrels on unglaciated hill country of the southeastern portion of the state (Bakken 1952). Climax forests support fewer fox squirrels than do subclimax, or secondary, forest types (Baumgartner 1943b). Typical fox squirrel habitat consists of woodlots one or more ha in size with about 50 trees/ha, mostly oaks. Many woodlots are inferior habitat and can support few, if any, fox squirrels in a food crisis. Such habitats support fox squirrels only seasonally or for one year out of every three or four. For example, stream vegetation consisting of elms (*Ulmus* sp.), ash (*Fraxinus* sp.), red maple (*Acer rubrum*), blackberry (*Rubus* sp.), buckeye (*Aesculus* sp.), and sycamore (*Platanus occidentalis*) are used mostly in the spring, while hickory stands lack spring food. Pure beech (*Fagus grandifolia*) or beech-maple stands provide mast (beechnuts) in abundance only in certain years.

In east Texas, fox squirrels inhabit oak-hickory ridges, while grays occupy dense river bottom woods and swamps. Farther west, grays are absent or restricted, and fox squirrels occupy both the timbered river bottoms and oak ridges (Baker 1944). Fox squirrels are more adapted to cutover woods and have increased their range westward into such cutover areas. However, the removal of hardwood species with the creation of pure pine (*Pinus* sp.) stands (timber stand improvement) is the main cause of fox squirrel habitat destruction (Goodrum 1940; Baker 1944; Smith 1970). On the other hand, fox squirrels often do occur in mixed woodlands of pines and hardwoods. In Alabama, they occur along watercourses, the shores of bayous and deep bald cypress (*Taxodium distichum*) swamps, and on upland dry pine stands (Bakken 1952). In southern Florida, fox squirrels occupy the pine hammocks, and range into mangroves and cypress stands (Moore 1957). In the coastal regions of northern Florida, Georgia, and the Carolinas, fox squirrels are found in long-leaf pine (*P. palustris*)-turkey oak (*Q. laevis*) habitats. These habitats are often typical pine savannahs with dispersed trees where crowns do not touch each other and with grass understory. Such habitat is maintained by frequent controlled burning. On Maryland's eastern shore, fox squirrels prefer old-growth loblolly pine (*P. taeda*) and also are found in deep deciduous swamps, usually close to pine woods (Dozier and Hall 1944).

Gray squirrels primarily inhabit extensive mature hardwood forests with dense undergrowth and abundant den cavities, but throughout their range their habitats are quite varied. Their range, as well as density, has expanded or contracted depending on changes in land use patterns wrought by people. They are generally found only in dense or mature stands of oak and hickory, while fox squirrels may occupy the same habitat and other wooded lands, especially where understory is minimal. In Florida, gray squirrels are confined to large live oak (*Q. virginiana*) hammocks, river swamps, and some bay heads.

Nixon et al. (1978) found that landscapes in central and northern Illinois must be at least 20 percent forested to possess even a few gray squirrels and at least 30 percent forested to support reasonable hunting. Forests with gray squirrels tend to be extensive, ungrazed, and predominantly saw timber with at least 15 tree cavities per ha. This dependence on den cavities was also noted by Bakken (1952), who stated that gray squirrels have a greater need for tree dens than do fox squirrels.

FOOD HABITS

Food preferences of fox and gray squirrels seem to be similar, but their different foraging habits result in some differences in consumption (Smith and Follmer 1972). Both species are heavily dependent on mast crops for caloric intake during the colder months. This is reflected in the relationship of squirrel abundance to acorn (*Quercus* sp.) crops (Burns et al. 1954; Nixon and McClain 1969). However, both consume a wide variety of foods and readily take advantage of unusual food sources.

Acorns, beechnuts, hickory nuts, walnuts, corn (*Zea mays*), and soybeans (*Glycine max*) are the main food source of fox squirrels. However, the list of food items of this species is long and extremely varied depending on geographic location and land use practices. Fox squirrels are opportunists and readily adapt to new and exotic foods.

In Illinois, fox squirrels rely heavily on hickories from late August through September. Pecans, black walnuts, osage orange (*Maclura pomifera*) fruits, and corn are also important fall foods. In early spring, elm buds and seeds are the most important food. In May and June, mulberries (*Morus* sp.) are heavily utilized.

By early summer, corn in the milk stage becomes a primary food (Brown and Yeager 1945). Fox squirrels in Illinois seemed more tolerant of a limited variety of food items than did gray squirrels, and in some areas with only one or two staple foods only fox squirrels occurred.

During summer in Kansas, berries and other fruits are high on the list of preferred foods. A variety of insects are also consumed by fox squirrels. Acorns of dwarf chinquapin oak (*Q. prinoides*) are also a favored food at this time. During winter, osage orange is a staple item supplemented with seeds of the Kentucky coffee tree (*Gymnocladus dioicus*) and honey locust (*Gleditsia triacanthus*), corn, wheat (*Triticum aestivum*), cottonwood (*Populus deltoides*) bark, ash seeds, and red cedar (*Juniperus virginianus*) berries. In spring, fox squirrels feed primarily on buds of elm, maple, and oaks but also on newly sprouting leaves and insect larvae (Packard 1956). Plant galls, fruit and bark of Russian olive (*Elaeagnus angustifolia*), and wild gourds are additional food items for Kansas fox squirrels (Bugbee and Riegel 1945).

Fox squirrels in Ohio prefer hickory nuts, acorns, corn, and black walnuts. They are absent where two or more of these mast species are missing. Fox squirrels also ate buckeyes, seeds and buds of maple and elm, hazelnuts (*Corylus* sp.), blackberries, tree bark, and small amounts of soil. After the cached food supply was exhausted in March, the squirrels fed chiefly on the buds and seeds of elm, maple, aspen (*Populus* sp.), and willow (*Salix* sp.), until midsummer (Baumgartner 1939). Ohio fox squirrels had the following order of food preference: white oak (*Q. alba*) acorns, black oak (*Q. velutina*) acorns, red oak (*Q. rubra*) acorns, walnuts, and corn (Baumgras 1944). Ofcarcik et al. (1973) in contrast found that red oak acorns were preferred over white oak.

East Texas fox squirrels prefer the acorns of bluejack oak (*Q. incana*), southern red oak (*Q. falcata*), and overcup oak (*Q. lyrata*). The least preferred foods are acorns of swamp chestnut oak (*Q. michauxii*) and overcup oak (Ofcarcik et al. 1973). In California fox squirrels feed on English walnuts (*J. regia*), oranges (*Citrus sinensis*), avocados (*Persea gratissima*), strawberries (*Fragaria* sp.), and tomatoes (*Lycopersicon esculentum*). In midwinter, they feed on eucalyptus seeds (Wolf and Roest 1971).

Michigan fox squirrels feed on a variety of foods throughout the year. Spring foods are mainly tree buds and flowers, insects, bird eggs, and seeds of red and silver (*A. saccharinum*) maples and elms. Summer foods include a variety of berries, plum and cherry (*Prunus* sp.) pits, fruits of basswood (*Tilia americana*), box elder (*Acer negundo*), black oak, hickory nuts, seeds of sugar (*A. saccharum*) and black (*A. nigrum*) maple, grains, insects, and unripe corn. Fall foods consist mainly of acorns, hickory nuts, beechnuts, walnuts, butternuts (*J. cinerea*), and hazelnuts. Caches of acorns and hickory nuts are heavily used in winter. Cahalane (1942) found that 10 percent were used by 1 January and by May less than 1 percent remained. Other foods cached during fall included corn, seeds of false buckwheat (*Polygonum scandens*), fruits of osage orange, basswood, pines, ash, pawpaw (*Asimina triloba*), blackberry, bittersweet (*Celastrus* sp.), plus rosehips (*Rosa* sp.), haws (*Crataegus* sp.), and wild grapes (*Vitis* sp.). Bark is eaten, especially from sugar maple but also from elm, red maple, and beech. During winter, isolated trees that had formerly been ignored may be investigated (Allen 1943).

Ordinarily water needs are provided by succulent plant materials, but during drought such materials may be scanty and surface water may become a necessity. In Michigan, lack of moisture for prolonged periods can be a serious mortality factor (Allen 1943).

Fruits, seeds, and bark of many trees are also important food items for gray squirrels. Fruits of bull bay magnolia (*Magnolia grandiflora*) are important on Everglade hammocks in Florida (Blair 1935). Elsewhere, gray squirrels favor seeds of hawthorn (Dambach 1942), cherry and apple (*Malus* sp.) pulp (Woods 1941), honey locust seeds, horse chestnuts (*Aesculus hippocastanum*), pussy willow catkins, and sycamore tree buds. Seeds of several species of maple are important foods (Robinson and Cowan 1954). Elm seeds are considered to be a favored food (Terres 1939; Judd 1955). Buds and bark of maple trees are also eaten (Nichols 1958). Scraping the inner bark of maples to get at the sweet sap can be extensive (Brenneman 1954). The bark of many trees is consumed, including chestnuts (*Castanea* sp.), chestnut oaks (*Q. prinus*), and white oaks.

Fungi are also an important gray squirrel food item, as are insects, especially larval and pupal stages (Britton 1933). Fungi in the diet of squirrels is mentioned by several authors. Squirrels will devour many species of mushroom. Fungi are considered to be the staple year-round food source of *Sciurus griseus* (Stienecker and Browning 1970). Possibly some of the bark consumed from dead or injured trees is eaten because of the fungi present (Gunter and Eleuterius 1971). Hamilton (1943) reported gray squirrels searching for and eating geometrid caterpillars from the bark of hickory trees, while Layne and Woolfenden (1958) found them searching automobile radiators in a parking lot for dragonflies, butterflies, grasshoppers, and other insects.

Calcium and other minerals are obtained by gray squirrels from bones, deer antlers, and turtle shells (Madson 1964). Analysis of 22 gray squirrel stomachs also revealed the presence of soil, 2.5 percent by volume in autumn, 5.0 percent in winter, and 6.7 percent in spring (Packard 1956). Soil is probably ingested intentionally, judging from the way captives sometimes eagerly devour it when a shovelful is thrown into their cage.

BEHAVIOR

Mating Behavior. Estrous female fox squirrels are pursued by sexually active males in a lengthy but somewhat leisurely chase, more or less in single file, over the forest floor, up and down tree trunks, and through tree branches. Such chases may last all day

with frequent pauses while the female stops to rest or feed. Occasionally, a male will approach the female, only to be driven away (Bakken 1952; McCloskey 1975). Presumably the dominant male becomes the individual closest to the female and finally the one to mate with her. Gray squirrels behave in the same manner. Moore (1968) reported males of both species pursuing a female gray squirrel in a mating chase.

Aggressive Behavior and Social Hierarchy. Fox and gray squirrels are relatively nonaggressive toward either their own species or the other species. During the colder months, they may share dens with a number of individuals of their own species depending on the size of the den cavity. When feeding on a concentrated food source such as at a corn bin or under an oak tree, individuals maintain an interval of 1.2 to 1.5 m. Pregnant or nursing females of both fox and gray squirrels, however, will aggressively drive off, by tooth chattering and lashing out with forepaws, other squirrels that happen to intrude on the tree containing a nest or den (McCloskey 1975).

When population density is seven or more squirrels per ha, social organization within the population becomes discernible. Both species exhibit dominance relationships such that a definite hierarchy exists (Flyger 1955; Pack et al. 1967). A dominant animal will chatter its teeth as a warning to a lower-ranking individual that may be approaching too closely. Status is determined by experience, size of the individual, and age; thus, older squirrels rank higher than younger squirrels. In studying two squirrel populations in Virginia, Pack et al. (1967) found that males and females were involved in a single hierarchy in each population. Males were dominant over females in 79 percent of the interactions observed. The ranking within the hierarchy was very stable. Dominance was maintained only during mating chases, at den trees, and at very concentrated food sources such as a bird feeder, a squirrel feeding station, or in captivity (Flyger 1955; Bakken 1959). Otherwise, squirrels usually stay well dispersed during their daily activities.

Escape Behavior. Frightened or pursued fox and gray squirrels utilize several strategies to elude enemies. If surprised while in a tree a fox squirrel may keep the tree trunk or a large branch between itself and its enemy. Gray squirrels are less likely to "play hide and seek" around a tree trunk and are more inclined to run off through the tree canopy. It is unusual for fox squirrels to flee through the forest canopy from tree to tree. Often, the squirrel's habit of remaining motionless will conceal its presence to an observer. If badly frightened, e.g., when a fox squirrel has been shot at, it may come down out of the tree and run over the ground to its den tree.

Vocalizations. The best-known tree squirrel sound is the alarm call, a series of rapid *kuk, kuk, kuks* uttered while the animal rapidly flicks its tail in the air. In general, fox squirrel barks seem "softer" than those of the gray squirrel. During the mating chase, males sometimes utter a short whine (Bakken 1952). When cornered, as in a trap, fox and gray squirrels will rapidly chatter their teeth. Juveniles and nestlings often scream loudly if removed from the nest or restrained. Such calls may cause the mother to approach.

Miscellaneous Behavior. Young fox and gray squirrels often engage in "play wrestling" with litter mates. Wild squirrels have been observed running and leaping about up onto tree trunks, over the forest floor, and rolling in the leaf litter. Sometimes pieces of wood are held in their forepaws and teeth. They frequently can be seen alternately wiping each side of the face on a branch. This behavior is probably not grooming but scent marking (Benson 1975).

When an excess of nuts is available in the fall, squirrels engage in caching them. Bakken (1959) noted that after hunger was satisfied, caching of the remaining food items usually followed. This behavior can occur any time of the year. Fox and gray squirrels cache at random and communal hoarding is common. Occasionally nuts are cached in a tree cavity, but most often they are stored singly underground (Madson 1964). The squirrel picks up a nut or acorn, hops over the ground a few meters, and digs a shallow hole in the earth or leaf litter with its forepaws. The nut is placed in the hole, the squirrel rakes leaves or soil over it with its forepaws, and then tamps it with its snout. Nuts are so well hidden that it is difficult for a person to find them even though the observer immediately searches the spot. During the winter, squirrels locate nuts by smell, even through 30 cm of snow (Cahalane 1942).

MORTALITY

Mortality Rates. Little attention has been paid to mortality rates of fox squirrels, but it is reasonable to assume that they are somewhat similar to those of gray squirrels. Barkalow et al. (1970) found that the mortality rate of gray squirrels was 50 percent per year. The mortality rate of gray squirrels based on trapping over a three-year period in a Baltimore, Maryland, suburb was 50 percent per 10-month period (Flyger 1956).

Longevity is similar in both species. Linduska (1947) trapped five fox squirrels in Michigan that were 4.5 years or older, one being over 6 years of age. Fox squirrels have survived 13 years in captivity. Twelve wild gray squirrels reached ages of over 6 years, and one female 12.5 years old was pregnant at her last capture (Barkalow and Soots 1975).

Mortality Factors. Predation seems to be a minor factor in the dynamics of fox squirrel populations. Fox squirrels have been reported as minor food items of Great Horned Owls (*Bubo virginianus*) (Errington 1932), Red-shouldered Hawks (*Buteo lineatus*) (English 1934), and bobcats (*Lynx rufus*) (Progulske 1955). The opening of a good den is too small to allow a raccoon (*Procyon lotor*) access to nestling squirrels. Large rattlesnakes (*Crotalus* sp.) occasionally may take a squirrel, although most dens are too high for them. Weasels (*Mustela* sp.), however, may be important predators because they are common in most squir-

rel habitat and can easily enter tree dens. Squirrels are more active in autumn and winter and therefore more susceptible to predation at that time.

A major cause of fox squirrel population decline has been a combination of mange mites (*Cnemidoptes* sp.), severe winter weather, and inadequate food (Allen 1943). Hunting is another major mortality factor throughout most of this species range.

Natural mortality seems to be the controlling factor with gray squirrel populations. Uhlig (1957) found a correlation between the fall population level and subsequent mortality the following year in West Virginia. In this state, hunters harvested 13 percent of the populations on state-owned lands but natural mortality was three or four times this amount (Uhlig 1956). Intensive hunting and trapping in Great Britain had no noticeable effect on squirrel densities (Shorten 1959). Carson (1957) also believed that heavy hunting in West Virginia had no effect on gray squirrel population levels.

Squirrels may occasionally lose their footing and fall from a tree branch but will right themselves in midair before reaching the ground. They seem to be unharmed by high falls. Injuries in squirrels heal quickly. Missing or abbreviated forepaws are found on rare occasions.

Automobiles take their toll of squirrels. Road kills among fox squirrels are probably proportionally higher than among gray squirrels, because fox squirrels prefer open habitats while gray squirrels are found in dense woods.

Tree squirrels are not as heavily affected by diseases as other wildlife. They are, however, susceptible to a number of ectoparasites and endoparasites (table 11.2). Ectoparasites appear to be a more serious problem. Fleas often are found in tree cavities or den boxes. The problem is reduced when the squirrels inhabit leaf nests (Packard 1956). Ticks, abundant in early summer, affect the ears and neck of squirrels. Chiggers are abundant in autumn. Larvae of the bot fly (*Cuterebra* sp.) are responsible for warbles (Allen 1952). The most common malady appears to be mange caused by a mite (*Cnemidoptes* sp.) that results in loss of hair (figure 11.4). This condition is most prevalent during winter, when fat reserves are their lowest.

Bacterial and Viral Diseases. California encephalitis virus has been reported from fox and gray squirrels in Ohio (Masterson et al. 1971) and Wisconsin (Moulton and Thompson 1971). Evidence of infection with western equine encephalitis in Colorado and California encephalitis in Wisconsin have been reported for fox squirrels (Lennette et al. 1956, Moulton and Thompson 1971).

An outbreak of plague (*Yersinia pestis*) occurred among urban fox squirrels in Denver, Colorado, in 1968. Individual plague-infected fox squirrels were also reported from Palo Alto, California, and Greely, Colorado (Hudson et al. 1971).

For the most part, gray squirrels seem to be relatively free of bacterial and viral agents (White et al. 1975). Gray squirrels are, on rare occasions, found

TABLE 11.2. Parasites of fox and gray squirrels

Parasite	Fox Squirrel	Gray Squirrel	Source
Protozoa			
Eimeria confusa		X	Joseph 1972
E. kniplingi	X		Levine and Ivens 1965
E. lancasterensis		X	Joseph 1972
E. ontarioensis		X	Joseph 1972
Hepatozoon griseisciuri		X	Clark 1958
Toxoplasma gondii		X	Walton and Walls 1964
Cestodes			
Bothriocephalus sciuri	X		Self and Esslinger 1955
Catenotaenia fasciolaris	X		Dobrovolny and Harbaugh 1934
C. pusilla		X	Rausch and Tiner 1948
Choanotaenia sciuricola	X		Harwood and Cooke 1949
Citellinema bifurcatum	X		Dikmans 1938
Cittotaenai pectinata		X	Rankin 1946
Cysticercus passeriformes	X		Baumgartner 1940
Hymenolepis diminuta	X	X	Rausch and Tiner 1948
H. nana		X	Olexik et al. 1969
Mesocestoides latus	X		Rausch and Tiner 1948
Multiceps serialis	X	X	Bonnal et al. 1933
Raillietina bakeri	X	X	Chandler 1942; Moore 1957
Taenia crassiceps		X	Freeman 1962
T. hydatigena	X	X	Meggitt 1924; Baylis 1939
T. pisiformis	X	X	Graham and Uhrich 1943; Brown and Yeager 1945
T. taeniaeformis	X		Harkema 1936
Nematodes			
Ascaris columnaris	X	X	Rausch and Tiner 1948; Tiner 1949
A. lumbricoides	X	X	Brown and Yeager 1945; Rausch and Tiner 1948
Bohmiella wilsoni	X	X	Lucker 1943; Rausch and Tiner 1948
Capillaria hepatica	X		McQuown 1954
Citellinema bifurcatum		X	Parker 1968
Dipetalonema interstitium		X	Price 1962
Dirofilariaeformia pulmoni		X	Davidson 1975
Enterobius sciuri	X	X	Rausch and Tiner 1948, Parker and Holliman 1971
Gongylonema pulchrum		X	Parker and Holliman 1971
Heligmodendrium hassalli	X	X	Rausch and Tiner 1948
Macracanthorhynchus hirudinaceus	X		Rausch and Tiner 1948
Moniliformis clarki	X	X	Chandler 1947
Nudacotyle sp.		X	Olexik et al. 1969
Physaloptera sp.		X	Rausch and Tiner 1948
P. massino	X		Morgan 1943
Rictularia coloradensis		X	Parker 1968
Strongyloides papillosus		X	Reiber and Byrd 1942
S. robustus	X	X	Chandler 1942; Parker and Holliman 1971
Trichinella spiralis	X		Zimmerman and Hubbard 1969
Trichostrongylus calcaratus		X	Rausch and Tiner 1948
T. calubriformes		X	Baylis 1934
T. retortaeformis		X	Cameron and Parnell 1933
Acarina			
Amblyoma americanum	X	X	Morlan 1952; Moore 1957
A. maculatum	X	X	Hixon 1940; Moore 1957
A. tuberculatum	X		Moore 1957
Androlaelaps glasgowi		X	Parker and Holliman 1971
Atricholaelaps glasgawi	X		Uhrich and Graham 1941
A. megaventralis	X		Moore 1957
Brueliai rotundata		X	Parker 1968
Cheladonta micheneri		X	Lipovsky et al. 1955

Parasite	Fox Squirrel	Gray Squirrel	Source
Dermacentor variabilis	X	X	Tugwell and Lancaster 1962
			Sonenshine and Stout 1971
Euschongastia jonesi		X	Loomis 1956
E. setosa		X	Loomis 1956
Eutrombicula			
alfredalugesi	X		Jenkins 1948
Haemaphysalis leporis-			
palustris	X	X	Bishopp and Trembley 1945
Haemogamasus ambulans		X	Clark 1958
Haemolaelaps			
megaventralis	X		Morlan 1952
Ixodes cookei	X	X	Bishopp and Trembley 1945
I. hearlei	X		Cooley and Kohls 1945
I. marxi		X	Bishopp and Trembley 1945
I. muris		X	Anastos 1947
Microtrombicula trisetica		X	Webb and Loomis 1971
Ornithonyssus bacoti		X	Morlan 1952
Sarcoptes sp.		X	Chapman 1938
Sarcoptes scabiei	X		Baumgartner 1940
Speleognathopsis sciuri		X	Clark 1960
Trombicula spp.	X		Loomis 1956
Trombicula alfredalugesi		X	Jenkins 1948
T. autumnalis		X	Loomis 1956
T. fitchi		X	Loomis 1954
T. gurneyi		X	Loomis 1955
T. splendens		X	Morlan 1952
T. sylvilagi		X	Kardos 1954
T. whartoni		X	Brennan and Wharton 1950
Walchia americana		X	Loomis 1956
Anoplura			
Enderleinellus longiceps	X	X	Mathewson and Hyland 1962;
			Kim 1966*a*, 1966*b*
Hoplopleura hesperomydis		X	Mathewson and Hyland 1962
H. sciuricola	X	X	Morlan 1952
Neohaematopinus			
antennatus		X	Keegan 1943
N. sciurinus	X	X	Harkema 1936; Johnson 1958
N. sciuri		X	Mathewson and Hyland 1962
Polyplax spinulosa		X	Morlan 1952
Siphonaptera			
Cediopsylla simplex		X	Holland and Benton 1968
Ceratophyllus fasciatus	X		Baumgartner 1940
C. gallinae		X	Burbutis 1956
Conorhinopsylla stanfordi		X	Fox 1940
Ctenocephalides felis	X	X	Morlan 1952; Mathewson and
			Hyland 1964
Echidnophaga gallinacea	X	X	Morlan 1952; Ellis 1955
Epidedia wenmauni		X	Main 1970
Hoplopsyllus glacialis		X	Layne 1971
H. offinis	X		Graham and Uhrich 1943
Leptopsylla segnis	X		Morlan 1952
Megabothris asio		X	Mathewson and Hyland 1964
Monopsyllus sciurorum		X	George 1954
M. vison		X	Woods and Larson 1969
Nosopsyllus fasciatus		X	Burbutis 1956
Opisocrostis bruneri		X	Fox 1940
Opisodasys robustus	X		Costa Lima and Hathaway 1946
Orchopeas howardii	X	X	Fox 1940; Holland and Benton
			1968
Pulex irritans		X	Shaftesbury 1934
Spilopsyllus cuniculi		X	George 1954
Diptera			
Cuterebra sp.	X	X	Parker and Holliman 1971

FIGURE 11.4. A gray squirrel (*Sciurus carolinensis*) infected with mange mite (*Cnemidoptes* sp.).

covered with skin tumors closely resembling the Shopes fibromas found on cottontail rabbits (*Sylvilagus* sp.) (Hirth et al. 1969; King et al. 1972). These tumors are caused by a pox virus and are transmitted by blood-sucking arthropods, possibly mosquitoes (Kilham 1954).

Shotts et al. (1975) isolated *Leptospira* from fox and gray squirrels in Florida. Wobeser (1969) reported a gray squirrel infected with *Clostridium tetani* in Ontario. Tularemia has been reported repeatedly from fox and gray squirrels (Davis et al. 1970), and *Coxiella burneti* has been reported from gray squirrels (Enright et al. 1971). Fungi of 19 genera were recovered from hair and skin scrapings and from toenails of Florida gray squirrels, including ringworm (*Trichophyton mentagrophytes*) (Lewis et al. 1975).

AGE DETERMINATION

In the field, age of tree squirrels is usually determined by sex characters. Males with pigmented, descended scrotums are considered adults 11 months old or older. Females with pigmented teats, indicating pregnancy or previous pregnancy, are also considered to be 11 months or older. This is not a desirable method of age determination for females, because older females that have not been bred would mistakenly be classified as subadults (Hoffman and Kirkpatrick 1959). The banding on guard hairs of the lateral rump region of the winter pelt of the gray squirrel can be used to distinguish nestlings, juveniles, subadults, and adults (Barrier and Barkalow 1967).

The dry weight of eye lenses can also be used for age determination. Fox squirrel eye lenses weighing 40 milligrams or more are probably 1.5 to 2.5 years, 46 mg lenses indicate 2.5 to 3.5 years, and 50 mg 3.5 years or older (Beale 1962).

ECONOMIC STATUS AND MANAGEMENT

Fox and gray squirrels are popular game animals east of the 100th meridian. The annual harvest in the United States is considerable. These two species rank among the country's top game animals (table 11.3).

Complaints of damage by fox squirrels are not common when compared to either the gray or the red (*Tamiasciurus* sp.) squirrel. Fox squirrels can become a nuisance around corn cribs or peel bark from trees. They can experimentally transmit spores of oak wilt fungus (Himelick and Curl 1955). In California, fox squirrels sometimes become a nuisance by eating English walnuts, oranges, avocados, strawberries, and tomatoes, but most farmers like to see the squirrels and are willing to tolerate the damage (Wolf and Roest 1971).

The most efficient method of control is to remove offending animals by shooting or trapping. Peanuts (*Arachis hypogaea*), hickory nuts, pecans, or ear corn are good trap baits. Repellents such as moth flakes in petroleum jelly are ineffective.

Gray squirrels are hunted to a greater extent than fox squirrels, and roughly 40,000,000 are harvested annually in the United States. The annual harvest is highest in the southern states. A conservative expenditure of five dollars per squirrel (license, shells, travel, etc.) would make the total annual expenditure by squirrel hunters about $200 million. Gray and fox squirrels,

TABLE 11.3. Numbers of fox and gray squirrels harvested in different parts of their range

Species	Annual Harvest	Location	Source
Fox squirrel	2,091,980	Illinois	Preno and Labisky (1971)
	947,054 (1942)	Illinois	Brown and Yeager (1945)
	1,425,530	Indiana	Allen (1952)
	1,000,000	Missouri	Christisen (1976)
	778,240	Ohio	Hicks (1938)
	645,779 (4-year average)	Michigan	Allen (1943)
Gray squirrel	2,500,000	Mississippi	Preno and Labisky (1971)
	1,150,000 (1955–56)	Florida	Beckwith (1959)
	556,000	Florida	Eichorn (1962)
	735,020	Illinois	Preno and Labisky (1971)
	437,760	Ohio	Hicks (1938)

especially the latter, are enjoyed by many people and these species are often the only wild animals that urban dwellers can readily see. The pelts of either species have no present market value except the tails, which have a very limited market in the manufacture of fishing lures.

Gray squirrels are frequently rated as nuisances in urban situations because of: (1) destructiveness to flower gardens, trees, and shrubs; (2) entry into buildings resulting in damage to structures and objects inside; (3) gnawing wires and lead-covered telephone cables; and (4) consuming food put out for birds. Such nuisance squirrels can best be eliminated by capture in live-traps and either killing the offenders or releasing them 10 or more kilometers from the capture site. Squirrels in buildings can often be driven away by *liberal* use of moth balls (napthaline) or moth crystals (paradichlorobenzene) in spaces occupied by the animals. After the nuisance animals have been eliminated all entrance holes should be closed by covering them with heavy 0.5-inch (1.3-cm) wire mesh.

Management for fox and gray squirrels has essentially consisted of adjustments in hunting seasons. The consensus among wildlife managers has been that hunting has relatively little impact on squirrel populations. As previously noted, Uhlig (1956) estimated that natural mortality was three to four times the hunting mortality. Overharvest of fox squirrels by hunting has been reported from small woodlots and public shooting areas in Ohio (Baumgartner 1943*b;* Nixon et al. 1974), in Michigan (Allen 1943), and in Indiana (Allen 1952). The small and widely separated woodlots are sometimes "shot out" and restocking does not occur immediately by dispersal because dispersing squirrels may not find these woodlots. Travel lanes of wooded fencerows would help in the restocking of such woodlots (Baumgartner 1943*b*). Differences in bag limits and length of season have a minor effect on squirrel numbers.

The opening date of squirrel season varies between states and is often a controversial issue. Some hunters want an early opening date because hunting is easier when squirrels are feeding in nut trees. After the nuts have fallen squirrels are harder to find. Other hunters object to the high percentage of nursing females shot during August and September. Flyger (1952) found that 55 percent of adult-sized female squirrels shot during the last half of September were either nursing or pregnant.

Delaying the opening date until after the first frost greatly reduces the percentage of warbles found in squirrels. Hunters often needlessly discard infested squirrels. An opening date of 15 October in the mid-Atlantic states may be a reasonable compromise; nursing females are rare and few squirrels are infested with warbles.

Habitat can be improved for fox squirrels by selective cutting to encourage nut-bearing trees and other food species, planting corn and soybeans, leaving overmature and large-crowned trees, and opening up the forest understory by burning or light grazing. Keeping wooded fencerows and breaking up forests

into small, 2- to 4-ha woodlots of irregular shapes also would promote fox squirrels. Winter feeding and providing nesting boxes may be of help under certain circumstances. However, the use of corn as supplemental winter food when tree seed crops fail is a nutritionally questionable practice (Havera and Nixon 1980).

Management of forested lands for the benefit of gray squirrels means increasing mast production by encouraging mature trees with large crowns. Diversity of food species is important so that a crop failure of one or more tree species will not result in a severe food shortage. Den sites seem to be more important for maintaining gray squirrels than for fox squirrels. Sanderson (1975) estimated that an average of one or two dens per hectare is necessary to maintain a density of 0.6 gray squirrel per hectare. The beneficial impact of artificial dens (figure 11.5) was demonstrated by Barkalow and Soots (1965) and by Burger (1969). Restocking of overharvested woodlots or unoccupied habitat has not been reported but may also be a practical management technique. Because most forest stands in the eastern hardwoods are approaching commerical maturity, foresters and wildlife managers must cooperate in the long-term planning of cuts over large areas of forested lands in order to maintain yields of both gray squirrels and timber (Roach 1974). Nixon et al. (1980) have recommended the use of small (<8 ha), narrow (<160 m), carefully located clear-cuts in forests where 40-60 percent of the stands are retained in a seed-producing age.

CURRENT RESEARCH AND MANAGEMENT NEEDS

The range of fox squirrel subspecies in the eastern states has been reduced greatly in the past 100 years, and one subspecies (*S. n. cinereus*) has been declared

FIGURE 11.5. Artificial dens such as this one are readily used by gray squirrels (*Sciurus carolinensis*) in lieu of natural cavities.

endangered. Information on habitat requirements and methods of restocking is needed if these squirrels are to be effectively managed.

Factors that control fox squirrel densities are poorly known and research on this aspect of squirrel biology would provide a better understanding of population dynamics. Such studies would include work on the effects of hunting, and the interrelationships of poor nutrition, mange, and inclement weather plus factors that control reproduction and movements of these squirrels.

One interesting line of research would be to discover the advantage (if any) of porphyria in fox squirrels. Possession of this trait is a substantial physiological burden on the animal. It seems unlikely that this condition would be retained unless it provided some benefit to the species. This species could also serve as an animal model for learning more about porphyria from a medical aspect.

Gray squirrels, and fox squirrels to a lesser extent, can provide greater recreation for sportsmen than they do at present. Studies of how this can best be accomplished would involve learning more about the behavior of hunters. Methods to reduce or eliminate damage to telephone cables would save the telephone companies and the public considerable sums of money.

OTHER *SCIURUS*

Tassel-Eared Squirrel. The tassel-eared squirrel (*Sciurus aberti*) is also called Abert Squirrel and Kaibab squirrel. Subspecies include: *S. a. aberti*, *S. a. barberi*, *S. a. chuscensis*, *S. a. durangi*, *S. a. ferreus*, *S. a. mimus*, *S. a. navajo*, and *S. a. phaeurus*. The Kaibab squirrel, formerly given specific status as *S. kaibabensis*, is now considered a subspecies, *S. a. kaibabensis.*

Hall and Kelson (1959) reported a total length of 463 to 584 mm for this species, tail length 195 to 255 mm, and hind foot length 65 to 80 mm. For 40 males and 26 females, total length was 482 and 483 mm, respectively; tail vertebral length 208 and 221 mm, respectively; and hind foot length 69 and 70 mm, respectively (Keith 1965). Hind foot length of *S. a. mimus* averaged 62.3 mm for females and 64.7 mm for males (Ramey and Nash 1976*a*).

Weights of this squirrel vary from a high in October with males and females averaging 715 and 690 g, respectively, to a corresponding 613 g and 600 g in April and May (Patton et al. 1976). Keith (1965) compiled weights of 313 adults and found that males averaged 589 g and females 602 g.

The most distinguishing characteristic of this species is the tufted ears. The tufts, which are especially noticeable in winter, are composed of black hairs that extend from the tips of the ears to a height of 2.5 cm.

Several color phases of this squirrel occur. The most common is gray above with a broad reddish brown band down the back. In some regions, this band is pale or absent. Underparts are white with a black lateral band between the forelegs and hindlegs. The tail is gray above and white below, often with a black tip. An animal in a melanistic phase appears black all over, including the tail. Ramey and Nash (1976*b*) reported that the melanistic phase predominates in northern Colorado. The gray phase is most common in the southern part of the state, and there are no intermediate forms. The subspecies *S. a. kaibabensis* is dark bodied above and below with a distinct white tail, giving the animal almost a striped skunklike appearance (Goldman 1928). The species molts twice yearly, as do fox and gray squirrels.

Tassel-eared squirrels are found in the ponderosa pine (*P. ponderosa*) forests in the foothills of the Rocky Mountains and the Colorado plateau in the United States (figure 11.6). They occur in Colorado, New Mexico, and Arizona plus a small portion of central Utah. In Mexico, scattered populations occur in the Sierra Madre Occidental as far south as southern Durango (Hall and Kelson 1959). The subspecies *kaibabensis* is confined to the Kaibab Plateau of Arizona on the north rim of the Grand Canyon.

The squirrel appears to be heterothermous. Body temperatures average 40.7° C but vary from 38.5° to 42.2° C (Patton et al. 1976). Golightly and Ohmart (1978) demonstrated that tassel-eared squirrels had body temperatures varying from 35.2° to 41.1° C, depending on behavior.

The reproductive anatomy of the tassel-eared squirrel is probably similar to that of the gray squirrel. Keith (1965) stated that mating was confined to the month of May in the vicinity of Flagstaff, Arizona. Several males congregate near the nest of an estrous female and participate in a one-day event for each female called a mating bout (Farentinos 1980). The most aggressive, dominant males (alphas) copulate first and most frequently with estrous females during such bouts.

FIGURE 11.6. Distribution of the tassel-eared squirrel (*Sciurus aberti*) and the western gray squirrel (*S. griseus*) in North America.

Young are born in May or June in Colorado. The gestation period ranges from 38 to 46 days (Keith 1965). Tassel-eared squirrels have one litter a year consisting of three to four young (Nash and Ramey 1970). The weight of newborn squirrels is 12 g. They are born hairless but with vibrissae and claws. At two weeks of age, they possess short hairs on the tail, sides, and back, but the belly is bare. Breeding age is probably similar to fox and gray squirrels.

Males appear to predominate in the population. Keith (1965) found 131 males to 100 females in a sample of 729. Farentinos (1972) found 62 percent of 60 animals were males.

The average home range of five Arizona tassel-eared squirrels was 7.3 ha, which was diminished to 2.0 ha under snow cover (Keith 1956). Nash and Ramey (1970) believed that males may treat a portion of this home range as a territory to be defended against other males, but Keith (1965) found no evidence of such behavior. Patton (1975a) found home ranges of three squirrels were 4.0, 12.2, and 34.4 ha. Hall (1972) found a home range of 11.8 ha for a female during a one-year period.

These squirrels are so dependent on ponderosa pine (especially seeds) that their abundance fluctuates in response to cycles of cone production of this tree. Squirrels on the Kaibab Plateau also have fluctuated noticeably over the years, depending upon ponderosa pine conditions (Rasmussen 1941; Hall 1972). Farentinos (1972) found that densities fluctuated from 3.3 to 5.6 per 10 ha. Tassel-eared squirrels consume seeds, inner bark (especially in winter), terminal buds, twigs, and flowers of ponderosa pine plus mushrooms that grow in the duff in the immediate vicinity of the trees (Keith 1965; Hall 1967; Nash and Ramey 1970). They also consume carrion, pinyon pine (*P. cembroides*) seeds, bones, acorns, and fleshy fungi. They frequently raid the spruce (*Picea* sp.) cone middens of *Tamiasciurus* (Hall 1972).

Tassel-eared squirrels are restricted to ponderosa pine stands at middle elevations where annual rainfall is less than 63.5 cm (Patton 1975b). Forests with open understories are preferred (Hall 1973). Optimum habitat consists of uneven-aged pines with the trees spaced in even-aged groups within the stand. Such stands should have densities of 496 to 618 trees per ha with average dbh (diameter at breast height, 1.4 m) of 28 to 33 cm. One or two Gambel oaks (*Q. gambelii*) in the 30 to 36 cm dbh size class should be included. Nest trees are usually pines with a crown comprising 35 to 55 percent of the total tree height and a diameter of 36 to 41 cm. Adjacent trees are of similar size with touching crowns. This provides escape routes (Patton 1975b).

Nests are built of twigs and leaves on limbs against the tree trunk of pines or Gambel oaks at heights of 9.1 to 15.2 m (Patton 1975b). Squirrels often take advantage of "witches brooms" caused by dwarf mistletoe (*Arceuthobium pusillum*) as locations for their nests (Farentinos 1972).

Farentinos (1974) described aggressive or excitement signals as tail fluffing, foot thumping, and tail flicking. Aggressive or dominant squirrels will face off subordinate individuals. Escape behavior and vocalizations of tassel-eared squirrels are similar to those of other squirrels. Farentinos (1974) characterized vocalizations as cluck, growl, bark, screech, adult squeal, and juvenile squeal. Activity periods extend from before sunrise to before sunset with rest periods in between.

Reported mortality factors are few. Reynolds (1963) believed that goshawks (*Accipiter gentilis*) may be a regulatory mechanism for stabilizing fluctuating populations of tassel-eared squirrels. Bailey and Niedrak (1965) stated that one of the goshawk's main foods in Colorado ponderosa pine forests was tassel-eared squirrels. The common flea (*Opisodasys robustus*) and mange are prevalent parasites in the spring (Keith 1965).

Age is determined as for fox and gray squirrels. The length and fullness of ear tufts is age related and can possibly be used to separate young animals from adults (Keith 1965).

Tassel-eared squirrel population density can be encouraged by providing optimum habitat conditions such as ponderosa pine stands of 28 to 76 cm dbh trees (Patton and Green 1970). Kaibab squirrels, which are of greatest immediate concern, can be transplanted to nearby suitable habitats. Controlled burning to reduce woody understory plants and to provide open, parklike conditions would be beneficial (Hall 1973).

Where tassel-eared squirrels become troublesome to ponderosa pines, they can be kept out by placing bands of 50-cm-wide sheets of aluminum around the trunks of isolated trees at a height of 1.5 to 2.1 m from the ground (Squillace 1953). Reynolds (1963) estimated that they cause an increment loss of 10 percent to ponderosa pines. Larson and Schubert (1970) found that over a 10-year period they reduced ponderosa pine production by one-fifth. Squillace (1953) stated that 60 to 89 percent of the cones produced in poor or fair seed years were consumed. Tassel-eared squirrels also support some hunting in Colorado, Arizona, and New Mexico. Pederson (1978) demonstrated that tassel-eared squirrels will readily utilize nesting boxes and erecting such boxes in submarginal habitat may encourage this species. Research and management needs include the development of methods for the reduction of damage to ponderosa pines and for censusing populations (Hall 1972). More information on the life history aspects of this species is also needed, especially on reproduction (Hall 1972).

Western Gray Squirrel. The western gray squirrel (*Sciurus griseus*) is also called the gray squirrel or the California gray squirrel. Subspecies include *S. g. anthonyi*, *S. g. griseus*, and *S. g. nigripes*.

This squirrel is gray above with white underparts. The dorsal and ventral coloration is sharply disjunct. The upper portions are slate gray grizzled with white, sometimes with a slight yellow brown cast. Ears are never tufted but are distinctly more furred in winter than in summer. The tail is gray with indistinct dark bands and edged with white. The subspecies *nigripes* has black hind feet. Eyes are surrounded by white rings. The western gray squirrel is 510 to 570 mm in total length, 265 to 290 mm in tail length, and 74 to 80 mm in hind foot length (Hall and Kelson 1959).

The western gray squirrel ranges from central Washington, western Oregon, the Sierras and Coastal Ranges of California to the northernmost Baja California. Johnson (1954) reported this species as occurring in western Nevada.

Ingles (1947) found the home range of western gray squirrels to vary from 0.1 to 0.3 ha for females, and 0.5 to 0.6 ha for males. Grinnell and Storer (1924) reported densities of the western gray squirrel as 0.2 per ha in foothills of California.

This species has one breeding season, extending from January to May, with perhaps two litters a year. A female examined on 30 October was nursing. Ross (1930) reported that litter sizes varied from two to four young. Neonates had the following dimensions: male 74.6 g, total length 205 mm, tail length 93 mm, hind foot length 34 mm; female 80.0 g, total length 225 mm, tail 95 mm, hind foot 37 mm.

The western gray squirrel inhabits forests of mixed hardwoods and evergreens and dry open hardwoods. They nest in tree cavities or among tree branches in structures made of twigs and leaves.

Fungi are a staple food, especially subterranean forms. However, acorns and pine mast are the most critical foods that provide the energy for wintering squirrels. Other foods are fruit of California bay (*Umbelluloria californica*) and leaves or stems of forbs (Stienecker and Browning 1970; Stienecker 1977). They also eat bark from tree branches.

In captivity, western gray squirrels exhibit vicious behavior toward each other (Ross 1930). Loud barking sounds are often uttered while thumping front feet. These are louder than calls of fox or gray squirrels. This species is active at any time of day but has long rest periods. They spend more time on the ground than does *S. carolinensis*. They bury nuts and acorns singly, as do fox and gray squirrels (Ingles 1947).

An adult in captivity lived 11 years (Ross 1930). Mean longevity in the wild is probably similar to that of fox and gray squirrels. Mange appears to be the chief malady of the western gray squirrel. In 1917 and 1926, mange resulted in a severe decline among these squirrels, which resulted in low densities for a number of years. Populations began to increase in the 1930s (Ross 1930; Moffitt 1931; Payne 1940). Mange may also have so weakened some individuals that they became more vulnerable to predators (Murie 1936). Parasites include mange mite (*Notoedres* sp.) and fleas (*Orchopias nepos, O. latens, O. dieteri, Opisodasys enoplus*) (Ingles 1947).

Management goals should be to provide den trees, especially favoring western sycamore (*P. racemosa*) and Fremont cottonwood (*P. fremontii*). Planting to increase food trees, such as silver maple, black walnut, mulberry, Monterey cypress (*Cupressus macrocarpa*), blue oak (*Q. douglasii*), and black oak (*Q. kelloggii*), is also recommended. This species is responsible for

damage to walnut, almond (*Prunus amygdalus*), and filbert (*Corylus maxima*) orchards (Bailey 1936). Young redwoods (*Sequoia sempervirens*) suffer when squirrels remove bark near the treetop. Large middens of sugar pine (*P. lambertiana*) cones accumulated by the squirrels may be a fire hazard. Damage to forest trees is probably not significant (Grinnel and Storer 1924). These squirrels are economically important as hunting recreation. Future research and management needs should focus on more details of the life history of the western gray squirrel.

This chapter is contribution no. 1091-AEL, Center for Environmental and Estuarine Studies, University of Maryland.

LITERATURE CITED

Allen, D. L. 1942. Populations and habits of the fox squirrel in Allegan County, Michigan. Am. Midl. Nat. 27:338–379.

———. 1943. Michigan fox squirrel management. Michigan Dept. Conserv., Game Div. Publ. 100. 404pp.

Allen, J. M. 1952. Gray and fox squirrel management in Indiana. Indiana Dept. Conserv. P-R Bull. 1. 112pp.

Anastos, G. 1947. Hosts of certain New York ticks. Psyche 54:178–180.

Bailey, A. M., and Niedrach, R. J. 1965. Birds of Colorado. 2 vols. Denver Mus. Nat. Hist., Denver, Colo. 595pp.

Bailey, V. 1936. The mammals and life zones of Oregon. North Am. Fauna 55. 416pp.

Baker, R. H. 1944. An ecological study of tree squirrels in eastern Texas. J. Mammal. 25:8–24.

Bakken, A. A. 1952. Interrelationships of *Sciurus carolinensis* (Gmelin) and *Sciurus niger* (Linnaeus) in mixed populations. Ph.D. Dissertation. Univ. Wisconsin, Madison. 188pp.

———. 1959. Behavior of gray squirrels. Proc. Annu. Conf. Southeastern Assoc. Fish Game Commissioners 13:393–407.

Barkalow, F. S., Jr. 1954. The status of the names *Sciurus niger cinereus* Linnaeus and *Sciurus niger vulpinus* Gmelin. J. Elisha Mitchell Sci. Soc. 70:19–26.

Barkalow, F. S., Jr.; Hamilton, R. B.; and Soots, R. F., Jr. 1970. The vital statistics of an unexploited gray squirrel population. J. Wildl. Manage. 34:489–500.

Barkalow, F. S., Jr., and Shorten, M. 1973. The world of the gray squirrel. J. B. Lippincott Co., Philadelphia. 160pp.

Barkalow, F. S., Jr., and Soots, R. F., Jr. 1965. An analysis of the effect of artificial nest boxes on a gray squirrel population. Trans. North Am. Wildl. Nat. Resour. Conf. 30:349–360.

———. 1975. Life span and reproductive longevity of the gray squirrel, *Sciurus c. carolinensis* Gmelin. J. Mammal. 56:522–524.

Barrier, M. J., and Barkalow, F. S., Jr. 1967. A rapid technique for aging gray squirrels in winter pelage. J. Wildl. Manage. 31:715–719.

Baumgartner, L. L. 1938. Population studies of the fox squirrel in Ohio. Trans. North Am. Wildl. Conf. 3:685–689.

———. 1939. Foods of the fox squirrel in Ohio. Trans. North Am. Wildl. Conf. 4:579–584.

———. 1940. The fox squirrel: its life history, habits, and management in Ohio. Ohio State Univ. Wildl. Res. Stn. Release 138. 257pp.

———. 1943a. Pelage studies of fox squirrels (*Sciurus niger rufiventer*). Am. Midl. Nat. 29:588–590.

_____. 1943*b*. Fox squirrels in Ohio. J. Wildl. Manage. 7:193-202.

Baumgras, P. 1944. Experimental feeding of captive fox squirrels. J. Wildl. Manage. 8:296-300.

Baylis, H. A. 1934. Miscellaneous notes on parasitic worms. Annu. Mag. Nat. Hist. 13:223-228.

_____. 1939. *Trichostrongylus colubriformis* from *S. carolinensis* in Oxfordshire, England. Annu. Mag. Nat. Hist. 18:796.

Beale, D. M. 1962. Growth of the eye lens in relation to age in fox squirrels. J. Wildl. Manage. 26:208-211.

Beckwith, S. L. 1959. Purple squirrels. Florida Wildl. 11:18-19.

Benson, B. N. 1975. Dominance relationships, mating behavior and scent marking in fox squirrels (*Sciurus niger rufiventer*). Ph.D. Dissertation. Southern Illinois Univ., Carbondale. 96pp.

Besnady, C. D. 1957. Bushytail business. Wisconsin Conserv. Bull. 22:17-19.

Bishopp, F. C., and Trembley, H. L. 1945. Distribution and hosts of certain North American ticks. J. Parasitol. 31:1-54.

Blair, W. F. 1935. The mammals of a Florida hammock. J. Mammal. 16:271-277.

Bolls, N. J., and Perfect, J. R. 1972. Summer resting metabolic rate of the gray squirrel. Physiol. Zool. 45:54-59.

Bonnal, G.; Joyeux, C. E.; and Bosch, P. 1933. Un cas de cenurose humaine dûà *Multiceps serialis* (Gervais). Bull Soc. Pathol. Exot. 26:1060-1071.

Brennan, J. M., and Wharton, G. W. 1950. Studies on North American chiggers. No. 3, The subgenus Neotrombicula. Am. Midl. Nat. 44:153-197.

Brenneman, W. S. 1954. Tree damage by squirrels: silviculturally significant. J. For. 52:604.

Britton, W. E. 1933. Injuries to trees by squirrels. Proc. Natl. Shade Tree Conf. 9:85-91.

Brown, L. G., and Yeager, L. E. 1945. Fox squirrels and gray squirrels in Illinois. Illinois Nat. Hist. Surv. Bull. 23:449-536.

Bugbee, R. E., and Riegel, A. 1945. Seasonal food choices of the fox squirrel in western Kansas. Trans. Kansas Acad. Sci. 48:199-203.

Burbutis, P. P. 1956. The Siphonaptera of New Jersey. New Jersey Agric. Exp. Stn. Bull. 782. 36pp.

Burger, G. V. 1969. Response of gray squirrels to nest boxes at Remington Farms, Maryland. J. Wildl. Manage. 33:796-801.

Burns, P. J.; Christisen, D. M.; and Nichols, J. M. 1954. Acorn production in the Missouri Ozarks. Univ. Missouri Agric. Exp. Stn. Bull. 611. 8pp.

Cahalane, V. H. 1942. Caching and recovery of food by the western fox squirrel. J. Wildl. Manage. 6:338-352.

Cameron, T. S. M., and Parnell, I. W. 1933. The internal parasites of land mammals in Scotland. Proc. R. Physiol. Soc. Edinburgh 22:133-154.

Carson, J. 1957. Squirrels! Are they invincible? West Virginia Conserv. 21:11-13.

Chandler, A. C. 1942. Helminths of tree squirrels in southeast Texas. J. Parasitol. 28:135-140.

_____. 1947. Notes on *Moniliformis clarki* in North American squirrels. J. Parasitol. 33:278-281.

Chapman, F. B. 1938. Summary of the Ohio gray squirrel investigation. Trans. North Am. Wildl. Conf. 3:677-684.

Chesemore, D. L. 1975. Ecology of fox and gray squirrels (*Sciurus niger* and *Sciurus carolinensis*) in Oklahoma.

Ph.D. Dissertation. Oklahoma State Univ., Stillwater. 370pp.

Christisen, D. M. 1976. Squirrel production and harvest. Missouri Dept. Conserv. Fed. Aid Proj. W-13-R-30. Missouri Dept. Conserv., Jefferson City. 35pp.

Clark, C. M. 1958. *Hepatozoon griseiscuri* n. sp., a new species of *Hepatozoon* from the grey squirrel (*Sciurus carolinensis* Gmelin, 1788) with studies on the life cycle. J. Parasitol. 44:52-63.

_____. 1960. Three new nasal mites (Acarina: Speleognathidae) from the gray squirrel, the common grackle, and the meadowlark in the United States. Proc. Helminth. Soc. Washington 27:103-110.

Cooley, R. A., and Kohls, G. M. 1945. The genus *Ixodes* in North America. Natl. Inst. Health Bull. 184. 246pp.

Cooper, G. F., and Robson, J. G. 1966. Directionally selective movement detectors in the retina of the grey squirrel. J. Physiol. London 186:116-117.

Cordes, C. L., and Barkalow, F. S., Jr. 1973. Home range and dispersal in a North Carolina gray squirrel population. Proc. Annu. Conf. Southeastern Assoc. Fish Game Commissioners 26:124-135.

Costa Lima, A., and Hathaway, C. R. 1946. Pulgas. Bibliografia, catalogo e animals por elas sugados. Monogr. Inst. Oswaldo Cruz 4. 522pp.

Dambach, C. A. 1942. Gray squirrel feeding on *Crataegus*. J. Mammal. 23:337.

Davidson, W. R. 1975. *Dirofilariaeformia pulmoni* sp. n. (Nematoda: Onchocercidae) from the eastern gray squirrel (*Sciurus carolinensis* Gmelin). J. Parasitol. 61:351-354.

Davis, J. W.; Karstad, L. H.; and Trainer, D. O. 1970. Infectious diseases of wild mammals. Iowa State Univ. Press, Ames. 364pp.

Dikmans, G. 1938. A consideration of the nematode genus *Citellinema* with a description of a new species, *Citellinema columbianum*. Proc. Helminth. Soc. Washington 5:55-58.

Dobrovolny, C. G., and Harbaugh, M. J. 1934. *Cysticercus fasciolaris* from the red squirrel. Trans. Am. Microscop. Soc. 53:67.

Doebel, J. H., and McGinnes, B. S. 1974. Home range and activity of a gray squirrel population. J. Wildl. Manage. 38:860-867.

Donohoe, R. W., and Beal, R. O. 1972. Squirrel behavior as determined by radiotelemetry. Ohio Dept. Nat. Resour. Fish Wildl. Rep. 2. 20pp.

Dozier, H. L., and Hall, H. E. 1944. Observations on the Bryant fox squirrel. Maryland Conserv. 21:2-7.

Eichhorn, R. 1962. Facts about Florida squirrels. Florida Wildl. 15:10-15.

Ellis, L. L., Jr. 1955. A survey of the ectoparasites of certain mammals in Oklahoma. Ecology 36:12-18.

English, P. F. 1934. Some observations on a pair of red-tailed hawks. Wilson Bull. 46:228-235.

Enright, J. B.; Franti, C. E.; Behymer, D. E.; Longhurst, W. M.; Dutson, V. J.; and Wright, M. E. 1971. *Coxiella burneti* in a wildlife-livestock environment: distribution of fever in wild animals. Am. J. Epidemiol. 94:79-90.

Errington, P. L. 1932. Food habits of southern Wisconsin raptors. Part 1, Owls. Condor 34:176-186.

Farentinos, R. C. 1972. Observations on the ecology of the tassel-eared squirrel. J. Wildl. Manage. 36:1234-1239.

_____. 1974. Social communication of the tassel-eared squirrel (*Sciurus aberti*): a descriptive analysis. Z. Tierpsychol. 34:441-558.

_____. 1980. Sexual solicitation of subordinate males by

female tassel-eared squirrels (*Sciurus aberti*). J. Mammal. 61:337–341.

Flyger, V. 1952. A study of the nest box habits and the breeding season of the gray squirrel (*Sciurus carolinensis bucotis*) in Maryland and Pennsylvania. M.S. Thesis. Pennsylvania State Univ., University Park. 59pp.

———. 1955. Implications of social behavior in gray squirrel management. Trans. North Am. Wildl. Conf. 20:381–389.

———. 1956. The social behavior and populations of the gray squirrel (*Sciurus carolinensis* Gmelin) in Maryland. Sc.D. Dissertation. Johns Hopkins Univ., Baltimore. 97pp.

———. 1959. A comparison of methods for estimating squirrel populations. J. Wildl. Manage. 23:220–223.

———. 1960. Movements and home range of the gray squirrel, *Sciurus carolinensis*, in two Maryland woodlots. Ecology 41:365–369.

———. 1969. The 1968 squirrel "migration" in the eastern United States. Proc. Northeast Fish Wildl. Conf. 26:69–79.

Flyger, V., and Levin, E. Y. 1977. Congenital erythropoietic porphyria: normal porphyria of fox squirrels (*Sciurus niger*). Am. J. Pathol. 87:269–272.

Fox, I. 1940. Fleas of eastern United States. Iowa State Coll. Press, Ames. 191pp.

Freeman, R. S. 1962. Studies on the biology of *Taenia crassiceps* (Zeder, 1800) Rudolphi, 1910 (Cestoda). Can. J. Zool. 40:969–990.

George, R. S. 1954. Siphonaptera from Gloucestershire. Entomol. Gazette 5:85–94.

Goldman, E. A. 1928. The Kaibab or white-tailed squirrel. J. Mammal. 9:127–129.

Golightly, R. T., Jr., and Ohmart, R. D. 1978. Heterothermy in free-ranging Abert's squirrels (*Sciurus aberti*). Ecology 59:897–909.

Goodrum, P. D. 1940. A population study of the gray squirrel in eastern Texas. Texas Agric. Exp. Stn. Bull. 591. 34pp.

Gouras, P. 1964. Duplex function in the grey squirrel's electroretinogram. Nature 203:767–768.

Graham, E., and Uhrich, J. 1943. Animal parasites of the fox squirrel in southeast Kansas. J. Parasitol. 29:159–160.

Grinnell, J., and Storer, T. I. 1924. Animal life in the Yosemite. Univ. California Press, Berkeley. 752pp.

Gunter, G., and Eleuterius, L. 1971. Bark-eating by the common gray squirrel following a hurricane. Am. Midl. Nat. 85:235.

Hall, E. R., and Kelson, K. R. 1959. The mammals of North America. Vol. 1. Ronald Press Co., New York. 546pp.

Hall, J. G. 1967. White tails and yellow pine. Natl. Parks (April). 3pp.

———. 1972. Kaibab squirrel report, 1971–1972. Rocky Mountain For. Range Exp. Stn., Tempe, Arizona. 13pp. Manuscript.

———. 1973. The Kaibab squirrel. Pages 18–21 *in* Symposium on rare and endangered wildlife of the southwestern United States. New Mexico Dept. Game and Fish, Santa Fe. 167pp.

Hamilton, W. J., Jr. 1943. Caterpillars as food of the gray squirrel. J. Mammal. 24:104.

Harkema, R. 1936. The parasites of some North Carolina rodents. Ecol. Monogr. 6:153–232.

Harwood, P. D., and Cooke, V. 1949. The helminths from a heavily parasitized fox squirrel, *Sciurus niger*. Ohio J. Sci. 49:146–148.

Havera, S. P., and Nixon, C. M. 1980. Winter feeding of fox and gray squirrel populations. J. Wildl. Manage. 44:41–55.

Hicks, L. E. 1938. The status of game mammals in Ohio. Trans. North Am. Wildl. Conf. 3:415–420.

Himelick, E. B., and Curl, E. A. 1955. Experimental transmission of oak wilt fungus by caged squirrels. Phytopathology 45:581–584.

Hirth, R. S.; Wyand, D. S.; Osborne, A. D.; and Burke, C. N. 1969. Epidermal changes caused by squirrel pox virus. J. Am. Vet. Med. Assoc. 155:1120–1125.

Hixon, H. 1940. Field biology and environmental relationships of the Gulf Coast tick in southern Georgia. J. Econ. Entomol. 33:179–189.

Hoffman, R. A., and Kirkpatrick, C. M. 1959. Current knowledge of tree squirrel reproductive cycles and development. Proc. Annu. Conf. Southeastern Assoc. Fish Game Commissioners 13:363–367.

Holland, G. P., and Benton, A. H. 1968. Siphonaptera from Pennsylvania mammals. Am. Midl. Nat. 80:252–261.

Hougart, B. 1975. Activity patterns of radio-tracked gray squirrels. M.S. Thesis. Univ. Maryland, College Park. 37pp.

Hudson, B. W.; Goldenberg, M. I.; McCluskie, J. D.; Larson, H. E.; McGuire, C. D.; Barnes, A. M.; and Poland, J. D. 1971. Serological and bacteriological investigations of an outbreak of plague in an urban tree squirrel population. Am. J. Trop. Med. Hyg. 20:255–263.

Ingles, L. G. 1947. Ecology and life history of the California gray squirrel. California Fish Game 33:139–158.

Jenkins, D. W. 1948. Trombiculid mites affecting man. I. Bionomics with reference to epidemiology in the United States. Am. J. Hyg. 48:22–35.

Johnson, N. K. 1954. New mammal records for Nevada. J. Mammal. 35:577–578.

Johnson, P. T. 1958. Type specimens of lice (Order Anoplura) in the United States National Museum. Proc. U.S. Natl. Mus. 108:39–49.

Jordan, J. S. 1971. Yield from an intensively hunted population of eastern fox squirrels. U.S.D.A. For. Serv. Res. Pap. NE-186. Northeast For. Serv. Exp. Stn., Upper Darby, Pa. 8pp.

Joseph, T. 1972. *Eimeria lancasterensis* Joseph, 1969 and *E. confusa* Joseph, 1969 from the grey squirrel, *Sciurus carolinensis*. J. Protozool. 19:143–150.

Judd, W. W. 1955. Gray squirrels feeding on samaras of elm. J. Mammal. 36:296.

Kardos, E. H. 1954. Biological and systematic studies on the subgenus Neotrombicula (genus *Trombicula*) in the central United States (Acarina, Trombiculidae). Univ. Kansas Sci. Bull. 36:69–123.

Keegan, H. L. 1943. Some host records from the parasitological collection of the State University of Iowa. Bull. Brooklyn Entomol. Soc. 38:54–57.

Keith, J. O. 1956. The Abert squirrel (*Sciurus aberti aberti*) and its relationship to the forests of Arizona. M.S. Thesis. Univ. Arizona, Tucson. 106pp.

———. 1965. The Abert squirrel and its dependence on ponderosa pine. Ecology 46:150–163.

Kilham, L. 1954. Metastasizing viral fibromas of gray squirrels: Pathogenesis and mosquito transmission. Am. J. Hyg. 61:55–63.

Kim, K. C. 1966a. The nymphal stages of three North American species of the genus *Enderleinellus* Fahrenholz. J. Med. Entomol. 2:327–330.

———. 1966b. The species of *Enderleinellus* (Anoplura, Hoplopleuridae) parasitic on the Sciurini and Tamiasciurini. J. Parasitol. 52:988–1024.

King, J. M.; Woolf, A.; and Shively, J. 1972. Naturally

occurring squirrel fibroma with involvement in internal organs. J. Wildl. Dis. 8:321–324.

Kline, P. D. 1965. Iowa squirrels: hunting statistics, sex and age ratios, and the influence of mast and agriculture. Proc. Iowa Acad. Sci. 71:216–227.

Larson, M. M., and Schubert, G. H. 1970. Cone crops of ponderosa pine in central Arizona including the influence of Abert squirrels. U.S.D.A. For. Serv. Res. Pap. RM-58:1–12. U.S.D.A. For. Serv., Fort Collins, Colo. 12pp.

Layne, J. N. 1971. Fleas (Siphonaptera) of Florida. Florida Entomol. 54:35–51.

Layne, J. N., and Woolfenden, G. E. 1958. Gray squirrels feeding on insects in car radiators. J. Mammal. 39:595–596.

Lennette, E. H.; Ota, M. I.; Dobbs, M. E.; and Browne, A. S. 1956. Isolation of western equine encephalomyelitis virus from naturally-infected squirrels in California. Am. J. Hyg. 64:276–280.

Levine, N. D., and Ivens, V. 1965. The coccidian parasites (Protozoa, Sporozoa) of rodents. Illinois Bio. Monogr. 33:1–365.

Lewis, E.; Hoff, G. L.; Bigler, W. J.; and Jefferies, M. B. 1975. Public health and the urban gray squirrel. J. Wildl. Dis. 11:502–504.

Linduska, J. P. 1947. Longevity of some Michigan farm game mammals. J. Mammal. 28:126–129.

Lipovsky, L. J.; Crossley, D. A., Jr.; and Loomis, R. B. 1955. A new genus of chigger mites (Acarina, Trombiculidae). J. Kansas Entomol. Soc. 28:136–143.

Longley, W. H. 1963. Minnesota gray and fox squirrels. Am. Midl. Nat. 69:82–98.

Loomis, R. B. 1954. A new subgenus and six new species of chigger mites (genus *Trombicula*) from the central United States. Univ. Kansas. Sci. Bull. 36:919–941.

————. 1955. *Trombicula gurneyi* Ewing and two new related chigger mites (Acarina, Trombiculidae). Univ. Kansas Sci. Bull. 37:251–267.

————. 1956. The chigger mites of Kansas (Acarina, Trombiculidae). Univ. Kansas Sci. Bull. 37:1195–1443.

Lucker, J. T. 1943. A new trichostrongylid nematode from the stomachs of American squirrels. J. Washington Acad. Sci. 33:75–79.

Lustig, L. W., and Flyger, V. 1975. Observations and suggested management practices for the endangered Delmarva fox squirrel. Proc. Annu. Conf. Southeastern Assoc. Fish Game Commissioners 29:433–440.

McCloskey, R. J. 1975. Description and analysis of the behavior of the fox squirrel in Iowa. Ph.D. Dissertation. Iowa State Univ., Ames. 238pp.

McQuown, A. L. 1954. *Capillaria hepatica*. Am. J. Clinical Pathol. 24:448–452.

Madson, J. 1964. Gray and fox squirrels. Olin Mathieson Chem. Corp., East Alton, Ill. 112pp.

Main, A. J., Jr. 1970. Distribution, seasonal abundance and host preference of fleas in New England. Proc. Entomol. Soc. Washington 72:73–89.

Masterson, R. A.; Stegmiller, H. W.; Parsons, M. A.; Croft, C. C.; and Spencer, C. B. 1971. California encephalitis: an endemic puzzle in Ohio. Health Lab. Sci. 8:89–96.

Mathewson, J. A., and Hyland, K. E., Jr. 1962. The ectoparasites of Rhode Island mammals. Part 2, A collection of Anoplura from nondomestic hosts. J. New York Entomol. Soc. 70:167–174.

————. 1964. The ectoparasites of Rhode Island mammals. Part 3, A collection of fleas from nondomestic hosts (Siphonaptera). J. Kansas Entomol. Soc. 37:157–163.

Meggitt, F. J. 1924. The cestodes of mammals. Edward Goldston, London. 282pp.

Moffitt, J. 1931. Diseases reducing tree squirrel populations in southern California. California Fish Game 17:338–339.

Moore, J. C. 1957. The natural history of the fox squirrel, *Sciurus niger shermani*. Bull. Am. Mus. Nat. Hist. 113:1–71.

————. 1968. Sympatric species of tree squirrels mix in mating chase. J. Mammal. 49:531–533.

Morgan, B. B. 1943. The Physaloptera (Nematoda) of rodents. Wasmann Collector 5:99–107.

Morlan, H. B. 1952. Host relationships and seasonal abundance of some southwest Georgia ectoparasites. Am. Midl. Nat. 48:74–93.

Moulton, D. W., and Thompson, W. H. 1971. California group virus infections in small, forest-dwelling mammals of Wisconsin: some ecological considerations. Am. J. Trop. Med. Hyg. 20:474–482.

Murie, A. 1936. A predator eliminates a sick animal. J. Mammal. 17:418.

Nash, D. J., and Ramey, C. A. 1970. The tassel-eared squirrel in Colorado. Occas. Publ. no 2. Dept. Zool., Colorado State Univ., Fort Collins. 6pp.

Nichols, J. T. 1958. Food habits and behavior of the gray squirrel. J. Mammal. 39:376–380.

Nixon, C. M.; Donohoe, R. W.; and Nash, T. 1974. Overharvest of fox squirrels from two woodlots in western Ohio. J. Wildl. Manage. 38:67–80.

Nixon, C. M.; Havera, S. P.; and Greenberg, R. E. 1978. Distribution and abundance of the gray squirrel in Illinois. Illinois Nat. Hist. Surv. Bio. Note 105:1–55.

Nixon, C. M., and McClain, M. W. 1969. Squirrel population decline following a late spring frost. J. Wildl. Manage. 33:353–357.

Nixon, C. M.; McClain, M. W.; and Donohoe, R. W. 1980. Effects of clear-cutting on gray squirrels. J. Wildl. Manage. 44:403–412.

Ofcarcik, R. P.; Burns, E. E.; and Teer, J. G. 1973. Acceptance of selected acorns by captive fox squirrels. Southwestern Nat. 17:349–355.

Olexik, W. A.; Perry, A. E.; and Wilhelm, W. E. 1969. Ectoparasites and helminth endoparasites of tree squirrels of southwest Tennessee. J. Tennessee Acad. Sci. 44:4–6.

Pack, J. C.; Mosby, H. S.; and Siegel, P. B. 1967. Influence of social hierarchy on gray squirrel behavior. J. Wildl. Manage. 31:720–728.

Packard, R. L. 1956. The tree squirrels of Kansas. Mus. Nat. Hist. State Bio. Surv., Univ. Kansas, Lawrence. 67pp.

Parker, J. C. 1968. Parasites of the gray squirrel in Virginia. J. Parasitol. 54:633–634.

Parker, J. C., and Holliman, R. B. 1971. Observations on parasites of gray squirrels during the 1968 emigration in North Carolina. J. Mammal. 52:437–441.

Patton, D. R. 1975a. Nest use and home range of three Abert squirrels as determined by radio tracking. U.S.D.A. For. Serv. Res. Note RM-281:1–3.

————. 1975b. Abert squirrel cover requirements in southwestern ponderosa pine. U.S.D.A. For. Serv. Res. Note RM-145:1–12.

Patton, D. R., and Green, W. 1970. Abert's squirrels prefer mature ponderosa pine. U.S.D.A. For. Serv. Res. Note RM-169:1–3.

Patton, D. R.; Ratcliff, T. D.; and Rogers, K. J. 1976. Weight and temperature of the Abert and Kaibab squirrels. Southwestern Nat. 21:236–238.

Payne, E. A. 1940. The return of the California gray squirrel. Yosemite Nature Notes 19:1–2.

Pederson, J. C. 1978. Use of artificial nest boxes by Abert's squirrels. Southwestern Nat. 23:681–709.

Preno, W. L., and Labisky, R. F. 1971. Abundance and harvest of doves, pheasants, bobwhites, squirrels and cottontails in Illinois, 1956-1969. Illinois Dept. Conserv. Tech. Bull. 4:1–76.

Price, D. L. 1962. Description of *Dipetalonema interstitium* n. sp. from the grey squirrel and *Dipetalonema llewellyni* n. sp. from the raccoon. Proc. Helminth. Soc. Washington 29:77–82.

Progulske, D. R. 1955. Game animals utilized as food by the bobcat in the southern Appalachians. J. Wildl. Manage. 19:249–253.

Ramey, C. A., and Nash, D. J. 1976a. Geographic variation in Abert's squirrel (*Sciurus aberti*). Southwestern Nat. 21:135–139.

———. 1976b. Coat color polymorphism of Abert's squirrel, *Sciurus aberti*, in Colorado. Southwestern Nat. 21:209–217.

Rankin, J. S., Jr. 1946. Helminth parasites of birds and mammals in western Massachusetts. Am. Midl. Nat. 35:756–768.

Rasmussen, D. I. 1941. Biotic communities of Kaibab Plateau, Arizona. Ecol. Monogr. 11:229–275.

Rausch, R., and Tiner, J. D. 1948. Studies on the parasitic helminths of the north central States. Part 1, Helminths of Sciuridae. Am. Midl. Nat. 39:728–747.

Redmond, H. R. 1953. Analysis of gray squirrel breeding studies and their relation to hunting season, gunning pressure and habitat conditions. Trans. North Am. Wildl. Conf. 18:378–389.

Reiber, R. J., and Byrd, E. E. 1942. Some nematodes from mammals of Reelfoot Lake in Tennessee. J. Tennessee Acad. Sci. 17:78–89.

Reynolds, H. G. 1963. Western goshawk takes Abert squirrel in Arizona. J. For. 61:839.

Roach, B. A. 1974. Scheduling timber cutting for sustained yield of wood products and wildlife. U.S.D.A. For. Serv. Tech. Rep. NE-14. 13pp.

Robinson, D. J., and Cowan, I. McT. 1954. An introduced population of the gray squirrel (*Sciurus carolinensis*, Gmelin) in British Columbia. Can. J. Zool. 32:261–282.

Ross, R. C. 1930. California Sciuridae in captivity. J. Mammal. 11:76–78.

Sanderson, H. R. 1975. Den-tree management for gray squirrels. Wildl. Soc. Bull. 3:125–131.

Schorger, A. W. 1949. Squirrels in early Wisconsin. Trans. Wisconsin Acad. Sci. Arts Letters 39:195–247.

Self, J. T., and Esslinger, J. H. 1955. A new species of bothriocephalid cestode, from the fox squirrel (*Sciurus niger* Linn.). J. Parasitol. 41:256–258.

Seton, E. T. 1920. Migrations of the gray squirrel (*Sciurus carolinensis*). J. Mammal. 1:53–58.

Shaftesbury, A. D. 1934. The Siphonaptera (fleas) of North Carolina with special reference to sex ratios. J. Elisha Mitchell Sci. Soc. 49:247–263.

Short, H. L., and Duke, W. B. 1971. Seasonal food consumption and body weights of captive tree squirrels. J. Wildl. Manage. 35:435–439.

Shorten, M. R. 1954. Squirrels. Collins, London. 212pp.

———. 1959. Squirrels in Britain. Proc. Annu. Conf. Southeastern Assoc. Fish Game Commissioners 13:375–378.

Shotts, E. B., Jr.; Andrews, C. L.; and Harvey, T. W. 1975. Leptospirosis in selected wild mammals of the Florida panhandle and southwestern Georgia, U.S.A. J. Am. Vet. Med. Assoc. 167:587–589.

Smith, C. C., and Follmer, D. 1972. Food preferences of squirrels. Ecology 53:82–91.

Smith, H. K. 1970. A method of analyzing fox squirrel stomach contents. Texas Parks Wildl. Dept. Tech. Serv. Bull. 3:1–75.

Smith, N. B. 1967. Some aspects of reproduction in the female gray squirrel, *Sciurus carolinensis carolinensis* Gmelin, in Wake County, North Carolina. M.S. Thesis. North Carolina State Univ., Raleigh. 92pp.

Sonenshine, D. E., and Stout, T. J. 1971. Ticks infesting medium-sized wild mammals in two forest localities in Virginia (Acarina:Ixodidae). J. Med. Entomol. 8:217–227.

Soots, R. F., Jr. 1964. An analysis of the effects of artificial nest boxes on gray squirrel populations. M.S. Thesis. North Carolina State Univ., Raleigh. 111pp.

Squillace, A. E. 1953. Effect of squirrels on the supply of ponderosa pine seed. Northern Rocky Mountain For. Range Exp. Stn. Res. Note 131:1–4. Mimeogr.

Stienecker, W., and Browning, B. M. 1970. Food habits of the western gray squirrel. California Fish Game 56:36–48.

Stienecker, W. E. 1977. Supplemental data on the food habits of the western gray squirrel. California Fish Game 63:11–21.

Terres, J. K. 1939. Gray squirrel utilization of elm. J. Wildl. Manage. 3:358–359.

Terrill, H. V. 1941. A preliminary study of the western fox squirrel, *Sciurus niger rufiventer* (Geoffroy), in Missouri. M.S. Thesis. Univ. Missouri, Columbia. 164pp.

Thoma, B. L., and Marshall, W. H. 1960. Squirrel weights and populations in a Minnesota woodlot. J. Mammal. 41:272–273.

Tiner, J. D. 1949. Preliminary observations on the life history of *Ascaris columnaris*. J. Parasitol. 35:13.

Tugwell, P., and Lancaster, J. L., Jr. 1962. Results of a tick-host study in northwest Arkansas. J. Kansas Entomol. Soc. 35:202–211.

Uhlig, H. G. 1955. Weights of adult gray squirrels. J. Mammal. 36:293–296.

———. 1956. Effect of legal restrictions and hunting on gray squirrel populations in West Virginia. Trans. North Am. Wildl. Conf. 21:330–338.

———. 1957. Gray squirrel populations in extensive forested areas of West Virginia. J. Wildl. Manage. 21:335–341.

Uhrich, J., and Graham, E. 1941. The animal parasites of the fox squirrel, *Sciurus niger rufiventer* and the gray squirrel, *Sciurus carolinensis carolinensis*. Anat. Rec. 81:65.

Walton, B. C., and Walls, K. W. 1964. Prevalence of toxoplasmosis in wild animals from Fort Stewart, Georgia, as indicated by serological tests and mouse inoculation. Am. J. Trop. Med. Hyg. 13:530–533.

Webb, J. P., Jr., and Loomis, R. B. 1971. The subgenus Scapuscutala of the genus *Microtrombicula* (Acarina: Trombiculidae) from North America. J. Med. Entomol. 8:319–329.

White, F. H.; Hoff, G. L.; Bigler, W. J.; and Buff, E. 1975. A microbiologic study of the urban gray squirrel. J. Am. Vet. Med. Assoc. 167:603–604.

Wobeser, G. 1969. Tetanus in a gray squirrel. Bull. Wildl. Dis. Assoc. 5:18–19.

Wolf, T. F., and Roest, A. I. 1971. The fox squirrel (*Sciurus niger*) in Ventura County. California Fish Game 37:219–220.

Woods, C. E., and Larson, O. R. 1969. North Dakota fleas. Part 2, Records from man and other mammals. Proc. North Dakota Acad. Sci. 23:28.

Woods, G. T. 1941. Mid-summer food of gray squirrels. J. Mammal. 22:321–322.

Zimmermann, W. J., and Hubbard, E. D. 1969. Trichiniasis in wildlife of Iowa. Am. J. Epidemiol. 90:84–92.

VAGN FLYGER, Department of Animal Science, University of Maryland, College Park, Maryland 20742.

J. EDWARD GATES, Appalachian Environmental Laboratory, Center for Environmental and Estuarine Studies, University of Maryland, Frostburg State College Campus, Frostburg, Maryland 21532.

12

Pine Squirrels

Tamiasciurus hudsonicus and *T. douglasii*

Vagn Flyger
J. Edward Gates

NOMENCLATURE

COMMON NAMES. Red squirrel, boomer, fairydiddle, piney, pine squirrel, spruce squirrel
SCIENTIFIC NAME. *Tamiasciurus hudsonicus*
SUBSPECIES. *T. h. abieticola, T. h. baileyi, T. h. columbiensis, T. h. dakotensis, T. h. dixiensis, T. h. fremonti, T. h. grahamensis, T. h. gymnicus, T. h. hudsonicus, T. h. kenaiensis, T. h. lanuginosus, T. h. laurentianus, T. h. loquax, T. h. lychnuchus, T. h. minnesota, T. h. mogollonensis, T. h. pallescens, T. h. petulans, T. h. picatus, T. h. preblei. T. h. regalis, T. h. richardsoni. T. h. streatori, T. h. ungavensis,* and *T. h. ventorum.*

COMMON NAMES. Douglas squirrel, pine squirrel, chickaree
SCIENTIFIC NAME. *Tamiasciurus douglasii*
SUBSPECIES. *T. d. albolimbatus, T. d. douglasii, T. d. mollipilosus,* and *T. d. mearnsi.*

DISTRIBUTION

Red squirrels are associated with the North American boreal forest from its northern edge in Alaska and Canada southward into the coniferous forest of the Rocky Mountains to southern Arizona and New Mexico (figure 12.1). In the east, they occur from eastern North Dakota to the Atlantic Ocean and southward almost to the 40th parallel, with a spur extending down the Appalachians to Georgia. On Mt. Washington, New Hampshire, they are found in the dwarfed balsam fir (*Abies balsamea*) close to the tree line, where these trees rarely, if ever, produce seeds. These squirrels are tame and perhaps rely on handouts from summer tourists. Whether or not they are there in the winter is not known.

Douglas squirrels are found from southwestern British Columbia southward through western Washington and Oregon, northern California, and southward in the Sierras to northern Baja California (see figure 12.1). The ranges of the two species are allopatric.

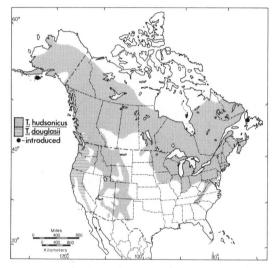

FIGURE 12.1. Distribution of the red squirrel (*Tamiasciurus hudsonicus*) and Douglas squirrel (*T. douglasii*) in North America.

DESCRIPTION

Red squirrels are distinctly reddish brown dorsally with white underparts, while Douglas squirrels are gray brown dorsally with yellow-tinted underparts. The degree of yellow tint varies. Both species have a pronounced lateral black stripe separating the dorsal coloration from the ventral. This line is most distinct in summer; it may almost disappear in winter on some individuals. Both species have slight ear tufts, which are most noticeable in winter but never approach the prominence of the tufts of the tassel-eared squirrel (*Sciurus aberti*). White eye rings are distinct in both species.

The tail of pine squirrels is smaller and flatter in proportion to body size than that of squirrels of the genus *Sciurus*. It lacks the distinct longitudinal bands of fox squirrels (*S. niger*) and gray squirrels (*S. carolinensis*). The tail hairs of red squirrels have yellowish tips bordered with black; Douglas squirrel tails have white tips bordered with a wider black band. Albinism and melanism are rarer among pine squirrels than among species of *Sciurus*. However, albinism has been reported in red squirrels (Wood 1965), as has melanism (Mengel and Jenkinson 1971).

In the wild, pine squirrels are easily recognized by their nervous scampering habits, small size, and noisy chatter. This chatter is so characteristic that their presence can be determined by sound alone. At close range, the body appears less robust and the head shorter and rounder than in the species of *Sciurus*. These are the most arboreal of all North American tree squirrels.

Adult red squirrel body measurements are: 270–385 millimeters (mm) total length, 92–158 mm tail length, and 35–57 mm hind foot length. They weigh from 145 to 260 grams (g). The corresponding measurements for Douglas squirrels are: 270–345 mm total length, 102–156 mm tail length, 45–55 mm hind foot length, and an average body weight of 255 g. Skulls of both species vary from 42 to 48 mm condylobasal length (figure 12.2). Those of males are slightly longer than those of females (Nellis 1969). There are no other sexual differences in size or coloration in either species. The dental formula for both species is 2/2, 0/0, 1 or 2/1, 3/3 = 24 or 26.

ANATOMY AND PHYSIOLOGY

The eyes of tamiasciurids are adapted to daytime vision. They have a yellow-colored lens, as in *Sciurus*, yet their eyes seem better adapted to night vision than people's eyes (Dippner and Armington 1971). Pine squirrels possess sebaceous, sudoriferous, and mucous glands in the mouth region (Quay 1965) that may have some function in scent marking.

Both species molt twice a year, with the winter pelt being brighter than that of summer. The spring molt occurs from the beginning of April to mid-August. It starts at the nose and progresses posteriorly (Nelson 1945). The fall molt begins on the hindquarters and progresses toward the head, as in *Sciurus* (Layne 1954).

Red squirrels, and probably Douglas squirrels, possess brown adipose tissue, with greater amounts present in winter than in summer (Aleksiuk 1970, 1971). It occurs most noticeably as a pod in each axilla. Brown fat permits pine squirrels to mobilize free fatty acids more quickly in cold weather than can rats (Muridae) or mice (Cricetidae) (Ferguson and Folk 1971). Adult metabolic rate is 143 to 168 kcal/kg and varies with ambient temperature and season (Grodzinski 1971). The 2N chromosome number is 46 (Nadler and Hoffmann 1970).

Pine squirrels differ so markedly from other tree squirrels in the structure of the male and female reproductive tracts that Mossman (1940) suggested that they may warrant reclassification into a separate family. Male pine squirrels possess rudimentary, almost microscopic, bacula (Layne 1952). In *Sciurus* the baculum is about 1 centimeter long. Also, in contrast to *Sciurus*,

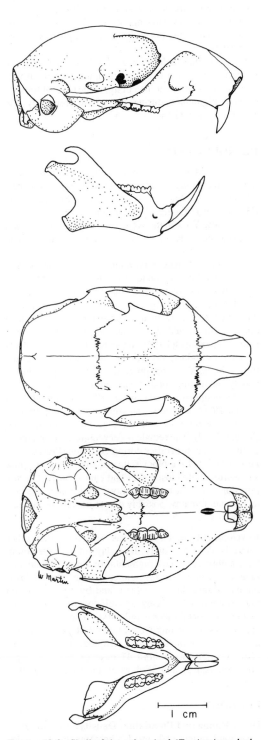

FIGURE 12.2. Skull of the red squirrel (*Tamiasciurus hudsonicus*). From top to bottom: lateral view of cranium, lateral view of mandible, dorsal view of cranium, ventral view of cranium, dorsal view of mandible. The skull of the Douglas squirrel is very similar.

male pine squirrels have minute Cowpers glands, no bulbar glands, and a very long, threadlike, symmetrical penis. The most striking difference in females is the coiled vagina, which is unique in pine squirrels (Mossman 1940).

As in other male sciurids, the testes descend into the testicles, which become pigmented and relatively hairless at about 10 months of age. The teats of females become pigmented during their first pregnancy and remain pigmented for life.

REPRODUCTION

Pine squirrels have only one litter a year throughout most of their range (Dolbeer 1973). Young are generally born from March through May in Saskatchewan (Davis 1969). Layne (1954) reported two litters per year in New York State. Kemp and Keith (1970) found no evidence of second litters among red squirrels in Canada.

When a female pine squirrel enters estrus, she permits neighboring males to enter her territory and mates with one or more of them (Smith 1965). The female is receptive for only one day (Smith 1968). After estrus, she again defends her territory from all other squirrels and raises the offspring by herself. Kemp and Keith (1970) suggested that red squirrels respond reproductively to the abundance of spruce flower buds, thus maximizing their ability to use a widely fluctuating food supply (spruce cones).

Pine squirrels breed when 10 to 12 months old (Millar 1970). One to seven young, averaging 7.5 g in weight, are born after 35 days gestation (Ferron and Prescott 1977). The female is very aggressive and intolerant toward conspecifics during gestation (Prescott and Ferron 1978). Like other sciurids, the young are pink and hairless with closed eyes and ears at birth. The sex ratios are equal at birth but more adult males are collected than females, possibly due to bias in the collecting methods.

Development of Young. Pine squirrels, like other squirrels, develop slowly. By 6 days of age, the young have a noticeable dark pigmentation on the dorsal surface just prior to the appearance of hair. By 16 days, the dark lateral line is distinct; and by 20 days, when they weigh just 17 g, they are well covered with brown hair on the dorsal surface and darker hairs on the head. At 28 days, the young weigh from 24 to 44 g. At this age, the lower incisors have erupted. The upper incisors do not appear until 41 days of age. The eyes open between 27 and 35 days of age (Svihla 1930).

ECOLOGY

Home Range and Population Density. Red squirrels have home ranges of 1.3 to 1.5 hectares (ha) for males and females, respectively (Davis 1969). Both sexes defend territories of 0.2 to 1.2 ha (Smith 1965). Population densities vary from 0.4 per ha in mixed jack pine (*Pinus banksiana*)-black spruce (*Picea mariana*) to 2.3 per ha in white spruce (*P. glauca*) forests (Davis

1969). Biomass was 640 g per ha on a college campus and 158 g per ha in a hemlock (*Tsuga canadensis*) grove (Mohr 1965). Population densities reach a high every three to four years in the Adirondacks of New York (Krull 1970).

Habitat. Evergreen forests are preferred by both species, but these squirrels often inhabit evergreen-hardwood mixtures and sometimes pure hardwood stands. In Arizona and Utah, pine squirrels are found in fir and spruce forests but never in pines (Bailey 1932; Long 1940; Burnett and Dickerman 1956). Lodgepole pine (*P. contorta*) forests are favored habitat in Colorado, where open forests and ponderosa pines (*P. ponderosa*) do not support pine squirrels (Hatt 1943). In Alaska, they are found mostly in white spruce forests, rarely in black spruce (Dice 1921). In the Great Slave Lake–Athabaska region of Canada, they occur in jack pine groves (Harper 1932). Petrides (1942) reported their use of fencerow habitats in New York. In the Great Smoky Mountains of North Carolina and other highlands of this state, red squirrels are found in evergreen as well as hardwood forests at the upper elevations (Komarek and Komarek 1938; Odum 1949). In the eastern United States, where red squirrels occupy hardwoods or hardwood-evergreen mixtures, they frequently overlap with fox or gray squirrels. Competition results for food and den sites.

Den Sites. Pine squirrels readily make use of tree hollows for dens but also will build on abandoned bird nests or make use of underground dens (Shaw 1936). Throughout much of the range of red squirrels, tree dens are rare and leaf nests are most common. Nests are large, bulky structures constructed of twigs and leaves. They are built in branches of evergreens at an average height of 5.3 meters (Rothwell 1979). Red squirrels show a preference for nest trees whose crowns are at least partially surrounded by, and often interlocking with, those of other trees. This situation offers protection from weather, provides several escape routes, and reduces foraging time (Rothwell 1979).

FOOD HABITS

Pine squirrels characteristically inhabit evergreen forests and are so dependent on these trees that in much of their range their population density fluctuates with the conifer cone crops. However, both species are opportunists. Their food habits are similar except that red squirrels are adapted to feeding on serotinous cones, while Douglas squirrels are not (Smith 1970). Most differences in food habits, where they exist between the two species, are due to geographical differences.

The diet of pine squirrels includes winter terminal buds of evergreens (pines, spruce, and juniper [*Juniperus virginiana*]) (Hatt 1930; MacNamara 1943; Abbott and Belig 1961). White spruce is such an important food that it influences the distribution and densities of red squirrels in the taiga of interior Alaska (Brink and Dean 1966). Box elder (*Acer negundo*) seeds are a preferred food (Burton 1930; Fox 1939), along with red pine (*P. resinosa*) seeds (Clarke 1939) and jack pine seeds (Smith and Aldous 1947). Other foods include: buds and sap of red (*Acer rubrum*) and sugar (*A. saccharum*) maple (Klugh 1918), seeds of Norway spruce (*P. abies*) (Yeager 1937), sap from gray birch (*Betula populifolia*) oozing from holes made by Yellow-bellied Sapsuckers (*Sphyrapicus varius*) (Kilham 1958), corn (*Zea mays*) (Linduska 1942), basswood seeds (*Tilia americana*) (Fox 1939), berries and bark in the winter diet (Bailey 1936), and the portions of lodgepole pine infested with dwarf mistletoe (*Arceuthobium pusillum*) (Baranyay 1968). The bark of staghorn sumac (*Rhus typhina*) and seeds of witch hazel (*Hamamelis virginiana*), *Cotoneaster,* and mockernut hickory (*Carya tomentosa*) are also consumed (McClelland 1948).

Mushrooms and other fungi are important foods, including *Amanita muscaria* (Klugh 1927). The diet is supplemented by bones (Coventry 1940), young birds (Adams 1939), young cottontails (*Sylvilagus* sp.) and gray squirrels (Hamilton 1934; Alkon 1962), and beef (Howard 1935). Bailey (1936) and Mailliard (1931) reported that conifer seeds and hazelnuts (*Corylus* sp.) were used by Douglas squirrels. Layne (1954) found that insects were an important summer food.

BEHAVIOR

Pine squirrels are constantly scurrying up and down tree trunks or on the ground, or are jumping from tree to tree. They are extremely aggressive toward conspecifics and other tree squirrels in defense of territories. Territories vary from 1 to 2 ha and are inversely proportional to available food (Smith 1968). Aggressive behavior is characterized by "scolding" chatter, threatening postures, rapid jerking of the tail, or actual attack with forepaws and teeth. These squirrels will also drive off birds and flying squirrels (*Glaucomys* sp.) (Preston 1948; Henshaw 1970); their periodic birdlike singing advertises occupancy of territories. When threatened, red squirrels hide in thick foliage, behind a branch or tree trunk, or quickly run to safety in a den cavity or leaf nest. If surprised on the ground, they run up the nearest tree and along adjacent branches. Rarely, they seek safety in a hole in the ground.

Vocalizations. Pine squirrels are much more vocal than other tree squirrels, especially in relation to territorial defense. Their repertoire varies from a *tcher, tcher* chatter to more musical calls. Young animals utter high-pitched "screams" when frightened. Douglas squirrels, chipmunks (*Eutamias* sp.), and ground squirrels (*Spermophilus* sp.) will respond to each other's alarm calls (Boyer 1943).

Caching. In contrast to other North American tree squirrels, the pine squirrels store food in one or more large caches. These may be in a hollow tree, an underground den, or even in the open at the base of a tree. Both species often locate large caches of cones in wet areas, seepages, springs, or depressions, where constant moisture keeps cones from opening. Such caches are not covered and seeds may remain viable for

years (Shaw 1936). Caches may be 0.3 to 0.6 m high and up to 2 m in diameter. They may represent accumulations from several years, often by a series of successive occupants of the territory. Cones of red pine are cached either singly or in twos or threes but not in large piles, as spruce cones are cached (Clarke 1939). Mushrooms are sometimes arranged on branches to dry, and, when dried, are cached in hollow trees or logs (Dice 1921; Cram 1924).

Activity Patterns. During cold weather, red squirrels in the northern portions of the range dig extensive tunnels in the snow; they are rarely seen in the trees (Pruitt 1960). They remain active all winter foraging in their tunnels. Where winters are less severe, they remain in trees and on the ground. During winter in New York, red squirrels sometimes remained in their nests for a day or two during bad weather but ordinarily they were most active during midday. In summer, they were least active during midday. Generally, when temperatures rise from 10° to 35° C activity decreases (Clarkson and Ferguson 1969).

Ordinarily, pine squirrels do not undergo extensive fall movements as gray squirrels do; however, Klugh (1927) stated that emigrations of red squirrels have occurred in the Adirondacks on several occasions.

MORTALITY

Red squirrels live up to nine years in captivity (Klugh 1927). In the wild, the natural mortality rate is highest during the first two years of life and among older adults. Five percent of a cohort live beyond five years of age and 0.4 percent live beyond eight years (Davis and Sealander 1971). Kemp and Keith (1970) estimated the average annual mortality to be 67 percent for red squirrels, with 34 percent among yearlings and 61 percent among older adults. They found that both intrauterine and neonatal mortality was low.

Predation. Layne (1954) considered predation a nonsignificant mortality factor for pine squirrel populations. No predator takes more than an occasional red or Douglas squirrel. Predators include Red-tailed Hawks (*Buteo jamaicensis*) (English 1934; Luttich et al. 1970), red fox (*Vulpes vulpes*) (Johnson 1970), and Bald Eagle (*Haliaeetus leucocephalus*). The pine marten (*Martes americana*) is able to overtake red squirrels in chases through tree branches. Nellis and Keith (1968) found that 1 percent of the diet of lynx (*Lynx lynx*) consisted of red squirrels. Seasonal trends were noted by Van Zyll de Jong (1966), with 2 percent of the diet of lynx in winter and 9 percent in summer composed of red squirrels.

Parasites, Diseases, and Viruses. In addition to a variety of parasites (table 12.1), the following infectious agents have been reported for red squirrels: tularemia (Francis 1937; Burroughs et al. 1945), *Haplosporangium* (Dowding 1947), *Adiaspiromycosis* (Dvorak et al. 1965), Silverwater virus from Alberta (Hoff et al. 1971), California encephalitis virus in Ohio

TABLE 12.1. Parasites of the red squirrel

Parasite	Source
Protozoa	
Eimeria tamiasciuri	Dorney 1963
E. toddi	Dorney 1963
Trypanosoma sp.	Dorney 1967
Cestodes	
Andrya primordealis	Mahrt and Chai 1972
Cladotaenia globifera	Leiby 1961
Fibricola nana	Rausch and Tiner 1948
Hymenolepis horrida	Schiller 1952
Paruterina candelabraria	Mahrt and Chai 1972
P. rauschi	Freeman 1957
Taenia crassiceps	Freeman 1962
T. mustelae	Mahrt and Chai 1972
T. rileyi	Mahrt and Chai 1972
Nematodes	
Capillaria americana	Lichtenfels and Haley 1968
Citellinema bifurcatum	Lichtenfels and Haley 1968
Heligmodendrium hassalli	Rausch and Tiner 1948
Sphacia thompsoni	Lichtenfels and Haley 1968
Strongyloides robustus	Lichtenfels and Haley 1968
Trichinella spiralis	Rausch 1962
Acarina	
Androlaelaps casalis	Whitaker and Pascal 1979
A. fahrenholzi	Whitaker and Pascal 1979
Dermacarus tamiasciuri	Whitaker and Pascal 1979
Dermacentor variabilis	Knipping et al. 1950
Euhaemogamasus ambulans	Keegan 1951
Euschoengastia peromysci	Shoemaker and Joy 1966
E. setosa	Shoemaker and Joy 1966
Haemogamasus alaskensis	Keegan 1951
H. ambulans	Allred and Beck 1966
H. reidi	Whitaker and Pascal 1979
Haemolaelaps glasgawi	Allred and Beck 1966
Ixodes angustus	Fay and Rausch 1969
I. marxi	Whitaker and Pascal 1979
Leptotrombidium peromysci	Whitaker and Pascal 1979
Miyatrombicula cynos	Whitaker and Pascal 1979
Neotrombicula whartoni	Whitaker and Pascal 1979
N. fitchii	Whitaker and Pascal 1979
Ornithodoros hermsi	Kohls et al. 1965
Orycteroxenus soricis	Whitaker and Pascal 1979
Trombicula fitchi	Shoemaker and Joy 1966
T. harperi	Lawrence et al. 1965
Walchia americana	Shoemaker and Joy 1966
Anoplura	
Enderleinellus longiceps	Kim 1966
E. nitzchi	Scholten et al. 1962
E. tamiasciuri	Kim 1966
Hoplopleura erratica	Whitaker and Pascal 1979
Neohaematopinus sciurincus	Mahrt and Chai 1972
N. semifasciatus	Whitaker and Pascal 1979
Siphonaptera	
Ctenophthalmus pseudagyrtes	Benton et al. 1971
Epitedia wenmanni	Main 1970
Foxella ignota	Wiseman 1955
Megabothris quirini	Buckner 1964
Monopsyllus vison	Benton et al. 1971
Opiscrostis bruneri	Benton et al. 1971

Parasite	Source
Opisodasys pseudarctomys	Benton et al. 1971
Orchopeas cadens	Benton et al. 1971
O. howardii	Holland and Benton 1968
O. leucopus	Holland and Benton 1968
Peromyscopsylla hesperomys	Benton et al. 1969
Tassopsylla octodecimdentata	Holland 1963
Diptera	
Cuterebra emasculator	Dury 1898

(Masterson et al. 1971), and Powassan virus (McLean 1963; McLean et al. 1968).

AGE DETERMINATION

Age can be determined either by weight of the eye lens or by the closure of epiphyseal sutures of the humerus or femur (Davis and Sealander 1971). Adults can be separated from subadults by the presence of a distinctly pigmented scrotum on adult males or pigmented teats on adult females. However, if an adult female did not become pregnant, she would be mistakenly classed as a subadult.

ECONOMIC STATUS AND MANAGEMENT

Pine squirrels are rarely hunted for sport. Between one million and three million are harvested annually in Canada for their fur. An economic value of one million dollars makes them the second or third most valuable fur in Canada (Kemp and Keith 1970). Another important aspect of pine squirrels is that they provide a source of conifer seeds for foresters. Seeds collected from large caches show higher germination rates than seeds from cones collected directly from trees (Wagg 1964). Cone caches are the most important source of seeds from western conifers for planting in the forest tree nurseries of some western states (Yeager 1937; Finley 1969).

Pine squirrel densities may be increased by cultivation of conifers. Suburban communities within their geographic range having extensive plantings of conifers usually have pine squirrels. In the northeastern United States, the establishment of conifers and the subsequent diminution of hardwoods frequently causes associated decreases in densities of gray squirrels and a predominance of red squirrels. Pine squirrel populations can also be enhanced by providing nesting boxes. These should be approximately 15 x 15 cm and 30 cm deep inside with a 4.5-cm-diameter hole in an upper corner. Boxes should be hung on tree trunks at a height of 6 m or more to discourage vandalism. The life of a box can be more than doubled and also made more acceptable to squirrels by cleaning out the accumulated nesting material once a year. Plantings or supplemental feeding are practical only on a small scale, such as when an urban dweller wishes to entice pine squirrels to his yard.

Sometimes pine squirrels, especially the red squirrel, can become troublesome. They sometimes enter buildings and cause damage, but not to the extent that gray squirrels do. Troublesome pine squirrels can be shot or trapped. Peanuts are a good trap bait for live traps. Such traps should have minimum inside dimensions of 8 x 8 x 30 cm. Peanut butter applied to the treadle of a commercial rat trap wired or tied to a tree trunk can be an effective method for removing unwanted pine squirrels. This technique is only practical where other animals are not likely to be killed. Trees can be protected from pine squirrels, if their canopies are separated, by the installation of sheet metal sleeves, as previously described for tassel-eared squirrels (Chapter 11).

The pine squirrel's habit of caching cones can seriously interfere with natural reseeding because in some cases the entire cone crop of forests may be collected (Finley 1969). Pulling (1924) believed that occasionally pine squirrels took 100 percent of white pine (*P. strobus*) cones in New York. However, Moore (1940) stated that squirrels played an insignificant role in conifer reforestation in Oregon and Washington. Red squirrels possibly retard the natural reseeding of jack pines in Minnesota (Smith and Aldous 1947) and may also reduce red pine seed production by eating the seeds before the cones mature (Roe 1948).

Much of the reported damage by red squirrels is to ponderosa pine. Squillace (1953) estimated that *Tamiasciurus* and *Sciurus* together may remove 60 to 80 percent of ponderosa pine cones in poor or fair seed years. Tree squirrels may be a major deterrent to the regeneration of this valuable tree species (Schubert 1953). Schmidt and Shearer (1971) estimated that red squirrels harvest about two-thirds of the mature ponderosa pine cones in some portions of the west and another 14 percent are removed before the cones are ripe. The squirrels' habit of nipping off branch tips with immature cones, in addition to reducing future cone crops, also severely prunes the trees (Adams 1955). Table mountain pine (*P. pungens*), a tree of the central Appalachians, may be so heavily pruned by red squirrels that growth and tree form are severely affected (Mollenhaur 1939).

Pine squirrels also damage trees by nipping off buds at the ends of branches and leaders. Trees affected include balsam fir and spruce in maritime Canada (Balch 1942), jack pines (Cheyney 1929), Japanese larch (*Larix kaempferi*) in Michigan (McCulloch 1937), and red pine in Minnesota (Schantz-Hansen 1945). In New York and Connecticut, they damage Norway spruce, white spruce, Scots pine (*P. sylvestris*), and white pine by eating buds (Hosley 1928; Hart 1936). Cook (1954) also reported that they damaged European larch (*L. decidua*). The damage caused by nipping off buds is almost entirely a winter activity and the severity of damage depends largely on snow depth and the duration of snow cover (Hart 1936).

Pine squirrels also damage trees by peeling bark from branches and the main trunk. In Alaska, such damage may be extensive to paper birch (*B. papyrifera*) stands, causing the deaths of many trees by completely girdling the trunks (Lutz 1956). In the black hills of South Dakota, red squirrels have caused damage by eating the inner bark and cambium of ponderosa pines (Pike 1934). They also strip bark from sapling jack pines (Schantz-Hansen 1945), which may hasten degradation by gall rusts (Lavallee and Bard 1973). Storer (1875) reported that red squirrels damaged fruit orchards in New England by eating ovaries of cherry (*Prunus* sp.) blossoms and destroyed pears (*Pyrus* sp.) by tearing the fruit apart to eat the seeds.

Damage to trees by pine squirrels by either pruning or bark peeling leaves open wounds that are subject to invasion by fungi and bacteria. Shigo (1964) reported cankers on red and sugar maples in New Hampshire as the result of gnawing by red squirrels to get sap in early spring.

CURRENT RESEARCH AND MANAGEMENT NEEDS

Additional research is needed to determine the role of pine squirrels in conifer reforestation in different parts of their range. In situations where damage by pine squirrels could be considerable, e.g., pine plantations and orchards, new and better techniques of reducing damage should be developed. In this regard, factors that influence pine squirrel reproductive success and survival in conifer, conifer-hardwood, and hardwood stands should be investigated for possible application. The mechanism allowing red squirrels, gray squirrels, fox squirrels, eastern chipmunks (*Tamias striatus*), flying squirrels (*Glaucomys volans, G. sabrinus*), or some combination of these species to coexist in certain areas would be an interesting line of research. More information is also needed on the life history aspects and ecology of the Douglas squirrel.

This chapter is contribution no. 1095-AEL, Center for Environmental and Estuarine Studies, University of Maryland.

LITERATURE CITED

Abbott, H. G., and Belig, W. H. 1961. Juniper seed: a winter food of red squirrels in Massachusetts. J. Mammal. 42:240–244.

Adams, L. 1939. Sierra chickaree eats young blue-fronted jays. Yosemite Nature Notes 18:93.

———. 1955. Pine squirrels reduce future crops of ponderosa pine cones. J. For. 53:35.

Aleksiuk, M. 1970. The occurrence of brown adipose tissue in the adult red squirrel (*Tamiasciurus hudsonicus*). Can. J. Zool. 48:188–189.

———. 1971. Seasonal dynamics of brown adipose tissue function in the red squirrel (*Tamiasciurus hudsonicus*). Comp. Biochem. 38:723–731.

Alkon, P. U. 1962. Red squirrel predation on nestling cottontail. New York Fish Game J. 9:142.

Allred, D. M., and Beck, D. E. 1966. Mites of Utah mammals. Brigham Young Univ. Sci. Bull. Bio. 8:1–123.

Bailey, V. 1931. Mammals of New Mexico. North Am. Fauna 53. 412pp.

———. 1936. The mammals and life zones of Oregon. North Am. Fauna 55. 416pp.

Balch, R. E. 1942. A note on squirrel damage to conifers. For. Chron. 18:42.

Baranyay, J. A. 1968. Squirrel feeding on dwarf mistletoe infections. Dept. Fisheries, Ottawa, Bi-Monthly Res. Notes 24:41–42.

Benton, A. H.; Larson, O. R.; and Van Huizen, B. A. 1971. Siphonaptera from Itaska State Park region. J. Minnesota Acad. Sci. 37:91–92.

Benton, A. H.; Tusker, H. L., Jr.; and Kelly, D. L. 1969. Siphonaptera from northern New York. J. New York Entomol. Soc. 77:193–198.

Boyer, R. H. 1943. Weasel versus squirrel in Sequoia National Park. J. Mammal. 24:99–100.

Brink, C. H., and Dean, F. C. 1966. Spruce seed as a food of red squirrels and flying squirrels in interior Alaska. J. Wildl. Manage. 30:503–512.

Buckner, C. H. 1964. Fleas (Siphonaptera) of Manitoba mammals. Can. Entomol. 96:850–856.

Burnett, F. L., and Dickerman, R. W. 1956. Type locality of the Mogollon red squirrel, *Tamiasciurus hudsonicus mogollonensis*. J. Mammal. 37:292–294.

Burroughs, A. L.; Holdenreid, R.; Longanecker, D. S.; and Meyer, K. F. 1945. A field study of latent tularemia in rodents with a list of all known naturally infected vertebrates. J. Infect. Dis. 76:115–119.

Burton, S. S. 1930. A new diet for the red squirrel. J. For. 28:233.

Cheyney, E. G. 1929. Damage to Norway and jack pine by red squirrels. J. For. 27:382–383.

Clarke, C. H. D. 1939. Some notes on hoarding and territorial behavior of the red squirrel, *Sciurus hudsonicus* (Erxleben). Can. Field Nat. 53:42–43.

Clarkson, D. P., and Ferguson, H. J. 1969. Effect of temperature upon activity in the red squirrel. Am. Zool. 9:1110.

Cook, D. B. 1954. Susceptibility of larch to red squirrel damage. J. For. 52:491–492.

Coventry, A. F. 1940. The eating of bone by squirrels. Science 92:128.

Cram, W. E. 1924. The red squirrel. J. Mammal. 5:37–41.

Davis, D. W. 1969. The behavior and population dynamics of the red squirrel, *Tamiasciurus hudsonicus,* in Saskatchewan. Ph.D. Dissertation. Univ. Arkansas, Fayetteville. 222pp.

Davis, D. W., and Sealander, J. A. 1971. Sex ratio and age structure in two red squirrel populations in northern Saskatchewan. Can. Field Nat. 85:303–308.

Dice, L. R. 1921. Notes on the mammals of interior Alaska. J. Mammal. 2:20–28.

Dippner, R., and Armington, J. 1971. A behavioral measure of dark adaptation in the American red squirrel. Psychonom. Sci. 24:43–45.

Dolbeer, R. A. 1973. Reproduction in the red squirrel (*Tamiasciurus hudsonicus*) in Colorado. J. Mammal. 54:536–540.

Dorney, R. S. 1963. Coccidiosis-incidence, epizootiology in two Wisconsin Sciuridae. Trans. North Am. Wildl. Nat. Resour. Conf. 28:207–215.

———. 1967. Incidence, taxonomic relationships and development of Lewisi-like trypanosomes in Wisconsin Sciuridae. J. Protozool. 14:425–428.

Dowding, E. S. 1947. *Haplosporangium* in Canadian rodents. Mycologia 39:372–373.

Dury, C. 1898. Squirrel bot fly (*Cuterebra emasculator*). J. Cincinnati Soc. Nat. Hist. 19:143.

Dvorak, J.; Otcenasek, M.; and Prokopic, J. 1965. The distribution of adiaspiromycosis. J. Hyg. Epidemiol. Microbio. Immunol. 9:510–514.

English, P. F. 1934. Some observations on a pair of red-tailed hawks. Wilson Bull. 46:228–235.

Fay, F. H., and Rausch, R. L. 1969. Parasitic organisms in the blood of arvicoline rodents in Alaska. J. Parasitol. 55:1258–1265.

Ferguson, J. H., and Folk, G. E., Jr. 1971. Effect of temperature and acclimation upon FFA levels in three separate species of rodents. Can. J. Zool. 49:303–305.

Ferron, J., and Prescott, J. 1977. Gestation, litter size, and number of litters of the red squirrel (*Tamiasciurus hudsonicus*) in Quebec. Can. Field Nat. 91:83–84.

Finley, R. B., Jr. 1969. Cone caches and middens of *Tamiasciurus* in the Rocky Mountain region. Univ. Kansas Mus. Nat. Hist. Misc. Publ. 51:233–273.

Fox, A. C. 1939. Red squirrels eat basswood and boxelder seeds. J. Mammal. 20:257.

Francis. E. 1937. Sources of infection and seasonal incidence of tularemia in man. Publ. Health Rep. 52:103–113.

Freeman, R. S. 1957. Life cycle and morphology of *Paruterina rauschi* n. sp. and *P. candelabraria* (Goeze, 1782) (Cestoda) from owls, and significance of plerocercoids in the order Cyclophyllidea. Can. J. Zool. 35:349–370.

———. 1962. Studies on the biology of *Taenia crassiceps* (Zeder, 1800) Rudolphi, 1910 (Cestoda). Can. J. Zool. 40:969–990.

Grodzinski, W. 1971. Food consumption of small mammals in the Alaskan taiga forest. Ann. Zool. Fenn. 8:133–136.

Hamilton, W. J., Jr. 1934. Red squirrel killing young cottontail and young gray squirrel. J. Mammal. 15:322.

Harper, F. 1932. Mammals of the Athabaska and Great Slave Lakes region. J. Mammal. 13:19–36.

Hart, A. C. 1936. Red squirrel damage to pine and spruce plantations. J. For. 34:729–730.

Hatt, R. T. 1930. The relation of mammals to the Harvard forest. Roosevelt Wildl. Bull. 5:625–671.

———. 1943. The pine squirrel in Colorado. J. Mammal. 24:311–345.

Henshaw, J. 1970. Conflict between red squirrels and gray jays. Can. Field Nat. 84:390–391.

Hoff, G. L.; Yuill, T. M.; Iversen, J. O.; and Hanson, R. P. 1971. Silverwater virus serology in snowshoe hares and other vertebrates. Am. J. Trop. Med. Hyg. 20:326–330.

Holland, G. P. 1963. Faunal affinities of the fleas (Siphonaptera) of Alaska: With an annotated list of species. Pacific Sci. Congr., Pacific Basin Biogeogr., Symp. 10:45–63.

Holland, G. P., and Benton, A. H. 1968. Siphonaptera from Pennsylvania mammals. Am. Midl. Nat. 80:252–261.

Hosley, N. W. 1928. Red squirrel damage to coniferous plantations and its relations to changing food habits. Ecology 9:43–48.

Howard, W. J. 1935. Apparently neutral relations of weasel and squirrel. J. Mammal. 16:322–326.

Johnson, W. J. 1969. Food habits of the Isle Royale red fox and population aspects of three of its principal prey species. Ph.D. Dissertation. Purdue Univ., Lafayette, Ind. 293pp.

Keegan, H. L. 1951. The mites of the subfamily Haemogamasinae (Acari: Laelaptidae). Proc. U.S. Natl. Mus. 101:203–268.

Kemp, G. A., and Keith, L. B. 1970. Dynamics and regulation of red squirrel (*Tamiasciurus hudsonicus*) populations. Ecology 51:763–779.

Kilham, L. 1958. Red squirrels feeding at sapsucker holes. J. Mammal. 39:596–597.

Kim, K. C. 1966. The species of *Enderleinellus* (Anopleura,

Hoplopleuridae) parasitic on the Sciurini and Tamiasciurini. J. Parasitol. 52:988-1024.

Klugh, A. B. 1918. The behavior of the red squirrel. Ottawa Nat. 32:9-12.

———. 1927. Ecology of the red squirrel. J. Mammal. 8:1-32.

Knipping, P. A.; Morgan, B. B.; and Jerome, D. R. 1950. Notes on the distribution of Wisconsin ticks. Trans. Wisconsin Acad. Sci. Arts Letters 60:185-197.

Kohls, G. M.; Sonenshine, D. E.; and Merritt, C. C. 1965. The systematics of the subfamily Ornithodorinae (Acarina: Argasidae). Part 2, Identification of the larvae of the Western Hemisphere and descriptions of three new species. Ann. Entomol. Soc. Am. 58:331-364.

Komarek, E. V., and Komarek, R. 1938. Mammals of the Great Smoky Mountains. Bull. Chicago Acad. Sci. 5:137-162.

Krull, J. N. 1970. Response of chipmunks and red squirrels to commercial clearcut logging. New York Fish Game J. 17:58-59.

Lavallee, A., and Bard, G. 1973. Observations sur deux rouilles-tumeurs du pingris (*Pinus banksiana*). Can. J. For. Res. 3:251-255.

Lawrence, W. H.; Hays, K. L.; and Graham, S. A. 1965. Arthropodous ectoparasites from some northern Michigan mammals. Occas. Pap. Mus. Zool. Univ. Michigan 639:1-7.

Layne, J. N. 1952. The os genitale of the red squirrel, *Tamiasciurus*. J. Mammal. 33:457-459.

———. 1954. The biology of the red squirrel, *Tamiasciurus hudsonicus loquax* (Bangs), in central New York. Ecol. Monogr. 24:227-267.

Leiby, P. D. 1961. Intestinal helminths of some Colorado mammals. J. Parasitol. 47:311.

Lichtenfels, J. R., and Haley, A. J. 1968. New host records of intestinal nematodes of Maryland rodents and suppression of *Capillaria bonnevillei* Grundmann and Frandsen, 1960 as a synonym of *C. americana* Reed, 1969. Proc. Helminth. Soc. Washington 35:206-211.

Linduska, J. P. 1942. Winter rodent populations in field-shocked corn. J. Wildl. Manage. 6:353-363.

Long, W. S. 1940. Notes on the life histories of some Utah mammals. J. Mammal. 21:170-180.

Luttich, S.; Rusch, D. H.; Meslow, E. C.; and Keith, L. B. 1970. Ecology of red-tailed hawk predation in Alberta. Ecology 51:190-203.

Lutz, H. J. 1956. Damage to paper birch by red squirrels in Alaska. J. For. 54:31-33.

McClelland, E. H. 1948. Notes on the red squirrel in Pittsburgh. J. Mammal. 29:409-412.

McCulloch, W. F. 1937. Red squirrels attack Japanese larch. J. For. 35:692-693.

McLean, D. W. 1963. Powassan virus isolations from ticks and squirrel blood. Fed. Proc. 22:323.

McLean, D. W.; Ladyman, S. R.; and Purvingood, K. W. 1968. Westward extension of Powassan virus prevalence. Can. Med. Assoc. J. 98:946-949.

MacNamara, C. 1943. An apparently unrecorded food of the red squirrel. Can. Field Nat. 57:107.

Mahrt, J. L., and Chai, S.-J. 1972. Parasites of red squirrels in Alberta, Canada. J. Parasitol. 58:639-640.

Mailliard, J. 1931. Redwood chickaree testing and storing hazel nuts. J. Mammal. 12:68-70.

Main, A. J., Jr. 1970. Distribution, seasonal abundance and host preference of fleas in New England. Proc. Entomol. Soc. Washington 72:73-89.

Masterson, R. A.; Stegmiller, H. W.; Parsons, M. A.; Croft, C. C.; and Spencer, C. B. 1971. California encephalitis: An endemic puzzle in Ohio. Health Lab. Sci. 8:89-96.

Mengel, R. M., and Jenkinson, M. A. 1971. A melanistic

specimen of the red squirrel. Am. Midl. Nat. 86:230-231.

Millar, J. S. 1970. The breeding season and reproductive cycle of the western red squirrel. Can. J. Zool. 48:471-473.

Mohr, C. O. 1965. Home area and comparative biomass of the North American red squirrel. Can. Field Nat. 79:162-171.

Mollenhauer, W., Jr. 1939. Table mountain pine: squirrel food or timber tree. J. For. 37:420-421.

Moore, A. W. 1940. Wild animal damage to seed and seedlings on cut-over Douglas fir lands of Oregon and Washington. U.S.D.A. Tech. Bull. 706:1-28.

Mossman, H. W. 1940. What is the red squirrel? Trans. Wisconsin Acad. Sci. Arts Letters 32:123-134.

Nadler, C. F., and Hoffmann, R. S. 1970. Chromosomes of some Asian and South American squirrels (Rodentia: Sciuridae). Experientia 26:1383-1386.

Nellis, C. H. 1969. Sex and age variation in red squirrel skulls from Missoula County, Montana. Can. Field Nat. 83:324-330.

Nellis, C. H., and Keith, L. B. 1968. Hunting activities and success of lynxes in Alberta. J. Wildl. Manage. 32:718-722.

Nelson, B. A. 1945. The spring molt of the northern red squirrel in Minnesota. J. Mammal. 26:397-400.

Odum, E. P. 1949. Small mammals of the highlands (North Carolina) plateau. J. Mammal. 30:179-192.

Petrides, G. A. 1942. Relation of hedgerows in winter to wildlife in central New York. J. Wildl. Manage. 6:261-280.

Pike, G. W. 1934. Girdling of ponderosa pine by squirrels. J. For. 32:98-99.

Prescott, J., and Ferron, J. 1978. Breeding and behavior development of the American red squirrel *Tamiasciurus hudsonicus* in captivity. Int. Zoo Yearb. 18:125-130.

Preston, F. W. 1948. Red squirrels and gray. J. Mammal. 29:297-298.

Pruitt, W. O., Jr. 1960. Animals in the snow. Sci. Am. 202:60-68.

Pulling, A. V. S. 1924. Small rodents and northeastern conifers. J. For. 22:813-814.

Quay, W. B. 1965. Comparative survey of the sebaceous and sudoriferous glands of the oral lips and angle in rodents. J. Mammal. 46:23-37.

Rausch, R., and Tiner, J. D. 1948. Studies on the parasitic helminths of the north central states. Part 1, Helminths of Sciuridae. Am. Midl. Nat. 39:728-747.

Rausch, R. L. 1962. Trichinellosis in the Arctic. Pages 80-86 in Proc. 1st Int. Conf. Trichinellosis, Warsaw.

Roe, E. I. 1948. Effects of red squirrels on red pine seed production in off years. J. For. 46:528-529.

Rothwell, R. 1979. Nest sites of red squirrels (*Tamiasciurus hudsonicus*) in the Laramie Range of southeastern Wyoming. J. Mammal. 60:404-405.

Schantz-Hansen, T. 1945. Red squirrel damage to mature red pine. J. For. 43:604-605.

Schiller, E. L. 1952. Studies on the helminth fauna of Alaska. Part 10, Morphological variation in *Hymenolepis horrida* (von Linstow, 1901) (Cestoda: Hymenolepididae). J. Parasitol. 38:554-568.

Schmidt, W. C., and Shearer, R. C. 1971. Ponderosa pine seed: for animals or trees. U.S. For. Serv. Res. Pap. INT-112:1-14.

Scholten, T. H.; Ronald, K.; and McLean, D. M. 1962. Parasite fauna of the Manitoulin Island region. Part 1, Arthropoda Parasitica. Can. J. Zool. 40:605-606.

Schubert, G. H. 1953. Ponderosa pine cone cutting by squirrels. J. For. 51:202.

Shaw, W. T. 1936. Moisture and its relation to the cone-storing habit of the western pine squirrel. J. Mammal. 17:337–349.

Shigo, A. L. 1964. A canker on maple caused by fungi infecting wounds made by the red squirrel. Plant Dis. Rep. 48:794–796.

Shoemaker, J. P., and Joy, S. L. 1966. *Trombicula fitchi* (Loomis 1954) in West Virginia. J. Parasitol. 52:1198.

Smith, C. C. 1965. Interspecific competition in the genus of tree squirrels, *Tamiasciurus*. Ph.D. Dissertation. Univ. Washington, Seattle. 306pp.

———. 1968. The adaptive nature of social organization in the genus of tree squirrels *Tamiasciurus*. Ecol. Monogr. 38:31–63.

———. 1970. The coevolution of pine squirrels (*Tamiasciurus*) and conifers. Ecol. Monogr. 40:349–371.

Smith, C. F., and Aldous, S. E. 1947. The influence of mammals and birds in retarding artificial and natural reseeding of coniferous forests in the United States. J. For. 45:361–369.

Squillace, A. E. 1953. Effect of squirrels on the supply of ponderosa pine seed. Northern Rocky Mountain For. Range Exp. Stn., Missoula, Montana. Res. Note 131:1–4.

Storer, F. H. 1875. Cherry blossoms destroyed by squirrels. Nature 13:26.

Svihla, R. D. 1930. Development of young red squirrels. J. Mammal. 11:79–80.

Van Zyll de Jong, C. G. 1966. Food habits of the lynx in Alberta and the Mackenzie District, N.W.T. Can. Field Nat. 80:18–23.

Wagg, J. W. B. 1964. Viability of white spruce seed from squirrel-cut cones. For. Chron. 40:98–110.

Whitaker, J. O., Jr., and Pascal, D. D., Jr. 1979. Ectoparasites of the red squirrel (*Tamiasciurus hudsonicus*) and the eastern chipmunk (*Tamias striatus*) from Indiana. J. Med. Entomol. 16:350–351.

Wiseman, J. S. 1955. The Siphonaptera (fleas) of Wyoming. Univ. Wyoming Publ. 19:1–28.

Wood, T. J. 1965. Albino red squirrel collected in Wood Buffalo Park. Blue Jay 23:90.

Yeager, L. E. 1937. Cone-piling by Michigan red squirrels. J. Mammal. 18:191–194.

Vagn Flyger, Department of Animal Science, University of Maryland, College Park, Maryland 20742.

J. Edward Gates, Appalachian Environmental Laboratory, Center for Environmental and Estuarine Studies, University of Maryland, Frostburg State College Campus, Frostburg, Maryland 21532.

13

Pocket Gophers

Geomyidae

Janis D. Chase
Walter E. Howard
James T. Roseberry

NOMENCLATURE

FAMILY GEOMYIDAE. Species featured in this chapter are: *Thomomys umbrinus,* southern pocket gopher; *Thomomys bottae,* Botta's pocket gopher; *Thomomys talpoides,* northern pocket gopher; *Geomys bursarius,* plains pocket gopher; and *Pappogeomys castanops,* yellow-faced pocket gopher.

The three genera of pocket gopher found in North America are the western pocket gopher (*Thomomys*), with eight species; the eastern pocket gopher (*Geomys*) with seven species; and the yellow-faced pocket gopher (*Pappogeomys*) (=*Cratogeomys*), with one species (Russell 1968; Jones et al. 1975). Five additional genera are recognized in Central America (Hall and Kelson 1959).

Because of the great variability among pocket gophers, about 30 species and more than 300 subspecies have been named, many of them since the revision of the family by Merriam (1906).

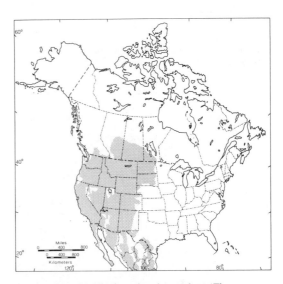

FIGURE 13.1. Distribution of pocket gophers (*Thomomys* species).

DISTRIBUTION

Pocket gophers (Geomyidae) are found only in the western hemisphere, where they occur from central Alberta south to Panama. Of the three genera of pocket gopher in the United States, the western pocket gopher (*Thomomys* sp.) occurs over most of western North America (Turner et al. 1973). *T. talpoides,* the Northern pocket gopher, has the widest distribution of all pocket gophers, extending from Central Alberta to northern New Mexico and Arizona. and from the western three-fourths of North Dakota and South Dakota to eastern Washington, Oregon, and northeastern California. The yellow-faced pocket gopher (*Pappogeomys castanops*) occurs from southeastern Colorado, eastern New Mexico, and western Oklahoma south through western Texas into Mexico. The eastern pocket gopher (*Geomys* sp.) is found on the plains and prairies and in the eastern Gulf states. There are no pocket gophers in the northeastern United States (figures 13.1–13.3).

Pocket gopher species are almost always allopatric in their distribution (Vaughan and Hansen 1964). The ecological separation of the three genera of pocket gopher found in the United States was discussed by Best (1973). When, on occasion, two species do occur in the same area, interspecific intolerances and competition are observed. Therefore, the similarity in ecological requirements of the different species of pocket gopher probably prevents sympatric distribution (Miller 1964; Vaughan 1967a). Reichman and Baker (1972) cite an example of *P. castanops* expanding its range by displacing two subspecies of *T. bottae.* Williams and Baker (1976) noted that *T. bottae* has a low level of vagility (ability to disseminate) compared with *P. castanops.*

DESCRIPTION

Pocket gophers are fossorial rodents that are morphologically well adapted for a burrowing existence (Hill 1937). The head is small and flattened, the

239

FIGURE 13.2. Distribution of the eastern pocket gophers (*Geomys* sp.).

FIGURE 13.3. Distribution of the yellow-faced pocket gopher (*Pappageomys castanops*).

neck is short without constriction, and the shoulders and forearms are broad, thick, and muscular (Hollinger 1916) (figure 13.4). The eyes are small, with eyelids so close fitting that even fine sand cannot penetrate them. Their eyesight is relatively inferior to that of other rodents (Grinnell 1923). The small, roundish ears, lying close to the head, are equipped with valves so that the pocket gopher can close them while digging (Cahalane 1947). The nearly naked, short tail of most pocket gophers, together with the long body guard hairs, aids in guiding the backward movements of gophers through the tunnel systems (Hoffmeister 1971).

The mouth of the pocket gopher is an opening with furred lips that, when closed, press firmly against

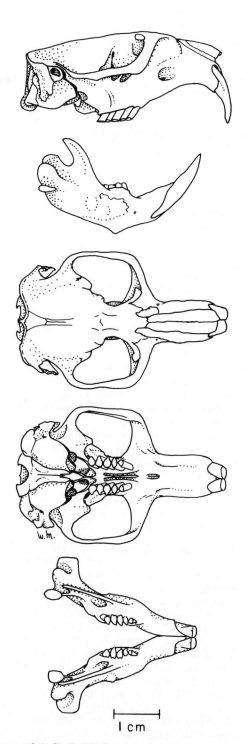

FIGURE 13.4. Skull of the Botta's pocket gopher (*Thomomys bottae*). From top to bottom: lateral view of cranium, lateral view of mandible, dorsal view of cranium, ventral view of cranium, dorsal view of mandible.

and extend behind the projecting incisors (Merrian 1906; Grinnell 1923). This adaptation enables pocket gophers to cut roots and use their incisors in digging without "eating" dirt, for the front teeth are exposed even when the mouth is closed (figure 13.5). The fur-lined cheek pouches on either side of the mouth are used in food-gathering activities (Grinnell 1923). These "pockets," which can be turned inside out like pants pockets, distinguish pocket gophers from other rodents, insectivores, reptiles, and amphibia locally called gophers. The vibrissae surrounding the mouth and nose area are extremely sensitive to touch and aid the pocket gopher in navigating its forward and backward movements in the dark burrow (Grinnell 1923; Cahalane 1947). The auditory ability of *T. talpoides* is comparatively less than that of most rodents; low vibrations are detected most readily (Grinnell 1923).

The large forefeet of the pocket gopher are encircled with a wristlet of long stiff hairs and are equipped with large sharp claws, while the hindfeet are smaller, with short toes and claws, and lack wristlets (Cahalane 1947). Most of the digging activity is accomplished by using the strong front claws, which consequently subjects the nails of the forefeet to great wear; in compensation, the middle three nails on the front feet grow faster than the other nails. Howard (1953a) found that the mean annual growth of the three center nails of the forefeet of *T. bottae* was approximately 140mm, about twice that of the rest of the nails.

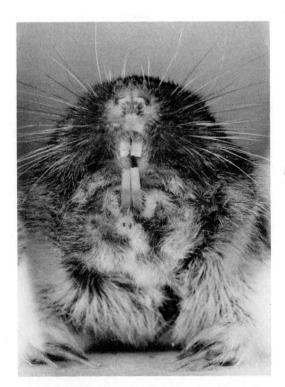

FIGURE 13.5. The lips of a pocket gopher close behind the incisors, enabling the animal to gnaw roots and excavate soil without getting dirt in its mouth.

Pelage. The fur of the pocket gopher is fine and soft, with the nap of the coat lying in one direction and close to the body (Cahalane 1947). The pelage color of pocket gophers is highly variable, depending on the subspecies; isolated colonies or families of pocket gophers with color aberrations are not uncommon (Storer and Gregory 1934). Most *Thomomys* are black to pale cream, usually nearly unicolor above and relatively paler below; *Geomys* are pale brown to black, usually paler below than above; and *Pappogeomys* are yellowish brown or buffy ochreous (Hall and Kelson 1959).

Molt lines are often conspicuous on pocket gophers. Morejohn and Howard (1956) noted that the molt lines of *T. bottae* indicated boundaries between summer and winter pelages. Occasionally, however, these molt lines were a demarcation between two portions of a similar pelage that molted at different times. Some animals possessed parts of three coats simultaneously. This on-and-off molting exhibited by some pocket gophers is quite different from the typical molt of nonfossorial mammals.

Dentition. As is typical of all rodents, pocket gophers have prominent chisellike incisors that are rootless and grow continuously from a persistent pulp throughout the life of the animal (Howard and Smith 1952). The upper incisors of *Thomomys* are relatively smooth, in contrast to the deeply center-grooved upper incisors of *Geomys* and *Pappogeomys* (Turner et al. 1973). Pocket gophers use their chisel-shaped incisors for cutting roots, stems, and tubers while foraging and burrowing. Bailey (1895) noted that pocket gophers are capable of a phenomenal rate of cutting action per minute, and that the animal may use its upper incisors to anchor its body in position while digging. Where soils are too hard or stony for the claws to be completely effective, the pocket gopher will often use its incisors to loosen the soil and rocks (Cahalane 1947; Howard and Smith 1952). Such abrasive actions subject the incisors to rapid wear and breakage, so the continuous replacement is essential.

The evolved growth rate of the incisors is largely a function of the nature of their use. Howard and Smith (1952) showed that the growth rate of all four incisors of *T. bottae* is 225 percent of that of domestic rabbits (*Oryctolagus cuniculus*), 226 percent of that of albino rats (*Rattus* sp.), 280 percent of that of guinea pigs (*Cavia porcellus*), and 374 percent of that of porcupines (*Erethizon dorsatum*). Small pocket gophers living in coarser and more compact soils had a slower extrusive growth than large pocket gophers from irrigated alfalfa fields (Miller 1958).

The lower incisors of the pocket gopher grow significantly faster than the upper incisors. Howard and Smith (1952) found that for the *T. bottae* the average annual growth rate of the lower incisors was 360mm, while it was 226mm for the upper incisors. Miller (1958) measured an annual growth of 445 mm for the lower incisor of one *T. talpoides*. The reasons for this unequal growth are not established. Howard and Smith (1952) hypothesized several reasons: (1) the lower jaw moves while the upper jaw remains stationary; (2) the lower jaw is subject to more abrasive actions; or (3) the greater amount of cutting and grating with the lower incisors creates more of a chance for chipping or breaking, leading to faster wear and regrowth.

Caged pocket gophers have been observed grating the lower and upper incisors, which probably helps keep the teeth chisel-sharp while preventing them from growing excessively long. It is therefore not necessary for pocket gophers continuously to gnaw on hard objects, as is required for most other rodents (Howard 1953a).

The dental formula for pocket gophers is i $\frac{1}{1}$, c $\frac{0}{0}$, p $\frac{1}{1}$, m $\frac{3}{3}$ = 20 total teeth.

Sexual Dimorphism. There is a marked sexual dimorphism in adult pocket gophers. Male pocket gophers continue to grow throughout their lifetime, whereas females grow very little after reaching sexual maturity (Tryon 1947; Miller 1952). The total body length of mature male pocket gophers of the genera discussed in this article ranges from 217 to 372 mm (Hall and Kelson 1959). Adult males of *T. talpoides* are about 10 percent heavier than adult females and 3 to 4 percent longer in total length (Hansen 1960; Hansen and Bear 1964; Hansen and Ward 1966). The mean average weight of *T. talpoides* from subalpine meadows in Utah was 91.4 ± 1.6 g for adult females and 104.4 ± 3.4 g for adult males (Anderson 1978). Turner et al. (1973) noted that body size of *T. talpoides* varied with locality, vegetative type, and altitude. Larger body sizes of the northern pocket gopher at higher elevations have been suggested as being a phenotypic response to food and soil depth (Tryon and Cunningham 1968). Howard and Childs (1959) showed that, for *T. bottae,* two-year-old males weigh about twice as much as two-year-old females (figure 13.6). At one year of age, males were already 50 percent heavier than females. Weight differences between the sexes at various ages also fluctuate markedly with the season. For comparative weight data on pocket gophers to be reliable, one must deduct the weight of the contents of the stomach, cecum, and intestine from the gross body weight. Howard and Childs (1959) found that these organ weights averaged 21.3 percent of gross body weight in pocket gophers trapped in alfalfa fields.

PHYSIOLOGY

Thermoregulation. Even though pocket gophers inhabit microenvironments that are relatively stable in light, temperature, respiratory gas concentrations, and humidity compared with the environments encountered by other terrestrial mammals (Gettinger 1975), their thermoregulation is facilitated by certain morphological and physiological adaptations. Increased peripheral blood circulation to the naked tail is a means of dissipating heat. Consequently, morphological adaptations of the ratio of body size to tail length are a result of local climatic variations, and these adaptations may

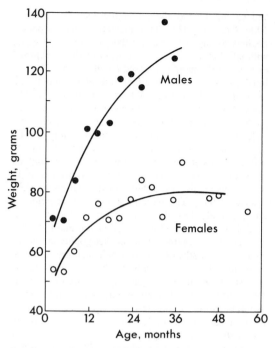

FIGURE 13.6. Female pocket gophers grow relatively little once mature, but males continue to grow throughout their life, as is illustrated for the Botta's pocket gopher (*Thomomys bottae*). From Howard and Childs 1959.

explain the ecological distribution of certain species of pocket gopher (McNab 1966; Bradley et al. 1974).

In a comparative study of thermoregulatory responses in the northern pocket gopher (*T. talpoides*) and the southern pocket gopher *T. umbrinus*), *T. talpoides* had a more variable regulatory response but a greater insulative ability. Northern pocket gophers had relatively larger bodies and shorter tails at higher altitudes than at lower elevations (Bradley et al. 1974). Gettinger (1975) found that thermoneutrality of *T. talpoides* extended from about 26.0° to 32.0° C. Body temperature was maintained at 36.9° C when ambient temperatures were between 1° and 30° C, but above 30° C the ability to dissipate heat was reduced and body temperature rose. Several animals died following one hour of exposure at ambient temperatures above 38° C. Howard and Childs (1959) observed that if *T. bottae* were left very long in live traps that were only lightly covered with soil, the animals succumbed from the heat when the temperature reached 38° C.

GENETICS

The many subspecies of pocket gopher may reflect the discontinuous distribution imposed on them by their fossorial existence, and may also indicate a great genetic plasticity. Among the attributes often associated with a fossorial life mode are disjunct distribution patterns resulting from patchy soil conditions and a closed social system of individual territories, favoring a high degree of both intrapopulation and interpopulation divergence in genetic systems (Patton and Yang 1977).

Solitary and subterranean in habit, limited in vagility, and narrow in habitat requirements, pocket gophers are inbred, and are interesting subjects for cytogenetic studies on speciation and evolutionary processes. For example, when considering the karyotypic variation represented by the chromosomal groups within *G. bursarius,* it becomes apparent that a taxonomic system based on only a few characters is unlikely to be valid (Hart 1978). Hart has found several karyotypes within this species, suggesting that reproductive compatibility between subspecies is doubtful.

There probably have been few changes in the genome of the northern pocket gopher (*T. talpoides*) since the late Pleistocene, most likely because of the constant environment offered by the subterranean tunnel system (Nevo et al. 1974). Analysis of the genome of several populations of this species in the southern Rockies has revealed little if any genetic variation. Thaeler (1974) examined four situations in western Colorado where two populations having different chromosome forms of the *Thomomys talpoides* complex have overlapping ranges and found three different results: population contacts at one place had no hybridization, two contacts had limited hybridization, and more or less unrestricted interbreeding occurred at the fourth contact between different populations. He concluded that karyotype differences in pocket gophers are frequently, though not always, indicative of species-level differences. Skull measurements from isolated populations of *T. talpoides* do not distinguish populations any better than certain external characteristics, including pelage color or number of teats (Tryon 1951). Patton et al. (1972) electrophoretically investigated allelic variations at 27 structural gene loci in five populations of *T. bottae* and one of *T. umbrinus,* thus it is apparent that divergence at the chromosomal and morphologic levels can occur without concomitant genetic differentiation.

REPRODUCTION

Age, Season, and Gestation. Breeding activity of pocket gophers begins at about one year of age, when the animal matures both behaviorally and physiologically. Between 6 months of age and when maturity is attained, there is little apparent external urogenital difference between sexes. With immature live-trapped individuals the best method of sex determination is to apply pressure inward to cause the penis (if any) to extrude (Howard and Childs 1959). The growth of the baculum and sex glands of male *T. talpoides* is complete at about one year of age (Hansen 1960). Sexual maturity of females can be determined by palpating to feel if the pubic symphysis is open. The female's pubic gap forms when the animal is between 9 and 12 months of age (Hansen 1960). Hisaw (1924, 1925) found that the female's pubic symphysis was permanently absorbed during the first breeding cycle through the action of the ovarian hormone relaxin.

Breeding among most pocket gopher species occurs in the spring of the year. In Colorado, pregnant *G. bursarius* were taken most frequently in April and May (Vaughan 1962) and pregnant *T. talpoides* were taken from the first of March through June, with peak activity occurring in mid-May (Hansen 1960; Turner et al. 1973). Pregnant *T. bottae* in the California foothills were taken from the middle of January to the middle of April. However, Miller (1946) showed that some female *T. bottae* in irrigated alfalfa fields in the Sacramento Valley were breeding at all times of the year, although it mostly occurred in spring with some in summer and almost none in fall. Breeding was variable in winter.

Howard and Childs (1959) found that male *T. bottae* were polygamous. In their field study the sex ratio of adults in different years varied from equality to a ratio of one male to four females. The home ranges of males did not overlap, but male home ranges overlapped with ranges of one to four females. Home ranges of females did not noticeably overlap.

During the breeding season there is a conspicuous enlargement of the testes and seminal vesicles of male pocket gophers. However, Miller (1946) found that the size and position of testes cannot always be used to assess or predict reproductive performance. Testes become larger as temperature decreases in winter and tend to be relatively small and scrotal in position in the hot summer. The gestation period, at least for the western pocket gopher (*Thomomys*), is about 18 or 19 days, as determined from *T. bottae* bred in captivity (Schramm 1961; Turner et al. 1973; Andersen 1978).

Number and Size of Litters. Most forest and rangeland pocket gophers reportedly produce only a single litter a year (Ellison and Aldous 1952; Howard and Childs 1959; Hansen 1960; Vaughan 1962), although both Cahalane (1947) and Hoffmeister (1971) suggested that more than a single litter per year is likely. Miller (1946) found *T. bottae* to have multiple litters.

The number of young per litter, as determined by counting the number of embryos or placental scars, varies with the species, size of the female, food availability, and altitude of habitat. Hansen (1960) reviewed the literature on the litter size of *T. talpoides,* and found that it ranged from 3.2 to 6.4. The differences were due more to locality than to yearly fluctuation. For the same species, Hansen and Ward (1966) found that litter size ranged from 3.7 to 4.3. For *T. bottae,* Miller (1946) found that the number of embryos averaged 5.8, whereas the number of placental scars was only 4.9. For the plains pocket gopher (*G. bursarius*). Vaughan (1962) found the number of embryos and scars to be 3.4 and 3.5, respectively.

Growth and Longevity. Newborn pocket gophers are completely altricial. Neonates of *T. talpoides* weigh between 2.8 and 4 g (Reid 1973; Andersen 1978). The cheek pouches are only slight furrows in the skin at birth, and the eyes are visible only as dark spots under the skin. Cheek pouches opened at 24 days, eyes and ears at 26 days; pouches were used to carry food at 39 days; siblings fought and had to be separated at 2 months; and molting to adult pelage was about half completed at 100 days (Andersen 1978). With most pocket gophers, weaning is thought to occur when the young are about 35 to 40 days old (Wight 1930; Howard and Childs 1959), although the young may remain with the mother for a while longer, not dispersing until they are about 2 months old (Cahalane 1947).

Pocket gophers have a maximum longevity of around 5 years. The oldest *T. talpoides* found in Colorado were two males, 4 and 5 years old (Hanson 1962; Reid 1973). Ingles (1952) reported 4 female *T. monticola* that were at least 4 years old when last trapped. Out of 330 *T. bottae* live-trapped by Howard and Childs (1959) in a 5-year field study, the oldest female was at least 4 years and 9 months old, and the oldest male 3 years of age. However, mean lifespan for that population of gophers was about 13.6 months for males and 18.3 months for females, with 96 percent being 2 years old or less when last trapped.

POPULATION DYNAMICS

Home Range. The home range of pocket gophers is practically synonymous with their territory, except during the breeding season. Howard and Childs (1959) observed that once *T. bottae* had established a territory it later made only minor shifts. For example, it shifted when crowded away by a larger animal or when claiming a portion of an abandoned tunnel system that had better food and soil conditions. The territories were of all shapes. Members of one sex often have a limited penetration of tunnels across the outermost boundaries of the other sex's territory. During winter and spring, when the population was stable, Howard and Childs (1959) found that based on trap-release data the home range of male *T. bottae* averaged about 250 m², with a maximum of 445 m². The home ranges of females were about half as large, 120 m², with the maximum being about 240 m². Tietjen et al. (1967), Hansen and Reid (1973), and Turner et al. (1973) reported comparable data for *T. talpoides*. In flood-irrigated alfalfa, however, Miller (1957) found the average home range of nine *T. bottae* to be much smaller, only 74 m², apparently because of the abundant food supply.

In a removal experiment, Howard and Childs (1959) maintained one live-trap set continuously in the same burrow, removing all animals captured. Between 10 March and 6 May, 18 gophers (16 individuals) were trapped, including 2 recaptures used in homing experiments. Of the 16 gophers captured, all 7 adults and 1 of the 9 subadults were marked animals from surrounding territories. Four of the marked adults had traveled 21 to 34 m beyond the boundary of their home territory (range), which had been established earlier by trap and release studies.

Movements. Tietjen et al. (1967) reported that the average distance between points of capture of live-trapped *T. talpoides* on the Black Mesa was 8.5 m, with a maximum distance for one individual of 42.7 m.

In the same area, Hansen (1962) found that the average distance between successive captures was greater for juveniles than for adults. Vaughan (1963) reported that *T. talpoides*, unlike *T. bottae*, can cross large expanses of otherwise unsuitable habitat by building tunnels in snow. Such snowpack tunnels were up to 45.7 m long (Turner et al. 1973).

When the snow melts, pocket gophers on Black Mesa often have to move to islands formed by Mima mounds (Hansen 1962). Ingles (1949) found that *T. monticola* living at the edges of mountain meadows shifted their home range to higher ground, then moved back in late spring. Vaughan (1963) released *T. bottae* and *T. talpoides* into different field areas and recorded the distances they moved. The longest movement was about 275 m for *T. bottae* and 790 m for *T. talpoides* during snow cover.

Young pocket gophers often disperse aboveground. Drift fences established to capture aboveground dispersing rodents in California caught 158 pocket gophers. As is typical with small rodents dispersing far away from their parental home range, roughly two-thirds (62 percent) were males (Howard 1961). Extensive dispersal movements of sexually maturing rodents are due to an "innate" trait. Such individuals showed no evidence of having been driven aboveground. Many subadults appear to live for some time in or near their natal burrow system. Other young adult dispersants no doubt moved off the study plot and were not recaptured, but six dispersed only between 90 and 122 m (Howard and Childs 1959). Other researchers have noted a spring and early-summer dispersal of subadults. Imler (1945) captured many pocket gophers in drift-fence funnel traps he established to take bullsnakes (*Pituophis melanoleucus sayi*).

In homing experiments to determine whether *T. bottae* returned underground or traveled across the ground surface, Howard and Childs (1959) established drift-fence-with-funnel traps. They intercepted gophers returning aboveground when released on the opposite side of the fence from where they lived. One animal was caught on the side where released and 2 were caught on the opposite side, but 8 presumably returned home by using existing burrow systems through territories of a number of other gophers. Thus, 10 of 11 released animals traveled mostly in existing burrow systems. This raised the question of whether some tunnels were more or less common property. One of these females returned home in existing burrow systems for a distance of 200 m without being caught in the drift-fence trap.

Density. The population density of pocket gophers is greatly influenced by the local climate, suitability of the soil, kind and amount of soil drainage, altitude, land use, and other habitat factors. Hansen and Bear (1964) found that densities of *T. talpoides* were greatest in the subalpine zone of southwest Colorado and lowest in the low shrub-grassland zone. Grazed rangelands of ponderosa pine (*Pine ponderosa*) and bunchgrass (*Agrophyron* sp.) supported only 9.9 *T. talpoides* per hectare, whereas ungrazed rangelands of similar habitat supported 22.2 per hectare (Turner et al. 1973). Reid noted that the rangelands that supported the largest pocket gopher populations were those with dark, friable, light-textured soils with large herbage yields, particularly if composed of succulent forbs having fleshy underground storage structures.

On the San Joaquin Experimental Range, in California, 99 percent of the herbaceous forage is composed of annual plants; more than half of it consists of broadleaf filaree (*Erodium botrys*), soft chess (*Bromus mollis*), and foxtail fescue (*Festuca megalura*). This habitat supported from 8 to 17 pocket gophers per hectare in different years. In contrast, Hansen (1965) reported a high of 183 *T. talpoides* per hectare in an exclosure of natural grass-forb range on the Black Mesa, but a high of only 47 per hectare on the free range, where the food supply was much less.

Hansen and Remmenga (1961) measured the average distances to closest neighbors to show density relationships, but concluded that it may be more feasible to obtain density estimates from trapping transects than by trapping plots of trapped-out *T. talpoides* populations. Relative densities also have been determined by counting mounds, expressed as number of mounds per hectare or kilometers of transect traveled. Numerous researchers have utilized this method (Phillips 1936; Richens 1965b; Reid et al. 1966; Hansen and Ward 1966; Julander et al. 1969; Barnes et al. 1970).

During peak burrowing activity, it is relatively easy to obtain a fairly good census of the population by counting the number of separate groups of fresh mounds that have been made within the last few days. Even though a pocket gopher can usually be trapped in any part of its burrow system, its current excavation activity usually remains in just one portion of the home range.

Howard (1961) developed a modified transect method that proved effective even when the density of pocket gophers was 20 or more per hectare. One hand-tally counter was used to record the number of premeasured paces taken, and another counter to record the number of fresh pocket gopher workings seen. The transects were any width that permitted good visibility of the mounds. With this method, whenever one or more fresh mounds are sighted within the transect, they are recorded as one, and no new count is made for the next five paces. It is easy to calculate in advance how many paces are necessary, depending upon the width of the transect, to indicate the size of the area covered. Howard (1961) found this a very effective method, making pocket gophers the easiest rodent to census, and the reliability was considerable.

Cycles. Northern pocket gophers are subject to intermittent fluctuations in population size (Hansen 1962; Hansen and Ward 1966; Tietjen et al. 1967; Julander et al. 1969). Aldous (1957) found high and low population levels at the same time in two separate areas in Utah. Abrupt population changes, in most cases, appear to be caused by a combination of extrinsic and unrecognized intrinsic factors. Snowmelt and resulting high water tables may be critical factors related to pocket gopher survival. In some places, extreme snowpacks have a disastrous effect on survival of the young of the year, changing the age-class structure drastically. Loss of the young of the year followed by additional loss of older pocket gophers before the subsequent breeding season appears to precipitate some declines (Turner et al. 1973). Population declines are occasionally marked. In the winter of 1958/59, Hansen (1962) found a die-off on the Black Mesa. Howard (1961) recorded a sudden decline in *T. bottae* in San Mateo County, California, where the drop in density in nine improved rangeland pastures ranged from 57 to 92 percent.

ECOLOGY

Habitat. In general, pocket gophers occupy a variety of habitat types, with some species occurring from high mountain meadows to lowland plains and rangelands. Widespread distribution of *T. talpoides,* for example, is attributable to a broad range of soil tolerance; soil factors play a critical role in limiting the distribution of all pocket gophers (Miller 1964; Hansen and Morris 1968).

Preferred soils are usually light in texture, very porous, and have good drainage. The closed burrow system of the pocket gopher makes it necessary for gas exchange to take place through the soil. The peat soils east of San Francisco, California, do not have pocket gophers, although pocket gophers are common in all other soils adjacent to the peat soil. Soils that are continuously wet and/or of a small particle size (such as clay) diffuse gas poorly and are generally avoided (Davis et al. 1938; Davis 1940; Kennerly 1964; Miller 1964; McNab 1966).

Soil depth may affect the local distribution of pocket gophers. Shallow burrows in soils less than 10 cm deep might result in a high number of cave-ins. Shallow soils also prevent pocket gophers from digging deeper to escape high air and ground temperatures. Consequently, the combination of soil depth and local climate influences the local distribution of pocket gophers (Kennerly 1964; McNab 1966; Turner et al. 1973).

The chemical properties of the soil have not been found directly to limit the distribution of pocket gophers (Davis et al. 1938; Kennerly 1964), but may indirectly affect the distribution through plant composition. Even though pocket gophers use a variety of plants for food, an absence of suitable vegetation may exclude the animal from otherwise tolerated soils (Turner et al. 1973).

Burrows. Pocket gophers excavate tunnels by loosening the earth with their long, curved front claws and, when necessary, their large, chisel-shaped incisors. The dirt is pushed backward under their body and then, by turning around within the diameter of their own body, they push the soil to the surface or into a burrow that is no longer used (Grinnell 1923). Although rocks are found in loosened soil brought to the surface, Hansen and Morris (1968) reported that *T. talpoides* usually avoided rocks larger than 2.5 cm.

The diameter of the tunnels, which ranges from 6.4 to 8.9 cm, is usually positively correlated with the size of the animal that dug it (Grinnell 1923; Davis et al. 1938; Smith 1948; Kennerly 1964). The depth of the main tunnel of a burrow system ranges from about 10 to 30 cm underground, depending on soil consistency (Grinnell 1923; Turner et al. 1973; Best 1973). Lateral tunnels frequently extend from the main tunnel and terminate at a plug near the surface. Some tunnels that are plugged at the main tunnel appear abandoned. These short side branches from the main tunnel are used for increased access to the surface for feeding, for storing food (Grinnell 1923; Hansen and Morris 1968), and for disposing of excavated soil. Also, there may be deep, vertical (often spiral) tunnels averaging 46 to 61 cm or more in depth. These connect the main tunnel system to the deeper system, where the enlarged breeding and nesting chambers are found (Tryon 1947). Nests in these areas are spherical, about 25 cm in diameter, and consist of dried grasses. Enlarged chambers found along the shallower main tunnel system are probably resting and feeding stations, since they contain no nesting material (Grinnell 1923; Turner et al. 1973).

Air temperatures of the main tunnels of *T. talpoides* in Colorado were measured from July to September and beneath the snowpack from February to April. The burrow temperatures ranged from 10° to 18° C in summer and from -0.5° to 2° C beneath the snowpack in winter (Turner et al. 1973).

In contrast to ground squirrels (*Spermophilus*) and most other rodents, pocket gophers usually maintain a closed burrow system. A pocket gopher surfaces to push up excavated soil or to explore for food in the immediate area around the burrow opening. Returning to the tunnel, the pocket gopher closes the opening with loosened soil, which forms a visible plug about 3 to 8 cm in diameter (Hansen and Morris 1968). In most cases the loosened soil and rocks from the burrow are the only materials brought to the surface. Old nest materials, rejected food, and fecal material all remain in the burrow system, usually in abandoned and plugged burrows or in currently unused chambers as previously noted. Mounds of excavated soil and earth plugs mark sites where the pocket gopher has surfaced. The mound is typically a fan-shaped pile of soil with an upraised surface marked by concentric rings. The mouth of the burrow is outlined as a circular plug of earth that is lower than the rest of the soil (Grinnell 1923). Those features distinguish it from the volcano-type mound made by moles.

In areas of shallow soils underlain by hardpan or a slowly permeable soil, pocket gophers create unlevel gound surfaces, or "Mima mounds," often regularly spaced and up to 1 m or so in height and 3 to 5 m in diameter (Arkley and Brown 1954; Branson et al. 1965; Ross et al. 1968. Common in western rangeland, these mounds were first attributed strictly to the action of gophers in 1940 (Scheffer and Kruckeberg 1966).

However, they can be formed by other fossorial mammals (Scheffer 1958). Pocket gophers also form isolated smaller mounds by depositing excavated soil under fallen brush, which provides them security from predators (Howard and Childs 1959).

In winter months, especially where the ground is frozen, the tunnel system of northern pocket gophers is extended by the formation of tunnels in the snowpack. The snow tunnels provide a convenient depository for excavated soil, which forms solid earth cores in the snowpack above the ground surface (Grinnell 1923; Hansen and Morris 1968). During the snowmelt, and before much rain, the cores remain intact as they sink to the surface of the ground, remaining as a visible sign of where a winter burrow of a pocket gopher had been. This method of forming tunnels in the snowpack is thought to allow pocket gophers to cross inhospitable environments as well as providing a means of easy winter travel (Marshall 1941). The snowpack tunnels also increase the vertical range in which the pocket gopher can forage in snow-covered months, making aboveground woody plants covered by snow available without danger of predation (Grinnell 1923).

The burrowing rates of pocket gophers appear to vary seasonally. Richens (1966) investigated the digging behavior of one pocket gopher in Utah for five months, during two of which the ground was frozen, and reported that it dug a total of 146 m of tunnels during the five-month period. The 161 mounds that were formed varied from 0 to 14 per day and the average amount of air-dry soil per mound was about 3 kg. He estimated that each pocket gopher moves about 1,130 kg of soil a year. Marshall (1941) noted that *T. talpoides* in Idaho could construct 152 cm of burrow per minute through the snow.

Pocket gophers live mostly in a sealed burrow system and are thus able to survive heavy rain and temporary flooding. Water can enter a tightly plugged pocket gopher burrow only from below. A rising ground water table may force a local population to move (Ingles 1949). Burrows are often inadequately plugged for withstanding the sudden rush of water in flood irrigation, but they withstand sprinkler irrigation very well. Water percolating through soil is prevented from entering the sealed burrows by the capillary attraction of the soil particles. Howard (unpublished data) demonstrated the ability of a pocket gopher to remain dry and to survive for several weeks when buried in a hardware-cloth cylinder just 70–100 mm under the bottom of a pond that was kept flooded. Ample food had been placed in the artificial tunnel, and a small tube provided air.

Burrow Associates. Pocket gopher burrows, especially abandoned ones, also serve as shelter for other animals. Vaughan (1961) reported 22 species of animals that utilized pocket gopher burrows in Colorado: reptiles and amphibian occupants include the tiger salamander (*Ambystoma tigrinum*), spadefoot toad (*Scaphiopus* sp.), ornate box turtle (*Terrapene ornata*), six-lined racerunner (*Cnemidophorus sex-*

lineatus), earless lizard (*Holbrook maculata*), gopher snake (*Pituophis catenifer*), and prairie rattlesnake (*Crotalus viridis*). He further noted that pocket gopher burrows might account for the local distributions of some of the reptiles and the tiger salamander. Mammals that use gopher burrows include the eastern mole (*Scalopus aquaticus*), desert cottontail (*Sylvilagus auduboni*), ground squirrels (*Spermophilus* sp.), Ord's kangaroo rat (*Dipodomys ordii*), deer mouse (*Peromyscus maniculatus*), meadow voles (*Microtus* sp.), and long-tailed weasel (*Mustela frenata*). Vertebrates found less frequently in pocket gopher burrows in Colorado include the plains toad (*Bufo cognatus*), burrowing owl (*Speotyto cunicularia*), northern grasshopper mouse (*Onychomys leucogaster*), and striped skunk (*Mephitis mephitis*).

FOOD HABITS

Pocket gophers are strict herbivores and use a wide variety of plants for food (Miller 1964). They are opportunistic feeders, collecting succulent food aboveground near the entrance of feed holes and feeding on roots or, when necessary, aboveground woody vegetation under the snow. A list of the common food items of *T. talpoides* is given in table 13.1. The Mazama pocket gopher (*T. mazama*) in south-central Oregon prefers perennial forbs during the summer, and throughout the rest of the year prefers the most succulent plants available (Burton and Black 1978). Its diet was mainly aboveground plant parts (40 percent forbs and 32 percent grasses), with the remainder being roots (24 percent) and woody plants (4 percent). Ponderosa pine (*Pinus ponderosa*) and other woody plants were eaten mostly in winter and constituted only a minor portion of the diet.

The food habits of *T. talpoides* have been deter-

TABLE 13.1. Representative common foods of *Thomomys talpoides*

Species	Food
Erigeion speciosus	aspen fleabane
Lupinus spp.	lupines
Chrysopsis villosa	hairy goldaster
Collomia linearis	slenderleaf gilia
Lathyrus leucanthus	aspen peavine
Taxacum offinale	common dandelion
Geranium fremontii	Fremont geranium
Potentilla pulcherrima	beauty cinquefoil
Achillea banulosa	western yarrow
Agosperis spp.	agoseris
Opuntia polyacamtha	plains prickly pear
Stipa comata	needle-and-thread grass
Sphaeralcea coccinea	scarlet globemallow
Agropyron smithii	bluestem wheat grass
Alfalfa spp.	alfalfa
Chrysothamus parryi	parry rabbitbrush

Source: Ward and Keith 1962; Vaughan 1967; Keith et al. 1973.

mined by examining underground food caches (Aldous 1945), by surveying clipped plants (Aldous 1951), and by examining stomach contents (Ward 1960, 1973; Ward and Keith 1962; Vaughan 1967). Those studies show that the northern pocket gopher changes its diet seasonally in response to plant availability and nutritional values. On the Black Mesa in Colorado, the diet of *T. talpoides* was 74 percent stem and leaf material in summer, with that proportion decreasing to 40 percent in October (Ward and Keith 1962). Ward (1960) reported that *T. talpoides* inhabiting alfalfa fields had a diet of 75.5 percent alfalfa stems and leaves in summer, whereas in winter the diet was 96.4 percent alfalfa roots.

The preferred food of *T. talpoides* is perennial forbs. Even where grasses were the dominant vegetative type, forbs still made up 93 percent of the diet, with grasses only 6 percent (Ward and Keith 1962). On some of Colorado's short-grass prairies, Vaughan (1967) found that the total yearly diet was 67 percent forbs and 30 percent grasses. On the Wasatch Plateau in Utah, Aldous (1951) surveyed clipped plants and found that 33 species of forb and 15 species of grass were utilized. The amount of grasses taken is largest during the the growing season of April to June, but even then they constitute only 15 percent by volume of the diet (Ward 1960).

Shrubs, if present, are not a major food source, usually being less than 3 percent of the yearly diet of *T. talpoides* (Ward and Keith 1962; Vaughan 1967). However, Ronco (1967) reported that roots of Engelmann spruce (*Picea engelmann*) and lodgepole pine (*Pinus contorta*) in Colorado were heavily clipped by northern pocket gophers during the winter months.

The water requirements of *T. talpoides* are satisfied at least in part, and most likely entirely, by the succulent vegetation in their diet (Bailey 1895; Tietjen et al. 1967). Vaughan (1967) reported that the succulent plains prickly pear (*Opuntia polyacantha*) was 49.9 percent of the yearly diet of *T. talpoides* on a short-grass prairie in Colorado, and was probably the major source of water. He also found that caged northern pocket gophers could survive well under laboratory conditions on a diet of dry alfalfa pellets and plains prickly pear without any available free water. However, after extensive feeding tests of caged northern pocket gophers, Teitjen et al. (1967) determined that succulent grasses and good food storage structures provided only a limited supportive diet. Howard and Childs (1959) showed that *T. bottae* could not survive in a laboratory on dry rolled oats unless it could also establish a burrow system in damp soil.

The northern pocket gopher stores plant material in food caches, usually located in side chambers. Observations on captive pocket gophers indicate that food caches are usually placed in sealed compartments and the short tunnels leading to them are also plugged. However, these food caches do not necessarily indicate that the plants were taken for food, as evidenced by the higher percentage of root material found in food caches than in stomach analyses (Aldous 1945, 1951;

Turner et al. 1973). The root of the common dandelion (*Taraxacum offinale*), where available, accounts for up to 99 percent of individual caches.

BEHAVIOR

Vocalization. It is not known whether pocket gophers vocalize in the wild, but in captivity they produce a grinding chatter of teeth and make aspirated sounds in combat, squeals of anger when annoyed, and a vocal expression of pain when bitten or caught in a trap (Scheffer 1931; Tryon 1947). Howard and Childs (1959) noted that *T. bottae* was usually silent in captivity but occasionally uttered soft murmurs and squeaks; otherwise, the principal sound produced was a clicking of teeth. In a laboratory housing several dozen *T. bottae,* the clicking of teeth by one animal would occasionally elicit clicking by many other animals.

Swimming Ability. Kennerly (1963) showed that *G. bursarius* were rather vigorous swimmers in placid water. Dispersants could cross narrow waterways, which could affect the gene flow of pocket gophers. With flood irrigation of alfalfa in the west, field workers commonly took a dog along to locate pocket gophers that were swimming or had already swum to a levee. In a study of the effects of floods on *T. bottae,* Williams (1976) found that most individuals were able to survive floods.

Best and Hart (1976) determined that if a flooded area were placid and shallow enough for the average pocket gopher (*Geomys*) to touch the bottom occasionally, the animal could cross more than 50 m of water. In contrast, a more xeric species, *Pappogeomys castanops,* had to be removed after only 130 seconds in water, only a third of the immersion that *G. pinetis* and *G. bursarius* could withstand.

Seasonal Activity. Pocket gophers are active throughout the year (Turner et al. 1973), although some species produce more mounds of excavated earth in the autumn, when the population is highest (Scheffer 1931; Crouch 1933; Miller 1948; Reid et al. 1966; Hickman and Brown 1973). Laycock (1957), observing *T. talpoides* in Jackson Hole, Wyoming, noted little if any surface activity in early summer and late August. Miller and Bond (1960) also found little activity by *T. talpoides* in June and July on the Black Mesa in Colorado but found a peak in activity by late August. They noted that the seasonal trend did not correlate with precipitation or soil moisture but appeared to be related to breeding and feeding behaviors. Miller (1948) reported that the digging rate of *T. bottae* was highest in spring and lowest in summer and early fall when the soil was hot and dry. However, activity rate rose after the first fall rain.

In the foothills of California, Howard and Childs (1959) noted very little surface activity by *T. bottae* from June through September. During this season the animals were known to have retreated to the deeper burrow systems, closing off the shallower ones, which were in hot, dry soil. It is suspected, although not proved, that some pocket gophers may go into short

periods of estivation at this season. In the laboratory, at any rate, they sleep soundly during this period (English 1932) and can sometimes be picked up and removed from their cage before they wake (Howard and Childs 1959).

Considering the diel cycle of pocket gophers, most observers report that they feed or burrow throughout the day. No particular time of the day is most fruitful for trapping them. Under laboratory conditions, Vaughan and Hansen (1961) recorded the daily activity rhythm of *G. bursarius.* They found the average activity period was 36 minutes and the average resting period 1 hour. The pocket gophers studied averaged 16.1 activity periods per day.

Aggression and Competition. Pocket gophers are the only small rodent we know that is so docile that it can be picked up by the tail without biting. Also, pocket gophers placed on the open pan of a dietary scale to be weighed usually will not jump off. Howard has maintained hundreds of *T. bottae* in open-topped cages having sides 30.5 cm high. The only occasion when any jumped out was when three young females, caged together, matured sexually. They began fighting, and two managed to jump out (Howard and Childs 1959).

Except during the breeding season, pocket gophers are highly territorial and extremely intolerant of each other, regardless of sex and age. Cahalane (1947) noted that intraspecific aggressive behavior usually begins with an outward display of grinding and chattering of teeth. When artificially confined together, fighting usually ensued within minutes or a day at the most. The heavier animal usually killed the other. Although English (1932) was able to cage *G. bursarius* together, Howard and Childs (1959) were unable to get more than one *T. bottae* to survive in a 7.8 m² glass-topped arena. In experimental bouts between two captive strangers, Anderson (1978) noted that some pocket gophers actually groomed each other. Once he left two males together for as long as 45 minutes without their fighting.

During the breeding season, plural occupancy of pocket gopher burrows by various combinations of adult males, adult females, and young is not uncommon (Hansen and Miller 1959; Miller and Bond 1960; Vaughan and Hansen 1964; Baker 1974). These intraspecific associations coincide with periods of surface inactivity. It is thought that the territorial boundaries of northern pocket gophers, *T. talpoides,* are relaxed during the breeding season (Miller and Bond 1960). Territories are then reestablished by September, remaining mutually exclusive again until the next spring. Other species of pocket gopher breed year round and consequently are more tolerant of plural occupancy throughout the year (Miller 1946).

Howard and Childs (1959) sometimes caught a male and a female *T. bottae* or two females from the same burrow on the same day during the breeding season. Some multiple catches were from two traps set side by side. Others were obtained by setting a second trap in those burrows that were plugged while a first animal had been temporarily removed from the plot to be examined. Vaughan (1962) found multiple occupancy highest in early March, when 42 percent of the *G. bursarius* trapped were multiple occupants. However, these traps had sometimes remained set for three days, which casts doubt on the multiple-occupancy figures because neighbors quickly enter another burrow system once the resident disappears. This was demonstrated dramatically when Reichman and Baker (1972) trapped a *Thomomys* from a burrow that had yielded a *Pappogeomys* only four hours earlier.

Evidently *T. bottae* and *T. talpoides* are not sympatric, at least not in Colorado. Both intraspecific and interspecific competition may account for mutually exclusive local distributions. Interspecific relations between *T. talpoides* and *T. bottae* have been studied by several investigators. Miller (1964) concluded that the eastern distribution of *T. talpoides* in Colorado is restricted by the competitive exclusion ability of *T. bottae*. Baker 1974, however, reported that *T. talpoides* is aggressively superior to *T. bottae,* so aggression is unlikely to be the lone mode of competitive interference. Vaughan and Hansen (1964) stated that the most important differences between these two species were the greater dispersal prowess of *T. talpoides*— facilitating immigration, emigration, and reproductive success at low population densities—and its slightly greater breadth of environmental toleration.

MORTALITY

Parasites. A wide variety of parasites are associated with pocket gophers. Table 13.2 lists those endoparasites and ectoparasites found on *T. talpoides*. The ectoparasites found on *T. talpoides* in Colorado include 2 species of louse, 1 species of flea, 1 species of tick, and 11 species of mite (Miller and Ward 1960). Botfly larvae (*Cuterebra* sp.) have also been found on this animal (Richens 1965a). In Indiana, Tuszynski and Whitaker (1972) found that all of the *Geomys bursarius* they trapped had lice, and 30.6 percent had fleas. Five species of mite were abundant. On *T. talpoides* examined by Tryon (1947) fleas were abundant,

lice were found sometimes in great numbers, and early stages of ticks were present on many animals. Howard and Childs (1959) found no ticks or fleas on *T. bottae,* but 1 species of louse and 3 species of mite were common, as were tapeworms. With regard to nematode infections, Todd et al. (1971) found that out of 46 *T. talpoides* in Wyoming 81 percent had *Ransomus rodentorum,* 65 percent had *Trichuris fossor,* and 39 percent had *Capillaria hepatica.* The number having cestodes was 46 percent *Paranoplocephala* sp. and 38 percent *Paranoplocephela variabilis.* So far, most studies of parasites of pocket gophers have been incidental to other aspects of their biology.

Predators. The effect of vertebrate predators on the density of a pocket gopher population is not well understood. Because pocket gophers are mostly fossorial, they are subject to predation mostly while aboveground. This occurs when feeding near a feed hole, pushing excavated earth into a mound, dispersing aboveground, or being temporarily aboveground because of snowmelt, flood irrigation, or through interaction with a more aggressive gopher. Several important predators of pocket gophers are given in table 13.3.

The principal regulatory factor determining the density of pocket gopher populations is the suitability of the habitat—the food, cover, soil type, and moisture. The Colorado Cooperative Pocket Gopher Project (1960) concluded that the rate of predation was not sufficient to influence the size of the gopher population. Hansen and Ward (1966) determined that weasel predation may slow pocket gopher increases but does not prevent large populations from developing. Pocket gophers are more important as a prey item to predators than predators are as a controlling factor of gophers. Competition between pocket gophers and other intrinsic factors are more important than predators in regulating pocket gopher numbers (Howard and Childs 1959). When Howard and Childs observed the interaction of rattlesnakes and pocket gophers in earth tunnels behind glass, it became apparent that a snake is too confined to strike while in a burrow. Even

TABLE 13.2. Endoparasites and ectoparasites reported to infest *Thomomys talpoides*

Endoparasite	Ectoparasite
Protozoa	Diptera (botfly)
Eimeria thomomysis	*Cuterebra* sp.
Eimeria fitzgeraldi	Acarina (mites and ticks)
Cestoda	*Haemogamasus ambulans*
Cysticerci	*Hirstionyssus geomydis*
Paranoplocephala infrequens	*Haemolaelaps geomys*
Paranoplocephala variabilis	*Ixodes sculptus*
Nematoda	Mallophaga (lice)
Ransomus rodentorum	*Geomydoecus thomoyus*
Longistriata vexillata	*Geomydoecus chapini*
Protospirura ascaroidea	Siphonaptera (flea)
Trichuris fossor	*Foxella ignota*
Capillaria hepatica	

SOURCE: Miller and Ward 1960; Richens 1965a; Todd et al. 1971.

TABLE 13.3. Common predators of pocket gophers

Common Name	Scientific Name
Coyotes	*Canis latrans*
Fox	*Vulpes* sp. and *Urocyon cinereoargenteus*
Bobcat	*Felis rufus*
Badger	*Taxidea taxus*
Skunk	*Mephitis* sp. and *Spilogale* sp.
Weasel	*Mustela* sp.
House cat	*Felis domesticus*
Red-tailed hawk	*Buteo jamaicensis*
Swainson's hawk	*Buteo swainsoni*
Ferruginous hawk	*Buteo regalis*
Goshawk	*Accipiter gentilis*
American kestrel	*Falco sparverius*
Great-horned owl	*Bubo virginianus*
Long-eared owl	*Asio otus*
Great gray owl	*Strix nebulosa*
Burrowing owl	*Speotyto cunicularia*
Barn owl	*Tyto alba*
Gopher and bull snakes	*Pituophis* sp.
Rattlesnakes	*Crotalus* sp.

SOURCE: Hisaw and Gloyd 1926; Tryon 1943, 1947; Imler 1945; Fitch et al. 1946; Evans and Emlen 1947; Fitch 1947; Craighead and Craighead 1956; Howard and Childs 1959; Vaughan 1961; Hansen and Ward 1966; Marti 1969; Fitzner et al. 1977.

though the pocket gopher gave way to the rattlesnake, it was not very frightened, and the pocket gopher repeatedly attempted to drive the snake away by pushing dirt toward it.

Avian predation appeared to be somewhat more efficient as a control measure. Fitzner et al. (1977) found that ferruginous hawks (*Buteo regalis*) most frequently preyed upon *T. talpoides*. Studies in Utah (Kimball et al. 1970) indicated a significant reduction of pocket gopher populations that were in close proximity to roosting sites of birds of prey.

RELATIONSHIP TO ENVIRONMENT

Pocket gophers have influenced the western rangelands of North America since the Pliocene. Their effect appears to vary with population size, nature of the habitat, season, and land use practices (Turner et al. 1973). The exposure of subsoil and rocks within the burrow to air, water, and solvents may hasten the weathering of the substratum. Subsoil brought to the surface is also weathered. The constant digging action counteracts the packing effect of grazers and makes the soil more porous, slowing spring run-off as well as providing more aeration for plant roots. The constant vertical cycling of the soil, with increased exposure and mineralization, increases soil fertility (Grinnell 1923). Grinnell (1923) also suggested that vegetation stored in the ground as well as the fecal materials present increased the humus content of the soil. In some pocket gopher mounds, rocks larger than 0.64 cm in diameter made up 40 percent of the excavated material (Hansen and Beck 1968). Rocks up to 2.5 cm in diameter are excavated (Hansen and Morris 1968). Branson et al. (1965) noted that digging in rocky soils decreased the bulk density and increased the soil volume.

Richens (1966) estimated that 74 pocket gophers located in one hectare could collectively move 93.7 tons of soil in one year.

Mielke (1977), assessing the role of fossorial rodents such as geomyids, concluded that pocket gophers were a dynamic force in determining the biochemical attributes of the North American prairie lands; the burrowing activities of such fossorial rodents may provide an explanation for the genesis of these prairie soils. He stated that the bison (*Bison bison*) grazed and trampled the dense prairie vegetation, accelerating forb growth, on which the gophers thrived. The gopher, in turn, worked the soil, thus increasing soil fertility and stimulating vegetative growth, increasing food for the bison.

Although the increased mixing and aerating of soils has been linked with accelerated soil erosion on overgrazed ranges (Ellison 1946), and with gulley formation in other soils (Longhurst 1957), constant soil displacement deepens the soil mulch and must be beneficial to soil formation and vegetative growth over time. However, it is not easy to evaluate how the soil formation in the northeastern United States, where there are no pocket gophers, might have been different if Geomyidae had been present. Uneaten vegetation, the prolific plant roots, the myriad of bacteria, protozoans, worms, crustaceans, arachnids, insects, and other small animals together may be able to be even more effective in soil formation if gophers are not present (Howard and Childs 1959).

Laycock and Richardson (1975) compared the conditions of the soils and the types of vegetation on areas grazed by sheep with those of two enclosures protected from sheep, one of which was kept free of *T. talpoides* for 31 years. Pocket gopher densities have often been found to increase on ranges in this area if depleted by overgrazing and thus to perpetuate large amounts of ephemeral and annual plants (Richens 1965*b*). In general, the pocket gophers tend to keep such ranges in poor condition even after grazing pressure has been reduced or eliminated. Annuals were scarce in the exclosures without sheep or pocket gophers for 31 years though they were maintained for this long in the sheep exclosure that had pocket gophers (Laycock and Richardson 1975). Where pocket gophers were present, noncapillary porosity, organic matter, total nitrogen, and total phosphorus were higher, but the increased soil fertility was not reflected in total peak standing crops. The increased fertility may have been offset by damage to vegetation by consumption and burrowing activities of the pocket gophers.

The presence of large populations of pocket gophers has often caused changes in the plant composition. Tietjen et al. (1967) found a marked drop in pocket gophers if forbs were chemically removed. Ellison and Aldous (1952) reported that where *T. talpoides* was present in Utah, the common dandelion (*Taxaxacum officinale*) markedly decreased and rhizomatous species, grasses, sedges, and forbs increased. As Turner et al. (1973) noted, pocket gophers often grow their own food by causing soil conditions that tend to favor continual growth of the plants they commonly eat.

Mima-mound formation (see the section "Burrows") also has a considerable effect on rangelands. Besides creating a microrelief that favors adult pocket gopher survival (Hansen 1962; Allgood and Gray 1974), Mima mounds support more vegetation than do the spaces between mounds, which have lost soil. McGinnies (1960) found that the Mima-mound soils in Colorado have a much higher content of organic matter and a greater volume of soil for water storage above shallow bedrock. Laycock (1958) investigated the revegetation of mounds and suggested that the mounds are microsites where early plant successional species (therophytes) grow continually, even in climax communities.

ECONOMIC STATUS AND MANAGEMENT

Pocket gophers are interesting animals to biologists and others, but can cause great economic loss to the farmer, rancher, and forester. They are frequently a considerable nuisance to the home gardner (Scheffer 1931). It takes only one animal to cause a ditchbank or dike to wash out or to cause land slips at a housing development. One or more pocket gophers can be costly in orchards, vineyards, and truck gardens; and, at least in California and the Pacific Northwest, they pose the principal animal-damage problem in reforestation (Capp 1976).

According to Ward (1973), where they are abundant, rangeland pocket gophers may reduce herbage yield 20 percent or more by harvesting and burying vegetation, and may be the primary cause of exposure of bare soil. The reduction of grasses and forbs by uncontrolled populations of pocket gophers results in

less plant material for livestock grazing, and on overgrazed rangeland the combination of pocket gophers and livestock can create severe erosion problems (Laycock and Richardson 1975). The formation of Mima-mounds may make range management difficult because of the transferral of so much of the soil into mounds. In California, pocket gophers compete with livestock for range forage more vigorously than California ground squirrels (*Spermophilus beecheyi*), especially because there are more pocket gophers than squirrels (Fitch and Bentley 1949). Pocket gophers can cause further deterioration and hamper the improvement of mountain rangeland and meadows already in poor range condition (Moore and Reid 1951; Julander et al. 1969). Their incessant gnawing can damage plastic irrigation pipe and buried electric cables (Howard 1953*b*; Connolly and Landstrom 1969). Pocket gophers may be beneficial to foresters by working the soil, but they also feed on tree roots, girdle stems, and, when under snow, damage stems and branches a meter or more high (Dingle 1956; Hooven 1971; Barnes 1973).

Information on methods of controlling pocket gophers (Tietjen 1973; Marsh and Howard 1978) has long been available from most western states and the federal government. For a review of pocket gopher control methods, see Barnes (1973). These methods include the use of Macabee, box, and other kill traps; the placing of poison baits in tunnels; and, as Miller (1954) demonstrated, the less effective method of inserting toxic gases and gas bombs into pocket gopher burrows. Since all entrances are usually plugged, the gases do not rapidly penetrate all the tunnels. The pocket gophers wall off a portion of their burrow system before lethal amounts of the gases can reach them.

The most practical and efficient method of controlling large numbers of pocket gophers is to use toxic bait. With a pointed stick, rod. or commercial gopher probe, a trowel, or a shovel, gopher tunnels can be located and poison bait applied. The opening into the tunnel must be completely closed with soil after baiting. Baiting consists of a spoonful of treated grain or cut vegetables inserted in the runway.

The effect of spraying ranges and vegetation with herbicides has been examined in several studies (Keith et al. 1959; Hansen and Ward 1966; Tietjen et al. 1967). The herbicide 2,4-D (2,4-dichlorophenoxyacetic acid) has been intensively investigated to determine its relationship to the vegetation and pocket gopher populations. At the proper dose, it is extremely effective in reducing perennial forbs within one year and usually resulted in an increase in grass yields. Keith et al. (1959) reported that forb production was reduced as much as 83 percent, while grass production increased 37 percent after spraying with 2,4-D. They further noted that the diet of the northern pocket gophers in the sprayed area consisted of increasing amounts of grasses and that the pocket gopher population after one year of treatment was reduced 87 percent. Tietjen et al. (1967) determined that the population reductions resulted from the marginal diet offered by the increased proportion of grasses present, and not from any toxic

FIGURE 13.7. Pocket gophers can be effectively controlled by using a machine to meter poison grain into an artificial burrow that intercepts some of the natural gopher burrow systems.

reaction to eating the herbicide-treated vegetation, nor from emigration from the sprayed areas. However, this method is no longer practical because of application costs and registration difficulties. Current federal government (Environmental Protection Agency) policies concerning animal-damaging toxicants have also restricted or discouraged the baiting of pocket gophers with toxicants such as strychnine and 1080 (sodium monofluoroacetate).

To modernize and mechanize pocket gopher control for alfalfa and other agricultural crops, Kepner and Howard (1960) developed the mechanical pocket gopher bait applicator, a means of placing toxic bait in artificially constructed burrows formed by pulling a "mole" or burrow-building device behind a tractor or other vehicle having a three-point hitch (figure 13.7). The burrow builder was later modified by the Colorado Cooperative Pocket Gopher Project for use in mountains and Colorado conditions (Ward and Hansen 1960), and similar devices are now widely used even in some forests (Marsh and Cummings 1968; Barnes et al. 1970). In spite of present knowledge of how to control pocket gophers, much more information is needed on their ecology (Howard and Ingles 1951) for an even better understanding of their true role in environments disturbed by people.

LITERATURE CITED

Aldous, C. M. 1945. Pocket gopher food caches in central Utah. J. Wildl. Manage. 9(4):327-328.

———. 1951. The feeding habits of the pocket gopher *Thomomys talpoides moseio*, on the high mountain ranges of central Utah. J. Mammal. 32(1):84-87.

———. 1957. Fluctuations in pocket gopher populations. J. Mammal. 38(2):266-267.

Allgood, F. P., and Gray, F. 1974. An ecological interpretation for the small mounds in landscapes of eastern Oklahoma. J. Environ. Qual. 3(1):37-41.

Anderson, D. C. 1978. Observations on reproduction, growth, and behavior of the northern pocket gopher (*Thomomy talpoides*). J. Mammal. 59(2):418-422.

Arkley, R. J., and Brown, H. C. 1954. The origin of Mima mound (hogwallow) microrelief in the far western states. Proc. Soil Sci. Soc. Am. 18:195-199.

Bailey, V. 1895. The pocket gophers of the United States. U.S.D.A., Div. Ornithol. and Mammal. Bull. 5. 47pp.

Baker, A. E. 1974. Interspecific aggressive behavior of pocket gophers *Thomomys bottae* and *T. talpoides* (Geomyidae: Rodentia). Ecology 55(3):671-673.

Barnes, V. G., Jr. 1973. Pocket gophers and reforestation in the Pacific Northwest: a problem analysis. U.S. Fish Wildl. Serv. Special Sci. Rep. Wildl. 155. 18pp.

Barnes, V. G., Jr.; Martin, P.; and Tietjen, H. P. 1970. Pocket gopher control on Oregon ponderosa pine plantations. J. For. 68(7):433-435.

Best, T. L. 1973. Ecological separation of three genera of pocket gophers (Geomyidae). Ecology 54(6)1131-1319.

Best, T. L., and Hart, E. B. 1976. Swimming ability of pocket gophers (Geomyidae). Texas J. Sci. 27(3):361-366.

Bradley, W. G.; Miller, J. S.; and Yousef, M. K. 1974. Thermoregulatory pattern in pocket gophers: desert and mountain. Physiol. Zool. 47(3):172-179.

Branson, F. A.; Miller, R. A.; and McQueen, T. S. 1965. Plant communities and soil moisture relationships near Denver, Colorado. Ecology 46(3):311-319.

Burton, D. H., and Black, H. C. 1978. Feeding habits of Mazama pocket gophers in south-central Oregon. J. Wildl. Manage. 42(2):383-390.

Cahalane, V. H. 1947. Mammals of North America. MacMillan Co., New York. 682pp.

Capp, J. C. 1976. Increasing pocket gopher problems in reforestation. Pages 221-228 *in* Proc. 7th Vertebr. Pest Conf., Univ. California, Davis.

Colorado Cooperative Pocket Gopher Project. 1960. Pocket gophers in Colorado. Colorado State Univ. Agr. Exp. Stn. Bull. 508-S. 26pp.

Connolly, R. A., and Landstrom, R. E. 1969. Gopher damage to buried electric cable materials. Am. Soc. for Testing and Materials, Materials Res. and Standards 9(12):13-18.

Craighead, J. J., and Craighead, F. C., Jr. 1956. Hawks, owls, and wildlife. Stackpole Co., Harrisburg, Pa., and Wildl. Manage. Inst., Washington, D.C. 443pp.

Crouch, W. E. 1933. Pocket gopher control. U.S.D.A., Farmer's Bull. 1709. 21pp.

Davis, W. B. 1940. Distribution and variation of pocket gophers (Genus *Geomys*) in the southwestern United States. Texas A & G Coll., Texas Agric. Exp. Stn. Bull. 590. 38pp.

Davis, W. B.; Ramsey, R. R.; and Arendale, J. M., Jr. 1938. Distribution of pocket gophers (*Geomys breviceps*) in relation to soils. J. Mammal. 19(4):412-418.

Dingle, R. W. 1956. Pocket gophers as a cause of mortality in eastern Washington pine plantation. J. For. 54(12):832-835.

Ellison, L. 1946. The pocket gopher in relation to soil erosion on mountain range. Ecology 27(2):101-114.

Ellison, L., and Aldous, C. M. 1952. Influence of pocket gophers on vegetation of subalpine grassland in central Utah. Ecology 33(2):177-186.

English, P. R. 1932. Some habits of the pocket gopher, *Geomys breviceps breviceps*. J. Mammal. 13:126–132.

Evans, F. C., and Emlen. J. T., Jr. 1947. Ecological notes on the prey selected by a barn owl. Condor 49(1):3–9.

Fitch, H. S. 1947. Predation by owls in the Sierran foothills of California. Condor 49(4):137–151.

Fitch, H. S., and Bentley, J. R. 1949. Use of California annual-type forage by range rodents. Ecology 30:306–321.

Fitch, H. S.; Swenson, F.; and Tillotson, D. F. 1946. Behavior and food habits of the red-tailed hawk. Condor 48(5):205–237.

Fitzner, R. E.; Berry, D.; Boyd, L. L.; and Riech, C. A. 1977. Nesting of ferruginous hawks (*Buteo regalis*) in Washington 1974–75. Condor 70(2):245–249.

Gettinger, R. D. 1975. Metabolism and thermoregulation of a fossorial rodent, the northern pocket gopher (*Thomomys talpoides*). Physiol. Zool. 48(4):311–322.

Grinnell, J. 1923. The burrowing rodents of California as agents in soil formation. J. Mammal. 4(3)137–149.

Hall, E. R., and Kelson, K. R. 1959. The mammals of North America. Vol. 1. Ronald Press Co., New York. 546pp.

Hansen, R. M. 1960. Age and reproductive characteristics of mountain pocket gophers in Colorado. J. Mammal. 41(3):323–335.

———. 1962. Movements and survival of *Thomomys talpoides* in a Mima-mound habitat. Ecology 43(1):151–154.

———. 1965. Pocket gopher density in an enclosure of native habitat. J. Mammal. 46(3):508–509.

Hansen, R. M., and Bear, G. D. 1964. Comparison of pocket gophers from alpine, subalpine and shrub grassland habitats. J. Mammal. 45(4):638–640.

Hansen, R. M., and Beck, R. F. 1968. Habitat of pocket gopher in Cochetopa Creek drainage, Colorado. Am. Midl. Nat. 79(1):103–117.

Hansen, R. M., and Miller, R. S. 1959. Observations on the plural occupancy of pocket gopher burrow systems. J. Mammal. 40(3):577–584.

Hansen, R. M., and Morris, M. J. 1968. Movement of rocks by northern pocket gophers. J. Mammal. 49(3):391–399.

Hansen, R. M., and Reid, V. H. 1973. Kinds and distribution of pocket gophers. Colorado State Univ. Agric. Exp. Stn. Tech. Bull. 554S:1–19.

Hansen, R. M., and Remmenga, E. E. 1961. Northeast neighbor concept applied to pocket gopher populations. Ecology 42(4):812–814.

Hansen, R. M., and Ward, A. L. 1966. Some relations of pocket gophers to rangelands on Grand Mesa, Colorado. Colorado State Univ. Agric. Exp. Stn. Tech. Bull. 88. 22pp.

Hart, E. B. 1978. Karyology and evolution of the plains pocket gopher, *Geomys bursarius*. Univ. Kansas. Mus. Nat. Hist. Occas. Pap. 71:1–20.

Hickman, G. C., and Brown, L. N. 1973. Pattern and rate of mound production in the southeastern pocket gopher (*Geomys pinetis*). J. Mammal. 54(4):971–975.

Hill, J. E. 1937. Morphology of the pocket gopher mammalian genus *Thomomys*. Univ. California Publ. Zool. 42:81–171.

Hisaw, F. L. 1924. The absorption of the pubic symphysis of the pocket gopher, *Geomys bursarius* (Shaw). Am. Nat. 58(654):93–96.

———. 1925. The influence of the ovary on the resorption of the pubic bones of the pocket gopher, *Geomys bursarius* (Shaw). J. Exp. Zool. 42:411–433.

Hisaw, F. L., and Gloyd, H. K. 1926. The bullsnake as a natural enemy of injurious rodents. J. Mammal. 7(3):200–205.

Hoffmeister, D. F. 1971. Mammals of Grand Canyon. Univ. Illinois Press, Urbana. 115pp.

Holliger, C. D. 1916. Anatomical adaptations in the thoracic limb of the California pocket gopher and other rodents. Univ. California Publ. Zool. 13:447–494.

Hooven, E. F. 1971. Pocket gopher damage on ponderosa pine plantations in southwestern Oregon. J. Wildl. Manage. 35(2):346–353.

Howard, W. E. 1953a. Growth rate of nails of adult pocket gophers. J. Mammal, 34(3):394–396.

———. 1953b. Tests of pocket gophers gnawing electric cables. J. Wildl. Manage. 17(3):296–300.

———. 1960. Innate and environmental dispersal of individual vertebrates. Am. Midl. Nat. 63(1):152–161.

———. 1961. A pocket gopher population crash. J. Mammal. 42(2):258–260.

Howard, W. E., and Childs, H. E., Jr. 1959. Ecology of pocket gophers with emphasis on *Thomomys bottae mewa*. Hilgardia 29(7):277–358.

Howard, W. E., and Ingles, L. G. 1951. Outline for an ecological life history of pocket gophers and other fossorial mammals. Ecology 32(3):537–544.

Howard, W. E., and Smith, M. E. 1952. Rate of extrusive growth of incisors of pocket gophers. J. Mammal. 33(4):485–487.

Imler, R. H. 1945. Bullsnakes and their control on a Nebraska wildlife refuge. J. Wildl. Manage. 9(4):265–273.

Ingles, L. G. 1949. Ground water and snow as factors affecting the seasonal distribution of pocket gophers, *Thomomys monticola*. J. Mammal. 30(4):343–350.

———. 1952. The ecology of the mountain pocket gopher, *Thomomys monticola*. Ecology 33(1)87–95.

Jones, J. K., Jr.; Carter, D. C.,; and Genoways, H. H. 1975. Revised checklist of North American mammals north of Mexico. Texas Tech. Univ. Mus. Occas. Pap. 28. 14pp.

Julander, O.; Low, J. B.; and Morris, O. W. 1969. Pocket gophers on seeded Utah mountain range. J. Range Mange. 22(5):325–329.

Keith, J. O.; Hansen, R. M.; and Ward, A. L. 1959. Effect of 2,4-D on abundance of foods of pocket gophers. J. Wildl. Manage. 23(2):137–145.

Kennerly, T. E., Jr. 1963. Gene flow and swimming ability in the pocket gopher. Southwestern Nat. 8:85.

———. 1964. Microenvironmental conditions of the pocket gopher mounds. Texas J. Sci. 16(4):395–441.

Kepner, R. A., and Howard, W. E. 1960. Gopher-bait applicator. Univ. California, California Agric. 14(3):7, 14.

Kimball, J.; Paulson, T. A.; and Savage, W. F. 1970. An observation of environmental rodent control. U.S. For. Serv. Range Impr. Notes 15(2):8–9.

Laycock, W. A. 1957. Seasonal periods of surface inactivity of the pocket gopher. J. Mammal. 38(1):132–133.

———. 1958. The initial pattern of revegetation of pocket gopher mounds. Ecology 39(2):346–351.

Laycock, W. A., and Richardson, B. Z. 1975. Long-term effects of pocket gopher control on vegetation and soils of a subalpine grassland. J. Range Manage. 28(6):458–462.

Longhurst, W. M. 1957. A history of squirrel burrow gully formation in relation to grazing. J. Range Manage. 10(4):182–184.

Marsh, R. E., and Cummings, M. W. 1968. Pocket gopher control with mechanical bait applicator. Univ. California Agric. Ext. Serv., AXT-261. 8pp.

Marsh, R. E., and Howard, W. E. 1978. Vertebrate pest

control manual: pocket gophers. Pest Control 46(3):30–34.

Marshall, W. H. 1941. *Thomomys* as burrowers in the snow. J. Mammal. 22(2):196–197.

Marti, C. D. 1969. Some comparisons of the feeding ecology of four owls in northcentral Colorado. Southwestern Nat. 14(2):163–170.

McGinnies, W. J. 1960. Effect of Mima-type microrelief on herbage production of five seeded grasses in Western Colorado. J. Range Manage. 13:231–234.

McNab, B. K. 1966. The metabolism of fossorial rodents: A study of convergence. Ecology 47(5):712–733.

Merriam, C. H. 1906. Monographic revision of the pocket gopher family Geomyidae, exclusive of the species of *Thomomys*. North Am. Fauna 8. 262pp.

Mielke, H. W. 1977. Mound building by pocket gophers (Geomyidae): their impact on soils and vegetation in North America. J. Biogeog. 4:171–180.

Miller, M. A. 1946. Reproductive rates and cycles in the pocket gophers. J. Mammal. 27(4):335–358.

———. 1948. Seasonal trends in burrowing of pocket gophers (*Thomomys*). J. Mammal. 29(1):38–44.

———. 1952. Size characteristics of the Sacramento Valley pocket gopher (*Thomomys bottae navus* Merriam). J. Mammal. 33(4):442–456.

———. 1954. Poison gas tests on gophers. Univ. California, California Agric. 8(10):7,14.

———. 1957. Burrows of the Sacramento Valley pocket gopher in flood-irrigated alfalfa fields. Hilgardia 26(8):431–452.

Miller, R. S. 1958. Rate of incisor growth in the mountain pocket gopher. J. Mammal. 39(3):380–385.

———. 1964. Ecology and distribution of pocket gophers (Geomyidae) in Colorado. Ecology 45(2):256–272.

Miller, R. S., and Bond, H. E. 1960. The summer burrowing activity of pocket gophers. J. Mammal. 41(4):469–475.

Miller, R. S., and Ward, P. A. 1960. Ectoparasites of pocket gophers from Colorado. Am. Midl. Nat. 64(2):382–391.

Moore, H. W., and Reid, E. H. 1951. The Dalles pocket gopher and its influence on the forage production of Oregon mountain meadows. U.S.D.A. Circ. 884. 36pp.

Morejohn, G. V., and Howard, W. E. 1956. Molt in the pocket gopher, *Thomomys bottae*. J. Mammal. 37(2):201–213.

Nevo, E.; Kim, Y. J.; Shaw, C. R., and Thaeler, C. S., Jr. 1974. Genetic variation, selection and speciation in *Thomomys talpoides* pocket gophers. Evolution 28(1):1–23.

Patton, J. L.; Selander, R. K.; and Smith, M. H. 1972. Genetic variations in hybridizing populations of gophers (Genus *Thomomys*). Syst. Zool. 21(3):263–270.

Patton, J. L., and Yang, S. Y. 1977. Genetic variations in *Thomomys bottae* pocket gophers: macrogeographic patterns. Evolution 31(4):697–720.

Phillips, P. 1936. The distribution of rodents in overgrazed and normal grasslands of central Oklahoma. Ecology 17(4):673–679.

Reichman, O. J., and Baker, R. J. 1972. Distribution and movements of two species of pocket gophers (Geomyidae) in an area of sympatry in the Davis Mountains, Texas. J. Mammal. 53(1):21–23.

Reid, V. H. 1973. Population biology of the north pocket gopher. Colorado State Univ. Exp. Stn. Bull. 554S:21–24.

Reid, V. H.; Hansen, R. M.; and Ward, A. L. 1966. Counting mounds and earth plugs to census mountain pocket gophers. J. Wildl. Manage. 30(2):327–334.

Richens, V. B. 1965a. Larvae of botfly in Northern pocket gopher. J. Mammal. 46(4):689–690.

———. 1965b. An evaluation of control of the Wasatch pocket gopher. J. Wildl. Manage. 29(3):413–425.

———. 1966. Notes on the digging activity of the northern pocket gopher. J. Mammal. 47(3):531–533.

Ronco, R. 1967. Lessons from artificial regeneration studies in a cut-over beetle-killed spruce stand in western Colorado. U.S. For. Serv. Res. Note RM-90. 8pp.

Ross, B. A.; Tester, J. R.; and Breckenridge, W. J. 1968. Ecology of mima-type mounds in northwestern Minnesota. Ecology 49(1):172–177.

Russell, R. J. 1968. Evolution and classification of the pocket gophers of the subfamily Geomyinae. Univ. Kansas Mus. Nat. Hist. Publ. 16(6):473–579.

Scheffer, T. H. 1931. Habits and economic status of the pocket gophers. U.S.D.A. Tech. Bull. 224. 26pp.

Scheffer, V. B. 1958. Do fossorial rodents originate Mima type microrelief? Am. Midl. Nat. 59(2):505–510.

Scheffer, V. B., and Kruckeberg, A. 1966. The Mima mounds. BioScience 16(11):800–801.

Schramm, P. 1961. Copulation and gestation in the pocket gopher. J. Mammal. 42(2):167–170.

Smith, D. F. 1948. A burrow of the pocket gopher (*Geomys bursarius*) in eastern Kansas. Trans. Kansas Acad. Sci. 51:313–315.

Storer, T. I., and Gregory, P. W. 1934. Color aberrations in the pocket gopher and their probable genetic explanation. J. Mammal. 15(4):300–312.

Thaeler, C. S., Jr. 1974. Four contacts between ranges of different chromosome forms of the *Thomomys talpoides* complex (Rodentia: Geomyidae). Syst. Zool. 23(3):343–354.

Tietjen, H. P. 1973. Control of pocket gophers. Colorado State Univ. Exp. Stn. Bull. 554S:73–81.

Tietjen, H. P.; Halvoran, C. H.; Hegdal, P. L.; and Johnson, A. M. 1967. 2,4-D herbicide, vegetation, and pocket gopher relationships; Black Mesa, Colorado. Ecology 48(4):634–643.

Todd, K. S.; Lepp, K. L.; and Arch, C. 1971. Endoparasites of the northern pocket gopher from Wyoming. J. Wildl. Dis. 7(2):100–104.

Tryon, C. A., Jr. 1943. The great gray owl as a predator of pocket gophers. Wilson Bull. 55(2):130–131.

———. 1947. The biology of the pocket gopher (*Thomomys talpoides*) in Montana. Montana Agric. Exp. Stn. Tech. Bull. 448. 3pp.

———. 1951. The use of skull measurements at the subspecific level in mammalian taxonomy, with special reference to *Thomomys talpoides*. J. Mammal. 32(3):313–318.

Tryon, C. A., and Cunningham, H. N. 1968. Characteristics of pocket gophers along an altitudinal transect. J. Mammal. 49(4):699–705.

Turner, G. T. 1973. Responses of mountain grassland vegetation to gopher control, reduced grazing, and herbicide. J. Range Manage. 22(6):377–383.

Turner, G. T.; Hansen, R. M.; Reid, V. H.; Tietjen, H. P.; and Ward, A. L. 1973. Pocket gophers and Colorado mountain rangeland. Colorado State Univ. Exp. Stn. Bull. 554S. 90pp.

Tuszynski, R. C., and Whitaker, J. O., Jr. 1972. External parasites of pocket gophers, *Geomys bursarius*, from Indiana. Am. Midl. Nat. 87(2):545–548.

Vaughan, T. A. 1961. Vertebrates inhabiting pocket gopher burrows in Colorado. J. Mammal. 42(2):171–174.

———. 1962. Reproduction in the plains pocket gopher in Colorado. J. Mammal. 43(1):1–13.

———. 1963. Movements made by two species of pocket gophers. Am. Midl. Nat. 69:367–372.

_____. 1967. Food habits of the northern pocket gopher on shortgrass prairie. Am. Midl. Nat. 77(1):176–189.

Vaughan, T. A., and Hansen, R. M. 1961. Activity rhythm of the plains pocket gopher. J. Mammal. 42(4):541–543.

_____. 1964. Experiments on interspecific competition between 2 species of pocket gophers. Am. Midl. Nat. 72(2):444–452.

Ward, A. L. 1960. Mountain pocket gopher food habits in Colorado. J. Wildl. Manage. 24(1):89–92.

_____. 1973. Food habits and competition. Colorado State Univ. Agric. Exp. Stn. Bull. 554S:43–49.

Ward, A. L., and Hansen, R. M. 1960. The burrow-builder and its use for control of pocket gophers. U.S. Fish Wildl. Serv. Spec. Sci. Rep. Wildl. 47. 7pp.

Ward, A. L., and Keith, J. O. 1962. Feeding habits of pocket gophers in mountain grasslands, Black Mesa, Colorado. Ecology 43(4):744–749.

Wight, H. M. 1930. Breeding habits and economic relations of the Dalles pocket gopher. J. Mammal. 11(1):40–48.

Williams, S. L. 1976. Effect of floods on *Thomomys bottae* in Texas. Southwestern Nat. 21(2):169–175.

Williams, S. L., and Baker, R. J. 1976. Vagility and local movement of pocket gophers (Geomyidae: Rodentia). Am. Midl. Nat. 96(2):303–316.

JANIS D. CHASE, Office of Fuels Conversion, Environmental Analysis Branch, Department of Energy, Washington, D.C. 20461.

WALTER E. HOWARD, Department of Wildlife and Fisheries Biology, University of California at Davis, Davis, California 95616.

JAMES T. ROSEBERRY, Bureau of Land Management, Box 1828, 2515 Warren Street, Cheyenne, Wyoming 82001.

14

Beaver

Castor canadensis   Edward P. Hill

NOMENCLATURE

COMMON NAMES. Beaver, Canadian beaver, el Castor, American beaver. Among some Canadian Indian tribes the local name means "little people."
SCIENTIFIC NAME. *Castor canadensis*
SUBSPECIES. *C. c. acadicus, C. c. baileyi, C. c. belugae, C. c. caecator, C. c. canadensis, C. c. carolinensis, C. c. concisor, C. c. duchesnei, C. c. frondator, C. c. idoneus, C. c. labradorensis, C. c. leucodontus, C. c. mexicanus, C. c. michiganensis, C. c. missouriensis, C. c. pallidus, C. c. phaeus, C. c. repentinus, C. c. rostralis, C. c. sagittatus, C. c. shastensis, C. c. subauratus, C. c. taylori,* and *C. c. texensis.*

Areas suggested as ranges of *C. c. canadensis, carolinensis, belugae, missouriensis,* and *acadicus* accounted for most of the pristine ranges of the beaver in North America (Hall and Kelson 1959) and are believed to be essentially unchanged. Propagation programs that followed extirpation in substantial portions of the range have altered these suggested subspecies ranges. Some subspecies may have disappeared or their gene pools may have been substantially diluted through introductions and subsequent mixing with other subspecies. Because subspecies are difficult to determine even with an animal in hand, subsequent discussions will be concerned with species.

The American beaver is one of two living species in the genus *Castor. Castor canadensis* is found in North America. *Castor fiber* was native to, and has been reestablished in, portions of Europe and Asia. Some taxonomists consider these two species to be so similar as to be conspecific. However, work by Lavrov and Orlov (1973) has shown distinct karyotypes and craniological differences between them, indicating that they are separate species.

Fossil remains of a giant beaver, *Casteroides,* and a number of closely related prehistoric mammals also have been found (Cahn 1932). The family dates back to the Oligocene, and is highly diversified in the Tertiary period in North America (Kowalski 1976). The genus dates to the Pleistocene (Garrison 1967) or late Tertiary (Schlosser 1902).

FIGURE 14.1. Distribution of the beaver (*Castor canadensis*) in North America. Modifications of the range map by Deems and Pursley (1978) include populations in Mexico, southern California, west-central Florida, and Delaware, and the absence of beaver in the north slope of Alaska.

DISTRIBUTION

Historical Range. Exploration and early development of the North American wilderness in pursuit of beaver from the 1790s through the early 1900s has been the subject of extensive writings (Bryce 1904; Davidson 1918; Innis 1930). Based on these descriptions, the pristine range of the beaver was believed to have included areas of North America where there was water and plant materials suitable for winter food. Seton (1900) estimated the populations in primitive times at 60,000,000, which seems reasonable. The northernmost geographic range was probably the mouth of the MacKenzie River, Northwest Territory, Canada. Beaver are believed to have been widespread in Alaska except the Arctic Slope from Point Hope east to the Canadian border (Hakala 1952). They occurred throughout the subarctic of mainland Canada to the tundra (Novakowski 1965) as well as Anticosti and

Queen Charlotte Islands (N. Van Nostrand and M. O'Brien personal communication 1980). Their original range in Mexico is difficult, if not impossible, to determine, but they were present on the tributaries of the Colorado River and the Rio Grande (Leopold 1959) as well as on coastal streams of the western Gulf of Mexico. They occurred only rarely in the arid portions of Arizona, Nevada, and southern California, except along the larger streams and rivers.

Present Range. Beaver are present throughout most of continental North America from the Atlantic to the Pacific (Nordstrom 1972; Deems and Pursley 1978; Jenkins and Busher 1979), in Alaska including the Alaskan peninsula (M. S. Boyce personal communication 1980), and Kodiak and adjacent islands (Hakala 1952) (figure 14.1).

Beaver have been observed 300 meters above the timberline in the Alaska Range. However, they have been unable to colonize the Alaskan and Canadian arctic tundra, perhaps because of the formation of thick winter ice that would limit their access to stored food (V. Van Ballenberger personal communication 1980). Wind action on Alaskan north slope frequently removes insulating snow cover and facilitates ice formation in excess of 1 m in thickness. Also, areas of tundra appear to lack essential woody plants for winter food and for lodge construction.

Populations have been reestablished along the Santa Anna and Colorado rivers in southern California and recent damage complaints have been received from as far south as Kern and San Bernadino counties (G. I. Gould personal communication 1980). Beaver are present 75.0 km southeast of Monterey, Nuevo Leon, Mexico, on the Rio Pilon (Bethal Landin, J. A. Jimenez Guzman, and A. Jimenez Guzman personal communication 1980). Leopold (1959) gave their range in Mexico as the tributaries of the Rio Grande and the Colorado River and the north-flowing streams of northern Chihuahua. He suggested that the shortage of appropriate foods was a factor in the absence of beaver in the mountain streams of Sierra Madre occidental. They are not found along the Louisiana coast or inland 80 to 160 km (Lowery 1974; G. Linscombe personal communication 1979). They are present in north Florida, the panhandle within 16 km of the mouth of the Swannee River, and wet areas along Otter Creek in Levy County, but are absent from the southern part of the state (L. E. Williams and J. R. Brady personal communication 1980). Although their range is expanding in South Carolina, they are presently limited primarily to the Savannah and Pee Dee river drainages (Woodward et al. 1976).

Introductions in Finland (Lahti and Helminen 1974) and Asian Kamchatka (Safonov 1979) have resulted in the establishment of viable populations on other continents.

DESCRIPTION

The beaver is the largest rodent in North America. Most adults weigh 16 to 31.5 kilograms and attain a total length of as much as 120 cm. They have heavily muscled bodies supported by large bones. The forelegs are shorter than the hind legs, which gives the animal greater height in its hips than in its shoulders when walking overland. Viewed dorsally, the beaver is short, thick, and broadest just anterior to its hips, tapering gradually toward its nose; its short, thick neck appears almost continuous with the shoulders and head. The most characteristic feature of a beaver is its dorsoventrally flattened, paddlelike tail, the unfurred portion of which in most adults varies in length from 230 to 323 mm and in width from 110 to 180 mm (Davis 1940). The distal three-quarters of the tail is covered with black, leathery, uncornified scales (Kowalski 1976) containing a few scattered, heavy hairs. A second prominent feature of the beaver is its incisors, which are generally orangish in color. In adults, the anterior surface of individual incisors is generally greater than 5 mm in width. This fact may be useful in distinguishing beaver damage from other rodent damage on the basis of toothmark width.

The beaver's ambulatory movement on land is an awkward waddle. In water however, its undisturbed swimming appears graceful and efficient, though slow and deliberate. Its hind legs and large hind feet, up to 200 mm in length, and webbed toes facilitate swimming. The short, heavily clawed front feet facilitate digging. Great dexterity in the forepaws enables the beaver to fold individual leaves into its mouth and to rotate small pencil-sized stems as it gnaws the cambium.

The ears are rounded, short (30 mm), and fleshy. They are situated to the rear and high on the head. The small eyes also are located high on the head, about midway between the nose and the base of the skull.

The fur of the beaver of eastern Canada is chestnut to dark brown or almost black in the more northern latitudes (Innis 1930). Throughout its range, the fur of the flanks, abdomen, and cheeks is usually shorter and lighter than the back fur. Pelt coloration is variable within populations; reddish, chestnut to almost black, and yellowish brown specimens may occur in the same watershed.

The guard hairs are about 10 times the diameter of the hairs comprising the underfur, and give the pelt a coarse appearance. Guard hairs attain greatest length (50 mm) and density along the back. The underfur is also longest on the back (25 mm) and has wavy individual hairs that give the pelt a downy softness. The underfur may be a dark gray to chestnut on the back and, like the guard hairs, becomes lighter on the sides and ventral areas. Unlike many furbearers, coloration of individual guard hairs of the beaver tends to be homogeneous throughout their length.

The two inside toes of each hind foot have movable, doubled, or split nails, which the beaver uses as combs for grooming the fur (Wilsson 1971). The beaver has closable nostrils, valvular ears, nictitating eye membranes, and lips that close behind the large incisors, all of which are well suited for its aquatic existence. The reproductive organs of both sexes are internal and lie anterior to and open into a common anal cloaca (Svendsen 1978) with castor and anal glands.

During periods of active lactation and when parturition is near, four pectoral mammae are discernable on the chest of the adult female.

A notable characteristic of beaver is the aroma from the paired castor glands (Svendsen 1978) located beneath the skin just anterior to the cloacal opening. Contents of the castor glands are deposited during scent marking (discussed in the section "Behavior"). Castor glands are used as a base aroma in perfume and in making trappers' lures.

Skeleton, Skull, and Dentition. The skeleton of the beaver, compared to that of animals of similar length, is quite massive. The skull and the mandible are heavy (figure 14.2), and provide a strong foundation to support the incisor teeth. A less rugged skull would be unable to withstand the physical stress and strain of jaw muscle contractions of sufficient strength to cut and chip hardwoods such as oak (*Quercus* spp.) and maple (*Acer* spp.). The braincase is narrow and there is a small infraorbital canal. A prominent rostrum is anterior to the massive zygomatic arch.

The skull in adults is large enough (120 to 148 mm condylbasal length) to reduce confusion with other rodents in North America. The skull of a juvenile may be similar in size to an adult nutria (*Myocastor coypus*), porcupine (*Erethizon dorsatum*), or mountain beaver (*Aplodontia rufa*), but differences are readily apparent on close examination. As in other amphibious mammals such as the otter, the acetabulum is shifted dorsally (Kowalski 1976). The caudal vertebrae are dorsoventrally flattened, and the multiple processes of the lumbar and thoracic vertebrae hamper removal of the backstrap muscle for use as food. The male beaver has a baculum that generally enlarges with age (Friley 1949). It is palpable and can be an aid in determining the sex of live beavers (Denney 1952) and unskinned carcasses. Osteological changes that take place with growth and development are described by Robertson and Shadle (1954).

The dental formula is i 1/1, c 0/0, p 1/1, m 3/3 = 20. The most significant aspect of the teeth is the size of the prominent orange incisors, which grow continuously and are sharpened by grinding the uppers against the lowers (Wilsson 1971). The hard enameled front surface serves as the cutting edge as beaver cut trees and peel bark. The cheek teeth are hypsodont and grow only through the deposition of cementum at the root base. Deciduous premolars are replaced at about 11 months of age by permanent molariform premolar teeth (Cook and Maunton 1954).

PHYSIOLOGY

Anatomically, morphologically, and ethologically, the beaver is more specialized for swimming than any other rodent (Wilsson 1971). Many of its anatomical specializations (such as its size; type and location of ears, eyes, and nose; legs of different size and function; and flattened tail) appear to have individually and collectively enhanced survival in wetlands with varied temperature ranges. However, its high degree of

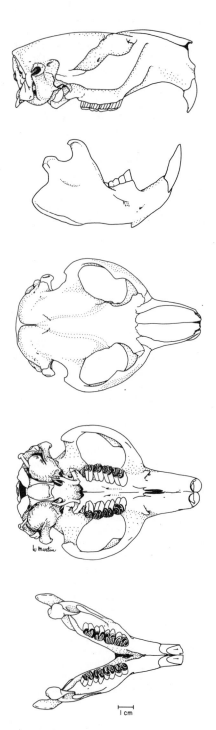

FIGURE 14.2. Skull of the beaver (*Castor canadensis*). From top to bottom: lateral view of cranium, lateral view of mandible, dorsal view of cranium, ventral view of cranium, dorsal view of mandible.

specialization had a serious drawback. Its inability to exist in other than aquatic habitats made it vulnerable to exploitation and caused its extirpation from vast areas as it was intensively sought by humans.

That the beaver persisted over such a wide range of North America is remarkable, particularly in view of the uncontrolled and relentless hunting and trapping pressure to which it was subjected by 1900. Its vulnerablity is a crucial factor that must be considered in population management.

Growth. Mean adult weight of beaver varies in different regions, depending on the extent of exploitation and quality of available food. Except in kits, which grow through the first winter (Novakowski 1965), a steplike pattern of growth appears to occur in northern latitudes, primarily between June and November during the first four years (Hakala 1952; Buckley and Libby 1955; Aleksiuk and Cowan 1969). Although it is frequently mentioned that beaver continue to grow throughout life, reports from the northern range limit indicate that growth terminates at age four (Novakowski 1965; Alekiuk and Cowan 1969; Gunson 1970). The largest beaver are apparently produced under the habitat conditions available at midcontinent (table 14.1).

Of approximately 1,450 specimens examined from previously unexploited populations in southeastern Alabama, only 3 females were as heavy as 27.2 kg: 2 weighed 27.2 kg and the other weighed 30.6 kg. Body weight appeared to plateau prior to five years of age, and diminished slightly after age nine. Maximum mean weight of 19.3 kg was attained at age seven, whereas all specimens nine and over average 18.6 kg. This region of the state generally has less fertile soils than the Black Belt and Tennessee Valley (Hill 1972b), where a 29.4-kg beaver has been reported (Howell 1921). This suggests that beaver at the southern range limit, like those at northern latitudes, grow until about age four or five. Exceptional individuals may continue growing at a reduced rate, but such growth appears slower at the southern and northern range limits than at midrange.

Beaver in Alabama appear to become almost sedentary after June. Specimens taken in early summer are visibly leaner than those taken in December after the accumulation of fat reserves. Pearson (1960)

and Aleksiuk and Cowan (1969) noted that body fat in the beaver was deposited in autumn in Prince Albert National Park, Saskatchewan, and the Mackenzie Delta, Northwest Territory, respectively. Energy demands would appear to be quite different at the range extremes, but utilization of reserves appears to reach similar levels by early summer.

Failure of beaver at the southeastern range limit to maintain extensive fat reserves is not due to lack of food, as it may be at the northern range limit where icing and freeze-up occurs (Aleksiuk and Cowan 1969). Woody vegetation is available throughout the winter months at the southern range limit, since the region is essentially ice free. Curtailment of food consumption, as described for captive beaver of a northern population (Pearson 1960), is believed to occur at the southern range limit with the onset of warming trends in February and March, perhaps associated with seasonal physiological changes (Potvin and Bovet 1975) or sunning behavior. Southern beaver are frequently observed lying on lodges during clear sunny days of late winter and early spring.

Digestion. Digestion in the beaver seems almost unique, enhanced by a prominent cardiogastric gland in the stomach (Kitts et al. 1957), glandular digestive areas elsewhere (Kowalski 1976), and a large trilobed cecum containing commensal microbiota (Nasset 1953; Currier et al. 1960). Year-round consumption of soft green excrement directly from the anus was described in captive European beavers (Wilsson 1971) and occurs in *C. canadensis* (H. Hodgden personal communcation 1980). This coprophagy, found in other rodents and lagomorphs, is believed to improve digestive efficiency. However, the beaver does not have any superior capacity for converting cellulose. Digestion of 32 and 33 percent of the available cellulose fed two beaver (Currier et al. 1960) is not unlike the digestive capability of a wide range of other nonruminant mammals. The apparent superiority of a beaver's digestive powers comes from its capability to gather and ingest the most nutritious portions of large amounts of very fibrous, woody plants.

Circulation. The heart is relatively small (McKean and Carlton 1977), consistent with the beaver's characteristic slow life style that is normally free of extended intensive exertion. The beaver has no unusual oxygen storage capacity (McKean and Carlton 1977) that enables it to remain underwater for comparatively long periods. However, changes occur in blood parameters, heart rhythm, and circulation to the extremities and body muscles that enable the beaver to make dives lasting up to 15 minutes (Irving and Orr 1935) without asphyxiation (Irving 1937; Clausen and Ersland 1968, 1970). Aleksiuk (1970) noted that "minute blood vessels permeate the entire tail, and a counterguard heat exchange system is present at the base." This specialized circulatory feature would help conserve heat energy in extremely cold water and radiate heat during hot weather. Normal blood parameters of the beaver were given by Kitts et al. (1958), Stevenson et al. (1959), and Currier et al. (1960).

TABLE 14.1. Weight, location, and source of reported large beaver

Weight (kg)	Location	Source
39.4, 36.5, 35.6	Wisconsin	Schorger (1953)
39.0	California	Grinnell et al. (1937)
38.0, 32.6	Washington	Anonymous (1955)
32.3	Colorado	Denney (1952)
31.7	Tennessee	Hill (unpublished data)
31.3	Colorado	Hay (1957)
30.6, 27.2	Alabama	Hill (unpublished data)
29.4	Alabama	Howell (1921)

REPRODUCTION

In the management of any wild or domestic animal, knowledge of reproductive parameters and annual recruitment is prerequisite to formulation of most decisions. Sex ratios, age of sexual maturity, pregnancy rates, litter size, and juvenile survival influence recruitment success, and knowledge of these parameters is basic to sound management.

Sexual Maturity. Although puberty may be reached several months prior to the first breeding, the important factor to be considered in population management is the age of the female at her first parturition. For the purposes of this discussion, *sexual maturity* will mean the age at breeding that results in the first litter. The first parturition in beaver normally occurs at age 3, but may occur near age 2 or 4 depending on the extent of population exploitation and environmental factors (Semyonoff 1953). Gunson (1970) believed that early sexual maturity in Saskatchewan beaver was enhanced by high-quality habitat; an estimated two-thirds to three-quarters of the 2 year olds were producing young. Parsons and Brown (1979) noted that reproduction ceased in yearlings where beaver were utilizing more than 40 percent of the potential sites. Payne (1975) found that 24 percent of the yearling females sampled in Newfoundland had bred. Four of 38 females in the 2- to 3-year-old age class from New Brunswick had placental scars, suggesting that they had produced litters at about age 2 (Nordstrom 1972). Lyons (1979) noted that in western Massachusetts 17 of 39 of the 1.5–2–year class had bred. In Ohio, 8 of 20 1.5- to 2-year-old females taken from exploited and expanding populations in early February had ovulated (Henry and Bookhout 1969). Boyce (1974) noted a tendency toward earlier sexual maturity in exploited populations in Alaska but found no indication of reproduction in 2 year olds. In northern Canada, Novakowski (1965) found first pregnancies in 3 of 21 females that were approaching their third birthday but no indications of conception in females that were almost 2. There were no indications of ovulation or pregnancy in 50 females 1.5 to 2 years old taken in Alabama, and only 16 of 65 female 2.5 to 3 year olds had ovulated. Similar ovulation rates were reported in small samples of 2- and 3-year-old females from an expanding population in South Carolina (Woodward 1977). This suggest a pattern of breeding and sexual maturity at age 1.5 to 2 at middle latitudes, provided food conditions are favorable and there is some exploitation or expansion. However, the pattern seems inconsistent at the northern and southern range limits, where breeding is delayed to 2.5 years of age or more.

Pregnancy Rates. Knowledge of pregnancy rates among age groups in a monogamous species such as the beaver increases accuracy in computing estimates of reproductive performance. Sex ratios (discussed in the section "Population Structure") are also important in determining pregnancy rates because the absence of sexually mature individuals of either sex in a colony precludes reproduction. Pregnancy rates vary among age groups (Henry and Bookhout 1969; Gunson 1970;

Payne 1975; Woodward 1977; Hill unpublished data); rates usually increase to about age 4 and remain high in the other age groups except in extremely old individuals. Pregnancy rates in 2.5- and 3.5-year groups are influenced by population structure within colonies and the extent of dispersal. These, in turn, are impacted by the extent of exploitation and the quality of habitat. Gunson (personal communication 1980) believed that there was less dispersal in good-quality habitats and therefore less breeding among young adults in colonies containing older dominant pairs. Habitat quality and extent of exploitation should be considered when using pregnancy rates to calculate reproductive performance. Sampling times should also be late enough to insure that 2.5- and 3.5-year groups have completed breeding, otherwise true pregnancy rates in these age groups may not be determined.

Breeding. Beavers are monogamous (Zharkov and Sokolov 1967; Boyce 1974; Brooks 1977; Fleming 1977; Lyons 1979) and produce only one litter per year. Although early researchers apparently mistook dancing or wrestling behavior for copulation, the act now believed to constitute breeding was described by Roth (1938), and Bradt (1940) and illustrated by Wilsson (1971). Breeding occurs between January and March in cold climates (Grasse and Putnam 1950; Hakala 1952; Hodgdon and Hunt 1953; Henry and Bookhout 1969; Bergerud and Miller 1977) as well as in California (Grinnell et al. 1937). However, Wilkinson (1962) reported December and January as the breeding period in Alabama, and back-aged fetuses indicated the occurrence of breeding in late November or early December in Alabama (Hill unpublished data), Mississippi (Thomason 1978), South Carolina (Woodward 1977), and Texas (Miller 1948). In cold climates, breeding takes place in water (Kowalski 1976), under the ice, or in bank dens or lodges (Wilsson 1971).

Wilsson (1971) reported that *C. fiber* remains in estrus 10 to 12 hours and has a second estrus in 14 days if not fertilized. He also reported 105 days as the gestation period for this species. Panfil (1964) reported 105–107 days. Hediger (1970) reported a 104–111– day gestation period for *C. canadensis*. Others suggest that gestation in the beaver in North America lasts 98–100 days (Dugmore 1914; Bergerud and Miller 1977; Woodward 1977).

Reproductive Performance. Placental scar counts (Hodgdon 1949), counts of developing embryos, and, with some limitations, counts of corpora lutea and corpora albicantia are useful indices of reproductive performance in the beaver (Provost 1958). Preimplantation losses from unfertilized ova or failure of fertilized ova to implant, and postimplantation losses, resulting in resorptions (Provost 1958, 1962), account for differences between ovulation rates and litter size. Intrauterine mortality amounted to 16 percent in 48 beaver from Ohio (Henry and Bookhout 1969) and almost 19 percent in 40 females from western Massachusetts (Lyons 1979). Rates of prenatal loss should be considered in obtaining precise estimates of litter size and annual productivity.

In areas where carcasses are available from fall or early winter trapping, counts of placental scars and persisting corpora albicantia corrected for current resorption rates and prenatal loss, respectively, provide an index of litter size the previous spring. Sources of error, such as regression or discoloration of implantation sites by some preservatives, and degeneration of corpora albicantia with the onset of the breeding season (Provost 1962) suggest that estimates of reproductive performance in fall are less accurate than those made at other times. Where trapping seasons overlap the breeding season, a combination of placental scar counts corrected for the past season resorption rates, and counts of corpora lutea of pregnancy corrected for current resorption rates, can provide information on litter size. However, early in gestation it is difficult to distinguish between corpora lutea of ovulation and corpora lutea of pregnancy. This is a potential source of error in estimating reproductive performance (Provost 1962). Also, embryo counts and ovulation rates of females trapped in January and February may not provide precise estimates of the extent of current-year breeding among the 1.5- and 2.5-year age groups. These age groups may breed later than adults (Grinnell et al. 1937), particularly at southern latitudes and if they disperse. Thus, winter samples may not reflect the true reproductive contribution that would be revealed if these age groups were sampled in May or June.

Where trapping seasons are set after the breeding season, counts of developing embryos corrected for current resorption rates will provide an accurate index of litter size among age groups. Resorption is negligible if embryos are in an advanced stage of development. Where possible, delaying the trapping season until breeding has occurred may lower the incidence of unbred females (Hunt 1952; Hodgdon and Hunt 1953) whose mates were trapped from the population before breeding. It also facilitates the measurement of current litter size through embryo counts.

Litter size throughout most of the range varies with the age of the female (Osborn 1949; Pearson 1960; Henry and Bookhout 1969; Gunson 1970; Payne 1975; Hodgdon 1978; Lyons 1979; Hill unpublished data) and nutritional plane (Pearson 1960). The latter appears to be influenced by restricted food intake due to ice at high altitudes (Rutherford 1955) and at the northern range limit, and by reduced food quality or availability as a result of plant species composition (Huey 1956) or excessive use.

A typical litter will most often contain three or four young, with a range of from one to nine. Gunson (1970) and Payne (1975) compiled litter size data from previous studies in eastern and western North America. Data from their compilations plus that contained in subsequent studies provide no consistent pattern of litter size related to single variables such as latitude or temperature (table 14.2). However, many agree that large litters are associated with better quality habitats, particularly those containing aspen.

Fetal and Kit Development. Fetal growth curves for gestation periods of 90 and 100 days were estimated

TABLE 14.2. Index of litter size in the beaver reported throughout its range in North America listed by increasing litter size

Litter Size	Type/Count	Location	Source
2.2	embryos	South Carolina	Woodward (1977)
2.5	embryos	Alabama	Wilkinson (1962)
2.7	embryos	New Mexico	Huey (1956)
2.7	embryos	California	Grinnell et al. (1937)
2.8	embryos	Newfoundland	Payne (1975)
2.8	ovulation	Alabama	Hill (unpublished data)
2.8	embryos	Alaska	Hakala (1952)
2.9	embryos	Wyoming	Osborn (1953)
3.0	embryos	Colorado	Rutherford (1955)
3.1	embryos	Alaska	Boyce (1974)
3.1	embryos	Alberta	Novakowski (1965)
3.4	embryos	Idaho	Leege and Williams (1967)
3.5	embryos	Saskatchewan	Gunson (1970)
3.6	embryos	Maine	Hodgdon and Hunt (1953)
3.6	embryos	Washington	Provost (1958)
3.7	embryos	Saskatchewan	Pearson (1960)
3.7	embryos	Michigan	Bradt (1938)
3.8	placental scars	Ohio	Henry and Bookhout (1969)
3.9	embryos	Michigan	Benson (1936)
4.0	embryos	New York	Johnson (1927)
4.1	embryos	Saskatchewan	Gunson (1970)
4.1	embryos	Vermont	Bond (1956)
4.2	embryos	North Dakota	Hammond (1943)
4.2	embryos	New Mexico	Huey (1956)
5.3	embryos	Wisconsin	Longley and Moyle (1963)
5.5	embryos	Pennsylvania	Brenner (1964)
6.3	embryos	Alberta	Gunson (1970)

by Provost (1958) and Woodward (1977), respectively. These may be useful for estimating peak periods of conception and parturition through the back-aging process (Hodgdon and Hunt 1953).

Newborn kits are reddish, cinnamon brown (Wilkinson 1962), dark chestnut, or almost black. At birth, kits weigh about 0.5 kg (Bailey 1927). Wilkinson (1962) reported two newborn kits that weighed 411 g each. Grinnell et al. (1937) reported the mean weight of a litter of four fetal kits near term as 485 g. Ten kits from three litters averaged 495 g (Bradt 1939). Shadle (1930) reported average weights of 335 g for six kits in a litter, but several kits died soon after parturition. Wilsson (1971) reported that seven *C. fiber* kits weighed from 0.3 to 0.7 kg at birth, and within 12 months attained weights of 5.0 to 13.0 kg.

Suckling behavior in *C. fiber* kits was observed by Wilsson (1971) and Zurowski et al. (1974). Suckling occurred as many as nine times daily, from 4 A.M. to 10 P.M. during the period of lactation, which lasted about 60 days past parturition. Kits normally suckled for 5–10 minutes, pressing the nipple rhythmically with their paws. Zurowski et al. (1974) described lactation and the chemical composition of milk and noted that the upper nipples were 50–75 percent less productive than the lower nipples. They also noted that kits took solid food at about one month of age, whereas Wilsson (1971) observed a kit eat a whole leaf at 11 days of age and begin to eat considerable quantities of solid food at about one month as suckling began to decrease.

ECOLOGY

Habitat. Beaver are presently found throughout most of their original range. Where they have not been extirpated, they occur commonly in large rivers, impoundments, and large lakes with relatively constant water levels, in protected areas of large lakes that have extensive wave action, streams, tributaries, and small seepages that have adequate flow for damming. Although they occur in streams containing steep, rocky, or bedrock bottoms, the destructiveness of high water conditions appears to prevent population densities typical of areas that are more suitable (Retzen et al. 1956). Beaver appear to prefer relatively flat terrain of fertile valleys that produce their preferred winter foods. They reach optimum population densities per unit of stream in such habitats. They thrive in small irregular pockets, heads, and drainage depressions at and above 3000 m where aspen is available. If unmolested, they cohabit with human populations.

Generally, good quality habitats containing an abundance of preferred and available foods support more beaver than those with less available foods or those that are of generally poorer quality. Habitat is known to influence or be related to litter size (Huey 1956) and is believed to influence age of dispersal and therefore age at first breeding (Gunson 1970) and colony size and composition.

Colony Size. Bradt (1938) defined the colony unit as "a group of beavers occupying a pond or stretch of

stream in common utilizing a common food supply, and maintaining a common dam or dams." A typical colony in midwinter contains the adult pair, two to four kits from the previous spring litter, two or three yearlings, and occasionally one or more that are about 2.5 years old. Gunson (1970) suggested that more beaver were maintained at colonies in higher quality habitats. The maximum number of individuals found in 57 colonies studied in Michigan was 12; more colonies contained 6 individuals, while the average colony contained 5.1 individuals (Bradt 1938). Swank (1949) and Woodward (1977) found an average of 5.3 and 5.5 individuals in colonies they trapped out in West Virginia and South Carolina, respectively. Hodgdon (1978) noted that a well-established population in Massachusetts had an average of 6.2 beaver. Wilkinson (1962) estimated 4.8 per colony in Alabama. Based on survey data from 22 states and provinces, Denney (1952) reported an overall average of 5.2 beaver per colony in the United States and an average of about 5 in Canada. However, Nordstrom (1972) reported 3.7 and 2.7 beaver per colony, respectively, in unexploited and exploited populations in New Brunswick. Gunson (1970) reported mean colony sizes of 4.1 and 2.8 in exploited populations in good and poor habitats, respectively, in Saskatchewan. These four studies suggest that smaller colonies may occur in Canada.

Although dispersal is believed an important population regulator in some populations (Gunson 1970), in most of the range trapping more than any other factors reduces the number of beaver per colony (Nordstrom 1972; Boyce 1974; Payne 1975). Population density estimates based on the colony unit should be made only with valid estimates of mean colony size, extent of single and pair colonies, and habitat quality, and with knowledge of the history of recent exploitation.

Age Structure. Population structure based on ages (Van Nostrand and Stephenson 1964) within trapped samples is subject to several sources of bias (Nordstrom 1972; Boyce 1974; Payne 1975). Except where food caches, lodges, and bank dens are trapped, most trapping schemes take kits and yearlings last. Trapping the entrances to living areas and cached food supplies will usually reduce sampling bias, but sets in those sites are unlawful in some areas. Many trappers prefer not to take kits and yearlings and will make sets that are more selective for larger, older beaver. Varied trapping techniques, intensity, and duration also will influence the composition of other age groups within the harvest. Intensive observational studies of unexploited colonies during spring and summer best reflect true numbers of kits and yearlings (Hodgdon 1978), but may underestimate numbers of dispersing two year olds. Estimates of pristine population structure may best be obtained by trapping out previously unexploited populations.

Percentages of kits and yearlings in trapped samples have been reported from several studies (Osborn 1949; Cook and Maunton 1954; Novakowski 1965; Larson 1967; Leege and Williams 1967; Henry and Bookhout 1969; Payne 1975; Woodward 1977; Hodgdon 1978; Hill unpublished data). Disregarding the various sources of trapping bias, the mean percentages in the kit, yearling, and two year old and older groups in these studies were 30.1, 22.8, and 47.1, respectively.

As mentioned earlier, habitat quality influences reproductive performance and colony composition. Gunson (1970) reported that the percentage of kits in intensively live- and kill-trapped populations varied from 51 percent (N=534) in habitat he classified as high quality to 30 percent (N=455) in that considered low quality.

Density Estimates. Beaver population density is usually expressed in colonies per unit of area or per length of stream. The former seems appropriate for regions of abundant precipitation, relatively flat terrain, and numerous lakes and drainages. In arid areas or areas of well-defined watercourses or rivers, density measurements might be more appropriately expressed as colonies per length of stream. Both expressions have been used. In Algonquin Park, Ontario, colony densities were 0.38 to 0.76 per km^2 (Voigt et al. 1976). Aleksiuk (1968) reported similar densities—0.39 colonies per km^2 from year to year on the Mackenzie Delta, Northwest Territory—as did Bergerud and Miller (1977) for Newfoundland. Nordstrom (1972) reported 1.0–1.2 colonies per stream km in unexploited areas versus 0.34 colonies per km of stream in areas that were exploited. Aerial surveys in 14 counties in New Brunswick between 1966 and 1970 revealed mean densities of 0.09–0.14 per km of stream (Nordstrom 1972). Boyce (1974) reported 0.35 to 0.48 lodges per km of stream in Alaska.

A density of approximately 1.24 colonies per km of stream was estimated as the saturation point for beaver in New York (Johnson 1927; Buckley 1950) and in Utah (Haas 1943). Novakowski (1965) suggested 0.4 colonies per km as the saturation point for beaver in northern Alberta. Work in Alabama on headwaters of four watersheds indicated that colony saturation in these selected areas approached 1.9 colonies per km of stream (Hill 1976).

Sex Ratios. Sex ratios in monogamous species like the beaver are particularly important because they influence pregnancy rates. Ratios reported from trapped populations have varied, perhaps because of bias inherent in varied trapping methods (Lyons 1979), small sample sizes, and because they may reflect true local sex imbalances (Henry and Bookhout 1969; Leege and Williams 1967). Where relatively large samples are considered, sex ratios appear relatively even. For example, Gunson (1970) reported 105:100 males to females (N=1,310) in Saskatchewan, Payne (1975) reported 93:100 (N=416) in Newfoundland, Bond (1956) reported 108:100 (N=2,164) in Vermont, and in Alabama a trapped sample contained 105:100 (N=977). The mean of the ratios reported in these four studies was 105:100 (N=4,867). Woodward (1977) compiled age-specific sex ratio data from 15 studies. When these ratios were averaged, they were 90.7:100 for adults and 111.4:100 for subadults. Collectively, however, the ratio was 98.5:100, or almost even. These differences may reflect a slight preponderance of males in kit and yearl-

ing age groups and trap susceptibility or other mortality in adult males, as many have noted. However, Gunson (1970) believed trapping to be selective for adult females. Payne (1980) found no difference in susceptibility of sexes to baited Conibear traps set under ice. This variability suggests the need for caution when averaging sex ratios, because important regional differences can be masked (J. Gunson personal communication) and because trap bias associated with sex and trapping methods are not well known. Some writers have hinted that the beaver had an inherent compensatory preponderance of males in newborn litters. However, the ratio in 671 unborn fetuses from six separate studies was 96:100.

BEHAVIOR

Movements. Other than routine movements within the territories of their individual colonies, the major movements of beaver have been classified into four categories by Bergerud and Miller (1977): (1) movement of the entire colony between ponds within its territory, (2) wandering of yearlings, (3) dispersal of two-year-old beaver to establish new colonies, and (4) miscellaneous movement of adults who likely lost their mates. Hodgdon (1978) combined the last three of these with family movements outside their respective territories, forming categories to cover immigration and emigration movements. The longest movements are consistently made by dispersing two year olds (Bradt 1947; Townsend 1953; Beer 1955) in response to an "innate tendency to leave their home colony" (Leege 1968; Bergerud and Miller 1977). In an intensive study of behavior in wild beaver populations in Massachusetts, Hodgdon (1978) observed no intrafamily aggressive behavior associated with dispersal. He noted that dispersing individuals occasionally returned to the home colony for short periods. Activity centers shifted within the colony territory from July to September with a period of premovement exploration. Families moved before parturition in early June. Pairs and single two year olds dispersed from April to September. Single animals usually dispersed before pairs (Hodgdon 1978). Based on studies of beaver in varied habitats in Saskatchewan, Gunson (1970) suggested that greater numbers of beaver were maintained in colonies in higher quality habitats. One of the regulatory mechanisms was the "apparent greater frequency of dispersal of immature animals from parent colonies in the low quality areas."

Distances moved and the time of movement and dispersal are important considerations in formulating management strategies. Documented movements of transplanted beaver have involved distances of 48 km (Knudsen and Hale 1965) to 238 km (Hibbard 1958). The mean distance traveled by 472 translocated and recaptured beaver was 7.4 km (Knudsen and Hale 1965). Beaver tagged and released at the point of capture moved 19 km (Leege 1968) to 241 km (Libby 1957).

Where icy conditions prevail, major dispersal and movements by families, pairs, and singles are confined to summer ice-free periods (Hodgdon 1978).

Dispersal and other beaver movements in ice-free areas are less restricted. Some dispersal occurs in late February and March. Other winter movements are greater because no food caches are prepared. Other indications are winter scent-making behavior and the high rate of pelt damage believed to be associated with territory defense. Both have been observed in populations in midwinter in Alabama (Wilkinson 1962), Arkansas (P. Dozier personal communication), Georgia, and western Oregon (R. Denney personal communication 1980).

Nightly movements by beaver about the colony territory are normally made by swimming, both on and beneath the water surface. Initial departure from the lodge or bank den is beneath the surface, but thereafter most movement is accomplished by swimming on the surface. Swimming is normally accomplished by alternate kicks of the hind legs. If chased at night with a motorboat and spotlight, beaver will swim alternately on and beneath the surface for distances of 10 to 60 m at a speed of about 5 km/hr.

If cornered or frightened, beaver are capable of swimming rapidly beneath the surface for short distances. The accelerated speed, estimated by Wilsson (1971) at in excess of 2 m per sec, is facilitated by undulating movements of the tail. Diving is described by Tevis (1950) and Wilsson (1971).

Beavers are crepuscular and nocturnal, following a normal 24-hour period in spring and summer at northern latitudes (Tevis 1950; Wilsson 1971; Hodgdon 1978; Lancia 1979) and during winter in ice-free areas. Winter activity periods of beaver near Quebec City, Quebec, displayed a "freerunning circadian rhythm of period length 26.25 to 18.0 h" (Potvin and Bovet 1975). Lancia (1979) noted that members of some colonies in Massachusetts displayed a free-running activity rhythm while others did not.

Ambulatory gait on land is similar to that of most quadrupeds, and beaver will gallop if frightened. They can walk upright in a bipedal fashion partially supported by the tail, while they carry mud or other materials held against the chest with chin and front legs (Seton 1900; Wilsson 1971; Hodgdon 1978).

Sound. Kit beaver emit a series of high-pitched whines if food is taken from them. Other vocalizations reported by Lighton (1933), Tevis (1950), Wilsson (1971), and Hodgdon (1978) vary from low- to high-pitched whines, whimpers, and whistles. Perhaps the most frequently heard sound produced by the beaver is the tail slap, a warning sound (Tevis 1950) described by Wilsson (1971) and Hodgdon (1978) and illustrated by Wilsson (1971).

Scent Marking. Mud and vegetational debris (Bollinger 1980) are placed in piles near the colony territorial boundaries and wetted with urine, castorium, and perhaps contents of the oil or anal glands (Svendsen 1978). This territorial marking appears to be limited to open water periods at northern latitudes (Nordstrom 1972; Hodgdon 1978), but the behavior occurs as early as February in California (Grinnell et al. 1937) and in December and January in Alabama, where scent mounds 35.5 cm in height occur.

Construction. The construction of lodges, dams, and food caches was described by Wilsson (1971), Hodgdon and Larson (1973), and Hodgdon (1978). Dam construction in Massachusetts occurred between April and June, and again from August through October. Lodge construction occurred between August 28 and September 25 and always before food cache construction. In many areas both lodges and bank dens are utilized. A number of situations has been observed in Alabama, where bank dens were converted to lodges as water levels increased with pond age. Bank dens are not used in some areas, such as Newfoundland (N. Payne personal communication 1980). Rocky conditions in that area and permafrost in other regions may preclude bank den construction.

In the Mackenzie River delta food cache construction begins during the last week of August (Aleksiuk 1970). It occurs progressively later with more moderate climates, and in areas free of extensive winter ice, it does not occur. In Alabama and presumably other ice-free areas, dam and lodge construction appears to be more active in fall, winter, and spring than in summer. Food cache construction is not frequently seen at low altitudes south of 38° N. latitude. It is irregular on the streams at lower elevations in Colorado (Yeager and Rutherford 1957). Food caches are seen in elevated areas of Arkansas, but not on the large rivers or lowland areas (P. Dozier personal communication 1980). Beaver may cache food on inland ponds at Land between the Lakes, Kentucky, but not usually in adjoining Kentucky or Barkley lakes, where wave action hampers ice formation (R. Nall personal communication 1978). Beaver do not cache food in the Willamette Valley and coastal regions of Oregon but they do in the eastern part of the state (R. Denney personal communication 1980).

In describing the construction of bank dens, Wilsson (1971) noted the existence of feeding chambers near the water within the main tunnel leading to the resting or sleeping area. Although an entrance feeding chamber was illustrated by Grinnell et al. (1937), I could find no reference to its use in the North American literature. The feeding chamber may be more closely associated with extreme cold northern latitudes, but its use is believed to occur in some southern populations as expelled peeled branches and the protruding tassel of corn stalks are frequently seen at bank den entrances in winter and summer, respectively, in Alabama. Other construction behavior, such as digging, pushing, and shoveling, is described by Wilsson (1971) and Hodgdon (1978).

Aggressive Interactions. I have observed beaver to lunge at a potential adversary on two occations: on one occasion, a beaver hunting dog was severely bitten in the back; I had to jump to avoid a bite on my right leg on the other occasion. Both attacks were preceded by hissing. Another form of behavior that may or may not be aggressive is tooth sharpening (Wilsson 1971). Recently captured adult beaver displayed hissing and tooth-sharpening behavior when approached. However, they also displayed tooth sharpening intermittently while peeling sweet gum (*Liquidambar styraciflua*) limbs after a few weeks in captivity. The deposition of foreign castoreum upwind from a lodge will often result in investigation, hissing, and tail slapping behavior as beaver emerge in the evening. This response to foreign castoreum was known among early trappers who used the castoreum to bait traps.

MORTALITY

The beaver is most vulnerable to predation by mammalian predators such as the coyote (*Canis latrans*) and the timber wolf (*Canis lupus*) (Young and Jackson 1951; Mech 1966) when it is away from water. Where these large predators occur, food shortages that require beaver to forage great distances from water cause greater exposure to predation. Food habit studies indicate that the coyote does not prey as heavily on the beaver as does the timber wolf, whose diet on Isle Royale was about 11 percent beaver (Mech 1966). In Algonquin Park, Ontario, the beaver gradually became the most important prey item (55 percent by frequency of occurrence in scats) over a nine-year period that deer declined (Voigt et al. 1976). Other potential mammalian predators that are believed to be of minor importance are bears (*Ursus* spp.), wolverines (*Gulo gulo*), river otters (*Lutra canadensis*), lynx (*Felis lynx*), bobcat (*Felis rufus*), and mink (*Mustela vison*) (Swank 1949; Semyonoff 1951; Hakala 1952; Gunson 1970; Boyce 1974).

Sudden snow melts in midwinter or violent spring breakups can raise water levels in streams and may destroy lodges and occupants or drown large numbers of beaver under the ice (Hakala 1952; Boyce 1974). Starvation at northern latitudes has also been noted as a mortality factor (Gunson 1970; Bergerud and Miller 1977).

Tularemia, often a water-borne disease caused by *Pasteurella tularense*, has caused epizootics that resulted in widespread mortality and decimation of beaver populations (Grasse and Putnam 1950; Stenlund 1953).

ECONOMIC STATUS

Historical Role. The involvement of the beaver in the exploration, development, and early economy of North America has been referred to as many times as there are separate accounts of the history of the era. Writers suggest that early trappers were concerned with meeting the needs of daily life and accumulating what fortune they could from the great wild land. Lured by trinkets and habits of western civilization, native inhabitants abandoned their historical use of the beaver only when needed and along with white trappers took beaver using whatever method they could, including shooting, trapping, and organized hunting parties without regulated seasons and as weather and pelt primeness permitted. Clearly the beaver was the most widely and intensively sought natural resource of the continent during the 1700s and 1800s (Bryce 1904; Davidson 1918; Innis 1930).

Because of the frequent moving that was characteristic of the early trappers' life style and the vastness of the continent over which they traveled and trapped,

they seemed unaware that they were destroying the beaver resource through excessive harvest in one place after another. Except in a few isolated pockets, most of the beaver population at middle and southern latitudes was depleted before 1900.

With growing public concern over the extinction of some species and the welfare of wildlife in general, state, provincial, and federal conservation agencies were established and given the responsibilities for management and perpetuation of wildlife resources. The beaver responded to the protection under new regulations that controlled times, methods, and number harvested. Live-trapping and restocking programs followed in the 1920s through the 1950s, and the beaver returned gradually to much of its original range. However, human populations increased and occupied most of the fertile land and in one area after another divergent attitudes developed among those that the beaver influenced. The persistence of these attitudes provides the stimulus for management and the sources of controversy that makes work with the beaver such a challenge today.

With the possible exception of the white-tailed deer (*Odocoileus virginianus*), the beaver has been the subject of more historical, popular, and scientific writing than any other mammal in North America. Much of the early writing was related to the role of the beaver in the early fur trade of Canada and the United States. However, with the emergence of governmental natural resources agencies and wildlife scientists, the trend in beaver literature turned toward technical publications on beaver status, restocking efforts, life history, damage control, and management. A major contribution toward a bibliography on the beaver was compiled by Yeager and Hay (1955). Denney (1952) summarized beaver management through 1948. Work by Wilsson (1971) on captive European beaver (*C. fiber*) provided substantial information that appears applicable to the North American beaver and is referred to frequently. Jenkins and Busher (1979) provided an updated summary of pertinent literature on life history and ecology of the species.

Value. The beaver is important economically, both as a pest species and as an income producer. Its worth is tied chiefly to the dollar value placed on its pelt. Fluctuating world demand for beaver fur correspondingly stimulates or depresses local markets and is the dominating force controlling the price of pelts and the income going to trappers. A discussion of the complexities of the economics of the fur industry is beyond the scope of this chapter, but knowledge of them can provide some understanding of instability in the demand for beaver (Fuchs 1957).

Negative economic aspects of the beaver are primarily associated with nuisance damage caused by its habitat modifications and food habits. Costs of handling nuisance complaints tend to be relatively stable except for inflationary changes, whereas income from beaver pelts follows fluctuating world markets. The magnitude of nuisance complaints frequently displays an inverse relationship to world demand for beaver fur. The response of trappers to lucrative fur prices stimulates trapping pressure, particularly on beaver populations that are free of harvest restrictions and are readily accessible near human habitation. Populations that are likely to cause problems are usually trapped first in response to price increases.

When raw, handled pelts (fleshed, stretched, and dried) are converted to money or used for exchange or in trade for goods, the beaver has realized value. When an animal such as the beaver with both pest and beneficial attributes attains value, it begins to receive favorable consideration from individual landowners for its continued existence. Landowners consider the beaver as part of their wealth in much the same way they value other marketable commodities and afford them a measure of protection. The commodity value of the annual harvest therein becomes a major point for defense of management and trapping of beaver.

Through the compilation of the fur harvest records at state, province, and national levels, knowledge of the beaver's annual value is provided. State and provincial governments vary widely in their attitudes toward and practice of monitoring or determining the annual beaver harvest. The royalties paid on beaver pelts in the Canadian provinces provide a reliable source for enumerating annual harvest in Canada. Pelt

tagging and sealing provides a similar side benefit where it is practiced in the United States; however, the lack of a reliable monitoring system at the national level and within several individual states leaves large gaps in the information available on past beaver harvests and their annual value.

Annual harvest statistics projected from mean prices paid for handled raw fur provide accurate estimates of the annual value to the trapper (table 14.3). Income in terms of employment and commercial endeavors derived by other user groups as beaver fur is processed into its final product contributes to the total economy and value of the beaver and should be considered in questions where the management, perpetuation, and consumptive use of the beaver are under review.

Voluntary responses from beaver trappers in Alabama to a questionnaire were compiled during a two-year study as an indication of the income potential of beaver trapping. Only individuals that trapped primarily for beaver and used 10 or more no. 330 Conibear traps were considered. These criteria reduced the sample size to 32 and 20 trappers, respectively, for the 1972/73 and 1973/74 fur seasons.

During the 1972/73 trapping season the average

TABLE 14.3. Annual reported harvest and value of beaver from Canada and the United States from the 1977/78 fur season back through the 1918/19 fur season

Fur Year	Canadian Sales[ac]			United States Sales[bc]			Total Continental Value
	Total Pelts	Average Price per Pelt	Total Value	Total Pelts[d]	Average Price per Pelt	Total Value	
1977/78				211,133	$10.93	$2,307,683.	
1976/77	404,625	$24.31	$9,836,433.	232,710 (45)	16.00	3,723,360.	$13,559,793.
1975/76	328,721	10.00	3,287,210.	188,329 (38)	6.00	1,129,974.	4,417,184.
1974/75	360,798	22.00	7,937,556.	166,282 (37)	11.00	1,829,102.	9,766,658.
1973/74	422,558	22.00	9,296,276.	168,975 (38)	13.50	2,281,162.	11,577,438.
1972/73	452,994	21.00	9,512,874.	228,200 (35)	14.00	3,194,800.	12,707,674.
1971/72	375,213	17.18	6,446,159.	161,409 (34)	12.00	1,936,908.	8,386,067.
1970/71	355,379	12.55	4,460,006.	100,049 (36)	10.00	1,000,490.	5,460,496.
1969/70	433,408	14.63	6,340,759.				
1968/69	437,875	18.40	8,056,900.				
1967/68	420,437	15.05	6,327,255.				
1966/67	371,533	12.73	4,729,615.				
1965/66	372,635	15.40	5,738,579.				
1964/65	415,261	11.81	4,904,232.				
1963/64	463,837	13.33	6,182,947.				
1962/63	436,708	12.48	5,450,115.				
1961/62	386,823	10.99	4,251,184.	145,030 (41)	8.87	1,286,416.	5,537,600.
1960/61	399,459	11.83	4,725,599.				
1959/60	344,766	13.73	4,733,637.	174,236 (32)			
1958/59	320,584	10.20	3,205,584.				
1957/58	341,674	10.45	3,570,493.				
1956/57	280,671	11.54	3,238,943.	158,346 (32)			
1955/56	282,036	12.10	3,412,635.	173,553 (31)			
1954/55	320,389	14.88	4,767,388.				
1953/54	242,453	10.50	2,545,756.	73,954 (18)			
1952/53	224,606	13.90	3,122,023.	143,012 (29)			
1951/52	222,932	14.91	3,323,916.	140,189 (31)			
1950/51	180,817	23.58	4,263,664.	130,030 (26)			
1949/50	157,416	20.99	3,304,161.	104,540 (27)			
1948/49	161,926	20.72	3,355,106.	76,330 (22)			

TABLE 14.3—*Continued*

Fur Year	Canadian Sales[a][c]			United States Sales[b][c]			Total Continental Value
	Total Pelts	Average Price per Pelt	Total Value	Total Pelts[d]	Average Price per Pelt	Total Value	
1947/48[e]	168,242	32.31	5,435,899.	112,700 (30)			
1946/47	127,622	29.46	3,759,744.	91,646 (21)			
1945/46	153,899	50.80	7,694,950.	108,690 (20)			
1944/45	128,999	36.33	4,686,533.	55,800 (17)			
1943/44	130,764	37.02	4,840,883.	34,899 (6)			
1942/43	102,241	29.96	3,063,140.	29,807 (4)			
1941/42	106,176	22.55	2,394,268.	5,555 (1)			
1940/41	90,123	23.03	2,075,532.	2,500 (1)			
1939/40	78,659	18.18	1,430,020.				
1938/39	64,086	15.38	985,642.				
1937/38	54,148	10.50	568,554.				
1936/37	55,759	12.54	699,217.				
1935/36	44,600	10.11	450,906.				
1934/35	50,175	8.23	412,940.				
1933/34	59,199	8.05	476,551.				
1932/33	71,799	9.74	699,322.				
1931/32	65,276	11.56	754,590.				
1930/31	51,313	14.77	757,893.				
1929/30	47,775	21.46	1,025,251.				
1928/29	57,043	26.61	1,517,914.				
1927/28	74,338	26.78	1,990,771.				
1926/27	100,364	22.85	2,293,317.				
1925/26	111,707	19.77	2,208,447.				
1924/25	151,913	20.22	3,071,680.				
1923/24	169,172	15.03	2,542,655.				
1922/23	175,275	14.04	2,460,861.				
1921/22	232,134	18.38	4,266,622.				
1920/21	164,656	16.31	2,685,539.				
1918/19	210,880	25.30	5,335,264.				

[a]Canadian harvest and mean price data through 1969/70 were provided by Statistics Canada, Animal Products Unit, Ottawa, Canada.

[b]From Ashbrook 1950, 1951, 1953, 1954; Janzen 1963; USDI 1956, 1957, 1961.

[c]Data from 1970/71 to present from Deems and Pursley (1978).

[d]Numbers in parentheses are number of states reporting.

[e]From Denny 1952; United States data includes Alaska.

beaver trapper operated 20 Conibear traps, trapped an average of 58 days, and caught 50 beaver, or an average of 0.86 beaver per day. The average price paid locally for beaver pelts was $7.89, which was less than was paid locally the previous two years and substantially less than the $21 average paid for beaver taken in Canada. The total average income per trapper from beaver including the sale of castor and oil glands and meat was $480.23, or $8.28 per day.

During the 1973/74 fur season, the average trapper operated 16 Conibear traps and, limited by extended periods of high water, trapped an average of 32.2 days. He caught 39.7 beaver, or 1.23 per day. The average price received per pelt was $7.46 and the total income from beaver was $288.51, or $8.96 per day. These amounts would be greater if calculated on a per-trip basis or if trappers had checked their traps daily rather then every other day.

Beaver trappers usually caught one river otter for every 8 to 10 beaver in addition to other fur bearers including mink and raccoon in water sets and bobcat and fox in land sets adjacent to beaver ponds. Exclusive of bounties paid by landowners, the total average income realized by beaver trappers for all the fur caught during the season ending February 1973 was $1,736.52, or $29.94 per day. The total average income per trapper from all fur caught during the 1974/75 fur season was $768.61, or $23.87 per day. After expenses, the daily profit was $27.30 per trapper in 1972/73 and $21.42 in 1973/74.

The income from the harvest of beaver in Canada in an equivalent effort would have yielded $1,050 (18.10 per day) and $873.40 (27.12 per day), respectively, during the 1972/73 and 1973/74 fur seasons from the sale of beaver pelts alone. Income from beaver by-products should have been adequate to cover costs or royalties. In addition, income from other furbearers taken in the process of beaver trapping would have made total income figures from beaver trapping in Canada more profitable.

Pelt Primeness. The winter coat of the beaver begins to develop as fat reserves are stored in October, the last area to become prime being the front shoulders. *Primeness,* a trade term describing quality and the amount of fur per unit of area, is typified by optimum fur length and a luxurious gloss or sheen, conditions fur farmers associate with good nutrition. Another indication of primeness is a thickening of the leather and the almost complete absence of bluish or dark coloration of the skin.

Primeness is an important consideration in determining the value of beaver pelts. Pelt size, the extent of damage, the quality of preparation, and, to a lesser extent, fur coloration influence pelt value. To obtain maximum income from a beaver resource, pelts should be harvested at the time of optimum primeness. The period of primeness in beaver, like that of other fur-bearers, varies from one part of the continent to another (Staines 1979). Huey (1956) indicated that the months of greatest primeness in New Mexico were December, January, and February, with a slight drop in February values due to the high incidence of cuts and damaged pelts. De Vos et al. (1959) reported that beaver in Ontario appeared to reach primeness about 15 October, as indicated by fully developed guard hairs and a clear hide. Staines (1979) reported January through April as months of primeness, with March the month of greatest primeness. In Alabama, most pelts are prime by late November. Yearly variations, believed to be associated with prolonged warm fall weather, have been noted. Huey (1956) also reported the influence of climatic factors on pelt primeness. Primeness diminished rapidly in late February in Alabama and in April in New Mexico (Huey 1956) due to "rubbing," a trade term describing shedding.

Trapping seasons are generally set to coincide with the time of pelt primeness, with some consideration in colder regions given to practical aspects of harvest. Denney (1952) reported varied trapping seasons. The earliest in Newfoundland opened 15 October (N. Payne personal communication 1980), and the latest in Yukon Territory and Ontario continued through 31 May. Trapping seasons in effect today vary with location, but most fall within these starting and ending dates (Hill 1975).

Pelt primeness is maintained longer at middle and northern latitudes than at the southern range extreme. "Springyness," a trade word describing loss of luster and sheen, or a bleached, flat, or singed appearance affecting the guard hairs, is the first indication of loss of primeness. Springyness at the southern range limit is associated with warm weather and may be delayed two or three weeks by late winter cold. Early springyness at the northern range limit is a more gradual change, perhaps a reflection of a lowered nutritional plane as food caches are used or sour in late winter.

Apart from primeness, color is another criterion for grading beaver pelts. Darker colors are considered best. Dark, approaching black pelts come from Quebec, Northwest Territory, western Canada, and Washington. Somewhat lighter pelts come from the northeastern United States. Even lighter colored pelts come from warmer areas of the western United States and from the plains areas, where alkaline waters may influence coloration. Beaver from the southeastern region have a chestnut reddish sheen that has been valued substantially below pelt colors of other regions (Denney 1952), probably due in part to poor handling techniques (Stains 1979). However, they are usually as valuable as pelts from California (Hensley and Fox 1948) and Arizona (Wire and Hatch 1943).

Beaver pelts are graded according to size by the total of the length from between the eyes to the tail plus the length from side to side on oval and round pelts. Variations in pelt measurements, sizes, and grades are also used locally. These may involve the sum of lengths from pelt edge to pelt edge across left front and right rear leg holes, and across right front and left rear leg holes. The sum of the measurements is matched to standard trade names and sizes for sorting oval or round pelts (table 14.4).

Size alone does not guarantee a high pelt value, as there may be differences in fur quality among pelt sizes. In the southeastern United States, medium, large medium, and large sizes are normally better furred than extra large and larger sizes. Fur thickness should be considered, as pelts are stretched to dry. Overstretched pelts tend to have thin fur, whereas understretching tends to thicken fur. To maximize fur quality in their various sizes, pelts should be stretched only to the lower limit of the appropriate size class.

Water Benefits. In many instances beaver are beneficial to people because their need for and manipulation of water results in water storage. Except during periods of high seasonal runoff and "cloudbursts," water is

TABLE 14.4. Standard trade names and sizes for oval or round beaver pelts

Trade Sizes	English Dimensions (in.)	Metric Dimensions (cm.)
Cub (or kit)	less than 40 in.	less than 101 cm.
Small	40–45	102–114
Small medium	45–50	114–127
Large medium	50–55	127–140
Large	55–60	140–152
X-large	60–65	152–165
XX-large (blanket)	65–70	165–178
XXX-large (super blanket)	greater than 70	greater than 178

effectively trapped and stored. The amount depends on the number of beaver dams. If dams are numerous and well distributed, they have the effect of holding most of the precipitation near where it falls or melts, thus filling the soil to saturation. It is then released gradually, moving by gravitational pull downslope and laterally through subterranean seepage. The water finds its way into feeder streams and subsequently to larger streams providing regulated flow that continues through dry periods. Water storage of this type is extremely valuable in sheltered mountainous areas of western North America, where precipitation occurs primarily as winter snow (Smith 1938; Houk 1942).

There are numerous instances and areas where beaver pond water has been useful to people for such purposes as fire fighting, irrigation, city water supplies (Wire and Hatch 1943), and livestock watering (Moore and Martin 1949). Finley (1937) placed a value of $300 on a western beaver for its water storage alone.

In addition to storing water and stabilizing stream flow, beaver manipulation of flowing water is valuable in elevating water tables (Finley 1937; Tappe 1942). Benefits include improved production of pasture, row crops, fruit, or wood products through moisture provided by subterranean irrigation in areas adjacent or downslope from ponds.

Benefits to Fish. Beaver ponds in the Southeast enhance conditions for populations of warm water fishes both in ponds and downstream (Hanson and Campbell 1963; Pullen 1967). They are generally beneficial to trout fisheries in mountainous and other areas of extremely cold waters, but are harmful where marginal water temperature in the eastern and north central United States become excessively warmed by beaver ponds (Knudsen 1962). Denney (1952) and Hakala (1952) summarized available literature on the benefits of beaver ponds to trout fisheries as: "(1) greater production of plankton and other microscopic organisms in ponds, thus serving as a rearing unit for small fish, (2) deeper water for protection in winter, (3) maintenance of continuous trout fishing in swift, rocky streams, (4) creation of trout habitat where none existed before, and (5) the warming of waters too cold for optimum trout growth."

Erosion Control. In the same fashion that beaver dams store water, they serve as basins for the entrapment of streambed silt and eroding soil. Their effectiveness within a watershed is dependent upon the number and distribution of dams. Sheet erosion, particularly on cultivated and excessively grazed mountainous areas, and gully erosion, resulting from poor logging, farming, and road construction practices, carry tons of soil downslope. Slit-laden waters slow as they pass through a series of beaver ponds and allow the heavier particles and some of the colloidals to settle out before flowing into larger streams. Following the establishment of aquatic and early successional plant species in the newly deposited sediment, conditions become favorable for the stabilization of the flood plain by more permanent woody vegetation.

Beaver are credited with stabilizing numerous eroding branch heads in the piedmont physiographic region of Alabama and Georgia following the small farm row-crop operations that were prominent in the early 1900s. It seems they could again function to correct one of mankind's abuses of the land if stocked and protected on the headwaters of streams draining strip-mined regions of North America. Acid mine waters also might undergo a degree of purification by pooling and downslope seepage through soil. More importantly, a series of beaver ponds high on a strip-mined watershed probably would provide some of the essentials for the establishment of early plant successional stages.

Benefits to Wildlife. Many species of wildlife are benefited as a result of alteration of the landscape by beaver for their own purposes (Johnson 1927; Rutherford 1955; Swank 1949; Knudsen 1962; Johnson 1967). Although habitat types that support beaver are extremely diverse from one area to another, in a given locality, habitats are often homogeneous, and without beaver are less attractive to many wild fauna. The major benefits of a beaver pond complex come from the creation of standing water, edge, and plant diversity, all in close proximity.

In the southeastern coastal plain, beaver activities inject a marsh dimension into otherwise homogeneous or mixed stands of water oak (*Quercus niger*), sweetbay (*Magnolia* spp.), sweetgum, or pine (*Pinus* spp.). Before the dead trees in a new pond have fallen, early successional stages including sedges (*Carex* spp.), pondweeds (*Potamogeton* spp.), cattails (*Typha* spp.), water lilies (*Nuphar* spp.), bullrush (*Scirpus* spp.), and *Sparganium* spp. become established in shallow areas and along the shore. Beaver may maintain dams and open water areas for decades if sprouting and second-growth sweetgum, willow, and other winter food species are adequate. They appear to make less use of areas where solid stands of alder (*Alnus serrulated*) become established in heavily silted ponds.

The open water areas provide habitat for birds (Lochmiller 1979), waterfowl (Hakala 1952; Beard 1953; Speake 1955; Arner 1963), other aquatic furbearers (Grasse 1951; Knudsen 1962; Anonymous 1967; Lauhachinda and Hill 1977), and fish (Rasmussen 1940; Huey 1956; Pullen 1967; Hanson and Campbell 1963). Areas opened by beaver cutting, and their edges, are attractive to regionally different assortments of rodents, lagomorphs, ungulates, predators, birds, and other wild fauna. Although the plant species may differ, there are similarities in the ways plants in different regions respond to the beaver's modification of habitat. These beneficial relationships are relatively constant; the habitat diversity of beaver ponds occurs in the upland aspen types of the Rocky Mountains, large river valleys of central Alaska, low-gradient floodplains of the midwestern states, and the sugar maple–ash–birch types of eastern North America. However, the year-round benefits to other wildlife resulting from beaver-created habitat diversity may be greater at the middle and lower latitudes than in the northward-flowing glaciated and lake-dotted drainages of Canada. In Canada, benefits may be negated by an abundance of water and extended periods of cold weather.

Esthetics. The esthetic and public relations value of a beaver pond complex are potentially immense. Because every schoolchild has heard of and can identify a beaver, it represents to many a starting point in the concept of "nature" or "wildlife." Denney (1952) mentioned the public sentiment related to the beaver's intelligence and engineering skills. The public can identify and relate to a peeled stick, the stump of a cut tree, a dam and the pool it holds, a lodge, or the sight of a beaver swimming on the surface in late afternoon. The beaver pond is an ideal area to observe a variety of wildlife activity. Wildlife agencies can accrue public relations benefits simply by identifying, marking with an explanatory sign, and maintaining strategically located and observable beaver ponds. To a segment of the public, activities and physical presence of a beaver pond can represent a bit of peaceful wilderness, often within the city limits. It is an ideal place to introduce and gain public support for management concepts.

Food Resource. Beaver meat is good to eat if properly processed. In portions of Canada, it has been as important a food resource as a fur resource. Among some groups in Alaska, it is the choice of available wild meats (M. Boyce personal communication 1979) and is used extensively for dog food. Nutritionally, it contains 20.3 percent protein and is comparable to most red meat protein sources. In a series of taste panel comparisons, "beaver burger" had slightly less overall acceptability than venison or beef burgers (Hill unpublished data). It was considered good barbequed, fried, or baked, the first two being preferred slightly over the last. It is used for food in some states and in the northern portions of Canada except in southern Ontario and Quebec and the Atlantic provinces. Substantial numbers of carcasses throughout North America are wasted, or are not used for human food. Based on current harvest figures (Deems and Pursley 1978) and an approximate yield of 3.7 kg of meat per carcass, the continental beaver population could provide 2,900 metric tons of meat annually. At $2.20 per kg the annual value would be approximately $6.4 million.

Recreation. An important benefit of the beaver is the recreation provided by tending a beaver trap line and processing the fur taken during the harvest. Beaver also provide an opportunity to learn and teach woodsmanship, animal track and sign reading, and pride associated with well-handled pelts. In addition, the beaver, through its habitat modifications, provides habitat for other furbearers that also provide trapping recreation. Some people removed from rural environments often do not understand relationships among wild animals and their habitats. These people may not accept consumptive use of any animal in such a way that causes its death. However, most people that have lived in close association with rural settings look at furbearer populations and their perpetuation rather than the fate of individuals. To these people, trapping is not only acceptable but also a valuable recreational endeavor, because it occurs in a natural setting and is a diversion from the confusion of the metropolitan sprawl. Most trappers generally agree that regardless of the species of furbearer they are attempting to take, the thrill and anticipation of tending a trap line is a very enjoyable and pleasant experience. Most look forward at the end of the summer to the first frost, the first ice, and the time for setting traps, knowing that they could make greater financial gains in any number of endeavors. Nevertheless, they will not engage in trapping without some profit incentive.

Another recreational pursuit associated with beaver pond habitat is waterfowl hunting. With relatively little expense, drainage devices can be installed in beaver dams to control water levels, thereby facilitating the seeding and production of foods attractive to waterfowl (Arner 1963).

Damage. Following the continent-wide restocking programs that took place during the 1950s, wildlife administrators began to receive increasing numbers of beaver damage complaints. Yeager and Rutherford (1957), commenting on beaver management in the Rocky Mountain Region, stated, "Western wildlife administrators, although appreciative of the beaver's potentialities for better hunting, fishing, and trapping are now faced with the problems of over-supply." Administrators experienced increasing damage, costly control, stream erosion, siltation and deteriorating environments where beavers had exhausted food supplies. In the southeast, beaver damage problems were noted in the late 1940s and increased to the point that in 1967 a symposium was devoted to the seriousness of problems and how to cope with them (anonymous 1967).

Beaver damage varies by type, magnitude, and region. Based on a survey of states and provinces, Denney (1952) listed the three most common beaver complaints in order of decreasing importance as flooding of roads, fields, and pastures; damage to timber by flooding and cutting; and damage to dikes, ditches, and dams. The first of these occurs range wide, the second is most serious in the southeastern United States, and the last is primarily a problem of arid western states. Where gravity flow irrigation systems are used in western states, beaver often plug ditches, make holes in dikes, and interfere with water control stuctures (Grinnell et al. 1937; Tappe 1942; Grasse and Putnam 1950; Huey 1956). Flooding of roads and highways is presently a serious problem in most of Canada (J. Gunson personal communication 1980).

Pastures and row-crop fields adjacent to streams or small drainages containing willows or other suitable food provide both access and other essentials for beaver, who dam the drainages, spreading water over the adjoining fields. Where corn is grown adjacent to beaver-inhabited streams, lakes, or wetlands, beaver damage can be expected. Beaver also damage ornamentals and shade trees near streams and lakeshores.

The importance of timber damage in the southeastern states was recognized soon after beaver were restocked (Moore and Martin 1949). Their populations increased dramatically during the 1950s, and timber damage now far exceeds other types of complaints (Smith 1964; Beshears 1967; Moore 1967; Godbee and Price 1975). Timber damages were reported by 67 percent of the respondents to landowner questionnaires in Alabama (Hill 1976) and 90 percent in South Carolina (Woodward et al. 1976). Recent estimates of beaver-caused timber damage (in millions of dollars) were 1.8 in Arkansas, 2.2 in Alabama, and 3.1 in Georgia (Hill 1976). Subsequent damage estimates exceeded $45 million in Georgia (Godbee and Price 1975) and $17 million per year in Mississippi (Arner and DuBose 1980).

The magnitude of the timber damage in the southeastern states is related to tree value and the relatively flat terrain along stream flood plains. A beaver dam 25 to 38 cm in height there inundates proportionally larger areas than it would in regions with steeper topography. If the root systems remain inundated for one or two growing seasons, a proportionally larger number of trees die.

Timber killed or damaged in the Southeast by beaver may be of low quality depending on past cutting practices. However, some stands in the larger dbh classes are of very high quality. Some stands may contain four or five logs per bole, and if killed, the financial losses are proportional. Beaver frequently gnaw bark from sawlog- and veneer-quality hardwoods such as sweetgum, ash, and sugarberry in large riverbottoms. Bite wounds may involve only a portion of the tree circumference; however, they subject the tree to diseases and subsequent rot (Toole and Krinard 1967). At injury sites, sweetgum trees exude storax, an aromatic balsam, to which beaver seem attracted. They gnaw such trees repeatedly.

Beaver damage to high-value timber resources of the Southeast was the major stimulus for meetings between forestry and conservation agencies to assess regional damage problems and to review possible beaver control programs (anonymous 1967). Mounting public pressure for control measures and the lack of trapper harvest associated with relatively poor regional pelt prices forced many conservation agencies to open their respective states to year-round harvest of the beaver. Legislative acts in Alabama, Tennessee, and Kentucky placed bounties on nuisance beaver populations. A number of regional studies also were done to evaluate control measures other than trapping (Cooper 1970; Williams 1971; Gordon and Arner 1976; Hill 1976).

Compared to benefits of beaver throughout their range, damage to trout fisheries is probably less important, but it can be serious where low-gradient or marginal trout streams become sufficiently warmed to be unproductive. Degradation of trout waters is important in areas of New York, Minnesota, Wisconsin, and other states, particularly if local trout streams are scarce. Hakala (1952) and Denney (1952) summarized the harmful effects of beaver ponds on trout as: (1) higher temperatures of water in beaver ponds on sluggish streams with subsequent reduction in available oxygen; (2) destruction of spawning areas and eggs by siltation; (3) reduction of certain species of aquatic insects used as food by trout; and (4) the barrier effect of dams to trout movement.

Beaver damage types as reported by landowner questionnaires were given by Hill (1976) and Woodward et al. (1976). Smith and Knudsen (1955) also noted numerous types of less frequently reported damage.

MANAGEMENT

Because of its commercial value, its potential for creating wetland wildlife habitat, and its potential as a pest, the beaver in some areas presents the greatest management challenge of any wildlife species. In areas where it is both beneficial and a nuisance, managers are faced with opposing situations. They must stimulate trapper support to provide adequate harvest and to assist with nuisance complaint problems, yet limit trapping intensity to prevent overharvest. Compounding this problem, beaver management strategies are influenced by public attitudes and participation, political pressures, national economics, agency funding levels, fuel availability, landownership, climatic and environmental conditions, human integrity, the animal's regional edaphic and ecologic role (Yeager and Rutherford 1957), and pelt price as controlled by styles and fashions of garments and hats. With all of these variables, management objectives differ with regions. In the Rocky Mountains the basic management objective is to provide range maintenance through control of population numbers. In the Southeast, objectives are to control density to avoid damage problems. In much of Canada and in areas of the northern United States, objectives are to maintain beaver numbers to provide sustained annual harvest.

Aside from state or provincial laws, and regions where beaver are taken on quota systems or used primarily for food, harvest pressure is controlled by the price paid for raw, handled pelts. Pelt price is the major unstable factor to which administrators must adjust beaver management programs. The magnitude of price fluctuations and the resultant management challenges are less severe in the Canadian provinces and northern and elevated parts of the United States than in the southern portion of the range. Depression of historically poorer prices of the Southwest (Wire and Hatch 1943) and the Southeast make beaver trapping a financial loss.

If prices decrease and remain depressed, there is a subsequent drop in trapper interest and effort, except where harvest quotas are set within bands, traplines, or area systems to take a minimum but not exceed a maximum number of beaver. Beaver populations then normally increase, damaging the habitat or spreading into areas where they become a nuisance. Conversely, if prices increase, trapper interest and resultant pressure increases and, except for areas under quota system management, the wildlife administrator must adjust strategies to guard against overharvest.

In areas where fur prices fluctuate widely and the beaver is a frequent source of nuisance complaints, a manager cannot support unlimited harvest during years of depressed prices and rigidly restrict harvest during years of high prices. Such a management strategy would jeopardize the confidence and support of resource user groups and those faced with nuisance problems. Management programs should attempt to strike a balance between enhancing consumptive use during periods of high prices and minimizing damage complaints during years of low prices.

Maintenance of the consumptive use aspect of beaver management assures the species of an economic as well as an esthetic status. This is essential, particularly at the southern range limit, if the beaver is to avoid being relegated through public or legislative action to a pest status. It would then become difficult to defend or maintain on public or private land. The beaver has no greater enemy than uninformed but well-meaning individuals who become caught up in protectionist movements to abolish trapping under the guise of "doing something for wildlife." Collectively these individuals can destroy what they claim to champion through rendering it valueless and forcing it into a pest status.

Public Attitudes. Opinions and attitudes of individuals, communities, and regions concerning the value of the beaver differ. They are formed and influenced by the benefits or expenses experienced as the respective entities interact with the beaver. Landownership, local options for private use of public land, and the way its use affects the use of adjacent private land have an impact on public attitudes toward the beaver. In general, the beaver is considered beneficial when not competing with humans for the use of land, and is viewed very positively throughout most of the nonagricultural areas of Canada, primarily because of its income potential. Opposing attitudes occur near the southern range limit where the fur value is substantially less and the damage to timber and interference with human land uses are much more severe. Intermediate between these two extremes are situations both east and west at the middle latitudes and the agricultural areas of Canada where the beaver is normally of good value but where it frequently conflicts with human land use practices. Where regional attitudes are generally favorable, there are occasionally individual landowners who are opposed to beaver. These situations add to the complexities of making management decisions for local beaver populations.

Local and regional strategies for management of the beaver are normally limited by human attitudes. An understanding of damage types, benefits, related costs and gains, and the effectiveness of information and education programs in influencing attitudes must exist before changes in management strategies are initiated.

Management Prerequisites. Strategies for management of beaver populations should be based first on an understanding of the beaver's vulnerability to overharvest. Its movement and behavior patterns are sedentary, routine, and easily learned (Wire and Hatch 1943). Also, a sign of active beaver is obvious to the novice observer, and even the inexperienced individual

wastes little or no time trapping or hunting unproductive areas. The relative ease with which beaver can be taken by hunting and trapping, combined with the fact that they are crowded into valleys (Yeager and Rutherford 1957), confined to watercourses, and can usually be extirpated through efforts on less than 10 percent of a geographic area, add to its vulnerability. The species' slow rate of reproduction and delayed sexual maturity preclude adequate reproduction to offset losses resulting from intensive annual harvest. Among furbearers, the beaver is one of the most vulnerable to trapping. This vulnerability enables people to eliminate the beaver locally and alter its habitat completely in one season of intensive trapping.

If an agency is to manage beaver populations, it must have flexibility and the authority to regulate the harvest (Bradt 1947) to fit changing harvest pressures and local beaver population needs. This authority is requisite to beaver management programs to insure perpetuation of the resource and provide for consumptive use. Such authority is highly variable among state and provincial governments (Denney 1952) and may be too restrictive to meet changing management objectives. In these cases, efforts should be initiated to gain public support for greater freedom to adjust harvest regulations to needs. Once management administrators have authority to regulate the beaver harvest, and recognize the need in beaver populations for both harvest and protection from overharvest, they can begin to address the various alternatives of beaver population management at local and regional levels.

It is always desirable to have the maximum amount of information available before undertaking the management of a beaver population. Yeager and Rutherford (1957) listed information necessary for beaver management in the Rocky Mountains as follows: "(1) determination of mangement units, (2) physical appraisal of area, (3) number of animals, (4) productivity, (5) degree of competition, (6) carrying capacity, (7) range trend, and (8) degree of harvest required to maintain stabilized populations and range conditions on areas suitable for beaver occupancy." Patric and Webb (1953) suggested that under intensive management, habitat and population inventories should be made to include determination of suitable habitat, its carrying capacity, beaver numbers, and their productivity. These information requirements may be appropriate for intensive management on the erodable soils of the Rocky Mountains.

In other regions beaver management similar to extensive beaver management described by Patric and Webb (1953) can be and is normally accomplished with less information. Population estimates compiled during the annual fall census (Gunson 1970) and annual harvest statistics obtained during pelt sealing or other methods of harvest monitoring are adequate information for managing beaver in most areas. A reasonably accurate count of the annual harvest is the minimum amount of information on which to base management decisions, even for states where the beaver is largely a nuisance. Without this minimum amount of information an agency has no measure of where it has been, much less where it intends to go in terms of beaver management, and would have difficulty defending its position as custodian of the resource.

The basic beaver management unit for most Rocky Mountain conditions is considered to be the stream and its tributaries (Yeager and Rutherford 1957). In most of Canada, the colony is the basic management unit within watershed, band, or trapline areas. Some provinces, however, base management on the number of lodges. In the management of beaver in South Dakota, Harris and Aldous (1946) treated colonies as the basic management unit. Beaver management in the southeastern United States, for the most part, involves no management unit and may or may not include a trapping season. Management seems to be centered on control, with no apparent safeguards to avoid overharvest and removal of beaver from extensive areas if extremely high prices prevail for three or four years.

Regional Management Philosophy. The philosophy for beaver management in the nonagricultural areas of Canada is totally different from that in most populated areas. The beaver is the major furbearer that for many years provided some economic stability for many forest-dwelling Canadians who, through choice or necessity, depended on the hunt for subsistence (Conn 1951). The beaver is considered mostly beneficial; its negative aspects are minor and within acceptable limits of multiple use on federal or Crown lands.

Close cooperation exists between the federal and provincial governments of Canada. Beaver management on provincial lands is administered by the provincial wildlife staff, which also handles nuisance problems on private land (Conn 1951). Except for nuisance complaints (Nordstrom 1972; Horstman 1979) confined primarily to the southern agricultural areas, beaver management in Canada centers on sustained annual income production from wise and consumptive use. In some areas, management is accomplished under limited provincial supervision with citizen participation in a remarkable system that incorporates private incentives to insure perpetuation of the resource (Conn 1951). In some provinces, the system involves registered traplines, fur management areas, or community areas or bands similar to the registered trapline system initiated in 1926 in British Columbia (Eklund 1946). Registered traplines or similar systems have been adopted in Manitoba and Quebec (Conn 1951), Ontario (de Vos et al. 1959), Saskatchewan (Gunson 1970), insular Newfoundland (Payne 1975), and Alberta (R. Butlin personal communication 1979). In addition, traditional traplines that are passed down from generation to generation exist in Yukon Territory, Northwest Territory, and Labrador (Payne 1975).

Traplines or trapping areas within natural boundaries such as watersheds, canoe routes, physiographic regions, or other divisions are assigned to communities, families, or individual trappers (Edwards and Cowan 1957; Payne 1975). Trapping areas average from about 57 (Payne 1975) to 70 km² (Edwards and Cowan 1957), varying among provinces and with demand for trapping opportunities. The trapper, family, or community retains the trapping rights within the bands, watersheds, or registered traplines unless they are revoked for failure to meet harvest quotas or for excessively trapping the area. Also, trappers have long-term harvest incentives to avoid overharvest for an immediate one-time gain. The provincial governments, and perhaps Dominion representatives and local fur management councils, assist with trapline or area selection and designation, and instruct trappers on basic census techniques and management and harvest concepts. In fall, trappers and provincial personnel make lodge counts, food cache counts, or other population estimates a basis for harvest quotas.

In areas where there are human concentrations and greater competition for trapping opportunities, aerial food cache or lodge counts also are made each fall. Based on the colony unit, harvest quotas are set. Varied regulations, including animal quotas per trapper, seasons, and methods of harvest, are employed to limit beaver harvest in specific zones.

Within the United States and some of the agricultural areas of southern Canada, beaver management philosophy is mixed. The problems arise as the local nuisance aspects of the beaver approach magnitudes that begin to negate the positive benefits. In some of the northern, sparsely populated states the beaver is considered beneficial despite the nuisance problems. In areas where beaver are economically beneficial, management philosophy normally centers on providing the maximum sustained income. Where beaver are both valued and considered a nuisance, management philosophies usually favor the beneficial aspects and reduce the nuisance aspects. It would be unfortunate indeed if the challenges of local beaver management were perceived to be so complex that state, province, or regional administrators choose alternatives such as eradication to management.

Harvest Objectives. Harvest objectives have to be varied to meet different conditions, but some harvest is necessary for sustained production in most habitats. No habitat can absorb continued and unrestricted beaver increase without deterioration and prolonged periods of unproductiveness. In some of the poor and marginal habitats that may be damaged by beaver and where they are a constant nuisance, colonies should be removed through intensive trapping (Yeager and Rutherford 1957; Payne 1975).

In areas where they are economically beneficial, beaver should be harvested in such a manner as to produce maximum sustained income. In Pennsylvania, for example, trappers far exceed the supply of beaver, and populations are maintained by stringent two-beaver-per-year bag limits and restrictive trapping techniques. Where beaver are both valuable and a nuisance, harvests should be adjusted to perpetuate the desired populations and hold nuisance problems to a minimum.

Because of their habitat modifications beaver constitute a threat to steep, erodable soils and irrigation systems in the mountains of Colorado, and are a source of nuisance complaints from private landowners. Despite being a valuable fur resource, beaver harvest is permitted year-round. Yeager and Rutherford (1957) suggested percentages of the population that should be harvested in a season to increase, decrease, or stabilize population levels in aspen and willow habitats below and above 1500 m. The practical mechanics of population reduction, through trapping colonies or individuals and its inherent tendency to be selective for adults cast some doubt that the percentages given would produce stabilized or increasing populations elsewhere. Denney (1952) suggested an annual harvest of one-third of the estimated population to maintain numbers or allow a slow increase. In the South, nuisance problems, damage to timber, and comparatively low fur value of the beaver have prompted the removal of harvest restrictions in Alabama, Arizona, Arkansas, Florida, Georgia, Mississippi, and Texas.

A harvest objective of one beaver per colony may not provide the maximum possible yield of beavers. Except where trapping is allowed at lodge or den entrances, trapping an individual colony selectively removes adults before other age groups. This practice tends to reduce the maximum reproductive potential of the colony, and in successive years, the number of harvestable beaver; trapping accomplished primarily in early winter may take the adult female or male before the female is bred. Gunson (1970) evaluated and recommended a rotational system of harvest, based on trapping colonies every two or three years with shorter rotations advisable in low-quality habitats. Boyce (1974) proposed trapping out some colonies for four or five years and allowing others to remain undisturbed. He suggested that such a practice would provide optimum colony dispersal sites, allow a rotation of vegetation use to permit regrowth in some areas while others are heavily utilized, and guard against removal of adult parous females, thereby maintaining a breeding population with an equal sex ratio. Payne (1975) suggested dividing traplines into sections and trapping within each section at three- or four-year intervals. He also suggested as another approach trapping only single and pair colonies. A determination of the most productive, practical, and profitable harvest schemes for management of beaver populations would seem an appropriate challenge for those desiring an extended study as well as for those engaged in exercises in mathematical modeling.

Beyond recreation, trappers are more concerned with the total sustained dollar income from the pelts handled than with numbers of pelts handled. Trappers generally prefer fewer large beaver to more small beaver that bring the same price. The selectivity of one-beaver-per-colony trapping for adults usually provides a greater proportion of larger-size beaver than other trapping schemes.

With mean colony sizes of 5 individuals (Denney 1952), a harvest quota of about 1 beaver per colony, or 20 percent of the stable population, approximates the 20–25 percent harvest level mentioned by Payne (1975) below which trappers would experience an economic loss. Trapping one-third of the colonies on a rotational system, as is done in Saskatchewan (Gunson

1970), provides for a sustained harvest rate of about 1.5 beavers per colony in good to fair habitats. Bradt (1947) suggested that one-third of the estimated population could be taken and still allow a slow increase. The colony seems to be the smallest convenient and practical unit on which to base management and harvest schemes at present. However, managers should be prepared to implement new harvest schemes if better techniques of determining colony size, age, and sex structure and selectively harvesting dispersing subadults are developed and prove economically advantageous on a sustained basis.

Program Funding. Beaver management programs in North America are normally a part of or subordinate to the fur resource programs of the individual states and provinces. Therefore, discussions of the quality and problems of local beaver management ultimately should include total fur program parameters. In some states or provinces, fur resources are a constant source of discussion or irritation to wildlife administrators, who frequently find they have funds to cover only the high-priority programs, and in many cases can or do not meet fur resource management responsibilities assigned under state or legislative acts. Ashbrook (1941) stated, "Almost every state has fur resources that are a source of income to some of its citizens. The methods of handling these resources are almost entirely haphazard, and in fact, few state game and conservation commissions have given sufficient, if any, serious thought to the matter." This statement was made nearly 40 years ago, yet despite many improvements and much progress, it has considerable validity today.

Lack of knowledge or interest, general apathy, or not wanting to "rock the boat" may be underlying motives. Reasons often given for failure to implement progressive fur resource programs are inadequate staff, insufficient funding, and fear of other influential resource user groups. The last is perhaps a valid yet weak excuse. The lack of staff and funding levels, however, can and should be overcome. If there is one segment of a conservation program that is capable of paying its own way, it is fur resources. The cash income incentives in trapping should be used to help support weak fur resource programs. "The most serious weakness in the conservation movement is its subordination of tangible to intangible values, of commercial to noncommercial, or if you please, of trapping to hunting. This subordination of fur resources is an incongruous viewpoint in a very materialistic world" (Ashbrook 1941). Most trappers and other fur resource user groups would favor better management programs and would voluntarily participate by supporting adequate license increases. Support of trappers can be gained by giving trapper organizations information on resource needs and on their capabilities to generate funding and influence the implementation of progressive programs based on sound biological principles. It involves a positive attitude and much work on the part of conservation administrators.

Damage Control. A number of potential beaver control measures have been evaluated, with varying measures of success. Large American alligators (*Alligator*

mississippiensis) in the Southeast were believed potentially capable of killing and eating adult beavers and were believed to take juveniles readily. Alligators were known to survive and grow well within the southern range limit of the beaver, but they were not effective in controlling beaver in three lakes where they were released (Hill et al. 1977). Also, extremely large alligators are known to have coexisted with beaver in ponds in Alabama for at least eight years. More importantly, alligators present a hazard to dogs and occasionally humans and are therefore undesirable as a control measure.

Efforts were directed toward development of poison bait substances such as silatrane, strychnine, gophacide, and sodium monofluoroacetate to control nuisance beaver (Cooper 1970; Williams 1971; Hill 1976). Poisons were found to be effective in killing penned beaver in Alabama. However, these compounds are not registered for that purpose, and the costs of distributing poisons would approximate those of a trapping operation while denying operators the benefit of both the meat and the pelts. More importantly, such programs are unproven on wild populations and would probably be opposed by the public.

Two chemosterilants were evaluated as reproductive inhibitors in wild beaver populations in Mississippi (Gordon and Arner 1976). Orally administered chemosterilants suppressed spermatogenesis, caused disruption of cells of the seminiferous tubules in males, and reduced ovulation and pregnancy in females. However, effective methods of treating wild beavers must be developed before the technique can be applied in practical management programs.

Trapping as an effective means of reducing beaver populations has been demonstrated continuously. The most prudent approach to animal damage problems is annual harvest, particularly where such harvest can involve citizen participants with no resultant public expense (Hill 1974, 1976). Trapping techniques for taking beaver vary with trapper preference and climatic conditions, the greatest differences occurring between areas that have extremely thick ice and those that are ice free. Harvest techniques that have been proven successful involve shooting, snaring, and trapping with either no. 3 or no. 4 leghold traps or no. 330 or no. 220 Conibear traps.

Shooting is a harvest technique employed in Canada following ice breakup, but it has application elsewhere. In situations where beaver have become serious pests, and where special permits can be used to hunt at night using spotlights and motorboats on large rivers or lakes, as many as 20 to 30 beaver can be taken per night. The procedure requires a scoped .22 caliber automatic rifle and involves driving the beaver to shore with a nose shot and killing it instantly with a neck or head shot as it clears the water. Beaver that sink when shot will usually float within an hour and can be recovered.

For special nuisance situations such as feeding area sets in ice-free areas, or for catching trap-shy beaver in open shallow water, no. 3 and no. 4 double-endspring leghold or jump traps with drowning devices are useful. They are also useful for taking beaver in baited sets under the ice at food caches (Canadian Trappers Federation 1977). At dam crossings, in runs, and in canals, leghold traps are generally less effective and more difficult to use than no. 330 or no. 220 conibear traps (Hill 1974).

Two or three conibear traps per colony, set for about two weeks during each of two years, effectively eliminated beaver in four Alabama watersheds (Hill 1976). Most of the adults and a few beaver three years and younger were caught the first year. Reasonably good, but less profitable, catches were made the second year, and during the interim period there was relatively little, if any, reproduction due to the removal of most of the adult females during the first year.

If it becomes necessary to remove beaver from an area, any of the techniques mentioned might be used. However, trapping efforts at southern latitudes are relatively unproductive in summer. If it is desirable to capture beaver alive for movement to other areas, live-traps or snares can be used. Although different live-traps are best suited for specific sets, those that have fewer moving parts to malfunction are more consistent and are preferred (Harris and Aldous 1946; Woodward 1977; Hodgdon 1978). Snares are preferable to live-traps because they are cheaper and can be more easily carried, handled, and set. An area of nuisance beavers can be saturated with snares and the catch per unit of effort and expense will frequently match other methods. Snare tie wires should be secured on land to avoid drowning the beaver, and snares should be checked early each morning to avoid death of a beaver due to hyperthermia from prolonged exposure to the sun.

A little-known but effective method of controlling nuisance beaver, particularly where it is desirable to drain water areas immediately, involves the use of trained dogs. Mixed breeds have been used during warm weather to remove beaver from pond complexes in southwest Alabama. Commencing at the lower end of drainages, dams are broken or blown with dynamite to allow pond areas to drain. Dogs are then introduced in lodges and exposed bank den entrances to flush out the beavers. Beaver in lodges can be exposed with less difficulty than those in bank dens. They frequently can be caught alive or may be shot.

Beaver frequently cut ornamentals and other valuable plants or trees near lakes or streams. Plant trunks can be protected by hardware cloth or similar wire mesh extending about 1 m above the ground. A diluted solution of creosote sprayed or painted on tree trunks has been found to reduce gnawing damage by beaver.

Census. Techniques suitable for census of wild animals over an extensive area, either through direct count or by indices, should be rapid. Although it is almost impossible to obtain absolute densities of beaver, relatively quick and accurate population estimates in elevated areas and northern latitudes can be obtained. These estimates include counting on foot or during aerial flights the number of food caches, lodges, or other indices of active colonies (Hay 1958; Gunson 1970; Nordstrom 1972; Payne 1975; Brown and Parson 1979) between leaf fall and permanent ice formation in September or October (Hay 1957). At the southern range limit, aerial survey estimates may be less accurate because winter foods are not cached and because beaver use both bank dens and lodges. Therefore, the only indices available for counting are dams, lodges, and pond areas (Beshears 1967). Hazards associated with low-level flight in fixed-wing aircraft in mountainous terrain or at high altitudes should be considered. Confining flights to early morning or late afternoon on relatively calm and cool days will help reduce hazardous conditions. Rotary-wing aircraft have greater maneuverability, but have similar flight limitations at high altitudes.

Interpretation of aerial photographs also can provide an index of population density (Dickinson 1971; Parson and Brown 1978). This technique is most applicable to drainages where beavers have been resident a number of years so that landscape modifications and impoundments show on photographs. Advantages of aerial photographs are that they provide a historical record and in some cases can show the quantities of timber lost to flooding. They also may be helpful in detecting population trends, trapper effectiveness, harvest or management needs, and food conditions. However, as with any aerial photo interpretation, knowledge of surface field conditions enhances the amount of information that can be obtained.

Harvest Monitoring. Monitoring the annual harvest of beaver can provide information requisite to beaver management. It establishes a system of data collection as a basis for determining population trends, and is a rapid means of determining the annual value of the beaver resource. Harvest monitoring can be accomplished in several ways. The payment of pelt royalties (as in some Canadian provinces) and pelt tagging or sealing systems are often used. Compulsory reporting systems by both fur buyers and trappers are also used, and provide a cross-check to detect some sources of bias (Todd and Geisbrecht 1979). Ideally, trapper fur-catch report forms should be designed to facilitate keypunch and computer analysis of fur harvest reports. Trapping licenses could be issued through the same system. By assigning each trapper a permanent license number, a variety of public relations, law enforcement, and related benefits could accrue (M. Beaudet personal communication 1980; M. Novak personal communication 1979).

Foods. Food is a necessity for the beaver. It is less important in some habitats than others, but should be considered in beaver management. In most areas, adult beaver eat a variety of woody and herbaceous plants and some vines. Utilization of herbaceous plants, particularly aquatics, row crops, grains, and succulent parts of woody plants, occurs primarily during the period of rapid growth before plant parts become dried and fibrous. The period of dependence on and utilization of cambium and noncorky bark of woody plants occurs primarily in winter months, when succulent foods are in short supply. Beaver eat the leaves, bark, buds, portions of branch ends, coppice sprouts, sap exudates, roots, and fruit of woody plants. Grinnell et al. (1937) and Wilkerson (1962) reported the use of acorns by beaver.

In the Mackenzie Delta, Northwest Territories, where the period of plant growth lasts from the third week in June to late August, the use of nonwoody vegetation was negligible (Aleksiuk 1970). Brenner (1964) reported that woody and succulent vegetation was used by beaver in northern Pennsylvania during periods of eight and four months, respectively; however, in warmer climates where herbaceous vegetation is available earlier in the spring and later in the fall, it is used for almost six months (Grinnell et al. 1937; Cooper 1970).

There is wide regional variation in the number and composition of woody plant species utilized. There are also variations in the species utilized among local physiographic regions (Denney 1952; Huey 1956) and among the colony territories within specific watersheds (Jenkins 1975). Woody plants used vary from as few as 3 at the northern range limit (Aleksiuk 1970) to as many as 22 species in Louisiana (Chabreck 1958) and 38 in South Carolina (Woodward 1977). Where available, the most preferred genera of woody food plant are *Populus* and *Salix,* the latter having greater capability for sustained use because of its second growth sprouting characteristics. Other genera that have been frequently reported as important are *Acer, Alnus,* and *Liquidambar* (Woodward 1977).

Availability of a winter food supply is the most important factor in determining range and distribution of the beaver. Aleksiuk (1970) reported heavy dependence on willow and that its removal caused a high degree of instability in beaver populations at the northern range limit. The dependence of beaver on aspen at high altitudes and at middle latitudes is freqently noted. Fire, logging operations, or other disturbances that reduce climax or advanced forest species to earlier deciduous succession stages including aspen or alder have a long-range beneficial effect on woody foods of the beaver (Slough and Sadleir 1977). With respect to manipulating foods in beaver management, Slough and Sadleir (1977) stated, "there are only a limited number of alternatives available for beaver-aspen management." These alternatives involve manipulating the food supply through regeneration and beaver numbers through harvest strategies; they are more applicable to regions where winter foods include only a few species. At the southern range limit, where winter foods are not cached and include many species, stimulation of the food supply is not important as a management concept and should be avoided when potential nuisance problems exist.

Intensive site preparation for pine plantations adjacent to some minor tributaries in the gulf coastal plain has been beneficial to beaver. When bottomland hardwoods were removed an abundance of beaver food was created adjacent to streams. The new growth of woody plants was followed by nuisance beaver problems that had not existed under food conditions of the closed canopy in previous stands.

Management Methods and Techniques. During some phases of management it may become necessary to restrain beaver in order to to palpate them to determine sex or pregnancy, to take measurements, or to ear tag or web mark them (Aldous 1940; Bradt 1947; Grasse and Putnam 1950). Beaver, particularly large ones, are capable of inflicting severe bites and their heavily clawed, powerful feet can cause bad scratches or lacerations to the hands of one attempting to hold them. In addition, some beaver may die of shock if physically restrained sufficiently long to attach ear tags. To overcome these problems, Whitelaw and Pengalley (1954) developed a bag with draw strings to manipulate beaver. Beaver larger than 13.5 kg were partially anesthetized by placing ether-soaked cotton over the respiratory openings in the head of the sack.

Other drugs have been evaluated for anesthetizing beaver (Allen 1965; Brooks and Dodge 1978; Lancia et al. 1978). For yearling and adult age classes, dosage rates of 10 to 13 mg/kg of ketamine hydrochloride plus 2.5 mg of acepromazine maleate were found to be a safe and effective immobilant that produced narcosis in less than 10 minutes without deleterious effects (Lancia et al. 1978). Doses of 100 mg of ketamine hydrochloride produced narcosis in kits; however, depending on the research needs, they can frequently be handled without anesthesia.

To facilitate individual study, identification and marking methods have been described using web punching (Aldous 1940), tail marking (Bradt 1947), ear tagging (Rasmussen and West 1943; Miller 1964), night-illuminating collars (Brooks and Dodge 1978), and radio telemetry techniques (Cooper 1970; Lancia and Dodge 1977). A modification of the ear-tagging technique described by Miller (1964) that employs reflective scotchlight-backed aluminum disks (1.6 cm in diameter) (Hill 1972*b*) attached with a no. 3 monel ear tag enables color marking sex or age groups within individual colonies to facilitate night study.

Characteristics of tooth eruption and annual cementum layers were found reliable in determining age in the beaver (Van Nostrand and Stephenson 1964; Larson and Van Nostrand 1968). This technique has essentially replaced earlier aging methods except that several body measurement criteria, including weight, tail measurement, and zygomatic breadth, have been useful for separating live beaver into four recognized age classes (Patric and Webb 1960). The nature of the cementum deposition reflects the major growth periods in summer and retarded growth in winter as described earlier. Teeth may be sectioned by grinding off the side of the root end of the molar teeth using a 150-grain circular stone turning at 1,800 to 4,000 rpm. This fine-grain stone facilitates gradually exposing transverse sectional views of teeth for frequent inspection during the grinding process. The first mandibular molars have been used in several studies, but were found less reliable than premolars for age determination on Alabama beaver. The lingual surface and occasionally the anterior edge of the mandibular premolar generally showed more distinct annuli. Molar annuli also are occasionally indistinct throughout the length of the tooth cementum. In these situations, other surfaces of the tooth can be ground or additional molars or premolars may be removed and several surfaces ground. Ages assigned (by two individuals) to teeth from 1,405 Alabama beaver differed by one, two,

three, four, and five years in 131, 15, 3, 2, and 1 specimens, respectively. Differences were reconciled by additional grinding and reexamination of the teeth, examination of other teeth, and comparisons with stained and mounted sections of decalcified teeth.

Landownership. Landownership patterns influence beaver management strategies that state and provincial agencies can administer within their boundaries. In most states and provinces, laws provide protection to private landowners from depredations and nuisance aspects of wild animals. When there is a legal conflict between maintaining depredating wildlife and protection of private property, judgments are usually in favor of protecting private property. It is unrealistic to expect private landowners to devote portions of their land to beaver production (Johnson 1927). However, natural resources agencies should be prepared to provide private landowners who value the beneficial aspects of beaver with the guidance to maintain populations at desired levels. Stoddard and Day (1969) discussed the use of private lands for public recreation. Wire and Hatch (1943) described methods of distributing income proceeds among participating parties.

Beaver damage draws less attention on public, federal, or Crown lands than on private property. Private citizens are quick to file damage complaints with the appropriate local agency for corrective action, particularly if control measures and harvest are rigidly regulated. In addition, there is generally public support for use of public land for wildlife. Yeager and Rutherford (1957) recommended involving forest managers and land management appraisers in the formulative and planning steps for beaver management on public lands in the Rocky Mountains.

Future Role in Soil and Water Management. The role of the beaver in the geological development of meadows in mountain valleys was alluded to by Ives (1942), but the immense value of their contribution toward protecting and conserving the continental soils is poorly understood and may never be fully appreciated. We should observe, study, and attempt to understand how we can best utilize the beaver and its actions to its maximum potential. Future generations may need to emulate their techniques of manipulating and storing water where it falls, particularly if we are forced to reclaim productive lands now enundated by large flood control impoundments.

Financial support for this chapter was provided in part by the Alabama Forestry Commission and the Game and Fish Division, Alabama Department of Conservation and Natural Resources, as Pittman-Robertson Project 44-IV, administered through the Alabama Cooperative Wildlife Research Unit, Auburn University, Auburn, Alabama. J. W. Lovett contributed substantially to the collection of data from Alabama. The author gratefully acknowledges the assistance of the following individuals who critically read portions or drafts of the manuscript or shared their knowledge or expertise on the beaver: M. S. Boyce, D. J. Cartwright, J. L. Dusi, J. R. Gunson, H. A. Hodgdon, E. L. McGraw, R. E. Mirarchi, N. F. Payne, E. E. Provost, and N. Van Nostrand.

LITERATURE CITED

Aldous, S. E. 1940. A method of marking beavers. J. Wildl. Manage. 4:145–148.

Aleksiuk, M. 1968. Scent-mound communication, territoriality, and population regulation in beaver (*Castor canadensis* Kuhl). J. Mammal. 49:759–762.

———. 1970. The function of the tail as a fat storage depot in the beaver (*Castor canandensis*). J. Mammal. 51:145–148.

Aleksiuk, M., and Cowan, I. 1969. Aspects of seasonal energy expenditure in the beaver (*Castor canandensis*). Can. J. Zool. 47:471–481.

Allen, K. E. 1965. Immobilizing beaver with succinylcholine chloride. Game Research in Ohio 3:215–220

Anonymous. 1955. Large beaver. Washington State Game Bull. 7:2.

Anonymous. 1967. Alabama beaver symposium. Alabama Dept. Conserv., Game and Fish Div., Montgomery. 51pp.

Arner, D. H. 1963. Production of duck food in beaver ponds. J. Wildl. Manage. 27:76–81.

Arner, D. H., and DuBose, J. S. 1980. The impact of the beaver on the environment and economics in the southeastern United States. Proc. Int. Wildl. Conf. 14. 13pp. In press.

Ashbrook, F. G. 1941. The position of fur resources in the scheme of wildlife management. Trans. North Am. Wildl. Conf. 6:326–331.

———. 1950. Annual fur catch of the United States. U.S.D.I. Wildl. Leafl. 315. 23pp.

———. 1951. Annual fur catch of the United States. U.S.D.I. Wildl. Leafl. 315. 25pp.

———. 1953. Annual fur catch of the United States. U.S.D.I. Wildl. Leafl. 346. 24pp.

———. 1954. Annual fur catch of the United States, U.S.D.I. Wildl. Leafl. 362. 24pp.

Bailey, V. 1926. Construction and operation of Biological Survey beaver trap. U.S.D.A. Misc. Circ. 69. 4pp.

———. 1927. Beaver habits and experiments in beaver culture. U.S.D.A. Tech. Bull. 21:1–40.

Beard, E. B. 1953. The importance of beaver in waterfowl management at the Seney National Wildlife Refuge. J. Wildl. Manage. 17:398–436.

Beer, J. R. 1955. Movements of tagged beaver. J. Wildl. Manage. 19:492–493.

Benson, S. B. 1936. Notes on the sex ratio and breeding of beaver in Michigan. Univ. Michigan Mus. Zool. Occas. Pap. 335. 6pp.

Bergerud, A. T., and Miller, D. R. 1977. Population dynamics of Newfoundland beaver. Can. J. Zool. 55:1480–1492.

Beshears, W. W. 1967. Status of the beaver in Alabama. Pages 2–6 *in* Alabama beaver symposium. Alabama Dept. Conserv. Game and Fish Div., Montgomery. 51pp.

Bollinger, S. 1980. Scent marking behavior of beaver (*Castor canadensis*). M.S. Thesis. Univ. Massachusetts, Amerst. 165pp.

Bond, C. F. 1956. Correlations between reproductive conditions and skull characteristics of beaver. J. Mammal. 37:506–512.

Boyce, M. S. 1974. Beaver population ecology in interior Alaska. M.S. Thesis. Univ. Alaska, Fairbanks. 161pp.

Bradt, G. W. 1938. A study of beaver colonies in Michigan. J. Mammal. 19:139–162.

———. 1939. Breeding habits of beaver. J. Mammal. 20:486–489.

———. 1940. Note on breeding of beaver. J. Mammal. 21:220–221.

———. 1947. Michigan beaver management. Michigan Dept. Conserv. Game Div. 56pp.

Brenner, F. J. 1964. Reproduction of the beaver in Crawford County, Pennsylvania. J. Wildl. Manage. 28:743–747.

Brooks, R. P. 1977. Induced sterility of the adult female beaver (*Castor canadensis*) and colony fecundity. M.S. Thesis. Univ. Massachusetts, Amherst. 90pp.

Brooks, R. P., and Dodge, W. E. 1978. A night indentification collar for beavers. J. Wildl. Manage 42:448–452.

Brown, M. K. and Parson, G. 1979. Reliability of fall aerial censuses in locating active beaver colonies in northern New York. Proc. Northeast Fish and Wildl. Conf. 36. In press.

Bryce G. 1904. The remarkable history of the Hudson Bay Company. Reprinted 1968 by Burt Franklin, New York, 501pp.

Buckley, J. L. 1950. The ecology and economics of the beaver (*Castor canadensis* Kuhl). Ph.D. Dissertation. State Univ. New York, Syracuse. 251pp.

Buckley, J. L., and Libby, W. L. 1955. Growth rates and age determination in Alaskan beaver. Trans. 20th North Am. Wildl. Conf. 20:495–507.

Cahn, A. R. 1932. Records and distribution of the fossil beaver, *Castoroides ohioensis*. J. Mammal. 13:229–241.

Canadian Trappers Federation. 1977. Canadian trappers manual. Can. Trappers Federation, North Bay, Ontario. Sections 1–18, looseleaf.

Chabreck, R. H. 1958. Beaver-forest relationship in St. Tammany Parish, Louisiana. J. Wildl. Manage. 22:179–183.

Clausen, G., and Ersland, A. 1968. The respiratory properties of the blood of two diving rodents, the beaver and the water mole. J. Respir. Physiol. 5:350–359.

———. 1970. Blood O_2 and acid-base changes in the beaver during submersion. J. Respir. Physiol. 11:104–112.

Conn, H. R. 1951. Federal aid in fur resources management in Canada. Trans. North Am. Wildl. Conf. 16:437–439.

Cook, A. H., and Maunton, E. R. 1954. A study of criteria for estimating the age of beavers. New York Fish Game J. 1:27–46.

Cooper, W. L. 1970. Preliminary investigations of certain beaver control methods in Alabama. M.S. Thesis. Auburn Univ., Auburn, Ala. 74pp.

Currier, A.; Kitts, W. D.; and Cowan, I. 1960. Cellulose digestion in the beaver (*Castor canadensis*). Can. J. Zool. 38:1109–1116.

Davidson, G. C. 1918. The North West Company. Russell & Russell Co., New York. 349pp.

Davis, W. B. 1940. Critical notes on the Texas beaver. J. Mammal. 21:84–86.

Deems, E. F. and Pursley, D. 1978. North American furbearers: their management, research and harvest status in 1976. Univ. Maryland Press, College Park. 165pp.

Denney, R. N. 1952. A summary of North American beaver management, 1946–1948. Current Rep. 28. Colorado Game and Fish Dept., Denver. 58pp.

de Vos, A.; Cringan, A. T.; Reynolds, J. K.; and Lumsden, H. G. 1959. Biological investigations of traplines in Northern Ontario. Ontario Dept. Lands and For. Wildlife Series 8. 62pp.

Dickinson, N. R. 1971. Aerial photographs as an aid in beaver management. New York Fish Game J. 18:57–61.

Dugmore, A. R. 1914. The romance of the beaver: being the history of the beaver in the Western hemisphere. W. Heinemann, London. 225pp.

Edwards, R. Y., and Cowan, I. M. 1957. Fur production of the boreal forest region of British Columbia. J. Wildl. Manage. 21:257–267.

Eklund, C. R. 1946. Fur resource management in British Columbia. J. Wildl. Manage. 10:29–33.

Finley, W. L. 1937. The beaver: conserver of soil and water. Trans. North Am. Wildl. Conf. 2:295–297.

Fleming, M. W. 1977. Induced sterility of adult beaver (*Castor canadensis*) and colony fecundity. M.S. Thesis. Univ. Massachusett, Amherst. 59pp.

Friley, C. E., Jr. 1949. Use of the baculum in age determination of Michigan beaver. J. Mammal. 30:261–265.

Fuchs, V. R. 1957. The economics of the fur industry. Columbia Univ. Press, New York. 168pp.

Garrison, G. C. 1967. Pollen stratigraphy and age of an early post-glacial beaver site near Columbus, Ohio. Ohio J. Sci. 67:96–105.

Godbee, J., and Price, T. 1975. Beaver damage survey. Georgia For. Comm. 24pp.

Gordon, K. L., and Arner, D. H. 1976. Preliminary study using chemosterilants for control of nuisance beaver. Proc. Southeast Assoc. Fish and Wildl. Agencies 30:463–465.

Grasse, J. E. 1951. Beaver ecology and management in the Rockies. J. For. 49:3–6.

Grasse, J. E., and Putnam, E. F. 1950. Beaver management and ecology in Wyoming. Bull. 6. Wyoming Game and Fish Comm. 52pp.

Grinnell, J.; Dixon, J. S.; and Linsdale, J. M. 1937. Fur bearing mammals of California, their natural history, systematic status, and relations to man. 2 vols. Univ. California Press, Berkeley.

Gunson, J. R. 1970. Dynamics of the beaver of Saskatchewan's northern forest. M.S. Thesis. Univ. Alberta, Edmonton. 122pp.

Haas, P. 1943. A study of beaver populations of some Utah streams. M.S. Thesis. Utah State Agric. Coll., Logan. 50pp.

Hakala, J. B. 1952. The life history and general ecology of the beaver (*Castor canadensis* Kuhl) in interior Alaska. M.S. Thesis. Univ. Alaska, Fairbanks. 181pp.

Hall, E. R., and Kelson, K. R. 1959. The mammals of North America. Vol. 2. Ronald Press Co., New York. 1083pp.

Hammond, M. C. 1943. Beaver on the Lower Souris Refuge. J. Wildl. Manage. 7:316–321.

Hanson, W. D., and Campbell, R. S. 1963. The effects of pool size and beaver activity on distribution and abundance of warm-water fishes in a north Missouri stream. Am. Midl. Nat. 69:136–149.

Harris, D., and Aldous, S. E. 1946. Beaver management in the Black Hills of South Dakota. J. Wildl. Manage. 10:348–353.

Hay, K. G. 1957. Record beaver litter for Colorado. J. Mammal. 38:268–269.

————. 1958. Beaver census methods in the Rocky Mountain region. J. Wildl. Manage. 22:395–402.

Hediger, H. 1970. The breeding behavior of the Canadian beaver (*Castor fiber canadensis*). Forma et Functio 2:336–351.

Henry, D. B., and Bookhout, T. A. 1969. Productivity of beavers in northeastern Ohio. J. Wildl. Manage. 33:927–932.

Hensley, A. L., and Fox, B. C. 1948. Experiments on the management of Colorado River beaver. California Fish and Game 34:115–131.

Hibbard, E. A. 1958. Movements of beaver transplanted in North Dakota. J. Wildl. Manage. 22:209–211.

Hill, D., Jr. 1975. Trapping laws and regulations. 2 vols. Woodstream Corp., Lititz, Pa.

Hill, E. P. 1972a. The cottontail rabbit in Alabama. Agric. Exp. Stn. Bull. 440. Auburn Univ., Auburn, Ala. 103pp.

————. 1972b. Litter size in Alabama cottontails as influenced by soil fertility. J. Wildl. Manage. 36:1199–1209.

————. 1974. Trapping beaver and processing their fur. Zoo-Ent. Dept. Series, Alabama Coop. Wildl. Res. Unit No. 1, Agric. Exp. Stn. Auburn Univ., Auburn, Ala. 10pp.

————. 1976. Control methods for nuisance beaver in the southeastern United States. Proc. 7th Vertebr. Pest Control Conf. Univ. California, Davis. 13pp.

Hill, E. P.; Lasher, D. N.; and Roper, R. B. 1977. A review of techniques for minimizing beaver and white-tailed deer damage in southern hardwoods. Pages 79–93 in Proc. 2nd Symp. Southern Hardwood. U.S. For. Serv. Southeast Region. 187pp.

Hodgdon, H. E. 1978. Social dynamics and behavior within an unexploited beaver (*Castor canadensis*) population. Ph.D. Dissertation. Univ. Massachusetts, Amherst. 292pp.

Hodgdon, K. W. 1949. Productivity data from placental scars in beavers. J. Wildl. Manage. 13:412–414.

Hodgdon, K. W., and Hunt, J. J. 1953. Beaver management in Maine. Game Div. Bull. 3. 102pp.

Hodgdon, K. W., and Larson, J. S. 1973. Some sexual differences in behavior with a colony of marked beavers (*Castor canadensis*). Anim. Behav. 21:147–152.

Horstman, L. P. 1979. Evaluation of beaver depredation and control in the Edmonton Fish and Wildlife Region, 1979. Alberta Energy and Nat. Res. Fish and Wildl. Div. 35pp.

Houk, I. E. 1942. When beaver aid irrigation. Sci. Am. 130:161.

Howell, A. H. 1921. A biological survey of Alabama. U.S.D.A. Bio. Surv. North Am. Fauna. 45. 88pp.

Huey, W. S. 1956. New Mexico beaver management. Bull. 4. New Mexico Dept. Game and Fish. 49pp.

Hunt, J. H. 1952. Some Maine beaver carcass study results. Proc. 8th Annu. Northeast Wildl. Conf. 8pp.

Innis, H. A. 1930. The fur trade in Canada. Rev. ed. 1962. Yale Univ. Press, New Haven. 446pp.

Irving, L. 1937. The respiration of beavers. J. Cell. Comp. Physiol. 9:437–451.

Irving, L., and Orr, M. D. 1935. The diving habits of the beaver. Science 82:569.

Ives, R. L. 1942. The beaver-meadow complex. J. Geomorph. 5:191–203.

Janzen, D. H. 1963. Annual fur catch in the United States. U.S.D.I. Wildl. Leafl. 452. 4pp.

Jenkins, S. H. 1975. Food selection by beavers: a multidimentional contingency table analysis. Oecologia 21:157–173.

Jenkins, S. H., and Busher, P. E. 1979. *Castor canadensis*. Mammal. Species. 120:1–8.

Johnson, A. S. 1967. Some beneficial aspects of beavers and beaver activities. Pages 12–17 in Alabama Beaver Symp. Alabama Dept. Conserv. Game and Fish Div., Montgomery. 51pp.

Johnson, C. E. 1927. The beaver of the Adirondacks: its economics and natural history. New York State Coll. For., Rossevelt Wildl. Bull. 4:500–641.

Kitts, W. D.; Bose, R. J.; Wood, A. J.; and Cowan, I. M. 1957. Preliminary observations on the digestive enzyme system of the beaver (*Castor canadensis*). Can. J. Zool. 35:449–452.

Kitts, W. D.; Robertson, M. C.; Stephenson, B.; and Cowan, I. M. 1958. The normal blood chemistry of the beaver (*Castor canadensis*). Section A: packed-cell volume, sedimentation rate, hemoglobin, erythrocyte diameter, and blood cell counts. Can. J. Zool. 36:279–283.

Knudsen, G. J. 1962. Relationship of beaver to forests, trout and wildlife in Wisconsin. Tech. Bull. 25, Wisconsin Conserv. Dept. 52pp.

Knudsen, G. J., and Hale, J. B. 1965. Movements of transplanted beavers in Wisconsin. J. Wildl. Manage. 29:685–688.

Kowalski, K. 1976. Mammals, an outline of theriology. Polis Sci. Publishers. 617pp.

Lahti, S., and Helminen, M. 1974. The beaver *Castor fiber* (L.) and *Castor canadensis* (Kuhl) in Finland. Acta Theriol. 19:177–189.

Lancia, R. A. 1979. Year-long activity patterns of radio-marked beaver (*Castor canadensis*). Ph.D. Dissertation. Univ. Massachusetts, Amherst. 126pp.

Lancia, R. A.; Brooks, R. P.; and Fleming, M. W. 1978. Ketamine hydrochloride as an immobilant and anesthetic for beaver. J. Wildl. Manage. 42:946–948.

Lancia, R. A., and Dodge, W. E. 1977. A telemetry system for continuously recording lodge use, and nocturnal and subnivean activity of beaver (*Castor canadensis*). Pages 86–92 *in* Proc. 1st Int. Conf. Wildl., Laramie, Wyoming. 159pp.

Larson, J. S. 1967. Age structure and sexual maturity within a western Maryland beaver (*Castor canadensis*) population. J. Mammal. 48:408–413.

Larson, J. S., and Van Nostrand, F. C. 1968. An evaluation of beaver aging techniques. J. Wildl. Manage. 32:99–103.

Lauhachinda, V., and Hill, E. P. 1977. Winter food habits of river otters from Alabama and Georgia. Proc. Annu. Conf. Southeast Assoc. Fish and Wildl. Agencies 31:246–253.

Lavrov, L. S., and Orlov, V. N. 1973. Karyotypes and taxonomy of modern beavers (Castor, Castoridae, Mammalia). Zool. Zhur. 52:734–742.

Leege, T. A. 1968. Natural movements of beavers of southeastern Idaho. J. Wildl. Manage. 32:973–976.

Leege, T. A., and Williams, R. M. 1967. Beaver productivity in Idaho. J. Wildl. Manage. 31:326–332.

Leighton, A. H. 1933. Notes on the relations of beavers to one another and to the muskrat. J. Mammal. 14:27–35.

Leopold, A. S. 1959. Wildlife of Mexico, the game birds and mammals. Univ. California Press, Berkley and Los Angeles, 568pp.

Libby, W. L. 1957. Observations on beaver movements in Alaska. J. Mammal. 38:269.

Lochmiller, R. L. 1979. Use of beaver ponds by southeastern woodpeckers in winter. J. Wildl. Manage. 43:263–266.

Longley, W. H., and Moyle, J. B. 1963. The beaver in Minnesota. Minn. Dept. Conserv., Div. Game and Fish. Tech. Bull. 6. 87pp.

Lowery, G. H., Jr. 1974. The mammals of Louisiana and its adjacent waters. Louisiana Wildl. and Fisheries Comm. Louisiana State Univ. Press, Baton Rouge. 565pp.

Lyons, P. J. 1979. Productivity and population structure of western Massachusetts beavers. Proc. Northeast Fish and Wildl. Conf. 36. In press.

McKean, T., and Carlton, C. 1977. Oxygen storage in beavers. J. Appl. Physiol. 42:545–547.

Mech, L. D. 1966. The wolves of Isle Royale. Fauna of the Natl. Parks of the United States Ser. 7. 210pp.

Miller, D. R. 1964. Colored plastic ear markers for beavers. J. Wildl. Manage. 28:859–861.

Miller, F. W. 1948. Early breeding of the Texas beaver. J. Mammal. 29:419.

Moore, G. C., and Martin, E. C. 1949. Status of the beaver in Alabama. Alabama Dept. Conserv., Montgomery. 30pp.

Moore, L. 1967. Beaver damage survey. Georgia For. Comm., Atlanta. 24pp.

Nasset, E. S. 1953. Gastric secretion in the beaver (*Castor canadensis*). J. Mammal. 34:204–209.

Nordstrom, W. R. 1972. Comparison of trapped and un-trapped beaver populations in New Brunswick. M.S. Thesis. Univ. New Brunswick, Fredericton. 104pp.

Novakowski, N. S. 1965. Population dynamics of a beaver population in northern latitudes. Ph.D. Dissertation. Univ. Saskatchewan, Saskatoon. 164pp.

Osborn, D. J. 1949. A study of age classes, reproduction and sex ratios of beaver in Wyoming. M.S. Thesis. Univ. Wyoming, Laramie. 105pp.

———. 1953. Age classes, reproduction and sex ratios of Wyoming beaver. J. Mammal. 34:27–44.

Panfil, J. 1964. Beaver: a vanishing animal in Poland. Center for Nature Protection of the Polish Acad. of Sci., Krakow. 64pp.

Parson, G. R., and Brown, M. K. 1978. An assessment of aerial photograph interpretation for recognizing potential beaver colony sites. Trans. Northeast Fish and Wildl. Conf. 35:181–184.

———. 1979. Yearling reproduction in beaver as related to population density in a portion of New York. Proc. Northeast Fish and Wildl. Conf. 36. In press.

Patric, E. F., and Webb, W. L. 1953. A preliminary report on intensive beaver management. North Am. Wildl Conf. 18:533–537.

———. 1960. An evaluation of three age determination criteria in live beavers. J. Wildl. Manage. 24:37–44.

Payne, N. F. 1975. Trapline management and population biology of Newfoundland beaver. Ph.D. Dissertation. Utah State Univ., Logan. 178pp.

———. 1980. Under-ice trap response and colony size, age, and sex structure of Newfoundland beaver. In preparation.

Pearson, A. M. 1960. A study of the growth and reproduction of the beaver (*Castor canadensis* Kuhl) correlated with the quality and quantity of some habitat factors. M.S. Thesis. Univ. British Columbia, Vancouver. 103pp.

Potvin, C. L., and Bovet, J. 1975. Annual cycle of patterns of activity rhythms in beaver colonies (*Castor canadensis*). J. Comp. Physiol. 98:243–356.

Provost, E. E. 1958. Studies on reproduction and population dynamics in beaver. Ph.D. Dissertation. State Coll. Washington, Pullman. 85pp.

———. 1962. Morphological characteristics of the beaver ovary. J. Wildl. Manage. 26:272–278.

Pullen, T. M., Jr. 1967. Some effects of beaver (*Castor canadensis*) and beaver pond management on the ecology and utilization of fish populations along warm-water streams in Georgia and South Carolina. Ph.D. Dissertation. Univ. Georgia, Athens. 84pp.

Rasmussen, D. I. 1940. Beaver-trout relationship in the Rocky Mountain region. Trans. North Am. Wildl. Conf. 5:257–263.

Rasmussen, D. I., and West, N. 1943. Experimental beaver transplanting in Utah. Trans. North Am. Wildl. Conf. 8:311–318.

Retzer, J. L.; Swope, H. W.; Remington, J. D.; and Rutherford, W. H. 1956. Suitability of physical factors for beaver management in the Rocky Mountains of Colorado. Fed. Aid Proj. W-83-R. Tech. Bull. 2. 32pp.

Robertson, R. A., and Shadle, A. R. 1954. Osteologic criteria of age in beavers. J. Mammal. 35:197–203.

Roth, A. R. 1938. Mating of beavers. J. Mammal. 19:108.

Rutherford, W. H. 1955. Wildlife and environmental relationships of beaver in Colorado forests. J. For. 53:803–806.

———. 1964. The beaver in Colorado: its biology, ecology, management, and economics. Colorado Game and Parks Dept., Game Res. Div. Tech. Bull. 17. 49pp.

Safonov, V. G. 1979. Experience of American beaver, *Castor canadensis,* introduction in Kamchatka. All Union

Research Inst. Game Manage. and Fur Farming U.S.S.R. Internat. Wildl. Congr. 14. English abstr.

Schlosser, M. 1902. Extinct beaver (*Castor neglectus*) from Tertiary of South Germany. Palaeontoligische Abh. 9:136.

Schorger, A. W. 1953. Large Wisconsin beaver. J. Mammal. 34:260–261.

Semyonoff, B. T. 1951. The river beaver in Archangel province. Pages 5–45 *in* Translations of Russian game reports. Vol. 1. Natl. Parks Br.-Can. Wildl. Serv., Ottawa. 109pp.

———. 1953. Beaver biology in winter in Archangel. Pages 71–92 *in* Biology of furbearers, English translation of Russian Game Reports, Vol. 1 translated by Can. Wildl. Serv., Ottawa.

Seton, E. T. 1900. Lives of game animals. 8 vols. Reissued 1953, by Charles T. Branford Co., Boston. 4:441–500. 949pp.

Shadle, A. R. 1930. An unusual case of parturition in beaver. J. Mammal. 11:483–485.

Slough, B. G., and Sadleir, R. M. F. S. 1977. A land capability classification for beaver. Can. J. Zool. 55:1324–1335.

Smith, A. E., and Knudsen, G. J. 1955. Beaver control in Wisconsin. Wisconsin Conserv. Bull. 20:21–31.

Smith, K. C. 1964. Beaver in Louisiana. Louisiana Wildl. and Fish. Comm. Wildl. Education Bull. 83. 8pp.

Smith, L. H. 1938. Beaver and its possibilities in water regulation. Univ. Idaho Bull. 32:85–88.

Speake, D. W. 1955. Seasonal abundance of waterfowl on sample areas in Lee County, Alabama. M.S. Thesis. Alabama Polytechnic Institute, Auburn. 110pp.

Stains, H. J. 1979. Primeness in North American furbearers. Wildl. Soc. Bull 7:120–124.

Stenlund, M. H. 1953. Report of Minnesota beaver die-off, 1951–52. J. Wildl. Manage. 17:376–377.

Stevenson, A. B.; Kitts, W. D.; Wood, A. J.; and Cowan, I. M. 1959. The normal blood chemistry of beaver (*Castor canadensis*). Section B: Blood glucose, total protein, albumin, globulin, fibrinogen, non-protein nitrogen, amino acid nitrogen, creatine, cholesterol, and volatile fatty acids. Can. J. Zool. 37:9–14.

Stoddard, C. H., and Day, A. M. 1969. Private lands for public recreation: is there a solution? Trans. North Am. Wildl. Conf. 34:186–196.

Svendsen, G. E. 1978. Castor and anal glands of the beaver (*Castor canadensis*). J. Mammal. 59:618–620.

Swank, W. P. 1949. Beaver ecology and management in West Virginia. Conserv. Comm. of West Virginia, Charleston. 65pp.

Tappe, D. T. 1942. The status of beavers in California. California Dept. Nat. Resour., Div. Fish and Game. Game bull. 3. 59pp.

Tevis, L., Jr. 1950. Summer behavior of a family of beavers in New York State. J. Mammal. 31:40–65.

Thomason, W. B., III. 1978. Seasonal patterns of beaver reproduction in east-central Mississippi. M.S. Thesis. Mississippi State Univ., Starkville. 28pp.

Todd, A. W., and Geisbrecht, L. C. 1979. A review of Alberta fur production and management, 1920–21 to 1977–78. Alberta Energy and Nat. Resour., Fish and Wildl. Div. 64pp.

Toole, E. R., and Krinard, R. M. 1967. Decay in beaver-damaged southern hardwoods. For. Sci. 13, no. 3 (September 1967):316–318.

Townsend, J. E. 1953. Beaver ecology in western Montana with special reference to movements. J. Mammal. 34:459–479.

U.S. Department of the Interior. 1956. Annual fur catch of the United States. U.S.D.I. Wildl. Leafl. 388. 3pp.

———. 1957. Annual fur catch of the United States. U.S.D.I. Wildl. Leafl. 398. 3pp.

———. 1961. Annual fur catch of the United States. U.S.D.I. Wildl. Leafl. 436. 3pp.

Van Nostrand, F. C., and Stephenson, A. B. 1964. Age determination for beavers by tooth development. J. Wildl. Manage. 28:430–434.

Voight, D. R.; Kolenosky, G. B.; and Pimlott, D. H. 1976. Changes in summer foods of wolves in central Ontario. J. Wildl. Manage. 40:663–668.

Whitelaw, C. J., and Pengalley, E. T. 1954. A method for handling live beaver. J. Wildl. Manage. 18:533–534.

Wilkinson, P. M. 1962. A life history of the beaver in east-central Alabama. M.S. Thesis. Auburn Univ., Auburn, Ala. 76pp.

Williams, R. K. 1971. Investigations of bait substances suitable for control of nuisance beaver. M.S. Thesis. Auburn Univ., Auburn, Ala. 69pp.

Wilsson, L. 1971. Observations and experiments on the ethology of the European beaver (*Castor fiber* L.) Swedish Sportsmen's Assoc. 8:266.

Wire, F. B., and Hatch, A. B. 1943. Administration of beaver in the western United States. J. Wildl. Manage. 7:81–92.

Woodward, D. K. 1977. Status and ecology of the beaver (*Castor canadensis carolinensis*) in South Carolina with emphasis on the Piedmont. M.S. Thesis. Clemson Univ., Clemson, S.C. 208pp.

Woodward, D. K.; Hair, J. D.; and Gaffney, B. P. 1976. Status of beaver in South Carolina as determined by a postal survey of landowners. Proc. 30th Annu. Conf. Southeastern Assoc. Fish and Wildl. Agencies 30:448–449.

Yeager, L. E., and Hay, K. G. 1955. A contribution toward a bibliography on the beaver. Colorado Dept. Game and Fish. Tech. Bull. 1. 103pp.

Yeager, L. E., and Rutherford, W. H. 1957. An ecological basis for beaver management in the Rocky Mountain region. Trans. North Am. Wildl. Conf. 22:269–299.

Young, S. P., and Jackson, H. T. 1951. The clever coyote. Univ. Nebraska Press, Lincoln. 411pp.

Zharkov, I. V., and Sokolov, V. E. 1967. The European beaver (*Castor fiber* L. 1758) in the Soviet Union. Castoria VI. Acta Theriol. 12:27–46.

Zurowski, W.; Kisza, J.; Kruk, A.; and Roskosz, A. 1974. Lactation and chemical composition of milk of the European beaver (*Castor fiber* L.) J. Mammal. 55:847–850.

EDWARD P. HILL, Mississippi Cooperative Fish and Wildlife Research Unit, Mississippi State University, Mississippi State, Mississippi 39762.

15

Muskrats

Ondatra zibethicus and *Neofiber alleni*

H. Randolph Perry, Jr.

NOMENCLATURE

COMMON NAME. Muskrat

SCIENTIFIC NAME. *Ondatra zibethicus*

The muskrat is a semiaquatic gnawing mammal (Rodentia: Cricetidae). The rodents constitute the largest mammalian order in numbers of species and individuals.

SUBSPECIES. Including the Newfoundland muskrat, formerly *Ondatra obscurus* (Pietsch 1970), there are 16 subspecies of *Ondatra zibethicus* in North America (Hall and Kelson 1959): *O. z. zibethicus* (eastern United States, southeastern Canada), *O. z. albus* (Manitoba and adjacent central Canada), *O. z. aquilonius* (Labrador, adjacent Ungava and Quebec), *O. z. bernardi* (Colorado River areas of southeastern California, southern Nevada, western Arizona, and Mexico), *O. z. cinnamominus* (Great Plains region), *O. z. goldmani* (southwestern Utah, northwest Arizona, southeastern Nevada), *O. z. macrodon* (middle Atlantic coast), *O. z. mergens* (northern Nevada, parts of adjacent states), *O. z. obscurus* (Newfoundland), *O. z. occipitalis* (coastal Oregon and Washington), *O. z. osoyoosensis* (Rocky Mountains, southwestern Canada), *O. z. pallidus* (south-central Arizona, west-central New Mexico), *O. z. ripensis* (southwestern Texas, southeastern New Mexico), *O. z. rivalicius* (southern Louisiana; coasts of Mississippi, western Alabama, and eastern Texas), *O. z. spatulatus* (northwestern North America), and *O. z. zalophus* (southern Alaska). Although morphologically similar, these subspecies vary widely as to population status, distribution, habits, and habitat; subspecific differences were discussed in detail by Errington (1963).

McMullen (1978) reported late Pleistocene fossil remains of *O. zibethicus* from Ellis County, Kansas. Similarly, fossil remains from Iberia and West Feliciana parishes, Louisiana, indicated that muskrats were present in the Pleistocene (Arata 1964; Domning 1969; Lowery 1974).

DISTRIBUTION

The muskrat is indigenous, common, and widely distributed throughout most of North America. It occurs

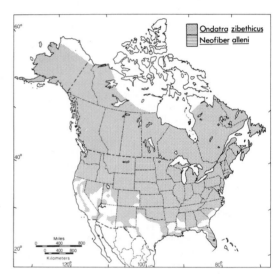

FIGURE 15.1. Distribution of the muskrat (*Ondatra zibethicus*) and the round-tailed muskrat (*Neofiber alleni*).

throughout the greater part of the continent from northern Mexico to northern Alaska and northern Canada (Dozier 1953; Lowery 1974) (figure 15.1). The muskrat is absent in Florida, most of Mexico, and parts of extreme northern Alaska and Canada. Errington (1963) extensively discussed the occupied North American range of the muskrat.

The muskrat is absent in many areas despite apparently suitable habitat, especially along the southeastern Atlantic coast (Dozier 1953; Wilson 1968). The nature of the absence of *O. zibethicus* from southern Georgia and Florida has been the subject of much discussion. Errington (1963) indicated that several authors believed the round-tailed muskrat (*Neofiber alleni*) took the place of *O. zibethicus* in Florida. However, tidal fluctuations, lack of expanses of habitat during catastrophic periods, and scarcity of certain vegetative communities may be factors that limit muskrat distribution in the southeast (Errington 1963). Apparently *O. zibethicus* did inhabit Florida during the Pleistocene period (Sherman 1952).

Numerous authors described range extensions of the muskrat on the Pacific coast, especially into California and Oregon (Hansen 1965; Beleich 1974; Wood 1974). Twining and Hensley (1943) discussed the history of muskrats in California and listed localities inhabited. Although it has been suggested that severe winters and lack of suitable areas of marsh might be factors that prevent muskrats from inhabiting tundra, muskrats have been harvested in the Northwest Territories at least 129 kilometers beyond the tree line (Stewart et al. 1975). Similarly, Noble and Wright (1977) observed muskrats on an Alaskan tundra lake 160 km west of the tree line.

Muskrat transplantations began after 1900. The early history of foreign and North American transplantation efforts was described by Storer (1937), along with some of the resulting problems, especially in irrigated regions. No muskrats were reportedly introduced into any of the United States or Canadian provinces or territories in 1976 (Deems and Pursley 1978).

DESCRIPTION

The muskrat is stocky, with a broad head, small ears, and small eyes; the neck is not distinctively restricted. Its legs are short and the tail is scaly, sparsely haired, and laterally flattened. Limbs are modified for aquatic life. The front feet are unwebbed, and have four toes, a rudimentary nailed thumb, and sharp claws. The five clawed toes of the hind feet are webbed proximally and have fringes of stiff hairs along the sides of the toes and feet.

The specific epithet refers to the typical musky odor during the breeding season. Paired glands, which enlarge during the mating period, emit a musky yellowish secretion through openings within the foreskin of the penis. This secretion mixes with the urine and is deposited throughout the muskrat's territory on defecation posts, lodge bases, stations along travel routes, and at the actual time of mating. Females have similar glands that are not as active as those of males (Schwartz and Schwartz 1959). Stevens and Erickson (1942) found that scent glands contain musk oil, a mixture of cyclopentadecanol and cycloheptadecanol and corresponding odiferous ketones. The scent remains viable long after exposure to air and serves as a means of advertisement (Errington 1963).

The pelage is soft and velvety. It consists of a thick, waterproof, soft underfur overlain by long, glossy guard hairs that practically conceal the fur dorsally and laterally. Dorsal coat color ranges from bright rusty red to dark brown (almost black). The fur is lighter on the sides, frequently with a brown, red, or yellow hue. Ventrally the coat is lighter in color, from whitish to broccoli brown, with various shades of rufous, cinnamon, and brown grading to a whitish throat. There are no apparent color differences between the sexes (Hall and Kelson 1959; Schwartz and Schwartz 1959).

Albinism, mutation, and pigment dilution cause individual color variations from white to black (Lowery 1974). Dozier (1948a) described the genetic basis of color mutations in the muskrat. The common black agouti coat (*ABC*) results from the joint action of three independent and dominant genes: *A*, a gene for the agouti, or wild-type, coat pattern; *B*, a gene that conditions production of black pigment as the darkest element of the coat; and *C*, a gene for the development of full color. A recessive mutation of the *C* gene results in some type of albinism. Dozier (1948a) classified several color mutations: a uniform black in Virginia and North Carolina marshes; a "fawn" color in the Chestertown, Maryland, area; a true albino with pink eyes and white fur; and a "Maryland white," where young individuals are maltese gray but white underneath. Four color phases have been observed in New York: normal brown, melanistic, albinistic, and "Maryland white" (Cook 1952). Detailed information on hair shape, patterns, and colors was provided by Cook (1952).

Dozier and Allen (1942) found that 54 percent of 9,895 muskrats trapped on the Blackwater National Wildlife Refuge in Maryland in 1941 were of the black color phase and 46 percent were brown. Further work on the refuge showed that 52 percent of the 1941–45 catch were of the black-and-tan phase and 48 percent were brown (Dozier et al. 1948). Sixty percent of the muskrats examined by Wilson (1956a) at Currituck County, North Carolina, fur sheds were brown and 40 percent were black.

Muskrats are 406 to 641 millimeters in total length; tail length ranges from 177 to 295 mm. Ranges of other body measurements are: hind foot length 63 to 92 mm, ear length 15 to 25 mm, skull length 60 to 70 mm, and skull width 38 to 44 mm (Hall and Kelson 1959; Schwartz and Schwartz 1959; Lowery 1974). The dental formula is 1/1, 0/0, 0/0, 3/3 = 16. The surfaces of the molars characteristically form a pattern of angled enamel folds surrounded by dentine. Characteristic of rodents, the single pair of upper and lower incisors is chisellike and separated from the cheek teeth by a diastema. The lips close behind the incisors, which allows underwater gnawing. The skull of the muskrat was described in detail by Gould and Kreiger (1948) (figure 15.2).

Muskrat weights vary geographically, but males average slightly heavier than females (Gould and Kreiger 1948) (table 15.1). As expected, muskrats tend to be larger (heavier) in northern latitudes. Maximum weights reported include 1,814 grams in Maryland (Dozier et al. 1948), 1,248 g in Delaware (Stearns and Goodwin 1941), and 1,956 g in North Carolina (Wilson 1956a). One notable reversal of the male:female weight trend was reported by Buss (1941): he found that, based on 589 animals trapped in Wisconsin during the fall, males averaged 42 g less than females.

PHYSIOLOGY

Smith (1938) stated that muskrats can suspend breathing when underwater for long periods; he described one muskrat that stayed submerged for 17 minutes, surfaced for not more than 3 seconds, then submerged for another 10 minutes. Respiratory adaptations for div-

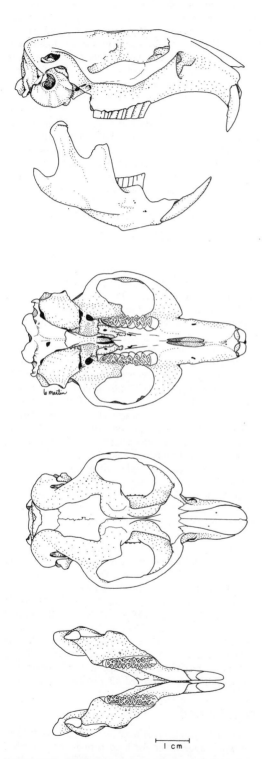

FIGURE 15.2. Skull of the muskrat (*Ondatra zibethicus*). From top to bottom: lateral view of cranium, lateral view of mandible, dorsal view of cranium, ventral view of cranium, dorsal view of mandible.

ing include a high tolerance of carbon dioxide; apparently carbon dioxide does not elicit the rapid internal responses for the escape from asphyxia during apnea generally characteristic of nondiving mammals (Errington 1963). However, Fairbanks and Kilgore (1978) found that postdive excess oxygen consumption (volume of oxygen consumed above what would have been consumed if the animal had not dived) was always greater than the calculated oxygen debt.

An increase in the metabolic rate and a decrease in whole-body insulation was observed by Fish (1977) for muskrats in water at 20–30° C, compared to those in air. He also reported that temperature of the appendages and whole-body insulation were inversely related. The higher thermal conductance of water was presumed responsible for the difference in physiological adjustment; apparently pelage insulation alone was insufficient to prevent general body cooling. Although the muskrat's pelage is well developed and waterproof, the tail and feet are essentially naked and provide radiator surfaces for excess heat produced during exercise and for heat exchange during high temperatures (Johansen 1962).

Although body cooling occurs during foraging activity in winter, it is retarded by periodic withdrawal from water (MacArthur 1979). Muskrats apparently avoid hypothermia by elevating the body temperature before entering water and by periodic rewarming within feeding shelters. The onset of diving bradycardia apparently occurs rapidly following submergence (Thornton et al. 1978). Submergence in cold water (as opposed to warm water) caused a significant drop in body temperature, and submergence in water at 2°, 20°, and 35° C suppressed heart rate to 24, 34, and 50 percent, respectively, of the resting heart rate (200 beats/minute). MacArthur and Aleksiuk (1979) found feeding shelters and push-ups were spaced to allow access to all points within the home range by short excursions, thus minimizing exposure in cold water. Huddling behavior is also an important overwinter survival mechanism of muskrats in northern latitudes (MacArthur 1977; MacArthur and Aleksiuk 1979).

Thompson and Baumann (1950) investigated the vitamin A content of muskrat livers to determine if a seasonal deficiency might influence population levels. Although amounts of vitamin A were usually higher in adult than juvenile muskrats, they concluded that it was unlikely that the population sampled could have suffered seriously from a deficiency. The composition of tryptic peptides of muskrat hemoglobin was presented by Genaux et al. (1976).

Stomachs (with contents) of 2,778 Delaware muskrats varied in volume from 2 to 130 cubic centimeters; average volumes for males and females were 22.34 and 23.04 cc, respectively (Stearns and Goodwin 1941). Without contents, stomach weights ranged from 0.2 to 3.4 g, with an average of 1.5 g.

REPRODUCTION

Anatomy. Superficially, the genitalia of male and female muskrats are similar. The prominent urinary papilla (clitoris) of the female lies anterior to the vagi-

TABLE 15.1. Selected muskrat weight data for North America by location, age, and sex

| Source | Location | N | Average Weight (g) | | | |
| | | | Immature | | Adult | |
			Male	Female	Male	Female
Donohoe 1961	Lake Erie marshes	1,782	1,096	1,031	1,299	1,257
Erickson 1963	central New York				1,153	1,181
					1,370[a]	1,323[a]
Dozier et al. 1948	Blackwater NWR, Maryland				1,030[b]	962[b]
Dozier 1945	Montezuma NWR, New York				1,644[b]	1,503[b]
Heit 1949	central New York	275			1,242[b]	1,202[b]
Wilson 1956a	Currituck County, North Carolina				1,102[c]	1,053[c]
Sather 1958	north-central Kansas	1,743	953	862	1,180	1,089
Stearns and Goodwin 1941	Delaware	2,778		907[d]		
O'Neil 1949	Louisiana			817–1,043[d]		

[a]Second-year adults.
[b]Age unknown.
[c]Nonkits greater than 624 g.
[d]Age and sex unknown.

nal opening and resembles the penis of the male. However, the perineum is furred in males but naked near the urethral papilla in females. Immature females have a completely naked perineum. Also, the anal to papilla distance is greater in males (Schwartz and Schwartz 1959; Taber 1969; Lowery 1974). A membrane covers the vaginal opening in unbred immature females. However, Chamberlain (1951) reported that in the fall and early winter a hymen may reform in adult females. Opening of the vaginal orifice of southern and central Wisconsin muskrats occurred in March, with partial reclosure in about one-third of the adults the last of October (Beer 1950).

Teats are absent on males. On females, three to four (rarely five) pairs of teats are present on the belly between the front and hind legs: two pairs are inguinal and two pectoral (Lowery 1974). Dozier (1942, 1953) described a technique to determine sex by palpation. In males the penis may be exposed or felt by stripping the urinary papilla posteriorly. Without palpation, females examined in the fall with vaginal openings sealed may be mistaken for males (Buss 1941). Sex determination of young muskrats was described by Baumgartner and Bellrose (1943).

Sather (1958) found that sex determination of young muskrats by the presence or absence of hair on the perineum was subject to error. However, very young muskrats could be sexed by the presence or absence of mammary glands and older animals by manipulation of the urinary sheath. Dried muskrat pelts may be sexed by the presence of mammae (Buss 1941).

Breeding Parameters. As early as the turn of the century, published accounts of the breeding of muskrats

varied widely (Lantz 1910). Reproductive data for the muskrat in North America are voluminous and are presented in practically every life history–ecological study published. Photoperiodism probably prepares muskrats for the breeding season, while the onset of mating may be triggered by a weather factor. Initiation of breeding and temperature are correlated (Olsen 1959b).

Muskrats are polyestrous (Schwartz and Schwartz 1959). Based on vaginal smears, Beer (1950) found the estrous cycle of southern and central Wisconsin muskrats averaged 28.7 days. Similarly, a 30-day estrous cycle with variations from 24 to 34 days was reported by Wilson (1955a). Initial research by McLeod and Bondar (1952) on Manitoba muskrats substantiated the findings by Beer (1950) that the estrous cycle was about 30 days. However, subsequent work by them produced quite variable results: estrous cycles ranged from 2 to 22 days (the mean for 136 cycles was 6.1 days). They interpreted the longer cycles reported earlier as abnormal cycles or pseudopregnancies induced by the experimental technique and not estrous cycles. Laboratory findings of McLeod and Bondar (1952) indicated that the female muskrat may go through three or four estrous periods before the male comes into sexual activity; similarly, the progressive decline in number of litters after the first litter appeared to be from a progressive decline in the sexual activity of the male with the approach of warmer weather. Therefore, the reproductive rate of the muskrat depends upon the sexual activity of the male.

Males and females are promiscuous or loosely monogamous (Errington 1963). All copulations in pens observed by Svihla and Svihla (1931) occurred when the muskrats were partially submerged in water. Similarly, Salinger (1950) described copulation of a pair of muskrats in an Idaho stream; copulation lasted about five minutes.

Errington (1937a) found that a postpartum estrus occurred in the muskrat. The interval was about a month between successive litters. The gestation period varied from an atypical minimum of 19 days to a more usual minimum of 22–23 days, with the ordinary period around 30 days. Similarly, Wilson (1955a) reported that the normal gestation period was 28–30 days.

Follicle maturation began in January for adult females in east Tennessee; sperm were present in adult males year-round (Schacher and Pelton 1975). After the breeding season had ceased, adrenal weights of adult male muskrats were at a peak. However, adrenal weights of nonpregnant adult females maximized during the early breeding season, coinciding with follicle maturation and early births. The maximum adrenal weight for pregnant females occurred in May (Schacher and Pelton 1976). Additional details of ovarian conditions during the reproductive cycle were presented by Enders (1939).

Beer and Meyer (1951) evaluated seasonal changes in endocrine organs and reproductive tracts. Gonadotropic activity of the pituitary was highest during spring and early summer and lowest in late and early winter. Seasonal metabolic functions of the muskrat were further evaluated by Aleksiuk and Frohlinger (1971).

Muskrat litter size and the number of young per adult female were found positively correlated with latitude (Boyce 1977). Additionally, he reported that the number of litters per year was inversely related to latitude. Breeding parameters vary greatly by geographical area (table 15.2). However, above 37° N latitude litter sizes generally average from four to seven, with three or fewer litters per year. South of the 37th parallel litters tend to be smaller (three to four), with more litters per year; breeding generally occurs year-round, except during severe winters, with reduced sexual activity in the deep south during the hotter months. Elsewhere, except in Canada and high elevations of the United States, breeding occurs from March through October, with March–June the peak period (Wilson 1955a).

The average number of embryos per female increased significantly from January to March (3.1 to 4.3, respectively) in Maryland (Harris 1952). Of 1,709 trapped females, from 93.3 to 95.9 percent of the adults had scars. Similarly, Beer and Truax (1950) reported that about 97 percent of adult females had bred, based on placental scar counts. In Mississippi, 13.3 percent fewer neonates occurred in the average litter than was indicated by embryo counts (Freeman 1945). Embryo resorption in muskrats is rare; Dozier (1947a) found only one incidence in over 15,000 female carcasses examined.

McLeod and Bondar (1952) reported that young were rarely born before 18 May in Manitoba. Within two weeks, depending upon breakup, a distinct rise in production of young occurred. Production then dropped until about one month later, when a second but lower upsurge in production occurred. Subsequent lesser peaks occurred at roughly monthly intervals through August. Olsen (1959b) found that average litter size varied, with peaks of litter production occurring at four-week intervals.

Young. Neonates are blind, nearly helpless, and have hair so scant that they appear almost naked (Errington 1939). At birth they weigh about 21 g and are 100 mm in total length. By one week of age they are covered with coarse gray brown fur (Schwartz and Schwartz 1959). Eyes open between 14 and 16 days, when young are about 80 g and 200 mm; after this, they may swim, dive, and climb (Errington 1939). Initially the young cling tightly to the mother's nipples. When the young are older, she carries them by the skin of the belly. The female can dive while carrying young, but while swimming she holds them above the water. Generally when the female leaves a nest containing naked young, they are well covered with shredded dry vegetation (Hensley and Twining 1946). The male rarely cares for the young.

Growth curves for typical, overfed, and underfed muskrats from 1 to 30 days of age were formulated by Errington (1939). Sather (1958) also provided measurements for newborn muskrats. Young may suckle up to 21 days, though 15–16 days is more common (Smith 1938). Weaning usually occurs in the fourth week (about 180 g and 285 mm) (Errington 1939). By this time the young are independent (figure 15.3) and their mother is usually ready to give birth again. The young may continue to utilize the nest or move to another area. If they remain and the female renests, she may construct a new chamber in the same house or bank (Schwartz and Schwartz 1959).

Young are referred to as "kits" when 70–90 days old (500–650 g, 400–480 mm). At 3.5 months of age muskrats are similar in size and appearance to small-medium adults. By 1 year of age they reach an average weight close to 1,100 g and a total length about 550 mm (Errington 1939).

Age at first breeding is variable. Generally, immatures reach puberty by one year. According to Errington (1940), Iowa muskrats did not reproduce during the same calendar year in which they were born; however, at eight to nine months of age larger males showed signs of approaching sexual maturity (Errington 1939). Schwartz and Schwartz (1959) stated that young born in early spring may breed in late summer but that their first breeding period usually occurred during the spring following their birth. However, female Louisiana muskrats may be pregnant at six to eight weeks of age (O'Neil and Linscombe 1976). In North Carolina the smallest muskrat with placental scars weighed 908 g (Wilson 1954a). Productivity was positively related to weight: an average of 6.6 young in the 907- to 964-g group to 12.4 young for females weighing 1,361 g or more.

ECOLOGY

Nest and Den. Muskrats that inhabit sloughs, levees, dikes, ditches, and shorelines generally den in banks if

FIGURE 15.3. Muskrats are blind and nearly helpless at birth, but are usually weaned in their fourth week.

suitable sites are available (Glass 1952; Schwartz and Schwartz 1959). Although they may prefer denning in high, firm banks to denning in houses (Dozier 1948*b*), muskrats inhabiting ponds with a fairly constant water level and abundant construction materials build houses (Glass 1952).

Stream-dwelling muskrats on the Athabasca-Peace Delta, Canada, were observed by Fuller (1951) using bank burrows for shelter throughout the year. Slough-dwelling muskrats used burrows in summer but constructed houses in the fall and push-ups in winter. Muskrats in marshes or in areas where banks are not readily available construct houses from aquatic vegetation.

Bank dens generally are 15 by 20 centimeters; depth varies from very superficial to over 1.5 meters. An air shaft to the surface may be constructed for dens in dense soil (Schwartz and Schwartz 1959). Dens made by burrowing are usually connected to water by a complex system of tunnels 12–15 cm in diameter. Tunnels may be 3 to 15 m long. Tunnel entrances are generally beneath the water level and surface openings are usually loosely plugged with vegetation. In farm ponds of Alabama the number of burrow entrances ranged from one to nine and averaged two; typical burrows began about 15 cm beneath the water surface. Nests were found in most of the major burrows excavated (Beshears 1951).

Earhart (1969) found that burrow construction in ponds was related to muskrat age; adults constructed the extensive, permanent breeding burrows and presumably young occupied small temporary burrows. Use of burrows by muskrats at Delta Marsh, Manitoba, appeared greatest in summer (MacArthur and Aleksiuk 1979).

The house provides a dry nest and stable temperatures. Nests are located above normal high tides and

provide insulation from summer heat and winter cold (O'Neil 1949). Although the maximum range in macroclimatic air temperature in Delta Marsh, Manitoba, Canada, was −39° to 34° C, microclimatic air temperature within lodges, burrows, and push-ups ranged from −9° to 30° C (MacArthur 1977; MacArthur and Aleksiuk 1979). Mean temperatures within occupied dwelling lodges and burrows in summer and winter were close to the thermoneutral zone of single, adult muskrats (10° to 25° C). The insulative value of dwellings was highest in winter; during summer, nest temperatures were similar to air temperatures. However, Cook (1952) reported that temperatures in a house in New York composed of cattails (*Typha* sp.) remained relatively cool despite high ambient temperatures.

Distinguishing among the feeding huts and platforms, push-ups, and various retreats that marsh muskrats may construct is frequently difficult. Houses or lodges are generally the largest, are more or less elliptical in shape, and have a rather lopsided top (figure 15.4). They are up to 2.5 m in diameter and vary in height from 0.4 to 1.3 m with a wall thickness of 0.1 to about 0.3 m (Dozier 1948*b*, 1953; O'Neil 1949; Schwartz and Schwartz 1959). One or two internal nest chambers are constructed. Individual family members may construct their own alcove, resulting in a multichambered house (Schwartz and Schwartz 1959). When a nest is present, it is above the water level with one or more plunge hole entrances to tunnels that lead to open water.

During periods of low water, canals are frequently constructed from the house to deeper water. Smith (1938) stated that muskrats prefer swimming to walking. They construct an elaborate system of canals in the marsh; these canals are from 15 to 30 cm wide and vary from muddy surface trails to ditches 30 cm or more deep. Debris, which is usually piled up along the

TABLE 15.2. Selected muskrat reproductive data for North America

Source	Location	Breeding Dates	Young per Litter	Litters per Year	Technique
Alexander 1951	Montezuma NWR, New York		5.6	2	
Arthur 1931, in Wilson 1954a	Louisiana		3.8	(14 young/season)	
Beer 1950	Wisconsin	April–June; occasional animals February–August			litters (N = 109)
Beer and Truax 1950	Wisconsin marsh		15.5 young/adult female		scars
Chamberlain 1951	Massachusetts marsh	mid-March–September	5.0	2.7	
Donohoe 1966	Lake Erie marshes	February–August (reduced April to August)		up to 3, based on breeding season	pregnancies
Dorney and Rusch 1953	Horicon Marsh, Wisconsin	highest-May, last September; secondary-June			
Erickson 1963	central New York ponds	May–July	6.3	1.5	
Errington 1937a	northeast Iowa				litters (N = 158)
Errington 1939	Iowa		6.5	2	litters (N = 163)
Errington 1943	Iowa		6.5		ovaries
Forbes and Enders 1940	Maryland	first cycle early February to mid-March			
Freeman 1945	Mississippi marsh	all year; greatest-late winter, early spring; least-July–August			
Fuller 1951	Athabasca-Peace Delta streams	initiated after ice in late April	7-10	2	
Gashwiler 1948	central and eastern Maine	March–July	5.4	2	
Gashwiler 1950b				2	
Harris 1952	Maryland (Dorchester Co.)	data collected January–March	3.9		embryos (N = 95)
Hensley and Twining 1946	California	data collected April–June, August–September	2.4	2	houses (N = 22)
Lay 1945b	Texas	practically year-round	4.7 (2-8)		
Louisiana Dept. of Conservation 1931	Louisiana	year-round except for late summer	4.0		

Reference	Location	Breeding season			Basis
Lowery 1974	Louisiana	year-round; highest-November, March; lowest-July, August found pregnancies last 10 days of March			pregnancies
Marshall 1937	Utah	mean date = 10 May			
Mathiak 1966	Horicon Marsh, Wisconsin				
McCann 1944	Minnesota				
McLeod and Bondar 1952	Manitoba	begins late March April-October	2	6	
Olsen 1959a	Delta Marsh, Manitoba		multiple	7.1-7.3	
O'Neil 1949	Louisiana marsh			3.2	litters
O'Neil and Linscombe 1976	Louisiana	year-round; highest-November, March; lowest-July, August	5-6 / 5-6	3.6-3.8 / 3-4	embryo counts
Reeves and Williams 1956	southeastern Idaho	May-August	1.6 / 2.4	7.0 / 7.3	N = 35 / N = 33
Sather 1958	north-central Wisconsin marshes and lakes	peaks-late April-early May, June, late August-early September	2.5	6.0-6.5 / 6.7-7.1 / 4.5-5.0	nestlings; adult scars { subadult and juvenile scars }; gross examination
Schacher and Pelton 1975	eastern Tennessee	late February and early March-mid-September; peak production-April-July	2.3	5.4	
Shanks and Arthur 1951	Missouri farm ponds		no more than 1		
Smith 1938	Maryland	most of year; intensive-mid-March-September	3	4-5	
Stewart and Bider 1974	southern Quebec collection ditches	peak-mid- to late April	2	6.3-6.6	
Svihla and Svihla 1931	Louisiana	all year; heaviest-November-April			
Wilson 1954a	North Carolina marshes	year-round except in cold winters; most-mid-February-May; small peak in August-September; heaviest-April-May	11 young/year	3.7 / (1-6)	placental scars (N = 194)
	Currituck, North Carolina		3 (11 young/female/year)	3.7	

ditches, is excavated with the front feet and cast up with the hind feet (Schwartz and Schwartz 1959).

Muskrat houses are typically made of natural materials present in the marsh or pond. Construction materials, which generally consist of roots and stems, are formed into conical piles; house walls are sometimes cemented with mud.

MacArthur and Aleksiuk (1979) found that lodges of the Delta Marsh in Manitoba were usually constructed of cattails (*Typha latifolia*) or bulrush (*Scirpus* spp.) interspersed with pondweed (*Potamogeton* spp.) and bottom detritus. Muskrats of a California marsh used tule (*Scirpus* sp.) stalks and roots and waterweed (*Elodea* sp.) for house construction (Hensley and Twining 1946). They also constructed open "porch nests" on the side of the house for loafing and sleeping.

Regarding muskrat use of vegetation in the Illinois River Valley, Bellrose and Brown (1941) concluded: (1) cattail communities supported more muskrat houses per unit area than other plant associations; (2) other plants utilized for house construction included marsh smartweed (*Polygonum* sp.), hornwort (*Ceratophyllum demersum*), and American lotus (*Nelumbo lutea*); and (3) muskrats used vegetation immediately surrounding the dwelling for construction purposes.

A hectare of good food habitat in South Dakota contained 2.8 houses (Aldous 1947). In North Carolina, undiked marsh frequently without water averaged 0.4 houses per ha; constantly flooded diked marsh averaged 1.6–2.2 houses per ha (Wilson 1955*b*).

House building began in early October and peaked in the first week of November in Massachusetts (Chamberlain 1951). Nicholson and Davis (1957) studied the duration of muskrat houses in marshes of the eastern shore of Maryland. The number built per month was constant during fall, increased about four-fold in February, and then practically stopped. Houses built from September to December lasted about 7.7 months; houses built in February persisted only 3.5 months.

Huenecke et al. (1958) investigated the accumulation of marsh gases in muskrat houses in Minnesota.

FIGURE 15.4. The muskrat house provides protection from predators and the elements as well as a suitable site for rearing young.

The only gas that accumulated in the houses was carbon dioxide. A gradual build-up of 5 to 7 percent occurred in February, followed by a sharp decline. They concluded that concentrations of this magnitude would have no effect on muskrats.

The feeding hut, or "feeder," is a place where muskrats bring in food for consumption on a small platform. The hut provides protection from the elements and predators and, during frozen conditions, ready access to oxygen. In marsh areas covered with water, muskrats construct feeding platforms or rafts by breaking down plants and supplementing them with stalks. Frequently, platforms are found on small, natural elevated areas of the marsh. Muskrats use these so that they may leave the water to eat (Smith 1938). Feeding platforms are temporary structures and are usually abandoned after a few days' use (Dozier 1948*b*, 1953).

The feeding hut is smaller than the house, roughly circular, and generally 30–41 cm above the water level. Several runs lead into the hut from beneath the ice or water level. MacArthur and Aleksiuk (1979) categorized winter lodges as dwelling or feeding shelters based primarily on external size. Winter lodges had greater external, nest chamber, and wall thickness dimensions than summer dwellings. Feeding huts characteristically had a reduced floor-to-water distance and contained rejected food items on the floor and in the plunge hole. Optimum water depth for lodges and feeding houses in Massachusetts was 41–46 cm (Chamberlain 1951).

A ratio of 2.7 huts per house was recorded in a central New York marsh Dozier (1948*b*). In Massachusetts, feeding houses were located at random and not correlated to lodges (Chamberlain 1951). Further south, where weather is less severe, feeding huts may be replaced by rather flimsy feeding shelters or feeding platforms (Louisiana Department of Conservation 1931). Muskrats in Oklahoma ponds only rarely constructed a feeding shelter, as is common with marsh muskrats (Glass 1952).

"Push-ups" are generally associated with frozen marshes and snow. They are made by muskrats cutting a 10- to 13-cm hole through ice and pushing up a 30- to 46-cm pile of fine fibrous roots or other vegetation. This pile of debris, forming an enclosed cavity, rests on top of the ice. The cavity serves as a rest site and may be used as a feeding area in severe weather. Push-ups are temporary and collapse as the ice melts. On streams, houses can be distinguished from push-ups because houses are built on a firm base or are otherwise anchored, contain a well-defined chamber, are usually on the periphery of a slough, and are constructed largely of emergent rather than submergent plants (Fuller 1951).

Muskrats may also use various shelters, such as hollow logs or stumps, piled vegetative debris, overhanging banks, or dens of other animals (MacArthur and Aleksiuk 1979). These retreats may be temporary and vary with local environmental conditions. For example, a floating blind was used by muskrats in Manitoba to rear young (Anderson 1969).

Waterfowl and other birds, reptiles, amphibians, spiders, insects, and mammals frequently utilize muskrat houses as resting, basking, or residence sites. Mathiak (1966) found 70 clutches of snapping turtle (*Chelydra serpentina*) eggs in muskrat houses on Horicon Marsh, Wisconsin, over a two-year period. Svihla and Svihla (1931) reported that numerous animals occupied muskrat houses in Louisiana, including ants; beetles; ground skinks (*Leiolopisma laterale*); frogs; narrow-mouthed toads (*Gastrophryne texensis*); Gulf coast toads (*Bufo valliceps*); and snakes, including the garter snake (*Thamnophis* sp.), kingsnake (*Lampropeltis getulus*), eastern cottonmouth (*Agkistrodon piscivorus*), and blue racer (*Coluber constrictor*). Muskrat houses in a California marsh were used as nesting sites by black terns (*Chlidonias niger*), Forster's terns (*Sterna forsteri*), Canada geese (*Branta canadensis*), ducks, and other birds (Hensley and Twining 1946).

Sex Ratios and Mortality of Young. The sex ratio tends to favor males (table 15.3). The degree of imbalance varies greatly but ranges from about 1:1 to 2:1, depending upon area, time of year, and method of evaluation.

The differential sex ratios found for stream-dwelling muskrats on the Athabasca-Peace Delta were attributed to differential trap mortality, fighting, and the greater tendency of males to wander (Fuller 1951). Similarly, Dozier and Allen (1942) indicated that sex ratios from trapping data may be biased because of the greater tendency of males to wander.

Differential mortality may be a contributing factor to sex ratio imbalance. Males outnumbered females in North Carolina in all age and weight groups, and a higher rate of mortality was indicated for adult females (Wilson 1956a). However, muskrats trapped at southwest Lake Erie had a male:female ratio of 87:100 for animals less than 907 g; other weight groups showed a preponderance of males (Anderson 1947). Similarly, Olsen (1959b) found proportionately more males with increasing litter age. The same relationship through the subadult age was observed by Sather (1958) for muskrats in north-central Wisconsin. A mortality factor acted on females between birth and winter and on males between their first and second winters. Females in Louisiana had a parasite infestation rate nearly four times that of males—15.65 percent versus 4.20 percent, respectively (O'Neil 1949).

Differential spring and summer mortality in Minnesota reduced the number of adult males and resulted in a sex ratio among adults of 83 males:100 females by fall; having a majority of immature males apparently compensated for natural losses within the adult male class during spring-summer (McCann 1944). Gashwiler (1950a) also suggested that males sustained disproportionately heavier mortality during the winter and spring periods.

Chamberlain (1951) stated that mortality of juvenile muskrats in a marsh near Concord, Massachusetts, based on 1949 trapping data, was 61 percent; 1950 data indicated that 73 percent of young were lost between birth and the fall harvest. Similarly, Schwartz and Schwartz (1959) reported that only about one-third of the young survived into the first winter. Sooter (1946) calculated that 56 percent of the young were lost between birth and the trapping season on Tule Lake Refuge, California. Mortality of juveniles from birth to 1 March was about 40 percent for muskrats on the Athabasca-Peace Delta (Fuller 1951). Mathiak (1966) found 87 percent mortality in muskrats during the first year of life.

In southern Quebec, a high rate of mortality occurred in muskrats under 6 weeks of age and during the first winter after birth; the number of young per breeding female declined from 5.0 in fall to 2.8 during the spring trapping season (Stewart and Bider 1974). Dorney and Rusch (1953) found that muskrat mortality that occurred between 6 and 24 days of age affected entire litters, whereas subsequent mortality occurred at random.

Subadult mortality rates in southeastern Idaho marshes ranged from 16 to 47 percent (Reeves and Williams 1956). A 62 percent mortality from birth to time of harvest occurred on the Montezuma National Wildlife Refuge, New York (Alexander 1951).

Density and Population-Habitat Interactions. Population density varies widely and depends upon such factors as phase of the population cycle, habitat type and condition, social pressures, competition, harvest, predation, and geographical area. Shoreline length is a more important factor than pond size in determining muskrat population levels; thus, ponds with more edge than larger lakes tend to support proportionally larger populations (Glass 1952). Newly constructed ponds may be populated within one to three years; population densities in new ponds of New York, Pennsylvania, and Maryland generally peaked in five to six years (Erickson 1966).

Den density averaged 6.9 per ha, or 7.9 dens per 1,000 m of shoreline, in ponds associated with strip mines in southern Illinois (Arata 1959a). Beshears (1951) estimated 2.8 muskrats per ha for a 23.9-ha Alabama farm pond.

Gashwiler (1948) found that muskrat density in a Maine marsh was much lower than that in some other states: the pretrapping population was estimated as 0.3–1.8 muskrats per ha. When open water was excluded from density calculations, pretrapping populations ranged from 1.1 to 3.9 muskrats per ha.

Good *Scirpus olneyi* marsh in Louisiana can support about 38 muskrats per ha; of these, about 19 per ha may be harvested safely (O'Neil 1949). Temporary densities may often be higher—up to 154 per ha—but such occurrences were attributed to a great influx of animals from surrounding areas to good areas.

Carrying capacity near Mafeking, Manitoba, ranged from 7.4 muskrats per ha for sedge (*Carex* spp.) habitat to 64.2 per ha for common reed (*Phragmites communis*) and sweetflag (*Acorus calamus*) habitats (Butler 1940). The highest winter muskrat density recorded during the Iowa investigations was about 86 per ha on cattail marsh, compared to a maximum of 37–49 per ha for *Scirpus* spp. marshes; in

TABLE 15.3. Selected muskrat sex ratio data for North America

Source	Location	Habitat	Date	N	How Obtained	Age	Ratio (Males/100 Females)
Aldous 1947	Sand Lake NWR, South Dakota	marsh	January		trapping		178
			late summer		trapping		113
Anderson 1947	southwestern Lake Erie, Ohio	marsh-bay	December–March	5,759	trapping		124
Arata 1959a	Illinois	strip-mine ponds			trapping		133
Beer and Truax 1950	various (literature review)		fall, winter, early spring	89,540	trapping		125
	Wisconsin		fall	18,832	trapping	immatures	134
	Wisconsin		fall	5,250	trapping	adults	98
	Wisconsin		fall	1,149		nestlings	121
Beshears 1951	Alabama	ponds	November–January	2,287	trapping		117
Buss 1941	Wisconsin		fall		trapping		222
Chamberlain 1951	Great Meadows NWR, Massachusetts	marsh	fall		trapping		163
							82
Donohoe 1961	Maryland	marsh	1942	9,091	trapping		more males
Dozier 1944	Maryland	marsh	1942	6,384	trapping	adults	142
Dozier 1945	Montezuma NWR, New York	marsh	January–March	239	trapping	kits	92
				3,919	trapping		122
Dozier 1950	Montezuma NWR, New York			>31,000	trapping		100
Dozier and Allen 1942	Blackwater NWR, Maryland	marsh	trapping season	9,304	trapping	adults	127
Dozier et al. 1948	Blackwater NWR, Maryland	tidal marsh	1941–45 trapping season	23,539			133
Erickson 1963	central New York	small water areas (potholes, ponds)	September	174	nestlings	nestlings	105
			July–October	279	trapping	immatures	118
			September	228	trapping	adults	93
				364	trapping		
Errington 1939	Iowa		September	878 in 162 litters	nestlings <2 weeks old	nestlings	119

Reference	Location	Habitat	Time of year	Number	Method	Young of year / age class	Mean
			late fall, early winter	584	trapping		119
Freeman 1945	Mississippi	marsh	November–January	3,324	trapping		127
			November–January	1,360	trapping		
Fuller 1951	Athabasca–Peace Delta, Canada		autumn	1,200	live and steel trapping	all	138
			winter			juveniles	146
						adult	122
Gashwiler 1950a	central and eastern Maine		summer	392	steel trapping (mostly) and live-trapping	young	146
			fall	681		subadults	144
			fall	270		adults	125
			spring	3,676		adults	124
Harris 1952	Maryland		April–May	275	trapping		108–178
Heit 1949	New York			68 in 10 litters	trapping		257
					litters	kits	117
Lay 1945b	Texas	dikes of banks	October–December	1,065	trapping		145
Marshall 1937	Utah		March–April	491			92
Mathiak 1966	Horicon Marsh, Wisconsin	marsh		3,364 in 460 litters	litters		153
				43,696			138
McCann 1944	Minnesota		fall–winter	12,307	trapping	immatures	132
			fall–winter			adults	101
Olsen 1959a	Delta Manitoba	marsh	March	1,094	trapping		122
			December	610	trapping		118
Reeves and Williams 1956	southeastern Idaho	marsh		69 litters	litters	litters	138
				4,963	trapping		143
Sather 1958	north-central Wisconsin		winter	4,567	trapping	adult	111
						subadults	149
						young	102
						young	96
Schwartz and Schwartz 1959	Missouri			224	34 litters		89–170
Stearns and Goodwin 1941	Delaware	tidewater marsh	December–March	2,778	trapping		157
Wilson 1956a	Currituck, North Carolina	marsh	winter	85	litters	embryos	118

less extreme cases, 49 per ha may be expected in cattails, as opposed to 25 muskrats per ha in *Scirpus* spp. (Errington 1948).

Maximum breeding density was reported by Schwartz and Schwartz (1959) as 5 pairs per ha. Breeding densities of 1 pair per 46 m of shoreline and between 2.5 and 5.0 pairs per ha in deeper areas were indicated as most desirable for fur production in Iowa marshes uniformly well grown with vegetation (Errington 1940). Breeding densities of 1 pair per 183–457 m of shoreline were optimal for open lakes. Ditches and medium-small streams intersecting food-rich agricultural lands may contain 6–8 breeding pairs per 1,609 m; such watercourses in pastures may support only 3–4 breeding pairs or fewer per 1,609 m. Similarly, Freeman (1945) stated that muskrat density was greater in northern Mississippi ponds than in flowing streams. Stewart and Bider (1974) found an average of 366 m of permanently watered collection ditch was needed to support each breeding female in southern Quebec.

Muskrats themselves may greatly influence their habitat and its carrying capacity. The significance of this influence may be partially reflected in muskrat population cycles. Cyclic patterns in muskrat reproduction, disease, and behavior were extensively discussed by Errington (1951, 1954a, 1963).

Factors responsible for cyclic changes in muskrat populations are not clear. O'Neil (1949) stated that parasitism was not related to cyclic population fluctuations of the Louisiana coast. He contended that food conditions formed the primary basis of population control and that muskrat population cycles and eat-outs were directly related.

The basic phases of muskrat cycle development in Louisiana are low muskrat numbers, development of a big food supply, overpopulation, range damage, and then starvation (Lowery 1974). Ten- to 14-year population cycles were reported on good *S. olneyi* marsh and longer cycles on less productive marsh types (O'Neil 1949). The cycles are local and independent and there is no set pattern for the entire brackish marsh (Lowery 1974). Muskrat populations of Athabasca-Peace Delta, Ottawa, undergo a marked annual cycle of increase and decrease with longer cycles of about 10 years (Fuller 1951).

Yearly production of muskrat pelts in Saskatchewan from 1915 to 1960 exhibited a 6-year rather than the expected 10-year cycle (Butler 1962). The highest catches occurred after 2 wetter-than-normal years in the prairie section. However, Elton and Nicholson (1942) and Butler (1953) found that the muskrat population followed a 10-year cycle in most sections of Canada. This periodicity was similar to, but out of phase with, the 10-year cycle of most other furbearers south of the tree line. Errington (1951, 1954a, 1963) also indicated that muskrats followed a 10-year cycle more than a 3- to 4-year cycle.

In Louisiana coastal marshes, Palmisano (1972b) reported that muskrat populations frequently increased rapidly for three to four years and then declined precipitously within a few months. Such fluctuations also affect nutria (*Myocastor coypus*), which surpassed the muskrat as the predominant furbearer in Louisiana in the early 1960s. Freshwater marshes are best for nutria production, whereas brackish marshes are superior for muskrat production. Most management of brackish marshes in Louisiana is aimed at production of Olney bulrush (*S. olneyi*) and muskrats. However, muskrat populations in brackish marshes fluctuate drastically and average fur production in poor years is below that of fresh areas. Therefore, out of every decade, well-managed marsh in Louisiana will exhibit only about three peak years of muskrat production.

During peak muskrat years, the average maximum fur revenue of the brackish marshes is more than three times that of the fresh type. When muskrat populations are low, however, trappers shift to nutria. Such cyclic production patterns are not exhibited in fresh and intermediate marsh types; consequently, muskrat populations in these areas vary little from year to year when compared to brackish areas (Palmisano 1972b).

Increasing and decreasing muskrat populations in northeastern Iowa were compared by Neal (1968). The decreasing population was associated with less dense vegetation and exhibited lighter average weight, a shorter breeding season, fewer and smaller litters, and a larger population size compared to the increasing population. Production (young/adult) was over 2.8 times greater in the increasing population. Errington et al. (1963) found that a build-up of muskrats at Goose Lake, Iowa, was probably due to underharvesting and "cyclic" changes in muskrat physiology; changes in breeding performance were also noted in this dense population.

Boyce (1978) found that morphological characters of adult muskrats were related to the annual precipitation pattern. Nutritive demands may exceed the supply for large individuals; therefore, selection in areas of low food resources favors smaller body size. The best areas for muskrats in Maryland have the normal water level at or near ground level (Smith 1938). A positive association between muskrat population levels and the availability of water areas was reported by Adams (1962); similar associations have been noted by O'Neil (1949), Dozier (1953), and Artimo (1960).

Muskrats examined during the trapping season from Blackwater National Wildlife Refuge, Maryland, had a rather small amount of fat, including those coming from good areas of *S. olneyi*. However, weights did increase progressively from 1 January to 15 February, after which they declined rapidly (Dozier et al. 1948); muskrats from deteriorated (eaten-out) marshes showed a progressive decrease in weight. Delta, Manitoba, muskrats reportedly have a rapid growth rate during summer; the rate showed a decline after fall with little or no growth from November to March (Olsen 1959a). Muskrats from river areas of Tennessee weighed more than those from creek areas—probably because of increased quality and quantity of food on the river (Schacher and Pelton 1976). During winter, adults inhabiting the river area contained more internal body fat than creek muskrats. Adult males had increased fat levels from spring through fall and decreased amounts from winter to spring. Nonpregnant females showed decreased body fat from winter through summer, whereas pregnant animals added fat from spring to summer.

Positive relationships were reported by Dozier and Allen (1942) and Dozier et al. (1948) between abundance and quality of food plants and muskrat weight and size. Similarly, Dozier (1945) found a positive relationship between muskrat weight and optimum pool water levels on Montezuma National Wildlife Refuge. However, an examination of the Montezuma data by Scheffer (1955) indicated no correlation between average weight of muskrats and the number of houses per hectare, or between weight and abundance on most of the trapping units. Nevertheless, on two of the units there was a strong association between low muskrat weight and low density. Scheffer (1955) agreed that muskrats tend to destroy their food supply and decrease in weight. Intolerance to crowding, strife, and occasional cannibalism may act more quickly than food resources to limit population expansion.

Shanks and Arthur (1952) and Greenwell (1948) found no correlation between pond muskrat populations and the type or kind of vegetation present. However, according to Glass (1952) the suitability of a pond for muskrats depends upon the availability of food plants.

Movement and Home Range. The area occupied by a muskrat depends upon the size, configuration, and diversity of aquatic habitat; social pressures; sex; age; season; and environmental conditions. Animals that inhabit the center of a marsh may have a circular home range, while those that occupy edge habitat usually have oblong ranges.

Beer and Meyer (1951) believed muskrat movements were caused by normal physiological cycles and emigrations from drought and population pressure. Errington (1963) reported that summer home ranges of marsh muskrats averaged approximately 61 m in diameter; river-dwelling muskrats reportedly may range along the banks up to 183 m and across most rivers (Schwartz and Schwartz 1959). Muskrats have shown the ability to home from distances up to 1,646 m (Errington 1963).

A muskrat may disperse to a new area in spring; after it becomes established in a new area, under normal conditions the area is occupied until the following spring (Schwartz and Schwartz 1959). Spring emigration began in north-central Wisconsin when the ice broke up (Sather 1958). Sather noted both cross-country emigration and within-marsh dispersal; females remained in their home territory while the males dispersed. Initial spring dispersal in Iowa muskrats began between 22 February and 24 March and was also influenced by snow and ice (Sprugel 1951). Distances traveled and rate of movement was variable. Most dispersal activity was over within 1.5 months and territories were then established.

Movement of muskrats inhabiting strip-mine ponds was greatest in February and in late summer

when low water caused den abandonment (Arata 1959a). Spring movement in central New York ponds was the most extensive and tended to be triggered by ice break-up; mean distance traveled was 224 m (Errington 1963). Almost all movement was from pond to pond.

Muskrats in ponds less than 0.4 ha were reportedly much more likely to move (primarily from pond to pond) than animals in larger ponds (Shanks and Arthur 1951, 1952). Muskrats inhabiting streams showed no indication of moving to ponds; movements tended to be along the streams with localized concentrations. Two migratory peaks—spring and fall—were found, with populations fairly stable during summer and winter. Initial fall movement may be from population pressure but pair formation may also be involved. By late fall most ponds contained a pair of muskrats.

Exploratory drifting out of a familiar area may result in muskrats moving as far as 34 km. Errington (1951) noted that under extreme population pressure, muskrats may be forced to make massive pioneering thrusts into totally uninhabitable habitats. Drifting is most common in August and September (Schwartz and Schwartz 1959).

MacArthur (1978) found that radio-tagged muskrats were within 15 m of their primary dwelling lodge during 50 percent or more of the position determinations. Most foraging occurred within a 5- to 10-m radius of a lodge or push-up and few muskrat movements exceeded 150 m. Similarly, the majority of movements of muskrats in an Illinois lake were within 15 m of the home den (Coon 1965).

Other studies of marked muskrats have yielded similar results. Only 5 percent of 1,579 tagged muskrats at Horicon Marsh, Wisconsin, moved farther than the adjacent trapping unit (Mathiak 1966). Likewise, about two-thirds of the kits recovered by Dorney and Rusch (1953) on the Horicon marsh had moved no more than 91 m from the last site handled. In South Dakota, Aldous (1947) found that 54 percent of marked muskrats had not moved from the last release site; only 15 percent moved more than 156 m. Similarly, for a marsh in central Maine, about 50 percent of the recaptures were within a 7.6-m radius of the original tagging site (Takos 1944). Less than 30 percent were beyond a 30.5-m radius and all immatures recaptured were within a 30.5-m radius. The maximum distance muskrats moved between points of capture on an Alabama pond was about 457 m (Beshears 1951).

Competition. Muskrats are not the only major herbivore inhabiting some marshes. The nutria coexists with the muskrat in the Gulf coastal marshes and in scattered marshes throughout the western half of the United States and on the Atlantic seaboard. Apparently the nutria feeds on coarser vegetation than does the muskrat and competition is not a serious problem; the two even get along well together in crowded cages (O'Neil 1949).

Gainey (1949) compared winter food habits of muskrat and nutria in captivity. Considerable similarity was revealed between the two animals in food preference but he stated that the degree of competition would depend upon whether nutria concentrated in the brackish marshes generally preferred by muskrats. Later studies showed that nutria generally preferred freshwater areas (Lowery 1974). Gainey (1949) indicated that, compared to muskrats, nutria generally preferred the larger plants and would eat more of the rougher, less palatable plant parts. On certain plants, such as alligatorweed (*Alternanthera philoxeroides*), nutria ate the leaves while muskrats ate the submerged part of the stem. Muskrats, more than nutria, were prone to eat rhizomes.

About 85 percent of the muskrats in Louisiana inhabit approximately one-quarter of Louisiana's marsh—that portion dominated by *S. olneyi*—during peak muskrat years (Lowery 1974). *S. olneyi* marsh has been gradually changing, and in places disappearing, since the 1940s from drought, hurricanes, industrialization, and eat-outs. The record high 8 million muskrat harvest of the 1945/46 season declined to 200,000 in 1964/65. Nutria catches increased from 436 in the 1943/44 season to over 1.5 million pelts in 1964/65. However, biologists could find only circumstantial evidence that the muskrat decline was related to the nutria increase. Nevertheless, nutria removal from an area inhabited by both species often resulted in a rise in the muskrat population (Evans 1970). In addition to possible competition for food where they exist together, muskrats may be harassed by nutria, and they may compete for general living space and elevated areas for retreat during high water (Lowery 1974).

Ecosystem Disruption. When muskrats become overpopulated, generally an eat-out occurs and the feeding area is ruined for several years (O'Neil 1949). An eat-out is a condition that results when muskrats populate an area to the extent that existing vegetation, including the soil-binding root systems, is consumed. Eat-outs may result in destruction of both food and cover (Ferrigno 1967; Wilson 1968). Lowery's (1974) description of eat-outs is especially revealing. When food is abundant muskrat populations increase rapidly, all available vegetation is eaten, and muskrats may dig down into the peaty marsh floor as deep as 50 cm to devour roots. As roots that bind marsh soils together are removed, the earth disintegrates into loose muck with decaying remains of plants floating in the ooze.

Lynch et al. (1947) found that marsh damage was inevitable when areas heavily populated by muskrats were undertrapped. Part of the decrease in vegetated areas of the marsh in Maryland was also attributed to marsh damage from previously high muskrat populations (Harris 1952).

Complete or partial eat-outs have differing effects on the marsh. The recovery of damaged marsh depends primarily on the quantity of seeds and propagative bodies surviving the eat-outs. Factors stimulating marsh recovery from eat-outs include low tides and dry weather, occurrence of annual weeds, development of flotage, and survival of original vegetation. High water

and salt-water intrusion retard recovery (Lynch et al. 1947).

On the positive side, muskrat eat-outs may be beneficial to certain wildlife species such as shorebirds, wading birds, and certain waterfowl. Muskrats at Ogden Bay Refuge, Utah, cleared sections of vegetation when building houses and runways. This benefited waterfowl by opening up dense stands of vegetation (Nelson 1954). Chamberlain (1951) stated that allowing muskrats to attain a high density on Great Meadows National Wildlife Refuge, Massachusetts, would improve conditions for waterfowl. However, on the negative side, wildlife that use eaten-out areas may be predisposed to diseases from stagnation (Lynch et al. 1947).

Marshes may be filled or ditched for drainage. However, ditching is not a recent practice; Bourn and Cottam (1950) indicated that by the end of 1938 about 90 percent (over 202,000 ha) of the total original acreage of tidewater marshland of the Atlantic coast from Maine to Virginia had been ditched. Marshland alteration may result in profound changes in faunal and floral components, usually in conjunction with changes in surface and subsurface hydrology. Destruction of marshes is no small problem. According to Ferrigno (1967), the only marshes in New Jersey that have escaped habitat destruction are those owned by state and federal governments or landowners interested in conservation.

On the lower Delaware Bay marshes, past mosquito control practices have resulted in extensive marsh destruction. These practices included the use of dikes, sluice boxes, and pumps to exclude tidewater from marshes in an effort to strand mosquito larvae before they emerged as adults. One example of the drastic marsh deterioration caused by such mosquito control practices was that of a previously important wet, tidal, cattail (*Typha angustifolia*)-cordgrass (*Spartina* sp.) marsh that was converted to a dry, red maple (*Acer rubrum*)-reed (*Phragmites* sp.) area. Dredging of the Delaware River channel and the diking in of salt hay agriculture areas also reportedly destroyed some important muskrat marshes (Ferrigno 1967).

Similarly, earlier data provided by Stearns et al. (1939, 1940) indicated that ditching for mosquito control in Delaware rapidly lowered the water table. Good muskrat food and house-building plants, such as *S. olneyi* and *Spartina cynosuroides*, were replaced by less desirable plants, such as *Hibiscus oculiroseus*, *Kosteletzkya virginica*, *Solidago sempervirens*, *Bidens trichosperma*, and *Aster novi-belgii*. As a result, muskrats moved.

Under certain conditions, drainage may cause damage to muskrat marshes by disruption of tidal action and removal of barriers or low ridges that trap waters required for high marsh productivity. Bourn and Cottam (1950) found that natural grass associations in ditched Delaware tidewater marshes were largely replaced by shrubby growths of groundsel baccharis (*Baccharis halimifolia*) and sumpweed (*Iva frutes-*

cens). Appreciable reductions in marsh value for muskrats and invertebrate populations also occurred. Likewise, reclamation (drainage) of wetland areas for agricultural purposes in southeastern Louisiana eliminated the aquatic environment and destroyed the endemic wildlife (Penfound and Schneidau 1945).

Numerous natural swamp areas that typically contain a meandering, sluggish stream have been channelized and straightened by private and governmental agencies, especially in the southeastern United States. Purported project benefits include flood control (especially for agricultural purposes), increased navigation, and mosquito control. Muskrat populations may be enhanced or impaired by channelization projects, but nonetheless a major ecological shift of unknown magnitude likely occurs (Perry 1969).

Studies by Gray and Arner (1977) indicated that muskrats were more numerous in the unchannelized segment than in either the old or newly channelized segments of the Luxapalila River in Mississippi and Alabama. On the positive side, many wetland reclamation areas that are abandoned (reflooded) provide excellent wildlife habitat (Penfound and Schneidau 1945). Also, extensive bank habitat is created by channelization projects. However, frequent and dynamic floods may occur as a result of water being rapidly shunted through the swamp system, the backwater areas that typically contain succulent food plants become more xeric, and overall water quality may be reduced.

General encroachment of urban sprawl on stream muskrat habitat has resulted in major ecological changes. When earthen ditches and canals are concreted, muskrat den sites and food sources are virtually eliminated.

FOOD HABITS

Muskrats are chiefly herbivorous (Bailey 1937; Dozier 1953; O'Neil and Linscombe 1976; and others). They eat various portions of aquatic plants, especially shoots, roots, bulbs, tubers, stems, and leaves (table 15.4). They commonly dig for food, frequently in sod on lake and pond bottoms (Bailey 1937; O'Neil and Linscombe 1976).

TABLE 15.4. Foods of muskrats of North America

Northeastern U.S.A. [a]
  Cattail (*Typha latifolia, T. angustifolia*)
  Burreed (*Sparganium eurycarpum*)
  Bulrush (*Scirpus americanus, S. acutus, S. fluviatilis*)
  Arrowhead (*Sagittaria latifolia*)
  Sweetflag (*Acorus calamus*)
  Duckweed (*Wolffia* sp. and *Lemna minor*)
  Pondweed (*Potamogeton natans*)
  Sedges (*Dulichium arundinaceum, Carex rostrata, C. lacustris, C. lasiocarpa*)
  Bayonet rush (*Juncus militaris*)
  Pickerelweed (*Pontederia cordata*)
  Freshwater clams, carp, crayfish, turtles, snails

TABLE 15.4—*Continued*

Mid-Atlantic U.S.A. [b]
  Bulrush (*Scirpus americanus, S. fluviatilis, S. olneyi, S. robustus*)
  Cattail (*Typha latifolia, T. angustifolia*)
  Cowlily (*Nuphar advena*)
  Pickerelweed (*Pontederia cordata*)
  Marshhay cordgrass (*Spartina patens*)
  Saltgrass (*Distichlis spicata*)
  Big cordgrass (*Spartina cynosuroides*)
  Common reed (*Phragmites communis*)
  Spikerush (*Eleocharis* sp.)
  Carp, mussels, turtles, blue crabs, dead birds, fish
Southeastern U.S.A. [c]
  Bulrush (*Scirpus americanus, S. robustus, S. olneyi*)
  Cattail (*Typha latifolia, T. angustifolia*)
  Arrowhead (*Sagittaria latifolia*)
  Waterlily (*Nymphaea odorata*)
  Big cordgrass (*Spartina cynosuroides*)
  Common reed (*Phragmites communis*)
  Crayfish, fish
North Central U.S.A. [d]
  Cattail (*Typha latifolia, T. angustifolia*)
  Bulrush (*Scirpus validus, S. fluviatilis, S. acutus*)
  Arrowhead (*Sagittaria latifolia*)
  Waterlily (*Nymphaea tuberosa*)
  Dry grasses (*Eragrostis, Panicum, Echinochloa, Bromus, Muhlenbergia, Agropyron, Elymus*)
  Corn (*Zea maize*)
  Common reed (*Phragmites communis*)
  Duckweed (*Lemna minor*)
  Burreed (*Sparganium eurycarpum*)
  Smartweed (*Polygonum* spp.)
  Willow (*Salix* spp.)
  Poplar (*Populus balsamifera*)
  Maple (*Acer* spp.)
  Shepherds purse (*Capsella bursa-pastoris*)
  Bluegrass (*Poa pratensis*)
  Goldenrod (*Solidago ulmifolia*)
  Barnyard grass (*Echinochloa crusgalli*)
  Beggartick (*Bidens* spp.)
  Hawthorn (*Crataegus* spp.)
  Love grass (*Eragrostis cilianensis*)
  Acorn (*Quercus* spp.)
  Aster (*Aster* spp.)
  Mint (Labiatae)
  Sweetclover (*Melilotus alba*)
  Giant ragweed (*Ambrosia trifida*)
  White clover (*Trifolium* sp.)
  Bluestem (*Andropogon virginicus*)
  Stonewort (*Chara* spp.)
  Pondweed (*Potamogeton foliosus*)
  Sedge (*Carex crinita, C. rostrata*)
  Pickerelweed (*Pontederia cordata*)
  Crayfish, frogs, turtles, fish, freshwater clams, snails, birds (young)
South Central U.S.A. [e]
  Bulrush (*Scirpus olneyi, S. robustus*)
  Cattail (*Typha latifolia, T. angustifolia*)
  Spikerush (*Eleocharis quadrangulata*)
  Arrowhead (*Sagittaria* sp.)
  Maidencane (*Panicum hemitomon*)
  Rice (*Oryza* sp.)
  Rush (*Juncus roemerianus, J. effusus*)
  Pickerelweed (*Pontederia cordata, P. lanceolata*)
  Naiad (*Najas* sp.)

TABLE 15.4—*Continued*

South Central U.S.A.
Lotus (*Nelumbo* sp.)
Johnsongrass (*Sorghum halepense*)
Bermuda grass (*Cynodon dactylon*)
Willow (*Salix* spp.)
Waterlily (*Nymphaea* spp.)
Pondweed (*Potamogeton* spp.)
Lizard's tail (*Saururus cernuus*)
Paspalum (*Paspalum* spp.)
Burreed (*Sparganium* sp.)
Smartweed (*Persicaria* spp.)
Wild millet (*Echinochloa* spp.)
Cordgrass (*Spartina* spp.)
Switchgrass (*Panicum virgatum*)
Alligatorweed (*Alternanthera philoxeroides*)
Crayfish, mussels, fish, freshwater clams, turtles
Western U.S.A. [f]
Cattail (*Typha latifolia, T. angustifolia*)
Bulrush (*Scirpus acutus*)
Clover (*Trifolium* sp.)
Arrowhead (*Sagittaria* sp.)
Canada [g]
Cattail (*Typha latifolia*)
Bulrush (*Scirpus validus*)
Pondweed (*Potamogeton* sp.)
Water horsetail (*Equisetum fluviatile*)
Reed (*Phragmites maximus*)

SOURCE: Adapted from Willner et al. 1975.
[a] From Johnson 1925; Enders 1931; Seamans 1941; Takos 1947; Dozier 1945, 1950; Gashwiler 1948; Anonymous 1950; Bednarik 1956; Alexander 1956.
[b] From Lantz 1910; LeCompte 1928; Bailey 1937; Smith 1938; Dozier et al. 1948; Cofer 1950; Harris 1952; Wilson 1956a.
[c] From Svihla and Svihla 1931; Freeman 1945; O'Neil 1949.
[d] From Errington 1937c, 1939, 1941; Hamerstrom and Blake 1939; Butler 1940; Bellrose and Low 1943; McCann 1944; Aldous 1947; McLeod 1948; Bellrose 1950; Sprugel 1951; Sather 1958; Arata 1959a, 1959b,; Schwartz and Schwartz 1959.
[e] From Louisiana Department of Conservation 1931; Svihla and Svihla 1931; Lay and O'Neil 1942; Freeman 1945; O'Neil 1949; Glass 1952.
[f] From Rawley et al. 1952; Borell and Ellis 1934.
[g] From Butler 1940; McLeod 1948; Fuller 1951.

Muskrat foods and feeding habits vary widely and depend somewhat on habitat, season, and availability (Butler 1940). Food is generally scarcer in winter and is often restricted to underground plant parts or to what can be reached under ice (Fuller 1951; Dozier 1953). Parts of the lodge or bedding may be eaten in winter and spring (Errington 1941). The muskrat does not store large quantities of food (Smith 1938; Errington 1941; Schwartz and Schwartz 1959). It depends upon its abilities to dive and while submerged obtain pieces of aquatic plants by gnawing (Dozier 1953). Water levels greatly influence feeding areas. Low water will force muskrats to move to more desirable areas where they may establish temporary homes.

Bailey (1937) described muskrats as dainty feeders. The front feet are employed to gather food and

transport it to the mouth. Frequently muskrats will rear on their hind legs, utilizing the tail for support. Food is generally eaten not in the nest but in some other area that affords protective cover. Much of the muskrat's food in summer is consumed at the water's edge (Lantz 1910). Muskrats are sometimes active during the day, but most feeding is done at night (Freeman 1945).

There are differences in the feeding of muskrats a few weeks old and adults: young feed more on bank vegetation than do the adults (Warwick 1940). No significant difference was found between stomach contents from males and from females in percentages of moisture, ether extract, crude protein, crude fiber, ash, and nitrogen-free extract (Stearns and Goodwin 1941). Analyses of the roots and tops of ten plants representative of the vegetation of the Delaware marsh where the muskrats were trapped indicated that crude protein and nitrogen-free extract for the roots were higher than the plant tops; the reverse was true for crude fiber.

Muskrat foods are more diverse in stream or canal systems than in marsh communities, probably because freshwater areas tend to be more diverse than saltwater areas. Indeed, 70–80 percent of the foods in some marshes may consist of one or two species (see O'Neil 1949), whereas in more diverse systems 8–10 species may be important.

Willner et al. (1975) concluded that bulrush (*Scirpus* sp.) and cattail (*Typha* sp.) were the most important muskrat foods throughout the United States and that cattail (*Typha latifolia*) was the most important in Canada. The utilization of bulrush and cattail for food by muskrats may be extremely high. O'Neil (1949) reported that 80 percent of the diet of muskrats in the brackish subdelta marshes of Louisiana was *S. olneyi*. Smith (1938) stated that 80 percent of the foods of muskrats in Maryland marshes consisted of *S. olneyi*, *S. americanus, T. latifolia,* and *T. angustifolia.* Common cattail is reportedly best suited to supply the needs of muskrats during winter because of its wide distribution, high nutritive quality of rhizomes, tolerance of a wide range of conditions, and preference by muskrats (Cook 1952). In spite of the obvious importance of bulrush and cattail to muskrats, in some areas and/or seasons the animals will utilize the most available species and to some extent any plant available (Smith 1938; Takos 1947; Sather 1958).

Muskrat food habits on one brackish and three freshwater areas in Maryland were evaluated by Willner et al. (1975). More than 30 vascular plants were identified from fragments from the stomachs of muskrats inhabiting freshwater habitats. Green algae constituted the most important group of plants consumed. Although vascular plants were unimportant individually, they made up 50 percent of the diet of muskrats in freshwater. Only nine plant species were in the diet of muskrats inhabiting brackish water areas; cattail (*T. angustifolia*) was the most significant plant in their diet.

Cultivated plants of which muskrats are fond include carrots, corn, raw peanuts, clover, alfalfa, soybeans, and apples (Dozier 1953). MacArthur and Aleksiuk (1979) found that muskrats heavily utilized field crops. Muskrats in captivity have been successfully fed apples, sweet potatoes, rolled oats, corn, fresh lettuce, wheat, rice, and oats (Svihla and Svihla 1931; Bailey 1937).

Although muskrats are largely vegetarians, in certain habitats or seasons, especially winter, they may rely upon animal food (Lantz 1910; Bailey 1937). Stearns and Goodwin (1941) studied muskrat food habits on a Delaware tidewater marsh by analyzing the nutritive values of stomach contents and raw plants. They concluded that animal matter was being consumed by muskrats in an appreciable but undetermined amount during winter. Lake and reservoir muskrats are opportunistic feeders and may feed more on animal matter than do marsh muskrats (O'Neil 1949). Similarly, Schwartz and Schwartz (1959) indicated that Missouri stream muskrats feed on a variety of fauna. Animals eaten include clams, crustaceans, mussels, snails, and young birds. Other studies have shown only limited feeding by muskrats upon animal matter (Bellrose 1950; Dozier 1953).

Sather (1958) suggested that utilization of animal matter may result from a shortage of preferred vegetation. Incidences of muskrats feeding on unretrieved waterfowl shot by hunters and on weaker muskrats were reported by Lantz (1910) and Errington (1941).

Some muskrat stomachs examined by Lantz (1910) contained over 63 g of moist food. Stomach contents (dry) of Delaware muskrats ranged from 0.1 to 17.7 g and averaged 2.4 g (Stearns and Goodwin 1941). Svihla (1931) found that the Louisiana muskrat consumes about one-third of its weight each day. At this consumption rate, one large muskrat can consume in one day about 929 cm^2 of *S. olneyi* marsh or about 1,858 cm^2 of any other marsh type (O'Neil 1949).

HABITAT

Muskrats generally prefer lentic or slightly lotic water containing vegetation. Such areas are found in the coastal marshes and in marshy areas around lakes, sloughs, streams, and rivers. However, muskrats are adaptable, and they inhabit a wide range of community types, including strip-mined ponds, ditches, canals, and pits.

Research by biologists of the Louisiana Department of Conservation (1931) indicated the importance of humus in determining muskrat abundance. Areas that supported the greatest numbers of muskrats contained peaty humus several centimeters thick. This humus was about 90 percent vegetable matter. Humus was considered important to muskrats because it provided a medium from which roots and burrows could be dug with ease.

Highest muskrat populations occur in Louisiana in brackish areas where fresh and salt water mix (O'Neil and Linscombe 1976). Louisiana has over 1.6 million ha of marsh, and most marsh types of the country are represented. However, over 80 percent of the muskrats are produced on about 405,000 ha of brackish marsh where Olney bulrush (*S. olneyi*) is dominant or subdominant (O'Neil 1949). *S. olneyi* makes up 90 percent of the Gulf coast muskrats' food supply. Advantages of *S. olneyi* marsh over other types of marsh are that it yields the highest weight per m^2 of edible material and it grows year-round (O'Neil 1949). The brackish marshes of the Texas coast, characterized by marshhay cordgrass or wiregrass (*Spartina patens*), saltgrass (*Distichlis spicata*), needle rush (*Juncus romerianus*), Olney bulrush (*Scirpus olneyi*), and saltmarsh bulrush (*S. robustus*), are also the most productive community for muskrats. The cordgrass is highly desirable for lodge building, but it must be controlled so it will not exclude food plants (Lay and O'Neil 1942).

Data presented by Palmisano (1972b) and Lynch et al. (1947) substantiate the importance of brackish marshes for muskrats of Gulf coastal marshes. Palmisano (1972b) reported that 72 percent of the muskrat houses counted on aerial surveys were in brackish marshes although this habitat type accounted for only 37 percent of the habitat within the survey area. Less than 20 percent of the muskrat houses were recorded outside the brackish marsh in normal years. These data indicate that muskrat density in the brackish marsh was twice the average density for the entire coastal marsh area. Saline marshes represented 12 percent of the habitat and contained 14 percent of the houses, whereas fresh and intermediate marshes accounted for 40 percent of the survey area and contained only 13 percent of the houses observed. The maximum yield of muskrats in the brackish marsh was 10 times (16 pelts/ha) the maximum harvest in the fresh marshes. However, during periods of drought stress in the preferred brackish marshes, saline, intermediate, and fresh marshes increase in importance because they are not as affected by dry weather (Palmisano 1972b).

Scirpus sp. is also an important habitat component for muskrats in certain areas along the Atlantic coast. On Blackwater National Wildlife Refuge, Maryland, Dozier et al. (1948) reported that the fresh to slightly brackish *S. olneyi–Typha* spp. marshes, adjacent to timbered areas, were best for muskrat production (harvest). Similarly, on Montezuma National Wildlife Refuge, New York, the abundant supply of blue flag cattail (*T. glauca*) was listed as the greatest factor contributing to the accelerated and continuous growth, huge pelt size, and record weight of muskrats (Dozier 1950). Wilson (1949) listed needle rush, sawgrass (*Cladium jamaicense*), and big cordgrass (*Spartina cynosuroides*) as undesirable muskrat vegetation in coastal North Carolina; *Scirpus-Typha* marshes were listed as desirable communities to support muskrats.

Hamerstrom and Blake (1939) found that ditches that were well shaped or that ran deep and swift were practically unused by muskrats. Slow-running streams with numerous aquatic plants had the highest muskrat populations in northern Mississippi; old stream runs and lakes were preferred over flowing water areas (Freeman 1945). An evaluation of the suitability of stream habitat for muskrats should consider that: (1) corn fields may enhance the attractiveness of stream environments to muskrats, (2) stream habitat for muskrats is dynamic and in a state of flux, (3) semipermanent retreats that provide muskrat security are important, and (4) preferences for various aspects of stream habitat may vary greatly during the year (Errington 1937c).

Muskrat harvest and population levels were compared by Nichols (1973) in marsh (mostly fresh) and swamp systems in Louisiana. No significant difference was found between relative abundance of muskrats in the swamp and marsh areas, but the mean number of muskrats caught per trapper was more than three times greater in the marsh region. However, there was no significant difference in the muskrat harvest per square kilometer between the swamp and marsh systems.

BEHAVIOR

Muskrats are generally nocturnal, although they are sometimes out during the day, especially in spring and early summer. About 80 percent of the muskrat's activities are nocturnal under normal conditions (O'Neil 1949). Most muskrats he studied were in beds from daylight to midafternoon, after which the adults were seldom in the nest. Survival of muskrats depends largely upon concealment; they are wary and spend a large amount of time underground in tunnels and burrows or in their nests and feeding shelters.

Muskrats are territorial (Errington 1943; Sather 1958). Shanks and Arthur (1951) found the summer range of stream-dwelling muskrats was equal to about one-half the distance between colonies. Similarly, Schwartz and Schwartz (1959) reported that houses are seldom closer than 8 m. The muskrat was described by Fuller (1951) as not being particularly tolerant of others of its species, even though it is semicolonial. Muskrats usually live alone or with their mates, but during winter, several muskrats may be denned or housed together (Schwartz and Schwartz 1959). In small ponds, adult males are rarely found with adult females and their young (Shanks and Arthur 1951).

On land muskrats move with an amble or slow hop and, unless alarmed, they enter water slowly. They swim with their hind feet while holding the front feet against the chin. The tail is trailed and, although not used in surface swimming, may be used as a rudder in turning. Swimming speed is about 3–5 km/hr. In underwater swimming, the tail is used extensively. Schwartz and Schwartz (1959) reported that muskrats may swim up to 46 m underwater.

Adults are usually silent (Smith 1938), but Svihla and Svihla (1931) and Schwartz and Schwartz (1959) reported that adult muskrats emit low squeaks, loud squeals, and snarls, and when cornered or fighting they chatter. Both sexes may emit a high-pitched "n-n-n-n" during the breeding season. According to Schwartz and Schwartz (1959), muskrats may warn of impending danger by slapping their tail on the water. The young have a squeaky cry or squeal and are comparatively noisy (Smith 1938). When the young call, the mother shows concern and runs to them at once. Frequently a mother may attempt to carry young away in her mouth.

Svihla and Svihla (1931) stated that muskrats they observed indulged in no games or play. They described a possible gesture of affection performed by both males and females that consisted of nibbling on each other's back and neck, especially during the breeding season. Hensley and Twining (1946) described a behavior whereby females use their front feet to groom and comb the backs and heads of suckling young.

Muskrats are very clean and spend a great deal of time washing and combing their fur. Excrement is usually deposited in water, and nests or houses are normally clean; pellets are sometimes found on a resting platform or feeding raft (Svihla and Svihla 1931). Although muskrats often defecate randomly throughout the marsh as they walk and swim, Smith (1938) described defecation "posts" where muskrats defecate repeatedly in definite spots. Most posts are on a slight elevation, such as the end of a log or piece of sod. One post may be used by more than one animal and any suitable item in the muskrat's habitat may be used as a defecation post. Generally 3–12 black, oval droppings are deposited at one time (Schwartz and Schwartz 1959).

When muskrats are injured they keep the wound clean by licking. They gnaw away infected flesh, and legs amputated by traps heal rapidly; one-, two-, and three-legged muskrats have been captured (Svihla and Svihla 1939).

Muskrats frequently jump when it is impossible to climb (Svihla and Svihla 1931). When caught or cornered they fight desperately, striking at their opponents. Their incisor teeth are long and sharp and capable of inflicting much damage. Muskrats encountered away from water sometimes are fierce and have been known to attack persons savagely without apparent provocation (Lantz 1910).

Muskrats readily fight each other, especially when food is scarce. According to Schwartz and Schwartz (1959), individuals are more tolerant of their own than the opposite sex, with females being most tolerant (Beer and Meyer 1951). Errington (1951) stated that population density was one of the most influential factors affecting muskrat behavior. During weaning, competition is keen when populations are dense, and older young may eat the newborn. Intraspecific strife among young muskrats was observed related to overpopulation (Sather 1958). A greater tolerance of crowding was noted during the "cyclic high" which was characterized by increased numbers of subadult and young-of-the-year breeders.

Muskrat activity is apparently related to weather and environmental factors. Hensley and Twining (1946) indicated that muskrats do less work during periods with a full moon; Svihla and Svihla (1931) reported that a very windy night or warm weather results in little muskrat activity. Signs of muskrat activity in the marsh are also scarce during the summer or when the weather is hot and dry (Lay 1945a). Water and cool weather are the most favorable factors for muskrat activity (Svihla and Svihla 1931). Muskrats are more active on cold, snappy, frosty nights and less active on warm, muggy nights (Louisiana Department of Conservation 1931). Other aspects of muskrat behavior were described by Cook (1952).

MORTALITY

Muskrats have a high reproductive potential and generally a short life span. However, one tagged Missouri muskrat reportedly lived four years in the wild (Schwartz and Schwartz 1959). Factors limiting muskrat populations include diseases, parasites, predators, accidents, climatic factors, food, intraspecific strife (fighting), and exploitation. These factors vary widely by area, season, and population level. However, adult survivorship and the percentage of adults in the population generally are inversely related to latitude (Boyce 1977). Basic aspects of juvenile mortality were discussed earlier in connection with the unbalanced sex ratios reported for muskrats.

Diseases and Parasites. Diseases of Maryland muskrats described by Smith (1938) include abscesses, septicemia (from *Chlamydia* sp.), coccidiosis (caused by a protozoan), leukemia, gallstones, and inflammation of the eye. Severe malocclusion of the incisors was described by Alexander and Dozier (1949). Mathiak (1966) reported pasteurellosis, hepatitis, uremia, pneumonia, and heart and liver degeneration in muskrats but indicated that these conditions had an insignificant effect on local populations; tularemia, from *Francisella tularensis* (=*Pasteurella tularensis*=*Bacterium tularense*), and Errington's disease were considered serious disease conditions.

Errington's disease, or hemorrhagic disease, of muskrats is highly infectious and various studies have been performed to determine the causative agent. Lord et al. (1956a, 1956b) described the pathological changes and etiology of the condition and found the bacterium *Clostridium* sp. common to cultures from diseased animals. Wobeser et al. (1978) diagnosed Tyzzer's disease (infection with *Bacillus piliformis*) in muskrats from a Saskatchewan marsh. Other investigators found similar infections of muskrats in other areas (Karstad et al. 1971; Chalmers and MacNeill 1977). Pathological and epizootiological similarities of Tyzzer's and Errington's diseases suggested to Wobeser et al. (1978) that these two diseases have a similar etiology.

Hockett (1968a, 1968b) isolated 19 genera of bacteria from muskrats. The most frequently found species (with percentage occurrence) were *Citrobacter freundii* (53 percent), *Enterobacter/Aerobacter* (57 percent), and *Proteus vulgaris* (40 percent). Cultural examinations of 200 muskrats by Friend and Muraschi (1963) yielded 28 isolates characteristic of *P. tularensis* (now *F. tularensis*).

A literature review by Jilek (1977) indicated 66 species of parasitic helminths in muskrats: 36 trematodes, 11 cestodes, 15 nematodes, and 4 acanthocephalans. General parasitological surveys of muskrats have been completed by Barker (1915), Chandler

(1941), Ameel (1942), Penn (1942), Rausch (1946), Edwards (1949), Meyer and Reilly (1950), Fuller (1951), Knight (1951), Abram (1968b), Gash and Hanna (1972), and Rice and Heck (1975). A detailed review of the earlier literature on diseases and parasites was provided by Takos (1940). Generally, percentage incidences in muskrats range from 78 to 93 for trematodes, from 0 to 41 for cestodes, and from 8 to 25 for nematodes. However, muskrat parasite burdens are variable and are frequently related to habitat, sex, and age of the host (Anderson and Beaudoin 1966; Abram 1968a; Cromer 1968).

The major trematodes (35 percent or greater incidences), including evaluations by Jilek (1977) and MacKinnon and Burt (1978), are *Echinochasmus schwartzi*, *Quinqueserialis quinqueserialis*, *Pseudodiscus zibethicus*, *Nudacotyle novica*, *Notocotylus filamentis*, *N. urbanensis*, *N. quinqueserialis*, *Echinostoma revolutum*, *Echinostomum coalitum*, and *Wardius zibethicus*. Important cestodes (10 percent or greater frequency) reported, including work by Byrd (1953), are *Hymenolepis evaginata*, *Taenia tenuicollis*, *T. opaca*, and *T. taeniaeformis*. Frequent nematodes (10 percent or greater frequency) are *Trichuris opaca*, *Rictularia ondatrae*, and *Capillaria hepatica*.

Penn and Martin (1941) reported that about 9 percent of 1,032 muskrats surveyed in Louisiana were infested with the pentastome nymph *Porocephalus crotali*. Other internal organisms that infect muskrats include the tapeworm (*Taenia crassicollis*), nematode (*Dirofilaria* sp.), and liver fluke (*Parametorchis* sp.) (Smith 1938), and two protozoans (*Giardia* sp. and probably *Trichomonas* sp.) (Penn 1942). Uncommon parasites of muskrats are discussed by Beckett and Gallicchio (1967).

Errington (1942) described a dermatomycosis (fungus disease of the skin), chiefly attributed to *Trichophyton mentagrophytes*, that affects young muskrats. The condition was present in 9.6 percent of the litters examined and had an apparent mortality rate of 91.8 percent for infected individuals. This ringworm disease, which is reportedly of zoonotic importance, was also described in muskrats by Dozier (1943).

Other maladies that may affect muskrats include yellow fat disease (nutritional panniculitis, steatitis), a dietary condition (Debbie 1968), and lumpy jaw (actinomycosis) (Dozier 1943).

Ectoparasites of muskrats include the mites *Tetragonyssus spiniger*, *Ichoronyssus spiniger*, and *Laelaps multispinosus*, and the larvae of the dipteran *Sarcophaga* sp. (Smith 1938; Penn 1942). Generally, percentage incidences for acarina range around 98.

Predation. Muskrats are heavily preyed upon but in better habitat their rate of production is high enough to prevent population damage. O'Neil (1949) listed 17 muskrat predators in descending order of effect: mink (*Mustela vison*), raccoon (*Procyon lotor*), barn owl (*Tyto alba*), barred owl (*Strix varia*), alligator (*Alligator mississippiensis*), ant, marsh hawk (*Circus cyaneus*), eastern cottonmouth (*Agkistrodon pis-*

civorus), bullfrog (*Rana catesbeiana*), garfish (*Lepisosteus* sp.), bowfin (*Amia calva*), snapping turtle (*Chelydra serpentina*), largemouth bass (*Micropterus salmoides*), crab (Decapoda), hog, house cat, and dog. The last 8 predators were supported by only limited scientific data.

Muskrats caught in traps may be damaged by marsh hawks, gulls (*Larus* spp.), raccoons, grackles (*Quiscalus* spp.), vultures (*Cathartes aura* and *Coragyps atratus*), muskrats, hogs, crabs, garfish, turtles, and crows (*Corvus* spp.) (Smith 1938; O'Neil 1949). Although O'Neil (1949) included ants as a predator of muskrats, studies in Louisiana by Newsom et al. (1976) failed to reveal any significant deleterious effect of fire ants (*Solenopsis invicta*) on whether or not muskrat houses were active or contained young.

Marsh hawks and raccoons were listed by Lay (1945a) as the common winter predators on muskrats in the Texas Jackson marsh; he also stated that losses due to fighting among muskrats may be large at times. Hawks, owls, coyotes (*Canis latrans*), foxes (*Urocyon cinereoargenteus* and *Vulpes vulpes*), dogs, raccoons, weasels (*Mustela frenata*), snakes, snapping turtles, and certain fishes were also listed as predators of lesser importance.

Predators noted by Sather (1958) on muskrats in north-central Kansas were mink and coyote; badger (*Taxidea taxus*) and raccoon were suspected. Muskrat predators reported in Maryland marshes include the bald eagle (*Haliaeetus leucocephalus*), marsh hawk, barred owl, and great horned owl (*Bubo virginianus*) (Smith 1938). He noted, however, that most birds of prey are scavengers and may not kill all the muskrats they feed upon. Additional predators of muskrats include weasels (*Mustela* sp.) and pickerels (*Essox* sp.) (Lantz 1910; Sather 1958).

In Missouri, Schwartz and Schwartz (1959) reported minks and humans as the most important muskrat predators. The predatory relationship between the mink and muskrat was addressed by Errington (1943, 1951, 1954b). The 1943 analysis led to several conclusions: (1) environmental conditions, intraspecific tolerance, and drought are important in predisposing muskrats to mink, and, with a few exceptions, minks had little influence upon the net mortality of muskrats; (2) when mink predation is severe, losses of muskrats from other causes proportionally diminish (a decrease of mortality from predation was largely offset by increased killing of young by older muskrats or from increased miscellaneous losses); (3) adults under crowded conditions not only kill young but prematurely stop breeding; and (4) little increase in muskrat trapping revenues could be realized by mink repression in north-central areas where muskrat trapping occurs in fall and winter, that is, both minks and muskrats should be managed on a sustained-yield basis. The 1954 evaluation was based on analyses of 13,176 mink scats. Of the 2,415 scats containing muskrats, 1,600 were believed to have resulted from scavenging on muskrats with hemorrhagic disease, 100 from miscellaneous scavenging, and only 674 from predation.

Therefore, mink selection pressure upon muskrats was not considered great.

Wilson (1954b) found that muskrats were a minor fall-winter food of mink and an unimportant fall-winter food of otters (*Lutra canadensis*). Similarly, Glass (1952) could not verify mink predation of muskrats in Oklahoma ponds.

Data that indicated that raccoons were an important muskrat predator in the Currituck marshes of North Carolina were presented by Wilson (1953). He documented extensive damage to muskrat houses by raccoons during the breeding season but practically no damage in December; apparently invasion of houses was influenced by the presence of young muskrats.

Raccoons seemed to be an unimportant factor in the destruction of muskrat houses in the marshes of the Maryland eastern shore (Nicholson and Davis 1957). However, Harris (1952) found that 50 percent of 1,892 muskrat structures in Maryland were disturbed by predators, largely raccoons. Muskrat remains occurred in 19.2 percent of 150 raccoon stomachs; 3 percent of 551 raccoon droppings collected in the marsh contained muskrat remains.

Errington and Scott (1945) documented heavy depredation of muskrats by red foxes under summer drought conditions in a marsh in north-central Iowa. Although the severe drought would have resulted in muskrat losses without fox depredation, the foxes seemed to have reduced muskrats in such a manner as to be noncompensatory. A possible net decrease of about 25 percent of the trappers' income was believed to have been caused by fox depredation. Similarly, of 17 red fox stomachs from Maryland examined by Harris (1952), 59 percent had muskrat remains and 56 percent of 132 fox droppings contained muskrat.

Accidents. Adverse or abnormal climate is an important factor limiting muskrat population growth since it affects salinity, pH, dissolved oxygen, water tables, and food plants (Ferrigno 1967). Floods, hurricanes, and drought are natural factors that cause the largest degree of muskrat mortality and habitat destruction (Lantz 1910; Svihla and Svihla 1931; Wilson 1968).

Muskrats in stream habitats seem especially vulnerable to sudden water rises. Consequently, floods from heavy rains during late winter and spring may be the most important factor limiting muskrat populations in northern Mississippi; such floods cause stream rises that fill muskrat burrows and drown the young (Freeman 1945). Errington (1937b) also pointed out that drowning is a mortality factor for muskrats due to rising water and suckling young being pulled underwater by their mothers.

Bellrose and Low (1943) found that muskrats flooded in the Illinois River Valley initially sought to remain on top of lodges, secondly on floating rafts of vegetative debris, and lastly on branches of willows (*Salix nigra*) and buttonbush (*Cephalanthus occidentalis*). During the first few days of displacement by flooding muskrats were lethargic. Emergency rations consisted of twigs and bark of buttonbush and willows, hornwort (*Ceratophyllum demersum*), duckweeds (*Lemna* spp.), river bulrush (*Scirpus fluviatilis*), arrowhead (*Sagittaria latifolia*), American lotus (*Nelumbo lutea*), and pickerelweed (*Pontederia cordata*). Intraspecific strife, as illustrated by fighting, occurred among the muskrats most exposed; death was frequently caused by muskrat-inflicted wounds and possibly drowning, disease, or exposure.

Equinoctial storms are more damaging than rainwater floods in marsh areas (Freeman 1945). O'Neil (1949) stated that Gulf coast muskrats were safe from storms as long as the tides did not raise water levels over the tops of marsh grasses. He indicated that a muskrat forced to remain in water until the fur is penetrated to the skin will soon die. Storms frequently cause high water to be blown into the marshes rapidly, giving muskrats little time to raise their houses or escape to higher ground. They may be buffeted by the rough water until they become exhausted and drown. Food is scarce because of high water, and muskrats driven from their homes are exposed to enemies; also, the food supply is often damaged by rough water (Freeman 1945).

Floods in Maryland marshes have caused many muskrats to drown but even more to succumb to disease, starvation, and lack of fresh water (because of the tide) (Smith 1938). During floods, food is scarce and of poor quality. As in streams, during quick rises of water levels in the marsh, muskrats, especially the young, may die from entrapment in houses or dens (Schwartz and Schwartz 1959). The Louisiana Department of Conservation (1931) described how life rafts were constructed as refuges for muskrats during the 1928 flood.

Low precipitation and the associated salinity increase were identified as the causes of the decline of muskrats on the Atlantic coast during the early 1940s (Dozier 1947b). Drought may concentrate muskrats and cause increased social pressures and higher rates of predation and disease. Smith (1938) described the effects of drought in Maryland: muskrats deserted their homes and followed receding water down the creeks and rivers; the water became stale and salt water from the bay intruded into the marsh and destroyed favorite foods. Errington (1938b) observed that a large proportion of muskrats in drought-inflicted habitats stayed in familiar range. Although they frequently suffered heavy mortality, they were usually better off than animals that attempted to go elsewhere. When conditions improve, moderately low, drought-stricken muskrat populations show greater rates of increase than do denser populations.

Muskrats subjected to low water conditions during summer are affected less seriously than during low water conditions in cold weather. Searching for food may become a special problem for muskrats during winter drought periods. Dry periods allow the lowered water to freeze through, thus sealing off food resources ordinarily available in unfrozen mud and water; ice entrapment may also be a problem (Schwartz and Schwartz 1959). Snowless winters in northern regions

may allow similar deep freezing of muskrat marshes. Lack of snow cover during winter was associated with heavy ice formation, which resulted in a freeze-out and heavy mortality in muskrat populations in northeastern North Dakota (Seabloom and Beer 1964). Similar effects of severe winters on muskrats in New Jersey were described by Pancoast (1937).

Productivity, more than survival, may be seriously affected by the unavailability of food during drought-stricken winters. Muskrats in most areas have a wide range of alternative foods, depending upon conditions. Errington (1941) attributed similar winter survival rates of well-fed and poorly fed muskrats to this feeding versatility. However, greater productivity was found in the well-fed population.

McEwan et al. (1974) found that heavy oiling (crude) increased the thermal conductance of muskrats by as much as 122 percent. To compensate for the loss of thermal insulation, oiled muskrats increased their dry-matter intake 2.5-fold. The investigators doubted that muskrats exposed to moderate quantities of oil could survive under natural conditions. Similarly, Wragg (1954) found that fuel oil had a persistent and cumulative wetting effect on muskrats.

Compensation. Although disease, parasitism, climatic factors, and accidents act directly on individual muskrats, the cumulative or actual influence of any or all of these factors may not significantly alter long-term population levels. A loss from one factor may be compensated by decreased losses from other factors (Errington 1951) or by corresponding increases in production (e.g., prolonged breeding, additional litters) (Errington 1948). Similarly, heavy losses at one season may be offset by lower losses at other times of the year.

Several examples of compensation have been reported. Muskrat drowning rates tend to rise under natural conditions as populations become denser (Errington 1937b, 1948). Similarly, Sather (1958) indicated that hemorrhagic disease appeared associated with overpopulation.

Penn and Martin (1941) related high parasitism to high muskrat populations in Louisiana. Preliminary data of O'Neil (1949) indicated that an inverse relationship existed between degree of parasitic infestation and muskrat density: high muskrat populations (greater than 25 per ha) had a percentage infestation rate of 5.7, while low densities (less than 25 muskrats per ha) had an infestation rate of 13.8 percent. However, he pointed out that since high populations may decline very rapidly, a significant lag in degree of infestation may occur because of the time necessary for parasites to mature. Additional details of intercompensatory adjustments and control of muskrat populations were provided by Errington (1963).

AGE DETERMINATION AND AGE RATIOS

Several techniques have been suggested for age determination in muskrats: the use of the appearance of internal reproductive organs, structure and weight of the baculum, size and structure of external reproductive organs, pattern of pelt primeness, length of the pelt, tail length, dental and skull characteristics, and eye lens weight. Additionally, Errington (1939) provided a growth curve for estimating the age of young muskrats. Generally, young muskrats are grayer on the back and paler on the sides than adults.

Age determination of male muskrats was described by Baumgartner and Bellrose (1943) and Schofield (1955). Males that have a dark penis over 5.15 mm in diameter with a blunt, round tip are adult, whereas males with a lighter red penis less than 5.15 mm in diameter with a knob-shaped tip are immature. Testes of adult males are 11 mm or greater in length, flattened, wrinkled, and discolored; testes of animals that have not bred are turgid and cream colored (Schwartz and Schwartz 1959).

Adult females have a thin or missing vaginal membrane, whereas immatures have a thick membrane (Baumgartner and Bellrose 1943; Schofield 1955). Uteri of adult females are thickened and contain placental scars whereas uteri from unbred muskrats are thin, transparent, and lack placental scars (Schwartz and Schwartz 1959). Methods suggested by Errington (1939), based upon the reproductive tract, and by Baumgartner and Bellrose (1943), based on appearance of the external genitalia, were found by Shanks (1948) to be valid but more time consuming than the pelt primeness method.

Elder and Shanks (1962) found that the distal end or distal processes of the baculum could be used to determine the age of Missouri muskrats with up to 97.6 percent accuracy during the harvest. In muskrats five to eight months old, the distal processes of the baculum are cartilagenous and calcification within the baculum has just begun; from three to six ossification centers develop and calcification is nearly complete in adult muskrats. Also, the baculum shaft is heavier and more rugose in adults than in juveniles.

Aging by molar teeth appearance was 100 percent accurate on 58 known-age skulls from Kansas (Sather 1958). In cleaned muskrat skulls less than nine months old, the upper end of the first fluting of the first upper molar extends deep into the bony socket, the end of the fluting is not visible, and the anterior edge of the first molar is straight; in skulls from adults this fluting groove does not reach the socket, it is visible, and the front edge of the first molar is slightly curved (Sather 1954). Olsen (1959a) further refined this technique by devising a method by which muskrats trapped during March and April could be aged. He found that fluting length and degree of root development could be employed to classify animals as adult, subadult, or juvenile with a low degree of overlap.

Brohn and Shanks (1948) reported that although the upper incisors of adult muskrats averaged 1.09 mm wider than those of subadults, the degree of overlap of incisor width of the two age classes was so high that this method could not be used to age Missouri muskrats accurately. Similarly, Elder and Shanks (1962) indi-

cated that dentition was not a valid criterion of age in Missouri muskrats.

Zygomatic breadth of freshly skinned carcasses, in conjunction with fluting of the first lower molar, was used by Alexander (1951, 1960a) to separate adults from subadults. A measurement greater than 40.6 mm indicated an adult muskrat on the Montezuma National Wildlife Refuge, New York. The zygomatic breadth technique (Alexander 1951; Schofield 1955; Dean 1957) may be subject to error unless measurements are standardized. A curve to correct measurements of zygomatic breadth from muskrat skulls prepared at different seasons or humidities was presented by Alexander (1960b).

Lay (1945a, 1945b) divided harvested muskrats of Texas into three age groups, based on either pelt or total length measurements: kits and mice with a total length up to 459 mm and a thin, papery pelt 195 mm long or less; subadults that were 460–509 mm long with thicker pelts 200–220 mm long; and adults that were larger than subadults. However, this technique was unreliable for Missouri muskrats (Shanks 1948). Similarly, pelt length and total body length were of little value in aging muskrats in north-central Nebraska because of overlap between age classes (Sather 1958).

Muskrats may be aged by the pattern of pelt primeness (Applegate and Predmore 1947; Shanks 1948). Pelts from sexually mature adults present an asymmetrical, mottled primeness pattern of dark and light areas, whereas the fleshy side of pelts from young are bilaterally symmetrical. The sexing and aging techniques of Shanks (1948) worked on Horicon, Wisconsin, muskrats with an error of less than 2 percent (Linde 1963). Pelts from adult females also exhibit teats, usually blackened and 1.6–4.8 mm in diameter, whereas pelts of immature females have usually unpigmented teats less than 1.6 mm in diameter (Schwartz and Schwartz 1959).

The age of live muskrats may be estimated by fitting a second-order polynomial to a set of measurements of tail length taken at various ages from an unknown-age animal (LeBoulenge 1977). Birth date is estimated as the time at which the tail length equals the mean tail length of muskrats at birth. LeBoulenge (1977) reported that this method gave unbiased estimates of age and was based on sounder assumptions than classical age estimation based on a growth curve calibrated from known-age muskrats. He also reported a method for aging dead muskrats by using a reference growth curve of crystalline lens weight.

Age ratios of muskrats vary widely with measurement techniques, age, season, and geographical area. Because of this variation, Mathiak (1966) questioned the value of age ratios as production indicators. Reported ratios range from 7 to 650 young per 100 adults (table 15.5). The large discrepancies in ratios between geographical areas are partly due to differences in aging techniques.

Lay (1945a) indicated that on a moderately trapped marsh, the standard age-group ratio was about 8 percent kits and mice, 28 percent subadults, and 64

TABLE 15.5. Selected age ratio data for muskrats trapped in North America

Source	Location	Ratio (young[a]/100 adults)
Alexander 1951	New York	560[b]
Alexander 1955	New York	264
Arata 1959*a*	Illinois	75
Beer and Truax 1950	Wisconsin	359
Dozier and Allen 1942	Maryland	7
Erickson 1963	New York	328
Freeman 1945	Mississippi	27–39[c]
Fuller 1951	Canada	43
Gashwiler 1950a	Maine	257
Harris 1952	Maryland	89–145[d]
McCann 1944	Minnesota	266
Reeves and Williams 1956	Idaho	386–403
Schwartz and Schwartz 1959	Missouri	350–650

[a] Younger than adult.
[b] Per 100 adult females.
[c] Greater than 12 weeks old.
[d] Young females per 100 adult females.

percent adults (56 young and subadults per 100 adults). Kits (≤624 g) comprised 6.6 percent of all trapped animals examined by Wilson (1956*a*) from Currituck County, North Carolina, fur sheds. However, age composition in the fall for muskrats on the Athabasca-Peace Delta may run as high as 733 young of the year per 100 adults (Fuller 1951). Beer and Truax (1950) found that percentage young in a Wisconsin marsh was inversely related to breeding population density, and that the best muskrat utilization of the habitat was obtained with a fall ratio of 400–500 young per 100 adults.

ECONOMIC STATUS

Value. The muskrat is the most valuable fur animal in North America (Harris 1952; Dozier 1953; O'Neil and Linscombe 1976). It leads all other North American wild furbearers in number caught and overall pelt value, and muskrat pelts are the most common fur on the market (Schwartz and Schwartz 1959; Palmisano 1972*a*; Deems and Pursley 1978).

During the 1975/76 season, over 8 million muskrats were harvested in North America, with a value exceeding $29 million (table 15.6). Data indicate an escalation of the value of the North American fur harvest, apparently because of increases in pelt value and number harvested, and improved harvest recording systems (Deems and Pursley 1978). The muskrat has been of major economic importance in the fur trade of North America since colonial times (Storer 1937) because of its wide distribution, abundance, and pelt value (O'Neil and Linscombe 1976).

Louisiana is an important muskrat harvesting area in terms of numbers harvested. However, before the nutria became the most important fur animal in Louisiana, even greater numbers of muskrats were harvested. For instance, over 8 million muskrats, with a total value exceeding $12.5 million, were harvested during the 1945/46 season in Louisiana. The muskrat appears to be making a comeback in Louisiana. During the 1978/79 season 445,525 pelts were sold for a value of $1.48 million; the meat was valued at approximately $1,600 (O'Neil and Linscombe 1976).

TABLE 15.6. Extent and value of the North American muskrat harvest, 1970–76 (6 seasons)

Season	Average Pelt Price	United States		Canada		Total	
		Number	Value	Number	Value	Number	Value
1970/71	$1.40	4,344,685	$6,082,559	1,475,284	$2,065,398	5,819,969	$ 8,147,957
1971/72	2.25	4,142,519	9,320,668	1,501,123	3,377,527	5,643,642	12,698,195
1972/73	3.00	4,612,494	13,837,482	1,457,928	4,373,784	6,070,422	18,211,266
1973/74	3.00	6,287,369	18,862,107	1,408,629	4,225,887	7,695,998	23,087,994
1974/75	3.00	5,528,894	16,586,682	1,805,805	5,417,415	7,334,699	22,004,097
1975/76	3.50	6,415,861	22,455,514	2,102,513	7,358,796	8,518,374	29,814,310

SOURCE: Extracted from tables of Deems and Pursley (1978).

Pelt Primeness. The primeness of muskrat fur is important because it influences the quality and value of the final pelt. Pelt value may also be affected by handling, stretching, skinning, and metabolic hormones. Hormones are controlled by such factors as food, water, and periodicity (Bissonnette and Wilson 1939; Bassett et al. 1944). Fur buyers and garment manufacturers are concerned with pelt primeness and pelt damage because these affect the value of the final garment.

Linde (1963) performed a detailed analysis of pelt patterns and primeness based on muskrats from Horicon Marsh, Wisconsin. He defined a pelt as being prime (of highest quality) when the fur is at its maximum length, density, and finest texture; when hairs have matured and no further pigment is being produced; and when the flesh of the pelt appears devoid of hair root pigmentation. He stated that hair growth occurs in a succession of wavelike patterns, each one complementing the previous until the fur is prime, and that certain forms of physical stress may suppress hair growth. Stains (1979) defined a prime pelt as one in which new guard hair and underfur have reached maximum growth.

The time of maximum primeness for a population is the period when the hair of the majority of the animals has reached maximum growth (Stains 1979). The degree of primeness can be ascertained by the amount of blue black streaking or spotting that occurs where new hairs have not yet emerged from their follicles (Dozier 1953).

Primeness varies geographically and by species: the fur of northern furbearers generally becomes prime earlier and remains prime later than that of southern animals. The muskrat is among the species that remain prime the longest (about four months). Generally the time of primeness for muskrats is from early December through March, with greatest primeness in March (Stains 1979).

Marketing. Dozier (1953) provided an excellent discussion of marketing muskrat pelts and O'Neil (1949) outlined some of the early relationships that existed in Louisiana among landowners, operators, and trappers (leasers). The skins are universally case handled and dried, generally on wire U- or V-shaped stretchers, with the flesh side out (Dozier 1953). The fur trade industry recognizes various types of muskrat pelt, such as: (1) the brown, or northern, (2) the black (a dark phase of brown), and (3) the Louisiana, or southern. Pelts are usually graded by size group (mice and kits, small, medium, large, X large, XX large, and XXX large) and season (fall, winter, and spring). Further grading is based on quality of color, taints, and pelt damage. Practically all pelts are tanned and dressed in the large centers of the fur trade, such as New York; Newark, New Jersey; St. Louis, Missouri; and Seattle, Washington (Dozier 1953).

Muskrat fur wears well and the skin takes dye readily and makes strong leather. The pelt is generally cut into several longitudinal strips that separate the sides from the belly and back. Muskrat fur is popular and blends well into any type of fashion. The fur is a favorite in Europe for linings and trimmings and for casual coats and jackets (O'Neil and Linscombe 1976). Muskrat fur is sometimes called river mink, red river seal, water mink, electric seal, or Hudson seal (Indiana Department of Conservation 1963).

Muskrat pelts from different geographical areas offer different advantages to garment manufacturers. For example, Louisiana's muskrats have a darker pelage than most northern muskrats. This eliminates the need for dyeing, a feature that appeals to furriers. The Louisiana muskrat pelt also has three distinct colors that provide material for three separate garments. For example, three coats of three colors can be made from about 90 pelts (Lowery 1974; O'Neil and Linscombe 1976). Louisiana muskrats also have a lower nap and shorter guard hair than northern muskrats. Thus, Louisiana furs drape more easily and gracefully when sewn together as a garment or coat. Northern muskrat pelts, in contrast, are larger and more uniform in color (Lowery 1974). Bailey (1937) stated that the best pelts from Maryland are the largest, darkest, and most valuable of any in the world.

Muskrats As Food. Muskrat meat has been considered as a food for stock poultry (Goff and Upp 1942) and for human consumption (Gowanloch 1943). Muskrats were esteemed as food by aborigines of North America and were eaten by early colonists, trappers, hunters, and voyageurs of the North and Northwest, and by American Indians (Lantz 1910). The flesh has a gamey flavor and carcasses are sold on the market in certain areas. Bailey (1937) stated that muskrats dressed to avoid a trace of musk and properly cooked and served have sweet, rich, and tender meat with the game flavor of wild duck. Detailed information on preparing muskrat meat for consumption is provided by Lantz (1910), Dozier (1943), and Dailey (1954). The muskrat was listed as a food resource by Deems and Pursley (1978) in 18 states and 11 provinces or territories. Muskrat meat is often referred to as marsh rabbit, water squirrel, marsh hare, hare, or Chesapeake terrapin (Cook 1952; Indiana Department of Conservation 1963).

Carcasses taken during the breeding season are frequently impregnated with musk, which renders the meat offensive to humans (Schwartz and Schwartz 1959). A second factor that prevents wider acceptance of muskrats as food may be its designation as "rat." Dried musk, however, was listed as an important component in perfumes, especially before synthetics were developed (Erickson and Stevens 1944; Taylor 1980), and in commercial trapping scents (Schwartz and Schwartz 1959).

Farming. During 1925–30, muskrat farming was promoted heavily in several states and many people were deceived into thinking such operations were highly profitable (Dozier 1953). Early muskrat ranching and farming operations have been described in detail (Louisiana Department of Conservation 1931). Field (1948) discussed muskrat farming and cited returns of 22 to 34 percent on investment. Annual yield varied with the quality of soil and skill of the farmer.

The production of about seven pelts per ha was considered good.

Dailey (1954) made the following points concerning muskrat farming: (1) practically every section of America, at least in places, is acceptable for muskrat rearing; (2) muskrats develop quickly and spring- and summer-born animals can be marketed the next spring (he advised against harvesting in fall and winter); (3) muskrat farming provides a quicker monetary return than the raising of any other furbearer because muskrats are prolific and can be cheaply maintained; (4) large numbers must be raised for farming to be successful, even though the muskrat is very self-sustaining and less capital is needed than for some other aspects of the fur industry; and (5) muskrats should be raised under conditions as natural as possible.

In a farming situation, a hectare should support 124–185 pairs and their progeny each year (Dailey 1954). However, Bradley and Cook (1951) indicated that muskrats cannot be raised successfully in a seminatural environment where fencing restricts movement. They found that 75 of 86 muskrats released in fenced 0.3- to 1.2-ha enclosures died as a result of wounds from fighting. Similarly, Dozier (1953) stated that although muskrats can be raised on a limited scale in pens, it is not profitable because of irregular reproduction under restraint and losses from fighting, poor sanitation, and handling.

The greatest muskrat-producing areas of North America were listed as Louisiana, Maryland, Delaware, New Jersey, Virginia, and parts of Canada (Dailey 1954). Today, muskrat farming in North America is limited, mainly because of low profitability.

Damage. Negative economic attributes of muskrats include burrowing and damage to dikes, ditches, ponds, and levees, and losses of crops. Glass (1952) described the methods by which muskrats damage ponds. Dams sodded with grass are frequently undercut by muskrats as much as 45 to 51 cm. When the water level drops, muskrat holes are expanded to keep pace with the water line. Holes constructed during periods of low water are expanded upward when the water level rises. Erosion problems from muskrat activities are worsened by wave action caused by high winds. Excavation activities may seriously weaken a dam. Trees growing on the dam may exacerbate the problem, because muskrats may tunnel along one or more of the roots. The most serious type of damage described was leakage, and damage was correlated to pond age and time inhabited by muskrats.

Muskrats especially relish corn, both the stalk and roasting ears. Damage to farm crops in Iowa by muskrats was reportedly confined mostly to cornfields close to watercourses (Errington 1938a). Lantz (1910) indicated that damage to crops by muskrats occurred only in limited areas, but noted the muskrat was a pest in rice plantations in parts of Louisiana because of its damage to embankments and plants. A 1969 survey indicated that damages by muskrats to Arkansas rice crops amounted to around $1 million annually, in addition to damages to the levees and pond banks of fish farmers (Sealander 1979). Conversely, muskrats are important in the control of emergent aquatic vegetation (Cook 1952).

MANAGEMENT

Management of systems to enhance muskrats centers around the manipulation of the harvest and specific treatments to provide water, food, cover, and protection. Muskrats are relatively easy to manage compared to other furbearers. They are prolific breeders and their habitat requirements are readily met.

Muskrat populations flourish in early seral stages (Penfound and Schneidau 1945), and many land management practices attempt to retard or set back plant succession. Treatment measures that benefit muskrats include burning, vegetative poisoning, planting, sodding, mowing, and water level and salinity control with ditches, weirs, and impoundments. These measures are frequently compatible and several may be employed simultaneously to maximize benefits, effectiveness, and efficiency.

Even with sophisticated management practices, however, muskrat populations cannot be stockpiled from year to year (Dozier 1953). Investigations in Wisconsin by Mathiak and Linde (1954) and Mathiak (1966) disproved the need for closed seasons in the hopes of producing bumper crops.

Two books have been written on marsh and muskrat management and they provide detailed descriptions and illustrations of specific management techniques. O'Neil (1949) addressed southern coastal areas and Errington (1961) detailed the practical management of muskrats in northern areas.

An overall view of muskrat research and management in North America was presented by Deems and Pursley (1978). No state, province, or territory granted total protection to the muskrat; 8 states had a muskrat hunting season; 47 of the 48 states where the muskrat was present had a trapping season and all of the 12 provinces or territories that contained the muskrat had a trapping season; 4 states (Arizona, Michigan, Texas, and Utah) allowed year-round harvesting; habitat management for the muskrat was being conducted in 15 states and 4 provinces or territories; and muskrat population inventories were conducted in 15 states and 7 provinces or territories.

Habitat Manipulation. Burning. Although Errington (1951) stated that a muskrat marsh may deteriorate from no apparent cause, oftentimes the culprit is advancing succession. In coastal marshes burning is an efficient tool suitable to enhance plant communities attractive to muskrats. Fire is a natural part of marshland ecology in many coastal areas. Burning is beneficial for muskrats because it sets back succession, provides optimum growing conditions for Olney bulrush (*Scirpus olneyi*), prevents the accumulation of roughs (which may cause damaging wildfires), and controls undesirable plants (Freeman 1945; O'Neil 1949; Dozier 1953; Wilson 1968).

From a management standpoint, *S. olneyi* is the most important food for muskrats in the Gulf coastal marshes. The greatest abundance of nonfoods consists of wiregrass or marshhay cordgrass (*Spartina patens*), needle rush (*Juncus roamerianus*), sawgrass (*Cladium jamaicense*), giant southern wildrice (*Zizaniopsis miliacea*), and maidencane panicum (*Panicum hemitomon*). *S. olneyi* is a subclimax species dependent upon suppression of climax vegetation.

Several methods of eradicating undesirable climax marsh vegetation were investigated by Wilson (1949). Burning prevented marsh build-up, and, in conjunction with flooding in needle rush–sawgrass marsh, resulted in 75–100 percent eradication of undesirable vegetation. Without proper burning, *S. olneyi* and other food plants are invaded by needle rush, sawgrass, and wiregrass (Wilson 1968).

Lay (1945*a*) reported that *S. olneyi* grows best during cool weather and grows little or none in summer. Therefore, there appears to be an advantage to burning *Spartina patens* in late summer to encourage *S. olneyi*. Burning in the spring, on the other hand, would encourage the grass rather than the preferred sedge. However, clean burning in summer seriously impaired muskrat marshes in Texas, especially if followed by drought or high tides (Lay and O'Neil 1942). Return of muskrats to the burned areas was hampered by a lack of house-building materials.

Burning wiregrass and saltgrass (*Distichlis spicata*) during normal water periods was not very effective in enhancing *Scirpus* spp. in Louisiana (Palmisano 1967). However, burning such marshes during fall with low water levels and then immediately flooding with 25 to 38 cm of water until late spring was successful in controlling unwanted grasses and in enhancing *S. robustus* (Babcock 1967; Palmisano 1967).

Specific burning dates recommended by other investigators vary by marsh type and geographical area. O'Neil (1949) stated that a normal three-cornered grass marsh in Louisiana should be burned between 10 October and 1 January with a 0- to 5-cm water level.

The best time to burn Olney bulrush in Mississippi is in November after the vegetation has died down. Burns should be made the day after a rain or after a high tide to prevent injuring *S. olneyi* roots or burning muskrat houses. Needle rush marsh that is suitable for Olney bulrush should be burned the last of February or the first of March. The marsh should be dry so the fire will be hot enough to injure needle rush roots. With suitable marsh, switchgrass panicum (*Panicum virgatum*) and Olney bulrush will begin to come in the second or third year (Freeman 1945).

Wilson (1968) found that in North Carolina burning could be done between November and January. However, Dozier (1953) suggested waiting until late February to burn Atlantic coast marshes to insure adequate muskrat cover during winter.

Specific burning techniques also vary by marsh type and area. For maximum production of muskrats, O'Neil (1949) recommended spot burning about 60 percent of Olney bulrush marshes between 1 and 15 October; then around 1 December about 20 percent of the remaining roughs should be burned. After the Olney bulrush gets a 20–25 cm start, the remaining 20 percent of the marsh should be burned. Burning at least every other year seems necessary (Lay 1945*a*).

The burning of marsh sections to provide cover and nest-building materials for muskrats was also recommended in Mississippi. A three-year rotation, one-third per year, was recommended for burning Olney bulrush. Needle rush areas should be burned every two years or two years in succession and then one year skipped (Freeman 1945).

In Texas, alternate strips should be burned every other spring in marshes that do not have rank vegetation, such as common reed (*Phragmites communis*) (Lay and O'Neil 1942). Advantages are the assurance of cover, a discouragement of muskrat movements to other areas, a conservation of marsh water supply, and provision of autumn vegetation too rank for geese.

WATER CONTROL. Water level manipulation is a principal muskrat management technique and the importance of proper water levels should not be underestimated. Changing water levels were found to (1) be more important than the type of marsh vegetation in determining muskrat population levels in the Illinois River valley (Bellrose and Brown 1941); (2) be the principal factor limiting muskrat populations in the marshes of Currituck Sound, North Carolina (Wilson 1949); (3) adversely affect muskrat populations in Illinois marshes (Bellrose 1950); and (4) prevent the establishment of muskrat food plants in certain areas of Louisiana (Moody 1950).

Prolonged flooding may kill desirable muskrat plants such as common cattail (*Typha latifolia*) (Errington 1948). When cattail marsh is destroyed, maximum muskrat densities that could otherwise be supported on Iowa marshes may be reduced by half. In Gulf coastal marshes, excessive water depths enhance undesirable plants such as cattail, bullwhip bulrush (*Scirpus californicus*), and southern wildrice (Lay and O'Neil 1942; Lay 1945*a*).

In contrast, water-deficient marshes in Mississippi are frequently dominated by needle rush, and needle rush replaces Olney bulrush when the water table is lowered (Freeman 1945). Water levels below the marsh floor also cause loss of peat through oxidation (Lay and O'Neil 1942; Lay 1945*a*).

Water levels also influence the physical well-being of muskrats. On areas of Montezuma National Wildlife Refuge where the water level had been lowered muskrat feeding beds froze, ice formation reached the impoundment bottom, and large muskrats lost weight (Friend et al. 1964). On areas where normal water levels were maintained, smaller weight losses occurred. Most of the weight lost was regained when the drained areas were reflooded. Disease and parasite infection were also greater in muskrats from the low-water areas.

Optimum water levels for muskrats vary geographically. Gulf coastal marsh water depths should be controlled year-round between 2 and 31 cm, and preferably between 15 and 20 cm (Lay and O'Neil 1942;

Lay 1945*a*). Palmisano (1967) recommended even more stringent water level control for the enhancement of *S. olneyi*: the maintenance of water levels at or slightly above the surface of the marsh and never more than 5 to 8 cm below the marsh floor. Water depth in *S. olneyi* marshes in Louisiana averaged 5.8 cm in August (Palmisano 1970). However, marsh water levels are significantly influenced by rainfall, tidal, and groundwater fluctuations.

Marsh water in Maine should be maintained 15 to 51 cm deep to improve habitat for muskrats (Gashwiler 1948). Such levels optimize muskrat usage of the entire marsh; increase desirable food, cover, and muskrat breeding sites; and minimize the effects of destructive floods. Other beneficial water management practices include the construction of potholes that have areas about 1 m deep to prevent solid freezing and the providence of shallow marsh areas for food production. Natural contours should be used to determine pothole spacing to reduce costs.

Salinity is also a factor controlling marsh plant associations (Lay and O'Neil 1942; Palmisano 1970). Marsh salinity, like water depth, is extremely variable, both seasonally and daily, and is affected by similar factors. Saline water blown over brackish and fresh marsh by storm tides will destroy stands of *S. olneyi* and *Typha* spp. (Wilson 1968). A salt level of 1 to 1.5 percent in the soil water was recommended by Lay and O'Neil (1942) as most desirable for muskrats. Similarly, Palmisano (1967) found *S. olneyi* growing in Louisiana where the percentage of soluble salts in the soil ranged from 1.0 to 1.7 (salinity 10 to 17 parts per thousand) with a pH of 4.1 to 7.9. Later investigations by Palmisano (1970) revealed that the total soil salts in Louisiana marshes where *S. olneyi* occurred averaged 0.9 percent in August. To enhance *S. olneyi*, Palmisano (1967) indicated that salinity must be maintained between 0.5 and 2.0 percent salt. Interestingly, however, a 50 percent reduction in germination of *S. olneyi* occurred at 0.4 percent salt.

Water control structures can help control salinity and ameliorate water deficiencies. A small change in water levels can completely change marsh vegetation (Wilson 1968). Consequently, the absence of water control structures may greatly reduce muskrat harvests (Sather 1958). Water levels may be controlled by diking and impounding, construction of weirs, and pumping. Where water levels are affected by tides, dikes and weirs are especially useful to stabilize water levels, stop erosion, and control the supply of muskrat food plants.

Diking and ditching, if done properly, provide numerous benefits to muskrat populations. Wilson (1953, 1955*b*) studied diked and undiked marsh in North Carolina. He found that: (1) marsh drained by ditches and subject to constant water level fluctuations constituted poor muskrat habitat; (2) year-round flooded marsh containing poor quality foods supported few muskrats; and (3) in marshes containing equal quantities of food and cover, muskrat populations were low in marshes where the water level was frequently below marsh level compared to the relatively high populations in marshes covered by an average of 5-15 cm of water.

Studies in Wisconsin by Anderson (1948) and Mathiak and Linde (1956) indicated that level ditches provided muskrats: (1) sufficient water to prevent solid freezing in winter, (2) access to food during the critical winter period, (3) escape and feeding areas during spring floods, (4) important muskrat food plants, (5) open water during summer drought periods, and (6) protection against housing freeze-outs (because houses on spoil banks are less likely to freeze than houses constructed in marsh). Trapper access to the marsh was also enhanced by the ditches. Similarly, levees in Texas marsh are useful for water level control until made ineffective by muskrats. Levees also provide habitat for muskrat burrowing, feeding, bedding, resting, and traveling (Lay 1945*a*).

Muskrat harvests and densities are generally greater on properly diked or ditched marshes. Wilson (1953, 1955*b*) found that constantly flooded, diked marsh averaged 4 to 5.5 times as many muskrat houses per hectare as undiked marsh frequently void of water, and that diked marsh provided about 6 times as many harvested muskrats as undiked marsh. In general, Atlantic coastal marshes managed with control structures can yield 3 to 5 times as many muskrats as undiked marsh (Wilson 1968). Donohoe (1961, 1966) also reported that marsh enclosures created by spoil bank construction increased muskrat densities, probably because of increased juvenile survival.

In leveed marshes, a high percentage of muskrats is trapped on levees. Lay (1945*a*) indicated that in the Jackson, Texas, marsh about 80 percent of muskrats were taken on levees spaced 274 m or more apart. He surmised that all trapped muskrats would have been caught on levees if the levees had been spaced every 183 m.

Likewise, the muskrat harvest from ditched plots in Wisconsin marsh was 4 to 10 times higher than the harvest in the surrounding bog (Mathiak and Linde 1956). Ditches with 15-m spacings provided the greatest harvest (37 per ha per yr); ditches spaced 122 m apart provided a harvest of only 14 muskrats per ha per year. However, annualized returns on the ditching investment were highest with the 61- and 122-m ditch spacings. The 61-m spacing was recommended because if gave a higher yield of muskrats per hectare than the 122-m spacing.

Ditching was recommended by Mathiak and Linde (1956) as being especially useful for providing deeper-water areas when flooding by dams or dikes is not possible. Ditches should not be dredged in a straight line so that boat travel will be safer during high winds. Spoil banks should be about 12 m long and staggered on alternate sides of the ditch to reduce the fire hazard. They should be seeded to grass as soon after dredging as possible. Marsh burning should not be done in a manner that would endanger spoil bank vegetation. To concentrate muskrats, a portion of a large marsh should be developed with 61-m spaced ditches rather than with scattered ditches throughout the entire area (Mathiak and Linde 1956).

For muskrat farms, ditches should be at least 1.2 m deep and not less than 2.4 m wide; when constructed in soft, flotable soil, ditches should be double width to allow for filling (Field 1948). Optimum ditch dimensions recommended by Mathiak and Linde (1956) were 1.5 m deep by 4 to 4.6 m wide. Ditches should remain productive for muskrats for about 6 to 10 years.

Ditches may be plowed, dynamited, or dug. Good water and burrowing conditions for muskrats may be provided by dynamiting ditches on hard *S. olneyi* marshes or in depressions to hold water at marsh level. Mathiak and Linde (1956) found that ditch dredging was superior to blasting because it was more economical, provided a spoil bank, and filled in less rapidly. Muskrats moved into ditched areas in the first few weeks following construction and the population increased for the next three years.

Filling of ditches from muskrat tunneling may be rapid. Levees in Texas marsh were made ineffective for water level control by muskrats in about four years.

MISCELLANEOUS LAND TREATMENT. *S. olneyi* can be introduced into new areas by planting seed or by transplanting rhizomes or rooted plants. Areas to be planted should be pretreated with a hard burn, preferrably in early spring (Freeman 1945), and flooded immediately after planting (Wilson 1968). Seed should be scattered when the area is wet or covered with water (Freeman 1945).

Sometimes desired shifts in vegetation can be made quicker by sodding *S. olneyi* and seeding *Typha* spp. (Wilson 1955c, 1968). Peat on the area planted should be at least 10 cm deep. Spring and summer is the best time to sod in North Carolina.

Three techniques of site preparation were evaluated in Louisiana by Ross (1972); tilling was best and burning was poorest. However, *S. olneyi* survival in the burned area was almost twice that which occurred in the area with no site preparation. The combination of burning and tilling was recommended by Palmisano (1967) and Soileau (1968) for reducing competition from wiregrass and saltgrass.

Ross (1972) reported that planting *S. olneyi* in December and January was best, with 100 percent monthly survival. Palmisano (1967) indicated that the planting of eight to nine rhizome nodes per m² in April will produce a dense stand by the end of the growing season in October. However, he stated that two to three nodes per m² is more than adequate for marsh revegetation; Ross (1972) suggested a node spacing of 1.8 m. Planting depth should be about 10 to 15 cm. Water should be maintained 5 to 10 cm above the soil surface for three to four weeks after planting. Best survival and growth of plantings occurred at 5 and 10 parts per thousand salinity (Ross 1972). To be successful, rhizome predation from geese, nutria, and muskrats must be prevented.

Mowing, disking, scalping, and herbicides are too expensive for large marshes but may be used on a small scale to enhance *S. olneyi* and *Typha* spp. (Wilson 1968). Economically, a combination of burning and chemical treatment may be the only site preparation measure affordable for the average landowner (Ross 1972). Competition from wiregrass and saltgrass can be reduced significantly by the use of a soil sterilant (5-bromo-3 sec butyl-6 methybduracil) (Soileau 1968). Wilson (1949) found that spraying with sodium arsenite resulted in vegetative mortality of 23 to 91 percent in unmowed and 12 to 70 percent in mowed plots. Mowing of needle rush-wiregrass-sawgrass areas followed by flooding also resulted in high mortality of climax vegetation.

Solid stands of Olney bulrush are more likely to sustain eat-outs than are mixed communities of climax plants (O'Neil 1949). Eaten-out areas should be rehabilitated by digging ditches approximately 76 cm² to enhance revegetation (Dozier 1953). To assist vegetative growth, Donohoe (1961) suggested the application of lime to bare areas of spoil banks where acidity approaches a pH of 4.0 to 5.0. If the eat-out area is immediately drained and conditions are favorable for plant growth, the marsh may be revegetated the winter following the eat-out. With poor growing conditions, which is generally the case in denuded areas, the eat-out area will return to climax rough (wiregrass in brackish marsh) in two to three years. A subclimax marsh can be produced with proper management in two to three additional years (O'Neil 1949).

Donohoe (1961) found that bluejoint reedgrass (*Calamagrostis canadensis*) was adequate to check erosion on spoil banks after three growing seasons with proper soil conditions. The planting of canarygrass (*Phalaris arundinacea*) was effective in checking erosion on spoil banks after one growing season. Mathiak and Linde (1956) suggested seeding new spoil banks with a mixture of canarygrass and sweetclover (*Melilotus officinalis*).

An ecosystem approach will yield the best results when deciding upon muskrat management strategies. Treatment effects on target and nontarget species, both plants and animals, should be evaluated. Specifically, basic life requisites and growth requirements of desired and undesired plants must be considered. For instance, Palmisano (1970) noted that *Spartina patens*, a major competitor of *S. olneyi*, occurs at a wider range of salinity and water level conditions than does *S. olneyi*. Therefore, it would be difficult to use only salinity and water level control to enhance *S. olneyi* at the expense of *Spartina patens*.

Excellent results may be obtained by using several treatment measures simultaneously to enhance habitat for muskrats. For example, ditching (or impounding) and burning complement one another well in the management of *S. olneyi*. Combinations of tilling and burning or burning and chemical treatment are effective for controlling climax vegetation, especially when coupled with water level and salinity control. When creating new marsh on clay soils, the lack of peat, which is desired by muskrats for burrowing, might considerably reduce habitat quality for muskrats (Lay and O'Neil 1942). The best management response during a muskrat disease outbreak may be to conduct detrimental management practices for muskrats to force the population down (Errington 1948).

Population Assessment. Proper management of marshes for muskrats includes a survey or census by which changes in the population can be assessed. Counts of muskrat houses may be used in establishing harvest regulations. Freeman (1945) suggested a survey of 20 percent of the area for occupied houses in order to determine suitable trapping pressure.

Dozier (1948b, 1953) evaluated the muskrat census method that employs house counts. Houses may be counted from the air, water, or ground over a specified area, band, or transect. Accuracy requires that the observer differentiate between unoccupied and occupied houses and between dwellings and feeding huts or shelters. House counts should be taken in early fall in order to plan trapping strategies adequately (Dozier 1948b).

Aerial censuses of muskrat houses permit coverage of a large area in a short time (figure 15.5). Dozier (1948b) stated that with good visibility larger houses are discernible from an altitude of 250 to 300 m within a 1.2-km radius. However, observations at 60 m or less may be necessary to accurately distinguish houses from other structures, depending upon local conditions. House counts vary seasonally. In the marshes of Louisiana they were highest in February when water levels were high, temperatures low, and spring breeding was under way (Palmisano 1972a).

Earlier work, summarized by Dozier (1948b), listed five muskrats per large house as a conversion factor for estimating muskrat populations from house counts. However, he proposed using the average number of muskrats per litter as a conversion factor, because during the fall and winter the parent muskrats and their last litter frequently occupy the same dwelling until the spring breeding season.

House counts from the ground or air, and the Lincoln index, were of questionable value in determining muskrat populations in north-central Kansas (Sather 1958). The most accurate population calculations were based on the number of young per adult female in the winter harvest and on the spring breeding-territory inventory. Winter muskrat populations are calculated by the inventory method on the basis of the number of breeding territories present during the previous breeding season and the number of young per adult female in the harvest (Errington 1943).

Capture-mark-recapture techniques may be employed to evaluate the size or trend of the muskrat population (Mathiak and Linde 1956). However, obtaining a large enough sample may be difficult, especially in summer, because of the muskrat's limited activities and mortality in traps from exposure (Dozier 1953). Hensley and Twining (1946) and Chamberlain (1951) provided additional details on the problems of live-trapping muskrats.

Drop-door live-traps, such as double-door, collapsible 15- x 15- x 61-cm traps of 2.5-cm gridded, 14-gauge, welded wire, are effective for capturing furbearers for biological investigation. Family-type live-traps may be useful for capturing muskrats for special purposes (Snead 1950).

Apples, carrots, cabbage, and potatoes have been used to attract muskrats into traps. Carrots and apples were superior baits in areas of central New York, with carrots far more durable in warm months and slightly more effective (Erickson 1963). The use of a commercially prepared scent, which produces an odor that attracts muskrats by appealing to the mating instinct, increased live-trapping success from an average of 25 to 45 percent (Williams 1951).

The use of unbaited live-traps in runways was very productive in catching muskrats in Louisiana marshes (Robicheaux 1978). Mathiak and Linde (1956) provided submerged aquatics within their live-traps for cover and placed the traps on floats in ditches. Erickson (1963) covered live-traps with vegetation to provide protection for captured animals and to increase trapping success.

FIGURE 15.5. During good weather conditions, trained observers can readily census muskrat houses from the air to provide a population or breeding index. (The house density here is about 25 per hectare.)

Muskrats may be captured within houses for studies that would not be detrimentally affected by system disruption (O'Neil 1949). Vegetation around the house should be burned to drive muskrats into the house. After blocking runs that may be used for escape, the house may be opened to obtain the muskrats.

Muskrats in live-traps may be carefully removed by the tail and placed head first into a wire cone for tagging, aging, sexing, and measurement (Erickson 1963). However, a noose attached to a pole (Robicheaux 1978) can be employed to remove and handle trapped muskrats safely. Serially numbered, size 1, monel metal tags may be applied to the ears of muskrats for identification (Aldous 1946; Snead 1950). Tags should be inserted to about half their length into the nonfleshy, dorsal portion of the ear to prevent infection (Erickson 1963). Tag sizes 1 and 3 may be similarly utilized on the webbing of the rear feet (Robicheaux 1978).

A 5- x 24-mm aluminum tag, inserted through two slits in the skin of the back, was described by Errington and Errington (1937) and Errington (1944). Hensley and Twining (1946) described a waterproof plastic marking button, which can be threaded through the loose skin of the back with a flexible steel needle, and the use of a no. 7 bird band around the base of the tail to mark adults. O'Neil (1949) recommended a band on the Achilles tendon. Evaluations by Takos (1943) of the back-slit, leg-ringing, and Achilles tendon methods of tagging muskrats indicated that the Achilles tendon (or Cook method) was best.

Muskrats may be toe clipped for identification purposes (Dorney and Rusch 1953). Most clipping should be done on the rear toes because the front feet are used extensively in feeding and burrowing. On kits up to five days of age, a toe is clipped with cuticle clippers as close to the foot as possible; after toe elongation, a toe is clipped off at the first joint. Ferric chloride may be used to cauterize and sterilize the wound and prevent excessive blood loss, especially in small kits.

Coon (1965) used radio transmitters that weighed about 43 g to monitor movements of muskrats. The transmitters were powered by a mercury battery. Several types of attachment collar and harness were tested, but a tight-fitting plastic neck collar was best. Intraabdominal FM transmitters were used by MacArthur (1978) to radiotrack muskrats in winter in Delta marsh, Manitoba.

For population estimation, subsequent week's (or day's) captures (depending on sample size) can be used to generate Schnabel, Schumacher, Bailey's Triple Catch, or Jolly-Seber statistics, depending upon the assumptions that can be met. Alternatively, trapper returns of tags, if close in time to tagging, can be used to obtain a Petersen population estimate. Among other assumptions required, these methods assume equal probability of capture, which is seldom obtainable but frequently testable. If enough data are available, segregation of data on the basis of age, sex, and habitat will likely reduce violation of the equal probability assumption. The frequency-of-capture method of population estimation may be used as a means to relax the restriction of assumption of equal capture probability if the frequency of capture data reasonably fit a described distribution (Caughley 1977). Animals per trap day, preferably segregated by age/sex status, will yield an index to detect changes over time or areas if standardized trapping methods are used. The effects of environmental factors on trap response should be considered to obtain an accurate index (Perry 1974).

Harvest. Regulation of the harvest is an important muskrat management measure. Trapping is a feasible way to regulate muskrat numbers (Cook 1952) and the steel leg-hold trap is a very effective tool to capture muskrats.

Underharvesting may result in extensive eat-outs that seriously reduce muskrat populations (Ferrigno 1967). Low populations of muskrats in some Louisiana marshes were attributed to previous overpopulation and undertrapping (O'Neil 1949). Mathiak (1966) reported that underharvesting was more of a problem than overharvesting at Horicon marsh, Wisconsin. He stated that seasons should be set so that an adequate harvest occurs on the more populated areas. Such areas apparently exist even in years of low populations.

Other factors that must be considered in establishing muskrat trapping seasons include weather conditions, trapper access to the marsh (which may be enhanced by burning), time of breeding, and geographical area. A long trapping season has the advantage of allowing for variations in weather and for overpopulations that may require thinning (Lay and O'Neil 1942). Sather (1958) found that the Kansas season should be (1) long enough to permit adequate harvesting, (2) during the time of year when pelts are of highest quality, and (3) closed during breeding territory establishment. He suggested a 15 December to 15 March season.

Highby (1941) suggested harvesting in early winter to reduce (1) muskrat mortality from freezeouts; (2) food shortage; (3) disease; (4) loss by emigration, mink predation, and mink trapping; and (5) damage to pelts from muskrat fighting. Fall trapping on small streams may deplete the population below carrying capacity; trapping was usually more difficult in spring than in fall but was more effective than winter trapping (Cook 1952).

Fall trapping in South Dakota was less effective than spring trapping or trapping near houses in winter (Aldous 1947). Gashwiler (1948) indicated the season should be based on zones of climatic lines because pelts taken in spring are worth 10 to 20 percent more than those taken in fall and a spring season results in the harvest of many pregnant females. Stewart and Bider (1974) suggested that a fall trapping season should be considered in southern Quebec. They noted that the number of muskrats available for harvest decreased as the season progressed: young per breeding female declined from 5 in the fall to 2.8 in spring.

Pelt quality, time of harvesting, and geographical area are interrelated. Pelts from muskrats caught in

Missouri in December were great enough in value to offset likely reduced catches that would result from the fall shuffle (Shanks and Arthur 1952). The maximum percentage of furs that were graded "tops" for muskrats harvested in Texas, occurred in January and the first half of February (Lay 1945*a*). Peak primeness was reached by muskrats on Blackwater National Wildlife Refuge, Maryland, during the last half of February (Dozier et al. 1948). Although muskrats had reached their maximum size by this time, trapping yielded the smallest number of pelts.

Populations on intensively managed marshes in New York should be reduced until the cattails have grown to a density that will sustain a relatively high muskrat population (Bradley and Cook 1951). Fifty percent of the muskrat population can be harvested without jeopardizing the population in Minnesota (McCann 1944), 60 to 65 percent in North Carolina (Wilson 1968), 66 percent along the Atlantic coast (Dozier 1953), 70 percent in New Jersey (Ferrigno 1967), and 75 percent in Wisconsin (Mathiak and Linde 1956). During years of high populations, 88 to 91 percent of the population from an area in Maine was trapped whereas about 80 percent was trapped during years of low populations (Gashwiler 1948). A 60 to 65 percent rate of harvest may seldom be accomplished where muskrat populations exceed 74 per ha in marsh. Consequently, such a density of muskrats often destroys the habitat (Wilson 1968).

The number of muskrats trapped on a marsh is determined by weather, skill and economic needs of the trapper, price of pelts, muskrat abundance, and geographical area (Harris 1952). Bailey (1937) indicated that yields of 7 to 10 muskrats per ha for Maryland marshes were good; yields of 15 to 20 per ha were not unusual on good, well-managed marsh. During peak populations in Maryland, 12 muskrats were trapped over extensive areas and up to 49 muskrats per ha have been recorded (Harris 1952). Wilson (1968) reported that productive *S. olneyi* marshes in New Jersey and Maryland yielded catches of 7 to 25 muskrats per ha and occasionally 37 per ha. Ten to 15 muskrats per ha may be harvested on the better marshes of North Carolina and catches exceeding 49 per ha were reported in Louisiana (Wilson 1968). In South Dakota, Aldous (1947) reported that 7.9 muskrats were taken per ha of good habitat.

Lay and O'Neil (1942) found a high degree of positive association between cold weather and muskrat trapping success. Catches on cold nights included approximately 6 percent more adults than those in warm weather (Lay 1945*a*).

The yield per house may be a better indicator of trapping efficiency than yield per hectare. Dozier et al. (1948) recommended that trapping should be started when muskrat density reaches 2.5 dwelling houses per ha under Atlantic coastal conditions. Intensive trapping should be undertaken to prevent eat-outs and reversion to open water and mud flats when populations reach 6.2 dwelling houses per ha. However, much variability exists in the exact number of harvestable muskrats represented by each house. An average of 2.6 to 2.9 muskrats was taken per house in South Dakota (Aldous 1947). Palmisano (1972*a*) found an average of 3.2 harvestable muskrats per house in Louisiana but recommended a harvest rate of 2.5 to 3.0 per house with a stable population. Potential catch estimates by other investigators range from a conservative 2 (Lay 1945*a*) to 5 per house (O'Neil 1949).

Freeman (1945) stated that good muskrat-producing marsh in Mississippi should yield a harvest of about 70 percent adults (>12 weeks old) and 30 percent juveniles. If the percentage of adults is higher, poor reproduction is indicated. If the number of kits and mice approaches the number of adults being caught or becomes greater, trapping for the season should be stopped. Similarly, Lay (1945*a*) indicated that trapping should be stopped when an abnormal proportion of kits and mice is caught. The ratio of kits to subadults to adults in a normal population was given as 8-28-64. A ratio of 20-40-40 indicates overtrapping. An abnormally high harvest of adults may indicate little or no breeding that season or poor survival of young; a low proportion of adults may indicate a close ratio or overtrapping in the preceding or current year (Dozier 1953).

An overview of trapping muskrats with steel leg-hold traps and handling the catch was provided by Freeman (1945), Lay (1945*a*), Cook (1952), Dozier (1953), and Wingard and Sharp (n.d.). No. 1 or no. 1 1/2 traps were suggested; sets discussed include the log, plant, underwater hole, slide, and bait. Illustrations of various sets for trapping muskrats were presented by Errington (1961).

Body-gripping traps are also effective for capturing muskrats. These traps are designed to kill and hold an animal by a blow to the neck or chest region, rather than just retaining an animal by its limb. Effective sets for body-gripping traps can be made in holes or runways and on rafters and poles. They may be used in water, under ice, or on the ground (Woodstream Corporation n.d.).

Several trapping schemes may be employed for muskrats. The whole territory may be encircled and then gradually worked inward, or the area may be divided and trapped on a rotational basis to reduce travel. If traps are concentrated too much, generally greater than one trap per five beds, muskrats may move (Lay 1945*a*). Traps should be moved continually, and, when success is low and no fresh sign is observed, a site should be at least temporarily abandoned. Fresh food cuttings and pellets are easily recognized, so traps can be placed where muskrat activity occurred the previous night. Daily checking of traps is essential, preferably in early morning. to prevent inhumane treatment, spoiled or tainted skins, and depredation.

Most catches in leg-hold traps should result in immediate drowning, but if not, a blow to the base of the skull should humanely kill a muskrat. Shooting is not recommended, because it may damage the pelt, nor is the handling of live muskrats. Skinning is initiated at the rear feet and proceeds forward. After skinning, the pelt is fleshed (removal of excess meat and fat) and stretched for drying (Louisiana Department of Conservation 1931) (figure 15.6). Compressed air may be utilized to enhance the skinning process and to improve pelt value (Dozier and Radway 1948; Dozier 1951).

Predator and Competitor Control. Almost every vertebrate predator in the marsh will prey upon muskrats (Wilson 1968). Marsh intensively managed for muskrats may justify predator control, especially for marsh hawks and raccoons (Lay 1945a).

Techniques tested by Wilson (1956b) for the control of raccoon depredation of muskrats involved using corn, eggs, persimmons, prunes, and sardines treated with strychnine. Control effectiveness was influenced by the attractiveness and lethal effect of the bait, raccoon population density in the area and in the surrounding habitat, and seasonal raccoon activity. Corn was the most effective bait and sardines were the least effective; however, 90 percent of the control effort was with corn. Control of raccoons was found necessary from March to November to maximize muskrat production. Raccoons usually repopulated the treated areas from surrounding habitat in May and June.

Although alligators eat muskrats, they may benefit muskrat populations by (1) forming holes that retain water during dry periods, (2) preventing pool water stagnation by their movements, and (3) scouring ditches, which reduces clogging by vegetation (Lay and O'Neil 1942).

Geese may feed upon favorite muskrat foods such as *S. olneyi* and *S. robustus*. The use of burned areas by snow geese (*Chen caerulescens*) may convert vegetated areas of the marsh to mud, resembling muskrat eat-outs. Muskrats must emigrate when goose eat-outs occur (O'Neil 1949). Bare areas form ponds and may

not recover for several years. To reduce competition between muskrats and geese, adjacent areas may be burned in autumn to produce succulent vegetation that attracts geese, and geese may be hunted on muskrat areas (Lay and O'Neil 1942).

Cattle in the marsh may disturb both muskrats and the habitat by feeding on *S. olneyi* and trampling beds and runs (Lay 1945a, Dozier 1953). Beds disturbed by cattle may not be rebuilt. In marsh heavily grazed by cattle, 5 to 10 percent of muskrat traps may be sprung by cattle (O'Neil 1949). Fencing to exclude cattle may be justified in the interest of muskrats and muskrat trapping (Lay and O'Neil 1942).

Damage Control. Muskrats may extensively damage ponds, impoundments, and ditches by tunneling into banks and dam areas. Burrowing activities are most pronounced from March through May (Erickson 1966). A high percentage of available ponds may be inhabited by muskrats. Forty-one of 89 Alabama ponds investigated by Beshears (1951) had muskrats; however, burrows were found in the dams of only 11 of these ponds.

Ponds can be constructed and maintained to minimize muskrat damage, although it is difficult to manipulate habitat to prevent muskrats from occupying suitable ponds (Shanks and Arthur 1951, 1952). Ponds in New York, Pennsylvania, and Maryland that were built to Soil Conservation Service specifications or the equivalent seldom were severely damaged by muskrats (Erickson 1966). Similarly, Cook (1957) found that some dikes made of heavy clay soils were not damaged by muskrats in spite of a substantial population.

Specific measures to minimize muskrat damage to ponds include constructing or providing: (1) a treeless dam of well-packed clay soil at least 7 m thick at the

FIGURE 15.6. After being cleaned of excess meat and fat, muskrat pelts are stretched for drying with the flesh side out.

water line; (2) vertical asbestos-cement sheets within the dam; (3) at least 0.9 m of dam freeboard, measured vertically from normal water level, to allow for denning activities; (4) a dam width minimum of 6.1 m at the water line; (5) a 3:1 dam slope on the pond side with steeper (2:1) pond banks to encourage burrowing in the steeper banks; (6) the dam and other burrow sites with a covering of 30 cm or more of sand to preclude effective construction of breeding burrows; and (7) a good sod cover over the tops of embankments. Additional measures to alleviate severe dam problems include providing: (1) dam riprap to about 1 m below the lowest possible water level; (2) selected pond margin cutbacks to allow for den excavation; (3) a sufficiently wide and sloped spillway to prevent water level increases of more than 15–20 cm during excessive rains; (4) for exclusion of livestock (by fencing) and farm machinery from embankment tops to prevent caving; (5) regular mowing of bank vegetation to increase the likelihood of discovering burrowing; and (6) riprap for problem areas (Nagel 1945; Glass 1952; Beshears and Haugen 1953; Erickson 1956, 1966; Brooks 1959; Schwartz and Schwartz 1959).

Trapping, poisoning, shooting, gassing, and water level manipulation have been used successfully to control muskrats in ponds. Erickson (1966) indicated that poisoning yields the cheapest effective control but has legal limitations. Muskrat control with 0.75 percent zinc phosphide bait in combination with cut carrots or sweet potatoes and corn oil was described by the U.S. Fish and Wildlife Service (1973).

Riprapping, which is expensive, and trapping appear to be the best means to control muskrat damage to ponds; gassing, poisoning, and shooting are not recommended (Erickson 1956). Damage to earthen dikes may be most practically controlled by using asbestos-cement sheets as vertical barriers within the dikes (Cook 1957).

Trapping may be the most economical and efficient method to control muskrats in farm ponds (Beshears 1951; Shanks and Arthur 1951; Beshears and Haugen 1953; Brooks 1959; Indiana Department of Natural Resources n.d.). Body-gripping traps can be used for muskrat control on dams and levees. Brooks (1959) described several pond-sets for muskrats using both leg-hold and body-gripping traps. Removal operations should be conducted after spring dispersal but before litters are born (Shanks and Arthur 1952).

Refuges and Restocking. Natural restocking of muskrat-vacant habitats is generally connected with spring breeding and dispersal but varies considerably. A system of refuges to insure adequate maintenance of breeding densities was suggested by Errington (1940) for states having late fall and winter trapping seasons. Suggested systems were 2.6-km² wintering refuges every 6 to 10 km for overtrapped streams and quarter-section refuges 2 to 3 km apart for extensive marsh areas. Highby (1941) also discussed a system of refuges for Minnesota whereby 0.8 km of stream would be selected every 10 km, and, for large marshes, a corner or a bay would be blocked off as a breeding

ground. Muskrat populations in farm ponds may be increased by establishing impoundments of at least 0.8 ha as refuges (Shanks and Arthur 1952). However, Mathiak and Linde (1954) found that the muskrat population failed to increase in a centrally located 38-ha refuge in Horicon marsh, Wisconsin. Muskrat movement from the refuge was insignificant.

Many past muskrat restocking operations have been successful, such as at Tule Lake, California, in 1930 and at Moyock, North Carolina, in 1936. Between 1915 and 1944, 50,000 wild muskrats were sold by one Maryland firm for restocking. Muskrats should be stocked only after careful study and all biological factors have been considered (Dozier 1953).

Stocking muskrats in bodies of water with high water level fluctuations is generally undesirable; stocking inland ponds having sufficient food plants may be practicable and desirable (Moody 1950). However, stocking may be unnecessary. Bradley and Cook (1951) found that muskrats moved from adjacent areas into newly created marsh developments in New York. Anderson (1969) described muskrat use of a floating blind to rear young and suggested that such structures might be used to increase density artificially.

CURRENT RESEARCH AND MANAGEMENT NEEDS

Informational needs on muskrats center around (1) impacts of land use changes, (2) effects of environmental contaminants, (3) effects of various muskrat harvest strategies on muskrat and land resources, and (4) population genetics. According to Deems and Pursley (1978), 15 states and 9 Canadian provinces or territories conducted some form of research on the muskrat during 1976.

Land uses, and the various treatments applied to lands, are, among other factors, dictated by the degree to which various profitable operations can be conducted. General and specific changes in land use are causing major shifts in the dynamics of local muskrat populations and in habitat use. Such changes include ditching; draining; dredging; filling; oil exploration; mining; road, urban, and industrial construction; and agricultural operations.

Drainage for development, commercial and agricultural operations, and mosquito and flood control are common practices that affect marshes and swamp (stream) wetlands. Although there are data on the negative effects of such modifications, the potential positive aspects of such land-treatment practices should receive more attention. That is, we need research designed to (1) solve conflicts between land treatment measures and muskrat populations, not just to document further that a problem exists; (2) discover alternative ways to implement development projects so that adverse resource impacts are minimized; and (3) formulate effective mitigation strategies for use when serious ecosystem disruptions must occur.

For instance, semidry marshes that are ditched properly can be efficiently managed for wildlife production rather than just to provide drainage for agricultural purposes (Mathiak and Linde 1956). Similarly, research on the Bombay Hook National Wildlife Refuge, Delaware, revealed that mosquito breeding on a tidal marsh may be effectively controlled by impoundment and proper swamp drainage (Bourn and Cottam 1950). Detrimental effects on wildlife habitat can be minimized if such a system is adequately constructed, operated, and maintained. Ferrigno (1967) stated that there is no reason why important tidal marshes low enough to be flooded by daily tides should be sprayed, because such marshes will not produce mosquitoes. Similarly, productive muskrat marshes in Delaware were found to be relatively unimportant from the standpoint of mosquito breeding (Stearns et al. 1940). Even in impoundments the best control for mosquitoes is proper water management.

The problem and effects of habitat shrinkage become exponential as the degree of habitat modification in an area increases. Consequently, research aimed at providing additional quantitative information on the effects of specific environmental modifications needs to be undertaken. Problems could most efficiently be addressed on a local level, but a diversity of biotic associations should be investigated. Demonstration areas should be constructed that illustrate how mosquito control, flood control, and wildlife enhancement can be integrated through marsh impoundment and/or water level control. These areas should be promoted and made open to the public to gain the political support necessary to fund such projects on a large scale.

Additional data need to be compiled on the effects of marsh management practices on the life history and ecology of muskrats. Common techniques readily available to landowners include ditching, impounding, and burning, but all the effects of these practices on muskrat populations and other fauna are not totally clear. In some areas only subjective evaluations of current management practices are available. More specific objective and quantitative information must be made available to land and resource managers if we are to maintain furbearer populations on a smaller and poorer quality land base.

Relatively little or no information is available on the long- or short-term effects of herbicides and pesticides on muskrats. Sprays, dusts, and systemics are commonly employed in agricultural and urban settings. Agricultural uses of chemicals are diverse and widespread; some communities bordering marsh or swamp areas are routinely fogged for insect pests, especially mosquitoes. Eventually both herbicides and pesticides enter the aquatic food chain and sometimes muskrats may be affected directly. The relationship between pesticide spraying and muskrat population changes was not clear in the Cumberland County–Delaware Bay area (Ferrigno 1967). Information is needed on the immediate toxic effects of pollutants on muskrats and on the long-term effects on muskrat life requisites. A data base for muskrats should be assembled on the immediate and long-term effects of commercial/industrial pollutants and point and nonpoint eutrophic discharges from urban and agricultural areas.

Coastal marsh landowners in several areas rely

upon the fur crop, of which the muskrat may be a significant contributor, to maintain their land as marsh. In essence, the fur crop is a major incentive for keeping land wet, as opposed to conversion to pasture or cultivation. Marsh use in these areas may be closely tied to the ability of landowners to harvest muskrats efficiently. The steel leg-hold trap is a primary method of harvesting furbearers. If the steel leg-hold trap is prohibited, many biologists believe that land use in many areas will quickly change from marsh to a more profitable system. For example, work by Mathiak (1966) indicated that low muskrat pelt prices greatly increased development of wetlands for other wildlife. Furbearers will not be the only populations affected by such land use shifts. Other resident and migratory marsh-dependent species, such as waterfowl, nongame birds, fishes, shellfishes, reptiles, and amphibians, will also be impacted. Even if use patterns result in the maintenance of marsh areas, the real impacts of nonharvested or underharvested furbearer populations on marsh ecology can only be hypothesized.

However, antitrapping groups are requiring that more humane trapping methods be investigated. Even early investigators recognized the unpleasantness of the steel trap. Bailey (1937) stated that about one-half of the muskrats trapped twisted off a leg or foot, and he believed trapping could be greatly improved. He experimented and developed quick-kill devices and traps for capturing animals alive. Gilbert (1976) also investigated the energy impact required to kill muskrats and other furbearers by blows to the neck or chest region to aid in the development of humane traps.

The modern steel trap may be more efficient than those used by Bailey (1937); however, another apparent problem with steel leg-hold traps is the infliction of injury or death to nontarget species. Gashwiler (1949) indicated that waterfowl mortality resulting from muskrat traps may be serious in certain areas. On six study areas in Maine, spring trapping resulted in 1 duck captured per 14.7 muskrats in 1946 and 1 per 17.7 in 1947. The total kill of ducks in muskrat traps for Maine in 1946 was estimated as 1,945, with an estimated 2,220 birds trapped and released alive but suffering from various degrees of injury.

Currently on the market are quick-kill, body-gripping traps and a variety of live-traps. An evaluation by Palmisano and Dupuie (1975) indicated that trappers were opposed to the use of body-gripping traps because of their high cost, weight, bulkiness, potential danger to the trapper and domestic animals, and the additional difficulty of skinning a cold animal. However, they found the body-gripping trap more effective than the leg-hold trap for capturing muskrats in the flooded marsh. About 6 percent of the muskrats were alive when removed from the body-gripping traps, whereas 20 percent were alive with the leg-hold traps.

Further comparisons between leg-hold and body-gripping traps were made by Linscombe (1976). He found no statistical difference in the number of muskrats captured with the two trap types in fresh or brackish marsh. More animals remained alive in the leg-hold than in the body-gripping traps. In brackish marsh the leg-hold trap was sprung without captures three times more than the body-gripping trap. However, the body-gripping trap captured more nontarget animals than the leg-hold trap.

Even with the availability of effective alternative trap types, the steel leg-hold trap is still the major muskrat trap in many areas. Possibly wider acceptance of more humane traps will be forthcoming with additional trapper education, as old traps require replacement, and as a new generation of trappers emerges. Trappers must also assume a greater responsibility. They should become more aware of and concerned about their public image, and the impacts of their use of the muskrat resource. They must treat the habitat and resources with respect. Research aimed at formulating new and effective methods of trapper education would be useful.

Manufacturers of traps are also making a concentrated effort to improve the humaneness of capturing furbearers. Such improvements include (1) slide locks on leg-hold traps to insure quick drowning of trapped animals, (2) the installation of an auxiliary guard on leg-hold traps, (3) increasing the holding power of body-gripping traps, and (4) development of positioning mounts for body-gripping traps so that they can be operated by a pan. The auxiliary guard moves high on the trapped animal's body to keep it pushed away from the trapped limb; thus, gnawing or twisting to escape should be minimized. The positioning mount allows the setting of body-gripping traps for quicker killing of captured animals and for a higher degree of selectivity (Woodstream Corporation n.d.).

Additional information on population genetics could provide a better understanding of how inheritance affects pelt value. Genetic and environmental effects should be experimentally separated to understand more fully parameters that control pelage condition. Such information may also be applied to selective breeding to improve pelt quality or enhance desired characteristics. In addition, since muskrats have an inherently high turnover rate, short generation time, and the ability to inhabit a wide range of habitat types, a population strain could possibly be selected that would thrive in certain polluted aquatic systems.

ROUND-TAILED MUSKRAT

The round-tailed muskrat (*Neofiber alleni*) is a common microtine rodent of Florida and parts of Georgia. The species is frequently referred to as prairie rat, water rat, or muskrat (Harper 1927).

Three subspecies of *Neofiber alleni* were recognized by Miller and Kellogg (1955) but five subspecies were listed in Hall and Kelson (1959). Burt (1954) stated that subspecific designations of *Neofiber alleni* are arbitrary and should be discarded. Fossils of *Neofiber* have been described from the Pleistocene (Sherman 1952).

The dorsal pelage of round-tailed muskrats is soft and lustrous. The soft underfur is gray and brown tipped. Long guard hairs are glossy tipped and brown.

Ventrally the soft, dense, grayish underfur is overlain by scattered, pale guard hairs (Schwartz 1953). Juveniles are lead gray (Birkenholz 1972).

Measurements of the round-tailed muskrat are: total length 285 to 381 mm, tail length 99 to 168 mm, and hind foot length 40 to 50 mm. Sexes are similar in size but adult males are slightly heavier (average 279.0 g) than females (average 262.0 g). The skull is similar to *Ondatra* but smaller. The body form, fur, ears, and eyes resemble *Ondatra* with slight differences: guard hairs produce a tuft above the tail, fringe hairs on hind foot margins are less developed, and the tail is terete and sparsely haired (Birkenholz 1972).

All *Neofiber* occur in either Florida or southeastern Georgia (Birkenholz 1972) (see figure 15.1), but once its range was more extensive (Schwartz 1953). Round-tailed muskrats are found throughout Florida except in the northeast, western panhandle, and upper east Gulf coast, where they are rare or absent (Paul 1967; Birkenholz 1972).

Neofiber may be restricted to its present range because of its adaptations to warm climatic conditions. The houses of round-tailed muskrats are constructed of fine-textured grass and would not be very strong, durable, or protective in areas where the marsh freezes for long periods (Birkenholz 1962, 1972). Their habitat consists of grassy marshes and prairies. Bangs (1899) and Chapman (1889) described the round-tailed muskrat as being common in salt savannahs, around the edges of saltwater pools, and in freshwater ponds and marshes of interior Florida. In Georgia it inhabits treeless wet areas and is more of a bog mammal than an aquatic mammal (Harper 1920, 1927; Golley 1962).

Populations of round-tailed muskrats in central Florida are highest in areas with water depths of 150 to 460 mm, a sand substrate, and dense stands of *Panicum hemitomon* and *Leersia hexandra*. Slightly deeper areas that also contain *Pontederia lanceolata* are also used (Birkenholz 1972). Water areas nearly completely filled in with sphagnum and other aquatic vegetation are preferred in the Okefenokee Swamp (Harper 1927).

The houses of *Neofiber* are dome shaped and nearly spherical, 180 to 610 mm in diameter. They are mainly constructed of tightly woven *P. hemitomon,* but *Pontederia* spp., *Sagittaria lancifolia, Ceratophyllum* spp., and *Polygonum* spp. are also used. The house base is on partly decayed vegetation and is partly supported by standing vegetation. The floor is slightly above the water level (Birkenholz 1962, 1972).

Houses are constructed by bending vegetation over a platform and incorporating additional vegetation into the walls. Interior chambers are approximately 100 mm in diameter and generally contain two plunge or exit holes. Houses are occupied by only one individual. The chambers are lined with grasses when young are present (Birkenholz 1962, 1972). Houses may be used up to six months (Birkenholz 1962) or from year to year (Porter 1953).

Nests of *Neofiber* in Okefenokee Swamp are occasionally built at the bases of solitary baldcypress (*Taxodium distichum*) trees (Golley 1962) or clumps of bushes. Nest foundations are on sphagnum rather than in water. The nests are anchored with larger prairie plants such as *Nymphae macrophylla, Erianthus saccharoides, P. hemitomon,* and *Cephalanthus occidentalis.* Fibrous roots may also be used (Harper 1920, 1927).

In the Everglades region of Florida, *Mariscus jamaicense, C. occidentalis,* and *Salix amphibia* are used as nest foundations (Porter 1953). Nests of *Neofiber* were found by Chapman (1889) and Bangs (1899) in stumps at the bases of black mangroves (*Avicennia germinans*).

As the water level rises, the floor of the nest is built higher and higher. If the level rises 0.9 to 1.2 m, new nests may be made using tree or bush limbs (Porter 1953). The houses of round-tailed muskrats may be used by turtles, frogs, snakes, other mammals, arthropods, and insects (Harper 1920, 1927; Porter 1953; Birkenholz 1962).

Breeding in *Neofiber* is positively related to favorable water and cover conditions (Birkenholz 1962, 1972; Paul 1967). Most reproduction takes place in late fall and early winter but breeding occurs year round. Duration of the estrous cycle appears to be 15 days with a gestation period of 26–29 days. The number of young per litter ranges from one to four (average 2.2). Four to six litters are produced each year, depending upon conditions.

Newborn are blind and almost hairless at birth (Hamilton 1956). Their average weight is 12 g. At 2 days after birth they are covered by short, coarse fur. Eyes are opened, all teeth are erupted, and the young can sustain themselves by 14 to 18 days. The young resemble adults by 1 month of age and they become sexually mature at 90 to 100 days of age (average weight 275 g). Males constitute about 56 percent of all age classes (Birkenholz 1962, 1963).

Round-tailed muskrats construct feeding platforms and shelters for feeding and defecation. These consist of a pad of vegetation, about 100 x 150 mm, elevated slightly above the water. Feeding platforms contain one to two plunge holes. Feeding shelters are platforms that contain a roof and are more common in late winter when protective cover is scarcer. Platforms may also be converted to houses. Depending upon population density, animals may share feeding platforms (Birkenholz 1962, 1972).

Grass structures that float on the surface of the water may also serve as feeding platforms (Golley 1962). During high water, platforms may be constructed on the limbs of trees. The size of the platform may vary with the abundance of food in the region (Chapman 1889).

Round-tailed muskrats are herbivorous and mostly utilize the grasses (*P. hemitomon, Mariscus* sp., *Sporobolus* sp., and *Echinochloa* sp.) (Birkenholz 1972). Other plants eaten include the stems of *Sagittaria graminea* and *Brasenia* spp., seeds of *Peltandra* sp. and *Iris* spp., and roots of *Anchistea virginica* (Harper 1927); and *Sagittaria lancifolia, Pontederia* sp., and *Nymphaea lutea* (Porter 1953).

Golley (1962) stated that crayfish and other invertebrates may be utilized by *Neofiber*, but this was not supported by analyses of stomach contents by Birkenholz (1963). Captive animals will eat lettuce, carrots, potatoes, and commercial rat pellets. Although they travel over a larger area, individuals feed in a circle about 9.1 m in diameter (Birkenholz 1962).

Populations on small areas in central Florida range from 250 to 300 per ha. Densities fluctuate greatly and are regulated primarily by water level fluctuations. Density on a larger marsh area (1,000 ha) averaged 50 per ha. Adults constitute the greatest percentage of the population from May through August (Birkenholz 1962, 1972).

Round-tailed muskrats are primarily nocturnal; peak activity occurs shortly after dark (Birkenholz 1972). They are very shy. Birkenholz (1962) found no evidence of stress, conflict, or territoriality beyond the limits of the house.

Occasionally the round-tailed muskrat will burrow into banks and will live in muck, especially during low water levels (Porter 1953; Paul 1967; Birkenholz 1972). Burrows and tunnels are usually filled with water and may be used for feeding. Above-ground runways may also be constructed (Chapman 1889; Harper 1927; Golley 1962).

Predators of *Neofiber* include cats and dogs (Porter 1953), barn owls (*Tyto alba*) (Schantz and Jenkins 1950), marsh hawks (*Circus cyaneus*), red-tailed hawks (*Buteo jamaicensis*), barred owls (*Strix varia*), bobcats (*Felis rufus*), and eastern cottonmouth (*Agkistrodon piscivorus*). They are normally preyed upon while away from their houses, mostly in late winter and early spring (Birkenholz 1962, 1963). Changes in water level may be the most important factor contributing to mortality of *Neofiber*. Population reductions up to 85 percent were recorded by Birkenholz (1962) after flooding. Mortality resulted from predation and from animals being hit by automobiles.

Internal parasites of the round-tailed muskrat include the cestodes *Paranoplocephala neofibrinus, Cittotaenia praecoquis,* and *Taenia lyncis;* the trematode *Quinqueserialis floridensis;* and the nematodes *Longistriata adunca* and *Litomosoides* sp. (Porter 1953; Birkenholz 1962, 1972). Birkenholz (1962) reported a helminth infestation rate of 41.3 percent: tapeworms 49.0 percent, flukes 13.1 percent, and roundworms 6.7 percent. Helminth prevalence varied from 16.7 percent for juveniles to 67.0 percent for adults. Mites infesting *Neofiber* include *Laelaps evansi, Haemolaelaps glasglowi* (=*Androlaelaps fahrenholzi*), and an undescribed species of the family Listrophoridae (Birkenholz 1972); and *Garmania bulbicola, Macrocheles* sp., *Trombicula (Eutrombicula) splendens,* and *Tyrophagus lintneri* (Porter 1953).

Porter (1953) described a condition of captive round-tailed muskrats in which slight breaks in the skin resulted in cysts swollen with pus. The condition was not observed in the wild and was attributed to a bacterial infection. Parasitism and disease do not appear to be important mortality factors for *Neofiber* (Birkenholz 1962).

The niches of *Neofiber* and *Ondatra* differ relative to vegetation used. The round-tailed muskrat uses fine-textured grasses or sedges for homesite construction, whereas *Ondatra* prefers coarser vegetation, such as bulrushes (*Scirpus* spp.) (Birkenholz 1962).

The burrowing and tunneling of the round-tailed muskrat may undermine canals and ditches and result in caving and bank sloughing. They may also damage sugarcane stalks and undermine the stools to the extent that they are damaged by high winds (Porter 1953). Round-tailed muskrats occurred in 40.3 and 80.4 percent of the fields in two sugarcane areas evaluated by Steffen (1978). Damage was listed as probably economically significant. Fields harvested by hand had a higher *Neofiber* occurrence rate than did fields harvested mechanically, probably because of increased cover in the hand-harvested fields. Potential control strategies include manipulation of field harvesting and husbandry practices (Steffen et al. in press).

Round-tailed muskrats may be trapped using no. 110 body-gripping or no. 0 underspring steel traps. Traps should be set in houses, covered feeding shelters, feeding pads, or runways. Animals may be captured alive using 10- x 10- x 20-cm hardware cloth traps with an inward-swinging wire door. Birkenholz (1962) found that trapping success varied from 10 percent in warm weather and in areas of dense cover to 50 percent after water level rises. Seasonally, the susceptibility of round-tailed muskrats to traps (captures per 100 trap nights) varied from 1.5 in October to 7.6 in March (Steffen 1978).

Sincere appreciation is expressed to all who assisted with the preparation of this chapter. Reviewers of all or a portion of the manuscript included J. D. Newsom, J. W. Carpenter, and J. C. Lewis of the U.S. Fish and Wildlife Service; J. T. Joanen, L. L. McNease, and R. G. Linscombe of the Louisiana Department of Wildlife and Fisheries; R. E. Noble, R. H. Chabreck, R. B. Hamilton, V. L. Wright, and T. R. Klei of Louisiana State University; and my wife, Linda. M. H. Chisholm of Louisiana State University, with early assistance from M. S. Miksa, obtained and compiled the literature sources and citations. D. G. Hewitt of the Louisiana Cooperative Wildlife Research Unit typed several drafts of the manuscripts; typing support was also provided by M. K. Conner, B. W. Nichols, and S. E. Fox. Proofreading assistance was provided by wildlife students of Louisiana State University: B. Bell, J. L. Hughes, J. A. Allen, P. M. Shealy, B. L. Shiflet, W. C. Matthews, Jr., R. P. Lanctot, A. L. Foote, J. A. Dugoni, J. C. Sumler, D. J. Riedlinger, P. A. Baer, and F. J. Zeringue II. K. A. Margel, J. A. Stegeman, and R. R. Gabel assisted with proofreading the galleys.

LITERATURE CITED

Abram, J. B., Jr. 1968a. Ecological factors influencing gastrointestinal helminths of the Maryland muskrats. Ph.D. Thesis. Oklahoma State Univ., Stillwater. 76pp.

———. 1968b. Some gastrointestinal helminths of *Ondatra zibethica*, the muskrat in Maryland. Proc. Helminthol. Soc. Washington 36:93–95.

Adams, A. W. 1962. 1961 aerial muskrat and beaver survey. North Dakota State Game and Fish Dept. Pittman-Robertson Div., Proj. W-67-R-2, Phase E, Furbearer Investigations, Job no. 1. 7pp.

Aldous, S. E. 1946. Live trapping and tagging muskrats. J. Wildl. Manage. 10:42–44.

———. 1947. Muskrat trapping on Sand Lake National Wildlife Refuge, South Dakota. J. Wildl. Manage. 11:77–90.

Aleksiuk, M., and Frohlinger, A. 1971. Seasonal metabolic organization in the muskrat (*Ondatra zibethica*). Part 1: Changes in growth, thyroid activity, brown adipose tissue, and organ weights in nature. Can. J. Zool. 49:1143–1154.

Alexander, M. M. 1951. The aging of muskrats on the Montezuma National Wildlife Refuge. J. Wildl. Manage. 15:175–186.

———. 1955. Variations in winter muskrat habitats and harvests. Am. Midl. Nat. 53:61–70.

———. 1956. The muskrat in New York State. State Univ. New York Coll. For. 15pp.

———. 1960a. Dentition as an aid in understanding age composition of muskrat populations. J. Mammal. 41:336–342.

———. 1960b. Shrinkage of muskrat skulls in relation to aging. J. Wildl. Manage. 24:326–329.

Alexander, M. M., and Dozier, H. L. 1949. An extreme case of malocclusion in the muskrat. Am. Midl. Nat. 42:252–254.

Ameel, D. J. 1942. Two larval cestodes from the muskrat. Trans. Am. Microscop. Soc. 61:267–271.

Anderson, D. L. 1969. The persistent use of a floating blind by muskrats. Am. Midl. Nat. 81:601–602.

Anderson, D. R., and Beaudoin, R. L. 1966. Host habitat and age as factors in the prevalence of intestinal parasites of the muskrat. Bull. Wildl. Dis. 2:70–76.

Anderson, J. M. 1947. Sex ratio and weights of southwestern Lake Erie muskrats. J. Mammal. 28:391–395.

Anderson, W. L. 1948. Level ditching to improve muskrat marshes. J. Wildl. Manage. 12:172–176.

Anonymous. 1950. Small marsh management for fur crops. New York Conserv. Dept. Info. Bull. 8. 11pp.

Applegate, V. C., and Predmore, H. E., Jr. 1947. Age classes and patterns of primeness in a fall collection of muskrat pelts. J. Wildl. Manage. 11:324–330.

Arata, A. A. 1959a. Ecology of muskrats in strip-mine ponds in southern Illinois. J. Wildl. Manage. 23:177–186.

———. 1959b. A quick method for gross analysis of muskrat stomach contents. J. Wildl. Manage. 23:116–117.

———. 1964. Fossil vertebrates from Avery Island. Appendix pages 69–72 *in* S. M. Gagliana. An archaeological survey of Avery Island. Coastal Studies Inst., Louisiana State Univ., Baton Rouge. 76pp.

Arthur, S. C. 1931. The fur animals of Louisiana. Louisiana Dept. Conserv. Bull. 18. 439pp.

Artimo, A. 1960. The dispersal and acclimatization of the muskrat, *Ondatra zibethicus* (L.), in Finland. Pap. Game Res. (Helsinki) 21. 101pp. Abstr.

Babcock, K. M. 1967. The influence of water depth and salinity on wiregrass and saltmarsh grass. M.S. Thesis. Louisiana State Univ., Baton Rouge. 109pp.

Bailey, V. 1937. The Maryland muskrat marshes. J. Mammal. 18:350–354.

Bangs, O. 1899. The land mammals of peninsular Florida and the coast region of Georgia. Proc. Boston Soc. Nat. Hist. 28:157–235.

Barker, F. D. 1915. Parasites of the American muskrat (*Fiber zibethicus*). J. Parasitol. 1:184–197.

Bassett, C. F.; Pearson, O. P.; and Wilke, F. 1944. The effect of artificially-increased length of day on molt, growth, and priming of silver fox pelts. J. Exp. Zool. 96:77–83.

Baumgartner, L. L., and Bellrose, F. C. 1943. Determination of sex and age in muskrats. J. Wildl. Manage. 7:77–81.

Beckett, J. V., and Gallicchio, V. 1967. A survey of helminths of the muskrat, *Ondatra zibethica,* in Portage County, Ohio. J. Parasitol. 53:1169–1172.

Bednarik, K. 1956. Muskrat in Ohio Lake Erie marshes. Ohio Dept. Nat. Resour., Columbus. 67pp.

Beer, J. R. 1950. The reproductive cycle of the muskrat in Wisconsin. J. Wildl. Manage. 14:151–156.

Beer, J. R., and Meyer, R. K. 1951. Seasonal changes in the endocrine organs and behavior patterns of the muskrat. J. Mammal. 32:173–191.

Beer, J. R., and Truax, W. 1950. Sex and age ratios in Wisconsin muskrats. J. Wildl. Manage. 14:323–331.

Beleich, V. C. 1974. Muskrats (*Ondatra zibethicus*) in Amargosa Canyon, Inyo and San Bernardino Counties, California. Murrelet 55:7–8.

Bellrose, F. C. 1950. The relationship of muskrat populations to various marsh and aquatic plants. J. Wildl. Manage. 14:299–315.

Bellrose, F. C., and Brown, L. G. 1941. The effect of fluctuating water levels on the muskrat population of the Illinois River Valley. J. Wildl. Manage. 5:206–212.

Bellrose, F. C., and Low, J. B. 1943. The influence of flood and low water levels on the survival of muskrats. J. Mammal. 24:173–188.

Beshears, W. W., Jr. 1951. Muskrats in relation to farm ponds. Proc. Southeast. Assoc. Game and Fish Commissioners. 8pp.

Beshears, W. W., Jr., and Haugen, A. O. 1953. Muskrats in your farm pond. Alabama Conserv. 25:4–5, 22.

Birkenholz, D. E. 1962. A study of the life history and ecology of the round-tailed muskrat (*Neofiber alleni* True) in north-central Florida. Ph.D. Thesis. Univ. Florida, Gainesville. 149pp.

———. 1963. A study of the life history and ecology of the round-tailed muskrat (*Neofiber alleni* True) in north-central Florida. Ecol. Monogr. 33:255–280.

———. 1972. *Neofiber alleni.* Mammal. Species 15. 4 pp.

Bissonnette, T. H., and Wilson, E. 1939. Shortening daylight periods between May 15 and September 12 and the pelt cycle of the mink. Science 89:418–419.

Borell, A. E., and Ellis, R. 1934. Mammals of the Ruby mountains region of north-eastern Nevada. J. Mammal. 15:12–45.

Bourn, W. S., and Cottam, C. 1950. Some biological effects of ditching tidewater marshes. U.S. Fish Wildl. Serv. Res. Rep. 19. 30pp.

Boyce, M. S. 1977. Life histories in variable environments: applications to geographic variation in the muskrat (*Ondatra zibethicus*). Ph.D. Thesis. Yale Univ., New Haven, Conn. 146pp.

———. 1978. Climatic variability and body size variation in the muskrats (*Ondatra zibethicus*) of North America. Oecologia 36:1–19.

Bradley, B. O., and Cook, A. H. 1951. Small marsh development in New York. Trans. North Am. Wildl. Conf. 16:251–265.

Brohn, A., and Shanks, C. E. 1948. Incisor width as an age criterion in muskrats. J. Wildl. Manage. 12:437–439.

Brooks, D. M. 1959. Muskrat damage-prevention and control. Res. Rep. Indiana Dept. Conserv., Indianapolis. 6pp.

Burt, W. H. 1954. The subspecies category in mammals. Syst. Zool. 3:99–104.

Buss, I. O. 1941. Sex ratios and weights of muskrats (*Ondatra zibethica*) from Wisconsin. J. Mammal. 22:403–406.

Butler, L. 1940. A quantitative study of muskrat food. Can. Field Nat. 54:37–40.

———. 1953. The nature of cycles in populations of Canadian mammals. Can. J. Zool. 31:242–262.

———. 1962. Periodicities in the annual muskrat population figures for the Province of Saskatchewan. Can. J. Zool. 40:1277–1286.

Byrd, M. A. 1953. High occurrence of *Taenia taeniaformis* in the muskrat. J. Wildl. Manage. 17:384–385.

Caughley, C. 1977. Analysis of vertebrate populations. John Wiley & Sons, New York. 234pp.

Chalmers, G. A., and MacNeill, A. C. 1977. Tyzzer's disease in wild-trapped muskrats in British Columbia. J. Wildl. Dis. 13:114–116.

Chamberlain, J. L. 1951. The life history and management of the muskrat on Great Meadows Refuge. M.S. Thesis. Univ. Massachusetts, Amherst. 68pp.

Chandler, A. C. 1941. Helminths of muskrats in southeast Texas. J. Parasitol. 27:175–181.

Chapman, F. M. 1889. On the habits of the round-tailed muskrat (*Neofiber alleni* True). Bull. Am. Mus. Nat. Hist. 11:119–122.

Cofer, H. P. 1950. Delaware furbearers. Delaware Game and Fish Comm., Dover. 10pp.

Cook, A. H. 1952. A study of the life history and management of the muskrat in New York State. Ph.D. Thesis. Cornell Univ., Ithaca, N.Y. 128pp.

———. 1957. Control of muskrat burrow damage in earthen dikes. New York Fish Game J. 4:213–218.

Coon, R. A. 1965. Daily movements and home range of the muskrat (*Ondatra zibethicus*). M.S. Thesis. Western Illinois Univ., Macomb. 75pp.

Cromer, J. I. 1968. Helminths of the muskrat in the lower peninsula of Michigan. M.S. Thesis. Univ. Michigan, Ann Arbor. 36pp.

Dailey, E. J. 1954. Practical muskrat raising. A. R. Harding, Columbus, O. 136pp.

Dean, F. C. 1957. Age criteria and kit growth of central New York muskrats. Ph.D. Thesis. State Univ. Coll. For., Syracuse, N.Y. 80pp.

Debbie, J. G. 1968. Yellow fat disease in muskrats. New York Fish Game J. 15:119–120.

Deems, E. F., Jr., and Pursley, D., eds. 1978. North American furbearers: their management, research and harvest status in 1976. Univ. Maryland Press, College Park. 171pp.

Domning, D. P. 1969. A list, bibliography, and index of the fossil vertebrates of Louisiana and Mississippi. Trans. Gulf Coast Assoc. Geol. Soc. 19:385–422.

Donohoe, R. W. 1961. Muskrat production in areas of controlled and uncontrolled water-level units. M.S. Thesis. Ohio State Univ., Columbus. 280pp.

———. 1966. Muskrat reproduction in areas of controlled and uncontrolled water-level units. J. Wildl. Manage. 30:320–326.

Dorney, R. S., and Rusch, A. J. 1953. Muskrat growth and litter production. Wisconsin Conserv. Dept., Tech. Wildl. Bull. 8. 32pp.

Dozier, H. L. 1942. Identification of sex in live muskrats. J. Wildl. Manage. 6:292–293.

———. 1943. Occurrence of ringworm disease and lumpy jaw in the muskrat in Maryland. J. Am. Vet. Med. Assoc. 102:451–453.

———. 1944. Color, sex ratios and weights of Maryland muskrats, part 2. J. Wildl. Manage. 8:165–169.

———. 1945. Sex ratio and weights of muskrats from the Montezuma National Wildlife Refuge. J. Wildl. Manage. 9:232–237.

———. 1947*a*. Resorption of embryos in the muskrat. J. Mammal. 28:398-399.

———. 1947*b*. Salinity as a factor in Atlantic coast tidewater muskrat production. Trans. North Am. Wildl. Conf. 12:398-420.

———. 1948*a*. Color mutations in the muskrat (*Ondatra z. macrodon*) and their inheritance. J. Mammal. 29:393-405.

———. 1948*b*. Estimating muskrat populations by house counts. U.S. Fish Wildl. Serv. Wildl. Leafl. 306. 17pp.

———. 1950. Muskrat trapping on the Montezuma National Wildlife Refuge, New York, 1943-1948. J. Wildl. Manage. 14:403-412.

———. 1951. Air inflation as an aid in pelting muskrats. J. Wildl. Manage. 15:199-205.

———. 1953. Muskrat production and management. U.S. Fish Wildl. Serv. Circ. 18. 42pp.

Dozier, H. L. and Allen, R. W. 1942. Color, sex ratios and weights of Maryland muskrats. J. Wildl. Manage. 6:294-300.

Dozier, H. L.; Markley, M. H.; and Llewellyn, L. M. 1948. Muskrat investigations on the Blackwater National Wildlife Refuge, Maryland, 1941-1945. J. Wildl. Manage. 12:177-190.

Dozier, H. L., and Radway, M. 1948. Pelting muskrats by air inflation. J. Wildl. Manage. 12:333-334.

Earhart, C. M. 1969. The influence of soil texture on the structure, durability, and occupancy of muskrat burrows in farm ponds. California Fish Game 55:179-196.

Edwards, R. L. 1949. Internal parasites of central New York muskrats (*Ondatra zibethica*). J. Parasitol. 35:547-548.

Elder, W. H., and Shanks, C. E. 1962. Age changes in tooth wear and morphology of the baculum in muskrats. J. Mammal. 43:144-150.

Elton, C., and Nicholson, M. 1942. Fluctuations in numbers of the muskrat (*Ondatra zibethica*) in Canada. J. Anim. Ecol. 11:96-126.

Enders, R. K. 1931. Muskrat propagation in Ohio. Ohio Dept. Agric. Bull. 19. 17pp.

———. 1939. The corpus luteum as an indicator of the breeding of muskrats. Trans. North Am. Wildl. Conf. 4:631-634.

Erickson, H. R. 1956. Muskrat damage to Pennsylvania farm ponds with emphasis on Indiana County. M.S. Thesis. Pennsylvania State Univ., University Park. 85pp.

———. 1963. Reproduction, growth, and movement of muskrats inhabiting small water areas in New York State. New York Fish Game J. 10:90-117.

———. 1966. Muskrat burrowing damage and control procedures in New York, Pennsylvania and Maryland. New York Fish Game J. 13:176-187.

Erickson, J. L. E., and Stevens, P. G. 1944. American musk from muskrats used in perfume manufacture. Louisiana Conserv. 2:3, 6, 8.

Errington, P. L. 1937*a*. The breeding season of the muskrat in northwest Iowa. J. Mammal. 18:333-337.

———. 1937*b*. Drowning as a cause of mortality in muskrats. J. Mammal. 18:497-500.

———. 1937*c*. Habitat requirements of stream-dwelling muskrats. Trans. North Am. Wildl. Conf. 2:411-416.

———. 1938*a*. Observations on muskrat damage to corn and other crops in central Iowa. J. Agric. Res. 57:415-421.

———. 1938*b*. Reaction of muskrat populations to drought. Ecology 20:168-186.

———. 1939. Observations on young muskrats in Iowa. J. Mammal. 20:465-478.

———. 1940. Natural restocking of muskrat-vacant habitats. J. Wildl. Manage. 4:173-185.

———. 1941. Versatility in feeding and population maintenance of the muskrat. J. Wildl. Manage. 5:68-69.

———. 1942. Observations on a fungus skin disease of Iowa muskrats. Am. J. Vet. Res. 3:195-201.

———. 1943. An analysis of mink predation upon muskrats in north-central United States. Iowa State Coll. Agric. Exp. Stn. Res. Bull. 320:798-924.

———. 1944. Additional studies on tagged young muskrats. J. Wildl. Manage. 8:300-306.

———. 1948. Environmental control for increasing muskrat production. Trans. North Am. Wildl. Conf. 13:596-609.

———. 1951. Concerning fluctuations in populations of the prolific and widely distributed muskrat. Am. Nat. 85:273-292.

———. 1954*a*. On the hazards of overemphasizing numerical fluctuations in studies of "cyclic" phenomena in muskrat populations. J. Wildl. Manage. 18:66-90.

———. 1954*b*. The special responsiveness of minks to epizootics in muskrat populations. Ecol. Monogr. 24:377-393.

———. 1961. Muskrats and marsh management. Stackpole Co., Harrisburg, Pa. 183pp.

———. 1963. Muskrat populations. Iowa State Univ. Press, Ames. 665pp.

Errington, P. L., and Errington, C. S. 1937. Experimental tagging of young muskrats for purposes of study. J. Wildl. Manage. 1:49-61.

Errington, P. L., and Scott, T. G. 1945. Reduction in productivity of muskrat pelts on an Iowa marsh through depredations of red foxes. J. Agric. Res. 71:137-148.

Errington, P. L.; Siglin, R. J.; and Clark, R. C. 1963. The decline of a muskrat population. J. Wildl. Manage. 27:1-8.

Evans. J. 1970. About nutria and their control. U.S.D.I. Bur. Sport Fish. Wildl. Resour. Publ. 86. 65pp.

Fairbanks, E. S., and Kilgore, D. L., Jr. 1978. Post-dive oxygen consumption of restrained and unrestrained muskrats (*Ondatra zibethica*). Comp. Biochem. Physiol. 59A:113-117.

Ferrigno, F. 1967. First in fur value: muskrats and their management. Part 2: research, management, and influences. New Jersey Outdoors 17:8, 13-19.

Field, W. H. 1948. Muskrats and muskrat farming. Wisconsin Conserv. (August). Pp. 9-12.

Fish, F. E. 1977. Thermoregulation in the muskrat (*Ondatra zibethicus*): the use of regional heterothermia. M.S. Thesis. Michigan State Univ., East Lansing. 39pp.

Forbes, T. R., and Enders, R. K. 1940. Observations on corpora lutea in the ovaries of Maryland muskrats collected during the winter months. J. Wildl. Manage. 4:169-172.

Freeman, R. M. 1945. Muskrats in Mississippi. Mississippi Game and Fish Comm., Jackson. 48pp.

Friend, M.; Cummings, G. E.; and Morse, J. S. 1964. Effect of changes in winter water levels on muskrat weights and harvest at the Montezuma National Wildlife Refuge. New York Fish Game J. 11:125-131.

Friend, M., and Muraschi, T. F. 1963. Montezuma muskrats and tularemia: an investigation of the incidence of tularemia among muskrats at the Montezuma National Wildlife Refuge. Northeast Fish Wildl. Conf. 19pp.

Fuller, W. A. 1951. Natural history and economic importance of the muskrat in the Athabasca-Peace Delta, Wood Buffalo Park. Can. Wildl. Serv. Wildl. Manage. Bull. Ser. 1., no. 2. 98pp.

Gainey, L. F. 1949. Comparative winter food habits of the

muskrat and nutria in captivity. M.S. Thesis. Louisiana State Univ., Baton Rouge. 41pp.

Gash, S. L., and Hanna, W. L. 1972. Occurrence of some helminth parasites in the muskrats, *Ondatra zibethicus*, from Crawford County, Kansas. Trans. Kansas Acad. Sci. 75:251–254.

Gashwiler, J. S. 1948. Maine muskrat investigations. Maine Dept. Inland Fish. Game Bull. 38pp.

_____. 1949. The effect of spring muskrat trapping on waterfowl in Maine. J. Wildl. Manage. 13:183–188.

_____. 1950*a*. Sex ratios and age classes of Maine muskrats. J. Wildl. Manage. 14:384–398.

_____. 1950*b*. A study of the reproductive capacity of Maine muskrats. J. Mammal. 31:180–185.

Genaux, C. T.; Ernst, K. U.; and Morrison, P. 1976. A comparison of the tryptic peptides of hemoglobin from two microtine genera: *Clethrionomys* and *Ondatra*. Biochem. Syst. and Ecol. 4:295–301.

Gilbert, F. F. 1976. Impact energy thresholds for anesthetized raccoons, mink, muskrats, and beavers. J. Wildl. Manage. 40:669–676.

Glass, B. P. 1952. Factors affecting the survival of the Plains muskrat *Ondatra zibethica cinnamomina* in Oklahoma. J. Wildl. Manage. 16:484–491.

Goff, O. E., and Upp, C. W. 1942. Dried muskrat meal for poultry. Pages 100–101 *in* Louisiana Agric. Exp. Stn. Annu. Rep., 1942–43. Baton Rouge, La. 142pp.

Golley, F. B. 1962. Mammals of Georgia. Univ. Georgia Press, Athens. 218pp.

Gould, H. N., and Kreiger, N. H. 1948. The skull of the Louisiana muskrat (*Ondatra zibethica rivalicia* Bangs). Part 1: The skull in advanced age. J. Mammal. 29:138–149.

Gowanloch, J. N. 1943. Louisiana muskrat industry as a source of human food. Trans. North Am. Wildl. Conf. 8:213–217.

Gray, M. H., and Arner, D. H. 1977. The effects of channelization on furbearers and furbearer habitat. Proc. Southeast. Assoc. Fish Wildl. Agencies 31:259–265.

Greenwell, G. A. 1948. Wildlife values of Missouri farm ponds. Trans. North Am. Wildl. Conf. 13:271–281.

Hall, E. R., and Kelson, K. R. 1959. The mammals of North America. Ronald Press Co., New York. 2 vols. 1241pp.

Hamerstrom, F. N., Jr., and Blake, J. 1939. Central Wisconsin muskrat study. Am. Midl. Nat. 21:514–520.

Hamilton, W. J., Jr. 1956. The young of *Neofiber alleni*. J. Mammal. 37:448–449.

Hansen, E. L. 1965. Muskrat distribution in south-central Oregon. J. Mammal. 46:669–671.

Harper, F. 1920. The Florida water-rat (*Neofiber alleni*) in the Okefinokee [*sic*] Swamp, Georgia. J. Mammal. 1:65–67.

_____. 1927. The mammals of the Okefinokee [*sic*] Swamp region of Georgia. Proc. Boston Soc. Nat. Hist. 38:191–396.

Harris, V. T. 1952. Muskrats on tidal marshes of Dorchester County. Maryland Board Nat. Resour., Chesapeake Bio. Lab., Dept. Res. Education Publ. 91. 36pp.

Heit, W. S. 1949. Muskrat weights and sex ratio in the Riverbend Marshes of Wayne County, New York. J. Mammal. 30:122–124.

Hensley, A. L., and Twining, H. 1946. Some early summer observations on muskrats in a northeastern California marsh. California Fish Game 32:171–181.

Highby, P. R. 1941. A management program for Minnesota muskrat. Proc. Minnesota Acad. Sci. 9:30–34.

Hockett, R. N. 1968*a*. Bacteria of the muskrat, *Ondatra zibethicus*. M.S. Thesis. Univ. Michigan, Ann Arbor. 62pp.

_____. 1968*b*. Normal flora of the muskrat, *Ondatra*

zibethicus (Linnaeus), and attempts to isolate *Pasteurella pseudotuberculosis* and *Listeria monocytogenes*. Ph.D. Thesis. Univ. Michigan, Ann Arbor. 62pp.

Huenecke, H. S.; Erickson, A. B.; and Marshall, W. H. 1958. Marsh gasses in muskrat houses in winter. J. Wildl. Manage. 22:240–245.

Indiana Department of Conservation. 1963. Life series: the muskrat. Div. Fish Game Leafl. 5. 2pp.

Indiana Department of Natural Resources. n.d. Muskrat control with Conibear traps. Div. Fish Wildl. Leafl. 2. 2pp.

Jilek, R. 1977. Trematode parasites of the muskrat, *Ondatra zibethicus*, in southern Illinois. Trans. Illinois State Acad. Sci. 70:105–107.

Johansen, K. 1962. Buoyancy and insulation in the muskrat. J. Mammal. 43:64–68.

Johnson, C. E. 1925. The muskrat in New York. Roosevelt Wildl. Bull. 3:193–320.

Karstad, L.; Lusis, P.; and Wright, D. 1971. Tyzzer's disease in muskrats. J. Wildl. Dis. 7:96–99.

Knight, I. M. 1951. Diseases and parasites of the muskrat (*Ondatra zibethica*) in British Columbia. Can. J. Zool. 29:188–214.

Lantz, D. E. 1910. The muskrat. U.S.D.A. Farmers' Bull. 396. 20pp.

Lay, D. W. 1945*a*. Muskrat investigations in Texas. J. Wildl. Manage. 9:56–76.

_____. 1945*b*. The problems of undertrapping in muskrat management. Trans. North Am. Wildl. Conf. 10:75–78.

Lay, D. W., and O'Neil, T. 1942. Muskrats on the Texas coast. J. Wildl. Manage. 6:301–312.

LeBoulenge, E. 1977. Two ageing methods for muskrats: live or dead animals. Acta Theriol. 22:509–520.

LeCompte, E. L. 1928. Muskrat industry of Maryland. Maryland Conserv. Dep., Annapolis. 32pp.

Linde, A. F. 1963. Muskrat pelt patterns and primeness. Wisconsin Conserv. Dept. Tech. Bull. 29. 66pp.

Linscombe, G. 1976. An evaluation of the no. 2 Victor and 220 Conibear traps in coastal Louisiana. Proc. Southeast. Assoc. Fish Wildl. Agencies 30:560–568.

Lord, G. H.; Todd, A.C.; and Kabat, C. 1956*a*. Studies on Errington's disease in muskrats. Part 1: Pathological changes. Am. J. Vet. Res. 17:303–306.

Lord, G. H.; Todd, A. C.; and Mathiak, H. 1956*b*. Studies on Errington's disease in muskrats. Part 2: Etiology. Am. J. Vet. Res. 17:307–310.

Louisiana Department of Conservation. 1931. The fur animals of Louisiana. Dept. Conserv. Bull. 18 (rev.). 444pp.

Lowery, G. H., Jr. 1974. The mammals of Louisiana and its adjacent waters. Louisiana State Univ. Press, Baton Rouge. 565pp.

Lynch, J. J.; O'Neil, T.; and Lay, D. W. 1947. Management significance of damage by geese and muskrats to Gulf Coast marshes. J. Wildl. Manage. 1:50–76.

MacArthur, R. A. 1977. Behavioral and physiological aspects of temperature regulation in the muskrat (*Ondatra zibethicus*). Ph.D. Thesis. Univ. Manitoba, Winnipeg. 168pp.

_____. 1978. Winter movements and home range of the muskrat. Can. Field Nat. 92:345–349.

_____. 1979. Seasonal patterns of body temperature and activity in free-ranging muskrats (*Ondatra zibethicus*). Can. J. Zool. 57:25–33.

MacArthur, R. A., and Aleksiuk, M. 1979. Seasonal microenvironments of the muskrat (*Ondatra zibethicus*) in a northern marsh. J. Mammal. 60:146–154.

McCann, L. J. 1944. Notes on growth, sex and age ratios, and suggested management of Minnesota muskrats. J. Mammal. 25:59–63.

McEwan, E. H.; Aitchison, N.; and Whitehead, P. E. 1974.

Energy metabolism of oiled muskrats. Can. J. Zool. 52:1057–1062.

MacKinnon, B. M., and Burt, M. D. B. 1978. Platyhelminth parasites of muskrats (*Ondatra zibethica*) in New Brunswick. Can. J. Zool. 56:350–354.

McLeod, J. A. 1948. Preliminary studies on muskrat biology in Manitoba. Trans. R. Soc. Can. 42. 14pp.

McLeod, J. A., and Bondar, G. F. 1952. Studies on the biology of the muskrat in Manitoba. Part 1: Oestrous cycle and breeding season. Can. J. Zool. 30:243–253.

McMullen, T. L. 1978. Mammals of the Duck Creek local fauna, late Pleistocene of Kansas. J. Mammal. 59:374–386.

Marshall, W. H. 1937. Muskrat sex-ratios in Utah. J. Mammal. 18:518–519.

Mathiak, H. A. 1966. Muskrat population studies at Horicon marsh. Wisconsin Conserv. Dept. Tech. Bull. 36. 56pp.

Mathiak, H. A., and Linde, A. F. 1954. Role of refuges in muskrat management. Wisconsin Conserv. Dept. Tech. Wildl. Bull. 10. 16pp.

———. 1956. Studies on level ditching for marsh management. Wisconsin Conserv. Dept. Tech. Wildl. Bull. 12. 49pp.

Meyer, M. C., and Reilly, J. R. 1950. Parasites of muskrats in Maine. Am. Midl. Nat. 44:467–477.

Miller, G. S., Jr., and Kellogg, R. 1955. List of North American recent mammals. U.S. Natl. Mus. Bull. 205. 954pp.

Moody, R. 1950. The possibilities of transplanting muskrats in East Baton Rouge, West Feliciana and Natchitoches Parishes, Louisiana. M.S. Thesis. Louisiana State Univ., Baton Rouge. 68pp.

Nagel, W. O. 1945. Controlling muskrat damage in ponds. Missouri Conserv. 6:10–11.

Neal, T. J. 1968. A comparison of two muskrat populations. Iowa State J. Sci. 43:193–210.

Nelson, N. F. 1954. Factors in the development and restoration of waterfowl habitat at Ogden Bay Refuge, Weber County, Utah. Pages 70–71 *in* Utah State Dept. Fish and Game. Publ. 6, Fed. Aid Div. 87pp.

Newsom, J. D.; Perry, H. R., Jr.; and Schilling, P. E. 1976. Fire ant-muskrat relationships in Louisiana coastal marshes. Proc. Southeast. Assoc. Fish Wildl. Commissioners 13:414–418.

Nichols, J. D. 1973. A survey of furbearer recources in the Atchafalaya River Basin, Louisiana. M.S. Thesis. Louisiana State Univ., Baton Rouge. 184pp.

Nicholson, W. R., and Davis, D. E. 1957. The duration of life of muskrat houses. Ecology 38:161–163.

Noble, R. E., and Wright, J. M. 1977. Nesting birds of the Shishmaref Inlet and Seward Peninsula, Alaska and the possible effects of reindeer herding and grazing on nesting birds. Alaska Coop. Wildl. Res. Unit Tech. Rep. 202pp.

Olsen, P. F. 1959*a*. Dental patterns as age indicators in muskrats. J. Wildl. Manage. 23:228–231.

———. 1959*b*. Muskrat breeding biology at Delta, Manitoba. J. Wildl. Manage. 23:40–53.

O'Neil, T. 1949. The muskrat in the Louisiana coastal marshes. Louisiana Dept. Wild Life and Fish., New Orleans. 152pp.

O'Neil, T., and Linscombe, G. 1976. The fur animals, the alligator, and the fur industry in Louisiana. Louisiana Wildl. and Fish. Comm. Wildl. Education Bull. 106. 66pp.

Palmisano, A. W., Jr. 1967. Ecology of *Scirpus olneyi* and *Scirpus robustus* in Louisiana coastal marshes. M.S. Thesis. Louisiana State Univ., Baton Rouge. 145pp.

———. 1970. Plant community-soil relationships in Louisiana coastal marshes. Ph.D. Thesis. Louisiana State Univ., Baton Rouge. 98pp.

———. 1972*a*. The distribution and abundance of muskrats (*Ondatra zibethicus*) in relation to vegetative types in Louisiana coastal marshes. Proc. Southeast. Assoc. Game and Fish Commissioners 26:160–177.

———. 1972*b*. Habitat preference of waterfowl and fur animals in the northern Gulf coast marshes. Pages 163–190 *in* R. H. Chabreck, ed. Proceedings of the marsh and estuary management symposium. Div. Continuing Education, Louisiana State Univ., Baton Rouge. 316pp.

Palmisano, A. W., and Dupuie, H. H. 1975. An evaluation of steel traps for taking fur animals in coastal Louisiana. Proc. Southeast. Assoc. Game and Fish Commissioners 29:342–347.

Pancoast, J. M. 1937. Exhibit "A": muskrat industry in southern New Jersey. Trans. North Am. Wildl. Conf. 2:527–530.

Paul, J. R. 1967. Round-tailed muskrat in west central Florida. Q. J. Florida Acad. Sci. 30:227–229.

Penfound, W. T., and Schneidau, J. D. 1945. The relation of land reclamation to aquatic wildlife resources in Southeastern Louisiana. Trans. North Am. Wildl. Conf. 10:308–318.

Penn, G. H., Jr. 1942. Parasitological survey of Louisiana muskrats. J. Parasitol. 28:348–349.

Penn, G. H., Jr., and Martin, E. C. 1941. The occurrence of porocephaliasis in the Louisiana muskrat. J. Wildl. Manage. 5:13–14.

Perry, H. R., Jr. 1974. Trap responses of the gray squirrel (*Sciurus carolinensis carolinensis*), raccoon (*Procyon lotor lotor*), and opossum (*Didelphis virginiana virginiana*) to environmental factors in two swamp watersheds of northeastern North Carolina. Ph.D. Thesis. North Carolina State Univ., Raleigh. 190pp.

Perry, R. 1969. Swamp drainage: a non-ecological approach to resource management. Wildl. North Carolina 33:20–23.

Pietsch, M. 1970. Vergleichende untersuchungen an Schädeln nordamerikanischer und europäischer Bisamratten (*Ondatra zibethicus* L. 1766). Z. Säugetierk. 35:257–288. (In German; English summary.)

Porter, R. P. 1953. A contribution to the life history of the water rat, *Neofiber alleni*. M.S. Thesis. Univ. Miami, Coral Gables. 84pp.

Rausch, R. L. 1946. Parasites of Ohio muskrats. J. Wildl. Manage. 10:70.

Rawley, E.; Low, J. B.; and Sharp, D. 1952. The muskrat: a farm crop. Utah State Agric. Coll. Ext. Circ. 168. 16pp.

Reeves, H. M., and Williams, R. M. 1956. Reproduction, size, and mortality in the Rocky Mountain muskrat. J. Mammal. 37:494–500.

Rice, E. W., and Heck, O. B. 1975. A survey of the gastrointestinal helminths of the muskrat, *Ondatra zibethicus*, collected from two localities in Ohio. Ohio J. Sci. 75:263–264.

Robicheaux, B. L. 1978. Ecological implications of variably spaced ditches on nutria in a brackish marsh, Rockefeller Refuge, Louisiana. M.S. Thesis. Louisiana State Univ., Baton Rouge. 50pp.

Ross, W. M. 1972. Methods of establishing natural and artificial stands of *Scirpus olneyi*. M.S. Thesis. Louisiana State Univ., Baton Rouge. 99pp.

Salinger, H. E. 1950. Mating of muskrats. J. Mammal. 31:97.

Sather, J. H. 1954. The dentition method of aging muskrats. Chicago Acad. Sci. Nat. Hist. Misc. 130. 3 pp.

———. 1958. Biology of the Great Plains muskrat in Nebraska. Wildl. Monogr. 2. 35pp.

Schacher, W. H., and Pelton, M. R. 1975. Productivity of

muskrats in East Tennessee. Proc. Southeast. Assoc. Game and Fish Commissioners 29:594-608.

———. 1976. Sex ratios, morphology, and condition parameters of muskrats in East Tennessee. Proc. Southeast. Assoc. Fish Wildl. Agencies. 30:660-666.

Schantz, V. S., and Jenkins, J. H. 1950. Extension of range of the round-tailed muskrat, *Neofiber alleni*. J. Mammal. 31:460-461.

Scheffer, V. B. 1955. Body size with relation to population density in mammals. J. Mammal. 36:493-515.

Schofield, R. D. 1955. Analysis of muskrat age determination methods and their application in Michigan. J. Wildl. Manage. 19:463-466.

Schwartz, A. 1953. A systematic study of the water rat (*Neofiber alleni*). Univ. Michigan Mus. Zool. Occas. Pap. 547. 27pp.

Schwartz, C. W., and Schwartz, E. R. 1959. The wild mammals of Missouri. Univ. Missouri Press and Missouri Conserv. Comm., Columbia. 341pp.

Seabloom, R. W., and Beer, J. R. 1964. Observations of a muskrat (*Ondatra zibethica cinnamominus*) population decline in North Dakota. Proc. North Dakota Acad. Sci. 17:66-70.

Sealander, J. A. 1979. A guide to Arkansas mammals. River Road Press, Conway, Ark. 313pp.

Seamans, R. 1941. Lake Champlain fur survey. Vermont Fish Game Serv. State Bull. 3-4. 33pp.

Shanks, C. E. 1948. The pelt-primeness method of aging muskrats. Am. Midl. Nat. 39:179-187.

Shanks, C. E., and Arthur, G. C. 1951. Movements and population dynamics of farm pond and stream muskrats in Missouri. Missouri Dept. Conserv., Columbia. 16pp.

———. 1952. Muskrat movements and population dynamics in Missouri farm ponds and streams. J. Wildl. Manage. 16:138-148.

Sherman, H. B. 1952. A list and bibliography of the mammals of Florida, living and extinct. Q. J. Florida Acad. Sci. 15:86-126.

Smith, F. R. 1938. Muskrat investigations in Dorchester County, Maryland, 1930-34. U.S.D.A. Circ. 474. 24pp.

Snead, I. E. 1950. A family type live trap, handling cage, and associated techniques for muskrats. J. Wildl. Manage. 14:67-79.

Soileau, D. M. 1968. Vegetative reinvasion of experimentally treated plots in a brackish marsh. M.S. Thesis. Louisiana State Univ., Baton Rouge. 75pp.

Sooter, C. A. 1946. Muskrats of the Tule Lake Refuge, California. J. Wildl. Manage. 10:68-70.

Sprugel, G., Jr. 1951. Spring dispersal and settling activities of central Iowa muskrats. Iowa State Coll. Sci. 26:71-84.

Stains, H. J. 1979. Primeness in North American furbearers. Wildl. Soc. Bull. 7:120-125.

Stearns, L. A., and Goodwin, M. W. 1941. Notes on the winter feeding of the muskrats in Delaware. J. Mammal. 5:1-12.

Stearns, L. A.; MacCreary, D.; and Daigh, F. C. 1939. Water and plant requirements of the muskrat on a Delaware tidewater marsh. New Jersey Mosquito Extermination Assoc. Annu. Conf. 26:213-221.

———. 1940. Effects of ditching for mosquito control on the muskrat population of a Delaware tidewater marsh. Univ. Delaware Agric. Exp. Stn. Bull. 225, Tech. no. 26. 55pp.

Steffen, D. E. 1978. The occurrence of and damage by the Florida water rat in Florida sugar cane production areas. M.S. Thesis. Virginia Polytechnic Inst. and State Univ., Blacksburg. 104pp.

Steffen, D. E.; Holler, N. R.; Lefebvre, L. W.; and Scanlon, P. F. In press. Factors affecting the occurrence and distribution of Florida water rats in sugarcane fields. J. Am. Soc. Sugar Cane Tech.

Stevens, P. G., and Erickson, J. L. E. 1942. The chemical constitution of the musk of the Louisiana muskrat. J. Am. Chem. Soc. 64:144-147.

Stewart, R. E. A.; Stephen, J. R.; and Brooks, R. J. 1975. Occurrence of muskrat, *Ondatra zibethicus albus,* in the District of Keewatin, Northwest Territories. J. Mammal. 56:507.

Stewart, R. W., and Bider, J. R. 1974. Reproduction and survival of ditch-dwelling muskrats in southern Quebec. Can. Field Nat. 88:429-436.

Storer, T. I. 1937. The muskrat as native and alien. J. Mammal. 18:443-460.

Svihla, A. 1931. The field biologist's report. Pages 278-287 *in* S. C. Arthur, compil. The fur animals of Louisiana. Louisiana Dept. Conserv., Bull. 18 (rev.), New Orleans. 439pp.

Svihla, A., and Svihla, R. D. 1931. The Louisiana muskrat. J. Mammal. 12:12-28.

Taber, R. D. 1969. Criteria of sex and age. Pages 325-401 *in* R. H. Giles, Jr., ed. Wildlife management techniques. Wildl. Soc., Washington, D.C. 623pp.

Takos, M. J. 1940. A review of the literature on diseases and parasites of the muskrat. Maine Coop. Wildl. Res. Unit, Orono. 14pp.

———. 1943. Trapping and banding muskrats. J. Wildl. Manage. 7:400-407.

———. 1944. Summer movements of banded muskrats. J. Wildl. Manage. 8:307-311.

———. 1947. A semi-quantitative study of muskrat food habits. J. Wildl. Manage. 11:331-339.

Taylor, M. 1980. Our valuable furbearers. Wildl. North Carolina 44:2-7.

Thompson, D. R., and Baumann, C. A. 1950. Vitamin A in pheasants, quail, and muskrats. J. Wildl. Manage. 14:42-49.

Thorton, R.; Gordon, C.; and Ferguson, J. H. 1978. Role of thermal stimuli in the diving response of the muskrat (*Ondatra zibethica*). Comp. Biochem. Physiol. 61A:369-370.

Twining, H., and Hensley, A. L. 1943. The distribution of muskrats in California. California Fish Game 29:64-78.

U.S. Fish and Wildlife Service. 1973. Muskrat and nutria control with zinc phosphide. Wildl. Leafl. 504. 2pp.

Warwick, T. 1940. A contribution to the ecology of the musk-rat (*Ondatra zibethica*) in the British Isles. Proc. Zool. Soc. London, Ser. A. 110:165-201.

Williams, R. M. 1951. The use of scent in live-trapping muskrats. J. Wildl. Manage. 15:117-118.

Willner, G. R.; Chapman, J. A.; and Goldsberry, J. R. 1975. A study and review of muskrat food habits with special reference to Maryland. Maryland Wildl. Adm. Publ. Wildl. Ecol. 1. 25pp.

Wilson, K. A. 1949. Investigations on the effects of controlled water levels upon muskrat production. Proc. Southeast. Assoc. Game and Fish Commissioners 3. 7pp.

———. 1953. Raccoon predation on muskrats near Currituck, North Carolina. J. Wildl. Manage. 17:113-119.

———. 1954*a*. Litter production of coastal North Carolina muskrats. Proc. Southeast. Assoc. Game and Fish Commissioners 8:13-19.

———. 1954*b*. The role of mink and river otter as muskrat predators in northeastern North Carolina. J. Wildl. Manage. 18:199-207.

———. 1955*a*. A compendium of the principal data on muskrat reproduction. Fed. Aid in Wildl. Restoration Proj. W-6-R. North Carolina Wildl. Resour. Comm., Raleigh. 11pp.

———. 1955*b*. Effects of water level control on muskrat populations. Job Completion Rep., Fed. Aid in Wildl. Restoration Proj. W-6-R. North Carolina Wildl. Resour. Comm., Raleigh. 9pp.

———. 1955*c*. Experimental marsh management near Currituck, North Carolina. Job Completion Rep., Fed. Aid in Wildl. Restoration Proj. W-6-R-15, Special Rep. 1. North Carolina Wildl. Resour. Comm., Raleigh. 27pp.

———. 1956*a*. Color, sex ratios, and weights of North Carolina muskrats. Fed. Aid in Wildl. Restoration Proj. W-6-R-15, Sp. North Carolina Wildl. Resour. Comm., Raleigh. 20pp.

———. 1956*b*. Control of raccoon predation on muskrats near Currituck, North Carolina. Proc. Southeast. Assoc. Game and Fish Commissioners 10:221–233.

———. 1968. Fur production on southeastern coastal marshes. Page 149–162 *in* J. D. Newsom, ed. Proceedings of the marsh and estuary management symposium. Div. Continuing Education, Louisiana State Univ., Baton Rouge. 250pp.

Wingard, R. G., and Sharp, W. M. n.d. Trapping and skinning muskrats. Pennsylvania State Univ. Coll. Agric. Ext. Serv. Circ. 478. 7pp.

Wobeser, G.; Hunter, D. B.; and Daoust, P. Y. 1978. Tyzzer's disease in muskrats: occurrence in free-living animals. J. Wildl. Dis. 14:325–328.

Wood, W. 1974. Muskrat origin, distribution, and range extension through the Coastal Areas of Del Norte Co. Calif. and Curry Co. Oregon. Murrelet 55:1–4.

Woodstream Corporation. n.d. Wildlife traps for predator control, fur trapping, and wildlife management. Woodstream Corp., Lititz, Pa. 6pp + suppl.

Wragg, L. E. 1954. The effect of D.D.T. and oil on muskrats. Can. Field Nat. 68:11–13.

H. RANDOLPH PERRY, JR., Louisiana Cooperative Wildlife Research Unit, School of Forestry and Wildlife Management, Louisiana State University, Baton Rouge, Louisiana 70803.

16

Voles
Microtus species

Murray L. Johnson
Sherry Johnson

NOMENCLATURE

COMMON NAMES. Voles, meadow mice. Microtines, arvicolines, and arvicolids are less restrictive terms. Various species may be referred to by specific common names, as noted in the species list. The French name *campagnol* and the German name *Wühlmaus* are used extensively in European literature. In this chapter, as a matter of convenience, the names *vole* and *meadow mouse* will refer to the genus *Microtus*.

SCIENTIFIC NAME. Genus *Microtus*

SPECIES NORTH OF MEXICO. *M. abbreviatus* (7 subspecies, including *M. miurus*) (Rausch 1974, 1977), insular vole. *M. californicus* (16 subspecies), California vole. *M. canicaudus* (no subspecies), gray-tailed vole. *M. chrotorrhinus* (no subspecies), rock vole. *M. longicaudus* (15 subspecies, including *M. coronarius*), long-tailed vole. *M. mexicanus* (12 subspecies, including *M. fulviventer*), Mexican vole. *M. montanus* (17 subspecies), montane vole. *M. ochrogaster* (6 subspecies, including *M. ludovicianus*), prairie vole. *M. oeconomus* (12 subspecies in North America), tundra vole. *M. oregoni* (4 subspecies), Oregon vole, creeping vole. *M. pennsylvanicus* (26 subspecies, including *M. breweri* and *M. nesophilus*), meadow vole. *M. pinetorum* (8 subspecies, including *M. parvulus* and *M. quasiater*; also currently considered a separate genus *Pitymys* by some authorities), pine vole. *M. richardsoni* (3 subspecies; also currently referred to the genus *Arvicola* by some authorities), Richardson's vole, water vole. *M. townsendii* (6 subspecies), Townsend's vole. *M. xanthognathus* (no subspecies), yellow-cheeked vole. (*Source:* Hall and Kelson 1959; Jones et al. 1975.)

OTHER ARVICOLIDS. The genus *Microtus* is varied and complex in its taxonomic grouping and relationships. Because of this complexity there is a useful justification for applying the restrictive family name Arvicolidae (Gray 1821) to voles of the genus *Microtus* and others closely related (Kretzoi 1962; Hibbard et al. 1978). According to several authorities, the family name Microtidae is preferable. Some people lump together the large, related families Cricetidae and Muridae.

The morphology of all the arvicolids is similar; a

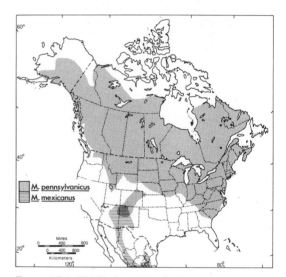

FIGURE 16.1. Distribution of the meadow vole (*Microtus pennsylvanicus*) and Mexican vole (*M. mexicanus*).

majority are small in size and several species or genera may occur in the same region. In some instances these factors combine to make identification difficult for any but a museum-based expert.

In addition to the genus *Microtus*, the following arvicolids are found in North America: *Arborimus albipes* (no subspecies), white-footed vole. *Arborimus longicaudus* (2 subspecies and an additional sibling species [Johnson 1973]), red tree vole. *Dicrostonyx torquatus* (11 forms, including *D. exsul*, several of which may be full species [Rausch 1977]), collared lemming. *Dicrostonyx hudsonius* (no subspecies), Labrador collared lemming. *Clethrionomys californicus* (3 subspecies), California red-backed vole. *Clethrionomys gapperi* (32 subspecies, including *occidentalis*), Gapper's red-backed vole. *Clethrionomys rutilus* (8 subspecies in North America), northern red-backed vole. *Lagurus curtatus* (6 subspecies), sagebrush vole. *Lemmus sibiricus* (6 subspecies in North America, including *L. nigripes*), brown lemming.

Neofiber alleni (5 subspecies), round-tailed muskrat. *Ondrata zibethica* (16 subspecies), muskrat. *Phenacomys intermedius* (9 subspecies), heather vole. *Synaptomys borealis* (9 subspecies), northern bog lemming. *Synaptomys cooperi* (6 subspecies), southern bog lemming.

The family Arvicolidae is a most interesting complex of species. Their nomenclature has never been stable; reasonable differences of opinion as to taxonomic usage exist. However, with modern tools, especially molecular biology, karyotyping, and statistical comparisons, and with improved paleontologic information, a better synthesis is becoming apparent. Understanding of basic relationships and evolutionary history is better today than ever before.

Arvicolids appear to have arisen from a "late-appearing" cricetid ancestor (Dawson 1967). They are first recognized from the late Miocene of North America and possibly arose here (Kurtén 1968). They have been reported in the Pliocene of Eurasia, but recent data correlated with potassium-argon 40 dating suggest an origin no more than 10 million years ago (Repenning 1979). The genus *Microtus* is an advanced

form and occurs in both the Old World and the New World, in the northern regions. The primary center of evolution is unknown, but could have occurred either in the central Eurasian region or in North America. Fossil records of modernlike or modern species of *Microtus* are known from North America, Asia, and Europe, dating back at least 2 million years (Repenning personal communication).

DISTRIBUTION

The genus *Microtus* has a circumboreal distribution. Many species occur throughout Europe and much of Asia (Corbet 1978).

In North America 19 species (table 16.1) are distributed from the northernmost reaches of the continent south into Mexico and Guatemala (figure 16.1). The highest species diversity is in the temperate regions and in the western part of the continent. Generally, in the southern regions the populations are isolated and occur only in the higher elevations.

Microtus pennsylvanicus has the most extensive range, from coast to coast and north to above the Arctic Circle. It is closely related to *Microtus agrestis*, a

TABLE 16.1. Distribution of subgenera of North American *Microtus,* by species

Subgenus	Species	Major Distribution[a]				
		N	W	M	E	S
Stenocranius	M. abbreviatus (incl. *miurus*)	x				
Microtus	M. californicus		x (P)			
	M. canicaudus		x (P)			
	M. chrotorrhinus	x			x	
	M. longicaudus (incl. *coronarius*)	x	x			
	M. mexicanus (incl. *fulviventer*)					x
	M. montanus		x			
	M. pennsylvanicus (incl. *breweri* and nesophilus)	x	x	x	x	
	M. oaxacensis					x
	M. oeconomus	x				
	M. townsendii		x (P)			
	M. xanthognathus	x				
Aulacomys	M. richardsoni		x			
Chilotus	M. oregoni		x (P)			
Pedomys	M. ochrogaster (incl. *ludovicianus*)			x		
Pitymys	M. pinetorum (incl. parvulus)				x	
	M. quasiater					x
Herpetomys	M. guatemalensis					x
Orthriomys	M. umbrosus					x
	Total 19 sp. (conservative)	6	8	2	3	5

[a]N = Far north, into permafrost regions
W = West of Mississippi River; P denotes Pacific states only
M = Midcontinent
E = East of the Mississippi River
S = Southern (Arizona and New Mexico and south)

FIGURE 16.2. Distribution of the tundra vole (*Microtus oeconomus*), creeping vole (*M. oregoni*), and prairie voles (*M. ochrogaster* and *M. ludovicianus*).

FIGURE 16.3. Distribution of the insular vole (*Microtus abbreviatus*), gray-tailed vole (*M. canicaudus*), and montane vole (*M. montanus*).

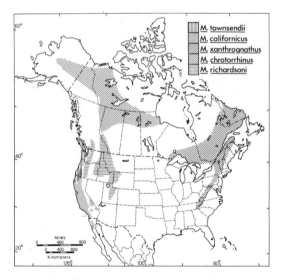

FIGURE 16.4. Distribution of the Townsend's vole (*Microtus townsendii*), California vole (*M. californicus*), yellow-cheeked vole (*M. xanthrognathus*), rock vole (*M. chrotorrhinus*), and water vole (*M. richardsoni*).

FIGURE 16.5. Distribution of the long-tailed vole (*Microtus longicaudus*) and pine voles (*M. pinetorum* and *M. parvulus*).

common vole of Europe and parts of Asia. It overlaps the general range of several other species of *Microtus,* and is separated from these by differences in ecological niche, especially noted in zones of contact, and by behavioral characteristics.

Other North American species have wide distributions. On a relative basis of size of range, as compared to *M. pennsylvanicus,* these include: *Microtus longicaudus,* 41 percent; *M. ochrogaster,* 30 percent; *M. pinetorum,* 24 percent; *M. montanus,* 16 percent; and *M. mexicanus,* 12 percent. These calculations do not take into account the frequently disjunct nature of the

ranges, especially that of *M. mexicanus.* They indicate a relative adaptive success within the North American ecosystems.

In the northern latitudes, two species exist: the circumboreal *Microtus oeconomus* in Alaska, Yukon Territory, and the Northwest Territories (figure 16.2); and *Microtus abbreviatus (miurus),* closely related to the Asiatic *M. gregalis,* with a discontinuous range in much of Alaska, Yukon Territory, and adjacent Northwest Territories (figure 16.3).

The remainder of the North American species are not as closely related to Eurasian counterparts. Except for *M. xanthognathus* (figure 16.4), they are more remotely (southerly) placed from Beringia, where Holarctic distribution occurred (figures 16.2–16.5). There is evidence of evolutionary differentiation, which may be interpreted as meaning more isolation from Eurasian genetic influence, over prolonged geologic time. The number of species in the west probably reflects isolating mechanisms of mountain systems, climatologic differences, and available ecological niches. This is seen throughout the family Arvicolidae, as well as in the genus *Microtus*.

To the south, in Mexico and Guatemala the presumed result of long continued isolation shows in the spectrum of more differentiated forms of *Microtus*. Of the five species occurring in the south, four subgenera are represented, two of which are unique (see table 16.1).

DESCRIPTION

General Morphology. Members of the North American genus *Microtus* are characterized by uniformity in morphology with little variation from species to species. They are generally adapted to subterranean, terrestrial, and, in some, semiamphibious life.

The body is stocky and rounded. The nose is blunt with vibrissae that are generally inconspicuous compared to many other rodents. The ears are small, rounded, and a little longer than the body fur, and are usually more obvious in this genus than in some other arvicolids. The eyes are small; they are relatively smaller in the fossorial species, *M. oregoni* and *M. pinetorum*. The legs are short. The tail is scantily haired, longer than the hind foot but shorter than the head and body length. This varies from species to species. It is relatively shorter in juveniles, achieving full length only in adults. The fur is moderately coarse in appearance and is various shades of brown to gray, lighter ventrally.

The size is variable within the genus. The largest two species are *M. richardsoni* and *M. xanthognathus*, both of which have adult weights that may exceed 120 g, comparable to the size of a half-grown Norway rat (*Rattus norvegicus*). The length of the head and body of these two species can exceed 180 mm. The smallest is *M. oregoni*, with a weight usually under 20 g.

Skull and Dentition. Skulls are usually angular in adults and have a "standard" microtine appearance (figure 16.6). The teeth are characteristic in the genus *Microtus* (figure 16.7). Incisors are typically rodent, open rooted and ever growing. The lower incisors, as in several other genera, pass from the lingual to the labial side of molars between the bases of the roots of m2 and m3. They ascend behind the molars to terminate within or near the condylar process (Hall and Kelson 1959). The molariform teeth are prismatic, hypsodont, and ever growing. The numbers and conformation of the prisms vary somewhat according to the species. As in other rodents of the suborder Myomorpha, the medial portion of the masseter muscle originating from the maxilla sends only a small slip of

muscle through the infraorbital canal to attach on the mandible. The dental formula is 1/1, 0/0, 0/0, 3/3 = 16.

Skeletal System. The skeleton shows no remarkable evolutionary changes. It consists of 7 cervical, 13 dorsal, and 6 lumbar vertebrae; the number of caudal vertebrae varies with the length of the tail. The hyoid bones are fused into a single, curved, unattached bone. There are well-developed clavicles joining the sternum with the acromion processes, helping to stabilize the shoulder joints. There are four functional digits on the forelimb, five on the hind limb. The radius and ulna are not fused; a supracondylar foramen is not present. The distal fibula is solidly fused with the tibia.

Pelage and Molts. Newborn voles have no external hair (except vibrissae) or dermal pigment. Hatfield (1935) found first dermal pigmentation at 8 hours of age in the California vole. This progresses from the top of the head posteriorly along the back. At 36 hours, pigment covers more than half of the back. At 72 hours, the back is completely pigmented and the hair on the back is 2 mm long. By day five, the hair over the rest of the body, exclusive of feet and tail, reaches 2 mm in length.

On day 9 or 10 the skin pigmentation begins to disappear, leaving the skin pink in color under the fur until the third week. This is the juvenile pelage. At the third week the postjuvenile molt begins with dark pigmentation all over the skin and new fur growth all over the body. At 35 days of age pigmentation is reduced to a dark dorsal streak. At about week eight the adult pelage begins to grow. In *M. californicus* this begins on the belly, progressing to the middorsal line. The molt line can be identified by this skin pigmentation, moving forward across the shoulders to the head region.

It is likely that the three molts—juvenile, postjuvenile and adult—occur in all species of *Microtus*. There may be some variation in timing and pattern, however.

Adult moltings are frequently seen in wild-caught and captive animals. Seasonal molts—winter and summer—appear to be of regular occurrence throughout the Arvicolidae. They are especially marked in the northern species. Hamilton (1938) described the summer molt of *M. pinetorum,* which replaced the dark winter pelage in May or June. The winter molt was noticeable in November or December.

The darkening of the skin is due to the increased melanin content in the hair follicles, which actively produce hair. There is also probably increased vascularity during this time.

Other Anatomy. Mammary glands are usually eight in number: two pairs of pectoral and two pairs of inguinal glands. This number may be reduced. In younger animals the pectoral mammae especially may not be fully developed, even during pregnancy (Howell 1924). Posterolateral (flank) glands and anal glands are usually present and are particularly noticeable in adult males.

The gastrointestinal system is similar to that of most other rodents (Vorontsov 1962, 1967). There is a simple stomach that is mostly lined by squamous epithelium, with a small pouch on the greater curvature, lined with glandular epithelium. The cecum is very large. This is apparently the site of symbiotic bacterial digestion, for no mammal has developed a cellulase (Grassé 1973; Hume 1978). The colon is small and forms the typical fecal pellets, with some storage enlargement in the rectum.

The kidneys of *Microtus* do not show the elongated renal papillae. It is presumed that there is no particular adaptation for water conservation.

Testes are in a shallow scrotum in a fully mature animal. There is characteristic baculum composed of a shaft and three separate digitate processes that become fully ossified at maturity. The uterus is bicornuate.

GENETICS AND PHYSIOLOGY

Chromosomes. A moderate amount of information is available regarding the chromosomes of North American mice of the genus *Microtus*. With the advent of newer techniques, especially banding (Cooper and Hsu 1972), much potential investigation of specific and general import remains to be done. Because of the close relationships of the Eurasian forms of *Microtus* and the importance of this active specialty, reports from many sources and regions have been examined and are summarized here (table 16.2). *Microtus agrestis* of the Old World and *M. chrotorrhinus* and *M. oregoni* of North America have diverse findings unique in karyology of mammals.

Karyotyping has been of taxonomic importance. Examples include the evidence for separation of *Microtus canicaudus* from *M. montanus* (Hsu and Johnson 1970), against combining *M. pennsylvanicus* with *M. agrestis* (Johnson and Ostenson 1959), and against including *Microtus richardsoni* in the genus *Arvicola* (Matthey 1969). This is a fertile field for investigation. We may expect further relationships within the genus and family to be clarified with additional research.

Molecular Biology. For more than 20 years increasingly sophisticated analytical procedures have allowed progressively better synthesis of molecular relationships (Johnson 1974). It is apparent that *Microtus* is similar to all mammals having near-identical molecules that involve all structures and all functions. In simple, smaller molecules there is absolute identity in different species, but as the complexity of the molecule and its function grows the identity gives way to homology. Polymorphism is frequently found, to give metabolic (or genetic?) choices in some of the molecular forms. Proteins in particular are suitable for examination. In the range of size from a few thousand dalton to about 60,000 dalton (as hemoglobin and albumin), and to longer conglomerates of the macromolecules, rich sources of information are available. The investigative surface has been barely scratched.

In a survey study related to taxonomy, 12 species and 9 additional subspecies of *Mictrotus* were exam-

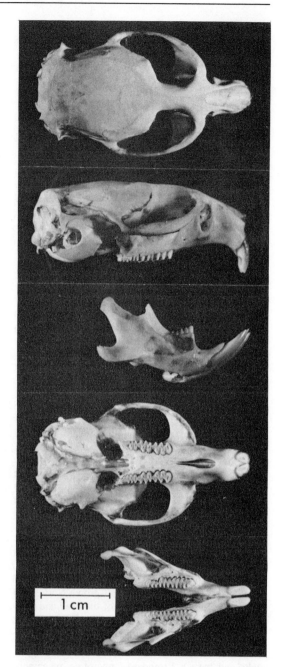

FIGURE 16.6. Skull of the meadow vole (*Microtus pennsylvanicus*). From top to bottom: lateral view of cranium, lateral view of mandible, dorsal view of cranium, ventral view of cranium, dorsal view of mandible.

ined by electrophoresis. Hemoglobins were found to be characteristic of the species, and identical migrations occurred in several species (Johnson 1974). A double major hemoglobin was found in *M. oeconomus,* and a unique hemoglobin was present in *M. longicaudus,* probably a combination of two nearly isoelectric molecular forms. Differentiation between species could frequently be determined by examination of the

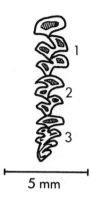

5 mm

FIGURE 16.7. Right upper toothrow of the montane vole (*Microtus montanus*), showing characteristic prisms.

hemoglobin and serum proteins: species pairs of *M. agrestis* and *M. pennsylvanicus*, and *M. montanus* and *M. canicaudus* were examples.

There may be markers in the spectrum of protein molecules. These can be very useful for identification of a population, especially when looking for evidence of gene exchange or biotype change. It has been found that proportions of genotypes within units, as shown by electrophoretic studies, can and do vary with time and place (Southern 1979). As an example, Krebs et al. (1976) found variants of leucine aminopeptidase genotypes in *M. townsendii*. Most of these kinds of studies require that a polymorphism exists; many can be demonstrated as a one-gene Mendelian trait and analyzed as such. This is an elegant system, when such genotypic variations can be found.

A review of the subject of biochemical relationships of voles, with additional data, was presented by Nadler et al. (1978). North American species of *Microtus* examined were *abbreviatus*, *ochrogaster*, *oeconomus*, *pennsylvanicus*, and *richardsoni*. These were compared with several Old World species and with members of the genus *Clethrionomys*. Holarctic identity of *M. oeconomus* was supported; and a close relationship of *Microtus* and *Clethrionomys* was suggested.

It must be noted that these kinds of studies equate allelic identity of the isoelectric electrophoretic bands. It is known, however, that a single band may contain more than one molecule form and that an isoelectric band most likely represents one molecular form in one species and a different molecular form in another. Apparent identity on electrophoresis is not, therefore, equivalent to genetic identity. Johnson (1977) felt that "serious error may occur in all such investigations when similar R_f in a gel does not reflect genetic identity," when R_f is the mobility of a protein. When two variants migrate at different rates, there is no ambiguity, and here lies the most powerful service of electrophoresis.

Wilhelm (1977) investigated 18 protein systems of *M. mexicanus*, 15 with confidence; 24 scorable loci were found, of which 14 segregated for more than one allele. He concluded that *M. mexicanus* was an "evo-lutionary conservative microtine," and that numerous techniques, including protein electrophoresis, should be used in reconstructing past biological events in systematic and evolutionary studies.

Temperature Regulation. *Microtus* are normal mammalian endotherms. Within the boreal regions the various species are well adapted to survive from arctic through temperate and into southern regions. They are active every day of the year, without hibernation or prolonged fasting.

Body temperatures are controlled within a narrow and normal range by metabolic activity, by regulating external heat loss or gain by various external maneuvers, and by anatomical characteristics. The most important of these are the use of burrows, runways, nests, and snow cover (the original "heat exchange" energy savers); adjusting of nocturnal or diurnal activity and huddling of family groups; and the globular body shape, with short ears and limbs, a relatively short tail, and a good insulating pelage, especially in the northern and high altitude species. All of these adaptations refer to the basic boreal origin of the arvicolids.

The small size of some species is disadvantageous for heat conservation. In this regard, *M. oregoni* and *M. pinetorum*, the two smallest species, are primarily subterranean in their existence.

Brown fat has been found to be an important thermogenic tissue in several mammals. Didrow and Hayward (1969) in a study on *Microtus pennsylvanicus* found a doubling in the weight of brown fat in winter, to a level of 2.3 percent of the total body weight. Thus, brown fat may function in winter survival. Torpor as a heat-conserving mechanism has not been described in *Microtus*.

Water Regulation. Arvicolids are not known to have any particular physiologic mechanism for water conservation, nor are they considered to be desert-adapted animals. Certain species, however, occur, occasionally in great numbers in regions considered to be arid (Krebs 1966; Rose and Gaines 1978). These include the sagebrush (*Artemisia* sp.) ecosystem (*M. montanus*) and the southwestern grasslands ecosystem (*M. californicus*).

Experience with these systems, however, has revealed that the simplistic view is probably most correct. Optimal habitat is that containing stable amounts of moisture and moisture-laden plant food, influenced either by permanent water sources including the water table, or by altitude and microclimate effects of geomorphology. From these stable ecosystems, populations increase at irregular intervals and seasons, as the plant communities expand from increased moisture.

High populations may thus be formed and the pest effect is again manifested, both in mankind's working domain and in natural areas. Retreat from the temporarily enhanced environments is inevitable as the herbivores denude the habitat.

Those species most restricted by ecological niches are least likely to become pests to humans. This restric-

TABLE 16.2. Chromosomes of the genus *Microtus*

Species	Diploid Number	Reference and Remarks
M. abbreviatus	54	Rausch and Rausch 1968; Rausch 1974, 1977. Includes *miurus*.
M. agrestis	50	(Europe.) Hsu and Benirschke 1969; Cooper and Hsu 1972. Unique karyotype: most of constitutive heterochromatin is confined to enormous sex chromosome. Some polymorphism has been reported.
M. arvalis	46	(Europe.) Hsu and Benirschke 1970, Zivovic et al. 1975. *M. epiroticus (subarvalis)* differentiated by karyotype.
M. californicus	54	Rausch 1964.
M. canicaudus	24	Hsu and Johnson 1970; Hsu and Benirschke 1971. Karyotype (sex chromosomes) differentiates from *M. montanus*.
M. chrotorrhinus	60	Meylan 1967. Identified giant sex chromosomes, similar to *M. agrestis*.
M. (Pitymys) duodecimcostatus	62	(Europe.) Matthey 1964.
M. epiroticus	54	(Europe.) Ruzic et al. 1975. Karyotype identified with *M. subarvalis*.
M. (Pitymys) felten	54	(Europe.) Zivkovic and Petrov 1974; Petrov and Rimsa 1976. Similar karyotype to *M. subepiroticus*.
M. gregalis	54	(Asia-Siberia.) Fedyk 1970.
M. guentheri	54	Matthey 1953, 1969. *M. iran* has the same diploid number. *M. socialis* considered conspecific by Corbet (1978), but see this list.
M. kikuchii	30	(Asia.) Makino 1950.
M. longicaudus	56–70	Hsu and Benirschke 1969; Matthey 1969; Judd and Cross 1980.
M. mexicanus	44	Matthey 1955; Wilhelm 1977.
M. middendorffi	50	(Asia-Siberia.) Matthey and Zimmerman 1961.
M. montanus	24	Hsu and Benirschke 1968.
M. montebelli	30	(Asia.) Hsu and Benirschke 1970.
M. (Pitymys) multiplex	48	(Europe.) Matthey 1964.
M. nivalis	54	(Europe.) Todorovic et al. 1971. Some populations apparently have a diploid number of 56.
M. ochrogaster	54	Rausch 1964.
M. oeconomus	30	Hsu and Benirschke 1970 (Europe); Makino 1950 (Asia); Rausch and Rausch 1968 (Alaska).
M. oregoni	17F 18M	Matthey 1958; Ohno et al. 1963; Hsu and Benirschke 1969. A unique sex determination system: in the somatic cells, the male has an XY construction and the female, XO; in the gonads the male has a YO and the female, XX.
M. pennsylvanicus	46	Matthey 1952; Johnson and Ostenson 1959; Hsu and Benirschke 1967. Karyotype differentiates from *M. agrestis*.
M. (Pitymys) pinetorum	62	Matthey 1964; Donati et al. 1977.
M. richardsoni	56	"There is no clue of close kinship with our European *Arvicola*" (R. Matthey, personal communication 5 August 1958). Matthey 1969.
M. socialis	62	(Eurasia) Matthey 1953, 1969.
M. (Pitymys) subterraneus	54	(Europe) Matthey 1964.
M. (Pitymys) tatricus	32	(Europe) Matthey 1964.
M. townsendii	50	Hsu and Benirschke 1971.
M. xanthognathus	54	Rausch and Rausch 1974.

tion frequently relates to the inability to adapt to conditions of lessened availability of water.

Digestion. The most abundant source of plant energy is in the cell wall portion (Parra 1978). This is a polysaccharide comprised of 3,000 or more glucose units. Mammals, specifically, break down the fiber. Chemically this is by hydrolysis of the B1-4 glycosidic bonds of cellulose (Hume 1978), and by symbiotes, bacteria and protozoa within the intestinal tract of the fiber-eating animals (fibrivores). These organisms live in the microenvironment, the microtine gut, where they are provided a never-ending supply of finely ground foodstuffs.

In the vole, compared to other fibrivores, considerably lower figures of fermentation products within the gastrointestinal tract have been found (McBee 1970). In terms of the capacity of wet contents, fermentation contents were 6.0 percent of body weight, calculated at 25 g/kg. This may be compared to cattle and the camel, where over 17 percent fermentation contents occur, calculated at 728 g/kg (Parra 1978).

The microbiology of the intestinal tract and utilization of the fermentive products of *Microtus* have been little studied. Some comparative information must, therefore, be used in suggesting what may occur during the digestive processes. Hind gut microbial populations apparently are under similar microconditions as those in the rumen of the larger herbivores, about which more is known. Parra (1978) referred to several species and indicated that physical constants of temperature, anaerobiosis, and buffering capacity are comparable. Herbivores evidently contain comparable numbers of bacteria and a spectrum of morphologically similar bacteria (*Streptococcus, Bacteroides,* gram-negative spore-forming bacilli, and unidentified cellulytic bacteria) in their expanded gut compartments. In the cecum of the vole, culture counts of about 300 x 10^9 per gram of dry matter were obtained (McBee 1970). However, symbiotic digestion may not be as efficient in the voles as in the larger herbivores (Keys and Van Soest 1970).

There are numerous gaps in the information about the nutritional system of *Microtus*:

1. What role do the flagellated protozoans play? Are they parasites that at times may produce harmful effects or are they true symbionts? Do they aid in digestion or produce byproducts utilizable to *Microtus,* or are they themselves eventually digested by the vole's system to add to the energy budget?

2. What nutritional role do microorganisms in general play?

3. What degree of coprophagy exists in nature that allows *Microtus* to utilize partially digested cellulose in some states of development or under certain environmental conditions?

4. Is there any significant biochemical breakdown of the celluloses in the upper gastrointestinal tract, and, if so, what microorganisms are involved?

5. What production of vitamins (especially B complex and K) takes place in the gastrointestinal tract? Is this required by the vole?

6. What microorganisms are found in the various species of *Microtus*?

7. Are there any valid biological comparisons among *Microtus,* other symbiont digestion, and humans? Are there breakdown products of certain ingested items that are potentially harmful to the host (e.g., toxic or carcinogenic)?

REPRODUCTION

Microtus in general have an efficient and potentially high rate of reproduction. This provides the potential for rapidly developing population irruption in years favorable to this phenomenon.

Anatomy. Mice can be identified as female externally by the close proximity of the urethral orifice to the anus. The vaginal introitus is in between. In young animals that have copulated, the vagina is visibly opened and "perforate." Slight traction anteriorly will usually reveal a vaginal mucous membrane. Immediately postpartum, the vagina will be dilated and sometimes blood stained.

The uterus is bicornuate (Raynaud 1969); there are two functional horns that open into a common lower segment, then through a single cervix into the vagina. Pregnancies occur simultaneously and equally in both horns.

In *M. agrestis, M. arvalis,* and the subgenus *Pitymys* a prostate gland is present in females; in *M.*

agrestis a tiny os clitoridis has been found (Raynaud 1969). It is likely that the same types of structures are present in most, if not all, arvicolid species. Urethral and preputial glands are also found in the female.

Males are usually easily identified externally by an anogenital distance of more than 8 mm, and in the adults by the shallow scrotum and descended testes. The external soft tissue micromorphology of the penis has some value as a taxonomic character. The enclosed baculum is comprised of a main shaft and three distal digital processes (Burt 1960; Wiseman 1967). *M. mexicanus* is unique in having rudimentary lateral processes (Anderson 1960; Wilhelm 1977). Accessory glands consist of the paired seminal vesicles, urethral glands, paired bulbourethral (Cowper's) glands, paired preputial glands, and the prostate. The prostate is a complex, spread-around gland comprised of a pair of large anterior lobes ("coagulation glands") associated with the seminal vesicle, and paired dorsal, lateral, and ventral lobes (Raynaud 1969). Sperm morphology, as related to taxonomy, has been analyzed in *M. mexicanus* by Wilhelm (1977). Some polymorphism between populations was found. It was suggested that differences might exist at the species level comparable with other groups of animals. This is a little-studied subject, but one with promise of clarifying relationships.

Reproductive Cycle. There is a remarkable uniformity of the reproductive patterns of the species of *Microtus*. Female voles reach reproductive maturity earlier than males. They may ovulate as early as three weeks of age, and become pregnant. Males, however, require more than six to eight weeks before spermatogenesis produces mature sperm, and, equally important, before hormonal influences stimulate appropriate mating behavior. As noted by Hamilton (1941), this discrepancy helps to prevent inbreeding, for the juvenile females are pregnant before their sibling brothers are in breeding condition. These early reproductive activities occur well before adult size is attained. Early reproductive maturity is largely responsible for the ability of the various species of *Microtus* to increase populations and expand their range; it allows an early beginning of population increase during an environmental cycle favorable to the vole. A mild mitigation of the effect of early breeding is the lower number of offspring in the first or second pregnancy (Negus et al. 1977).

The short gestation period of 20–23 days is another most important factor in producing a high population level of *Microtus*. In a theoretical calculation, Negus et al. (1977) showed that a beginning population of 100 pairs of mice per 40 acres could create a total density of 8,900 between April and September.

Ovulation may be spontaneous in *Microtus,* but there is continued and increasing information that induced ovulation occurs (Breed 1967; Cross 1972; Rose and Gains 1978). Certainly there are cyclic changes in preparation for ovulation, as demonstrated by repeated vaginal smears.

Behavior and Copulation. Mating behavior occurs when a female is receptive to a male; at other times agonistic activities occur. During the receptive period a female may actively repulse and severely wound certain males and accept others. Males vary in their aggressiveness for mating, and, in general, it is the female that determines mating success. After a few exploratory contacts between male and female, the female will pursue the male, smelling the genital area and then in turn being followed by the male.

Copulation follows shortly and is done in the stereotyped fashion, male over the back of the female. The act takes only a few seconds, with rapid thrusting by the male, and it may be repeated several times. The male may attempt further mountings, which the female will aggressively repulse. Both male and female then groom themselves, especially the genital area.

A copulation plug consisting of congealed mucoid substance forms within the vagina and lasts about two days (Hamilton 1941).

Pregnancy and Parturition. Implantation occurs on either one or both sides of the bicornuate uterus. Preimplantation mortality may occur, possibly from genetic or nutritional causes. Absorption of implanted embryos regularly occurs; embryo counts done early in pregnancy are higher than the number of young born. It is assumed that at least some of these phenomena are related to incompatible unions (such as hybridization) and recognition of defective embryos. The preimplantation losses in various species may vary from 5 to 10 percent; postimplantation losses are 1.5 to 7 percent (Rose and Gaines 1978).

Parturition is usually fast and simple. Hamilton (1941) described the process as beginning usually in the morning in captive *M. pennsylvanicus.* The mouse assumes a sitting posture, and licks her external genitalia. She moves about in a slow and "seemingly laborious" fashion. Contractions are evident and presentation is either breech or cephalic. The birth is frequently aided by the female with gentle traction by her teeth, and may take only four or five seconds. Intervals between births are usually a few minutes, but may be as long as seven hours. Little discomfort is evident.

Death of the mother or fetus occasionally occurs in late pregnancy or during parturition. This occurs in cases of hybridization between sibling species of the red tree vole (*Arborimus*). This phenomenon is assumed to be part of the reproductive barrier between similar species.

The placenta is expelled after each young and is usually eaten by the mother. The umbilical cord breaks spontaneously; the young is cleaned by maternal licking immediately after birth. Nursing may begin before the entire litter is delivered.

Birth weight in *M. pennsylvanicus* ranged from 1.6 to 3 g, averaging 2.1 g (Hamilton 1941). This range of size in viable young is not unusual. Birth weights in various species are in general correlated with the size of the adults. For instance, in *M. townsendii,* a species about twice as large, a litter of seven ranged from 3.9 to 5.4 g.

Early Development. Young are born in a nest. They are pink and have closed eyes and ears. The only external hair is the mystacial bristles. In the pine vole (*M. pinetorum*) hair appears on day 5 or 6; the incisors erupt at the same time. The ear pinnae open at 8 days and the eyes open at 9 to 12 days. Wandering begins at about two weeks and weaning at three weeks (Benton 1955).

Most species of *Microtus* follow this general pattern, although the timing may vary somewhat. The weaning time in particular may occur as early as two weeks, and the young mice are gradually shifted to a diet of succulent vegetation while still nursing. Hatfield (1935) suggested that weaning might be variable as needs of the animals occurred. Thus, captive situations would not necessarily correlate with those of the wild.

Reproductive Potential. Bailey (1924) reported one captive female *M. pennsylvanicus* that produced 17 consecutive litters in one year (25 May–20 May), for a total of 83 young. One of her young, from the 25 May litter, produced 13 litters, totaling 78 young, before she was a year old. Thus, *Microtus* may be one of the most prolific mammals. There are species differences, and *M. pinetorum* "stands out for its conservation of reproductiveness" (Schadler and Butterstein 1979).

There are many kinds of natural control, which are incompletely understood. One of the most intriguing has been investigated by Negus and coworkers for the past decade and a half (Negus and Pinter 1965; Negus et al. 1977; Berger et al. 1977; Negus and Berger 1979). This related to (*a*) the stimulation of breeding by natural plant "estrogens," now isolated and "identified" (but not released), and (*b*) the inhibition of reproduction by naturally occurring phenolic plant compounds. How these operate in the natural ecosystem is not clear.

As a final statement regarding the reproductive phenomena of *Microtus,* Bailey (1924:534), with the insight of field experience, stated: "More than any group of small mammals, meadow mice hold the key to balance of natural adjustment for a large portion of our native bird and mammal population. They also have a vivid lesson for us in the struggle for life."

ECOLOGY

The ecology of the genus *Microtus* is intimately tied into grasslands. This relates to the conflict between man and mouse: mankind is invading many natural ecosystems, and in creating its own agroecologic systems is in a competitive situation with *Microtus*.

Evolution of the Grasses and Herbivores. During the Tertiary, and especially following the Eocene, grasses and mammals underwent rapid evolution and diversification (Durham 1974; Raven and Axelrod 1974; Dawson 1967). As the plants provided an abundant energy source, herbivores expanded in number and evolved into the ecological niches. The Perissodactyla and Artiodactyla both evolved in the Eocene, about 40–55 million years ago. The grass-eating marsupials probably arose 25–38 million years ago (Hume 1978). The Holarctic microtine rodents probably have a history no older than 10 million years, with rapid evolution and diversification for the past 6 million years (Repenning 1979). Thus, the various portions of the grassland niches filled.

Grasses became the most important of all flowering plants in a modern, economic sense, because of their nutritious grains and soil-forming function. Thus, they developed into the principal sustenance of mankind, of domestic grazing livestock, and of wild herbivores, including *Microtus*.

Two major adaptive characteristics (the masticatory system and the digestive system) have evolved in *Microtus* that have placed this genus in competition with humans. Both adaptations are related to the ability to eat grasses, including cereals, upon which both humans and voles subsist.

The first adaptive modification for a fibrivore was the ability to bite and masticate the tough celluloses. The toughness of grass is compounded into abrasiveness by the presence of silica cells or bodies that are in the leaf epidermis interspersed among the other cells (Pohl 1974). This is particularly critical in the small mammals, especially in *Microtus,* with their tiny teeth. The solution to this ecological problem was ever-growing teeth. The incisors and the molariform teeth are ever growing in *Microtus*. As they are abraded and worn down, they grow, providing a perfect solution to the problem. The wearing pattern provides both sharpness for the cutting function of the incisors and a ridged rasping surface for the grinding molars. The specialization of ever-growing teeth is of modern origin. The concept of evolution into the grass habitat fits well with this statement.

The second important adaptive modification was the physiologic need to digest complex carbohydrates (celluloses and hemicelluloses) contained in the fibrous portion of grasses. Most omnivores and carnivores are unable to digest these celluloses (Loosli 1974; Grassé 1973). Herbivores have evolved strategies that allow them to utilize this part of the energy stores (see the section "Genetics and Physiology").

The central theme of this critical cellulose digestion is gut fermentation by microorganisms. This may be further divided into foregut (stomach) fermentation (with the anatomic evolution of large sacculated chambers, the rumen reticulum) and hind gut fermentation (also with the development of large chambers, the cecum and proximal colon) (Hume 1978). The Artiodactyla and Marsupialia, in general, possess the foregut adaptation. As a rule, the Perissodactyla, Lagomorpha, and Rodentia have the cecum-proximal colon systems. Experimental data on many species are lacking, but correlation with the better-studied domestic animals provides clear evidence for these hypotheses (Eisenberg 1978; Vorontsov 1962, 1967).

Ecological Energetics. The damage done by mice to agriculture or to the environment may be considerable. Evaluating such damage has usually been on an empiric basis. How much damage is done? How many mice can be tolerated before the economic balance shifts to the debit side? How important are these small mammal pests?

These and many other questions demand answers. This involves the concepts of the "food chain," "energy web," "bioenergetics," or similar terms. The concepts are strictly in conformity with physical laws of thermodynamics; that is, energy can be changed from one form to another (first law) and there is a loss of energy no matter what transfer occurs (second law). Energy from the sun promotes photosynthetic plant activity. Small mammals that constitute an important component of primary consumers eat the plants. Plant foods are recycled by normal metabolic processes. The mice either are eaten by predators (the secondary consumers) or eventually die and are cycled by microorganisms into elemental parts available to the plants.

Competition between *Microtus* and humans exists in several portions of this simplified system. Especially during periods of high population, *Microtus* may eat, and thus remove from mankind's use, significant amounts of plant foods upon which humans directly depend. Cereal grains, root crops, and fruit trees are of particular importance. Even during times of moderate mouse populations there is a significant drain within the ecosystem that may be considered unacceptable. Add to this the competition with other primary consumers that are in mankind's reservation—domestic livestock and wild herbivores that he hunts or enjoys. The trophic impact of the same biomass of small mammals is considerably more than that of large herbivores, because of the physiologic needs for energy in a small-sized animal.

In forming a model of a mammalian energy budget, Grodzinski and Wunder (1975) included the following factors that are important considerations:

1. Body size. The smaller size of *Microtus* requires higher proportionate intake.

2. Environmental and nest temperature. The metabolism of the animal is described as a function of temperature.

3. Nest insulation. Subnivean or underground nests are very effective in preventing energy loss.

4. Huddling. Limited information suggests that this may be a general function in conserving energy during cold weather.

5. Sex. In the male, mating searches are significant; in the female pregnancy and lactation may increase metabolic cost by 50–70 percent.

6. Activity (needs for food gathering, in particular).

7. Light/dark. This is related to activity cycles; five to eight times as much energy may be needed during activity cycles.

8. Season. Acclimation produces modification in the animal stimulated by seasonal changes in the field. Heat conservation is apparently the most important modification.

Ecosystems and Niches. There are numerous well-recognized ecosystems in North America. *Microtus* occurs in most of these. Several species of *Microtus*

TABLE 16.3. Niche preference of *Microtus* as it relates to competitive and noncompetitive overlap and pest index

No.	Microtus Species	Niche Preference	Competitive Overlap (by reference number)	Noncompetitive Overlap (by reference number)	Pest Index[a]	Reference and Remarks
1	*abbreviatus*	Arctic or alpine tundra, areas grown to shrubs, burns	9	15	3	Rausch 1964.
2	*californicus*	Grasslands of California, annual	5	10, 14	1	Bailey 1936; Krebs 1966.
3	*canicaudus*	Grass-dominated meadows, pastures	5	10	2	Bailey 1936
4	*chrotorrhinus*	Rocky outcrops in deciduous forests, disturbed areas.	11	2, 12	3	Timm et al. 1977; Kirkland 1977
5	*longicaudus*	Rocky, grass-dominated wide range	2, 3, 6, 7, 11, 13, 14	10	3	Overlap agricultural areas but no reports of damage.
6	*mexicanus*	Pinyon-juniper zone, yellow pine forested highlands, grasses	5, 7		2	Findley and Jones 1962
7	*montanus*	Grass-dominated meadows agroecosystem, wide range	2?, 5, 10	10?	1	Serious irruptions reported (Piper 1909; Oregon State College 1959).
8	*ochrogaster*	Wet and dry grasslands agro-ecosystem	11	12	1	Martin 1956
9	*oeconomus*	Grass-dominated tundra, arctic and subarctic	5, 11, 15		1	May affect productivity of tundra ecosystem. Carries echinococcus (Rausch 1979).
10	*oregoni*	Grass-dominated meadow and forest, burns		2, 3, 4, 7, 13, 14	3	Gashwiler 1972
11	*pennsylvanicus*	Grass-dominated old field, marsh, bog mats	4, 7, 9?	12	2	Getz 1961
12	*pinetorum*	Dense grass, forbs, and brush; orchards, forests		4, 8, 11	1	Major damage to fruit orchards (Goertz 1971).
13	*richardsoni*	Grass-dominated waterside subalpine	5, 7	10	2	Overlaps agroecosystem (Racey 1960), affects alpine and subalpine ecosystem (Johnson and Johnson 1979).
14	*townsendii*	Grass-dominated meadow marshes, agroecosystem	3, 5, 7	2, 10	2	Overlaps agroecosystem. Carrier of giardiasis (Sheffield 1979).
15	*xanthognathus*	Grass-dominated taiga with microrelief, waterside.	9	1	3	Douglas 1977

[a]Most harmful is 1.

may exist in a single ecosystem, and when this occurs partitioning of niches can usually be recognized. Not infrequently, partitioning separates sympatric species; there have been questions of particular adaptations of these populations to certain microenvironments, or the possibility of exclusion by dominance of one species over another. Trapping experience has indicated considerable overlap, with more than one species being collected at a single trap station. The truth of the local situation can be found only by intensive study, with saturation trapping and sampling of various niches within the ecosystem. Even with this information, if the investigation is only during a single season, the conclusions may be confounded in subsequent years by population peaks, asynchronous cycles between species, subtle or drastic climatic changes. Also, other mammal or bird species may exert definite and recognizable pressures.

The observation is, therefore, that the following be investigated before making firm conclusions upon the ecological relationships of *Microtus* in any particular environment: (1) samples should be obtained from a sufficiently wide area to examine all niches; (2) numbers should be adequate for valid comparisons; (3) the conditions of the environment should be stable within the parameters of natural fluctuations; and (4) the data base should extend over each season during several consecutive years.

Table 16.3 summarizes the niche preferences of various species of North American *Microtus*. Central American forms, except *M. mexicanus,* are omitted because of lack of data. The niche preferences also will indicate those species that are potential or actual pests, and a ''pest index'' (on a value of one to three, one is most harmful) can be applied to each species.

The role of competition in determining apparent niche preference requires more study. In a balanced system of stable populations, the communities of plants and animals appear to coexist. However, underlying are the dynamic ebb and flow of population numbers, resource limitation, behavioral characteristics, and multiple other factors that confound attempts at modeling and prediction of even the most simple system. Studies such as Randall's (1977) on *Microtus montanus* and *M. longicaudus* indicated some of the subtleties that exist.

Runways, Nests, and Burrows. Most species of *Microtus* construct networks of well-kept runways on the surface of the ground. Fresh cuttings indicate the species and parts of the plants that are being eaten. Areas of broadening and a tamped-down appearance of the ground show that extra time is spent in certain spots. Concentration of scats is usually found; this points to some degree of sanitation. In a busy area, however, fecal pellets may be found scattered throughout. Some reingestion occurs, but this appears minimal under most circumstances. Main roads and secondary highways may be identified by gross examination, and the exploratory efforts away from the obvious trail can be identified.

All species appear to burrow to some degree. Several species, such as the pine vole and the Oregon vole, are almost wholly subterranean. Others, such as the long-tailed vole, show little evidence of trail building.

Nest building is well developed. The nests are used as nurseries, resting areas, and protection against environmental excesses. These are subterranean in placement, in an enlarged nest chamber, and constructed with dried grass and plant fibers. They are frequently placed under rocks or logs. In mankind's domain, boards, fence posts, hay bales, and brush piles are normal attractants. Subnivean nests are frequently found in snow country, usually placed against some natural bulwark.

It is likely that most of the time is spent in the nest, once a home range is established. Barbour (1963) found that in February, one *M. ochrogaster* spent only 175 minutes out of 24 hours away from the nest. These 175 minutes encompassed 15 activity periods, with an average duration of 11.7 minutes, during daylight hours.

Competition. This ecologic interaction is defined here as two types of organism competing for the same resource. It primarily relates to food, but also to mating, water, and space. Predation is not considered in this definition.

Microtus, being a small mammal, is restricted by the continual need for food, water, and protection from the environment, including predators. It cannot travel long distances safely, and the more stable populations are those with continuous acceptable habitat.

Interspecific Competition, *Microtus/Microtus.* This is one of the more interesting relationships in an ecologic sense, because various species of *Microtus* either have identical niches or there is a broad overlap of two or more species of *Microtus* (see table 16.3). Exacting data need to be obtained.

In cases of identical niche occupation by species of *Microtus,* with overlap, there is some apparent give and take. One species may, in general, be dominant and thus maintain its same niche, while the other may alter its preferred microhabitat, even over extensive areas. The factors involved in this arrangement are not completely known: the temptation is simplistically to state a dominance/submission condition. This does not look at the real community as it has evolved, and does not consider that a more complicated behavioral complex is at work. Randall (1977) uses the term *withdrawal,* which may better describe the mechanisms of interreaction in part.

In many cases where several species of *Microtus* occur in overlapping areas, a definite niche difference can be demonstrated. These are presumed to be situations of coevolution and a dynamic competition situation may be, in fact, continuously at work.

Home Range. *Microtus* in general are found to conform to the general principles of home range and are considered to be territorial. Because individuals are attracted to and live in a particular ecological niche, partitioning is required so that each animal will have its share of the resource. This is accomplished by establishing the home range. Home range is defined as an area used by the animal in food gathering, mating, and rearing its young. Home ranges are usually irregular because of land form and vegetation characteristics, and overlapping is common. Within the home range, for territorial species there is a smaller region that an individual will actively defend against conspecifics. In the home range or territory a single mature animal may exist. Female voles have smaller home ranges. Males invade female territories at breeding times and may be accepted by estrous females for mating purposes. The female and litter then occupy the home range until dispersal. Progressive litters may overlap, so two generations occasionally may be present in a single home range at one time. During the active breeding season dispersal generally occurs as the young reach sexual maturity or when a new litter is produced. The young mice then must find their own home ranges and territories. However, as the breeding cycle wanes, especially as winter approaches, the juveniles may stay in a group until the next breeding season. There is some evidence to suggest that monogamy may occur in some species of *Microtus* (*californicus* and *ochrogaster*) and that a social system of small groups centered around one adult pair may direct the home range behavior (Lidicker 1979).

The mouse has obvious advantages provided by a home range. It has all the necessities of life close at hand, without subjecting itself to predation outside of a familiar, protected area. Dispersal is advantageous for certain individuals and helps mitigate the total predation of a colony. This also extends the population to the next suitable habitat. In *Microtus*, population cycling complicates this simplistic balance. In most, if not all species, there are dynamic shifts from times of much unexploited area to times of saturation. At these times, behavioral and genetic changes occur that may be evolved mechanisms especially beneficial for dispersal of the species (Krebs et al. 1976).

Determination of home range is by observation of individuals' locations and by plotting on a map. In general with *Microtus*, live-trapping and release has been the optimal method. Radio tagging, including the use of radioisotopes (Harvey and Barbour 1965), has also been used and can be very effective. Methods of calculation vary, and, with experience, fewer data points may be used to secure a satisfactory calculation of the home range area. However, the more points of data obtained, the more understanding of total range, times, patterns of use, and atypical explorations will be possible.

Cycles. Three general categories of *Microtus* populations are of basic scientific and management interest: (1) low population, (2) high population, and (3) irruptions. Dramatic changes of numbers of mice have evoked great biologic interest. The economic relationship to high populations and irruptions of the voles is

an important consideration (see the section ''Economic Status and Management'').

Cycling of populations (both regular, seasonal in type and less regular, irruptive in type) is a nearly universal biological phenomenon in all forms of life. Many cycles go unnoticed, but those of several species of *Microtus* may be in direct conflict with agricultural pursuits. These, thus, become obvious during the high points of the cycle. The herbivorous diet, the wide distribution of the species, the high reproductive rate, and diurnal activity all bring these dramatic increases in numbers of meadow mice to the attention of the farmer.

This is illustrated by one population of *M. pennsylvanicus* reported by Golley (1961). In May, the density was very low. There was an excess of food and cover and little competition. Birth rates were high and the population increased rapidly throughout summer and fall. Breeding ceased in January and did not reoccur until March. Immigration continued to increase the density through February. Mortality was highest in the summer and least in the winter, and was considered to be related to the age composition of the population and a protective snow cover. Mortality rates (61 percent) were highest in postnestling juveniles (11 to 20 g) and were similar (58 percent) in young adults (21-30 g). A mortality rate of 53 percent persisted in the older age groups. The heaviest weight class (over 51 g) averaged over 104 days old and had a mortality rate of 100 percent with an expectation of life of 12.8 days. In the nestlings (0 to 10 g) the mortality rate was lowest (50 percent).

Arvicolids have been well studied in both Eurasia and North America. They are good laboratory subjects; they can be kept satisfactorily in natural and controlled environments. But there is still not a generally accepted reason for cycling in *Microtus*.

To test the hypothesis that behavioral interactions are responsible for the decline, Boonstra (1977) created a low density by removal of *M. townsendii* during and before population declines in order to increase survival. The life expectancy of males was 7 weeks longer on the removal grid than on the control grid, and in females it was 4 weeks longer. The study also revealed the expectation of life to be 21.3 weeks for males and 19.6 weeks for females. These kinds of studies generally clarify some of the potential factors involved in population regulation, but point out the difficulty in proving the cause-and-effect relationship.

In North America, population peaks occur at intervals of about 4 years, not in catastrophic numbers, and not at all regularly. The spectacular mouse irruptions, however, are at longer intervals and do not occur in all species. These population explosions are preceded by a year of high population. The period of swarming during an irruption is rarely longer than a single year. Typically, subsidence is dramatic, during the winter and spring following the maximum population. Then follow several years of population depression. In cases of widespread irruptions, there may be an irregularity of subsiding in different areas; some pockets of high population density may remain. Irruptions have been recorded no more often than every 8 to 10 years in any one area. The major biological benefit to the species would appear to be that of forcing maximal dispersal, to give a spread of range and colonizing of previously uninhabited areas. Additional benefits may exist, such as protection of the individual simply by being a member of a large group or selection for the genotype of better fitness under stress; the hypotheses of group selection and natural selection can each be supported by these kinds of considerations.

FOOD HABITS

Several methods of studying food habits have been applied to the voles. Direct examination of cuttings and food caches, stomach and intestinal contents, and fecal pellets has been used. In reviewing many of these reports, it is apparent that (*a*) agricultural products used for food by humans and domestic animals are also prime food for *Microtus*, hence, it is a pest; (*b*) there is a large spectrum of food items used by most species of *Microtus*; (*c*) the food items may change in relative proportions and species from season to season; (*d*) the food items may change from year to year; (*e*) food items may differ from one part of the range to another; (*f*) there may be subtle differences (niche differences?) between sympatric species; and (*g*) much reported about food habits is inconclusive—the state of the art is in need of more extensive evaluation.

The amount of edible energy sources varies with the ecological niche and is directly related to the microhabitat requirement of *Microtus* species. In forests, the estimated available food for herbivorous grazing rodents in only 4–13 percent of the plant biomass, but in the grasslands, including cultivated fields, nearly 100 percent is considered potential food. In all ecosystems, the corridor and edge effect of streams, lakes, and other natural openings and breaks must be considered. In agricultural areas, the intermittency of land use by plowing and crop rotation, and the potential highway effect (upon native mice) of roads, drainage, and other changes all provide energy sources and potentially may affect the distribution and spread of populations of *Microtus*. Modern highway rights-of-way have been found to be an excellent source of food for many species of animals (Schmidley and Wilkins 1977). *M. ochrogaster* populations were greater within rights-of-way (Michael 1976). Roads may act as avenues of dispersal for individuals and populations because of this.

The role of food preference is important, especially during periods of high population (Batzli and Pitelka 1970). It was found that during periods of densities of 387 *M. californicus* per ha, the major food plants, where the mice existed, contributed 85 percent less volume to vegetation than was present in the exclusion control area, and the seed crop was reduced by 70 percent in the grazed area. These kinds of studies indicate a much more complicated interrelationship

with the ecological community than has been considered in the past.

The levels of digestibility and assimilation, as well as the total daily input, must be known in computing ecologic impact. Information for grazing herbivores, including voles, indicates a daily intake of 30–35 percent of body weight per 24 hours (Ognev 1950), and a relatively low digestibility/assimilation rate of 65–67 percent, 2–3 percent energy cost as urine, and about 30 percent as fecal loss. Seed eaters have a much higher assimilation coefficient of 88–90 percent (Grodzinski and Wunder 1975).

To complete the involvement of the meadow mice in the energy chain and at the next trophic level, it is apparent from many studies that *Microtus* may be the most important as well as the preferred item by many of the secondary consumers. Both avian and mammalian predators are involved. Pearson (1964) recovered 3,782 *M. californicus* from carnivore scats on a 14.5-ha study plot. This was a "catastrophic reduction" in which the meadow mice were nearly annihilated by the predator cadre, consisting of feral cats (*Felis catus*), gray foxes (*Urocyon cinereoargenteus*), raccoons (*Procyon lotor*), striped skunks (*Mephitis mephitis*), and spotted skunks (*Spilogale putorius*).

Agriculture Relationships. Directly related to potential pest effect and the food habits of *Microtus* are agricultural practices, including the selection of plant species placed in the agroecosystem. This is further related to seasons and food availability and population numbers.

A general vegetable and cultivated plant classification (Warid 1974) indicates a broad spectrum of potentially damaged crops:

1. Root vegetables (carrots, beets, turnips, sweet potatoes). These are frequently used by voles in small home and truck gardens, especially when easy access is provided by mole (*Scapanus* sp.) tunnels.

2. Stem vegetables (asparagus and kohlrabi). Not usually included in the vole diet.

3. Tuber vegetables (potato, yam). As item 1, these may become an item of vole food.

4. Leaf and leafstalk vegetables (cabbage, lettuce, celery, spinach). These are naturally exploited items, and are frequently used as captive dietary items.

5. Immature inflorescent vegetables (artichoke, broccoli, cauliflower). Not usually included in the vole diet.

6. Fruits used as vegetables.
   a. Immature (bean, okra, sweet corn).
   b. Mature (melon, pumpkin, tomato).
Not usually included in the voles' diets, but under the stress of high populations some items may be included.

7. Fruit trees (bark and root systems). May be seriously used, even killed, by overwintering populations, especially *Microtus pinetorum*, *M. ochrogaster*, and *M. montanus*.

8. Pasture, grasslands, hay crops, and grain. These plant foods used by humans and domestic herbivorous animals can be directly transposed from the

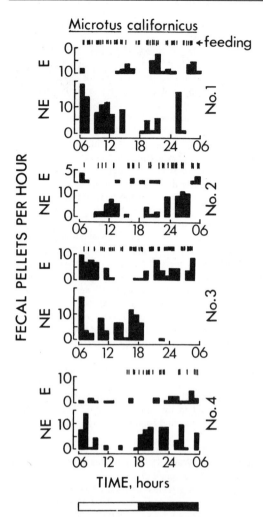

FIGURE 16.8. Daily temporal pattern of coprophagy in the California vole (*Microtus californicus*): individual records for four animals for a 24-hour observation period. Vertical bars show the number of pellets eaten (E) and not eaten (NE) in each hour. The single vertical lines indicate quarter-hour periods during which feeding occurred. The black and white horizontal bar at the bottom represents the daily light-dark cycle. Animals 1 and 2 received lettuce only; animals 3 and 4 received lab chow and water. From Kenagy and Hoyt 1980.

natural food of *Microtus*. There is usually some continuous drain upon the food energy produced. These plant species may become the primary food item for *Microtus*, especially during population highs, and may be dramatically destroyed during population explosions.

Small gardens, as compared to those of larger area, are at increased risk associated with ecotone, because of the nearness of the pests and their natural homes.

Underground Storage. In 1804, Lewis and Clark (cited by Bailey 1920) found and recorded the Indian custom of raiding the underground stores of native

mice for food. Bailey (1920) reported that both Indians and whites greatly prized these stores. He identified the foods as, among others, the beans produced on the underground shoots of a trifoliate bean vine (*Falcata comosa*) and artichokes as tubers of a wild sunflower (*Helianthus tuberosa*). The mouse was identified as *Microtus pennsylvanicus*.

Microtus abbreviatus also stores root parts in the arctic, where they have long been used by the Eskimos. In September 1959, we collected these voles and found a gallon-sized food cache. Whenever the mouse's storehouse is raided, Eskimo natives leave a bit of seal oil on a leaf so that next year another cache will be made. Seal oil was traditionally the most prized item the Eskimos extracted from their environment.

It is likely that other species of *Microtus*, including *M. oeconomus*, have evolved this old fashioned but admirable trait to aid in winter survival, though Ognev (1950) states that the Eurasian form has not.

Reingestion of Feces. Coprophagy occurs in many species of mammal. It has been ascribed as an evolved trait to the need for vitamins produced in the gastrointestinal tract, and to the need to use the original feedings more efficiently, after partial digestion by symbiotic organisms in the lower bowel. Kenagy and Hoyt (1980), working with *M. californicus*, and Ouellette and Heisinger (1980), observing *M. pennsylvanicus*, gave quantitative data regarding this interesting feeding behavior.

M. californicus reingests about one-quarter of its feces in a series of rhythmic, short-term alternations of one to several hours duration between reingestion and nonreingestion (figure 16.8). This correlates with the day and night foraging pattern. *M. pennsylvanicus* reingests primarily in the period of inactivity following the cessation of feeding, with some overlap, and in a rhythmic manner. If possible, these observations should be enlarged to include wild conditions of diet, and to quantify during various seasons and under various kinds of physiologic stress.

In general: (1) *Microtus* species feed primarily on green succulent vegetation; (2) they may be selective plant feeders (Zimmerman 1965) or general plant feeders (Whitaker and Martin 1977); (3) they frequently eat items other than green plants, including roots, bark, fungi, insects, and meat; (4) there are species differences in food habits; and (5) coprophagy occurs as part of the feeding behavior.

BEHAVIOR

Knowledge of field behavior of voles is important in preventing damage and protecting agricultural lands. Likewise, during population surges or in analyzing and avoiding public or animal health problems the responses of mice to control measures should be understood.

Microtus are active all times of the year and may be out day or night. These generalizations may be modified by differences imposed by species, season, habitat, cover, temperature, and other factors. There is typically intermittent activity. They must eat at frequent intervals, for the volume is large in proportion to the energy budget realized. It is not unusual to see them during the daytime. Catches on trap lines indicate that nocturnal activity in general predominates, and this is verified by instrumented studies. Dawn and dusk are periods of maximum activity for most species, but there are many mitigating factors, as noted. Getz (1961) found less diurnal activity during hot summer days.

This mouse is adaptable to human society, this being a part of its evolved behavior. Vernon Bailey (1924:525) (one of the great and experienced small-mammal biologists with the old U.S. Biological Survey) delightfully stated his views on *M. pennsylvanicus:* "While not highly endowed with even rodent intelligence, they are quick to adapt themselves to the food supply that is available, to the best possible means of protection from enemies, and to make the best of what, from the point of view of even a squirrel, might seem a very humble existence."

Locomotion. These animals generally move at a run. This may be intermittent and appears to be related to an inherent need for some protection, as they have spurts of speed over open areas and may pause in sheltered spots. There is none of the bounding, jumping progression that most other mammals exhibit. In captivity, there is little of the stereotyped captive pacing. As long as they have food, water, and space they are restful, though some species and individuals (as with other animals) are hyperactive.

Most voles climb well, especially such species as *M. longicaudus*. Trees and bushes present no difficulty to many voles.

Most *Microtus* species swim well. Several occupy well-known amphibious niches, especially *M. richardsoni*. *M. pennsylvanicus* was reported to swim up to 1120 m in cold salt water in Newfoundland (Riewe and Pruitt 1968). *M. townsendii* and *M. longicaudus* are semiaquatic in some habitats. *M. californicus* was observed on one occasion to swim a 12.2-m slough, diving to a depth of over 1.2 m and swimming 6.7 m underwater (Fisler 1961).

Vocalization. At times, much high squeaky sound communicating is heard in captive groups, which occasionally can be detected in the field. Young animals especially appear to be more vocal. Newborn nestlings have a barely audible, breathless chirping that increases in vigor day by day. Older animals interact with a variety of neutral to agonistic chirps, squeals, and little growls. They are most vocal during altercations and indeed this may be accompanied by considerable bloody wounding. Chattering of teeth is frequently heard as they interact.

Senses. There are no particularly developed senses, nor is there any lack of usual mammalian sensitivities in *Microtus*. Touch is well developed, even in neonates (Hatfield 1935). Heat and cold likewise give responses. Contact with a warm object stimulates nuzzling in the newborn. Vision is well developed for short

distances and in response to shadows and movements; before the eyes open there is a phototactic response. Hearing is probably the best-developed sense. "Screeping" (imitating the distress sound) produces immediate response in most young and adult animals. The ear canals do not open until about day eight; though some sound reaction can be elicited before this time, good reactions to usual sounds do not occur until several days later. Olfaction is also relatively keen and mice can be attracted by particular odors and respond predictably under experimental conditions.

MORTALITY

Avian Predators. The genus *Microtus* occurs in a variety of habitats, most of which are relatively open, grassy regions. These areas are ideal for hunting by avian predators. Many food habit studies of predatory birds have been done, and mice of the genus *Microtus* are the most common food item reported.

Population cycles of *Microtus* greatly influence the abundance of the hunting birds; during periods of high population large numbers of birds may invade the region. Usual avian predators are hawks and owls.

There are numerous reports to indicate the relationship of avian predators. In southern Idaho, 66.3 percent of the total prey and 80.0 percent of the biomass of the barn owl (*Tyto alba*) was *Microtus montanus* (Roth and Powers 1979); for the great horned owl (*Bubo virginianus*), the figures were 52.9 percent and 48.2 percent, respectively. In the same region, Sonnenberg and Powers (1976) found that *Microtus montanus* was also the preferred prey species (53.7 percent) of the long-eared owl (*Asio otus*). Maser et al. (1970) in central Oregon, on a yearly basis, found that *Microtus montanus* made up 13.4 percent of the prey species of the great horned owl, 3.6 percent of the long-eared owl, and 9.8 percent of the short-eared owl. This was in an area of low population of *Microtus*. Fitzner and Fitzner (1975) reported that *Microtus montanus* made up 65.5 percent and *Microtus longicaudus* 8.0 percent of winter food items of the short-eared owl in the Palouse prairies of eastern Washington. Seidensticker (1970) reported 8 percent occurrence of *Microtus* sp. among 28 kinds of food item during the nesting season of red-tailed hawks (*Buteo jamaicensis*) in Montana. Rageot (1957) found that the barred owl (*Strix varia*) took *M. pennsylvanicus* and *M. pinetorum* in Virginia.

Bent (1937, 1938) cited numerous examples of hawks and owls feeding upon *Microtus* in many parts of their ranges. Almost all species of raptor take *Microtus*. Other birds not usually considered predators of mice may actively catch and eat *Microtus*, especially during times of mouse plagues. Sea gulls may be attracted in great numbers. Piper (1909*a*) showed a photograph of gulls hunting mice in an alfalfa field during a mouse plague in Nevada. Craighead (1959) identified excessively high counts of California and ring-billed gulls (*Larus californicus* and *L. delawarensis*) during an Oregon mouse irruption. *Microtus montanus* was the vole species involved in these two reports. Pellet analysis indicated that the diet of the California and ringed-billed gulls was 99 percent *Microtus*. Gulls may move long distances from their usual residence areas of the large inland lake and coastal regions.

Other species of bird are observed predators of *Microtus*, especially during periods of mouse irruption. This list includes the Northern shrike (*Larius borealis*) and magpie (*Pica pica*) (Craighead 1959), crow (*Corvus brachyrhynchos*), raven (*C. corvax*), white-necked raven (*C. cryptoleucus*) (Bent 1946), and great blue heron (*Ardea herodias*) and bittern (*Botaurus lentiginosus*) (Piper 1909*b*; Jewett et al. 1953).

Mammalian Predators. Natural mammalian enemies of *Microtus* include the full spectrum of predators (Maser and Storm 1970). Those that eat voles as a major item of food are short-tailed shrews (*Blarina brevicauda*) (Eadie 1952), badgers (*Taxidea taxus*), coyotes (*Canis latrans*), foxes (*Vulpes vulpes*), skunks (*Mephitis*), weasels (*Mustela* sp.), and wildcats (*Felis rufus*) (Piper 1909*b*), as well as other carnivores. As a general rule, predators are opportunists; during periods of low population they will shift their diet to other items. During high populations they may eat little but the readily available *Microtus*.

Other Predators. Other vertebrate predators are well known to take *Microtus*, either in accidental encounters or, as in the case of snakes, by evolved deliberate hunting strategies. We have observed the following to take or contain ingested voles: trout (*Salmo* sp.), Pacific giant salamander (*Dicamptodon ensatus*), garter snake (*Thamnophis* sp.), western yellow-bellied racer (*Coluber constrictor*), gopher snake (*Pituophis melanoleucas*), rattlesnake (*Crotalus viridis*), and rubber boa (*Charina bottae*).

Parasites. Fleas (*Siphonaptera*) are well-known ectoparasites occurring on a variety of animals. Numerous species have been reported from *Microtus*. They tend to be host specific, at least to the genus level of the host. However, because they are so mobile, they frequently occur on other genera. If an animal dies the fleas soon leave and may temporarily engage the next mammal that comes along. For this reason predators often have "rodent fleas" upon them. Because *Microtus* runways are used by many other small mammals, there also is a good chance of temporary exchanges of fleas between species and genera.

Nests may harbor species of flea not readily found on examination of the host because of the protected and stable environment. According to Hubbard (1947), many more fleas may be found in nests than on animals. However, compared to other small mammals, *Microtus* nests seldom have many fleas in them; the mice, too, often are entirely without them.

Fleas are important as vectors of disease, being able to ingest infected blood and then inject it into an uninfected animal. The medical import for humans is related to plague, tularemia, and possible flea allergies.

Other ectoparasites (mites, ticks, lice, and others) are less well known. Reports are fragmentary and scattered and there are many lapses of information related to their biology. As part of the ecological community in which *Microtus* exists, there is much to be learned.

Because information is scattered over many publications and throughout much time, our listing of ectoparasites (table 16.4) is not to be considered complete. It will, however, be helpful as an indicator of the potential of this subject and provide a start for more complete studies.

TABLE 16.4. Ectoparasites of *Microtus*

Fleas
  Amataraens penicilliger (1)
  Atyphloceras multidentatus (1)
  Catallagia charlottensis (1)
  Catallagia dacenkoi (1)
  Catallagia dicipiens (2)
  Corrodopsylla curvata
  Ctenophthalmus pseudalgyrtes
  Delotelis telegmi (2)
  Doratopsylla blarina (2)
  Doratopsylla jellisoni (2)
  Epitedia jordani (1)
  Epitedia scapani
  Epitedia stewarti
  Epitedia wenmani (1)
  Hystrichopsylla gigas (1)
  Hystrichopsylla orophila
  Malaraeus (Amalaraeus) penicilliger
  Malaraeus telchinum (2)
  Malaraeus doddsi (1)
  Megabothris abantis (1)
  Megabothris asio (2)
  Megabothris calcarifer (1)
  Megabothris groenlandicus (1)
  Megabothris quirini (1)
  Monopsyllus ciliatus (2)
  Monopsyllus wagneri (2)
  Opisodasys keeni (2)
  Peromyscopsylla catatina
  Peromyscopsylla hesperomys (1)
  Peromyscopsylla ostsibirica (1)
  Peromyscopsylla ravalliensis (1)
  Peromyscopsylla selensis (1)
  Stenoponia americana
Mites
  Androlaelaps fahrenholzi
  Haemogamasus alaskensis
  Haemogamasus ambulans
  Haemogamasus barberi
  Haemolaelaps glasgowi
  Demacarus sp.
  Dermacarus hypudaei
  Laelaps alaskensis
  Laelaps clethrionomydis
  Laelaps kochi
  Laelaps microti
  Laelaps muris
  Mycoptes sp.
  Mycoptes muscuinus
  Neotrombicula harperi
  Neotrombicula microti

TABLE 16.4—*Continued*

Pygmephorus sp.
Radfordi lemnina
Ticks
  Ixodes angustus (1)
  Ixodes dammini
  Ixodes muris
  Dermocentor andersoni (1) (nymph stage)
Lice
  Anopleura sp.
  Hoplopleura acanthopus
Others
  Bot fly (*Cuterebra* sp.)
  Nest beetle (*Leptinus testaceus*)

SOURCE: Hubbard 1947; Judd 1950; Benton 1955; Wilson 1957; Golley 1963; Rausch and Rausch 1968; Buech et al. 1977; Timm et al. 1977; Haas and Barrett 1978; Healy 1979; McDaniel 1979; Whitaker 1979.

NOTE: Those commonly found are marked (1), those considered accidental or unusual are marked (2). Unmarked species are indeterminate from the literature at hand.

PARASITIC WORMS. Helminths characteristically found in *Microtus* spp. belong to genera that are represented in rodents of various families. In members of the family Arvicolidae, helminths frequently exhibit little host specificity at the generic level, and some species may occur in two or more rodent genera. Qualitative differences in the helminth faunas of *Microtus* spp. seem usually to be attributable to ecologic factors that influence the distribution of various parasite-host assemblages. Nonetheless, voles of some species differ in degree of susceptibility to infection by certain helminths.

As indicated by both diversity of species and numbers of individuals, cestodes and nematodes predominate among helminths occurring in *Microtus* spp. Trematodes are less common. Acanthocephalans are generally rare and probably accidental in occurrence, although Benton (1955) reported *Moniliformes clarki* as common in *Microtus pinetorum* in New York. Most of the helminths of voles are the adult (sexually reproducing) stage and inhabit the lumen of the alimentary canal; those in the larval stage usually localize in the liver or in other organs.

Helminths for which voles serve as the final host may have direct cycles (some nematodes), or intermediate hosts are involved (cycles of cestodes, trematodes, and some nematodes). Voles themselves serve as the intermediate host of some helminths, the most important of which are cestodes that occur in the strobilar stage either in carnivores (canids or mustelids) or in hawks or owls. Completion of the cycles of these cestodes is favored by the predator-prey relationship existing between the respective final and intermediate hosts.

The most common cestodes in *Microtus* spp. represent the subfamily Anoplocephalinae, family Anoplocephalidae. As far as is known, oribatid mites serve exclusively as intermediate hosts of these cestodes. Voles become infected by the incidental ingestion of the mites while feeding. Also relatively common are cestodes of the family Hymenolepididae, of which insects serve as intermediate hosts. These cestodes are placed in the genus *Hymenolepis*. Additional genera are recognized by helminthologists in the Soviet Union.

The common anoplocephaline cestodes of voles represent the genera *Anoplocephaloides*, *Paranoplocephala*, and *Andrya*. The genera *Paranoplocephala* and *Andrya* are distinguished by differences in the early development of the uterus (cf. Rausch 1976). The most common species of *Anoplocephaloides* is *A. troeschi*, which inhabits the terminal part of the ileum, immediately above the ileocecal junction. Another species, perhaps *A. variabilis*, is less common; it is found more anteriorly in the small intestine. *Paranoplocephala macrocephala* is a common cestode in *Microtus* ssp. in North America. This species has been reported also from voles in Eurasia, but it appears now that these cestodes represent distinct species in the Palearctic. *P. omphalodes* is a Holarctic species, widely distributed in Eurasia but occurring in only northwestern North America, where its range corresponds to that of voles of comparatively recent Palearctic origin (e.g., *Microtus oeconomus*). *Andrya primordialis* is a rather common cestode in *Microtus* spp. in North America. Other species apparently referable to *Andrya* occur in voles in Eurasia. For detailed information concerning cestodes of *Microtus* spp. in Eurasia, see Ryzhikov et al. (1978).

Voles and other small rodents serve as intermediate hosts for several cestodes of the family Taeniidae. These represent the genera *Taenia* and *Echinococcus*. All of the species considered here are Holarctic, and some occur most commonly in the Arctic and Subarctic. *T. crassiceps* and *T. polyacantha* are parasites of foxes (*Vulpes* and *Alopex*). Their larval stages, which reproduce asexually in the intermediate host, are found in peritoneal cavities or in tissues of voles. *T. mustelae* and *T. martis* are host-specific parasites in the strobilar stage of mustelids. The larvae are found characteristically in the liver of the intermediate host, but other tissues or organs may be involved. In inhabited areas, the larval *T. taeniaeformis* is often found in *Microtus* spp., in which it forms a large cyst at the hepatic surface. The adult cestode is a common parasite of domestic cats. Larvae of other taeniid species may sometimes occur in voles, but are not common. Additional information concerning the taxonomy of *Taenia* spp. is available in the revision by Verster (1969).

Microtus sp., as well as rodents of other genera, serve as the intermediate host of *Echinococcus multilocularis*, of which the strobilar stage occurs typically in foxes (*Vulpes* and *Alopex*). Normal development of the adult cestode also takes place in domestic dogs and cats, if they capture and eat infected rodents. The larval cestode develops in the liver of the intermediate host, often causing extreme enlargement of the organ. With massive infections, movement of the rodent is hampered, probably making them more vulnerable to predation. *E. multilocularis* is of particular interest in that the larval stage also develops in the human liver, causing a severe and frequently fatal disease (alveolar hydatid disease). The cestode is a common parasite of foxes in arctic and subarctic regions, where the disease occurs also in indigenous human populations (e.g., northeastern Siberia and Alaska). *E. multilocularis* was first observed in the contiguous United States in 1964, when it was found in foxes in North Dakota. Since that time, the cestode has been found in rodents or foxes in five additional states and in three provinces of Canada (Rausch 1979). The first human case was reported recently, in a resident of Minnesota (Gamble et al. 1979). Because of the general presence of suitable hosts, continuing spread of this cestode in the United States is anticipated.

The families Paruterinidae and Dilepididae include cestodes whose larval stages occur in voles and other small mammals. Certain species of *Paruterina* are host-specific parasites of owls, whereas cestodes of the dilepidid genus *Cladotaenia* occur only in birds of the order Falconiformes. Both have cysticercoidlike larvae that are found in the liver or other organs of voles and other smaller rodents that are preyed upon by hawks and owls.

The helminth fauna of *Microtus* spp. includes a large component of nematodes, of which relatively few occur commonly in these rodents in North America. Nematodes of the genus *Heligmosomoides* family Heligmosomidae, inhabit the small intestine of voles and other rodents throughout the Holarctic. The cycle is direct, involving the ingestion of the embryonated egg and subsequent development taking place in the small intestine. The taxonomy and distributional history of the heligmosomes have been discussed in detail by Durette-Desset (1971). One species of *Heligmosomoides*, *H. polygyrus* (formerly known as *Nematospiroides dubius*), has been intensively studied under laboratory conditions.

Nematodes of the oxyurid genus *Syphacia* inhabit the cecum of voles and other rodents. *S. nigeriana* is a Holarctic parasite of *Microtus* spp. (Quentin 1971). In earlier publications, this nematode was usually reported erroneously under the name *S. obvelata*, which is a parasite of murine rodents (cf. Rausch and Tiner 1949). *S. nigeriana* is one of the most common parasites of voles in North America, often occurring in large numbers in the cecum of the host. The cycle is direct. More common nematodes that require intermediate hosts represent the genera *Pterygodermatites* (Rictulariidae) and *Mastophorus* (Spiruridae). *P. microti* is found in the small intestine of *Microtus* spp. in northwestern North America. *M. muris* is a Holarctic species that occurs widely in rodents. It is a large nematode that typically inhabits the stomach of the host. Insects serve as intermediate hosts for both of these nematodes. *Capillaria* spp. occurs in many species of rodent, showing high organ specificity but low host specificity (Dunaway et al. 1968). It has been reported in several species of *Microtus* as a gastric worm; in laboratory rat experiments *Capillaria* ssp. can induce squamous cell carcinoma.

As was remarked earlier, *Microtus* spp. harbor comparatively few trematodes. A netocotylid, *Quinqueserialis quinqueserialis,* is common in voles in wet habitat, where the metacercariae encyst on emergent vegetation. The trematodes inhabit the cecum of the final host, and may be very numerous. *Mediogonimus ovilacus* (Plagiorchiidae) is a parasite of *M. pennsylvanicus.* It occurs in the bile ducts of the host. Voles occupying wet habitat may also become infected by a schistosome, *Schistosomatium douthitti,* the adults of which occur in the veins of the portal circulation. This trematode is of unusual interest because its cercarial stage is a cause of dermatitis in humans. Trematodes occurring in *Microtus* spp. in Eurasia have been discussed by Ryzhikov et al. (1978).

Diseases. One important deleterious effect of *Microtus* on humans is the transmission of several diseases. Some of these are fatal to humans and to *Microtus* and other native fauna. Some are little known and their potential effects as lethal diseases on mouse and man cannot be stated. Others are better known, including cystic hydatid disease, tularemia, and plague.

DISEASES DUE TO MICROORGANISMS. Tularemia occurs as an endemic disease in several regions of the west and far north. This is a serious illness in humans and has been reported in association with rodents in Europe, Asiatic portions of the U.S.S.R., and the United States. Water supplies may be involved (Osgood et al. 1959). It has been reported from *Microtus,* among other mammals, especially during periods of high population. Kartman et al. (1959) reported an infection rate of 43 in 126 *Microtus montanus* collected in Oregon. The role that this disease may play as an effective agent of destruction of the mice during a mouse irruption is not clear, but dead mice were positive for tularemia organisms. Live voles challenged with known dosages of *Pasteurella tularensis* showed a high susceptibility to the disease. However, the oral route of infection, assumed to be the method of natural infection, required 1,000 to 100,000 times as many organisms as a subcutaneous injection. The suspect route of infection during the winter (when there are few ectoparasites) was cannibalism.

Rausch et al. (1968) isolated *P. tularensis* from *M. oeconomus.* The isolate was found to resemble most closely the less virulent Eurasian strains. They suggested that this strain of organism was responsible for a relatively high rate of subclinical tularemia in humans in northern and western Alaska, as indicated by serological tests.

An interesting potential for spread to domestic water supplies has been reported by Craighead (1959): not only may infected mice contaminate the water directly, but they may do so secondarily when caught by hawks or eaten as scavenged dead animals by gulls. The possibility of being carried by birds, with resultant long-range spread, exists in these situations.

Plague bacillus, *Yersinia pestis,* has been reported intermittently from *Microtus.* Isolates from *M. californicus* have been found repeatedly in California (Hubbard 1947). Kartman et al. (1959) reported the isolation of the plague bacillus from *M. montanus* in eight instances during the Oregon irruption of 1958. No epizootic or human infection resulted.

Additional examination for disease other than tularemia and plague, including brucellosis and leptospirosis, potentially hazardous to humans was negative during the 1957–58 Oregon irruption of *Microtus* (Osgood et al. 1959).

Organisms of various kinds have been reported to occur in *Microtus*. In general, these reports are not the result of long continued research, and in-depth quantitative data are not available. Most, but not all, organisms are assumed to be symbiotic in nature and of little significance to the hosts or to the public health of humans and animals. This may not be the case, however, and a great amount of research remains to be done. There appears to be increasing surfacing and recognition of disease of humans and animals related to those "unusual" infections.

Many of these microorganisms (including protozoans) are poorly understood; their relationships and taxonomy are controversial and much empiricism exists in the literature. *Bartonella, Haemobartonella,* and *Grahamella* (Bartonellaceae) are closely related to rickettsiae organisms, but are more bacterial in life style. Several protozoans have been reported from *Microtus*, including amoebae, *Coccidia* ssp., *Hepatozoan* ssp., *Hexamita* ssp., *Giardia* ssp., piroplasms, *Babesia* ssp., *Spirochaeta* ssp., *Trichomonas* ssp., and trypanosomes.

Several parasites have been reported from the blood of *Microtus*. Fay and Rausch (1969) reported in *M. oeconomus* from Alaska a Grahamella-like organism (Bartonellaceae), a trypanosome, *T. microti* (Trypanosomatidae), and two morphologically similar but biologically different strains of piroplasm (Theileriidae). A piroplasm was also found in *M. pennsylvanicus*. These parasites were found during a period of high population.

Babesia, a widespread piroplasm, has recently been diagnosed in human cases in New England (Healy 1979). It appears to be much more serious than was previously thought. *Babesia microti* is the specific organism. The dominant vector is the deer tick, *Ixodes dammini*, which is an indiscriminant feeder; both larvae and nymph stages may bite and inject mice, deer, or humans. Several deaths have been reported in splenectomized persons, probably related to their comprised immunologic system. This disease has a potential of being very widespread, related to its occurrence in smaller mammals, including *Microtus*. Serologic examination in New England revealed that 11 of 577 (1.9 percent) humans routinely tested had a positive titer, but 10 out of 133 (7.5 percent) humans with a tick bite or a fever of at least three days were positive.

Some flagellated protozoans are intestinal parasites. Their distribution, nomenclature, and importance require further study. One, *Giardia* sp., indistinguishable from human *G. lamblia*, has been recovered from *Microtus townsendii* intestine and fecal pellets in our laboratories (Sheffield 1979). *G. lamblia* is the infecting organism of giardiasis, a serious world-wide diarrheal disease. Recently it has been found in several domestic water supplies in the United States from coast to coast, associated with outbreaks. Our isolate was in such a situation, the watershed for Camas, Washington. Giardiasis has also been recently implicated as a more widespread disease, and the term *backpacker's diarrhea* has been applied. Analysis of the ecological niches and behavior of several species of *Microtus* suggests positive correlation to this public health problem that should be further investigated. *Giardia* sp. have been found in *Microtus californicus* and *M. pennsylvanicus* in this country and *Microtus (Pitymys) savii* in Europe, not associated with disease outbreak (Sheffield 1979).

No virus diseases have implicated North American *Microtus* with certainty. There has, however, been little investigation in this field. In Europe, there have been most interesting reports of rodent variants of rabies virus from *Microtus arvalis,* as well as *Clethrionomys glareolus* and *Apodemus flavicollis,* all common species (Arata 1975). These rodent variants are not linked to rabies in other species.

AGE DETERMINATION

Research involving anatomy, physiology, population structure, and reproduction generally requires that the age of individuals be determined. Many evaluations require only that a general age status of "immature" or "mature" be obtained, but because of the great influence of age on many functions, more accurate determinations have been sought (Hoffmeister and Getz 1968).

All vole species achieve reproductive maturity at the early age of a few weeks. However, many have a growth period that proceeds over a period of many months. In general, maximum size is reached between 2 and 10 months of age; few voles survive more than 2 years.

If one has experience, or has available a large number of individuals, it is possible to segregate: (1) the *juvenile,* by pelage (duller and darker), size, and virgin reproductive condition; (2) the *subadult,* which is larger and shows evidence of breeding (perforated vagina, corpora lutea of pregnancy, enlarged testes); (3) the *mature adult,* which has achieved a general maximum body size by weight and measurement; and (4) the *old adult,* which may show scruffy tail and pelage, deteriorating conditions of bones and teeth, and frequently a weight decline.

Size. This can be used only after sufficient evaluative experience has been achieved. In large series of animals, weight alone or length of head and body can be suitably accurate. Live-trapped animals may immediately have abnormally low weights unless optimal care in frequent trap checking is done. Captive animals frequently are overweight. Under some natural conditions, seasonal or individual obesity may occur to skew this method. Body weight alone was used in studies by Didow and Hayward (1969) on *M. pennsylvanicus.*

Bone size can be quantitated and may be useful, especially related to certain of the long bones. This method must be standardized by species and population (subspecies) before being useful.

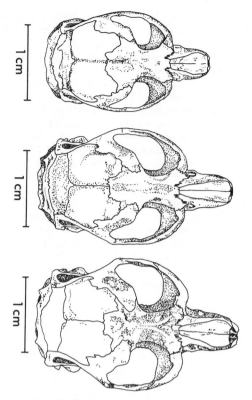

FIGURE 16.9. Skull of the montane vole (*Microtus montanus*), showing changes with age. From top to bottom: juvenile, young adult, and old adult.

Reproductive System. Examination of the reproductive system gives information primarily regarding sexual maturity, not age, but is useful. In the female, there may be gross evidence of perforation of the vagina or recent parturition, or internal corpora lutea, pregnancy, or uterine scars. The simultaneous presence of scars and an active pregnancy indicates age beyond the first pregnancy (generally three weeks first conception and three weeks gestation period, or six weeks minimum age). Evidence of active lactation or enlargement of the teats, postweaning, can be found. In the male, the best external evidence is descent and enlargement of the testes. In older males, darkening and some hair loss of the scrotal skin occurs. Internally, the enlargement of the testes and the turgid enlargement of the seminal vesicles indicate full maturity; measurement of the testis may be used to indicate this relative physiologic age. Microscopic examination of the testis and epididymis and of the ovary, will show mature sperm or ova, respectively. This accurately times sexual maturity (Didow and Hayward 1969). The baculum of the male can be cleared and stained, and can accurately age juveniles and younger males by the progressive ossification of the digital processes as well as by size.

Skeleton. In addition to skeletal growth, there is one marked qualitative change at sexual maturity

of the females. This is the opening of the symphysis pubis, which occurs during the development of the first pregnancy. This is of limited use but can be helpful in certain studies. Closure of epiphyses is another method useful within certain age groups and requiring known-age controls. Various bones may close the epiphysis at different ages. Myers (1978) used a regression formula upon data obtained from x-rays of live animals and obtained a good congruence with controls.

Skull. The general size and conformation of the skull is helpful (figure 16.9). The rounded, softer skull of the juvenile becomes angular and hard. There is gradual closing of the suture lines of the skull bones; several remain open during the lifetime of the vole, but there is a reasonable progressive closure, different for each suture. This method has not proven to be as accurate as it theoretically should be, due to the wide range of times of closure, but the potential exists.

Irruption of the molariform teeth occurs at an early age, before weaning; it is useful only at this restricted time. Because the teeth are all ever growing, they do not wear down; the crown pattern is usually as good in a superannuated mouse as in a juvenile. Howell (1924) provided an excellent, detailed anatomic study of age variation in *M. montanus*.

Eye Lens and Lens Proteins. This is a method widely used in small mammals, with apparent good results (e.g., Birney et al. 1975). Total lens weight and determination of the soluble lens protein appear to be best for the first four months, after which examination of the insoluble proteins is better.

Because any method of age determination measures biological changes occurring over a finite time, it is a dangerous practice to interpolate the results of others into a given situation without critical examination. In practice, some experience must be obtained with the technique decided upon. Known-age specimens should be examined to give confidence in the method. Factors of natural variation within a population, and population (or subspecies/species) differences, must be considered.

ECONOMIC STATUS AND MANAGEMENT

Within the temperate and boreal zones of Europe, Asia, and North America, mankind has evolved with an extensive supportive agricultural base. Also within the most productive and most populous regions, the meadow vole, *Microtus*, has evolved and prospered. There have always been problems and conflicts (Lantz 1907; Bailey 1924; Elton 1942; Myllimäki 1979). In North America, as in the Old World, the ecologic communities existed with *Microtus* as an integral part; the vole has continued to compete with and provide misery for humans.

Mankind has invaded ecosystems that are best for living and providing water, food, and a good environment for agricultural and hunting or fishing pursuits. Almost all original settlements were selected for their optimal ecological niche by humans. Generally, this also was prime environment for *Microtus*, i.e., river bottom land, prairies, meadows, fertile forest areas. Space was lost and competitive lines drawn. Continuous habitat was broken up into islands of lesser size, but it was not all bad for the voles, for irrigation and forest clearing opened up large new areas. Food for domestic grazing animals provided a steady source for *Microtus*, too, if not overused. New species of grass and other forage and food species were introduced. Fence rows and highway rights-of-way provided greater access for the moving and migration of individuals or populations of small mammals, particularly beneficial for *Microtus* (Schmidley and Wilkins 1977).

A new habitat recently categorized as the agroecosystem was created (Loucks 1977). Now in the modern context of economic accountability and systems analyses a different kind of evaluation of the effects of *Microtus* in the bioenergetics of agriculture is possible. If not new, at least it is more accurate and when properly done has some revealing information for the farmer.

Beyond the consideration of humans and domestic animals, voles compete in a more natural and evolved way with other species. This is especially related to the food resource. The larger grazing herbivores are the big game animals. Except in special preserves and national parks, they are managed and do not overpopulate, and there is a dynamic balance of resource. During periods of irruption, however, the biomass of voles may become extremely large, and there may be wide areas of wildlife habitat where links in the energy chain temporarily drop out.

This deleterious effect of voles upon the game animals approaches the theoretical and has not been reported. We have, however, observed the effects upon the habitat during irruption of *M. oeconomus* in the North and *M. montanus* and *M. richardsoni* in the West. It is likely that partitioning of the resources, especially the browsing potential of the larger animals, may compensate the browsing to a degree in this competitive ground-plant overkill.

The competition between the smaller herbivores and small mammals of other species has not been critically studied. Serious competitive problems have seldom been noted. As described above in relationship to the big game animals, it would be only during the periods of vole irruption (or at a time of severe climatic disaster, such as drought) that a critical situation might develop in particular regions.

Bioenergetics and Agricultural Management. Looking at the future in this context, it is obvious from statistics at hand that resources have limitations. Management (e.g., grass-fed versus grain-fed livestock, as a well-studied problem) will have to be carefully investigated and considered. Pimentel et al. (1980) stated that in the United States 135 metric tons of grain (about 10 times the amount consumed by the human population) is fed to livestock annually. A system of "grass only" as now envisioned would reduce the input of energy about 60 percent and land resources 8 percent. It would also reduce animal protein production in the United States by about half. Evaluation of

the resources for feeding domestic herbivores requires a thorough understanding of the biologic and natural forces at work and the competition within the natural ecosystem.

Damage to Grassland and Field Crops. In areas used for pasture and for the raising of hay and grain and various surface and root crops, there is frequently a continual drain on the productivity caused by voles. There is also measurable damage to natural vegetation used by either grazing domestic animals or the native larger herbivores. This damage is frequently little noticed during most years, abundantly evident during periods of peak populations, but especially dramatic during the years of "mouse plagues."

These irruptions are not continuous, nor are they countrywide. They rise and fall "locally like the swells of the ocean" (Bailey 1924:523). They are usually composed of a single species, although sympatric species of other small mammals, including *Microtus*, may show high population numbers at the same time and experience a die-off at the same time.

Damage consists of directly eating the succulent crowns of clover and other grasses, so whole fields may be ruined. Growing grass, hay, and grains (wheat, oats, barley, sage, and buckwheat) are attacked by directly eating early sprouts, but more important, by cutting mature stalks. During the storage phase, severe damage may be done by eating the most nutritious portions.

Much injury is done to planted seeds in open gardens, hotbeds, cold frames, or even greenhouses (Lantz 1907). Bulbs and growing plants are eaten or damaged. Early peas, cabbage, celery, and other surface crops, and potatoes and other root crops (such as beets, turnips, carrots, parsnips, and sweet potatoes) may be destroyed. Vegetables stored on the ground or in pits may also be drastically attacked during periods of high population. Melon and cantaloupe plants may be badly injured by eating of the roots, especially when preparatory plowing has not been done.

Damage to Small Fruits. Blackberries, raspberries, grapes, currants, gooseberries, and strawberries may be badly damaged; whole plantations may be ruined (Lantz 1907). Plantings that are kept less tidily, when canes are uncut and where mulching is done, are especially subject to these attacks.

Damage to Trees. Severe and at times devastating harm is done to nursery stock, to fruit orchards, and, in some cases, to trees in the eastern deciduous forests (Lantz 1907). Tree damage consists of debarking under the cover of snow. Destruction of roots is associated whenever extensive damage has been reported, with an irruption of mouse populations. It also has a positive correlation with long, hard winters.

The species of *Microtus* involved depends upon the region. The pine vole, *M. pinetorum,* the meadow vole, *M. pennsylvanicus,* and the montane vole, *M. montanus,* are the species most seriously involved in tree damage reports (Piper 1909b; Garlough and Spencer 1944; Sartz 1970; Baumgartner 1978; Myllymäki 1979). Tree species include orchard trees, shade trees, common deciduous forest trees in eastern forests, and several coniferous trees in young plantations of silvicultural operations. In this latter group, the preferred trees were Scotch pine (*Pinus sylvestris*), Douglas fir (*Pseudotsuga menziesii*), Austrian pine (*Pinus nigra*), Norway spruce (*Picea abies*), and European larch (*Larix decidua*). Red (*Pinus resinosa*), white (*Pinus strobus*), and jack (*Pinus banksiana*) pine, white cedar (*Thuja occidentalis*), balsam fir (*Abies balsamea*), and several larch (*Larix*) species were secondarily attacked. White spruce (*Picea glauca*) was hardly touched except by *M. ochrogaster* (Sartz 1970). In eastern North American forests, locust (*Robinia pseudoacacia*), black cherry (*Prunus serotina*), oak (*Quercus* ssp.), chestnut (*Castanea dentata*), and hickory (*Carya* ssp.) plantations have been particularly at risk (Lantz 1907).

Tree damage is not necessarily complete. Root damage may be occult and go unrecognized. Aboveground debarking is more noticeable. This may completely encircle the tree and cause death, or it may be partial and the tree may survive. Such partial damage results in lowered production and promotes secondary decay and insect depredation.

Population Irruptions. Population irruptions are variously referred to as "mouse years" and "mouse plagues." They have been well documented in the Old World for centuries (Bailey 1924), but were not officially recognized as a problem in North America until 1907–8.

Historically, an irruption of *Microtus montanus* in 1907–8 in Nevada, Utah, and northeast California was the most serious recorded in the United States up to that time (Piper 1909*a*). Numbers of mice were estimated (probably overestimated) at 20,000 to 30,000 per hectare. Fields were honeycombed by holes up to 60,000 per hectare. During the summer cropping season, one-third of the area's alfalfa crops, three-fourths of the potatoes, and much of the root crops (beets and carrots) were ruined. In the fall, green food disappeared and the roots of alfalfa and orchard and shade trees were extensively damaged until the dramatic abatement occurred by the following summer. It was noted the regions of damage were essentially reclaimed arid lands, which provided ideal enrichment for this irruptive phenomenon.

In 1948–49, an irruption of the montane vole occurred in eastern Washington and was investigated on the experimental agricultural plot at Orondo. As the snow cover left, the degree of population density became apparent. A large biomass of mice could easily be obtained. Examination of the sample of mice obtained indicated an actively increasing population. Examination of the experimental stone fruit tree plots showed that in three plots containing no ground cover, there was no mouse damage; three plots where some cover was maintained suffered 15.8 percent losses. Of special interest was one plot where 38 of 294 trees were girdled where ground cover was present, and only

1 of 182 trees in the adjoining clean-cultivated section was girdled.

An irruption of *Microtus montanus* was documented in eastern Oregon in 1957–58 (Oregon State College 1959). This was widespread in adjoining northern California and Nevada and Idaho. It was the most severe outbreak in the memory of long-time residents.

The 1957–58 winter was mild and open; population numbers rose to a peak at the end of the year and began to drop in February. Maximum counts of populations in the agricultural areas ranged from 500 to 10,000 mice per ha. Severe damage resulted in many fields at a level of 500 to 1,200 mice per ha. There was an irregularity of die-off: in some areas complete disappearance occurred, in others a build-up continued into July and August of 1958.

Damage was to several kinds of crops. Alfalfa, grain fields, and grass fields were damaged. Some fields harvested in 1957 were not worth cutting in 1958. Hundreds of hectares of ordinarily high-yield crops were not harvested. Red clover seed and grain in some areas were not harvested because the mice had eaten the seed heads. Potato fields and cellars suffered damage up to 50 percent. Cattlemen had to move cattle from natural meadows many weeks before the usual time.

Grass had better ability to regrow in areas of severe infestation, but much clover was removed. In late December and January, there was an apparent movement of the mice into brushland-type habitat. Serious damage in rangeland occurred in bitterbrush (*Purshia tridentata*), rabbit brush (*Chrysothamnus viscidiflorus*), wild cherry (*Prunus emarginata*), wild plum (*Prunus subcordata*), and greasewood (*Sarcobatus vermiculatus*).

Additional potential damage by tunneling in irrigation ditches was averted by heavy poisoning. Other damage occurred in baled hay by tunneling, chewing, and fouling of the lower two or three layers. In numerous cases, 5- to 15-year-old apple and other fruit trees were completely girdled and their supporting roots cut.

Management. A most effective means of control of mice in agricultural areas is eradication of local populations of pest animals. This means primarily control of roadside areas, fence lines, and ditches. Such control entails either frequent cultivation, application of herbicides, and/or grass cutting, in the agricultural areas themselves and in the surrounding borders. This may be a severe economic drain on the landholder and farmer. Because it requires working a large amount of unproductive land, it may not be economically feasible. Further, the word *control* must be clearly understood. To people involved in vertebrate pest control, the aim is not to exterminate all members of a species. The "purpose is to reduce the numbers of these animals to tolerable densities only in areas where they are pests" (California Vertebrate Pest Control Committee 1964:2).

In most situations of "mouse plagues" it has been

documented that there is a build-up of population the year prior to the irruption. Thus, control measures may be taken prior to the population irruption. Once the population cycle is in its ascent, it has been impossible to control the development of these populations.

What may a farmer do on a year-to-year basis to produce an ecosystem that is maximally effective? (1) Eradicate the ground cover ("clean" farming). (2) Protect natural enemies (see the section "Mortality"). (3) Control directly. In initiating control, the farmer must be absolutely certain that action is necessary, and here frequently consultation may be obtained and the following evaluated: (*a*) how much damage has already occurred; (*b*) how much damage is anticipated without control; (*c*) what were the economic factors of cost versus control and its benefits; (*d*) what is the effect of a control program on the ecosystems and on non-target animals.

In general, properly timed cultivation and controlled fires have some effect; trapping and fencing for *Microtus* are not effective, except locally. Poisons have evolved as the most efficient (and most destructive) management tool. Repellents are attractive in theory, but not yet satisfactorily effective (Myllymäki 1979).

Methods of Poisoning. In the past, poison grain and numerous poison items, including arsenic, zinc, phosphorus, and thallium, have been utilized (Lantz 1907). These have gradually been refined. It is imperative to consider that the best results (and least harm) occur when the application is done by an experienced person. Hand baiting is the most flexible kind of poisoning and is better suited to most small operations, especially when the soils and moisture conditions are not suitable for mechanical trail building. It is also best for supplemental baiting during the winter season when the ground is frozen, and for local control when this is required. The best time of the year for application is in the fall. Zinc phosphide–treated oats, groats, or soft, white wheat is used. For winter and spring baiting, zinc phosphide–treated apples or carrot cuttings are preferred. Small amounts of about 4 grams (1 teaspoonful) of poison bait in active, covered runways are applied and additional bait is placed in dense cover near the trunks of trees or on fence lines. Mechanical baiting can be done using a "mouse trail builder." This is best suited for large orchards and large fields. The same materials—zinc phosphide–treated oats, groats, and soft, white wheat—are used. Approximately 75 kg of poison bait per ha is applied at a depth of 5–10 cm (Baumgartner 1978). In orchards a single trail is made on each side of the trees, as close to the trunks as possible, along the orchard borders, and as near to other crops and adjoining areas as possible. The soil should be moist so that the artificial runways remain intact until the mice discover the bait. The area should remain undisturbed for several days following the application.

Ground Sprays. Endrin is used where baiting methods cannot be used, or may be used to supplement the mechanical baiting. It may also be used in inaccessible areas. Endrin, however, is not recommended for general use, and ordinarily must be registered because of the potential for ecosystem damage. In the state of Washington, for instance, registration permits the use in restricted areas only. Endrin is used at the rate of about 1 liter per 400 liters of water (Baumgartner 1978). The cover adjacent to the orchards and within the orchards should be thoroughly wet with about 2,800 to 3,300 liters of spray per hectare and the application should be done as soon after the harvest as possible. Spraying of the ground cover should be especially thorough around the bases of trees, along the borders, and in strips down the tree rows. Basically, the ground should be dry and also should be mowed short, if possible, to give maximum spray penetration. Endrin is extremely toxic and must be handled with care during all phases of mixing and application. All pets and persons should be kept out of sprayed areas for a 30-day period. Signs must be placed around the edge of the orchard, and sprayed windfall fruit cannot be eaten. Sprayed cover crops cannot be grazed or fed to livestock.

County, state, and federal agencies should be consulted in all problem cases. The legality of any control measure, especially the use of plant and animal poisons, should be carefully investigated. International and interstate coordination of efforts is desirable.

As a final note, it cannot be stressed too heavily that the total ecosystem must be considered. The risk of all of the poisoning methods to the nontarget species should be evaluated and the basic ethics of consideration for nature must be applied.

CAPTIVE MAINTENANCE

Captive colonies of *Microtus* are maintained to check response to control measures, or to study the susceptibility, carrying ability, and transmission of disease, either to other animals or to humans.

Most species of *Microtus* adapt very well to captivity. They appear to do equally well in small standard laboratory cages or other modified caging. Maintenance of health, increased longevity, breeding in captivity, and rearing of the young are the prime criteria for determining the adaptability of a species. By these criteria *Microtus* does well. They are hardy, their food is readily available, they are prolific breeders, and they are easily transported. For these reasons, *Microtus* has been well studied for their own merits or as a laboratory animal to investigate other biological problems. Over much of North America (see the section "Distribution"), one or several species of *Microtus* are readily available.

There are a few disadvantages, related to some other laboratory or native mammals. Most species are hydrophilic, meaning an increased output of urine and saturation soiling of the cage litter or drainage pan. Likewise, the fecal output is usually large. Many species are independent and irritable. Compared to laboratory mice or rats, they do not handle easily; they can bite savagely.

Food. Most species do best if they are given root vegetables, especially carrots, and greens daily; this may range from grass cuttings to a more standardized supermarket fare. Lab Chow (Purina) and other laboratory food commercially available can be a part of the standard fare. Rabbit pellets, alfalfa, or other hay makes good food and the hay is fine nesting material. Food is best supplied in some form of manger up off the floor level. This conserves food and prevents waste and soiling.

Water must be supplied *ad lib* and should be present at all times. The mice will drink quite well from laboratory fountains or drinking bottles. If water is supplied in dishes, most species, especially hydrophiles, will selectively defecate in the water.

Social Structure. Colonies may be maintained with several breeding animals of either sex present or a harem arrangement, if the enclosure is large enough. This may have some advantages under certain conditions and may be productive. However, controlled breeding is better. This not only identifies the genetic background but, in general, it allows for better survival. It also demands better record keeping.

Miscellaneous. Ambient room temperature is usually satisfactory. Colonies may be kept out of doors in all but the coldest months. Nest boxes may be helpful in handling and providing a secure microenvironment for the subjects.

Litter must be changed at frequent intervals to avoid parasites and disease problems. Especially in wild-caught animals, build-ups of potentially noxious organisms can be prevented by good sanitation.

Anesthesia is frequently needed for certain manipulations: blood drawing, tissue biopsy, etc. Ether, in general, works well and is safe for the animal, but it is explosive. Nembutal and others can be used for longer-acting sessions.

Blood samples may be obtained by heart puncture, lower cervical vessel puncture, or ocular plexus puncture at low mortality/morbidity rates.

The general principles of laboratory animal care should be observed and the legal restrictions adhered to. Most important is to individualize the species if setting up a laboratory colony. None will do well without daily humane care by a person who enjoys the job and likes animals. And under these circumstances, bites are few!

The section on parasitic worms in this chapter was provided by Dr. Robert L. Rausch.

LITERATURE CITED

Anderson, S. 1960. The baculum in microtine rodents. Univ. Kansas Publ., Mus. Nat. Hist. 12:181–216.

Arata, A. A. 1975. The importance of small mammals in public health. Chapter 14.3 (pages 349–359) *in* F. B. Golley et al., eds. Small mammals: their productivity and population dynamics. IBP5. Cambridge Univ. Press, Cambridge. 451pp.

Bailey, V. 1920. Identity of the bean mouse of Lewis and Clark. J. Mammal. 1:70–72.

———. 1924. Breeding, feeding and other life habits of meadow mice (*Microtus*). J. Agric. Res. 27(8):523–536.

———. 1936. The mammals and life zones of Oregon. North Am. Fauna 55. 416pp.

Barbour, R. W. 1963. *Microtus:* a simple method of recording time spent in the nest. Science 141:41.

Batzli, G. A., and Pitelka, F. A. 1970. Influence of meadow mouse population on California grassland. Ecology 51:1027–1039.

Baumgartner, D. M. 1978. Animal damage control in the Pacific Northwest. EM 3908, Rev. Coop. Extension Serv., Washington State Univ., Pullman. 92pp.

Bent, A. C. 1937. Life histories of North American birds of prey. Part 1: order Falconiformes. U.S.N.M. Bull. 167. Smithsonian Institution Press, Washington, D.C. 409pp.

———. 1938. Life histories of North American birds of prey. Part 2: order Falconiformes and Strigiformes. U.S.N.M. Bull. 170. Smithsonian Institute Press, Washington, D.C. 482pp.

———. 1946. Life histories of North American jays, crows, and titmice. U.S.N.M. Bull. 191. Smithsonian Institution Press, Washington, D.C. 493pp.

Benton, A. H. 1955. Observations on the life history of the northern pine mouse. J. Mammal. 36(1):52–62.

Berger, P. J.; Sanders, E. H.; Gardner, P. D.; and Negus, N. C. 1977. Phenolic plant compounds functioning as reproductive inhibitors in *Microtus montanus*. Science 195:575–577.

Birney, E. C.; Jenness, R.; and Baird, D. D. 1975. Eye lens proteins as criteria of age in cotton rats. J. Wildl. Manage. 39(4):718–728.

Boonstra, R. 1977. Effect of conspecifics on survival during population declines in *Microtus townsendii*. J. Anim. Ecol. 46:835–851.

Boulière, F. 1975. Mammals, small and large: the ecological implications of size. Pages 1–9 *in* F. B. Golley, K. Petrusewicz, and L. Ryszkowski, eds. Small mammals: their productivity and population dynamics. Cambridge University Press, Cambridge. 451pp.

Breed, W. G. 1967. Ovulation in the genus *Microtus*. Nature 214:826.

Buech, R. R.; Timm, R. M.; and Siderits, K. 1977. A second population of rock voles, *Microtus chrotorrhinus*, in Minnesota with comments on habitat. Can. Field Nat. 91:413–414.

Burt, W. H. 1960. Bacula of North American mammals. Misc. Publ. Mus. Zool. Univ. Michigan No. 113. 76pp.

California Vertebrate Pest Control Committee. 1964. Proceedings of Second Vertebrate Pest Control Conference, Anaheim, Calif. Coll. of Agriculture, Univ. California, Davis. 160pp.

Cooper, J. E. K., and Hsu, T. C. 1972. The C-band and G-band of *Microtus agrestis* chromosomes. Cytogenetics (Basel) 11:295–304.

Corbet, G. B. 1978. The mammals of the Palearctic region. Br. Mus. (Nat. Hist.) and Cornell Univ. Press, London and Ithaca. 314pp.

Craighead, J. J. 1959. Predation by hawks, owls and gulls. Pages 35–42 *in* Oregon State College, The Oregon meadow mouse irruption of 1957–1958. Special Rep. Fed. Coop. Extension Serv., Corvallis. 88pp.

Cross, P. C. 1972. Observation on the induction of ovulation in *Microtus montanus*. J. Mammal. 53(1):210–212.

Dawson, M. R. 1967. Fossil history of the families of recent mammals. Pages 12–53 *in* S. Anderson and J. K. Jones, Jr., eds. Recent mammals of the world. Ronald Press Co., New York. 453pp.

Didow, L. A., and Hayward, J. S. 1969. Seasonal variations in the mass and composition of brown adipose tissue in the meadow vole, *Microtus pennsylvanicus*. Can. J. Zool. 47(4):547–555.

Donati, N. L.; Beck, M. L.; and Price, P. K. 1977. Chromosomal study of *Microtus pinetorum*. J. Tennessee Acad. Sci. 52(2):73. Abstr.

Douglas, R. J. 1977. Population dynamics, home range and habitat association of the yellow cheeked vole, *Microtus xanthognathus*, in the Northwest Territories. Can. Field Nat. 91:237–247.

Dunaway, P. B.; Cosgrove, G. E.; and Story, J. D. 1968. *Capillaria* and *Trypanosoma* infestations in *Microtus ochrogaster*. J. Wildl. Dis. 4:18–20.

Durette-Desset, M. C. 1971. Essai de classification des nématodes héligmosomes: corrélations avec la paléobiogéographie des hôtes. Mém. Mus. Nat. d'Hist. Naturelle 69, Ser. A, Zool. 126pp.

Durham, J. W. 1974. Tertiary period. Encyclopaedia Britannica, Macropedia 18:151–160.

Eadie, W. R. 1952. Shrew predation and vole populations on a localized area. J. Mammal. 33(2):185–189.

Eisenberg, J. F. 1978. The evolution of arboreal herbivores in the class mammalia. Pages 135–152 *in* G. G. Montgomery, ed. The ecology of arboreal folivores. Smithsonian Inst. Press, Washington, D.C. 574pp.

Elton, C. 1942. Voles, mice and lemmings: problems in population dynamics. Clarendon Press, London. 496pp.

Fay, F. H., and Rausch, R. L. 1969. Parasitic organisms in the blood of Arvicoline rodents in Alaska. J. Parasitol. 55(6):1258–1265.

Fedyk, S. 1970. Chromosomes of *Microtus (Stenocranius) gregalis major* (Ognev, 1923) and phylogenetic connections between sub-arctic representatives of the genus *Microtus* Schrank, 1798. Acta Theriol. 15:143–152.

Findley, J. S., and Jones, C. J. 1962. Distribution and variation of voles of the genus *Microtus* in New Mexico and adjacent areas. J. Mammal. 43(2):154–166.

Fisler, G. F. 1961. Behavior of salt marsh *Microtus* during winter high tides. J. Mammal. 42(1):37–43.

Fitzner, R. E., and Fitzner, J. N. 1975. Winter food habits of short-eared owls in the Palouse prairie. Murrelet 56(2):2–4.

Gamble, W. B.; Segal, M.; Schantz, P. M.; and Rausch, R. L. 1979. Alveolar hydatid disease in Minnesota: first human case acquired in the contiguous United States. J. Am. Med. Assoc. 241:904–907.

Garlough, F. E., and Spencer, D. A. 1944. Control of destructive mice. U.S. Fish Wildl. Serv. Conserv. Bull. 36. 37pp.

Gashwiler, J. S. 1972. Life history notes on the Oregon vole, *Microtus oregoni*. J. Mammal. 53(3):558–569.

Getz, L. 1961. Factors influencing the local distribution of *Microtus* and *Synaptomys* in Southern Michigan. Ecology 42(1):110–119.

Goertz, J. W. 1971. An ecological study of *Microtus pinetorum* in Oklahoma. Am. Midl. Nat. 86(1):1–12.

Golley, F. B. 1961. Interaction of natality, mortality and movement during one annual cycle in a *Microtus* population. Am. Midl. Nat. 66(1):152–159.

———. 1963. A contribution to a bibliography of mice of the genus *Microtus*. U.S. Atomic Energy Comm. T1D-18274, Biology and Medicine. Office of Technical Services, Dept. of Commerce, Washington, D.C. 114pp.

Golley, F. B.; Petrusewicz, K.; and Ryszkowski, L., eds. 1975. Small mammals: their productivity and population dynamics. Int. Bio. Programme 5. Cambridge Univ. Press, Cambridge. 451pp.

Grassé, P. P., ed. 1969. Traité de zoologie. Tome 16, fascicule 6: Mamelles, appareil génital, gamétogenèse, fécondation, gestation. Masson et Cie, Paris. 1027pp.

———. 1973. Traité de zoologie. Tome 16, Fascicule 5, Vol 1: Splanchnologie. Masson et Cie, Paris. 1063pp.

Gray, J. E. 1821. On the natural arrangement of vertebros animals. Medic. Repos. 15:296–310.

Grodzinski, W., and Wunder, B. A. 1975. Ecological energetics of small mammals. Pages 173–204 *in* F. B. Golley, K. Petrusewicz, and L. Ryszkowski, eds. Small mammals: their productivity and population dynamics. Int. Bio. Programme 5. Cambridge Univ. Press, Cambridge. 451pp.

Haas, G. E.; Barrett, R. E.; and Wilson, N. 1978. Siphonaptera from mammals in Alaska. Can. J. Zool. 56(2):333–338.

Hall, E. R., and Kelson, K. R. 1959. The mammals of North America. 2 vols. Ronald Press Co., New York. 1083pp.

Hamilton, W. J., Jr. 1938. Life history notes on the northern pine mouse. J. Mammal. 19(2):163–170.

———. 1941. Reproduction of the field mouse *Microtus pennsylvanicus* (Ord). Cornell Univ. Agric. Exp. St. Memoir 237:1–23.

Harvey, M. J., and Barbour, R. W. 1965. Home range of *Microtus ochrogaster* as determined by a modified minimum area method. J. Mammal. 46(3):398–402.

Hatfield, D. M. 1935. A natural history study of *Microtus californicus*. J. Mammal. 16(4):261–271.

Healy, G. R. 1979. *Babesia* infections in man. Hospital Practice (June 1979), pp. 107–116.

Hibbard, C. W.; Zakrzewski, R. J.; Eshelman, R. E.; Edmund, G.; Griggs, C. D.; and Griggs, C. 1978. Mammals from the Kanapolis local fauna, Pleistocene (Yarmouth) of Ellsworth County, Kansas. Contrib. Mus. Paleont. Univ. Michigan 25(2):11–44.

Hoffmeister, D. F., and Getz, L. 1968. Growth and age classes in the prairie vole, *Microtus ochrogaster*. Growth 32:57–69.

Howell, A. B. 1924. Individual and age variations in *Microtus montanus yosemite*. J. Agric. Res. 28(10):977–1015.

Hsu, T. C. and Benirschke, K. 1967–71. An atlas of mammalian chromosomes. Springer-Verlag, New York.

Hsu, T. C., and Johnson, M. L. 1970. Cytological distinction between *Microtus montanus* and *Microtus canicaudus*. J. Mammal. 51(4):824–826.

Hubbard, C. A. 1947. Fleas of western North America. Iowa State Coll. Press, Ames. 533pp.

Hume, I. D. 1978. Evolution of the Macropodidae digestive system. Aust. Mammal. 2:37–42.

Jewett, S. G.; Taylor, W. P.; Shaw, W. T.; and Aldrich, J. W. 1953. Birds of Washington State. Univ. Washington Press, Seattle. 767pp.

Johnson, G. B. 1977. Assessing electrophoretic similarity: the problem of hidden heterogeneity. Ann. Rev. Ecol. Syst. 8:309–328.

Johnson, M. L. 1973. Characters of the heather vole, *Phenacomys* and the red tree vole, *Arborimus*. J. Mammal. 54(1):239–244.

———. 1974. Mammals. Pages 1–87 *in* C. A. Wright, ed. Biochemical and immunologic taxonomy of animals. Academic Press, London. 490pp.

Johnson, M. L., and Johnson, S. 1979. Plague of *Microtus*: possible adverse affect on *Thomomys*. Pap. presented at

59th Annu. Meet. Am. Soc. Mammal., Corvallis, Oreg. Abstr.

Johnson, M. L., and Ostenson, B. T. 1959. Comments on the nomenclature of some mammals of the Pacific Northwest. J. Mammal. 40:571–577.

Johnson, M. L., and Wicks, M. J. 1959. Serum protein electrophoresis in mammals: taxonomic implications. Syst. Zool. 8(2):88–95.

Jones, J. K., Jr.; Carter, D. C.; and Genoways, H. H. 1975. Revised checklist of North American mammals north of Mexico. Occas. Pap. Mus. Texas Tech. Univ. 28:1–14.

Judd, S. R., and Cross, S. P. 1980. Chromosomal variation in Microtus longicaudus. Murrelet 61(1):2–5.

Judd, W. W. 1950. Mammal host records of Acarina and Insecta from the vicinity of Hamilton, Ontario. J. Mammal. 31(3):357–358.

Kartman, L.; Prince, F. M.; and Quan, S. F. 1959. Epizootiologic aspects. Pages 43–54 in Oregon State College. The Oregon meadow mouse irruption of 1957–1958. Special Rep. Fed. Coop. Extension Serv., Corvallis. 88pp.

Kenagy, G. J., and Hoyt, D. F. 1980. Reingestion of feces in rodents and its daily rhythmicity. Oecologia (Berlin) 44:403–409.

Keys, J. E., and Van Soest, P. J. 1970. Digestibility of forages by the meadow vole (Microtus pennsylvanicus). J. Dairy Sci. 53:1502–1508.

Kirkland, G. L., Jr. 1977. The rock vole Microtus chrotorrhinus (Miller) in West Virginia. Ann. Carnegie Mus. 46(5):45–53.

Krebs, C. J. 1966. Demographic changes in fluctuating populations of Microtus californicus. Ecol. Monogr. 36:239–73.

Krebs, C. J.; Wingate, I.; Leduc, J.; Redfield, J.; Taitt, M.; and Hilborn, R. 1976. Microtus population biology: dispersal in fluctuating populations. Can. J. Zool. 54(1):79–95.

Kretzoi, M. 1962. Arvicolidae oder Microtidae? Vertebrata Hungarica 4:171–175.

Kurtén, B. 1968. Pleistocene mammals of Europe. Aldine Publ. Co., Chicago. 317pp.

Lantz, D. E. 1907. An economic study of field mice (genus Microtus). U.S.D.A. Bio. Survey Bull. 31:1–64.

Lidicker, W. Z., Jr. 1979. Analysis of two freely-growing enclosed populations of the California vole. J. Mammal. 60(3):447–466.

Loosli, J. K. 1974. Animal feed. Encyclopaedia Brittanica 1:908–911.

Loucks, O. L. 1977. Emergence of research on agroecosystems. Ann. Rev. Ecol. Syst. 8:173–192.

McBee, R. H. 1970. Metabolic contributions of the cecal flora. Am. J. Clinical Nutrition 23:1514–1518.

McDaniel, B. 1979. Host records of ectoparasites from small mammals of South Dakota. Southwestern Nat. 24(4):689–691.

Makino, S. 1950. Studies on murine chromosomes. Part 6: Morphology of the sex chromosomes in two species of Microtus. Annot. Zool. Japan 23:63–68.

Martin, E. P. 1956. A population study of the prairie vole (Microtus ochrogaster) in northeastern Kansas. Univ. Kansas Publ. Mus. Nat. Hist. 8(6):361–416.

Maser, C.; Hammer, E. W.; and Anderson, S. H. 1970. Comparative food habits of three owl species in central Oregon. Murrelet 51(3):29–33.

Maser, C., and Storm, R. M. 1970. A key to Microtinae of the Pacific Northwest (Oregon, Washington, Idaho). Oregon State Univ. Book Stores, Corvallis. 162pp.

Matthey, R. 1952. Chromosomes de Muridae (Microtinae et Cricetinae). Chromosoma 5:113–138.

———. 1953. Les chromosomes des Muridae. Rev. suisse Zool. 60:225–283.

———. 1955. Nouveaux documents sur les chromosomes des Muridae. Rev. suisse Zool. 62:163–206.

———. 1958. Un nouveau type de détermination chromosomique du sexe chez les mammifères, Ellobius lutescens Th. et Microtus (Chilotus) oregoni Bachm. (Muridés-microtinés). Experientia 7:240.

———. 1964. La formule chromosomique et la position systématique de Pitymys tatricus Kratochvil (Rodentia-Microtinae). Z. Säugetierk. 29(4):235–242.

———. 1969. Les chromosomes et l'évolution chromosomique des mammifères. Pages 855–909 in P. P. Grassé, ed. Traité de zoologie. Tome 16, fascicule 6: Mammelles, apparcil génital, gâmétogenèse, fécondation, gestation. Masson et Cie, Paris. 1027pp.

Matthey, R., and Zimmerman, K. 1961. La position systématique de Microtus middendorffi Poliakov, Rev. suisse Zool. 68:63–72.

Meylan, A. 1967. Karyotype and giant sex chromosomes of Microtus chrotorrhinus (Miller) (Mammalia: Rodentia). Can. J. Genet. Cytol. 9:700–703.

Michael, E. D. 1976. Effects of highways on wildlife. West Virginia Univ. Agric. Exp. Stn. Rep.

Montgomery, G. G., ed. 1978. The ecology of arboreal folivores. Smithsonian Inst. Press, Washington, D.C. 574pp.

Myers, P. 1978. A method for determining the age of living small mammals. J. Zool. Soc. London 186:551–556.

Myllymäki, A. 1979. Importance of small mammals as pests in agriculture and stored products. Pages 239–279 in D. M. Stoddard, ed. Ecology of small mammals. Chapman and Hall, London. 386pp.

Nadler, C. F.; Zhurkevich, N. M.; Hoffman, R. S.; Kozlowskii, A. I.; Deutsch, L.; and Nadler, C. F., Jr. 1978. Biochemical relationships of the Holarctic vole genera (Clethrionomys, Microtus and Arvicola. Rodentia: Arvicolinae). Can. Zool. 56(7):1564–1575.

Negus, N. C., and Berger, P. J. 1979. Plant compounds: a component of Microtine cycles? Annu. Meet. Am. Soc. Mammal. Abstr.

Negus, N. C.; Berger, P. J.; and Forsland, L. 1977. Reproductive strategy of Microtus montanus. J. Mammal. 58(3):347–353.

Negus, N. C., and Pinter, A. J. 1965. Litter sizes of Microtus montanus in the laboratory. J. Mammal. 46(3):434–445.

Ognev, S. I. 1950. Mammals of the U.S.S.R. and adjacent countries. Translated from Russian by Israel Program for Scientific Translations, Jerusalem. 1964. 626pp.

Ohno, S.; Jainchill, J.; and Stenius, C. 1963. The creeping vole (Microtus oregoni) as a gonosomic mosaic. Part 1: The OY/XY constitution of the male. Cytogenics 2:232–239.

Oregon State College. 1959. The Oregon meadow mouse irruption of 1957–1958. Special Rep. Fed. Coop. Extension Serv., Corvallis. 88pp.

Osgood, S. B.; Holmes, M. A.; Runte, V.; and Brandon, G. R. 1959. Epidemiological aspects. Pages 61–69 in Oregon State College. The Oregon meadow mouse irruption of 1957–1958. Special Rep. Fed. Coop. Extension Serv., Corvallis. 88pp.

Ouellette, D. E., and Heisinger, J. F. 1980. Reingestion of feces by Microtus pennsylvanicus. J. Mammal. 61:366–368.

Parra, R. 1978. Comparison of foregut and hindgut fermentation in herbivores. Pages 205–229 in G. G. Montgomery, ed. The ecology of arboreal folivores. Smithsonian Inst. Press, Washington, D.C. 574pp.

Pearson, O. P. 1964. Carnivore-mouse predation: an example

of its intensity and bioenergetics. J. Mammal. 45(2):177–188.

Petrov, B., and Rimsa, D. 1976. Uber die arteigenstandigkeit de Kleinwühlmaus, *Pitymys felteni*. Senckenberg. Biol. 57:1–10.

Pimentel, D.; Oltenacu, P. A.; Nesheim, M. C.; Krummel, J.; Allen, M. S.; and Chick, S. 1980. The potential for grass fed livestock: resource constraints. Science 207:843–848.

Piper, S. E. 1909a. Mouse plagues, their control and prevention. Yearb. Dept. Agric. (1908), pp. 301–310.

———. 1909b. The Nevada mouse plague of 1907-8. U.S.D.A. Farmers Bull. 352. 23pp.

Pohl, R. W. 1974. Poales. Encyclopaedia Britannica, Macropedia, 14:584–595.

Quentin, J. C. 1971. Morphologie comparée des structures céphaliques et génitales des Oxyures du genre *Syphacia*. Ann. Parasit. Hum. Comp. 46:15–60.

Racey, K. 1960. Notes relative to the fluctuation in numbers of *Microtus richardsoni richardsoni* about Alta Lake and Pemberton Valley, B.C. Murrelet 41(1):13–14.

Rageot, R. H. 1957. Predation on small mammals in the Dismal Swamp, Virginia. J. Mammal. 38(2):281.

Randall, J. A. 1977. Habitat preference and interspecific dominance as mechanisms of habitat segregation between two sympatric species of *Microtus*. Ph.D. Thesis. Washington State Univ. 51pp.

Rausch, R. L. 1964. The specific status of the narrow-skulled vole (sub-genus *Stenocranius* Kashchenko) in North America. Z. Säugetierk. 29(4):343–358.

———. 1974. On the zoogeography of some Beringian mammals. Proc. 1st Int. Theriol. Congr., 6–12 June 1974, Moscow.

———. 1976. The genera *Paranoplocephala* Lühe, 1910 and *Anoplocephaloides*, Baer, 1973. Ann. Parasit. Hum. Comp. 51:513–562.

———. 1977. O zoogeografii nekotorykh Beringiiskikh mlekopitayushchikh, Voprosi teriologii. Nauka, Moscow. Pp. 162–177.

———. 1979. Taeniidae. Supplement to Chapter 53 *in* W. T. Hubbert, W. F. McCulloch, and P. R. Schnurrenberger, eds. Diseases transmitted from animals to man. Chas. C. Thomas, Publ., Iowa Univ. Press. In press.

Rausch, R. L.; Huntley, B. E.; and Bridgens, J. G. 1968. Notes on *Pastuerella tularensis* isolated from a vole, *Microtus oeconomus* Pallas, in Alaska. Can. J. Microbiol. 15(1):47–55.

Rausch, R. L., and Rausch, V. R. 1968. On the biology and systematic position, *Microtus abbreviatus* Misser, a vole endemic to the St. Matthew Islands, Bering Sea. Z. Säugetierk. 33(2):65–99.

Rausch, R. L., and Tiner, J. D. 1949. Studies on the parasitic helminths of the North Central states. Part 2: Helminths of voles (*Microtus* spp.). Preliminary rep. Am. Midl. Nat. 41:665–694.

Rausch, V. R., and Rausch, R. L. 1974. The chromosomal complement of the yellow-cheeked vole, *Microtus xanthognathus* (Leach). Can. J. Genet. Cytol. 16:267–272.

Raven, P. H., and Axelrod, D. I. 1974. Angiosperm biogeography and past continental movements. Ann. Missouri Bot. Gardens 61(3):539–637.

Raynaud, A. 1969. Les organes génitaux de mammifères. Pages 149–636 *in* P. P. Grassé, ed. Traité de zoologie. Tome 16, fascicule 6: Mammelles, appareil génital, gamétogenèse, fécondation, gestation. Masson et Cie, Paris. 1027pp.

Repenning, C. A. 1979. The new mammalian biochronology. Annu. Meet. Geol. Soc. Am., San Diego, Calif. Abstr.

Riewe, R. R., and Pruitt, W. O., Jr. 1968. Migrations, homing, and home ranges of insular populations of the meadow vole in Newfoundland. Pap. presented at Annu. Meet. Am. Soc. Mammal., Fort Collins, Colo. (Title only.)

Roscoe, B., and Majka, C. 1976. First records of the rock vole (*Microtus chrotorrhinus*) and the Gaspé shrew (*Sorex gaspensis*) from Nova Scotia and a second record of the Thompson's pygmy shrew (*Microsorex thompsoni*) from Cape Breton Island. Can. Field Nat. 90:497–498.

Rose, R. K., and Gaines, M. S. 1978. The reproductive cycle of *Microtus ochrogaster* in eastern Kansas. Ecol. Monogr. 48:21–42.

Roth, D., and Powers, L. R. 1979. Comparative feeding and roosting habits of three sympatric owls in southwestern Idaho. Murrelet 60(1):12–15.

Ruzić, A.; Petrov, B.; Zivković, S.; and Rimsa, D. 1975. On the species independence of the 54-chromosome vole *Microtus epiroticus*, Ondrias, 1966 (Mammalia, Rodentia), its distribution, ecology and importance as a pest in the west part of the Balkan Peninsula. J. For. Sci. Agric. Res. (Belgrade) 104:153–160.

Ryzhikov, K. M.; Gvozdev, E. V.; Tokobaev, M. M.; Shaldybev, L. S.; Matsasberidze, G. V.; Nerkusheva, I. V.; Nadtochii, E. V.; Khokhlova, I. G.; and Sharpilo, L. D. 1978. Orpredelitel' gel'mintov gryzunov fauny SSSR. Tsestody i trematody. Nauka, Moscow. 231pp.

———. 1979. Opredelitel' gel'mintov gryzunov fauny SSSR. Nematody i akantotsefaly. Nauka, Moscow. 276pp.

Sartz, R. S. 1970. Mouse damage to young plantations in southwestern Wisconsin. J. For. 68:88–89.

Schadler, M. H., and Butterstein, G. M. 1979. Reproduction in the pine vole, *Microtus pinetorum*. J. Mammal. 60(4):841–844.

Schmidly, D. J., and Wilkins, K. T. 1977. Composition of small mammal populations on highway rights-of-way in east Texas. Rep. of study 2-8-76-197. Natl. Tech. Inform. Serv. PB-275089. U.S. Fed. Highway Adm. and Texas State Dept. Highways and Public Transportation. 96pp.

Schrank, F. P. 1798. Fauna Boica, Durchgedachte Gesch. der in Bayern Einheimisch. u. Zahmen Thiere. Bd. 1, Abth, 1, Nürnberg.

Seidensticker, J. C. 1970. Food of nesting red-tailed hawks in south-central Montana. Murrelet 51(3):38–40.

Sheffield, S. 1979. Intestinal sarcomastigophorans in *Peromyscus maniculatus* from southwestern Washington. M.S. Thesis. Univ. Puget Sound, Tacoma, Wash. 74pp.

Sonnenberg, E. L., and Powers, L. R. 1976. Notes on the food habits of long-eared owls in southwestern Idaho. Murrelet 57:63–64.

Southern, H. N. 1979. Population processes in small mammals. Pages 63–101 *in* D. M. Stoddard ed. Ecology of small mammals. Chapman and Hall, London. 386pp.

Stoddard, D. M., ed. 1979. Ecology of small mammals. Chapman and Hall, London. 386pp.

Storer, T. I., and Usinger, R. 1963. Sierra Nevada natural history. U. California Press, Berkeley. 374pp.

Timm, R. M.; Heaney, L. M.; and Baird, D. D. 1977. Natural history of rock voles (*Microtus chrotorrhinus*) in Minnesota. Can. Field Nat. 91:177–181.

Todorović, M.; Soldatović, B.; and Savić, I. Karyotype of the species *Microtus nivalis* Martins, 1842 (Rodentia) from Sar Planina. Arhiv Bioloskih Nauk (Beograd) 23:7–9.

Vaughn, T. A. 1978. Mammalogy. 2nd ed. W. B. Saunders Co., Philadelphia. 522pp.

Verster, A. 1969. A taxonomic revision of the genus *Taenia* Linnaeus 1758 s. str. Onderstepoort. J. Vet. Res. 36:3–58.

Vorontsov, N. N. 1962. The ways of food specialization and evolution of the alimentary tract. Pages 360–377 *in* Proc. Int. Symp. Methods of Theriol. Invest., Prague.

———. 1967. Evolution of the alimentary system of Myomorph rodents. Publ. House "Nauka" Siberian Branch, Novosibirsk. 240pp. (In Russian, with English summary and title.)

Warid, W. A. 1974. Vegetables and vegetable farming. Encyclopaedia Britannica, Macropaedia 19:43–53.

Whitaker, J. O. 1979. Origin and evolution of the external parasite fauna of western jumping mice, genus *Zapus*. Am. Midl. Nat. 101(1):49–60.

Whitaker, J. O., Jr., and Martin, R. L. 1977. Food habits of *Microtus chrotorrhinus* from New Hampshire, New York, Labrador and Quebec. J. Mammal. 58:99–100.

Wilhelm, D. E. 1977. Zoogeographic and evolutionary relationships of selected populations of *Microtus mexicanus*. Ph.D. Thesis. Texas Tech. Univ. 85pp.

Wilson, N. 1957. Some ectoparasites from Indiana mammals. J. Mammal. 38:281–282.

Wiseman, V. S. 1957. A morphological study of the baculum in some microtine rodents. M.A. Thesis. Univ. Puget Sound. 47pp.

Zimmerman, E. G. 1965. A comparison of habitat and food of two species of *Microtus*. J. Mammal. 46:605–612.

Zivković, S., and Petrov, B. 1974. Record of a vole of the *Microtus arvalis* group with 54 chromosomes (*Microtus subarvalis* Myer, Orlov et Skholl, 1972) in Yugoslavia and comparison of its karyotype with that of *Pitymys felteni* Malec et Storch, 1963 (Rodentia, Mammalia). Genetika 6:283–288.

Zivović, S.; Rimsa, D.; Ruzić, A.; and Petrov, B. 1975. Cytogenetic characteristics, taxonomic status and distribution of the voles with 46 and 54 chromosomes of the *Microtus arvalis* group in Yugoslavia (Rodentia, Mammalia). Archiv. Bioloskih Nauka 26:123–134.

MURRAY L. JOHNSON, Puget Sound Museum of Natural History, University of Puget Sound, Tacoma, Washington 98416.

SHERRY JOHNSON, Puget Sound Museum of Natural History, University of Puget Sound, Tacoma, Washington 98416.

17

Porcupine

Erethizon dorsatum

Wendell E. Dodge

NOMENCLATURE

COMMON NAMES. Porcupine, porky, hedgehog, quillpig, porc epic, stachelschwein, pricklepig, quiller

SCIENTIFIC NAME. *Erethizon dorsatum*
SUBSPECIES. *E. d. dorsatum, E. d. myops, E. d. nigrescens, E. d. bruneri, E. d. epixanthum, E. d. couesi,* and *E. d. picinum.*

The porcupine is the second largest native North American rodent (the beaver, *Castor canadensis,* is larger) and *Erethizon* is one of four New World genera of the family Erethizontidae. Its closest relative, the Mexican porcupine (*Coendou* sp.), is uncommon and found only in the high mountain ranges of Mexico and northern South America. There are seven described subspecies of *Erethizon* (Anderson and Rand 1943). Although the geological age of *E. dorsatum* is complicated by lack of fossil evidence, the porcupine's first known occurrence is in the middle Pleistocene (Hibbard and Mooser 1963). Considerable disagreement exists regarding the porcupine's taxonomic position, affinities, and evolutionary dispersal. Wood (1955) suggested that erethizontids' and hystricids' only affinity is possession of quills. Moody and Doniger (1956) indicated that perhaps erethizontids should be accorded their own suborder. Woods (1973) placed the Erethizontidae in the suborder Hystrichongnatha.

DISTRIBUTION

With the exception of a few widely separated areas of unsuitable habitat, such as plains, drier desert areas, coastal islands, and the southeastern coastal plain, the porcupine's range includes nearly half of the United States, nearly all of Canada and Alaska, and part of northwestern Mexico (figure 17.1). Fossil evidence of Erethizon is found in Alabama (Barkalow 1961), Florida (White 1970), Missouri (Parmalee 1971), West Virginia (Guilday and Hamilton 1973), and other areas where no extant porcupines are presently found or apparently existed during recorded history. No evidence

FIGURE 17.1. Distribution of the porcupine (*Erethizon dorsatum*) in North America.

is available that these animals were ever present in the southern Mississippi basin.

DESCRIPTION

Uniquely coated, short, obese appearing, bowlegged, pigeon-toed when walking, and obviously myopic are terms often used to describe the porcupine's appearance. It has a relatively small face, blunt muzzle, shoe-button eyes, no discernable neck, and a short, muscular, clublike tail. The feet are bearlike with naked soles and possess long prominent claws (four front, five rear). The forefeet are capable of grasping and holding by opposition of the claws and foot pad. The animal often sits upright on its hind legs and tail, especially when feeding.

It is difficult to misidentify this animal in North America, with the possible exception of *Coendou* in Mexico. There have been reports of confusion with bear cubs when viewed at some distance or high in a tree.

Adult porcupines normally weigh 4 to 6 kilograms and some have been reported to weigh more than 17 kg. Total length of adults ranges from 60 to 100 centimeters, and tail length 17.5 to 22.5 cm. Height at the shoulder rarely exceeds 30 cm but an adult standing on its hind legs can often reach 110 cm or more.

Pelage. The Erethizontidae are the only North American mammal possessing hairs modified as quills. These are distributed dorsally from in front of the eyes (figure 17.2) to nearly the tip of the tail. Quills vary in length and diameter but the longest are over the shoulders and the thickest are on the lateral and dorsal surfaces of the tail. The entire quill area of the back is apparently controlled by a muscular sheath so that all quills can be erected when the animal is alarmed. The undersurfaces of the porcupine's body lack quills but are densely furred. The underpart and tip of the tail are covered with stiff bristles that provide assistance in ascending and descending trees.

Porcupine quills are unique defensive weapons with a spongy interior and are sharply pointed on the distal end. This sharp end, usually black or dark colored, is covered with shingle-lapped barbules (figure 17.3) that can only be seen under magnification. When the quill is embedded in tissue, these barbules expand and, as muscle fibers contract, the quill is pulled deeper into the tissue. The proximal end is also pointed but only to provide follicle attachment.

Porcupine quills have proven fatal to many predators and inquisitive animals including dogs, owls, cattle, horses, and even people (McDade and Crandall 1958). The human death was an indirect result of eating a porcupine meat sandwich containing a quill.

Porcupines throughout their range are generally dark colored, but as in other mammal species, many variations are found. The western yellow-haired subspecies (*E. d. epixanthum*) appears yellow to gold at a distance due to its yellowish gold guard hairs. Winter coats with dense underfur impart a dark color to most porcupines. Young animals are generally dark until the following spring, when the immature underfur is shed at one year of age. Chestnut, cream (figure 17.4), brown, gray, and albino porcupines are occasionally seen. Prepartum and postpartum females usually exhibit a dark brown halo around each of the six mammae.

Skull and Dentition. The skull of the porcupine (figure 17.5) contains 35 bones (Swena and Ashley 1956) but is essentially divided into two sections: facial and cranial. Most obvious and distinguishing of the facial portion is the unusually large infraorbital foramen that permits passage and insertion in the premaxillary area of a large portion of the deep masseter muscle. Other predominant features are the large incisor and molar teeth, the lateral curving zygomatic arches, and the large orbital fossae. Older animals possess a pronounced sagittal crest. Other skull characteristics that unmistakably separate the porcupine from any other North American mammal (with the exception of *Coendou*) are the hystricid angular process on the mandible and the open pterygoid fossa (Woods 1973).

The cheek teeth are flat crowned, subhypsodont, and rooted, and possess a folded-ridged enamel pattern. The deciduous premolar is replaced by a permanent tooth at about two years of age (Dodge 1967) and is useful for age estimation of live animals. The large extrusive incisors are dark to light orange colored on the anterior surface. The dental formula is: (I 1/1, C 0/0, P 1/1, M 3/3) × 2 = 20.

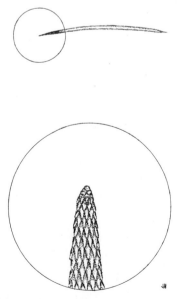

FIGURE 17.2. The porcupine (*Erethizon dorsatum*) has quills over most of its body, with the exception of the nose, ventral body, and foot pads.

FIGURE 17.3. Enlarged drawing of the distal tip of a porcupine (*Erethizon dorsatum*) quill, illustrating the shingle-lapped barbules.

FIGURE 17.4. A cream-colored male porcupine (*Erethizon dorsatum*) captured in Massachusetts.

Other Skeletal Features. Unique and perhaps a useful taxonomic characteristic is the fusion of the second and third cervical vertebrae of *Erethizon*. This feature is also present in other erethizontids (Ray 1958). The proximal ends of the tibia and fibula are also fused, as are the malleus and incus of the inner ear (see figure 17.5).

Digestive Tract. Although the digestive tract is not particularly unusual, certain features merit discussion. The mouth of the porcupine is similar to that of the beaver in that the hairy integument of the upper lips continues into the mouth. Lateral opposition of these lips posterior to the incisors effectively separates the anterior and posterior portions of the mouth cavity. In the beaver, this feature apparently permits the animal to feed underwater. The porcupine occasionally feeds on aquatic plants but generally only wades in shallow water.

Another distinct feature is the large parotid gland that extends from the base of the ear to shoulder level and ventromedially to the submaxillary glands, thus forming a nearly complete collar.

The small intestine, excluding the cecum, is about half the total six to seven meters of intestinal tract. The cecum is a large, sacculated organ with a lumen capacity of one liter or more and ending in a J-shaped tail similar to that of the domestic rabbit. Fecal pellets begin to form in the transverse portion of the colon and become firm as they reach the level of the cecal tail.

The liver differs from that of most other rodents, as it possesses a sixth lobe (unnamed) in the lessor peritoneal cavity that forms the right margin of the foramen epiploicum and is attached to the lesser omentum. There is a bile duct in the adult but no gall bladder, although Struthers (1941) reported that rudiments are seen during early ontogeny.

REPRODUCTION

Reproduction and breeding behavior of the porcupine has been the subject of much discussion. While some aspects are unique and others remain unanswered, the porcupine closely parallels many other species of mammal.

Anatomy. Perotta (1959) described the female porcupine's reproductive tract as being a Y-shaped and transitional duplex-bicornuate structure. The uterine horns have a spiral lumen and, although externally they appear joined, are separated by an internal septum. The vagina is relatively long and joins with the urethra to form a urogenital sinus that is protected by densely haired labial folds ventral and anterior to the anal opening. The labia also have numerous long, light-colored vibrissae that are apparently tactile. The genital prominence appears similar in both sexes but sex can be determined easily by palpation of the folded penis or scrotal testes during the breeding season.

The ovary is almond shaped, about 5×10 millimeters in size during anestrus, and suspended by a typical round ligament and mesovarium to the horns and dorsal body wall, respectively. Mossman and Duke (1973) stated that only a partial bursa is present. The oviducts are extremely long, complexly convoluted structures, possessing a distinct funnel and typical fimbria.

Although the relationship to breeding and pregnancy has not been confirmed, the female porcupine exhibits a thick vaginal membrane that completely seals the vaginal opening. It is not established whether this is perforated by the penis during copulation or if it is physiologically perforated during estrus. It is present in postpartum females but its presence in prepartum females has not been confirmed.

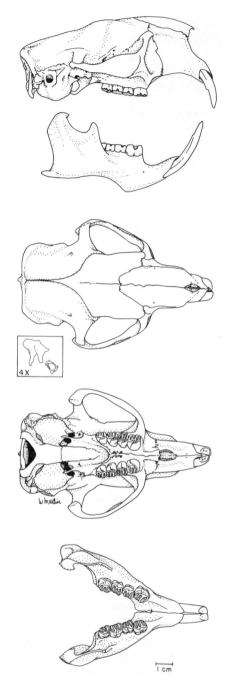

FIGURE 17.5. Skull of the porcupine (*Erethizon dorsatum*) (1x). From top to bottom: lateral view of cranium, lateral view of mandible, dorsal view of cranium (inset: inner ear ossicles, 4x), ventral view of cranium, dorsal view of mandible. (UMA 3391, February 1968, Glenwood, Wash., male.)

The male reproductive tract was described by Mirand and Shadle (1953); it is uncommon in that the penis normally lies folded with the distal end pointed posterior in a penial sheath within the genital prominence. It is unusual, as erection apparently requires a complex muscular contraction rather than the usual venous engorgement. The associated prostate and Cowper's glands are exceptionally large. The testes are typical elongate glands with a complex convoluted seminiferous tubule and large caput epididymis. The testes descend from the abdominal cavity through a large inguinal foramen during June or July and remain scrotal through the breeding season.

Female Reproductive Cycle. The female porcupine generally does not exhibit ovarian activity until about one year of age; ovulation occurs in the second fall at one and a half years old. Although the basic ovarian activity is similar to that of most mammals, there are noteworthy differences.

The ovary containing the corpus of pregnancy presents a confusing picture when compared with most other mammals. Mossman and Judeas (1949) presented a detailed description of the formation and maintenance of accessory corpora lutea. Many developing ovarian follicles normally vacuolate with only one becoming the corpus luteum of pregnancy. However, a number of follicles become accessory corpora lutea that persist into the next gestation. Even a few nonovulating ovaries possessed luteinized follicles, but these disappeared after implantation. It must be presumed that these accessory corpora play a role in supporting the fertilized ovum and fetus and may be related to the abnormally long gestation period, precocial young, and low fetal mortality in porcupines.

It is thought that ovulation in the porcupine is unilateral. Unilateral ovulation in many other mammals normally alternates between ovaries during succeeding estrous periods (Nalbandov 1958). Ovulation is almost always from the right ovary and the embryo develops in the right horn (Dodge 1967). This may be because of the large cecum occupying the left quadrant, which may obstruct pregnancies in the left uterine horn. This phenomenon is similar to implants in the right horn of cattle, which, although not fully substantiated, are thought to be due to the influence of the rumen that occupies the left quadrant of the abdominal cavity.

Breeding. Porcupines are normally solitary animals except during the breeding season and in winter denning concentrations. A number of males may be found scattered around adult females during the breeding season of September through November when estrus normally occurs. Heat apparently lasts 8 to 12 hours, during which time a receptive female may even be aggressive toward an unattentive male by assuming the male's role. The high-pitched screams of fighting males and nonreceptive females have probably resulted in numerous reports of the presence of wild felids.

Precopulatory behavior of the male is usually meticulous and time consuming. Males expend considerable time making careful olfactory examinations of the ground, the base and trunks of trees, projecting rocks, roots, or other likely female urination sites. Males vigorously rub their genitoanal area over any of these female sites.

Although observed attempts at mating in trees

have not been successful, the male usually coaxes the female to the ground, where he will rear on hind legs and tail while emitting low vocal "grunts." He then proceeds to spray the female with bursts of urine from a rapidly erecting penis and, after wrestling, chases, vocalization, and more urine showers, coitus is effected. It is performed, as in most mammals, with the male taking the active role and, contrary to folklore, mounting from the rear. The receptive female elevates her hindquarters and arches her tail over her back, providing the male a platform for his forepaws or chest. The male then permits his forelegs to hang free. Coital contact is brief, with a violent ejaculation, and afterward the male drops back to groom and clean. Further matings may ensue until one of the pair climbs a tree and ends the contact by hostile screaming and lunging. A vaginal plug is formed shortly after mating and the female is then no longer receptive.

Males are extremely hostile during the breeding season and violent battles have been witnessed where both combatants have received vicious bites and hundreds of foreign quills. A captive male received fatal wounds from another male during a contest over a receptive female.

Although Mossman and Duke (1973) stated that the female is monestrous with one ovum, unconfirmed evidence indicates that porcupines may be seasonally polyestrous and recycle 25–30 days after a sterile mating. Newborns found as late as August indicate that some breeding may occur as late as January.

Gestation and Parturition. For a rodent, the porcupine has an exceptionally long gestation period of 205 to 215 days (Shadle 1951), which approximates that of the Cervidae. Most rodents give birth after a short gestation of a few weeks.

Parturition usually occurs in late April or early May, but some births occur as late as August. One pup, or "porcupette," is born (multiple births have been discussed but never authenticated) weighing 400–500 grams, eyes open, incisor and premolar teeth erupted, quills present and functional when dry, mobile, and often displaying defensive activities. Although the young are capable of eating vegetation within a week, they generally remain with the female throughout the summer. Lactating females have been observed as late as September.

ECOLOGY

Porcupines are found in varied habitats throughout the North American continent from the open tundra of Canada and Alaska to the desert chapparal and rangelands of Northern Mexico. Although not abundant, they are found in the humid Pacific coastal areas of Washington and Oregon (Dodge 1966; Dodge and Canutt 1969) and eastward through the northern United States and Canada to the Atlantic seaboard. Porcupines are found in Pennsylvania but are rare to absent in the Atlantic states from Connecticut south.

It is evident that the porcupine is essentially cosmopolitan and not restricted to any specific plant or edaphic community. Porcupines are normally not gregarious except during winter denning in the colder habitats where rockfall caves, hollow trees, abandoned buildings, and brush piles may contain a number of individuals of both sexes. Some animals are independent of any particular den site and often remain in trees for long periods during the winter. One abandoned house in New Hampshire provided dens for six porcupines—two in the chimney and one in each of the four bake ovens (Dodge 1967).

Population densities are extremely variable throughout the porcupine's range and, especially in the colder areas, change markedly from winter to summer because of high denning concentrations. Taylor (1935) reported 0.77 per square kilometer in Arizona, and Kelker (1943) reported 9.5 per km^2 in northeastern Wisconsin. Densities tend to shift with food availability. Higher populations are found near maturing agricultural crops such as corn, alfalfa, and other legumes or, during dry periods, in riparian-associated habitats.

FOOD HABITS

Porcupines are primarily vegetarians with a winter diet that is almost exclusively cambium, phloem, and foliage of woody shrubs and sapling to saw log trees. Although the porcupine feeds on nearly all species of trees within its range, the principal winter foods in northern and western forests are conifers and hardwoods. Eastern hemlock (*Tsuga canadensis*) cambium, phloem, and foliage is the principal winter food in the Northeast (Dodge 1967) and the lake states (Brander 1973), and western hemlock (*T. heterophylla*) is second to Douglas fir (*Pseudotsuga menziesii*) in the Washington Cascades (Dodge 1970). Ponderosa pine (*Pinus ponderosa*) is the principal winter food in the transition zone of Oregon (Dodge and Canutt 1969) and where it occurs in the porcupine's range. The shiny, white stems and limbs resulting from fresh porcupine girdling are obvious winter indicators of the porcupine's presence in western forests.

Other common conifers fed upon by porcupines are: lodgepole pine (*P. contorta*), western white pine (*P. monticola*), eastern white pine (*P. strobus*), piñon pine (*P. edulis* and *P. monophylla*), and limber and bristlecone pine (*P. flexilis* and *P. aristata*). The larches (*Larix* spp.), especially in New England, many of the spruces (*Picea* spp.), and firs (*Abies* spp.) are also fed upon by porcupines. Jones (1973) reported that spruce, balsam fir (*A. balsamea*), and white pine were preferred in Nova Scotia when compared to availability. Gill and Cordes (1972) reported that localized destruction of limber pine in the southern Alberta foothills may warrant active control measures against porcupines.

Many species of hardwood are eaten by porcupines but sugar maple (*Acer saccharum*), red maple (*A. rubrum*), oak (*Quercus* spp.), beech (*Fagus grandifolia*), and birch (*Betula* spp.) are preferred. Sweet birch (*B. lenta*) is often severely deformed or killed near winter den sites in New England (Dodge 1967). Ornamentals, such as silver maple (*A. saccharinum*), boxelder (*A. negundo*), butternut (*Juglans cinerea*), American elm (*Ulmus americana*), basswood (*Tilia* spp.), and willow (*Salix* spp.) are often eaten. Apple (*Malus* spp.) and cherry (*Prunus* spp.) in orchards are also severely damaged by winter feeding porcupines.

Woody shrubs too numerous to list are also winter foods of the porcupine. Most obvious of these are blackberry and raspberry canes (*Rubus* spp.) in late winter and spring (Dodge 1967). Elderberry (*Sambucus* sp.) damaged by porcupines, because of the shiny, white stems, is an excellent roadside indicator of porcupine presence in the western Cascades (Dodge 1970). Taylor (1935) reported that spiny buffaloberry (*Shepherdia argentea*) damage in North Dakota is another good indicator of porcupine presence. Taylor (1935) reported that porcupines in Arizona consumed large quantities of dwarf mistletoe (*Arceuthobium* sp.). He expressed some concern that the porcupine may spread mistletoe in the Southwest.

Food habits of the porcupine in the spring and summer shift from aboreal to nearly exclusive ground feeding. Porcupines do, however, feed on buds, catkins, and new leaves on aspen (*Populus* spp.), elm, and birch. Ground vegetation includes numerous herbaceous forbs, grasses, and succulent riparian or wetland plants. Some of the more commonly sought species are: sedges (*Carex* spp.), violets (*Viola* spp.), dandelion (*Taraxacum* spp.), clovers (*Trifolium* spp.), alfalfa (*Medicago* spp.), and common grains (Dodge 1967). Taylor (1935) reported large quantities of buckbrush (*Ceanothus fendleri*) eaten by porcupines in Arizona. Agricultural crops, especially ripening corn, are selected during summer and fall.

The porcupine spends considerable time in wetlands and riparian habitats where succulents and other green vegetation are available. It consumes large quantities of aquatic plants such as water lilies (*Nymphea* spp.), arrowhead (*Sagittaria* spp.), pondweeds (*Potamogeton* spp.), and aquatic liverworts (*Riccia* spp.) (Dodge 1967). They are not adverse to swimming and are often observed in shallow water feeding on the above aquatics. Dean (1950) observed a porcupine swim to floating water lilies, which it cut and brought to shore to consume. Porcupines in the Washington Cascades frequented late winter open seeps, where they ate sprouts of skunk cabbage (*Lysichitum americanum*) and other emerging succulents (Dodge and Barnes 1975).

Porcupines continue to feed on ground vegetation throughout summer and fall until frosts destroy available plants. In New England and other areas, where oak and other mast is found, they spend considerable time clipping mast-bearing branches to obtain the acorns. The ground under oaks where porcupines feed is littered with clipped branches.

BEHAVIOR

Vocalization. Porcupines, like most wild mammals, are not overly vocal but are capable of a variety of calls and screeches. Postpartum females communicate with their young during spring and summer by subdued grunts and whines (somewhat similar to a mourning dove coo) and occasional clicking of the teeth. Females are very responsive to imitations of this call and it can be useful for live capture.

The fall breeding season is characterized by numerous vocalizations, especially from male porcupines: low grunts, ascending and descending scale whines, and shrill screeches. The last are probably responsible for reports of wild felids, as they can be heard from considerable distances.

Imitations of the up-and-down scale whine can be useful for live-capturing porcupines, particularly males, in the fall. Males will readily descend trees, emerge from dens, and move considerable distances to investigate the imitated call. Females generally exhibit no interest or a negative response by clicking the teeth and moving away.

Quill Removal. Porcupines are amazingly dexterous at removing foreign quills received during encounters with hostile males or unreceptive females. They will sit upright on the tail and hind legs and methodically extract the offending quill from the face, neck, or ventral body surface by using the incisors and the semiprehensile forefeet. Shadle (1955) also described this activity among captive porcupines.

Alarm and Defense. Although the porcupine's vision is poor, the auditory and olfactory senses are relatively acute. Wild porcupines will often stop, rear on their hind legs (like a bear), and carefully smell their surroundings. They are more sensitive to low-frequency sounds such as footsteps and rustling leaves. If slightly alarmed, they will remain unmoving but a threshold alarm will cause them to erect their quills and retreat to a den, tree, or protective shelter where only their back and tail are presented. Cornered porcupines, in addition to quill erection, will lower their head and shoulders and raise the tail for flailing the attacker. The tail is heavily muscled and can drive the thick quills into the attacker's flesh 20 mm or more.

Tail flailing is apparently an innate defensive mechanism, as it is one of a newborn's first reactions after parturition. Juveniles will often "play" and assume a defensive attitude of quill erection, spinning, and tail flailing. Solitary immatures will often respond to gusts of wind in the same manner.

Movements and Home Range. The home ranges of porcupines vary with the climate and habitat. In some areas of cold climate the winter range is severely restricted to a few hectares, whereas summer and winter ranges may be essentially identical in warmer locations. Linear daily movements during the summer range from 111–129 meters (Marshall et al. 1962) to 249 m (Dodge and Barnes 1975). Winter movements in northern habitats are usually restricted to feeding areas near heavily used den sites. Monthly movements averaged 1500 m and females exceeded males by 50 m. One subadult male moved 31 km in 66 days. This was an unusual move but numerous individuals did make 8–10 km moves when shifting home ranges.

Home range areas were also dependent on climate and habitat and varied from a few hectares in winter to over 100 ha in summer (Dodge and Barnes 1975).

Sex Ratios. Most reported sex ratios tend to favor females. Spencer (1949) reported 30 males per 100 females, Gensch (1946) reported 86 males per 100 females, and Dodge (1967) found 92 males per 100 females. These sex ratios are not unusual for a polygamous species.

Age Determination. Porcupines are relatively long-lived rodents but are difficult to age accurately after sexual maturity. Age determination based on cementum annuli has not been adequately evaluated because known-age specimens beyond a few years are not available. Wild porcupines probably do not normally exceed 5–7 years but Brander (1971) reported a recaptured porcupine living in the wild 10.1 years. A male porcupine, originally tagged in 1971 as a 3 year old by Kelly (1973), was recaptured in good condition by the author in 1978. Shadle (1951) and Burge (1966) reported that captive porcupines live to be at least 10 years of age.

Live young-of-the-year porcupines can be easily distinguished by their generally dark, woolly pelage and less than 4 kg weight. Any porcupine over 5 kg in the fall can be considered an adult animal and sexually mature.

Tooth eruption, replacement, and wear provides a more definitive technique for aging but is naturally restricted to anesthetized animals or skulls with at least one side of the mandible intact. Tooth eruption and replacement has proven reliable for animals examined in New England, Colorado, Washington, and Oregon (Dodge 1967) through 22 months of age. The tooth eruption and replacement criteria are:

1. Newborn to 3 weeks—incisors (I) and premolar (P) erupted.
2. 1 to 4 months—three cheek teeth (P_1, M_1, M_2) erupted.
3. 5 to 11 months—little wear on cheek teeth, M_3 visible but no wear (figure 17.6A).
4. 12 to 13 months—M_3 attains level of M_2, no wear. Deciduous P_1 can be distinguished from permanent P_1 by presence of double-pronged sulcus in lingual center of tooth that becomes twin "lakes" with wear. Permanent P_1 possesses only one central sulcus.
5. 14 to 21 months—M_3 shows some wear, premolar with excessive wear (figure 17.6B) and lump on ramus under P_1 (figures 17.7A,B).
6. 22 to 26 months—permanent P_1 erupted, with little or no wear (figure 17.8A).
7. 26+ months—the maxillary lump disappears, and M_1 and M_2 show irregular wear.

Aging by wear beyond 26+ months becomes difficult and is complicated by local food habits, because wear patterns appear older due to unusual amounts of abrasive materials in more arid sandy areas. This would be typical of porcupines in the more arid areas east of the Washington Cascades, in contrast to those from the coastal plains. Old, debilitated porcupines may be found with the occlusal tooth surfaces deeply cupped, pulp cavities exposed, and molar teeth dark colored in various stages of decay (figure 17.8B). Such tooth wear leads to rapid debilitation and death.

Capture and Handling. Porcupines are easily captured, either by hand or with various types of live or steel-jaw traps. If needed for research, live-traps or hand capture are more suitable, as steel traps, even with offset and padded jaws, often cause broken legs and lacerations.

Live box, wire mesh, or corral-type traps can be quite successful. Apple baits, particularly Red Delicious, are preferred. Fetid scents, such as those used for coyotes, are also useful and trap sites near dens or active feeding areas are desirable. Porcupines will follow the same trail on nightly feeding forages and a visible trail is usually evident in leaves, vegetation, or snow.

Hand capture is a useful method of taking large numbers of porcupines unharmed for research purposes. Once the animal is located, it should be quietly approached and the tip of its tail grasped firmly between the thumb and fingers. The free hand can then quickly reach under the tail base and, by simultaneously smoothing the quills and grasping, remove the animal or administer anesthesia by hypodermic or a nose cone. Once captured, the porcupine's tail provides an excellent "handle," and the animal can thus be transported alive for long distances. Adult animals are generally incapable of curling up to bite but younger, more acrobatic porcupines can. Porcupines in trees can be captured as above but it is more convenient to have an associate on the ground who can tail-catch the porcupine after it has been driven down by gentle taps on the head and nose.

Dogs, trained to locate and bay porcupines without encounter, are extremely valuable for research or control. The author and an assistant with a dog captured 17 live porcupines in three hours using the above techniques.

Steel traps can be used as blind sets in consistently used dens or trails without bait, but these may be hazardous to nontarget species. Local laws may prohibit the use of leg-hold traps and therefore state conservation agencies should be consulted.

Evidence of porcupine presence is usually easy to detect. Snow trails are obvious. Trails from winter dens to feeding areas are usually deep troughs about 25 cm wide and often marked with urine spots, scattered fecal pellets (18–20 mm × 10–13 mm), and shed quills. The typical pigeon-toed track with obvious claws, dimpled foot pads, and often an obvious tail drag, easily identify the porcupine. Telltale signs of quills and droppings are usually present where the porcupine spends any time feeding or resting. Consistently used dens possess a characteristic musky odor and fecal deposits are obvious. Hollow trees, ledge caves, and other heavily used dens may have piles of fecal material often a meter or more in depth. Porcupine gnawing or girdling can usually be identified by the size of the incisor marks, size of discarded bark, and diameter of clipped limbs. These signs, in addition to fecal material and quills, usually readily identify the

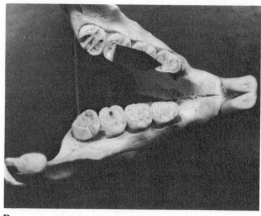

A **B**

FIGURE 17.6. A, jaw of a 5–11-month-old porcupine (*Erethizon dorsatum*), with third molar visible but not occlusal. B, double-pronged central sulcus of deciduous premolar, showing characteristics of 14–21-month-old jaw.

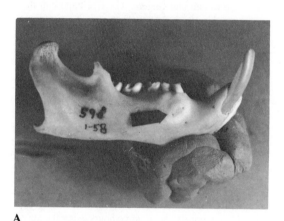
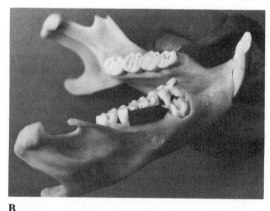

A **B**

FIGURE 17.7. A, lump on the maxillary ramus of a porcupine (*Erethizon dorsatum*), indicative of erupting permanent premolar at 20–21 months. B, part of the ramus cut away to reveal the erupting permanent premolar at 20–21 months.

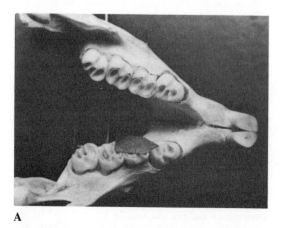
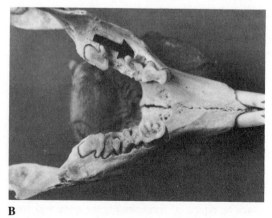

A **B**

FIGURE 17.8. A, newly erupted permanent premolar of a porcupine (*Erethizon dorsatum*), showing single central sulcus and no wear at 22–26 months. B, jaw of old, debilitated animal, showing decay, exposed pulp cavities, and loose sockets.

362

porcupine, although gray squirrel (*Sciurus* spp.) girdling in Douglas fir and maples may initially be confusing.

Activity. Porcupines are generally nocturnal with some crepuscular activity. They exhibit little diurnal movement other than shifting position in a tree or den, scratching, and stretching. Telemetric monitoring of porcupines during 24-hour periods showed most movement between 2100 and 0700 hours (Dodge and Barnes 1975). Nocturnal summer activity was generally spent foraging on the ground. Summer days were usually spent in trees or large shrubs. Winter feeding was mostly foliage and phloem material, again at night. Porcupines were in some type of den during winter days, although a few animals utilized trees for feeding and resting for long periods. Porcupines may be seen actively feeding during daylight hours but this is relatively uncommon.

MORTALITY

Predation. Mankind, especially in settled areas, is the porcupine's most lethal enemy. Bounty systems, vermin control programs, and highway kills eliminate many porcupines annually. In spite of a century of bounties, poisoning, and other control measures, the porcupine is still a fairly common inhabitant of much of rural North America. It is probably most susceptible to death by predation during the first few weeks of life.

Of the porcupine's natural predators, the fisher (*Martes pennanti*) is probably the deadliest and has been reintroduced in many areas for biological control. The fisher, usually the male (Kelly 1977), will harass a porcupine and slash its facial area until it is rendered helpless. The fisher usually consumes nearly the entire carcass, leaving only a skin case.

Other carnivores, such as coyotes (*Canis latrans*), cougars (*Felis concolor*), bobcats (*F. rufus*) and the wolf (*C. lupus*) (Stenlund 1955), have been reported to have porcupine remains in their stomach contents. The great horned owl (*Bubo virginianus*) takes porcupines as prey, but these are probably young-of-the-year porcupines. Eifrig (1909) reported capturing a great horned owl with 56 porcupine quills throughout its body. Jonkel (1968) reported an unusual incident of finding porcupine quills in the lips and facial skin of a captured polar bear (*Ursus maritimus*).

Parasites. The mange mite (*Sarcoptes* sp.) is probably the most serious parasite of the porcupine, especially in the northeastern United States. This mite, thought to be the same as that infecting wild foxes (Dodge 1958; Payne and O'Meara 1958), soon renders a porcupine completely debilitated and it is doubtful that a porcupine ever recovers from a mange infection.

The porcupine is seasonally infested with lice (*Trichodectes setosus*) in the winter and ticks (*Ixodes* sp. and *Dermacentor* sp.) in the summer, but these apparently cause little problem other than discomfort.

The porcupine hosts an amazing number of internal parasites, but does not appear to be unduly impaired. Hart (1957) counted 378 tapeworm (*Mo-*

necocestus sp.) scoleces in one animal. A new species, *Monecocestus giganticus,* was recently reported by Buhler (1970). Olsen and Tolman (1951) estimated that one porcupine they examined contained over 30,000 nematodes (*Wellcomia evaginata*), the most common and abundant endoparasite found in the porcupine. Highby (1943) identified and described the life cycle of a hairlike filarid (*Dipetalonema arbuta*) found in the porcupine's abdominal cavity. These filarids are common but not found in large numbers. Microfilaria in circulating blood are likely to be an early life stage of this parasite. Choquette et al. (1973) reported *Schistosomatium douthitti* in the portal system of a porcupine in eastern Ontario that apparently caused granuloma and necrosis of the liver.

None of these endoparasites, with the possible exception of *S. douthitti,* regardless of numbers, seems adversely to affect the host. Most of these parasite species are acquired during the first summer and fecal pellets contain hundreds of eggs and cestode sections. The pellets are often found strung together by pieces of parasite debris.

Salkin et al. (1976) reported a cutaneous infection of a porcupine with *Auriobasidium pullulans.* This is a fungal infection and a first report for wild or domestic animals.

Diseases. Few diseases have been reported for the porcupine. Hull (1930) and Kuhns et al. (1953) reported tularemia (*Francisella tularensis*) in the porcupine, and the latter authors reported a serial transmission from porcupines through dogs to humans.

Porcupines occasionally contract an infection similar to the human cold with sniffles; a green, mucous nasal discharge; and respiratory complications. This may result in the death of the animal.

Papillomas, resembling heavy black warts, are often found on nearly all adult porcupines, especially on the upper lips, but it is doubtful that these produce fatalities. Some are extensive and generally depilated. Taylor (1935) reported finding a number of animals with large tumors in various sites that may have been cancerous. Dodge (1967) found a liposarcoma in a uterine horn that was initially mistaken for a developing fetus.

Symbionts. Although the porcupine utilizes cellulose, only one or possibly two flagellate protozoans, tentatively identified as *Monocercomonoides* (Dodge 1967), have been found in the intestines. Balows and Jennison (1949) described a thermophylic cellulose-decomposing bacteria of sufficient numbers to aid digestion and Johnson and McBee (1967) reported that cecal fermentation contributed 16–33 percent of the porcupine's maintenance energy requirements.

Accidents and Injuries. In addition to highway accidents, the porcupine is also subject to injuries including broken or lost toes; broken legs, hips, and ribs; broken incisors; eye and ear injuries; and hernias. Most of these are probably the result of falling from trees but foot and leg injuries may be due to steel trap encounters. One animal, exhibiting constant sinestral movements, had a packet of white pine needles encapsulated in the right auditory canal. The tympanic membrane and inner ear ossicles had been destroyed but there was no indication of infection. Another animal had a 10-cm-long, 1.25-cm-diameter pine limb completely encapsulated in the dorsal abdominal wall. If a porcupine lives to old age (7-10 years), molar tooth wear, malocclusion, and subsequent jaw infections are extremely debilitating and often result in death.

ECONOMIC STATUS AND MANAGEMENT

The porcupine gained its unpopularity mainly because of its feeding habits. Its economic impact on forest or timber crops, especially conifers, is well documented. Although essentially a vegetarian, it occasionally gnaws on inanimate objects that often are of economic importance. Damage to buildings and signs, especially those made of plywood, represents substantial losses in some areas. Apparently plywood bonding glue is attractive to porcupines; extensive surface areas are gnawed away to expose the glue line. Other inanimate objects such as vehicular synthetic fuel and hydraulic lines, tires, steering wheels, seat coverings, shift knobs, and electrical wiring have all been gnawed on. Aluminum canoe seats and gunwales, canoe paddles, axe handles, and other implements associated with human perspiration have been damaged. New England sugarbush owners are often plagued by porcupines gnawing plastic sap lines that gravity feed sap to the sugarhouse. Ski resorts often report damaged electrical and communication cables and some have found the plastic shock bumpers on lift towers damaged by porcupines. Power companies are sometimes troubled by outages due to porcupines gnawing on cables or monitoring lines.

Numerous economic surveys have been conducted relative to porcupine timber damage with variable conclusions depending on the area and forest crop involved. Lawrence (1957) estimated that one porcupine is capable of inflicting $6,000 of damage to the forest industry. Van Deusen and Meyers (1962) reported that thinning tends to produce timber stands more favorable to porcupines and felt that control should be anticipated in those stands. With current high-yield forestry practices of chemical thinning, aerial fertilization, and even-aged stand management, porcupine damage can inflict considerable economic loss and require control. In western Oregon, for example, most damage occurs in thinned stands 10-30 years old; dominant trees are selected by porcupines (Dodge and Canutt, 1969). Gill (1972), in his investigation of porcupine winter habitat preferences in Alberta, reported that excessive damage in limber pine stands may warrant control measures. Generally, most foresters feel that timber damage by porcupines is relatively insignificant or can be tolerated. In New England, porcupine winter habitat and feeding is generally in precipitous ledge or rock-fall areas and timber in those sites is usually low grade and not economically feasible to harvest. Therefore, direct control may be justified in areas of localized damage where economic losses are significant but widespread control and bounty systems are no longer recommended.

Many states, possessing a forest industry within the porcupine's range, have had bounty systems or subsidized porcupine control programs. New Hampshire, the last state to have a statewide porcupine bounty system, repealed that law in 1979. In spite of bounties, poisoning campaigns, and hunting and trapping contests, the porcupine is still well represented throughout its range.

Toxic materials, such as the strychnine salt block (Spencer 1950) and sodium arsenite apples (Faulkner and Dodge 1962), have been used with some success but recent legislation has removed these and other poisons from use.

Many repellents have been evaluated (Welch 1954; Dodge 1958) but none is currently recommended for porcupines. Common wood preservatives such as those containing copper napthanate and pentachlorophenol have proven effective as a porcupine repellent when applied to exterior plywoods but none is registered as an animal repellent.

Mechanical barriers, such as electric fences (Spencer 1948) and individual tree guards using metal flashing, are effective in curtailing localized damage but are not economically feasible for large areas. A simple 75-cm band of aluminum flashing will effectively prevent porcupines from climbing ornamental or fruit trees providing these are attached above the snow line.

Natural population increase and reintroduction of fisher (*Martes pennanti*) has apparently been a useful biological control method because of the fisher's effectiveness as a predator on porcupines. Cook and Hamilton (1957) reported a decline in porcupine populations in New York as fisher populations increased. Many states, including Michigan, Wisconsin, Montana, Idaho, Oregon, and California, have successfully reintroduced the fisher.

Natural forest succession, to closed canopy stands where understory vegetation is suppressed, can reduce porcupine populations. Also, forest cutting such as "high grading" or selection of better-quality hemlock near long-used den sites is an effective control measure in the Northeast, where porcupines depend on hemlock during the winter and thus the absence of hemlock is a limiting factor.

With changing legislation regarding use of toxicants, traps, and firearms, one should consult state conservation agencies when a porcupine damage problem exists. Some agencies, such as state conservation, fish and wildlife, county and state extension services, and U.S. Fish and Wildlife, Wildlife Assistance Offices, will provide information and advice, and some may loan live-traps for porcupine control.

LITERATURE CITED

Anderson, R. M., and Rand A. L. 1943. Variation in the porcupine (Genus *Erethizon*) in Canada. Can. J. Res., Sect. D, 21:292-309.

Balows, A., and Jennison, M. W. 1949. Thermophilic, cellulose-decomposing bacteria from the porcupine. J. Bacteriol. 57:135.

Barkalow, F. F., Jr. 1961. The porcupine and fisher in Alabama archeological sites. J. Mammal. 42:544–545.

Brander, R. B. 1971. Longevity in wild porcupines. J. Mammal. 52:835.

———. 1973. Life history notes on the porcupine in a hardwood-hemlock forest in upper Michigan. Pap. Michigan Acad. Sci., Arts, and Letters 5:425–433.

Buhler, G. A. 1970. *Monecocestus giganticus* sp.h. (Cestoda: Anoplocephalidae) from the porcupine *Erethizon dorsatum* L. (Rodentia). Proc. Helmin. Soc. Washington 37:243–245.

Burge, B. L. 1966. Vaginal casts passed by captive porcupine. J. Mammal. 47:713–714.

Choquette, L. P. E.; Broughton, E.; and Gibson, G. G. 1973. *Schistosomatium douthitti* (Cort, 1914) Price, 1927 in a porcupine (*Erethizon dorsatum*) in eastern Ontario, Canada. Can. J. Zool. 51:1317.

Cook, D. B., and Hamilton, W. J. 1957. The forest, the fisher and the porcupine. J. For. 55:719–722.

Dean, H. J. 1950. Porcupine swims for food. J. Mammal. 31:94.

Dodge, W. E. 1958. Investigations concerning the repellency and toxicity of various compounds for control of porcupine, *Erethizon d. dorsatum*. M.S. Thesis. Univ. Massachusetts, Amherst. 50pp.

———. 1966. Status of porcupine in western Washington. Washington For. Prot. Assoc., Seattle. 6pp.

———. 1967. Life history and biology of the porcupine (*Erethizon dorsatum*) in western Massachusetts. Ph.D. Thesis. Univ. Massachusetts, Amherst. 167pp.

———. 1970. The porcupine in western Washington. Washington For. Prot. Assoc., Seattle. 6pp.

Dodge, W. E., and Barnes, V. G., Jr. 1975. Movements, home range, and control of porcupines in western Washington. Wildl. Leaf. 508. U.S. Fish Wildl. Serv., Washington, D.C. 7pp.

Dodge, W. E., and Canutt, P. R. 1969. A review of the status of the porcupine (*Erethizon dorsatum epixanthum*) in western Oregon. Bur. Sport Fisheries and Wildl. and U.S. For. Serv., Portland, Oreg. 25pp.

Eifrig, G. 1909. Winter birds of New Ontario and other notes on northern birds. Auk 26:46–59.

Faulkner, C. E., and Dodge, W. E. 1962. Control of the porcupine in New England. J. For. 60:36–37.

Gensch, R. H. 1946. Observations on the porcupine in northern Wisconsin and northern Michigan. U.S. Fish Wildl. Serv. Prog. rep. 16pp. Manuscript.

Gill, D., and Cordes, L. D. 1972. Winter habitat preference of porcupines in the southern Alberta foothills. Can. Field Nat. 86:349–355.

Guilday, J. E., and Hamilton, H. W. 1973. The late Pleistocene small mammals of Eagle Cave, Pendleton County, West Virginia. Ann Carnegie Mus. 44:45–58.

Hart, J. E. 1957. Helminth studies on eleven species of mammals native to New England. M.S. Thesis. Univ. Massachusetts, Amherst. 116pp.

Hibbard, C. W., and Mooser, O. 1963. A porcupine from the Pleistocene of Aquascalientes, Mexico. Contrib. Mus. Paleont., Univ. Michigan 18:245–250.

Highby, P. R. 1943. Mosquito vectors and larval development of *Dipetalonema arbuta* Highby (Nematoda) from the porcupine, *Erethizon dorsatum*. J. Parasitol. 29:243–252.

Hull, T. G. 1930. Diseases transmitted from animals to man. C. C. Thomas. Springfield, Ill. 350pp.

Johnson, J. L., and McBee, R. H. 1967. The porcupine cecal fermentation. *Erethizon dorsatum epixanthum*. J. Nutr. 91:540–546.

Jones, M. S. 1973. Ecology and life history of the porcupine in Nova Scotia. M.S. Thesis. Acadia Univ., Wolfville. 142pp.

Jonkel, C. J. 1968. A polar bear and porcupine encounter. Can. Field Nat. 82:222.

Kelker, G. H. 1943. A winter wildlife census in northeastern Wisconsin. J. Wildl. Manage. 7:133–141.

Kelly, G. M. 1973. The biology of an isolated porcupine population. M.S. Thesis. Univ. Massachusetts, Amherst. 48pp.

———. 1977. Fisher (*Martes pennanti*) biology in the White Mountain National Forest and adjacent areas. Ph.D. Thesis. Univ. Massachusetts, Amherst. 178pp.

Kuhn, E.; Houtz, C. S.; and Axley, A. 1953. Tularemia from a porcupine. Rocky Mtn. Med. J. 50:736.

Lawrence, W. H. 1957. Porcupine control: a problem analysis. For. Res. Notes. Weyerhaeuser Co., Centralia, Wash. 43pp.

McDade, H. C., and Crandall, W. B. 1958. Perforation of the gastrointestinal tract by an unusual foreign body—a porcupine quill. New England J. Med. 258:746–747.

Marshall, W. H.; Gullion, G. W.; and Schwab, R. B. 1962. Early summer activities of porcupines as determined by radio-positioning techniques. J. Wildl. Manage. 26:75–79.

Mirand, E. A., and Shadle, A. R. 1953. Gross anatomy of the male reproductive system of the porcupine. J. Mammal. 34:210–220.

Moody, P. A., and Doniger, D. E. 1956. Serological light on porcupine relationships. Evolution 10:47–55.

Mossman, H. W., and Duke, K. L. 1973. Comparative morphology of the mammalian ovary. Univ. Wisconsin Press, Madison. 461pp.

Mossman, H. W., and Judeas, I. 1949. Accessory corpora lutea, lutein cell origin, and the ovarian cycle in the Canadian porcupine. Am. J. Anat. 85:1–40.

Nalbandov, A. V. 1958. Reproductive physiology. W. H. Freeman & Co., San Francisco. 217pp.

Olsen, O. W., and Tolman, C. D. 1951. Wellcomia evaginata (Smitt 1908) (Osyuridae: Nematoda) of porcupines in mule deer, *Odocoileus hemionus*, in Colorado. Proc. Helm. Soc. Washington 18:120–122.

Parmalee, P. W. 1971. Fisher and porcupine remains from cave deposits in Missouri. Trans. Illinois State Acad. Sci. 64:225–229.

Payne, D. D., and O'Meara, D. C. 1958. *Sarcoptes scabei* infestation of a porcupine. J. Wildl. Manage. 22:321–322.

Perotta, C. A. 1959. Fetal membranes of the Canadian porcupine (*Erethizon dorsatum*). Am. J. Anat. 104:35–59.

Ray, C. E. 1958. Fusion of the cervical vertebrae in the Erethizontidae and Dinomyidae. Breviora 97:1–11.

Salkin, I. F.; Gordon, M. A.; and Stone, W. B. 1976. Cutaneous infection of a porcupine (*Erethizon dorsatum*) by *Auriobasidium pullulans*. Sabouraudia 14:47–49.

Shadle, A. R. 1951. Laboratory copulations and gestations of porcupine, *Erethizon dorsatum*. J. Mammal. 32:219–221.

———. 1955. Removal of foreign quills by porcupines. J. Mammal 36:463–465.

Spencer, D. A. 1948. An electric fence for use in checking porcupine and other mammalian crop depredations. J. Wildl. Manage. 12:110–111.

———. 1949. Porcupine problems on the Nicolet National

Forest. U.S. Fish Wildl. Serv. Spec. Rep. 14pp. Manuscript.

——. 1950. Porcupines, rambling pincushions. Natl. Geogr. 98:247–264.

Stenlund, M. H. 1955. A field study of the timber wolf (*Canis lupus*) on the Superior National Forest, Minnesota. Minnesota Dept. Conserv. Tech. Bull. 4:1–55.

Struthers, P. H. 1941. The prenatal development of the pancreatic and exta-hepatic ducts in the Canadian porcupine (*Erethizon dorsatum*). Anat. Rec. 81:143–161.

Swena, R., and Ashley, L. M. 1956. Osteology of the common porcupine *Erethizon dorsatum*. Publ. Dept. Bio. Sci. and Bio. Stat., Walla Walla Coll. 18:1–26.

Taylor, W. P. 1935. Ecology and life history of the porcupine (*Erethizon epixanthum*) as related to the forests of Arizona and the southwestern United States. Univ. Arizona Bull. 6:1–177.

Van Deusen, J. L., and Meyers, C. A. 1962. Porcupine damage in immature stands of ponderosa pine in the Black Hills. J. For. 6:811–813.

Welch, J. F. 1954. Rodent control: A review of chemical repellents for rodents. J. Agric. and Food Chem. 2:142–149.

White, J. A. 1970. Late Cenozoic porcupines (Mammalia, Erethizontidae) of North America. Am. Mus. Novit. 2421:1–15.

Wood, A. E. 1955. A revised classification of rodents. J. Mammal. 36:165–187.

Woods, C. A. 1973. *Erethizon dorsatum*. Mammal. Species 29:1–6.

WENDELL E. DODGE, Massachusetts Cooperative Wildlife Research Unit, University of Massachusetts, Amherst, Massachusetts 01003.

IV

Cetacea

18

Bottlenose Dolphin

Tursiops truncatus

and Other Toothed Cetaceans

Stephen Leatherwood

Randall R. Reeves

NOMENCLATURE

COMMON NAMES. Bottlenose(d) dolphin or porpoise, black porpoise (used indiscriminately by some U.S. fishermen in the eastern tropical Pacific to refer to this and other species of small, basically black cetaceans); common porpoise, gray porpoise (British); Delfín naríz de botella (Spanish).
SCIENTIFIC NAME. *Tursiops truncatus*

Systematics. The suborder Odontoceti of the order Cetacea is subdivided into 6 families, as follows (Mead 1975*a*, appendix 1): Physeteridae, including 3 species, all of which inhabit waters bordering North America (see "Sperm Whale" section below); Monodontidae, the arctic narwhal (*Monodon monoceros*) and the white whale, or beluga (*Delphinapterus leucas*); Ziphiidae, including 5 or 6 genera and at least 18 species worldwide; Delphinidae, including 12 genera and 20 recognized species known to occur in the North Pacific and/or the North Atlantic; Phocoenidae, including two species confined to the North Pacific, the Dall's porpoise (*Phocoenoides dalli*) and the Gulf of California harbor porpoise, or cochito (*Phocoena sinus*), as well as the more widely distributed harbor porpoise (*Phocoena phocoena*) of the Northern Hemisphere; and Platanistidae, the river dolphins of South America and the Indo-Pacific.

The bottlenose dolphin (figure 18.1) is a midsized delphinid and, as the archetypal dolphin, possesses many of the features commonly associated with this group. A perceived problem, frequently raised by laymen, concerns the distinction between dolphins and porpoises. Most cetologists use the terms interchangeably and do not ascribe special meaning to them. However, for clarity in this chapter we refer to members of the family Delphinidae as dolphins and to members of the family Phocoenidae as porpoises. Delphinids usually have two or more fused vertebrae, a well-defined beak, and peglike or conical teeth in each jaw, numbering 20 or more in individuals of smaller species. (*Feresa attenuata, Globicephala* spp., *Orcinus orca*,

Pseudorca crassidens, and *Grampus griseus* are exceptional among the delphinids in having, usually, fewer than 13 teeth.) Phocoenids, by contrast, have a free atlas and axis, a poorly defined or at least inconspicuous beak, and usually fewer than 25 pairs of laterally flattened, spade-shaped teeth in each jaw.

Although many subspecific and at least 20 specific names have been attributed to the genus *Tursiops* (Hershkovitz 1966), only three are commonly used to refer to bottlenose dolphins in North American waters: *T. truncatus* (Montagu 1821), *T. gilli* (= *gillii*) (Dall 1873), and *T. nuuanu* (Andrews 1911). The systematics and taxonomy of bottlenose dolphins off South Africa (Ross 1977) and in the eastern North Pacific (Walker 1980) have been reviewed recently. It continues to be true, however, that "a taxonomic revision of this popularly known and widely distributed group of dolphins is urgently needed" (Hershkovitz 1966). Pending such a review, *T. truncatus* is the scientific name most generally accepted for bottlenose dolphins found in North American waters (Mitchell 1975*b*).

In all areas where bottlenose dolphin systematics have been studied, there appear to be two ecotypes, a coastal form and an offshore form. The latter includes residents near oceanic and coastal islands (Walker 1975; Mitchell 1975*b*; Leatherwood and Reeves 1978). Differences between the ecotypes are subtle and even such nominal species as *T. gilli* and *T. nuuanu* are so much alike that they may not be readily distinguished in the field (Van Gelder 1960). Proposed species and ecotypes have been defined principally on the basis of skull differences, such as size of the mandibular condyles (Dall 1873; True 1889) and teeth (Van Gelder 1960). Participants in a recent international review of the biology of small cetaceans concluded that there probably is only one species of *Tursiops*, with sharply defined geographic races, varying in body size and tooth proportions, and distributed differentially relative to sea temperature and depth (Mitchell 1975*b*).

In comparing offshore and coastal specimens from the eastern North Pacific, Walker (1980) demon-

FIGURE 18.1. Captive bottlenose dolphins (*Tursiops truncatus*) performing at Sea World, San Diego, Calif. The animal low in the foreground is from the Pacific; the other two are from the Atlantic.

FIGURE 18.2. Approximate distribution of the bottlenose dolphin (*Tursiops truncatus*) in North American waters.

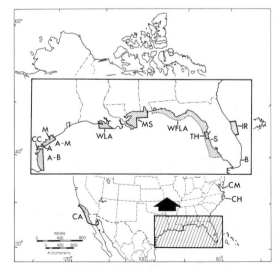

FIGURE 18.3. Principal areas where the bottlenose dolphin (*Tursiops truncatus*) has been studied and/or collected.

370

strated modal differences in skull measurements. He found clear separations among three eastern North Pacific stocks in size at sexual maturity, tooth diameter, parasite loads, and feeding habits, however. Electrophoretic examination of variability at a number of gene loci revealed sharp differences between Atlantic and Pacific specimens and suggested disjunct offshore-nearshore populations in the Atlantic (Duffield et al. 1979). Preliminary data also suggest differences in some basic hemoglobin and red blood cell characteristics from coastal/estuarine versus offshore bottlenose dolphins in various areas (Duffield et al. in preparation). It is not within the scope of this chapter to address the complex and still unanswered questions about the systematics and taxonomy of bottlenose dolphins. Quantitative comparisons of large series of specimens, together with more detailed regional studies of the limits of variability within populations (after Ross 1977 and Walker 1980), will be necessary before most questions are resolved.

DISTRIBUTION

Bottlenose dolphins are abundant worldwide in tropical and warm temperate waters, and they occasionally stray into colder areas (Mitchell 1975*b*).

In the western North Atlantic, bottlenose dolphins have been reliably reported from as far north as Nova Scotia and southern Greenland, presumably in or near intrusions of temperate water under influence of the Gulf Stream (Leatherwood et al. 1976; National Marine Fisheries Service 1978). Northernmost records are from offshore. These dolphins are better known from the New York Bight southward, where they begin to occur closer to the coast. They are found almost continuously southward from New Jersey along the Atlantic, Gulf of Mexico, and Caribbean coasts, including the continental shelf and oceanic islands (figures 18.2 and 18.3).

North Atlantic and North Pacific populations appear to be effectively isolated from each other, judging by accumulation of gene differences (Duffield et al. 1979) and geographic distribution. In the North Atlantic there appear to be different populations, coastal and offshore, between which there is limited gene interchange (Duffield et al. 1979; Duffield 1980).

In the eastern North Pacific, bottlenose dolphins have been alleged to occur "infrequently . . . in offshore currents, perhaps as far north as southern Oregon" (National Marine Fisheries Service 1978). However, there are no published records north of San Francisco and only two north of Point Conception. A single skull, estimated to have been in the water for 50 to 100 years, was snagged by a fisherman near the Richmond–San Rafael Bridge, inside San Francisco Bay, in 1958 (Orr 1963). The type of *T. gilli* is a specimen "from Monterey" brought to Dall by Scammon (Dall 1873). Bottlenose dolphins are currently known only from south of Point Conception (Leatherwood et al. in press). Outside tropical waters they are encountered principally in the coastal zone, less frequently to the edge of the continental shelf and

beyond. The nearshore ecotype is distributed continuously from southern Los Angeles County south along the Mexican coast, including the Gulf of California. Bottlenose dolphins are found offshore throughout much of the tropical eastern Pacific, including the oceanic islands of the Revillagigedos, Cocos, Clipperton, and Hawaii (figure 18.2). We know that bottlenose dolphins occur near shore off Central America as well.

DESCRIPTION

External Appearance. Most bottlenose dolphins have a short, thick beak, a robust, almost chunky head and trunk, and a body tapering rapidly behind the dorsal fin. Slimmer individuals with a longer beak are often seen, sometimes even in the same group with stockier individuals. There is much variability in individual appearance.

It is worth noting that hybridization between bottlenose dolphins and Risso's dolphins (*Grampus griseus*) (M. Nishiwaki 1980, personal communication) and bottlenose dolphins and rough-toothed dolphins (*Steno bredanensis*) (Dohl et al. 1974) has occurred in captivity, and at least one instance of suspected hybridization between free-ranging bottlenose and Risso's dolphins is known (Fraser 1940).

The flippers are convex and smoothly curved on the forward margin and pointed on the tip. The flukes are sinuously curved on the rear margin and separated by a deep notch. Their combined width is equal to 10–13 percent of total body length (Ridgway 1972). The dorsal fin, located near midbody, varies in shape from falcate to subtriangular, reaching a maximum height of about 35 cm. There appear to be significant age-related changes in body proportions (Tomilin 1967).

Coloration. Bottlenose dolphins are not strikingly marked. In a comparison of delphinid pigmentation patterns as possible indicators of evolutionary relationships among genera, Mitchell (1970) found the pattern of *Tursiops* to be among the most generalized. Perrin (1972), using discrete pattern component analysis to describe the markings of representative delphinids, illustrated the *Tursiops* pattern as principally a simple cape and cape overlay. Dominant tones are lead gray, charcoal, and brownish gray. Markings, including brushed striping from the eye to the rostrum and from the blowhole to the apex of the melon, an eye to flipper stripe, and a small genital blaze, are subtle and barely visible on most animals. In some individuals the cape system is obscured and the dorsum appears black. In others, interplay of cape and cape overlay components produces a more complex pattern of dark back, lighter sides, and still lighter belly (Walker 1980). The belly is often pink. The eye generally is dark brown.

Color variants occur. They include albinos and Himalayan-patterned, cinnamon or buff individuals (Gray 1964; Caldwell and Caldwell 1972). Many of these anomalously colored animals have red or pink eyes. Some individuals are spotted, and others have and 650 kg (National Marine Fisheries Service 1978), given for other regions have not been verified off North America. At birth, bottlenose dolphins are about 0.9–1.3 m long and weigh 9–12 kg (Gunter 1942; McBride and Kritzler 1951; Harrison et al. 1972; Sergeant et al. 1973; Ridgway and Benirschke 1977). Newborn calves are an estimated 10–15 percent of their mother's weight (Slijper 1966).

Skull and Skeleton. General features of the *Tursiops* skull include: a rostrum length of more than 2.2 times its breadth, a greatest length of less than one-half of the condylobasal length, and the mandibular symphysis less than one-fifth of the mandibular length (Nishiwaki 1972; Green 1972) (figure 18.4). The teeth are relatively large, 3–4 cm in length and 10–11 mm in diameter. There are eighteen to twenty-six teeth in each row (Dall 1873; True 1889; Ridgway 1972; Leatherwood et al. 1976).

The flippers have 5 digits; the carpals are separate bones; and, especially in the middle 3 digits, the metacarpals may be indistinguishable from proximal phalanges (Green 1972). Vertebral counts are reported as 63–66, comprised of 7 cervical, 12–14 thoracic, 15–17 lumbar, and 26–30 caudal vertebrae (Tomilin 1967; Green 1972; Nishiwaki 1972). The cervical vertebrae fuse with "no apparent pattern" except that all seven are rarely ankylosed (Fischer 1881). The neck is shortened, with from 2 to all the caudal vertebrae fused to form a single osseus unit (Green 1972).

REPRODUCTION

Reproductive parameters and behavior of bottlenose dolphins are poorly known. Most available information on life history derives from materials and observations of Florida specimens, principally captives. Data on reproductive condition continue to be correlated with age using uncertain techniques of age determination (Hui 1977; Perrin and Myrick in press; see also the section "Age Determination"). There have been few studies on behavior of free-ranging animals (e.g., see Würsig and Würsig 1977, 1980; Leatherwood 1977; Shane and Schmidly 1978). Generalizations on behavior of natural populations from studies of captive colonies should continue to be regarded with caution (see Ridgway and Benirschke 1977). For informed management of bottlenose dolphin populations, detailed studies of reproductive parameters and age-sex structure of "population(s)" and stocks are needed.

Free-ranging bottlenose dolphins live an estimated 25–30 years or more (Ridgway 1968, 1972; Sergeant et al. 1973) and natural mortality rates are thought to be low (National Marine Fisheries Service 1978). Assuming deposition of one dentinal layer per year, males are sexually mature by 10–13 years of age and 245–260 cm of length, and females by 5–12 years of age and 220–235 cm of length (Harrison et al. 1972; Sergeant et al. 1973; Mitchell 1975*b*; Odell 1975; Ridgway and Benirschke 1977). Experience with captive breeding colonies indicates that older males are responsible for most conceptions (D. Duf-

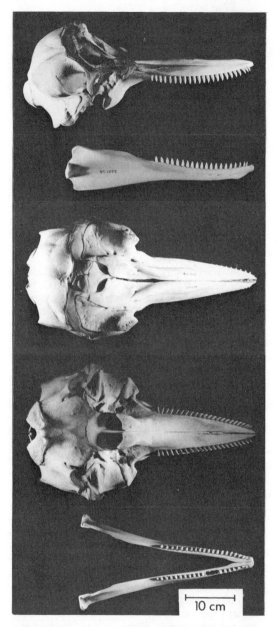

FIGURE 18.4. Skull of an Atlantic bottlenose dolphin (*Tursiops truncatus*). From top to bottom: lateral view of cranium, lateral view of mandible, dorsal view of cranium, ventral view of cranium, dorsal view of mandible.

diffuse pigmentation in the cape (Caldwell and Caldwell 1972) or an extension of the dorsal field overlay onto the ventrum (Ross 1977). Most geographic variation in pigmentation is in the dorsal field system overlaying a basic general pattern (Perrin 1972).

Size. Bottlenose dolphins off North America grow to at least 3 m and 275 kg (Ridgway 1968; Mitchell 1975*b*). Males are somewhat longer and heavier than females. The larger sizes, to 4 m or more (Gaskin 1972) field 1980, personal communication), which suggests that social maturity in males may be as important a factor in reproduction as sexual maturity.

Sexual activity may be initiated by either sex. It includes chasing, stroking with flippers or flukes, head butting, and vocalizations. Copulation is accomplished by ventral-ventral swimming (McBride and Hebb 1948; Tavolga and Essapian 1957; Caldwell and Caldwell 1972; contributors to Ridgway and Benirschke 1977). Intromission is brief, usually lasting less than one minute, and there are no behavioral indications of ejaculation.

Males appear to be sexually active year-round, although seasonal peaks in hormonal activity have been postulated (Harrison and Ridgway 1971). Ovulation has been reported as spontaneous (Slijper 1966) and as induced (Harrison et al. 1972). In captive animals, ovulation could not be induced by artificial vaginal stimulation, and spontaneous ovulation was observed to occur in the absence of males (Kirby and Sawyer-Steffan 1979). Corpora albicantia persist in the cetacean ovary, and in bottlenose dolphins often more corpora are found than might reasonably be derived from pregnancy (Harrison 1977).

Female reproductive organs include a narrow vaginal entrance and a spermathecal recess capable of holding up to 10 ml of fluid. These modifications are thought to prevent sea water from washing away semen following copulation (Green 1972, 1977). As in other cetaceans, the uterus is bicornate and implantation and development almost always occur in the left horn.

Gestation is about 12 months (Ridgway 1968; Ridgway and Benirschke 1977). Labor may last from 30 minutes to well over one hour, though most signs are subtle. In normal births less than 6 minutes elapse between the appearance of the umbilicus and parturition. Stillbirths are protracted (Ridgway 1968). Birth is tail first (figure 18.5) and the young are precocious. A second adult is often present at birth and participates in support and attention to the calf. Although generally called a "midwife" or "auntie" and thought to play a positive role in the birth process, an attending female may actually be competing with the mother for the calf. In captive colonies other adults sometimes must be removed if the calf is to survive (Ridgway and Benirschke 1977).

In the wild, the calving interval is probably two years (Mitchell 1975b). A February–May breeding and calving season has been postulated for bottlenose dolphins in the southeastern United States (Odell 1975). There is substantial evidence for a fall rut as well (Harrison and Ridgway 1971; Ridgway and Benirschke 1977). Seasonal differences in the number of calves observed in free-ranging populations (Asper and Odell 1980) (table 18.1) support the hypothesis of year-round breeding and calving with peaks in spring and fall.

ECOLOGY

Home Range. In all areas where it has been studied, the nearshore ecotype appears to occur in discrete units with limited, overlapping home ranges (table 18.2). Evidence of home range comes mainly from long-term studies of artificially marked (Irvine et al. 1979a) or naturally distinguishable (Caldwell 1955; Würsig and Würsig 1977; Shane and Schmidly 1978; Würsig 1978) individuals. Individuals that could be recognized by natural marks comprised about 12 percent of a Sarasota, Florida, population (loc. cit.) and 9 percent of an Aransas, Texas, population (Shane 1977; Shane and Schmidly 1978). Combinations of freeze brands, button tags, streamers, and radio tags have been applied to over 150 bottlenose dolphins in Sarasota Bay (Irvine et al. 1979a) and Indian River (Asper and Odell 1980).

In addition to the well-studied populations in

Marine Studios, Marineland, Fla.

FIGURE 18.5. Dolphins are born tail first, so the blowholes are late to emerge. Young dolphins are precocious, but they are closely tended by an adult for the first months of life.

TABLE 18.1. Some estimates of percentages of calves of the year in "populations" of bottlenose dolphins in coastal waters of the United States

Area[a] Code	Area Description	Data Source	Period Covered	Observed Percentage of Calves[b]	Remarks
A	Aransas Pass area, Texas	Shane (1977)	June 1975–May 1977	7.61	x̄ for 15 months
		Shane (1977)	February 1976	3.65	low monthly mean
		Shane (1977)	May 1976	12.92	high monthly mean
S	Tampa Bay south to Big Pass, Florida	Irvine et al. (1979a)	Spring	to 14.0	
TH	Tampa to Hudson, Florida	Leatherwood and Show (1980)	November 1979	2.72–3.93	Considered exceptionally low because observers not experienced with classification criteria.
CC	Corpus Christi Bay, Texas	Leatherwood and Show (1980)	September 1979	6.25–8.59	
AB	Aransas Pass–Brownsville, Texas	Leatherwood and Show (1980)	September 1979	5.00–8.00	
IR	Indian and Banana rivers, Florida	Leatherwood and Show (1980)	November 1979	10.27–15.58	
AM	Aransas-Matagorda	Barham et al. (1980)	March 1978	9.40–10.7	
IR	Indian and Banana rivers	Leatherwood (1979)	August 1977	8.10–10.10	
MS	Mississippi Sound and adjacent waters of eastern Louisiana	Leatherwood (1977); Leatherwood et al. (1978a)	July 1974	7.7–12.4	
MS-LA	Mississippi Sound to western Louisiana	Leatherwood (1977); Leatherwood et al. (1978a)	July 1975	7.9–11.4	
WFLA	western Florida 28°32′N to 25°08′N	Irvine et al. (1980)	September 1979	5.7[c]	Study covered July–December
WFLA	Sarasota Bay, Florida	Irvine et al. (1979a)	yearly high	to 11.0	
IR	Indian and Banana rivers	Asper and Odell (1980)	August–December 1979	8.1	

[a]See figure 18.3 and table 18.2.

[b]The reported ranges are positively identified calves (the low value) + animals tentatively identified as "yearlings" but possibly actually larger calves of the year = total possible calves (the high value).

[c]This figure is the high month.

Florida and Texas, there is some evidence of home range or residency in Mississippi Sound (Leatherwood and Platter 1975) and off southern California (Leatherwood and Reeves 1978). Animals of the offshore form appear to be resident around some of the oceanic islands, such as Clarion in the Revillagigedos Islands off Mexico, and Clipperton, nearly 2,000 km west of Costa Rica.

Migration. Long-distance seasonal migrations apparently are undertaken primarily by offshore bottlenose dolphins, whose diet is heavily dependent on highly migratory species of fish and squid. Along the Outer Banks of North Carolina, "resident" coastal animals are supplemented by large influxes of southbound fall and northbound spring migrants passing through the area (True 1891). Presence of the stalked barnacle *Xenobalanus* cf. *X. globicipitis* has been noted on the tails of northward-migrating animals in late April and May, but not on southward-migrating animals in fall or winter (Mead 1975b). There is a marked increase in bottlenose dolphin abundance during summer in Calibogue Sound, Georgia, and at the

mouth of the Savannah River (Hogan 1975), and a corresponding increase inshore during winter off eastern Florida (Leatherwood 1979; Leatherwood and Show 1980). There are more than twice as many bottlenose dolphins in the Aransas Pass area during midwinter as there are at other seasons (Shane 1977; Shane and Schmidly 1978).

Bottlenose dolphins of the offshore ecotype extend their range northward to the area of the Northern Channel Islands off California principally during summer and early fall (Leatherwood and Reeves 1978).

Habitat Preferences. Bottlenose dolphins exploit an impressive range of habitats. The nearshore form occurs occasionally in freshwater rivers (True 1890), including the Mississippi as far as 160 miles upstream (Gunter 1954). It is found more frequently in river mouths, bays, lagoons, and estuarine complexes, and in virtually any shallow marine region (0.5–20 m deep) (Irvine et al. 1979a; Leatherwood et al. 1976, in press; Odell and Reynolds 1980). In a detailed study off Argentina, coastal bottlenose dolphins were observed

TABLE 18.2. Principal areas of North America for which data are available on "populations" of bottlenose dolphins and the most recent or complete sources

Area Code[a]	Area Description	Principal Sources of Data on Bottlenose Dolphin "Populations"
EASTROPAC	The portions of the eastern tropical Pacific in which purse-seine fishing activity and research on yellowfin tuna are conducted, referred to as the Commission Yellowfin Regulatory Area (CYRA)	National Marine Fisheries Service, Southwest Fisheries Center and Inter-American Tropical Tuna Commission.
CA	California and Baja California waters	Norris and Prescott (1961); Walker (1975); Leatherwood (1975); Walker (1980).
STEX	The area off southern Texas bounded by about 27°02′N, 95°09′W; 27°02′N, 97°22′W; 26°02′N, 94°56′W; and 26°02′N, 97°09′W	U.S. Fish and Wildlife Service (1980).
A-B	The waters between Aransas Pass and Brownsville, Texas, including Laguna Madre and Gulf coast waters to ten fathoms	Leatherwood and Show (1980).
CC	The waters of Corpus Christi and Nueces bays, Texas	Leatherwood and Show (1980).
NTEX	The area off central Texas bounded by about 27°48′N, 94°10′W; 28°26′N, 96°17′W; 27°N, 95°40′W; and 27°39′N, 97°11′W	U.S. Fish and Wildlife Service (1980).
A	The Aransas Pass area of southern Texas	Shane (1977); Shane and Schmidly (1978); Barham et al. (1980); Shane (1980).
M	The Pass Cavallo area of southern Matagorda Bay, Texas	Gruber (1979); Barham et al. (1980).
A-M	The waters inside the barrier islands between Aransas Pass and the junction of the intracoastal waterway and northern Matagorda Bay	Barham et al. (1980).
WLA	Atchafalaya Bay and adjacent coastal waters to ten fathoms	Leatherwood et al. (1978a).
MS	Mississippi, Chandeleur, and Breton sounds, and adjacent shallow marshlands and Gulf of Mexico	Leatherwood (1975); Leatherwood and Platter (1975); Leatherwood (1977); Leatherwood et al. (1978a).
WFLA	Thirteen study blocks of coastal waters between about Sanibel Island, Florida, and the Alabama-Florida border	Odell and Reynolds (1980).
TH	The waters from mainland shore to 20 nm seaward from Passe-au-Grille, the northern entrance to Tampa Bay, to Hudson, Florida	Leatherwood and Show (1980).
NFLA	The area off northwestern Florida bounded by about 28°N, 82°52′W; 28°N, 85°05′W; 27°N, 82°24′W; and 27°N, 84°41′W	U.S. Fish and Wildlife Service (1980).
S	Coastal and inland marine waters extending south from the southern edge of Tampa Bay 40 km to Big Pass, Florida, and offshore to about 3 km	Irvine and Wells (1972); Irvine et al. (1979a); Wells et al. (1980).
SFLA	The area off southwestern Florida bounded by about 26°11′N, 81°43′W; 26°11′N, 83°54′W; 25°11′N, 81°43′W; and 25°11′N, 83°55′W	U.S. Fish and Wildlife Service (1980).
E	The waters of Everglades National Park, South Florida	Odell (1979).
B	The waters of Biscayne Bay, near Miami, Florida	Odell (1976a, 1976b).
IR	The intracoastal waterway and connected waters of Indian and Banana rivers, eastern Florida, from about 28°47′N to 27°10′N.	Leatherwood (1979); Leatherwood and Show (1980); Asper and Odell (1980).
CH	Cape Hatteras, North Carolina	True (1885, 1891); Townsend (1914); Mead (1975b); Mitchell (1975a).
CM	Cape May, New Jersey	True (1891); Mead (1975b); Mitchell (1975a).

[a]See figure 18.3.

in water 2–39 m deep, apparently preferring depths of 2–6 m (Würsig and Würsig 1977). Along the Gulf of Mexico and Atlantic coasts of North America, they penetrate shallow grassy flats, using deeper channels for transit (Irvine et al. 1979a; Leatherwood 1979). Passes between open ocean and enclosed bays or lagoons are often centers of abundance, and the dolphins use intracoastal waterways and other deep channels to gain access to productive shallows (Shane 1977; Irvine et al. 1979a; Barham et al. 1980; Leatherwood and Show 1980).

The offshore ecotype is well known on the margins of some coastal (Leatherwood and Reeves 1978) and oceanic islands (Rice 1960) and atolls (Ely and Clapp 1973). It is also encountered in the open ocean (National Marine Fisheries Service 1978).

FIGURE 18.6. Bottlenose dolphins (*Tursiops truncatus*) were caught by fisheries at Cape May, N.J. (Photo from ca. 1907.)

Abundance and Trends in Population Size. There is no reliable estimate of the total number of bottlenose dolphins in North American waters. Until recently, reports on abundance were largely the subjective impressions of collectors, commercial fishermen, and other observers in areas where the animals are most accessible. Comments on early abundance off New Jersey and North Carolina are based on cumulative catch figures for 19th and early 20th century shore-based net fisheries (Mitchell 1975a; Mead 1975b) (figure 18.6). The absence of more rigorous estimates of present abundance is due both to the lack of focused research and to the unavailability of reliable methods of conducting surveys and interpreting their results.

Since 1974, largely in response to management considerations imposed by federal legislation (see the "Management" section), density and population estimates have been made using low-altitude aerial surveys in selected areas. Techniques for censusing bottlenose dolphins and analyzing survey data, similar initially to those used in other wildlife population studies (e.g., Seber 1973), have evolved through field application and experimentation (Leatherwood and Platter 1975; Leatherwood et al. 1978a; Leatherwood and Show 1980). Quadrat, strip, and line transect models have been used, and variables such as observer experience, season, time of day, survey altitude, and techniques used to measure distance have been tested. Problems remain, particularly those of estimating the probability of detection and correcting for environmental conditions. Our own work in the Indian and Banana rivers in 1979 and 1980, described in Leatherwood and Show (1980), resulted in good agreement between aerial estimates and attempted 100 percent surface counts. This suggests that aerial survey methods currently in use can give acceptable, conservative population estimates for enclosed areas (see table 18.2 and 18.3).

Dolphin density has been observed to be highest in shallow marshlands (Leatherwood and Platter 1975; Leatherwood and Show 1980) and in passes joining enclosed bays or lagoons with open Atlantic or Gulf waters (Irvine et al. 1979a; Barham et al. 1980). Much lower densities have been reported for deeper waters seaward of about one nautical mile (Odell and Reynolds 1980).

Claims have been made about dramatic changes in local dolphin populations caused by human activities. Unfortunately, there is little documentation for these statements, and we find it difficult to judge their validity. Bottlenose dolphins were present in San Diego Bay through the early 1960s (Norris and Prescott 1961), then reportedly were absent for nearly a decade. The apparent decline has been attributed to severe pollution of the bay (Food and Agriculture Organization 1978). Water quality in the bay has improved markedly since the early 1970s (Food and Agriculture Organization 1978), and bottlenose dolphins can now be seen in or near the bay at any season. A downward trend for the population off Texas has been reported (Gunter 1954). The population in Biscayne Bay, Florida, was at one time large enough to support a local live-capture fishery (Gray 1964). However, it declined to as few as 12 individuals by the mid-1950s, perhaps due to increased boat traffic or a pollution-linked decline in food availability (Odell 1976a; Food and Agriculture Organization 1978). A net fishery for "porpoises" off the east end of Long Island, New York, during the late 18th century probably exploited bottlenose dolphins and "may have contributed significantly to the lack of that species along those shores today" (Mead 1975b).

TABLE 18.3. Some estimates of density and total population of bottlenose dolphins in coastal waters of the United States and in the eastern Tropical Pacific

Area Code[a]	Location	Time Period	Reference	Dolphins km²	Dolphins nm²	Estimated Total Population	95 Percent Limits
CH	Cape Hatteras, North Carolina	1884	Mitchell (1975a)			13,740–17,000[b]	
EASTROPAC	Eastern Tropical Pacific	1979	Holt and Powers (1980)			40,200[c]	
MS	Mississippi coast (sounds)	July 1974	Leatherwood et al. (1978a)	0.14		1342	495–2189
	Mississippi coast (marshes)	July 1974	Leatherwood et al. (1978a)	0.13		438	144–732
	Mississippi coast (sounds)	July 1975	Leatherwood et al. (1978a)	0.09		879	511–1247
LA	Louisiana coast	July 1975	Leatherwood et al. (1978a)	0.09		897	436–1358
FLA	Florida Gulf panhandle	(averages from year-round surveys, 1977)	Odell and Reynolds (1980)	0.12[d]		527	$1.85 \times 10^{-3\,e}$
	Florida Gulf peninsula		Odell and Reynolds (1980)	0.06		787	$1.14 \times 10^{-4\,f}$
A-M	Texas: inland marine waters from Aransas Pass to Matagorda	March 1978	Barham et al. (1980)	0.75		1319	± 189[f]
ĪR	Florida: Indian and Banana rivers	August 1977	Leatherwood (1979)	0.52		438	311–565
LM	Texas: Gulf coastal waters off Padre and Mustang islands	September 1979	Leatherwood and Show (1980)	0.31		300	226–374
LM	Texas: Corpus Christi Bay and Aransas Pass	September 1979	Leatherwood and Show (1980)	1.02[g]		103[h]	67–139
TA	Florida Gulf Coast, Tampa to Hudson	September 1979	Leatherwood and Show (1980)	0.52	1.35	1190	964–1416
		November 1979	Leatherwood and Show (1980)	0.61	1.59	1397	1291–1503
IR	Florida: Indian and Banana rivers	November 1979	Leatherwood and Show (1980)	0.89	2.31	303[h]	785[e]
		January 1980	Leatherwood and Show (1980)	1.22	3.18	597[h]	

[a] See figure 18.3 and table 18.1.

[b] Based on cumulative catch records for years 1884–90.

[c] This is only a partial estimate, obtained from aerial surveys of 25 percent of the Commission Yellowfin Regulatory Area (CYRA) of the Inter-American Tropical Tuna Commission.

[d] Derived from their table 10 by computing the product of mean herd size (5.43) and mean herd density (0.497).

[e] Reported as $S^2_{\bar{x}}$.

[f] Reported as standard error of mean, not 95 percent confidence limits.

[g] This figure is somewhat misleading. The majority of the bay is not heavily populated. Almost all animals were observed in or immediately adjacent to Aransas Pass and the Intracoastal Waterway.

[h] Alternative estimates, using more conservative assumptions (Leatherwood 1979), are: LM, September 11 ± 5.2; IR, November 222 ± 33; IR, January 214 ± 42.

Reasons for long-term changes in distribution or abundance are not always obvious. Seasonal movements of odontocetes appear superficially to be related to changes in surface water temperature (Gaskin 1968; Norris and Prescott 1961; Leatherwood and Walker 1979). Similarly, long-term trends in occurrence of an odontocete species near its distributional boundary may be related to long-term drift in water temperature or other oceanographic characteristics (Leatherwood et al. 1980). Even low-level hunting pressure can have a significant impact on local stocks (Food and Agriculture Organization 1978).

Herd Size. In aerial surveys, bottlenose dolphin "herds" have been defined in short-term observations as groups found within close proximity of one another that converge and maintain coordinated movement while being chased (Leatherwood and Platter 1975) (table 18.4). While that definition may be adequate for censusing purposes, it probably does not reflect the actual scope of a given dolphin's interactions. Its herd is likely to include various groups and subgroups that coalesce or disperse, with regular mixing and exchange of individuals (Asper and Odell 1980). The groups of dolphins observed during the long-term studies in Sarasota Bay were all regarded as members of a single "superherd" (Irvine et al. 1979a). It is necessary to apply the term *herd* to cetaceans with some caution, as their aquatic environment and sophisticated communication system may allow them to maintain herd integrity over great geographic distances (Payne and Webb 1971).

The size of dolphin aggregations depends on water depth and complexity of habitat. Groupings tend to be smaller in marshlands and estuaries, where acoustic contact and coordinated activities would be difficult (Irvine et al. 1979a; Wells et al. 1980). Inshore herds of as many as 200 (Leatherwood and Platter 1975) are exceptional and probably represent convergence on an area of food abundance. Offshore herds of 500 or more are seen occasionally (Leatherwood and Reeves 1978). Large inshore aggregations of 1,000 dolphins have been reported from South Africa (Saayman and Tayler 1973).

Herd Composition. There is evidence of a limited degree of segregation within nearshore herds of bottlenose dolphins, based on age and sex (Irvine et al. 1979a; Wells et al. 1980). Adult males rarely associate with subadult males, the latter usually remaining in bachelor groups or with one or two adult females. Females accompanied by calves associate with one another and occasionally with other age and sex classes. There does not appear to be a strong, sustained separation by class, but rather a high degree of mixing.

During most aerial surveys of bottlenose dolphins, observers have estimated the number of calves and yearlings present (see table 18.1). These estimates have been used as a crude index of productivity. Aerial photographs of groups, particularly when taken after dropping known-diameter discs near the animals, can be used to calculate at least relative lengths of dolphins (Whitehead and Payne 1976). As age-length keys be-

come available, such photogrammetric techniques might be used to describe herd composition more precisely.

FOOD HABITS

Bottlenose dolphins are often described as catholic or opportunistic feeders (Caldwell and Caldwell 1972). The list of known prey includes a wide variety of fish, particularly littoral and sublittoral forms, and invertebrates such as squid, clams, and crabs (Leatherwood et al. 1978b; Walker 1980) (table 18.5). Primary reported prey in North American waters are mullet (*Mugil* spp.), croaker (*Micropogon* spp.), spot (*Leiostomus* spp.), sea trout or weakfish (*Cynoscion* spp.) and whiting (*Menticirrhus* sp.) in most coastal areas, and squid (*Loligo* spp.) in offshore areas. Differences in food habits between coastal and offshore animals sometimes are reflected in the degree of tooth wear (Ross 1977; Walker 1980).

In shallow habitat, individuals commonly chase prey independently, swimming at high burst speeds along erratic courses, "pinwheeling" at the termination of the chase (Lawrence and Schevill 1954). They also often swim on their sides or upside down, positions thought to improve echolocation by reducing interference from surface echoes (Leatherwood 1975). Feeding bottlenose dolphins often leave long, erratic mud trails punctuated by plumes of sediment where pinwheeling or acceleration has occurred. In some shallow areas we have observed bottlenose dolphins feeding along mud lines created by surf, tides, or currents near steep dropoffs.

Bottlenose dolphins also hunt and feed cooperatively, especially in deep, open water (Wells et al. 1980), where there is a greater tendency of fish to school (Williams 1964). Such cooperation can involve groups holding a school of fish at bay while individuals alternately feed (Caldwell and Caldwell 1972) or several dolphins driving fish into shallow water (Hoese 1971). According to the density, distribution, and behavior of prey, coordinated activity probably improves the chances of dolphins finding and capturing it. Thus, their feeding strategies are in large part determined by the type of habitat they occupy (Irvine et al. 1979a; Würsig and Würsig 1977, 1980). We have observed that bottlenose dolphins exhibit a greater variety of feeding modes in areas with diverse habitats, such as Indian River, Mississippi Sound, and portions of the Texas coast, than they do in areas with a more uniform habitat, such as both coasts of Florida outside the barrier islands and beaches.

Some bottlenose dolphins living near human population centers have learned to exploit certain human activities. They feed on fish attracted to stationary vessels or platforms, churned up by boat propellers or working shrimp nets, netted or hooked by sport or commercial fishermen, or discarded overboard by shrimpers (Leatherwood 1975). Several individuals returned on three successive nights to feed on flying fish (*Exocoetus* sp.) near a vessel anchored at Clarion Island (Ljungblad et al. 1977). They took live and dead

TABLE 18.4. Some estimates of sizes of "herds" of bottlenose dolphins in coastal waters of the United States

Area Code[a]	Location Description (Habitat-Type)	Number of Herds	$\bar{X}$ Dolphins/ Group	Range	$S_{\bar{x}}$	Reference Source
CH	Cape Hatteras, North Carolina (open coast)		12–several hundred	12–several hundred		Townsend (1914)
TEX	Texas (bays)		5–10			Gunter (1942)
HI	Hawaii (within 1 km of reefs)		1–50	1–50		Rice (1960)
CA	California (bays and coastal waters)		20 <15			Evans & Bastian (1969); Leatherwood & Reeves (1978)
FL	Florida (inshore)		12			Caldwell & Caldwell (1972)
MS	Mississippi Sound (open bay, 1974)		25.4			Leatherwood & Platter (1975); Leatherwood et al. (1978a)
MS	Mississippi Sound (1975)		10.5			Leatherwood & Platter (1975); Leatherwood et al. (1978a)
GU	northern Gulf of Mexico (marshland)		16.7	1–66	8.2	Leatherwood & Platter (1975); Leatherwood et al. (1978a)
GU	northern Gulf of Mexico (sounds and islands)		28.9	1–175	8.7	Leatherwood & Platter, (1975); Leatherwood et al. (1978a)
B	Biscayne Bay, Florida (open bay)		8	3–13		Odell (1976a)
E	Everglades National Park (shallow bays, many small islands)		2.95	1–25		Odell (1979)
S	Sarasota, Florida (inshore gulf, bays, flats, channels, passes)	736	5.31	1–40	14.27	Irvine et al. (1979a); Wells et al. (1980)
A	Aransas Pass, Texas (bays, channels, passes)		3–4			Shane (1977); Shane and Schmidly (1978)
WFLA	Florida, west coast (offshore)	326	4.66	1–50		Odell & Reynolds (1980)
NFLA	Gulf of Mexico, Sarasota, Florida (open ocean)	31	2.87			Odell & Reynolds (1980)
WLA	Atchafalaya Bay, Louisiana (open bay)		22.4			Leatherwood et al. (1978a)
SFLA	southern Florida (August 1979)	5	3.2		2.17	U.S. Fish & Wildlife Service (1980)
NFLA	northern Florida (August 1979)	14	3.5		3.7	U.S. Fish & Wildlife Service (1980)
NTEX	northern Texas (August 1979)	10	17.5		13.9	U.S. Fish & Wildlife Service (1980)
SFLA	southern Florida (November 1979)	10	2.5		2.27	U.S. Fish & Wildlife Service (1980)
NFLA	northern Florida (November 1979)	10	17.7		18.22	U.S. Fish & Wildlife Service (1980)
STEX	southern Texas (November 1979)	10	11.7		14.68	U.S. Fish & Wildlife Service (1980)
NTEX	northern Texas (November 1979)	7	3.7		3.3	U.S. Fish & Wildlife Service (1980)
WFLA	portion of western Florida between 28° 32′N and 25° 08′W	431	3.21	1–72	SE ± 0.26	Irvine et al. (1980)

[a]See figure 18.3 and table 18.1.

TABLE 18.5. Summaries of published accounts of food and feeding behavior of bottlenose dolphins off North America

Reference	Summary of Report
True (1885)	Common gunnard (*Prionothus carolinus*) in stomachs of two females examined at the Cape May fishery.
Townsend (1914)	Second-hand report of fishermen that animals taken in the Hatteras fishery fed primarily on squeteague or weakfish.
Gunter (1938)	Animals feeding behind working shrimp boats.
Harris (1938)	Two photographs and a brief description of feeding on striped mullet (*Mugil cephalus*) off St. Augustine, Florida, including the dolphins catching their prey in midair.
Kemp (1949)	Squid (*Loligo* sp.), shrimp (*Penaeus* sp.), crabs, and nine fish species from stomachs of ten animals collected off Texas. Striped mullet, hardhead catfish (*Galeichthys felis*), and ribbonfish (*Trichiurus lepturus*) were most numerous.
Gunter (1942)	Report that animals in northern Gulf of Mexico fed most commonly on striped mullet, but also took fish of eleven other species and shrimp; comments on further feeding behavior.
Gunter (1951)	Second-hand report from E. I. McIlhenny of whole shrimp in stomachs.
Gunter (1954)	A comprehensive review of knowledge to date.
Norris and Prescott (1961)	Stomach contents of a single animal plus anecdotal records of feeding in association with human activities.
Hoese (1971)	Feeding in shallow *Spartina* marshes of Georgia coast.
Caldwell and Caldwell (1972)	Anecdotal accounts of feeding behavior and a review of knowledge.
Leatherwood (1975)	Seven feeding patterns observed in the Gulf of Mexico and off Nayarit, Mexico, and California. Stomach contents of one coastal animal were primarily menhaden (*Brevooria* spp.), striped mullet, silver eels (probably *Trichiurus lepturus*), and shrimp (*Penaeus* spp.).
Ljungblad et al. (1977)	Feeding on flying fish (*Exocetus* sp.) near a vessel anchored at Clarion Island, Mexico, demonstrating ability to distinguish between normal and experimentally altered fish.
Shane (1977)	Six types of feeding behavior in or near Aransas Pass, Texas. Frequent association with shrimp boats. Refusal to eat small sharks and hardhead catfish but acceptance of spot (*Leiostomus xanthurrhus*), moonfish (*Vomer setapinnis*), and whiting (*Menticirrhus* spp.)
Irvine et al. (1979a)	Speculation that observed changes in dolphin distribution in southern Sarasota Bay related to movements of prey species and observation that feeding strategies vary according to depth and differ from those of pelagic small cetaceans.
Leatherwood et al. (1978b)	Ten fish species (predominantly Atlantic croakers, seatrout, and spot) plus some fragments of mollusks and arthropods in stomachs of animals from the northeastern United States. Similarities in stomach contents of stranded and netted animals. A calf apparently fully weaned prior to twelve months of age.
Leatherwood (1979)	Feeding behavior in Indian River, Florida, is similar to that in the Gulf of Mexico.
Gruber (1979)	Shrimpboat-dolphin association in Pass Cavallo, Texas. Feeding observed as early as 0545 and continued past 2200. Following the boat seemed preferable to eating trash fish except after last trawl of day. Late spring to early fall, most dolphins fed in association with shrimpboats.
Shallenberger (1979)	Bottlenose dolphins stealing or damaging skipjack tuna (*Katsuwonus pelamis*), other bait fish, and commercially valuable fish from longlines, trollers, and bottom-fishing rigs off Hawaii.
Irvine et al. (1979a)	Opinions of fishermen off western Florida that when mullet are not plentiful, local dolphins in Sarasota Bay eat any available fish, including catfish.

SOURCE: Modified from Leatherwood (1975) and Wells et al. (1980).

fish, including some fish altered to affect their acoustic characteristics. The most extensive and complex association between bottlenose dolphins and fishermen has developed with the shrimp fisheries in the Gulf of Mexico and Gulf of California (Gunter 1942; Leatherwood 1975; Gruber 1979). The dolphins apparently can distinguish acoustically between transiting and working vessels, often converging on a shrimp boat shortly after the nets are deployed (Gunter 1951; Norris and Prescott 1961).

Captive bottlenose dolphins eat an average of 6–7 kg of food per day (Mitchell 1975b). However, Sergeant's (1969) estimated consumption rate of 4–6 percent of body weight per day would mean that a large adult eats as much as 16.5 kg per day in the wild. Smaller individuals probably consume up to 10 percent

of their body weight per day (Ridgway 1972), and captive individuals of any size may consume from 8 to 12 percent of their body weight per day for several months (E. D. Asper 1980 personal communication).

Newborn bottlenose dolphins begin sucking within 4 to 24 hours of birth (McBride and Kritzler 1951; Tavolga and Essapian 1957). Nursing periods may last from a few seconds to over 30 minutes and vary from 3–4 per hour (Gurevich 1977) to 7 per day (McBride and Kritzler 1951). Gross composition of the milk is 71.4 percent water, 28.6 percent total solids, 16.7 percent fat, 9.6 percent protein, and 0.8 percent ash (Eichelberger et al. 1940).

The duration of calf dependency varies. One calf in captivity showed interest in fish at 2 months of age (Gurevich 1977); another took pieces of fish at 3 ½ months (Tavolga 1966). Captive calves usually begin to accept fish and squid regularly at 4–6 months of age, although nursing often continues for 18 months or more (McBride and Kritzler 1951; Caldwell and Caldwell 1972; Gurevich 1977; Prescott 1977; Saayman and Tayler 1977). A free-ranging juvenile, estimated to be less than a year old, was seen eating fish (Leatherwood 1977). The stomach of a 177-cm male, thought to be somewhat less than a year old, contained fish but no milk (Leatherwood et al. 1978*b*).

NATURAL MORTALITY

Sharks are a constant presence in the bottlenose dolphin's environment. Evidence exists of predation, such as direct observation of attacks on dolphins and discovery of dolphin remains in shark stomachs, and of unsuccessful attacks, such as healed shark-bite scars on live animals (Wood et al. 1970). Sharks no doubt take a toll of sick, injured, and unwary bottlenose dolphins. However, there is increasing evidence that sharks and dolphins of many species travel together and peacefully coexist (Leatherwood et al. 1972). We do not wish to perpetuate the popular view of sharks and porpoises as the dogs and cats of the sea, or to reinforce the false sense of security from sharks enjoyed by sailors and bathers when dolphins are present.

We are aware of no documented accounts of predation on bottlenose dolphins by killer whales (*Orcinus orca*). However, we assume that it occurs, and there is some indirect evidence for it (Würsig and Würsig 1980).

AGE DETERMINATION

Examination of cyclic dentine deposits in teeth has been used to estimate age in various odontocetes, including sperm whales (*Physeter macrocephalus*) (Ohsumi et al. 1962), bottlenose whales (*Hyperoodon ampullatus*) (Christensen 1973), white whales (Sergeant 1973), common dolphins (*Delphinus delphis*) (Hui 1977), striped dolphins (*Stenella coeruleoalba*) (Kasuya 1972, 1976), spotted dolphins (*S. attenuata*) (Kasuya 1976; Perrin et al. 1976), dusky dolphins (*Lagenorhynchus obscurus*) (Best 1976*a*), and bottlenose dolphins (Sergeant 1959; Sergeant et al. 1973; Hui 1978). Use of the largest tooth available is preferable for aging bottlenose dolphins.

With prohibitions on collection imposed by the U.S. Marine Mammal Protection Act of 1972, beach-cast specimens and captive animals have become important sources of material for research. A simple technique of applying block anesthesia to the jaw of a living dolphin allows removal of a tooth with minimal trauma (Ridgway et al. 1975). Using this technique, teeth have been removed from approximately 100 bottlenose dolphins captured, marked, and released in Indian River (Asper and Odell 1979, 1980). If required routinely of capture operations, such a procedure provides useful data on population structure without requiring the animals to be killed.

A recent workshop on odontocete age determination produced an important summary of the state of the art (Perrin and Myrick in press). The best preparation of teeth from smaller species appears to be a decalcification-staining technique. Although many positive statements have been published concerning odontocete ages, significant problems of reading and interpretation of growth zones are unresolved. Oxytetracycline marking studies are in progress and should eventually provide a breakthrough in our understanding of the significance of growth zones in odontocete hard tissue.

Other methods, including use of mandibular layers in white whales (Brodie 1969) and aspartic acid accumulation levels in teeth and lens nuclei of other marine mammals (Bada et al. 1977), have been considered. However, refinement of techniques of age determination through use of growth zones in teeth presently seems to be the most promising approach.

ECONOMIC RELATIONSHIPS

Direct Fisheries. Bottlenose dolphins have been hunted with nets, hand-held harpoons, and rifles (Mitchell 1975*a*). The principal products of such fisheries have been bait, oil, and leather. In North America, the most important fishery was at Cape Hatteras, North Carolina (Mead 1975*b* and associated references). It began in 1797 and continued sporadically until 1929. Using seines deployed from small boats, fishermen forced the dolphins onto the beach. The main objects of the Hatteras dolphin fishery were jaw oil, body oil, and leather. Catch data are confusing, but it is clear that thousands were captured in exceptional years. A "porpoise" fishery at Cape May, New Jersey, used specially designed nets towed by a steamer to catch bottlenose dolphins in 1884–85 (see figure 18.6). Another poorly documented "porpoise" fishery on Long Island, New York, may have exploited bottlenose dolphins in the late 18th century.

A small harpoon fishery may once have existed along Florida's east coast (Caldwell and Caldwell 1972). During the 19th century 8–10 dolphins per day were taken with lances in the Gulf of Mexico, apparently coincidental with a shark fishery based in Tampa Bay, Florida (Stearnes 1887). Small numbers were killed for leather off Texas and Louisiana (Gunter 1938, 1951, 1954).

Presently, some bottlenose dolphins are killed off mainland Mexico and Baja California for food (National Marine Fisheries Service 1978), and we have noted the use of bottlenose dolphins as shark bait off Baja California, in the Gulf of California, and off the west coasts of Guatemala and Costa Rica. Bottlenose dolphins are still hunted along with other small cetaceans at Saint Vincent (Caldwell et al. 1971; Caldwell and Caldwell 1975*a*) and Saint Lucia (Gaskin and Smith 1977) in the Caribbean Sea. Despite legal protection, some small delphinids, mainly common dolphins but including bottlenose dolphins, are taken occasionally with hand harpoons by Venezuelan fishermen for use as shark bait (Caldwell and Caldwell 1971; G. di Sciara 1979 personal communication).

Conflicts with Fisheries. Commercial fishermen in the Indian and Banana rivers of Florida complain that bottlenose dolphins are a major nuisance, sometimes injuring fishermen and causing an estimated $441,000 worth of damage per year to mackerel longlines and trammel nets as of 1976 (Cato and Prochanska 1976). The fishermen have requested federal help in controlling dolphin populations, but doubt remains as to whether large sharks or dolphins are causing the damage (Leatherwood 1979). Attempts to discourage dolphins by projecting aversive sounds into the water near lines, nets, and boats had little effect beyond initially eliciting startled responses (Caldwell and Caldwell 1975b). This is not surprising, considering that captive bottlenose dolphins habituate to loud, repetitive sounds in their environment (Wood 1973).

Although the relationship between bottlenose dolphins and fishermen in the Gulf of Mexico appears to be generally amicable (Shane 1977), some shooting occurs (Irvine et al. 1979a; Schmidly and Shane 1978). Occasionally, dolphins rip segments of shrimp net as they try to remove trapped fish, or they blunder into a tow or handling line and do minor damage while struggling to get free. Beach seines off Louisiana are sometimes damaged by dolphins attempting to remove gilled fish from the seaward side (Leatherwood 1975).

Underwater photographs have confirmed that bottlenose dolphins steal caught fish from longlines in Hawaii (Iverson 1975). They also take skipjack tuna (*Katsuowmus pelamis*) from the lines of fishermen and cause damage to the catch and gear of troll, bottom, and night jig fishermen in Hawaii (Shallenberger 1979). Attempts to resolve these conflicts have met with limited success. Quinine-injected fish have been used experimentally to determine if dolphins would find such bait distasteful. Fishermen have wrapped several turns of wire around bait fish to change the character of the echolocation signal received from them by marauding dolphins. In frustration, some fishermen have used poisoned bait with the intention of killing problem dolphins.

Bottlenose dolphins, like many other small cetaceans, die after becoming fouled in fishing gear. Small numbers are killed in shrimp nets and tending lines off Texas (Gunter 1942) and in the Mississippi River delta (Leatherwood unpublished data). The few beach seines still operated along the Outer Banks of North Carolina occasionally entangle bottlenose dolphins accidentally (J. G. Mead 1980 personal communication). As many as 50 per year are accidentally netted in a shark and sea bass fishery in the Gulf of California, primarily near San Felipe, Mexico (National Marine Fisheries Service 1973, 1978; Mitchell 1975a). Small numbers die annually when they become entrapped in menhaden seines or entangled in float lines of crab or lobster traps in the Gulf of Mexico (Leatherwood unpublished data). The purse seine fishery for tuna in the eastern tropical Pacific kills some bottlenose dolphins, although other cetacean species are involved to a much greater extent (see "The Dolphin-Tuna Problem"). No bottlenose dolphins were reported in the catch between 1959 and 1973. However, with more reliable reporting since 1974 the following estimates of the kill have been made: 1974, 103; 1975, 198; 1976, 26; 1977, 47; 1978, 149 (Smith 1979).

A widely publicized problem involving bottlenose dolphins outside North American waters is the large recent kills of small cetaceans by Japanese fishermen in the western North Pacific (Jones 1980; Perrin 1980). These slaughters, which included large numbers of false killer whales (*Pseudorca crassidens*) and melon-headed whales (*Peponocephala electra*) as well as bottlenose dolphins, have been authorized by Japanese local governments as control measures required to protect commercial fish stocks. However, "while the potential for significant competition between fishermen and marine mammals exists and such competition has been demonstrated in some cases, in nearly all such situations the relationship is only asserted to exist, rather than demonstrated to exist" (International Whaling Commission 1979). The significance of pollution and human overexploitation of fish stocks is too often conveniently ignored.

Live-Capture Fishery. The live-capture fishery for bottlenose dolphins in U.S. waters is the oldest sustained fishery of its kind in the world. Bottlenose dolphins were first reported to have been maintained in captivity in 1861, when a stranded individual was held briefly at Boston's Aquarial Gardens (Wyman 1863). The first ones known to have been captured specifically for display were taken at Cape Hatteras, North Carolina, in 1914 for the New York Aquarium (Townsend 1914). The display industry developed rapidly after the opening of Marine Studios (now Marineland of Florida) in Saint Augustine in 1938 (Lawrence and Schevill 1954; Caldwell and Caldwell 1972). Since that time bottlenose dolphins, most of them caught in U.S. waters, have been supplied to research institutions, zoos, aquariums, and marine parks in many countries (Gray 1964; Ridgway 1972).

Techniques for live-capture of smaller cetaceans—killer whales, pilot whales, white whales, and smaller dolphins and porpoises, including bottlenose dolphins—have been reviewed and illustrated (Ridgway 1972; Wood 1973; Asper 1975; Walker 1975; Goldsberry et al. 1978). Formerly, bottlenose dolphins were caught with a "tail grabber" as they rode a vessel's bow wave (Gray 1964). The tail grabber is a spring-loaded clamp mounted on a pole and attached by a line to the collecting vessel or to a large, brightly colored float. It is no longer used because it is considered ineffective and inhumane. For capture in deep water, it has been replaced by the breakaway hoop net (figure 18.7), which allows a collector to net a single dolphin as it rides a vessel's bow wave. However, the vast majority of bottlenose dolphins are collected in shallow water, usually less than about 1.5 m deep, using large-mesh seine nets. Small groups of selected animals are herded or followed to an appropriate location, where they are encircled by boats

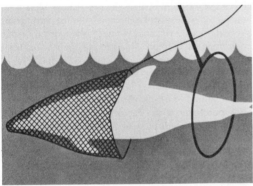

FIGURE 18.7. A breakaway hoop-net with a bag attached to the ring by means of tape. B, illustration showing how dolphins are collected with this net as they ride a vessel's bow.

FIGURE 18.8. Capturing dolphins with a seine net, showing the inner circle completed and the boat in position.

paying out net behind them. The circumference of the net circle is gradually reduced until the animals become entangled in the net and can be handled by swimmers (figure 18.8). Sometimes the net is dragged onto the beach and the animals are stranded.

Mortality of coastal bottlenose dolphins during capture has not been well documented. It might have been as high as 40 percent for pelagic animals captured off California during the years from 1960 through 1972 (Walker 1975). Capture techniques and the skill of collectors have improved, so mortality directly related to capture procedures is now unusual. Similarly, the chances of a bottlenose dolphin's surviving in captivity have improved significantly during the last decade. We believe this to be due in part to legislation requiring prospective dolphin owners to obtain permits and to maintain animals in approved facilities using tested medical and husbandry practices (National Marine Fisheries Service 1978). As of 20 August 1976, at least 286 bottlenose dolphins were living in institutions in the United States alone (Cornell and Asper 1978). The number in aquariums outside the United States is not known.

The success of bottlenose dolphins in captivity has been attributed to their natural occurrence in shallow, turbid, and sometimes noisy and polluted waters near human population centers. They are, in a sense, preadapted to the shallow confines of a tank, the noise, and the regular, close contact with people (Lawrence and Shevill 1954; Ridgway 1972; Odell 1975). Captivity appears to be far less stressful for coastal bottlenose dolphins than it is for oceanic dolphins.

Statistics on the live-capture fishery are difficult to reconstruct. The Florida Department of Natural Resources documented collections in Florida from the mid-1950s through 1972. They made available to us the detailed records for the years 1966 through 1972, including estimates of the value of each animal at the time of sale (table 18.6). Since the passage of the Marine Mammal Protection Act of 1972, collections of bottlenose dolphins in the United States, and by U.S. citizens in foreign waters, have been documented by the National Marine Fisheries Service (table 18.7). Collections in Florida prior to 1966 and in other areas prior to 1972 must be reconstructed from a few published sources and from anecdotal accounts of collectors and researchers familiar with each region (table 18.8). We believe these records are incomplete for most areas and periods for which no official records were maintained, and we recognize that poaching has occurred even when the fisheries have been monitored. Therefore, we must consider as conservative our estimate that by 20 March 1980 over 1,500 bottlenose dolphins had been removed from the wild off the United States, Mexico, and the Bahamas for the purposes of captive display or research.

Although the population of bottlenose dolphins in aggregate appears sufficiently large to support the current levels of take, there is concern that "such numbers [taken] may represent a significant impact on the level of local stocks" (Food and Agriculture Organization 1978). Current research programs sponsored by the

TABLE 18.6. Summary of bottlenose dolphin collecting in Florida, 1967–73

Year	Total Captures	Total Deaths[a]	Total Sold	Total Value[b]	Average Price/Dolphin[b]
1967[c]	39	2	31	$ 14,250	$ 460
1968	172	38	92	52,700	573
1969	177	60	144	59,750	415
1970	215	55	135	67,355	459
1971	172	27	148	145,900	986
1972	214	23	162	147,348	910
1973[d]	0	31	23	53,700	2,335
Total	989	236	735	$541,003	

SOURCE: Prepared from unpublished data on Florida commercial marine landings, held by the Florida Department of Natural Resources, Tallahassee, Florida.

[a]Includes animals that died in captivity as well as during collection and transport.

[b]In U.S. dollars at the date of purchase.

[c]October, November, and December only.

[d]Animals collected prior to the effective date of the Marine Mammal Protection Act in December 1972.

National Marine Fisheries Service emphasize determination of population size (Leatherwood and Show 1980) and discreteness (Asper and Odell 1980; Duffield 1980).

The economic value of the bottlenose dolphin live-capture fishery is difficult to estimate with any reliability. Actual values of dolphins at the time of sale are available only for Florida for the years 1967–72 (see table 18.6). From interviews with collectors or officials at institutions making purchases, and from our own unpublished data, we have conservatively estimated values of bottlenose dolphins at time of sale for all years possible and have interpolated the values for intermediate years for which no estimates were possible. The results are shown in tables 18.7 and 18.8. These figures are not complete and are presented only

TABLE 18.7. Total number of bottlenose dolphins removed from the wild under permits from the U.S. National Marine Fisheries Service, 1 January 1973–30 December 1979

Year	Total Removals	Estimated Average Value[a,b]	Estimated Total Value[b]
1973	14	$1,775	$24,850
1974	21	2,015	42,315
1975	36	2,675	96,300
1976	28	3,450[c]	96,600
1977	49	4,255	208,495
1978	48	5,263[c]	252,624
1979	38	6,270	238,260
Total	234		$956,744

SOURCE: Provided to the authors by the U.S. National Marine Fisheries Service, Washington, D.C., April 1980.

[a]The mean of estimates by eight or more collectors or individuals at institutions making purchases. Equal values are assumed for animals sold and for animals retained for use by collectors making capture.

[b]In U.S. dollars at the time of sale.

[c]Inadequate estimates available. This value is the mean of the two adjacent years.

to show the approximate magnitude of the fishery at various times. In addition to the incomplete collection records, our attempts to characterize the economics of the bottlenose dolphin live-capture fishery were confounded by: widespread reluctance of most participants to reveal prices, even anonymously; the indeterminable proportions of dolphins in different value categories (an animal of prespecified sex, age, size, and level of training guaranteed for some period of time beyond shipment to the customer is obviously more expensive than an animal of unspecified characteristics delivered to the purchaser at the capture site); and the need to apply a single multiplier to annual total numbers of dolphins collected in divergent localities with different regional economies. Projecting the figures in tables 18.6–18.8 to the total economic value of bottlenose dolphins off North America at the time of sale, the value is between $2.0 and 2.5 million (U.S.), depending on the assumptions. We resisted presenting a single figure because available data in no way reflect secondary economic impacts of bottlenose dolphins (for example, on gate receipts at aquaria and on sales of souvenirs, photos, films, movies, television programs, magazine articles, and whale-watching trips) or some recent increases in value of the animals. Linehan (1979), for example, reported that a trained bottlenose dolphin recently sold for $23,000.

Indirect Impact of Human Activities. As conspicuous, high-order predators, bottlenose dolphins may be good indicators of the health and vitality of the inshore marine ecosystem. Definitive research has not been done to test the effects of chemical pollution and harassment on bottlenose dolphins and other cetaceans. Nevertheless, it is reasonable to expect changes in behavior, distribution and movement, and reproductive success as quality of the coastal environment deteriorates.

Noise levels in the water column can have a serious impact on sound-sensitive cetaceans (Myrberg 1978). Bottlenose dolphins are vocal animals (Evans 1973) that probably depend on acoustic cues for long-

TABLE 18.8. Estimates of number of bottlenose dolphins live-captured and removed from the wild off North America, Mexico, and Bahamas, 1938–80

Year(s)	Estimated Minimum Value/Dolphin	Number of Bottlenose Dolphins Live-Captured by Areas								
		Hawaii	California	Mexico	All Areas, Principally Florida and North Carolina	Florida	Texas	Mississippi	Open Ocean	Bahamas
1938										
1956										
1957	200[a]		1[b]							
1958	230[a]		3[b]							
1959	259[a]									
1960	288[a]		20–25[c]		200[d]					
1961	317[a]									
1962	345[a]									
1963	374[a]									
1964	398[a]		20–25[c]							
1965	419[a]									
1966	440[a]									
1967	460[e]			40–60[f]		31[g]				
1968	573[e]					92[g]				
1969	415[e]		18[h]			144[g]				
1970	459[e]	35[i]				135[g]				
1971	986[e]					148[g]				
1972	910[e]					162[g]				
1973	1,775[j]									21[l]
1974	2,015[j]									
1975	2,675[j]			18–25[k]						
1976	3,450[j]									
1977	4,255[j]		9[m]			130[n]	33[n]	66[n]	8[n]	
1978	5,263[j]									
1979	6,270[j]									
1980	[j]									

[a] Prices increased from about $200 in 1957 to $450 in 1967; intermediate years are interpolated.
[b] Norris and Prescott (1961)
[c] Authors' estimates based on conversation with J. Prescott, W. A. Walker, F. G. Wood, S. H. Ridgway, and W. E. Evans.
[d] Gray (1964); Caldwell and Caldwell (1972)
[e] See table 18.6
[f] Leatherwood (unpublished data)
[g] See table 18.6
[h] Walker (1975)
[i] Shallenberger (1979)
[j] See table 18.7
[k] J. C. Sweeney (1980 personal communication)
[l] By a U.S. collector for transport to Europe (Linehan 1979).
[m] See table 18.7
[n] See table 18.7

range maintenance of social bonds, as do other cetaceans (Payne and Webb 1971), and for locating concentrations of prey (Ljungblad et al. 1977; Würsig and Würsig 1979). Changes in noise and reverberation levels in captive enclosures have altered whistle contours in common dolphins (Norris in preparation). Free-ranging dolphins subjected to high and sustained noise levels at certain frequencies may be similarly affected.

The fact that bottlenose dolphins persist as they do in heavily trafficked ship channels and among sprawling shoreline developments testifies to their basic compatability with humans. Like the white-tailed deer

(*Odocoileus virginianus*) and the coyote (*Canis latrans*), bottlenose dolphins might be adaptable enough to expand their range and develop new feeding strategies as a result of human use and manipulation of the environment. Near Pass Cavallo, Texas, "shrimping activities may have influenced the activity cycles, daily and seasonal movements and social interactions of the dolphins" (Gruber 1979), so accustomed have they become to feeding in association with the shrimpers. We have observed bottlenose dolphins intensively using the dredged passes and channels of the Laguna Madre complex in southern Texas. It is our impression that human efforts to reduce salinity and improve navigation in this waterway have made new habitat and new sources of food available to the local coastal bottlenose dolphin population.

MANAGEMENT

The abundance and wide distribution of bottlenose dolphins in North American waters make their management a less urgent concern than that of depleted or endangered species (Food and Agriculture Organization 1978). The only active fishery of any consequence in North America is the live-capture industry, which involves bottlenose dolphins off eastern Florida, in the Gulf of Mexico, and occasionally off southern California. Dolphin mortality from accidental fouling in fishing gear and from direct retaliatory gestures by fishermen needs to be monitored, but at present such mortality does not appear to be overtaxing any part of the population.

The U.S. Marine Mammal Protection Act of 1972 (MMPA) imposed a moratorium on the taking of bottlenose dolphins and other marine mammals, except under special permits. Permits for scientific research and public display are issued by the Department of Commerce's National Marine Fisheries Service (NMFS), following consultation with the Marine Mammal Commission (MMC) and its Committee of Scientific Advisors. In evaluating permit applications, responsible agencies are required by law to consider: The maintenance of optimum sustainable populations; the esthetic and recreational as well as economic significance of marine mammals; and the need to maintain the health and stability of ecosystems of which the animals are a part.

Management of the bottlenose dolphin live-capture fishery under the MMPA has been of an interim nature, pending the results of directed research programs designed to address management-related questions. It is generally agreed that the total bottlenose dolphin population is large enough to fill existing demands for research and display. However, collecting activities have focused historically on a few limited areas. Local coastal stocks appear to be effectively isolated from each other (Irvine et al. 1979a; Asper and Odell 1980; Duffield 1980) and from offshore populations (Duffield et al. 1979, in preparation). Thus, it is feared that concentration of collecting activities in specific areas already may have contributed to a decline in some local stocks. Based on what is known of population size, net recruitment rate, and magnitude of past removals, the NMFS and MMC have identified local areas for which a provisional prohibition on live-capture activities is desirable and other areas for which a controlled take is justified. In general, authorized removals are not to exceed 2 percent per year of the minimum estimated population in areas for which such estimates are available. This conservative approach to management appears to be adequate, pending results of ongoing research. Much can be learned about social structure, age, growth, maturation, naturally occurring disease, blood characteristics, and pregnancy rates from rigorous adherence to carefully planned sampling schedules during live-capture activities (Asper and Odell 1979, 1980). The combination of conservative federal management, industry cooperation, and sustained research should ensure that bottlenose dolphins remain an important esthetic resource in coastal waters of the southern United States, and a valuable research, entertainment, and educational resource in institutions around the world.

The Mexican federal government is considering policies and regulations that would be patterned after U.S. precedents, requiring permits for scientific research and live-capture and the confiscation by federal officials of cetaceans inadvertently caught in nets (M. Esqueda, Mexican Fisheries Commission, San Diego, 1980, personal communication). Presently bottlenose dolphins and other cetaceans have no formal protection in Mexican waters.

SPERM WHALE

DESCRIPTION

The sperm whale (*Physeter macrocephalus*) is the largest of the toothed whales. The family to which it is assigned includes only two other species: *Kogia breviceps*, the pygmy sperm whale, and *K. simus*, the dwarf sperm whale. All three have a strongly developed corpus adiposum, which forms a frontal protuberance of the head; a markedly asymmetric skull; an underslung lower jaw; and functional teeth only in the lower jaw.

The sperm whale has a disproportionately large head (to one-third the body length); a rounded, fleshy dorsal hump; and fused second to seventh cervical vertebrae (Tomilin 1967). The S-shaped blowhole is at the front of the head, to the left of the midline; consequently, the blow angles forward and to the left. The two *Kogia* species, both much smaller than the sperm whale, have a proportionately smaller head (to only one-sixth the body length), a falcate dolphinlike dorsal fin, and completely fused cervical vertebrae (Handley 1966).

Sexual dimorphism in size is more exaggerated in the sperm whale than in any other living cetacean (Best 1979). Physically mature males, at 15.8 m, are 1.44 times as long and, at 43.5 tons, 3.22 times as heavy as physically mature females. Few bulls grow longer than 18.5 m (Berzin 1972). The heaviest whale ever weighed whole was a 13.3-m male sperm whale caught off Durban, South Africa, in 1969 (Gambell 1970): it weighed about 31.5 tons.

The skin of the sperm whale, posterior to the head, has a shriveled or wrinkled appearance. Color is generally slate gray or brownish with a pale, almost white area around the corners of the mouth. The body is generally paler on the abdomen, but the dorsal and ventral surfaces of the flukes and flippers are the same slate gray or brown as the rest of the body (figure 18.9). White or piebald animals have been reported occasionally throughout whaling history (Gilmore 1969; Berzin 1972).

FIGURE 18.9. A group of mass-stranded sperm whales (*Physeter macrocephalus*) near Florence, Oregon. Sperm whales are the only great whales known to mass strand.

FIGURE 18.10. Skull and skeleton of a sperm whale (*Physeter macrocephalus*) on display at Whalers Village, Maui, Hawaii.

The large head dominates the sperm whale's appearance. The lower jaws each contain about 25 conical teeth, to 10 cm in diameter and often badly worn in larger, older individuals. These mandibular teeth fit snugly into sockets in the upper jaws. The upper jaws normally contain only vestigial, unerupted teeth.

The shape of the skull is surprising, given the external appearance of the head. It resembles a chariot with the gondola turned backward (figure 18.10). The remaining bulk of the head consists primarily of the spermaceti organ and its related structures (Clarke 1978a). The spermaceti organ is a nearly hollow cavity filled with oil (Clarke 1978b). Most of the spermaceti oil is contained in what the whalers called the case, a mass of parallel cablelike ligaments. Below the case is a fat, fibrous region known as the junk. There is marked asymmetry in the development of the nasal passages. The left one is rather simple. The right passage forms a complicated pathway within the head that links the concave anterior wall of the skull with a membranous vestibule near the anterior end of the head, at the junction of the case and the junk.

Several functions have been ascribed to the complex of nasal passages, spermaceti organ, and associated veins, arteries, and other organs. Clarke (1978c) regarded it as a mechanism to control buoyancy by alternate heating and cooling of the oil through the range of physiologically possible temperatures. However, Ridgway (1971) found this argument unconvincing.

Norris and Harvey (1972) believed these structures to function in sound production. The two air-filled sacs may act as acoustic mirrors, reflecting sounds back and forth through the oil-filled case and finally releasing them anteriorly as a rapid series of clicks. By this view, the spermaceti organ would be a sound generating, focusing, and reverberating chamber responsible for the loud clicks characteristic of the sperm whale. There is some field evidence in support of this hypothesis (Mohl et al. 1976). Time intervals between sperm whale clicks were used to estimate the length of the hypothesized sound mechanism in the head of a sperm whale of measured head, skull, and body length. Estimated lengths were identical to measured lengths.

A third hypothesis is that the spermaceti organ assists in the evacuation of the lungs prior to deep diving (Schenkkan and Purves 1973). It might also help absorb nitrogen during deep dives, thereby offsetting the dangers of nitrogen narcosis.

The sperm whale has the largest brain of any cetacean (Pilleri and Gihr 1970): it can weigh as much as 9.23 kg (Berzin 1972).

DISTRIBUTION AND MIGRATION

Sperm whales are distributed in oceanic water worldwide (Townsend 1935; Berzin 1972). Their distribution is not random, but shows definite preferences for continental margins (usually along or seaward of the 100-fathom contour), sea mounts, and, in general, areas of upwelling where food is abundant.

Yankee whalers of the 19th century were among the first to recognize the haunts and habits of the sperm whale (Beale 1839). As their activities spread, they managed to identify most of the currently known sperm whale grounds. Maury (1851, 1853) and Clark (1887a) produced charts showing the locations of the sperm whaling grounds, based on the lore of whaler informants and on actual catch records of the American whaling fleet (see Bannister and Mitchell 1980 for reproduction of charts and full references). Townsend (1935) summarized the catches of 53,877 whales caught on more than 1,600 Yankee whaling voyages. Of the total, 36,909 were sperm whales. The location and season for all the catches are evident on the Townsend charts (see also Gilmore 1959).

The major grounds in the North Atlantic, according to Townsend (1935), consisted of the Western Ground, centering at 31° N, 50° W in the midocean Sargasso region; the Western Islands Ground, around the Azores; the Southern Ground (33°–40° N, 60°–75° W), northwest of Bermuda; the Charleston Ground (28°–33° N, 67°–78° W), south of the Southern Ground and southeast of Cape Hatteras; the Commodore Morris Ground (47°–51° N, 20°–25° W); the San Antonio Ground, around the Cape Verde Islands; the Cornell Ground (4° N, 22° W), near Saint Paul Island; and a midocean whaling ground known as the Twelve-Forty (12° N, 40° W), between the West Indies and Cape Verde Islands. The more northern grounds were used only in summer; those at intermediate latitudes generally provided whaling opportunities in spring and fall as well as in summer. Only near the equator did sperm whaling occur year-round.

In the North Pacific, the major whaling grounds were the Japan Ground (28°–35° N, 150°–179° E), the Coast of Japan Ground (34°–40° N, 142°–149° E), a region around the Bonin Islands southeast of southern Japan, the Hawaiian Ground, the Kodiak or Northwest Ground in the Gulf of Alaska, and along the southern half of Baja California and central Mexico (Townsend 1935; Gilmore 1959). The grounds known as Panama, Galapagos, Off Shore, and On the Line formed a nearly continuous hunting region from the coast of Ecuador across almost the entire Pacific, and provided whaling at all seasons.

The Maury and Townsend charts, as reviewed by Bannister and Mitchell (1980), showed that during the peak of the breeding season—April and May—there are discontinuities in sperm whale occurrence among many of the major North Pacific grounds as well as the minor Hawaiian Ground. There probably are, then, at least three, and possibly more, separate stocks of sperm whales in the North Pacific.

Centers of whale distribution change seasonally to some extent, as animals move between their equatorial wintering areas and their high-latitude summering grounds (Best 1969b). Some mixing among the stocks is thought to occur in summer. Only large adult males, however, reach the higher latitudes, occurring anywhere between 70° N and 70° S. The seasonal shifts in distribution of sperm whales are not necessarily comparable to the long-distance migrations of some baleen whales, for "there is little evidence for a periodicity in feeding and the length of pregnancy is not geared to a twelve-month cycle" (Best 1969b). Also, a considerable number of sperm whales can be found in equatorial regions throughout the year. Some mature females wander north of latitude 50° N, but most females and young remain inside the equatorial belt bordered by latitude 40° in either hemisphere.

SOCIAL ORGANIZATION

Knowledge of the social organization of sperm whales is fundamental to understanding sperm whale reproduction. The considerably larger size of males is presumably associated with a greater metabolic requirement, an ability to dive more deeply and for longer periods, a predisposition to feed on larger prey found at greater depths, and a need to migrate to higher latitudes and colder water to find such favored prey. The interplay of these characteristics helps explain why females are restricted mainly to areas between the northern and southern subtropical convergences, while males migrate to higher latitudes as they grow larger.

Females mature sexually at approximately 9 years of age, males at 25–27 years (Best 1970). Fetal sex ratios are equal. If it is assumed that adult mortality is the same for both sexes, there would be an abundance of immature and maturing males in the population (Best 1969a). Subadults comprise about 77 percent of the male population, 38.5 percent of the total population (Best 1979). The ratio of sexually mature males to sexually mature females is thus 1:2.6.

The basic social unit appears to be the nursery school, or mixed school of adult females plus their calves and some juveniles of both sexes (Best 1979). As the young animals age, they form separate schools. Such segregation probably begins shortly after weaning. The males tend to segregate into bachelor schools. These can be subdivided according to age and size into small, medium, and large bachelors. The mean age at which males depart from the mixed schools off South Africa may be as low as 4–5 years, when they are 7.6–7.9 m long (Best 1970). Segregation by male sperm whales from mixed or all-female schools is complete at an age of 15 years and a length of 10.7 m.

The mixed schools appear to be long-term, cohesive units led by one or two older females and with which the adult males associate only briefly during the peak of the breeding season (Best 1970). This brief association occurs during midsummer while the majority (75–90 percent) of adult males are in high latitudes and the females are still between the subtropical convergences. This ensures geographic separation between the majority of sexually mature males and the reproductive females. Replacement of a breeding bull under these circumstances could be difficult during a given breeding season.

The term *harem* needs clarification when applied to sperm whales, for it is often used to describe their social structure. The mixed schools seem to be led by older females and to have adult males in attendance only during a part of the breeding season. Ralls (1976) suggested the terms *extended matricentric family* or *extended mother family* for this type of social organization, which is also characteristic of elephants. There is no evidence that the females are dominant to the bulls when the latter are present. Mature bulls may circulate among different mixed schools during a given breeding season (Best and Butterworth 1980).

Sexually mature males fight with one another, as evidenced by the common occurrence of parallel tooth rakes on their skin (Best 1979). Judging by the spacing between these tooth rakes and the interdental spacing of male sperm whales at different ages, the majority of the combatants are over 14.6 m long. Although it is difficult to prove, males probably have a higher natural mortality rate than females (Ralls et al. 1980).

REPRODUCTION

The reproductive anatomy of the sperm whale is similar to that of other cetaceans (Slijper 1966; Berzin 1972). The corpora albicantia on the ovaries never completely disappear, leaving a permanent record of the number of ovulations (Best 1967).

Gestation probably lasts 14–16 months (Best 1968, 1974). The interval between pregnancies averages five years (Best 1979). Thus, only about 20 percent of the mature females are receptive in a given year. There is disagreement about whether the sperm whale is monestrous (Ohsumi 1965; Clarke et al. 1980b) or polyestrous (Best 1968; Gambell 1972). Also, it is not certain whether ovulation is induced (provoked) (Clarke et al. 1980a, 1980b) or spontaneous (Tormosov 1976; Best and Butterworth 1980). Although the empirical evidence for a density-dependent adjustment in the length of the sperm whale's breeding cycle is equivocal, a calving interval of four rather than five years is applied to the computer model currently used by the International Whaling Commission for yield calculations (Bannister 1980: table 2).

The breeding season is protracted, lasting as long as nine months off Durban, South Africa, for example (Gambell 1972). Nevertheless, 80 percent of conceptions occur there during the austral summer (November to January). Such a long breeding season may improve the chances for conception, compensating to some extent for the geographic separation of the majority of breeding bulls from receptive females.

At birth, sperm whales are about 4 m long and weigh about 750 kg (Best 1968). Few births have been observed (Gambell et al. 1973). Twinning is rare, with an incidence of about 0.5 percent (Gambell 1972).

FOOD HABITS

Sperm whales apparently use their impressive diving capabilities to feed in deep water. They have become entangled in submarine cables as deep as 1,134 m (Heezen 1957). Dives of 488 m were "usual" for sperm whales being tracked with sonar in the Indian Ocean (Rice 1978). Bulls have been followed with an array of hydrophones to a depth of nearly 2,500 m.

Two bull sperm whales were caught off Durban, South Africa, after an 80-minute dive (Clarke 1976). One of them had two individuals of a bottom-dwelling species of shark in its stomach. The water depth at the point of capture was 3,193 m. Thus, these bulls must have dived to depths of more than 3,000 m.

Squid (e.g., *Moroteuthis, Gonatus*) are the principal food of sperm whales. More than 28,000 individual squid were taken from the stomach of one whale (Akimushkin 1955). The stomach of a sperm whale killed in the eastern North Pacific contained 19 large *Moroteuthis robusta*, averaging about 16 kg, making a total of about 300 kg of squid (Caldwell et al. 1966). In the stomachs of 277 males and 79 females taken off South Africa, 130 kg was the maximum amount of food found in an individual's stomach; the average was far less (Best 1979).

Deepwater bottom fish are the second most common item in the diet (Caldwell et al. 1966; Rice 1978). Rag fish (*Acrotus willoughbyi*) and rockfish (*Sebastodes* sp.) are known prey off British Columbia (Pike and MacAskie 1969).

Very little is known about the sperm whale's

method of finding or capturing its prey. As there is little or no light in the depths where the sperm whale hunts, its acoustic sense is assumed to be of paramount importance (but see Watkins 1977, 1980). Some well-fed sperm whales have been caught and found to have deformed jaws and badly worn teeth (Caldwell et al. 1966). Gaskin (1967) speculated that the biolumines-cent mucus from caught squid smeared onto the sperm whale's teeth and mouth may act as a lure.

BEHAVIOR

Although scientists have timed sperm whale dives of more than an hour, 19th-century whalers claimed to have observed dives lasting up to 90 minutes (Caldwell et al. 1966). These prolonged dives are followed by series of 60–70 blows.

Compared to some of the baleen whales, sperm whales are slow swimmers. Aerial whale spotters working off South Africa regarded swimming speeds of 11 km/hr as fast for sperm whales (Gambell 1968). They considered 5–6 km/hr to be average. Watkins and Schevill (1975), using acoustic tracking techniques, estimated speeds of 2–3 km/hr for sperm whales.

There are numerous accounts of epimeletic, or care-giving, behavior in sperm whales (Caldwell and Caldwell 1966). Adult males sometimes approach struck females. Adult females frequently "stand by" near injured companions, and a cow, or sometimes even entire nursery schools, will often remain with an injured calf until it dies or escapes the whalers.

Yankee whalers were aware of sperm whale sounds that could be heard in air or through the bottom of their boats, but these sounds were not monitored by hydrophone or tape recorded until the 1950s (Worth-ington and Schevill 1957). Known sperm whale sounds are impulsive, consisting principally of clicks at a va-riety of repetition rates (Backus and Schevill 1966; Watkins and Schevill 1977a, 1977b). The function of these clicks is not clear, although they probably are used for echolocation and communication. Watkins and Schevill (1977a) described stereotyped, repetitive patterns of sperm whale clicks, which they termed *codas*. The codas, which appear to be unique for each whale, were heard only from whales underwater, when whales met, and from whales that were close together (Watkins 1977). When a coda is in progress, nearby whales become or remain silent. These circumstances strongly suggest that the codas may allow individual acoustic identification.

Sperm whales are the only large whales known to mass strand. Seven instances were summarized by Gilmore (1959) and 17 by Van Heel (1962). At least three factors have been identified that might precipitate such mass strandings: strong social cohesion with a leadership hierarchy, a tendency to panic in unfamiliar and stressful situations, and lack of adaptation to shal-low water (Gilmore 1959).

There are only five authenticated cases of sperm whales attacking whale ships (Martin 1973, 1976), but there are numerous records of their attacking whale boats. The first and most famous reported attack on a whale ship resulted in the sinking of the *Essex* on 29 November 1820 in the mid-Pacific near the equator. All reported attacks apparently were made by single, old, "rogue" bulls. Such bulls seem prepared to attack anything in the water. They are assumed to be search-ing for a harem, recent losers in battles for a harem, or simply individuals having recently passed their prime. Whatever their reasons, such attacks are rare, much more so than popular accounts imply.

ECONOMICS AND HISTORY OF WHALING

Yankee sperm whaling began in the North Atlantic in the early 18th century and had expanded outside the Atlantic by the 1780s (Starbuck 1878, Stackpole 1953). The *John R. Manta* made the last successful whaling voyage from New Bedford in 1925, taking 300 barrels of sperm whale oil from the Atlantic in less than four months (Hegarty 1959). Thus, sperm whales drew American seafarers to far corners of the globe during more than two centuries. The techniques of open-boat, hand-harpoon whaling, as practiced by the sperm whalers (Brown 1887), were transferred to local inhabitants of certain oceanic islands. Consequently, sperm whales are still hunted in the old manner at the Azores (Clarke 1954; Compton-Bishop et al. 1979), Madeira (Maul and Sergeant 1977), and to a lesser extent Saint Vincent in the Caribbean (Rathjen and Sullivan 1970; Caldwell et al. 1971).

Pelagic sperm whaling did not begin in the central Pacific until 1818, and the Coast of Japan Ground was opened in 1820 (Starbuck 1878). Since then, the North Pacific has been a major center of sperm whaling. An estimated 60,842 sperm whales were taken in the North Pacific between 1800 and 1909 (Best 1976b).

Petroleum, discovered in Pennsylvania in 1864, provided an alternative fuel for lighting homes and city streets. The price of sperm whale oil reached a high of $2.55 per gallon in 1866, but then steadily declined, never exceeding $1.00 per gallon after 1882 (Starbuck 1878; Hegarty 1959). This fall in oil prices removed much of the incentive that had driven the American sperm whale fishery. The Civil War also contributed to the industry's demise, as many Yankee whalers were sunk by Confederate cannons (e.g., see various con-temporary accounts in the *Whalemen's Shipping List*, published weekly in New Bedford). Between 1800 and 1909 an estimated 255,440 sperm whales were killed by Yankee whalers (Best 1976b).

Modern whaling is usually considered to have begun in 1864 with Svend Føyn's twin invention of the explosive harpoon and harpoon cannon and his innova-tion of mounting them on fast, steam-powered catcher vessels (Mackintosh 1965). The main result of these developments was that the larger and faster-swimming rorquals, primarily the blue whale (*Balaenoptera mus-culus*), could now be harvested instead of, or in addition to, sperm whales. Prior to World War II, the annual catch of sperm whales world-wide rarely exceeded 6,000 (Mackintosh 1965: figure 19). However, after the war, as the stocks of large rorquals became depleted, the at-tention of the highly efficient whaling industry began to focus once again on the sperm whale. By 1967 sperm whales comprised more than half of the total reported whale catch world-wide (*International Whaling Statistics,* published annually in Sandefjord, Norway). Until 1979, when the ban on pelagic whaling imposed by the International Whaling Commission went into effect (Asgeirsson 1980), sperm whales continued to be the principal target of the international whaling industry, accounting for 70–80 percent of the annual catch between 1975/76 and 1977/78.

Nearly 269,000 sperm whales (76 percent males) were killed in the North Pacific between 1910 and 1976 (Ohsumi 1980). The peak kill in one year, just over 16,000, occurred in 1968. The vast majority of the sperm whales killed in the North Pacific in the modern era were taken by the Japanese, operating principally from shore stations, and the Soviets, operating principally from pelagic factory ships.

Modern shore-based whaling began on the west coast of North America in 1905 (Rice and Wolman 1971). Land stations operated along the coasts of Alaska, British Columbia, Washington, Oregon, California, and Baja California (Kellogg 1931; Scheffer and Slipp 1948; *International Whaling Statistics*). Sperm whales were among the species captured. By 1971 only the station at Richmond, California, remained in operation, and it closed for economic considerations and in anticipation of federal legislation to ban commercial whaling in the United States.

The California shore stations took sperm whales between 1956 and 1961 only when baleen whales were not available (Rice 1974). After 1962, when the take was 30 percent sperm whales, the percentage increased steadily, reaching nearly 90 percent in 1970. The availability of adult male sperm whales had declined significantly by 1970, and the average length of legal-sized males in the catch off California declined from 13.4 m in 1956–60 to 12.1 m in 1968–70.

Canadian shore stations operated in British Columbia from 1905 to 1967, taking more than 5,000 sperm whales (Pike and MacAskie 1969). In most years, fin whales (*Balaenoptera physalus*) and mature male sperm whales predominated in the catch.

Eastern Canadian coastal sperm whaling had a somewhat longer history: from 1898 to 1972 in Newfoundland and from 1964 to 1972 in Nova Scotia (Mitchell 1974; Jonsgård 1977). It was ancillary to whaling for balaenopterids. The cumulative reported catch of sperm whales (all males) was only 424 in Newfoundland and 109 in Nova Scotia (Jonsgård 1977).

Sperm whale products consist primarily of industrial oils and oil derivatives (Berzin 1972; Food and Agriculture Organization 1978). The oil is especially useful to industry because of its lubricating properties and resistance to high temperature and pressure. It has wide application in cosmetics. Sperm whale meat is eaten in Japan but is regarded as inferior in quality and distasteful compared to meat from baleen whales (Brownell and Omura 1980). It contains high levels of mercury. Other products include ivory items, called scrimshaw, made from the teeth; leather made from the skin; animal foods, fertilizers, and stock feed additives made from the meat, bone, and blood meal; and ambergris, a valuable perfume fixative sometimes found in the intestines (Clarke 1954).

Substitutes are now available for nearly all of the commercially important products taken from sperm whales (Food and Agriculture Organization 1978). Oil from the bean of the jojoba plant (*Simmondsia chinensis*), a desert shrub indigenous to the American Southwest, is virtually identical chemically to sperm whale oil (National Academy of Sciences 1975; Vietmeyer 1975). It is hoped that a significant part of the world demand for sperm whale oil will eventually be met by jojoba oil.

MANAGEMENT

The management of sperm whales is necessarily an international concern, as their distribution is pelagic and their movements are extensive. On the North American continent there is at present no management of sperm whales aside from total protection. However, the United States, Canada, and Mexico are members of the International Whaling Commission (IWC), the primary international forum for development of management procedures. Although whaling for sperm whales by floating factories was banned by the IWC in 1979 (Asgeirsson 1980), some shore-based whaling continues in the Northern Hemisphere at Iceland, Spain, Japan, the Azores, Madeira, western Greenland, and the Lesser Antilles in the Caribbean.

Quotas by regional stock unit, minimum size limits on whales that can be killed, and closed seasons by area are set by the IWC. Member nations are allowed to consider such measures for 90 days before becoming bound by them. Objections lodged within 90 days exempt a nation from a given regulation. However, even after 90 days have passed, enforcement is the sole responsibility of the governments whose citizens are involved in the fishery. When two or more countries fish the same stock, quotas are apportioned through bilateral or multilateral agreements. An observer scheme, consisting largely of exchanges of personnel among whaling countries, is in effect. Infractions are reported to the IWC, but there is no legal recourse for imposing penalties.

According to the New Management Procedures introduced in 1976, there are three classifications of whale stocks (Rindal 1977): initial management stock (IMS), which may be reduced in a controlled manner to achieve maximum sustainable yield (MSY) levels or optimum levels as these are determined; sustained management stocks (SMS), which should be maintained at or near MSY levels and then at optimum levels as these are determined; and protection stocks (PS), which are below the level of SMS and should be fully protected.

As of 1980, sperm whale stocks were classified as follows (Asgeirsson 1980): In the North Atlantic a single stock, listed provisionally as an SMS, was recognized, with an annual quota of 273 (for 1980 only) to be divided between Spain and Iceland. In the North

Pacific the "eastern stock" was listed as a PS, while the "western stock" hunted off Japan was listed as an SMS with a catch limit of 1350. Stocks north of the equator in the Indian Ocean were unclassified, with a quota of zero assigned pending more information. The Southern Hemisphere has nine management divisions. In all but two, sperm whales were accorded full protection in 1980. The division 1 stock off eastern South America was classified provisionally as an SMS, with a quota of 30, the division 9 stock off western South America as an SMS, with a quota of 550.

The complex social structure and breeding behavior of sperm whales complicate management (Best 1974, 1979). Historically, management policies have reinforced the tendency of whalers to hunt selectively for larger individuals, particularly in the Southern Hemisphere pelagic fishery, where quotas were set at much higher levels for males than for females (Bollen 1979: table 2). A maximum length limit of 45 ft (13.7 m) was set in 1976 for the Southern Hemisphere north of latitude 40° S during October to January. This was a measure designed to protect large, socially mature males ("harem masters") on the breeding grounds (Mitchell 1977a). Such protection should be extended to all the sperm whale's mating and calving grounds during the entire breeding season to prevent a reduction in pregnancy and calf survival rates (Mitchell and Kozicki 1978).

The impossibility, or at least impracticality, of censusing sperm whales has meant that most population assessment has been based on the catch per unit of effort (CPUE). However, CPUE data—resting, chasing, towing, and searching times—are difficult to quantify and interpret. There are major differences in hunting technology and procedures used by the whalers of different nations and even by different fleets of the same nation. The shift in whaling emphasis from high latitudes in the early postwar years to middle and low latitudes in recent years has had a major effect on the CPUE. The large males found alone or in small groups at high latitudes require different hunting strategies from the harem or bachelor herds pursued at lower latitudes. Improved understanding of the factors influencing whale distribution reduces the required searching time. Satellite imagery and colorimetry tests now help the whalers locate concentrations of whales, and aircraft spotters and sonar assist in the chase. These technological innovations tend to mask trends in whale abundance on the whaling grounds. Also, during some phases of modern sperm whaling, sperm whales have been but one of several target species, and it is difficult to compare the CPUE of a fishery directed exclusively at one species with that of a multispecies fishery.

Several other methods of population assessment and monitoring have been attempted. Sightings data for the whaling grounds are not independent of whaling effort. Consequently, they are inappropriate for extrapolation to population estimates. Many sperm whales have been tagged with stainless steel darts (Discovery marks) (Mackintosh 1965). However, in no area has a sufficiently large sample been tagged to allow population estimates from tag recovery rates.

There is uncertainty about "initial" population levels and about the biological response of sperm whales to exploitation. Models presently used by the IWC's Scientific Committee to predict the consequences of various harvest strategies assume that major sperm whale populations had recovered fully by the end of World War II. Considering the long history of whaling for sperm whales and the protracted nature of their reproductive cycle, such an assumption may be optimistic. Also, there is no conclusive evidence of a density-dependent response in the form of higher pregnancy and recruitment rates.

As a valuable and historically abundant renewable resource, the world population of sperm whales can be conserved only through "continual public surveillance and control of the physical yield and the condition of the stocks" (Clark 1973). Otherwise, the inherent combined effects of common-property competitive exploitation and private-property maximization of profits will ensure overexploitation and invite the risk of biological extinction. Single-species management is no longer adequate, given mankind's intrusion upon natural ecosystems at many different levels (May et al. 1979). For instance, in the Southern Ocean a decrease in the sperm whale population may cause a decrease in the amount of krill available for human use because squid, the main prey of sperm whales, are important krill predators. Theoretically, the largest sustainable krill yield is attained when sperm whales are unexploited. As the depleted Antarctic stocks of baleen whales will themselves require a large standing stock of krill to support their recovery, they too have a stake in the sperm whale's continuing ability to function as a squid predator. Such complex interactions challenge human analytical skills as well as political and moral will.

OTHER TOOTHED CETACEANS

SMALL ARCTIC WHALES: NARWHALS AND BELUGAS

Among the odontocetes, there are two arctic specialists: the narwhal (*Monodon monoceros*) and the white whale, or beluga (*Delphinapterus leucas*). They have no Southern Hemisphere counterparts, although the Irrawaddy dolphin (*Orcaella brevirostris*) of the Indo-Pacific bears a close resemblance to the beluga, suggestive of a phyletic relationship (Mitchell 1975a).

Narwhals and belugas are both suited to living in close proximity to pack ice. The absence of a dorsal fin and the presence of a thick dermis and substantial subcutaneous blubber layer are usually viewed as arctic adaptations, as is the all-white coloration of adult belugas. Sergeant (1978) considered narwhals and belugas to be ecologically allopatric, the former preferring deep basins and fjords, the latter coastal margins and river mouths (in summer). It is not unusual, however, to find the two species together, especially in winter when availability of open water is reduced (Brown 1868; Vibe 1950). Neither species is obliged to make regular, long-distance latitudinal migrations; the critical factor governing their movements appears to be ice cover.

Dentition differs strikingly between these two Recent monodontids. Narwhal embryos can have as many as six pairs of dental papillae in the maxillae (Eales 1950). All but the anterior pair fail to develop. The two developed teeth normally remain in the jaw in females; the left tooth in males grows through the gum and becomes an impressive, sinistrally spiraled tusk, 2–3 m long. The tusk is believed to be employed in aggressive encounters between males (Silverman and Dunbar 1980). Belugas have 8–11 pairs of conical teeth in both jaws.

Both narwhals and belugas are sexually dimorphic; males grow consistently larger than females. Also, in both species a significant change in coloration takes place as the animals age. Neonates are basically gray to brown. Young narwhals become uniformly black, then develop white patches on the ventrum and eventually on the sides (Reeves and Tracey 1980; Hay and Mansfield in press). Adults are white ventrally, mottled on the sides, and dark dorsally. Old individuals are almost completely white. Juvenile belugas are uniform gray to bluish gray (Sergeant 1973). At some time after attainment of sexual maturity they become all white (Brodie 1971). Besides lacking a dorsal fin, both species have upcurled flippers as adults (Vladykov 1943), and both have a convex trailing margin on the flukes.

Although both have a circumpolar distribution, the beluga is more widely distributed and apparently more numerous overall than the narwhal. The largest narwhal concentration in the world is centered in Baffin Bay and Lancaster Sound (Mansfield et al. 1975; Reeves and Tracey 1980), numbering upward of 20,000 whales (Davis et al. 1978). Smaller numbers occur in western Hudson Strait and northern Hudson Bay/southern Foxe Basin, especially Repulse Bay and Frozen Strait (Soper 1944; Sergeant 1978). Narwhals are uncommon in the Beaufort and Chukchi seas, and virtually absent from the Bering Sea (Geist et al. 1960; Smith 1977; Reeves 1978).

In North American waters there are at least five centers of beluga abundance (Sergeant and Brodie 1975; Perrin 1980: table 1). An isolated stock numbering 200–500 inhabits Cook Inlet and nearby waters of the Gulf of Alaska (Harrison and Hall 1978; Perrin 1980: table 1). A much larger population, perhaps consisting of several noninterbreeding stocks, is found from Bristol Bay to the eastern Beaufort Sea (Perrin 1980: table 1). At least 4,000 animals congregate in the Mackenzie River estuary in summer (Fraker 1980). The majority probably migrates along the northern Alaska coast eastward in spring and westward in fall.

In the eastern North American arctic there are about 5,000–10,000 belugas along the west coast of Hudson Bay (Sergeant 1973). They summer in es-

tuaries between the Nelson River and Eskimo Point, and some probably winter in the northwest part of the bay. Others may migrate into or through Hudson Strait. Some overwintering also occurs in James Bay (Jonkel 1969). A relict, badly depleted population numbering only about 350 or fewer inhabits the Saint Lawrence estuary (Pippard and Malcolm 1978). Summer populations in eastern Hudson Bay, Ungava Bay, and Hudson Strait are heavily exploited (Perrin 1980: appendix 3). There is a large population in the Lancaster Sound region (Sergeant and Brodie 1975). A census in Cumberland Sound, Baffin Island, revealed a summer population of about 750 there in 1967 (Brodie 1971), but this has declined since then (Kemper 1980). The Cumberland Sound and Saint Lawrence populations each numbered several thousand prior to commercial exploitation (Mitchell 1974; Mitchell and Reeves in press). Small groups and individuals occasionally wander as far south as New Jersey (Reeves and Katona 1980). The only stock for which movement data exist is in western Hudson Bay, where several animals tagged in the Seal River (Sergeant and Brodie 1969) were recovered at Whale Cove and Repulse Bay (Sergeant 1973).

Because of their unique ivory appendage, narwhals have acquired a special kind of commercial importance. Their spiraled tusk was a valued commodity of trade from very early times, and it still brings a good price to native hunters (Kemper 1980; Mitchell and Reeves in press). The price of narwhal ivory paid to hunters increased from less than $1.00 per pound in 1961 to more than $45.00 per pound in 1980 (Mitchell and Reeves in press). Wholesalers sell narwhal tusks by length rather than weight, and tusks are graded according to their condition, i.e., whether they are broken or intact. Although the character of the narwhal ivory market is not well known, most of the demand is believed to come from a desire to possess the tusks as curios. Narwhal ivory may still have some superstitious value in the Far East, as it did during the 18th and 19th centuries (see references in Mitchell and Reeves in press).

European whalers hunting for bowhead whales (*Balaena mysticetus*) took narwhals only as a pastime; catches of 100 in Davis Strait were considered exceptional (Gray, quoted in Buckland 1890). Whaling by natives for subsistence and trade purposes has at times been intensive. A white resident of the Canadian arctic claimed that 2,800 narwhals were taken in and near Eclipse Sound, off the north coast of Baffin Island, in one year (William Duval, quoted in Soper 1928). Mitchell and Reeves (in press) included this exceptional catch in their estimate of nearly 11,000 narwhals killed in Canada and Greenland, combined, between 1914–15 and 1923–24. Some narwhal hides were exported to France in the early 20th century, where they were used for leather to make fine gloves. The landed and reported catch by North American natives has averaged about 300 in recent years (Kemper 1980) (figure 18.11). Correcting the landed and reported catch to account for animals shot but lost to the hunters and for

FIGURE 18.11. A narwhal (*Monodon monoceros*) killed by native hunters in the eastern Canadian arctic.

animals landed but not reported, and adding the Greenlandic catch, total annual removals from the narwhal population(s) in Canadian and western Greenland waters may exceed 1,500 (Kemper 1980). A domestic quota of 402, allocated by village, was established in Canada in 1977 (Kemper 1980), but it has not been strictly enforced (Finley et al. 1980).

White whales have been of more commercial significance than narwhals. European whalers took belugas frequently to fill their holds when bowheads were unavailable (Lubbock 1937; Mitchell and Reeves in press). In addition, major shore-based fisheries have been conducted in the Soviet arctic (Kleinenberg et al. 1969), at Spitsbergen (Lønø and Øynes 1961), and in Canada. The Saint Lawrence fishery is well documented (Vladykov 1944, 1946), as is the fishery at Churchill in western Hudson Bay (Doan and Douglas 1953; Sergeant 1968). Products included blubber oil used in the manufacture of soaps and cosmetics, hides used for making leather bootlaces, and head and jaw oil used as a lubricant. Some of the meat from animals taken at Churchill was supplied to mink (*Mustela vison*) farms. The meat and muktuk (the thick dermis and subcutaneous fat) from animals killed in a short-lived (1961–70) net fishery at Whale Cove were canned and marketed for human consumption (Sergeant and Brodie 1975). However, when mercury levels in excess of 0.5 ppm wet weight were found in the meat, this aspect of the fishery was stopped.

An Alaskan fishery based in Anchorage was established in 1916, and another in Nome in 1920 (Tønnessen 1967). The former depended on animals from the Cook Inlet population, the latter on those in Norton Sound. Both fisheries closed in 1921 because of economic considerations. The catches amounted to 9, 42, 41, 236, and 50 whales from 1917 to 1921. A brief commercial net fishery accounted for about 100 belugas in the Cook Inlet area during the 1930s (Klinkhart 1966). Some sport hunting of belugas took place in Cook Inlet during the 1960s.

The beluga was one of the first cetaceans to be displayed alive in captivity in North America. A young male, apparently caught in 1861 in the Saint Lawrence, lived for nearly two years in Boston's Aquarial Gardens (Wyman 1863). It was trained to tow a small car ridden by a woman around its tank, and lived peaceably with a sturgeon, a small shark, and a bottlenose dolphin. Other white whales taken in eastern Canadian waters were shipped alive to England (Lee 1878), and more recently live-captures have been made in Alaska (Ray 1961, 1962; Sergeant 1973; Leatherwood unpublished data) and western Hudson Bay (Ricciuti 1967; Hewlett 1978; Anonymous 1979).

Since 1966 the live-capture fishery at Churchill, Manitoba, has involved the driving of whales into shallow water, where individuals are lassoed by divers (Asper 1975; Farquhar 1978; anonymous 1979) (figure 18.12). This fishery presently supplies most of the North American, Japanese, and European demand for captive white whales (Leatherwood unpublished data). From the mid-1960s through 1973, small-scale "sport whaling" was conducted from Churchill. Thrill seekers, under supervision of local Indians, were allowed to harpoon white whales in the Churchill River. Attempts to maintain narwhals in captivity have been unsuccessful (Newman 1971).

The known prey of narwhals consists mainly of arctic cod (*Boreogadus saida*), Greenland halibut (*Reinhardtius hippoglossoides*), cephalopods (*Gonatus fabricii*), and shrimp (*Pasiphaea tarda*) (Mansfield et al. 1975; Hay and Mansfield in press). Belugas are more versatile in their food habits. In western Hudson Bay they feed on capelin (*Mallotus villosus*), river fish such as cisco (*Leucichthys artedi*) and pike (*Esox lucius*), marine worms (*Nereis* sp.), and squid (Sergeant 1968). Farther north they rely on decapod crustaceans, arctic char (*Salvelinus alpinus*), Greenland cod (*Gadus ogac*), and arctic cod. In the Saint Lawrence the main part of their diet consists of capelin, sand lance (*Ammodytes americanus*), marine worms (*Nereis* sp.), and squid (Vladykov 1946). Five species of salmon (*Oncorhynchus* spp.) are included in the diet of Alaskan belugas (Klinkhart 1966).

In the Kvichak River, Alaska, predation by belugas on red salmon smolt (*Oncorhynchus nerka*) brought them into conflict with Bristol Bay's lucrative commercial salmon fishery during the 1950s (Fish and Vania 1971). Charges of dynamite were dropped into the water to repel herds of belugas ascending the Kvichak River (Klinkhart 1966). Later, recorded killer whale vocalizations broadcast underwater were used to drive belugas away from the fishing grounds. Fluctuations in commercial finfish landings—mainly cod (*Gadus morhua*) and Atlantic salmon (*Salmo salar*)—in the Saint Lawrence during the 1930s prompted the creation of a bounty system for eradicating belugas there (Vladykov 1944). Subsequent studies of white whale feeding in the Saint Lawrence demonstrated that cod are not a major prey item and that salmon form an insignificant part of their diet (Vladykov 1946). The periodic decline in availability of these commercially important species of fish was almost certainly a result of natural cyclic changes or of human overfishing

FIGURE 18.12. Procedures for live capture and transport of beluga whales (*Delphinapterus leucas*) near Churchill, Manitoba. Top: outboard-power canoes herd the whales into shallow water; middle: the whale is transported to temporary holding facilities; bottom: the whale is loaded for long-distance transport.

rather than of excessive natural predation. The bounty lasted from 1932 to 1939 (not including 1936). A total of 1,897 bounties was paid in the period 1932–37. To our knowledge, this is the only instance in North America in which an official government bounty has been used to control a cetacean population. According to Grenfell (1934), the Canadian government also sponsored the bombing of Saint Lawrence belugas to protect the cod fishery in the late 1920s or early 1930s.

Belugas have been captured in North America by a variety of techniques. In the Saint Lawrence specially constructed weirs were once used to entrap belugas that entered at high tide and became stranded at low tide (Vladykov 1944). There and elsewhere they have been driven by motorboats close to shore where they could be stranded or netted and shot (e.g., Low 1929; Soper 1928). Northern natives once used kayaks and harpoons to hunt both belugas and narwhals (Vibe 1950), but today most such hunting is done from motor-driven canoes and with rifles (Reeves 1976; Hunt 1979; Fraker 1980; Kemper 1980). Modern methods are wasteful because many shot animals are not recovered (Fraker 1980), particularly at the ice edge hunt for narwhals (Kapel 1977; Finley et al. 1980). At the same time, motorized transportation and high-powered rifles allow hunters to cover a wider area and to interact with a larger segment of a given whale population, even though human settlement has become more concentrated and sedentary (e.g., Treude 1975, 1977). The International Whaling Commission's Subcommittee on Small Cetaceans has recommended that the aboriginal hunt for arctic monodontids be added to the commission's schedule so that it will be subject to regulation (Perrin 1980). The badly depleted beluga population in Cumberland Sound requires immediate protection if its survival is to be assured (Mitchell and Reeves in press). The small Cook Inlet and Saint Lawrence stocks need to be monitored closely. The latter was given explicit protection from all forms of hunting in 1979 under Canada's Fisheries Act (Pippard 1980).

Killer whales prey on both narwhals and belugas, although most evidence for successful predation is circumstantial or hearsay (Kleinenberg et al. 1969). Polar bears occasionally manage to catch them at narrow openings in the ice (Munn 1932; Freeman 1973; Heyland and Hay 1976; Hay and Mansfield in press). Records of walruses (*Odobenus rosmarus*) attacking narwhals resting at the surface are exceptional (Gray 1939).

A major hazard of living year-round in the arctic is ice entrapment. This phenomenon, which is most frequent off West Greenland, where it is called *savssat,* occurs when narwhals and/or belugas fail to exit from a bay or fjord before ice forms across its mouth in the autumn (Porsild 1918). As the ice cover spreads, the breathing space for the whales becomes more limited; mass mortality from crushing, starvation, suffocation, or predation may result. *Savssats* are a boon to hunters. The most recent major occurrence of this kind in North America was in southern Gulf of Boothia in the fall of 1979, when natives killed more than 100 narwhals, most of them females and calves of the year (Mitchell in press). Records of *savssats* in the eastern North American arctic were reviewed by Mitchell and Reeves (in press). There is some evidence for long-term survival of trapped whales that are not molested by hunters (Freeman 1968; but also see Hill 1967).

KILLER WHALE

The genus *Orcinus* includes one species (*Orcinus orca*). Killer whales are locally common in shelf waters of North America, particularly in the Pacific Northwest. They apparently are unique among the Cetacea in that they regularly prey on warm-blooded vertebrates. However, several species of pelagic "blackfish"—the false killer whale (*Pseudorca crassidens*), the pilot whales (*Globicephala* spp.), the pygmy killer whale (*Feresa attenuata*), and the melon-headed whale (*Peponocephala electra*)—are aggressive and are believed under some circumstances to attack and kill cetaceans (Perryman and Foster 1980). Their predatory habit, together with scattered records of killer whale "attacks" on vessels (di Sciara 1977), has burdened killer whales until recently with a reputation as creatures dangerous to humans.

The killer whale's appearance is very striking. Coloration is black with well-defined white markings. There is a distinctive white oval postocular patch; the throat and belly are white. A white zone contiguous with the midventral whiteness encroaches onto each side, originating at the umbilicus and oriented posteriorly. Variability in the size and shape of these white zones and the gray saddle behind the dorsal fin can be used to identify regional stocks (Evans and Yablokov 1978). Dentition consists of 10–12 large conical teeth per row. The flippers are broad and paddlelike. Males, which grow larger (to 9.4 m) than females (to 8.2 m), have a tall (to 1.8 m), erect dorsal fin; the female's dorsal fin is falcate and no more than a meter high.

Whaling for killer whales in North American waters has been modest—certainly in comparison to catches made off Japan, Iceland, and Norway (Mitchell 1975*a*). The Japanese (Nishiwaki and Handa 1958) and Norwegian (Jonsgård and Lyshoel 1970) killer whale fisheries for meat and oil have no parallel in North America. A single killer whale can yield 1.5 tons of oil and up to 2 tons of meat for animal food (Yablokov and Bel'kovich 1968).

Understandably, their predaceous nature has made killer whales unpopular in many fishing communities. They have been, and occasionally still are, shot by fishermen in coastal North America (Scheffer and Slipp 1948; Balcomb 1978). The U.S. Navy was enlisted by disgruntled Icelanders during the 1950s to destroy killer whales off their coast with machine guns, rockets, and depth charges (see Mitchell 1975*a*). Control measures of this kind are still being contemplated in Iceland to protect the herring fishery (Food and Agriculture Organization 1978). The Norwegian killer whale fishery for oil and animal food is presently subsidized and justified as a control measure to protect inshore herring stocks (Perrin 1980).

The commercial importance of killer whales in North America today rests primarily in their value as display animals. Their metamorphosis from villain to darling actually occurred very recently. In 1964 a collecting team from the Vancouver Public Aquarium attempted to capture a killer whale by harpooning it from shore (Newman and McGeer 1966). Their intention was to kill it and use the carcass as a model for constructing a replica. However, the whale survived the harpooning. Its captors then decided to treat its wounds and attempt to maintain it alive in captivity. They found that "the most astonishing aspect of the behavior [of this whale] was the complete lack of ferocity or aggressiveness." Not only did the killer whale in captivity fail to live up to its reputation for pugnacity, but the autopsy of this animal, which died after somewhat less than three months of confinement, revealed a brain weighing 6.45 kg, initiating speculation about the species' "intelligence." The relationship between mankind and the killer whale would never be the same.

In 1965, when fishermen at Namu, British Columbia, accidentally netted a large male killer whale, it was purchased by the Seattle Marine Aquarium for $8,000 (Griffin 1966) and towed 700 km in a floating pen to Seattle (Griffin and Goldsberry 1968). During the ensuing year "Namu" proved docile and trainable, and it delighted thousands of spectators. As a direct result of this whale's popularity, the live-capture of killer whales became a rewarding enterprise.

Between 1964 and 1973, 43 killer whales were captured in inside waters of southwestern British Columbia; of these, 1 died during capture and 20 escaped or were released (Bigg and Wolman 1975). In and adjacent to Puget Sound, Washington, 220 were captured from 1962 to 1973; of these, 11 died during capture and 181 were freed or escaped. Most of the whales were taken by encirclement with gillnets. Entire pods, including one consisting of 80 individuals, were trapped and sorted. Whales longer than 6 m were generally released; young, weaned individuals 3.5–4.7 m long were selected preferentially.

Killer whales apparently eat any palatable form of animal life in the oceans, from seabirds and herring to the great whales. Recently a pod of killer whales was filmed off Baja California attacking a blue whale (*Balaenoptera musculus*) (Tarpy 1979). They had long been known to prey on gray whales (*Eschrichtius robustus*) (Scammon 1874; Baldridge 1972), humpback whales (*Megaptera novaeangliae*) (Packard 1903; Mead 1961), sei whales (*B. borealis*) (Nishiwaki and Handa 1958), and minke whales (*B. acutorostrata*) (Hancock 1965), but this was the first well-documented account of predation on the largest of the balaenopterids. The degree to which killer whales depend on large mammals for food varies greatly among regions and may also vary seasonally. Soviet studies in the Southern Hemisphere have revealed that cetaceans (including sperm whales but mainly minke whales) and pinnipeds are an important part of the diet (Yukhov et al. 1975; Shevchenko 1975). By contrast, killer whales in the eastern North Atlantic prey heavily on herring (Jonsgård and Lyshoel 1970), and those off

Japan prey on various fish species (Nishiwaki and Handa 1958). In the eastern North Pacific, remains of cetaceans and pinnipeds were found in 6 of 10 killer whale stomachs (Rice 1968), and there are numerous observations of predation on pinnipeds and small cetaceans off California and Mexico (Norris and Prescott 1961). In Washington waters killer whales are mainly piscivorous (Balcomb 1978); salmon (*Oncorhynchus* spp.) are regarded by fishermen as an important part of the killer whale's diet there (Scheffer and Slipp 1948). Shevchenko (1975) reported an instance of what he called cannibalism: two killer whales feeding on a third killer whale.

A unique program of research on killer whales has been conducted in Washington and British Columbia. Using a technique of photodocumentation pioneered by M. A. Bigg and colleagues of the Pacific Biological Station in Nanaimo, B.C., Balcomb (1978) and Chandler et al. (1977) recorded a long series of encounters with recognized individuals and pods. Patterns of notching in the trailing edge of the dorsal fin and of dorsal pigmentation at the base of the dorsal fin are the primary keys for individual recognition. These studies have demonstrated that pods consisting of 1–40 whales found in Puget Sound and environs have great stability over time. The best known of these—J pod—is resident in Puget Sound and consisted of 3 adult males, 8 adult females, and 6 juveniles as of 1977. The whales are often seen hunting cooperatively.

Balcomb (1978) described "rutting" behavior among large males in Puget Sound in September, when sexual activity there peaks. Adult males become "arrogant and disinclined to yield right of way to other whales and small boats." An adequate sample of specimens has not been examined to give details of life history. Calves, about 2.4 m long, are born after at least 12 months of gestation (Mitchell 1975b). Sexual maturity in females off Norway is attained at lengths of about 4.9 m (Jonsgård and Lyshoel 1970).

BEAKED WHALES

Four genera of beaked whales are represented in coastal waters of North America (Moore 1968; Rice 1977). The most striking feature shared by them is reduced dentition: maxillary teeth are absent or relictual and there are no more than four teeth in the mandibles. Baird's beaked whale (*Berardius bairdii*) has two pairs of teeth positioned toward the front of the mouth; the northern bottlenose whale (*Hyperoodon ampullatus*) usually has a single pair of apical mandibular teeth, as does Cuvier's beaked whale (*Ziphius cavirostris*); and the various species of *Mesoplodon* have one pair of laterally compressed, tusklike teeth positioned anywhere from the apex of the mandibles to behind the mandibular symphysis, according to species. In general, the teeth are functional only in adult males; *Berardius* is an exception.

The beaked whales are distributed primarily along or seaward of the edge of the continental shelf and are believed to feed primarily on cephalopods and deepsea fishes (Mitchell 1975b). Very little is known about any of them. Most information about Cuvier's beaked whale and the mesoplodonts comes from experience with stranded specimens (Mitchell 1968; Moore 1966, 1968). The two largest Northern Hemisphere species, Baird's beaked whale and the northern bottlenose whale, have been exploited commercially and are, as a result, somewhat better known. There are no estimates of present population size for any of the beaked whales in North American waters.

Baird's beaked whale is by far the largest of the family, reaching lengths of 11.8 m (males) and 12.8 m (females) (Mitchell 1975b). It occurs only in the North Pacific, mainly between southern California and the Pribilof Islands of Alaska on the east side and between the Sea of Okhotsk or Kamchatka and southeastern Japan on the west. A sizable shore-based Japanese commercial fishery for meat and oil peaked in 1952, when a catch of 322 Baird's beaked whales was reported (Ohsumi 1975a). This fishery continues at a reduced level (Balcomb and Goebel 1977).

In the eastern North Pacific there has been no significant exploitation. Pods of 10–20 Baird's beaked whales, mainly males, were seen in summer on the whaling grounds off Vancouver Island, British Columbia, during the 1950s and 1960s (Pike and MacAskie 1969). Only 25 were caught by shore whalers there between 1950 and 1966. This species was also casually exploited by the California shore whaling industry: 15 were taken there between 1956 and 1970 (Rice 1974). Most were taken off California in July and October. As in British Columbia, a high proportion consisted of males. At present there is no reason to believe that Baird's beaked whale poses a management problem off western North America.

In contrast, the status of the northern bottlenose whale has become a controversial international management issue. This species is confined to the North Atlantic, where it has been hunted commercially since 1877 (Gray 1941). Scottish, and later Norwegian, sealers and whalers caught bottlenose whales on an opportunistic basis until the 1890s, when bottlenose whaling developed into a separate industry in Norway (Christensen 1975). During the 1890s the annual catch averaged about 2,500. The fishery declined steadily thereafter, reaching insignificant levels by the late 1920s. Another episode of Norwegian whaling in the North Atlantic, aimed principally at minke whales, began in the 1920s. Bottlenose whales were taken in this fishery, along with killer whales and long-finned pilot whales (*Globicephala melaena*), to supplement minke catches. The highest catch of bottlenose whales during this period was 692 in 1965. In the early 1970s the toothed whale portion of this fishery became unprofitable and was virtually discontinued. The products of the Norwegian bottlenose fishery were oil and animal food (Foote 1975). A short-lived Canadian bottlenose fishery based in Nova Scotia took 87 whales between 1962 and 1967 (Mitchell 1974).

Bottlenose whales appear to be capable of longer dives than any other cetacean; submergences lasting up to two hours have been reported for harpooned individuals (Benjaminsen and Christensen 1979). Their principal prey is squid (*Gonatus fabricii*), and they rarely are found in water less than 1,000 m deep. There is a striking degree of sexual dimorphism. Males grow to lengths of 9.0 m, females to about 7.5 m. Adult males also are more robust than adult females, and they develop an extremely bulbous melon, supported by large maxillary crests. Mitchell and Kozicki (1975) speculated that these serve an acoustic function. Bottlenose whales are long lived. Assuming that laminae in the dentine are formed annually, maximum age is at least 37 years (Benjaminsen and Christensen 1979). Length at birth is about 3 m. Mating and births occur mainly in April. Both sexes probably reach sexual maturity some time after 7 years of age.

The known southern limit of bottlenose whale distribution off North America is at the latitude of Rhode Island (Mitchell and Kozicki 1975). Wintering grounds may be along and seaward of the continental slope from Cape Cod to the Grand Bank. Although winter distribution is not well documented, sightings in February (MacLaren Atlantic Ltd. 1977) and March (MacLaren Marex Inc. 1979) along the ice edge in southern Davis Strait suggest that at least part of the population remains in arctic waters year-round. They are present at the mouth of Hudson Strait and near Frobisher Bay in spring, and remain there through summer (Smith et al. 1980). Several hundred have been seen off West Greenland at about 64° N latitude in spring and early summer (Benjaminsen and Christensen 1979). The Gully, a deepwater area off Nova Scotia influenced by the Gulf Stream, seems to contain bottlenose whales year-round. Occasional strandings occur in the Bay of Fundy and Gulf of Saint Lawrence. Northern bottlenose whales have been caught as much as 10 nm into the pack ice off Labrador (Benjaminsen and Christensen 1979).

Because there was no tagging program in conjunction with the bottlenose fisheries, the question of stock identity remains open. It seems likely, based on the progressively westward expansion of the Norwegian fishery documented by Christensen (1975) and Mitchell (1977b), that there are local stocks that were sequentially overhunted. Mitchell (1977b) used cumulative catch analysis to estimate population size for the entire North Atlantic prior to the 1890s as at least 28,376. A downward trend in a population of this size should be expected from catches of 2,000–3,000 per year as occurred during the 1890s.

There are several aspects of bottlenose whale behavior that have special significance for management (Mitchell 1977b). One is that these whales approach vessels, apparently out of curiosity. This inevitably affects the manner in which they are hunted, adding a passive dimension to the "chase." It also means that shipboard censusing is suspect, as "seeking" behavior tends to inflate counts beyond representative values. In the early Norwegian fishery, vessels were sometimes outfitted with harpoon guns at the stern and waist as well as the bow, thus affording an opportunity to shoot whales as they approached a stationary vessel from several different angles. Another aspect is called "standing by," or epimeletic behavior. Bottlenose whales occur in pods of 4–10 animals, and individuals

generally do not abandon wounded companions. Several whales, then, sometimes even entire pods, can be taken almost simultaneously.

The question of why the bottlenose fishery ended in the early 1970s remains problematic. Christensen (1975) attributed its decline to the lowered prices paid for animal foods. Mitchell (1977b), on the other hand, emphasized the possibility that the westward expansion of the fishery and decline in catch were due to stock depletion. The northern bottlenose whale is presently listed as "vulnerable" in the International Union for the Conservation of Nature and Natural Resources (IUCN) Red Data Book. Also, it was the second "small" cetacean (after the minke whale) to be included in the IWC schedule. In 1977, and annually thereafter, the species has been designated a protected stock by the IWC. Therefore, it cannot legally be hunted by member nations. Bottlenose whales are now effectively protected, and stocks should be recovering.

PILOT WHALES

These extremely abundant whales are common in shelf waters of North America. There are at least two well-differentiated species of pilot whale in the North Atlantic—an antitropical long-finned form (*Globicephala melaena*) and a tropical short-finned form (*G. macrorhynchus*) (van Bree 1971). A third putative form (*G. scammonii*) has been attributed to the eastern North Pacific, but it is properly referred to *G. macrorhynchus* pending a review of the genus (see Mitchell 1975b).

All pilot whales have a bulbous forehead ("pothead") and a broad-based, low-profile dorsal fin positioned unusually far forward on the body. The flippers are long and sickle shaped; the caudal peduncle is strongly keeled. They are sexually dimorphic. Males are much larger (to 6 m long, versus 5 m in females) and have a thicker dorsal fin and a more robust melon. Length at birth is about 1.4–1.8 m. Short-finned pilot whales are generally somewhat smaller than long-finned pilot whales. However, the two species are difficult to tell apart on the basis of external features (Sergeant 1962a). Both are generally black to brown overall, with varying amounts of light gray on the back behind the dorsal fin, on the ventrum, and on the head.

Fisheries for pilot whales have existed in many parts of the world (Mitchell 1975a). Perhaps the oldest and best known of these is at the Faeroe Islands, where large herds have been driven close to shore and killed for local meat consumption since at least as long ago as 1584 (Williamson 1948). Mitchell (1975a), after a review of published information on this fishery, concluded: "This set of data surely represents one of the longest runs of whaling statistics available anywhere in the world, and further indicates that the shore-driving techniques as practiced by the Faeroe islanders have not overtaxed the Ca'aing [pilot] whale populations."

Unfortunately, the drive fishery conducted off eastern Newfoundland between 1951 and 1972 showed little of the resilience characteristic of the Faeroese fishery. It was established in 1947 to supply meat for ranch mink (Sergeant 1963). The capture technique involved the deployment of several catcher boats that would "slowly guide them [the whales] shorewards, like dogs herding sheep" (Sergeant 1962*b*). Once deep inside a bay, fishermen in motor launches and rowboats would take over, forming a crescent and driving the whales into very shallow water. There the whales were lanced, hauled ashore by power wagon or tractor, and butchered. Occasionally some pilot whales were harpooned from the catcher boats.

The highest landed catch at Newfoundland occurred in 1956, when 9,794 were taken (Mercer 1975). The average for the period 1950–59 was 3,906 per year; for 1960–69, 1,491. Catches were insignificant by the time commercial whaling on the Atlantic coast of Canada was suspended, on 11 December 1972. Studying catch composition and life history of the exploited population through 1959, Sergeant (1962*b*) found no evidence that pilot whales had been overhunted. However, Mitchell (1974) and Mercer (1975) concluded that the population had been severely reduced by the time hunting ceased.

Another major drive fishery existed at Cape Cod from the mid-1700s to the 1920s (Clark 1887*b*; Mitchell 1975*a*). A catch of about 1,400 in November 1884 was believed at the time to have been the biggest single drive on record (anonymous 1885). However, the mean annual catch during the late 1800s may have been on the order of 2,000–3,000 whales (Mitchell 1975*a*). Vessels from the New England sperm whaling fleet occasionally went to the Grand Bank to hunt pilot whales in the 19th century (Clark 1887*b*). Nineteenth-century whalers often encountered, and hunted, pilot whales, the catches being reported as "blackfish" (see various accounts in the *Whalemen's Shipping List*). Interestingly enough, it was during a pilot whale hunt that the Hatteras Ground for sperm whaling was discovered in 1837 (Clark 1887*b*).

Short-finned pilot whales are exploited off eastern North America only in the Lesser Antilles, particularly at the islands of Saint Vincent and Saint Lucia (Caldwell and Caldwell 1975*a*). Using relatively primitive, open-boat, hand-harpooning techniques, the islanders take up to several hundred pilot whales and a few hundred dolphins of various species each year for local consumption.

Off the Pacific coast the only significant recent exploitation of the short-finned pilot whale has been by the aquatic display industry (Norris and Prescott 1961). Pilot whales are among the most difficult of the small odontocetes to live-capture because they do not bow ride (Walker 1975). A breakaway hoop net with a long handle is used. The collector is positioned at the front of a bow extension, and he nets one animal at a time. A total of 33 pilot whales was taken by Marineland of the Pacific from 1966 to 1972.

Long-finned pilot whales in the western North Atlantic feed almost exclusively on short-finned squid (*Illex illecebrosus*); they take cod and Greenland halibut (or turbot) (*Reinhardtius hippoglossoides*) when squid are unavailable (Sergeant 1962*b*; Mercer 1967). Mercer (1975), using 50,000 as an estimate of initial population size for the pilot whales hunted off Newfoundland (after Mitchell 1974), calculated that approximately 605,900–908,850 metric tons of squid were consumed annually by the whales. The inshore appearance of long-finned pilot whales at Newfoundland coincides with the arrival of concentrations of squid in late summer and fall. A similar pattern exists in southern California, where squid spawning in winter attract pilot whales onto the continental shelf (Norris and Prescott 1961).

Pilot whales are exceedingly gregarious; herds of several hundred are not unusual. The degree of cohesion within a herd seems to vary with activity—individuals disperse to feed, then coalesce to rest or flee (Brown and Norris 1956; Sergeant 1962*b*). Their tendency to remain together and act in concert makes pilot whales vulnerable to driving fisheries and to mass stranding. Pilot whale strandings, often involving 100 or more animals, are common in North America (Hall et al. 1971; Geraci and St. Aubin 1977). The reasons for this phenomenon are still not known. Among the possibilities discussed are: (1) feeding in dangerously shallow water; (2) acoustical confusion; (3) certain hydrological and meteorological conditions; and (4) unusual disturbances such as underwater explosions. Although press reports have referred to excessive middle ear nematode infestation as a possible cause of mass stranding, this explanation has not yet gained wide acceptance among cetologists.

Public enthusiasm for cetaceans has led to efforts, usually futile, to refloat stranded animals. An interesting effort of this kind occurred in February 1977 (Irvine et al. 1979*b*). At least one pilot whale of a group of 175–200, tagged with spaghetti and roto tags at a stranding site near Jacksonville, Florida, restranded near Charleston, South Carolina, about 220 km (straight line distance) to the northeast six days later. It was examined and refloated, apparently successfully. Because of the high frequency of mass strandings in pilot whales and the large number of animals involved, stranding mortality may be an important factor in regulating their abundance.

An unrestrained pilot whale was trained to retrieve objects from the ocean floor at depths to 550 m, made exploratory dives to 667 m, and remained submerged for periods of nearly 15 minutes (Bowers and Henderson 1972).

Understanding of life history is based mainly on the large sample of long-finned pilot whales studied by Sergeant (1962*b*) at Newfoundland. Although births occur throughout the year, there is a peak in summer. Assuming, as Sergeant did, that two growth layers in the dentine represent 1 year, females become sexually mature at about 6–7 years of age, males at about 12. Gestation may last as long as 15 ½–16 months and lactation about 22–22 ½ months. Males apparently have a significantly higher mortality rate than females.

HARBOR PORPOISES AND DALL'S PORPOISES

The family of "true porpoises," Phocoenidae, is represented in North American waters by three species. Dall's porpoise (*Phocoenoides dalli*) is found only in the temperate to subarctic North Pacific, primarily in deep coastal canyons and offshore, from Baja California north to the central Bering Sea (Cowan 1944; Mitchell 1975b). The harbor porpoise (*Phocoena phocoena*) is locally abundant in temperate waters of the Northern Hemisphere, principally in shallow shelf waters (Gaskin et al. 1974). Significant morphological differences indicate that the eastern Pacific and western Atlantic populations are discrete from each other (Yurick 1977). Along the Pacific coast of North America harbor porpoises are common from the Bering Sea to central California, with stragglers reaching southern California (Leatherwood and Reeves 1978) and the Mackenzie River Delta (Van Bree et al. 1977). Along the Atlantic coast they are found between Cape Cod and Labrador, with individuals reaching North Carolina and Baffin Island in some years (Mitchell 1975b). The third species, the cochito or Gulf of California harbor porpoise (*Phocoena sinus*), was described only in 1958 (Norris and McFarland 1958). It is currently confined to the upper Gulf of California (Brownell 1976). Scammon (1874) left open the possibility that the species was formerly more widely distributed: "our observation proves that they are found as far south as Banderas Bay, and about the mouth of the Pigento River, on the coast of Mexico (which estuary is in latitude 20° 30′)." He could have been referring to either of the eastern North Pacific species in the genus *Phocoena*.

These three porpoises are the smallest cetaceans in North American waters. Male Dall's porpoises grow to lengths of 2.1 m; females are slightly smaller. Harbor porpoises rarely grow to lengths of 1.8 m and weights of 90 kg. Dall's porpoises have a striking black-and-white pigmentation pattern. Some adult males have an exaggerated keel on the caudal peduncle. The head is disproportionately small. Both species of harbor porpoise are basically gray, lighter ventrally, with the ventral lightness intruding high onto the sides anterior to the dorsal fin. The dorsal fin of the three phocoenids is generally in the shape of a modified triangle. They do not have a well-defined beak.

Dall's porpoises occur primarily in water at least 100 fathoms deep (Morejohn 1979). They avoid shallow coastal areas and generally are seen near shore only where deep channels or canyons, with ready access to the open sea, are present. Regular increases in numbers of Dall's porpoises in Monterey Bay and southern California in winter-spring months suggest they migrate (Leatherwood and Reeves 1978). Dall's porpoises are rarely found in water warmer than 18° C (Norris and Prescott 1961). Judging by stomach contents, they feed at great depths. Major prey species in the eastern North Pacific include hake (*Merluccius productus*), herring (*Clupea harengus*), juvenile rockfish (*Sebastes* sp.), anchovy (*Engraulis mordax*), and squid (*Loligo opalescens* and *Gonatus* sp.). As nocturnal feeders (Ridgway 1966), Dall's porpoises probably prey on organisms associated with the deep scattering layer.

With the exception of subsistence hunting by Indians (Scheffer and Slipp 1948), Dall's porpoises have not been exploited by humans off North America. A few have been taken alive for research (Ridgway 1966; Wood 1973) and public display (Walker 1975), but survival has been poor.

The Japanese pelagic salmon fishery in the North Pacific, using drift gill nets, has killed large numbers of Dall's porpoises since 1962 (Ohsumi 1975b; Kasuya 1978). There is much uncertainty about the magnitude of this incidental by-catch. It has been viewed, in combination with an intensive direct harpoon fishery off the coast of Japan (Kasuya 1978), as a major conservation and management problem (Mitchell 1975a). A research program at the U.S. Department of Commerce's Northwest and Alaska Fisheries Center in Seattle began in 1978. It is designed to assess the salmon fishery's impact on porpoise populations and to develop techniques to reduce mortality (Perrin 1980).

Dall's porpoises are regarded as fast swimmers, capable of exceeding 16 knots (Morejohn 1979). They make a very distinctive cone-shaped splash as they surface. It is not unusual for Dall's porpoises to bow ride, but they do it erratically and often quickly lose interest. A peculiar swimming formation, with animals arranged single file at even intervals, has been observed (Norris and Prescott 1961).

Dall's porpoises breed year-round off western North America (Morejohn 1979), but most calves off Japan are born in August and September (Kasuya 1978). Gestation has been estimated as 11.4 months, and nursing may last 2 years (Kasuya 1978). The calving interval is probably 3 years. Their only known predator is the killer whale (Rice 1968; Pike and MacAskie 1969; Barr and Barr 1972).

The more widely distributed harbor porpoise has been subjected to heavy inadvertent catching in gill nets and weirs in many parts of its range (Mitchell 1975a, 1975b; Prescott and Fiorelli 1980). An international gill net fishery for salmon off West Greenland, for instance, caught an estimated 1,500 harbor porpoises in 1972 (Lear and Christensen 1975). This was in addition to the incidental kill by Greenlandic salmon gillnetters and the direct kill by local fishermen for food—estimated as approximately 1,000 per year (Kapel 1975).

There has been no major direct commercial fishery for harbor porpoises in North America. However, Clark (1887b) mentioned the existence of a short-lived net fishery for "porpoises," probably *P. phocoena,* at Salem, Massachusetts, in the 1740s. According to Laurin (1976), they are still hunted with shotguns in the Saint Lawrence for food, but the annual catch is small, probably fewer than 100. Fishing families in New Brunswick and Maine still use harbor porpoises for meat occasionally (Mitchell 1975b; Prescott and Fiorelli 1980).

The Passamaquoddy and MicMac Indians of the Canadian Maritimes and Maine hunted harbor porpoises from canoes, killing them with guns and lances, in a small commercial fishery during the 19th century (Gilpin 1878; Leighton 1937). Some were taken in Passamaquoddy Bay for mink food (Fisher and Harrison 1970). The people on the Passamaquoddy reservation near Eastport, Maine, still relish porpoise meat, and a small-scale domestic fishery continues there (Prescott and Fiorelli 1980). Unfortunately, the Maine Indians apparently were not recognized as exploiters of marine mammals by drafters of the Marine Mammal Protection Act of 1972, and their continued hunting may be technically in violation of federal law. Judging by the estimate by Gaskin (1977), roughly 4,000 for the size of the porpoise population in the approaches to the Bay of Fundy, it is unlikely that overexploitation is occurring. However, active management of this fishery, with careful monitoring of the porpoise population, is desirable. In the Pacific Northwest, Indians have hunted harbor porpoises for meat (Scheffer and Slipp 1948), but commercial exploitation is not known to have occurred.

Harbor porpoises are not important to the aquarium display industry, although they have been kept and trained by some institutions in Europe, usually having been caught by accident (Andersen 1978). Four captured alive in a weir in Passamaquoddy Bay were transferred to the New York Aquarium in 1965 (Sergeant et al. 1970). A few harbor porpoises captured in weirs in New Brunswick have been tagged and released, but there have been few resightings or tag returns (Gaskin et al. 1975).

Because they are inconspicuous and do not occur in the large aggregations characteristic of many species of oceanic dolphins, seasonal movements of harbor porpoises have proven difficult to determine. However, there is considerable evidence that their numbers increase inshore in spring or early summer, at least in the Bay of Fundy and Gulf of Maine (Gaskin 1977). There may be an inshore-offshore migration that coincides with the movements of prey.

The diet of harbor porpoises consists primarily of pelagic, nonspiny schooling fishes, but they prey on a wide variety of species including polychaete worms and cephalopods (Smith and Gaskin 1974). Some prey species—e.g., herring, cod, and mackerel (*Scomber scombrus*)—are of economic importance to mankind. Given its unusually high metabolic rate (Kanwisher and Sundnes 1965), an individual harbor porpoise must consume relatively large quantities of food often.

Harbor porpoises are vulnerable to predation by killer whales (Chandler et al. 1977) as well as large sharks (Arnold 1972). High levels of pollutants—chlorinated hydrocarbon insecticides, polychlorinated biphenyls (PCB), and dieldrin—have been found in tissue of harbor porpoises in the Bay of Fundy region (Gaskin et al. 1971, 1976; Gaskin and Smith 1979). As an "abundant and important component in the upper part of the North Atlantic inshore food wed," the harbor porpoise may be a good indicator of environmental contamination. A catastrophic decline in inshore populations in western Europe may be due, at least in part, to environmental degradation (Perrin 1980).

Although age determination techniques have not been refined, harbor porpoises are believed to live for no more than about 13 years, with 8–9 years being the average life span (Gaskin and Blair 1977; Gaskin and Smith 1979). Newborn calves are about 25 percent of their mother's weight and are usually 80–90 cm long (Fisher and Harrison 1970; Gaskin et al. 1974). Birth and mating usually occur in summer. The interval between births is probably more than 8 months (Mitchell 1975*b*).

The Gulf of California harbor porpoise, known locally as the cochito or vaquita, is one of the most poorly known cetaceans. Only a few specimens have been examined, and living animals have rarely been observed. Only 21 confirmed records existed as of 1976 (Brownell 1976). Because of its limited range and low numbers, the Gulf of California harbor porpoise is probably highly susceptible to overexploitation and environmental disturbances. Incidental entanglement in association with shrimp trawling and the gill net fishery for totoaba (*Cynoscion macdonaldi*) has resulted in an unknown level of mortality for this porpoise (Norris and Prescott 1961). The net fishery was in operation by the late 1940s. In the early 1970s as many as 10 porpoises are known to have been taken in a single day at San Felipe (Brownell 1976). The total incidental catch may have been in the hundreds annually. Because of a drastic decline in the totoaba population, the fishery was closed by the Mexican government in 1975 (Flanagan and Hendrickson 1976). However, continued gillnetting for other sciaenids and for sharks may mean the Gulf of California harbor porpoise is still taken incidentally.

THE DOLPHIN-TUNA PROBLEM

Although the eastern tropical Pacific tuna purse seine fishery is not conducted in shelf waters of North America, the extensive involvement of North American fishermen, the fishery's commercial importance, and its significance to marine mammal management are reasons enough to include some mention of the fishery in this chapter. The international high-seas tuna fishery presents a high-profile and complex resource management problem (Joseph and Greenough 1979).

During the late 1950s high-seas tuna fishermen began to exploit the relationship between herds of dolphins—primarily spinner dolphins (*Stenella longirostris*), spotted dolphins (*S. attenuata*), and common dolphins (*Delphinus delphis*)—and tuna—primarily yellowfin (*Thunnus albacares*) (Perrin 1969, 1970; Green et al. 1971). A strong bond exists between the mammals and the fish, such that successful encirclement and pursing of a dolphin school often ensures a good catch of tuna. The dolphin-tuna complex is sighted, often at great distances, by the presence of bird activity or surface disturbance caused by the dolphins. Motor skiffs launched from the seiner chase and herd the dolphins until the purse seine is set around them. The net is then gathered to the vessel until the tuna caught in association with the dolphins can be brailed aboard.

The chase is exhausting and hazardous for the dolphins. Once netted many suffocate as they panic and become entangled; others exhibit passive behavior and are prematurely given up for dead and pursed in the net

TABLE 18.9. Estimated total numbers of porpoises killed (including those seriously injured), by stock, for U.S. and non-U.S. fleets purse-seine fishing for yellowfin (principally) and skipjack tuna, 1959–80

Species and Stock	1959	1960	1961	1962	1963	1964	1965	1966	1967	1968	1969
Spotted											
Coastal	0	0	0	0	0	0	0	0	0	0	0
No. offshore	82,539	409,908	460,063	189,485	204,382	332,646	361,061	324,197	216,252	187,628	382,843
S. offshore	0	0	0	0	0	0	0	0	0	0	0
Spinner											
Eastern	30,657	152,251	170,880	70,380	75,913	123,554	134,108	120,416	30,322	69,690	125,791
N. whitebelly	0	0	0	0	0	0	0	0	0	0	16,408
S. whitebelly	0	0	0	0	0	0	0	0	0	0	0
Common											
N. tropical	0	0	0	0	0	0	0	0	0	0	0
C. tropical	4,092	20,322	22,808	9,394	10,132	16,491	17,900	16,072	10,721	9,302	18,980
Striped											
N. tropical	0	0	0	0	0	0	0	0	0	0	0
C. tropical	0	0	0	0	0	0	0	0	0	0	0
S. tropical	0	0	0	0	0	0	0	0	0	0	0
Bottlenose	0	0	0	0	0	0	0	0	0	0	0
Fraser's	0	0	0	0	0	0	0	0	0	0	0
Rough-toothed	0	0	0	0	0	0	0	0	0	0	0
Risso's	0	0	0	0	0	0	0	0	0	0	0
Total	117,288	582,481	653,751	269,259	290,427	472,691	513,069	460,685	307,295	266,620	544,022

Year

Species and Stock	1970	1971	1972	1973	1974	1975	1976	1977	1978	1979	1980
Spotted											
Coastal	0	0	0	4,452	24	0	0	167	0	97	0
N. offshore	371,328	184,326	298,154	131,863	95,643	105,564	47,460	22,549	19,241	4,148	4,744
S. offshore	0	0	0	17,782	0	2,376	22,957	379	1,038		
Spinner											
Eastern	122,008	60,564	97,965	32,250	26,088	45,301	8,700	5,041	2,270	767	355
N. whitebelly	15,914	7,900	12,778	33,163	47,794	34,355	20,485	4,554	4,215	640	1,264
S. whitebelly	0	0	0	16,614	0	2,333	17,843	489	146		
Common											
N. tropical	0	0	0	7,485	0	9	457	428	505	4,557	348
C. tropical	18,409	9,138	14,781	21,299	4,822	3,134	7,067	17,470	2,417		
Striped											
N. tropical	0	0	0	0	36	127	133	18	0	247	82
C. tropical	0	0	0	65	163	1,056	154	32	517		
S. tropical	0	0	0	0	0	0	2,230	103	15		
Bottlenose	0	0	0	0	103	198	26	47	149	7	0
Fraser's	0	0	0	0	9	4	671	76	0		
Rough-toothed	0	0	0	0	0	0	0	0	0	98[a]	86[a]
Risso's	0	0	0	0	0	0	39	0	0		
Total	527,659	261,928	423,678	264,973	174,682	194,457	128,222	51,353	30,513	10,554	6,879[b]

SOURCE: Data for the years 1959–78 are taken from Smith (1979), table 15, p. 94. Those for 1979 and 1980 are from the report of the Porpoise Rescue Foundation, San Diego, California, 5 June 1980, table 1.

NOTE: Totals reflect only stocks listed in this table.

[a] Also include Pacific white-sided dolphins (*Lagenorhynchus obliquidens*), short-finned pilot whales, and unidentified animals.

[b] Projection based on incomplete returns.

(Coe and Stuntz 1980). From several hundred thousand to over a half million dolphins were killed or seriously injured annually by the multinational fishery until the late 1970s, when the mortality level declined to less than 100,000 per year (Smith 1979). Such high mortality occurred despite the introduction of two important dolphin-saving techniques developed by American fishermen relatively early in the fishery (Barham et al. 1977). "Backdown" takes place after most of the net has been gathered; it causes the far end of the corkline, near where the dolphins generally congregate, to submerge, spilling the dolphins out of the net. The Medina Panel is a stretch of netting with smaller-than-usual (5-cm) mesh situated in the backdown area. It is designed to deflect dolphins whose beaks or flippers would become caught in netting of standard (approximately 11-cm) mesh. Specially designed "aprons" and "chutes" made with even smaller mesh webbing have been developed and tested, and they show promise for reducing dolphin mortality further.

Perhaps the most favorable thing that can be said about this wasteful fishery, aside from its obvious nutritional and economic importance to mankind, is that it has afforded scientists an unprecedented opportunity to study pelagic dolphin populations. Research has been based at the Southwest Fisheries Center of the National Marine Fisheries Service, Department of Commerce, in La Jolla, California. Two lines of inquiry have been followed. One is an effort to establish the status of stocks and to judge the fishery's impact on the dolphins. The other is the search for new fishing technology that would facilitate a decrease in dolphin mortality without reducing tuna landings. Incidental to these practical pursuits has been the generation of basic information on the biology and ecology of the animals.

Management of the fishery must be attendant to the existence of discrete stocks within exploited dolphin species. Distinctions between stocks have been made primarily on the basis of demonstrable morphological differences (e.g., Perrin 1975). A limited tagging program has been conducted to elucidate dolphin home ranges and seasonal movements (Perrin et al. 1979).

The current understanding of status of stocks is reflected in the report of a workshop held at La Jolla in August 1979 (Smith 1979). Fourteen eastern tropical Pacific stocks (population units having limited genetic exchange with adjacent units of the same species) of spinner, spotted, common, and striped (*S. coeruleoalba*) dolphins were considered. Of these, the northern offshore spotted and eastern spinner dolphin stocks were of greatest concern, having been reduced to levels well below the "optimum sustainable population" (OSP) range.

The Marine Mammal Protection Act of 1972 (MMPA) set as a management goal the maintenance of marine mammal populations at levels within the OSP range. The existing definition of this concept, as published in the *Federal Register* and interpreted in the workshop report, sets the OSP range between the population level that is the largest supportable within the ecosystem and the population level that results in maximum net productivity. The level of maximum net productivity for these pelagic dolphins is believed to range between 65 and 80 percent of the equilibrium unharvested population level (i.e., carrying capacity or largest supportable population).

Administrative hearings were held early in 1980 to reassess the status of stocks and to consider management measures. The interpretation of available evidence by government scientists was that strict adherence to the Marine Mammal Protection Act and the interests of conservation require that fishing on some dolphin stocks be closed. The fishing industry, buoyed by recent significant decreases in total dolphins killed and in kill rates per set (table 18.9), opposes any further regulation. Currently, the industry's intention to transfer more and more vessels to foreign registry (not an idle threat, as shown by a number of recent reregistrations) poses a serious obstacle to successful management of the fishery. Such transfers remove oversight responsibility from the U.S. government and give it to the Interamerican Tropical Tuna Commission and member governments for whom protection of dolphin stocks is a low priority. If the trend toward evasion of conservation regulations persists, the outlook for the dolphin stocks, as well as the tuna, deteriorates.

One of the more interesting—and distressing— biological findings is that the expected density-dependent response to exploitation, an increase in gross annual reproductive rate, has not been shown for the badly depleted eastern spinner stock (Perrin and Henderson 1979). This is thought to be due at least in part to a dearth of sexually mature males in the population, a factor resulting from "age- or maturity-selective" fishing mortality.

Kim Goodrich assisted in compilation of information and in checking of references. Giuseppe di Sciara shared information and reviewed portions of the manuscript. Sam H. Ridgway, Forrest G. Wood, and William A. Walker criticized portions of the manuscript and generously provided us with expert counsel over the years. Edward Mitchell and James G. Mead made available to us their extensive libraries of marine mammal literature, and this made our work easier. We especially thank Ronn Storro-Patterson, who contributed to this chapter by writing a first draft of the sperm whale section.

LITERATURE CITED

Akimushkin, I. I. 1955. [The feeding of the cachalot.] Compt. Rende., Acad. Sci. USSR 101:1139–1140. [In Russian; not seen.]

Andrews, R. C. 1911. Descriptions of an apparently new porpoise of the genus *Tursiops* with remarks upon a skull of *Tursiops gilli* Dall. Bull. Am. Mus. Nat. Hist. 30:233–237.

Andersen, S. H. 1978. Experiences with harbour porpoises, *Phocoena phocoena*, in captivity: mortality, autopsy findings, and influence of the captive environment. Aquatic Mammal. 6(2):39–49.

Anonymous. 1885. Capture of blackfish. Bull. U.S. Fish Comm. 5:91–92.

Anonymous. 1979. Of whalers and white whales. Manitoba

Dept. Mines, Nat. Resourc. and Environ. Conserv. Comment (July), pp. 1–16.

Arnold, P. W. 1972. Predation on harbor porpoise, *Phocoena phocoena*, by a white shark, *Carcharodon carcharias*. J. Fish. Res. Board Can. 29:1213–1214.

Asgeirsson, T. 1980. Chairman's report of the thirty-first annual meeting. Rep. Int. Whaling Comm. 30:25–41.

Asper, E. D. 1975. Techniques of live capture of smaller Cetacea. J. Fish. Res. Board Can. 32:1191–1196.

Asper, E. D., and Odell, D. K. 1979. Collection of biological data from bottlenose dolphins, *Tursiops truncatus*, during live-capture fishing operations. Page 1 *in* Abstr. Presentations at 3rd Bienn. Conf. Bio. Marine Mammals, 7–11 October 1979, Seattle, Wash. 64pp.

———. 1980. *Tursiops truncatus* studies; bottlenose dolphin local herd monitoring: capture, marking, collection of biological data, and follow up observations of marked animals. Tech. Rep. no. 80-122. Hubbs Sea World Research Institute, San Diego, Calif. 107pp.

Backus, R. H., and Schevill, W. E. 1966. *Physeter* clicks. Pages 510–528 *in* K. S. Norris, ed. Whales, dolphins and porpoises. Univ. California Press, Berkeley and Los Angeles. 789pp.

Bada, J. L.; Brown, S.; and Masters, P. M. 1977. Aspartic acid racemization in the teeth and lens nucleus of marine mammals: use in age determinations. Page 48 *in* Proc. 2nd Conf. Biol. Marine Mammals, 12–15 December 1977, San Diego, Calif. 88pp. Abstr.

Balcomb, K. C., III. 1978. Orca survey, 1977. Marine Mammal Div., U.S. Natl. Marine Fish. Serv. 10pp. Manuscript.

Balcomb, K. C., III, and Goebel, C. A. 1977. Some information on a *Berardius bairdii* fishery in Japan. Rep. Int. Whaling Comm. 27:485–486.

Baldridge, A. 1972. Killer whales attack and eat a gray whale. J. Mammal. 53:898–900.

Bannister, J., convenor. 1980. Report of the Sub-Committee on Sperm Whales. Rep. Int. Whaling Comm. 30:79–96.

Bannister, J., and Mitchell, E. 1980. North Pacific sperm whale stock identity: distributional evidence from Maury and Townsend charts. Rep. Int. Whaling Comm. (special issue 2), pp. 219–230.

Barham, E. G.; Sweeney, J. C.: Leatherwood, S.; Beggs, R. K.; and Barham, C. L. 1980. Aerial census of bottlenosed dolphins (*Tursiops truncatus*) in a region of the Texas coast. Fish. Bull. 77:585–595.

Barham, E. G.; Taguchi, W. K.: and Reilly, S. B. 1977. Porpoise rescue methods in the yellowfin purse seine fishery and the importance of Medina Panel mesh size. Marine Fish. Rev. 39(5):1–10.

Barr, N., and Barr, L. 1972. An observation of killer whale predation on a dall porpoise. Can. Field-Nat. 86:170–171.

Beale, T. 1839. The natural history of the sperm whale. 2nd ed. John Van Voorst; London. 393pp.

Benjaminsen, T., and Christensen, I. 1979. The natural history of the bottlenose whale, *Hyperoodon ampullatus* (Forster). Pages 143–164 *in* H. E. Winn and B. L. Olla, eds. Behavior of marine animals: current perspectives in research. Vol. 3: Cetacea. Plenum Press, New York and London. 438pp.

Berzin, A. A. 1972. The sperm whale. A. V. Yablokov, ed. Pishchevaya Promyshlennost, Moscow. Transl. Israel Program for Sci. Transl., Jerusalem, 1972. 394pp.

Best, P. B. 1967. The sperm whale (*Physeter catodon*) off the west coast of South Africa. 1. Ovarian changes and their significance. Republic of South Africa, Dept. Industries, Div. Sea Fish. Invest. Rep. no. 61:1–27.

———. 1968. The sperm whale (*Physeter catodon*) off the west coast of South Africa. 2. Reproduction in the female. Republic of South Africa, Dept. Industries, Div. Sea Fish. Invest. Rep. no. 66:1–32.

———. 1969a. The sperm whale (*Physeter catodon*) off the west coast of South Africa. 3. Reproduction in the male. Republic of South Africa, Dept. Industries, Div. Sea Fish. Invest. Rep. no. 72:1–20.

———. 1969b. The sperm whale (*Physeter catodon*) off the west coast of South Africa. 4. Distribution and movements. Republic of South Africa, Dept. Industries, Div. Sea Fish. Invest. Rep. no. 78:1–12.

———. 1970. The sperm whale (*Physeter catodon*) off the west coast of South Africa. Part 5: Age, growth and mortality. Republic of South Africa, Dept. Industries, Div. Sea Fish. Invest. Rep. no. 79:1–27.

———. 1974. The biology of the sperm whale as it relates to stock management. Pages 257–293 *in* W. E. Schevill, ed. The whale problem: a status report. Harvard Univ. Press, Cambridge, Mass. 419pp.

———. 1976a. Tetracycline marking and the rate of growth layer formation in the teeth of a dolphin (*Lagenorhynchus obscurus*). South African J. Sci. 72:216–218.

———. 1976b. A review of world sperm whale stocks. Advisory Com. on Marine Resourc. Res., Sci. Consultation on Marine Mammals, ACMRR/MM/SC/8. Rev. 1 (Limited). 106pp.

———. 1979. Social organization in sperm whales, *Physeter macrocephalus*. Pages 227–289 *in* H. E. Winn and B. L. Olla, eds. Behavior of marine animals: current perspectives in Research. Vol. 3: Cetacea. Plenum Press, New York and London. 438pp.

Best, P. B., and Butterworth, D. S. 1980. Timing of oestrus within sperm whale schools. Rep. Int. Whaling Comm. (special issue 2), pp. 137–40.

Bigg, M. A., and Wolman, A. A. 1975. Live-capture killer whale (*Orcinus orca*) fishery, British Columbia and Washington, 1962–73. J. Fish. Res. Board Can. 32:1213–1221.

Bollen, A. G. 1979. International Whaling Commission report. Rep. Int. Whaling Comm. 29:7–20.

Bowers, C. A., and Henderson, R. S. 1972. Project Deep Ops: deep object recovery with pilot and killer whales. Tech. Publ. no. 306. Naval Undersea Center, San Diego, Calif. 86pp.

Brodie, P. F. 1969. Mandibular layering in *Delphinapterus leucas* and age determination. Nature 221 (5184): 956–958.

———. 1971. A reconsideration of aspects of growth, reproduction, and behavior of the white whale (*Delphinapterus leucas*), with reference to the Cumberland Sound, Baffin Island, population. J. Fish. Res. Board Can. 28:1309–1318.

Brown, D. H., and Norris, K. S. 1956. Observation of captive and wild cetaceans. J. Mammal. 37:311–326.

Brown, J. T. 1887. The whalemen, vessels and boats, apparatus, and methods of the whale fishery. Pages 218–293 *in* G. B. Goode, ed. The fisheries and fishery industries of the United States. Section 5: History and methods of the fisheries. Government Printing Office, Washington, D.C. Vol. 2. 881pp.

Brown, R. 1868. Notes on the history and geographical relations of the Cetacea frequenting Davis Strait and Baffin's Bay. Proc. Zool. Soc. London 35:533–556.

Brownell, R. L., Jr. 1976. Status of the cochito, *Phocoena sinus*, in the Gulf of California. Working Document, Meet. Ad Hoc Consultants Group on Small Cetaceans and Sirenians (Ad Hoc Group 2). Working Party on Marine Mammals, Advisory Com. of Experts on Marine

Resourc. Res. (ACMRR), Food and Agric. Organization (FAO), United Nations. ACMRR/MMII/47. 9pp.

Brownell R. L., and Omura, H. 1980. Whale meat in the Japanese diet. Science 208:976.

Buckland, F. 1890. Notes and jottings from animal life. Smith, Elder & Co., London. 414pp.

Caldwell, D. K. 1955. Evidence of home range of an Atlantic bottlenose dolphin. J. Mammal. 36:304–305.

Caldwell, D. K., and Caldwell, M. C. 1971. Porpoise fisheries in the southern Caribbean: present utilizations and future potentials. Pages 195–206 in Proc. 23rd Annu. Sess. Gulf and Caribbean Fish. Inst.

_____. 1972. The world of the bottlenosed dolphin. J. B. Lippincott Co., Philadelphia and New York. 157pp.

_____. 1975a. Dolphin and small whale fisheries of the Caribbean and West Indies: occurrence, history, and catch statistics—with special reference to the Lesser Antillean Island of St. Vincent. J. Fish. Res. Board Can. 32:1105–1110.

_____. 1975b. Dolphins and fisheries. Pages 28–29 in A report on the Sea Grant program. State Univ. System of Florida.

Caldwell, D. K.; Caldwell, M. C.; Rathjen, W. F.; and Sullivan, J. R. 1971. Cetaceans from the Lesser Antillean Island of St. Vincent. Fish. Bull. 69(2):303–312.

Caldwell, D. K.; Caldwell, M. C.; and Rice, D. W. 1966. Behavior of the sperm whale, *Physeter catodon* L. Pages 677–717 in K. S. Norris, ed. *Whales, dolphins, and porpoises.* Univ. California Press, Berkeley and Los Angeles. 789pp.

Caldwell, M. C., and Caldwell, D. K. 1966. Epimeletic (care-giving) behavior in Cetacea. Pages 755–789 in K. S. Norris, ed. *Whales, dolphins, and porpoises.* Univ. California Press, Berkeley and Los Angeles. 789pp.

Cato, J. C., and Prochanska, F. J. 1976. Porpoise attacking hooked fish and injuring Florida fishermen. Natl. Fisherman (January), pp. 3-B, 16-B.

Chandler, R.; Goebel, C.; and Balcomb, K. 1977. Who is that killer whale?: a new key to whale watching. Pacific Search 11(7): 25–35.

Christensen, I. 1973. Age determination, age distribution and growth of bottlenose whales, *Hyperoodon ampullatus* (Forster), in the Labrador Sea. Norw. J. Zool. 21:331–340.

_____. 1975. Preliminary report on the Norwegian fishery for small whales: expansion of Norwegian whaling to Arctic and northwest Atlantic waters, and Norwegian investigations of the biology of small whales. J. Fish. Res. Board Can. 32:1083–1094.

Clark, A. H. 1887a. Map of the world on Mercator's projection showing the extent and distribution of the present and abandoned whaling grounds. Plate 183 in G. B. Goode, ed. The fisheries and fishery industries of the United States. Section 5: History and methods of the fisheries. Government Printing Office, Washington, D.C.

_____. 1887b. The blackfish and porpoise fisheries. Pages 297–310 in G. B. Goode, ed. The fisheries and fishery industries of the United States. Section 5: History and methods of the fisheries. Vol. 2. Government Printing Office, Washington, D.C. 881 pp.

Clark, C. W. 1973. The economics of overexploitation. Science 181:630–634.

Clarke, M. R. 1976. Observation on sperm whale diving. J. Marine Bio. Assoc. (U.K.) 56(1):809–810.

_____. 1978a. Structure and proportions of the spermaceti organ in the sperm whale. J. Marine Bio. Assoc. (U.K.) 58(1):1–17.

_____. 1978b. Physical properties of spermaceti oil in the sperm whale. J. Marine Bio. Assoc. (U.K.) 58(1):19–26.

_____. 1978c. Buoyancy control as a function of the spermaceti organ in the sperm whale. J. Marine Bio. Assoc. (U.K.) 58(1):27–71.

Clarke, R. 1954. Open boat whaling in the Azores: the history and present methods of a relic industry. Discovery Rep. 26:281–354.

Clarke, R.; Aguayo, A. L.; and Paliza, O. 1980a. Pregnancy rates of sperm whales in the Southeast Pacific between 1959 and 1962, and a comparison with those from Paita, Perú between 1975 and 1977. Rep. Int. Whaling Comm. (special issue 2), pp. 151–158.

Clarke, R.; Paliza, O.; and Aguayo, A. L. 1980b. Some parameters and an estimate of the exploited stock of sperm whales in the Southeast Pacific between 1959 and 1961. Rep. Int. Whaling Comm. 30:289–305.

Coe, J. M., and Stuntz, W. E. 1980. Passive behavior by the spotted dolphin, *Stenella attenuata*, in tuna purse seine nets. U.S. Fish. Bull. 78(2):535–537.

Compton-Bishop, Q. M.; Gordon, J. C. D.; Allen, P. LeG.; and Rotton, N. 1979. The report of the Cambridge Azores Expedition, 1979. Cambridge Univ., Cambridge, England. 46pp.

Cornell, L. H., and Asper, E. D. 1978. A census of captive marine mammals in North America. Int. Zoo Yearb. 18:220–224.

Cowan, I. M. 1944. The Dall porpoise, *Phocoenoides dalli* (True), of the Northern Pacific Ocean. J. Mammal. 25(3):295–306.

Dall, W. H. 1873. Preliminary descriptions of three new species of Cetacea from the coast of California. Proc. California Acad. Sci. 5:12–13.

Davis, R. A.; Richardson, W. J.; Johnson, S. R.; and Renaud, W. E. 1978. Status of the Lancaster Sound narwhal population in 1976. Rep. Int. Whaling Comm. 28:209–215.

Di Sciara, G. N. 1977. A killer whale (*Orcinus orca* L.) attacks and sinks a sailing boat. Natura-Soc. Ital. Sci. Nat., Musco civ. Stor. nat. e Acquario civ. (Milan) 68 (3–4):218–220.

Doan, K. H., and Douglas, C. W. 1953. Beluga of the Churchill region of Hudson Bay. Bull. Fish. Res. Board Can. 98:1–27.

Dohl, T. P.; Norris, K. S.; and Kang, I. 1974. A porpoise hybrid: *Tursiops* x *Steno.* J. Mammal. 55(1):217–221.

Duffield, D. 1980. Electrophoretic comparison of genetic variability in *Tursiops*. Appendix 1 (pages 1–12) in E. D. Asper and D. K. Odell. *Tursiops truncatus* studies; bottlenose dolphin local herd monitoring: capture, marking, collection of biological data, and follow up observations of marked animals. Hubbs Sea World Research Institute, Tech. Rep. no. 80–122. San Diego, Calif. 107pp.

Duffield, D.; Ridgway, S. H.; and Cornell, L. C. In preparation. Comparison of hemoglobins and red blood cell parameters in onshore versus offshore *Tursiops*.

Duffield, D.; Ridgway, S. H.; Sparkes, M. E.; and Sparkes, R. 1979. Evidence for population differentiation in *Tursiops*. Page 14 in Abstr. from Presentations at 3rd Bienn. Conf. Bio. Marine Mammals, 7–11 October 1979, Seattle, Wash. 64pp.

Eales, N. B. 1950. The skull of the foetal narwhal, *Monodon monoceros* L. Phil. Trans. R. Soc. London 235:1–33.

Eichelberger, L.; Fetcher, E. S.; Geiling, E. M. K.; and Vos, B. J. 1940. The composition of dolphin milk. J. Bio. Chem. 134:171.

Ely, C. A., and Clapp, R. B. 1973. The natural history of Laysan Island, Northwestern Hawaiian Islands. Atoll Res. Bull. 171:1–361.

Evans, W. E. 1973. Echolocation by marine delphinids and one species of fresh-water dolphin. J. Acoust. Soc. Am. 54(1):191–199.

Evans, W. E., and Bastian, J. 1969. Marine mammal communication: social and ecological factors. Pages 425–475 in H. T. Andersen, ed. The biology of marine mammals. Academic Press, New York. 511pp.

Evans, W. E., and Yablokov, A. B. 1978. Intraspecific variation of the colour pattern of the killer whale (*Orcinus orca*). Pages 102–114 in B. E. Sokolov and A. B. Yablokov, eds. Advances in pinniped and cetacean research. USSR Academy of Sciences. 143pp. (In Russian; English abstract.)

Farquhar, A. 1978. Beluga blues. North Nord 25(2):30–33.

Finley, K. J.; Davis, R. A.; and Silverman, H. B. 1980. Aspects of the narwhal hunt in the eastern Canadian Arctic. Rep. Int. Whaling Comm. 30:459–464.

Fischer, P. 1881. Cétacés du Sud-Ouest de la France. Extrait des Actes de la Société Linnéenne de Bordeaux. Vol. 35. 220pp.

Fish, J. F., and Vania, J. S. 1971. Killer whale, *Orcinus orca,* sounds repel white whales, *Delphinapterus leucas.* Fish. Bull. 69:531–535.

Fisher, H. D., and Harrison, R. J. 1970. Reproduction in the common porpoise (*Phocoena phocoena*) of the North Atlantic. J. Zool. London 161:471–486.

Flanagan, C. A., and Hendrickson, J. R. 1976. Observations on the commercial fishery and reproductive biology of the totoaba, *Cynoscion macdonaldi,* in the northern Gulf of California. Fish. Bull. 74:531–544.

Food and Agriculture Organization of United Nations. 1978. Mammals in the sea. Report of the FAO advisory committee on marine resources research, working party on marine mammals. FAO Fish. Ser. no. 5. Food and Agriculture Organization of United Nations, Rome. Vol. 1. 264pp.

Foote. D. C. 1975. Investigation of small whale hunting in northern Norway, 1964. J. Fish. Res. Board Can. 32:1163–1189.

Fraker, M. A. 1980. Status and harvest of the Mackenzie stock of white whales (*Delphinapterus leucas*). Rep. Int. Whaling Comm. 30:451–458.

Fraser, F. C. 1940. Three anomalous dolphins from Blacksod Bay, Ireland. Proc. R. Irish Acad. Vol. 45, sec. B, no. 17, pp. 413–455.

Freeman, M. M. R. 1968. Winter observations on beluga (*Delphinapterus leucas*) in Jones Sound, N. W. T. Can. Field Nat. 82:276–286.

———. 1973. Polar bear predation on beluga in the Canadian Arctic. Arctic 26:162–3.

Gambell, R. 1968. Aerial observations of sperm whale behaviour. Norsk Hvalfangst-Tidende 57:126–138.

———. 1970. Weight of a sperm whale, whole and in parts. South African J. Sci. 66(7):225–227.

———. 1972. Sperm whales off Durban. Discovery Rep. 35:199–358.

Gambell, R.; Lockyer, C.; and Ross, G. J. B. 1973. Observations on the birth of a sperm whale calf. South African J. Sci. 69:147–148.

Gaskin, D. E. 1967. Luminescence in a squid *Morotheuthis* sp. (probably *ingens* Smith), and a possible feeding mechanism in the sperm whale *Physeter catodon* L. Tuatara 15(2):86–88.

———. 1968. Distribution of Delphinidae (Cetacea) in relation to sea surface temperatures off eastern and southern New Zealand. N.Z. J. Marine and Freshwater Res. 2:527–534.

———. 1972. Whales, dolphins and seals, with special reference to the New Zealand region. Heinemann Educational Books, London. 200pp.

———. 1977. Harbour porpoise, *Phocoena phocoena* (L.) in the western approaches to the Bay of Fundy 1969-75. Rep. Int. Whaling Comm. 27:487–492.

Gaskin, D. E.; Arnold, P. W.; and Blair, B. A. 1974. *Phocoena phocoena.* Mammalian Species no. 42:1–8.

Gaskin, D. E., and Blair, B. A. 1977. Age determination of the harbour porpoise, *Phocoena phocoena* (L.), in the western North Atlantic. Can. J. Zool. 55:(1):18–30.

Gaskin, D. E.; Holdrinet, M.; and Frank, R. 1971. Organochlorine pesticide residues in harbour porpoises from the Bay of Fundy. Nature 233:499–500.

———. 1976. DDT residues in blubber of harbour porpoise, *Phocoena phocoena* (L.), from eastern Canadian waters during the five year period 1969–1973. ACMRR/MM/SC/96. Advisory Com. on Marine Resourc. Res., Food and Agric. Organizations of United Nations, Sci. Consultation on Marine Mammals, Bergen, Norway. 11pp.

Gaskin, D. E., and Smith, G. J. D. 1977. The small whale fishery of St. Lucia, W. I. Rep. Int. Whaling Comm. 27:493.

———. 1979. Observations on marine mammals, birds and environmental conditions in the Head Harbour region of the Bay of Fundy. Pages 69–86 in D. J. Scarratt, ed. Evaluation of recent data relative to potential oil spills in the Passamaquoddy area. Techn. Rep. no. 901. Fish. and Marine Serv. (Canada). 107pp.

Gaskin, D. E.; Smith, G. J. D.; and Watson, A. P. 1975. Preliminary study of movements of harbor porpoises (*Phocoena phocoena*) in the Bay of Fundy using radiotelemetry. Can. J. Zool. 53:1466–1471.

Geist, O. W.; Buckley, J. L.; and Manville, R. H. 1960. Alaskan records of the narwhal. J. Mammal. 41:250–253.

Geraci, J. R., and St. Aubin, D. J. 1977. Mass stranding of the long-finned pilot whale, *Globicephala melaena,* on Sable Island, Nova Scotia. J. Fish. Res. Board Can. 34:2196–2199.

Gilmore, R. M. 1959. On the mass strandings of sperm whales. Pacific Nat. 1(9–10):9–16.

———. 1969. Mocha Dick, or the white whale of the Pacific: a leaf from a manuscript journal by J. N. Reynolds, Esq. Oceans 1(4):65–80.

Gilpin, J. B. 1878. On the smaller cetaceans inhabiting the Bay of Fundy and shores of Nova Scotia. Proc. Nova Scotia Inst. Nat. Sci. 4:21–34.

Goldsberry, D. G.; Asper, E. D.; and Cornell, L. H. 1978. A live capture technique for the killer whale (Orcinus orca). Aquatic mammals 6(3):91–95.

Gray, R. W. 1939. The walrus. Naturalist (London) 991:201–207.

———. 1941. The bottlenose whale. Naturalist (London) 791:129–132.

Gray, W. B. 1964. Porpoise tales. A. S. Barnes & Co., New York. 111pp.

Green, R. E.; Perrin, W. F.; and Petrich, B. P. 1971. The American tuna purse seine fishery. Pages 182–194 in H. Kristjonsson, ed. Modern fishing gear of the world. Fishing News (Books), London. Vol. 3.

Green, R. F. 1972. Observations on the anatomy of some cetaceans and pinnipeds. Pages 247–297 in S. H. Ridgway, ed. Mammals of the sea: biology and medicine. Charles Thomas Publishers, Springfield, Ill. 812pp.

———. 1977. Gross anatomy of the reproductive organs in

dolphins. Pages 185–194 *in* S. H. Ridgway and K. Benirschke, eds. Breeding dolphins: present status, suggestions for the future. U.S. Marine Mammal Comm. Rep. no. MMC-76/07. NTIS no. PB-273673. U.S. Dept. Commerce, Natl. Tech. Information Serv., Arlington, Va. 308pp.

Grenfell, W. 1934. The romance of Labrador. MacMillan Co., New York. 329pp.

Griffin, E. I. 1966. Making friends with a killer whale. Natl. Geogr. 129:418–446.

Griffin, E. I., and Goldsberry, D. G. 1968. Notes on the capture, care and feeding of the killer whale *Orcinus orca* at Seattle Aquarium. Int. Zoo. Yearb. 8:206–208.

Gruber, J. 1979. Aspects of the population biology of *Tursiops truncatus* in the Pass Cavallo area of Texas, with respect to the dolphin-shrimpboat association. Page 26 *in* Abstr. Presentations at 3rd Bienn. Conf. Bio. Marine Mammals, 7–11 October 1979, Seattle, Wash. 64pp.

Gunter, G. 1938. Seasonal variations in abundance of certain estuarine and marine fishes in Louisiana, with particular reference to life histories. Ecol. Monogr. 8:313–346.

———. 1942. Contributions to the natural history of the bottle-nose dolphin, *Tursiops truncatus* (Montague), on the Texas coast, with particular reference to food habits. J. Mammal. 23:267–276.

———. 1951. Consumption of shrimp by the bottle-nosed dolphin. J. Mammal. 32:465–466.

———. 1954. Mammals of the Gulf of Mexico. *In* P. S. Galtsoff, coordinator. Gulf of Mexico, its origin, waters, and marine life. U.S. Fish Wildl. Serv., Fish. Bull. 55:543–551.

Gurevich, V. S. 1977. Post-natal behavior of an Atlantic bottlenosed dolphin calf (*Tursiops truncatus,* Montagu) born at Sea World. Pages 168–184 *in* S. H. Ridgway and K. Benirschke, eds. Breeding dolphins: present status, suggestions for the future. U.S. Marine Mammal Comm. Rep. no. MMC-76/07. NTIS no. PB-273673. U.S. Dept. Commerce, Natl. Tech. Information Serv., Arlington, Va. 308pp.

Hall, J. D.; Gilmartin, W. G.; and Mattson, J. L. 1971. Investigation of a Pacific pilot whale stranding on San Clemente Island. J. Wildl. Dis. 7:324–327.

Hancock, D. 1965. Killer whales kill and eat a minke whale. J. Mammal. 46:341–343.

Handley, C. O., Jr. 1966. A synopsis of the genus *Kogia* (pygmy sperm whales). Pages 62–69 *in* K. S. Norris, ed. Whales, dolphins, and porpoises. Univ. California Press, Berkeley and Los Angeles. 789pp.

Harris, J. C. 1938. Porpoises feeding. Life Magazine 5(12):67.

Harrison, C. S., and Hall, J. D. 1978. Alaskan distribution of the beluga whale, *Delphinapterus leucas.* Can. Field-Nat. 92:235–241.

Harrison, R. J. 1977. Ovarian appearances and histology in *Tursiops truncatus.* Pages 195–204 *in* S. H. Ridgway and K. Benirschke, eds. Breeding dolphins: present status, suggestions for the future. U.S. Marine Mammal Comm. Rep. no. MMC-76/07. NTIS no. PB-273673. U.S. Dept. Commerce, Natl. Tech. Information Serv., Arlington, Va. 308pp.

Harrison, R. J.; Brownell, R. L., Jr.; and Boice, R. C. 1972. Reproduction and gonadal appearance in some odontocetes. Pages 362–429 *in* R. J. Harrison, ed. Functional anatomy of marine mammals. Academic Press, New York. Vol. 1. 451pp.

Harrison, R. J., and Ridgway, S. H. 1971. Gonodal activity in some bottlenose dolphins (*Tursiops truncatus*). J. Zool. London 165:355–366.

Hay, K. A., and Mansfield, A. W. In press. The narwhal, *Monodon monoceros* Linnaeus. 1758. *In* S. H. Ridgway and R. J. Harrison, eds. Handbook of marine mammals. Academic Press, London.

Heezen, B. C. 1957. Whales entangled in deep sea cables. Deep-Sea Res. 4:105–115.

Hegarty, R. B. 1959. Returns of whaling vessels sailing from American ports: a continuation of Alexander Starbuck's "History of the American whale fishery." Old Dartmouth Historical Soc. and Whaling Mus., New Bedford, Mass. 58pp.

Hershkovitz, P. 1966. Catalog of living whales. U.S. Natl. Mus. Bull. 246. 259pp.

Hewlett, S. I. 1978. It's a boy! At the Vancouver Aquarium. Animal Kingdom 81(2):15–27.

Heyland, J. D., and Hay, K. 1976. An attack by a polar bear on a juvenile beluga. Arctic 29:56–57.

Hill, R. M. 1967. Observations on beluga whales trapped by ice in Eskimo Lakes, winter 1966/67. Inuvik Res. Lab. Rep. 13pp. Manuscript.

Hoese, J. D. 1971. Dolphin feeding out of water in a salt marsh. J. Mammal. 52:222–223.

Hogan, T. 1975. Movements and behavior of the bottlenosed dolphin in the Savannah River mouth areas. Graduate School of Oceanography, Univ. Rhode Island, Narragansett. 16pp. Manuscript.

Holt, R. S., and Powers, J. E. 1980. Abundance estimation of dolphin stocks involved in the eastern tropical Pacific yellowfin tuna fishery determined from aerial and ship surveys. Nat. Marine Fish. Serv., Southwest Fish. Center, La Jolla, Calif. Manuscript.

Hui, C. A. 1977. Growth and physical indices of maturity in the common dolphin, *Delphinus delphis.* Pages 231–260 *in* S. H. Ridgway and K. Benirschke, eds. Breeding dolphins: present status, suggestions for the future. U.S. Marine Mammal Comm. Rep. no. MMC-76/07. NTIS no. PB-273673. U.S. Dept. Commerce, Natl. Tech. Information Serv., Arlington, Va. 308pp.

———. 1978. Reliability of using dentin layers for age determination in *Tursiops truncatus.* Final Rep., Contract MMC-155. U.S. Marine Mammal Comm., Washington, D.C. 25pp.

Hunt, W. J. 1979. Domestic whaling in the Mackenzie estuary, Northwest Territories. Fish. and Marine Serv. (Canada) Tech. Rep. 769:14pp.

Husson, A. M., and Holthuis, L. B. 1974. *Physeter macrocephalus* Linnaeus, 1758, the valid name for the sperm whale. Zool. Meded. 48:205–217.

International Whaling Commission. 1979. Report of the Subcommittee on Small Cetaceans. Rep. Int. Whaling Comm. 29:87–89.

Irvine, A. B.; Caffin, J. E.; and Kochman, H. I. 1980. Aerial surveys for manatees and dolphins in western Florida. U.S. Dept. Interior, Natl. Fish and Wildl. Lab., Rep. to Bur. Land Manage. Region, New Orleans, La. Manuscript.

Irvine, A. B.; Scott, M. D.; Wells, R. S.; Kaufmann, J. H.; and Evans, W. E. 1979*a*. A study of the movements and activities of the Atlantic bottlenosed dolphin, *Tursiops truncatus,* including an evaluation of tagging techniques. Final Rep., U.S. Marine Mammal Comm., Washington, D. C. Natl. Tech. Information Serv. PB 298042. 54pp.

Irvine, A. B.; Scott, M. D.; Wells, R. S.; and Mead, J. G. 1979*b*. Stranding of the pilot whale, *Globicephala macrorhynchus,* in Florida and South Carolina. Fish. Bull. 77:511–513.

Irvine, A. B., and Wells, R. S. 1972. Results of attempts to tag Atlantic bottlenose dolphins, *Tursiops truncatus.* Cetology 13:1–5.

Iverson, R. T. B. 1975. Bottlenose dolphins stealing fish from Hawaiian fishermen's lines. Rep., Southwest Fish. Center, Honolulu Lab., Natl. Marine Fish. Serv., NOAA. 12pp. Manuscript.

Jones, H. 1980. Why the dolphins died. Int. Wildl. 10(5):4–11.

Jonkel, C. J. 1969. White whales wintering in James Bay. J. Fish. Res. Board Can. 26:2205–2207.

Jonsgård, Å. 1977. Tables showing the catch of small whales (including minke whales) caught by Norwegians in the period 1938–75, and large whales caught in different North Atlantic waters in the period 1868–1975. Rep. Int. Whaling Comm. 27:413–426.

Jonsgård, Å., and Lyshoel, P. B. 1970. A contribution to the knowledge of the biology of the killer whale, *Orcinus orca* (L.). Nytt Mag. Zool. 18:41–48.

Joseph, J., and Greenough, J. W. 1979. International management of tuna, porpoises, and billfish: biological, legal, and political aspects. Univ. Washington Press, Seattle and London. 253pp.

Kanwiser, J., and Sundnes, G. 1965. Physiology of a small cetacean. Hvalradets Skrifter 48:45–53.

Kapel, F. O. 1975. Preliminary notes on the occurrence and exploitation of smaller Cetacea in Greenland. J. Fish. Res. Board Can. 32:1079–1082.

———. 1977. Catch of belugas, narwhals and harbour porpoises in Greenland, 1954–75, by year, month and region. Rep. Int. Whaling Comm. 27:507–520.

Kasuya, T. 1972. Growth and reproduction of *Stenella coeruleoalba* based on age determination by means of dentinal growth layers. Sci. Rep. Whales Res. Inst. (Tokyo) 24:57–79.

———. 1976. Reconsideration of life history parameters of the spotted and striped dolphins based on cemental layers. Sci. Rep. Whales Res. Inst. (Tokyo) 28:73–106.

———. 1978. The life history of Dall's porpoise with special reference to the stock off the Pacific coast of Japan. Sci. Rep. Whales Res. Inst. (Tokyo) 30:1–63.

Kellogg, R. 1931. Whaling statistics for the Pacific coast of North America. J. Mammal. 12(1):73–77.

Kemp, R. J. 1949. Report on stomach analysis from June 1, 1949–August 31, 1949. Annu. Rep. Marine Lab. Texas Fish Game, Oyster Comm., for fiscal year 1948–1949, pp. 111–112, 126–127.

Kemper, J. B. 1980. History of use of narwhal and beluga by Inuit in the Canadian eastern arctic including changes in hunting methods and regulations. Rep. Int. Whaling Comm. 30:481–492.

Kirby, V. L., and Sawyer-Steffan, J. E. 1979. Reproductive hormones and induced ovulation in captive Atlantic bottlenose dolphins, *Tursiops truncatus*. Page 64 *in* Abstr. Presentations at 3rd Bienn. Conf. Bio. Marine Mammals, 7–11 October 1979, Seattle, Wash. 64pp.

Kleinenberg, S. E.; Yablokov, A. V.; Bel'kovich, V. M.; and Tarasevich, M. N. 1969. Beluga (*Delphinapterus leucas*): investigation of the species. Transl. by Israel Program for Sci. Transl., Jerusalem. 376pp. (Russian publication in Moscow, 1964.)

Klinkhart, E. G. 1966. The beluga whale in Alaska. Fed. Aid in Wildl. Restoration Proj. Rep. Covering invest. Completed by 31 December 1965. Vol. 7: Proj. W-6-R and W-14-R. State of Alaska, Dept. Fish and Game, Juneau. 11pp.

Laurin, J. 1976. Preliminary study of the distribution, hunting and incidental catch of harbour porpoise, *Phocoena phocoena* L. in the Gulf and estuary of the St. Lawrence. Advisory Com. Marine Resour. Res., Food and Agric. Organization of United Nations, Sci. Consultation on Marine Mammals, Bergen, Norway. ACMRR/MM/SC/93. 14pp.

Lawrence, B., and Schevill, W. E. 1954. *Tursiops* as an experimental subject. J. Mammal. 35(2):225–232.

Lear, W. H., and Christensen, O. 1975. By-catches of harbour porpoise (*Phocoena phocoena*) in salmon drift nets at West Greenland in 1972. J. Fish. Res. Board Can. 32:1223–1228.

Leatherwood, J. S. 1979. Aerial survey of populations of the bottlenosed dolphin, *Tursiops truncatus*, and the West Indian manatee, *Trichechus manatus*, in the Indian and Banana Rivers, Florida. Fish. Bull. 77:47–59.

Leatherwood, J. S.; Perrin, W. F.; Garvie, R. L.; and La-Grange, J. C. 1972. Observations of sharks attacking porpoises (*Stenella* spp. and *Delphinus* cf. *D. delphis*). Naval Undersea Center, Tech. Note no. 908. San Diego, Calif. 7pp.

Leatherwood, J. S., and Platter, M. F. 1975. Aerial assessment of bottlenose dolphins off Alabama, Mississippi and Louisiana. Pages 49–86 *in* D. K. Odell, D. B. Siniff, and G. H. Waring, eds. *Tursiops truncatus* assessment workshop. Final Rep., U.S. Marine Mammal Comm. Contract MM5AC021 to Univ. Miami.

Leatherwood, S. 1975. Some observations of feeding behavior of bottlenosed dolphins (*Tursiops truncatus*) in the northern Gulf of Mexico and (*Tursiops* cf *T. gilli*) off southern California, Baja California, and Nayarit, Mexico. Marine Fish. Rev. 37(9):10–16.

———. 1977. Some preliminary impressions of the numbers and social behavior of free-swimming bottlenosed dolphin calves (*Tursiops truncatus*) in the northern Gulf of Mexico. Pages 143–167 *in* S. H. Ridgway and K. Benirschke, eds. Breeding dolphins: present status, suggestions for the future. U.S. Marine Mammal Comm. Rep. no. MMC-76/07. NTIS no. PB-273673. U.S. Dept. Commerce, Natl. Tech. Information Serv., Arlington, Va. 308pp.

Leatherwood, S.; Caldwell, D. K.; and Winn, H. E. 1976. The whales, dolphins and porpoises of the western north atlantic: a guide to their identification. NOAA Tech. Rep., NMFS Circ. 396. 176pp.

Leatherwood, S.; Deerman, M. W.; and Potter, C. W. 1978*b*. Food and reproductive status of nine *Tursiops truncatus* from the northeastern United States coast. Cetology, no. 28, pp. 1–5

Leatherwood, S.; Gilbert, J. R.; and Chapman, D. G. 1978*a*. An evaluation of some techniques for aerial censuses of bottlenose dolphins. J. Wildl. Manage. 42:239–250.

Leatherwood, S.; Perrin, W. F.; Kirby, V. L.; Hubbs, C. L.; and Dahlheim, M. 1980. Distribution and movements of Risso's dolphin (*Grampus griseus*) in the eastern North Pacific. Fish. Bull. 77(4):951–963.

Leatherwood, S., and Reeves, R. 1978. Porpoises and dolphins. Pages 97–111 *in* D. Haley, ed. Marine mammals of eastern North Pacific and arctic waters. Pacific Search Press, Seattle. 257pp.

Leatherwood, S.; Reeves, R. R.; Perrin, W. F.; and Evans, W. E. In press. The whales, dolphins, and porpoises of eastern North Pacific and arctic waters: a guide to their identification. NOAA Tech. Rep., NMFS Circ.

Leatherwood, S., and Show, I. T. 1980. Development of systematic procedures for estimating sizes of "populations" of bottlenose dolphins. Contract Rep. to NMFS, SEFC, Miami, Fla. 98pp.

Leatherwood, S., and Walker, W. A. 1979. The northern right whale dolphin, *Lissodelphis borealis* Peale, in the eastern North Pacific. Pages 85–141 *in* H. E. Winn and B. L. Olla, eds. Behavior of marine animals: current perspectives in research. Vol. 3: Cetaceans. Plenum Press, New York and London. 438pp.

Lee, H. 1878. The white whale. R. K. Burt & Co., Printers, London. 16pp.

Leighton, A. H. 1937. The twilight of the Indian porpoise hunters. Nat. Hist. 40(1):410–416, 458.

Linehan, E. J. 1979. The trouble with dolphins. Natl. Geogr. 155(4):506–540.

Ljungblad, D. K.; Leatherwood, S.; Johnson, R. A.; Mitchell, E. D.; and Awbrey, F. T. 1977. Echolocation signals of wild Pacific bottlenose dolphins, *Tursiops* sp. Page 36 in Proc. 2nd Conf. Bio. Marine Mammals, 12–15 December 1977, San Diego, Calif. 88pp. Abstr.

Lønø, Ø., and Øynes, P. 1961. White whale fishery at Spitzbergen. Norsk Hvalfangst-tidende 50:267–287.

Low, A. P. 1929. Extracts from reports on the District of Ungava or New Quebec. 3rd ed. Province of Quebec, Canada. Dept. Highways and Mines, Bur. Mines. R. Paradis, Quebec. 210pp.

Lubbock, B. 1937. The arctic whalers. Brown, Son & Ferguson, Glasgow. 483pp.

McBride, A. F., and Hebb, D. O. 1948. Behavior of the captive bottlenosed dolphin, *Tursiops truncatus*. J. Comp. Physiol. Psychol. 41(2):111–123.

McBride, A. F., and Kritzler, H. 1951. Observations on pregnancy, parturition and postnatal behavior in the bottlenose dolphin. J. Mammal. 32:251–266.

MacLaren Atlantic, Ltd. 1977. Report on Cruise 77-1, February 1977: environmental aspects of the cruise to Davis Strait and the Labrador coast for Imperial Oil Ltd., Aquitaine Co. of Canada Ltd. and Canada—Cities Services Ltd. Arctic Petroleum Operator's Assoc. Rep., Series no. 127. Manuscript.

MacLaren Marex, Inc. 1979. Report on aerial surveys of marine mammals and birds in southern Davis Strait and eastern Hudson Strait in March, 1978 for Esso Resources Canada Ltd., Aquitaine Co. of Canada Ltd. and Canada Cities Services Ltd. Arctic Petroleum Operators Assoc., Proj. no. 146. Manuscript.

Mackintosh, N. A. 1965. The stocks of whales. Fishing News (Books) Ltd., London. 232pp.

Mansfield, A. W.; Smith, T. G.; and Beck, B. 1975. The narwhal, *Monodon monoceros*, in eastern Canadian waters. J. Fish. Res. Board Can. 32:1041–1046.

Martin, K. R. 1973. Yankee whalemen and the enigma of the avenging whale. Mankind 3(11):54–61.

————. 1976. A whale attack on the whaleship Joseph Maxwell, 1845. Extracts 27:1–4.

Maul, G. E., and Sergeant, D. E. 1977. New cetacean records from Madeira. Bocagiana 43:1–8.

Maury, M. F. 1851. Whale chart. Natl. Observatory, U.S. Bur. Ordnance and Hydrography, Washington, D.C. 1p.

————. 1853. A chart showing the favorite resort of the sperm and right whale by M. F. Maury, L. L. D., Lieut., U.S. Navy. Constructed from Maury's whale chart of the world, by R. H. Wyman, Lieut., U.S.N., by authority of Comm. Charles Morris, U.S.N., Chief, Bureau Ordnance and Hydrography, Washington, D.C. 1p.

May, R. M.; Beddington, J. R.; Clark, C. W.; Holt, S. J.; and Laws, R. M. 1979. Management of multispecies fisheries. Science 205:267–277.

Mead, J. G. 1975a. Anatomy of the external nasal passages and facial complex in the Delphinidae (Mammalia: Cetacea). Smithsonian Contributions to Zool., no. 207. 72pp.

————. 1975b. Preliminary report on the former net fisheries for *Tursiops truncatus* in the western North Atlantic. J. Fish Res. Board Can. 32(7):1155–1162.

Mead, T. 1961. The killers of Eden. The story of the killer whales of Twofold Bay. Angus & Robertson, London. 22pp.

Mercer, M. C. 1967. Wintering of pilot whales, *Globicephala melaena*, in Newfoundland inshore waters. J. Fish. Res. Board Can. 24:2481–2484.

————. 1975. Modified Leslie DeLury population models of the long-finned pilot whale (*Globicephala melaena*) and annual production of the short-finned squid (*Illex illecebrosus*) based upon their interaction at Newfoundland. J. Fish. Res. Board Can. 32:1145–1154.

Mitchell, E. D. 1968. Northeast Pacific stranding distribution and seasonality of Cuvier's beaked whale *Ziphius cavirostris*. Can. J. Zool. 46:265–279.

————. 1970. Pigmentation pattern evolution in delphinid cetaceans: an essay in adaptive coloration. Can. J. Zool. 48(4):717–740.

————. 1974. Present status of northwest Atlantic fin and other whale stocks. Pages 108–169 in W. E. Schevill, ed. The whale problem: a status report. Harvard Univ. Press, Cambridge, Mass. 419pp.

————. 1975a. Porpoise, dolphin and small whale fisheries of the world: Status and problems. Int. Union for Conserv. of Nature and Nat. Resour., Morges, Switzerland. IUCN Monograph no. 3:129.

————. 1977a. Sperm whale maximum length limit: proposed protection of "harem masters." Rep. Int. Whaling Comm. 27:224–225.

————. 1977b. Evidence that the northern bottlenose whale is depleted. Rep. Int. Whaling Comm. 27:195–203.

————. In press. Canada progress report on cetacean research June 1979–May 1980. Rep. Int. Whaling Comm.

Mitchell, E. D., ed. 1975b. Report of the meeting on smaller cetaceans, Montreal, 1–11 April 1974. J. Fish. Res. Board Can. 32:923–925.

Mitchell, E. D., and Kozicki, V. M. 1975. Autumn stranding of a northern bottlenose whale (*Hyperoodon ampullatus*) in the Bay of Fundy, Nova Scotia. J. Fish Res. Board Can. 32:1019–1040.

————. 1978. Sperm whale regional closed seasons: proposed protection during mating and calving. Rep. Int. Whaling Comm. 28:195–198.

Mitchell, E., and Reeves, R. R. In press. Catch history and cumulative catch estimates of initial population size of cetaceans in the eastern Canadian arctic. Rep. Int. Whaling Comm.

Mohl, B.; Larsen, E.; and Amundin, M. 1976. Sperm whale size determination: outlines of an acoustic approach. Food and Agric. Organization of United Nations, Sci. consultation on marine mammals, 31 August-9 Sept 1976, Bergen, Norway. ACMRR/MM/SC/84. 4pp.

Montagu, G. 1821. Description of a species of *Delphinus* which appears to be new. Mem. Wernerian Nat. Hist. Soc. 3:75–82.

Moore, J. C. 1966. Diagnoses and distributions of beaked whales of the genus *Mesoplodon* known from North American waters. Pages 32–61 in K. S. Norris, ed. Whales, dolphins, and porpoises. Univ. California Press, Berkeley and Los Angeles. 789pp.

————. 1968. Relationships among the living genera of beaked whales with classifications, diagnoses and keys. Fieldiana: Zool. 53:209–298.

Morejohn, G. V. 1979. The natural history of Dall's porpoise in the North Pacific ocean. Pages 45–83 in H. E. Winn and B. L. Olla, eds. Behavior of marine animals: current perspectives in research. Vol. 3: Cetaceans. Plenum Press, New York and London, 438pp.

Munn, H. T. 1932. Prairie trails and arctic by-ways. Hurst and Blackett, Ltd., London. 288pp.

Myrberg, A. A., Jr. 1978. Ocean noise and the behavior of

marine animals: relationships and implications. Pages 169–208 in J. L. Fletcher and R. G. Busnel, eds. The effects of noise on wildlife. Academic Press, New York. 305pp.

National Academy of Sciences. 1975. Products from Jojoba: a promising new crop for arid lands. Com. Jojoba Utilization, Office of Chemistry and Chemical Tech., Assembly of Mathematical and Physical Sci., Washington, D.C. 30pp.

National Marine Fisheries Service. 1973. Pages 20564–20601 in Administration of the Marine Mammal Protection Act of 1972, 21 December 1972, to 21 June 1973: Rep. of Secretary of Commerce. Fed. Register (Washington) 38 (147), Wednesday, 1 August 1973.

———. 1978. The Marine Mammal Protection Act of 1972. Annu. Rept. Natl. Marine Fish Serv. U.S. Dept. Commerce, Washington, D.C. 183pp.

Newman, M. 1971. Capturing narwhals for the Vancouver Public Aquarium 1970. Polar Record 15:922–923.

Newman, M. A., and McGeer, P. L. 1966. The capture and care of a killer whale, *Orcinus orca,* in British Columbia. Zoologica 51(5):59–70.

Nishiwaki, M. 1972. General biology. Pages 1–204 in S. H. Ridgway, ed. Mammals of the sea: biology and medicine. Charles C. Thomas, Springfield, Ill. 812pp.

Nishiwaki, M., and Handa, C. 1958. Killer whales caught in the coastal waters off Japan for recent 10 years. Sci. Rep. Whales Res. Inst. (Tokyo) 13:85–96.

Norris, J. N. In preparation. The effects of tank reverberation on whistle vocalizations of captive common dolphins, *Delphinus delphis.* Hubbs Sea World Res. Institute, San Diego, Calif. 11pp.

Norris, K. S., and Harvey, G. W. 1972. A theory for the function of the spermaceti organ of the sperm whale (*Physeter catodon* L.). Pages 397–417 in S. R. Galler, K. Schmidt-Koenig, G. J. Jacobs, and R. E. Bellebille, eds. Animal orientation and navigation. Natl. Aeronautics and Space Admin., Washington, D.C. 606pp.

Norris, K. S., and McFarland, W. N. 1958. A new porpoise of the genus *Phocoena* from the Gulf of California. J. Mammal. 39:22–39.

Norris, K. S., and Prescott, J. H. 1961. Observations on Pacific cetaceans of Californian and Mexican waters. Univ. California Publ. Zool. 63(4):291–402.

Odell, D. K. 1975. Status and aspects of the life history of the bottlenosed dolphin, *Tursiops truncatus,* in Florida. J. Fish. Res. Board Can. 32:1055–1058.

———. 1976a. Distribution and abundance of marine mammals in south Florida: preliminary results. Pages 203–212 in Biscayne Bay: past/present/future, proceedings of a symposium. Univ. Miami Sea Grant Program, Spec. Rept. no. 5.

———. 1976b. A preliminary study of the ecology and population biology of the bottlenosed dolphin in southeast Florida. Contract Rep. for 10 June 1974–9 June 1975, U.S. Marine Mammal Comm. Contract no. MM4AC003.

———. 1979. Distribution and abundance of marine mammals in the waters of the Everglades National Park. Pages 673–681 in R. M. Linn, ed. Proc. 1st Conf. Sci. Res. Waters of Everglades Natl. Park, 9–12 November 1976, New Orleans, Vol. 1. U.S. Dept. Interior, Natl. Parks Serv., Trans. and Proc., Series no. 5.

Odell, D. K., and Reynolds, J. E. 1980. Abundance of the bottlenose dolphin, *Tursiops truncatus,* on the west coast of Florida. U.S. Dept. Commerce, NTIS Publ. PB80-197650: 47pp.

Ohsumi, S. 1965. Reproduction of the sperm whale in the north-west Pacific. Sci. Rep. Whales Res. Inst. (Tokyo) 19:1–35.

———. 1975a. Review of Japanese small-type whaling. J. Fish. Res. Board Can. 32:1111–1121.

———. 1975b. Incidental catch of cetaceans with salmon gillnets. J. Fish. Res. Board Can. 32:1229–1235.

———. 1980. Catches of sperm whales by modern whaling in the North Pacific. Rep. Int. Whaling Comm. (special issue 2), pp. 11–18.

Ohsumi, S.; Kasuya, T.; and Nishiwaki, M. 1962. Accumulation rate of dentinal growth layers in the maxillary tooth of the sperm whale. Sci. Rep. Whales Res. Inst. (Tokyo) 17:17–36.

Orr, R. T. 1963. A northern record for the Pacific bottlenose dolphin. J. Mammal. 44(3):424.

Packard, W. 1903. The young ice whalers. Houghton, Mifflin and Co., Boston. 397pp.

Payne, R., and Webb, D. 1971. Orientation by means of long range acoustic signaling in baleen whales. Ann. New York Acad. Sci., pp. 100–141.

Perrin, W. F. 1969. Using porpoise to catch tuna. World Fishing 18(6):42–45.

———. 1970. Porpoise fishing. Pages 45–48 in Proc. 6th Annu. Conf. Bio. Sonar and Diving Mammals, 17–18 October 1969, Stanford Res. Institute, Menlo Park, Calif.

———. 1972. Color patterns of spinner porpoises (*Stenella c.f. S. longirostris*) of the eastern Pacific and Hawaii, with comments on delphinid pigmentation. Fish. Bull. 70(3):983–1003.

———. 1975. Distribution and differentiation of populations of dolphins of the genus *Stenella* in the eastern tropical Pacific. J. Fish. Res. Board Can. 32:1059–1067.

Perrin, W. F., chairman. 1980. Report of the Sub-Committee on Small Cetaceans. Rep. Int. Whaling Comm. 30:111–128.

Perrin, W. F.; Coe, J. M.; and Zweifel, J. R. 1976. Growth and reproduction of the spotted porpoise, *Stenella attenuata,* in the offshore eastern tropical Pacific. Fish. Bull. 74:229–269.

Perrin, W. F., and Henderson, J. R. 1979. Growth and reproductive rates in two populations of spinner dolphins, *Stenella longirostris,* with different histories of exploitation. Admin. Rep. LJ-79-29. Southwest Fish. Center, Natl. Marine Fish. Serv., La Jolla, Calif. 11pp.

Perrin, W. F.; Evans, W. E.; and Holts, D. B. 1979. Movements of pelagic dolphins (*Stenella* spp.) in the eastern tropical Pacific as indicated by results of tagging, with summary of tagging operations, 1969–76. NOAA Tech. Rep. NMFS SSRF-737. 14pp.

Perrin, W. F., and Myrick, A. C., Jr. In press. Age determination of toothed cetaceans and sirenians. International Whaling Commission, Cambridge.

Perryman, W. L., and Foster, T. C. 1980. Preliminary report on predation by small whales, mainly the false killer whale, *Pseudorca crassidens,* on dolphins (*Stenella* spp. and *Delphinus delphis*) in the eastern tropical Pacific. Admin. Rep. LJ-80-05. Southwest Fish. Center, Natl. Marine Fish. Serv., La Jolla, Calif. 9pp.

Pike, G. C., and MacAskie, I. B. 1969. Marine mammals of British Columbia. Fish. Res. Board Can. Bull. 171:1–54.

Pilleri, G., and Gihr, M. 1970. The central nervous system of the mysticete and odontocete whales. Pages 89–127 in G. Pilleri, ed. Investigations on Cetacea. Brain Anatomy Inst., Berne. Vol. 2. 296pp.

Pippard, L. 1980. Looking after our own: the white whales of the St. Lawrence need protection. Nat. Can. 9(1):39–45.

Pippard, L., and Malcolm, H. 1978. White whales (*Del-*

phinapterus leucas): observations on their distribution, population and critical habitats in the St. Lawrence and Saguenay rivers. Proj. C1632, Contract 76-190. Dept. Indian and Northern Affairs, Parks Canada. 160pp. Manuscript.

Porsild, M. P. 1918. On "savssats": a crowding of arctic animals at holes in the sea ice. Geogr. Rev. 6:215–228.

Prescott, J. H. 1977. Comments on captive births of *Tursiops truncatus* at Marineland of the Pacific (1957-1972). Pages 71–76 in S. H. Ridgway and K. Benirschke, eds. Breeding dolphins: present status, suggestions for the future. U.S. Marine Mammal Comm. Rep. no. MMC-76/07. NTIS no. PB-273673. U.S. Dept. Commerce, Natl. Tech. Information Serv., Arlington, Va. 308pp.

Prescott, J. H., and Fiorelli, P. M. 1980. Review of the harbor porpoise (*Phocoena phocoena*) in the U.S. northwest Atlantic. Final rep. to U.S. Marine Mammal Comm. in Fulfillment of Contract MM8AC016. Natl. Tech. Information Serv., U.S. Dept. Commerce, Springfield, Va. 64pp.

Ralls, K. 1976. Mammals in which females are larger than males. Q. Rev. Bio. 51:245–276.

Ralls, K.; Brownell, R. L.; and Ballou, J. 1980. Differential mortality by sex and age in mammals, with specific reference to the sperm whale. Rep. Int. Whaling Comm. (special issue 2), pp. 233–243.

Rathjen, W. F., and Sullivan, J. R. 1970. West Indian whaling. Sea Frontiers 16(3):130–137.

Ray, C. 1961. White whales for the aquarium. Anim. Kingdom 64(6):162–170.

———. 1962. Three whales that flew. Natl. Geogr. 121(3):346–359.

Reeves, R. R. 1976. The narwhal: another endangered species. Can. Geogr. J. 92(3):12–19.

———. 1978. The narwhal: the arctic's unicorn. Alaska 44(9):10–11, 88–89.

Reeves, R. R., and Katona, S. 1980. Extralimital records of white whales (*Delphinapterus leucas*) in eastern North American waters. Can. Field-Nat. 94(3):239–247.

Reeves, R. R., and Tracey, S. 1980. *Monodon monoceros*. Am. Soc. Mammal., Mammalian Species no. 127, pp. 1–7.

Ricciuti, E. R. 1967. Quest for the white whale: the New York Aquarium brings back two belugas. Anim. Kingdom 70(5):148–156.

Rice, D. W. 1960. Distribution of the bottlenose dolphin in the Leeward Hawaiian Islands. J. Mammal. 41:407–408.

———. 1968. Stomach contents and feeding behavior of killer whales in the eastern North Pacific. Norsk Hvalfangst-tidende 57:35–38.

———. 1974. Whales and whale research in the eastern North Pacific. Pages 170–195 in W. E. Schevill, ed. The whale problem: a status report. Harvard Univ. Press, Cambridge, Mass. 419pp.

———. 1977. A list of the marine mammals of the world. Natl. Marine Fish. Serv., NOAA Tech. Rep. SSRF-711. 15pp.

———. 1978. Sperm whales. Pages 82–87 in D. Haley, ed. Marine mammals of eastern North Pacific and arctic waters. Pacific Search Press, Seattle, 257pp.

Rice, D. W., and Wolman, A. A. 1971. The life history and ecology of the gray whale (*Eschrichtius robustus*). Am. Soc. Mammal. Special Pub. no. 3. 142pp.

Ridgway, S. H. 1966. Dall porpoise, *Phocoenoides dalli* True: observations in captivity and at sea. Norsk Hvalfangst-tidende 55(5):97–110.

———. 1968. The bottlenose dolphin in biomedical research. Pages 387–446 in W. I. Gay, ed. Methods of animal experimentation. Academic Press, New York. 469pp.

———. 1971. Buoyancy regulation in deep diving whales. Nature 232(5306):133–134.

———. 1972. Homeostasis in the aquatic environment. Pages 590–747 in S. H. Ridgway, ed. Mammals of the sea: biology and medicine. Charles C Thomas, Springfield, Ill. 812pp.

Ridgway, S. H., and Benirschke, K. W., eds. 1977. Breeding dolphins: present status, suggestions for the future. U.S. Marine Mammal Comm. Rep. no. MMC-76/07. NTIS no. PB273673. U.S. Dept. Commerce, Natl. Tech. Information Serv., Arlington, Va. 308pp.

Ridgway, S. H.; Green, R. F.; and Sweeney, J. L. 1975. Mandibular anesthesia and tooth extraction in the bottlenose dolphin. J. Wildl. Dis. 11:415–418.

Rindal, I. 1977.Chairman's report of twenty-seventh meeting. Rep. Int. Whaling Comm. 27:6–15.

Ross, G. J. B. 1977. The taxonomy of bottlenosed dolphins *Tursiops* species in South African waters, with notes on their biology. Ann. Cape Prov. Mus. (Nat. Hist.) 11:135–194.

Saayman, G. S., and Tayler, C. K. 1973. Social organisation in inshore dolphins (*Tursiops aduncus* and *sousa*) in the Indian Ocean. J. Mammal. 54(4):993–996.

———. 1977. Observations on the sexual behavior of Indian Ocean bottlenosed dolphins (*Tursiops aduncus*). Pages 113–129 in S. H. Ridgway and K. Benirschke, eds. Breeding dolphins: present status, suggestions for the future. U.S. Marine Mammal Comm. Rep. no. MMC-76/07. NTIS no. PB-273673. U.S. Dept. Commerce, National Tech. Information Serv., Arlington, Va. 308pp.

Scammon, C. M. 1874. The marine mammals of the northwestern coast of North America, described and illustrated together with an account of the American whale fishery. Carmany & Co., San Francisco, 320pp.

Scheffer, V. B., and Slipp, J. W. 1948. The whales and dolphins of Washington State with a key to the cetaceans of the west coast of North America. Am. Midl. Nat. 39:257–337.

Schenkkan, E. J., and Purves, P. E. 1973. The comparative anatomy of the nasal tract and the function of the spermaceti organ in the *Physeteridae* (Mammalia, Odontoceti). Bijdr. Dierkd. 43(1):93–112.

Schmidly, D. J., and Shane, S. H. 1978. A biological assessment of the cetacean fauna of the Texas coast. NTIS no. PB-281763. Natl. Tech. Information Serv., Washington D.C. 38pp.

Seber, G. A. F. 1973. The estimation of animal abundance and related parameters. Griffin, Inc., London. 499pp.

Sergeant, D. E. 1959. Age determination of odontocete whales from dentinal growth layers. Norsk Hvalfganst-tidende 6:273–288.

———. 1962*a*. On the external characters of the blackfish or pilot whales (genus *Globicephala*). J. Mammal. 43:395–413.

———. 1962*b*. The biology of the pilot whale or pothead whale *Globicephala melaena* (Traill) in Newfoundland waters. Fish. Res. Board Can., Bull. no. 132. 84pp.

———. 1963. Minke whales, *Balaenoptera acutorostrata* Lacépède, of the western North Atlantic. J. Fish. Res. Board Can. 20:1489–1504.

———. 1968. Whales. Pages 388–396 in C. S. Beals, ed. Science, history and Hudson Bay. Dept. Energy, Mines and Resources, Ottawa. 501pp.

———. 1969. Feeding rates of Cetacea. Tisk Dir. Skr. Der. Hav. Unders. 15:246.

———. 1973. Biology of white whales (*Delphinapterus*

leucas) in western Hudson Bay. J. Fish. Res. Board Can. 30:1065–1090.

――――. 1978. [Ecological isolation in some Cetacea.] Pages 20–34 *in* Novae v Izuchenii Kitoobraznykh i Lastonogykh [Recent advances in the study of whales and seals]. A. N. Severtsov Inst. of Evolutionary Morphol. and Ecol. of Anim. Acad. Sci. of U.S.S.R., Moscow.

Sergeant, D. E., and Brodie, P. F. 1969. Tagging white whales in the Canadian Arctic. J. Fish. Res. Board Can. 25:2201–2205.

――――. 1975. Identity, abundance, and present status of populations of white whales, *Delphinapterus leucas,* in North America. J. Fish. Res. Board Can. 32:1047–1054.

Sergeant, D. E.; Caldwell, D. K.; and Caldwell, M. C. 1973. Age, growth, and maturity of bottlenosed dolphin (*Tursiops truncatus*) from northeast Florida. J. Fish. Res. Board Can. 30:1009–1011.

Sergeant, D. E.; Mansfield, A. W.; and Beck, B. 1970. Inshore records of Cetacea for eastern Canada, 1949–68. J. Fish. Res. Board Can. 27(11):1903–1915.

Shallenberger, E. W. 1979. The status of Hawaiian cetaceans. Final Contract Rep., Contract no. MM7AC028. U.S. Marine Mammal Comm., Washington, D.C. 103pp.

Shane, S. 1977. Population biology of *Tursiops truncatus* in Texas. Page 57 *in* Proc. 2nd Annu. Conf. Bio. and Conserv. Marine Mammals, 12–15 December 1977, San Diego, Calif. 88pp.

Shane, S. H. 1980. Occurrence, movements and distribution of bottlenose dolphin, *Tursiops truncatus,* in southern Texas. Fish. Bull. 78(3):593–601.

Shane, S. H., and Schmidly, D. J. 1978. Population biology of Atlantic bottlenosed dolphins, *Tursiops truncatus,* in the Aransas Pass area of Texas. Contract Rep. to U.S. Marine Mammal Comm., Natl. Tech. Information Serv., Washington, D.C.

Shevchenko, V. I. 1975. Kharakter Vzaimootnoshenii kasatok i drugikh kitoobraznykh. Pages 173–174 *in* G. B. Agarkov, et al., eds. Morskie mlekopitaiushchie. Chast' 2. Materialy VI vsesoyuznogo soveshchaniya (Kiev, Oktyabr' 1975 g.) Izdatel'stvo ''Naukova Dumka,'' Kiev. [Transl. as: The nature of the interrelationships between killer whales and other cetaceans. Can. Dept. Environment, Fish. and Marine Serv., Transl. Series no. 3839. 1976. 4pp.]

Silverman, H., and Dunbar, M. J. 1980. Aggressive tusk use by the narwhal (*Monodon monoceros* L.). Nature 284:57–58.

Slijper, E. J. 1966. Functional morphology of the reproductive system in Cetacea. Pages 278–319 *in* K. S. Norris, ed. Whales, dolphins, and porpoises. Univ. California Press, Berkeley and Los Angeles. 789pp.

Smith, G. J. D., and Gaskin, D. E. 1974. The diet of harbor porpoises (*Phocoena phocoena* [L.]) in coastal waters of eastern Canada with special reference to the Bay of Fundy. Can. J. Zool. 52:777–782.

Smith, T., ed. 1979. Report of the Status of Porpoise Stocks Workshop, 27–31 August 1979. Admin. Rep. no. LJ-79-41. Southwest Fish. Center, Natl. Marine Fish. Serv. La Jolla, Calif. 121pp.

Smith, T. G. 1977. The occurrence of a narwhal (*Monodon monoceros*) in Prince Albert Sound, western Victoria Island, Northwest Territories. Can. Field.-Nat. 91:299.

Smith, T. G.; Hammill, M. H.: Doidge, D. W.; Cartier, T.; and Sleno, G. A. 1980. Marine mammal studies in southeastern Baffin Island. Can. Manuscript Rep. of Fish. and Aquatic Sci. no. 1552. 70pp.

Soper, J. D. 1928. A faunal investigation of southern Baffin Island. Natl. Mus. Can., Bio. Series no. 15, Bull. no. 53, pp. 1–143.

――――. 1944. The mammals of southern Baffin Island, Northwest Territories, Canada. J. Mammal. 25:221–254.

Stackpole, E. A. 1953. The sea-hunters: the New England whale men during two centuries, 1635–1835. J. B. Lippincott, Philadelphia and New York. 510pp.

Starbuck, A. 1878. History of the American whale fishery from its earliest inception to the year 1876. U.S. Comm. Fish and Fish., Part 4, Rep., 1875–1876, appendix A (reprinted in 1964 by Argosy-Antiquarian Ltd., New York). 2 vols. 779pp.

Stearnes, S. 1887. Fisheries of the Gulf of Mexico. Page 533–587 *in* G. B. Goode, ed. The fisheries and fishery industries of the United States. Sect. 2. A geographical review of the fisheries industries and fishing communities for the year 1880. 787pp. Part 15: Fisheries of the Gulf of Mexico.

Tarpy, C. 1979. Killer whale attack. Natl. Geogr. 155(4):542–545.

Tavolga, M. C. 1966. Behavior of the bottlenosed dolphin (*Tursiops truncatus*): social interactions in a captive colony. Pages 718–730 *in* K. S. Norris, ed. Whales, dolphins, and porpoises. Univ. California Press, Berkeley and Los Angeles. 789pp.

Tavolga, M. C., and Essapian, F. S. 1957. The behavior of the bottlenosed dolphin (*Tursiops truncatus*): mating, pregnancy, parturition and mother-infant behavior. Zoologica 42(2):11–31.

Tomilin, A. G. 1967. Mammals of the U.S.S.R. and adjacent countries. Vol. 9: Cetacea (Kitoobrazne). Israel Program for Sci. Transl., Jerusalem. 717pp. (Originally published in Russian, 1957.)

Tønnessen, J. N. 1967. Den moderne hvalfangst histoire: opprinnelse og utvikling. [History of modern whaling: foundation and development. Vol. 2: Global whaling, 1883–1924, part I, 1883–1914.] Norwegian Whaling Association, Sandefjord. 619pp. (Pages 129–173 transl. by R. Mathisen, 1973.)

Tormosov, D. D. 1976. On ovulation in sperm whale females. Rep. and Papers of Sci. Com. of the Comm., 1975. Int. Whaling Comm., London. Pages 360–365.

Townsend, C. H. 1914. The porpoise in captivity. Zoologica (New York) 1:289–299.

――――. 1935. The distribution of certain whales as shown by logbook records of American whaleships. Zoologica 19(1):1–50.

Treude, E. 1975. Studies in settlement development and evolution of the economy in the eastern central Canadian Arctic. Musk-Ox 16:53–66.

――――. 1977. Pond Inlet, northern Baffin Island: the structure of an Eskimo resource area. Polar Geogr. 1:95–122.

True, F. W. 1885. The bottle-nose dolphin, *Tursiops tursio,* as seen at Cape May, New Jersey. Science 5:338–339.

――――. 1889. Contributions to the natural history of the cetaceans: a review of the family Delphinidae. Bull. U.S. Natl. Mus. 36:1–191.

――――. 1891. Observations on the life history of the bottlenose porpoise. Proc. U.S. Natl. Mus. 13(1890):197–203.

U.S. Fish and Wildlife Service. 1980. Pilot study of marine mammals, birds and turtles in OCS areas of the southeastern United States including waters of the Gulf of Mexico and Atlantic Ocean. Draft Contract Rep. to Bur. Land Manage. in MOUAA551-MU9-19. 56pp.

Van Bree, P. J. H. 1971. On *Globicephala sieboldii* Gray, 1846, and other species of pilot whales (notes on Cetacea, Delphinoidea III). Beaufortia 19:79–87.

Van Bree, P. J. H.: Sergeant, D. E.; and Hoek, W. 1977. A harbour porpoise *Phocoena phocoena* (Linnaeus, 1758), from the Mackenzie River delta, Northwest Territories, Canada (notes on Cetacea, Delphinoidea VIII). Beaufortia 26:99–105.

Van Gelder, R. G. 1960. Results of the Puritan-American Museum of Natural History expedition to western Mexico. 10. Marine mammals from the coasts of Baja California and the Tres Marías Islands, Mexico. Am. Mus. Nov. 1992:1–27.

Van Heel, W. H. D. 1962. Sound and Cetacea. Netherlands J. Sea Res. 1(4):407–507.

Vibe, C. 1950. The marine mammals and the marine fauna in the Thule district (northwest Greenland) with observations on ice conditions in 1939–1941. Medd. Grønland 150(6):1–115.

Vietmeyer, N. D. 1975. Can a whale find life in the desert? Audubon 77(5):105–106.

Vladykov, V. D. 1943. A modification of the pectoral fins in the beluga from the St. Lawrence River. Studies on aquatic mammals, 2. Contributions de l'Institut de Biologie de l'université de Montréal, no. 11. (Extrait du Naturaliste Canadien 70:23–40.)

_____. 1944. Chasse, biologie et valeur économique du marsouin blanc ou béluga (*Delphinapterus leucas*) du fleuve et du golfe Saint-Laurent. Études sur les mammifères aquatiques, 3. Contribution du Département des Pêcheries Québec, no. 14. 182pp.

_____. 1946. Nourriture du marsouin blanc ou béluga (*Delphinapterus leucas*) du fleuve Saint-Laurent. Études sur les mammifères aquatiques, 4. Contribution du Département des Pêcheries Québec, no. 17. 129pp.

Walker, W. A. 1975. Review of the live-capture fishery for smaller cetaceans taken in southern California waters for public display, 1966–1973. J. Fish. Res. Board Can. 32:1197–1211.

_____. 1980. Geographic variation in morphology and biology of bottlenose dolphins (*Tursiops*) in the eastern North Pacific. Contract Rep., Contract no. 03-7-208-35238. Southwest Fish. Center, Natl. Marine Fish. Serv. La Jolla, Calif. 13pp.

Watkins, W. A. 1977. Acoustic behavior of sperm whales. Oceanus 20(2):50–58.

_____. 1980. Acoustics and the behavior of sperm whales. Pages 283–290 in R. G. Busnel and J. F. Fish, eds. Animal sonar systems. Plenum Publishing Corp., New York.

Watkins, W. A., and Schevill, W. E. 1975. Sperm whales (*Physeter catodon*) react to pingers. Deep-Sea Res. 22:123–129.

_____. 1977a. Sperm whale codas. J. Acoust. Soc. Am. 62(6):1485–1490.

_____. 1977b. Spatial distribution of *Physeter catodon* (sperm whales) underwater. Deep-Sea Res. 24:693–699.

Wells, R. S.; Irvine, A. B.; and Scott, M. D. 1980. The social ecology of inshore odontocetes. Pages 263–317 in L. M. Herman, ed. Cetacean behavior.: Mechanisms and functions. John Wiley & Sons, New York. 463pp.

Whitehead, H., and Payne, R. 1976. New techniques for assessing populations of right whales without killing them. Paper presented at Sci. Consultation on Marine Mammals, Bergen, Norway, 21 August–9 September 1976. 45pp.

Williams G. C. 1964. Measurement of consociation among fishes and comments on the evolution of schooling. Michigan State Mus. Publ. Bio. Sci. 2:351–383.

Williamson, K. 1948. The Atlantic islands: a study of the Faeroe life and scene. Collins St. James's Place, London. 360pp.

Wood, F. G., Jr. 1973. Marine mammals and man: the navy's porpoises and sea lions. R. B. Luce, Inc., Washington and New York. 264pp.

Wood, F. G., Jr.; Caldwell, D. K.; and Caldwell, M. C. 1970. Behavioral interaction between porpoises and sharks. Pages 264–277 in G. Pilleri, ed. Investigations on Cetacea. Brain Anatomy Inst., Berne. Vol. 2. 296pp.

Worthington, L. V., and Schevill, W. E. 1957. Underwater sounds heard from sperm whales. Nature 180(4580):291.

Würsig, B. 1978. Occurrence and group organization of Atlantic bottlenose porpoises (*Tursiops truncatus*) in an Argentine Bay. Bio. Bull. 154(2):348–359.

Würsig, B., and Würsig, M. 1977. The photographic determination of group size, composition, and stability of coastal porpoises (*Tursiops truncatus*). Science 198(4318):755–756.

_____. 1979. Day and night of the dolphin. Nat. Hist. 88(3):60–67.

_____. 1980. Behavior and ecology of the bottlenose dolphin, *Tursiops truncatus*, in the South Atlantic. Fish. Bull. 77(2):399–412.

Wyman, J. 1863. Description of a "white fish" or "white whale" (*Beluga borealis* Lesson). Boston J. Nat. Hist. 7:603–612.

Yablokov, A. V., and Bel'kovich, V. M. 1968. Cetaceans of the arctic: prospects of their proper utilization and conservation. Problems of the North 11:199.

Yukhov, V. L.; Vinogradova, E. K.; and Medvedev, L. P. 1975. Ob'ekty pitaniya kosatok (*Orcinus orca* L.) v Antarktike i sopredel'nykh vodakh. Pages 183–185 in G. B. Agarkov et al., eds. Morskie mlekopitaiushchie. Chast' 2. Materialy VI vsesoyuznogo soveshchaniya (Kiev, Oktyabr' 1975 g.). Izdatel'stvo "Naukova Dumka," Kiev. [Transl. as: The diet of killer whales (*Orcinus orca* L.) in the Antarctic and adjacent waters. Can. Dept. Environment, Fish. and Marine Serv., Transl. Ser. no. 3844. 1976. 5pp.]

Yurick, D. B. 1977. Populations, subpopulations, and zoogeography of the harbour porpoise, *Phocoena phocoena* (L.). M.S. thesis. Univ. Guelph, Guelph, Ontario. 148pp.

STEPHEN LEATHERWOOD, Hubbs Sea World Research Institute, P. O. Box 13101, Orlando, Florida 32859.

RANDALL R. REEVES, National Fish and Wildlife Laboratory, National Museum of Natural History, Washington, D.C. 20560.

19

Baleen Whales
Eubalaena glacialis and Allies

Randall R. Reeves
Robert L. Brownell, Jr.

NOMENCLATURE

COMMON NAMES. Right whale, Nördkaper, Biscay(an) whale, black right whale, black whale, northwest whale, North Atlantic right whale, North Pacific right whale, southern right whale, Sletbag, Sarda, Sarde
SCIENTIFIC NAME. *Eubalaena glacialis*

Systematics. The systematics of right whales are unresolved. The distinction between the temperate region right whale(s) (*Eubalaena*) and the arctic region bowhead whale (*Balaena mysticetus*) was not firmly established until the late 19th century (Eschricht and Reinhardt 1866), long after most stocks of both genera had been hunted to near extinction (Allen 1908). There is no longer any doubt as to their specific identities, although some authorities, including Rice (1977*a*), consider them to belong to a single genus, *Balaena*. Proposed divisions within the genus *Eubalaena* have been based more on assumed geographic isolation than on well-established morphological differences. Allen (1908), who considered the lack of genetic interchange as adequate grounds for separating species, recognized four "fairly well founded" species of *Eubalaena*—one in each of the four major ocean basins of temperate latitudes. Omura (1958), Klumov (1962), and Omura et al. (1969) found no appreciable differences in morphology between North Pacific and North Atlantic right whales. The incidence of callosities along the upper edge of the lower lips may be greater in Southern Hemisphere right whales than in those from the Northern Hemisphere (Best 1970). Muller (1954) found marked differences in the shape and arrangement of the alisphenoid between Northern and Southern Hemisphere right whales. Tomilin (1967) recognized a single species, *E. glacialis*, and proposed three subspecies: *E. g. glacialis* in the North Atlantic, *E. g. sieboldii* in the North Pacific, and *E. g. australis* in the Southern Hemisphere.

Considering the difficulty of preserving and handling the bulky carcasses of right whales, the shortage of available comparative material, and the fully protected status of these animals, taxonomic ambivalence is likely to persist for some time. We are of the opinion that, until a larger series of specimens is examined,

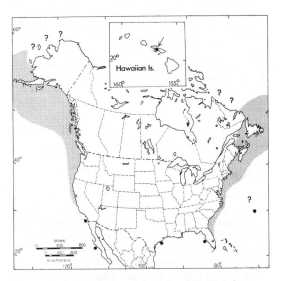

FIGURE 19.1. Distribution of the right whale (*Eubalaena glacialis*) in coastal North America. Dots indicate records outside "normal" range; question marks indicate uncertainty about occurrence in the area so marked. (Based on published accounts.)

Tomilin's view should prevail, and only the species *E. glacialis* should be recognized in the genus.

Other Baleen Whales. The suborder *Mysticeti* encompasses three extant families: Balaenidae, the right whales, including *Eubalaena* as well as *Balaena mysticetus*, the Greenland or bowhead whale, and *Caperea marginata*, the pygmy right whale; Balaenopteridae, the groove-throated, streamlined rorquals, including *Balaenoptera musculus*, the blue whale, *B. physalus*, the fin whale, *B. borealis*, the sei whale, *B. edeni*, the Bryde's whale, *B. acutorostrata*, the minke whale, and *Megaptera novaeangliae*, the humpback whale; and Eschrichtidae, including only *Eschrichtius robustus*, the gray whale. All of these except the pygmy right whale inhabit waters off North America. The gray whale of the North Pacific is the only Northern Hemisphere mysticete presently confined to a single ocean

415

basin, having been extirpated from the North Atlantic within historic times (Fraser 1970; Mitchell and Mead 1977).

ORIGINS

Although cetaceans clearly derived from land-dwelling mammalian (probably ungulate) ancestors, the published fossil record is inadequate for describing certain stages in their early evolution. There is some controversy, but Cetacea generally are considered monophyletic, although the identity of the common ancestor of mysticetes (baleen whales) and odontocetes (toothed whales) as well as the timing and nature of their divergence remain speculative (Van Valen 1968; Barnes and Mitchell 1978). Fully recognizable mysticetes first appeared in the Oligocene, between 37 million and 22 million years ago (Whitmore and Sanders 1976). The Balaenidae might have already become differentiated as a family before the Early Miocene in the Southern Hemisphere (Cabrera 1926; Kellogg 1928). The earliest apparent balaenids known from the North Pacific are Middle Miocene (Barnes 1976). The immediate ancestors of modern balaenids apparently evolved during the Pliocene, perhaps 4 to 5 million years ago (Kellogg 1928). They were present in both the temperate Atlantic (Van Beneden 1880) and the Pacific (Barnes 1976) at that time.

DISTRIBUTION

Historically, right whales inhabited cold- and warm-temperate marine waters (as defined by Briggs 1974) worldwide (figure 19.1 gives North American distribution). Now they are extirpated or severely depleted throughout their range, although local populations in the Southern Hemisphere (Best 1970, 1974; Cawthorn 1978; Mermoz 1980) and the western North Atlantic (Reeves et al. 1978) appear to have made a modest recovery in numbers. Townsend (1935), after examining 19th-century whaling logbooks, concluded that the region between approximately 30° N and 30° S is normally devoid of right whales, ensuring the separation of Northern Hemisphere and Southern Hemisphere populations. An earlier study by Maury (1851) had shown that while right whales in the Northern Hemisphere remain primarily north of 30° N latitude, those in the Southern Hemisphere penetrate the subtropics, occasionally reaching 10° S latitude. Recent evidence evaluated by Clarke (1965) indicates that right whales in both the Atlantic and Pacific wander inside Townsend's "vacant tropical belt," but probably not far enough or regularly enough to allow significant mixing of northern and southern populations. The possibilities for effective genetic interchange are further lessened by the fact that the breeding seasons above and below the equator appear to be out of phase.

Near the west coast of North America, the right whale's principal summering grounds are in the Gulf of Alaska and the southeastern Bering Sea. These areas constituted the "Kodiak Ground" of 19th-century whaling (Clark 1887). The region between 55-58° N

latitude and 140-152° W longitude was a prime hunting area. Nasu (1960) reported a sighting of two right whales in the Chukchi Sea in August, but few right whales reach as far north as St. Lawrence Island (Townsend 1935), and no details were provided by Nasu to eliminate the possibility that the whales seen were actually bowheads. The normal winter range was from Washington State south to northern California, with stragglers occasionally reaching southern California and northern Baja California (Rice and Fiscus 1968). The southernmost record is of a pair seen near Punta Abreojos, Baja California, at 26°39' N latitude. None had been reported reliably from Hawaii during the 20th century (Gilmore 1956) until spring of 1979, when one was observed there (Herman et al. 1980; Rowntree et al. 1980). Although at one time the right whale may have been fairly abundant off western North America (Scammon 1874), it is now very rare there, numbering "only a few individuals" (Rice 1974).

Right whales were at one time abundant along Kamchatka and in the Sea of Okhotsk (Clark 1887), and they are still found in these areas (Klumov 1962; Omura et al. 1969; Berzin and Kuz'min 1975). However, they apparently have almost disappeared from the Sea of Japan (Nishiwaki and Hasegawa 1969). According to Klumov (1962), the Asian and American stocks are discrete from each other.

Along the east coast of North America, right whales are fairly common during spring in Cape Cod waters (Watkins and Schevill 1976, 1980), as they were historically (Allen 1916). During summer they can be found in the New York Bight (Reeves 1975), throughout much of the Gulf of Maine (Reeves et al. 1978), in the lower Bay of Fundy (Arnold and Gaskin 1972), and near Nova Scotia and Newfoundland (Sergeant 1966; Mitchell 1974; Sutcliffe and Brodie 1977). The presence of right whales in winter from Cape Cod (Allen 1916; Watkins and Schevill 1976, 1980) south to the east coast of Florida (Moore 1953; Layne 1965) and into the Gulf of Mexico (Moore and Clark 1963; Schmidly et al. 1972) is well documented. Right whales are rarely encountered in spring near Bermuda (Payne and McVay 1971). Watkins and Schevill (1980) reported seeing as many as 70-100 different right whales on a single day near Cape Cod. These figures serve as a minimum estimate for the size of this small stock.

A presumably separate stock in the eastern North Atlantic, distributed seasonally from the Bay of Biscay to the Norwegian Sea and the coast of Iceland (Allen 1908; Thompson 1918), is almost extinct. Two captures at Madeira and a single sighting near Ireland serve as the only recent published evidence of its continued existence (Maul and Sergeant 1977).

DESCRIPTION

Like other baleen whales, the right whale possesses striking adaptations enabling it to live in a marine environment (Kellogg 1928). The two nostrils, called spiracles or blowholes, are situated on top of the head

FIGURE 19.2. A 41-ft (12.5-m) right whale (*Eubalaena glacialis*) at the Coal Harbour, British Columbia, whaling station, May 1951. Note the long baleen, fleshy tongue, serrated upper margin of the lower lip, and white patch on the side behind the flipper. The eye is clearly seen just above and slightly behind the corner of the mouth. The chunky appearance, absence of a dorsal fin, and broad flipper are all characteristic of the right whale. (See Pike and MacAski 1969.)

and well behind the snout. Forelimbs are paddlelike flippers, and hind limbs are reduced to small bones resting in the ventral musculature. Genitalia and mammary glands are concealed inside folds on the abdomen. The tail is well developed and muscular, ending in a pair of horizontally flattened flukes that, when thrust vertically, afford the principal means of locomotion. Highly specialized feeding structures called baleen plates, formed of keratin, are rooted in the gums of the upper jaw. These allow zooplankton—and schooling fish in the case of rorquals—to be sifted and trapped in the mouth.

External Appearance. Right whales are stout compared to the balaenopterids and the gray whale (figure 19.2). Their girth in front of the flippers is often more than half of body length (Thompson 1918; Omura et al. 1969). The mean ratio of maximum measured girth to total body length in 10 specimens was 0.73, with the highest ratio 0.89 in one large (17.1 m) male (Omura et al. 1969). A North Atlantic specimen, a 14-m male, had a girth (manner of measurement unspecified) equal to its length, looking "almost as round 'as a ball' when it lay on the beach" (Collett 1909). The right whale's fatness is attested to by the fact that it is the heaviest species of whale relative to length, with the possible exception of the closely related bowhead (Lockyer 1976). The head of the right whale occupies about one-fourth to one-third of total body length, the proportion increasing markedly with age (Allen 1908). The shape of the head is very distinctive, with a narrow rostrum, arched in side view, that widens anteriorly and is "pinched" or enfolded on either side by strongly bowed lower lips. The upper margin of the lower lips is usually scalloped. The eye, "approxi-

mately the size of an orange" (Payne 1976), rests near and just above the corner of the mouth. The ear is "a mere hole or pit externally, about large enough to admit the end of a parlor match" (Allen 1916).

The paired blowholes appear as two slits resting on a crest well behind the tip of the snout. They are oriented so as to form an incompletely closed V, diverging posteriorly, and their spout normally consists of two divergent columns of vapor.

The back is smooth, lacking the dorsal fin characteristic of balaenopterids and the irregular ridges of the gray whale. There are no grooves or pleats on the ventrum. The flippers, or forelimbs, are long (up to 1.7 m) and broadly splayed, in contrast to the narrow flippers of the balaenopterids. Right whale flukes are broad (up to 6 m wide) and distinctly V shaped along the smooth rear margin.

Size. As in other baleen whales, females are larger than males (Ralls 1976). The largest specimen measured in the North Atlantic was reported as 18 m long (Norman and Fraser 1937); in the North Pacific, an 18.3-m female has been reported (Klumov 1962). Right whales in the North Pacific seem to grow larger than those in the Southern Hemisphere and the North Atlantic (Omura 1958), although inadequate sampling or the lack of adherence to standard measurement techniques may account for some of the discrepancies in size.

A 17.4-m female examined by Klumov (1962) weighed 106.5 metric tons. The largest male examined by Omura et al. (1969) weighed 78.5 metric tons; it was 16.4 m long. They plotted body length against body weight and developed the formula $W = 0.01238(L)^{3.065}$, where W = weight in metric tons and

L = length in meters. Lockyer (1976) revised this formula to account for fluid loss before weighing. Her formula is $W = 0.0132(L)^{3.06}$.

Coloration. For ease of distinguishing it from the Greenland right, or bowhead, whale, which has a white chin, *Eubalaena* is frequently referred to as the *black* right whale. This is something of a misnomer, as many right whales are not completely black. Of 50 examined off the Hebrides in the early 20th century, 10 had white markings on the belly (Collett 1909). Omura et al. (1969) observed that only 2 of 13 taken in the North Pacific were entirely black. Eleven had at least a white abdominal patch; 4 of these had extensive white markings on the belly, sides, or throat. Klumov (1962) regarded white ventral patches as normal markings for right whales. Such white areas are well demarcated from their black surroundings, and they are not superficial, for the whiteness continues through the entire thickness of the skin (Brimley 1894, quoted in Allen 1908:321–322). Many right whales in the Southern Hemisphere also have irregular white markings on various parts of the body (Matthews 1938; Clarke 1965), and photos have been published of almost completely white individuals off Argentina (Payne 1974, 1976) and South Africa (Best 1970). White patches are present on fetuses and neonates, indicating that their character is determined genetically rather than by scarring (Omura et al. 1969). White markings should not be confused with gray areas (''mottling'') that are present on the black skin of most right whales, evidently caused by uneven sloughing of epidermis.

Many white linear scars are present on the bodies of some right whales. Collett (1909) suspected they result from the whale's rubbing along the bottom in its search for food; Omura (1958) attributed them to the teeth of killer whales. Donnelly (1969) and Payne (1976) suggested that the origin of scarring may be the scraping of one whale's rough-edged callosities against the sensitive skin of another.

Callosities. One of the most conspicuous and distinctive external features of the right whale is the series of raised, wartlike excrescences, usually called *callosities,* that adorn much of the head. Because the tissue of these excrescences is much the same color as the whale's skin, their white, orange, or pinkish hue is attributed primarily to the cyamid crustaceans that almost always live on them. There is generally a large oval or kidney-shaped callosity on the anterior half of the rostrum; this is called ''the bonnet.'' Other callosities, most of them much smaller than the bonnet, are variably arranged elsewhere on the rostrum. There is usually a large callosity over each eye, reminiscent of an eyebrow. Another large callosity is present on either side of the lower jaw, roughly on a line with the bonnet, and it is often followed by a string of smaller callosities. Some animals, particularly those in the Southern Hemisphere (Best 1970), have callosities along the scalloped upper edge of the lower lips.

There is considerable speculation regarding the origin and nature of callosities. Ridewood (1901) concluded that the structure of callosities is similar to that of normal skin except for a thickening of the outer cornified layers. He felt that ''the cornified layers fail to rub off at the normal rate, but remain and accumulate to produce a hard mass, projecting above the general surface of the epidermis as a kind of corn.''

The callosities are associated with clusters of hairs and heavy infestations of cyamids, or ''whale lice''. How, and indeed whether, either of these features is functionally involved in the creation or maintenance of callosities is unclear. The presence of callosities on fetuses and suckling calves convinced Matthews (1938) and Omura et al. (1969) that they are congenital. As for their function, Payne (1976) surmised that they could serve as splash deflectors; they also may serve as louse attractants, localizing irritation (Matthews 1938). Their horny texture might make them rather effective weapons, inflicting painful abrasions on the skin of competitors or predators. Payne and Dorsey (1979) demonstrated for a population of right whales in the Southern Hemisphere that males have more and larger callosities than females, and they interpreted the higher incidence of scrape marks on males as evidence that the callosities are used for intraspecific aggression.

Payne (1976) determined that: ''On every right whale the number, size, shape, and placement of callosities are unique, making it possible for us—and presumably the whales too—to tell individuals apart on sight.'' Callosity patterns, constituting as they do a natural ''mark,'' afford a promising research tool. Photographs of recognized individuals can document repeat encounters. Payne and Dorsey (1979) reportedly were able to study the movements and behavior of more than 500 individual right whales in Argentine waters over a period of eight years, using 20,000 photographs. A useful schematic presentation of callosity patterns found on 11 specimens from the North Pacific was given by Omura et al. (1969) (figure 19.3).

Hair. The baleen whales are almost devoid of hair, although most retain at least a few bristles somewhere on the head. Large clusters of short (0.5 to 1.0 cm), grayish hairs are usually found at the tip of the rostrum and on the chin of the right whale (Andrews 1908; Allen 1916; Matthews 1938), and smaller patches of hair are sometimes present near the blowholes (Omura et al. 1969). Allen (1916) pointed out that a single long bristle, greater than 2 cm, emerged from the prominent callosities on either side of the chin in one whale; Matthews (1938) noted that ''callosities occupy just those positions where hair groups would be found in other whales.'' Indeed, Omura et al. (1969) managed to find hairs in and near many of the callosities, particularly the bonnet. In fetuses they discovered hairs at each site where a callosity could be expected to develop.

Baleen. The right whale has longer baleen than any species other than the bowhead, which has baleen about twice as long as that of the right whale (figure 19.4). Most plates are dark gray to black, sometimes

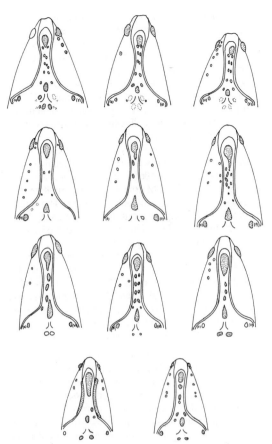

FIGURE 19.3. Dorsal view of the heads of 11 right whales (*Eubalaena glacialis*) from the North Pacific, showing variability and distinctiveness of callosity patterns. (After Omura et al. 1969, fig. 3.)

FIGURE 19.4. Baleen from two male right whales (*Eubalaena glacialis*) killed in the North Pacific. The photo shows the baleen in place, rooted in the upper jaw.

bluish tinged, with all-black fringes (Omura et al. 1969). Some of the anteriormost plates can be light colored (Andrews 1908; Collett 1909; Klumov 1962), and in living whales all the plates frequently are lighter underwater (see, for example, the photo in Payne 1974). Maximum length of a single plate is about 2.7 m, measured from the gumline along the lateral edge to the tip. Williamson (1973) suggested standard measurement techniques and terminology relating to baleen. The longest plates are situated near the middle of the mouth. Average number of plates is about 250 on either side (Allen 1916); Omura et al. (1969) found a mean of 229 per side in a series of 23 specimens from the North Pacific.

Compared to other mysticetes, right whales (and bowheads) have supple, finely fringed baleen (Nemoto 1970). At the front of the mouth their two rows of baleen are separated by an open space, whereas in the balaenopterids the two rows are connected by a series of hairlike structures (Watkins and Schevill 1976). Nemoto (1970) and Pivorunas (1976) described right whales as true filter feeders, or "skimmers," in contrast to the balaenopterids, which normally "swallow" rather than "skim" their prey. Right whales are less versatile feeders than most rorquals (see the section "Food Habits").

Skull and Skeleton. Telescoping—the progressively posterior migration of the nares to the vertex of the skull and/or the anterior displacement of the occipital shield—is expressed in the skulls of modern mysticetes, including the right whale (Miller 1923) (figure 19.5). The most conspicuous features of the right whale skull are its long, outwardly bowed mandibles and the almost equally long and strongly arched maxillae (see True 1904; Allen 1908). The shape of the bony shield of the supraoccipital is distinctive in the right whale. The brain of the right whale weighs from 2.4 to 3.1 kg, representing 0.0038 to 0.0058 percent of body weight (Omura et al. 1969); its small size and relatively low degree of encephalization were noted by Pilleri (1964), who compared the right whale's brain to those of other mysticetes.

The following features of the right whale skeleton distinguish it from those of balaenopterids (Allen 1916): the presence of five fingers in the hand instead of four, a greater number of ribs (14–15) that have a double articulation with the vertebrae, the complete or nearly complete fusion and anterioposterior compression of the cervical vertebrae (see figure 1 in Omura et al. 1971), the relatively large vestigial femur bones, and the absence of a distinct coronoid process of the mandible.

REPRODUCTION

Anatomy. The male sex organs of right whales follow the normal cetacean pattern, described in detail by Slijper (1966). The testes are permanently abdominal; there is no scrotum. Right whales appear to be exceptional among the mysticetes in the enormous size of their testes: Omura et al. (1969) reported one testis 201 cm long, 78 cm in diameter, and weighing 525 kg. In contrast, the largest pair of gray whale testes weighed by Rice and Wolman (1971) were 67.5 kg, and those of blue whales are "only" about 90 kg (Slijper 1966).

The darkly pigmented penis is long, slender, and retractile, reaching a length of 215–270 cm with a girth at the base of 90–110 cm (Omura et al. 1969) (figure 19.6). Turner (1913) mistakenly attributed an os penis to the right whale; cetaceans, unlike pinnipeds, have no bacula. The glans penis is nearly rectangular in cross-section, rather than round, as in the balaenopterids. In a 17-m specimen, the center of the genital opening was 1.6 m posterior to the umbilicus and 2.4 m anterior to the anus (Omura et al. 1969).

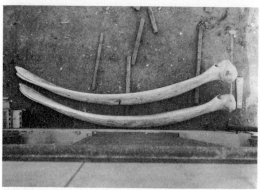

FIGURE 19.5. Skull of a right whale (*Eubalaena glacialis*). From top to bottom: lateral view of cranium, view of two left mandibles, view of occipital area.

The female's ovaries occur in about the same position as the male's testes. In mature female mysticetes, follicles in different stages of development protrude from various points along the surface of the ovaries, "resembling a bunch of grapes" (Slijper 1966). The largest right whale ovary examined by Omura et al. (1969) weighed 6.3 kg. A corpus luteum is formed at each ovulation; these eventually degenerate to permanent corpora albicantia, visible as scars on the surface of the ovary.

Like other cetaceans, right whales have a bicornuate uterus. In most mysticetes, the fetus can develop in either horn (Slijper 1966), and this is probably also true of right whales. Matthews (1938) found several corpora in both ovaries of one specimen. The female genitalia share a single ventral slit with the anus. At the front of this groove is a unilobular clitoris, projecting into the genital slit. A vaginal band (hymen), consisting of fibrous connective tissue that obstructs the vaginal orifice, was found in one immature specimen (Omura 1958) but has not been reported in others (Matthews 1938; Omura et al. 1969). The two mammae are housed in short slits on either side of the genital groove (figure 19.7). In most cetaceans these slits are vestigial in males and are found just anterior to the anus.

Reproductive Cycle. Most evidence suggests that mysticetes, and possibly many odontocetes, are spontaneous ovulators, with sexual activity conditioned primarily by day length (Slijper 1966). Matthews (1938) concluded that the female right whale's sexual season "consists of about three dioestrous cycles and four ovulations." This is not consistent with observations of most other baleen whales that seemingly are seasonally monestrous, with the peak of estrus occurring at the tropical end of their annual migration (Laws 1961; Slijper 1966). Rice and Wolman (1971) speculated that gray whales occasionally might ovulate twice, or even three times, in a single breeding season. They concluded: "Potential polyestry would be of considerable selective advantage in a species that can produce no more than one offspring every 2 years, that does not form permanent pair bonds, and that may be so widely dispersed that a male might not be available when the female first comes into estrus." This statement certainly could apply to right whales, even though their predictable seasonal concentration in bays in the Southern Hemisphere should help make receptive females easy to locate.

The existence of a male rut in right whales has not been proven empirically. One specimen displayed active spermatogenesis in late August (Omura et al. 1969). Male gray whales have a marked seasonal cycle, with a peak of spermatogenetic activity in late autumn and early winter (Rice and Wolman 1971). Conclusions about the sexual cycles of male and female right whales must await better documentation.

Breeding Season. Some courting activity apparently occurs in the Northern Hemisphere in summer (Collett 1909; Reeves personal observations in the Bay of Fundy in August). It is also observed regularly in winter and spring in the Southern Hemisphere (Donnelly 1969; Saayman and Tayler 1973; Cummings et al. 1974; Payne 1976). Most effective mating probably occurs in December and January in the North Pacific (Klumov 1962). Fetuses 1 to 1.5 m long have been found in June and July in the North Atlantic (Collett 1909). Fetuses 2.2 and 2.7 m long were reported in August in the North Pacific (Omura et al. 1969). Term fetuses and newborn right whales, which are 4.5–6 m long (Tomilin 1967), have been seen in the North Atlantic in winter (Manigault 1885; Allen 1916; Layne 1965; Caldwell and Caldwell 1971); and young calves, some apparently newborn, are present near Cape Cod between March and May (Watkins and Schevill 1980). Probable births have been observed off South Africa in late June and October, i.e., during the austral winter and early spring (Best 1970), and term fetuses have been removed from females killed off South Africa in July (Committee for Whaling Statistics 1938). It seems, then, that the peak period for both effective mating and calving is winter, or perhaps more accurately late autumn through early spring. This conclusion is consistent with the general pattern among mysticetes (Rice and Wolman 1971; Yablokov et al. 1974). The gestation period is probably about 11–12 months (Klumov 1962).

Birth. There is no record of multiple births of right whales, although the rare occurrence of multiple embryos in other mysticetes (Harrison 1969) suggests that right whales might occasionally bear more than one young. In this regard, a female accompanied by two young is reported to have entered the Bay of San Sebastian, northern Spain, in 1854 (Markham 1881), and Scammon (1874) noted that "twins have been observed." Caudal presentation is normal, judging by the few whale births that have been witnessed and by the position of the fetus inside the uterus in other mysticetes (Slijper 1966). Best (1970) described what was probably the birth of a right whale: the mother turned on her back and slapped the water surface with her flippers, shortly after which a small tail and peduncle were seen protruding in the area of the genital slit. The umbilical cord ruptures during birth.

Lactation. The duration of lactation is unknown. Most authors have estimated it as 6–7 months (Matthews 1938; Klumov 1962; Omura et al. 1969), although Best (1970) assumed it to be "a little less than 11–12 months." Weaning in cetaceans is usually gradual and prolonged (Rice and Wolman 1971). A known yearling was observed in close association with its mother off Cape Cod (Watkins and Schevill 1980). Klumov (1962, as interpreted in Klumov 1963), found a small quantity of copepods mixed with a large amount of milk in the stomach of a right whale killed in late July in the North Pacific. An 11.35-m individual killed in August appeared to be fully weaned.

One of Matthews's specimens was in anestrus, but a lactating animal had a developing follicle in one ovary. He guessed that a short period of anestrus may follow lactation. In some baleen whales, notably the humpback (Chittleborough 1958) and minke whales (Mitchell 1975c), simultaneous lactation and preg-

nancy is not uncommon. The frequency of this phenomenon, however, may be density dependent or age specific within a population. Though right whales may be physiologically capable of annual reproduction (Tomilin 1967 cites some evidence for it), a two-year cycle is more likely (Donnelly 1969). Payne (1976) and Watkins and Schevill (1980) have, in fact, resighted known individual female right whales off

FIGURE 19.6. Penis of a 17.1-m right whale (*Eubalaena glacialis*) taken in the North Pacific.

FIGURE 19.7. Genital groove and mammary slits of a 16.1-m right whale (*Eubalaena glacialis*) taken in the North Pacific.

TABLE 19.1. Food of right whales, based on stomach content analysis or plankton tows in the immediate vicinity of feeding whales

Location	Species	Source
South of Kodiak Island, Okhotsk Sea, coast of Japan	*Calanus plumchrus*	Klumov 1962, 1963; Omura et al. 1969
Okhotsk Sea	*Metridia* sp.	Omura et al. 1969
North of eastern Aleutian Islands	*Calanus cristatus*	Klumov 1962, 1963; Omura et al. 1969
Coast of Japan	*Euphausia pacifica* (larvae)	Omura 1958
"Cape Cod waters"	*Calanus finmarchicus*[a]	Watkins and Schevill 1976
"Cape Cod waters"	"Juvenile euphausiids"[a]	Watkins and Schevill 1976
Off Iceland and the Hebrides	*Thysanoessa inermis*	Collett 1909; Klumov 1962
North Atlantic	*Clione limacina*	Klumov 1962
North Atlantic	*Limacina helicina*	Klumov 1962
South Georgia Island, South Atlantic	*Euphausia superba*	Matthews 1938
Coast of Patagonia, Argentina	Grimothea (postlarvae) of *Munida gregaria*	Matthews 1932
Southern hemisphere	*Munida gregaria*	Klumov 1962, 1963
North Pacific	*Calanus pacificus* (unconfirmed)	Klumov 1962, 1963
North Pacific	*Calanus glacialis* (unconfirmed)	Klumov 1962, 1963

[a]Based on plankton tows

Argentina and Cape Cod, respectively, that calved one year and apparently waited two years before calving again. This suggests that in some cases there is an interbirth interval of three years. If this relatively wide interval is usual, it would help explain the slowness of the species' recovery from overexploitation.

Growth. On the basis of their small sample of North Pacific right whales, Omura et al. (1969) concluded that right whale calves nearly double their length during the period of lactation. This phenomenal growth rate can be achieved because the female's milk is rich in lipids (50 percent fat in some cetaceans) and high in protein (Harrison 1969). The growth rate slows markedly after the first year. Males probably become sexually mature at a body length of 14.5–15.5 m, and females at 15–16 m. These sizes might roughly correspond to an age of about 10 years (Omura et al. 1969). If right whales in the Northern Hemisphere are larger than their Southern Hemisphere counterparts (Omura 1958), then this is the reverse of the situation for other baleen whales.

FOOD HABITS

Right whales subsist almost exclusively on copepods and euphausiids (Klumov 1963) (table 19.1). Allen (1916) and Tomilin (1967) rejected the notion perpetuated by an early author that herring (*Clupea harengus*) were found in the stomach of a specimen killed near the Shetlands. No fish have been reliably reported in the diet of right whales.

Collett (1909) found only euphausiids about 1.25 cm long—probably *Boreophausia* (=*Thysanoëssa*)

inermis—in animals killed off Iceland and the Hebrides. Allen (1916) regarded *Th. inermis* and *Calanus finmarchicus* as staples in New England waters. In the North Pacific *C. plumchrus, C. finmarchicus,* and *C. cristatus* are consumed along with small quantities of *Euphausia pacifica* (Omura 1958) and *Parathemisto japonica* (Klumov 1962). Watkins and Schevill (1976) made plankton tows in the vicinity of feeding right whales near Cape Cod and verified the presence of large swarms of *C. finmarchicus* and juvenile euphausiids. Whalers often referred to the reddish zooplankton on which right and other baleen whales feed as "krill," and to the copepods that are the principal prey of right whales as "brit."

At times, right whales can be seen "skimming" along the surface with their rostrum exposed in air, mouth agape (Watkins and Schevill 1976) (figure 19.8). The triangular opening at the front of the mouth between the two diverging rows of baleen admits water, which escapes out the sides of the mouth after passing through the baleen. Particles of food accumulate against the bristles that line the inner edges of the baleen plates. The feeding whale pauses at intervals, closes its mouth, and sweeps its tongue across the baleen bristles to collect the adhering plankton (Payne 1974).

Feeding right whales follow an erratic path that, when examined from an aerial perspective, can be seen to follow "discrete slicks or patches of plankton" (Watkins and Schevill 1976). They appear to remain unerringly on a course through the densest concentration of plankton in a given area (Watkins and Schevill 1979). Although surface skimming is, for obvious rea-

FIGURE 19.8. This extraordinary view of a surface-feeding right whale (*Eubalaena glacialis*), near Cape Cod, Mass., shows clearly the arrangement of the long baleen plates. The animal is swimming from right to left in the photo, with its mouth open.

sons, the most frequently observed and reported mode of feeding, Watkins and Schevill (1976) regarded it as exceptional. More often the whales find their prey below, but within 10 m of, the surface. The same authors recorded sounds of right whales remaining at depths greater than 10 m, concluding that the whales swim "at particular depths on particular days." The preferred depths, which are probably feeding strata, sometimes coincide with thermoclines and may correspond to vertical migrations of zooplankton.

Some baleen whales essentially fast during the winter when they move to less productive low latitudes (Mackintosh 1965; Brodie 1975). However, right whales reportedly feed frequently in their winter calving and breeding grounds off Argentina (Payne 1974). Because they seem to be mainly confined to temperate latitudes throughout the year, right whales may avoid the extremes of "feast or famine" experienced by species whose annual migrations take them from the tropics to the polar regions and back again. However, Thompson (1918), based on a small but apparently representative sample of right whales killed off Scotland in the early 20th century, argued in favor of a feast-famine cycle. He found ratios of girth to length in June 1909 to average 45 percent; in July 1908, 75 percent in females and 82 percent in males. These data suggest that whales arriving early in the season were appreciably leaner than those caught in midsummer. If girth is a good measure of condition, and if the catch was not strongly biased in either sampling period, then a rather dramatic seasonal change in condition is indicated.

Nemoto (1970) made a series of interesting hypothetical calculations concerning the amount of food ingested by right whales. A right whale with its mouth wide open could have a filtering area of baleen of 13.5 m². Swimming through a plankton concentration of 4,000 mg/m at a speed of 6 km/hr, the whale

could harvest 324 kg of food per hour. Because plankton is not often that concentrated, and the effective filtering area of baleen may be significantly less than 13.5 m², the 324 kg/hr figure is probably much greater than normal.

Klumov (1963) estimated that a large right whale's stomach could accommodate 2–3 tons of food. The maximum volume recovered from one stomach was 150 l of *C. plumchrus*. Klumov also estimated that an average of 35–40 g of food per kg of the whale's live weight is required for a 24-hour period.

Sei whales have dietary preferences similar to those of right whales, and the two species may compete for the same resources (Klumov 1963; Nemoto 1970; Mitchell 1975b; Kawamura 1978). Watkins and Schevill (1979) observed right whales feeding in a plankton-fish complex at close quarters with sei, fin, and humpback whales. The lone sei whale devoted all its attention to the plankton, as did the right whales, while the fin whales and humpbacks worked only the fish school. According to Omura et al. (1969) right whales compete with fin whales for copepods in the southeastern Bering Sea and with sei whales for copepods in the Gulf of Alaska. Overhunting of right whales may have given an advantage to their balaenopterid competitors, which could help explain the failure of right whales to increase rapidly during their long period of protection (Mitchell 1975b; Kawamura 1978). Many marine birds and pelagic fishes also prey on copepods, and they also might be regarded as competitors of right whales.

BEHAVIOR

Migration. Scammon (1874) was puzzled by the apparent failure of right whales in the North Pacific to congregate in protected "nursery" grounds during winter, as do their Southern Hemisphere counterparts. The same thing can be said of right whales in the western North Atlantic, where a well-defined center of winter distribution has yet to be identified. Payne (1974) noted that Cape Cod Bay and Golfo San Jose are at roughly the same latitude, though in opposite hemispheres, and that they have many physical similarities. Cape Cod Bay seems, at least superficially, to be a nearly ideal winter nursery for right whales.

Reeves et al. (1978) reviewed evidence for a right whale migration along the east coast of North America. This evidence is equivocal, as seasonal changes in sighting conditions and variability of observer effort may influence the regularity of sightings (Watkins and Schevill 1980). Most reported occurrences from Cape Hatteras south have been in late winter or early spring. Although right whales are present in winter as far north as Cape Cod, they are not commonly seen there from late December to mid-February. As the whales wintering at lower latitudes move north toward richer feeding banks, an apparent peak of abundance occurs in the New York Bight and the Cape Cod area, usually sometime between mid-February and early May. Most of the northward-migrating population has probably cleared Cape Cod by early June, and right whales are uncommon off southern New England between mid-June and early October (Allen 1916). Most individuals spend the summer feeding at relatively high latitudes. By the second half of October the population begins to shift southward again, with numbers near Cape Cod building to a winter peak in December. A similar pattern of north-south shifts in distribution is evident off the coasts of Europe (Thompson 1918) and Japan (Omura et al. 1969).

In general, the right whale's migration does not appear as orderly as that of the gray whale or the humpback whale. Its strategy for meeting energy demands may involve a comparatively wide dispersal of individuals in seasons other than summer. Calving and mating activity in the Northern Hemisphere must take place either well offshore or in scattered, relatively isolated areas.

There is good evidence for a marked age and sex segregation within migrating populations of other baleen species (Laws 1961; Rice and Wolman 1971), and this may also apply to right whales. Layne (1965), for instance, hypothesized from very limited sighting data in eastern Florida that adult males and females unaccompanied by young begin their northward trek earlier in the spring than postparturient females, or that the latter might have a stronger tendency to move inshore with their young. Whales wintering off Peninsula Valdez, Argentina, appear to be separated into "herds" composed either of females with young or of consorting adults (Payne 1972). Off South Africa calving females arrive at inshore areas ahead of courting adults, and the two groups remain apart (Donnelly 1969). This kind of spacing is very much in evidence at the breeding grounds of gray whales in Baja California (Scammon 1874; Norris et al. 1977; Swartz and Jones, 1978). Records from European right whale fisheries in the early 20th century indicate a fairly even mix of males and females in the catches during summer months (Collett 1909; Thompson 1918). This is in striking contrast to the situation in South African bays, where females predominated in the catch (Eschricht and Reinhardt 1866; Best 1970).

Group Size. Right whales are only moderately gregarious. Scammon (1874) wrote that single animals were frequently encountered and that pairs or trios were not uncommon. By the end of the whaling season (apparently late summer or early autumn), however, they were "scattered over the surface of the water as far as the eye can discern from the mast-head." Large concentrations on the feeding grounds would, of course, be expected if the planktonic prey were patchily distributed. Thus, these associations may indicate little more than an adventitious type of "herding."

Recent observations in the North Atlantic (Watkins and Schevill 1976) are consistent with those of Scammon (1874) a century earlier in the North Pacific. Watkins and Schevill (1976) found single animals, pairs, and groups of three to eight to be most common, but they also referred to occasional "loose aggregations" of "up to thirty animals within a few square kilometers." Even in their wintering grounds at low latitudes, right whales off North America seem to be most often encountered alone, in pairs, or in very small groups (Moore and Clark 1963; Rice and Fiscus 1968).

Individuals are believed to remain in a given area for several days, or even weeks at a time (Arnold and Gaskin 1972; Watkins and Schevill 1976). This can lead to overestimation of population size. For instance, off Nova Scotia whalers reported regular sightings of right whales during certain periods. Mitchell (1975*b*) cautioned that these probably represented resightings of the same few individuals rather than a dramatic increase in total numbers.

Nurturant Behavior. Whalers sometimes took advantage of the female right whale's inclination to protect or rescue her calf by harpooning the calf first and counting on the mother to remain nearby (Allen 1916). Female right whales interpose themselves between their calves and any potential source of danger, including other whales as well as boats or divers (Payne and Payne 1971) and aircraft (Watkins and Schevill 1980). Similar behavior has been observed in other species of cetacean (Caldwell and Caldwell 1966).

Response to Humans. One characteristic of the right whale that made it especially attractive to early whalers was its approachability. Rice and Fiscus (1968) described an encounter off Baja California in which a pair of right whales, hard to approach at first, eventually allowed the vessel to come within a few meters of them. As the vessel came nearer, one of the whales turned on its side and exposed its lip and eye. The observers felt that the animal was examining them visually. A strikingly similar incident was described by Watkins and Schevill (1976), but this time the vessel was not under way and the whale was feeding:

> One of these surface feeding whales slowed and approached the side of our drifting 12-meter boat head-on, stopping within 2 m of the gunwale. Its mouth still open, the head was progressively raised until only the bottom of the lower jaw remained submerged with even the forward lip edge above water. The whale remained in this position for 8 to 10 sec with an eye visible just above the surface. Then it slowly closed its mouth and turned as it backed away and submerged.

What appears to be in-air visual reconnaissance by right whales has also been reported (Gilmore 1956; Donnelly 1969). Payne (1976) illustrated the manner in which right whales flex their backs in order to focus both eyes on objects in front of them.

Scammon (1874) reported that right whales in the North Pacific had become "very wild and difficult to get near to" after many years of intensive hunting pressure. We have observed that right whales sometimes react to low-level aircraft overflights by surfacing at irregular intervals and by remaining under the surface until the aircraft has passed overhead. However, careful approach has allowed Watkins and Schevill (1979) to make useful aerial observations of apparently undisturbed behavior.

The powerful tail and flukes are apparently the primary means of defense. Early open-boat whalers showed considerable respect for the right whale's ability to retaliate with flailing, thrashing flukes lifted high above the surface (Eschricht and Reinhardt 1866; Scammon 1874). Holder (1883) recounted a capture off South Carolina during which one stroke of the struggling whale's flukes unhinged a capture vessel's cabin door. During an apparent attack by five killer whales (*Orcinus orca*), two adult right whales slapped the surface with flukes and flippers while rolling and twisting in tight maneuvers (Cummings et al. 1971). After a 25-min struggle, the attack ended with the right whales apparently left unharmed.

Right whales are by nature placid and gentle, permitting extremely close approach by divers (Payne 1972). Divers may actually touch them, both above and below the surface. Once when a diver plunged into the water next to a whale, the animal violently swung its head from side to side, apparently making a threat gesture. However, no evidence of outright aggression was observed. While filming an interacting (courting) group of three right whales, Saayman and Tayler (1973) were approached closely by one individual, an action they interpreted as "an unmistakable challenge." It could as easily have been an expression of curiosity. Gilmore (1956) interpreted the breaching behavior of a right whale observed off southern California as a display meant to intimidate small boats nearby (for other interpretations of breaching, see below).

"Play," "Rest," and "Greeting." One of the first things noticed about the right whales in Golfo San Jose, Argentina, after their "discovery" in 1969, was their tendency to headstand for long periods, during which the tail and flukes extend well above the surface (Cummings et al. 1972). Although at first believed to indicate some form of bottom feeding, this behavior has been observed in water deeper than 25 m, where the whale's head would have to be well above the bottom (Cummings et al. 1974). After seeing whales headstand continuously with eyes closed for about 20 minutes, Payne (1972) concluded that it must be a rest posture. Later, he decided that some of the headstanding represented efforts by females to avoid the attentions of male suitors (Payne 1976). He also guessed that some headstanding right whales are actually sailing. They lift their tails so that the flukes are set at right angles to the wind and then drift as fast and as far as the wind will take them.

Breaching and slapping of the surface with flippers or flukes are common antics of right whales (Matthews 1938; Gilmore 1956; Clarke 1965; Payne 1976). Although they are frequently interpreted as play, Payne (1976) considered some of this breaching and slapping to be a form of communication. Noting that ambient noise in the water increases enough during storms to interfere with the whales' low-frequency communication signals, he suggested that the resounding smacks and smashes resulting from aerial display, which can be heard at considerable distances, might help to maintain herd integrity. Donnelly (1969) described a spectacular series of leaps by a courting pair, the female breaching 10 times in succession and the male 30.

Right whales are frequently seen in the company of other cetaceans, particularly bottlenose dolphins (*Tursiops truncatus*) in Florida (Layne 1965). A very interesting instance of interaction was reported by Rowntree et al. (1980), who saw a right whale "schooling" with a pod of three to four humpback whales in Hawaii. Also in Hawaii, but during a different sighting of what was probably the same right whale, a solitary humpback made "frequent contact with the right whale in ways suggesting courtship" (Baker et al. 1979; see also Herman et al. 1980). Payne (1974) described "playful" interactions between southern right whales, sea lions (*Otaria flavescens*), and dolphins. Although the whales' movements were more deliberate than those of the smaller animals, the whales sometimes appeared to take a vigorous and active role in sustaining episodes of play: "For example, we have seen a subadult whale breach away from porpoises in several successive leaps, but then swim slowly back toward them. Once in their midst again, it stopped to roll about in the water, while 'flippering' and displaying various other forms of underwater playfulness frequently observed between mothers and young." Payne also found the whales to be playful and curious toward floating seaweed, buoys, and tide gauges. Right whales off the U.S. east coast also have been seen "playing" with buoys (Reeves 1975).

Right whales have remarkably flexible bodies. Collett (1909) reported that they can nearly touch their head to their flukes, although the trauma of harpooning may be required to induce a whale to assume such an unlikely posture. Undisturbed right whales, however, sometimes arch their backs in what has been interpreted as a greeting ceremony. Payne and Payne (1971) watched several times as a whale basking at the surface was approached from the rear by a submerged whale. Each time, the basking whale flexed its spine so that its back became curved both laterally and vertically while the other whale passed by below.

Diving. Among the baleen whales, right whales are not thought to be capable of particularly deep or prolonged dives. Doi (1974) calculated mean duration of dives by right whales in the Antarctic as 3.7 minutes, shorter than that of any other large whale. Best (1970) watched one right whale surface 19 times, with an average dive duration of 55 seconds. The longest dive timed by Cummings et al. (1972) lasted for about 8 minutes. They observed one individual to swim 4 km without diving. Best claimed that females with calves remain at the surface for longer periods, dive less often, and remain submerged for shorter periods than adults unaccompanied by young.

Watkins and Schevill (1976) found that feeding right whales surface at intervals of 4 to 6 minutes, regardless of the depth at which their prey may be located. They remain at the surface for 30 seconds or more, during which they blow 2 to 6 times or more.

In spite of their reputation, right whales can submerge for prolonged periods. According to Scammon (1874), the usual pattern in the North Pacific is to surface for 7–9 blows and then to submerge for 12–15 min; Collett (1909) gave 10–20 min as the average interval between respiration sequences. Best (1970) described aerial observations suggesting that "occasionally individuals submerge for a long period and remain virtually motionless close to if not on the sea floor." Payne (1976) recorded a 25-min dive by a male waiting for a reluctant female to make herself available for copulation. Such behavior could have an important bearing on surveys, especially high-speed overflights, as a whale indulging in a prolonged dive would very likely be overlooked. Watkins and Schevill (1980), who have timed right whale dives well in excess of 20 minutes, noted that whales sometimes disappear suddenly and move off underwater. Occasionally they blow underwater, making no visible spout. Right whales, like humpbacks and sperm whales (*Physeter macrocephalus*), often lift their flukes as they begin a "terminal" dive, that is, a dive following a series of respirations.

Swimming. Right whales are slow swimmers; it is this characteristic that helped make them the "right" species to hunt from rowboats. Cummings et al. (1972) estimated maximum swimming speed to be about 8 knots (14.8 km/hr), noting that a speed of 10 knots (18.5 km/hr) was adequate for overtaking them. Feeding right whales maintain a swimming speed of about 3 knots (5.5 km/hr) (Watkins and Schevill 1976).

Mating. Copulation by right whales has been observed several times in the Southern Hemisphere, including at least one instance in which the process was recorded on film underwater (Cummings et al. 1974). Courting episodes last as long as five hours, with animals rolling and twisting at the surface, "exposing all parts of their bodies, including flippers, flukes, belly, back, and head" (Cummings et al. 1971). An instance of sexual activity between a single pair of right whales lasted *at least* from dawn to dusk (Donnelly 1969). Payne (1972) described mating behavior this way: "courtship involved much stroking and hugging of a female by the males competing for her. She would avoid her suitors by rolling onto her back and lying at the surface. But eventually she had to right herself to breathe, and the males would grab hasty breaths and dive beneath her, swimming upside down and pushing and shoving each other." The "normal" mating position is with the female upright at the surface and the male ventral side up, beneath her (Payne 1976). Collett (1909) reported an encounter off the Hebrides in which a female was seen lying on her back between two males, each of whose penis was extended. According to Payne (1972) they "hold each other with their flippers, belly to belly, while mating." Although most accounts at least imply that the male is the more active partner, Saayman and Tayler (1973) observed what they interpreted as interchangeability of roles during

right whale courtship. Cummings et al. (1974) were impressed by the absence of overt aggression among males seeking the favor of a single female.

Vocalization. Although right whales have an extensive and fairly complex repertoire of underwater sounds, they are not known to construct "songs" as the humpback whale does (Payne and Payne 1971). Furthermore, none of the sounds recorded from them is suggestive of echolocation (Watkins and Schevill 1976); the use of this system of orientation and navigation by mysticetes remains unproven.

The first record of right whale underwater sounds, a low-frequency "moan" heard off Martha's Vineyard, was published with a spectrograph and phonographic disc (Schevill and Watkins 1962). Such moans are heard sporadically from feeding right whales in the Cape Cod region and are believed to be communicative (Watkins and Schevill 1976). Their fundamental frequency is generally well below 400 Hz. The vocalizations, recorded in air, of a neonate stranded alive in Florida were pulsed and ranged in fundamental frequency from 100 to 750 Hz (Caldwell and Caldwell 1971).

The sounds of southern right whales assembled in the vicinity of Peninsula Valdez, Argentina, where they nurse their young, mate, and do some feeding, have been studied intensively. The principal frequency range is 50 Hz to 500 Hz, sometimes reaching 1.5 kHz (Payne and Payne 1971) and rarely exceeding 2.2 kHz (Cummings et al. 1974). Each sound's frequency range tends to be narrow, but the harmonic structure of a given sound can be complex. Cummings et al. (1974) reported that "bellowing" and "moans of rising pitch" were especially common among mating whales. Vocal activity increases at night (Payne and Payne 1971). Sound playback experiments have demonstrated that right whales can differentiate between conspecific sounds and other sounds (Clark and Clark 1980).

The mechanism by which many right whale sounds are produced is unknown. Cummings et al. (1974) correlated some of them with the emission of bubbles from the blowhole and determined that low-frequency pulse trains can be caused by subsurface exhalation. A low-level, pulsed "rattle," with frequencies between 2 and 4 kHz, was produced by right whales feeding near the surface with a portion of their baleen exposed (Watkins and Schevill 1976). "Baleen rattle" appears to be caused by "the physical movement of unloaded baleen plates as their free ends trip on a wavelet or are suddenly freed by a trough." This clicking sound can be heard both in air and over an underwater sound system, and it is considered to be wholly adventitious rather than purposeful.

MORTALITY

Conflicts with Human Activities. Right whales are caught occasionally in fish nets and traps, and such incidents may result in injury or death (Reeves et al. 1978). In at least one area (Tristan da Cunha in the

South Atlantic) they have incurred the hostility of fishermen because of damage to crawfish nets and the hazard they were seen to present to navigation (Elliott 1953). The right whale's coastal distribution brings it into contact with a large volume and variety of potential contaminants, but the effects of industrial pollution on fecundity and calf survival are not known. A right whale beached on Long Island, New York, showed evidence of having been struck by a large propeller, which severed its caudal peduncle (James G. Mead personal communication). Thus, ship collisions may contribute to right whale mortality.

Parasites. Like other cetaceans, right whales host a number of endoparasites, although their level of infestation appears to be relatively low (Klumov 1963). This may be related to their fish-free diet, because predatory fishes are frequently intermediate hosts for parasitic helminths. However, it may simply reflect limited sampling. Dailey and Brownell (1972), in a comprehensive checklist of parasites known to infest marine mammals, mentioned only three intestinal helminths for right whales: one cestode (*Priapocephalus grandis*) and two acanthocephalans (*Bolbosoma brevicolle* and *B. turbinella*). Yablokov et al. (1974) removed *B. brevicolle* from the list and added eight more species, including a cestode that occurs in blubber (*Phyllobothrium delphini*) and a nematode found in kidneys (*Crassicauda costata*). Of the 10 species attributed to right whales, 9 are also known to infest balaenopterids.

Right whales host an impressive array of epizoites. "Whale lice," amphipods of the genus *Cyamus*, are usually abundant on the callosities, in the genital region, and along the lips (Tomilin 1967). Wounds also often attract concentrations of whale lice. The life history and ecology of cyamids living on the callosities of right whales are promising avenues for further research. Allen's (1916) sketches and the photographs of Omura et al. (1969) and Payne (1972, 1976) illustrate these small crustaceans. Films of diatoms are occasionally found on the skin of right whales (Omura et al. 1969).

Cirriped barnacles are present on the callosities of some right whales in the Southern Hemisphere (Pilsbry 1916). Right whales appear to be specific hosts for *Tubicinella major* (Newman and Ross 1976). The "barnacles" on a callosity pictured in Payne (1976), however, are sharp outgrowths of skin protruding through the mass of cyamids; such irregularities on the surface of the callosities should not be mistaken for barnacles. As Gray (1866) noted from firsthand observation of *Tubicinella, in situ,* these barnacles sink into the whale's epidermis without altering to a noticeable degree the structure of the skin surrounding the attachment site. We have found no definite evidence that barnacles occur on right whales in the Northern Hemisphere, although Pilsbry (1916) mentioned an early account of *Tubicinella* found on a stranded whale in the Faeroe Islands, and Eschricht and Reinhardt (1866) presented some evidence to the effect that cirripeds occur on right whales in the North Atlantic.

Predation. Presumably the killer whale, *Orcinus orca,* is the only predator of right whales. Allen (1916) referred to unconfirmed accounts in which several killer whales would combine and "appear to be trying to bite the lips and tongue" of a right whale. Cummings et al. (1972) gave a firsthand description of an incident off Argentina in which a pod of killer whales apparently launched an unsuccessful attack on two right whales, and Omura (1958) published a photograph of scarring on a right whale's flank that he judged to have been inflicted by a killer whale. Evidence for successful predation on balaenids by killer whales is scarce, and most of what there is comes from observations made long ago when bowheads and right whales were more abundant (Mitchell and Reeves in press). Given the extensive documentation for killer whale attacks on other baleen species, many of which are faster swimmers than right whales (Tomilin 1967; Shevchenko 1975; Tarpy 1979), there is no reason to think right whales would be immune from predation.

AGE DETERMINATION

Among various techniques for age determination reviewed by Jonsgård (1969), three are especially promising. The most reliable method for balaenopterids involves the counting of growth layers—alternating light and dark laminae—in waxy plugs found in the external auditory meatus (Purves 1955). These plugs are present in right whales. However, they are soft—like "wet mud both before and after fixation in 10% solution of formalin"—and difficult to remove without distortion and breakage (Omura et al. 1969). Omura et al. (1969) found only one set of earplugs in their sample of 13 right whales to be "readable" (containing 12 dark striations). They concluded that the ear plugs of right whales are of limited value in assessing age, due both to their soft texture and to the difficulty of reading them.

A second method of age determination of whales is the counting of corpora albicantia in the ovaries of adult females. If the rate of ovulation is known, then these permanently visible scars will indicate the number of years elapsed since the attainment of puberty (Laws 1961). This method of age determination has not been applied on a significant scale to right whales.

The third method, introduced by Ruud (1940, 1945), relies on seasonal changes in the thickness of baleen plates. Such changes are reflected in transverse ridges that can be detected—and counted—on the surface of the plates. Omura et al. (1969) found these ridges, or steps, to be present on right whale baleen, and they counted a maximum of 12+ on plates removed from two sexually mature males. Taking into account neonatal growth, these individuals are believed to have been at least 16 years old. It is assumed that each ridge represents one year's growth. It must be remembered, however, that after a certain age, the ridges most distal to the gum probably begin to wear away.

Yablokov and Andreyeva (1965) examined transverse sections of baleen from several species, including the right whale. They discovered growth zones in the tubules that compose the medullar layer of the plates. Although these bands probably are laid down on an annual or otherwise regular basis, no information is available concerning the usefulness of applying this aging technique to right whales.

ECONOMICS

History of Whaling. Whaling by Europeans near the east coast of North America probably began in the 16th century, when Basques are known to have been hunting right whales and/or bowheads in the Gulf of St. Lawrence and Strait of Belle Isle (Eschricht and Reinhardt 1866; Barkham 1977, 1978). Unfortunately, "records of this industry are for the most part buried in obscurity, or have been destroyed, and such as are now known contain no descriptions of whales" (True 1904). The animals were attacked with hand-held harpoons and lances, thrust from aboard oar-driven boats, a technique pioneered by the enterprising Basques who had developed a commercial right whale fishery along the shores of Europe by the 12th century and possibly much earlier (Markham 1881; Allen 1908).

Statements of early New England colonists indicate that native peoples may have hunted right whales, at least casually, before the arrival of Europeans (see Waymouth 1605, in True 1904:20–21). Also, the abundance of commercially desirable whales, essentially right whales, sometimes may have influenced the colonists' decisions about where to settle (see de Vries 1631, in True 1904:24–26). Upon settling, the colonists quickly established a shore-based whaling industry centered on the abundant and approachable right whale (Allen 1916). By the end of the 17th century the right whale population along the east coast of North America had apparently diminished considerably. However, sporadic whaling continued throughout much of New England (Allen 1916), New York (Edwards and Rattray 1932), and the Carolinas (Holder 1883; Stick 1958) until well into the 20th century. Shore whalers in Newfoundland killed single right whales in 1937 and 1951 (Sergeant 1966), and a young individual, the last of its kind known to have been killed intentionally in U.S. waters, was taken by fishermen in Florida in 1935 (Moore 1953).

On the west coast of North America, aboriginal whalers probably took right whales at least occasionally (Rice and Fiscus 1968), and Mitchell (1979) suggested that Aleuts in particular killed large numbers of right whales with lances tipped with aconite poison. The first right whale killed by commercial whalers off northwestern North America was taken on the Kodiak Ground in 1835 (Scammon 1874). After this time, the species' exploitation in the North Pacific was intensive. Many American vessels engaged in sperm whaling spent the summer hunting right whales in the Gulf of Alaska and Bering Sea, then passed the winter off California or Mexico chasing sperm and gray whales

(Henderson 1972). Because right whales were considered the most desirable species from an economic point of view, it is reasonable to suppose that the whalers pursued any that were encountered by chance during winter. According to Scammon (1874), "before harpoon or bomb guns came into general use, the whalemen of the North-western Coast made such havoc among these colossal animals [right whales] . . . as to have nearly extirpated them." There is no record of right whales being taken—or even seen—by shore whalers in Washington State between 1911 and 1925 (Scheffer and Slipp 1948). A few were killed by modern shore-based whale catchers in the 1920s in British Columbia and Alaska (Kellogg 1931), and the last one known to have been killed by west coast whalers was caught near Vancouver Island in 1951 (Pike and MacAskie 1969). Commercial shore whalers of the western United States caught very few right whales. Few records exist for kills along the coast of California, the last one off the Farallon Islands in 1924 (Gilmore 1956). During the 1950s and 1960s Japan and the Soviet Union authorized the taking, for scientific purposes, of a small sample of right whales in the North Pacific (Omura 1958; Klumov 1962; Omura et al. 1969).

Products. Right whales were hunted primarily for oil and baleen (often referred to as whalebone or just "bone"). Whale oil was valued as an illuminant before the advent of petroleum, and it was used to a minor extent in the process of leather tanning (Clark 1887; Tressler 1923). Some was also used in the manufacture of candles, soap, and lard substitutes. The uses of whale oil and its economic importance were reviewed by Stevenson (1904) and Brandt (1940). The oil fields of Pennsylvania began production soon after the Civil War, and petroleum quickly "entered into active, relentless competition with whale oil," driving the price downward from a high of $1.45 per gallon in 1865 (Starbuck 1878:660).

Most of the oil was rendered from the right whale's thick layer of body blubber, more than 30 cm in fat individuals. Oil from the lower lips was also often saved (Brown 1887). Dudley (1726) wrote of extracting "near twenty [barrels of oil] out of the Tongue." Average oil yield is about 120–130 barrels (Scammon 1874; Allen 1916). References to large right whales yielding as much as 280 barrels (Scammon 1874) are suspect, as they may have been bowheads (e.g., Goode 1884:26). A barrel is equivalent to 31.5 American gallons (119.2 liters).

Right whale baleen was used in the early days of whaling to stiffen the clothing of fashionable women (Allen 1916). It is somewhat coarser than that obtained from bowheads, and hence was of slightly less value, frequently being made into whips and canes in addition to the well-known corset stays (Clark 1887). The price of baleen crept above one dollar per pound in 1863, just in time to offset the decline in oil prices that was by then imminent (Starbuck 1878). Baleen maintained its commercial importance until 1907, when the market for it deteriorated due to the emergence of spring steel as a much cheaper and readily available substitute (Bockstoce 1977). The quantity of baleen that can be obtained from a right whale is problematical. Tomilin (1967) gave 600 kg as a maximum. Omura et al. (1969) reported far greater weights, ranging as high as 940 to 1,015 kg in two North Pacific specimens. According to Dudley (1726), whose experience was limited to the North Atlantic, "A good large Whale has yielded a thousand weight of Bone." Collett (1909) claimed that no more than 250–330 kg of baleen could be expected from a full-grown right whale in the North Atlantic. Best (1970) found the mean baleen yield from right whales captured in the 19th century off South Africa to be in close agreement with Collett's figures, and suggested that the plates weighed by Omura et al. probably had not been cleaned of their gum before weighing.

The use of whales for human and animal food does not appear to have been of particular interest to North American commercial whalers. In Japan and Norway, however, both maritime nations with a whaling tradition, the meat of baleen whales has always been in demand (Foote 1975; anonymous 1978). As top-order predators that feed at a low trophic level, right whales are perhaps among the more efficient converters of protein in the sea. Their depleted status and low reproductive rate, however, make their future nutritional value of questionable importance.

Whale Watching. Interest in cetaceans and other marine mammals developed among some western nations during the 1970s to the extent that the nonconsumptive (nonlethal) commercial importance of certain species approached or exceeded their industrial value (anonymous 1978). The whale-watching business in southern California is now a well-established, multimillion dollar enterprise. Any proposed, sustained yield harvesting of gray whales might be commercially insignificant in comparison. Right whales, of course, are too scarce in the North Pacific to be of significance to tourism. However, in eastern North America, particularly in New England, they have become, together with humpback and fin whales, the backbone of a rapidly expanding and profitable branch of tourism.

MANAGEMENT

Right whales were protected from commercial hunting in 1935, when a League of Nations agreement called the Convention for the Regulation of Whaling took effect (Mackintosh 1965). This convention probably made little difference in populations of right whales, not only because they were already badly depleted but also because the Second World War quickly made such international agreements ineffectual (Scarff 1977). However, the convention did establish the principle of international regulation of the whaling industry, and it specifically prevented, in principle at least, the opportunistic capture of right whales by hunters at sea in pursuit of more abundant species.

After the war, a new International Convention for the Regulation of Whaling was signed in 1946. The International Whaling Commission (IWC) was empowered to implement the convention, and it remains the dominant, though often controversial, voice of international whale management (McVay 1974; McHugh 1974). The commission has always considered the right whale a species warranting protection from commercial whaling. It has no enforcement powers, but recently member governments have tended to accept and follow the commission's recommendations regarding kill quotas, size and sex limitations, seasonal or regional closures, and the importance of international observers on whaling vessels. Among the more significant advances in recent years is the IWC's recognition of the need for stock, rather than species, management. Most of the cosmopolitan large whale species appear to occur as discrete populations that require management by area or stock unit.

In North America several laws protecting right whales and other species have been instituted during the past decade. In 1970, most commercially attractive and endangered species—right, bowhead, gray, humpback, blue, fin, sei, and sperm—were placed, and remain, on the U.S. Department of the Interior's Endangered Species List. This action also made the importation of whale products illegal. In 1972 the U.S. Marine Mammal Protection Act was passed. It forbade not only the purposeful killing or molestation of all marine mammals, but it specifically made harassment illegal as well. The issue of whale harassment by well-intentioned but thoughtless tourists has become a major concern of the Marine Mammal Commission, an independent advisory agency created by the 1972 act to monitor its enforcement by the executive branch. In Canada there presently is no commercial hunting of whales. Mexican policy has been to resist strongly any overtures from parties interested in the resumption of whaling along the coast of Baja California.

An important aspect of whale management is the native or subsistence exemption that is part of virtually all relevant national and international legislation (Mitchell and Reeves 1980). Bowhead and gray whales are hunted by aborigines in northern Alaska (Maher 1960; Maher and Wilimovsky 1963; Wolman and Rice 1979; Marquette 1979); humpbacks, minkes, and occasionally fin whales are taken by West Greenlanders (Kapel 1979); and some humpbacks are killed by whalers in the West Indies (Adams 1975; Winn et al. 1975). In 1977 the IWC Scientific Committee recommended that the Eskimo hunt for bowheads be ended, as available evidence suggested that even the relatively low harvest (less than 100 animals struck per year) could be driving the species toward extinction (Allen 1978). Currently, the hunt continues, but under close scrutiny by the IWC, the U.S. National Marine Fisheries Service, and the Alaska Eskimo Whaling Commission, a consortium of Eskimo whaling captains. Unquestionably, the issue of Alaskan aboriginal whaling for the critically endangered bowhead is the most serious whale conservation and management problem in North America and the world today (Mitchell and Reeves 1980).

Protection from overhunting clearly is but one element of an effective baleen whale management policy. Several species, including the right whale, appear to be in immediate danger of extirpation from some areas. Their "critical habitat" needs to be identified and maintained such that it remains (or becomes) suitable to their energy and reproduction requirements. The provincial legislature of Chubut Province, Argentina, made Golfo San Jose "a permanent sanctuary for right whales" in 1974 (Payne 1976). Similar designations have been made for Scammon's Lagoon, a national gray whale sanctuary in Baja California (Brownell 1977), and certain parts of Maui County, Hawaii, where a substantial humpback whale population congregates in winter (Hudnall 1978). Consideration has been given to the creation of a National Marine Sanctuary for humpback whales in Hawaii (Hudnall 1977, 1978; Norris and Reeves 1978). Habitat protection for the large baleen whales is made difficult by ignorance of their ecological needs and by their tendency to range over great distances, which almost always involves crossing international boundaries.

CURRENT RESEARCH AND MANAGEMENT NEEDS

Continued total protection from all types of hunting will be required during the foreseeable future if the right whale is to effect a recovery to levels approaching those that existed before commercial whaling began. If stocks do recover, it may be necessary to weigh the esthetic and commercial importance of whale watching against the value of sustainable-yield harvesting. Considering tourist appeal, the tendency of right whales to approach shore and behave with animation near the surface puts them on a par with gray whales and humpback whales. Furthermore, there is presently no market for baleen, and the returns from the meat, oil, and bonemeal may not even cover the costs of harvesting and processing right whales. A persistent problem in whale management has been the continued illegal "opportunistic" capture of depleted species after a fishery has redirected its effort to the capture of more abundant species. The right whales killed recently at Madeira are an unfortunate example (Maul and Sergeant 1977). The existence of the whale fishery there hinges primarily on the hunt for sperm whales (Clarke 1954). The occasional right whale that comes within range apparently is pursued only as a diversion.

Feeding grounds can be affected by overfishing or pollution; death and debilitation of whales can result from accidental entrapment in fishing gear; noise and turbulence from offshore drilling and marine traffic may eventually influence behavior in a negative manner. Even overly zealous whale watchers could pose a threat, inadvertently disrupting vital activities by their close, noisy, and persistent approaches. Finally, competition with other baleen whales, in particular the sei whale, for food could inhibit the right whale's recovery. The dynamics among predators and prey in systems where multispecies overexploitation has occurred are extremely complex. The equilibrium levels of abundance that existed before exploitation may be impossible to reestablish (May et al. 1979).

Taxonomic studies as well as every conceivable attempt to improve our understanding of the biology and ecology of the right whale are desirable. Capture for scientific purposes, however, should be discouraged as long as populations appear to be in a depressed state. Specimens occasionally wash ashore, and it is critical that these be examined promptly and that parts be preserved whenever feasible.

Because it is unrealistic to expect objects buried in the muscle of protected animals to be recovered, conventional mark-recapture techniques involving implantation and recovery of numbered stainless steel darts are inadvisable for right whales. Also, there are special problems with the immobilization and capture of cetaceans, whose slightly negative buoyancy can cause them to sink when narcotized and whose breathing reflex is not as automatic as those of terrestrial mammals (Schevill et al. 1967). Schevill and Watkins (1966) attached radio transmitters to two right whales but were unable to track the animals. Since then, most development and experimentation with radio tagging has been applied to other cetacean species (Ray et al. 1978; Mate and Harvey 1979; Watkins 1979; Watkins et al. 1980). The technological advances resulting from such work (reviewed by Hobbs in press) will enhance future attempts to learn about right whale movements and behavior.

The cataloguing of photographs of callosity patterns is a possible nonlethal means of reidentifying individual animals over time (Whitehead and Payne 1976). Unfortunately, the most useful photographs for such cataloging are of the head in dorsal view, and these can best be acquired by aerial photography, an extremely expensive endeavor. Also, it has yet to be demonstrated that the patterns remain stable over long periods of time.

One of the most critical questions pertaining to right whales is simply, "How many are there?" The only Northern Hemisphere stock that appears to be large enough and sufficiently concentrated for censusing is that in the western North Atlantic. In our experience, attempts to count the whales from aircraft and ships deployed intensively in space and time have met with limited success. Extrapolations from direct counts are difficult to justify in light of ignorance about such things as daily activity patterns, the nature of whale distribution (random or clumped), and the time spent at the surface. Considerable basic behavioral research will be necessary before meaningful population estimates can be made.

Areas of seasonal concentration like the protected nursery in Golfo San Jose, Argentina, offer opportunities for protection of critical habitat, for tourism, and for much-needed scientific observation. No specific place in North America has been identified as a counterpart to Golfo San Jose, although waters near Cape Cod show some promise in this regard. Our own recent observations suggest that inshore waters bordering southeast New Brunswick may provide good opportunities for close and sustained observation of right whales in summer.

OTHER BALEEN WHALES

BLUE WHALE

The blue whale is the largest animal that ever lived. Populations occur along both coasts of North America, and have experienced periods of heavy commercial exploitation. After the development of steam-powered catcher boats and explosive harpoons in the late 19th century, the blue whale, because of its huge size and large oil yield, became the whale of choice within the industry (Mackintosh 1956). It is interesting to note that in 1946, when the International Whaling Commission was created, whaling quotas were given in blue whale units: 1 blue whale = 2 fin whales = 2.5 humpback whales = 6 sei whales. Inherent in this scheme, which remained in force until the 1971–72 season, was an incentive to harvest the largest whales available, and it insured that blue whales would continue to be taken whenever they were encountered.

Like other *Balaenoptera* spp., the blue whale is long and streamlined, in contrast to the chunkier balaenids and the humpback whale. Its distinctive features include: mottled blue gray coloration; a strikingly small dorsal fin located in the posterior third of the body; a flat, U-shaped rostrum; and a tall, dense blow. A yellowish film of diatoms, often present on the ventrum, prompted the whaler term *sulfurbottom* for the species. Maximum length in the eastern North Pacific is about 26 m (Pike and MacAskie 1969). Weights of blue whales caught in the Antarctic (which are larger than their Northern Hemisphere counterparts) averaged about 80 metric tons (Lockyer 1976).

The blue whale is, and apparently always has been, rather rare off the east coast of North America (Mitchell 1974). Less than 100 per year were killed in the Gulf of St. Lawrence during the second decade of this century; close to 250 per year were taken off Newfoundland and Labrador from 1903 to 1905. The total harvest in this region decreased to less than 50 per year through 1951; and 360 were taken in Davis Strait by Norwegian pelagic whalers between 1922 and 1934. These comparatively small catches had a severe effect on the population, which probably was never more than about 1,100 whales. Present estimates of the blue whale population in the western North Atlantic are in the low hundreds. Although a single blue whale has been reported from the coast of New Jersey (True 1904) and several unconfirmed records exist for the Texas coast (Schmidly and Melcher 1974), the majority of the population appears to remain on the Nova Scotian shelf and north, or well offshore. The small number visiting the inner Gulf of St. Lawrence in late summer and fall helps to support a small, local whale-watching industry (Reeves 1976).

The blue whale population in the eastern North Pacific is one of the few in the world that was not seriously damaged by overexploitation (Rice 1974). A total of 989 was caught in coastal waters of Baja California by Norwegian floating factories during the first third of this century, and California shore-based whalers took 48 between 1958 and 1965. Some were also taken at the northern end of their range, including

571 in British Columbia from 1905 to 1965 (Pike and MacAskie 1969) and 13 by Washington shore whalers from 1911 to 1922 (Scheffer and Slipp 1948). Although stock relationships are obscure, Rice (1974) postulated an extensive migration from winter grounds near Baja California, passing far offshore of central California, to summer grounds off Vancouver Island and throughout the Gulf of Alaska. Returning whales pass near the coast of central California in late September and October, with an influx in Baja California occurring in October. Rice (1974) could not account for the whales' whereabouts between November and January, but Wade and Friedrichsen (1979) reported blue whales from January to March off Costa Rica. They concluded that this might represent the population's center of winter distribution, with individuals reaching the coast of Baja California again in February–April. The several thousand blue whales remaining in the eastern North Pacific might become commercially valuable as tourist attractions (Rice 1974).

Pike and MacAskie (1969) asserted that blue whales in British Columbian waters remain clear of "marginal or inland seas." However, Rice (1974) found them off Baja California within 3 km of shore in water only 50–200 m deep. Blue whales are not noticeably gregarious. They are normally found alone or in small groups of up to three. Several such groups are sometimes found within a few kilometers of one another (Wade and Friedrichsen 1979).

Blue whales have short, coarse, black baleen, with which they entrap the euphausiids on which they feed almost exclusively (Nemoto 1970). Evidence from the Southern Hemisphere suggests that blue whales feed very little in winter. Rice (1974) believed, however, that they eat pelagic red crabs (*Pleuroncodes planipes*) near Baja California during this season. The high productivity of the Costa Rican Dome could be one of the reasons that blue whales come to this area in winter (Wade and Friedrichsen 1979).

Most blue whale calves are probably conceived and born at low latitudes in winter (Mackintosh 1965). They are thought to suckle for about six to seven months. Individual females probably give birth at two- or three-year intervals. Despite their bulk, blue whales are vulnerable to killer whale attacks (Tarpy 1979) and to ice entrapment and consequent starvation or suffocation (Mitchell 1977).

FIN WHALE

The fin whale is the most cosmopolitan and most abundant large baleen whale. Because it is second in size only to the blue whale, whalers in most areas began hunting it in earnest when the stocks of blues began to decrease from overhunting. The history of modern shore whaling for this species off Newfoundland and Labrador dates to the turn of the century (690 were taken in 1904 alone), with the last major episode occurring between 1966 and 1972, when an average of 341 per year was killed (Jonsgård 1977). Off Nova Scotia an average of 174 per year was caught between 1964 and 1972. Although commercial whaling in Canada was terminated after the 1972 season, a small catch quota of 90 per year is still allocated by the IWC for the stock formerly hunted off Newfoundland.

Fin whales are charcoal gray on the back, although their skin frequently has a dusky or greenish sheen. Some also have a very distinctive light gray or whitish chevron behind the neck and a light blaze on the right side of the neck. The head is asymmetrically pigmented, the right lower lip being white and the left gray. The dorsal fin is typically low in profile, set well behind the middle of the back, and strongly falcate, or recurved.

Mitchell (1974) conducted an extensive study of fin whales in the western North Atlantic that included sampling of landed whales and shipboard observations. He demonstrated that the major summer feeding grounds lay between shore and the 1,000-fathom curve, from 41° to 57° N latitude. A limited migration probably occurs in shelf waters from Cape Cod in June and July north to Labrador. Sergeant (1977), after reviewing available information on fin whales throughout the North Atlantic (see especially Jonsgård 1966), concluded that they exist as many discrete, local stocks that undertake relatively short seasonal movements in order to find food. They are present in winter on the Continental Shelf as far south as Cape Lookout, North Carolina, and the extent of northward movement by these animals is unknown. Fin whales penetrate far inside the Gulf of St. Lawrence in summer but are driven out by ice in winter. Sightings on the Nova Scotian shelf from December to May could be either of migrants from the gulf or of a resident stock. The best available estimate of stock size in the 1960s, based on Mitchell's summer cruise data, is about 7,200 for Newfoundland and Nova Scotia combined, with about 5,000 comprising the northeast Newfoundland component (Rørvik 1977).

Rice (1974) had difficulty determining the winter distribution of fin whales in the eastern North Pacific and concluded that many of them probably move far offshore. A peak of abundance occurs off central California in late May to early June, and there is another influx later in the summer. The southern limit appears to be the tip of Baja California (Cabo San Lucas); a possibly isolated population inhabits the central part of the Gulf of California. Shore-based whalers in California took about 70 fin whales per year from 1956 to 1970 (Rice 1974). Their Washington counterparts caught 602 between 1911 and 1925, mostly in spring and fall (Scheffer and Slipp 1948). Pike and MacAskie (1969) deemed fin whales the most abundant baleen species in British Columbian waters, where a few thousand were killed by coastal whalers between 1905 and 1967. Some younger animals appear to summer there, while adults seem only to pass through en route to and from northern feeding grounds.

Fin whales are fairly broad in their food preferences. Copepods, euphausiids, and schooling fishes, particularly capelin (*Mallotus villosus*) and herring, are important parts of their diet in southeastern Canada (Sergeant 1977). Krill (*Euphausia pacifica*) and anchovies (*Engraulis mordax*) are important prey off California (Rice 1963). Fish are particularly attractive in their spawning phases, when they presumably are easier to catch (Sergeant 1977). Competition for fishery resources between fin (and other) whales and humans is likely to be recognized in the future as a factor influencing decisions about whale management.

While feeding near the surface, fin whales sometimes turn onto one side, causing a pectoral fin and a fluke to extend into the air (Watkins and Schevill 1979). They are methodical in their movements, rarely breaching and almost never lifting their flukes clear of the water as they dive. Their breathing pattern, except when feeding, is highly regular, consisting of several blows at intervals of about 20 seconds followed by submergences of 5–10 minutes (Mackintosh 1965). Finbacks are relatively gregarious, often found traveling in groups of 7–10; 20 or more have been seen feeding together on a single dense school of fish near Cape Cod (Watkins and Schevill 1979).

The reproductive cycle of fin whales is fairly well understood in both Northern and Southern hemispheres (Rice 1963; Mackintosh 1965; Mitchell 1974, 1978). Sexual activity peaks in late autumn and winter, and calves are born after a gestation period of about 11 months. Weaning occurs 6–7 months after birth. The interbirth interval is usually two years. Because of uncertainties in age determination methodology, age at sexual maturity is not known. The difficulty of estimating age at sexual maturity is compounded by density-dependent effects in exploited populations. Both sexes become reproductively mature at a length of 17–18 m.

SEI WHALE

The common name of the sei whale was coined by Norwegian mariners who noticed long ago that it arrived on their coast with the *seje*, or coalfish (*Pollachius virens*) (Andrews 1916). Sei whales are medium-sized balaenopterids (with a maximum length of about 18 m and weight of 30 tons). They were hunted off Norway, among the British Isles, and near Iceland during the 19th and early 20th centuries, but were not reported reliably in catches off New England and Newfoundland until the first decade of the 20th century. Andrews (1916) discovered as recently as 1912 that a summer coastal fishery off Japan had been exploiting sei whales for some time. He remained unconvinced of their regular occurrence off western North America. However, British Columbian and Californian whalers, having reduced local populations of larger species, made sizable catches of sei whales in the 1960s, thus demonstrating that sei whales do occur, with some regularity, off western North America (Pike and MacAskie 1969; Rice 1974).

In many respects the sei whale appears to be a slightly smaller version of the fin whale. Indeed, the two are frequently confused. The best way to distinguish them is by paying close attention to the color of the right lower lip: it is white in fin whales, gray in sei whales. The head is symmetrically pigmented, as in all balaenopterids except the fin whale. The dorsal fin of the sei whale is generally higher and more erect than that of the blue whale or the fin whale, and is strongly falcate. The baleen of sei whales has finer fringes than that of any other species of *Balaenoptera*—about 0.1

mm in diameter at the base of the bristle, compared to about 0.3 mm or more for other species (Mead 1977). The sei whale and minke whale have shorter ventral grooves than other members of the genus. They end about midway between the flipper and the umbilicus rather than reaching the umbilicus.

The distribution of sei whales in North American waters is not well known. A population may be present in the Caribbean Sea and Gulf of Mexico during much of the year (Mead 1977). There is also limited evidence for a seasonal north-south migration between South Carolina and Massachusetts. Mitchell and Chapman (1977) tentatively identified two stocks: one found in the Labrador Sea early in June and migrating north along the coasts of Labrador, West Greenland, and possibly Iceland later in the season; the other remaining along the continental slope of the United States in winter and arriving at Georges Bank, Northeast Channel, and Browns Bank by mid- or late June. Some exploitation of the northern stock occurred from land stations in Labrador and northeast Newfoundland during the early 20th century. Between 1966 and 1972 several hundred whales, apparently from the southern stock, were killed in summer on the Nova Scotia shelf (Mitchell 1974). There are believed to be no more than several thousand sei whales in the western North Atlantic today (Gambell 1977).

Off western North America sei whales are known to range from the Revillagigedo Islands off Mexico at 18° 30' N latitude north to the Gulf of Alaska but rarely passing beyond the Aleutian Islands (Rice 1974). They are thought to winter well offshore, becoming common off central California only in late summer and early autumn. Tag recoveries have demonstrated that at least some individuals move seasonally from California to British Columbia. In California, 284 sei whales were caught between 1959 and 1970, and well over 2,000 were taken in British Columbia during the same period (Pike and MacAskie 1969). The existing exploitable population in the eastern North Pacific is estimated at 7,000–8,000 whales (Gambell 1977).

The sei whale is considered one of the most versatile feeders among the baleen whales, showing a strong preference for copepods and euphausiids but also feeding on schooling fishes, amphipods, and squids (Nemoto 1970; Nemoto and Kawamura 1977). It can alternate between "skimming" and "swallowing" in its pursuit of food. Such versatility may have allowed the sei whale to expand its range and population size as the numbers of right, blue, and fin whales were reduced by overhunting (Mitchell 1975*b;* Nemoto and Kawamura 1977; Kawamura 1978).

The calving season off California extends from September to March, although most calving is completed by the end of November (Rice 1977*b*). The female reproductive cycle includes a gestation period of approximately 12.7 months, 9 months of lactation, and 14.3 months before breeding again. This results in an average interbirth interval of three years. A two-year cycle is thought to be usual off Norway, with a well-defined peak of births in late autumn and winter (Jonsgård and Darling 1977). A definite calving peak occurs during winter in the Nova Scotian population (Mitchell 1974).

Various authorities have indicated that the mass occurrence of sei whales in different areas is episodic and unpredictable, causing substantial year-to-year fluctuations in the catches (Andrews 1916; Rice 1974; Jonsgård and Darling 1977). In the Antarctic at least, a high degree of segregation exists within the migrating sei whale population, as pregnant females reach high latitudes in advance of the others (Lockyer 1977). Pods of three to six individuals are not unusual in the North Pacific (Rice 1979; Brownell unpublished data).

Sei whales are more heavily infested with parasitic helminths than are other mysticetes (Rice 1974). Their skin is often badly scarred as a result of bites by ectoparasites (including the copepod *Penella* sp.), warm-water sharks, and lampreys (Andrews 1916; Shevchenko 1977). An unexplained disease that causes the shedding of baleen and its replacement by abnormal papillomalike growths was observed in 7 percent of the sei whales taken off central California (Rice 1974, 1977*b*).

BRYDE'S WHALE

Although it has been heavily exploited in the western Pacific by Japan and the Soviet Union during recent years (Ohsumi 1977*a*), the tropical and subtropical Bryde's whale has not been hunted to any significant degree off North America (Rice 1974). It is generally absent from water cooler than 18.0°C and so is excluded from most U.S. and Canadian coastal regions. Mead (1977) found several historical records of sei and fin whales in the southeastern United States and the Caribbean Sea to have been misidentified Bryde's whales. He guessed that Bryde's whales are resident in the Gulf of Mexico and the Caribbean Sea, with only rare strays reaching the Atlantic coast as far north as Chesapeake Bay.

In the Pacific, Rice (1977*b*) viewed the Bryde's whale as a permanent resident of coastal waters between about 26° 21' N latitude in western Baja California and the Trés Marías Islands at 21° N latitude. It is also known to range well into the Gulf of California (Rice 1974). A sighting off La Jolla is the only record of the Bryde's whale's occurrence off California (Morejohn and Rice 1973). Some Bryde's whales have been caught northwest of the Hawaiian archipelago (Ohsumi 1977*b*). The exploitable population of Bryde's whales in the North Pacific east of 150° E longitude is estimated at more than 9,000.

Whaling for Bryde's whales off North America has been limited to small catches off western Mexico by floating factories in the 1920s and 1930s (Rice 1977*b*). Substantial catches have been made in recent years by Japanese and Soviet whalers in the pelagic North Pacific as far north as 40° N latitude and including Hawaiian waters to some extent (Ohsumi 1977*a*). Historical whaling records generally do not discriminate between sei and Bryde's whales.

In appearance the Bryde's whale differs very little from the sei whale, and it resembles the fin whale except for its symmetrical head coloration. The best way to distinguish them at open sea is by noting the three prominent longitudinal ridges on the rostrum of the Bryde's whale; all other balaenopterids have a single median ridge, the two auxiliary ridges being either faint or nonexistent (Mead 1977). The Bryde's whale's dorsal fin is intermediate in relative size and shape between those of the fin whale and the sei whale, and its surfacing behavior is variable enough to make it difficult to distinguish from the other two species (Rice 1979). It is a solitary species, only occasionally found in pairs.

Bryde's whales are known to feed on red crabs off southern Baja California, and they probably eat anchovies (*Engraulis mordax*) in the same area (Rice 1977*b*). Whales caught in the pelagic whaling grounds of the western North Pacific feed mainly on euphausiids and fishes (Ohsumi 1977*a*).

MINKE WHALE

The minke, or little piked, whale is the smallest balaenopterid, with a maximum length of about 10 m. Consequently, it has been of little commercial significance. Some native tribes along the Washington and British Columbia coasts are known to have hunted minkes (Scheffer and Slipp 1948). However, the species was almost never taken by commercial whalers in Alaska, where minkes are very abundant, or British Columbia (Pike and MacAskie 1969). A small commercial fishery for minke whales was established at northeast Newfoundland in 1947, with the principal aim after 1954 of supplying mink (*Mustela vison*) farms with fresh meat (Sergeant 1963). The average catch by this fishery from 1955 to 1969 was 35 whales per year (Mitchell 1974). Japanese and Korean coastal fisheries capture minke whales for human consumption, as do local fisheries in West Greenland and Iceland and a pelagic commercial Norwegian fishery in the North Atlantic (Mitchell 1975*a*).

Minke whales are distributed across virtually the entire North American continental shelf, including major estuaries. They occur from Revillagigedo Islands off western Mexico to the Chukchi Sea in the Pacific (Scammon 1874; Scattergood 1949; Rice 1974) and from the Gulf of Mexico and Caribbean Sea to Ungava Bay and Hudson Strait in the Atlantic (Sergeant 1963; Mitchell 1974; Winn and Perkins 1976). Marked seasonal north-south migrations have been postulated, although there appear to be resident populations in some areas.

The prey of minke whales includes fish, especially capelin, sand lance (*Ammodytes* spp.), and anchovy, and planktonic euphausiids and copepods (Mitchell 1975*c*). They appear to follow spawning capelin into Newfoundland bays in late spring and early summer, then move offshore or northward as the fish relocate (Mitchell and Kozicki 1975). The potential for conflicts is considerable between an uncropped minke population, as well as humpback and fin whales, and a major commercial fishery for capelin in the northwest Atlantic (anonymous 1978; Perkins and Beamish 1979). In addition to such direct competition

with humans for marine resources, minkes are susceptible to accidental capture in salmon traps (Scattergood 1949), trawl netting (Mitchell 1975*c*), and cod traps and gillnets (Perkins and Beamish 1979).

All individuals in the Northern Hemisphere have a very conspicuous broad white band on each pectoral flipper. This feature is often the most reliable key to identification of the species (Mitchell 1975*c*). The body is black dorsally and white ventrally, with a light zone extending onto the sides anterior to the dorsal fin. The short baleen plates are cream white, although some are streaked with dark gray or black.

Minke whales are believed to be exceptional among baleen whales in that females frequently calve in successive years (Mitchell 1975*c*). Pairing in the western North Atlantic occurs between December and May, and gestation lasts 10–10.5 months. Neonates are about 2.4–2.8 m long and are probably weaned before 6 months of age.

Minke whales are generally solitary, and only infrequently found in pairs or small groups. Their tendency to approach vessels led Mitchell (1974) to caution against population estimates based on shipboard sighting data. The spout of minke whales is low and often not visible. They breach more often than their congeners.

The minke is the only baleen species, except the gray whale, to have been maintained successfully in captivity. One was kept alive for nearly three months in Japan's Mito Natural Aquarium (Kimura and Nemoto 1956).

HUMPBACK WHALE

The slow-moving humpback's habits of remaining close to coasts during migration and clustering near islands in subtropical regions during winter have made it vulnerable to overexploitation throughout its range (Mackintosh 1965). Along both coasts of North America the humpback has been a major target of commercial whaling (Kellogg 1931; Scheffer and Slipp 1948; Pike and MacAskie 1969; Rice 1974; Mitchell 1974). It was the mainstay of shore whaling in British Columbia, where 500–1,000 were caught annually before 1913; in Washington State, where 252 were killed in one season (1915); and off California, where between 100 and 200 were taken annually from 1956 to 1959. In addition to these shore-based catches, pelagic floating factories took large numbers around the eastern Aleutians and south of the Alaska Peninsula in the early 1960s. The International Whaling Commission gave protected status to humpbacks in the North Pacific in 1965. They have not been hunted commercially in the Northern Hemisphere since then.

Commercial hunting of humpbacks in the western North Atlantic was centered at Newfoundland and Labrador, where shore stations took as many as 287 in one year (1903). At Bermuda and the West Indies, 19th-century American sperm whalers and modern 20th-century Norwegian whalers killed humpbacks, at least opportunistically (Townsend 1935; Ingebrigtsen

1929). Catches in the north decreased dramatically after 1905, and humpbacks have been protected from commercial exploitation in the North Atlantic since 1955. Two small fisheries persist in the western North Atlantic, however. One is in West Greenland, where an average of 9 humpbacks per year was caught between 1973 and 1976 (Kapel 1979), and the other at the island of Bequia in the West Indies, where 0–6 whales, primarily females and calves, are landed each year (Adams 1975; Winn et al. 1975). In addition to these subsistence catches, a total of 41 humpbacks was taken by Canada for research between 1969 and 1971 (Mitchell 1973). An increasing number of humpbacks run afoul of fishing gear, and die as a result, in inshore areas of Newfoundland each summer (Perkins and Beamish 1979; Lien et al. 1979; Lien and Merdsoy 1979).

Although humpback populations have been studied off North American coasts, their status remains uncertain. The species is broadly distributed in cool temperate waters during summer, ranging from Point Conception north to the Chukchi Sea in the Pacific (Rice 1978), and from Cape Cod north to Labrador and Disko Bay in the Atlantic (Mitchell 1973). The winter grounds are in tropical or subtropical waters. It is here that the concentrated distribution of the animals facilitates censusing. In the Pacific, a small number winters along the coast of Baja California and mainland Mexico, including the Gulf of California and some offshore islands (Rice 1978). An apparently larger population reaches Hawaiian waters, where it has become a major tourist attraction (Hudnall 1977, 1978; Norris and Reeves 1978). Estimates of this stock's size range from 200–250 (Herman and Antinoja 1977) to 410–590 (Rice and Wolman 1978). In the Atlantic, humpbacks winter on shallow banks in the Caribbean Sea. There is some disagreement about stock relationships, but most population estimates in the western North Atlantic are between 800 and 1,500 (Sergeant 1966; Mitchell 1973; Winn et al. 1975; Balcomb and Nichols 1978).

Humpbacks grow to a length of about 16.5 m. They are stockier in build than species of *Balaenoptera* but less chunky that the balaenids. Wide grooves are present on the ventrum, extending posteriorly to the umbilicus. Among their most conspicuous external features are irregular rows of knobs, reminiscent of rivets, on the head; extremely long, flexible pectoral flippers; a dorsal fin that varies in shape from low and humped to erect and falcate, positioned far back on the body; and deeply notched flukes with a serrated rear margin. The dorsal surface is generally dark gray or black; the ventral surface is often at least partly white. Although the flippers of North Atlantic humpbacks are usually all white, a high percentage of North Pacific humpbacks have flippers that are black dorsally (Pike 1953).

Like gray whales, humpbacks are highly migratory. One of the more interesting approaches to determining their patterns of movement is the use of photographs that allow recognition of individuals. A catalogue of photographs of the undersides of the tail flukes, usually distinctively marked, has been com-piled for Atlantic humpbacks (Kraus and Katona 1977). This has been used to analyze individual movements (Katona et al. 1979; Balcomb and Nichols 1978). A similar effort is under way in the eastern North Pacific (Wolman 1978). Matched fluke and dorsal fin photos have demonstrated a relationship between the wintering population near the Trés Marías Islands off southern Mexico and the summering population in Frederick Sound, southeastern Alaska (Lawton et al. 1979).

The diet of humpbacks consists mainly of krill (euphausiids) and schooling fishes—mostly herring, capelin, and sand lances in the North Atlantic, and herring, anchovies, and sardines in the North Pacific (Mitchell 1973; Wolman 1978). Their feeding behavior, much of which occurs at the surface, is more spectacular and more frequently observed than that of any other baleen whale. They are sympatric and perhaps competitive with fin whales; we have seen humpbacks and fin whales feeding in close proximity to each other on numerous occasions. Watkins and Schevill (1979), after watching a humpback and several fin whales feeding on the same school of fish, wrote: "Feeding rushes by the humpback contrasted sharply with the smoother approaches of the finbacks, and the schools of fish were more scattered by a humpback rush than by a finback pass." Surely one of the most remarkable forms of whale behavior yet described is the humpback's "bubble net feeding." A Norwegian whaler (Ingebrigtsen 1929) postulated that a humpback would swim in a circle beneath a swarm of krill, emitting a wall of bubbles to herd them into a tight ball. The whale would then lunge vertically from under the mass of crustaceans, mouth open, and engulf its prey. Observations in recent years in southeastern Alaska (Wolman 1978; Jurasz and Jurasz 1979) and off Cape Cod (anonymous 1979) have confirmed the accuracy of this description. This method of feeding is but one of several displayed by the demonstrative, acrobatic humpback. Of all baleen species, the humpback is probably the most entertaining to human observers. The whale-watching industry in the Cape Cod area depends on the breaching, spyhopping (raising the head vertically above the surface), tail-lobbing, and ship-approaching antics of humpbacks to entertain its patrons. In Newfoundland, where a sharp decline in offshore capelin stocks is believed to have forced humpbacks inshore in search of food, the conflict between whales and cod fishermen caused by whale damage to fishing gear has soured the relationship between whales and people and resulted in calls for a resumption of whaling (Lien and Merdsoy 1979).

The sounds made by "singing" humpback whales at their southern wintering grounds have been well studied by scientists and are of great interest to the public. Their haunting underwater "melodies" have been sold as record albums and incorporated into the works of human musicians. Payne and McVay (1971) demonstrated that these sounds, which span a broad range of frequencies, are songs in a technical sense, similar to bird songs. Remarkably, humpbacks in different geographical areas have different songs, and these songs change from year to year (Winn and Winn 1978; Payne 1979). The singers, or callers, are believed to be solitary males, but this has yet to be proven. In addition to their complex assortment of underwater vocalizations, humpbacks have been known to make wheezing sounds in air as they blow, apparently purposefully (Watkins 1967).

Most humpback calves are born in winter after a gestation period of about a year. Lactation continues for approximately 11 months (Chittleborough 1958). Some females become pregnant soon after giving birth, thereby calving in successive years, although more often the calving interval is two years.

GRAY WHALE

The gray whale now exists only in the North Pacific, where two largely separate populations occur—one migrating between the Sea of Okhotsk and Korea (Andrews 1914), the other between the Bering and Chukchi seas and western Mexico (Rice and Wolman 1971). There is a long history of coastal whaling in Japan. Modern commercial pursuit of gray whales in Asia began around the turn of the 20th century, and at least 1,474 were killed off Korea between 1910 and 1933. The Korean stock is now very near extinction (Brownell and Chun 1977). The eastern, or California, stock was first subjected to heavy commercial exploitation beginning in 1846; whalers invaded the lagoons along the outer coast of Baja California where the species retires in winter (Scammon 1874). This hunting included both the American pelagic fleet and a series of shore stations in Mexico and California. Whalers hunting for the preferred right and bowhead whales on northern grounds in summer took grays only occasionally, and many spent the winter season pursuing gray whales at lower latitudes. The cumulative catch between 1846 and 1900 was estimated to be about 9,600 (Henderson 1972).

Modern whaling by pelagic floating factories was directed at gray whales for the first time in 1914, when a Norwegian vessel took 19 off Baja California (Rice and Wolman 1971). Subsequently, these whales were hunted by Norway from 1924–25 to 1928–29 outside the breeding lagoons; by the United States from 1927 to 1937 off California; and by Japan and the U.S.S.R. from 1933 to 1946 in the Bering and Chukchi seas. In 1947 the species became fully protected from commercial whaling. Since then, 326—316 in California (Rice and Wolman 1971) and 10 in British Columbia (Pike and MacAskie 1969)—have been killed for scientific purposes. Between 150 and 200 have been taken in most years by Soviet catcher boats on behalf of Siberian aborigines (Brownell 1977). A few are also killed by Alaskan Eskimos (Wolman and Rice 1979).

The California stock probably numbered 15,000–20,000 whales before the onset of commercial whaling (Henderson 1972; Ohsumi 1976). Rice and Wolman (1971) judged the present population to be stable and estimated it to be 9,000 to 13,000 whales. Counts made at Unimak Pass in Alaska during fall of 1977 and 1978, however, suggest that 15,099

($\pm$2,341) may be a better estimate (Rugh and Braham 1979).

Between late May and October the majority of the gray whale population is scattered across the shallow shelf waters of the Bering Sea (Rice and Wolman 1971). A significant number passes through the Bering Strait, with stragglers reaching as far east along the North Slope of Alaska as Barter Island and as far west as Wrangel Island, off the coast of Siberia. They infrequently penetrate broken pack ice. From October to January, the whales that have spent the summer in arctic waters begin one of the longest migrations of any baleen whale, a 18,000-km round trip. They move southward, funneling through Unimak Pass, and follow the coast to Baja California and mainland Mexico.

Rice and Wolman (1971) demonstrated that the migratory stream is led by late-term pregnant females; the peak of migration consists primarily of adult males and females; and immature animals lag behind the rest of the population. Some gray whales are present in British Columbian waters throughout the year, which suggests that the entire population may not complete the long-distance migration each year (Hatler and Darling 1974). The concentrated nature of the fall migration and its proximity to shore provide an unmatched spectacle for shore-based whale watchers.

There are five known calving sites currently used by gray whales: Ojo de Liebre Lagoon (Scammon Lagoon) and adjacent Guerrero Negro Lagoon (Black Warrior Lagoon); San Ignacio Lagoon; Magdalena Bay and adjacent waters; the open coast south of Yavaros, Sonora, Mexico; and Reforma Bay, Sinaloa, Mexico. Most on-water whale-watching activity takes place in San Ignacio Lagoon since Scammon Lagoon, formerly the center of such activity, was made a Mexican National Whale Refuge in 1974 (Brownell 1977). Whale watching at Scammon Lagoon is now limited to observation from shore.

Female gray whales generally grow to lengths of about 14 m, and males 13 m (Rice and Wolman 1971). They are about 4.9 m long at birth. Their bodies are mottled gray, although the background can vary from light gray to almost black. The skin hosts large numbers of barnacles (*Cryptolepas rhachianecti*), especially on the rostrum, flippers, and caudal peduncle, and three species of cyamid, or "whale lice." Gray whales are in some ways morphologically intermediate between balaenopterids and balaenids. The rostrum is arched and tapered anteriorly. There are two to four deep grooves about 2m long on the throat. Gray whales have no dorsal fin, but they do have a rounded hump followed by a knuckled, or saw-toothed, ridge along the spine.

The diet of gray whales consists primarily of benthic gammaridean amphipods, which are ingested in huge quantities on the whales' northern feeding grounds (Rice and Wolman 1971). Very little feeding is believed to take place during migration or at the winter grounds. However, feeding behavior occurs in the lagoons occasionally, so the gray whale's "fast" may be less strict than is commonly asserted (Norris et al. 1977). Kasuya and Rice (1970) speculated that asymmetric wear on the baleen plates and unequal colonization by barnacles on the right and left sides of the gray whale's jaw result from prey-bearing sediment being scooped and filtered mostly with the right anterior plates. The role of suction in feeding by gray whales, as well as the nature of their diet, was discussed by Ray and Schevill (1974).

This species is relatively gregarious, with migrating pods of up to 16 animals sometimes observed. Gray whales on their southward migration generally travel at speeds of 4–5 knots (7–9 km/hr). They surface every 3–5 min to breath 3–5 times. While heading north, they apparently travel more slowly (Pike 1962; Leatherwood 1974). Breaching and spyhopping are common forms of behavior during migration, and at the calving and mating grounds. The aggressive behavior of provoked gray whales is well documented by whalers' accounts (Scammon 1874), and more recently by the experiences of researchers who chanced to separate females from their calves (Norris et al. 1977). Their reputation as "devilfish" may be exaggerated, based as it is primarily on experience in the peculiar milieu of lagoon whaling and on attempts to take young calves from the sides of solicitous females. "Friendly" behavior toward people has become almost commonplace at San Ignacio Lagoon, where several whales regularly solicit attention from tourists in small vessels (Swartz and Cummings 1978; Swartz and Jones 1978). These whales allow themselves not only to be approached closely but to be scratched and rubbed by human hands. One gray whale was captured shortly after birth, maintained in a California oceanarium for one year, and returned, apparently successfully, to the wild (Evans 1974).

Although sexual activity has been observed in many parts of the gray whale's range, most conceptions occur during the southward migration in late fall and early winter (Rice and Wolman 1971). Gestation lasts about 13 months, so births generally take place at the southern extremity of the gray whale's range. Calves are weaned on the feeding grounds in the north after about 7 months.

BOWHEAD WHALE

The bowhead, the closest relative of the right whale, became the object of intensive commercial hunting during the early 17th century in the Spitsbergen area (Scoresby 1820). Despite its circumpolar distribution, the species was overlooked in arctic waters of North America until early in the 18th century, after the Spitsbergen stock had begun to show signs of serious depletion. Once the "Davis Straits fishery" began, it quickly grew to replace in importance the declining "Greenland fishery" (Lubbock 1937). By the late 19th century the stock of whales inhabiting the eastern Canadian arctic and western Greenland had been reduced to a very low level (Ross 1979). A short, final episode of commercial whaling in this region, undertaken principally by the Scots, accounted for at least 688 bowheads in Hudson Bay and Foxe Basin between 1860 and 1915 (Ross 1974). Since 1913, when the last Dundee whaling ship left Scotland, bowheads have been unexploited in the eastern arctic

except for sporadic attempts at capture by native hunters (Mansfield 1971; Mitchell and Reeves in press).

The last phase of arctic whaling began in 1848 when an American whaler, Captain Thomas Roys, worked his vessel through the Bering Strait and into the Chukchi Sea, where he found bowheads to be extremely abundant (Scammon 1874). The western arctic fishery flourished for over 60 years, as whalemen strove to satisfy the demand for whale oil to light homes and streets, and baleen to stiffen fashionable women's garments (Bockstoce 1977). When the market for baleen finally collapsed in 1907, the few bowheads that remained in the Bering, Chukchi, and Beaufort seas were left in relative peace.

Because of its strictly arctic distribution and the intense, protracted nature of its exploitation, the bowhead is probably the most critically endangered of the baleen whales. The Spitsbergen stock, which is believed to be genetically isolated from other bowhead populations, is very near extinction (Reeves 1980). The population(s) in Davis Strait/Baffin Bay and Hudson Bay/Foxe Basin probably now numbers no more than a few hundred (Allen 1978; Davis and Koski 1980). The third major stock, in the western arctic of North America, is certainly the largest today. However, its continued pursuit by Alaskan Eskimos is viewed as a threat to its survival (Allen 1978; Mitchell and Reeves 1980). This aboriginal fishery has been the focus of impassioned controversy in recent years, since the annual catch at Barrow, the leading whaling village, increased from an average of 5.8 between 1928 and 1960 (Maher and Wilimovsky 1963) to 11 between 1973 and 1976, with a struck-but-lost rate higher than 100 percent of the landed catch (Marquette 1977, 1979; Durham 1979). Beginning in 1977 the IWC Scientific Committee recommended a zero quota for Alaska's native whalers as the only safe course for ensuring stock, and perhaps species, survival (Allen 1978 and subsequent IWC Scientific Committee reports).

In appearance, bowheads closely resemble right whales. However, they lack the callosities on the head, and most individuals have a very distinctive cream white chin. Some also have a light gray to cream zone around the caudal peduncle, just anterior to the flukes. In addition, bowheads have a more prominent crest at the site of the blowholes, and their baleen is much longer than that of right whales. Although the maximum body length of 17 m is similar to the right whale's, a large bowhead yields considerably more oil and much more baleen than a right whale of comparable length.

The distribution of bowheads in North American waters is fairly well understood for all but the winter months. In the eastern arctic they range from the northeast corner of Labrador, along the east and north coasts of Baffin Island, into many of the sounds and channels adjacent to Baffin Bay, to northern Foxe Basin and Hudson Bay and throughout Hudson Strait (Mansfield 1971). There is some evidence for wintering in Hudson Bay and Foxe Basin, thus the possibility of an isolated, resident population there (Ross 1974). In the Pacific Northwest they are believed to winter along the pack ice front in the southwest Bering Sea, pressing north through Bering Strait and east through nearshore leads in the spring ice along the northern coast of Alaska (Braham et al. 1980). The main summer feeding grounds appear to be east of Barrow, including the vicinity of Banks and Herschell islands, Amundsen Gulf, and the Mackenzie River Delta (Fraker et al. 1978). In autumn the return migration takes the whales at least as far west as Wrangel Island, where most seem to veer southwestward, preceding the advancing ice back through the Bering Strait.

The spring migration of bowheads usually consists of solitary animals or small groups. In summer and fall the population appears to be more concentrated, with up to 50 whales occasionally being seen in close proximity. The spring migration past Point Hope and Point Barrow usually occurs in several "waves," the first comprised mainly of subadults and the last of adults (Marquette 1977; Braham et al. 1980). Swimming speed of migrating bowheads is about 2.6–3.2 knots (Braham et al. 1979). Bowheads can submerge for periods of at least 28 minutes, and they may use this capability to move long distances between openings in heavy ice (McVay 1973). There are reports of bowheads during fall migration breaking through newly forming ice to breathe. Bowheads are demonstrative at the surface during spring migration, breaching often and frequently slapping the surface with their flukes. An unbroken series of 39 breaches by one animal has been observed (Braham et al. 1979).

The feeding habits of bowheads have not been well studied, although it is generally assumed that they prey primarily on zooplankton. Johnson et al. (1966) found polychaete worms, gastropod mollusks, echinoidean echinoderms, and crustaceans in one specimen's stomach. Lowry et al. (1978) reported that two specimens caught near Barrow had 90 percent euphausiids, 7 percent gammarids, and 3 percent hyperiids, by volume, in their stomachs. A specimen caught at Barrow in spring had mainly copepods in its stomach (Marquette 1979). Bowheads clearly feed on benthic organisms at times, although their usual feeding behavior is probably similar to that of right whales.

Important though it is for enlightened management of the Eskimo hunt, information on bowhead life history is almost entirely lacking. Marquette (1977) reviewed the scattered and frequently inconsistent evidence and concluded, tentatively, that the western arctic herd mates and calves mainly in April and May, during spring migration. Gestation may last about 12 months, and weaning probably occurs some time in the first year of an individual's life. It is unlikely that females breed more than once every two years, although Marquette (1977) reported one instance of a pregnant female taken with a newborn calf. Length at birth is 3.0–4.3 m. Males probably reach sexual maturity at 11–12 m, females at slightly over 12 m.

The authors wish to thank Dr. Lawrence G. Barnes, of the Los Angeles County Museum of Natural History; Dr. Peter Best, of the Sea Fisheries Branch, South Africa; and Dr. James G. Mead, of the Smithsonian Institution, for critical reviews of all or portions of the manuscript. Dr. Hideo Omura, of the Whales Research Institute, Tokyo, and Dr. W. A. Watkins, of the Woods Hole (Mass.) Oceanographic Institution, generously provided photographs and shared information.

LITERATURE CITED

Adams, J. E. 1975. Primitive whaling in the West Indies. Sea Frontiers 21(5):303–313.

Allen, G. M. 1916. The whalebone whales of New England. Mem., Boston Soc. Nat. Hist. 8(2):107–322.

Allen, J. A. 1908. The North Atlantic right whale and its near allies. Bull. Am. Mus. Nat. Hist. 24:277–329.

Allen, K. R., chairman. 1978. Report of the Scientific Committee. Rep. Int. Whaling Comm. 28:38–92.

Andrews, R. C. 1908. Notes upon the external and internal anatomy of *Balaena glacialis* Bonn. Bull. Am. Mus. Nat. Hist. 24:171–182.

———. 1909. Further notes on *Eubalaena glacialis* (Bonn.). Bull. Am. Mus. Nat. Hist. 26:273–275.

———. 1914. Monographs of the Pacific Cetacea. Part 1: The California gray whale (*Rhachianectes glaucus* Cope). Mem. Am. Mus. Nat. Hist. (New Ser.) 1:227–287.

———. 1916. Monographs of the Pacific Cetacea. Part 2: The sei whale (*Balaenoptera borealis* Lesson). Mem. Am. Mus. Nat. Hist. (New Ser.) 1:289–388.

Anonymous. 1978. Mammals in the seas. Vol. 1. Rep. of FAO Advisory Com. on Marine Resour., Working Party on Marine Mammals. Food and Agric. Organization (United Nations) Fish. Ser. no. 5, Vol. 1. 264pp.

Anonymous. 1979. A whale of a survey. Maritimes (Univ. Rhode Island, Graduate School of Oceanography) 23(4):1–5.

Arnold, P. W., and Gaskin, D. E. 1972. Sight records of right whales (*Eubalaena glacialis*) and finback whales (*Balaenoptera physalus*) from the lower Bay of Fundy. J. Fish. Res. Board Can. 29:1477–1478.

Baker, S.; Forestell, P. H.; Antinoja, R. C.; and Herman, L. M. 1979. Interactions of the Hawaiian humpback whale, *Megaptera novaeangliae*, with the right whale, *Balaena glacialis*, and odontocete cetaceans. Page 2 *in* Abstr. 3rd Biennial Conf. Bio. Marine Mammals, Seattle, Wash. 64pp.

Balcomb, K. C., and Nichols, G. 1978. Western North Atlantic humpback whales. Rep. Int. Whaling Comm. 28:159–164.

Barkham, S. 1977. The first will and testament. Geogr. Mag. 49(9):574–581.

———. 1978. The Basques: filling a gap in our history between Jacques Cartier and Champlain. Can. Geogr. J. 96(1):8–19.

Barnes, L. 1976. Outline of eastern North Pacific fossil cetacean assemblages. Systematic Zool. 25:321–343.

Barnes, L., and Mitchell, E. D. 1978. Cetacea. Pages 582–602 *in* V. J. Maglio and H. B. S. Cooke, eds. Evolution of African mammals. Harvard Univ. Press, Cambridge, Mass. 641pp.

Berzin, A., and Kuz'min, A. A. 1975. Gray and right whales of the Sea of Okhotsk. Pages 30–32 *in* Material of 6th All-Union Meeting on Marine Mammals, Izdatelstuo Naukova Dumka, Kieve, October 1975.

Best, P. B. 1970. Exploitation and recovery of right whales *Eubalaena australis* off the Cape Province. Invest. Rep., Div. Sea Fish. South Africa 80:1–20.

———. 1974. Status of the whale populations off the west coast of South Africa, and current research. Pages

53–81. *in* W. E. Schevill, ed. The whale problem. Harvard Univ. Press, Cambridge, Mass. 419pp.

Bockstoce, J. R. 1977. Steam whaling in the western Arctic. Old Dartmouth Hist. Soc., New Bedford, Mass. 127pp.

Braham, H.; Fraker, M.; and Krogman, B. 1980. Spring migration of the western arctic population of bowhead whales. Marine Fish. Rev. 42:36–46.

Braham, H.; Krogman, B.; Leatherwood, S.; Marquette, W.; Rugh, D.; Tillman, M.; Johnson, J.; and Carroll, G. 1979. Preliminary report of the 1978 spring bowhead whale research program results. Rep. Int. Whaling Comm. 29:291–306.

Brandt, K. 1940. Whale oil: an economic analysis. Food Res. Inst., Stanford Univ., Calif. 264pp.

Briggs, J. C. 1974. Marine zoogeography. McGraw-Hill, New York. 475pp.

Brodie, P. F. 1975. Cetacean energetics, an overview of intraspecific size variation. Ecology 56:152–161.

Brown, J. T. 1887. The whalemen, vessels and boats, apparatus, and methods of the whale fishery. Pages 218–293 *in* G. B. Goode, ed. The fisheries and fishery industries of the United States. Sec. 5, Vol. 2, Part 15.

Brownell, R. L., Jr. 1977. Current status of the gray whale. Rep. Int. Whaling Comm. 27:209–211.

Brownell, R. L., Jr., and Chun, C. 1977. Probable existence of the Korean stock of the gray whale (*Eschrichtius robustus*). J. Mammal. 58:237–239.

Cabrera, A. 1926. Cetaceos fosiles del Muséo de la Plata. Rev. de Mus. de la Plata 29:363–411.

Caldwell, D. K., and Caldwell, M. C. 1971. Sounds produced by two rare cetaceans stranded in Florida. Cetology 4:1–6.

Caldwell, M. C., and Caldwell, D. K. 1966. Epimeletic (care-giving) behavior in Cetacea. Pages 755–789 *in* K. S. Norris, ed. Whales, dolphins, and porpoises. Univ. California Press, Berkeley, Calif. 789pp.

Cawthorn, M. W. 1978. Whale research in New Zealand. Rep. Int. Whaling Comm. 28:109–114.

Chittleborough, R. G. 1958. The breeding cycle of the female humpback whale, *Megaptera nodosa* (Bonnaterre). Aust. J. Marine and Freshwater Res. 9:1–18.

Clark, A. H. 1887. History and present condition of the (whale) fishery. Pages 3–218 *in* G. B. Goode, ed. The fisheries and fishery industries of the United States. Sec. 5, Vol. 2, Part 15.

Clark, C. W., and Clark, J. M. 1980. Sound playback experiments with southern right whales (*Eubalaena australis*). Science 207:663–665.

Clarke, R. 1954. Open boat whaling in the Azores. Discovery Rep. 26:281–354.

———. 1965. Southern right whales on the coast of Chile. Norsk Hvalfangst-tidende 54:121–128.

Collett, R. 1909. A few notes on the whale *Balaena glacialis* and its capture in recent years in the North Atlantic by Norwegian whalers. Proc. Zool. Soc. London 1909:91–98.

Committee for Whaling Statistics, eds. 1938. International whaling statistics. Vol. 11. Oslo, Norway. 36pp.

Cummings, W. C.; Fish, J. F.; and Thompson, P. O. 1972. Sound production and other behavior of southern right whales, *Eubalaena glacialis*. Trans. San Diego Soc. Nat. Hist. 17:1–14.

Cummings, W. C.; Fish, J. F.; Thompson, P. O.; and Jehl, J. R., Jr. 1971. Bioacoustics of marine mammals off Argentina: R/V *Hero* cruise 71-3. Antarctic J. United States 6:266–268.

Cummings, W. C.; Thompson, P. O.; and Fish, J. F. 1974. Behavior of southern right whales: R/V *Hero* cruise 72-3. Antarctic J. United States 9:33–37.

Dailey, M. D., and Brownell, R. L., Jr. 1972. A checklist of marine mammal parasites. Pages 528–589 *in* S. H. Ridgway, ed. Mammals of the sea: biology and medicine. Chas. C. Thomas, Springfield, Ill. 812pp.

Davis, R. A., and Koski, W. R. 1980. Recent observations of the bowhead whale in the Eastern Canadian High Arctic. Rep. Int. Whaling Comm. 30:439–444.

Doi, T. 1974. Further development of whale sighting theory. Pages 359–368 *in* W. E. Schevill, ed. The whale problem. Harvard Univ. Press, Cambridge, Mass. 419pp.

Donnelly, B. G. 1969. Further observations on the southern right whale, *Eubalaena australis,* in South African waters. J. Reprod. Fert., Suppl. 6:347–352.

Dudley, P. 1726. An essay upon the natural history of whales, with a particular account of the ambergris found in the Sperma Ceti whale. Philos. Trans. London 33:256–269.

Durham, F. E. 1979. The catch of bowhead whales (*Balaena mysticetus*) by Eskimos, with emphasis on the Western Arctic. Contrib. Sci. Nat. Hist. Mus. Los Angeles County 314:1–14.

Edwards, E. J., and Rattray, J. E. 1932. "Whale off": the story of American shore whaling. Coward McCann, New York. 285pp.

Elliott, H. F. I. 1953. The fauna of Tristan da Cunha. Oryx 2(1):41–53.

Eschricht, D. F., and Reinhardt, J. 1866. On the Greenland right-whale (*Balaena mysticetus,* Linn.). Pages 1–150 *in* W. H. Flower, ed. Recent memoirs on the Cetacea. Ray Soc., London. 312pp.

Evans, W. E., ed. 1974. The California gray whale. Pap. presented at California Gray Whale Workshop, Univ. California, San Diego, Scripps Inst. Oceanogr., 21–22 August 1972. Marine Fish. Rev. 36(4):1–64.

Foote, D. C. 1975. Investigations of small whale hunting in northern Norway, 1964. J. Fish. Res. Board Can. 32:1163–1189.

Fraker, M. A.; Sergeant, D. E.; and Hoek, W. 1978. Bowhead and white whales in the southern Beaufort Sea. Beaufort Sea Proj. Dept. Fish. and Environ. (Can.), Tech. Rep. 4:1–114.

Fraser, F. C. 1970. An early 17th century record of the Californian gray whale in Icelandic waters. Invest. on Cetacea 2:13–20.

Gambell, R., ed. 1977. Report of the special meeting of the Scientific Committee on Sei and Bryde's Whales, La Jolla, Calif., 3–13 Dec. 1974. Rep. Int. Whaling Comm., Special Issue 1:1–9.

Gilmore, R. M. 1956. Rare right whale visits California. Pacific Discovery 9(4):20–25.

Goode, G. B. 1884. The fisheries and fishery industries of the United States. Section 1: Natural history of useful aquatic animals. U.S. Comm. Fish and Fisheries. Gov. Printing Office, Washington, D.C. 895pp.

Gray, J. E. 1866. Catalogue of seals and whales in the British Museum. London. 402pp.

Harrison, R. J. 1969. Reproduction and reproductive organs. Pages 253–348 *in* H. T. Andersen, ed. The biology of marine mammals. Academic Press, New York and London. 511pp.

Hatler, D. F., and Darling, J. D. 1974. Recent observations of the gray whale in British Columbia. Can. Field Nat. 88:449–459.

Henderson, D. A. 1972. Men and whales at Scammons Lagoon. Dawson's Book Shop, Los Angeles. 313pp.

Herman, L. M., and Antinoja, R. C. 1977. Humpback whales in the Hawaiian breeding waters: population and pod characteristics. Sci. Rep., Whales Res. Inst. (Japan) 29:59–85.

Herman, L. M.; Baker, C. S.; Forestell, P. H.; and Antinoja, R. C. 1980. Right whale *Balaena glacialis* sightings near Hawaii: a clue to the wintering grounds? Mar. Ecol. Prog. Ser. 2:271–275.

Hobbs, L. J. In press. Tags on whales, dolphins and porpoises. Appendix A *in* S. Leatherwood, R. R. Reeves, W. F. Perrin, and W. E. Evans, eds. Whales, dolphins and porpoises of eastern North Pacific and adjacent Arctic waters. NMFS Tech. Rep.

Holder, J. B. 1883. The Atlantic right whales. Bull. Am. Mus. Nat. Hist. 1:99–137.

Hudnall, J. 1977. In the company of great whales. Audubon 79(3):62–73.

——. 1978. A report on the general behavior of humpback whales near Hawaii and the need for creation of a whale park. Oceans 11(2):8–15.

Ingebrigtsen, A. 1929. Whales caught in the North Atlantic and other seas. Cons. int. Expl. Mer, Rapp. Proc.-verb. Reun. 56:1–26.

Johnson, M. L.; Fiscus, C. H.; Ostenson, B. T.; and Barbour, M. L. 1966. Marine mammals. Pages 877–924 *in* N. J. Wilimovsky, ed. Environment of the Cape Thompson region, Alaska. U.S. Atomic Energy Comm. 1,250pp.

Jonsgård, A. 1966. Biology of the North Atlantic fin whale *Balaenoptera physalus* (L.): taxonomy, distribution, migration and food. Hvalradets Skrifter 49:1–62.

——. 1969. Age determination of marine mammals. Pages 1–30 *in* H. T. Andersen, ed. The biology of marine mammals. Academic Press, New York and London. 511pp.

——. 1977. Tables showing the catch of small whales (including minke whales) caught by Norwegians in the period 1938–75, and large whales caught in different North Atlantic waters in the period 1968–1975. Rep. Int. Whaling Comm. 27:413–426.

Jonsgård, A., and Darling, K. 1977. On the biology of the eastern North Atlantic sei whale, *Balaenoptera borealis* Lesson. Rep. Int. Whaling Comm., Special Issue 1:124–129.

Jurasz, C. M., and Jurasz, V. P. 1979. Feeding modes of the humpback whale, *Megaptera novaeangliae*, in southeast Alaska. Sci. Rep., Whales Res. Inst. (Japan) 31:69–83.

Kapel, F. O. 1979. Exploitation of large whales in West Greenland in the twentieth century. Rep. Int. Whaling Comm. 29:197–214.

Kasuya, T., and Rice, D. W. 1970. Notes on baleen plates and on arrangement of parasitic barnacles of gray whales. Sci. Rep., Whales Res. Inst. (Japan) 22:39–43.

Katona, S.; Baxter, B.; Brazier, O.; Kraus, S.; Perkins, J.; and Whitehead, H. 1979. Identification of humpback whales by fluke photographs. Pages 33–44 *in* H. E. Winn and B. L. Olla, eds. Behavior of marine animals. Vol. 3: Cetaceans. Plenum Press, New York and London. 438pp.

Kawamura, A. 1978. An interim consideration on a possible interspecific relation in southern baleen whales from the viewpoint of their food habits. Rep. Int. Whaling comm. 28:411–419.

Kellogg, R. 1928. The history of whales: their adaptation to life in the water. Q. Rev. Bio. 3:29–76, 174–208.

——. 1931. Whaling statistics for the Pacific coast of North America. J. Mammal. 12:73–77.

Kimura, S., and Nemoto, T. 1956. Note on a minke whale kept alive in aquarium. Sci. Rep., Whales Res. Inst. (Japan) 11:181–189.

Klumov, S. K. 1962. Gladkie (Yaponskie) kity Tikhogo Okeana. [The right whales in the Pacific Ocean.] Trudy Inst. Oceanol. 58:202–297. (Russian with English summary.)

——. 1963. Pitanie i gelmintofauna usatykh kitov (Mystacoceti) v osnovnykh promyslovykh raionakh mirovogo okeana. [Food and helminth fauna of whalebone whales (Mystacoceti) in the main whaling regions of the world ocean.] Trudy Instituta Okeanologii 71:94–194. (Fish. Res. Board Can. Trans. Ser. no. 589, 1965.)

Kraus, S., and Katona, S., eds. 1977. Humpback whales in the western North Atlantic: a catalogue of identified individuals. Coll. Atlantic, Bar Harbor, Maine. 26pp.

Laws, R. M. 1961. Reproduction, growth and age of southern fin whales. Discovery Rep. 31:327–486.

Lawton, W.; Rice, D.; Wolman, A.; and Winn, H. 1979. Occurrence of southeastern Alaskan humpback whales, *Megaptera novaeangliae* in Mexican coastal waters. Page 35 *in* Abstr. 3rd Biennial Conf. Bio. Marine Mammals, Seattle, Wash. 64pp.

Layne, J. N. 1965. Observations on marine mammals in Florida waters. Florida State Mus. Bull. 9(4):131–181.

Leatherwood, J. S. 1974. Aerial observations of migratory gray whales, *Eschrichtius robustus*, off southern California, 1969–72. Marine Fish. Rev. 36(4):45–49.

Lien, J., and Merdsoy, B. 1979. The humpback is not over the hump. Nat. Hist. 88(6):46–49.

Lien, J.; Perkins, J.; Merdsoy, B.; and Johnson, S. 1979. Baleen whale collisions with inshore fishing gear in Newfoundland. Page 37 *in* Abstr. 3rd Biennial Conf. Bio. Marine Mammals, Seattle, Wash. 64pp.

Lockyer, C. 1976. Body weights of some species of large whales. J. Con. Explor. Mer. 36(3):259–273.

——. 1977. Some possible factors affecting age distribution of the catch of sei whales in the Antarctic. Rep. Int. Whaling Comm., Special Issue 1:63–70.

Lowry, L. F.; Frost, K. J.; and Burns, J. J. 1978. Food of ringed seals and bowhead whales near Point Barrow, Alaska. Can. Field-Nat. 92(1):67–70.

Lubbock, B. 1937. The arctic whalers. Brown, Son and Ferguson, Glasgow. 483pp.

McHugh, J. H. 1974. The role and history of the International Whaling Commission. Pages 305–335 *in* W. E. Schevill, ed. The whale problem. Harvard Univ. Press, Cambridge, Mass. 419pp.

Mackintosh, N. A. 1965. The stocks of whales. Fishing News (Books), London. 232pp.

McVay, S., 1973. Stalking the arctic whale. Am. Sci. 61(1):24–37.

——. 1974. Reflections on the management of whales. Pages 369–382 *in* W. E. Schevill, ed. The whale problem. Harvard Univ. Press, Cambridge, Mass. 419pp.

Maher, W. J. 1960. Recent records of the California grey whale (*Eschrichtius robustus*) along the north coast of Alaska. Arctic 13:257–265.

Maher, W. J., and Wilimovsky, N. J. 1963. Annual catch of bowhead whales by Eskimos at Point Barrow, Alaska, 1928–1960. J. Mammal. 44:16–20.

Manigault, G. E. 1885. The black whale captured in Charleston Harbor, January, 1880. Proc. Elliott Soc. (September 1885); pp. 98–104.

Mansfield, A. W. 1971. Occurrence of the bowhead or Greenland whale (*Balaena mysticetus*) in Canadian arctic waters. J. Fish. Res. Board Can. 28:1873–1875.

Markham, C. R. 1881. On the whale-fishery of the Basque Provinces of Spain. Proc. Zool. Soc. London (1881), pp. 969–976.

Marquette, W. 1977. The 1976 catch of bowhead whales (*Balaena mysticetus*) by Alaskan Eskimos, with a review of the fishery, 1973–1976, and a biological summary of the species. Processed Rep., U.S. Dept. Commerce, Northwest and Alaska Fish. Center, Seattle, Wash. 80pp.

———. 1979. The 1977 catch of bowhead whales (*Balaena mysticetus*) by Alaskan Eskimos. Rep. Int. Whaling Comm. 29:281–289.

Mate, B. R., and Harvey, J. 1979. A new radio tag for large whales. Page 39 *in* Abstr. 3rd Biennial Conf. Bio. Marine Mammals, Seattle, Wash. 64pp.

Matthews, L. H. 1932. Lobster krill: anomuran crustaceans that are the food of whales. Discovery Rep. 5:467–484.

———. 1938. Notes on the southern right whale, *Eubalaena australis*. Discovery Rep. 17:169–182.

Maul, G. E., and Sergeant, D. E. 1977. New cetacean records from Madeira. Bocagiana 43:1–8.

Maury, M. F. 1851. Whale chart. (Preliminary sketch.) Constructed by Lts. Leigh, Herndon, and Fleming and Pd. Midn. Jackson. Series F. Natl. Observatory, Washington, D.C. 1 sheet.

May, R. M.; Beddington, J. R.; Clark, C. W.; Holt, S. J.; and Laws, R. M. 1979. Management of multispecies fisheries. Science 205:267–277.

Mead, J. G. 1977. Records of sei and Bryde's whales from the Atlantic coast of the United States, the Gulf of Mexico, and the Caribbean. Rep. Int. Whaling Comm., Special Issue 1:113–116.

Mermoz, J. F. 1980. Preliminary report on the southern right whale in the southwestern Atlantic. Rep. Int. Whaling Comm. 30:183–186.

Miller, G. S. 1923. The telescoping of the cetacean skull. Smithsonian Misc. Collections 76(5):1–71.

Mitchell, E. D. 1973. Draft report on humpback whales taken under special scientific permit by eastern Canadian land stations, 1969–1971. Rep. Int. Whaling Comm. 23:138–154.

———. 1974. Present status of northwest Atlantic fin and other whale stocks. Pages 108–169 *in* W. E. Schevill, ed. The whale problem. Harvard Univ. Press, Cambridge, Mass. 419pp.

———. 1975*a*. Porpoise, dolphin and small whale fisheries of the world. Monogr. no. 3. Int. Union for Conserv. Nature and Nat. Resourc. (IUCN), Morges, Switzerland. 129pp.

———. 1975*b*. Trophic relationships and competition for food in northwest Atlantic whales. Proc. Can. Soc. Zool. Annu. Meet., pp. 123–133.

———. 1977. Canadian progress report on whale research, June 1975 to May 1976. Rep. Int. Whaling Comm. 27:73–85.

———. 1978. Finner whales. Pages 36–45 *in* D. Haley, ed. Marine mammals of eastern North Pacific and arctic waters. Pacific Search Press, Seattle, Wash. 256pp.

———. 1979. Comments on magnitude of early catch of East Pacific gray whale (*Eschrichtius robustus*). Rep. Int. Whaling Comm. 29:307–314.

Mitchell, E. D., ed. 1975*c*. Review of biology and fisheries for smaller cetaceans. (Rep. and Pap. from meet. of Subcom. on Smaller Cetaceans, Int. Whaling Comm., Montreal, 1–11 April 1974.) J. Fish. Res. Board Can. 32:875–1242.

Mitchell, E. D., and Chapman, D. G. 1977. Preliminary assessment of stocks of northwest Atlantic sei whales (*Balaenoptera borealis*). Rep. Int. Whaling Comm., Special Issue 1:117–120.

Mitchell, E. D., and Kozicki, V. M. 1975. Supplementary information on minke whale (*Balaenoptera acutorostrata*) from Newfoundland fishery. J. Fish. Res. Board Can. 32:985–994.

Mitchell, E. D., and Mead, J. G. 1977. History of the gray whale in the Atlantic Ocean. Page 11 *in* Abstr. Proc. 2nd Conf. Bio. Marine Mammals, San Diego, Calif. 88pp.

Mitchell, E. D., and Reeves, R. R. 1980. The Alaska bowhead problem: a commentary. Arctic 33(4):686–723.

———. In press. Factors affecting abundance of bowhead whales (*Balaena mysticetus*) in the eastern Arctic of North America, 1915–1980. Biological Conservation.

Moore, J. C. 1953. Distribution of marine mammals to Florida waters. Am. Midl. Nat. 49:117–158.

Moore, J. C., and Clark, E. 1963. Discovery of right whales in the Gulf of Mexico. Science 141:269.

Morejohn, G. V., and Rice, D. W. 1973. First record of Bryde's whale (*Balaenoptera edeni*) off California. California Fish Game 59:313–315.

Muller, J. 1954. Observations on the orbital region of the skull of the Mystacoceti. Zoologische Mededelingen 32, no. 23, pp. 279–290.

Nasu, K. 1960. Oceanographic investigation in the Chukchi Sea during the summer of 1958. Sci. Rep. Whales Res. Inst. (Japan) 15:143–157.

Nemoto, T. 1970. Feeding pattern of baleen whales in the oceans. Pages 241–252 *in* J. H. Steele, ed. Marine food chains. Univ. Calif. Press, Berkeley and Los Angeles. 552pp.

Nemoto, T., and Kawamura, A. 1977. Characteristics of food habits and distribution of baleen whales with special reference to the abundance of North Pacific sei and Bryde's whales. Rep. Int. Whaling Comm., Special Issue 1:80–87.

Newman, W. A., and Ross, A. 1976. Revision of the balanomorph barnacles; including a catalog of the species. San Diego Soc. Nat. Hist. Mem. 9:1–108.

Nishiwaki, M., and Hasegawa, Y. 1969. The discovery of the right whale skull in the Kisagata Shell Bed. Sci. Rep. Whales Res. Inst. (Japan) 21:79–84.

Norman, J. R., and Fraser, F. C. 1937. Giant fishes, whales and dolphins. Putnam, London. 376pp.

Norris, K. S.; Goodman, R. M.; Villa-Ramirez, B.; and Hobbs, L. 1977. Behavior of California gray whale, *Eschrichtius robustus*, in southern Baja California, Mexico. Fish. Bull. 75:159–172.

Norris, K. S., and Reeves, R. R., eds. 1978. Report on a workshop on problems related to humpback whales (*Megaptera novaeangliae*) in Hawaii. Contract Rep., Marine Mammal Comm., Natl. Tech. Information Serv. PB-280 794. 90pp.

Ohsumi, S. 1976. Population assessment of the California gray whale. Rep. Int. Whaling Comm. 26:350–359.

———. 1977*a*. Bryde's whales in the pelagic whaling ground of the North Pacific. Rep. Int. Whaling Comm., Special Issue 1:140–150.

———. 1977*b*. Further assessment of population of Bryde's whales in the North Pacific. Rep. Int. Whaling Comm. 27:156–160.

Omura, H. 1958. North Pacific right whale. Sci. Rep. Whales Res. Inst. (Japan) 13:1–52.

Omura, H.; Nishiwaki, M.; and Kasuya, T. 1971. Further studies on two skeletons of the black right whale in the North Pacific. Sci. Rep. Whales Res. Inst. (Japan) 23:71–81.

Omura, H.; Ohsumi, S.; Nemoto, T.; Nasu, K.; and Kasuya, T. 1969. Black right whales in the North Pacific. Sci. Rep. Whales Res. Inst. (Japan) 21:1–78.

Payne, R. 1972. Swimming with Patagonia's right whales. Natl. Geogr. 142:576–587.

———. 1974. A playground for whales. Animal Kingdom 77(2):7–12.

———. 1976. At home with right whales. Natl. Geogr. 149:322–339.

———. 1979. Humpbacks: their mysterious songs. Natl. Geogr. 155:18–25.

Payne, R. S., and Dorsey, E. M. 1979. Sexual dimorphism in the callosities of southern right whales. Page 46 *in* Abstr. 3rd Biennial Conf. Bio. Marine Mammals, Seattle, Wash. 64pp.

Payne, R., and McVay, S. 1971. Songs of humpback whales. Science 173:585–597.

Payne, R. and Payne, K. 1971. Underwater sounds of right whales. Zoologica 56:159–165.

Perkins, J. S., and Beamish, P. C. 1979. Net entanglements of baleen whales in the inshore fishery of Newfoundland. J. Fish. Res. Board Can. 36:521–528.

Pike, G. C. 1953. Color pattern of humpback whales from the coast of British Columbia. J. Fish. Res. Board Can. 10:320–325.

———. 1962. Migration and feeding of gray whale (*Eschrichtius gibbosus*). J. Fish. Res. Board Can. 19:815–838.

Pike, G. C., and MacAskie, I. B. 1969. Marine mammals of British Columbia. Bull. Fish. Res. Board Can. 171:1–54.

Pilleri, G. 1964. Morphologie des gehirnes des "southern right whale," *Eubalaena australis* Desmoulins 1822 (*Cetacea, Mysticeti, Balaenidae*). Acta Zool. 46:245–272.

Pilsbry, H. A. 1916. The sessile barnacles (Cirripedia) contained in the collections of the U.S. National Museum, including a monograph of the American species. Bull. U.S. Natl. Mus., 93:1–361.

Pivorunas, A. 1976. A mathematical consideration on the function of baleen plates and their fringes. Sci. Rep. Whales Res. Inst. (Japan) 28:37–55.

Purves, P. E. 1955. The wax plug in the external auditory meatus of the Mysticeti. Discovery Rep. 27:293–302.

Ralls, K. 1976. Mammals in which females are larger than males. Q. Rev. Bio. 51(2):245–276.

Ray, G. C.; Mitchell, E. D.; Wartzok, D.; Kozicki, V. M.; and Maiefski, R. 1978. Radio tracking of a fin whale (*Balaenoptera physalus*). Science 202:521–524.

Ray, G. C., and Schevill, W. E. 1974. Feeding of a captive gray whale, *Eschrichtius robustus*. Marine Fish. Rev. 36(4):31–38.

Reeves, R. R. 1975. The right whale. Conservationist 30(1):32–33, 45.

———. 1976. White whales of the St. Lawrence. Can. Geogr. J. 92(2):12–19.

———. 1980. Spitsbergen bowhead stock: a short review. Marine Fish. Rev. 42:65–69.

Reeves, R. R.; Mead, J. G.; and Katona, S. 1978. The right whale, *Eubalaena glacialis,* in the western North Atlantic. Rep. Int. Whaling Comm. 28:303–312.

Rice, D. W. 1963. Progress report on biological studies of the larger Cetacea in the waters off California. Norsk Hvalfangst-tidende 7:181–187.

———. 1974. Whales and whale research in the eastern North Pacific. Pages 170–195 *in* W. E. Schevill, ed. The whale problem. Harvard Univ. Press, Cambridge, Mass. 419pp.

———. 1977*a*. A list of the marine mammals of the world. U.S. Dept. Commerce, NOAA Tech. Rep. NMFS SSRF-711:1–15.

———. 1977*b*. Synopsis of biological data on the sei whale and Bryde's whale in the eastern North Pacific. Rep. Int. Whaling Comm., Special Issue 1:92–97.

———. 1978. The humpback whale in the North Pacific: distribution, exploitation, and numbers. Pages 29–44 *in* K. S. Norris and R. R. Reeves, eds. Report on a Workshop on Problems Related to Humpback Whales (*Megaptera novaeangliae*) in Hawaii. Contract Rep., Marine Mammal Comm., Natl. Tech. Information Serv. PB-280 794. 90pp.

———. 1979. Bryde's whales in the equatorial eastern Pacific. Rep. Int. Whaling Comm. 29:321–324.

Rice, D. W., and Fiscus, C. H. 1968. Right whales in the southeastern North Pacific. Norsk Hvalfangst-tidende 57:105–107.

Rice, D. W., and Wolman, A. A. 1971. The life history and ecology of the gray whale (*Eschrichtius robustus*). Am. Soc. Mammal., Special Publ. no. 3. 142pp.

———. 1978. Humpback whale census in Hawaiian waters, February 1977. Pages 45–53 *in* K. S. Norris and R. R. Reeves, eds. Report on a Workshop on Problems Related to Humpback Whales (*Megaptera novaeangliae*) in Hawaii. Contract Rep., Marine Mammal Comm., Natl. Tech. Information Serv. PB-280 794. 90pp.

Ridewood, W. G. 1901. On the structure of the horny excrescence, known as the "bonnet," of the southern right whale (*Balaena australis*). Proc. Zool. Soc. London 1 (1901):44–47.

Rørvik, C., chairman. 1977. Report of the Working Group on North Atlantic Whales. Rep. Int. Whaling Comm. 27:369–387.

Ross, W. G. 1974. Distribution, migration, and depletion of bowhead whales in Hudson Bay, 1860 to 1915. Arctic and Alpine Res. 6(1):85–98.

———. 1979. The annual catch of Greenland (bowhead) whales in waters north of Canada, 1719–1915: a preliminary compilation. Arctic 32:91–121.

Rowntree, V.; Darling, J.; Silber, G.; and Ferrari, M. 1980. Rare sighting of a right whale (*Eubalaena glacialis*) in Hawaii. Can. J. Zool. 58:309–312.

Rugh, D. J., and Braham, H. W. 1979. California gray whale (*Eschrichtius robustus*) fall migration through Unimak Pass, Alaska, 1977: a preliminary report. Rep. Int. Whaling Comm. 29:315–320.

Ruud, J. T. 1940. The surface structure of the baleen plates as a possible clue to age in whales. Hvalradets Skrifter 23:1–24.

———. 1945. Further studies of the structure of baleen plates and their application to age determination. Hvalradets Skrifter 29:1–69.

Saayman, G. S., and Tayler, C. K. 1973. Some behaviour patterns of the southern right whale *Eubalaena australis*. Z. Saukgetierk. 38:172–183.

Scammon, C. M. 1874. The marine mammals of the northwestern coast of North America, described and illustrated; together with an account of the American whale-fishery. Carmany and Co., San Francisco. 320pp.

Scarff, J. E. 1977. The international management of whales, dolphins, and porpoises: an interdisciplinary assessment. Ecol. Law Q. 6:323–427, 574–638.

Scattergood, L. W. 1949. Notes on the little piked whale. Murrelet 30:3–16.

Scheffer, V. B., and Slipp, J. W. 1948. The whales and dolphins of Washington State with a key to the cetaceans of the west coast of North America. Am. Midl. Nat. 39(2):257–337.

Schevill, W. E.; Ray, C.; Kenyon, K. W.; Orr, R. T.; and Van Gelder, R. G. 1967. Immobilizing drugs lethal to swimming mammals. Science 157:630–631.

Schevill, W. E., and Watkins, W. A. 1962. Whale and porpoise voices: a phonograph record. Woods Hole Oceanogr. Inst., Woods Hole, Mass. Disc + 24pp.

_____. 1966. Radio-tagging of whales. Tech. Rep. 66–17. Woods Hole Oceanogr. Inst., Woods Hole, Mass. 15pp.

Schmidly, D. J.; Martin, C. O.; and Collins, G. F. 1972. First occurrence of a black right whale (*Balaena glacialis*) along the Texas coast. Southwestern Nat. 17:214–215.

Schmidly, D. J., and Melcher, B. A. 1974. Annotated checklist and key to the cetaceans of Texas waters. Southwestern Nat. 18:453–464.

Scoresby, W., Jr. 1820. An account of the arctic regions, with a history and description of the northern whale fishery. Constable and Co., Edinburgh. 2 vols. 1125pp.

Sergeant, D. E. 1963. Minke whales, *Balaenoptera acutorostrata* Lacépède, of the western North Atlantic. J. Fish. Res. Board Can. 20:1489–1504.

_____. 1966. Populations of large whale species in the western North Atlantic with special reference to the fin whale. Fish. Res. Board Can., Arctic Bio. Stn. Circ. 9:1–13.

_____. 1977. Stocks of fin whales *Balaenoptera physalus* L. in the North Atlantic Ocean. Rep. Int. Whaling Comm. 27:460–473.

Shevchenko, V. I. 1975. [Nature of correlations between killer whales and other cetaceans.] Pages 173–175 *in* Morsk Mlekopitayushchiye, Kiev. (Transl. from Russian by U.S. Joint Publ. Res. Serv., L/6049).

_____. 1977. Application of white scars to the study of the location and migrations of sei whale populations in Area III of the Antarctic. Rep. Int. Whaling Comm., Special Issue 1:130–134.

Slijper, E. J. 1966. Functional morphology of the reproductive system in Cetacea. Pages 277–319 *in* K. S. Norris, ed. Whales, dolphins, and porpoises. Univ. California Press, Berkeley. 789pp.

Starbuck, A. 1878. History of the American whale fishery; from its earliest inception to the year 1876. Rep. U.S. Comm. Fish and Fish., part 4. 767pp.

Stevenson, C. H. 1904. Aquatic products in arts and industries: fish oils, fats, and waxes; fertilizers from aquatic products. Fish and Fish. Rep. Comm., part 28 (for 1902): 177–279, pl. 10–25.

Stick, D. 1958. The Outer Banks of North Carolina, 1584–1958. Univ. North Carolina Press, Chapel Hill, N.C. 352pp.

Sutcliffe, W. H., Jr., and Brodie, P. F. 1977. Whale distribution in Nova Scotia waters. Fish. and Marine Serv. (Canada), Tech. Rep. no. 722. 83pp.

Swartz, S. L., and Cummings, W. C. 1978. Gray whales, *Eschrichtius robustus,* in Laguna San Ignacio, Baja California, Mexico. Contract Rep., U.S. Marine Mammal Comm., Natl. Tech. Information Serv., PB-276 319. 44pp.

Swartz, S. L., and Jones, M. L. 1978. The evaluation of human activities on gray whales, *Eschrichtius robustus,* in Laguna San Ignacio, Baja California, Mexico. Contract Rep., U.S. Marine Mammal Comm., Natl. Tech. Information Serv., PB-289 737. 34pp.

Tarpy, C. 1979. Killer whale attack. Natl. Geogr. 155:542–545.

Thompson, D. W. 1918. On whales landed at the Scottish whaling stations, especially during the years 1908–1914. Scottish Nat. 81:197–208.

Tomilin, A. G. 1967. Mammals of the U.S.S.R. and adjacent countries, Vol. 9: Cetacea. Israel Program for Sci. Transl., Jerusalem. 717pp. (First published in Russian, 1957.)

Townsend, C. H. 1935. The distribution of certain whales as shown by logbook records of American whaleships. Zoologica 19:1–50.

Tressler, D. K. 1923. Marine products of commerce. Chemical Catalog Co., New York. 762pp.

True, F. W. 1904. The whalebone whales of the western North Atlantic. Smithsonian Contributions to Knowledge 33:1–332.

Turner, W. 1913. The right whale of the North Atlantic, *Balaena biscayensis:* its skeleton described and compared with that of the Greenland right whale, *B. mysticetus.* Trans. R. Soc. Edinburgh 48:889–992.

Van Beneden, M. P. -J. 1880. Description des ossements fossiles des environs d'anvers. Vol. 4, part 2: Cetacés; genres Balaenula, Balaena, et Balaenotus. Ann. Mus. R. d'hist. nat. belgique (ser. paleontol.), Brussels. 82pp.

Van Valen, L. 1968. Monophyly or diphyly in the origin of whales. Evolution 22:37–41.

Wade, L. S., and Friedrichsen, G. L. 1979. Recent sightings of the blue whale, *Balaenoptera musculus,* in the northeastern tropical Pacific. Fish. Bull. 76:915–919.

Watkins, W. A. 1967. Air-borne sounds of the humpback whale, *Megaptera novaeangliae.* J. Mammal. 48:573–578.

_____. 1979. A projectile point for penetrating whale blubber. Deep-Sea Res. 26A:1301–1308.

Watkins, W. A., and Schevill, W. E. 1976. Right whale feeding and baleen rattle. J. Mammal. 57:58–66.

_____. 1979. Aerial observation of feeding behavior in four baleen whales: *Eubalaena glacialis, Balaenoptera borealis, Megaptera novaeangliae,* and *Balaenoptera physalus.* J. Mammal. 60:155–163.

_____. 1980. Right whales (*Eubalaena glacialis*) in Cape Cod waters. Manuscript.

Watkins, W. A.; Wartzok, D.; Martin, H. B., III; and Maiefski, R. R. 1980. A radio whale tag. Pages 227–241 *in* F. P. Diemer, F. J. Vernberg, and D. Z. Mirkes, eds. Advanced concepts in ocean measurements for marine biology. Belle W. Baruch Library in Marine Sci. no. 10. Univ. South Carolina Press, Columbia, S.C.

Whitehead, H., and Payne, R. 1976. New techniques for assessing populations of right whales without killing them. Advisory Com. Marine Resour. Res., Sci. Consultation on Marine Mammals, Bergen, Norway. Food and Agric. Organization (United Nations), ACMRR/MM/SC/79. 23pp.

Whitmore, F. C., Jr., and Sanders, A. E. 1976. Review of the Oligocene Cetacea. Systematic Zool. 25:304–320.

Williamson, G. 1973. Counting and measuring baleen and ventral grooves of whales. Sci. Rep., Whales Res. Inst. (Japan) 25:279–292.

Winn, H. E.; Edel, R. K.; and Taruski, A. G. 1975. Population estimate of the humpback whale (*Megaptera novaeangliae*) in the West Indies by visual and acoustic techniques. J. Fish. Res. Board Can. 32:499–506.

Winn, H. E., and Perkins, P. J. 1976. Distribution and sounds of the minke whale, with a review of mysticete sounds. Cetology 19:1–12.

Winn, H. E., and Winn, L. K. 1978. The song of the humpback whale *Megaptera novaeangliae* in the West Indies. Marine Bio. 47:97–114.

Wolman, A. A. 1978. Humpback whale. Pages 47–53 *in* D. Haley, ed. Marine mammals of eastern North Pacific and arctic waters. Pacific Search Press, Seattle, Wash. 256pp.

Wolman, A. A., and Rice, D. W. 1979. Current status of the gray whale. Rep. Int. Whaling Comm. 29:275–279.

Yablokov, A. V., and Andreyeva, T. V. 1965. Age determination in baleen whales. Nature 205:412–413.

Yablokov, A. V.; Bel'kovich, V. M.; and Borisov, V. I. 1974. Whales and dolphins. Joint Pub. Res. Serv., Arlington, Va. 2 parts. 402pp. (Transl. of Russian Kity i Del'finy. 1972. Nauka, Moscow. 472pp.)

RANDALL R. REEVES, National Fish and Wildlife Laboratory, National Museum of Natural History, Washington, D.C. 20560.

ROBERT L. BROWNELL, JR., National Fish and Wildlife Laboratory, National Museum of Natural History, Washington, D.C. 20560.

V

Carnivora

20

Coyote

Canis latrans   Marc Bekoff

NOMENCLATURE

COMMON NAMES. Coyote, brush wolf, prairie wolf, Heul wolf, Steppenwolf, lobo, American jackal. The word *coyote* means "barking dog" and is taken from the Aztec word *coyotl*.

The coyote is one of eight recognized species in the genus *Canis*. By the late Pliocene, the ancestral coyote, *Canis lepophagus*, was widespread in North America. The Eastern coyote, *Canis latrans* var., formerly referred to as the New England canid, appears to be a stable subspecies, having predominantly coyote ancestry with some introgression of wolf and dog genes (Bekoff et al. 1975; Hilton 1978; Lawrence and Bossert 1969, 1975; Silver and Silver 1969).

SCIENTIFIC NAME. *Canis latrans*

SUBSPECIES. There are 19 recognized subspecies of *C. latrans*. However, due to increased mobility of coyotes, the integrity of individual subspecies and their taxonomic utility is questionable (Nowak 1978). In chronological order, the subspecies are: *C. l. latrans, C. l. ochropus, C. l. cagottis, C. l. frustror, C. l. lestes, C. l. mearnsi, C. l. microdon, C. l. peninsulae, C. l. vigilis, C. l. clepticus, C. l. impavidus, C. l. goldmani, C. l. texensis, C. l. jamesi, C. l. dickeyi, C. l. incolatus, C. l. hondurensis, C. l. thamnos,* and *C. l. umpquensis.*

DISTRIBUTION

Coyotes are nearctic canids. They occupy many diverse habitats between about 10° north latitude (Costa Rica) and 70° north latitude (northern Alaska). They are found throughout the continental United States and in many areas of Canada (figure 20.1). In some states, such as Florida (Cunningham and Dunford 1970) and Georgia (Fisher 1975), it appears as if coyotes have been transplanted by humans. Urban coyotes are also observed in large cities (Andelt 1977; Gill 1965; Time 1975).

DESCRIPTION

Coyotes are often confused with other canids, such as gray wolves (*C. lupus*), red wolves (*C. rufus*), and

FIGURE 20.1. Distribution of the coyote (*Canis latrans*). T = probable transplants by people. Note that in Bekoff (1977*a*), the distribution of coyotes in Quebec was inadvertently omitted.

domestic dogs (*C. familiaris*), with all of whom they can successfully interbreed and produce fertile hybrids. However, coyotes can be differentiated from the above species (although overlap is not uncommon) using serologic parameters, dental characteristics, cranial measurements, neuroanatomical features, diameter of the nose pad, diameter of the hindfoot pad, ear length, track size, stride length, pelage, and behavior (for reviews, see Bekoff 1977*a*; Elder and Hayden 1977; and Lawrence and Bossert 1967). For example, coyotes are typically smaller than gray wolves (see table 20.1), so the nose pad (about 25 mm in diameter) and hindfoot pads (less than 32 mm) are correspondingly smaller. The track length and stride of the coyote are shorter than those of gray and red wolves (Murie 1954; Riley and McBride 1975). The coyote brain is also anatomically different from that of gray wolves (Radinsky 1973). The wolf has a dimple in the middle of the coronal gyrus, whereas the coyote does not (see also

447

TABLE 20.1. Some representative mean coyote weights (kg) from a variety of locales

| | Adults | | Juveniles | | |
Study	Males	Females	Males	Females	State/Province
Andrews and Boggess 1978	13.4	11.4			Iowa
Berg and Chesness 1978	12–13	11–12	10–11	10	Minnesota
Gier 1968	14.1	11.8			Kansas
Hawthorne 1971	11.2	9.8			California (northeast)
Daniel 1973	16.8	13.6			Texas
Richens and Hugie 1974[a]	15.8	13.7			Maine
Boggess and Henderson 1978	13.1	11.0			Kansas
Bowen 1978	12.1	11.5			Alberta
Litvaitis 1978	14.7[b]	12.1			Oklahoma

[a]Eastern coyote: Hilton (1978) reports a range of 15.8–18.1 kg. He notes that the average weight of Eastern coyotes is less than 18 kg and that authenticated reports of individuals exceeding 22 kg are uncommon.

Atkins and Dillon 1971 and Atkins 1978). Coyotes may be differentiated from dogs using the ratio of palatal width (distance between the inner margins of the alveoli of the upper first molars) to the length of the upper molar toothrow (from the anterior margin of the aveolus of the first premolar to the posterior margin of the last molar alveolus) (Howard 1949). If the tooth row is 3.1 times the palatal width the specimen is a coyote; if the ratio is less than 2.7 the specimen is a dog. This method is about 95 percent reliable.

The coyote has a relatively larger braincase than *C. lupus* (Mech 1974), and there is no overlap when comparing large coyotes to small wolves (*lupus*) in zygomatic breadth (greatest distance across zygomata), greatest length of the skull, or bite ratio (width across the outer edges of the alveoli of the anterior lobes of the upper carnassials divided by the length of the upper molar toothrow) (Paradiso and Nowak 1971). *C. latrans* is usually smaller in size than *C rufus* and there is almost no overlap between them in greatest length of the skull. *Rufus* also has a more pronounced sagittal crest than *latrans*. Multivariate techniques have clearly shown that coyotes, wolves, and dogs can be differentiated anatomically (Elder and Hayden 1977; Lawrence and Bossert 1967) and behaviorally (Bekoff et al. 1975, Bekoff 1978a) and provide for more rigorous analyses than do univariate methods.

Size and Weight. Coyotes are about 1–1.5 m in length; the tail is about 400 mm long. Size varies with geographic locale and subspecies (Hall and Kelson 1959; Jackson 1951). Adult males are usually heavier and larger than adult females (table 20.1).

Pelage. The banded nature of coyote hair is responsible for the appearance of the blended color, gray mixed with a reddish tint. Coyotes show great variation in color, ranging from almost pure gray to rufous. Melanistic coyotes are rare (Gipson 1976; Mahan 1978; Van Wormer 1964; Young 1951). Texture and color of the fur also varies geographically. In northern

subspecies the hair is longer and coarser. In desert habitats, coyotes tend to be fulvous in color, while those at higher latitudes are more gray and black (Jackson 1951). The belly and throat are paler than the rest of the body. Course guard hairs are acuminate in the proximal region, crenate medially, and flattened distally. They are about 50–90 mm in length; in the mane they tend to be longer (80–110 mm). The fine underfur (up to 50 mm long) has coronal scales (Adorjan and Kolenosky 1969; Ogle and Farris 1973). The summer coat is shorter than the winter coat. Coyote hair may be differentiated from hair of dogs and red foxes (*Vulpes vulpes*) by the number, order, and color of the bands, the cross-sectional transluscence and shape, and the circular scale pattern (Hilton and Kutscha 1978). Coyote hairs typically are coarser, longer, larger in diameter, and rougher and stiffer.

The coyote's fur is similar in insulative value to that of the gray wolf (Ogle and Farris 1973). The critical temperature of *latrans* is −10° C (Shield 1972). When wearing the shorter summer coat, there is a decrease of about 87 percent in thermal conductivity (Ogle and Farris 1973). There is usually one main molt between late spring and autumn. About 50 mm down from the base of the tail there is an oval tail gland (Hildebrand 1952).

Age Determination. Coyote age can be estimated by counting dental cementum annuli (Linhart and Knowlton 1967; Nellis et al. 1978). Roberts (1978) pointed out that there is variation in age determination from different teeth and suggested using canines in age determination. Eye lens weight, baculum weight, and thermal contraction of tail tendons can also be used to estimate age accurately. Ages of young coyotes can be estimated from weight, body length, and length of the hindfoot (Barnum et al. 1979; Bekoff and Jamieson 1975; Gier 1968). The regression equation for the weight of hand-reared pups (0–30 days of age) is $y = 0.2685 + 0.197x$; for coyotes 30–154 days the

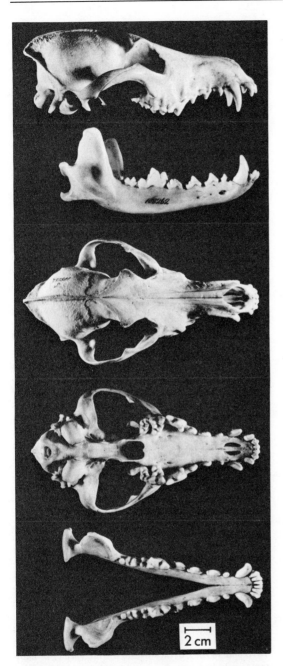

FIGURE 20.2. Skull of the coyote (*Canis latrans*). From top to bottom: lateral view of cranium, lateral view of mandible, dorsal view of cranium, ventral view of cranium, dorsal view of mandible.

equation is y=0.5049 + 0.0469x (Barnum et al. 1979). The regression equations to predict the weight of known-age coyotes are y=−13.57 + 50.59x (0–30 days) and y=11.386 + 21.11x (30–154 days). The correlations between weight and age for 0–30 days and 30–154 days are 0.999 and 0.995, respectively (Barnum et al. 1979).

Skull and Dentition. Males show greater development of the sagittal crest than females. The dental formula is i 3/3, c 1/1, p 4/4, m 2/3 = 42 (figure 20.2). The skull of a mature male is about 180–205 mm long from the tip of the premaxilla to the posterior rim of the coronal crest (Gier 1968) and weighs between 170 and 210 gm.

PHYSIOLOGY

Central Nervous System. Although the cerebrum and cerebellum share many common features with other canids, there are some interspecific differences (see Atkins 1978 and Atkins and Dillon 1971 for review). With respect to cerebellar morphology, coyotes may be distinguished from all other *Canis* studied to date as follows: the anterior lobe is more than one-half the total width, the parafloccular process is relatively prominent, the vermian lobule reaches its greatest size, there are fewer and larger posterior hemispheric folia, the posterior ventral parafloccular limb is reduced in size, and there is a broad vermian twist (Atkins 1978). The remainder of the central nervous system, the brain stem and spinal cord, does not show much variation when compared to the domestic dog.

Adrenals. Coyote adrenals are similar in structure to those of most other canids (Heinrich 1972). In both males and females, the left adrenal is heavier than the right, and the adrenals of females tend to be heavier than those of males.

Audition and Vision. The region of maximal sensitivity to auditory stimuli is 100–30,000 Hz with a limit of approximately 80 kHz (Petersen, Heaton, and Wruble 1969). The retina is duplex and has a preponderance of rods. The absolute scotopic (rod) threshold is about 1.4 foot-candles and the adaptation curve shows distinct rod-cone breaks (Horn and Lehner 1975).

REPRODUCTION

Because of the great interest in coyote control, a good deal is known about basic reproductive biology of coyotes.

Genetics and Hybridization. The coyote has 38 pairs of chromosomes (Wurster and Benirschke 1968). The autosomes are acrocentric or telocentric and the sex chromosomes are submetacentric (Hsu and Benirschke 1967, cited by Mech 1974). Fertile hybrids have been produced by matings of coyotes with domestic dogs (Kennelly and Roberts 1969; Mengel 1971; Silver and Silver 1969; Young 1951), red and gray wolves (Kolenosky 1971; Paradiso and Nowak 1971; Riley and McBride 1975; Young 1951), and golden jackals (*C. aureus;* Seitz 1965). Coyote-dog hybrids exhibit decreased fecundity (Gipson et al. 1975; Mengel 1971).

Anatomy and Physiology. There are no detailed reports of the gross or microscopic anatomy of coyote reproductive systems. With the exception of the seasonality of breeding and associated changes in reproductive anatomy and physiology, there appear to be only minor differences, if any, between coyotes and domestic dogs (Kennelly 1978). Berg and Chesness (1978) found no correlation between carcass weight and ovarian weight (n=105, r=0.218, p<0.05).

Detailed accounts of spermatogenesis and the estrous cycle are provided by Kennelly (1972, 1978). Proestrus lasts about two to three months and estrus up to ten days, depending on locale (Hamlett 1939; Kennelly 1978). Copulation ends with the copulatory "tie," during which time (up to 25 min) the male's penis is locked in the female's vagina (Grandage 1972). Juvenile males and females are able to breed (see below); juvenile females may ovulate less than adult females.

Both males and females show annual cyclic changes in reproductive anatomy and physiology. Females are seasonally monestrous, showing one period of "heat" per year, usually between January and March, depending on geographic locale (Gier 1968; Hamlett 1939; Kennelly 1978).

Pair Bonding. The dynamics of heterosexual pair bonding are not known for wild coyotes in any detail. Data on captive individuals that do not appear to differ significantly from wild coyotes (Bekoff and Wells unpublished data) indicate that courtship may begin as long as two to three months before successful copulation (Bekoff and Diamond 1976). Associated changes in behavior, especially increases in scent marking observed at the beginning of the breeding season, have been described by Bekoff and Diamond (1976) and by Wells and Bekoff (1981). During early stages of courtship the male becomes increasingly attracted to the female's urine and feces. When the female is ready to copulate she will tolerate mounting attempts by the male and will flag her tail to one side. When tieing, the male steps over the female's back and the couple remain locked at 180° for a period of up to 25 minutes. The same pair may breed from year to year but not necessarily for life.

Pregnancy Rate, Gestation, and Litter Size. The percentage of females that breeds in a given year varies with local conditions (Gier 1968; Gipson et al. 1975; Knowlton 1972). Food supply is usually the prime factor; in "good" years, more females, especially yearlings, breed (Gier 1968). Usually, about 60–90 percent of adult females and 0–70 percent of female yearlings produce litters. The greatest annual variation in the number of breeding females is due to the number of juveniles that become sexually mature (Kennelly 1978; Knowlton 1972). Adults may show more placental scars than yearlings; however, Nellis and Keith (1976) reported that the difference was not statistically significant. Gier (1975) estimated that the number of young born was about 80 percent of the ovulated ova and Knowlton (1972) estimated that about 87 percent of inplants were represented by viable ova. Gipson et al. (1975) reported the mean number of ova per breeding female to be 6.2, with 4.5 (73 percent) becoming implanted. Between about 85 percent (Hamlett 1939) and

92 percent (Asdell 1964) of embryos develop into viable young.

Gestation lasts approximately 63 days. Average litter size is 6, but it is known that litter size can be affected by population density. Knowlton (1972) reported average litter sizes of 4.3 at high densities and 6.9 at low densities. Food availability can also affect litter size. In years of high rodent density, mean litter size is higher (5.8–6.2) than in years of reduced densities of rodents (4.4–5.1) (Gier 1968). The sex ratio is about 1:1, though it *may* vary with the level of control to which a local population is subjected (Berg and Chesness 1978; Kleiman and Brady 1978; Knowlton 1972).

Development. Young are born blind and helpless in a den. Birth weight is about 240–275 gm and the length of the body from the tip of the head to the base of the tail is about 160 mm (Bekoff and Jamieson 1975; Gier 1968). Young are cared for by the mother and possibly by other "helpers," usually siblings from a previous year (Bekoff and Wells 1980). The presumed father and other males may also help to rear the young by providing food as the young grow older. Young are weaned at about 5–7 weeks of age (Snow 1967) and begin to eat solid food at about 3 weeks, when the caregivers regurgitate semisolid food. Between birth and week 8 the average weight increase per week is about 0.31 kg. The pups reach adult weight at about 9 months of age. Eyes open at about 14 days. Teeth erupt on the average as follows: upper canines, day 14; lower canines and upper incisors, day 15; and lower incisors, day 16 (Bekoff and Jamieson 1975). The young are able to urinate and defecate on their own by 2–3 weeks. They emerge from the den at about 3 weeks.

ECOLOGY

More is known about the ecology of coyotes than, perhaps, any other carnivore. Much of the information has been collected because of control and management interests. Yet, the marked lack of success of most control programs (see below) suggests the strong possibilities that little of the information is useful, that it has not been correctly used, or that more data are needed. It is fortunate that more federal money is being put into behavioral/ecological research (U.S.D.I. 1978: 48): there has been about a fivefold increase between 1970 and 1978. Coyotes live in a wide variety of habitats, including grasslands, deserts, and mountains. They do not compete well with larger carnivores such as wolves (Mech 1966, 1974) and pumas (Young 1951) and do not tolerate foxes (*Vulpes* spp.; *Urocyon cinereoargenteus*) or bobcats (*Felis rufus*) (Young 1951).

Dens. Coyotes den in a wide variety of places, including brush-covered slopes, steep banks, rock ledges, thickets, and hollow logs. Dens of other animals (e.g., badgers, *Taxidea taxus*) are frequently used. Dens may have more than one entrance and often there are many interconnecting tunnels. The same den may be used

TABLE 20.2. Distances (km) of coyote movements in 10 different studies

Study	Adults		Juveniles		Both sexes
	Males	Females	Males	Females	
Garlough 1940					42.9
Robinson and Cummings 1951	12.6	17.8			16.8
Young 1951	36.2				
Young 1951	45.3	40.0			
Robinson and Grand 1958	45.6	34.2			40.6
Hawthorne 1971	6.4	7.6	5.2	6.4	
Chesness 1972	10.1		6.4	6.6	
Gipson and Sealander 1972	20.5	8.16		7.4	
Nellis 1975	6.0	7.0	6.0	26.0	
Nellis and Keith 1976	5.5	6.6	28.2 (5.8)[a]	31.5 (23.5)[b]	
Andrews and Boggess 1978	30.2	31.2[c]			

[a]male pups
[b]female pups
[c]excludes data for 1 female that moved a record distance of 323.2 km.

from year to year. Den sharing occurs only rarely (Camenzind 1978; Nellis and Keith 1976). Movement of pups from one den area to another is very common. The reasons for these moves are not known but disturbance and possibly infestation by parasites may be factors. Most moves are over relatively short distances; however, moves over 2–4 km are not uncommon (Bekoff and Wells unpublished data). Van Wormer (1964) reported that a male coyote moved four pups, one at a time, a distance of 8.0 km.

Population Age Structure and Mortality. Coyotes of different ages have different mortality rates, and, furthermore, mortality, of course, depends on the level of control to which a population is exposed. Pups and individuals less than one year of age tend to have the highest mortality rate (68 percent, Nellis and Keith 1976; 67 percent, Robinson and Cummings 1951). For individuals over 1 year of age, mortality rates vary geographically (36–42 percent, Nellis and Keith 1976; 45 percent, Bowen 1978; 39 percent, Andrews and Boggess 1978; 40 percent, Knowlton 1972). Mathwig (1973) found greatest life expectancy for coyotes in Iowa at 1 ½ years of age and lowest life expectancy at 5 ½ years. Knowlton (1972) reported relatively high survival in coyotes between 4 and 8 years of age. About 70–75 percent of coyote populations are between 1 and 4 years of age (see Mathwig's [1973] summary of seven studies and Berg and Chesness 1978:239). To maintain population stability, a net survival of about 33–38 percent seems to be necessary (Knowlton 1972; Nellis and Keith 1976).

Coyotes in captivity may live as long as 18 years (Young 1951), but in wild populations life expectancy is considerably shorter. Maximum ages of wild individuals are 13.5 (Nellis and Keith 1976) and 14.5 years (Knowlton 1972).

Activity and Movements. Coyotes may be active throughout the day, but they tend to be more active during the early morning and around sunset. Gipson and Sealander (1972) showed a principal activity peak at sunset and a minor peak at daybreak in Arkansas. In summer, they found that animals were more active during the day. In Jackson Hole, Wyoming, coyotes show sporadic activity throughout the night (Wells and Bekoff unpublished data). Seasonal activity patterns are also obvious, especially during winter months, when there is a change in food base. For example, coyotes living in Grand Teton National Park (in the area of Blacktail Butte) rest more during winter months, when they are dependent primarily on ungulate carrion for food, than during other seasons, when they feed mainly on small rodents (Bekoff and Wells 1981). Hunting attempts also decrease during winter months, when rodents are relatively inaccessible due to snow-covered ground and because the major rodent food items (Uinta ground squirrels, *Spermophilus armatus*) are hibernating. Coyotes living in packs rest more and travel less during winter months than do animals living as mated pairs or alone. Large energy saving has been shown for a pregnant female living in a pack than for a female living alone with her mate (Bekoff and Wells 1981). However, more data are needed to establish whether or not pack-living females are better off reproductively because of the energy saved during pregnancy.

Coyotes, like wolves, undergo movements within "territories" and home ranges, disperse from the natal site, and also make long migrations. Although there

TABLE 20.3. Mean coyote home range sizes (km²) from some representative studies

Study and Location	Adults		Juveniles & Pups		Comments
	Males	Females	Males	Females	
Chesness and Bremicker 1974 (Minnesota)	41.9	10.0			There was considerable overlap between home ranges of males and no evidence of territoriality; for females there was little overlap and they appeared to be territorial.
Berg and Chesness, 1978 (Minnesota)	68.0	16.0			Same comments as for Chesness and Bremicker 1974, above.
Gipson and Sealander 1972 (Arkansas)	32.8	13.1		11.8	Home ranges tended to be smaller than those reported for western coyotes; unable to determine if the animals were territorial.
Hibler 1977 (Utah)	17.8	20.2			
Bowen 1978 (Alberta)	13.9[a] 14.9[c] 14.2[d]	10.3[a] 16.3[c] 13.3[d]	5–8[b]		Coyotes were territorial; they appeared to mark boundaries with urine. (See also Bowen 1980.)
Litvaitis 1978 (Oklahoma)	15.0	27.9		21.3[e]	Adjacent male-female and female-female home ranges overlapped.
Bekoff and Wells unpublished (Wyoming)	31.33	35.83			Pack members were territorial (see Wells and Bekoff 1981).

NOTE: In addition to there being large geographical differences, individual variability has been found in most studies that is related to sex, age, and social organization.
[a]summer, no difference between sexes
[b]males and females
[c]winter, no difference between sexes
[d]overall, no difference between sexes
[e]yearling females; home range for pups was 0.4 km²

are no consistent sex differences in dispersal distance, in some cases female pups may move farther than male pups (Nellis and Keith 1976). Dispersal by juveniles usually occurs during autumn and early winter, though some individuals do not disperse during the first year and remain to provide care for future siblings. Dispersal, in most cases, occurs randomly in all directions, and pups will move more than 80–160 km. Berg and Chesness (1978) reported mean dispersal distances of 48 km that occurred at a mean rate of about 11 km/wk. Data from coyote movements in 10 studies are presented in table 20.2.

TABLE 20.4. Coyote densities (individuals/km²) in different geographic areas and seasons

Study	Location	Density
Gier 1968	Kansas	0.8[a]
Knowlton 1972	Texas	0.9;[a] 1.5–2.3[b]
Chesness and Bremicker 1974	Minnesota	0.2–0.4[a]
Nellis and Keith 1976	Alberta	0.1–0.6[c]
Bowen 1978	Alberta	0.46;[a] 0.35[d]

[a]postwhelping
[b]fall
[c]winter
[d]late winter

Home Range and Territory. Home range size, or the area used on a regular basis but not actively defended, varies geographically and seasonally, and within populations there is a lot of individual variability as well (Bekoff and Wells 1980; Bowen 1978; Edwards 1975; Gipson and Sealander 1972; Springer 1977). Home range size is influenced by social organization. Coyotes living in packs who defend ungulate carrion during the winter have much smaller, compressed home ranges ($\bar{X}=14.3$ km²) than coyotes living in pairs or alone (30.1 km²; Bekoff and Wells 1980). The most solid evidence demonstrating that coyotes actively defend well-defined territorial boundaries is provided by Bekoff and Wells (1980), Bowen (1978), and Wells and Bekoff (1981). Typically, only pack members defend territories; pairs of coyotes and solitary individuals do not (Bekoff and Wells 1980). (See tables 20.3 and 20.4.)

FOOD HABITS

Coyotes eat a wide variety of food items. Gipson (1974) found regional and seasonal differences in food habits of coyotes living in Arkansas. Overall, the most common food item from 168 coyote stomachs were poultry (34 percent occurrence), persimmons (23 percent), insects (11 percent), rodents (9 percent), songbirds (8 percent), cattle (7 percent), rabbits (7 percent), deer (5 percent), woodchucks (4 percent), goats (4 percent), and watermelon (4 percent). Korschgen (1973), in an extensive analysis of coyote food habits in Missouri, also found seasonal differences. For example, rabbits were found in 57 percent of coyote stomachs in winter and in 14.8 percent in spring. On the other hand, carrion was found in 15.6 percent of stomachs in winter and in 37.0 percent in spring. Furthermore, Korschgen (1973) concluded that coyotes did not prey heavily on livestock. Indeed, most researchers agree that livestock and wild ungulates are represented in coyote stomachs and scats as carrion (Bekoff and Wells 1980; Korschgen 1973; Murie 1935; Murie 1951; Ozga and Harger 1966; Weaver 1977); actual predation is rare (Hamlin and Schweitzer 1979; Henderson 1977). In the Jackson Hole area of Wyoming, the percentage of coyote stomachs containing ungulate meat from scavenging carrion may increase as much as threefold in winter months when compared to summer (Weaver 1977; see also Houston 1978). For other studies of food habits, see Sperry (1933a, 1933b), Young (1951), Gier (1968), and Holte (1978).

BEHAVIOR

Direct observation of coyote behavior in the wild is difficult due to the elusive nature of the species. However, good progress has been made recently (Bekoff and Wells 1980, 1981; Bowen 1978; Camenzind 1978; Wells and Bekoff 1981). Detailed longitudinal studies of identified individuals of known parentage and genetic relationship are still forthcoming; one such study has been in progress in Grand Teton National Park for the past four years (Bekoff and Wells study in progress) and will continue for at least the next three years. Data on behavioral development are presented in Bekoff (1972, 1974, 1977b, 1978). Coyotes show early development of aggressive behavior when compared to wolves and most domestic dogs; they will engage in serious fights when they are only three to five weeks of age. The significance of the early development of rank relationships within litters to later behavior is still unknown (Bekoff 1977b). General behavioral patterns (postures, gestures, tail movements, facial expressions, vocalizations) are described and discussed in Bekoff (1972, 1974, 1978), Fox (1970), Kleiman (1966), Lehner (1978a, 1978b), and Wells and Bekoff (1981). Gait, stance, ear and tail position, and retraction of the lips to expose the teeth are all very important in social communication and may vary independently or together, depending on individual "mood." Comparative reviews are presented in Kleiman and Eisenberg (1973) and Kleiman and Brady (1978). Here, I shall briefly consider social organization and predatory behavior.

Social Organization. Generally, coyotes are less social than wolves. The basic social unit is the adult heterosexual pair (Bekoff and Wells 1980; Bowen 1978). Camenzind (1978) recognized four "levels" of organization, ranging from solitary individuals (nomads) to "packs," that occupied the same geographical area, maintained social hierarchies, and often fed and denned together (see also Bowen 1978). Coyote packs resemble wolf packs (Bekoff and Wells

1980; Bowen 1978) and Bowen concluded that the differences between coyotes and wolves were quantitative and not qualitative.

Just as with all other aspects of coyote biology, there is a considerable amount of variability in observed social organization (Bekoff and Wells 1980, 1981; Bowen 1978). In many areas, solitary individuals are most frequently observed (Bekoff and Hill unpublished data for Rocky Mountain National Park; Berg and Chesness 1978) outside of the breeding season. In other areas, such as Jackson Hole, Wyoming (Bekoff and Wells 1980, 1981; Camenzind 1978), and Jasper, Alberta (Bowen 1978), groups of coyotes are very frequently observed. One factor that seems to be of utmost importance in affecting coyote sociality is prey size (Bekoff and Wells 1980; Bowen 1978). In populations where the major prey items throughout the year are small rodents, coyotes tend to be solitary. In populations where large animals are available (e.g., elk, deer) either as live individuals or as carrion, large groups of coyotes are formed. But coyote groups are not necessarily formed for taking down large prey, though coyotes will occasionally hunt as a pair or a group (Hamlin and Schweitzer 1979; Henderson 1977). Rather, cooperative group defense appears to be the major selective force favoring increased sociality (Bekoff and Wells 1980; Berger 1978; Bowen 1978; Lamprecht 1978). Yet it has been observed that one bald eagle (*Haliaeetus leucocephalus*) could prevent five coyotes from feeding on cow carrion (Wells and Bekoff 1978). It is important to note that coyotes tend to be more social during winter, when carrion is a very important food resource (Bekoff and Wells 1980; Bowen 1978; Camenzind 1978). It is also possible that the level of control to which a population is exposed can affect sociality.

Predatory Behavior. Very few detailed data are available concerning the predatory behavior of wild coyotes. Predatory sequences may be divided into at least six components: search, orientation, stalk, pounce, head thrust/close search into ground cover, and rush (Wells and Bekoff unpublished data). Pouncing is used mostly for capturing small microtine rodents, while the rush is used most frequently on larger animals, such as Uinta ground squirrels. It has been shown experimentally that coyotes depend on various senses to locate prey. In order of decreasing importance they are vision, audition, and olfaction (Wells and Lehner 1978), though these "priorities" may change depending on environmental conditions (Wells 1978). Developmental data on predatory behavior are present in Vincent and Bekoff (1978).

Because of the impact of coyotes and other predators (including dogs [Boggess et al. 1978]) on the livestock industry (see Control and Management), it is shocking that so little is known about the predatory habits of wild coyotes on domestic livestock. Connolly et al. (1976) studied the sheep-killing behavior of captive coyotes. The results were very interesting, though inconclusive. Coyotes killed sheep (in confinement) only in 20 of 38 tests (52.6 percent). Mean latency to attack was very long (47 min), and there was considerably variability (standard deviation [sd] = 48 min). Mean killing time, likewise, was rather long (13 min), and again considerable variability was evident (sd = 13). Needless to say, it would be highly unlikely that this inefficient predatory behavior would have evolved due to natural selection. However, when one considers the behavior of sheep, animals that have been bred for their food- and wool-producing abilities with a concomitant loss of virtually all defensive behavior, it is understandable that the coyote would "take its time" in killing them. Defensive behavior by sheep in the above study deterred coyotes only 31.6 percent of the time. In most instances, coyotes attacked by biting the throat and sheep died of suffocation. Gluesing (1977) found no behavioral differences between lambs caught by coyotes and those that escaped predation. He concluded that the probability of being captured was related more to the lamb's being on the periphery of the bedground than to behavior. Rock (1978) found that coyote-killed lambs weighed less than the pasture average. Ogle (1971) found it was possible to distinguish to a certain degree between deer killed by coyotes and those fed on as carrion. The most important criterion for adult deer kills was the presence of large patches of hide leading to the carcass. Other general data on coyote predatory behavior are reported in Bekoff (1977a) and Kauffeld (1977).

MORTALITY

Coyotes are inflicted with a wide variety of parasites and diseases. These include fleas (the most common external parasite), various ticks, lice, cestodes, roundworms, intestinal worms, hookworms, heartworms, whipworms, pinworms, thorny-headed worms, lungworms, and coccidia fungus. Coyotes also carry tularemia, rabies distemper, and bubonic plague and may suffer from aortic aneurysms, cardiovascular diseases, mange, and cancer (for reviews and taxonomic analyses of parasites, see Bekoff 1977a and Gier et al. 1978).

CONTROL AND MANAGEMENT

From civilized man's selfish point of view, predators are commonly looked upon as pests or outlaws with almost every hand raised against them. In fairness to these animals, it should always be kept in mind that their destructive habits cannot be due to any criminal intent, but are due wholly to their efforts to gain a livelihood by the only means that nature has provided through untold ages of evolution. From this just point of view anyone, except perhaps those who have recently suffered heavy losses from their depredations, must be impressed by the skill with which they carry on the fight for existence, which with the exception of the coyote, is a steadily losing battle. . . . Predatory animals which once lived solely on native game and rodents very promptly learned that the clumsy and in most in-

stances stupid domestic animals introduced in their territory by man were a more certain and easier prey than the wild things that were always on guard against them, thus, they followed the lines of least resistance in gaining a livelihood through the same mental processes that man would have used under similar conditions. (Young 1951:225)

It occurred to me early on in my association with wolves that I was distrustful of science. Not because it was unimaginative, though I think that this is a charge that can be made against wildlife biology, but because it is narrow. I encountered what seemed to be eminently rational explanations for why wolves did some of the things they did, only to find wildlife biologists ignoring these ideas . . . it was admittedly taking quite a leap to extrapolate from the behavior of captive animals to include those in the wild. (Lopez 1978:77)

The coyote is a victim of success. By taking advantage of poorly devised domestication practices that left most livestock virtually defenseless against predation, coyotes have established a reputation of being fierce and ruthless predators. They have even narrow-mindedly been referred to as being a dispensable species (Shelton 1973). The lack of success of control programs is, in fact, rather easy to understand and does not necessarily involve pitting one federal agency against another (Howard 1974). Rather, the major problem is that most control programs have not been successful because of the basic failure to gather information on predators under field conditions. Instead of sitting back and marveling at the phenomenal success of coyotes despite large-scale control programs, it would be more appropriate to recognize that large-scale control programs that are based on inadequate foundations are no different from other research endeavors. Both are doomed to fail, and they do (Bekoff 1979a).

Economics of Coyote Predation. A good deal of money is spent on predator control. In 1971, coyote control cost the United States government about $8 million (Amory 1973). In California from 1973–74, the cost of governmental control was about $1,015,810 (Nesse et al. 1976). It has been estimated that in 1969, the state of Arizona spent $157,603 to eliminate coyotes, who did approximately $42,211 worth of damage (Cole 1970). There is no doubt that coyotes do prey on various livestock and poultry and have an economic impact. For all practical purposes, some control is necessary.

In 16 studies reviewed by Sterner and Shumake (1978a: 301–3), coyotes were responsible for 82 percent of all sheep losses due to predators. But it is important to stress that only a few flocks typically showed sizeable losses (Sterner and Shumake 1978a). It is also important to consider the fact that coyote predation is *not* the major cause of losses in almost all cases (Bekoff 1979b; USDI 1978) (table 20.5). There is little evidence to support the timeworn notion that coyote predation is a *primary* limiting factor on popu-

TABLE 20.5. Some comparative data for sheep losses to predators, including coyotes

Study	Location	Sheep Losses			
		All Predators	Coyotes	Disease	Other
Cummings 1972	California	$3,691,000[a]			
Early and Brewer 1973	Idaho	$ 180,750 (E)[b]	$ 120,780 (E)	$261,150 (E)	$281,182 (E)
		$ 274,833 (L)[b]	$ 214,947 (L)	$593,081 (L)	$416,745 (L)
DeLorenzo and Howard 1975	New Mexico[c]	53/116 (46%)			44/116 (38%)[d]
Nesse, Longhurst, and Howard 1976	California	$1,677,775	$1,375,775		

[a] livestock and poultry
[b] E = ewe; L = lamb
[c] no control, no adults lost
[d] 19 losses were not accounted for in the categories listed above

lations of big game or domestic livestock (for review, see Bekoff 1977a:5 and Bekoff 1979b). Let us consider some of the more detailed information available from studies of predation on sheep in Idaho. Early et al. (1974) found that 77 percent of the total economic loss of sheep in 1970–71 was due to disease (43 percent) and unspecified causes (31 percent), and only 23 percent was due to predators. Early and Brewer (1973) studied losses of ewes and lambs in Idaho during 1970–71 and presented comparative figures in terms of dollar losses (table 20.5). It is clear that disease and other losses far outweighed losses to coyotes and other predators. For example, the total loss of lambs equaled $1,499,607. Losses to all predators accounted for 18.3 percent, while losses to coyotes were only 14.3 percent. Obviously, steps must be taken to reduce losses to causes other than predators (Henderson et al. 1977). One way that losses can be reduced is by removing carrion, since coyotes are attracted to animal remains and may then try to kill live individuals (Gier 1968:23; Gipson 1974; Todd and Keith 1976). In addition, removal of remains may reduce disease transmission.

Control Methods. Control methods have recently been reviewed extensively by Sterner and Shumake (1978a) and Wade (1978). Two basic approaches are *preventative* (application of methods beforehand to alleviate expected damage) and *corrective* (reducing damage after it begins). Either approach can employ lethal or nonlethal methods. For all procedures, safety, selectivity, and cost efficiency must be considered.

Both lethal and nonlethal approaches can be divided into chemical and nonchemical procedures. Nonlethal chemical methods include drug-induced aversions (aversive conditioning), repellents, and chemosterilants. The use of lithium chloride (LiCl), for example, to produce aversions to sheep (e.g., Gustavson et al. 1974) is a very attractive idea but research in this area has for the most part been methodologically sloppy and the results very inconclusive (Bekoff 1975; Conover, Francik, and Miller 1979; Griffiths et al. 1978; Lehner 1976; Olsen and Lehner 1978; Sterner and Shumake 1978a, 1978b; Wade 1978; but see Cor-

nell and Cornely 1979), despite overzealous interpretations appearing in popular literature. Furthermore, even if these methods can stop consumption of prey, they must also stop killing of prey as well. The search for effective repellents has not been successful because most also have bad effects on sheep (Lehner et al. 1976). With respect to chemosterilants (e.g., diethylstilbestrol), both regulation and distribution have been problematic (Balser 1964; Kennelly 1969; Stellflug, Gates, and Saller 1978). Nonlethal nonchemical procedures include sheep confinement and herding, the use of guard dogs, and the use of exclusion fences. It should be noted that coyotes are quite adept at crossing a wide variety of fences (Thompson 1978). However, Gates et al. (1978) have developed what appears to be a coyoteproof fence (see also DeCalesta and Cropsey 1978; Thompson 1979).

Lethal chemical approaches include M-44 registers (a tube containing a sodium cyanide capsule) and toxic sheep collars (Connelly, Griffiths, and Savarie 1978). Nonchemical lethal methods include trapping and shooting (from airplanes and snowmobiles). Whether or not control programs that remove selected individuals (Knowlton 1972) can be effective remains to be determined. We know still too little about the relationship between individual differences in behavior and predatory behavior (Bekoff 1978).

In summary, coyote control has been relatively ineffective. It is easy to understand the lack of patience by those suffering economic losses. However, instead of continuing to use inefficient methods it might be best to begin by gathering relevant basic data on wild populations of coyotes that may have some bearing on their predatory tendencies as related to livestock and poultry. We also need to know more about coyote behavioral ecology and population dynamics in general. Put simply, it is difficult to manage and control something that is not well understood. Better methods are also needed for assessing damage (Sterner and Shumake 1978a). The latest large-scale assessment of coyote damage in the west (U.S.D.I. 1978) is based on only a 23.45 percent return rate for a mailed questionnaire.

Some of the research reported in this chapter was supported by PHS grant MH29571 and NSF grants 27616 and 23463 to M. Bekoff. Ms. Dion McMain kindly typed the manuscript. The editors of this volume provided very useful comments on an earlier draft of this chapter.

LITERATURE CITED

Adorjan, A. S., and Kolenosky, G. B. 1969. A manual for the identification of hairs of selected Ontario mammals. Ontario Dept. Lands For. Res. Rep. 64pp.

Amory, C. 1973. Little brother of the wolf. Am. Way 6:18–20.

Andelt, W. F. 1977. Ecology of an urban coyote. Proc. Nebraska Acad. Sci. 87:5.

Andrews, R. D., and Boggess, E. K. 1978. Ecology of coyotes in Iowa. Pages 249–265 in M. Bekoff, ed. Coyotes: biology, behavior, and management. Academic Press, New York. 384pp.

Asdell, S. A. 1964. Patterns of mammalian reproduction. Cornell Univ. Press, Ithaca, N.Y. 670pp.

Atkins, D. L. 1978. Evolution and morphology of the coyote brain. Pages 17–35 in M. Bekoff, ed. Coyotes: biology, behavior and management. Academic Press, New York. 384pp.

Atkins, D. L., and Dillon, L. S. 1971. Evolution of the cerebellum in the genus Canis. J. Mammal. 52:96–107.

Balser, D. S. 1964. Management of predator populations with antifertility agents. J. Wildl. Manage. 28:353–358.

Barnum, D. A.; Green, J. S.; Flinders, J. T.; and Gates, N. L. 1979. Nutritional levels and growth rates of hand-reared coyote pups. J. Mammal. 60:820–823.

Bekoff, M. 1972. An ethological study of the development of social interaction in the genus Canis: a dyadic analysis. Ph.D. Dissertation. Washington Univ., St. Louis. 164pp.

_____. 1974. Social play and play-soliciting by infant canids. Am. Zool. 14:323–341.

_____. 1975. Predation and aversive conditioning in coyotes. Science 187:1006.

_____. 1977a. Canis latrans. Mammal. Species 79:1–9.

_____. 1977b. Mammalian dispersal and ontogeny of individual behavioral phenotypes. Am. Nat. 111:715–732.

_____. 1978. Behavioral development in coyotes and eastern coyotes. Pages 97–126 in M. Bekoff, ed. Coyotes: biology, behavior and management. Academic Press, New York. 384pp.

_____. 1979a. Coyote control: the impossible dream? BioScience 29:4.

_____. 1979b. Coyote damage assessment in the west: review of a report. BioScience 29:754.

Bekoff M., and Diamond, J. 1976. Precopulatory and copulatory behavior in coyotes. J. Mammal. 57:372–375.

Bekoff, M.; Hill, H. L.; and Mitton, J. B. 1975. Behavioral taxonomy in canids by discriminant function analysis. Science 190:1223–1225.

Bekoff, M., and Jamieson, R. 1975. Physical development in coyotes (Canis latrans) with a comparison to other canids. J. Mammal. 56:685–692.

Bekoff, M., and Wells, M. C. 1980. Social ecology and behavior of coyotes. Sci. Am. 242:130–148.

_____. 1981. Behavioral budgeting by wild coyotes: the influence of food resources and social organization. Anim. Behav. (in press).

Berg, W. E., and Chesness, R. A. 1978. Ecology of coyotes in northern Minnesota. Pages 229–247 in M. Bekoff, ed. Coyotes: biology, behavior and management. Academic Press, New York. 384pp.

Berger, J. 1979. "Predator harassment" as a defensive strategy. Am. Midl. Nat. 102:197–199.

Boggess, E. K.; Andrews, R. D.; and Bishop, R. A. 1978. Domestic animal losses to coyotes and dogs in Iowa. J. Wildl. Manage. 42:362–372.

Boggess, E. K.; and Henderson, F. R. 1977. Regional weights of Kansas coyotes. Trans. Kansas Acad. Sci. 80:79–80.

Bowen, W. D. 1978. Social organization of the coyote in relation to prey size. Ph.D. Dissertation. Univ. of British Columbia, Vancouver. 230pp.

Bowen, W. D. Scent marking in coyotes. Can. J. Zool. 58:473–480.

Camenzind, F. J. 1978. Behavioral ecology of coyotes on the National Elk refuge, Jackson, Wyoming. Pages 267–294 in M. Bekoff, ed. Coyotes: biology, behavior and management. Academic Press, New York. 384pp.

Chesness, R. A. 1972. Home range and territoriality of coyotes in north-central Minnesota. 34th Midwest Fish Wildl. Conf., Des Moines.

Chesness, R. A., and Bremicker, T. P. 1974. Home range, territoriality, and sociability of coyotes in north-central Minnesota. Coyote Res. Workshop, Denver.

Cole, G. A. 1970. Notes on the cost of coyote meat in Arizona. J. Arizona Acad. Sci. 6:2,85.

Connolly, G. E.; Griffiths, R. E., Jr.; and Savarie, P. J. 1978. Toxic collar for control of sheep-killing coyotes: a progress report. Pages 197–205 in Proc. 8th Vertebr. Pest Conf.

Connolly, C. E.; Timm, R. M.; Howard, W. E.; and Longhurst, W. M. 1976. Sheep killing behavior of captive coyotes. J. Wildl. Manage. 40:400–407.

Conover, M. R.; Francik, J. G.; and Miller, D. E. 1979. Aversive conditioning in coyotes: a reply. J. Wildl. Manage. 43:209–211.

Cornell, D., and Cornely, J. E. 1979. Aversive conditioning of campground coyotes in Joshua Tree National Monument. Wildl. Soc. Bull. 7:129–131.

Cummings, M. W. 1972. Predator management in relation to domestic animals. Nat. Resour. Forum, Ukiah, Calif.

Cunningham, V. D., and Dunford, R. D. 1970. Recent coyote record from Florida. Q. J. Florida Acad. Sci. 33:279–280.

Daniel, W. S. 1973. Investigation of factors contributing to subnormal fawn production and herd growth patterns. Texas Parks and Wildl. Dept. Job Prog. Rep. no. 10.

DeCalesta, D. S., and Cropsey, M. G. 1978. Field test of a coyote-proof fence. Wildl. Soc. Bull. 6:256–259.

DeLorenzo, D. G., and Howard, W. E. 1975. Evaluation of sheep losses on a range lambing operation without predator control in southeastern New Mexico. U.S. Fish and Wildl. Serv., Denver. Prog. Rep.

Early, J. O., and Brewer, G. R. 1973. Preliminary analysis of range and sheep losses in Idaho. Agric. Econ. Ext. Dept., Univ. Idaho.

Early, J. O.; Roetheli, J. C.; and Brewer, G. R. 1974. An economic study of predation in the Idaho range sheep industry, 1970–71 production cycle. Idaho Agric. Res., Univ. Idaho. Prog. Rep. no. 182.

Edwards, L. L. 1975. Home range of the coyote in southern Idaho. M. S. Thesis. Idaho State Univ., Pocatello. 36pp.

Elder, W. H., and Hayden, C. M. 1977. Use of discriminant function in taxonomic determination of canids from Missouri. J. Mammal. 21:17–24.

Fisher, J. 1975. The plains dog moves east. Nat. Wildl. 13:14–16.

Fox, M. W. 1970. A comparative study of the development of facial expressions in canids: wolf, coyote, foxes. Behavior 36:49–73.

Garlough, F. E. 1940. Study of the migratory habits of coyotes. U.S. Bur. of Sport Fish and Wildl., Denver. 5pp.

Gates, N. L.: Rich, J. E.; Godtel, D. D.; and Hulet, C. V. 1978. Development and evaluation of anti-coyote electric fencing. J. Range Manag. 31:151–153.

Gier, H. T. 1968. Coyotes in Kansas. Kansas Agric. Exp. Stn., Kansas State Univ. 118pp.

———. 1975. Ecology and social behavior of the coyote. Pages 247–262 in M. W. Fox, ed. The wild canids. Van Nostrand Reinhold, New York. 508pp.

Gier, H. T.; Kruckenberg, S. M.; and Marler, R. J. 1978. Parasites and diseases of coyotes. Pages 37–71 in M. Bekoff, ed. Coyotes: biology, behavior and management. Academic Press, New York. 384pp.

Gill, D. A. 1965. Coyote and urban man: a geographic analysis of the relationship between the coyote and man in Los Angeles. M.A. Thesis. Univ. of California, Los Angeles. 114pp.

Gipson, P. S. 1974. Food habits of coyotes in Arkansas. J. Wildl. Manage. 38:848–853.

———. 1976. Melanistic *Canis* in Arkansas. Southwest Nat. 21:124–126.

Gipson, P. S.; Gipson, I.K.: and Sealander, J. A. 1975. Reproductive biology of wild *Canis* in Arkansas. J. Mammal. 56:605–612.

Gipson, P. S.; and Sealander, J. A. 1972. Home range and activity of the coyote (*Canis latrans frustror*) in Arkansas. Proc. Annu. Conf. Southeast Assoc. Fish Game Commissioners 26:82–95.

Gluesing, E. A. 1977. Sheep behavior and vulnerability to coyote predation. Ph.D. Dissertation. Utah State Univ., Logan. 120pp.

Grandage, J. 1972. The erect dog penis: a paradox of flexible rigidity. Vet. Rec. 91:141–147.

Griffiths, R. E., Jr.; Connolly, G. E.; Burns, R. J.; and Sterner, R. T. 1978. Coyotes, sheep, and lithium chloride. Pages 190–196 in Proc. 8th Vertebr. Pest Conf.

Gustavson, C. R.; Garcia, J.; Hankins, W. G.; and Rusinak, K. W. 1974. Coyote predation control by aversive conditioning. Science 184:581–583.

Hall, E. R., and Kelson, K. R. 1959. Mammals of North America. Ronald Press, New York. 2: 547–1083.

Hamlett, G. W. D. 1939. The reproductive cycle of the coyote. U.S.D.A. Tech. Bull. 616:1–11.

Hamlin, K. L., and Schweitzer, L. L. 1979. Cooperation by coyote pairs attacking mule deer fawns. J. Mammal. 60:849–850.

Hawthorne, V. M. 1971. Coyote movements in Sagehen Creek Basin, northeastern California. California Fish Game 57:154–161.

———. 1972. Coyote food habits in Sagehen Creek Basin, northeastern California. California Fish Game 58:4–12.

Heinrich, D. 1972. Vergleichende Untersuchungen an Nebennieren einger Arten der Familie Canidae Gray 1821. Z. Wiss. Zool. Leipzig 185:122–192.

Henderson, F. R.; Boggess, E. K. and Brown, B. A. 1977. Understanding the coyote. Coop. Ext. Serv. Kansas State Univ. Rep. no. C-578.

Henderson, R. E. 1977. A winter study of coyote predation on white-tailed deer in the Miller drainage, Montana. M. S. Thesis. Univ. Montana, Missoula.

Hibler, S. J. 1977. Coyote movement patterns with emphasis on home range characteristics. M.S. Thesis. Utah State Univ., Logan. 84pp.

Hildebrand, M. 1952. The integument in Canidae. J. Mammal. 33:419–428.

Hilton, H. 1978. Systematics and ecology of the eastern coyote. Pages 209–228 in M. Bekoff, ed. Coyotes: biology, behavior and management. Academic Press, New York. 384pp.

Hilton, H., and Kutscha, N. P. 1978. Distinguishing characteristics of the hairs of eastern coyote, domestic dog, red fox and bobcat in Main. Am. Midl. Nat. 100:223–227.

Holle, D. 1978. Food habits of coyotes in an area of high fawn mortality. Proc. Oklahoma Acad. Sci. 58:11–15.

Horn, S. W., and Lihner, P. N. 1975. Scotopic sensitivity in the coyote (*Canis latrans*). J. Comp. Physiol. Psychol. 89:1070–1076.

Houston, D. B. 1978. Elk as winter-spring food for carnivores in northern Yellowstone National Park. J. Appl. Ecol. 15:653–661.

Howard, W. E. 1949. A means to distinguish skulls of coyotes and domestic dogs. J. Mammal. 30:169–171.

———. 1974. Predator control: whose responsibility? BioScience 24:359–363.

Jackson, H. H. T. 1951. Part 2, Classification of the races of coyotes. Pages 227–341 in S. P. Young and H. H. T. Jackson. The clever coyote. Wildl. Manage. Inst., Washington, D.C. 411pp.

Johnson, M. K., and Hansen, R. M. 1979. Coyote food habits on the Idaho National Engineering Laboratory. J. Wildl. Manage. 43:951–956.

Kauffeld, J. D. 1977. Availability of natural prey and its relationship to coyote predation on domestic sheep. M. S. Thesis. Univ. Nevada, Reno. 144pp.

Kennelly, J. J. 1969. The effect of mestranol on canine reproduction. Bio. Reprod. 1:282–288.

———. 1972. Coyote reproduction. Part 1, The duration of the spermatogenic cycle and epididymal sperm transport. J. Reprod. Fert. 31:163–170.

———. 1978. Coyote reproduction. Pages 73–93 in M. Bekoff, ed. Coyotes: biology, behavior and management. Academic Press, New York. 384pp.

Kennelly, J. J., and Roberts, J. D. 1969. Fertility of coyote-dog hybrids. J. Mammal. 50:830–831.

Kleiman, D. G. 1966. Scent marking in Canidae. Symp. Zool. Soc. London 18:167–177.

Kleiman, D. G., and Brady, C. A. 1978. Coyote behavior in the context of recent canid research: problems and perspectives. Pages 163–188 in M. Bekoff, ed. Coyotes: biology, behavior and management. Academic Press, New York. 384pp.

Kleiman, D. G., and Eisenberg, J. F. 1973. Comparison of canid and felid social systems from an evolutionary perspective. Anim. Behav. 21:637–659.

Knowlton, F. F. 1972. Preliminary interpretations of coyote population mechanics with some management implications. J. Wildl. Manage. 36:369–382.

Kolenosky, G. B. 1971. Hybridization between wolf and coyote. J. Mammal. 52:446–449.

Korschgen, L. J. 1973. Food habits of coyotes in north-central Missouri. Missouri Dept. Conserv. Fed. Aid Proj. no. W-13-R-27.

Lamprecht, J. 1978. The relationship between food competition and foraging size group in some larger carnivores. Z. Tierpsychol. 46:337–343.

Lawrence, B., and Bossert, W. H. 1967. Multiple character analysis of *Canis lupus, latrans* and *familiaris*, with a discussion of the relationships of *Canis niger*. Am. Zool. 7:223–232.

———. 1969. The cranial evidence of hybridization in New England *Canis*. Breviora 330:1–13.

———. 1975. Relationships of North American *Canis* shown by a multiple character analysis of selected populations. Pages 73–96 in M. W. Fox, ed. The wild canids- Van Nostrand Reinhold, New York. 508pp.

Lehner, P. N. 1976. Coyote behavior: implications for management. Wildl. Soc. Bull. 4:120–126.

———. 1978a. Coyote communication. Pages 127–162 in M. Bekoff, ed. Coyotes: biology, behavior and management. Academic Press, New York, 384pp.

———. 1978b. Coyote vocalizations: a lexicon and comparisons with other canids. Anim. Behav. 26:712–722.

Lehner, P. N.; Krumm, R.; and Cringan, A. 1976. Tests for olfactory repellents for coyotes and dogs. J. Wildl. Manage. 40:145–150.

Linhart, S.; and Knowlton, F. 1967. Determining age of coyotes by tooth cementum layers. J. Wildl. Manage. 31:362–365.

Litvaitis, J. A. 1978. Movements and habitat use of coyotes on the Wichita Mountains National Wildlife Refuge. M.S. Thesis. Oklahoma State Univ., Stillwater. 70pp.

Lopez, B. H. 1978. Of wolves and men. Scribners, New York. 309pp.

Mahan, B. R. 1977. Comparison of coyote and coyote x dog hybrid food habits in southeastern Nebraska. Prairie Nat. 9:50–52.

———. 1978. Occurrence of melanistic canids in Nebraska. Proc. Nebraska Acad. Sci. 5:121–122.

Mathwig, H. J. 1973. Food and population characteristics of Iowa coyotes. Iowa State J. Res. 47:167–189.

Mech. L. D. 1966. The wolves of Isle Royale. U.S. Nat. Park Serv. Fauna Ser. 7:1–210.

———. 1974. Canis lupus. Mammal. Species 37:1–6.

Mengel, R. M. 1971. A study of coyote-dog hybrids and implications concerning hybridization in Canis. J. Mammal. 52:316–336.

Murie, A. 1940. Ecology of the coyote in the Yellowstone. U.S. Nat. Park Serv. Fauna Ser. 4:1–206.

Murie, O. J. 1935. Food habits of the coyote in Jackson Hole. U.S.D.A. Circ. 362:1–24.

———. 1951. The elk of North America. Wildl. Manage. Inst. Washington, D.C.

Nellis, C. H. 1975. Population dynamics and ecology of coyotes in central Alberta. Coyote Res. Newsl. 3:5.

Nellis, C. H., and Keith, L. B. 1976. Population dynamics of coyotes in central Alberta, 1964–68. J. Wildl. Manage. 40:380–399.

Nellis, C. H.; Wetmore, S. P.; and Keith, L. B. 1978. Age-related characteristics of coyote canines. J. Wildl. Manage. 42:680–683.

Nesse, G. E.; Longhurst, W. M.; and Howard, W. E. 1976. Predation and the sheep industry in California. Univ. California, Davis, Coop. Ext. Bull. 1878.

Nowak, R. 1978. Evolution and taxonomy of coyotes and related Canis. Pages 3–16 in M. Bekoff, ed. Coyotes: biology, behavior and management. Academic Press, New York. 384pp.

Ogle, T. F. 1971. Predator-prey relationships between coyotes and white-tailed deer. Northwest Sci. 45:213–218.

Ogle, T. F., and Farris, E. M. 1973. Aspects of cold adaptation in the coyote. Northwest Sci. 47:70–74.

Olsen, A., and Lehner, P. N. 1978. Conditioned avoidance of prey in coyotes. J. Wildl. Manage. 42:676–679.

Paradiso, J. L., and Nowak, R. 1971. A report on the taxonomic status and distribution of the red wolf. U.S. Fish Wildl. Serv. Spec. Sci. Rep. 145:1–36.

Petersen, E. A.; Heaton, W. C.; and Wruble, S. D. 1969. Levels of auditory response in fissiped carnivores. J. Mammal. 50:566–587.

Richens, V. B., and Hugie, R. D. 1974. Distribution, taxonomic status, and characteristics of coyotes in Main. J. Wildl. Manage. 38:447–454.

Riley, G. A., and McBride, R. T. 1975. A survey of the red wolf (Canis rufus). Pages 263–277 in M. W. Fox, ed. The wild canids. Van Nostrand Reinhold, New York. 508pp.

Roberts, J. D. 1978. Variation in coyote age determination from annuli in different teeth. J. Wildl. Manage. 42:454–456.

Robinson, W. B., and Cummings, M. W. 1971. Movements of coyotes from and to Yellowstone National Park. Spec. Wildl. Sci. Rep. 11.

Robinson, W. B., and Grand, E. F. 1958. Comparative movements of bobcats and coyotes as disclosed by tagging. J. Wildl. Manage. 22:117–122.

Rock, T. W. 1978. An evaluation of seasonal coyote control techniques and sheep losses in Saskatchewan. M. S. Thesis. Univ. Nevada, Reno. 54pp.

Seitz, A. 1965. Früchtbare Kreuzungen Goldschakal x Coyote und veziprok Coyote x Goldschakal; erste früchtabre Ruckkreuzung. Zool. Gart. 31:174–183.

Shelton, M. 1973. Some myths concerning the coyote as a livestock predator. BioScience 23:719–720.

Shield, J. 1972. Acclimation and energy metabolism of the dingo, Canis dingo, and the coyote, Canis latrans. J. Zool. London 168:483–501.

Silver, H., and Silver, W. T. 1969. Growth and behavior of the coyote-like canid of northern New England with observations on canid hybrids. Wildl. Monogr. 17:1–41.

Snow, C. J. 1967. Some observations on the behavioral and morphological development of coyote pups. Am. Zool. 7:353–355.

Sperry, C. C. 1933a. Autumn food habits of coyotes: a report of progress. J. Mammal. 14:216–220.

———. 1933b. Winter food habits of coyotes: a report of progress. J. Mammal. 15:286–290.

Springer, J. T. 1977. Movement patterns of coyotes in south-central Washington as determined by radio telemetry. Ph.D. Dissertation. Washington State Univ., Pullman. 109pp.

Stellflug, J. N.; Gates, N. L.; and Sasser, R. G. 1978. Reproductive inhibitors for coyote population control: developments and current status. Pages 185–189 in Proc. 8th Vertebr. Pest Conf.

Sterner, R. T., and Shumake, S. A. 1978a. Coyote damage-control research: a review and analysis. Pages 297–325 in M. Bekoff, ed. Coyotes: biology, behavior and management. Academic Press, New York. 384pp.

Sterner, R. T., and Shumake, S. A. 1978b. Bait-induced prey aversions in predators: some methodological issues. Behav. Bio. 22:565–566.

Thompson, B. C. 1978. Fence-crossing behavior exhibited by coyotes. Wildl. Soc. Bull. 6:14–17.

———. 1979. Evaluation of wire fences for coyote control. J. Range Manage. 32:457–461.

Time. 11 August 1975. P. 44.

Todd, A. W., and Keith, L. B. 1976. Responses of coyotes to winter reductions in agricultural carrion. Alberta Rec., Parks and Wildl., Fish and Game Div., Wildl. Tech. Bull. no. 5. 32pp.

U.S. Department of the Interior. 1978. Predator damage in the west: a study of coyote management alternatives. U.S. Fish Wildl. Serv. Spec. Sci. Rep.

Utsler, H. W. 1974. An evaluation of potential criteria for estimating age of coyotes. M.A. Thesis. Oklahoma State Univ. 28pp.

Vincent, L., and Bekoff, M. 1978. Quantitative analyses of the ontogeny of predatory behavior in coyotes, Canis latrans. Anim. Behav. 26:225–231.

Wade. D. A. 1978. Coyote damage: a survey of its nature and

scope, control measures and their application. Pages 347–368 *in* M. Bekoff, ed. Coyotes: biology, behavior and management. Academic Press, New York. 384pp.

Weaver, J. L. 1977. Coyote–food base relationships in Jackson Hole, Wyoming. M.S. Thesis. Utah State Univ., Logan. 88pp.

Wells, M. C. 1978. Coyote senses in predation: environmental influences on their relative use. Behav. Proc. 3:149–158.

Wells, M. C., and Bekoff, M. 1978. Coyote–bald eagle interactions at carrion. J. Mammal. 59:886–887.

_____. 1981. An observational study of scent marking by wild coyotes. Anim. Behav. 29:332–350.

Wells, M. C., and Lehner, P. N. 1978. The relative importance of the distance senses in coyote predatory behavior. Anim. Behav. 26:251–258.

Van Wormer, J. 1964. The world of the coyote. Lippincott, Philadelphia. 150pp.

Wurster, D. H., and Benirschke, K. 1968. Comparative cytogenetic studies in the order Carnivora. Chromosoma 24:336–382.

Young, S. P. 1951. Part 1, Its history, life habits, economic status, and control. Pages 1–226 *in* S. P. Young and H. H. T. Jackson. The clever coyote. Wildl. Manage. Inst., Washington, D.C. 411pp.

MARC BEKOFF, University of Colorado, Department of Environmental, Population and Organismic Biology, Behavioral Biology Group, Boulder, Colorado 80309.

21

Wolves

Canis lupus and Allies

John L. Paradiso
Ronald M. Nowak

NOMENCLATURE

COMMON NAMES. Gray wolf, timber wolf
SCIENTIFIC NAME. *Canis lupus*
SUBSPECIES. For North America, Hall and Kelson (1959) listed 24 subspecies of *C. lupus,* none described more recently than 1943. It is doubtful whether a systematist revising the wolves today would designate so many subspecies, and possibly none would be recognized. As yet, however, there has been no formal synonymization of any of these subspecific names, and they are still used regularly in literature and in conservation programs. Moreover, it has even been suggested that one subspecies (*C. lupus lycaon*) actually is separable into two distinct entities in Ontario: a larger, "boreal" type in the north and a smaller, "Algonquin" type in the southeast (Kolenosky and Standfield 1975).

Our own studies, involving mainly cranial morphology, indicate that the gray wolf varies in gradual clines over much of North America, and that there are few meaningful places to draw lines separating one kind from another. On the average, the largest skulls of gray wolves are from the northwestern part of the continent and the smallest are from Mexico and southeastern Canada. A relatively abrupt transition in size seems to occur between specimens taken in the main part of Alaska, the Yukon, the mainland Northwest Territories, interior British Columbia, and Alberta, and those taken in the panhandle of southeastern Alaska, coastal British Columbia, and the western conterminous United States. Skulls from the northern arctic islands usually are relatively much broader than those from elsewhere.

EVOLUTION

The gray wolf is a member of the Canidae, or dog family, which is part of the order Carnivora. It is generally considered to be among the most morphologically primitive of the living carnivores, and, along with the coyotes (*C. latrans*), usually is placed at the beginning of systematic treatments of the order. It is probable, however, that foxes of the genus *Vulpes* represent a more primitive group. Indeed, the genus *Canis* seems to have originated from foxlike ancestors in the early to middle Pliocene.

It is not known when the wolf line became distinct from those of other members of *Canis,* but conceivably this could have occurred by the late Pliocene. Early in the subsequent period, the Pleistocene, there apparently was extensive development and diversification of the wolf line in North America, possibly in response to the extinction of the borophagines, a group of massive dogs that had been present through the Pliocene. The first clearly identifible wolf was *Canis edwardii,* known from a few fossils collected in the southwestern part of the continent. The red wolf (*Canis rufus*), another small, relatively primitive species, and perhaps a direct descendant of *C. edwardii,* also arose in the early Pleistocene, and continued to occupy the southeastern quarter of North America until modern times. Meanwhile, still another branch of this stock of small wolves entered the Old World, where it seems eventually to have evolved into the large gray wolf (*Canis lupus*).

The first of the larger wolves to appear in North America was *Canis armbrusteri,* known from extensive mid-Pleistocene material in Maryland and Florida and which probably occurred all across the continent. In the late Pleistocene the dire wolf (*Canis dirus*) appeared from southern Canada to South America. The largest member of the genus ever to exist, it may have evolved from *Canis armbrusteri,* or developed in South America from the earlier stock of small wolves. Its disappearance about 8,000 years ago probably was associated with the sudden extinction of many of the large herbivorous mammals upon which it preyed. It may also have lost in competition with *C. lupus,* which had invaded North America from Eurasia.

DISTRIBUTION

When European settlement of North America began, the gray wolf occupied the whole continent except the southeastern coastal plain, Baja California, the coastal lowlands of Mexico, and that region south of Estado de

460

FIGURE 21.1. Past and present distribution of the gray wolf (*Canis lupus*).

FIGURE 21.2. History of the distribution of the red wolf (*Canis rufus*).

Oaxaca, Mexico (figure 21.1). There are no precise records for most of the state of California; the wolf may have been there, but was eliminated at an early time, in association with the development of a major livestock industry by the Spanish in the eighteenth century. The species inhabited the Island of Newfoundland, Vancouver Island, the islands of the Canadian arctic, several coastal parts of Greenland, and most of the islands of southeastern Alaska. It is not known to have ocurred in the West Indies, on the Queen Charlotte Islands, on Kodiak Island, or on Admiralty, Baranof, and Chichagof islands off southeastern Alaska. In historical time the gray wolf occupied all of Eurasia except the tropical forests of southeast Asia. Its range included Ireland, Great Britain, Sicily, Sakhalin, and Japan, but not Ceylon, Formosa, the Philippines, or the East Indies. Human persecution and habitat modification have eliminated the wolf in many parts of the Old World, especially in Europe, and the species has been entirely extirpated in Ireland, Great Britain, and Japan.

The domestic dog (*Canis familiaris*) probably originated over 15,000 years ago from some South Asian population of the gray wolf, and subsequently spread throughout the world in association with people. Some authorities consider the domestic dog to be only a subspecies of *C. lupus*. The wild dingo (*Canis familiaris dingo*) of Australia apparently descended from dogs introduced by the aboriginal human population.

The red wolf once inhabited the region from central Texas and central Oklahoma to the Atlantic, and from the Gulf of Mexico north to southeastern Kansas, southern Missouri, the Ohio Valley, and southern Pennsylvania. Lawrence and Bossert (1967, 1975) questioned the specific separation of *C. rufus* from *C. lupus*. For evidence supporting continued recognition of the red wolf as a distinct species, for documentation

of its range (figure 21.2), and for more details on the entire above discussion, see Nowak (1979). A brief discussion of the red wolf is presented separately at the close of this chapter.

DESCRIPTION

Externally the gray wolf resembles a large domestic dog of an unspecialized breed, such as a German shepherd, but usually differs in having relatively longer legs, larger feet, and a narrower chest (Banfield 1974). In addition, the wolf's face can be distinguished by its wide tufts of hair that project down and outward from below the ears (Mech 1970). If a long tail is present on a domestic dog, it generally curls upward posteriorly, but a wolf's tail is straight. Adult wolves, except for some melanistic individuals, have white fur around the mouth, but dogs usually have black fur in this area.

The total length of North American specimens of *C. lupus* usually is about 1,300 to 1,800 mm, of which approximately one-fourth is tail length. Shoulder height is about 700 to 800 mm. Among the adults of any one region, males are usually, but not always, larger than females. Mech (1974) stated that males weigh from 20 to 80 kg and females from 18 to 55 kg. Average weight for males is about 30 kg in southeastern Canada and Mexico and 45 kg in northwestern Canada and Alaska. For more discussion on weight in *C. lupus,* see Mech (1970, 1974) and Young and Goldman (1944).

Wolves are digitigrade, walking so that only the toes touch the ground. There are five toes on the front foot, the first being only rudimentary and not reaching the ground but having a well-developed dew claw. The hind foot has four toes. The claws are nonretractile, blunt, and nearly straight. Young (1944) reported that wolf tracks in the Rocky Mountains averaged 90 mm in length and 70 mm in width for the front foot, and 82 mm in length and 64 mm in width for the hind foot. In comparison with those of large dogs, tracks of wolves are more elongated, have the front two toe prints closer together, and show the marks of the two front toenails more prominently.

Pelage. The pelage of wolves consists of long, coarse guard hair, mostly measuring 60 to 100 mm, and much shorter, softer underfur (Young and Goldman 1944; Mech 1974). The fur is considerably longer and denser in northern populations. Dorsal hairs generally are longer and darker than those of the underparts. The longest hairs of all, measuring 120 to 150 mm, are found in the mane, a special erectile part of the pelage that extends along the center of the back from the neck to behind the shoulders. Wolves usually have one long annual molt, beginning in late spring when the old coat is shed. Simultaneously, the new, short summer coat develops, which grows through the fall and winter.

The coloration of wolves is so highly variable that it is generally of little value in ascertaining the geographic origin of the specimens. Over much of its range the "gray" wolf may vary in color from pure white to coal black. The usual color is not gray, but is basically light tan or cream mixed with brown, black, and white. Much of the black is concentrated on the back, the forehead tends to be brown, and the lower parts of the head and body are whitish. Light-colored or all-white wolves predominate in much of the northern arctic, but black individuals also are present there. Dark-colored or all-black wolves are relatively common in Alaska and the interior of western Canada. Distinctly white or black individuals seem to have been less common in the conterminous United States, although white wolves were reported from the Great Plains and black wolves from some of the eastern forest areas. Standfield (1970) stated that, in Ontario, wolves to the north of Lake Superior varied in pelage from white to black but wolves to the east and southeast of Lakes Superior and Huron were invariably gray brown.

Certain specialized hairs are present in the pelage of wolves. Elongated whiskers, or vibrissae, on the muzzle are organs of touch. A group of stiff hairs surrounds the precaudal gland located on the back about 70 mm above the base of the tail. These hairs usually are tipped with black, even in animals that otherwise are completely white (Mech 1970).

Skull and Dentition. The skull of a gray wolf usually has a greatest length of 230–290 mm and a zygomatic width of 120–150 mm. Recently, the largest skulls of *C. lupus* on record, one measuring 305 mm in greatest length, were found in Alberta (Gunson and Nowak 1979). A wolf skull has an elongated rostrum, a broadly spreading zygomata, a heavily ossified braincase, and usually a pronounced sagittal crest (figure 21.3). A skull of *C. familiaris* of equivalent size can readily be distinguished by its much more massive, steeply rising, frontal region (one usual result of which is a higher orbital angle; see Mech 1970), and its relatively smaller teeth. The normal dental formula for all

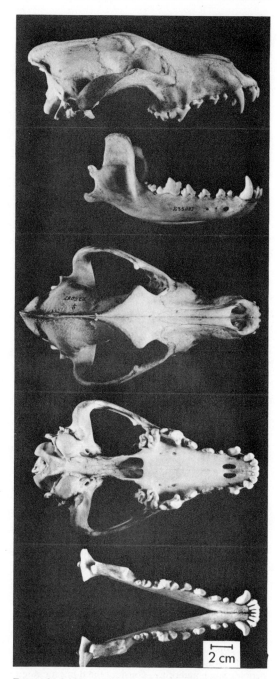

FIGURE 21.3. Skull of the gray wolf (*Canis lupus*). From top to bottom: lateral view of cranium, lateral view of mandible, dorsal view of cranium, ventral view of cranium, dorsal view of mandible.

members of the genus *Canis* is (3/3, 1/1, 4/4, 2/3) x 2 = 42. The incisors are relatively small, and the canines are large, with an exposed dorsoventral length of about 26 mm in *C. lupus*. The fourth upper premolar and the first lower molar form the carnassials.

The molars of wolves retain a flattened or chewing surface, but not to the same extent as in the coyote

(*C. latrans*), which depends more on vegetable matter in its diet.

The red wolf resembles the gray wolf in most respects, but is smaller in average size. Total length usually is about 1,300–1,600 mm, and weight usually 20–35 kg for males and 16–25 kg for females. It reportedly has relatively longer legs, larger ears, and shorter fur. Its color is not really red, as in a red fox (*Vulpes vulpes*); it is much like that of most *C. lupus,* though perhaps sometimes with a stronger rufous tinge to the flanks and limbs. Some gray wolves, however, also are reddish. A dark-colored or black phase of *C. rufus* apparently was locally common in the heavily forested parts of the range of the species.

PHYSIOLOGY AND GENETICS

The internal anatomy of the gray wolf is not known to differ substantially from that of domestic dogs as described by Miller et al. (1964). The digestive system of the gray wolf was discussed in detail by Mech (1970), who commented on its efficiency in absorbing large amounts of meat while ridding itself of indigestible matter such as hair and bone. He also observed that malnutrition probably is not generally a direct threat to the survival of the wolf.

Serological and karyological studies have not yet disclosed a reliable means of distinguishing among *C. lupus, C. rufus, C. latrans,* and *C. familiaris.* All have a diploid chromosome number of 78, and the chromosomes appear identical in each species (Chiarelli 1975). Hybridization appears to occur relatively readily in the genus *Canis.* Viable hybrids have been reported between *C. lupus* and *C. familiaris, C. lupus* and *C. latrans,* and *C. rufus* and *C. latrans* (Gray 1972; Nowak 1979). Large-scale hybridization in the wild has occurred between the gray wolf and the coyote in southeastern Canada, and between the red wolf and the coyote in the south-central United States, and has resulted in the modification of populations of *Canis* over large areas.

REPRODUCTION

Mech (1974) has summarized the breeding data on wolves, and most of the following is based on his discussion. Wolves gain sexual maturity in their second year, but often do not breed until their third year. It is commonly thought that wolves mate for life, and in captivity wolves do demonstrate strong and long-lasting mate preferences. The receptive period of the female may be anytime between January and April, depending on the latitude. Courtship takes place between pack members or between lone wolves that pair during the mating season. The female is in estrus for 5 to 7 days and blood may flow from the vulva for a few days to a few weeks before estrus. Copulation is in the typical canid fashion, with the bulbous base of the male's penis locking into the female's vaginal sphincter, the tie lasting anywhere up to 30 minutes. The gestation period is 63 days. An average of 6 young (1 to 11) are born blind and helpless, usually in a shel-tered place in a hole, rock crevice, hollow log, or overturned stump.

Development of the Pups. The female stays near the young for several months, while the male and other pack members hunt and feed them. The pups' eyes open at 11 to 15 days, and weaning takes place at about five weeks. After about eight weeks, the pups are moved up to an aboveground nest (the "rendezvous site"), where they romp and play over an area up to 0.4 hectare in size. The pups continue their development over the summer and may join the adults of the pack in their travels by October, at which time they are almost of adult size. Adult teeth replace milk teeth in the 16th to 26th month.

From the third week of their lives, wolf pups begin appearing outside the den, romping and playing (Young and Goldman 1944). This marks the beginning of their period of socialization, during which they develop behavior patterns and emotional attachments to places and other wolves (Scott 1967). Play fighting and agonistic behavior during this period eventually help establish the dominance relations that the wolves will develop later.

The period of socialization is also important to the formation of emotional bonds. These emotional attachments are the basis for the formation and continuation of the pack. This process takes place during a period when members of a litter begin following one another and acting as a group (Mech 1970).

Feeding behavior of pups begins to change at this time. At first, the young are forced to nurse while standing, follow the mother around the den, and eat the food regurgitated by adults. Feeding on these semiliquid disgourgements eventually leads to weaning, which is a more gentle process in wolves than in domestic dogs (Ginsburg 1965). Also during the period of socialization, wolf pups learn to run, climb, jump, and play in adult patterns, and the beginnings of predatory behavior can be observed. Mech (1970) reported that his captive male pup snapped "viciously" at raw meat offered to him when he was 34 days old, and chewing and tugging at soft objects was apparent before this age. By the 10th week of life, a wolf pup will menacingly shake mops, and will even chase small animals.

Based on his observations of wolf pups, Mech (1970) felt that the species does not have an inborn tendency to kill, but rather is born with certain behavior patterns that allow it to learn to kill. Apparently, both imitation of killing behavior of the adults and association of killing with eating are important steps in the learning process. This learning process continues on into what Scott and Fuller (1965) call the juvenile period, from about the 12th week to sexual maturity. During this period, the wolf pups do some limited mouse hunting (Murie 1944), gradually replace the milk teeth with permanent dentition, and grow to adulthood.

At about the middle of the juvenile period, 10 months of age, the pups are the size of adults and are participating in the hunt, but the learning of when, where, and how to hunt appears to be a continuing process throughout the life of each animal (Mech 1970). The maximum life expectancy of wolves is 16 years, although a 10-year-old wolf can be regarded as a very old animal.

ECOLOGY

Gray wolves do not seem particular about habitat. They originally occurred in arctic tundra, taiga, plains or steppes, savannahs, and hardwood, softwood, and mixed forests. Examination of their former distribution reveals that gray wolves occupied nearly the entire land surface of the two northern continents (Mech 1970). Young and Goldman (1944) doubted that any other wild land animal ever had a greater range. (It now seems likely that one did; if all Pleistocene lions represented *Panthera leo,* as thought by Hemmer [1974], that species would have occurred throughout Africa, Eurasia, North America, and South America.)

The only major terrestrial barriers to wolves appear to have been the hot, dense forests of southeast Asia and the neotropics, and the hot, dry deserts of northern Africa and Baja California. Wolves even have freely crossed pack ice to occupy Greenland and all the larger islands of the Canadian arctic. Rather than on land form, climate, or vegetation, the presence of wolves seems to have depended on the availability of suitable prey.

Dens. Unlike many mammals, wolves do not make regular use of shelters. A den is constructed only for a female to give birth and care for her young for about two months. Dens usually are located on slopes, ridges, or other high ground, and near a source of water. A typical den is a hole with an entrance 0.36–0.63 meters in diameter, and a tunnel extending 1.3–4.5 m into the ground. There may be several entrances, and the tunnel often curves upward, downward, or sideways. At the end of the tunnel is a nesting chamber that measures about 1.5 m long, 1.2 m wide, and 0.9 m high. No bedding material is used. Wolves may also den in such places as abandoned beaver lodges, hollow logs, rock crevices, or merely surface depressions (Mech 1970, 1974; Stephenson 1974).

Dens may be reused year after year, but occasionally the young are shifted from one den to another during a single season. When the pups are about 8 to 10 weeks old the use of a den ceases, and the young are then brought to what is called a rendezvous area. In the Great Lakes region such sites usually are near a pond or bog, and consist of a system of trails, beds, and areas in which vegetation has been flattened, presumably by playing of the pups. Reported sizes of rendezvous areas vary from about 0.4 to 1 km long. The pups remain in these areas while the pack hunts. A succession of rendezvous sites are used, generally until the fall, when the young have grown sufficiently to accompany the adults on their travels. The reported period of occupation of each site varies from 6 to 59 days (Joslin 1967; Mech 1970, 1974; Van Ballenberghe et al. 1975; Peterson 1977).

Seasonal Activity. The wolf generally has two main types of wanderings: (1) those that center around the den and the pups from April to late fall, and (2) those that take place during the rest of the year, when the animals roam widely. Generally, there is a day/night activity pattern in the summer, with the animals staying close to the dens during the day and wandering out to hunt during the late afternoon or early evening; they usually return to the den by morning. Perhaps the primary reason for night hunting during the summer is the wolves' sensitivity to heat. The animals generally spend their days trying to keep cool, seeking out shady areas and seldom staying in direct sunlight for even short periods. Wolves pant a great deal in hot weather and quickly become overheated even with slight exertion; their preference for cooler nighttime travels therefore is readily understood (Mech 1970).

In summer, in some areas, such as the tundra, wolves may travel as much as 30 km from the den site to obtain food, a round trip of 60 km or more; in the forested areas, distances traveled are usually far less. This is probably because the food supply is more widely and unevenly distributed on the tundra. Wolves wander along waterways, dirt roads, game, cattle and sheep trails, ridges, and shorelines during their regular hunting activities. When pups are young, the movements of the adults are shorter and less frequent than later in the year. In May, June, and July, the average distance traveled during the hunt may be as little as 1.5 km, whereas later the animals may range out to 3 or 4 km a day. Of course, animals without pups or packs without reproducing members are not restricted as closely to dens and wander more widely than packs with pups.

In the fall, wolf pups abandon their rendezvous sites and join the adults on hunting forays. Thus, the pack no longer needs to restrict its wandering to small areas near the dens. During the winter then, the packs wander freely and widely over vast areas, tracking their prey wherever it may flee. Generally, winter hunting is nocturnal, but often the wolves may travel for days on end, resting only at periodic intervals to recover their strength. In winter, wolves use game trails, roads, ridges, and sometimes even highways to avoid the deep snows that would hamper them in any hunting or moving activity.

In winter, movements of wolves are surprisingly lengthy and rapid. In Ontario, seven wolves moved 65 km in no more than 20 hours and during that time killed and ate one large and one small deer. In Minnesota, a pack traveled 56 km overnight on a chain of frozen lakes, and in Alaska, a pack moved 72 km during no more than 24 hours. Mech (1970) reported that on Isle Royale, the fastest long-distance move he observed was 72 km in 24 hours, mostly along well-established shore routes. The greatest movement of a wolf on record covered a straight line distance of 670 km during an 81-day period in central Canada (Van Camp and Gluckie 1979).

In northern areas, where the primary source of food for wolves is caribou (*Rangifer tarandus*), the wolves appear to be migratory and to follow the caribou as those animals travel throughout the year. Kelsall (1968) stated that in the Canadian tundra the tracks of wolf packs have been followed in a straight line for distances of 160 km one way, moving seasonally from tundra to forest or vice versa. Sometimes they followed caribou trails and at other times they headed for the caribou herds with "uncanny accuracy from directions not used by the caribou." In Alaska, wolves are known to accompany migrating caribou herds. After making a kill, they remain with the carcass until it is devoured, after which they may rapidly travel 40 to 65 km to catch up with the moving herd. On the other hand, in areas where prey is not migratory and is numerous, wolves will remain in a specific area the year around.

In areas with high mountains and heavy snowfalls, such as in the Rocky Mountains, wolves may move down from higher elevations into the valleys during winter. This shift to lower elevations, however, does not appear to be an actual migration on the part of the wolves. Probably it is a response to the difficulty they experience in maneuvering in the deep snow, and the fact that most game on which they feed has migrated to lower elevations. During this winter period when the wolves are in the valleys at the lower elevations their movements may be more restricted than usual and their entire range may be no more than 26 km² per wolf. In the spring and summer, when the snow begins to melt and deer and other game move up to their summer ranges at higher elevations, the wolves follow them and resume their wider-ranging hunting activities until they are again restricted by the birth of pups (Mech 1970).

FOOD HABITS

The wolf is a meat-eating animal, and its entire digestive system is adapted to a carnivorous diet. Typically, it consistently feeds on large prey such as deer or caribou rather than on smaller animals such as rabbits (*Sylvilagus* sp.). The large size of the wolf itself, combined with its habit of traveling in packs, makes it perfectly adapted to feed on larger species of prey.

Studies by Murie (1944) on Mount McKinley in Alaska, Cowan (1947) in the Rocky Mountains of Canada, Thompson (1952) in Wisconsin, Mech (1966) on Isle Royale in Michigan, Pimlott et al (1969) in Algonquin National Park, Ontario, and Van Ballenberghe et al. (1979) in northeastern Minnesota, showed that 59 to 96 percent of the food items consumed by wolves were animals the size of beavers (*Castor canadensis*) or larger. The most frequently taken prey were white-tailed deer (*Odocoileus virginianus*), mule deer (*Odocoileus hemionus*), moose (*Alces alces*), caribou (*Rangifer tarandus*), Dall sheep (*Ovis dalli*), bighorn sheep (*Ovis canadensis*), and beaver.

Cowan (1947) found that in the Canadian Rockies, 80 percent of the food consumed by wolves was big game and only 18 percent consisted of rodents. In northern Wisconsin, Thompson (1952) reported that 97 percent of 425 wolf scats collected were comprised of deer remains, while only 9 percent of the scats contained snowshoe hares (*Lepus townsendii*). In northern Minnesota, Stenlund (1955) found that white-tailed deer comprised 95.5 percent of the total volume of wolf stomach contents. He also found that small animals were more often consumed in summer than in winter. Pimlott et al. (1969) reported that deer were the primary prey of wolves in Algonquin Park, while moose and beaver were of lesser importance. Moose were the only prey for wolves on Isle Royale in Michigan. Mech (1966) determined that the wolf and moose populations were in dynamic equilibrium, with wolves culling out the weak and infirm moose and thus stimulating moose reproduction. In areas where caribou are abundant, Banfield (1954), Kelly (1954), Kelsall (1960), and Kuyt (1972) found that the wolves feed almost exclusively on the caribou. Kuyt also found that wolves prey on arctic fox (*Alopex lagopus*), red fox (*V. vulpes*), arctic hare (*Lepus arcticus*), arctic squirrel (*Spermophilus undulatus*), microtine rodents, birds, eggs, fish, and insects, but to a much lesser extent than they prey on caribou.

According to Mech (1970), domestic animals usually eaten by wolves in North America include cattle, sheep, horses, swine, dogs, and cats. He stated that these species have evolved under constant protection by humans and that they are unable to protect themselves well. Therefore, wherever they occur in the vicinity of wolves they fall easier prey than animals that evolved with the ability to protect and defend themselves.

The teeth of wolves are designed to tear and cut large chunks of meat and to crush and crack bone. Wolves bolt their food and make little attempt at chewing. The size of pieces of prey swallowed by a wolf is impressive. In Alaska, one wolf's stomach contained a caribou ear, tongue, lip, two kidneys, liver, and windpipe, plus hair and large chunks of meat (Kelly 1954). Each wolf can consume almost 9 kg of meat at a feeding (Mech 1970). The food is digested quickly, so the animal probably eats several times a day when large amounts of food are available. Mech (1970) several times saw packs of 15 or 16 animals on Isle Royale consume all the edible parts of a moose calf weighing about 135 kg within 24 hours. On one occasion he saw the pack finish about half of a mature moose in less than 2 hours, which meant that each wolf consumed about 9 kg of meat in 1.5 hours. Mech (1970) estimated that a healthy, active wolf would need to consume a minimum of 1.7 kg of meat a day in order to maintain itself. Average reported consumption rates were 2.5 to 6.3 kg of moose per wolf per day (Mech 1974).

Although wolves can eat enormous amounts of food in very short periods at frequent intervals, the species is also well adapted to go for long periods, sometimes several days, without food. Mech (1966) reported that a pack on Isle Royale once went at least 95 hours without eating anything except hair and bones that they might have gleaned from old kills. Young and Goldman (1944) mention a wild male wolf that, when kept in captivity, did not eat anything for 7 days, but on the 8th day gorged itself. In the Soviet Union, it has been reported that a wild wolf went without food for 17 days, the longest recorded fast for a wolf (Makridin 1962). In the light of the evidence, Mech (1974) thought that wolves can probably fast for two weeks or more while searching about for suitable prey, and then gorge themselves on enormous quantities in order to prepare for another period of fasting. The benefit of such a digestive system is obvious in a large predatory animal like the wolf.

Hunting. Wolves spend almost their entire waking time either eating or hunting. After finishing a meal, wolves begin almost immediately to search for new prey. Mech (1970) found that on Isle Royale the average distance traveled by wolves between kills was 36 km. The pack could show interest in a new kill within 35 minutes of abandoning the old one. They are apparently continually ready to hunt and have no special hunting grounds or behavior. The animals simply travel around and whenever they find a potential prey they attack it. In some regions, they may have to travel through vast areas where game is scarce, but they move through these areas rapidly and spend most of their time wandering and hunting in areas with prey.

Wolves have three main methods of locating prey: direct scenting, chance encounter, and tracking (Mech 1970). Of these three, direct scenting is the most often used. In 51 hunts on Isle Royale in which hunting methods could be determined, wolves used direct scenting in 42 cases. Usually the wolves could scent moose when within 300 m downwind, but on one occasion a cow and its calves were scented about 3 km away. Observations have shown that deer may also be located by odor. When the wolf packs are traveling and detect the scent of prey, the leaders stop and all animals stand alert, looking toward the source of the scent. Then they veer abruptly and head directly toward the prey.

Wolves locate prey through chance encounter less often than by direct scent. Most chance encounters would take place only in areas where prey species are very abundant. Thus, in areas inhabited by deer, which normally occur in high densities, chance encounters would be especially important (Mech 1970). Mech observed several hunts in Minnesota where deer were located in this fashion. Evidence suggests that chance encounters play a major role with wolves that prey on Dall sheep in Mount McKinley Park (Murie 1944). Banfield (1954) reported that on the Canadian tundra, wolves "patrol" an area and flush out caribou at close range.

In 9 of 51 hunts observed by Mech (1970) on Isle Royale, tracking played the major role in obtaining the prey. Observations indicate that wolves use this method only when the tracks are very fresh; older tracks are generally ignored.

Once the wolves have located suitable prey, either by direct scenting, chance encounter, or tracking, they stalk it, usually approaching the animal in the same fashion. That is, they slowly close the gap between themselves and the prey, becoming more excited and quickening their pace the closer they get. They wag their tails and peer straight ahead, but they continue to show restraint and move quietly and cautiously. In this way the wolves may move to within 30 m of the prey without being observed (Crisler 1956; Mech 1970). Once the quarry has spotted the stalking wolves, it may either stand its ground and fight or bolt and try to escape. Usually, only very large prey such as moose or bison will stand and fight, in which case they often fend off the attackers. In fact, wolves seem quite hesitant about attacking any animal that does not attempt to run to safety (Mech 1970). If the prey bolts, however, the wolves immediately give chase in a maneuver that Mech terms the *rush*. This is the most crucial part of the hunt, and often the attacked animal succeeds in escaping without the wolves even getting close to it. If, however, the wolves do overtake the fleeing prey, they attack by biting on the rump, flanks, and shoulders; rarely, if ever, do wolves hamstring any animal they attack. Occasionally, wolves pursue their quarry many kilometers before giving up or launching their final attack. Usually chases are much shorter, generally 1 km or less (Mech 1970).

Upon overtaking and killing the quarry, all the wolves in the pack immediately begin feeding, and within a short time are finished and ready to hunt again. The only exception seems to be when female wolves are restricted to dens during the pupping season. In such cases, all the wolves in the pack apparently help feed the female and young. Murie (1944) reported on at least three wolves that carried food back to a mother in Mount McKinley Park. They took some of it directly to the den, but cached the majority of it a considerable distance away, some as far as 0.8 km distant. On three nights, an adult female other than the mother stayed at the den while the mother accompanied the pack on the hunt. The males in this pack also showed much interest in the pups.

Predator-Prey Relations. The effects of wolves on their prey fall into three categories: (1) culling of inferior animals; (2) the control or partial control of prey populations; and (3) the stimulation of productivity in herds of prey (Mech 1970).

There is little question that wolves play a major beneficial role in removing sick and inferior animals from a herd. Wolf predation generally is selective, resulting in the removal of very young, very old, sick, wounded, crippled, and other infirm individuals. These animals contribute little to herd dynamics and their removal would increase the amount of food, space, and cover for the more productive members of the herd. The diseases of hoofed animals are numerous. Internal and external parasites also plague the prey of the wolf. In some cases, wolf predation might help reduce parasites and diseases by culling the individuals that carry them.

Concerning the wolf's control or partial control of prey populations, the evidence is less clear. The problem is an extremely complex one and is discussed in detail by Stenlund (1955), Pimlott (1967), and Mech (1970). Many factors are involved and interrelated, including prey density, predator density, weather,

TABLE 21.1. Density per square kilometer for various North American wolf populations

Location	Area (approx. km²)	Density of Wolves (approx. km²/wolf)	Authority
Isle Royale, Michigan	546	18–26	Mech 1966; Jordan et al. 1967
Algonquin Park, Ontario	2,600	26	Pimlott et al. 1969
Ontario	26,000	260–500	Pimlott et al. 1969
Minnesota	6,475	26	Olson 1938
Minnesota	8,660	44	Stenlund 1955
Minnesota	1,865	24	Van Ballenberghe et al. 1975
North Central Brooks Range	9,120	170–320	Stephenson 1975
Mt. McKinley Natl. Park, Alaska	5,200	130	Murie 1944
Mt. McKinley Natl. Park, Alaska	3,900	85	Haber 1968
Unit 13, Alaska	52,000	130	Rausch 1967
Tanana Flats, Alaska	18,200	90	Stephenson 1977
Southeast Alaska	19,500	65–100	Atwell et al. 1963
Saskatchewan		104–216	Banfield 1951
Northwest Territories	1,248,000	155–312	Kelsall 1957
Western Canada	1,500	229	Carbyn 1974
Baffin Island	4,680	312	Clark 1971

available browse, and other mortality factors. Studies of wolves on Isle Royale, in Superior National Forest, and elsewhere indicate that they do not play a major part. However, Pimlott (1967) and Mech (1970) believe that wolf predation may have been the main limiting factor on most, if not all, big game species before people so greatly disturbed the habitat.

Prey herds with an adequate food supply that have had old, sick, and debilitated members removed by wolf predation could be expected to reproduce most vigorously. The only place, however, where such a situation has been examined closely is on Isle Royale, Michigan. There, moose inhabited the island for decades before the wolf arrived, so comparison can be made on reproduction prior to and after the advent of the wolf. Mech (1970) reported that the only figure for which enough data are available on moose reproduction on Isle Royale is the twinning rate. This is probably the most sensitive indicator of productivity in a moose herd. Before wolves arrived on Isle Royale, about 1949, very few twin calves were observed. In 1929 and 1930, only 6 percent of 53 cows had twins. However, in 1959, after the wolves had cropped the moose herd for 10 years, the twinning rate was about 38 percent. This rate is much higher than that for any other moose population in North America. Thus, although the data pertain to only a single population and to only one index of fertility, it seems likely that wolves do stimulate productivity in herds of larger prey.

POPULATION DYNAMICS

One wolf per 26 km² constitutes a high density; much lower densities are common over large areas (Pimlott 1967). Pimlott felt that one wolf per 26 km² represents the saturation point beyond which wolf populations cannot exist. However, Kuyt (1972) reported that wolf densities can compress in winter in some parts of Northwest Territories (Mackenzie) to about one wolf per 10 km². Further, Van Ballenberghe (1974), working in Superior National Forest, found a 550-km² area of wolf pack territories in which the densities reached an average of about one per 13.8 km². In these instances, however, it should be noted that prey densities were extremely high; caribou averaged 68 per about 2.5 km² in their Canadian wintering areas. Deer may have averaged as high as about 166 per 2.6 km² in one of their winter yards in Superior National Forest. Thus, as Mech (1974) concluded, while it is true that average wolf densities do not exceed about one wolf per 26 km² (and are usually far lower than this), during certain periods of exceptionally high prey concentrations, wolf densities may almost double. The highest density ever reported for wolves was about one per 8 km², on Coronation Island, Alaska (Merriam 1964), and the lowest was one wolf per 260 to 500 km², in Ontario (Pimlott et al. 1969). Table 21.1 shows the density per square kilometer for various North American wolf populations.

Home Range Size. The term *home range*, as generally used by wolf biologists, means the area of land either enclosed by the runways of a particular pack or available to use by the pack, given its usual travel habits (Mech 1970). In the summer, tundra wolves are known to range as far as 30 to 35 km from the den. Mech (1970) speculates that if the den is in the center of a pack's range and the wolves forage in several directions, the area of the range could be as great as 3,100 km². If, however, the den is only on one edge of the range, the area would approximate only 390 to 780 km². Although home ranges must vary widely, it is evident that on the tundra most of them must be relatively extensive.

Little is known about the summer ranges of wolves in the forested regions, but evidence indicates that they are considerably smaller than in the tundra. Studies by Joslin (cited in Mech 1970) indicated that the range of one pack in southern Ontario was at least 20 km² in one year and 18 km² in another. Mech felt that these figures were probably too low and that the actual range was far larger. Joslin also believed that the total summer range of one Ontario pack that he studied may have been as large as 65 km² (from Mech 1970).

Wolves have a larger home range in winter than in summer. Various studies indicate a range of from about 94 km² for a pack of 2 animals (in Minnesota) to about 13,000 km² for a pack of 10 animals (in Alaska) (Mech 1970). Reduced to a km²/wolf basis, the range would be from about 47 km² per wolf to 1,300 km² per wolf. There is much room for skepticism with regard to published figures of wolf home ranges; such ranges may actually be larger than the published figures indicate.

BEHAVIOR

Social Life and the Pack. Wolves are highly social animals and almost always live in packs. The pack is the basic unit in the social structure and consists of a group of individual wolves that hunt, feed, travel, and rest together. Members of a pack are loosely associated with each other, but the bonds of attachments of individuals living within the pack are strong. Pack size is highly variable. The highest number of wolves recorded in a pack was 36 animals observed by Rausch in south-central Alaska (Mech 1967). Other large packs of 20–21 animals have been observed on occasion in Alaska. Mech (1970) spent several weeks tracking a pack of wolves on Isle Royale, Michigan, that numbered 15–16 animals. However, most wolves associate in packs of 8 animals or less.

The social composition of wolf packs is not clear. It has generally been assumed that the pack contains a breeding pair and their pups, yet many packs contain several adults in breeding condition. Mech (1970) listed the following basic facts about wolf packs:

1. Populations of wolves consist of packs occupying adjacent and sometimes overlapping regions of range.
2. Most packs contain fewer than eight members.
3. Temporary associations of two or more packs sometimes occur, forming a very large group (this is quite rare, however).
4. Several instances are known of packs chasing non-members away.
5. In some cases, a pack has accepted one member of a different group and rejected another member.
6. Strong bonds are needed to hold a pack together; if there were no bonds, each wolf would go its own separate way.
7. Most packs include a pair of breeding adults, pups, and extra adults that may also breed.

It appears that packs are held together by strong bonds of affection. These bonds might develop when two lone adult wolves come together to mate. The courtship of wolves is a lengthy ritual. During this period, the pair may develop an attachment for each other that extends beyond the breeding period and keeps them together as the nucleus of a pack. Pups born to the pair develop the bonds of affection for each other at an early age, generally before five months of age (Mech 1974). Thus, wolf packs may be composed entirely of related individuals. Mech (1970) presented the following theory of pack formation, which is consistent with present knowledge of pack composition and behavior, and explains conflicting observations: the basic component of the pack is the breeding pair, and a pack is first formed when a lone male and female mate in late winter. The pack is added to by the first litter (often as many as six pups), which in order to learn to hunt must stay with the parents at least through the first winter. Because the pups do not mature until they are two years old, there would be no sexual conflicts during the next breeding season when the original pair mate again. As the first pups advance into their third year, rivalries may develop between them and the original pair, or strong sexual attachments may develop between them and strange wolves, so that some of them leave the pack to form new packs. In some instances, however, littermate matings might occur and these wolves would stay with the original pack and increase its size.

The next stage at which a pack might break up is during the third denning period, three or four weeks before the original pair give birth to their third litter (Mech 1970). Each mated pair might separate from the pack, dig its own den, and raise its own young apart from the rest of the pack. This might develop new bonds and help to break old ones, thus forming a new pack nucleus. When this new pack met with the old one, it would recognize its former pack mates. The new pack might travel together with the old temporarily, but soon would separate to develop its own travel routes and hunting areas, particularly in times of food shortages.

Dominance and Leadership. Within any pack of wolves, there is a strict, ridgidly enforced social structure based on dominance. There are two dominance orders within each pack, a male order and a female order. Dominance within these orders is linear in nature, with the highest-ranking male being dominant over the next highest ranking, and so on down the order to the least dominant male. A similar order is found among the females. Generally, the highest-ranking male, known as the alpha male, is dominant over all other animals in the pack, males and females, and is the recognized leader. The most dominant female, known as the alpha female, is dominant over all females in the pack, but is subordinate to the alpha male. She may, however, be dominant over some of the other males in the pack but generally males dominate females. The alpha male and female usually are the original founding members of the pack.

The dominance shown by the alpha animals and other high-ranking wolves was described by Mech (1970) as a kind of "forceful initiative." When a situation does not require initiative, dominance may not be shown, for example, when a pack is resting. However, dominance comes into play when the pack is feeding, mating, seeking favored space, encountering strange wolves, or in other such active or competitive situations.

There are two main aspects of dominance in a wolf pack: privilege and leadership (Mech 1970). Privilege involves the dominant wolf's taking the initiative in any competitive situation and claiming whatever it desires. Thus, the dominant wolf has the first choice of food, bedding sites, and mates. Usually none of the other members of a pack disputes the claims of a dominant wolf, although two animals closely related in rank may compete. There are some exceptions to the privileges of a dominant animal, however, notably when a female gives birth to young. A normally subordinate female, after its pups are born, may take an aggressive attitude toward a dominant male and drive it away. Schoenberner (1965) stated that the only time an alpha male is dominated by a female is when the female is caring for a newborn young.

Leadership is probably the most important aspect of pack dominance. Obviously no pack could exist as a unit if each individual member decided for itself when to rest, hunt, or seek refuge. Some members of the pack must take the initiative in these activities, and dominance appears to be the major factor in determining leadership. Generally, in any pack, the alpha male is the most highly motivated animal and serves as leader of the pack. Leadership is evidenced in wolves when they are attacking prey, traveling overland, or waking from a long nap (Mech 1970). Highly dominant males, other than the alpha male, may at times play a leadership role in the pack, but females rarely do.

Leadership also involves a guarding function in the pack. Murie (1944) described a case where the leader of a pack remained alert one day while other wolves rested, and then suddenly led an attack on a strange wolf that approached. On another occasion, this same wolf was the most aggressive in defending the pack's denning area from an invading grizzly bear (*Ursus arctos*).

As yet, the manner in which an alpha male controls the pack is not known. It could be an autocratic type of leadership in which the leader dictates the activity and the other members follow the lead without protest. Or, the leadership could be democratic, with the alpha male taking its cue from the behavior of other pack members. A third possibility is that the leadership could combine elements of both autocracy and democracy.

Territoriality. Mech (1970) accepts the definition of an animal's territory as being the area that the animal will defend against individuals of the same species. Defense of this area is the main difference between "territory" and "home range." With wolves, it appears that packs are territorial, at least to the extent that their territories include most of their hunting and traveling areas. Information on territoriality in wolves is scanty and somewhat conflicting. Data on the Isle Royale population shows that the largest pack used the entire island in winter, but concentrated its activities in about one-half to one-third of the area (Mech 1966). Two smaller packs confined their movements to the part of the island used least by the large pack. Most information on territories of the Isle Royale packs pertains to winter, but limited summer observations suggest that the packs are spaced about as they are in winter. In Ontario, Joslin (1966) found summer ranges of packs to be separate, with no overlap of ranges between packs. However, Van Ballenberghe et al. (1975) felt that territories would generally be entirely discrete only when minimum territory size is approached, probably because of the greater ease in patrolling small territories.

Howling and Communication. Wolves communicate with each other in a great variety of ways. They may posture or position themselves in various poses to indicate aggression or subservience. They whimper, growl, or utter other sounds in response to fear, pleasure, or pain. The ears may be held erect or flattened against the head, the tail wagged or held stiff, hairs erected or laid flat, all depending on the mood or specific situation in which the animal finds itself.

The wolf whimper is a high, soft, plaintive sound similar to that uttered by domestic dogs; Joslin (1966) considered it to be a submissive or friendly greeting sound. Growling, on the other hand, seems to be uttered primarily in aggressive situations and is a threatening, unfriendly sound. The barking of wolves is deep, guttural, and coarse (Mech 1970). It is apparently basically an alarm call given when other animals impose on a pack's territory (Joslin 1966).

The most commonly heard wolf sound is the howl, a long, low, mournful moan (Mech 1970). A wolf pack may howl at any time of day and any time of year. A single wolf may have a howling session (Joslin 1966). When a pack performs, one wolf begins the howling, and after its first or second howl other animals join in. Each animal starts by itself, beginning with long, low howls and working up to a series of shorter, higher ones. Such a howling session lasts 85 seconds on the average, and is often followed by a repeat performance. All the functions of howling are not yet known. Crisler (1956) believed that howling, like a community sing, is a happy, social occasion. An important function of howling seems to be as an aid in assembling the pack (Mech 1970). The advantage of a method of assembling scattered pack members is obvious, and it is easy to see how this function could have evolved from social gratification. Mech (1970) believed that much of the howling of wolves actually represents the assembling of pack members after a chase. Often wolves get separated during the hunt, and on such occasions one animal may climb a ridge and howl to attract other pack members to it. Other possible adaptive functions of howling are to advertise and maintain territories (Harrington and Mech 1979), to identify individuals, and to supply information about behavior, such as whether the animal is lying down, walking slowly, or pacing (Mech 1970).

One of the most important expression centers for the wolf is the head. In general, teeth bared, open mouth, wrinkled and swollen forehead, and ears erect and pointed forward indicate a full threat by a dominant wolf. Subordination is indicated by a closed mouth, a smooth forehead, eyes closed to slits, and ears drawn back and held flat against the head.

Another important communicating organ for the wolf is the tail. A threatening wolf raises the tail above the level of the back; a submissive wolf holds the tail low, sometimes tucked between the legs. Loose, free tail wagging usually indicates friendliness, while tight, abrupt wagging of the tail tip can often be associated with aggression.

MORTALITY

Wolves are subject to a great many diseases, of which parasites, both internal and external, are a major cause. Some of the endoparasites that affect wolves are various species of fluke, tapeworm, roundworm, and thorny-headed worm. Lice, tongueworms, fleas, ticks, mange mites, mosquitoes, deerflies, horseflies, black flies, and stable flies are some of the external parasites that attack wolves.

Rabies is the most important disease in wild populations of wolves (Mech 1970). Whether rabies is important in control of wolves is uncertain. Cowan (1949) felt that rabies might play such a role in limiting the number of foxes and possibly wolves as well. Rausch (1958), however, concluded that there is simply not enough known about canine rabies to determine whether it is important in the control of boreal populations of carnivores.

Distemper is another viral canine disease that may occur among young wolves in captivity (Gross 1948). However, Rausch (1958) and Mech (1970) report no cases known to them of distemper in wild populations of wolves. Other diseases that have been reported in captive wolves range from liver and thyroid cancer to bladder stones and chronic nephritis, but the prevalence of these disorders in wild populations is not known.

Other factors that may play significant roles in wolf mortality and possibly in control of populations include injury and accidents, malnutrition, social stress, and persecution and exploitation by humans.

ECONOMIC STATUS

Wherever the wolf occurs, it has generally been feared, hated, and destroyed; people seem always to have been the most relentless and determined foe of the wolf whenever the two species occurred together. Generally, people have killed wolves for three reasons: (1) wolf fur is durable and warm and has been sought for clothing in some areas; (2) the wolf is a large, aggressive animal and people generally have feared for their own personal safety when living in proximity to wolves; and (3) the wolf is a known predator on domestic livestock in some situations.

By far, hatred of the wolf as a predator on livestock has been the major factor in human persecution of the species. In general, wolf fur has never been widely appreciated as a major article of clothing. Some American Indian tribes used wolf fur to make shoes, caps, and robes, but deer and bison hides were gener-

ally of much greater importance to them. Europeans and American settlers used wolf fur as trimming on their garments, and because of its durability, warmth, and frost-free qualities, this is its primary use today. However, persecution for its hide has been, and continues to be, only a minor factor in the decline of the wolf.

Of more importance has been the fear that the wolf is a serious threat to human safety. At one time the wolf was so dreaded in some parts of Europe that merely mentioning its name was considered a crime (Ricciuti 1978). It is believed that as many as 3,000 people have been killed by wolves in Europe over the past 500 years. Many terrible tales are told in most European countries of wolves attacking women and children and terrorizing whole villages during hard winters. It seems almost certain that wolves in Europe did pose a threat to human safety in some areas. However, rabies may have been the primary factor in recorded wolf attacks; most experts (Novikov 1956, Rutter and Pimlott 1968) concede that very few attacks by nonrabid animals probably occurred.

When European settlers came to the New World, they brought their fear of the wolf with them, and naturally sought to extirpate wolves whenever they were encountered. Yet, wolves in the New World apparently were different from those in the Old World, and genuine records of unprovoked attacks on humans here are very rare indeed. In fact, Mech (1970) stated that there is no acceptable evidence to support any claim that healthy wild wolves in North America are dangerous to people. The only scientifically documented case of a North American wolf attacking a man was reported by Peterson (1947): a wolf pulled a man from a railroad "speeder" and continued to attack him for about 25 minutes. Although this wolf was not tested for rabies, its extremely unusual behavior would strongly indicate that it had the disease.

There are numerous hearsay accounts of attacks by wolves, but none has ever been verified. On the contrary, there is strong evidence that the North American wolf is harmless to humans. Numerous field researchers have worked closely with the species, and none has ever reported being attacked or threatened. In fact, nearly all accounts show that North American wolves are shy animals that usually try to avoid people as much as possible.

There is no question that the wolf, wherever it occurs, can be a serious predator on domestic animals. Young (in Young and Goldman 1944) stated that there is overwhelming evidence that wolves prey upon cattle, and he listed numerous cases of such predation. He cited Joseph Neal, a stockman and conservationist from Meeker, Colorado, as stating that the history of the wolf in the West has been a chronicle of the struggle for supremacy between it and the livestock industry, with the success or failure of the livestock business depending upon the outcome. Even Theodore Roosevelt, judging from his experience on a cattle ranch in North Dakota, commented that the wolf, wherever it exists in numbers, is a veritable scourge to stockmen. Equally, it is reported that wolves played havoc with sheep flocks. As early as 1790, wolves were serious drawbacks to sheep raising in New York (Young and Goldman 1944), and in Pennsylvania a woolen goods manufacturer failed in business because wolves destroyed his sheep in large numbers.

MANAGEMENT

History of Decline. Altogether, the role of the wolf as a predator on livestock may have been exaggerated, but it was still serious enough that settlers in North America started very early to attempt to eradicate the species on this continent. Nowak (1974) documented how these attempts to extirpate the wolf led to its decline in the Western Hemisphere. With the arrival of English settlers along the Atlantic Coast of North America, the range of the wolf began to shrink. The extreme hatred responsible for the early extermination of wolves on the British Isles was carried into the New World. By the 1880s the species had been wiped out all along the east coast of the United States, as well as in the Ohio Valley and the eastern plains. By 1914 the last wolves had been killed in Newfoundland, Canada, south of the St. Lawrence River, and in New England, New York, the Appalachians, the southern peninsula of Michigan, and much of the Great Plains region. In the following year the United States Bureau of Biological Survey began a program aimed at controlling wolf depredations on domestic stock. Partly as a result of this campaign, wolves had nearly disappeared from the western United States by 1944.

The decline in the gray wolf's range from 1944 to 1974 was much less than that of the two previous 30-year periods. The greatest recent loss seems to have occurred in Mexico, and the number of wolves moving from that country into the southwestern United States has been reduced. Otherwise, resident populations have disappeared in the Oregon Cascades, northern Wisconsin, possibly the upper peninsula of Michigan, and the Bruce Peninsula of Ontario, and on some arctic islands. Partially offsetting these losses, however, has been the reestablishment or rediscovery of wolves in certain areas, such as the Kenai Peninsula of Alaska, Glacier National Park and adjacent parts of the Rocky Mountains, and Isle Royale.

Apparently, the most critical factor in the shrinking distribution of the wolf was the spread of domestic livestock. Where cattle and sheep were raised, wolves were exterminated. Where the livestock industry was not significant, wolves were persecuted, but not to the same extent, and their numbers sometimes had a chance to rebound. The profitable limits of livestock range seem to have been reached by the 1940s, and relatively little change in the wolf's distribution has occurred since then. The one major exception is in Mexico, where wolves continue to occupy grazing lands and are under heavy human pressure. In certain other areas, such as Alberta, Ontario, and Minnesota, fairly stable lines of demarcation have existed for years between wolf range and agricultural lands. The notion that the wolf is being pushed back steadily each year does not seem altogether correct.

Within their primary range, however, wolves certainly have been intensively hunted in the last 30 years, and some populations were at least temporarily reduced. Government control programs, usually involving mass poisoning, were widespread in the 1940s and 1950s in Alaska and western Canada. Such programs were carried out mostly for the avowed purposes of protecting wild ungulates or halting the spread of rabies. Private hunting and trapping, especially from aircraft, also are thought to have reduced populations in some areas. Nonetheless, most populations seem to have remained viable, and with the general curtailment of control programs by the early 1960s, wolf numbers began to recover. Presently there is a general consensus among wildlife officials that wolves exist in safe numbers throughout most of the range they still occupy.

Current Status of the Wolf in North America. Although evaluation of numerical status would seem to be useful in determining the condition of a wolf population, such an approach presents problems. Wolves usually inhabit remote terrain, are difficult to locate, and move over great distances. There is still much to be learned about their movements, population structure, and spacing mechanisms. Estimates of numbers frequently are given, but, interestingly, those authorities most reluctant to provide estimates include some persons who have studied the wolf the longest. Even if reliable estimates were available, they would not necessarily be an effective measure of status. Numbers or population densities in a given area at a given time must be seen in a relative manner. To assess human impact on particular wolf populations, current numbers would have to be compared to those of primeval times. There is no certain way of knowing how many wolves inhabited the various regions of North America before the arrival of Caucasian mankind. Presumably, population densities approached those now found in relatively undisturbed areas where both wolves and the original kinds of prey species are fully protected. The only authority to calculate an estimate of primeval numbers was Seton (1925), who thought that the continent once had about 2 million wolves, with densities as great as 1 per 2.6 km^2, and that by 1908 about 200,000 still survived. These figures seem remarkably high in the light of present knowledge. As previously discussed, densities do not generally exceed 1 wolf per 26 km^2 even in protected areas with abundant prey, and most populations have much lower densities. One major region in which densities conceivably could have been higher is the prairie of the western United States, formerly with vast herds of bison and other large ungulates. Early travelers reported easily seeing many wolves.

Present populations densities may also be looked at relative to carrying capacity. Even if wolves exist at a lower average density than originally, they might still approach maximum levels with regard to availability of prey species. This situation may now hold in parts of northern Canada where caribou numbers are far smaller than before the introduction of firearms (see chapter 47).

Distribution, numbers, and population densities provide one idea of the status of wolf populations, and, perhaps unfortunately, most available data deal with these subjects. There are, however, other factors that must be considered in assessing the status of the wolf. The human attitudes that caused the persecution of wolves for centuries still exist to a large extent. Morever, mankind has demonstrated the technical ability to destroy wolf populations if that is desired. During the control programs of the 1940s and 1950s, poisoning and aerial gunning killed thousands of wolves in the remote tundra and taiga regions. Wolves, especially those of the far north, remain vulnerable to such activities.

Even without human dislike and direct hunting pressure, wolf populations may be under several potential threats. Economic developments now are taking place in all regions where wolves still exist. The most critical of these operations is oil and gas exploration, which is penetrating even the most remote parts of the arctic. In addition, water diversion projects, mining, and road construction are increasing in northern wilderness areas. Although such activities seldom directly affect wolves, increased human presence could mean more potential hunters and more harassment by aircraft and snowmachines. Ungulate herds, on which wolves depend, could be disrupted and their movements hindered. Some observers think that entire northern ecosystems are endangered, and, if so, then inclusive wolf populations would be threatened.

Estimates of Current Wolf Numbers. The following figures are from Nowak 1974.

Alaska: conservationists familiar with the species suggest there are between 5,000 and 10,000 wolves in the state.

Northwest Territories: 2,000 to 5,000 (Cahalane 1964).

Yukon Territory: 2,000 to 5,000 (director, Yukon Game Department).

British Columbia: 2,500 to 5,000 (Cahalane 1964).

Alberta: 3,550 in the 1965–66 period (Stelfox 1969).

Saskatchewan: 1,500 to 2,500 (Cahalane 1964).

Manitoba: 1,500 to 2,000 (government of Manitoba).

Ontario: 10,000 to 15,000 (Standfield 1970).

Quebec: 1,500 to 3,000 (Cahalane 1964).

Labrador: several hundreds (Cahalane 1964).

Michigan: upper peninsula, 6 wolves between 1971 and 1973 (Hendrickson et al. 1973); Isle Royale, "Wolf numbers continue to fluctuate above and below approximately 20 animals as they have for more than 10 years" (Mech and Rausch 1973).

Minnesota: 1,000 to 2,000 (Mech 1977).

Wyoming: Yellowstone National Park, 10 to 15 animals (Cole 1971); Shoshone National Forest, 9 in 1972; Teton National Forest, 4 in 1972.

Occasionally in recent years, wolves have been reported from or killed in Idaho, Oregon, Washington, Montana, North Dakota, South Dakota, Arizona, New Mexico, and Texas, and there may be small populations of wolves in each of these states. Else-

where in the United States, wolves were extripated by the 1930s.

Mexico: no population estimates, but wolves are found in small, scattered groups in Chihuahua, Sonora, Coahuila, Durango, and Zacatecas.

Conservation Measures. Internationally, commercial traffic in wolves and wolf products is controlled by the Convention on International Trade in Endangered Species of Wild Fauna and Flora; the wolf is on appendix II of that convention. An appendix II species is one that is not necessarily threatened with extinction but that may become so unless trade is strictly regulated to avoid utilization incompatible with its survival. Permits are available for such trade where warranted.

Additionally, the wolf in the continental United States (except Alaska and Minnesota) and in Mexico is classified as an "endangered" species under the U.S. Endangered Species Act of 1973. This means that interstate trade, export, import, and take of wolves are severely regulated by the federal government. In Minnesota, the wolf has the federal classification of a "threatened" species and take and trade are also regulated, but a limited take of the species by authorized personnel is permitted in areas where wolves cause severe depredations of livestock.

In Alaska, there is no federal classification, but the wolf is classified by the state as a "big game" animal and its killing is regulated by season, limit, locality, and method. Regulations in Alaska have varied considerably over the years depending on wolf densities and other factors established by state biologists.

In Canada, the wolf is classified as "big game" in British Columbia; "predatory animal" in Northwest Territories, Yukon, and Manitoba; and "fur-bearing animal" in Alberta; it receives no classification in Saskatchewan, Ontario, and Quebec. With certain reservations, classification as "predatory animal" and "fur-bearing animal," as well as no official classification, means that wolves may be taken at any time and place, and in unlimited numbers. The "big game" classification in British Columbia allows the province to set rules and regulations governing the take of wolves.

Under the U.S. Endangered Species Act of 1973, the federal government has also determined specific areas in the lower 48 states to be "critical habitat" for the wolf. These areas are Isle Royale, Michigan, and about 26,000 km² of woodland in northeastern Minnesota. An official designation of "critical habitat" by the federal government, pursuant to the Endangered Species Act of 1973, prohibits all federal agencies from undertaking, authorizing, or financing any activities within the critical habitat area that might destroy the habitat or modify it in such a way as to prove detrimental to the survival of the species for which it is critical.

At present, the wolf is legally protected in Mexico, and cannot be taken anywhere, except under special permit issued by the federal director general of wildlife. This regulation, however, is difficult to enforce and is not generally applied; in some areas, local stockmen apparently actively engage in efforts to eliminate the species.

THE RED WOLF

The red wolf (*Canis rufus*) is a species that has often been confused with both the gray wolf and the coyote (*Canis latrans*). In most of its range it resembles the gray wolf in color but is smaller, weighing from 18 to 34 kg, with a narrower physique and shorter fur.

The red wolf was formerly distributed from southern Pennsylvania to Florida and west to central Texas (Nowak 1979). Human persecution over the years caused a steady contraction of the species' range. Meanwhile, the more prolific coyote pushed into the range of the red wolf from the west and the north, its way opened by the elimination of the larger red wolf and environmental disruption that proved advantageous to its survival. These expanding coyote populations interbred with and eventually absorbed the scattered remnant red wolves, and by 1970 the only pure red wolves were found along the Gulf Coast of Texas and Louisiana. Eventually, even these populations succumbed to the unique process of genetic erosion, so, to the best of our knowledge, the red wolf is now extinct in the wild. About two dozen red wolves, however, were removed from the wild before the end, and were taken to breeding facilities in Tacoma, Washington. It is hoped that this small group will form a breeding nucleus that will keep the species alive and eventually provide animals for reintroductions into the wild.

Less is known about the life history of the red wolf than about that of the gray wolf or coyote. Its prey was mostly smaller than that of the gray wolf, primarily consisting of rabbits and rodents. Its home range was also smaller, about 30 km². Pairs established territories, mated in winter, and produced four or five young in the spring (Paradiso and Nowak 1972).

LITERATURE CITED

Atwell, G.; Garceau, P.; and Rausch, R. A. 1963. Wolf investigations. Alaska Fed. Aid Wildl. Rest. Proj. W-6-R-3, Work Plan K. Juneau. 28pp.

Banfield, A. W. F. 1951. Populations and movements of the Saskatchewan timber wolf *Canis lupus knightii* in Prince Albert National Park. Saskatchewan, 1947–1951. Wildl. Manage. Bull. Ser. 1, no 4. 24pp.

———. 1954. Preliminary investigation of the barren ground caribou. Part 2, Life history, ecology and utilization. Can. Wildl. Serv., Wildl. Manage. Bull. Ser. 1, no. 10B. 112pp.

———. 1974. The mammals of Canada. Univ. Toronto Press, Toronto. 438pp.

Cahalane, V. H. 1964. A preliminary study of distribution and numbers of cougar, grizzly and wolf in North America. New York Zool. Soc. 12pp.

Carbyn, L. N. 1974. Wolf predation and behavioral interactions with elk and other ungulates in an area of high prey diversity. Can. Wildl. Serv. Rep., Edmonton. 238pp. Manuscript.

Chiarelli, A. B. 1975. The chromosomes of the Canidae. *In* M. W. Fox, ed. The wild canids: their systematics, behavioral ecology and evolution. Van Nostrand Reinhold, New York. 508pp.

Clark, K. R. F. 1971. Food habits and behaviour of the timber wolf on Central Baffin Island. Ph. D. Thesis. Univ. Toronto. 223pp.

Cole, G. F. 1971. Yellowstone wolves (*Canis lupus irremotus*). Yellowstone Natl. Park, Res. Note 4. 6pp.

Cowan, I. M. 1947. The timber wolf in the Rocky Mountain national parks of Canada. Can. J. Res. 24:139–174.

———. 1949. Rabies as a possible population control of Arctic Canidae. J. Mammal. 30:396–98.

Crisler, L. 1956. Observations of wolves hunting caribou. J. Mammal. 37:337–346.

Ginsburg, B. E. 1965. Coaction of genetical and nongentical factors influencing sexual behavior. Pages 53–75 *in* F. Beach, ed. Sex and behavior. John Wiley & Sons, New York. 592pp.

Gray, A. P. 1972. Mammalian hybrids. Commonwealth Agric. Bur., Slough, England. 262pp.

Gross, L. J. 1948. Species susceptibility to the virus Carre and feline enteritis. Am. J. Vet. Res. 9:65–86.

Gunson, J. R., and Nowak, R. M. 1979. Largest gray wolf skulls found in Alberta. Can. Field Nat. 93:308–309.

Haber, G. C. 1968. The social structure and behavior of an Alaskan wolf population. M.S. Thesis. Northern Michigan Univ., Marquette. 198pp.

Hall, E. R., and Kelson, K. R. 1959. The mammals of North America. Ronald Press Co., New York. 2:547–1083.

Harrington, F. H., and Mech, L. D. 1979. Wolf howling and its role in territory maintenance. Behavior 68:207–249.

Hemmer, H. 1974. Untersuchungen zur Stammesgeschichte der Pantherkatzen (Pantherinae). Part 3, Zur Artgeschichte des Lowen *Panthera (Panthera) leo* (Linnaeus 1758). Veroff. Zool. Staatssaml. Munich 17:167–280.

Hendrickson, J.; Robinson, W. L.; and Mech, L. D. 1975. Status of the wolf in Michigan, 1973. Am. Midl. Nat. 94:226–232.

Jordan, P. A.; Shelton, P. C.; and Allen, D. L. 1967. Numbers, turnover and social structure of the Isle Royale wolf population. Am. Zool. 7:233–252.

Joslin, P. W. B. 1966. Summer activities of two timber wolf (*Canis lupus*) packs in Algonquin Park. M.S. Thesis. Univ. Toronto. 99pp.

———. 1967. Movements and home sites of timber wolves in Algonquin Park. Am. Zool. 7:279–288.

Kelly, M. W. 1954. Observations afield on Alaskan wolves. Proc. Alaska Sci. Conf. 5:35.

Kelsall, J. P. 1957. Continued barren ground caribou studies. Can. Wildl. Serv., Wildl. Manage. Bull. Ser. 1, no. 12. 148pp.

———. 1960. Co-operative studies of barren-ground caribou, 1957–1958. Can. Wildl. Serv., Wildl. Manage. Bull. Ser. 1, no. 15.

———. 1968. The migratory barren ground caribou of Canada. Can. Wildl. Serv., Queen's Printer, Ottawa. 340pp.

Kolenosky, G. B., and Standfield, R. O. 1975. Morphological and ecological variation among gray wolves (*Canis lupus*) of Ontario, Canada. Pages 62–72 *in* M. W. Fox, ed. The wild canids/their systematics, behavioral ecology and evolution. Van Nostrand Reinhold, New York. 508pp.

Kuyt, E. 1962. Movements of young wolves in the Northwest Territories of Canada. J. Mammal. 43:270–271.

———. 1972. Food habits and ecology of wolves on barren-ground caribou range in the Northwest Territories. Can. Wildl. Serv., Rep. Ser., no. 21. 36pp.

Lawrence, B., and Bossert, W. H. 1967. Multiple character analysis of *Canis lupus, latrans,* and *familiaris,* with a discussion of the relationships of *Canis niger.* Am. Zool. 7:223–232.

———. 1975. Relationships of North American *Canis* shown by a multiple character analysis of selected populations. Pages 73–86 *in* M. W. Fox, ed. The wild canids/their systematics, behavioral ecology and evolution. Van Nostrand Reinhold, New York. 508pp.

Makridin, V. P. 1962. The wolf in Yamal north. Zool. Zhur. 41(9):1413–1417. Translated by Peter Lent.

Mech, L. D. 1966. The wolves of Isle Royale. U.S. Natl. Park Serv. Fauna Ser. 7. 210pp.

———. 1970. The wolf. Nat. Hist. Press. Garden City, New York. 384pp.

———. 1974. A new profile for the wolf. Nat. Hist. 83:26–31.

———. 1977. A recovery plan for the eastern timber wolf. Natl. Parks and Conserv. 50:17–21.

Mech, L. D., and Rausch, R. L. 1973. The status of the wolf in the United States, 1973. 13pp. Manuscript.

Merriam, H. R. 1964. The wolves of Coronation Island. Proc. Alaska Sci. Conf. 15:27–32.

Miller, M. E.; Christensen, G. C.; and Evans, H. E. 1964. Anatomy of the dog. W. B. Saunders Co., Philadelphia. 941pp.

Murie, A. 1944. The wolves of Mount McKinley. U.S. Nat. Park Serv. Fauna Ser. no. 5. 238pp.

Novikov, G. A. 1956. Carnivorous mammals of the U.S.S.R. Zool. Inst. Acad. Sci. U.S.S.R., Moscow. 284pp.

Nowak, R. M. 1974. The grey wolf in North America; a preliminary report. Rep. for U.S. Fish Wildl. Serv. 255pp. Manuscript.

———. 1979. North American Quaternary *Canis.* Univ. Kansas, Mus. Nat. Hist. Monogr. 6. 154pp.

Olson, S. F. 1938. A study in predatory relationship with particular reference to the wolf. Sci. Month. 66:323–336.

Paradiso, J. L., and Nowak, R. M. 1972. *Canis rufus.* Mammal. Species, Am. Soc. Mamm., no. 22. 4pp.

Peterson, R. L. 1947. A record of a timber wolf attacking a man. J. Mammal. 28:294–295.

———. 1976. The role of wolf predation in a moose population decline. 1st Conf. Sci. Res. in the Natl. Parks, New Orleans, La. Manuscript.

———. 1977. Wolf ecology and prey relationships on Isle Royale. U.S. Natl. Park Ser. Monogr. Ser. no. 11. 210pp.

Pimlott, D. H. 1967. Wolf predation and ungulate populations. Am. Zool. 7:267–278.

Pimlott, D. H.; Shannon, J. A.; and Kolenosky, G. B. 1969. The ecology of the timber wolf in Algonquin Park. Ontario Dept. Lands For. 92pp.

Rausch, R. L. 1958. Some observations on rabies in Alaska, with special reference to wild canidae. J. Wildl. Manage. 22:246–260.

———. 1967. Some aspects of the population ecology of wolves, Alaska. Am. Zool. 7:253–265.

Ricciuti, E. R. 1978. Dogs of war. Int. Wildl. 8:36–40.

Rutter, R. J., and Pimlott, D. H. 1968. The world of the wolf. J. P. Lippincott Co., Philadelphia. 202pp.

Schoenberner, D. 1965. Observations on the reproductive biology of the wolf. Z. Saugetierk. 30:171–178.

Scott, J. P. 1967. The evolution of social behavior in dogs and wolves. Am. Zool. 7:373–381.

Scott, J. P., and Fuller, J. L. 1965. Genetics and the social behavior of the dog. Univ. Chicago Press, Chicago. 468pp.

Seton, E. T. 1925. Lives of game animals. Vol. 1, part 1. Charles T. Branford. Boston.

Standfield, R. O. 1970. A history of timber wolf (*Canis lupus*) populations Ontario, Canada. Trans. Int. Congr. Game Bio. 9:505–512.

Stelfox, J. G. 1969. Wolves in Alberta: A history, 1880–1969. Alberta Lands—For.—Parks—Wildl. 12:18–27.

Stenlund, M. H. 1955. A field study of the timber wolf (*Canis lupus*) on the Superior National Forest, Minnesota. Minn. Dept. Conserv. Tech. Bull. 4. 55pp.

Stephenson, R. O. 1974. Characteristics of wolf den sites. Alaska Dept. Fish Game, Fed. Aid in Wildl. Rest. Final Rep. 27pp.

———. 1975. Wolf report. Alaska Fed. Aid. Wildl. Rest. Prog. Rep. Proj. W-17-6 and 17, Juneau. 11pp.

———. 1977. Wolf report. Alaska Fed. Aid Wildl. Rest. Prog. Rep. Proj. W-17-7, Juneau. 18pp.

Thompson, D. Q. 1952. Travel, range and food habits of timber wolves in Wisconsin. J. Mammal. 33:429–442.

Van Ballenberghe, V. 1974. Wolf management in Minnesota: An endangered species case history. Trans. North Am. Wildl. Nat. Res. Conf. 39:313–320.

Van Ballenberghe, V.; Erickson, A. W.; and Byman, D. 1975. Ecology of the timber wolf in Minnesota. Wildl. Monogr. 43. 43pp.

Van Camp, J., and Gluckie, R. 1979. A record long-distance move by a wolf (*Canis lupus*). J. Mammal. 60:236–237.

Young, S. P., and Goldman, E. A. 1944. The wolves of North America. Am. Wildl. Inst., Washington, D. C. 385pp.

JOHN L. PARADISO, Office of Endangered Species, U.S. Fish and Wildlife Service, Washington, D.C. 20240.

RONALD M. NOWAK, Office of Endangered Species, U.S. Fish and Wildlife Service, Washington, D.C. 20240.

22

Foxes
Vulpes vulpes and Allies

David E. Samuel
Brad B. Nelson

NOMENCLATURE

COMMON NAMES. Red fox: black fox, cross fox, silver fox. Gray fox. Kit fox. Swift fox.

SCIENTIFIC NAME AND SUBSPECIES. Red fox: *Vulpes vulpes*. Subspecies: *V. v. fulva* (eastern United States), *V. v. rubricosa* (southern Quebec, Nova Scotia), *V. v. regalis* (plains, United States and Canada), *V. v. macroura* (Rocky Mountains), *V. v. necator* (California, Nevada), *V. v. abietorum* (western Canada), *V. v. alascensis* (Alaska), *V. v. harrimani* (Kodiak Island), *V. v. kenaiensis* (Kenai Peninsula), and *V. v. cascadensis* (northwest, coastal United States).

Gray fox: *Urocyon cinereoargenteus*. Subspecies: *U. c. borealis* (New England), *U. c. cinereoargenteus* (eastern United States), *U. c. floridanus* (Gulf states), *U. c. ocythous* (central plains states), *U. c. scottii* (southwestern United States), *U. c. townsendi* (California, Oregon), and *U. c. californicus* (southern California).

Kit fox: *Vulpes macrotis*. Subspecies: *V. m. macrotis* (southwest corner of California), *V. m. mutica* (San Joaquin Valley), *V. m. nevadensis* (Nevada), *V. m. arsipus* (southern California, Arizona), and *V. m. neomexicana* (New Mexico, Mexico). Confusion exists here. Recent literature (Waithman and Roest 1977) listed *mutica* as a separate subspecies and synonymized *arsipus* with *macrotis*.

Swift fox: *Vulpes velox*. Subspecies: *V. v. hebes* (northern plains states) and *V. v. velox* (southern plains states)

DISTRIBUTION

There is some question whether the red fox was native to North America. Churcher (1959) believed the red fox was native to North America north of latitude 40° N, but was scarce or absent in most of the vast hardwood forests where gray foxes were common. Others believe that the North American red fox originated from the European red fox, which was introduced into the southeastern section of the United States around 1750. It may have interbred with the scarce

indigenous population to produce a hybrid population (Godin 1977). It was also introduced into New England (Waters 1964). Current distribution is shown in figure 22.1. Red foxes are also native to Europe and Asia, and have been introduced into Australia (Ryan 1976).

The native gray fox occurred over much of eastern North America (Churcher 1959). Fossils from the late Post-Wisconsin period were found in Pennsylvania (Guilday and Bender 1958). Gray foxes apparently disappeared in New England, as did the red fox, until this past century (Grayce 1957).

Dorf (1959) suggested that the range of *Vulpes* contracted northward during a warm Hypsithermal period (5000–2000 B.C.), then moved southward as conditions cooled. During the same warm period, *Urocyon* moved north and east, then its range contracted southward as the weather cooled. He further suggested that other shifts in range may have occurred during a second warm period (1000–1300 A.D.) and recent reentry of *Urocyon* into New England may be related to a warming trend (since 1850).

Gray foxes are found in 45 to 50 states, being absent in the northern Rocky Mountains and Canada (figure 22.2). One insular species is found on islands off the California coast and *Urocyon* subspecies are found south through Mexico and Central America.

DESCRIPTION

Foxes belong to the family Canidae in the order Carnivora. The red fox weighs 3–7 kg, with long, pointed, erect ears and an elongated, pointed muzzle. Adults are about 1 meter in length, with a 32-cm tail that is used for balance and for keeping the face warm while sleeping. Storm et al. (1976) found that the total length of red foxes ranged from 827 to 1097 mm, tail length from 291 to 455 mm, hind foot length from 124 to 182 mm, and ear length from 65 to 102 mm. They also presented detailed skull measurements of juvenile, subadult, and adult red foxes. The white tip on the tail distinguishes the red fox from the kit fox, swift fox, and gray fox. The dense, soft pelage of the upper body is reddish yellow and darkest on the shoulders and

FIGURE 22.1. Current distribution of the red fox (*Vulpes vulpes*).

FIGURE 22.2. Current distribution of the gray fox (*Urocyon cinereoargenteus*).

back. The moderately long legs have black stockings and the belly and chin are white. The ears are often trimmed in black, and there are varying amounts of black on the tail. Females have eight mammae. The forefoot has five toes, whereas the hind foot has four toes, each with a long, nonretractile claw.

Several color phases of red foxes are found, especially in cold regions: black, silver, cross (dark cross on shoulders), bastard (bluish gray), and Samson fox (no guard hairs). The silver fox is black with white-tipped guard hairs giving it a silvery appearance.

Foxes have a 2-cm-long subcaudal gland on the upper portion of the tail that exudes a "foxy" odor. The rostrum is narrow and the teeth are relatively widely spaced.

The gray fox is slightly smaller than the red fox. It weighs 4–5 kg, and is 60 cm in head and body length plus a 30-cm tail. There is little color variation in gray foxes. They have a "pepper and salt" coat, a median black stripe on the black-tipped tail, and black-tipped ears. They have six mammae and a rather large (11 cm long) musk gland on top of the tail. This gland may function in individual identification (Rue 1968). The temporal ridges of the skull are separated anteriorly but connect posteriorly to form a distinctive U shape. A distinct "step" is found on the ventral border of the dentary, midway between the interior border and the coronoid process and the tip of the angular process (Hall and Kelson 1959).

In both the red and the gray fox, the incisors are not prominently lobed. The dorsal surface of the post-orbital process is concave and the dental formula is 3/3, 1/1, 4/4, 2/3 = 42 (figure 22.3).

PHYSIOLOGY

During the first three and a half months of life, many changes occur in the pelage of red fox pups. At first they have a grayish brown fur that is fine and silky.

Then the fur becomes more woolly and turns light yellow around the neck and head during the second week. Conversely, gray fox pups are nearly hairless at birth and have a dark skin. Red foxes open their eyes at 8–9 days and begin to walk at 3 weeks, while gray fox pups open their eyes at 10–12 days. At 4 weeks of age, the ears of the red fox are black and become erect. The pelage becomes a pale yellowish brown at 7 weeks and by 14 weeks the guard hair appears and the pups take on a reddish brown color. Black guard hairs appear on the tail, legs, and ears. A gradual molt begins in late August or September (Linhart 1968).

Adult red foxes develop a thick, dark red pelage during late September and October. This coat lasts through the winter until February or March, when the thick guard hair is replaced by the thin, lighter summer pelage.

Newborn red fox pups range from 71 to 119 g in weight (Storm and Ables 1966). Full growth is attained in 6 months and by October juveniles are similar in size to adults (Storm et al. 1976). Juvenile males are usually larger than juvenile females.

REPRODUCTION

Female red and gray foxes are monestrous (Rowlands and Parkes 1935). Breeding takes place from December to March in red foxes and from January to April in gray foxes. However, the onset of breeding varies in different parts of their range, earlier in the south, later in the north. Sheldon (1949) found that 76 percent of red fox matings in New York occurred between late December and end of March. Layne and McKeon (1956) found that 83 percent of matings occurred between mid-January and early February in New York. For the gray fox, 71 percent of the breeding occurred in late February to mid-March (Sheldon 1949), while Layne and McKeon (1956) noted that 52.5 percent took place in March. Latitude does play a role in the breeding season in *Urocyon*. In southern Illinois, the main breeding season extended from late January through February (Layne 1958), while in Alabama the peak is even earlier (Sullivan and Haugen 1956).

In Illinois and Iowa, Storm et al. (1976) calculated conception dates for 20 red fox vixens by backdating 52 days (gestation period) from known birth dates, and by aging fetuses from growth curves prepared by Layne and McKeon (1956). Three vixens bred between 13 and 29 December, 9 bred between 1 and 15 January, and 8 bred between 16 and 31 January. The peak of *Vulpes* breeding season occurs the third or fourth week of January (Sheldon 1949; Lloyd and Englund 1973; Storm et al. 1976). Wood (1958) noted that the peak for gray fox breeding was early February in Florida, while Follman (1973) suggested it was March in Illinois. The gestation period is between 51 and 54 days for *Vulpes,* 53 days for *Urocyon* (Sheldon 1949).

Estrus lasts 1 to 6 days (Asdell 1964) and females may breed at 10 months of age. However, not all females breed their first year. Storm et al. (1976) found that 5 percent of red fox vixens were unsuccessful in breeding, whereas Layne and McKeon (1956) found that 11.5 percent were barren. Wood (1958) noted that 8 percent of gray fox females failed to produce young in their first breeding season. Sheldon (1949) reported that 3.3 percent of all adult females did not breed in New York, while Wood (1958) found that 6.4 percent did not breed in Georgia and Florida. Failure of a female to produce young may be attributed to no mate, failure of the eggs to be fertilized, or embryo mortality (Layne and McKeon 1956).

In red and gray foxes spermatozoa are formed in October and November (Sullivan 1956; Lloyd and Englund 1973; Storm et al. 1976). Most young males are capable of breeding their first year, even though their os penis is not completely ossified (Storm et al. 1976). Joffre (1976) reported that puberty occurs in young male red foxes in January. Testis weight reaches a maximum from December to January and then decreases until May, after which a period of quiescence is observed until September. During this quiescent period, spermatogenic and androgenic activity decrease. Layne (1958) reported a similar curve for gray foxes, but the quiescent period does not begin until late June. An enlarged prostate gland is an indication of reproductive status of gray foxes.

Variations in the female reproductive organs of red and gray foxes were studied by Layne and McKeon (1956) and Layne (1958). The uterine horns of young of the year vixens are smooth, firm, and creamy white in color. The mesometrium usually contains a considerable amount of fat and the uterine blood vessels are inconspicuous. The uterine horns of adult foxes are much larger and flattened, usually with conspicuous placental scars. The uterine surface is rough and discolored. Uterine blood vessels are prominent and little fat is found in the mesometrium.

During pregnancy a mucous plug is formed in the cervix, and the vagina becomes flaccid and enlarged. As

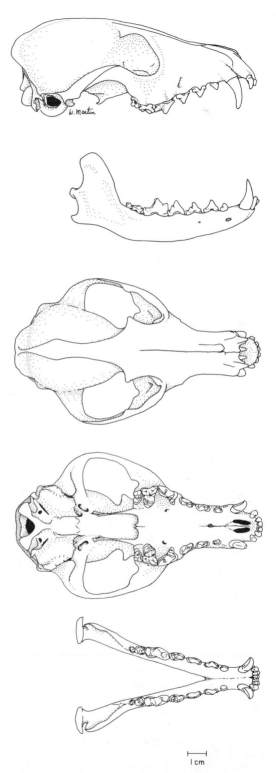

FIGURE 22.3. Skull of the red fox (*Vulpes vulpes*). From top to bottom: lateral view of cranium, lateral view of mandible, dorsal view of cranium, ventral view of cranium, dorsal view of mandible.

TABLE 22.1. Red and gray fox mean litter size as reported in the literature

Reference	Location	Method[a]	Mean Litter Size
Red Fox			
Sheldon (1949)	New York	PS, E	5.4
Switzenberg (1950)	Upper Michigan	D	4.2
	Lower Michigan	D	5.4
Hoffman and Kirkpatrick (1954)	Indiana	PS, E	4.3–6.8
Layne and McKeon (1956)	New York	PS	5.4
McIntosh (1963)	Australia	PS	4.3
		E	3.8
		D	3.8
Stanley (1963)	Kansas	D	4.5
Storm et al. (1976)	Illinois, Iowa	PS	7.1
		E	6.8
		D	4.2
Ryan (1976)	Australia	PS	3.7
		E	4.0
Insley (1977)	England	D	4.0
Pils and Martin (1978)[b]	Wisconsin	PS, E, D	5.6
Gray Fox			
Wood (1958)	Georgia, Florida	PS, E	4.6
Sheldon (1949)	New York	PS, E	3.7
Layne and McKeon (1956)	New York	PS	4.5
Layne (1958)	Illinois	E	4.4
Layne (1958)	Illinois	PS, E	3.8

NOTE: Comparisons can be unclear due to the different methods used in these studies.

[a]Abbreviations for method of determination are: E = embryos of fetuses, D = number of young found in den, PS = placental scars (adapted from Storm et al. 1976: 21).

[b]Refer to pp. 12–18 of Pils and Martin (1978) for a discussion of the problems in using placental scars to obtain estimates of litter sizes.

the fetuses develop, the uterine walls become stretched and thin (Layne and McKeon 1956).

Following parturition, the walls of the uterus become thick and wrinkled. Uterine blood vessels and placental scars are conspicuous. Cutting open a uterine horn will clearly show the placental scars as broad bands of villi encircling the interior wall of the horn (Layne and McKeon 1956).

Mean litter size has been calculated by different variables including placental scar counts, embryos or fetus counts, and number of young found in the den (table 22.1). There are biases to all three methods. If you are counting young in the den, you may not excavate all the pups or you may encounter communal denning (Hoffman and Kirkpatrick 1954). Using placental scars and embryo counts to obtain litter sizes is biased due to intrauterine mortality (Englund 1970; Allen 1975). Pils and Marten (1978) excavated 27 red fox dens and found a mean litter size of 5.2, while their placental scar and embryo counts were 5.5 and 6.4, respectively. However, for 70 litters examined they found no significant differences in those mean litter sizes. In general, the mean litter size is 5 for red fox and 4 for gray fox.

Newly born pups remain at the den for the first month of life. Red fox parents may move the pups from one den to another as many as three times before they are six weeks old (Scott 1943; Sheldon 1949;

Storm et al. 1976). Litters are sometimes split, with half the litter remaining in one den and half in another.

Once red fox pups are 10 weeks old, they may travel short distances from the den without being accompanied by a parent. Beginning about the 12th week, pups (or juveniles, as they may now be considered) begin to explore different parts of their parents' home range during the daylight hours. By mid-September or early October they begin to disperse. Males usually disperse before females (Storm et al. 1976).

ECOLOGY

Sex and Age Ratios. Sex ratios of red foxes have been estimated using many techniques with varying results. The maximum production of young of monogamous species occurs when the sex ratio is 50/50 (Dasmann 1964). Various authors (Sheldon 1949; Layne 1958; Fairley 1970; Storm et al. 1976) reported relatively equal ratios of male and female fetuses for red and gray foxes. Layne and McKeon (1956), however, noted 61 percent male red fox fetuses. Further, Storm et al. (1976) found a significantly higher proportion of male red fox pups from excavated dens. In the fall there is a greater percentage of males, suggesting higher vulnerability of males to hunting and trapping (Sheldon 1950; Richards and

Hine 1953; Friend and Linhart 1964). Storm et al. (1976) suggested that if the sex ratio of 50:50 is real, then females suffer a higher mortality than males during gestation, parturition, or in their very early days.

Home Range. The home range of foxes has been calculated mainly by radiotelemetry techniques. In Wisconsin, Ables (1969) reported home range sizes of 57.5 to 161.9 ha for seven red foxes radio collared in diverse habitat. He found a much larger home range of 5.12 km² for an adult male in a less diverse farming area. In Illinois, Storm (1965) found that one adult male red fox had a home range 3.1 km long by 2.2 km wide. Fox home ranges tend to be about twice as long as they are wide (Follman 1973) and are often influenced by habitat features (Storm 1965; Ables 1974). Schofield (1960) followed tracks in the snow and estimated red fox home ranges to be 1.6–2.4 km in radius.

Lord (1961) reported the home range of gray foxes to be 3.2 km². Follman (1973) and Yearsley and Samuel (1980) used telemetry and found home ranges to vary from 75–185 ha. Overlap in the home ranges of gray foxes was observed in both studies.

Factors that may affect the size of a fox home range include the abundance of food, the degree of intraspecific and interspecific competition, the type and diversity of habitat, and the presence of natural physical barriers such as rivers and lakes. Fuller (1978) suggested that differences in sizes of gray fox home ranges were related to quality and quantity of habitat rather than weather, age, and breeding season. However, Follman (1973) noted that female red foxes reduced movements in March and female gray foxes did the same in April, indicating that the birth of pups had occurred.

Habitat. Red foxes prefer diverse habitats consisting of intermixed cropland, rolling farmland, brush, pastures, mixed hardwood stands, and edges of open areas that provide suitable hunting grounds (MacGregor 1942; Eadie 1943; Cook and Hamilton 1944; Ables 1974). They select areas of greatest diversity and use edges heavily (Ables 1974). Dense forests are undesirable. Pils and Martin (1978) found that juvenile red foxes favored unpicked cornfields in Wisconsin, while adults utilized a wide variety of cover types. In winter adults favored marshes, retired croplands, upland hardwoods, and strip cover.

Red foxes may also inhabit suburban areas, particularly parks, golf courses, cemeteries, and large gardens. Obtaining a daytime hiding place seems to be an important factor in determining a suitable suburb for red foxes (Harris 1977).

Gray foxes are similar to red foxes in that they prefer a diversity of fields and woods rather than a large tract of homogeneous habitat (Wood et al. 1958). The basic difference between the two species is that gray foxes prefer woodlands more than red foxes do (Kozicky 1943; Follman 1973; Failor 1974). In Illinois, gray foxes selected early successional stage woodlands in all seasons except summer. Red foxes utilized woods in the winter (Follman 1973).

In California, Fuller (1978) noted that gray foxes used riparian and old field habitats more than expected. Peterson et al. (1977) found that the best gray fox habitat in Wisconsin consisted of areas having 12 percent or greater slope within an area of at least 64 ha and a change in elevation of at least 12 m. Such habitat was about 30 percent forests and 40 percent farmed land, with much interspersion maximizing the amount of edge present.

Dens. Red foxes may dig their own dens or more often use an abandoned woodchuck (*Marmota* sp.) or badger (*Taxidea taxus*) burrow. The same den may be used for many generations (Stanley 1963), with burrows being added each year (Pils and Martin 1978). In the Midwest red foxes preferred strip cover for den sites. Pils and Martin (1978) believed cover type was the most important factor influencing den selection, followed by human disturbance, water, and length of use. Gray foxes utilized woodpiles, rocky outcrops, hollow trees, or brushpiles for dens (Trapp and Hallberg 1975).

Pils and Martin (1978) found five communal red fox dens representing 11 percent of all dens found in southern Wisconsin during 1972–75. The reasons for such dens are not understood, although Tullar et al. (1976) suggested that disturbance at one den may cause a vixen to move her pups to another.

FOOD HABITS

The food habits of foxes are well documented, a result of early studies done when predators were thought to be killing large numbers of poultry and wild game. Foxes feed on a wide variety of animal and plant materials depending mainly on the availability of the food source. Small mammals, birds, fruits, and insects comprise the bulk of the foxes' diet.

Red foxes often store food in caches. By digging a hole in loose dirt or sand a red fox can bury a mouse or part of a rabbit for later consumption. In Michigan, Murie (1936) found that 40.3 percent of the food items taken by red foxes were cached. However, in Kansas, Stanley (1963) reported very few instances of food being cached. He attributed this infrequent use of caches to the relative abundance of food in the study area (making caches unnecessary) and the density of animals likely to raid caches, such as skunks, opossums, dogs, coyotes, owls, hawls, and crows. MacDonald (1977) reported that red foxes will continue hunting after they have filled their stomachs and, conversely, will store food in caches when they are still hungry. Caching is necessary when prey becomes scarce for red foxes because they must feed on a regular basis.

Adult red foxes consumed 2.25 kg of prey/week, while 12-week-old pups required 1.90 kg/week (Sargeant 1978). This could put a high demand on prey, especially during the denning season.

The food habits of the red fox have been studied in many different locations using techniques such as stomach analysis, scat analysis, analyses of bones dis-

TABLE 22.2. Summary of red fox food habits as reported in the literature

Principal Foods	Location	Number and Kind of Samples	Investigator
Mice, rabbits	New York, New England	206 stomachs	Hamilton (1935)
Mice, rabbits woodchucks	New Hampshire	313 scats	Eadie (1943)
Rabbits, mice	Virginia	15 stomachs	Nelson (1933)
Deer carcasses, mice, rabbits	Michigan	trails	Schofield (1960)
Hares, rabbits, mice	Missouri	1006 stomachs	Korschgen (1959)
Mice, rabbits	Minnesota	34 stomachs	Hatfield (1939)
Mice, rabbits	Wisconsin	25 stomachs	Errington (1935)
Mice, rabbits, birds	Iowa	40 stomachs	Errington (1963)
Rabbits, cottontails, mice	Kansas	69 scats	Stanley (1963)
Rabbits, sheep carrion, mice	Victoria, Australia	967 stomachs	Coman (1973)
Small mammals, rabbits, birds	Wisconsin	1020 scats 132 stomachs 58 dens	Pils and Martin (1978)
Rabbits, sheep carrion, mice	New South Wales, Australia	899 stomachs	Croft and Hone (1978)
Snowshoe hares, red squirrels, deer, mice	Isle Royale, Michigan	448 scats	Johnson (1970)

carded at active dens, and observations of dead prey along red fox trails. In New York and New England, Hamilton (1935) found that mice were the main food item eaten by red foxes. The meadow mouse (*Microtus pennsylvanicus*) occurred most often in the 206 red fox stomachs examined. It was not uncommon to find three or four meadow mice in a single stomach, and one stomach contained eight meadow mice. Mice occurred in 40 percent of the stomachs examined by Hamilton (1935); rabbits occurred in 27.2 percent of the stomachs. Nelson (1933) in Virginia and Eadie (1943) in New Hampshire also found mice and rabbits comprising the bulk of the diet (table 22.2).

Seasonal variations are prominent in the diet of red foxes. Korschgen (1959) found that rabbits constituted a larger percentage of the diet during winter than during summer, while rats and mice were eaten most often in the spring. In Michigan, Schofield (1960) reported that deer carrion comprised the bulk of the winter food of the red fox. Insects and plant foods were heavily utilized in the summer. Others noted a similar change in diet from animal matter in winter to insects and fruit in the summer (Eadie 1943; Cook and Hamilton 1944).

Besides rabbits and mice, the red fox may also eat squirrels, young oppossums, raccoons, skunks, housecats, dogs, woodchucks, weasels, mink, muskrats, shrews, moles, porcupines, pocket gophers, songbirds, crows, pheasants, quail, grouse, ducks, turkeys, chickens, geese, woodcock, hawks, owls, bird eggs, turtles, turtle eggs, and insects. Plant foods such as grasses, sedges, nuts, berries, pears, apples, grapes, and other fruits, and corn, wheat, and many grains are eaten by the red fox. Livestock such as pigs, sheep, calves, and

goats have also been taken. Foxes will eat carrion, which may partially account for the great diversity of animals found during stomach and scat analysis. MacDonald (1977) reported an incidence of cannabalism in the red fox.

Gray fox food habits are similar to but less known than those of the red fox. First, availability plays a major role in what is consumed. Secondly, animal matter is most important in winter, insects and fruits in summer (Wood 1958; Yearsley 1976).

In a review of gray fox food habits in the eastern United States, cottontails were the major food, with rodents second (Trapp and Hallberg 1975). English and Bennett (1942) found that cottontails made up 64 percent of the winter diet of *Urocyon*. Yoho and Henry (1972) examined 24 stomachs from foxes collected in Tennessee from January to June and found cottontails to be most important, with arthropods and small mammals next. Some of the summer-fall fruits consumed are huckleberries, wild grapes, apples, hawthorns, and elderberries (Kozicky 1943; Glover 1949).

Turkowski (1969) compared gray fox foods taken in the Lower and Upper Sonoran life zones of the southwestern United States. In both zones the foxes were opportunistic. Mammals occurred in 65 percent of the stomachs of animals collected in the Lower Sonoran zones and in 57 percent of the animals from the Upper Sonoran zones. Arthropods were found in 57 percent of stomachs collected in Lower Sonoran, 53 percent in Upper Sonoran. Plants were found in 39 percent of stomachs from Lower Sonoran, 40 percent from Upper Sonoran. Important plants were mesquite beans (Lower Sonoran) and juniper berries (Upper Sonoran).

BEHAVIOR

During the breeding season the male and female mate and remain in one area until the pups are raised. Males defend this territory against intruding foxes while the female gives birth to her pups. The male brings food to the vixen until the pups can be left alone for short periods of time. Then both the male and the female hunt for food, with the vixen returning to nurse the pups during the day. Both parents remain with the pups until dispersal in the fall (Sheldon 1949). Dispersal is defined as a straight-line movement of more than 8 km between first and last captures (Phillips et al. 1972; Storm et al. 1976; Pils and Martin 1978).

In the Midwest, subadult red foxes begin dispersal in October, males dispersing earlier than females (Storm et al. 1976; Pils and Martin 1978). A higher percentage (88 percent) of tagged subadult males than females (58 percent) were recovered more than 8 km from their dens, while 83 percent of the adult males and 56 percent of adult females dispersed (Pils and Martin 1978). Foxes move greater distances from October to December than later in the winter and spring, with males moving greater distances than females.

In the Midwest, red foxes disperse in a north or northeast direction (Arnold and Schofield 1956; Storm et al. 1976; Pils and Martin 1978). Storm et al. (1976) suggested that this is not a reflection of movement but rather that hunters in the north have better snow conditions, thus increasing the chance of killing a marked fox. Pils and Martin (1978) felt that an interstate highway located south of their Wisconsin study area may have discouraged foxes from moving in that direction.

The male red fox assumes responsibility for territory defense. Scent markings play a major role in defining the territory. Not only does this allow nonresidents to assess the status of the area but it also reinforces the male's familiarity with its own home range (Preston 1975).

Preston (1975) introduced nonresident male red foxes into established home ranges of resident male-female pairs. He observed 75 encounters between residents and intruders. Thirty-five culminated in agonistic display, 37 in chase, and 3 in physical contact.

The voice of the red fox varies from short, sharp yaps to long howls and screeches (Godin 1977). Gray foxes are most commonly heard barking in February, during breeding, and in late summer when pups are being taught to hunt (Rue 1968).

Interspecific relationships between foxes and other animals have been documented. Merriam (1966) reported several instances of woodchucks and foxes living together in the same den. Other interspecific relationships include foxes playing with mountain sheep (*Ovis* sp) in Alaska, and Newfoundland foxes have been observed leaping and nibbling at the muzzles of caribou (*Rangifer tarandus*) (Godin 1977).

MORTALITY

The three most prevalent causes of death for red foxes in Illinois and Iowa were hunting, trapping, and road kills. Pils and Martin (1978) noted the same major

causes for mortality of red foxes in Wisconsin. Hunting and trapping mortality is also affected by snow conditions, with more snow making red foxes more visible, especially in open farm country (Storm et al. 1976). Highway mortality is much greater for juvenile red foxes than for adults. In June when juveniles are wandering short distances from the den, such mortalities begin. The frequency of highway deaths increases through the summer as juveniles move further from the den.

The proportion of foxes killed by farm machines is believed to be small, but again is more frequent in juveniles than adults. Phillips et al. (1972) reported that juvenile red foxes were 1.2 times more vulnerable to hunters or trappers than adults and 1.5 times more vulnerable than adults to all forms of mortality.

Lloyd and Englund (1973) found that for red foxes 9 percent of ovulated eggs failed to implant or were lost before becoming visible. Ten to 20 percent of the red fox fetuses were resorbed.

Little is known about mortality in gray foxes; however, Layne (1958) reported natal mortality of about 32 percent.

Total longevity for red and gray foxes is about five years and four years, respectively (Wood 1958; Storm et al. 1976).

Parasites and Diseases. Mange (*Sarcoptes scabiei*) occurs in red fox populations throughout North America and Europe (Ross and Fairley 1969; Trainer and Hale 1969; Stone et al. 1972). Sarcoptic mange is density dependent and caused by mites that irritate the skin. The symptoms are flaking and cracking of the skin as well as loss of hair. As the condition persists, flaking skin areas develop lesions and the skin becomes crusted and wrinkled. After two or three months the mites will invade the eyelids, causing the fox to squint. This leads to scratching and biting and in most cases the fox will die. Often infested foxes will lose fear of humans, making them highly likely to be killed (Trainer and Hale 1969). Tullar (1979) noted that 45 percent of the red foxes found dead in New York had mange. Pils and Martin (1978) reported only a 7 percent mange-related mortality for Wisconsin red foxes, but suggested that it was really much higher, as carcasses were hard to find.

Apparently pups are infected in the den and will die before the fall season. Tullar et al. (1976) indicated that communal denning further compounds the problem and leads to high spring mortality of pups. Smith (1978) found that 67 percent of red foxes in New Brunswick and Nova Scotia had mange mites. Other external parasites included many species of flea and two species of tick, *Dermacentor variables* (Stanley 1963) and *Ixodes ricinas* (Ross and Fairley 1969).

Two viral diseases that occur in foxes are rabies and distemper. During an outbreak of distemper in New York, Monson and Stone (1976) examined 133 red foxes and found distemper in 30.8 percent. One hundred and thirty-one gray foxes were examined and 64.1 percent had distemper (Monson and Stone 1976). Distemper differs from mange in that animals can recover and survive future infection (Tullar 1979). Gray foxes are important in the spread of enzootic rabies to domestic animals (Jennings et al. 1960). In Florida, where high populations were reduced by "fox encephalitis" (disease of distemper-hepatitis complex), the spread of rabies was not reduced. Not surprisingly, the outbreak spread most rapidly during the January and February mating time and during the fall dispersal period for the gray fox. The behavior of rabid foxes conforms to two general patterns (Jennings et al. 1960). In one type the fox appears suddenly and makes an aggressive attack, while in the other the fox is visibly weak; it has usually been found by dogs or people and offers little struggle. Bites are not as common when the second type of behavior occurs.

Storm et al. (1976) reported a low incidence of rabies in Iowa red foxes, only 1-3 positive cases per year, while Schmurrenberger and Martin (1970) reported 10 cases per year in Illinois. Rabies does not appear to be a major mortality factor; however, in some cases it has been known to regulate fox numbers (Parker et al. 1957).

The heartworm, *Dirofilaria immitis,* also occurs in red foxes. This is of particular significance because of its possible transmission to domestic dogs (Monson et al. 1973). The larval worm is transmitted by mosquitoes.

In 40 of 438 pairs of kidneys examined by Ross and Fairley (1969) there was chronic nephritis due to an infection of leptospira. This indicates that the red fox may be a reservoir of leptospirosis, which can be transmitted to other wild and domestic animals.

Other internal parasites of the red fox reported by Stiles and Baker (1935) were *Alaria arisaemoides, Cryptocotyle lingua, Opisthorchis conjunctus, Parametorchis intermedius, Pseudamiphistomum conus,* Ancylostomidae, *Capillaria plica, Eucoleus aerophilus, Toxascaris limbata, Toxicara canis, T. cati, T. marginata, Trichuris vulpis, Uncinaria polaris,* and *U. stenocephala.* Stanley (1963) found the dog tapeworm (*Taenia pisiformis*), the stomach worm (*Physaloptera rara*), an intestinal worm (*Toxascaris leonina*), and the dog hookworm (*Ancylostoma caninum*) in 24.6 percent of the foxes he examined. The most common nematode was *Capillaria aerophila,* which occurred in 67.2 percent of the foxes. The only cestode found was *Taenia crassiceps,* occurring 50.8 percent of the time.

Of 366 red foxes examined in Northern Ireland, 64 percent were infected with the nematode *Uncinaria stemocephala* and 35 percent contained *Toxocara canis* and *T. leonina* (Ross and Fairley 1969).

AGE DETERMINATION

Many different techniques have been developed to determine the age of foxes. Some of these techniques can be used only to separate juveniles from adults, while other methods can separate individuals into more precise age classes.

Sullivan and Haugen (1956) used x-rays of forefeet to separate juveniles from adults. The distal epiphyses of the radius and ulna ossified at about eight to nine months.

Churcher (1960) examined the skulls of 188 known-aged captive red foxes and 1,288 wild specimens. The most reliable skull criteria for determining age was the closure of the basioccipital-basisphenoid suture during the first year of life. The presphenoid-basisphenoid suture closes in the first and second years, and the palatal portion of the premaxillar-maxillary suture closes during years four to six. He also observed changes in the shape of the postorbital processes with age. Juvenile foxes had pointed postorbital processes that became blunt and rounded with age. The lateral anterior process of the nasals also varied in shape from pointed in juveniles to blunt and rounded in adults. Harris (1978) found the basioccipital-basisphenoid suture to be the most useful in determining age and believed that there was too much variability in other cranial characteristics accurately to determine age.

Young of the year may be separated from adults by the degree of ossification of the proximal epiphysis of the humerus. The epiphysis closes at the age of 9 to 9½ months in red foxes (Reilly and Curren 1961). Pils and Martin (1978) aged red fox pups in days by measuring the right hind foot length and fitting the data to a growth curve (following the technique of Johnson et al. 1975).

Eye lens weights have been used to separate juvenile red foxes (Friend and Linhart 1964; Phillips 1970; Ryan 1976) and gray foxes (Lord 1961) from adults. Eyes were removed from freshly killed foxes and placed in 10 percent buffered formalin. The lenses were separated from the eyeball and dried in a drying oven for three to four days at 80° C, then weighed. The juvenile-adult separation range is between 203 and 213 milligrams (Friend and Linhart 1964).

One of the most reliable and accurate techniques for determining the age of foxes is to count the annular cementum rings of the teeth. This method has been used by Monson et al. (1973), Grue and Jensen (1973), Storm et al. (1976), Geiger et al. (1977), and Harris (1977, 1978). Canines, premolars, or molars are placed in a decalcifying solution until they become soft. Using a cryostat, the root portion of the tooth is frozen in bedding medium, then sections are stained and placed under a microscope to count the number of rings.

Length and weight of the baculum may be used to age male gray foxes, but not red foxes (Petrides 1950). Baculum weight provides less overlap for red foxes than baculum length, but such differentiation is more pronounced in gray foxes.

ECONOMIC STATUS AND MANAGEMENT

There have been few management programs aimed specifically at foxes (Pils and Martin 1978). Most "management" was a result of predator control utilizing bounties. Trippensee (1953:111) noted that "little

TABLE 22.3. The four most productive states for red and gray foxes based on numbers trapped

Red fox

Season	Minnesota	North Dakota	South Dakota	Wisconsin	Dollar Value of Pelt	Total in United States
1971–72	54,000	22,779	no records	26,496	$17.00	151,818
1972–73	55,000	37,827	no records	22,295	25.00	187,701
1973–74	46,000	61,556	30,222	33,827	36.00	283,044
1974–75	49,000	32,872	26,838	21,322	30.00	237,494
1975–76	55,000	38,938	23,825	22,165	30.00	272,017
Average	51,800	38,794	26,961	25,221		

Gray fox

Season	Texas	Ohio	Indiana	Pennsylvania	Dollar Value of Pelt	Total in United States
1971–72	no records	6,873	2,812	6,873	$ 3.50	29,334
1972–73	7,518	12,380	7,639	5,328	14.00	60,476
1973–74	21,428	18,972	9,654	9,362	15.00	108,863
1974–75	35,973	19,174	10,411	11,401	13.00	146,584
1975–76	25,077	23,200	17,041	12,742	17.00	163,458
Average	22,499	16,123	9,511	9,141		

SOURCE: Adapted from Deems and Pursley 1978
NOTE: Some states missing; either did not respond or did not keep records.

attention, except a sensible hunting take and limited trapping, is needed to maintain a good fox population.''

The use of bounties to "control" numbers of foxes was inconsistent. In Wisconsin, for example, bounty existed in 1923, was repealed in 1931, reintroduced in 1946, dropped in 1957, reintroduced in 1959, and finally dropped in 1963 (Pils and Martin 1978). Such erratic activity is an indication of how little we know about bounties as regulators of fox populations. Even though isolated counties still pay fox bounties, this is not considered a management technique (Pils and Martin 1978).

Management today is variable from one state to the next. Although most management involves harvest regulations, some habitat management has been suggested. Scott and Klimstra (1955) suggested planting trees with fruit available to foxes, thus reducing utilization of other prey. Yearsley and Samuel (1980) suggested that maximizing edge would aid utilization of reclaimed surface mines by gray foxes in the east. Optimum gray fox habitat can support 10 to 15 foxes per 256 ha. Any type of tree harvest may benefit red foxes, and regeneration should be in linear and irregular shapes to maximize edge effect. Cutting down pines and converting to a variety of mast- and fruit-producing hardwoods benefit foxes (anonymous 1971).

Fox harvest regulations are variable (Pils and Martin 1978). Bounties and predator control programs were designed to control fox populations, while other regulations existed to increase fox populations because of their economic value. In 1976, the red fox was totally protected, with no hunting or trapping season in 5 of 48 states where it was found (Arkansas, California, Delaware, Florida, and Tennessee). Thirty-four states had hunting seasons on the red fox in 1976, while 35 states and all 12 Canadian provinces and territories had a trapping season on the red fox. Laws limiting the number of red foxes one can harvest are enforced in 26 states and 7 provinces and territories. Colorado, Michigan, and Missouri practice some form of habitat management for the red fox. Populations were inventoried in 11 states and 3 provinces (Deems and Pursley 1978).

The major problem in setting harvest regulations stems from the fact that we have poor indexes to fox populations. With the relatively recent increase of pelt prices, accompanied by declining fox populations in some areas (Petersen et al. 1977; Pils and Martin 1978), the need for estimations of numbers and subsequent harvest regulations is even more important. In fact, Pils and Martin (1978) suggested that the increase in fur prices and declining fox populations provided the catalyst for recent initiation of fox seasons in the Midwest.

The total numbers of reds and grays trapped has steadily increased from 1971–72 to 1975–76 (table 22.3). The value of pelts has also taken a drastic jump (figures 22.4 and 22.5). Even in states such as West

Virginia where red and gray foxes are not considered major fur bearers (when compared to those states in table 22.3), based on sale prices they accounted for $1,200,000 in 1978-79, with an average price of $54-55 for both species (J. R. Hill personal communication). This increase in prices, and thus trapping effort, further complicates efforts to estimate fox populations.

Annual harvest may not be an indicator of abundance because harvest would vary with pelt prices (Seagers 1944). Increasing pelt prices in Wisconsin brought more intensive harvest and resulted in a declining red fox population (figure 22.5). Petersen et al. (1977) found a significant correlation between pelt prices and harvest of gray foxes in Wisconsin (figure 22.4). They also reported a decrease in harvest numbers during the 1960s and 1970s. In Missouri, inflated market values of pelts was positively correlated with trapper and fur trader numbers (Erickson and Sampson 1978). Thus, pelt value does cause changes in the harvest and is another reason that fox harvest may not be a good index to fox populations. New York is attempting to get population indexes on foxes by surveys of tracks on roads and counting reoccupied dens (Tullar 1979). Use of scent post stations is under investigation in the East. It may be a technique that will give an index to fox populations over the years (Linhart and Knowlton 1975).

Fox trappers are motivated by profits, but fox hunters are not. Fox hunters and trappers are major causes of mortality for red foxes in some areas (figure 22.6). Petersen et al. (1977) noted that trappers took over 40 percent of the gray foxes in November while hunter kills remained relatively stable for November, December, and January (figure 22.7). They found a significant correlation between gray fox harvest and red fox and raccoon harvest, and pointed out that gray foxes were incidental targets.

Foxes were once viewed as harmful predators consuming large numbers of pheasants and rabbits. Poultry farmers viewed foxes as dangerous thieves that would rob their chicken coops. Today, however, the image of the fox has changed from harmful predator to valuable fur bearer. Ecologists emphasized the importance of foxes in the biological food chain, especially their role in controlling population explosions of rodents and rabbits. Poultry raisers realized that restriction of poultry to pens or houses at night greatly decreased fox predation. Orchard growers recognized foxes as controllers of small rodents that eat the bark and roots of fruit trees.

Red fox pelt prices continued to rise with the increased demand for long-haired furs. In 1963 the price paid for a red fox pelt varied from $0.50 to $1.50 (Stanley 1963). In 1979 the price of red fox pelts rose above $50, with a similar increase for gray fox pelts. The esthetic value of watching a fox or tracking it through the snow must not be overlooked (Pils and Martin 1978). Photography of predators can be a year-round activity that requires skill and patience. Photographing pups at active dens can be a rewarding and satisfying experience, as can simple observation of fox signs in the wild.

CURRENT RESEARCH AND MANAGEMENT NEEDS

The ideal management plan would be one that attempts to maximize fox harvest and still achieve a balance between this predator and its prey species. However, most states have far too little data available to allow the establishment of regulations that can achieve this balance. Rapidly rising fur prices and the once-common bounty system are factors that further complicate efforts to determine population densities from harvest.

Wisconsin has made major strides in developing harvest regulations. Iowa, Indiana, Illinois, and Missouri have also considered this problem. Petersen et al. (1977) and Pils and Martin (1978) made the following management suggestions for gray foxes in Wisconsin, most of which should apply to other states. The gray fox population has stabilized at a low level, so Petersen et al. (1977) suggested a shortened statewide trapping season. Pils and Martin (1978) suggested shortening the Wisconsin season for red foxes, eliminating February, to prevent trapping at the den site during reproductive activities, and October. They added that the trapping season should be opened concurrently for muskrats and mink, to reduce trapping of nontarget fur bearers. They further suggested that the season might

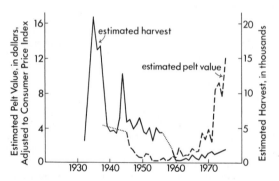

FIGURE 22.4. Relationships of harvest and pelt value of gray foxes (*Urocyon cinereoargenteus*) in Wisconsin, 1932-75. Modified from Petersen et al. 1977.

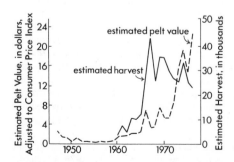

FIGURE 22.5. Estimated harvest and pelt value of red foxes (*Vulpes vulpes*) in Wisconsin, 1947-76. Modified from Pils and Martin 1978.

be shortened to the months of December and January to reduce deer hunter pressure on red foxes. Thus, a delay until 1 December might improve snow conditions. Finally, Pils and Martin (1978) and Petersen et al. (1977) suggested that better estimates of harvest could be made if a separate license for all fox hunters and trappers were provided, and if there were a mandatory registration of pelts. When red foxes cause problems for domestic fowl or game species on state wildlife lands, removing pups from dens and transferring them to other areas has proven beneficial (Pils and Martin 1978).

The most intensive management practices suggested involved the construction of artificial dens and the raising of foxes for release in areas of low density (Everett 1952). It is doubtful if this is a practical or realistic possibility, however.

Other states have reacted to declining red fox populations. Missouri closed the trapping season on red foxes in 1977, until pelt prices return to a lower level (Sampson 1977), and California closed its season in 1974 due to low population levels (Gray 1975). If pelt prices remain high, other states will have to reduce harvest to protect their fox populations.

In summary, the major research need is to obtain regional population estimates or indexes. Perhaps the best method for this is to develop permanent scent station transects on a state-by-state or regional basis. Further research in this area is fully warranted.

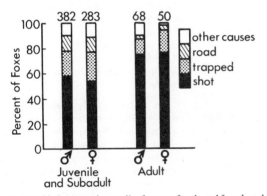

FIGURE 22.6. Annual mortality factors of male and female red foxes (*Vulpes vulpes*) in Iowa and Illinois. Adapted from Storm et al. 1976.

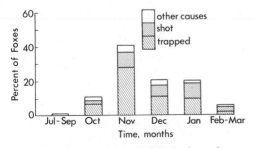

FIGURE 22.7. Causes and season of death of gray foxes (*Urocyon cinereoargenteus*) purchased by fur buyers in Wisconsin, 1975–76. Adapted from Petersen et al. 1977.

THE KIT FOX AND THE SWIFT FOX

Taxonomic Status. The kit fox (*Vulpes macrotis*) is a narrowly specialized animal adapted for life on the deserts of western United States and Mexico (Egoscue 1962). There are eight subspecies of kit fox (Hall and Kelson 1959); however, Waitham and Roest (1977) studied five of these subspecies and amended the classification as follows: the San Joaquin kit fox, *Vulpes m. mutica*, is recognized as a distinct subspecies, while *arsipus*, *devia*, and *tenuirostris* are synonymized with *macrotis*. Based on this work there are now five distinct subspecies. Hall (1946) and Durrant (1952) suggested that *nevadensis* is not distinct from *arsipus;* if this is true then there are only four distinct subspecies.

The common names *swift fox* and *kit fox* have been used interchangeably for both *Vulpes macrotis* (the kit fox) and *Vulpes velox* (the swift fox). This is probably because the general outward appearances are similar to *V. velox,* but possessing shorter ears and a broader skull and being a slightly larger size than *V. macrotis*. Indeed, some scientists have questioned whether the species distinction is genuine. Hall and Kelson (1959:860) noted that the swift fox "may be only subspecifically distinct from the kit fox." Snow (1973) reported that specific status for swift and kit foxes is justified. Hall (1981) lists both groups as *Vulpes velox*.

Distribution. The distribution of both species is quite interesting and that of *Vulpes macrotis* is important because one subspecies, *Vulpes m. mutica*, was given "rare" status by the U.S. Fish and Wildlife Service and the state of California.

The former range of *V. m. mutica* was the plains and foothills of San Joaquin Valley before white settlers arrived. Today, theses animals are greatly restricted to southern portions of the valley and north along the eastern and western part of the valley. They remain on the valley floor only in the southwestern portion (Laughrin 1970). Grinnell et al. (1937) attributed the population decline to trapping and coyote poisoning efforts. In 1969 the population was estimated to be between 1,000 and 3,000 (Laughrin 1970), or about 1 fox per 256 ha. Egoscue (1956) estimated the population of *Vulpes m. nevadensis* at one pair per 922 ha in Tooele County, Utah.

Vulpes m. neomexicana is found in southeast Arizona and most of New Mexico. The swift fox is found in eastern New Mexico, eastern and northern Colorado, parts of Wyoming and Montana, northwestern Texas, western Kansas, western Oklahoma, all of Nebraska, and South and North Dakota (figure 22.8). However, for the past 30–40 years the swift fox has had a much more limited range than was previously believed. For example, the last confirmed sighting of the swift fox in Montana was reported by Bailey and Bailey (1918), and Hoffman et al. (1969) thought the animals to be extirpated in Montana. In North Dakota there were no sightings for 1915 to 1970, in South Dakota there were no sightings from 1914 to 1966, and in Nebraska there were none from 1901 to 1953 (Jones

1964; Pfeifer and Hibbard 1970; Hillman and Sharps 1978).

Recent reports indicate that the swift fox is returning to its original range. Moore and Martin (1980) reported a specimen taken in 1978 in Montana, and other recent sightings have been made in North Dakota (Pfeifer and Hibbard 1970), South Dakota (Van Ballenberghe 1975; Hillman and Sharps 1978), Wyoming (Long 1965), Nebraska (Blus et al. 1967), New Mexico, and Colorado (Robinson 1961). The causes of the original extirpation are human related. The swift fox was apparently easily trapped, and Johnson (1969) documented many thousands being trapped between 1835 and 1838. As the prairies were settled the swift fox declined due to trapping, hunting, predator control, and rodent control programs as well as to the loss of native prairie (Bailey 1926). Later, Robinson (1953) suggested that species similar to the kit fox were vulnerable to "coyote getters" and to the poison 1080, used for coyote control. Thus, the swift fox populations were severely depleted from the early 1800s thru the 1950s. However, they have recently rebounded and apparently are increasing in numbers and returning to their original range.

Description. The kit fox has exceptionally large ears and a body length of 40 centimeters. It weighs 2–3 kg and has a tail length of 25–30 cm. The dorsal coloration varies from a pale gray and rust to a buffy yellow; the belly is whitish and the tail is tipped with black (Burt and Grossenheider 1976).

The swift fox is very similar, but is slightly larger (2.5 to 3.5 kg) and a more buffy yellow color.

Growth, Development, and Reproduction. The kit fox is the smallest fox native to North America. The average weight is 1 kg. Breeding activity begins in late December through January and February for *Vulpes m. mutica* and for *V. m. nevadensis*. Egoscue (1962) felt that *V. m. nevadensis* may pair for life, while Morrell (1972) noted that only one of seven pairs of *V. m. mutica* had the same members over two breeding seasons. They are probably monogamous (Egoscue 1956, 1962), but he did observe two lactating females using the same den with one male. They did not successfully breed their first year (Morrell 1972).

The gestation period is unknown, but Egoscue (1962) assumed that it was the same as that of the red fox, 49–55 days. Newborn *V. m. nevadensis* have red fur on the head, neck, and sides, while the back, thighs, and tail are gray brown (Egoscue 1966). The litter size is four to five (Egoscue 1956, 1962; Morrell 1972). Young are born in February or March and emerge from the den in one month. In early May, puppy fur is replaced by adult pelage, beginning on the forehead then progressing middorsally and to the sides. The molt is complete by mid-June (Egoscue 1962). At one month of age the coloration is brown black on the back and hips and reddish on the front shoulders and legs, with a dark muzzle and a grayish black tail. The undersides are white (Egoscue 1956).

The growth, development, and reproduction of the swift fox is apparently quite similar to that of the kit

FIGURE 22.8. Distribution of the kit fox (*Vulpes macrotis*) and swift fox (*Vulpes velox*).

fox. Size of adults, coloration, breeding period, gestation, and litter size are also similar. However, breeding may occur later in the northern regions than in the South (Hillman and Sharps 1978). Kilgore (1969) suggested that swift foxes probably mate for life, are monestrous, and have one litter annually. Pups remain in natal dens until early June, then move to temporary dens in the nearby vicinity. Young remain with family groups until late August. Kilgore (1969) and Hillman and Sharps (1978) both pointed out that this strong family group association at a den site is unique among foxes and other canids.

Ecology. The kit fox is ecologically adapted to the desert shrub biome, while the swift fox lives in a grassland prairie environment. Both species have been affected by loss of habitat, but the swift fox probably exists on what once was marginal habitat due to prairie loss through modern agricultural practices (Snow 1973). He noted that a suitable den is a critical habitat component for kit foxes, as they use dens throughout the year.

Morrell (1972) found that San Joaquin Valley kit fox dens belong to specific family groups. He found that large dens are frequently found in association with abandoned ground squirrel mounds. Most dens are located on flat ground in fairly open areas; they average two openings per den. Eighty percent of the dens in a Utah study area were located on sparsely vegetated shadscale flats (Egoscue 1962). Natal dens are selected in January and tend to be located in groups of as many as 8–10 dens in 1–2 ha (Egoscue 1956). However, these dens are located at least 3 km from each other. Morrell (1972) noted that for *Vulpes m. mutica*, smaller dens were utilized during the breeding season. In September and October female foxes would reoccupy the larger winter dens (called brood dens). The male joined the female in October and they remained in the brood den until May or June. Unpaired foxes used the smaller dens all year around. Kit foxes utilized various dens during the summer, but were more sedentary during the winter months. This use of more than one den created many vacant dens, which were frequently utilized by burrowing owls.

Swift foxes have been characterized as the most subterranean of native foxes (Seton 1929). They utilize dens throughout the year, and the dens are found in short-grass pasture lands (Cutter 1958b; Kilgore 1969).

Food Habits. Kit foxes in western Utah depend on black-tailed jackrabbits (*Lepus californicus*), cottontails (*Sylvilagus* spp.), and nocturnal rodents (Egoscue 1962, 1975). Those in the San Joaquin Valley depend on kangaroo rats (*Dipodomys* sp.) in the spring and cottontails during spring and summer (Morrell 1972). *Vulpes m. mutica* also consumes gophers and a few birds when available, and grasses may play an important role in the diet (Morrell 1972). Kit foxes in Utah consume some birds as well as deer mice and kangaroo rats (Egoscue 1962).

Egoscue (1975) noted the first nonbreeding adult female kit fox two years after a population decline of jackrabbits. The decline continued into the third year and most adult females foxes did not breed. Four years from the initial rabbit decline the fox population was one-half of the original. Also, as rabbits declined the kit fox did not shift home ranges to areas where secondary prey were plentiful.

Adult kit foxes in captivity ate an average of 175 grams of fresh meat per day. Egoscue (1962) estimated that a total family requirement for the first 64 days following birth of a litter was 44,605 grams.

Cutter (1958a) noted that the main food items for swift foxes in Texas were jackrabbits and cottontails. Kilgore (1969) noted that cottontails and black-tailed jackrabbits made up the largest proportion of the swift fox diet in the Oklahoma panhandle. Hillman and Sharps (1978) found that prairie dogs (*Cynomys* sp.) are important during the spring and summer in South Dakota. Other items of prey for South Dakota swift foxes included thirteen-lined ground squirrels (*Spermophilus tridecemlineatus*), cottontail rabbits (*Sylvilagus floridanus*), white-tailed jackrabbits (*Lepus townsendii*), deer mice (*Peromyscus* sp.), and voles (*Microtus* spp.). Birds included mourning doves (*Zenaida macroura*) and western meadowlarks (*Sturnella neglecta*) (Hillman and Sharps 1978).

Behavior. Kit foxes are basically nocturnal, emerging at sunset to hunt. They hunt sporadically throughout the night, with different family groups occasionally hunting the same area, but not at the same time. This led Morrell (1972) to suggest that no specific hunting territory is defended. He also indicated much overlap in kit fox home ranges.

The kit fox utilizes at least five vocalizations: a sharp bark given as a warning, a ''burp'' given by a trapped adult, a hacking growl as a warning, a soft growl given when a trapped fox was approached, and a lonesome call given when mates or pups and parents were separated (Egoscue 1962; Morrell 1972).

Mortality. The San Joaquin kit fox is suffering from a decreasing habitat. From 1960 to 1970 there was a 34 percent reduction in the native habitat of this fox (Laughrin 1970). But vermin hunters and indiscriminate shooting were judged to be the major mortality factors for the San Joaquin subspecies (Morrell 1972). Egoscue (1962) indicated that vehicles were a major problem for kit foxes in Utah. Morrell (1972) also felt this was a problem in California, especially when there were low rodent populations leading to more animals feeding on road kills.

Vulpes m. arsipus suffered from shooting and from disturbance by motorcycle and dune buggy enthusiasts (Laughrin 1970). There is some evidence that coyotes occasionally attempt to capture the kit fox (Seton 1937; Egoscue 1962).

Today as in years past, it is human activities that present the greatest problems for the swift fox. Trapping is the major problem in South Dakota but automobiles, shooting, and free-ranging dogs also are problems (Kilgore 1969; Hillman and Sharps 1978). Rodent control programs may be detrimental to kit foxes as well (Snow 1973). Recent higher fur prices for red foxes and coyotes could lead to fewer swift foxes, which are very susceptible to trapping (Van Ballenberghe 1975). Hillman and Sharps (1978) also pointed out that coalmining is being proposed in areas of the northern plains that may have swift foxes.

PARASITES. Kit foxes often are heavily infested with fleas. Morrell (1972) noted large populations of *Echinonorhoya gallinacia* on *V. m. mutica*. Two flea species found on swift foxes in Oklahoma were *Pulex irritans* and *P. simulans* (Kilgore 1969). Four species of flea were found on kit foxes in Utah: *P. irritans*, *Meringis parker*, *Monopsyllus wagneri wagneri*, and *Thrassis bacchi gladiolis* (Egoscue 1962). The most important tick found on Utah kit foxes was *Ixodes texanus*; *I. kingi* and *Dermacentor parumapterus* were also found. *I. kingi* was found on Oklahoma kit foxes (Kilgore 1969). Egoscue (1962) suggested that heavy flea infestations may be the reason kit foxes change den sites often.

Other swift fox parasites found include protozoans (*Coccidium bigeminum, Isospora bigemina, I. felis*), flatworms (*Dipylidium caninum, Taenia multiceps*), and roundworms (*Physaloptera* spp., *Ancylostoma caninum, Toxocara* spp., *Toxicara canis*, and *Uncinaria* spp.) (Kilgore 1969).

Age Structure and Sex Ratios. Egoscue (1962) noted a slightly higher number of male adult kit foxes in the 1950s in Utah, and more male than female pups (23 of 36). During a jackrabbit decline the adult sex ratio went from 46 percent males in 1966 to 62 percent in 1967, 56 percent in 1968, and 50 percent in 1969, with the high 1967 count reflecting no need for breeding because food supplies were low (Egoscue 1975). The sex ratios of pups during the same period varied from a high of 67 percent males in 1968 to a low of 27 percent in 1969, the preponderance of male pups suggesting an overcrowded population relative to available food supplies. Egoscue (1975) noted that the average age of kit foxes during a period of low available food supply was 1.96 years.

Current Research and Management Needs. As mentioned, human-related problems have led to the decline of kit and swift foxes. To date there have been no management practices for either species. except the full protection given *Vulpes m. mutica* and endangered species in California, and *Vulpes m. nevadensis* in Oregon (Snow 1973). Military reservations may be playing a role in the comeback of the kit fox, as grazing is controlled and predator control is excluded (Snow 1973). Snow (1973) and Hillman and Sharps (1978) have made species and habitat management recommendations. These include: surveying BLM lands to determine presence and numbers of kit and swift fox; learning more about the distribution and status of subspecies, habitat and requirements, home range, and populations dynamics so that adequate management plans can be drawn; studying the effects of rodent control programs on kit and swift foxes; determining the impact of illegal and legal hunting; and determining the effects of intense public use on lands occupied by kit and swift foxes.

LITERATURE CITED

Ables, E. D. 1969. Activity studies of red foxes in southern Wisconsin. J. Wildl. Manage. 33:145–153.

———. 1974. Ecology of the red fox in North America. Pages 148–163 *in* M. W. Fox, ed. The wild canids. Van Nostrand Reinhold Co., New York. 508pp.

Allen, S. H. 1975. The influence of age and other factors on red fox reproduction. Pap. presented at 37th Midwest Wildl. Conf., Toronto, Canada. 10pp.

Anonymous. 1971. Wildlife habitat management handbook. U.S.D.A. For. Serv., Washington, D.C., 189pp.

Arnold, D. A., and Schofield, R. D. 1956. Home range and dispersal of Michigan red foxes. Pap. Michigan. Acad. Sci. Arts and Letters 41:91–97.

Asdell, S. A. 1964. Patterns of mammalian reproduction. 2nd ed. Cornell Univ. Press, Ithaca, N.Y. 670pp.

Bailey, V. 1926. A biological survey of North Dakota. North Am. Fauna 49. 226pp.

Bailey, V., and Bailey, F. M. 1918. Wild animals of Glacier National Park. Natl. Park Serv., Washington, D.C. 210pp.

Blus, L. J.; Sherman, G. R.; and Henderson, J. D. 1967. A noteworthy record of the swift fox in McPherson County, Nebraska. J. Mammal. 48:471–472.

Burt, W. H., and Grossenheider, R. P. 1976. A field guide to the mammals. 3rd ed. Houghton Mifflin Co., Boston. 289pp.

Churcher, C. S. 1959. The specific status of the new world red fox. J. Mammal. 40:513–520.

———. 1960. Cranial variation in the North American red fox. J. Mammal. 41:349–360.

Coman, B. J. 1973. The diet of red foxes *Vulpes vulpes* L. in Victoria. Aust. J. Zool. 21:391–401.

Cook, D. B., and Hamilton, W. J., Jr. 1944. The ecological relationship of red fox food in eastern New York. Ecology 24:94–104.

Croft, J. D., and Hone, L. J. 1978. The stomach contents of foxes, *Vulpes vulpes*, collected in New South Wales. Aust. Wildl. Rev. 5:85–92.

Cutter, W. L. 1958*a*. Food habits of the swift fox in northern Texas. J. Mammal. 39:527–532.

———. 1958*b*. Denning of the swift fox in northern Texas. J. Mammal. 39:70–74.

Dasmann, R. F. 1964. Wildlife biology. John Wiley and Sons, Inc., New York. 231pp.

Deems, E. F., and Pursley, D. 1978. North American furbearers: their management, research and harvest status in 1976. Univ. Maryland Press, College Park, Md. 171pp.

Dorf, E. 1959. Climatic changes of the past and present. Contrib. Mus. Paleontol. Univ. Mich. 13:199.

Durrant, S. D. 1952. Mammals of Utah, taxonomy and distribution. Univ. Kansas Publ., Mus. Nat. Hist. 6. 549pp.

Eadie, W. R. 1943. Food of the red fox in southern New Hampshire. J. Wildl. Manage. 7:74–77.

Egoscue, H. J. 1956. Preliminary studies of the kit fox in Utah. J. Mammal. 37:351–357.

———. 1962. Ecology and life history of the kit fox in Tooele County, Utah. Ecology 43:481–497.

———. 1966. Description of a new born kit fox. Southwestern Nat. 11:501–502.

———. 1975. Population dynamics of the kit fox in western Utah. Bull. Southern California Acad. Sci. 74:122–127.

English, P. F., and Bennett, L. J. 1942. Red fox food habits study in Pennsylvania. Pennsylvania Game News 12:6–7, 22.

Englund, J. 1970. Some aspects or reproduction and mortality rates in Swedish foxes (*Vulpes vulpes*), 1961–63 and 1966–69. Viltrevy 8:1–82.

Erickson, D. W., and Sampson, F. W. 1978. Impact of market dynamics on Missouri's furbearer harvest system. 32nd Southeast Fish Wildl. Conf. In press.

Errington, P. L. 1935. Food habits of mid-west foxes. J. Mammal. 16:192–200.

———. 1963. Muskrat populations. Iowa State Univ. Press, Ames. 665pp.

Everett, E. E. 1952. Restocking fox. *In* The hunter's horn. May. 16pp.

Failor, P. L. 1974. Pennsylvania trapping and predator control methods. Pennsylvania Game Comm., Harrisburg, Pa. 92pp.

Fairley, J. S. 1970. The food, reproduction, form, growth, and development of the fox *Vulpes vulpes* (L.) in northeast Ireland. Proc. R. Irish Acad. 69, Section B, no. 5:103–137.

Follman, E. H. 1973. Comparative ecology and behavior of red and gray foxes. Doctoral Dissertation. Southern Illinois Univ. 193pp.

Friend, M., and Linhart, S. B. 1964. Use of the eye lens as an indicator of age in the red fox. New York Fish Game J. 11:58–66.

Fuller, T. K. 1978. Variable home-range sizes of female gray foxes. J. Mammal. 59:446–449.

Geiger, V. G.; Bromel, J.; and Habermehl, K. H. 1977. KonKordanz verschiedener methoden der altersbestimmung beim Rot fuchs (*Vulpes vulpes* L. 1958). Sonderdruck aus Bd. 23:57–64.

Glover, F. A. 1949. Fox foods on West Virginia wild turkey range. J. Mammal. 30:78–79.

Godin, A. J. 1977. Wild mammals of New England. Johns Hopkins Univ. Press, Baltimore. 304pp.

Gray, R. L. 1975. Sacramento Valley redfox survey, 1975. California Dept. Fish and Game. Prog. Rep. W-54-R. Job II-1.2. 5pp. Mimeogr.

Grayce, R. L. 1957. Checklist of New England mammals. Bull. Massachusetts Audubon Soc. 41:15–24.

Grinnell, J.; Dixon, J. S.; and Linsdale, J. M. 1937. Furbearing mammals of California. Vol. 2. Univ. California Press, Berkeley. 400pp.

Grue, H., and Jensen, B. 1973. Annular structures in canine

tooth cementum in red foxes (*Vulpes vulpes* L.) of known age. Danish Rev. Game Bio. 8:1–12.

Guilday, J. E., and Bender, M. S. 1958. A recent fissure deposit in Bedford County, Pennsylvania. Ann. Carnegie Mus. 35:127–138.

Hall, E. R. 1946. Mammals of Nevada. Univ. California Press, Berkeley. 710pp.

———. 1981. The mammals of North America. 2nd ed. 2 vols. John Wiley and Sons, New York. 1181pp.

Hall, E. R., and Kelson, K. R. 1959. The mammals of North America. Ronald Press Co., New York. Vol. 2. 536pp.

Hamilton, W. J., Jr. 1935. Notes on food of red foxes in New York and New England. J. Mammal. 16:16–21.

Harris, S. 1977. Distribution, habitat utilization and age structure of a suburban fox (*V. vulpes*) population. Mammal. Rev. 7:25–29.

———. 1978. Age determination in the red fox (*Vulpes vulpes*): an evaluation of technique efficiency as applied to a sample of suburban foxes. J. Zool. London 184:91–117.

Hatfield, D. M. 1939. Winter foods habitat of foxes in Minnesota. J. Mammal. 29:202–206.

Hillman, C. N., and Sharps, J. C. 1978. Return of swift fox to northern great plains. Proc. South Dakota Acad. Sci. 57:154–162.

Hoffman, R. A., and Kirkpatrick, C. M. 1954. Red fox weights and reproduction in Tippecanoe County, Indiana. J. Mammal. 55:504–509.

Hoffman, R. S.; Wright, P. L.; and Newby, F. E. 1969. The distribution of some mammals in Montana. J. Mammal. 50:579–604.

Insley, H. 1977. An estimate of the population density of the red fox (*Vulpes vulpes*) in the New Forest, Hampshire. J. Zool. London 183:549–553.

Jennings, W. L.; Schneider, N. J.; Lewis, A. L.; and Scatterday, J. E. 1960. Fox rabies in Florida. J. Wildl. Manage. 24:171–179.

Joffre, M. 1976. Puberty and seasonal sexual cycle of the wild male fox (*Vulpes vulpes*). Ann. Bio. Anim. Biochem. Biophys. 16:503–520.

Johnson, D. H.; Sargeant, A. B.; and Allen, S. H. 1975. Fitting Richards curve to data of diverse origins. Growth 39:315–330.

Johnson, D. R. 1969. Returns of the American Fur Company, 1835–1839. J. Mammal. 50:836–839.

Johnson, W. J. 1970. Food habits of the red fox in Isle Royale National Park, Lake Superior. Am. Midl. Nat. 84:568–572.

Jones, J. K. 1964. Distribution and taxonomy of mammals of Nebraska. Univ. Kansas Mus. Nat. Hist. Publ. 16. 356pp.

Kilgore, D. L., Jr. 1969. An ecological study of the swift fox (*Vulpes velox*) in the Oklahoma panhandle. Am. Midl. Nat. 81:512–534.

Korschgen, L. J. 1959. Food habits of the red fox in Missouri. J. Wildl. Manage. 23:168–176.

Kozicky, E. L. 1943. Food habits of foxes in wild turkey territory. Pennsylvania Game News 14:8–9, 28.

Laughrin, K. 1970. San Joaquin kit fox: its distribution and abundance. California Dept. Fish and Game, Admin. Rep. 70–72. 20pp.

Layne, J. N. 1958. Reproductive characteristics of the gray fox in Southern Illinois. J. Wildl. Manage. 22:157–163.

Layne, J. N., and McKeon, W. H. 1956. Some aspects of red fox and gray fox reproduction in New York. New York Fish Game J. 3:44–74.

Linhart, S. B. 1959. Sex ratios of the red fox and gray fox in New York. New York Fish Game J. 6:116–117.

———. 1968. Dentition and pelage in the juvenile red fox (*Vulpes vulpes*). J. Mammal. 49:526–528.

Linhart, S. B., and Knowlton, F. F. 1975. Determining the relative abundance of coyotes by scent station lines. Wildl. Soc. Bull. 3:119–124.

Lloyd, H. G., and Englund, J. 1973. The reproductive cycle of the red fox in Europe. J. Reprod. Fert. Sullp. 19:119–130.

Long, C. A. 1965. The mammals of Wyoming. Univ. Kansas Publ., Mus. Nat. Hist. 14:493–758.

Lord, R. D. 1961. A population study of the gray fox. Am. Midl. Nat. 66:87–109.

MacDonald, D. W. 1977. On food preference in the red fox. Mammal. Rev. 7:7–23.

MacGregor, A. E. 1942. Late fall and winter foods of foxes in central Massachusetts. J. Wildl. Manage. 6:221–224.

McIntosh, D. L. 1963. Reproduction and growth of the fox in the Camberra District. C.S.I.R.O. Wildl. Res. 8:132–141.

Merriam, H. G. 1966. Temporal distribution of woodchuck interburrow movements. J. Mammal. 47:103–110.

Monson, R. A., and Stone, W. B. 1976. Canine distemper in wild carnivores in New York. New York Fish Game J. 23:149–154.

Monson, R. A.; Stone, W. B.; and Parks, F. 1973. Aging red foxes (*Vulpes fulva*) by counting the annular cementum rings of their teeth. Fish Game J. 29:54–61.

Moore, R. E., and Martin, N. S. 1980. A recent record of the swift fox (*Vulpes velox*) in Montana. J. Mammal. 61:161.

Morrell, S. 1972. Life history of the San Joaquin kit fox. California Fish Game 58:162–174.

Murie, A. 1936. Following fox trails. Univ. Mich. Mus. Zool. Misc. Publ. 32. 45pp.

Nelson, A. L. 1933. A preliminary report on the winter food of Virginia foxes. J. Mammal. 14:40–43.

Parker, R. L.; Kelly, J. W.; Cheatum, E. L.; and Dean, D. J. 1957. Fox population densities in relation to rabies. New York Fish Game J. 4:219–228.

Petersen, L. R.; Martin, M. A.; and Pils, C. M. 1977. Status of gray foxes in Wisconsin, 1975. Wisconsin Dept. Nat. Resour. Rep. 94. 17pp.

Petrides, G. A. 1950. The determination of sex and age ratios in fur animals. Am. Midl. Nat. 43:355–382.

Pfeifer, W. K., and Hibbard, E. A. 1970. A recent record of swift fox (*Vulpes velox*) in North Dakota. J. Mammal. 51:835.

Phillips, R. L. 1970. Age ratios of Iowa foxes. J. Wildl. Manage. 34:52–56.

Phillips, R. L.; Andrews, R. D.,; Storm, G. L.; and Bishop, R. A. 1972. Dispersal and mortality of red foxes. J. Wildl. Manage. 36:237–248.

Pils, C. M., and Martin, M. A. 1978. Population dynamics, predator-prey relationships and management of the red fox in Wisconsin. Wisconsin Dept. Nat. Resour. Rep. 105. 56pp.

Preston, E. M. 1975. Home range defense in the red fox (*Vulpes vulpes* L.) J. Mammal. 56:645–652.

Reilly, J. R., and Curren, W. 1961. Evaluation of certain techniques for judging the age of red foxes (*Vulpes fulva*). New York Fish Game J. 8:122–129.

Richards, S. H., and Hine, R. L. 1953. Wisconsin fox populations. Wisconsin Conserv. Dept. Tech. Wildl. Bull. Note. 78pp.

Robinson, W. P. 1953. Population trend of predators and fur animals in 1080 station areas. J. Mammal. 34:220–227.

———. 1961. Population changes of carnivores in some coyote-control areas. J. Mammal. 42:510–515.

Ross, J. G., and Fairley, J. S. 1969. Studies of disease in the red fox (*Vulpes vulpes*) in northern Ireland. J. Zool. 157:375–381.

Rowlands, I. W., and Parkes, A. S. 1935. The reproductive proceeds of certain mammals. Part 8: Reproduction in foxes (*Vulpes* spp.). Proc. Zool. Soc. London. Pages 823–841.

Rue, L. L. 1968. Sportsman's guide to game animals. Harper and Row, New York. 655pp.

Ryan, G. E. 1976. Observations of the reproduction and age structure of the fox, *Vulpes vulpes* L., in New South Wales. Aust. Wildl. Res. 3:11–20.

Sampson, F. 1977. Status of Missouri's red fox. Missouri Conserv. 37:26–29.

Sargeant, A. B. 1978. Red fox prey demands and implications to prairie duck production. J. Wildl. Manage. 42:520–527.

Schmurrenberger, P. R., and Martin, R. J. 1970. Rabies in Illinois foxes. J. Am. Vet. Med. Assoc. 157:1331–1335.

Schofield, R. D. 1960. A thousand miles of fox trails in Michigan's ruffed grouse range. J. Wildl. Manage. 24:432–434.

Scott, T. G. 1943. Some food coactions of the northern plains red fox. Ecol. Monogr. 13:427–479.

Scott, T. G., and Klimstra, W. D. 1955. Red foxes and a declining prey population. Southern Illinois Univ. Monogr. Ser., No. 1. 123pp.

Seagers, C. B. 1944. The red fox in New York. New York Conserv. Dept. Educ. Bull. 85pp.

Seton, E. T. 1929. Lives of game animals. Doubleday and Co., Inc., Garden City, N.Y. 746pp.

Sheldon, W. G. 1949. Reproductive behavior of foxes in New York state. J. Mammal. 30:236–246.

———. 1950. Denning habits and home range of red foxes in New York state. J. Wildl. Manage. 14:33–42.

Smith, H. J. 1978. Parasites of red foxes in New Brunswick and Nova Scotia. J. Wildl. Dis. 14:366–370.

Snow, C. 1973. Habitat management series for endangered species. Rep. no. 6, San Joaquin Kit Fox, B.L.M. Tech. Note, Denver Serv. Center. 24pp.

Stanley, W. C. 1963. Habits of the red fox in northeastern Kansas. Univ. Kansas Mus. Nat. Hist. Misc. Pub. 34:1–31.

Stiles, C. W., and Baker, C. E. 1935. Key-catalogue of parasites reported for carnivora with their possible health importance. U.S. Natl. Inst. Health Bull. 163:913–1223.

Stone, W. B.; Parks, E.; Wever, B. L.; and Parks, F. J. 1972. Experimental transfer of sarcoptic mange from red foxes and wild canids to captive wildlife and domestic animals. New York Fish Game J. 19:1–11.

Storm, G. L. 1965. Movements and activities of foxes as determined by radio tracking. J. Wildl. Manage. 29:1–12.

Storm, G. L., and Ables, E. D. 1966. Notes on newborn and fullterm wild red foxes. J. Mammal. 47:116–118.

Storm, G. L.; Andrews, R. D.; Phillips, R. L.; Bishop, R. A.; Siniff, D. B.; and Testes, J. R. 1976. Morphology, reproduction, dispersal and mortality of midwestern red fox populations. Wildl. Monogr. 49:1–82.

Sullivan, E. G. 1956. Gray fox reproduction, denning, range and weights in Alabama. J. Mammal. 37:346–351.

Sullivan, E. G., and Haugen, A. D. 1956. Age determination of foxes by x-ray of forefeet. J. Wildl. Manage. 29:210–212.

Switzenberg, D. F. 1950. Breeding productivity in Michigan red foxes. J. Mammal. 31:194–195.

Trainer, D. O., and Hale, J. B. 1969. Sarcoptic mange in red foxes and coyotes of Wisconsin. Bull. Wildl. Dis. Assoc. 5:387–391.

Trapp, G. R., and Hallberg, D. L. 1975. Ecology of the gray fox (*Urocyon cinereoargenteus*): a review. Pages 164–178 *in* M. W. Fox, ed. The wild canids. Van Reinhold Co., New York. 508pp.

Trippensee, R. E. 1953. Wildlife management, furbearers, waterfowl, and fish. Vol. 2. McGraw-Hill Book Co., New York. 572pp.

Tullar, B. F. 1979. The management of foxes in New York state. Conservationist. Pages 33–36.

Tullar, B. F.; Berchielli, K. T.; and Saggese, E. P. 1976. Some implications of communal denning and pup adoption among red foxes in New York. New York Fish Game J. 23:92–94.

Turkowski, F. J. 1969. Food habits and behavior of the gray fox (*Urocyon cinereoargenteus*) in the lower and upper sonoran life zones of Southwestern United States. Doctoral Dissertation. Arizona State University. 136pp.

Van Ballenberghe, V. 1975. Recent records of the swift fox (*Vulpes velox*) in South Dakota. J. Mammal. 56:525.

Waithman, J., and Roest, A. 1977. A taxonomic study of the kit fox, *Vulpes macrotis*. J. Mammal. 58:157–164.

Waters, J. H. 1964. Red and gray fox from New England archeological sites. J. Mammal. 45:307–308.

Wood, J. E. 1958. Age structure and productivity of a gray fox population. J. Mammal. 39:74–86.

Wood, J. E.; Davis, D. E.; and Komarek, E. V. 1958. The distribution of fox populations in relation to vegetation in Southern Georgia. Ecology 39:160–162.

Yearsley, E. F. 1976. Use of reclaimed surface mines by foxes in Preston County, West Virginia. Masters Thesis. West Virginia Univ. 81pp.

Yearsley, E. F., and Samuel, D. E. 1980. Use of reclaimed surface mines by foxes in West Virginia. J. Wildl. Manage. 44:729–734.

Yoho, N. S., and Henry, V. G. 1972. Foods of the gray fox (*Urocyon cinereoargenteus*) on European wild hog (*Sus scrofa*) range in east Tennessee. J. Tennessee Acad. Sci. 47:77–78.

David E. Samuel, Division of Forestry, West Virginia University, Morgantown, West Virginia 26506.

Brad B. Nelson, Department of the Interior, Bureau of Land Management, Laramie, Wyoming.

23

Arctic Fox
Alopex lagopus

Larry Underwood
James A. Mosher

NOMENCLATURE

COMMON NAMES. Arctic fox, white fox, blue fox
SCIENTIFIC NAME. *Alopex lagopus*

The genus *Alopex* is thought to be intermediate within the family Canidae between the genus *Canis* and the genus *Vulpes*. For a time, *Alopex* was regarded as a subgenus of *Vulpes;* however, modern taxonomists recognize *Alopex* as a separate genus. The Corsac fox, *A. corsac,* was once considered congeneric (Grzimek 1975). Today, however, the only species included in the genus is *A. lagopus.*

In North America five subspecies are recognized (Hall 1981): *A. l. groenlandicus* in Greenland; *A. l. hallensis* on St. Matthew Island in the Bering Sea; *A. l. pribilofensis* on the Pribilof Islands, Bering Sea; *A. l. lagopus* in northern Alaska and Canada northwest of Hudson's Bay; and *A. l. ungava* in northeastern Canada. Additional subspecies are recognized in Eurasia.

DISTRIBUTION

The arctic fox is a circumpolar inhabitant of the high-latitude lands of the Northern Hemisphere (figure 23.1). Along with the musk ox (*Ovibos moschatus*), arctic hare (*Lepus arcticus*), collared lemming (*Dicrostonyx torquatus*), and polar bear (*Ursus maritimus*), it is virtually restricted to the arctic biome. Excellent descriptions of this biome in North America are given in Bee and Hall (1956), Britton (1957), and Bliss et al. (1973).

In Alaska, the arctic fox is commonly found in coastal regions north of the Kuskokwim River delta (Chesemore 1968a, 1975) and throughout the northern part of the state north of the Brooks Range. It is particularly common in the arctic coastal plain (Bee and Hall 1956). The range of common occurrence in Canada continues eastward, and includes all of the Canadian Archipelago, the westward and eastward shores of Hudson's Bay north of James Bay, and northern Quebec. In Greenland, the arctic fox is common along the coast (Vibe 1967).

The distribution of arctic foxes in western Europe

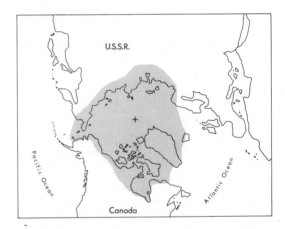

FIGURE 23.1. Distribution of the arctic fox (*Alopex lagopus*).

appears to be not as common as in former times. In the Pleistocene, the species apparently inhabited all of Europe (Chesemore 1975). Pedersen and Larsen (1971) reported that white foxes are only infrequently observed in northern Norway. In the U.S.S.R., the arctic fox is found throughout the Far North (Tchirkova 1968), but is more common toward the coast. It also inhabits the islands of the North Atlantic and the Bering Sea.

Mankind has influenced the distribution of Arctic foxes in Alaska. Originally, they did not occur in the central Aleutian Islands but were restricted to the westernmost and easternmost islands (Seton 1929). They also apparently occupied the Pribilof Islands (Lembkey and Lucas 1902). However, in the 1900s through the 1930s, fur trappers transplanted the blue color phase to certain Aleutian Islands, where they persist today (Murie 1959; Berns 1969; Burris and McKnight 1973; Carnahan 1979).

There are considerable differences in seasonal distribution. In summer, the bulk of the population is restricted to the breeding grounds. In the U.S.S.R., dens are rarely encountered south of 67° N latitude (Dementyeff 1958), which corresponds roughly with the southern boundary of the tundra (Bannikov 1970).

In North America, breeding is most common in the coastal plain (Chesemore 1975; Eberhardt 1977) and the coastal regions of continental Canada and the High Arctic islands (MacPherson 1969).

During winter the population disperses widely from the breeding grounds. The bulk of the population appears to move toward the coast and many individuals travel far out onto the pack ice (Bannikov 1970; Banfield 1977). Arctic foxes have been reported within 140 kilometers of the true north pole, and within 80 km of the northern pole of inaccessibility, that is, the point in the arctic pack ice furthest from land (Underwood 1971). Less frequently, significant numbers of arctic foxes appear to travel south of the breeding ranges and at times encroach upon the subarctic lands (Dementyeff 1958; Banfield 1977). The extreme southern ranges in North America appear to be the Kenai Peninsula of Alaska (Seton 1929), southern Manitoba (Wrigley and Hatch 1976), and the mouth of the St. Lawrence River (Cameron 1950). Vibe (1967) reported arctic foxes occasionally found in the middle of the Greenland ice cap.

There are significant differences in the distribution of the two color phases (Fetherston 1947; Chesemore 1969; Banfield 1977). The blue phase is largely insular, predominating in the islands of the North Atlantic, the Bering Sea, and the Aleutians. The white phase is found predominantly in continental areas. In rare instances, blue foxes appear in a white population (Chesemore 1975), most commonly in eastern Canada. Occurrences of white foxes among the Greenland blue population are more frequent (Braestrup 1941; Vibe 1967). White individuals apparently come across Davis Strait from Canada with some regularity— approximately every four years. Some individuals may survive a few years and even breed, but they do not become established.

DESCRIPTION

The arctic fox is somewhat smaller than the red fox, *Vulpes vulpes*. In winter, the luxuriant fur gives a look of plumpness that is deceptive. The legs and muzzle of the arctic fox are short. However, the tail is long, about 400 millimeter in length, and brushlike. The ears are rounded and inconspicuous in the winter fur. The feet are thickly furred on the pads. The size of an adult ranges from 1.4 to 3.2 kilograms (Chesemore 1975), with a total length of about 1 meter. Seton (1929) reported exceptional individuals as large as 9.4 kg. A colony of captive, well-fed arctic foxes maintained in northern Alaska averaged 4.1 kg for a year (Underwood 1971).

There are two color phases of the arctic fox based on winter pelage. In the white color phase, the animal is typically pure white except for the black tip of the nose and yellowish eyes. A few individuals have some dark guard hairs along the flanks or a few dark hairs at the tip of the tail and on the ears. However, throughout much of arctic North America, individuals with even these slight deviations from pure white are unusual. The blue phase is a dark to dull slate gray. Vibe (1967)

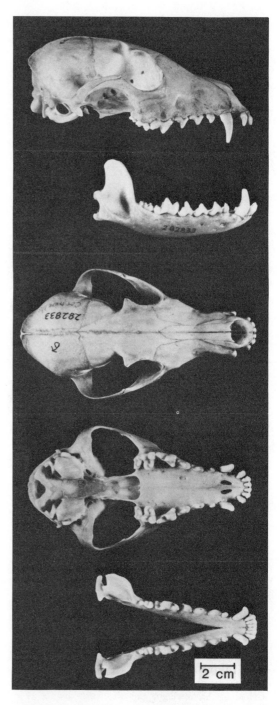

FIGURE 23.2. Skull of the arctic fox (*Alopex lagopus*). From top to bottom: lateral view of cranium, lateral view of mandible, dorsal view of cranium, ventral view of cranium, dorsal view of mandible.

differentiated two size groups among the blue fox. Those of the Bering Sea islands averaged larger than the white foxes of continental Alaska and Canada, while the blue foxes of Greenland were smaller. Pelage color is genetically controlled, with white resulting from an autosomal recessive gene (Johansen 1960).

There is only one summer coat color. In both the blue and white phases, winter fur is replaced with a much shorter fur, gray in color along the back and sides, with the underside and flanks cream to fawn colored.

There are few morphological differences between the sexes. Generally, however, males are somewhat larger than females, with a significantly longer tail (Chesemore 1967).

Skull and Dentition. As in all canids, the rostrum is narrow. The postorbital processes in the arctic fox are concave dorsally and the interorbital region is elevated. Condylobasal length is about 120 mm. The carnassial teeth are well developed, and the dental formula is 3/3, 1/1, 4/4, 2/3 = 42 (figure 23.2).

PHYSIOLOGY

General Adaptations. The arctic fox is a relatively small, nonmigratory carnivore chronically exposed to the arctic environment. In order to withstand the conditions of this rather rigorous area, the species must possess special adaptations. Scholander et al. (1950*a*) felt that the arctic fox was the most highly adapted of the arctic forms. The species is well adapted to arctic conditions, with reduced limb size; a short snout; short, rounded ears; dense winter fur (Chesemore 1975); and, in continental populations, the white color of winter pelage. Furred soles of the feet provide the animal with additional warmth and traction on the snow (Bee and Hall 1956). The arctic fox also possesses an acute olfactory sense, which allows it to smell lemming nests under several centimenters of snow (Banfield 1977). Henshaw et al. (1972) found that the arctic fox, like the wolf (*Canis lupus*), is able to maintain foot temperatures at just above the freezing point when standing on extremely cold snow, thus minimizing heat loss. Additional adaptations of the thermal qualities of the fur, reproduction, and energy requirements are discussed below.

Many significant physiological aspects of the arctic fox, such as reproduction, seasonal movements, and pelage growth, are closely attuned to environmental events, particularly changes in photoperiod and average ambient temperature (Underwood 1975). The role of hormone mediators in regulating and timing these events has not been determined.

Molting and Pelage. The pelage of the arctic fox is seasonally dimorphic. Changes occur in color and fur length in the white phase, and fur length in the blue phase. Fur length measurements taken in winter are very nearly double those in the summer (figure 23.3). Considerable variations in fur length are observed between various regions of the body (Underwood 1971). In all seasons, fur lengths are considerably greater on

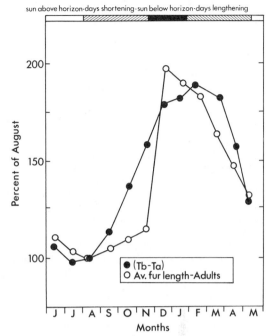

Day Lengths at Barrow, Alaska

sun above horizon·days shortening·sun below horizon·days lengthening

FIGURE 23.3. Changes in average fur length and body-to-ambient temperature (Tb-Ta) of the arctic fox (*Alopex lagopus*), expressed as percentages of August values. Adapted from Underwood and Reynolds in press.

the trunk than on the extremities, with the exception of the tail. Fur lengths are least on the muzzle, the dorsal surface of the paws, and the front of the legs. Fur lengths are considerably less on the belly than on the back. They are longer on the anterior portions of the trunk than on the posterior portions.

These differences appear to be related to the animal's need to conserve heat in a hostile, cold environment (Underwood and Reynolds 1980). Site-specific variations in fur length reflect the animal's surface conditions while resting, when the animal typically lies in a curled position. Poorly furred sites, such as the lower extremities and the muzzle, are protected by curling, with the tail and body hair. The skin temperature of a curling fox can be as great as 22.7° C above the ambient temperature (Follmann 1977). When an animal is not resting, the scantily furred areas are exposed and could serve as heat dissipators.

Cold tolerance of the arctic fox in winter is extreme. Scholander et al. (1950*b*) noted that arctic foxes, along with many other arctic species, have to maintain a body-to-ambient air temperature gradient of approximately 100° C for many weeks. They calculated that an arctic fox with a lower critical temperature in excess of −40° C could compensate for an air temperature of −70° C by increasing the heat production only 37 percent over basal values. Irving and Krog (1954) observed that an arctic fox was able to maintain body temperature for one hour in an environmental temperature of −80° C. Certainly a most important adaptation of the arctic fox, and the one on which their survival in winter may very well depend, is the insulative quality of their fur. Indeed, this may be the primary adaptation of the arctic fox.

Considerable variation, both individual and geographical, is observed in fall molting. Chesemore (1970) reported that in Barrow, Alaska, the belly fur turns white first, followed by other areas of the body. However, Underwood (1971) observed white fur to appear first on the head and back regions and spread dorsocaudally to the tail and ventrum. In general, the fall molt starts in late August to late September. Fur lengths reach their maximum in December in northern Alaska (Underwood 1971). Chesemore (1970) reported that the fall molt may extend only to mid-October on Baffin Island, Canada; into November on the southeastern coast of the Taimyr Peninsula in the U.S.S.R.; and well into December on Novaya Zemlya, U.S.S.R.

There may be fundamentally important differences between the fall molting pattern of foxes in Greenland and that of foxes in Canada (Vibe 1967). In the blue foxes of Greenland, the winter fur occurs in early October and November, well before the onset of cold and heavy snow. White color phase foxes from Canada, which periodically invade Greenland, develop their winter fur later in the season. Because of this, Vibe (1967) believed that the white foxes do not survive to breed in Greenland. Braestrup (1941) considered the white fox of Canada and the blue fox of Greenland as physiological races with distinct genetic differences and separate breeding seasons. MacPherson (1969), however, attributed the periodic disappearance of white foxes in Greenland to homing, which allows them to return to Canada during the breeding season. Regardless, for distinct populations, the growth of fur may be correlated with the development of winter cold. Thus, the growth of fur compensates for the decreased winter temperatures and increased winter winds.

The spring molt seems to be less variable, generally occurring from March through May. This molt begins as an obscure dark patch in the hip region. In the latter stages of the molt, white fur is shed in large patches from the flanks, back, and legs. Small patches of white fur may be seen on some individuals into early August.

Fur lengths in the arctic fox vary in apparent response to seasonal changes in temperature. Indeed, there appears to be a striking parallel between average fur lengths and seasonal changes in northern Alaska (see figure 23.3) (Underwood 1971). As a result, the lower critical temperature changes with the season. Underwood (1971) found a seasonal shift in lower critical temperature from in excess of $-40°$ C in winter to approximately 0 to $-10°$ C in summer. Summer pelage with less effective thermal qualities may facilitate heat loss during the warmer season.

Little work has been done to determine the mechanisms controlling molting in the arctic fox, but photoperiod and ambient temperature both may be important (Underwood and Reynolds 1980). The fall molt

may be more endogenously controlled than the spring molt.

REPRODUCTION

Arctic foxes are monogamous and the vixens are monestrous, with estrus lasting from 12 to 14 days (Dementyeff 1958; Banfield 1977). Individual variation in breeding dates can be attributed to a variety of factors, including the nutritional state of the foxes. Generally, however, breeding takes place sometime after January, usually in March or April (Fay 1973). Gestation lasts approximately 52 days (Chesemore 1975). Arctic foxes may mate for life (Seton 1929; Dementyeff 1958). Natal dens are dug or expanded in Canada toward the end of March (Banfield 1977).

Litter Size. The number of young produced varies considerably from year to year. The correlation between arctic fox breeding success and lemming density is most evident (Speller 1972). Low lemming density between breeding seasons may delay estrus and increase prenatal mortality, with insufficient nutrition having its greatest effect on reproduction in yearling females. Insufficient food during the latter stages of pregnancy could cause abortion, or reabsorption of all or part of the embryos (Tchirkova 1958a; Speller 1972). However, MacPherson (1969) found no variation in average placental count from year to year or between vixens of various ages, so postnatal pup mortality may be a more important factor in determining breeding success. MacPherson (1969) found a mean placental count of 10.6 and an average weaned litter size of 6.7. He attributed the differences to sibling aggression in which pups killed littermates. The cause of such aggression was attributed to insufficient food being provided to the pups. Berns (1969) reported an average litter size of 2.8 on Rat Island, Alaska, in an area where the food supply consisted only of arthropods, kelp fly pupae, and a few birds. Specifically, lemmings were not available prey. The average litter size in the U.S.S.R. varies from 8 to 12 (with extremes of 20–22) during years of excess prey availability and 3 to 5 in bad years (Bannikov 1970). Chesemore (1975) reported that the number of pups in a litter ranged up to 25, with 6.3 being the approximate mean derived from numerous sources.

Development. Several aspects of arctic fox reproduction show specialized adaptations. The average litter size is higher for arctic foxes than for temperate zones species. A comparison of embryo and placental scar counts showed that arctic foxes average 10.6 pups at birth, while red foxes average 5.5 pups and gray foxes 4.1 (MacPherson 1969). Relatively few arctic fox vixens breed in their first or second year, also in contrast to temperate zone foxes.

Arctic foxes are sexually mature at the age of 9–10 months, and some take part in the next breeding season (Banfield 1977). MacPherson (1969) found that one-third of the vixens at one and a half years of age had placental scars, while 85 percent of the three-plus year olds had placental scars. One and two year olds breed in their first year if they are from a large cohort, and in the second year if they are from a small cohort. Such an arrangement has adaptive value. Animals born in a small cohort were probably born in a year of food shortage. Because years of food shortage are usually followed by other years of food shortage, it would be selectively advantageous if the females of that cohort did not breed in the next year. Animals born in a large cohort are born in a year when food is abundant. Quite often this is followed by a year in which food is also relatively abundant and it would be advantageous for those individuals to breed when they are one year old.

Precocious development of hierarchical aggression within the litter may be adaptive. This phenomenon seems to be expressed only in years when food is limited, and presumably tends to favor the most highly developed, aggressive, largest individuals among the litter. Bee and Hall (1956) and Underwood (1971) reported that arctic foxes appear to be more active during the summer months, when breeding activities are intense, than foxes in the more temperate regions.

ECOLOGY

Ecological Role. The arctic fox is a top predator throughout its range. The effect of arctic foxes on the abundance of birds can be devastating (Mayfield 1976), especially if the arctic fox population is high and the lemming population is low. In Canada, a single pair of breeding arctic foxes and their litter consumed an average of 1,800 lemmings during a 90-day denning season (Speller 1972). MacInnes and Misra (1972) failed to observe any severe fox damage to nesting birds during a 9-year period in northern Canada. They felt that significant damage by foxes is relatively infrequent and highly localized. However, others feel that the effects of arctic fox predation on birds may be more significant. The distribution of many arctic nesting birds may be related to their adaptations to arctic fox predation (Speller 1969). Ross's geese (*Chen rossii*), other geese, and swans nest on islands in Canada, perhaps to avoid arctic foxes. Foxes may set an upper limit on the number of nesting guillemots (*Cepphus spp.*) and eiders (*Somateria spp.*) on Bear Island in the Barents Sea south of Svalbard (Bertram and Leek 1938). Riewe (1977) reported that five captive foxes, between 1 May and 30 September, consumed the equivalent of 2,000 28-gram (fresh weight) lemmings, 2,000 30-gm snow buntings (*Plectrophenax nivalis*), and 92 3.8-kg arctic hares. In his study, foxes had a devastating effect on nesting birds in the lowland and adjacent areas.

The presence of a fox den on the tundra radically alters the plant community around the burrow. It replaces the typically dry tundra community with a lush, grassy cover that stands out in bold contrast to the other vegetation beyond the influence of fox activity. These effects are due to the addition of organic materials, including food remains and droppings, aeration, mixing of the soil, and disturbance due to digging.

Dens. Dens range from single burrows to large, conspicuous structures used and expanded year after year. Some dens may be used by foxes for 50 to 60 years in the U.S.S.R. (Dementyeff 1958) or as long as 330 years in Canada (MacPherson 1969). MacPherson classified dens as follows: a youthful den has few burrows and no characteristic vegetation; a mature den has good mats of vegetation with no collapsing of burrows; an old den has large vegetative mats with grasses dominant and extensive collapsing burrows; a senile den is inactive, with extensive collapsing.

Typically, on the surface, dens appear to be circular in shape, covering from 3 to 100 m², and even up to 250 m² (Chesemore 1967; Shibanoff 1958). The number of entrances varies from 1 to 50 in northern Alaska (Chesemore 1969; Underwood 1975). As many as 45 entrances have been reported in the U.S.S.R. (Shibanoff 1958), while MacPherson (1967) reported over 100 entrances in a den in Canada. The average number of entrances ranges from 7 to 13. Strecker et al. (1952) excavated a den near Barter Island, Alaska, and found a nest chamber 1 m underground with 6 to 8 openings. Dens in the region were elevated above the normal ground surface about 1.5 m.

Den sites appear to be chosen for their small accumulations of winter snow, good exposure to spring sun, protection from severe summer winds, and elevation above water and frost lines (MacPherson 1969). Dens are usually located in areas with deep active layers, that is, thawed soils in summer greater than 1 m deep above the permafrost. Thus, mounds 1–4 m high, low hills, and ridges are preferred sites (Shibanoff 1958; Chesemore 1967, 1969; Voilochnikov 1968; Underwood 1975; Eberhardt 1976). Stability of the surface and depth of permafrost may be the most important factors. Although MacPherson (1969) found no dens among boulders in northern Canada, in areas where preferred habitat is nonexistent, such as islands in the Bering Sea, dens occur in alpine fellfields and boulder fields (Speller 1972; Fay 1973), and among rocky cliffs (Braestrup 1941). Most dens are south facing, on slopes with a 48–90 percent grade. Generally, fewer than 10 percent face north (Chesemore 1968b; Underwood 1975). Dens on north and east slopes in the U.S.S.R. become snow-free one and a half months later than those on south slopes (Dementyeff 1958).

Most dens are found in sandy soils, when available, with clay generally being avoided. Dens have been found in conspicuous groupings separated by only a few meters (Chesemore 1969, 1975), indicating that where habitat is favorable exploitation can be extensive. The number of occupied dens, however, does not seem to be limited by habitat, at least in Canada (MacPherson 1969). The placement of dens with relation to lakes and rivers apparently varies with the locality (Chesemore 1967). The proportion of dens occupied in any given year is a function of the food supply.

Home Range and Seasonal Movements. The arctic fox is a highly mobile species. McEwen (1951) described four types of movements: (1) local, general daily travels; (2) sporadic, unpredictable occurrences of individuals many kilometers from their normal range; (3) seasonal migrations; and (4) periodic migrations—many foxes traveling long distances in one sustained direction. Males appear to travel farther than females (Vibe 1967; MacPherson 1969).

Daily movements primarily involve localized searches for food. The size of breeding and home ranges may vary with the abundance of food, and, to some extent, the geographic location (Bannikov 1970). Home range areas may vary by a factor of six, depending on food abundance. Speller (1972) reported a mean hunting range of 2.9 km^2 in Canada. Underwood (1975) presented circumstantial evidence that a fox may have obtained food for a den from more than 13 km away. Whenever food is limited, more substantial movements are required. A localized shortage may force one or a small number of foxes to seek out new resources, even in what is generally a good food year. In a poor year, larger numbers of foxes may have to wander.

Several studies describe the sporadic movements of arctic foxes. MacPherson (1968) reported a fox that was captured in arctic Canada, was tagged, and then escaped in Ontario. It was apparently recaptured two years later at Hudson's Bay, having traveled a straight-line distance of 1,120 km in two years. In Wainwright, Alaska, an Eskimo trapped a white fox bearing a Russian neck tag (Chesemore 1969). Eberhardt and Hansen (1978) report the results of a tagging study documenting seven long-distance movements of 129–945 km. Some individuals averaged 24 km in a straight line per day. Travel occurred both in years of food abundance and during food scarcity, principally scarcity of lemmings. Northcott (1975) reported an arctic fox trapped and marked in Newfoundland, approximately 2,400 km from the normal geographic range, that survived one year and traveled 107 km straight-line distance.

In Alaska, there are two seasonal movements (Chesemore 1969, 1975), a seaward movement in fall and early winter, probably triggered by food scarcity, and the reverse in late winter and early spring. The percentage of migrants making round trips remains unknown (Wrigley and Hatch 1976).

The age structure of the wandering portion of the population changes from year to year (Dementyeff 1958; MacPherson 1969). In years of whelp abundance (which implies that food is abundant), adults are scarce in the trapping returns, probably because they remain at the breeding sites. The whelps wander in this situation. When food is scarce, pup production is down and the adults cannot stay at the den sites. In this situation they wander, and the number of older foxes trapped increases.

Extensive migrations of arctic foxes occur in the U.S.S.R., involving large numbers of individuals (Dementyeff 1958). Autumn migrations occur along shore. Old, male foxes migrate first. There is less movement during good food years. Migrations may occur over hundreds, possibly thousands, of kilometers. According to Bannikov (1970), up to two-thirds of

the entire arctic fox population in the U.S.S.R. leaves the tundra. Approximately one-half travels southward toward the forest, while the others travel northward onto the pack ice. Few, if any, individuals return.

Migrations of arctic foxes in North America either occur in small magnitude compared to Eurasia or perhaps just have not been as well recorded (Chesemore 1975; Wrigley and Hatch 1976). Chesemore (1968b) found no evidence of North American migrations. Large-scale migrations into the forest zone occur only when numbers are exceptionally high and disease becomes prevalent (Elton 1931). Such movements occur chiefly in winter and spring.

Population Cycles. Arctic fox populations are subject to great fluctuations in density on a regional basis. For instance, before trapping, the entire U.S.S.R. population of arctic foxes averages approximately 130,000 animals (Bannikov 1970), including approximately 12,000 breeding females (Smirov 1968). However, there may be a tenfold fluctuation in animals from one year to the next. Fur harvest data from Alaska indicate that periodic peaks in population density occur approximately every 3.6 years, and lows every 3.9 years (Chesemore 1967). The population of white foxes in Greenland reaches a climax every 4 years, as it does in Canada (Braestrup 1941). Arctic fox numbers appear to be causally related to food abundance, particularly the abundance of small rodents, including the lemming (Chesemore 1967). Years of abundant fox harvest follow peaks in small mammal density by 1–2 years (Chitty 1950; MacPherson 1969). Other factors, including local reproduction and migration (Shibanoff 1958), disease, parasites, predation, competition, migration, human economic activity, and weather (Tchirkova 1958a), are all of secondary importance. However, they do affect arctic fox population numbers. Arctic fox population cycles may depend ultimately on climate, and proximally on food (Tchirkova 1958b). Vibe (1967) felt that white arctic foxes in Greenland are favored by severe, cold winters, while the blue foxes thrive in warm, mild winters.

FOOD HABITS

Lemmings (*Lemmus sibiricus* and *Dicrostonyx torquatus*) are primary food sources for arctic foxes (Braestrup 1941; Chesemore 1968b). In regions of the tundra where other small rodents occur, these play a similar role. For example, throughout much of the U.S.S.R. (Shibanoff 1958; Voilochnikov 1968) and on St. Lawrence Island (Stephenson 1970; Fay 1973), the tundra vole (*Microtus oeconomus*) is important. Only in a few areas where products of the sea and access to nesting birds are consistently available can arctic foxes survive without small rodents (Braestrup 1941; Vibe 1969). Such foxes are almost exclusively insular and are of the blue color phase. In Greenland, where both color phases occur, white foxes are dependent on lemmings and blue foxes are dependent on products of the sea.

Even when strongly dependent on them, it is

doubtful that any foxes can survive strictly on small rodents (Riewe 1977). Arctic foxes are opportunistic in their choice of secondary food sources. Birds and their products are extremely important. Adult birds (including both those associated and those not associated with nests), nestlings, fledglings, and eggs are all taken when available (Seton 1929; Dementyeff 1958; MacPherson 1969; Stephenson 1970; Fay 1973; Eberhardt 1977).

Carrion is perhaps the next most important food source for arctic foxes. Caribou (*Rangifer tarandus*) remains and various marine mammals are of particular importance in winter. Schiller (1954) described arctic foxes feeding in walrus (*Odobenus rosmarus*) carcasses by burrowing into the abdomen and feeding on the inside of the unfrozen but putrifying carcass. Mullen and Pitelka (1972) reported that arctic foxes were extremely effective in locating lemming carcasses during winter. By the following spring during a lemming population low only 4 of 188 were not consumed. All 155 carcasses set out in a moderate lemming year were consumed.

Numerous miscellaneous food items are eaten, including fish (Dementyeff 1958), frogs, mollusks, crustaceans (Shibanoff 1958; Eberhardt 1976, 1977), and mummified human skin (Murie 1959). This last author relates Steller's observations of foxes attacking sick and dead humans in the Aleutian Islands in the 1700s. Foxes will also occasionally eat nonfood items (Seton 1929; Underwood 1975). Up to 14 percent of stomach samples surveyed near Prudhoe Bay, Alaska, showed nondigestible items. Arctic foxes, especially pups, may occasionally be cannibalistic (Shibanoff 1958; Murie 1959; MacPherson 1969; Chesemore 1975).

There are significant seasonal differences in the diet of the arctic fox. Lemmings, other small rodents, and birds are much more readily available in summer than in winter. In winter, lemmings continue to be an important food item and carrion becomes important. Stephenson (1970) found that on St. Lawrence Island, Alaska, carrion, though readily available, was seldom used in summer. Thus, prey items were adequate for the population of foxes in summer. Carrion became important to the population during midwinter and late winter. Trappers report that marine mammal bait is not effective until after October. Dementyeff (1958) found that during winter with food available, foxes were indifferent to trap bait. This suggests that a preference for live food persists even when it is not usually available.

During spring, arctic foxes on the pack ice become active predators of seals by digging out pups in their lairs (Stirling and Smith 1975; Smith 1976; Riewe 1977). A keen olfactory sense apparently allows the arctic fox to locate subnivian seal lairs through snow depths of over 150 cm. Up to 26 percent of the seal dens sampled had been preyed upon. Calculations of daily energy requirements suggest that newborn seal pups would provide 30.2 to 45.2 fox-days of maintenance energy. Seal pups almost weaned would provide 227 to 341 fox-days of maintenance energy (Smith 1976). Such an energy source may be extremely impor-

tant to maintain at the den sites arctic foxes that have migrated off the pack ice, until snow melts, lemmings become more readily available, and migratory birds return to the tundra.

Chesemore (1968*b*) found no evidence of caching by arctic foxes, although a number of other authorities reported that these foxes regularly cache food during the summer months (Braestrup 1941; MacInnes and Misra 1972). Arctic foxes on the pack ice stay close to polar bears and feed on the seal remains after a bear has made a kill (Seton 1929; Bannikov 1970). They occasionally follow wolves for the same reason (MacPherson 1969).

The effects of arctic fox predation on tundra communities can be significant (Speller 1972). Food consumption for an average litter of 10.6 whelps is 127 kg, the equivalent of 2,400 lemmings. This is an average of 60 lemmings per day.

Energy Requirements. A few studies have been conducted to determine the energy requirements of arctic foxes. In these studies, captive foxes experienced arctic conditions (Underwood 1971; Speller 1972) or conditions typical of southern Canada. Underwood (1971) found significant differences in the amount of food required seasonally; energy demands were significantly higher in summer than in winter. The amount voluntarily ingested ranged from a high of 370 kilocalories per kg body weight per day in July to a minimum of 63 kcal per kg body weight per day in January. Four arctic foxes maintained in southern Canada subsisted on an average intake of 125 kcal per kg body weight per day, with no decrease or increase in body weight (Riewe 1977). Two captive arctic fox whelps maintained at Aberdeen Lake, Northwest Territories, ingested an average of 490 kcal per kg body weight per day during July and August (Speller 1972).

As noted, the amount of food required by the arctic fox shifts seasonally, being highest in summer and lowest in winter. Ecologically, this is to their advantage. Energy requirements are lowest during the season when energy sources are most limited (winter). Conversely, in summer, when activity states and food requirements are highest because of reproduction, and average ambient temperatures are above freezing, food sources are most abundant.

When food is readily available, both wild and captive arctic foxes accumulate a considerable amount of body fat (Underwood 1971). Fat is deposited primarily around the visceral organs and two areas of the trunk. The most extensive is a large area along the back, from the base of the skull to the base of the tail, coming laterally onto the sides. The second area is a patch running down the ventral midline. These deposits may be as great as 1 cm thick and may weigh as much as 1.8 kg. With an estimated caloric content of 16,200 kcal, they could maintain caged arctic foxes at moderate levels of activity for 14 days (Underwood 1971) to 30 days (Riewe 1977). Well-fed arctic foxes maintain their body weight consistently throughout the year, so the accumulation of fat does not seem to be related to the season. However, in the case of limited food availability, the fat reserve is used as an energy source until additional food sources can be located.

BEHAVIOR

Activity. Except during the breeding season, arctic foxes are primarily solitary (Banfield 1977). They tend to congregate in areas where food is available. They are active, nervous, and unsuspecting around human habitation until they are molested, and then they become more wary. They can become great camp thieves. Arctic foxes are highly tolerant of, and are attracted by, human activity (Eberhardt 1976). They learn to follow biologists in the field (Wrigley and Hatch 1976). Individuals became tame and tolerant of camp activities associated with oilfield developments at Prudhoe Bay, Alaska (Eberhardt 1977). These foxes use unskirted buildings for shelter and den sites. At present, they cause only minor problems.

Kavanau and Ramos (1975) tested the activity cycles of one arctic fox. Under a 24-hour light cycle, this individual was mainly nocturnal, outdoors and indoors, summer and winter. Other studies verify nocturnal tendencies (Seton 1929). Activity cycles peak at midnight in summer; some animals had two periods of activity, peaking at midnight and noon (Folk 1964).

Vocalizations vary from a kind of yelp, a loud, high-pitched bark, and a rasping cry to a purr (Seton 1929). Arctic foxes scream when fighting, but do not howl (Banfield 1977).

Seton (1929) stated that arctic foxes "never voluntarily enter the water or attempt to swim," because waters in the arctic are too cold. Dementyeff (1958), however, reported that foxes voluntarily swim across rivers, bays, and inlets. Banfield (1977) also reported that they swim readily and float high in the water, and Murie (1959) found that arctic foxes commonly swim between close islands in the Aleutians.

Territoriality. Arctic foxes may be territorial during the breeding season. MacPherson (1969) reported a distinct tendency for denning foxes to keep their distance from each other, with the minimum distance between dens approximately 1.6 km. An arctic fox was observed on St. Matthew Island driving away others from a reindeer carcass on which they were feeding. Never more than one fed on the carcass at a time (Chesemore 1975). Arctic foxes also exhibited antagonistic behavior toward one another at a garbage pit near a campsite on the same island. Bedard (1967, as reported in Stephenson 1970) noted that the periphery of a large cirque on St. Lawrence Island, Alaska, was divided up among five or six foxes, which frequently engaged in territorial clashes.

Den Site Behavior. Both parents care for the young (Chesemore 1967; MacPherson 1969). For the first few days after the birth of the pups, females almost never leave them (Dementyeff 1958; Speller 1972). During this time, the male brings food to the den for the female. Both hunt after the first two weeks, when the whelps are consuming lemmings (Speller 1972). Even then the female continues to stay relatively close to the den and visits it often, while the male ranges farther afield. Males try to draw off intruders. Females maintain a close relationship to whelps and probably have a greater concern for their feeding than do the males. As the female begins to hunt, the male's hunting effort is reduced, which releases more time for territorial defense. Female hunts are frequent, short in duration, and close to the den, that is, similar to male hunts earlier in the season. In Canada, all hunting is conducted between 1600 and 1000 hours, perhaps related to lemming activity (Speller 1972). During a year when the food supply was quite limited, adult foxes were observed to visit the den relatively infrequently (Underwood 1975). The adults were defensive of the den early in the season (in June). It appeared that they became less attached to the den as the litter matured. Litters are occasionally moved from one den to another during summer (Underwood 1975; Eberhardt 1977).

Young foxes emerge from the den at two to four weeks of age (Tchirkova 1958a; Dementyeff 1958; Banfield 1977). Typical pup behavior includes exploring the den site, pouncing on grass clumps and other objects, and tossing objects into the air. Group play rarely involves more than two individuals, and consists mainly of chasing and wrestling. The latter is preceded by a stiff tail display, that is, the tail upright, the back arched, and a sideways walk, followed with a quick pounce by the displayer. Wrestling consists of rolling and biting, especially on the back of the neck and rarely on the tail and legs.

This behavior leads to excursions away from the den, which increase in duration and distance. At the end of summer, the dens are abandoned. Males abandon the dens first in late August, followed by the females some weeks later. After the parents leave the den site the young are on their own. In normal years this takes place in July on Rat Island, Alaska (Berns 1969), mid-August to September in Canada and the United States (Chesemore 1967, 1968a) and the U.S.S.R. (Shibanoff 1958). Tchirkova (1958a) stated that dispersal takes place in the U.S.S.R. in August in poor food years, and in September in good food years. MacPherson (1969) believed that in exceptional years the older animals do not disperse from the den site but the young animals do. This is why trappers take a higher proportion of young animals than older animals during years of increased food availability.

Hunting Behavior. Arctic foxes carefully follow the progress of lemmings under the snow, either by sound or smell (Banfield 1977). Suddenly, they pounce and dig rapidly with their forefeet. In Iceland, they are accused of killing sheep. Arctic foxes apparently travel considerable distances to secure food, traveling at a sustained, steady lope. Adult arctic foxes run with tail stiffly extended. Active hunting can be categorized into various intensities (Speller 1972): wide sweeps with brief stops; a slow lope with narrow sweeps and brief digging; and walking, stalking, digging, and/or watching lemming holes with frequent kills. Capture behavior includes digging, remaining in a semicrouch with tail extended, stalking, lunging, dashing, pouncing, and leaping. Shibanoff (1958) found freshly killed lemmings that were hidden as a food reserve. Several other authors also refer to caching behavior by arctic foxes (Seton 1929; Braestrup 1941; Vibe 1967).

MORTALITY

Premature den abandonment and sibling aggression may be the leading causes of pup mortality among arctic foxes. MacPherson (1969) found that from 0 to 57 percent of the dens were abandoned in Canada during a four-year period. He found that weaned litter size and the occurrence of den abandonment were responsible for variability in the production of young. These factors were causally related to food scarcity, particularly a scarcity of lemmings. In areas where both blue and white color phases occur, such as Greenland, one litter may have both white and blue pups. There may be an occasional gray pup, although the gray phase seems to be weak and not fit for survival (Vibe 1967).

Most arctic foxes over six months old die in winter due to trapping or starvation (Vibe 1967; MacPherson 1969; Bannikov 1970; Speller 1972). The ratio of young to old trapped animals may be as great as 30:1, indicating rather low survival rates (Smirov 1968). In winter, survival is precarious and the mortality rate may be severe. In summer, life is comparatively easy for adults. Although little direct evidence exists, food scarcity appears to be the most important factor in whelp mortality. The differences between placental scar counts and mean litter size vary from year to year. Little information is available on mortality of cubs after weaning and before the trapping season.

Wolves may be the chief predator of the arctic fox (Banfield 1977). Occasionally, a grizzly bear (*Ursus arctos*) digs out a fox den, probably in search of ground squirrels (*Spermophilus* sp.). Other probable natural predators include red foxes (*Vulpus vulpes*), wolverines (*Gulo gulo*), snowy owls (*Nyctea scandiaca*), large hawks, eagles, and jaegers (*Stercorarius sp.*) (Bee and Hall 1956; Chesemore 1967, 1975; Berns 1969). Red foxes harry arctic foxes in areas of sympatry. Bannikov (1970) reported that arctic foxes have few natural enemies in the U.S.S.R., but that

snowy owls take young. Around villages, domestic dogs are dangerous to foxes.

Epizootic rabies may be a principal cause of natural mortality (Fay 1973; Chesemore 1975). Disease may act in conjunction with a lack of food to reduce arctic fox density, especially in winter (Speller 1972). In the U.S.S.R., during a year with high rabies incidence, only 12 to 13 percent of young animals survive, whereas in normal years 60 to 70 percent survive (Syuzymova 1968).

Avian predators, red foxes, wolves, wolverines, least weasels (*Mustela nivalis*), ermine (*Mustela erminea*), and dogs actively compete with white foxes for food, especially in winter. A general northward advance of the red fox has occurred in the U.S.S.R. and has replaced the white fox wherever their two ranges are sympatric (Chesemore 1967). Similar observations were made by MacPherson (1969) in Manitoba. No competition between red and arctic foxes was observed by Eberhardt (1977) in Alaska, however.

MacPherson (1969) constructed a tentative life table based on catch statistics of arctic foxes in Canada (table 23.1). The age composition changes drastically from one year to the next. A heavy catch of whelps one year is followed the next year by a heavy catch of yearlings. Thus, there is no demonstrable correlation between ranked cohort size and harvest size. The life table shows that the highest mortality occurs in whelps, and only a few individuals survive beyond 4 years of age. The maximum life expectancy is approximately 8 to 10 years (Banfield 1977). Smirov (1968) found 3, 9-year-old individuals among 684 individuals for which age was determined.

Diseases. Epizootics of rabieslike diseases have been recorded in the far north for over a century, and have been called rabies, arctic dog disease, rabidity, polar madness, arctic nervous disease, and fits (Crandell 1975). The scarcity of Negri bodies is a feature of these diseases in arctic regions. Epizootics of rabies are widespread among arctic and red fox populations throughout the arctic seaboard (Elton 1931; Braestrup 1941; Cowan 1949; Vibe 1957; Syuzymova 1968; Bitsch and Knox 1971; Banfield 1977). Rabies is a common disease of arctic foxes throughout Alaska. It

TABLE 23.1. Life table for the arctic fox in the Northwest Territories

Age (years)	Number Dying in Age Interval out of 1,000 Born	Number Surviving at Beginning of Age Interval out of 1,000 Born	Mortality Rate per 1,000 Alive at Beginning of Age Interval	Number Alive between Ages x and x + 1	Expectation of Life, or Mean Life Remaining to Those Attaining Age Interval (years)
0 –0.5	609	1,000	609.0	695.5	1.30
0.5–1.5	135	391	345.3	323.5	1.54
1.5–2.5	176	256	687.5	168.0	1.09
2.5–3.5	27	80	337.5	66.5	1.39
3.5–4.5	17	53	320.8	44.5	0.50
4.5+	36	36	1,000.0	18.0	unknown

SOURCE: MacPherson 1969. Based on data from reproductive tracts, den surveys, weaned litter counts, and harvest samples. Reproduced by permission of the minister of supply and services, Canada.

was confirmed in all dead foxes found and tested in arctic Alaska between 1949 and 1957 (Rausch 1958). The frequency of rabies incidence varies from year to year and is related to the population cycle. High instances occur during population highs and subside as the population decreases. However, rabies has been diagnosed in arctic foxes at times of low population density (Rausch 1958). Infectivity varies from 0.7 to 1 percent during population lows, and increases to over 20 percent during population highs (Syuzymova 1968). The virus has been recovered from the brains of four foxes showing no signs of illness (Crandell 1975), suggesting that arctic foxes may be prolonged carriers of rabies (Bannikov 1970). Age structure of the population is important, with rabies spreading mainly among young foxes. Most cases are observed during the colder months of the year in the U.S.S.R., Alaska, and Canada. In Greenland, most diseased foxes are observed in March and April (Crandell 1975). Mass movements of foxes following significant declines in lemming population numbers favor the spread of rabies and may contribute to population decreases of arctic foxes (Rausch 1958). Arctic foxes may be the main vector transmitting rabies to sled dogs in the far north (Crandell 1975). However, apparently no cases of rabies have been reported in humans, even among those bitten by rabid foxes (Bannikov 1970).

Woloszyn et al. (1973) reported that farm-reared arctic foxes suffered from trichophytosis, a ringworm fungus. The morbidity index ranged from 2.9 to 27.5 percent, with young animals being most sensitive. Under natural conditions, arctic foxes, especially the young, are vulnerable to leptospirosis (Bannikov 1970). Brucellosis, in both farm and wild foxes, has been reported in the U.S.S.R. (Pinigin et al. 1970). The disease is obtained from eating reindeer wastes and scavenging dead reindeer. Brucellosis was also found in lemmings.

Parasites. The cestode *Echinococcus multilocularis* occurs throughout the holarctic tundra zone as a parasite of the arctic fox, the definitive host, and several species of small rodent (Rausch 1967; Banfield 1977). In North America, the known range of *E. multilocularis* at high latitudes corresponds closely to that of the arctic fox. It is a potential pathogen to humans because it attacks the liver and can be fatal. This tapeworm cyst inhabits the small intestine of the arctic fox—from 10,000 to 25,000 per fox on St. Lawrence Island (Fay 1973). The proportion of animals harboring the cestodes ranges from 40 to 100 percent. Variations seem to be closely correlated with fluctuations in the abundance of both arctic fox and rodent hosts, being most numerous when host numbers are high. They are not reported from Greenland.

Numerous other helminths are found in the arctic fox. In the U.S.S.R., 58 species of worm, including 11 nematodes, 10 cestodes, 7 trematodes, and 2 Acanthocephala have been identified in the arctic fox (Bannikov 1970). Individual species and genus names were not given.

Fay and Williamson (1962) found the following helminths in Pribilof Island foxes: the hookworm (*Uncinaria stenocephala*), the fluke (*Maritrema afanassjewi*), the ascarid nematode (*Toxascaris leonina*), and the tapeworms *Mesocestoides karbyi*, *Taenia polyacantha*, and *T. crassiceps*. The infection by the three species of Taeniid cestode is dependent on the fox's having consumed microtine rodents harboring larvae. Webster (1974) reported the occurrence of immature tapeworms, *T. polyacantha*, in arctic fox specimens in Canada. Eaton and Secord (1979) found that 48 out of 50 foxes showed one or more parasitic species in the small intestine. These included *T. crassiceps* (78 percent occurrence) and *Toxascaris leonina* (60 percent occurrence). They did not find *Toxocara canis* or hookworms. Chesemore (1967) found nematode parasites in 88 percent of arctic fox specimens observed.

AGE DETERMINATION

Although it is possible to separate young of the previous year from older foxes on the basis of tooth pulp cavity size, further subdivisions are not possible (Grue and Jensen 1976). Annual incremental lines apparently form in the cementum layer that may distinguish age up to five or six years.

ECONOMIC STATUS

The arctic fox is a valuable resource. It is easily trapped. In Canada alone, 10,000 to 80,000 arctic fox pelts go to market each year (Mayfield 1976). The price paid per pelt has varied between $6.00 and $75.00 in recent years. Many Eskimos are economically dependent upon the fur of the arctic fox. Smith and Taylor (1977) reported an average harvest of 27,378 arctic foxes per year in 42 settlements in Canada between 1962 and 1971. Fox pelt cash value at auction has been between $50,000 and $2 million annually (MacPherson 1969).

The trapping industry is one of the main sectors of the economy in the Far North of the U.S.S.R. (Geller and Skrobov 1968). The arctic fox supplies 90 to 95 percent of the total furs trapped there (Bannikov 1970). The actual mean annual yield of arctic foxes comes to 0.8 pelts per 1,000 ha of tundra. The proportion of arctic foxes is currently 43.7 percent of the total harvest, more than two times that of any other species. The arctic area of the U.S.S.R. yields between 72,600 and 107,000 arctic fox pelts per annum. Bannikov (1970) reported an annual average harvest of 83,000. The harvest accounts for no more than one-third of the annual population increase.

Trapping data from Alaska are widely scattered and poorly analyzed. The number of arctic fox pelts sold on the open market has averaged 2,000 per year between 1970–71 and 1978–79 (Van Ballenberghe, Alaska Dept. Fish and Game, personal communication). During these years, the average price paid per pelt has increased from $17.00 in 1970–71 to $45.00 in 1978–79. The fox harvest could be considerably greater (Grauvogel, Alaska Dept. Fish and Game, personal communication). Fox farming operations in the

Aleutians declined in the 1930s because of economics (Chesemore 1975). Interest in arctic fox farming activities is reviving, particularly among certain Native groups.

MANAGEMENT

Aboriginal peoples, with the exception of the Aleuts (Carnahan 1979), made extensive use of the arctic fox for clothing and food (Spencer 1959). Many arctic Natives still trap arctic foxes. The situation in Canada may be typical. There are fewer trappers now in Canada than there were ten years ago. However, for those that still trap, the arctic fox continues to be important to Native subsistence (MacPherson 1970). It seems unlikely that the arctic fox habitat will be affected by human activities for several decades. If active management becomes necessary, breeding adults should be protected in spring, before whelping. Late spring trapping should be discouraged because it may decrease population densities the following year. The value of live-trapping young arctic foxes and keeping them until furs become prime should be studied. This method is being tested in the U.S.S.R. (Bannikov 1970). In other experiments, low-quality fish, remains of slaughtered reindeer, and marine mammals are put out at 600 feeding stations. Each station attracts about 30 foxes, which can be trapped when prime.

Development activities at Prudhoe Bay, Alaska, may have both beneficial and detrimental effects on arctic foxes (Vaughan 1979). Garbage may supply an alternate food source, leading to increased population. However, this also increases the chance for rabies in arctic foxes and contacts with humans. Eberhardt (1977) found no evidence of rabies or other fox diseases in his studies in northern Alaska. The potential for a severe epizootic is greatly increased because of tameness, availability of artificial food, and migratory habits of the animal, however. On St. Lawrence Island, the risk of human exposure to *E. multilocularis,* the causative agent of alveolar hydatid disease, which can be fatal in humans, seems unusually high because of the greater abundance of arctic foxes (Fay and Williamson 1962; Fay 1973). Contamination comes through handling and processing fox furs.

Transplanted arctic foxes have had a detrimental effect on wildlife in the Aleutians. On some of the smaller Aleutian Islands, nesting birds have almost been eliminated by the arctic fox. They are especially effective in catching Aleutian Canada geese (*Branta canadensis*), whiskered auklet (*Aethia pygmaea*), and Cassin's auklet (*Ptychoramphus aleuticus*). Predations by foxes seems to be related to Aleutian Canada goose extinction, at least on Amchitka Island, and has disturbed the growth of the sea otter (*Enhyda lutris*) population (Carnahan 1979). The U.S. Fish and Wildlife Service began a program to exterminate blue foxes on certain Aleutian Islands in the early 1950s. Poisoning was used, but has been successful only on Amchitka. Lembkey and Lucas (1902) reported that in order to increase population densities of foxes, only males were trapped on certain Aleutian Islands. However, the sex ratio stayed the same. This may have been the first attempt to manage arctic fox populations.

CURRENT RESEARCH AND MANAGEMENT NEEDS

Most of the research needs concerning the arctic fox involve applied questions. In Canada, research needs include: (1) the role of large carnivores and people in providing food for winter-scavenging arctic foxes; (2) the variability in hunting range as a function of lemming population; (3) the degree and patterns of dispersal; and (4) the effects of red foxes on arctic foxes. Red fox range has been expanding in northern Canada for at least 35 years (MacPherson 1970). Perhaps the necessary studies could best be mounted under the aegis of a long-term integrated productivity project. In the U.S.S.R., efforts are being made to develop reliable criteria for compiling population forecasts and estimating harvests (Geller and Skrobov 1968). The study of migration routes and other factors that determine the number of arctic foxes in different regions is far from adequate. Smirov (1968) felt that the number of animals dying from rabies should be studied. Current studies in Alaska are examining the effects of human development projects on arctic fox numbers, and the possible interactions between sympatric red and arctic fox populations. The use of sterilized red foxes as an agent to reduce arctic fox numbers in the Aleutian Islands is also being explored.

LITERATURE CITED

Banfield, A. W. F. 1977. The mammals of Canada. Univ. Toronto Press, Toronto. 438pp.

Bannikov, A. G. 1970. Arctic fox in the U.S.S.R. Pages 121–130 *in* W. A. Fuller and P. G. Kern, eds. Conference on Productivity and Conservation in Northern Circumpolar Lands, Morges: proceedings. Int. Union for Conserv. of Nature and Nat. Resour., New Ser., Publ. 16. 344pp.

Bedard, J. H. 1967. Ecological segregation among plankton-feeding alcidae (*Aethia* and *Cyclorrhynchus*). Ph.D. Thesis. Univ. British Columbia.

Bee, J. W., and Hall, E. R. 1956. Mammals of northern Alaska on the arctic slope. Univ. Kansas Mus. Nat. Hist. Misc. Publ. 8. 309pp.

Berns, V. D. 1969. Notes on the blue fox of Rat Island, Alaska. Can. Field Nat. 83:404–405.

Bertram, G., and Leek, D. 1938. Notes on the animal ecology of Bear Island. J. Anim. Ecol. 7:27–52.

Bitsch, V., and Knox, B. 1971. On pseudorabies in carnivores in Denmark. Part 2: The blue fox (*Alopex lagopus*). Acta Vet. Scandinavica 12:285–292.

Bliss, L. C.; Courtin, G. M.; Pottie, D. L.; Riewe, R. R.; Whitfield, P. W. A.; and Widdon, P. 1973. Arctic tundra ecosystems. Pages 359–399 *in* R. F. Johnston, ed. Annual review of ecology and systematics. Annu. Rev., Inc., Palo Alto. 424pp.

Braestrup, F. W. 1941. A study on the arctic fox in Greenland: immigrations, fluctuations in numbers based mainly on trading statistics. I Kommission Hos. C. A. Reitzels Forlag, Copenhagen. 98pp. (Reprinted from Meddelelser om Gronland. Bd. 131).

Britton, M. E. 1957. Vegetation of the arctic tundra. Pages

26–61 *in* H. P. Hansen, ed. Arctic biology. Oregon State Univ. Press, Corvallis. 134pp.

Burris, O. E., and McKnight, D. E. 1973. Game transplants in Alaska. Div. Game, Alaska Dept. Fish Game Tech. Bull. 4. 57pp.

Cameron, A. W. 1950. Arctic fox on Cape Breton. Can. Field Nat. 64:154.

Carnahan, J. 1979. Fox farming in the Aleutians. Pages 76–96 *in* D. L. Spencer, C. M. Naske, and J. Carnahan, eds. National wildlife refuges of Alaska: a historical perspective. Arctic Environ. Inform. and Data Center, Univ. Alaska, Anchorage. 183pp.

Chesemore, D. L. 1967. Ecology of the arctic fox in northern and western Alaska. M.S. Thesis. Univ. Alaska, College. 148pp.

———. 1968*a*. Distribution and movements of white foxes in northern and western Alaska. Can. Zool. 46:849–854.

———. 1968*b*. Notes on the food habits of arctic foxes in northern Alaska. Can. J. Zool. 46:1127–1130.

———. 1969. Den ecology of the arctic fox in northern Alaska. Can. J. Zool. 47:121–129.

———. 1970. Notes on the pelage and priming sequence of arctic foxes in northern Alaska. J. Mammal. 51:156–159.

———. 1975. Ecology of the arctic fox *Alopex lagopus* in North America: a review. Pages 143–163 *in* M. W. Fox, ed. The wild canids. Van Nostrand Reinhold Co., New York. 508pp.

Chitty, H. 1950. Canadian arctic wildlife enquiry, 1943–49, with a summary of results since 1933. J. Anim. Ecol. 19:180–193.

Cowan, I. M. 1949. Rabies as a possible population control of arctic Canidae. J. Mammal. 30:396–398.

Crandell, R. A. 1975. Arctic fox rabies. Pages 23–40 *in* G. Baer, ed. The natural history of rabies. Vol. 2. Academic Press, New York. 387pp.

Dementyeff, N. E. 1958. Biology of the arctic fox in the Bolshezemelskaya tundra. Pages 166–181 *in* Translations of Russian game reports. Vol. 3. Can. Wildl. Serv., Ottawa. (Translation of: Dement'yev, N. I. 1955. K biologii pestsa Bol'shezemel'skoy tundry. Voprosy Biologii Pushnykh Zverey 14:123ff.)

Eaton, R. D. P., and Secord, D. C. 1979. Some intestinal parasites of arctic fox, Banks Island, N.W.T. Can. J. Comp. Med. 43:229–230.

Eberhardt, L. E., and Hanson, W. C. 1978. Long-distance movements of arctic foxes tagged in northern Alaska. Can. Field Nat. 92:386–389.

Eberhardt, W. L. 1976. The biology of arctic and red foxes on the North Slope. Pages 238–239 *in* G. C. West, ed. Science in Alaska. Proc. 27th Alaska Sci. Conf., Fairbanks. Alaska Div., Am. Assoc. Advancement of Sci. 125pp.

———. 1977. The biology of arctic and red foxes on the North Slope. M.S. Thesis. Univ. Alaska, Fairbanks. 125pp.

Elton, C. 1931. Epidemics among sledge dogs in the Canadian arctic and their relation to disease in the arctic fox. Can. Res. 5:673–692.

Fay, F. H. 1973. The ecology of *Echinococcus multilocularis* Leuckart, 1863 (*Cestoda: Taeniidae*), on St. Lawrence Island, Alaska. Ann. Parasitol. 48:523–542.

Fay, F. H., and Williamson, F. S. L. 1962. *Echinococcus multilocularis* Leuckart, 1863, and other helminths of foxes on the Pribilof Islands. Studies on the helminth fauna of Alaska 39. Can. J. Zool. 40:767–772.

Fetherston, K. 1947. Geographic variation in the incidence of occurrence of the blue phase of the arctic fox in Canada. Can. Field Nat. 61:15–18.

Folk, G. E. 1964. Daily physiological rhythms of carnivores

exposed to extreme changes in arctic daylight. Fed. Proc. 23:1221–1228.

Follmann, E. H. 1977. The thermal significance of the curled posture during rest in arctic foxes and wolves. Page 78 *in* Science Information Exchange in Alaska. Tech. Session Pap. 28th Alaska Sci. Conf., Anchorage. Vol. 4. Alaska Div., Am. Assoc. Advancement of Sci.

Geller, M. K., and Skrobov, V. D. 1968. Methods of developing and increasing the productivity of the fur industry in the far north. Problems of the North 11:19–45.

Grzimek, B. 1975. Grzimek's animal life encyclopedia. Vol. 12: Mammals, book 3. Van Nostrand Reinhold Co., New York. 657pp.

Grue, H., and Jensen, B. 1976. Annual cementum structures in canine teeth in arctic foxes (*Alopex lagopus* L.) from Greenland and Denmark. Danish Rev. Game Bio. 10:3–12.

Hall, E. R. 1981. The mammals of North America. 2nd ed. 2 vols. John Wiley and Sons, New York. 1181pp.

Henshaw, R. E.; Underwood, L. S.; and Casey, T. M. 1972. Peripheral thermoregulation: foot temperature in two arctic canines. Science 175:988–990.

Irving, L., and Krog, J. 1954. Body temperature of arctic and subarctic birds and mammals. J. Appl. Physiol. 6:667–680.

Johansen, I. 1960. Inheritance of the color phase in ranch-bred blue foxes. Hereditas 4:753–766.

Kavanau, J. L., and Ramos, J. 1975. Influences of light on activity and phasing of carnivores. Am. Nat. 109:391–418.

Lembkey, W. I., and Lucas, F. A. 1902. Blue fox trapping on the Pribilof Islands. Science 16:216–218.

McEwen, E. H. 1951. Literature review of the arctic foxes. M.A. Thesis. Univ. Toronto, Toronto. n.p.

MacInnes, C. D., and Misra, R. K. 1972. Predation on Canada goose nests at McConnell River, Northwest Territories. J. Wildl. Manage. 36:414–422.

MacPherson, A. H. 1968. Apparent recovery of translocated arctic fox. Can. Field Nat. 82:287–289.

———. 1969. The dynamics of Canadian arctic fox populations. Wildl. Serv., Can. Dept. Indian Affairs and Northern Development, Ottawa. Rep. ser. no. 8. 52pp.

———. 1970. Situation report on Canadian arctic fox research. Pages 130–132 *in* W. A. Fuller and P. G. Kern, eds. Conference on productivity and conservation in Northern Circumpolar Lands, Morges: proceedings. Int. Union for Conserv. of Nature and Nat. Resour., New Ser., Publ. 16. 344pp.

Mayfield, H. F. 1976. Of arctic foxes and birds and men. Audubon 78:2–23.

Mullen, D. A., and Pitelka, F. A. 1972. Efficiency of water scavengers in the arctic. Arctic 25:225–231.

Murie, O. J. 1959. Fauna of the Aleutian Islands and Alaska Peninsula. N. Am. Fauna 61. 405pp.

Northcott, T. 1975. Long-distance movement of an arctic fox in Newfoundland. Can. Field Nat. 89:464–465.

Pedersen, J. A., and Larsen, T. M. 1971. Fjellrev, *Alopex lagopus*, pa Fosenhalvoya. Fauna 24:187–188. (English summary.)

Pinigin, A. F.; Zabrodin, V. A.; and Nikulina, V. I. 1970. Brucellosis in the arctic fox. Krolikovodstvo I Zverovodstvo (Moscow) 5:39–40. (English translation.)

Rausch, R. 1958. Some observations on rabies in Alaska, with special reference to wild Canidae. J. Wildl. Manage. 22:246–260.

———. 1967. On the ecology and distribution of *Echinococcus* spp. (*Cestoda: Taeniidae*), and characteristics of their development in the intermediate host. Ann. Parasitol. Humaine et Comparée 42:19–63.

Riewe, R. R. 1977. Mammalian carnivores utilizing Truelove Lowland. Pages 493–501 *in* L. C. Bliss, ed., Truelove Lowland, Devon Island, Canada: a high arctic ecosystem. Univ. Alberta Press, Edmonton. 714pp.

Schiller, E. L. 1954. Unusual walrus mortality on St. Lawrence Island, Alaska. J. Mammal. 35:203–209.

Scholander, P. F.; Hoek, R.; Walters, V.; and Irving, L. 1950*a*. Adaptation to cold in arctic and tropical mammals and birds in relation to body temperature, insulation and basal metabolic rate. Bio. Bull. 99:259–271.

Scholander, P. F.; Walters, V.; Hoek, R.; and Irving, L. 1950*b*. Body insulation of arctic and tropical mammals and birds in relation to body temperature, insulation and basal metabolic rate. Bio. Bull. 99:225–236.

Seton, E. T. 1929. Lives of game animals. Vol. 1, pt. 2. Doubleday, Born & Co., Inc., New York. 639pp.

Shibanoff, S. V. 1958. Dynamics of arctic fox numbers in relation to breeding, food and migration conditions. Pages 5–28 *in* Translations of Russian game reports. Vol. 3. Can. Wildl. Serv., Ottawa. (Translation of: Shibanov, S. V. 1951. Dinamika chislennosti pestsa v svyazi s usloviyami razmnizheniya, pitaniya i migratsiyami. Voprosy Biologii Pushnykh Zverey i Tekhniki Okhotnich'yego Promysla 11:57ff.)

Smirov, V. S. 1968. Analysis of arctic fox population dynamics and methods of increasing the arctic fox harvest. Problems of the North 11:81–101.

Smith, T. G. 1976. Predation of ringed seal pups (*Phoca hispida*) by the arctic fox (*Alopex lagopus*). Can. J. Zool. 54:1610–1616.

Smith, T. G., and Taylor, D. 1977. Notes on marine mammal, fox and polar bear harvests in the Northwest Territories, 1940 to 1972. Tech. rep. 694. Fisheries and Marine Ser., Environment Can., Ste. Anne de Bellevue, Quebec. 37pp.

Speller, S. W. 1969. Arctic fox attacks on molting Canada geese. Can. Field Nat. 83:62.

———. 1972. Food ecology and hunting behavior of denning arctic foxes at Aberdeen Lake, Northwest Territories. Ph.D. Thesis. Univ. Saskatchewan, Saskatoon. 145pp.

Spencer, R. F. 1959. The north Alaskan Eskimo: a study in ecology and society. Smithsonian Inst., U.S. Bur. Am. Ethnol. Bull. 171. Reprinted by Dover Publ., 1976. 490pp.

Stephenson, R. O. 1970. A study of the summer food habits of the arctic fox on St. Lawrence Island, Alaska. M.S. Thesis. Univ. Alaska, College. 76pp.

Stirling, I., and Smith, T. G. 1975. Interrelationships of Arctic Ocean mammals in the sea ice habitat. Pages II-129–II-136 *in* Circumpolar Conference on Northern Ecology, Ottawa: proceedings. Nat. Res. Council of Canada. IX-7p.

Strecker, R. L.; Ryser, F. A.; Tietz, W. J.; and Morrison, P. R. 1952. Notes on mammals from Alaska. J. Mammal. 33:476–480.

Syuzyumova, L. M. 1968. Epizootiology of rabies among arctic foxes on the Yamal Peninsula. Problems of the North 11:113–121.

Tchirkova, A. F. 1958*a*. Experiments in mass visual census and forecasting harvest of arctic foxes, 1944–49. Pages 101–165 *in* Translations of Russian game reports. Vol. 3. Can. Wildl. Serv., Ottawa. (Translated from Russian.)

———. 1958*b*. A preliminary method of forecasting changes in numbers of arctic foxes. Pages 29–49 *in* Translations of Russian game reports. Vol. 3. Can. Wildl. Serv., Ottawa. (Translated from Russian.)

———. 1968. The relationship between arctic fox and red fox in the far north. Problems of the North 11:129–131.

Underwood, L. S. 1971. The bioenergetics of the arctic fox (*Alopex lagopus* L.). Ph.D. Thesis. Pennsylvania State Univ., State College, Pa. 85pp.

———. 1975. Notes on the arctic fox (*Alopex lagopus*) in the Prudhoe Bay area of Alaska. Pages 145–149 *in* J. Brown, ed. Ecological investigations of the tundra biome in the Prudhoe Bay region, Alaska. Univ. Alaska, Fairbanks, Bio. Pap., Special Rep. no. 2. 215pp.

Underwood, L. S., and Reynolds, P. 1980. Photoperiod and fur lengths in the arctic fox (*Alopex lagopus* L.). Inter. Biometeorol. 24(1):39–48.

Vaughan, B. E. 1979. Pacific Northwest Laboratory annual report for 1978 to the DOE assistant secretary for the environment. Pt. 2: Supplement ecological sciences. Pages 12.5–12.10.

Vibe, C. 1967. Arctic animals in relation to climatic fluctuations. C. A. Reitzels Forlag, Copenhagen. Meddelelser on Gronland. Bd. 170, no. 5. 227pp.

Voilochnikov, A. T. 1968. Types and characteristics of tundra hunting ranges in the vicinity of the Vashutkiny Lakes. Problems of the North 11:73–79.

Woloszyn, S.; Kamyszek, F.; Andrychiewicz, J.; and Krukowski, W. 1973. Clinics and control of trichophytosis in rearing foxes. Medycyna Weterynaryjna 29:73–77. (English summary.)

Webster, W. A. 1974. Records of cestodes in varying lemmings and an arctic fox from Bathurst Island, Northwest Territories. Can. J. Zool. 52:1425–1426.

Wrigley, R. E., and Hatch, D. R. M. 1976. Arctic fox migrations in Manitoba. Arctic 29:147–158.

Larry Underwood, Arctic Environmental Information and Data Center, 707 A Street, Anchorage, Alaska 99501.

James A. Mosher, Appalachian Environmental Laboratory, Center for Environmental and Estuarine Studies, University of Maryland, Frostburg State College Campus, Frostburg, Maryland 21532.

24

Black Bear

Ursus americanus

Michael R. Pelton

NOMENCLATURE

COMMON NAME. Black bear
SCIENTIFIC NAME. *Ursus americanus*
SUBSPECIES. *U. a. altifrontalis, U. a. amblyceps, U. a. americanus, U. a. californiensis, U. a. carlottae, U. a. cinnamomum, U. a. emmonsii, U. a. eremicus, U. a. floridanus, U. a. hamiltoni, U. a. kermodei, U. a. luteolus, U. a. machetes, U. a. perniger, U. a. pugnax,* and *U. a. vancouveri* (Hall 1981).

DISTRIBUTION

Black bears are the most common and widely distributed of the three ursids in North America. The primitive range of *U. americanus* covered the forested areas of North America, including Mexico. Black bears are now found primarily in less settled, forested regions in at least 23 states and all the Canadian provinces and territories (figure 24.1). Their status and density vary considerably within the existing range. In some western states and many Canadian provinces the species is almost relegated to pest status, with thousands being harvested annually. Cowan (1972) reported that probably 25,000 to 30,000 black bears are harvested each year in North America. The species has been extirpated from many midwestern and eastern states, including Alabama, Kentucky, Ohio, and Illinois. Other states—Florida, Georgia, Massachusetts, and South Carolina—report only remnant numbers surviving in enclaves of inaccessible habitat.

DESCRIPTION

Black bears are plantigrade, pentadactyl, and have short, curved nonretractable claws. Average weights range from 40 to 70 kilograms for adult females and from 60 to 140 kg for adult males. An occasional adult male will exceed 250–300 kg. Body length ranges from 1 to 2 meters. Black bears are generally full grown at four years of age. The eyes are small; the ears are small, rounded, and erect; the tail is short. Fur is uniform in color except for a brown muzzle; an occasional white blaze occurs on the chest. The black color

FIGURE 24.1. Distribution of the black bear (*Ursus americanus*).

phase is most prevalent in the East and the brown phase most prevalent in the West. Unique white and blue phases occur on the Pacific coast in the Northwest. The dental formula is 3/3, 1/1, 4/4, 2/3 = 42. The first three premolars of each jaw are usually rudimentary (figure 24.2). Dentition is bunodont and reflects the black bear's omnivorous food habits.

REPRODUCTION

Female black bears become sexually mature at three to five years of age. Occasional reports indicate some individuals in estrus as young as two years (Hamilton 1978). Other reports indicate that some females do not enter first estrus until five to seven years old (Rogers 1977). Rausch (1961) suggested that an increase in age of reproductive maturity may occur as one goes north in latitude. Nutrition also may have an impact on the age of reproductive maturity and subsequent fecundity of females. Years of poor berry or acorn production

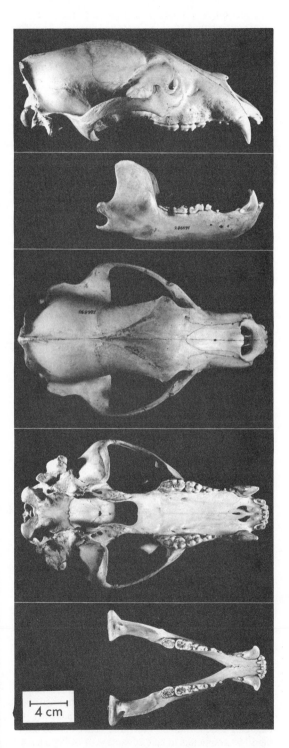

FIGURE 24.2. Skull of the black bear (*Ursus americanus*). From top to bottom, lateral view of cranium, lateral view of mandible, dorsal view of cranium, ventral view of cranium, dorsal view of mandible.

result in delayed first estrus, decreased litter sizes, and increased incidence of barren females.

Breeding occurs in summer and peaks in the latter part of June and July. However, females in estrus have been observed as early as late May and as late as mid-August (Knudsen 1961; Jonkel and Cowan 1971). Black bears have a seasonally constant estrus; females remain in estrus until bred or until the ovarian follicles begin to degenerate. Black bears are induced ovulators (Wimsatt 1963; Erickson and Nellor 1964); ovulation occurs only as a result of coital stimulation.

The gestation period for black bears is seven to eight months (Wimsatt 1963). However, most fetal development occurs only during the last six to eight weeks. Delayed implantation occurs in the black bear; blastocysts float free in the uterus and do not implant until late November or early December. Only minimal cell differentiation occurs prior to implantation (Wimsatt 1963).

Cubs are born in winter dens at the end of January or the beginning of February. The young are altricial—helpless, hairless, and eyes closed at birth—and 200 mm in length. The normal litter size is two, but three and four cubs are not uncommon (Erickson and Nellor 1976; Jonkel and Cowan 1971). Parturient females emerge from dens from late March to early May. Cubs stay with their mother through summer and fall and den with them the second winter. Young disperse as yearlings in spring or summer, prior to the female's period of estrus. Thus, the adult female normally breeds every other year. There are indications that females will occasionally skip a year between reproductive cycles (Jonkel and Cowen 1971). Because of the relatively low biological potential of the species, changes in the various factors influencing reproduction—age of reproductive maturity, number cubs produced, frequency of successful breeding and litter production—can significantly alter population stability. Thus, low biological potential becomes an increasingly important consideration for management.

ECOLOGY

Population Density. Because of their generally sparse numbers, characteristic shy and secretive nature, and inaccessible habitat, black bears are difficult to census. Researchers have used a variety of methods to census black bears. The use of bear signs includes attempts by Spencer (1955) to make a statewide census in Maine using tracks and other bear signs observed on cruise lines. Pelton (1972) indicated that the incidence of scats along established index routes might reflect relative densities of black bears. However, Matthews (1977) discussed the complex variables affecting scat incidence, such as season, altitude, weather, and ecological site.

Harvest data have been used in a variety of ways. Poelker and Hartwell (1973) in Washington, Carpenter (1973) in Virginia, and Spencer (1955) in Maine, in computing statewide estimates of black bears, used methods based on mean annual harvest figures. These authors assumed that the total harvest is a known percentage of the population. Carpenter (1973) assumed that the annual kill in Virginia was 20 percent of the population. Spencer (1955) arrived at an annual harvest percentage based on the total harvest and the percentage of young animals in the harvest. Erickson and Petrides (1964) used the ratio of marked to unmarked bears in the harvest of Michigan black bears to determine estimates using the Lincoln index.

Removal techniques are not applicable in areas where populations are not hunted or sample sizes are too small. Thus, direct counts of bears have been used to determine density. Bray and Barnes (1967) made direct counts of black bears along roadsides and backcountry areas in Yellowstone National Park. Hornocker (1962) believed that direct counts of bears were reliable where there were adequate open areas for observations to be made. However, in many parts of North America where the vegetation is dense and often continuous, the technique is not feasible. The roadside census technique often is biased due to human traffic and activities (Hayne 1949). Mark-reobserve estimates in park areas are often biased because of the influence of the differential observability of panhandler versus wild bears (Marcum 1974).

Several studies have utilized multiple recapture methods to estimate black bear density. Kemp (1972) used the Lincoln and Schnabel methods to estimate black bear populations in Canada; the size of the study area was quite small and this enabled him to engage in a very intensive trapping program. Hornocker (1962) and Troyer and Hensel (1964) were able to correlate estimates obtained by the Schnabel method with direct counts of grizzly bears in the Northwest. However, too often the results based on capture-recapture data are not reliable because samples are too small and the data are biased due to nonrandom sampling, loss of marks, and unequal vulnerability of the individuals in the population. Pelton and Marcum (1977) reported on the use of radioisotope feces tags as a technique for estimating population density of black bears using the Schnabel method; they felt that several of the biases inherent in the technique were removed or lessened. Eagar (1977) compared radioisotope feces tagging techniques (Schnabel and Schumacher-Eschmeyer) with capture-recapture techniques (between-year Petersen and triple catch) and a minimum backdating technique in Tennessee; he reported no significant differences among any of the estimates. Piekielek and Burton (1975) used the Lincoln index as well as the technique derived by Edwards and Eberhardt (1967) to estimate population densities of black bears in California. The above variability of methods for censusing black bears may, in part, explain the highly variable population estimates reported from various parts of North America. Reported estimates range from one bear per 1.2 km² (Kemp 1976) to one per 14.3 km² (Spencer 1955).

Because of the limitation of time, money, and personnel most agencies presently rely on trends in harvest data for assessing populations, rather than trying to make accurate population estimates. Recent advances in population modeling now enable biologists to predict harvest levels and population trends by using data on natality, mortality, and sex and age ratios. Although black bears enjoy a higher reproductive rate than the other two species of bear in North America, they are nonetheless a "K"-selected species. As such, the margin of error for making miscalculations in management (such as too much harvest pressure) is narrow. Consequently, the need to arrive at more accurate estimates of population density is important.

Sex Ratios. The sex ratio of black bears at birth is essentially 50:50. However, sex ratios determined from older animals are biased strongly toward males. Most of this bias results from the sampling procedures used. Males tend to be more aggressive and bold than females, thus increasing their chances of contact with people. Males also exhibit greater mobility and home range size, and thus greater vulnerability than females. As a result, sex ratios based on hunter harvest are predominantly male. Nuisance bears trapped at campgrounds and garbage dumps exhibit an even greater predominance of males than females. Sampling a wild bear population utilizing a variety of techniques normally reveals a sex ratio of about 50:50. Because one adult male is capable of breeding with a number of females, a greater mortality of males probably would not detrimentally affect the population.

Age Ratios. Black bears are most accurately aged using the tooth section, cementum annuli technique (Willey 1974); with this method animals are placed in 1-year age classes. The average age of males in a healthy black bear population ranges from 3 to 5 years, while females average 5 to 8 years old (figure 24.3). The more restricted movements, smaller home range, and less aggressive behavior of females contribute to their longer life span. Few black bears in a population ever reach 10 years of age. However, an occasional animal will survive to 15–20 years of age. Changes in age ratios in a population may reflect the relative health of that population if other parameters such as population density are also known. As with sex ratios, constructing age structures from harvested bears can be biased and may contribute to a misinterpretation of the actual age structure. Factors such as selectivity by hunters, laws prohibiting shooting of smaller bears, differential dispersal of young animals, and timing of hunting seasons all affect the availability to hunters of various sex and age classes of bears.

Home Range and Movements. The size and shape of a black bear's home range is determined by the capability of an area to provide the animal's annual needs (Hamilton 1978). Home ranges also may vary considerably depending on such factors as sex, age, season, and population density. Because these factors change, so also does the home range size for an individual. For example, black bears can respond dramatically to changing food resources; individuals have moved more than 160 kilometers to take advantage of isolated pockets of available food (Rogers 1977).

Concentrations of soft mast (Piekielek and Burton 1975), hard mast (Sauer et al. 1969), or artificial food

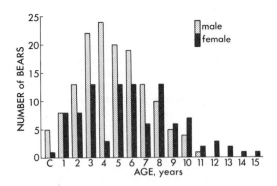

FIGURE 24.3. Age structure of male and female black bears (*Ursus americanus*) captured in Great Smoky Mountains National Park. From Johnson and Pelton 1980*c*.

sources (Rogers 1977) provide the stimulus for extensive movements and consequent range expansion, at least temporarily. These movements occur particularly during the late summer and fall when foraging activities increase. There are conflicting reports of average home range sizes of various sex and age groups of bears (table 24.1); some of these differences can be attributed to the variables noted above. However, the methods of data collection and analysis also contribute to the differences in home range sizes. Consistent among past studies, though, is that the home range size of adult male bears is typically three to eight times larger than that of adult females. Contributing to these movements is the social pressure exerted by larger, older adult males on younger animals, causing them to disperse from the areas occupied by older animals.

The great mobility of black bears, particularly males, also results in a significantly increased mortality rate. Their movements often take them beyond the confines of familiar areas and, in many instances, into closer contact with people and their activities, thus making them more vulnerable.

Habitat Requirements. Black bears are very adaptable. Populations are maintained surprisingly well in the presence of humans if the bears are not overharvested. However, in most instances, if habitat areas of relative refuge are not available, local populations succumb to the intolerances of humans. Throughout its range, a prime black bear habitat is characterized by relatively inaccessible terrain, thick understory vegetation, and abundant sources of food in the form of shrub or tree-borne soft or hard mast. As the pressures of an expanding human population decrease existing habitat, quality food and cover become more critical. If bears are forced to leave relatively protected enclaves to forage on less protected sites, increased mortality rates result in a declining population. Eventually, only remnant populations remain, as now occur in states such as Florida, Georgia, Massachusetts, and South Carolina.

In the Southwest, prime black bear habitat is restricted to vegetated, mountainous areas ranging from 900 to 3,000 m in elevation. Habitats consist mostly of chapparal and pinyon-juniper woodland sites (Waddell 1979). Bears occasionally move out of the chapparal into more open sites and feed on prickly pear cactus (*Opuntia* sp.).

There are at least two distinct, prime habitat types in the Southeast. Black bears in the southern Appalachian Mountains survive in a predominantly oak-hickory and mixed mesophytic forest. Understory consists of such food plants as blueberry (*Vaccinium* sp.), huckleberry (*Gaylussacia* sp.), raspberry, and blackberry (*Rubus* sp.). Laurel (*Kalmia latifolia*) and rhododendron (*Rhododendron* sp.) provide additional thick cover. In the coastal areas of the Southeast, bears inhabit a mixture of flatwoods, bays, and swampy hardwood sites; black gum (*Nyssa sylvatica*) and cypress (*Taxodium dictichum*) are common overstory species on wetter sites. On drier sites, pine (*Pinus* sp.) and oak (*Quercus* sp.) predominate. The understory consists primarily of dense thickets of evergreen, woody species, and greenbriar (*Smilax* sp.). Holly (*Ilex* sp.), greenbriar, gallberry (*Ilex coriacea*), arrow-arum (*Peltandra virginica*), and huckleberry are

TABLE 24.1. Home range sizes of black bears

Area	Study Period	Number of Bears		Mean Home Range Size (km²)		Reference[a]
		Male	Female	Male	Female	
Great Smoky Mountains National Park	1976–77	8	12	42[b]	15[b]	Garshelis 1978
Northeastern Pennsylvania	1974–75	5	8	196[b]	38[b]	Alt et al. 1976
Washington[c]	1973–75	5	6	5[d]	2.3[d]	Lindzey and Meslow 1977
Idaho	1973–74	2	7	112.1[e]	49.8[e]	Amstrup and Beecham 1976

[a]All home ranges obtained through radiotelemetry.
[b]95 percent confidence ellipse.
[c]Island population.
[d]Convex polygon.
[e]Minimum area method.

representative of the understory food species of this habitat. Black gum seed and some acorns provide additional food (Hamilton 1978).

In the Northeast, prime habitat consists of a forest canopy of hardwoods such as beech (*Fagus*), maple (*Acer*), and birch (*Betula*) and coniferous species including red spruce (*Picea rubens*) and balsam fir (*Abies balsamea*). Swampy habitat areas are mainly white cedar (*Thuja occidentalis*). Raspberries and blueberries are common understory food plants (Hugie 1974). Abandoned apple orchards also play an important role as a source of food. Corn crops and oak-hickory mast are also common sources of food in some sections of the Northeast; small, thick swampy areas provide excellent refuge cover.

Along the Pacific coast, redwood (*Sequoia sempervirens*), sitka spruce (*Picea sitchensis*), and hemlocks (*Tsuga* sp.) predominate as overstory cover. Ponderosa pine (*Pinus ponderosa*), lodgepole pine (*Pinus contorta*), and Douglas fir (*Pseudotsuga taxifolia*) are common on the drier sites. Within these forest types are early successional areas important for black bears, such as brushfields, wet and dry meadows, high tidelands, riparian areas, and a variety of mast-producing hardwood species (Lawrence 1979).

The spruce-fir forest dominates much of the range of the black bear in the Rockies. Important nonforested areas are wet meadows, riparian area, avalanche chutes, roadsites, burns, sidehill parks, and subalpine ridgetops (Kemp 1979).

Winter Denning. Winter dormancy is one of the more interesting and important aspects of the biology of black bears. Black bears, the least predacious of the North American carnivores, circumvent food shortages and severe weather conditions by becoming dormant. They apparently exhibit a dormant period even in the southern extremities of their range in Florida and Arizona. Bears in the northern part of their range take advantage of heavy snowfall for concealment and insulation; bears in the Southeast may select cavities high above ground in large trees (Johnson and Pelton 1980c), or simply construct a crude nest on the ground in dense thickets of briars and/or low shrubs (Hamilton and Marchinton 1977). Secure dens are important because birth and early maternal care of cubs is limited to winter dens. As one progresses southward, black bears appear to be less lethargic and can be easily aroused from their dens. This may be related more to den selection and climate relationships than to differences in physiology and denning behavior, however. After increased movements and intense foraging in the fall, black bears usually return to their spring and summer home ranges to den (Garshelis and Pelton 1980). Denning is preceded and followed by a period of decreased movements and activities (Johnson and Pelton 1979). This transition into and out of denning, which may last as long as one month, may be a physiological and behavioral adaptation through which bears must adjust their digestive system to a lengthy period of quiescence. Bears may enter dens between October and early January depending on latitude, available food,

sex and age, and local weather conditions. Adult females generally den first, followed by subadults and adult males (Johnson and Pelton 1980a). Unless disturbed because of weather, people, or other animals, they tend to stay in one den until emergence, from mid-March to early May. Females with cubs usually are the last to leave the den. Foot pads may be shed during denning, and new pads formed during the post-denning recovery period. Factors that cause shedding of old pads are unknown at present (Rogers 1974). Also, on emergence, black bears generally defecate a fecal plug that has blocked the gastrointestinal tract since they entered the den in the fall.

Unique physiological adaptations of black bears to winter dormancy have resulted in a variety of classifications of this behavior. Body temperature is reduced only 7–8°C, metabolism is reduced 50–60 percent, heart rate drops from 40–50 bpm to 8–19 bpm, and a weight loss of 20–27 percent (adipose tissue only) occurs; the animals do not eat, drink, defecate, or urinate during the entire denning period (Hock 1960; Nelson et al. 1973; Folk et al. 1977). A typical mammalian hibernator reduces its body temperature, heart rate, and metabolism until body temperature is within 1°C of the ambient temperature. A regular awakening every few days to eat, drink, defecate, and urinate is also typical, but when in hibernation the animal can be handled and even removed from the den without awakening (Hock 1960; Folk et al. 1977). On the contrary, black bears can be easily aroused and will react to a disturbance (Jonkel and Cowan 1971; Folk et al. 1972).

Deviation from the norm established for "true" hibernators has resulted in such phrases as "dormancy" (Matson 1946), "ecological hibernators" (Morrison 1960), and "carnivorean lethargy" (Hock 1960) to describe the denning behavior of bears. Further physiological evidence, particularly concerning the electrocardiogram of bears (Folk et al. 1977) and their metabolic and excretory mechanisms (Lundberg et al. 1976), indicates that the "hibernator" designation applies to bears. The winter weight loss of bears (20–27 percent) versus smaller hibernators (25–30 percent) (Hock 1960; Kayser 1961) indicates that the adaptations of bears are equivalent or even superior to those of other hibernators, since a relatively high body temperature enables black bears to remain somewhat alert and care for the young in the winter den. This ability to react to disturbances is important to an animal as large as a black bear, since complete concealment in a den is not usually possible.

FOOD HABITS

Throughout their range in North America, black bears consume primarily grasses and forbs in spring, soft mast in the form of shrub and tree-borne fruits in summer, and a mixture of hard and soft mast in fall. However, the availability of different food types varies regionally. Only a small portion of their diet consists of animal matter, and then primarily in the form of colonial insects and beetles (figure 24.4). Most vertebrates

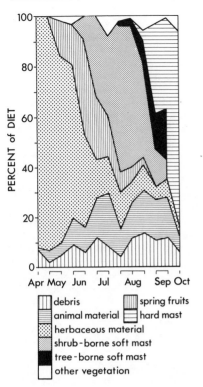

FIGURE 24.4. Typical food habits for black bears (*Ursus americanus*), illustrating the high percentage of plant materials in their diet. Bears also feed on garbage from early June until the middle of August. From Eagle and Pelton 1980.

are consumed in the form of carrion. Black bears are not active predators and only feed on vertebrates if the opportunity exists. The diet of black bears is high in carbohydrates and low in proteins and fats. Consequently, they generally prefer foods with high protein or fat content, thus their propensity for the food and garbage of people. Bears feeding on a protein-rich food source show significant weight gains and enhanced fecundity (Rogers 1976).

Spring, after the bears' emergence from winter dens, is a period of relative food scarcity. Bears tend to lose weight during this period and continue to subsist partly off of body fat stored during the preceding fall. They take advantage of any succulent and protein-rich foods available; however, these are typically not in sufficient quantity to maintain body weight.

As summer approaches, a variety of berry crops become available. Summer is generally a period of abundant and diverse foods for black bears, enabling them to recover from the energy deficits of winter and spring. Late summer and fall are critical periods for black bears. It is during this period that fat stores must be increased. Extensive foraging may occur and animals travel great distances to take advantage of available food supplies. Weight gains of more than 1 kg daily have been recorded during this period (Jonkel and Cowan 1971). However, it is also a period of greater vulnerability to mortality. The extensive movements increase the chances of traveling into unfamiliar habitats and subsequent contact with people. During years of abundant fall foods, bears tend to forage longer before returning to their spring-summer ranges and entering winter dens.

BEHAVIOR

Black bears are generally crepuscular, although breeding and feeding activities may alter this pattern seasonally (figure 24.5) (Garshelis and Pelton 1980). Where human food or garbage is available, individuals may become distinctly diurnal (on roadsides) or nocturnal (in campgrounds). Nuisance activities are usually associated with sources of artificial food and the very opportunistic feeding behaviors of black bears. Activities are depressed at temperatures above 25°C or below freezing. As with many other mammalian species, they tend to be most active after the passage of a low pressure weather front (figure 24.6) (Garshelis and Pelton 1980). During periods of inactivity, black bears utilize bed sites in forest habitat; these sites generally consist of a simple shallow depression in the forest leaf litter.

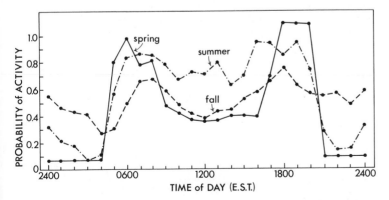

FIGURE 24.5. Seasonal differences in the daily activity patterns of black bears (*Ursus americanus*) in Great Smoky Mountains National Park. From Garshelis and Pelton 1980.

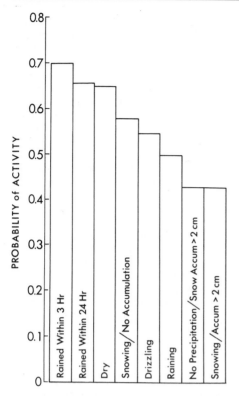

FIGURE 24.6. Relationship between precipitation and activity of black bears (*Ursus americanus*) in Great Smoky Mountains National Park. From Garshelis and Pelton 1980.

FIGURE 24.7. Mark trees are indicators of the presence of black bears (*Ursus americanus*) in an area.

Feeding activities intensify as the summer progresses. Bears tend to leave abundant sign of their presence in an area. Scats on game trails, rotten logs broken open, opened yellowjacket nests, matted paths through berry patches, and broken limbs and climb marks on a variety of mast-bearing plants are all signs of their intensified summer and fall feeding activities. The amount and ease of observation of such signs indicate their potential as an index of bear abundance or activity.

"Mark trees" are also evident in parts of a black bear's range (figure 24.7). Why bears mark objects is still open to question. Marks are usually made by biting or clawing at 1.5–2.0 m above the ground on conifer or deciduous trees (Burst and Pelton 1980). Marked trees are usually located along game trails, ridge tops, abandoned roads, or hiking trails. Although the function of marking in the natural history of black bears is unknown, the incidence of marking peaks during the breeding season (midsummer), indicating a possible relationship to breeding behavior. The ritualistic nature of the behavior, its intensity, and its defined location suggest that it may be associated with some important aspect of the social structure of the population.

Black bears are normally solitary animals except for female groups (adult female and cubs), breeding pairs in summer, and congregations at feeding sites. Territories are established by adult females during the summer (Rogers 1977). Temporal spacing is exhibited

by individuals at other times of the year and is likely maintained through a dominance hierarchy system.

The highly evolved family behavioral relationships probably are the result of the slow maturation of cubs and the high degree of learning associated with the family organization. Black bears possess a high level of intelligence and exhibit a high degree of curiosity and exploratory behaviors (Bacon and Burghart 1976; Pruitt 1976). Although black bears are generally characterized as shy and secretive animals toward humans, they exhibit a much wider array of intraspecific and interspecific behaviors than originally thought, exhibiting agonistic behavior toward conspecifics as well as people. Threat behaviors as well defined by such characteristics as ear position, body postures, and vocalizations. Tate and Pelton (1980) delineated seven types of aggression by black bears toward people; these include noncontact behaviors such as a low moan vocalization, blow vocalizations coupled with lip extensions and head oriented downward, and charge. Quadripedal and bipedal swat and snapping or biting were two types of aggressive contact behavior. During the above study, 624 aggressive acts of bears were recorded toward visitors to the Great Smoky Mountains National Park; only 37 (5.9 percent) resulted in contact. Thus, from the standpoint of actual physical contact, black bears are less aggressive than the other North American ursids. Defensively, black bears have had the security of a dense habitat in which to escape. Conversely, grizzly bears (*U. arctos*) and polar bears (*U. maritimus*) evolved on relatively open plains or tundra habitat and were forced to "stand and defend" against antagonists (Herrero 1979).

MORTALITY

Although a number of neoplastic, Rickettsial, viral, bacterial, and traumatic diseases have been reported for black bears (table 24.2), none appears to contribute greatly to the natural regulation of black bear populations. Black bears also exhibit a relatively low prevalence and diversity of parasites compared to other mammalian species. The literature reports 25 genera and 37 species of endoparasite and 8 genera and 12 species of ectoparasite (table 24.3). As in other mammals, a greater prevalence and variety of parasites have been detected in the southeastern United States, where combined higher temperatures and humidity provide a more favorable environment. In this region, 22 different species of parasite have been recovered from bears (Crum 1977).

MANAGEMENT

The development and/or improvement of four techniques during the 1960s and 1970s have enabled biologists to collect data on bear populations to an extent never before realized: (1) Aldrich spring-activated snares now make it possible for bears to be captured efficiently and in relatively inaccessible locations (Johnson and Pelton 1980b); (2) new immobilization drugs allow for the safety of the bear and the biologist (drugs such as ketamine hydrochloride [Hugie et al. 1976] or M-99 Etorphine [Beeman et al. 1974] exhibit wide tolerances and relatively short latency periods); (3) refinements in the tooth section method presently provide a high degree of accuracy in aging; (4) more sophisticated telemetry systems allow researchers to monitor movements and activities of large numbers of animals for long periods of time. Also, computer programs are now available that are capable of efficiently handling the large amounts of data generated through biotelemetry. All of the above techniques provide the necessary means by which wildlife biologists can piece together comprehensive

TABLE 24.2. Major diseases of black bears

Neoplastic
  Liposarcoma and unidentified tumors
  (Rausch 1961; King et al. 1960)
Rickettsial
  Elokomin fluke fever (Farrell et al. 1973)
Viral
  Rabies (reported in Ursidae) (Schoening 1956)
Bacterial
  Bronchiectasis (King et al. 1960)
  Bronchopneumonia (Rausch 1961)
  Dental caries (Rausch 1961)
  Osteomyelitis (Rausch 1961)
  Peridontal disease (Erickson 1965)

TABLE 24.3. Parasites of black bears

Endoparasite
  Protoza
    3 genera and 4 species
  Trematoda
    1 genus and species (*Nanophyetus salmincola*)
  Cestoda
    4 genera and 10 species
  Acanthocephala
    1 genus and species (*Macracanthorhynchus ingens*)
  Nematoda
    16 genera and 20 species
Ectoparasite
  Mallophaga
    1 genus and species (*Trichodectes pinguis*)
  Siphonaptera
    2 genera and 3 species
  Acarina
    Ixoides
      3 genera and 6 species
    Sarcoptiformes
      2 genera and 2 species

SOURCE: Summarized from Hamilton 1978.

life history information to develop better management plans.

Steps in management are as varied as the status of the species, which ranges from pest to threatened in North America. Where there are large expanses of forested area and relatively sparse human populations, such as in Canada and the northwestern United States, harvest regulations are very liberal, sometimes amounting to a nearly year-round season. However, small residual enclaves may receive total protection or a very restricted harvest. To date, management of the species has consisted of regulations on harvesting in order to sustain populations at certain levels. To achieve this, hunting and/or trapping seasons, bag limits, and the relative amount of protection are adjusted in response to the assessed status of these populations based on harvest trends and/or research results. For example, an adjustment that has been made by many resource agencies in recent years is to move the bear hunting season back to late fall. This adjustment was made after telemetry studies revealed that females were less vulnerable to hunting the closer it was to the denning period (Johnson and Pelton 1979). In some cases the amount of protection is mandated by the land rights where the bear population exists. For instance, due to the policy of nonhunting in United States national parks, bears are afforded complete protection under the law.

Wherever the species exists, if allowed to do so, it has a tendency to adapt to the presence of people. Management strategies designed to deal with human-bear interactions are relatively new. The comparatively high intelligence of the species and the emotions it evokes in people combine to present a singular dilemma for responsible resource agencies. Regulations dealing with separating people and their food or garbage from bears are often difficult to enforce, particularly where high densities of both may interact. Transplant efforts offer only a partial solution because a high percentage of the transplants result in bears either returning to their original capture area or being killed soon after release in unfamiliar habitat. A homing success rate greater than 50 percent was achieved by translocated bears in Yosemite National Park (Harms 1977). Beeman and Pelton (1976) found similar homing success rates in the Great Smoky Mountains National Park. This success rate was found to be negatively correlated with the distance bears were moved (that is, the further the animal was moved the less chance there was that it would return). Furthermore, intensive trapping in areas near release sites has resulted in very few recaptures of the relocated bears. This suggests that a large portion of the bears that are not successful in homing are likely to succumb to some form of mortality. However, most managers still prefer this method to the more drastic option of destroying problem bears.

Much hope has been placed in the future use of aversive stimuli to deal with problem bears rather than shooting or translocating animals. Aversive conditioning employs any of a number of materials that will evoke a negative response in the animal when it is involved in conflict with humans. For instance, lithium chloride, a strong emetic, placed in a specific food may make a bear sick and keep it from eating that particular food again. Shock from an electric fence may keep a bear from trying to get to beehives. However, neither of the above examples has been completely effective in separating bears from human food or garbage. Research is being carried out to assess the effectiveness of these and other aversive stimuli in dealing with human-bear conflicts. Ultimately, however, the most effective management is aimed at the human side of the problem: making unnatural food sources unattainable to the black bear. Given this species' strength and intelligence, this is a formidable task, but it offers a better long-range solution to a persistent problem.

CURRENT RESEARCH NEEDS

Although the development of new techniques and refinements of existing ones have allowed researchers to accumulate much valuable information on black bear populations, there are still many significant gaps in our knowledge about this species. We are only beginning to understand the social structure of bear populations, the mechanisms controlling certain behaviors, and the implications of such behaviors on population regulation. We are aware only in general terms of the possible factors affecting natality and mortality. Through intensive field efforts using radiotelemetry, enough long-term data are accumulating on a series of individual animals to calculate age-specific natality and mortality rates. The methods of censusing currently used by agencies responsible for managing black bears are crude at best. Inexpensive, precise, and accurate ways of censusing bears on a regular basis are unknown. As future pressures demand that we know with greater certainty the number of animals per unit area that we are trying to manage, we will have to intensify

our efforts toward developing more reliable population estimators. If bears and people are to coexist in the future, we must also develop ways to decrease or eliminate the negative interactions that occur. We know very little about the dynamics of that segment of a bear population that interacts with people, or human food and/or garbage. More research is needed in the area of proper handling of translocated bears, better techniques for dealing with troublesome bears, and especially more innovative methods of separating bears from sources of human food and/or garbage. Land-use practices will ultimately decide the fate of black bears throughout their range. Information on black bear populations must be integrated with land-use agencies (for example, U.S.F.S., B.L.M., N.P.S., private companies) to insure survival of the species.

The black bear resource is in competition with other natural resources such as timber, mining, and commercial development. Black bears have important economic, esthetic, recreational, social, and scientific values. Black bear management must be based on sound research so this resource can compete with other demands on the land.

LITERATURE CITED

Alt, G. L.; Alt, F. W.; and Lindzey, J. S. 1976. Home range and activity patterns of black bears in northeastern Pennsylvania. Trans. Northeast Sect. Wildl. Soc. 33:45–56.

Amstrup, S. C., and Beecham, J. 1976. Activity patterns of radiocollared black bears to Idaho. J. Wildl. Manage. 40:340–348.

Bacon, E. S., and Burghardt, G. M. 1976. Learning and color discrimination in the American black bear. Pages 27–36 in M. R. Pelton, J. W. Lentfer, and G. E. Folk, Jr., eds. Bears: their biology and management. 3rd Int. Conf. Bear Res. and Manage. IUCN Publ. New Series no. 40. 467pp.

Barnes, V. G., and Bray, O. E. 1967. Population characteristics and activities of black bears in Yellowstone National Park. Colorado Coop. Wildl. Res. Unit and Colo. Colorado State Univ., Fort Collins. 199pp.

Beeman, L. E., and Pelton, M. R. 1976. Homing of black bears in the Great Smoky Mountains National Park. Pages 87–95 in M. R. Pelton, J. W. Lentfer, and G. E. Folk, Jr., eds. Bears: their biology and management. 3rd Int. Conf. Bear Res. and Manage. IUCN Publ. New Series no. 40. 467pp.

Beeman, L. E.; Pelton, M. R.; and Marcum, L. C. 1974. Use of M-99 Etorphine for immobilizing black bears. J. Wildl. Manage. 38(3):568–569.

Burst, T. L., and Pelton, M. R. 1980. Black bear mark trees in the Smoky Mountains. 5th Int. Conf. Bear Res. and Manage., Madison, Wis. in press.

Carpenter, M. 1973. The black bear in Virginia. Virginia Comm. Game and Inland Fish, Richmond. 22pp.

Cowan, I. McT. 1972. The status and conservation of bears. (Ursidae) of the world, 1970. Pages 343–367 in S. Herrero, ed. Bears: their biology and management. Int. Conf. Bear Res. and Manage. IUCN Publ. New Series no. 23. 371pp.

Crum, J. M. 1977. Some parasites of black bears (Ursus americanus) in the southeastern United States. M.S. Thesis. Univ. Georgia, Athens. 76pp.

Eagar, D. C. 1977. Radioisotope feces tagging as a population estimator of black bear (*Ursus americanus*) density in the Great Smoky Mountains National Park. M.S. Thesis. Univ. Tennessee, Knoxville. 89pp.

Edwards, W. R., and Eberhardt, L. L. 1967. Estimating cottontail abundance from live trapping data. J. Wildl. Manage. 33:87–96.

Erickson, A. W. 1965. The black bear in Alaska: its ecology and management. Alaska Dept. Fish and Game. Fed. Aid in Wildl. Restor. Dept. Prog. W-6-R-5, Work Plan F. 19pp.

Erickson, A. W., and Nellor, J. E. 1964. Breeding biology of the black bear. Part 1, pages 1–45 *in* A. W. Erickson, J. Nellor, and G. A. Petrides. The black bear in Michigan. Michigan State Agric. Exp. Stn. Res. Bull. 4. 102pp.

Erickson, A. W., and Petrides, G. A. 1964. Population structure, movements, and mortality of tagged black bears in Michigan. Part 2, pages 46–67, *in* A. W. Erickson, J. Nellor, and G. A. Petrides. The black bear in Michigan. Michigan State Agric. Exp. Stn. Res. Bull. 4. 102pp.

Farrell, R. K.; Leader, R. W.; and Johnston, S. O. 1973. Differentiation of salmon poisoning disease and Elokomin fluke fever: studies with the black bear (*Ursus americanus*). Am. J. Vet. Res. 34(7):919–922.

Folk, G. E.; Folk, M. A.; and Minor, J. J. 1972. Physiological condition of three species of bears in winter dens. Pages 107–124 *in* S. Herrero, ed. Bears: their biology and management. IUCN New Ser. Publ. 23. Int. Union for Conserv. Nature and Nat. Resour., Morges, Switzerland. 371pp.

Folk, G. E.; Hunt, J. M.; and Folk, M. A. 1977. Hibernating bears: further evidence and bioenergetics. Proc. 4th Int. Conf. Bear Res. and Manage., Kalispell, Mont. in press.

Garshelis, D. L. 1978. Movement, ecology and activity behavior of black bears in the Great Smoky Mountains National Park. M.S. Thesis. Univ. Tennessee, Knoxville. 117pp.

Garshelis, D. L., and Pelton, M. R. 1980. Activity of black bears in the Great Smoky Mountains National Park. J. Mammal. 61:8–19.

Hall, E. R. 1981. The mammals of North America. 2nd ed. 2 vols. John Wiley and Sons, New York. 1181pp.

Hamilton, R. J. 1978. Ecology of the black bear in southeastern North Carolina. M.S. Thesis. Univ. Georgia, Athens. 214pp.

Hamilton, R. J., and Marchinton, R. L. 1977. Denning activity of black bears in the coastal plain of North Carolina. Proc. 4th Int. Conf. Bear Res. and Manage., Kalispell, Mont. in press.

Harms, D. 1977. Black bear management in Yosemite National Park. Proc. 4th Int. Conf. Bear Res. and Manage., Kalispell, Mont. in press.

Hayne, D. W. 1949. An examination of the strip census method for estimating animal populations. J. Wildl. Manage. 13:145–157.

Herrero, S. M. 1979. Black bears: the grizzly's replacement. Pages 179–195 *in* D. Burk, ed. The black bear in modern North America. Boone and Crockett Club. Amwell Press, Clinton, N.Y. 299pp.

Hock, R. J. 1960. Seasonal variations in physiologic functions of arctic ground squirrels and black bears. Harvard Univ. Mus. Comp. Zool. Bull. 124:155–173.

Hornocker, M. G. 1962. Population characteristics and social reproductive behavior of the grizzly bear in Yellowstone National Park. M.S. Thesis. Univ. Montana, Missoula. 94pp.

Hugie, R. H. 1974. Habitat of the black bear in Maine. Pages 151–157 *in* M. R. Pelton and R. H. Conley, eds. Second eastern workshop on black bear management and research. Gatlinburg, Tenn. 242pp.

Hugie, R. H.; Landry, J.; and Hermes, J. 1976. The use of Ketamine hydrochloride as an anesthetic in black bears in Maine. Trans. Northeast Fish and Wildl. Conf. 33:83–86.

Johnson, K. G., and Pelton, M. R. 1979. Denning behavior of black bears in the Great Smoky Mountains National Park. Annu. Conf. Southeast Fish and Wildl. Agencies. 33. in press.

———. 1980*a*. Environmental relationships and the denning period of black bears in Tennessee. J. Mammal. 61. in press.

———. 1980*b*. Prebaiting and snaring techniques for black bears. Wildl. Soc. Bull. 8:46–54.

———. 1980*c*. Selection and availability of dens for black bears in Tennessee. J. Wildl. Manage. 44. in press.

Jonkel, C. J., and Cowan, I. McT. 1971. The black bear in the spruce-fir forest. Wildl. Monogr. 27:1–57.

Kayser, C. 1961. The physiology of natural hibernation. Pergammon Press., New York. 325pp.

Kemp, G. A. 1972. Black bear population dynamics at Cold Lake, Alberta, 1968–1970. Pages 26–31 *in* S. Herrero, ed. Bears: their biology and management. Int. Conf. Bear Res. and Manage. IUCN Publ. no. 23. 371pp.

———. 1976. The dynamics and regulation of black bear populations in Northern Alberta. Pages 191–197 *in* M. R. Pelton, J. W. Lentfer, and G. E. Folk, Jr., eds. Bears: their biology and management. 3rd Internat. Conf. on Bear Res. and Manage. IUCN Publ. New Series no. 40. 467pp.

———. 1979. The Rocky Mountain working group. Pages 217–236 *in* D. Burk, ed. The black bear in modern North America. Boone and Crockett Club. Amwell Press, Clinton, N.Y. 299pp.

King, J. M.; Black, H. C.; and Hewitt, O. M. 1960. Pathology, parasitology, and hematology of the black bear in New York. New York Fish Game J. 7(2):99–111.

Knudsen, G. J. 1961. We learn about bears. Wisconsin Conserv. Bull. 27:13–15.

Lawrence, W. 1979. Working group reports: Pacific working group: habitat management and land use practices. Part 3, pages 196–201 *in* D. Burk, ed. The black bear in modern North America. Boone and Crockett Club. Amwell Press, Clinton, N.Y. 299pp.

Lindzey, F. G., and Meslow, E. C. 1977. Home range and habitat use by black bears in southwestern Washington. J. Wildl. Manage. 41:413–425.

Lundberg, D. A.; Nelson, R. A.; Wahner, H. W.; and Jones, J. P. 1976. Protein metabolism in the black bear before and during hibernation. Mayo Clinic Proc. 51:716–722.

Marcum, L. C. 1974. An evaluation of radioactive feces-tagging as a technique for determining population densities of the black bear in the Great Smoky Mountains National Park. M.S. Thesis. Univ. Tennessee, Knoxville. 95pp.

Matson, J. R. 1946. Notes on the dormancy of the black bear. J. Mammal. 27:203–212.

Matthews, S. J. K. 1977. The seasonal, altitudinal, and vegetational incidence of black bear scats in the Great Smoky Mountains National Park. M.S. Thesis. Univ. Tennessee, Knoxville. 89pp.

Morrison, P. 1960. Some interrelations between weight and hibernation function. Harvard Univ. Mus. Comp. Zool. Bull. 124:75–91.

Nelson, R. A.; Wahner, H. W.; Jones, J. D.; Ellefson, R. D.; and Zollman, P. E. 1973. Metabolism of bears before, during, and after winter sleep. Am. J. Physiol. 224:491–496.

Pelton, M. R. 1972. Use of foot trail travellers in the Great Smoky Mountains National Park to estimate black bear

(*Ursus americanus*) activity. Pages 36–43 *in* S. Herrero, ed. Bears: their biology and management. Int. Conf. Bear Res. and Manage. IUCN Publ. New Series 23. 371pp.

Pelton, M. R., and Marcum, L. C. 1977. The potential use of radio-isotopes for determining densities of black bears and other carnivores. Pages 221–236 *in* R. L. Phillips and C. Jonkel, eds. Proc. 1975. Predator Symp., Montana For. and Conserv. Exp. Stn., Univ. Montana, Missoula. 268pp.

Piekielek, W., and Burton, T. S. 1975. A black bear population study in northern California. California Fish Game J. 61:4–25.

Poelker, R. J., and Hartwell, H. D. 1973. Black bear of Washington. Washington State Game Dept. Bio. Bull. 18. 180pp.

Pruitt, C. H. 1976. Play and agonistic behavior in captive black bears. Pages 79–86 *in* M. R. Pelton, J. W. Lentfer, and G. E. Folk, Jr., eds. Bears: their biology and management. 3rd Int. Conf. Bear Res. and Manage. IUCN Publ. New Series no. 40. 467pp.

Rausch, R. L. 1961. Notes on the black bear in Alaska, with particular reference to dentition and growth. Z. Saugertierk. 26:65–128.

Rogers, L. L. 1974. Shedding of foot pads by black bears during denning. J. Mammal. 55:672–674.

———. 1976. Effects of mast and berry crop failures on survival, growth, and reproductive success of black bears. Trans North Am. Wildl. and Nat. Resour. Conf. 41:432–438.

———. 1977. Movements and social relationships of black bears in northeastern Minnesota. Ph.D. Dissertation. Univ. Minnesota, St. Paul. 194pp.

Sauer, P. R.; Free, S. L.; and Browne, S. D. 1969. Movement of tagged black bears in the Adirondacks. New York Fish Game J. 16:205–223.

Schoening, H. W. 1956. Rabies. Pages 195–202 *in* A. Stefferud, ed. Animal diseases: U.S.D.A. yearbook. Gov. Printing Office, Washington, D.C. 591pp.

Spencer, H. E. 1955. The black bear and its status in Maine. Maine Dept. Inland Fish and Game, Game Div. Bull. no. 4. 55pp.

Tate, J., and Pelton, M. R. 1980. Human-bear interactions in the Smoky Mountains: focus on ursid aggression. 5th Int. Conf. Bear Res. and Manage., Madison, Wis. in press.

Troyer, W. A., and Hensel, R. J. 1964. Structure and distribution of a Kodiak bear population. J. Wildl. Manage. 28:769–772.

Waddell, T. 1979. State and provincial status reports: Arizona. Pages 33–37 *in* A. LeCount, ed. First western black bear workshop. Tempe, Ariz. 339pp.

Willey, C. H. 1974. Aging black bears from first premolar tooth sections. J. Wildl. Manage. 38:97–100.

Wimsatt, W. A. 1963. Delayed implantation in the ursidae, with particular reference to the black bear. Pages 49–76 *in* A. C. Ender, ed. Delayed implantation. Univ. Chicago Press, Chicago. 316pp.

MICHAEL R. PELTON, Department of Forestry, Wildlife, and Fisheries, University of Tennessee, Knoxville, Tennessee 37916.

25

Grizzly Bear

Ursus arctos

John J. Craighead
John A. Mitchell

COMMON NAMES. Grizzly bear, grisly bear, range bear, roach-back, smut-face, griz, Old Ephraim, Moccasin Joe, great white bear, silvertip, white bear
SCIENTIFIC NAME. *Ursus arctos*

The species is distributed widely throughout the Palearctic and Nearctic across a variety of habitats. Local variations in body size, skull structure, pelage color, and other morphological characteristics were utilized by early taxonomists as specific and subspecific classification criteria. This resulted in early taxonomic schemes that have defied accurate interpretation. The most noteworthy early effort to classify the brown and grizzly bears of North America was that of C. Hart Merriam. His work, produced over a period of about 20 years, culminated in a comprehensive taxonomy of brown and grizzly bears embracing 87 different species in North America alone (Merriam 1918). Although accepted as authoritative by Hall and Kelson (1955), Merriam's classification has been largely discarded in favor of the single holarctic species concept established by the works of Couterier (1954), Rausch (1953, 1963), and Kurtén (1968).

SUBSPECIES. Rausch (1953, 1963), on the basis of skull structure, body size, and coloration, suggested that *Ursus arctos* on the North American continent and its adjacent islands is comprised of three subspecies: *U. a. horribilis* Ord, to include all brown and grizzly bears of continental North America; *U. a. middendorffi* Merriam, to include brown bears of the Alaskan Islands of Kodiak, Afognak, and Shuyak; and *U. a. gyas* Merriam, to include brown bears confined to the Alaskan peninsula.

Although *U. a. gyas* is no longer considered a distinct subspecies (Rausch 1963), the taxa *U. a. horribilis* and *U. a. middendorffi* are recognized by most current workers. The grizzly bear is considered a genetically strong variant of the classical brown bear phenotype of *U. a. horribilis*.

The family Ursidae originated in Europe early in the Miocene epoch as a derivative of the Miacidae, a family of small, carnivorous, tree-climbing mammals (Simpson 1945). Subsequent phylogenetic develop-

FIGURE 25.1. Present and past distribution of the grizzly bear (*Ursus arctos*). After Rausch 1963.

ment of the ursids has been well documented in the works of Thenius (1959) and Kurtén (1968). A thorough review of bear evolution that relates environmental selective pressures in postglacial North America with behavioral, ecological, morphological, and physiological adaptations that, today, constitute distinct differences between black bears and grizzly/brown bears was done by Herrero (1972). Also, Martinka (1976) briefly reviewed the phylogeny of bears.

At least three distinct evolutionary lines emerged from the earliest ursid progenitors. Of these, only one was of major importance in the origin of modern day bears. Divergence of this major evolutionary line during the early Pliocene gave rise to forms considered representative of the two extant genera of bears, *Ursus* and *Tremarctos*.

The Auvergne bear, *Ursus minimus* Devèze and Bouillet, has been identified from remains in Europe dating to the latter phases of the Pliocene some 4 to 6 million years ago. Among the most primitive of fossil *Ursus* spp., it was relatively small and similar in structure to the modern Asiatic black bear. The bear was a forest dweller and, despite persisting relatively unchanged into the Pleistocene, provided the early progenitors to the Etruscan bear, *Ursus etruscus* Cuvier.

The Etruscan bear was well established in Europe and Asia by the early Pleistocene 2 to 3 million years ago. The repeated cycles of continental glaciation that shaped the Northern Hemisphere throughout the time of the mid-Pleistocene provided selective pressures that spurred adaptive radiation in the Etruscan bear population nucleus. All extant species of *Ursus* had been derived by the late Pleistocene, with the polar bear, *Ursus maritimus* (*Thalarctos maritimus*), being most recently derived as an offshoot of the basic grizzly bear stock (*U. arctos*).

The grizzly/brown bear group, the black bear group, and the great cave bears (*U. spelaeus*) all were derived from Etruscan bear stock in Asia. The cave bears of Europe were extinct by recent times. The black bear group, now represented in Asia by *U. thiebetanus*, had radiated via the Aleutian land bridge into North America by the middle of the Pleistocene to provide the black bear (*U. americanus*) lineage. The grizzly/brown bear group did not cross into North America until the end of the Pleistocene epoch and appears to have been confined to the northern reaches of Alaska until the withdrawal of the continental ice flows. *Ursus arctos* then expanded its range southward to inhabit land area from the northern limits of North America well into Mexico.

DISTRIBUTION

Historical Range. From early records and paleontological finds, it is clear that *Ursus arctos horribilis* was once native to a far more extensive area of North America than it now inhabits (Roosevelt 1907; Wright 1909; Dobie 1950; Storer and Tevis 1955; Stebler 1972; Schneider 1977). The historical distribution of the bear was best summarized by Rausch (1963), although many other workers provided similar descriptions compiled from the literature (figure 25.1). It should be noted that, according to current documentation, the historical distribution described by Rausch is more accurate than that reported in Hall and Kelson (1959). However, discoveries of skulls in southern Ontario (Peterson 1965) and on the northern coast of Labrador (Spiess 1976; Spiess and Cox 1977) suggest that the range historically may have extended across the breadth of North America (figure 25.1). Guilday (1968) documented the presence of *U. a. horribilis* in the vicinity of what is now Ohio and Kentucky.

From the beginning of the European invasion of North America, the continental range of the grizzly bear receded, especially from the south and east. Bears were killed out of fear, for food, or to protect livestock. The natural habitat over which grizzlies ranged widely was often eliminated. The early and rapid extinction of populations from most of Mexico and from the central and southwestern United States and California suggests that many were weak populations dynamically, and of marginal status in the community structure.

Current Range. Grizzly bears are present, and even common, throughout much of the current range (figure 25.1). Populations continue to thrive in the remote, largely unsettled areas of Alaska and northwestern Canada (Pearson 1972, 1975, 1976; Reynolds 1979; Reynolds et al. 1976; Hamer et al. 1977, 1978, 1979). Within the contiguous 48 states populations are more sporadic, particularly at the western extremes of the range. Populations in the Yellowstone ecosystem have declined in recent years (Craighead et al. 1974; Craighead 1980b); various population units in the vicinity of the continental divide and north through Montana appear to be stable or declining slowly. Sightings of grizzlies in the Selway-Bitterroot drainages separating Idaho and Montana indicate only the presence of isolated animals. Layser (1978) presented evidence that grizzlies survive, although probably marginally, in the Selkirk Mountains of northern Idaho and northeastern Washington. A small, but probably viable, population inhabits the Cabinet Range, the Yaak River area, and adjoining forests in eastern Idaho and northwestern Montana. There is also a remote possibility that a small remnant population survives far to the south in the San Juan Wilderness of Colorado.

Because the grizzly bear requires large areas for its natural ranging habits, continued human competition will undoubtedly further reduce its range. It is important to note that much of the habitat that supported the grizzly bear in its historic range still exists and that the range of the animal could easily be extended by transplantation. The technology necessary for range extension is available, but the socioeconomic and sociopolitical conditions necessary for support of such a venture are not.

DESCRIPTION

A considerable mystique has long been associated with the grizzly bear. This, combined with the confused state of early taxonomic distinctions, provided for many early descriptions of the bear of only incidental scientific value (Allen 1814; James 1823; Fremont 1843; Coues 1893; Roosevelt 1907; Wright 1909; Mills 1919; Holzworth 1930; Dobie 1950; Hubbard 1960; Haynes and Haynes 1966). Observations on the morphology were superficial, subjective, and often sensationalized. Most natural history and behavioral descriptions were oriented toward hunting techniques, Indian mythology, or the bear's vaunted aggressive tendencies. Nevertheless, the early writings concerning the grizzly bear are enjoyable reading and have contributed to popular legends that endure even today.

Although well known to the American Indians and most probably encountered by the 1540 Coronado expedition to the seven cities of Cibola (now west-central New Mexico), the first known record of the grizzly bear is that of Sebastian Vizcaino (Storer and Tevis 1955). In 1602 while camped at the site of

Monterey, he observed bears feeding on the carcass of a beached whale. Because black bears were not native to that area, these could only have been grizzlies. The next record of a grizzly bear was that of Henry Kelsey, an Englishman employed by the Hudson Bay Company in Canada, when he wrote on 20 August 1691 of encountering a "silver hair'd" bear (Schneider 1977). Although bears undoubtedly were observed by other travelers moving west of the Mississippi River, it was not until the mounting of the Lewis and Clark expedition more than a century later that data of some value were collected. The first type-specimen was collected, as were occasional feet, claws, teeth, and skulls. Measurements and general morphological descriptions of specimens killed or observed were often recorded. The explorers killed at least 43 grizzly bears (Burroughs 1961), a number that journals of the expedition do not correlate clearly with a need for food (Allen 1814; Coues 1893; DeVoto 1953).

George Ord, credited with the first scientific naming of the grizzly bear (*Ursus horribilis* Ord) in Guthrie (1815), actually had little first-hand knowledge of the bears (Storer and Tevis 1955). His species description was obtained indirectly from information of the Lewis and Clark expedition published by Brackenridge (1814).

The earliest scientific descriptions of the grizzly bear based on adequate specimen numbers were those of Swainson, Baird, and Elliot from the arctic, western United States, and British Columbia, respectively (Storer and Tevis 1955). C. Hart Merriam (1918) was the next important contributor to a scientific description of the grizzly bear. As discussed earlier, his efforts were comprehensive, but eventually were of limited systematic value.

Description of the grizzly bear must be approached with the understanding that it is a genetic variant within a subspecies with the large brown bears. Moreover, any description must account for considerable variations in size, color, and morphology between populations.

General Morphology and Structure. Members of *Ursus arctos horribilis* are larger and more heavily built than most other ursids, with relatively short tails and ears, and with the four limbs of approximately equal length tapering to large feet structured for plantigrade locomotion. Features that distinguish the subspecies include a large hump of muscle overlying the scapulae, unusually long foreclaws, characteristic skull and dental structure, and, at least in some specimens, the color and appearance of the pelage (figure 25.2).

The feet of the grizzly bear are cushioned by heavy plantar and digital pads of fibrous connective tissue covered by cornified epidermis (Storer and Tevis 1955; Ewer 1973). The major plantar pad of the forefoot is somewhat rectangular and is wider than it is long. The distal extremity of the pisiformis serves as the "heel" of the forefoot and is capped by an oval pad. Each of the five digits of the forefoot also has a small oval pad. The plantar surface of the hindfoot is comprised of a single, triangular pad that extends posteriad over the calcaneus to form the heel. A small oval pad surfaces each of the five digits. While there are minor differences in pad conformations (Wright 1909; Holzworth 1930), the feet of grizzly bears do not differ substantially from those of black bears except for being larger and lacking interpedal hair. The claws, however, differ considerably and are the feature that often permits distinction of bear tracks.

The foreclaws of grizzly bears are heavier, longer, broader, and straighter than those of black bears. Measurements along the external curvature of claws from four grizzly bear skins yielded value extremes for the claws of the forefeet ranging from 62 to 83 mm and for claws of the hindfeet ranging from 25 to 59 mm (Storer and Tevis 1955). Foreclaws of black bears seldom exceed 51 mm in length and usually do not produce track markings, as those of grizzlies routinely do.

The pelage of grizzly bears consists of an underfur of very fine hairs overlaid with coarse, long guard hairs that are more densely distributed in some bodily regions than in others. Often the bears have a full, thick mane, or roach, of guard hairs from the skull to the shoulders. Vibrissae appear to be present only vestigially. Most of the underfur is shed during the late spring and summer and replaced between August and October, depending on the climate of the locale (Holzworth 1930; Ewer 1973). Scholander et al. (1950) found that grizzly bear winter fur was an excellent insulator. When compared with the winter furs of seven other North American mammals, its insulating capacity was exceeded only by the pelts of the arctic fox (*Alopex lagopus*) and timber wolf (*Canis lupus*).

The colors of the grizzly bear pelage are extremely variable, but not so much, perhaps, as they were prior to the virtual extinction of the animal south of Yellowstone Park, Wyoming. The journals of the Lewis and Clark expedition (Allen 1814), as well as many other sources (Roosevelt 1907; Wright 1909; Holzworth 1930; Dobie 1950; Storer and Tevis 1955), note specimens with pelages of white, black, gray, or various shades of brown, tan, yellow, cream, or red. These specimens also often had the silvering or "frosting" of the guard hairs characteristic of the "grizzly" grizzly bear. In general, grizzly bears are colored a dark to blondish brown and may have silver- or blond-tipped guard hairs. Some specimens may even appear to be broadly striped dorsally or laterally. Color appears to be partially related to age and to annual replacement of pelage; old males are normally dark brown to red brown with few silver-tipped guard hairs. As in many large mammals, the new pelage is darker and richer in color than the old pelage. Some cubs and yearlings may exhibit a white or cream neck V-patch that disappears with age.

The axial and appendicular skeleton of the grizzly bear is similar to that of the black bear except, perhaps, that the hind legs are longer relative to the forelegs in the grizzly. The bone structure of both species is relatively massive and there is no fusion of leg bones as in

FIGURE 25.2. Female grizzly bear (*Ursus arctos*) and two one-year-old offspring.

some other carnivores. An interesting anatomical adaptation for climbing is the large flangelike postscapular fossae on the upper part of the posterior scapular margins (Ewer 1973). The subscapulares minor arise from the lateral and mesial surfaces of these flanges and insert on the heads of the humeri. These muscles play a direct role in resisting the pull of the humeri away from their glenoid articulations when a bear pulls its body up by its forelimbs. Despite having evolved foreclaws adapted to digging rather than to climbing, the grizzly bear retains this skeletomuscular arrangement in common with the black bear.

Size and Weight. Much of the early American natural history and adventure literature already cited includes reports of bears of gargantuan proportions. Some may have been Alaskan brown bears, while others may have been subjectively inflated by the observer. Few reliable records of grizzly bear measurements from specimens taken prior to the 20th century survive. It has been suggested that the race now extinct in California (designated *Ursus arctos californicus* currently and *U. magister* by Merriam 1918) was somewhat larger than other races south of Alaska (Storer and Tevis 1955). Leopold (1959) described a race of brown bears from the mountains of northern Mexico that, according to Rausch (1963), qualifies as the smallest form of *U. a. horribilis* in North America. In general, size and weight of adult grizzly bears are highly variable between populations (Rausch 1953, 1963). Within populations, adults are found to vary over a wide range of dimensions and weight relative to genotypic expression, gender, age, circumstances of habitat, and season of year.

Commonly referenced sources of information on mammals do not agree on any range of measurements for grizzly bears. In part, this is because of the paucity of verifiable data and the failure of systematics adequately to define the heterogeneity of the species. Weights and dimensions obtained from animals kept in captivity usually are not representative of their free-ranging counterparts. Such animals commonly are relatively inactive and obese.

Burt and Grossenheider (1964) list the following parameters for the grizzly bear: length of head and body 1.8–2.2 m, height at shoulders 0.9–1.1 m, weight 147–386 kg. Walker et al. (1964) refer to a maximum weight of almost 363 kg and a maximum length of more than 2.5 m. Storer and Tevis (1955) included reports for California grizzlies, many of dubious reliability, estimating total length to be 3.2 m and weight in excess of 544 kg.

Reliable data on measurements of grizzly bears exist for populations in the Brooks Range of Alaska (Rausch 1963), the Yukon Territory (Pearson 1975), and the Yellowstone ecosystem (Craighead and Craighead 1973b). Weights of large adult animals from the Yellowstone ecosystem ranged from 158 to

TABLE 25.1. Mean weights and measurements of grizzly bears

Sex and Age	Weight	Foot Dimensions (cm) Right Front	Left Front	Right Rear	Left Rear	Length (cm)	Claw Base (cm) Left	Right	Neck Circumference (cm)	Interocular Length (cm)	Nose to eye Length (cm)	Ear Length (cm)
Males												
under 9 mos.	(34) 31.6	(31) 7.9 × 8.8	(30) 7.8 × 8.8	(30) 14.4 × 8.3	(31) 14.2 × 8.2	(29) 99.3	(23) 1.43	(23) 1.44	(27) 38.1	(30) 6.5	(30) 10.0	(28) 8.9
17–21 mos.	(39) 68.0	(37) 10.6 × 11.5	(22) 10.1 × 11.5	(22) 19.0 × 11.0	(22) 18.9 × 10.7	(22) 131.8	(18) 1.82	(18) 1.85	(19) 51.2	(19) 7.9	(20) 13.4	(18) 11.4
29–33 mos.	(16) 110.8	(14) 11.7 × 12.8	(14) 11.7 × 13.0	(14) 21.3 × 12.1	(14) 21.3 × 12.2	(12) 159.0	(8) 1.84	(8) 1.85	(13) 61.9	(13) 8.7	(13) 15.2	(13) 12.2
41–45 mos.	(16) 124.6	(10) 12.1 × 13.8	(10) 11.9 × 13.7	(10) 23.0 × 13.0	(10) 22.7 × 13.1	(10) 165.2	(6) 2.04	(6) 2.07	(8) 65.9	(9) 9.6	(9) 16.6	(9) 11.9
53–57 mos.	(5) 152.7	(5) 12.0 × 14.0	(5) 12.4 × 13.9	(5) 23.8 × 12.9	(5) 23.0 × 12.8	(5) 177.2	(5) 2.11	(5) 2.12	(5) 69.3	(5) 9.3	(5) 16.9	(5) 12.5
65 + mos.	(33) 245.0	(25) 13.8 × 15.4	(25) 13.8 × 15.1	(28) 25.0 × 14.6	(28) 24.7 × 14.5	(24) 195.4	(15) 2.29	(15) 2.31	(19) 85.3	(18) 11.0	(19) 17.7	(19) 12.8
Females												
under 9 mos.	(17) 26.9	(17) 8.0 × 8.5	(17) 7.8 × 8.4	(17) 13.8 × 8.0	(17) 13.9 × 8.0	(16) 94.7	(15) 1.40	(14) 1.35	(14) 38.0	(15) 6.4	(15) 9.6	(15) 9.1
17–21 mos.	(19) 57.6	(18) 9.8 × 10.7	(18) 10.0 × 10.6	(18) 18.2 × 10.1	(18) 18.0 × 10.0	(15) 128.1	(13) 1.74	(13) 1.74	(15) 49.9	(15) 7.5	(15) 12.6	(15) 10.7
29–33 mos.	(22) 83.8	(19) 11.1 × 11.7	(19) 11.1 × 11.6	(19) 19.8 × 11.2	(19) 19.8 × 11.2	(16) 144.6	(11) 1.75	(12) 1.73	(13) 57.4	(13) 8.3	(14) 13.6	(13) 10.9
41–45 mos.	(7) 125.4	(6) 11.4 × 12.5	(6) 11.0 × 12.8	(6) 21.0 × 11.9	(6) 20.8 × 11.7	(6) 162.1	(5) 1.90	(5) 1.94	(6) 63.5	(6) 8.7	(6) 15.1	(6) 10.4
53–57 mos.	(4) 132.4	(4) 11.4 × 12.8	(4) 11.4 × 12.8	(4) 20.5 × 12.2	(4) 20.3 × 11.6	(4) 166.4	(3) 1.88	(3) 1.85	(4) 66.7	(4) 9.0	(4) 15.1	(4) 12.0
65 + mos.	(72) 152.0	(42) 11.8 × 13.0	(42) 11.7 × 12.9	(45) 21.2 × 12.0	(45) 21.3 × 12.0	(41) 172.2	(19) 2.05	(19) 2.05	(34) 69.4	(35) 9.2	(36) 15.6	(34) 12.9

NOTE: The size of the sample for each category is given in parentheses.

204 kg for females, and from 363 to 500 kg for males. Average weights and physical dimensions by age classes of Yellowstone grizzly bears are given in table 25.1 (J. J. Craighead unpublished data).

Skull and Dentition. The skull of the grizzly bear is highly variable in its size and configuration. Rausch (1953, 1963) presented an intensive evaluation of 357 skulls from 26 regions of North America, which served as the basis for his distinction of subspecies *U. a. horribilis* and *U. a. middendorffi*. The structure of the skull is characteristically massive (figure 25.3). However, tremendous variations exist within and between *U. a. horribilis* populations in such commonly evaluated parameters as condylobasal length, zygomatic width, frontal profile, rostral length, sagittal crest development, length and width of palate, length and form of mandibular ramus, and length of maxillary tooth row. Variations in mean condylobasal length were reported by Rausch (1963) to exist on a clinal gradient, mean lengths increasing from south to northwest. Zavatsky (1976), in an attempt to correlate age with general skull development and morphology, carefully described a skull series from 43 Russian brown bears of age classes assigned according to tooth cementum layers. The specific skull characteristics used to distinguish the 11 age classes would not seem to apply well to the heterogeneous *U. arctos* of North America.

The dental structure of *U. a. horribilis* is generally distinguishable from that of *U. americanus*. For both species, the dental formula is 3/3, 1/1, 4/4, 3/2 = 42. However, the length of the third upper molar of the adult grizzly bear is seldom less than 38 mm, while that of the adult black bear does not attain 31 mm (Storer and Tevis 1955). If teeth are badly worn or fractured, differentiation often is not easy.

Karyotype. All ursids, with the exception of *Tremarctos* sp., have a diploid chromosome number of 74 (Ewer 1973). If centric fusion of chromosomes has accompanied evolution of the families of carnivores from their miacid progenitors, then this number, and the chromosome numbers of some canid species, are primitive to most other carnivores (Wurster and Benirschke 1968). If, as Todd (1970) suggested, karyotypic fission is characteristic of evolutionarily progressive species, the majority of carnivores have retained the primitive condition and the ursids and canids are advanced.

PHYSIOLOGY

In general, bears exhibit the basic systemic physiology common to most mammals. Few studies have defined physiological characteristics specific to *Ursus arctos horribilis*. Jenness et al. (1972) presented a comparative analysis of total solids, fats, lactose, casein, whey proteins, and various minerals and vitamins found in the milk of four bear species. Milk specimens analyzed from eight wild grizzly females differed greatly from the milk of two females confined to zoos and, to a lesser degree, from milk specimens of *U. americanus*,

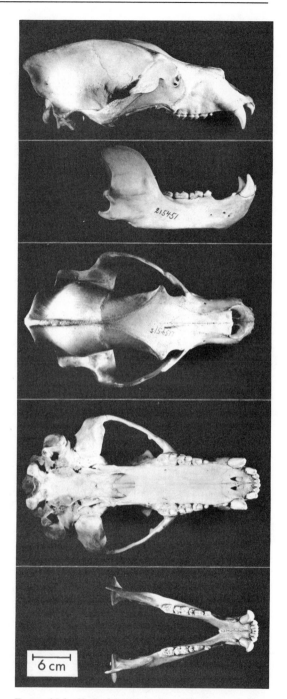

FIGURE 25.3. Skull of the grizzly bear (*Ursus arctos*). From top to bottom: lateral view of cranium, lateral view of mandible, dorsal view of cranium, ventral view of cranium, dorsal view of mandible.

U. maritimus, and other *U. arctos* subspecies. Gel electrophoresis of milk caseins revealed differences in protein composition between samples from the different bear species. There was inadequate resolution for determining if differences were species related or simply due to polymorphism within a species, however.

Pearson and Halloran (1972) reported consistent anisocytosis of erythrocytes in blood samples from 22 *U. arctos* of southwestern Yukon Territory. On repeated sampling, they documented statistically significant decreases in erythrocyte count and increases in basic erythrocyte indices from spring to summer, with some evidence of trend reversal in autumn. They speculated that a relationship existed with the hibernation cycle, but the adaptive value of such a relationship is not obvious.

Studies of hibernation physiology and related cyclic phenomena comprise the bulk of literature treating bear physiology. Nelson et al. (1980) presented a thorough review of current literature. Folk et al. (1967, 1972, 1976) have studied intensively the cardiac cycle of *U. a. horribilis* in Alaska. They found that bears required at least two weeks to enter deep "winter sleep" and, unless disturbed, did not normally waken until spontaneously aroused in the spring. Body temperature did not decrease appreciably (no more than 5° C on the average) as in many other hibernators. A very distinct bradycardia was observed—25-43 percent of normal summer heart rates—but correlative data on respiratory rate, blood pressure, and cardiac output were not obtained. Folk et al. (1972) postulated that the bradycardia is associated with circulatory shunting, which transforms the bears into "heart-lung-brain" preparations and provides other organ systems only marginal support.

Grizzly bears generally do not feed, urinate, or defecate during "winter sleep." According to Folk et al. (1976), this qualifies as a true state of "hibernation" more highly evolved than that observed in small mammals. The latter exhibit bouts of hypothermic torpor periodically interrupted by arousal for imbibition, feeding, and excretion.

The reduction in cardiac rate is paralleled in black bears by a reduction in oxygen consumption (Hock 1960). The lowered oxygen consumption is reflected in the relatively low respiratory quotient values of 0.7–0.85 for bears during cold exposure (Folk et al. 1972). Such respiratory quotients are quite appropriate for energy metabolism based primarily on fats (South and House 1967). Although grizzlies are known to develop extensive fat deposits prior to winter denning, the exact sites and mechanics of deposition and relative quantities have not been well documented. Attempts to measure fat utilization by weighing animals before and after hibernation are often invalid because of unmeasured variations in water weight.

Most work on hibernation metabolism has been on black bears (Brown et al. 1971; Nelson et al. 1971, 1973, 1975). Black bears maintain constant fluid levels in the blood and tissues by retaining the metabolic water of fat catabolism. Blood levels of total protein, urea, and uric acid remain relatively constant, but creatinine concentrations increase. Urea is formed by normal pathways during hibernation. The deamination of amino acids for energy is reduced in favor of lipid utilization. Any urea resulting is recycled via a deamination and recombination with glycerol from lipolysis to form alanine. Thus, lean body weight is preserved and uremia avoided (Nelson et al. 1980). It is assumed that metabolic processes in the hibernating grizzly bear are very similar to those of the black bear. Folk et al. (1976) presented data comparing urine volumes and compositions from a grizzly bear during and following hibernation, which compare well with blood and urine analyses from the black bear.

REPRODUCTION

Relatively little was known about the reproductive biology of grizzly bears prior to the development and use of immobilizing drugs to capture, individually color mark, and radio-instrument specific animals for study over extended periods of time (Craighead et al. 1960, 1963; Craighead and Craighead 1969, 1972, 1973a).

Prior to the use of the definitive field techniques, the bulk of information on ursine reproductive biology concerned the European brown bear (*U. a. arctos*), the polar bear, and the American black bear. Studies by Rausch (1961), Wimsatt (1963), and Erickson et al. (1964) contributed greatly to knowledge of the reproductive biology of the American black bear. Dittrich and Kronberger (1962) reviewed the reproductive biology of the European brown bear.

Murie (1944) reported on breeding dates for grizzly bears in Alaska. Erickson et al. (1968) and Hensel et al. (1969) described breeding biology and discussed reproduction in the Alaskan brown bear.

In the past two decades intensive research has focused on all aspects of the grizzly bear's reproductive biology. Early work was focused on the southern interior ecotype of brown bear inhabiting Wyoming, Montana, and Canada (Craighead et al. 1960, 1961, 1963; Craighead and Craighead 1967, 1969; Hornocker 1962). Knight (1975) and Servheen and Lee (1979) have current investigations under way. Reproductive biology for the ecotype inhabiting northern British Columbia, Yukon, Northwest Territory, and Alaska was treated by Pearson (1975), while that for the ecotype found at the northern extreme of the range on the north slope of the Brooks Range, Alaska, was well documented by Reynolds (1979). Reproduction in brown bears of coastal Alaska was described by Glenn et al. (1976).

The reproductive tract of the female grizzly bear is similar to that of the black bear (Erickson et al. 1964; Kordek and Lindzey 1980). Pearson (1975) described the gross anatomy of reproductive tracts of male and female grizzlies. The size of the uterus varies with the stage of the reproductive cycle and age of the animal. Changes in gross anatomy of the testes also occur.

In the Yellowstone ecosystem, the mating season may begin as early as mid-May and terminate in mid-July. As the season progresses, the vulva enlarges twofold or more, retracting to nonbreeding size in July. Ovaries increase in size with attainment of sexual maturity. Placental scars are present and readily visible in properly prepared specimens. Placental scars, as well as size and coloration of mammae, are indicative of postreproductive history.

Young female grizzly bears mate in Yellowstone National Park from 26 May to 9 July, a period of 45 days (Craighead et al. 1969). The earliest mating recorded was a 7.5-year-old female that copulated twice in one afternoon. Three females were seen to mate on 28 May. The latest mating was recorded on 9 July; other late mating dates were 1 July and 6 July. Records covering a 6-year period showed that during seasons in which mating began early, it terminated early, and vice versa. The periods over which copulation annually occurred proved remarkably similar, averaging 26 days per mating season (figure 25.4).

Precopulatory and postcopulatory behavior were noted as early as 14 May and as late as 15 July, respectively. A period of estrous behavior persisted for approximately 62 days. Dittrich and Kronberger (1962) reported a mating season of approximately 72 days (end of April to mid-July) for captive European brown bears.

From 1962 to 1967, 49 copulations were observed in Yellowstone (table 25.2). They show that 80 percent of all copulations occurred in June, with a preponderance during the first two weeks. Twelve percent occurred in late May and 8 percent in early July.

Copulation. Copulation by grizzly bears is vigorous and prolonged. The copulatory act and related overt behavior vary considerably with the age of the female, her responsiveness, and the number of males vying for her (table 25.3). Several females were observed to copulate more than once in a single day.

The length of a successful copulation varies greatly. Normally a minimum of 10 minutes is required; the maximum time recorded was 60 minutes. The breeding histories clearly show that the female grizzly will mate with a number of males (table 25.3). It is not uncommon for females to accept two males in

TABLE 25.2. Frequency of copulations occurring between 26 May and 9 July (1961–67)

Date of Breeding	Number of Copulations	Percentage of Copulations
26–31 May	6	12.2
1–15 June	23	46.9
16–30 June	16	32.7
1– 8 July	4	8.2
Total	49	100.0

one day. During the period 13–15 June, female #29 copulated with four different males; female #200 showed similar behavior.

Grizzly bears in Yellowstone Park and vicinity are definitely polygamous. This is probably the case wherever they congregate or are sufficiently abundant to allow a female in estrus to meet more than one adult male. Pairing normally occurs only for short periods of time and the maintenance of this bond is dependent on the ability of the male to defend his female against contenders. A postestrous female quickly loses the attention of her mate if another estrous female appears. Although pairing was observed among the grizzlies in Yellowstone, it was not normally a partnership that lasted throughout the mating season. It may be a tenuous, short-term arrangement for the convenience of mating or, in a few instances, a partnership that persists throughout the period of estrus. Neither the male nor the female in these partnerships remains unresponsive to the sexual condition or sexual advances of other bears.

Duration of Estrus. Females in estrus are readily detected by the number and behavior of the male or males they attract. Preestrous and postestrous periods are characterized by complete lack of sexual interest by both male and female.

Some females may experience relatively brief es-

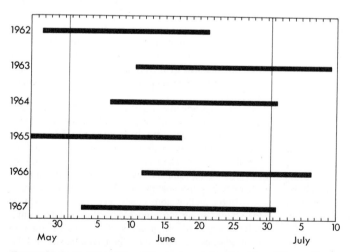

FIGURE 25.4. Duration of the mating season, defined by dates of first and last copulations of female grizzly bears (*Ursus arctos*).

TABLE 25.3. Copulation behavior of female grizzlies (1961–67)

Designation of Female	Total Number of Copulations Observed (All Seasons)	Number of Males Involved	Maximum Number of Copulations Observed during a Breeding Season	Number of Males Involved
40	13	11	6	5
29	7	5	7	5
200	7	5	4	3
101	5	3	5	3
109	5	4	4	3
81	4	3	2	2
6	2	2	2	2
15	2	2	1	1
10	2	2	1	1
96	1	1	1	1
187	1	1	1	1
Totals	49	39	34	27

trous periods. For example, in 1962 a female in Yellowstone, #6, was observed to be in estrus from 6 to 13 June (table 25.4). Her heat period, which lasted eight days, probably began on 6 June. She was then observed for eight consecutive days after 13 June, during which she attracted no males and exhibited no mating behavior. When observed again on 29 and 30 June and 2 and 4 July, she still attracted no males. This female was 3.5 years old and experiencing estrus for the first time.

Females #200 and #29, both entering first estrus

at the age of 3.5 years, had relatively short heat periods (see table 25.4). Female #40 exhibited a short estrous period in 1965 at the age of 7.5 years, but had considerably longer periods in 1963 and 1967.

In 1966, female #200 was in estrus over a period of 15 days. Because she was seen on only 6 days of the mating season and was observed to mate on 3 of these, it is quite probable that the estrous period exceeded that observed (see table 25.4). Female #29 was in estrus for 17 days in 1964 and, when first observed on 7 June, had attracted two large males that showed interest but

TABLE 25.4. Durations of estrus (1961–67)

Designation of Female	Age of Female	Inclusive Dates of Estrus	Observed Period of Estrus (Days)	Result of Breeding (Number of Cubs)
15	3½	6/28 1961	1	none
15	4½	6/11–6/13 1962	3	2
15	6½	6/26 1964	— (26)	none
15	9½	6/27 1967	1	
96	4½	5/28 1962	1	2
200	3½	6/10–6/13 1965	4	none
a 200	4½	6/22–7/6 1966	15	2
187	4½	6/20 1967	1	
29	3½	6/15–6/20 1962	6	none
a 29	5½	6/7–6/23 1964	17	2
a 40	5½	6/12–6/27 1963	16	2
a 40	7½	5/26–5/30 1965	5	2
a 40	9½	6/3–6/27 1967	25	
132	5½	6/21 1967	1	
a 6	3½	6/6–6/13 1962	8	none
6	5½	6/23–6/30 1964	did not come into estrus	none
a 101	8½	6/5–6/17 1965	13	1
109	4½	6/10 1965	1	none
109	5½	6/12 1966	1	none
a 109	6½	6/5–7/1 1967	27	
81	5½	5/28 1962	1	none
81	6½	6/11 1963	1	none
81	7½	6/22–6/30 1964	9	1
81	10½	6/9 1967	1	
10	6½	7/9 1963	1	none
10	7½	7/1 1964	1	none

[a]Most accurately established estrous periods

did not attempt to mate with her. That she was just entering estrus was confirmed 2 days later when she copulated, the first of six copulations. Female #40 exhibited estrous periods of 16, 5, and 25 days duration in 1963, 1965, and 1967, respectively (see table 25.4).

The longest estrous periods recorded were 26 and 27 days. With the exception of one, all females exhibiting long periods of estrus whelped cubs the following year. No 3.5 year olds showed estrous periods exceeding 8 days. Dittrich and Kronberger (1962) reported captive European brown bears in estrus for two to five weeks.

Age at Puberty. Female grizzlies under 3.5 years of age exhibited no heterosexual behavior. Some 3.5-year-old females copulated long, vigorously, and frequently. However, no 3.5-year-old females whelped following this mating activity.

The age at which female black bears become sexually mature is reported to vary with latitude and individual growth rates (Rausch 1961). Erickson et al. (1964, citing Baker 1912) and Rausch (1961) suggest 3.5 years for captive black bears.

Observations in Yellowstone show that female grizzlies produce a first litter at 5.5 years of age. However, some females may whelp for the first time considerably later than this. Among 15 females, 7 had their first litter at 5.5 years of age, 2 at 6.5, 4 at 7.5, and 1 each at 8.5 and 9.5 years of age. It is possible that the first observed litters were the result not of first pregnancies, but of later ones. Also, marked females whose first litters were born when the mothers were 6 to 9 years old may have been pregnant at an earlier age, and lost embryos before birth or cubs soon after it. This would mean that 53 percent of the females studied failed to produce offspring from conceptions occurring before the time they produced their first observed litter. This seems highly unlikely, but cannot be completely discounted.

Estrous and Anestrous Periods. Young females may breed in alternate years or every third year. Older bears often show greater intervals between breeding. The sequences, by years, when individual females either were in estrus or were anestrous were determined for seven bears (figure 25.5). The anestrous condition normally accompanies lactation, but it may occur at other times. Seven females showed continuous anestrous periods of one, two, or three years between their first active mating season and their first established pregnancy.

It is evident that there is much variability in the breeding pattern. Four out of seven females exhibited "false estrus" at 3.5 years, five exhibited anestrus with no mating at age 4.5, while two mated successfully for the first time at 4.5 years. Four females exhibited estrous periods, presumably without ovulation and characterized by unsuccessful breeding.

Wimsatt (1963) offered strong, but not conclusive, evidence that ovulation in the black bear may be coitus induced. There is evidence that the female grizzly is an induced ovulator, but conclusive histolog-

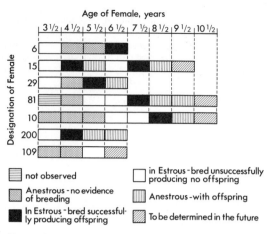

FIGURE 25.5. Estrous condition of seven female grizzly bears (*Ursus arctos*), observed in successive breeding seasons.

ical evidence is lacking. It is intriguing to consider the possibility that the mechanism for inducing ovulation by coital stimulus may develop slowly and, thus, may vary considerably with the age of the individual bear. This could explain effectively why some female grizzlies between the ages of 4.5 and 8.5 years apparently do not ovulate following mating.

Mating Interval and Estrous Cycle. Hansson (1947) provided strong quantitative evidence that the female mink (*Mustela vison*) has several estrous cycles during the same mating season. The female permits copulation at periods of follicular maturity. The time between development of mature follicles and subsequent matings was termed the *mating interval*. Dittrich and Kronberger (1962) studied the mating behavior of the European brown bear in the Zoological Gardens of Leipzig and found that the females allowed numerous copulations. The periods of copulation were followed by days without copulation, implying a mating interval similar to that described for mink. These pauses in the mating of the European brown bear might be short or long and occurred in the presence of males. Prell (cited by Rausch 1961) reported an interval of variable length between a "pseudoestrus" and true estrus in the brown bear and polar bear. Schneider (cited by Rausch 1961) concluded that "pseudoestrus does not occur in the polar bear, but rather a true estrus of long duration."

Observations suggest that the female grizzly has two estrous cycles during the same mating season (Craighead et al. 1969). A female comes in estrus and is receptive and attractive to males. Copulation occurs for a period of days (figure 25.6), after which the female is no longer receptive and the male is not attracted to her. Following a period varying from 4 to 18 days, the female again is receptive and once again attracts males. Thus, an interval exists during which mating does not occur and is probably coincident with follicular development following ovulation. It appears to be quite similar, if not identical, to the "mating interval" in mink described by Hansson (1947).

Collectively, data on durations of the two mating

periods show that each probably does not exceed 10 days, and that there may be little or no difference between the duration of the two periods (figure 25.6). Each cycle begins and terminates rather abruptly, probably following ovulation induced by mating. It is of interest that the inclusive dates of the "mating interval" all occur within the month of June, when 80 percent of all copulations were recorded.

Delayed Implantation. Hamlett (1935) postulated the occurrence of delayed implantation in the black bear, European brown bear, grizzly bear, and polar bear. Dittrich and Kronberger (1962) later provided evidence for delayed implantation in the European brown bear and the Himalayan black bear (*Selenarctos thibetanus*) and Wimsatt (1963) provided conclusive proof of the phenomenon in the black bear.

There is clear evidence that discontinuous embryonic development also occurs in the grizzly bear (Craighead et al. 1969). A female mated during June 1965, at the age of 6.5 years, and was killed on 27 July of the same year. Free blastocysts were flushed from the uterine horns and fixed for future histological study. A second female mated on 18 June 1967, at the age of 14.5 years. When she was killed 50 days later, unimplanted blastocysts were recovered. The long interval between presumed time of ovulation and recovery of the unimplanted blastocysts is evidence of developmental arrest. Erickson et al. (1964) discounted the possibility of a delayed ovulation in the black bear because of the early formation of corpora lutea following mating. Wimsatt (1963) likewise ruled out this possibility for the black bear. Large corpora present in the ovaries of the female specimens mentioned above ruled out delayed ovulation in the grizzly bear as well. The period in years between the earliest recorded pregnancy (age 4.5) and the age at which a first pregnancy actually occurs in a bear is the prepregnancy period. Among 30 females with reproductive histories, this period ranged from 0 to 4 years.

Craighead et al. (1974) recorded the age at first pregnancy for 16 of 30 marked females. Eleven of these (69 percent) first became pregnant at 4.5 years of age, 1 at 5.5, 3 at 6.5, and 1 at 8.5 years of age. Although younger females copulated, none became pregnant before she was 4.5 years old (Craighead et al. 1969). The average age of first pregnancy was 5.2 years for 16 females. The age at which female grizzlies attain sexual maturity varies widely in other bear populations. Age of sexual maturity ranged in the eastern Brooks Range, Alaska, from 6.5 to 12.5 years (Reynolds 1976) and in the Yukon Territory from 6.5 to 7.5 years (Pearson 1976). Brown bears on the Alaska Peninsula showed an age range of 3.5 to 6.5 years for first pregnancy (Glenn et al. 1976).

Reproductive Cycles. The chronology of events occurring in a reproductive cycle varies with the cycle length. The length of a cycle is dependent on when the female weans and how soon thereafter she comes into estrus.

In a two-year cycle, the female becomes pregnant in June or July, whelps the following February, suckles cubs through summer and winter, weans them as yearlings in the spring, and then comes into estrus, breeds, and becomes pregnant following weaning. In a three-year cycle, the female becomes pregnant, whelps cubs, attends them as yearlings, dens with them, weans them as two-year olds soon after leaving the den, and then comes into estrus and breeds to begin another cycle. In a four-year cycle, the female follows the three-year cycle, but after weaning two year olds, she either remains anestrous or comes into estrus and is not fertilized. She is bred the following year and becomes pregnant. In longer cycles, the female may remain anestrous or for various reasons fail to produce cubs.

Precise data on reproductive cycles require that marked animals be recognized and observed over a period of years. Reproductive cycles of 19 marked females were calculated from known pregnancies (Craighead et al. 1974). The number of cycles per female varied from 1 to 3 and totaled 33 for all 19 animals during a cumulative reproductive period of 99 years. The reproductive cycle varied from 2 to 5 years. Of the 33 cycles, 9 lasted 2 years; 16, 3 years; 7, 4 years; and 1, 5-years. Three-year cycles were more prevalent than 2-year cycles (64 to 36 percent). For some females, the reproductive period consisted of a single reproductive cycle, but for others it included 2 or more cycles.

An average reproductive cycle of 3.00 years was obtained when the total of reproductive periods in years for all 19 females (99) was divided by the total number of cycles (33). This parameter was then refined by including prepregnancy data. For example, among 19 females recorded, 5 were older than 4.5 years at first pregnancy. The average reproductive cycle of 3.00 was adjusted for the 11 years that these females were not pregnant. With this adjustment (99 years + 11 = 110/33 = 3.33), the average reproductive cycle becomes 3.33 years.

By assuming when each of 30 females would become pregnant following her last litter, it was possible

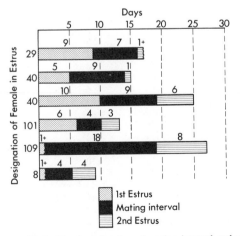

FIGURE 25.6. Observed duration of mating interval and estrous cycles of six female grizzly bears (*Ursus arctos*).

to use a larger number of cycles and reproductive years to compute reproductive rate. With longer reproductive histories to examine, changes occurred in average length of the reproductive cycle for individual females.

A sample of 30 marked females yielded 68 reproductive cycles, a cumulative reproductive period of 231 years, and an average reproductive cycle of 3.40 years. Data used to calculate average reproductive cycles for marked females in four data samples yielded values of 3.33, 3.21, 3.26, and 3.40 years, indicating the range occurring in this parameter with changes in sampling. They also indicate that a representative reproductive rate for a population of long-lived animals can be obtained only from a relatively large sample of animals over an extended period of time, because the accuracy of this biological parameter is dependent on an accurate measurement of cycle length.

Litter Sizes. Thirty marked females produced 68 litters. Among these, 9 were 1-cub litters, 38 were 2-cub litters, 18 were 3-cub litters, and 3 were 4-cub litters. Fifty-six and 26 percent were 2- and 3-cub litters, respectively. Reynolds (1979) reported litter sizes ranging from 1 to 3 per female with a mean litter size of 2.08, determined from 50 offspring of 17 marked females and 7 unmarked identifiable females in the Brooks Range, Alaska. Mean litter size probably reflects the nutritional quality of food available to bears in different regions. Average litter size in the Yellowstone ecosystem was 2.24 (Craighead et al. 1976a). A number of authors have reported mean litter size; the validity of the parameter in many instances is questionable because samples were small, yearling lit-

ters were included with cub litters, or data were not from marked animals.

Reproductive Rates. In a sample of 19 females, reproductive rates for individuals ranged from a low of 0.33 to a high of 1.50 (Craighead et al. 1974). The low represented two cubs produced in two cycles totaling six years, whereas the high resulted from six cubs produced in two cycles totaling four years. Of the 19 females, 4 exhibited reproductive rates of 1.00 or higher (table 25.5). The average rate for all 19 females was 0.70.

Individual bears showed highly variable reproductive cycles and reproductive rates. Because these biological parameters are so important for evaluating the status of a population and for computing reproductive rates for population units, more research effort should be directed toward isolating the factors causing such variability.

In a sample involving 30 females, a reproductive cycle of 3.40 years and a composite reproductive period of 231 years gave a reproductive rate of 0.66.

Maximum and minimum reproductive rates for individual females or a group of females are useful because they indicate the potential of a population to grow or to decline (table 25.6). A population exhibiting compensatory reproduction following a population decline should contain females with high reproductive rates. Similarly, a declining population under environmental stresses could be expected to have females with low reproductive rates.

Although the data indicate that one female exhibited a reproductive rate of 1.50 during a six-year

TABLE 25.5. Reproductive rates of 19 marked female grizzly bears (33 reproductive cycles), 1959–62

Bear Number	Age Marked	Number of Reproductive Cycles in Years				Total Cycles	Reproductive Period in Years	Date of Last Known Pregnancy	Number of Cubs	Reproductive Rates
		2	3	4	5					
5	1.5		1			1	3	1967	2	0.667
7	12.5		1	1		2	7	1965	6	0.857
10	2.5			1		1	4	1969	2	0.500
15	1.5		2			2	6	1968	3	0.500
34	14.5	1			1	2	7	1967	4	0.571
40	1.5	2				2	4	1967	4	1.000
42	5.5	1	2			3	8	1969	6	0.750
65	adult	2				2	4	1963	6	1.500
84	adult			1		1	4	1964	3	0.750
96	3.5		2			2	6	1968	6	1.000
101	4.5	1		1		2	6	1969	2	0.333
120	12.5			1		1	4	1964	2	0.500
125	5.5		3			3	9	1970	8	0.889
128	10.5	1	2			3	8	1969	10	1.250
144	0.5			1		1	4	1970	2	0.500
150	4.5			1		1	4	1966	3	0.750
172	11.5	1	1			2	5	1967	4	0.800
173	2.5		1			1	3	1969	2	0.667
175B	adult		1			1	3	1962	2	0.667
Totals		9	16	7	1	33	99		77	

period, it is highly unlikely that she could sustain this throughout her entire reproductive life (see table 25.6). Data suggest that a maximum for several females averaged 1.17, or about 1 cub per adult female per year. A reproductive rate of this magnitude for a population of females would indicate a potential for that population to grow if mortality was not excessive.

The minimum reproductive rate recorded for an individual female was 0.29; however, this was for only one reproductive cycle and was not considered representative. Minimum rates were calculated using methods employed for samples 1, 2, and 3 (table 25.6). The reproductive rate for female #120 averaged 0.36 over an 11-year period. The average of four females in samples 2 and 3 was 0.50. Therefore, an average minimum reproductive rate among marked females was approximately half the maximum. A rate of this magnitude among female grizzlies in Yellowstone would clearly indicate a declining population, even if human-caused mortalities were kept to a minimum (Craighead et al. 1974).

Reproductive Rate for Yellowstone Ecosystem Population. The reproductive cycle of 3.40 and rate of 0.66 are average parameters for 30 marked females observed over a 12-year period. To obtain a reproductive rate that would more accurately represent the entire population of grizzly bears inhabiting Yellowstone National Park and adjacent areas, the sample size was increased by including data from an additional 25 marked females. These had been omitted from reproductive cycle calculations because of observational discontinuities, but provided data valid for calculating litter size. Total data gave a long-term reproductive rate of 0.63 for the population. This long-term rate, derived from annual counts of 55 marked and recognizable females with litters over a 12-year period, is considered to be the most accurate long-term average rate for the population between 1959 and 1970. Reproductive rates summarized for several other grizzly bear populations are 0.66 for the Alaska Peninsula (Glenn et

al. 1976) and 0.45 and 0.51 for the eastern and western portions of the Brooks Range, respectively (Reynolds 1976, 1978).

Before valid comparisons can be made and conclusions drawn between the reproductive status of these populations and those of the Yellowstone population and other populations currently under study, the reproductive rate data for each population must be quantitatively comparable. This degree of precision

Reproductive Longevity. Pearson (1975) mentioned a female 24.5 years of age with a cub and Reynolds (1978) reported three females that produced cubs at 17.5, 21.5 (or 22.5), and 25.5 years of age. In Yellowstone, one female 14.5 years old when marked produced her last litter of two cubs at the age of 22.5, and weaned them when she was 24.5 (Craighead et al. 1974). Two other females produced litters when they were 19.5 years old and two when 17.5. The greatest age recorded for a female was 25 years. Therefore, reproductive longevity approximates physical longevity, most adult females producing offspring as long as they live. Although the minimum breeding age in Yellowstone is 4.5 years, a female cub born into the population required an average of 6.3 years to whelp her first litter. With an average reproductive cycle of 3.40 years and 2.24 the average litter size, a 25-year-old female could experience 6 reproductive cycles and produce 13 cubs.

Presumably, flexibility of these biological parameters should enable the species to adjust to environmental factors that affect the population favorably or unfavorably. However, for a long-lived species exhibiting delayed maturity, compensatory reproductive processes (increases in litter size, decreases in length of reproductive cycle, and/or higher survivorship rates for subadult bears) would act slowly. On the other hand, population-regulating mechanisms (infanticides from aggressive males and hormonal activity regulating the intervals between estrus in females) are factors that can offset compensatory processes. Infanticide was low in the Yellowstone population (eight instances) but may be higher in other populations (Pearson 1975; Reynolds 1978). The great variability in the length and sequences of reproductive cycles could be important in regulating reproduction, but it will be difficult to draw conclusions from this information until similar data are obtained from other populations and norms established.

TABLE 25.6. Maximum and minimum reproductive rates as illustrated by certain grizzly bears for which more than one reproductive cycle was observed

Bear Number	Sample 1 (19)	Sample 2 (19)	Sample 3 (24)
40	1.000	1.000	1.000
65	1.500	1.500	1.500
96	1.000	0.889	0.889
128	1.250	1.300	1.300
Average	1.188	1.172	1.172
101	0.333	0.500	0.500
120	0.500	0.364	0.364
10	0.500	0.571	0.571
15	0.500	0.556	0.556
Average	0.458	0.498	0.498

ECOLOGY

Habitat. Various aspects of grizzly bear habitat south of Canada have been described by Shaffer (1971), Craighead and Craighead (1972), Sumner and Craighead (1973), Varney et al. (1974), Mealey (1975, 1976), Roop (1975), U.S.D.A. Forest Service (1975), Pearson (1975), and Craighead et al. (1976*b*). This literature deals with surveys, establishment of criteria for evaluating habitat, habitat typing and mapping techniques, distribution and occurrence of plant foods, and relating food habits of grizzlies to habitat types and generalized vegetation complexes.

Some recent studies delineate critical habitat (Craighead 1980*b*) or describe and/or evaluate specific forest and range habitat types (Hamer et al. 1977, 1978, 1979; Hechtel 1979; Servheen and Lee 1979; Schallenberger and Jonkel 1980; Sterling Miller personal communication). Craighead and Scaggs (1979), Craighead and Sumner (1980), and Scaggs (1979) addressed the problem of developing a standarized system for describing, evaluating, and rating habitat types within climatic zones. Craighead (1980*a*) utilized multispectral imagery with computer assistance to map and evaluate grizzly bear habitat and to develop an ecospectral vegetation classification.

The grizzly has been able to survive in North America only where spacious habitat has insulated it from excessive human-caused mortality. Its habitat has traditionally been protected by rugged physiography or inaccessibility. These factors alone, however, are no longer effective. Populations in the contiguous 48 states have survived through the past decade primarily because suitable habitat was preserved by the Wilderness Act of 1964, which established a National Wilderness Preservation System. This system now includes much of the spacious, mountainous habitat where grizzly bears are found south of the Canadian border and where they presumably can survive in the future. The grizzly is not threatened in Canada or Alaska, primarily because large expanses of wilderness habitat still exist, unmodified by human development. Habitat in the contiguous 48 states is largely confined to three grizzly bear "ecosystems"—the Yellowstone, the Selway-Bitterroot, and the Bob Marshall–Lincoln-Scapegoat. In at least one, and probably two, of the three ecosystems, grizzly bears occur as geographically and genetically isolated populations. In the third, the Bob Marshall–Lincoln-Scapegoat, the population can be reinforced genetically and numerically by movement and interchange of individual bears from adjacent occupied habitat in Canada.

The purpose of the Endangered Species Act of 1973 is to perpetuate threatened and endangered species and, where possible, to extend their populations. On 1 September 1975, grizzly bears (*Ursus arctos horribilis*) were listed as "threatened" south of the Canadian border. With this designation, all U.S. federal agencies were required to conduct their land management programs to prevent destruction or adverse modification of critical grizzly bear habitat. Critical habitat determination involved delineating an area essential for the survival and recovery of the species. Federal rules published 22 April 1975 defined critical habitat as that necessary to provide for: nutritional and spatial needs of the species; specialized sites for breeding, reproduction, and shelter; and other specific physical, seasonal, and behavioral requirements.

Most of the range currently occupied by the grizzly bear has been proposed as critical habitat through professional and agency recommendations. A proposed rule by federal authorities delineating critical grizzly bear habitat in the contiguous 48 states was published in the *Federal Register*, 6 November 1976. This was followed by public hearings (U.S. Senate Hearings 1977). The extensive land areas proposed by the U.S. Fish and Wildlife Service as habitat for the grizzly bear's survival total about 5.3 million hectares and consist of four discreet parcels, as follows:

(1) the region where Wyoming, Montana, and Idaho come together in Yellowstone National Park and adjacent areas including parts of Custer, Shoshone, Teton, Targhee, Beaverhead, and Gallatin national forests and part of Grand Teton National Park;

(2) northwestern Montana in Glacier National Park, the Bob Marshall Wilderness Area, and most of the Flathead National Forest and adjacent areas, including parts of the Lewis and Clark, Helena, and Lolo national forests and small parts of the Blackfeet and Flathead Indian reservations;

(3) extreme northwestern Montana and northern Idaho in the Cabinet Mountains, mostly in the Kootenai, Kaniksu, and Lolo national forests; and

(4) extreme northern Idaho and northeastern Washington in the Selkirk Range, mostly in the Kaniksu National Forest.

The enactment of the Endangered Species Act of 1973 also spurred habitat studies in Canada and Alaska.

A vegetation/landtype classification of grizzly bear habitat in the Scapegoat Wilderness, Montana, was developed by Craighead and Scaggs (1979) for the grass-shrublands of the alpine, subalpine, and temperate climatic zones. It was based on the ecoclass methods of Daubenmire (1952), Peterken (1970), and Corliss et al. (1973). Twelve land units (habitat units) were delineated and described in the alpine zone, as were five landtypes in the subalpine zone. Forest habitat types of both the subalpine and temperate climatic zones were grouped as xeric, mesic, or hydric types. Eight major forest habitat types (Pfister et al. 1977) included within these groupings were sampled for ground cover and described in terms of grizzly bear food plants. The habitat type/land type classification allowed measurement and quantification of bear food plants on a comparative basis.

The most important habitat units of the alpine zone, based on the percentage abundance of food plants, were the Alpine Meadow, Alpine Meadow Krummholz, Glacial Cirque Basin, and Mountain Massif, all of which showed an abundance of bear food plants in excess of 50 percent of the total ground vegetation (figure 25.7). Landtypes in the subalpine zone with the greatest abundance of food plants were fire-caused Seral Stages, Dry Forb Grasslands, Snowslides, and Ridgetop Glades, all of which showed an abundance of bear food plants in excess of 50 percent of the total ground cover.

Forest habitat types of the subalpine zone had high potential as plant energy sources for grizzly bears. Those with the greatest abundance of food plants (in excess of 60 percent of total ground cover) were *Abies lasiocarpa/Luzula hitchcockii-Vaccinium scoparium* and *Abies lasiocarpa-Pinus albicaulis/Vaccinium scoparium*. The poorest was *Abies lasiocarpa/Luzula hitchcockii-Menziesia ferruginea.* The presence of *Pinus albicaulis* made the subalpine zone unique as an energy source for grizzly bears.

Ecological landtypes of the temperate zone showed greater variation as energy sources than their equivalents in the subalpine zone. Seral stages (burns) and Dry Forb Grasslands showed the highest potential, based on food plant abundances exceeding 70 percent of the total ground cover.

The forest habitat types of the temperate zone exhibited the highest food plant potential of all vegetation units measured. Those with the greatest abundance of undergrowth food plants were *Abies lasiocarpa/ Xerophyllum tenax* (*Vaccinium globulare* phase), *Abies lasiocarpa/Xerophyllum tenax* (*Vaccinium scoparium* phase), and *Pseudotsuga menziesii/ Calamagrostis rubescens* habitat types (figure 25.8). Food plant abundance values for each of these habitat types exceeded 80 percent.

In potential energy sources for the grizzly bear, the subalpine zone rated highest, the temperate zone second, and the alpine zone third. The resources of all three zones are essential to the grizzly within the Bob Marshall–Lincoln-Scapegoat Wilderness areas. This is probably true for the other large wilderness ecosystems supporting populations of grizzly bears.

Those portions of the grizzly bears' total environment that contain preferred food plants in greatest abundance are critical to the bears' welfare. Some of these—for example, seepage areas where *Equisetum arvense* grows in heavy mats—are small in size. Others—for example, the *Abies lasiocarpa/Luzula hitchcockii* habitat type (*Vaccinium scoparium* phase) where grouse whortleberry may average 50 percent of the forest undergrowth—are quite large. Some critical food source areas are at high altitudes, including the semivegetated talus that supports *Claytonia megarhiza* and the glacial cirque basins with *Lomatium cous.* Others—for example, the sedge marshes and *Abies lasiocarpa/Xerophyllum tenax* habitat type (*Vaccinium scoparium* phase)—lie near the lower altitudinal limits

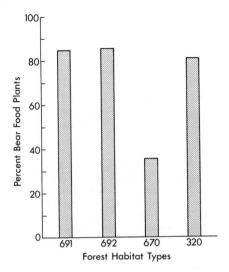

FIGURE 25.8. Grizzly bear (*Ursus arctos*) food plant abundance by forest habitat type of the temperate climatic zone. 691 = *Abies lasiocarpa/Xerophyllum tenax–Vaccinium globulare;* 692 = *Abies lasiocarpa/Xerophyllum tenax–Vaccinium scoparium;* 670 = *Abies lasiocarpa/Menziesia ferruginea;* 320 = *Pseudotsuga menziesii/Calamagrostis rubescens.*

of the bears' wilderness environment. High-altitude areas provide succulent vegetation in early spring; the lower areas, where 85 percent of the ground cover may be plants eaten by grizzlies, are a veritable storehouse of various plant foods.

Pearson (1975) provided a general description of bear habitat in the Yukon Territory; Mealey (1976) surveyed and evaluated the importance of specific forest habitat types west of Yellowstone National Park; Hamer et al. (1977, 1978) described a number of forest and vegetation types for Banff National Park. None of these researchers quantifies these in terms of percentage of food plants or evaluates or rates them as energy

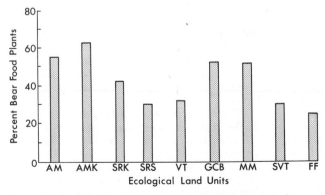

FIGURE 25.7. Grizzly bear (*Ursus arctos*) food plant abundance by ecological land unit of the alpine climatic zone. AM = alpine meadow, AMK = alpine meadow Krummholz, SRK = slab rock Krummholz, SRS = slab rock steps, VT = vegetated talus, GCB = glacial cirque basin, MM = mountain massif, SVT = semivegetated talus, FF = Fellfield.

sources for the grizzly. Scaggs (1979) described forest and vegetation types in the Selway-Bitterroot Wilderness area of Montana and Idaho using the same classification system employed by Craighead and Scaggs (1979) for the Scapegoat Wilderness, Montana. The use of specific forest habitat types and vegetation complexes by radio-instrumented bears was documented by Hechtel (1979), Servheen and Lee (1979), Schallenberger and Jonkel (1980), and Craighead (1980b).

Grizzly bears require spacious habitat, characterized by great diversity. The species thrives best when its habitat is isolated from humans and their activities. Although the grizzly bear is essentially a wilderness animal, it can, and does, adapt to the presence of humans; however, it cannot and has not adapted to mankind's intensive use and modification of its habitat. Mankind must adapt to the grizzly, a tolerance that may not yet have been attained.

Where grizzly and mankind are competing for the same habitat, human-caused bear deaths rise. The bear has a low reproductive rate that does not offset heavy and persistant human-caused mortality. Precautions must be taken to keep such mortalities to a minimum. These appear to be an even greater threat to the grizzly throughout its range than is direct modification of the habitat.

Ranges and Movement. The species' omnivorous feeding habits, complex population and social interactions, winter denning, and aggressive intraspecific and interspecific behavior require extensive movement throughout a spacious habitat. How a population unit moves and interacts within a large geographic area primarily depends on the spatial and temporal distribution of food. There appear to be at least two distinct types of bear populations as characterized by their movements: those populations that inhabit an ecosystem where concentrations of salmon or refuse attract them to feed communally, and those populations where no massive concentration of food exists. The hierarchical relationships that develop at communal sites have been described for the brown bear by Stonorov and Stokes (1972), Egbert and Stokes (1976), and Luque and Stokes (1976). Hierarchical behavior in grizzly bears has also been observed (Hornocker 1962; J. J. Craighead in preparation).

Both the grizzly and the brown bear establish traditional movements to exploit dependable sources of high-calorie foods. These food sources are generally seasonal and available for a period of only several months. They represent a long-established, seasonal pattern for some bear populations, attracting and holding large aggregations of bears for prolonged periods. In Yellowstone, as many as 137 individuals were observed in a single evening (Craighead et al. 1971) and at the McNeil River, Alaska, approximately 50 in a similar period of time (Larry Aumiller personal communication). Where such conditions occur, whether "natural" (figure 25.9) or human-induced (figure 25.10), they influence the daily and seasonal movements, as well as the spatial requirements, of

many members of the population. These population concentration sites can be characterized as population activity centers, or "ecocenters." They may best be visualized as ecological magnets that attract and hold high densities, not only of bears, but also of many other omnivorous species such as ravens, gulls, magpies, coyotes, and raccoons.

Specific information on movements of grizzly bears to and from ecocenters, and information on how much movements affected the development, size, and configuration of home ranges, were obtained in the Yellowstone ecosystem. This was accomplished by directly observing color-marked individuals, by recording the place of capture and locality of death of marked bears, and by monitoring radio-instrumented animals (Craighead et al. 1961; Craighead 1976, 1980b).

Movement data from color-marked grizzly bears within the park or immediately adjacent to it at Gardiner and West Yellowstone, Montana, showed that the bears moved extensively throughout a 2,023,500-ha ecosystem (figure 25.11) that could be considered as critical habitat for the population (Craighead 1980b). Bears marked at sites where they congregated to feed at open pit garbage dumps in Yellowstone Park were observed and recognized in four national forests. Twelve grizzly bears marked at Trout Creek (the geographic center of Yellowstone Park) were observed in the Shosone National Forest. The maximum recorded movement was 74 km. Similarly, other bears marked at Trout Creek were observed in the Gallatin (13), Targhee (2), and Teton (6) national forests. Maximum movements of 70, 72, 78, and 87 km were recorded. Among bears marked at Rabbit Creek approximately 27.5 km to the southwest of Trout Creek (figure 25.11), five were observed in Targhee National Forest and one in Teton National Forest. Of all bears marked in Yellowstone National Park but sighted outside, 67 percent were in the Shoshone and Gallatin national forests, an indication of the considerable spatial needs of the species.

Information obtained from kill records of marked animals provided further evidence for extensive movement of individual animals and for the vast spatial requirements of the entire population. Of 277 color-marked grizzly bears, there were 137 known mortalities: 79 were killed within Yellowstone Park, while 58 were killed outside the park. The deaths of 35 (15.6 percent) of 224 bears marked at or near the geographic center of Yellowstone National Park were recorded in the five adjoining national forests (figure 25.12).

The movement of grizzly bears from the original marking sites (ecocenters) was massive and extensive. The seven greatest airline distances from locality of marking to site of death ranged from 72 to 86 km and averaged 80 km.

Census data (Craighead et al. 1960, 1974; Craighead and Craighead 1967) showed that much of the movement between Yellowstone National Park and the national forests was seasonal. The summer aggregation of grizzly bears at Trout Creek (Craighead et al. 1971) also supported strongly the concept of major

FIGURE 25.9. Grizzly bears (*Ursus arctos*) congregated on the salmon runs at McNeil River, Alaska.

FIGURE 25.10. Grizzly bears (*Ursus arctos*) congregated at the Trout Creek dump, Yellowstone Park.

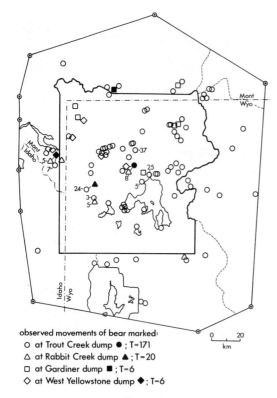

observed movements of bear marked:
○ at Trout Creek dump ● ; T=171
△ at Rabbit Creek dump ▲ ; T=20
□ at Gardiner dump ■ ; T=6
◇ at West Yellowstone dump ◆ ; T=6

FIGURE 25.11. Individual natural movements of grizzly bears (*Ursus arctos*) in the Yellowstone ecosystem, 1959–74.

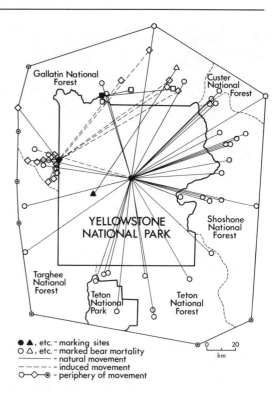

● ▲ , etc. – marking sites
○ △ , etc. – marked bear mortality
—————— – natural movement
— — — — – induced movement
○—◇—◉ – periphery of movement

FIGURE 25.12. Movements of marked grizzly bears (*Ursus arctos*) from site of marking to localities of death in the Yellowstone ecosystem, 1959–74.

seasonal movement. Berns and Hensel (1972) reported established movement patterns among brown bears on Kodiak Island, Alaska. Similar movement to and from ecocenters occurs among brown bears of the Alaska Peninsula (see figure 25.9) at the McNeil River, at Katmai National Monument, and at other sites in coastal Alaska. Some individual bears within the Yellowstone ecosystem centralized their year-round activities near the food source, or ecocenter (see figure 25.10) and were not observed to move large distances (Craighead 1980*b*). Such individuals tended to have small, discreet home ranges, viz., females #40, #150, and #39 (see table 25.6). Other bears, such as #37 and #14, exhibited home ranges encompassing much larger geographic areas and including seasonal migratory "corridors" to and from ecocenters. Such ecocenters are characterized by a unique concentration of readily available high-protein food. The large aggregation of bears that is attracted has developed a high order of social interaction expressed as a linear hierarchy.

Home ranges are usually defined to be areas within which individuals meet all of their biological requirements. These requirements may be met for individual or family units within small core areas or centers of activity, or they may require extensive movement to range peripheries. Many home ranges delineated in Yellowstone contained well-defined seasonal ranges, some separated by long migratory cor-

ridors (Craighead et al. 1974). Reynolds et al. (1976) postulated migratory corridors between seasonal ranges of grizzlies in the Brooks Range, Alaska. Home ranges of the bears in Yellowstone varied greatly in area, depending on the sex and age of the animal, seasonal and annual food availability, foraging ability, reproductive condition of females, and other factors (table 25.7). Equally important was the influence that ecocenters exerted on the movement of most bears. Adult female #7 exhibited a home range of 275 km² during the spring, summer, and fall of 1963. She had three yearlings and traveled extensively, seeking food at the Trout Creek garbage dump (an ecocenter) and in the grass-shrub parklands and forest habitat types of the subalpine zone, with occasional foraging treks into the alpine zone. After weaning her offspring she remained anestrous until the following spring, during which time she exhibited a much smaller home range.

Female #150 with three cubs had a home range during 1963 of only 70 km². She regularly visited an open pit garbage dump and also occupied a core area nearby where meadow mice (*Microtus* spp.) and pocket gophers (*Thomomys talpoides*) were abundant. Sedges, grasses, and the starchy, onionlike bulbs of *Melica spectabilis* were also very abundant. This family met its nutritional requirements within a much more limited space than did female #7 and her three yearlings. Females #101, #39, and #187 had relatively small home ranges that overlapped one another. The

TABLE 25.7. Grizzly bear home ranges, Yellowstone ecosystem

Bear Designation	Sex	Age	Total Years Color Marked (1959–70)	Consecutive Years Radio-tracked (1963–68)	Mode of Detection[a]	Days Radiolocated[b] (1963–68)	Range (km²)
7	F	adult	10	1	R	44	275
150	F	adult	9	1	R	33	70
40	F	subadult–adult	9	6	R	400	78
101	F	adult	11	2	R	125	111
39	F	adult	12	1	R	51	57
187	F	adult	6	2	R	98	104
202	M	yearling–2 year old	2	2	R	174	324
158	M	yearling	1	1	R	51	57
37	M	subadult	3	1	C		1217
14	M	adult	10	1	R & C	41	2600

[a]R = radio-tracking; C = color ear tags.
[b]Radio-located refers to a radio fix, a series of bearings or a radio signal indicating the presence of the instrumented animal.

three females shared the ecocenter as well as much adjacent habitat. However, each utilized discreet centers of activity, seasonal foraging areas, and denning sites within her home range (Craighead 1980*b*).

Many bears in the Yellowstone population moved 65 to 90 km, or more, between denning areas or early spring foraging sites and ecocenters of localized food abundance. Such movement patterns become traditional and affect the size of home ranges, as well as the way in which bears utilize the space and the resources within those ranges. The home range concept implies that each animal is limited to a definable area from which it seldom ventures. When applied to a large, long-lived omnivore such as the grizzly bear, the concept has limited interpretive value. A range representing the spatial requirements of an individual for a period of years or for its lifetime is needed. For example, bear #16, a 340-kg male, had a radio-defined range of only 31 km² within Yellowstone National Park during the fall of 1964 and his home range that year did not greatly exceed this. Data from recaptures and sightings over a period of years, however, indicated that the life range of this male probably exceeded 2,600 km² (see table 25.7). Home and seasonal ranges of grizzlies in the Yukon are discussed by Pearson (1976); in the Brooks Range, Alaska, by Reynolds (1976, 1979); in Yellowstone National Park after 1970 by Knight et al. (1978) and Judd and Knight (1980); and in western Montana by Rockwell et al. (1978), Servheen and Lee (1979), and Schallenberger and Jonkel (1980). Summer ranging of brown bears in the alpine zone of Kodiak Island is described by Atwell et al. (1980). Information from these studies confirms the larger spatial requirements of males versus females, the utilization of seasonal ranges within the total home range, and the presence of discreet activity centers. Home range calculations for the brown bear on Kodiak Island are presented by Berns et al. (1980).

A life range can be defined as an area that provides the biological requirements of an individual bear for all or most of its lifetime. For females, this includes the requirements for bearing and raising offspring. Female

#40 (see table 25.7) was radiotracked for eight consecutive years from 1961 through 1968 (Craighead 1976, 1980*b*). She was instrumented at the age of 2.5 years and shot when 10.5 years old. Her life range, smaller than home ranges of most females, encompassed an area of only 78 km². Her core areas remained basically the same year after year, none exceeding a square mile. Her seasonal and home ranges, however, varied considerably. During 1961 and 1962, her summer-fall range as a subadult did not exceed 21 km². In 1963, at the age of 4.5 years, she used an area of 21 km² during the summer, was observed breeding, and became pregnant. In 1964 she produced two cubs (one of which died) and had a fall range of 40 km². She entered her den on 10 November with her cub. In 1965 she weaned her yearling and mated; she was radiotracked for 106 days beginning 28 June and is known to have entered a den on 11 November. Her home range was 52 km². Accompanied by two new cubs in 1966, her summer and fall range was 19 km². She dug a new den and wintered with her cubs. In the spring of 1967 she weaned the cubs and bred. During the fall of 1967 she ranged within an area of 29 km². Her den was not located, but she emerged in 1968 with three cubs and occupied a home range of 57 km². She was shot in 1969 at the age of 10.5 years. The life range of female #40 was small because it encompassed the Trout Creek dump, an ecocenter, where she satisfied many of her nutritional needs. This food source supplemented her "natural" food intake and that of her offspring, thereby reducing her foraging activities and her spatial requirements. Nevertheless, she made frequent and extensive seasonal movements to feed on winter-killed elk (*Cervus elaphus*) and bison (*Bison bison*) in the riparian communities and the sagebrush-bunchgrass habitat types. She also ate *Vaccinium* berries in the subalpine fir-huckleberry and dwarf whortleberry habitat types, both of which were well represented within her life range. In fall she traveled to the ridges for whitebark pine nuts (*Pinus albicaulis*) in the subalpine fir-whitebark pine forest types and hunted *Microtus* spp. in the sagebrush-bunchgrass parklands. Her life range contained seasonal foraging areas, travel corridors, denning and escape sites, and activity centers. It lay entirely within the subalpine climatic zone.

The life range of female #101 exhibited a similar developmental pattern throughout an 11-year period (see table 25.7). During this time, she raised three litters (Craighead et al. 1974) and showed area requirements linked closely to her reproductive condition and family responsibilities. Among the many biological needs that must be satisfied within a home range, food abundance and availability appear to be more important than all others in determining the size, character, and configuration of a life range.

The extent and duration of movements within a home or a life range are generally responses to specific needs of an animal. In general, distances moved by males radiotracked in the Yellowstone ecosystem were greater than those moved by females. Radiofixes made every 12 hours revealed that female #150 averaged an airline distance of 3.7 km over 13 separate movements. Female #7 averaged 5.1 km for 12 distinct movements and a young adult male averaged 11.5 km for 8 movements. The greatest distance covered by the young male in a 12-hour period, 16 km, was recorded on four separate occasions. He frequently moved 11 to 16 airline kilometers daily within a home range of 435 km². On one occasion, he traversed 3,122-m Mt. Washburn about the 2,750 m level and crossed the Grand Canyon of the Yellowstone River five times, traveling 93 airline kilometers over extremely rough terrain during an 8-day foraging period. The ground distance was estimated to be three times the airline distance. Data on the movements of color-marked grizzlies in the park supported the observation of radioed animals. On an average, males moved greater distances than did female bears, and subadults averaged distances per move about equal to those of adult bears. However, averages for male subadults exceeded those for either male or female adult bears, while average distances moved by female subadults were much less than those observed for adults or for male subadults. This is explained, perhaps, by the fact that, within the population as a whole, less aggression was directed toward females than toward males; this was especially evident with regard to subadult females. Nevertheless, there were instances where both male and female subadults established home ranges within their mother's home range, indicating considerable spatial tolerance and compatibility in some individual adult females or unusual assertiveness by their offspring.

Grizzly bears in Yellowstone made daily (24-hour) movements from feeding sites to bedding sites. Some of these movements were only a few kilometers, while others were 10 to 11 kilometers. During the summer, a high percentage of the daytime bedding occurred from 1000 to 1600 hours. Movements to feed and to bed occurred at all times of the day and night. Maximum movements were recorded in late afternoon and evening, while minimal travel occurred during midday and the middle of the night. Comparative movements of male and female adult and subadult bears in the Yukon, Brooks range, Alaska, and Kodiak Island have been

recorded by Pearson (1975), Reynolds (1979), and Berns et al. (1980), respectively.

All grizzlies of both sexes were radiotracked in Yellowstone on prehibernation "treks" to locate suitable denning sites and to initiate digging of dens. For some individuals, treks began as early as 3 September (Craighead and Craighead 1972). Prior to entering dens for winter sleep, some individuals made as many as four trips from activity centers within their home ranges to their dens. The distances traveled from summer to fall foraging areas to dens or denning sites varied greatly among individual animals. The minimum distance recorded was 3 km and the maximum 25.6 km, although some treks were known to be greatly in excess of this maximum. Final movements to hibernate were generally shorter than the prehibernation movements and often were more direct and rapid. Movements associated with locating a site, digging the den, preparing it for winter, and finally entering it were numerous and closely related to fall foraging movements.

Dispersal from summering areas to fall foraging areas was common to all bears radiotracked. In some instances, the winter den was located within the fall range. In other cases, the den was located many kilometers away. Male #76 and female #96 moved 32 and 64 km, respectively, from summer to fall foraging areas. In general, movement from summer to fall foraging sites occurred in September and often was very abrupt and rapid. Movement to these areas within home ranges was partly a response to food availability, but also was due to a need to be near the winter denning sites while preparing the den. Some dens were not completed until mid-November. Final den entry did not occur for adult females #120 and #101 until as late as 21 November. Some grizzlies, such as #164, moved directly from a fall foraging area to enter a winter den; other bears, such as females #40, #101, and #34, spent many days at, or in the vicinity of, their dens prior to entering them for the winter.

Movement from the winter dens to spring foraging areas was less complex. Some animals simply moved to the nearest snow-free foraging site or to the carcass of a winter-killed elk or bison. Females #40, #101, and #34, for example, moved distances between 6.5 and 13 km within the subalpine zone. For a number of animals, the movements exceeded 50 or 60 airline kilometers and involved several days to several weeks of leisurely traveling, during which the individual or family unit descended from the subalpine zone, traversed river valleys in the temperate zone, and returned to their former spring and summer ranges in the subalpine zone many kilometers from the den sites. Some individuals were first attracted to emerging sedges and grasses, others to carrion, and still others to rodent populations exposed by the receding snow.

Den-related movements of grizzlies have been reported for the Yukon by Pearson (1975); for the Brooks Range by Reynolds et al. (1976); for Yellowstone after 1970 by Knight et al. (1978); and for western Montana by Schallenberger and Jonkel (1980).

Movements to seasonal food sources can be extensive. In Yellowstone it was not uncommon for individual animals of either sex to travel 16 to 30 airline kilometers to feed on the seeds of whitebark pine. Similar movements were recorded in the Lincoln-Scapegoat and southern Bob Marshall wilderness areas of Montana (Sumner and Craighead 1973). In some instances, movements transected three climatic zones. Some individual animals moved many miles to feed on moths (Noctuidae), biscuitroot (*Lomatium cous*), and spring beauty (*Claytonia megarhiza*) in the alpine zone of the Scapegoat Wilderness.

Grizzlies congregate in relatively small numbers at unusually rich or extensive sources of pine nuts, berries, insects, forbs, and other green vegetation, and at carrion. Movements to such sites were repeated annually by some bears and appeared to involve a learning process (Craighead and Sumner 1980). Movements to food sources may be direct and rapid. One adult male traveled 14 km in a single afternoon. Movements to such food sources should not be confused with those to ecocenters. The food attraction is much more limited and dispersed and does not draw and hold animals in large seasonal aggregations.

Movement to carrion is normally rapid and directed by scent. Many bears were observed to move 5 to 12 km daily to and from carcasses. A subadult male in Yellowstone traveled an airline distance of 30 km to a carcass in 36 hours. On the other hand, an adult sow with three yearlings required approximately 60 hours to locate a carcass less than 3 airline km away. Similar daily movements to carcasses were observed in the Scapegoat Wilderness (Sumner and Craighead 1973).

Movements of bears to fall foraging areas were normally sudden and swift in Yellowstone. A female with siblings moved 24 km overnight. Another family unit traveled 19 km in 48 hours, while a lone adult female made a continuous move of 64 km in less than 36 hours. An adult male averaged 1 km per hour in a 24-km move to a fall foraging site, while another traveled 32 airline km in a similar type movement. A yearling male covered an airline distance of 88 km in 20 days. The most rapid movement to a winter den was 25.6 km in 12 hours from a fall foraging area. In nearly all instances where speed of travel was documented, the terrain traversed was rough; the airline distances recorded can be at least doubled to obtain approximate ground distances per unit of time.

Induced movements resulting from the release of a grizzly at some distance from the place of capture averaged greater for adult males than for adult females and greater for subadult males than for subadult females. Among 145 releases of grizzly bears within Yellowstone National Park at varying distances from the campgrounds and developed areas where they were captured, 68 percent returned to the same or another campground following release. As the following examples illustrate, the homing instinct of grizzlies is strong. Cub #78 was orphaned when his mother was captured and shipped to a zoo in 1961. The orphan began entering campgrounds and traveling the highways. Captured in the Lake Campground on 10 July, transported across Yellowstone Lake, and released on Promontory Point, a large peninsula extending into the

lake from the south, he returned to his old haunts at the north end of Yellowstone Lake 7 days later. To accomplish this, he traveled due south 9.7 airline km, east 6.4 km, north 24.1 km, and then west 4.8 km to Pelican Campground—a total airline distance of about 45 km. Actual ground travel was probably more than double this distance.

A two-year-old male bear, #38, captured at Lake Campground, was also boated to Promontory Point and released. Four days later he was back at the Lake Campground, after traveling essentially the same route used by bear #78.

Female #170 and her two cubs were captured just outside the northern border of Yellowstone at Gardiner, Montana. When they were released in Hayden Valley, the geographic center of Yellowstone National Park, the female was color marked and fitted with a radio transmitter. She traveled 50 airline km in 62 hours to return to her home range. The ground distance determined by radio fixes was approximately 80 km, not taking into consideration elevational movements. When again trapped, transported, and released, she returned a distance of 85 airline km.

Movements following transport and release have been recorded by Craighead and Craighead (1967, 1972), Pearson (1975), Reynolds (1979), and Servheen and Lee (1979).

The overlap of home and seasonal ranges of a large number of animals and the extensive travel to and from food sources, daytime retreats, and denning sites were not characterized by territorial defense. The social order inherent in grizzly and brown bear populations precludes the need for holding and defending a well-defined territory. In Yellowstone, grizzly bears did not defend activity centers, seasonal ranges, or their dens from other bears. Aggressive adults defended kills and choice feeding sites until their hunger was appeased, after which other bears shared the food and the site. For example, the carcass of an adult male bison was first defended by an alpha male, but, over a period of several days, more and more bears utilized the food source. Eventually, 23 animals attended the carcass at one time and shared it with only infrequent confrontations.

In summary, the extensive movements of grizzly bears is probably directly related to the absence of defended territories and the functioning of a social linear hierarchy that permits freedom of travel and maximum exploitation of rich food sources.

Dens. The denning tendency is well developed in the brown/grizzly bear group in northern latitudes. Earliest evidences of the Ursidae are cave associated (Kurtén 1968) and imply that natural shelters, at least, were utilized by European/Asian progenitors common to both the *Ursus americanus* and the *U. arctos* lines. As in other hibernating mammals, the adaptive value of winter denning by bears relates to survival during inclement weather conditions. Reduced food supply during winter, together with decreased mobility and the bear's increased energy needs for thermoregulation, have represented a real threat to its survival. The evolution of denning and associated behavior has been the biologic response. The strength of the behavioral mechanism is evident in orphaned cubs that were recorded to dig dens and hibernate successfully in Yellowstone Park.

As discussed earlier in the chapter, hibernation physiology of bears differs from that of most other hibernators in that bears do not assume a state of hypothermic torpor. A den aids in reducing the energy necessary to maintain body temperatures at levels only slightly lower than those maintained during warmer seasons of the year (Hock 1960; Folk et al. 1972, 1976; Craighead et al. 1976a). The period of denning coincides with the period of most inclement weather as well as with the length of gestation. Young, conceived between late spring and early summer, are born in midwinter in the comparative safety and isolation of the den.

Denning behavior, as observed in brown/grizzly bears, coincides with general time frames that relate to regional climates and latitudes. Dates of entry and emergence in a particular population vary in response to weather conditions from year to year. However, grizzly bears inhabiting the contiguous United States generally locate sites and excavate dens between early September and mid-November, enter dens between mid-October and mid-November, and emerge between late March and early May (Craighead and Craighead 1966, 1972; Knight 1975; Werner et al. 1978; Servheen and Lee 1979). Approximately the same chronology has been recorded for grizzlies farther north in the Banff National Park, Canada (Hamer et al. 1977; Vroom et al. 1980), and in the southern Yukon Territory (Pearson 1975), as well as for brown bear populations on Kodiak Island (Berns et al. 1980) and the Alaska Peninsula (Glenn and Miller 1980). Grizzly bears observed in the Brooks Range of northern Alaska entered dens throughout October, on an average somewhat earlier than more southerly populations (Reynolds et al. 1976; Reynolds 1979). Observations on emergence dates from the Brooks Range bears were not reported. Work by Harding (1976) on grizzly bears inhabiting Richards Island off the coast of the Northwest Territories revealed even earlier entry into and later emergence from the winter dens. Entry occurred from late September through mid-October, with emergence from late April through early May.

Factors that govern denning behavior are not as yet clearly understood. Several studies have noted that grizzly bears commonly become increasingly lethargic as winter weather becomes more inclement and finally move to and enter dens, some years, during heavy snowstorms (Craighead and Craighead 1966, 1972; Reynolds et al. 1976). Other workers have observed such responses only in some animals (Servheen and Lee 1979), or not at all (Pearson 1975). All workers have suggested that some factor(s) other than weather conditions provides the critical denning stimulus.

In general, adult male bears remain active longer and emerge from dens earlier than do other sex or age classes (Craighead and Craighead 1972; Pearson 1975; Reynolds 1979). Females with newborn cubs are usually the last to leave the denning areas in the spring.

The physiography of grizzly bear denning sites, including associated habitat types, and the physiognomy of the dens themselves have been studied extensively in Yellowstone Park (Craighead and Craighead 1966, 1969, 1972, 1973a; Knight et al. 1978), in northern Montana (Werner et al. 1978; Servheen and Lee 1979; Schallenberger and Jonkel 1980), in Banff National Park (Hamer et al. 1977; Vroom et al. 1980), in Yukon Territory (Pearson 1975, 1976), in northern Alaska (Reynolds et al. 1976; Reynolds 1979), and on Richards Island off the coast of the Northwest Territories (Harding 1976). Similar data have been reported for dens and denning sites on Kodiak Island (Lentfer et al. 1972; Glenn and Miller 1980). Because the data reported by these workers are detailed and extensive, in the interest of brevity only an overview will be presented. For more detailed treatments of denning topics, the reader is directed to the individual papers.

The ranges of elevation within which discrete grizzly bear populations den are variable relative to latitude. Most sites in the continental interior are located in the upper reaches of the subalpine biogeoclimatic zone. Habitat types characteristic of the subalpine zone vary over the range of the grizzly and selection of the denning sites seemingly relates to the seasonal temperature extremes characteristic of the zone. Just as the elevation of the subalpine zone is progressively lower with increasing latitude, so also are the ranges of elevation within which most dens are located. Ranges of elevation within which dens are most common decline from a high of 2,024–2926 m in Yellowstone Park (Craighead and Craighead 1972; Knight et al. 1978) to a low of 270–1,280 m in northern Alaska (Reynolds et al. 1976; Reynolds 1979). Where large bodies of water are in close proximity or the topography is in low relief, denning elevation is not so clearly related to temperature zonation. Brown bears on Kodiak Island and along the Alaska Peninsula were reported to den at elevations ranging between 31 and 1,006 m (Lentfer et al. 1972). A subsequent study of denning only on Kodiak Island reported elevations ranging between 487 and 670 m (Berns et al. 1980). On Richards Island, an area of low relief, grizzly bears were observed to den primarily in river or lake banks (Harding 1976).

Dens of both grizzly and brown bears have been observed in terrain sloped between 0° and 75°. However, the majority of dens have been reported from slopes of 30°–45°. Steep slopes, along with the porous soils into which the dens are generally excavated, provide relatively easy digging and good drainage of rainwater and snowmelt away from the denning chamber. In deep snow country they support snow cornices that may act as insulation for the den, as well as a physical barrier to any intruder.

The orientation of den openings varies within populations and from one population to another. A majority of den openings for a particular bear population commonly are found in slopes oriented toward some particular quadrant. Charting of seasonal wind directions indicates that the slopes most favored for dens are leeward of prevailing winter winds in the area. Such orientation would insure accumulation of heavy snow burdens to provide insulation to the dens. Those den openings not situated to the apparent leeward of the prevailing winds often are found oriented to local topography such that wind eddying provides heavy snow deposition (Reynolds 1979). Selection and construction of a suitable den appears to be a learning process, the sophistication of the den increasing with the age of the animal (Craighead and Craighead 1972).

Though grizzly bears are known to den in natural caves (Knight et al. 1978; Reynolds et al. 1976) and, in one instance, in a hollow tree (Knight et al. 1978), the majority of grizzly bear dens and all brown bear dens reported have been excavated. Den entrances are bare or may be enclosed by brush. Ideally, the dens are constructed such that they enclose a space of very minimal air movement. Tunnels and chambers are commonly excavated within the root systems of trees and shrubs or beneath large boulders or rock strata. This imparts structural strength to the top of the den and reduces the threat of cave-ins during midwinter thaws. Most bears apparently excavate new dens each year, but there is indirect evidence and suspicion on the part of many observers that dens are reused year after year.

The physical measurements of dens are probably determined most by the age and, thus, the physical excavating ability of individual bears. This translates into a volume of enclosed airspace that must be warmed by the hibernating bear. Accordingly, chambers are generally just large enough to permit minimal stretching and change of position by the bear. Tunnels often lead horizontally directly into the chamber, although chambers may open at right angles to the tunnel and the tunnel may angle up or down. The floor of the chamber is sometimes lower than the floor of the tunnel, but more often is shelved above the tunnel floor. The latter construction would provide a "well" system in which cold air would sump. The chambers are usually longer than wide, with ceilings higher than those of the tunnels. The chambers of Yellowstone dens were usually lined with nests of grass and rootlets or tree boughs (Craighead and Craighead 1972, 1973a), but this was not always the case in other geographic areas. Such nests appear to relate to the age and sex of the bear, being more common among adults and/or females. Concise physical parameters of dens throughout the ranges of both grizzly and brown bears have been reported in the papers cited earlier.

Population Statistics. Estimates of grizzly bear numbers have been, until recent years, largely educated guesses. Storer and Tevis (1955) estimated that in California there were once 10,000 grizzlies; all had vanished by 1924. Grizzlies are notoriously difficult to census, and thus density figures for large geographic areas are often of limited value. Over the past two decades, greatly improved field techniques have enabled researchers to count members of population units and of small segments of those units more accurately.

During a 12-year period in the Yellowstone ecosystem, 264 grizzly bears were captured, individually color marked (Craighead et al. 1960), and returned to the population. Approximately 41 censuses of 3.5 hours each were made each year from 1959 through 1970 at five localities throughout Yellowstone Park. This enumeration of individually recognizable animals provided the population characteristics for deriving a mathematical model of this population (Craighead et al. 1974). The model was then used to estimate the size of the grizzly population and to predict future rates of growth or decline. The most probable estimate of numbers was 222 animals for the year 1959, with an upper bound for the population of 309 and a lower of 172. This provided a most probable density of one bear per 80.3 km² in an area of 20,200 km². Densities for small units of the ecosystem were much higher.

Reynolds (1978), working with a population of marked animals in the Brooks Range, Alaska, estimated 121 animals for a 5,180-km² area, a density of

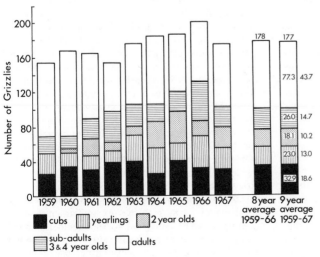

FIGURE 25.13. Age structure of the Yellowstone grizzly bear (*Ursus arctos*) population, 1959–67.

one bear per 42.8 km². By using densities derived on smaller areas of intensive study and extrapolating the data to broad areas, a population of 420 animals was estimated. Estimated densities within this large area were one bear/778 km² in the coastal plain, one bear/91 km² in the low foothills, one bear/52–130 km² in the range, one bear/130 km² in the high foothills, and one bear/259 km² in the mountains. Density figures for other areas have been reported by Mundy and Flook (1973), Martinka (1974), and Troyer and Hensel (1964).

Sex and Age Composition. The age structure of the Yellowstone grizzly bear population was determined for the years 1959–67 (figure 25.13). The average age composition was 18.6 percent cubs, 13.0 percent yearlings, 10.2 percent 2 year olds, 14.7 percent 3 and 4 year olds, and 43.7 percent adults. A further breakdown of the adult age structure was obtained by randomly capturing and determining the age of 52 adults (27 males and 25 females). Fourth premolars extracted from each captured adult before release were sectioned and cementum layers counted to determine age (Scheffer 1950; Craighead et al. 1970). The sample of 52 aged adults was increased to 60 by including 8 known-age adult members of the population. This adult age structure and the age structure of animals from 0.5 to 5.5 years old (see figure 25.13) were then combined and applied to an average population level of 177 animals in order to construct an age- and sex-specific life table (table 25.8).

The age- and sex-specific survivorship rates provided the basic data for determining the number of animals in each age and sex class that would die and/or survive from one year to the next. These are presented as mortality probabilities (Qx) and as survivorship probabilities (Px). The number of cubs born each year can be predicted by counting the number of adult females in the population each year and applying the proper reproductive rate and sex ratio (Craighead et al. 1974).

Age structure for the grizzly bear population in Yellowstone is compared with those for the Brooks Range, Alaska, and with a brown bear population at McNeil River, Alaska (table 25.9). Construction of reliable population age structures requires a number of years of consecutive data; therefore, the comparisons between populations and the conclusions drawn from table 25.9 must be considered tentative. The Yellowstone and McNeil River populations exhibit higher proportions of cubs than those in the Brooks Range. This is probably directly related to the much greater abundance of high-protein food annually available to them.

The Yellowstone population was increasing between 1959 and 1967 (see figure 25.13). The age structures for the eastern and western Brooks Range show low proportions of cubs and suggest that the populations either are declining slowly or are in equilibrium.

The McNeil River population shows a proportion of cubs lower than that for Yellowstone, but higher than that for the Brooks Range. The low percentages of yearlings and two year olds in the McNeil population would, in itself, indicate a declining population with heavy first- and second-year mortality. However, the very high percentage of three and four year olds suggests that the low percentages recorded for yearlings and for two year olds is not due to mortality. These age classes do not frequent the concentration area at the falls and therefore, are not recorded until they return as three and four year olds. At this age they can compete more favorably with the large number of adults present. The population appears to be increasing.

Age structures derived in terms of live animals in a population may appear quite different from those constructed from death statistics for the same population. Use of mortality statistics for the construction of a life table requires unrealistic assumptions that are difficult to reconcile.

Sex Ratios. Sex ratios are essential for understanding the dynamics of a population. Especially important is the ratio of males to females born each year. The cumulative cub sex ratio in Yellowstone from a sample of 78 cubs was 0.59 males to 0.41 females. This may have resulted from differential mortality of females in utero or immediately postpartum, or from sampling procedure. The ratio of males to females for yearlings and for two, three, and four year olds is shown in table 25.8. Among 577 observations of adult grizzlies, most recognized as individuals, 53.7 percent were females and 46.3 percent were males. A differential sex mortality was operative among adults, probably because of selective hunting of males and higher mortality caused by greater movement.

In the Brooks Range, Alaska, the sex ratio of a marked population was 39.8 percent males to 60.2 percent females. The sex ratio of cubs and yearlings was equal. Pearson (1975) reported a sex ratio of 68 males to 32 females among captured animals, but provided no data on the sex ratio of cubs.

Mortality and Survivorship Rates. Mortality in the Yellowstone population (Craighead et al. 1974) was measured in two ways: first by changes in sex-age structure from year to year, and second by verifying and recording actual deaths. Mortality and survivorship rates for the population were obtained by using age structures, sex ratios, and census data described earlier to construct an age- and sex-specific life table for the period 1959–67 (see table 25.8). Data for this 9-year period were used, rather than data for a longer period of time, because new management procedures greatly increased the annual death rate of the population after the summer of 1967. The survivorship rates for the 1959–67 period characterized a population in stable age distribution. The age structure data (see figure 25.13) were converted to an age- and sex-specific structure by applying the sex ratios and then smoothing this to the form shown in table 25.8. Mortality and survival expressed through the sex-age structure of the population were converted to the number annually dying and the number annually surviving in a population of 178 animals. The subadult age classes (0.5 to 4.5 years) represent 9-year averages for the population;

TABLE 25.8. Age- and sex-specific life table for the Yellowstone grizzly bear population, 1959–67

Males

Age	Number in Age Class	Number of Males	Number Dying in Age Class (Dx)	Number Surviving per Thousand (Lx)	Survivorship Rate (Px)	Mortality Rate (Qx)
0.5	33.0	19.5	5.0	1000.	0.7436	0.2564
1.5	23.0	14.5	4.6	744.	0.6828	0.3172
2.5	18.0	9.9	1.4	508.	0.8586	0.1414
3.5	14.0	8.5	1.5	436.	0.8235	0.1765
4.5	12.0	7.0	3.4	359.	0.5143	0.4857
5.5	7.7	3.6	0.2	185.	0.9444	0.0556
6.5	7.4	3.4	0.2	174.	0.9412	0.0588
7.5	7.0	3.2	0.1	164.	0.9688	0.0313
8.5	6.8	3.1	0.1	159.	0.9677	0.0323
9.5	6.6	3.0	0.1	154.	0.9667	0.0333
10.5	6.3	2.9	0.1	149.	0.9655	0.0345
11.5	6.1	2.8	0.1	144.	0.9643	0.0357
12.5	5.8	2.7	0.3	138.	0.8889	0.1111
13.5	5.2	2.4	0.3	123.	0.8750	0.1250
14.5	4.5	2.1	0.5	108.	0.7619	0.2381
15.5	3.5	1.6	0.4	82.	0.7500	0.2500
16.5	2.6	1.2	0.2	62.	0.8333	0.1667
17.5	2.2	1.0	0.2	51.	0.8000	0.2000
18.5	1.7	0.8	0.2	41.	0.7500	0.2500
19.5	1.4	0.6	0.1	31.	0.8333	0.1667
20.5	1.1	0.5	0.1	26.	0.8000	0.2000
21.5	0.8	0.4	0.1	21.	0.7500	0.2500
22.5	0.6	0.3	0.1	15.	0.6667	0.3333
23.5	0.4	0.2	0.1	10.	0.5000	0.5000
24.5	0.2	0.1	0.1	5.	0.5000	0.5000
25.5	0.1	0.1	0.1	3.	0.0000	1.0000
Total	178.0	95.4	19.6			

Females

Age	Number in Age Class	Number of Females	Number Dying in Age Class (Dx)	Number Surviving per Thousand (Lx)	Survivorship Rate (Px)	Mortality Rate (Qx)
0.5	33.0	13.5	5.0	1000.	0.6296	0.3704
1.5	23.0	8.5	0.4	630.	0.9529	0.0471
2.5	18.0	8.1	2.6	600.	0.6790	0.3210
3.5	14.0	5.5	0.5	407.	0.9091	0.0909
4.5	12.0	5.0	0.9	370.	0.8200	0.1800
5.5	7.7	4.1	0.1	304.	0.9756	0.0244
6.5	7.4	4.0	0.2	296.	0.9500	0.0500
7.5	7.0	3.8	0.1	281.	0.9737	0.0263
8.5	6.8	3.7	0.1	274.	0.9730	0.0270
9.5	6.6	3.6	0.2	267.	0.9444	0.0556
10.5	6.3	3.4	0.1	252.	0.9706	0.0294
11.5	6.1	3.3	0.2	244.	0.9394	0.0606
12.5	5.8	3.1	0.3	230.	0.9032	0.0968
13.5	5.2	2.8	0.4	207.	0.8571	0.1429
14.5	4.5	2.4	0.5	178.	0.7917	0.2083
15.5	3.5	1.9	0.5	141.	0.7368	0.2632
16.5	2.6	1.4	0.2	104.	0.8571	0.1429
17.5	2.2	1.2	0.3	89.	0.7500	0.2500
18.5	1.7	0.9	0.1	67.	0.8889	0.1111
19.5	1.4	0.8	0.2	59.	0.7500	0.2500
20.5	1.1	0.6	0.2	44.	0.6667	0.3333
21.5	0.8	0.4	0.1	30.	0.7500	0.2500
22.5	0.6	0.3	0.1	22.	0.6667	0.3333
23.5	0.4	0.2	0.1	15.	0.5000	0.5000
24.5	0.2	0.1	0.1	7.	0.5000	0.5000
25.5	0.1	0.1	0.1	4.	0.0000	1.0000
Total	178.0	82.7	13.6			

TABLE 25.9. Comparison of age cohorts of grizzly bears in four populations

Location	Number of Years Data Base	Percentage of Cubs	Percentage of Yearlings	Percentage of 2 Year Olds	Percentage of 3 and 4 Year Olds	Percentage of 5 Year Olds and Older	Status of Population
Yellowstone Park (Craighead et al. 1974)	9	18.6	13.0	10.2	14.7	43.7	increasing
Eastern Brooks Range (Reynolds 1976)	3	7.9	10.9	10.9	5.0	65.3	declining[a]
Western Brooks Range (Reynolds 1978)	2	10.8	9.5	10.8	9.5	50.0	unknown
McNeil River (Christopher Smith, Alaska Game and Fish 1980, personal communication)	5	13.2	8.2	6.9	19.1	52.5	increasing

[a]Based on reproduction and age distribution data.

the adult age classes (5.5 to 25.5 years) represent one-time samples of 60 adults as described earlier (see column 2 of table 25.8).

The age-specific mortality represents death from all causes (see table 25.8). Among these deaths, some were known and recorded; others were unknown and unrecorded, except as they were reflected in the age structure. Each year from 1959 through 1973, all known grizzly bear deaths were recorded. Because it was difficult to obtain the precise ages of these animals, they were grouped into three classes: subadults, adults, and a class of unknown sex and age. In general, the adult and subadult classes represent the reproductive and nonreproductive periods in the life of a female grizzly bear. From 1959 through 1967, a total of 170 known deaths occurred, an average of 18.9 bears per year, or a 10.6 percent known mortality in an average annual population of 178 animals. A total of 189 known deaths occurred from 1968 through 1973 (an average of 31.5 bears per year), with maximum deaths of 53 and 48 grizzlies in 1970 and 1971, respectively. Known deaths for the 15-year period (1959–73) thus totaled 359. Deaths of adult and subadult females alone increased from 39.8 percent (51/128) during 1959–67 to 44.7 percent (71/159) for the 1968–73 period (Craighead 1980b).

The mortality percentages by sex and age among the 359 known deaths show the adult and subadult deaths to be equal at 40.7 percent. Forty-six percent of all deaths were males, 34 percent were females, and 20 percent were of unknown sex. In all probability, the differential sex mortality has led to the unbalanced adult sex ratio of 46.3 percent males to 53.7 percent females noted previously. The preponderance of males to females in the subadult age structure does not reflect the differential male mortality among subadults. This may be due to sampling error.

Survivorship calculations and calculations of yearly increments based on reproductive rates provided the basis for describing the way grizzly bears enter and leave age classes from year to year. Simulation runs were made for three cases: the upper and lower bounds on the population, and the most probable case. The latter showed the ecosystem population increasing from 222 animals in 1959 to 245 in 1967, then declining to 136 animals in 1974 (Craighead et al. 1974).

Varied estimates of grizzly bear numbers utilizing the Craighead data have been made for the Yellowstone population by others (Cowan 1974; McCullough 1978; Shaffer 1978). Disparities have arisen primarily because of differences in simulation models and the problem of evaluating the relative strength of biological parameters used in the models by writers unfamiliar with the field conditions.

The Interagency Grizzly Bear Study Team has estimated 300–350 animals in the population from "inductive inference" each year since 1974, yet postulated that a population of this size should have recovered from the excessive 1970, 1971, and 1972 mortality (Roop 1980). However, field data does not indicate recovery. No scientific population estimates have been offered and the status of the population remains unanswered after summary displacement of one long-term research effort with another.

The threatened status of the grizzly focuses attention on the viability or survivability of grizzly bear populations. From biological parameters (Craighead et al. 1974) and from habitat variables, Shaffer (1978) gave a theoretical analysis of survivability. He concluded that populations of fewer than 30–70 bears occupying less than 2,500–7,400 km² have less than a 95 percent chance of surviving 100 years. The ease with which grizzlies can be baited and killed, the difficulties of detecting a wide range of illegal deaths, the threats to habitat security, the problems of making accurate censuses, and the longevity of the species are all crucial factors tending to mask detection of population declines. The lack of current scientific population information for the Yellowstone grizzlies leaves no alternative but use of stringent protective measures.

FOOD HABITS

John Muir said of the grizzly, "to him almost everything is food except granite." Recent quantitative studies of the food and feeding habits of grizzlies, as well as the casual observations of early explorers and naturalists, tend to confirm his statement. Grizzly bears are omnivorous, feeding on an extremely broad range of food items.

Early observers reported grizzly bears feeding on beached whales, acorns, and cultivated corn (Storer and Tevis 1955). The bears competed directly with Native Americans for such plants as blue camas (*Camassia quamash*), *Lomatium cous* and other species of biscuitroot, yampa (*Perideridia gairdneri*), the berries of *Vaccinium* spp., and the nuts of *Pinus albicaulis* and other nut-bearing pines.

Between 50 and 60 percent of the grizzly diet may be animal life varying in size from ants and moths to elk (*Cervus elaphus*) and bison (*Bison bison*). The grizzly is both directly and indirectly dependent on the plant base. Feeding behavior suggests that the grizzly prefers high-protein animal food but readily takes plant foods lower in protein when the former is unavailable.

Like other North American bears, grizzlies are attracted to garbage and refuse dumps, large and small, visiting them periodically during the foraging season. Foraging at open pit garbage dumps has been documented by Hornocker (1962), Craighead and Craighead (1967), and Cole (1972). The universal attraction of garbage dumps and carrion disposal sites is evidenced by the large number of animals captured and marked by all bear research biologists at such sites. Similarly, brown bears form aggregations to feed on salmon (Stonorov and Stokes 1972). Studies of feeding habits show clearly that grizzly bears are attracted to large and persistent energy sources, both natural and "artificial," and visit such ecosystems seasonally and annually. High-energy food sources attract both bears and humans and are generally closely associated with human activities; despite the solitary nature of the grizzly bear, this has tended to bring bears and mankind in close association.

The use and availability of food plants are readily quantified through direct observation, fecal analysis, and measurements of plant abundance. Food habit analyses have been made by Tisch (1961), Mundy (1963), Shaffer (1971), Russell (1971), Sumner and Craighead (1973), Mealey (1975), Hamer et al. (1977, 1978, 1979), Husby et al. (1977), Husby and McMurray (1978), Hechtel (1979), Servheen and Lee (1979), Craighead and Sumner (1980), and Schallenberger and Jonkel (1980).

Craighead and Sumner (1980) utilized a number of parameters to evaluate the plant food and feeding habits of grizzly bears in the Scapegoat Wilderness, Montana. An importance value percentage (IVP) of food plants identified in scats was calculated for a number of food items to permit direct comparison between food plant usage and food plant abundance (Sumner and Craighead 1973). The IVPs of food plants in the Scapegoat Wilderness were ranked for use in describing the dietary importance to the grizzly bear of individual food plants. Scat analysis indicated four major plant energy sources in the alpine and subalpine zones: graminales, forbs, berries, and pine nuts with IVPs of 29.7, 37.6, 12.5, and 20.4, respectively.

IVP values for specific plants varied from 20.4 for pine nuts (*Pinus albicaulis*) to 0.1 for several forb species. A positive correlation was found between grizzly bear use of grasses (Gramineae) and their relative abundance values in the grass-shrublands of the alpine and subalpine zones. The sedges (Cyperaceae) were not consumed in relation to their relative abundance values. The high IVPs of specific forbs such as *Lomatium cous* and *Claytonia megarhiza* indicated that preference and a high order of selectivity, rather than relative abundance, determined the extent to which they were utilized by grizzlies.

Energy values were determined for the more important food plants. Available energy of specific food plants varied from a low of 1.91 kcal/g in the roots of *Veratrum viride* to 3.99 kcal/g in whitebark pine nuts (*Pinus albicaulis*). Specific energy values were then related to each plant's abundance, distribution, and seasonal and annual availability.

Among the four major energy sources utilized by grizzlies, the graminales and forbs were chiefly spring and summer foods, berries were almost exclusively summer food, and pine nuts were primarily fall food (except during years of exceptional seed production, when they were consumed in spring as well). The grasses, a highly stable energy source available during the entire foraging season, served as a "survival ration" to carry the bears through periods when other energy sources were low.

To quantify further the relative value to grizzly bears of food plants and food plant groupings, food plant value percentage (FPV) was calculated. This value incorporated five distinct values, strictly comparable for each plant (table 25.10). Based on the FPVs calculated for each food plant, it was concluded that most important, in order of ranking, for the grizzly were: Gramineae, *Pinus albicaulis, Vaccinium* spp., Cyperaceae, *Lomatium cous, Shepherdia canadensis,*

Claytonia megarhiza, Fragaria spp., and *Arctostaphylos uva-ursi.* Gramineae and Cyperaceae exhibited high food plant value percentages, but individual species of grasses and sedges could not be rated.

If this method of rating plant foods from a composite of values (see table 25.10) were adopted in other studies, more precise comparisons could be made between the food habits of bears inhabiting different biogeographical areas. At present, this is not possible.

Comparison of the importance value percentages (a single component of FPV) for major food plant groupings can be made between the Scapegoat and the Yellowstone ecosystems (table 25.11). The IVP for graminales in Yellowstone was twice that for the Scapegoat Wilderness, Montana. The values for forbs were sixfold greater in Scapegoat than in Yellowstone. However, graminales and forbs, both low-calorie food plant groups, when considered together showed almost identical values of 67.3 for Scapegoat and 67.1 for Yellowstone. The IVPs for the high-calorie groupings, berries and nuts, were also nearly identical. That the values presented for the two areas (widely separated in time and distance) would so closely match suggests that, in general, the abundant, highly dependable, low-calorie food plants represent about 2/3 of the grizzlies' vegetable diet. The less abundant, less dependable, high-calorie food plants comprise the remainder. Abundance and availability, rather than energy values, may well determine the grizzlies' long-term utilization of plant foods.

The wide assortment of plant species used as food by the grizzly is becoming increasingly evident. Mealey (1975) listed approximately 25 species for the Yellowstone area without specifically identifying grasses and sedges. J. J. Craighead (in preparation) recorded over 35 species utilized in the same area between 1959 and 1969 prior to the closure of the open pit garbage dumps.

Craighead and Sumner (1980) listed 68 species and plant categories (genera and families) as bear food plants in the Scapegoat Wilderness, Montana, between 1972 and 1978. Servheen and Lee (1979) recorded approximately 36 plant species for the Mission Mountains, Montana; Husby and McMurray (1978), 74 for northwestern Montana; and Hamer et al. (1978), about 41 for Banff National Park, Canada. There are undoubtedly well over 200 plant species whose seeds, fruits, foliage, flower heads, stems, roots, tubers, and root stocks are eaten by grizzlies within their North American range.

The utilization of any specific food plant by a population of grizzly bears is usually dependent upon the relative temporal and spatial abundance and availability of other food plants, as well as upon energy expended for acquisition relative to energy provided by the food. Thus, a plant or a plant group heavily utilized in one area may be recorded as lightly utilized in another. Nevertheless, sufficient information is available to indicate the most important and basic sources of food from the great variety of plants used by grizzlies in specific geographic areas.

Mealey (1975) listed the following as major plant

TABLE 25.10. Calculation of composite food plant values (FPV) for the Scapegoat study area from a series of five comparable food plant evaluations

	Importance Value % (IVP)	Preference Value % (PVP)	Random Abundance Value %	Climatic Zone Occurrence	Energy Value % (EVP)	Food Plant Value % (FPV)
Berries						
Vaccinium spp.	5.4	18.7	12.8	3.0	2.8	42.7
Shepherdia canadensis	3.5	10.4	0.8	2.0	2.7	19.4
Fragaria virginiana	2.0	2.6	0.8	3.0	2.5	10.9
Arctostaphylos uva-ursi	1.6	2.1	1.6	3.0	2.4	10.7
Plant group	12.5	33.8	16.0	11.0	10.4	83.7
Mean values	3.1	8.5	4.0	2.8	2.6	20.9
Nuts						
Pinus albicaulis	20.4	17.6	7.9	1.0	3.3	50.2
Berries and nuts combined						
Plant group	32.9	51.4	23.9	12.0	13.7	133.9
Mean values	6.9	10.3	4.8	2.4	2.7	26.8
Forbs						
Claytonia megarhiza	5.5	5.7	0.1	1.0	2.1	14.4
Lomatium cous	5.3	12.8	0.3	2.0	2.2	22.6
Equisetum arvense	2.3	0.9	0.3	3.0	2.2	8.7
Claytonia lanceolata	1.2	2.1	0.2	2.0	3.3	8.8
Polygonum spp.	0.9	2.0	0.4	2.0	2.0	7.3
Erythronium grandiflorum	0.5	2.1	0.5	3.0	3.0	9.1
Heracleum lanatum	0.1		0.3	3.0	2.1	5.5
Cirsium scariosum	0.1		T	3.0	2.5	5.6
Hedysarum spp.	0.1		0.5	2.0	1.7	4.3
Plant group	16.0	25.6	2.6	21.0	21.1	86.3
Mean values	1.8	2.9	0.3	2.3	2.3	9.6
Graminales						
Gramineae	25.9	7.0	16.6	3.0	1.8	54.3
Cyperaceae	3.8	4.6	12.3	3.0	1.9	25.6
Plant group	29.7	11.6	28.9	6.0	3.7	79.9
Mean values	14.9	5.8	14.5	3.0	1.9	40.0
Sum of food plant parameters	78.6	88.6	55.4	39.0	38.5	300.1

NOTE: The FPV percentages are the sums of the five plant values preceding them.

food sources for grizzlies in Yellowstone National Park, Wyoming: Gramineae/Cyperaceae, *Claytonia lanceolata, Cirsium scariosum, Perideridia gairdneri, Lomatium cous, Vaccinium scoparium, Equisetum arvense,* and *Pinus albicaulis.* Pearson (1975), working in southwestern Yukon Territory, Canada, recorded *Hedysarum alpinum, Shepherdia canadensis,*

TABLE 25.11. Comparison of importance value percentages for major food plant groups in the Scapegoat ecosystem to those for the Yellowstone ecosystem

Plant Food Group	Scapegoat Importance Value Percentages	Yellowstone Importance Value Percentages	Average (kcal/g)
Graminales	29.7 ⎫	60.8 ⎫	2.56
	⎬ 67.3	⎬ 67.1	
Forbs	37.6 ⎭	6.3 ⎭	2.81
Berries	12.5	12.0	3.21
Nuts	20.4	20.8	3.99

NOTE: Number of scats analyzed in: Yellowstone = 487 (J. J. Craighead 1968–70), Scapegoat = 282 (1972–76)

Gramineae, and *Salix* spp. as important sources of food. For grizzlies in the Mission Mountains, Montana, Servheen and Lee (1979) recorded Graminoids, *Amelanchier alnifolia, Equisetum arvense, Osmorhiza occidentalis, Prunus* spp. (domestic), *Taraxacum* spp., *Heracleum lanatum, Trifolium repens,* and *Malus* spp. (domestic) as major plant foods.

Husby and McMurray (1978), working in northwestern Montana, found the following to be important: *Vaccinium globulare,* species of Umbellifereae, Gramineae/Cyperaceae, *Equisetum* spp., *Arctostaphylos uva-ursi, Shepherdia canadensis, Amelanchier alnifolia,* and Formicidae. Hamer et al. (1978), working in Banff National Park, Canada, recorded as important: *Hedysarum* spp., *Equisetum* spp., Gramineae/Cyperaceae, *Heracleum lanatum, Rumex* spp., *Shepherdia canadensis, Vaccinium* spp., and *Arctostaphylos uva-ursi.*

The range of food plants available to grizzly bears and their omnivorous feeding habits does not necessarily ensure an adequate food supply from year to year. During years of widespread failure of such preferred food as *Vaccinium* berries and/or pine nuts, grizzlies generally must travel more, enlarge their home ranges, visit man-made sources of food more frequently, and exhibit greater aggressiveness in defense of their food sources. When berries and nuts are scarce, grizzlies sustain themselves with green vegetation (grasses, sedges, and forbs), but generally will lose weight because these foods are not completely digested. Grizzlies feeding primarily on green vegetation in spring fail to gain weight, but those securing high-protein food such as carcasses, the young of big game species, or various man-derived food sources maintain or gain weight. When pine nuts are abundant, grizzlies gain weight rapidly from this high-energy plant food (3.99 kcals/g). A young adult male killed early in the spring following an exceptionally good pine nut season had 14 cm of fat over the rump. The excellent condition of individual Yellowstone bears captured and weighed in September and October correlated well with good crops of pine nuts. Similarly, grizzlies gained weight rapidly in those summers when berry crops were good.

Grizzlies exhibit different metabolic stages (exhibited in terms of nutritional status) that are associated with seasonal changes. Nelson et al. (1980) described four metabolic stages for the black bear: (1) hibernation, or winter sleep, (2) transition, or hypophagia, (3) normal activity, and (4) hyperphagia. Craighead and Sumner (1980) determined that these metabolic stages in the grizzly are closely attuned to plant and animal phenology and can be observed and documented in the behavior and activity of a bear population.

In spring when adult grizzlies leave their winter dens, they eat sparingly for several weeks (stage 2). Their movements are generally slow and deliberate. During this transition stage from hibernation to normal activity, they continue to metabolize body fat. As food becomes increasingly available, the bears' food consumption increases. Observations of feeding behavior and weight records taken in Yellowstone suggest that losses in body weight during April and May may exceed gains as grizzlies continue to utilize body fat (J. J. Craighead in preparation). By June, grizzlies are on a normal feeding regime (stage 3) involving a wide range of foods, but they still exhibit little or no gain in body weight. Not until late July and August are there noticeable increases in body weight associated with the seasonal increase in food quality and availability.

From mid-July through September a maximum amount of food (energy) is present from both plant and animal sources. Bears spend much of their time feeding (stage 4), and gains in body weight are substantial. Among 28 individual grizzlies captured and weighed periodically in Yellowstone, a two-year-old female showed an average weight gain of 1.65 kg/day over a 24-day period from mid-July to mid-August; a yearling male, 0.97 kg/day over a similar time span; and one adult female, 1.13 kg/day over a 26-day period. Bears for which weights were averaged over longer time spans of 111, 114, and 118 days showed gains of 0.41, 0.24, and 0.46 kg/day, respectively. In adults, the rapid weight gains are due largely to fat deposition, but in subadults, lean body mass also increases. The aver-

age annual increase in weight of yearlings was 145 percent for males and 130 percent for females.

As winter nears, metabolic changes occur that prepare the grizzly for winter sleep (stage 1). Among well-fed members of a population, feeding activity decreases in mid-October; some of these animals exhibit a state of lethargy before entering winter dens (Craighead and Craighead 1972). Those animals not so well fed may continue to feed up to the time they enter their dens for winter sleep. In Yellowstone, for example, color-marked animals were observed that moved almost daily from den areas to feed on elk carcasses. They terminated feeding only when heavy snow storms finally confined them to the dens.

In the northern rockies of the United States, grizzlies hibernate from October/November to March/April, a period when both plant and animal foods are unavailable. Normally they remain in the dens throughout the winter (Craighead and Craighead 1972). However, several instances were recorded in Yellowstone of adult grizzlies leaving dens in midwinter when ambient temperatures rose and mild weather prevailed for five to six days. There was no evidence that grizzlies fed while on these excursions away from their dens. While in the den, grizzlies metabolize stored body fat (Folk et al. 1972). This requires no intake of free water and produces no wastes requiring defecation or urinary excretion. However, water is expelled through respiration. Body fat remains the sole ultimate energy and water source until late March or April (Nelson et al. 1980). At this time, all members of a population except females with cubs will normally leave the dens.

The transition from fat to carbohydrate/protein metabolism (stage 2) takes place slowly, in association with behavioral and activity patterns and changes in physical conditions. By mid-May to mid-July, the bears have again become active, exploiting all of the energy sources available to them. At this time, adult females come into estrus and the larger, more aggressive males breed them (Craighead et al. 1969). Agonistic behavior is common among adult males; many severe encounters occur during the mating period. It is a time of great energy expenditure by all members of a population. The relatively low energy intake and high energy utilization is reflected in the nutritional level of the population. Body weights of individual animals reach an annual low.

The six- to seven-month period from den emergence to return is, in general, one of preparing for hibernation. The entire year is defined in this cyclic phenomenon of metabolic stages that dictates the bears' behavioral patterns, especially those associated with foraging and feeding.

Most of the grizzlies' foraging movements are deliberate. Information obtained during 10 years of monitoring color-marked or radio-collared grizzlies of all ages and both sexes in the Yellowstone ecosystem (Craighead 1980b, in preparation) showed that individual grizzlies do not normally move randomly or aimlessly throughout their large home ranges, feeding opportunistically; rather, the bears are attuned to the plant phenology. Their activities are associated with the emergence and maturation of plants.

From the Yellowstone study, and that in the Scapegoat, a general pattern of movement and activity for securing food emerged for populations south of the Canadian border. Some bears leave their hibernation dens as early as March, traveling when the snow is crusted or keeping to the bare south-facing ridges. They move from the subalpine zone where they have denned to the lower subalpine and the temperate zones where snow is light or absent. By late April to mid-May, many of the mature bears, and most of the subadults, have moved from winter dens to the lower altitudes. Females with cubs of the year may emerge from late April to late May. They also tend to move to lower altitudes. At this time, overwintering rodents such as voles (*Microtus* spp.), deer mice (*Peromyscus* spp.), and pocket gophers (*Thomomys talpoides*) are consumed. High overwintering populations of these rodents occur periodically. At such times, they are especially vulnerable as the snow cover melts. A female and three yearlings were observed to feed for several weeks on *Microtus* spp. During this time, these rodents constituted a significant portion of the total diet of this family group. When big game is abundant, grizzlies move to the winter ranges of these ungulates and feed on winter-killed animals or prey on those in a state of advanced malnutrition. Grizzly bear predation on big game species is generally greatest from mid-April to mid-May. At the periphery of the wilderness, the bears may kill livestock, feed on carrion, or routinely visit livestock disposal sites common on most large ranches. Often, more than one grizzly may feed on a carcass. Craighead and Sumner (1980) reported 172 grizzly bear sightings on 118 big game carcasses over a 13-year period in Yellowstone. Carcasses were usually visited before the snow had melted. Sometimes as many as 6–7 individual grizzlies utilized a carcass, and there were instances in which carcasses were periodically revisited for 10 to 15 days. One grizzly returned to a carcass at least nine times during a 15-day period. Grizzlies were readily attracted to carcasses distributed through three climatic zones and over a 259-km² area in the Scapegoat Wilderness of Montana (Sumner and Craighead 1973).

Where food is abundant and concentrated, aggregations of bears occur and a social order is operative (Hornocker 1962; Craighead and Craighead 1971; Craighead 1980b). The social hierarchy serves to increase foraging efficiency by allowing large numbers of a population to share a common food source. In Yellowstone, 23 grizzly bears were recorded feeding on a bison carcass and large aggregations in excess of 80 grizzlies per evening were documented at open pit garbage dumps (Craighead and Craighead 1971). Grizzlies supplement an early spring meat diet with the early emerging sedges and grasses. At this time of year they frequently forsake the relative safety of the large national forest and wilderness areas to forage on emerging grasses, sedges, and forbs in the temperate zone. Individual bears may remain at low elevations, utilizing plant foods for several weeks or more. However, as big game species leave winter ranges and move to higher elevations, the bears tend to follow the same pattern, feeding primarily on grasses and forbs. If carrion or other high-protein food is not available, they sustain themselves almost exclusively on the plant resource. Adult males, the subadults of both sexes, and females without offspring are generally solitary foragers. Females with cubs, yearlings, or two year olds forage as family groups. A female with cubs may form a close bond with a similar age family, and they then travel and forage as a unit.

In early June elk begin dropping their calves in the temperate and subalpine parklands of northwestern Wyoming and western Montana. Calving sites tend to be traditional, the elk returning to them year after year (Craighead et al. 1972b). Grizzlies whose home ranges encompass these calving areas appear to locate elk by scent and follow them as they migrate to these areas. In some instances, individual bears apparently recall the locations from past experience. Calves are vulnerable to grizzlies for a relatively short period of time. Soon after calving, the cows and their offspring move to higher elevations, their movements determined by the recession of snow and the emergence of plants. Grizzlies follow the same general pattern, so by July they are feeding on the grasses, sedges, and forbs in both the subalpine and alpine zones.

From late June through July, the alpine zone is used extensively as a source of *Lomatium cous,* *Claytonia megarhiza,* and other succulent and nutritious tubers, bulbs, and greens. Insects become important items of diet during this period. Grizzlies seem to have a craving for such insects as moths, beetles, ants, and even earthworms that is partially, but not entirely, related to their high protein content.

As August approaches, the berries of *Vaccinium scoparium* and *V. globulare* begin to ripen in the temperate zone and those of *Shepherdia canadensis* in the subalpine. Grizzlies traveling within large, but undefended, home ranges move to lower elevations to utilize this energy resource, which, in years of peak abundance, is exploited until snow covers the subalpine country. When berries are abundant, bears tend to utilize this food source almost exclusively and gain weight rapidly. In years when berry crops are poor, the greens help alleviate the energy shortage; however, bears do not gain weight on this diet. At such times the nuts of the whitebark and limber pines (*Pinus albicaulis* and *P. flexilis*) become a critical energy source. Grizzlies will move to the extremities of their home ranges to feed on pine nuts and will utilize them through September and October and, in some instances, until mid-November. Radiotracked grizzlies were observed to move over 80 km to feed on whitebark pine nuts. In the Yellowstone ecosystem, and in the Scapegoat study area as well, the nuts of whitebark pine provided the high-energy diet necessary for the grizzly to enter hibernation with a heavy layer of stored fat. Bumper pine nut crops occurred twice throughout Yellowstone over a 12-year period and twice over a 7-year period in Scapegoat. This ideal situation never occurred uniformly throughout the Yellowstone ecosystem, but did occasionally occur within specific home ranges of individual bears.

Stored fat is vital to the bears' survival. During the long period of hibernation (a winter sleep of approximately five to six months), it is the bears' only energy source. Although most grizzlies leave their dens with sufficient body fat to carry them through the lean months of spring, females with cubs reach a lower nutritional level because energy reserves are expended to give birth to young and to produce milk. Lactating females may not show renewed fat deposition until late August or September. The degree of fat deposition in fall may influence the estrous cycle, and thereby determine whether a female will wean her cubs as yearlings or carry them through another year (J. J. Craighead in preparation). When both berry and pine nut crops peak, grizzlies fare exceedingly well.

Grizzlies locate and learn to use specific locales where plant and animal foods are most abundant. The more productive sites become centers of activity within home ranges. In the course of a long life span, such areas become well known to individual bears. These may be large or small and at high or low elevations. Whether they support many or few bear food plants, they are all parts of larger vegetation units that the grizzly utilizes throughout the year with an uncanny sense of its biological needs and a knowledge of where it can meet its dietary requirements.

MORTALITY

An accurate measurement of mortality is essential for formulating long-range management goals and for annually evaluating hunting success. Holding the annual kill to a predetermined quota has been the basic management tool employed for both brown and grizzly bears. Human-caused mortalities can be categorized as hunting and nonhunting. The former data are quite accurate and relatively easy to obtain, but the latter are subject to inaccuracies because of the difficulty of detecting and verifying them. Deaths in both categories can be substantial and, therefore, data on both are necessary for making precise management recommendations. This is especially true where the species is threatened. In that respect, it is revealing that the basic brown bear management goal in southeastern Alaska where bears are abundant is to maintain a high-quality hunting experience. In the lower 48 states where the grizzly is threatened, the primary goal is recovery. Brown bear management in southeastern Alaska has been thoroughly reviewed by Johnson (1980). Hunting statistics and bear mortalities for northwestern and south-central Montana have been summarized by Greer (1980). The effect of heavy human-caused mortalities on the Yellowstone grizzly population was analyzed by Craighead et al. (1974); strong agency reaction and public concern resulted at that time (Craighead 1979).

An update of human-caused mortalities over the past two decades is revealing in its management impli-

cations for the Yellowstone population (table 25.12). Grizzly bear mortalities are summarized for the 11-year period 1959 to 1969 (Craighead et al. 1974, 1980b) and for the 10-year period from 1970 to 1979 (Knight unpublished data). Over the 11-year period prior to closure of the open pit dumps (ecocenters) in 1969–70, grizzly bear deaths averaged 19.4 bears per year. During the 10 years following elimination of the ecocenters, deaths averaged 19.0 bears per year. For the four critical years following closure of the ecocenters (1970–73), Knight's records show 14 fewer deaths than were recorded by Craighead et al. (1974). To avoid possible controversy, the lower death statistics have been employed in table 25.12; however, it should be noted that inclusion of those deaths indicates a total mortality of 204 and an annual mean of 20.4 bear deaths for the 10-year period following closure of the ecocenters. Also, use here of Knight's mortality data for the 1970–79 period does not apply to the mortality statistics reported for the same period as a basis for evaluation of the Yellowstone grizzly population discussed earlier (Craighead et al. 1974). Regardless of which set of data is used, it is evident that the mortality rate rose dramatically during the first 4 years following closure, and then gradually leveled off. If we assume that the level of sampling has been comparable (and we believe it has been), then one must conclude from table 25.12 that the percentage of nonhunting deaths, both inside and outside the park, increased nearly threefold in the decade following closure of the ecocenters. This can be attributed primarily to nutritional stress and dispersion (Craighead 1980b), which greatly increased the incidence of bear-human conflicts.

The percentage of hunter kills decreased in the latter decade from 36.4 percent to 22.6 percent, but this was due entirely to a partial hunting ban imposed by Montana and Wyoming in 1975. Although it is difficult to judge from the total of all bear deaths during the decade, the ban appears to have been effective in reducing the total of human-caused deaths. Relative mortality due to bear control within the park dropped from 45.8 percent to 25.7 percent, reflecting a concerted effort by park officials to reduce and/or to show a reduction in this cause of death concomitant with curtailment of hunter kills. Because of the consistently large number of nonhunting deaths occurring annually in the area around Yellowstone National Park (51.6 percent), the mean mortality for the 1970–79 decade equaled that of the previous 11-year period. This can only be viewed as a serious threat to the integrity of the population when analyzed in context with a decline in reproductive rate from 0.66 to 0.56 (Craighead et al. 1974; Knight personal communication 1980) and direct observations that show a 70 to 80 percent decrease in grizzly bear use of winter-killed elk and bison (Craighead and Sumner in preparation). Data presented in table 25.12 should eventually be incorporated into computer-modeled population analyses, but certain conclusions relevant to management do not require such sophisticated treatment. The hunting ban in Montana, Wyoming, and Idaho must continue; efforts to curtail nonhunting deaths, especially attributable to

TABLE 25.12. Known grizzly bear mortalities by year in Yellowstone National Park and adjoining areas, 1959-79

Year	Area adjacent to YNP				YNP		Total		Total without Hunting	
	Nonhunting		Hunting							
	No.	%	No.	%	No.	%	No.	%	No.	%
1959	0	0	4	5.1	8	8.2	12	5.6	8	5.9
1960	2	5.3	14	18.0	8	8.2	24	11.2	10	7.4
1961	7	18.4	5	6.4	9	9.2	21	9.8	16	11.8
1962	1	2.6	4	5.1	10	10.2	15	7.0	11	8.1
1963	1	2.6	5	6.4	9	9.2	15	7.0	10	7.4
1964	1	2.6	3	3.8	8	8.2	12	5.6	9	6.6
1965	1	2.6	7	9.0	7	7.1	15	7.0	8	5.9
1966	7	18.4	2	2.6	4	4.1	13	6.1	11	8.1
1967	8	21.0	24	30.8	11	11.2	43	20.1	19	14.0
1968	6	15.8	3	3.8	12	12.2	21	9.8	18	13.2
1969	4	10.5	7	9.0	12	12.2	23	10.8	16	11.8
Total	38	99.8	78	100.0	98	100.0	214	100.0	136	100.2
Percentage of 11-Year Total	17.8		36.4		45.8		11 yr. $\bar{x} = 19.4$		11 yr. $\bar{x} = 12.4$	
1970	10	10.2	13	30.2	20	40.8	43	22.6	30	20.4
1971	23	23.5	13	30.2	6	12.2	42	22.1	29	19.7
1972	11	11.2	4	9.3	9	18.4	24	12.6	20	13.6
1973	14	14.3	6	14.2	2	4.1	22	11.6	16	10.9
1974	5	5.1	7	16.3	2	4.1	14	7.4	7	4.8
1975	4	4.1			0	0	4	2.1	4	2.7
1976	3	3.1			3	6.1	6	3.2	6	4.1
1977	12	12.2			4	8.2	16	8.4	16	10.9
1978	7	7.1			2	4.1	9	4.7	9	6.1
1979	9	9.2			1	2.0	10	5.3	10	6.8
Total	98	100.0	43	100.2	49	100.2	190	100.0	147	100.1
Percentage of 10-Year Total	51.6		22.6		25.7		10 yr. $\bar{x} = 19.0$		10 yr. $\bar{x} = 14.7$	

SOURCE: 1959-69, Craighead 1980; 1970-79, Knight unpublished data.

poaching and illegal bear controls, must be greatly intensified; and the death rate within Yellowstone Park itself must continue depressed. Preliminary analyses indicate that, for recovery, the total annual human-caused deaths within the ecosystem must be held to a number considerably fewer than the mean death toll of 10 per year recorded between 1975 and 1979. To accomplish this will require changes in livestock, logging, and recreation competition within the ecosystem (Craighead 1980*b*) as well as enactment of the recommendations above. To effect these changes on the scale and intensity needed for recovery will require interagency recognition of the critical nature of the problem and cooperative interagency action. The lesson that the Yellowstone situation offers to management is that positive corrective action, based on solid research, cannot be delayed a full decade. Management must follow rapidly on the heels of research and, indeed, be concommitant with it.

Parasites and Disease. Most of the literature on North American ursine parasites concerns helminths. Rogers and Rogers (1976) provided a good review of parasites known from bears around the world.

Only two trematode species have been reported from *Ursus arctos horribilis*. Worley et al. (1976) found *Echinostoma revolutum* in the intestines of 2 of 31 Montana grizzlies. Schlegel et al. (1968) reported

Nanophyetus salmincola from Alaskan brown bears. Salmonid fishes serve as intermediate hosts for *N. salmincola*. Bears are infected when fishes, especially salmon, containing the metacercariae are ingested. *Nanophyetus salmincola* is well known to veterinarians as the vector of *Neorickettsia helminthoeca*, a bacterium that causes the highly lethal "salmon poisoning disease" in canids. Although ursids are apparently refractile to infection with *N. helminthoeca*, a different, uncharacterized rickettsia also carried by the fluke has been shown experimentally to cause Elokomin fever in black bears (Rogers and Rogers 1976). Presumably, this could also infect grizzly bears.

Tapeworms found in grizzly bears include species of *Diphyllobothrium*, a pseudophyllidean cestode. Infections are most likely incurred when bears eat fish containing the tapeworm pleurocercoids. Choquette et al. (1969) collected *Diphyllobothrium* from 3 of 21 grizzlies in northwestern Canada and tentatively identified the species as *D. ursi*. This species was provisionally described by Rausch (1954); however, it has not been consistently distinguished by many researchers from the much more common *D. latum*. Worley et al. (1976) reported *Diphyllobothrium* spp. from 16 of 66 grizzly bears, but did not determine the species. Interestingly, all 16 infected animals were from the Yellowstone ecosystem of Montana and Wyoming.

The only cyclophyllidean tapeworms reported in grizzlies are *Taenia* spp. Choquette et al. (1969) found *T. krabbei* in 2 of 21 bears in northwestern Canada. Worley et al. (1976) reported *Taenia* sp. from 14 of 66 grizzlies in Montana, but, again, did not determine the species. Although *Echinococcus* spp. have not been reported from grizzlies, the geographic distributions and natural intermediate hosts of the hydatid worms would imply that grizzlies are exposed via their natural prey. It seems likely that *Echinococcus* spp., and other cyclophyllideans common to feral mammals, will be reported from grizzly bears with continued work.

Of all helminths, nematode species are those most commonly found in bears. *Baylisascaris transfuga* was reported from the intestines of 16 of 21 grizzlies in northwestern Canada (Choquette et al. 1969) and 53 of 70 grizzlies in Montana (Worley et al. 1976). A hookworm, *Uncinaria* (=*Dochmoides*) *yukonensis,* was found in 10 of 21 grizzlies in northwestern Canada (Choquette et al. 1969); Worley et al. (1976) reported 12 of 69 Montana grizzlies infected with *Uncinaria* sp. Olsen (1968) described a new species of hookworm, *U. rauschi,* from both black and grizzly bears in Alaska. Rausch (1961; cited in Rogers and Rogers 1976) found *U. yukonensis* in Alaskan brown bears.

Choquette et al. (1969) observed the mosquito-borne, filarial nematode *Dirofilaria ursi* in 3 of 27 grizzlies in northwestern Canada; Worley et al. (1976) reported it from 2 of 13 Montana grizzlies. Rausch (1961, in Rogers and Rogers 1976) stated that *D. ursi* was observed quite commonly in Alaskan brown bears.

As a host-inspecific parasite of many mammals, including humans and bears, *Trichinella spiralis* is of major concern in contexts of public health and wildlife management. All species of *Ursus* have been found to host the nematode. Larvae encysted in the flesh of the bear, if not destroyed by cooking, are infective to humans. Infections appear to be maintained in wild bear populations more through cannibalism and feeding on the carcasses of other carnivores than through feeding on refuse at garbage disposal sites (Worley et al. 1974). *Trichinella spiralis* has been reported from 10 of 20 Alaskan grizzly bears (Rausch et al. 1956), from 21 of 24 grizzlies in northwestern Canada (Choquette et al. 1969), and from 103 of 141 grizzlies in Montana (Worley et al. 1976). The last group also noted that larval density, in terms of average larval cysts per gram of tissue, was highest in the tongue, followed by the femoral muscle, the masseter, and the diaphragm.

Few arthropod parasites have been reported from grizzlies. The fleas that appear to be native to grizzlies are all *Chaetopsylla* spp. Holland (1949) reported *C. setosa* from grizzlies in British Columbia and *C. tuberculaticeps ursi* from grizzlies in parts of western Canada and in Alaska. Worley et al. (1976) found *Chaetopsylla sp.* on one of three Montana grizzly bears. The single tick species reported from the grizzly bear is *Dermacentor andersoni* (Rogers and Rogers 1976).

Exactly what role is played by protozoan parasites in grizzly bear populations is undetermined. There is, likewise, virtually no knowledge of diseases of bacterial, fungal, or viral etiology. It is likely that this paucity of information is related more to a lack of investigation than to unusual disease resistance in bears.

Although *Eimeria ursi* and *Isospora fonsecai* have been reported from *Ursus arctos* in the USSR and other coccidia have been found in North American *U. americanus,* no protozoan of any kind is reported from North American *U. arctos.* Worley et al. (1976) noted coccidian oocysts in the feces of grizzly bears, but they did not identify them or examine the intestinal tissues for sporozoites.

Although grizzly bears are known to show symptoms of gastrointestinal and respiratory illness, etiology has seldom been researched. As mentioned earlier, grizzly bears are undoubtedly exposed to the rickettsia that causes Elokomin fever experimentally in black bears and naturally in other mammals. Neiland (1975) found that a very high percentage of grizzlies in the Brooks Range of Alaska had antibodies to *Brucella suis* type 4, the agent of rangiferine brucellosis in caribou. Discernible antibody titres in such a large portion of the bear population indicate a high degree of exposure to *B. suis* through predation on infected caribou and suggest that brucellosis might be of importance in the dynamics of some grizzly bear populations. Heddleston (1976) reported a positive isolate of *Pasteurella multocida* from a bear (species unstated). This bacterium is widely distributed in North American birds and mammals, including, most likely, the grizzly bear.

Agents of disease, whether enzootic or explosively epizootic, can have a powerful effect on the status of an animal population. As regards the grizzly bear, it is clear that extensive work is necessary to develop even an elementary understanding of the health dynamics.

MANAGEMENT

Study Techniques. As has been shown, grizzly and brown bears generally have extensive spatial needs and tend to range almost continuously. This mobility, together with the animals' large size, secretiveness, and potential aggressiveness, has made scientific study difficult. As little as three decades ago, scientifically definitive data concerning the bears and their habits were lacking. Population enumeration based simply on counting tracks or recording sightings of bears not individually identifiable, practices all too common even today, were inaccurate and misleading. Current knowledge of grizzly and brown bears has been amassed through use of innovative study methods and application of inventive new technologies.

To study individual bears and to mark them distinctively requires that they be subdued with minimal injury. Animals are either baited into culvert traps constructed of steel bars and spiral pipe (often on trailer frames), trapped with baited snares, or approached and shot with propulsive syringe darts. A muscle-relaxing drug or anesthetic is administered intramuscularly by means of either a heavy syringe mounted on a long rod ("jab stick") or a gas-propelled dart fired from a rifle (developed by Crockford et al. 1958). Although pentobarbital sodium, a potent general anesthetic, had been used in earlier work on black bears (Erickson 1957; Black 1958), the first efforts to immobilize brown bears (Troyer personal communication 1960) and grizzly bears (Craighead et al. 1960) were based on the fact-acting muscle relaxant succinylcholine chloride (Sucostrin). This drug blocks nervous transmission at the myoneural junction by competitively inhibiting acetylcholine and is degraded by cholinesterase only very gradually. For small to medium grizzly bears, optimal dosage was about 1 mg per 1.41 kg body weight. This dosage was found to prolong immobilization in larger bears, however, and had to be modified to account for age and amount of body fat (Craighead et al. 1960). Other drug preparations reported in recent literature include phencyclidine hydrochloride (Sernylan) alone (Craighead et al. 1964, 1969, 1972c; Pearson 1975, 1976; Reynolds 1979; Glenn and Miller 1980) or in combination with promazine hydrochloride (Sparine) (Joslin et al. 1977; Servheen and Lee 1979; Schallenberger and Jonkel 1980) and ketamine hydrochloride (Ketaset) in combination with acepromazine or promazine hydrochloride (Joslin et al. 1977). Phencyclidine hydrochloride appears to be the preferred immobilizing agent for use on larger bears, while ketamine hydrochloride is becoming more common for use with smaller grizzlies and for black bears.

Once the bear has been immobilized, primary physical data can be collected. Morphometry, body weight, breeding condition, general physiological characteristics, and age can be determined. The age of a bear, especially important in constructing life tables and determining reproductive longevity for a population, is accurately determined from preparations of an extracted tooth. The technique, originated for study of the Pinnipedia by Scheffer (1950), entails decalcifying and cross-sectioning the tooth and staining the sections to define annuli in the cementum. The annuli occur as a result of seasonal variation in the rate of cementum deposition; their number relates to the age of a specimen. Successful applications of the technique to third molars, fourth premolars, and first premolars have been reported by Craighead et al. (1970), Pearson (1975), and Reynolds (1978). The extraction of fourth premolars from live members of a population before release permitted age determination necessary for constructing a life table (Craighead et al. 1974).

While immobilized, the bear may be marked in some manner such that it is individually identifiable while roaming free. Marking of grizzly bears is necessary to obtain accurate biological data. Color-coded, plastic ear tags, in conjunction with tatoos, were first used to study the Yellowstone grizzly bears during the late 1950s (Craighead et al. 1960). This marking technique has since become a common practice in population work.

Of all technical innovations, the radio-transmitter collar and tuned directional receiving antenna have probably proven most valuable in documenting the biology and life history of grizzly/brown bears. The method was first applied when Yellowstone bears were radioinstrumented in 1961 (Craighead et al. 1963) and the population monitored for the next decade (Craighead and Craighead 1965, 1969, 1971, 1973a, 1974). Many of the current data on movements, space requirements, activity centers, nocturnal activity, reproductive biology, denning ecology, and food habits throughout the range of grizzly/brown bears have been obtained by adoption of this technique. Application was widened and further improved through use of orbiting satellites to collect data transmitted by radio collars and implanted sensors (Buechner et al. 1971; Craighead et al. 1971, 1972a). Radiotracking, mandatory reporting by hunters, scat analysis, and aerial surveys have provided the methodology upon which current management depends.

Modality-specific thermistors coupled with microtransmitters have revolutionized in situ physiological studies in bears. Radio receivers, coupled with appropriate signal transducers, are usually used in recording data (Folk 1967; Folk et al. 1972, 1976; Craighead et al. 1972c). The technology for recording data by satellite has been demonstrated (Craighead et al. 1971). Termed *biotelemetry*, the process involves implanting a thermistor sensitive to the desired modality within the body of the bear. The microtransmitter provides for remote recording of data via land-based or satellite radio receivers.

Understanding the ways in which bears depend on and utilize their habitat requires a thorough understanding of the physical, botanical, and faunal characteristics of that habitat. Through indirect evidence and direct observation, the feeding behavior of grizzly/brown bears has been documented in many parts of their natural range. The seasonal importance of food plants, carrion, and prey species has been assessed and in depth chemical analysis of many food items to determine nutritional values has been performed. Although useful, such information alone is inadequate for evaluating comprehensively the potential of a spacious wilderness habitat. The distribution and availability of the plant food base and the bear's ecological efficiency in utilizing food items must be understood. A vital, new technology developed during the 1970s provides the means quantitatively to evaluate and to rate relative habitat structure for very large biogeographic areas. Such an evaluation was recently completed for grizzly bear habitat in the Lincoln-Scapegoat Wilderness in Montana and extrapolated to an adjoining 5,200-km² area in the Bob Marshall Wilderness (Craighead et al. 1976b; Craighead and Scaggs 1979; Craighead 1980a; Craighead and Sumner 1980). First, a holistic description of the vegetation composing the grizzly bear habitat must be organized quantitatively into a type map demarcated according to zones of elevation. There are many methods in the literature for typing vegetation/land systems that could be adapted to develop habitat classification systems. In those studies cited above, forests were classified and mapped according to the forest habitat types of Daubenmire and Daubenmire (1968) and Pfister et al. (1977), while the vegetation/landtype classification was developed for the grass-shrublands of the alpine, subalpine, and temperate zones in terms of the "ecoclass method" of Daubenmire (1952), Peterken (1970), and Corliss et al. (1973).

The data derived from type mapping and from vegetation sampling allow vegetation complexes to be quantified with regard to bear food plants on a comparative basis. This information is then converted to a computer-enhanced simulation using satellite imagery. In the Scapegoat Wilderness study, the polar-orbiting LANDSAT-1 was the source of the high-altitude photographic frames (images) depicting 177-x-177-km areas. A frame is a record of spectral energy reflected from the earth's surfaces. It is composed of over 6×10^6 "pixels," each of which is a record of the brightness level of a 0.453-ha unit on the surface. The frame can be computer oriented and analyzed, pixel by pixel, for spectral value. When the vegetation characteristics of grouped pixels of similar spectral values are supplied, a user-interactive computer can be employed to identify and map all other portions of the frame having those same spectral values. Spectral values ("signatures") unique to specific vegetation groupings or complexes can then be color coded on a computer thematic map. Thus, an ecospectral classification of vegetation is constructed from a purely ecological classification, using satellite multispectral imagery and computer assistance. The resulting thematic map and summary statistical read-outs are checked in the field to develop the level of veracity (ground-truth data) and to perfect further the signature separations for the major vegetation habitat components (complexes). Also, the spectral signatures recorded for known vegetation/landform associations can be computer extrapolated directly to large unmapped geographic areas having comparable habitat structure. The final computer statistics and thematic map, corrected and verified, can then be used as an extremely valuable tool in designing bear management programs, estimating population levels, and monitoring habitat changes. Multispectral imagery mapping has unlimited potential for all aspects of wilderness, game, and forest management in any part of the world.

General Status. The grizzly bear presents a unique management problem among North American mammals because of its aggressive behavior and space requirements. The earliest management methods consisted of eliminating offending animals. Only within the last decade have serious efforts been made to manage grizzlies utilizing scientific information and interagency cooperation. The Wilderness Act of 1964 effectively prevented adverse modification of millions of acres of grizzly bear habitat in the lower 48 states. Thus, the most serious threat to the grizzly within the last 20 years has not been habitat destruction, but rather human-caused deaths. With a low reproductive rate, a history of competing with mankind for space and resources, and a propensity to attack humans occasionally, the grizzly has suffered heavy mortality. Current management must, of course, preserve existing

habitat, but equally critical is the need to reduce human-caused bear deaths throughout the range of this animal so that the birth rate equals or exceeds the mortality rate. The task is rendered yet more difficult by the need for verifiable birth and death statistics throughout millions of acres of rugged wilderness country.

Jurisdictional problems have proven especially troublesome. The bear's habitat transcends national park, national forest, and state boundaries. Management philosophies of the land agencies have varied widely, as have also their specific management objectives. Enactment of the Endangered Species Act of 1973, and resulting federal rules defining critical habitat, stimulated greater interagency cooperation, increased standardization of management objectives for development of detailed management guidelines, and synthesized a common philosophy of preservation rather than exploitation. That the grizzly bear is seriously threatened in the lower 48 states is now well established, if not well accepted.

Detailed guidelines for managing grizzly and brown bears under a wide range of habitat and jurisdictional conditions are currently being formulated by agencies responsible for their welfare (Habitat Management Guidelines for Grizzly Bears of the Greater Yellowstone Area 1976). In the lower 48 states, a recovery plan to restore the grizzly bear to nonthreatened status is being compiled through the U.S. Fish and Wildlife Service by a recovery plan leader and a group of knowledgeable biologists and administrators. Critical habitat has been delineated for the Yellowstone region (Craighead 1980b) and current investigations (discussed earlier in the text) are in progress for other regions. Recovery of the grizzly in the Yellowstone region and in two other regional areas where suitable habitat exists will require time. These distinct areas support either viable or remnant populations, and are composed of one or more "ecosystems" (ecosystem defined as a large biogeographic area supporting a common ecological vegetation classification).

The Yellowstone Ecosystem (Wyoming, Montana, and Idaho) of approximately 2.2 million hectares supports a population variously estimated at 130 to 350 grizzlies. Based on long-term population parameters (Craighead et al. 1974) and current death statistics and reproductive rates (Knight personal communication 1980), as well as on a sharp decline in utilization of winter-killed elk and bison (J. J. Craighead in preparation), the Yellowstone population could lie closer to the lower than to the higher estimate. The failure of the Interagency Grizzly Bear Study Team to make a scientific population estimate after 10 years of field effort has seriously delayed management and has created widespread public concern for the survival of the bears.

The Western Montana Region (including at least two distinct ecosystems) of over 2 million hectares is confluent with Canadian habitat with which it shares many bears. The population has not been enumerated, but preliminary data on female-cub ratios and long-term kill statistics (Greer 1980) suggest a downward trend.

The Selway-Bitterroot Region (Idaho and Montana) of over 1.2 million hectares is not yet well defined, but probably supported a viable grizzly population historically. Recent observations suggest the presence of a small resident population, but whether it is viable is unknown.

Three other regions, each of sufficient size to support viable populations in the future, are the Cabinet-Yaak in northwestern Montana and northeastern Idaho, the Selkirk Mountains in northeastern Washington and northwestern Idaho, and the northern Cascades from north-central Washington to the Canadian border. Grizzly bears are rarely observed in these areas and population structure, density, and relative numbers are unknown. Whether viable populations exist is also unknown, but if viable, then interchange with larger population centers in Canada is probably essential to their welfare and survival.

Thus, only two large population centers exist in the United States, excluding Alaska. In both of these, the Yellowstone and the western Montana region, the populations are in trouble and need precise management to lower the death rate and to maintain present habitat conditions.

In Canada and Alaska, grizzly and brown bears still have adequate habitat and present populations have not been seriously threatened. However, increased logging, mining, recreation, and energy development are not compatible with continued survival of the bears. Problems are rapidly emerging and will continue to mount in the future. Fortunately, research has accelerated to meet the challenge. As in the lower 48 states, management goals should include minimizing the death rate and preserving habitat.

Throughout North America, both research and management efforts should be focused on the largest wilderness areas of prime habitat. Space and solitude are essential for maintaining grizzly bears in perpetuity. Nonwilderness areas adjacent to wilderness must be managed as critical habitat and, where feasible, reclassified as wilderness. In nonwilderness areas, grizzlies have but short-term security. Eventually, intensified resource use will displace them. Maintaining large wilderness areas of prime habitat inviolate to energy exploitation is essential to the future of grizzly/brown bears throughout their range in North America. The threat of mining and energy exploration in wilderness areas will abate considerably in 1983 by virtue of provisions established by the Wilderness Act of 1964. The next two years, however, will be a period of great pressure for development of wilderness resources.

Bear-Human Relations. The problem of managing grizzlies and people has been most acute in American and Canadian national parks where bear-human encounters and fatal maulings have increased over the past decade. The causes for these are not well understood and solutions have been hampered by agency fears of litigation. Some of the problems have been addressed by Craighead and Craighead (1967, 1972), Cole (1972), Cowan (1972), Herrero (1972), Craighead (1973), and Martinka (1976). With the listing of the grizzly as "threatened" south of the Canadian border on 1 September 1975, the problem of how to manage bears and humans in national parks became even more acute. Fatal maulings receive national publicity. The culprit bears, and frequently other bears, are killed. Though public sentiment is aroused both for and against them, each new media-exploited incident damages further the general support for grizzlies. Investigations of incidents have too often been "in house," thereby creating a credibility gap. The cause or causes proffered in explanation of specific attacks have varied widely. Such improbable stimuli as severe thunderstorms, forest fires, perfume, cosmetics, and menstruating women have been suggested.

Most attacks can be grouped into two categories: incidents in which bears, especially females with offspring or bears defending food sources, have been startled or approached too closely; and incidents involving animals conditioned to humans from close association, generally in national park or monument campgrounds. Solutions in the first case include increased public education concerning grizzlies and their behavior and more intensive patrolling of potentially high-risk areas and trails. In spite of the best management efforts, there undoubtedly will always be some attacks by unconditioned grizzlies to the extent that they are codominant with mankind in the wilds. The risk is very low from this type of grizzly. By far the greatest danger is from man-conditioned grizzlies—those that have lost their fear and respect for humans. Such animals are attracted to human-associated scents and have learned by conditioning that these scents lead to food-rewarding experiences. Dominant and aggressive, man-conditioned grizzlies behave as aggressively toward humans as they do toward subordinate bears. There are a wide range of situations in which man-conditioned bears have attacked humans: sometimes in the campgrounds and developed areas where most of the conditioning has occurred; other times in backcountry, miles from the conditioning centers. There is no simple or sure solution for preventing this type of attack, but certain protective measures are logical. Campgrounds and developed areas must be fully sanitized. Once there is suspicion or evidence that a grizzly has become man-conditioned, the animal should be closely monitored. Radiocollaring of such bears provides an excellent surveillance mechanism. However, the technique must be used with moderation and with judgment as to when monitoring is no longer effective or justifiable. If confrontations continue, the animal must be eliminated; transport and release have not proven effective.

Man conditioning of bears is basically a result of failing to manage bears and people properly in the same environment. Human injuries or deaths can be judged preventable, thereby making the responsible agency subject to litigation. The risk of attack from man-conditioned bears can be greatly reduced by providing funds for national park rangers well trained in bear management. Patrolling campgrounds and trails and monitoring suspected animals with the same expertise and fervor directed to patrolling park highways for errant drivers would surely reduce the risk of bear-man incidents. Problem animals could be controlled before a serious accident could occur. This preferred type of situation will be much more easily accomplished once the grizzly bear populations have recovered and stabilized and the species is removed from threatened status.

Other alternatives for reducing human-bear confrontations are reduction of visitor use, protection of large areas from human visitation, or great reduction in the number of bears. Although the last alternative is not biologically acceptable, it has been seriously considered. It would certainly severely threaten the survival of the species in its natural range. A biologically sound and feasible solution is to effect recovery of threatened populations. Then, if sanitation and other management procedures fail to prevent man conditioning, the subsequent elimination of rogue animals presents a minimal threat to restored bear populations.

To effect recovery of the grizzly bear populations in the lower 48 states, it will be necessary to conduct certain types of ongoing, management-oriented research and to apply the findings rapidly. There is justifiable concern that agencies may have overresponded to the plight of a threatened species with a surfeit of research. Certainly, habitat should be defined, described, rated, and mapped for all areas inhabited by grizzlies. However, it is highly questionable whether each population unit requires intensive study, and restudy, to document denning, home ranges, population structures, and reproductive biology. Radioinstrumenting of large numbers of bears to obtain biological parameters, already well documented elsewhere, should be reevaluated. Capturing and marking places stress on a population. It can be justified to obtain initially the basic biological information essential to a better understanding of the species. However, it becomes increasingly difficult to justify such measures for each of numerous population units inhabiting basically similar biogeographic environments. Handling of bears should be kept to a minimum with marking and monitoring techniques used only to meet specific and essential research and management objectives. The large-scale marking and radioinstrumenting that characterizes much of the current applied research can hardly be justified when used as a continuous monitoring and data-gathering technique in the ongoing management process.

Finally, if the prognosis for a population unit is for a slow decline based on long-term data, or nonviable based on very low bear densities, then the population should be declared endangered until recovery is documented.

For their years of dedication and hard work in the field of wildlife biology, we extend special thanks and credit to Dr. Frank C. Craighead, Jr., Mr. Jay Sumner, Dr. Maurice Hornocker, Dr. Robert Ruff, Mr. Joel Varney, Mr. Derek Craighead, and Mr. Harry Reynolds III. Without their expertise and single-minded devotion in the field and in the laboratory, much of the current knowledge of the grizzly bear, and of many other wild species, would not be available.

LITERATURE CITED

Allen, P., ed. 1814. History of the expedition under the command of captains Lewis and Clark, to the sources of the Missouri, thence across the Rocky Mountains, and down the River Columbia to the Pacific Ocean; performed during the years 1804-5-6 by order of the government of the United States. 2 vols. Bradford & Inskeep, Philadelphia. 992pp.

Atwell, G. C.; Boone, D. L.; Gustafson, J.; and Berns, V. D. 1980. Brown bear summer use of alpine habitat on the Kodiak National Wildlife Refuge. Pages 297-305 in C. J. Martinka and K. L. McArthur, eds. Bears: their biology and management. Bear Bio. Assoc. Conf. Ser. no. 3. 375pp.

Berns, V. D.; Atwell, G. C.; and Boone, D. L. 1980. Brown bear movements and habitat use at Karluk Lake, Kodiak Island. Pages 293-296 in C. J. Martinka and K. L. McArthur, eds. Bears: their biology and management. Bear Bio. Assoc. Conf. Ser. no. 3. 375pp.

Berns, V. D., and Hensel, R. J. 1972. Radio-tracking brown bears on Kodiak Island. Pages 19-25 in S. Herrero, ed. Bears: their biology and management. IUCN New Ser. 23 (Morges). 371pp.

Black, H. C. 1958. Black bear research in New York. North Am. Wildl. Conf. Trans. 23:443-461.

Brackenridge, H. M. 1814. Views of Louisiana; together with a journal of a voyage up the Missouri River, in 1811. Pittsburgh. 304pp.

Brown, D. C.; Mulhausen, R. O.; Andrew, D. J.; and Seal, U. S. 1971. Renal function in anesthetized dormant and active bears. Am. J. Physiol. 220(1):293-298.

Buechner, H. K.; Craighead, F. C., Jr.; Craighead, J. J.; and Cote, C. E. 1971. Satellites for research on free-roaming animals. Bioscience 21(24):1201-1205.

Burroughs, R. D., ed. 1961. The natural history of the Lewis and Clark expedition. Michigan State Univ. Press, Ann Arbor. 340pp.

Burt, W. H., and Grossenheider, R. P. 1964. A field guide to the mammals. 2nd ed. Houghton-Mifflin Co., Boston. 284pp.

Choquette, L. P. E.; Gibson, G. G.; and Pearson, A. M. 1969. Helminths of the grizzly bear, Ursus arctos L., in northern Canada. Can. J. Zool. 47:167-170.

Cole, G. F. 1972. Preservation and management of grizzly bears in Yellowstone National Park. Pages 274-288 in S. Herrero, ed. Bears: their biology and management. IUCN New Ser. 23 (Morges). 371pp.

Corliss, J. C.; Pfister, R. D.; Buttery, R. F.; Hall, F. C.; Mueggler, W. F.; On, D.; Phillips, R. W.; Platts, W. S.; and Reid, J. E. 1973. Ecoclass: a method for classifying ecosystems. INT Missoula, For. Serv. Lab, USDA, For. Serv., Missoula, Mont. 52pp.

Coues, E. 1893. History of the expedition under the command of Lewis and Clark. 4 vols. New York.

Couterier, M. A. J. 1954. L'ours brun, Ursus arctos L. Marcel Couterier, Grenoble, Isére, France. 905pp.

Cowan, I. McT. 1972. The status and conservation of bears (Ursidae) of the world, 1970. Pages 343-367 in S. Herrero, ed. Bears: their biology and management. IUCN New Ser. 23 (Morges). 371pp.

———, chairman. 1974. Report of committee on the Yellowstone grizzlies. Natl. Acad. Sci., Washington, D.C. 61pp.

Craighead, F. C., Jr. 1973. They're killing Yellowstone's grizzlies. Nat. Wildl. 11(6):4–8, 17.

———. 1976. Grizzly bear ranges and movement as determined by radiotracking. Pages 97–109 in M. Pelton, J. Lentfer, and G. Folk, Eds. Bears: their biology and management. IUCN New Ser. 40 (Morges). 467pp.

———. 1979. Track of the grizzly. Sierra Club Books, San Francisco. 261pp.

Craighead, F. C., Jr., and Craighead, J. J. 1966. Trailing Yellowstone's grizzlies by radio. Natl. Geogr. (August), pp. 252–267.

———. 1969. Radiotracking of grizzly bears in Yellowstone National Park, Wyoming, 1964. Natl. Geogr. Soc. Res. Rep., 1964 Proj. Washington, D.C. Pp. 35–43.

———. 1972. Grizzly bear prehibernation and denning activities as determined by radiotracking. Wildl. Monogr. no. 32. 35pp.

———. 1973a. Radiotracking of grizzly bears and elk in Yellowstone National Park, Wyoming, 1966. Natl. Geogr. Soc. Res. Rep., 1966 Proj. Washington, D.C. Pp. 33–48.

———. 1973b. Tuning in on the grizzly. Pages 34–49 in W. H. Nault, exec. ed. Science year. World Book Sci. Annu. Field Enterprises Educ. Corp., Chicago. 443pp.

Craighead, F. C., Jr.; Craighead, J. J.; Cote, C. E.; and Buechner, H. K. 1972a. Satellite and ground radiotracking of elk. Pages 99–111 in Animal orientation and navigation: a symposium. NASA Spec. Pub. 262. Supt. of Documents, Washington, D.C. 606pp.

Craighead, F. C., Jr.; Craighead, J. J.; and Davies, R. S. 1963. Radiotracking of grizzly bears. Pages 133–148 in L. E. Slater, ed. Biotelemetry: the use of telemetry in animal behavior and physiology in relation to ecological problems. MacMillan Co., New York. 372pp.

Craighead, J. J. 1980a. Grizzly bear habitat analysis. Sect. 3: LANDSAT-1 multispectral imagery and computer analysis of grizzly bear habitat. Wildl.-Wildlands Inst., Univ. Montana, Missoula. 275pp.

———. 1980b. A proposed delineation of critical grizzly bear habitat in the Yellowstone region. Bear Bio. Assoc. Monogr. Ser. no. 1. 20pp.

Craighead, J. J.; Atwell, G.; and O'Gara, B. W. 1972b. Elk migrations in and near Yellowstone National Park. Wildl. Monogr. no. 19. 48pp.

Craighead, J. J., and Craighead, F. C., Jr. 1967. Management of bears in Yellowstone National Park. Montana Coop. Wildl. Res. Unit, Univ. Montana, Missoula. 118pp. Mimeogr.

———. 1971. Grizzly bear-man relationships in Yellowstone National Park. Bioscience 21(16):845–857.

Craighead, J. J.; Craighead, F. C., Jr.; and Hornocker, M. G. 1961. An ecological study of the grizzly bear. 3rd Annu. Rep. (summary of work accomplished, 1959-1961), Subproject 2.9. Montana Coop. Wildl. Res. Unit, Univ. Montana, Missoula. 59pp.

———. 1964. An ecological study of the grizzly bear. 5th Annu. Rep. (summary of work accomplished, 1959-1963). Montana Coop. Wildl. Res. Unit, Univ. Montana, Missoula. 33pp.

Craighead, J. J.; Craighead, F. C., Jr.; and McCutchen, H. E. 1970. Age determination of grizzly bears from fourth premolar tooth sections. J. Wildl. Manage. 34(2):353–363.

Craighead, J. J.; Craighead, F. C., Jr.; and Sumner, J. S. 1976a. Reproductive cycles and rates in the grizzly bear, *Ursus arctos horribilis,* of the Yellowstone ecosystem. Pages 337–356 in M. Pelton, J. Lentfer, and G. Folk, eds. Bears: their biology and management. IUCN New Ser. 40 (Morges). 467pp.

Craighead, J. J.; Craighead, F. C., Jr.; Varney, J. R.; and Cote, C. E. 1971. Satellite monitoring of black bear. Bioscience 21(24):1206–1212.

Craighead, J. J.; Hornocker, M. G.; and Craighead, F. C., Jr. 1969. Reproductive biology of young female grizzly bears. J. Reprod. Fert., Suppl. 6:447–475.

Craighead, J. J.; Hornocker, M. G.; Woodgerd, W.; and Craighead, F. C., Jr. 1960. Trapping, immobilizing and color-marking grizzly bears. Trans. 25th North Am. Wildl. Conf. Pp. 347–363.

Craighead, J. J., and Scaggs, G. B. 1979. Grizzly bear habitat analysis. Sect. 1: Vegetation description of grizzly bear habitat in the Scapegoat Wilderness (Ground Truth). Wildl.-Wildlands Inst., Univ. Montana, Missoula. 158pp.

Craighead, J. J., and Sumner, J. S. 1980. Grizzly bear habitat analysis. Sect. 2: Evaluation of grizzly bear food plants, food categories and habitat. Wildl.-Wildlands Inst., Univ. Montana, Missoula. 161pp.

Craighead, J. J.; Sumner, J. S.; and Varney, J. R. 1976b. Mapping grizzly bear habitat using LANDSAT multispectral imagery and computer-assisted technology. Interim Rep. Montana Coop. Wildl. Res. Unit, Univ. Montana, Missoula. 129pp.

Craighead, J. J.; Varney, J. R.; and Craighead, F. C., Jr. 1974. A population analysis of the Yellowstone grizzly bears. Bull. 40. For. and Conserv. Exp. Stn., Sch. For., Univ. Montana, Missoula. 20pp.

Craighead, J. J.; Varney, J. R.; Craighead, F. C., Jr.; and Sumner, J. S. 1972c. Telemetry experiments with a hibernating black bear. Montana Coop. Wildl. Res. Unit, Univ. Montana, Missoula. 32pp.

Crockford, A. J.; Hayes, F. A.; Jenkins, J. H.; and Feurt, S. D. 1957. Nicotine salicylate for capturing deer. J. Wildl. Manage. 21:213–220.

Daubenmire, R. 1952. Forest vegetation of northern Idaho and adjacent Washington and its bearing on concepts of vegetation classification. Ecol. Monogr. no. 22:301–330.

Daubenmire, R., and Daubenmire, J. B. 1968. Forest vegetation of eastern Washington and northern Idaho. Washington Agric. Exp. Stn. Tech. Bull. 60. 104pp.

DeVoto, B., ed. 1953. The journals of Lewis and Clark. Houghton-Mifflin Co., Boston. 504pp.

Dittrich, L., and Kronberger, H. 1962. Biologische-Anatomische Untersuchungen über die Fortplanzungsbiologie des Braunbären (*Ursus arctos* L.) und anderer Urisden in Gefongenschaft. A. Säugetierk. 28(3). 129pp.

Dobie, J. F. 1950. The Ben Lilly legend. Little, Brown & Co., Boston. 237pp.

Egbert, A. L., and Stokes, A. W. 1976. The social behavior of brown bears on an Alaskan salmon stream. Pages 41–56 in M. Pelton, J. Lentfer, and G. Folk, eds. Bear: their biology and management. IUCN New Ser. 40 (Morges). 467pp.

Erickson, A. W. 1957. Techniques for live-trapping and handling black bears. North Am. Wildl. Conf. Trans. 22:520–543.

Erickson, A. W.; Moosman, H. W.; Hensel, R. J.; and Troyer, W. A. 1968. The breeding biology of the male brown bear (*Ursus arctos*). Zoologica 53:85–105.

Erickson, A. W.; Nellor, J. E.; and Petrides, G. 1964. The black bear in Michigan. Michigan State Univ. Res. Bull. 4. 102pp.

Ewer, R. F. 1973. The carnivores. Cornell Univ. Press, Ithaca, N.Y. 494pp.

Folk, G. E., Jr. 1967. Physiological observations of subarctic bears under winter den conditions. Pages 75–85 in K. C. Fisher and F. E. South, eds. Mammalian hibernation. Oliver & Boyd, Edinburgh. 535pp.

Folk, G. E., Jr.; Folk, M. A.; and Minor, J. J. 1972. Physiological condition of three species of bears in winter dens. Pages 221–231 in S. Herrero, ed. Bears: their biology and management. IUCN New Ser. 23 (Morges). 371pp.

Folk, G. E., Jr.; Larson, A.; and Folk, M. A. 1976. Physiology of hibernating bears. Pages 373–380 in M. Pelton, J. Lentfer, and G. Folk, eds. Bears: their biology and management. IUCN New Ser. 40 (Morges). 467pp.

Fremont, J. C. 1843. A report on an exploration of the country lying between the Missouri River and the Rocky Mountains on the line of Kansas and Great Platte rivers. Report to the Senate, 28th Congr., 2nd sess. 693pp.

Glenn, L. P.; Lentfer, J. W.; Faro, J. B.; and Miller, L. H. 1976. Reproductive biology of female brown bears, McNeil River, Alaska. Pages 381–390 in M. Pelton, J. Lentfer, and E. Folk, eds. Bears: their biology and management. IUCN New Ser. 40 (Morges). 467pp.

Glenn, L. P., and Miller, L. H. 1980. Seasonal movements of an Alaska Peninsula brown bear population. Pages 307–312 in C. J. Martinka and K. L. McArthur, eds. Bears: their biology and management. Bear Bio. Assoc. Conf. Ser. no. 3. 375pp.

Greer, K. R. 1976. Managing Montana's grizzlies for the grizzlies. Pages 177–189 in M. Pelton, J. Lentfer, and G. Folk, eds. Bears: their biology and management. IUCN New Ser. 40 (Morges). 467pp.

———. 1980. Grizzly bear studies during 1979. Wildl. Invest. Lab. Proj. W-120-R-11, Work Plan 5, Study no. L-1.1. Montana State Univ., Bozeman. 22pp.

Guilday, J. E. 1968. Grizzly bears from eastern North America. Am. Midl. Nat. 79(1):247–250.

Guthrie, W. 1815. A new geographical, historical, and commercial grammar; and present state of the several kingdoms of the world. 2nd Amer. ed., improved. 2 vols. Johnson & Warner, Philadelphia. 652pp.

Habitat management guidelines for grizzly bears of the greater Yellowstone area. 1976. Steering Comm., Interagency Study Team, Montana State Univ., Bozeman. 20pp.

Hall, E. R., and Kelson, K. R. 1959. The mammals of North America. Ronald Press Co., New York. Vol. 2. 536pp.

Hamer, D.; Herrero, S. M.; and Ogilvie, R. T. 1977. Ecological studies of the Banff National Park grizzly bear, Cuthead/Wigmore region, 1976. Parks Can. Contract WR 34-76. 239pp.

———. 1978. Ecological studies of the Banff National Park grizzly bear, Cuthead/Wigmore region, 1977. Parks Can. Contract WR 35-77. 50pp.

Hamer, D.; Herrero, S. M.; Ogilvie, R. T.; and Toth, T. 1979. Ecological studies of the Banff National Park grizzly bear, Cuthead/Wigmore region, 1978. Parks Can. Contract WR 96-78. 86pp.

Hamlett, G. W. D. 1935. Delayed implantation and discontinuous development in mammals. Q. Rev. Bio. 10:432.

Hansson, A. 1947. The physiology of reproduction in mink (Mustela vison, schreb.) with special reference to delayed implantation. Acta Zool. 28:1–136.

Harding, L. E. 1976. Den-site characteristics of arctic coastal grizzly bears (Ursus arctos L.) on Richards Island, Northwest Territories, Canada. Can. J. Zool. 54(8):1357–1363.

Haynes, B. D., and Haynes, E., eds. 1966. The grizzly bear: portraits from life. Univ. Oklahoma Press, Norman. 386pp.

Hechtel, J. 1979. Behavioral ecology of a barren-ground grizzly bear female and her young in the National Petroleum Reserve, Alaska. Prelim. Rep. Montana Coop. Wildl. Res. Unit, Univ. Montana, Missoula. 11pp.

Heddleston, K. L. 1976. Physiologic characteristics of 1268 cultures of Pasteurella multocida. Am. J. Vet. Res. 37(6):745–747.

Hensel, R. S.; Troyer, W. A.; and Erickson, A. W. 1969. Reproduction in the female brown bear. J. Wildl. Manage. 33:357–365.

Herrero, S. M. 1972. Aspects of evolution and adaptation in American black bears (Ursus americanus Pallas) and brown and grizzly bears (U. arctos Linné) of North America. Pages 221–231 in S. Herrero, ed. Bears: their biology and management. IUCN New Ser. 23 (Morges). 371pp.

Hock, R. J. 1960. Seasonal variation in physiological functions of arctic ground squirrels and black bears. Pages 155–171 in C. P. Lyman and A. R. Dawe, eds. Mammalian hibernation. Harvard Univ., Cambridge. 549pp.

Holland, G. P. 1949. The Siphonaptera of Canada. Can. Dept. Agric. Publ. 817, Tech. Bull. 70. 306pp.

Holtzworth, J. M. 1930. The wild grizzlies of Alaska. G. P. Putnam's Sons, New York. 417pp.

Hornocker, M. G. 1962. Population characteristics and social and reproductive behavior of the grizzly bear in Yellowstone National Park. M.S. Thesis. Montana State Univ., Bozeman. 158pp.

Hubbard, W. P. 1960. Notorious grizzly bears. Sage Books (Alan Swallow, Denver). 205pp.

Husby, P., and McMurray, N. 1978. Seasonal food habits of grizzly bears (Ursus arctos horribilis Ord) in Northwestern Montana. Pages 109–134 in Annu. Rep. 3. Border Grizzly Proj. Sch. For., Univ. Montana, Missoula. 256pp.

Husby, P.; Mealey, S. P.; and Jonkel, C. 1977. Seasonal food habits of grizzly bears (Ursus arctos horribilis Ord) in Northwestern Montana. Pages 103–116 in Annu. Rep. 2. Border grizzly proj. Sch. For., Univ. Montana, Missoula. 134pp.

James, E. 1823. Account of an expedition from Pittsburgh to the Rocky Mountains, 1819-20. H. C. Carey & I. Lea, Philadelphia. Vol. 2. 442pp.

Jenness, R.; Erickson, A. W.; and Craighead, J. J. 1972. Some comparative aspects of milk from four species of bears. J. Mammal. 53(1):34–47.

Johnson, L. 1980. Brown bear management in Southeastern Alaska. Pages 263–270 in C. J. Martinka and K. L. McArthur, eds. Bears: their biology and management. Bear Bio. Assoc. Conf. Ser. no. 3. 375pp.

Joslin, G.; McMurray, N.; Werner, T.; Kiser, S.; and Jonkel, C. 1977. Grizzly bear response to habitat disturbance. Pages 39–98 in Subproject no. 4, Annu. Rep., Border Grizzly Proj. Sch. For., Univ. Montana, Missoula. 134pp.

Judd, S. L., and Knight, R. R. 1980. Movements of radio-instrumented grizzly bears within the Yellowstone area. Pages 359–367 in C. J. Martinka and K. L. McArthur, eds. Bears: their biology and management. Bear Bio. Assoc. Conf. Ser. no. 3. 375pp.

Knight, R. R. 1975. Yellowstone grizzly bear investigations: annual report of the Interagency Study Team, 1975. Misc. Rep. no. 9. Montana State Univ., Bozeman. 46pp.

Knight, R. R.; Basile, J.; Greer, K.; Judd, S.; Oldenburg, L.; and Roop, L. 1978. Yellowstone grizzly bear investigations: annual report of the Interagency Study Team, 1977. Natl. Park Serv., U.S.D.I. 107pp.

Kordek, W. S., and Lindzey, J. S. 1980. Preliminary analysis of female reproductive tracts from Pennsylvania black bears. Pages 159–161 *in* C. J. Martinka and K. L. McArthur, eds. Bears: their biology and management. Bear Bio. Assoc. Conf. Ser. no. 3. 375pp.

Kurtén, B. 1968. Pleistocene mammals of Europe. World Nat. Ser. Weidenfeld & Nicolson, London. 317pp.

Layser, E. F. 1978. Grizzly bears in the southern Selkirk Mountains. Northwest Sci. 52(2):77–91.

Lentfer, J. W.; Hensel, R. J.; Miller, L. H.; Glenn, L. P.; and Berns, V. D. 1972. Remarks on denning habits of Alaska brown bears. Pages 125–132 *in* S. Herrero, ed. Bears: their biology and management. IUCN New Ser. 23 (Morges). 371pp.

Leopold, A. S. 1959. Wildlife of Mexico: the game birds and mammals. Univ. California Press, Berkeley. 568pp.

Luque, M. H., and Stokes, A. W. 1976. Fishing behavior of Alaska brown bears. Pages 71–78 *in* M. Pelton, J. Lentfer, and G. Folk, eds. Bears: their biology and management. IUCN New Ser. 40 (Morges). 467pp.

McCullough, D. R. 1978. Population dynamics of the Yellowstone grizzly bear. Pap. presented at Symp. Pop. Dynam. of Large Mammals. Utah State Univ., Logan. 55 pp.

Martinka, C. J. 1974. Population characteristics of grizzly bears in Glacier National Park, Montana. J. Mammal. 55(1):21–29.

———. 1976. Ecological role and management of grizzly bears in Glacier National Park, Montana. Pages 147–156 *in* M. Pelton, J. Lentfer, and G. Folk, eds. Bears: their biology and management. IUCN New Ser. 40 (Morges). 467pp.

Mealey, S. P. 1975. The natural food habitats of free-ranging grizzly bears in Yellowstone National Park, 1973–1974. M.S. Thesis. Montana State Univ., Bozeman. 158pp.

———. 1976. A survey for grizzly bear habitat on the Mount Hebgen winter sports special use application site and adjacent areas. Contract Rep. Ski Yellowstone, Inc., West Yellowstone, Mont. 22pp.

Merriam, C. H. 1918. Review of the grizzly and big brown bears of North America (genus *Ursus*) with description of a new genus *Vetularctos*. North Am. Fauna 41:1–36.

Mills, E. A. 1919. The grizzly. Houghton-Mifflin Co., New York. 289pp.

Mundy, K. R. D. 1963. Ecology of the grizzly bear (*Ursus arctos* L.) in Glacier National Park, British Columbia. M.S. Thesis. Univ. Alberta, Edmonton. 103pp.

Mundy, K. R. D., and Flook, D. R. 1973. Background for managing grizzly bears in the national parks of Canada. Can. Wildl. Serv. Rep. Ser. 22:1–36.

Murie, A. M. 1944. The values of Mount McKinley. Fauna Natl. Parks U.S. no. 5:1–238.

Neiland, K. A. 1975. Further observations on rangiferine brucellosis in Alaskan carnivores. J. Wildl. Dis. 11(1):45–53.

Nelson, R. A.; Folk, G. E.; Pfeiffer, E. W.; Craighead, J. J.; Jonkel, C. J.; and Wellik, D. M. 1980. Behavior and biochemical adaptation of black, grizzly and polar bears. *In* E. C. Meslow, ed. Bears: their biology and management. Proc. 5th Int. Conf. Bear Res. Manage. Madison, Wis. (in preparation).

Nelson, R. A.; Jones, J. D.; Wahner, H. W.; McGill, D. B.; and Code, C. F. 1975. Nitrogen metabolism in bears. Mayo Clinic Proc. 50:141–146.

Nelson, R. A.; Wahner, H. W.; and Jones, J. D. 1971. Urea and urine volume in bears. Physiologist 14:201.

———. 1973. Metabolism of bears before, during and after winter sleep. Am. J. Physiol. 224(2):491–496.

Olsen, O. W. 1968. *Uncinaria rauschi* (Strongyloidea: Nematoda), a new species of hookworms from Alaskan bears. Can. J. Zool. 46:1113–1117.

Pearson, A. M. 1972. Population characteristics of the Northern Interior grizzly in the Yukon Territory, Canada. Pages 32–35 *in* S. Herrero, ed. Bears: their biology and management. IUCN New Ser. 23 (Morges). 371pp.

———. 1975. The northern interior grizzly bear *Ursus arctos* L. Can. Wildl. Serv. Rep. Ser. 34. 84pp.

———. 1976. Population characteristics of the arctic mountain grizzly bear. Pages 247–260 *in* M. Pelton, J. Lentfer, and G. Folk, eds. Bears: their biology and management. IUCN New Ser. 40 (Morges). 467pp.

Pearson, A. M., and Halloran, D. W. 1972. Hematology of the brown bear *Ursus arctos* from southwestern Yukon Territory, Canada. Can. J. Zool. 50(3):279–286.

Peterken, G. R. 1970. Guide to check sheet for IBP (Int. Bio. Program) areas; including a classification of vegetation for general purposes by F. R. Fosberg. IBP Handb. 4. Blackwell Sci. Publ., Oxford. 133pp.

Peterson, R. L. 1965. A well-preserved grizzly bear skull recovered from a late glacial deposit near Lake Simco, Ontario. Nature 208(5016):1233–1234.

Pfister, R. D.; Kovalchik, B. L.; Arno, S. F.; and Presby, R. C. 1977. Forest habitat types of Montana. Intermountain For. and Range Exp. Stn. Rep. U.S.D.A. For. Serv., Ogden, Utah. 174pp.

Rausch, R. L. 1953. On the status of some arctic mammals. Arctic 6:91–148.

———. 1954. Studies on the helminth fauna of Alaska. Part 21: Taxonomy, morphological variation, and ecology of *Diphyllobothrium ursi* n. sp. provis. on Kodiak Island. J. Parasitol. 40:540–563.

———. 1961. Notes on the black bear, *Ursus americanus* Pallas, in Alaska, with particular reference to dentition and growth. Z. Säugetierk. Bd. 26, H. 2:65–128.

———. 1963. Geographic variation in size in North American brown bears, *Ursus arctos* L., as indicated by condylobasal length. Can. J. Zool. 41:33–45.

Rausch, R. L.; Babero, B. B.; Rausch, R. V.; and Schiller, E. L. 1956. Studies on the helminthic fauna of Alaska. Part 27: The occurrence of larvae of *Trichinella spiralis* in Alaskan mammals. J. Parasitol. 42:259–271.

Reynolds, H. V. 1976. North slope grizzly bear studies. Alaska Fed. Aid Wildl. Rest. Rep. Proj. W-17-6 and 7, Jobs 4.8R-4.11R. 25pp.

———. 1978. Structure, status, reproductive biology, movement, distribution, and habitat utilization of a grizzly bear population in NPR-A. Proj. 105 C Studies (Work Group 3). Alaska Dept. Fish and Game. 41 pp.

———. 1979. Structure, status, reproductive biology, movement, distribution, and habitat utilization of a grizzly bear population in NPR-A (Western Brooks Range, Alaska). Final Rep. to Alaska Dept. Fish and Game, Proj. 105 C (Work Group 3). 41pp.

Reynolds, H. V.; Curatolo, J. A.; and Quimby, R. 1976. Denning ecology of grizzly bears in northeastern Alaska. Pages 403–409 *in* M. Pelton, J. Lentfer, and G. Folk, eds. Bears: their biology and management. IUCN New Ser. 40 (Morges). 467pp.

Rockwell, S. K.; Perry, J. L.; Haroldson, M.; and Jonkel, C. 1978. Vegetation studies of disturbed grizzly habitat. Pages 17–68 *in* Subproj. no. 2, Annu. Rep. Border Grizzly Proj. Sch. For., Univ. Montana, Missoula.

Rogers, L. L., and Rogers, S. M. 1976. Parasites of bears: a review. Pages 411–430 *in* M. Pelton, J. Lentfer, and G. Folk, eds. Bears: their biology and management. IUCN New Ser. 40 (Morges). 467pp.

Roop, L. J. 1975. Grizzly bear progress report W-87-R-1. Wyoming Fish and Game Dept. 60pp.

———. 1980. The Yellowstone grizzly bear: a review of past and present population estimates. Yellowstone grizzly bear investigations: Annu. Rep. of Interagency Study Team, 1978–79. U.S.D.I. Natl. Parks Serv. 91pp.

Roosevelt, T. 1907. Hunting the grisly and other sketches. Part 2: The wilderness hunter. Current Literature Publishing Co., New York. 247pp.

Russell, R. H. 1971. Summer and autumn food habits of island and mainland populations of polar bears: a comparative study. M.S. Thesis. Univ. Alberta, Edmonton. 87pp.

Scaggs, G. B. 1979. Vegetation description of potential grizzly bear habitat in the Selway-Bitterroot Wilderness Area, Montana, and Idaho. M.S. Thesis. Univ. Montana, Missoula. 148pp.

Schallenberger, A., and Jonkel, C. J. 1980. Rocky Mountain East front grizzly studies, 1979 annual report. Spec. Rep. 39. Border Grizzly Proj., Sch. For., Univ. Montana, Missoula. 207pp.

Scheffer, V. B. 1950. Growth layers on the teeth of Pinnepedia as an indication of age. Science 112(2907):309–311.

Schlegel, M. W.; Knapp, S. E.; and Millemann, R. E. 1968. Salmon poisoning disease. Part 5: Definitive hosts of the trematode vector, *Nanophyetus salmincola*. J. Parasitol. 54:770–774.

Schneider, B. 1977. Where the grizzly walks. Mountain Press Publ. Co., Missoula, Mont. 191pp.

Scholander, P. F.; Walters, V.; Hock, R.; and Irving, L. 1950. Body insulation of some arctic and tropical mammals and birds. Bio. Bull. 99:225–236.

Servheen, C., and Lee, L. C. 1979. Mission Mountains grizzly bear studies: an interim report 1976–78. Border Grizzly Proj., Montana For. Conserv. Exp. Stn., Univ. Montana, Missoula. 299pp.

Shaffer, M. L. 1978. Determining minimum viable population size: a case study of the grizzly bear (*Ursus arctos* L.). Ph.D. Diss. Duke Univ., Durham, N.C. 190pp.

Shaffer, S. C. 1971. Some ecological relationships of grizzly bears and black bears of the Apgar Mountains in Glacier National Park, Montana. M.S. Thesis. Univ. Montana, Missoula. 134pp.

Simpson, G. G. 1945. The principles of classification and a classification of mammals. Bull. Am. Mus. Nat. Hist. 85. 349pp.

South, F. E., and House, W. A. 1967. Energy metabolism in hibernation. Pages 305–324 in K. C. Fisher and F. E. South, eds. Mammalian hibernation. Oliver & Boyd, Edinburgh. 535pp.

Spiess, A. 1976. Labrador grizzly (*Ursus arctos* L.): first skeletal evidence. J. Mammal. 57(4):787–790.

Spiess, A., and Cox, S. 1977. Discovery of the skull of a grizzly bear in Labrador. Arctic 29(4):194–200.

Stebler, A. M. 1972. Conservation of the grizzly: ecologic and cultural considerations. Pages 297–303 in S. Herrero, ed. Bears: their biology and management. IUCN New Ser. 23 (Morges). 371pp.

Stonorov, D., and Stokes, A. W. 1972. Social behavior of the Alaskan brown bear. Pages 232–242 in S. Herrero, ed. Bears: their biology and management. IUCN New Ser. 23 (Morges). 371pp.

Storer, T. I., and Tevis, L. P. 1955. California grizzly. Univ. Nebraska Press, Lincoln. 335pp.

Sumner, J. S., and Craighead, J. J. 1973. Grizzly bear habitat survey in the Scapegoat Wilderness, Montana. Montana Coop. Wildl. Res. Unit, Univ. Montana, Missoula. 49pp.

Thenius, E. 1959. Ursidenphylogenese und biostratigraphie. Zeitschr. Säugertierk. 24: 78–84.

Tisch, E. L. 1961. Seasonal food habits of the black bear in the Whitefish Range of Northwestern Montana. M.S. Thesis. Univ. Montana, Missoula. 108pp.

Todd, N. B. 1970. Karyotypic fissioning and canid phylogeny. J. Theoret. Bio. 26:445–480.

Troyer, W. A., and Hensel, R. J. 1964. Structure and distribution of a Kodiak bear population. J. Wildl. Manage. 28(4):769–772.

U.S.D.A. Forest Service. 1975. Criteria for grizzly bear critical habitat identification: a state of the art compendium. U.S.D.A., For. Serv., Reg. 1, Missoula, Mont. 18pp.

U.S. Senate Committee on Appropriations. 1977. Proposed critical habitat area for grizzly bears. Special hearing before Comm. on Appropriations, 94th Congr., 2nd sess. U.S. Gov. Printing Off., Washington, D.C. 209pp.

Varney, J. R.; Craighead, J. J.; and Sumner, J. S. 1974. An evaluation of the use of ERTS-1 satellite imagery for grizzly bear habitat analysis. Paper no. E-11, 3rd Earth Res. Tech. Sat. Symp. 1:1653–1670.

Vroom, G. W.; Herrero, S.; and Ogilvie, R. T. 1980. The ecology of winter den sites of grizzly bears in Banff National Park, Alberta. Pages 321–330 in C. J. Martinka and K. L. McArthur, eds. Bears: their biology and management. Bear Bio. Assoc. Conf. Ser. no. 3. 375pp.

Walker, E. P.; Warnick, F.; Lange, K. I.; Uible, H. E.; Hamlet, S. E.; Davis, M. A.; and Wright, P. F. 1964. Mammals of the world. Johns Hopkins Press, Baltimore. Vol. 2. 953pp.

Werner, T.; Gillespie, D.; and Jonkel, C. 1978. Grizzly and black bear dens in the Border Grizzly Area. Pages 173–213 in Subproj. no. 9, Annu. Rep. no. 3, Border Grizzly Proj. Sch. For., Univ. Montana, Missoula. 256pp.

Wimsatt, A. W. 1963. Delayed implantation in the Ursidae, with particular reference to the black bear (*Ursus americanus* Pallas). Pages 49–76 in E. C. Enders, ed. Delayed implantations. Univ. Chicago Press. 318pp.

Worley, D. E.; Fox, J. C.; and Winters, J. B. 1974. Prevalence and distribution of *Trichinella spiralis* in carnivorous mammals in the United States Northern Rocky Mountain Region. Pages 597–602 in C. W. Kim, ed. Trichinellosis. Proc. 3rd Int. Conf. Trichinellosis. Intext Educ. Pub., New York. 658pp.

Worley, D. E.; Fox, J. C.; Winters, J. B.; Jacobson, R. H.; and Greer, K. R. 1976. Helminth and arthropod parasites of grizzly and black bears in Montana and adjacent areas. Pages 455–464 in M. Pelton, J. Lentfer, and G. Folk, eds. Bears: their biology and management. IUCN New Ser. 40 (Morges). 467pp.

Wright, W. H. 1909. The grizzly bear. Univ. Nebraska Press, Lincoln (1977). 274pp.

Wurster, D. H., and Benirschke, K. 1968. Comparative cytogenetic studies in the order Carnivora. Chromosoma (Berlin) 24:336–382.

Zavatsky, B. P. 1976. The use of the skull in age determination of the brown bear. Pages 275–279 in M. Pelton, J. Lentfer, and G. Folk, eds. Bears: their biology and management. IUCN New Ser. 40 (Morges). 467pp.

JOHN J. CRAIGHEAD, Wildlife-Wildlands Institute, University of Montana, Missoula, Montana 59812.

JOHN A. MITCHELL, Wildlife-Wildlands Institute, University of Montana, Missoula, Montana 59812.

26

Polar Bear

Ursus maritimus   Jack W. Lentfer

COMMON NAME. Polar bear
SCIENTIFIC NAME. *Ursus maritimus*

The first designation of polar bears as a distinct species
was *Ursus maritimus*, by Phipps (1774). The generic
and subgeneric names *Thalassarctos, Thalarctos,* and
Thalatarctos were applied later. The authority on
bears, Erdbrink (1953), approved the designation
Ursus (Thalarctos) maritimus. Thenius (1953) agreed,
citing the frequently successful breeding between
brown bears (*U. arctos*) and polar bears in captivity
to support his case. More recent workers (Kurten 1964;
Harington 1966; Manning 1971; Wilson 1976) have
concluded that the most appropriate scientific name is
Ursus maritimus Phipps.

Comparative studies of skulls and teeth of bears
by Erdbrink (1953) and Thenius (1953) indicated that
polar bears and brown bears stemmed from a common
ancestor, *Ursus etruscus*, in the early Pleistocene.
Kurten (1964) summarized most of the present knowledge on polar bear evolution.

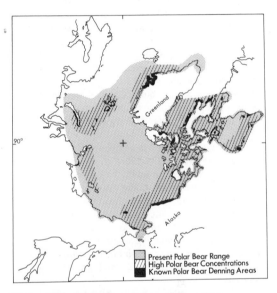

FIGURE 26.1. Distribution of the polar bear (*Ursus
maritimus*).

DISTRIBUTION

Polar bears occur only in the Northern Hemisphere,
nearly always in association with sea ice (figure 26.1).
Locations for six apparently discrete population centers
are: Wrangel Island–western Alaska, northern Alaska,
northern Canadian archipelago, Greenland, Spitsbergen–Franz Josef Land, and central Siberia (Parovschikov 1964; Uspenski 1965; Vibe 1967; Lentfer
1974*a*; Stirling et al. 1975; Jonkel et al. 1976). Discrete subpopulations also exist within the Canadian
arctic archipelago and James and Hudson bays (Stirling
et al. 1975, 1977, 1978). In general, bears are most
abundant around the perimeter of the polar basin for
200–300 kilometers offshore from land masses. They
do occur throughout the polar basin, however, and
have been recorded as far north as 88° N latitude
(Papanin 1939). In some areas bears make extensive
north-south migrations related to the southern edge of
the drifting pack ice. They have been recorded as far
south as St. Matthew Island and the Pribilof Islands

(Ray 1971) in the Bering Sea, James Bay (Jonkel et al.
1976) and Newfoundland (Smith et al. 1975) in
Canada, and Iceland in the North Atlantic. In summer,
polar bears are concentrated along the southern portion
of the drifting pack ice or in bays that retain ice. With
the onset of winter, bears move south with the ice and
concentrate along certain coastlines for denning and
feeding. Climatic changes, especially heating and cooling trends, affect polar bear habitat and thereby affect
distribution (Vibe 1967).

DESCRIPTION

Polar bears are similar to brown bears in size and
weight but are generally more elongated and less
robust. Adult males weigh 350 to 650 kilograms and
adult females weigh 150 to 300 kg. Weight and size
vary geographically with a cline of increasing skull
size from the Franz Josef Land–Spitsbergen area west

557

to the Chukchi Sea (Manning 1971). Newborn cubs weigh 600 to 700 g, are blind, do not have teeth erupted, and are covered with hair about 5 millimeters long; hair density on the back is about 650 hairs per cm² (Blix and Lentfer 1979).

Bears are completely furred other than on the nose and foot pads. The pelt has dense underfur and guard hairs of intermediate length. Fur is pure white after bears have shed in late summer and often attains a yellowish shade from staining and oxidation of seal oil.

The dental formula is 3/3, 1/1, 4/4, 2/3 = 42. Incisors have no specialized form. Canines are well developed. First premolars are rudimentary. Polar bear cheek teeth, unlike the teeth of other bears, are more suited for biting than grinding, an adaptation reflecting the evolution from an omnivorous to a carnivorous diet (figure 26.2). Canines of male polar bears have a greater basal diameter and are more robust than those of females (Kurten 1955), and upper and lower molar tooth rows of males are longer than those of females, with an overlap of about 5 percent (Larsen 1971). Tooth cementum annulations, as found in brown bears and black bears, *U. americanus* (Rausch 1961; Sauer et al. 1966; Stoneberg and Jonkel 1966; Craighead et al. 1970), are present but not as well defined in polar bears, perhaps because activity patterns differ, particularly as related to winter denning (Hensel and Sorensen 1980).

The age of a bear can be determined in two ways. Manning (1964) described how to assign skulls to age classes (young through adult) based on the closing of sutures between skull bones. Hensel and Sorensen (1980), from marking studies and recapture of known-age animals, described how to assign ages to live bears. This entailed tentative age assignment in the field based on reproductive status, body size, and tooth replacement or wear, combined with subsequent cementum annulation counts and growth regression analysis.

Polar bear milk has a fairly high fat content, about 31 percent (Cook et al. 1970). Other components of the milk were described by Baker et al. (1963a, 1963b, 1967), Cook et al. (1970), and Jenness et al. (1972). Lee et al. (1972) described baseline blood parameters.

Polar bear liver contains 15,000–30,000 units per g of vitamin A (Rodahl and Moore 1943; Rodahl 1949; Lewis and Lentfer 1967; Russell 1967). Quantity does not seem to vary with age or sex of the animal. The vitamin A content is high enough that liver is toxic to humans if eaten.

PHYSIOLOGY

Studies by Oritsland (1970), Oritsland et al. (1974), and Oritsland and Lavigne (1976) indicated that polar bears use fat and pelt insulation, heat dissipation through foot pads, and panting for thermoregulation. Blix and Lentfer (1979) suggested that newborn cubs depend on closeness of the mother and protection of a snow den to maintain body heat. A high metabolic rate and highly effective fur insulation are important for thermoregulation when cubs leave the winter maternity

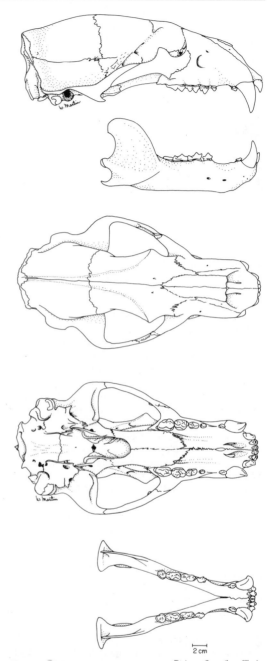

FIGURE 26.2. Skull of the polar bear (*Ursus maritimus*). From top to bottom: lateral view of cranium, lateral view of mandible, dorsal view of cranium, ventral view of cranium, dorsal view of mandible.

den at three months of age (Blix and Lentfer 1979). Cardiac physiology at rest and during exercise has been studied by Hock (1968), Folk et al. (1970, 1973), and Oritsland et al. (1977). During deep sleep, irregular heart activity (sinus arrhythmia) was observed. For instrumented animals, disturbance of sleep can probably be monitored by recording heart activity. The heart rate also provides an index for different activities, including sleeping, lying, standing, walking, and rapid walking (Oritsland et al. 1977).

REPRODUCTION

Spermatogenesis occurs between February and May, and possibly into June (Erickson 1962; Lentfer and Miller 1969; Lono 1970). Turgidity in vulvas, indicative of estrus, has been observed in females in Alaska between 21 March and 10 May. It should be noted that this condition may have occurred outside these dates but opportunities to make observations were limited (Lentfer et al. 1980). North of Alaska there is some segregation in the population, with mature males further offshore in early winter than females and younger males. Males start moving south, presumably for breeding, in late March (Lentfer et al. 1980). They locate estrous females by following their tracks. Bears paired for breeding have been observed in Alaska between 21 March and 10 May (Lentfer et al. 1980) and in Spitsbergen between 8 March and 20 June (Lono 1970). From histological examination of testes and ovaries from Spitsbergen bears, Lono (1970) concluded that breeding continues through mid-July. Implantation is apparently delayed, and the gestation period (conception to parturition) is therefore relatively long, 195 to 265 days (Uspenski 1977). Pregnant females seek out denning areas in late October and November and form maternity dens, generally in drifted snow (Harington 1968; Jonkel et al. 1972; Lentfer and Hensel 1980). Cubs are born in December and January. The number of young per litter varies from one to three. Estimates of the average litter size vary between 1.58 and 1.87 (Lono 1970; Stirling et al. 1975, 1977; Lentfer et al. 1980).

In most denning areas females and cubs break out of dens in late March and early April, but emergence is earlier in Hudson Bay, where most bears break out the first two weeks of March (Stirling et al. 1977). Cubs weigh 8–12 kg at the time of emergence (figure 26.3). Bears use the den for several days after breaking out, probably to allow cubs to become acclimated to outside temperatures before leaving the den completely. In Spitsbergen, Thomassen and Hanson (1979) observed 25 dens from the time bears broke out until they permanently left them. The time between breakout and desertion averaged 14 days and ranged from 5 to 28 days.

The age of first successful breeding for females averages 5.4 years and ranges from 3 to 7 years in the Alaskan sector of the Beaufort Sea (Lentfer et al. 1980). Stirling et al. (1975) stated that most female polar bears in the Canadian Beaufort Sea were not sexually mature until 5 years old. In the eastern Canadian arctic, however, the age of first breeding by females appears to be 1 year earlier (Stirling et al. 1978, 1979). The reason for this difference is not apparent. Maximum breeding age has not been clearly determined, but reproductively active females 21 years old have been reported (Stirling et al. 1975; Lentfer et al. 1980).

The presence of mature sperm in testes and

FIGURE 26.3. Polar bear (*Ursus maritimus*) adult female and young.

epididymides indicates that minimum and maximum ages at which males may be capable of breeding are 3 and 19 years, respectively (Lentfer and Miller 1969). Although the presence of sperm indicates breeding capability, it does not show that bears as young as 3 and as old as 19 are successful breeders. An understanding of the significance of breeding by young and old animals requires the study of social interactions and behavior. Males paired with females when captured during the breeding season ranged in age from 3 to 11 years (Lentfer et al. 1980).

The average length of breeding interval, or time between successive parturitions, is 3.6 years in the Alaskan portion of the Beaufort Sea (Lentfer et al. 1980). Young normally remain with the mother for 28 months after birth, and the female breeds again at about the time of separation (Lentfer et al. 1980). From studies in the Canadian section of the Beaufort Sea, Stirling et al. (1975) cited two instances of 3 year olds still with the females but stated that a 3-year breeding cycle is probably the most common. Lono (1970) stated that in Spitsbergen, young normally separate from the female at 17 months of age, and a 2-year breeding cycle is the most common.

Polar bears have been successfully bred and reared in captivity (Wemmer 1974; Nunley 1977), but success depends on protection from visual and audio disturbances, confinement to a small area, and a heated den.

Denning. In October and November pregnant females seek denning areas; exact times and locations depend on ice movements and freezing. Core denning areas occur on the Siberian coastline and offshore islands, Spitsbergen, Greenland, the high Canadian arctic, and Hudson Bay (Uspenski and Chernyavski 1965;

Harington 1968; Jonkel et al. 1972; Uspenski and Kistchinski 1972; Larsen 1976; Stirling et al. 1977). Dens are more sparsely located in Alaska (Lentfer and Hensel 1980) and parts of Canada (Stirling et al. 1975, 1978). Some denning also occurs on drifting pack ice (Lentfer 1975).

Dens are dug into drifted snow and are further formed as snow drifts over them. Dens range from single chambers with short tunnels to complex structures with several chambers. Bears may dig another chamber if the chamber in use forms an ice layer (Lentfer and Hensel 1980). A typical denning chamber is 80 cm high and 140–180 cm in diameter (Uspenski and Kistchinski 1972; Lentfer and Hensel 1980). Often dens are on slopes of 20–40° (Uspenski 1977) or on coastal banks or river banks (Harington 1968; Lentfer and Hensel 1980). Most dens are within 8 km of the coast, although a few are as far inland as 48 km (Harington 1968; Uspenski 1977; Lentfer and Hensel 1980).

Another type of den is a temporary shelter formed in the snow, particularly by females with young after they have deserted a maternity den. Bears on the Manitoba and Ontario coasts of Hudson Bay form earth dens in the summer (Kolenosky and Stanfield 1966; Doutt 1967; Jonkel et al. 1972). The basic types are surface pits, shallow dens, and deep burrows. All appear to be primarily for thermoregulation, but each type may be used later for shelter, protection from insects, protection from other bears, and winter dens. Bears have also been reported to use snow dens in the summer on Baffin and Bylot islands in the high Canadian arctic. Summer snow dens are of three types: shallow surface pits, shallow dens, and deep burrows (Schweinsburg 1979).

Winter denning for extended periods by bears other than parturient females has been reported occasionally in Canada (Van de Velde 1957, 1971; Harington 1968), in northern Taimyr, U.S.S.R. (Uspenski and Chernyavski 1965), and in Greenland (Freuchen 1935). There is no evidence of extended winter denning by other than parturient females in Alaska (Lentfer and Hensel 1980).

Population Characteristics. Pedersen (1945) suggested that polar bears are all part of one circumpolar population. However, subsequent studies involving individually marked bears indicated that polar bears form relatively discrete subpopulations (Jonkel 1967; Stirling and Jonkel 1972; Larsen 1972; Lentfer 1974a; Stirling et al. 1975; Uspenski 1977). Differences in skull sizes of bears from different areas also indicate that they form different subpopulations (Manning 1971; Wilson 1976). Differences in mercury levels in tissues of polar bears north and west of Alaska further suggest different subpopulations in these areas (Lentfer 1976). Serum differences as revealed by electrophoretic techniques occur between Spitsbergen and Alaskan bears (T. Larsen personal communication).

Various workers have estimated polar bear population size and abundance in various regions and throughout their range. Although most estimates have been based on limited data and broad assumptions, some have gained credence by being quoted without qualifying statements on the quality of the data.

Tovey and Scott (1957) obtained bear sighting data from Alaskan airplane hunting guides in the spring of 1957. By assuming that all bears within a 0.42-km-wide strip were reported and that bear density within 125 km of the coastline was constant throughout the polar basin, they calculated a density figure of 1 bear per 96 km² and a world population of 17,000 in this 125-km-wide strip. They stated that results of these calculations were highly speculative.

Scott et al. (1959), using the same technique as Tovey and Scott (1957) and with two more years of data (1956 and 1958) from airplane hunting guides, calculated a density figure of 1 bear per 83 km² and 2,500 bears in a 125-km-wide strip adjacent to the Alaska coast where bears occur, and 19,000 bears in a 125-km-wide strip around the rim of the polar basin. They did not estimate the number of bears occurring further than 125 km offshore.

Uspenski and Chernyavski (1965) estimated the number of denning females in specific areas in the Soviet arctic at 665, and then projected the total number of denning females in the polar basin to be 1,000–2,000. They extrapolated the total world population at 5,000–10,000 on the basis that pregnant females comprised 20 percent of the total population. They stated that their den estimates were highly speculative, and they did not estimate the number of females denning on sea ice. The number of females estimated for Wrangel Island may have been low; later work (Uspenski and Kistchinski 1972) resulted in a greater estimate of 180–200 denning females on Wrangel Island in the winter of 1969/70 and the conclusion that den numbers were underestimated from ground counts.

Harington (1964) estimated 6,000–7,000 bears for the Canadian arctic on the basis of aerial censuses in Ontario and Manitoba in conjunction with knowledge of the variations in bear productivity in different regions of the Canadian arctic, the method used by Scott et al. (1959), and an approximation of the method used by Uspenski and Chernyavski (1965). Harington stated that his method was only an informed guess and that the world polar bear population was probably well over 10,000.

Uspenski and Shilnikov (1969) estimated a world population of 10,000–15,000 polar bears and a Soviet population of 5,000–7,000 animals based on 58 bears seen on ice survey flights in the springs of 1962, 1967, and 1968. Survey planes flew 100–500 m above the ice at 270–290 km per hour. The width of the strip observed was estimated to be 500–750 m. Density in the Barents Sea area was estimated to be 1 bear per 1,080 km² in 1957 and 1 bear per 819 km² in 1968. These densities were about 10 times lower than Norwegian air survey estimates for the same general area (Larsen 1972). Extensive polar bear work from helicopters and light fixed-wing aircraft in Canada (Stirling et al. 1975, 1978) and Alaska (Lentfer et al. 1980) indicates that the Soviet figures are extremely low.

Larsen (1972), from observations of bears from sealing vessels in summer pack ice in the Spitsbergen region, estimated 1 bear per 66 km² in 1967 and 1 bear per 59 km² in 1968 and suggested a population of 1,500–1,900 for the Spitsbergen region. He stated that the use of ships in the summer pack ice was the best method yet available to estimate polar bear densities. He also stated that the great variation in polar bear population estimates obtained by both air and ship censuses must be regarded as an indicator of their inaccuracy. Larsen (1972) used his estimate for Spitsbergen; the most recent estimates for Alaska, Canada, and Russia; and his own minimum estimate of 1,000 for Greenland, and concluded that the world population of polar bears probably was closer to 20,000 than to lower figures previously presented.

Stirling et al. (1975, 1978) estimated the number of polar bears in the western Canadian arctic in a study area broadly defined as the Beaufort Sea east of 140° W longitude and south of 78° N latitude including Amundsen Gulf, M'Clure Strait, Banks Island, and western Victoria Island. Using capture-recapture data and a Seber-Jolly technique (Seber 1973), they estimated 1,000–1,700 bears, or roughly 1 bear per 37–52 km² during a period when bear numbers were undergoing large fluctuations due to natural causes.

Lentfer et al. (1980) used single-season mark-recapture data and a Seber-Jolly procedure (Seber 1973) to obtain an estimate of 320 bears in a 12,100-km² area (1 bear per 38 km²) between 17 March and 28 April. There was much movement of bears through the area and estimates of the number in the area in any three-day period ranged from 15 to 57 (1 bear per 212–807 km²).

Hearings on a proposal to waive the moratorium on taking polar bears imposed by the Marine Mammal Protection Act of 1972 resulted in several estimates of the size of Alaskan polar bear populations. The conservative estimate finally adopted was 5,700, with approximately one-third of these in the northern stock and two-thirds in the western stock (Schreiner 1979).

Adult males (age six and above) comprise 12 percent of the Alaskan population (Lentfer et al. 1980), 18 percent of the western Canadian arctic population (Stirling et al. 1975), and 17 percent of the Hudson Bay population (Stirling et al. 1977). Adult females comprise 26 percent, 19 percent, and 17 percent, respectively, of these three populations. Litter members (cubs, yearlings, and two year olds) comprise 32 percent and 26 percent of the Alaskan and western Canadian populations, respectively.

Age-specific rates of survival are not precisely known. For males, estimates of annual survival rates were 0.97 for ages 2–6 and 0.84 for older than 6. For females, estimates were 0.99 for ages 2–8 and 0.84 for older than 8 (Lentfer et al. 1980). In Alaska, Lentfer et al. (1980) suggested 25 years as the maximum age for polar bears, and in Canada, Stirling et al. (1975) suggested 25–30 years.

The mean number of young produced per adult female per year is calculated by dividing mean litter size by breeding interval. Using the best estimates of a litter size of 1.65 and a breeding interval of 3.6 years, Lentfer et al. (1980) in Alaska estimated the number of young produced per year per breeding age female to be 0.46. In Canada, Stirling et al. (1975, 1978) estimated the average litter size at 1.7 and the average reproductive interval at 3.1 years, an average annual rate of reproduction of 0.54 young per adult female.

FOOD HABITS

Polar bears feed primarily on ringed seals (*Phoca hispida*) (figure 26.4) and secondarily on bearded seals (*Erignathus barbatus*) (Stirling and Smith 1975). They also eat harp seals (*Phoca groenlandica*) and hooded seals (*Cystophora cristata*), and scavenge on carcasses of walrus (*Odobenus rosmarus*), beluga whales (*Delphinapterus leucas*), narwhales (*Monodon monoceros*), and bowhead whales (*Balaena mysticetus*). Bears have been reported to kill walrus (Kiliaan and Stirling 1978) and attack beluga whales (Degerbol and Freuchen 1935; Freeman 1973; Heyland and Hay 1976). Polar bears occasionally eat small mammals, birds, eggs, and vegetation when other food is not available (Russell 1975). The arctic fox (*Alopex lagopus*), which spends the winter on sea ice, feeds almost exclusively on remains of seals killed by bears.

Bears catch seals in several ways. In late April and May they break into ringed seal pupping dens formed on top of the ice and under the snow. In years of high seal productivity, pups constitute at least 50 percent of the seals killed by polar bears in the western Canadian arctic during the spring (Stirling and Archibald 1977). During the rest of the year, seals are taken most commonly by waiting at a breathing hole or the edge of open water. Bears also stalk seals that are hauled out on the ice in late spring and summer, though stalking is used less than waiting at holes or open water (Stirling 1974).

MORTALITY

Natural mortality factors and diseases in polar bears are poorly understood. Intraspecific strife occurs in the form of fighting between adult males during the breed-

FIGURE 26.4. Ringed seal (*Phoca hispida*), the principal food item of the polar bear (*Ursus maritimus*).

ing season, as evidenced by tracks in the snow and wounds and scars. This is not a significant mortality factor, however. Adult males may also feed on cubs, but this is not common. Alaskan workers found only one case of predation on cubs in a study that involved following tracks of approximately 700 bears that were tagged and many others that could not be located for tagging. The one case occurred north of Point Barrow on 13 April 1976, during a period when heavy ice conditions made seals difficult for bears to obtain. An adult male had followed a female and two cubs approximately 3 km and then killed and nearly completely consumed both cubs. Tracks indicated that the female had not aggressively defended the cubs (Alaska Department of Fish and Game data on file).

The only parasite that has been found in a significant number of animals is *Trichinella spiralis*. Roth (1950) reported that 28 percent of 112 bears in Greenland were positive for *Trichinella* larvae, and Lentfer (1976a) reported 64 percent of 292 Alaskan bears to be positive.

Lentfer (1976a) examined polar bear tissues collected from 1967 through 1972 for organochlorinated hydrocarbons, polychlorinated biphenyl (PCB), and mercury. Fat samples from nearly all bears examined ($N=94$) contained PCB and organochlorinated hydrocarbons including the DDT group, hexachlorobenzene, dieldrin, and endrin. However, organochlorinated hydrocarbons were at such low levels that they probably have a minimal effect on bears. The mean PCB level was relatively low (1.9 ppm in fat tissue) compared to levels, apparently nonlethal, reported in some other mammals. All liver samples examined for mercury contained low levels, with bears to the north of Alaska having significantly higher levels than bears to the west.

MANAGEMENT

The five countries with polar bears under their jurisdiction (the U.S.S.R., Norway, Greenland, Canada, and the United States) have somewhat different management philosophies. The U.S.S.R. stopped all hunting in 1956; the only animals removed from the population are a few cubs taken each year for zoos. Wrangel Island, an important denning area, received special protection by being designated a national park in 1978 (Uspenski personal communication).

Bears in the Spitsbergen area are managed by Norway. Until recent years, Norwegian sealers killed bears as predators, Spitsbergen trappers baited set-guns to obtain hides to sell, and trophy hunters from boats took bears in the summer. Between 200 and 400 bears were taken annually through 1968 (Lono 1970). In 1971, the Norwegian government stopped set-gun hunting and in 1973 enacted a five-year moratorium on killing bears because recovery of marked animals indicated fewer bears in the population than previously had been believed. The moratorium has since been extended indefinitely because Norway interprets the 1973 Oslo Agreement on Conservation of Polar Bears as prohibiting the type of hunting that could occur in Spitsbergen (T. Larsen personal communication).

Greenland was governed by Denmark until attaining home rule in May 1979. Killing polar bears in Greenland was limited to Eskimos or long-time residents for subsistence and sale of skins. The average annual kill was 125–150 bears. The new Greenland government probably will not be concerned with polar bear management or make regulation changes for a number of years (Vibe personal communication).

In Canada, polar bears have traditionally been taken by Eskimos for subsistence and for sale of skins. Harvests are regulated by quotas in hunting districts, and the annual take is about 700. The Northwest Territories, where the greatest number of bears are killed, has a regulation allowing an outside sport hunter to purchase the right to kill one of the bears of a hunting district's quota. Hunting must be by dog team and with an Eskimo guide. Relatively few bears are taken by sport hunters because snow machines are replacing dog teams, and guiding services are still in a developmental stage. Also, some Eskimos who could act as guides think that during a given hunting period they can make more money by killing several bears and selling their skins than by guiding a single sport hunter. Average prices per polar bear hide received in 1977–78 were $723 by Canadian hunters and $907 by fur auction houses (Smith 1979). Canada has an active research program related to management and considers the annual kill of 700 not excessive.

In Alaska, the only location where the United States has polar bears under its jurisdiction, management practices and authorities have changed over the years (Lentfer 1976b). Traditionally, polar bears were important in the subsistence economy and culture of Alaskan Eskimos. After commercial whaling began in the 1850s, the sale and barter of skins became especially important. Skins were used for robes and clothing, and meat was used for food. Eskimo ceremonies and dances were related to the taking of bears, and a hunter's prestige was enhanced considerably by success in killing bears. Prior to the late 1940s, nearly all Alaskan polar bear hunting was by Eskimos with dog teams, with a harvest of about 120 animals per year (Brooks and Lentfer 1966).

Trophy hunting of polar bears with the aid of aircraft began in the late 1940s. Most hunters took bears with the aid of a relatively few pilot/guides operating from Eskimo coastal villages. The use of airplanes for hunting continued after Alaska became a state and assumed game management authority in 1960. State hunting regulations became more restrictive as pilot/guides became more efficient in taking bears and more people desired to hunt. Between 1961 and 1972, an average of 260 polar bears was taken annually (Lentfer 1976b). The value of guided polar bear hunting to the Alaska economy during this period was approximately $450,000 annually (Brooks and Lentfer 1966).

In response to public sentiment and to control illegal killing by a few pilot/guides, Alaska stopped polar bear hunting with aircraft in 1972. Regulations were adopted to provide only for more acceptable sport hunting from the ground.

Polar bear population status from 1959 through 1976 was monitored in several ways, with not all methods being used for this entire period. Hunters were required to present hides and skulls to state Game Division representatives for examination so that sex, age, kill location, and other information could be obtained. Specimens, including reproductive organs, tissue samples, and blood, were obtained from hunters and provided information on breeding biology, pollutant and contaminant levels, parasites, and blood characteristics. Hunting guides were asked to report on the number and location of bears seen on hunting flights, which provided comparative density and population composition figures from year to year. Intensive research, which started in 1966, provided data from year to year on comparative densities, distribution, population composition, movements, population discreteness, breeding biology, and reproductive rates.

There was no basis for accurate estimation of population size and rates of reproduction and natural mortality during the period prior to 1972, when bears could be managed. The basis for management was maintenance of harvests at what were judged to be moderate levels, protection of females with young, and a close monitoring of population status as described above. Changes in population status that might indicate overharvest resulted in more restrictive regulations.

Polar bears were included as marine mammals in the United States Marine Mammal Protection Act of 1972. This act transferred management authority from the state to the federal government and enacted a moratorium on the taking of marine mammals by anyone other than Alaskan natives, who could take them without restriction provided waste did not occur. The previous state management program had imposed bag limits on polar bear subsistence hunters and provided complete protection for females with young.

Shortly after the Marine Mammal Act was passed, the state of Alaska requested return of management authority for certain species, including polar bears, as provided for in the act if certain conditions were met. The proposed state management program would provide for subsistence hunting and for recreational hunting, both to be done from the ground only. Females with young, and bears in dens would receive complete protection, and there would be quotas by areas and bag limits for all hunters.

The state of Alaska's request for return of management was approved by the federal government after a review period of six years, but with stipulations that were unsuitable to the state. As of 1980, all state management and research programs for marine mammals, including polar bears, had stopped. Federal management programs had not been implemented, although they were mandated by the Marine Mammal Act, and federal research programs were minimal.

Management is further complicated by the native exemption in the Marine Mammal Act, which allows

Alaskan natives to take marine mammals without restriction provided taking is not done in a wasteful manner. A court ruling (*People of Togiak* v. *United States of America,* Civil Action no. 77–0264, 1979) states that the native exemption cannot be waived. Therefore, natives cannot be regulated along with other users as part of a management program unless the population being managed is declared depleted. However, the constitution of the state of Alaska prohibits preferential management of resources for a single ethnic group; therefore, the state cannot accept management under present interpretation of the Marine Mammal Act. Polar bears cannot be considered depleted and the federal government also cannot regulate native take. An amendment to the act deleting or modifying the native exemption appears to be the only alternative for a sound management program. The act affects Alaska more than any other state and a member of Alaska's congressional delegation would be the logical one to propose an amendment. Before proposing a change, however, lawmakers would likely consider desires of Alaskan natives who would be restricted and who have much political influence. There is little incentive from other states to do away with the native exemption, and a broad environmentalist/preservationist faction would probably oppose opening the act to amendment because of concern that other parts of the act might be liberalized.

The five nations with polar bears under their jurisdiction provided for international management of polar bears in the 1973 Oslo Agreement on Conservation of Polar Bears (Lentfer 1974*b*). The agreement is based on the premise that nations have the ability to manage populations on and adjacent to their coasts. It creates a *de facto* "high seas" sanctuary for bears by not allowing them to be taken with aircraft, large motorized boats, or in areas where they have not been taken by traditional means in the past. It does provide for hunting by nationals of the five polar bear countries in their respective countries. The agreement states that nations shall protect the ecosystems of which polar bears are a part and emphasizes the need for protection of habitat components such as denning and feeding areas and migration routes. The agreement also states that countries shall conduct national research, coordinate management and research for populations that occur in more than one area of national jurisdiction, and exchange research results and harvest data. Resolutions appended to the agreement request the establishment of an international hide-marking system to control traffic in illegal hides, the protection of cubs and females with cubs, and the prohibition of hunting in denning areas when bears are moving into them or are in dens.

The cooperative aspects of the Oslo agreement relating mainly to data exchange are being followed quite well because an international Polar Bear Specialist Group is continuing to function as it did before the agreement. The United States has not enacted regulations to implement the agreement. One area of conflict involves protection of females with cubs. The native exemption to the Marine Mammal Protection Act allows natives to take bears without restriction, and females with young are taken. A resolution appended to the Polar Bear Agreement requests parties to the agreement to prohibit this, but in the United States no regulations have been written and no enforcement occurs.

Another international agreement that includes polar bears is the Convention on International Trade in Endangered Species of Wild Fauna and Flora, which came into effect in July 1975 and which more than 50 countries have signed. In each member nation a permit is required to export certain animals or parts thereof, including polar bears. A permanent record of exports is maintained by each member country.

Polar bear management and research is also considered in an international context by the Polar Bear Specialist Group affiliated with the International Union for the Conservation of Nature. The Polar Bear Group is composed of biologists specializing in polar bear research and management from each of the countries with jurisdiction over polar bears. It meets every two years to discuss current and proposed management actions and research findings and needs.

Future Management Needs. As increasing demands for petrochemicals and other resources result in increased human activity in the arctic, there will be more contacts with polar bears and effects on habitat and behavior. Conflicts between bears and humans are more common in areas where bears cannot be hunted. Management agencies should be aware of potential conflicts and habitat disturbances and establish protective measures as human developments occur.

A number of protective measures can be taken. Human activity should be reduced along the coast in polar bear denning areas during the late October–early November period, when bears come ashore to den, and also from late December through mid-April, when disturbance could cause bears to desert their dens after the cubs are born. Seismic lines, pipelines, and roads should be routed perpendicular to the coast rather than parallel and adjacent to it to reduce disturbances of bears and obstructions of their movements. Specific proposals for development, including plans for removal of snow from drift areas for roads and pads for drilling platforms and other structures, should be reviewed by wildlife specialists to minimize the impact on denning bears. No-activity zones approximately 3 km in diameter should be established around active polar bear dens.

Persons who may encounter polar bears should be indoctrinated on how to avoid conflicts. Camps to support petrochemical exploration and extraction should be established inland rather than on routes that bears normally travel along the coast. Studies to develop scaring devices and deterrents to keep bears away from camps should be continued (Wooldridge 1980; Wooldridge and Belton 1980). Garbage should be incinerated properly. On encountering a bear, persons should make noise to frighten it and stand their ground or back slowly away instead of running. A bear that comes into

a camp or settlement should not be allowed to become habituated and lose fear of humans, but should be thoroughly frightened and, if possible, driven several kilometers away with a snow machine or helicopter.

In polar bear areas there should be one-time-only seismic exploration on public lands, accomplished by treating information from seismic surveys as public property and making it available to all who might wish to evaluate oil potential on public lands. In denning areas near shore, seismic exploration should be conducted from boats during summer rather than from shorefast ice during late winter.

Drilling for and transporting oil should be done so as to minimize potential for spills into arctic marine waters. Oil could affect bears by contaminating fur and reducing its insulating value, by causing physiological damage from ingestion, and by affecting organisms in the food chain that supports bears. Problems associated with arctic marine drilling are especially critical because moving ice can break oil lines beneath drilling platforms and between wells and storage facilities. Transport ships are subject to ice damage. There are no effective methods for cleaning up oil that collects beneath the ice or is transported by wind and currents through extensive open-water lead systems in the ice.

RESEARCH

Much effort has been devoted to polar bear research since 1965, when representatives of the five countries with polar bears met in Fairbanks, Alaska, to report on the status of bears in their countries and to define informational needs. The research activity receiving the greatest emphasis has been a mark-recapture program conducted by each of the five countries. Approximately 2,000 bears were captured and marked between 1966 and 1977. Capturing and recapturing provide data on distribution, movements, breeding biology, reproductive rates, mortality rates, and estimates of abundance. Information from marking has been supplemented by a limited amount of radiotracking, now done by satellite (Kolz et al. 1978). Studies have also been directed at the delineation of critical habitat areas, especially those used for denning; productivity of denning areas, especially in Canada; physiology related mostly to thermoregulation and bioenergetics; levels of environmental contaminants; food habits and feeding behavior; estimates of population size in specific areas; and predictions of what effects disturbances of different magnitudes might have on populations.

Many of these same types of studies are continuing. Future research should also be directed toward gaining a better understanding of the relationship of bears to their habitat. This should include studies of arctic marine food chains, sea ice, ocean currents, and the effects of habitat disturbances on polar bears.

LITERATURE CITED

Baker, B. E.; Harington, C. R.; and Symes, A. L. 1963*a*. Polar bear milk. Part 1: Gross composition and fat constitution. Can. J. Zool. 41:1035–1039.

Baker, B. E.; Hatcher, V. E.; and Harington, C. R. 1967. Polar bear milk 3: Gel-electrophoretic studies of protein fractions isolated from polar bear milk and human milk. Can. J. Zool. 45:1205–1210.

Baker, B. E.; Huang, F. H.; and Harington, C. R. 1963b. The carbohydrate content of polar bear milk caesin. Biochem. Biophys. Res. Commun. 13:227–230.

Blix, A. S., and Lentfer, J. W. 1979. Modes of thermal protection in polar bear cubs: at birth and on emergence from the den. Am. J. Physiol. 236:R67–R74.

Brooks, J. W., and Lentfer, J. W. 1966. U.S. delegation brief. Pages 44–45 in Proc. 1st Int. Polar Bear Meet. U.S. Dept. Int. and Univ. Alaska.

Cook, H. W.; Lentfer, J. W.; Pearson, A. M.; and Baker, B. E. 1970. Polar bear milk. Part 4: Gross composition. fatty acid, and mineral constitution. Can. J. Zool. 48:217–219.

Craighead, J. J.; Craighead, F. C., Jr.; and McCutchen, H. E. 1970. Age determination of grizzly bears from fourth premolar sections. J. Wildl. Manage. 34:353–363.

Degerbol, M., and Freuchen, P. 1935. Mammals. Rep. 5th Thule Exped., 1921–1924. Vol. 2(4–5):109.

Doutt, J. K. 1967. Polar bear dens on the Twin Islands, James Bay, Canada. J. Mammal. 48:468–471.

Erdbrink, D. 1953. A review of fossil and recent bears of the Old World with remarks on their phylogeny based upon their dentition. Ph.D. Thesis. University of Utrecht. 597pp.

Erickson, A. W. 1962. Bear investigations. Alaska Dept. Fish Game Fed. Aid in Wildl. Restor. Seg. Rep. Proj. W-6-R-3. 8pp.

Folk, G. E.; Berberich, J. J.; and Sanders, D. K. 1973. Bradycardia of the polar bear. Arctic 26:78–79.

Folk, G. E.; Bremer, M.; and Sanders, D. K. 1970. Cardiac physiology of polar bears in winter dens. Arctic 23:130.

Freeman, M. F. 1973. Polar bear predation on beluga in the Canadian Arctic. Arctic 26:162–163.

Freuchen, P. 1935. Mammals, Part 2. Field notes and biological observations. Rep. 5th Thule Exped., 1921–1924. Vol. 2(4–5):68–278.

Harington, C. R. 1964. Polar bears and their present status. Can. Audubon 26:4–11.

———. 1966. Canadian Wildlife Service brief. Pages 9–15 in Proc. 1st Polar Bear Meet. U.S. Dept. Int. and Univ. Alaska.

———. 1968. Denning habits of the polar bear (*Ursus maritimus* Phipps). Can. Wildl. Serv. Rep. Ser. 5. 33pp.

Hensel, R. J., and Sorensen, F. E. 1980. Age determination of live polar bears. Pages 93–100 in C. J. Martinka and K. L. McArthur, eds. Bears: their biology and management. Government Printing Office, Washington, D.C.

Heyland, J. D., and Hay, K. 1976. An attack by a polar bear on a juvenile beluga. Arctic 29:56–57.

Hock, J. J. 1968. Some physical measurements of polar bears. Z. Saugtierk. 33:57–62.

Jenness, P.; Erickson, A. W.; and Craighead, J. J. 1972. Some comparative aspects of milk from four species of bears. J. Mammal. 53:39–47.

Jonkel, C. J. 1967. Life history, ecology, and biology of the polar bear: autumn 1966 studies. Can. Wildl. Serv. Prog. Notes 1. 18pp.

Jonkel, C. J.; Kolenosky, G. B.; Robertson, R. J.; and Russell, R. H. 1972. Further notes on polar bear denning habits. Pages 142–185 in S. Herrero, ed. Bears: their biology and management. IUCN Publ. New Ser. 23.

Jonkel, C. J.; Smith, P.; Stirling, I.; and Kolenosky, G. B. 1976. Notes on the present status of the polar bear in James Bay and the Belcher Islands. Can. Wildl. Serv. Occas. Pap. 26. 42pp.

Kiliaan, H. P. L., and Stirling, I. 1978. Observations on overwintering walruses in the eastern Canadian high arctic. J. Mammal. 59:197–200.

Kolenosky, G. B., and Stanfield, R. O. 1966. Polar bear of Canada. Animals 8:528–531.

Kolz, A. L.; Lentfer, J. W.; and Fallek, H. G. 1978. Polar bear tracking via satellite. 15th Rocky Mountain Bioeng. Symp. Ames, Iowa.

Kurten, B. 1955. Sex dimorphism and size trends in the cave bear, *Ursus spelaeus* Rosenmuller and Heinroth. Acta Zool. Fennica 90:1–48.

———. 1964. The evolution of the polar bear, *U. maritimus* Phipps. Acta Zool. Fennica 108:3–30.

Larsen, T. 1971. Sexual dimorphism in the molar rows of the polar bear. J. Wildl. Manage. 35:374–377.

———. 1972. Air and ship census of polar bears in Svalbard. J. Wildl. Manage. 36:562–570.

———. 1976. Polar bear den surveys in Svalbard, 1972 and 1973. Pages 199–208 in M. R. Pelton, J. W. Lentfer, and G. E. Folk, Jr., eds. Bears: their biology and management. IUCN Publ. New Ser. 40.

Lee, J.; Ronald, K.; and Oritsland, N. A. 1972. Some blood values of wild polar bears. J. Wildl. Manage. 41:520–526.

Lentfer, J. W. 1974a. Discreteness of Alaskan polar bear populations. Proc. Int. Congr. Game Bio. 11:323–329.

———. 1974b. Agreement on conservation of polar bears. Polar Rec. 17:327–330.

———. 1975. Polar bear denning on drifting sea ice. J. Mammal. 56:716–718.

———. 1976a. Environmental contaminants and parasites in polar bears. Alaska Dept. Fish Game Fed. Aid in Fish and Wildl. Res. Final Rep. Proj. 5.5R. 22pp.

———. 1976b. Polar bear management in Alaska. Pages 209–213 in M. R. Pelton, J. W. Lentfer, and G. E. Folk, Jr., eds. Bears: their biology and management. IUCN Publ. New Ser. 40.

Lentfer, J. W., and Hensel, R. J. 1980. Alaskan polar bear denning. Pages 101–108 in C. J. Martinka and K. L. McArthur, eds. Bears: their biology and management. Government Printing Office, Washington, D.C.

Lentfer, J. W.; Hensel, R. J.; Gilbert, J. R.; and Sorensen, F. E. 1980. Population characteristics of Alaskan polar bears. Pages 109–115 in C. J. Martinka and K. L. McArthur, eds. Bears: their biology and management. Government Printing Office, Washington, D.C.

Lentfer, J. W., and Miller, L. H. 1969. Report on 1968 polar bear studies. Alaska Dept. Fish Game Fed. Aid in Wildl. Restor. Seg. Rep. Proj. W-15-R-3 and W-17-1. 29pp.

Lewis, R. W., and Lentfer, J. W. 1967. The vitamin A content of polar bear liver: range and variability. Comp. Biochem. Physiol. 22:923–926.

Lono, O. 1970. The polar bear (*Ursus maritimus* Phipps) in the Svalband area. Norsk Polarinst. Skrifter 149:103pp.

Manning, T. H. 1964. Age determination in the polar bear *Ursus maritimus* Phipps. Can. Wildl. Serv. Occas. Pap. 5. 12p.

———. 1971. Geographical variation in the polar bear *Ursus maritimus* Phipps. Can. Wildl. Rep. Ser. 13. 27pp.

Nunley, L. 1977. Successful rearing of the polar bears. Int. Zoo. Yearb. 17:161–164.

Oritsland, N. A. 1970. Temperature regulation of the polar bear (*Thalarctos maritimus*). Comp. Biochem. Physiol. 37:225–233.

Oritsland, N. A., and Lavigne, D. M. 1976. Radiative surface temperatures of exercising polar bears. Comp. Biochem. Physiol. 53A:327–330.

Oritsland, N. A.; Lentfer, J. W.; and Ronald, K. 1974.

Radiative surface temperatures of the polar bear. J. Mammal. 55:459–461.

Oritsland, N. A.; Stallman, P. K.; and Jonkel, C. J. 1977. Polar bears: heart activity during rest and exercise. Comp. Biochem. Physiol. 57:139–141.

Papanin, I. 1939. Life on an ice-floe. Hutchinson, London. 240pp.

Parovschikov, V. J. 1964. Breeding of the polar bear on the Franz Josef archipeligo. Bull. Moscow Soc. Nat. 69:127–129. (In Russian.)

Pedersen, A. 1945. Der Eisbar. Verbreitung and Lebensweise. E. Brunn & Co., Copenhagen. 166pp.

Phipps, C. J. 1774. A voyage towards the north pole. J. Nourse, London. 253pp.

Rausch, R. L. 1961. Notes on the black bear, *Ursus americanus,* in Alaska with particular reference to dentition and growth. Z. Saugetierk. 26:65–128.

Ray, C. E. 1971. Polar bear and mammoth on the Pribilof Islands. Arctic 24:9–19.

Rodahl, K. 1949. The toxic effect of polar bear liver. Norsk Polarinst. Skrifter 92. 70pp.

Rodahl, K., and Moore, T. 1943. The vitamin A content and toxicity of bear and seal liver. Biochem. J. 37:166–168.

Roth, H. 1950. Nouvelles expériences sur la trichonose avec considérations spéciales sur son existence dans les régions arctiques. Rep. 18th Sess. Off. Int. Epizooties. 24pp.

Russel, R. H. 1975. The food habits of polar bears of James Bay and southwest Hudson Bay in summer and autumn. Arctic 28:117–129.

Russell, R. E. 1967. Vitamin A content of polar bear liver. Toxicon. 5:61–62.

Sauer, P. R.; Free, S.; and Browne, S. 1966. Age determination in black bears from canine tooth sections. New York Fish Game J. 13:125–139.

Schreiner, K. M. 1979. Federal Register 44(8):2540–2546.

Schweinsburg, R. E. 1979. Summer snow dens used by polar bears in the Canadian high Arctic. Arctic 32:165–169.

Scott, R. F.; Kenyon, K. W.; Buckley, J. L.; and Olson, S. T. 1959. Status and management of the polar bear and Pacific walrus. Trans. North Am. Wildl. Conf. 24:366–373.

Seber, G. A. F. 1973. The estimation of animal abundance. Hefner Press, New York. 506pp.

Smith, P.; Stirling, I.; Jonkel, C.; and Juniper, I. 1975. Notes on the present status of the polar bear (*Ursus maritimus*) in Ungava Bay and northern Laborador. Can. Wildl. Serv. Prog. Note 53. 8pp.

Smith, P. A. 1979. Resumé of the trade in polar bear skins in Canada, 1977–78. Can. Wildl. Serv. Prog. Note 3. 6pp.

Stirling, I. 1974. Midsummer observations on the behavior of wild polar bears. Can. J. Zool. 52:1191–1198.

Stirling, I.; Andriashek, D.; Latour, P.; and Calvert, W. 1975. Distribution and abundance of polar bears in the eastern Beaufort Sea. Final Rep. to Beaufort Sea Proj. Fisheries and Marine Serv., Dept. of Environ., Victoria, B.C. 59pp.

Stirling, I., and Archibald, W. R. 1977. Aspects of predation of seals by polar bears. J. Fish. Res. Bd. Can. 34:1126–1129.

Stirling, I., and Jonkel, C. J. 1972. The great white bears. Nature Can. 1:15–18.

Stirling, I.; Jonkel, C.; Smith, P.; Robertson, P.; and Cross, D. 1977. The ecology of the polar bear along the western coast of Hudson Bay. Can. Wildl. Serv. Occas. Pap. 33. 69pp.

Stirling, I.; Kiliaan, H. P. L.; Calvert, W.; and Andriashek,

D. 1979. Population ecology studies of polar bears in the area of southeastern and south Baffin Island and northern Labrador. Prog. rep. to Can. Wildl. Serv.; Esso Resour. Can. Ltd.; Aquitaine Can., Ltd.; Can. Cities Serv., Ltd.; and North West Territories Fish Wildl. Serv. 88pp.

Stirling, I.; Schweinsburg, R. E.; Calvert, W.; and Kiliaan, H. P. L. 1978. Population ecology of the polar bear along the proposed Arctic Islands gas pipeline route. Final Rep. to Environ. Manage. Serv., Can. Dept. Environ. 93pp.

Stoneberg, R. P., and Jonkel, C. J. 1966. Age determination of black bears by cementum layers. J. Wildl. Manage. 30:411–414.

Thenius, F. 1953. Zur Analyse des Gebisses des Eisbären, *Ursus* (Thalarctos) *maritimus* Phipps. Saugetierk. Mitt. pp. 1–7.

Thomassen, J., and Hansson, R. 1979. Ethological studies of the polar bear. Univ. Trondheim. 9pp. Manuscript.

Tovey, P. E., and Scott, R. F. 1957. A preliminary report on the status of polar bear in Alaska. Presented at Alaska Sci. Conf. 11pp. Mimeogr.

Uspenski, S. M. 1965. Distribution, number, and preservation of the white polar bear in the Arctic. Bull. Moscow Soc. Nat. 70:18–24. (English summary.)

———. 1977. The polar bear. Nauka, Moscow. 107pp. (English translation by Can. Wildl. Serv.)

Uspenski, S. M., and Chernyavski, F. B. 1965. "Maternity home" of polar bears. Priroda 4:81–86.

Uspenski, S. M., and Kistchinski, A. A. 1972. New data on the winter ecology of the polar bear (*Ursus maritimus* Phipps) on Wrangel Island. Pages 181–197 *in* S. Herrero, ed. Bears: their biology and management. IUCN Publ. New Ser. 23.

Uspenski, S. M., and Shilnikov, V. I. 1969. Distribution and the numbers of polar bears in the Arctic according to data of aerial ice surveys. Pages 89–102 *in* The polar bear and its conservation in the Soviet Arctic. Hydrometerol. Publ. House, Leningrad. 184pp.

Van de Velde, F. 1957. Nanuk, king of the arctic beasts. Eskimo 45:4–15.

———. 1971. Bear stories. Eskimo, New Ser. 1:7–11.

Vibe, C. 1967. Arctic animals in relation to climate fluctuations. Medd. om Gronl. 170(5). 227pp.

Wemmer, C. 1974. Design for polar bear maternity dens. Int. Zool. Yearb. 14:222–223.

Wilson, D. E. 1976. Cranial variation in polar bears. Pages 447–453 *in* M. R. Pelton, J. W. Lentfer, and G. E. Folk, Jr., eds. Bears: their biology and management. IUCN Publ. New Ser. 40.

Wooldridge, D. R. 1980. Chemical aversion conditioning of polar and black bears. Pages 167–173 *in* C. J. Martinka and K. L. McArthur, eds. Bears: their biology and management. Government Printing Office, Washington, D.C.

Wooldridge, D. R., and Belton, P. 1980. Natural and synthesized aggressive sounds as polar bear repellents. Pages 85–91 *in* C. J. Martinka and K. L. McArthur, eds. Bears: their biology and management. Government Printing Office, Washington, D.C.

JACK W. LENTFER, Alaska Department of Fish and Game, 230 South Franklin, Juneau, Alaska 99801.

27

Raccoon and Allies

Procyon lotor and Allies

John H. Kaufmann

NOMENCLATURE

COMMON NAMES. Raccoon, 'coon
SCIENTIFIC NAME. *Procyon lotor*
SUBSPECIES. *Procyon l. lotor, P. l. maritimus, P. l. solutus, P. l. litoreus, P. l. elucus, P. l. marinus, P. l. inesperatus, P. l. auspicatus, P. l. incautus, P. l. varius, P. l. megalodous, P. l. fuscipes, P. l. hirtus, P. l. excelsus, P. l. vancouverensis, P. l. pacificus, P. l. psora, P. l. pallidus, P. l. mexicanus, P. l. grinnelli, P. l. hernandezii, P. l. shufeldti, P. l. dickeyi, P. l. crassidens,* and *P. l. pumilus.*

In addition to these subspecies, Hall and Kelson (1959) recognized five closely related insular species in the Caribbean and off the west coast of Mexico. The status of some of these insular species is questionable, and indeed the entire species needs an up-to-date revision.

Other Procyonids. The northern raccoon belongs to the family Procyonidae, which also includes the crab-eating raccoon (*P. cancrivorous*) in Central and South America; the ringtails (*Bassariscus*) in North and Central America; the coatis (*Nasua* and *Nasuella*) in North, Central, and South America; the kinkajou (*Potos*) and olingo (*Bassariscyon*) in Central and South America; and probably the pandas (*Ailurus* and *Ailuropoda*) in Asia.

Among the Carnivora the Procyonidae seem most closely related to the Canidae, from which they probably diverged in the Oligocene or late Eocene. The earliest known procyonid fossils are from the Oligocene; *Procyon* has been traced back to the Pliocene, and *P. lotor* to the Pleistocene (Hollister 1916; Goldman 1950; Arata and Hutchison 1964; Stains 1967; Ewer 1973).

DISTRIBUTION

The distribution of the northern raccoon extends across Canada from Nova Scotia to British Columbia, throughout the United States except for portions of the northern Rocky Mountains and Great Basin, and south throughout Mexico and Central America (figure 27.1). Prior to the 1950s raccoons apparently were absent

FIGURE 27.1. Present distribution and subspecies of the northern raccoon (*Procyon lotor*). Modified from Hall and Kelson 1959.

from western Wyoming and western Montana. In recent years they have become common in parts of western Montana, but have been seen only rarely in western Wyoming (personal communications from M. Boyce, C. Martinka, B. O'Gara, and R. Siglin). Raccoons have also expanded their range northward in Canada in the last 30 years, reaching latitudes of 55°9'N in Ontario (Simkin 1966), 56°15'N in Manitoba (Lynch 1971), and 54°9'N in Saskatchewan (Houston and Houston 1973). There are two records (1920, 1930) from Wood Buffalo National Park at about 58°45'N latitude in Alberta (Soper 1942). Though occasional individuals have wandered to these latitudes, persistent populations are restricted to the southern parts of the provinces. Scheffer (1947) reported that raccoons transplanted from Indiana were thriving on two islands off the southeast coast of Alaska.

The raccoon was introduced into Russia in 1936. By 1955 commercial fur trapping was in progress, and

567

by 1964 the population was estimated at 40,000–45,000 (Redford 1962; Aliev and Sanderson 1966). Raccoons were also introduced into Germany in 1934, increased to 4000–5000 by 1965, and now occupy most of West Germany north of the river Main (Aliev and Sanderson 1966; Roeben 1975). There are several records from France (de Beaufort 1968).

DESCRIPTION

The body is stocky, with a broad head, pointed snout, and bushy tail. Adults range in total length from 600 to 1050 mm, including a tail of 200 to 400 mm. The feet are pentadactyl and plantigrade, with prominent nonretractile claws and naked soles. The long, thin, flexible fingers are opposable to some degree, very sensitive to tactile stimulation, and capable of delicate manipulations. The hind foot is about 80 to 140 mm long. The prominent, pointed ears are about 40 to 65 mm long (Pocock 1921; Goldman 1950; Whitney and Underwood 1952; Hall and Kelson 1959).

Adults typically vary in weight from about 3.6 kg to 9 kg, with males outweighing females by 10 to 15 percent and northern specimens generally weighing more than those from the south (Grinnell et al. 1937; Caldwell 1963; Johnson 1970). A few raccoons reach weights of up to 18 kg; the heaviest weights on record are 25.4 kg (Wood 1922) and 28.3 kg (Scott 1951). Both were of very fat males taken in the late autumn.

Pelage. A raccoon is easily recognized by its black mask on a whitish face, and the four to seven dark rings on the tail. The rings are poorly defined and sometimes interrupted below. The general body color is gray to black above, depending on the relative number of black- and white-tipped guard hairs, and paler below. The stiff guard hairs are banded, but the short, fine underfur is uniformly gray or brownish. Many raccoons have variable amounts of yellow or brown in the pelt, and occasional cinnamon and albino individuals have been reported. Fur farmers once selectively bred certain color variations (Whitney and Underwood 1952). The sexes are similar, and juveniles resemble adults.

The annual molt begins early in the spring and lasts about three months, beginning on the head and proceeding posteriorly. New fur continues to grow in throughout the summer, though the molt (loss of old hair) is completed by early summer. In the autumn the pelt thickens and pigment is lost from the proximal portions of the individual hairs. This priming, which begins ventrally and proceeds dorsally and posteriorally, takes about six weeks (Bissonette and Csech 1937; Stuewer 1942).

Skull and Dentition. The skull is broad and rounded (figure 27.2). The sagittal crest is variable in development or absent, the palate extends well beyond the last molars, and the large bullae are highly inflated on the inner side and compressed laterally, with a short central auditory canal (Goldman 1950).

The dentition is heavy, with broad, high-cusped molars and carnassials adapted for crushing rather than

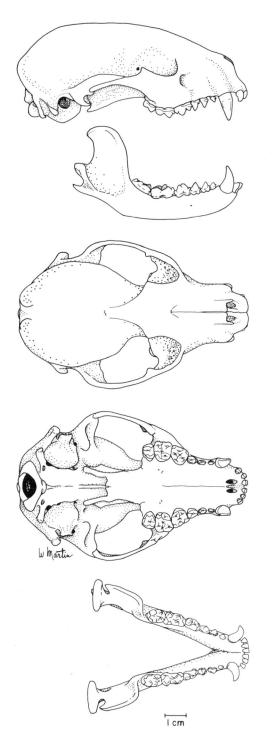

FIGURE 27.2. Skull of the northern raccoon (*Procyon lotor*). From top to bottom: lateral view of cranium, lateral view of mandible, dorsal view of cranium, ventral view of cranium, dorsal view of mandible.

cutting. The upper carnassial has a hypocone added, and on the lower the talonid is enlarged and broadened. The dental formula is 3/3, 1/1, 4/4, 2/2 = 40.

Karyotype. The diploid number of chromosomes is 38. Thirty autosomes are metacentric and six are autocentric. The metacentric X chromosome is medium in size compared to the autosomes, and the metacentric Y chromosome is relatively small (Wurster and Benirschke 1968).

PHYSIOLOGY

Temperature, Metabolism, and Body Weight. The resting body temperature of two eight-month-old, wild-caught Maine raccoons was 38.1°C, with extremes of 37.4°C and 38.7°C (Folk et al. 1957). These animals were kept in a room with a temperature of 20°C. Exposure to an ambient temperature of 5°C for three days had no effect on their body temperature. A 7.37 kg Florida raccoon had a basal rate of metabolism of 0.32 cc O_2/g hr, which is 86 percent of the value predicted by Kleiber's (1932) relation for placental mammals (personal communication from B. McNab). The heart rate of a raccoon maintained at an ambient temperature of 24°C was 200 per minute, with a P-R interval of 0.06 s and a Q-T interval of 0.16 s (Wilber 1955). This animal, which was anesthetized with Nembutal, had a rectal temperature of 37°C.

From the central United States north into Canada, raccoons typically undergo a variable period of winter dormancy that should not be confused with the deep torpor of true hibernation. The animals are easily roused, and the abdominal temperature remains above 35°C (Thorkelson and Maxwell 1974). Iowa raccoons sleeping outside on cold winter nights showed no drop in either temperature or heart rate (Folk et al. 1968). The period of winter sleep, which may last as long as four months, is sometimes broken during spells of warm weather. Because little if any food is consumed, however, the animals must rely on fat accumulated in the autumn to supply their energy needs. According to Whitney and Underwood (1952), these fat reserves account for 20 to 30 percent or more of the raccoon's autumn weight. In northern areas the winter weight loss may be 50 percent of the autumn weight in juveniles, yearlings, and adults (Mech et al. 1968). In Alabama, where raccoons are active all year, the winter weight loss of adults averaged 16 to 17 percent, with a maximum loss of 32 percent (Johnson 1970).

Central Nervous System. Because of their sensitive and dextrous forepaws, raccoons have been popular subjects for studies of sensory and motor function. They have four times as many sensory receptors in the glabrous forepaw skin as in the hindpaw (Zollman and Winkelmann 1962), and more than the closely related coati (Barker and Welker 1969). Compared to most other mammals, raccoons have relatively large numbers of cells responsive to ventral forepaw stimulation at the first order (Pubols et al. 1965), medullary (Johnson et al. 1968), thalamic (Welker and Johnson 1965), and cerebral neocortical levels of the dorsal column medial lemniscal system (Welker and Campos 1963; Welker and Seidenstein 1959; Welker et al. 1964).

The area of the raccoon's cerebral neocortex that controls forepaw motor function is only one-third the size of the forepaw sensory area, but considerably better developed than the motor areas of domestic dogs, *Canis familiaris,* and cats, *Felis catus* (Buxton and Goodman 1967; Hardin et al. 1968). This is consistent with the raccoon's abilities. The hand is an extremely sensitive sensory organ, but its motor capability, while superior to that of other nonprimates, is less well developed.

REPRODUCTION

Anatomy. The male and female reproductive tracts are generally like those of other members of the Carnivora. The ovary is completely surrounded by the *bursa ovarii,* which is intact except for a small slit on one side. The uterus differs from the bipartite uterus found in domestic dogs and cats in that, after the horns join externally to form the single uterine body, a septum keeps the lumina separate to a point near the cervix (Sanderson and Nalbandov 1973). Evidence of transuterine migration of ova was found by Sanderson and Nalbandov (1973) and by Llewellyn and Enders (1954a).

The placenta is zonary and deciduate, with two distinct types of organization (Creed and Biggers 1963). The complete annulus has an endotheliochorial labyrinth, whereas the hemophagous organ, which is attached to the annulus, is haemochorial. This peculiar organ, which regresses by term, phagocytoses maternal blood cells and may aid in fetal nutrition by digesting complex substances.

Females may have an *os clitoridis* up to 19 mm long (Rinker 1944), though Sanderson (1950) found one in only 1 of 100 females examined, and Sanderson and Nalbandov (1973) found one in only 1 of 50 females examined. Normally three pairs of teats are present: thoracic, abdominal, and inguinal.

As in other members of the Canoidea, the erectile tissue in the male's penis is reduced and replaced by a baculum. Penile spines of cornified epithelium are present on the glans (Zarrow and Clark 1968). The baculum is relatively large, ranging from 92 to 111.5 mm long. The distal third is bent downward at an angle of 90° to 135°, and the apex is expanded into a pair of condylelike lobes separated by a deep notch (Pocock 1921; Didier 1950; Burt 1960). The accessory glands include the prostate and ampullae; the Cowper's glands and seminal vesicles are absent (Sanderson and Nalbandov 1973).

Physiology. Up to 60 percent of wild and captive females first mate as juveniles, producing litters when they are one year old (Stuewer 1943b; Pope 1944; Petrides 1950; Wood 1955; Sanderson 1960; Johnson 1970; Cowan 1973). Captive juvenile males may father litters (Pope 1944; Petrides 1950), but wild males typically do not breed until their second year. Wild males may be physiologically mature at one year, but usually have no opportunity to breed because they come into breeding condition after the adult males, so the females are already pregnant (Sanderson and Nalbandov 1973). Yearling males may sire some of the litters born late in the season.

The testes were nearly always found in the scrotum, even in immature males, by Sanderson and Nalbandov (1973). Testes weights varied seasonally, reaching a maximum of about 7 g in December and a minimum of 2 to 3 g in June, July, and August. Although in a large sample males with sperm in the epididymis were found in every month, individuals had periods of three to four months, usually from June to October, when they had no sperm present. The weights of ovaries in parous females also varied seasonally, reaching peaks in November and April and a minimum in July.

That increasing photoperiod plays an important role in initiating reproduction was demonstrated by Bissonnette and Csech (1937). Males and females exposed to artificially increased day length came into breeding condition in December, rather than two to four months later, as did the controls. The time of onset of estrus in the penned controls was not affected by temperature or amount of snow. Late snows or low temperatures may restrict the movements of wild males, however, and thus delay mating (Stains 1956).

The approach of estrus is marked by swelling and reddening of the vulva. One or two weeks later the female becomes receptive to copulation for three to six days, and it is another three to four weeks before the vulva returns to its normal appearance (Stuewer 1943b; Whitney and Underwood 1952).

There is confusion in the literature concerning ovulation in raccoons. Whitney and Underwood (1952) stated that raccoons are induced ovulators. Llewellyn and Enders (1954b) agreed, claiming that four unbred females formed well-developed follicles but no corpora lutea. Although this conclusion has been frequently quoted, Sanderson and Nalbandov (1973) reported that three females kept isolated from males underwent spontaneous ovulation and pseudopregnancy, with corpora lutea forming and persisting for at least 75 days. Morris (1975) kept six females completely isolated from males during the breeding season, and two others were isolated except for visual, auditory, and olfactory stimuli. None of the first six ovulated, but one of the latter two did.

Zarrow and Clark (1968) showed that cervical stimulation may produce ovulation and pseudopregnancy in the laboratory rat (*Rattus norvegicus*), which is normally a spontaneous ovulator. Ewer (1973) suggested that the difference between spontaneous and induced ovulation may be one of degree only, and that "the fact that some carnivores do ultimately ovulate without copulation does not preclude the existence of some normal triggering effect of the stimuli resulting from mating." This may explain the seemingly contradictory observations made on raccoons.

Adult females that fail to become pregnant during their first estrus in the spring may have a second cycle two to four months later, and thus produce unusually late litters (Stuewer 1943b; Whitney and Underwood 1952; Stains 1956; Sanderson 1961a; Johnson 1970). Yearling females that fail to conceive during their first cycle probably do not breed until the next year (Stuewer 1943a; Sanderson and Nalbandov 1973). One litter per year is the rule, but if her litter is lost early the female may mate again and have a second litter (Whitney and Underwood 1952; Dellinger 1954; Sanderson and Nalbandov 1973). Some females that were experimentally brought into estrus early produced a second litter after weaning the first (Bissonnette and Csech 1938, 1939).

Mating. Near the northern limits of their range (Minnesota, North Dakota, Manitoba) raccoons mate from February until June, with peak activity in March (Schneider et al. 1971; Cowan 1973; Fritzell 1978a). Most litters are thus born in May, though some are born as late as September. Throughout most of their range in North America raccoons mate from January to March, with a peak in February, and most litters are born in April. Lehman (1968) reported a litter in Indiana that was apparently conceived in July and born in September. Late litters probably have little chance of surviving the coming winter, which may help to limit the species' spread northward (Dorney 1953). In the extreme southeastern United States mating typically occurs later than it does farther north, but more often continues later into the summer. In South Carolina (Cunningham 1962), Georgia (McKeever 1958), and Louisiana (Cagle 1949) most raccoons mate in March and bear their litters in May. In Alabama mating occurs from March until June or later, with the peak in April, and most litters are born in June (Johnson 1970). In Florida most mating takes place from December through August (Caldwell 1963). Some young are probably born throughout the year in South Carolina, Florida, and Alabama.

Though males typically mate with several females each spring, Whitney and Underwood (1952) presented evidence for some sort of pair bond between individual males and females. These authors claimed that females form bonds with familiar males about one month before mating, but will also mate with other males at the peak of estrus. They reported two cases in which a male denned with a female all winter and copulated with her early in the spring. Mating takes place immediately after the start of spring activity, and may be followed by another brief period of inactivity (Hamilton 1936; Bissonnette and Csech 1937).

Copulation may last an hour or more, with vaginal penetration maintained with the aid of the decurved baculum (Whitney and Underwood 1952). The only copulation reported from the field took place at 1405 hours in bright sunshine in a marsh (Goldman 1950). When sighted, the male had already mounted and was alternately thrusting and pausing with his head resting on the female's back. The female was passive until she terminated the copulation after a half hour by turning her head and baring her teeth.

Litters. For the last few days before giving birth a female occupies the den where her litter will be born, chewing and scratching wood from the den cavity into a rude bed. No other nest is built. At this time she becomes more aggressive toward males and other females. During the night following birth the female usually stays in the den, and her movements are greatly reduced for the next several nights. She soon resumes her nightly foraging trips, spending each day in the den with her young (Bissonnette and Csech 1937; Whitney and Underwood 1952; Sanderson 1961a; Ellis 1964; Schneider et al. 1971).

Gestation usually lasts from 63 to 65 days, with reported extremes of 54 and 70 days (Gander 1928; Goldman 1950; Whitney and Underwood 1952). Litters of one to eight have been reported, with mean litter sizes ranging from two to five (Asdell 1964). Though there is some tendency for southern females to have smaller litters, factors other than latitude are probably involved and there are no really convincing geographical correlations with litter size (Llewellyn 1952; Johnson 1970). The sex ratios reported for newborn litters have not varied significantly from 1:1.

A 21-day-old fetus has a crown-rump length of 24 mm (Llewellyn 1953). It has reached 45 mm at 35 days, 65 mm at 46 days, and 95 mm at birth (63 days). At birth a raccoon weighs about 60 to 75 g (Hamilton 1936; Stuewer 1943b). Hamilton reported the following average weights from two infants in a captive litter: 196 g at 7 days, 454 g at 19 days, 567 g at 30 days, 681 g at 40 days, and 908 g at 50 days. By autumn a juvenile may weight up to 7 kg, but full growth is not achieved until the second year.

The eyes and ear canals are closed at birth. Both usually open after 18 to 24 days (Hamilton 1936; Stuewer 1943b; Whitney and Underwood 1952), but Welker (1959) reported three infants whose eyes opened at 30 to 32 days.

The schedule of tooth eruption was worked out by Montgomery (1964). The deciduous first, second, and third incisors and canines are in place at 1 month; the deciduous second, third, and fourth premolars at 1.5 months; the deciduous first premolars and permanent first incisors at 2 months; the permanent second incisors and first molars at 2.5 months; the permanent third incisors at 3 months; and the permanent canines at 3.5 months.

At birth a raccoon is covered with hair, though the mask and tail rings are represented only by sparsely haired, pigmented skin (Montgomery 1968). The mask is fully haired at two weeks, and the tail rings at three weeks. The back, sparsely furred at birth, is well covered by the end of the first week. Guard hairs appear at six weeks, and adultlike pelage is complete when the first molt begins at age seven weeks.

Infants less than 3 weeks old squirm actively and chitter, but they cannot support their weight with their legs (Hamilton 1936; Montgomery 1969). Captives begin walking in the fourth to sixth week, and by the end of the seventh week can walk, run, and climb. At about this time there is a sharp increase in activity, and the infants begin leaving the den. Wild infants first leave the den and begin following their mother when she forages when they are 8 to 12 weeks old (Stuewer 1943b; Schneider et al. 1971).

During their stay in the den, infant raccoons are brought no solid food and consume only milk. The female sits up to nurse, and often holds one or more young up to the anterior teats with her forepaws (Whitney and Underwood 1952). Weaning begins when the young leave the den and begin to forage for themselves. Most are weaned by the time they are 16 weeks old (Stuewer 1943b; Montgomery 1969; Schneider et al. 1971), but some may continue to nurse occasionally for several months more (Scheffer 1950; Sharp and Sharp 1956; Montgomery 1969).

Females that are disturbed often move their litters to new dens (Stuewer 1943b). They usually grasp the infants with their teeth around the neck, but sometimes around the middle. The female may also move her litter from the tree den to a surface location when they begin to move freely in and out of the den. This may prevent their falling, and also permit them to begin obtaining their own food. Montgomery et al. (1970) reported three litters that were moved to the ground when 45, 47, and 63 days old.

Postden Development. Schneider et al. (1971) used radio telemetry to follow several females and their litters during the months after leaving the den. The raccoons' behavior was variable, but a typical sequence was as follows. The infants were born in a hollow tree in May and moved to ground beds during their 7th to 9th week. During the 10th week they began trying to follow their mother on her nightly excursions, but gave up within 100 m and returned to the bed. The mother made brief visits to them during the night. In the 11th week, in August, the mother began taking the cubs on brief trips. Within a week the family was traveling and bedding together. This continued through the 20th week, into October. During this period the young were weaned and began to leave their mother occasionally, bedding alone or with their siblings. In November, as freezing nights became more frequent, the entire family began sleeping together again in hollow trees. Finally, when the first permanent snow fell in late November, they denned for the winter in the same or nearby trees.

ECOLOGY

Throughout their range raccoons are found almost everywhere that water is available. They are most abundant in hardwood swamps, mangroves, flood plain forests, and fresh and salt marshes. They are also common in mesic hardwood stands, in cultivated and abandoned farmlands, and in suburban residential areas. They are relatively scarce in dry upland woodlands, especially where pines are mixed with hardwoods, and few are found in southern pine forests. Raccoons also tend to avoid large open fields, and where they have moved onto the prairies of the northern United States and southern Canada they favor buildings, woodlots, and wetlands. In deserts they are restricted to watercourses, and they are seldom found in mountains above 2000 m.

Numerous investigators have recorded age and sex ratios and computed densities from large samples of raccoons killed by hunters and trappers or live trapped on study areas. Age ratios vary widely, with juveniles generally in a minority in declining populations and high proportions of juveniles in growing populations. For example, Keefe (1953) reported ratios of 48 percent juveniles to 52 percent adults in a low population in Missouri and 67 percent juveniles to 33 percent adults in a high population. The lowest proportions of juveniles have been reported in the southeastern United States: 20 to 23 percent in Georgia (Cunningham 1962), 32 percent in Alabama (Johnson 1970), and 37 percent in Florida (Caldwell 1963).

In most studies captured males have outnumbered females by up to two to one. In some cases this may be because males move around more than females, especially in the breeding season. In a number of other studies the sex ratio was about equal, and in a few growing populations females have outnumbered males. Dellinger (1954) found 64 percent males to 36 percent females in a low population in Missouri, but only 48 percent males in a high population. The lowest proportion of males, 40 percent, was recorded in a Missouri marsh in January by Twichell and Dill (1949). This was also the densest population of raccoons ever recorded—about 400 per km².

Low densities generally range up to 5 raccoons per km², including 0.5 to 1.0 per km² (Fritzell 1978a) and 1.5 to 3.2 per km² (Cowan 1973) on the prairies of North Dakota and Manitoba, respectively. Higher densities, up to 20 raccoons per km², have been recorded by many investigators from bottomlands and marshes in the midwestern and eastern United States (e.g., Yeager and Rennels 1943; Butterfield 1944a; Dorney 1954; Urban 1970; VanDruff 1971). Johnson (1970) reported 49 per km² in a beaver swamp in Alabama, and Hoffmann and Gottschang (1977) reported 68.7 per km² in a residential suburb in Ohio.

Activity. Although there is much individual and seasonal variation in daily activity cycles, raccoons are typically active from approximately sunset to sunrise. The peak of feeding activity is generally before midnight (Sharp and Sharp 1956; Ellis 1964; Berner and Gysel 1967; Bider et al. 1968; Sunquist et al. 1969; Urban 1970; Schneider et al. 1971). Activity rarely begins more than one hour before sunset, but the return to the daytime resting site is occasionally delayed for several hours after sunrise. The most extensive daytime activity has been reported from coastal marshes (Mosby 1947; Ivey 1948), where the raccoons fed on mollusks and crustaceans at low tide, and from beach areas, where they visited fresh-water wells in the middle of the day during dry spells (Moore 1953).

Seasonal activity cycles vary with latitude. Where subfreezing temperatures and permanent snow cover prevail during the winter, raccoons typically sleep for several months in hollow trees or other sheltered sites. In the northern part of the United States inactivity usually lasts from late November or December until February, March, or early April. Most authors agree that the appearance of permanent snow cover is more important than low temperatures in initiating dormancy

(Bissonnette and Csech 1937; Stuewer 1943a; Sharp and Sharp 1956; Mech et al. 1966). Later in the winter, however, one to three days of temperatures above freezing may bring raccoons out to forage even in deep snow (Hamilton 1936; Steuwer 1943a; Whitney and Underwood 1952; Cabalka et al. 1953). In the southern states and in California raccoons are active throughout the winter except for occasional periods of several days when the temperature is unusually low (Grinnell et al. 1937; Mosby 1947; Cunningham 1962; Caldwell 1963; Johnson 1970).

Shelter. Raccoons use a variety of shelters for different purposes. The most commonly used dens for winter sleep are in hollow trees. Tree dens may be in any hollow limb or trunk of sufficient size. Den cavities examined by Stuewer (1943a) averaged 29 by 36 cm, and were mostly from 3 to 12 m above the ground. Stains (1961) found that the temperature in a tree cavity varied only from −4°C to 24°C while the outside temperature varied from −8°C to 28.5°C.

Ground burrows dug by foxes (*Urocyon cinereoargenteus* and *Vulpes vulpes*), groundhogs (*Marmota monax*), skunks (*Mephitis mephitis*), and badgers (*Taxidea taxus*) are also used, especially in areas where hollow trees are scarce. One abandoned fox burrow used by raccoons was both warmer in winter and cooler in summer than the best-insulated tree cavity examined, and had an average temperature fluctuation only one-fifth that of the tree den (Berner and Gysel 1967). Other winter den sites are in rock crevices and caves, drains, abandoned buildings, and brush piles, and on the ground in swamps under clumps of cedar. Muskrat (*Ondatra zibethicus*) houses are used rarely in marshes where hollow trees are scarce. Urban (1970) found a raccoon with all four feet frozen in a muskrat house in January.

After emerging from her winter sleep and mating, a pregnant female chooses a different den in which to have her litter. In most areas, hollow trees are again the most popular choice. Underground burrows are also used, and in marshes may be used more often than trees. Various authors have reported litters raised in rock crevices, caves and abandoned mine shafts, brush and slab piles, sawdust piles, muskrat houses, wood duck (*Aix sponsa*) boxes, and even in a domed magpie (*Pica* sp.) nest. Schneider et al. (1971) found that once the cubs leave the den where they were born it is not used again that year.

All of the sites already mentioned, plus a number of others offering less protection, are used for daytime sleep during the warmer months. In marshes, swamps, and open fields the most common sleeping site is often on the ground in herbaceous vegetation (Cabalka et al. 1953; Ellis 1964; Mech et al. 1966; Schnell 1970; Cowen 1973). Usually no nest is prepared, but at high tide raccoons in salt marshes build flat platforms of *Spartina* and *Juncus* as much as 1.6 km from dry land (Ivey 1948). Raccoons also rest during the day on bare tree limbs and mashed-down gray squirrel (*Sciurus carolinensis*) nests, and in clumps of Spanish moss (*Tillandsia usneoides*).

Sleeping sites and dens may be anywhere in the home range. Even dens where litters are born may be on the periphery of the female's range. Dens of all kinds, however, are located near water. The average distance to water in four studies was 67 to 140 m, with maximum distances of 180 to 800 m (Giles 1942; Stuewer 1943*a*; Cabalka et al. 1953; Schneider et al. 1971). Raccoons may shift their sleeping sites daily during most of the year, though some sites are used more than others (Mech et al. 1966; Frampton and Webb 1974). Even the winter den is sometimes changed—one female used three dens in the same winter (Schneider et al. 1971).

Home Range and Movements. There is great variation in the home range sizes reported for raccoons. This is caused by differences in the sex and age of the raccoons reported, and by differences in population level, habitat quality, season, length of study, and methods of obtaining and analyzing data. Most of the maximum home range diameters that have been reported fall between 1 and 3 km, with a few of up to 6.4 km (Stuewer 1943*a*; Butterfield 1944*a*; Cunningham 1962; Turkowski and Mech 1968; Sunquist et al. 1969; Urban 1970; Johnson 1970; Schneider et al. 1971). Hoffman and Gottschang (1977) reported range diameters of only 0.3 to 0.7 km in a dense suburban population, while Fritzell (1978*b*) found ranges up to 10 km across on the North Dakota prairies. Similarly, the areas calculated for raccoon home ranges vary from less than 5 ha in the Ohio suburbs to almost 5000 ha in North Dakota, with most falling in the 40- to 100-ha range elsewhere.

As pointed out by Johnson (1970), raccoons seem to restrict most of their short-term movements to relatively small, shifting areas within a larger area of general familiarity. Some parts of the larger area, especially those away from watercourses, are visited rarely if at all. The raccoons tracked by Schneider et al. (1971) visited all parts of their seasonal home ranges within any two-week period, but spent more time in marshes, swamps, and oak woods than in bogs and open fields. Relatively long movements outside of the usual home range are made to frequent temporary food sources such as cornfields or fruit trees. Grinnell et al. (1937) reported that two raccoons in California regularly made a trip of 5.6 km each way to visit a plum orchard.

Adult males generally seem to have larger home ranges than adult females, and may temporarily expand their ranges to visit several females during the mating period. Females with young greatly restrict their movements during the first few weeks after their litters are born, and juveniles occupy their mother's home range for at least the first few months after leaving the den. As juveniles become more independent in the fall, winter, or following spring their home ranges expand. Dispersal of juveniles or yearlings may further complicate matters.

There is abundant evidence that some juvenile males and females disperse from their natal home range in the fall or winter (Stuewer 1943*a*; Butterfield 1944*a*; Urban 1970). In the northern United States and Canada dispersal may be delayed until spring or summer of the second year (Schneider et al. 1971; Cowan 1973; Fritzell 1978*a*). In some cases these one-way movements cover only a few kilometers, but occasionally they are of remarkable length. Priewert (1961) reported that a Minnesota male tagged as a juvenile was recaptured three years later 264 km to the north, and Lynch (1967) told of a Manitoba male that was tagged as a juvenile in May and killed in November, 164 days later, 253 km to the northwest.

Fritzell (1978*a*) found that in North Dakota only yearling males disperse. Nine "predispersers" stayed within 5 km of their first capture site for 10 to 50 days, and then left. Ten "dispersers" were recorded as they passed through, and five "postdispersers" remained after being newly caught on the study area. The main dispersal period was May to June, but some left as late as September. One yearling moved 23.5 km between June and November, and most dispersers probably moved more than 20 km at a rate of 0.3 to 10 km per night.

Wild raccoons transplanted to unfamiliar territory provide no evidence of homing ability, though they often wander long distances in various directions. Transplants in Arkansas and South Carolina were killed or recaptured up to 240 and 288 km, respectively, from the points of release (Giles 1943; R. Wood in discussion following Butterfield 1944*a*; Nelson 1955; Frampton and Webb 1974). Pen-reared raccoons are more likely to remain near where they are released (Stuewer 1943*a*).

FOOD HABITS

The gut, like that of most carnivores, is unspecialized. There is no caecum and the intestine is relatively short (Mitchell 1905). The salivary glands are well developed (Neseni 1938), contrary to suggestions made to explain the raccoon's fabled food-washing behavior.

A survey of the many studies of raccoon food habits yields two obvious generalizations. First, raccoons are omnivorous and opportunistic. Hundreds of species of plant and animal food have been recorded from scats and stomachs, and the relative proportions of different foods vary with season and locality. Second, in most habitats plants are generally more important than animals in the raccoon's diet.

The plant foods eaten by raccoons fall into several main categories. The largest of these in number of species is fleshy fruits. Wild grapes (*Vitis* spp.), cherries (*Prunus* spp.), apples (*Malus* spp.), persimmons (*Diospyros* spp.), and berries of all kinds are eaten whenever they are available. Cultivated fruits such as peaches (*Prunus persica*), plums (*Prunus augustifolia*), figs (*Ficus carica*), citrus fruits (*Citrus* spp.), and watermelons (*Citrullus vulgaris*) are taken on occasion. Nuts, especially acorns (*Quercus* spp.), are important seasonal foods. Beech nuts (*Fagus grandifolia*), hickory nuts (*Carya* spp.), pecans (*Carya illinoensis*), and walnuts (*Juglans* spp.) are also eaten. Corn (*Zea mays*) is the most important item in the diet in some localities, and raccoons take lesser quantities of other grains such as wheat (*Triticum* spp.), oats (*Avena sativa*), millet (*Setaria italica*), and sorghum (*Sorghum vulgare*). Weed seeds, buds, fungi, and grasses and other herbaceous growth round out the list.

Among the animals eaten, more invertebrates than vertebrates are consumed. The most important animal food is crayfish (*Cambarus* spp., *Astacus* spp.). Insects, especially grasshoppers, beetles, caterpillars, and hymenopterans, constitute the next most important category. Marshes and beaches provide crabs, shrimp, oysters, mussels, echinoderms, and marine annelids, whereas snails, slugs, and earthworms are taken in terrestrial habitats.

Most classes of vertebrates are eaten. Among mammals, rodents are the most important, including cricetids, gophers, ground squirrels, and tree squirrels. Young muskrats are sometimes eaten in the spring, while adults may be taken from traps or as carrion. Cottontails and other rabbits in the genus *Sylvilagus,* shrews, and moles are also commonly eaten, and even jackrabbits (*Lepus* spp.), small raccoons, and mink (*Mustela vison*) have been recorded. The last three species are taken at least sometimes as carrion; raccoons also eat carrion from deer, cows, and horses. Garbage is a common element of the diet around farms and towns.

Raccoons eat small numbers of passerine birds and woodpeckers, and a few pheasants (*Phasianus colchicus*) and quail (mostly *Colinus virginianus*). Occasionally they also take ducks, coots, and gallinules, though the waterfowl are most often taken as cripples or carrion during the hunting season. Even crippled or dead geese (*Branta canadensis*) may be important food items on refuges where hunting is allowed (Yeager and Elder 1945). A few unusual bird species are known to have fallen prey to raccoons, including short-eared owls, *Asio flammeus* (Grinnell et al. 1937), great blue herons, *Ardea herodias* (Lopinot 1951), and double-crested cormorants, *Phalacrocorax auritus* (Yeager and Rennels 1943). Some raccoons eat bird eggs, including those of pheasants, quail, turkeys (*Meleagris gallopavo*), ducks, and shore birds. One was even seen robbing the nest of a red-tailed hawk (*Buteo jamaicensis*) 15 m above the ground (Grinnell et al. 1937).

Reptiles are relatively unimportant in the raccoon's diet, though turtles and especially their eggs suffer in some localities. Freshwater genera preyed on include *Pseudemys, Kinosternon, Chelydra,* and *Trionyx* (Erickson and Scudder 1947; Cagle 1949). The eggs of marine turtles are dug up and eaten in large numbers on some nesting beaches (Worth and Smith 1976; Burger 1977; Davis and Whiting 1977). In some southeastern marshes raccoons open the nests and eat the eggs of alligators (*Alligator mississippiensis*). Raccoons eat a few snakes, mostly *Natrix* and *Thamnophis,* and an occasional lizard.

Frogs are not as important in the diet as one might expect, given the raccoon's proclivity for aquatic foraging. Salamanders are rarely eaten.

Fishes are often taken in small numbers, and may temporarily become important food items when they are easily caught in drying pools. Centrarchids and cyprinids seem to be the most commonly caught, and gars, suckers, catfish, trout, pickerel, shad, eels, and perch have all been found in scats and stomachs.

Despite the great variety of foods eaten in different localities, raccoons in most habitats throughout their range follow the same general pattern of seasonal diet changes. Only in the spring do most raccoons eat more animal than plant food. Crayfish are the most important food at this time, followed by insects and small vertebrates (Stuewer 1943a; Baker et al. 1945; Llewellyn and Uhler 1952; Dorney 1954). In most areas more vertebrates are eaten in the spring than at any other season. In some parts of Iowa corn provides most of the spring diet (Giles 1940; Cabalka et al. 1953). Acorns are also an important food early in the spring before other foods are available.

During the summer raccoons in most habitats eat primarily fruits, with corn also important in some areas (Giles 1940; Hamilton 1940, 1951; Stuewer 1943a; Llewellyn and Uhler 1952; Cabalka et al. 1953; Baker et al. 1945; Sonenshine and Winslow 1972). The most important animal foods are crayfish, followed by insects and small vertebrates. Dearborn (1932) found that crayfish made up 59 percent of the summer diet in one area of Michigan, Tyson (1950) reported that raccoons on beaches on the Washington coast ate marine invertebrates almost exclusively, and Dorney (1954) found that muskrats and crayfish provided most of the diet in a Wisconsin marsh.

In the autumn, plants, especially fruits and corn, continue to be more important than animals in most areas (Giles 1939, 1940; Stuewer 1943a; Yeager and Rennels 1943; Baker et al. 1945; Llewellyn and Uhler 1952; Cabalka et al. 1953; Tester 1953). Dorney (1954) still found animals, led by crayfish and fish, to be more important on the marsh, and Yeager and Elder (1945) found that 1000 to 2000 crippled geese made up most of the diet on one refuge.

Acorns become the most important food in winter, supplemented by what little corn and fruit remain (Hamilton 1936; Stuewer 1943a; Baker et al. 1945; Llewellyn and Uhler 1952). A variety of invertebrates and small vertebrates is also important during this season, and in Dorney's (1954) study continued to provide the bulk of the diet. In addition, crippled waterfowl and trapped muskrats added significantly to the winter diet on the marsh.

Looking at the overall diet for the entire year, the great majority of investigators have found that plants are more important than animals. Of the authors already mentioned, only Dorney (1954) and Grinnell et al. (1937) disagreed. In addition, Wood (1954), Caldwell (1963), Johnson (1970), and Cowan (1973) all found that plants made up most of the annual diet.

BEHAVIOR

Senses, Intelligence. The raccoon's most studied sensory mode is its delicate tactile sense concentrated in the forepaws. Cole (1912) commented on the well-developed sense of touch, and Davis (1907) described the raccoon's dexterity at catching flying insects, picking up minute morsels, and opening a variety of fastenings. Davis (1907), and Whitney and Underwood (1952) also described object manipulation as a prominent part of the play behavior of juveniles. The raccoon's powers of tactile discrimination of form and size were studied more systematically by Thorgeson (1958) and Rensch and Dücker (1963). The latter found that raccoons did as well as humans in distinguishing between spheres with a difference in diameter of only 0.53 percent (2.51 versus 2.64 cm), and between objects of different shape and texture.

Several observers of raccoons have noted their fine sense of hearing and their alertness to strange sounds (Cole 1912; Tevis 1947; Whitney and Underwood 1952; Sharp and Sharp 1956). Peterson et al. (1960) tested the levels of auditory response in a number of fissiped carnivores, and found that raccoons have an upper limit of 85 kHz, with the upper level of the high sensitivity range at 50 kHz. These limits are lower than those of the ringtail (100 and 70 kHz) and coati (95 and 60 kHz), which are more specialized for hunting small prey. Visual acuity is similarly keen (Johnson 1957; Johnson and Michels 1958). Consistent with their nocturnal habits, raccoons are color blind (Davis 1907; Michels et al. 1960), but have excellent night vision. A well-developed tapetum behind the retina produces at night at bright red eyeshine with flashes of green from some angles.

The raccoon's alertness, curiosity, and intelligence are celebrated in Indian folklore and in the stories of the early settlers. Davis (1907) and Cole (1907) both found that raccoons quickly learned to open various fastening devices to get food. Davis reported that they retained the solutions for up to a year or more without practice. Cole, who tested them successfully on combinations of up to seven fastenings, found raccoons midway between domestic cats and Rhesus monkeys (*Macaca mulatta*) in speed of learning. Shell and Riopelle (1957) found raccoons comparable to cats and poorer than primates in their rate of learning in visual discrimination problems, and Johnson (1957) found that raccoons were superior in learning set formation to any nonhuman species except the higher monkeys and apes. Johnson (1970) pointed out that individual raccoons in the wild have from time to time learned to take advantage of new food sources, and that this behavior has been copied by other individuals and passed on through cultural inheritance to succeeding generations. The examples quoted included eating eggs, opening and eating melons after broken ones were left in the field, and killing sheep (*Ovis aries*).

Locomotion. The raccoon's normal gait is a slow, waddling walk, but when pressed it can run rapidly for considerable distances. Raccoons are excellent climbers, aided by their grasping forepaws and sharp claws, and readily ascend trees to feed or seek shelter. Procyonids are among the few mammals that can descend vertical trunks head first. Raccoons do this by rotating the hind foot 180° at the subtalar joint, whereas coatis rotate it at the transverse tarsal joint (McClearn 1977). Ringtails achieve rotation of the hind limb at the hip-femur, calcaneum-talus, and intratarsal joints (Trapp 1972). Raccoons take readily to water when feeding and traveling, and are strong swimmers. They regularly swim between keys in the Florida everglades and cross rivers and lakes up to 300 m wide.

Feeding and Defecation. Foraging raccoons move quickly for short periods, then spend long periods in small areas, especially where shallow water is present (Urban 1970). The jaws and especially the forepaws are used to catch and manipulate food items. On land raccoons use their delicate hands to explore every hole and cranny, and crayfish and other aquatic animals are located by feeling under rocks and in crevices. Small items are eaten where found, but large crayfish are carried to the bank or to a boulder (Tevis 1947). Live prey is sometimes rubbed or rolled under the forepaws before it is eaten. Raccoons also use their forepaws to dig crayfish and fiddler crabs from burrows in marshes (Moore 1953; Schoonover and Marshall 1951). All food is carefully and finely chewed.

Near the beach at Cape Sable, Florida, Moore (1953) found seven wells up to 50 cm deep that had been dug by raccoons during the dry season. Well-used trails led to these wells, and raccoons visited them throughout the day. They also lapped dew from leaves when fresh water was scarce. Moore speculated that the holes may have been dug to obtain crabs or crayfish, and then were revisited when fresh water seeped into them.

The raccoon's Latin name, *lotor*, means "the washer" and refers to its familiar habit of dousing food in water before eating it. A critical evaluation of the evidence, however, reveals that only captives douse their food, and that this behavior has nothing to do with cleanliness or moistening the food (Lyall-Watson 1963). Clean foods and dirty, wet and dry, are all doused with equal frequency. Dabbling is the fixed motor pattern used in searching for aquatic prey in the wild, and "washing" food is simply a substitute for this normal behavior, which has no other outlet in captivity. Ewer (1973) has further suggested that raccoons may enjoy their monotonous captive diet more if they go through the motions of catching it first.

Raccoons in the wild often deposit single scats wherever they forage, but collections of dung have been widely reported (Giles 1939, 1942; Yeager and Rennels 1943; Tevis 1947; Moore 1953; Tester 1953; Caldwell 1963). Some of these piles are of considerable size and seem to indicate use over a long period. Favorite sites are on logs, stumps, and rocks, but accumulations have also been found on the ground beside raccoon trails, on mounds of soil, and up in trees on large horizontal limbs. At least some of these dung heaps seem to be used by a number of individuals.

Communication. My notes on raccoon vocalizations are generally consistent with the scattered comments of Davis (1907), Cole (1912), Hamilton (1936), Stuewer

(1943a), Tevis (1947), and Sharp and Sharp (1956). Like the young of many other mammals, raccoons have an all-purpose juvenile distress call. Used when they are hungry or separated from their mother, it has been variously described as high-pitched "chittering," "winnowing," and a "quavering purr." Cole states that juveniles may indicate contentment with a quieter, lower-pitched purr. A similar "twittering" or "purring" is used by the mother to maintain contact with her young. Davis stated that mothers also make a short, explosive, almost inaudible "mm" sound to warn the young. Hissing and a sharp "bark" or "snort" apparently express fear or defensive threat, and are uttered by an individual in the head-down posture or as it snaps. Aggressively threatening and fighting raccoons give a variety of fierce growls, snarls, and squeals.

Several visual displays have also been described. General agitation is indicated by tail lashing, and threat by baring the teeth, laying the ears back, raising the shoulder hackles, arching the back, and raising the tail (Tevis 1947; Sharp and Sharp 1956; Barash 1974). The last three movements tend to exaggerate the animal's size. Submissiveness is indicated by pressing the head, body, and tail to the ground and retreating. Often a mixture of aggressive and defensive threat is seen, with the head thrust forward but lowered to the ground between the forepaws, and the back sharply arched.

Little is known about olfactory communication. Pocock (1921) mentioned a pair of anal glands, and I have observed anal sniffing of one raccoon by another, presumably as a means of recognition. Perhaps collections of dung function as home range markers.

Social Organization. The most common social group among raccoons consists of a mother and her young of the year. In some areas the juveniles become independent during the fall or winter. They may den for the winter alone or with siblings, but their mother typically dens alone (Preble 1940; Stuewer 1943a; Schoonover 1950; Whitney and Underwood 1952; Sharp and Sharp 1956). Farther north, the family dens together for the winter and breaks up the following spring (Sunquist 1967; Schneider et al. 1971; Cowan 1973). In North Dakota yearling males dispersed in the spring and summer of their second year, but females did not disperse and remained tolerant of other females and males (Fritzell 1977, 1978a). The most common groups recorded, except for females with their young of the year, were pairs of yearlings. These pairs gradually separated during the spring.

Although most groups found in winter dens consist of a female and her cubs, or groups of two or three siblings, other combinations are sometimes found. Whitney and Underwood (1952) stated that an adult male and female sometimes denned together, and they found up to eight individuals in one winter den. Twichell and Dill (1949) found up to nine individuals in winter dens, and several groups of two or more included both adult males and adult females. About one-half of the females denned alone, but only one-fifth of the males did. The largest group ever found in a winter den was 23: 7 juvenile males, 7 juvenile females, 1 juvenile of undetermined sex, 3 yearling or adult males, and 5 parous females (Mech and Turkowski 1966). It was suggested that these may all have been the descendants of a single female.

The only groups that have been reported other than family groups and winter denning groups are temporary feeding aggregations. Tevis (1947) saw up to nine raccoons at one time feeding along a stream in summer. The family groups and individuals in these aggregations generally avoided close contact. During the winter in Nebraska, Sharp and Sharp (1956) observed nightly aggregations at a feeding station. Seventy-five to 80 percent came in groups of two or more, though only females and their cubs also left together. Three or more females with their cubs often fed in an area 2 m in diameter. As the mating period approached in February, the adults began fighting more and finally stopped coming.

With the exceptions already noted, adult raccoons tend to be solitary. All investigators agree, however, that the home ranges of adult females overlap broadly and there is no evidence of any form of territoriality (Stuewer 1943a; Urban 1970; Johnson 1970; VanDruff 1971; Fritzell 1978a). Fritzell reported that even though the home ranges of two parous females overlapped, no encounters were noted during the 47 days when they were simultaneously monitored. In fact, no encounters were noted between adult females and any other yearling or adult raccoons even though their ranges freely overlapped those of both males and females.

The home ranges of adult males also overlap broadly at times, and most authors could find no evidence of territoriality (Stuewer 1943a; Urban 1970; VanDruff 1971). Johnson (1970) suggested, however, that dominant adult males may defend temporary, local feeding territories. After simultaneously monitoring the movements of several pairs of adult males, Fritzell (1978a) reported that the monthly home ranges of most males overlapped less than 10 percent. Furthermore, in over 80 percent of the simultaneous locations, adjacent adult males were more than 2 km apart. Only three encounters were noted; one was hostile, one was indeterminate, and in the third the two males slept and traveled together for two successive days and nights. The available data do not rule out the possibility of some form of territoriality between adult males. They may defend local areas temporarily, maintain areas of more or less exclusive use through mutual avoidance, or have priority of access to the resources (food, females, etc.) in a given area through social dominance (Kaufmann 1971).

It is generally agreed that the home ranges of adult males and females overlap each other freely and that each male may mate with several females each breeding season (Stuewer 1943a; Whitney and Underwood 1952; Mech et al. 1966; Johnson 1970; Fritzell 1977, 1978a). No associations have been reported between males and females after the mating period, however, except for one observation made by Tevis (1947). In the summer, when most females had their cubs with them, he observed a male and a female foraging together. They maintained vocal communication, rubbed against each other, and stayed within 15 m of each other. The male was attentive to the female and drove off other raccoons that approached her. It seems likely that this was a female who had failed to breed earlier and was experiencing a late estrus.

Raccoons are evidently capable of individual recognition and the formation of dominance hierarchies both in captivity and in the wild (Moore 1953; Tevis 1947; Sharp and Sharp 1956; Johnson 1970; Barash 1974). Barash found that "neighbors" (trapped 1 to 3 km apart) formed dominant-subordinate relations more quickly and had fewer fights than "strangers" (trapped more than 8 km apart). The sum of the available evidence indicates that adult raccoons of both sexes have overlapping home ranges, but that they usually remain apart through mutual avoidance except briefly during the mating period. Dominance relations between neighbors may play a role in resource allocation, especially among males.

MORTALITY

The principal causes of mortality in raccoons are the activities of man (hunting, trapping, automobiles), and malnutrition and related effects in the late winter and early spring (Stuewer 1943*a*; Whitney and Underwood 1952; Sanderson 1960; Mech et al. 1968; Johnson 1970; Cowan 1973). Juveniles are especially susceptible to starvation because they have less body tissue to lose during the winter. Because malnutrition decreases resistance to disease, parasites, and predators in raccoons of all ages, the proximate cause of death may be any or a combination of these. Well-fed raccoons can carry a heavy parasite load without apparent ill effects, and diseases alone are rarely significant in population control except at high densities.

Raccoons may live up to 16 years in the wild (Haugen 1954; Johnson 1970), but most die during their first 2 years. According to Sanderson (1951*a*, 1960), only 1 raccoon in 100 attains an age of 7 years in the wild. He calculated an adult mortality rate of 56 percent and an average longevity of 1.8 years in Missouri, with a turnover period of 7.4 years (turnover = number of years required for an age class to decrease to zero). Cowan (1973) calculated the same average longevity and turnover period for raccoons in Manitoba, and estimated that the annual mortality rate may be as high as 60 percent for yearlings and over 50 percent for the entire population. He also estimated that the prenatal mortality rate was 67 percent for yearling females but only 4 percent for adult females. In Alabama, where winter mortality is less severe and hunting and trapping pressures are less than farther north, Johnson (1970) reported an average longevity of 3.1 years and a turnover period of 10 years. He believed that yearlings suffered the highest rate of mortality, with juveniles next, and found no evidence for extreme prenatal mortality.

Predators. The list of reported raccoon predators includes pumas (*Felis concolor*), bobcats (*Felis rufus*),

wolves (*Canis lupus*), coyotes (*Canis latrans*), foxes (*Vulpes vulpes* and *Urocyon cinereoargenteus*), fishers (*Martes pennanti*), great horned owls (*Bubo virginianus*), and alligators (Whitney and Underwood 1952; Sanderson 1960; Johnson 1970; Lowery 1974). These species probably never killed many raccoons, however, and most of them are now absent or rare where raccoons live.

Parasites and Diseases. The literature on raccoon parasites and diseases is voluminous. The ever-increasing list of internal parasites includes dozens of species of nematode, cestode, trematode, and acanthocephalan. Whitney and Underwood (1952) discussed the symptoms and treatment for some of these. Raccoons are known to harbor externally a variety of ticks, fleas, and sucking lice. Stains (1956) listed all of the parasites reported for raccoons up to that time. Chandler (1942), Harkema and Miller (1964), and Schultz (1962) also made major studies of raccoon parasites, and extensive bibliographies on parasites and diseases were compiled by Halloran (1955) and Sanderson et al. (1967). According to Johnson (1970) the most harmful parasites are the nematodes *Gnathostoma procyonis*, which produce extensive lesions in the stomach and other tissues, and *Crenosoma goblei*. The latter is the most common lungworm in raccoons, and along with distemper is a major cause of chronic respiratory infection, especially in malnourished animals.

The only diseases likely to have significant impact on raccoon populations are canine distemper and rabies (Johnson 1970). Distemper, which, according to Gorham (1966), is a significant factor in controlling wild carnivore populations, is widespread in raccoons and has been diagnosed in several epizootics (Robinson et al. 1957; Habermann et al. 1958). Raccoons are also susceptible to feline distemper, or infectious enteritis (Whitney and Underwood 1952). Although apparently less common in raccoons than canine distemper, rabies sometimes causes epizootics. In 1969–72 raccoons accounted for 68 percent of 440 reported cases of rabies in Florida (Bigler et al. 1973). Rabies does not appear to spread readily from raccoons to other species, and rabid raccoons are often passive and unaggressive. Other viral diseases to which raccoons are susceptible include St. Louis encephalitis, eastern equine encephalitis, and fox encephalitis (Johnson 1970). Johnson summarized the literature on other raccoon diseases, which include the fungal disease histoplasmosis and the protozoon diseases trypanosomiasis (Chagas's disease), coccidiosis, and toxoplasmosis. In addition, raccoon populations are reservoirs for a number of bacterial diseases, including tularemia, tuberculosis, listeriosis, and leptospirosis. The high incidence of *Leptospira* in raccoon populations represents a potential public health problem in some areas.

AGE DETERMINATION

Males. Juvenile males can be told from adults by the condition of the baculum (Sanderson 1950, 1961*c*; Petrides 1950; Johnson 1970). The baculum of a juvenile is relatively small, straight, and porous at the base, has a cartilaginous tip, weighs less than 1.2 g, and is less than 90 mm long. With age the base enlarges and becomes more curved. The weight may increase to 3 g in yearlings, but any male with a baculum weighing less than 2.5 g is almost certainly immature.

The penile orifice is small in juveniles, making it impossible to extrude the penis from the sheath (Sanderson 1961c).

Females. Placental scars are present in the uteri of parous females (Sanderson 1950). The teats of parous females and those that have undergone pseudopregnancy are dark and 5 to 7 mm long, whereas those of nulliparous females are unpigmented and only 1 to 3 mm long (Stuewer 1943a; Sanderson 1950; Petrides 1950).

Both Sexes. Embryos can be aged by measuring the crown-rump length (Llewellyn 1953), and the length of the ear and hind foot are useful in aging infants up to four months old (Stuewer 1943a).

The weight of the eye lens, hardened in formalin and oven dried, can be used to determine the age in months up to one year, when it will weigh about 110 mg (Sanderson 1961b). Freezing the lens for more than one day, or allowing decomposition for more than two days, causes it to lose weight, producing inaccurate results (Montgomery 1963). Lens weight can also be used to distinguish between yearlings and adults, though with less accuracy (Johnson 1970; Cowan 1973). Measuring the total nitrogen content of the lens is more expensive and no more accurate than weighing it (Grau et al. 1970).

Several dental characteristics are useful in aging raccoons. The schedule of tooth eruption is accurate up to 110 days (Montgomery 1964; see "Litters"). Counting annuli in the dental cementum is useful up to the age of four years (Grau et al. 1970; Johnson 1970), and the size of the pulp cavity is useful in distinguishing among raccoons that are one, two, or more than two years old (Johnson 1970). Tooth wear is helpful in telling three-year-old animals from younger ones (Stuewer 1943a; Petrides 1950; Johnson 1970).

Skeletal characters can also be used. X-rays of the distal ends of the radius and ulna reveal that juveniles have broad epiphyses, whereas yearlings and adults have epiphyses that are narrow or fused with the shaft. Yearlings and adults cannot reliably be separated because the epiphyses in some adults are not completely fused (Petrides 1950, 1959; Sanderson 1961c). Sundell (1956) found that, while fusion of the epiphyses is convenient, the most accurate basis for age estimation is the maximum length of the humerus, radius, and ulna. The degree of closure of the cranial sutures is useful for determining relative age classes in males up to five years old (Grau et al. 1970).

Total body weight is the most convenient age indicator, but also the least reliable. It is fairly accurate for the first six months if reference curves are constructed using local, known-aged animals (Sanderson 1961c). Body weight may also be helpful, especially for males, in distinguishing between juveniles and older animals in the autumn (Stuewer 1943a; Sanderson 1950; Dellinger 1954; Johnson 1970). It is important to develop local criteria, however, as fall juveniles in Alabama weighed only 2.7 to 3.2 kg, compared to 6 to 8 kg in Missouri. Because of the considerable variation in growth rates in different localities, it is best to develop local criteria for aging by any of the above methods.

ECONOMIC STATUS AND MANAGEMENT

Economic Status. Though the demand for its pelt has fluctuated, the raccoon has long been an important fur species (Sanderson 1951b; U.S. Department of Commerce, Business and Defense Services Administration 1966; Hubert 1977, 1978; Deems and Pursley 1978; personal communication from G. Sanderson 1978). In the 1920s, when the demand was high, the prices paid for pelts averaged $5 to $6 each, with maximum of $15. As raccoon fur became less popular the price dropped in the 1930s, rose temporarily during World War II to the 1920s level, and then dropped to an average of $1 to $3 per pelt in the 1950s and 1960s. As other furs became unavailable the price rose rapidly in the 1970s to the current level of $20 to $30 per pelt. Most raccoon pelts from the United States are exported to Canada and Europe, especially West Germany. The skins are often sheared and dyed, and sold as imitation mink, otter, and seal.

The number of raccoons killed for their fur in the United States averaged less than 300,000 per year in the 1930s, climbed slowly in the early 1940s, and then rose rapidly from the mid-1940s to the mid-1950s despite dropping prices. From the mid-1950s to the early 1970s the harvest averaged 1 to 2 million annually. In the 1974–75 season the kill was 2.9 million, and in 1975–76 it was 3.2 million. The harvest in Canada climbed from 21,014 in the 1970–71 season to 78,836 in the 1975–76 season.

Raccoon fur farming was popular in the 1920s and 1930s, but proved unprofitable and was abandoned.

The raccoon is also a popular game species, primarily hunted at night with trained dogs. Relatively small numbers of live raccoons are sold for use in field dog trials and as pets, and to raccoon hunting clubs for put-and-take hunting. A few are killed and sold for meat (Stuewer 1943a; Sanderson 1960; Caldwell 1963).

Occasionally raccoons are accused of causing agricultural damage. They may be a nuisance in orchards, vineyards, melon patches, cornfields, peanut fields, and chicken yards. Raccoons also eat the eggs of upland game birds, waterfowl, and sea turtles, and prey on young muskrats. These depredations are sometimes regarded as undesirable, especially in areas managed primarily for one of the prey species. Thus, Wilson (1953) regarded raccoons as responsible for the decline of muskrat populations in North Carolina coastal marshes, and Johnson and Rauber (1971) reported that raccoons destroyed large numbers of shore bird and sea turtle nests. Raccoons are sometimes regarded as serious threats to wood ducks (Sanderson

1960; Bellrose et al. 1964), and marsh nesting waterfowl (Llewellyn and Webster 1960; Urban 1970). In most cases, however, raccoon damage to crops and game species is inconsequential, temporary, or very local and often caused by only one or a few individuals.

Roeben (1975) stated that introduced raccoons pose a serious threat to the native fauna of West Germany, and Johnson (1970) warned that transplanted raccoons may spread parasites and diseases such as rabies.

Management. Most efforts at raccoon management are aimed at maintaining or increasing the population. Raising and releasing raccoons, once popular with state game departments, was found to be both expensive and ineffective, and has been discontinued (Stuewer 1941, 1943a; Berard 1951; Woehler 1957). Trapping wild raccoons from dense populations and transplanting them to areas where they are scarce has apparently increased some local populations (Jones 1946). The animals often do not stay where they are released, however, and the practice is generally not recommended (Giles 1943; Butterfield 1944a; Kellner 1954; Nelson 1955).

Laws restricting raccoon hunting to the late fall and winter are useful in protecting populations in heavily hunted areas. It is important to open the season after the juveniles are mature enough to care for themselves, and to close it before the breeding season begins. Refuges may also help to maintain breeding populations.

Probably the most effective means of perpetuating raccoon populations is habitat improvement and protection (Stuewer 1943a, 1948; Butterfield 1944b, 1950; Berard 1951). Small woodlands in agricultural areas should be protected from intense fire, cutting, and grazing. Wild fruits should be encouraged, and nut trees—especially oaks and beeches—preserved. Streams, swamps, marshes, and beaver colonies should be protected from destruction and pollution, and ponds and marshes can be constructed near woodlands. Den trees and potential den trees—those without cavities as yet, but damaged or diseased—should be given special protection. Where natural dens are scarce it may be helpful to put up den boxes in woodlands, especially near water.

When the need is amply proven, damage control is usually best accomplished through the selective removal of offenders by trapping or shooting. Electric fences are effective for small plots such as gardens and chicken yards. Properly designed and spaced nest boxes will reduce predation on wood ducks (Llewellyn and Webster 1960; Eaton 1966; Bellrose et al. 1964). In special circumstances it may be desirable to reduce high raccoon populations through carefully controlled poisoning. Strychnine-treated eggs placed in dummy nests were found effective by Llewellyn and Webster (1960). Wilson (1957) successfully used strychnine-treated corn in a muskrat marsh, and Johnson and Rauber (1971) achieved a rapid population decrease with corn treated with Fumarin, an anticoagulant. Poisoning should be considered only as a method of last resort in special cases, and indiscriminate killing should be prevented through careful design and placement of the bait stations and careful choice of the baits and poisons.

CURRENT RESEARCH AND MANAGEMENT NEEDS

Further research is needed on such basic topics as social structure and interactions, individual behavior, and reproductive physiology. Considering their intelligence and adaptability, raccoons have been strangely ignored by ethologists. Comparative studies of behavioral and population ecology in widely differing habitats would also be illuminating. On a more applied level, strenuous efforts should be made to monitor local raccoon population trends in view of the increasingly heavy demand for fur. Measures should be taken to ensure the continuing success of the species through habitat protection and controls on hunting and trapping based on current population data.

OTHER PROCYONIDS

Ringtail. The ringtail, *Bassariscus astutus* (Lichenstein), is also called ringtailed cat, miner's cat, civet cat, and cacomistle (from the Aztec). Its range extends from southwestern Oregon south and east through California, southern Nevada, southern Utah, southern Colorado, Arizona, New Mexico, and Texas to Louisiana and central Mexico. It has recently been reported from Arkansas (Sealander and Gipson 1972). According to Willey and Richards (1974), ringtails in Colorado have steadily declined in recent years, and are threatened with extirpation there.

Ringtails are smaller and more slender than raccoons. The face is pointed, with relatively large eyes and ears, and the tail is bushy and approximately as long as the body. The semiplantigrade feet have hairy soles and naked pads, and the five toes bear semiretractile claws. The dental formula is identical to that of the raccoon, but the canines are more rounded, the incisors have small secondary lobes, and the molars and premolars are doglike, with sharp cusps. Total length ranges from 630 to 810 mm, the tail from 305 to 438 mm, the hind foot from 57 to 77 mm, and the ear from 45 to 54 mm. The weight ranges from 750 to 1100 g. Females are slightly smaller than males.

The general color is tan with black-tipped guard hairs dorsally, and yellowish white below. The eyes are ringed conspicuously with white and there are white spots in front of the ears. The tail has six to nine (usually seven or eight) black bands alternating with white, and a black tip.

Whether or not ringtails mate permanently is unknown, but pairs have been reported together at all seasons. Birth occurs from April to July, usually in May or June, after a gestation period of about eight weeks. One captive female that ate her litter in May immediately mated again and bore a second litter two months later. Newborn young are fuzzy and blind and weigh about 28 g. The eyes open at about 22 to 34 days (usually about 30), and the teeth begin to appear in the fourth or fifth week. The infants begin to walk in the fifth to sixth week, and to climb in the eighth week (Richardson 1942; Bailey 1974; Toweill and Toweill 1978; J. Kaufmann unpublished; I. Poglayen unpublished).

The father is kept away from the young by the mother for at least the first three weeks, then both parents may bring food to them. The young begin foraging with their mother in their second month and are usually weaned at about three to five months. They may also forage with their father, though this is apparently not a usual occurrence. The young become increasingly independent during the late summer and fall, but have been seen with their mother as late as December (Grinnell, Dixon and Linsdale 1937; Toweill 1976; Lemoine 1977).

Ringtails live in a wide variety of rocky and wooded habitats from sea level to 2800 m. They are found in dense riparian growth, montane evergreen forests, oak woodlands, pinyon-juniper, chaparral, and deserts. Although they are capable of considerable urine concentration (Richards 1976), they are seldom found more than one-half mile from water in the wild. Reported densities range from 0.08 per km² to 3.9 per km² (Grinnell et al. 1937; Taylor 1954; Trapp 1978).

More nocturnal than raccoons, ringtails are only occasionally active at dawn and dusk, and rarely during the day. Daytime rest sites and dens for reproduction are located in caves; rock crevices; burrows dug by other animals; brush piles; hollow trunks, roots, and limbs; and buildings. Monthly home ranges of males and females in Zion National Park were from 49 to 233 ha, with a mean of 136 ha (Trapp 1978). Seasonal home ranges there may be much larger, as some individuals used entirely different home ranges in different months. Toweill (1976) reported home ranges of 16 to 52 ha in Texas, during periods of three to six months. Males had larger home ranges than females.

Ringtails are omnivorous, but consume a higher proportion of animal matter than do raccoons. Mammals, arthropods, and fruits are the principal foods, with each of these being most important in different studies (Wood 1954; Taylor 1954; Davis 1960; Lemoine 1977; Toweill and Teer 1977; Trapp 1978). Passerine birds and lizards round out the diet. The mammals eaten are chiefly cricetid rodents, plus pocket gophers, squirrels, and cottontails, and carrion from sheep, cattle, deer, and jackrabbits. Grasshoppers, crickets, beetles, and arachnids make up the arthropod contribution. The fruits most commonly eaten are juniper (*Juniperus* spp.), hackberry (*Celtis* spp.), persimmon, oak, and prickly pear (*Opuntia*).

Ringtails are agile runners and climbers, at home on the ground, in trees and cliffs, and in holes and burrows. Their locomotory adaptations include ricocheting, chimney stemming, power leaps, and claustrophilia (Trapp 1972). Described vocalizations include whimpers and squeaks by infants, chittering (juvenile distress, copulating females), chirps (contact call), barks (alarm, defensive threat), and aggressive

hisses, grunts, growls, snarls, and squeals (Bailey 1974; Richards 1976; Toweill and Toweill 1978; J. Kaufmann unpublished). The conspicuously ringed tail is fluffed out in excitement or agitation, and is sometimes arched over the back. Urine is rubbed on the ground or on raised objects as a home range marker, and a strong anal musk is released when the animal is alarmed. The fragmentary information available on social structure indicates that there is sometimes much overlap in the home ranges of both males and females, with perhaps less overlap between males (Trapp 1978). Toweill (1976) found no overlap, however, in the home ranges of the females he studied; home ranges of males overlapped those of females and may have overlapped those of other males. Both Trapp and Lemoine (1977) found latrine areas with accumulations of scats.

Known predators on ringtails are great horned owls, snakes, and domestic dogs and cats. A few are killed by automobiles, and many more are killed by trappers, often in traps set for more valuable fur bearers. Ringtail fur is thin and of poor quality, and is used only as trim on cheap coats. The price of pelts has generally ranged from $0.50 to $5.00. The ringtail cannot legitimately be regarded as a game, fur, or pest species, and should be given full legal protection.

Coati. Coatis, *Nasua narica* (Linnaeus), are often indiscriminately called coatimundis, though the latter term properly refers only to solitary adult males. Coatis range throughout Central America from Panama to Mexico, and barely across the border into the southwestern United States. In the United States they breed from the Animas Mountains in southwestern New Mexico to the Baboquivari Mountains in southcentral Arizona, and north to the Gila River (Kaufmann et al. 1976). Coatis have lived in the United States for over 80 years, and there is no evidence that they are currently expanding their breeding range northward.

Slightly smaller than raccoons, coatis have a slender, mobile snout and a long, slender tail that is often carried erect. Typical measurements are: total length 850 to 1340 mm, tail 420 to 680 mm, hind foot 95 to 122 mm, and ear 38 to 44 mm. The usual weight is from 4 to 6 kg, though adult males are sometimes heavier. The feet are plantigrade, naked on the sole, and strongly clawed on each of the five toes. The dental formula is the same as that of the raccoon, but the canines are more bladelike, and the premolars and molars have comparatively high crowns with sharp cusps. The usual color is chocolate brown with a yellowish wash over the chest and shoulders, a white snout, and broken white eye rings. Individuals vary in color, independent of age or sex, from pale sandy brown to reddish to almost black. The tail is indistinctly ringed.

Coatis have a matriarchal social system. Adult males are solitary, whereas females and males less than two years old live in bands (Kaufmann 1962). Most bands have fewer than 20 individuals, but larger bands have been reliably reported from Arizona. An adult male is allowed to accompany a band for only about one month during the mating period. During this time he is completely subordinate to the females, but drives off other males that approach the band. In the United States most mating occurs in April, the bands break up before the young are born in June, and the females with their new litters regather with the yearlings of both sexes in August.

Coatis are most abundant in wet, lowland tropical forests, where home ranges of up to 50 ha and densities of up to 42 per km² have been reported (Kaufmann 1962). In the drier, less productive oak and pine woodlands of our southwestern mountains, bands and solitary males have home ranges of up to 300 ha and densities of less than 10 per km² (Lanning 1976; Kaufmann et al. 1976). Moreover, they are seminomadic in this habitat, and may leave an area entirely after a few months. Unlike other procyonids, coatis are chiefly diurnal. Though they sleep in trees and bear their young in tree nests in tropical forests, rocky ledges and caves are typically used in the United States.

Like other procyonids, coatis are highly opportunistic omnivores. Insects, arachnids, and other small terrestrial invertebrates are their principal foods. Coatis also eat whatever fruits are available. In the United States these include juniper, madrone (*Arbutus arizonica*), manzanita (*Arctostaphylos* spp.), oaks, and prickly pear. Small vertebrates, chiefly lizards, are a minor part of the diet.

Only red-tailed hawks, golden eagles (*Aquila chrysaëtos*), and humans have been seen to kill coatis in the United States, though cats and large snakes prey on them in the tropics. Coatis are susceptible to rabies, and an epizootic diagnosed as canine distemper caused a population crash among Arizona coatis in 1960–61 (Risser 1963). Recovery has been slow and irregular, and the United States population still has not regained the high levels of the late 1950s. Coatis are neither a game nor a pest species in this country, and their fur is virtually worthless. Nevertheless, many have been carelessly or thoughtlessly killed by trappers and hunters, and in government "predator" control campaigns. They should receive complete legal protection as an unusual and scarce member of our natural fauna.

LITERATURE CITED

Aliev, F. F., and Sanderson, G. C. 1966. Distribution and status of the raccoon in the Soviet Union. J. Wildl. Manage. 30:497–502.

Arata, A. A., and Hutchison, J. H. 1964. The raccoon (*Procyon*) in the Pleistocene of North America. Tulane Stud. Geol. 2:21–27.

Asdell, S. A. 1964. Patterns of mammalian reproduction. 2nd ed. Cornell Univ. Press, Ithaca. 670pp.

Bailey, E. P. 1974. Notes on the development, mating behavior, and vocalization of captive ringtails. Southwestern Nat. 19:117–119.

Baker, R. H.; Newman, C. C.; and Wilke, F. 1945. Food habits of the raccoon in eastern Texas. J. Wildl. Manage. 9:45–48.

Barash, D. P. 1974. Neighbor recognition in two "solitary" carnivores: the raccoon (*Procyon lotor*) and the red fox (*Vulpes vulpes*). Science 185:794–796.

Barker, D. J., and Welker, W. I. 1969. Receptive fields of first-order somatic sensory neurons innervating rhinarium in coati and raccoon. Brain Res. 14:367–386.

Bellrose, F. C.; Johnson, K. L.; and Meyers, T. U. 1964. Relative value of natural cavities and nesting houses for wood ducks. J. Wildl. Manage. 28:661–676.

Berard, E. V. 1951. Basic study of the raccoon. Pages 1–53, *appendix to* S. McKeever, W. G. Frum, and E. V. Berard, eds. A survey of West Virginia mammals. Cons. Comm. West Virginia, Final Rep., P-R Proj. 22-R. 126pp. + 53pp. appendix.

Berner, A., and Gysel, L. W. 1967. Raccoon use of large tree cavities and ground burrows. J. Wildl. Manage. 31:706–714.

Bider, J. R.; Thibault, P.; and Sarrazin, R. 1968. Schemes dynamiques spatiotemporels de l'activité de *Procyon lotor* en relation avec le comportement. Mammalia 32:137–163.

Bigler, W. J.; McLean, R. G.; and Trevino, H. A. 1973. Epizootic aspects of raccoon rabies in Florida. Am. J. Epidemiol. 98:326–355.

Bissonnette, T. H., and Csech, A. G. 1937. Modification of mammalian sexual cycles. Part 7, Fertile matings of raccoons in December instead of February induced by increasing daily periods of light. Proc. R. Soc. London, ser. B, vol. 122, no. 827:246–254.

————. 1938. Sexual photo-periodicity of raccoons on low protein diet and second litters in the same breeding season. J. Mammal. 19:342–348.

————. 1939. A third year of modified breeding behavior with raccoons. Ecology 20:156–162.

Burger, J. 1977. Determinants of hatching success in diamond-backed terrapins, *Malaclemmys terrapin*. Am. Midl. Nat. 97:444–464.

Burt, W. H. 1960. Bacula of North America mammals. Univ. Michigan Mus. Zool. Misc. Publ. 113:1–76.

Butterfield, R. T. 1944*a*. Populations, hunting pressure, and movement of Ohio raccoons. Trans. North Am. Wildl. Conf. 9:337–344.

————. 1944*b*. Raccoon management. Ohio Conserv. Bull. 8(3):20–21.

————. 1950. The buying of den trees for raccoon management. J. Wildl. Manage. 14:244–246.

Buxton, D. F., and Goodman, D. C. 1967. Motor function and the corticospinal tracts in the dog and raccoon. J. Comp. Neurol. 129:341–360.

Cabalka, J. L.; Costa, R. R.; and Hendrickson, G. O. 1953. Ecology of the raccoon in central Iowa. Proc. Iowa Acad. Sci. 60:616–620.

Cagle, F. R. 1949. Notes on the raccoon, *Procyon lotor megalodous* Lowery. J. Mammal. 30:45–47.

Caldwell, J. A. 1963. An investigation of raccoons in north-central Florida. M.S. Thesis. Univ. Florida. 107pp.

Chandler, A. C. 1942. The helminths of raccoons in east Texas. J. Parasitol. 28:255–268.

Cole, L. W. 1907. Concerning the intelligence of raccoons. J. Comp. Neurol. Psychol. 17:211–262.

————. 1912. Observations of the senses and instincts of the raccoon. J. Anim. Behav. 2:299–309.

Cowan, W. F. 1973. Ecology and life history of the raccoon (*Procyon lotor hirtus* Nelson and Goldman) in the northern part of its range. Ph.D. Thesis. Univ. North Dakota. 176pp.

Creed, R. F. S., and Biggers, J. D. 1963. Development of the raccoon placenta. Am. J. Anat. 113:417–446.

Cunningham, E. R. 1962. A study of the eastern raccoon *Procyon lotor* (L.) on the Atomic Energy Commission Savannah River Plant. M.S. Thesis. Univ. Georgia. 55pp.

Davis, G. E., and Whiting, M. C. 1977. Loggerhead sea

turtle nesting in Everglades National Park, Florida. Herpetologica 33:18–28.

Davis, H. B. 1907. The raccoon: a study in intelligence. Am. J. Psychol. 18:447–489.

Davis, W. B. 1960. The mammals of Texas. Texas Game and Fish Comm., Austin. Bull. 41.

Dearborn, N. 1932. Foods of some predatory fur-bearing animals in Michigan. Univ. Michigan Bull. School For. Conserv. 1:1–52.

de Beaufort, F. 1968. Apparition de raton-laveur, *Procyon lotor* (L.) en France. Mammalia 32:307.

Deems, E. F., Jr., and Pursley, D. 1978. North American furbearers: their management, research and harvest in 1976. Int. Assoc. Fish Wildl. Agencies, in cooperation with the Maryland Dept. Nat. Resour., Wildl. Adm. 171pp.

Dellinger, G. P. 1954. Breeding season, productivity, and population trends of raccoons in Missouri. M.A. Thesis. Univ. Missouri, Columbia. 86pp.

Dídier, R. 1950. Etude systématique de l'os pénien des mammifères: Procyonidés, Ursidés. Mammalia 14:78–94.

Dorney, R. S. 1953. Some unusual juvenile raccoon weights. J. Mammal. 34:122–123.

———. 1954. Ecology of marsh raccoons. J. Wildl. Manage. 18:217–225.

Eaton, R. L. 1966. Protecting metal wood duck houses from raccoons. J. Wildl. Manage. 30:428–430.

Ellis, R. J. 1964. Tracking raccoons by radio. J. Wildl. Manage. 28:363–368.

Erickson, A. B., and Scudder, H. I. 1947. The raccoon as a predator of turtles. J. Mammal. 28:406–407.

Ewer, R. F. 1973. The carnivores. Cornell Univ. Press, Ithaca. 494pp.

Folk, G. E., Jr.; Coady, K. B.; and Folk, M. A. 1968. Physiological observations on raccoons in winter. Proc. Iowa Acad. Sci. 75:301–305.

Folk, G. E., Jr.; Dodge, C. H.; and Folk, M. A. 1957. Resting body temperatures of raccoons and domestic rabbits. Proc. Anim. Care Panel 7(4):253–258.

Frampton, J. E., and Webb, L. G. 1974. Preliminary report on the movement and fate of raccoons released in unfamiliar territory. Pages 170–183 *in* Proc. 27th Annu. Conf. Southeast Assoc. Fish Game Commissioners.

Fritzell, E. K. 1977. Dissolution of raccoon sibling bonds. J. Mammal. 58:427–428.

———. 1978a. Aspects of raccoon (*Procyon lotor*) social organization. Can. J. Zool. 56:260–271.

———. 1978b. Habitat use by prairie raccoons during the waterfowl breeding season. J. Wildl. Manage. 42:118–127.

Gander, F. F. 1928. Period of gestation in some American mammals. J. Mammal. 9:75.

Giles, L. W. 1939. Fall food habits of the raccoon in central Iowa. J. Mammal. 20:68–70.

———. 1940. Food habits of the raccoon in eastern Iowa. J. Wildl. Manage. 4:375–382.

———. 1942. Utilization of rocky exposures for dens and escape cover by raccoons. Am. Midl. Nat. 27:171–176.

———. 1943. Evidences of raccoon mobility obtained by tagging. J. Wildl. Manage. 7:235.

Goldman, E. A. 1950. Raccoons of North and Middle America. North Am. Fauna 60. 153pp.

Gorham, J. R. 1966. The epizootiology of distemper. J. Am. Vet. Med. Assoc. 149:610–622.

Grau, G. A.; Sanderson, G. C.; and Rogers, J. P. 1970. Age determination of raccoons. J. Wildl. Manage. 34:364–372.

Grinnell, J.; Dixon, J. S.; and Linsdale, J. M. 1937. Fur-bearing mammals of California. Univ. California Press, Berkeley. 2 vol.

Habermann, R. T.; Herman, C. M.; and Williams, F. P. 1958. Distemper in raccoons and foxes suspected of having rabies. J. Am. Vet. Med. Assoc. 132:31–35.

Hall, E. R., and Kelson, K. R. 1959. The mammals of North America. Ronald Press Co., New York. 2 vol.

Halloran, P. O. 1955. A bibliography of references to diseases of wild animals and birds. Am. J. Vet. Res. 16, no. 61, pt. 2. 465pp.

Hamilton, W. J., Jr. 1936. The food and breeding habits of the raccoon. Ohio J. Sci. 36:131–140.

———. 1940. The summer food of minks and raccoons on the Montezuma Marsh, New York. J. Wildl. Manage. 4:80–84.

———. 1951. Warm weather foods of the raccoon in New York State. J. Mammal. 32:341.

Hardin, W. B., Jr.; Arumugasamy, N.; and Jameson, H. D. 1968. Pattern of localization in "pre-central" motor cortex of raccoon. Brain Res. 11:611–627.

Harkema, R., and Miller, G. C. 1964. Helminth parasites of the raccoon, *Procyon lotor,* in the southeastern United States. J. Parasitol. 50:60–66.

Haugen, O. L. 1954. Longevity of the raccoon in the wild. J. Mammal. 35:439.

Hoffmann, C. O., and Gottschang, J. L. 1977. Numbers, distribution, and movements of a raccoon population in a suburban residential community. J. Mammal. 58:623–636.

Hollister, N. 1916. The genera and subgenera of raccoons and their allies. Proc. U.S. Nat. Mus. 49, no. 2100:143–150.

Houston, C. S., and Houston, M. I. 1973. A history of raccoons in Saskatchewan. Blue Jay 31:103–104.

Hubert, G. F., Jr. 1977. Fur harvest survey, 1976–77. Illinois Dept. Conserv., Job Completion Rep., Surveys and Investigations Projects no. W-49-R(24). 8pp.

———. 1978. Fur harvest survey, 1977–78. Illinois Dept. Conserv., Job Completion Rep., Surveys and Investigations Projects no. W-49-R(25). 14pp.

Ivey, D. W. R. 1948. The raccoon in the salt marshes of northeastern Florida. J. Mammal. 29:290–291.

Johnson, A. S. 1970. Biology of the raccoon (*Procyon lotor varius* Nelson and Goldman) in Alabama. Auburn Univ. Agric. Exp. Stn. Bull. 402. 148pp.

Johnson, E. F., and Rauber, E. L. 1971. Control of raccoons with rodenticides: a field test. Pages 277–281 *in* Proc. 24th Annu. Conf. Southeast Assoc. Fish Game Commissioners.

Johnson, J. I., Jr. 1957. Studies of visual discrimination by raccoons. Ph.D. Thesis. Purdue Univ. 111pp.

Johnson, J. I., Jr., and Michels, K. M. 1958. Discrimination of small intervals and objects by raccoons. Anim. Behav. 6:164–170.

Johnson, J. I., Jr.; Welker, W. I.; and Pubols, B. H., Jr. 1968. Somatotropic organization of raccoon dorsal column nuclei. J. Comp. Neurol. 132:1–44.

Jones, G. 1946. The raccoon in Oklahoma. Oklahoma Game and Fish News 2:4–5, 7.

Kaufmann, J. H. 1962. Ecology and social behavior of the coati, *Nasua narica,* on Barro Colorado Island, Panama. Univ. California Publ. Zool. 60:95–222.

———. 1971. Is territoriality definable? Pages 36–40 *in* A. H. Esser, ed. Behavior and environment: the use of space by animals and men. Plenum Press, New York. 411pp.

Kaufmann, J. H.; Lanning, D. V.; and Poole, S. E. 1976.

Current status and distribution of the coati in the United States. J. Mammal. 57:621–637.

Keefe, J. 1953. Knee deep in coons. Missouri Conserv. 14:10–11.

Kellner, W. C. 1954. The raccoon in southwestern Virginia. Virginia Wildl. 15:16–17, 21.

Kleiber, M. 1932. Body size and metabolism. Hilgardia 6:315–353.

Lanning, D. V. 1976. Density and movements of the coati in Arizona. J. Mammal. 57:609–611.

Lehman, L. E. 1968. September birth of raccoons in Indiana. J. Mammal. 49:126–127.

Lemoine, J. 1977. Some aspects of ecology and behavior of ringtails (*Bassariscus astutus*) in St. Helena, California. M.S. Thesis. Antioch College. 56pp.

Llewellyn, L. M. 1952. Geographic variation in raccoon litter size. Pages 1–7 *in* Proc. 8th Annu. Northeast Wildl. Conf.

———. 1953. Growth rate of the raccoon fetus. J. Wildl. Manage. 17:320–321.

Llewellyn, L. M., and Enders, R. K. 1954*a*. Trans-uterine migration in the raccoon. J. Mammal. 35:439.

———. 1954*b*. Ovulation in the raccoon. J. Mammal. 35:440.

Llewellyn, L. M., and Uhler, F. M. 1952. The foods of fur animals of the Patuxent Research Refuge, Maryland. Am. Midl. Nat. 48:193–203.

Llewellyn, L. M., and Webster, C. G. 1960. Raccoon predation on waterfowl. Trans. North Am. Wildl. Conf. 25:180–185.

Lopinot, A. C. 1951. Raccoon predation on the great blue heron, *Ardea herodias*. Auk 68:235.

Lowery, G. H., Jr. 1974. The mammals of Louisiana and its adjacent waters. Louisiana State Univ. Press, Baton Rouge. 565pp.

Lyall-Watson, M. 1963. A critical re-examination of food "washing" in the raccoon (*Procyon lotor* Linn.). Proc. Zool. Soc. London 141:371–393.

Lynch, G. M. 1967. Long-range movement of a raccoon in Manitoba. J. Mammal. 48:659–660.

———. 1971. Raccoons increasing in Manitoba. J. Mammal. 52:621–622.

McClearn, D. 1977. Morphological and behavioral adaptations for arboreality in the raccoon family, Procyonidae. Am. Zool. 17:975.

McKeever, S. 1958. Reproduction in the raccoon in the southeastern United States. J. Wildl. Manage. 22:211.

Mech, L. D.; Barnes, D. M.; and Tester, J. R. 1968. Seasonal weight changes, mortality, and population structure of raccoons in Minnesota. J. Mammal. 49:63–73.

Mech, L. D.; Tester, J. R.; and Warner, D. W. 1966. Fall daytime resting habits of raccoons as determined by telemetry. J. Mammal. 47:450–466.

Mech, L. D., and Turkowski, F. J. 1966. Twenty-three raccoons in one winter den. J. Mammal. 47:529–530.

Michels, K. M.; Fischer, B. E., and Johnson, J. I., Jr. 1960. Raccoon performance on color discrimination problems. J. Comp. Physiol. Psychol. 53:379–380.

Mitchell, P. C. 1905. On the intestinal tract of mammals. Trans. Zool. Soc. London 17:437–536.

Montgomery, G. G. 1963. Freezing, decomposition, and raccoon lens weights. J. Wildl. Manage. 27:481–483.

———. 1964. Tooth eruption in preweaned raccoons. J. Wildl. Manage. 28:582–584.

———. 1968. Pelage development of young raccoons. J. Mammal. 49:142–145.

———. 1969. Weaning of captive raccoons. J. Wildl. Manage. 33:154–159.

Montgomery, G. G.; Lang, J. W.; and Sunquist, M. E. 1970. A raccoon moves her young. J. Mammal. 51:202–203.

Moore, J. C. 1953. Raccoon parade. Everglades Nat. Hist. 1:119–126.

Morris, J. 1975. Ovulation in raccoons and note on reproductive physiology of free-ranging raccoons. Trans. Missouri Acad. Sci. 7:261–262.

Mosby, H. S. 1947. The raccoon. Virginia Wildl. 8(2):8–9, 20.

Nelson, F. P. 1955. The place of stocking in game management. South Carolina Wildl. 2:2–3.

Neseni, R. 1938. Beitrag zur Ernährung und Verdaaung des Waschbären. Z. Säugetierk 13:77.

Peterson, F. A.; Heaton, W. C.; and Wruble, S. D. 1960. Levels of auditory response in fissiped carnivores. J. Mammal. 50:566–578.

Petrides, G. A. 1950. The determination of sex and age ratios in fur animals. Am. Midl. Mat. 43:355–382.

———. 1959. Age ratios in raccoons. J. Mammal. 40:249.

Pocock, R. I. 1921. The external characters and classification of the Procyonidae. Proc. Zool. Soc. London pp. 389–422.

Pope, C. H. 1944. Attainment of sexual maturity in raccoons. J. Mammal. 25:91.

Preble, N. A. 1940. The status and management of raccoon in central Ohio. M.S. Thesis. Ohio State Univ.

Priewert, F. W. 1961. Record of an extensive movement by a raccoon. J. Mammal. 42:113.

Pubols, B. H., Jr.; Welker, W. I.; and Johnson, J. I., Jr. 1965. Somatic sensory representation of forelimb in dorsal root fibers of raccoon, coatimundi, and cat. J. Neurophysiol. 28:312–341.

Redford, P. 1962. Raccoon in the U.S.S.R. J. Mammal. 43:541–542.

Rensch, B., and Dücker, G. 1963. Haptisches Lern und Unterscheidungsvermögen bei einem Waschbären. Z. Tierpsychol. 20:608–615.

Richards, R. E. 1976. The distribution, water balance and vocalization of the ringtail, *Bassariscus astutus*. Ph.D. Thesis. Univ. Northern Colorado. 104pp.

Richardson, W. B. 1942. Ring-tailed cats: their growth and development. J. Mammal. 23:17–26.

Rinker, G. C. 1944. *Os clitoridis* from the raccoon. J. Mammal. 25:91–92.

Risser, A. C., Jr. 1963. A study of the coati mundi *Nasua narica* in southern Arizona. M.S. Thesis. Univ. Arizona. 76pp.

Robinson, V. B.; Newberne, J. W.; and Brooks, D. M. 1957. Distemper in the American raccoon (*Procyon lotor*). J. Am. Vet. Med. Assoc. 131:276–278.

Roeben, P. 1975. Zur Ausbreitung des Waschbaeren, *Procyon lotor* (Linne 1758), und des Marderhundes, *Nyctereutes procyonoides* (Gray 1834), in des Bundesrepublik Deutschland. Saügetierk. Mitt. 23:93–101.

Sanderson, G. C. 1950. Methods of measuring productivity in raccoons. J. Wildl. Manage. 14:389–402.

———. 1951*a*. Breeding habits and a history of the Missouri raccoon population from 1941 to 1948. Trans. North Am. Wildl. Conf. 16:445–460.

———. 1951*b*. The status of the raccoon in Iowa for the past twenty years as revealed by fur reports. Proc. Iowa Acad. Sci. 58:527–531.

———. 1960. Raccoon values—positive and negative. Illinois Wildl. 16:3–6.

———. 1961*a*. The reproductive cycle and related phenomena in the raccoon. Ph.D. Thesis Univ. Illinois. 165pp.

————. 1961*b*. The lens as an indicator of age in the raccoon. Am. Midl. Nat. 65:481–485.

————. 1961*c*. Techniques for determining age of raccoons. Illinois Nat. Hist. Surv. Bio. Notes 45. 16pp.

Sanderson, G. C.; Mech, L. D.; and Schnell, J. H. 1967. A contribution to a bibliography of the raccoon (*Procyon lotor*). U.S. Atomic Energy Commission, Contract AT (11-1)-1332 (Document COO-1332). 51pp. Mimeogr.

Sanderson, G. C., and Nalbandov, A. V. 1973. The reproductive cycle of the raccoon in Illinois. Illinois Nat. Hist. Surv. Bull. 31:29–85.

Scheffer, V. B. 1947. Raccoons transplanted in Alaska. J. Wildl. Manage. 11:350–351.

————. 1950. Notes on the raccoon in southwest Washington. J. Mammal. 31:444–448.

Schneider, D. G.; Mech, L. D.; and Tester, J. R. 1971. Movements of female raccoons and their young as determined by radio-tracking. Anim. Behav. Monogr. 4:1–43.

Schnell, J. H. 1970. Rest site selection by radio-tagged raccoons. J. Minnesota Acad. Sci. 36:83–88.

Schoonover, L. J. 1950. A study of raccoon (*Procyon lotor hirtus* Nelson and Goldman) in north-central Minnesota. M.S. Thesis. Univ. Minnesota.

Schoonover, L. J., and Marshall, W. H. 1951. Food habits of the raccoon (*Procyon lotor hirtus*) in north-central Minnesota. J. Mammal. 32:422–428.

Schultz, A. L. 1962. A survey of parasites of the raccoon, *Procyon lotor*, in southeastern Michigan. M.S. Thesis. Univ. Michigan.

Scott, W. E. 1951. Wisconsin's first prairie spotted skunk, and other notes. J. Mammal. 32:363.

Sealander, J. A., and Gipson, P. S. 1972. Range extension of ringtail cat into Arkansas. Southwestern Nat. 16:458–459.

Sharp, W. M., and Sharp, L. H. 1956. Nocturnal movements and behavior of wild raccoons at a winter feeding station. J. Mammal. 37:170–177.

Shell, W. F., and Riopelle, A. J. 1957. Multiple discrimination learning in raccoons. J. Comp. Physiol. Psychol. 50:585–587.

Simkin, D. W. 1966. Extralimital occurrences of raccoons in Ontario. Can. Field Nat. 80:144–146.

Sonenshine, D. E., and Winslow, E. L. 1972. Contrasts in distribution of raccoons in two Virginia localities. J. Wildl. Manage. 36:838–847.

Soper, J. D. 1942. Mammals of Wood Buffalo Park, northern Alberta and District of Mackenzie. J. Mammal. 23:119–145.

Stains, H. J. 1956. The raccoon in Kansas: natural history, management, and economic importance. Univ. Kansas Mus. Nat. Hist. and State Bio. Surv. Misc. Publ. 10. 76pp.

————. 1961. Comparison of temperatures inside and outside two tree dens used by raccoons. Ecology 42:410–413.

————. 1967. Carnivores and pinnipeds. Pages 325–354 *in* S. Anderson and J. K. Jones, Jr., eds. Recent mammals of the world. Ronald Press Co., New York. 453pp.

Stuewer, F. W. 1941. 'Coon stocking not for Michigan. Michigan Conserv. 10:3, 11.

————. 1942. Studies of molting and priming of the fur of the eastern raccoon. J. Mammal. 23:399–404.

————. 1943*a*. Raccoons: their habits and management in Michigan. Ecol. Monogr. 13:203–257.

————. 1943*b*. Reproduction of raccoons in Michigan. J. Wildl. Manage. 7:60–73.

————. 1948. Artificial dens for raccoons. J. Wildl. Manage. 12:296–301.

Sundell, R. A. 1956. Age determination studies of the raccoon (*Procyon lotor* Linnaeus). M.S. Thesis. Michigan State Univ., East Lansing.

Sunquist, M. E. 1967. Effects of fire on raccoon behavior. J. Mammal. 48:673–674.

Sunquist, M. E.; Montgomery, G. G.; and Storm, G. L. 1969. Movements of a blind raccoon. J. Mammal. 50:145–147.

Taylor, W. P. 1954. Food habits and notes on the life history of the ring-tailed cat in Texas. J. Mammal. 35:55–63.

Tester, J. R. 1953. Fall food habits of the raccoon in the South Platte Valley of northeastern Colorado. J. Mammal. 34:500–502.

Tevis, L., Jr. 1947. Summer activities of California raccoons. J. Mammal. 28:323–332.

Thorgeson, H. L. 1958. Studies of tactual discrimination by raccoons. Ph.D. Thesis. Purdue Univ. 67pp.

Thorkelson, J., and Maxwell, R. K. 1974. Design and testing of a heat transfer model of a raccoon (*Procyon lotor*) in a closed tree den. Ecology 55:29–39.

Toweill, D. E. 1976. Movements of ringtails in Texas' Edwards Plateau region. M.S. Thesis. Texas A & M Univ., College Station. 75pp.

Toweill, D. E., and Teer, J. G. 1977. Food habits of ringtails in the Edwards Plateau region of Texas. J. Mammal. 58:660–663.

Toweill, D. E., and Toweill, D. B. 1978. Growth and development of captive ringtails. Carnivore 1(3):46–53.

Trapp, G. R. 1972. Some anatomical and behavioral adaptations of ringtails, *Bassariscus astutus*. J. Mammal. 53:547–557.

————. 1978. Comparative behavioral ecology of the ringtail and gray fox in southwestern Utah. Carnivore 1:3–32.

Twichell, A. R., and Dill, H. H. 1949. One hundred raccoons from one hundred and two acres. J. Mammal. 30:130–133.

Turkowski, F. J., and Mech, L. D. 1968. Radio-tracking the movements of a young male raccoon. J. Minnesota Acad. Sci. 35:33–38.

Tyson, E. L. 1950. Summer food habits of the raccoon in Washington. J. Mammal. 31:448–449.

Urban, D. 1970. Raccoon populations, movement patterns, and predation on a managed waterfowl marsh. J. Wildl. Manage. 34:372–382.

U.S. Department of Commerce, Business and Defense Services Administration. Fur facts and figures: a survey of the United States fur industry. Gov. Printing Off., Washington, D.C. 20pp.

VanDruff, L. W. 1971. The ecology of the raccoon and opossum, with emphasis on their role as waterfowl nest predators. Ph.D. Thesis. Cornell Univ. 151pp.

Welker, W. I. 1959. Genesis of exploratory and play behavior in infant raccoons. Psychol. Rep. 5:764.

Welker, W. I., and Campos, G. B. 1963. Physiological significance of sulci in somatic sensory cerebral cortex in mammals of the family Procyonidae. J. Comp. Neurol. 120:19–36.

Welker, W. I., and Johnson, J. I., Jr. 1965. Correlation between nuclear morphology and somatotropic organization in ventro-basal complex of the raccoon's thalamus. J. Anat. 99:761–790.

Welker, W. I.; Johnson, J. I., Jr.; and Pubols, B. H., Jr. 1964. Some morphological and physiological characteristics of the somatic sensory system in raccoons. Am. Zool. 4:75–96.

Welker, W. I., and Seidenstein, S. 1959. Somatic sensory representation in the cerebral cortex of the raccoon (*Procyon lotor*). J. Comp. Neurol. 111:469–509.

Whitney, L. F., and Underwood, A. B. 1952. The raccoon. Practical Science Publ. Co., Orange, Conn. 177pp.

Wilber, C. G. 1955. Electrocardiogram of the raccoon. J. Mammal. 36:283–284.

Willey, R. B., and Richards, R. E. 1974. The ringtail (*Bassariscus astutus*): vocal repertoire and Colorado distribution. J. Colorado-Wyoming Acad. Sci. 7:58.

Wilson, K. A. 1953. Raccoon predation on muskrats near Currituck, North Carolina. J. Wildl. Manage. 17:113–119.

———. 1957. Control of raccoon predation on muskrats near Currituck, North Carolina. Pages 221–233 in Proc. 10th Annu. Conf. Southeast Assoc. Game and Fish Commissioners.

Woehler, E. E. 1957. How about raccoon stocking? Wisconsin Conserv. Bull. 22:12–14.

Wood, J. E. 1954. Food habits of furbearers in the upland post oak region of Texas. J. Mammal. 35:406–414.

———. 1955. Notes on reproduction and rate of increase of raccoons in the post oak region of Texas. J. Wildl. Manage. 19:409–410.

Wood, N. A. 1922. The mammals of Washtenaw, County, Michigan. Univ. Mich. Mus. Zool. Occ. Pap. 123:1–23.

Worth, D. F., and Smith, J. B. 1976. Marine turtle nesting on Hutchison Island, Florida, in 1973. Florida Marine Res. Publ. 18. Florida Dept. Nat. Resour., Marine Research Lab., St. Petersburg, Fla. 17pp.

Wurster, D. H., and Benirschke, K. 1968. Comparative cytogenetic studies in the order Carnivora. Chromosoma Berl. 24:336–382.

Yeager, L. E., and Elder, W. H. 1945. Pre- and post-hunting season foods of raccoons on an Illinois goose refuge. J. Wildl. Manage. 9:48–56.

Yeager, L. E., and Rennels, R. G. 1943. Fur yield and autumn foods of the raccoon in Illinois river bottom lands. J. Wildl. Manage. 7:45–60.

Zarrow, M. X., and Clark, J. H. 1968. Ovulation following vaginal stimulation in a spontaneous ovulator and its implications. J. Endocrinol. 40:343–352.

Zollman, P. E., and Winkelmann, R. K. 1962. The sensory innervation of the common North American raccoon (*Procyon lotor*). J. Comp. Neurol. 119:149–157.

JOHN H. KAUFMANN, Department of Zoology, University of Florida, Gainesville, Florida 32611.

28

Fisher

Martes pennanti

Marjorie A. Strickland
Carman W. Douglas
Milan Novak
Nadine P. Hunziger

NOMENCLATURE

COMMON NAMES. Fisher, wejack, black cat, Pennant's marten, pekan. The most commonly used name, fisher, may have come from early immigrants who noted its similarity to the European polecat, whose other names include fitchet, fitche, or fitchew (Brander and Books 1973).

SCIENTIFIC NAME. *Martes pennanti*

SUBSPECIES. *M. p. pennanti, M. p. pacifica,* and *M. p. columbiana.* Hagmier (1959) concluded that none of the subspecies is separable on the basis of pelage or skull characteristics, and questioned this classification.

DISTRIBUTION

The original range of this exclusively North American mustelid was extremely well documented by Hagmeier (1956), to whose work the reader is referred.

Fisher are members of the Canadian and Transition Life Zones (Merriam 1898), and their northern limit coincides fairly well with the 60° F July isotherm (Meterorological Branch 1962). The northern limit shown by Hall and Kelson (1959) for Ontario and Quebec either was in error or was influenced by very rare occurrences such as the one taken in Labrador in 1978. Fisher are very rare in the Northwest Territories, the Yukon, and the northwest portion of British Columbia (figure 28.1). The original southern range has been much reduced through overharvest, anthropogenic factors, or both. Figure 28.1 also shows the approximate extirpation dates of original populations and recent live trapping and transfer programs.

DESCRIPTION

Fisher have long bodies with short legs; dark fur; small, rounded ears; and a long bushy tail. The head is wedge shaped, with a face that tapers to a narrow black nose. Males are larger, heavier, and more muscular than females, particularly in the neck and shoulders.

Fisher are digitigrade and their feet are large, presumably to assist them in walking on snow. They have five toes on all feet with sharp, curved, unsheathed

FIGURE 28.1. Distribution of the fisher (*Martes pennanti*).

claws. There are pads on each toe and central pads on each foot. The feet are heavily furred in winter, almost obscuring the pads.

Size and Weight. Adult male fisher usually weigh 3.5 to 5.5 kg but may, in exceptional cases, weigh over 9 kg (Blanchard 1964). Adult females weigh 2.0 to 2.5 kg. Sexual dimorphism is apparent in body length as well. The total length of males is 90 to 120 cm and of females is 75 to 95 cm, of which about one-third is tail (Powell 1977).

By mid-October, young born the previous spring are virtually adult length, although their mean weight, especially in males, is less than that of adults (table 28.1). Males gain up to 5 percent in length and 15 to 40 percent in weight between six months of age and adulthood. Because the epiphyses of the long bones are ossified in females by seven months of age, their growth is complete then. In males, the epiphyses often are not ossified until ten months of age (Dagg et al. 1975).

TABLE 28.1. Weights of unpelted fisher by sex and age

Author	Mean Weight (kg ± 1SD)			
	Adult Males	Juvenile Males	Adult Females	Juvenile Females
Coulter[a]	4.85 ± 0.13	3.44 ± 0.11	2.40 ± 0.06	2.0 ± 0.07
(1966)	(N = 36)	(N = 15)	(N = 32)	(N = 12)
Kelly	4.63 ± 0.76	4.03 ± 0.41	2.10 ± 0.14	2.26 ± 0.22
(1977)	(N = 9)	(N = 11)	(N = 8)	(N = 3)
Strickland	4.72 ± 0.66	3.88 ± 0.56	2.43 ± 0.14	2.14 ± 0.27
(1978)	(N = 15)	(N = 20)	(N = 8)	(N = 12)

[a]original data in oz.

Anatomy. Leach (1977*a*, 1977*b*) and Leach and de Kleer (1978) give detailed descriptions of the postcranial osteology and musculature of fisher and their comparison with marten.

Pelage. The fur varies between sexes, seasons, and individuals. Females have darker, silkier fur than that of males, which is coarser and more grizzled, especially around the head and shoulders. This grizzling is caused by tricolored guard hairs that have a light subterminal band. In both sexes, the fur becomes darker toward the rump and tail. The ventral surface is dark brown with variable-sized white patches on the chest and around the genitals. Animals less than a year old are usually darker than adults (Coulter 1966).

Spring and summer pelage is lighter colored and less dense than winter pelage, due to molting. A new coat begins to grow in September and is complete in November, so that in early winter, the fur is dense, glossy, and prime. By late January, the fur is past peak primeness. Pelts of males have an obvious preputial orifice.

There are four inguinal mammae. These are distinct in both sexes at birth (Coulter 1966). Their location is evident on the flesh side of pelts from parous females (see "Age Determination" section).

Scent Glands. As is common in carnivores and pronounced in mustelids, both sexes have two lateral anal sacs that contain a musky-smelling, viscous liquid. The brown fluid is deposited by fisher at scent stations, especially during the breeding season. Coulter (1966) did not observe either penned fisher or those caught in traps releasing fluid when frightened or hurt, although we have noticed it to some extent. Powell (1977) reported scent glands on the soles of the hind feet.

Fat Deposits. Among winter-caught animals, males almost always contain more fat in relation to body size than females. In males, fat deposits are common in the inguinal region as well as on the rump, shoulders, and occasionally over the entire body surface. In females, subcutaneous fat deposits are generally restricted to the inguinal region. Internally, fat storage begins on the mesenteries and kidneys, and in very fat animals the ovaries and kidneys are completely encased in fat and deposits partially fill the body cavity. The animals with the least amount of fat are adult females.

Skull and Dentition. Skull length in males is 112–130 mm, females 95–106 mm. Skull width in males is 62–84 mm, females 52–61 mm (Peterson 1966). Among the mustelids, only fisher, marten (*Martes americana*), and wolverine (*Gulo gulo*) have four upper and lower premolars. Fisher can be distinguished from marten by their relatively greater skull length (more than 95 mm) and from wolverine by their relatively smaller skull length (less than 130 mm) (Peterson 1966). Figures 28.2 and 28.3 illustrate the differences in skull features between males and females of various ages. Figure 28.4 shows details of the skull of a young female. The dental formula is: 3/3, 1/1, 4/4, 1/2 = 38.

Coulter (1966) reported that canines, but not premolars or incisors, erupted by 62 days after birth. There are no reliable reports as to when these presumably deciduous teeth are replaced by permanent ones. We have found slightly over 6 percent of young of the year and 4 percent of older fisher, caught by trappers in southern Ontario from November to February, do not have all four canines fully erupted. These unerupted canines have deposits of cementum. A few animals have skull anomalies that may prevent the canines from erupting, but most with unerupted canines are normal. In some cases, lost canines may be replaced. R. D. Leonard (personal communication 1978) reported one radio-collared adult female fisher broke a canine in a live trap. When collected two years later, it had an unerupted replacement canine.

REPRODUCTION

Anatomy. The ovaries are completely encapsulated in a bursa and the oviduct encircles the ovary. The uterus has a common corpus uteri, which allows embryos developing in one horn to migrate to the other horn. There is no os clitoridis.

Males have a baculum. Paired testes are carried externally in a scrotal sac. Testis size and weight increase in all age groups with the approach of the breeding season. Males of all ages had abundant sperm in March (Coulter 1966; Strickland 1978).

Fisher have delayed implantation. A fertilized egg starts its development as it travels down the oviduct. Reaching the uterus in the blastocyst stage, it then

FIGURE 28.2. Skulls and bacula of male fisher (*Martes pennanti*), showing variations with age. Note the development of the sagittal crest and baculum and the fusion of the sutures with age. Numerals denote cementum annuli. The juvenile at the extreme left was captured in November; the one adjacent was captured in February.

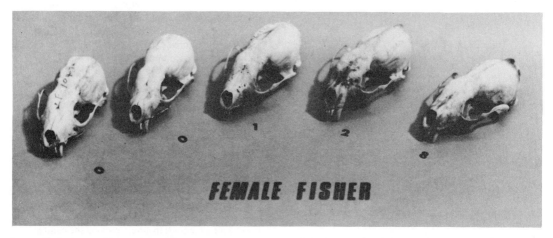

FIGURE 28.3. Skulls of female fisher (*Martes pennanti*), showing variations with age. Note the sutures and minimal development of the sagittal crest. The juvenile at the extreme left was captured in November; the one adjacent was captured in February.

becomes inactive, the metabolic rate falls, and cell division ceases (Ewer 1973). The unimplanted blastocyst remains quiescent for a period of ten to eleven months before implantation and normal development resumes.

The blastocyst can be flushed from the excised fresh uterus by syringing with water. It is a clear, transparent sphere, approximately 1 mm in diameter, surrounded with a thick zona pellucida. The inner cell mass is plainly visible. Enders and Pearson (1943) give cell counts of 798, 807, and 844 in three unimplanted blastocysts.

Breeding. Most early information comes from fur farmers who reported that matings of Ontario fisher occurred between March 26 and April 23 (Hodgson 1937; Douglas 1943). Twenty-six matings in British Columbia occurred between April 5 and 27. Females come into estrus six to eight days after the young are born, and remain in heat for two or three days (Hall 1942). Hodgson (1937) noted that females "will sometimes mate a second time in fourteen days, leading one to think that she has two or more heat periods, each lasting about two days." No other authors reported this.

All reported that the female is the dominant individual at breeding time. Copulation takes place with the male astride the back of the female, forepaws encircling her body behind her shoulders, and lasts about one hour (Hall 1942). Males are highly polygamous at fur farms, although they can be bred both monogamously and polygamously (Hodgson 1937; Douglas 1943). Coulter (1966) believed that fisher are most likely polygamous in the wild.

Female fisher are sexually mature at one year of age and breed then, producing their first litter at two years of age. They have one litter per year, followed immediately by breeding (Asdell 1946; Coulter 1966; Eadie and Hamilton 1958; Hall 1942; Wright and Coulter 1967; Strickland 1978). However, the authors and others (G. R. Parsons and M. K. Brown, and P. Dwyer personal communication) have recently observed a possible estrus in both juveniles and adults in September.

The age at which males are fully sexually mature is not known. Smears taken in February and early March showed that fisher at 10 and 11 months of age were producing sperm but it is not known if they were effective breeders. Douglas (1943) reported no success in mating males and females both 12 months of age. Because females can become pregnant at 12 months of age, the lack of success was probably due to the inadequacy of the male. It may also have been the result of captive conditions.

Induced ovulation has been demonstrated in several species of *Mustela* and *Martes,* and the occurrence of a large baculum may be associated with ensuring that the stimulation provided during copulation is sufficient to cause ovulation (Ewer 1973). This is one reason why young males, who have a smaller baculum, may be less effective breeders.

Gestation. Gestation varies from 327 to 358 days with a mean of 352 (Hall 1942; Douglas 1943). The period of active pregnancy, following implantation, is reported to be from 30 days (Coulter 1966) to about 2 months (Hamilton and Cook 1955), but this has not been firmly established.

Wright (in Enders 1963) reported inducing early implantation in several mustelids by increasing the photoperiod. He concluded that this was the most important factor controlling the time of implantation. However, we found small embryos at about the same stage of development in females caught in southern Ontario on January 30 and February 28. Dwyer (personal communication 1979) reported an early parturition date (February 27) in a female brought into artificial light conditions in late December.

Parturition Dates. Parturition is most common in March but varies from early March to mid-April (table 28.2).

Litter Size. All of the available evidence indicates that fisher produce an average of 3 young per adult female. Data on actual or potential numbers of young have been obtained by counting neonates, embryos, blastocysts, corpora lutea, placental scars, and corpora albicantia (table 28.3). Most accurate, but available only from captive animals, is the count of young born alive, which gives a mean for 60 litters of 2.9. Eadie and Hamilton (1958) found 2.7 corpora lutea per female in 28 pairs of hand-sectioned ovaries, which may not give accurate counts because of the small size of the ovary. We found a mean of 2.3 corpora lutea in 21 pairs of hand-sectioned ovaries. The same ovaries

had a mean of 3.2 corpora lutea when sectioned with a microtome.

Wright and Coulter (1967) found a mean of 3.0 unimplanted blastocysts when they serially sectioned the uterus and examined it microscopically. Flushing the uterus to obtain blastocysts, as was done by Hamilton and Cook (1955), Kelly (1977), and Strickland (1978), gave lower blastocyst counts, due, no doubt, to the poor condition of the carcasses.

Coulter (1966) found a mean of 2.9 placental scars in 27 females, compared to 3.4 embryos and

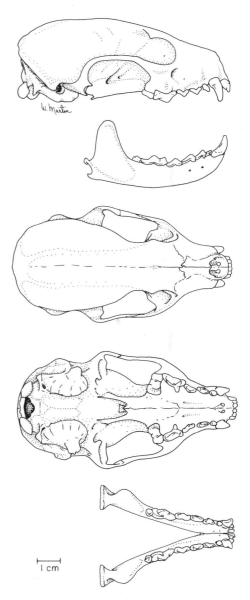

FIGURE 28.4. Skull of the fisher (*Martes pennanti*). From top to bottom: lateral view of cranium, lateral view of mandible, dorsal view of cranium, ventral view of cranium, dorsal view of mandible.

TABLE 28.2. Parturition dates of fisher

Author	Location	Number of Litters	Parturition Dates
Coulter (1966)	Maine	2	March 3 and March 20
Douglas (1943)	Manitoba		mid-March to mid-April
Hall (1942)	British Columbia	22	March 23 to April 7
Leonard (personal communication)	Manitoba	2	March 28 and early April
Parsons (personal communication)	Vermont	1	April 15
Powell (1977)	Michigan	2	late March

corpora lutea. He noted difficulty in detecting implantation sites unless the material is fresh. Clearing the uteri did not make the scars more visible.

We found neither placental scars nor corpora albicantia in females less than 24 months of age. Although the number of corpora albicantia per female increased with age, they could not be related to past productivity because of the varying rates at which they regressed.

We found adult females were sometimes barren in areas where overharvesting had depressed the number of breeding males. In years when the sex ratio of trapped adults (greater than 12 months of age) was more than 2 females to each male, 6 to 8 percent of the females were barren. Coulter (1966), who reported 1.3 females per male in 168 adults, found no barren females. However, New York, where the sex ratio of breeders was nearly 1:1, had 5 to 15 percent of the adult females barren annually (Parsons personal communication).

We found that, in 2 percent of the females examined, there were more blastocysts or embryos than corpora lutea, indicating polyovular release or polyembryonic development.

Growth and Development of Young. Coulter (1966) gave a detailed report on the growth and development of three fisher born in captivity on March 3. They were altricial, sparsely furred, and weighed about 40 g. Growth was rapid (5 to 10 g per day), so that by 96 days (June 5) they weighed over 800 g. The young were relatively helpless until about the sixth or seventh week. They started crawling in the eighth week, when their eyes opened (53 days). They were walking during the ninth week and climbing during the tenth week. The adult did not take meat to the young until they were 62 days old, by which time the deciduous canine teeth had erupted. Nursing continued until the 114th day (June 23). The young were 125 days old before they learned to kill prey effectively. The female and her young separate sometime between late summer and the period of first snows. Adult males apparently take no part in rearing the young.

Differences in size and weight between sexes is not evident up to four months of age. This may be an adaptation to permit the survival and growth of females because, by the third month, litter mates are very aggressive in competing for food (Coulter 1966).

ECOLOGY

Habitat. A fisher's movements are governed by the availability of food, topography, cover, location of dens, and weather conditions. Food is probably the

TABLE 28.3. Litter sizes of fisher

		Number of Litters Examined		Mean Litter Size	Author
Young born		29		2.9	Hodgson (1937)
		26		2.7	Hall (1942)
		2		3.0	Coulter (1966)
		1		2.0	Powell (1977)
		2		2.5	Hamilton & Cook (1955)
	Total	60	Mean	2.9	
Embryos		11		3.4	Coulter (1966)
		32		3.2	Strickland (1978)
	Total	43	Mean	3.2	
Blastocysts		11		3.0	Wright & Coulter (1967)
		7		2.7	Hamilton & Cook (1955)
		60		2.4	Strickland (1978)
		12		2.2	Kelly (1977)
	Total	90	Mean	2.5	
Corpora lutea		300		3.3	Strickland (1978)
		44		3.3	Coulter (1966)
		12		3.7	Kelly (1977)
	Total	356	Mean	3.3	
Placental scars		27		2.9	Coulter (1966)

most important factor. This may explain why there are so many diverse reports of preferred fisher habitat: around swamps, especially if they are among large timber (Seton 1953); spruce forests (Hamilton 1943); heavier coniferous forests, especially damper areas (Rand 1948); mixed stands as well as coniferous stands; cedar swamps (de Vos 1952); "green timber" (Cook and Hamilton 1957); deep unbroken tracts of virgin forest (Mathiessen 1959); young forest stands that followed cutting, burns, or agricultural use (Coulter 1960); mixed forests of coniferous and deciduous trees (Cowan and Ginguet 1960); wetland and softwood-hardwood areas (Kelly 1977); open, upland hardwood where porcupines den; and lowland spruce-fir, spruce-aspen, and alder stands that are hare habitat (Powell 1978).

In general, fisher are adaptable and their habitat is sympatric with their available prey species. All authors agree that fisher avoid open places with no overhead cover, such as fields, roads, burns, and open bogs. They run when crossing open spaces and frequently minimize the amount of open distance to be crossed.

Dens. Fisher dens appear to be of two types; temporary dens used for cover and protection, and nesting dens used for whelping (Coulter 1966; de Vos 1952; Powell 1977; Hamilton and Cook 1955). The temporary dens may be located under logs, brush piles, or tree roots, in hollow trees or ground burrows, or under the snow. Temporary dens are seldom used for more than two or three days. Fisher often den near a food supply or after a heavy snowfall. Nesting dens occur most often high in hollow trees, although dens in rocky ledges may also be utilized.

Density and Home Range. Little definitive work has been done on either density or home range of fisher and most of the information available is based on harvest statistics and trapper reports.

Maximum densities of approximately one fisher per 2.59 km² of suitable habitat were reported by Hamilton and Cook (1955) in New York. Coulter (1966) in Maine found densities of one fisher per 2.59–11.65 km² in intensive tracking studies but noted that it is doubtful if densities of this magnitude can be sustained.

In New Hampshire, Kelly (1977) by telemetry calculated densities of one animal per 3.9–7.5 km². He also determined mean yearly home ranges of about 1,500 ha for females (both adult and subadult), 1,971 ha for adult males, and 2,563 ha for subadult males. There was much overlap of ranges in all sex and age categories.

Coulter (1966) and Hamilton and Cook (1955) doubted that fisher populations have regular cycles. However, Bulmer (1974, 1975) found that a 10-year cycle (9.63 years) existed in fisher harvest records from 1751 to 1969. It followed the snowshoe hare-lynx cycle by about 1.2 years and was most pronounced in midwestern Canada.

There is an indication in some areas of an inverse relationship between population levels of fisher and marten (de Vos 1952; Strickland 1978). This may be

not due to competition for food between these species (Clem 1975) but a reflection of changes in habitat.

FOOD HABITS

Because fisher are opportunistic feeders their diet varies with the prey available. Predominant foods are: porcupines (*Erethizon dorsatum*); snowshoe hares (*Lepus americanus*); small rodents; and carrion, especially deer (*Odocoileus* sp), moose (*Alces alces*), and beaver (*Castor canadensis*). They also consume, when available, birds and their eggs, insects, reptiles, amphibians, and various fruits and nuts (Seton 1953; de Vos 1951*a*, 1952; Quick 1953; Hamilton and Cook 1955; Coulter 1966; Clem 1977; Powell 1977; Brown and Will 1979).

The belief that porcupines are an essential part of the fisher's diet may not be valid, since fisher have been known to occur in areas completely devoid of this prey species. However, their ability successfully to prey upon porcupines makes available an otherwise generally unexploited food source.

Scat analysis tends to overestimate the significance of small mammals in the diet due to the higher ratio of indigestible residues (Davidson et al. 1978). Analysis of digestive tracts of trapped animals may also give an inaccurate estimate of the diet because of the use of baits in trapping. Variation between animals in the length of time between their last meal and capture and between capture and death affects the amount of material remaining in the digestive tract, as does the differential digestibility of the materials eaten. Differences in fluctuations in food habits are difficult to assess without simultaneous indices of food abundance and other ecological data for the area.

Clem (1977) and Coulter (1966) found no significant difference in overall diet between male and female fisher in spite of their size difference. Rosenzweig (1966) noted that fisher eat larger prey than marten and have greater diversity in their diet.

To estimate how much food must be eaten to replace the daily energy expended, Powell (1977, 1979*a*) devised models for the nonreproductive aspects of the winter energy budget of free-living fisher. He estimated daily expenditures of about 86 kcal/kg for a 2.4-kg female to 94 kcal/kg for a 5.1-kg male, both of which were deemed to be active about 30 percent of the time. He also estimated the caloric value of various foods and concluded that, depending on weight, free-living fisher will require a daily average of about one-quarter of a snowshoe hare, one-twentieth of a porcupine, one-quarter kg of deer carrion, one squirrel, or a dozen mice. Optimal foraging strategies based on search time and energy available from the kill showed that a fisher should hunt for porcupines only if it will take about four times as long to find a hare.

Using caged female fisher, Davidson et al. (1978) determined the net energy expended for maintenance and limited physical activity to be 162 kcal/kg $W^{0.75}$/day and calculated the amounts of four prey species that would be required to equal the energy expended. He found that gross energy intake was highest on diets of deer and coturnix quails (*Coturnix coturnix*), but that hares were best for conversion of nitrogen to body tissue. Deer carrion would provide the greatest energy balance for wild fisher, as it requires no energy expenditure for pursuit, capture, or killing, although it requires search energy.

BEHAVIOR

As a hunter, the fisher is an opportunist. It does not lie in wait or pursue prey for long distances. Powell and Brander (1975) described two different foraging behaviors in fisher. When foraging for hares the hunting pattern appears disorganized, with frequent changes of direction that flush the hares from cover and make them vulnerable to a rush attack. This pattern is typical of weasels (*Mustela* spp.). Fisher exhibit different behavior in porcupine habitat, traveling long distances in one direction. They travel on more or less regular circuits inspecting most of the porcupine dens along the way. If a porcupine is encountered, an attack may occur.

A porcupine is usually attacked on the ground, with the fisher repeatedly circling it and attempting to bite its face. The porcupine attempts to keep its back toward the fisher and its face close to a tree trunk and, when possible, charges backward with flailing tail. During these backward charges, the porcupine exposes its face, which the fisher attacks until the porcupine is so weakened that the fisher can inflict a mortal wound. The whole process may take 30 minutes or more. The fisher does not always escape porcupine quills. They are commonly found in trapped fisher, quite undissolved, but cause no inflammation or festering even when inside the body cavity.

When porcupines are encountered in a tree, the fisher will climb the tree and attempt to force the porcupine to the end of a branch. If it falls to the ground, it is open to attack. If the porcupine is on the branch with its back toward the trunk, it is relatively safe.

The fisher begins to eat the porcupine on the ventral side, consuming the heart, lungs, and liver first, and eventually consumes everything except the skin, large bones, feet, and intestines.

Fisher kill all prey except porcupines by biting at the back of the neck. Also in typical weasel fashion, they use their body and feet to assist in handling prey (Powell 1977).

Senses. Hamilton and Cook (1955) reported that the olfactory rather than the visual sense is used in following trails. Powell (1977) stated that while fisher use sight or sound to locate prey, olfaction seems most important when choosing places to inspect for prey, particularly in locating porcupine dens. He noted that fisher respond to prey movement as a stimulus for attack.

Activity. The fisher is generally a solitary traveler during much of the winter. Travel is in rough circuits varying in size, according to food availability, from 10 to 30 km in diameter (de Vos 1951). These circuits are used repeatedly at intervals of 4 to 12 days. Radio-

collared animals moved a mean straight-line distance in 24 hours of 1.3 km for adult females and 2.7 km for adult males. For subadults of both sexes, the mean was 2.1 km/24 hours (Kelly 1977).

By mid-March, when the breeding season is at its peak, tracks of fisher often appear together and there is frequent circling and crossing (Coulter 1966). This difference in pattern is important when evaluating track intersects, which are sometimes used as an index of population density.

Fisher are most active at night, although daytime activity is not uncommon. Diurnally, they are most active at sunrise and sunset (Kelly 1977). Females have a smaller home range in the summer than in the winter, probably due to restrictions of dependent young. Adult males displayed the greatest mobility, subadults of both sexes moved lesser distances, and adult females were least mobile (Kelly 1977). According to Powell (1977), daily activity is confined to a number of active periods interspersed with sleeping periods. He also suggested that fisher might not be complete homeotherms. Their body temperature drops while hunting and then they seek shelter to curl in a ball to raise body temperature. This could explain why fisher are inactive during extreme cold and why activity increases during the summer. Coulter (1966) found that his penned fisher were inactive for as long as 48 hours after gorging on a meal of meat.

Fisher are mostly terrestrial but they do climb trees readily and well. Coulter (1966) observed a fisher jumping from one tree to another. Females are more arboreal than males, because of their smaller size (Powell 1977). The generalized skeleton and forelimb musculature of fisher preadapts them to arboreality (Leach 1977a). They descend trees head first with their hind limbs rotated outward and caudally, somewhat after the nature of a squirrel (D. M. Leach personal communication 1979). Fisher will swim but do not generally do so (de Vos 1952). In deep snow their usual means of travel is a walk, while on firmer ground fisher use a lope or gallop (Pittaway 1978). Fisher, when annoyed, growl or emit hissing coughs or clucks.

MORTALITY

Prenatal Mortality. The mean size of 60 newborn litters is 2.9 and the mean number of corpora lutea of pregnancy is 3.3 (see table 28.3). The difference between these can be regarded, for all practical purposes, as intrauterine mortality and amounts to 12 percent of the ova shed.

The sex ratio of 163 newborn fisher as reported by various authors is 45:55 (table 28.4), which is not significantly different from a 50:50 ratio. We examined 33 late stage fetuses, which had a sex ratio of 33:67, again not significantly different from 50:50. We also determined the sex of 66 embryos, too small for external sexing, using the sex chromatin method of Moore (1966) and found an exact 50:50 ratio. These data may suggest some differential mortality in various stages of prenatal development but sample sizes are not large and more evidence is needed.

TABLE 28.4. Sex ratios of newborn fisher

Author	Sex Composition of Litters	
	Males	Females
Coulter (1966)	2	3
de Vos (1952)	44	40
Hall (1942)	13	20
Hodgson (1937)	11	19
Leonard (personal communication)	1	3
Parsons (personal communication)	0	3
Dwyer (personal communication)	2	2
Total	73 (45%)	90 (55%)

Neonatal Mortality. The extent of postpartum mortality or whether it is sex biased in wild fisher is unknown. It is known that within a few days of giving birth, the female leaves the naked kits in the den, perhaps for several hours, while she searches for a male and mates. Hodgson (1937) described the near demise of one kit from this desertion.

Other Natural Mortality. Fisher are at the top of the food pyramid and there is no evidence that any other animal except humans preys extensively upon adult fisher. It is probable that mortality in kits from chilling, intraspecific fighting, and predation from such animals as hawks (*Accipiter, Buteo, Falco*), owls (*Bubo*), red foxes (*Vulpes vulpes*), bobcats (*Felis rufus*), lynx (*F. lynx*), and back bears (*Ursus americanus*) is higher than that of adults (Coulter 1966).

Porcupine quills may cause some deaths in fisher but apparently the mortality factor is not large. There are reports from trappers that quills sometimes kill fisher (de Vos 1952; Hamilton and Cook 1955). (See "Behavior.")

Information on other natural mortality in fisher is sparse. Evidence of some mortality factor, albeit unknown, that operates on males between their first and second trapping season (i.e., between 10 and 19 months) but not on females is indicated from the sex ratio of our sample of trapped animals. The sex ratio of 1,171 fisher less than 12 months of age is 51:49, while that of 476 older animals shows a distinct dominance of females: 31:69. This added mortality in males may be due to intraspecific competition or strife. However, the imbalance in the sex ratio of trapped animals may be due to a number of other factors and may not necessarily represent the true ratio of the population.

Parasites and Diseases. Parasites and diseases do not appear to be a serious mortality factor in fisher. Parasitism is frequent but the level of infestation is low. Nematodes are the most frequently encountered parasite and include *Capillaria mustelarum, Physaloptera* sp., *Arthrocephalus lotoris, Molineus patens, Dracunculus insignis, Uncinaria stenocephala, Dioctophyme renale, Ascaris mustelarum, Ascaris devosi, Crenosoma petrowi, Soboliphyme baturini, Sobolevingylus* sp., and *Trilobostrongylus bioccai*. The cestodes *Mesocestoides variabilis* and *Taenia* sp. have also been found in fisher, as have the trematodes *Alaria* sp. and *Metorchis conjunctus*. Ectoparasites reported include the tick (*Ixodes cookei*), the flea (*Oropsylla arctomys*), and the mite (*Sarcoptes scabei*) (Coulter 1966; Craig and Borecky 1976; Dick and Leonard 1979; Fyvie 1964; Hamilton and Cook 1955; Meyer and Chitwood 1951; Morgan 1943; O'Meara et al. 1960). Fisher from our study have been found positive for trichinosis (5.6 percent), leptospirosis (5.5. percent), Aleutian disease (2.5 percent), and toxoplasmosis (41 percent).

Trapping Mortality. The main mortality factor in fisher is trapping. This includes both animals killed by traps and those wounded when escaping from traps. Being easy to trap and readily baited into an area, fisher can be quickly overharvested unless they are managed properly.

Sex Ratios in Harvest. Sex ratios of 2,784 trapped fisher reported by a number of investigators show a significant predominance of females: 45:55 (table 28.5). This imbalance in the sex ratio of trapped animals can be due to a difference between the sexes in: (1) their natural mortality following birth, (2) their trappability or ability to escape from traps (which can vary with the season and the age of the animal), or (3) selection by trappers for the most valuable animal. Increased trapping effort increases the likelihood of capturing the less susceptible individuals, so that in heavily trapped areas, the sample should more nearly approximate the population (Kelly 1977; Strickland 1977).

Differences in trappability of males and females are suggested by Coulter (1966) and Strickland (1978). Both found an increase in the percentage of males in the harvest after January. This is probably due to the different activity patterns between the sexes during late winter, when the females are heavy with embryos or confined by nursing young and the males are "trotting the trails" with the approach of breeding season.

Males are more likely to escape from traps (Coul-

TABLE 28.5. Sex ratios of trapped fisher

Source	Percentage Males	Percentage Females	N
de Vos (1952)	49	51	505
Quick (1953)	32	68	25
Hamilton & Cook (1955)	39	61	69
Coulter (1966)	45	55	367
Kelly (1977)	42	58	171
Strickland (1978)	45	55	1,647
Total	45	55	2,784

ter 1966) because of their greater strength, thus decreasing their numbers in the harvest.

Trappers have been known to try to select females by track size, as their fur is more valuable. This selection is often unreliable due to the overlap in foot size. Coulter (1966) found hind foot size ranged from 10.1 to 13.5 cm for males and 8.6 to 12.1 cm for females.

Where quotas are strictly enforced, trappers may sell the females and discard the less valuable males so that there is no record of their capture (Douglas 1953). This is a potential bias of which an investigator should be aware.

Age Composition of Harvest. Information on age composition of harvested animals is limited because accurate aging by cementum layers is a recent technique.

Coulter (1966) estimated the age of 223 fisher by skull, baculum, and ovarian characteristics and found 36 percent juveniles and 64 percent adults. He noted that the proportion of juveniles was higher in populations that were increasing rapidly. He speculated that a fisher population cannot maintain itself if the harvest exceeds 30 percent of the population.

Kelly (1977) estimated the age of 202 animals by cementum layers, and reported 50 to 55 percent juveniles. Both males and females reached age 7. He calculated that yearly adult mortality rate averaged 39 percent.

Our cementum aging extends over six years, with a total sample of 1,648 fisher. We have consistently found 70 to 80 percent of our harvest sample to be juveniles. Only 1 percent of the males were older than 4 years but 9 percent of the females were older than age 4. Our oldest specimen was a female aged 14 years. Mortality rates of adult fisher ranged, in various years, from 22 to 60 percent in males and 21 and 31 percent in females.

In our sample, the ratio of young fisher (less than 12 months old) to adult females (more than 30 months old, and therefore old enough to have been mothers of the young) is 5.6:1 (table 28.6). This is almost twice as many young in the harvest as might be expected from the reproductive rate of 2.9 young per female (see table 28.3). What is more, this ratio is higher in November and December than in late winter. The high proportion of young animals harvested early in the season can be used as a management method to select young nonbreeding animals. The reason more young are caught early is not known. It may be that they are inexperienced and thus more vulnerable to traps or that, following dispersal of the family group, they are traveling more widely than the resident animals and thus are more likely to encounter traps.

AGE DETERMINATION

Fisher show several changes in skeletal characteristics associated with age, including fusion of sutures, development of the sagittal crest and the suprafabellar tubercle, and changes in the conformation and size of the baculum. Age can also be determined by counting the annuli in the dental cementum (Strickland 1978; Stone et al. 1975; Kelly 1977). Boise (1975) examined various skull measurements in adults and juveniles and found a significant difference between the means of a number of them. Especially useful in both sexes was zygomatic width. Overlap, however, precluded an absolute index of age.

The zygomatic-temporal, nasomaxillary, and nasofrontal sutures remain unfused in both males and females throughout their first year. The sagittal crest in males begins to form by November of their first year and often is well developed by February although it does not overhang the supraoccipital bones as it does in adult males (see figure 28.2). Development of the sagittal crest in females is slower than in males, the temporal ridges often remaining separated for all of their first year and sometimes longer (see figure 28.3). Any female that has a sagittal crest is an adult, but not all adult females have crests. Crests of females never approach the size of those in males. The lack of any crest in some adult females and the presence of a crest in some late-caught juvenile males makes skull palpation an unreliable aging method.

Bacula of adult (older than 12 months) fisher weigh more than 1,500 mg and do not change significantly with age or month of capture. Bacula of young males (less than 12 months) increase in weight during their first winter but, at least until February, weigh less than 1,500 mg (Coulter 1966; Strickland 1977). Leach (in preparation) reports that judging development of the suprafabellar tubercle is a reliable and rapid method of distinguishing juvenile fisher from adults, at least in males.

Annuli are not visible in the cementum of canines or premolars of fisher during their first winter but appear in animals classified as adult by other characteristics (figure 28.5). Females pregnant for the first time (without corpora albicantia or placental scars) had one annulus; those that have had more than one pregnancy (and have corpora albicantia and placental scars) had two or more annuli (Kelly 1977; Strickland 1977).

Various pelt measurements (body length, weight of pelt, width behind ears) have been examined as a method of separating fisher less than 12 months of age from those older than 12 months. Mean values differ significantly in males (but not in females) between the age groups, but overlap is too great to give a useful aging method (Strickland 1978).

The development of the mammae are an indication of age. Mammae can be felt readily in early winter on the fur side of the pelts of parous females and to a lesser extent in nulliparous animals. In the pelts of

TABLE 28.6. Ratio of harvested fisher < 12 months old to adult females (> 30 months old)

Period When Trapped	Young: Adult Female
November and December	7.7:1.0
January and February	2.9:1.0
All months (November to February)	5.6:1.0

SOURCE: Strickland 1978

FIGURE 28.5. Sagittal section of the root of a lower canine of a fisher (*Martes pennanti*), showing three cementum annuli. The animal was captured on 1 January.

young females that have not yet bred, the mammae are often difficult to feel.

Preliminary studies show that "canine drop" in fisher as described by Nellis et al. (1978) for coyotes (*Canis latrans*) promises to be a useful technique for separating young fisher from old ones.

ECONOMIC STATUS

Never an abundant animal and perhaps in part because of this rarity, fisher have long held a premier position in the fur market (Innis 1962). Most valued in the trade are black, silky pelts of high gloss; usually these are females. Recent developments in dyeing processes permit the more grizzled pelts of males to be rendered more acceptable and valuable. Originally used mostly as luxury linings and trims, fisher fur is now made into complete garments.

Historically occurring in 38 administrative units and still occurring in 30 units, fisher were harvested in ten provinces and territories and six states during the decade 1967–77. In that period, recorded annual pelt production varied from 8,094 to 18,402, with an annual mean of 12,790 (table 28.7). At the same time, mean Canadian price rose from $13.81 to $114.76 per pelt, and the price continues to rise. At all times, much higher prices have been realized for exceptional pelts, and in 1979 prices in excess of $300 were received for outstanding skins or grade lots of excellent and especially well matched pelts.

Generally, female pelts bring 50 percent more than male pelts in the market. Juvenile females are worth about 10 percent less than adult females; conversely juvenile males are worth 10 percent more than adult males (Strickland 1978).

MANAGEMENT

As noted by Coulter (1966), at least four groups of people are concerned about the control or harvest of fisher: (1) trappers, who view them as a source of

income; (2) foresters and landowners, who see them as a means of controlling porcupines; (3) naturalists, who wish to preserve native fauna; and (4) some sportsmen and farmers, who may view them as predators. Not all of these interests are compatable, but all must be considered by the administrations responsible for fisher management.

Management methods for fisher fall into these categories:
1. Regulation of harvest by licences, seasons, and quotas.

The most common method of management is administrative control of trapping by varying length of seasons (or closing them), by licensing of trappers, and by regulating the number of fisher caught by setting quotas (allowable harvest per trapper).

Closure of trapping for fisher has been enforced at one time or another by most administrations and is probably responsible for the reestablishment of fisher in a number of areas. Coulter (1966) suggested alternating areas open to trapping and suggested that a closure of two to three years would be adequate to replenish a population. By using a registered trapline system wherein each trapper has a specific assigned

TABLE 28.7. Annual harvest of fisher by country, state, and province over a ten-year period

Season	Canada	United States	Total
1967–68	5,450	2,644	8,094
1968–69	7,684	3,051	10,735
1969–70	8,042	3,222	11,264
1970–71	6,442	5,008	11,450
1971–72	8,149	3,640	11,789
1972–73	13,837	4,565	18,402
1973–74	12,400	4,197	16,597
1974–75	10,224	3,989	14,213
1975–76	8,683	4,104	12,787
1976–77	9,669	2,902	12,571

Mean annual harvest, 1967–68 to 1976–77

British Columbia	1,014
Alberta	1,592
Saskatchewan	512
Manitoba	820
Ontario	3,112
Quebec	1,767
New Brunswick	198
Nova Scotia	(64)
N.W. Territories	33
Yukon	4
Canada	9,116
Maine	1,910
Massachusetts	(11)
New Hampshire	694
New York	1,027
Vermont	(318)
West Virginia	(2)
United States	3,962
North America	13,078

NOTE: Figures in brackets show areas in which trapping was for only part of decade.

area, harvest can be controlled by individual area and whole sections need not be closed.

All administrations that have an open season limit the duration. The length of the season varies from one to five months, often starting about 1 November. As explained previously, young of the year are overrepresented in the harvest in the early months of the season. This means that the harvest from November and December will contain a preponderance of young animals, and the adult breeding stock will largely escape capture.

Because fisher are readily trapped, they can easily be overharvested, even in a short, early season. Permissible harvest per trapper (quota) has been used by a number of administrations. Unfortunately, information on population density in an area is often inadequate, so quotas must be set on what information is available— usually harvests of past years, total area trapped, the opinions of trappers and game officials, and age and sex distribution of the harvest.

The aim of quota setting is sustained yield. However, there is a fine line between quotas low enough to maintain the population or allow it to increase to carrying capacity and those high enough to keep the population below the nuisance level. Fisher may be a nuisance to trappers when they get into traps set for other animals (e.g., fox) or when they rob traps in which other furbearers are captured. The use of poison as a harvest method or for the control of other species has had a marked effect upon the abundance of fisher in some areas.

2. Refuges.

The creation of refuges, where trapping of all furbearers is prohibited, has proven effective in some areas by providing foci for population spread (de Vos 1951b, 1952). In several southern areas, relict populations owe their existence to such sanctuaries or to closed seasons of long standing.

3. Restocking.

Restocking of fisher by live trapping and transfer has been attempted in a number of areas with varying amounts of success (see figure 28.1). A most successful program resulted in the reestablishment of a viable fisher population in our study area in southern Ontario. Ninety-seven fisher (37 males, 60 females) were transplanted into an area of about 13,600 km² during a six-year period (1957 to 1963). Within nine years, 730 fisher were harvested annually from this area, and the population continues to thrive and produce sizable harvests.

4. Trapper and public education.

Keeping trappers and the general public informed about the purposes and goals of management programs helps to ensure their cooperation and active support.

5. Selective harvesting

To some extent, harvest can be made selective by limiting the trapping period; early season trapping gives a preponderance of young animals (Strickland 1978). Greater selectivity may be possible in the future by live-trapping fisher and determining sex and age. Breeding stock could then be released unharmed and only those animals surplus to the population harvested.

6. Habitat management

By and large, habitat management is not undertaken because there is not sufficient knowledge to practice it intelligently. Forest cover of some type is required, however (see habitat section). Kelly (1977) discussed ways of increasing fisher habitat in mixed forest by silvicultural practices that produce shade during reforestation (e.g., shelterwood and east-west oriented strip cuts), thus encouraging the growth of shade-tolerant softwood. Alder wetlands can be increased by regulating beaver trapping to increase the number of beaver colonies.

7. Feeding

Many trappers distribute carcasses of beaver and other species as food for fisher. This is a way of increasing the carrying capacity of the range.

CURRENT RESEARCH AND MANAGEMENT NEEDS

From the management point of view, research is needed that will permit accurate estimation of present population density and predictions for the future. Population modeling appears to offer promise (Strickland 1978; Powell 1979b).

Information on habitat preferences and food requirements that relates to the carrying capacity of an area is also vital, as is investigation of interspecific competition. Very little is known about home range sizes and movements of fisher at different densities and in different habitats.

For both males and females, whether the age of sexual maturity varies geographically must be established. The possibility of a September estrus should be confirmed. The effect of altered photoperiod on ranch-raised animals may be significant. It is helpful to know whether sex ratios differ at various stages of embryonic development and if they vary with population density or habitat suitability.

Exploration of methods of selective harvesting of fisher should be undertaken, especially of ways for the trapper accurately to determine the age of the animal while it is in a live-trap so that he can select the age and sex groups he wishes to harvest. Information on the basis and applicability of cycles to fisher populations would be useful.

LITERATURE CITED

Asdell, S. A. 1946. Patterns of mammalian reproduction. Comstock Publishing Co., Ithaca, N. Y. 420pp.

Blanchard, H. 1964. Weight of a large fisher. J. Mammal. 45:487–488.

Boise, C. M. 1975. Skull measurements as criteria for aging fishers. New York Fish Game J. 22(1):32–37.

Brander, R. B., and Books, D. J. 1973. Return of the fisher. Nat. Hist. 82(1):52–57.

Brown, M. K., and Will, G. 1979. Food habits of the fisher in northern New York. New York Fish Game J. 26(1):87–92.

Bulmer, M. G. 1974. A statistical analysis of the 10 year cycle in Canada. J. Anim. Ecol. 43(3):701–708.

———. 1975. Phase relations in the ten year cycle. J. Anim. Ecol. 44(2):609–621.

Clem, M. 1975. Interspecific relationships of fisher and marten in Ontario during winter. Pages 165-182 *in* Proc. 1975 Predator Symp. Univ. Montana, Missoula.

———. 1977. Food habits, weight changes and habitat use of fisher (*Martes pennanti*) during winter. M.S. Thesis. Univ. of Guelph, Ontario. 49pp.

Cook, D. B., and Hamilton, W. J. Jr. 1957. The forest, the fisher and the porcupine. J. For. 55:719-722.

Coulter, M. W. 1960. The status and distribution of fisher in Maine. J. Mammal. 41:1-9.

———. 1966. Ecology and management of fisher in Maine. Ph.D. Thesis. State Univ. Coll. of For., Syracuse Univ. 183pp.

Cowan, I. McT., and Ginguet, C. J. 1960. The mammals of British Columbia. Pages 304-307 *in* British Columbia Prov. Dept. Educ. and Dept. Rec. and Con. Handb. 11.

Craig, R. E., and Borecky, R. A. 1976. Metastrongyles (Nematoda: Metastrongyloidea) of fisher (*Martes pennanti*) from Ontario. Can. J. Zool. 54(5):806-807.

Dagg, A. I.; Leach, D.; and Summer-Smith, G. 1975. Fusion of the distal femoral epiphysis in male and female marten and fisher. Can. J. Zool. 53(1):1514-1518.

Davidson, R. P.; Mautz, W.; Hayes, H.; and Holter, J. B. 1978. The efficiency of food utilization and energy requirements of captive female fishers. J. Wildl. Manage. 42(4):811-821.

de Vos, A. 1951*a*. Recent findings in fisher and marten ecology and management. Trans. North Am. Wildl. Conf. 16:498-505.

———. 1951*b*. Overflow and dispersal of marten and fisher from wildlife refuges. J. Wildl. Manage. 15(2):164-175.

———. 1952. Ecology and management of fisher and marten in Ontario. Tech. Bull. Ontario Dept. Lands For., Wildl. Ser. 1. 90pp.

Dick, T. A., and Leonard, R. D. 1979. Helminth parasites of fisher *Martes pennanti* (Erxleben) from Manitoba, Canada. J. Wild. Dis. 15(3):409-412.

Douglas, C. W. 1953. Fluctuations in sex ratios of fisher and marten during the 1952-53 trapping season, White River District. Ontario Dept. Lands For. Fish Wildl. Manage. Rep. 14. 14pp.

Douglas, W. O. 1943. Fisher farming has arrived. Am. Fur Breeder 16(2):18-20.

Eadie, W. R., and Hamilton, W. J. Jr. 1958. Reproduction in the fisher in New York. New York Fish Game J. 5:77-83.

Enders, A. C. 1963. Delayed implantation. Univ. Chicago Press, Chicago, Ill. 318pp.

Enders, R. K., and Pearson, O. P. 1943. The blastocyst of the fisher. Anat. Rec. 85:285-287.

Ewer, R. F. 1973. The carnivores. Weidenfeld & Nicolson, London. 494pp.

Fyvie, Audrey. 1964. Manual of common parasites, diseases and anomalies of wildlife in Ontario. Publ. Ontario Dept. Lands For. 102pp.

Hagmeier, E. M. 1956. Distribution of marten and fisher in North America. Can. Field Nat. 70:149-168.

Hall, E. R. 1942. Gestation period in the fisher with recommendations for the animal's protection in California. California Fish Game 28:143-147.

Hall, E. R., and Kelson, K. R. 1959. The mammals of North America. Ronald Press Co. Vol. 2, pp. 547-1083.

Hamilton, W. J., Jr. 1943. Mammals of Eastern United States. Comstock Publishing Co., Ithaca, N. Y. 432pp.

Hamilton, W. J., Jr. and Cook, A. H. 1955. The biology and management of the fisher in New York. New York Fish Game J. 2:13-35.

Hodgson, R. G. 1937. Fisher farming. Fur Trade J. Can. 103pp.

Innis, H. A. 1962. The fur trade in Canada. Univ. Toronto Press, Toronto. 446pp.

Kelly, G. M. 1977. Fisher (*Martes pennanti*) biology in the White Mountain National Forest and adjacent areas. Ph.D. Thesis. Univ. Massachusetts, Amherst. 178pp.

Leach, D. 1977*a*. The descriptive and comparative post cranial osteology of marten (*Martes americana* Turton) and fisher (*Martes pennanti* Erxleben): The appendicular skeleton. Can. J. Zool. 55:199-214.

———. 1977*b*. The forelimb musculature of marten (*Martes americana* Turton) and fisher (*Martes pennanti* Erxleben). Can. J. Zool. 55(1):31-41.

Leach, D., and de Kleer, V. S. 1978. The descriptive and comparative postcranial osteology of marten (*Martes americana* Turton) and fisher (*Martes pennanti* Erxleben): the axial skeleton). Can. J. Zool. 56(5):1180-1191.

Matthiessen, P. 1959. Wildlife in America. Viking Press, New York. 304pp.

Merriam, C. H. 1898. Life zones and crop zones of the United States. U.S.D.A. Bull. 10. Government Printing Office, Washington. 79pp.

Meteorological Branch. 1962. The climate of Canada. Info. Canada, Ottawa. 74pp.

Meyer, M., and Chitwood, B. G. 1951. Helminthes from fisher (*Martes pennanti*) in Maine. J. Parasitol. 37:320-321.

Moore, K. L. 1966. The sex chromatin. W. B. Saunders Co., Philadelphia. 275pp.

Morgan, B. B. 1943. New host records of Nematodes from Mustelidae (Carnivora). J. Parasitol. 29(2):158-159.

Nellis, C. H.; Wetmore, S. P.; and Keith, L. B. 1978. Age-related characteristics of coyote canines. J. Wildl. Manage. 42(3):680-683.

O'Meara, D. C.; Payne, D. D.; and Witter, J. F. 1960. Sarcoptes infestation of a fisher. J. Wildl. Manage. 24:399.

Parsons, G. R.; Brown, M. K.; and Will, G. B. 1978. Determining the sex of fisher from the lower canine teeth. New York Fish Game J. 25(1):42-44.

Peterson, R. L. 1966. The mammals of Eastern Canada. Oxford Univ. Press, Toronto. 465pp.

Pittaway, R. J. 1978. Observations on the behavior of the fisher (*Martes pennanti*) in Algonquin Park, Ontario. Le Naturaliste canadien 105:487-489.

Powell, R. A. 1977. Hunting behavior, ecological energetics, and predator-prey community stability of the fisher (*Martes pennanti*). Ph.D. Thesis. Univ. Chicago. 132pp.

———. 1978. A comparison of fisher and weasel hunting behavior. Carnivore 1:28-34.

———. 1979*a*. Ecological energetics and foraging strategies of the fisher (*Martes pennanti*). J. Anim. Ecol. 48:195-212.

———. 1979*b*. Fishers, population models and trapping. Wildl. Soc. Bull. 7(3):149-154.

Powell, R. A., and Brander, R. B. 1975. Adaptations of fishers and porcupines to their predator prey systems. Pages 45-53 *in* Proc. 1975 Predator Symp. Univ. Montana, Missoula.

Quick, H. F. 1953. Wolverine, fisher and marten studies in a wilderness region. Trans. North Am. Wildl. Conf. 18:512-533.

Rand, A. L. 1948. Mammals of the Eastern Rockies and Western Plains of Canada. Nat. Mus. Can. Bull. 108, Bio. Ser. 35. 237pp.

Rosenzweig, M. L. 1966. Community structure in sympatric carnivora. J. Mammal. 47(4):602-612.

Seton, E. T. 1953. Lives of game animals. Chas. T. Branford Co., Boston, Mass. Vol. 2, pt. 2. 367pp.

Stone, W. B.; Clauson, A. S.; Slingerlands, D. E.; and Weber, B. L. 1975. Use of Romanowsky stains to prepare tooth sections for aging mammals. New York Fish Game J. 22(2):156–158.

Strickland, M. A. 1977, 1978. Fisher and marten study. Ontario Ministry Nat. Res., Algonquin Region Prog. Rep. 4 and 5. 93 and 106 pp. Manuscripts.

Wright, P. L., and Coulter, M. W. 1967. Reproduction and growth in Maine fishers. J. Wildl. Manage. 31(1):70–86.

MARJORIE A. STRICKLAND, Ontario Ministry of Natural Resources, 7 Bay Street, Parry Sound, Ontario.

CARMAN W. DOUGLAS, Ontario Ministry of Natural Resources, Box 9000, Huntsville, Ontario.

MILAN NOVAK, Ontario Ministry of Natural Resources, Wildlife Branch, Queen's Park, Toronto, Ontario.

NADINE P. HUNZIGER, Ontario Ministry of Natural Resources, 7 Bay Street, Parry Sound, Ontario.

29

Marten

Martes americana

Marjorie A. Strickland
Carman W. Douglas
Milan Novak
Nadine P. Hunziger

NOMENCLATURE

COMMON NAMES. American marten, pine marten (or martin), pussy marten, marten cat, American sable, Canadian sable, saple, Hudson Bay Sable

SCIENTIFIC NAME. *Martes americana*. Of five holarctic species of *Martes*, only *M. americana* occurs in North America.

SUBSPECIES. Some authors have subdivided this species into *americana* or *caurina* subspecies groups, as described by Hagmeier (1961). The *americana* group consists of *M. a. americana, M. a. brumalis, M. a. atrata, M. a. abieticola, M. a. actuosa, M. a. kenaiensis,* and *M. a. abietinoides;* and the *caurina* group consists of *M. a. caurina, M. a. humboldtensis, M. a. vancouverensis, M. a. nesophila, M. a. vulpina, M. a. origenes,* and *M. a. sierrae.* However, Hagmeier concluded that ''little is to be gained and much to be lost by continuing to divide the species into subspecies.'' We shall follow Hagmeier's lead and consider all North American marten to be one species, and not distinguish among possible subspecies.

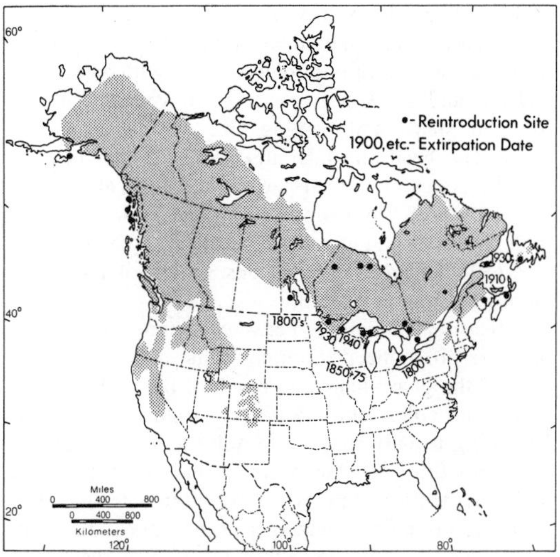

FIGURE 29.1. Distribution of the marten (*Martes americana*).

DISTRIBUTION

The excellent work of Hagmeier (1956) documented the primordial range of marten in North America. Marten are members of Merriam's (1898) Canadian and Hudsonian life zones, and their northern limit closely approximates the 50°F July isotherm (Meteorlogical Branch 1962; Environmental Data Services 1968). This limit is fairly expressed as north to the limit of trees and includes mountain habitat (figure 29.1). Throughout the original southern limits, there has been considerable range loss due to anthropogenic factors. Marten are easily baited and trapped. This has led to overharvest if not near extirpation in many settled or readily accessible areas and can render recovery difficult.

In the more remote sections of their range, particularly in the north and west, marten limits compare favorably with the primordial range. Protection in sanctuaries or by closed seasons is responsible for the continued existence of some populations. In some mountain areas that have high populations of marten,

their greatest protection results from their inaccessibility due to excessively rugged terrain or very deep snow and extreme cold during the trapping period.

DESCRIPTION

The marten is the size of a small house cat but, being a member of the weasel family, it is longer and more slender, with a bushy tail and a sharply pointed face. It is digitigrade, with five toes on each foot. The feet, including toe pads, are furred, especially in winter. The sharp claws are semiretractable, but not sheathed.

Size and Weight. Marten vary in size in different localities (Hagmeier 1961). Adult size is attained in about three months (Brassard and Bernard 1939). Males are usually larger and heavier than females.

In our study area of central Ontario, winter-caught male marten had a total length of 513 to 659 mm. Females were 465 to 572 mm in total length. Tail length averaged about 150 mm in both sexes.

599

The winter weights of our unpelted males varied from 563 g to 990 g; those of females were 400 to 605 g. The unfleshed pelt was about 20 percent of this total weight. Newby and Hawley (1954) found some indication of seasonal weight variation, with lower weights in winter.

These total length measurements agree reasonably well with those given for marten by de Vos (1952) and Peterson (1966) for Ontario, Gunderson and Beer (1953) for Minnesota, and Burt (1948) for Michigan. Our weights, however, were slightly less than theirs, due, perhaps, to the fact that the animals had been frozen for some weeks before weighing.

Anatomy. A detailed description of the postcranial osteology and musculature of marten is given by Leach (1977a, 1977b) and Leach and de Kleer (1978), who also compared the anatomy of marten to that of fisher.

Pelage. The color of marten varies markedly with locality, season, and individual. When the fur is prime, the most usual color is a golden brown with darker legs and tail and a distinctive orange or yellow patch of irregular shape and varying size on the throat and chest. The head and face are usually lighter in color and the edges of the ears are white. Individuals may vary, however, from almost black to yellow. In the fur trade, a totally yellow marten is called a "canary." De Vos (1952) gave a detailed description of color variations in Ontario marten.

Seasonal variations between summer and winter pelts are pronounced. Summer pelts, due to the shedding of the guard hairs and much of the underfur, are lighter in color, with a faded, rough appearance. Bleaching usually starts in March (de Vos 1952).

Growth of new fur begins at the tip of the tail in early September and is nearly complete by mid-October, giving the marten a prime appearance with long, dense underfur and a thin covering of guard hairs (Markley and Bassett 1942). One occurrence of "Samson" condition in marten is reported by Libby (1961) in which the animal had no guard hairs except at the tip of its tail. The entire pelage consisted of soft underfur.

Pelts can be distinguished on the basis of sex by the preputial orifice of males or by the direction of growth of the underfur at the posterior end of the abdominal gland. The direction of fur growth reverses at this point in males but not in females.

There are six inguinal mammae in both males and females.

Scent Glands. The marten has a pair of anal scent glands. In addition, both sexes have abdominal scent glands. These occur along the midventral line, extending anteriorly from the preputial orifice for about 50 mm; they are in the same position in females. On this area, the hair is short and often stained and worn. Both sexes have been observed dragging their bellies over logs and limbs, especially during the breeding season, depositing scent (Markley and Bassett 1942). Hall (1926) gave a detailed description of these glands.

Skull and Dentition. Adult male skull lengths range from 80 to 90 mm, females from 69 to 76 mm. Skull widths of males range from 46–53 mm, while those of females are from 38 to 46 mm (Peterson 1966). A marten can be separated from the similar-sized mink (*Mustela vison*) because it has 4 upper and lower premolars rather than 3. It can be separated from the fisher (*Martes pennanti*), which has the same dental formula, by its smaller skull length (less than 90 mm). The dental formula is: 3/3, 1/1, 4/4, 1/2 = 38. Details of the skull are shown in figure 29.2.

Fat Deposits. Among winter-caught marten, males usually contain more fat in relation to body size than do females. Subcutaneous fat in both sexes is confined to the inguinal region. Fat storage occurs on the mesenteries and around the kidneys. In general, marten are less fat than fisher examined from the same areas and trapping period (Strickland 1978).

REPRODUCTION

Anatomy. The ovary is completely encapsulated in a bursa. The uterus has a common corpus uteri, which allows embryos to migrate from one horn to the other.

Enders and Leekley (1941) detailed the changes in the vulva with approaching estrus. They observed that copulation takes place at the time of maximum swelling of the vulva. Vulvar swelling may last for a month or more and does not regress following early copulation.

In sexually mature males, the testes enlarge "to the size of hazel nuts" (Markley and Bassett 1942) in late June and maintain this size until September. The penis, too, shows an increase in hardness and size.

Induced ovulation has been demonstrated in several species of *Mustela* and *Martes*. The occurrence of a large baculum and penis may be associated with ensuring that the stimulation provided during copulation is sufficient to cause ovulation (Ewer 1973). Because baculum weight increases with age (Strickland 1976), older males may be more effective breeders.

Marten have delayed implantation. The fertilized egg develops to the blastocyst stage in the uterus, then becomes inactive for about 190 to 250 days until implantation occurs and normal development resumes (Hamilton 1943). The blastocyst can be flushed from the uterus by syringing with water. It is a clear, transparent sphere 500 to 900 microns in diameter. It is surrounded with a thick zona pellucida and has an inner cell count of 300 to 400 cells (Marshall and Enders 1942).

Age of Sexual Maturity. Although adult size is attained by marten by three months of age, sexual maturity in both males and females is not achieved until at least 15 months of age. There is considerable uncertainty in the literature about when sexual maturity in marten is attained, especially in wild females. This is because of the uncertainty of determining age in female marten. The best evidence, therefore, comes from fur farms where the age of the animals is known. The breeding age of captive animals, however, may differ from that in the wild. Age at sexual maturity also may vary with locality.

Markley and Bassett (1942) at the experimental fur farm in New York had ample evidence that at least some males and females were sexually active at 15 months of age, although neither sex had a successful mating until 27 months of age. Ashbrook and Hanson (1930), also in New York, observed three instances of females less than 4 months of age mating, although none produced a litter. Successful matings at 15 months of age in both males and females were reported by Yerbury (1947) and Orsborn (1953) on fur farms in western Canada. Ritchie (in Lensink 1953) believed "ranch marten first breed when two years old, though rarely when only one year old and occasionally not until they are three years old." However, Ritchie's marten from western Canada, when transported to California, normally bred when only one year old.

Jonkel and Weckwirth (1963) studied the age of sexual maturity of wild marten in Montana. They tagged and released a number of young marten and recaptured one male and three females as yearlings. The male, recaptured in July, was in active spermatogenesis. The three females, captured after the breeding season, were not pregnant.

In our study of trapper-caught wild marten, we found that nearly 80 percent of the 18–20–month-old females (determined from cementum annuli) had corpora lutea of pregnancy, indicating that they were pregnant. W. R. Archibald (personal communication 1979) reported the same incidence of pregnancy in marten in the Yukon at 15 months of age. We have no information on testicular development in males, as none of our specimens was captured during the breeding season. Mean baculum weight, however, nearly doubled between the ages of 12 and 24 months, suggesting that sexual maturity occurs during that period (see "Age Determination").

Females breed successfully until 15 years of age in captivity (Ritchie 1953; Yerbury 1947). In our study, pregnant females 12 years of age were found.

Breeding Behavior. Because observations of wild marten are rare, information on breeding behavior is largely from captive animals (Ashbrook and Hanson 1930; Markley and Bassett 1942; Brassard and Bernard 1939; Orsborn 1953).

During the estrous period, the female squats and urinates frequently and rubs her abdominal gland on stones and other objects. Marten communicate not only by scent but also by a range of vocalizations, which were detailed by Belan et al. (1978). These include huffs, growls, screams, and a chuckle that apparently is a "love call." During courtship, which may last 15 days, there is much playing and wrestling.

Copulation usually occurs on the ground. The male holds the female with his teeth by the skin on the back of her neck, and may drag her around for 30 minutes or more before coitus occurs. The vulva is exposed voluntarily. Matings may last as long as 90 minutes.

Two or three matings may occur in one day and the female may mate many times during one breeding season, not necessarily with the same male. Males and

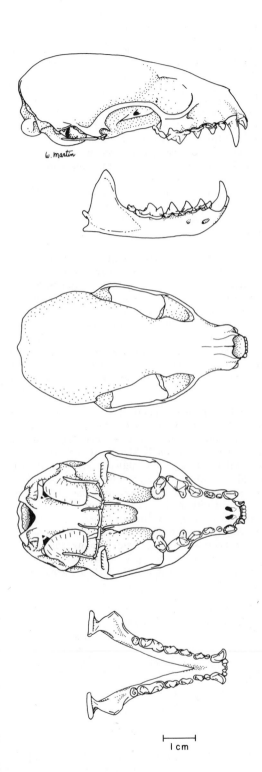

FIGURE 29.2. Skull of the marten (*Martes americana*). From top to bottom: lateral view of cranium, lateral view of mandible, dorsal view of cranium, ventral view of cranium, dorsal view of mandible.

females are polygamous. Female marten may have two or more periods of heat, separated by a quiescent interval of a few days.

Markley and Bassett (1942) reported that females became antagonistic toward each other during the mating season, demonstrated by bitter or even fatal fighting. Certain males were also rapacious during the mating season and caused injuries to females as well as to other males (Francis 1958).

Date of Breeding. Marten mate in midsummer, usually July or August. Markley and Bassett (1942) in New York State reported that extreme limits of the breeding season of captive marten ranged from July 10 to September 2. The lengths of the breeding season varied from 24 to 46 days. Most of their animals mated in July. Ritchie (1953) recorded the success of matings on a fur farm in Alberta during a 24-year period. He found the extreme limits to be June 18 to August 21. Lensink (1953) believed wild marten in Alaska mated in late June and July. Because of multiple matings, the exact date of a successful mating is unknown.

Gestation. The gestation period in marten is 220 to 276 days and can be quite variable even among the same group of ranch animals. Ashbrook and Hanson (1930), who first determined the breeding season of marten, reported that gestation lasted 259 to 275 days. Markley and Bassett (1942) agreed, and gave two instances of single, fertile matings that had gestation periods of 261 and 276 days. Brassard and Bernard (1939) reported gestation times of 220 to 230 days in five litters. Ritchie (1953) reported 17 parturitions that resulted from matings on 1 or 2 consecutive days. They had gestation times that ranged from 228 to 276 days.

The period of active pregnancy was determined by Jonkel and Weckwirth (1963) through a series of weekly laparotomies. They found that the time from implantation to parturition was approximately 27 days.

Enders and Pearson (1943) induced early implantation in marten by increasing the photoperiod, beginning 21 September. They obtained embryos that by 4 December were as developed as those of wild marten in late March. They also noted that the large percentage (up to 88 percent) of barren females among captive

marten may be due in part to loss of blastocysts prior to or at implantation, and that early illumination may increase the percentage implanted.

Parturition Dates. Parturition is most common in April but varies from mid-March to late April (table 29.1).

Litter Size. Marten produce an average of slightly less than three young per female. Adult females produce one litter per year. The most reliable data on litter sizes are from fur farmers (table 29.1).

In order to determine reproductive rates, we counted corpora lutea, blastocysts, and corpora albicantia from trapper-caught animals and found a mean of 2.7 blastocysts and 3.5 corpora lutea per pregnant female (table 29.2). However, the lower number of blastocysts relative to corpora lutea may be due to their disintegration because of poor preservation of the carcasses.

No corpora albicantia were found in females less than 24 months old. In older animals, corpora albicantia were found in various stages of regression, but we were not able to determine their age or how many had disappeared entirely. Using these structures as a measure of past productivity is not reliable.

Placental scars generally were not evident in either fixed or unfixed uteri. The uteri examined were from trapper-caught animals, which had been dead for a long time, and the scars may have disappeared as a result.

Growth and Development of Young. Brassard and Bernard (1939) gave a detailed account of the growth and development of a litter of three male and one female marten born on 28 March. At birth, the young weighed about 28 g, were altricial, and were sparsely furred. Growth was rapid, and by 4 weeks of age the males weighed over 200 g each, the female 173 g. This sexual disparity in size, evident at an early age, continued to maturity. A similar disparity in size was found by Remington (1952). A sharp decline in weight occurred at 6 weeks of age, with weaning, and again at 90 days, when the young were separated from the mother. Adult size was attained in about three months.

TABLE 29.1. Parturition dates and litter sizes of marten from fur farms

Author	Location	Parturition Dates	Number of Litters	Total Number of Young	Mean Litter Size	Range
Ashbrook and Hanson (1930)	New York	15–23 April	2	6	3.0	3
Brassard and Bernard (1939)	Quebec	28 March–17 April	5	11	2.2	1–3
Markley and Bassett (1942)	New York	6–27 April	20	52	2.6	1–4
Ritchie (1953)	Alberta	12 March–29 April	99	287	2.9	1–5
Yerbury (1947)	British Columbia	18 March–end April	10	31	3.1	?
Total			136	387	2.85	1–5

TABLE 29.2. Mean reproductive rate of Ontario marten

Age When Captured	Mean Number of Corpora Lutea per Pregnant Female	Mean Number of Blastocysts per Female	Mean Number of Corpora Albicantia per Female
18–20 months	3.2	2.3	0
30–32 months	3.6	2.9	2.0
> 40 months	3.7	2.9	2.3
all ages	3.5	2.7	

SOURCE: Strickland 1978

By 3 weeks of age, males were easily recognized by the small ridge (raphe) that runs from the anus to the penis. The ears and eyes were closed at birth. The ears opened at about 24 days, the eyes at about 39 days. By 46 days, the young were able to crawl out of the nest and by 7 weeks were quite active. Permanent dentition was complete on both jaws by 18 weeks.

The male apparently has no part in rearing the young. The kits leave their mother and disperse in late summer or early autumn.

The sex ratio of the combined litters reported by Ritchie (1953), Brassard and Bernard (1939), and Markley and Bassett (1942) is 181 males to 161 females (53:47), not significantly different from a 50:50 ratio.

ECOLOGY

Habitat. Although the marten is usually considered an inhabitant of the climax forest, recent studies have shown it to be quite adaptable to a variety of forest habitats (Soutiere 1978; Gebo 1953). In general, however, the marten prefers mature conifer or mixed forest stands.

In Maine, Soutiere (1978) found that, in undisturbed forest, marten preferred softwood-dominated mixed stands and used hardwood stands less than expected. This was also found by Francis and Stephenson (1972) in Ontario. In both summer and winter, marten preferred forest types with a high proportion of conifer, in this case hemlock (*Tsuga canadensis*).

In the western states, Clark and Campbell (1977) and Koehler et al. (1975) found that marten used a variety of forest communities. They occurred most frequently in mature spruce-fir forest, especially during the winter. They suggested this was related to the greater abundance of *Clethrionomys* on these sites, and to reduced snow accumulation. In summer, meadows may be utilized more if food is more available there.

The effects of disturbance (cutting or fire) on marten populations were studied by Soutiere (1978), Koehler et al. (1975), and Clark and Campbell (1977). The effects varied with the size and intensity of the disturbance. In Idaho, Koehler et al. (1975) found loss of cover caused a xeric condition that supported deer mice and chipmunks, not preferred foods of marten. Low-intensity fires or selective logging on mesic sites where canopy cover was maintained at 30 percent did not adversely affect marten habitat.

Soutiere (1978) also found that partial timber harvesting had little effect on density of marten populations. He concluded that marten were not restricted to wilderness spruce-fir forests. Harvesting methods that maintained a residual stand with a basal area of 20–25 m²/ha in pole stage and older trees provided adequate marten habitat. Clear cutting, however, reduced use by marten for up to 15 years.

The use of habitat is naturally related to the foods available there. Lightly disturbed mixed and conifer stands are good sites for *Clethrionomys*, staple food for marten.

Dens. Marten dens are of two main types: dens at considerable height from the ground in hollow trees and lined with grass, leaves, and moss (Seton 1953; Hamilton 1943); and dens on or under the ground in rock piles, hollow logs, tree roots, or under the snow (van Zyll de Jong 1969). Marten use squirrel nests for dens (Gebo 1975). Large dead trees with woodpecker holes are important nesting sites for marten (Clark and Campbell 1977). Masters (1980) reported daytime resting sites 1.7 m or more above the ground in live trees.

Home Range. Minimum home range size is about 2–3 km² for males and about 1 km² for females (table 29.3).

Mech and Rogers (1977) monitored radio-collared marten in Minnesota and found much larger home

TABLE 29.3. Home range size of marten using capture-recapture methods

Author	Location	Home Range Area (km²) Males	Home Range Area (km²) Females
Hawley and Newby (1957)	Montana	2.4	0.7
Francis and Stephenson (1972)	Ontario	3.6 (2.0–4.8)	1.1 (0.5–2.4)
Clark and Campbell (1977)[a]	Wyoming	2.2	0.8
Soutiere (1978)	Maine	0.1–4.4	0.1–2.3

[a]Clark and Campbell (1977) used telemetry as well as recapture methods to determine range size.

range sizes of 10 to 20 km² for males and 4.3 km² for females.

Francis and Stephenson (1972) determined "foraging" ranges (that area in which activity was centered for a week or more on a continuous basis) to be 1.7 km² for males and 0.8 km² for females. Newby and Hawley (1954) found minimum foraging areas of 1.5 km² for males and 0.3 km² for females. Males shifted foraging ranges more often than females but usually retained some overlap with the former range. Females stayed in a relatively small area for long periods of time (Francis and Stephenson 1972).

In most of these studies males had ranges two to three times the size of females. Differences in calculated range size between studies may be due to different methods and to differences in forest types. Soutiere (1978) noted that the size of a home range depends on the animal's energy requirements and the food resources available, so, in part, it reflects habitat quality. He found that males had larger home ranges in clear-cut areas. Population density of marten was thought to be another factor affecting home range size.

Francis and Stephenson (1972) found an overlap of the home ranges of males and females but a spatial or temporal separation of resident animals of the same sex. Immature marten did not remain as permanent residents and probably do not establish fixed ranges until they are sexually mature, and then only if vacant areas are available. Francis and Stephenson (1972) found the majority of immature marten were first captured in August as the family groups were dispersing. Clark and Campbell (1977) also found that males tend territorially to exclude other males but not females. Their marten population remained stable during the early summer but transient, immature animals appeared by mid-August. They postulated that marten populations were organized around the territories of the adult males. In spring and summer, adult males were able to hold territories because food was abundant. This territory was maintained against other males. By tolerating adult females and young (especially their own), however, adult males provided an area of reduced competition for the growth and development of the young and maximized the possibility of reproductive encounters.

Hawley and Newby (1957) also found the home ranges of adult males to be discrete but to overlap ranges of one or more females. Juveniles of either sex were also tolerated. Resident marten remained on the same home range for periods as long as 819 days with no seasonal movement observed. Juveniles may disperse long distances. One juvenile male was recaptured 40 km from its original capture site.

Fluctuations in Population Density. In Montana Hawley and Newby (1957) and Weckwirth and Hawley (1962) found that the marten population was continually changing due to the movement of both transients and residents. A population decline closely followed a decrease in microtine rodent density. "Loss of female marten in the resident population, a reduction in reproductive success, and failure of juvenile marten to remain on the study area were indications of a period of overpopulation apparently caused in part by a food shortage that reduced the carrying capacity of the area. The loss of marten from the area brought the population more in line with the carrying capacity and the marten population began to recover with the addition of both juveniles and adults" (Weckwirth and Hawley 1962:73). Overcrowding occurred when the population density of marten was more than about 1.2 residents per km².

In Ontario Francis and Stephenson (1972) found a resident density of 0.6 animals per km², comprised of equal numbers of males and females. Other adults with part of their range in the study area and immatures, which appeared in August, increased the density to 1.2–1.9 per km². This represented a population surplus that could be harvested.

Clark and Campbell (1976) found that their marten population in Wyoming was "fluid, with animals coming and going throughout most of the year." Only during June and July was their population stable.

In Maine, Soutiere (1978) studied the effects of timber harvesting on marten populations. He found that while the density of adult resident animals averaged 1.2 per km² in both undisturbed and partially disturbed forests, it dropped to 0.4 animals per km² in commercial clear cuts. In the clear-cut areas, home range size was larger and there were proportionally fewer immature and transient marten.

Sex Ratio. The sex ratio of trapper-caught marten is usually unbalanced toward males. Often, two or three males are captured for each female. Certain authors report a greater imbalance in the early part of the trapping season and more equal numbers later in the year (table 29.4).

Population levels and habitat conditions may influence sex ratios in the adult resident population. Francis and Stephenson (1972) and Newby and Hawley (1954) found the sex ratio of resident adult animals to be approximately 1 to 1. However, Hawley and Newby (1957) and Weckwirth and Hawley (1962) reported that the equal ratio (8 males to 9 females) originally found on their study area later shifted predominantly to males (11 males to 3 females). They attributed this to the disappearance of females and immatures during a period of reduced food supply. Both Soutiere (1978) and Clark and Campbell (1977) found twice as many adult males as females resident in their study areas.

It seems unlikely that the sex ratio of harvested marten represents the true biological character of populations. As previously noted, harvests usually have a preponderance of males; marten sex ratio at birth is 1 to 1, and data from live-trapping studies often show equal numbers of resident males and females. Yeager (1950) suggested this inconsistency was due to the larger foraging area of the male as compared to the female, which would increase his chance of encountering a trap. Studies have confirmed the differential size of foraging range between sexes (Francis and Stephenson 1972; Newby and Hawley 1954). Also, live-trapping studies found differential recapture rates for males and

TABLE 29.4. Seasonal differences in the sex ratios of trapper-caught marten

Author	Location	Percentage Males[a]		
		Trapped October to December	Trapped January to March	Trapped all Season[b]
Yeager	Colorado	65	55	62
(1950)		(405–218)	(146–122)	(729–454)
de Vos	Ontario	65	50	55
(1952)		(62–34)	(85–85)	(147–119)
Reynolds	Ontario	66	62	65
(1953)		(471–244)	(352–215)	(823–439)
Quick	British	82		44
(1953)	Columbia	(32–7)		(71–92)
Lensink	Alaska	54	49	62
(1953)		(184–157)	(64–66)	(810–506)
Strickland	Ontario	73	75	69
(1975)		(126–46)	(44–14)	(292–129)
Total		64	57	62
		(1280–706)	(691–502)	(2872–1739)

[a]Figures in brackets are sample sizes (males-females)
[b]Discrepancies between totals in this column versus totals in the two preceding columns are because for many specimens exact capture date was unknown.

females. Francis and Stephenson (1972) found that males were captured 1.64 times as often as females; Newby and Hawley (1954) reported that males were caught 2.09 times as often as females. These ratios are very similar to the male:female ratios of harvested animals.

The increase in females taken by trappers in January and February probably is related to the decreasing number of males in the population due to trapping. (Yeager 1950). Also, the scarcity of food in midwinter and the increased energetic demand of females entering active pregnancy may cause the females to range more widely, increasing their chance of encountering a trap. Another possible reason for the increased number of females in the late-season capture is that a trapper, where harvest is limited by quota, might declare the more valuable males first. The females would be declared only after the probability of catching more males was past.

Of trapped animals, males predominate until four years of age, after which the sexes are captured more equally (table 29.5). Because marten have a neonatal sex ratio of one to one and have similar longevity, yet more males than females are captured at every age up to four years, it follows that females must have a higher natural mortality than males. Otherwise females would soon outnumber males in the population but studies cited above show either equal or male-dominated populations. This natural mortality no doubt varies temporally and spatially according to food abundance and general carrying capacity and may explain much of the variation between studies.

Age Distribution of Harvest. We determined the age by cementum annuli of 1300 marten caught by trappers from 1973 to 1978. About 60 percent were less than 12 months of age.

Marten are relatively long lived. Both Markley and Bassett (1942) and Ritchie (1953) reported marten living at least 15 years in captivity. We have found as many as 13 annuli in wild marten. Longevity of males and females is similar. During the period of this study, the mean age of adults increased during the study period. The mean age of adults increased from 2.1 to 2.9 years for males and from 2.2 to 3.4 years for females. This was a statistically significant difference for females only. At the same time, both the total harvest and the population size in the study area were increasing (Strickland 1978).

The ratio of young marten (less than 12 months old) to adult females (more than 30 months old, that is,

TABLE 29.5. Sex ratios of Ontario marten by age (Strickland 1978)

Age Estimated by Cementum Annuli (Years)	Males	Females	Percentage Males
0	272	121	69
1	99	34	74
2	63	26	71
3	24	11	68
4	11	10	52
5	9	4	
6	4	3	
7	0	5	
8	3	3	
9	2 } 24	3 } 23	51
10	4	1	
11	0	2	
12	2	1	
13	0	1	
Total	493	225	69

females that could have been mothers of those young) is 7.6 to 1 (table 29.6). This is more than 2.5 times as many young in the harvest as might be expected considering the reproductive rate of 2.8 young/female (see table 1). Also, this ratio is higher in the early months of trapping than later in the season. Whether this is due to the fact that the young, soon after dispersal, are inexperienced and thus vulnerable, or to the fact that they travel more than the older residents is not known. The large percentage of young harvested early in the season can be used as a management method to select for young, nonbreeding animals.

Francis and Stephenson (1972) found that 18 of the 35 marten they live trapped were immature. All were captured in late summer and coincided with the dispersal period. Hawley and Newby (1957) also found evidence of dispersal of young, especially the males, in late summer. They too noted a change in the ratio of young animals to adult females (older than 12 months). As their population declined, this ratio declined due to lowered reproduction or decreased survival of the young.

Soutiere (1978) found that timber harvesting affected the age structure of a marten population. The number of immature animals varied inversely with the amount of disturbance: 23 percent in undistrubed forests, 19 percent in partially disturbed, and 11 percent in clear-cut areas. This is contrary to what might be expected, since the immatures unable to compete with resident adults are more likely to move to unoccupied areas that also may be less favorable ecologically (Weckwirth and Hawley 1962; Lensink 1953). Soutiere, however, does not report the degree of saturation in the undisturbed area, which would be important in this consideration.

FOOD HABITS

Marten food items fit into four general categories: small mammals, birds, insects, and fruits. Food habits of marten have been extensively studied and are similar in Maine (Soutiere 1978), Ontario (Francis and Stephenson 1972; Clem 1975), British Columbia (Cowan and MacKay 1950; Quick 1955), Alaska (Lensink 1953; Lensink et al. 1955), Idaho (Koehler et al. 1975), Wyoming (Clark and Campbell 1977), and Montana (Weckwirth and Hawley 1962; Marshall 1946).

Various species of mice, especially the redbacked vole (*Clethrionomys gapperi*), are the staple

TABLE 29.6. Ratio of juvenile marten to adult females (>30 months) in Ontario harvest

Period When Trapped	Juveniles/Adult Females
November	10.4
December	6.8
January and February	4.1
Total	7.6

SOURCE: Strickland 1978

food of marten in all seasons but are most important in winter. The meadow vole (*Microtus pennsylvanicus*) is also a common and perhaps preferred food. Francis and Stephenson (1972) found that marten selected *Microtus* sp. but made little use of the more abundant deer mouse (*Peromyscus* sp.), either because they did not prefer it or were not able to catch it.

Birds and their eggs are important foods in June and July, when they are most vulnerable. Fruits and berries, as they become more abundant in late summer, make up a large part of the diet. Most utilized are the low-growing species, such as blueberries (*Vaccinium* sp.), which are most available to the marten. Insects are also a common food in summer.

The occurrence of chipmunks (*Tamias striatus, Eutamias* sp.), various species of squirrel, and snowshoe hares (*Lepus americanus*) in the diet of marten varies, but generally they are less important than mice. Cowan and MacKay (1950) found that hares were underrepresented in marten scats relative to their availability. They concluded that any cyclic trend in marten populations was not due to their dependence on hares.

The marten is an opportunist and takes a wide variety of foods, especially when the usual or preferred diet is not availabe. Carrion apparently is more important in the winter diet, perhaps because it is more available. A number of factors affect the use of a particular food. Preference and availability, as well as availability of other foods, are all important and probably account for variation in the utilization of certain foods seasonally and geographically.

Quick (1955) believed that marten are versatile enough to survive despite extreme changes in prey populations. However, Weckwirth and Hawley (1962) found that although the marten utilized a wide variety of foods only a few were staple items in their diet and that the abundance of small mammals was directly associated with the abundance of marten. Changes in small mammal densities were sufficient to affect the carrying capacity of the area for marten. Because adult females and juveniles have the highest energy requirements, they are most affected by a food shortage. Clem (1975) speculated that fisher and marten compete for foods from December through February.

BEHAVIOR

Marten are largely solitary except during the breeding season. Fur farmers report that marten can be kept together amicably during most of the year (Ashbrook and Hanson 1930; Markley and Bassett 1942; Yerbury 1947). However, social tolerance in wild marten is exhibited between males and females only in the breeding season and between females and young only during the period of juvenile dependency. Aggressive behavior between two juvenile males in the wild, as well as many marten with wounds and scars around the head and shoulders, were reported by Hawley and Newby (1957).

Curious and thus easily trapped, marten are active throughout the year (Halvorson 1961). They may den during heavy rains or extremely low temperatures (Clark and Campbell 1977; Lensink 1953). Although they are purported to be primarily arboreal, most of their time is spent traveling and hunting on the ground. Their movements are quick and agile. They are able to swim, even underwater (de Vos 1952; Mech and Rogers 1977).

By radio tracking two males and one female, Clark and Campbell (1977) found that the males were most active in the evening, while the female was active throughout the day, perhaps because of greater energy requirements. Marten are also active at night (Remington 1952). The availability of food is probably the most important factor affecting their daily range and activity. Marten have erratic travel routes, crossing and recrossing their own tracks. They hunt from one stump, windfall, or brush pile to the next and search out every hole and crevice. They often use fallen logs for runways (Francis and Stephenson 1972; Clark and Campbell 1977).

Although marten travel on top of the snow, much of their actual hunting is done beneath the snow's surface. They gain access to this level by tunneling down near fallen logs or tree stumps that project above the snow (Clark and Campbell 1977; Koehler et al. 1975).

Marten do not appear to hunt in openings much greater than 100 m across in winter, but may do so in summer if food and cover are available (Koehler and Hornocker 1977). However, Soutiere (1978) found that marten in Maine crossed large openings (200 m) in winter, and while they traveled these more directly than in the forest, they did investigate windfalls and slash that protruded through the snow cover. He felt that winter use of open areas appeared to be related not to overhead cover but to availability of prey and to noninhibiting snow depth.

Marten probably kill small animals by means of a neck bite (Remington 1952).

MORTALITY

Although actual evidence is limited, there are scattered reports of marten being preyed upon by coyote (*Canis latrans*), fisher, red fox (*Vulpes vulpes*), lynx (*Felis lynx*), cougar (*Felis concolor*), eagles (*Aquila chrysaetus, Haliaeetus leucocephalus*), and great horned owls (*Bubo virginianus*) (Marshall 1951*a;* de Vos 1952; Quick 1953; Williams 1957). Natural enemies are few, however, and climbing ability removes the marten from most terrestrial predators.

Intraspecific and interspecific competition for food is undoubtedly a mortality factor, especially in late winter (Marshall 1951*a;* Clem 1975). De Vos (1951*b,* 1952) noted that on many traplines the population level of marten and fisher were inversely related. We have found the same thing in our study area. Although fisher and marten often occur in the same area, some sort of competition between these two species is suspected. It is not known if this competition is most pronounced relative to food or habitat. Actual predation is probably rare (de Vos 1952).

Because female marten are smaller than males

from an early age (Brassard and Bernard 1939), they may be less able to compete with litter mates for food, thus increasing their mortality. Hawley and Newby (1957) noted that, in their live-trapping studies, females were more subject to trap death than males. They speculated that females may be more subject to mortality from other factors also. Some of their dead females, although they apparently died of starvation, had mean weights greater than those of the average female, indicating that the energy balance of females normally may be near the critical level. Energy requirements of adult females and juveniles are higher, so during periods of food scarcity these would be more prone to starvation (Weckwirth and Hawley 1962) (see "Sex Ratio").

People are the greatest enemy of the marten both directly, through trapping, and indirectly, through reduction of suitable habitat.

Parasites and Diseases. A number of ectoparasites of marten have been reported. De Vos (1952, 1957) found the tick *Ixodes cookei*, and the fleas *Monopsyllus vison* and *Megabothris atrox*. Holland (1950) listed nine species of flea in marten. Cowan (1954) reported a marten infested with mites (*Listrophorus mustelae*).

Endoparasites reported by Holmes (1963) were the helminths *Alaria taxideae, Taenia martis*, and *T. mustelae*. R. J. Cawthorn (unpublished data) and R. C. Anderson (unpublished data) list *Capillaria* spp., *Capillaria aerophila, Physoloptera* spp., *Uncinaria* spp., *Uncinaria stenocephala, Crenosoma* spp., *Creonosoma petrowi, Sobolevigylus* spp., and *Ascaris devosi* from marten in Ontario. *Trichinella* spp. has been identified in 5 percent of the marten examined from our study area. We have also found 3 percent of the animals with *Dioctophyme renale* and 0.5 percent with the Guinea worm (*Dracunculus insignis*).

Marten from our study area were tested for several diseases with the following results: Toxoplasmosis incidence rate 11 percent; Aleutian disease incidence rate 1.4 percent; and Leptospirosis incidence 0 percent. None of these parasites or diseases has been noted as a significant mortality factor in marten.

AGE DETERMINATION

A number of characteristics have been used to determine the ages of marten.

Body Size. Young are full grown by their third month (Brassard and Bernard 1939). The epiphyses of the long bones of both males and females are ossified before November of their first year (Dagg et al. 1975), which negates this feature as a technique for separating adults from juveniles during the trapping season.

Pelt Characteristics. The softer texture of the juvenile pelt is useful in separating young from mature animals until about late September (Newby and Hawley 1954; Francis and Stephenson 1972). Females that have suckled young have conspicuous mammae, while the mammae of immature females are difficult to ob-

TABLE 29.7. Measurements of cleaned marten skulls

Item	Sex and age group			
	Juvenile Male	Adult Male	Juvenile Female	Adult Female
Mean total length (cm)	7.8	7.9	7.1	7.1
SD	1.4	1.5	1.3	1.0
N	29	12	32	9
Range	7.6–8.0	7.6–8.0	6.8–7.5	6.8–7.3
Mean zygomatic width (cm)	4.0	4.5	3.7	3.8
SD	1.2	2.3	0.9	1.9
N	29	12	32	9
Range	3.7–4.6	3.9–4.8	3.4–3.9	3.4–4.0

SOURCE: Strickland 1978

serve (Newby and Hawley 1954; Lensink 1953; Strickland 1978).

We examined the amount of wear and stain on the fur of the abdominal scent gland and found that, while, in general, older animals had more wear, we could not reliably divide the marten into age groups by this method. Also, the weight, lengths, and widths of pelts were measured and found not to be useful in separating young from mature marten (Strickland 1978).

Skull Characteristics. Fusion of most skull sutures in marten occurs relatively early, but the nasofrontal sutures usually remain open for most of the first year and sometimes longer. Skull measurements of total length and zygomatic width, while useful for separating males from females, will not divide the age groups (table 29.7).

The development of the sagittal crest and its use as an aging technique were discussed by Marshall (1951b). He described immature males as having no crest or one less than 20 mm long; mature males had a crest longer than 20 mm. In females, those lacking a sagittal crest were classified as immature while those with sagittal crests were considered adult. We have found some exceptions to his system, however, especially in females. Females with a crest are invariably adult, but not all adults have developed crests. We have found that a number of adult females have borne

young yet have no crest or even a coalescence of the temporal ridges. The same problem is evident in males, but to a lesser extent. In general, males and females that have developed a crest are adult but those without a crest may be either immature or adult. This is particularly important when determining the age of animals from skull palpation. Newby and Hawley (1954) noted that age determination of young marten was uncertain in the period between attainment of adult size and the development of fully adult characteristics.

Baculum Weight. As noted by Marshall (1951b), the baculum of a marten less than 12 months old was less than 100 mg. Our study largely confirmed this. We also found that the mean baculum weight continued to increase with age to about 200 mg (table 29.8). There was enough overlap between age classes, however, to preclude this as a precise aging technique. The conformation of the baculum also changes with age; adults have a more bulbous proximal end.

Suprafabellar Tubercle. Leach (in preparation) found that noting the presence of suprafabellar tubercle was a rapid and reliable method of distinguishing adult marten.

Tooth Development and Wear. Because adult dentition is attained by 18 weeks, tooth replacement is not

TABLE 29.8. Baculum weights of Ontario marten

Age from Cementum Annuli (Years)	Mean Weight (mg)	SD	N	Range
0	67.0	17.96	76	40.0–140.00
1	132.7	25.64	49	80.0–190.00
2	150.1	27.09	40	110.00–220.00
3	170.0	28.28	11	130.00–220.00
4	171.8	24.83	11	130.00–220.00
5	186.7	39.33	6	110.00–220.00
6	178.3	20.41	6	150.00–210.00
7	140		1	
8	196.7	22.36	9	170.00–230.00
9	200.0		1	
10	183.3	20.82	3	160.00–200.00

SOURCE: Strickland 1976

useful for age determination after that time. Tooth wear is not a useful aging tool (Marshall 1951*b;* Strickland 1975). We have done some preliminary studies on "canine drop" (Nellis et al. 1978), which shows some promise as a method of separating young from old animals.

Cementum Aging. We have sectioned and stained the canines and premolars of several hundred trapper-caught marten and examined the cementum for annuli. We have had no known-age animals. However, we have related the number of cementum lines to the reproductive state of the females and to various skeletal characteristics. We believe that cementum layers offer a reliable way to determine age in marten. Cementum aging has also been used for marten by Klevezal and Kleinenberg (1967), Stone et. al. (1975), and Brown (1980).

ECONOMIC STATUS

Not as valuable as its relative the European sable (*Martes zibellina*), the marten nonetheless is an important part of the fur economy of North America, particularly in the northern part of its range. The pelt of a marten, due to its light weight (a consideration when packed out from a trapline) and its relative ease of preparation, represents a very large monetary return for the labor expended when compared with, for example, beaver. This, coupled with its ease of capture, has often led to severe overharvest or extirpation. The important position of marten in the fur trade was documented by Innes (1962).

Used in the manufacture of complete coats, capes, or jackets, marten pelts are also used as trim and have been used as linings for luxury garments.

Historically occurring in 38 administrative units and still present in 30, marten were harvested in 10 provinces and territories and 9 states during the seasons 1967–77. During that period, recorded annual production varied from 50,733 (1967–68) to 130,237 (1976–77) (table 29.9). Generally, the harvest has continued to increase across most of its range. During that decade, mean Canadian price has risen from $6.41 to $17.67, with much higher price for outstanding pelts. In 1979 top Canadian prices of $126.00 were noted.

Pelts of adult male marten bring the best price, juvenile males bringing about 10 percent less. Pelts of adult females are worth about half that of adult males, juvenile females about 25 percent less than that.

MANAGEMENT

Marten are important for both economic and esthetic reasons, and several techniques have been used to enhance or control their numbers.

Regulation of the Harvest. Administrative controls of trapping such as licensing trappers, varying the length or timing of seasons, and setting quotas to limit the harvest are common methods of management.

Trapping may be limited to November and December, when the pelts are most prime. This also re-

TABLE 29.9. Annual marten harvest, 1967–68 to 1976–77

Season	Canada	United States	Total
1967–68	42,168	8,565	50,733
1968–69	64,391	7,216	71,607
1969–70	66,563	11,066	77,629
1970–71	52,096	8,806	60,902
1971–72	55,746	8,912	64,658
1972–73	61,104	9,935	71,039
1973–74	61,851	18,841	80,692
1974–75	47,790	13,092	60,882
1975–76	53,467	13,787	67,254
1976–77	102,987	27,250	130,237

Mean annual harvest

British Columbia	10,939
Alberta	2,061
Saskatchewan	553
Manitoba	681
Ontario	24,968
Quebec	9,971
New Brunswick	663
Labrador	77
N.W. Territories	9,327
Yukon	1,577
Canada	60,816
Alaska	11,220
Colorado	167
Idaho	490
Maine	262
Montana	645
Oregon	34
Utah	4
Washington	83
Wyoming	100
United States	12,747
North America	73,563

moves animals surplus to the winter carrying capacity of the range. Both Yeager (1950) and Quick (1956) found that early-season catches had a preponderance of males and younger age classes. We found that early-season trapping selected for young nonbreeding animals but we did not find a variation in the percentage of males in the harvest for various months (see tables 4 and 6).

Closing the season entirely may be necessary, and has been used by various administrations, when the population is at a low density. This is largely a corrective measure used to permit recovery of the population. Sometimes alternating open and closed seasons or alternating the sections of the trapline utilized can be effective in sustaining a population.

Marten are easily trapped and can be overharvested even in a short, early season, especially where the trapping pressure is heavy. To regulate harvest in heavily trapped areas some sort of quota is necessary. Information on which to base quotas is often inadequate, so what data are available must be used. These include harvest per area in past years and sex and age ratios of the current harvest. The opinions of trappers and game officers are also useful in determining population trends. The use of the registered trapline system, in which each trapper has a specific assigned area, permits adjustment of quotas by individual area. Quick (1956) suggested that at least 40 percent of the preseason population should be harvested. The ideal is to trap all of the young except those necessary to counter mortality in the older groups, that is, sustained yield.

Refuges. Refuges provide a reservoir from which animals disperse into the surrounding areas. De Vos (1951*a*) detailed the dispersal of marten from the Chapleau Game Preserve in Ontario. Algonquin Park is another reservoir for marten in Ontario (Strickland 1978). Van Zyll de Jong (1969) strongly recommended that refuges be established in Manitoba. Lensink (1953) noted that insufficient information about various characteristics of dispersal was available to determine the most effective size and spacing for refuges. He also noted that small, untrapped areas exist even in heavily trapped areas and that these might serve as natural refuges.

Restocking. Reintroduction of marten to parts of their former range has been undertaken by a number of administrations (see figure 29.1). Restocking of marten was undertaken in part of our study area in southern Ontario in 1957 to 1963. Two hundred and forty-nine marten (155 males and 94 females) were live-trapped in Algonquin Park and released in a 13,600 km² area west of the park. These animals have increased and are presently producing a sustained harvest of 300 to 400 animals annually.

Selective Trapping. To some extent, sex and age groups can be selectively harvested by modifying the time and intensity of trapping (Quick 1956; Yeager 1950; Strickland 1978). Greater selectivity may be possible by using some type of live-trap so that breeding stock can be released and surplus animals retained.

Feeding. The carrying capacity of the winter range can be increased by distributing carcasses of beaver (*Castor canadensis*) and other species as food for marten. Many Ontario trappers now do this.

Habitat Management. A number of recent studies have investigated the effects of disturbance (fire and timber harvesting) on marten populations (see "Habitat"). They generally agree that the effect depends on the intensity of the disturbance and that low-intensity fires and selective logging usually have little adverse effect on marten habitat. Clear cutting is incompatible with marten, as it eliminates resting sites, hunting sites, and overhead cover and causes alteration to the forest floor, which affects the preferred prey species of the marten. Effects of clear cuts could be minimized if they are kept small.

Wildfire and logging, if not too extensive, may actually improve habitat for marten. These increase shrub growth, which, in turn, produces fruits and promotes increased population densities of small mammals, both important foods of marten. Grazing, however, may cause serious depletion of the habitat in some areas (Yeager 1950; Koehler and Hornocker 1977).

Mech and Rogers (1977) noted that marten populations in both Ontario and Minnesota declined when areas were heavily logged in the 1920s and 1930s. In fact, the disappearance of the marten from much of its range has been attributed to the effects of fire and logging (Seton 1953; Yeager 1950; Koehler et al. 1975; Clark and Campbell 1977; Soutiere 1978). There is a need for land managers to consider the habitat requirements of marten in their plans for forest management.

CURRENT RESEARCH AND MANAGEMENT NEEDS

Much information is available about certain aspects of marten biology but often it is difficult to compare because it has been collected at different times, places, and stages in the population cycles.

More data on known-age animals will enhance our knowledge of age at sexual maturity of both males and females. Is age at sexual maturity affected by latitude? How much can the reproductive cycle be advanced by the use of artificial lighting on ranches? Would such lighting also affect the reproductive rate?

Most important to management are techniques that will give accurate estimations of density, the potential carrying capacity of an area, and ways to predict future population trends. Computer modeling may prove useful in this endeavor.

The establishment of a reliable method of age determination that could be used by trappers on live-trapped animals would permit a selective harvest.

Other interesting questions that could be pursued are: What are the energy requirements of marten by age, sex, season, and latitude? What are the comparative nutritional values of known marten foods? Some areas have larger, darker marten; is this due to genetic polymorphism or to local environmental conditions? What influences marten abundance and cycles? Why is the marten population across North America now at an all-time high?

LITERATURE CITED

Ashbrook, F. G., and Hanson, K. B. 1930. The normal breeding season and gestation period of martens. U.S. D. A. Circ. 107. 6pp.

Belan, I.; Lehner, P. N.; and Clark, T. W. 1978. Vocalizations of the American pine marten, *Martes americana*. J. Mammal. 59(4):871–874.

Brassard, J. A., and Bernard, R. 1939. Observations on breeding and development of marten. Can. Field Nat. 53(2):15–21.

Brown, M. K. 1980. Pine marten: the sprite of the northern forests. Conservationist 34(4):15–17.

Burt, W. H. 1948. The mammals of Michigan. Univ. Michigan Press, Ann Arbor. 288pp.

Clark, T. W., and Campbell, T. M. 1976. Population organization and regulatory mechanisms of pine martens in Grand Teton National Park, Wyoming. 1st Conf. Res. in Nat. Parks, New Orleans, La. 9pp.

———. 1977. Short-term effects of timber harvests on pine marten behavior and ecology. Idaho State Univ., Pocatella. 60pp. Manuscript.

Clem, M. 1975. Interspecific relationships of fisher and martens in Ontario during winter. Proc. Predator Symp. Univ. Montana, Missoula. 268pp.

Cowan, I. McT. 1954. An instance of scabies in the marten (*Martes americana*) J. Wildl. Manage. 19(4):499.

Cowan, I. McT., and MacKay, R. H. 1950. Food habits of the marten in the Rocky Mountain region of Canada. Can. Field Nat. 64(3):100–104.

Dagg, A. I.; Leach, D.; and Summer-Smith, G. 1975. Fusion of the distal femoral epiphysis in male and female marten and fisher. Can. J. Zool. 53(1):1514–1518.

de Vos, A. 1951a. Overflow and dispersal of marten and fisher from wildlife refuges. J. Wildl. Manage. 15(2):164–175.

———. 1951b. Recent findings in fisher and marten ecology and management. Trans. North Am. Wildl. Conf. 16:498–505.

———. 1952. Ecology and management of fisher and marten in Ontario. Tech. Bull. Ontario Dept. Lands For., Wildl. Ser. 1. 90pp.

———. 1957. Pregnancy and parasites of marten. J. Mammal. 38(3):42.

Enders, R. K., and Leekley, J. R. 1941. Cyclic changes in the vulva of the marten (*Martes americana*). Anat. Rec. 79(1):1–5.

Enders, R. K., and Pearson, O. P. 1943. Shortening gestation by inducing early implantation with increased light in the marten. Am. Fur Breeder (January). P. 18.

Environmental Data Service. 1968. Climatic atlas of the United States. U.S. Dept. Commerce, Washington. Reprinted 1977 by National Oceanic and Atmospheric Administration.

Ewer, R. F. 1973. The carnivores. Weidenfeld & Nicolson, London. 494pp.

Francis, G. R. 1958. Ecological studies of marten (*Martes americana*) in Algonquin Park, Ontario. M.S. Thesis. Univ. British Columbia. 74pp.

Francis, G. R., and Stephenson, A. B. 1972. Marten ranges and food habits in Algonquin Provincial Park, Ontario. Ontario Ministry Nat. Resour. Res. Rep. Wildl. 91. 53pp.

Gebo, T. 1953. The pine marten (*Martes americana*) in the Adirondacks: distribution and habitat affinities. Adirondack Ecol. Centre, State Univ. of New York, Coll. of Environ. Sci. and For. 56pp.

Gunderson, H. L., and Beer, J. R. 1953. The mammals of Minnesota. Univ. Minnesota Press, St. Paul. 196pp.

Hagmeier, E. M. 1956. Distribution of marten and fisher in North America. Can. Field Nat. 70(4):149–168.

———. 1961. Variation and relationships in North American marten. Can. Field Nat. 75(3):122–138.

Hall, E. R. 1926. The abdominal skin gland of *Martes*. J. Mammal. 17:227–229.

Halvorson, C. H. 1961. Curiosity of a marten. J. Mammal. 42(1):111–112.

Hamilton, W. J., Jr. 1943. Mammals of Eastern United States. Comstock Publishing Co., Ithaca, N. Y. 432pp.

Hawley, V. D., and Newby, F. E. 1957. Marten home ranges and population fluctuations. J. Mammal. 38(2):174–184.

Holland, G. P. 1950. The siphonaptera of Canada. Sci. Serv. Div. Ent. Livest. Insect Lab., Kamloops, B.C. 306pp.

Holmes, J. C. 1963. Helminth parasites of pine marten, *Martes americana*, from the district of MacKenzie. Can. J. Zool. 41:333.

Innis, H. A. 1962. The fur trade in Canada. Univ. Toronto Press, Toronto. 446pp.

Jonkel, C. J., and Weckwirth, R. P. 1963. Sexual maturity and implantation of blastocysts in the pine marten. J. Wildl. Manage. 27(1):93–98.

Klevezal, G. A., and Kleinenberg, S. E. 1967. Age determination of mammals from annual layers of teeth and bones. TT 69–55033. Israel Programme for Sci. Transl., Jerusalem. 128pp.

Koehler, G. M., and Hornocker, M. G. 1977. Fire effects on marten habitat in Selway-Bitterroot wilderness. J. Wildl. Manage. 41(3):500–505.

Koehler, G. M.; Moore, W. R.; and Taylor, A. R. 1975. Preserving the pine marten: management guidelines for western forests. Western Midlands 2(3):31–36.

Leach, D. 1977a. The descriptive and comparative postcranial osteology of Marten (*Martes americana* Turton) and fisher (*Martes pennanti* Erxleben): the appendicular skeleton. Can. J. Zool. 55:199–214.

———. 1977b. The forelimb musculature of marten (*Martes americana* Turton) and fisher (*Martes pennanti* Erxleben). Can. J. Zool. 55(1):31–41.

Leach, D., and de Kleer, V. S. 1978. The descriptive and comparative postcranial osteology of marten (*Martes americana* Turton) and fisher (*Martes pennanti* Erxleben): the axial skeleton. Can. J. Zool. 56(5):1180–1191.

Lensink, C. J. 1953. An investigation of the marten in interior Alaska. M.S. Thesis. Univ. Alaska, Fairbanks. 89pp.

Lensink, C. J.; Skoog, R. O.; and Buckley, J. L. 1955. Food habits of marten in interior Alaska and their significance. J. Wildl. Manage. 19(3):364–368.

Libby, W. L. 1961. Occurrence of "Samson" condition in marten. J. Mammal. 42(1):112.

Markley, M. H., and Bassett, C. F. 1942. Habits of captive marten. Am. Midl. Nat. 28:605–616.

Marshall, W. H. 1946. The winter food habits of the pine marten in Montana. J. Mammal. 27(1):83–84.

———. 1951a. Pine marten as a forest product. J. For. 49(2):889–905.

———. 1951b. An age determination method for pine marten. J. Wildl. Manage. 15(3):276–283.

Marshall, W. H., and Enders, R. K. 1942. The blastocyst of the marten. (*Martes*). Anat. Rec. 84(9):307–310.

Masters, R. D. 1980. Daytime resting sites of two Adirondack martens. J. Mammal. 61(1):157.

Mech, L. D., and Rogers, L. L. 1977. Status, distributions and movements of martens in northeastern Minnesota. U.S.D.A. For. Serv. Res. Pap. NC-143. 7pp.

Merriam, C. H. 1898. Life zones and crop zones of United States. U.S.D.A. Bull. 10. 79pp.

Meteorological Branch. 1962. The climate of Canada. Information Canada, Ottawa. 74pp.

Nellis, C. H.; Wetmore, S. P.; and Keith, L. B. 1978. Age-related characteristics of coyote canines. J. Wildl. Manage. 42(3):680–683.

Newby, F. E., and Hawley, V. D. 1954. Progress on a marten live-trapping study. Trans. North Am. Wildl. Conf. 19:452–462.

Orsborn, E. V. 1953. More on marten raising. Fur Trade J. Can. 31:14.

Peterson, R. L. 1966. The mammals of Eastern Canada. Oxford Univ. Press, Toronto. 465pp.

Quick, H. F. 1953. Wolverine, fisher and marten studies in a wilderness region. Trans. North Am. Wildl. Conf. 18:512–533.

———. 1955. Food habits of marten (*Martes americana*) in northern British Columbia. Can. Field Nat. 69(4):144–147.

———. 1956. Effects of exploitation on a marten population. J. Wildl. Manage. 20(3):267–274.

Remington, J. D. 1952. Food habits, growth and behavior of two captive pine martens. J. Mammal. 33(1):66–70.

Reynolds, K. 1953. Sex ratios of marten and fisher in Ontario. Ontario Dept. Lands For. Div. Fish Wildl. Prog. Rep. 4pp.

Ritchie, J. W. 1953. Raising marten for twenty-four years. Fur Trade J. Can. 30:10.

Schupback, T. A. 1977. History, status and management of the pine marten in the Upper Peninsula of Michigan. M.S.F. Thesis. Michigan Technol. Univ., Houghton. 79pp.

Seton, E. T. 1953. Lives of game animals. Chas. T. Branford Co., Boston. Vol. 2, pt. 2. 367pp.

Soutiere, E. C. 1978. The effects of timber harvesting on the marten. M.S. Thesis. Univ. Maine, Orono. 62pp.

Stone, W. B.; Clauson, A. S.; Slingerlands, D. E.; and Weber, B. L. 1975. Use of Romanowsky stains to prepare tooth section for aging mammals. New York Fish Game J. 22(2):157–158.

Strickland, M. A. 1975, 1976, 1978. Fisher and marten study. Ontario Ministry Nat. Resour. Algonquin Region Prog. Rep. 3 to 5. 92, 93, and 106 pp. Manuscript.

van Zyll de Jong, C. G. 1969. The restoration of marten in Manitoba: an evaluation. Manitoba Wildl. Branch Rep. Manuscript.

Weckwirth, R. P., and Hawley, V. D. 1962. Marten food habits and population fluctuations in Montana. J. Wildl. Manage. 26(1):55–74.

Williams, T. R. 1957. Marten and hawk harass snowshoe hare. J. Mammal. 38(4):517–518.

Yeager, L. E. 1950. Implications of some harvest and habitat factors on pine marten management. Trans. North Am. Wildl. Conf. 15:319–335.

Yerbury, H. 1947. Raising marten in captivity. Fur Trade J. Can. 25:14.

MARJORIE A. STRICKLAND, Ontario Ministry of Natural Resources, 7 Bay Street, Parry Sound, Ontario.

CARMAN W. DOUGLAS, Ontario Ministry of Natural Resources, Box 9000, Huntsville, Ontario.

MILAN NOVAK, Ontario Ministry of Natural Resources, Wildlife Branch, Queen's Park, Toronto, Ontario.

NADINE P. HUNZIGER, Ontario Ministry of Natural Resources, 7 Bay Street, Parry Sound, Ontario.

30

Weasels

Mustela species

Gerald E. Svendsen

NOMENCLATURE

Weasels and the black-footed ferret are small, terrestrial carnivores of the family Mustelidae. This family contains 25 genera and nearly 70 species that occupy terrestrial, fresh-water, and salt-water habitats throughout the world except for Australia, the Antarctic, and most oceanic islands. The genus *Mustela* within which the North American weasels and the black-footed ferret are placed, contains about 15 living species. The present classification separates the weasels and the black-footed ferret into two subgenera.

    genus *Mustela*
        subgenus *Putorius*
            Mustela nigripes, black-footed ferret
        subgenus *Mustela*
            Mustela erminea, ermine, or stoat
            Mustela frenata, long-tailed weasel
            Mustela nivalis, least weasel

Up until 1890, the names of European weasels were applied to the weasels of the New World. For the next 50 years, the American weasels were generally regarded to be specifically distinct from counterparts in the Old World and there was a proliferation of newly named species in the New World (see Hall 1951 for a review of taxonomy and nomenclature). Later, however, Allen (1933) argued that the least weasel occurred in both North America and the Old World and should be considered as a single species, *Mustela nivalis (rixosa)*. Hall (1944) then emphasized the circumpolar distribution of the ermine and recognized only a single species, *Mustela erminea*, in both the Old World and the New World. At present, there is continuing controversy over whether or not there is a single species of least weasel in both the Old World and the New World or whether there are two species, *M. nivalis* in western Europe and *M. rixosa* in northeastern Eurasia and North America (Siivonen 1967; R. S. Hoffmann personal communication). Hall (1951), in his revision of American weasels, recognized four species: *M. erminea* and *M. nivalis* limited to North America, *M. frenata* from North America through Central America and into South America, and *M. africana* in tropical South America. Hall and Kelson (1959) recognized 20 subspecies of *M. erminea*, 4

subspecies of *M. nivalis*, and 35 subspecies of *M. frenata* in North America.

DISTRIBUTION

The genus *Mustela* occurs throughout North America (figures 30.1–30.4). *M. erminea* is a boreal, circumpolar species having the most widespread distribution of any species in the family Mustelidae (figure 30.1). In the Nearctic, it is found from the arctic south to northern California, Nevada, New Mexico, and Colorado in the West, to the upper Great Lakes states in the Midwest, and to Pennsylvania in the East. *M. erminea* occupies a variety of boreal habitats from agricultural lowlands, woodlands, and meadows, to montane habitats at elevations of 3,000–4,000 meters. It avoids dense coniferous forests and deserts.

M. nivalis also occupies a circumpolar distribution. In North America, it is found from Alaska southward to Montana and Nebraska in the West, through Wisconsin and Minesota and into Illinois, Indiana, and Ohio in the Midwest, and south into western Pennsylvania and through the Appalachians into North Carolina in the East (figure 30.2). It is not known from upper New England, Iceland, Greenland, or the arctic islands. *M. nivalis* appears to be sporadically distributed or uncommon throughout much of its range (Hatt 1940). It inhabits marshy areas, meadows, cultivated fields, brushy areas, and open woods (Jackson 1961). In Indiana it occurs primarily in open, cultivated fields (Mumford 1969), whereas in Wisconsin it prefers high marsh areas (Beer 1950).

M. frenata occurs from southern Canada throughout all of the United States and Mexico, southward through all of Central America and into northern South America (figure 30.3). It inhabits all life zones from alpine-arctic to tropical, except for deserts. The availability of water in summer appears to limit distribution (Hall 1951). Favored habitats include brushland and open timber, brushy field borders, grasslands along creeks and lakes, and swamps.

M. nigripes was once distributed over the grassy prairies from southern Alberta and Saskatchewan south to Texas and Arizona, but was never abundant (figure

FIGURE 30.1. Distribution of the ermine (*Mustela erminea*). Adapted from Hall and Kelson 1959.

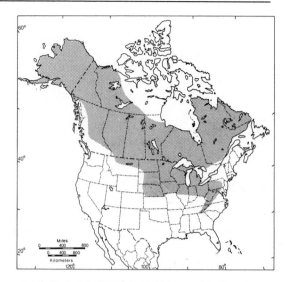

FIGURE 30.2. Distribution of the least weasel (*Mustela nivalis*). Adapted from Hall and Kelson 1959.

FIGURE 30.3. Distribution of the long-tailed weasel (*Mustela frenata*).

FIGURE 30.4. Historical (adapted from Hall and Kelson 1959) and present distribution of the black-footed ferret (*Mustela nigripes*).

30.4). The geographic range of the black-footed ferret coincides with the range of the prairie dog (*Cynomys* sp.). This secretive mustelid spends most of its life underground in the borrows of prairie dog towns.

DESCRIPTION

Weasels and the black-footed ferret are characterized by a long slender body. The short limbs bear five digits with nonretractible, curved claws. Ears are short and rounded. The pelage of weasels living in northern latitudes turns white during the winter but the pelage of weasels living in southern latitudes may remain brown

or become a mixture of brown and white. Males are generally larger than females.

Skull. The typical skull shape of weasels and black-footed ferret has a slight facial angle, a long braincase, a short rostrum, and a greatly inflated typmanic bullae with the paraoccipital processes closely appressed to the bullae (figure 30.5). The skull of the adult *M. frenata* is more angular and the postorbital processes are more pointed than in the skull of *M. erminea*. The skull of *M. erminea* is elongated with a smoothly rounded braincase. The saggital crest is weakly developed and the frontal region is not inflated but is flat. The zygomata are slender, weak, and moderately arched. The rostrum is short, the auditory bullae are large and flattened, and there is no auditory meatus. The skull of *M. nivalis* is a miniature of that of *M. erminea*. The braincase is narrower through the mastoid process, and has a more distinct saggital ridge and a more prominent postorbital process. The skull of *M. nigripes* is large and massive compared to the weasels, with the mastoid process projecting and angular.

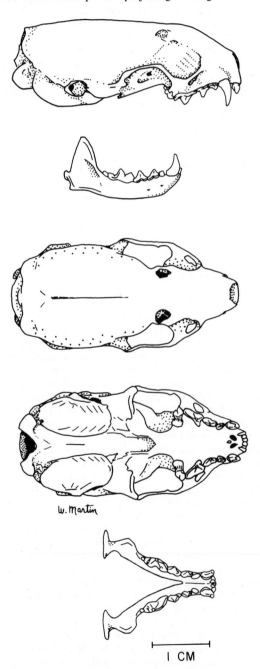

1 CM

FIGURE 30.5. Skull of the long-tailed weasel (*Mustela frenata*). From top to bottom: lateral view of cranium, lateral view of mandible, dorsal view of cranium, ventral view of cranium, dorsal view of mandible.

Dentition. The dental formula is 3/3, 1/1, 3/3, 1/2 = 34. The typical permanent dentition is highly specialized for a diet of flesh. The degree of specialization is second only to the Felidae. The first lower molar is highly specialized, the metaconid is completely suppressed, and the taloid forms an elevated blade for cutting food rather than a basin for crushing. The upper blade of the fourth upper premolar and the first lower molar form a shearing surface. Other modifications of the dentition for a diet of flesh include reduction in size of the second lower molar, the small size of the inner lobe of the upper molar, and the absence of the last two upper molars and third lower molar. The teeth vary little among the species except in size.

The high degree of specialization for a diet of flesh is as evident in the deciduous dentition as it is in the permanent dentition. The upper carnassial of the deciduous dentition is more highly sectorial than the permanent tooth, suggesting that the deciduous dentition is more specialized for a diet of flesh than the permanent dentition. Hamilton (1933) found that milk teeth are used for eating solid food as soon as the carnassials are in place, at about three weeks after birth. The deciduous teeth are used for almost two months before being replaced by permanent teeth in *M. frenata* (Hall 1951).

Species. *Mustela frenata* is the largest weasel, with a total length of 300–350 mm. The tail is 40–70 percent of the head and body length and it has a distinct black tip. Males are about 10–15 percent larger than the females in most populations. The upper parts are brown in summer. The underparts are whitish, tinged with yellowish or buffy brown from the chin to the inguinal region. The tail is uniformly brown except for the black tip. Populations in Florida and the southwestern United States sometimes have white or yellowish facial markings. The winter pelage in northern populations is normally entirely white, sometimes tinged with yellow, except for the black tip of the tail. *M. frenata* is separable from *M. erminea* by size in regions where the two species are sympatric. The tail is more than 44 percent of the head and body length in *M. frenata*, whereas it is about 40 percent in *M. erminea*. When the sex is known, the greatest breadth of skull will distinguish the two weasels. *M. frenata* is always larger than *M. erminea*.

M. erminea is a medium-sized to small weasel. The total length is 225–340 mm in males and 190–290 mm in females. The tail has a distinct black tip and is 30–45 percent of the head and body length. In summer pelage, the upper parts are brown and the underparts are whitish from the chin to the inguinal region but sometimes are interrupted by the brown of the upper parts encircling the body in the abdominal region. The winter pelage is white, with a black-tipped tail. Soles of the feet are densely haired in winter and only a small area of pads is exposed in summer. Males are about 30 percent larger than females.

M. nivalis is a tiny weasel. The total length is less than 250 mm in males and 225 mm in females. The tail length is 25 percent or less of the head and body length and lacks a distinct black tip. The summer pelage color of the upper parts is a rich chocolate brown extending well down on each side onto the ventral parts, sometimes meeting midventrally. Underparts are white, or rarely the same brown as the upper parts. The normal winter pelage is entirely white; rarely the summer pelage is maintained throughout winter. The small size, length of tail, and absence of a black tip to the tail readily distinguish *M. nivalis* from *M. erminea* and *M. frenata*. Also, the fur of *M. nivalis* will fluoresce under ultraviolet light, producing a lavender color, whereas the fur of the other two species remains dull brown under ultraviolet light (Latham 1953).

M. nigripes is larger than any weasel. Total length is 500–600 mm, with the tail about one-third the length of the head and body. The ears are relatively large but rounded. Males are 10 percent larger than females. The pelage is buffy, becoming lighter on the face, throat, and ventrum and brownish over the middle of the back. The feet and the end of the tail are black and a black mask occurs across the face and eyes. The pelage is slightly paler in winter. The skull is large and massive, broad between the orbits, and constricted behind the postorbital processes.

ANATOMY

Weasels and the black-footed ferret are characterized by a long, slender body that apparently is adapted to allow them to enter the hiding places of rodents, their main prey. There is an elongation of the vertebral column compared to mammals with a body shape like a raccoon (*Procyon lotor*) (Hall 1951). The different species of weasel do not vary in the number of vertebrae except for the caudal vertebrae, which are fewest in the short-tailed species and most numerous in the long-tailed species. There is also some geographical variation in the number of caudal vertebrae within each species.

The structure of the baculum varies among the weasels and ferret. It is a useful characteristic in distinguishing *M. nivalis* from *M. erminea* and indicating maturity in *M. frenata*. Male *M. frenata* reach adult body size in three to four months yet do not reach sexual maturity until about one year of age. The baculum remains small during the year of immaturity but develops to adult proportions when the animals become sexually active (Wright 1951).

External characters vary according to geographical distribution, season, age, and gender. Measurements of both body size and skull vary with age (Hamilton 1933) and the skull of the female is lighter and more slender than that of the male (Hall 1951). Individual variation occurs in molt patterns, skull measurements, and dentition. Pelage differences are seen between seasons and among populations at different latitudes. Body the size and the shape of the skull show geographical differences. Rosenzweig (1966) statistically analyzed variation in the body size of mammalian carnivores including weasels with respect to latitude, temperature, community production, competition, and size of available prey. He concluded

that temperature and latitude appeared to be likely causes of change in body size. Bergmann's rule states that "races from cooler climates tend to be larger in species of warm-blooded vertebrates than races of the same species in warmer climates" (in Mayr 1966:319). The usual explanation for this rule is that large animals expend less energy for thermoregulation because of the small surface-to-volume ratio, thus it is physiologically more economical for large animals to live in cold climates. Scholander (1955) criticized this interpretation and McNab (1971) found no consistent pattern among length, latitude, and food habits for either sex of *M. nigripes, M. frenata, M. erminea,* or *M. nivalis.* Only *M. nivalis* and *M. erminea* exhibited increased body length at high latitudes. Variation in the size of weasels and the ferret was concluded to be due to the size of available prey and the interaction among sympatric species of weasel.

PHYSIOLOGY

Metabolism. The elongated body shape of weasels and the ferret results in a higher surface-to-volume ratio than standard-shaped mammals of the same weight. The metabolic rates of adult male *M. frenata* are greater than the metabolic rates of adult females. At 32° C, the thermoneutral zone, the metabolic rate of resting, postabsortive males, is 136 ± 0.12 kcal/hour (Brown and Lasiewski 1972). Metabolism was inversely related to ambient temperature below 35° C, and mean body temperature did not vary with ambient temperature. Weasels lost almost twice as much heat to the environment as comparably sized woodrats (*Neotoma* sp.). Evaporative heat loss accounted for about 14 percent of the minimal metabolic coefficient. The higher heat loss in weasels was attributed largely to their greater surface area, shorter pelage, and inability to achieve a spherical resting posture. Weasels are unable to minimize their heat loss at low ambient temperatures by rolling into a nearly round ball assumed by most mammals. Instead, they coil their elongated bodies into a flattened disc, a shape with considerably more surface area exposed than a sphere. Compensation of heat loss through the evolution of greater insulation qualities of the pelage has not occurred (Brown and Lasiewski 1972).

M. nivalis in the arctic has a metabolic rate two to three times higher than would be expected from the values for basal metabolic rate and body size derived from the standard "mouse to elephant curve." In cold experiments, weasels were able to mobilize heat production nearly four times greater than the basal rate. These are as high a metabolic rate as an mammal can maintain for long periods of time in a small cage. Weasels in Wisconsin have a much less elevated metabolic rate, indicating that arctic weasels have a greatly increased resting metabolic rate compared with the southern form (Scholander et al. 1950). Parer and Metcalfe (1967) postulated that in animals less than about 1 kg there is an increased oxygen need by the tissues. This is reflected in an inverse relationship between capillary density and body weight (Schmidt-

Nielsen and Pennycuik 1961). Iversen (1972) repudiated this and suggested that the high basal metabolic rate observed in weasels is due to a physiological adjustment to high metabolic rates in the smaller species of mustelids.

Molt. The summer pelage of all American weasels is brown with white or whitish underparts. Winter whitening occurs in the northern ranges. Increase in day length initiates molting and growth of white hair, and decrease in day length causes molt and growth of brown hair (Bissonnette and Bailey 1944). The white pelage is commonly termed *ermine* in the fur trade. Weasels in the southern part of their range in North America remain in the brown pelage during both molts; intermediate populations may be characterized by mottling of white and brown in the pelage. Seasonal pelage change is controlled by the pituitary gland (Rust 1965; Rust and Meyer 1968). The pineal gland produces melantonin, which initiates changes in the central nervous system and causes molting and growth of the white winter pelage (Rust and Meyer 1969). The seasonal pelage cycle is controlled by light (Bissonnette and Bailey 1944), which acts to mediate the activity of the pituitary (Lyman 1942).

The onset of the molt to the brown pelage was delayed in a cold room at 20° F, whereas at 70° F brown hairs appeared as soon as the white hairs were shed (Rust 1962). Temperature was a modifying factor affecting the speed and nature of the pelage change but it was not the major factor responsible for spring pelage change. Weasels in the north have thicker pelage than those in the south. The soles of the feet are least hairy in regions of high average temperature and hairest in regions of low average temperature. The degree of hairiness is determined by the diameter and length of individual hairs and the number of hairs per dermal surface (Hall 1951).

M. nigripes also molts twice a year but dichromism is not evident. Summer pelage is sleek and short, whereas winter pelage is only slightly longer and darker. A captive ferret was shedding in June but no abrupt seasonal change in pelage followed (Progulske 1969).

Fall molt of weasels begins on the ventral surface and moves along the flanks, encroaching on the facial region and limbs. The last area to whiten is the dorsum: a few patches of old brown fur on the shoulders, neck, and head are the last to remain. The molt is completed in a little over three weeks (Hamilton 1933). The spring molt begins on the dorsum and head and spreads over the body and legs; it is not necessarily symmetrical in its progress. Small patches of white in front of the ears may be the last remains of the winter pelage. Each individual weasel has a slightly different pattern of molt, and the pattern will vary in the same weasel from year to year.

REPRODUCTION

Many mustelids exhibit an unusually long period between conception and parturition because of delayed implantation (Hamlett 1935; Wright 1942). Both *M. frenata* and *M. erminea* have delayed implantation, whereas *M. nivalis* and *M. nigripes* do not (table 30.1).

Breeding and Gestation. Male *M. frenata* attain adult body weight in three to four months, but do not become sexually mature until about one year of age. The bacula of sexually immature *M. frenata* weigh 14–29 milligrams. Those of sexually mature individuals weigh 53–100 mg (Petrides 1950). The earliest sign of spermatogenic activity occurs in March, coincident with the beginning of the spring molt. Testes reach a maximum volume seven to eight times larger than the inactive size, and do not regress until August or September. Testes are inactive throughout the winter (Wright 1948a). Female *M. frenata* are fully grown and sexually mature at three to four months of age.

TABLE 30.1. Summary of reproductive data for the North American weasels and black-footed ferret

Species	Age of Sexual Maturity	Breeding Season	Gestation	Litter Size	Number of Litters per Year	Reference
M. frenata	male, 1 yr female, 3–4 months	July–August	approx. 278 days, 27 days implantation	6–9	1	Hamilton 1933; Wright 1942, 1948a, 1948b; Svendsen unpublished data
M. erminea	male, 1 yr female, 3–4 months	July–August	approx. 270 days, 21–28 days implantation	6–9	1	Hamilton 1933; Svendsen unpublished data
M. nivalis	male and female, 3–4 months	all year	35 days	3–6	2–3	Deanesly 1944; Heidt et al. 1968; Heidt 1970
M. nigripes	male and female, 1 year	spring	approx. 41 days	4–5	1	Hillman 1968a

NOTE: All species are polygamous.

Sanderson (1949) reported that a female came into heat (estrus) at the age of nine weeks. Both adult and young-of-the-year females come into heat from June through August. Heat is indicated by a swollen, doughnut-shaped vulva and females are receptive for 3–4 days from the time of first mating. The swollen vulva of successfully mated females shows regression within 4–5 days, whereas the vulva of unmated females remains swollen for several weeks. Estrus does not occur during lactation; females that suckle young come into heat 55–104 days after parturition, while females that give birth and do not suckle come into heat 36–71 days after parturition (Wright 1948a).

Development. The first signs of pregnancy in *M. frenata* occur 50–80 hours after copulation: the two-cell stage of 70–85 hours, four-cell stage of 81–99 hours, eight-cell stage at 4–8 days, and morula at 11 days. The early blastula stage occurs at 15 days and at this time the blastula is positioned in the upper end of the uterus. The full blastula with early stages of cell differentiation remains quiescent until spring, then implantation occurs and further development takes place. Active but not implanted blastocysts were recovered 251 days after mating and 27 days before birth (Wright 1948b). Parturition occurs in April and May and a single litter is produced each year.

Male *M. erminea* reach 85 percent of their adult weight by July but do not become sexually mature until the following May or June. Spermatogenesis begins with the initiation of the spring molt. Testes enlarge to about 13 times the resting size by April and then begin to regress after July (Deanesly 1944). Treatment with melatonin causes regression of testes, indicating that the pineal gland is involved in the regulation of seasonal reproduction (Rust and Meyer 1969). Female *M. erminea* also reach 85 percent of their adult weight by four months of age, but, unlike males, females are sexually mature during their first summer of life. Spontaneous ovulation may begin in June and repeat monthly unless the female is bred. Nine to 10 ova are shed each ovulation (Hamilton 1933). Delayed implantation accounts for all but 21–28 days of the 10-month gestation period. Parturition occurs in April and May and a single litter is produced each year.

The reproductive cycle of *M. nivalis* is quite different from that of *M. frenata* and *M. erminea*. *M. nivalis* produces up to three litters per year and delayed implantation does not occur. Pregnancies are recorded in most months of the year but peak numbers of pregnancies occur in spring and midwinter. Males from the spring litter grow quickly and produce active sperm by fall. Females may breed in the year of their birth. Males and females of the fall litter do not become sexually mature until the following spring (Southern 1964). The gestation period is about 35 days (Heidt et al. 1968).

The reproductive biology of *M. nigripes* is poorly documented but there does not seem to be delayed implantation. Mating takes place in early spring and the young are born after about 41 days gestation. This reproductive pattern resembles that of the European ferret (*M. putorius*) and the closely related Siberian polecat (*M. eversmanni*). Young black-footed ferrets were observed in early July in South Dakota and attained adult size in August (Hillman 1968a). Parturition apparently occurred in May.

Neonates. Weasels are both blind and helpless but postnatal development is rapid and they attain adult size in 4–6 weeks depending on the species (table 30.2). *M. nivalis* is the smallest weasel and it attains adult size and pelage earliest and becomes self-sufficient in 4–5 weeks. The eyes and ears of young *M. nivalis* open at 26–29 days, whereas the eyes and ears of *M. erminea* and *M. frenata* do not open until 5 weeks of age. Permanent dentition is complete by 40–42 days in *M. nivalis* but not until 75 days in *M. frenata* (Hamilton 1933). In general, self-sufficiency of the large two species of weasel is attained 2–3 weeks later than in *M. nivalis*.

Young *M. erminea* can be distinguished from the other species by a prominent dark-colored mane extending from the forehead to the shoulders that is evident shortly after birth and persists up to six weeks of age (Hamilton 1933; Svendsen unpublished data). Both *M. erminea* and *M. frenata* exhibit a sexual disparity in body size as adults. This disparity is evident by two weeks of age in *M. frenata* but not until four weeks of age in *M. erminea*.

ECOLOGY

Activity. Weasels and the ferret are active in both winter and summer; they do not hibernate. Weasels are commonly thought to be nocturnal but evidence does not support this. In Pennsylvania, the peak activity was at dusk in winter (Glover 1942). In Iowa, the peak winter activity occurred at sunrise (Polder 1968), and Fitzgerald (1977) observed both hunting and traveling during the day. In a montane habitat in Colorado, *M. frenata* was active throughout the day during the summer, with highest activity occurring in the late afternoon and morning (Svendsen unpublished data). *M. erminea* alternates periods of activity and rest over a 24-hour period but appears to be more diurnal in summer and more nocturnal in winter (Erlinge 1977b). *M. nivalis* also remains active throughout 24-hour periods (Ognev 1962). King (1975) caught more *M. nivalis* during the day than at night, and 98 percent of all *M. frenata* live-trapped in a study of Colorado were caught between dawn and dusk (Svendsen unpublished data).

M. nigripes is mainly nocturnal (Clark 1976) and appears aboveground for only a few minutes every few days (Hillman 1968b). Activity of a captive black-footed ferret was influenced by both season and weather. It fed and dug tunnels mostly at night. During the days, it retreated to a hollow tile pipe, and during cold and snow it emerged daily to feed and romp but spent most of the time in the nest box (Progulske 1969). Most aboveground activity by black-footed ferrets is recorded during late summer, when the young are old enough to be moving about on their own (Hillman 1968b).

TABLE 30.2. Postnatal development of three species of weasel, *M. nivalis*, *M. erminea*, and *M. frenata*

Age	*M. nivalis*	*M. erminea*	*M. frenata*
Birth	1–1.5 g; total length 44 mm, tail 5 mm, hind foot 3 mm; naked except for sparse, long, white hair; high-pitched squeak vocalization; can control the forelimbs but not the hind-limbs	1.5–2.0 g; total length 51 mm; naked except for fine white hair, can lift itself on forelimbs, movement is uncoordinated	3 g; total length 56 mm, tail 13 mm, hindfoot 7 mm; flesh colored with long, white hair; high-pitched vocalizations
1 week	4 g; pigmentation begins on dorsum, remaining hair white; vocalizations are short and lower pitched	5 g; total length 70 mm, tail 10 mm, hind foot 7.5 mm; no sexual dimorphism in size; forepaws have dark brown claws; body darker on dorsum, hair on neck and shoulders very long	5–9 g; total length 92 mm, tail 21 mm, hind foot 11 mm; fine white fuzzy hair over body, all hair of same length
2 weeks	8 g; brown hairs on dorsum; crawls in a straight line; vocalizations consist of a hiss and a chirp	10 g; total length 90 mm, tail 13 mm, hind foot 10 mm; heavy brown mane from forehead to shoulders; tip of tail black, claws and tip of nose brown	sexual disparity in size, female 13.5 g, total length 89 mm, male 17 g, total length 100 mm; ears pigmented; can stand but not walk
3 weeks	pelage brown and white of adult; carnassials and incisors erupt; takes solid food and nursing declines; aggressive in play; eyes open at 26–29 days	16 g; total length 115 mm, tail 16 mm, hind foot 12 mm; canines and carnassials erupt; dark mane persists	males 27 g, total length 150 mm, females 21 g, total length 127 mm; canines and carnassials erupt, feeds on meat; crawls outside the nest; dorsum gray, tail black-tipped
4 weeks	weaned, able to kill mice; active outside of the nest, aggressive in play and reproductive behavior	sexual disparity in size first evident, males 25 g, females 22 g; upper incisors erupt; mane partly obscured by surrounding fur	males 39 g, total length 211 mm, tail 49 mm, hind foot 28 mm; females 31 g, total length 196 mm, tail 47 mm, hind foot 24 mm; walks long distances from the nest; upper incisors erupt; both trill and squeal vocalizations
5 weeks	trill vocalization appears	eyes and ears open; mane still visible but adult pelage and coloration nearly complete; males average 4–5 g heavier than females, females 28 g, males 32 g	eyes open; fights over food and consumes nearly its own weight daily, weaned; pelage nearing adult summer coloration; males 10–15 g heavier than females.
6 weeks	deciduous teeth replaced by permanent teeth	females 37 g, males 55 g, males weight as much as their mothers but are smaller in body dimensions; no trace of the mane	males 81 g, total length 240 mm; females 62 g, total length 215 mm; long guard hairs appear

SOURCE: Hamilton 1933; Hartman 1964; Heidt et al. 1968; Sanderson 1949; Svendsen unpublished data

In laboratory experiments, both *M. frenata* and *M. nivalis* preferred the brightest illumination available (500–900 lux) (Kavanau 1969). This preference was related to the visual needs of these carnivores for greater color, contrast, pattern, and intensity discrimination. In later studies it was determined that the visual systems were best adapted to dim light but suitable for daylight. Complete darkness inhibited running activity (Kavanau and Ramos 1975).

Movement. Movement is related primarily to hunting and seeking mates. Glover (1942) found that male *M. frenata* traveled farther at night than did females in Pennsylvania. Males traveled an average distance of 214 m (18–775), whereas females traveled 105 m (6–430) per night. Weasels in open timber traveled farther per trip than did those in brushland and dense stands of trees. Trails lead from one rodent den to another; dense vegetation with cover was used more than open areas and the same route was not used two nights in a row. In Iowa fields, the average cruising radius was 125 m from a central den, the maximum was 225 m (Polderboer et al. 1941). In Idaho, *M. frenata* and *M. erminea* followed circuits and did not use central dens. *M. frenata* followed 7- to 12-day circuits and the smaller *M. erminea* 10- to 15-day circuits. The same general route was followed on each circuit (Musgrove 1951). Where snow is deep, weasels burrow beneath the snow to hunt (Fitzgerald 1977). They also will use snowshoe hare (*Lepus americanus*) runways (Keith and Meslow 1966) and pocket gopher (Geomyidae) burrows as foraging routes (Vaughan 1961).

A large male *M. erminea* holds the record for the longest known distance traveled by a weasel crossing 35 airline km in 7 months (Burns 1964). Peak long-distance movements seem to occur in spring, coincident with mating (Erlinge 1977*b*), and primarily involves males. Movement of *M. nigripes* is unknown but probably nondispersal movements are limited by the size of the prairie dog towns inhabited.

Home Range. Home range size varies with habitat, population density, season, sex, food availability, and species. *M. nivalis* is the smallest weasel and it has the smallest home range; *M. frenata* is the largest weasel and it has the largest home range. *M. nivalis* males use 7–15 hectares, females 1–4 ha. Home ranges are smaller in young pine plantations (1–5 ha) than in mature deciduous woodland (7–15 ha), and home ranges generally are 65–85 percent smaller in habitats with the most food (King 1975). *M. erminea* is larger than *M. nivalis* and has a larger home range. Males use an average of 34 ha and females 7.4 ha based on snow tracking (Nyholm 1959). Using radio-collared animals, Erlinge (1977*b*) determined home ranges to be 2–3 ha for females and 8–13 ha for males in late autumn. In late summer, females used 4–10 ha and males 15 ha. Males increased the area of use considerably in spring, coincident with mating. When small rodents were scarce, home ranges were two to three times larger than during periods with food abundance. The largest weasel, *M. frenata*, has home ranges of 12–16 ha, and males have larger home ranges than do females in summer (Svendsen unpublished data). Estimates of *M. nigripes* home range have not been made but it is probably determined in part by the burrow structure and density of prairie dogs within a prairie dog town.

Population Dynamics. Weasel population densities vary with season, food availability, and species. In favorable habitat, maximum densities of *M. nivalis* may reach 25 per km², *M. erminea* 8 per km², and *M. frenata* 6–7 per km² (Jackson 1961; Quick 1944; Glover 1942; King 1975). Population densities fluctuate considerably with the year-to-year changes in small mammal abundances (MacLean et al. 1974; Fitzgerald 1977), and there are great differences in densities in different habitats and parts of the distribution (Glover 1942; Polderboer et al. 1941; King 1975). In central and western New York *M. frenata* and *M. erminea* were found in equal numbers, in northern New York *M. erminea* was the most common, and in southern New York *M. frenata* was most common (Hamilton 1933).

Sex Ratios. King (1975) reviewed the sex ratios of trapped *M. nivalis* from 12 studies. Males usually outnumbered females by about 3 to 1, a ratio similar to those reported for trapped *M. frenata* and *M. erminea* (Hamilton 1933; Jackson 1961). The sex ratio at birth is 1:1 and King (1975) found no evidence of differential mortality. Thus, the observed imbalance probably was due to sampling bias.

Social Structure. The social organization of weasel populations is that of a solitary existence except during the breeding season. Residents maintain intrasexual territories and transients do not. Males defend territories against males, and females defend against females, but the territory of an individual of one sex may overlap with the territory of a member of the other sex (Lockie 1966; Erlinge 1977*a*). Male *M. nivalis* and *M. erminea* have nearly nonoverlapping territories in winter; overlap was greatest during the summer but even while there was overlap, exclusive areas of the territory were maintained. Females of both species occupy smaller territories than do males. Transients regularly moved through the area occupied by residents. Even though a female territory was completely within the boundaries of a male territory, the male seldom ventured into the female territory except during the breeding season (Erlinge 1977*a*). When a male did enter a female territory during the nonbreeding season, he used different parts of it (Erlinge 1975). In the region of overlap between two male weasels, avoidance characterized the behavior of residents toward one another; each occupied a different portion of the area at any given time (King 1975).

Dominance relationships play an important part in maintaining territories in Sweden. *M. erminea* males and females exhibited avoidance and defensive actions (mutual avoidance, retreat, flee, escape, submissive behaviors) more frequently than offensive actions (approach, thrust and threat vocalizations, chase, nest occupation, prey robbery) in paired encounters. Adult males were dominant to juvenile males and nonbreeding females. Pregnant females were the most aggressive and dominant or of equal dominance to adult males. Adult males were ranked in order of dominance based on body size, age, and weight of anal glands. Adult females did not exhibit this relationship. Established males and females were dominant to introduced individuals, but established nonbreeding females and juvenile males were not able to maintain dominance over an introduced male. Established animals exhibited threat behavior, causing the introduced animal to flee. Important factors in the spatial organization of *M. erminea* are solitary habits maintained by avoidance and threat display (Erlinge 1977*a*).

Territories of *M. nivalis* have well-defined boundaries. Animals holding territories come into breeding condition from March to August. Nonterritory-holding animals are highly variable as to when they enter breeding condition and sometimes do not breed. According to Lockie (1966), *M. nivalis* females in Scotland are always subdominant to males and the females live on the male's territory at its pleasure. When food is scarce, resident males treat females as trespassing males. This results in poor breeding and high female mortality. Lockie (1966) felt that before a territorial system could be established, a certain minimum threshold density is necessary. Erlinge (1974) failed to find a minimum density necessary for territory formation but did detect seasonal changes in the pattern of territoriality. In summer and autumn, males were territorial, but in spring the males moved about and stayed in one area for only a short time. Prey density and individual qualities were important in territory formation. Some individuals failed to maintain a territory at any time of the year.

The size of territories varies greatly between individuals and with the season (King 1975; Erlinge 1974, 1977*a*). Vacated territories are most likely to be occupied by adjacent residents even though nonresidents are present. Not all available habitat is occupied. Those habitats lacking food and cover are avoided (Moors 1975). During the breeding season, males travel widely and well-defined male territories are not maintained (Erlinge 1974, 1977*a*; Lockie 1966; Moors 1975); however, females maintain well-defined territories during the breeding season (Erlinge 1974; Svendsen unpublished data).

The social organization of *M. nigripes* appears to be much the same as that of weasels. Adult ferrets of the same sex have not been reported to inhabit the same prairied dog town. Adults of the opposite sex are found in the same town but not in the same burrow system (Hillman 1968*a*, 1968*b*).

FOOD HABITS AND HUNTING BEHAVIOR

Between 50 and 80 percent of the yearly food intake of weasels consists of small mammals, especially rodents. Other foods vary in proportion depending on the season, availability, and individual preferences. The largest weasels generally take larger-sized prey than do smaller weasels. Based on scat analysis, Erlinge (1974) determined that *M. nivalis* in Sweden used rodents as the staple food item; they represented 80 percent of the total prey eaten. The vole *Microtus agrestis* was the primary prey and other small rodents and hares (*Lepus* sp.) were of secondary importance. Shrews (Soricidae), birds, and lizards were taken only rarely. In autumn and winter, rodents represented 94 percent of all food items. However, in spring and summer, as rodent populations declined, the proportion of rodents taken as food also declined and the use of lagomorphs increased. Shrews were taken at a low rate in captive studies and weasels showed little interest in them when hunting. Male *M. nivalis* are twice as large as females and there is evidence of food segregation between them. During food scarcity, males were able to shift to larger prey (i.e., lagomorphs and the larger water vole, *Arvicola terrestris*), whereas the smaller vole, *M. agrestis,* remained the primary prey of females. Differences in hunting behavior were also seen. The smaller females were better suited for entering rodent burrows. They spent more time hunting in rodent tunnels, while males spent more time hunting aboveground. Overall, small rodent populations were exploited in relation to their numbers. The low rate of predation on birds was attributed to the low density of breeding birds in the study area. Brugge (1977) also reported that male *M. erminea* and *M. nivalis* preyed upon larger animals than did females.

Small mammals were predominant as prey items of *M. erminea* in New York. The proportions of prey were: 36 percent voles (*Microtus* sp.) 16 percent unidentified mice, 15 percent short-tailed shrews (*Blarina brevicauda*), 11 percent deer mice (*Peromys-*

cus sp.), 9 percent cottontail rabbits (*Sylvilagus* sp.), 4 percent rats (*Rattus* sp.), and 4 percent eastern chipmunks (*Tamias striatus*). Songbirds represented 2 percent of all items, and grasshoppers, crickets, and frogs were consumed in the fall (Hamilton 1933). Winter food items determined from stomach analysis in the Northwest Territories, Canada, included 55 percent mammals, 13 percent fish, 7 percent amphibia, 4 percent birds, 4 percent insects, and 14 percent vegetation. Of the mammals, prey included 18 percent meadow jumping mice (*Zapus hudsonius*), 16 percent deer mice (*Peromyscus maniculatus*), 7 percent meadow voles (*Microtus pennsylvanicus*), and 3 percent vagrant shrews (*Sorex vagrans*). Diets of a comparison population of weasels in Alberta consisted of 66 percent mammals, 6 percent birds, 6 percent insects, and 15 percent vegetation. Of the mammals in the diet, 30 percent were jumping mice, 18 percent deer mice, 12 percent meadow voles, and 3 percent vagrant shrews. At both sites, the jumping mouse appeared in the diets at rates higher than anticipated, especially since these mice were hibernating at the time of the study. Weasels live under snow when the air temperature falls below $-13°$ C (Kraft 1966) and probably entered the hibernacula of *Z. hudsonius* and preyed upon the mice during these periods of subnivean existence (Northcott 1971).

Rodents, shrews, and young cottontail rabbits were major prey items for *M. frenata* in New York State. Of 163 scats analyzed, 34 percent of the prey were voles (*Microtus* sp.), 17 percent cottontail rabbits (*Sylvilagus* sp.), 17 percent unidentified mice, 11 percent deer mice (*Peromyscus* sp.), 10 percent rats (*Rattus* sp.), 6 percent short-tailed shrews (*B. brevicauda*), 3 percent squirrels (*Sciurus* sp.), 1 percent eastern chipmunk (*T. striatus*), 1 percent star-nosed moles (*Condylura cristata*), and 1 percent muskrats (*Ondatra zibethicus*) (Hamilton 1933). Voles were most common in diets of *M. frenata* in Colorado, with deer mice and chipmunks next in order of abundance (Quick 1951). Moles (Talpidae), deer mice, and harvest mice (*Reithrodontomys* sp.) made up 75 percent of the prey of *M. frenata* in Iowa (Polderboer et al. 1941). Rabbits were taken primarily when they were a few weeks old. Songbirds were rarely taken by *M. frenata*. Errington (1936) reported that 13-lined ground squirrels (*Spermophilus tridecemlineatus*) bore the brunt of the predation pressure by one weasel family but cottontail rabbits (*Sylvilagus floridanus*), voles (*Microtus* sp.), deer mice (*Peromyscus* sp.), birds, and insects were also taken. Among the insects preyed upon were ground beetles, ants, grasshoppers, and blowflies. The loss of three blue-winged teal (*Anas discors*) nests was attributed to *M. frenata* in Manitoba. Captive weasels made small, paired conical punctures in eggs that were identical to those found in the eggs of plundered teal nests (Teer 1964). Stoddard (1931) included weasels as predators on bobwhite (*Colinus virginianus*) eggs based on the fact that they fed on them in captivity. Bump et al. (1947) ranked weasels second to foxes (*Vulpes vulpes* and *Urocyon cinereoargenteus*) as predators on ruffed grouse (*Bonasa umbellus*) eggs, estimating that about 10 percent of all grouse nests were destroyed by weasels.

During periods of rodent scarcity, *M. frenata* and *M. erminea* both may shift to alternative prey that includes poultry and other domestic fowl (Jackson 1961). *M. frenata* have preyed upon three-day-old pigs (Polderboer 1948). *M. nivalis* are not known to prey on domestic animals, however. At normal prey densities, weasels occupy the same buildings as poultry, do not molest them, and effectively reduce the numbers of rats and mice.

Shrews were found to be part of the normal diet of weasels in some studies (Hamilton 1933; Aldous and Manweiler 1942; Jackson 1961; Ognev 1965) but were uncommon in other studies (Day 1968; Erlinge 1974). Individual preferences may reflect these differences to some degree. A single captive female *M. nivalis* had a distinct taste for shrews and preferred shrews to voles if offered them both (Erlinge 1974). Jackson (1961) reported that *M. erminea* preyed on shrews mainly in winter.

The impact of weasels on populations of microtines has been demonstrated by MacLean et al. (1974) and Fitzgerald (1977). Predation pressure by *M. nivalis* during the declining phase of the population cycle of a collared lemming (*Dicrostonyx groenlandicus*) population was important in sustaining the amplitude of the cycle. In a California study on the impact of weasels on the vole cycle, predation was related to the population density of voles. *M. frenata* and *M. erminea* preyed on voles under the snow and then used the vole nests for dens. Both species of weasel varied in population density over the course of the study. *M. erminea* was most common. In winter, both species preyed entirely on the voles and winter food habits did not change with changing vole density. The percentage of occupied vole nests varied from 5 to 54 percent of the vole population. Predator pressure was heaviest when voles were at their lowest densities, lightest as voles began to peak in density. At low vole densities, all losses between autumn and spring were accounted for by predation, while at high vole densities there was less mortality due to predation (Fitzgerald 1977). These studies support the hypothesis of Pearson (1971) that predation during and after the crash of microtine populations is responsible for timing and amplitude of microtine cycles.

It is generally assumed that weasels utilize living prey and do not scavenge. However, *M. nivalis* scavenged on frozen carcasses of brown lemmings (*Lemmus pibiricus*) that were systematically placed under the snow in various tundra habitats (Mullen and Pitelka 1972).

Weasels have evolved a highly successful prey capture and killing behavior consisting of a rapid dash to the prey, biting the prey at the base of the neck, and entwining the prey with the body and legs. In this manner, the prey is securely held even though the bite may be changed to a better vantage point for the kill. Weasels do not suck blood, nor do they subsist on blood. After the prey is dead, weasels may devour the entire body or select certain parts and leave the rest. It is not uncommon for a well-fed weasel to devour only the viscera, muscle, and brains, and leave the feet, tail, and skin.

The killing and feeding behaviors of *M. nivalis* were detailed by Heidt (1972). The killing behavior is stereotyped: weasels seize the prey at the nape of the neck and bite through the base of the skull and/or the throat. The initial bite may be on any part of the prey's body in order to gain a hold and leverage for the neck bite. Weasels manipulate and position the prey with their feet, and wrap the body about the prey to provide further leverage. Mice were commonly dropped after they had quit struggling. If more than one mouse was present, the weasel would drop the first after it quit struggling, then catch and kill the second before returning to the first. The time elapsed per kill ranged from 10 to 60 seconds. Attack stimulus was the movement of the mouse. Weasels often passed within centimeters of a still mouse without seeing it.

Killing appears to be innate (Heidt 1970). Young *M. nivalis* that were separated from their mother before their eyes had opened could kill mice at 50–60 days of age with no previous experience. However, training by the mother caused the young to be more efficient in capturing and killing mice at an earlier age of 40–45 days (Heidt et al. 1968).

Weasels hunt by traveling through the habitat in a "random search" manner (Ewer 1973), investigating tunnels, nests, and potential hiding places of rodents as they encounter them. Cues for finding prey are not known, but sight, sound, and odors are all potential channels for prey identification. *M. erminea* was observed pursuing the flight sounds of grasshoppers and responding to them in a manner that indicated that it was hunting the grasshoppers (Willey 1970). The visual cues of a moving prey elicit an immediate response and attack in *M. nivalis* (Heidt 1972) and *M. frenata* (Svendsen unpublished data), but stationary prey are not readily seen. In confined conditions, weasels pass within centimeters of still prey and do not notice them, indicating less reliance on the sense of smell in searching for prey than on other cues.

Prairie dogs are generally assumed to be the principle food of the black-footed ferret. Scats recovered from burrows occupied by a mother and four young ferrets contained 18 percent mouse remains and 82 percent prairie dog remains (Sheets and Linder 1969). A captive ferret readily ate freshly shot prairie dogs, ground squirrels (*Spermophilus* sp.), small rodents, birds, and cottontail rabbits. It was fond of fresh fish, and also ate calf liver, hamburger, fat pork, milk, and bread (Aldous 1940). A black-footed ferret lived for several days under a wooden sidewalk in Hays, Kansas, where it killed rats.

Like other members of the weasel family, the black-footed ferret kills prey by attacking the neck and base of the skull. In a confined cage, a black-footed ferret stalked each prairie dog from the rear, lunged for the hind end of the animal, and then struck the base of the skull. When the black-footed ferret bit, a fierce fight followed until the prairie dog was killed. Prairie dogs kicked loose soil at the attacker. Under natural conditions, prairie dogs are probably pursued into their burrows, killed within the confines of the burrow or nest, and devoured. Like the weasels, the black-footed ferret evidently has poor distance vision; it had to move close to the prairie dog or other small mammal prey before noticing it (Progulske 1969).

Caching of food is reported among all weasels. When weasels encounter a local abundance of food in excess of what they can consume, they will store the unused food and return to eat it later. The site of the cache may be a side burrow off the main home burrow or a site located near the kill. A female *M. frenata* was observed hunting in an alpine meadow in Colorado, where it found two nests of one- to two-week-old golden-mantled ground squirrels (*Spermophilus lateralis*). It killed the young ground squirrels in the burrow and then carried them to an old pocket gopher burrow, where they were cached. There were nine young ground squirrels cached in all. After removing all the young squirrels, the weasel returned to the burrow where they were cached and remained for 30 minutes. It then emerged and crossed the meadow to a creek and then returned to the cache. This was repeated 50 minutes later. The next morning it returned to the food cache and was live-trapped and marked. The weasel was lactating and had a burrow containing young about 175 m away. Over the next two weeks it was observed to cache in the same pocket gopher burrow on three occasions (Svendsen unpublished data). Caching behavior is adaptive when a locally abundant food source is discovered. Prey can be quickly killed and then used later, thus conserving energy used in foraging. Other accounts of caching by weasels include immense numbers of rats dragged together and piled in a compact heap beneath the floorboards of a barn. Piles of 100 or more rats and mice are reported (Seton 1929).

Weasels require a constant supply of drinking water. They take only a little at a time but drink frequently. *M. frenata* drinks about 25 cc daily (Hamilton 1933).

BEHAVIOR

Weasels have the reputation of being objectionable, blood-thirsty, wandering demons of carnage. This view is expressed in the quote of Dr. Elliot Coues:

A glance at the physiognamy of the weasel, would suffice to betray their character. The teeth are almost the highest known raptorial character; the jaws are worked by enormous masses of muscles covering all sides of the skull. The forehead is low, and the nose is sharp; the eyes are small, penetrating, cunning, and glitter with an angry green light. There is something peculiar, moreover, in the way that this fierce face surmounts a body extraordinarily wiry, lithe, and muscular. It ends a remarkable long and slender neck, in such a way that it may be held at right

angles with the axis of the latter. When the creature is glancing forward, swaying from one side to the other, we catch the likeness in a moment—it is the image of a serpent. (Seton 1929: 599)

Seton goes on to single out weasels alone among the animals that seem to revel in slaughter for its own sake and to find unholy joy in the horrors of a dying squeak, final quiver, and wholesale extermination.

These anthropomorphisms are of little value in understanding weasel behavior, however. Instead, weasel behavior should be interpreted as a highly specialized and adaptive carnivorous way of life that is a result of a long and successful evolutionary process. Weasels are not angry, cunning, or wanton killers. Instead, they are efficient predators with behavioral, anatomical, physiological, and sensory adaptations that allow them to survive as small carnivores. Their quick actions and curious nature are evolutionary adaptations useful in hunting and finding prey as well as avoiding being taken as prey. Their fearless and pugnacious nature is a necessary behavioral adaptation that allows the weasel to attack, capture, and kill prey that may be much larger than they. The "wanton slaughter" of prey is a very adaptive and efficient way of securing food. When a locally abundant and easily captured prey is located, some is eaten immediately and the rest is cached for future use. The long wiry, serpentlike body enables weasels to enter burrows and hiding places of prey and aids in prey capture and killing. Weasels are not bloodsuckers, but rather are like most carnivores that lap up the blood flowing from the wounds of a freshly killed prey as part of the feeding pattern. This enables them to use all of the nutrients of the kill. Finally, the nonsocial behavior is an adaptation to the solitary way of life. When viewed from an evolutionary viewpoint, weasels are a remarkable group.

Locomotion. The traveling gait of weasels and the black-footed ferret is a slow gallop or series of jumps; frequently they walk. *M. nivalis* moves at a slow pace, about 8–10 km per hr; the larger species are capable of moving faster when pressed but probably travel at about the same pace (Jackson 1961). Weasels are capable of climbing trees in pursuit of prey but are not adept in doing so. They will also take to the water but are slow swimmers.

Reproductive Behavior. Mating behavior is prolonged in weasels and the black-footed ferret. Copulation may last three hours or more and occurs repeatly over several days. There is little difference in mating behavior among the species. Although a female in heat allows a male to approach, vigorous struggling and fighting prior to copulation is typical. However, the brief struggle and resistance to the advances of the male ends abruptly when the male grabs the female by the scruff of the neck with his teeth. If the female continues to struggle, the male holds the female down by the neck until she is subdued, then clasps her lower abdomen with the forelegs and arches his back over her posterior. Pelvic thrusts occur in the upright position or from the side. The female remains rather passive during coitus. Bursts of pelvic thrusts alternate with periods of rest. At the end of coitus, the female breaks away (Wright 1948a; Hartman 1964; Heidt et al. 1968).

Interactions among Individuals. Young weasels are playful and spend a great deal of time in playfighting and in reproductive behavior (Hamilton 1933, Heidt et al. 1968). However, adults live a solitary existence. Interactions among adults occur primarily during mating, otherwise individuals avoid one another. Complex social behaviors characterizing the social carnivores are not well developed in this group. When two strange *M. erminea* were placed together simultaneously, the result was avoidance. When one animal was established and then a second animal was added, the intruder was met with threat behavior that caused the intruder to escape. Threat display through scent marking, visual, and acoustic signals were involved in territorial defense (Erlinge 1977c).

Like other mustelids, weasels and the black-footed ferret produce a pungent odor from their anal glands. When irritated, they discharge the odor, which can be detected at some distance. *M. frenata* has been observed rubbing and dragging its body over surfaces, possibly to leave scent from its anal glands (Jackson 1961). As in other mammals, this odor probably serves to identify individuals just as visual features do in humans. However, knowledge of scent communication in weasels is not sufficient to make succinct statements on either the role or the information content of the scent.

Vocalizations. Vocalizations of weasels can be grouped into three types: a trill, a screech, and a squeal. The trill is a common vocalization accompanying several behaviors. It occurs when animals are investigating their surroundings, when in play, in conjunction with mating, and when hunting. The trill is composed of a series of short-duration, rapidly occurring, low-frequency calls. The smallest weasel, *M. nivalis,* has the highest frequency trill; the larger weasels, *M. erminea* and *M. frenata,* produce trills at a lower frequency. When a weasel is suddenly disturbed, the screech is elicited. This vocalization is accompanied by an open-mouthed gape and a sudden lunge. The probable function is a startle effect in the defensive or threat display. The screech is composed of complex harmonics and white noise caused by forcefully expelling air. The squeal is elicited under stress, possibly accompanying pain (Heidt et al. 1968; Huff and Price 1968; Svendsen 1976). Vocalizations of the black-footed ferret are not as well known. A chattering scold was made by an adult female toward men trying to capture one of its young (Aldous 1940). When people were near a captive female, it chattered constantly in a staccato voice, and between bursts of six to seven loud chirps it emitted low hissing sounds. Chatter was related to excitement and was made in response to unusual conditions, such as a bird flying overhead or people nearby (Progulske 1969).

Behavior toward Other Animals. Weasels are well known for their boldness and courage. It is said that a weasel will face any animal, including humans. There are several reports of weasels attacking people; almost always this occurred when someone way trying to take freshly killed prey away from a weasel (Wight 1932; Oehler 1944). On occasions when someone has tried to grab a live weasel, the power and tenacious grip of the jaws were quickly evident in the bite. A bleached skull of a weasel was found with a "death grip" on the throat of a shot bald eagle (*Haliaeetus leucocephalus*). It was surmized that the eagle had captured the weasel and in its struggle the weasel had latched onto the eagle and died (Seton 1929).

MORTALITY

Longevity in the wild is not well documented for weasels or the black-footed ferret. Jackson (1961) reported the life span of *M. erminea* to be 4–6 years in the wild, potentially up to 10 years. King (1975) determined that resident *M. nivalis* lived on her study area about 1 year and the mean age of death of 171 trapped weasels was 11 months. Marked adult *M. frenata* were resident on a Colorado site for up to 3 years (Svendsen unpublished data). Natural mortality of weasels is a result of several factors interacting simultaneously: disease, parasites, nutrition, population stress, and predation.

Predation. Black-footed ferrets have many potential predators, including badgers (*Taxidea taxus*), coyotes (*Canis latrans*), bobcats (*Felis rufus*), rattlesnakes (*Crotalus* sp.), eagles, hawks, and owls. A great horned owl (*Bubo virginianus*) was observed diving at an adult ferret (Hillman 1968a). Sperry (1941) reported the remains of black-footed ferrets in three coyote stomachs. The greatest cause of death of black-footed ferrets, however, has been human activity in control of prairie dogs. The widespread use of sodium fluoroacetate (1080) to poison prairie dogs has resulted in secondary poisoning of black-footed ferrets that fed on them. However, probably the most dramatic source of mortality results from the large-scale eradication of prairie dogs from the range of the black-footed ferret, thus greatly reducing their food supply.

Predators of weasels are also numerous. There are records of weasels being eaten by rattlesnakes (*Crotalus* sp.) and blacksnakes (*Elaphe obsoleta*), snowy owls (*Nyctea scandiaca*), great horned owls, barred owls (*Strix varia*), rough-legged hawks (*Buteo lagopus*), goshawks (*Accipiter gentilis*), red foxes, gray foxes, and domestic cats (Errington et al. 1940; Hamilton 1933; Handley 1949; Jackson 1961). The small *M. nivalis* is even prey for the larger *M. frenata* (Polderboer et al. 1941). The erratic hunting pattern and black tip of the tail have been postulated as adaptations to reduce losses to other predators (Powell 1973). He suggested that a black tip to the tail of the very short-tailed *M. nivalis* would improve the hawk's chances of catching this small weasel because the black tip would direct the predator's attention to the short tail and body. Hence, *M. nivalis* has no black tip to its tail. Conversely, the black tip on the longer-tailed *M. erminea* and *M. frenata* reduces the chances of being captured by a hawk. The predator's attention is directed toward the black tip and not at the body. This is especially true when the weasels are in their white winter coat on a background of snow.

Diseases and Parasites. The incidence of disease and parasites of weasels and the black-footed ferret are poorly known. External parasites of the weasels include ticks, (*Dermacentor variabilis, Ixodes cookei*), fleas (*Ceratophyllus vison, C. fasciatus, Nearctopsylla brooksi, Neotrichodectes mephitidis*), and mites (*Lutrilichus canadensis*). Reported internal parasites include nematodes (*Trichinella spiralis, Dracunculus medinensis, Molineus* spp., *Physaloptera maxillaris, Filaroides martis, Capillaria mustelorum, Skrjabingylus nasicola*), the cestode *Taenia taeniaformis,* and trematodes (*Alaria mustelae, A. taxideae*) (Goble 1942; Jackson 1961; Schmidt 1965; Soleman and Warner 1969; Fain et al. 1974). The nematode *Dracunculus medinensis* invades the frontal sinuses, causing swelling and perforations of the skull in the supraorbital region (Dougherty and Hall 1955; van Soest et al. 1972). Seton (1929) reported that weasels held in captivity had a tendency to develop fits and die. Disease probably plays an important role in the population biology of weasels and black-footed ferret, but this factor has not been studied in detail. Canine distemper is known to affect black-footed ferrets. Parker (1934) reported that the black-footed ferret and weasels were highly susceptible to tularemia (*Pasteurella tularensis*).

ECONOMIC STATUS AND MANAGEMENT

Overall, weasels are more of an asset than a liability. They destroy quantities of rats and mice that otherwise would eat and damage additional crops and produce.

This asset is partially counterbalanced by the fact that weasels occasionally kill beneficial animals and game species. The killing of domestic poultry may come only after the rat population around a farmyard is diminished. In fact, rats may have destroyed more poults than the weasel. In most cases, a farmer lives in harmony with weasels on the farm for years without realizing that they are even there until they kill a chicken. Quick (1951) estimated that Gunnison County, Colorado, an area of 10,240 km², was inhabited by 8,000 weasels. These were killing more than 30,000 small mammals per day, or 10,000,000 per year in that county alone. In New York State, a conservative estimate of the number of rats and mice killed per year by weasels was 60,000,000 mice and several million rats (Hamilton 1933).

The white winter pelage of *M. erminea* and *M. frenata* is valued in the fur trade as "ermine." Select winter pelts of *M. frenata* bring more than smaller pelts of *M. erminea*. Pelts of *M. nivalis* have practically no commercial value. Prices for pelts vary greatly. The total number of weasel pelts sold in the United States and Canada between 1970 and 1976 ranged from a low of 43,876 to 1971–72 to a high of 102,090 in 1974–75 (table 30.3). The average price per pelt during this time was $0.58 and the average yearly income from weasel pelts was $43,377.

The relationship of weasels to humans dates back to antiquity. Indian superstition held that if one captured a *M. nivalis*, it was regarded as a piece of good fortune and one was destined to have great wealth and power (Seton 1929). Ermine skins decorated the headdress and ceremonial clothing of many Indian tribes in North America and stuffed skins are found among ceremonial relics (Clark 1976; Peterson and Berg 1954).

The black-footed ferret was never numerous enough to pose a serious threat to domestic stock or to become economically feasible for trade on the fur market. Overall, it benefited the early settlers by helping to

TABLE 30.3. Number and price of weasel pelts bought in the United States and Canada during 1970 to 1976

Date	Number of Skins	Average Price per Pelt	Total Value
Mustela frenata			
1970/71	47,693	0.60	$28,615.80
1971/72	22,717	0.50	$11,358.50
1972/73	61,819	0.50	$30,909.50
1973/74	38,002	0.50	$19,001.00
1974/75	50,616	0.50	$25,308.00
1975/76	38,521	0.90	$34,668.90
Mustela erminea			
1970/71	8,949	0.60	$ 5,369.40
1971/72	21,159	0.50	$10,579.50
1972/73	29,907	0.50	$14,953.50
1973/74	30,316	0.50	$15,158.00
1974/75	51,474	0.50	$25,737.00
1975/76	42,895	0.90	$38,605.50

SOURCE: Deems and Pursley 1978

keep populations of prairie dogs from expanding on grazing lands. Ironically, mankind has driven the black-footed ferret to near extinction by reducing the habitat and number of prairie dog towns to levels where an effective breeding population of ferrets may not now exist. Various Indian groups—Sioux, Blackfoot, Crow, Cheyenne, and Pawnee—used the black-footed ferret in a variety of ways. Blackfoot Indians in Montana used ferret hides as pendants on their chief's headdress, while the Crow Indians of Wyoming and Montana used ferrets in their sacred tobacco society. There is evidence that Indians also used ferrets for food (Clark 1976). Stuffed ferret skins were included in ceremonial relics of the Crow Indians, suggesting that the Indians were aware of the rarity of the species and attached a mystic significance to it (Peterson and Berg 1954).

The black-footed ferret is listed as an endangered species and efforts are being made to protect critical habitat where ferrets have been sighted over the past few years. A captive breeding effort at the Patuxent Wildlife Research Center, U.S. Fish and Wildlife Service, proved to be unsuccessful because the black-footed ferret did not breed in captivity. The only hope for the future of the black-footed ferret is that identification of critical habitat and an increase in the number and size of prairie dog towns within that habitat have not come too late for recovery of this unique mammal. The most intensely studied area is in Mellette County, South Dakota, where ferrets have been observed at 14 black-tailed prairie dog (*Cynomys ludovicianus*) towns between 1964 and 1974 (Hillman 1968*b*; Hillman et al. 1979). Prairie dog towns in this area were chiefly on ranches and Indian lands. Several landowners were sympathetic to the study and withheld treatment of towns that were inhabited by ferrets; there was no prairie dog control program on Indian lands. Whereas prairie dog towns increased in numbers during the study, there was no evidence that ferrets had increased (Hillman et al. 1979).

LITERATURE CITED

Aldous, S. E. 1940. Notes on a black-footed ferret raised in captivity. J. Mammal. 21:23–26.

Aldous, S. E., and Manweiler, J. 1942. The winter food habits of the short-tailed weasel in northern Minnesota. J. Mammal. 23:250–255.

Allen, G. M. 1933. The least weasel: a circumboreal species. J. Mammal. 14:316–319.

Beer, J. R. 1950. Least weasel in Wisconsin. J. Mammal. 31:146–149.

Bissonnette, T. H., and Bailey, E. E. 1944. Experimental modification and control of molts and changes of coat-color in weasels by controlled lighting. Ann. New York Acad. Sci. 45:221–260.

Brown, J. H., and Lasiewski, R. C. 1972. Metabolism of weasels: the cost of being long and thin. Ecology 53:939–943.

Brugge, T. 1977. Prooidierkeuze van wezel, Hermelijn en Bunzing in relatie tot geslacht en lichaamsgrootte [Prey selection of weasel, stoat and polecat in relation to sex and size]. Lutra 19:39–49.

Bump, G.; Darrow, R. W.; Edminster, F. C.; and Crissey,

W. F. 1947. The ruffed grouse: life history, propagation, management. New York State Conserv. Dept., Albany. 915pp.

Burns, J. J. 1964. Movements of a tagged weasel in Alaska. Murrelet 45:10.

Clark, T. W. 1976. The black-footed ferret. Oryx 13:275–280.

Day, M. G. 1968. Food habits of British stoats (*Mustela erminea*) and weasels (*Mustela nivalis*). J. Zool London. 150:485–497.

Deanesly, R. 1944. The reproductive cycle of the female weasel (*Mustela nivalis*). Proc. Zool. Soc. London. 114:339–349.

Deems, E. F., and Pursley, D., eds. 1978. North American furbearers: their management, research and harvest status in 1976. Int. Assoc. Fish Wildl. Agencies. 171pp.

Dougherty, E. C., and Hall, E. R. 1955. The biological relationships between American weasel (genus *Mustela*) and nematodes of the genus *Skrjabingylus* Petrov, 1927 (Nematoda: *Metastrongylidae*), the causative organisms of certain lesions in weasel skulls. Rev. Iberica Parasitol. 531–569.

Erlinge, S. 1974. Distribution, territoriality and numbers of the weasel *Mustela nivalis* in relation to prey abundance. Oikos 25:308–314.

———. 1975. Feeding habits of the weasel *Mustela nivalis* in relation to prey abundance. Oikos 26:378–384.

———. 1977*a*. Spacing strategy in stoat *Mustela erminea*. Oikos 28:32–42.

———. 1977*b*. Home range utilization and movements of the stoat, *Mustela erminea*. pages 31–42 *in* Proc. 13th Congr. Game Bio.

———. 1977*c*. Agonistic behavior and dominance in stoats (*Mustela erminea* L.). Z. Tierpsychol. 44:375–388.

Errington, P. L. 1936. Food habits of a weasel family. J. Mammal. 17:406–407.

Errington, P. L.; Hammerstrom, F.; and Hammerstrom, F. N., Jr. 1940. The great horned owl and its prey in north-central United States. Iowa Agric. Exp. St. Bull. 277:757–850.

Ewer, R. F. 1973. The carnivores. Cornell Univ. Press, Ithaca, N.Y. 494pp.

Fain, A.; Lukoschus, F. S.; Kok, N. J. J.; and Clulow, F. V. 1974. A key to the genus *Lutrilichus* Fain and description of a new species from the ermine, *Mustela erminea*, in Canada (Acarina: Sarcoptiformes). Can. J. Zool. 52:941–944.

Fitzgerald, B. M. 1977. Weasel predation on a cyclic population of the montane vole (*Microtus montanus*) in California. J. Anim. Ecol. 46:367–397.

Glover, F. A. 1942. A population study of weasels in Pennsylvania. M.S. Thesis. Pennsylvania State Univ., Univ. Park. 210pp.

Goble, F. C. 1942. The Guinea-worm in a Boneparte weasel. J. Mammal. 23:221.

Hall, E. R. 1944. Classification of the ermines of eastern Siberia. Proc. California Acad. Sci. 23 (ser. 4):555–560.

———. 1951. American weasels. Univ. Kansas Mus. Nat. Hist. Publ. 4:1–466.

Hall, E. R., and Kelson, K. R. 1959. The mammals of North America. Ronald Press, New York. 2 vols.

Hamilton, W. J., Jr. 1933. The weasels of New York. Am. Midl. Nat. 14:289–337.

Hamlett, G. W. D. 1935. Delayed implantation and discontinuous development in mammals. Q. Rev. Bio. 10:432–447.

Handley, C. O., Jr. 1949. Least weasel, prey of barn owl. J. Mammal. 30:431.

Hartman, L. 1964. The behavior and breeding of captive

weasels (*Mustela nivalis* L.). New Zealand J. Sci. 7:147–156.

Hatt, R. T. 1940. The least weasel in Michigan. J. Mammal. 21:412–416.

Heidt, G. A. 1970. The least weasel, *Mustela nivalis* L.: developmental biology in comparison with other North American *Mustela*. Mich. State Univ. Mus. Nat. Hist. Publ. 4:227–282.

———. 1972. Anatomical and behavioral aspects of killing and feeding by the least weasel, *Mustela nivalis* L. Proc. Arkansas Acad. Sci. 26:53–54.

Heidt, G. A.; Petersen, M. K.; and Kirkland, G. L., Jr. 1968. Mating behavior and development of least weasels (*Mustela nivalis*) in captivity. J. Mammal. 49:413–419.

Hillman, C. N. 1968*a*. Life history and ecology of the black-footed ferret in the wild. M.S. Thesis. South Dakota State Univ., Brookings. 28pp.

———. 1968*b*. Field observations of black-footed ferrets in South Dakota. Trans. North Am. Wildl. Nat. Res. Conf. 33:433–443.

Hillman, C. N.; Linder, R. L.; and Dahlgren, R. B. 1979. Prairie dog distribution areas inhabited by black-footed ferrets. Am. Midl. Nat. 102:185–187.

Huff, J. N., and Price, E. O. 1968. Vocalizations of the least weasel, *Mustela nivalis*. J. Mammal. 49:548–550.

Iversen, J. A. 1972. Basal energy metabolism of mustelids. J. Comp. Physiol. 81:341–344.

Jackson, H. H. T. 1961. Mammals of Wisconsin. Univ. Wisconsin Press, Madison. 504pp.

Kavanau, J. L. 1969. Influences of light on activity of small mammals. Ecology 50:548–557.

Kavanau, J. L., and Ramos, J. 1975. Influences of light on activity and phasing of carnivores. Am. Nat. 109:391–418.

Keith, L. B., and Meslow, E. C. 1966. Animals using runways in common with snowshoe hares. J. Mammal. 47:541.

King, C. M. 1975. The home range of the weasel (*Mustela nivalis*) in an English woodland. J. Anim. Ecol. 44:639–668.

Kraft, V. A. 1966. Effect of temperature on the mobility of the ermine in winter. Zool. Zhur. 45:148–150.

Latham, R. 1953. Simple method for identification of least weasel. J. Mammal. 34:385.

Lockie, J. D. 1966. Territory in small carnivores. Symp. Zool. Soc. London 18:143–165.

Lyman, C. P. 1942. Control of coat color in the varying hare by daily illumination. Proc. New England Zool. Club 19:75–78.

MacLean, S. F., Jr.; Fitzgerald, B. M.; and Pitelka, F. A. 1974. Population cycles in arctic lemmings: winter reproduction and predation by weasels. Arctic and Alpine Res. 6:1–12.

McNab, B. K. 1971. On the ecological significance of Bergmann's rule. Ecology 52:845–854.

Mayr, E. 1966. Animal species and evolution. Belknap Press, Cambridge, 796pp.

Moors, P. J. 1975. The annual energy budget of a weasel (*Mustela nivalis* L.) population in farmland. Ph.D. Dissertation. Univ. Aberdeen, Scotland. 126pp.

Mullen, D. A., and F. A. Pitelka, 1972. Efficiency of winter scavengers in the arctic. Arctic 25:225–231.

Mumford, R. E. 1969. Distribution of mammals of Indiana. Indiana Acad. Sci., Indianapolis. 114pp.

Musgrove, B. F. 1951. Weasel foraging patterns in the Robinson Lake area, Idaho. Murrelet 32:8–11.

Northcott, T. H. 1971. Winter predation of *Mustela erminea* in northern Canada. Arctic 24:142–144.

Nyholm, E. S. 1959. Kärpästä ja lumikosta ja niiden talvisista elinpiireistä. [Stoats and weasels and their winter habitat]. Suomen Riista 13:106–116.

Oehler, C. 1944. Notes on the temperament of the New York weasel. J. Mammal. 25:198.

Ognev. S. I. 1962. Mammals of U.S.S.R. and adjacent countries. Israel Program for Sci. Transl. Vol. 3, 641pp. (Orig. Publ. in Russian, 1935).

Parer, J. T., and Metcalfe, J. 1967. Oxygen transport by blood in relation to body size. Nature (London) 215:653–654.

Parker, R. R. 1934. Recent studies of tick-borne diseases made at the United States Public Health Service Laboratory at Hamilton, Montana. Proc. 5th Pacific Sci. Congr. B5:3367.

Pearson, O. P. 1971. Additional measurements of the impact of carnivores on California voles (*Microtus californicus*). J. Mammal. 52:41–49.

Peterson, L. A., and Berg, E. D. 1954. Black-footed ferrets used as ceremonial objects by Montana Indians. J. Mammal. 35:593–594.

Petrides, G. A. 1950. The determination of sex and age ratios in fur animals. Am. Midl. Nat. 43:355–388.

Polder, E. 1968. Spotted skunk and weasel populations den and cover usage by northeast Iowa. Iowa Acad. Sci. 75:142–146.

Polderboer, E. B. 1948. Predation on the domestic pig by the long-tailed weasel. J. Mammal. 29:295–296.

Polderboer, E. B.; Kuhn, L. W.; and Hendrickson, G. O. 1941. Winter and spring habits of weasels in central Iowa. J. Wildl. Manage. 5:115–119.

Powell, R. A. 1973. A model for raptor predation on weasels. J. Mammal. 54:259–263.

Progulske, D. R. 1969. Observations of a penned, wild-captured black-footed ferret. J. Mammal. 50:619–621.

Quick, H. F. 1944. Habits and economics of New York weasel in Michigan. J. Wildl. Manage. 8:71–78.

———. 1951. Notes on the ecology of weasels in Gunnison County, Colorado. J. Mammal. 32:281–290.

Rosenzweig, M. L. 1966. Community structure in sympatric carnivora. J. Mammal. 47:602–612.

Rust, C. C. 1962. Temperature as a modifying factor in the spring pelage change of short-tailed weasels. J. Mammal. 43:323–328.

———. 1965. Hormonal control of pelage cycles in the short-tailed weasel (*Mustela erminea bangsi*). Gen. Comp. Endocrin. 5:222–231.

Rust, C. C., and Meyer, R. K. 1968. Effect of pituitary autograft on hair color in the short-tailed weasel. Gen. Comp. Endocrin. 11:548–551.

———. 1969. Hair color, molt, and testis size in male, short-tailed weasels treated with melatonin. Science 165:921–922.

Sanderson, G. C. 1949. Growth and behavior of a litter of captive long-tailed weasels. J. Mammal. 30:412–415.

Schmidt, G. 1965. *Molineus mustelae* sp. m. (Nematoda: Trichostrongylidae) from the long-tailed weasel in Montana and *Molineus chabaudi* nom. n., with the key to the species of *Molineus*. J. Parasitol. 51:164–168.

Schmidt-Nielsen, K., and Pennycuik, K. 1961. Capillary density in mammals in relation to body size and oxygen consumption. Am. J. Physiol. 200:746–750.

Scholander, P. R. 1955. Evolution of climatic adaptation in homeotherms. Evolution 9:15–26.

Scholander, P. R.; Walters, V.; Hock, R.; and Irving, L. 1950. Body insulation of some arctic and tropical mammals and birds. Bio. Bull. 99:225–236.

Seton, E. T. 1929. Lives of game animals. Doubleday & Co., Inc., Garden City, N.Y. 746pp.

Sheets, R. G., and Linder, R. L. 1969. Food habits of the

black-footed ferret (*Mustela nigripes*) in South Dakota. Proc. South Dakota Acad. Sci. 48:58–61.

Siivonen, L. 1967. Pohjolan nisäkkäät [Mammals of northern Europe]. Ursa Major Publishers, Helsinki. 181pp.

Soleman, G. B., and Warner, G. S. 1969. *Trichinella spiralis* in mammals of Mt. Lake, Va. J. Parasitol. 55:730–732.

Southern, H. N. 1964. The handbook of British mammals. Blackwell Sci. Publ., Oxford. 465pp.

Sperry, C. C. 1941. Food habits of the coyote. U.S. Fish Wildl. Serv. Wild. Res. Bull. 4. 70pp.

Stoddard, H. L. 1931. The bobwhite quail: its habits, preservation and increase. Chas. Scribner's Sons, New York 559pp.

Svendsen, G. E. 1976. Vocalizations of the long-tailed weasel (*Mustela frenata*). J. Mammal. 57:398–399.

Teer, J. G. 1964. Predation by long-tailed weasels on eggs of blue-winged teal. J. Wildl. Manage. 28:404–406.

van Soest, R. W. M.; van der Land, J.; and van Bree, P. J. H. 1972. *Skrjabingylus nasicola* (Nematoda) in skulls of *Mustela erminea* and *Mustela nivalis* (Mammalia) from the Netherlands. Beaufortia 20:85–97.

Vaughan, T. A. 1961. Vertebrates inhabiting pocket gopher burrows in Colorado. J. Mammal. 42:171–174.

Wight, H. M. 1932. A weasel attacks a man. J. Mammal. 13:163–164.

Willey, R. 1970. Sound location of insects by the dwarf weasel. Am. Midl. Nat. 84:563–564.

Wright, P. L. 1942. Delayed implantation in the long-tailed weasel (*Mustela frenata*), the short-tailed weasel (*Mustela cicognaniz*) and the marten (*Martes americana*). Anat. Rec. 83:341–349.

———. 1948a. Breeding habits of captive long-tailed weasels (*Mustela frenata*). Am. Midl. Nat. 39:338–344.

———. 1948b. Preimplantation stages in the long-tailed weasel (*Mustela frenata*). Anat. Rec. 100:593–603.

———. 1951. Development of the baculum of the long-tailed weasel. Proc. Sco. Exp. Bio. Med. 75:820–822.

GERALD E. SVENDSEN, Department of Zoology, Ohio University, Athens, Ohio 45701.

31

Mink

Mustela vison

Greg Linscombe
Noel Kinler
R. J. Aulerich

NOMENCLATURE

COMMON NAMES. Mink, American mink, *belette* (French colonist of Louisiana), *ching-woose se* (Chippewa), common mink, *iskixpa* (Biloxi), Mississippi Valley mink, *n'pshikwa* (Potawatomi), *toni* (Choctaws), vison, water weasel (Arthur 1928; Jackson 1961; Lowery 1974).

SCIENTIFIC NAME. *Mustela vison*

SUBSPECIES. *M. v. aestuarina*, *M. v. aniakensis*, *M. v. energumenos*, *M. v. evagor*, *M. v. evergladensis*, *M. v. ingens*, *M. v. lacustris*, *M. v. letifera*, *M. v. lowii*, *M. v. lutensis*, *M. v. melampeplus*, *M. v. mink*, *M. v. nesolestes*, *M. v. vison*, and *M. v. vulgivaga*.

DISTRIBUTION

Mink are distributed throughout most of North America (figure 31.1). They occur in all the United States except Arizona. They inhabit all of Canada south of the treeline except for Anticosti Island and the Queen Charlotte Islands and have been introduced into Newfoundland (Banfield 1974).

FIGURE 31.1. Distribution of the mink (*Mustela vison*).

DESCRIPTION

Mink are typically weasellike in shape, possessing a long, thin neck and body; short, sturdy legs; a pointed flattened face; and small ears. The bushy tail comprises approximately one-third to one-half of the animal's total length. Males are usually 10 percent larger in size and weight than females (Jackson 1961). The feet are characterized by semiwebbed toes and are fully furred except for pads on the toes and the soles of their feet (Lowery 1974).

Size and Weight. Weights of adult male mink range from 0.9 to 1.6 kilograms; adult females weigh 0.7 to 1.1 kg. Body measurements for adult males are: total length 580–700 mm, tail length 190–230 mm, and hind foot length 68–80 mm. Adult females body measurements are: total length 460–575 mm, tail length 150–190 mm, and hind foot length 60–70 mm (Jackson 1961). The mink is found throughout most of North America; thus, body weights and measurements show variation from one portion of its range to another.

Pelage. The fur of the mink is dark chestnut brown, with the exception of a few white spots on the chin, throat, chest, abdomen, and anal region. These white spots can be utilized to identify individual specimens. The ventral surface tends to be slightly lighter than the dorsum. The tail is similar to the body in color but gets progressively darker toward the tip. The pelage is soft and lustrous, with thick underfur and oily guard hairs that tend to waterproof the animal's coat (Lowery 1974).

Scent Glands. As do all mustelids, mink possess a pair of anal scent glands that emit a liquid with a very strong, musky odor. This liquid may be discharged when the animal is excited or aggravated. They will also emit this liquid during any conflict with conspecifics (Jackson 1961). The scent is especially strong during the breeding season.

629

Skull and Dentition. The mink skull is distinguished by the presence of 34 teeth, the extension of the palate beyond the last molar, the dumbbell shape of the last molar, and the large auditory bullae (figure 31.2). The deciduous teeth erupt between the 16th and 49th days following birth and remain for a short period of time. They are replaced by the permanent teeth, which erupt between 44 and 71 days of age. The dental formula is 3/3, 1/1, 3/3, 1/2 = 34.

PHYSIOLOGY

Body Weight and Temperature. Mink show pronounced sexual dimorphism, with males usually weighing 1.5 to 1.8 times more than females. Mink body weights also vary significantly among subspecies or geographic races. Males of some western races weigh up to 2.3 kg, while large males of certain eastern subspecies may weigh only 1.4 kg (Harding 1934).

The normal body temperature of adult mink is 39.5° to 40.5° C (Berestov 1971). However, as Kennedy (1951) pointed out, body temperature may rise several degrees when these animals are handled, due to their nervous temperament.

Cardiovascular, Hematological, and Pulmonary Parameters. Ringer et al. (1974) reported the following cardiovascular values for mink. The mean blood pressure averages 153 mm Hg, with a pulse pressure of 34.8 mm Hg. Blood constitutes 7.2 percent of the body weight. Because of body size differences, the average heart rate of females (300 beats per minute) is faster than that of males (250 beats per minute). The respiration rate ranges from 18.8 to 20.2 per minute.

Hemoglobin and hematocrit values and red blood cell counts increase steadily from weaning to maturity. Average values for adults range 18–20 gram percent hemoglobin, 47–60 percent hematocrit, and $7.5–9 \times 10^6$ red blood cells per mm³ (Kennedy 1935; Kubin and Mason 1948; Rotenberg and Jorgensen 1971; Fletch and Karstad 1971; Asher et al. 1976).

Urine Values. Mink urine is normally greenish yellow in color and slightly cloudy, with a distinct musky odor. The pH ranges from 6.8 to 7.5 and the specific gravity varies from 1.018 to 1.036. The amount excreted is in direct proportion to body weight and in adults varies from 35 to 60 ml per day (Kubin and Mason 1948).

Pelage Cycle. The furring cycle (spring and fall molts), as in other species such as the ferret and weasel (*Mustela frenata*), is controlled by the hypophysis (Rust et al. 1965) in response to changes in photoperiod. The cycle is independent of environmental temperature (Bissonnette and Wilson 1939). However, the mechanism whereby light influences the furring cycle is poorly understood (Duby and Travis 1972). It has been suggested (Rust and Meyer 1961) that the pineal may be involved in controlling pituitary function.

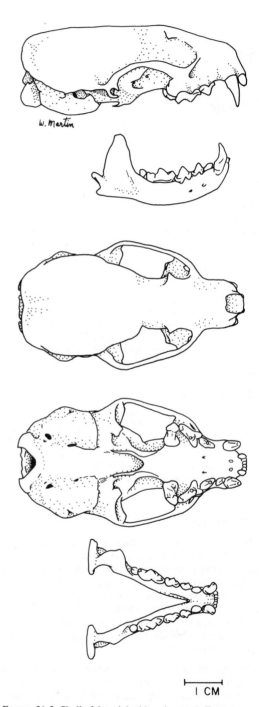

W. Martin

⊢ I CM

FIGURE 31.2. Skull of the mink (*Mustela vison*). From top to bottom: lateral view of cranium, lateral view of mandible, dorsal view of cranium, ventral view of cranium, dorsal view of mandible.

GENETICS

The chromosomes of the mink were described by Lande (1957), Humphrey and Spencer (1959), and Fredga (1961). The technique used by Humphrey and Spencer (1959) utilized only testicular material from adult male minks from a fur farm in Oregon and from wild mink in Canada. The tissue culture technique utilized by Fredga (1961) has allowed the study of chromosome morphology in somatic tissues from both sexes of any age. Both of these studies showed that the number of chromosomes for the mink is 2N = 30. These results are in agreement with the findings of Lande (1957). However, they are not in agreement with the results of Shackelford and Wipf (1947), who determined a chromosome number of 2N = 28. Fredga (1961) found no variation in chromosome number and morphology between different sex and age groups or between the different tissue materials and in cultures.

The chromosome pairs have the following centromere positions: four median, six submedian, three subterminal, and one terminal (Fredga 1961). The X sex chromosome has a submedian centromere position and the Y sex chromosome has a median-submedian centromere position and is the smallest of all chromosomes. There is one pair of chromosomes that have a secondary constriction.

The fur color of the geographic races of wild mink varies from light brown to almost black. The factors that determine the degree of brownness in wild-type mink are, however, genetically distinct from those responsible for the brown mutant color phases commonly produced on commercial mink farms (anonymous 1955). According to Pastirnac (1974), the wild-type color is conditioned by 20 alleles (14 dominant and 6 recessive), of which at least 5 are linked with other genes on the same chromosome.

REPRODUCTION

Since mink emerged as valuable fur bearers during the 1920s, there has been considerable interest in their husbandry. As a result, much of the present knowledge of mink biology has been obtained through research conducted with farm-raised mink, where good records and large numbers of animals are available for study. The results of many of these studies are applicable to mink in the natural state, because farm-raised mink have been reared commercially for a relatively short period of time and represent an interbreeding of almost all geographic subspecies of *M. vison*.

Anatomy. The gross anatomy of the mink reproductive system is similar to that of other carnivores. The reproductive organs vary seasonally in size and functional activity in response to variations in the length of daylight (Hammond 1951; Holcomb et al. 1962; Aulerich et al. 1963; Duby and Travis 1972).

The ovaries are "bean shaped" and range in mean weight from 0.3 g, during anestrus, to 0.65 g, during the mating season (March). There is no difference in size between the right ovary and the left (Enders 1952).

Although primary follicles are present throughout the year, the increase in the size of the ovaries is mainly due to an increase in the number and growth of the follicles.

Corpora lutea, formed after ovulation, vary considerably in size but their presence does not suppress the development of primary follicles (Venge 1973). Following parturition the corpora lutea undergo a change in character but persist, undiminished in size, for up to five weeks (Enders 1952).

The oviduct, approximately 3 cm in length, is firmly attached to the ovarian capsule. Fimbria are well developed during estrus and the mucosa is convoluted. The tubouterine junction is marked by a smooth muscle sphincter (Hansson 1947).

The uterus is bicornate. The uterine horns fuse proximal to the cervix to form a rudimentary corpus that is about one-third the length of the horns. During anestrus, the horns of nulliparous females are thread-like and approximately 3 cm in length, while the horns of parous mink are thicker and longer (Enders 1952). As the breeding season approaches, the uterus elongates and the walls become thickened. By two weeks after parturition, the external appearance and size of the uterus have returned to those of the anestrus state (Kolpovsky 1979). Placental scars are not considered to be a reliable indicator of the breeding history (Elder 1952).

The cervix has a well-developed circular sphincter and the cervical canal opens into the vagina through a dorsally projecting papilla (Hansson 1947). The muscular vagina averages 3.4 cm in length from the vestibule to the cervix. Both longitudinal and transverse folds are present in the mucosa. A well-developed transverse fold lies across the dorsal wall just posterior to the cervix. As estrus approaches, there is increased circular muscle development in the cervical portion and distension of the anterior portion of the vagina (Enders 1952).

As with many other species, changes in the vaginal epithelium of mink can be correlated with the estrus cycle. Vaginal smears taken from mink throughout the year show the following cellular structure (Hansson 1947; Holcomb et al. 1962):

1. Proestrus—Large, cornified and small, rounded epithelial cells with clearly stainable nuclei (February).
2. Estrus—Large, irregular, cornified epithelial cells with faintly visible nuclei (March).
3. Late estrus—Small, autolysed epithelial cells with clearly stainable nuclei and a few polymorphonuclear leukocytes (late March).
4. Postestrus—Primarily polymorphonuclear leukocytes (early April).
5. Anestrus—Lack of cells and cellular detritus of various kinds (late April–January).

The vulva shows a modest progressive swelling and darkening (from off-white to red) as the breeding season approaches. The swelling, which is much less than in the ferret (*Martes nigripes*) or marten (*Martes americana*), persists throughout estrus and can be correlated with mating performance (Travis et al. 1978).

Six to eight mammary glands are present but may not all be functional. Hansson (1947) characterized the placenta of mink as a zonodiscoidal type, which closely resembles that of the ferret.

The penis of the adult mink is approximately 5.6 cm long and is enclosed in a sheath (Enders 1952). Its shape is determined by a baculum, which is triangular in cross-section and recurved at the tip (Elder 1951; Lechleitner 1954; Long and Frank 1968; and Paul 1968). The urethra is contained in a longitudinal groove on the ventral surface of the baculum. The glans portion of the penis constitutes the distal one-third of the organ. During erection, the baculum protrudes beyond the glans so that the urethral orifice opens ventral and anterior to the recurved tip (Enders 1952).

The testes are permanently scrotal. There is little change in the size and functional activity of the testes and epididymides of adult mink from June to November. From mid-November through early March there is a progressive increase in size and spermatogenic activity (Lundh 1961; Onstad 1967; Tiba et al. 1968); by the mating season (March) testicular weights average about 0.124 percent of the body weight (Bostrom et al. 1968). After April these organs regress in size and functional activity.

The accessory glands of the male consist of a distinct ampulla and a prostate that completely surrounds the urethra. Seminal vesicles and bulbourethral glands are absent.

Breeding. Mink are generally solitary, unsociable animals. Males and females associate only for brief periods during the mating season, which occurs from late February to early April (Enders 1952; Hodgson 1958; Travis and Schaible 1961). There is some variation in the length of the mating season of different subspecies, which may be due to genetic adaptation to the climate. Mink are sexually mature at 10 months of age and are fecund for 7 or more years (Enders 1952).

As the breeding season approaches, the general physical activity of the animals increases (Marshall 1936). The males travel widely, seeking females for mating. When sexually excited, both emit vocalizations that resemble "chuckling." The mating act is frequently quite vigorous. The male grasps the female by the back of the neck and, with the aid of his hind feet and body, attempts to position her for copulation. Scabs and numerous white guard hairs are often found on the dorsal neck surface of females that have recently bred. Usually after a struggle of varying duration, providing the female is receptive, she will allow intromission. When the female is unreceptive, a violent struggle may ensue. If she is successful in breaking away from the male, the encounter usually ends in a fight.

Mink frequently exhibit prolonged copulation. Periods of activity followed by periods of rest lasting for several hours are not uncommon. There are, however, no intumescent bodies, as in the Canidae, and intromission can be interrupted at any time. During prolonged copulations, the male may ejaculate several times.

Female mink remain in estrus throughout the breeding season and have receptive periods at 7- to 10-day intervals (Venge 1959). The female may allow several matings during these receptive periods. Ovulation is not spontaneous but is induced by the stimulation associated with the mating act. Ovulation occurs 33 to 72 hours after coition (Venge 1959). When females remate, a second ovulation capable of producing fertilizable ova occurs only when the interval between matings exceeds six days (Hansson 1947). Breeding studies, involving multiple matings with various color phases of farm mink, have demonstrated that both superfetation and superfecundation are common (Caven 1944; Shackleford 1952; Johansson and Venge 1951) and, according to Enders (1952), occur under natural conditions.

Gestation. The duration of pregnancy in mink averages 51 days but may vary from 40 to 75 days due to variation in the duration of the preimplanation period (Hansson 1947; Enders 1952). Following ovulation, fertilized ova develop to the blastula stage. They then enter into a "more or less" quiescent state of variable duration before uterine implantation and further development take place. The duration of the delay in implantation is less in females mated late in the breeding season and appears to be influenced by increasing photoperiods (Pearson and Enders 1944). In females that accept multiple matings, both ovulation and fertilization may occur during the delay period (Enders and Enders 1963).

Parturition. The young are born 28–30 days after implantation (Enders 1952). Whelping usually occurs between the last week in April and the middle of May.

Litter size is from one to eight or more, averaging about four young at birth. Hansson (1947) noted there is a trend toward larger litters following shorter gestation periods. He suggested that intrauterine mortality prior to implantation may be greater during long gestation periods.

Newborns. Neonatal mink are altricial and weigh 8–10 grams. At whelping, light-colored hairs are present on the dorsal surface of the body. The first deciduous teeth erupt between the second and third weeks (Aulerich and Swindler 1968). The eyes open around three weeks of age (Travis and Schaible 1961), after which the young begin to consume solid food. By 35 days of age, the young are fully homothermic (Kostron and Kukla 1970). They grow rapidly, attaining 40 percent of their adult body weight and 60 percent of their body length by seven weeks of age. The kits are very playful, constantly squealing and attacking one another until the litter breaks up in early fall.

ECOLOGY

Habitat. An examination of fur harvest records for North America (Deems and Pursley 1978) shows that Louisiana, Minnesota, and Wisconsin produce the most wild mink pelts. In Canada, Saskatchewan and Manitoba top the list. The common denominator is the amount of wetland habitat. Mink inhabit wetland areas of all kinds, including banks of rivers, streams, lakes, ditches, and other waterways, and swamps, marshes, and backwater areas.

Marshall (1936), studying winter activities of the mink in Michigan, found 50 percent of mink tracks in types of hydrophytic successions, 37 percent in bushy and timbered areas, and 13 percent in "sedge" and "cattail" types. The observations in xerophytic successions were all within 100 feet of the Huron River.

Burns (1964) described high, medium, and low mink densities on the tundra of the Yukon-Kuskokwim Delta. The highest density occurred in the low, swampy terrain surrounding the largest bodies of water in the area, and in the extensively interconnected water system with large concentrations of blackfish (*Dallia pectoralis*) and whitefish (*Coregonus* spp.). In Louisiana, Arthur (1931) indicated that the highest density of mink occurred in coastal marshes, followed by deep cypress-tupelo swamp and then backwater bottomland hardwood areas in central and northern portions of the state.

Work on habitat selection by mink in Sweden indicated changes in summer and winter preference related to food supply (Erlinge 1972). The lakes densely populated with mink were generally small and most were oligotrophic with stony shores and sparse vegetation. Most of these lakes were rich in crayfish. Mink were also common along sections of streams surrounded by marshes, providing a good supply of fish and mammals.

The distribution of mink in some cases is definitely related to habitat preference; however, the presence of den sites can also be extremely important. Schladweiler and Storm (1969), working in Minnesota, found that one mink family within a 31-ha area used 20 different den sites. Marshall (1936) found that 2 females used only 2 dens regularly during the winter. Males did not use the same dens repeatedly, but used those most convenient.

An Idaho study indicated that old beaver (*Castor canadensis*) lodges are sometimes utilized as denning locations (Jack Whitman, Idaho Cooperative Wildlife Unit, personal communication 1980). Errington (1961) suggested that muskrat bank burrows or houses provide den sites for mink. In Louisiana, peak mink production in coastal marsh areas has coincided with peak muskrat production. This does not necessarily indicate increased numbers but may result from more intensive trapping (Palmisano 1971). On the other hand, if availability of dens can be a limiting factor, then an increase in muskrat houses may provide additional den sites. Gerell (1970), working in Sweden with telemetry techniques, found that the greatest intensity of use of a home range by mink occurred in areas with suitable dens.

Population Density. In Louisiana, peak mink production occurred in the cypress-tupelo swamps following logging operations during the early part of this century (Palmisano 1971). Numbers have declined dramatically from that time.

St. Amant (1959) reported good cypress-tupelo areas producing one pelt per 2.4–4 ha. The average swamp produced about one pelt per 10–12 ha in southern Louisiana. A catch of one pelt per 24.3–60.7 ha was representative of better-drained bottomlands. Peak mink production on a 31,984-ha tract of brackish marsh in coastal Louisiana occurred in 1940. Approximately 16.8 pelts per 405 ha (1,000 acres) were harvested during that year. Maximum production in other areas of brackish marsh has yielded 10 mink per 405 ha. Peak production in fresh marsh areas has been approximately 15 per 405 ha (Palmisano 1971).

Mitchell (1961), using mark and recapture data from a population along a Montana river, estimated 21.9 mink per 259 ha (640 acres) in 1957, and 8.5 in 1958. Marshall (1936) estimated 1.5 females per 259 ha, with a 1:1 sex ratio after heavy trapping (in Michigan).

Activity. Gerell (1969) found that male mink were mainly nocturnal at all seasons, with the level of activity increasing with the length of the night and decreasing temperature. In Michigan, Marshall (1936) found that both sexes most frequently moved about in the winter from dawn to dusk, with some activity during the middle of the day. Mink were inactive during periods of low temperature following snowfall. Gerell (1969) found that male activity peaked with peaks in prey activity. During pregnancy, the female mink showed a very low activity level, which increased during the litter period and was primarily diurnal. There was a tendancy toward the same activity during similar environmental conditions. A free-living ranch mink under study showed activity related to farm feeding time rather than environmental conditions. However, adaptation seemed to be taking place, with the animal feeding in relationship to prey availability.

Home Range and Movement. Using telemetry data, Gerell (1970) calculated home ranges for adult males to average 2,630 meters in stream length and vary from 1,800 to 5,000 m. Juvenile males had an average range of 1,230 m of stream length, but varied from 1,050 to 1,400 m. Adult females averaged 1,850 m and varied from 1,000 to 2,800 m. These figures show that adult male mink have a home range twice as large as that of the juvenile male. However, when less frequented areas (5 percent use intensity) are excluded, the ranges of the adult and juvenile males are very similar. Although the home range of females is smaller than that of males, females appeared to use a greater portion of their range more intensely. Gerell (1970) also found a definite tendency for activity to be concentrated in a more restricted portion of the home range during winter. Marshall (1936) found that males covered a much larger territory, and females utilized a more restricted area. Ritchey and Edwards (1956) summarized their movement data from live-trapping in British Columbia much the same way. Juveniles were more restricted in their travels than adult males, and adult females were the most sedentary.

Gerell (1970) noted the great variety of habitats utilized by mink, which indicated a definite adaptability. Males at times exploited areas far from water, probably due to lack of adequate food in the more aquatic areas. Differences found in the intensity of use of portions of the home range were best explained by the distribution of food and hunting places. Gerell (1970) felt that males spend about 90 percent of their active time hunting. As mentioned earlier by several researchers, suitable den sites could be important in determining use activity. He also felt that use intensity was influenced by territorial behavior. Radio-tracking and trapping research in Idaho seems to question the meaning of ''territorial.'' Mink in the Idaho study were not strictly territorial; they did not actively defend any portion of their home ranges from conspecifics of the same sex (Jack Whitman, Idaho Cooperative Wildlife Research Unit, personal communication 1980).

The largest-scale movements in the Swedish mink population were dispersal of the juveniles beginning in July. One juvenile male moved 21 km in 27 days, an average of approximately 800 m per day. The longest distance Gerell (1970) recorded was a juvenile male moving 45 km. Juveniles increased their movements as they increased in size and age. Most juveniles left the study area; however, some losses were compensated for the following spring as animals from adjacent areas moved in.

The most obvious seasonal movements were mainly associated with males and the mating season. Although yearly variations occurred, the mating season appeared to start the first week in March. Along a Montana river, six juvenile males averaged a move of 18.1 km each (Mitchell 1961). This occurred during a population build-up. However, the majority of the movements for juvenile and adult males did not exceed 4.8 km. The home range for a female captured 7 times was 7.8 ha and was 20.4 ha for another female captured 10 times. The first female was located in an area of excellent cover with a small irrigation ditch running through it, whereas the second area could have been marginal because of limited cover from heavy grazing.

FOOD HABITS

The digestive system of the mink was studied by Kainer (1954a, 1954b). The size and location of the stomach vary with the amount of material contained. Stomach weight averaged 3.8 ± 0.8 grams, with a range of 2.5 to 9.8 g from 77 laboratory animals whose body weights ranged from 400 to 1,450 g. The stomach averaged 0.52 percent of the body weight. When the stomach was fully expanded, it contained from 40 to 70 cc of material. The cecum is absent in the mink. The ratio of intestinal length to body length had a mean of 3.97:1 with a range of 3.60:1 to 4.38:1. Passage of food through the digestive system is very rapid. Undigested material may appear in scats only an hour after ingestion (Waller 1962). Various types of food may have different retention rates in the mink digestive system. Hard, indigestible materials are passed quickly from the stomach to the intestine, while soft, fleshy food items remain in the stomach for a longer period of time (Sealander 1943b; Wilson 1954).

The food habits of mink have been studied by Errington (1943, 1954), Sealander (1943b), Wilson (1954), Korschgen (1958), Waller (1962), Erlinge (1969), and Eberhardt (1973), through stomach, intestinal, and scat analyses. Stomachs and intestinal tracts are usually obtained from hunters, trappers, or fur houses. Because mink generally deposit feces in latrines near a den site, collecting large numbers of scats is relatively easy. The mink is a carnivore—ingestion of plant debris is incidental to feeding of other prey items (Sealander 1943b; Waller 1962). Mink utilize a diversified array of prey items and will feed on any animal material that they can find and kill (table 31.1).

The mink is well adapted for hunting both aquatic and terrestrial prey. Common food items found in the diet of mink include mammals, fishes, birds, amphibians, crustaceans, insects, and reptiles. No one individual food item seems to be consistently more important in the mink diet. The importance of each prey item varies with the location and the time of year in which the stomachs and scats were collected.

TABLE 31.1. Frequency of foods in mink stomachs collected during winter months in Missouri and Michigan

	Source			
	Korschgen 1958		Sealander 1943b	
Food	Percentage Occurrence	Percentage Volume	Percentage Occurrence	Percentage Volume
Mice & rats	22.6	23.9	15	11
Rabbits	5.9	10.2	15	14
Muskrats	1.1	1.3	36	31
Frogs	25.5	24.9	23	20
Fish	30.9	19.9	11	6
Crayfish	19.9	9.3	6	2
Birds	5.9	5.6	13	15
Invertebrates	4.0	0.4		
Snakes	0.3	0.1	2	2
Sample size		372		102

TABLE 31.2. Comparison of foods eaten by mink collected from an Iowa marsh and streamside habitat in Sweden

Food	Winter		Spring		Summer		Fall		Total	
	a	b	a	b	a	b	a	b	a	b
Mammals	11.1[c]	49.6	12.5	34.0		50.2	20.0	47.7	9.5	46.0
Birds	14.8	48.0	12.5	71.6	66.6	32.2	17.8	40.5	22.4	44.8
Fish	95.5	5.7	96.7	11.1	33.2	14.9	79.9	2.3	60.2	10.5
Frogs		9.8		9.2		22.8		30.0	1.9	19.6
Crustaceans		2.8		9.4		22.3		10.8		14.9
Insects		2.8		4.4		12.5		8.1	4.0	8.7
Reptiles				1.2		1.3		0.7		1.0
Sample size	27	246	32	521	18	1030	45	444	122	2241

[a]Erlinge (1969)
[b]Waller (1962)
[c]percentage occurrence in scats

Waller (1962), working in an Iowa marsh, found mink predation of mammals lowest in the spring and the highest in the summer (table 31.2). Some of the most common mammals preyed upon included deer mice (*Peromyscus* spp.), meadow mice (*Microtus* sp.), muskrats (*Ondatra zibethicus*), and cottontail rabbits (*Sylvilagus floridanus*).

Fish comprise a major portion of the food eaten by mink. Korschgen (1958) found an increased usage of fish as they became more available due to decreasing water levels in Missouri. Burbot (*Lota lota*), killifish (Cyprinodontidae), pike (*Esox lucius*), perch (*Perca* spp.), sunfish (Centrarchidae), catfish (Ameiuridae), trout (*Salmo trutta*), and eel (*Anguilla bostoniensis*) are all consumed by mink.

Frogs are frequently consumed by mink. Mink predation on frogs decreased during years of drought when many small streams and ponds were dry (Korschgen 1958). Waller (1962) found the highest use of frogs during summer and fall months, when they were most abundant and vulnerable to mink predation. The frogs most frequently found included the green frog (*Rana clamitans*), leopard frog (*R. pipiens*), bull frog (*R. catesbeiana*), and spring peeper (*Hyla crucifer*).

Crayfish (*Cambarus* sp.) are another prey item frequently utilized by mink. Korschgen (1958) stated that mink predation on crayfish increased during drought periods, which made crayfish more available to mink. Waller (1962) found crayfish most abundant in scats collected during summer and fall months and in areas where crayfish were most plentiful.

Blackbirds, waterfowl, songbirds, and eggs are eaten by mink. Waller (1962) reported high usage of birds throughout the year, with highest frequency of occurrence in scats collected during winter and spring. Sargeant et al. (1973) and Eberhardt (1973) investigated the effect of mink predation on waterfowl on prairie wetlands. They concluded that mink prey on all parts of the waterfowl population. Especially vulnerable to mink predation were incubating females and released hand-reared ducklings.

Sealander (1943b), working in Michigan, reported that the diet of male and female mink varied with respect to the size of the prey items consumed by each group. Males consumed significantly more muskrats (P<0.01) than did females, while females consumed more small mammals (P<0.05) than did males. No other significant differences were found in other food items utilized by mink. The smaller size of the female mink probably accounts for these differences in prey usage.

Prey selection and feeding behavior of mink were studied by Errington (1943, 1954) and Waller (1962) in Iowa. Errington (1943) determined that any factor that caused instability in the muskrat population (drought, competition, or strangeness to environment) increased the vulnerability of the muskrat to mink predation. Waller (1962) found that although mink diets are characterized by diversity, mink may exhibit a responsiveness to a prey item that is highly available. The responsiveness of the mink to available prey results in a diet that is extremely variable by season and location.

The variability in the diet of mink is suggested by comparing foods utilized by mink in two different habitats (see table 31.2). Waller (1962) collected scats from a fresh marsh in Iowa; Erlinge (1969) collected scats from mink habitat along a stream in Sweden. In Sweden the most readily available prey was fish and waterfowl. The resultant diet was dominated by these two prey categories.

The Iowa marsh provided habitat for a variety of prey species. The mink's diet was also characterized by a variety of prey items. Also of interest is the high usage of specific prey items at certain times of the year. The results of these studies show that the opportunistic feeding behavior of the mink results in a varied diet.

MORTALITY

Predation. Mink do not appear to suffer significant mortality due to predators other than humans. They may occasionally fall victim to fisher (*Martes pennanti*), red fox (*Vulpes vulpes*), gray fox (*Urocyon cinereoargenteus*), bobcat (*Felis rufus*), lynx (*Felis canadensis*), wolf (*Canis lupus*), alligator (*Alligator*

TABLE 31.3. Parasites of the mink

Parasite	Study Location	Source
Trematoda		
Alaria freundi		Erickson 1946
Alaria mustelae	Minnesota	Erickson 1946
	Oregon	Senger & Neiland 1955
	North Carolina	Miller & Harkema 1964
Baschkirouitrema incrassatum	North Carolina	Miller & Harkema 1964
Brachylaemus virginianus	Louisiana	Lumsden & Zischke 1961
Dicrocoelium dendriticum		Stiles & Baker 1935
Distoma sp.	Maryland	Stiles & Baker 1935
Enhydridiplostomum alarioides	North Carolina	Miller & Harkema 1964
Euparyphium beaveri	Canada	Law & Kennedy 1932
	Minnesota	Erickson 1946
	Louisiana	Lumsden & Zischke 1961
	North Carolina	Miller & Harkema 1964
Euparyphium inerne	Canada	Law & Kennedy 1932
Euryhelmis monorchis	Minnesota	Erickson 1946
Euryhelmis pacificus	Oregon	Senger & Neiland 1955
Euryhelmis squamula	Oregon	Senger & Neiland 1955
	North Carolina	Miller & Harkema 1964
Fibricola cratera	Louisiana	Lumsden & Zischke 1961
Fibricola lucida	Louisiana	Lumsden & Zischke 1961
Metagonimoides oregonensis	Oregon	Senger & Neiland 1955
	North Carolina	Miller & Harkema 1964
Metorchis conjunctus	Canada	Law & Kennedy 1932
	Minnesota	Erickson 1946
Nanophyetus salmincola	Oregon	Senger & Neiland 1955
Paragonimus kellicotti	Michigan	Sealander 1943*a*
	Minnesota	Erickson 1946
Paragonimus rudis	North Carolina	Miller & Harkema 1964
Paragonimus sp.		Erickson 1946
Plagiorchis proximus	Canada	Swales 1933
Procyotrema marsupiformis	North Carolina	Miller & Harkema 1964
Rhopalias macracanthus	Louisiana	Lumsden & Zischke 1961
Sellacotyle mustelae	Minnesota	Erickson 1946
	North Carolina	Miller & Harkema 1964
Sellacotyle vitelosa	Louisiana	Sogandares-Bernal 1961
Cestoda		
Mesocestoides litteratus		Stiles & Baker 1935
Pseudophyllidea	North Carolina	Miller & Harkema 1964
Taenia mustelae	North Carolina	Miller & Harkema 1964
Taenia tenuicollis	Minnesota	Erickson 1946
	Oregon	Senger & Neiland 1955
Nematoda		
Ascaris spp.	Canada	Law & Kennedy 1932

mississippiensis), and the great horned owl (*Bubo virginianus*).

Parasites and Diseases. The mink is a host for a large number of internal parasites (table 31.3). The external parasites that have been found to occur on mink include five mites (*Ixodes cruciaris, I. hexagonus, I. spinosa, I. ricinus, I. kingii*), nine fleas (*Ceratophyllus oculatus, C. vison, C. acomantis, Hystrichopsylla dippiei, Nearctopsylla hyrtaci, Oropsylla arctomys, Orchopeas howardii, Nosopsyllus fasciatus, N. genalis*), and two lice (*Lipearus dissimilis, Trichodectes retuses*) (Gorham and Griffiths 1952; Jackson 1961; Lowery 1974).

Gorham and Griffiths (1952) investigated the diseases common to ranch mink, which in some instances may also be applicable to wild mink. The following diseases were noted: botulism, anthrax, abscesses, salmonella, streptococcus, acute infections of skin and subcutaneous tissues of the head, septicemia, sinusitis, tularemia, tuberculosis, nonspecific enteritis, pneumonia, distemper, enteritis, urinary calculi, steatitis nursing sickness, pregnancy disease, nutritional anemia, Chastek's paralysis, rickets, gastroenteritis, fatty change of the liver, and hydrocephalus. These diseases have not been known seriously to affect wild mink populations.

Prenatal Mortality. Data relating to prenatal mortality in wild mink are limited. Studies with ranch mink (Hansson 1947; Enders 1952; Venge 1973) indicated that losses occur throughout the gestation period

TABLE 31.3 (*continued*)

Parasite	Study Location	Source
Capillaria mustelorum	North Carolina	Miller & Harkema 1964
Capillaria spp.	Canada	Law & Kennedy 1932
	Canada	Swales 1933
	Minnesota	Erickson 1946
Dioctophyma gigas		Stiles & Baker 1935
Dioctophyma renale	Canada	Law & Kennedy 1932
	Canada	Swales 1933
	Michigan	Sealander 1943*a*
	Minnesota	Erickson 1946
	Louisiana	Schacher & Faust 1956
	North Carolina	Miller & Harkema 1964
Dracunculus insignis		Gorham & Griffiths 1952
Dracunculus medinensis	Minnesota	Erickson 1946
Epomidistomum sp.	North Carolina	Miller & Harkema 1964
Filaria dentata		Stiles & Baker 1935
Filaria muscularis		Stiles & Baker 1935
Filaria sp.	Canada	Law & Kennedy 1932
		Swales 1933
		Stiles & Baker 1935
Filaroides bronchialis	Michigan	Sealander 1943*a*
Filaroides martis	Canada	Law & Kennedy 1932
	Canada	Swales 1933
	New York	Goble & Cook 1942
	Minnesota	Erickson 1946
	North Carolina	Miller & Harkema 1964
Gnathostoma spinigerum		Stiles & Baker 1935
Heterakis isolonche	North Carolina	Stiles & Baker 1935
Molineus patens	North Carolina	Miller & Harkema 1964
Molineus sp.	Minnesota	Erickson 1946
Physaloptera maxillaris		Erickson 1946
Physaloptera spp.	Michigan	Sealander 1943*a*
	Minnesota	Erickson 1946
Seurocyrnea sp.	North Carolina	Miller & Harkema 1964
Skrjabingylus nasicola	New York	Goble & Cook 1942
	Michigan	Sealander 1943*a*
Strongyloides sp.	Canada	Law & Kennedy 1932
	Canada	Swales 1933
Trichinella spiralis		Erickson 1946
Acanthocephalans		
Centrorhynchus conspectus	North Carolina	Miller & Harkema 1964
Macracanthorhynchus igens	North Carolina	Miller & Harkema 1964
Sporozoa		
Isospora bigemina	Michigan	Sealander 1943*a*

and amount to 50–60 percent of the zygotes. Hansson (1947) found that 83.7 percent of ovulated ova implant; 50.2 percent of these result in kits. These figures, however, refer only to females that ovulated, had implanted fetuses, and whelped, respectively. Therefore, actual fertility would be lower due to the absence of ovulation, fertilization, and implantation or to total resorption of fetuses by some females.

Possible causes of prenatal mortality in mink are: (1) reduced sperm viability due to the long interval between mating and fertilization; (2) high estrogen levels (resulting from the continued growth of ovarian follicles), which in other species prevent nidation; (3) unequal development of ova and blastocysts due to remating; (4) delayed implantation; and (5) variation among females in the ability of the uterus to support implanted blastocysts (Enders 1952).

Environmental Contaminants. Residues from environmental pollutants, such as mercury and halogenated hydrocarbon compounds (PCBs, DDT, DDE, dieldrin, and others), constitute potential biohazards for mink.

MERCURY. Mercury pollution of aquatic environments is widespread (Cumbie 1975; Wobeser et al. 1976*a*) and in areas of known contamination is suspected to be an unrecognized cause of mink mortality (Wobeser and Swift 1976). Mercury poisoning has been reported in wild mink from the consumption of contaminated fish (Wobeser and Swift 1976) and seabirds (Borg 1975).

Feeding trials have shown that mink are quite sensitive to methyl mercury (Aulerich et al. 1974; Wobeser et al. 1976b), phenylmercuric acetate (Borst and van Lieshout 1977), and magnesium-bromalkylmercuric chloride (Ahman and Kull 1962) but comparatively tolerant of mercury in an inorganic form (Aulerich et al. 1974).

Clinical signs of mercuralism in mink include anorexia, weight loss, incoordination, tremors, ataxia, paralysis, paroxymal convulsions, and high-pitched vocalizations as well as the typical limb-crossing phenomenon described in other species (Aulerich et al. 1974). The latency period varies inversely with the amount of mercury consumed. Kirk (1971) reported that mink could survive on a diet that contained 0.5-ppm mercury (fish source) but that 1 ppm could prove fatal within two months. Feeding 5-ppm methyl mercury to mink produced clinical signs of poisoning within 25 days, with deaths occurring between the 30th and 37th days (Aulerich et al. 1974).

POLYCHLORINATED BIPHENYLS. Polychlorinated biphenyls (PCBs) are a series of compounds that have had wide industrial use. Although no longer manufactured in the United States, they have become major pollutants of many aquatic and terrestrial ecosystems. As with other halogenated hydrocarbon compounds, PCBs have low biodegradation rates. They tend to concentrate in fatty tissues of animals as these substances move through the food chain.

Mink are extremely sensitive to PCBs. Residues of these compounds in fresh-water fish used for feeding ranch mink have caused mortality and reproductive problems on commercial mink farms (Aulerich et al. 1973). As little as 0.64 ppm of PCB (from contaminated meat) in the diet of mink for 160 days caused nearly complete reproductive failure, while 3.57-ppm dietary PCB was lethal to adult mink (Platonow and Karstad 1973).

O'Shea et al. (1980) found PCB residues in wild mink from western Maryland comparable with the residue levels reported by Platonow and Karstad (1973) to cause reproductive failure in ranch mink. These results suggest that wild mink populations in areas of PCB pollution may be experiencing serious reproductive complications.

Clinical signs of PCB poisoning in mink consist of anorexia, bloody stools, fatty infiltration and degeneration of the liver and kidneys, hemorrhagic gastric ulcers, brain edema, and increased hepatic cytochrome P450 levels (Platonow and Karstad 1973; Aulerich and Ringer 1977; Jensen et al. 1977).

PESTICIDES. Chlorinated hydrocarbon pesticides, such as DDT, DDE, and dieldrin, have had wide agricultural use and because of their persistent nature have become troublesome pollutants. These pesticide residues in Great Lakes fish were previously suspected of causing reproductive failure and mortality in ranch mink (Hartsough 1965; anonymous 1966a; anonymous 1966b). Feeding studies have shown, however, that mink can survive and reproduce when fed diets that contain 100 ppm of DDT, DDE, or DDD for prolonged periods (Aulerich and Ringer 1970; Duby 1970). Feeding 2.5-ppm dieldrin was toxic to adult mink, especially if the animals were stressed, but it did not impair reproduction when fed at 5 ppm during gestation (Aulerich and Ringer 1970).

The levels of pesticides used in these feeding studies are considerably higher than the residue levels one would expect to find in typical prey species of mink. The relatively low levels of organochlorine residues that have been reported from tissues of wild mink (Franson et al. 1974; Frank et al. 1979; O'Shea et al. 1980) suggest that some margin of safety exists concerning these compounds.

Neonatal Mortality. Burns (1964), working in the Yukon-Kuskokwim Delta in Alaska, stated that dry, warm weather during the whelping season resulted in better kit survival. He felt that this climatic condition may be the most important factor affecting fall population levels prior to the trapping season. This factor may be of lesser importance in areas not subject to climates as severe as those occurring in Alaska.

Trapping Mortality. Mink harvest fluctuates widely from year to year (Deems and Pursley 1978). Harvest data for mink taken in Louisiana from 1940–41 through 1978–79 (compiled by authors) showed no significant correlation (r = 0.3) between the number of mink harvested and the average price received per pelt. Because price and harvest levels are not significantly correlated, and if one assumes that trapping pressure is relatively constant, then harvest levels can indicate changes in the population level. However, the extent to which trapping mortality influences population levels is not known.

Age and Sex Ratios of the Harvest. Data available on age ratios of trapped mink are limited. Croxton (1960) reported a juvenile-to-adult ratio of 1.4:1 in mink trapped in coastal southeastern Alaska. Other researchers also found high juvenile-to-adult ratios in harvested mink. In Montana, Lechleitner (1954) reported an approximate juvenile-to-adult ratio of 4:1 from a sample of 280 carcasses collected from trappers. Petrides (1950), working in Ohio, found a juvenile-to-adult ratio of 3.14:1.

Greer (1956) in Montana noted that juveniles make up a high percentage of the early harvest and that the majority of the harvest is taken early in the season. These figures may indicate either differential vulnerability to trapping, high breeding success, an age structure skewed toward young animals, or a combination of these factors. Mitchell (1961) concluded that a nearly complete population turnover occurred in a three-year period in Montana.

Research concerning sex ratios of harvested mink has yielded inconsistent results. Burns (1964) found a predominance of males to females (145:100) in the harvest from the Yukon-Kuskokwim Delta of Alaska, which was significantly different from a 1:1 sex ratio. He stated that more males are caught early in the season due to their wide-ranging movements; more females are caught late in the season when searching for a late winter food supply. In years of low abundance of mink, trappers tend to stop earlier in the season, thus leaving a high percentage of the females for breeding.

Croxton (1960) reported an 81:100 ratio of males to females in mink trapped in coastal southeast Alaska. This ratio also differs significantly from the anticipated 1:1 sex ratio. He felt that these data may have resulted from an uneven sex ratio in the population, trapping methods selective for females, or a nonrepresentative sample of the harvest.

Petrides (1950) found an essentially even (1:1) sex ratio in 249 wild mink in Ohio. Due to the variability in the sex ratios of harvested mink, no general conclusions can be drawn.

SEX AND AGE DETERMINATION

In live mink, the sexes can easily be distinguished by external genitalia and by the larger size of males. Petrides (1950) determined that dried pelts can be easily sexed. The presence of a penis scar on the dried pelts is indicative of a male pelt, while the absence of this scar is indicative of a female. A secondary factor for female pelts is the presence of mammary nipples.

The sex of mink may be accurately determined with the use of several skull measurements. Hall (1951) and Burns (1964) utilized the basilar length and the length of tooth rows to distinguish accurately between males and females. The basilar length measurement yielded approximately 99 percent accuracy—measurements of 63 mm or more distinguished males. The length of tooth rows was of equal accuracy, with male skulls yielding measurements of 25.0 mm or greater. Burns (1964) also studied the depth of the skull at the anterior margin of the basioccipital as a means to determine sex from mink skulls. This measurement is not as accurate as the others because of a slight overlap between sexes, and because the skull "shrinks" as age increases. Birney and Fleharty (1966) determined that condyle-premaxilla length provided 95 percent accuracy in sexing mink skulls for animals 18 months of age or older.

Baculum weight and development, femur morphology, and skull and pelt characteristics are indexes used in determining the age of mink. The baculum of the mink is J shaped with a sharp crook at the distal end. It is characterized by a urethral groove that extends along the distal half of the bone and deepens as it approaches the tip of the bone. This urethral groove gives the distal portion of the bone the shape of an inverted V in cross-section. The proximal end of the bone has a cross-sectional shape of a triangle with the apex on the dorsal side of the baculum (Elder 1951).

The development of the baculum provides valid criteria for separating male mink into adult and juvenile age classes (Elder 1951; Lechleitner 1954). The bacula of adult mink are heavier, more massive, and have a longer and rougher proximal end. In addition, the bacula of adults have a ridge formed at the proximal end at the point of attachment of the corpus cavernosum. Bacula from juvenile animals do not develop this ridge until the animals reach sexual maturity (Lechleitner 1954).

Elder (1951) and Lechleitner (1954) discussed the use of baculum weight as a method of age determination of mink. There is some overlap in weights at the 95 percent confidence limit. Lechleitner (1954) gave the following mean bacula weights for ranch and wild mink: ranch mink—adults 395 mg ± 90.0 mg, juveniles 203 mg ± 25.4 mg; wild mink—adults 398.1 mg ± 97.0 mg, juveniles 172.0 mg ± 34.2 mg. Because the mean bacula weights overlap at the 95 percent level, this method should be used in conjunction with baculum conformation to estimate the age of the animal.

Mink can be aged by changes in the epiphyseal closure of the femur. The epiphyseal closure in young mink has a spongy appearance, while on adults the bone becomes dense and smooth. Adults also have a calcium deposit, called the suprasesamoid tubercle, at the gastrocnemius muscle. Age determination is based on the appearance of the epiphyseal closure and on the presence or absence of the suprasesamoid tubercle (Lechleitner 1954).

From a series of 20 measurements Lechleitner (1954) concluded that mink skulls were unreliable for determining age classes. Greer (1957) found that the presence (adult) or absence (juvenile) of the jugalsquamosal suture could be used for age determination of mink.

Dried pelts of female mink can be aged by the size and color of the nipples. Pelts from adults have dark nipples, which are usually more than 1 mm in diameter. Nipples on juvenile pelts are light in color, flattened, and less than 1 mm in diameter. Pelts of juvenile male mink can be distinguished from adult male pelts by their small size and thin skin (Petrides 1950).

ECONOMIC STATUS

The prestige long attached to owning mink garments has perpetuated one of the greatest demands for a particular type of fur. The influence of mink ranching also has maintained the demand for this fur. The ability of the ranching industry to produce a relatively large supply of perfectly matched pelts, as compared to wild fur harvest, has created a stabilizing effect on the market.

During the period of 1953 to 1966, ranch mink production in North America increased from 1,000,000 to 11,000,000 pelts. During the same interval, world mink production increased from 2,500,000 to approximately 22,000,000 pelts. Mink historically were regarded as a luxury fur and, prior to this expansion of the ranching industry, were confined to fur coats at high prices. The tremendous increase in pelt production necessitated greater diversification and broader consumer outlets. Prentice (1976) estimated that 95 percent of mink were used for jackets, stoles, and trimmings of fur and cloth coats. By 1975, many women in the moderate income bracket had become potential purchasers of mink in some form.

Between 1967 and 1974, North America's mink

farm production decreased by 65 percent, because of the increased world supply of ranch mink and lower production costs in Russia and Europe. While the United States and Canada were liquidating herds, Russia and Europe expanded production. Russia has since become the largest single mink producer in the world. The U.S. Department of Agriculture reported about 7,200 American mink ranches in the mid-1960s and only 1,221 in 1974. The introduction of "fun furs" revived consumer interest for fur garments in general in the early 1970s, and reversed the downward trend in mink ranching (Prentice 1976). The total mink production of the top seven countries in 1975–76 was estimated at over 21,000,000 pelts, or 88 percent of the total world production. These countries included: the Soviet Union—8,700,000; Finland—3,200,000; the United States—3,000,000; Denmark—2,960,000; Sweden—1,200,000; Norway—1,130,000; and Canada—900,000.

Over a 22-year period from the 1956/57 season to 1977/78, an average of 256,373 wild mink pelts were taken yearly in the United States (computed from U.S. Department of Interior, Fish and Wildlife Service 1957–1971 and Deems and Pursley 1978). Considering the size of the world mink ranch production and the rather insignificant U.S. harvest of wild mink, it is difficult to believe that supplies of ranch mink could influence prices for wild mink.

Some fur marketing experts believe that the markets for wild mink are completely different from and to some extent unrelated to those for the many color shades of ranch mink (Alex Shieff, Sales Manager, Ontario Trappers Association, North Bay, Ontario, personal communication 1980).

During the 1960/61 and 1961/62 seasons, mink represented 25.1 percent of the total value of the wild fur harvest in North America and accounted for approximately 3 percent of the pelts (Williams 1966). However, during the 1970/71 through 1976/77 seasons, mink represented about 2 percent of the harvest and only 3 percent ($2.4 million) of the value in the United States (Deems and Pursley 1978).

An examination of the value of long-haired pelts versus mink pelts in Minnesota and North Dakota during the 1960–62 seasons compared to the 1976/77 season reveals why the overall value of mink dropped so drastically during the later period. In Minnesota during the 1960/61 and 1961/62 seasons, mink accounted for 67 percent of the total value of the fur harvest in the state (Williams 1966). However, by 1976/77, mink value in Minnesota accounted for only 6 percent, while red fox (*Vulpes vulpes*) was 36 percent and raccoon (*Procyon lotor*) made up 35 percent (computed from Deems and Pursley 1978). North Dakota is another good example. From 1960 to 1962, the mink value made up 64 percent of the total value of the fur harvest. By 1976/77, however, mink accounted for only 2 percent, with red fox at 31 percent, raccoon making up 28 percent, and coyote (*Canis latrans*) supplying 24 percent (computed from Deems and Pursley 1978). The tremendous interest in "fun furs" (mainly long haired) during the 1970s created spectacular prices and in-

creased the harvest of animals such as the fox, coyote, and raccoon.

The ranking of the top five states in wild mink production from 1970/71 through 1976/77 was: Louisiana—36,056; Minnesota—34,057; Wisconsin—24,267; Iowa—17,743; and Ohio—11,978. The average price per pelt during this time period was: 1970/71—$7.50; 71/72—$9.50; 72/73—$12.00; 73/74—$11.50; 74/75—$8.00; 75/76—$10.00; and 76/77—$14.00. The average price paid in the different states varies significantly depending on the local marketing system and the quality of the fur. Bachrach (1953) discussed in detail the characteristics of pelt quality in various regions of North America. Pelt quality is dependent upon pelt size, color, texture, and density of the fur fiber. The eastern Canadian peltries are considered by Bachrach (1953) to be the highest quality of mink on the North American continent.

MANAGEMENT

Management of mink populations, as with most fur animals, generally consists of three factors: harvest regulations, harvest monitoring, and habitat maintenance.

Regulation of the Harvest. Forty-seven states and all Canadian provinces conduct limited trapping seasons on mink (Deems and Pursley 1978). Season lengths and dates vary tremendously from area to area and state to state. In many states, the mink season coincides with seasons for other fur bearers. In some regions, temperatures or pelt priming are major considerations. In addition to trapping seasons, five states (Arkansas, Florida, Indiana, Kentucky, and Michigan) have a limited hunting season. In Canada, only Nova Scotia reported conducting a hunting season for mink. Eighteen states and five provinces reported having special regulations for mink trapping (Deems and Pursley 1978).

Many wildlife managers in Canada have concluded that an imposed quota on registered trap lines maintains a more stable annual mink production. Without this quota, overtrapping may cause lowered annual production (Dick Standom, Ontario Fish and Wildlife Service, Winnepeg, Ontario, personal communication). Such areas in Canada have a relatively lower density than many of the areas known to produce large numbers of mink in the United States. Overtrapping, which could result without quotas, could lower population density to a level where recovery would be very slow.

Trapping techniques vary greatly from the cold northern areas to the warmer south. Generally, the Conibear trap used with bait is the most common in ice and snow country. In the warmer areas, the leghold is the most common trap used.

Monitoring the Harvest. The most basic and essential data necessary to monitor the harvest are naturally the total annual harvest figure. Also, data on distribution of the harvest and trappers are extremely important in order to detect possible population density changes or trapping pressure trends. Naturally, the management agency must have the capability of separating trappers from other wildlife consumers. Many states and provinces presently have these monitoring capabilities. Some agencies have now begun collecting mink carcasses to develop baseline data on the harvested segment of the mink population.

Maintenance of Habitat. Undoubtedly more important than any regulation or monitoring data is the maintenance of mink habitat. Enhancing or maintaining habitat for many forms of wildlife probably benefits mink. This is most certainly true for state or provincial agencies creating, enhancing, or maintaining waterfowl habitat. All data on mink demonstrate the importance of aquatic systems. Loss of such habitat would pose a great threat to the mink populations of North America.

CURRENT RESEARCH NEEDS

In 1976, seven universities or wildlife departments were conducting research on mink reproduction, food habits, disease, and habitat (Deems and Pursley 1978). The generally secure status of most mink populations probably influences its priority in research planning. Species with lower densities or limited habitat, or species that are easily overharvested, rank higher than mink in research priority planning.

Considering the overall knowledge of this species in North America, some areas lacking data are obvious. As in the case of many of the fur animals, better techniques are needed to determine population density. Much additional work is needed concerning population structure through long-term studies of exploited and unexploited populations in order to measure the impact of harvest. Such research would probably reveal explanations of what appears to be drastic fluctuations in population density in areas of North America producing relatively large annual harvests.

Telemetry studies in these areas of high density could produce valuable data on movement, habitat preference, and the impact of habitat alteration, and would assist in producing information on population density and social structure.

The severe impact of PCBs, pesticides, and heavy metals on mink, particularly on reproduction, warrants continuous monitoring in areas with high contaminant levels. Areas of prime mink habitat or production should be checked periodically for the presence and level of pollutants.

LITERATURE CITED

Ahman, G., and Kull, K. E. 1962. Prövning av magnesiumbromalkylkvicksilverklorid i foder til mink. Int. rep. S.P.R., Inst. f. Husdj. Otf. O. Vard Palsdjursavdelningen. 5pp.

Anonymous. 1955. Types of mink. Sect. 1. Pages 23–42 *in* The blue book of fur farming. Editoral Service Co., Milwaukee, Wis.

Anonymous. 1966a. Lake pollution a threat to Canadian mink herds. U.S. Fur Rancher 45:19.

Anonymous. 1966b. Mink made sterile. Fur Can. 31:6.

Arthur, S. C. 1928. The fur animals of Louisiana. Louisiana Dept. Conserv., Bull. 18:1–433.

————. 1931. The fur animals of Louisiana. Louisiana Dept. Conserv., Bull. 18(revised):1–444.

Asher, S. J.; Aulerich, R. J.; Ringer, R. K.; and Kitchen, H. 1976. Seasonal and age variations which occur in the blood parameters of ranch mink. U.S. Fur Rancher 56:4, 6, 9.

Aulerich, B. J.; Holcomb, L.; Ringer, R. K.; and Schaible, P. J. 1963. Influence of photoperiod on reproduction in mink. Michigan Agric. Exp. Stn. Q. Bull. 46:132–138.

Aulerich, R. J., and Ringer, R. K. 1970. Some effects of chlorinated hydrocarbon pesticides on mink. Am. Fur Breeder 43:10–11.

————. 1977. Current status of PCB toxicity to mink, and effect on their reproduction. Arch. Environ. Contam. Toxicol. 6:279–292.

Aulerich, R. J.; Ringer, R. K.; and Iwamoto, S. 1973. Reproductive failure and mortality in mink fed on Great Lakes fish. J. Reprod. Fert. Suppl. 19:365–376.

————. 1974. Effects of dietary mercury on mink. Arch. Environ. Contam. Toxicol. 2:43–51.

Aulerich, R. J., and Swindler, D. R. 1968. The dentition of the mink (*Mustela vison*). J. Mammal. 49:488–494.

Bachrach, M. 1953. Fur: a practical treatise. Prentice-Hall, Inc., New York. 660pp.

Banfield, A. W. F. 1974. The mammals of Canada. Univ. Toronto Press, Toronto. 438pp.

Berestov, V. 1971. Biochemistry and blood morphology of fur bearing animals. Agric. Res. Serv. U.S. Dept. Agric. and Nat. Sci. Foundation, Washington, D.C. 348pp. (Translated from the Russian Biokhimiya I Morfologiya Krovi Pushnykh Zverei.)

Birney, E. C., and Fleharty, E. D. 1966. Age and sex comparisons of wild mink. Trans. Kansas Acad. Sci. 69:139–145.

Bissonnette, T. H., and Wilson, E. 1939. Shortened daylight periods between May 15 and September 12 and the pelt cycle of the mink. Science 89:418–419.

Borg, K. 1975. Den svenska faunan och miljogifterna. Pages 19–25 *in* Symp. vid jubileet Svensk Veterinarmedicin 200 ar. Skara.

Borst, G. H. A., and van Lieshout, C. G. 1977. Phenylmercuric acetate intoxication in mink. Scientifur 1:46–48.

Bostrom, R. E.; Aulerich, R. J.; Ringer, R. K.; and Schaible, P. J. 1968. Seasonal changes in the testes and epididymides of the ranch mink (*Mustela vison*). Michigan Agric. Exp. Stn. Q. Bull. 50:538–558.

Burns, J. J. 1964. The ecology, economics, and management of mink in the Yukon-Koskokwim Delta, Alaska. M.S. Thesis. Univ. Alaska, Fairbanks. 114pp.

Cavan, D. V. 1944. Two fathers—one mother. Am. Fur Breeder 17:34.

Croxton, L. W. 1960. A southern Alaska mink study. M.S. Thesis. Univ. Alaska, Fairbanks. 74pp.

Cumbie, P. M. 1975. Mercury levels in Georgia otter, mink and freshwater fish. Bull. Environ. Contam. Toxicol. 14:193–196.

Deems, E. F., Jr., and Pursley, D., eds. 1978. North American furbearers: their management, research and harvest status in 1976. Univ. Maryland Press, College Park. 171pp.

Duby, R. T. 1970. Pesticides vs. reproduction still a puzzle. Am. Fur Breeder 43:15.

Duby, R. T., and Travis, H. F. 1972. Photoperiodic control of fur growth and reproduction in the mink (*Mustela vison*). J. Exp. Zool. 182:217–225.

Eberhardt, R. T. 1973. Some aspects of mink-waterfowl relationships on prairie wetlands. Prairie Nat. 5:17–19.

Elder, W. H. 1951. The baculum as an age criterion in mink. J. Mammal. 32:43–50.

————. 1952. Failure of placental scars to reveal breeding history in mink. J. Wildl. Manage. 16:110.

Enders, R. K. 1952. Reproduction in the mink. Proc. Am. Phil. Soc. 96:691–755.

Enders, R. K., and Enders, A. C. 1963. Morphology of the female reproductive tract during delayed implantation in the mink. Pages 129–139 *in* A. C. Enders, ed. Delayed implantation. Univ. Chicago Press, Chicago. 318pp.

Erickson, A. B. 1946. Incidence of worm parasites in Minnesota Mustelidae and host lists and keys to North American species. Am. Midl. Nat. 36:494–509.

Erlinge, S. 1969. Food habits of the otter *Lutra lutra* L. and mink *Mustela vison* Schreber in a trout water in southern Sweden. Oikos 20:1–7.

————. 1972. Interspecific relations between otter *Lutra lutra* and mink *Mustela vison* in Sweden. Oikos 23:327–335.

Errington, P. L. 1943. An analysis of mink predation upon muskrats in north-central United States. Res. Bull. 320. Iowa Agric. Exp. Stn. Pp. 797–924.

————. 1954. Special responsiveness of minks to epizootics in muskrat populations. Ecol. Monogr. 24:377–393.

————. 1961. Muskrats and marsh management. Stockpole Co., Harrisburg, Pa. 183pp.

Fletch, S. M., and Karstad, L. H. 1971. Blood parameters of healthy mink. Can. J. Comp. Med. 36:275–281.

Frank, R.; Holdrinet, M. V. H.; and Suda, P. 1979. Organochlorine and mercury residues in wild mammals in southern Ontario, Canada, 1973–74. Bull. Environ. Contam. Toxicol. 22:500–507.

Franson, J. C.; Dahm, P. A.; and Wing, L. D. 1974. Chlorinated hydrocarbon insecticide residues in adipose, liver, and brain samples from Iowa mink. Bull. Environ. Contam. Toxicol. 11:379–385.

Fredga, K. 1961. The chromosomes of the mink. J. Heredity 52:90–94.

Gerell, R. 1969. Activity patterns of the mink *Mustela vison* Schreber in southern Sweden. Oikos 20:451–460.

————. 1970. Home ranges and movements of the mink in southern Sweden. Oikos 21:160–173.

Goble, F. C., and Cook, A. H. 1942. Notes on nematodes from the lungs and frontal sinuses of New York furbearers. J. Parasitol. 28:451–455.

Gorham, J. R., and Griffiths, H. J. 1952. Diseases and parasites of minks. U.S. Dept. Agric., Farmer's Bull. 2050 41pp.

Greer, K. R. 1956. Mink age and sex ratios. Montana Fish Game Dept. Pages 34–54 *in* Wildl. Rest. Div., Annu. Rep. Mimeogr.

————. 1957. Some osteological characters of known-age ranch mink. J. Mammal. 38:319–330.

Hall, E. R. 1951. American weasels. Univ. Kansas Mus. Nat. Hist. 4:1–466.

Hammond, J., Jr. 1951. Control by light of reproduction in ferrets and mink. Nature 167:150–151.

Hansson, A. 1947. The physiology of reproduction in mink (*Mustela vison* Schreb.) with special reference to delayed implantation. Acta Zool. 28:1–136.

Harding, A. R. 1934. Mink trapping. A. R. Harding, Columbus, Ohio. 171pp.

Hartsough, G. R. 1965. Great Lakes fish now suspect as mink food. Am. Fur Breeder 38:25.

Hodgson, R. G. 1958. The mink book. Fur Trade J. Can., Toronto. 271pp.

Holcomb, L. C.; Schaible, P. J.; and Ringer, R. K. 1962. The effects of varied lighting regimes on reproduction in mink. Michigan Agric Exp. Stn. Q. Bull 44:666–678.

Humphrey, D. G., and Spencer, N. 1959. Chromosome number in mink. J. Heredity 50:245–247.

Jackson, H. H. T. 1961. Mammals of Wisconsin. Univ. Wisconsin Press, Madison. 504pp.

Jensen, S.; Kihlstrom, J. E.; Olsson, M.; Lundberg, C.; and Orberg, J. 1977. Effect of PCB and DDT on mink (*Mustela vison*) during the reproductive season, Ambio 6:23.

Johansson, I., and Venge, O. 1951. Relation of the mating interval to the occurrence of superfetation in the mink. Acta Zool. 32:255–258.

Kainer, R. A. 1954a. Gross anatomy of the digestive system of the mink. Part 1: Headgut and foregut. Am. J. Vet. Res. 15:82–90.

————. 1954b. Gross anatomy of the digestive system of the mink. Part 2: Midgut and the hindgut. Am. J. Vet. Res. 15:91–97.

Kennedy, A. H. 1935. Cytology of the blood of normal mink and raccoon. Part 2: The numbers of the blood elements in normal mink. Can. J. Res. 12:484–494.

————. 1951. The mink in health and disease. Fur Trade J. Can., Toronto. 307pp.

Kirk, R. J. 1971. Fish meal, higher cereal levels perform well. U.S. Fur Rancher 50:4.

Kolpovsky, V. M. 1979. Postnatal changes of uterus and ovaries in *Mustela vison* (Carnivora, Mustelidae). Zool. Zhur. 58:409–418.

Korschgen, L. T. 1958. December food habits of mink in Missouri. J. Mammal. 39:521–527.

Kostron, K., and Kukla, F. 1970. Changes of thermoregulation in mink kits within the 45 days of ontogenesis. Acta Universitatis Agriculturae, Facultas Agronomica, Sbornik Vysoké skoly zemedelské v Brne (rada A) 18:461–469.

Kubin, R., and Mason, M. 1948. Normal blood and urine values for mink. Cornell Vet. 38:79–85.

Lande, O. 1957. The chromosomes of the mink. Hereditas 43:578–582.

Law, R. G., and Kennedy, A. H. 1932. Parasites of furbearing animals. Dept. Game and Fisheries, Ontario, Bull. 4. 30pp.

Lechleitner, R. R. 1954. Age criteria in mink, *Mustela vison*. J. Mammal. 35:496–503.

Long, C. A., and Frank, T. 1968. Morphometric variation and function in the baculum, with comments on correlation of parts. J. Mammal. 49:32–43.

Lowery, G. H., Jr. 1974. The mammals of Louisiana and its adjacent waters. Louisiana State Univ. Press, Baton Rouge. 565pp.

Lumsden, R. D., and Zischke, J. A. 1961. Seven trematodes from small mammals in Louisiana. Tulane Stud. Zool. 9:89–98.

Lundh, E. 1961. Testikelutveckling och cykliska forandrigar i spermiebildningen hos mink. Var Palsdjur 14:380–383.

Marshall, W. H. 1936. A study of the winter activity of the mink. J. Mammal. 17:382–392.

Miller, G. C., and Harkema, R. 1964. Studies on helminths of North Carolina vertebrates. Part 5, Parasite of the mink, *Mustela vison* Schreber. J. Parasitol. 50:715–720.

Mitchell, J. L. 1961. Mink movements and populations on a Montana River. J. Wildl. Manage. 25:48–54.

Onstad, O. 1967. Studies on postnatal testicular changes, semen quality, and anomalies of reproductive organs in the mink. Acta Endocrinol. Suppl. 117. 117pp.

O'Shea, T. J.; Askins, G. R.; Chapman, J. A.; and Kaiser, T. E. 1980. Polychlorinated biphenyls in a wild mink population. Proc. Worldwide Furbearer Conf., Frostburg, Md. In press.

Palmisano, A. W. 1971. Louisiana's fur industry. Commercial Wildl. Work Unit Rep. of Louisiana Wildl. and Fisheries Comm. to U.S. Army Corps of Eng., New Orleans District. Mimeogr.

Pastirnac, N. 1974. Genetic elements of colour in mink. Rev. Zooteh. Med. Vet. 24:43–46.

Paul, J. R. 1968. Baculum development in mink. Trans. Illinois Acad. Sci. 61:308–309.

Pearson, O. P., and Enders, R. K. 1944. Duration of pregnancy in certain Mustelids. J. Exp. Zool. 95:21–35.

Petrides, G. A. 1950. The determination of sex and age ratios in fur animals. Am. Midl. Nat. 43:355–382.

Platonow, N. I., and Karstad, L. H. 1973. Dietary effects of polychlorinated biphenyls on mink. Can. J. Comp. Med. 37:391–400.

Prentice, A. C. 1976. A candid view of the fur industry. Clay Publishing Co., Ltd., Bewdley, Ontario. 319pp.

Ringer, R. K.; Aulerich, R. J.; Pittman, R.; and Cogger, E. A. 1974. Cardiac output, blood pressure, blood volume and other cardiovascular parameters in mink. J. Anim. Sci. 38:121–123.

Ritchey, R. W., and Edwards, R. Y. 1956. Live trapping mink in British Columbia. J. Mammal. 37:114–116.

Rotenberg, S., and Jorgensen, G. 1971. Some haematological indices in mink. Nord. Vet. Med. 23:361–366.

Rust, C. C., and Meyer, R. K. 1961. Hair color, molt and testes size in male, short-tailed weasels treated with melatonin. Science 165:921–922.

Rust, C. C.: Shackelford, R. M.; and Meyer, R. K. 1965. Hormonal control of pelage cycles in the mink. J. Mammal. 46:549–565.

St. Amant, L. S. 1959. Louisiana wildlife inventory and management plan. Louisiana Wildl. and Fisheries Comm., New Orleans. 329pp.

Sargeant, A. B.; Swanson, G. A.; and Doty, H. A. 1973. Selective predation by mink, *Mustela vison* on waterfowl. Am. Midl. Nat. 89:208–214.

Schacher, J. F., and Faust, E. C. 1956. Occurrence of *Dioctophyma renale* in Louisiana with remarks on the size of infertile eggs of this species. J. Parasitol. 42:533–535.

Schladweiler, J. L., and Storm, G. L. 1969. Den-use by mink. J. Wildl. Manage. 33:1025–1026.

Sealander, J. A. 1943*a*. Notes on some parasites of the mink in southern Michigan. J. Parasitol. 29:361–362.

———. 1943*b*. Winter food habits of mink in southern Michigan. J. Wildl. Manage. 7:411–417.

Senger, C. M., and Neiland, K. A. 1955. Helminth parasites of some furbearers of Oregon. J. Parasitol. 41:637–638.

Shackleford, R. M. 1952. Superfetation in the ranch mink. Am. Midl. Nat. 86:311–319.

Shackleford, R. M., and Wipf, L. 1947. Chromosomes of the mink. Proc. Natl. Acad. Sci. 33:44–46.

Sogandares-Bernal, F. 1961. *Sellacotyle vitellosa*, a new troglo trematid trematode from the mink in Louisiana. J. Parasitol. 47:911–912.

Stiles, C. W., and Baker, C. G. 1935. Key-catalogue of parasites reported for Carnivora (cats, dogs, bears, etc.) with their possible health importance. Natl. Inst. Health Bull. 163:913–1223.

Swales, W. E. 1933. A review of Canadian helminthology. Can. J. Res. 8:468–482.

Tiba, T.; Ishikawa, T.; and Murakami, A. 1968. Histologische Utersuchung der Kimetik der Spermatogenese bein Mink (*Mustela vison*). Part 7: Samenepithelzyklus in der Paarungszeit. Jap. J. Vet. Res. 16:73–85.

Travis, H. F.; and Schaible, P. J. 1961. Fundamentals of mink ranching. Circ. Bull. 229, Coop. Ext. Serv., Michigan State Univ., East Lansing, Mi. 101pp.

Travis, H. F.; Pilbeam, T. E.; Gardner, W. J.; and Cole, R. S. 1978. Relationship of vulvar swelling to estrus in mink. J. Anim. Sci. 46:219–224.

U.S. Fish Wildl. Serv. Bur. Sport Fish. and Wildl. 1957. Fur catch in the United States, 1956. Wildl. Leafl. 388. Washington, D.C. 3pp.

———. 1958. Fur catch in the United States, 1957. Wildl. Leafl. 398. Washington, D.C. 3pp.

———. 1959. Fur catch in the United States, 1958. Wildl. Leafl. 410. Washington, D.C. 3pp.

———. 1960. Fur catch in the United States, 1959. Wildl. Leafl. 424. Washington, D.C. 3 pp.

———. 1961. Fur catch in the United States, 1960. Wildl. Leafl. 436. Washington, D.C. 3pp.

———. 1962. Fur catch in the United States, 1961. Wildl. Leafl. 444. Washington, D.C. 3pp.

———. 1963. Fur catch in the United States, 1962. Wildl. Leafl. 452. Washington, D.C. 4pp.

———. 1964. Fur catch in the United States, 1963. Wildl. Leafl. 460. Washington, D.C. 4pp.

———. 1965. Fur catch in the United States, 1964. Wildl. Leafl. 471. Washington, D.C. 4 pp.

———. 1966. Fur catch in the United States, 1965. Wildl. Leafl. 474. Washington, D.C. 4pp.

———. 1967. Fur catch in the United States, 1966. Wildl. Leafl. 478. Washington, D.C. 4pp.

———. 1968. Fur catch in the United States, 1967. Wildl. Leafl. 482. Washington, D.C. 4 pp.

———. 1969. Fur catch in the United States, 1968. Wildl. Leafl. 488. Washington, D.C. 4pp.

———. 1970. Fur catch in the United States, 1969. Wildl. Leafl. 493. Washington, D.C. 4pp.

———. 1971. Fur catch in the United States, 1970. Wildl. Leafl. 499. Washington, D.C. 4pp.

Venge, O. 1959. Reproduction in the fox and mink. Anim. Breed. Abstr. 27:129–145.

———. 1973. Reproduction in the mink. Pages 95–146 *in* 1973 yearb., R. Vet. and Agric. Univ., Copenhagen.

Waller, D. W. 1962. Feeding behavior of minks at some Iowa marshes. M.S. Thesis. Iowa State Univ. 90pp.

Williams, C. E. 1966. A map of the wild fur harvest: Anglo-American. J. Geogr. 65:377–383.

Wilson, K. A. 1954. Mink and otter as muskrat predators in northeastern North Carolina. J. Wildl. Manage. 18:199–207.

Wobeser, G.; Nielsen, N. O.; and Schiefer, B. 1976*a*. Mercury and mink. Part 1: The use of mercury contaminated fish as a food for ranch mink. Can. J. Comp. Med. 40:30–33.

———. 1976*b*. Mercury and mink. Part 2: Experimental methyl mercury intoxication. Can. J. Comp. Med. 40:34–45.

Wobeser, G., and Swift, M. 1976. Mercury poisoning in a wild mink. J. Wildl. Dis. 12:335–340.

GREG LINSCOMBE, Louisiana Department of Wildlife and Fisheries, Route 4, Box 78, Darnell Road, New Iberia, Louisiana 70560.

NOEL KINLER, Louisiana Department of Wildlife and Fisheries, Route 4, Box 78, Darnell Road, New Iberia, Louisiana 70560.

R. J. AULERICH, Fur Animal Project, Department of Animal Science, Michigan State University, East Lansing, Michigan 48824.

32

Wolverine

Gulo gulo

Don E. Wilson

NOMENCLATURE

COMMON NAMES. Wolverine, glutton, Vielfrass, Veel-varaat
SCIENTIFIC NAME. *Gulo gulo*
SUBSPECIES. *G. g. gulo* (Old World wolverines); *G. g. luscus* (New World wolverines).

The taxonomic arrangement presented here is that of Kurten and Rausch (1959), who followed Degerbol (1935) in regarding Old and New World wolverines as conspecific rather than as two species. An alternative arrangement was presented by Hall and Kelson (1959), wherein the New World wolverines were considered a distinct species (*Gulo luscus*) with four recognized subspecies (*katschemakensis, luscus, luteus, vancouverensis*).

DISTRIBUTION

In the Old World, wolverines are found from Scandinavia across the taiga and forest-tundra zones of Eastern Europe and Asia. Extensive distributional records for the U.S.S.R. were given by Ognev (1935) and Stroganov (1969). In North America, wolverines once were distributed widely across the northern part of the continent, southward to roughly the 38th parallel. Southern range extensions were probably limited to montane boreal regions, with conspicuous gaps in the Great Basin and Great Plains areas.

Today, wolverines are uncommon in the contiguous United States and only slightly more common in some parts of Canada and Alaska. An interesting account of historical records with documentation of their disappearance from parts of North America was prepared by Allen (1942). In Canada, they have become rare in the east, and the southern limits of their range have receded in the prairie provinces (van Zyll de Jong 1975) (figure 32.1).

Wolverine numbers apparently declined steadily in the United States beginning in the latter half of the last century. However, in recent years, they seem to have made a comeback in several western states.

FIGURE 32.1. Distribution of the wolverine (*Gulo gulo*).

DESCRIPTION

Size and Weight. The wolverine is one of the largest members of the family Mustelidae. Adults range in size from 650 to 1,050 mm in length of head and body (Stroganov 1969). Tail length ranges from 170 to 260 mm, and height at the shoulders from 355 to 432 mm. Weights range from 14 to 27.5 kg (Walker et al. 1975). Some sexual dimorphism is apparent; males average larger than females (table 32.1).

The external appearance of the wolverine is more like that of badgers and skunks than like other members of the family Mustelidae, such as weasels (*Mustela* sp.) or otters (*Enhydra* and *Lutra*). The heavy body and lumbering gait may give a false impression of clumsiness. The head and tail are carried lower than the somewhat arched back. Although the head resembles the genus *Martes,* it seems broader and more rounded. The jaws are almost canidlike. The distal 1.3 cm (nose pad) of the snout is naked. The eyes are small and wide set. The ears are short, rounded,

Table 32.1. Cranial measurements (mm) of male and female wolverines from Alaska

Character	Female			Male		
	N	Mean	Range	N	Mean	Range
Greatest skull length	9	147.6	(139.3–161.4)	11	159.6	(139.8–170.3)
Zygomatic breadth	9	94.9	(90.7–103.8)	8	104.3	(100.3–107.3)
Interorbital breadth	10	32.9	(30.5– 36.0)	9	32.7	(30.9– 34.8)
Mastoid breadth	9	83.0	(71.3– 95.2)	9	88.7	(80.1– 96.1)
Postpalatal length	9	54.5	(52.7– 58.0)	10	55.7	(52.3– 58.6)
Maxillary toothrow length	10	49.2	(47.3– 53.4)	11	51.1	(42.7– 54.1)
Mandibular length	10	96.5	(93.5–105.3)	11	101.9	(92.1–109.4)
Mandibular toothrow length	10	59.8	(57.0– 64.6)	11	62.8	(59.3– 66.2)

and well furred. Whiskers and head bristles are few and short.

Pelage. The underfur is coarse, kinky, woolly, and about 2–3 cm long. The guard hairs are about 10 cm long, giving an overall shaggy appearance. The fur becomes shorter anteriorly, and is thick and close on the head. The drooping tail is especially shaggy, with hairs measuring 15–20 cm. According to Grinnell et al. (1937), there is a single molt each year occurring between August and December. They described specimens taken in July that had worn pelage and were beginning to replace caudal guard hairs.

The background color is blackish brown, with a pale reddish or pale brown stripe extending from behind the shoulder to the rump on each side. These bands join, forming a contrasting, though not always sharply defined, patch across the base of the tail. There is often a somewhat paler stripe on each side of the head from the eye to the ear. There are light-colored patches on the throat and between the forelegs, as well as irregular ones on the venter. The legs, feet, and most of the tail are quite dark.

The legs are stocky and powerful, and the feet large. The palms and soles are densely haired with small bare pads at the balls and bases of the digits. Digit four is the longest on the front feet, and digit three the longest on the hind feet. The claws are pale, sharply curved, and about 2–3 cm long.

Glands. Like other members of the family, wolverines possess anal glands, which secrete a yellowish brown fluid from lateral papillae located just inside the anus. According to Coues (1877), the glands are about the size of a walnut and ''the scent is foetid in a high degree.'' The scent and urine are deposited on objects, usually near food left by the animal, a habit thought by Jackson (1961) to signal ownership. Ewer (1973) said

the anal sac secretion is used only when the animal is alarmed, and that the ventral gland is used for marking.

Hall (1926) described the ventral gland of *Martes*, and noted that both sexes of wolverine have similar glandular areas. Wolverines use marking trees much as bears do, chewing and biting on them, thus scarring trees in a characteristic fashion (Ewer 1973). Pulliainen and Ovaskainen (1975) detailed 26 marking places visited by a wolverine during a single night. The majority were on small trees, but boulders and the branches of a larger tree were also marked, apparently from the anal glands.

Tracks. Although doglike, wolverine tracks are wider, show the fifth toe, the hairiness of the sole, and the characteristic division of the sole pads of the front foot (Jackson 1961). The belly may leave a drag mark in heavy snow. Although tracks in the snow may suggest otherwise, wolverines are digitigrade (Seton 1929).

Skull. Wolverine skulls are massive in comparison to other mustelids. In profile, the skull is convex, sloping down anteriorly and posteriorly from a high point just behind the orbits. The zygomatic arches are heavy, laminar, and horizontal with only a slight arch. The arches diverge from anterior to posterior. The relatively small infraorbital foramina are situated above the anterior border of P^4. The paraoccipital and mastoid are heavily developed adjacent to the auditory bullae. The bullae are inflated only on the medial side, with a tabular meatus extending laterally. They diverge somewhat posteriorly, resulting in a wedge-shaped basioccipital. The palate is roughly triangular, with straight sides and a moderately U-shaped posterior emargination. The pterygoids are broad basally, but extend in long, narrow, hamular processes. The lambdoidal crests are well developed and terminate in prominent mastoids. The sagittal crest develops with age and diverges anteriorly to form the supraorbital projections. The braincase is trapezoidal, more like the badger (*Taxidea taxus*) than the oval shape of *Mustela* (figure 32.2).

Dentition. The dental formula is 3/3, 1/1, 4/4 1/2 = 38. The upper molar is displaced medially from the toothrow. It is the normal mustelid "dumbbell" shape, with a medial constriction separating lingual and labial cusps. The premolars form a size-graded series from the large carnassial, P^4, down to the small peglike P^1. The canines are broad at the base, somewhat blunt, and canted slightly forward. The lateral pair of incisors are the largest, and occlude laterally with the lower canines. The remaining four incisors form an even row across the front, and each has a pair of divaricating ridges on the posterior surface.

The lower middle pair of incisors are displaced posteriorly from the inner, and slightly larger, outer pair, although the displacement makes the teeth appear smaller when viewed from the front. This arrangement allows added bulk, and presumably strength, with no increase in space. The lower canines are shorter and more strongly curved than the uppers. They occlude with the outer upper incisors. The first lower premolar is small, medially displaced from the toothrow, and sits against the canine. The next three premolars are size graded, with P_4 the largest. The M_1, which with P^4 forms the carnassial pair, is the largest member of the mandibular toothrow. The last lower molar (M_2) is small and subcircular.

Baculum. The baculum of adult males is 80–90 mm long and round or oval in shape, with an enlarged base and a dorsally curved distal end. The tip is expanded into three prongs (Wright and Rausch 1955).

PHYSIOLOGY

In a study of metabolic rate changes during growth, Iversen (1972) found that the basal metabolic rate increased in proportion to the 1.41 power of body weight ($W^{1.41}$) during the first 2.5 months of life. Most other mammals show an increase proportional to simple body weight ($W^{1.0}$). The difference was attributed to more rapid development of high energy-producing tissues (liver, heart, brain. and kidney) in wolverines. After 2.5 months of age (3 kg body weight), o̶̶̶gen consumption increased almost linearly with the ̶.64 power of body weight ($W^{0.64}$). The timing of the̶̶̶eak was assumed to coincide with weaning, as it does in most other mammals. This in turn may reflect a change in food composition resulting in increased carbohydrate intake. There may also be a relationship with improved thermoregulatory ability as the mother leaves the young for longer periods after weaning.

GENETICS

The diploid number is 42. The autosomes consist of 24 metacentrics and submetacentrics, and 16 acrocentrics. The X chromosome is metacentric and the Y is ac-

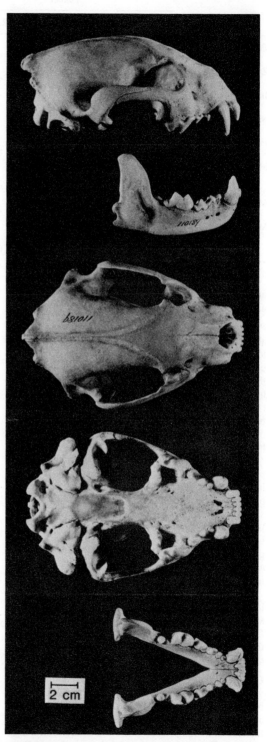

FIGURE 32.2. Skull of the wolverine (*Gulo gulo*). From top to bottom: lateral view of cranium, lateral view of mandible, dorsal view of cranium, ventral view of cranium, dorsal view of mandible.

rocentric (Hsu and Benirschke 1970). For additional information, see Fredga (1967) and Wurster and Benirschke (1968).

REPRODUCTION

Rausch and Pearson (1972) provided a summary of reproductive information. Male wolverines are not in breeding condition in late fall and early winter; spermatogenesis begins in late winter and continues through early spring. Yearling males may be sexually mature, although Liskop et al. (in press) found that males over three years old in British Columbia were sexually mature. Some females mature at 12–15 months and produce their first litter when two years old. Breeding occurs in the summer and unimplanted blastocysts are known from October through January. Implantation may occur as early as December or as late as March. Parturition occurs as early as January and as late as April, but most young are born in February or March.

There seems to be only two published records of the length of the gestation period, both from captive animals. A female in the Copenhagen Zoo gave birth following 215 days of gestation and another in the Dakota Zoo gave birth after 272 days (Mehrer 1976). The period of active gestation is only 30–40 days.

Wolverines are probably polygamous. Corpora lutea counts, placental scars, and analysis of uterine contents for fetuses all suggested a litter size averaging 3.5 in Alaska (Rausch and Pearson 1972). The range of *in utero* litters was 1–6. In Lapland, litter size averaged 2.5 for 161 litters in dens (Pulliainen 1968) while in British Columbia the mean litter size was 2.6 based on five reproductive tracts (Liskop et al. in press). According to Ewer (1973), wolverines breed no more frequently than every other year, possibly owing to a lengthy period of postweaning maternal care. However, Liskop et al. (in press) found that most mature females were reproductively active. The females have four abdominal and four inguinal mammae (Jackson 1961).

The placenta was described in detail by Wislocki and Amoroso (1956:98–99):

It is a zonary placenta with a large central hematoma which has its closest affinities to the placentas of other mustelids, particularly the martin [sic], otter and badger. The placental labyrinth is characterized by greatly hypertrophied endothelial cells which line the large maternal capillaries, and by the prominence of a basement membrane which completes the wall of the maternal vessels. In two other mustelids, namely the ferret and *Zorilla*, the maternal vessels are similarly constructed.

Pulliainen (1968) described 31 dens found by hunters in Lapland. Most were on the fells, especially in ravines, although 6 were found on spruce and pine peat bogs. The dens are dug beneath the snow for distances up to 40 m. The young are normally in a shallow, unlined pit on the ground. In Siberia, they also den in caves, among boulders and tree roots, and in broken wood and accumulations of dry twigs (Stroganov 1969). Additional accounts of dens in rocks may be found in Youngman (1975) and Harper (1956).

Sex Ratios. Of 37 kits taken from dens in Finland, Pulliainen (1968) found 15 males and 22 females. Rausch and Pearson (1972) found 6 males and 8 females of 14 fetuses from Alaska. These figures are not significantly different from a 1:1 ratio.

Captures tend to be biased toward males because of the males' tendency to roam further than females. Rausch and Pearson (1972) listed a male:female ratio of 1.64:1 for Alaskan catch statistics of 554 animals. They also found that for kits the ratio was 1.28:1 and for animals above age class 5 the ratio was 1:1. This means that intermediate-aged animals were predominantly males.

Growth and Development. Three kits born in captivity were fully covered with white fur, and had unopened eyes and unerupted teeth. They averaged 84 g in weight, 131 cm in crown-rump length, 29 mm in tail length, 22 mm in hindfoot length, and 4 mm in ear length (Mehrer 1976). Young animals grow rapidly, and leave the dens in April and May. The young are transported by the mother by the usual carnivore method of carrying them by the scruff of the neck (Ewer 1973). They are weaned beginning at about 7–8 weeks (Myhre and Myrberget 1975). Adult size is often reached by early winter. The baculum undergoes a growth spurt at about 1 year of age, and some males begin producing sperm at 14–15 months of age (Rausch and Pearson 1972).

Woods (1944) surveyed zoos for records on captive animals, and reported several individuals that had lived for 15 years. According to Jackson (1961), longevity in the wild is 8–10 years, with a possible maximum of 18.

ECOLOGY

Habitat. Wolverines are primarily found in boreal forests, but there are many reports from open tundra areas in the far north. They are particularly fond of marshy areas, and are most at home in regions with snow on the ground during winter. The snow makes it easier for them to obtain large prey. They are found at a wide variety of elevations, and in the southern parts of their range are limited to montane areas. In California they are found in Douglas fir, mixed conifer, and lodgepole pine forest types (Schempf and White 1977). These authors listed 1,300 ft (400 m) as the lowest elevational record for wolverines in California.

Activity. The wolverine is primarily nocturnal, but may be active during daylight as well. Krott (1960) reported a continuous activity cycle of alternate 3–4 h periods of activity and sleep. This cycle may be disrupted by unusually inclement weather, when they tend to spend more time sleeping. Hunger, however, may lead to extended periods of activity. They are nonmigratory, and do not hibernate.

According to Jackson (1961) wolverines do not disperse far from their birth site, but adults may wander over relatively large areas. They are able to cover distances of up to 65 km without rest if pursued. Krott (1960) determined that an eight-month-old male once covered 33 km in a single night. Dixon (1938) indicated that some individuals travel regularly over the same routes. In Montana, evidence from recent range extensions suggested that wide-ranging males are the first to invade new territory (Newby and McDougal 1964).

Males appear to be territorial, excluding other males but permitting females to enter (Ewer 1973). Of 20 observations reported by Bee and Hall (1956), 17 were of solitary animals, 2 of pairs, and 1 of three animals together. Females are also mutually intolerant, but have smaller territories than males, and more than one female may occur within a male's territory (Krott 1959b). Males have territories as large as 2,000 km², and females about 400–500 km². Krott (1959b) suggested that availability of denning sites and food supply were important determinants of territory size.

FOOD HABITS

Myhre and Myrberget (1975) studied stomach contents of 121 wolverines from Norway. The most important winter food was reindeer, followed in order by moose (*Alces*), roe deer (*Capreolus*), fox (*Vulpes*), hare (*Lepus*), small rodents, birds, and plants. Other food items eaten in Norway (Myrberget 1968; Myrberget et al. 1969) included vole (*Microtus*), lemming (*Lemmus*), sheep (*Ovis*), and eggs. In Russia, they also prey heavily on reindeer (*Rangifer*) and other cervids. Ognev (1935) reported them feeding on fish, frogs, and geese along marshy rivers and lakes. Jackson (1961) suggested that they feed mainly on rabbits (*Sylvilagus*), beavers (*Castor*), squirrels (*Sciurus*), chipmunks (*Eutamias*), and mice, as well as grouse and waterfowl. In California, Grinnell et al. (1937) listed marmots (*Marmota*), carrion, gophers (*Thomomys*), and mice as most important. Rausch (1959) reported moose and caribou to be important prey items in Alaska. Rausch and Pearson (1972) added walrus (*Odobenus*), seals, and whales as important carrion items in the diet. They also pointed out that berries were eaten in the summer. In Sweden, Krott (1959b, 1960) found that in addition to the wide variety of previously mentioned small animals, wolverines fed extensively on wasp larvae and berries of all types in the summer.

Predatory Behavior. Although much of the literature documents the importance of large animals in the diet of the wolverine, these accounts often do not differentiate between prey items actually killed by the wolverine and those scavenged as carrion. It is likely that a considerable percentage of the larger-sized animals are taken as carrion. It is apparently not uncommon to find wolverine footprints accompanying those of other carnivores such as bears, foxes, and wolves, suggesting that they may feed extensively on kills made by other animals (Ognev 1935).

Nevertheless, wolverines are capable of killing large animals, especially when deep snow hinders the prey. Burkholder (1962) observed a wolverine attack-

ing a medium-sized male caribou in Alaska. The wolverine repeatedly attacked the caribou from the rear, and once from the front. It held on even when swung from the ground by the caribou's attempts to escape. In attacking large animals, wolverines often leap on the animal's back and kill it by biting the neck (Ewer 1973). Smaller prey may be killed by a blow from the forepaw, but a characteristic head bite always follows (Krott 1959a).

Ognev (1935) cited stories of as many as 20 foxes and 100 ptarmigan in wolverine caches stored under the ice and snow. Krott (1960) described this habit of burying food caches in some detail. In addition, they place large pieces of carrion in the forks of tree branches. All caches are marked with scent or urine or both.

Drinking. Although wolverines drink by lapping, as do all carnivores, they make treading movements with the forepaws as they lap (Ewer 1973). This movement was interpreted by Krott (1959b) as a response to drinking in marshy areas where the treading may push down vegetation and cause the water to puddle. The movement is even made by captive animals, often resulting in difficulty in drinking from a dish without upsetting it. Ewer (1973) suggested that the movement might be derived from the infantile milk tread, which stimulates milk flow from the mother's mammary glands.

MORTALITY

Predators. Apparently, people are the only important enemy of the wolverine. Grinnell et al. (1937) provided a lengthy account of a wolverine killed by the quills of a porcupine that it had eaten. Burkholder (1962) found the body of a wolverine near a caribou carcass. Tracks in the area showed that the wolverine was probably killed by a pack of wolves. Boles (1977) reported two other instances of wolverines killed by wolves; one wolverine was unable to escape because it was caught in a trap. In none of the reports was the wolverine eaten.

Parasites. Erickson (1946) listed flukes (*Opisthorchis felineus*), tapeworms (*Bothriocephalus* sp.; *Taenia twitchelli*), and roundworms (*Dioctophyme renale; Soboliphyme baturini*) from wolverines. Rausch (1959) found helminths in 86 percent of 80 wolverines examined from Alaska. He listed seven species, four of which (*Alaria* sp., *Mesocestoides kirbyi*, *Ascaris devosi*, and *Physaloptera torquata*) were recorded for the first time from wolverines. A fifth, *Molineus patens*, was recorded for the first time from a North American wolverine. The remaining two were *Taenia twitchelli* and *Trichenella spiralis*. Williams and Dade (1976) recorded the heartworm *Dirofilaria immitis* from a captive specimen.

AGE DETERMINATION

Rausch and Pearson (1972) presented data on aging wolverines by a variety of methods including tooth cementum deposition, eye lens weight, ossification of long bones and skull sutures, and weights of ossa bacula. They found that only cementum deposition provided reliable estimates beyond one year of age. They aged 779 wolverines, placing them in 14 age categories that roughly corresponded to annual increments. Unfortunately, the annuli were not always completely legible, causing some difficulty. Rausch and Pearson obtained the most consistent materials for animals less than 28 months old. Class 0 canines (0–15 months) have an open root, thin dentine layer, and relatively little cementum. Class 1 canines (16–28 months) have a closed root, increased dentine layers, a large pulp cavity foramen, and a distinct dark layer in the cementum between two clear areas.

ECONOMIC STATUS

Wolverine fur is reputed to be valuable as trim for parkas because of its moisture resistance. Quick (1952) pointed out that frost does form on wolverine fur, but unlike wolf or coyote fur, it can be brushed off easily. He believed that the fur was valued for its beauty and rarity as much as for its function.

Rausch and Pearson (1972) summarized records from Alaskan fur dealers dating back to 1978. These records show high harvests during 1918–20, 1927–29, 1947–50, and 1964–68. Pelts apparently are now in demand especially for garment trim. Table 32.2 summarizes recent harvest records from Alaska and Montana.

Although wolverines feed on reindeer (see above), their negative impact on livestock is probably slight. They are an economic nuisance to fur trappers because of their well-documented tendencies to rob traplines. They are also well known for their ability to destroy food caches of field workers. In Stefansson's (1929:524–525) book on life with the Eskimos, Dr. R. M. Anderson provided a good illustration of their strength and perseverance:

> Hardly any kind of caches can be made strong enough to keep out a Wolverine if he has plenty of time to work undisturbed; for the animal is strong enough to roll away heavy stones and logs, gnaw through timbers, climb to elevated caches, and excavate buried goods. The most nearly Wolverine-proof cache I have seen was constructed by an Indian near Great Bear Lake. It was constructed by finding four trees in suitable position to form upright posts at the cor-

ners of the cache and cutting them off ten or twelve feet from the ground. The posts were notched on the inner sides to support horizontal beams, and logs laid across to form a floor, projecting two or three feet beyond each end. When filled up, the cache was roofed with a layer of heavy green logs three or four deep, too heavy for a Wolverine to move and too deep to gnaw through if he succeeded in getting on top. The uprights are stripped of bark and made as smooth as possible. If a Wolverine succeeds in climbing the upright posts, the projecting ends of the floor timbers prevent him from getting around to the top of the cache. Having no foothold, he cannot work at the bottom or sides of the cache, and consequently one thickness of timber suffices for these.

Any animal that causes people to go to such lengths to intrude on his territory deserves our respect and protection rather than the bounties and destruction they have faced in the past. Fortunately, wolverines now enjoy some form of protection in most states, and it is hoped that their numbers will continue to increase.

The status of the wolverine in North America is summarized below.

California. Jones (1950, 1955) recorded 26 recent occurrences of wolverines in California, most from the northern Sierra Nevada. Cunningham (1959) summarized observations from Yosemite National Park. Ruth (1954) reported on an observation from Squaw Valley. Nowak (1973) noted apparent range expansions into Shasta and Trinity counties and from near Gasquet in the northwestern part of the state. Yocom (1973) listed a dozen recent records from the Siskiyou, Klamath, Salmon, Trinity, Scott, and Scott Bar mountains. The current range seems to be from Del Norte and Trinity counties eastward through Siskiyou and Shasta counties, and southward through the Sierra Nevada to Tulare County (Schempf and White 1977). Numbers appear to be increasing slightly in California.

Colorado. Armstrong (1972) listed numerous old records from the western half of the state, but was able to locate only a single specimen, from Clear Creek County. He stated that wolverines were rare or possibly extirpated from the state. However, Nowak (1973) reported a kill from a few miles south of Denver in 1965, and mentioned other recent sight records. A variety of sight records was summarized by Field and Feltner (1974). Wolverines were almost extinct in Colorado earlier in the century, but perhaps their numbers are slowly increasing today.

Idaho. Early numbers were summarized by Davis (1939), who believed that wolverines were extinct or at best limited to the more inaccessible mountains in the central part of the state. Pengelly (1951) reported the first known specimen, an adult male trapped near Clark's Fort and Johnson Peak in Bonner County. He also reported several sight records from the state. Nowak (1973) said several kills had been made in the central mountains of Idaho. A wolverine was captured in 1964 in Flat Creek. Wolverines were never common in the state, and they do not appear to be increasing.

TABLE 32.2. Recent harvest records of wolverines from Alaska and Montana

Year	Alaska	Montana	Average Price ($)
1971–1972	548	23	90.00
1972–1973	946	31	100.00
1973–1974	1037	15	125.00
1974–1975	805		150.00
1975–1976	984	40	100.00
1976–1977	939	58	182.00

SOURCE: Deems and Pursley 1978

Indiana. Lyon (1936) listed early records from Noble and Knox counties, but no recent records are known. The wolverine is almost certainly extirpated in Indiana.

Iowa. Haugen (1961) listed a record from Tama County, thought to be accidental. Bowles (1975) suggested it was more likely natural. The scarcity of records from states even to the north of Iowa suggests that wolverines are not likely to occur in the state today.

Maine. There are no recent records, but Palmer (1937) mentioned unverified early reports. Although wolverines probably did occur in Maine at one time, they are extirpated there now.

Michigan. Although Burt (1946) mapped early unverified records, there are no confirmed records of wolverines from Michigan, the Wolverine State.

Minnesota. Birney (1974) summarized twentieth-century records from Minnesota, and showed that the last unquestionable record was from Itaska County in 1899. Wolverines probably do not occur in the state today, although occasional wanderers from Canada might be encountered.

Montana. Although apparently near extirpation in the state by 1920, wolverines made a significant comeback along the continental divide between 1940 and 1953 (Newby and Wright 1955). Newby and McDougal (1964) reported a continued increase and expansion into southwestern Montana during the next decade. They suggested that reestablishment occurred through dispersal from Canada and Glacier National Park, and subsequent breeding. Nowak (1973) reported more than 200 animals taken by fur trappers during the 1960s. Montana has the highest wolverine population of any of the contiguous United States today.

Nebraska. Jones (1964) reported two definite records: near Gering, Scotts Bluff County, and near Chimney Rock (the former a specimen, the latter an observation). There is a skeleton in the National Museum of Natural History (no. 21493) labeled "Nebraska," but the accession records show only that it was purchased from Professor H. A. Ward in 1884. It is highly unlikely that wolverines occur in the state today.

Nevada. The only published record from Nevada is that of a cranium found in a cave 11.3 km S Baker, White Pine County, as reported by Barker and Best (1976). Despite this recent finding, it seems doubtful that wolverines occur in the state at present.

New Hampshire. Jackson (1922) reported a pair from the Diamond region east of the Connecticut Lakes. As elsewhere in New England, wolverines are extirpated in the state today.

New Mexico. Although Bailey (1932) included wolverines in his fauna of the state based on Coues's (1877) remarks, they were not included in the most recent faunal list (Findley et al. 1975) due to an absence of specimens or verified observations. Seton (1931) reported an Indian legend that he believed indicated that wolverines had formerly occurred in the Sangre de Cristo mountains. Thus, it is uncertain whether wolverines ever occurred in the state, and they do not today.

New York. Merriam (1882) reported one from Rensselaer County, apparently based on an early Audubon record. Wolverines no longer occur in the state.

North Dakota. Bailey (1926) summarized records, all of which seem to be based on hearsay from early trappers. They were probably never common in the state and do not occur there now.

Ohio. On 8 December 1943, a wolverine was shot east of Beaver Lake, near the Mahoning-Columbiana County line. The animal was mounted and put on display at the Mill Creek Park Museum. If accurate, this is the only record for the state of Ohio (Anonymous 1944, not seen; from Natl. Fish Wildl. Lab. files). It seems unlikely that wolverines occur in Ohio today.

Oregon. Bailey (1936) reported wolverines to be rare in Oregon, and listed records for Mount Hood and the McKenzie Valley. Olterman and Verts (1972) and Nowak (1973) listed recent records from near Three Fingered Jack Mountain in the north-central part of the state. An individual was trapped on Steen's mountain in Harney County in 1974 (G. Feldhamer personal communication). Although still uncommon, wolverines may be increasing in numbers in the Cascades, as they apparently are in Washington and California.

Pennsylvania. Allen (1942) listed an early record from near Great Salt Lick in Portage Township. As with other eastern states, wolverines probably no longer occur in Pennsylvania.

South Dakota. Turner (1974) reported a record from near Timber Lake, in Dewey County, but listed wolverines as being of uncertain status now. It is unlikely that any remain in the state.

Utah. Durrant (1952) summarized records. He said that wolverines were never common in Utah and were now possibly extirpated. A lack of recent sightings supports his supposition.

Vermont. Although wolverines no longer occur in the state, Bachman apparently saw three skins from Vermont in about 1811 (Audubon and Bachman 1849).

Washington. Dalquest (1948) mapped records from the Cascade Mountains. Nowak (1973) reported six kills from the 1960s and other sightings. All twentieth-century records were summarized by Johnson (1977). His data showed an increase in numbers during 1960–75. Thus, wolverines appear to be increasing in numbers in Washington, as they are in other northwestern states.

Wisconsin. Jackson (1954, 1961) recorded only a single specimen from the state, from Bogie's Cave, near Gotham. He listed numerous "accurate" records, and stated that wolverines probably occurred throughout the state before 1870. They apparently no longer occur in Wisconsin.

Wyoming. Long (1965) summarized recent sight records, mainly restricted to mountains in the northwestern part of the state. The only specimen is a skull from Yellowstone National Park in the National Museum of Natural History. Nowak (1973) reported a kill from near Boulder in west-central Wyoming. The increase in numbers seen in Montana apparently has not yet extended southward into Wyoming.

CURRENT RESEARCH AND MANAGEMENT NEEDS

Most of our knowledge of wolverines is based on observations by trappers and other interested amateur naturalists. Field studies on solitary carnivores are quite difficult, and we still have much to learn about wolverines. Ecological and behavioral data based on studies of natural populations are sorely needed. Population structure, demography, and density in various habitat types will have to be elucidated before good management plans can be formulated.

It is axiomatic that a management plan directed toward wolverines should consider the possible effects on other species, such as predators, competitors, and prey items. Assuming that the goal of such a plan is to maintain some optimum population size of wolverines, it is possible that the best way to manage this species is to do nothing. Wolverines would probably flourish if sufficiently large tracts of proper habitat were set aside and protected.

Unfortunately, it is not always possible to provide totally protected refuges in areas where it may be possible to maintain wolverine populations. Protection from human predation should be provided in all areas of the contiguous United States. The development and maintenance of harvest regulations in those parts of Alaska and Canada where sizable wolverine populations remain should help to ensure their survival.

LITERATURE CITED

Allen, G. M. 1942. Extinct and vanishing mammals of the western hemisphere with the marine species of all the oceans. Am. Com. Int. Wild Life Protection, Spec. Publ. 11. 620pp.

Anonymous. 1944. Ohio Conserv. Bull. 8:10.

Armstrong, D. M. 1972. Distribution of mammals in Colorado. Univ. Kansas Mus. Nat. Hist. Monogr. 3. 415pp.

Audubon, J. J., and Bachman, J. 1849. The quadrupeds of North America. G. R. Lockwood, New York. Vol. 1. 383pp.

Bailey, V. 1926. A biological survey of North Dakota. North Am. Fauna 49. 226pp.

———. 1932. Mammals of New Mexico. North Am. Fauna 53. 412pp.

———. 1936. The mammals and life zones of Oregon. North Am. Fauna 55. 416pp.

Barker, M. S., Jr., and Best, T. L. 1976. The wolverine (*Gulo luscus*) in Nevada. Southwestern Nat. 21:133.

Bee, J. W., and Hall, E. R. 1956. Mammals of northern Alaska on the Arctic Slope. Univ. Kansas Mus. Nat. Hist. Misc. Publ. 8. 309pp.

Birney, E. C. 1974. Twentieth century records of wolverine in Minnesota. Loon 46:78–81.

Boles, B. K. 1977. Predation of wolves on wolverines. Can. Field Nat. 91:68–69.

Bowles, J. B. 1975. Distribution and biogeography of mammals of Iowa. Texas Tech. Univ. Mus. Spec. Publ. 9. 184pp.

Burkholder, B. L. 1962. Observations concerning wolverine. J. Mammal. 43:263–264.

Burt, W. H. 1946. The mammals of Michigan. Univ. Michigan Press, Ann Arbor. 288pp.

Coues, E. 1877. Fur-bearing animals: a monograph of North American Mustelidae. U.S. Geol. Surv. Terr. Misc. Publ. 8. 348pp.

Cunningham, J. D. 1959. The wolverine and fisher in the Yosemite region. J. Mammal. 40:614–615.

Dalquest, W. W. 1948. Mammals of Washington. Univ. Kansas Mus. Nat. Hist. Publ. 2. 444pp.

Davis, W. B. 1939. The recent mammals of Idaho. Caxton Printers, Ltd., Caldwell, Id. 400pp.

Deems, E. F., Jr., and Pursley, D. 1978. North American furbearers. Int. Assoc. Fish Wildl. Agencies. 171pp.

Degerbol, M. 1935. Report of the mammals collected by the fifth Thule expedition to Arctic North America. Part 1: Systematic Notes. Vol. 2, no. 4. Gyldendal, Copenhagen. 67pp.

Dixon, J. S. 1938. Birds and mammals of Mount McKinley National Park, Alaska. Fauna Natl. Parks U.S. 3. 236pp.

Durrant, S. D. 1952. Mammals of Utah. Univ. Kansas Mus. Nat. Hist. Publ. 6. 549pp.

Erickson, A. B. 1946. Incidence of worm parasites in Minnesota Mustelidae and host lists and keys to North American species. Am. Midl. Nat. 36:494–509.

Ewer, R. F. 1973. The carnivores. Cornell Univ. Press, Ithaca, N.Y. 494pp.

Field, R. J., and Feltner, G. 1974. Wolverine. Colorado Outdoors 23:1–6.

Findley, J. S.; Harris, A. H.; Wilson, D. E.; and Jones, C. 1975. Mammals of New Mexico. Univ. New Mexico Press, Albuquerque. 360pp.

Fredga, K. 1967. Comparative chromosome studies of the family Mustelidae (Carnivora, Mammalia). Hereditas 57:295.

Grinnell, J.; Dixon, J. S.; and Linsdale, J. M. 1937. Fur-bearing mammals of California. Univ. California Press, Berkeley. Vol. 1. 375pp.

Hall, E. R. 1926. The abdominal skin gland of Martes. J. Mammal. 7:227–229.

Hall, E. R., and Kelson, K. R. 1959. The mammals of North America. Ronald Press Co., New York. 2:547–1083.

Harper, F. 1956. The mammals of Keewatin. Univ. Kansas Mus. Nat. Hist. Misc. Publ. 12. 94pp.

Haugen A. O. 1961. Wolverine in Iowa. J. Mammal. 42:546–547.

Hsu, T. C., and Benirschke, K. 1970. An atlas of mammalian chromosomes. Folio 184. Springer-Verlag, New York. Vol. 4, folios 151–200.

Iversen, J. A. 1972. Basal metabolic rate of wolverines during growth. Norwegian J. Zool. 20:317–322.

Jackson, C. F. 1922. Notes on New Hampshire mammals. J. Mammal. 3:13–15.

Jackson, H. H. T. 1954. Wolverine (*Gulo luscus*) specimens from Wisconsin. J. Mammal. 35:245.

———. 1961. Mammals of Wisconsin. Univ. Wisconsin Press, Madison. 504pp.

Johnson, R. E. 1977. An historical analysis of wolverine abundance and distribution in Washington. Murrelet 58:13–16.

Jones, F. L. 1950. Recent records of the wolverine (*Gulo luscus luteus*) in California. California Fish Game 36:320–322.

———. 1955. Records of southern wolverine, *Gulo luscus luteus,* in California. J. Mammal. 36:569.

Jones, J. K., Jr. 1964. Distribution and taxonomy of mammals of Nebraska. Univ. Kansas Mus. Nat. Hist. Publ. 16. 356pp.

Krott, P. 1959*a*. Der Vielfrass (*Gulo gulo* L. 1758). Monogr. Wildsauget. 13. 159pp.

———. 1959*b*. Demon of the north. Knopf, New York. 260pp.

———. 1960. Ways of the wolverine. Nat. Hist. 69:16–29.

Kurten, B., and Rausch, R. 1959. A comparison between Alaskan and Fennoscandian wolverine (*Gulo gulo* Linnaeus). Acta Arctica, Fasc. 11:5–20.

Liskop, K. S.; Sadleir, R. M. F. S.; and Saunder, B. P. In press. Reproduction and harvest of wolverine (*Gulo gulo*) in British Columbia. Proc. Worldwide Furbearer Conf. J. A. Chapman and D. Pursley, ed.

Long, C. A. 1965. The mammals of Wyoming. Univ. Kansas Mus. Nat. Hist. Publ. 14:493–758.

Lyon, M. W. 1936. Mammals of Indiana. Am. Midl. Nat. 17. 384pp.

Mehrer, C. F. 1976. Gestation period in the wolverine, *Gulo gulo*. J. Mammal. 57:570.

Merriam, C. H. 1882. The vertebrates of the Adirondack region, northeastern New York. Trans. Linnaean Soc., New York. 106pp.

Myhre, R., and Myrberget, S. 1975. Diet of wolverines (*Gulo gulo*) in Norway. J. Mammal. 56:752–757.

Myrberget, S. 1968. The breeding den of the wolverine, *Gulo gulo*. Fauna 21:108–115. (In Norwegian; English summary.)

Myrberget, S.; Groven, B.; and Myhre, R. 1969. Tracking wolverine, *Gulo gulo,* in the Jotunheim Mountains, south Norway, 1965–1968. Fauna 22:237–252. (In Norwegian; English summary.)

Newby, F. E., and McDougal, J. J. 1964. Range extension of the wolverine in Montana. J. Mammal. 45:485–487.

Newby, F. E., and Wright, P. L. 1955. Distribution and status of the wolverine in Montana. J. Mammal. 36:248–253.

Nowak, R. M. 1973. Return of the wolverine. Natl. Parks Conserv. 47:20–23.

Ognev, S. I. 1935. Mammals of U.S.S.R. and adjacent countries. Vol. 3: Carnivora (Fissipedia and Finnepedia). Transl. from Russian by Israel Program for Sci. Transl., Jerusalem, Israel, 1962. 641pp.

Olterman, J. H., and Verts, B. J. 1972. Endangered plants and animals of Oregon. Vol. 4: Mammals. Oregon State Univ. Agric. Exp. Stn. Spec. Rep. 364. 47pp.

Palmer, R. S. 1937. Mammals of Maine. B.A. Thesis. Univ. Maine, Orono, 181pp.

Pengelly, W. L. 1951. Recent records of wolverines in Idaho. J. Mammal. 32:224–225.

Pulliainen, E. 1968. Breeding biology of the wolverine (Gulo gulo L.) in Finland. Ann. Zool. Fennici 5:338–344.

Pulliainen, E., and Ovaskainen, P. 1975. Territory marking by a wolverine (*Gulo gulo*) in northeastern Lapland. Ann. Zool. Fennici 12:268–270.

Quick, H. F. 1952. Some characteristics of wolverine fur. J. Mammal. 33:492–493.

Rausch, R. 1959. Studies on the helminth fauna of Alaska. Vol. 36, Parasites of the wolverine, *Gulo gulo* L., with observations on the biology of *Taenia twitchelli* Schwartz, 1924. J. Parasitol. 45:465–484.

Rausch, R. A., and Pearson, A. M. 1972. Notes on the wolverine in Alaska and the Yukon Territory. J. Wildl. Manage. 36:249–268.

Ruth, F. S. 1954. Wolverine seen in Squaw Valley, California. J. Mammal. 35:594–595.

Schempf, P. F., and White, M. 1977. Status of six furbearer populations in the mountains of Northern California. U.S.D.A., For. Serv., California Region. 51pp.

Seton, E. T. 1929. Lives of game animals. Doubleday, Doran & Co., Inc., Garden City, N.Y. Vol. 2, pt. 2. 746pp.

———. 1931. Two records for New Mexico. J. Mammal. 12:166.

Stefansson, V. 1929. My life with the Eskimo. Macmillan, New York. 538pp.

Stroganov, S. U. 1969. Carnivorous mammals of Siberia. Israel Program for Sci. Transl., Jerusalem, Israel. 522pp.

Turner, R. W. 1974. Mammals of the Black Hills of South Dakota and Wyoming. Univ. Kansas Mus. Nat. Hist. Misc. Publ. 60. 178pp.

van Zyll de Jong, C. G. 1975. The distribution and abundance of the wolverine (*Gulo gulo*) in Canada. Can. Field Nat. 89:431–437.

Walker, E. P., et al. 1975. Mammals of the world, ed. J. L. Paradiso. 3rd ed. Johns Hopkins Univ. Press, Baltimore. 2:645–1500.

Williams, J. F., and Dade, A. W. 1976. *Dirofilaria immitis* infection in a wolverine. J. Parasitol. 62:174–175.

Wislocki, G. B., and Amoroso, E. C. 1956. The placenta of the wolverine (*Gulo gulo luscus* Linnaeus). Bull. Mus. Comp. Zool. Harvard 114:91–100.

Woods, G. T. 1944. Longevity of captive wolverines. Am. Midl. Nat. 31:505.

Wright, P. L., and Rausch, R. 1955. Reproduction in the wolverine, *Gulo gulo*. J. Mammal. 36:346–355.

Wurster, D. H., and Benirschke, K. 1968. Comparative cytogenetic studies in the order *Carnivora*. Chromosoma 24:336–382.

Yocom, C. F. 1973. Wolverine records in the Pacific Coastal states and new records for northern California. California Fish Game 59:207–209.

Youngman, P. M. 1975. Mammals of the Yukon Territory. Natl. Mus. Nat. Sci. Canada Publ. Zool. 10. 192pp.

Don E. Wilson, U.S. Fish and Wildlife Service, National Fish and Wildlife Laboratory, National Museum of Natural History, Washington, D.C. 20560.

33

Badger
Taxidea taxus

Frederick G. Lindzey

NOMENCLATURE

COMMON NAMES. Badger, North American badger
SCIENTIFIC NAME. *Taxidea taxus*
SUBSPECIES. *T. t. taxus, T. t. jeffersonii, T. t. jacksoni,* and *T. t. berlandieri.*

DISTRIBUTION

The current range of the badger extends from the northern part of Alberta, Canada, to central Mexico and eastward from the Pacific coast to a line running roughly from east Texas to the central lake states (figure 33.1). Recent records suggest, however, an eastward extension of their range (Nugent and Choate 1970). Badgers are generally associated with treeless regions, prairies, parklands, and cold desert areas. Altitudinally, their range extends from below sea level to over 3,600 m. The Rocky Mountains and Grand Canyon are geographic features associated with the distribution of western subspecies (Long 1973).

FIGURE 33.1. Distribution of the badger (*Taxidea taxus*).

DESCRIPTION

Many unique physical characteristics of the badger, which adapt it for preying on fossorial rodents, make it readily identifiable. The body of the badger is depressed, with short stout legs having long recurved foreclaws and short shovellike hind claws (figure 33.2). The external auditory meatus is small and rounded; eyes are small with a nictitating membrane present (Long 1973). Badgers have eight mammae: two inguinal, two abdominal, and four pectoral. The skin of the badger is loose, particularly across the front, shoulders, and back.

Size and Weight. Body measurements reported by Long (1973) were: total length, 60–73 cm; tail length, 10.5–13.5 cm; hind foot length, 9.5–12.8 cm for sexes combined. Similar measurements recorded by Messick (1981) in southwestern Idaho for male and female badgers were: total length, 73.9 cm and 70.8 cm; body length, 59.9 cm and 57.8 cm; hind foot length, 10.7 cm and 10.3 cm.

Adult males weigh an average of 26 percent more than adult females (Wright 1966). Average weights for adult male and female badgers were 8.4 and 6.4 kg, respectively, in South Dakota (Wright 1969) and 8.7 and 7.1 kg, respectively, in northern Utah and southern Idaho (Lindzey 1971). Young badgers grow rapidly until July, and they generally enter their first fall close to adult size. Body weight is least during the winter and early spring. Large males may exceed 11.5 kg.

Pelage. Color may vary from yellowish brown to silver gray on the dorsal surface and from a light cream to buff ventrally. The feet are black to dark brown. The sides of the face are white with a black triangular patch anterior to the ears. A white medial stripe extends dorsally from the nose. Both the width and continuity of this stripe are variable, but in *T. t. berlandieri* the stripe continues for the length of the body and terminates at the base of the tail (Long 1973). The amount of guard hairs present is variable and apparently results in the local terms *hair* and *fur* badger.

653

FIGURE 33.2. Male badger (*Taxidea taxus*), seven months old, in southern Idaho. Note the facial characteristics and markings, the claws on the front feet, and the excavated soil.

Skull and Dentition. The skull is wedge shaped, being broader posterior (figure 33.3). Auditory and mastoid bullae are large. Skulls vary from 11.3 to 14.1 cm in length (Long 1973). Even skulls of old males often do not have a prominant sagittal crest (Wright 1969). The adult dental formula is 3/3, 1/1, 3/3, 1/2 = 34. The formula for deciduous dentition is 1/0 or 2/0, 1/1, 3/3, with no deciduous molars and a total of 18 to 20 deciduous teeth (Long 1973). The triangular upper molar in skulls of adults is an important diagnostic feature for this species. Long (1965) provided a detailed discussion of jaw articulation and wear on canines. Long and Long (1965) discussed dental abnormalities in old badgers.

PHYSIOLOGY

On the basis of energy trials, Jense (1968) estimated that for maintenance, an 8.3-kg badger would require 82.6 g dry weight of cottontail rabbit (about 22 percent of an adult rabbit) or 72.3 g of ground squirrel per day. He noted, however, that these estimates were probably below the intake that would be required for maintenance in active badgers. He also found that maintenance requirements of juveniles were as much as 62 percent more than those of adult badgers.

Based on maintenance requirements estimated by Jense (45 kcal/kg/day), Messick (1981) determined that an 8-kg badger would need to consume 1.2–1.4 Townsend ground squirrels per day. He assumed, however, that a free-ranging badger would have energy requirements 1.8 times greater. Thus, intake would actually need to be 2.3 ground squirrels per day.

Based on observations and energy requirements determined from monitored heart rate, Lampe (1976) determined the energy costs of various activities. He estimated that it cost a badger actively searching for prey 25.2 kcal per hr^{-1} and 0.55 kcal per liter^{-1} to displace soil while digging. The cost of resting while underground was, however, only 7.2 kcal per hr^{-1}. A single pursuit lasting 34 minutes required expenditure of energy at the rate of 38.4 kcal per hr^{-1}. On the basis of a diel activity schedule of 10 hours of activity and 14 hours of rest (east-central Minnesota), Lampe estimated a daily energy requirement, exclusive of energy costs of pursuit activity, of 418 kcal per day^{-1}. Employing a digestive efficiency of 85.6 percent, which varies inversely with intake and time spent in each activity and their metabolic costs, a 6.5-kg badger was found to require a minimum of 1.7 pocket gophers per day for maintenance.

Harlow (1981) determined that a badger, under a regime of *ad libitum* food, would spend an average of 15 percent of its total metabolism on activity, 20 percent on food processing, and 65 percent on maintenance. Assimilation efficiencies of badgers fed dry dog food averaged only 66 percent. This value, however, was low in comparison to efficiencies of captive badgers fed more natural foods (Jense 1968; Lampe 1976).

To investigate the strategies used by badgers to exist during periods of food shortages, Harlow (1978) monitored the responses of badgers to 30-day fast periods. Total metabolism decreased by 54 percent during the period, of which part could be explained by the lack of energy required for food processing and a decrease in body weight. Reductions of activity metabolism by 37 percent and maintenance metabolism by 44 percent together accounted for slightly more than half of the reduction in total metabolism. Harlow concluded that these reductions were adaptive and actually increased a badger's chance of surviving periods of food deprivation. He calculated that if a badger continued to metabolize at rates observed when food was abundant, it would deplete its body stores and die at a time 26 percent earlier than a badger experiencing reduced metabolism.

Badgers also react to periods of food scarcity by increasing the efficiency with which they use ingested foods (Harlow 1981). Assimilation efficiency before a 30-day fast was 66.6 percent; after the fast period it was 74.7 percent. The amount of food consumed by badgers before and after the trial did not differ, but passage time increased from 520 to 614 minutes. Caloric content of feces collected during the trial increased after day 10 of the fast, suggesting that the stomach may have acted as a storage organ for 5 to 10 days.

Winter activity of captive badgers indicated they avoided temperatures below −15°C by remaining in the moderated microclimate of their dens (Harlow 1981). While in dens during cold periods, badgers underwent bradycardia and diurnal hypothermia. Harlow noted only an average of 24 percent depletion of fat reserves during the winter period.

REPRODUCTION

Badgers reportedly have a promiscuous breeding pattern, with breeding occurring in late July and August (Wright 1966; Messick 1981). Badgers exhibit delayed implantation. Development of the blastocysts is arrested until they are implanted in February (Hamlett 1932; Wright 1966). Parturition occurs in late March or early April, and litter size ranges between one and five. A minority of juvenile females, aged 4–5 months, may breed but males do not attain full spermatogenesis until 14 months of age (Wright 1966, 1969). Messick (1981) noted increased fecundity with age of females and Lindzey (1971) observed that 38 percent of yearling females did not ovulate; all older females did.

Male badgers were found to be aspermatic in late winter but to have fully active testes for three months during the summer, from late May or June to August (Wright 1969). Although males are capable of breeding for three months, the breeding season appears considerably shorter and is apparently determined by the timing of estrus in the females.

Anatomy of Reproductive Tracts. The reproductive tract of the female badger and active testes of the male are considerably larger in relation to body size than those of other North American mustelids (Wright 1966, 1969). Swelling of the vulva is noticeable in estrous females (Messick 1981). Ovaries are encapsulated, and the oviduct totally encircles the ovary and forms a part of the capsule. The oviduct hypertrophies after mating. The uterus is bicornuate, with horns as long as 60 mm and up to 5 mm in diameter in the early stages of pregnancy. Following implantation of the ova, corpora lutea increase to nearly twice their inactive size (Wright 1966).

Corpora albacantia persist for a relatively short time and may disappear by early June (Wright 1966; Messick 1981). Placental scars frequently used as indicators of young carried to term may be visible for over a year, but their persistence is highly variable (Wright 1966; Messick 1981).

Both the os clitoris and the baculum are present in badgers, although the os clitoris is not externally visible and requires clearing to be found. The baculum

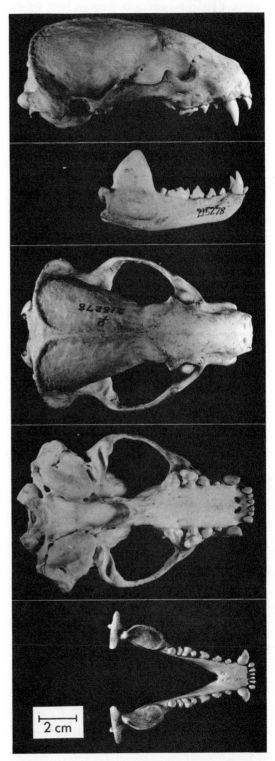

FIGURE 33.3. Skull of the badger (*Taxidea taxus*). From top to bottom: lateral view of cranium, lateral view of mandible, dorsal view of cranium, ventral view of cranium, dorsal view of mandible.

is large, and its size and ossification are of value in separating juveniles from adults (Petrides 1950; Wright 1966; Lindzey 1971).

Testes size varies with the stage of spermatogenesis. Wright (1966) found the size of the testes began to increase in late March, developed fully by late May or June, and began to regress by late August. Testes were only one-eighth as large during the inactive period as during the breeding season. Paired testes weights during the breeding season averaged 43.9 g (Wright 1966).

ECOLOGY

Density. Both Seton (1929) and Lindzey (1971) reported a minimum estimate of badger density of one per 2.6 km². However, Messick (1981) estimated a density of between two and six resident badgers per km² from intensive trapping efforts on a 5-km² area in southwestern Idaho. He noted that densities may be even greater during the period when juveniles are dispersing.

Sex and Age Structure. The sex ratio of trapped badgers was essentially 1:1 (Messick 1981). Samples from populations in southwestern Idaho and northern Utah and southeastern Idaho were composed of 31.1 and 35 percent juveniles, respectively. The oldest badgers collected by Lindzey (1971), Crowe and Strickland (1976), and Messick (1981) were 8, 13, and 14 years of age, respectively. A badger in the National Zoological Park in Washington, D.C., lived 15 years and 5 months (Jackson 1961). Both age and sex ratio should be viewed with caution until the relative vulnerability of various sex and age classes to collection methods is determined.

Mortality. Deaths of badgers in many areas appear largely caused by people, but deaths caused by other factors may easily go undetected. Larger predators, such as bears (*Ursus spp.*), coyotes (*Canis latrans*), wolves (*C. lupus*), and cougars (*Felis concolor*), occasionally kill badgers. Messick (1981) listed the major causes of badger deaths in southwestern Idaho as automobiles, farmers, indescriminate shooting, and fur trappers. Lindzey (1971) reported finding a male badger dead in a den after it had been observed consuming a portion of a horse carcass that had been treated with sodium fluoroacetate ("1080"), a poison used at the time in coyote control programs. Deaths induced by other badgers may be greatest in the juvenile age class. Two of four young badger found dead in a burrow were partially consumed, apparently by another badger (Lindzey 1971). Messick (1981) reported finding wounds on juvenile badgers that had apparently been inflicted by other badgers. Neither the frequency nor the influence of such deaths on badger populations has been determined.

Home Range. Home range size of badgers varies considerably within and between regions of their geographic distribution. Size differences undoubtedly occur because of variable habitat characteristics and prey densities, but also may result from patterns of intraspecific interactions. A radio-collared female badger monitored in Minnesota used a total area of 850 ha (Sargeant and Warner 1972). Another female monitored in the same area a number of years later, however, utilized a 1,700-ha home range (Lampe and Sovada in press). Two male and three female badgers radio-tracked in northern Utah by Lindzey (1978) occupied home ranges that averaged 583 ha and 237 ha in size, respectively Although home ranges of males were twice as large as those of females, the males were not followed during most of the breeding season and their annual home ranges may have been even larger. In southwestern Idaho, male and female badgers two years of age and older occupied home ranges of 240 and 160 ha in size (Messick 1981). One-year-old badgers in this same locale had home ranges of 60 and 80 ha, respectively, for males and females. Messick felt that juveniles (less than one year old) were dispersing during their first year of life and that the area they occupied could not accurately be termed a home range.

Home ranges of male badgers overlap those of several females and home ranges of females may overlap one another (Lindzey 1978; Messick 1981). Home ranges of adult males, however, have not been found to overlap. Transient juveniles may be found temporarily in the established home ranges of adult residents.

The locations of home ranges of adults appear to vary little from year to year (Lindzey 1978; Messick 1981). Shape of home ranges may vary from elongate to nearly circular in response to topographic features.

The amount and portion of the home range used may vary seasonally. Summer, fall, and winter ranges of a female badger in Minnesota were 725, 53, and 2 ha, respectively (Sargeant and Warner 1972). Lampe and Sovada (in press) monitored another female badger in the same area and noted a similar reduction in area used in the fall. They noted, however, that there was little overlap of the seasonally selected portions of the home range. Badgers used areas of their home range more frequently during different seasons in Idaho and Utah as well, but the seasonally emphasized areas were not exclusive of each other (Lindzey 1978; Messick 1981).

Movement of an adult female with dependent young is restricted by the need to return to the natal den. A female may, however, shift her hunting area by moving her young from one den to another (Lindzey 1971; Messick 1981; Lampe and Sovada in press). One female used over half of her 210-ha home range and three different dens while caring for her dependent young (Lindzey 1971).

Activity. Badgers are principally nocturnal, foraging at night and remaining underground during daylight hours. Although all age classes of badger are occasionally active during the day, young of the year tend to be the most diurnal (Messick 1981).

Badgers do not hibernate, but rather react to cold weather and reduced prey availability by reducing their above-ground activity. Badgers rarely remain in a den for more than a day without emerging during the warmer months, but extended habitations are common during the winter and late fall. Extended habitations were significantly more common in Utah between November and May than during the remainder of the year (Lindzey 1978). Extended habitations of 2–12 days are common, and Messick (1981) reported observing a badger that apparently did not move for 38 days. Harlow (1979) monitored the activity of penned badgers and observed periods of up to three weeks without emergence. He found that even when badgers were active during periods of cold weather, their above-ground activity coincided with the warmest periods during the night.

Den Use. Dens play a central role in the ecology of the badger, functioning as sites for diurnal inactivity, food storage, and parturition, and as foci for foraging. Dens are variable in characteristics because most are dug in pursuit of prey. Generally, they have only a single, often eliptical entrance. Soil excavated during formation of the den is piled at the entrance. When a den is occupied, particularly during cold weather, the tunnel is often partially plugged with loose soil. Scats are frequently found in the mound of soil at the entrance and in the den itself.

Old dens are common over the range of the badger and persist for varying lengths of time depending on weather and density of livestock. Lindzey (1978) found an average of 1.6 dens with open entrances per hectare in northern Utah. These dens were heavily used by badgers as diurnal resting sites, with only 15 percent of all dens used by badgers being dug immediately before their use (Lindzey 1978). Some dens were reused on numerous occasions by the same badger, which suggests a knowledge of the location of the den. Dens are used only infrequently on successive days following a night's foraging.

The den in which the female gives birth to and raises her young is structurally more complex than other dens and reflects the needs of the family group. In Utah, natal dens had the following characteristics in common: (1) a main tunnel that branched into two secondary tunnels that later rejoined; (2) dead-end side tunnels that projected from the main tunnel, secondary tunnels, and chambers; (3) pockets less than 15 cm in length in the sides of tunnels and chambers; (4) shallow excavations in the floors of tunnels; and (5) chambers. Branching of the main tunnel presumably allowed badgers to pass one another in the burrow system. Pockets and shallow excavations were filled with feces and covered with soil. Dead-end tunnels contained both soil and feces. The soil mound at the entrance to these dens generally was comparatively larger than those at the entrance to other dens, and badger hair was mixed throughout the soil, presumably because the construction of these dens coincided with the spring molt in females with young (Lindzey 1976). A natal den excavated by Lampe (personal communication) in Minnesota had similar characteristics to tbose noted above,

but it contained no scats. Although Palmer (1954) and Jackson (1961) reported finding chambers lined with grass, no nest material was observed by Lindzey (1976).

Sociality. Badgers are solitary, with the exception of breeding pairs, family groups, and occasional short-term association between siblings after family breakup (figure 33.4). Although Messick (1981) cites an incidence of badgers fighting, the absence of scars on the vast majority of individuals examined suggested that physical encounters are infrequent. The dispersion of home ranges, however, suggests that there is some interaction among members. Lindzey (1978) found that 44 percent of dens reused by badgers were located on the periphery of home ranges, perhaps indicating a respect for boundaries of home ranges of other badgers in the vicinity. Additionally, three dens were known to be used at different times by more than one badger. In each case, the den was on the periphery of the occupant's home range.

Although it is not known how badgers communicate to establish and maintain observed dispersion patterns, it seems likely that scent plays a dominant role. Badgers commonly explore each den or old dig site they encounter. In drier areas in particular, the habit of defecating and perhaps urinating in dens or at their entrance, because they are shaded and cooler, provides for extended duration of scent. Additionally, Lampe (personal communication) observed a badger rubbing its abdominal gland on the soil mound at a den entrance. Because of the central role of dens in badger behavior, they are a likely location for the accumulation and olfactory exchange of information.

The association between a female and her young lasts about 10 to 12 weeks. Young are apparently fed solid foods by the female during the 5- to 6-week lactation period (Lampe and Sovada in press). Cubs seldom come above ground on their own until 4 to 5 weeks of age. Although they are capable of moving short distances at this age, they tire easily (Lampe and Sovada in press). Before this time, the female probably carries the young if she moves to a different den. Lampe and Sovada observed a female badger that began to remain away for 30- to 32-hour periods from the den containing her young by late May. The timing of family breakup appears variable and may be characterized by a progressively looser association among family members. A family group of badgers monitored by Messick (1981) remained spatially close until 10 June. However, Hetlet (1968) captured four badgers at one den that he felt were a family group on 10 and 11 July. A radio-collared female badger monitored by Lindzey (1976) remained spatially close to her young until at least 10 June, and he did not observe independent young until 10 July.

Movement of the young after family breakup is erratic and more extensive than adults (Messick 1981). Their movements are frequently through areas devoid of other badgers in what would seem unsuitable habitat. Juvenile badgers occasionally remain near

FIGURE 33.4. Young male badgers (*Taxidea taxus*), about 2.5 months old, in southern Idaho. Note the variation in markings.

their mother's home range but most disperse. One juvenile female dispersed 52.1 km from the natal range (Messick 1981).

FOOD HABITS

The North American badger is uniquely adapted for capturing fossorial prey. However, it is flexible and readily takes advantage of local abundances of more terrestrial mammals (table 33.1) as well as avian, reptilian, and insect species. Insects, birds, and reptiles are apparently taken opportunistically (Errington 1937; Snead and Hendrickson 1942). Much of the vegetative material ingested by badgers is probably taken incidentally to the consumption of animal matter. Jense (1968), however, noted an increase in vegetable matter in the badger's diet in the fall. Corn was found in 10 of 17 stomachs of badgers collected in the fall in Iowa; four of these stomachs were completely filled with corn (Lampe personal communication).

Although ground squirrels (*Spermophilus* spp.) are the major food in many locales (Jense 1968; Snead and Hendrickson 1942; Messick 1981), smaller rodents dominate the diet in other areas (Lindzey 1971; Dearborn 1932). Composition of the diet changes seasonally and between years in locales as the phenology and abundance of prey species influence their availability to badgers (Jense 1968; Snead and Hendrickson 1942; Messick 1981). Badgers are known to eat carrion and will cache food items, frequently in old dens (Snead and Hendrickson 1942; Lindzey 1971).

Messick (1981) found that young badgers consumed more arthropods and birds and fewer mammals than adult badgers. He felt the difference in diet might

be partially accounted for by differences in movement of the relatively sedentary adults and dispersing juveniles.

Predation. Foraging patterns undoubtedly differ among regions of the badger's range, as they are adjusted to provide efficiency under varying habitat and prey regimes. Behaviors involved in the search phase of foraging appear to be more general than the often specific behaviors exhibited in the capture of certain species. Badgers tracked in the snow moved from one old den or digging site to another, thoroughly investigating each (Lindzey 1971; Messick 1981). Observations and trapping success during snowless periods suggest that this is a common pattern of movement throughout the year. The value of such a foraging pattern is implied by the frequency with which these sites are used by prey species. Hetlet (1968) noted that badgers often returned to old den sites and that of 112 dens examined, 5.4 percent were occupied by badgers, 16.0 percent by Richardsons ground squirrels (*S. richardsonii*), and 1.8 percent by least chipmunks (*Eutamias minimus*). Species such as cottontail rabbits (*Sylvilagus* spp.), which would generally be unavailable to badgers, occupy old dens and occur in the badger's diet. Snead and Hendrickson (1942) discussed the increased vulnerability to badger predation of species using badger dens in Iowa.

Knopf and Balph (1969) described the selective predation by badgers on family groups of Uinta ground squirrels (*S. armatus*). Over a 30-day period badgers excavated the burrows of seven family groups while not disturbing adjacent burrows that contained only a single squirrel. Although the disturbance caused by

TABLE 33.1. Mammalian prey of the North American badger and life style of prey

	Life Style				
Species of Prey	Burrowing	Burrow User	Terrestrial	Arboreal	Reference
Lagomorpha					
Sylvilagus floridanus		X	X		Bailey 1931
Lepus townsendii		X	X		Snead and Hendrickson 1942
L. californicus		X	X		Lindzey 1971
Rodentia					
Eutamias minimus		X	X		Lindzey 1971
E. amoenus		X	X		Broadbook 1970
Marmota monax	X		X		Snead and Hendrickson 1942
M. flaviventris	X		X		Verbeek 1965
Spermophilus townsendii	X		X		Hall 1946
S. richardsonii	X		X		Seton 1929
S. armatus	X		X		Balph 1961
S. beldingi	X		X		Grinnell et al. 1937
S. columbianus	X		X		Manville 1959
S. tridecemlineatus	X		X		Seton 1929
S. franklinii	X		X		Errington 1937
S. beecheyi	X		X		Howell 1924
Cynomys ludovicianus	X		X		Bailey 1931
Tamiasciurus hudsonicus				X	Dearborn 1932
Thomomys talpoides	X				Criddle 1930
T. umbrinus	X				Grinnell et al. 1937
Geomys bursarius	X				Bailey 1888
Pappogeomys castanops	X				Baker 1956
Perognathus parvus	X		X		Lindzey 1971
Dipodomys sp.	X		X		Lindzey 1971
Reithrodontomys megalotis		X	X		Snead and Hendrickson 1942
Peromyscus (maniculatus or leucopus)		X	X		Dearborn 1932
Neotoma sp.			X		Lindzey 1971
Microtus (pennsylvanicus or ochrogaster)			X		Bailey 1931
M. montanus			X		Lindzey 1971
Lagurus curtatus			X		Lindzey 1971
Synaptomys cooperi			X		Dearborn 1932
Mus musculus			X		Snead and Hendrickson 1942
Zapus hudsonicus			X		Jense 1968
Carnivora					
Canis latrans			X		Young 1951
Mephitis mephitis		X	X		Bailey 1929

NOTE: The earliest record of predation for each species is cited from Lampe 1976.

young around the burrow entrance gave it a unique appearance that was detectable by badgers, it seems likely that distinctive olfactory clues were present as well. Badgers captured the squirrels by plugging all entrances but one, then excavated the remaining entrance. Messick (1981) observed a similar pattern of entrance plugging when badgers preyed on Townsend ground squirrels (*S. townsendii*). Balph (1961) described underground concealment by badgers as a method of predation on Uinta ground squirrels. After plugging a second entrance, the badger apparently dug to a point close to the burrow entrance and waited for squirrels to enter the open entrance.

Lampe (1976) felt that the technique exhibited by badgers to capture pocket gophers (*Geomys bursarius*) was used to avoid digging out the entire burrow. The technique entailed the penetration of the burrow system at various points. A comparison of the olfactory clues at each point by running between points determined the approximate location of the gopher. Extensive digging was then begun to capture the gopher. From the examination of 30 sites, Lampe determined that badgers were successful at capturing pocket gophers in 73 percent of attempts. He felt that badgers employed a similar technique to capture ground squirrels.

Despite limited olfactory or visual clues, badgers prey on hibernating mammals. The technique used to capture these prey generally includes only a single excavation (Lampe 1976).

PARASITES

The badger is host to numerous endoparasites including nematodes, trematodes, and cestodes (table 33.2).

Ectoparasites include the tick (*Dermacentor variabilis*) (Ellis 1955) and fleas (*Pulex irritans, Thrassis acamantis,* and *Echidnophaga gallinacea*) (Hubbard 1947). Badgers are apparently susceptible to both tularemia and rabies. Hetlet (1968) found antibodies to the plague organism *Yersinia pestis* in five of six sera tested, and he postulated that badgers may be instru-mental in the transmission of plague among rodent colonies. Messick et al. (in press) proposed the use of seralogical testing of badgers to monitor the dynamics of plague in Townsend ground squirrel populations. Seventy-nine percent of 294 sera tested between 1975 and 1977 were positive.

TABLE 33.2. Helminth parasites from the badger in North America

Helminth	Geographic Location and Reference
Trematode	
Alaria (Paralaria) taxideae Swanson and Erickson 1946	Minnesota (Swanson and Erickson 1946; Erickson 1946).
Euparyphium melis (Schrank 1788) Dietz 1909	North Dakota (present work).
Euparyphium sp.	Minnesota (Erickson 1946).
Cestodes	
Atriotaenia (Ershovia) procyonis (Chandler 1942) Spassky 1951	North Dakota (present work). Wyoming (Keppner 1969*b*).
Mesocestoides carnivoricolus Grundmann 1956	Utah (Grundmann 1956, 1958).
Monordotaenia taxidiensis (Skinker 1935) Little 1967	Colorado (Leiby 1961). Montana (Skinker 1935). North Dakota (Pederson and Leiby 1969; present work). Wisconsin (Rausch 1947). Wyoming (Honess 1937; Keppner 1967).
Nematode	
Ancylostoma caninum (Ercolani 1859) Hall 1913	Arizona (Hannum 1942).
Ancylostoma taxideae Kalkan and Hansen 1966	Kansas (Kalkan and Hansen 1966). North Dakota (present work).
Angiocaulus gubernaculatus (Dougherty 1946) Schultz 1951	California (Dougherty 1946).
Ascaris columnaris Leidy 1856	Colorado (Leiby 1961). Minnesota (Erickson 1946). North Dakota (present work). Wisconsin (Morgan 1943).
Ascaris sp.	Kansas (Worley 1961).
Filaria martis Gmelin 1790[a]	Mexico (Cabellero y C. 1948). North Dakota (present work). Wyoming (Keppner 1969*a*).
Filaria taxideae Keppner 1969	
Molineus felineus Cameron 1923	Utah (Grundmann 1957).
Molineus mustelae Schmidt 1965	Wyoming (Keppner 1969*b*).
Molineus patens (Dujardin 1845) Petrov 1928	Minnesota (Erickson 1946). North Dakota (present work).
Monopetalonema? eremita Leidy 1886	Wyoming (Leidy 1886)
Physaloptera maxillaris Molin 1860	Minnesota (Erickson 1946). Unknown (Morgan 1941*a*).
Physaloptera torquata Leidy 1886	Arizona (Hannum 1942). California (Morgan 1942). Illinois (Morgan 1941*b*, 1942). Minnesota (Erickson 1946). Montana (Ehlers 1931). North Dakota (present work). Pennsylvania (Leidy 1886; Walton 1927; Canavan 1931). Wisconsin (Morgan 1941*b*, 1942, 1943). Wyoming (Leidy 1886).
Trichinella spiralis (Owen 1835) Railliet 1895	Wyoming or New York (Herman and Goss 1940).

SOURCE: Leiby et al. 1971
[a]Keppner (1969*a*) states that nematodes reported as *F. martis* by Worley (1961) should be considered conspecific with *F. taxideae* and also questions the identity of *F. martis* of Caballero g C. (1946).

AGE DETERMINATION

Wright (1966) used dry eye-lens weights to separate adult from juvenile badgers, but later found (Wright 1969) that this separation could be made more easily on the basis of cranial suture coalescence. Nearly all cranial sutures are closed by one year of age. Juvenile and adult males are easily separated on the basis of testes and bacula weights. Even during the anestrous season, adult testes weighed more than those of juveniles (Wright 1969). Average weights of bacula from juveniles and adults reported by Wright (1969) were 1.5 g and 4.2 g, respectively. Bacula weights reported by Lindzey (1971) were 1.1 g and 4.6 g for juveniles and adults, respectively.

The presence of bands in the cementum layer of canines and in the mandible were originally reported by Wright (1969). Lindzey (1971) and Crowe and Strickland (1975) investigated the relation between these bands and age and concluded that they could be used to determine age of badgers. Because the first annulus is not deposited until the badger's second summer or fall, the count of annuli must be increased by one to arrive at age in years.

The amount of wear on canines caused by the upper and lower canines rubbing on one another (Long 1965) and the presence of dental abnormalities most frequent in older age animals (Long and Long 1965) may allow discrimination between young and old animals.

CURRENT RESEARCH AND MANAGEMENT NEEDS

Badgers are generally offered little protection across their range. While this policy may have been excusable in the past, it seems likely that in the face of increasing demands for natural furs and changing land uses more active management programs will be needed in the future. Historically, the badger has played only a minor role in the North American fur trade. Demand and prices have generally been low and the pelt is relatively difficult to prepare. Many of the badgers sold were caught in traps set for other animals and many others caught were simply killed and discarded. Attempts to raise badgers commercially for their fur proved uneconomical.

Unfortunately, the benefit provided agriculture by the badger's consuming rodents is outweighed in many instances by the problems caused by its diggings. Excavations made by badgers have caused broken legs in cattle and horses, led to the loss of valuable water from earthen irrigation structures, and made mowing and plowing of fields more difficult.

Changing land use patterns undoubtedly will influence populations on a local basis. Clearing of timbered areas in the eastern part of its geographic range may prove beneficial to the badger, while clearing of sagebrush in the cold desert region for dryland farming or cattle grazing may prove detrimental. Both the species composition of small rodents and biomass available to the badger will change as habitats are al-

tered. More intensive agricultural practices in the prairie regions and rodent control programs will adversely affect the badger.

Increased demand for furs will probably result in increased trapping pressure on the badger. Because they seldom go undetected and are relatively easy to trap, the intensity of trapping and population levels may need to be monitored carefully.

In the future, we will need to know more about the relationship between badgers and their prey populations to assess the impact of changing land uses. The role of badgers in epizootics similarly requires further study. Although it is likely that badgers will benefit from increased emphasis on "nongame" programs, steps need to be taken to insure them adequate protection.

H. J. Harlow, R. P. Lampe, and J. P. Messick all provided recent information for inclusion in this chapter. T. R. McCabe provided criticism of the manuscript.

LITERATURE CITED

Bailey, V. 1888. Trip through parts of Minnesota and Dakota, 1887. Rep. Ornithol. Mammal. Dept. Agric. 433pp.
———. 1929. Mammals of Sherburne County, Minnesota. J. Mammal. 10:153–164.
———. 1931. Mammals of New Mexico. North Am. Fauna 53. 412pp.
Baker, R. H. 1956. Mammals of Coahuila, Mexico. Univ. Kansas Publ., Mus. Nat. Hist., 9:125–335.
Balph, D. F. 1961. Underground concealment as a method of predation. J. Mammal. 42(3):423–424.
Broadbook, H. E. 1970. Populations of the yellow-pine chipmunk, *Eutamias amoenus*. Am. Midl. Nat. 83:472–488.
Caballero y C., E. 1948. *Filaria martis* Gmelin, 1790 en mamiferos de Neuvo Leon y consideraciones sobre las especies del genero *Filaria* Müller, 1787. Rev. Soc. Mexican Hist. Nat. 9:257–261.
Canavan, W. P. N. 1931. Nematode parasites of vertebrates in the Philadelphia Zoological Gardens and vicinity. II. J. Parasitol. 23:196–229.
Criddle, S. 1930. The prairie pocket gopher, *Thomomys talpoides*. J. Mammal. 11(1):265–280.
Crowe, D. M., and Strickland, M. D. 1975. Dental annulation in the American badger. J. Mammal. 56(1):269–272.
Dearborn, N. 1932. Foods of some predatory fur-bearing animals in Michigan. Univ. Michigan School For. Conserv. Bull. 1:1–52.
Dougherty, E. C. 1946. The genus *Aclurostrongylus* Cameron, 1927 (Nematoda: Metastrongylidae), and its relatives; with descriptions of *Parafilaroides*, gen. nov., and *Angiostrongylus gubernaculatus*, sp. nov. Proc. Helmin. Soc. Washington 13:16–25.
Ehlers, G. H. 1931. The authelmintic treatment of infestations of the badger with spirurids (*Physaloptera* sp.), J. Am. Vet. Med. Assoc. 31:79–87.
Ellis, L. L., Jr. 1955. A survey of ectoparasites of certain mammals in Oklahoma. Ecology 36(1):12–18.
Erickson, A. B. 1946. Incidence of worm parasites in Minnesota Mustelidae and host lists and keys to North American species. Am. Midl. Nat. 36:494–509.

Errington, P. L. 1937. Summer food habits of the badger in northwestern Iowa. J. Mammal. 18(2):213–216.

Grinnell, J.; Dixon, J. S.; and Linsdale, J. M. 1937. Furbearing mammals of California. Univ. Calif. Press, Berkeley. 375pp.

Grundmann, A. W. 1956. A new tapeworm, *Mesocestoides carnivoricolus,* from carnivores of the Great Salt Lake Desert region of Utah. Proc. Helmin. Soc. Washington 23:26–28.

_____. 1957. Nematode parasites of mammals of the Great Salt Lake Desert region of Utah. J. Parasitol. 43:105–112.

_____. 1958. Cestodes of mammals from the Great Salt Lake Desert region of Utah. J. Parasitol. 44:425–429.

Hall, E. R. 1946. Mammals of Nevada. Univ. California Press, Berkeley. 710pp.

Hamlett, G. W. 1932. Observations on the embryology of the badger. Anat. Rec. 53:283–301.

Hannum, C. A. 1942. Nematode parasites of Arizona vertebrates. Ph.D. Thesis. Univ. Washington, Seattle. Pp. 66–67, 90–91.

Harlow, H. J. 1979. A photocell monitor to measure winter activity of confined badgers. J. Wildl. Manage. 43:997–1001.

_____. 1980. Influence of the burrow on cold adaptations and energy requirements of the American badger. Ph.D. Thesis. Univ. Wyoming, Laramie.

_____. 1981. Effect of fasting on rate of food passage and assimilation efficiencies in badgers. J. Mammal. 62:173–177.

Herman, C. M., and Goss, L. J. 1940. Trichinosis in an American badger, *Taxidea taxus taxus.* J. Parasitol. 26:157.

Hetlet, L. A. 1968. Observations on a group of badgers in South Park, Colorado. M.S. Thesis. Colorado State Univ., Fort Collins. 30pp.

Honess, R. F. 1937. Un nouveau cestode: *Fossor angertrudae* n.g., n. sp. du blaireau d'Amérique *Taxidea taxus taxus* (Schreber 1778). Ann. Parasit. 16:363–366.

Howell, A. B. 1924. The mammals of Mammoth, Mono County, California. J. Mammal. 5(1):25–36.

Hubbard, C. A. 1947. Fleas of western North America. Iowa State College Press, Ames. 533pp.

Jackson, H. H. T. 1961. Mammals of Wisconsin. Univ. Wisconsin Press, Madison. 504pp.

Jense, G. K. 1968. Food habits and energy utilization of badgers. M.S. Thesis. South Dakota State Univ., Brookings. 39pp.

Kalkan, A., and Hansen, M. F. 1966. *Ancylostoma taxideae* sp. n. from the American badger, *Taxidea taxus taxus.* J. Parasitol. 52:291–294.

Keppner, E. J. 1967. *Fossor taxidiensis* (Skinker, 1935) n. comb. with a note on the genus *Fossor* Honness, 1937 (Cestoda: Taeniidae). Trans. Am. Microscop. Soc. 86:157–158.

_____. 1969a. *Filaria taxidea* n. sp. (Filarioidea: Filariidae) from the badger, *Taxidea taxus taxus* from Wyoming. Trans. Am. Microscop. Soc. 88:581–588.

_____. 1969b. Occurrence of *Atriotaenia procyonis* and *Miloncus mustelae* in the badger, *Taxidea taxus* (Schreber, 1778), in Wyoming. J. Parasitol. 55:1161.

Knopf, F. L., and Balph, D. F. 1969. Badgers plug burrows to confine prey. J. Mammal. 50(3):635–636.

Lampe, R. P. 1976. Aspects of the predatory strategy of the North American badger (*Taxidea taxus*). Ph.D. Thesis. Univ. Minnesota, St. Paul 102pp.

Lampe, R. P., and Sovada, M. A. In press. Seasonal variation in home range of a female badger (*Taxidea taxus*). J. Mammal.

Leiby, P. D. 1961. Intestinal helminths of some Colorado mammals. J. Parasitol. 47:311.

Leiby, P. D.; Sitzmann, P. J.; and Kritsky, D. C. 1971. Studies on helminths of North Dakota. Parasites of the badger *Taxidea taxus* (Schreber). Proc. Helmin. Soc. Washington 38:225–228.

Leidy, J. 1886. Notices of nematoid worms. Proc. Philadelphia Acad. Nat. Sci. 38:308–313.

Lindzey, F. G. 1971. Ecology of badgers in Curlew Valley, Utah and Idaho with emphasis on movement and activity patterns. M.S. Thesis. Utah State Univ., Logan. 50pp.

_____. 1976. Characteristics of the natal den of the badger. Northwest Sci. 50(3):178–180.

_____. 1978. Movement patterns of badgers in northwestern Utah. J. Wildl. Manage. 42(2):418–422.

Long, C. A. 1965. Functional aspects of the jaw-articulation in the North American badger, with comments on adaptiveness of tooth wear. Trans. Kansas Acad. Sci. 68:156–162.

_____. 1973. *Taxidea Taxus.* Mammalian Species 26. Am. Soc. Mammal. 4pp.

Long, C. A., and Long, C. F. 1965. Dental abnormalities in North American badgers, genus Taxus. Trans. Kansas Acad. Sci. 68(1)145–154. Ecology 36(1):12–18.

Manville, R. H. 1959. The columbian ground squirrel in northwestern Montana. J. Mammal. 40(1):26–45.

Messick, J. P. 1981. Ecology of the badger in southwestern Idaho. Ph.D. Thesis. Univ. Idaho, Moscow. 127pp.

Messick, J. P.; Smith, G. W.; and Barnes, A. M. In press. Serological testing of badgers to monitor the presence of plague in Townsend ground squirrel populations in southwestern Idaho. J. Wildl. Dis.

Morgan, B. B. 1941a. A summary of the Physalopterinae (Nematoda) of North America. Proc. Helmin. Soc. Washington 8:28–30.

_____. 1941b. Additional notes on North American Physalopterinae (Nematoda). Proc. Helmin. Soc. Washington 8:63–64.

_____. 1942. The Physalopterinae (Nematoda) of North American vertebrates. Summ. Doctoral Diss. Univ. Wisconsin 6:88–91.

_____. 1943. New host records of nematodes from *Mustelidae* (Carnivora). J. Parasitol. 29:158–159.

Nugent, R. F., and Choate, J. R. 1970. Eastward dispersal of the badger, *Taxidea taxus,* into the northeastern United States. J. Mammal. 51(3):626–627.

Palmer, R. S. 1954. The mammal guide. Doubleday & Co. Inc., Garden City, N.Y. 384pp.

Pederson, E. D., and Leiby, P. D. 1969. Studies on the biology of *Monordotaenia taxidiensis,* a taeniid cestode of the badger. J. Parasitol. 55:759–765.

Petrides, G. A. 1950. The determination of sex and age ratios in fur animals. Am. Midl. Nat. 43(2):355–382.

Rausch, R. 1947. A redescription of *Taenia taxidiensis* Skinker, 1935. Proc. Helmin. Soc. Washington 14:73–75.

Sargeant, A. B., and Warner, D. W. 1972. Movements and denning habits of a badger. J. Mammal. 53(1):207–210.

Seton, E. T. 1929. Lives of game animals. Charles Branford Co., Boston, Mass. 949pp.

Skinker, M. S. 1935. Two new species of tapeworms from carnivores and a redescription of *Taenia laticollis* Rudolphi, 1819. Proc. U.S. Natl. Mus. 83:211–220.

Snead, E., and Hendrickson, G. O. 1942. Food habits of the badger in Iowa. J. Mammal. 23(1):380–391.

Swanson, G., and Erickson, A. B. 1946. *Alaria taxideae* n. sp., from the badger and other mustelids. J. Parasitol. 32:17–19.

Verbeek, N. A. 1965. Predation by badger on yellow-bellied marmots in Wyoming. J. Mammal. 46(2):506.

Walton, A. 1927. A revision of the nematodes of the Leidy collection. Proc. Philadelphia Acad. Nat. Sci. 79:49–163.

Worley, D. E. 1961. The occurrence of *Filaria martis* Gmelin, 1790, in the striped skunk and badger in Kansas. J. Parasitol. 47:9–11.

Wright, P. L. 1966. Observations on the reproductive cycle of the American badger (*Taxidea taxus*). *In* Comparative biology of reproduction in mammals. Symp. Zool. Soc. London 15:37–45.

———. 1969. The reproductive cycle of the male American badger (*Taxidea taxus*). J. Reprod. Fert. Suppl. 6:435–445.

Young, S. P. 1951. The clever coyote. Stackpole Co., Harrisburg, Pa. 226pp.

FREDERICK G. LINDZEY, Utah Cooperative Wildlife Research Unit, Utah State University, Logan, Utah 84322.

34

Spotted and Hog-Nosed Skunks

Spilogale putorius and Allies

Walter E. Howard

Rex E. Marsh

NOMENCLATURE

COMMON NAME. Eastern spotted skunk
SCIENTIFIC NAME. *Spilogale putorius*. *Spilogale* is derived from the Greek *spilos,* or spot, and *gale,* or weasel. *Putorius* is Latin for "stinker."

The prairie spotted skunk, *Spilogale interrupta,* has been reclassified to subspecific status as *S. putorius interrupta.*

COMMON NAME. Western spotted skunk
SCIENTIFIC NAME. *Spilogale gracilis.* Van Gelder (1959), in his taxonomic revision of spotted skunks, did not recognize the species *S. gracilis,* but considered all spotted skunks in the United States to be monotypic (*S. putorius*). However, Hall and Kelson (1959) and Jones et al. (1975) recognized two species, *S. putorius* and *S. gracilis.*

COMMON NAMES. Hog-nosed skunk, rooter skunk, rooter
SCIENTIFIC NAME. *Conepatus mesoleucus. Conepatus* comes from the native Mexican name *Conepate* or *Conepatl.* The species name, *mesoleucus,* is from the Greek *mesos,* or middle, and *leucos,* or white, referring to its unbroken white mantle.

COMMON NAMES. Eastern hog-nosed skunk, white-backed skunk, white-tailed skunk, Texan skunk, badger skunk, conepate
SCIENTIFIC NAME. *Conepatus leuconotus*

Other common names for *Spilogale* include civet, polecat, hydrophoby cat, marten, skunk, weasel skunk, tree skunk, four-way skunk, four-striped skunk, little spotted skunk, sachet kitty, and black marten.

SUBSPECIES. Hall and Kelson (1959) list 14 subspecies of these two species of spotted skunk north of Mexico, and 6 subspecies of hog-nosed skunk. Both genera have additional representatives in Mexico and Central America.

DISTRIBUTION

Spotted Skunk. *Spilogale putorius,* the eastern spotted skunk, ranges from northeastern Mexico through the Great Plains to the Canadian border, and throughout the southeastern United States (figure 34.1). Baker and Baker (1975) noted that eastern spotted skunks are found not only in open lowlands but also in the mountainous country of western and southern North America. In Mexico, Baker and Baker observed the skunks at altitudes of about 2,415 meters (8,000 feet).

The range of *Spilogale gracilis,* the western spotted skunk, is the western states, extending from Mexico to southwestern British Columbia, but in only small parts of Montana and North Dakota. To the east, the range meets the western distribution of the eastern spotted skunk (see figure 1). One western spotted skunk was captured at an altitude of 2,560 m (8,400 feet) in the Sierra Nevada of California (Orr 1943).

Hog-Nosed Skunk. *Conepatus mesoleucus,* the hog-nosed skunk, ranges over the southwestern United States and Mexico, including parts of Arizona, New Mexico, Colorado, the Oklahoma panhandle, and the lower half of Texas (figure 34.2). The genus is generally distributed in South America and is believed to be spreading northward. The white-furred area of the back continues broadly onto the tail, except where this species intergrades with the eastern hog-nosed skunk (*C. leuconotus*). In the eastern hog-nosed skunk, the white area is only narrowly continuous from the back to the tail. Some observers consider them one species.

Conepatus leuconotus, the eastern hog-nosed skunk, ranges over the southern tip of Texas and the east coast of Mexico (see figure 34.2).

DESCRIPTION

Many workers consider both genera (*Spilogale* and *Conepatus*) to be monotypic, although we will treat them here as separate genera, with each having more than one species. The spotted skunks (*Spilogale* sp.) are the smallest skunks; adult males weigh about 700 grams and females 450 g. Their habits are weasel-

FIGURE 34.1. Approximate distribution of two species of spotted skunk (*Spilogale putorius* and *S. gracilis*). Data from Hall and Kelson 1959.

FIGURE 34.2. Approximate distribution of two species of hog-nosed skunk (*Conepatus leuconotus* and *C. mesoleucus*). Data from Hall and Kelson 1959.

like. They have jetblack pelage with white spots and four to six broken stripes, while the much larger striped skunks (*Mephitis* sp.) and hog-nosed skunks (*Conepatus* sp.) may have just one, two, or no stripes. Adult hog-nosed skunks, depending on their age and amount of stored body fat, weigh between 1,100 and 4,500 g (Davis 1945). Spotted skunks are more nocturnal, quick, and alert than all of the larger skunks. Other differences are that spotted skunks climb trees and have finer and denser fur, tail hairs unicolored to the base, and a more pungent scent. Spotted skunks are easily excited and Crabb (1941*b*) considered them difficult to tame.

Female hog-nosed skunks (*Conepatus*) have 6 to 8 mammae, in contrast to female common striped skunks (*Mephitis*), which have 10 to 14 mammae.

Both sexes have similar coloration. The patchwork pattern of black and white, especially of spotted skunks, makes them nearly invisible at night, even in moonlight, if they hold still. Color patterns of various subspecies of spotted skunk differ primarily in the width of the white stripes in relation to intervening black areas (Davis 1945).

In most *C. mesoleucus*, the white area of the back continues broadly onto the tail, while in *C. leuconotus* the white stripe on the back is much narrower, wedge shaped rather than truncate on the head, and reduced in width or absent on the rump. With *C. mesoleucus*, the long, bushy tail is all white except for a few black hairs scattered underneath, while in *C. leuconotus*, the upper side of the tail is white but the underside is black toward the basal half and white toward the tip (Davis 1945).

White portions of the skin of spotted skunks (*Spilogale*) frequently are tinged with yellow, especially around the anterior ends of the shoulder stripes.

This may be caused by the rubbing of wax from the ears or by urine (Van Gelder 1959).

Spilogale are the only skunks having a broad, triangular nose patch and more than four white body stripes. *Conepatus* have no white on the nose, and either a single broad white stripe or a pair of dorsal white stripes. *Mephitis* have a thin white stripe medially on the nose and either a broad stripe of mixed black and white hairs, two lateral white stripes, or a combination of both dorsal and lateral stripes (Van Gelder 1959).

The eastern spotted skunk (*Spilogale putorius*) has small white spots in front of each ear and one on the forehead. The ear spots often join the middle of six white stripes on the back. The lower pair of stripes begins on the back of the foreleg. The western spotted skunk (*S. gracilis*) resembles its eastern relative; however, white markings are more extensive. The black and white stripes on the upper back are nearly equal in width; the dorsal pair of white stripes begins between the ears or just behind them; the white area on the face is large. extending nearly from the nose pad to a line behind the eyes and covering more than half of the area between the eyes; the underside of the tail is white for nearly half its length, and the tip is extensively white (Davis 1945).

To determine how to age *S. putorius,* Mead (1967) used the weights of dried eye lens and baculum, obliteration of cranial sutures, ankylosis of the epiphysis to the diaphysis of long bones, and presence or absence of placental scars on females. Lens weight proved to be the most useful measurement for distinguishing among juveniles, subadults, and adults.

Claws. The five claws on the forefeet of all skunks are adapted for digging. On the forefeet of members of the largest of the three genera, *Conepatus,* they are quite long (20 mm or more)—about three times as long as the five claws on the hind feet. The claws of the members of the genera *Mephitis* (considered middle sized) are about 10 mm long, twice the length of those on the hind feet. With the little *Spilogale,* forefeet claws, which are used in climbing and digging, are about 7 mm long, a little more than twice that of the hind feet. Claws on all skunks are highly curved.

Dentition. Hog-nosed skunks (*Conepatus*) differ in many ways from both *Spilogale* and *Mephitis* in cranial characters and dentition. The dentition of *Conepatus* is 3/3, 1/1, 2/3, 1/2 = 32, while *Spilogale* or *Mephitis* usually have an additional upper premolar and hence may have a total of 34 teeth (figure 34.3).

Sexual Dimorphism. Sexual dimorphism seems to occur in all skunks, with males larger than females. In *Spilogale* the males are about 7 percent larger than females in cranial measurements, and 10 percent larger in length of the head and body and the hind foot (Van Gelder 1959).

SCENT

All skunks can be objectionable when they use their principal means of defense: the spraying of musk. The capacity of each of the two anal scent glands is about 15 cc for the larger striped skunk (*Mephitis* sp.) (Blackman 1911). In the spotted skunk, anal fluid looks like skim milk mixed with some curds of cream, although the basic color varies from white to light yellow or greenish yellow, usually with white curds (Crabb 1948). The active ingredient is a sulphide called mercaptan.

For defense, the spotted skunk usually stamps or pats its front feet rapidly or tries to bluff by handstanding to look much larger, using musk as a last resort.

Crabb (1948) found that spotted skunks caught in his tagging chute could, even though they could not lift their tails, swing sideways, expose the anal sphincter, and discharge all over their tails.

Spotted skunks make interesting pets, but only when captured while young. Because of rabies, it is illegal in some states to possess or capture for sale any skunk. They can be easily "deskunked" with a pair of sharp scissors by snipping off the ends of the two musk ducts that open just inside the vent on each side. This will cause the ends of the ducts to seal over and prevent emission of musk (Davis 1945). Bailey (1937) described this operation.

Amundson (1950) provided some useful information about the musk spraying of both spotted (*Spilogale*) and striped (*Mephitis*) skunks. Skunks can expel their strongly acid musk spray with considerable accuracy for several meters and do not exhaust their supply in one expulsion. Most authors agree that skunks cannot readily expel their musk if they are held by their tails or are unable to get their tails erect. Dice (1921) pointed out that a skunk held off the ground by one hind foot in a trap cannot discharge. If the animal

is trapped by a front foot and can get both hind feet on the trap, however, it can discharge scent. Being sprayed by a skunk can be nauseating, but will not cause more than momentary blindness if the fluid gets in the eyes.

Live-trapped skunks can be moved or transported in live-traps with little hazard of their spraying if they are not shaken or jarred. An additional precaution is to

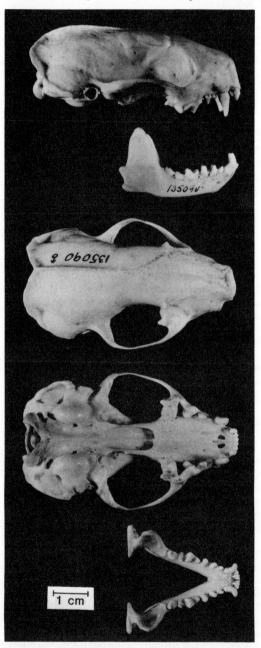

FIGURE 34.3. Skull of the eastern spotted skunk (*Spilogale putorius*). From top to bottom: lateral view of cranium, lateral view of mandible, dorsal view of cranium, ventral view of cranium, dorsal view of mandible.

cover the trap with a sack or tarp. Crabb (1948) found that spotted skunks are easy to keep in captivity and do not use their musk; none, however, would tolerate handling. The removal of and counteraction of skunk odors will be discussed in the "Economic Status and Management" section.

REPRODUCTION

All skunks are born blind and helpless; their eyes do not open for about one month. Young *S. putorius* can emit musk when only about 46 days old (Davis 1945). They are weaned at about 54 days of age and are almost as large as adults by three months of age. Some spotted skunks are thought to be polygamous, others monogamous. *S. putorius,* at least, is known to be a polyestrous, spontaneous ovulator upon copulation and has estrous cycles occurring in September through January (Greensides and Mead 1973). Males do not assist in the care of the young.

Conepatus females probably have small litters because the females have only three pairs of mammae—one pair inquinal and two pairs pectoral (Bailey 1931). Meager records indicate that litter size is only about two to four young. The gestation period is 42 days (Hall and Kelson 1959). The mean litter size of *Spilogale* is about four (three to six).

Spotted skunks exhibit delayed implantation, as do other mustelids. For 12 spotted skunks from Oregon (presumably *S. gracilis latifrons* [formerly *S. putorius*]), Foreman and Mead (1973) found that the duration of postimplantation ranged from 28 to 31 days. Morphological interaction of trophoblasts and uterus during the preimplantation period and implantation has been described (Sinha and Mead 1976). Mead et al. (1979) reported on the changes in uterine protein synthesis during delayed implantation and its regulation by hormones.

According to Mead (1968*b*), most adult male *S. gracilis* have spermatozoa in their testes and epididymides by June. Some juvenile males mature sexually by September. Western spotted skunk females (*S. gracilis*) come into heat in September and most are bred by the first week in October, although the southern subspecies may breed as early as July and some may produce two litters in the same year. Copulation occurs in September and both adult females and the year's young are bred. After the blastocyst is formed, the embryos float freely in the uterine lumen for about 180 to 200 days. Nidation occurs in April and most litters are born in May after gestating for 210 to 230 days.

In contrast, Mead (1968*a*) pointed out that for the eastern spotted skunk (*S. putorius*) females come into heat in late March and nearly all are bred by the end of April. Some young male *S. putorius* are known to reach sexual maturity when five months old, as evidenced by testes and epididymides with mature sperm, and breed in September (Mead 1967).

According to Crabb (1948), the sex ratio of three complete litters (N=16) of *S. putorius* averaged 130:100 males to females.

ECOLOGY

Habitat. The eastern spotted skunk (*S. putorius*) has a strong propensity to spend much of its life in or near farmyards. It moves to fields only when the food supply of insects or rodents is better there than around human developments. It likes good cover, such as along fences, embankments, gullies, willows and hedgerows, rock outcrops, barns, and outbuildings (Crabb 1948). This skunk primarily inhabits woods and tallgrass prairies, preferring rock canyons and outcrops when available (Davis 1945).

Henderson (1976) reported that in Kansas the striped skunk is generally more tolerant of humans than is the spotted skunk. The latter species prefers hillsides that are dry, somewhat rocky, and more brushy than wooded. When found in buildings, spotted skunks will be in those only intermittently occupied, such as summer cabins.

Dens. Almost any location will serve as a den for spotted skunks if there is safe accessibility to it along a fence row or through other cover, and if it also provides protection from predators, weather, and light. Den sites include rock crevices, granaries, barns, spaces under outbuildings, drains, or piles of hay. They occasionally use hollow trees, logs, and stumps, and frequently use burrows made by other animals. Sometimes they dig their own burrows.

In northeast Iowa, *Spilogale* and long-tailed weasels (*Mustela frenata*) both showed a strong preference for burrows of the Franklin ground squirrel (*Spermophilus franklinii*) and the pocket gopher (*Geomys bursarius*) (Polder 1968). Both animals favored similar vegetation cover. When corn was shocked and left standing in late winter, both species used the shocks as temporary dens to feed on the concentrations of meadow mice (*Microtus* sp.) that formed.

Spotted skunks normally make nests of grass or hay. Occasionally, no nest is constructed when they live in attics or walls of a building. Dens distributed over an area seem to belong not to an individual skunk or kin group (if such exists), but to a whole population. An exception is a mother and her brood during the breeding season. Spotted skunks move about from farmyard to farmyard and from den to den as suits their inclination. Several individuals may use the same den, either at different times or simultaneously. If harassed by dogs or smoked out by humans, they abandon their dens readily.

Signs. According to Crabb (1948), one of the best signs of the spotted skunk is its scats. Generally, spotted skunks defecate indiscriminately (except in their nests) in their runways, usually in inaccessible places, or under buildings, log piles, and haystacks. Scats also are likely to be found in cow paths, in dry washes in fields, and along fence lines. When skunks use barns, their scats may be found all along the runways, which usually follow rafters, poles, and braces. When they use piles of firewood, scats may appear anywhere. In attics and walls long used by spotted skunks, scat accumulations may be 2 to 5 cm deep.

Most scats of spotted skunks measure a little over 1 cm in diameter; however, Crabb (1941a) found an unusually large scat 17 mm long that contained insect remains. The age of scats in the field cannot be determined after about two weeks. Spotted skunks are less inclined than striped skunks (*Mephitis* sp.) to make so-called latrines (Crabb 1941a).

Musk odor is an excellent field sign. With a little experience and a reasonably good nose, one can differentiate between the musks of striped and spotted skunks. Spotted skunks also often leave hairs about den entrances or other holes they squeeze through in buildings; the hairs can be confirmed with a magnifying glass (Crabb 1948). The key developed by Stains (1958) is useful in the identification of hairs, or hairs can simply be compared with the hairs from specimens in hand.

Tracks made in mud or snow are useful signs. Both the hind feet and the forefeet of skunks have five toes, although the fifth toe may not imprint in some instances. Imprints of the long, curved claws of the forefeet are highly visible, but the heels may not be. Tracks are of several types, depending on the animal's activity. While the animal is slowly hunting, the hind feet are often placed exactly in the print of the front feet, with a spacing of 13 or 14 cm. The bounding gait leaves tracks similar in placement to those of weasels, with the front and hind feet working in pairs. The space between tracks varies with speed of progress and animal size, but generally ranges between 23 and 36 cm. In the galloping gait the track pattern resembles that of the striped skunk, whose hind feet tracks do not register exactly with those of the front feet (Crabb 1948).

Population Dynamics. We are aware of few studies on the home range, population density, fluctuations, and degree of territoriality of spotted and hog-nosed skunks. Crabb (1948) believed the eastern spotted skunk did not occupy territories. These skunks are somewhat nomadic and apparently do not defend their home ranges. He observed no defense of a specific domain. During a 39-month study, Crabb (1948) handled 238 spotted skunks in his 22.5-km² (14 mi²) study area. His estimate of the density was 8.8 skunks per km² (5.6 mi²).

In Kansas and elsewhere on the Great Plains, the long-term population density of *S. putorius* has fluctuated markedly (Choate et al. 1974). Choate et al. (1974) speculated that this commensal species increased as humans settled this region. It declined at the onset of the Great Depression, which coincided with a rapid decline in the number of farms (favorable habitats), so the spotted skunk is not common today in Kansas or elsewhere in the Great Plains.

Polder (1968) stated that spotted skunks were uncommon in northeastern Iowa before 1900, but have been common since about 1925. However, he noted that much spotted skunk habitat has disappeared since 1957, when his observations concluded. Farms have been enlarged, fences and windbreak groves removed,

gullies filled in, and continuous corn or corn-soybean rotation increasingly practiced.

FOOD HABITS

The food of both *Spilogale* and *Conepatus* contains many beetles, worms, crickets, grasshoppers, grubs, carrion, rodents, young rabbits, bird eggs, frogs, crayfish, lizards, and some fruit. Polder (1968) observed that *Spilogale* did not appear to store food but frequently pilfered weasel (*Mustela* sp.) caches, and that the large bones of rabbits, chickens, and carrion of pheasants dragged to the nest were not eaten. Feathers were pulled out and scattered about (Crabb 1948).

Based on an analysis of 834 scats of *Spilogale putorius* collected over one year, Crabb (1941a) provided a good picture of the seasonal diet of the eastern spotted skunk in southeastern Iowa. Mammals provided 90 percent of winter foods, with the eastern cottontail (*Sylvilagus floridanus*) present in almost 51 percent of the scats. There was an unusually dense population of cottontails in that area at that time, however. Meadow mice (*Microtus pennsylvanicus* and *M. ochrogaster*) were the next most frequent mammals eaten, appearing in 19 percent of the scats. Rats (*Rattus norvegicus*) appeared in the greatest numbers in winter, being present in over 9 percent of the scats. The next most important food item was corn (*Zea mays*), which exceeded all other items except cottontails and occurred in over 25 percent of the scats. Even though there are not many insects in winter, they were found in nearly 15 percent of the scats.

The main differences in spring diet were an increase in arthropods (present in 48 percent of the scats), and a decline of plant material (11 percent). In summer, mammal remains were found in only 11 percent of the scats, and arthropods were found in 92 percent. In fall, mammals increased to 58 percent; arthropods were still numerous at 80 percent.

In Iowa, the diet of the spotted skunk (*Spilogale putorius*) differs considerably from that of the striped skunk (*M. m. mesomelas*). Striped skunks eat about twice as many insects (grasshoppers, crickets, and white grubs) as *Spilogale,* but spotted skunks eat about four times as much mammalian material—mostly meadow mice, but also cottontails. Selko (1937), however, concluded that food availability may have strongly influenced these diets. We assume, but do not know, that the diet of *Conepatus* sp. is much more like that of *Mephitis* sp. than that of *Spilogale* sp., because the larger skunks appear to feed more on insects.

Van Gelder (1953) described how a captive female *S. putorius* was able to eat hens' eggs that were too large for the animal to bite. After giving up on biting, she propelled the egg underneath her and gave it a quick kick backward. She repeated this act until the egg broke. She then ate the egg contents and some of the shell.

Bailey (1931) noted that even where *C. mesoleucus* was common in the Lower Sonoran area of New Mexico, the animals were not often trapped, as they seemed to prefer natural foods of beetles and larvae. Their peculiar long, flexible nose apparently is adapted for capturing ground beetles in shallow burrows, grasshoppers, crickets, and ripe prickly pear fruit. Even though the skunks will eat fresh or stale meat and occasionally capture poultry or small game, their main food is insects (figure 34.4).

BEHAVIOR

Johnson (1921) was not the first to report on the "handstand" behavior of the spotted skunk. However, after observing one animal stand on its front feet with hind feet in the air many times as he pursued it through a field, he described the behavior in some detail. It walked several meters on its hands. This appears to be a defensive mechanism that allows the skunk to face a tall adversary and arch its back almost into a semicircle, so that its anal sphincter can direct a discharge of scent into an intruder's face. This posture enables this little skunk to see where it aims its spray (Sherwood 1931). However, the handstand posture may well frighten off potential enemies, for the skunk presents a more frightening threat when it stands on its hands. Sherwood (1931) provided an excellent photograph of a *Spilogale* that used the handstand posture when it was approached by a house cat.

Manaro (1961) noted that *S. putorius* is so nocturnal that it is rarely ever encountered in moonlight. One night when the moon was full, for example, no skunks were trapped until after the moon set at 3:30 A.M., when nine were captured between 3:30 and sunrise at 6:10 A.M. In the dark, spotted skunks often can be located by directing a flashlight beam at the noise, made by the patting of their front feet, an easily heard warning behavior (Polder 1968).

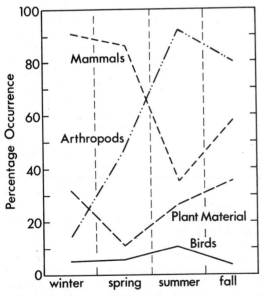

FIGURE 34.4. Percentage of food items found in stomach and digestive tract of the hog-nosed skunk (*Conepatus mesoleucus*) in Texas. In summer, snails were 5 percent of the diet. Data from Bailey 1931.

Contrary to earlier views, it is recognized now that skunks, including spotted and hog-nosed skunks, do not truly hibernate. However, they may go into short periods of inactivity or winter sleep during bad weather to help conserve their body fat.

MORTALITY

Humans are the main mortality factor of both *Spilogale* and *Conepatus*. Animals of both genera are frequently run over by automobiles. In some localities skunk populations are controlled to reduce the potential exposure of humans, pets, and livestock to rabies. Skunks are often trapped, shot, or poisoned when found around farm buildings because of their odor (especially when harassed by dogs), their habit of making dens beneath or in buildings, and concerns about their predation on poultry. Fur trappers take many skunks; striped skunks are taken in far greater numbers than spotted skunks. Some skunks are trapped accidentally in traps intended for more valuable fur bearers. Insecticides may also have been a minor factor in skunk mortality in the past, when chlorinated hydrocarbon compounds such as dieldrin were used widely to fight insect pests such as the Japanese beetle (*Papillia japonica*). Scott et al. (1959), however, recovered only one striped skunk and no spotted skunks in their extensive search for all affected animals.

Data are meager on natural predation of skunks. All skunks have an effective defense against predators. Errington et al. (1940) found *Spilogale* present in only 0.02 percent of 4,815 great horned owl (*Bubo virginianus*) pellets. Schwartz and Schwartz (1959) indicated that spotted skunks were preyed upon to some extent by great horned owls, bobcats (*Felis rufus*), and domestic dogs and cats. Because dogs and house cats occasionally kill spotted skunks, it seems likely that other predators such as foxes, bobcats, and coyotes (*Canis latrans*) also may prey on them. However, we know of no quantified data on the natural fate of either *Spilogale* or *Conepatus*. Henderson (1976) stated that a dead skunk is a good bait for luring a coyote into a buried trap.

Parasites. *Spilogale* and *Conepatus* probably carry many kinds of flea and perhaps also lice, mites, and ticks, as well as a variety of endoparasites. Tables 34.1 and 34.2 show examples of ectoparasites and endoparasites recorded from *Spilogale*. One account for *Conepatus* listed a new roundworm, *Filaria mastis* (Tiner 1946).

Zoonoses. Histoplasmosis (*Histoplasma capsulatum*) was isolated in cultures from five eastern spotted skunks (*S. putorius*) trapped in widely separated locations in southwestern Georgia, but no significant relationship was found between skunks and human zoonoses (Emmons et al. 1949). Other skunk diseases that might affect humans include microfilaria, listeriosis, mastitis, distemper, Q fever, and, most importantly, rabies.

Rabies virus (*Formida inexerabilus*) is the most serious threat skunks provide to humans. In a report on

TABLE 34.1. Some ectoparasites of the spotted skunk (*Spilogale* sp.)

Flea
Echidnophaga gallinacea,
Tropical hen flea
Pulex simulans
Human flea
Diamanus montanus
Ground squirrel flea
Holoplopsyllus anomalus
Ground squirrel flea
Ctenosephalides felis
Cat flea
Foxella ignota
Pocket gopher flea
Louse
Neotrichodectes mephitidis
Skunk louse
Dermacentor variabilis
Tick
Ornithodoros sp.
Tick

SOURCE: Mead 1963

all cases of rabies in pet skunks (the striped skunk is probably the most common pet) reported to the Center for Disease Control from 1958 to 1973 (Hattwick et al. 1973), the 32 positive cases and 99 possibly rabid skunks were all striped skunks. In the United States, since 1960 skunks have become the most frequently reported rabid field animal and the most common source of human rabies.

The first report of skunk rabies in North America was in Baja California in 1826, where residents said the spotted skunk sometimes entered houses at night, bit people, and gave them hydrophobia. The more recent general outbreak of wildlife rabies in the United States began in 1940 in the Southeast, with the gray fox (*Urocyon cinereoargenteus*) as the epidemic host, and in 1953, when rabies was identified in insectivorous bats. Johnson (1959) believed that the spotted skunk is involved in the long-term ecology of rabies virus in the western United States, although there is also much evidence to incriminate the long-tailed weasel (*M. frenata*). In 12 north-central states, the epizootic skunk rabies, which started in the early 1950s, peaked in 1958; 782 laboratory cases were reported (Parker 1961). Presumably these cases involved mostly striped skunks.

Skunks that are active in the daytime, especially if they are bold and aggressive, should be suspected of being rabid.

If it is necessary to kill a suspect animal, care should be taken to avoid serious damage to the head; damage would occur if the animal were shot in the head. A sharp blow at the base of the skull or across the spine with a heavy stick is sufficient to kill without making the brain useless for diagnostic purposes. Handle the animal with gloves.

"Ammon's horn," or the hypothalmus of the brain, is used to detect rabies in the laboratory; care must be used in removing this nerve center. Removal is best left to a veterinarian or trained individual. Untrained persons should do no more than sever the head carefully and turn it over to the professional. It is desirable to refrigerate (not freeze) the specimen, although the virus can be detected in specimens that are partially decomposed. Infective saliva when thoroughly dried (exposed to sunlight) is rendered harmless. The possibility of contracting rabies in ways other than by biting concerns humane societies and others who find it necessary to confine rabid animals. Pens or kennels where rabid animals have been impounded may be cleaned before occupancy by other animals by washing with a Lysol solution or other materials of equivalent phenol coefficient, and then allowing them to dry thoroughly (Maynard 1965).

ECONOMIC STATUS AND MANAGEMENT

Fur Harvest. The pelts of the hog-nosed skunks are of little commercial value because the fur is short and coarse (Bailey 1931). Hog-nosed skunk pelts are inferior and never command as high a price as the pelts of spotted or striped skunks. But spotted skunks represent an insignificant part of the U.S. fur harvest. Table 34.3 compares the number of spotted skunks taken in 1975/76 with the number of other common fur bearers. The spotted skunk harvest was only 0.1 percent of the muskrat harvest, 0.2 percent of the raccoon harvest, and 8.8 percent of the striped skunk harvest. The value of striped skunk pelts taken that year was only 0.02 percent of the total raccoon harvest.

The commercial value of furs fluctuates from year to year and is based primarily on demand linked to fashions in fur. The value of fur pelts is determined largely by their quality, which includes their color, durability, and size, and the length and density of guard hairs and fur fibers.

In the early 1900s, skunk fur farmers invariably used the larger striped skunk as breeding stock. It is not very attractive to raise either spotted or hog-nosed skunks for their fur. Skunk farming had a relatively short history and was not very popular even during its peak. In addition to the fact that skunk fur is not very valuable compared with most other fur bearers, many other factors tend to make skunk farming unprofitable. These include the cost of building adequate enclosures, having to feed skunks to maturity, the threat of rabies, the cost of maintenance, and objections by neighbors.

TABLE 34.2. Some endoparasites of the spotted skunk (*Spilogale* sp.)

Endoparasite	Source
Tapeworm	
Oochoristica pedunculata	Mead (1963)
Oochoristica wallacei	Chandler (1952)
Oochoristica oklahomensis	Peery (1939)
Mesocestoides corti	Mead (1963)
Roundworm	
Skrjabingylus chitwoodorum	Mead (1963), Tiner (1946)
Capillaria hepatica	Layne and Winegarner (1971)
Physaloptera maxillaris	Tiner (1946)

TABLE 34.3. Number, average price, and total value of pelts of spotted skunk (*Spilogale* sp.) and five other common genera of fur bearer, taken in the United States, 1975/76

Fur Bearer	Number of Pelts	Average Price per Pelt	Total Value
Spotted skunk	6,834	$ 2.00	$ 13,668
Striped skunk	77,654	1.50	111,481
Raccoon	3,232,159	19.00	61,411,021
Muskrat	6,415,861	3.50	22,455,513
Red fox	272,017	30.00	8,160,510
Coyote	175,734	45.00	7,908,030

SOURCE: Deems and Pursley 1978

Skunk management involves two opposing approaches depending on the region and situation: to increase skunk population density as desirable fur bearers, or to decrease their numbers when they become pests. The two aspects are not as incompatible as they might appear. Even when skunks are reduced as pests, the fur value from those taken need not be lost if they are trapped when the fur is in prime condition.

Management for increasing spotted skunk populations or sustaining a fur bearer resource is done primarily through game laws and regulations. Colorado and Missouri were the only two states with a habitat management program for the benefit of the spotted skunk in 1976 (Deems and Pursley 1978). The spotted skunk, which is classed as a fur bearer in some states and as a nonprotected mammal in others, was listed as present in 33 states and in British Columbia. It received total protection only in Kentucky and was open for hunting in 15 states and trapping in 25 states. There was year-round harvesting in 20 states and limited harvesting in 12 states in 1976.

It is often implied that spotted skunks are highly beneficial to the farmer because they eat insects and rodent pests, yet there is no good evidence that they reduce any pest species to tolerable or economically acceptable levels at which no other control methods would be necessary. But even if they are not clearly beneficial, skunks should not be destroyed without good reason. Humans should be encouraged to be tolerant simply because skunks are interesting animals and are part of our diverse animal heritage.

If spotted skunks present a threat to human or animal health, cause damage to poultry, crops, or ornamental plants, or become intolerable nuisances, individual skunks may have to be eliminated or local populations reduced substantially. Population reduction sometimes is conducted where rabies is a great threat (Humphrey 1967).

Removing Musk. The musk odor of skunks can be neutralized in various ways. Clothing may be deodorized by washing several times with soap or detergent; household ammonia should be added to the wash. Dogs commonly get sprayed in their first encounter with skunks, and they and other pets can be deodorized by saturating them with tomato juice and then bathing.

Neutroleum-alpha is probably one of the most useful chemicals for alleviating skunk scent. About 15

cc of the water-soluble form in a water bath can be used to decontaminate dogs and humans (Cummings 1965). It can also be used to scrub basements, garages, floors, walls, and outdoor furniture sprayed by skunks. At a higher concentration (15 cc to 1 liter of water), it can be sprayed on soil in a contaminated area. Hospital supply houses may be the best local source of this material.

Damage and Control. Spotted skunks are implicated most often in damage and nuisance problems. However, both spotted and hog-nosed skunks can do considerable damage to poultry once they develop a taste for eggs or chicks. Because of their offensive odor, the most common complaint occurs when they take up residence in or beneath buildings used by humans, or in various farm buildings. They also have been known to raid beehives and dig in vegetable gardens. Their habit of digging in lawns and golf courses in search of grubs and other soil invertebrates is a trait that puts them in a pest category. The fact that skunks are major carriers of rabies is the paramount reason for controlling spotted skunks.

Before reducing any skunk populations, check on the legality of it with officials of the local fish and game or conservation agency. Even where reduction is regulated, legal provisions normally permit removal of skunks if a health threat exists or if damage occurs (Marsh and Howard 1978).

As with many other vertebrate pests, the best solution to skunk problems under buildings is to screen or block the animals out. All entrances or openings in the foundations of homes and outbuildings should be sealed. Spaces under porches, stairs, and mobile homes should be closed off.

Once a skunk makes its home under a building, it is a little more difficult to determine whether the animal has left before the entrance is sealed. This sometimes can be done by sprinkling a smooth 22-mm (⅛-inch) layer of flour on the ground at the suspected entrance, and then examining the patch for skunk tracks soon after dark. If tracks lead out of the entrance, the opening usually then can be closed off safely.

If the number of skunks in the den is not certain, the tracking patch can be supplemented by placing over the opening a section of 13-mm ½-inch) mesh hardware cloth that is hinged at the top and left loose on the other three sides. It must be larger than the opening so that it cannot swing inward. The skunks will push it open to leave but cannot reenter.

To repel skunks from an attic or beneath a building, for example, use a generous amount of moth balls (paradichlorobenzene) or naphthalene crystals. Paradichlorobenzene is used widely but is not registered by the Environmental Protection Agency. Henderson (1976) suggested using about 450 g for burrows and 5 to 10 times that amount for attics or crawl spaces under average houses. It should be placed on shallow trays for easy removal, or hung from sills or rafters in coarse-mesh cloth sacks to keep it out of the reach of children and pets. Some people may object to the odor of paradichlorobenzene or naphthalene.

Fences can also exclude skunks from landscaped areas, backyard gardens, and schoolyards. One-meter-high wire fences with 2.5-cm hexagonal mesh, extending about 15 cm down below the ground surface and 15 cm horizontally in an L shape beneath the surface will discourage most skunks. Spotted skunks will rarely scale such a fence.

When skunks dig in lawns and golf courses in search of grubs and insects, controlling the insects will often solve the problem, though perhaps not immediately. To many, this indirect control may be more acceptable than controlling the skunks themselves.

Live-catch box traps constructed of wood or wire mesh probably offer the best method of removing skunks from under or around buildings. The traps should be about 20 × 20 × 61 cm, although smaller ones will work occasionally. Traps should be placed where animals enter a building or in trails they are known to use. Bait can be fish (canned or fresh), fish-flavored cat food, raw or cooked bacon, or chicken parts.

Skunks are relatively easy to trap, and, if the trap is handled with minimum shaking and disturbance, can be carried in the trap to a suitable location for disposal. With live-catch traps constructed of wire mesh, a piece of tarp or burlap placed carefully over the trap before moving it will keep the skunk in darkness and make it less likely to release its scent. But there is no assurance that a trapped skunk will not spray when it is trapped or moved while alive.

Chloroform or ether can be used to subdue or kill trapped skunks, or the trap can be immersed in water to destroy the animal. In regions where rabies is prevalent, trapped skunks should not be kept as pets or released in a different locality because of the possibility of spreading rabies.

Skunks can be caught with no. 1 or no. 1½ leghold traps at the entrance to dens or in regularly used trails. The trap should be set and concealed by covering it. In some situations traps can be left unbaited or baited just beyond the trap or between two trail-set traps. Fetid scents are often used to attract skunks to the traps. The same baits suggested for live-catch box traps also can be used for leghold traps.

If a trap is set at the opening of a den or building, the trap stake should be driven into the ground the full length of the trap chain away from the opening. That will keep the trapped animal out of the den, so that it can be removed without tugging on the animal and exciting it into a defensive attitude.

Because skunks are nocturnal, they can be spotlighted and shot at night for control in some situations, assuming that the shooting can be done legally and safely.

Because normally only a few skunks are involved in infestations beneath or around buildings, exclusion and trapping should be the control methods. In past years, however, skunks in rural areas were controlled with acute poisons such as strychnine, particularly where rabies epidemics threatened. Except under special authorization, strychnine is no longer registered for such use. Anticoagulants have been found effective in recent years, although they are not registered for the control of skunks on farms (Brooks and Peck 1969). Toxicants are rarely warranted for the control of a few skunks on farms.

Dens found away from buildings can be fumigated with the same fumigants used for pest ground squirrels (*Spermophilus* sp.) and woodchucks (*Marmota* sp.). Under an occupied building, however, fumigation can be unsafe. Rarely is fumigation needed in ridding a premise of a few skunks.

Keeping poultry penned and making henhouses or laying houses as skunk-proof as possible will do much to prevent chicken and egg losses. Because high numbers of rodents will attract skunks and other predators, good rodent control will make a farm less attractive to skunks. The elimination of denning sites by removing scrap lumber piles and old unused farm machinery may help make the habitat less desirable. Such habitat modifications are indicative of a well-groomed farmyard. Modern clean farming practices also destroy much skunk cover and many denning sites.

LITERATURE CITED

Amundson, R. 1950. Striped and spotted skunks. Wildl. North Carolina 14 (June):4–7.

Bailey, V. 1931. Mammals of New Mexico. North Am. Fauna 53. 412pp.

———. 1937. Deodorizing skunks. J. Mammal. 18:481–482.

Baker, R. H., and Baker, M. W. 1975. Montane habitat used by the spotted skunk (*Spilogale putorius*) in Mexico. J. Mammal. 56:671–673.

Blackman, M. W. 1911. The anal glands of *Mephitis mephitica*. Anat. Rec. 5:491–504.

Brooks, J. E., and Peck, T. D., eds. 1969. Community pest and related vector control. Pest Control Operators of Calif., Los Angeles. 224pp.

Chandler, A. C. 1952. Two new species of *Oochoristica* from Minnesota skunks. Am. Midl. Nat. 48:69–73.

Choate, J. R.; Fleharty, E. D.; and Little, R. J. 1974. Status of the spotted skunk, *Spilogale putorius*, in Kansas. Trans. Kansas Acad. Sci. 76:226–233.

Crabb, W. D. 1941a. Food habits of the prairie spotted skunk in southeastern Iowa. J. Mammal. 22:349–364.

———. 1941b. A technique for trapping and tagging spotted skunks. J. Wildl. Manage. 5:371–374.

———. 1944. Growth, development and seasonal weights of spotted skunks. J. Mammal. 25:213–221.

———. 1948. The ecology and management of the prairie spotted skunk in Iowa. Ecol. Monogr. 18:201–232.

Cummings, M. W. 1965. Skunks and their control. One-Sheet Answers no. 126. Univ. California Extension Serv., Davis. 2pp.

Davis, W. B. 1945. Texas skunks. Texas Game and Fish 3:8–11, 25–26.

Deems, E. F., Jr., and Pursley, D. 1978. North American furbearers: their management, research and harvest status in 1976. Int. Assoc. Fish Wildl. Agencies and Maryland Dept. Nat. Res., College Park, Md. 171pp.

Dice, L. R. 1921. Erroneous ideas concerning skunks. J. Mammal. 2:38.

Emmons, C. W.; Morlan, D. H. B.; and Hill, E. L. 1949. Histoplasmosis in rats and skunks in Georgia. Public Health Rep. 64:1423–1430.

Errington, P. L.; Hamerstrom, F.; and Hamerstrom, F. N., Jr. 1940. The great horned owl and its prey in north-central United States. Iowa Agric. Exp. Stn. Res. Bull. 227:757–850.

Foresman, K. R., and Mead, R. A. 1973. Duration of post-implantation in a subspecies of the spotted skunk (*Spilogale putorius*). J. Mammal. 54:521–523.

Greensides, R. D., and Mead, R. A. 1973. Ovulation in the spotted skunk (*Spilogale putorius latifrons*). Bio. Reprod. 8:576–584.

Hall, E. R., and Kelson, K. R. 1959. The mammals of North America. Ronald Press Co., New York. Vol. 11, pp. 929–943.

Hattwick, M. A. W.; Marcuse, E. K.; Britt, M. R.; Zehmer, R. B.; Currier, R. W.; and Elledge, W. N. 1973. Skunk rabies: the risk to man—or never trust a skunk. Am. J. Public Health 63:1080–1085.

Henderson, F. R. 1976. How to handle problem skunks. Pages 35–38 *in* Proc. 2nd Great Plains Wildl. Damage Control Workshop, 9–11 December 1975, Manhattan, Kans.

Humphrey, G. L. 1967. The current status of wild animal rabies in California. Pages 19–30 *in* Proc. 3rd Vertebr. Pest Conf. 7, 8, and 9 March 1967, San Francisco, Calif.

Johnson, C. E. 1921. The "hand-stand" habit of the spotted skunk. J. Mammal. 2:87–89.

Johnson, H. N. 1959. The role of spotted skunks in rabies. Proc. U.S. Livest. Sanit. Assoc. 63:267–274.

Jones, J. K., Jr.; Carter, D. C.; and Genoways, H. H. 1975. Revised checklist of North American mammals north of Mexico. Occas. Pap. Mus., Texas Tech Univ. 28:1–14.

Layne, J. N., and Winegarner, C. E. 1971. Occurrence of *Capillaria hepatica* (Nematoda: Trichuridae) in the spotted skunk in Florida. J. Wildl. Dis. 7:256–257.

Manaro, A. J. 1961. Observations on the behavior of the spotted skunk in Florida. Q. J. Florida Acad. Sci. 24:59–63.

Marsh, R. E., and Howard, W. E. 1978. Vertebrate control manual: skunks. Pest Control 46:23, 30, 31.

Maynard, R. P. 1965. The biology and control of skunks in California. California Vector Views 12:17–20.

Mead, R. A. 1963. Some aspects of parasitism in skunks of the Sacramento Valley of California. Am. Midl. Nat. 70:164–167.

———. 1967. Age determination in the spotted skunk. J. Mammal. 48:606–616.

———. 1968*a*. Reproduction in eastern forms of the spotted skunk (genus *Spilogale*). J. Zool. 156:119–136.

———. 1968*b*. Reproduction in western forms of the spotted skunk (genus *Spilogale*). J. Mammal. 49:373–390.

Mead, R. A.; Rourke, A. W.; and Swannock, A. 1979. Changes in uterine protein synthesis during delayed implantation in the western spotted skunk and its regulation by hormones. Bio. Reprod. 21:39–46.

Orr, R. T. 1943. Altitudinal record for the spotted skunk in California. J. Mammal. 24:270.

Parker, R. L. 1961. Rabies in skunks in the north-central states. Proc. U.S. Livest. Sanit. Assoc. 65:273–280.

Peery, H. J. 1939. A new unarmed tapeworm from the spotted skunk. J. Parasitol. 25:487–490.

Polder, E. 1968. Spotted skunk and weasel populations den and cover usage by northeast Iowa. Iowa Acad. Sci. 75:142–146.

Schwartz, C. W., and Schwartz, E. R. 1959. The wild mammals of Missouri. Univ. Missouri Press and Missouri Conserv. Comm., Kansas City. 341pp.

Scott, T. G.; Willis, Y. L.; and Ellis, J. A. 1959. Some effects of field application of dieldrin on wildlife. J. Wildl. Manage. 23:409–427.

Selko, L. F. 1937. Food habits of Iowa skunks in the fall of 1936. J. Wildl. Manage. 1:70–76.

Sherwood, W. E. 1931. The ways of *Spilogale*. Nature 17:224–267.

Sinha, A. A., and Mead, R. A. 1976. Morphological changes in the trophoblast, uterus and corpus luteum during delayed implantation and implantation in the western spotted skunk. Am. J. Anat. 145:331–356.

Stains, H. J. 1958. Field key to guard hair of middle western furbearers. J. Wildl. Manage. 22:95–97.

Tiner, J. D. 1946. Some helminth parasites of skunks in Texas. J. Mammal. 27:82–83.

Van Gelder, R. G. 1953. The egg-opening technique of a spotted skunk. J. Mammal. 34:255–256.

———. 1959. A taxonomic revision of the spotted skunks (Genus *Spilogale*). Bull. Am. Mus. Nat. Hist. 117 (Art. 5):229–392.

WALTER E. HOWARD, Wildlife and Fisheries Biology, University of California, Davis, California 95616.

REX E. MARSH, Wildlife and Fisheries Biology, University of California, Davis, California 95616.

35

Striped
and Hooded Skunks

Mephitis mephitis and Allies

Alfred J. Godin

NOMENCLATURE

COMMON NAMES. Striped skunk, large striped skunk, skunk, polecat

SCIENTIFIC NAME. *Mephitis mephitis*

The meaning of *Mephitis* is pestilent or bad odor, referring to the characteristic obnoxious odor given off by the scent glands.

SUBSPECIES. Hall and Kelson (1959) recognized 13 subspecies of striped skunk occurring in North America: *M. m. avia, M. m. elongata, M. m. estor, M. m. holzneri, M. m. hudsonica, M. m. major, M. m. mephitis, M. m. mesomelas, M. m. nigra, M. m. notata, M. m. occidentalis, M. m. spissigrada,* and *M. m. varians.*

The subfamily of Mephitinae comprises the skunks. These stout carnivores with long, bushy tails are characterized by their highly developed musk glands, which can project a nauseous liquid up to a few meters. The coloration of black and white may comprise contrasting spots (*Spilogale*) or stripes (*Mephitis* and *Conepatus*).

The three extant genera of the Mephitinae include seven species of hog-nosed skunk (*Conepatus*), two species of spotted skunk (*Spilogale*), and one species of striped skunk and hooded skunk (*Mephitis*). Stains (1967) reported that the geologic range of skunks in North America is from the Early Oligocene to Recent.

DISTRIBUTION

The striped skunk occurs throughout southern Canada from Nova Scotia, the Hudson Bay country, and British Columbia to the United States (except the desert areas of the southwest) and extreme northern Mexico (figure 35.1).

DESCRIPTION

The striped skunk is known for its typical black and white striped pelage pattern and the unique odor of its musk. The animal is about the size of a house cat (*Felis catus*) but has a relatively small, somewhat triangular-shaped head tapering to a bulbous nose pad.

FIGURE 35.1. Distribution of the striped skunk (*Mephitis mephitis*).

The eyes are black and beady. The nictitating membrane is absent, a unique feature among carnivores (Smythe 1961). The ears are small and rounded. The body is robust, with a wide rump. The legs are short, somewhat shorter in front than in the rear. The feet are plantigrade; the soles of the feet are almost completely naked. The forefeet have five long, curved claws for digging; the five claws of the hindfeet are shorter and straighter. The tail is long and bushy. Females may have from 10 to 14 mammae; the usual number is 12.

As with all skunks, the striped skunk has a pair of well-developed oval scent glands, one on each side of the anus, embedded in a mass of sphincter muscles. These muscles can compress one or both glands so forcefully that the secretion can be ejected several meters. Each gland has a single duct leading to a prominent nipplelike papilla that can be protruded from the anus. Either separately or together, the glands can discharge a malodorous yellowish, oily, phosphorescent, nauseating musk sometimes containing creamy, yellowish curds.

The musk is a sulfur-alcohol compound known chemically as butylmercaptan. Taken internally in considerable doses, butylmercaptan produces unconsciousness, lowers the body temperature and blood pressure, slows the pulse, and acts as a depressant of the central nervous system (Jackson 1961).

The pungent musk causes an intense burning sensation when sprayed into the eyes but, contrary to popular belief, does not cause permanent ill-effects.

Sexual Dimorphism. Adult males are somewhat larger than adult females. The total length of adults varies from 520 to 770 mm, tail length is 170 mm to 400 mm, and hindfoot is 55 to 85 mm. Weights of adults range from 1.8 to 4.5 kg. Fat individuals sometimes weigh up to 5.5 kg.

Of 197 adult male and female striped skunks from Illinois, the mean weight of males is 2.6 kg and of females 2.0 kg. The mean monthly weights of the adult females increases substantially from August through October. Females gain in April and May, probably because of pregnancy, but they average only 114 g heavier in July than in March. The failure of the females to gain weight in June and July is probably caused by nursing and caring for the young (Verts 1967).

Both sexes of striped skunk lose weight during winter. In New York, weight losses were 13.8 percent of the summer weight for males and 38.0 percent for females (Hamilton 1937). In Michigan, weight losses were 36.3 percent for males and 31.6 percent for females (Allen 1939). In Illinois, weight losses were 47.7 percent for males and 55.1 percent for females (Verts 1967). In Minnesota, striped skunks lose 55–65 percent of their body weight during winter (Houseknecht 1969). In Manitoba, Canada, the weights of striped skunks increase in October and November, remain stable in December and January, decrease in February and March, and increase in April (Aleksiuk and Stewart 1977).

Measurements of skulls from 60 male and 53 female adult striped skunks from Illinois varied in total skull length from 69.5 to 88.0 mm, zygomatic breadth from 42.0 to 54.5 mm, and length of toothrow from 26.0 to 32.2 mm (Verts 1967). Measurements of 17 skulls from adult striped skunks from Louisiana varied in zygomatic breadth from 31.3 to 45.7 mm and length of maxillary toothrow from 19.7 to 22.7 mm (Lowery 1974).

Pelage. The pelage of the striped skunk is composed of underfur of soft wavy hairs 25.4 to 30.5 mm long, intermixed with glisterning guard hairs 38.1 to 76.2 mm long. The bases of the underfur within the black areas are dark grayish and the tips are black. The hairs forming the characteristic white stripe are uniformly white and are usually slightly longer than adjacent black hairs. The skin beneath the black hairs is grayish black, and beneath the white hairs is pinkish (Verts 1967).

The color of striped skunks is mostly glossy black but a thin white stripe runs from the nose to the back of the forehead and a broad white stripe extends from the crown of the head over the nape, branching at the shoulders, and continues posteriorly along the upper sides to the rump or onto the tip of the tail. The hairs of the tail are white at the bases. A white stripe may also occur on the outside of each front leg. Some individuals have a white patch on the chest. Some striped skunks' "stars" lack white stripes on the back. Immatures are striped more or less like adults.

The sexes are colored alike and show no noticeable seasonal change. Cream colored, brown, and all-black specimens sometimes occur, while silver and albino individuals are rare. Stains and Stuckley (1960) reported a striped skunk with brachial-antebrachial stripes each about 1.3 mm long on the lateral surface of each front leg.

In Illinois, molting of striped skunks begins in April when the underfur is shed in large wads of matted fur. These wads appear first in the scapular region and later in the lumbar and pelvic regions. Molting of the guard hairs begins in July, at which time both underfur and guard hairs are replaced. The direction of the molt is from anterior to posterior and is completed in early September. Females begin to molt about 2 weeks before males. Juveniles apparently do not molt until they are 11 to 12 months old (Verts 1967).

Skull and Dentition. The skull of the striped skunk is long, relatively narrow, highly arched, and deepest in the frontal region; the rostrum is deep and truncate. The auditory bullae are slightly inflated; the infraorbital foramina are small, narrow, and sometimes divided by thin septa. The posterior border of the short palate is nearly in line with the posterior borders of the upper molars. The squarish mandible has a distinct posterior notch (figure 35.2).

The teeth are relatively heavy, and the upper molars are somewhat dumbbell shaped. The upper rear molars are larger than the carnassials. The dental formula is 3/3, 1/1, 3/3, 1/2 = 34.

REPRODUCTION

Anatomy and Physiology. The female striped skunk reproductive organs are composed of a urogenital sinus, a vagina, a bipartite-type uterus, paired oviducts, ovaries, and uterine horns. The vagina is joined by the urethra, forming a urogenital sinus that opens to the exterior. There appears to be no os clitoris (Verts 1967). Leach and Conaway (1963) reported on the histological development of ovaries of 60 striped skunks 3 days to 11 weeks old. They found both monovular and polyovular follicles in ovaries from striped skunks 3 to 9 weeks old, and first observed degeneration of both types of follicle in skunks 11 weeks old. Verts (1967) reported that the polyovular follicles probably did not persist until the skunk's first breeding season.

Female striped skunks, at about five months old,

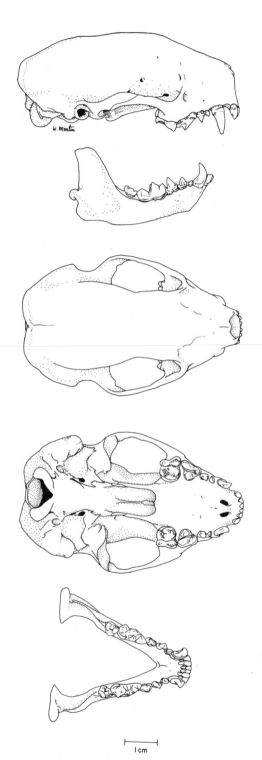

have ovaries that contain numerous primary follicles in the cortex. However, no bulging of the largest follicles occurs on the surfaces of the ovaries. At about seven months of age, blood vessels in the ovaries are much more numerous than in ovaries of skunks about five months old. Females about eight and a half months old have follicles that cause a slight bulging of the surfaces of the ovaries. In late February just prior to estrus, females about nine months old have much enlarged follicles, which cause pronounced bulging of the surfaces of the ovaries (Verts 1967).

The reproductive organs of male striped skunks consist of a somewhat fibrous penis, a urogenital sinus, a prostate gland, paired vas deferens, epididymides, and scrotal testes. A thin, slightly curved os baculum is present. The size of the testes and epididymides best expresses seasonal changes in the reproductive organs of male striped skunks. The mean monthly weights of testes of juveniles indicate that the size of the testes increases as a straight-line function of age at least until December, then increases at a decelerating rate until a peak is reached in March, after which the average size of the testes declines. Testes in adult males are smallest in August (Verts 1967).

Breeding Season. Striped skunks are polygamous and normally breed once a year during the brief breeding season, which occurs mainly in February or March throughout most of their range.

In Canada, the breeding season occurs in late February and early March (Banfield 1974). In Texas, it begins in February or March (Davis 1960). In Illinois, the breeding season is restricted to a relatively short period in late February and early March (Verts 1967).

Occasionally a female striped skunk will produce two litters in one year (Seton 1929). A captive female produced two litters, on 16 May and on 28 July (Shadle 1953).

Female striped skunks that have not bred during their first estrous cycle may have a second cycle about four weeks later (Seton 1929). One female striped skunk that was found dead had bred on 29 March, 31 days later than the earliest breeding date in Illinois. This late breeding date possibly represented that the female had bred during a second estrous cycle (Verts 1967). Although Shadle (1953) suggested that lactation inhibited the next sexual cycle, Parks (1967) stated that lactation did not seem to inhibit copulation or pregnancy.

Ernst (1965) observed the reproductive behavior of a docile 10-month-old de-scented captive male striped skunk. The animal altered his nocturnal activities in late January and nervously prowled about and examined every hiding place, sniffing loudly as he went. The young male became aggressive and alternately stamped his feet and charged at the attendant near the cage. The aggressive behavior period lasted 36 days and ended in early March. After the period of aggressive behavior, the male resumed his docile behavior.

Mating. Little is known about the mating behavior of striped skunks in the wild. Wight (1931) observed mating behavior by a captive pair of striped skunks. In

FIGURE 35.2. Skull of the striped skunk (*Mephitis mephitis*). From top to bottom: lateral view of cranium, lateral view of mandible, dorsal view of cranium, ventral view of cranium, dorsal view of mandible.

essence, when the estrous female was placed in a cage with the male, he approached her from the rear and smelled and licked her vulva. The male then grasped the female by the skin of the neck with his teeth and mounted her. The male made several rapid copulatory movements until intromission was achieved. The female remained passive until intromission. At intromission, the female often attempted to bite the male. Copulation lasted about one minute. After separation each animal acted indifferently toward the other. Neither animal vocalized during mating.

Captive males that have successfully bred one female usually immediately attempt to mount another female, and ignore the first female for several hours. Captive females that have bred usually fight viciously with males attempting to mount, but unbred females not in estrus usually do not fight with males attempting to mount and exhibit a receptive posture (Verts 1967).

Gestation Period. The gestation period is variable, reported to be from 66 to 75 days (Shadle 1953), but ordinarily it is 62 to 66 days (Verts 1967). The kits are born about the middle of May.

Parturition. Preparturition behavior of wild striped skunks is little known. In captive striped skunks Wight (1931) observed a female carrying nesting material 17 days before the litter was born. Shadle (1956) found that a female made a nest of shredded newspapers within two hours after giving birth to its litter. Verts (1967) observed that pregnant captive female striped skunks became unusually nervous and aggressive about one week before parturition. Even the slightest noise caused the females to rush from their nest boxes and stamp their front feet and make hissing noises. Before giving birth the females usually removed nearly all the straw from their nest boxes. The newborn were allowed to lie on the wooden floor of the nest box.

Litter Sizes. Female striped skunks produce 4 kits in their first litter and 6 to 8 in subsequent litters (Seton 1929). The average litter size usually consists of 5 or 7 kits, but may be 2 to 10 (Banfield 1974; Lowery 1974).

Growth, Development, and Longevity. At birth the kits are thinly furred, wrinkled, and blind. The mean weight of 12 striped skunks within 24 hours of birth was 33.4 g (range 26.7–41.4 g), and the mean total length was 131.4 mm (range 124–137 mm). The black and white patterns show distinctly on the pink skin, the whitish claws are well developed, and the sexes can be distinguished. The eyes open at about two to four weeks of age. About the time the eyes open, the young can assume the defensive posture and can discharge fluid from the scent glands (Verts 1967). Female striped skunks at first nurse their young by sprawling over them; as they become older, the female lies on her side. Weaning occurs when the kits are six to eight weeks old (Burns 1953).

The kits grow rapidly and follow the female on hunting trips when they are about two months old, often keeping close behind her in a long single file along a trail. Soon thereafter the kits usually become independent and disperse.

Striped skunks can breed during the spring following their birth (Verts 1967). Longevity ranges from 2 to 3½ years in the wild (Linduska 1947; Verts 1967). Captive striped skunks may live as long as 10 years (Schwartz and Schwartz 1959).

ECOLOGY

Striped skunks occur from sea level to over 4,000 m elevation, but rarely above about 2,000 m (Nelson 1918; Grinnell et al. 1937). They inhabit a variety of habitat types: rolling hayfields, fencerows, edges of woodlots, wooded ravines, patches of brush, rocky outcrops, recesses under stone walls or fences, and drainage ditches where water is available. They are also encountered under porches of houses and vacant buildings, in culverts, at the bases of trees, beneath the small branches of fallen trees, at rest sites in sedges (*Carex* spp.), and near dumps. Striped skunks may dig their own burrows, but they prefer to use natural cavities among rocks, caves, and holes under fallen logs and old stumps.

Most often striped skunks retreat to abandoned burrows of such mammals as nine-banded armadillos (*Dasypus novemcinctus*), badgers (*Taxidea taxus*), woodchucks (*Marmota monax*), red foxes (*Vulpes vulpes*), and muskrats (*Ondatra zibethicus*) (Selko 1938a; Allen and Shapton 1942; Jones 1950). Slopes seem to be the preferred den sites, probably because drainage is good. In Michigan, most striped skunk dens were on slopes of 5 to 10 percent or greater (Scott and Selko 1939). In Illinois, the mean slope was 5.3 percent for skunk dens (Verts 1967).

Where skunks dig their own dens the burrows are seldom very long or deep. Most burrows are from 1.8 to 6.1 m long, but an occasional burrow may extend 15 m or more. Burrows generally entend 0.9 to 1.2 m underground, and those less than 50 cm underground are apt to be summer dens without nests. The burrow ends in one to three special chambers 30 to 38 cm in diameter. The chambers may be lined with a bushel of mixed leaves, hay, and grass. In cold weather the natal nest material is used to plug the entrances. A den may have from 1 to 5 well-hidden entrances with openings about 20 cm in circumference, which are found more or less on the sunny side of slopes. The mean number of entrances ranges from 1.3 to 2.2 per den (Selko 1938a; Allen and Shapton 1942; Verts 1967).

Movements. Striped skunks normally do not roam far, though individuals may wander a considerable distance. Allen and Shapton (1942) tracked five striped skunks in February, and recorded that the longest straight-line distance of skunk tracks ranged from 548 m to 2,416 m. They believed that sexual activity contributed to these relatively extensive movements. Juvenile striped skunks appear to have a strong attraction to specific feeding grounds. They usually travel directly to their feeding grounds, frequently by the same routes, for several consecutive nights. Movements of juveniles averaged 39 m in late summer and autumn; after the breeding season, movements of adult males averaged about 0.4 to 0.8 km more. The range of pregnant females is about 0.4 km from their dens (Verts 1967). Both adult and juvenile males tend to move greater distances than females during late summer and autumn (Storm 1972).

Communal Denning. Striped skunks may den alone or in communal dens in any combination of sexes and ages, and the combination can vary from day to day. There may be a pair of adults, a female with her litter of the past season, or a male with several females.

In Minnesota, the males utilize aboveground rest sites throughout spring, summer, and early autumn, but the females utilize dens during parturition and nursing. In late June, there is a shift to aboveground nest sites in swamps and marshes when the juveniles begin to travel with the female. During late autumn both sexes begin to use underground dens more frequently, and much of the communal denning occurs during winter. Most skunks remained in a single den for 75–100 days (Houseknecht 1969).

Striped skunks sometimes winter in the same burrows with, but usually in different chambers from, opossums (*Didelphis virginiana*), woodchucks, and cottontail rabbits (*Sylvilagus* spp.).

FOOD HABITS

Striped skunks are omnivorous. They feed mainly on a variety of insects such as grasshoppers and crickets (Orthoptera), beetles (Coleoptera), and bees and wasps (Hymenoptera). They eat spiders (Arachnida), earthworms (*Lumbricus terrestris*), snails (Gastropoda), clams (Pelecypoda), crayfish (*Astacus* spp., *Combarus* spp.), salamanders (Urodela), lizards and toads (Bufonidae), frogs (Ranidae and Hylidae), eggs of turtles (Chelonia), snakes (Colubridae), and minnows (Cyprinidae). Striped skunks also prey on field mice (*Peromyscus* spp.), jumping mice (*Napaeozapus* spp., *Zapus* spp.), house mice (*Mus musculus*), voles (*Microtus* spp.), rats (*Neotoma* spp., *Rattus* spp.), bats (Vespertilionidae), moles (Talpidae), shrews (Soricidae), ground squirrels and chipmunks (Sciuridae), other small rodents, rabbits (Leporidae), an occasional bird and its eggs, carrion, and garbage. They are fond of numerous species of berry including blackberry (*Rubus allegheniensis*), black cherry (*Prunus serotina*), blueberries (*Vaccinium* spp.), ground cherry (*Physalis heterophylla*), gooseberries (*Ribes* spp.), raspberries (*Rubus* spp.), persimmons (*Diospyros* spp.), grasses, nuts, roots, grains, corn (*Zea mays*), and other vegetation.

Striped skunks are too slow to chase fast-moving prey for capture. They often hunt much as do domestic cats, by lying in wait or slowly stalking their intended victim. Striped skunks usually catch beetles and grasshoppers in summer twilight by springing on them with the front feet. After dark the skunks depend upon their senses of smell and hearing to locate prey. Verts (1967) observed a striped skunk, attached with a radio transmitter, canter for a distance of between 2.7 and 9.1 m in a cornfield in which the corn was about 25 cm tall. The skunk turned toward the prey (probably a cara-

bid beetle) at a right angle to the former direction of travel, made one or two bounces, and pounced like a house cat on the prey. The skunk apparently trapped the prey beneath its front feet and ate it immediately.

Striped skunks capture much of their prey by digging in the ground for insects and invertebrates or tearing apart ground nests of small mammals, often leaving the ground with small conical holes. They are known to scratch the front of beehives to induce bees to leave, and the bees are then caught and eaten. The skunks eat the larvae and probably the honey stored by the bees. Apparently striped skunks do not mind bee stings. Striped skunks may roll hairy caterpillars or toads on the ground with the forepaws to remove hair or skin poison before eating them.

Striped skunks can open an egg by propelling it by the front feet between the hind feet until it strikes a hard object and is broken (King 1944). When eating an egg, striped skunks bite off one end and lick out the contents, and leave the shell more or less together at the nest. Striped skunks crush the eggshell more than do spotted skunks, and their work can thus be identified (Schwartz and Schwartz 1959). Striped skunks may return to their own droppings to eat beetles feeding thereon (Chapman 1946).

Insects and mammals appear to be the preferred foods of striped skunks during spring and summer, and animal and plant matter during autumn and winter. In New York, fruit was the most important food of skunks in autumn and winter and was second in importance in spring and summer (Hamilton 1936). The diet of striped skunks in Maryland was 60 to 90 percent animals (Llewellyn and Uhler 1952). In Illinois, the frequency of occurrence of foods of animal origin and of plant origin were about equal during all seasons. Beetles and grasshoppers occurred with the greatest frequency, and carabid and scarabid beetles were the most frequently encountered beetles, possibly because of their nocturnal habits (Verts 1967). In Canada, the autumn and winter diet consisted of about 26.7 percent by bulk fruit, 19.6 percent small mammals, 13.5 percent carrion, 11.3 percent grains and nuts, 10.8 percent grasses, leaves, and buds, and 6.8 percent insects. During spring and summer the diet consisted of 43.0 percent insects, 27.5 percent fruits, 16.2 percent small mammals, and 8.7 percent grains (Banfield 1974).

Captive striped skunks eat both types of tenebrionid beetles, those with defensive chemical secretions and those without. Before eating beetles with defensive secretions the skunks roll the beetles in the ground, during which time the beetles discharge their foul secretions. When the supply of secretions is presumably exhausted, the beetles are consumed. The skunks appear to require an odor cue to recognize a beetle. Once the skunks recognize the beetle by odor, they use visual cues in choosing the largest beetle, regardless of whether or not it has defensive secretions (Slobodchikoff 1978).

Several biologists have noted predation by skunks on domestic and game species. Dice (1926) reported that striped skunks occasionally preyed on young domestic cats. Kalmbach (1938) observed that striped skunks destroyed 30.4 percent of 351 duck nests in North Dakota. In Pennsylvania, Beule (1941) reported that 4 of 24 cottontail rabbits nests were destroyed by skunks. Stoddard and Komarek (1941) reported that skunks destroyed many nests of bobwhite quail (*Colinus virginianus*) in the southeastern United States. In the Dakotas, Kimball (1948) found that badgers and striped skunks destroyed 55.7 percent of 350 ring-necked pheasant (*Phasianus colchicus*) nests. Davis (1960) reported that striped skunks preyed on small birds during spring and summer in Texas.

BEHAVIOR

The senses of sight and hearing are poor to fair. Skunks occasionally utter low growls, grunts, snarls, churring, short squeals, shrill screeches, or hissing noises. Captive juveniles are much more vociferous than adults, and pregnant or lactating females utter hissing noises whenever they are disturbed.

Striped skunks are not sociable animals, though as noted above several skunks of both sexes den together in winter. Adult males are usually intolerant of other males at all times of the year but may den together in winter (Seton 1929; Allen and Shapton 1942). Striped skunks are rather docile animals, and will respond to gentle treatment. But when provoked they will face their adversaries, give warnings of displeasure, and assume a defensive posture before discharging (scenting) their musk. The warnings include arching of the back, stamping the stiffened front legs rapidly on the ground, and shuffling backward. They may walk on their front feet for a short distance with the tail high in the air, and may click their teeth, growl, or hiss.

Just before scenting, a skunk raises its flaired plume tail and quickly bends into a U-shaped position with both the head and the rear facing the adversary. It scents swiftly. A striped skunk does not scatter the secretion with the tail, as is commonly believed. When discharging scent the skunk can aim behind, to either side of, or in front of itself by changing the direction of the aim of the nipples, so that the jet of musk covers an arch of 30 to 45°. By this means the striped skunk greatly increases the probability of hitting the adversary with the musk. No urine is released during scenting (Verts 1967).

The skunk discharges its musk either as an atomized spray of almost invisible droplets or as a short stream of rain-sized drops. The musk may be discharged accurately to a distance of about 3 m and somewhat less accurately for as much as 5 m. The smell can be detected for much greater distances.

Striped skunks generally avoid spraying themselves and will refrain from spraying if their tails are held tightly over the anus or if the anus is held tightly against the ground. A striped skunk can be handled without danger of biting or spraying if it is grasped around the neck with one hand and by the tail with the other. The animal is turned on its back with the tail pointed away and downwind, and the back prevented from arching. A skunk can spray while suspended by the tail.

Striped skunks are not agile and normally walk slowly at about 1.6 km per hour. The running speeds of two striped skunks were about 9.7 km per hour for 137.2 m and between 12.8 and 14.5 km per hour for 91.4 km (MacLulich 1936). The running speed of striped skunks for short distances is about 16.1 km per hour (Verts 1967).

Striped skunks are mainly terrestrial and are not climbers, though juvenile captive skunks have climbed on the woven-wire sides of their cages (Verts 1967). Striped skunks usually avoid water but will swim when necessary (Cole 1921). They are able to swim for at least 463 minutes (7.7 hours) in water at 23° C (Wilber and Weidenbacher 1961). A juvenile skunk, when released, traveled rapidly to a stream about 182.8 m away, drank, went into the water, and swam downstream for about 45.7 m (Verts 1967).

Daily and Seasonal Activities. Striped skunks are crepuscular or nocturnal but are sometimes active during daylight hours. They usually emerge about sunset to hunt. Newly weaned skunks appear to spend the daylight hours in cornfields (Verts 1967). Striped skunks extensively utilize hayfields, pastures, fence-rows, and waterways as daytime resting sites (Storm 1972). In areas where rabies in skunks is prevalent, a skunk seen during the daytime is likely to be rabid (Parker 1962).

Juvenile striped skunks in Illinois apparently become active between 1800 and 1900 h during August through October, remain active throughout the night, and return to their dens or resting sites from 0500 to 0600 h. Activity of the juveniles did not appear to be affected by rain or strong winds, but on very cold nights the skunks occasionally retired to their dens after being active for two to four hours (Verts 1967).

During late summer, striped skunks acquire much body fat, and by late autumn and colder weather they spend most of the time in the underground den. They do not hibernate but merely go into a winter sleep, since their rate of metabolism does not correspond to lower ambient temperatures.

In the northern portion of their range, striped skunks apparently remain more or less inactive for extended periods during winter (Hamilton 1937; Selko 1938*b;* Jones 1939; Terres 1940). In Texas, striped skunks are active throughout the year, though possibly less active in summer than in winter (Davis 1951).

There appeared to be no close correlation between ambient temperature and resumption of activity by skunks in spring (Smith 1931). At lower temperatures male striped skunks were more active than females, but low temperatures did not prevent skunks from being active, especially during the breeding season (Hamilton 1937). Striped skunks remained active in January and February, when the mean monthly ambient temperatures were −16.8° C and −12.7° C, respectively (Selko 1938*b*). Females remained in their dens in winter but males made periodic sorties aboveground (Allen 1939). Temperatures below freezing reduced, but did not prevent, activity of skunks (Jones 1939). A captive skunk was noted as "hibernating" when the mean ambient temperature was as high as 3° C (Terres 1940). During February and March skunks were active 7 to 23 nights. The mean minimum temperature of 7 nights on which the animals were active was −10.8° C, and on the nights when the skunks were inactive the mean minimum temperature was −9.4° C. There was no significant difference between the numbers of nights on which skunks were active when minimum temperatures were above and below −9.3° C. Snow cover on nights when the skunks were active ranged from 12.7 to 25.4 cm, and some skunks appeared to have been trapped in their dens by crusted snow (Verts 1967).

A combination of factors such as temperature, sudden changes in temperature, snow cover, crusted snow over dens, hunger, and sexual drive probably determine the times and duration of activity periods in striped skunks. Aleksiuk and Stewart (1977) found no evidence of natural hypothermia of four captive striped skunks held outdoors in midwinter near Winnipeg, Manitoba, Canada. Both food intake and physical activity of the skunks showed a strong correlation with ambient temperature. The skunks were largely nocturnal in October, November, and April, less so from December to March (inclusive). They hypothesized that the characteristics of winter dormancy of striped skunks would be more fully expressed in the absence of a winter food supply.

MORTALITY

Humans are the main predator of striped skunks. Coyotes (*Canis latrans*), foxes (*Vulpes vulpes* and *Urocyon cinereoargenteus*), badgers, lynx (*Felis canadensis*), bobcats (*Felis rufus*), mountain lions (*Felis concolor*), fishers (*Martes pennanti*), golden eagles (*Aquila chrysaetos*), great horned owls (*Bubo virginianus*), and barred owls (*Strix varia*) are known to prey on striped skunks, but they are usually near starvation when they do so (Bent 1938; Cahalane 1961; Young and Jackson 1951; Hall 1955; Young 1958; Banfield 1974). Most dogs (*Canis familiaris*) avoid skunks after an initial encounter. Some dogs, however, seem to become persistent skunk killers. Many young nursing skunks die from starvation when adult lactating females are killed by predators or automobiles and farm marchines. In addition, some skunks die from starvation when they are forced to retire to their winter dens earlier than usual, or when individuals are forced to remain in their winter dens during extremely prolonged winters under deep snow.

Insecticides, including dieldrin, used to control insects injurious to field crops are suspected of being toxic to skunks (Scott et al. 1959). Scott et al (1959) stated, "Wildlife species show differential vulnerability to poisoning by dieldrin. This seems to result from characteristic behavior and the degree of inherent resistance to the poison. Warm-blooded vertebrates which spend a good deal of time on the ground and include insects in the diet, especially during the season when the poison is applied, are highly vulnerable." They included the striped skunk in this group of species.

Parasites. Striped skunks are parasitized by many ectoparasites and endoparasites. Stegman (1939) believed that fleas, mites, and lice of striped skunks increased materially in numbers during the winter, when the skunks remain within the dens, since the ectoparasites all complete their life cycle on the skunks and within the den. The ectoparasites of striped skunks are listed in table 35.1.

Babero (1960) and Verts (1967) summarized the many studies concerning endoparasites of striped skunks. Endoparasites include 3 protozoans, 6 acanthocephalans, 10 cestodes, 25 nematodes, and 9 trematodes. Verts (1967) reported botfly larvae (*Cuterebra* spp.) from two striped skunks collected in Illinois. Additional endoparasites have been reported by Webster (1967), Dyer (1969a, 1969b), Webster and Casey (1970), Bemrick and Schlotthauer (1971), Kirkland (1975), and Cawthorn and Anderson (1976a, 1976b).

Among 79 striped skunks collected in New York, those examined in late summer and autumn were more heavily infested with endoparasites than those examined in late winter and spring. The seasonal difference in rates of infestation may be attributed to starvation of the parasites when the digestive tracts of the skunks were empty during their period of winter dormancy (Stegman 1939). In addition, rates of infestation of skunks collected in Illinois appeared to decline in midsummer and to increase rapidly in early autumn (Verts 1967). Verts (1967) could not explain the decline in the number of intestinal parasites during midsummer on the basis of known skunk habits or behavior.

Disease. Striped skunks are susceptible to many diseases, especially leptospirosis (*Leptospira* spp.) and rabies. Leptospirosis is probably the world's most

TABLE 35.1. Ectoparasites of striped skunks

Species	Source
Flea	
Cediopsylla simplex	Jackson (1961)
Chaetopsylla lotoris	Verts (1967)
Ctenophthalmus felis	Jackson (1961)
Ctenophthalmus pseudagyrtes	Verts (1967)
Megabothris asio	Verts (1967)
Nosophyllus fasciatus	Lowery (1974)
Opiscocrotis bruneri	Verts (1967)
Orchopeas howardi	Verts (1967)
Oropsylla arctomys	Jackson (1961)
Pulex irritans	Jackson (1961)
Lice	
Trichodextes mephitis	Jackson (1961)
Mites	
Hirstionyssus staffordi	Verts (1967)
Tick	
Amblyomma americana	Jackson (1961)
Dermacentor variabilis	Jackson (1961)
Ixodes cookei	Verts (1967)
Ixodes hexagonus	Lowery (1974)
Ixodes kingi	Jackson (1961)

TABLE 35.2. Occurrences of Leptospirosis in striped skunks

Number of Skunks	Percentage Occurrence	Geographic Location	Source
286	60.1	Louisiana	Roth (1961)
650	57.4	Louisiana	Roth et al. (1963)
106	34.9	Illinois	Verts (1967)
No data	33.3	Pennsylvania	Clark (1962)
10	20.0	eastern Canada	McKiel et al. (1961)
430	15.8	Georgia	Gorman et al. (1962)
132	13.6	Georgia	McKeever et al. (1958a)

widespread zoonosis. It occurs particularly when infested animals, urine, water, mud, and humans come together. The bacteria of this disease are frequently shed by dogs, rodents, cattle, swine, and wildlife. The bacteria enter humans through abrasions or cuts in the skin, or through the intact membranes of the mouth, nose, or eyes. They produce fever, muscular pains, and jaundice. Tabel (1970) reported that the kidneys of striped skunks chronically infested with *Leptospira pomona* contained a conspicuous number of plasma cells. The occurrences of leptospirosis in striped skunks are listed in table 35.2.

Rabies, or hydrophobia, one of the most important zoonoses, is a fatal viral disease affecting the central nervous system of mammals, including humans. It is caused by the filterable virus *Formido ineroxabilis*. There are two kinds of rabies: the furious form and the dumb form. In the furious form, the infected animal is aggressive. It snaps at objects placed before it, and will attack any animal and people. In the dumb form of the disease, the infected animal is not vicious. It has no tendency to bite or roam. It is not excitable. In fact, it may be the opposite.

In essence, rabies is transmitted to humans by the bite of an animal that excretes the virus with its saliva. Rabies is the only viral disease in which specific vaccine and serum treatment can successfully be applied after infection. Rabies is prevalent in many parts of the range of striped skunks. The incidence of rabies in striped skunks varies yearly.

The U.S. Public Health Service, Center for Disease Control (CDC), Rabies Surveillance Annual Report, 1978, summarized that of the laboratory-confirmed cases of rabies, skunks constituted 59.6 percent in wildlife in 1977. The states that reported over 100 cases in skunks were Minnesota (260), Texas (257), California (247), Oklahoma (188), and South Dakota (105). Thirteen states reported increases in the number of skunk cases in comparison with the preceding year, with Minnesota (+115), Oklahoma (+66), Texam (+52), and California (+37) reporting the greatest increases. Sixteen states reported decreases in comparison with 1976 cases, with Montana (−35) and Arkansas (−28) showing the greatest decreases.

Verts (1967) noted that skunks infected with rabies were most frequently reported in the North Central states, Texas, and California. Martin and Leonard (1949) reported that in the Midwest, the distribution of rabies in skunks appeared to conform fairly closely to the distribution of corn-producing areas. Richards (1957) found that in North Dakota, 90 percent of the skunks with rabies did not scent; many did not scent while dying. Verts and Storm (1966) indicated that rabies was not often mutually transmitted between foxes and striped skunks in Illinois. Verts (1967) noticed that captive rabid skunks were usually quiet for relatively long periods unless stimulated by movements, loud noise, or strong light. Usually, the irritated animals appeared completely exhausted before they ceased their efforts to reach the sources of irritation, often lying on their sides and gasping for several minutes from exertion. Verts (1967) stated that during copulation or attempted copulation male striped skunks almost invariably bite the females and that females often reciprocated. Houseknecht (1969) implied that these activities would seem to provide a way for direct transmission of leptospirosis and rabies within the winter den.

Burkel et al. (1970) found that the fluorescent antibody technique correctly identified rabies virus longer than either of the other two diagnostic rabies detection techniques (microscopic examination for Negri bodies, and mouse inoculation used in rabies detection in road-killed skunks). Jacobson et al. (1970) improved a technique for handling striped skunks in rabies investigation.

Striped skunks are subject to other diseases, some of common occurrence, others rare. Since population levels of striped skunks appear to fluctuate widely from unknown causes, the following is a discussion of some of the diseases reported in striped skunks.

Pulmonary aspergillosis is a respiratory disease caused by the fungus *Aspergillosis fumigatus*. Cheesy nodules, from pinpoint size to about 13 mm across, occur in the lungs and air sacks of the infected animal. Durant and Doll (1939) isolated aspergillosis from a juvenile captive striped skunk in Missouri. They suspected that the skunk became infected in the basement, where moldy grain was kept. Verts (1967) stated that the prevalence of aspergillosis among wild skunks is unknown, but in damp weather, moldy nesting material could be a source of aspergillosis in skunks.

Pleuritis, or inflammation of the lining membranes of the chest cavity, often causes death. The lungs become compressed and covered with adhesions. Some of the bacteria that have been isolated from this pus include Streptococcus and Micrococcus species. Junod and Beydek (1945) reported a pleuritis condition in a wild skunk in Ohio, but did not report on organisms involved in the infection.

Histoplasmosis is an infection characterized by diarrhea, chronic cough, weakness, anemia, and general emaciation. The disease is caused by the fungus *Histoplasmoa capsulatum*. The mortality rate in the acute cases is high. No definite proof has been found to indicate that histoplasmosis is transmitted from one animal to another. Emmons et al. (1955) isolated the disease in 2 of 18 striped skunks in Virginia. Menges et al. (1955) reported a striped skunk infected with histoplasmosis in Kansas.

Listeriosis is an infectious disease that affects animals and humans. The disease is caused by a bacterial organism (*Listeria monocytogenes*). Listeriosis causes changes in the cells of the nervous system. It may be considered an insidious disease; the body may harbor the organisms for a long period before the disease appears in clinical form. Little success has been reported for treatment of listeriosis. The disease has been isolated from wild striped skunks in North Dakota (Bolin et al. 1955) and California (Osebold et al. 1957).

Canine distemper is an acute infection disease caused by a filterable virus usually complicated by many bacterial secondary invaders. In infected animals, the pelage roughens. The appetite decreases. Usually inflammation of the lining of the nasal cavities begins one to two days after the first fever. Death occurs when any group of vital muscles is affected by paralysis. Helmboldt and Jungherr (1955) diagnosed canine distemper in a striped skunk that exhibited symptoms of rabies but was found negative by mouse inoculation for rabies.

Canine hepatitis is a viral disease that affects primarily the endothelial and liver cells. It often includes symptoms of acute shock, followed by coma and death. Infectious canine hepatitis is difficult to differentiate from leptospirosis and canine distemper. Karstad et al. (1975) reported hepatitis in skunks.

Q-fever is a rickettsial disease. It causes no signs of illness in animals studied in the United States and only occasionally produces clinical signs of infection in humans. Verts (1967) indicated that skunks were not important reservoirs of Q-fever.

Tularemia is caused by the bacterium *Pasteurella tularensis*. Ticks or insects that have fed on infected animals may transmit tularemia to skunks. This disease is transmissible to humans. McKeever et al. (1958b) believed that striped skunks were one of the main mammalian reservoirs of tularemia in Georgia.

Chagas' disease is caused by a group of parasitic hemoflagellates (*Trypanosoma* spp.) that are blood inhabiting. These parasites cause some of the most serious diseases of humans and livestock, such as African sleeping sickness. Norman et al. (1959) found 1.1 percent of 306 striped skunks taken in southern Georgia and northern Florida to be infected with *Trypanosoma cruzi*-like hemoflagellates.

Although renal lesions have not been reported as a major problem in striped skunks, Crowell et al. (1977) found inflammation in 74 percent of the renal tissue from 100 striped skunks from Louisiana.

AGE AND SEX DETERMINATION

There is no single criterion for determining ages of striped skunks. Schwartz and Schwartz (1959) suggested using the length and shape of the os baculum of the penis to distinguish males less than one year old from older males, and the color, length, and diameter of teats to distinguish young females from parous females. In males less than one year old the os baculum is less than 1.9 cm long, slender, irregularly curved, and without an enlarged basal portion. The os baculum of adult males is about 2.3 cm long, somewhat stouter and less curved, with an enlarged basal portion.

In cased skins or pelts the teats of adult females are at least 2 mm in diameter and 2.5 mm long, usually dark; they are usually flesh colored and under 1 mm long in juveniles (Petrides 1950). Verts (1967) determined ages within limits of 1 month to 19 months by using a combination of criteria of normal parturition dates, the closure of the epiphyseal cartilage of the radius and ulna bones, and the dry weight of the eye lens. He found that closure of the epiphyses was completed at about 8 to 9 months of age. In summer and autumn, young-of-the-year striped skunks could be distinguished from older skunks by this criterion alone. He noted that the dry weight of the eye lens also appeared to increase with age and that lenses of males appeared to be slightly heavier than those of females of the same age.

Casey and Webster (1975) determined age and sex of 351 females and 399 males, from Ontario, Manitoba, and Quebec, Canada. In determining the age, they counted the number of annular layers in the tooth cementum. Sex was determined by examining the hippocampal neurons for sex chromatin. Of 750 skunks, 376 were less than one year of age. The oldest skunks were between five and six years old.

ECONOMIC STATUS

The chief economic value of striped skunks lies in their destruction of injurious animal life. Their value may be more than what is normally credited to them. For example, striped skunks consume a prodigious amount of insects injurious to agriculture, such as army worms (*Cirphis unipuncta*), cutworms (*Chorizagrotis* spp.), tobacco bud worms (*Heliothis virescens*), hop-plant bugs (*Taedia hawleyi*), Colorado potato beetles (*Leptinotarsa decemlineata*), the larvae and adults of scarab beetles (Scarabaeidae), May beetles (*Phyllophaga fusca*), June beetles (*Cotinis nitida*), grasshoppers, cicadas (Cicadidae), crickets (Gryllidae), sphinx moths (Sphingidae), and squash bugs (Coreidae). Striped skunks are excellent mousers. They prey on mice and rats that infest barns.

The amount and distribution of white in striped fur pelts varies considerably among individuals. The fur industry grades skunk pelts according to the amount of white in the pelage. In the best grade, or no. 1 "star," the white may be absent or restricted to the head and neck. The no. 2 "short stripe" pelts have white stripes that do not extend beyond the middle of the body. No. 3 "narrow stripe" pelts have long, narrow stripes, and no. 4 "broad stripe" pelts are broad striped. Northern pelts are more valuable than southern pelts because the fur is finer and the black color more intense (Lantz 1923; Bachrach 1953).

As a source of fur, striped skunks are an important asset to the fur industry. In the cooler parts of their range their economic value as producers of fur entitles striped skunks to protection. It is difficult to obtain complete and reliable statistics as to the monetary income derived from the annual skunk fur. However, the pelt brings trappers a substantial income annually. Thousands of skunks are trapped annually for their fur. The pelts are durable, thick, rich, and glossy, and are used for coats, trimmings, jackets, muffs, and scarfs. In preparing a garment, the white portion is cut out and the pelt sewed together without it, or the white portion may be dyed. The tails are used for manufacturing brushes.

Some authors (Merriam 1884; Seton 1929) have declared that the flesh of skunks is good, if properly prepared, and that it was a source of food among North American Indians and trappers.

The liabilities of striped skunks are many, although usually less serious than those of some other species of wildlife. The chief cause for concern is the odor. The charge is often made that skunks kill poultry and destroy eggs. A skunk usually takes a single bird at a time, but once it forms the habit, it may repeatedly visit the poultry yard. Such a guilty individual should be disposed of. Upon investigation, however, it is frequently found that the real culprits are weasels (*Mustela* spp.) or mink (*Mustela vison*) or even rats (*Rattus* spp.), all of which are adept climbers and kill far more ruthlessly.

Their fondness for insects causes skunks to make depredations upon beehives. Skunks occasionally dig holes in lawns and meadows for grubs and other invertebrates.

Occasionally, striped skunks may be important disease vectors. Diseases such as rabies or distemper may be transmitted through skunk populations to wild and domestic animals.

Where striped skunks are numerous, people find them a nuisance. A main complaint against skunks is their tendency to den under the floors of houses and other buildings, and other places of human habitation. Sometimes several striped skunks den near houses and other buildings, and the penetrating skunk odor becomes highly obnoxious.

Striped skunks sometimes do considerable damage to sweet corn when it is in the milky stage. However, it is necessary to determine by close inspection which of the many wild mammals cause depredation to corn. Skunks, raccoons (*Procyon lotor*), and woodchucks (*Marmota monax*) can strip the husks from the entire ear of corn. Skunks and raccoons shred the husks

as they pull them away with their pointed canine teeth. Striped skunks usually make more, finer shreds and limit their damage to the corn near the ground. Raccoons are stronger and frequently break off an entire ear, often carrying it away. Woodchucks cut the husks smoothly with their incisors.

MANAGEMENT

It seems from the liabilities of striped skunks that they do more damage than good. That is not true. The instances reported above are the more unusual and those that occur least frequently. In spite of the fact that striped skunks do damage at times, they are more beneficial than harmful.

Because of their economic value, skunks should be prevented from establishing themselves where they will become a nuisance rather than being killed needlessly. If they are numerous, seal all openings through which any small mammal might enter the foundations of porches, garages, and buildings. Striped skunks are not efficient burrowers. They can often be discouraged from denning and be driven away by making den sites difficult to find. If they are already beneath buildings, a method of elimination is to seal off with suitable materials, such as cement, boards, or wire netting, all openings except one. Sprinkle a flour patch at the opening and examine it after dark. If trail signs indicate that the skunk(s) has left the den, the remaining opening should be closed off immediately. In the spring of the year, beware of a litter of young left behind in a den.

When skunks are found under buildings, they can sometimes be driven away by the use of repellents. About 2 kg of naphthalene moth flakes sprinkled around in the den is satisfactory for this purpose.

Striped skunks can be easily excluded from poultry houses by closing all doors and other openings each night. Properly built skunk-proof fencing will keep skunks out of open chicken yards. The range fence is a 0.9-m wire netting fence, 0.6 m of which should be above and 0.3 m below the surface. Bend outwardly at right angles 15 cm of the part below the surface and bury it 15 cm deep. When the skunk starts digging down along the vertical wire fence, it will become discouraged and will stop digging when it strikes the horizontal flange.

Skunks damaging lawns can be indirectly managed by ridding the lawns of grubs and other insects. It is suggested that information on the control of lawn insects be obtained from a county agricultural agent, the state university, state agricultural experiment station, or the state fish and game department for advice.

As a precaution against depredation by skunks on bees, beehives should be raised at least 1 m above the ground, or a temporary skunk-proof fence placed around groups of beehives.

Striped skunks are easily trapped. When they must be removed from an area, they should preferably be live-trapped by a baited box or wire-cage trap covered with burlap. Skunks seldom return if they are released 3 km away.

Poisoning of skunks is not recommended, because of their close association with humans and domestic animals. Also, many states protect skunks most of the year but usually permit an open season when the skunk fur is prime. If skunks are injurious to property, the owner or occupant can usually obtain a permit to control them. Persons wishing to take skunks should familiarize themselves with local and state ordinances. Local game wardens or the state game department will furnish this information.

Skunk odors on pets, clothing, under buildings, etc., may be removed by use of a deodorant, neutroleum alpha, which may be purchased from hospital supply houses or pest control operators.

CURRENT RESEARCH NEEDS

More information is needed on the communal denning behavior, genetics, reproductive biology, and distribution of striped skunks.

NOMENCLATURE

COMMON NAMES. Hooded skunk, white-sided skunk, southern skunk
SCIENTIFIC NAME. *Mephitis macroura*
SUBSPECIES. Hall and Kelson (1959) recognized five subspecies of hooded skunk occurring in North America: *Mephitis m. eximius, M. m. macroura, M. m. milleri, M. m. richardsoni,* and *M. m. vittata.*

DISTRIBUTION

The hooded skunk is a southern skunk. It occurs from southwestern Texas, southwestern New Mexico, and southern Arizona south through Mexico and into Nicaragua (figure 35.3). Packard (1965) reported that

FIGURE 35.3. Distribution of the hooded skunk (*Mephitis macroura*).

an adult female *M. m. milleri* was taken near Royalty, Ward County, Texas. This specimen extended the known geographic range approximately 121 km northeast of the Mount Livermore, Jeff Davis County, Texas location previously reported by Blair (1940). Packard (1974) also reported that a second adult female *M. m. milleri* was obtained 16 km northwest of Guaymas, Sonora, Mexico, in the San Carlos Bay area. This specimen represented an extension of the geographic range 112.6 km southwest of that previously reported by Hall and Kelson (1959).

DESCRIPTION

The hooded skunk, when compared with the striped skunk, is a small, slender animal with a very long tail. It differs from *M. mephitis* in having a ruff of long hairs on the neck and back of the head that usually spreads out into a hood or cape; by the tail, which is usually as long as the body; and by the dorsal white stripe rarely divided into a V. The hooded skunk differs from the hognosed skunk (*Conepatus mesoleucus*) in having a smaller size, smaller snout, much longer tail, and finer fur.

Sexual Dimorphism. The total length of adult male hooded skunks ranges from 558 to 790 mm, tail length 357 to 400 mm, and hindfoot length 60 to 68 mm (Bailey 1932; Hall and Kelson 1959; Davis 1960; Hubbard 1972). An adult male collected in July weighed 965 g and an adult female with three embryos, trapped in June, weighed 1,212 g (Davis and Lukins 1958).

Pelage. The hooded skunk has two color patterns: the white-backed pattern and the black-backed pattern, with intermediate variants. In the white-backed pattern, the back and top of the tail are mainly white and mixed with some black hairs, often with a narrow white stripe along each side behind the shoulder, and the belly is black or mottled with white. In the black-backed pattern, the back and the tail are chiefly black, and the white is reduced to a narrow lateral stripe along each side of the body, instead of being narrowly separated on the back of the animal. Females have 10 mammae (Cahalane 1961).

Skull and Dentition. In the skull, the hooded skunk is distinguished from the striped skunk by larger tympanic bullae. The dental formula is the same as that of the striped skunk.

REPRODUCTION

Bailey (1932) recorded five embryos in a female hooded skunk. Davis and Lukens (1958) reported that a female collected on 23 June at Colotlipa, Mexico, contained three embryos 35 mm in rump-crown length.

ECOLOGY

The hooded skunk occurs in a variety of habitats from sea level to 2,440 m altitude (Hubbard 1972). Hooded skunks are uncommon, but are widespread in Guerrero, Mexico, below elevations of 1,830 m (Davis and Lukens 1958). The species inhabits mainly low-elevation desert but has been taken in high-elevation ponderosa pine (*Pinus ponderosa*) forest (Findley et al. 1975). Hooded skunks appear to be more abundant in riparian vegetation along permanent watercourses and near water in forested or shrubby uplands than in arid areas of the desert plain (Baker 1956). Hooded skunks have been observed foraging along streams, where they frequented the rocky ledges and tangles of streamside vegetation for safety. Occasionally they resorted to burrows in the banks and washes. A hooded skunk was trapped in a heavy stand of willows (*Salix* spp.) along the sandy bank of Tornillo Creek, Brewster County, Texas. The animal had been feeding in the vicinity with hog-nosed skunks (Davis 1960).

Hooded skunks have been reported as occurring in desert, grassland, and riparian habitats of the northern region of the Chihuahuan Desert of the northern portion of the Mexico Plateau (Findley and Caire 1974), from the desert and grassland habitats of the southern portion of the Chihuahuan Desert (Packard 1974), and from the mesquite (*Prosopis* spp.) grassland and high Montane pine-oak (*Pinus-Quercus* spp.) woodland habitats in the Chihuahuan Desert region (Schmidly 1974).

BEHAVIOR

The behavior of the hooded skunk is little known. Reed and Carr (1949) chased a large hooded skunk for 0.4 km in the semiarboreal Sonoran desert, about 24.1 km west of Tucson, Pima County, Arizona, about midafternoon. The skunk took refuge under a jumping cholla (*Opuntia fulgida*) where a woodrat (*Neotoma* spp.) had its nest. The skunk picked its way warily, but seemingly surely, between the dried cholla piled upon the ground until it got in the middle of the woodrat nest. Then the skunk put its nose to the ground and pushed down through the heaped cholla into one of the burrows below. They stated: "This incident would indicate that woodrats, even though covering their nests with cholla segments, are not immune to predatory skunks."

FOOD HABITS

Bailey (1932) reported that the stomach of a hooded skunk trapped near a river in New Mexico was filled with fragments of large black beetles, but the stomachs of others were empty, except for bark and sticks, which the skunks had chewed up while in the traps.

Regarding hunting behavior of hooded skunks, "The stomachs taken in mild weather usually contain insects, principally those obtained by digging. In settled areas these animals show a preference for fields and trails in the lower valleys. Here they search intensively the nooks and corners, where numerous shallow diggings indicate that ground-inhabiting insects, including beetles have been detected and unearthed" (anonymous 1939).

MORTALITY

Erickson (1946) reported the nematode *Physaloptera maxillaris* from a hooded skunk.

CURRENT RESEARCH NEEDS

Little is known concerning the biology, ecology, and distribution of the hooded skunk. Research needs include the physiology, genetics, food habits, reproductive biology, and habitat requirements of this species.

LITERATURE CITED

Aleksuik, M., and Stewart, A. P. 1977. Food intake, weight changes and activity of confined skunks (*Mephitis mephitis*) in winter. Am. Midl. Nat. 98:331–332.

Allen, D. L. 1939. Winter habits of Michigan skunks. J. Wildl. Manage. 3:212–228.

Allen, D. L., and Shapton, W. W. 1942. An ecological study of winter dens, with special reference to the eastern skunk. Ecology 23:59–68.

Anonymous. 1939. Hooded skunk. Nature 32:272.

Babero, B. A. 1960. A survey of parasitism in skunks, *Mephitis mephitis*, in Louisiana, with observations on pathological damages due to helminthiasis. J. Parasitol. 46:26–27.

Bachrach, M. 1953. Fur: a practical threatise. 3rd ed. Prentice-Hall, Inc., New York. 660pp.

Bailey, V. 1932. Mammals of New Mexico. North Am. Fauna 53:1–412.

Baker, R. H. 1956. Mammals of Coahuila, Mexico. Univ. Kansas Publ. Nat. Hist. 9:125–335.

Banfield, A. W. F. 1974. The mammals of Canada. Univ. Toronto Press, Toronto and Buffalo. 438pp.

Bemrick, E. J., and Schlotthauer, J. C. 1971. *Paragonimus kellicotti* (Ward, 1908) in a Minnesota skunk (*Mephitis mephitis*). J. Wildl. Dis. 7:36.

Bent, A. C. 1938. Life Histories of North American birds of prey. Part 2. U.S. Natl. Mus. Bull. 170:1–482.

Beule, J. D. 1941. Cottontail nesting-study in Pennsylvania. Trans. North Am. Wildl. Conf. 5: 320–328.

Blair, W. F. 1940. A contribution to the ecology and faunal relationships of the mammals of the Davis Mountain Region, southwestern Texas. Misc. Publ. Mus. Zool., Univ. Michigan 46:1–39.

Bolin, F. M.; Turn, J.; Richards, S. H.; and Eveleth, D. F. 1955. Listeriosis of a skunk. Bimonthly Bull. North Dakota Agric. Exp. Stn. 18:49–50.

Burkel, M. D.; Andrews, M. F.; Meslow, E. C. 1970. Rabies detection in road-killed skunks (*Mephitis mephitis*). J. Wildl. Dis. 6:496–499.

Burns, E. 1953. The sex life of wild animals. A North American study. Rinehart & Co., Inc., New York. 290pp.

Cahalane, V. H. 1961. Mammals of North America. Macmillan Co., New York. 682pp.

Casey, G. A., and Webster, W. A. 1975. Age and sex determination of striped skunk (*Mephitis mephitis*) from Ontario, Manitoba, and Quebec. Can. J. Zool. 53:223–226.

Cawthorn, R. J., and Anderson, R. C. 1976a. Development of *Physaloptera maxillaris* (Nematoda: Physalopterioidea) in skunk (*Mephitis mephitis*) and the role of paratenic and other hosts in its life cycle. Can. J. Zool. 54:313–323.

_____. 1976b. Seasonal population changes of *Physaloptera maxillaris* (Nematoda: Physalopteriodea) in striped skunk (*Mephitis mephitis*). Can. J. Zool. 54:522–525.

Chapman, F. B. 1946. An interesting feeding habit of skunks. J. Mammal. 27:397.

Clark, L. G. 1962. Leptospirosis in Pennsylvania: a progress report. Proc. U.S. Livest. Sanit. Assoc. 65:140–146.

Cole, H. E. 1921. A swimming skunk. Wisconsin Conserv. 3:6.

Crowell, W. A.; Stuart, B. P.; and Adams, W. V. 1977. Renal lesions in striped skunks (*Mephitis mephitis*) from Louisiana. J. Wildl. Dis. 13:300–303.

Davis, W. B. 1951. Texas skunks. Texas Game and Fish 9:18–21, 31.

———. 1960. The mammals of Texas. Texas Game and Fish Comm. Bull. 27:1–252.

Davis, W. B., and Lukens, P. W. 1958. Mammals of the Mexican state of Guerrero, exclusive of Chiroptera and Rodentia. J. Mammal. 39:347–367.

Dice, L. R. 1926. Skunk eats kittens. J. Mammal. 7:131.

Durant, A. J., and Doll, E. R. 1939. Pulmonary aspergillosis in a skunk. J. Am. Vet. Med. Assoc. 95:645–646.

Dyer, W. G. 1969a. Helminths of the striped skunk, *Mephitis mephitis*, North America. Am. Midl. Nat. 82:601–605.

———. 1969b. *Skrjabingylus chitwoodorum* (Nenatoda: Pseudaliidae) from *Mephitis mephitis* in north central North Dakota. Bull. Wildl. Dis. Assoc. 5:140.

Emmons, C. W.; Rowley, D. A.; Olson, B. J.; Mattern, C. F. T.; Bell, J. A.; Powell, E.; and Marcey, E. A. 1955. Histoplasmosis: Proved occurrence of inapparent infection in dogs, cats and other animals. Am. J. Hyg. 61:40–44.

Erickson, A. B. 1946. Incidence of worm parasites in Minnesota Mustelidae and host list and keys to North American species. Am. Midl. Nat. 36:494–509.

Ernst, G. H. 1965. Rutting activities in a captive striped skunk. J. Mammal. 46:702–203.

Findley, J. S., and Caire, W. 1974. The status of mammals in the northern region of the Chihuahuan Desert. Pages 127–139 *in* R. H. Wauer and D. H. Riskind, eds. Transactions of the symposium on the biological resources of the Chihuahuan Desert Region, United States and Mexico. U.S. Natl. Park Serv. Trans. and Proc. Ser., no. 3, 1977.

Findley, J. S.; Harris, A. H.; Wilson, D. E.; and Jones, C. 1975. Mammals of New Mexico. Univ. New Mexico Press, Albuquergue. 360pp.

Gorman, G. W.; McKrever, S.; and Grimes, R. D. 1962. Leptospirosis in wild mammals from southwestern Georgia. Am. J. Trop. Med. and Hyg. 11:518–524.

Grinnell, J.; Dixon, J. S.; and Linsdale, J. M. 1937. Furbearing mammals of California: their natural history, systematic status, and relations to man. Univ. California Press, Berkeley. Vol. 1. 375pp.

Hall, E. R. 1955. Handbook of mammals of Kansas. Univ. Kansas Mus. Nat. Hist. Misc. Publ. 7. 303pp.

Hall, E. R., and Kelson, K. R. 1959. The mammals of North America. Ronald Press, New York. 2 vols. 1162pp.

Hamilton, W. J., Jr. 1936. Seasonal food of skunks in New York. J. Mammal. 17:240–246.

———. 1937. Winter activity of the skunk. Ecology 18:326–327.

Hemboldt, C. F., and Jungherr, E. L. 1955. Distemper complex in wild carnivores simulating rabies. Am. J. Vet. Res. 16:463–469.

Houseknecht, C. R. 1969. Denning habits of the striped skunk and the exposure potential for disease. Bull. Wildl. Dis. Assoc. 5:302–306.

Hubbard, J. P. 1972. Hooded skunk on the Mongollon Plateau, New Mexico. Southwestern Nat. 16:458.

Jackson, H. T. J. 1961. Mammals of Wisconsin. Univ. Wisconsin Press, Madison. 504pp.

Jacobson, J. O.; Meslow, E. C.; and Andrews, M. F. 1970. An improved technique for handling striped skunks in disease investigations. J. Wild. Dis. 6:510–512.

Jones, F. H. 1950. Natural history of the striped skunk in northeastern Kansas. M.A. Thesis. Univ. Kansas, Lawrence. 38pp.

Jones, H. W., Jr. 1939. Winter studies of skunks in Pennsylvania. J. Mammal. 20:254–256.

Junod, F. L., and Bezdek, H. 1945. Pleuritis in wild skunk. J. Mammal. 26:309–310.

Kalmbach, E. R. 1938. A comparative study of nesting waterfowl on Lower Souris Refuge, 1936–1937. Trans. N. Am. Wildl. Conf. 3:610–623.

Karstad, L.; Ramsen, B.; Berry, T. J.; and Binn, L. N. 1975. Hepatitis in skunks caused by the virus of infectious canine hepatitis. J. Wildl. Dis. 11:494–496.

Kimball, J. W. 1948. Pheasant population characteristics and trends in the Dakotas. Trans. North Am. Wildl. Conf. 13:291–311.

King, D. C. 1944. Skunk and egg. Reader's Digest 44:85.

Kirkland, G. L., Jr. 1975. Parasitosis of the striped skunk (*Mephitis mephitis*) in Pennsylvania by the nasal nematode (*Skrjabingylus chitwoodorun*). Proc. Pennsylvania Acad. Sci. 49:51–53.

Lantz, D. E. 1923. Economic value of North American skunks. U.S. Dept. Agric. Farmers's Bull. 587:1–24.

Leach, B. J., and Conaway, C. H. 1963. The origin and fate of polyovular follicles in the striped skunk. J. Mammal. 44:67–74.

Linduska, J. P. 1947. Longevity of some Michigan farm game mammals. J. Mammal. 28:126–129.

Llewellyn, L. M., and Uhler, F. M. 1952. The foods of fur animals of the Patuxent Research Refuge, Maryland. Am. Midl. Nat. 48:193–203.

Lowery, G. H., Jr. 1974. The mammals of Louisiana and its adjacent waters. Louisiana State Univ. Press, Baton Rouge. 565pp.

McKeever, S.; Gorman, G. W.; Chapman, J. F.; Galton, M. M.; and Powers, D. K. 1958a. Incidence of leptospirosis in wild mammals from southwestern Georgia, with a report of new hosts for six serotypes of leptospires. Am. J. Trop. Med. Hyg. 7:646–655.

McKeever, S.; Schubert, J. H.; Moody, M. D.; and Chapman, J. F. 1958b. Natural occurrence of tularemia in marsupials, carnivores, lagomorphs, and large rodents in southwestern Georgia and northwestern Florida. J. Infect. Dis. 103:120–126.

McKiel, J. A.; Cousineau, J. G.; and Hall, R. R. 1961. Leptospirosis in wild animals in eastern Canada with particular attention to the disease in rats. Can. J. Comp. Med. Vet. Sci. 25:15–18.

MacLulich, D. A. 1936. Running speeds of skunk and European hare. Can. Field Nat. 50:92.

Martin, J. H., and Leonard, W. H. 1949. Principals of field crop production. Macmillan Co., New York. 1176pp.

Menges, R. W.; Habermann, R. T.; and Stains, H. J. 1955. A distemper-like disease in raccoons and isolation of *Histoplasma capsulatum* and *Haplosporangium parvum*. Trans. Kansas Acad. Sci. 58:58–67.

Merrian, C. H. 1884. The mammals of the Adirondacks. Henry Holt Co., New York. 516pp.

Nelson, E. W. 1918. Smaller North American mammals. An intimate study of the smaller wild animals of North America by the foremost authorities. Natl. Geogr. Soc. Mag. 33:477–479.

Norman, L.; Brooke, M. M.; Allain, D. S.; and Gorman, G. W. 1959. Morphology and virulence of *Trypanosoma cruzi*-like hemoflagellates isolated from wild mammals in Georgia and Florida. J. Parasitol. 45:457–463.

Osebold, J. W.; Shultz, G.; and Jameson, W. W., Jr. 1957. An epizootiological study of listeriosis. J. Am. Vet. Med. Assoc. 130:471–475.

Packard, R. L. 1965. Range extension of the hooded skunk in Texas and Mexico. J. Mammal. 46:102.

———. 1974. Mammals of the southern Chihuahuan Desert: an inventory. Pages 141–153 *in* R. H. Wauer and D. H. Riskind, eds. Transactions of the symposium on the biological resources of the Chihuahuan Desert Region, United States and Mexico. U.S. Natl. Park Serv. Trans. and Proc. Ser., no. 3. 1977.

Parker, R. L. 1962. Rabies in skunks in the North-Central States. Proc. U.S. Livest. Sanit. Assoc. 65:273–280.

Parks, E. 1967. Second litters in the striped skunk. New York Fish Game J. 14:208–209.

Petrides, G. A. 1950. The determination of sex and age ratios in fur animals. Am. Midl. Nat. 43:355–382.

Reed, C. A., and Carr, W. H. 1949. Use of cactus as protection by hooded skunk. J. Mammal. 30:79–80.

Richards, S. 1957. Rabies in North Dakota wildlife. North Dakota Outdoors 20:4–5, 16.

Roth, E. E. 1961. Leptospirosis in striped skunks. M. S. Thesis. Texas A. & M. Coll., College Station. 58pp.

Roth, E. E.; Adams, W. V.; Sanford, G. E., Jr.; Greer, B.; Newman, K.; Moore, M.; Mayeux, P.; and Linder, D. 1963. The bacteriologic and serologic incidence of leptospirosis among striped skunks in Louisiana. Zoonoses Res. 2:13–39.

Schmidly, D. J. 1974. Factors governing the distribution of mammals in the Chihuahuan Desert Region. Pages 163–192 *in* R. H. Wauer and D. H. Riskind, eds. Transactions of the symposium on the biological resources of the Chihuahuan Desert Region, United States and Mexico. U.S. Natl. Park Serv. Trans. and Proc. Ser., no. 3. 1977.

Schwartz, C. W., and Schwartz, E. R. 1959. The wild mammals of Missouri. Univ. Missouri Press and Missouri Conserv. Comm., Columbia. 341pp.

Scott, T. G., and Selko, L. F. 1939. A census of red foxes and striped skunks in Clay and Boone counties, Iowa. J. Wildl. Manage. 3:92–98.

Scott, T. G.; Willis, Y. L.; and Ellis, J. A. 1959. Some effects of field application of dieldrin on wildlife. J. Wildl. Manage. 23:409–427.

Selko, L. F. 1938a. Notes on the den ecology of the striped skunk in Iowa. Am. Midl. Nat. 20:455–463.

———. 1938b. Hibernation of the striped skunk in Iowa. J. Mammal. 19:320–324.

Seton, E. T. 1929. Lives of game animals. Doubleday, Doran & Co., Garden City, N. Y. 949pp.

Shadle, A. R. 1953. Captive striped skunk produces two litters. J. Wildl. Manage. 17:388–389.

———. 1956. Parturition in a skunk, *Mephitis mephitis hudsonica*. J. Mammal. 37:112–113.

Slobodchikoff, C. N. 1978. Experimental studies of teneb-rionid beetle predation by skunks. Behaviour 66:313–322.

Smith, W. P. 1931. Calendar of disappearance and emergence of some hibernating mammals at Wells River, Vermont. J. Mammal. 12:78–79.

Smythe, R. H. 1961. Animal vision: what animals see. Charles C. Thomas, Springfield, Ill. 250pp.

Stains, H. J. 1967. Carnivores and pinnipeds. Pages 325–365 *in* S. Anerson and J. K. Jones, Jr., eds. Recent mammals of the world. Ronald Press Co., New York.

Stains, H. J., and Stuckley, D. 1960. Brachial-antebrachial stripes on a striped skunk. J. Mammal. 41:139.

Stegeman, L. C. 1939. Some parasites and pathological conditions of the skunk (*Mephitis mephitis nigra*) in central New York. J. Mammal. 20:493–496.

Stoddard, H. L., and Komarck, E. V. 1941. Predator control in southeastern quail management. Trans. North Am. Wildl. Conf. 6:288–293.

Storm, G. L. 1972. Daytime retreats and movements of skunks on farmlands in Illinois. J. Wildl. Manage. 36:31–45.

Storm, G. L., and Verts, B. J. 1966. Movements of a striped skunk infected with rabies. J. Mammal. 47:705–708.

Tabel, H. 1970. Local production of antibodies to *Leptospira pomana* in kidneys of chronically infected skunks (*Mephitis mephitis*). J. Wildl. Dis. 6(4):299–304.

Terres, J. K. 1940. Notes on the winter activity of a captive skunk. J. Mammal. 21:216–217.

U.S. Public Health Service, Center for Disease Control. 1978. Rabies Surveillance Report, Annual Summary, 1977. (Issued September 1978.)

Verts, B. J. 1967. The biology of the striped skunk. Univ. Illinois Press, Urbana. 218pp.

Verts, B. J., and Storm, G. L. 1966. A local study of prevalence of rabies among foxes and striped skunks. J. Wildl. Manage. 30:419–21.

Webster, W. A. 1967. *Filaroides mephitis* N. sp. (Metastrongyloidea: Filariodidae) from the lungs of eastern Canadian skunks. Can. J. Zool. 45:145–147.

Webster, W. A., and Casey, G. A. 1970. The occurrence of *Drancunulus insignis* (Leidy, 1858) Chandler, 1942, in a skunk from Ontario, Canada. J. Wildl. Dis. 6:71.

Wight, H. M. 1931. Reproduction in the eastern skunk (*Mephitis mephitis nigra*). J. Mammal. 12:42–47.

Wilber, C. G., and Weidenbacher, G. H. 1961. Swimming capacity of some wild mammals. J. Mammal. 42:428–429.

Young, S. P. 1958. The bobcat of North America. Stackpole Co., Harrisburg, Pa. 153pp.

Young, S. P., and Jackson, H. H. T. 1951. The clever coyote. Stackpole Co., Harrisburg, Pa. 411pp.

ALFRED J. GODIN, U.S. Fish and Wildlife Service, P.O. Box 1684, Federal Building, 402 East State Street, Trenton, New Jersey 08607.

36

River Otter

Lutra canadensis

Dale E. Toweill

James E. Tabor

NOMENCLATURE

COMMON NAMES. Northern river otter, Canadian otter, land otter, fish otter
SCIENTIFIC NAME. *Lutra canadensis*
SUBSPECIES. *L. c. brevipilosus, L. c. canadensis, L. c. chimo, L. c. degener, L. c. evexa, L. c. extera, L. c. interior, L. c. kodiacensis, L. c. lataxina, L. c. nexa, L. c. optiva, L. c. pacifica, L. c. periclyzomae, L. c. preblei, L. c. sonora, L. c. texensis, L. c. vaga, L. c. vancouverensis,* and *L. c. yukonensis.*

The 19 subspecies above are those listed by Hall and Kelson (1959), who also show the Prince of Wales Island otter (*Lutra mira*) and the southern river otter (*Lutra annectens*) as North American forms. A taxonomic revision of the Lutrinae (the taxonomic subfamily that includes all forms of otter) has been done by van Zyll de Jong (1972), who suggests consolidating the 19 subspecies above and the Prince of Wales Island form into 7 subspecies of *Lutra canadensis* but leaving the southern river otter as a separate species.

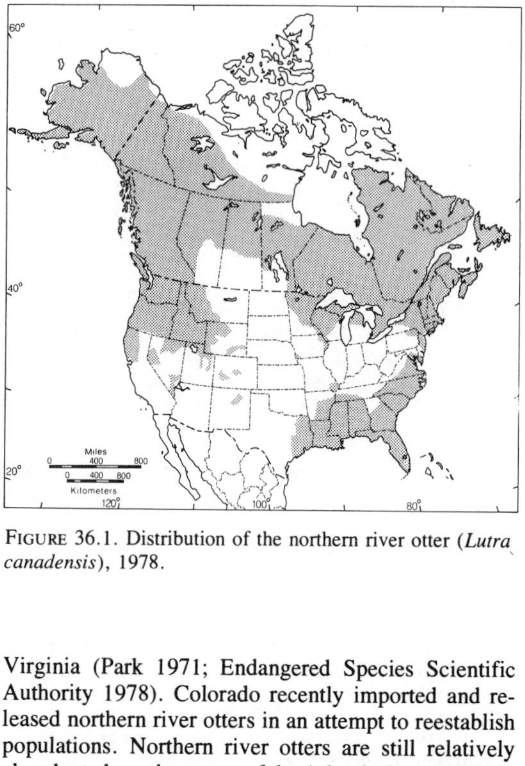

FIGURE 36.1. Distribution of the northern river otter (*Lutra canadensis*), 1978.

DISTRIBUTION

The northern river otter was found historically over much of the North American continent (Hall and Kelson 1959). Along with the beaver (*Castor canadensis*) and the timber wolf (*Canis lupus*), it occupied one of the largest geographic areas of any North American mammal, an area estimated by Anderson (1977) to encompass 2×10^7 km². Present distribution of the northern river otter extends from 25° north latitude in Florida to beyond 70° north latitude in Alaska (figure 36.1), and from eastern Newfoundland to the Aleutian Islands.

Northern river otters were found in all major waterways of the United States and Canada until at least the eighteenth century. Settlement and attendant changes in habitat, and perhaps overharvest from some portions of their range, resulted in their extirpation from some areas. Otters are extirpated or rare in Arizona, Colorado, Indiana, Iowa, Kansas, Kentucky, Nebraska, New Mexico, North Dakota, Ohio, Oklahoma, South Dakota, Tennessee, Utah, and West

Virginia (Park 1971; Endangered Species Scientific Authority 1978). Colorado recently imported and released northern river otters in an attempt to reestablish populations. Northern river otters are still relatively abundant along the coasts of the Atlantic Ocean and the Gulf of Mexico, throughout the Pacific Northwest and Great Lakes states, and across most of Canada and Alaska. Populations were listed as stable or increasing in 1978 in Alabama, Alaska, Arkansas, Connecticut, Delaware, Florida, Georgia, Idaho, Louisiana, Maryland, Massachusetts, Michigan, Minnesota, Mississippi, Montana, New Hampshire, New York, North Carolina, Oregon, Rhode Island, Texas, Vermont, Washington, and Wisconsin (Endangered Species Scientific Authority 1978). There is recent evidence that otters are currently expanding their range northward in Alaska (Magoun and Valkenburg 1977).

Because northern river otters are adapted to existence in freshwater habitats, the primary barriers to otter dispersal have been considered to be arid areas,

FIGURE 36.2. The northern river otter (*Lutra canadensis*), in typical habitat.

mountain ranges, glaciated areas, and salt water straits (Pohle 1920). Thus, the arid southwestern portion of North America may function as a barrier between northern and southern forms, and the North Atlantic Ocean and the Bering Sea between northern river otters and European and Asian forms. Glaciers and sea ice limit northward range. The degree to which mountain ranges and salt-water straits serve as barriers to otter dispersal has perhaps been overstated, however. There are numerous accounts of northern river otters making extensive overland movement, and Laughlin (1955), Morejohn (1969), and Magoun and Valkenburg (1977) have reported apparent range extension of northern river otters across mountain ranges. Similarly, otters along both the Atlantic and Pacific coasts make extensive use of estuarine areas such as the Chesapeake Bay (Mowbray et al. 1979), and forms along the northern Pacific coast from Washington northward through the Alexander Archipelago inhabit brackish and marine environments including the Strait of Juan de Fuca and the exposed outer coast (R. Hirschi personal communication 1979).

DESCRIPTION

In general body conformation, the northern river otter resembles a long cylinder that reaches its greatest diameter in the thoracic region (Tarasoff et al. 1972). The head is rather blunt, small, and somewhat flattened. It is characterized by a bulbous nose on the end of a short muzzle, small rounded ears set well back, and eyes set high on the head and closer to the nose than to the ears. The neck is thick and cylindrical. Legs are short and stocky, and the feet are pentadactyl and plantigrade, with interdigital webs. The tail is relatively long, thick, and pointed (figure 36.2).

Size and Weight. Adult northern river otters measure 915 to 1270 mm in length, with the tail accounting for just over one-third of the total. Weight ranges from 5.0 to 13.7 kg (Harris 1968). The largest of the North American otters, the Prince of Wales Island otter, occurs in southeastern Alaska, while the smallest form, *Lutra c. degener*, is found in Newfoundland; a clinal decrease in size is not evident, however, from west to east (van Zyll de Jong 1972). There is some evidence for a clinal decrease from north to south in the size of forms occurring along the Pacific Coast.

Sexual dimorphism in size is evident among all subspecies of northern river otter, with females being smaller than the males (Harris 1968; van Zyll de Jong 1972).

Pelage. Northern river otters are characterized by a relatively short but very dense fur. It ranges in color from a rich, dark chocolate brown to a pale chestnut dorsally and light brown to silver gray ventrally. Differences in length and density of the fur are related to climate, with northern forms having the longest and most dense pelage (van Zyll de Jong 1972). Similarly, western and southern forms tend to be lighter in color than northern and eastern forms.

Skull, Dentition, and Skeleton. The skull is depressed dorsoventrally and the dorsal edge of the profile presents nearly a straight line (figure 36.3). The rostrum is short and blunt. The point of maximum constriction is well forward, just posterior to the broad, flat interorbital area; behind this constriction, the cranium bulges to nearly the posterior margin of the skull. The auditory bullae are flattened.

Dentition, while less massive than that of the sea otter (*Enhydra lutra*), is heavy when compared with most other Mustelids. Teeth are adapted for crushing (Grinnell et al. 1937). The dental formula is: 3/3, 1/1, 4/3, 1/2 = 36. Supernumerary premolars have been reported (Dearden 1954).

Northern river otters have 14 rib-bearing vertebrae and normally 52 vertebrae in total—7 cervical, 14 thoracic, 6 lumbar, 3 sacral, and 22 caudal (van Zyll de

Jong 1972). Fisher (1942) discussed the osteology and myology of northern river otters.

Karyotype. The diploid (2n) number of chromosomes is 38. Thirteen pairs of autosomes are metacentric or submetacentric, while six pairs are acrocentric or sub-acrocentric (Wurster and Benirschke 1968).

PHYSIOLOGY

Few studies have been published concerning the physiology of northern river otters. Iversen (1972) reported that the basal metabolic rate of otters and other mustelids weighing 1 kg or more could be expressed by the equation $M = 84.6\ W^{0.78}\ (+\ 0.15)$, where M is the basal metabolic rate in kcal/day and W is the body weight in kg. This is about 20 percent higher than expected from the mammalian standard curve described by $M = 70\ W^{0.75}$ (Iversen 1972).

Comparative studies have indicated that northern river otters exhibit typical terrestrial locomotory patterns while at the same time showing adaptations of the feet, tail, and fur typical of more highly aquatic mammals (Tarasoff et al. 1972). Modifications of the respiratory system for aquatic life, including an increase in relative lung size, have also been described (Tarasoff and Kooyman 1973a, 1973b).

REPRODUCTION

Anatomy. Female northern river otters have a typically Mustelidlike bicornuate uterus, similar in morphology to that described by Sinha et al. (1966) for the sea otter. The uterine horns of a multiparous adult from New York were 12 mm in diameter and were 90 mm long; the body of the uterus of this animal was 30 mm long (Hamilton and Eadie 1964). Uteri of nonparous otters in their first year are nonvascular, translucent, and light in color; vascularization increases and uterine walls become darker with increasing age and growth of the uterus (Tabor 1974). Uterine measurements are useful in determining the age of otters up to three years (Tabor 1974; Lauhachinda 1978).

Ovaries of immature river otters and one-year-old animals are smaller than those of adult animals and lack enlarging follicles and corpora lutea (Tabor 1974; Lauhachinda 1978). Otters less than two years of age may show developing follicles generally less than 0.25 mm in diameter (Tabor 1974).

Adult female northern river otters may develop an os clitoridis, the female counterpart to the male os baculum (Scheffer 1939). The os clitoridis is a cartilagenous structure in females less than two years old, but may ossify in older individuals (Lauhachinda 1978).

Testes of adult male northern river otters fluctuate seasonally in size depending on the level of development and sperm production. Testes of otters from Oregon 31 months old and older averaged 37.8 x 24.4 mm in length and weighed an average of 12.5 g prior to the peak of breeding activity (Tabor 1974). As is typical of most Mustelidae, adult male river otters normally ex-

hibit a well-developed os baculum. This bone ranges from 56.6 to 102.9 mm in length and from 0.72 to 8.43 g in weight. The baculum increases in length until an otter reaches about three years of age; weight increases until about six years of age (Stephenson 1977). In appearance, the os baculum of northern river otters is shaped like a thin S with a sharply distinct, rough proximal end, a smooth shaft, and a slightly enlarged distal end with a V-shaped urethral groove on its dorsal surface.

Physiology. Northern river otters normally do not reach sexual maturity until two years of age (Liers 1951b; Hamilton and Eadie 1964; Tabor and Wight 1977; Lauhachinda 1978), although a few females may breed at 15 months (Liers 1958; Mowbray et al. 1979).

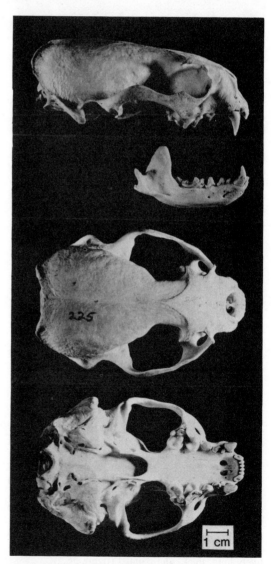

FIGURE 36.3. Skull of the northern river otter (*Lutra canadensis*). From top to bottom: lateral view of cranium, lateral view of mandible, dorsal view of cranium, ventral view of cranium.

The ovary and uterus do not reach normal adult size until an otter is two years old (Hamilton and Eadie 1964; Tabor 1974; Lauhachinda 1978).

Males begin to produce sperm at about two years of age. Hamilton and Eadie (1964) reported that no otter examined at 20 to 21 months of age in New York as producing sperm but that sperm production did occur in otters as young as 23 to 24 months of age. Tabor (1974) reported that 32 percent of the males 19 to 22 months of age in Oregon were producing sperm. Testes descend into the scrotum in November (Liers 1951b).

Although males become sexually mature at about two years of age, they may not become successful breeders until they reach five to seven years (Liers 1951b).

Breeding. Northern river otters apparently breed in late winter or early spring, with the breeding season spread over a period of three months or longer. In the northern Mackenzie River District breeding occurs from March through May (Macfarlane 1905), in Michigan from January through May with a peak in March or April (Hooper and Ostenson 1949), in Minnesota from December through early April (Liers 1951b), and in March or April in New York (Hamilton and Eadie 1964). The estrous period of an individual female extends 42 to 46 days, with peaks in the period of receptiveness occurring about every 6 days (Liers 1951b).

Copulation normally occurs in the water (Liers 1951b; McDaniel 1963) but may also occur on land. Liers (1951b:5) described copulation among otter as follows:

> The male approaches the female from the rear, holds the female by the scruff of the neck with his teeth, and bends the posterior part of his body around and below the broad tail of the female. Breeding is very vigorous when contact is made. The vigorous periods are interspersed with periods of rest. In two that I timed the copulatory process lasted 16 minutes and 24 minutes, respectively.... A pair may copulate several times in successive days. The female caterwauls when breeding.

Receptive female otters may advertise their condition by marking at scent stations or haul-outs (Erlinge 1968b). Liers (1951b) reported that otter hunters sometimes locate such areas and that they may kill two or three males following a female in estrus. He reported that male otters are normally solitary and do not form pair bonds with females, although they may fight to prevent other males from approaching a female in estrus. Home range of a solitary male may overlap that of one or more females (Erlinge 1967).

Adult female northern river otters apparently breed each year in Oregon, based on a pregnancy rate near 100 percent (Tabor and Wight 1977). However, of otter examined in Maryland, only 4 of 16 animals with regressing corpora lutea (indicating pregancy the previous year) also exhibited active corpora lutea (Mowbray et al. 1979); Lauhachinda (1978) concluded

that alternate- year breeding apparently occurred in Alabama and Georgia, based on a 50 percent pregnancy rate and the finding that nonpregnant adults of sufficient age had typically had litters the previous year. One captive female in Minnesota bred successfully only once each two to three years over four successive litters (Liers 1951b).

Gestation Period and Prenatal Growth. There is much confusion in the literature regarding the length of gestation in the northern river otter. Liers (1951b, 1958) recorded 288 to 375 days gestation for eight litters of captive animals. By contrast, the gestation period for the European river otter is 61 to 63 days (Ognev 1931). The extreme length of the gestation period for the northern river otter is due to a process called "delayed implantation," a phenomenon common to most Mustelids, wherein the development of the blastocyst is arrested for a period before it implants into the uterine wall. Mechanisms for this phenomenon are not known, although experiments with other Mustelids indicate that it may be related to photoperiod (Pearson and Enders 1944). The exact duration of neither the inactive (unimplanted) nor the active (implanted) stage of pregnancy is known. However, Kenyon (1969), using a method developed by Huggett and Widdas (1951), estimated that the active period of pregnancy lasted about 50 days.

Hamilton and Eadie (1964) reported that northern river otters bred in March or April in New York, but that development of the blastocyst was apparently arrested until January or February. No embryo was recovered from otters collected prior to 11 February, although 10 of 12 females examined had bred, as indicated by the presence of either blastocysts or corpora lutea. Implanted embryos were recovered during February, March, and April. They concluded that the unimplanted stage of development lasted approximately 240 to 285 days (Hamilton and Eadie 1964). Lauhachinda (1978) reported implantation dates ranging from 19 November to 8 January for otters from Alabama and Georgia, while Tabor (1974) found no implanted embryo prior to 3 February in Oregon.

The average ovulation rate for northern river otters has been reported to range from 2.40 to 3.02 (table 36.1). In one study (Lauhachinda 1978), mean numbers of corpora lutea were reported to increase slightly (from 2.4 to 3.3 per female) from age 3 to 6 and then decline to an average of 3.0 for otters aged 7 to 15.

Embryonic growth proceeds rapidly once implantation is complete (Huggett and Widdas 1951).

Parturition and Growth. Parturition may occur from November (Liers 1951b) through May (Bailey 1936; Hooper and Ostenson 1949; Mowbray et al. 1979) or possibly even later (Grinnell et al. 1937). Most authors agree that the peak time for parturition occurs in March or April (Coues 1877; Grinnell et al. 1937; Hooper and Ostenson 1949; Hamilton and Eadie 1964; Tabor and Wight 1977; Mowbray et al. 1979), although there are many reports of litters being born in January or February (Hooper and Ostenson 1949; Liers 1951b, 1958; Crandall 1964). This wide variation in the timing of

TABLE 36.1. Mean numbers of corpora lutea and implanted embryos reported for pregnant northern river otters two years of age and older

State	Period of Collection	Corpora Lutea			Implanted Embryos			Source
		Number of Females	Mean	Range	Number of Females	Mean	Range	
New York	November–April	15	2.40[a]	1–5	9	2.11[a]	1–3	Hamilton and Eadie 1964
Oregon	November–February	43	3.02	2–4	4	2.75	2–3	Tabor and Wight 1977
Alabama and Georgia	November–February	59	2.9	1–4	48	2.6	1–4	Lauhachinda 1978
Maryland	January–March	31	2.74	1–4	22	2.73	1–4	Mowbray et al. 1979

[a]Calculated from data presented in Hamilton and Eadie (1964), table 2.

parturition has long been attributed to differences in latitude of the respective populations. However, Harris (1968) has reviewed much of the available data and has shown that such a belief is unfounded. As pointed out by Mowbray et al. (1979), parturition dates may not be closely synchronized even within a given population.

Litter sizes range from one to six (Hooper and Ostenson 1949), with two to four young produced most commonly. Mean litter sized based on fetal counts has been reported as 2.29 (Hamilton and Eadie 1964); 2.75 (Tabor and Wight 1977), and 2.73 (Mowbray et al. 1979). Intrauterine mortality levels apparently are low (Hamilton and Eadie 1964; Tabor and Wight 1977); Mowbray et al. (1979) suggested that it may reach 9 percent for the Maryland population they studied.

Prior to parturition, the female normally retires alone to a natal den. The young are born as the female stands upright on all four feet; the process may require three to eight hours (Liers 1951b). Newborn otters weight about 132 g and are about 275 mm in total length. Although blind and helpless, the auditory canals of the ears are open, and neonates are fully furred and appear as smaller replicas of adult otters (Liers 1951b). Newborn otters are without erupted teeth (Hamilton and Eadie 1964).

Otter milk has a high content of both fat and protein and is low in carbohydrates. Analysis shows the following relative proportions of components: water 62.0 percent, fat 24.0 percent, protein 11.0 percent, carbohydrates 0.1 percent, and ash 0.75 percent (Ben-Shaul 1962). Young otters grow rapidly on this diet. Liers (1951a) reported that an otter pup that weighed 187 g at 7 to 10 days of age weighed 454 g 10 days later.

Although helpless at birth, young otters begin to open their eyes by the age of 21 to 35 days, and by 25 to 42 days they begin playing with each other and with their mother (Liers 1951a; Harris 1968). The pups may be introduced to the water by the age of 48 days and may venture out of the den on their own by the age of 59 to 70 days (Liers 1951b; Harris 1968). By 49 days, young began to use a specific, localized area for defecation. At the age of 63 to 76 days they begin eating solid food, although weaning does not occur until about 91 days (Liers 1951b; Harris 1968).

ECOLOGY

Habitat. As indicated by their vast geographic range, northern river otters have adapted to a wide variety of aquatic habitats, from marine environments to high mountain lakes. Northern river otters are generally most abundant along food-rich coastal areas, such as the lower portions of streams and rivers and in estuaries, and in areas having extensive nonpolluted waterways and minimal impact by humans (Wilson 1959; Tabor and Wight 1977; Mowbray et al. 1979). R. Hirschi (personal communication 1979) found high densities of northern river otters not only in the estuarine areas of coastal Washington but also throughout the marine areas of the San Juan Archipelago and the Strait of Juan de Fuca. Northern river otters are also relatively abundant in major nonpolluted river systems and in the lakes and tributaries that feed them. They are scarce, however, in heavily settled areas, particularly if the waterways are polluted, and in food-poor mountain streams. Mowbray et al. (1979) reported that no northern river otter occurred in waters altered by acidic mine drainages in Maryland. Little work has been done in evaluating the range of water quality that otters will tolerate; most investigators have confined themselves to describing the land type adjacent to the aquatic otter habitat. Because northern river otters are almost entirely aquatic mammals and because of the tremendous variety of possible land types throughout their vast range, it seems probable that a survey of water quality would do much in clarifying the value of particular areas for otter habitat.

Sex Ratios. Sex ratios of northern river otters generally do not differ significantly from 1:1, although the number of males in most samples (usually trapper-caught otters) is generally greater than the number of females (Wilson 1959; McDaniel 1963; Hamilton and Eadie 1964; Tabor 1974; Lauhachinda 1978; Mowbray et al. 1979). The observed preponderance of males in most samples is believed to be due to their greater vulnerability to harvest, because they range more widely than females (Lauhachinda 1978).

Density. Few researchers have determined numerical estimates for northern river otter population densities.

Seton (1926) felt that otter populations might approach five animals per 40 mi^2 (approximately 5/100 km^2) in favorable habitat in Ontario. Intensive studies of European otter territoriality in Sweden indicated otter densities of one per 0.7 to 1.0 km^2 of water, or one otter per 2.0 to 3.0 km of lakeshore or 5.0 km of stream (Erlinge 1968*b*). Recent work with the northern river otter in Idaho has indicated that otter densities there approach one animal for each 2 to 3 km of *straight-line* distance of waterway (i.e., meanders and tributaries not included) in favorable habitat. This later figure was derived by dividing the study area size by the estimated population based on social structure: one family group—an adult female and two or three pups—and one to three subadults or nonbreeding adults would occupy about 15 km of waterway; in addition, one breeding adult male would be found for each 20 to 30 km of waterway (Melquist and Hornocker 1979).

Activity. Although northern river otters may be active at any time of day, most activity occurs from dawn to midmorning and during the evening (Melquist and Hornocker 1979). The peak of feeding activity apparently occurs from dawn to midmorning.

Seasonal activity patterns of the northern river otter are not well documented. Otters remain active throughout the year, even in areas where winter snowfall levels are high and ice seals most rivers and lakes. Along the Washington coast, R. Hirschi (personal communication 1979) reported that otter activity was significantly greater in estuarine areas during the spring and in freshwater areas during the fall.

Shelter and Dens. Northern river otters do not excavate their own dens, but rather use dens dug by other animals or natural shelters. Audubon (in Coues 1877) described an otten den he examined in the hollow trunk of a large tree; Yeager (1938) described similar dens. Also used are vacated beaver or nutria (*Myocastor coypus*) dens, hollow logs, log jams or drift piles, jumbles of loose rock, abandoned or unused boathouses, and duck blinds. Otters will on occasion build a nestlike structure in aquatic vegetation (Grinnell et al. 1937; Liers 1951*a*).

Home Range and Movements. Little is known concerning either the size of northern river otter home ranges or the extent of their movements. The best information available comes from studies of the European river otter in Sweden. There, otters have well-defined home ranges, each containing a number of areas of relatively intensive use—dens, rolling areas, slides, feeding places, sprainting areas (defecation sites), haul-outs, and runways (Erlinge 1967). Family groups consisting of an adult female and her young were found to utilize an area about 7 km in diameter during a year, with the diameter increasing from 3 to 4 km to the full width as the pups mature (Erlinge 1967). Male otters were found to occupy home ranges much more extensive and variable in size than those of females. Home range areas of males averaged about 15 km in width but were highly variable in length, owing

to individual differences, topography, and occurrence of other otters (especially males). Male otters commonly traveled extensively: the mean distance recorded was 9 to 10 km per night, but distances of up to 16 km were recorded (Erlinge 1967). Otters maintained territories within their home ranges (Erlinge 1968*b*). Territories were delineated by marks and signs (scent posts and sprainting areas, for example). Direct conflicts were rare. The area within territories was used almost exclusively by the defending otter. Females established territories excluding other females and family groups, and males established territories excluding other males; territories of female and male otters overlapped and were of unequal size, with male territories larger (Erlinge 1968*b*). The primary significance of territoriality for family groups involved securing feeding areas, while territoriality in males served to secure a breeding area. Transient otters could pass through maintained territories. Northern river otters may behave in a similar manner, although Melquist and Hornocker (1979) did not find evidence of a rigid social structure. According to W. E. Melquist (personal communication 1979), otter population density is probably an important factor in determining whether territoriality is exhibited. Little evidence of territoriality among northern river otters was found in Idaho.

Little is known of the mechanisms of dispersal among northern river otters. Family groups may begin to disassociate about three months after the pups are weaned (Melquist and Hornocker 1979).

FOOD HABITS

Food habits of northern river otters have been studied in many portions of their widespread range. Almost without exception, the bulk of the river otter's diet is composed of fish, with crustaceans (primarily crayfish), amphibians, insects, birds, and mammals comprising lesser portions (table 36.2). However, the importance of food items other than fish is not to be underestimated; Liers (1951*b*) has pointed out that captive otters fed a basic diet of live fish often did poorly.

Northern river otters have rather high metabolic rates for land mammals (Iversen 1972) and an efficient digestive system. Otters previously fed bland foods passed exoskeletal remains of crayfish about one hour after feeding (Liers 1951*b*). Harris (1968) found that otters in captivity required about 700 to 900 g of prepared food. In controlled feeding experiments, Erlinge (1968*a*) reported that captive European otters "satisfied their hunger" after eating 900 to 1000 g of live food. Toweill (unpublished data) recorded similar volumes of food in moderately distended northern river otter stomachs containing food.

Northern river otters take a tremendous variety of food items. Although a wide variety of fish species appear in the diet, certain patterns of fish vulnerability to otter predation are evident in almost all studies. The most important is that fish are preyed on in direct proportion to their availability (i.e., occurrence and density) and in inverse proportion to their swimming

TABLE 36.2. Frequency of occurrence of river otter food categories based on major otter food habits studies

State	Type of Sample	Sample Size	Period of Collection	Frequency of Occurrence of Food Category							Source
				Fish	Crustacean	Insect	Amphibian	Bird	Mammal	Other	
Michigan	stomach	173	March–April	a	35.3	13.3	16.2	b	b	4.1[c]	Lagler and Ostenson 1942
Michigan	intestine	220	March–April	d	59.2	31.8	25.4			1.0[c]	Lagler and Ostenson 1942
North Carolina	combination[e]	85[e]	December–February	91	39	6		3	1	2	Wilson 1954
Michigan	stomach	75	March–April	f	22.2	13.0	16.7			b	Ryder 1955
Montana	scat	1374	January–December	93.2	g	41.2[g]	18.4	5.2	6.1		Greer 1955
New York	digestive tract	141	October–April	70.2	34.7	13.5	24.8	0.7	4.3	0.4	Hamilton 1961
Florida	stomach	63[h]	December–March	i							McDaniel 1963
Massachusetts	scats	517	January–December	92	46	j	1.5	1	3	b	Sheldon and Toll 1964
Wisconsin	stomach	131	January–December	81	31	8	13		5	b	Knudsen and Hale 1968
Wisconsin	intestine	260	January–December	85	40	19	2		3		Knudsen and Hale 1968
Michigan	stomach	28	February–April	79	29	7	11		0	b	Knudsen and Hale 1968
Michigan	intestine	41	February–April	81	59	22	10		0	b	Knudsen and Hale 1968
Minnesota	stomach	12	February–April	100	17	33	17		0		Knudsen and Hale 1968
Minnesota	intestine	29	February–April	90	34	38	0		0		Knudsen and Hale 1968
Oregon	stomach	44	November–February	86	20	k	9	9	b	b	Toweill 1974
Oregon	intestine	75	November–February	80	27	k	12	7	b	b	Toweill 1974
California	scats	120	January–December	29	98	8[l]	0	38	7	15[m]	Grenfell 1974
Alabama and Georgia	digestive tracts	315	November–February	83.2	62.5	9.8	5.4	0.3	0.9	b	Lauhachinda 1978
Washington	scats	254	January–December	93	46	b	b	3	b	b	Tabor (unpublished)

[a] Data presented as: game and pan fishes, 29.5; forage fishes, 56.6; fish remains, 37.6. [b] Present. [c] Vertebrates other than fish or amphibians. [d] Data presented as: game and pan fishes, 47.3; forage fishes, 76.1; fish remains, 53.2. [e] Based on 24 digestive tracts and 61 scats. [f] Data presented as: game and pan fishes, 40.7; forage fishes, 55.5; fish remains, 27.8. [g] Category "Invertebrates" includes freshwater shrimp (11.0) and millipede (0.7). [h] Percentages calculated on the basis of 63 stomachs, although only 18 of these contained food. [i] Data presented as: rough fish, 17.4; game fish, 6.3. [j] "Invertebrates" occurred in 56 percent of the sample, but included crustaceans. [k] Present in 25 percent of total sample. [l] Assumed from data presented; categories combined in data. [m] Insects included in category "Invertebrates."

ability (Ryder 1955; Erlinge 1968a; Toweill 1974). Otters tend to capture the first fish they encounter that is not able to escape capture efforts. Although such a pattern of selection would appear obvious, it contains three general concepts important to the understanding of otter prey selection: (1) otters do not select a particular species of fish when hunting, (2) slow-swimming species of fish are more vulnerable than fast-swimming species, and (3) injured or weakened fish are more vulnerable to otter predation than healthy, vigorous fish. Controlled feeding studies of captive European otters confirmed these general statements (Erlinge 1968a). In these studies, too, otters tended to select larger fish, which were less maneuverable and less able to find effective hiding cover than smaller fish, except when the smaller fish were much more abundant.

In practical terms, these patterns imply that abundant, slow-swimming fish species will be selected as food by otters more often than their abundance in the water would indicate. Fish in this category include suckers (*Catostomus* sp.) and redhorses (*Moxostoma* sp.) of the family Catostomidae; carp (*Cyprinus* sp.), chubs (*Semotilus* sp.), daces (*Rhinichthys* sp.), shiners (*Notropis* sp.), and squawfishes (*Ptychocheilus* sp.) of the family Cyprinidae; and bullheads and catfishes (*Ictalurus* sp.) of the family Ictaluridae. Also important would be fish species that are often abundant and found in large schools—the sunfishes (*Lepomis* sp.), darters (*Etheostoma* sp.), and perches (*Perca* sp.)—and those bottom-dwelling species that are particularly susceptible to otters because of their habit of remaining immobile until a potential predator is quite close—for example, mudminnows (*Umbra limi*) and sculpins (*Cottus* sp.). Fast-swimming fish species—for example, trout (*Salmo* sp.) and pike (*Esox* sp.)—would be taken by otters in lesser numbers than their abundance in the water might suggest. In general, these patterns have been reported by most researchers (Lagler and Ostenson 1942; Wilson 1954; Greer 1955; Ryder 1955; Hamilton 1961; Sheldon and Toll 1964; Knudsen and Hale 1968; Field 1970; Toweill 1974; and Lauhachinda 1978).

Exceptions to these general patterns may occur whenever a particular population of fish is, for some reason, especially vulnerable. For example, Toweill (1974) reported a high incidence of coho salmon (*Oncorhynchus kisutch*) in the diet of otters in western Oregon during the period of salmon spawning migration. It is likely that many of these fish were taken as carrion or weak " spawned out" fish; others undoubtedly became vulnerable as they entered small streams or pools for spawning. Grinnell et al. (1937) reported that otters ate dead salmon in California.

Crayfish (*Cambarus* sp., *Pacifasticus* sp., and others) comprise a major portion of the otter diet throughout North America (see table 36.2); in one study, crayfish comprised the bulk of the otter diet (Grenfell 1974). A variety of crabs are also eaten by otters in estuarine areas (Wilson 1954; Toweill 1974).

Reptiles and amphibians, particularly frogs (*Rana* sp.), are commonly eaten by otters (see table 36.2). Although turtles may be abundant in otter habitat and sometimes are consumed (Stophlet 1947; Liers 1951b), many researchers have found no evidence of turtles in the otter diet in areas where turtles were abundant (Greer 1955; Grenfell 1974; Lauhachinda 1978; Toweill 1974).

Waterfowl and rails (Rallidae) comprise an important part of the otter diet in the Pacific coast states (Grenfell 1974; Toweill 1974) and in many other regions (Lagler and Ostenson 1942; Wilson 1954; Greer 1955; Banko 1960; Meyerriecks 1963; Knudsen and Hale 1968; Lauhachinda 1978). Although an unknown percentage of those taken may have represented hunter-crippled birds or carrion (most studies were conducted during or shortly after the waterfowl hunting season), there are several accounts of northern river otters observed actively hunting and killing healthy individuals (Cahn 1937; Meyerriecks 1963; Grenfell 1974). Otters also take a number of other bird species occasionally (see table 36.2); Hamilton (1961) gave a secondhand account of signs indicating that an otter had captured a "partridge" on dry land. A number of accounts of northern river otter predation on glaucous-winged gull (*Larus glaucescens*) chicks in nesting colonies along the north Pacific coast have appeared recently (Kennedy 1968; Hayward et al. 1975; Foottit and Butler 1977; Verbeek and Morgan 1978). Verbeek and Morgan suggested that this is a recent phenomenon, and one that may have a serious impact on gull populations in local areas.

A variety of mammals have been reported in the otter diet. Although Field (1970) presents evidence of northern river otters actively hunting and capturing small mammals in the snow and chasing mammals up to the size of snowshoe hares (*Lepus americanus*), the incidence of mammal remains in otter food habits studies is uniformly low.

Freshwater mussels (*Anodonta californiensis*), freshwater periwinkles (*Oxytrema silicula*), and unidentified clams and snails have been reported in the otter diet (see table 36.2) but apparently are not important food items.

A great variety of insects have been recorded in otter food habits studies, but as these are staples in most fish diets and normally occur with fish remains in the otter diets, the importance of these to the otter is not well understood. It is known that otters do prey on some of the larger diving beetles and other large insects that they encounter.

Several other kinds of food, including some plant parts—blueberries (*Vaccinium* sp.) and rose hips (*Rosa californica*)—have been reported (see table 36.2).

BEHAVIOR

Northern river otters are highly intelligent. They are extremely curious animals and may readily be trained to perform a wide variety of activities, and, of course, their inclination to make a "game" out of almost any activity is almost legendary (Harris 1968; Park 1971). Much of their active time is spent exploring new surroundings or objects, often in the form of apparent play. Northern river otters have been taught to retrieve objects from land and water, to capture and retrieve fish, and to hunt other animals (Harris 1968; Park 1971). Liers (1951*b*) reported that one captive otter learned to retrieve waterfowl by watching a Labrador retriever perform.

The memory of river otters is exceptional. Liers (1951*a*) reported an apparent reunion between two otters—supposedly a mother and pup—after they were separated a minimum of three weeks. Harris (1968) recites an account of a European otter that was trained to operate a self-feeding device in its cage. The device was then dismantled for 26 months. It was reassembled in the otter's absence, yet immediately upon returning to its cage, the otter went to the device and operated it successfully.

Senses. The tactile senses of northern river otter are highly developed. In a study of sensory specialization of otter brains, Radinsky (1968) found that the coronal gyrus of the brain was enlarged, suggesting highly developed receptor fields in the head, probably associated with the numerous and stout facial vibrissae. It has been suggested that prey moving in turbid water may be detected through sensations received by the vibrissae (Harris 1968). An indication of the manual dexterity of northern river otters is suggested by an account of an otter's manipulating a small lead pellet underwater, fondling it in its paws, dropping it, and retrieving it from the bottom (Park 1971). The pellet was the size of a "no. 6 shot," or less than 3 mm in diameter.

Visual senses are not acute in the otter. Otters are nearsighted, an adaptation for underwater vision, but apparently can detect movement at considerable distances.

Auditory senses seem to be quite well developed. Harris (1968) reports that the hearing of all otters is exceptionally acute, and a further suggestion of this is found in the variety of noises otters make for communication.

Little is known about the sense of smell, although indications that it is acute come from the fact that otters regularly communicate by means of scent marking. Northern river otters have also been observed playing "hide and seek" with a newly dead fish in a manner that suggests that the fish was located primarily by smell (Park 1971). Likewise, little is known concerning taste discrimination in otters.

Locomotion. Locomotion patterns of northern river otters consist of walking, running, and bounding on land and a thrust-recovery movement of the limbs and flexure of the posterior portion of the body and the tail vertically in the water (Tarasoff et al. 1972).

Terrestrial locomotion occurs in the pattern typical of terrestrial carnivores, with alternate movements of the opposite forelimbs and hind limbs (Tarasoff et al. 1972). When walking, limbs are moved parallel to the long axis of the body, which is held rigid with the head and neck outstretched; the distal portion of the tail may touch the ground. The same pattern of limb movement occurs in running, but the tail is held higher

above the ground and in a slight arch. Bounding, the most rapid form of movement on land (Liers 1951*a*), is characterized by the body being held in an arched position and the tail held stiffly. Both forefeet are lifted from the ground simultaneously but one strikes the ground slightly ahead of the other; both rear feet are moved simultaneously. The tail is used as a balancing organ in all terrestrial movements.

The principal mode of aquatic locomotion involves thurst-recovery movements of one or more of the limbs. Typically, the movement consists of "paddling" with one or both hind limbs alternated with a period of gliding; forelimbs are used principally in turning movements. During rapid swimming, the caudal portion of the body and the tail are moved rapidly in a vertical direction (Liers 1951*b*; Tarasoff et al. 1972). Top speed in the water is 6 to 7 mph (Harris 1968).

Body Care. The most common means of body care in otters is rubbing and rolling, whether in sand, grass, snow, or whatever else is available and relatively dry. This activity is commonly associated with considerable scratching, and both activities apparently serve to clean the animal's fur and thereby maintain its insulative qualities, as well as to dry the otter quickly after its emergence from the water. Areas used for this activity, called "rolling sites," "scrapes," "haul-outs," or "landings" (Mowbray et al. 1979; Melquist and Hornocker 1979), are among the most common evidence of otter activity.

Allogrooming does occur to some extent among otters, but is not common.

Otters typically have particular "toilets" (Greer 1955) near regular landings for defecation purposes, although single scats may be deposited near rolling areas, scent posts, or on logs or points extending out into the water (Mowbray et al. 1979; Melquist and Hornocker 1979).

Foraging and Feeding. Otters typically forage by diving and chasing fish or by digging in the substrate of ponds or streams (Liers 1951*b*). Other methods documented include "stalking" aquatic birds by approaching them underwater and seizing them from below (Meyerriecks 1963; Grenfell 1974), stalking birds and mammals on land (Hamilton 1961; Field 1970), and raiding bird nests for eggs and nestlings (Hayward et al. 1975; Verbeek and Morgan 1978). Otters may take waterfowl or fish as carrion (Grinnell et al. 1937; Toweill 1974) but showed no inclination to feed on white-tailed deer carrion (Field 1970).

Although river otters may consume a wide variety of foods, observations by Kruuk and Hewson (1978) indicate that successful foraging is not always easy. European otters observed foraging in a marine environment were successful in slightly less than 20 percent of all foraging dives; mean duration of a successful dive was 15.9 seconds, while mean duration of an unsuccessful dive was 24.8 seconds. Of 120 observed dives, the longest lasted 49.0 seconds. These otters were feeding on the bottom, in 2 to 3 m of water.

Play. Northern river otters have a reputation for being extremely playful animals. Intelligent, quick, and highly active, otters seem to go about the serious business of life in an exuberant manner. In captivity, they often engage in repetitive actions, such as sliding on mud banks or in the snow, that would seem to have little or no survival value for wild otters. Such "play" is poorly documented for wild otters and may merely be a behavioral response to captive conditions. It appears so commonly among captive individuals, however, that the concept of the "fun-loving otter" is firmly entrenched in otter literature. As Harris (1968:6–7) noted of the captive otter in general:

> its playfulness is perhaps its most marked characteristic and has been noticed by innumerable naturalists from very early times. *'Alacris ad ludos est'* wrote Albertus in the thirteenth century. With many animals, their play activities can be recognizably related to some other 'earnest' occupation, but with otters the urge to play, and the method of playing, appear in the main to be unconnected with reproductive, territorial, or prey/predator processes. Otters will play alone, with each other, with other animals, or with people. An otter alone will choose a small pebble, carry it to the water, swim out and drop it; before it can reach the bottom the otter will dive and come up underneath the stone and catch it on the flat top of its head. It will then engage in a series of underwater acrobatics—continuing to balance the stone on its head. Otters together will wrestle, play tag and duck each other in the water, and communally form slides down steep muddy banks and spend hours sliding. An otter will happily play hide-and-seek with a dog in a pile of straw....
>
> Amongst the bedding of . . . otters . . . it is common to find small shells and rounded pebbles which they have carried in as playthings.

Communication. Although the northern river otter is less vocal than most other members of the subfamily Lutrinae (Harris 1968), a variety of vocalizations have been reported. Liers (1951*b*) reported a shrill chirp, a soft "chuckle," a scream, and finally a caterwaul, used only by the female and only during copulation. Harris (1968) reported a "low grunting noise." W. E. Melquist (personal communication 1979) reported that adult otters exhibit a "grunt" or "cough" used when the animals are startled or feel threatened and in response to "chirps" uttered by pups. At higher intensities of threat, a "growl" may be uttered. Otter pups use the birdlike "chirp" as an apparent contact sound whose volume may depend on the intensity of the desire to renew contact with the mother or a sibling. In addition, pups may use a series of low murmuring grunts in soliciting the mother, they may cry or squeal in distress, and they utter a growl when threatened.

Few visual displays have been recorded. With their short muzzle and ears and their more or less uniform coat of hair, otters are poorly adapted for visual display-based communication. A "threat face" characterized by pulling the ears back and a gape display is used.

Olfaction apparently plays a major role in otter communication. Northern river otters possess anal scent glands, and scent may be released from these glands in times of fear or rage. Otters also maintain "scent posts" throughout their territory (Grinnell et al. 1937; Mowbray et al. 1976; Mowbray et al. 1979; Melquist and Hornocker 1979). Mowbray et al. (1979) described scent posts as sites 1 to 2 m² with digging and scratching sites but no food remains, scats, or beds. In addition to scent posts, they described other areas used by otters as haul-outs, bedding sites, rolling sites, scrapes, dens, and diggings. Apparently, any of these sites may be used for scent deposition. Erlinge (1967, 1968*b*), working with European otters in Sweden, reported that scented excrements were deposited at very obvious places in the otter's route of travel. Dens, rolling places, slides, runways, and haul-outs were mentioned specifically; also mentioned were such sites as elevated or projecting points of land, mouths of ditches and streams, and openings in the streamside vegetation.

Marking activities include not only deposition of scented excrement but also scratching together mounds of soil and debris (which may reach one dm in height) or twisting tufts of grass together, either of which may have scent deposits or spraints (excrement) deposited on top.

Such olfactory signals play an important role in demarcation and maintenance of otter home ranges and territories (Erlinge 1967, 1968*b*). They serve not only as boundary markers but also as signals that allow transient otters to pass through an occupied territory without encountering the territorial "owner," that is, the scent posts keep transients informed of the location of other otters. Thus, a means of avoiding direct conflicts is provided.

Social Behavior. Adult female northern river otters are devoted parents, and much time is spent teaching vital skills to the pups. The females not only introduce their young to water and teach them to swim, but also capture and release food so that the young can develop foraging skills (Liers 1951*b*). Females aggressively protect their young from potential danger, which accounts for most reported attacks upon humans (anonymous 1977; anonymous 1979).

Little is known of the social structure of northern river otter populations. Individuals of different ages and sex may freely interact, although adult males are normally solitary. Adult females are normally accompanied by their pups of the previous breeding season (Melquist and Hornocker 1979). Female otters with pups and adult male otters may have relatively well defined home ranges. A study of European otters indicated that 30 to 40 percent of otters in fresh-water habitats were territory holders. Approximately an equal number were transients (largely subadults that had not yet established a territory), and the remainder of the population was comprised of pups still with their mother in her home range (Erlinge 1968*b*). Melquist and Hornocker (1979) found little evidence of territoriality in studies of northern river otters in Idaho.

MORTALITY

Northern river otters are subject to few natural predators, and parasitism and disease are not known to have severe impacts on otter populations. Malnourishment is probably not a limiting factor in most areas, although it could limit some populations in the far north, particularly in icebound waters. By far the most serious causes of mortality among river otters are the activities of people, through harvest of otters for pelts and, far more importantly, through destruction and modification of required habitat.

Predators. Bobcats (*Felis rufus*), dogs (*Canis familiaris*), coyotes (*Canis latrans*), foxes, and American alligators (*Alligator mississippiensis*) have all been reported to kill northern river otters (Seton 1926; Grinnell et al. 1937; Young 1958; Vallentine et al. 1972; Mowbray et al. 1979). In addition, it is likely that other predators, including cougars (*Felis concolor*), wolves (*Canis lupus*), black bears (*Ursus americanus*), American crocodiles (*Crocodylus acutus*), and perhaps some of the large raptors also kill otters on occasion. No predator has been shown to have a serious impact on otter populations, and most

TABLE 36.3. Helminth parasites reported from northern river otters. (Numbered sources are listed below and refer to citations listed in literature cited.)

Species	Body Location	Geographic Region	Source
Trematodes			
Alaria canis	Subcutaneous fat	Ontario	9, 17
Baschkirovitrema incrassatum	Stomach, intestine	Alabama, Georgia, New York, North Carolina	7, 10, 14, 19
Crepidostomum cooperi	Stomach	Alabama, Georgia	10
Enhydridiplostomum alarioides	Stomach, intestine	Alabama, Georgia, North Carolina	7, 10, 14, 19
Euparyphium melis		Michigan	4
Nanophytes salmincola			20
Telorchis spp.	Small intestine	Alabama	7
Cestodes			
Diphyllobothrium mansonoides			13
Ligula intestinalis	Proglottids in feces	Montana	8
Proteocephalus perplexus	Stomach	Alabama	10
Schistocephalus solidus		Newfoundland	22
Nematodes			
Capillaria plica	Urinary bladder	North Carolina	14
Crenosoma goblei	Respiratory tract	North Carolina	14
Dirofilaria lutrae	Subcutaneous tissue	Florida, Louisiana	16
Dracunculus insignis	Connective tissue	Alabama, Florida	10, 11
Dracunculus lutrae	Subcutaneous tissue	Ontario	6
Dracunculus sp.	Subcutaneous tissue	New York	5
Eustrongyloides spp.		Maryland	1
Filaroides canadensis	Nodules in lungs	Ontario	2
Gnathostoma miyazakii	Kidneys	Alabama, Ontario	3, 10, 14
Physaloptera sp.	Stomach, intestine	Alabama	10, 15
Spinitectus gracilis	Large in intestine	Alabama	7, 10
Skrjabingylus lutrae	Frontal sinuses	Ontario	9
Strongyloides lutrae	Intestines	Alabama, Louisiana	7, 10, 12
Unknown microfilaria		Michigan	18
Acanthocephalans			
Acanthocephalus spp.	Large intestine	Alabama	7
Leptorhynchoides spp.	Stomach	Alabama	10
Metachinorhynchus lateralis		Newfoundland	22
Neoechinorhynchus spp.	Stomach	Alabama	10
Paracanthocephalus rauschi		Alaska	21
Pomphorhynchus spp.	Large intestine	Alabama	7

[1]Abram and Lichtenfels 1974
[2]Anderson 1963
[3]Anderson 1964
[4]Beaver 1941
[5]Cheatum and Cook 1948
[6]Crichton and Beverly-Burton 1973
[7]Fleming et al. 1977
[8]Greer 1955
[9]Lankester and Crichton 1972
[10]Lauhachinda 1978
[11]Layne et al. 1960

[12]Little 1966
[13]McIntosh 1937
[14]Miller and Harkema 1968
[15]Morgan 1941
[16]Orihel 1965
[17]Pearson 1956
[18]Rothenbacher 1962
[19]Sawyer 1961
[20]Schlegel et al. 1968
[21]Schmidt 1969
[22]Smith and Threlfel 1973

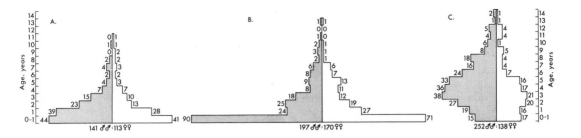

FIGURE 36.4. Age and sex pyramids for three widely disjunct northern river otter (*Lutra canadensis*) populations. A, Oregon (Tabor and Wright 1977); B, Maryland (Mowbray et al. 1979); C, Alabama-Georgia (Lauhachinda 1978).

predation is probably directed toward young animals and those adults that are encountered by a predator while away from water.

Parasites and Diseases. A variety of internal parasites have been reported for northern river otters (table 36.3). Of these, two roundworms—*Strongyloides lutrae,* a small roundworm that lives in the mucosa of the small intestine and that completes its life cycle in the lungs, and *Gnathostoma miyazakii,* which lives in the collecting tubules of the kidneys—may cause serious pathological damage. The former could cause lesions during its migratory stage, while the latter could cause blockage and partial loss of kidney function (Fleming et al. 1977; Lauhachinda 1978).

Little is known of diseases of free-ranging northern river otters. They are apparently susceptible to canine distemper, jaundice, hepatitis, and perhaps feline panleucopenia. Pneumonia was implicated in the deaths of 13 of 29 otters of all species that died in the London Zoo (Harris 1968). Otters are also susceptible to human tuberculosis (Borg 1964).

Human Impacts. The most readily apparent human impact on northern river otter populations results from trappers harvesting otters for their fur. However, life table analyses of river otter populations in Oregon (Tabor and Wight 1977) and Maryland (Mowbray et al. 1979) indicated that under existing harvest levels otter populations were approximately stable, while a decline in numbers and decrease in average longevity was noted for otter populations in Alabama and Georgia (Lauhachinda 1978) (figure 36.4).

Northern river otters have been extirpated from nine states and one Canadian province within recent times; without exception, the primary cause of the decline in otter numbers was listed as habitat destruction (Deems and Pursley 1978). Such destruction may take any of a number of forms. For example, the disappearance of otters from West Virginia and parts of Tennessee and Kentucky was attributed to increased acidity of ground water due to mining operations (Lauhachinda 1978). Other causes of habitat destruction include development of waterways for economic or recreational purposes, destruction of riparian habitat for homesites or farmland, or declines in water quality resulting from such things as increased siltation or introduction of pesticide residues into the water as a result of intensive farming operations.

Residues of pesticides including mercury (Cumbie 1975), DDT and its metabolites, and Mirex (Hill and Lovett 1976) have been reported from northern river otter tissues.

AGE DETERMINATION

A number of physical characters have been found useful in determining the age of northern river otters. For males, these include growth characteristics of the baculum, developement of the testes, and production of spermatozoa. Development of the female reproductive tract is useful for determining age of females. Both sexes may be aged to some degree by use of body size, skull characteristics, dental characters (including tooth eruption patterns and the presence of annuli in tooth cementum), eye lens weight, and skeletal characters.

Males. Length, weight, and volume of bacula have been used to determine the age of male otters (Hooper and Ostenson 1949; Friley 1949; Stephenson 1977; Lauhachinda 1978). Bacula of northern river otters increase rapidly in length until the otter is three to five years of age, and in weight and volume until an age of five to six years. However, there is considerable overlap in older age classes of otters, limiting the usefulness of this technique for otters older than juveniles and yearlings (Stephenson 1977).

Size of the testes is useful in distinguishing yearling river otters from older animals (Hamilton and Eadie 1964; Tabor 1974), and production of spermatozoa indicates that otters are approximately two years of age or older.

Females. The uterus of juvenile female river otters is very small, transluscent, and nonvascular, with horns measuring 1 to 5 mm in diameter. The horns increase in size and degree of vascularization as otters become older. Dimensions and weight of the ovaries may also be used to separate juvenile and yearling otters from older animals (Hamilton and Eadie 1964; Tabor 1974).

Both Sexes. Body size may be used as an indication of age for embryos (Hamilton and Eadie 1964); although no known-age series is available, Kenyon (1969) presented a hypothetical equation for fetal growth rate.

Body size of male otters is greater than that of females of the same age, and relative sizes of individual otters of different subspecies differ, limiting the usefulness of body size as an age criterion except within local areas. Stephenson (1977) presented body size information for Ontario otters. Skull measurements show similar patterns of variation, but patterns of suture closure in skull bones are useful in grouping otters into juvenile, yearling, and adult age groups (Hooper and Ostenson 1949; Hamilton and Eadie 1964; Stephenson 1977).

Patterns of tooth eruption are useful in determining the age of northern river otter pups.

The most reliable and useful age determination technique for use with northern river otters results from the annual deposition of dark-staining bands of tooth cementum, resulting from an annual period of slowed development. Tabor (1974), the first to use this technique with river otters, found that the initial band was deposited in spring or summer at approximately one year of age and that the number of bands equaled the otter's age in years. Stephenson (1977) concluded that the initial band was deposited during the summer and also found that the number of bands equaled the otter's age in years; complete agreement was found between cementum annuli counts and age of five known-age otters. Up to 14 cementum annuli have been reported in wild-caught otters (Lauhachinda 1978).

Eye lens weight as an indicator of age was explored by Lauhachinda (1978) and was found to be highly correlated with age. Differences in eye lens weight between male and female otters were noted but were not significant.

Closure of the epiphyses of the long bones is useful in grouping river otters into juvenile, yearling, and adult age classes (Hamilton and Eadie 1964).

ECONOMIC STATUS AND MANAGEMENT

Economic Status. Because of its thick, beautiful, and very durable fur, northern river otters have been an economically important furbearer species since Europeans first arrived in North America. In 1976, 27 of the 49 continental states and 11 Canadian provinces and territories allowed trappers to harvest otters for their fur. During the period 1970 through 1976, the total harvest of northern river otters in the United States ranged from a low of about 11,000 in the 1970–71 season to a high of almost 19,000 in 1972–73. Canadian harvest totals ranged from about 15,000 in 1970–71 to a high of over 18,000 in 1972–73 (Deems and Pursley 1978). Total annual raw pelt values for northern river otters in North America ranged from $600,000 in 1970–71 to almost $3,000,000 in 1976–77 (Pursley 1978). Harvest levels were largely influenced by prices paid to trappers for the pelts, which in turn are dictated by fashion and availability (Pursley 1978). Most northern river otter pelts were eventually sold to furriers in Central Europe, and were used in the garment industry.

Because of their large size and specialized requirements, raising of northern river otters on "fur farms" was never a profitable venture. A few animals have been maintained as pets by interested individuals, but training otters for fishing or other activities, as is practiced in some parts of Europe and Asia, never became popular in North America.

Northern river otters have often been accused of causing serious depredations on game fish, particularly trout. Food habits studies, while indicating that otters do occasionally take trout, have all indicated that the bulk of the otter diet consists of nongame fish species. Otters are beneficial to game fish populations in many instances because they remove nongame fish that would otherwise compete with game fish for food (Toweill 1974; Lauhachinda 1978, and many others). River otters may cause severe depredations in and around fish hatcheries on occasion.

Northern river otters have also been accused of damaging the fur resources of certain areas by their supposed depredations on beaver (Green 1932) or muskrats (*Ondatra zibethicus*) (Wilson 1954). In each case where studies of such supposed predation were conducted, otter predation on other furbearers has been found to be extremely unusual.

Management. Management of northern river otters may include any of five major elements: regulation of harvest, monitoring of harvest and/or populations, total protection, reintroduction into areas where the population has been depleted, and protection of habitat. In 1978, no state or Canadian province allowed unregulated taking of otters (Endangered Species Scientific Authority 1978), partially in response to the species' being listed in appendix 2 (species not currently threatened with extinction, but may become threatened unless trade in them is subject to strict regulation) of the Convention on International Trade in Endangered Species of Wild Flora and Fauna.

Because river otters were listed as an appendix 2 species, the Endangered Species Scientific Authority of the United States has required each state to submit data on river otter populations within its boundaries, and has taken the lead in setting export quotas on the number of pelts from each state that may enter international commerce. Quotas may be revised annually on the basis of information received, and most states have begun issuing tags that must be affixed to each pelt harvested, thereby providing a vehicle for monitoring harvest—and, indirectly, population—trends. Data presented for the 1977 season indicated that otter populations were stable or expanding in most states that allowed harvest, although local overharvest was reported in Maine (Endangered Species Scientific Authority 1978).

Because otters are extremely difficult to count and most index techniques are of questionable accuracy, most states estimate otter populations on the basis of general habitat surveys, trapper or furbuyer reports, and the consensus of field personnel of the state management agency. Harvest quotas thus are based more often on the number harvested in previous years than on goals related to population size and viability. Since most trappers harvest otters incidental to trapping other species, particularly beaver, this generally results in conservative management policy. Policy is made even more conservative in those states that impose a limit on the number of otters any one trapper may take during the season, which theoretically limits those individuals who are competent trappers and might otherwise choose to trap specifically for otters. Tabor (1974) reported that 33 percent of the total otter harvest in Oregon was accounted for by trappers who harvested more than four otters per season. He also reported that although most otter trapping was incidental to beaver trapping efforts, otters were not normally caught in beaver trap sets. Most otters (79 percent) were caught in traps set specifically for otters. Lauhachinda (1978) also reported that a large portion of the otters he obtained were taken by a few trappers.

Although management policies tend to be conservative, northern river otter populations are susceptible to overharvest because of the fact that they do travel extensively and in the restricted avenues provided by watercourses. Thus, the impact of even a single knowledgeable trapper at some point along a given watercourse may severely affect local populations. For this reason, until adequate indices to river otter population levels can be developed and until otter habitat requirements are clearly understood, conservative policies are essential to assure the maintenance of healthy and viable populations.

CURRENT RESEARCH AND MANAGEMENT NEEDS

Much additional research concerning the relationship of the northern river otter to its environment needs to be done. Particular concerns are habitat requirements in terms of water quality and tolerances to industrial and agricultural pollutants, responses to mechanical alterations of otter habitat, and responses to differing levels and kinds of human disturbance. Additional studies in the area of population structure and recruitment need to be conducted, in order to gain insight into the resiliency of populations subjected to harvest. Studies of social structure within otter populations and the effects of social structure on otter movements are needed, so that the manager may be able to predict the effect an impact at a particular point, as through habitat modification, may have on the population within a given area. All of the above may be needed in order to develop a reliable index to population levels, which is necessary at all levels of management.

Although the above-mentioned items are of particular importance, a number of other areas of river otter biology also need attention. For example, little is known of the physiological mechanisms otters have developed in response to living in an aquatic environment or of river otter ethology. The phenomenon of delayed implantation and its survival value is poorly understood, as is the role of otters in disease transmission—a potential area of considerable concern, since otters move widely through natural waterways. This list could be extended at some length. Suffice it to say that much work needs to be done to understand completely the biology of the northern river otter.

The authors extend their sincere thanks to W. E. Melquist, for sharing his insights into otter ecology and behavior and for offering valuable suggestions during the preparation of this material, and to R. Hirschi, for sharing his knowledge of otter ecology in western Washington. Ed Park generously contributed an excellent photograph. Throughout this effort, the volume editors, J. A. Chapman and G. A. Feldhamer, assisted in every way possible.

LITERATURE CITED

Abram, J. B., and Lichtenfels, J. R. 1974. Larval *Eustrongyloides* sp. (Nematoda: Dioctophymatoidae) from otter, *Lutra canadensis,* in Maryland. Proc. Helmin. Soc. Washington 41(2):253.

Anderson, R. C. 1963. Further studies on the taxonomy of *Metastrongylis* (Nematoda: Metastrongyloidae) of Mustelidae in Ontario. Can. J. Zool. 41(1):801–809.

_____. 1964 *Gnathostoma miyazakii* n. sp. from the otter (*Lutra canadensis*) with comments on *G. sociale* (Leidy, 1858) of mink (*Mustela vison*). Can. J. Zool. 42(2):249–254.

Anderson, S. 1977. Geographic ranges of North American terrestrial mammals. Am. Mus. Novit. 2629. 15pp.

Anonymous. 1977. Ornery otter attacks Salmon sailors. Lewiston Tribune, Lewiston, Idaho, 19 August 1977.

Anonymous. 1979. Swimming boy, 5, attacked by otter. Portland Oregonian, Portland, Oregon, 17 August 1979.

Bailey, V. 1936. The mammals and life zones of Oregon. North Am. Fauna 55. 416pp.

Banko, W. E. 1960. The trumpeter swan. North Am. Fauna 63. 214pp.

Beaver, P. C. 1941. Studies on the life history of *Euparyphium melis* (Trematoda: Echinostomidae). J. Parasitol. 27(1):35–44.

Ben-Shaul, D. M. 1962. The composition of the milk of wild animals. Int. Zoo Yearb. 4:333–342.

Borg, K. 1964. Human tuberculosis in an otter (*Lutra l. lutra*). Int. Symp. Dis. Zoo Anim. 5:89–90.

Cahn, A. R. 1937. The mammals of the Quetico Provincial Park of Ontario. J. Mammal. 18(1):19–30.

Cheatum, E. L., and Cook, A. H. 1948. On the occurrence of the North American guinea worm in mink, otter, raccoon, and skunk in New York state. Cornell Vet. 38:421–423.

Coues, E. 1877. Fur-bearing animals; a monograph of North American Mustelidae. U.S. Geol. Surv. Terr. Misc. Publ. 8:1–348.

Crandall, L. S. 1964. The management of wild animals in captivity. Univ. Chicago Press, Chicao. 761pp.

Crichton, V. F. J., and Beverley-Burton, M. 1973. *Dracunculus lutrae* n. sp. (Nematoda: Dracunculoidea) from the otter, *Lutra canadensis,* in Ontario, Canada. Can. J. Zool. 51(5):521–528.

Cumbie, P. M. 1975. Mercury levels in Georgia otter, mink and freshwater fish. Bull. Environ. Contamination and Toxicol. 14(2):193–196.

Dearden, L. C. 1954. Extra premolars in the river otter. J. Mammal. 35(1):125–126.

Deems, E. F., Jr., and Pursely, D. 1978. North American furbearers: their management, research and harvest status in 1976. Int. Assoc. Fish Wildl. Agencies and Univ. Maryland, College Park. 171pp.

Endangered Species Scientific Authority. 1978. Export of bobcat, lynx, river otter, and American ginseng. Fed. Register 43(52):11082–11093.

Erlinge, S. 1967. Home range of the otter *Lutra lutra* in southern Sweden. Oikos 18(2):186–209.

_____. 1968a. Food studies on captive otters *Lutra lutra* L. Oikos 19(2):259–270.

_____. 1968b. Territoriality of the otter, *Lutra lutra* L. Oikos 19(1):81–98.

Field, R. J. 1970. Winter habits of the river otter (*Lutra canadensis*) in Michigan. Michigan Acad. 3(1):49–58.

Fisher, E. M. 1942. Osteology and myology of the California river otter. Stanford Univ. Press, Palo Alto, Calif. 65pp.

Fleming, W. J.; Dixon, C. F.; and Lovett, J. W. 1977. Helminth parasites of river otters (*Lutra canadensis*) from southeastern Alabama. Proc. Helmin. Soc. Washington 44(2):131–135.

Foottit, R. G., and Butler, R. W. 1977. Predation on nesting glaucous-winged gulls by river otter. Can. Field Nat. 91(2):189–190.

Friley, C. E. 1949. Age determination, by use of the baculum, in the river otter (*Lutra c. canadensis*) Schreber. J. Mammal. 30(2):102–110.

Green, H. U. 1932. Observations of the occurrence of the otter in Manitoba in relation to beaver life. Can. Field Nat. 46 pp.204–206.

Greer, K. R. 1955. Yearly food habits of the river otter in the Thompson Lakes region, northwestern Montana, as indicated by scat analysis. Am. Midl. Nat. 54(2):299–313.

Grenfell, W. E., Jr. 1974. Food habits of the river otter in Suisin Marsh, central California. M.S. Thesis. California State Univ., Sacramento. 43pp.

Grinnell, J.; Dixon, J. S.; and Linsdale, J. M. 1937. Fur-bearing mammals of California. Vol. 1. Univ. California Press, Berkeley. 375pp.

Hall, E. R., and Kelson, K. R. 1959. The mammals of North America. Ronald Press Co., New York. 1083pp.

Hamilton, W. J., Jr. 1961. Late fall, winter, and early spring foods of 141 otters from New York. New York Fish Game J. 8(2):106–109.

Hamilton, W. J., Jr., and Eadie, W. R. 1964. Reproduction in the otter, *Lutra candensis*. J. Mammal. 45(2):242–252.

Harris, C. J. 1968. Otters: a study of the Recent Lutrinae. Weidenfield & Nicolson, London. 397pp.

Hayward, J. L., Jr.; Amlaner, C. J., Jr.; Gillett, W. H.; and Stout, J., F. 1975. Predation on nesting gulls by a river otter in Washington state. Murrelet 56(2):9–10.

Hill, E. P., and Lovett, J. W. 1976. Pesticide residues in beaver and river otter from Alabama. Proc. Southeast Assoc. Fish Game Commissioners 29:365–369.

Hooper, E. T., and Ostenson, B. T. 1949. Age groups in Michigan otter. Univ. Michigan, Ann Arbor, Mus. Zool. Occas. Pap. 518. 22pp.

Huggett, A. St. G., and Widdas, W. F. 1951. The relationship between mammalian foetal weight and conception age. J. Physiol. 114(3):306–317.

Iversen, J. A. 1972. Basal energy metabolism of Mustelids. J. Comp. Physiol. 81(4):341–344.

Kennedy, K. 1968. River otter feeding on glaucous-winged gull. Blue Jay 26:109.

Kenyon, K. W. 1969. The sea otter in the eastern Pacific Ocean. North Am. Fauna 68. 352pp.

Knudsen, K. F., and Hale, J. B. 1968. Food habits of otters in the Great Lakes region. J. Wildl. Manage. 32(1):89–93.

Kruuk, H., and Hewson, R. 1978. Spacing and foraging of otters (*Lutra lutra*) in a marine habitat. J. Zool. London 185(2):205–212.

Lagler, K. F., and Ostenson, B. T. 1942. Early spring food of the otter in Michigan. J. Wildl. Manage. 6(3):244–254.

Lankester, M. W., and Crichton, V. J. 1972. *Skrjabingylus lutrae* n. sp. (Nematoda: Metastrongyloidae) from otter (*Lutra canadensis*). Can. J. Zool. 50(3):337–340.

Laughlin, J. 1955. River otter noted east of Sierran crest. California Fish Game 41(2):189.

Lauhachinda, V. 1978. Life history of the river otter in Alabama with emphasis on food habits. Ph.D. Dissertation. Auburn Univ., Auburn, Ala. 169pp.

Layne, J. N.; Birkenholz, D. E.; and Griffo, J. V. 1960. Records of *Dracunculus insignis* (Leidy, 1858) from raccoons in Florida. J. Parasitol. 46(6):658.

Liers, E. E. 1951a. My friends the land otters. Nat. Hist. 60(7):320–326.

_____. 1951b. Notes on the river otter (*Lutra canadensis*). J. Mammal. 32(1):1–9.

_____. 1958. Early breeding in the river otter. J. Mammal. 39(3):438–439.

Little, M. D. 1966. Seven new species of *Strongyloides* (Nematoda) from Louisiana. J. Parasitol. 52(1):85–97.

McDaniel, J. C. 1963. Otter population study. Proc. Southeast Assoc. Fish Game Commissioners 17:163–168.

Macfarlane, R. 1905. Notes on mammals collected and observed in the northern Mackenzie River District. Proc. U.S. Natl. Mus. 23:716–717.

McIntosh, A. 1937. New host records for *Diphyllobothrium mansonoides* Muller, 1935. J. Parasitol. 23(3):313–315.

Magoun, A. J., and Valkenburg, P. 1977. The river otter (*Lutra canadensis*) on the north slope of the Brooks Range, Alaska. Can. Field Nat. 91(3):303–305.

Melquist, W. E., and Hornocker, M. G. 1979. Methods and techniques for studying and censusing river otter populations. For., Wildl. and Range Exp. Stn., Tech. Rep. 8. Univ. Idaho, Moscow. 17pp.

Meyerriecks, A. J. 1963. Florida otter preys on common gallinule. J. Mammal. 44(3):425–426.

Miller, G. C., and Harkema, R. 1968. Helminths of some wild mammals in the southeastern United States. Proc. Helmin. Soc. Washington 35(2):118–125.

Morejohn, G. V. 1969. Evidence of river otter feeding on freshwater mussels and range extension. California Fish Game 55(1):83–85.

Morgan, B. 1941. Additional notes on North American Physalopterinae (Nematoda). Proc. Helmin. Soc. Washington 8(2):63–64.

Mowbray, E. E.; Pursley, D.; and Chapman, J. A. 1979. The status, population characteristics and harvest of the river otter in Maryland. Maryland Wildl. Admin., Publ. Wildl. Ecol. 2. 16pp.

Mowbray, E. E., Jr.; Chapman, J. A.; and Goldsberry, J. R. 1976. Preliminary observations on otter distribution and habitat preferences in Maryland with descriptions of otter field sign. Northeast Fish Wildl. Conf. 33:125–131.

Ognev, S. I. 1931. Mammals of eastern Europe and northern Asia. Vol. 2, Carnivora (Fissipedia). Transl. from Russian by Israel Prog. Sci. Transl., 1962. 590pp.

Orihel, T. C. 1965. *Dirofilaria lutrae* n. sp. (Nematoda: Filarioidea) from otters in the southeast United States. J. Parasitol. 51(3):409–413.

Park, E. 1971. The world of the otter. J. B. Lippincott & Co., Philadelphia and New York. 159pp.

Pearson, J. C. 1956. Studies on the life cycles and morphology of the larval stages of *Alaria arisaemoides* Augustine and Uribe, 1927 and *Alaria canis* LaRue and Fallis, 1936 (Trematoda: Diplostomidae). Can. J. Zool. 34(4):295–387.

Pearson, O. P., and Enders, R. K. 1944. Duration of pregnancy in certain Mustelids. J. Exp. Zool. 95(1):21–35.

Pohle, H. 1920. Die Unterfamilie der Lutrinae. Arch. Naturgesch. Jahrg. 85, 1919, abt. A., vol. 9, pp. 1–247.

Pursley, D. 1978. Economic values of furbearers in North America. Proc. Western Assoc. Fish Wildl. Agencies 58:123–140.

Radinsky, L. B. 1968. Evolution of somatic sensory specialization in otter brains. J. Comp. Neurol. 134(4):495–506.

Rothenbacher, H. 1962. Acute microfilariasis of unknown etiology in an otter. Michigan State Univ. Vet. 23:43–45.

Ryder, R. A. 1955. Fish predation by the otter in Michigan, J. Wildl. Manage. 19(4):497–498.

Sawyer, T. K. 1961. The American otter, *Lutra canadensis vaga,* as a host for two species of trematodes previously unreported from North America. Proc. Helmin. Soc. Washington 28(2):175–176.

Scheffer, V. B. 1939. The os clitorides of the Pacific otter. Murrelet 20(1):20–21.

Schlegel, M. W.; Knapp, S. E.; and Millemann, R. E. 1968. "Salmon poisoning" disease. Part 5: Definitive hosts of the trematode vector, *Nanophyetus* salmincola. J. Parasitol. 54(4):770–774.

Schmidt, G. D. 1969. *Paracanthocephalus rauschi* n. sp. (Acanthocephala: Paracanthocephalidae) from grayling, *Thymallus arcticus* (Pallas), in Alaska. Can. J. Zool. 47(3):383–385.

Seton, E. T. 1926. Lives of game animals. Doubleday, Doran & Co., New York. Vol. 2. 671pp.

Sheldon, W. G., and Toll, W. G. 1964. Feeding habits of the river otter in a reservoir in central Massachusetts. J. Mammal. 45(3):449–455.

Sinha, A. A.; Conaway, C. H.; and Kenyon, K. W. 1966. Reproduction in the female sea otter. J. Wildl. Manage. 39(1):121–130.

Smith F. R., and Threlfel, W. 1973. Helminths of some mammals from Newfoundland. Am. Midl. Nat. 90(1):215–218.

Stephenson, A. B. 1977. Age determination and morphological variation of Ontario otters. Can. J. Zool. 55(10):1577–1583.

Stophlet, J. J. 1947. Florida otters eat large terrapin. J. Mammal. 28(2):183.

Tabor, J. E. 1974. Productivity, survival, and population status of river otter in western Oregon. MS. Thesis. Oregon State Univ., Corvallis. 62pp.

Tabor, J. E., and Wight, H. M. 1977. Population status of river otter in western Oregon. J. Wildl. Manage. 41(4):692–699.

Tarasoff, F. J.; Basaillon, A.; Pierard, J.; and Whitt, A. P. 1972. Locomotory patterns and external morphology of the river otter, sea otter, and harp seal (Mammalia). Can. J. Zool. 50(7):915–929.

Tarasoff, F. J., and Kooyman, G. L. 1973a. Observations on the anatomy of the respiratory system of the river otter, sea otter, and harp seal. Part 1, The topography, weight, and measurements of the lungs. Can. J. Zool. 51(2):163–170.

———. 1973b. Observations on the anatomy of the respiratory system of the river otter, sea otter, and harp seal. Part 2, Trachea and bronchial tree. Can. J. Zool. 51(2):171–177.

Toweill, D. E. 1974. Winter food habits of river otters in western Oregon. J. Wildl. Manage. 38(1):107–111.

Vallentine, J. M., Jr; Walther, J. R.; McCartney, K. M.; and Ivy, L. M. 1972. Alligator diets on the Sabine National Wildlife Refuge, Louisiana. J. Wildl. Manage. 36(3):809–815.

van Zyll de Jong, C. G. 1972. A systematic review of the Nearctic and Neotropical river otters (Genus *Lutra,* Mustelidae, Carnivora). R. Ontario Mus., Life Sci. Contr. 80. 104pp.

Verbeek, N. A. M., and Morgan, J. L. 1978. River otter predation on glaucous-winged gulls on Mandarta Island, British Columbia. Murrelet 59(3):92–95.

Wilson, K. A. 1954. The role of mink and otter as muskrat predators in northeastern North Carolina. J. Wildl. Manage. 18(2):199–207.

———. 1959. The otter in North Carolina. Proc. Southeast Assoc. Fish Game Commissioners 13:267–277.

Wurster, D. H., and Benirschke, K. 1968. Comparative cytogenetic studies in the order *Carnivora.* Chromosoma (Berl.) 24:336–382.

Yeager, L. E. 1938. Otters of the Delta hardwood region of Mississippi. J. Mammal. 19(2):195–201.

Young, S. P. 1958. The bobcat of North America. Stackpole Co., Harrisburg, Pa., and Wildl. Manage. Inst. Washington, D. C. 193pp.

DALE E. TOWEILL, Research and Development Section, Oregon Department of Fish and Wildlife, P.O. Box 8, Hines, Oregon 97738.

JAMES E. TABOR, Washington Department of Game, P.O. Box 1237, Ephrata, Washington 98823.

37

Sea Otter

Enhydra lutris

Karl W. Kenyon

NOMENCLATURE

COMMON NAME. Sea otter
SCIENTIFIC NAME. *Enhydra lutris*
SUBSPECIES. *E. l. nereis, E. l. lutris,* and *E. l. gracilis.*

The taxonomic status of the subspecies *nereis* is controversial, and was named on the basis of parts of one specimen. Scheffer and Wilke (1950) reviewed all available skeletal material from Alaska and California otters. They concluded that the very slight differences between these geographically separated animals were insignificant and that *nereis* was not a valid subspecies. Roest (1973) reviewed additional skeletal material from California, Prince William Sound, and the Aleutians. He concluded that the small statistical differences were clinal and also suggested eliminating *nereis.* Davis and Lidicker (1975), however, proposed that this subspecific status be retained; they believed that differences in body size and behavior exist. These differences, however have never been scientifically demonstrated. Roest (1979) studied 15 additional skulls taken prior to 1909 and added these data to his previous 248 specimens. He concluded, "all sea otters from the Pacific Coast of North America represent a single, clinally varying population, *Enhydra lutris lutris.*" I agree that the best scientific work supports Roest's conclusion.

DISTRIBUTION AND NUMBERS

The sea otter is found only in the northern Pacific Ocean. Prior to 1740 it ranged from Morrow Hermoso Bay, Baja California, Mexico, along the west coast of the United States, Canada, Alaska, the Aleutian and Commander islands, the Kamchatka Peninsula, and the Kuril Islands to the northern islands of Japan. Its historic northern range in the Bering Sea (the Pribilof Islands) appears to be limited by the southern limit of ice. An unusual advance of ice caused mortality of otters at Urup Island in the Kurils (Nikolaev 1965). Sea otters trapped by a sudden freeze-up in Bristol Bay in 1971 and again in 1972 were unable to dive for food. Some of these died in their attempt to escape to open water by crossing the Alaskan Peninsula (Schneider and Faro

FIGURE 37.1. Distribution of the sea otter (*Enhydra lutris*), encompassing the ranges in 1740, 1911, and 1980. The width of the range has been greatly exaggerated: the animal seldom ranges more than 2 km offshore. A, Pribilof Islands; B, Alaska Peninsula; C, Kodiak area; D, Prince William Sound; E, southeastern Alaska (Yakobi, Chicagof, Baranof, and Prince of Wales islands were release sites); F, Queen Charlotte Islands; G, Vancouver Island; H, Washington coast; I, Oregon coast; J, Monterey area; K, San Benito Island.

1975). Rarely otters travel far north, perhaps while resting on retreating ice or borne by currents that flow in the spring from the Bering Sea into the Chukchi Sea. Two sea otters were taken on the northern Siberian Coast of the East Siberian Sea in recent years (Zimushko et al. 1968). Also in 1977, a sea otter was found at Savoonga, St. Lawrence Island, far north of the species' usual range in the Bering Sea. (F. H. Fay personal communication). Undoubtedly, otters far north of their usual range would perish in winter ice if they could not retreat to ice-free waters.

During the 18th and 19th centuries the sea otter was hunted intensively for its rich, soft fur. Hunters were so persistent that the species was extirpated from many parts of its range. Only a few "seed" populations remained by 1911, when the species was given protection through inclusion in an international treaty primarily protecting the northern fur seal (*Callorhinus ursinus*). For years sea otter populations increased slowly. In the late 1940s, though, a number of Aleutian Island populations had reached a stage of exponential growth. Today the Alaska population numbers well over 100,000 (Alaska Department of Fish and Game 1973) and the species has reoccupied most of its former range (figure 37.1). In order to expedite the repopulation of vacant habitat, various government agencies cooperated in translocating otters from densely populated areas to vacant parts of the former range. From 1965 to 1972 a total of 708 sea otters was captured in Alaska. Among these, 467 were taken to unoccupied parts of the former range in Alaska. The areas of release included southeastern Alaska near Chicagof Island and the Pribilof Islands. The remaining 241 were taken by air and released on the west coast of Vancouver Island, B. C., on the Washington coast, and on the Oregon coast (Jameson et al. 1981). The survival of these translocated populations has been observed and studied intermittantly. Among the 89 otters liberated on the west coast of Vancouver Island, a total of 70 was recorded in 1977. Mothers with pups were observed, indicating that this colony was established (Bigg and MacAskie 1978). Further studies of the ecology and behavior of the Vancouver Island colony were conducted in 1978. It was found that reproduction was good but the total otters counted was 67. These counts indicate that at present the population is probably about stable (Morris et al. 1979). On the Washington coast, reproduction has been observed and in 1978 a total of 18 otters was counted (Jameson et al. 1981). On the coast of Oregon, the population declined from 23 animals observed in the early 1970s to only 4 in 1978. Probably this colony and one on the Pribilof Islands, where few otters have been seen in recent years, will not survive (Jameson et al. 1981).

The small seed colony that survived the early period of exploitation on the rugged coast of California was first reported by Bryant (1915). That this population was well established was recorded on 19 March 1938 (Bolin 1938), when it numbered about 100 animals. The colony grew slowly, spreading along the coast, and by 1957, 638 otters were counted (Boolootian 1961). More recent counts have ranged from 1,146 in 1973 to a high of 1,561 in 1976. The 1979 count was down to 1,119 otters. Whether this figure indicates a population decline or simply a low count because of bad weather conditions can only be ascertained during future surveys. The official estimate for the total population on the California coast is 1,357 to 1,537 (anonymous 1979).

DESCRIPTION

Size. The sea otter is the largest member of the family Mustelidae. A large adult male sea otter may weigh 45 kilograms; however, most weigh 27 to 38 kg. A large adult female may weigh 32.6 kg, but most weigh from 16 to 27 kg. The total length of the adult male is 148 centimeters and of the female 140 cm. Pups at birth normally weigh about 2–2.3 kg and measure about 60 cm in total length (Kenyon 1969).

Pelage and Molt. The sea otter lacks a blubber layer (which protects the seals and whales from their cold environment) and is therefore entirely dependent for warmth on an insulating blanket of air trapped among its tightly packed fur fibers. These fibers may number approximately 800 million hairs (Scheffer in Kenyon 1969). Because the otter's survival depends on the insulation of its fur coat, the molt is diffuse, taking place during the entire year. Although the molt is continuous, more fur fibers are shed in midsummer than in midwinter. The pelage color varies from nearly black to various shades of brown. Light or blond-colored animals are rare. Although most sea otter pelts look quite similar, it is difficult to match exactly even 3 or 4 skins out of 100 (J. S. Vania personal communication). Guard hairs vary in color with individuals, from white to black.

Glands. Unlike most other mustelids, the sea otter does not possess functional anal scent glands.

Feet. The front paws are highly mobile and adapted to picking up food items from the sea bottom and holding them during eating. The claws are rectractile and may be extended when the otter grasps slippery prey such as a fish. The hind flippers are broadly flattened. They are adapted to propelling the otter on the surface by alternate strokes while it floats on its back. Because of this method of swimming, the outer toes are elongated and the species is very clumsy on land.

Skull and Dentition. The rostrum is very short and much wider than long. The post-orbital region is only moderately constricted and the palate extends beyond the last molars.

The dentition differs from that of other carnivores in being adapted to crushing rather than cutting food material. The teeth are bunodont for crushing the shells of mollusks and crustaceans, the otters' primary food; there are no carnassials. The dental formula of the adult is: 3/2, 1/1, 3/3, 1/2 = 32. Four incisors in the mandible of the sea otter are unique among carnivores. The musculature and joints of the skull are also adapted to the sea otter's food habits (figure 37.2).

Ears. The sea otter's ears resemble those of seals more than they do its close relative the river otter (*Lutra canadensis*). The ears are short and somewhat curled or "valvelike." While the otter is at the surface the ears usually are held erect. When it is beneath the surface they are held pointing sharply downward. Presumably in this position water is prevented from entering the ear.

PHYSIOLOGY

Apparently the fur-air insulating mechanism of the sea otter is less efficient than the insulating blubber of whales and seals. This is indicated by the observation that metabolism in the sea otter is much more rapid than in other marine mammals. The sea otter consumes food at the rate of approximately 20–25 percent of its body weight per day. Food containing red dye passed through an otter in about three hours (Kenyon 1969). One of the first problems encountered when sea otters were held in captivity was that they died of hypothermia if their fur became soiled with food scraps or excrement so that the "air blanket" was lost. Recent studies have quantitatively expressed the effect of fur soiling on metabolism. Costa and Kooyman (1979) found that the standard metabolic rate of sea otters was 12.0 ml O_2/kg-min. An otter that was coated with crude oil on 20 percent of its body surface for an eight-day period increased its metabolic rate to 35.1 ml O_2/kg-min. This rate was 127 percent above the normal (or control) rate. Otters caught in an oil spill would certainly die of hypothermia very quickly after becoming coated with oil.

In a study of comparative mammalian respiratory mechanics, Leith (1976) found that the sea otter has an exceptionally large lung capacity for its size. In conjunction with the otter's high metabolic rate and food consumption, the organs associated with digestion, such as the liver and kidneys, are proportionately very large (Kenyon 1969).

REPRODUCTION

Sex Ratio. The neonatal sex ratio is approximately 1 to 1. During periods of hardship, males may have a higher mortality rate than females. Thus, among 481 adults taken at Amchitka Island, 33 percent were males and 67 percent were females, indicating that females outnumbered males by a 2 to 1 ratio. It is possible, however, that this ratio is biased by the fact that areas frequented primarily by females were more extensive and accessible than locations frequented primarily by males (Kenyon 1969).

Pupping Season. In areas of relatively warm climate, newborn pups are seen in every month of the year. In the Aleutians a peak in the pupping season occurs in spring and early summer (Kenyon 1969). On the coast of California the peak in pupping is February (Vandevere 1979). In Prince William Sound, the most northerly sea otter breeding area, winter weather is violent and pups are not born during winter months (A. M. Johnson personal communication). It is indicated that in areas of high population density, sea otters breed every other year. In areas of lower density, they may breed in consecutive years (Johnson and Jameson 1979; Vandevere 1979). More data on the successful weaning of the pups of mothers seen in consecutive years with new young are needed before annual reproduction is established as the norm. Histological studies of the male reproductive tracts indicate that some males are capable of reproduction in all seasons. No breeding among juveniles has been observed.

Mating. Males frequently visit areas occupied primarily by females to search for animals in estrus. Copulation takes place in the water and may last for 20 min-

utes or more. The mated pair may remain together for as long as three days, after which the female deserts the male (Kenyon 1969). Among captive otters, females that fail to become pregnant reenter estrus at intervals of about five weeks.

Gestation Period. Histological studies of reproductive tracts show that the sea otter, like other mustelids and some seals, undergoes a variable period of delayed implantation. A pup born at the Seattle Aquarium on 16 May 1979 apparently had a gestation period of seven and a half months, based on observation of copulations (Sheng 1979). Additional studies are necessary to improve our understanding of this aspect of reproduction.

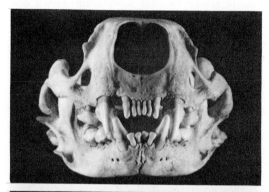

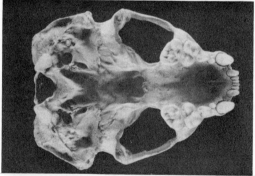

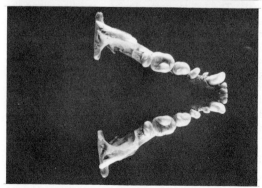

FIGURE 37.2. Skull of the sea otter (*Enhydra lutris*). From top to bottom: anterior view of cranium and mandible, ventral view of cranium, dorsal view of mandible.

Litter Size. Although twin fetuses have been found, no mother has ever been observed with more than one pup. It is doubtful that a mother otter could care for twins. The newborn sea otter requires almost constant care. It is completely helpless and is carried on the mother's chest at all times, except when she is diving for food.

Birth and Maturation of Young. An observation of a mother with a newborn pup on land, and fetal membranes and blood beside the pool of captives, indicate that birth takes place on land (Cecil Brosseau personal communication). Some observations indicate that the pup may be born in the water (Nightingale 1980). Further observations are needed, but at present I believe that pups may be born either on land or in the water. At birth the pup is unable to swim or dive. Observations of one pup born in captivity at the Seattle Aquarium indicated that it was first able to swim weakly at the age of two weeks; by four weeks it was attempting shallow dives (Sheng 1979). By 9 months of age, this pup weighed 22.3 kg. At 11 months it was nearly the size of its mother and continued to nurse, but with reduced frequency. The mother discouraged nursing by rolling over and over. Although the pup was able to gather the easily available food, its mother continued to hand food to it. The pup still slept in close association with its mother, placing its head on the mother's chest or abdomen while the two floated on their backs. The mother remained very possessive of the pup (age 11 months) and tried to restrain it if it attempted to leave her side (J. Styers personal communication).

ECOLOGY

Habitat. The sea otter is limited to a narrow band of coastal water in most areas. Aerial surveys indicate that the majority of otters occur within the 30-fathom depth curve. Otters prefer to feed in shallow water when food is available there. When a dense otter population depletes food resources, the animals must move to offshore areas of forage. The maximum depth at which a sea otter has been taken was 97 m (Newby 1975). Apparently, sea otters do not inhabit inland waterways, such as Puget Sound (Kenyon 1969), and there is no record of a sea otter voluntarily entering fresh water.

Interrelationships. When sea otters exploit invertebrate populations (e.g. abalones and sea urchins), the macroalgae increase and dominate the habitat. This condition is accompanied by an increase in fish populations. Conversely, when humans exploit the sea otter an expanding herbivore population dominates the ecosystem, creating an alternate stable state community (Simenstad et al. 1978).

The near extirpation of sea otters on the California coast during the 18th and 19th centuries made it possible for an abalone population to "explode" and flourish and become the basis for an abalone fishery. Now that the sea otter is again entering areas from which it was long absent, considerable controversy has developed.

It is impossible for a peak population of sea otters and a commercial fishery to utilize the same food resources and to coexist in the same area. Where peak populations of sea otters exist and herbivorous invertebrates are greatly reduced, the otters spend 55 percent of their time in foraging activity. By contrast, in areas of sparse otter populations, only 17 percent of their time is spent foraging (Estes and Jameson 1979). In addition, where a peak otter population exists and fish are more abundant (because of the macroalgae), the diet of otters may consist of 60 percent fish. In a sparsely populated area, however, where herbivorous invertebrates are abundant the otter's food intake consists entirely of these invertebrates (Estes and Jameson 1979).

Whether sea otters or abalones and sea urchins will dominate any given geographical area of marine environment is a matter that may be resolved on the basis of human value judgments.

Movements. Recent finds indicate that otters may move long distances, probably in response to food availability. Benech (1979) found that seasonal fluctuations in population density occurred during winter and spring, when approximately 200 otters entered the observation area. During the summer and fall the population declined to about 50. Of 53 otters tagged in 1978 and 1979, Jameson (1979) found that adult males traveled nearly 96 km from the Piedras Blancas area to the Pecho Rock–Pismo Beach area, California. They returned to the Piedras Blancas area in the late spring and early summer, apparently in search of estrous females. Previous studies in the Aleutians, primarily involving tagged females, indicated that their home range included approximately 16 km of coastal habitat (Kenyon 1969).

FOOD HABITS

Food preferences have been studied both in the wild and among captives. In California a serious conflict exists between commercial and sport fishermen and those who would afford the sea otter complete protection. Observations in the wild indicate that abalones (*Haliotus* sp.) are preferred food when available. Jameson (1979) found that after abalone and urchins (*Stronglylocentrotus,* sp.) became restricted to crevice refuges, areas that were free of predation, the otters shifted to crabs. Benech (1979) found that feeding otters took 54 percent abalones, 16 percent crabs, and 30 percent other species. Also, during a six-year observation period (1973–78) the sea urchin density was reduced to 1 percent of that found before otters entered the study habitat. In captivity an attempt was made to find the food most preferred by otters. They were fed squid (*Loligo opalescens*), rockfish filets (*Sebastes* sp.), crabs, and various species of clam. Crabs were the most favored food (Nightingale et al. 1979).

There is no question that the voracious appetite of the sea otter exerts severe pressure on the herbivorous invertebrates of the near-shore environment and that this pressure can alter an entire ecosystem. For exam-

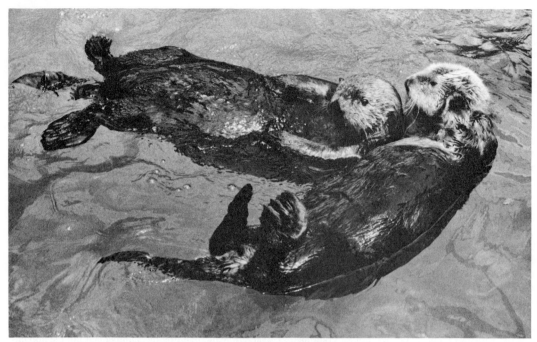

FIGURE 37.3. Male and female sea otter (*Enhydra lutris*) floating on their backs.

ple, archeological excavations of kitchen middens in the Aleutian Islands indicate that during past centuries sea otters were greatly reduced (probably because of their warm fur) and populations of sea urchins flourished. These furnished a considerable food resource for primitive mankind. In more recent years, after the decimation of the aboriginal human population, sea otter populations there recovered. Now the habitat is dominated by sea otters and herbivorous invertebrates are greatly reduced.

BEHAVIOR

Territoriality was not strongly expressed among adult male otters at Amchitka Island (Kenyon 1969). However, A. M. Johnson (personal communication) found that adult male otters will remain in a limited area, from which they exclude other males but welcome females, expressing rather strong territorialism.

In general, males and females choose areas in which they are segregated to a considerable degree. Females accompanied by young particularly stay away from areas occupied primarily by adult males. Adult males may gather on beaches in the Aleutians, where they rest and sleep in close proximity. Males searching for estrous females enter female areas (figure 37.3).

MORTALITY

Eagles. At Amchitka Island there is a substantial population of both bald eagles (*Haliaeetus leucocephalus*) and sea otters. Otter remains are often found in eagle nests. During the period of 1969 through 1973, 114 nests were checked for food, and

parts of 100 otters were found. Sea otter remains constituted from 9 to 18 percent of the prey remains in nests. Although there are three observations of eagles taking living sea otter pups, this predation has little effect on the population. Most of the sea otter remains found in nests are probably scavenged from dead animals found on beaches (Sherrod et al. 1975).

Sharks. Along the California coast, sharks undoubtedly kill a considerable number of otters. Shark teeth have been found inbedded in the bones of otters washed ashore on beaches (Orr 1959). More recent information shows shark tooth remains in the bones of sea otters originally presumed to be killed by the propellers of outboard motors. In certain areas it appears that shark predation may be a significant cause of mortality.

Killer Whale. A Soviet biologist reports that killer whales (*Orcinus orca*) may take sea otters (Nikolaev 1965). I have seen killer whales in close association with sea otters on several occasions, but have never seen any indication of predation.

For a discussion of parasites and diseases, see Kenyon (1969:271–278).

ECONOMIC STATUS AND MANAGEMENT

Alaska. The policy of complete protection of the sea otter was maintained by the federal government until the state of Alaska took over the management of marine mammal resources in 1958. It was recognized by the federal government that otter populations were large enough to support an annual harvest of otter pelts.

However, it was believed that if otters were harvested, patrol of their extensive marine habitat would be prohibitively expensive. It was decided, therefore, that a commercial harvest would not be attempted. After statehood, the Alaska Department of Fish and Game took the first modern harvest of sea otter pelts in 1962. A more serious effort was undertaken from 1967 to 1971; the total take was 2,933 skins. During five auctions held annually from 1968 through 1972, 2,717 skins were sold for a total of $364,550, an average price of $134.17 per skin (V. B. Scheffer and J. S. Vania personal communication). After 1972, Alaska decided against further harvesting or sale of skins. An editorial in *Otter Raft* (published by Friends of the Sea Otter, Box 221220, Carmel, California 93922) stated, "The State of Alaska received a poor return from its attempt to auction 500 freshly dressed sea otter skins this February at the Seattle Fur Exchange. The approximate cost for killing and dressing each otter pelt is $40. Because the skins brought only $100, $110, and $120, and because the future demand for sea otter furs is uncertain, the State Legislature is unwilling to appropriate $40,000 to hire Aleut hunters to obtain additional skins for future auctions" (anonymous 1972).

The question arises: why should the pelt that was considered "the most valuable fur in the world" during the 18th and 19th centuries fail to be in demand in the mid-20th century? (At the turn of the century, skins sold for more than $1,000 each [Kenyon 1969: 41–42].) I asked several officials in the fur industry, particularly at the Seattle Fur Exchange, where the auctions took place, for their opinions. The following reasons were proposed: (1) the fur was not advertised widely enough to create a sufficient demand; (2) it has been off the market for so long (60 years) that people do not know what it is; (3) the fur is too heavy for most modern uses; (4) preservationists have encouraged the belief that the sea otter remains an endangered species (only in California is the sea otter classified as "threatened"); and (5) there is a growing feeling among the increasingly urban population, encouraged by many moving pictures "humanizing" wildlife, that it is wrong to take wild animals for their skins.

It is a moot question whether or not the sea otter will again be used as an article of commerce. Certainly the present Alaskan population would indefinitely support an annual harvest of several thousand skins. Officials of the fur industry believe that if sufficient effort were invested in advertising, a substantial market for sea otter fur would again develop.

California. On the California coast, strong feelings concerning sea otters have developed between two opposing groups, Friends of the Sea Otter, and Friends of the Abalone. The latter group encourages the formation of a coastal zone management plan that would exclude sea otters from certain areas set aside as abalone fishing grounds. The Friends of the Sea Otter, however, are opposed to any exploitive management of the sea otter until the species has completely reoccupied its aboriginal range. This group contends that the sea otter is a natural part of an ecosystem that

evolved over millions of years. The members feel that an ecosystem containing the sea otter as an integral part would be more productive and useful to mankind than one managed for a single resource, such as the abalone or even the sea urchin, both of which have considerable commercial value. The Friends of the Sea Otter have presented a position paper stating that (1) they support the natural spread of sea otter populations and there should be no management that would include range restriction; (2) in order to establish additional populations, otters should be translocated from areas of abundance to new areas (this program would insure that a future oil spill along the California coast would not completely annihilate the sea otter population); (3) they advocate the continued protection of the sea otter in California under the Endangered Species Act of 1973, as a threatened species, and that protection under the Marine Mammal Protection Act of 1972 should be continued; and (4) they recommend that both state and federal wardens continue and increase their efforts to protect the sea otter from any possible human harassment or killing (Davis 1979). I consider that the sea otter would be a very easily managed mammal. It is highly visible and the number within any given area could be quite easily controlled. On the other hand, now that the shipping of petroleum is greatly expanded, primarily around California, the prospect of a spill and subsequent serious damage to the sea otter population is a real threat—thus the status as a "threatened species." Also, attitudes have changed greatly in the past 20 years. Many people believe that just because we can manage something is no reason that we should. Others would rather see sea otters than eat abalones.

Trapping and Censusing. Sea otters are taken for translocation by means of gill nets. Censuses are made by boat, by aircraft, and also from shore using telescopes and binoculars.

CURRENT RESEARCH AND MANAGEMENT NEEDS

At present the U.S. Fish and Wildlife Service maintains a research station at Piedras Blancas, in the heart of the California sea otter range. This branch of the National Bird and Mammal Laboratories is currently staffed by biologists studying the behavior and ecological relationships of the sea otter. Valuable information is being found and this study will be continued. Additional studies, in the Aleutians and in Prince William Sound, are in progress by state and federal wildlife biologists. In California many university professors, students, and state employees are involved in observational and experimental ecological studies.

LITERATURE CITED

Alaska Department of Fish and Game. 1973. Alaska's wildlife and habitat. Anchorage, Alaska. 144 pp.+ 563 maps.
Anonymous. 1972. Fur prices down. Otter Raft 7:2.
Anonymous. 1979. DFG sea otter census 1979: "Holding steady" or losing ground? Otter Raft 22:6.

Benech, S. V. 1979. Sea otter observations and activities on the southern frontier (1973–1978). Otter Raft 22:7.

Bigg, M. A., and MacAskie, I. B. 1978. Sea otters established in British Columbia. J. Mammal. 59:874–876.

Bolin, R. L. 1938. Reappearance of the southern sea otter along the California coast. J. Mammal. 19:301–303.

Boolootian, R. A. 1961. The distribution of the California sea otter. California Fish Game 47:287–292.

Bryant, H. C. 1915. Sea otter near Point Sur. California Dept. Fish Game 1:134.

Costa, D., and Kooyman, G, 1979. Effects of crude oil contamination on the sea otter's ability to thermoregulate. Page 10 in 3rd Abstr. Biennial Conf. Bio. Marine Mammals, 7–11 October 1979.

Davis, B. S. 1979. Friends of the Sea Otter; position paper on management perspectives. Otter Raft 22:3.

Davis, J., and Lidicker, W. Z., Jr. 1975. The taxonomic status of the southern sea otter. Proc. California Acad. Sci., Ser. 4, 40:429–437.

Estes, J. A., and Jameson, R. J. 1979. Relationships among species interaction, food selection, and activity of sea otters in the Aleutian Islands. Page 17 in 3rd Abstr. Biennial Conf. Bio. Marine Mammals, 7–11 October 1979.

Jameson, R. J. 1979. An interview with Ron Jameson, otter studies at Piedras Blancas, by Margaret Owings. Otter Raft 22:8

Jameson, R. J.; Kenyon, K. W.; Johnson, A. M.; and Wight, H. W. 1981. The history and status of translocated sea otter populations. J. Wildl. Manage. In press.

Johnson, A., and Jameson, R. 1979. Evidence of annual reproduction among sea otters. Page 31 in 3rd Abstr. Biennial Conf. Bio. Marine Mammals, 7–11 October 1979.

Kenyon, D. W. 1969. The sea otter in the eastern Pacific Ocean. North Am. Fauna 68. 352pp.

Leith, D. E. 1976. Comparative mammalian respiratory mechanics. Physiologist 19:485–510.

Morris, R.; Ellis, D.; Emerson, B.; and Norton, S. 1979. Assessment of the B. C. otter transplants, 1978; including data on stocks of invertebrates and macrophytic algae. Rep. to Ecol. Reserves Unit, B.C. Ministry of Environment. 98pp. Mimeogr.

Newby, T. C. 1975. A sea otter (Enhydra lutris) food dive record. Murrelet 56:7.

Nightingale, J. 1980. Field notes (an account of birth and health of a sea otter pup born 16 May 1979). Cetus, Whale Mus. 2:3–4.

Nightingale, J. W.; Timmis, C.; and Styers, J. 1979. Food preference studies with captive sea otters (Enhydra lutris) using a comparison matrix. Page 43 in 3rd Abstr. Biennial Conf. Bio. Marine Mammals, 7–11 October 1979.

Nikolaev, A. M. 1965. On the feeding of the Kuril sea otter and some aspects of their behavior during the period of ice. Pages 231–236 in E. N. Pavlovskii et al., eds. Marine mammals. 315pp. Transl. Nancy McRoy, National Marine Mammal Laboratory, Seattle, Washington, April 1966.

Orr, R. T. 1959. Sharks as enemies of sea otters. J. Mammal. 40:617.

Roest, A. I. 1973. Subspecies of the sea otter, Enhydra lutris. Los Angeles City Mus., Contrib. Sci. 252:1–17.

———. 1979. a reevaluation of sea otter taxonomy. Page 14 in Abstr. Sea Otter Workshop, 23–25 August 1979, Santa Barbara, Calif.

Scheffer, V. B., and Wilke, F. 1950. Validity of the subspecies Enhydra lutris nereis the southern sea otter. J. Washington Acad. Sci. 40:269–272.

Schneider, K. B., and Faro, J. B. 1975. Effects of sea ice on sea otters (Enhydra lutris). J. Mammal. 56:91–101.

Sheng, S. 1979. The otters steal the show. Seattle Aquarium J. 3:4–5.

Sherrod, S. K; Estes, J. A.; and White, C. M. 1975. Depredation of sea otter pups by bald eagles at Amchitka Island, Alaska. J. Mammal. 56:701.

Simenstad, C. A.; Estes, J. A.; and Kenyon, K. W. 1978. Aleuts, sea otters, and alternate stable-state communities. Science 200:403–410.

Vandevere, J. E. 1979. The dependency period for Enhydra lutris nereis. Page 15 in Abstr. Sea Otter Workshop, 23–25 August 1979, Santa Barbara, Calif.

Zimushko, V. V.; Fedoseev, G. A.; and Shustov, A. P. 1968. A sea otter in the Arctic. Priroda (Moscow), (1968), p. 104.

KARL W. KENYON, 11990 Lakeside Place, N. E., Seattle, Washington 98125.

38

Mountain Lion

Felis concolor

Kenneth R. Dixon

Kenneth R. Dixon

NOMENCLATURE

COMMON NAMES. Mountain lion, puma, cougar, panther, painter, catamount

SCIENTIFIC NAME. *Felis concolor*

SUBSPECIES NORTH OF MEXICO. *F. c. couguar, F. c. missoulensis, F. c. hippolestes, F. c. oregonensis, F. c. vancouverensis, F. c. olympus, F. c. californica, F. c. kaibabensis, F. c. browni, F. c. improcera, F. c. azteca, F. c. stanleyana,* and *F. c. coryi.*

The mountain lion has existed in its present form since the early Pleistocene. Of the extant North American felids, it is second in size only to the jaguar (*Felis onca*).

DISTRIBUTION

The mountain lion occurs only in the western Hemisphere and once had one of the most extensive ranges of any terrestrial mammal. Although still covering over 100° latitude from the Straits of Magellan to the Canadian Yukon, there has been a significant reduction in the mountain lion distribution as a result of the extirpation of the species from much of its former range (figure 38.1). In North America, substantial mountain lion populations occur in suitable habitat in the western United States and Canada, although ranges have been reduced in these areas also. The distribution of the mountain lion ranges from sea level to elevations over 3,000 meters. The mountain lion is restricted largely to the habitat of its primary prey—the open forests of the white-tailed deer (*Odocoileus virginianus*) and the mule deer (*O. hemionus*).

The mountain lion may be expanding its range in some areas, holding steady in others, and declining in still others depending on the level of hunting pressure and prey availability. In western North America, the bounty on mountain lions was removed beginning in 1958 with British Columbia and ending in 1970 with Arizona. Without the bounty, and with increasing deer populations, the mountain lion has increased its range in many western states and provinces. Recent sightings, which may indicate either range extensions or reestablishment of former range, have been reported

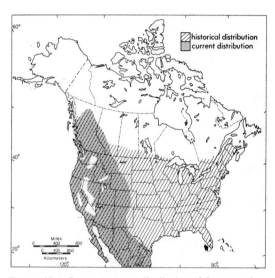

FIGURE 38.1. Past and present distribution of the mountain lion (*Felis concolor*).

from eastern Alberta (Nowak 1976), southern Yukon (Weddle 1965; Youngman 1968), southern and east-central Saskatchewan (Maliepaard 1971; Riome 1973), southern Manitoba and northern Minnesota (Nero and Wrigley 1977), central Montana (Greer 1979, 1980), eastern Wyoming (Roop 1971), and the Black Hills of South Dakota (Turner 1971). A decline in the Nevada mountain lion population in the early 1970s (Greenley 1971) has been reversed with the establishment of a hunting quota system, and the population is now approaching its carrying capacity (Tsukamoto 1980).

Areas where mountain lion populations, and subsequently their range, are decreasing include south-central and southeastern British Columbia (Spalding 1971; Dewar and Dewar 1976), northern Utah (Nowak 1976), and the trans-Pecos area and southern brush country of Texas (Russell 1971). The status of the lion in Texas is of particular concern, since it is believed (e.g., Russell 1971) that many of the state's lions migrate from Chihuahua, Mexico, an area in which the mountain lion's range also is decreasing (Anderson 1972).

In eastern North America the only recognized population of mountain lions is found in southern Florida. However, reports of sightings of lions are made every year from many other eastern states and provinces. Reliable sightings are most numerous from the Appalachian Mountains, south-central and southeastern Canada, the Ozark Mountains of Arkansas and Missouri, southeastern Arkansas, and northern Louisiana. Recent specimens were taken in northwest Louisiana near Keithville in 1965 (Goertz and Abegg 1966), in southeastern Arkansas near Hamburg in 1969 (Noble 1971), in eastern Oklahoma near Checota in 1968 (Lewis 1969) and between Stringtown and Redden in 1975 (Nowak 1976), in eastern Tennessee near Pikeville in 1971 (Nowak 1976), and in northwestern Pennsylvania near Edinboro in 1967 (Doutt 1969). It is not known whether these sightings and kills represent remnants of former populations, transient lions migrating from disjunct populations, or lions escaped from captivity. In any case, there is not sufficient evidence at present to assume that viable populations exist in eastern North America except in southern Florida.

DESCRIPTION

Mountain lions vary considerably in size and weight. Adult males average about 2.4 meters in length and range in weight from 80 to 91 kilograms. Adult females average slightly less than 2 m and weigh from 34 to 80 kg.

Pelage. The adult mountain lion's fur generally is uniformly tawny. However, the color ranges from a slate gray (sometimes referred to as "blue") to a rufous brown (sometimes referred to as "red"). The underside of the body is whitish, as is the upper lip. The back of the ears and tip of the tail are blackish brown.

Kittens have a pattern of blackish brown spots and a dark-ringed tail. This pattern gradually fades until it is replaced by a tawny coat by the end of the first year.

Anatomy. The mountain lion is one of the most highly specialized carnivores. Like most felids, it has anatomical features for an almost exclusively predatory diet.

The rostrum of the skull is short and rounded (figure 38.2). The sagittal crest and lambdoidal ridge are well developed. The mandible articulates with the mandibular fossa of the temporal bone. The power for the lion's jaw comes primarily from the two largest muscles of the head, the temporalis and the masseter. The temporalis is the largest and strongest muscle, originating in the temporal fossa with insertion over the entire coronoid process of the mandible. The masseter originates from the zygomatic arch and inserts over the coronoid fossa. Working with these muscles to elevate the mandible are the two pterygoideus muscles, with origins on the pterygoid fossa and insertions near the angular process of the mandible.

Vibrissae are found on the cheek and upper lip, and above the upper eyelid. The vibrissae on the upper

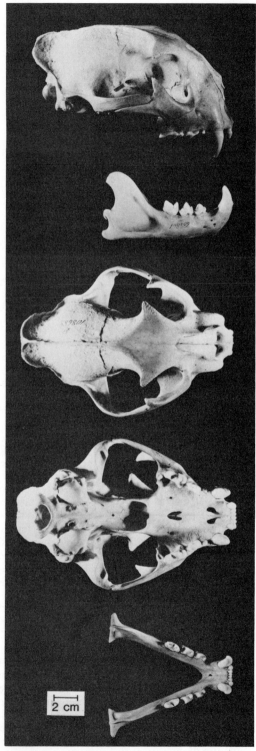

FIGURE 38.2. Skull of the mountain lion (*Felis concolor*). From top to bottom: lateral view of cranium, lateral view of mandible, dorsal view of cranium, ventral view of cranium, dorsal view of mandible.

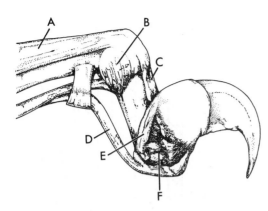

FIGURE 38.3. Retractile claw mechanism of the mountain lion (*Felis concolor*). A, extensor expansion; B, middle interphalangeal joint; C, extensor tendon; D, flexor digitorum profundus tendon; E, lateral dorsal elastic ligament; F, distal interphalangeal joint. Redrawn from Gonyea and Ashworth 1975.

lip and cheek are erected by the angular head muscles and retracted by the caninus muscle.

Like most mammals, the mountain lion is diphyodont, with the adult having 30 teeth—16 in the upper jaw and 14 in the lower jaw. The cheek teeth are secondont with well-developed carnassial teeth. The dental formula is 3/3, 1/1, 3/2, 1/1 = 30.

Mountain lions, like all felids, are digitigrade, with five digits in the forepaw (manus) and four in the hindpaw (pes). Each of the distal phalanges supports a terminal, ensheathed, retractile claw (figure 38.3). The claws are not used for locomotion as in most other carnivores, but function mainly for grasping prey. Retraction is made possible by the angular articulation of the distal phalanx with the second phalanx and the attachment of dorsal elastic ligaments and forearm muscles (Gonyea and Ashworth 1975).

PHYSIOLOGY

Blood. Mountain lions have hemoglobins in their blood in which the beta chain terminus is substituted with an acetyl group. The acetylated and nonacetylated components occur in approximately equal proportions in the mountain lion; however, the acetylated component may comprise as much as 90 percent of the total hemoglobin in the lion (*Felis leo*), tiger (*F. tigris*), and snow leopard (*F. uncia*) (Taketa et al. 1972). Similar acetylated beta chain amino termini in domestic cat hemoglobins are insensitive to the modifying influence of organic phosphates on oxygen affinity, whereas unsubstituted beta chain amino termini are sensitive (Taketa et al. 1971). The significance of the occurrence of variable proportions of acetylated and nonacetylated hemoglobins in the blood of different feline species from the physiological and evolutionary point of view is not known (Taketa et al. 1972).

Vision. There are several adaptations in the vision of mountain lions, including both diurnal and nocturnal vision and stereoscopic vision, that make them effective predators. Nocturnal vision requires a large pupil

to gather light. This entails a large lens, increasing its focal length so that its curvature also has to be increased to bring the focused image back onto the retina. Ordinarily an increase in curvature produces a smaller but brighter image; therefore, the whole eye is enlarged to increase the size of the image. The sensitivity to light also is increased by the presence of a tapetum, which reflects the light back through the visual cells of the eye and by dark adaptation involving neural reorganization of the retina (Barlow et al. 1957). The adaptations for nocturnal behavior require increased protection for the lion's retina in daylight; hence, the pupil contracts to a vertical slit or "cat's-eye" in bright light.

Stereoscopic vision requires a large area of overlapping visual fields of the two eyes. In cats, the binocular field is 130° with a total visual field of 287°. Stereoscopic vision or depth perception enables the mountain lion to attack its prey with extreme accuracy but functions only within a distance of 15 to 24 m (Tansley 1965). Depth perception is possible to some extent without binocular vision by the use of perspective, image size, and parallax. A male mountain lion killed in California was blind in one eye and otherwise appeared to be healthy, indicating that monocular vision was sufficient in this case (McLean 1954).

The perception of movement obviously is important for any predator that feeds on moving prey. In fact, movement of prey may be a necessary stimulus for attack behavior (Hubel 1959). This may be why prey species tend to "freeze" after detecting a predator.

REPRODUCTION

Male mountain lions are attracted to females early in estrus and more than one male may accompany a female until it is receptive to mating. If the males come in contact, they may engage in aggressive behavior and fighting during this time. Mating, which may induce ovulation, is by the dominant male. Copulation usually lasts less than one minute. An estrous cycle in mountain lions lasts approximately 23 days, with the estrus usually lasting 8 days (Rabb 1959). Periods of estrus up to 11 days have been reported, however (Young and Goldman 1946; Rabb 1959).

Mountain lions first breed when about two and one-half years old (Young and Goldman 1946; Rabb 1959). Those born during the summer peak (figure 38.4) first breed during the winter following their second birthday. Variation from this pattern has been reported (Seidensticker et al. 1973), with individual cases determined by the season of the female's own birth and its condition when sexually mature.

Once lions are sexually mature they are capable of breeding throughout the year and successful litters can be produced any month of the year (Bruce 1922; Musgrave 1926; Hibben 1937; Young and Goldman 1946). There is a peak in litter production during the summer, however (see figure 38.4) (Robinette et al. 1961).

Following a gestation period of 82–98 days, a litter of one to six young is produced. The frequency of various litter sizes was reported by Robinette et al. (1961) (figure 38.5). Prenatal mortality of 15 percent, including resorption of embryos, also has been reported (McLean 1954; Robinette et al. 1961). The young are weaned when two to three months old and remain with their mother until nearly 2 years old (Merriam 1884; Bruce 1922; Hibben 1937). Individuals produce litters usually every other year, but may breed in consecutive years under optimal conditions (Robinette et al. 1961). A captive mountain lion produced seven litters in 16 years (Conklin 1884), although its productivity declined with age (figure 38.6).

Mountain lions provide rather simple den sites in which to have their litters, with only enough cover to keep out heavy rain and the hot sun. In rough terrain they usually find a shallow nook on the face of a cliff or rock outcrop. In less mountainous areas, dens are

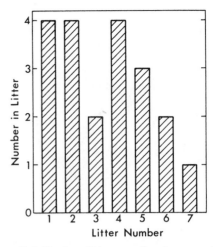

FIGURE 38.6. Number of kittens per litter in consecutive litters of a mountain lion (*Felis concolor*) over a 16-year period. From data of Conklin 1884.

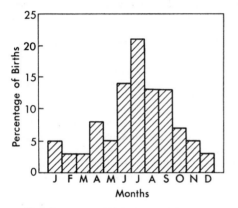

FIGURE 38.4. Percentage of births by month for 145 mountain lion (*Felis concolor*) litters in Utah and Nevada. From Robinette et al. 1961.

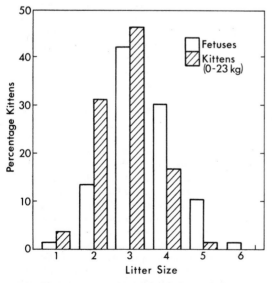

FIGURE 38.5. Frequencies of litter sizes in mountain lions (*Felis concolor*). Clear bars indicate the frequencies of numbers of prenatal young (N = 66); hatched bars indicate the frequencies of postnatal young up to 22.7 kg (N = 131). From Robinette et al. 1961.

located in dense thickets or under fallen logs. Little bedding is used in lion dens, although Young and Goldman (1946) report the finding of a female's soft belly hair in one bed. Mills (1923) reported that one female used the same den for several years.

ECOLOGY

Habitat. Mountain lion habitat essentially is that of their primary prey—mule deer in western North America and white-tailed deer in eastern North America. Typical lion habitat in western North America is open woodland such as oak (*Quercus* spp.) scrub, pinyon pine (*Pinus cembroides*), juniper (*Juniperus* spp.), mountain mahogany (*Cercocarpus ledifolius*), snowbush (*Ceanothus velutinus*), and manzanita (*Arctostaphylos* spp.) associations. Within these habitat types, lions tend to prefer rocky cliffs, ledges, or other areas that provide cover. The presently reduced mountain lion populations in eastern North America make it difficult to determine the preferred habitat. Historically, the evidence suggests that the mountain lion and the white-tailed deer shared the same basic habitat types. These habitats primarily were open areas in oak stands, marshes and swamps, and small prairies with an abundance of "edge" (Severinghaus and Cheatum 1956). Likewise, early reports of mountain lion habitat emphasized areas along watercourses and swamps, particularly the native bamboo (*Arundinaria gigantea*) association known as canebrakes (Nuttall 1821; Audubon and Bachman 1851; Wailes 1854; Hallock 1877).

Home Range. Rather than a true territory that is actively defended, lions have a land tenure system (Seidensticker et al. 1973) in which home ranges are maintained by resident lions but not transient lions. The home range consists of a first-order home area, used primarily for resting, and a much larger area used for hunting.

TABLE 38.1. Seasonal home ranges (km²) of male and female mountain lions in Idaho

Mountain Lion	Winter–Spring	Fall–Summer	Yearly Total
Males (N=1)	120.5	293.0	453.0
Females			
with kittens (N=2)	91.2	110.0	196.5
without kittens (N=2)	128.8	185.0	339.5

SOURCE: Seidensticker et al. 1973

Home range size in the mountain lion varies by sex and age of the lion, season of the year, and spatial pattern and density of the lion's prey. Seasonal and sex differences in home range size (table 38.1) were reported by Seidensticker et al. (1973) on the Idaho Primitive Area. Different spatial patterns of prey probably account for the differences in home range size reported from three different states (table 38.2).

The size of female home ranges varies according to reproductive status. Adult females without young and females with newborn young have the smallest home ranges. As the age and food requirements of the kittens increase, the home range of the parent female also increases (Seidensticker et al. 1973).

Studies on the Idaho Primitive Area (Hornocker 1969; Seidensticker et al. 1973) showed that resident lions maintained contiguous but fairly distinct home ranges in winter and summer, determined largely by the movements of their most important prey—mule deer and elk (*Cervus elaphus*). Similar results were found in California, except where weather conditions and nonmigratory prey precluded seasonal variation in range occupancy (Sitton and Weaver 1977). In Idaho there was little overlap of resident male home ranges; however, each male home range overlapped more than one female home range. Home ranges of resident females overlapped, sometimes completely (Seidensticker et al. 1973). However, different results were found in Nevada (Ashman 1977) and California (Sitton and Weaver 1977): home ranges of males overlapped and those of females did not.

The differences in home range size and overlap found in the studies in Idaho, Nevada, and California may be due partly to the method of calculating the

TABLE 38.2. Annual home ranges of male and female mountain lions in California, Idaho, and Nevada

State	Home Range Size (km²)	
	Male	Female
California[a]	194.4	103.7
Idaho[b]	453.0	268.0
Nevada[c]	575.4	203.0

[a]Sitton and Weaver 1977; California Department of Fish and Game 1973
[b]Seidensticker et al. 1973
[c]Ashman 1979

home range. The method of connecting outlying telemetry locations used in these studies does not distinguish between the first-order home range and the total area used for hunting. These methods also tend to include areas not used by the lions (Dixon and Chapman 1980). Determining the amount of overlap, or even whether or not the home ranges overlap, is not as important as identifying areas that are used with different degrees of intensity and the correlation of range use with habitat or prey density. Also, overlap may occur in space but not in time (Hornocker 1969; Leyhausen 1979).

Hornocker (1969) suggested that home range boundaries are maintained by mutual avoidance as opposed to active defense. He found that transient lions are tolerated by resident lions and that there was no evidence of fighting. This form of home range maintenance through avoidance behavior probably serves as a mechanism to limit population density (Hornocker 1969, 1970) and increase predation success rates (Lamprecht 1978).

The use of lion scrapes appears to function as signals to other lions and to mark home range boundaries. First described by Darwin (1839), the scrape (or scratch) is a collection of pine needles, leaves, or dirt scraped into a pile with the hind feet. Urine or feces often are deposited on the pile. Seidensticker et al. (1973) found that resident males scraped more often at home range boundaries than at the center. Scrapes usually are located along travelways under a tree or along the edge of a cliff or ridge (Hibben 1937). Although most scrapes are made by resident males, transient males and females without kittens also make scrapes (Musgrave 1926; Seidensticker et al. 1973).

Similar olfactory marking behavior has been observed in free-ranging domestic cats (Leyhausen 1965) and may serve to identify wolf (*Canis lupus*) pack territory boundaries (Mech and Peters 1977).

Mountain lions, like domestic cats, will scratch tree trunks. However, this behavior probably is used only to sharpen their claws and not as a means of communication (True 1889; Seidensticker et al. 1973).

Movements. The close association between mountain lions and deer extends to their seasonal movements. In much of western North America, mule deer migrate to lower elevations in the winter and higher elevations in the summer. These seasonal movements also are made by mountain lions as they follow their major source of prey. In the Kaibab region of Arizona, the mule deer is the most important prey species in the Ponderosa pine (*Pinus ponderosa*) forest (Rasmussen 1941; Russo 1970). The deer herd migrates from lower elevations into the pine forest beginning in April (figure 38.7). From June to August, the deer are found at even higher elevations in the spruce-fir (*Picea engelmanni–Abies concolor*) community. By September or October, most of the herd again is found in the pine forest. The mountain lions follow the migrating deer herd in and out of the pine forest. Within this community, lions are found primarily near canyon rims or other rugged terrain (Rasmussen 1941).

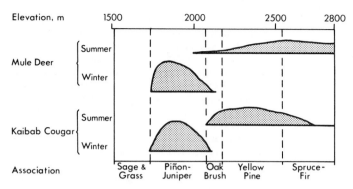

FIGURE 38.7. Elevational distribution of mountain lions (*Felis concolor*) and mule deer (*Odocoileus hemionus*) on the Kaibab Plateau, Arizona. Adapted from Rasmussen 1941.

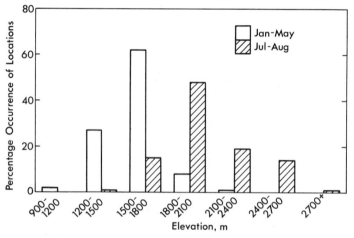

FIGURE 38.8. Elevation distribution of mountain lion (*Felis concolor*) activity on winter (January-May) and summer (July-August) ranges. From Seidensticker et al. 1973.

By late October, the deer herd begins migrating to lower elevations, where they spend the winter—primarily in the pinyon-juniper community. The mountain lions follow the herd to its lower elevations and remain there until the herd again migrates upward in the spring.

A similar pattern of elevational movement was found on the Idaho Primitive Area (Seidensticker et al. 1973) and in California (Sitton and Weaver 1977). In Idaho, the winter range of mule deer and elk is restricted largely to elevations below 1,800 m and is dominated by ponderosa pine and Douglas fir (*Pseudotsuga menziesii*). Normally, both resident and transient lions also restrict their movements to these lower elevations. (figure 38.8). However, resident lions occasionally crossed high mountain passes filled with snow and transient lions moved considerable distances at high elevations between winter ranges (Seidensticker et al. 1973). Beginning in May, the deer and elk begin migrating to higher elevations and areas dominated by spruce (*Picea engelmanii*) and fir (*Abies lasocarpa*). From June to August resident female lions spend most of their time on their summer range; similarly, resident males occupy their summer range from July to August (see figure 38.8). Beginning in September and extending into November, the ungulates and lions again migrate down to their winter range.

FOOD HABITS

In North America the mountain lion depends almost exclusively on deer for its food, although other species of big game and small mammals are eaten depending on local abundance. Food habits studies in western North America show a remarkable consistency in the frequency of occurrence of the major food items in the mountain lion's diet (table 38.3). Only three items—mule deer, porcupine (*Erethizon dorsatum*), and grass—comprise from 86 to 100 percent of the diet. Elk also are taken where they and the mountain lion occur together. In Oregon (Toweill and Meslow 1977), elk comprised 11 percent of the lion's diet; in Washington 9 percent (Schwartz and Mitchell 1945); in

TABLE 38.3. Percentage occurrence of food items of mountain lions in western North America

Food Item	Utah and Nevada[a]		Oregon[b]	British Columbia[c]	Arizona[d]	New Mexico[e]	California[f]
	Winter	Summer					
Mule deer	75.1	63.5	72.2	78.0	42.6	61.7	
White-tailed deer					32.4	26.5	
Total	75.1	63.5	72.2	78.0	75.0	88.3	79.1
Porcupine	14.7	19.2	16.7	11.9	9.2	4.1	
Grass or vegetation	5.3	9.6	11.1	3.4	2.8	1.5	7.0
Domestic sheep	4.1	15.4					
Elk		1.9	11.1				
Domestic cattle	0.4			3.4	2.8	1.0	2.3
Skunk (*Mephitis mephitis*)	0.4					0.5	2.3
Beaver (*Castor canadensis*)	1.2	7.7		5.1			
Lagomorphs	3.7[g]	7.7[g]		1.7[h]	11.1[i]	3.6[g]	
Miscellaneous	1.2[j]	3.8[k]		8.5[l]	0.9[m]	1.0[n]	9.2[o]

[a]Robinette et al. 1959. Winter analysis is from stomach samples (232), summer from scats (52).
[b]Toweill and Meslow 1977. Winter analysis from stomach samples (18).
[c]Spalding and Lesowski 1971. Winter analysis from stomach samples (59).
[d]Hibben 1937. Winter and summer analysis from stomach (15) and scat (103) samples.
[e]Hibben 1937. Winter and summer analysis from scat samples (196).
[f]Dixon 1925. Winter and summer analysis from stomach samples (43).
[g]Cottontail rabbit (*Sylvilagus* spp.), black-tailed jackrabbit (*Lepus californicus*).
[h]Snowshoe hare.
[i]Black-tailed jackrabbit.
[j]Horse, dog, woodrat (*Neotoma* spp.).
[k]Pocket mouse (*Perognathus* spp.), ground squirrel (*Spermophilus* spp).
[l]Carrion, coyote (*Canis latrans*), mountain lion, ruffed grouse (*Bonasa umbellus*), horse.
[m]Badger (*Taxidea taxus*).
[n]Gray fox (*Urocyon cinereoargenteus*).
[o]Raccoon (*Procyon lotor*), cat, hog.

Utah and Nevada 2 percent (Robinette et al. 1959); and in Idaho 24 percent (Hornocker 1970). In one area of British Columbia, moose (*Alces alces*) contributed a significant part of the lion's diet, as did snowshoe hares (*Lepus americanus*) during a peak hare population (Spalding and Lesowski 1971). White-tailed deer and porcupine were the principal foods of the mountain lion in eastern North America (Merriam 1884; Wright 1959).

In temperate parts of the mountain lion's range, small mammals (other than porcupine) represent a minor, suboptimal part of the lion's diet and probably are taken opportunistically. Porcupines are taken wherever they occur in the mountain lion's range and undoubtedly are a preferred food item, although the reason for this preference has not been established. In more tropical areas, the lion's diet is more variable. In Florida, mountain lions prey on wild hogs, raccoons, and armadillos (*Dasypus novemcinctus*) in addition to white-tailed deer. In the southwestern United States, peccaries (*Dicotyles tajacu*) can be an important part of the lion's diet.

Prey Selection. Prey selection refers to the occurrence of a relatively greater proportion of a given segment of the prey population in the predator's diet and does not connote any conscious behavior on the part of either prey or predator. Several studies have shown mountain

lions to select a greater proportion of mule deer fawns and mature bucks (table 38.4). The same pattern was observed in lion-killed elk in Idaho (Hornocker 1970) and in lion-killed mule deer in Colorado (Dixon 1967) and Nevada (Ashman 1977). Spalding and Lesowski (1971) found selection for mule deer nine years or older but selection against yearlings.

Mountain lions are known to prey on both healthy animals and animals in poor physical condition. Whether there is selection of unhealthy animals is not known.

Several reasons have been proposed for the selection of old male and young prey animals: (1) bucks are less wary during the mating season, (2) bucks are found more often in the preferred habitat of the mountain lion, and (3) the very young and very old are less vigorous and therefore more vulnerable to predation. All of these reasons could contribute to prey selection. If bucks are found more often in preferred habitat, this could be the reason the habitat is preferred by the lions. Selection on this basis would be due to increased availability and not necessarily an increase in vulnerability. Less vigorous animals and less wary males would be more vulnerable to successful stalking, attack, or both (see the section "Behavior").

The selection of old males and the very young in the lion's prey is the same strategy predicted by dif-

TABLE 38.4. Index of age and sex selection of prey by mountain lions (proportion of lion-killed mule deer divided by proportion of normal populations of mule deer)

	Utah and Nevada[a]	British Columbia[b]	Idaho[c]
Adult male	1.75	2.00	2.69
Adult female	0.83	0.64	0.46
Fawns	0.89	1.15	1.59

NOTE: Values greater than 1.0 indicate selection for a given age/sex class and values less than 1.0 selection against.
[a]Robinette et al. 1959
[b]Spalding and Lesowski 1971
[c]Hornocker 1970

ferent optimization models to maximize the quantity of prey consumed by predators and at the same time maximize productivity of the prey population (MacArthur 1960; Bellman and Kalaba 1960; Beddington and Taylor 1973). This is because the selected animals are those that contribute the least to the reproductive effort of the prey population. In addition, the removal of the prey animals increases the productivity of the prey population by reducing competition in the remaining population. This is true whether the prey are healthy or in poor physical condition.

Predation on Livestock. Unfortunately, the benefit to native prey populations of predation is not accrued by domestic prey. Even though livestock are a small proportion of the lion's diet (see table 38.3), any predation on livestock represents an economic loss to the livestock owner. As livestock numbers increase, the number of incidents of livestock predation increases proportionately. In a study involving Arizona cattle ranches (Shaw 1977), cattle occurred as often as 34 percent of the time in mountain lion diets. In Utah and Nevada the frequency of occurrence of domestic sheep in the mountain lion's diet reached 15 percent in the summer (see table 38.3). In the case of sheep, losses usually are greater than are indicated by stomach or scat

TABLE 38.5. Age and sex of depredating mountain lions in California

	Sex		
Age	Male	Female	Total
1	0	1	1
2	0	2	2
3	5	0	5
4	4	1	5
5	0	1	1
6	1	0	1
7+	2	2	4
Total	12	7	19

SOURCE: Sitton 1978

analysis, since mountain lions frequently kill several sheep at a time (see the "Behavior" section).

Although mountain lions of both sexes and all ages will prey on livestock, the probability of depredation is greater in some lion age and sex classes than in others. Of 19 lions taken on depredation permits in California, 12 were males and 7 were females (table 38.5). Of the males, the largest number were three to four years old. The three- and four-year-old males are at the age where they are likely to be transients seeking a home range. Young males, which have a dispersal range greater than that of females, are more likely to come into contact with livestock. Condition was determined on 18 depredating lions (11 males and 7 females): 9 of the male lions were in good to excellent health, and 1 three year old and 1 four year old were in fair condition. One of the females over seven years old was in poor condition: she had broken teeth and was impaled by a large number of porcupine quills, some penetrating the lungs. The second female over seven years old was pregnant and in good condition but also had worn and broken teeth (Sitton 1978).

The evidence suggests the hypothesis that a depredating mountain lion is most likely to be either a transient male (a young one recently on its own or an old one displaced from its resident status) or an old lion (male or female) in poor physical condition. In Colorado, a sheep-depredating lion was an old male that had toes missing from one front paw (Dixon 1967). In California a female with three of her front toes missing killed four sheep and was eating one of them when a hunter approached. The lion made no attempt to run away, and was killed (McLean 1954). In another case of sheep predation in Colorado, the lion had an upper canine missing and a healed, complete fracture of the left humerus (Dixon 1967).

BEHAVIOR

Mating Behavior. The mating posture of felids differs from that of most mammals in that the female lies down or crouches instead of standing. The male stands over the female, supported primarily by his forepaws while his hindpaws scrape the ground. The male often controls the female by gripping the back of the neck with his jaws. This copulation bite differs from the killing bite in that muscles controlling the closing of the jaw normally are inhibited (Leyhausen 1979). The unusual mating posture may be an adaptation that prevents any stimulus that would cause the male to kill the female (Ewer 1974).

Predatory Behavior. The complete predatory behavior cycle of the mountain lions, like that of most predators, can be divided into several phases: search, approach, pursuit, capture, and ingestion. Between feeding cycles there is a digestive pause. Then, when satiety falls below a certain threshold, the cycle is repeated. Each phase of the feeding cycle is regulated by the set of external and internal stimuli attending the particular situation. However, ethological and ecological studies have provided some information on a predator's response under given conditions for each phase. The conditions determine the rate at which different prey species are eaten by mountain lions. This rate can be expressed as the frequency of encounters (E) times the probability that the encountered prey is eaten (Pe) (deRuiter 1967). The frequency of encounter depends on the prey density and the mountain lion's searching behavior. The probability of eating (Pe) can be expressed as the product of individual probabilities that the encountered prey will be approached successfully (Pa), the probability that the prey will be captured if approached (Pc), and the probability that the prey will be ingested if captured (Pi):

$$Pe = Pa \times Pc \times Pi.$$

The search phase of the feeding cycle is characterized by extensive movement, zigzagging back and forth across the home range, avoiding large open areas, and taking advantage of whatever cover is available (Koford 1946; Seidensticker et al. 1973). As hunger increases, the tendency to roam increases, with the search largely restricted to areas where food previously was found (deRuiter 1967; Royama 1970; Leyhausen 1979).

The approach phase involves the encounter of potential prey and the detection and recognition as a prey item. After a prey item is detected (most probably by eyesight), the mountain lion begins its approach, usually referred to as stalking. Movement of the prey may increase the probability of detection and may release attack behavior in the mountain lion. Once a mountain lion begins its approach, the probability of success (defined by the initiation of an attack) likely is not very high, although it is hard to measure because of the difficulty of observing lions in the wild (however, see Koford 1946).

Capture, although undoubtedly influenced by hunger, may be influenced to some degree by other stimuli (Leyhausen 1979). The usual method of attacking deer is by grasping the shoulders and neck with the front paws, with the retractile claws extended, and digging the hind claws into the flanks. The deer then is killed by biting the back of the neck, forcing apart the vertebrae and breaking the spinal cord. In larger prey such as elk, the neck may be broken by pulling the head down and back, breaking it directly or in a fall (Cunningham 1971). The orientation of the killing bite toward the nape develops in two phases. The first phase is inherent and directs the bite to the indentation of the neck. The second is the final positioning of the nape bite and has to be learned (Leyhausen 1965, 1979). Based on tracks in the snow, Hornocker (1970) estimated the probability of successful attack as 0.82. Few direct observations of mountain lion captures have been reported (however, see Wade 1929). Allen (1950) reported observing a captive mountain lion killing a white-tailed deer. Direct estimation of the probability of a successful attack would have to be made under controlled conditions.

After making a kill, the mountain lion often carries or drags the prey under the cover of a bush or tree before eating. The mountain lion is capable of moving a carcass weighing several times its own weight. Musgrave (1926) reported that a mountain lion dragged a horse carcass weighing 360 to 450 kg, 8 to 10 m. Similar feats of strength have been reported of the African lion (Guggisburg 1963).

Mountain lions begin feeding on their larger prey usually by opening up the abdominal cavity and ingesting the liver, heart, and lungs. The probability of ingestion for preferred prey should be close to 1.0 unless the lion is disturbed from the kill. Mountain lions, as most felids, assume a crouched position when feeding. Using their carnassials as shears, they cut the meat into pieces that are swallowed whole. The shearing action is guided by the two masticating muscles, the masseter and the pterygoideus (Becht 1953). Experiments on the feeding behavior of domestic cats and the ocelot (*Felis pardalis*) indicate that the primary stimuli for cutting and eating are tactile; the vibrissae are stimulated by the direction of the hairs on the prey (Leyhausen 1979).

Often after a mountain lion has fed, it will cover the carcass with litter to conceal it from scavengers. It then may return to feed on the carcass several more times.

Surplus Killing. Mountain lions occasionally kill more prey than they eat, particularly domestic sheep. Near DeBeque, Colorado, a lion killed 9 domestic sheep out of a herd of about 1,000. However, it fed on only the hind quarters of the last sheep killed (Dixon 1967). A lion had killed sheep in the same area several months earlier; however, fresh lion tracks had been observed near the same herd at other times when none of the sheep was killed.

Surplus killing of wild prey by mountain lions has rarely been recorded. Shaw (1977) described three instances of mule deer does and fawns killed within a short period of time. Mills (1922) reported that a mountain lion killed nine bighorn sheep (*Ovis canadensis*) at one time and ate only part of one of them. Whether this occurrence was caused by the same factors as those accounting for surplus killing of domestic sheep is not known.

Surplus killing also has been reported in other predators. Kruuck (1972) reported that a leopard (*Felis pardus*) killed 17 domestic goats and ate only parts of 2. African lions in one case killed as many as 50 goats and, in another case, 12 cows (Guggisberg 1963). In a third case, a pride of lions killed 51 ostriches (*Struthio camelus*), leaving most of the carcasses untouched. Prey capture and prey ingestion behavior largely are independent, the stimuli to kill a prey outweighing stimuli to consume it. Leyhausen (1979) described several experiments with various cat species in which both hungry and satiated individuals continued to kill prey as long as it was presented rather than eat those already killed.

Thus, surplus killing occurs when several prey individuals are available at once. The stimulus for prey capture continues as long as prey are available or until the killing mood is exhausted. The fact that few domestic prey are eaten could be the result of the lack of appropriate tactile stimuli for cutting and eating. For example, the wool of a sheep would not produce the same stimulus as the hairs of normal prey, such as deer.

Social Behavior. Although mountain lions essentially are solitary animals, a social structure related to reproduction exists. Adult males and adult females without kittens are found together primarily during mating. Adult males also are found with adult females with large kittens. This probably is related to mating also, since a female's estrus may coincide with her separation from her kittens. Adult males do not associate with other males and solitary females do not associate with other solitary females (Seidensticker et al. 1973). Females with young kittens will not tolerate the presence of adult males, for males are likely to kill the young (see the section ''Mortality'').

Kittens remain with the female parent until late in their second winter. In two cases on the Idaho Primitive Area, the female left the kittens at a kill and did not return (Seidensticker et al. 1973). Following separation, the kittens remained together for a few days before they separated and moved out of the area. Once the kittens left, they were never recaptured on the study area. Such dispersal of young appears to be certain, even when there are vacant home ranges available. In Nevada, the distance two juvenile males dispersed were 40 and 45 km; whereas a juvenile female moved only a few kilometers before establishing a home range (Ashman 1975). According to Seidensticker et al. (1973), dispersal likely is a mechanism to prevent inbreeding.

MORTALITY

Natural Mortality. The mortality pattern of mountain lions seems to follow that of most mammals, with the highest mortality rates in the pre-reproductive young and oldest age classes (Caughley 1966), although age-specific mortality rates have not been determined. Some mortality occurs soon after birth of the litter (see figure 38.5) and may be due to food limitations. Adult males are known to kill young kittens and sometimes eat them (Young and Goldman 1946; Robinette et al. 1961; Hornocker 1970). Young kittens (without the protection of the mother lion) particularly are susceptible to all mortality sources. Robinette et al. (1959) reported a case of three of four kittens dying of wounds from porcupine quills sustained trying to kill a porcupine after their mother was killed by hunters.

Adult lions may be killed by encounters with both prey species and other lions. Once a mountain lion has attacked its prey and is clinging to its back, it may be dislodged by the prey running under low branches, by colliding with trees, or in falls. In some cases the lion is killed directly from injuries sustained in collision with trees or limbs. In other cases, it may be killed by attacks with hooves or antlers of deer or elk after having been stunned by a collision (Seton 1927; Gashwiler and Robinette 1957; Hornocker 1970). Usually adult lions will attempt to avoid each other. However, occa-

sionally fights occur that may result in the death of one of the lions (Bruce 1925; McLean 1954; Lesowski 1963; Sitton and Weaver 1977).

Mortality undoubtedly increases in older animals as they become senescent; next to hunting, deaths attributed to old age probably represent the most significant source of mortality. Hornocker (1970) described the deterioration of a mature male lion in Idaho, including extreme tooth wear and loss of body weight. McLean (1954) cited a report of a lion skull found in California also with extreme tooth wear. This deterioration of condition affects the lion's ability to kill effectively, ultimately resulting in its starvation (Young and Goldman 1946). This weakened condition has been cited as one reason for mountain lions preying on livestock (see the ''Food Habits'' section).

Hunting. One of the largest sources of mortality in mountain lions is hunting— either as game animals or as livestock depredators. Poaching probably contributes little to the mortality in mountain lion populations because of the difficulty of capturing lions without dogs and the low economic value of lion pelts.

Accidents. Road-killed mountain lions comprise the largest number of accidental deaths, although drownings in drainage canals in California also have been reported (Macgregor 1976).

Parasites. Mountain lions usually are only lightly infested with ectoparasites. By being highly mobile and spending little time in dens that contain only small amounts of litter, lions are able to keep ectoparasites to a minimum. Those ectoparasites that have been reported include ticks, fleas, lice, and mites (table 38.6).

Endoparasites frequently are found in mountain lion tissues, although in most cases parasite loads are probably at subclinical levels. Fifty to 55 percent of the mountain lions examined from Montana, Idaho, and Wyoming were infected with the nematode *Trichinella spiralis* (Winters 1969; Worley et al. 1974). Other internal parasites found in mountain lions include tapeworms and flukes (table 38.7).

Fecal samples from a mountain lion in New Mexico contained oocysts of the sporozoan *Isospora felis* and eggs of the ascarid *Toxocara* sp. (Marchiondo et al. 1976). These oocysts and eggs probably developed in the lion, although they could have been ingested by feeding on the intermediate host.

Toxoplasmosis was found in a road-killed mountain lion in southern California (Sitton and Weaver 1977). A Montana mountain lion developed toxoplasma oocysts after being fed mice infected with toxoplasma cysts (Miller et al. 1972). Toxoplasmalike oocysts also were found in feces from a Montana mountain lion (Marchiondo et al. 1976).

The occurrence of the trematode *Nanophyetus salmincola* often is associated with the disease ''salmon poisoning'' in canids and black bears *(Ursus americanus)*. The disease is caused by a rickettsialike organism transmitted by the metacercariae of *N. salmincola*. The occurrence (Kistner et al. 1979) of an estimated 650,000 adult trematodes in a wild female

TABLE 38.6. Known ectoparasites of the mountain lion

Common Name	Order	Species	Source
Fleas	Siphonaptera	*Arctopsylla setosa* (Roths.) 1906	Young and Goldman 1946
Ticks	Acarina	*Dermacentor variabilis*	Young and Goldman 1946
		D. cyaniventris[a]	Young and Goldman 1946
		Ixodes ricinus	Young and Goldman 1946
		I. cookei	Young and Goldman 1946
		Amblyomma cajennense[b]	Young and Goldman 1946
		Boophilus microplus	Young and Goldman 1946
Mites	Acarina		McLean 1954
Lice	Mallophaga	*Trichodectes felis*[c]	Young and Goldman 1946
		Felicola felis (Werneck)	Hopkins 1949

[a]Reported from South America.
[b]Reported from South America in mountain lions but found in other hosts in Texas.
[c]Reported from Brazil.

mountain lion kitten was the first case of infection reported in mountain lions and the first case of pathogenicity attributed to the adult trematode. Kistner et al. (1979) believed the kitten was an orphan that, because of its starving condition, fed on dying salmon. The massive infection of trematodes caused diarrhea and stimulated thickening of the intestinal wall, resulting in impaired nutrient absorption and malnutrition.

Disease. The only disease know seriously to affect mountain lions in the wild is rabies. This disease may account for unprovoked attacks by mountain lions on human beings (Storer 1923). Russell (1978) suggested that feline panleukopenia (distemper) may be a source of mortality; however, to date this disease in mountain lions is known only from animals in zoos.

AGE DETERMINATION

Methods of determining the age of mountain lions have not been too successful. The tooth-annulation method in particular has been difficult to develop as a result of the poor staining ability of lion teeth. However, Thomas (1977) found that using toluidine blue as a stain gave the best results. Greer (1972) has developed an aging technique based on the degree of suture closure in the lion skull. This method, which groups lions into 10 age classes, has been used in mountain lion studies in Montana (Greer 1974). Tooth wear and general appearance have been used for aging in Colorado (Currier et al. 1977). A combination of tooth eruption, wear, and staining was used in California (Sitton and Weaver 1977).

CENSUS TECHNIQUES

Several different techniques have been used to estimate the size of mountain lion populations. These methods include census techniques and indexes of population size. In California, Koford (1978) used a set of road sand transects to count lion tracks and obtain a popula-

TABLE 38.7. Known endoparasites of the mountain lion

Phylum	Class	Species	Source
Protozoa		*Isospora felis*	Marchiondo et al. 1976
		Toxoplasma gondii[a]	Sitton and Weaver 1977
Aschelminthes (roundworms)	Nematoda (nematodes)	*Toxocara* sp.	Marchiondo et al. 1976
		Metathelazia californica[b]	Skinker 1931
		Trichinella spiralis	Winters 1969
		Toxascaris leonina	Sitton and Weaver 1977
		Filaroides striatum (Mol.) 1858	Young and Goldman 1946
		Spirocerca sp.	Sitton 1978
Platyhelminthes (flatworms)	Cestoidea (tapeworms)	*Taenia* (Echinococcus) *oligarthus* (Diesing) 1813[c]	Young and Goldman 1946
		T. lyncis	Skinker 1935
		Echinococcus granulosus	Hall 1920
	Trematoda (flukes)	*Alaria* (Alaria) *marcianae* (LaRue 1917) Walton 1950[d]	Fischthal and Martin 1977
		Nanophyetus salmincola	Kistner et al. 1979

[a]Uncertain taxonomic status.
[b]Possibly *Vogeloides felis* (see Pence and Stone 1977).
[c]Reported from Brazil.
[d]Reported from Paraguay.

tion index. A separate California study (Sitton and Weaver 1977) used a combination of track counts and animal captures to obtain a minimum population estimate. Lion tracks also were used in Nevada (Ashman 1977, 1979), but the tracks were located in snow by aerial and ground searches. Also in Nevada (Ashman 1979), a population index based on the use and nonuse of lion scratch sites was employed.

An intensive capture program designed to provide a complete census of the lion population in a given area was used in Idaho (Hornocker 1970). A mark-recapture method of estimating population size has been used in a Colorado study (Currier et al. 1977).

ECONOMIC STATUS AND MANAGEMENT

Caughley (1977) summarized wildlife management as consisting of three basic approaches: control, sustained yield harvesting, and conservation. Each of these approaches has been exercised in the past management of the mountain lion. In fact, depending on the circumstances, all three approaches are being used today.

Control. Historically, the "management" approach to the mountain lion has been one of control—or more accurately, extermination. This resulted from the threat—real or imagined—to livestock as well as to human life. The image of the mountain lion as a blood-thirsty miscreant persisted in popular journals, largely as a result of the lack of any substantive data. For example, in the account of Lord Southesk's "Adventures on the Saskatchewan" (anonymous 1875:674) is the following: "Marking out a small party of hunters or travelers, it [the mountain lion] will follow them secretly for days, and watch by their camp at night, till at last it discovers one of their number resting a little separate from his companions. Then, when all is dark and silent, the insidious puma glides in, and the sleeper knows but a short awakening when its fangs are buried in his throat."

The early control of mountain lions was done primarily by private landowners to protect their livestock. To encourage the eradication of all predators, bounty laws were passed by the colonial assemblies—first to eliminate wolves, then mountain lions and other "noxious" species. As the livestock industry grew in western North America, bounties were placed on predators by states, counties, towns, and livestock associations. The official attitude of many western states in the early part of this century was given by the California State Lion Hunter, Jay Bruce, "the mountain lion is of practically no value as a fur bearer, game animal, or source of food, but is simply a liability which probably costs the state a thousand dollars a year in deer meat alone" (Bruce 1925:2). Only from 1958 to 1970 have bounty laws been repealed in the western states and provinces.

The policy of predator extermination extended even to the federal level. The national park system was established, beginning with Yellowstone National Park in 1872, for the protection of wildlife, but this protection was not extended to predators. F. A. Boutelle, Acting Superintendent of Yellowstone National Park in 1890, stated, "I am more than ever convinced that the bear and puma do a great deal of mischief and ought to be reduced in numbers" (Boutelle 1890:6).

The National Park Service was created by Congress in 1916 to administer the national parks. Its policy was "to conserve the scenery and the natural and historic objects and the wild life therein and to provide for the enjoyment of the same in such manner and by such means as will leave them unimpaired for the enjoyment of future generations." Again predators were excluded from this policy and a separate policy of extermination was in force in practically all the national parks from 1916 to 1925. The list of "predator" species marked for extermination by the National Park Service, with the aid of trappers from the Department of Agriculture Biological Survey, included the wolf, mountain lion, lynx *(Felis lynx),* fisher *(Martes pennanti),* fox *(Vulpes vulpes),* coyote, wolverine *(Gulo gulo),* bobcat *(Felis rufus),* ground squirrels, mice, gopher *(Geomyidae),* marten *(Martes americana),* badger, weasel *(Mustela* sp.), mink *(Mustela vison),* and gray squirrel *(Sciurus* sp.). The methods of extermination included poisoning, shooting, trapping, exploding shells, and dynamite (Cahalane 1939). According to S. T. Mather, Director of the National Park Service in 1923, "a campaign of extermination waged against mountain lions and coyotes has shown beneficial results" (Mather 1923:64).

Public outcries and the concerns expressed by several scientific societies led to a gradual softening of official National Park Service policy toward predators. H. M. Albright, Director of the National Park Service in 1931, stated that the policy of the service was (in part) that "predatory animals are to be considered as an integral part of the wild life protected within national parks, and no widespread campaigns of destruction are to be countenanced" (Albright 1931:186).

The decision to use the control option in predator management today is recognized as requiring: a combination of esthetic, social, economic, ecological, political, and administrative considerations; the coordination of local, state, and federal agencies; and that it be based on adequate data (Berryman 1972). In mountain lions, the control option is limited to domestic animal protection and, in some areas, the possible prevention of attack on human beings. Control measures should be limited to those individual animals identified as causing the depredations (Berryman 1972). This precludes sport hunting and indiscriminant hunting by professional hunters as a control option.

If depredations by mountain lions are occurring in an area, three options are available: (1) the lion can be killed, (2) it can be captured and translocated, or (3) no action can be taken. The first option is predicated on the assumption that the offending lion is likely to cause further depredations. However, evidence suggests that predation on livestock is opportunistic rather than a recurring phenomenon. None of the lions captured and released following depredations in California was involved in further incidents of depredation (Sitton 1978). If the decision is made to kill the lion, several methods are available:

a. The livestock owner is permitted to kill the lion. The disadvantage of this method is that it relies on the judgment and ability of usually untrained individuals.

b. The livestock owner is permitted to kill the lion (usually with the assistance of a professional lion hunter) after the depredation has been confirmed by authorized personnel. This is the method being implemented by the California Department of Fish and Game (Sitton and Weaver 1977).

c. Lions are captured by state or federal professional trappers. This method has the advantage that a decision to translocate the lion can be made at the time of capture.

d. The sport hunter, chosen from a list of applicants, can kill the lion with the aid of a professional guide. This method is being used by the Colorado Division of Wildlife (Russell 1978).

The second option (translocation), being nondestructive, provides for a number of further alternatives. If the lion is in poor condition, rehabilitation to minimize the chances of further depredation can be attempted before relocation. Rather than releasing the lion into a different area, it could be given to a zoo or other facility—possibly for inclusion in a program of artifical propagation. In British Columbia, lions found close to civilized areas and in good physical condition are translocated; only those lions that are threatening human beings or livestock are destroyed (Dewar and Dewar 1976).

Problems associated with the relocation option include the need to locate prospective release sites, the relatively high cost, and the possibility of mixing gene pools. The last problem exists only when lions are moved great distances, since dispersal distances of 250 km are normal. This concern may be unfounded in any case, because little is known of genetic differences among lion populations.

The third option (no action) has the advantage of no direct cost and allows for the possibility that further depredation by the offending lion would not occur. If further depredation did occur, options one and two are still available. The third option has much support from the public (Sitton and Weaver 1977; Kellert 1979). As part of a no-control option, a means of recovering the financial loss has been suggested—including insurance, tax credits, or compensation by the state. These means should be considered separately from the decision to control mountain lions, since the control decision is made after the financial loss from lion depredation has occurred.

Sustained-Yield Harvesting. Sustained-yield harvesting through regulated sport hunting is practiced by 12 western states and provinces, although in 1 state (Texas) mountain lion hunting is unregulated. Ordinarily, harvesting a population can proceed at a rate that balances the annual rate of productivity in that popula-

tion. In the harvest of mountain lion populations, particular consideration is needed of the significant role dispersal plays in maintaining equilibrium populations. A given population contributes mountain lions to and receives them from adjacent populations through dispersal. As mountain lions are removed from a population, vacant home ranges are filled by those dispersing from nearby populations. Isolated populations likely would have difficulty sustaining themselves unless they were within the migrating range of dispersing individuals. Therefore, it is necessary to harvest lions at a rate that is low enough to provide for dispersal. If the density of an isolated population falls below the minimum viable level, it will tend toward extirpation (Seidensticker et al. 1973).

Harvesting mountain lions can create some additional problems. Since mountain lions breed any time of the year, setting hunting seasons does not give them the protection it would give to seasonally breeding species. Killing a female lion at any time during the 18 to 24 months that it may be caring for kittens reduces their chance of survival (see the "Mortality" section). Since the killing bite of the lion takes time to develop (Bogue and Ferrari 1974), orphaned kittens—if they survive—possibly could turn to preying on livestock.

The loss of habitat is one of the greatest threats to mountain lion populations. To ensure a sustained harvest, a population requires a stable habitat. Any permanent loss of habitat will cause a reduction in the population. This particularly is the case with the loss of deer and elk winter range. This habitat is being eliminated in parts of British Columbia, Montana, and Utah as a result of land development for residential and recreational purposes. The planned development of oil shale deposits in Wyoming, Utah, and Colorado could eliminate large sections of lion habitat in those states. Large areas of habitat also would be lost with the planned development of the MX missile system, which would spread over 25,600 km^2 in Utah and Nevada.

Not only are large tracts of habitat necessary to maintain individual populations of mountain lions, but corridors that connect these tracts are required for the dispersal of mountain lions between populations.

Conservation. Caughley (1977:168) described conservation as "the treatment of a small or declining population to raise its density." As applied to mountain lions and other predators, conservation has gained much public support in recent years. This is particularly true where the species may be threatened or endangered. However, even where the mountain lion exists in large numbers, a large segment of the public is opposed to hunting only for sport or trophies (Sitton and Weaver 1977; Kellert 1979).

In western North America, mountain lion populations can increase by the natural processes of reproduction and dispersal provided there is sufficient habitat and an optimal harvest rate. In areas adjacent to established populations, such as southeastern Alberta, the southern parts of Saskatchewan, Manitoba, Ontario, and Quebec, and the eastern parts of Montana, Wyoming, Colorado, and New Mexico, dispersal should provide for increases in lion populations where suitable habitat occurs.

In eastern North America, where the mountain lion has been largely absent in this century, there has been increasing interest in restoring the species to its former range (Frome 1979; Downing 1979). Aside from the esthetics of having a species restored as part of a whole ecosystem, there are several ecological reasons as well (some of which have been described above). As pointed out by Talbot (1978:307) "predators play a key role in maintaining ecosystem integrity in terms of species and genetic composition, ecosystem functions, and long term stability."

A first step in the restoration of mountain lions in eastern North America is to determine their present status. This involves the establishment of a clearinghouse to investigate reports of sightings and tracks (Wright 1959). One center (the Florida Panther Record Clearinghouse of the Florida Game and Fresh Water Fish Commission in Gainesville, Florida) has been established to investigate reports in Florida (Belden 1977) and a second (U.S. Fish and Wildlife Service, Department of Forestry, Clemson University, Clemson, South Carolina) to investigate reports in the southern Appalachians (Downing 1979). A third center has been proposed for New York (Brocke 1979).

However, the mountain lion is not likely to make a comeback—even with complete protection—unless several lions are reintroduced into suitable habitat at one time, since there is little evidence of natural reestablishment of former ranges by resident lions. Reintroduction of lions in eastern North America can be from western stock or from artificial propagation of eastern stock if enough individuals can be located. At present the only breeding population of mountain lions in the East is found in Florida. The use of western stock for introduction in the East might be objectionable if genetic differences between eastern and western populations are found. If no other eastern populations can be verified, then genetic mixing is not a problem and the question might be one of having western lions in the East or none at all.

Artificial propagation programs can be divided into two types: the ecoethological approach and the physiogically induced approach (Sadleir 1974). The ecoethological approach involves creating conditions in captivity that duplicate as much as possible those of the lion's natural habitat and allow the natural production of offspring. The physiologically induced approach involves such techniques as artificial insemination, hormonally induced ovulation, and the preservation and transport of semen.

Whether reintroductions are from western populations or from artificial propagation, this management option needs to consider the impact of having mountain lion populations restored to eastern North America. Although the idea is ecologically sound and esthetically satisfying, social problems still exist. Even though animal husbandry practices have changed considerably, some livestock depredation could be expected to occur; and there is the threat, however slight, of attacks on human beings. To be successful, a mountain lion restoration project would have to have wide public support. This would require an extensive information and education program and a greater understanding of interactions among lions, livestock, and human beings.

CURRENT RESEARCH AND MANAGEMENT NEEDS

Although the mountain lion is one species for which it is most difficult to obtain data, much additional information is required for their effective management. These needs range from the systematic status of the species to a complete management plan for the recovery of the endangered eastern mountain lion:

1. The systematic status of the mountain lion today largely is that described by Goldman in 1946 (Young and Goldman 1946). His classification was based on "geographic variation in size, weight, color, and minor details of structure," with intergrades throughout the mountain lion's range. Museum specimens, both those used by Goldman and more recent collections, should be reevaluated by modern numerical methods. In addition, other techniques (such as electrophoresis of tissue proteins) should be used.

2. The development of aging and census techniques should be continued, as this information is necessary for the interpretation of behavioral and ecological phenomena.

3. Artificial propagation techniques need additional study. Both the ecoethological and physiologically induced approaches require further research to provide maximal production of young.

4. Predation on both livestock and natural prey is not fully understood. A more complete profile of those lions involved in livestock predation is needed. This profile includes age, sex, and status relative to the land tenure social system. Whether there is a difference in the type of lion involved in single or repeated depredation episodes needs to be determined. Although there have been several food habits studies (see "Food Habits" section), few of the hypotheses for the selection of deer fawns and older males have been studied.

5. Many of the recent studies on mountain lions have included the estimation of home ranges. However, studies are still needed to determine the spatial and temporal patterns of activity and how they relate to prey activity patterns and habitat physiognomy.

6. A recovery plan needs to be developed for the eastern mountain lion. This is not to recommend that mountain lions be reintroduced in eastern North American. However, such introductions should be based on a management plan that includes ecological, social, political, and economic factors. The research described above would be highly beneficial in the development of such a plan.

I would like to thank D. Ashman, K. R. Greer, and D. E. Worley for their helpful comments on the manuscript; J. Peters and N. R. DelDuca for providing invaluable discussions on mountain lion behavior; and C. N. Dixon for assistance in the literature search. This chapter is contribution no. 1100-AEL, Center for Environmental and Estuarine Studies, University of Maryland.

LITERATURE CITED

Albright, H. M. 1931. The National Park Service's policy on predatory mammals. J. Mammal. 12:185–186.

Allen, R. 1950. Notes on the Florida panther, *Felis concolor coryi* Bangs. J. Mammal. 31:279–280.

Anderson, S. 1972. Mammals of Chihuahua: taxonomy and distribution. Bull. Am. Mus. Nat. Hist. 148(2):149–410.

Anonymous. 1875. Among the Rocky Mountains. Appletons' J. 13 (323):673–677.

Ashman, D. 1975. Mountain lion study. Nevada Outdoors and Wildl. Rev. 9:12–14.

———. 1977. Mountain lion investigations. Job Performance Rep., Project W-48-8. Nevada Dept. Fish Game, Reno, Nevada. 11pp.

———. 1979. Mountain lion investigations. Job Performance Rep., Project W-48-10. Nevada Dept. Wildl., Reno, Nevada. 16pp.

Audubon, J. J., and Bachman, J. 1851. The quadrupeds of North America. Vol. 2. V. G. Audubon, New York.

Barlow, H. B.; Fitzhugh, R.; and Kuffler, S. W. 1957. Change of organization on the receptive fields of the cat's retina during dark adaptation. J. Physiol. 137:338–354.

Becht, G. 1953. Comparative biologic-anatomical researches on mastication in some mammals. Proc. Koninkl. Ned. Akad. van Wetenschappen (Amsterdam). Series C: Biological and Medical Sciences 56:508–527.

Beddington, J. R., and Taylor, D. B. 1973. Optimum age specific harvesting of a population. Biometrics 29:801–809.

Belden, C. 1977. If you see a panther. Florida Wildl. (September–October), pp. 31–43.

Bellman, R., and Kalaba, R. 1960. Some mathematical aspects of optimal predation in ecology and boviculture. Proc. Nat. Acad. Sci. 46:718–720.

Berryman, J. H. 1972. The principles of predator control. J. Wildl. Manage. 36:395–400.

Bogue, G., and Ferrari, M. 1974. The predatory "training" of captive reared pumas. Pages 35–45 in R. L. Eaton, ed. The world's cats. Vol. 3, No. 1: Contributions to status, management and conservation. Dept. Zool. Univ. Washington, Seattle. 95pp.

Boutelle, F. A. 1890. Report of the superintendent of the Yellowstone National Park to the secretary of the Interior. Government Printing office, Washington, D.C.

Brocke, R. H. 1979. Have we closed the "barn door" too late for the eastern puma? Pages 4–5 in R. L. Downing, ed. Eastern Cougar Newsletter (January 1979). U. S. Fish Wildl. Serv., Dept. For., Clemson Univ., Clemson, S.C. 7pp.

Bruce, J. 1922. The why and how of mountain lion hunting in California. California Fish Game 8:108–114.

———. 1925. The problem of mountain lion control in California. California Fish Game 11:1–17.

Cahalane, V. H. 1939. The evolution of predator control policy in the national parks. J. Wildl. Manage. 3:229–237.

California Department of Fish and Game. 1973. Report to the 1973 legislature on progress of the mountain lion study. 4pp. Mimeo.

Caughley, G. 1966. Mortality patterns in mammals. Ecology 47:906–918.

———.1977. Analysis of vertebrate populations. Wiley, New York, 234pp.

Conklin, W. A. 1884. Footnote, p. 34, in C. H. Merriam 1884. The mammals of the Adirondack Region, northeastern New York. L. S. Foster Press, New York, 316pp.

Cunningham, E. B. 1971. A cougar kills an elk. Can. Field Nat. 85:253–254.

Currier, M. J. P.; Sherriff, S. L.; and Russell, K. R. 1977. Mountain lion population and harvest near Canon City, Colorado, 1974–1977. Spec. Rep. 42, Colorado Div. Wildl., Colorado Coop. Wildl. Res. Unit, Colorado State Univ., Fort Collins. 12pp.

Darwin, C. R. 1839. Journal of researches into the geology and natural history of the various countries visited by H. M. S. Beagle, under the command of Captain FitzRoy, R. N., from 1832 to 1836. H. Colburn, London. 615pp.

Dewar, P., and Dewar, P. 1976. The status and management of the puma in British Columbia. Pages 4–19 in R. L. Eaton, ed. The world's cats. Vol. 3, No. 1: Contributions to status, management and conservation. Dept. Zool., Univ. Washington, Seattle. 95pp.

Dixon, J. 1925. Food predilections of predatory and fur-bearing mammals. J. Mammal. 6:34–46.

Dixon, K. R. 1967. Mountain lion predation on big game and livestock in Colorado. Job Completion Rep. Proj. W-38-R-21, Colorado Game, Fish Parks Dept., Fort Collins, Colo. 23pp.

Dixon, K. R.,and Chapman, J. A. 1980. Harmonic mean measure of animal activity areas. Ecology 61:1040–1044.

Doutt, J. K. 1969. Mountain lions in Pennsylvania? Am. Midl. Nat. 82:281–285.

Downing, R. L., ed. 1979. Eastern Cougar Newsletter (May 1979). U.S. Fish Wildl. Serv., Dept. For., Clemson Univ., Clemson, S.C. 8pp.

Ewer, R. F. 1974. Viverrid behavior and the evolution of reproductive behavior in the Felidae. Pages 90–101 in R. L. Eaton, ed. The world's cats. Vol. 2: Biology, behavior and management of reproduction. Feline Res. Group, Woodland Park Zoo, Seattle, Wash. 260pp.

Fischthal, J. H., and Martin, R. L. 1977. Alaria (Alaria) marcianae (LaRue 1917) Walton 1950 (Trematoda: Diplostomatidae) from a mountain lion *Felis concolor acrocodia* Goldman, from Paraguay. J. Parasitol. 63:202.

Frome, M. 1979. Panthers wanted: alive, back East where they belong. Smithsonian 10:82–88.

Gashwiler, J. S., and Robinette, W. L., 1957. Accidental fatalities of the Utah cougar. J. Mammal. 38:123–126.

Goertz, J. W., and Abegg, R. 1966. Pumas in Louisiana. J. Mammal. 47:727.

Gonyea, W., and Ashworth, R. 1975. The form and function of retractile claws in the Felidae and other representative carnivores. J. Morphol. 145:229–238.

Greenley, J. C. 1971. The status of the mountain lion in Nevada, 1971. Pages 103–106 in S. E. Jorgensen and L. D. Mech, eds. Proceedings of a symposium on the native cats of North America: their status and management. U.S. Fish Wildl. Serv., Twin Cities, Minn. 139pp.

Greer, K. R. 1979. Mountain lion studies, 1978–79. Job Prog. Rep., Proj. W-120-R-10, Montana Dept. Fish Game, Henena, Mont. 14pp.

———.1980. Mountain lion collections, 1979–80. Job Prog. Rep., Proj. W-120-R-11, Montana Dept. Fish Game, Henena, Mont. 8pp.

Guggisberg, C. A. W. 1963. Simba: the life of the lion. Chilton, New York 309pp.

Hall, M. C. 1920. The adult taenioid cestodes of dogs and cats, and of related carnivores in North America. Proc. U.S. Natl. Mus. 55:1–94.

Hallock, C. 1877. The sportsman's gazetteer and general guide: game animals, birds, and fishes of North America; their habits and various methods of capture. For. and Stream Publ. Co., New York 688pp.

Hibben, F. C. 1937. A preliminary study of the mountain lion (*Felis oregonensis* sp.). Univ. New Mexico Bull. 318. 59pp.

Hopkins, G. H. E. 1949. The host associations of the lice of mammals. Proc. Zool. Soc. London 119:387–604.

Hornocker, M. G. 1969. Winter territoriality in mountain lions. J. Wildl. Manage. 33:457–464.

————. 1970. An analysis of mountain lion predation upon mule deer and elk in the Idaho Primitive Area. Wildl. Monogr. 21. 39pp.

Hubel, D. H. 1959. Single unit activity in striate cortex of unrestrained cats. J. Physiol. 147:226–238.

Kellert, S. 1979. Public attitudes toward critical wildlife and natural habitat issues. National Technical Information Service, U.S. Dept. of Commerce, Springfield, Va. 138pp.

Kistner, T. P.; Wyse, D.; and Schmitz, J. A. 1979. Pathogenicity attributed to massive infection of *Nanophyetus salmincola* in a cougar. J. Wildl. Dis. 15:419–420.

Koford, C. B. 1946. A California mountain lion observed stalking. J. Mammal. 27:274–275.

————. 1978. The welfare of the puma in California, 1976. Carnivore 1 (Part 1):92–96.

Kruuk, H. 1972. Surplus killing by carnivores. J. Zool. London 166:233–244.

Lamprecht, J. 1978. The relationship between food competition and foraging group size in some larger carnivores. Z. Tierpsychol. 46:337–343.

Lesowski, J. 1963. Two observations of cougar cannibalism. J. Mammal. 44:586.

Lewis, J. C. 1969. Evidence of mountain lions in the Ozarks and adjacent areas, 1948-1968. J. Mammal. 50:371–372.

Leyhausen, P. 1965. The communal organization of solitary mammals. Symp. Zool. Soc. London 14:249–263.

————. 1979. Cat behavior: the predatory and social behavior of domestic and wild cats. Garland STPM Press, New York. 340 pp. Translated by B. A. Tomkin.

MacArthur, R. H. 1960. On the relation between reproductive value and optimal predation. Proc. Nat. Acad. Sci. 46:143–145.

Macgregor, W. G. 1976. The status of the puma in California. Pages 28–35 in R. L. Eaton, ed. The world's cats. Vol. 3, no. 1: Contributions to status, management and conservation. Dept. Zool. Univ. Washington, Seattle. 95pp.

McLean, D. D. 1954. Mountain lions in California. California Fish Game 40:147–166.

Maliepaard, H. S. 1971. A report on wildcats in Saskatchewan. Pages 37–44 in S. E. Jorgensen and L. D. Mech, eds. Proceedings of a symposium on the native cats of North America: their status and management. U.S. Fish Wildl. Serv. Twin Cities, Minn. 139pp.

Marchiondo, A. A.; Kiszynski, D. W.; and Maupin, G. O. 1976. Prevalence of antibodies to toxoplasma-gondii in wild and domestic animals of New Mexico, Arizona and Colorado, U.S.A. J. Wildl. Dis. 12:226–232.

Mather, S. T. 1923. Report of the director of the National Park Service to the Secretary of the Interior. Gov. Printing Off., Washington, D.C. 198pp.

Mech, L. D., and Peters, R. P. 1977. The study of chemical communication in free-ranging mammals. Pages 321–332 in D. Müller-Schwarze and M. M. Mozell, eds. Chemical signals in vertebrates. Plenum Press, New York. 609pp.

Merriam, C. H. 1884. The mammals of the Adirondack Region, northeastern New York L. S. Foster Press, New York 316pp.

Miller, N. L.; Frenke, J. K.; and Dubey, J. P. 1972. Oral infections with toxoplasma cysts and oocysts in felines, other mammals, and in birds. J. Parasitol. 58:928–937.

Mills, E. A. 1922. Watched by wild animals. Doubleday, Page & Co., Garden City, N. Y. 243pp.

————. 1923. Wild animal homesteads. Doubleday, Page & Co., Garden City, N. Y. 259pp.

Musgrave, M. E. 1926. Some habits of mountain lions in Arizona. J. Mammal. 7:282–285.

Nero, R. W., and Wrigley, R. E. 1977. Status and habits of the cougar in Manitoba. Can. Field Nat. 91:28–40.

Noble, R. E. 1971. A recent record of the puma *Felis concolor* in Arkansas. Southwest Nat. 16:209.

Nowak, R. M. 1976. The cougar in the United States and Canada. New York Zool. Soc. and U.S. Fish Wildl. Serv. Off. of Endangered Species, Washington, D.C. 190pp.

Nuttall, T. 1821. A journal of travels into the Arkansas territory during the year 1819. T. H. Palmer, Philadelphia. 296pp.

Pence, D. B., and Stone, J. E. 1977. Lungworms (Nematoda: Pneumospiruridae) from West Texas carnivores. J. Parasitol. 63:979–991.

Rabb, G. G. 1959. Reproductive and vocal behavior in captive pumas. J. Mammal. 49:616–617.

Rasmussen, D. I. 1941. Biotic communities of Kaibab Plateau, Arizona. Ecol. Monogr. 11:229–275.

Riome, S. D. 1973. Evidence of cougars near Nipawin, Saskatchewan. Blue Jay 31:100–102.

Robinette, W. L.; Gashwiler, J. S.; and Morris, O. W. 1959. Food habits of the cougar in Utah and Nevada. J. Wildl. Manage. 23:261–273.

————. 1961. Notes on cougar productivity and life history. J. Mammal. 42:204–217.

Roop, L. 1971. The Wyoming lion situation. Wyoming Wildl. 35(12):16–21.

Royama, T. 1970. Factors governing the hunting behavior and selection of food by the great tit (*Parus major* L.). J. Anim. Ecol. 39:619–668.

Ruiter, L. de. 1967. Feeding behavior of vertebrates in the natural environment. Pages 97–116 in C. F. Code, ed. Handbook of Physiology. Sect. 6:Alimentary canal. Vol. 1: Control of food and water intake Am. Physiol. Soc., Washington, D.C. 459pp.

Russell, D. N. 1971. History and status of the felids of Texas. Pages 53–58. in S. E. Jorgensen and L. D. Mech eds. Proceeding of a symposium on the native cats of North America: their status and management. U.S. Fish Wildl. Serv., Twin Cities, Minn. 139pp.

Russell, K. R. 1978. Mountain lion. Pages 207–225 in J. L. Schmidt and D. L. Gilbert, eds. Big game of North America. Stackpole Books, Harrisburg, Pa.

Russo, J. P. 1970. The Kaibab north deer herd: its history, problems and management. Wildl. Bull. 7. Arizona Game Fish Dept., Phoenix, Arizona.195pp.

Sadleir, R. M. F. S. 1974. Research priorities in reproduction (discussion led by R. M. F. S. Sadleir). Pages 247–258 in R. L. Eaton, ed. The world's cats. Vol. 2: Biology, behavior, and management of reproduction. Feline Res. Group, Woodland Park Zoo, Seattle, Wash. 260pp.

Schwartz, J. E., II, and Mitchell, G. E. 1945. The Roosevelt elk on the Olympic Peninsula, Washington. J. Wildl. Manage. 9:295–322.

Seidensticker, J. C., IV; Hornocker, M. G.; Wiles, W. V.; and Messick, J. P. 1973. Mountain lion social organization in the Idaho Primitive Area. Wildl. Monogr. 35. 60pp.

Seton, E. T. 1927. Lives of game animals. Vol. 3. Double-day, Page & Co., Garden City, N. Y.

Severinghaus, C. W., and Cheatum, E. L. 1956. Life and times of the white-tailed deer. Pages 57–186 *in* W. P. Taylor ed. The Deer of North America. Stackpole Co., Harrisburg, Pa. 668pp.

Shaw, H. G. 1977. Impact of mountain lion on mule deer and cattle in northwestern Arizona. Pages 17–32 *in* R. L. Phillips and C. Jonkel, eds. Proceedings of the 1975 Predator Symposium. For. and Conserv. Exp. Stn., Univ. Montana, Missoula.

Sitton, L. W. 1978. Mountain lion predation on livestock in California. Presented at Annu. Fisheries, Wildl. Soc. Conf., Hyatt Lake Tahoe, 4 February 1978.

Sitton, L. W., and Weaver, R. A. 1977. California mountain lion investigations with recommendations for management. California Dept. Fish Game. 35pp.

Skinker, M. S. 1931. Three new parasite nematode worms. Proc. U. S. Natl. Mus. 79:1–9.

———. 1935. Two new species of tapeworms from carnivores and a redescription of *Taenia laticolis rudolphi*, 1819. Proc. U.S. Natl. Mus. 83:211–220.

Spalding, D. J. 1971. The wildcats of British Columbia. Pages 59–67 *in* S.E. Jorgensen and L. D. Mech, eds. Proceedings of a symposium on the native cats of North America: their status and management. U.S. Fish Wildl. Serv., Twin Cities, Minn. 139pp.

Spalding, D. J., and Lesowski, J. 1971. Winter food of the cougar in south-central British Columbia. J. Wildl. Manage. 35:378–381.

Storer, T. I. 1923. Rabies in a mountain lion. California Fish Game 9:45–48.

Taketa, F.; Attermeier, M. H., and Mauk, A. G. 1972. Acetylated hemoglobins in feline blood. J. Bio. Chem. 247:33–35.

Taketa, F.; Mauk, A. G.; and Lessard, J. L. 1971. Chain amino termini of the cat hemoglobins and the response to 2,3-diphosphoglycerate and adenosine triphosphate. J. Bio. Chem. 246:4471–4476.

Talbot, L. M. 1978. The role of predators in ecosystem management. Pages 307–321 *in* M. W. Holdgate and M. J. Woodman, eds. The breakdown and restoration of ecosystems. Plenum Press, New York. 496pp.

Tansley, K. 1965. Vision in vertebrates. Chapman & Hall, London. 132pp.

Thomas, D. C. 1977. Metachromatic staining of dental cementum for mammalian age determination. J. Wildl. Manage. 41:207–210.

Toweill, D. E., and Meslow, E. C. 1977. Food habits of cougars in Oregon. J. Wildl. Manage. 41:576–578.

True, F. W. 1889. The puma, or American lion: *Felis concolor* of Linneus. Rep. of Nat. Mus. 1889, pp. 591–608.

Tsukamoto, G. 1980. Trophy big game investigations and hunting season recommendations. Job Prog. Rep., Proj. W-48-11, Nevada Dept. Wildl., Reno, Nevada.

Turner, R. W. 1971. Mammals of the Black Hills of South Dakota and Wyoming. Ph.D. Dissertation. Univ. Kansas. 361pp.

Wade, J. G. 1929. Mountain lion seen killing a doe. California Fish Game 15:73–75.

Wailes, B. L. C. 1854. Report on the agriculture and geology of Mississippi. Lippincott, Grambo, & Co., for E. Barksdale, State Printer. 317pp.

Weddle, F. 1965. The ghost cats of the Yukon. Defenders of Wildl. News 40:53–54.

Winters, J. B. 1969. Trichiniasis in Montana mountain lions. Bull. Wildl. Dis. Assoc. 5:400.

Worley, D. E.; Fox, J. C.; and Winters, J. B. 1974. Prevalence and distribution of *Trichinella spiralis* in carnivorous mammals in the United States northern Rocky Mountain region. Pages 597–602 *in* C. W. Kim, ed. Proc. 3rd Int. Conf. on Trichinellosis, 1972, Miami Beach, Fla. Intext Educational Publ., New York. 658pp.

Wright, B. S. 1959. The ghost of North America: the story of the Eastern panther. Vantage Press, New York. 140pp.

Young, S. P., and Goldman, E. A. 1946. The puma, mysterious American cat. Dover Publications, Inc., New York. 358pp.

Youngman, P. M. 1968. Notes on mammals of southeastern Yukon Territory and adjacent Mackenzie District. Natl. Mus. Can. Bull. 223, pp. 70–86.

Kenneth R. Dixon, Appalachian Environmental Laboratory, Center for Environmental and Estuarine Studies, University of Maryland, Frostburg State College Campus, Frostburg, Maryland 21532.

39

Bobcat and Lynx

Felis rufus and *F. lynx*

Chet M. McCord
James E. Cardoza

NOMENCLATURE

COMMON NAMES.

Bobcat: wildcat, bay lynx, catamount, cat o' the mountain, barred bobcat, pallid bobcat, red lynx, cat lynx, *chat sauvage* (French Canadian), *chat sauvage de la nouvelle cosae* (French Canadian), *loup-cervier* (French Canadian), *pichou* (French Canadian), *gato monte* (Mexican)

Lynx: Canada lynx, gray wildcat, gray lynx, lynx, link, lucivee, *loup-cervier* (French Canadian), *pichu* (French Canadian), *lynx boréal* (French), *Nordluchs* (German), *Luchs* (German)

SCIENTIFIC NAMES.

Felis rufus (bobcat).

Felis lynx (lynx).

The generic and specific identities of the wild cats have been much debated, and several classification schemes are in use. An authoritative revision of Felidae is required in order to clarify these relationships.

Based on comparisons of skull characters, Kurten and Rausch (1959) tentatively considered North American and Eurasian lynxes to be conspecific, although they did not consider the geographically intermediate Siberian forms. Van Gelder (1977) argued that the ability to hybridize negated the generic separation of *Lynx* and *Felis*. Jones et al. (1975) and Corbet (1978) concurred in listing the binomial of the bobcat and lynx as *Felis rufus* and *Felis lynx,* respectively.

SUBSPECIES.

Bobcat subspecies have been listed by Seton (1929), Peterson and Downing (1952), Miller and Kellogg (1955), Young (1958), Hall and Kelson (1959), and Samson (1981). However, the 11 to 14 subspecies described to date comprise few realistically distinguishable taxons that have any real biological or managerial significance. Peterson and Downing (1952) found that pelage color and markings were too variable to distinguish the eastern races with certainty. "Type" specimens as described by Seton (1929) displayed similar nondiscrete variability. Multivariate statistical analyses of a variety of skull measurements have been presumed useful in distinguishing subspecies. How-

ever, measurements frequently overlap and have meaning only in large samples, thus being ineffective in the subspecific assignment of individual specimens. Subspecific differentiations also lose meaning in contiguous populations that lack geographical barriers. Exceptions seem to be *F. r. fasciatus,* separated from *F. r. pallescens* by the Cascade Range and the endangered *F. r. escuinapae* of central Mexico. Management of bobcats, then, does not mean subspecies-oriented management but rather population management. Much more work needs to be done before subspecies are a valid management consideration in such a plastic species as *Felis rufus.*

Similarly, the taxonomic relationships of the lynx are not well defined, and certain subspecies may well be invalid (Corbet 1978). The so-called Sardinian lynx is apparently referable to the wildcat *Felis silvestris* (Guggisberg 1975). The eight presumptive lynx subspecies and their respective approximate range are as follows: *F. l. lynx,* boreal Europe, Asia, and the Carpathian Mountains; *F. l. canadensis,* boreal North America; *F. l. pardina,* Iberian peninsula; *F. l. isabellina,* Mountains of central and eastern Asia; *F. l. subsolanus,* Newfoundland; *F. l. sardiniae,* Sardinia; *F. l. kozlovi,* Buryatskaya, U.S.S.R.; *F. l. stroganovi,* Lake Baikal region, U.S.S.R.

The lynx, which has drawn the attention of researchers in several countries, is now generally recognized as a single circumpolar species comprised of several geographically defined subspecies. Keeping this in mind, biologists and managers should be attentive to the detailed corpus of literature from Eurasian sources. These investigations, in many instances, bear a direct relationship to North American situations and will be used, when appropriate, in this chapter.

DISTRIBUTION

The historic range of the bobcat was given by Hall and Kelson (1959). However, there is some question as to how common the animal once was in those midwestern areas where it is absent today. The current range is shown in figure 39.1. The reduction in range from historic times appears to have been primarily

728

FIGURE 39.1. Distribution of the bobcat (*Felis rufus*).

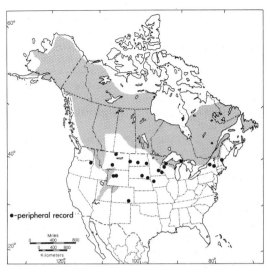

FIGURE 39.2. Distribution of the lynx (*Felis lynx*) in North America in 1977. Peripheral records since 1950 are indicated by a dot. From Seagears 1952; Erickson 1955*b;* Doll et al. 1957; Adams 1963; Long 1965; Coggins 1969; Hoffman et al. 1969; Rasmussen 1969; Hunt 1974; Fountain 1976; Scott 1977; and Gunderson 1978.

in the northern part of the original range and was associated with intensive agriculture or removal of forests in those areas lacking rugged or rocky mountainous terrain or extensive bogs and swamps.

The lynx is Holarctic in distribution, ranging across the boreal regions of North America (figure 39.2) and Eurasia. The species remains widespread in northern areas but has receded from much of its former range in the United States and western Europe.

In Canada, the lynx is found from Newfoundland, Labrador, and Quebec west to central British Columbia and the Yukon (Banfield 1974). It is probably extirpated from New Brunswick, lower Nova Scotia, and Prince Edward Island, and is absent from unforested areas of the Ungava Peninsula and the Northwest Territories (Rand 1944; Van Zyll de Jong 1971; Banfield 1974).

The lynx is also found in mainland Alaska (except the Seward Peninsula and the panhandle) north to the tree line (Bee and Hall 1956; Manville and Young 1965) and in the contiguous United States in northern New England (Godin 1977), parts of the Lake States (Gunderson 1978), the Pacific Northwest (Rust 1946; Ingles 1965; Hoffman et al. 1969; Nellis 1971), and the Rocky Mountains south to Utah (Durrant 1952). South and east of this area, the range of the lynx has changed substantially from the pristine distribution. Peripheral records outside occupied range may reflect the dispersal of the species, but in many areas, social and environmental changes may preclude viable reestablishment. Peripheral records since 1950 are shown in figure 39.2.

Matjuschkin (1978) described in detail distributional changes of the lynx in Eurasia. Currently, the species is resident in limited areas of Scandinavia, Poland, Czechoslovakia, Yugoslavia, Rumania, and Greece (Guggisberg 1975). A disjunct population (*F. l. pardina*) in southern Spain is now severely reduced in numbers and distribution and considered endangered (Goodwin et al. 1978). Stragglers occur rarely in the western Pyrenees of France (Navarre 1976). Lynxes also occur in parts of Iraq, Iran (Guggisberg 1975), Tibet, upper India (Prater 1971), Mongolia, Manchuria, and western and northwestern China (Tate 1947). Lynxes are found in the Soviet Union in timbered areas from the Kola Peninsula south to the northern Ukraine and the Mordovian A.S.S.R. and west through the southern Urals, the Altai, and western Siberia (excluding Kamchatka), the Transbaikal region, and on Sakhalin Island. An isolated population is found in the Carpathian Mountains in the southwest Ukraine and in the mountain forests of the Ciscaucasus and Transcaucasus (Novikov 1962; Ognev 1962).

DESCRIPTION

General Body Form. The bobcat appears about twice the size of a domestic cat (*Felis catus*), due to the bobcat's large-boned structure, particularly in the legs and head. The hind legs of the bobcat, especially the femur, tend to be proportionally longer and the tail proportionally shorter than that of the domestic cat. Generally, the bobcat is more muscular, more compact, and better adapted for springing than the domestic cat (Kelson 1946).

The more abrupt facial plane of the bobcat features a slightly deeper jaw, which is adapted for a wider opening of the mouth. The eyes are prominent, with elliptical pupils, and the sides of the face support a ruff of cheek whiskers. The ears are prominent and pointed, with a tuft of black hair at the tip. The dorsal surface of the ear is black with a central white spot.

The bobcat and lynx are digitigrade with sharp, retractile claws. There are four toes on the hind feet, the front feet are larger and show only four toes in the track, but there is a fifth toe that is raised. If tracked in snow over 1 cm deep the only imprint seen will be of the hind feet because these are placed in the impressions created by the front feet. The pads are relatively large and well defined in the track and the toes tend to be anterior to the pad.

The lynx is a medium-sized, robust, short-bodied cat with long legs; large, spreading, well-furred paws; prominent tufted ears; a flared facial ruff; and a short blunt tail. Its bone structure resembles that of a bobcat.

Pelage. The bobcat's general coloration is yellowish brown or reddish brown, variously streaked or spotted with black or dark brown. The guard hairs are black tipped. The underparts are white with black spots and there are several black bars along the insides of the forelegs. The tail is colored above like the coat, with the addition of several dark bands that become more distinct at the tip. The underside of the tail is whitish overall.

Bobcat fur is dense, short, and very soft. No true color phases occur, but melanism has been reported (Ulmer 1941). Pelts in the northwestern portion of the range tend to be more colorful and the spots more distinct than those of eastern or southern bobcats. Those animals from the more arid regions of the southwest tend toward pallidity; however, there is tremendous individual variation within any region. In bobcats there appear to be two annual molts, commencing in the spring and autumn. The summer coat is shorter and frequently more rufous than the longer and grayer winter pelage.

In the lynx, as in the bobcat, the sexes are colored similarly, although individual variation may be apparent. Adult North American lynxes in prime winter pelage have long thick fluffy fur with the upper body parts generally grizzled grayish brown mixed with buff or pale brown (Durrant 1952; Jackson 1961). The tricolored guard hairs are white basally and black terminally, separated by a darker middle band. The top of the head is brownish, with the ears buffy brown externally with a central white spot and the tufts and margins black. The facial ruff and throat are a mixture of grayish white, black, and brown. Underparts, feet, and legs are grayish white or buffy white, sometimes sprinkled with brown or blackish brown spots, particularly on the insides of the legs. The tail is brownish or pale buffy white with a completely black tip. Immature lynxes are pale buff, spotted or streaked with brown or black. Adult pelage is presumed to be attained the first winter (Jackson 1961).

The lynx's worn spring pelage is ragged, paler, and more buffy than the winter coat, with the fresh summer pelage darker and browner than the grizzled winter fur (Saunders 1961). There is apparently a single annual molt beginning in late spring (Jackson 1961), with the differences between summer and winter coloration resulting from the gradual replacement of the long, fine, gray-tipped guard hairs. The molt is generally complete by late autumn or early winter.

Color variations occur rarely in lynxes. Jones (1923) described "drab blue," "fawn yellow," "light brown" and "tabby" colorations. Cahalane (1947) referred to a bluish gray "dilute" phase. These unusual color phases probably represent mutations or changes in gene loci (Little 1958).

Other subspecies closely resemble the North American lynx in general appearance, although geographical variations in size are apparent. Bangs (1897), who described the Newfoundland lynx in summer pelage, referred to it as darker and richer than the mainland form, with black and hazel back and sides and irregularly spotted wood brown underparts. Saunders (1961, 1964) identified the winter coat of the Newfoundland lynx as brown with a silvery gray cast, with dull white, black-spotted underparts. Novikov (1962) described the Soviet Union lynx as variable in color, ranging from "smokey straw" to "rust red" with irregular spotting on the back and limbs. Spanish lynxes are more heavily spotted, while those of eastern Asia are indistinctly spotted or unspotted (Guggisberg 1975).

Size. Bobcats and lynxes are dimorphic in size, with males generally larger than females. Weights and physical dimensions also vary geographically. Some putative subspecies have been described merely on the basis of variations in average size. Male bobcats are generally about 33 percent larger in body weight than females. Adult males range from about 7.5 to 25.8 kg (Young 1958). Adult females range in weight from about 3.8 to over 15.0 kg. Berg (1981) reported a mean weight of 6.8 and 5.7 kg for fall-caught male and female young of the year.

Comparisons between mean weights for different regions should be made with great caution due to differences in the methods used to separate male from female and adult from juvenile animals (Foote 1945; Pollack 1949; Progulske 1952; Peterson 1966). Many early studies depended on trappers, hunters, and wardens for sex determination. These identifications often proved inaccurate. Some techniques for age determination also tended to lump large juveniles with adults. Therefore, these data should be used only to illustrate the consistency of sexual dimorphism.

Crowe (1975a) found that female bobcats averaged 9 kg at maturity, compared to 12 kg for males. The period of maximum growth was between four and seven months for females, while males grew at a faster rate, continuing until eight months. Lembeck (1978) believed that there may be a seasonal fluctuation in adults, with maximum weights in May and June and minimum weights in fall and early winter.

The consistency of sexual dimorphism is shown in table 39.1. Averaging the means from these studies yields standard measurements of 869 for total length, 148 for tail length, 170 for hind foot length, and 66 mm for ear length, for males. Respective measurements for females were 786, 137, 155, and 66 mm. The height at the shoulder varies between 500 and 600 mm.

TABLE 39.1. Standard measurements (mm) of bobcats (*Felis rufus*) from several regions of North America

Area	Sample Size	Total Length Range	Total Length Average	Tail Length Range	Tail Length Average	Hind Foot Length Range	Hind Foot Length Average	Ear Length Range	Ear Length Average	Reference
Canada										Banfield
Male	(?)	800–876	827	130–170	157	165–185	170	65–85	77	1974
Female		710–804	768	127–165	143	146–170	156	60–76	70	
Eastern Canada										Peterson
Male	(?)	853–1,252	950	119–190	145	170–223	185	60–75	66	1966
Female		745–950	829			155–190	172			
New Hampshire										Pollack
Male	(47)	691–972	865	108–173	155	137–196	177			1949
Female	(34)	708–869	805	110–171	145	144–181	165			
Massachusetts										Pollack
Male	(14)	731–948	846	130–163	148	160–187	173			1949
Female	(18)	670–828	786	117–161	137	140–170	158			
Michigan										Erickson
Male	(32)	851–1,041	935	140–201	160	152–201	175	53–81	71	1955a
Female	(24)	759–1,092	850	132–157	145	147–183	163	64–80	70	
Minnesota	(29)	562–1,022	902	117–174	148	136–193	168			Rollings 1945
North Carolina										Progulske
Male	(37)	610–940	823	113–178	138	137–178	159	37–70	55	1952
Female	(51)	610–864	728	90–152	121	127–165	145	32–70	52	
Virginia										Progulske
Male	(15)	475–1,016	839	110–155	132	135–165	155	49–78	61	1952
Female	(7)	660–787	736	114–142	127	141–146	143	53–65	59	
Louisiana										Lowery
Male	(3)	834–900	868	135–156	147	160–175	168	62–72	68	1974
Female	(3)	738–834	786	130–154	141	125–161	141	53–70	62	
Average male			869		148		170		66	
Average female			786		137		155		63	

In North America, adult male lynxes (N=24) from Alaska ranged from 710 to 850 mm in total length, with adult females (N=186) slightly smaller at 670–820 mm (Nava 1970). Newfoundland lynxes (Saunders 1961) were larger, with adult males (N=96) varying from 737 to 1067 mm overall and females (N=89) measuring 762–965 mm. Novikov (1962) reported total lengths of 820–1,050 mm for lynxes in the Soviet Union.

Weights of adult lynxes also vary according to sex, subspecies, and geographic location. Nava (1970) reported skinned weights of males (N=20) ranging from 6.8 to 14.1 kg and females (N=140) from 4.5 to 10.4 kg. In Newfoundland, Saunders (1961) weighed males (N=93) tallying 6.4–11.8 kg and females (N=91) from 5.0 to 11.8 kg. Lynxes are heavier in Sweden, averaging 17.9 kg for adult males (N=13) and 16.8 kg for adult females (N=10) (Haglund 1966). Novikov (1962) reported weights of 8.0 to 15.0 kg (sexes combined) for lynxes in the U.S.S.R. Extreme weights of 19.9 kg for Pennsylvania (Rhoads 1903) and 32.0 kg for central Russia (Ognev 1962) have been reported.

Skull and Skeleton. The skull of the lynx (*F. l. canadensis*) is short and broad. It is similar to (but smaller than) that of the cougar (*Felis concolor*), and has only two premolars on each side of the upper jaw

(figure 39.3). The postorbital processes of the frontals are small and the posterior palatine foramina are located near the palatine rim. A weak sagittal crest is present in older individuals. Lynx skulls differ from those of bobcats by a greater interorbital breadth (>30 mm), a posteriorly widened presphenoid (> 6 mm), longer upper carnassials (> 16 mm), and anterior condyloid foramina separate from the foramen lacerum posterius (Jackson 1961). Size is not a distinguishing character. Bobcat skulls range in total length from 108 to 143 mm, with zygomatic breadths of 75–107 mm (Jackson 1961; Lowery 1974). Alaskan lynxes display skull lengths of 118–140 mm and zygomatic breadths of 85–100 mm (Nava 1970). Soviet Union lynxes are generally larger, attaining skull lengths of up to 153 mm (Ognev 1962).

The dental formula for the adult bobcat and lynx is: 3/3, 1/1, 2/2, 1/1 = 28. A two-month-old lynx kitten described by Saunders (1961) had a dental formula of 3/3, 1/1, 2/2, 0/0 = 24.

The appendicular skeleton of the bobcat and lynx has been incompletely described. Mandal and Talukder (1975) studied the morphology of the girdles and limb bones of the lynx and caracal (*Felis caracal*). They explained the longer presacral and shorter postsacral ilium and the elongated tibiae in the lynx as modifications for its springing habits. The bobcat's structure is presumed to have a similar function.

PHYSIOLOGY AND MORPHOLOGY

The bobcat has two anal sacs or glands that are about 1 cm in diameter and can be discharged voluntarily and independently of each other (Bailey 1972). The secretions apparently vary with sex and age. Immature bobcats have a thin, milky substance; adult females have a thicker yellowish secretion. Adult males show a very thick brownish substance that is granular in comparison with the smooth texture of the immatures and adult females.

The physiology of the lynx has been studied little. Its thick, fluffy fur and broad, well-furred paws are presumed to be thermoregulatory and structural adaptations to a frigid snowy environment. Jackson (1961) stated that the lynx's resistance to cold is "phenomenal." Banfield (1974) reported that it beds down only in the most severe weather. The lynx, however, has poor endurance and tires easily after a chase (Seton 1929; Jackson 1961; Ognev 1962).

Lynxes have well-developed eyesight and hearing; however, their sense of smell is not well developed (Lindemann 1955; Saunders 1963b; Banfield 1974).

Stewart (1973) analyzed the fat content of bone marrow in 51 Ontario lynxes and found no significant difference between sexes or among months of capture (November–March). Yearlings, however, did have significantly lower amounts of fat than older animals, suggesting greater growth requirements and less well-developed hunting skills in yearlings than in adults. Brand and Keith (1979) rated renal and subcutaneous fat deposits in lynxes trapped in Alberta. Renal fat levels were affected by age, and subcutaneous fat levels by sex. Snowshoe hare (*Lepus americanus*) abundance and season also affected fat levels. Generally, the greatest fat levels were recorded during years of intermediate hare abundance and the lowest fat levels during years of hare scarcity.

The bobcat's energy requirements are not well understood. Kight (1962) kept a 4.97-kg wild-trapped bobcat and fed it one rabbit (averaging 553 g) every other day. At the end of 28 days, the bobcat had lost 0.54 kg of body weight. Golley et al. (1965), in feeding eight bobcats three different diets, found that food consumption averaged 110, 112, 113 g/kg of bobcat body weight on chicken, rabbit, and deer diets, respectively. Any variability in food intake appeared to be due to differences in the quantity of food available or the age of the test animal, rather than the type of diet. Golley et al. (1965) showed that 91 percent of the food intake was assimilated. They concluded that bobcats were able to maintain their body weight on widely varying quantities of food and that the utilization of consumed energy is relatively stable. Thus, a bobcat can increase its hunting effort on a low level of food intake and then, when prey is captured, assimilate large quantities of food. Adult bobcats appear better able to maintain body weight on a lower level of food intake than do young animals.

Saunders (1963a) calculated the food consumption rate of two captive adult lynxes at 18.6 kg/lynx/month (620 g/lynx/day), or about 170–200 hares per

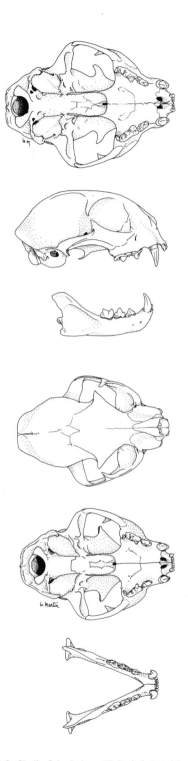

FIGURE 39.3. Skull of the bobcat (*Felis rufus*) and lynx (*F. lynx*). From top to bottom: ventral view of lynx cranium, lateral view of bobcat cranium and mandible, dorsal view of cranium, ventral view of cranium, dorsal view of mandible.

year. Nellis et al. (1972) reported that a captive juvenile lynx consumed about 370 g/day. Brand et al. (1976) reported an average daily consumption rate of 540–960 g/day based on tracking studies over a five-winter period. Consumption rates during periods of low hare density were less than in periods of high density and may not have been sufficient to maintain lynxes in "good condition." This nutritional stressing may have caused lynxes to become increasingly susceptible to trapping or other human-related mortality. Females apparently conceive regardless of hare densities; however, some minimum hare density may be required for females to raise young successfully (Nellis et al. 1972). Insufficient nutrition may result in starvation losses of kittens and/or decreased conception rates in yearlings (Brand et al. 1976; Brand and Keith 1979).

GENETICS

The karyotypes of the lynx were described by Hsu and Benirschke (1974). No noticeable difference was discerned between the karyotype of the lynx and that of the domestic cat. The complement of 38 chromosomes consisted of 32 metacentric and submetacentric and 4 acrocentric autosomes and 2 submetacentric sex chromosomes. The karyotype of the bobcat is similar to that of the lynx (Hsu and Benirschke 1970).

Gray (1972) reported a case of male bobcat X female domestic cat hybridization.

REPRODUCTION

Anatomy and Description. In the bobcat, the ovary is completely enclosed in the bursa ovarii, while that of the lynx is partially encapsulated (Saunders 1961). Ovarian histology and morphology in the bobcat were described by Duke (1949) and Mossman and Duke (1973). They described features important for management purposes, such as the stage of ovarian development and the number and size of the Graafian follicles and corpora lutea. The follicles are numerous in the anestrous stage and measure less than 1 mm. In preestrus, a few of the primary follicles enlarge to a diameter of 1 mm. In estrus, a single ovary may have up to seven follicles ranging in diameter from 2 to 8 mm (Crowe 1975a). An adult female bobcat will usually average five to six mature follicles for both ovaries (Erickson 1955a).

The bobcat and lynx have bipartite uteri that become swollen and turgid at the onset of estrus. Pollack (1949) and Erickson (1955a), working with bobcats, found that uterine diameters of about 1.5 to 4.0 mm were typical of the nonturgid condition and diameters between 5 to 10 mm were indicators of the turgid estrous state. Saunders (1961) found that uterine horns in lynxes during anestrus averaged 4.1 mm in diameter and 131 mm in length. However, uterine size varies with the size of the animal and the state of development of the uterus. There are one to three intermediate stages of uterine development (Erickson 1955a; Saunders 1961).

The development and persistence of the corpora lutea in the bobcat and lynx are unique and have generated much confusing terminology. In most mammals, the corpus luteum begins to degenerate late in pregnancy or during lactation to form the corpus albicans. Generally, corpora albicantia completely disappear within one year and in some species much sooner (Mossman and Duke 1973). However, Duke (1949) observed that corpora lutea in bobcats did not degenerate into corpora albicantia but that some luteal cells did remain. He thereby described two types of corpora lutea, based on cellular make-up, with the most recently formed containing both epithelial cells and large, heavily vacuolated cells. The second type of corpora lutea contained large cells with an acidophilic pigment. Duke (1949) believed these corpora to have derived from previous ovulations. The early, two-celled corpora and the later single-celled bodies have been referred to as "pale" and "dark" corpora lutea (Pollack 1949). The degree of pigmentation in the later, or "dark" corpora lutea was used by Progulske (1952) to distinguish corpora lutea of a previous cycle (medium dark color) from those of two or more cycles earlier (dark color). Erickson (1955a) also used "pale" and "dark" terminology to describe two classes of corpora lutea. Crowe (1975a) suggested the term *LBPC* (luteal bodies of previous cycles) instead of "corpora albicantia" or "dark corpora lutea," both of which referred to the single-celled luteal body.

Corpora lutea at various stages of development have been described in the lynx by Saunders (1961). In Newfoundland, active corpora lutea are largest (7–11 mm) during late March to late April and may be distinguished from nonfunctional corpora lutea by their lighter or yellowish pigmentation and generally larger size. Older, darker luteal bodies have been described as corpora albicantia (Saunders 1961; Nava 1970; Stewart 1973) but are probably derived from mature atretic follicles of previous seasons and should be called corpora lutea (Brand and Keith 1979), or LBPC.

To alleviate this nomenclatorial confusion, we recommend that those yellowish bodies derived from the most recent ovulation (which have been called "two-celled corpora lutea," "pale corpora lutea," and "corpora lutea") be called simply "corpora lutea." Those persistent luteal bodies that are variable in color (but generally a dark grayish brown) and derived from earlier cycles ("single-celled corpora lutea," "dark corpora lutea," and "corpora albicantia") should be called "LBPC" (Crowe 1975a). We have used these terms in the sections below.

Ovulation Rates. Ovulation rates for bobcats have been reported as 3.6 ova for the first ovulation and 5.5 thereafter in Michigan (Erickson 1955a). Average ovulation rates for all age classes have been reported as 4.8 in Utah (Gashwiler et al. 1961), 3.4 in Wyoming (Crowe 1975a), and 4.2 in Arkansas (Fritts and Sealander 1978). Brand and Keith (1979) reported that in lynxes, during a period of hare abundance, the combined corpora lutea–LBPC count per ovulating female averaged 5.1 ± 0.3. Fritts and Sealander (1978) and Brand and Keith (1979) cautioned that the determination of ovulation rates may be influenced by multiple ovulation during the same breeding season and by changes in the percentage of females ovulating in the sample. Asdell (1946) reported the bobcat to be an induced ovulator similar to the domestic cat. Duke (1949), however, stated that spontaneous ovulation could not be ruled out. Crowe (1975a) and Fritts and Sealander (1978) suggested that spontaneous ovulation is more likely because all primiparous females displayed from 3 to 12 LBPCs, which is well above the average for one ovulation. They believed that bobcats may cycle up to three times per season if they are not fertilized. Lynxes have been assumed to be induced ovulators (Saunders 1961; Nava 1970).

The LBPCs probably persist for the life of the animal, their number increases with each ovulation, and the size of the ovary increases as a result. Erickson (1955a), working with bobcats, found ovarian volumes to average 0.40 ml for juveniles, 0.88 ml for first-year animals, and 0.66 ml for older adults. Crowe's (1975a) bobcat studies showed a positive correlation between the number of LBPCs and age and between ovarian volume and age. The number of LBPCs ranged from 3 to 59 for age classes 1 to 11 and the ovarian volume for the same age classes varied from 0.8 to 3.5 ml in anestrus ovaries (Crowe 1975a). In lynxes in Alberta, Brand and Keith (1979) showed a rising trend in luteal body (combined corpora lutea and LBPC) counts with age, rising from about 5.5 per yearling female to 11.2 for four-year-old females. The ovaries of multiparous female lynxes average 14.6×8.6×8.7 mm, with an approximate oviduct length of 24–26 mm (Saunders 1961). Nulliparous females display smaller, smooth ovaries (averaging 10.4×5.5×5.2 mm) with numerous small tertiary follicles.

The uterus also contains scars from the implantation sites of the embryo. The placentation of the bobcat and lynx is endotheliochorial, deciduate, and zonary and the scars can be seen most readily as dark or opaque areas when a fresh uterus is viewed against a background light source (Erickson 1955a; Brand and Keith 1979). Scars may also sometimes be detected by passing the uterus through the fingers and noting the swelling at each scar site. Gashwiler et al. (1961) found that placental scars in bobcat were not discernible in all uteri, and, similarly, placental scars in lynxes rapidly regress and may be faint or difficult to distinguish (Nellis et al. 1972; Stewart 1973).

Transmigration of ova from one uterine horn to the other may occur. In lynxes, Nava (1970) found 150 cases where the placental scar count in one uterine horn was greater than the corpora lutea count in the corresponding ovary. Fritts and Sealander (1978) found three cases of transuterine migration of zygotes in bobcats.

Age of Puberty. The female bobcat appears to breed between its first and second years of life. Crowe (1975a) found that females that had gone through one reproductive season showed from 3 to 12 LBPCs in the ovary. Pollack (1950) and Erickson (1955a) reported bobcat breeding during their first reproductive season, based on the presence of corpora lutea. Fritts and Sea-

lander (1978) agreed that yearling animals bred, but believed that this probably took place later in the year than for adult animals. Neither Poelker (1977) nor Berg (1981) found corpora lutea in animals less than one year old.

Male bobcats show little evidence of spermatogenesis until their second year of life. There may be some seasonal and annual cycles of testicular development and regression. Crowe (1975a) found no mature spermatozoa in the testes or epididymides of 29 juvenile males. He also found that the growth curve of the testes was sigmoid, with the greatest growth rate corresponding to the beginning of spermatogenesis. Testes volume was positively correlated with body weight. Fritts and Sealander (1978) also found a lack of sexual activity in juvenile males, and found that all males over one year of age were fecund during the breeding season, as shown by sperm in both the testes and epididymides. Crowe (1975a) found that the mean testis volume in July and September was significantly less than the volume (2.31 ± 0.49 ml) for January through May. This decrease in volume was apparently due to a reduction in the size of the seminiferous tubules. Few spermatozoa were found in the tubules or epididymides of mature males during the summer months of July and August. Sperm production appeared to start in September and October, with mature males fecund until some time in the summer months. At no time was a total absence of spermatozoa observed. Duke (1954) and Fritts and Sealander (1978) believed that male bobcats were fertile the entire year.

The age of puberty in the lynx appears to vary and may be influenced by prey abundance. Saunders (1961), during a period of snowshoe hare scarcity, suggested that female lynxes did not attain sexual maturity until 22–23 months of age. Nava (1970), based on the presence of placental scars in 53 percent of his sample of yearlings, concluded that Alaskan lynxes normally breed at one year of age. However, his data also showed a significantly higher percentage of females with luteal bodies (corpora lutea and LBPC) in areas of hare abundance than in areas of hare scarcity. This indicated the possibility of an alternate-year breeding cycle under certain conditions of food availability.

Reproductive activity in the male lynx has not been fully described. Testes may range from 0.5 to 4.5 g in weight (Saunders 1961), with those 1.5 g or less probably representing immature animals. In Alaska, Nava (1970) demonstrated that mean testes size increased with age. Those of kittens averaged 8.5 mm in total length and 1.1 cc in volume, yearlings 19 mm and 1.9 cc, and adults 21 mm and 2.2 cc. Evidence of reproductive status can be determined by the presence or absence of sperm in the epididymis and seminiferous tubules. The age of sexual maturity in male lynxes is uncertain, but Saunders (1961) believed males incapable of breeding during their first year of life. Likewise, Stewart (1973) found no male kittens that appeared sexually mature.

Breeding Season. Seton (1929) reported that bobcats breed in February and March. Duke (1954) plotted the frequency of embryonic litters by month and concluded that the breeding season extended from February through April. Erickson (1955*a*) examined the condition of bobcat uteri and ovaries and the extent of blood clots in the urine and determined a period of January through May. Gashwiler et al. (1961) backdated embryonic litters and found that bobcats may breed from January to July or later. They found two female bobcats in March with chewed necks, indicating neck grips by their male consorts. Crowe (1975*a*) used the condition of the follicles, corpora lutea, and fetus to backdate to the ovulation dates and determine a mid-January to early February mating onset and a July or later termination.

Crowe (1975*a*) determined that the mean ovulation date was 15 March, but in a polyestrous animal recycling would probably occur in 44 days (about 28 April), and again if conception did not occur (about 11 June). This, of course, assumes polyestrous cycling three times annually, with spontaneous ovulation. McCord (1974*b*) observed signs in the snow of bobcat courtship behavior in early February and March in Massachusetts. Fritts and Sealander (1978), in one of the few reproductive studies from southern states, determined that the bobcat mating period extended from December to March. They used ovarian and uterine condition and backdated embryonic litters. They believed the bobcat in Arkansas capable of breeding throughout the year. Miller (1980) found breeding from February through July in Alabama, with peaks in March and April. He used measurements of reproductive tracts, number and size of follicles, plus backdating tooth replacement in kittens to determine the reproductive season.

In the lynx, Saunders (1961) determined that mating occurred in Newfoundland between 11 March and 7 April, peaking about the third week in March. The mating period in Alaska is similar, ranging from March to early April (Nava 1970). Nellis et al. (1972) suggested that in Alberta mating took place primarily in April and May. Eurasian lynxes, however, reportedly breed in February and March (Novikov 1962; Haglund 1966).

Variations in lynx and bobcat breeding seasons are probably influenced by latitude, longitude, altitude, climate, photoperiod, and perhaps prey availability. Bobcats in the southern portion of their range probably begin breeding activities earlier than those in the north.

Gestation Period. The gestation period in bobcats has been variously cited as 50 to 70 days, with about 62 days the most commonly reported (Palmer 1954; Jackson 1961; Peterson 1966; Ewer 1973; Lowery 1974; Guggisberg 1975).

The gestation period in the lynx is about 9 weeks. Seton (1929) reported a range of 60–65 days and Hemmer (1976) 61 days in North America. Saunders (1961) assumed a range of 63–70 days and a mode of 70 days in Newfoundland. Matjuschkin (1978) reported a period of 67–74 days in the U.S.S.R. and Haglund (1966) 63–73 days for captive European lynxes.

Parturition Dates. Parturition dates in bobcats have been determined from occasional field observations and by extrapolation from reproductive tracts. Gashwiler et al. (1961) found that birth dates for 13 newborn bobcat litters ranged from March through July, with most births occurring in April and May. Crowe (1975*a*) found that births ranged from 15 April to 1 September, with 59 percent falling between 15 May and 15 June. Fritts and Sealander (1978) examined tooth eruption on 8 kittens collected in fall and found birth dates ranging from March to September, with a peak from March to May. They cautioned, however, that tooth eruption rates may vary with genetic and nutritional variations in individual animals. Zezulak (1981) observed area reductions of 75 to 90 percent in female bobcat home ranges during February and March, behavior that is indicative of parturition. This behavior pattern was also noted by Bailey (1972) and Lembeck (1978).

Lynx litters are dropped from about late May to early June, peaking around 23–24 May in Newfoundland. In the Soviet Union, most young are born around the second half of May (Matjuschkin 1978).

Litter Size. Various methods have been used to determine litter sizes in lynx and bobcat populations. Generally, follicular maturation, corpora lutea counts, blastocyst counts, placental scar and embryo counts, and field observations of litters produce a descending estimate of litter size. Primarily, corpora lutea counts may be used as indicators of ovulation rates, but inflated figures may be obtained if the population appears to be polyestrous and ovulating spontaneously. Placental scar counts can be used as evidence of implantation and litter size. Observations of newborn young provide accurate assessment of actual recruitment to the population. Estimates of bobcat and lynx litter sizes for several regions of North America are presented in table 39.2. Litters from yearlings are normally smaller than those from older animals (Poelker 1977; Brand and Keith 1979).

Reproductive Rates. Lynxes and bobcats are generally considered to have only one litter per year. However, speculation persists regarding two litters in the bobcat, particularly in the southern portion of its range (Fritts and Sealander 1978). Nevertheless, in nine zoos that have successfully bred bobcats, all reported only one litter annually (Crowe 1975*a*). Speculation concerning two litters has been stimulated by high LBPC counts and by occasional fall litters. The high LBPC count probably is due to spontaneous ovulation in a polyestrous animal that may cycle up to three times per year. The occurrence of litters in August–September may be explained by the animal's being impregnated on its third cycle. Late litters also may be due to juvenile animals cycling late in the year of their first breeding season (Erickson 1955*a*; Crowe 1975*a*; Berg 1981) or to females that lost their litter at an early age and then recycled (Fritts and Sealander 1978).

Two bobcat litters per year are possible because an animal could conceive in mid-January, give birth 62 days later in mid-March, nurse the young for 2 months (Young 1958) until mid-May, then recycle and breed again while the first litter is still dependent. This is not

TABLE 39.2. Estimates of litter size in bobcat (*Felis rufus*) and lynx (*F. lynx*) for several regions of North America

| | Productivity Estimator | | | | | | | | | | |
| | Corpora Lutea | | Placental Scars | | Embryo Counts | | Litter Observed | | Combined | | |
Region	Range	Average	Range	Average	Range	Average	Range	Average	Range	Average	Reference
Bobcat											
California								2.7 (N = 9)			Zezulak 1981
Washington										2.6 (N = 171)	Poelker 1977
Idaho							1–5 (N = 16)	2.8			Bailey 1974
Wyoming		3.4	3–5								Crowe 1975a, 1975b
Utah		4.8		3.9 (4)		3.2 (N = 365) 3.2 (3)	1–6 (N = 47)	3.5 (4)		2.8 (3)	Gashwiler et al. 1961
Arkansas					1–8						Fritts and Sealander 1978
Minnesota			1–4	2.5							Berg 1981
Michigan				3.2						2.6	Erickson 1955a
Massachusetts									2–4	2.0	Pollack and Sheldon 1951
Lynx											
Alaska				4.5 (N = 493)						4.0	Nava 1970
Alberta	4.3–11.9[a] (N = 261)								3.4–4.6 (N = 207)		Brand and Keith 1979
Newfoundland	0.0–4.7[b]	2.9									Saunders 1961
Ontario										3.4	Stewart 1973

NOTE: Averages are means, except where the mode is expressed in parentheses.
[a] Combined corpora lutea and LBPC counts.
[b] LBPC.

inconceivable; Mathews (1941) found that the Scottish wildcat (*Felis silvestris*) produced two and rarely three litters per year.

Breeding Synchrony. The question of induced versus spontaneous ovulation in the bobcat (and presumably the lynx) is still open, with the most recent studies circumstantially tending toward spontaneous ovulation. Asdell (1946) merely assumed induced ovulation, based upon evidence from the domestic cat. Duke (1949) remained undecided on the question. Crowe (1975*a*) and Fritts and Sealander (1978) both believed that the bobcat was a spontaneous ovulator cycling up to three times per year. However, Scott and Lloyd-Jacob (1955) pointed out that the domestic cat may not always be an induced ovulator. Perry (1972) noted that there is no clear demarcation between induced and spontaneous ovulation. The bobcat may ovulate spontaneously at times and yet coitus may induce or hasten ovulation at other times (Crowe 1975*a*).

Description and Development of the Young. The growth of fetal bobcats and lynxes is not well documented, and much information has been derived from the embryology of the domestic cat. Crowe (1975*a*), for example, backdated the time of ovulation in bobcats from the size and development of the embryo (crown–rump measurements) using the pattern developed for the domestic cat. Similarly, Scott (1976) discussed the factors affecting the development of young felids by using the domestic cat as a model.

At birth, bobcat kittens have their eyes closed and commence suckling immediately while being licked dry by the mother. The newborn kittens display mottled or moderately spotted fur and more distinct facial markings than adults. The female generally stays with the young for two days, sustaining herself by consuming the placenta, feces, and any stillborn kittens. A well-fed kitten will gain about 10 g per day (Scott 1976). By the end of its fourth week it will have developed sufficient locomotory function and sensory perception to begin exploring its surroundings. By this time, milk alone is insufficient for the kitten's needs and solid food is required to sustain weight increases. The mother's milk output is gradually reduced, ceasing by about the seventh or eighth week. Gashwiler et al. (1961) reported that three-day-old kittens weighed 128 g, while Young (1958) cited 283–340 g as the common birth weight. This range is similar to that reported by Pollack (1950), Palmer (1954), and Banfield (1974). However, Peterson (1966) reported a broader range of 200 to 800 g.

Bobcat kittens' eyes open between 3 and 11 days (Grinnell et al. 1937; Pollack 1950; Young 1958), with 10 days being most common.

Newborn lynx kittens have closed eyes, folded ears, and well-developed grayish buffy pelage with dark longitudinal streaking on the back, flanks, and limbs (Merriam 1886). They measure between 158 mm and 163 mm in length and weigh from 197 to 211 g (Saunders 1961). They are altricial, with the eyes opening between 12 (Novikov 1962) and 17 days (Lindemann 1955; Wayre 1969). Weaning is accomplished

by 12 weeks and the natal pelage is lost between 60 and 300 days.

Growth Rates. Growth rates of bobcat young vary greatly, depending on food supply. Crowe (1975*a*) developed a growth curve and equation based on the actual weight of wild-caught animals and their estimated ages. The growth curves for both the male and the female start at about 4.5 kg at 200 days, with the male's growth peaking at about 11 kg and 500 days and the female's peaking at about 8 kg and 700 days. Colby (1974) periodically weighed two females and a male captive bobcat. The male and female curves were equal until about day 540, when the male grew from less than 9 kg to over 13 kg in about 60 days. The female's growth leveled off at about 8–9 kg. In both studies, the male attained maximum weight in about 500 days. The primary difference between the studies was that females leveled off in weight at about day 340 in Colby's study, compared to about day 650 in Crowe's study. The main increase in body size of the male bobcat appears to take place in the fall of its second year.

Bailey (1972:26) reported that in bobcats "kitten survival appears to be related to the food supply." His observations indicated that "most [58] kittens captured in 1969 and 1970 were captured in the fall and winter, indicating good survival. In 1971, no kittens were captured then despite much trapping effort" (p. 10). On his study area in Idaho, rabbit populations were increasing in 1969, peaked in 1970, and declined about 86 percent in 1971. Similarly, survival and productivity of the lynx are related to prey populations (Brand et al. 1976; Brand and Keith 1979).

ECOLOGY

Social Organization and Territoriality. Etkin (1964:21) defined territoriality as "any behavior on the part of an animal which tends to confine the movements of the animal to a particular territory." The animal may actively defend the area or there may be other responses such as the "mutual avoidance reaction" (Hornocker 1969*b*). This reaction, described by Leyhausen and Wolff (1959) in free-roaming domestic cats, involves animals simply moving away from an area in use by a conspecific, thus avoiding contact. Recognition of another animal's presence is generally believed to be through visual and olfactory means.

Older male and female bobcats usually have a territory or home range that is fairly well defined yet that varies in size depending on prey density, sex, season, and climate. Resident animals, with some exceptions, confine their movements and activities to these areas. McCord (unpublished data) has observed that resident bobcats utilize the best habitat areas, especially ledges or other activity centers within their home range. Other residents may also use these core areas on a mutual basis. However, transient animals are almost always excluded from these areas.

Bailey (1972:14) described transient bobcats as animals "that had not permanently settled in an area."

Transients are basically a floating population that coexists with the residents. General characteristics of transient populations include a predominance of young or sexually immature animals, erratic or long-range movements, and utilization of less desirable habitat areas. Bailey (1972) mentioned that transient bobcats appeared to be sexually immature, with smaller body sizes and lower weights as compared to residents. He believed that residents prevented transients from raising young; however, the mechanisms for this were unclear.

McCord (unpublished data), in the Quabbin Reservation in Massachusetts, studied three bobcats believed to be transients. The behavior of a young female was typical. When released, the animal moved about 1.6 km; it was found in the same place several days in a row on a hillside. McCord became convinced that the animal was dead. However, when he moved in to retrieve the carcass the animal flushed and moved 4.8 km to an area not used by resident cats. Again, it remained in a small area for about a week, then moved 6.4 km to a small hilltop. This hilltop was near two much-used ledge areas. However, this female avoided them. Again, after a few days, the animal moved another 6.4 km, where it stayed close to a fresh deer kill. The bobcat was again flushed and it again moved several kilometers. During all this time, this female never used the preferred habitat areas used by the resident bobcats. This stop-and-go behavior was very different from that of residents. Residents moved short or long distances within their home ranges, generally on a daily basis.

Aggressive behavior between resident and transient bobcats was not observed by McCord (unpublished data). Bailey (1972) and Miller and Speake (1978b) also reported that adults were apparently tolerant of juveniles within their home range.

Transient bobcats appear to be primarily dispersing juveniles. However, Lembeck (1978) reported that residents occasionally displayed transient behavior, especially when sick or dying. Dispersal of young bobcats may take place anywhere between fall and late winter. McCord (unpublished data) captured a juvenile female (3.4 kg) on 25 October. Signs around the trap indicated that another, larger animal was also present. The young bobcat was placed in a holding cage at the rear of the trap. After one night, the juvenile was radio-collared and released. The next night, an adult female was caught and released at the same location. The two animals stayed in the same area until 2 November, when both were located outside the study area. The two bobcats were not accompanying each other. This was the only instance in two seasons of tracking in which the old female was found outside of its home range. The young female subsequently began moving 4.8–8.0 km per night westward. The animal was observed once, and appeared to be traveling alone. After crossing the Connecticut River and a major interstate highway, the juvenile continued westward into the Berkshire Mountains, where radio contact was lost about 40 km west of the original capture site.

Hamilton (1942) believed that juvenile bobcats dispersed during their first fall. Erickson (1955a), however, believed that bobcat litters in Michigan may not disperse until the spring, based on six observations of two to four bobcats traveling together in the period from December through February. Erickson (1955a:27) stated, "In several of the above instances there were evidences of a larger cat followed by smaller individuals. When travelling conditions were suitable, the smaller tracks were usually found superimposed on those of the larger individual." Erickson (1955a) also reported an instance where the family break-up may have been violent. He speculated that, because the breeding season was about to begin, the mother had fought the offspring to force them to leave.

The transient population appears quickly to fill any voids that appear due to the death or removal of a resident animal. Bailey (1972) reported three instances where resident females died and their home ranges were taken over by one or two other females that subsequently raised young. These "new" females were believed to have come from the transient population, because the known residents did not move in. Zezulak and Schwab (1980) found a bobcat with a well-defined home range that abruptly abandoned its area. The vacant range was reoccupied within 11 days by another bobcat that used 90 percent of the previously occupied area. Two resident males died in Miller's (1980) study and their home ranges were taken over in one case by a transient male and in the other by an adjacent resident that simply expanded his home range.

The evidence to date indicates that bobcat social structure tends to follow general guidelines, with resident males and females occupying fixed home ranges and a transient population associated with the resident population. Females tend not to overlap ranges of other females but normally overlap those of males. However, the ranges of male bobcats are likely to be overlapped by those of other males. Males may also move more often and have larger home ranges than females. Nevertheless, bobcats appear quite adaptable: female ranges have overlapped each other and some apparent resident bobcats have even shifted home ranges altogether.

The behavior of some bobcats that seem to behave like residents for a few months and then move long distances for no apparent reason suggests that there could be a third, or nomadic, segment of the population. These bobcats may tend to be older than the normal transients. They may be able to establish temporary ranges where they have access to key areas such as ledges and good feeding sites, whereas the transients seem to be excluded from these key areas. The mechanism that excludes the animals and that stimulates movement may be olfactory or related to marking behavior. The nomadic animal may require a greater stimulus for movement than transients, such as an actual encounter with the resident animal. Zezulak and Schwab (1980) analyzed interactions between bobcats in areas of home range overlap and found that older or larger animals tended to initiate more interactions and that the larger home ranges were violated most often.

There appears to be a dominance hierarchy fostered by a land tenure system implied in many studies of bobcats. This may function most effectively on the submissive transients and less effectively on older nomadic animals that have not found a suitable home range.

Lynxes are solitary animals, normally associating only for reproduction and rearing of the young. Hardy's (1907) anecdotal references to "droves" of 7 and 11 lynxes are not substantiated by current evidence.

The young lynxes are believed to remain with the mother through their first winter or until the start of the next mating period (Seton 1929; Jackson 1961; Saunders 1961). Lindemann (1955) reported this period as about 10 months. Ognev (1962) reported that Russian lynx families are disbanded during their second year. Saunders (1961) recorded a family group remaining intact through April and suggested that those females accompanied by young may not breed annually.

Adult male lynxes travel briefly with females during the mating period, and males may also later tolerate family groups within their home range (Saunders 1963b; Haglund 1966). Adult females may be less tolerant of other females than are males (Berrie 1973). Home ranges of lynxes may well overlap (Nellis et al.

1972; Brand et al. 1976; Matjuschkin 1978), suggesting that mutual avoidance behavior serves to separate lynxes temporally and spatially (Keith 1974). Mech (1980), however, studying an expanding lynx population in Minnesota, found that ranges of females often overlapped, while those of males did not. Male ranges also were found to overlap those of females only slightly, even when adjacent. Communal hunting, probably involving both adult and young lynxes, may occur infrequently (Barash 1971).

Home Range. Home range estimates have varied from 0.6 to 201.0 km² for bobcats and from 10 to 243 km² for lynxes, depending on sex, age, population density, prey density, and survey method (table 39.3). Females generally display smaller home ranges than do males and appear to stay in an area several days, using their range intensively (Saunders 1961; Bailey 1974). Males may move about more frequently in a larger area and may have ranges divided into two or more distinct segments separated by up to 11 km (Lembeck 1978).

Home ranges of male and female bobcats may overlap but females rarely trespass on ranges of other females. Zezulak (1981), in California, however, re-

TABLE 39.3. Home range estimates for bobcat (*Felis rufus*) and lynx (*F. lynx*) for several regions of North America and Eurasia

Region	Range Estimate[a] (km²)	Sex (N)	Reference
Bobcat			
California	0.6–4.4 (2.1)	F (3)	Lembeck 1978
	0.9–6.4 (3.0)	M (9)	
California	26–59 (43)	F (4)	Zezulak 1981
	39–95 (73)	M (3)	
California	17.5	F (1)	Zezulak and Schwab 1980
	4.7–53.6 (28)	M (6)	
Washington	6.5–15.5	F ⎱ (21)	Brittell 1979
	3.9–8.4	M ⎰	
Idaho	(19.3)	F (8)	Bailey 1974
	(42.1)	M (4)	
	6.5–107.9	MF (12)	
Minnesota	15–92 (38)	AdF (6)	Berg 1981
	13–201 (62)	AdM (16)	
Massachusetts	26–31	MF (3)	McCord 1977
South Carolina	2.5	F (1)	Marshall and Jenkins 1966
	3.5–4.6 (3.9)	M (3)	
Alabama	(1.5)	F (6)	Miller 1980
	(3.0)	M (6)	
Louisiana	0.9–1.2 (1.0)	F (3)	Hall and Newsom 1976
	3.4–7.3 (4.9)	M (3)	
Lynx			
Alaska	12.8–25.5	F	Berrie 1973
	14–25	M	
Alberta	12.4–23.1	family group (2)	Brand et al. 1976
	11.9–49.5	adult (6)	
Minnesota[b]	51–122	F	Mech 1980
	145–243	M	
Newfoundland	15.5	F (1)	Saunders 1963b
	(19.4)	M (2)	
Soviet Union	1–25	U	Novikov 1962

[a] Means in parentheses.
[b] Expanding population.

ported three of four radio-collared females overlapping other females by up to 36 percent of home range area. Seasonal range differences may also occur. He reported that winter ranges of male bobcats in California were up to 41 percent smaller than summer ranges, with females showing reductions of up to 70 percent over the same period. Nellis et al. (1972) tracked an adult male lynx whose range fluctuated from 18.0 km² in late winter to 49.2 km² the following winter.

Home range estimates may be affected by the survey method. Berrie (1973) cautioned that insufficient telemetry fixes and erratic lynx movements may have affected some estimates. Mech (1980) stated that snow-tracking may bias movement and home range studies because of errors in identifying sexes and individuals and by restricting studies to a limited area and season. McCord felt that toe-clipping was essential to maintain the true identity of an individual bobcat being followed in the snow.

Movement and Activities. The first detailed study of bobcat movements was conducted by Robinson and Grand (1958), who tagged 81 bobcats and recovered 48 in a 27-month period in Montana. The mean recapture distance was 6.6 km, with males averaging 8.5 km and females 5.1 km between captures. Seventy-nine percent of the bobcats were captured within 8 km of the initial trap site; 92 percent were recaptured within 16 km (Robinson and Grand 1958).

Early bobcat investigators determined daily movements from snow-tracking. Rollings (1945) had five cases of 24-hour activity periods where the bobcats traveled from 4.8 to 11.3 km, averaging 8.8 km/day. Considering the same five cases, the shortest straight-line distance between points (cruising radius) was only 1.6–6.0 km (Rollings 1945). Erickson (1955a) similarly tracked four bobcats that displayed daily movements varying from 0.8 to 11.7 km and averaging 5.6 km. McCord monitored bobcats for 11 daily activity periods in Massachusetts, and found an average movement of 4.9 km per period (1.1–18.5 km).

The first substantive bobcat telemetry study was conducted by Marshall and Jenkins (1966), who calculated the average daily movements of an adult male and an adult female as 4.8 and 2.6 km, respectively. Differential movements have also been documented by Bailey (1972, 1974), who showed significant differences between male and female movements over distances greater than 1.6 km. Differences between average daily radio fixes of males (average distance traveled, 1.8 km) and females (1.2 km) were not statistically significant, however. Hall and Newsom (1976) showed that the average minimum distance traveled was 4.4 km for male bobcats and 2.9 km for females. Berg (1981) detected a 61 percent greater movement between weekly locations for males than for females.

The rate of movement was also greater for males (369–465 m/hr) than for females (192–224 m/hr) in a Louisiana study (Hall and Newsom 1976). Bailey (1974) believed that females normally used their smaller home ranges more intensively than males.

Females tended to stay in one locality for a number of days, while males moved about their home range without any apparent pattern. He also noted more movements for both sexes over 1.6 km in the spring and summer than in the fall and winter.

Average movements of marked lynxes have ranged up to 19.5 km between recaptures. In Newfoundland, 30 lynxes of both sexes averaged 5.0 km (0.0–16.1) and 88 days (1–430) from initial to final capture (Saunders 1963b), while 14 Alaskan animals ranged about 4.0 km (0.0–19.5) and 45 days (1–447) between recaptures (Berrie 1973).

Long-range movements have also been recorded. Saunders (1963b) noted an adult male lynx that moved 103 km in 587 days. Nellis and Wetmore (1969) observed an injured adult male that traveled 164 km in 163 days. This unusual movement was believed due to stresses induced by trapping and tracking by the investigators. Mech (1977) documented an adult female that dispersed from Minnesota to Ontario, 483 km in 807 days.

The total average daily distance (cruising distance) traveled by a lynx has varied from 5.0 km in Newfoundland (Saunders 1963b) to 11.0 km in central Russia (Iurgenson 1955) to 19.2 km in Sweden (Haglund 1966). In Alberta, Nellis and Keith (1968) noted a shift in average cruising distance from 4.8 km/day in 1965–66 to 8.8 km/day in 1967. These increased movements were believed to be related to hunting conditions, exclusive of prey numbers that increased during the study period.

The shortest distance between lynx bedding sites (cruising radius) averaged 1.0–1.8 km/day in Alberta (Nellis and Keith 1968), 4.2 km/day in Newfoundland (Saunders 1963b), and 7.6 km/day in Sweden (Haglund 1966). The reasons for higher diel movements in Europe are unknown.

The bobcat is best described as crepuscular rather than nocturnal. Marshall and Jenkins (1966) found that bobcats were most active from about three hours before sunset to midnight and from one hour before to about four hours after sunrise. Male and female diel activity levels are shown in figure 39.4.

Zezulak (1981) found that bobcat activity levels peaked at dawn and dusk in California. Cats were active throughout the day, but night movements averaged 2.5 km compared to 1.7 km in the daytime. There were no significant differences, however, between average distances traveled in the dawn (1.0 km) and dusk (1.6 km) periods. Also, contrary to most other studies, he could not relate mean bobcat activity levels to sex, age, and home range size. Zezulak (1981) suggested that individual variation was the major factor in electing different levels of activity.

In another California study, Zezulak and Schwab (1980) found some seasonal differences in bobcat activity levels between winter and spring periods. The crespuscular activity pattern (peak activity from 0400 to 1000 hrs and 1600 to 2200 hrs) characterized the winter season, while the bobcats became more nocturnal (1800 to 0600 hours) in the spring. The lack of daytime activity in the spring seemed to be related to

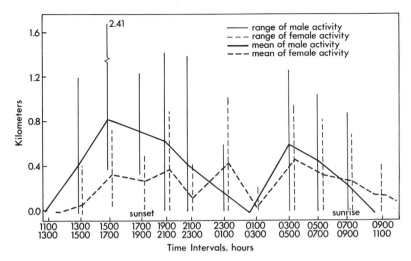

FIGURE 39.4. Rates of movement for male and female bobcats (*Felis rufus*) on Thistlethwaite Game Management Area, Louisiana. After Hall and Newson 1976.

temperature in that there was a decrease in bobcat activity above 26° C. It was unclear if the reduced activity levels derived from the high temperatures or from reduced prey activity; however, low temperatures in more northern areas may, conversely, increase bobcat daytime activity. Regardless of the season, Zezulak and Schwab (1980) found that bobcats were active about 12 hours per day.

Population Density. The density of resident bobcats can be determined in small tracts with extensive effort because of the stability of their home range structure. The greatest difficulty is in trapping and marking all the residents. Lembeck (1978) suspected, and later confirmed, the presence of an untrapped resident by the home range patterns of other females. He believed that there was an inverse relationship between home range size and density. Bailey (1974) also was able to detect the presence of uncaptured animals on his study area. Bobcat density estimates for several regions of North America are given in table 39.4.

Transient and nomadic bobcats are more difficult to locate, but such animals may be identified by their age and behavior, as described above. Their numbers may vary greatly on a yearly basis, yet McCord (unpublished data) found that his study area (54 km²) generally had one or two transients superimposed on a resident population of six animals. Transients appeared essential to the stability of the resident population due to their ability to fill quickly any vacated range. Rollings (1945) estimated catch rates on a known area and then extrapolated these statewide. His estimates were not dissimilar from estimates derived by telemetry nearly 40 years later (Berg 1981).

Miller and Speake (1978a) marked 13 bobcats on a 20.2-km² area, and estimated, with the Lincoln index, a population of 48 bobcats (2.4 per km²).

Gould (1978) developed bobcat population density estimates in California by using trapper catch rates

as an index to density, and compared these with catch rates from areas of known population size. Scent stations and track transects have also been used to estimate population status. However, all of these techniques are only indices that must be tested first in areas of known populations.

The major fluctuations in population levels from one fall to the next are apparently related to the survival of the kittens and nonresidents. Bailey (1974) caught few bobcat kittens after a crash in the rabbit population. However, the resident population appeared to remain stable.

Lynx densities have been observed to fluctuate numerically according to changes in hare densities (Nellis et al. 1972; Brand et al. 1976). Lynx numbers on a 148-km² area in central Alberta decreased from 11 (7.4/100 km²) in 1964–65 to a low of 3 (2.0/100 km²) in 1966–67, then rebounded to 13 (8.8/100 km²) in 1971–72. These high and low points followed within a year after equivalent changes in hare populations (Brand et al. 1976).

Lynx densities in the Soviet Union are greatest in the mixed forests of the central mountains (Iurgenson 1955), averaging 5.0/100 km². Over a 20-year period, lynx densities fluctuated from 1.7 to 5.6 animals per 100 km², with the incidence of lynx distribution closely related to the availability of hares (*Lepus timidus*).

Population Fluctuation (Cycles). Periodic numerical fluctuations have been demonstrated for several vertebrates since this phenomenon was first described (Seton 1911). These fluctuations, or cycles, are defined as patterns "of numerical change in which time intervals between successive highs and successive lows are significantly less variable and hence more predictable than in auto-correlated random fluctuations" (Keith 1974:19). The causes of cycles remain incompletely understood and several speculative causes have been

TABLE 39.4. Bobcat (*Felis rufus*) density estimates for several regions of North America

Region/Locality	Density (Bobcats/km²)	Remarks	Reference
North			
Massachusetts	0.10–0.13	late winter	McCord 1977
Minnesota	0.04–0.06	adults only	Berg 1981
New York	0.06	late winter	Fox 1980
South			
Alabama	0.77–1.16	total estimate	Miller and Speake 1978a
South Carolina	0.13–0.19	total estimate	Kight 1962
South Carolina	0.58	total estimate	Provost et al. 1973
Virginia	0.09–0.18	total estimate	Progulske 1952
West			
Arizona	0.28	total estimate	Jones 1977
San Diego County, California	1.27–1.53	residents only	Lembeck 1978
northeastern California	0.05	adults only	Zezulak 1981
Mojave Desert, California	0.10–0.19	captures only	Zezulak and Schwab 1980
Idaho	0.05	adults only	Bailey 1972
Texas	0.58–2.74	removal studies	Brownlee 1977

advanced, including chance, meteorological events, overpopulation, predator-prey interactions, and human influence (Grange 1949; Moran 1949, 1953; Cole 1954; Lack 1954; Gilpin 1973; Weinstein 1977). A current conceptual model of the "10-year" hare cycle involves synchronic interactions among hare populations, habitat and food supply, continental weather trends, and other demographic and environmental conditions (Keith 1974). Predator cycles apparently reflect the influence of fluctuations in prey populations rather than the direct effects of some environmental condition (Bulmer 1974).

Lynx harvests for the period 1735–1950 have been compiled from the fur returns of the North West and Hudson's Bay companies and the Dominion Bureau of Statistics (Elton and Nicholson 1942; Wing 1953). These meticulous analyses indicate a continuous, persistent, and regular fluctuation in the lynx harvest in boreal North America, with an average periodicity between peak years (1752–1935) of 9.6 years. Cycles in lynx populations are demonstrably associated with the "10-year" cycle in snowshoe hares (*Lepus americanus*) (Keith 1963, 1974). Numerical responses of lynx populations have followed within 1–2 years of similar changes in hare numbers (Brand et al. 1976; Brand and Keith 1979). Declines in hare populations appear to affect lynx numbers by diminishing recruitment through mortality of kittens and reduced conception in yearlings. Dietary changes and decreased consumption rates indicate that lynxes sustain a negative energy balance during periods of hare scarcity, probably resulting in stress from inadequate nutrition, intraspecific aggression, and increased susceptibility to human-related mortality. Cyclic phenomena have not been demonstrated in bobcats.

Lynxes are also susceptible to arhythmic irruptive population surges of variable duration and amplitude. Substantial causal evidence is lacking; however, speculative reasons include lack of food, disease, fire,

and habitat disruption (Gunderson 1978). During irruptions, lynxes may undergo long-range dispersal movements and enter atypical habitat, including cities, steppes, prairie, and tundra. Aggressive behavior and lack of fear of people may also be evident. In recent years, lynxes have irrupted in 1962–63 and 1971–72 in the north central United States and adjacent Canada (Adams 1963; Mech 1973). These irruptions may have survival value by allowing potential reoccupation of suitable habitat and by effecting gene flow between isolated populations (Gunderson 1978). Mech's (1980) studies in northeastern Minnesota, however, indicated that dispersing lynxes occupied very large (51–243 km²) home ranges, showed little productivity, and were susceptible to human-initiated mortality (50 percent of tagged animals retrieved). Sex ratios were initially equal, but became skewed to females about two years postirruption, presumably due to the males' greater vulnerability to harvesting.

HABITAT AND DENNING

Habitat. The bobcat is adapted to a wide variety of habitat types, from swamps to deserts and mountain ranges. The only habitat type that is not used is agricultural land that is so extensive as to eliminate rocky ledges, swamps, and forested tracts. Typical bobcat habitat in the north is generally broken country, including swamps, bogs, conifer stands, and rocky ledges, while in the south bobcats are common in mixed forest and agricultural areas that have a high proportion of early successional stages. Rollings (1945) believed that prey abundance, protection from severe weather, availability of rest areas, dense cover, and freedom from disturbance were the key factors in bobcat habitat selection.

The reasons why bobcats chose certain cover types during the winter were studied by McCord (1974a) on a reservation in Massachusetts. The most highly selected cover was a mixture of roads, spruce plantations, cliffs, and hemlock-hardwood forest. The use of secondary roads and spruce plantations was a matter of energy conservation through the avoidance of deep snow. Bobcats rested and utilized this cover where the least amount of energy would be expended in meeting the energetic requirements of hunting. Bobcats fed on white-tailed deer (*Odocoileus virginianus*) that were primarily killed in the hemlock-hardwood forest, but utilized the spruce plantations to satisfy an apparent physiological requirement to hunt even though food remained available in carcasses of previously killed deer.

Ledges were the most critical terrain. They appeared to be activity centers that also provided protective cover from weather and harassment. Courtship activities were always centered around ledges. McCord (1974*b*) and Bailey (1972) noted that courtship sequences either began or ended in ledge areas; ledges seemed to function as gathering places for the otherwise solitary cats. The importance of ledge areas in Massachusetts to bobcats is demonstrated by the distribution of these cats in the state. The historic range of the bobcat has always been the western two-thirds of the state, where the ledges are fairly common. In the east, where there are no ledges, bobcats were scarce or absent.

Ledges appear to be the most important terrain feature in bobcat habitat in the northern portion of the range, with the only satisfactory replacement being coniferous bogs and swamps. Berg (1981), in a telemetry study of bobcats, found that preferred habitat in Minnesota consisted of areas of black spruce (*Picea mariana*), white cedar (*Thuja occidentalis*), and balsam fir (*Abies balsamifera*) interspersed with quaking aspen (*Populus tremuloides*). No significant seasonal shifts in habitat occurred. Rollings (1945), also in Minnesota, found that bobcat winter habitat was primarily coniferous swamps. In the south, Miller and Speake (1978*b*) found that farming and logging provided food and cover for bobcat prey and that consequently the bobcat showed little aversion to human dwellings or equipment. Hall and Newsom (1976), in the hardwood bottomlands of Louisiana, found that the midsuccessional stages on cutover areas characterized by saplings, vines, and briars were the centers of bobcat activity. The replacement for the ledges and conifer swamps of the north appeared to be thickets of dense briar and palmetto.

The western half of the bobcat's range includes desert, semiarid lands, and mountains from the timber line to below sea level. Grippi (n.d.) found bobcats most common in Fresno Valley, California, between 610 and 1220 m elevation, with the preferred cover types in the eastern portion of the county including woodland-grass, pine, chaparral, and woodland-hardwood.

Lynxes are typically associated with extensive tracts of dense boreal forest interspersed with rocky outcrops, bogs, and thickets. Deep snow and extremely low temperatures may be characteristic of these areas.

In Alaska, lynxes were found in rolling hilly terrain ranging from 300 to 1075 m in elevation and one-half forested with approximately equal portions of coniferous and deciduous forest (Berrie 1973). The coniferous area comprised stands of black spruce underlain by mosses and white spruce (*P. glauca*) with an understory of alder (*Alnus* sp.) and willow (*Salix* sp.). White birch (*Betula papyrifera*), quaking aspen, and cottonwood were major components of the deciduous forest. Mixed types of white spruce–white birch and white spruce–aspen were found along roads. Unforested tundra, comprising one-half the study area, was rarely used by lynxes.

Lynx habitat in western Newfoundland was characterized by second-growth stands of balsam fir, black and white spruce, and white and yellow birch (*Betula alleghaniensis*) interspersed with bogs and shrub barrens (Saunders 1961). Lambkill (*Kalmia angustifolia*) and blueberry (*Vaccinium* sp.) were common on barren areas, with alder predominant on riverbanks and wet areas. Lynxes typically were most active in the forested areas.

Lynxes in central Alberta inhabited areas of mixed habitat approximately 48 percent forest, 33 percent pasture or cropland, 8 percent bog, and 11 percent brushland, marsh, and open water. Typical forest species included spruce, aspen, and balsam fir. Bog species included black spruce, tamarack (*Larix laricina*), bog birch (*Betula glandulosa*), and willow.

In the Soviet Union, lynxes are found chiefly in old-growth taiga, though sometimes on mixed or deciduous forests and wooded steppes, from sea level to about 2,500 m (Iurgenson 1955; Novikov 1962).

Dens and Shelters. The importance to bobcats of rock piles or broken rocky ledges for dens is well documented. The animals also use these areas for refuge, breeding, raising young, and shelter. In California, small rocky areas about 30 m above the desert floor were often used as sanctuaries and denning and resting sites (Zezulak and Schwab 1980). During periods of heavy rain or high temperatures, bobcats used these areas almost exclusively. Bailey (1974) also noted the importance of rock piles and caves for rearing young and refuge in severe weather.

The bobcat also uses brush piles, hollow trees, and logs as rest areas and natal dens (Gashwiler et al. 1961). McCord (unpublished data) has frequently found bobcat beds under low-hanging conifer boughs. Zezulak and Schwab (1980) noted bobcats resting under bushes and next to fallen Joshua trees (*Yucca brevifolia*).

Lynxes give birth in a den in a hollow log, stump, or clump of timber (Seton 1929; Jackson 1961). Murie (1961) found a den in a dense spruce windfall in Alaska. Berrie (1973) reported a den site with five kittens in a tangle of spruce roots washed up on a brook shore. Saunders (1963*b*) reported two instances of adult lynxes sheltering in caves, but did not observe the use of dens by females traveling with kittens. Lynxes frequently bed during daylight hours in thickets near game trails. Beds used for several hours may have tufts

FOOD HABITS

Predators tend to take prey within certain size limits. The bobcat most frequently kills prey weighing between 700 g and 5.5 kg (Rosenzweig 1966). This size range includes rabbits, large rodents, and opossum-sized mammals. The second most frequent prey size group taken comprised those larger than 5.5 kg (beaver and ungulates), followed closely by the 150–700 g category (most squirrels, rats, pocket gophers) and finally those weighing less than 150 g (mice, shrews, voles).

A review of food habit studies shows that the bobcat, like most predators, is opportunistic and will attempt to take almost anything available, including insects, fishes, reptiles, amphibians, birds, and mammals. Mammalian prey, however, is the most important group, with the bobcat best adapted for the taking of leporids. Of 24 areas where bobcat food habits were studied, leporids predominated in 16, were in the top 2 in 21 studies, and were in the top 3 in 23 (Sweeney and Poelker 1977; Miller and Speake 1978a; Berg 1981).

The cottontail rabbit (Sylvilagus sp.) appears to be the principal prey of the bobcat throughout its range. Primary exceptions occur in the northern and eastern parts of the bobcat's range from Minnesota to New England, where the white-tailed deer and snowshoe hare gain in importance. The cotton rat (Sigmodon spp.) may be more important than the cottontail from Florida to Louisiana. In the west, other rodents, especially the woodrat (Neotoma spp.), may be important prey items. If cottontails are abundant, however, they will be an important food source.

Bailey (1972) demonstrated selectivity for cottontails before and after a rabbit population decline in southern Idaho. Roadside counts showed that black-tailed jackrabbits (Lepus californicus) were the most abundant leporid and prey species (92 percent), followed by cottontails. Before the decline, cottontails occurred in 96 percent of 132 feces and 90 percent of 290 prey remains examined. Then, after an 86 percent drop in the cottontail population, 65 percent of the feces (N=55) and 85 percent of the prey remains (N=102) still contained cottontails. Although these cottontails represented only 6 percent of the available leporid population, they were found in 50 percent of all feces and 29 percent of all prey remains. At the same time, the proportions of birds and of rodents in the bobcat diet increased from 6 to 22 percent and from 2 to 14 percent, respectively.

The importance of deer in the diet of bobcats has always been a controversial topic. Many authors (Dearborn 1932; Hamilton and Hunter 1939; Rollings 1945; Pollack 1951a; Erickson 1955a) have attributed its occurrence to scavenging after the deer hunting season. However, the bobcat's ability to kill healthy deer has been well documented (Young 1928; Dixon 1934; Marston 1942; Smith 1945; Dill 1947; McCord 1974a; and others). In McCord's study, deer were plentiful (but alternative prey scarce) and it was not uncommon

for a bobcat to have several of its own kills available simultaneously. These bobcats visited carrion 16 times and cached deer 17 times. They fed on carrion twice (18 percent of visits), as opposed to 79 percent of the time on cached deer. Beasom and Moore (1977) found that deer occurred in the bobcat's diet primarily during periods of low density of alternate prey (rabbits and cotton rats).

The seasonality of bobcat predation on deer is not well defined; however, Erickson (1955a) and Miller and Speake (1978a) both showed deer to be utilized more in winter than in other seasons. This question of seasonality is frequently related to fawn mortality in wild ungulates. Late summer seems to be the time when most fawns are killed, although there is some evidence that selection for fawns continues into the winter. Guthery and Beasom (1977) showed that predator control in south Texas yielded a 70 percent increase in fawn production during one year. Beale and Smith (1973) showed that bobcat predation from June through August was the major cause of mortality (61 percent of losses) of pronghorn antelope (Antilocapra americana) calves in a five-year study. Linsdale and Tomich (1953) reported several attempts on fawns by bobcats during the summer. In Pennsylvania, however, fawn white-tailed deer were killed most often during the winter, when the bobcats preyed almost entirely on deer (Matson 1948). Bobcats in Oregon accounted for 20 percent of the total fawn predation and 10 percent of the total fawn mortality during both summer and winter months (Trainer 1975).

Lynxes are almost exclusively carnivorous and are dependent on the snowshoe hare as a primary food source (Saunders 1963a; Van Zyll de Jong 1966a; Stewart 1973; Brand et al. 1976; More 1976). Lynxes are capable, however, of responding functionally to changes in hare availability. In Alberta, winter tracking studies disclosed a diet of 43 percent hare in a period of hare scarcity and 100 percent in periods of hare abundance (Brand et al. 1976). Based on scat and digestive tract analyses, it can be said that hare comprise 65 percent of the lynx's diet in spring, 52–91 percent in summer, 52–62 percent in fall, and 61–85 percent in winter.

Other foods such as mice, squirrels, grouse, and ptarmigan may be important seasonally or when hares are unavailable. However, in order for alternate foods to contribute meaningfully to lynx nutrition, they must either be present in greater numbers than standard prey or else comprise a greater percentage of the available biomass (Brand et al. 1976). This diversification of food habits is particularly evident in summer. Predation on domestic animals is uncommon.

In Scandinavia, lynxes feed on hare (L. timidus), red fox (Vulpes vulpes), voles and lemmings (Microtidae), shrews (Sorex sp.), domestic cats, carrion, and several species of grouse (Haglund 1966; Pulliainen and Hyypia 1975). Cervids may also be seasonally important. Predation on large mammals may be difficult to ascertain, because scat and stomach analyses rarely differentiate between remains derived from carrion and those from direct predation. Hardy

(1907) and Seton (1929) related several instances of lynx attacks on white-tailed deer. Novikov (1962) reported that roe deer (*Capreolus capreolus*) were a predominant (59 percent) summer food in the Soviet Altai region. Haglund (1966) reported lynx predation on roe deer and reindeer (*Rangifer tarandus*) in Sweden. Reindeer were an essential component of the lynx's winter diet. Attacks on adult moose (*Alces alces*) are almost unknown, although predation on calves may occur (Saunders 1963*a*; Haglund 1974).

Bergerud (1971) studied the population dynamics of caribou (*Rangifer tarandus*) in Newfoundland and found that lynx predation on calves was a major limiting factor in herd growth between 1900 and 1967. The high incidence of lynx predation in Newfoundland, as compared to mainland Canada, probably relates to the defense behavior of caribou, which have adapted for open attack by wolves (*Canis lupus*). Wolves were extirpated from Newfoundland about 1911, while lynxes were rare or absent until the introduction of the snowshoe hare in the late 1800s (Bergerud 1971). Thus, caribou may not have developed defense mechanisms against the lynx's springing attack.

Grass and other vegetation have frequently appeared in bobcat and lynx scats and digestive tracts (Fritts 1973; Stewart 1973; and others). Much of this material is probably taken incidentally at trap sites or while feeding on prey. However, captive lynxes have readily eaten green grass (Saunders 1963*a*). Kight (1962) found that grass was intentionally eaten and constituted 11 percent of the bobcat's diet in South Carolina. Fritts (1973) suggested that this behavior may be purgative rather than nutritional.

BEHAVIOR

Mating Behavior. Gashwiler et al. (1961:78), in March, trapped two female bobcats that "had been chewed around the neck" and that displayed swollen genitalia. The chewed necks were probably an indication of the typical "neck grip" that is a common precopulatory behavior in felids (Antonius 1943; Schaller 1972; Eaton and Craig 1973). Colby (1974) observed bobcats under laboratory conditions and reported that the male grasped the nape of the female exactly as do domestic cats.

The neck grip was also hypothesized as the reason for a small tuft of bobcat hair found by McCord (1974*b*) in three instances of courtship and copulation in wild populations. Interpreting behavior from tracks in the snow, he described copulation sites as areas ranging in diameter from 99 to 114 cm. These contained body impressions in half the areas and tracks plus the tuft of hair in the other half.

McCord (1974*b*) described an elaborate sequence of "pursuit" or "running encounters," "bumping," and "ambushes" which occurred before and after a copulation. The intensity of courtship and copulation decreased after four to five copulation sites. The pair bond appeared to last longer than required just for breeding, however, as the animals appeared to hunt cooperatively and feed on a deer carcass before enter-

ing another ledge. It also appeared that other males not involved with active courtship were also present, and followed along without interfering in the activities. Although the mating system of bobcats has not been described, the evidence points strongly to polygamy, which is the typical system of the asocial felids (Eaton 1976).

Telemetry studies have also shed some light on the behavior of bobcats during the breeding season. McCord (1974*b*) reported a female bobcat that was followed by three other cats. Signs of courtship were evident, and two days later the same female and a marked male mated while another unmarked animal followed along unobtrusively. The presence of several males around a female in estrus raises a question regarding the behavior of males during the breeding season. Lembeck (1978) reported that four of seven males increased their home ranges during the breeding season. He also noted that most males and females were scarred around the face and ears; two males taken in May had recent cuts that appeared to be the result of fights with other bobcats. McCord (1974*b*), however, saw no signs of antagonism or fighting between males that were following an estrous female.

Zezulak (1981) could not rule out extreme aggression being possible in the breeding season. A male bobcat was observed eating another that had died of bites in the throat region, and Erickson (1955*a*) shot a large male that had split ears, gashes, and bleeding feet.

The mating behavior of the lynx has not been fully described. A number of males may court a single female, with fights sometimes occurring among competing males (Hardy 1907; Matjuschkin 1978). Mating may be preceded by periods of yowling (Wayre 1969), with the partially crouched mating posture showing some divergent or vestigial characteristics (Ewer 1968; Lindemann, in Leyhausen 1979).

Lynxes, like certain other felids, display Flehmen, a behavior involving scenting or tasting the urine of conspecifics, followed by a stylized scowl or grimace. This behavior may last about 10 seconds in juveniles, but up to a couple of minutes (perhaps involving multiple sniffs) in adults (Lindemann 1955). Flehmen often occurs in conjunction with sexual activity, but its exact significance is uncertain. However, Verberne (1970) has shown that the urine of receptive female cats elicits a more intense response in males than does that of nonreceptive females. Flehmen may provide a sexual stimulus by strengthening the sensory capabilities of the Jacobson's organ (Ewer 1968; Verberne 1970).

Marking Behavior. The social structure of bobcats appears to be maintained, in part, by a complex system of scent marking involving urine, feces, the anal glands, and enhancements of the excretions by scraping with the feet. Rollings (1945) described "stretching trees"—dry, barkless snags that the bobcat uses by standing on its hind legs and scratching with its front feet. He also observed that feces were usually deposited off a trail at elevated spots and might be covered or uncovered. Pollack (1951b:57) reported that "a female cat will squat to urinate and then usually covers the urine while the male always backs up to an object." Observing captive animals, Erickson (1955a) found that this was not the case. He found that 22 of 44 scat groups in the wild were covered, and that most were along the animal's trail. Young (1958) reported that female bobcats will urinate from an upright position, shooting a jetlike stream backward against a nearby object.

Bailey (1972) was the first to attempt to compile the basic components and concepts regarding bobcat scent marking. Much of our knowledge is preliminary because of the difficulties in learning an area and its bobcat population well, in order to evaluate interactions among animals. Snow tracking with toe-clipped animals is essential once the home ranges of the animals have been established through telemetry.

Bobcats may cover their feces with snow or other material, leave them uncovered on an unprepared site, or leave them uncovered on a prepared site such as a scrape. Scrapes are made with the hind feet and appear somewhat rectangular (25–30 cm × 10–15 cm). Additionally, the anal glands may be used to communicate additional information and the feces may be deposited in a specific location that may have meaning to other animals. These locations have been variously termed "scent posts" (Young 1958), "toilets" (Hawbaker 1965), "depositories" (Kight 1962), "marking locations" (Bailey 1972).

Bobcat kittens, when first emerging from the den, cover their feces. Older kittens and adults usually cover their feces, but a captive 16-month-old female left its feces uncovered significantly more often than a captive male of similar age (Bailey 1972). Adult bobcats tend to cover their feces in random locations while traveling or hunting; uncovered feces and scrapes are more frequent around dens and well-used trails. Bailey (1972) found no evidence of communal marking locations, with most (80 percent) being within the home ranges of marked females or areas used by both males and females. He also found that females marked more intensively when they had mobile, but defenseless, young. Zezulak (1981) found similar behavior when an adult female established at least 10 marking locations around its den.

In trailing bobcats for 117 km in Massachusetts, McCord (unpublished data) found 23 feces, of which 48 percent were covered, 21 percent uncovered on unprepared sites, 17 percent in scrapes, and 13 percent unrecorded as to status. Bailey (1972) found 60 percent covered and 40 percent exposed. McCord found five scrapes with no urine or feces in them and observed only females scraping. Immature and adult animals of both sexes were known sometimes to cover and sometimes to leave the feces uncovered.

Urination sites can be classified as "squirt backward," "squat," (which can be covered or uncovered, but involves only a small amount of urine), and "voiding," which incorporates a squatting position and a large volume of urine. Voided sites may also be covered or uncovered.

Because bobcats did not repeatedly urinate in the same places or marking locations, Bailey (1972) believed that urination probably had limited value for demarcating areas. He reported that the frequency of urine marking for two bobcats was three times in 1.6 km and three times in 0.4 km. Urine marking probably functioned more in helping bobcats to avoid each other than in marking individual areas.

There were 217 urination sites in the 117 km of bobcat trails followed by McCord (unpublished data). Over 89 percent of these urinations were the "squirt backward" type, with 4 percent "squats," 2 percent "voided," and 3 percent scrapes with urine detected. Seven scrapes without urine were made in association with the above; all scrapes were made by an adult female. The 217 urination sites were found on 35 different trails averaging 3.3 km in length; however, over 62 percent of the urination sites were found on only three of the trails. The frequency of urination on these trails was 5.2 urinations/km (range, 3.9 to 9.6), compared to only 0.9 urinations/km for the other 32 bobcat trails followed.

These data show that bobcats generally urine mark only about once per kilometer. However, at certain times the frequency is much greater. Two of the high-frequency trails were made by unknown cats, but the other was an adult female with a well-defined home range. This female was traveling outside its home range for the first time in the study and was moving through the home range of the adjacent adult female. This was the only documented overlap by either cat. These bobcats' home ranges did bound on two widely separated ledges that were used by both females, but never at the same time. The intruding female returned from its excursion, entered one of the ledge areas, and then circled the area, urine marking frequently before entering the ledge.

In Massachusetts, urine marking was used to delineate bobcat territories (McCord unpublished data). In fact, feces were never seen in the manner reported in the western United States. Only two marking locations were known in the study area. Only one dropping and three urinations were documented in these two locations that bordered home ranges and frequently used travel routes in the 117 km of trails followed.

Anal gland secretions may also function as markers, in conjunction with urine and feces. Apparently, the jet of urine in the "squirt backward" posture can be directed across the anal glands in order to pick up the glands' secretions. Bailey (1972) reported that captive bobcats of both sexes sometimes backed up and rubbed a small amount of the secretion on him. He also reported that, on one occasion, the anal gland was discharged on the snow covering the cat's feces. McCord (1974b) reported "trample urination sites," which were apparently anal gland secretions pressed into the snow during a courtship sequence. Eight such sites were found, all of which were associated with courtship.

The development of marking behavior is probably a continuation from the kitten to the transient, and finally to the adult. Bailey (1972) believed that the covered feces, voided urinations, and front foot scrapes associated with these were simply elimination methods, while other excretions and signs had more specific functions in scent marking. Scrapes appeared to have some visual meaning; McCord (unpublished data) found only females scraping. Feces seem to be a longer-lasting sign of the animal's presence as compared to urinations.

Lynxes may urinate frequently on stumps, bushes, and other sites along a trail, sometimes as often as 27–32 times per km (Saunders 1963b). Scats are deposited at random and are covered by kittens, but not by adults. These excretions have been described as territorial markers (Lindemann 1955; Matjuschkin 1978). However, Leyhausen (1979) suggested that his observations contraindicate any warning function and that other communication behavior may be involved. Marking behavior in the lynx has not been fully investigated.

Hunting Behavior. The bobcat hunts basically three sizes of prey: mice, rabbit-sized animals, and deer. The method of hunting these varies greatly, but generally prey may be attacked either when moving or when stationary. Moving prey—such as rodents, rabbits, and squirrels—are hunted by bobcats that frequently stop, sit and wait, or crouch, or from "hunting beds" or "lookouts" (Rollings 1945; Marshall and Jenkins 1966). McCord (1974a) found that the number of beds, sittings, and standing stops was three times higher per km of trail in areas where prey was abundant than in areas where prey was scarce.

Felids, which depend heavily on their sight and hearing to locate prey, are basically stalkers and move in as close to the prey as possible before making a short dash to capture it. When attacking moving prey, however, the bobcat does move into an advantageous position, waiting for the prey to come to him before making an attempt. Hall and Newsom (1976:436) stated that Louisiana bobcats "would often stop and sit in the road, peering intently into the roadside vegetation. Sometimes this position would be held for 5 or 10 minutes before the animal would move. If the object of investigation seemed to be a potential food item, a sitting bobcat would assume a crouched position, often followed by a pounce into the roadside cover." Marshall and Jenkins (1966) reported an instance of similar behavior for a bobcat that killed a cotton rat along the roadside. The bobcat took 13 minutes to move one meter before it trapped the rat with its forefeet.

The hunting bed or "lookout" is often found where bobcats hunt rabbits or hares. The bobcat crouches in an area where game is plentiful and merely waits. The cat will rotate this position, presumably to view a different area. This eventually results in a circular bed showing front paw prints along the edges. Bobcats spend considerable time in these beds, as evidenced by the hair frozen into the packed snow. Attempts at prey are sometimes made directly from the bed, or stalking may be required depending on prey movements. Generally, the bobcat is within 10 m of its prey before a quick rushing attack is made.

When hunting prey that is generally stationary—like bedded deer—the bobcat moves more frequently. Marston (1942), Erickson (1955a), and McCord (1974a) all found that the bobcat usually attempted to take bedded deer. McCord (1974a) noted that six of nine attempts at deer were on bedded animals. McCord (unpublished data) found where a bobcat chased a deer for 70 m and Stiles (1961) reported where a bobcat had leaped from an overhanging tree onto a deer's back.

Whether deer are killed by bobcats through strangulation, as suggested by Linsdale and Tomich (1953), or through hemorrhaging or other means, is still uncertain. McCord (unpublished data) found that bobcat-killed deer were bitten in three general areas: the throat, base of the skull, and chest. The chest area at the base of the throat is rich in blood vessels, and sometimes a few bite marks were found there in addition to bites in other spots. The base of the skull was more frequently bitten, and, on some deer, was the only area of attack. Apparently, the bobcat is able to reach the spine in the region of the atlas and axis. The throat, however, was almost always a focus of attack. Generally, the area just behind the angle of the jaw was punctured many times. Matson (1948) found that the trachea had been cut or pulled apart on several deer killed by bobcats.

The bobcat is able to bite very rapidly. McCord (unpublished data) was attempting to feed a five-month-old captive kitten when the animal fell on his hand and started to bite. Despite a reflexive withdrawal of the hand, the cat bit to the bone three to four times. This biting speed presumably enables an adult bobcat to pulverize the throat, including the major blood vessels and the trachea, in a matter of seconds.

The hunting technique of the lynx may include following game trails, concentrating movements in prey activity areas, and lying in wait in "hunting beds" overlooking areas of prey activity (Brand et al. 1976). Once located, prey is stalked and pursued or pounced upon from a place of concealment. The reliability of the different techniques may be related to prey abundance. Concentrating movements in prey "pockets" is probably most advantageous when overall prey populations are low (Berrie 1973; Brand et al. 1976).

A bobcat sometimes caches or covers its prey with whatever material is available. The animal's forefeet are used to scrape snow or leaves over a carcass. Sniffing of the site usually accompanies this covering behavior. Cached deer generally have large amounts of hair removed from the carcass mixed with the covering material (figure 39.5).

Caching of prey occurs inconsistently in lynxes. Seton (1929) believed this habit to be well developed. Nellis and Keith (1968) found that in 1965–66, 6 of 11 (54 percent) hares taken in 418 km of trailing were cached and six old caches were revisited. Pulliainen and Hyypia (1975) reported that lynxes returned to old kills in 10 of 34 instances. Haglund (1966), however, indicated that lynxes display weak caching behavior, with small animals covered carelessly and larger prey merely abandoned. Lynxes commonly returned to roe deer kills at least once but rarely returned at all to reindeer carcasses.

Lynx hunting success rates for all prey species ranged from 21 percent in Alberta (Brand et al. 1976) to 40 percent in Finland (Pulliainen and Hyypia 1975). Hunting success may depend heavily on the bearing strength of the snow (Haglund 1966; Nellis and Keith 1968), with crusty or compacted snow permitting a quicker pursuit.

Cooperative hunting in bobcats has been reported by McCord (1974b:80). A pair of bobcats, which were tracked in snow, had gone through courtship and copulation prior to entering a series of spruce plantations where snowshoe hares were common. In the last four areas, "the cats moved through the plantation about 10–15 meters apart and appeared to alternate stopping as the other moved forward 15–20 meters."

Cooperative hunting efforts also have been recorded in lynxes. Saunders (1963a) tracked two lynxes that unsuccessfully attacked a group of caribou in Newfoundland. Barash (1971), in Montana, observed a hunt involving two adults and a nonparticipating juvenile lynx. Haglund (1966) described both "training" hunts involving adults and young and cooperative hunts between two adults. Similarly, Matjuschkin (1978) recorded communal hunting of large prey by lynxes in the U.S.S.R.

Play. Play behavior is, basically, a precursor of adult patterns in which neural responses develop prior to the need or opportunity to use them (Ewer 1968). Leyhausen (1979) described and categorized the development of these play patterns in felids. Lindemann (1955) described combative and prey-catching behavior in lynx kittens as indicative of the major role of these activities at maturity. Saunders (1963b) also observed mock chasing and fighting among his captive young lynxes. Similar play behavior in captive bobcats was described by Young (1958).

Voice. Lindemann (1955) discussed the ontogeny of vocalizations in young lynxes, including sounds indicating hunger, discomfort, fear, and satisfaction. Certain of these sounds may change as the animal matures; the nature and function of lynx vocalizations should be clarified. Adults may purr and mew when tending the young and growl and hiss when disturbed or threatened. During mating, female lynxes may utter a loud yowl (Wayre 1969) or "raucous loud meow" (Matjuschkin 1978:99) followed by a deep growl or a muted purr from the male. The so-called scream or appalling cry (Seton 1929; Jackson 1961) attributed to lynxes seems distorted by exaggeration. The bobcat has a similar repertoire of vocalizations (Young 1958).

Weather Relations. Snow depth appears to be a major factor in limiting the bobcat's northward range, although the length of time the snow remains soft after a storm is also important. McCord (unpublished data) observed bobcats traveling through snow 38 cm deep, although they frequently would bound in efforts to continue moving. McCord (1974a) reported significant differences in snow avoidance behavior, such as walking on plowed roads, snowmobile paths, deer trails, and logs when the snow exceeded 15 cm in depth. Marston (1942), Matson (1948), and Bailey (1972) also observed that bobcat movements were restricted in deep snow.

The importance of snow depth in bobcat and lynx distribution was demonstrated on Cape Breton Island, Nova Scotia. Before 1955 the island had only a lynx population, but with the building of a causeway to the mainland, the intervening strait began to freeze, thus allowing bobcats access probably for the first time (J. Parker personal communication). The bobcat dramatically increased in the lowlands where snow depths were minimal, but did not establish populations in the northern highlands section where snow depths are much greater. Lynxes are now found only in the highlands. Apparently, the difference in snow depths between the highlands and lowlands is the only factor to explain the distribution of the two species.

Even though deep snow may limit bobcat movements, it is perhaps more important in the restrictions it places on their ability to obtain food. Golley et

FIGURE 39.5. White-tailed deer (*Odocoileus virginianus*) cached by a bobcat (*Felis rufus*).

al. (1965) found that the bobcat can maintain its body weight on an alternating feast-famine regimen; however, there must be a point in northern areas where the snow stays powdery longer than the average famine period that a bobcat can survive. Such a break point was apparently reached in northern Minnesota, where during six winters there were three separate periods when the snow was loose and deep and the weather was persistently cold for periods of four to five weeks. Petraborg and Gunvalson (1962:430) described the first period in January 1950 as follows: "At this time bobcats were frequently seen along highways and were found raiding farmers' chicken houses. More bobcats than usual were killed along highways with clubs and guns. Those examined by game personnel were . . . emaciated and their stomachs empty."

In 1953, the second period caused bobcats to enter plowed roadways and to refuse to yield to pulp haulers, who either had to run over them or kill them. Again, in 1954, similar snow conditions existed. Petraborg and Gunvalson (1962:430) reported that "animals were being clubbed to death by loggers and killed by cars on roadways and there were many reports of animals coming into settlements, crawling under porches, sneaking into outbuildings, and preying on house cats."

This need for food during the period after a deep snow may also explain why deer seem to be more important in the bobcat's diet in the northern part of the range. A cached deer would be able to supply food for a bobcat that could not hunt effectively but that could travel to and from the carcass. Also, any extension of the bobcat's range in the north may relate to man's creation of travel routes (snowmobile paths, roads) that could be used in periods of deep snow as easy travel routes for the bobcats. This assumes, of course, that such routes also intersect areas where bobcat prey is readily available.

The relative compactness or "bearing strength" of snow is important to lynxes also (Haglund 1966; Nellis and Keith 1968). North American lynxes may be better adapted to snow than those in Eurasia. The weight load for Alberta lynxes ranged from 34 to 38 g/cm² of foot surface area (Van Zyll de Jong 1966*a*), compared to 34–60 g/cm² in the U.S.S.R. (Iurgenson 1955).

Other Behavior. Bobcat-water relationships have generally been considered to be one of necessity rather than affinity. Young (1958:24) stated, "The bobcat will take to water if necessary, particularly if crowded by dogs." However, Yoakum (1964) described the behavior of semitame bobcats that played in water and caught fish. Lynxes are also capable swimmers (Seton 1929).

Bailey (1936) and Ognev (1962) allege that, in rare instances, desperate or starving lynxes have attacked people. These reports should be treated with caution. In two instances, however, lynxes have pounced upon men from trees, perhaps indicating confusion with natural prey (Hardy 1907; Hancock et al. 1976). Confusion and stress during irruptions may also precipitate encounters (Santy 1964).

MORTALITY AND LIMITING FACTORS

Mortality. Mortality can be categorized as prenatal, postnatal (kittens and juveniles), and adult. Prenatal mortality estimates are made from comparisons between the number of eggs ovulated (corpora lutea) and the number of fertilized and implanted ova (placental scars), or between the numbers of embryos and live births. Spontaneous ovulations could cause inflated corpora lutea counts in the former estimate. Fritts and Sealander (1978) found that only 59 percent of bobcat ova were fertilized and implanted. To date, however, there are few studies of *in utero* fatalities and stillbirths in both the bobcat and the lynx.

Kitten mortality is defined as those deaths occurring between birth and the withdrawing of maternal support. The most important factor is the availability of food. Bailey (1972) found few bobcat kittens after a crash in prey populations. He speculated that during periods of food scarcity the female would have to expend more energy to secure prey and would therefore be consuming most of its catch solely to maintain its own physical condition. The starvation of kittens appears to be the major factor; however, reductions in birth rates may also occur. Bailey (1972) found two resident female bobcats that either had lost their kittens early or did not have any due to prey scarcity. Similarly, Zezulak and Schwab (1980) found that radio-collared female bobcats did not reproduce, and no kittens were captured. They believed that the primary reason for this reproductive failure was related not to food supply, but instead to some population regulatory mechanism related to the high density of bobcats and the incidence of conspecific interactions. Miller (1980) speculated that in high-density populations with abundant prey, failure to breed, perhaps due to delayed maturity in females, affected population levels more than kitten survival.

Juvenile mortality occurs between the withdrawal of maternal support and the establishment of a home range territory (including participation in reproductive activities). This period coincides with the transient or nomadic phase described earlier. Although most previous studies on bobcat population mortality have combined kitten and juvenile mortality, our definitions are more suited to describing mortality mechanisms rather than the total number of young entering the next year's population. Both methods are valid and useful, however.

The major factor in juvenile mortality also appears to be food supply, but, instead of depending on the mother to bring food, the juvenile must depend on his own hunting skills. Crowe (1975*b*) believed that, in bobcats, the attainment of hunting proficiency was mostly a matter of experience. If inexperienced juveniles are successful only 10 percent of the time and prey is scarce, the length of time between catches may become too great. The length of time the bobcat remains inexperienced puts it in double jeopardy as winter approaches and food supplies decline. If, on the other hand, food supplies are abundant, the juvenile can become skillful more quickly and, therefore, better able to deal with reduced prey levels.

Observations of juvenile mortality are uncommon; however, Zezulak (1981) reported that two of three radio-collared juvenile bobcats died of malnutrition and parasitism in a population that was generally protected from hunting and trapping. He also reported an adult male bobcat eating another bobcat, which it had apparently killed, as evidenced by wounds in the victim's neck.

Crowe (1975b) estimated survival rates for juvenile bobcats and found that the rates varied from 0 to a high of 71 percent. This estimate is really a combination of prenatal and postnatal survivorship, as defined above. The methodology involves the extrapolation of population structure from a known sample of juveniles. For example, 30 juveniles (based on canine root closure) are identified in a sample of 100 bobcats, leaving 70 adults. Assuming a 50:50 sex ratio, then 35 are adult females. If the average litter size was determined to be 2.79 young/female, the expected production for the 35 females would be 98 young. However, the sample showed only 30 young. Thus, the survival rate is estimated by dividing the observed number of young by the expected (30/98), yielding 0.31, or 31 percent, survival.

Crowe (1975b) found no evidence of differential mortality between juvenile male and female bobcats. However, Fritts and Sealander (1978), attempting to explain the imbalanced sex ratio (169M:100F) in their sample of 180 bobcats, speculated on a higher female mortality in both kittens and juveniles. They believed that competition among kittens would favor males and that females would have greater difficulty feeding themselves as juveniles. This suspected sex-differential mortality has not yet been demonstrated, however.

Adult mortality in an unexploited bobcat population was first documented by Bailey (1972). He found a maximum of 3 resident adults (in a population of 35 resident bobcats) in three years that died of causes unrelated to people. This amounts to about 3 percent annual mortality. Lembeck (1978), on an unexploited area, recorded 13 bobcat deaths from disease, starvation, predation, human influence, and unknown causes. Nine of the 13 deaths arose from natural causes, representing a greater rate of natural adult mortality in a denser population. Deaths of radio-collared adults in an exploited population in Minnesota (Berg 1981) showed 14 from human causes, 2 from porcupine (*Erethizon dorsatum*) quills, and 1 from unknown causes. Petraborg and Gunvalson (1962) recorded widespread starvation losses of bobcat adults.

Crowe (1975b) reported a 67 percent survival rate among adult bobcats in an exploited population. Survivorship (S) for each age group was calculated by the formula $S = pX/pX - 1$, where pX is the number of individuals in age class X. A linear regression was used to calculate a constant survival rate for all age classes. Fredrickson and Rice (1981), using a time-specific life table, showed 60 percent survival in an exploited population in South Dakota.

Demographic fluctuations in lynx populations have followed similar changes in snowshoe hare populations. During periods of hare scarcity, lynxes may undergo a negative energy balance, with resultant decreases in conception rates and increases in kitten mortality. In Alberta, the adjusted ratio of kittens in a trapped sample dropped from 66 percent during 1971–72, when hares were abundant, to 3 percent in 1973–76, when hares were scarce (Brand and Keith 1979). Berrie (1973) also related kitten survival to lynx population growth. Causes of kitten mortality may include both starvation and intraspecific aggression; losses of adults are also starvation related, including increased susceptibility to trapping and other human contact (Brand et al. 1976).

Trapping pressure has been frequently implicated in changes in lynx numbers and distribution (Butler 1942; De Vos and Matel 1952; Nellis 1971; Van Zyll de Jong 1971). This may be critical when lynxes are at or near a cyclic low (Berrie 1973; Brand and Keith 1979). Nellis et al. (1972) reported that 3 of 9 marked lynxes in 1964–65 were killed within one year, while in 1965–68 only 1 of 8 were recovered. They attributed the higher mortality in 1964–65 to decreased prey populations and severe weather conditions. Brand and Keith (1979) recorded average winter trapping mortality rates of 10 percent (of fall lynx populations) in 1964–67 and 17–29 percent in 1973–75. Trapping mortality was believed to be density dependent, directly related to pelt price, and additive to nontrapping mortality. Trapping and nontrapping mortality did not appear to be related, however, and lynxes would presumably decline during hare lows in the absence of trapping, though to a lesser degree than when continuously trapped.

Sex and Age Ratios and Survivorship. Sex ratios for bobcats and lynxes show a wide variation, ranging in bobcats from 40M:100F (Foote 1945) to 199M:100F (Young 1958), with just as much variability in reasons explaining these differences (table 39.5). The most common viewpoint explaining male predominance is that males are more vulnerable to harvest because of their larger home ranges and greater movements, which presumably increase the chance of their encountering a trapper or hunter. Harvest seasons that include part of the breeding season have also been used to explain a male-dominated sex ratio because of the increased activity of sexually aroused males. Lembeck (1978) reported a higher average catch rate per individual for male bobcats than for females, which illustrates this bias. However, Lembeck was really reporting the recapture rate for 15 animals, with 3 of those accounting for over one-half (18 of 31) of the recaptures.

One major source of variation in reporting sex ratios is misidentification. Vermont bobcat harvests consistently showed a take of 65–80 percent females for three years until warden and field personnel reports were verified by internal examination. Subsequently, in 1979–80, Vermont biologists found a 17 percent error rate in field reports, resulting in an actual ratio (53 percent females) nearing equality (W. Cottrell personal communication). Notably, the most widely

TABLE 39.5. Reported sex ratios for bobcats (*Felis rufus*) and lynxes (*F. lynx*) for several areas of North America

Males/ 100 Females	Sample Size	Method of Sexing	Area	Reference
Bobcat				
40	351	unknown	Vermont	Foote 1945
105	180	internal	Massachusetts	Pollack 1949
87	144	internal	Virginia	Progulske 1952
154	6496	bounty records	Michigan	Erickson 1955*a*
119	103	internal	Michigan	Erickson 1955*a*
108	150	unknown	New Mexico	Young 1958
126	8703	state records	Arizona	Young 1958
299	315	trappers	Oregon	Young 1958
111	792	government hunters	Utah	Gashwiler et al. 1961
128	28,423	government hunters	unknown	Gashwiler et al. 1961
91	66	sexed by author	Idaho	Bailey 1972
101	161	internal	Wyoming	Crowe 1974
170	180	internal	Arkansas	Fritts and Sealander 1978
210	16	sexed by author	California	Lembeck 1978
110	761	internal	Washington	Brittell 1978, 1979
137	2443	field personnel	Colorado	Donoho et al. 1979
129	190	internal	Minnesota	Henderson 1981
64	1381	field personnel	Wisconsin	Klepinger et al. 1981
132 juvenile	102	internal	Minnesota	Berg 1981
117 adult	113	internal	Minnesota	Berg 1981
101	213	internal	Alabama	Miller 1980
Lynx				
93	286	reports	Newfoundland	Saunders 1961
137	896	internal	Alaska	Berrie 1973
100 kittens	30	internal	Ontario	Stewart 1973
525 yearlings	25	internal	Ontario	Stewart 1973
153 all ages	96	internal	Ontario	Stewart 1973
92	974	internal	Alberta	Brand and Keith 1979

skewed ratio reported was also from Vermont (Foote 1945). The most frequent error in sex determination appears to be identifying males as females. Fredrickson and Rice (personal communication) originally reported a 50M:100F ratio, but after verifying field checks with internal examination found a nearly equal sex ratio. Some errors may also be made in describing females as males, and, although adults are sometimes misidentified, juveniles are more frequently a source of error. These errors in field reports make it imperative that sex ratios be derived solely from internal examination by qualified personnel. It is also noteworthy that Henderson (1981) found such differences in sexing between pelt examination and internal checks that she determined that bobcats could not be reliably sexed by pelt characters.

The theory that male bobcats are more vulnerable than females due to larger home ranges or more intense breeding season activity is also suspect. Colorado has a bobcat hunting and trapping season extending from October through March. This covers the period before and during the breeding season. If males were vulnerable, the sex ratio should favor males later in the harvest season. However, a chi-square analysis of the data of Donoho et al. (1979) of over 2,300 males and females taken by week and month of harvest showed no significant difference. Similar analyses were done by month for 960 bobcats taken in the 1977/78 season with simi-

lar results. Assuming consistent sexing errors by field personnel, these data show no support for increased male vulnerability due to the onset of breeding. Erickson (1955*a*) and Gashwiler et al. (1961) showed some changes in sex ratios over a yearly period, but neither of these showed any real variation.

Differences in the size of the home range and movement patterns between males and females can be viewed two ways. Male bobcats generally have larger home ranges than females and therefore move greater distances in covering their range, while females use their smaller ranges more intensively (Bailey 1972). The probability of a trapper setting in a male bobcat's range may thus be greater, but the probability of an animal's encountering a trap once set would be greater for the female.

Selectivity for sex based on the methods of harvest also may be a source of variability in bobcat sex ratios. Trapping and hunting with dogs are the most successful methods of taking bobcats and usually account for the majority of the harvest. Brittell (1978, 1979) presented comparisons between sex and type of harvest in Oregon. These data (table 39.6) show no significant difference in sex ratios between hunting and trapping take. Brittell's sex ratios were not significantly different when tested separately against an assumed 50:50 sex ratio. The evidence does not generally support the assumption that skewed sex ratios in

TABLE 39.6. Number of male and female bobcats (*Felis rufus*) taken by hunting with dogs and trapping during 1976–79 in western Washington

Years	Sex	Method of Harvest		Totals	χ^2 (1 d.f.)
		Trapped	Hunted		
1976–77	Male	43	49	92	0.5491
	Female	43	41	84	
		86	90	176	
1977–78	Male	43	33	76	1.3108
	Female	42	23	65	
		85	56	141	
1978–79	Male	76	101	177	0.5979
	Female	74	85	159	
		150	186	366	
1976–79	Male	162	183	345	1.6109
	Female	159	149	308	
		321	332	653	

SOURCE: Brittell 1978, 1979.

bobcats are caused by the increased vulnerability of a particular sex due to the time of year or season or the harvest method. In fact, it appears as if samples collected through hunting or trapping seasons are accurately reflecting the sex ratios in the population. Lembeck (1978) provided a field test of this tenet in studies at the El Capitan Reservoir. Resident bobcats, studied by telemetry, before the trapping season showed a sex ratio of 201M:100F, while that of those captured in the study area during a time corresponding to the permitted trapping season was also 201M:100F. Thus, sex ratios may be adequately defined by combining trapper and hunter takes.

Bailey (1974) mentioned that bobcat kittens were more easily trapped and recaptured than were adults. However, other studies do not support that statement. Robinson and Grand (1958) initially captured and tagged a sample consisting of 80 percent adults and 20 percent kittens, with recapture rates of almost equivalent percentages. Crowe (1975b) implied no age selectivity due to trapping.

Chi-square tests were used to detect differences in age structure between the male and female samples in Crowe's (1975b), Fritts and Sealander's (1978), and Brittell's (1979) data (table 39.7). Generally, age classes 0–1 to 6+ were tested, except that Crowe (1975b) used age classes 0–1 to 5+. None of the tests was significant, showing no differences in age structure due to sex in these studies. If either hunting or trapping tended to select for specific age classes, it would not be valid to combine samples of animals taken by different methods. Brittell's (1978, 1979) work provides the opportunity to test age and sex selectivity in samples taken only by trapping and only by hunting. Table 39.8 shows these data; neither the trapped nor the hunted sample showed any selectivity for age groups based on sex. This means that the sex of the bobcat has no impact on the age structure derived from

samples collected by hunting alone or trapping alone. However, it does not answer the question about the advisability of combining samples of animals taken by the two methods and extrapolating this to an estimate of the population age structure.

Since there were no differences in age structure by sex in trapped or hunted bobcat samples, the sexes were combined. A chi-square test of the age structure by method of harvest showed a significant difference ($\chi^2 = 26.46$, 6 d.f.). The percentage of bobcats in each age class shows that hunting tends to take a higher proportion of juveniles (0–1 age class) than trapping, which shows a general trend toward older animals. These findings mean that, although trapped and hunted animals can be combined to assess sex ratios, bobcat age structure analyses should preferably be done on samples collected by similar techniques.

In lynxes, Stewart (1973) found equal sex ratios in all age classes except 1–2 years, in which there was an excess of males. This difference was believed due to dispersal behavior by the young males. Berrie (1973), however, recorded 14 females and 7 males (50M:100F) in a live-trapped sample. This disparity was similarly attributed to sex-related differences in movements and behavior.

Berrie's (1973) survivorship studies of Alaskan lynxes demonstrated that females lived longer than males from 1964/65 through 1967/68, but reversed this relationship in the three seasons thereafter. This indicated, as expected, a greater proportion of females than males during an expanding population. Brand and Keith (1979), during a hare decline in Alberta, found no significant difference from equality (92M:100F) in a trapped sample of 974 lynxes.

Stewart (1973) found that kittens comprised from 5 to 35 percent of the total on his three Ontario study areas in 1971/72. These differences may have reflected variable trapping pressure and a differential vulnerabil-

TABLE 39.7. Age distribution, by sex, of bobcats (*Felis rufus*) taken in five states

| | Wyoming (Crowe 1974) | | | Minnesota (Berg 1981) | | | South Dakota (Fredrickson and Rice 1981) | | | Washington | | | | | | Arkansas (Fritts and Sealander 1978) | | |
| | | | | | | | | | | 1977–78 (Brittell 1978) | | | 1978–79 (Brittell 1979) | | | | | |
Age Class	M	F	Total	M	F	Total	M	F	Total	M	F	Total	M	F	Total	M	F	Total
0–1	29	29	58	49	39	88	21	38	59	19	11	30	48	38	86	20	11	31
1–2	21	23	44	18	12	30	8	15	23	33	22	55	57	41	98	18	11	29
2–3	14	14	28	9	12	21	2	7	9	22	16	38	36	38	74	19	10	29
3–4	6	5	11	3	5	8	2	5	7	3	6	9	21	11	32	16	12	28
4–5	3	1	4	4	6	10	0	4	4	2	8	10	6	7	13	14	5	19
5–6	1	3	4	4	0	4	0	2	2	3	0	3	3	7	10	11	4	15
6–7	1	2	3	1	0	1	1	2	3	1	1	2	3	5	8	3	4	7
7–8	2	1	3	1	2	3	0	1	1	1	2	3	4	3	7	2	4	6
8–9	2	0	2	1	0	1	0	0	0	0	0	0	1	2	3	1	0	1
9–10	0	2	2	3	0	3	0	1	1	1	2	3	1	3	4	1	3	4
10+	1	1	2	0	0	0	0	0	0	0	0	0	0	4	4	4	0	5
Total	80	81	161	93	76	169	34	75	109	85	68	153	180	159	339	110	64	174

ity to trapping on the part of kittens. Berrie's (1973) investigations of trapped Alaskan lynxes indicated that the kitten cohort declined from 1964/65 through 1966/67. It increased thereafter from 3 percent in 1967/68 to 30 percent in 1969/70. Adults, on the other hand, increased and decreased inversely to kittens. These data suggested a decrease in the lynx population from 1964/65 to 1966/67 and an increase thereafter.

In Alberta, the adjusted percentage of kittens in trapped samples decreased from 66 percent in 1971/72 to 3 percent in 1973/76 (Brand and Keith 1979). This decline in recruitment was accompanied by a shift toward older age classes, with the mean age of trapped lynxes rising from 1.6 in 1971/72 to 3.6 in 1975/76. The diminished recruitment was attributed to the mortality of kittens resulting from severe decreases in available food. Age distributions in trapped samples did not seem to reflect actual population structure, due to disparities between modeled and trapped popula-

TABLE 39.8. Age structure of male and female bobcats (*Felis rufus*) taken in western Washington grouped by method of harvest

| | Trapped | | Hunted | |
Age Class	Male	Female	Male	Female
0–1	21	20	41	26
1–2	37	29	48	34
2–3	38	31	19	22
3–4	10	7	13	10
4–5	3	8	5	7
5–6	5	7	1	0
6+	5	14	6	7
Totals	119	116	133	106

SOURCE: Brittell 1978, 1979.

tions, and field surveys and trapped populations. The disparate representation of kittens in trapped samples was possibly due to dependence on their mother for food. True age distributions were estimated from yearling ratios, which should reflect the kitten component of the previous year (Brand and Keith 1979:840).

Predators. Bobcats are not commonly preyed upon. Kittens may be taken by foxes, owls, and adult male bobcats (Young 1958; Crowe 1975b). Young (1958) and Lembeck (1978) also reported predation on adults by mountain lions (*Felis concolor*) and coyotes (*Canis latrans*).

Pulliainen (1965) claimed that wolves were the most important natural enemy of the lynx in Finland. He attributed increases in lynx numbers and population density after the 1880s to a rapid decline in wolf numbers a few years previously. Matjuschkin (1978) demonstrated the same relationship in the Lake Baikal region, U.S.S.R.

Sheldon (in Murie 1961) and Matjuschkin (1978) cited instances where lynxes abandoned their prey to a wolverine (*Gulo gulo*). Rausch and Pearson (1972) found lynx remains in a wolverine scat in Alaska, as did Matjuschkin (1978) in the U.S.S.R. It is uncertain, however, if this represented direct predation or scavenging. In Alaska, Berrie (1973) found a livetrapped lynx that had been killed by a wolverine.

Feral dogs were reported to kill lynxes in Newfoundland (Saunders 1961). The larger Eurasian lynx, however, is capable of fierce resistance against dogs (Ognev 1962). Bears (*Ursus* sp.) are powerful enough to kill a lynx, although there is no record of their doing so. The Siberian tiger (*Panthera tigris*) may rarely kill a lynx (Matjuschkin 1978).

The bobcat and lynx may also be killed or injured by prey animals. The defense behavior of large ungulates toward bobcats has been observed in deer and bighorn sheep (*Ovis canadensis*). Young (1958) reported an instance where a bobcat was injured by a white-tailed deer. Linsdale and Tomich (1953) reported that mule deer located bobcats by scenting and sight, with does being particularly intolerant of the cat's presence. Numerous observations were made of doe mule deer chasing and treeing bobcats. Most chases did not extend beyond about 90 meters. The deer lost interest after the bobcat treed, allowing it to come down soon after the deer wandered off. Hornocker (1969a) observed three ewe bighorns advance abreast and chase away a bobcat that had pursued a lamb in their group.

Matjuschkin (1978) reported that about 13 percent of 285 lynxes taken in the Ural Mountains had old injuries to the skull and legs, presumably as a result of struggles with prey.

Competition. Competition between coyotes and bobcats has long been discussed but little studied. The general attitude has been that coyotes are more competitive than bobcats and will cause a decrease in their populations. Robinson and Grand (1958) noted that bobcats had increased over much of the west, particularly where coyotes had been eradicated. Bailey (1972) provided detailed information from Idaho, where coyotes outnumbered bobcats. Coyotes showed more versatility in their food habits and took proportionately more jackrabbits than did bobcats. He believed the difference in the jackrabbit capture rate between the two predators was due to their different habitat preferences and hunting methods. The bobcat, which stalked or still-hunted, tended toward rocky terrain, while the coyote, a chaser, worked the more open areas. The natal dens and rest areas of the two predators were also different, with the coyote enlarging badger (*Taxidea taxus*) holes and finding shelter in open areas, while the bobcat used rock piles and caves. Bailey (1972) concluded that there was little competition between coyotes and bobcats and that both species could exist on the study area sympatrically.

Habitat Changes. The influence of humans on bobcat habitat does not seem excessively great. Bobcats have been able to maintain their historic range in Massachusetts—one of the four most densely populated states in the nation—even though there is scattered urbanization throughout the cat's range. The key to this is the presence of ledges, which are critical habitat features for the bobcat. If these ledges are present, and not completely surrounded by urban sprawl, the bobcat has proved adaptable enough to utilize whatever other natural habitat remains. Miller and Speake (1978b) found that instrumented bobcats intensively used recently logged areas, because logging and farming practices provided food and cover for prey species. These bobcats showed little or no aversion to human dwellings or equipment; in fact, one frequently rested within 61 meters of an occupied dwelling. Resting bobcats responded to human disturbances, motor vehicles, and logging activities by moving a short distance and resuming their rest (Miller and Speake 1978c). Hall and Newsom (1976) found similar responses by bobcats to human activity. Incidental mortality related to human activities was noted by Bailey (1974), who reported six kittens that were electrocuted after climbing power transmission lines.

Habitat changes hold potential for impacting the lynx. Declines in lynx numbers in several areas of North America have been attributed to these causes (De Vos and Matel 1952; Jordahl 1956). Agricultural land clearing was implicated in the loss of lynx habitat in Europe up through the 1940s. Iurgenson (1955) regarded the diminishing area of forested mountains as the primary cause of lynx scarcity in the Central Zone, U.S.S.R. Matjuschkin (1978), however, found little evidence of human impact on lynxes in most of Asia, but considered that human actions and available habitat played a major role in defining the southern and western boundaries of lynx range in Russia.

Increased timber harvests and a shorter rotation period may result in a proliferation of successional or second-growth stages that will benefit hare populations and also the lynx. Increased hare populations, however, may result in unacceptable damage to seedlings and nursery stock, with resultant intensive hare control efforts and a potentially associated decline in lynxes (Nellis 1971). Forest fire control may also adversely affect lynxes by permitting forest maturation and consequent reduction of hare habitat (Burris 1971).

The expanding road networks associated with oil and mineral development, a swelling human population, and the growing popularity and availability of snowmobiles may indirectly affect lynx populations by increasing accessibility by trappers and decreasing the time necessary for trapline coverage (Berrie 1973).

Parasites and Diseases. Bobcat and lynx populations have not succumbed to epizootics or die-offs due to heavy parasitic infestations. They may be less vulnerable to this type of mortality due to their solitary nature and their propensity to change denning and resting areas frequently.

Documentation of diseases in wild populations of these felids is particularly scarce. There are 12 infectious diseases that have been shown to produce lesions or antibodies in wild bobcats (M. Placke personal communication). Panleukopenia, rabies, and leptospirosis are known to have produced lesions, and feline infectious peritonitis, rhinotracheitis, feline infectious anemia, brucellosis, feline leukemia, vesicular stomatitis, salmonellosis, toxoplasmosis, and sarcocystitis have been documented through serum antibody tests. Lembeck (1978) mentions that 2 of 31 bobcats on his study area died of panleukopenia (feline distemper); Povey and Davis (1977) reported this disease in captive lynxes. Additionally, lynxes are susceptible to rabies (Lewis 1975; Matjuschkin 1978), pasturellosis, and the mycoplasma *Acheleplasma laidlawii* (Langford 1974), which may also occur in bobcat.

The internal parasites of lynx include tapeworms, flukes, roundworms, and spiny-headed worms (table 39.9). In Alberta and MacKenzie, Van Zyll de Jong (1966b) found the most common parasite to be *Taenia*

TABLE 39.9. Endoparasites of the lynx (*F. lynx*) in North America

Parasite	Site in Host	Locality	Reference
Tapeworms (Cestoda)			
Taenia sp. [*T. monostephanos?*]	small intestine	British Columbia	Adams 1966
T. laticollis	small intestine	Alberta/MacKenzie	Van Zyll de Jong 1966*b*
T. rileyi	small intestine	Alberta/MacKenzie	Van Zyll de Jong 1966*b*
T. rileyi	small intestine	Minnesota	Van Zyll de Jong 1966*b*
T. pisiformis	small intestine	Alberta/MacKenzie	Van Zyll de Jong 1966*b*
T. krabbei	small intestine	North America	Fyvie and Addison 1979
Multiceps sp. [*T. multiceps*]	small intestine	Alberta/MacKenzie	Van Zyll de Jong 1966*b*
Flukes (Trematoda)			
Alaria sp.	small intestine	Alberta/MacKenzie	Van Zyll de Jong 1966*b*
Roundworms (Nematoda)			
Toxascaris leonina	small intestine	Alberta/MacKenzie	Van Zyll de Jong 1966*b*
Toxocara cati	small intestine	Alberta/MacKenzie	Van Zyll de Jong 1966*b*
Cylicospirura subaquealis	stomach	Alberta/MacKenzie	Van Zyll de Jong 1966*b*
C. felineus	stomach	Alberta	Pence et al. 1978
Physaloptera praeputialis	stomach	Alberta/MacKenzie	Van Zyll de Jong 1966*b*
Troglostrongylus wilsoni	lungs	Alberta/MacKenzie	Van Zyll de Jong 1966*b*
Spiny-headed Worms (Acanthocephala)			
Oncicola canis		Alaska	Schmidt 1968

laticollis, with a frequency of occurrence of 96 percent and a mean infection rate of 30 parasites/host. The incidence of this cestode was believed related to the lynx's preying on the snowshoe hare, the presumed intermediate host. The most common roundworm was *Toxascaris leonina*, found in 92 percent of lynxes examined, with an average of 14 parasites/host. Also in Alberta, Pence et al. (1978) noted that the nematode *Cylicospirura felineus* was well entrenched in the lynx population, infesting 73 percent of their sample. Matjuschkin (1978) listed some parasites of lynxes in the Soviet Union.

Rollings (1945) was one of the first workers to describe the parasites of the gastrointestinal tract of bobcats. Seventy-two percent of 50 bobcats had 1 of 13 different helminth parasites, of which *Toxocara cati, Taenia taeniaformis,* and *T. pisiformis* were the most frequent. The bobcat was the definitive host for all 13 parasites listed, with deer, mice, hares, and rabbits known as alternative hosts for several of the parasites. Pollack (1949) found *Taenia lyncis* most frequent in 100 northeastern bobcats, while Buttrey (1974) found *Toxocara cati* most frequent in Tennessee. Young (1958) listed several other minor endoparasites.

Examination of the lungs of 24 bobcats in Virginia showed that 23 harbored adult *Anafilaroides rostratus* and 22 had *Troglostrongylus wilsoni* (Klewer 1958). Little et al. (1971) noted different parasite loads in bobcats from western and southern Texas, with *Ancylostoma caninum* and *Toxascaris leonina* the most common species. Orihel and Ash (1964) found *Dirofilaria striata* microfilariae in the peripheral blood vessels and adult worms in the thigh muscles of bobcats.

Ectoparasites are relatively uncommon in the lynx and bobcat, perhaps due to their irregular denning and bedding behavior. Six species of flea and one louse (probably of accidental occurrence) were recorded for lynxes in North America (Van Zyll de Jong 1966*b*). In bobcats, Bailey (1974) identified the cat flea *Ctenocephalides felis* and Pollack (1949) found the rabbit fleas *Cediopsylla simplex* and *Hoplosyllus glacialis*. Young (1958) mentioned three other fleas rarely found on bobcats. The mange mite *Notoedres cati* was reported on bobcats by Pollack (1949), Penner and Parke (1954), and Young (1958).

Treatment for parasites may be effective in captivity. Kumar et al. (1975) treated a captive lynx for lungworm (*Troglostrongylus* sp.) infestation using the antihelminthic drug Mebendazole. An initial dose of 100 mg/day for four days had no effect. Seven weeks later, a regimen of 400 mg/day for five days was imposed. This dosage was double that recommended for dogs (*Canis familiaris*) of similar weight. No ill effects were observed and after three weeks the lynx showed no traces of nematode infestation.

AGE DETERMINATION

The most accurate methods currently used to determine the age of bobcats and lynxes are to examine tooth replacement in kittens, the canine teeth for an open apical foramen in juveniles, and count cementum annuli in adults. Saunders (1961, 1964) found that in the lynx the root of the canine tooth displayed a foramen that closed at 13 to 18 months of age. Crowe (1972) believed the same to be true in the bobcat. Berg (1981) radiographed bobcat canines and used the open pulp cavity to identify juveniles. Mahan (1981) took a transverse section of the canine and compared the diameter of the pulp cavity to the thickness of the cementum layer to develop a ratio that was used in separating juvenile, one-year-old, and adult bobcats.

Many studies have employed the open apical foramen to identify juveniles, but for older age classes they used the deposition of cementum annuli in stained and sectioned canines (Berrie 1973; Crowe 1975a). This process requires the assumption that the first cementum layer is deposited during the animal's second winter and that each subsequent annulus represents incremental growth over a one-year period (Nellis et al. 1972; Crowe 1972; Stewart 1973). Verification of the incremental deposition of annuli has been limited to work by Nellis et al. (1972), who retrapped five animals of known minimum age showing the minimum number of expected annuli. Brand and Keith (1979) found that the deposition of lynx annuli resembled that of other species that had been verified with known-age animals.

Crowe (1972) found that boiling teeth did not influence age structure. Teeth were removed from the jaw with a saw, decalcified in formic acid–formalin, sectioned on a cryostat, and stained with Paragon stain. Annuli definition was enhanced by using a green filter in the microscope illuminator.

The age of bobcats or lynxes can thus be segregated into young of the year (which have an open apical foramen), one to two year olds (which have a closed foramen), and those over two years old, depending on the number of annuli plus one year. Fritts and Sealander (1978) found that on teeth that have an open apical foramen occasionally there is a dark staining layer that is deposited the first year. They believed it probably occurred only on bobcats born very early in the year, which means that in their first winter there is considerable deposition of cementum. Fritts and Sealander (1978) believed that teeth from Arkansas and other southern states should be examined from the root tip instead of the side of the tooth. Estimates of the age in days of kittens around one year old can be made by several methods. Crowe (1975a) used tooth replacement to estimate bobcat ages up to 240 days. This eruption pattern was believed similar to that described for lynxes; Saunders (1961) found that canine teeth continued to grow through the first year of life. Crowe (1975a) used the length (L), the diameter at the gum line (D), and the diameter of the root apical foramen (FD) to calculate an age index [AI = (L + D − FD) 1000]. Crowe (1975a) also used the least squares equation [$\log_{10}Y = 1.9038 + 0.001514$ (X), where X is the age index and Y the age in days] to calculate ages up to about one year.

Crowe (1975a) derived regression equations for body weight in grams (Y) and age in days (X):

$$Y = 8976/(1 + 2.79e^{-.0056x})$$

for females, and

$$Y = 11652/(1 + 22.13e^{-.0135x})$$

for males.

Although body weight does vary on an individual basis, it is a factor that has some use in breaking out broad classes such as "adult" or "juvenile." Erickson (1955a) used weight in conjunction with reproductive status to differentiate juvenile from adult bobcats. All females over 7.26 kg were considered adult, as were all males over 8.16 kg. Lembeck (1978), however, cautioned against too much reliance on weight in aging because of seasonal variations he observed in resident adult bobcats.

Before the advent of age determination by cementum annuli, aging techniques merely grouped animals in age categories. Saunders (1961, 1964) investigated the sequence of tooth replacement and the growth, development, and ossification of the skull and humeri in the Newfoundland lynx. Kittens (younger than one year) possessed deciduous or erupting permanent dentition, lyre-shaped temporal ridges, an indistinct lambdoidal ridge, and incompletely ossified humeri. Adult characteristics (completed canine root development, permanent lambdoidal ridge, and fully ossified humeri) were established by the end of the second year in females and slightly later in males. Sagittal crest development was presumed to be indeterminate. Yearlings displayed characteristics intermediate between kittens and adults, particularly regarding the development of the canines and humeri.

Conley (1968) and Nava (1970) classified bobcats and lynxes into three age classes based on longbone development. Kittens displayed unattached epiphyses on the radius and ulna, and yearlings showed a line of demarcation between the epiphysis and the longbone. In adults, the epiphyseal plate had disappeared and the bone narrowed. Nava (1970) attempted to use eye lens weight and canine measurements as indices of age in lynxes; however, only the kitten class could be reliably distinguished. Conley (1968) also had little success with eye lenses, pelts, and most body measurements except total length. The most effective method was a series of skull measurements with which Conley (1968) believed he could distinguish bobcat kittens, young adults, and adults.

Stewart (1973) attempted to correlate physical growth characteristics and cementum annuli deposition in the lynx. Eye lens weight, humerus length, and the ratio of interior to exterior postorbital constriction were determined to be strong indicators of growth, but reliable age indicators only for kittens. No significant correlation was found between these three growth characteristics and annuli deposition.

ECONOMIC STATUS

Depredations. Bobcats occasionally prey upon livestock. Young (1958) reported that two bobcats killed 34 lambs in 48 hours in Nevada. Nielson and Curle (cited in Powell 1971) determined that mountain lions and bobcats killed 4,995 lambs and ewes in Utah during 1969. Gashwiler et al. (1960:228) alleged that bobcats "often hunt around lambing grounds" but documented sheep remains in only 1 of 53 (1.9 percent) bobcat stomachs. Nass (1977) found that only 1 of 222 ewe losses to predators in 1973–75 in Idaho was attributable to bobcats.

Lynxes rarely constitute a threat to domestic animals. Saunders's (1963a) food habit studies showed little evidence of depredation in Newfoundland, although several reports were received. These reports indicated that some attacks on poultry and sheep did occur in untended and unfenced areas adjacent to lynx habitat. Gunderson (1978) reported a few lynx attacks on chickens during a midcontinent lynx irruption. Matjuschkin (1978) stated that the Eurasian lynx occasionally attacked sheep, goats, calves, and foals in the U.S.S.R. with rare instances of a single animal killing 30 sheep and 12 cows in one night. Attacks generally took place in unprotected areas.

Fur Harvests and Value. The bobcat and lynx have been commercially exploited for fur purposes since the settlement of North America. About 514 bobcats were taken annually in Nova Scotia between 1910 and 1934 (Rand 1944), and in Colorado 12,631 bobcat were taken by fur trappers between 1939 and 1959 (Sandfort and Tully 1971). During the six seasons from 1970 to 1976, bobcat harvests in the United States averaged 21,860 (9,432–35,990) animals per season. The take in Canada during the same period averaged 5,860 (3,408–17,354) annually (calculated from Deems and Pursley 1978). The lynx take for Canada averaged about 23,400 annually between 1850 and 1899 and about 15,900 from 1920 to 1940 (calculated from Elton and Nicholson 1942). Recently, in the six seasons from 1970 to 1976, Canadian lynx harvests have varied from about 13,000 to 53,000 per season. The take in Alaska in 1916 was about 21,000 lynxes (Burris 1971), as compared to about 1,400–9,000 animals between 1970/71 and 1974/75. In the contiguous United States, lynx harvests have ranged between about 130 and 975 animals annually between 1970 and 1975.

Pelt prices have displayed considerable variability over the years according to demand, availability, and fashion. For bobcats, prices averaged $6.00–10.63 in Nova Scotia between 1931 and 1940 (Rand 1944) and varied from about $1.42 per pelt in Colorado in 1949/50 to an average of 47¢ per pelt in Mississippi during 1950/51 (Young 1958). Recently, however, the demand for and consequent value of bobcat and lynx pelts have increased dramatically. Average pelt prices for bobcats in the United States and Canada climbed from about $11.25 in 1970/71 to about $110.00 in 1975/76. For lynxes, prices increased from $35 in 1971/72 to $142 in 1972/73 in Ontario alone (Stewart 1973), and North American averages increased from about $27 in 1970/71 to about $216 in 1975/76 (Deems and Pursley 1978). In 1979, Canadian auction prices ranged from $25 to $172 for bobcats and from $30 to $480 for lynxes, depending on size and quality (Ontario Trappers Association 1979).

Lynx and bobcat pelts are used primarily in natural colors for coats, jackets, hats, and trimming (Ford 1971). Exports of bobcats and lynxes from the United States are believed to be between 80 and 90 percent of the total harvest, although there may be substantial domestic use of bobcat furs from some regions.

Lynx harvests in the Soviet Union ranged between 1,000 and 7,000 annually prior to 1917. Lynxes are now managed on a zoned basis, with harvests averaging between 3,800 and 5,800 annually (Matjuschkin 1978), comprising about 2 percent of the total fur harvest. As in North America, pelt prices have risen sharply in the past few years and lynxes now command premium prices. In 1971, small/medium Siberian lynxes sold at $240 in Seattle (Ford 1971).

Harvest figures elsewhere in Eurasia are largely lacking. In Finland, special lynx licenses may be issued, and harvests averaged about 11 animals per year between 1965 and 1971 (Pulliainen 1974).

Lynx flesh is alleged to be quite tasty (Seton 1929; Matjuschkin 1978), although the animal is not taken (at least in North America) for food purposes. Young (1958), however, found that bobcat meat was tough and unpalatable.

MANAGEMENT

Status. Bobcats appear to be well capable of dealing with human influence on the environment. Their populations are doing well in the United States (Jenkins 1971; Miller and Speake 1978*b*), except in areas of intensive farming and dense human populations, such as in the Midwest and along the central Atlantic coast in the area of Delaware and New Jersey. There is one endangered subspecies, *F. r. escuinapae*, from central Mexico; however, in Canada the species seems to be expanding its range into many areas that previously supported only lynxes. Generally, lynx populations throughout their range are considered to be safe (Paradiso 1972:33), except for the endangered Spanish subspecies (*F. l. pardina*).

Lynxes in Canada have apparently begun to recover from a downward trend evident from about 1910 through the 1950s (Elton and Nicholson 1942; De Vos and Matel 1952) and the species' status is now regarded as "good" in that area (Van Zyll de Jong 1971). A similar pattern is noted in the northwestern United States (Nellis 1971). However, in most of the contiguous United States, the lynx's range has shifted northward from pristine times (Cahalane 1947), and lynxes are now rare outside the northern-tier states adjacent to Canada. Berrie (1973) reported that the status of the lynx in Alaska is generally favorable, though he expressed concern about the effects of industrial development and increased human settlement and mobility on lynx populations.

Lynxes are afforded protection as a game species in most of the Soviet Union and populations are reasonably secure except where drastic habitat modifications have occurred (Matjuschkin 1978). In the absence of persecution, lynxes have succeeded in reoccupying some former suitable habitat (Guggisberg 1975), including areas of northern and eastern Europe, since the 1940s. Haglund (1974) reported that lynxes were increasing in Fennoscandia, and Cop (1977) documented a successful introduction in Yugoslavia.

In North America, the bobcat and the lynx are managed by the state and provincial governments. Export of pelts out of the country, however, is monitored and controlled by the federal government under provisions of the Convention on International Trade in Endangered Species of Wild Flora and Fauna (CITES).

Management History. The first management procedure affecting the lynx and bobcat was an attempt to control their numbers and distribution by year-round taking and bounty-oriented harvest incentives. Even though these drastic control measures were ultimately unsuccessful, they illustrated, for the bobcat at least, that in most areas these cats could withstand extreme harvest regulations. Such bounty programs began to be eliminated in the 1960s through the mid-1970s, and today both species are recognized as "game" or "furbearers" with protection usually afforded through closed seasons during their breeding seasons and restrictions on the method of take. In the United States in 1971, for example, lynxes were protected in 9 states and unprotected in 12, of which 4 paid bounties (Faulkner 1971). By 1976, lynxes had partial or total protection in 14 states and a year-round open season (with special regulations) in 1 other (Deems and Pursley 1978), with bounties apparently eliminated. Presently, Alaska, Minnesota, and Montana are approved for export of lynx pelts by the Management Authority under CITES (45 FR 64520-64537, 29 September 1980). Lynxes are regulated as furbearers in most of Canada (Van Zyll de Jong 1971), except Nova Scotia and New Brunswick, where a year-round closed season prevails.

In 1976, 47 states listed the bobcat as present (Deems and Pursley 1978); 13 of those had closed seasons, 23 had regulated seasons, 7 had open seasons with no harvest regulations, and 4 had no established seasons (National Wildlife Federation unpublished data). In 1980, the number of states with closed seasons dropped to 11 and all 37 of the states that allowed harvests were approved for export by the Management Authority under CITES (45 FR 64520-64537, 29 September 1980). Only 1 state did not have regulatory authority, and that power had apparently been recently removed by the state's legislature. All eight Canadian provinces where the bobcat is present have regulated harvests.

Research and management of the bobcat and lynx have been greatly accelerated in the past 10 years by a combination of three factors: (1) increased awareness of the value of predators, (2) skyrocketing pelt prices, and (3) the listing of these felids on appendix II of CITES, which requires that signatory states justify their management before pelt export can be approved. This last action was the result of political pressure rather than biological need and resulted in the requirement for some states to develop more sophisticated management systems than was necessary given the abundance of the animals and their habitat.

A group of biologists working with the bobcat, lynx, and river otter (*Lutra canadensis*) was convened by the Scientific Authority under CITES to develop a set of minimum requirements for biological and management programs that a state would have to meet before export of those species would be allowed.

The minimum requirements for biological information are: (1) population trend information, the method of determination to be a matter of state choice, (2) information on total harvest of the species, (3) information on the distribution of the harvest, and (4) habitat evaluation (Mech 1978). The minimum requirements for a management program are that: (1) there should be a controlled harvest, the methods and seasons to be a matter of state choice, (2) all pelts should be registered and marked, and (3) harvest level objectives should be determined annually (Mech 1978). These requirements certainly encompass the basic management principles, but they were really developed only as indicators of adequate state programs and they should be used only for that purpose.

The recognition of these species as useful resources and the infusion of federal monies for research on the bobcat and lynx have stimulated more research in the past 10 years than at any time in the past. New information and techniques are beginning to be developed and published rapidly, and managers should keep abreast.

Management Philosophy and Methods of Assessment. To manage a cat population one must first determine the objectives, develop indirect indices to population status, and manipulate, when possible, those factors that influence the status to achieve the desired result. Management is defined as the purposeful manipulation of factors that affect a species, and the management of secretive creatures like the bobcat and lynx is possible only by indirect assessments (indices) of their status.

Today, statewide or province-wide objectives invariably include protection and perpetuation of this natural resource and usually include utilization of the resource for recreational and commercial purposes. Control programs are also used in localized areas when the predatory nature of the cats conflicts with other interests that are deemed of higher priority, such as when they cause excessive mortality in pronghorn fawns (Beale and Smith 1973).

The use of indirect indices has always been an integral part of wildlife management. The opinions of early wildlife managers who became very familiar with local populations were the first type of index. In the United States opinions of sportsmen and naturalists as voiced in public hearings or random surveys continue to be a source of status information. This type of index has proven over the years to be sufficient, when the resource base is large, to insure the maintenance of a population that is capable of reestablishing itself relatively quickly if and when overharvest occurs. With few exceptions, however, bobcat and lynx management today requires some additional and more sophisticated indices.

There are five major areas where researchers and managers focus to develop reliable indices for management purposes: range, habitat, population size and structure, and harvest assessments. The spatial distribution or range of a species within any management authority's jurisdiction can be determined and monitored by the following methods: (1) distribution of harvest locations; (2) surveys of experienced sportsmen, naturalists, conservation agency personnel; (3) radio-collaring individuals near the periphery of the range; and (4) observations of signs. If the traditional range of the cats begins to shift, managers and researchers are alerted to the problem area and the causes can be investigated.

Changes in the amount and distribution of the necessary habitat components for the bobcat and lynx can also be used as an index. For example, certain types of broken rock ledges are known to be important to bobcats in Massachusetts. The location and use of these areas can be used as an index. Similarly, the abundance of prey species is a habitat component that can be used as an index, e.g., the snowshoe hare cycle as an index to lynx abundance. Methods of assessing habitat components are as follows: (1) define habitat components through tracking the animals in the snow or with radio-telemetry; (2) identify and monitor food habits through stomach, scat, or tracking studies; (3) identify and monitor areas of habitat through either ground checks by agency personnel or surveys of user groups, or by remote sensing with aerial photography or satellite imagery; and (4) monitor prey abundance.

The use of population structure indices has received much attention in recent years, with emphasis on sex and age structure, recruitment, and mortality rates. Sex ratio data are best derived by examination of reproductive tracts in harvested samples, while age ratios are obtained by sectioning the teeth of such samples. The only alternative to the use of harvest data is intensive efforts to capture the animals outside the regular season as part of a research project. These projects generally use similar capture methods and will provide useful data but too frequently they are limited to a small area and small samples due to the high manpower requirements involved.

Recruitment is the product of natality, survival rates, and immigration. Again, carcass collections from harvests provide the reproductive tracts from which natality rates can be estimated. Telemetry studies can give estimates of the percentage of females breeding and of the survival of the young. Otherwise, survival rates are directly estimated through field observations (primarily of kittens) and tagging studies, or indirectly through age structure analyses (Crowe 1975b). Immigration has not been studied well but telemetry is the best method available for such work.

Mortality rates after the first year are best estimated through analysis of age structures derived from sectioning canine teeth (Gould 1981). Telemetry can show the type and importance of different mortality factors but the duration of radio-collars is not sufficient presently to provide population mortality rates.

A population size assessment is simply an index repeated periodically to provide an indication of whether the population is growing, decreasing, or stable. This should not be confused with a census that is a total count or tally over a specific area at a specific time (Overton and Davis 1969). There are presently no census methods for the bobcat or lynx. However, estimates of the total population size can best be derived by determining the density through intensive study (including telemetry) on representative habitats and then multiplying those density estimates by the amount of habitat within the bobcat range. Gould (1978) used this method in California, where he had 17 habitat types averaging 22,062 km² each and density estimates from study areas within or similar to each habitat. Similar statewide population estimates have been done by Progulske (1952), McCord (1977), and Berg (1981). It should be remembered that these estimates are costly and take several years to complete, rendering them useless as an index or trend indicator except in something like 25-year increments.

The first annual population index used was the total harvest. This is a useful index but should be tempered with some measure of yearly effort. Zezulak (1981) used the capture per trap night and felt that this was a good indicator of trend. Klepinger et al. (1981) found that harvest declines in Wisconsin corresponded to declines noted in track counts from road transects. Track counts have received much attention recently at scent stations and on line transects. Although Grippi (n.d.) found that scent stations appeared to give useful trends, Zezulak (1981) found that they were probably the least effective method for indices. He felt that bobcats were not greatly attracted by the scent and that the stations were really only monitoring stations for tracks of animals that normally passed by the area. Therefore, he felt that transects were much more useful as indicators of population trend.

Surveys of hunters and trappers and other knowledgeable people have also been used successfully as population size indicators. Gould (1977a) interviewed knowledgeable federal and state game and land management agency personnel and experienced hunters and trappers in order to obtain status information and found the major problem with this technique to be the number and distribution of the observers. Gould (1977b) also found that correction factors had to be developed in order to reduce the high number of bobcat kills reported in the general survey of all types of hunters and their take.

Harvest method assessment for the bobcat and lynx should be divided into hunting and trapping methods and should also document the exact method, location, and date of harvest. The most common technique is a random survey of users after the season (Gould 1981), but for specimens that must be tagged, a check station interview is most useful. Field bag checks have limited use due to the low density of hunters. The recapture of marked animals in a population of harvested cats can provide an estimate of the proportion taken by each harvest method.

The key elements to be ascertained before management practices are modified are: (1) the direction of change of the population; (2) the rate of change, and (3) the factors causing or influencing the change. The rate of change is generally quite easy to evaluate from the magnitude and speed of change in the population indices. Crowe (1975b) developed a formula to determine if a bobcat population was increasing, decreasing, or stable. Using the parameters of adult survival (Sa), number of female young per female (F), and juvenile survival (Sy), Crowe derived the parameter lambda (λ) as: $\lambda = Sa + F(Sy)$. This parameter indicates that, for every cat in the population in a given year, there will be λ animals the following year. If λ is 1.0, the population is stable, while values above and below 1.0 indicate an increasing or decreasing population, respectively. The rate of change is also indicated by the amount of deviation from the norm of 1.0. The methods used to derive Sa, F, and Sy are described in previous sections.

Beyond the obvious population indicators, such as changes in population size or range reductions and expansions, there are several other ways to evaluate what is happening in a cat population. Age structure determined from a harvested sample has been used by a number of biologists to assess the "condition" of furbearer populations. However, Caughley (1974) demonstrated that age structure alone should not be used to monitor the status of a harvested population. The same age structure changes could result from increased/decreased mortality, increased/decreased natality, or some combination of the two. Therefore, it is necessary to monitor additional population parameters such as indices of abundance (catch per unit effort, transect track counts, etc.) and productivity (analysis of reproductive tracts) to obtain a more comprehensive indication of population status. The age structure in a healthy population should show around 15 percent young of the year, with a gradual decline in numbers per age class from the 1.5 year olds up to about 12.5 years, with roughly 35 percent of the animals in the maximum breeding age category, and few animals over 12.5 years old. Lembeck and Gould (1981) stated that in heavily harvested populations less than 30 percent of the bobcats were yearlings or older individuals, more than 30 percent were young of the year, and very few were older animals. Lembeck and Gould also noted that the sex ratios were about equal, and profiled a lightly harvested population as having more than 35 percent of the harvest comprising breeding-age females, less than 20 percent young of the year, a "noticeable segment" older than 6.5 years, and sex ratios of more than 1.2 males per female. Gould (1978:1) stated, "Healthy harvested populations appear to have . . . 15–30 percent young-of-the-year, 30–40 percent breeding age females, . . . some older individuals and . . . the sex ratio fairly well male dominated." In Alabama, Miller (1980) found that the 0–1 age class for three consecutive years comprised about 26 percent of the harvested population. This percentage approximated the 25–30 percent harvest rate for those years, and the population size remained relatively stable on the intensively studied area, as determined from the scent stations indices and presence of sign.

A word of caution is necessary about being too strict in applying the above percentages and relationships from California or Alabama to different areas of the country. Although the principles involved are sound, different seasons, harvest methods, or natural population characteristics may cause the values to vary. For example, in Massachusetts the minimum late winter population estimate is about 500 bobcats with a yearly harvest of 30, and hunting with hounds is the primary harvest method. Under these conditions where the harvest has little impact on the population and the primary harvest method is biased toward juveniles, it is not uncommon to have 50 percent juveniles in the take without any adverse impact because of low total harvest. The manager therefore must use these data as guidelines and apply them realistically to the local area. There is ultimately no substitute for monitoring the indices and seeing how they respond over the years as regulations are modified and natural conditions change. As more information is obtained, models can become more sophisticated and predict with greater accuracy the consequence of natural and man-made changes.

The factors causing the change in population status or behavior may be obvious from the data, or research may be necessary to identify causes and remedies. In either case there may not be a direct method of dealing with the problem, so instead management decisions may have to be made which increase the population's ability to handle a problem. For example, previous sections of this chapter have documented the close cyclic relationship between the lynx and the snowshoe hare, with declines in lynx numbers consequent upon similar declines in hares. Because trapping mortality appears to be additive to natural causes (Van Zyll de Jong 1971; Brand and Keith 1979), if trapping mortality is excessive after the hare population crashes it may result in local extirpations of lynx populations. Berrie (1973) expressed concern that intensive trapping could result in the constriction of lynx range and a decrease in cyclic amplitude. He suggested a closed season for up to three years during a lynx population low. Also, Brand and Keith's (1979) management strategies suggest curtailment of trapping for three to four years commencing the second year after the peak in fur harvest returns. Similarly, changes in natural food levels may best be initially handled by additional hunting and trapping to reduce surplus animals and lessen losses to domestic livestock. Bailey (1972) also suggested that sheep losses may be reduced by moving the animals away from rocky or brushy terrain at night. Beale and Smith (1973) similarly suggested placement of watering devices for pronghorns away from broken terrain to reduce predation by bobcats.

Introductions. Introductions have been used in a few instances to restore lynxes to vacant habitat. Cop (1977) released 6 Czechoslovakian lynxes in Slovenia, Yugoslavia, in 1973. Breeding was observed the first year and dispersal up to 45 km the second year. By 1976, the estimated population was 20–35 lynxes in an area of about 60,000 ha. Reintroductions have also been attempted in Germany, Switzerland, and Italy (Guggisberg 1975; Cop 1977).

Reintroduction of bobcats is currently under way in New Jersey. To date, 15 bobcats (7 males and 8 females) have been captured in Maine and released in northwestern New Jersey. One adult male moved 157 km north and was killed in New York, while another was found dead at the release site six months later. The third known mortality was a road kill eight months after release and 16 km from the site (R. C. Lund personal communication). Most of the animals appear to have established residence in the areas around their release sites.

Immobilization. The restraint and handling of the bobcat and lynx have been facilitated by the use of immobilizing chemicals, administered primarily by injection. Drugs that have been used on the bobcat and lynx include succinylcholine hydrochloride (SCC) (Saunders 1963*b*), tiletamine hydrochloride/ zolazepam HCl (D. Phillips and J. P. Sundberg unpublished data), phencyclidine HCl (PHC) (Bailey 1971), PHC-promazine HCl (Mech 1980; Berg 1981), ketamine HCl (KH) (Beck 1976; Zezulak and Schwab 1980), and atropine-KH-acepromazine (Jones 1977). Of the above, SCC is now infrequently used due to dosage variations and a narrow tolerance range in most species, and commercial production of PHC has been discontinued. Sex-differentiated responses with PHC may also occur (Berrie 1972). KH is now the preferred drug for immobilization of the bobcat and lynx, with dosages ranging from 5.5 to 21 mg/kg body weight. KH produces cataleptoid anesthesia accompanied by deep analgesia, with a wide margin of safety. It does not relax skeletal muscles and permits normal pharyngolaryngeal reflexes. However, convulsive seizures and salivation may occur as side effects. Used in conjunction with acepromazine, muscle relaxation is induced, but hyperthermia and apnea may occur at larger dosages. Excessive salivation may be reduced by use of atropine. Xylazine hydrochloride (Addison and Kolenosky 1979) or diazepam may prove effective in minimizing body rigidity or convulsions induced by KH anesthesia.

CURRENT RESEARCH AND MANAGEMENT NEEDS

The single greatest research need appears to be long-term, intensive studies of a bobcat or lynx population with known sex and age structure, reproductive activities, home ranges, habitat use, food habits, trends in prey species, and interactions with other predators. With these factors known and monitored in the population, different indexing methods such as scent stations or track transects should be applied to evaluate their sensitivity and perhaps develop methods to evaluate density from different indices. The area should then be subjected to varying harvest levels to evaluate the impact of harvest on reproduction, sex and age structure, home range establishment, etc. Other research needs include the following:

1. Evaluating mortality rates of juveniles after they leave parental care, and the importance of dispersal in maintaining stable populations.
2. Studying bobcats with known reproductive histories so that the rate of change from corpora lutea to LPBC can be described more accurately and more precise reproductive histories can be obtained from kill samples.

3. Tooth aging evaluations for animals less than one year old using criteria other than the open apical foramen. This will reduce any error in these important age classes.
4. Comparison of exploited and unexploited populations in different regions of the country with special reference to age and sex structure, recruitment, home range establishment, and mortality factors.
5. Evaluation of the second litter phenomenon in the southern portion of the range.
6. Reintroduction studies focusing on the best sex and age of the animals to be released.
7. Snow tracking studies of known sex and age animals to determine urination, defecation, and other behavior patterns that could be indicative of the sex, age, and social status of the individual or give some indication of population status.

We acknowledge the cooperation of S. D. Miller for his critical review of the manuscript, and of J. M. Holland and E. C. Horwitz for editorial comments and assistance with translations.

LITERATURE CITED

Adams, A. W. 1963. The lynx explosion. North Dakota Outdoors 26(5):20–24.

Adams, J. R. 1966. Taenids in lynx from British Columbia, with a comment on *Taenia monostephanos* Von Linstow and other taenids with a single crown of hooks. Pages 480–481 *in* A. Corradetti, ed. Proc. 1st Int. Congr. Parasitol. 2 vols. Pergamon Press, New York. 1,118 pp.

Addison, E. M., and Kolenosky, G. B. 1979. Use of ketamine hydrochloride and xylazine hydrochloride to immobilize black bears. J. Wildl. Dis. 15:253–258.

Antonius. O. 1943. Nachtrag zu "Symbolhandlungen und Verwandtes bei Saugetieren." Z. Tierpsychol. 1:259–289. (In German.)

Asdell, S. A. 1946. Patterns of mammalian reproduction. Comstock Publ. Co., Ithaca, N.Y. 437 pp.

Bailey, T. N. 1971. Immobilization of bobcats, coyotes, and badgers with phencyclidine hydrochloride. J. Wildl. Manage. 35:847–849.

———. 1972. Ecology of bobcats with special reference to social organization. Ph.D. Diss. Univ. Idaho. 82 pp.

———. 1974. Social organization in a bobcat population. J. Wildl. Manage. 38:435–446.

Bailey, V. 1936. The mammals and life zones of Oregon. North Am. Fauna 55:1–416.

Banfield, A. W. F. 1974. The mammals of Canada. Natl. Mus. Can. and Univ. Toronto Press, Toronto. 438pp.

Bangs, O. 1897. Notes on the lynxes of eastern North America, with descriptions of two new species. Proc. Bio. Soc. Washington 11:47–51.

Barash, D. P. 1971. Cooperative hunting in the lynx. J. Mammal. 52:480.

Beale, D. M., and Smith, A. D. 1973. Mortality of pronghorn antelope fawns in western Utah. J. Wildl. Manage. 37:343–352.

Beasom, S. L., and Moore, R. A. 1977. Bobcat food habit response to a change in prey abundance. Southwest Nat. 21:451–457.

Beck, C. C. 1976. Vetalar: a unique cataleptoid anesthetic agent for multispecies usage. J. Zoo Anim. Med. 7(3):11–38.

Bee, J. W., and Hall, E. R. 1956. Mammals of northern Alaska on the Arctic slope. Univ. Kansas Mus. Nat. Hist. Misc. Pub. 8:1–309.

Berg, W. E. 1981. Ecology of bobcats in northern Minnesota. Pages 55–61 *in* Blum, L. G., and Escherich, P. C., eds. Bobcat research conference proceedings. Natl. Wildl. Fed. Sci. Tech. Ser. no. 6. 137pp.

Bergerud, A. T. 1971. The population dynamics of Newfoundland caribou. Wildl. Monogr. 25:1–55.

Berrie, P. M. 1972. Sex differences in response to phencyclidine hydrochloride in lynx. J. Wildl. Manage. 36:994–996.

———. 1973. Ecology and status of the lynx in interior Alaska. Worlds Cats 1:4–41.

Brand, C. J., and Keith, L. B. 1979. Lynx demography during a snowshoe hare decline in Alberta. J. Wildl. Manage. 43:827–849.

Brand, C. J.; Keith, L. B.; and Fischer, C. A. 1976. Lynx responses to changing snowshoe hare densities in central Alberta. J. Wildl. Manage. 40:416–428.

Brittell, J. D. 1978. Analysis of current bobcat harvest. Washington Game Dept., Rep. Fed. Aid Wildl. Rest. Proj. W-84-R, Study II-2. 9pp.

———. 1979. Analysis of current bobcat harvest. Washington Game Dept., Rep. Fed. Aid Wildl. Rest. Proj. W-84-R, Study II-2. 3pp.

Brownlee, W. C. 1977. Status of the bobcat (*Lynx rufus*) in Texas. Texas Parks Wildl. Dept., Special rep. 30pp.

Bulmer, M. G. 1974. A statistical analysis of the 10-year cycle in Canada. J. Anim. Ecol. 43:701–718.

Burris, O. 1971. Lynx management in Alaska. Pages 30–33 *in* S. E. Jorgenson and L. D. Mech, eds. Proc. symp. native cats of North Am. U.S. Fish and Wildl. Serv., Twin Cities, Minn. 139pp.

Butler, L. 1942. Fur cycles and conservation. Trans. North Am. Wildl. Conf. 7:463–472.

Buttrey, G. W. 1974. Food habits and distribution of the bobcat, *Lynx rufus rufus* (Schreber), on the Catoosa Wildlife Management Area. M.S. Thesis. Tennessee Tech. Univ. 64 pp.

Cahalane, V. H. 1947. Mammals of North America. MacMillan Co., New York. 682pp.

Caughley, G. 1974. Interpretation of age ratios. J. Wildl. Manage. 38:557–562.

Colby, E. D. 1974. Artificially induced estrus in wild and domestic felids. World's Cats 2:126–147.

Cole, L. C. 1954. Some features of random population cycles. J. Wildl. Manage. 18:1–24.

Conley, R. H. 1968. An investigation of some techniques for determining age of bobcats (*Lynx rufus*) in the Southeast. M.S. Thesis. Univ. Georgia. 44pp.

Cop, J. 1977. The results of lynx introductions into Kocevsko, Slovenia, Yugoslavia. Int. Congr. Game Bio. 13:372–376.

Corbet, G. B. 1978. The mammals of the Palearctic region: a taxonomic review. Br. Mus. Nat. Hist./Cornell Univ. Press, London and Ithaca, N.Y. 314pp.

Crowe, D. M. 1972. The presence of annuli in bobcat tooth cementum layers. J. Wildl. Manage. 36:1330–1332.

———. 1974. Some aspects of reproduction and population dynamics of bobcats in Wyoming. Ph.D. Diss. Univ. Wyoming. 191pp.

———. 1975a. Aspects of aging, growth, and reproduction of bobcats from Wyoming. J. Mammal. 56:177–198.

———. 1975b. A model for exploited bobcat populations in Wyoming. J. Wildl. Manage. 39:408–415.

Dearborn, N. 1932. Foods of some predatory fur-bearing animals in Michigan. School For. Conserv., Univ. Michigan, Ann Arbor, Bull. 1. 52pp.

Deems, E. F., Jr., and Pursley, D., eds. 1978. North American furbearers: their management, research, and harvest status in 1976. Int. Assoc. Fish. Wildl. Agencies and Maryland Dept. Nat. Res. 157pp.

De Vos, A., and Matel, S. E. 1952. The status of the lynx in Canada, 1920–1952. J. For. 50:742–745.

Dill, H. H. 1947. Bobcat preys on deer. J. Mammal. 28:63.

Dixon, J. S. 1934. A study of the life history and food habits of the mule deer in California. California Fish and Game 20:229–272.

Donoho, H. S.; Riffel, H. D.; and Tully, R. J. 1979. Colorado small game, furbearer and varmint harvest, 1978. Colorado Div. Wildl., Rep. Fed. Aid Wildl. Rest. Proj. W-121-R. 256pp.

Duke, K. L. 1949. Some notes on the histology of the ovary of the bobcat (*Lynx*) with special reference to the corpora lutea. Anat. Rec. 103:111–132.

———. 1954. Reproduction in the bobcat, *Lynx rufus*. Anat. Rec. 120:816–817.

Durrant, S. D. 1952. Mammals of Utah. Univ. Kansas Publ. Mus. Nat. Hist. 6:1–549.

Eaton, R. L. 1976. The evolution of sociality in the Felidae. World's Cats 3(2):95–142.

Eaton, R. L., and Craig, S. J. 1973. Captive management and mating behavior of the cheetah. World's Cats 1:217–254.

Elton, C., and Nicholson, M. 1942. The ten-year cycle in numbers of the lynx in Canada. J. Anim. Ecol. 11:215–244.

Erickson, A. W. 1955a. An ecological study of the bobcat in Michigan. M.S. Thesis. Michigan State Univ. 133pp.

———. 1955b. A recent record of lynx in Michigan. J. Mammal. 36:132–133.

Etkin. W., ed. 1964. Cooperation and competition in social behavior. Pages 1–34 *in* Social behavior and organization among vertebrates. Univ. Chicago Press, Chicago. 307pp.

Ewer, R. F. 1968. Ethology of mammals. Elek Science, London. 418pp.

———. 1973. The carnivores. Cornell Univ. Press, Ithaca, N.Y. 494pp.

Faulkner, C. E. 1971. The legal status of the wildcats in the United States. Pages 124–125 *in* S. E. Jorgenson and L. D. Mech, eds. Proc. symp. native cats of North Am. U.S. Fish and Wildl. Serv., Twin Cities, Minn. 139pp.

Foote, L. E. 1945. Sex ratio and weights of Vermont bobcats in autumn and winter. J. Wildl. Manage. 9:326–327.

Ford, H. S. 1971. The economic value of the wildcats of North America as fur animals. Pages 121–122 *in* S. E. Jorgenson and L. D. Mech, eds. Proc. symp. native cats of North Am. U.S. Fish and Wildl. Serv., Twin Cities, Minn. 139pp.

Fox, L. 1980. Biology, ecology, and range of the bobcat in New York and its inferred interaction with lynx in the Adirondack Park. New York Dept. Environ. Conserv., Rep. Fed. Aid Proj. E-3-1, Job XII-1.

Fredrickson, L. F., and Rice, L. A. 1981. Bobcat management survey study in South Dakota, 1977–79. Pages 32–36 *in* Blum, L. G., and Escherich, P. C., eds. Bobcat research conference proceedings. Natl. Wildl. Fed. Sci. Tech. Ser. no. 6. 137pp.

Fritts, S. H. 1973. Age, food habits, and reproduction of the bobcat (*Lynx rufus*) in Arkansas. M.S. Thesis. Univ. Arkansas. 80pp.

Fritts, S. H., and Sealander, J. A. 1978. Reproductive biology and population characteristics of bobcats (*Lynx rufus*) in Arkansas. J. Mammal. 59:347–353.

Fyvie, A., and Addison, E. M. 1979. Manual of common

parasites, diseases, and anomalies of wildlife in Ontario. 2nd ed. Ontario Ministry Nat. Res. 120pp.

Gashwiler, J. S., Robinette, W. L.; and Morris, O. W. 1960. Foods of bobcats in Utah and eastern Nevada. J. Wildl. Manage. 24:226–229.

———. 1961. Breeding habits of bobcats in Utah. J. Mammal. 42:76–84.

Gilpin, M. E. 1973. Do hares eat lynx? Am. Nat. 197:727–730.

Godin, A. J. 1977. Wild mammals of New England. Johns Hopkins Press, Baltimore. 304pp.

Golley, F. B.; Petrides, G. A.; Rauber, E. L.; and Jenkins, J. H. 1965. Food intake and assimilation by bobcats under laboratory conditions. J. Wildl. Manage. 29:442–447.

Goodwin, H. A.; Holloway, C. W.; and Thornback, J. 1978. Red data book. Part 1: Mammalia. Int. Union Conserv. Nature, Morges, Switzerland. Paged separately.

Gould, G. I. 1977*a*. Bobcat distribution in northeastern California. California Dept. Fish and Game. Rep. Fed. Aid Wildl. Rest. Proj. W-54-R, Job IV-1.2. 12pp.

———. 1977*b*. Estimated hunter take of bobcat in California during 1976. California Dept. Fish and Game. Rep. Fed. Aid Wildl. Rest. Proj. W-54-R, Job IV-1.0. 10pp.

———. 1978. Bobcat study and survey. California Dept. Fish and Game, Rep. Fed. Aid Wildl. Rest. Proj. W-54-R, Job IV-1.6. 8pp.

———. 1979. Bobcat study and survey. California Dept. Fish and Game. Rep. Fed. Aid Wildl. Rest. Proj. W-54-R, Job IV-1.6. 8pp.

———. 1981. Techniques used in assessing bobcat populations and harvests in California. Pages 40–41 *in* Blum, L. G., and Escherich, P. C., eds. Bobcat research conference proceedings. Natl. Wildl. Fed. Sci. Tech. Ser. no. 6. 137pp.

Grange, W. B. 1949. The way to game abundance. Chas. Scribner's Sons, New York. 365pp.

Gray, A. P. 1972. Mammalian hybrids: a check-list with bibliography. Commonwealth Bur. Anim. Breeding and Genetics, Edinburg, Tech. Comm. 10 (Rev.). 262pp.

Grinnell, J.; Dixon, J. S.; and Linsdale, J. M. 1937. Fur-bearing mammals of California. Univ. California Press, Berkley. 2 vols.

Grippi, R. n.d. Bobcat distribution and abundance in Fresno County, California. California Dept. Fish and Game, Rep. Fed. Aid Wildl. Rest. Proj. W-54-R, Job IV-1.1. 16pp.

Guggisberg, C. A. W. 1975. Wild cats of the world. Taplinger Publ. Co., New York. 328pp.

Gunderson, H. L. 1978. A midcontinent irruption of Canada lynx, 1962–63. Prairie Nat. 10:71–80.

Guthery, F. S., and Beasom, S. L. 1977. Responses of game and nongame wildlife to predator control in south Texas. J. Range Manage. 30:404–409.

Haglund, B. 1966. [Winter habits of the lynx and wolverine as revealed by tracking in the snow.] Viltrevy 4:81–310. (In Swedish, with English summary.)

———. 1974. Moose relations with predators in Sweden, with special reference to bear and wolverine. Nat. Can. 101:457–466.

Hall, E. R., and Kelson, K. R. 1959. The mammals of North America. 2 vols. Ronald Press, New York. 1,088pp.

Hall, H. T., and Newsom, J. D. 1976. Summer home ranges and movements of bobcats in bottomland hardwoods of southern Louisiana. Proc. Annu. Conf. Southeast Assoc. Fish Wildl. Agencies 30:427–436.

Hamilton, W. J. 1942. Mammals of eastern United States. Comstock Publ. Co., Ithaca, N.Y. 432pp.

Hamilton, W. J., and Hunter, R. P. 1939. Fall and winter

food habits of Vermont bobcats. J. Wildl. Manage. 3:99–103.

Hancock, J. A.; Mercer, W. E.; and Northcott, T. H. 1976. Lynx attack on man carrying hares in Newfoundland. Can. Field Nat. 90:46–47.

Hardy, M. 1907. Canada lynx and wildcat. For. Stream 68:1010–1011.

Hawbaker, S. S. 1965. Trapping North American furbearers. Kuertz Bros., Fort Louden, Pa. 352pp.

Hemmer, H. 1976. Gestation period and postnatal development in felids. World's Cats 3(2):143–165.

Henderson, C. L. 1981. Bobcat (*Lynx rufus*) distribution, management, and harvest analysis in Minnesota, 1977–79. Pages 27–31 *in* Blum, L. G., and Escherich, P. C., eds. Bobcat research conference proceedings. Natl. Wildl. Fed. Sci. Tech. Ser. no. 6. 137pp.

Hoffman, R. S.; Wright, P. L.; and Newly, F. E. 1969. The distribution of some mammals in Montana. Part 1: Mammals other than bats. J. Mammal. 50:579–604.

Hornocker, M. G. 1969*a*. Defensive behavior in female bighorn sheep. J. Mammal. 50:128.

———. 1969*b*. Winter territoriality in mountain lions. J. Wildl. Manage. 33:457–464.

Hsu, T. C., and Benirschke, K. 1970. An atlas of mammalian chromosomes. Vol. 4, Folio 187: *Lynx rufus*.

———. 1974. An atlas of mammalian chromosomes. Vol. 8, Folio 385: *Felis lynx*.

Ingles, L. G. 1965. Mammals of the Pacific states. Stanford Univ. Press, Stanford, Calif. 506pp.

Iurgenson, P. B. 1955. [Ecology of the lynx in forests of the central zone of the USSR.] Zool. Zh. 34:609–620. (In Russian.)

Jackson, H. H. T. 1961. Mammals of Wisconsin. Univ. Wisconsin Press, Madison. 504pp.

Jenkins, J. H. 1971. The status and management of the bobcat and cougar in the southeastern states. Pages 87–91 *in* S. E. Jorgensen and L. D. Mech, eds. Proc. symp. native cats of North Am. U.S. Fish and Wildl. Serv., Twin Cities, Minn. 139pp.

Jones, J. H. 1977. Density and seasonal food habits of bobcats on the Three Bar Wildlife Area, Arizona. M.S. Thesis. Univ. Arizona. 48pp.

Jones, J. K., Jr.; Carter, D. C.; and Genoways, H. H. 1975. Revised checklist of North American mammals north of Mexico. Occas. Pap. Mus. Texas Tech. Univ. 28:1–14.

Jones, S. V. H. 1923. Color variations in wild animals. J. Mammal. 4:172–177.

Jordahl, H. C. 1956. Canada lynx. Wisconsin Conserv. Bull. 21(11):22–26.

Keith, L. B. 1963. Wildlife's ten-year cycle. Univ. Wisconsin Press, Madison. 201pp.

———. 1974. Some features of population dynamics in mammals. Int. Congr. Game Bio. 11:17–58.

Kelson, K. R. 1946. Notes on the comparative osteology of the bobcat and the house cat. J. Mammal. 27:255–264.

Kight, J. 1962. An ecological study of the bobcat (*Lynx rufus* Schreber) in west-central South Carolina. M.S. Thesis. Univ. Georgia. 52pp.

Klepinger, K. E.; Creed, W. A.; and Ashbrenner, J. E. 1981. Monitoring bobcat harvest and populations in Wisconsin. Pages 23–26 *in* Blum, L. G., and Escherich, P. C., eds. Bobcat research conference proceedings. Natl. Wildl. Fed. Sci. Tech. Ser. no. 6. 137pp.

Klewer, H. L. 1958. The incidence of helminth lung parasites of *Lynx rufus rufus* (Schreber) and the life cycle of *Anafilaroides rostratus* Gerichter, 1949. J. Parasitol. 44:29.

Kumar, V.; Mortelmans, J.; Vercruysse, J.; and Ceulemans, F. 1975. Chemotherapy of helminthiasis among wild mammals. Part 1: Lungworm infection of *Felis canadensis*. Acta Zool. Path. Antverpiensa 61:85–89.

Kurten, B., and Rausch, R. 1959. Biometric comparisons between North American and European mammals. Part 2: A comparison between the northern lynxes of Fennoscandia and Alaska. Acta Arctica 11:21–45.

Lack, D. 1954. Cyclic mortality. J. Wildl. Manage. 18:25–37.

Langford, E. V. 1974. *Acholeplasma laidlawii* and *Pasturella multocida* isolates from the pneumonic lung of a lynx. J. Wildl. Dis. 10:420–422.

Lembeck, M. 1978. Bobcat study, San Diego County, California. California Dept. Fish and Game, Rep. Fed. Aid Nongame Wildl. Invest. Proj. E-W-2, Study IV-1.7. 22pp.

Lembeck, M., and Gould, G. I., Jr. 1981. Dynamics of harvested and unharvested bobcat populations in California. Pages 53–54 *in* Blum, L. G., and Escherich, P. C., eds. Bobcat research conference proceedings. Natl. Wildl. Fed. Sci. Tech. Ser. no. 6. 137pp.

Lewis, J. C. 1975. Control of rabies among terrestrial wildlife by population reduction. Pages 243–259 *in* G. M. Baer, ed. The natural history of rabies. Vol. 2. Academic Press, New York. 387pp.

Leyhausen, P. 1979. Cat behavior. Garland STPM Press, New York. 340pp.

Leyhausen, P., and Wolff, R. 1959. Das Revier einer Hauskatze. Z. Tierpsychol. 16:666–670. (In German.)

Lindemann, W. 1955 [The early development of lynx and the wildcat.] Behaviour 8:1–45. (In German, English summary.)

Linsdale, J. M., and Tomich, P. Q. 1953. A herd of mule deer. Univ. California Press, Berkley. 567pp.

Little, C. C. 1958. Coat color genes in rodents and carnivores. Q. Rev. Bio. 33:103–137.

Little, J. W.; Smith, J. P.; Knowlton, F. F.; and Bell, R. R. 1971. Incidence and geographic distribution of some nematodes in Texas bobcats. Texas J. Sci. 22:403–407.

Lowery, G. H., Jr. 1974. The mammals of Louisiana and its adjacent waters. Louisiana State Univ. Press, Baton Rouge. 564pp.

McCord, C. M. 1974a. Selection of winter habitat by bobcats (*Lynx rufus*) on the Quabbin Reservation, Massachusetts. J. Mammal. 55:428–437.

———. 1974b. Courtship behavior in free-ranging bobcats. World's Cats 2:76–87.

———. 1977. The bobcat in Massachusetts. Massachusetts Wildl. 28(5):2–8.

Mahan, C. J. 1981. Age determination of bobcats (*Lynx rufus*) by means of canine pulp cavity ratios. Pages 126–129 *in* Blum, L. G., and Escherich, P. C., eds. Bobcat research conference proceedings. Natl. Wildl. Fed. Sci. Tech. Ser. no. 6. 137pp.

Mandal, A. K., and Talukder, S. K. 1975. Skeletal differences in the appendicular skeleton of the lynx and the caracal (Felidae: Carnivora) in relation to ecology. Anat. Anz. 137:447–453.

Manville, R. H., and Young, S. P. 1965. Distribution of Alaskan mammals. U.S. Bur. Sport Fish and Wildl., Washington, D.C., Circ. 211. 74pp.

Marshall, A. D., and Jenkins, J. H. 1966. Movements and home ranges of bobcats as determined by radio-tracking in the upper coastal plain of west-central South Carolina. Proc. Annu. Conf. Southeast Assoc. Game and Fish Comm. 20:206–214.

Marston, M. A. 1942. Winter relations of bobcats to white-tailed deer in Maine. J. Wildl. Manage. 6:328–337.

Mathews, J. H. 1941. Reproduction in the Scottish wild cat, *Felis silvestris*. Proc. Zool. Soc. London, ser. B, 111:59–77.

Matjuschkin, E. N. 1978. Der Luchs. A. Ziemsen, Wittenberg Lutherstadt, G.D.R. 160pp. (In German.)

Matson, J. R. 1948. Cats kill deer. J. Mammal. 29:69–70.

Mech, L. D. 1973. Canadian lynx invasion of Minnesota. Bio. Conserv. 5:151–152.

———. 1977. Record movement of a Canadian lynx. J. Mammal. 58:676–677.

———. 1978. Report of the working group on bobcat, lynx, and river otter. Natl. Sci. Foundation, Washington, D.C. 15pp.

———. 1980. Age, sex, reproduction and spatial organization of lynxes colonizing northeastern Minnesota. J. Mammal. 61:261–267.

Merriam, C. H. 1886. Description of a newly born lynx. Bull. Nat. Hist. Soc. New Brunswick 5:10–13.

Miller, G. S., Jr., and Kellogg, R. 1955. List of North American recent mammals. U.S. Natl. Mus. Bull. 205:1–954.

Miller, S. D. 1980. The ecology of the bobcat in southern Alabama. Ph.D. Diss. Auburn Univ., Auburn, Ala. 156pp.

Miller, S. D., and Speake, D. W. 1978a. Prey utilization by bobcats on quail plantations in south Alabama. Proc. Annu. Conf. Southeast Assoc. Fish and Wildl. Agencies 32:100–111.

———. 1978b. Status of the bobcat: an endangered species? Pages 145–153 *in* R. Odom and L. Landers, eds. Proc. Rare and Endangered Wildl. Symp., Athens, Ga. 184pp.

———. 1978c. Use of motion-sensitive transmitters to study felid ecology. Pages 163–166 *in* H. J. Klewe and H. P. Kimmich, eds. Proc. Int. Conf. Biotelemetry 4, Garmisch, West Germany.

Moran, P. A. P. 1949. The statistical analysis of the sunspot and lynx cycles. J. Anim. Ecol. 18:115–116.

———. 1953. The statistical analysis of the Canadian lynx cycle. Austr. J. Zool. 1:291–298.

More, G. 1976. Some winter food habits of lynx in the southern MacKenzie district, NWT. Can. Field Nat. 90:499–500.

Mossman, H. W., and Duke, K. L. 1973. Comparative morphology of the mammalian ovary. Univ. Wisconsin Press, Madison. 461pp.

Murie, A. 1961. A naturalist in Alaska. Devin-Adair Co., New York. 302pp.

Nass, R. D. 1977. Mortality associated with sheep operations in Idaho. J. Range Manage. 30:253–258.

Nava, J. A., Jr. 1970. The reproductive biology of the Alaska lynx. M.S. Thesis. Univ. Alaska. 141pp.

Navarre, J. 1976. Observations récentes sur le lynx dans les Pyrénées occidentales. Mammalia 40:518–519. (In French.)

Nellis, C. H. 1971. The lynx in the northwest. Pages 24–33 *in* S. E. Jorgenson and L. D. Mech, eds. Proc. symp. native cats of North Am. U.S. Fish and Wildl. Serv., Twin Cities, Minn. 139pp.

Nellis, C. H., and Keith, L. B. 1968. Hunting activities and success of lynxes in Alberta. J. Wildl. Manage. 32:718–722.

Nellis, C. H., and Wetmore, S. P. 1969. Long-range movement of lynx in Alberta. J. Mammal. 50:640.

Nellis, C. H.; Wetmore, S. P.; and Keith, L. B. 1972. Lynx-prey interactions in central Alberta. J. Wildl. Manage. 36:320–329.

Novikov, G. A. 1962. Carnivorous mammals of the fauna of the USSR. Israel Prog. Sci. Transl., Jerusalem, and Natl. Sci. Found., Washington, D.C. 284pp.

Ognev, S. I. 1962. Mammals of the USSR and adjacent countries. Part 3: Fissipedia and Pinnipedia. Israel Prog. Sci. Transl., Jerusalem, and Natl. Sci. Found., Washington, D.C. 641pp.

Ontario Trappers Association. 1979. Fur sales service, market bulletin. Can. Trapper 8(2):9–12.

Orihel, T. C., and Ash, L. R. 1964. Occurrence of *Dirofilaria striata* in the bobcat (*Lynx rufus*) in Louisiana with observations on its larval development. J. Parasitol. 50:590–591.

Overton, W. S., and Davis, D. E. 1969. Estimating the numbers of animals in wildlife populations. Pages 403–455 *in* R. H. Giles, Jr., ed. Wildlife management techniques. 3rd ed. Wildl. Soc., Washington, D.C. 623pp.

Palmer, E. L. 1954. Palmer's fieldbook of mammals. E. P. Dutton, New York. 321pp.

Paradiso, J. L. 1972. Status report on cats (Felidae) of the world, 1971. U.S. Fish and Wildl. Serv. Spec. Sci. Rep. Wildl. 157:1–43.

Pence, D. B.; Samoil, H. P.; and Stone, J. E. 1978. Spirocercid stomach worms (Nematoda: Spirocercidae) from wild felids in North America. Can. J. Zool. 56:1032–1042.

Penner, L. R., and Parke, W. N. 1954. Notoedric mange in the bobcat (*Lynx rufus*). J. Mammal. 35:458.

Perry, J. S. 1972. The ovarian cycle of mammals. Hafner Publ. Co., New York. 219pp.

Peterson, R. L. 1966. The mammals of eastern Canada. Oxford Univ. Press, Toronto. 465pp.

Peterson, R. L., and Downing, S. C. 1952. Notes on the bobcats (*Lynx rufus*) of eastern North America, with the description of a new race. Contrib. R. Ontario Mus. 33. 23pp.

Petraborg, W. H., and Gunvalson, V. E. 1962. Observations on bobcat mortality and bobcat predation on deer. J. Mammal. 43:430–431.

Poelker, R. J. 1977. Food habits, age, and reproductive status of collected bobcats. Washington Game Dept., Rep. Fed. Aid Wildl. Rest. Proj. W-84-R, Study II-4. 3pp.

Pollack, E. M. 1949. The ecology of the bobcat (*Lynx rufus rufus* Schreber) in the New England states. M.S. Thesis. Univ. Massachusetts. 120pp.

———. 1950. Breeding habits of the bobcat in northeastern United States. J. Mammal. 31:327–330.

———. 1951*a*. Food habits of the bobcat in the New England states. J. Wildl. Manage. 15:209–213.

———. 1951*b*. Observations on New England bobcats. J. Mammal. 32:356–358.

Pollack, E. M., and Sheldon, W. G. 1951. The bobcat in Massachusetts. Massachusetts Div. Fish. and Game, Boston. 24pp.

Povey, R. C., and Davis, E. V. 1977. Panleucopenia and respiratory virus infection in wild felids. World's Cats 3(3):120–128.

Powell, J. L. 1971. Problems posed to the livestock industry by felids. Pages 127–131 *in* S. E. Jorgenson and L. D. Mech, eds. Proc. symp. native cats of North Am. U.S. Fish and Wildl. Serv., Twin Cities, Minn. 139pp.

Prater, S. H. 1971. The book of Indian animals. 3rd ed. Bombay Nat. Hist. Soc., Bombay, India. 324pp.

Progulske, D. R. 1952. The bobcat and its relation to prey species in Virginia. M.S. Thesis. Virginia Polytechnic Inst., Blacksburg. 135pp.

Provost, E. E.; Nelson, C. A.; and Marshall, D. A. 1973. Population dynamics behavior in the bobcat. World's Cats 1:42–67.

Pulliainen, E. 1965. Studies of the wolf in Finland. Ann. Zool. Fenn. 2:215–259.

———. 1974. [The number of bear (*Ursus arctos*) and lynx (*Lynx lynx*) killed in Finland in 1971.] Suomen Riista 25:106–108. (In Finnish, with English summary.)

Pulliainen, E., and Hyypia, V. 1975. [Winter food and feeding habits of lynxes (*Lynx lynx*) in southeastern Finland.] Suomen Riista 26:60–63. (In Finnish, with English summary.)

Rand, A. L. 1944. The recent status of Nova Scotia furbearers. Can. Field Nat. 58:85–96.

Rausch, R. A., and Pearson, A. M. 1972. Notes on the wolverine in Alaska and the Yukon territory. J. Wildl. Manage. 36:249–268.

Rhoads, S. N. 1903. The mammals of Pennsylvania and New Jersey. Privately printed, Philadelphia. 266pp.

Robinson, W. B., and Grand, E. F. 1958. Comparative movements of bobcats and coyotes as disclosed by tagging. J. Wildl. Manage. 22:117–122.

Rollings, C. T. 1945. Habits, foods, and parasites of the bobcat in Minnesota. J. Wildl. Manage. 9:131–145.

Rosenzweig, M. L. 1966. Community structure in sympatric Carnivora. J. Mammal. 47:602–612.

Rust, H. J. 1946. Mammals of northern Idaho. J. Mammal. 27:308–327.

Samson, F. B. 1981. Multivariate analysis of cranial characters among bobcats with a preliminary discussion of the number of subspecies. Pages 80–86 *in* Blum, L. G., and Escherich, P. C., eds. Bobcat research conference proceedings. Natl. Wildl. Fed. Sci. Tech. Ser. no. 6. 137pp.

Sandfort, W. W., and Tully, R. J. 1971. Status and management of the mountain lion and bobcat in Colorado. Pages 73–85 *in* S. E. Jorgenson and L. D. Mech, eds. Proc. symp. native cats of North Am. U.S. Fish and Wildl. Serv., Twin Cities, Minn. 139pp.

Santy, D. 1964. Some interesting lynx observations. Blue Jay 22(1):35.

Saunders, J. K. 1961. The biology of the Newfoundland lynx. Ph.D. Diss. Cornell Univ. 114pp. (Univ. Microfilms, Ann Arbor, Mich. Libr. Congr. Card no. 62-00965.)

———. 1963*a*. Food habits of the lynx in Newfoundland. J. Wildl. Manage. 27:384–390.

———. 1963*b*. Movements and activities of the lynx in Newfoundland. J. Wildl. Manage. 27:390–400.

———. 1964. Physical characteristics of the Newfoundland lynx. J. Mammal. 45:36–47.

Schaller, G. B. 1972. The Serengeti lion: a study of predator-prey relations. Univ. Chicago Press, Chicago. 480pp.

Schmidt, G. D. 1968. *Oncicola canis* from *Felis* (*Lynx*) in Alaska. J. Parasitol. 54:930.

Scott, P. P. 1976. Diet and other factors affecting the development of young felids. World's Cats 3(2):166–177.

Scott, P. P., and Lloyd-Jacob, M. A. 1955. Some interesting features in the reproductive cycle of the cat. Stud. Fertil. 7:123–129.

Seton, E. T. 1911. The arctic prairies. Chas. Scribner's Sons, New York. 415pp.

———. 1929. Lives of game animals. Vol. 1, part 1: Cats, wolves, and foxes. Chas. Scribner's Sons, New York. 337pp.

Smith, B. E. 1945. Wildcat predation on deer. J. Mammal. 26:439–440.

Stewart, R. R. 1973. Age distributions, reproductive biology, and food habits of Canada lynx in Ontario. M.S. Thesis. Univ. Guelph. 62pp.

Stiles, V. D. 1961. Observations on the behavior of the bobcat (*Lynx rufus rufus*) in the Quabbin Reservation. Dept. For. Wildl. Manage., Univ. Massachusetts, Amherst. 80pp. Manuscript.

Sweeney, S., and Poelker, R. J. 1977. A contribution toward an annotated bibliography on the bobcat and Canada lynx. Washington Game Dept., Completion Rep. Fed. Aid Wildl. Rest. Proj. W-84-R, Studies II.1 and III.1. 55pp.

Tate, G. H. H. 1947. Mammals of eastern Asia. MacMillan Co., New York. 366pp.

Trainer, C. 1975. Direct causes of mortality in mule deer fawns during summer and winter periods on Steen's Mountain, Oregon: a progress report. Proc. Annu. Conf. Western Assoc. State Game and Fish Comm. 55:163–170.

Ulmer, F. A., Jr. 1941. Melanism in the Felidae, with special reference to the genus *Lynx*. J. Mammal. 22:285–288.

Van Gelder, R. G. 1977. Mammalian hybrids and generic limits. Am. Mus. Novit. 2635:1–25.

Van Zyll de Jong, C. G. 1966a. Food habits of the lynx in Alberta and the MacKenzie District, NWT. Can. Field Nat. 80:18–23.

———. 1966b. Parasites of the Canada lynx. Can. J. Zool. 44:499–509.

———. 1971. The status and management of the Canada lynx in Canada. Pages 16—19 *in* S. E. Jorgenson and L. D. Mech, eds. Proc. symp. native cats of North Am. U.S. Fish and Wildl. Serv., Washington, D.C. 139pp.

Verberne, G. 1970. [Observations and experiments on *Flehmen* in various felid species.] Z. Tierpsychol. 27:807–827. (In German, with English summary.)

Wayre, P. 1969. Breeding the European lynx at the Norfolk wildlife park. Int. Zoo Yearb. 9:95–96.

Weinstein, M. S. 1977. Hares, lynx, and trappers. Am. Nat. 111:806–808.

Wing, L. W. 1953. A composite index of lynx abundance. J. Cycle Res. 2:21–24.

Yoakum, J. 1964. Observations on bobcat-water relationships. J. Mammal. 45:477–479.

Young, S. P. 1928. Bobcat kills deer. J. Mammal. 9:64–65.

———. 1958. The bobcat of North America. Stackpole Co., Harrisburg, Pa., and Wildl. Manage. Inst., Washington, D.C. 193pp.

Zezulak, D. S. 1981. Northeastern California bobcat study. California Dept. Fish and Game Rep. Fed. Aid Wildl. Rest. Proj. W-54-R-12, job. IV-3. 19pp.

Zezulak, D. S., and Schwab, R. G. 1980. Bobcat biology in a Mojave Desert community. California Dept. Fish and Game Rep. Fed. Aid Wildl. Rest. Proj. W-54-R-12, job IV-4.

CHET M. MCCORD, Division of Fisheries and Wildlife, Field Headquarters, Westboro, Massachusetts 01581.

JAMES E. CARDOZA, Division of Fisheries and Wildlife, Field Headquarters, Westboro, Massachusetts 01581.

VI

Pinnipedia and Sirenia

40

Seals

Phocidae, Otariidae, and Odobenidae

Keith Ronald
Jane Selley
Pamela Healey

NOMENCLATURE

ORDER. Pinnipedia
SUPERFAMILY. Phocoidea
FAMILY. Phocidae
SUBFAMILY. Phocinae
TRIBE. Phocini
GENUS. *Phoca*
  SPECIES. *Phoca vitulina*
  COMMON NAMES. Harbor seal, spotted seal, common seal
    SUBSPECIES. *P. v. vitulina, P. v. concolor, P. v. richardii, P. v. kurilensis, P. v. largha, P. v. mellonae.*
  SPECIES. *Phoca groenlandica*
  COMMON NAMES. Greenland seal, harp seal, saddleback seal

GENUS. *Pusa*
  SPECIES. *Pusa hispida*
  COMMON NAMES. Ringed seal, floe rat
    SUBSPECIES. *P. h. hispida, P. h. krascheninikovi, P. h. ochotensis, P. h. botnica, P. h. saimensis, P. h. ladogensis.*

GENUS. *Histriophoca*
  SPECIES. *Histriophoca fasciata*
  COMMON NAMES. Banded seal, ribbon seal

GENUS. *Halichoerus*
  SPECIES. *Halichoerus grypus*
  COMMON NAME. Gray seal

TRIBE. Erignathini
GENUS. *Erignathus*
  SPECIES. *Erignathus barbatus*
  COMMON NAME. Bearded seal
    SUBSPECIES. *E. b. barbatus, E. b. nauticus.*

TRIBE. Monachini
GENUS. *Monachus*
  SPECIES. *Monachus schauinslandi*
  COMMON NAMES. Laysan seal, Hawaiian monk seal

SUBFAMILY. Cystophorinae
GENUS. *Cystophora*

SPECIES. *Cystophora cristata*
COMMON NAMES. Hooded seal, crested seal, bladder nose seal

GENUS. *Mirounga*
  SPECIES. *Mirounga angustirostris*
  COMMON NAME. Northern elephant seal

SUPERFAMILY. Otarioidea
FAMILY. Otariidae
SUBFAMILY. Otariinae
GENUS. *Eumetopias*
  SPECIES. *Eumetopias jubatus*
  COMMON NAMES. Northern sea lion, Steller's sea lion

GENUS. *Zalophus*
  SPECIES. *Zalophus californianus*
  COMMON NAME. California sea lion
    SUBSPECIES. *Z. c. californianus, Z. c. japonicus, Z. c. wollebaeki.*

GENUS. *Callorhinus*
  SPECIES. *Callorhinus ursinus*
  COMMON NAMES. Northern fur seal, Pribilof fur seal, Alaska fur seal

SUBFAMILY. Arctocephalinae
GENUS. *Arctocephalus*
  SPECIES. *Arctocephalus townsendi*
  COMMON NAME. Guadalupe fur seal

FAMILY. Odobenidae
GENUS. *Odobenus*
  SPECIES. *Odobenus rosmarus*
    SUBSPECIES. *O. r. rosmarus*
    COMMON NAME. Atlantic walrus
    SUBSPECIES. *O. r. divergens*
    COMMON NAME. Pacific walrus

GENERAL CHARACTERISTICS

Within the immediate vicinity of continental North America there are 14 different species of pinnipeds (seals, sea lions, and the walrus) whose distributions are within the political jurisdiction of the United States and Canada. Because of the number of species, certain biological and physiological topics that are pertinent to these mammals are limited to discussion in general terms. Individual characteristics of the species and distribution ranges will be discussed separately.

Because management of these animals is complex, the topic is divided into those systems through which management operates, the policies that exist in the two countries, and the current status of each species. The species with few exceptions have been the center of considerable public interest, and at times even marked and bitter controversy. We hope that in the future the emotional aspects of management can be minimized and the basic necessity of obtaining enough scientific knowledge not hindered, to ensure an adequate conservation strategy for the Pinnipedia. The scope of this topic is so great, and the available literature so vast in quantity, that reference has been made to only some key papers and ideas. Some of the material in this chapter is based upon original research, hence, it is unreferenced and should be considered as work carried out by the University of Guelph. We refer those interested in further details to Ronald et al. (1976).

DESCRIPTION

Generally, the pinniped form is adapted to the aquatic and usually cold environment in which these animals exist. Their large size has evolved mainly in response to this cold environment (Scheffer 1958). The entire body is streamlined, from the smooth, rounded head down the trunk, which has no sharp protuberances, to the small tail tucked between the hind flippers (Scheffer 1958; Harrison and King 1965). The shortened humerus and femur are swimming modifications that allow the limbs to be almost completely withdrawn into the body. Only the flippers project. This not only results in a streamlined body, it also brings the source of power through muscle insertion closer to the body. A connecting web of skin between the digits creates an increased surface area for moving water (Harrison and King 1965). The head is sometimes flattened—an advantage for diving—and the external ears are reduced or absent; large eyes are situated well forward and sometimes rather close together. Although the neck is thick and muscular, it is very flexible. External genitalia and mammary teats are almost always withdrawn beneath or level with the body surface. A thick layer of subcutaneous fat or blubber smooths the contours of the body (Scheffer 1958; Harrison and King 1965). The epidermis is thick and tough, the hairs flattened and arranged in clusters, the oil glands large and numerous, and the dermis highly vascular (Scheffer 1958).

Although seals are found in all concentrations of water, there has been little internal change to ensure ionic regulation. The basic mammalian system, therefore, apparently has enough scope to accommodate moderate changes in the aquatic medium of seals.

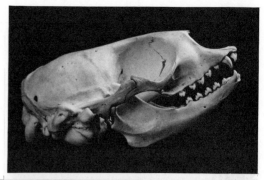

FIGURE 40.1. Skull of a young phocid. Top: lateral view of cranium and mandible. Bottom: dorsal view of cranium. Note that the sutures are open.

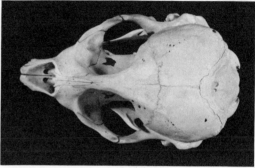

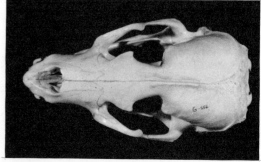

FIGURE 40.2. Skull of an adult otariid. Top: lateral view of cranium and mandible. Bottom: dorsal view of cranium.

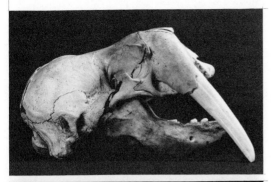

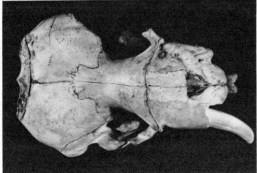

FIGURE 40.3. Skull of a young odobenid. Top: lateral view of cranium and mandible. Bottom: dorsal view of cranium.

Each of the three pinniped families, Phocidae, Otariidae, and Odobenidae, has its own distinguishing characteristics. North American phocids have no external ear pinnae, their testes are internal, and their mammae have two teats, with the exception of the bearded seal (*Erignathus barbatus*), which has four. Phocids have a dense single layer of hair, plus a thick layer of blubber under the skin to prevent heat loss (Maxwell 1967). The flippers are furred, and the large hind flippers face backward. These provide the means for aquatic propulsion by rhythmic lateral movements (Howell 1970). Locomotion is performed on land by a caterpillarlike flexion of the body. The hind flippers are elevated and the foreflippers are used for traction and negotiating rough surfaces or for swimming. The nails are the same size on all five digits (King 1964).

The postorbital processes of the skull are absent, as is the alisphenoid canal. The interorbital region of phocids is narrowly constricted relative to otariids and the walrus. The canines are elongated, and postcanine teeth usually have three or more cusps. The dental formula is 2-3/1-2, 1/1, 4-6/4-5 = 26-36 (figure 40.1).

The Otariidae, or eared seals, have prominent vestigial ear pinnae. This family is comprised of fur seals, which are distinguished by their dense, two-layered fur made up of long, coarse guard hairs and a fine, thick undercoat, and sea lions, which have a sparser and coarser coat. Like phocids, sea lions lack underfur (Maxwell 1967; Food and Agriculture Organization of the United Nations, Advisory Committee

on Marine Resources Research 1976, which will be referred to throughout the text as FAO Adv. Comm. 1976). The testes of otariids are scrotal and their mammae have four teats. Both the foreflippers and hind flippers are used in limited terrestrial locomotion (Howell 1970). The hind limbs can be rotated forward on land, enabling the animal not only to stand on four legs but also even to manage a kind of gallop. Most of the power for swimming comes from the forelimbs; the hind limbs are rarely active (Maxwell 1967). Their foreflippers have five rudimentary nails. The tail is distinct and free from the body. The tip of the tongue is notched and the dentition is unspecialized (King 1964).

In contrast to that of the phocids, the otariid skull has a postorbital process and an alisphenoid canal. The interorbital region is relatively wide. The first and second incisors are small; the third incisor is caninelike. Canines are large, conical, and recurved; the cheek teeth have one main cusp. The dental formula is 3/2, 1/1, 4/4, 1-3/1 = 34-38 (figure 40.2).

The Odobenidae, or walrus (*Odobenus rosmarus*), has no external ear pinnae, its testes are internal, and its mammae have four teats. Walruses can also rotate their hind limbs forward and terrestrial locomotion is similar to that of the Otariidae. When swimming, the hind flippers move from side to side while the foreflippers are used alternately (Howell 1970). The foreflippers have five small distinct nails. The tail of the walrus is enclosed in a web or skin. The tip of its tongue is rounded. Its integument is nearly bare. For warmth the walrus depends upon its large quantities of blubber (Maxwell 1967).

The upper canines are extremely enlarged, and form ever-growing tusks. The remainder of the dentition is smaller in size, similar, and forms a continuous row. The dental formula of adults is variable, but is usually 1/0, 1/1, 3/3, 0/0 = 18. (Figure 40.3.)

PHYSIOLOGY

Growth. In an ideal environment every animal has the potential to reach its maximum size. Growth is a feedback mechanism that reflects the variables in the environment. These include weather, food availability, herd density, and disease. Growth rates reflect changes in the environment, the herd, and the individual.

In seals, weights and girths change seasonally because of reproductive stress, availability of food, molting, and other factors. They are, therefore, not a reliable measure of size. Also, these measures are difficult to obtain in the larger species. A pinniped's size is usually expressed in length (Laws 1959) (Table 40.1). Girth measurements are valuable as an expression of blubber thickness.

SPECIES-SPECIFIC GROWTH PATTERN. Compared to other mammals, most seals receive a relatively short period of maternal care, because they are born in an advanced state of development. Neonates of the different species are all different weights at birth in relation to their mother's weight, but some seals accelerate prenatal growth to accommodate for the difference in

TABLE 40.1. Growth parameters of pinnipeds

Pinniped	Birth Weight (KG)	Birth Length (M)	Lactation	Weaned Weight (KG)	Weaned Length (M)	Increase during Lactation	Length at Sexual Maturity (M)	Length at Physical Maturity (M)	Length at Sexual Maturity As % of Final Length	Female Adult Weight (KG)	Weight of Pup As % of Mother's Weight	
											Birth	Weaned
Harbor seal (*Phoca vitulina*)	10[a]	0.82[a]	5–6 weeks[a]	24[a]	0.97[a]	2.6X[b]	1.55 M[a] 1.40 F[a]	1.61 M[a] 1.48 F[a]	87.5[c]	89[b]	11.2[b]	28.6[b]
Ringed seal (*Pusa hispida*)	4[a]	0.65[a]	8–10 weeks[a]	16[a]	0.80[a]	2.7X[b]	1.23 M[a] 1.17 F[a]	1.38 M[a] 1.35 F[a]	86.0[c]	68[b]	6.7[b]	18.0[b]
Ribbon seal (*Histriophoca fasciata*)	8.4[d]	80.5[d]	4–6 weeks[b]	21.3[a]		2.4X[b]		mean length 1.7 M[e]	92.4[a]	81[b]	13.6[b]	32.4[b]
Harp seal (*Phoca groenlandica*)	11[a]	0.90[a]	1.5 weeks[a]	33[a]	1.04[a]		1.70 M[a]	1.76 M[a]	92.3[c]	130[f]	8.0[f]	27.0[f]
Gray seal (*Halichoerus grypus*)	14[a]	0.89[a]	2–3 weeks[a]	40[a]	1.07[a]		2.00 M[a] 1.85 F[a]	2.40 M[a] 2.02 F[a]	83			
Bearded seal (*Erignathus barbatus*)	45[a]	1.2[a]	1.7–2.2 weeks[g]	90[a]	1.52[a]	2.5X[b]	2.2 M[a] 2.10 F[a]	2.30 M[a] 2.30 F[a]	80.6[c]	270[b]	12.3[b] 12.9[b]	31.2[b]
Hawaiian monk seal (*Monachus schauinslandi*)	16[a]	1.0[a]	5 weeks[a]	50[a]	1.38[a]			2–3[a]				
Hooded seal (*Cystophora cristata*)	12[a]	0.91[a]	1–2 weeks[a]		1.20[a]		2.36 M[a] 2.02 F[a]					
Northern elephant seal (*Mirounga angustirostris*)	36 M[g] 31.5 F[a]	1.53 M[g] 1.47 F[j]	4 weeks[h]					4.5 M[a] 3.6 F[a]				
Northern sea lion (*Eumetopias jubatus*)	12[a]	0.98[a]	12 weeks[a]	20[a]				3.20 M[a] 2.70 F[a]				
California sea lion (*Zalophus californianus*)	5–6[g]	0.75[g]	20–48 weeks[g]					2.2 M[b] 1.8 F[a]				
Guadalupe fur seal (*Arctocephalus townsendi*)			extended maternal care[i]					1.9 M[i] 1.4 F[i]				
Northern fur seal (*Callorhinus ursinus*)	5[a]	0.64[a]	12 weeks[a]	13[a]			1.4 M[a] 1.10 F[a]	2.00 M[a] 1.30 F[a]	88.2[c]	800[b]	4.3[b] 5.3[b]	
Walrus (*Odobenus rosmarus*)	55[a]	1.2[a]	100 weeks[a]	36[a]	2.00[a]	8.8X[b]	2.6 M[a] 2.4 F[a]	3.00 M[a] 2.60 F[a]	89.8[c]			41.9[b]

SOURCES:
[a] Bryden 1972
[b] Irving 1972
[c] Laws 1956
[d] Burns 1971
[e] Popov 1976
[f] Innes, Stewart, Lavigne, personal communication
[g] Burns 1978, DeLong 1978
[h] FAO, Adv. Comm. 1976
[i] Fleischer 1978a

size. They do not prolong gestation as land mammals do (Laws 1959). In fact, the period of active growth during gestation is the same for all pinnipeds except the walrus.

Recent studies on neonatal growth in harp seals illustrate some of the growth patterns of newborn seals. There are two critical periods in the harp seal's early life; the first few hours after birth, and the time period after weaning. Immediately after birth, the pup must cope with a harsh environment that taxes its thermoregulatory capacities, because it lacks a well-developed insulating blubber layer (Blix et al. 1975). Nonshivering thermogenesis (see the section "Thermoregulation") helps it to survive initially. Within minutes or hours of birth, the mother's fat-rich milk provides the energy necessary for the enormous weight gain experienced during lactation. All seals experience this rapid postnatal weight gain, but the rate is species specific. The harp seal pup gains approximately 2.5 kg per day for nine days. Newborn otariids, large at birth, nurse for several months. The majority of the weight gain in the harp seal is deposited as a subcutaneous layer of fat; the rest is gained in the core body mass. The sculp (skin and blubber weight) is 60 percent of the total body weight at the gray coat stage (Stewart and Lavigne 1980).

A weaned pup loses the energy gained from its mother's milk. This loss is partially compensated for by accumulation of external energy by the fur in the form of solar radiation. Delayed weaning corresponds to delay in dentition development. This is aptly demonstrated in the walrus, whose tusks, which are needed to assume adult dietary habits, are not developed until after the one- or two-year period of nursing (Laws 1959). Phocids, however, shed their milk teeth *in utero* and are born with adult dentition.

As the newly weaned pup learns to forage in the aquatic environment, further stress is placed on its energy requirements and it loses body weight. In the harp seal, it is lost from the body core, not from the subcutaneous fat reserves as is often suggested. These reserves are critical for the prevention of heat loss. Juvenile mortality at this time might be significant and should be taken into account in management strategies (Stewart and Lavigne 1980).

Once this postweaning weight loss is overcome, the nutritionally independent seal grows steadily. Final body length is attained at approximately five years of age in the harp seal (Innes et al. 1980). A pinniped generally reaches sexual maturity, but not necessarily reproductive activity, at about 86 percent of its final length.

Some species demonstrate sexual dimorphism, the male being larger than the female. This is mainly characteristic of otariids; however, the northern elephant seal is also sexually dimorphic. Attainment of this larger size by the male occurs during a postpubertal growth spurt (Laws 1959; Bryden 1972), although some species show sexual size differentiation as early as one month of age (Payne 1979; Stewart and Lavigne 1980). This later growth period accounts for the time lapse between sexual maturity and reproductive activity (Bryden 1972). Data on the growth of individual species is given in table 40.1.

DENSITY-DEPENDENT INFLUENCE ON GROWTH. Changes in per capita food consumption can result in significant differences in growth rates. This can have important ecological consequences regarding population parameters and the impact of the population on its ecosystem. An increase in stock size may result in increased competition for food both intraspecifically and interspecifically, which in turn causes a decrease in per capita consumption. A large population will consume less on a per capita basis. This decreased food consumption results in a slower growth rate and delayed attainment of sexual maturity. Per capita food consumption may change as a result of either increased population size or interspecific competition. Growth rates, age of maturation, and age of first whelping significantly affect the energy requirements of the entire population over time and correspondingly affect the ecosystem (Innes et al. 1980).

Molting. Periodic growth and replacement of the pelage of mammals are necessary if individuals are to survive and function efficiently. Pinnipeds renew their coats annually by molting (Ling 1970).

The pelage of seals performs three functions: waterproofing, streamlining, and insulating. While the adult pelage is hydrodynamic in function (Ling and Button 1975), the coat of the newborn seal is more important as an insulator (Ling 1974). In all pinnipeds the overhairs are flattened. Phocids have few if any underfur or secondary hairs. Otariids possess a thick, water-repellent underfur that insulates against heat loss in both air and water (Ling 1970).

Molting cannot be considered as an independent process. It merely marks the end of a complex pelage cycle and is the outward sign of more basic subcutaneous phenomena (Ling 1970). This pelage cycle is a morphogenetic process occurring within the annual cycle and is influenced not only by environmental factors but also by the physiology and behavior of the individual (Ling 1974). The direct proximate stimulus of light acting through neuroendocrine pathways determines the pelage cycle and regulates it with respect to season. Such factors as temperature, behavior, and nutritional and reproductive status modify the influence of photoperiod (Ling 1970; Ronald et al. 1970). The pelage cycle is independent of the reproductive cycle (Ling 1974), and molting usually occurs between parturition/lactation and implantation of the new blastocyst (Ling 1970). Molting is also affected by the ability of the seal to store energy sources in a subcutaneous layer of fat (Ling 1974).

In some pinnipeds the first molt occurs *in utero* shortly before birth, as in the harbor seal (Ling 1970) and probably Steller's sea lion (Ling and Button 1975). It is possible that prenatal molting of this first pelage allows the newborn of this species to enter the water and swim sooner than others (Mohr 1950). In ice-breeding pinnipeds, a rapid postnatal molt is typical, and allows them to live in the aquatic environment (Ling and Button 1975). Scheffer (1958) correlated color patterns of newborn seals with ancestral habitats. He suggested that the prenatal molt is an adaptation by species that now inhabit climatic zones other than those in which they originated. In each family, the adult pelage of seals is always different from that of newborns. Among the Phocidae, only the harbor and hooded seals molt *in utero*, while the others shed their first pelage within a few weeks of birth. Otariid pups have dark, wettable body fur. This is shed a few months after birth and replaced by a more adultlike pelage. The light coat of a newborn odobenid is molted after a few weeks, and after a delay of some months the growth of the adult pelage is slowly accomplished (Ling 1974).

Although pinnipeds molt annually, only the Phocidae replace their entire coat each time. In northern elephant, hooded, and monk seals, molting is rapid. The hairs are shed along with large sheets of cornified epidermis (Ling 1970). The northern fur seal replaces only 75 percent of its guard hair each year, while 25 percent of the underfur fibers of mature males and 35 percent of that of mature females is renewed annually (Scheffer and Johnson 1963).

In order to molt, pinnipeds must modify their behavior to maintain the necessary skin temperatures. Accordingly, additional rest is essential for heightened epidermal mitotic activity and for high tissue temperatures of around 37° C (Feltz and Fay 1966). Ling (1970) gave a detailed description of cellular activity during the hair cycle.

Thermoregulation. Many factors affect heat balance in animals. From a purely physical aspect, it is possible to produce a heat balance model for a homeothermic animal using simple heat transfer principles (Øritsland 1978). However, there are physiological, behavioral, and environmental complexities, many of which are only now being studied in humans and marine mammals.

The different species of seals share some of the same variables affecting thermoregulation. All seals must be able to maintain a uniform body temperature in two very different media, air and water. Because water has a greater thermal capacity than air, it is a more effective cooling agent. However, the different species of seal live throughout a range of habitats and climates from the arctic to the subtropics, which imposes different stresses. For example, seals such as the northern fur seal migrate annually from temperate waters to more northerly latitudes. Other pinnipeds, like the walrus, live a gregarious existence, crowded together by the thousands in rookeries where they are able to either benefit or suffer from their neighbor's thermoregulatory capacities.

Diving behavior places some demand on the animal's metabolism, and the need to conserve heat becomes critical, a factor that is compensated for by certain circulatory adaptations (see the section "Diving Physiology"). Disposing of heat can also be a problem. Because sweating is not an appropriate method for marine mammals (Hemplemen and Lockwood 1978), the seal has made specific accommodations in its flippers.

Phocids are generally the smallest pinnipeds, while otariids are medium to large in size, as is the walrus. Adult phocids have thick blubber, their newborn little or none. The walrus has a thicker layer than phocids to accommodate for its lack of fur, but it also has a very thick, rugose hide. The otariids have little or no blubber, as they do not need the excessive insulation, but rely on their fur for its insulative properties.

Pelage is important in establishing the heat balance of pinnipeds, except in the walrus. Its insulative properties (otariids have air trapped between the layers, whereas phocids normally have wettable hair), its color (often white in newborn phocids unless the initial layer has been shed *in utero*, gray to black in otariid young), and its structure are complexities that interact with solar and other environmental conditions (Øritsland and Ronald 1978a).

Coloration does not appear as important as the presence of air vacuoles and hair pigmentation in determining transmittance values. However, arctic fur allows more solar heating at skin level (producing a "greenhouse effect") than nonarctic fur (Øritsland 1978). In phocids, transmittance of radiation through the hairs is hindered only by pigments, because the hair is solid with no medulla and has a very transparent cuticle and cortex, in contrast to that of otarids. In mammals, hair transmittance is very important in solar heating (Øritsland et al. 1978). At skin level, heating is a function primarily of fur depth, coupled with hair transmittance and coat reflectance (Øritsland and Ronald 1978b).

Environmental conditions such as wind, radiation, and precipitation affect metabolism, but to what extent is still not known. Øritsland and Ronald (1977) produced a computer simulation model that attempted to include all the variables involved in heat balance in harp seals (see Øritsland 1978). Some of the factors studied were ambient temperature, solar radiation (absorbance, reflectance, and transmittance), wind chill, and precipitation (rain, hail, and snow), which can cause severe cold stress in neonates.

Because of their diving habit, and often extreme environmental conditions, seals have certain physiological adaptions for thermoregulation. Heat loss in harp seals is fairly localized, so expired heat loss and convective heat loss from the body are less than 25 percent of the metabolic heat production (Gallivan 1977). During a dive the seal's general metabolic rate drops (Blix 1976). However, the seal still maintains heat balance without increased muscle activity or metabolic rate—a temperature of 13° C for the harbor seal and 5° C for the gray seal. The harp seal can even maintain its body temperature while remaining motionless in the water, possibly because of a higher metabolic rate. Otariids, such as the California sea lion, that do not have the insulative properties of phocids must swim at 0° C to balance heat loss (Hempleman and Lockwood 1978). Water temperature normally does not exert a significant influence on respiration or diving patterns in harp seals. Measurements of metabolism and respiration in relation to water temperature showed no difference in metabolic rate, core temperature, ventilation, gas exchange, or diving pattern over a range of water temperature from 1.8 to 28.2°C. The cooling power of the aquatic environment always greatly exceeded the heat production of the seal, and internal conduction rather than external convection governed heat loss from body surfaces (Gallivan and Ronald 1979). The insulative properties are such that the core temperature of a seal in ice water conditions extends only to a depth of 5 cm from the skin surface (Hempleman and Lockwood 1978).

Primary sources of heat loss are the head and flippers (Gallivan 1977). The flippers are not insulated, but heat loss is controlled by the parallel structure of the veins and arteries, which allows for a countercurrent blood flow (Hempleman and Lockwood 1978). Phocids have a system of arteriovenous anastomoses in the superficial layer of the dermis (occurring in greater densities in harp seals than in hooded seals), which also seems to act as a temperature regulatory device (Bryden 1978; Blix et al. 1979).

During a dive the seal shunts its blood supply away from the surface areas and internal organs to supply those areas such as the heart and brain which are critically in need of oxygen. Some heat loss is prevented in this way and the large venous plexus in the neck is embedded in large deposits of brown adipose tissue, a condition also found on the pericardium and around the kidneys. It is speculated that this "venous plexus–brown fat complex might function as a high-efficiency tubular heat exchanger" (Blix et al. 1975; Blix 1976:176).

As harp seal pups are born they leave the 37° C warmth of the uterus (Grav et al. 1974) for a climate where ambient temperatures can be as low as −15° C, combined with winds of up to 30 ms⁻¹. Their white coat is wet and they have no blubber, so they shiver vigorously for over an hour until their coat is dry (Blix et al. 1979). There are also physiological adaptations that produce heat and help them to survive this critical period. They appear to have a high tolerance to low body core temperatures under cold conditions (Øritsland and Ronald 1978b), and the presence of loosely coupled mitochondria in brown fat and in dark muscle fibers allows for a process called nonshivering thermogenesis (George and Ronald 1973, 1975; Øritsland and Ronald 1978a; Blix et al. 1979). The "brown fat" actually appears to have an ultrastructure between brown and white fat, which might indicate the transformation to blubber which will take place during lactation. Glycogen stores are also believed to be depleted during this process (Blix et al. 1979). Hyperthermia, for the newborn exposed to prolonged direct sunlight, can be as severe a problem as cold conditions. Pups appear able to alleviate the condition up to ambient 15±1° C, by heat dissipation through the flippers, because they are unable to lower their metabolism sufficiently (Øritsland and Ronald 1978b).

Vision. The large, well-developed eyes of pinnipeds are an adaptation to both aquatic and terrestrial environments and are indicative of the importance of visually guided behavior (Walls 1942; Hobson 1966; Schusterman 1972). The ability to see well both above and below water under a variety of light intensities is essential to the survival of these animals. They use their vision for finding landmarks when migrating, establishing territories, or hauling out to give birth, for avoiding obstacles and predators, for exploring the ocean floor, for locating prey, and for recognizing conspecific individuals (Schusterman 1975). Behavior is visually guided but because each species' feeding and social habits differ, so do their perceptual capabilities (Schusterman 1972).

Anatomically, the seal eye is adapted in numerous ways to function in the two different optical media. Lavigne et al. (1977) believe that the seal retina is duplex or has two types of photoreceptors (rods and cones) that allow the seal to function both on land and underwater during both day and night. This theory is supported by the examination of various aspects of seal vision, such as visual acuity, spectral sensitivity, pupilomotor response, and critical flicker frequency. The retina's cone receptors permit detailed vision under high light intensities, while a rod-dominated retina with its increased sensitivity gives effective perception underwater, especially in low luminance. It is possible that rod and cone receptors function in both surface and underwater vision and that the transition from cone to rod vision occurs when background light is decreased. The presence of a Purkinje shift (Purkinje 1825) in the seal eye supports the idea of a duplex retina and at least two types of photopigment with different absorption spectra, one functioning under photopic and the other under scotopic conditions. The pupil responses, which give evidence of the different physiological responses of both rods and cones, occur at lower light intensities in seals than in humans and point to a more sensitive, duplex retina. The critical flicker frequency response contour of the seal shows a shift in function from one type of photoreceptor to another, again indicating a duplex retina (Lavigne et al. 1977).

Other anatomical adaptations include a highly developed tapetum (retinal reflecting layer), which provides increased visual sensitivity, especially underwater (Nagy and Ronald 1975; Lavigne et al. 1977), and a large, spherical lens similar to that in fish (Walls 1942; Munz 1971; Jamieson and Fisher 1972). This allows for sufficient accommodation underwater, in the absence of the corneal refracting surface (the refractive indices of the cornea and the water are the same), to focus an image on the retina (Lavigne et al. 1977; Schusterman 1972).

Seal visual pigments and the shape of the pupil are also important adaptations. Some pinnipeds are nocturnal, opportunistic feeders and will approach their prey from below as it is silhouetted against the surface light above. Under relatively clear coastal water conditions some seals can recognize small food items such as herring or sardines even under a cloudy sky at a depth of 200 m. Under ideal conditions in the open ocean the same species can be seen from depths slightly greater than 1 km (Schusterman 1975). The absorption spectra of rod visual pigments tend to correlate with the spectral distribution of radiant energy in the underwater environment, thus maximizing contrast for silhouetting prey species against the ambient surface light (Lavigne et al. 1977). The type of visual pigment each seal has is determined by its photic environment. For example, the spectral region of greatest intensity in temperate and polar seas is shifted toward greener wavelengths when compared to the blue of tropical oceans. Thus, the harp seal would have a greener visual pigment adaptation, while a tropical species such as the southern elephant seal would have a bluer visual pigment (Lavigne et al. 1977). In pinniped visual pigments, then, there is an adaptation to the underwater environment which facilitates feeding.

The pupil, which is narrow and "teardrop in shape" (Jamieson and Fisher 1972; Lavigne and Ronald 1972), performs an important function. Its size varies with the ambient light intensity, independent of whether the seal is on the surface or underwater (Lavigne et al. 1977). In air, the corneal astigmatism inherent in seals is effectively reduced when the pupil closes to a narrow vertical slit, which acts as a pinhole providing the eye with a huge depth of focus in that meridian. Under dim light this astigmatism causes a loss of visual acuity, making the eye strongly myopic (Schusterman 1972; Lavigne et al. 1977).

Finally, dark adaptation in the seal is important because of the animal's transition from high light intensities on land or ice during sunny days to much lower levels of light underwater, at night, or during the dark polar winter. Its narrow pupil reduces the amount of light energy reaching the retina, thus minimizing the effects of prolonged exposure to bright light, such as the sun on ice and snow, on subsequent dark adaptations (Lavigne et al. 1977). In humans, exposure to sunlight can cause temporary and cumulative effects on night vision (Hecht et al. 1948). Although the harp seal initially adapts quickly to dark, it takes 30–40 minutes for maximum sensitivity to be reached. This rate of adaptation could affect feeding habits, limiting the diving depth during daylight hours and accounting for the common occurrence of night feeding among seals when the low illumination in air allows for sufficient time to dark-adapt prior to the dive (Lavigne et al. 1977).

Hearing. Pinnipeds use their sense of hearing in a variety of ways: to communicate with conspecific individuals, find prey, avoid predators, and navigate by means of ice and shallow water wave noises (Terhune and Ronald 1974).

The seal is capable of hearing both in air and underwater, although in the latter environment its hearing is more sensitive. The range of hearing extends from 0.7 to 32 kHz in air and from 1 to 100 kHz underwater, with the best frequencies being 17–25 kHz (Terhune and Ronald 1974). The hearing capabilities of seals and humans have similar features, these being pitch discrimination abilities (Møhl 1968b), the critical ratios (influence of background noises) (Terhune and Ronald 1971), and the ability to localize sounds (Møhl 1964; Terhune 1973).

The seal ear is fully adapted to the aquatic environment (Møhl 1964, 1968a; Terhune and Ronald 1971, 1972). Many of its anatomical adaptations are described here, but for a more detailed description, see Ramprashad (1975). Water is prevented from entering the ear when the seal submerges by the loss of pinna in the outer ear of phocids, and by changes in the auricular muscles that open and close the outer ear. There is both an internal constriction of the membranous part of the outer ear at the internal pinna and a subsequent closing of the external orifice. Adaptations to accommodate the increased pressure during a dive are an elastic, cartilaginous outer ear canal and the presence of large blood sinuses within the wall of the outer ear canal. When engorged with blood, these sinuses could reduce the volume of the outer ear canal by acting as a pressure-regulating device, and they might also form a fluid wall around the outer ear canal, thus preventing its complete collapse during diving (Ramprashad 1975).

Adaptations within the middle ear are associated with pressure regulation. The presence of cavernous tissue within the middle ear mucosa (Tandler 1899) and the thick medial wall of the auditory tube are the most prominent features (Møhl 1968a; Kooyman et al. 1970). During a dive the distension of this cavernous tissue would reduce the pressure difference between the middle ear and the nasopharynx, thus facilitating muscular opening of the lateral wall of the auditory tube by its associated muscles (Møhl 1968a; Ramprashad et al. 1973). Air at ambient pressure within the nasopharynx would accomplish pressure regulation in a manner similar to that of other mammals (Møhl 1968a; Ramprashad et al. 1973).

No specific aquatic adaptations were found with the seal's vestibular system. However, the size of the macula utriculi and the presence of the crista neglecta (not found in terrestrial mammals) are unique features. The large macula is probably important for body orientation underwater due to the stimulation of the otolithic receptors in response to gravity. The crista neglecta acts as an extrasensory area that may be of importance in rotational acceleration of the seal's body (Ramprashad et al. 1972). The morphological features of the phocid cochlea are similar to those of other high-frequency hearing mammals. A major difference within the phocid cochlea is the position and size of the round window, which may be significant to the capability to hear at high frequencies (Ramprashad 1975).

In air, hearing results from sound conducted through the meatal orifice or, if this is blocked, via the superficial tissues ventral to the orifice; these latter are important in underwater hearing. Thus, there is a possibility of two parallel inputs that merge in front of the middle ear complex, the underwater input being best adapted (Møhl and Ronald 1975). Because the external ear is closed underwater, hearing cannot be accomplished by conventional air conduction mechanisms. The blood sinuses of the outer ear may aid in the transmission of sound energy to the tympanic membrane and hence to the cochlea via the ossicular chain. Sound energy may also be transmitted through the cavernous tissue of the middle mucosa (Ramprashad 1975). This feature may explain the seal's acute directional hearing capabilities underwater.

Mankind is continually introducing new sounds into the seal's underwater world that are a great deal louder than the natural ones. This "noise pollution" could adversely affect the seal's hearing and subsequent behavior as these animals make efforts to avoid such disturbances.

Echolocation and Phonation. Echolocation can be defined as the use of the echo of an animal's phonation by that animal to determine the location, distance, or characteristics of the echoing object (Norris 1969). Although some research has been done that would indicate that pinnipeds can echolocate, nothing definitive has yet emerged from these experiments. Norris (1969) supports the probability of an echolocation skill in seals, although the degree of development and extent of use likely vary with the habitat and the individual animal.

Strongly supporting the probability of an echolocation skill is the fact that pinnipeds, particularly those in the icy polar seas, tend to be very vocal and as yet few alternative functions for these sound emissions have been found. However, Ray et al. (1969) stated that the frequency-modulated calls of bearded seal males in breeding condition are perhaps used in establishing territories. Also, the underwater belllike sound of the walrus is believed to be associated with sexual activity. This "song," produced by adult males, probably has territorial and courtship functions (Ray and Watkins 1975). Although little is known of the functions of pinniped phonations, many species' sounds have been recorded, and the most common vocalizations are faint clicks (Norris 1969). Phonations characteristic of a particular species include the growls and barks of *Zalophus* (Schusterman 1967) and the bell sound of the walrus (Schevill et al. 1966), made with the help of its pharyngeal pouches (Fay 1960).

For further details on phonation, see the behavior sections under individual species.

Circulatory System. HEART. Diving animals have a larger blood volume than nondiving animals, to allow for an increased oxygen capacity during a dive (Elsner 1969; Vallyathan et al. 1969; Ronald 1970; Andersen 1977; Drabek 1977). However, the seal's heart is not proportionally larger than those of the terrestrial animals (Slipjer 1962; de Kleer 1972; Drabek 1977). In fact, Bryden (1972) calculated that, in phocid species, heart weights were on the average 0.68 percent of body weight. In form the bifid heart has a very deep cleft, a fetal characteristic that persists until nearly the adult stage (Murie 1970; de Kleer 1972). It is shorter and broader than that of terrestrial carnivores, such as the dog (Muller 1940; Drabek 1977). Drabek (1975) saw the broad form as a specific adaption to diving because of the extreme hydrostatic pressure encountered in deep dives. Since the flexible ribs allow the thorax to be highly collapsible, and since the diaphragm is at an oblique angle, the bifid heart is more advantageous because it is less subject to deformation (Harrison and Kooyman 1966; Drabek 1977). Placement of the heart in the thoracic

cavity is symmetrical, a condition that, along with the even distribution of the lungs, favors stability (de Kleer 1972).

The large aortic bulb is the most important external feature of the heart. Its size is directly correlated to the individual species' habitat and needs. It is functionally significant during a dive because its extreme elasticity allows and ensures blood perfusion of the coronary system and brain during diastole. It may also be important for storage during diastole, ensuring that blood pressure does not increase too much (Drabek 1977).

The seal's vagus nerve is more prominent than that in domestic animals, and unlike other animals it has no vagal escape mechanism for rest during prolonged vagal stimulation (de Kleer 1972). Such stimulation occurs in a dive and an escape mechanism could spell death for the seal, who might never come out of the dive. Resting time between heart beats is highly irregular in the seal, even during bradycardia.

In general, seals have long, ventricular chambers that are narrower than those of any fissiped family yet examined. The Weddell seal, a deep diver, has the longest and narrowest right ventricle of those seals so far examined; the leopard seal, a shallow diver, has a broad structure (Drabek 1977). The ventricular wall is thicker on the left than on the right, and the mass of the right is a lesser percentage of the total heart weight. Speculation about the significance of this is that a "thinner walled and a less massive right ventricle, with a long, narrow chamber, would be better designed for dilation in response to pulmonary resistance during diving" (Drabek 1977:223). The harp seal's right ventricle is extremely dilated during a diving reflex when the pulmonary vascular resistance is greatly increased. This fact would explain the compensatory mechanism of the alteration of the ventricular structure (Drabek 1977).

VENOUS SYSTEM. Most adaptions in the venous system are for the habit of diving. In pinnipeds the jugular system is not highly developed (Harrison and Tomlinson 1956). The blood drains from the cranium by two hypocondylar veins that join to form a large sinus (King 1964). The vessels leaving this sinus on either side of the spinal cord join together to form the major venous exit from the brain, the extradural vein (Ronald et al. 1977). The extradural vein also communicates with almost every other branch of the venous system (King 1964). Anteriorly there are branches from the dorsal muscular and intercostal veins. More of the dorsal intercostal veins are connected with plexuses to the abdominal walls and in the renal plexus. Posteriorly the extradural vein receives branches from the renal and pelvic plexuses, the dorsal muscles, and the veins of the abdominal walls (King 1964).

There are vast systems of these venous plexuses in the pinniped's neck, pericardium, and abdominal wall adjacent to the kidney, and one exists as a stellate renal plexus. In the harp seal these are imbedded in functional brown adipose tissue containing mitochondria that behave like loosely coupled mitochondria, sustaining a high rate of heat production. It is speculated that this brown adipose tissue–venous plexus complex functions as a high-efficiency tubular heat exchange system that warms venous blood returning to the already cooled core after a dive. This is probably effected by a reflex activation of the sympathetic innervation (Blix et al. 1975).

The vascular system of the harp seal has been fully detailed (de Kleer 1972, 1975; St. Pierre 1972, 1974; Ronald et al. 1977).

Diving Physiology. Because of its habitat, a seal is forced to submerge without breathing for protracted periods for protection, for food, for escape from predators, and for locating new breathing holes. While the lung oxygen capacity of seals is comparable to that of humans (5l/kg of body weight), seals display a number of physiological adaptations for homeostatic compensation during dives of which humans are certainly incapable (Irving 1966; Ronald et al. 1977; Hempleman and Lockwood 1978). For example, many pinniped species are capable of surviving dives lasting 20–30 minutes, and as long as 60 minutes in the Weddell (Kooyman 1966) and baikal seals.

These adaptations can best be described after the mechanism of diving and the circulatory changes examined.

As the seal's face is submersed, the trigeminal nerve receptors around the nose are stimulated, causing a reflex apnoea that is reinforced by the collapsed lungs from a predive expiration. The apnoea then causes a reduced activity in the pulmonary vagal reflex, which in turn results in bradycardia and vasoconstriction. As arterialhypoxia and hypercapnia develop, arterial chemoreceptors are stimulated. Their products provide a reflex reinforcement for cardioinhibition, vasoconstriction, and humoral responses (Angell James and De Burgh Daly 1972).

All these mechanisms turn the circulatory system of the seal into a heart-brain system, so that oxygen may be conserved, and therefore supplied to those tissues that are relatively sensitive to even short periods of hypoxia and rationed to the others.

Hol et al. (1975) found that the redistribution of blood during a dive was a result of the vena caval sphincter, a function independent of the dive reflex. This sphincter acted as a mediastinal bypass, protecting the anterior caval vein and the thermoregulatory flipper veins against venous stasis, as well as a possible creation of large venous reservoirs. In fact, studies show that reservoir formation in the hepatic sinuses and the abdominal section of the posterior vena cava occurred both before and up to 40 sec after constriction of the posterior vena caval sphincter. The tissues not receiving blood are almost completely ischaemic. This condition eliminates normal *vis a tergo,* the force that moves extensive volumes of oxygenated blood held in the posterior venae cavae (PVC), splanchnic circulation, and many tributaries of the venae cavae into the heart-brain circulation. At the same time, the oxygen-depleted blood is removed, allowing passage of oxygenated blood from the hepatic sinus to the heart-brain circulation (Ronald et al. 1977). Meanwhile, the

cervical vertebral venous system, the major connection between the extradural intervertebral vein and the anterior venae cavae (AVC), is closed during a dive, thus increasing the resistance to the flow of blood from the EIV into the AVC.

In the splanchnic system, the blood is apparently transferred to the PVC by a profound peristaltic venoconstriction. Muscle pumping aids the movement of blood into the diving circulation from the tributary veins of the vena cava. The hepatic sinus expands on diving, to accommodate the large volumes of blood being moved out of other veins. The caval sphincter appears to contract, retaining the blood engorging the hepatic sinus (Ronald et al. 1977).

During a dive, the heart is supplied with blood from both venae cavae. On postsystolic expansion of the heart, a stream of blood enters the heart from the AVC. Synchronously, the caval sphincter appears to open briefly, allowing a bolus of blood to pass from the hepatic sinus to the heart. The ratio of blood from the AVC to the PVC appears to be regulated by the occasional prevention of the opening of the caval sphincter on expansion of the heart and the occlusion of the cervical vertebral venous system, with a resultant increase in resistance to blood flow from the extradural intravertebral vein (EIV) to the AVC (Ronald et al. 1977).

It is felt that circulation through muscles is shut off early in the dive, and that lactic acid remains in the muscles until the end of the dive (Scholander 1940). Lactic dehydrogenase levels are also high in the blood. Vallyathan et al. (1969) suggest that it could be oxidized to pyruvate, enter the liver for conversion to glycogen, or proceed to the muscles, where via aerobic muscular activity it becomes part of the Krebs cycle or forms lactate anaerobically (George and Ronald 1975).

The act of diving imposes stress on the seal different from the stress of the terrestrial environment. The animal must be able to withstand greater pressure, must conserve oxygen, and yet must have energy for muscular action and thermoregulation. Adaptations that accommodate this diving take the form of morphological, physiological, and biochemical changes. They are also part of an integrated system, so they function synonymously.

One physiological compensation that all diving mammals make during water submergence (Irving 1966) is the variability of cardiac rhythms (Casson and Ronald 1975). In the harp seal, the normal in-air heart rate measured by an electrocardiogram is 55 beats per minute, but rates may range between 37.5 and 150.0 bpm (de Kleer 1975). Rates as low as 2 bpm have been recorded in air from a relaxed seal.

A normal circulatory response to the asphyxia encountered during a dive is bradycardia, the process of decreasing the heart's rate of beating. This can reduce the heart rate during a dive to about 8–12 bpm. A following tachycardia (increase in heart beats) has raised it to as much as 200–250 bpm. When the face of the seal is submerged, the heart rate drops immediately, and as apnoea occurs a further slowing results from a response to blood gas changes. Upon resurfacing, the heart rate and cardiac output are greater than predive rates, though they gradually return to normal (Casson and Ronald 1975; Hempleman and Lockwood 1978). It is also important to note that there are extended apneic pauses in the normal respiratory pattern of the harp seal, a condition that is reflected in the irregular occurrence of bradycardia even while on land (Dykes 1974). Dykes (1974) speculates that a maximum diving bradycardia may be established by the exhalation in the first 15–30 sec after immersion. He also suggests that the anticipatory onset of bradycardia seconds before a dive, alluded to by Casson (1971), may be not part of the dive reflex but rather an autonomic reaction to threatening stimuli.

The contrast between cardiac function and breathing in surface swimming and diving has been observed in unrestrained immature captive harp seals. The heart rate averaged 122 bpm while swimming on the surface, but only 50 bpm during the dive. In seals trained to dive on command, the diving heart rates were 34.9–43.0 bpm, as opposed to resting rates of 125.5 bpm when the seal was floating on the surface (Casson and Ronald 1975). There would seem to be, therefore, rates for psychological diving, i.e., restrained seal exposed to an enforced dive when the bradycardia is profound (10 bpm) and the physiological diving with more limited bradycardia (40 bpm).

Diving imposes certain metabolic problems that the seal has compensated for in three ways: it is able to store large amounts of oxygen, it has the ability to utilize these reserves economically, and it has biochemical adaptions that improve anaerobic metabolism (Blix 1976). Not only do some seals possess twice the blood volume of many terrestrial mammals (Ronald et al. 1977), but their blood has high hematocrit values (percentage of blood volume that consists of red blood cells), which increase its total oxygen-carrying capacity (Hempleman and Lockwood 1978). Oxygen is released to the tissues from the hemoglobin as the partial pressure of O_2 in the surrounding medium drops, and this release is aided by the presence of carbon dioxide, because of the Bohr effect (Hempleman and Lockwood 1978).

In air, seals ventilate their lungs more fully with each breath than do land mammals, in order to remove the excess O_2 quickly, so that blood and body fluids can more quickly be completely reoxygenated after a dive (Hempleman and Lockwood 1978). Not only is the oxygen utilized during a dive stored in the blood, but it is also bound to high concentrations of myoglobin in the seal muscle. In addition, when lung oxygen supplies are insufficient for the consumption requirements of a dive, the metabolic rate of the seal falls, conserving the O_2 stores in the blood (Ronald et al. 1977; Hempleman and Lockwood 1978). During long dives an oxygen debt is built up when the stores are insufficient to last. This debt is repaid when the seal surfaces and resumes regular breathing. The urge to surface and breathe can be resisted longer by seals, mainly because these mammals are less sensitive to CO_2 and have a larger capacity (hemoglobin level) to absorb excess CO_2 (Hempleman and Lockwood 1978).

Oxygen conservation results not only from the above-mentioned metabolic response but also from the restriction of circulation to tissues that are capable of sustaining short periods of hypoxia (Ronald et al. 1977). Filtration in the kidneys ceases during a dive, and waste from the tissues not receiving blood is held until circulation returns to normal (Hempleman and Lockwood 1978). The heart and brain, which are incapable of prolonged anaerobic respiration, continue to receive an adequate supply (Ronald et al. 1977; Hempleman and Lockwood 1978). This redistribution of blood has the advantage of saving the O_2 reserves for those tissues easily damaged by hypoxia but also the disadvantage of causing a dramatic decrease in heat production due to the change from aerobic to anaerobic metabolism in the deprived tissues. This cooling is reinforced by the increased circulation to the poorly insulated brain (Irving 1938). As a result, deep body temperature drops during a dive (Scholander 1942) and the venous return must be warmed in order to minimize the cooling of the central body core (Hempleman and Lockwood 1978). To accomplish this, the venous return is shunted through the venous plexus, which is imbedded in brown adipose tissue with its thermogenic potential (Hol 1975).

When a seal dives deeply, the increased hydrostatic pressure could increase the risk of rupture in the gas-filled spaces as well as the rate that nitrogen enters the blood, a condition that on resurfacing could cause the "bends." Anatomical adaptions that accommodate this increase with depth in hydrostatic pressure are: flexible lungs and chest wall (flexible rib attachments); the absence of connections between the thoracic wall and the lungs; strong, cartilaginous main airways; the presence of retia mirabilia lining the sinuses and the thoracic wall region; and a reduction of sinuses in the skull (excluding those of the middle ear) (Hempleman and Lockwood 1978). The rete expand with the blood during a dive and partially compensate for the change in volume that occurs as the lungs collapse. Nitrogen is prevented from penetrating into the blood not only by the exclusion of gas from the lungs but also by the large amounts of fat in which it is soluble. Thus, nitrogen is absorbed into lipid bodies rather than into more critical areas. Excess nitrogen is also accommodated in the fatty foam in sinuses and other air spaces. Sometimes seals exhale just prior to diving, which also limits the amount of available gas (Hempleman and Lockwood 1978).

REPRODUCTION

Ancestral pinnipeds slowly evolved a dependence on marine resources. There was, therefore, less motivation to return either inland or upriver. Social interaction and migrations were oriented along coastal regions. Present reproductive physiology and behavior were developed in response to these changes, and to seasonal climatic variations and food availability. Individual communication was difficult during the periods of dispersal throughout the sea, so there was

TABLE 40.2. Reproductive parameters of pinnipeds

Species	Age to Maturity (Years)		Pregnancy Rate[a]	Pupping Season	Lactation	Mating	Gestation[b]	Delayed Implantation	Comments
	Female	Male							
Harbor seal	2-5 (FAO Adv. Comm. 1976)	3-6 (FAO Adv. Comm. 1976)		duration is 1.5-2 months Alaska: early March to April Washington: early March to May Mexico: early February to March Western Atlantic: March-June (FAO Adv. Comm. 1976)	2 weeks after pups weaned (FAO Adv. Comm. 1976)	30 days (Boulva and McLaren 1979)	10.5-11 months (FAO Adv. Comm. 1976)	2 months (FAO Adv. Comm. 1976)	
Ringed seal	6-8 (Burns 1978)	7-9 (Burns 1978)	85% ovulation rate: +90% (FAO Adv. Comm. 1976)	late March to late April (FAO Adv. Comm. 1976)	4-6 weeks (depends on stability of ice where birth lair built) (Burns 1978)	4-6 weeks postparturition (FAO Adv. Comm. 1976)	11 months (Burns 1978)	3.5 months (Burns 1978)	Newborn ringed seals stay in the birth lair until they are weaned; if early ice break-up destroys the lairs, they are abandoned and usually starve (Burns 1978). Annual reproductive rate is 26% (Popov 1976).
Ribbon seal	4 (Burns 1971)	5 (Burns 1971)	85% (FAO Adv. Comm. 1976)	late March to mid-April (FAO Adv. Comm. 1976)	4-6 weeks (Irving 1972)	late April to early May (FAO Adv. Comm. 1976)	10.5-11 months (FAO Adv. Comm. 1976)	probability of some period of delayed implantation (FAO Adv. Comm. 1976)	
Harp seal	northwestern Atlantic	4.5	about 94% in 1979, compared with 85% in 1952	Front: early March (Sergeant 1975) Gulf of St. Lawrence 15 February to 15 March (King 1964)	2-4 weeks (Terhune et al. 1979)	occurs at end of 2nd week after parturition (Sivertsen 1941)	11.5 months (FAO Adv. Comm. 1976)	4.5 months (FAO Adv. Comm. 1976)	The maturity ogive of the harp seal appears to have responded to density pressures within the herd, so the female probably reached a mean whelping age in 1979 of 4.5 years, as compared to a mean age of 6.2 years in 1952. Fertility rate and mean age at maturity appear to be density dependent.
Gray seal	4-7 (Hewer 1964). In some colonies females mature as late as 9; males mature physically, but not socially, at the same time as females (Platt et al. 1974; Mansfield 1978).	10 (FAO Adv. Comm. 1976)	85% for 6+ year females (Mansfield and Beck 1977; Mansfield 1978)	late December-February peak: mid-January (Bruemmer 1979; Mansfield 1978)	weaned at 2 weeks (Mansfield and Beck 1977)	occurs about 7 weeks after lactation ends (FAO Adv. Comm. 1976)	11.5 months (FAO Adv. Comm. 1976)	3 months (FAO Adv. Comm. 1976)	Ovulation appears to be spontaneous, as in the southern elephant seal (Mansfield 1978).
Bearded seal	5-6 (Burns 1978). Age higher in Pacific seal, perhaps because of higher rate of exploitation (Burns 1978).	6-7 (Burns 1978)	83% (Burns 1978)	Bering Strait: mid-March-1st week in May; later further north (Burns 1978)	pups born on ice nursed 1.7-2.2 weeks (Burns 1978)	see comments	11 months (Burns 1978)	2 months (Burns 1978)	Banfield (1974) states that there is no postpartum estrus in this species, so some females therefore do not ovulate until early June when males are not potent. These same females would mate the following year and therefore pup once every second year. If this is true it is unique in phocids. According to Burns (1967), the reproduction rate or annual growth of the population as a whole is 22-25%.

Hawaiian monk seal	3 possibly (FAO Adv. Comm. 1976)	see comments	late December to July, with a peak from late March to May (Kenyon 1978b)	6 weeks (FAO Adv. Comm. 1976)	unknown though courting behavior has been observed between early March and July (King 1964)	unknown	see comments	Pregnancy rate is 56% of females over 2 years with 34% of these breeding in both seasons, 32% in the first year only, and 34% in the second year only (FAO Adv. Comm. 1976).
Hooded seal	3 (first pup at 4) (FAO Adv. Comm. 1976)	95% (FAO Adv. Comm. 1976)	second half of March (FAO Adv. Comm. 1976)	1.7 weeks (Sergeant 1976)	occurs toward end of lactation (Sergeant 1976)	11.7 months (FAO Adv. Comm. 1976)	more than 4 months (FAO Adv. Comm. 1976)	
Northern elephant seal	Give birth at 3–5 at 9 or 10 (FAO Adv. Comm. 1976); 4–5 socially mature	95% cows reproductively active until death (FAO Adv. Comm. 1976)	December–January and February (FAO Adv. Comm. 1976)	about 4 weeks (Delong 1978)	at time of weaning when female comes into estrus (Delong 1978)	11.3 months (FAO Adv. Comm. 1976)	3 months (FAO Adv. Comm. 1976)	
Northern sea lion	4–5; 5–7 socially and physically competitive at 7–9 (FAO Adv. Comm. 1976)	see comments	mid-May to late June (FAO Adv. Comm. 1976)	32–44 weeks (occasionally a female will nurse both a newborn and a yearling) (FAO Adv. Comm. 1976)	8–14 days after parturition (FAO Adv. Comm. 1976)	12 months (FAO Adv. Comm. 1976)	3 months (FAO Adv. Comm. 1976)	Pregnancy rate is variable according to crowding and age distribution. Can be as high as 85% for 7–19-year-old females in a low-density population (FAO Adv. Comm. 1976).
California sea lion	unknown	unknown	Mexico and California: late May to late June Galapagos Islands: October–December (FAO Adv. Comm. 1976)	20–48 weeks (FAO Adv. Comm. 1976)	14 days after parturition (FAO Adv. Comm. 1976)	1.9 months (FAO Adv. Comm. 1976)	unknown	
Guadalupe fur seal		in 1977 56% of females observed breeding (Fleischer 1978b)	late June or July (Fleischer 1978a)	see comments	1 week after parturition (Fleischer 1978a)		see comments	Little is known about the reproduction of this species. Pups do receive extended maternal care, though no data on weaning time are available. There is almost certainly some period of delayed implantation (Fleischer 1978a).
Northern fur seal	4–5 peak breeding between 7 and 15 (FAO Adv. Comm. 1976); 5–6 do not breed until 12–15	60% (FAO Adv. Comm. 1976)	late June to early August, with a peak in early July (Fiscus 1978)	16 weeks (Fiscus 1978)	5–10 days after parturition (Fiscus 1978)	1.95 months (FAO Adv. Comm. 1976)	3.25–4 months (FAO Adv. Comm. 1976)	
Walrus	7; 6 (Reeves 1978)	see comments	mid-April to mid-June (FAO Adv. Comm. 1976)	minimum of 78 weeks (Reeves 1978)	February and March (FAO Adv. Comm. 1976)	380 days (Reeves 1978)	3 months (Reeves 1978)	This species has a low level of productivity; 80% of females give birth every 2 years and 15% every 3 years (FAO Adv. Comm. 1976).

[a] Incidence of pregnancy in adult females.
[b] Includes delayed implantation stage.

TABLE 40.3. Dietary range of North American pinnipeds

	Harbor Seal	Ringed Seal	Ribbon Seal	Harp Seal	Gray Seal	Bearded Seal
INVERTEBRATES						
Phylum Coelenterata						
Phylum Annelida						
Phylum Mollusca						
Class Gastropoda (abalone)						
Class Pelecypoda (clams)						X
Class Cephalopoda (squid and octopus)	X		X		X	X
Phylum Arthropoda						
Class Crustacea						
Subclass Malacostraca						
Order Decapoda (crabs, crayfish, shrimp)	X	X	X	X	X	X
Phylum Echinodermata						
Class Asteroidea (starfish)	X					
Class Holothuroidea (sea cucumber)						
Class Echinoidea (sand dollars)						
VERTEBRATES						
Phylum Chordata						
Subphylum Urochordata (tunicates)						X
Subphylum Vertebrata						
Class Agnatha						
Family Petromyzonidae (lamprey)						
Class Chondrichthyes (shark)						
Family Squalidae (dogfish shark)						
Family Rajidae (skate)	X				X	
Family Chimaeridae (ratfish)						
Class Osteichthyes						
Family Anguillidae (eels)						X
Family Clupeidea (herring, sardine, alewife)	X			X	X	
Family Engraulidae (anchovy)						
Family Salmonidae (salmon)	X				X	
Family Osmeridae (capelin, smelt)	X		X	X	X	
Family Myctophidae (lantern fish)	X					
Family Gadidae (codfish, haddock, hake, pollock)	X	X	X	X	X	X
Family Zoarcidae (eelpouts)			X			
Family Scomberescocidae (saury)						
Family Embiotochidae (sea perch)						
Family Stichaeidae (pricklebacks)			X			
Family Amodytidae (sandlance)	X					
Family Scombridae (mackerel)	X				X	
Family Scorpaenidae (rockfish, redfish)	X					
Family Anoplopomatidae (sablefish)						
Family Hexagrammidae (greenlings)						
Family Cotridae (sculpins)	X		X			X
Family Cyclopteridae (snailfish, lumpfish)			X			
Family Pleuronectidae (plaice, halibut, flounder)	X				X	X
Family Bothidae (flounder)	X				X	X

strong selection pressure to return to the same sites to reproduce. Access to land was obstructed by ice, which forced breeding on ice. This resulted in selection pressures for short, synchronized suckling and mating periods, a less rigid social organization, and monogamy in some instances (Stirling 1975). Delayed implantation, an evolutionary response to this prob-

lem, resulted in synchronized periods in one location each year (Maxwell 1967). An impregnated seal experiences both a period of delayed implantation and a "true gestation," or actual fetal growth period (Maxwell 1967). Most biologists, however, include this period of delay in their estimates of gestation. Because habitats are different, species reproductive behavior is

Hawaiian Monk Seal	Hooded Seal	Northern Elephant Seal	Stellars Sea Lion	California Sea Lion	Guadalupe Fur Seal	Northern Fur Seal	Walrus
			X				
			X				X
				X			
							X
X	X	X	X	X		X	
X			X				
							X
			X				
							X
			X			X	
		X					
		X					
		X					
		X		X			
		X	X			X	
						X	
			X			X	
X	X					X	
	X	X	X	X		X	
						X	
			X				X
						X	X
						X	
	X	X	X	X		X	
			X				
			X				
			X				
	X		X	X		X	
	X		X	X		X	

variable, as is individual maturity, gestation length, and delayed implantation periods (Table 40.2).

ECOLOGY AND FOOD HABITS

Pinnipeds are carnivores and feed on fish and invertebrates, which are supported by a multilevel trophic food chain. Seals live in areas of the ocean where upwelling of currents occurs, bringing nutrients to the surface that nourish their food source. Because they require large quantities of food near their breeding and hauling-out sites, these sites are located near those regions in the ocean of high food productivity. These areas include the eastern sides of ocean basins and areas adjacent to high-latitude land masses (Lipps and Mitchell 1976).

Feeding behavior and food will be discussed in detail for each individual species. Table 40.3 gives the dietary range of the pinnipeds in North America.

MORTALITY

Mankind is the primary cause of death in several species of pinniped. The other common causes of mortality are: parasitism, predation, trauma or infectious agents, premature weaning, early abandonment, or an unsuccessful struggle against the stresses imposed by the marine environment (Sweeney and Geraci 1979), such as heavy precipitation on unprotected neonates. Stress, a physiological and biochemical response to changing environmental demands, weakens the animal and causes it to be more susceptible to disease and parasitic infections (Sweeney and Geraci 1979).

AGE DETERMINATION

Laws (1962) outlined the various methods used in determining the age of a pinniped. The distinctive coat of the newborn indicates a very young seal, and in some species, pelage distinguishes a subadult from an adult. Other aging characteristics include the degree of scarring in polygamous species, tooth size in the walrus, color, growth or wear of the vibrissae, and general body proportions. The degree of suture closure demonstrates skull development; however, a study of the epiphyseal fusion of other parts of the skeleton, such as the manus or pelvic bones, will result in only a rough approximation of the age of a seal (Sumner-Smith et al. 1972). The number of ovarian scars and the study of laminations in certain bones have limited value. By far the most accurate method of age determination of seals is examination of the annual increments of dentine which form rings in the teeth. When these dentine layers are not distinguishable, similar growth layers in the cementum can be used. These visible annular interruptions in the deposition of cement and dentine are caused by fasting periods (McLaren 1958; Kenyon and Fiscus 1963). They are accurate indicators of age in most species of pinniped, with the exception of the bearded seal, whose teeth begin to deteriorate at an earlier age (Banfield 1974). Baculum length is a good indicator of age in this species. Development of the os penis is stimulated by the increased production of androgens during the breeding season throughout the animal's life (Burns 1967). Age can also be determined in *E. barbatus* by examination of its foreclaws, which show annual rings. During the breeding season a light band forms and during the subsequent molt a ridge appears (Banfield 1974). The ringed seal can also be aged by this method. Its claws are marked by alternating light and dark bands in the spring and early summer, respectively. Aging by claws is useful only up to 10 years of age in the ringed seal, however. Beyond this age, claws become too worn (McLaren 1958).

NOMENCLATURE, DISTRIBUTION, AND GENERAL CHARACTERISTICS OF INDIVIDUAL SPECIES

PHOCIDS IN NORTH AMERICA

The following species of phocid have been reported as resident in the waters surrounding continental North America: *Phoca vitulina, Pusa hispida, Histriophoca fasciata, Phoca groenlandica, Halichoerus grypus, Erignathus barbatus, Monachus schauinslandi, Cystophora cristata,* and *Mirounga angustirostris.*

HARBOR SEAL

Phoca vitulina

NOMENCLATURE

COMMON NAMES. Harbor (sometimes spelled harbour) seal, spotted seal, common seal
SUBSPECIES. *P. v. vitulina, P. v. concolor, P. v. richardii, P. v. kurilensis, P. v. largha,* and *P. v. mellonae.*

DISTRIBUTION

Only four subspecies of *Phoca vitulina* are found in North American waters. *P. v. concolor* occurs from Maine north to Ellesmere Island and west to western Hudson Bay as well as on the coast of Greenland (Mansfield 1967a). *P. v. richardii,* the coastal seal, occurs in more temperate, ice-free waters extending south from the Bering Sea to Cedros Island off Baja California, while *P. v. largha,* the ice-breeding seal, is found from the southern bay of the Chukchi Sea through the Bering Strait, Bering Sea, and coast of the U.S.S.R. to the coast of China (Newby 1978). *P. v. mellonae* is a landlocked group of seals found only in upper and lower Seal Lakes on the east side of Hudson Bay (King 1964) (figure 40.4).

A sedentary species, the harbor seal spends most of its time offshore in winter, but may make some short, seasonal migrations away from the occurrence of fast ice to areas of open water, since it does not maintain breathing holes (Boulva 1976).

The world population of approximately 600,000–900,000 is in equilibrium, although it is decreasing around human habitation (Scheffer 1977). In 1973, the North Pacific population consisted of ±750,000 seals, mostly *P. v. richardii.*

Since the U.S. Marine Mammal Protection Act of 1972, the Washington State stock has been increasing; it was 5,500 in 1977. In British Columbia, there are approximately 35,000 harbor seals (Newby 1978). Although *Phoca vitulina* once penetrated the rivers and lakes of eastern Canada, exploitation and displacement by humans had reduced its population to about 12,700

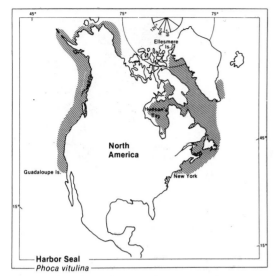

FIGURE 40.4. Distribution of the harbor seal (*Phoca vitulina*).

seals in semi-isolated groups in the waters south of Labrador by 1973 (Boulva and McLaren 1979). An overall decline of 4 percent a year occurred between 1950 and 1970 in eastern Canada (Boulva and McLaren 1979).

Habitat. The harbor seal is an adaptable species with a widely ranged tolerance of temperatures and water salinities (Boulva 1976) and can therefore occupy an extensive variety of environments (Scheffer 1977). In eastern North America its habitat consists of inlets, islets, reefs, and sandbars, which may be of importance for breeding (Boulva and McLaren 1979). In the Pacific it is found in areas of tidal mud flats, sand bars, shoals, river deltas, estuaries, bays, coastal rocks, and offshore islets (Johnson and Jeffries 1977) and it is commonly seen some distance inland. These seals tend to select a protected location with unobstructed access to water and at low tide are seen on sandspits, coastal rocks, and small reefs (Newby 1978).

FIGURE 40.5. The harbor seal (*Phoca vitulina*).

DESCRIPTION

The harbor seal has a short, heavy body and a round-smooth head. Valvular nostrils and large, convex eyes are dorsally situated (figure 40.5). Its face, which is slightly canine in appearance, has flattened, beaded mystacial vibrissae on each side of the muzzle (Banfield 1974). There is some sexual dimorphism in size, with adult males averaging 1.6 m in length and 87.6 kg and adult females about 1.5 m in length and 64.8 kg (FAO Adv. Comm. 1976). Newborn pups are at least 0.75 m long and 9 kg in weight (Fay et al. 1979). Coat color and pattern vary considerably and pups are born with an adult-type pelage (FAO Adv. Comm. 1976; Fay et al. 1979), their lanugo being shed in utero. The coat pattern is basically a mottle of dark spots on a lighter ground, but in some animals the spots coalesce, particularly on the back (FAO Adv. Comm. 1976). Color ranges from nearly black with scattered, whitish rings to pale and spotted with some whitish rings on the back (Fay et al. 1979). The pelage consists of an overcoat of stiff hairs 11 mm long and an undercoat of sparse, curly hairs 5 mm long (Banfield 1974). The harbor seal lives an average of 40 years (FAO Adv. Comm. 1976).

FOOD HABITS

Harbor seals are generally close inshore, shallow water feeders with catholic tastes (FAO Adv. Comm. 1976). Their food base consists of pelagic, demersal, anadromic, and catadromic fishes, cephalopods, and Crustacea. Dietary items of economic importance to humans are gadoids, clupeids, pleuronectids, and salmonids (FAO Adv. Comm. 1976). They swallow small fish whole underwater, but take the larger ones to the surface, where they eat them in pieces, usually leaving the heads. Because these seals are opportunistic feeders there is a great variation in meal size, although they generally have one large meal per day (Boulva and McLaren, 1979).

BEHAVIOR

Harbor seals tend to occur in small groups of about 30–80 animals, although larger groups are found in areas where food is plentiful (FAO Adv. Comm. 1976). Although gregarious on land, these seals have no developed social structure and in the water they tend to disperse and forage for food alone (Banfield 1974; FAO Adv. Comm. 1976). In Atlantic Canada, haul-out behavior is related to the tides and the weather; seals are seen resting on land at low tide, reentering the water as the tide returns (Johnson and Jeffries 1977). Few seals haul out if high winds are causing rough seas and almost no seals stay ashore except for females with pups and some seals in molt (Boulva and McLaren 1979). In winter when wind chill temperatures are below −15° C these pinnipeds stay in the water; when the inlets become frozen they stay away from the mainland, as they do not maintain breathing holes (Boulva and McLaren 1979). Haul-out sites must have immediate access to deep water and some protection from human activities and other disturbances and most commonly consist of offshore rocks and islets, high-banked grassy areas, log booms, and anchored beach floats (Johnson and Jeffries 1977). Harbor seals are very wary and will retreat to the water immediately if alarmed. They have an odd behavior trait that consists of clapping their foreflippers on their chests, causing a large splash. Vocalizations include a variety of grunts and growls and a high-pitched yapping bark usually heard on land; while in the water, they emit snorts and snuffles (Banfield 1974). *P. vitulina* do not form harems and are probably promiscuous (Bigg 1969; Bonner 1972). Of the two Pacific subspecies, *P. v. largna* is monogamous and forms family groups during the breeding season, while *P. v. richardii* is polygamous and females form nursery areas for the two-week period of lactation (Newby 1978).

MORTALITY

Newborn harbor seals are preyed on by sea eagles (*Haliaeëtus*), golden eagles (*Aquila chrysaëtos*), and foxes (*Vulpes*); adults are attacked by sharks, killer whales (*Orcinus orca*), bears, and walruses (FAO Adv. Comm. 1976). Predation on pups by polar bears (*Ursus maritimus*) and arctic foxes (*Alopex lagopus*) is heavy; that by killer whales and walruses is incidental.

In eastern Canada there is a high natural mortality rate for *P. vitulina* of 17.5 percent/year (postweaning), possibly as a result of shark predation and overexposed breeding sites, which make these seals particularly vulnerable to hunting (Boulva and McLaren 1979). The three major causes of pup mortality are stillbirth, desertion by the mother, and shark kills (Boulva and McLaren 1979). Human disturbance can cause desertion of the pup by its mother. Other causes of death are shooting, underwater blasting, propeller wounds, internal hemorrhage, and infection (Johnson and Jeffries 1977). The major causes of mortality can be attributed to man-related incidents, either intentional or by disturbance.

RINGED SEAL

Pusa hispida

NOMENCLATURE

SCIENTIFIC NAME. The scientific name is derived from the Greek word *phoca,* meaning seal, and the Latin word *hispida,* meaning rough or bristly, which refers to the stiff hairs of this seal's coat (Burns 1978).

SUBSPECIES. *P. h. hispida,* on all the arctic coasts of Europe and North America; *P. h. krascheninikovi* [*sic*], of the northern Bering Sea; *P. h. ochotensis,* of the Okhotsk Sea; *P. h. botnica,* of the Baltic Sea; *P. h. saimensis,* of Lake Saimaa; and *P. h. ladogensis,* of Lake Ladoga (King 1964).

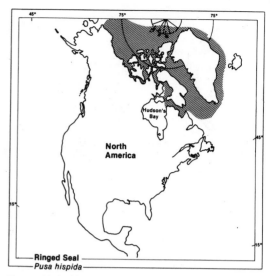

FIGURE 40.6. Distribution of the ringed seal (*Pusa hispida*).

DISTRIBUTION

The range of the ringed seal is circumpolar, from the southern edge of the pack ice to the north pole, with some animals in James and Hudson's bays and distinctive forms in Lake Baikal and the Caspian Sea (FAO Adv. Comm. 1976). Of the two North American subspecies, *P. h. krascheninikovi* [*sic*], occurs in the fast ice gulfs and bays of the Bering Sea. When the ice disappears, the seals remain sedentary along the coastal inlets of the Chukotsk and Kamchatka peninsula and the Commander Islands (Popov 1976) (figure 40.6). *P. h. hispida* is found along the arctic coast of North America and as far south as Labrador in the east (King 1964). Although there is some seasonal movement in the Beaufort Sea, little is known about their migrations (FAO Adv. Comm. 1976). The ringed seal is possibly the most abundant arctic seal. Because of the vast area of its occurrence, however, it is impossible to obtain an accurate population estimate, but one recent estimate is in excess of 5 million (FAO Adv. Comm. 1976).

Habitat. The ringed seal maintains a year-round association with ice, occupying the landfast or shore ice in the winter and migrating with the annual advance and retreat of the ice pack in other seasons (Burns 1978). Adults tend to remain on the stable, inshore ice, while subadults are found further offshore in areas of shifting but relatively stable ice (FAO Adv. Comm. 1976).

DESCRIPTION

The smallest North American pinniped is the ringed seal, *P. hispida.* It is similar in appearance to the harbor seal, although its head is rounder and its snout more pointed, which gives its face a feline character (Banfield 1974) (figure 40.7). There is some sexual

FIGURE 40.7. The ringed seal (*Pusa hispida*).

dimorphism in size, adult males being slightly larger than females (FAO Adv. Comm 1976). There are also some size differences and color distinctions between the North Pacific subspecies, *P. h. krascheninikovi* [*sic*], and the northeast Atlantic and arctic subspecies, *P. h. hispida*. The former are about 1.4 m in length and weigh about 63–66 kg, while the latter are 1.2–1.5 m in length and up to 80 kg in weight. Pups are similar in size, being about 0.62 m long and weighing 5 kg. Weights show much seasonal variation owing to changes in the proportion of blubber (FAO Adv. Comm. 1976). This layer of blubber reaches a maximum of 40 percent body weight in late autumn and a minimum of 23 percent during the spring fast (Banfield 1974). The North Pacific ringed seal is gray brown, sometimes with a light greenish yellow tint and light strips along the back and sides forming irregular rings. *P. h. hispida* adults vary from brown to gray with an olive shading dorsally, to dark gray and almost black. The ventral surface is light gray with silver shading. Light-colored veins of color are interlaced on the main background, forming the net or lace design that produces oval rings. The embryonic coat is fluffy and white, but changes to the short, adult-colored pelage after the first molt (Popov 1976). The life span of the ringed seal is believed to be about 40 years (FAO Adv. Comm 1976).

FOOD HABITS

Depending on the season and the area, the ringed seal consumes at least 72 food organisms (McLaren 1958). In deeper, offshore waters polar cod and zooplankton are most commonly eaten, while in other areas small fish and shrimp form the diet. Major prey species include amphids, mysids, euphausiids, schooling fish such as saffron and polar cod, capelin, and sand lance (Burns 1978). Patterns of feeding suggest that food is not a limiting factor in the ringed seal's distribution and abundance (McLaren 1958). A relaxation of feeding in the early spring is followed by intense fasting in June and early July, when most of the seals are basking on fast ice. Feeding is resumed and blubber restored after the departure from the ice (McLaren 1958).

BEHAVIOR

The ringed seal, *P. hispida,* is a solitary animal although it is occasionally observed in loosely dispersed groups (Banfield 1974). Vocalizations include whining

and moaning sounds when on land and a threatening growl when trapped (Banfield 1974). We have observed a ringed seal slapping the water with its flipper as we approached. Its two most distinctive behavioral traits are the habits of using the strong foreflipper claws to maintain breathing holes in thick, stable ice and to excavate lairs in the snow for resting and pupping (Burns 1978). The haul-out lair is single chambered and round, while the birth lair contains an extensive network of tunnels made by the pup (Smith and Stirling 1975). The use of such lairs results from a need for protection from predators and cold. Newborn pups accumulate blubber slowly and thus depend on the lair for thermal protection (Smith and Stirling 1975). Ringed seals do not normally haul out on land in Alaskan waters but remain on the fast ice (Burns 1978).

MORTALITY

Enemies of adult ringed seals, apart from mankind, are the polar bear, killer whale, and arctic fox, while young pups are sometimes eaten by walruses (King 1964).

RIBBON SEAL

Histriophoca fasciata

NOMENCLATURE

COMMON NAMES. Ribbon seal, banded seal
SCIENTIFIC NAME. The scientific name of this seal describes its appearance. *Histriophoca* is derived from the Latin word *Histro,* meaning a stage player, and the Greek word *Phoca,* meaning seal. *Fasciata* comes from the Latin *fascia,* meaning band or ribbon (Burns 1978).

DISTRIBUTION

H. fasciata occurs in the Bering Sea and the Sea of Okhotsk as well as in the southern Chukotsk Sea and the northern Sea of Japan. In the Bering Sea, its primary hauling-out grounds are in the Anadyr Gulf and adjacent southeastern St. Lawrence Gulf and on the ice massifs near St. Mathew Island as well as in the Bering Strait (Popov 1976) (figure 40.8). During an unusual winter freeze-up, ribbon seals were seen moving overland at Cape Prince of Wales, Alaska (King 1964). They are generally associated with the spring and winter ice front in the Bering Sea and range to only 150 km north of its southern periphery (Fay 1974). Ribbon seal migration consists of passive movement on the ice during the breeding season and a shift to solid ice areas preceding the molting period in the spring (Popov 1976). Population estimates in the Bering Sea were 80,000–90,000 in 1964 and 60,000 in 1969, with about 133,000 in the Sea of Okhotsk in May of 1969 (Popov 1976).

FIGURE 40.8. Distribution of the ribbon seal (*Histriophoca fasciata*).

Habitat. The ribbon seal tends to remain in areas with open water or very thin ice, possibly because of its feeding habits or because it is incapable of maintaining breathing holes (Fay 1974). It usually hauls out on firm pack ice, far from shore, where cracks and leads are found. Its choice of habitat is obviously influenced by the availability of food and the ice conditions (Popov 1976). In the summer this seal stays in ice-free waters near, but usually not in, the permanent ice pack (Fay 1974), which it utilizes for parturition, lactation, breeding, and molting (Popov 1976). The large, heavy claws of the ribbon seal, typical of northern phocids, are an adaption to its icy habitat (Fay 1974).

DESCRIPTION

When compared to other Bering Sea phocids, the ribbon seal is slender and medium in size (figure 40.9). Its eyes are wide in diameter (Burns 1971). There is little sexual dimorphism in size, with adults attaining a maximum length of 1.9 m and a mean length of 1.7 m. Adults may attain a weight of up to 100 kg, with the average about 70–80 kg (Popov 1976). Newborns are 0.80–0.90 m in length and weigh about 9 kg (Popov 1976). Blubber and skin account for 27–35 percent of the total body weight (Burns 1978). The pelage shows a striking degree of sexual dimorphism. Two distinctive pelage stages occur that are connected with age (Burns 1971; Burns 1978). Adult males are dark brown

FIGURE 40.9. The ribbon seal (*Histriophoca fasciata*).

to black with four white or yellowish ribbonlike bands, 10–12 cm in width, circling the neck and the hind end of the body and forming a large ring around each forefilipper that extends from the shoulder to the midtrunk area. The female is lighter, with less distinctive markings (King 1964; Popov 1976). Newborns are white, and they molt after four weeks to a pale yellow pelage that lacks ribbons. After the first molt, pups become dark gray dorsally and lighter ventrally (Popov 1976). Full adult color is distinct by age three (Burns 1978). Ribbon seals probably live for 22–25 years (FAO Adv. Comm. 1976).

FOOD HABITS

H. fasciata is capable of diving 200 m for its food. During the breeding and molting periods this species subsists mainly on crustaceans plus some fish and cephalophods; young seals feed mainly on crustaceans. Ribbon seals of the Bering Sea have a more diverse diet than those of the Sea of Okhotsk, consuming 14 species of crustaceans, mainly *Pandalus gonionus*, *Temisto* sp., *Pandalopsis* sp., *Enalus gaimardi*, and *Amphipodae*, as well as 10 species of fish, such as polar cod, lumpfish, navaga, and capelin (Popov 1976). From March to June the major prey items are pollock, capelin, and eelpouts, with pricklebacks, snailfish, sculpin, saffron cod, shrimps, octopus, and squid also being taken (Burns 1978).

BEHAVIOR

H. fasciata is a solitary animal and forms groups only during the breeding season. However, it is often observed in widely dispersed aggregations during the spring molt (Fay 974).

MORTALITY

Ribbon seal pups are not subject to much predation by polar bears and arctic foxes because they are not born in their normal range. However, during their time at sea, they are preyed on by sharks and killer whales (Burns 1978).

HARP SEAL

Phoca groenlandica

NOMENCLATURE

COMMON NAMES. Greenland seal, harp seal, saddleback seal (King 1964)

DISTRIBUTION

Three separate stocks of *P. groenlandica* exist in the following breeding grounds: the White Sea, the Grønland Sea between Jan Mayen and Svalbard, and the northwest Atlantic. This latter population is divided during the breeding season into two major groups, one near the Magdalen Islands and the other off the coast of

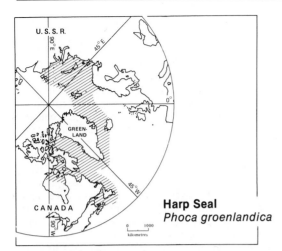

FIGURE 40.10. Distribution of the harp seal (*Phoca groen-landica*).

Newfoundland and Labrador. During the summer the immature harp seals (bedlamers), in particular, are found on the west coast of Greenland (especially near Disko Island) and north to Upiinavik at 73° N (FAO Adv. Comm. 1976). The adults move to the cooler waters surrounding the Canadian arctic archipelago and occur in the Thule area, in Jones Sound, in Land-caster Sound, along the east coast of Baffin Island, in Hudson Strait, and in northern Hudson Bay around Southhampton Island (FAO Adv. Comm. 1976; Lavigne 1978) (figure 40.10).

As the ice recedes each spring, harp seals migrate north along the east coast of Canada from the breeding grounds and in the fall they return ahead of the arctic pack ice, reaching Labrador and the Gulf of St. Lawrence in late December or early January (Lavigne 1978).

The world population of harp seals is thought to be between 1.25 and 2.0 million, with 200,000–500,000 in the White Sea, another 100,000 around Jan Mayen Island, and between 1 and 1.5 million in the western Atlantic. The Canadian population produces 300,000–350,000 young a year.

Habitat. The harp seal's habitat is the drifting pack ice of the North Atlantic. In summer this seal utilizes open water and loose ice and during the spring it is found on more solid ice in the Gulf of St. Lawrence or on the front off Newfoundland-Labrador on the thick winter ice at the edge of the advancing ice sheet. Harp seals use channels or leads that penetrate deep into the ice cover to gain access to breeding sites. Because these leads are also important places from which seals can mount the ice, their location often determines the haul-out sites. Harp seals inhabit rough, hummocky ice at least 0.25 m thick (Ronald et al. 1976).

DESCRIPTION

The harp seal has a small head and short neck with limited mobility (figure 40.11), a condition that is re-lated to its mode of propulsion as well as to its dietary habits (Bisaillon et al. 1976). Large eyes dominate the skull (King 1972), and the face has a few beaded whis-kers (King 1964). The paddlelike fin feet, both front and hind, are covered on each surface with fur (Green 1972). On each of the five digits are nails of equal size, and the third digit of the forelimbs is longer than the first or second (King 1964). There is no cartilaginous extension beyond the bony framework of the digits in the forelimbs, as there is in those of the hind flippers (Green 1972).

The average adult female harp seal measures from 1.7 m to 1.8 m in length; the male from 1.7 m to 1.9 m in length (Smirnov 1924). Newborn whitecoats mea-sure from 0.97 m to 1.08 m, and the molting pups from 0.90 to 1.2 m in length. The newborn harp seal pups weigh on the average about 11.8 kg at birth, but in-crease their weight rapidly to about 22.8 kg within four to five days. Their weight continues to increase at the same rapid rate to a maximum of 40 kg at weaning. Adult male harp seals average 135 kg and females about 119.7 kg (Sivertsen 1941). There are circannial weight changes in all age groups (Ronald et al. 1969).

The newborn "white coat," so named from its first, preweaning pelage color, molts through several color changes before reaching adulthood. The clarity of the color of the background fur, spots, and the dis-tinguishing harp on the back is dependent on both the medium (air or water) in which the animal is observed and whether the coat is wet or dry. A full-grown adult with a defined harp on its dorsal surface, if observed in water, appears as a silver animal with a deep jet black marking. The same animal seen wet in air would ap-pear as generally dark gray in color with markings that are indistinct and dull. An adult that has a dry coat and is observed in air would appear to have a dark choco-late brown or black harp on a gray background tinged with brown. Maturity of the pelage differs between males and females, so males may develop the harp pattern earlier, but it is usually distinct by age 7. There is such individual variation in markings among adult females, however, that only generalizations may be made. The harp begins to form in the females by age 5

FIGURE 40.11. The harp seal (*Phoca groenlandica*).

to 8 and usually closes by age 15. Spots may persist until age 19. Usually a more light-faced seal with a dark harp is a female, but a definitely dark-faced seal, while probably male, could also be female. Harp seals have a life span of 30 years or more (FAO Adv. Comm. 1976).

FOOD HABITS

Pelagic fish, especially capelin, and some benthic fish form the major part of the harp seal's diet (Sergeant 1973). Both regional and seasonal variations in feeding habits occur. For example, harp seals feed mainly on capelin and a few euphausids off western Greenland, while in the colder waters of northwestern Greenland they consume various crustaceans and polar cod (Kapel 1973). Weaned pups feed chiefly on *Euphausiacea,* changing to pelagic fish at one year. During the winter and summer *P. groenlandica* feed intensively, but during the spring and fall migrations and also during the whelping and molting periods little or nothing is eaten by adults (Mansfield 1967*a*; Sergeant 1973). Pregnant females are found in midwinter in the best feeding grounds and remain separated from other harp seals during lactation, feeding on decapod crustacea.

Small animals are sucked in, tail first, but larger ones are bitten. Harp seals can dive to a depth of 200–400 m for food. As individual seals consume about 2 percent of their body weight per day in food, their intake would be about 800–900 kg annually, allowing for 65 days of fasting. The northwest Atlantic harp seal herd therefore consumes substantial amounts of capelin annually in eastern Canada and western Greenland.

BEHAVIOR

Harp seals are very gregarious, except that old males tend to separate themselves from the herd. Although their terrestrial locomotion appears limited, they are agile and powerful swimmers, capable of speeds up to 5 m/sec. Because of their social nature the many sounds that these very vocal seals emit probably have a communicative function (Møhl et al. 1975). Fifteen underwater phonations have been recorded, but the only known air vocalizations are a hoarse hissing made by lactating females, distressed adults, and pups (Maxwell 1967; Møhl et al. 1975) and the wailing of the pups, which is similar to the cry of a human infant (Terhune and Ronald 1970).

In April and May, adults and immature one year olds haul out to molt, forming dense aggregations. These molting groups are even larger than the breeding colonies and may be in groups of tens of thousands of seals.

Breeding sites occur near water leads and natural or seal-made holes in the ice, which are used for exit and breathing and which the seals attempt to maintain. These holes, 60–90 cm at the top and widening toward the base, are often communal, with up to 40 seals occurring in a pool around the hole (Ronald et al. 1976). The females form "whelping patches" the size and shape of which change as the ice floes drift (Anderson personal communication).

When pupping, these seals tend to stay close to a lead and near females with pups of the same age (Dorofeev 1939). Males usually keep together in groups of about 12 (Maxwell 1967). Pups nurse five or six short times a day soon after birth and change to fewer but longer periods later in the nine days of lactation. However, there is not necessarily any correlation between length of time and quantity of milk obtained (Lavigne personal communication). The females leave the pups occasionally during the first week. After this they will return to the water, coming out periodically to feed their pups and thermoregulate. If threatened while in the water a few females will defend their pups, but most will swim away. If threatened on the ice, however, some will defend their pups vigorously while others will "play possum," as do the pups (Ronald et al. 1970). On clear windless days males and females are seen basking in the sun, but they tend to spend more time in the water in snowy and foggy weather (Ronald et al. 1976). Before mating, males will often fight, using their teeth and flippers (Popov 1966).

MORTALITY

Predation by polar bears, Greenland sharks, and killer whales on harp seals is low. Mortality rates from parasitism and disease are also incidental (FAO Adv. Comm. 1976).

GRAY SEAL

Halichoerus grypus

NOMENCLATURE

COMMON NAMES. Gray seal (King 1964), horsehead seal; Eskimo names include hodge and apa (Mansfield and Beck 1977)

DISTRIBUTION

Three stocks of gray seals (western Atlantic, eastern Atlantic, and Baltic) are thought to be distinguished by geographical isolation and differences in breeding season (FAO Adv. Comm. 1976). In North America there is a well-established population in eastern Canada with breeding colonies at the Magdalen Islands (Deadman Island), on Amet Island in Northumberland Strait, on Sable Island, on Point Michaud on the east coast of Cape Breton Island, and on the fast ice along the western Cape Breton shore from the Strait of Canso to Inverness (figure 40.12). During the summer gray seals disperse around the Gulf of St. Lawrence and coasts of Newfoundland and the maritime provinces, ranging from Hebron in Labrador to Nantucket Island. The only gray seal breeding grounds in the United States are on the islands of Muskeget and Tuckernuck off Nantucket (Mansfield and Beck 1977). The total world population is thought to be about 88,000–99,000 (FAO Adv. Comm. 1976).

Grey Seal
Halichoerus grypus

1976). After three weeks, pups molt their white lanugo and become blue gray dorsally and paler ventrally. With successive annual molts they gradually achieve the adult color and pattern. An adult coat color varies greatly, with many shades of dark and light gray, brown, and silver (King 1964). Adults are generally darker dorsally, shading lighter ventrally, and males tend to be darker than females (FAO Adv. Comm. 1976). Records exist of a 46-year-old female (Bonner 1971), and a male gray seal 26 years old has been found (Platt et al. 1974).

FIGURE 40.13. The gray seal (*Halichoerus grypus*).

There is limited evidence of migration in adult gray seals, although their breeding and feeding grounds are usually distinct. However, pups tend to disperse from their birthplace to all parts of eastern North America from New Jersey to northern Labrador. Recent tagging studies indicate that migration speeds reach up to 50 km/day. Most seals probably return to their birthplace as breeding adults, though some may move to other colonies (Mansfield and Beck 1977).

Since 1962, there has been a marked increase in gray seals in eastern Canada, and in 1979 this population was estimated at possibly over 43,000, with about 10,000 of these being young of the year (Gray and Beck 1979).

Habitat. The gray seal is essentially a coastal species, moving out to sea only to feed. It is generally associated with heavily indented, rocky coasts where there are small islands and reefs, but has also been observed in some estuaries and lagoons where sandbanks are found (Mansfield and Beck 1977). We have observed it breeding on ice.

DESCRIPTION

The gray seal is a moderately large pinniped whose long, broad, straight snout gives it an equine appearance (figure 40.13). Its flexible forelimbs and long, slender claws make this seal the most adept on land of all phocids (Banfield 1974). There is marked sexual dimorphism in coat pattern, size, and appearance (FAO Adv. Comm. 1976). More massive than the female, the adult male has a swollen, wrinkled neck and a broad, slightly "roman" snout. Older males acquire a prominent sagittal crest (Banfield 1974). Males average 2 m in length and 233 kg in weight, while females are about 1.8 m in length and weigh about half as much as the males (FAO Adv. Comm. 1976; Mansfield and Beck 1977). Newborn pups are 0.90–1.0 m in length and weigh about 14.5 kg (FAO Adv. Comm.

FOOD HABITS

Essentially a coastal species, *H. grypus* does spend part of its time at sea feeding on shallow, offshore fishing banks. It feeds primarily on bottom fish but takes several kinds of demersal, anadromous, and pelagic species during their inshore migrations. Occasionally the gray seal will eat invertebrates, particularly squid and epibenthic crustaceans such as crabs and shrimp, but rarely lobsters. The most important food species available all year are benthic skates and flounders, while herring, cod, squid, and mackerel become important when they begin their inshore migrations in the spring (Mansfield and Beck 1977). Gray seals fast during the breeding season—females for at least two weeks and adult males even longer. Feeding is minimal during February, March, and April when important food species such as herring, cod, and mackerel are not abundant inshore. Gray seals eat about 3 percent of their body weight daily. The Canadian gray seal stock eats approximately 47,083 t of food a year. These seals "compete" for food with harbor seals, some porpoises and larger cetaceans, and migrating harp and hooded seals early in the year. The quantities of food they consume are not considered to present significant competition to commercial fisheries (Mansfield and Beck 1977), although they are a nuisance factor to the fishermen on a local basis. To this end they are bountied in Canada, though this has proven to be ineffective.

BEHAVIOR

The gray seal is very gregarious, feeding in groups and hauling out to breed and molt in dense colonies. It dominates the smaller harbor seal, taking over the best hauling-out spots at low tide. It drives off intruders with a variety of vocalization calls from hisses and snarls to short barks and mournful hoots (Banfield

1974). Gray seal males tend to be polygamous and fiercely territorial. A harem is usually made up of two or three females and one male, but most females lie widely dispersed, each with a male in close attendance. Older females often return to specific spots where they bore pups in previous years (Bruemmer 1979). The gray seal, like other seals, has been observed sleeping on the bottom in shallow water and apparently surfaces to breathe without awakening.

MORTALITY

Aside from predation by humans, there is only the insignificant loss of gray seals by killer whales (FAO Adv. Comm. 1976).

BEARDED SEAL

Erignathus barbatus

NOMENCLATURE

COMMON NAMES. Bearded seal (King 1964), square flipper (because of the unusual shape of its foreflippers) (Banfield 1974); Eskimo names are *mukluk* and *oogruk* (Burns 1967).
SCIENTIFIC NAME. *Erignathus,* derived from Greek, refers to this animal's rather deep jaw, while *barbata* is from the Latin word *barba,* meaning beard, and refers to its numerous vibrissae.
SUBSPECIES. *E. b. barbatus,* in the eastern arctic and subarctic; *E. b. nauticus,* in the western arctic and subarctic (Burns 1967).

DISTRIBUTION

The bearded seal is found in the circumpolar region of the moving pack ice and is distributed southward around the globe to James Bay, in the Sea of Okhotsk, the White Sea, and occasionally Hokkaido, Scotland, Norway, and the Gulf of St. Lawrence (FAO Adv. Comm. 1976). The range of the subspecies *E. b. barbatus* includes suitable habitat from the Laptev Sea westward to the central Canadian arctic archipelago, and the subspecies *E. b. nauticus* occurs from the central Canadian arctic archipelago westward to the Laptev Sea (Burns 1967). In North America the largest concentrations of bearded seals are found in the Bering Sea on the inshore ice of the St. Lawrence, St. Mathew, and Hall islands (Popov 1976) (figure 40.14). In the Bering and Chukchi seas area, they stay in the region of the Bering-Chukchi platform, where they can reach the bottom to feed. Their seasonal movements and distribution patterns are primarily the result of ice conditions overlying this platform.

From January to April, *E. barbatus* is sparsely distributed throughout the Chukchi and northern Bering seas. A northward migration, which follows the ice edge to the Bering Strait area, begins in April. The southward fall migration is concurrent with the southward movement of sea ice (Burns 1967).

The Bering Sea population numbers approxi-

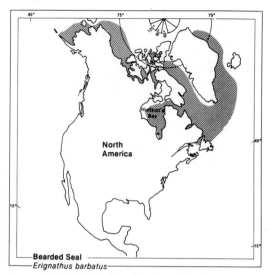

FIGURE 40.14. Distribution of the bearded seal (*Erignathus barbatus*).

mately 250,000 (Popov 1976) out of the world population of 500,000 (FAO Adv. Comm. 1976). A large group in the neighboring Sea of Okhotsk comprises about 200,000 animals.

Habitat. A seal's habitat is based on food resources and breeding and hauling-out sites. The bearded seal, *E. barbatus* is a benthic or bottom feeder and therefore is restricted to relatively shallow water. Such regions have heavy offshore ice that is in motion because it is influenced by winds, currents, and coastal features and therefore is often known as a "flaw zone" (Burns 1978). Animals remain in these open water situations until their breathing holes are destroyed by shifting ice (Burns 1967).

DESCRIPTION

The bearded seal is characterized by a conspicuous "moustache" composed of long, regular vibrissae that tend to curl at the tips when dry (Banfield 1974) (figure 40.15). Its foreflippers are an unusual shape, with the digits being almost equal in length. Other distinguishing features are its large eyes, prominent ear orifices, and thickened neck (Banfield 1974). Males and females are similar in size and can weigh up to and over 300 kg. The mean length is 2.2 m and maximum 2.6 m. Newborn pups are 1.2–1.4 m in length and weigh 27–35 kg (Popov 1976). The pelage of the pup is dark gray brown, with whitish dorsal blotches and

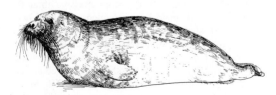

FIGURE 40.15. The bearded seal (*Erignathus barbatus*).

forelimbs. Adults are pale grayish to buff with a slightly darker saddle and are sometimes a rusty color around the head and neck (Fay et al. 1979). This brown facial and upper body color is thought to be a result of mud stains caused by their benthic feeding habits (FAO Adv. Comm. 1976).

FOOD HABITS

The bearded seal is a benthic feeder, consuming mainly epibenthic organisms—those that live on the surface of the sea floor. It also feeds on organisms that live in the bottom sediments. Its major prey items are a variety of crabs, hermit crabs, shrimp, clams, benthic fish, and schooling demersal (near bottom dwelling) fish. The feeding habits of this pinniped restrict it to waters of less than 200 m, such as the Bering-Chukchi continental shelf (Burns 1978). Although bearded seals consume a wide variety of food items, a relatively small number comprise the bulk of the diet. They do not utilize clams in any large quantities except during the summer, in spite of their year-round abundance (Burns 1967). The average volume of food found in a single bearded seal stomach was 854 ml, including sand, pebbles, parasites, and food items. Certain food items may aid in the removal of stomach parasites (Burns 1967).

BEHAVIOR

Erignathus barbatus is nongregarious and is rarely observed in large numbers (Banfield 1974). When basking, it remains widely distributed and faced away from others. Old males are often seen fighting on the ice floes, using their foreflippers as weapons; frequent observations of scarred seals of both sexes further confirm this pinniped's unsociableness (Burns 1967). The bearded seal is a curious animal but when surprised it becomes immobilized with fear and is easy prey for a polar bear or Eskimo (Maxwell 1967). During the spring breeding season, adult males make a loud, distinct, and highly characteristic noise like a whistle underwater which is thought to be connected with courtship behavior (Burns 1967).

MORTALITY

The major predators of the bearded seal are the polar bear and humans (Burns 1978); occasional deaths are attributed to killer whales (FAO Adv. Comm. 1976).

HAWAIIAN MONK SEAL

Monachus schauinslandi

NOMENCLATURE

COMMON NAMES. Hawaiian monk seal, Laysan seal (King 1964)

DISTRIBUTION

The range of the greatly endangered Hawaiian monk seal includes the atolls and islands of the Hawaiian

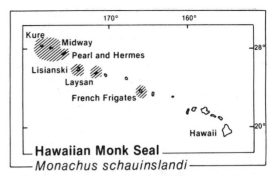

FIGURE 40.16. Distribution of the Hawaiian monk seal (*Monachus schauinslandi*).

leeward chain, a group that stretches northwest of the main Hawaiian Islands (King 1964) (figure 40.16). It breeds regularly in the five outermost atolls: French Frigate Shoals, Laysan Island, Lisianski Island, Pearl and Hermes Reef, and Kure Atoll (Kenyon 1978*b*).

In recent years there have been some births at Necker Island, which suggests there may be a new breeding group. Population counts in 1976 and 1977 were 695 and 625 seals, respectively, and total population estimates are not much more than 1,000 (Kenyon 1978*b*).

Habitat. The critical habitat of the Hawaiian monk seal, which includes environmental features essential to successful reproduction and survival, consists of dry coral sand beaches bordered by sheltering vegetation and shallow, calm water. Here the young can learn to swim and the seal can thermoregulate during the heat of the day in wet sand or shallow water. On cold, windy nights they can seek shelter and warmth on the dry sand and under the vegetation. They feed in the surrounding atoll lagoons and shallows (Kenyon 1976). This type of habitat is essentially undisturbed by man and his activities because of its isolation. This factor has enabled the seal to survive after an intense period of overexploitation (Kenyon 1973*b*).

DESCRIPTION

Monk seals, the most primitive living seals (Kenyon 1978*b*), are the only phocids to live permanently in tropical waters (figure 40.17). The Hawaiian monk seal has evolved in remote oceanic islands, as a trusting and tame animal. This is a detriment to its survival, since human disturbance now threatens its existence. Physiologically, the seal has not made great adapta-

FIGURE 40.17. The Hawaiian monk seal (*Monachus schauinslandi*).

tions to the warm climate, the blubber layer being much the same thickness as that of other phocids. However, the pelage of the pup is less woolly and more silky than that of its northern relatives (King 1964). *Monachus schauinslandi* exhibits sexual dimorphism in size, with females growing slightly larger than males as adults and also varying more in weight (King 1964). A typical adult male weighs up to 173 kg and is 2.1 m in length, while females are up to 273 kg and 2.3 m in length. Newborn pups are about 1 m long and weigh 16–17 kg (FAO Adv. Comm. 1976). The adult pelage is light silvery gray ventrally and slate gray dorsally, with males tending to be darker than females. At three to five weeks, the pups lose their soft, black birth coat; their new pelage is silvery blue dorsally, shading to silvery white ventrally (King 1964).

The life span of this species is unknown, although a female aged 11 years and a male aged 20 years have been recorded (FAO Adv. Comm. 1976).

FOOD HABITS

Hawaiian monk seals feed on bottom-inhabiting forms such as octopus and crayfish as well as a variety of reef fish. Eels, which are numerous in the atolls, are an important food item (Kenyon and Rice 1959). These seals also feed on lobster, although to what degree is not known (Green-Hammond 1980). Because monk seals may travel many miles at sea from one island to another, they probably feed in the open sea as well as in the shallow lagoons, where they have been observed eating at all hours. Feeding dives last about 10–15 minutes (Kenyon 1978*b*).

BEHAVIOR

Hawaiian monk seals spend most of their time in the relatively shallow water around the reefs they inhabit, or basking in the sun on the sandy beaches. Nocturnal in their feeding activities, monk seals are most likely to be seen on land in the afternoon, though some are in the water at all times of the day (Kenyon and Rice 1959). Because of its isolated environment, the Hawaiian monk seal evolved free of terrestrial enemies and is therefore innately docile and easily approached by humans, a trait that could be important in its survival (Kenyon 1978*b*). A characteristic habit of the monk seal is its tendancy to roll and root on coral sand beaches until a 15-cm depression is formed. These "sand wallows" are always evident on beaches frequented by this species and the seals resting in them have faces and eyes caked with sand from using their heads to burrow (Kenyon and Rice 1959).

MORTALITY

Human disturbance of the Hawaiian monk seal during the nursing period may affect the survival of the neonates. Greater pup mortality has been observed on islands occupied by humans. Shark attack is an important cause of death, and seals are often seen with healed scars from shark bites. In 1977 and 1978, 22 seals died from ciguatera poisoning (Kenyon 1973).

HOODED SEAL
Cystophora cristata

NOMENCLATURE

COMMON NAMES. Hooded seal, crested seal, bladder-nose seal (King 1964)

DISTRIBUTION

The population is comprised of one stock, which inhabits the North Atlantic from the Gulf of St. Lawrence, Newfoundland, and Labrador in the west (figure 40.18), to the eastern limit near Novaya Zemlya and Kanin Peninsula in the Barents Sea (Popov 1976). The population is concentrated in two areas during the breeding season, Jan Mayen Island and Newfoundland, with a possible recently "rediscovered" third concentration in Davis Strait (Sergeant 1976), although this could well be the northernmost part of the Newfoundland group. By July and August the seals have migrated to their molting grounds east of Greenland in Denmark Strait and also from 72 to 74° N (Sergeant 1976). Hooded seals move from Newfoundland to south, southwest, and southeastern Greenland for the summer (Sergeant 1976). The world population is thought to be 500,000–600,000, with 80–90 percent in the Jan Mayen area and the remainder near Newfoundland (Popov 1976).

Habitat. A pelagic animal, *C. cristata* inhabits areas of open sea and drifting shore ice (Popov 1976). Its habitat is limited by the fact that it dives deeply for its food, which consists of larger food organisms than that of the harp seal, whose range is similar (Sergeant 1976). The only time this species remains sedentary is during its breeding and molting periods, when it is concentrated on two areas of drift ice (Popov 1976).

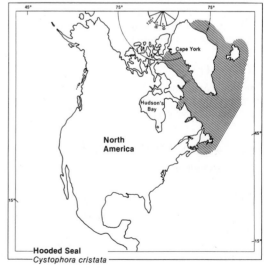

FIGURE 40.18. Distribution of the hooded seal (*Cystophora cristata*).

FIGURE 40.19. The hooded seal (*Cystophora cristata*).

DESCRIPTION

The most distinguishing feature of the hooded seal is the male proboscis (figure 40.19). This ''hood,'' or nasal sac, is actually the greatly enlarged skin of the snout, which is inflated in times of anger or excitement (Dunbar 1949). Mature males, who are not very tolerant of one another, may in anger also extrude a fiery red bladder through the nostril that is formed by the inflation of a very elastic portion of the internasal septum (Dunbar 1949). Sexual dimorphism is evident: males measure up to 2.8 m in length and weigh over 300 kg, while females are up to 2.3 m long and 160 kg in weight. Newborn pups are between 1.0 and 1.5 m long and weigh 22–25 kg (Popov 1976). The adult pelage is silvery gray with scattered black blotches and spots (Fay et al. 1979), and tends to be lighter ventrally (Dunbar 1949). The embryonic coat is shed *in utero,* so pups are dark gray with silver and blue tints on the back, sides, and dorsal sides of the back flippers; the chest and ventral surface are white. Following postpartum molt at one year, adult markings appear (Popov 1976). A hooded seal lives for approximately 20 years (Sergeant 1976).

FOOD HABITS

Because of their deep diving capabilities, their food is mainly squid, redfish, capelin, and polar cod (FAO Adv. Comm. 1976). In the Barents Sea, bottom fishes like halibut, cod, redfish, and flounder are taken (Popov 1976). Pups initially feed near the ice edge on crustaceans and squid (Popov 1976), and this seal's nutritional habits have no large effect on any commercial fishing interests (FAO Adv. Comm. 1976).

BEHAVIOR

Hooded seals spend most of the time swimming in deep water and are very active (Maxwell 1967), but when hauled out on land to whelp or molt, they form ''families'' consisting of a female, its pup, and one or several competing males (FAO Adv. Comm. 1976). Females are protective and will actively defend their pups. The male hooded seal is often maligned for its supposed aggressiveness. Such aggression only occurs toward humans, when they stand upright near the female and neonate. The seal's vocal repertoire is limited to a few sounds (Terhune and Ronald 1973). The head inflation and nasal bladder extrusion by the male are thought to be behavioral displays (Maxwell 1967).

MORTALITY

Major causes of death in hooded seals include predation by polar bears at whelping sites, predation by Greenland sharks in the molting areas, heartworm infection (FAO Adv. Comm. 1976), and humans.

ELEPHANT SEAL
Mirounga angustirostris

NOMENCLATURE

COMMON NAMES. Elephant seal, northern elephant seal (King 1964)

DISTRIBUTION

The breeding range of the northern elephant seal is restricted to the offshore islands from central Baja California and Mexico to central California, or from Isla Cedros, Mexico, to Point Reyes, California (figure 40.20). There is no mass migration of these seals, but a widespread dispersion in a northward direction occurs as soon as the pups leave their birthplace at about three months of age (FAO Adv. Comm. 1976).

The largest groupings of this seal are at Guadalupe Island (15,000–20,000), San Benito Island (5,000–10,000), and San Miguel Island. A colony was established in the southeast Farallon Islands in 1972 and 60 pups were born there in 1976. The Ano Nuevo Island population was reestablished in 1961 and numbers 2,000 (Delong 1978), while a new rookery has begun on the mainland at Ano Nuevo Point, where 16 females gave birth in 1977. This initiation of breeding at Ano Nuevo Point is a reflection of the continued growth and recovery of the elephant seal population and is possibly the result of overcrowding on Ano

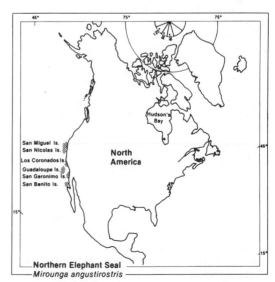

FIGURE 40.20. Distribution of the elephant seal (*Mirounga angustirostris*).

FIGURE 40.21. The elephant seal (*Mirounga angustirostris*).

Nuevo Island, which caused the move to the point on the mainland (LeBoeuf and Panken 1977). Other rookeries are found at Isla Cedros, Los Coronados, San Nicolas Island, and Point Reyes Peninsula (FAO Adv. Comm. 1976). Individuals are seen as far north as Alaska and up to 240 km from shore. The total population is about 45,000 (FAO Adv. Comm. 1976).

Habitat. The northern elephant seal spends most of its time at sea, but hauls out on sandy beaches to breed. *M. angustirostris* has been seen to cohabit peacefully in areas with the California sea lion, Steller's sea lion, Pacific harbor seal, and northern fur seal. However, this species breeds at a time of year when the others are at sea (FAO Adv. Comm. 1976). Gulls and cormorants are usually found in close association with these seals, walking among them or alighting on their backs. The birds are more alert to the approach of danger and act as sentinels for the elephant seals (Bartholomew 1952).

DESCRIPTION

Elephant seals are the largest of all pinnipeds. They are characterized by a very long proboscis, which, when relaxed, hangs over the mouth (figure 40.21). This is fully developed in the male only and is used as an instrument of vocalization when erected to threaten other seals. Another male trait is the rugose, corrugated neck shield, which also becomes more pronounced with age. There is considerable sexual dimorphism in size, but none in coat pattern. Adult males are 4.5 m in length and weigh 1.8–2.7 t, while females are 3.6 m in length and about 0.7 t. Newborn male pups are about 1.53 m in length and 36.0 kg and females are approximately 1.47 m in length and 31.5 kg. Elephant seals live about 14 years (FAO Adv. Comm. 1976).

FOOD HABITS

A deep diver, the northern elephant seal can reach a depth of 200 m and tends to feed well offshore (50 km) (DeLong 1978). Bottom-dwelling marine life such as skates (Rajidae), rays (Dasyatidae, Myliobatidae), ratfish (*Hydrolagus colliei*), small sharks (order

Chlamydoselachiformes), squid (order Decapoda), and Pacific hake (*Merluccius productus*) are part of their diet (Huey 1930; Morejohn and Baltz 1970).

BEHAVIOR

A solitary animal, this species spends most of the year at sea. When hauled out for the annual breeding and molting periods, these seals are gregarious and maintain an orderly social structure. Retention of social position is very important and results in fierce competition between males, who deal crushing blows to each other with their heads and necks (Bartholomew 1952). They are polygamous in their mating behavior, the males forming harems. Vocalizations include snorts, sneezes, grunts, and yawns. The most characteristic sound of the males is a prolonged loud snort that is produced with the help of the arched proboscis, pointed down the animal's throat (Maxwell 1967). Females and subadults make a vomiting cough when threatened (Bartholomew 1952). The animal is indifferent to the approach of humans but will rapidly retreat to the water if frightened (FAO Adv. Comm. 1976). The placid nature of these seals, apart from male aggression during the breeding season, is remarkable and must have evolved from hundreds of years of existence in remote areas where they were the supreme land mammals. On land they remain either asleep or inert and will periodically suspend breathing for about 5 min. To gain relief from dry skin and surface parasites they often engage in sand-flipping behavior, throwing sand over their backs with their foreflippers (Maxwell 1967). This may provide thermoregulatory relief.

MORTALITY

Predators of the northern elephant seal include killer whales and sharks (FAO Adv. Comm. 1976).

OTARIIDS IN NORTH AMERICA

The following species of otariid inhabit the waters adjacent to continental North America: *Eumetopias jubatus*, *Zalophus californianus*, *Arctocephalus townsendi*, and *Callorhinus ursinus*.

STELLER'S SEA LION

Eumetopias jubatus

NOMENCLATURE

COMMON NAMES. Steller's sea lion, northern sea lion (King 1964)

DISTRIBUTION

In the eastern Pacific, to the southern extent of their range (figure 40.22), adult males undergo a distinct

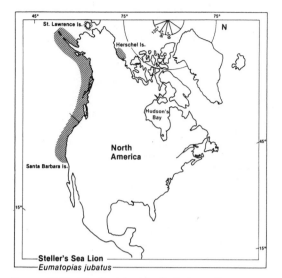

FIGURE 40.22. Distribution of the Steller's sea lion (*Eumetopias jubatus*).

in the Bering Sea in the summer (Gentry and Withrow 1978).

DESCRIPTION

Typical of other otariids, *Eumetopias jubatus* has large naked flippers with small nails (figure 40.23). It possesses four mammary teats (Fay et al. 1979). The female is slim, but the male has massive forequarters (Banfield 1974) and a thick, muscular neck with a substantial mane of long, coarse hairs (Maxwell 1967). Its head is bearlike, with a straight muzzle and small, stiff, pointed ear pinnae. Long (50 cm), stiff, pale mystacial vibrissae are present (Banfield 1974). The eyes of this species are very unusual because there is a circle of white around the outer edges of the iris (Maxwell 1967). As already mentioned, there is sexual dimorphism in size and appearance, adult males reaching lengths of 3 m and weights of 900 kg while females grow up to 2 m and 300 kg. Newborn pups are 1 m long and weigh 16–23 kg and are dark brown until the age of 4–6 months. Subadults are silver to light brown when wet (FAO Adv. Comm. 1976). The pelage is short and coarse and lacks an undercoat. Adults are tawny to brownish in color (Fay et al. 1979). This seal lives to an average age of 23 years (FAO Adv. Comm. 1976).

FOOD HABITS

Northern sea lions feed all year long, except during the breeding season, when the males in the harem starve themselves for two or three months. Their food consists of a wide variety of invertebrate marine life and fishes such as coelenterates, sand dollars, worms, and mollusks as well as cod, herring, halibut, and salmon (Banfield 1974).

BEHAVIOR

Steller's sea lions are highly gregarious, crowding together on breeding rookeries, swimming, and hauling out in groups the rest of the year. They haul out in sunny, calm weather but stay at sea when it is rough and stormy. They have been observed at play in the water but on land appear to be very quarrelsome. Although this is a polygamous species, only the larger

postbreeding seasonal northward migration. Some females may also participate in this northward movement during the winter (FAO Adv. Comm. 1976). On the other hand, when the far northern breeding aggregations disperse in August, the males of these rookeries leave immediately in a general southward migration from Alaska. The newborn pups disperse among the islands of the North Pacific.

The world population is 250,000 (Gentry and Withrow 1978), with fewer than 100 in the Channel Islands, 1,600–2,000 and stable on Ano Nuevo Island, 5,000–7,000 and stable in California; 2,000 in Oregon, 600 in Washington, and 5,000 in British Columbia (2,500 around Vancouver Island, 2,000 in the area of Cape St. James and the Queen Charlottes, and 500 along the northern British Columbia coast) (FAO Adv. Comm. 1976). In Alaska, Steller's sea lions number in excess of 200,000. The Gulf of Alaska populations are stable, but the eastern Aleutian Island population (10,000 in 1967) (Maxwell 1967) has experienced a drastic decline. The California population has been declining since the 1920s, with a sharp decrease since the late 1960s (Gentry and Withrow 1978).

Habitat. A nongregarious species, the animal is rarely seen in large numbers except occasionally in the summer when hauled out on beaches. During the winter this sea lion prefers shallow waters free of fast ice near the coast and may also be observed on gravel beaches and ice flows that are not too far out to sea (King 1964).

E. jubatus breeds along the west coast of North America from San Miguel in the California Channel Islands, northwest to the Gulf of Alaska, along the Alaska Peninsula, and throughout the Aleutian and Pribilof islands. Off Asia it is found in the Kuril Islands, Kamchatka, and the islands in the Okhotsk Sea. Some animals move as far north as St. Lawrence Island

FIGURE 40.23. The Steller's sea lion (*Eumetopias jubatus*).

and older sea lion males have their own harem (Banfield 1974). They arrive at the breeding grounds first, and they establish and defend their territories with the use of many threat displays (Gentry and Withrow 1978). Rarely, if ever, do they leave their territories during the breeding period and as a result may fast for up to 60 days. Females are more gregarious and do not show any attachment to one specific male or territory. They usually wean their pups in the first year, but occasionally suckle a yearling (Gentry and Withrow 1978). There is a definite social structure in breeding grounds, with barren females, bachelor males, and yearlings remaining at a distance (Banfield 1974). This species is normally very wary of humans and will hurry to the water as soon as an intruder is seen, except during the breeding season, when the females lose their shyness and will defend their pups (Maxwell 1967).

MORTALITY

The mortality rate of Steller's sea lions is 42.9 percent in the Farallon Islands off central California. Premature stillbirths here account for 35.7 percent of all births (Ainley et al. 1977).

CALIFORNIA SEA LION

Zalophus californianus

This species is probably the most widely known sea lion in the world, performing in shows, circuses, and zoos.

DISTRIBUTION

Some subadult and adult male California sea lions undergo a northward postbreeding migration and are seen as far north as Bull Harbour, Vancouver Island (51° N), during the winter. This is 1,000 km north of the northernmost rookery at San Miguel Island, California (34° N). Females have been reported as far north as Ano Nuevo Island (37° N). In the Gulf of California, males, females, and young are found throughout the entire breeding range all year, although some adult males and subadults move south following the breeding season along mainland Mexico to Manzanillo (19° N) and some have been sighted at 23° N (Los Frailes) (Mate 1976) (figure 40.24).

In North America as of 1975 there were 8,500 of these sea lions in Mexican waters, 20,000 in the Galapagos Islands, 15,000 on the Pacific side of the Baja, 20,000 in the Channel Islands, 14,000 in northern California, 2,500 in Oregon, 500 in Washington State, and 1,000 in British Columbia; as of 1965 there were over 16,000 in Cedros, Benitos, and Guadalupe islands (FAO Adv. Comm. 1976). The total North American population is estimated at 100,000–125,000 (Mate 1978).

Habitat. Pupping and breeding is carried out on island-based rookeries consisting of sandy beach areas (Mate 1978), although some rookeries in the Gulf of

California Sea Lion
Zalophus californianus

California are semiaquatic or fully aquatic but are always close to the mainland or an island (Mate 1976).

Competition with the northern sea lion for food, habitat, and other resources may significantly affect the distribution of both species. There are indications of mutual shifts in breeding range, short periods of cohabitation, use of similar hauling-out grounds, and similar prey species (FAO Adv. Comm. 1976).

The North American subspecies, *Z. c. californianus,* occurs from the tip of Baja, California (23° N), throughout the Sea of Cortez and north in the eastern Pacific to San Miguel Island, California (34° N), with a non-breeding range of 19–51° N. *Z. c. wollebaeki* occurs in the Galapagos Islands and *Z. c. japonicus* in the Sea of Japan (Mate 1976).

DESCRIPTION

The California sea lion is more slender than Steller's sea lion and does not have the overtly thickened neck of that species. The ear pinnae are stiff but inconspicu-

FIGURE 40.25. The California sea lion (*Zalophus californianus*).

ous and the mystacial vibrissae are long and stiff (Banfield 1974) (figure 40.25). At the age of five, the males develop a distinctive sagittal crest on top of the head resulting from the growth of this portion of the skull. This crest becomes increasingly prominent during the animal's adult life and is often more striking in older sea lions because the normally dark hair turns to light brown or tan in this area (Mate 1978). Males are noticeably larger than females, being 2.2 m in length and weighing 275 kg, while the females are 1.8 m in length and weigh 91 kg. Newborn pups are 0.75 m and 5–6 kg (FAO Adv. Comm. 1976). The pelage is short, dense, and chocolate brown in color (Fay et al. 1979). *Z. californianus* has a lifespan of 17 years (FAO Adv. Comm. 1976).

FOOD HABITS

California sea lions are shallow water, opportunistic, day and night feeders. They compete with the northern sea lion and other near-shore pinnipeds for the same food items (Mate 1976), such as squid, octopus, abalone, and a variety of fishes like herring, sardines, rockfish, hake, and ratfish. These sea lions bring their prey to the surface, bite through the neck, snap off the head with a powerful shake, and quickly swallow it "head first." They eat commercial fish in nets and on lines, damaging the fishing gear in the process (Banfield 1974).

BEHAVIOR

California sea lions are very gregarious, though wary of intruders (Banfield 1974). In the water they are very playful, chasing one another and leaping out of the water. They are also active on land, mainly because of parasitic infections that irritate their skin (Maxwell 1967). Vocalizations include a honking bark by the males, a quavering howl by the females, and a bleating sound by the pups. A distinctive behavioral trait of the sea lions is their ability to catch objects in the air with their teeth, and this has been exploited in animals in captivity (Banfield 1974). California sea lions are polygamous, forming loosely organized harems. The males patrol the waters off their beach territories and make threatening gestures toward intruders. In the water the males appear to be very protective toward the pups (Maxwell 1967).

MORTALITY

Predators of the California sea lion include sharks and killer whales. Pup mortality is caused by drowning in rough seas and injuries received on crowded rookeries (Mate 1976). In recent years there has been an apparent increase in the number of premature births in this species, possibly due to chemical residues or a bacterium (FAO Adv. Comm. 1976). PCBs and DDT metabolites have been found to be two to eight times higher in the blubber and liver tissues of aborting females and their pups than in the tissues of full-term animals. A correlation has been observed between premature pupping and high tissue organochlorine levels. These compounds can interfere with the reproductive processes directly by causing hormonal imbalances and or indirectly by immunosuppression resulting in increased susceptibility to disease. High PCB concentrations cause abnormal liver function, which may also be responsible for imbalances in heavy and trace elements in the premature parturient females (Gilmartin et al. 1975). *Leptospira pomona*, a bacterium found in California sea lions in 1972, is known to cause abortion in many domestic mammalian species. This virus, called the San Miguel Sea Lion Virus (SMSV), has been detected in sea lions giving premature birth (Gilmartin et al. 1975).

GUADALUPE FUR SEAL

Arctocephalus townsendi

NOMENCLATURE

SUBSPECIES. Two subspecies have been previously recognized: *A. p. philippii* and *A. p. townsendi*, the former being found on the Juan Fernandez Islands and the latter on Guadalupe Island (Scheffer 1958, King 1964) However, more recent opinions distinguished two separate species under the same genus, these being *A. philippii*, or the Juan Fernandez fur seal, and *A. townsendi*, or the Guadalupe fur seal (Repenning et al 1971). The more northern species is described here (figure 40.26).

DISTRIBUTION

The only known breeding colony of *A. townsendi* is on the east shore of Guadalupe Island off lower California. They range at least 500 km north and 270 km east of this rookery and haul out regularly on San Miguel Island (FAO Adv. Comm. 1976). A 1977 census of these seals on Guadalupe Island indicated 1,073 individuals (Fleischer 1978a). Current estimates of the population are 1,300–1,500 animals and its status is considered to be vulnerable, although recent censuses indicate a gradual increase. Furthermore, the area occupied by these seals on Guadalupe Island is continually expanding, indicating a probable increase in numbers (Fleischer 1978a).

Habitat. The Guadalupe fur seal prefers a rocky habitat. The volcanic caves along the east side of Guadalupe Islands provide shelter from prevailing winds, launching spots from which to swim in hot weather, and suitable places to breed (Fleischer 1978b).

DESCRIPTION

A. townsendi, the Guadalupe fur seal, is distinguished by its large head and extremely long, fleshy snout and pointed muzzle (Repenning et al. 1971; Fleischer 1978b) (figure 40.27). The ear pinnae are scroll shaped, and capable of limited movement, and the

FIGURE 40.26. Distribution of the Guadalupe fur seal (*Arctocephalus townsendi*).

muscles allow the pinna to be closed tightly when the animal is submerged. This seal is characterized by its large flippers, which are very dark in color and hairless to the area of the metacarpals (Fleischer 1978a). Sexual dimorphism occurs, with males measuring about 1.9 m in length and weighing 159 kg, while females are approximately 1.4 m in length and 45 kg (Fleischer 1978b). Very little is known about the young of this species. Adult male Guadalupe fur seals are dark brown with lighter hair on the neck, shoulders, and breast and possess a heavy mane. Females vary more in color, with shades of gray, brown, and cream on various areas of their body. Pups are dark dorsally and gray brown ventrally with lighter faces (Fleischer 1978a).

FOOD HABITS

Nothing is known about the feeding habits of the Guadalupe fur seal.

BEHAVIOR

A. townsendi is a polygamous species, the males forming territories during the breeding season. A typical

FIGURE 40.27. The Guadalupe fur seal (*Arctocephalus townsendi*).

territory consists of one adult territorial male, perhaps several nonterritorial fringe males (Fleischer 1978a), and two or three, but occasionally as many as six or eight, females with pups (Fleischer 1978b). A great deal of aggression has been observed between territorial males during the breeding season (Fleischer 1978a). Although the nursery areas are very noisy, seal vocalizations are limited and probably have some important social functions. Adult males growl, bark, whimper, and cough, while pregnant females are very quiet. Mothers are heard calling to their pups as a means of locating them. A distinctive behavioral trait of the Guadalupe fur seal is its enjoyment of floating and grooming itself in the water, much like the sea otter. This grooming allows the water to penetrate the seal's thick fur, giving a cooling effect.

Also related to this seal's thermoregulation is its habit of floating: with one flipper extended above the water, on its back with its snout above the surface, or upside down with its hind flippers in the air (Fleischer 1978a).

NORTHERN FUR SEAL

Callorhinus ursinus

NOMENCLATURE

COMMON NAMES. Northern fur seal (Scheffer 1958), Pribilof fur seal, Alaska fur seal (King 1964)

DISTRIBUTION

C. ursinus ranges the subarctic waters of the North Pacific Ocean, Bering and Okhotsk Seas, and the Sea of Japan (Fiscus 1978) (figure 40.28). It occurs as far south as San Miguel Island, a common home for this seal, the northern and California sea lions, and the Guadalupe fur seal. There is competition for space, and interspecific reproductive behavior has been observed between *Callorhinus* and *Zalophus*. Occasionally there is interaction between *Callorhinus* and *Arctocephalus* males (De Long 1975).

The main herd leaves the North Pacific and Bering Sea in October and migrates south down the North American coast as far as San Francisco (Fiscus 1978). Females and young of both sexes make the longest migration, while adult males tend to remain north near the Aleutian chain and the Gulf of Alaska (FAO Adv. Comm. 1976). The northward migration from California begins in March. The migration patterns of fur seals follow the movement of schools of prey species whose abundance influences the length of time they may stay in an area (Fiscus 1978).

The world population is almost 2 million, with 1.3–1.4 million in the Pribilof Islands (United States), 265,000 in the Commander Islands (U.S.S.R.), 165,000 on Robben Island (U.S.S.R.), 33,000 in the Kuril Islands (U.S.S.R.), and about 2,000 on San Miguel Island (United States) (FAO Adv. Comm. 1976; Fiscus 1978).

FIGURE 40.28. Distribution of the northern fur seal (*Callorhinus ursinus*).

Habitat. Apart from the breeding period when northern fur seals are hauled out on rocky island beaches, these animals spend most of the time at sea. They are usually seen offshore along the continental slope, and in areas where the topography causes upwellings of nutrient-rich water and an abundance of prey species (Fiscus 1978). They range through waters with a surface temperature of −1 to +15° C and are most abundant in waters of 8 to 12° C, probably because of food availability (Baker et al. 1970).

DESCRIPTION

The northern fur seal has a small head with large eyes; a short, pointed snout; a high forehead; and moderately long, stiff, slender ear pinnae (Banfield 1974) (figure 40.29). The ears are tightly rolled cylinders, each having a wax-coated orifice to prevent the entrance of water. Its most distinctive features are its thick, waterproof underfur and its unusually large flippers (Fiscus 1978). There is great sexual dimorphism, with males having a massive body and neck and females

FIGURE 40.29. The northern fur seal (*Callorhinus ursinus*).

being more gracefully proportioned (Banfield 1974). Adult males are 2.13 m in length and weigh 181.3 kg and females are 1.4 m in length and 43–50 kg, while newborn males are 0.66 m in length and 5.4 kg and females 0.63 m in length and 4.5 kg (FAO Adv. Comm. 1976). Pelage color varies between the sexes. Males are dark brown, except for the mane, which has a grayish tinge, while females are slate gray dorsally and a lighter reddish gray ventrally. The throat of both sexes is lighter while the rest of the coat and the underfur is a chestnut color (Maxwell 1967). The hair and fur fibers of its thick coat number 56,900/cm² (Fiscus 1978). The northern fur seal lives for 20 years or more (FAO Adv. Comm. 1976).

FOOD HABITS

Northern fur seals obtain some of their food in deep water and have been recorded to a depth of 80 m. They tend to feed during the evening, night, and early morning, sleeping during the day. The most common food items in their diet are squid, herring, pollack, and lantern fish, although their prey species vary with the season, the seal's age, and the area (FAO Adv. Comm. 1976). Fur seals will swallow small fish whole (25 cm) while underwater, but will bring larger prey to the surface to be torn into chunks by shaking before being eaten (Fiscus 1978). This seal is believed to consume about 10 percent of its body weight per day; if this is so, the total annual consumption by the Pribilof stock is 965,000 t. Major changes in the Bering Sea ecosystem as a result of intensive commercial fishing may have a profound effect on this species (FAO Adv. Comm. 1976).

BEHAVIOR

Usually this species swims alone, although groups of two or three are not unusual (Fiscus 1978). On land the seals are very gregarious, and a complex social structure exists on the breeding rookery, where territorial harems are the rule and subadults remain on the outskirts. They spend the day resting, preening, or swimming, feeding at night. They are able to sleep either on land or in water (Banfield 1974). A characteristic pose when sleeping at sea is the "jughandle position," where the seal lies on its back, with the hind flippers folded forward and held by a foreflipper (Fiscus 1978). Hindflipper waving is said to be a thermoregulatory device (Banfield 1974). Vocalizations consist of loud coughs, roars, barks, and bleating sounds (Maxwell 1967).

MORTALITY

Mortality rates of northern fur seal pups since 1963 have varied from 5 to 12 percent. Major causes of death are hookworm infection and malnutrition, with injuries, congenital defects, and bacterial infections killing some. Hookworms cause severe anemia and subsequent death. Other parasites probably kill a few, and some pups starve when their mother is killed at sea. Violent weather and the inability to obtain adequate food cause mortality during the seal's first year at sea. Up to 85 percent die before the age of three in some year classes. Predators of northern fur seals include killer whales and great white sharks. Because this seal as a behavioral trait will put its head through looplike objects floating at sea, scraps of synthetic netting can be dangerous. If caught around the animal's head or neck, they can impede feeding or cut deeply, causing crippling infection or death (Baker et al. 1970).

ODOBENIDS IN NORTH AMERICA

WALRUS

Odobenus rosmarus

NOMENCLATURE

SUBSPECIES. Two subspecies are recognized: *O. r. rosmarus*, the Atlantic walrus, and *O. r. divergens*, the Pacific walrus (Reeves 1978; Scheffer 1958). Both are found in North American waters.

DISTRIBUTION

A few thousand Atlantic walruses occur in the areas of the east coast of Greenland, Franz Josef Land, and the Barents and Kara seas, and about 10,000 occur in the eastern Canadian arctic and western Greenland (FAO Adv. Comm. 1976). The Pacific walrus is found in the areas of the Bering and Chukchi seas, numbering 3,000 in 1954, and in the Laptev Sea, where they numbered 140,000 in 1976 (FAO Adv. Comm. 1976) (figure 40.30). There has been a great diminution in the range of the walrus in recent history. It used to occur as far south as New York City and southwest Britain, but has deserted much of its former habitat as a result of human activities rather than as a result of natural factors (Reeves 1978). Although migration occurs, not all animals move every year. Atlantic walruses move north as the ice edge retreats in the summer and move south as the ice advances in October. Pacific walruses also move north in the spring, migrating on floating ice (King 1964).

Habitat. Major requirements for walrus survival include adequate hauling-out platforms of land or ice, and suitable adjacent feeding banks. The range and distribution are determined by the conformation of the shore and the temperature (Reeves 1978). In the summer, walruses haul out on all available kinds of habitat, such as cobble beaches, rock beaches, and large boulders (Miller 1976).

DESCRIPTION

The walrus is instantly recognizable from other pinnipeds by its long tusks (figure 40.31). The main differ-

FIGURE 40.30. Distribution of the walrus (*Odobenus rosmarus*).

ence between the Atlantic and Pacific odobenid subspecies is their tusk size and shape, those of the Atlantic walrus being shorter, slimmer, and straighter (Reeves 1978). In general, among pinnipeds, walruses are second in size to the elephant seal and have a small square head with a broad muzzle that bears a heavy, bristly moustache (Banfield 1974). Their eyes are small and frequently bloodshot. They possess no external ears, but the meatus is protected by a fold of skin (King 1964). The large, massive neck of the mature males is often covered with coarse tubercles, which is a secondary sexual characteristic (Banfield 1974). Located in the neck are a pair of air sacs or pharyngeal pouches that can be inflated and used as buoys in the water when the walrus is sleeping or wounded (King 1964; Kenyon 1978a). Muscular whisker pads, a highly vaulted mouth, and tusks are all adaptations to their feeding habitat. The tusks, which are large and conspicuous in both sexes but are usually more slender in females, are used as an aid when hauling out onto the pack ice and also as a means of propulsion as the animal feeds on the bottom (Kenyon 1978a). A single tusk may reach a length of 1 m and a weight of 5 kg (King 1964).

Adult males are larger than the females, measur-

FIGURE 40.31. The walrus (*Odobenus rosmarus*).

ing 3–3.6 m in length and weighing from 1.2 to 1.6 t. Females, on the other hand, are from 2.5 to 2.6 m in length and weigh from 0.75 to 1.0 t (FAO Adv. Comm. 1976). As the animal increases in age, so does the thickness of its skin, reaching 2.5–5.08 cm in adults and as much as 6.35 cm on the neck of a mature male (King 1964). Blubber accounts for one-third of the total body weight in the walrus (Maxwell 1967). Walruses have a scanty coat of reddish brown hair, although old animals may be nearly naked (King 1964). When the animal is in the water, the skin appears whitish in color, but when resting in the sunlight, blood circulation increases and the skin turns pink (Kenyon 1978a). Pups are a slate gray color, and change to the reddish brown after the first molt (King 1964). Walruses live to be about 40 years old (FAO Adv. Comm. 1976).

FOOD HABITS

Tusk development delays weaning in this species until the animal is between one and two years old. Adult walruses use their snouts to root on the bottom, with their vibrissae helping to locate food. A powerful sucking action, resulting from a highly vaulted mouth working together in the tongue, (Kenyon 1978a) extracts worms and tunicates or pulls clams and snails from their shells (Reeves 1978). This feeding process is an important means of recycling nutrients by stirring up the sediment on the ocean floor and makes walruses the "earthworms" of the subpolar seas (Reeves 1978). Their diet consists primarily of bivalve mollusks such as clams and mussels, though fish and decapod crustaceans are occasionally taken (Reeves 1978). Walruses will sometimes eat seal flesh or blubber (ringed and bearded), narwhals (*Monodon monoceros*), and belugas (*Delphinapterus leucas*) when invertebrates are unavailable (Reeves 1978). The capacity of the walrus stomach is 4–5 l, which is small for such a large mammal (Maxwell 1967). These pinnipeds feed in the early morning and haul out for the rest of the day, occasionally fasting for up to a week during pleasant weather (Banfield 1974). They feed extensively in shallow water in the autumn until the bays freeze over, at which point they try to keep breathing holes open by bunting the new, soft ice. Eventually they retreat to the shore leads (one mile offshore in winter) (Banfield 1974). Fasting occurs during the spring mating season and males feed very little during the summer (Banfield 1974). The 25,000 Atlantic walruses consume 463,570 t of benthic organisms a year, or 1.270 t a day (Reeves 1978).

BEHAVIOR

Walruses are very gregarious, feeding, hauling out, and migrating in densely packed groups (Banfield 1974). In cool weather they lie in passive body contact on land, the dominant walruses (ones with long, unbroken tusks) taking the best positions while the most subordinate ones lie on the periphery of the herd. This body contact provides warmth and aids the molting

process (Miller 1976). Individual bickering occurs (Banfield 1974), but if any one member is attacked the rest of the herd will defend it fiercely (Maxwell 1967). On land these animals are usually placid, but males, especially if they are wounded, have been known to attack boats (Banfield 1974). A sentinel warns the entire herd of possible danger with a low whistling bellow that sends the walruses into the sea (Maxwell 1967). In addition to helping the walrus climb out of the water onto the ice, the tusks may also have a role in the social structure of the herd by establishing dominance (Reeves 1978). The distinctive male gong or belllike vocalization (Schevill et al. 1966) is thought to be associated with courtship display, as are songlike responses (Ray and Watkins 1975). It is believed that walruses may navigate and orient by echolocation (Reeves 1978). Walruses do not have territories or harems. Mating occurs during the spring migration. Mothers will defend their young and will often clasp their young calves to themselves with their foreflippers to protect them. If the calf becomes tired in the water, the mother will sometimes carry it on her back (McClung 1978).

MORTALITY

Nonhuman predation is not a major cause of death for walruses, as the adults make a formidable adversary for both killer whales and polar bears. *Odobenus rosmarus* appears to be relatively free of parasitism and disease. Death of young by crushing during a stampede to the sea and as a result of interspecific aggression among males does occur (Reeves 1978).

PINNIPED MANAGEMENT IN NORTH AMERICA

Pinniped management in North America is often complicated. It may involve two or more countries, active legislation, and people with political, economic, scientific, and other interests. This section makes no attempt to deal with the topic in its entirety, but we hope it will serve as an introduction.

Who Is Involved. Marine mammals do not recognize international boundaries, so any discussion of the management and conservation of seals, sea lions, and walruses in North America must encompass the management policies of the governments involved, namely, the United States and Canada. In order to do this it is important to understand the laws and agencies from which these policies arise. It is also important to be aware of some of the international agreements that have bearing on the policies regarding these animals.

Like all furbearing animals, some seals have long been used as a source of wearing apparel for either practical or esthetic reasons. Recent interest groups, intent on preserving animals from this type of exploitation, have become part of seal management. Further complications arise because both seals and humans compete for the same food source, so the rights of commercial fishermen become important considerations.

Unable to obtain enough fuel on land, mankind is now exploring the seas for oil, particularly in the northern climes. The transport of such commodities poses certain direct problems. The livelihood of the North American Inuit, as well as a group of other native peoples, the Aleuts, has been intimately tied to marine life for many centuries. Preserving the right to maintain this tie has been an important consideration that is founded in the laws of each country. In Canada certain people are also permitted to hunt seals as a means of income, and these people have taken worldwide criticism and have faced on-the-job problems that most industries have never had to face.

Man is a curious animal, intent on seeking answers, so scientists have been studying the seal, puzzling over its existence and making recommendations that often run counter to the opinions of other groups involved. Finally, government civil servants have been charged with the responsibility of applying the laws of the land and sea, weighing the interests of all, and then attempting to produce a satisfactory plan of administration.

Historical Background. There is a complicated cast of characters who figure in the management of pinnipeds. Therefore, it may be helpful to define where they fit in the system and how the system works. Enough has been written on the historical exploitation of some species of seal in North America for most people to realize that between 1800 and 1960 many seal populations became depleted to levels that threatened extinction. People have acted naïvely, taking a natural resource without thought that in the use lay a responsibility for allowing replenishment. Over the past two decades people have come to see this, and to realize that payment for their actions is finally coming due.

To avoid the redundancy of those writers who have recounted the history of the great slaughters, suffice it to say that management of seals began after World War II in a very minor way and that by the late 1960s and early 1970s legislation began to appear in a way that meant an improvement in the life expectancy of the seal. Most of this legislation, which is examined in the next section, was part of a larger development of man's understanding of the environment, and of his use or misuse of its riches.

UNITED STATES

Legislation

Four pieces of legislation are the foundations of the U.S. government pinniped management program. The first is the National Environmental Policy Act, 1969. The impetus for this act came from the Committee on Merchant Marine and Fisheries, who saw many conflicts and overlapping research projects regarding the environment and little cooperation among the departments involved. The purpose of the act was to declare a national policy, by promoting preventative measures to protect the environment and the biosphere for the health and welfare of mankind. In order to carry out the creation of an understanding of the ecological

system and natural resources of the United States, the act established the Council of Environmental Quality.

The second piece of U.S. legislation important to pinnipeds is the Marine Mammal Protection Act, 1972 (MMPA). This act recognizes that certain marine mammal species have become depleted because of human activities, and as a result seeks to protect these animals or to take action to ensure that they remain a significant functioning element in the ecosystem. This might require measures to replenish depleted stocks and the protection of "areas of significance" for such stocks. Efforts were to be made to widen the sphere of ecological knowledge and population dynamics through international cooperation. Following the premise of preservation and protection comes the necessity of preserving the continuing balance of the ecosystem. The MMPA recognizes the esthetic and economic value of marine mammals and thus directs that the protection and conservation afforded to them must be conducted with the most sound resource management policies and that the primary objectives of such management should be "to maintain the health and stability of the marine ecosystem."

Major provisions are:

1. A moratorium is declared on the taking and importation of marine mammals and marine mammal products; permits are available for purposes of scientific research or public display.
2. During commercial fishing operations, incidental kill of marine mammals may be allowed under permit, provided that the method and equipment used meet certain requirements. The "immediate goal," however, in this instance would be to reduce such takings to "insignificant levels approaching zero mortality." The secretary of the treasury could ban the import of any commercial fish from countries that in catching such products did not meet the U.S. standards of incidental kill, and could demand reasonable proof from foreign governments that such standards were met.
3. Taking and importing may be waived by the secretary if such a recommendation is made by the best scientific advice with sound principles of resource protection and conservation. However, products from nations whose criteria do not meet the aims of this act may not be imported, even for processing.
4. Endangered species may not be taken except with written permission for research purposes.
5. Indians, Aleuts, or Eskimos who dwell on the coast of the North Pacific Ocean or Arctic Ocean are exempt from the provisions of this act if the taking of marine mammals is for subsistance purposes or for creating or selling authentic native handicrafts. Such taking must not be wasteful and if the species is endangered certain regulations may be made about their take by native peoples.
6. Pelagic hunting is banned.
7. International treaties, conventions, and agreements made before this act was effective are still viable. (Certain recommendations and exceptions

are discussed in the section "International Agreements").

8. Persons and vessels are banned from taking marine mammals or using ports for importing, and they may not offer any such animals or their products for sale.

9. It is unlawful to import any marine mammal that at the time of taking, is pregnant, nursing, less than eight months old, taken from a depleted species, or taken in an inhumane manner.

10. The secretaries of the departments involved are charged with making regulations about the taking and importing to ensure that such taking will not disadvantage the species or stocks. These regulations are to be made with the best scientific advice and evidence and in consultation with the Marine Mammal Commission. Consideration must be given to what effects the regulations will have on: (*a*) existing and future levels of marine mammal species and population stocks; (*b*) existing international treaty and agreement obligations; (*c*) the marine ecosystem and related environmental considerations; (*d*) the conservation, development, and utilization of fishery resources; and (*e*) the economic and technological feasibility of implementation.

11. The secretaries of the Department of Commerce and the Department of the Interior must report annually through the *Federal Register* on the current status of all marine mammal species and stocks, and on any actions taken on their behalf.

12. Permits for the taking or import of marine mammals may be granted by the secretary, subject to several stipulations.

There are severe monetary fines and prison terms for infractions against this act.

Three agencies act as administrative bodies for this act. The Department of Commerce, in which the National Oceanic and Atmosphere Administration operates, has jurisdiction over all pinnipeds except the walrus. The Department of the Interior has jurisdiction over the walrus. Title II of the act established an independent body, the Marine Mammal Commission, whose responsibility is to develop and review information, actions, and policies to achieve the objectives set forth in the MMPA.

The third important piece of legislation is the Endangered Species Act, 1973 (ESA). Although the United States was previously involved in international agreements about endangered flora and fauna, this act provides national recognition that such species exist in a state of extinction or near extinction because of economic growth and development and the lack of adequate conservation measures to protect such animals. The act recognizes that such endangered species are part of an ecosystem, and that ecosystems are in need of protection and preservation as well. The secretary of the interior is charged with the responsibility of using any viable method to bring threatened populations to the point where they are no longer considered to be endangered (Green-Hammond 1980).

The fourth piece of legislation is the Fishery Conservation and Management Act, 1976 (FCMA). This act was a response to the need for a program of conservation and management to protect the fish stocks in the U.S. waters from overfishing and to develop those fisheries that were not being utilized. Administration of the act was delegated to the Department of Commerce, and the secretary has jurisdiction over the Fishery Conservation Zone established by the act. This zone coincides with the U.S. territorial waters of 200 nautical miles from the coast.

Conservation and management were defined in the act as measures designed to avoid irreversible or long-term effects on the marine environment and measures useful in rebuilding, restoring, or maintaining any fishery resource and the marine environment. Fishery resources include stocks and their habitats. The act specifies that these fishery resources are to be utilized to get the optimum yield (OY) but prevent overfishing. In order to implement this policy of management and conservation, the act required the development of Fishery Management Plans, which were to be the responsibility of the new agencies established by the FCMA, namely, eight regional fishery management councils (Green-Hammond 1980).

Administrative Bodies

In the United States there are four separate government bodies that have jurisdiction, input, or advisory capabilities directly legislated to them with regard to the management of the seal and walrus populations and their environment. The Departments of Commerce and Interior and the Marine Mammal Commission are, however, the primary management agencies. Their responsibilities for marine mammals are set out in the MMPA, which may be regarded as the cornerstone of conservation and management of marine mammals in the United States. The other agency that has some impact on marine mammal populations is the Council on Environmental Quality.

It is important to note that while each of these bodies may in fact work together on certain aspects of seal management, they are legislatively separate departments or agencies. The Marine Mammal Commission and the Council on Environmental Quality may act in a watchdog capacity, reporting to the Congress that the two departments charged with administering the management policies are either fulfilling or not fulfilling their duties. As objective observers, they offer another perspective to many areas and may point out areas of need in research or management.

The Department of Commerce contains the National Oceanic and Atmospheric Administration (NOAA) and its subdivision National Marine Fisheries Service (NMFS). All seal populations in the United States are effectively managed by NOAA and NMFS. NOAA's responsibilities with regard to marine mammals are defined by: Coastal Zone Management Act, 1972, Marine Mammal Protection Act, 1972, Marine Protection, Research and Sanctuaries Act, 1972; Endangered Species Act, 1973; and Fishery Conservation and Management Act, 1976.

The Fishery Conservation and Management Act, 1976 (FCMA), created eight regional fishery management councils under the auspices of the Department of Commerce whose sphere of interests directly affects marine mammals and particularly seal populations, although seals are specifically NMFS management concern. Their districts are: New England, mid-Atlantic, South Atlantic, Gulf of Mexico, Caribbean, Pacific, North Pacific, and Western Pacific. Their jurisdiction extends outside of state waters, but within the fishery zone created by FCMA. Each council consists of voting members—those appointed by the secretary of commerce, those appointed by the governor, and the regional director of NMFS—and nonvoting members—representatives of the U.S. Fish and Wildlife Service, the Coast Guard, the Marine Fishery Commission, and the State Department. Each council has its own scientific and statistics committee (Green-Hammond 1980).

Certain species of seal come within the jurisdiction of each council, and two of those councils presently have fishery management plans that could have broadly based effects on seal populations. The breakdown of the relevant groups is as follows (Green-Hammond 1980):

North Pacific Council:
  Northern fur seal, bearded seal, Steller's sea lion, walrus, ribbon seal, ringed seal, largha (*Phoca vitulina largha*) harbor seal, hooded seal, northern elephant seal
  Fishery Management Plan: Bering Sea Groundfish Plan
Pacific Council:
  Northern seal, Steller's sea lion, northern elephant seal, harbor seal, California sea lion, Guadalupe fur seal
New England Council:
  Gray seal, harbor seal, hooded seal, harp seal
Mid-Atlantic Council:
  Harbor seal, hooded seal, gray seal, harp seal
South Atlantic Council:
  Harbor seal, hooded seal
Western Pacific Council:
  Northern seal, Hawaiian monk seal
  Fishery Management Plan:
  Spiny lobster.

The impact that the fishery management plans will have on seals has been reviewed by Green-Hammond (1980), and will be discussed later.

The Department of the Interior contains the U.S. Fish and Wildlife Service (U.S.F.W.S.). The U.S.F.W.S. has held its present structure and sphere of influence since 1974, when the old Bureau of Commercial Fisheries was transferred by Congress to the Department of Commerce. The Bureau of Sport Fisheries was incorporated into the new U.S. Fish and Wildlife Service. Accordingly, it maintains a responsibility for wild birds, mammals (except certain marine mammals), inland sport fisheries, and specific fisheries research activities. Specifically, it has jurisdiction over the walrus. Its objective is to ensure that the people of the United States enjoy the maximum opportunity and benefit from the fish and wildlife resources under a conservation and management program (U.S. Office of Federal Register 1979).

By virtue of Title II of the Marine Mammal Protection Act, 1972, the Marine Mammal Commission was established as an independent body. Members have the responsibility for developing and reviewing information, actions, and policies to achieve the objectives set forth in the act. The commission consists of three presidentially appointed commissioners, who appoint a nine-member committee of scientific advisers on marine mammals, composed of scientists knowledgeable in marine ecology and mammalian affairs (U.S. Marine Mammal Commission 1979).

The Council on Environmental Quality has as its objective to formulate and recommend policies to promote the improvement of the quality of the environment (U.S. Office of Federal Register 1979). In 1970 additional responsibilities were added. The council may conduct studies and research related to ecology. The three council members are directly appointed by the president. They make recommendations to the president on national environmental policies, monitor changing trends in the environment, and appraise any government policies affecting the environment (U.S. Office of Federal Register 1979).

CANADA. Canadian sealing differs from that of the United States. In Canada, seals are still part of a national fishery industry. In the United States, only the fur seal is harvested, and that is bound by international treaties. Although treaties do come into play in the harvest of the harp seal, legally, the sealers of the eastern Canadian provinces are commercial sealers who depend for a living on sealing. The vast difference between the legislation of the two countries is that in the United States, lawmakers have responded to various groups who sought to protect all marine mammals. The laws of a country are born out of the needs of the people and to this date the people of Canada have not seen fit to call a moratorium on sealing. The seal herds continue to be a renewable resource using the best government and nongovernment scientific knowledge available.

Legislation
One major piece of legislation, the Fisheries Act, 1970, and its corollary, the Seal Protection Regulations, 1966 (which are amended annually), govern seal management in Canada. The Fisheries Act delegates to the minister of fisheries and oceans various provisions regarding fisheries in Canada (at the inception of the act it was the minister of fisheries and forestry). With regard to seal fisheries, the act clearly provides protection for both the seals and the sealers: "No one shall with boat or vessel or in any other way during the time of fishing for seals knowingly or willfully disturb, impede or interfere with any seal fishery or prevent or impede the shoals of seals from coming into such fishery or knowingly or willfully frighten such shoals" (Fish Act. Amended List 1979).

The act also provides for enforcement of any regulations or provisions made under the act. Provision is made for the minister to issue any regulations necessary for carrying out the provisions of the act.

Details of seal management are outlined in the Seal Protection Regulations. These regulations detail every aspect of seal hunting and protection of seal herds, from season opening and closing dates to types of instruments approved to animals permitted to be taken. It would be redundant to repeat the provisions in detail, so major areas of concern and some important specific provisions are noted. (Reference is made here to the Seal Protection Regulation, Amended List, 30 March 1979.)

Provisions include:

1. An annual amendment sets out which seals may or may not be hunted, announces a quota on each species per person or group, and specifies in which areas the seals may be hunted and in what manner they may be taken. It is useful to note that in the harp seal hunt in 1980 the harvest on large vessels could not consist of over 5 percent yearlings or older, and that only 5 percent of hooded seals taken by large vessels may be adult females.
2. There is a clear definition of who may hunt seals, as well as of those people or vessels that may not. Generally speaking, those residents, native Eskimos or Indians, and those persons holding a sport sealing licence are permitted to hunt. Provision is also made for scientific research.
3. Methods of transportation to and from the hunt areas are indicated, i.e., types and lengths of vessels and aircraft, safety factors, equipment specifications, etc. Permits for such vessels are granted entirely at the discretion of the minister of fisheries and oceans. Specifications are made about the distance at which aircraft may land from the seal herds.
4. Observers and nonparticipants of the hunt are carefully regulated by permit only, and application of such permits must give definite information on purpose, equipment, and length of stay.
5. Season opening and closing dates may be changed by ministerial decree annually.
6. Hunting techniques are very clearly spelled out, with strict definitions of instruments, methods of humane slaughter, and regional particularities.
7. There are a great many requirements for permits of sealing licenses, each one clearly stated.
8. The regulations detail the means to judge when an animal is dead, and specifies methods and timing of skinning the dead animal.
9. The regulations grant permission for the killing of gray seals in the areas where bounties are allowed, except during breeding season.
10. Certain areas in the eastern coastal waters are detailed as areas where seals may not be taken.
11. Tagging, marking, and moving of live seals is forbidden without written permit.
12. Time periods in the day when seals may be hunted in the Gulf of St. Lawrence and the Front off Labrador are set out.
13. The vessel master is in charge of everyone on his boat with regard to the provisions of these regulations.
14. License prices are stated.

Please consult the amended regulations from which these provisions were taken, which state, ''all persons making use of this consolidation are reminded that it has no official sanction. . . . [See] Part II of the Canada Gazette for purposes of interpreting and applying the regulations'' (Fish. Reg. amended 1980).

Administrative Bodies

In Canada, there is one department or ministry that has a legislative mandate to manage pinnipeds, the Ministry of Fisheries and Oceans. There is an advisory committee as well, but, unlike the U.S. watchdog agency, the Canadian contingent reports to the ministry who created it.

The duties of the Ministry of Fisheries and Oceans include all matters in the area of renewable resources that are designated through the Parliament of Canada which are not by law assigned to another department. Pinnipeds are part of the fisheries resources. The primary objective of the management program is to guarantee maximum economic and social benefit to Canada from the use of fisheries and other aquatic living resources and to preserve and maintain these resources in a healthy, productive state.

Within the coastal areas a director general of fisheries management oversees field operations, which are managed and conducted from regional and field locations, usually in the form of biological stations. Legislative jurisdiction belongs exclusively to the federal government, but some provinces have accepted administrative responsibility for fisheries. When Canada announced the extension of the offshore 200-mile boundary in 1977, it became necessary to create a new forum for the management of the Atlantic fisheries resources. The Canadian Atlantic Fisheries Scientific Advisory Committee (CAFSAC) is a vehicle for discussion and assessment of methodology and advice to management within the department of Fisheries and Oceans and development of programs of management. CAFSAC reports to the Atlantic Fisheries Management Committee.

In 1971 observers at the annual seal hunt in the Gulf of St. Lawrence, Canada, felt that there was a need for a vehicle that would offer a direct and effective channel to the then minister of fisheries regarding sealing recommendations. These observers were from such diverse fields as the Canadian Federation of Humane Societies, university scientists, and representatives of the International Society for the Protection of Animals. As a result of the meeting, a committee was formed to advise the minister on all matters concerning seals and sealing, the Committee on Seals and Sealing (COSS).

This committee of six persons includes three scientists and three nonscientists. It functions in an advisory capacity and reports directly to the minister of the Department of Fisheries and Oceans. Recommendations are based on investigations into the socioeconomic, biological, ecological, and humane aspects of the seal hunts in Canada as well as any international aspects of sealing that affect Canada.

The objectives of the committee, as outlined in its terms of reference, are: (1) to advise on quota changes before the harp and hooded seal hunt, and over a long time period to observe the different phases of the hunt; (2) to evaluate the size and composition of the herd with any suitable techniques, and to carry out and evaluate the methods of statistical analysis and make recommendations on changes if any; and (3) to investigate hunting methods, and carry out research and necessary changes. Although COSS has mainly concerned itself with harbor, gray, harp, and hooded seals, it is able to and has made recommendations on any species.

INTERNATIONAL AGREEMENTS. Both Canada and the United States have unilateral and multilateral agreements regarding marine mammals, as well as membership in many international conventions whose ratification entails conservation and protection of and research into either the animals or their habitat.

North Pacific Fur Seal Convention, 1957
By 1911, pelagic sealing had seriously depleted the North Pacific fur seal herds, and a meeting in Washington involving the United States, Great Britain (acting for Canada), Japan, and the Soviet Union resulted in a convention that protected the seal and banned pelagic sealing. The United States and the Soviet Union would conduct their own hunts, but would each provide Japan and Great Britain with 15 percent of the take. The United States banned commerical sealing for three years. The herds increased so remarkably that by 1941 Japan broke the agreement, claiming concern for the interference of the seals with its commercial fisheries concerns. From 1942 until 1957, the seals were protected by a provisional agreement between Canada and the United States. In 1957, a new North Pacific Fur Seal Convention was convened by Canada, Japan, the Soviet Union, and the United States. Research and management of the North Pacific fur seal is coordinated by the Fur Seal Commission, a body provided by the convention, consisting of representatives of the four governments. Research programs and harvesting rates are recommended by a standing scientific committee on an annual basis. The convention is still effective and provides Canada and Japan with 15 percent of the annual commercial take on the Pribilofs, which is conducted by the United States, and on the Commander Islands and Robben Island, conducted by the Soviet Union. To effect the convention on the domestic level, the Fur Seal Act of 1966 was enacted; it renewed the authority of the secretary of the interior to adminster management of the Pribilofs as federal territory. In 1970 that authority was transferred to the Department of Commerce.

Under the MMPA the secretary of commerce, through the secretary of state, is charged with the responsibility of initiating multilateral or unilateral agreements with nations to protect or conserve marine mammals, and where the United States is specifically involved in multilateral agreements such as the North Pacific Fur Seal Convention, to initiate amendments if necessary that comply with the policies and purposes of the act. At the time the act was enacted the secretary of commerce was to ensure, in conference with the Marine Mammal Commission, that the North Pacific fur seals were at their optimum sustainable populations and to review the North Pacific Fur Seal convention to ensure that it was consistent with the MMPA. If the findings were, for example, that the fur seal was presently not at its maximum sustainable population, the secretary of state could then initiate negotiations with other parties of the convention to protect the herds.

Northwest Atlantic Fisheries Organization (NAFO)
In 1967 the International Commission for the Northwest Atlantic Fisheries (ICNAF) was established. The organization's members were from those countries that had coastal states in the area and whose interests were in the species of fish, seal, and other marine mammals within the waters of the northwest Atlantic Ocean. This formation effectively brought the management of the northwest Atlantic population of harp and hooded seals in Canada within ICNAF's area of concern. These seals migrate within the boundaries of Canada and Denmark (Greenland). Two other countries (Norway and the U.S.S.R.) were interested in the population of the Northern Atlantic. A Standing Committee on Research and Statistics (STACRES), was created, which provided in part the panel A (seals) personnel who were charged with making annual recommendations to the commission about seal management.

In 1976, Canada (and other nations) extended the zone of fisheries jurisdiction to 200 nautical miles from the coast. This meant that there were changes in the management authority for seals. The harp seal populations were residing within the districts of Canada and Denmark (Greenland). This required a restructuring of ICNAF, which was accordingly replaced by a new convention, which established the Northwest Atlantic Fisheries Advisory Committee (CAFSAC) is the Cana-200-nautical-mile extension of the fisheries jurisdiction of the member states (which meant that NAFO had no jurisdiction with these areas). STACRES was replaced with its counterpart, the Scientific Council of NAFO, which still provides scientific advice on harp and hooded seal management. The Canadian Atlantic Fisheries Advisory Committee (CAFSAC) is the Canadian subsidiary of this council. It is a formalized forum for biological advice on Canadian domestic fisheries in the 200-mile limit, including stocks not previously dealt with by ICNAF. Total allowable catches are now negotiated annually between Canada and the European Economic Community (representing Denmark). In northern Canada and Greenland this effectively amounts to a very small, locally important but relatively insignificant harvest by native peoples. (Agreement was reached between Canada and Norway on their mutual fisheries relations in 1976.)

Norway is the only country that presently hunts seals within Canadian boundaries. After World War I, a Canada-Norway Sealing Commission was established which acted as a research and management consultative body. A bilateral treaty was signed in 1971 regarding sealing and conservation of the Northwest Atlantic seal herds. In 1975, this agreement was amended to include hooded and bearded seals and the walrus. When both countries extended their fishing zones to 200 miles, a new agreement was made to encompass mutual fisheries relations.

INTERNATIONAL CONVENTION PROTOCOLS. Canada and the United States also subscribe to several international conventions that are concerned with the environment, as well as to several international conventions whose interests are marine life and its habitats. These are briefly listed below, with the exception of IUCN, which is discussed more fully. More detailed information can be found in the United Nations Environment Programme, Register of International Conventions and Protocols in the Field of Environment (1977), and its supplement (1978).

The International Union for the Conservation of Nature and Natural Resources (IUCN)
Founded in 1948 and stationed in Gland, Switzerland, IUCN is an independent international body whose membership is composed of member states, government bodies, private institutions, and international organizations irrespective of political or social systems. Its objective is to preserve, on a world-wide basis, the natural environment and its resources, which are rapidly being altered by man's urban and industrial development. It maintains active conservation programs, in cooperation with such agencies as UNESCO and FAO. The World Wildlife Fund (WWF), also located in Gland, is the basic fundraiser for IUCN, whose scientific and technical advice it receives. WWF allocates its funds to projects, either directly or through IUCN publicity programs, particularly focusing on the education of young people. The Species Survival Commission (SSC) of IUCN is concerned with the preservation and prevention of extinction of plants and animals. It also publishes the Red Data book, which provides data on the status of endangered species. Both Canada and the United States have national branches of WWF.

Canada (C) and/or the United States (USA) have signed the following international conventions that are directly or indirectly relevant to pinnipeds. Country and date of entry are indicated.

1. Convention on Nature Protection and Wildlife Preservation in the Western Hemisphere (USA, 1942).
2. International Convention for the High Seas Fisheries of the North Pacific Ocean (C and USA, 1953).
3. International Convention for the Prevention of Pollution of the Sea by Oil (C, 1958; USA, 1961).
4. Convention on the Continental Shelf (C, 1970; USA, 1964).
5. Convention on fishing and conservation of the living resources of the High Seas (USA, 1966).
6. Convention on High Seas (USA, 1962).
7. The Antarctic Treaty (USA, 1961).
8. Amendments to the International Convention for the Prevention of Pollution of the Sea by Oil, 1954 (C, 1942; USA, 1973).
9. International Convention relating to Intervention on the High Seas in Cases of Oil Pollution Casualties (USA, 1975).
10. Amendments to the International Convention for the Prevention of Pollution of the Sea by Oil, 1954, concerning Tank Arrangements and Limitation of Tank Size (C, 1974).
11. Convention on International Trade in Endangered Species of Wild Fauna and Flora (CITES) (C and USA, 1975).
12. Convention for the Conservation of Antarctic Seals (USA, 1976).
13. Convention for the International Council for the Exploration of the Sea (amending protocol 1975) (C, 1968; USA, 1973).
14. Treaty on the Prohibition of the Emplacement of Nuclear Weapons and Other Weapons of Mass Destruction on the Sea-Bed and the Ocean Floor and in the Subsoil Thereof (C and USA, 1972).

Policy and Application. How man manages any group of animals, wild or domestic, is dependent on his idea of what place in his life those animals will assume. Whether he decides to utilize the animal for food or clothing, to preserve and protect it for esthetic or biological reasons, or to keep it as a companion, he must formulate that idea of utilization into a policy either formally stated or clearly understood. Ideas of "conservation [are complex issues] not solely related to policies of federal and provincial government for the encouragement of restraint and curtailment of demand, but more importantly to choices freely made by the public at large on life styles and on social goals, as well as to economic considerations" (National Energy Board, Canada 1977: 1-67 to 1-68). The policies governing seal management in the United States and Canada differ in one very important way. The United States has made a policy statement that is definitely philosophical in nature, firmly stated and defined in legislation, and part of a much larger policy that affords protection to the environment. Canadian policy on pinnipeds is not philosophical in nature. It is somewhat more difficult to ascertain what Canadian policy is, because, unlike the Marine Mammal Protection Act, the Canadian Seal Fisheries Regulations do not state moral values about marine mammals. Nevertheless, the policy is there, but it is formulated in a more abstract form. The Fisheries Act and its seal regulations are rules and regulations that apply to a policy that is understood as a basic premise. In other words, if there are rules about hunting seals or protecting seals, the people as a whole believe in hunting or protecting seals. Canadian policy is usually made known by statements by the government in power, or the philosophy of policy often exists in such forms as commission reports. For example, when the Mackenzie Valley Pipeline Commission (Berger 1977) made its recommendations to halt a decision for at least 10 years to settle native claims, it was adopted as policy by the Canadian government (National Energy Board 1977). The philosophy of the policy is clearly stated.

Essentially, the United States and Canada share the same philosophy about pinnipeds, in other words, that the herds must be protected. Canadians have chosen a middle-of-the-road position that allows a small segment of their population to supplement their income from regulated hunting of harp and hooded seals. The United States has chosen the extreme position that does not allow any hunting, except by native peoples or by terms dictated by previous commitments such as those governing the Pribilofs.

If a policy is society's ideal, achievement of that ideal, or application of it, depends on the people's understanding of the concept, the correct methods for applying it, and the basic premise that the policy is correct or workable. It is in this area that the breakdown sometimes occurs. This is an interesting time in the history of seal management, because both the United States and Canada are in the process of discarding old methods, and new ones have not been formulated to deal with the myriad of problems facing the marine populations as a whole. In short, it is a period of transition. The policies are defined, but the means to carry them out are not.

Many of the difficulties are ones that are common to all periods of change. People find it difficult to keep up with new thinking, clinging instead to familiar methods and patterns. Although they realize that there is a new policy, often it is not clearly defined in terms of what changes it will mean for them. "Changes normally proceed more slowly than some elements in society wish" (Nat. Energy Board 1977: 1–68), and it is a matter of fact that the rate of change in all areas of our lives over the past few decades has been unequaled in the history of mankind (Toffler 1976). Scientific discoveries, resulting implications, different philosophies, accelerated growth in all areas, and increasing economic demands have all been progressing at an astonishing rate, so it is small wonder that even biologists have had a difficult time keeping pace. A single item, the computer, has caused major changes whose effects have still not manifested themselves today. Although it was predicted in the late 1960s that by 1980 management decisions would be made from a wide computer data base, a recent survey reveals that major management decisions in all spheres of society are still being made by individuals at one or two top levels, with no indication that such data are widely used (George 1979). The data may be there, but interpretation is often difficult, ignored, or not relevant to those people in decision-making positions.

Upper management decisions, particularly in government areas, are often a function of territoriality, with the unwritten law being "stay out of my area and I'll stay out of yours." Others remain isolated because of administrative duties, a common and often unavoidable situation, and more often than not many government decisions are politically motivated. Compounded with these realities are the present complexities of the

data, new opinions and theories. Managerial personnel must then rely on trusted and informed advisers. In the United States, administrators are faced with a stringent law, and a profusion of powerful and knowledgeable interest groups. Marine mammal management for them demands a tightrope act of balance between interpretation of the law and the watchdogs.

For those at the application level of pinniped management, civil servant and scientist alike, it is now becoming apparent they have not always changed their methods to agree with the laws. This is particularly true in the United States, where a major policy shift was made a decade ago (Green-Hammond 1980). In Canada, other factors, such as the failure of certain methods that have resulted in the collapse of certain fish stocks, have acted as an impetus in this regard. While the legislation in the United States is a multispecies-oriented philosophy, single species strategies, often disguised to be multidimensional, have been used for a decade in both pinniped and fisheries management.

The fate of the seal is in fact intimately tied to fisheries management. Although seals are effectively protected from excessive hunting by man, they are at the top of the food chain, the foundations of which are being utilized and eroded. All of these levels are in danger from man's exploitation, oil exploration and transportation, as well as pollution.

As we enter the 1980s it is apparent that management of pinnipeds, or any other marine animal, can no longer be spoken of in isolation. The ecosystem in which seals, sea lions, and the walrus live is being utilized consistently and intensively by mankind, and it is an environment that is interdependent on all levels. It is also a world that is largely unfamiliar to terrestrial mankind.

In spite of this, urban man and his needs have reached into every corner of the earth. Heavy metal and organochlorine contamination and pollution have plagued the oceans for a couple of decades. Compounding these two serious problems is the practice of fossil fuel exploration in the seas. It is quite feasible for a person living in Toronto, Canada, to drive a car with gas obtained from the Mackenzie Delta. Exploration for that gas could have unknown and irreversible effects on the populations of seals and people inhabiting that area.

Current Trends in Management Methods Because the policies of pinniped management in Canada and the United States and the concepts that are the basis for these policies were formulated in the late 1960s and early 1970s, and because a decade later some of the management methods have proven outdated, it is critical to understand some of the old concepts. Pinniped and fisheries management has traditionally been based on a theory of maximum sustainable yield (MSY). This strategy is still in use, but is undergoing continual change.

MSY is defined as "the greatest harvest that can be taken from a self-regenerating stock of animals year after year, while maintaining constant average size of the stock'' (Holt and Talbot 1978). This procedure has been an integral part of the methods used to harvest and preserve seal and fish stocks, but it has been overused; it was never intended to be the ''sole conceptual basis for management'' (Green-Hammond 1980: G-5). It is a strategy that is based on a few assumptions, which have been summarized by Green-Hammond (1980):

1. A small population increases rapidly, until the stock size increases, then the rate slows till the rate reaches 0, at which point losses to the herd are exactly replaced by additions; this point is known as the carrying capacity.
2. Unexploited populations are assumed to be at carrying capacity, i.e., a size that the environment can support.
3. Birth and survival rate relations are assumed to be density dependent, i.e., rates are higher when the population is lower.
4. Any population can have a high rate of birth and death turnover or a correspondingly low turnover. A high rate indicates food for predators, which are part of the ecosystem and therefore an indication replacement rate for a population, i.e., high rate = more production for population.
5. Stock density is the standing population per unit area. When density falls below carrying capacity, birth rate and survival rate increase to exceed death rate. Recruitment to an age or size class is dependent on birth rate and the rate of survival to the given age or size class. If recruits outnumber losses, this surplus can be harvested, which means the population remains below carrying capacity. If that surplus is unharvested it will increase the standing population size until the recruitment and death rates are equal. ''The population level at which the surplus of recruits over deaths for the total population is greatest, is the population level producing MSY.'' Another assumption here is that environmental conditions are constant.
6. Equilibrium yield (EY) or replacement yield is the level of harvesting that removes the surplus of births over deaths without changing population size. This may be less than, equal to, or more than MSY, but in reality, on a conceptual basis, could be constant only for one given carrying capacity.

For ''r''-selected species, MSY is assumed to be between 40 and 60 percent of carrying capacity (Holt and Talbot 1978); for ''K''-selected species, such as marine mammals, it may be as high as 60 percent (Green-Hammond 1980). In very broad terms, the environment of an organism may be one in which competition is almost nonexistent, or one in which it is much greater and where density effects are important. Organisms living in the less competitive situation are generally termed r-selected, in that they direct most of their energy to reproducing as often and as many offspring as possible. However, those organisms living in a competitive environment, or K-selected species, direct most of their energies to competition and their maintenance. It is also important to understand that this theory is used only for modeling purposes, and that in

reality organisms do not strictly conform to either category (Pianka 1978).

There are many faults with these MSY assumptions, many of which are simply naive, many of which are just too simplistic and not testable (Holt and Talbot 1978; Green-Hammond 1980). Green-Hammond summarizes:

1. The strategy is single-species oriented, and the stocks are not self-contained.
2. The environment and the natural mortality rate (M) do not remain constant.
3. This constant M assumption does not calculate a time lag for predator populations to adjust to changes in prey density, or for ways in which this will be apparent, e.g., older age of first reproduction in seals.
4. There is a strain in any area where more than one stock is being fished by the MSY concept.
5. The concept has tunnel vision, ignoring trophic level relationships, symbiotic or commensal relationships, environmental changes, and human influences (Holt and Talbot 1978).
6. Stocks are not always at ideal carrying capacity before exploitation.
7. Many of the variables underlying the MSY theory, such as carrying capacity, do not remain constant during exploitation.
8. Most of the methods used for the MSY calculations are recruitment, growth, natural mortality, fish mortality, and fishing effort (Ricker 1975); and most of this information comes from fishing. If there is no exploitation, information stops.
9. Collection of data has been difficult, and wrong calculations have been made, resulting in overexploitation.

Perhaps the most challenging conclusion that can be drawn from this problem is that there is no alternative that deals with the environment as a whole (Gulland 1976). Fisheries management plans may have considerable impact on seal populations. The time is ripe for innovative people and methods.

New methods, or changes in policies, can come from within the government, from interest groups such as Greenpeace or Monitor, Inc., from watchdog agencies such as COSS or the MMC, or from university scientists.

In the United States there are some very powerful groups who have considerable knowledge, wealth, and lobbying abilities, who monitor all management decisions about marine animals. Some of these groups have been instrumental in making some of the very early fundamental policy changes about marine mammals. In Canada, such groups have not had such far-reaching impact. Seal hunting is a right afforded by law to certain groups of people in Canada; dissenters and observers have been ineffective in changing that existing law, but certainly have forced humane slaughtering techniques to be adopted (Lavigne 1978).

In the Seal Protection Regulations the minister of fisheries and oceans is granted the right to implement any measure he sees fit to protect the sealers or the seals. The opinions of dissenters have always been welcomed in open forums, but management decisions have been changed only through the structured channels of government. Although world attention has focused on the seal hunt for many years, the emotionalism of the issue often exceeds reason or fact. One interest group that has a definite interest in sealing and that has also been effective in presenting its view to the Canadian and American people was the prohunt delegation lead by Premier Moores of Newfoundland. There was also a group of Newfoundlanders known as the Mummers, who, in dramatic format, presented the role that seals play in their lives.

Watchdog agencies in Canada have only advisory capabilities. Although they are invited to make recommendations, their abilities to change policy must be viewed as limited. The U.S. watchdog agencies have an impact that is somewhat more effective, if less direct, as they do not report to the body whom they are reviewing.

University scientists are usually not politically motivated, and much of the basic research is done at this level. It is encouraging to see their development of new methods of pinniped management, which have been intensive and refreshingly innovative.

In order to illustrate the effectiveness of new management methods in certain situations, two examples of current management situations are detailed. The Fisheries Management plan encompasses a wide scale of variables that are addressed in an overall management plan.

FISHERIES MANAGEMENT PLANS. The terms of the FMCA require that Fisheries Management Plans be drawn up and approved before any new fisheries industry is implemented. Green-Hammond (1980) examined two such plans that could have direct impact on pinniped populations. Her object was to examine their feasibility within the context of those theories prescribed by the FCMA, the MMPA, and the ESA to see if the plans were compatible with those intentions. Her conclusion was that all three legislative documents mandate an ecosystem-level perspective as an approach to fisheries management, but that the plans were not consistent with this theory in their practical application or considerations.

The first plan examined was the Bering Sea Groundfish Plan, 1978. The Bering Sea is a highly productive area for fish, marine mammals, and birds. The North Pacific Council proposed a plan to harvest all groundfish except herring and Pacific halibut. Domestic take is minimal, being only 1 percent, but the groundfishery will constitute about 2 percent of the entire world marine catch. The plan covers 33 percent of all fisheries under final or draft management plans in the United States.

In order to model the fish populations in this area, a computer simulation model (DYNUMES III: Dynamic Numerical Marine Ecosystem model) has been developed (Laevastu and Favorite, cited by Green-Hammond 1980).

The plan takes into consideration the variables of growth, recruitment, production, and mortality of individual fish and other fish. But it serves as a good example of why some new concepts are ineffective. The whole concept of management as outlined in this plan was single species, MSY being calculated separately for each species, so that no provisions were made for interconnections between the species that inhabit a common ecosystem. Any attempt at multispecifics was organizational, not conceptual.

Mention of marine mammals is restricted to their impact on the fisheries through consumption of fish caught in fishing gear, and the adverse affect on consumption of the fish stocks as a whole. The reverse of this impact was ignored. Fish consumption by predators is calculated, but there are no provisions for the impact on the marine mammals, for example, from a changed food availability. In fact, these data were available in the DYNUMES model, but "maintenance of optimum sustainable populations or maximum net productivity of marine mammals were not addressed in the plan" (Green-Hammond 1980:38).

There was no anticipation for recovery of deleted stocks except from immigration. The optimum yield levels were representative of very intense fishing. Such levels could have been greatly lowered without harming the very small fraction of the industry that accounted for domestic fisheries. As Green-Hammond pointed out, justification for such reductions could have been an ecological consideration, namely, reducing the considerable risk of adverse impacts on the groundfish and marine ecosystem. Unfortunately, economic returns are often more important, even when other evidence is pertinent. For example, intense groundfish harvesting in the 1970s in the Bering Sea is thought to have been responsible for recent reductions in the northern fur seal populations, a fact not mentioned in this plan (Green-Hammond 1980).

The second plan examined was the Spiny Lobster Draft Plan, 1978 (Western Pacific Council). If this plan is approved, it could have an impact on the endangered monk seal, although what the impact might be is unknown. This seal does feed on lobster, although its dietary importance to the seal is unknown. The aim of the plan was stated to be minimization of the environmental and ecological impacts of this fishery, especially on the monk seal. Prohibition of fishing in shallow water was a provision intended to protect the spawning area of the lobster; whether it will also protect the monk seal and its coastal habitat is uncertain.

Although the council suggested that the impact of fishery on monk seals be monitored and it recognized the risk factor involved, it will accept no responsibility for monitoring, and has adopted a "wait and see" attitude (Green-Hammond 1980).

POPULATION CENSUS. New concepts and new methods of determining the population of the northwestern Atlantic harp seal population have been developed in the last decade. The Canadian government revises the total allowable catch (TAC) of harp seals annually based on the best population estimates of the size and condition of the herd. Accurate estimates, therefore, are important. Methods of obtaining such figures have ranged from eyeball "guestimates" to survivorship indices (Sergeant 1969; Benjaminsen and Øritsland 1975), sequential population analyses (Lett and Benjaminsen 1977), and other visual methods relying on a systematic fixed method of estimation (Lett and Benjaminsen 1977). A new technique that utilizes the absorbance of ultraviolet radiation by the hair of a polar species has provided some interesting data. The method also encompasses an aerial survey and ground truthing, which result in a high level of accuracy (Lavigne 1975). The white-coated pup, which was not seen in previous methods, can therefore be photographed, even against the white snow and ice conditions. There have been disagreements over certain aspects of such censusing. Some scientists believe that the whole herd was not photographed in certain years. However, the method has never been disputed, and by and large, the population of the northwestern Atlantic harp seal has been studied in the past three or four years to such an extent that there is "more and better data for [them] than for most exploited marine stocks, seals, whales or fish" (Lavigne 1978:386). Such new techniques stimulate a review of older, more established techniques.

One of the issues that has arisen from the collection of such data has been the development of computer models, which attempt to accommodate all variables. The resulting disputes over which models are better and more reliable have been active and at times prolonged. The first model (Allen 1975) still stands undisputed, however. The most recent and thorough discussion of modeling and other management methods for seal populations can be found in the documentation of the IUCN-sponsored workshop held at the University of Guelph in 1979 (IUCN 1980).

This situation also serves to illustrate the changing role or evolution of the biologist who is involved in management situations. The computer simulation experiments are highly sophisticated, complex, and still in the innovation stages of development. As a result, there are very few persons conversant with the technicalities of the entire situation. The new biologist must be part statistician, part mathematician, part political scientist, and part biologist in the traditional sense. There is an imminent danger in the growth of a small elitist group whose knowledge of such things as population figures and condition of the herd is not disseminated. Their wisdom is vital for decisions, and the responsibility and trust that must be placed in their hands is enormous. The realities of the situation demand liaison people who can brief other managerial persons in simpler terms about the problems or theories behind such complex models. Compound this with a situation like that of the Bering Sea, where oil exploration is a possibility, fish exploitation is intense, and seals are still being harvested, and the need for managerial biologists with very specialized skills becomes paramount.

INTERNATIONAL MANAGEMENT. A response to world-wide pollution and international industrialization cannot be made using only methods that are local in scale. Certain issues and concerns are important nationally and can therefore be handled at that level. But in a situation where the problem is an international one, local or national remedies are often akin to brush fire fighting.

In 1980, the IUCN released a document entitled "World Conservation Strategy" that stressed the need for a global conservation strategy for management of shared resources (IUCN 1980). Its recommendations are worthy of mention here.

As Green-Hammond (1980) stated, economics is more often than not the impetus behind exploitation of fisheries. But "human beings in their quest for economic development and enjoyment of the riches of nature, must come to terms with the reality of resource limitation and the carrying capacities of the ecosystems, and must take account of the needs of future generations. This is the message of conservation" (IUCN 1980, Preamble and Guide, foreword).

If marine resources are understood to be part of an interdependent ecosystem, then we must understand that global interrelatedness is also a reality and that mankind must adopt an international management strategy and responsibility for our shared resources. The IUCN document defines shared resources as an ecosystem and its species that are shared by two or more states. This includes species like seals, which move between states and ecosystems. It also includes species that either depend on or are affected either by events, such as in coastal ecosystems, or by other animals, such as those associated with fisheries and migratory species. Proposals have been made accordingly for regional strategies of management in international river basins and seas. As demonstrated in the section on structures and agencies, the law is the most effective form of action for implementing management strategies. It also points out that two vehicles for such action that are potentially powerful and effective are CITES (see the section "International Agreements") and the Migratory Species Convention, although neither country has signed the latter convention.

There are 40 multilateral conventions that deal directly with management of living resources, but conservation is not their primary purpose. The suggestion is made that they be reviewed to strengthen regional or global approaches and to ensure that conservation is integral to such agreements.

Most coastal nations recently extended their territory to include a 200-mile limit. The time is ripe to renegotiate international agreements regarding regional fisheries commissions, to ensure that such areas will be managed as ecological entities.

The IUCN report endorses the view expressed by Green-Hammond (1980) that there is a critical need to protect the fisheries that support the ecosystem, especially on an international level. It cautions that ecologically sound management is needed to deal with the pollution and contamination that exist from one coastal area to another. Improved or new bilateral and multilateral management agreements are needed to ensure sustainable management of marine resources. Special recommendations are made for some regions that are critical habitats, such as the arctic.

This area would take a very long time to recover from any damage, and should be considered a priority by its member nations. IUCN suggests mapping of critical ecological areas, marine and terrestrial, long-term management guidelines, and protected areas. Measures should be taken, including joint research, to improve protection of migratory species breeding within the arctic and wintering inside or outside the region. Studies are needed on the impact of fisheries and other economic activities in the northern seas on ecosystems and nontarget species. The successful Agreement on Conservation of Polar Bears could serve as a model for a developing agreement among the arctic nations on the conservation of the region's vital biological resources.

Pinnipeds do cross open ocean and, as the IUCN report states, the open ocean and its living resources may be exploited by anyone. It is a critical habitat to some species, and other species who live in coastal areas travel through it. Such species should be regarded as common resources and special provision for both groups is needed in a formal international agreement, which at present is not in existence.

The report cautions against deep-sea mining, including oil exploitation, until the effects of such actions are better understood.

Summary of Seal Management. The complexities of seal management have only been introduced here. The rate of change is so great that even as this material is read, much of it will have become outdated.

One fact becomes apparent in reviewing management philosophies, plans, and methods: man does not accept responsibility for his actions. If risk factors, which are rarely built into management plans, are acknowledged, as in the Spiny Lobster Plan, the body utilizing the resource still does not want to take responsibility for either monitoring the situation or the consequences of poor planning (Green-Hammond 1980).

Management of marine resources can no longer be the sole responsibility of those persons working directly in that field. Persons utilizing those resources on personal and economic levels must also accept responsibility. IUCN suggests that, for example, industry be encouraged to analyze the resource base and to work in cooperation with governments and other commercial sectors of society to ensure that the base is being utilized sustainably. Furthermore, the other parts of that ecology that are part of the larger ecosystem must be similarly protected. Informed utilization may breed responsibility.

Scientists are often caught in the middle. Faced with a data base that is deficient in many aspects, they are called upon by bureaucrats who like to make management decisions based on the certainty of numbers. However, many assumptions are not empirically testable and the data in fact are far from ideal. The danger,

of course, is that selection can be made of data that support only one view, with the fact often being overlooked that there is no such thing as the certainty of numbers. Lavigne (1978) cautioned that future management decisions must be mindful of future events that cannot be predicted, such as climatic change and man-made interferences that could have chain reactions in the ecosystem. Security in data collections by the most sophisticated methods from any stock is a false concept (Lavigne 1978).

Recalling the time lag between policy and application, it is interesting to note a statement by Toffler (1970:5) more than a decade ago, which is applicable to seal management: "The rate of change has implications quite apart from, and sometimes more important than, the directions of change." He suggests "that there must be balance not merely between rates of change in different sectors, but between the pace of environmental change and the limited pace of human response." Such daring to slow the rate of change and economic development was illustrated in the landmark decision by the Canadian government to halt the Mackenzie River pipeline until other factors had been settled and studied.

North American seals are in no extreme danger at present, but as the transition period theory is at hand, it would be a good time to reassess our priorities, and to be mindful of not only where we are going, and how we are getting there, but also how fast we are going to travel before we know the risks involved. Sometimes common sense is overlooked in the haste to develop.

CURRENT RESEARCH AND MANAGEMENT NEEDS

An overview of the current status of seals in North America clearly points out that environmental conditions are of prime concern. The two most common problems facing seals are heavy metal contamination and pollution, and oil exploration. It is therefore important to review this situation, and to be aware that, as with all marine life, the seal's habitats need protection and continued monitoring of contamination (Johnson and Jeffries 1977).

Heavy Metal Contamination. Heavy metal contamination in the environment has been a concern since the 1960s. Such contaminants as mercury, DDT, dieldrin, selenium, and PCBs are taken up by various organisms within the food web, but are found in their heaviest concentrations in such terminal predators as seals and large pelagic fish (Mansfield and Sergeant 1973), and ultimately humans. Because DDT and PCBs are known to affect steroid reproductive hormones, thereby often prolonging estrus and decreasing successful implantation, the implications are that heavy metals could be the cause of reproductive failure in northern seals and sea lions (Holden 1978).

The highest concentrations of mercury are found in the liver of wild animals. (Mansfield and Sergeant 1973; Holden 1978). Deaths at high exposure levels, however, are normally associated with renal rather than hepatic failure (Ronald and Tessaro 1976). There is a wide range of contamination levels between species, and there is an age and sex correlation as well. For example, Addison (1973) found that lactating females showed lower organochlorine levels than other adult females, indicating that the contaminants are probably passed on to the pup through the milk. Variations of mercury levels also correspond to diet; harp seals who eat small pelagic fish and crustacea had lower levels than gray or harbor seals, who feed on large pelagic and benthic fish and cephalopods (Sergeant and Armstrong 1973).

Recent studies indicate that seals may not suffer ill effects from some of these contaminants, depending, of course, on concentrations, because of certain protective mechanisms (Holden 1978). Selenium occurs at higher levels in seals than in other animals, and shows a positive correlation with levels of mercury (Koeman et al. 1972, 1975); this may be similar to the situation in humans, where dietary selenium has a protective effect against heavy metal toxicities.

It appears that seals are able to demethylate mercury (Freeman and Horne 1973; Ronald and Tessaro 1976; Holden 1978), perhaps with an enzyme system (Freeman and Horne 1973). Evidence of demethylation has been found in kidneys and livers and in low amounts in the small intestine of the harp seal (Ronald and Tessaro 1976).

Although high concentrations of several organochlorine compounds have been found in the lipid of seals and other marine mammals (Holden 1978), there does not seem to be any evidence of adverse effects. However, Newby (1978) indicated that PCB concentrations between 0.6 and 3 ppm cause reproductive dysfunction. Such a link may be the cause of the abnormally high rate of abortion in California sea lions (Holden 1978).

Urbanization, Oil Exploration, Drilling, and Spills. Much of the habitat of pinnipeds, especially that of the northern hemisphere species, is presently being threatened by man and his activities. Incidental human disturbance, offshore oil drilling, the concentration of world industry in the northern hemisphere, and the increased human population may ultimately result in direct overexploitation of seals, environmental contamination, or intense competition for fish (FAO Adv. Comm. 1976).

Offshore oil drillings and transportation of fossil fuels and their by-products pose the most serious ecological threats to seals. These bring not only human disturbance but also the possibility of oil spills. Until recently, oil had not caused any serious effects on populations of marine mammals, possibly because seals are highly mobile most of the year and are capable of avoiding heavily polluted areas. However, any breeding, and therefore sedentary, population could be affected in a harmful way (Hyland et al. 1977). Such a situation could have disastrous consequences on the neonates.

Adverse effects of an oil spill in the Arctic Ocean, the habitat of many pinnipeds, would tend to be heightened and prolonged compared to those of spills in temperate waters (Percy 1977). Biodegradation in this climate is slow and complex, and clean-up operations in inclement weather are impossible. Harmful side effects of an oil spill on seals might possibly include the following situations. The rich algae on and under the ice surface in sea-ice habitat would be subject to severe contamination, and pinnipeds are part of the food chain based on this vegetation. Large quantities of oil could collect as slicks in the breathing holes, leads, and breeding areas, and oil tends to settle along the shoreline, which would affect animals hauling out (Percy 1977). This could result in ingestion of toxic oil droplets during grooming, loss of thermal insulation from coating, and irritation of the eyes and exposed mucous membranes (Hyland and Schneider 1977), areas that show extreme sensitivity (Percy 1977). Solar heating of the skin has been shown to increase in the presence of oil, which may be significant during haul-out. Under conditions of natural stress, such as molting, seals are particularly sensitive to an oil spill. Physiological repercussions seem to be kidney damage and behavioral changes (Percy 1977).

The transportation of liquid natural gas from the High Arctic to terminals produces a peculiar problem. There is reason to believe that in an area such as the Davis Strait the power requirements of the vessels would produce an acoustic barrage of such intensity as possibly to modify seal behavior and movements (Mohl personal communication).

Harbor Seal (*Phoca vitulina*). The harbor seal is entirely protected throughout its Pacific and Atlantic range. It is hunted to a small extent only by native peoples, which results in minimal impact on arctic seal populations. Since the inception of the Marine Mammal Protection Act, the Pacific harbor seal population has gradually increased; it currently seems to be in a high and stable condition (Newby 1978).

Economically, the common seal often represents the cornerstone of the coastal native economy. The flesh is used as food for humans and dogs, the liver being especially high in vitamin A (Banfield 1974). The animal is primarily hunted for skins, which produce a high-quality fur. The manufacture of trinkets and handicrafts is of minor importance. *P. vitulina* presents a popular tourist attraction in aquaria and zoos, and is also the seal most commonly used in experimental research (FAO Adv. Comm. 1976).

Harbor seals are not popular with commercial or sport fishermen, however, because they damage fishing gear and gill nets, often becoming entangled themselves. They also feed on salmon during upriver spawning periods (Banfield 1974), herring, and trout (Johnson and Jeffries 1977). If conflicts occur between this species and commercial fishermen, however, a provision in the MMPA allows for such individuals to be taken by a fisherman with a government permit (Newby 1978). The Canadian populations are also protected, but for many years the Canadian government allowed a bounty system on this species for commercial fishermen on both of the coasts (Banfield 1974).

Man has interfered and to some extent is still interfering with the harbor seal's habitat. Some traditional hauling-out grounds, such as the Nisqually Delta, Washington State, where 200 seals used to breed over 30 years ago, have been abandoned because of hunting pressures. Presently, waterways, such as San Francisco Bay, that are frequented by harbor seals are being taken over by man (Newby 1978).

Because *P. vitulina* is a shallow water dweller it is more affected by pollution than other species (FAO Adv. Comm. 1976). On Gertrude Island, Washington, PCB levels in harbor seal blubber and liver tissues were found to be as high as 400 ppm. A high incidence of birth defects was noted in this population of seals (Newby 1978).

Between November 1979 and May 1980, 437 dead harbor seals were recovered from an area between Cape Cod and southern Maine. The cause of the fatalities has been diagnosed as an influenza virus, possibly complicated by a species of mycoplasm that manifested itself as pneumonia. The outbreak originally occurred around Cape Cod, but the virus migrated with the seals 500–600 km up the coast (Geraci personal communication).

Current management needs exist in the area of taxonomic information based on morphometry, development of better stock assessment methods, and the study of haul-out behavior to assist with on-land counts (FAO Adv. Comm. 1976). In Washington State the population is stable and, in order to keep the status quo, an active research program should be encouraged. All locations that are currently used as haul-out sites must be evaluated and possibly designated as critical habitats. Because there is conflict between this seal and commercial fishermen, a need exists for the recognition and accurate documentation of this problem.

Ringed Seal (*Pusa hispida*). The ringed seal is the most important seal species to the Eskimo—the one relied upon most heavily for fur, meat, and other commercial products (Alaska Dept. Fish and Game 1976; FAO Adv. Comm. 1976). Its stocks are presently underexploited, the average annual take in Alaska being about 9,000–13,000. The U.S.S.R. also harvests this seal in the White Sea, with a limit of 3,500 per year. These totals are well within the limits for the productivity of ringed seals (Alaska Dept. Fish and Game 1976; Popov 1976). Hunting is actually controlled by regional climatic conditions and the availability of *P. hispida* (Alaska Dept. Fish and Game 1976). Because the ringed seal has had such little contact with humans it is very abundant and its present numbers are probably at the original levels. In Canada this species forms the basis of a very limited sport (12 licenses per year).

The only potential for a significant threat to this seal is presented by the gas and oil exploration on the outer continental shelf (Burns 1978). Recent investigations on indirect and direct effects of such an industry in the arctic point to a definite threat to the Beaufort Sea population of ringed seals. Surface contact with oil

would cause eye and kidney damage, stress, and interference in thermoregulation, especially for pups (Geraci and Smith 1976). A population of ringed seals near the Norman Wells has been tested for tolerance of crude oil. It was found that hydrocarbons are absorbed rapidly, from both ingestion and immersion. Although the liver appears to be the main detoxification and excretion organ, no serious lesions developed from the amounts tested. Results indicate that, in limited exposure situations, accumulations occur in body tissues and fluids that can be excreted, but more research is needed to establish the limits of safety (Englehardt et al. 1977).

The mere presence, however, of a large offshore oil field may affect the seasonal pattern of movement of ringed seals and reduce their ability to survive adverse natural conditions (Geraci and Smith 1976). Also, chronic low-volume releases of petrochemicals and the occasional major oil spill may have detrimental effects on the food of ringed seals. The eggs and larvae of the two codfish and various kinds of zooplankton are highly susceptible to petrochemical pollution (Burns 1978).

Current research on ringed seals encompasses population dynamics, censusing methods, interspecific relationships on the pack ice, caloric requirements, heavy metal and toxic chemical uptake and pollution, and physiology. Areas of need include improved censusing techniques and the study of female reproductive cycles, migrations and dispersal, the influence of environmental factors on reproduction and ecological relationships with other phocids (FAO Adv. Comm. 1976), and polar bears, for which the seal is a prime food source.

Ribbon Seal (*Histriophoca fasciata*). The ribbon seal is important economically to the residents of northern Alaska and represents part of the subsistence base of all the Eskimo settlements along the Alaskan coast from Kuskokuim Bay to Demarkation Point. It provides food and fiber for everyday use and is a source of income through pelt sales. In comparison with ringed, bearded, and harbor seals, however, it comprises an insignificant part of the Alaskan native harvest. On the other hand the U.S.S.R. is believed to have overexploited this species in recent years (Alaska Dept. Fish and Game 1976). Presently the United States allows harvesting of the ribbon seal for native take by permit only. The U.S.S.R. permits 3,500 to be taken from the Sea of Okhotsk and 3,000 from the Bering Sea. There is an unofficial agreement between the two countries to cooperate on research (FAO Adv. Comm. 1976).

The only threat to the ribbon seal population exists in the anticipated continental shelf development near their wintering and breeding areas in Bristol Bay and Saint George Basin. Although this will not result in direct seal mortality, it may reduce their food sources (Burns 1978).

Ribbon seals remain the least known of the arctic seals. There is a need for information on numbers, reproductive cycle, general biology, seasonal distribution, and migration routes (FAO Adv. Comm. 1976).

Harp Seal (*Phoca groenlandica*). The harp seal is the basis for income for local residents of the Gulf of St. Lawrence and Newfoundland-Labrador as well as west Greenland and aboriginals of the Canadian High Arctic. In fact, it is hunted every day of the year along its 6,000–8,000-km cyclic migration route.

The seal is used as a food (approximately 10,000 taken) and fiber source in the north, with limited return as a furbearing species. It is on the ice of the Gulf of St. Lawrence and Newfoundland-Labrador that the greatest exploitation occurs. Approximately 170,000 are permitted to be taken; about 80 percent of these are young of the year. The population is probably greater than 1 million, and the Canadian government policy is to allow the population to increase to 1.6 million. There have been conflicting views as to whether the herd can in fact attain such an optimum level under present management strategies.

The controversy over the open air and publicly visited killing of this species does not need to be relived. Such controversy may, however, have overshadowed the conservation of more endangered species.

The return from the hunt is around $5.5 million to Canada. A large proportion of the secondary and tertiary industries is carried out at high profit elsewhere, indeed, even in countries that have been leaders in protesting the seal hunt.

The harp seal is now receiving a great deal of management attention, and it is probably one of the best studied pinnipeds with respect to its biology. This does not mean, however, that it is completely understood; even though scientists are aware of such activities as its phonation, audiogram, visual acuity, dive responses, and biocidal residue levels, there is a great deal left to understand about its biology.

Four principles must be considered in the management of this and other pinniped species.

1. Is the present hunt at a proportionate level for the herd size? There is a great deal of evidence to indicate that the proportion taken is not excessive.

2. If the seal is hunted, is it killed humanely? There is no reputable evidence that the legal club and hakapik are not efficient (if brutal in appearance) weapons. A new firearm currently being developed will probably be as efficient but less offensive in connotation and therefore more acceptable.

3. If seals are harvested, is the maximum use being made of the resource? There is evidence that Canada is not receiving a maximum return from the meat or even from the pelt, as greater profit may be made in the final processing than in the collecting, fat removal, and primary pelt processing.

4. Does the seal deserve special status? Some persons differentiate between appealing animals, such as seals, and those that are less endearing, such as cockroaches. If enough persons make such emotional divisions it may lead to political decisions that are not

necessarily based on scientific fact. Nevertheless, it is the right of the public to express opinions on this issue.

Gray Seal (*Halichoerus grypus*). In New England, human occupation and erosion by storms have depleted the haul-out areas (Gilbert 1977) of the gray seal, resulting in a reduced population of the species there. In most of its North American range *H. grypus* is said to be undisturbed by man, as its island and pack ice colonies are rarely visited (Mansfield and Beck 1977). There may be some reason for doubting such a statement at the present time.

The main concern with management of this seal is competition for food source. Although it competes with the harbor seal for food (FAO Adv. Comm. 1976), its real competition is with humans. The gray seal has a considerable effect on inshore fisheries, especially those of mackerel, herring, cod, and salmon, because it damages nets and gear and eats or mutilates the trapped fish. It is also the primary host of the codworm, *Terranova decipiens*, which infests the flesh of groundfish, especially cod, thereby decreasing their commercial value. In 1967, the Conservation and Protection Branch of the Canadian Department of Fisheries began culling these seals (mostly molted pups) in an attempt to reduce the population and thereby limit competition with commercial fisheries. Despite an annual take of about 800 animals (maximum 2,300 taken in 1975), the population continues to expand. In April 1976, the Canadian government placed a bounty on the gray seal. By 1977 over 600 seals had been taken in this manner, and there were probably an equal number lost through sinking (Mansfield and Beck 1977). There is no justification, other than local income return to fishermen for such a bounty.

Biologists observed an oil spill in Wales. It occurred at the beginning of the gray seal breeding season and resulted in oiling of some pups and an obvious interference with the mother-pup relationship. These pups did not gain weight as rapidly as unoiled pups (Davis and Anderson 1976). To date, no such oil spill has occurred in North American waters to threaten gray seals. PCBs, methylmercury, arsenic, cadmium, and selenium, however, have been found in *H. grypus* in levels that may have a significant effect on their physiology and that already appear to have altered their steroid hormone metabolism (Freeman et al. 1975). The gray seal is on a collision course with man's needs and the next seal confrontation will probably arise over the activities in regard to the very viable colonies on Sable Island. Attempts are being made to control the population using antifertility drugs.

Long-term studies are needed on range, distribution, migration patterns and corridors, whelping sites, competition, and predation.

Bearded Seal (*Erignathus barbatus*). The bearded seal is vitally important in the economy of the coastal Eskimo because of its large size (Burns 1967; Banfield 1974), the high quality of its meat, its blubber, and its strong durable skin (Burns 1967). The meat often poses a health risk, because it may harbor the nematode *Trichinella,* a parasite causing trichinosis if meat is eaten uncooked. The liver of this seal contains high levels of vitamin A, so overindulgence could result in poisoning (Banfield 1974).

E. barbatus is often shot, a method that results in many animals being sunk and therefore wasted. Two types of hunting are practiced, namely, subsistence hunting along both the Siberian and Alaskan coasts, and commercial sealing by the U.S.S.R. and Japan. It is estimated that half this annual harvest is lost to sinkage. Annual take is dependent on the seasonal availability of the migrating seal. In 1967 the bearded seal was still abundant in its range, but a need existed for improvement and development of hunting techniques (Burns 1967).

Future management problems may occur because of some of the prey species of this seal, namely, pandalid and crangonid shrimps and eithode crabs, which are also presently of interest to man (FAO Adv. Comm. 1976). Although to date no serious competition for these food resources has arisen, prospects for developing a clam fishery in the bearded seal's range may present future problems (Burns 1978). The habitat of this seal is also potentially threatened by extensive offshore oil exploration and production, interisland pipeline construction, and the shipment of oil by tanker in arctic waters (FAO Adv. Comm. 1976). These may affect either the seal itself or its food web, which is comprised mainly of invertebrates with highly susceptible larval stages (Burns 1978).

Monk Seal (*Monachus schauinslandi*). Because the monk seal is a particularly sensitive pinniped, it is remarkable that it survived the period of intensive marine mammal exploitation during the 18th and 19th centuries, when it might have been eliminated. It probably survived because its habitat is so isolated and its low population levels and solitary habitats were not appealing commercially. A shy animal, it has disappeared from its former habitats, which are now inhabited by people. A definite correlation exists between the presence of humans and the absence of monk seals. An increase in pup mortality has been observed when nursing mothers are disturbed (Kenyon 1978*b*). When humans intrude on their habitat, these seals desert favored beaches and haul out on isolated shifting sand pits. Here the young are exposed to extremes of wind and tidal conditions and are also nearer to deep water that is frequented by sharks.

When human activity became extensive on Midway Atoll during World War II, the monk seal population began to decline; by 1968 it had disappeared (Kenyon 1978*b*). However, if protected and undisturbed a seal population can begin repopulating ancestral breeding grounds. The Laysan Islands have been repopulated in just this manner (Kenyon 1973*b*).

In 1976 the U.S. government declared the monk seal an endangered species. As a result, the Hawaiian Islands National Wildlife Refuge was created to encompass all of the leeward Hawaiian Islands except

Midway and Kure atolls, which comprise a separately protected area. The breeding population at Midway atoll had disappeared between 1958 and 1968, primarily because the nursing females were disturbed by the personnel at a large naval base. In 1976 the U.S. National Marine Fisheries Service and the U.S. Fish and Wildlife Service began an intensive study. The results of that survey indicated that populations were declining because of human disturbance of nursing females in all areas and that minor contributions were shark attack and ciguatera poisoning. In the French Frigate shoals the population seems to be increasing (Kenyon 1978*b*).

Fishing interests in the leeward Hawaiian Islands are being promoted by local state officials, and could pose a direct and serious threat to monk seals, who might become entangled in the fishing gear. This industry would place these seals and humans in direct competition for food, and nursing females would also be disturbed by fishermen. It is doubtful whether the monk seal could survive this type of pressure. In the Caribbean, the *M. tropicalis* species of monk seal is almost certainly extinct, and the primary cause of its early disappearance is attributed to fishermen (Kenyon 1978*b*). The remaining monk seal, *M. monachus,* is in an endangered condition in the Mediterranean and northeast Atlantic.

Recommendations have been made to close the Hawaiian Islands National Wildlife Refuge breeding and pupping grounds to all humans. A detailed study program has been proposed, as has a public education plan for Kure and Midway atolls (Kenyon 1978*b*).

Hooded Seal (*Cystophora cristata*). In the Newfoundland-Labrador area the hooded seal has been hunted annually since the early 18th century. Stock assessment was not attempted until the early 1960s. In 1961 the government introduced opening and closing dates for the hunt. Adult females have been protected on the whelping patch since 1965. ICNAF recommended quotas on the species in 1971, which were then put into effect. At present the stocks are thought to be increasing, since it is probable that they have been exploited at their sustainable yield levels since the early 1960s. For example, in 1974, the total allowable catch was 15,000 seals, but only about 12,000 seals were actually taken. The maximum was harvested only during the 1975 season. During the last three years the proportion of adult seals allowed in the catch has been decreased progressively to 5 percent.

In the Jan Mayen breeding area the hooded seal was exploited by the Norwegians until 1961, when the U.S.S.R. and Norway agreed that since the population was decreasing, management measures were in order. They therefore do not allow the taking of adult and immature animals at the molting grounds in Denmark Strait or a spring harvest season.

There is speculation that there has been an increasing availability of *C. cristata* over the last 30 years because of the existence of a reserve in the Davis Strait. It has been observed that the hunt for this seal increased as the availability of harp seals decreased

(FAO Adv. Comm. 1976). This hunt is associated with the same vessels carrying out the harp seal hunt, with the hooded seal being the species of choice.

C. cristata is prized for its pelts, which are of particularly high quality, especially the young bluebacks. The only other commodity that enters the commercial market is oil. In 1976 hooded seals contributed about 40 percent of the total land value in the Norwegian catch of harp and hooded seals (FAO Adv. Comm. 1976). At present, the hooded seal has no large effect on any commercial fish species (FAO Adv. Comm. 1976).

Elephant Seal (*Mirounga angustirostris*). During the early 19th century the northern elephant seal was abundant along the entire coast from Point Reyes to Cabo San Lazaro. For at least 40 years at the beginning of the 19th century, this seal was exploited continually, primarily for its blubber, which entered the market as oil. However, by 1860, after intensive slaughter, this oil was no longer a valuable commodity. By 1869 the species was considered to be extinct, but the occasional individual was sighted on San Benito and Guadalupe Island until 1880. It was not until 1892 that more were seen, and from that time until 1930 the species bred on Guadalupe in very small numbers. The Mexican government granted protection to the 264 seals living on the island in 1922. By 1932, the animal began a slow dispersal along the Baja coast of California. The United States also protected the elephant seal, so it increased to 13,000 by 1957. Thus, the population had tripled in twenty years. In the 1970s four new colonies were established, but only two colonies, namely, those on Guadalupe and San Benito, had reached equilibrium by 1977 (Le Bouef and Panken 1977).

Despite the rapid renewal in growth, the population of elephant seals, once so abundant, has suffered irreparable damage, as the current generation is lacking genetic variability and adaptability. Many gene forms were lost from the gene pool and there was a greater reduction in the number of heterozygotes. Any such survivor of a severely depleted population contains only a small fraction of the total genetic variability of the parent population (Le Bouef and Panken 1977).

Tourists do not have a chance to disturb the rookeries, but the elephant seals do have a great appeal for the public. Many people come to see them every year off the California mainland at Ano Neuvo Point (FAO Adv. Comm. 1976). Human presence does not appear to bother male elephant seals, but if the females are disturbed prior to parturition they have been known to enter the water and give birth elsewhere. Fortunately, females do not abandon their pups even if disturbed (Le Boeuf and Panken 1977).

Offshore oil exploration could disturb rookeries (FAO Adv. Comm. 1976) and any oil spills could also be harmful, although elephant seal pups exposed to an oil spil on San Miguel Island survived and later dispersed normally (Hyland et al. 1977).

Northern Sea Lion (*Eumetopias jubatus*). The current Alaskan population of 200,000 northern sea lions

is approaching maximum carrying capacity for the herd and in fact may be in a state that is conducive to disease. The legalities involving the management of marine mammal populations by Alaska as provided by the MMPA and the ensuing disagreement and subsequent bequeathing of all responsibilities back to the federal government in 1979 (see "Walrus" section) have delayed application of sound management policies to this population. Japan, which has a similarly high population, allows a bounty of 1,000 yen per animal on the Steller's sea lion without permits being required (FAO Adv. Comm. 1976).

Steller's sea lions are serious competitors to commercial fishermen (Banfield 1974), although some scientists point out that no commercially valuable fish is a major item in their diet (Gentry and Withrow 1978). Nonetheless, the sea lions destroy not only the fish but the nets, traps, and other gear (Banfield 1974). The Canadian government allowed a control program in British Columbia in 1959–60 that consisted of an intensive program to reduce the population because of commerical fishing conflicts, resulting in a high in 1956 of 12,000 to a low of 4,000 by 1969 (Banfield 1974) in this area. Unintentional harassment occurs to varying degrees by sport and commercial fishermen, divers, photographers, and tourists, although the consequences of this are unknown (FAO Adv. Comm. 1976).

Significant levels of heavy metals and chlorinated hydrocarbons have been found in sea lions in the southern extent of their eastern Pacific range, but the effects of these pollutants are unknown (FAO Adv. Comm. 1976).

California Sea Lion (*Zalophus californianus*). Although exploitation of the California sea lion occurred in the 19th century, it was not as intense as that for the Guadalupe fur seal or the northern elephant seal. From 1860 to 1888 there was a trade in seal oil. Later a trade developed in the manufacture of glue from their hides (Banfield 1974). The dried "trimmings" were sold in China as medicinal cure-alls. A minor reason for taking these animals was to make dog food. In the 1920s a small but lucrative trade developed in capturing them for zoos, aquaria, traveling shows, etc. (Banfield 1974; Mate 1978).

Present populations inhabit the entire range of former times. The animal is afforded complete protection in the United States and for some reason this seal has increased much more than the Steller's sea lion, which in the 1930s was the more abundant. Whether such circumstances as food or space have changed to favor this animal is not fully understood (Mate 1978).

Mexico offers the species protection, although small numbers are taken. In Canada, on the coast of British Columbia, there is no management program as such but fishermen are allowed to take a sea lion if it continually interferes in their fishing area.

Because of reduced fishery resources from overfishing and the destruction of fish spawning habitat due to poor forest practices, hydroelectric power dams, and pollution, California sea lions are now using upriver areas to feed, resulting in competition with humans, especially sportfishermen (Mate 1978). Complaints have also been made that this seal damages fishing gear and netted fish (FAO Adv. Comm. 1976).

The nearshore habitat of the California sea lion makes it a target for human disturbance, and such interference is suspected to have caused some colonies to move their breeding rookeries to more aquatic sites (FAO Adv. Comm. 1976). Proximity to industrialization also exposes them to toxic wastes through their prey species. This sea lion spends much of the year in the waters off southern California, where the marine organisms are highly contaminated with PCBs and DDT and its metabolites from the Los Angeles sewage discharge (Buhler et al. 1975). High concentrations of chlorinated hydrocarbons have been found in the seal's tissues and are believed to cause abortions and increased mortality in newborns (Mate 1978).

Guadalupe Fur Seal (*Arctocephalus townsendi*). Historically the Guadalupe fur seal ranged from the Farallon Islands off San Francisco to its southern limits near the Channel Islands and Guadalupe (McClung 1978). A massive and unrelenting slaughter of this species in the 18th century almost exterminated the animal by 1834. In fact, it was considered extinct until 1894, when one individual was taken from the coast of Baja California. There were thought to be 7 seals on Guadalupe Island at that time (Fleischer 1978*b*). In 1928, a very small herd was rediscovered there, but no others were sighted for 20 years (McClung 1978; Fleischer 1978*b*). In 1949 there was a report of 1 male Guadalupe fur seal living on San Nicolas Island (McClung 1978). By 1954 there was a small colony on Guadalupe Island and today the species numbers 500. The island is now a wildlife sanctuary under Mexican protection and has been since 1922. There is presently a proposal to make it part of a Mexican National Park in which visitors would be strictly controlled. In U.S. waters the seal is protected by the Fur Seal Act of 1966, but Mexican law does afford the seal complete protection (Kenyon 1973*a*). This is a particularly satisfactory history of conservation and we hope it will be practiced elsewhere.

Fur Seal (*Callorhinus ursinus*). The story of the fur seal, like that of the harp seal, is representative of how the governments of the United States and Canada manage a population of seals that has traditionally been hunted for its commercial value. The fur seal hunt in the Pribilofs is well controlled and one that meets the criteria set down in the MMPA. However, it is important to look briefly at the history of this seal and its early exploitation. (This section is mainly taken from a U.S. Department of Commerce publication on Pribilof fur seals.) It is relatively easy to manage a species of seal for which there is no demand but a multilateral agreement ties this seal to demands of the past.

Full-scale exploitation of the northern fur seal began in earnest in 1786, and by 1787 an overwhelming 2.5 million skins had been taken. It is not difficult to imagine that the population was reaching dangerously low proportions by 1834. From 1835 until 1867 only males on land could be taken, and the killing of females was forbidden. The result was an increase in the population of this species, although we know that at this time sealers were turning their interests to other species of seal as well.

In 1867 the United States acquired Alaska and the Pribilofs and until 1869 sealing in this area was very disorganized, so the Pribilof fur seal once more suffered a purge of 329,000 animals. In 1870 the first of two 20-year leases of sealing privileges was awarded. In the next 19 years, a total of 1,854,029 skins was officially shipped from these islands. From 1890 to 1909, during the second lease period, 342,651 animals were taken. Basically then, this period from 1869 to 1911 was a time when there were few restrictions on the taking of fur seals. Historically, it stands as the zenith of the fur sealing trade when pelagic sealing was also practiced regularly. Most of the seals taken were females. After this intensive slaughter, it is not surprising that by 1909 there were only 200,000–300,000 seals remaining on the Pribilof Islands.

Finally, by 1911, the sealing nations began to realize that management was an integral part of resource utilization. The North Pacific Fur Seal Convention was held and formalized (the history of that treaty was discussed in the section "International Agreements.")

The current policy regarding these seals ensures that only those seals not needed as replacements for breeding stock are taken and that the harvest is carried out as humanely as possible. The fur seal is polygamous, and presently only young males are taken. The few females harvested are incidental and accidental. Since 1973, the United States has declared St. George Island a research study area, where no seals can be harvested and where the growth and behavior of an unharvested population is being compared with those variants of the harvested population on St. Paul's Island. Such programs ensure practical management of the herd. In 1977 it was reported that there had been an unusually high recent mortality among young seals. There is speculation that the reduced availability of food caused by an intense trawl fishing of pollock near the Pribilof rookeries may be a cause. The Bering Sea food chain consists of plankton, pollock, and fur seals, and of these, man harvests two. Nursing females may be traveling further to find food, and young seals may be forced to search out their own food supply sooner than normal. The other problem is the usual one when seals and humans compete, in that seals become entangled in the nets or gear (Marine Mammal News 1977).

As was pointed out earlier, there are many persons involved in fur seal management. In terms of the MMPA, there is certain admiration for the way in which the government is applying the laws. Such groups as Greenpeace, however, have voiced clear objection to the annual harvest of 30,000 seals on the Pribilofs, claiming that justification for such killings cannot be made for either economic or environmental reasons. The U.S. government maintains, on the other hand, that a controlled harvest of these seals, with the pelts divided among the nations that make up the Fur Seal Commission, is the only logical way to prevent a seal hunt on the high seas, where pelagic hunting would mean a loss of at least half the catch through sinkage. The U.S. Congress, in keeping with its responsibility to the MMPA, has voiced objection to the Canadian harp seal hunt. This attitude is hard to understand with the continued harvest of the northern fur seal.

In August 1979 a bill was introduced into the U.S. Congress calling for the termination of the Interim Convention on the Conservation of the North Pacific Fur Seal and an end to the harvest. It also recommended the necessary changes with member nations to the treaty, creation of the Pribilof Island as a wildlife refuge and marine sanctuary, and a buffer around the 200-mile limit in this area. In September 1979, the continuation of the existing fur seal harvest under the present international agreement was endorsed by Congress.

The Pribilof fur seal faces problems other than the hunt. Man-made debris, such as scraps of fish nets, twine, and plastic wrapping bands, is having an increasingly serious effect on seals, who ingest these items or become entangled in them. Approximately 3,500 fur seals are accidently caught each year in high seas gill nets (FAO Adv. Comm. 1976) or in net fragments. This type of injury often means that the animals die of starvation (Fiscus 1978).

An oil spill would have a particularly detrimental effect on this species. A small amount of crude oil on the fur increases thermal conductance, which results in increased heat loss, and the seal must consequently increase its metabolic rate to maintain body temperature. Oil encountered at sea in any amount will probably harm fur seals by rendering their dense underfur ineffective as an insulator (Fiscus 1978).

Walrus (*Odobenus rosmarus*). Because the walrus has tusks, its valuable ivory has been marketed in a similar manner to seal fur. The use of the walrus as a natural resource of Alaskan native peoples has posed some of the most difficult management problems under the MMPA. The state of Alaska has returned the power of jurisdiction to the federal government, namely, the Department of the Interior.

The Pacific walrus was hunted by whalers and sealers in the 19th century who often abandoned the slaughtered animals after taking only their tusks. Klondike gold seekers used the walrus as an object of sport killing, similarily wasting the carcass. By 1900 the walrus was virtually extinct south of Nunivak Island. From 1650 to 1960 over 3.5 million animals were killed on this coast. In 1962 another 12,000 animals were taken by Eskimos, and many of these were lost through sinking. In contrast to the early days of proliferation, by 1950 the west coast walrus numbered only 40,000–50,000 (McClung 1978).

The Walrus Islands are a group of seven islands in northeast Bristol Bay, Alaska, and they are the only regular summer hauling-out grounds of the Pacific walrus. In 1960 they became the Walrus Islands State Game Sanctuary. By the mid-1970s, as a result of this protection, the population began to increase, and the animals started to return to their ancestral hauling-out grounds on the Pribilof Islands, a place they had inhabited before the slaughters of the previous century. The Marine Mammal Protection Act of 1972 banned trophy hunting of the walrus, and permitted the taking of this species by native peoples for subsistence reasons only (Kenyon 1978a). The state of Alaska by law prohibits the sale of uncarved ivory in order to prevent killing for ivory only (Kenyon 1978a).

Eskimos are now claiming traditional hunting rights. Before the MMPA in 1972, hunting permits were issued to residents dependent on the walrus for food to hunt north of Newenham and to take up to five adult cows or subadults and an unlimited number of adult males annually. The other licensed hunters were permitted to take only one male a year. Since 1972, only the Eskimos, Aleuts, and Indians have been allowed to take walruses for subsistence, with no restrictions being placed on numbers. The state of Alaska has placed restrictions on the site of the taking, however, the Walrus Island State Game sanctuary and another state game management unit in Bristol Bay being prohibited areas. In April 1977, the Alaskan native people sued the U.S. Department of the Interior for allowing the state of Alaska to forbid them the right to hunt on these traditional hunting grounds. Their argument was that it was necessary for them to hunt in these areas to fulfill their essential dietary and economic needs. Government officials working in the area stated that they believed that the overriding motivation for this action by the native people was not subsistence but ivory. In April 1979, a U.S. district court judge ruled that Alaskan natives could hunt walruses despite the state regulations against it. Alaska had fought for control of the management of marine mammals under the provisions of the MMPA and this right had been granted in April 1976 by Congress. However, after this judgment about the walrus, the state of Alaska, in June 1979, returned the responsibility for all marine mammal management to the federal government.

Because other problems affect the walrus population, namely, the clam industry, this action not only negated Alaska's walrus management program but also put the animals in possible danger if the federal government did not respond quickly and effectively. The U.S. Fish and Wildlife Service, under the auspices of the Department of the Interior, accordingly issued an emergency statement effective 8 July 1979 ruling that the federal restraints laid down under the MMPA be reinstated and that henceforth all recreational hunting, importation, and other activities that the state had permitted now were prohibited. Federal permits are now required to take walruses. Natives are allowed to take walruses for subsistence reasons or to create authentic handicraft articles or clothing. There is still conflict concerning the walrus in the Walrus Islands State

Game Sanctuary, which the state is still attempting to protect from the native hunters. Alaska has offered to allow natives to take walruses from nearby Twin Island, but the natives claim that the number of walruses is inadequate.

A far more serious threat to the Pacific walrus population exists in the proposed clam dredging industry in the Bering Sea. With the depletion of the clam resources of the western North Atlantic, proposals have been made to transfer the industry to the rich Bering Sea feeding grounds of the walrus. The development of such an operation could cause irreversible damage to the entire Bering Sea ecosystem, but particularly to the walrus, which depends primarily on clams for food, consuming about 27 kg per day (Marine Mammal News 1977; Kenyon 1978*a*). Another major problem, the effects of which are unknown, would be the disturbance of the sediment by dredging and the resulting resuspension of undesirable elements (Marine Mammal News 1977).

Furthermore, harassment by humans presents many problems resulting from extensive exploration for oil and minerals throughout the walrus's range. A major oil spill would be very serious (FAO Adv. Comm. 1976).

The Atlantic walrus is mainly a Canadian management problem. A brief look at exploitation of the Atlantic walrus during the 17th to 19th centuries shows the eradication of that animal from some of its important centers of abundance, such as Bear Island, from the height of its abundance in 1604 to near extinction by 1613 (McClung 1978). Other areas of depletion were most of Svalbard, Sable Island, and the Gulf of St. Lawrence (Reeves 1978). The Baffin Island herds were left untouched until the 1920s, when 175,000 animals were taken between 1925 and 1931. After this time hunting was restricted to the native peoples. The other countries, such as Norway and the U.S.S.R. that had traditionally taken part in the walrus hunt, have already passed restrictive legislation to protect these animals. There are two main populations, the Kara Sea to east Greenland region and from west Greenland to Canadian waters and land. Since 1956 Greenland (Denmark) has allowed hunting by permanent residents only with strict regulations (McClung 1978). In 1952 Norway passed the Norwegian Walrus Decree which forbade hunting for any purpose, and since 1973 it has created many marine sanctuaries encompassing habitats of the walrus. Canada permits subsistence hunting for Eskimos only, a condition that has been in effect since 1928, and the quota is still seven animals per year per family. Hides may not be exported, nor may unworked ivory. In May 1974 the Canadian government established a national park in some of the critical habitats of the walrus (Reeves 1978). The most threatening condition today for the walrus is still human encroachment, accidental or otherwise, on the hauling-out grounds. Behavioral research is needed to determine the effects of such trespassing. Investigation is also needed to determine if there might be any future conflict if man harvests the walrus invertebrate food

supply in the Canadian eastern arctic (Reeves 1978), as he is doing in the Pacific.

The economics of walrus consumption by humans are varied. Utilization varies by region, but on the average only 35 percent is consumed as human food, the rest being used as dog food (FAO Adv. Comm. 1976). The skin is still used for the tips of billiard cues in Greenland (Reeves 1978). Although there is still trade with the Hudson's Bay Company for ivory, much is sold privately. Recent quotes for raw ivory in Canada ranged from $44 to $55 kg (Reeves 1978). Traditional uses were much less wasteful, almost every part of the animal being used for subsistence purposes from food to wearing apparel, rope, tools, and fuel (Alaska Dept. of Fish and Game 1976; Reeves 1978). Life has changed for the Eskimo, other items having replaced traditional ones. One thing that cannot be taken away or replaced is the part that the hunt of the walrus takes in the culture of these people. The hunt still provides a man with pride, self-sufficiency, and a connection with his past.

As with any document that encompasses such a range and depth of material, the authors cannot claim sole credit for all the labor. Many persons aided us in our searches, our questions, our typing and editing, and our financial concerns. We therefore thank the following people, each of whom contributed in a unique way: W. Aaron, F. W. H. Beamish, C. Blondin, S. Brown, J. Dougan, J. Gallivan, P. Greenaway, R. Hofman, S. Innes, D. M. Lavigne, T. Loughlin, A. W. Mansfield, M. C. Mercer, R. V. Miller, P. L. J. Montreuil, N. A. Øritsland, M. Snowdon, R. E. Stewart, and H. Wilson.

LITERATURE CITED

Addison, R. F. 1973. Organochlorine residues in Canadian arctic marine mammals. Int. Counc. Explor. Sea, Counc. Meet., Marine Mammals Comm., N:4. 7pp.

Ainley, D. G.; Huber, H. R.; and Henderson, R. P. 1977. Studies of marine mammals at the Farallon Islands, California, 1975–76. U.S. Marine Mamm. Comm. 75/02. 32pp.

Alaska Department of Fish and Game. 1976. Ice inhabiting phocid seals. *In* Symp. Sci. consultation on marine mammals. Food and Agric. Organ. of U.N., Bergen, Norway, 13 August–9 September 1976.

Allen, J. A. 1880. History of North American pinnipeds: a monograph of the walruses, sea-lions, sea-bears, and seals of North America. Gov. Printing Office, Washington, D.C. 785pp.

Allen, R. L. 1975. A life table for harp seals in the northwest Atlantic. Pages 303–311 *in* K. Ronald and A. W. Mansfield, eds. Biology of the seal. Int. Counc. Explor. Sea, Rapp. & P.-V. Reun. 169. 557pp.

Angell James, J. E., and De Burgh Daly, M. 1972. Some mechanisms involved in the cardiovascular adaptations to diving. Pages 313–341 *in* M. A. Seligh and A. G. MacDonald, eds. Symposium: effects of pressure on organisms. Proc. 26th Symp. Soc. Exper. Biol., Wales, 6–10 September 1971. Academic Press, New York.

Baker, R. C.; Wilke, F.; and Baltzo, C. H. 1970. The northern fur seal. Reprint ed. U.S. Fish. Wildl. Serv., Circ. 336. 19pp.

Banfield, A. W. F. 1974. The mammals of Canada. Univ. Toronto Press, Toronto. 438pp.

Bartholomew, G. A. 1952. Reproductive and social behavior of the northern elephant seal. Univ. California Publ. Zool. 47:369–472, pls. 38–57.

Beddington, J. R., and Lavigne, D. M. 1979. International workshop on biology and management of northwest Atlantic harp seals. Proc. Symp. Univ. Guelph, Guelph, Ontario, 3-6 December 1979. World Wildl. Fund, Morges, Switzerland.

Benjaminsen, T., and Øritsland, T. 1975. The survival of year-classes and estimates of production and sustainable yield of northwest Atlantic harp seals. Int. Comm. Northwest Atlantic Fish., Res. Doc. 75/121, Ser. no. 3625.

Berger, T. R. 1977. Northern frontier, northern homeland; the report of the Mackenzie Valley pipeline inquiry. Vol. 1. Supply and Services, Ottawa. 213pp.

Bigg, M. A. 1969. The harbour seal in British Columbia. Fish. Res. Board Can. Bull. 172. 33pp.

Bisaillon, A.; Picard, I.; and La Rivière, N. 1976. Le segment cervical des carnivores (Mammalia: Carnivora) adaptés à la vie aquatique. Can. J. Zool. 54:431–436.

Blix, A. S. 1976. Metabolic consequences of submersion asphyxia in mammals and birds. Biochem. Soc. Symp. 41:169–178.

Blix, A. S.; Grav, H. J.: and Ronald, K. 1975. Brown adipose tissue and the significance of the venous plexuses in pinnipeds. Acta Physiol. Scand. 94:133–135.

_____. 1979. Some aspects of temperature regulation in newborn harp seal pups. Am. J. Physiol. 236:188–197.

Bonner, W. N. 1971. An aged grey seal (*Halichoerus grypus*). J. Zool. 164:261–262.

_____. 1972. The grey seal and common seal in European waters. Oceanogr. Marine Bio. 10:461–507.

Boulva, J. 1976. *Phoca vitulina concolor*. *In* Symposium: scientific consultation on marine mammals. Food and Agric. Advis. Comm. Mar. Resour. Res., Mar. Mammal, Sci. Consult. Organ. of U.N., Bergen, Norway, 13 August-9 September 1976.

Boulva, J., and McLaren, I. A. 1979. Biology of the harbour seal, *Phoca vitulina*, in Eastern Canada. Fish. Res. Board Can. Bull. 200. 24 pp.

Bowen, W. D.; Capstick, C. K.; and Sergeant, D. 1980. Changes in harp seal reproductive parameters: another look. Manuscript.

Bruemmer, F. 1979. The homecoming: the grey seals of Sable. Nat. Can. 8:48–53.

Bryden, M. M. 1972. Growth and development of marine mammals. Pages 2–79 *in* R. J. Harrison, ed. Functional anatomy of marine mammals. Vol. 1. Academic Press, New York. 451pp.

_____. 1979. Arteriovenus anastomoses in the skin of seals. Part 3: The harp seal *Pagophilus groenlandicus* and the hood seal *Cystophora cristata* (Pinnipedia: Phocidae). Aquat. Mammal 6:67–75.

Buhler, D. H.; Claeys, R. R.; and Mate, B. R. 1975. Heavy metal and chlorinated hydrocarbon residues in California sea lions (*Zalophus californianus californianus*). Fish. Res. Board Can. J. 32:2391–2397.

Burns, J. J. 1967. The Pacific bearded seal. Annu. Proj. Segment Rep. 8, Fed. Aid Wildl. Restor. Proj. W-6-R, W-14-R. Alaska Dept. Fish Game, Juneau. 66pp.

_____. 1971. Biology of the ribbon seal, *Histriophoca fasciata*, in the Bering Sea. Page 135 *in* Symposium: adaptation for northern life. Proc. 22nd Alaskan Sci. Conf., College, Alaska, 17–19 August 1971. Abstr.

_____. 1978. Ice seals. Pages 192–205 in D. Haley, ed.

Marine mammals of Eastern North Pacific and arctic waters. Pacific Search Press, Seattle. 256pp.

Burns, J. J., and Fay, F. H. 1970. Comparative morphology of the skull of the ribbon seal, *Histriophoca fasciata,* with remarks on systematics of Phocidae. J. Zool. London 161:363–394.

Casson, D. M., and Ronald, K. 1975. The harp seal (*Pagophilus groenlandicus* Erxleben 1777). Part 14: Cardiac arrythmias. Comp. Biochem. Physiol. 50A:307–314.

Coish, C. 1979. Season of the seal. Breakwater, St. John's, Nfld. 296pp.

Davis, J. E., and Anderson, S. S. 1976. Effects of oil pollution on breeding grey seals. Marine Bio. Bull. 7(6):115–118.

de Kleer, V. S. 1972. The antomy of the heart and the electrocardiogram of *Pagophilus groenlandicus*. M.S. Thesis. Univ. Guelph. 126pp.

DeLong, R. L. 1975. Interspectific reproductive behavior among four otariids at San Miguel Island, California. Page 13 *in* Symposium: conference on the biology and conservation of marine mammals. Univ. California, Santa Cruz, 4–7 December 1975. Abstr. 62pp.

_____. 1978. Northern elephant seal. Pages 206–211 *in* D. Haley, ed. Marine mammals of Eastern North Pacific and arctic waters. Pacific Search Press, Seattle. 256pp.

Dorofeev, S. V. 1939. (The influence of ice conditions on the behavior of harp seals.) Zool. Zh. 18:748–761. (In Russian.)

Dunbar, M. J. 1949. The Pinnipedia of the arctic and subarctic. Fish. Res. Board Can. Bull. 85:1–22.

Engelhardt, F. R.; Geraci, J. R.; and Smith, T. G. 1977. Uptake and clearance of petroleum hydrocarbons in the ringed seal, *Phoca hispida*. Fish. Res. Board Can. J. 34:1143–1147.

Fay, F. H. 1960. Structure and function of the pharyngeal pouches of the walrus (*Odobenus rosmarus* L.). Mammalia 24:362–371.

_____. 1974. The role of ice in the ecology of marine mammals of the Bering Sea. Alaska Univ., Inst. Marine Sci. Occas. Pap. 2:383–397.

Fay, F. H.; Shults, L. M.; and Dieterich, R. A. 1979. A field manual of procedures for postmortem examination of Alaskan marine mammals. Univ. Alaska, Inst. Marine Sci., Rep. 79-1. Univ. Alaska, Inst. Arctic Bio., Occas. Publ. Northern Life, 3. 51pp.

Feltz, E. T., and Fay, F. H. 1966. Thermal requirements *in vitro* of epidermal cells from seals. Cryobiology 3:261–264.

Fiscus, C. H. 1978. Northern fur seal. Pages 152–159 *in* D. Haley, ed. Marine mammals of Eastern North Pacific and arctic waters. Pacific Search Press, Seattle. 256pp.

Fisher, H. D. 1954. Studies on reproduction in the harp seal *Phoca groenlandica* Erxleben in the northwest Atlantic. Ph.D. Thesis. McGill Univ. 109pp.

Fleischer, L. A. 1978*a*. The distribution, abundance, and population characteristics of the Guadalupe fur seal, *Arctocephalus townsendi* (Meriam, 1897). M.S. Thesis. Univ. Washington. 93pp.

_____. 1978*b*. Guadalupe fur seal. Pages 160–165. *in* D. Haley, ed. Marine mammals of Eastern North Pacific and arctic waters. Pacific Search Press, Seattle. 256pp.

Food and Agriculture Organization of the United Nations, Advisory Committee on Marine Resources Research. 1976. Mammals in the seas. Ad Hoc Group III on seals and marine otters. Draft Rep. *In* Symposium: scientific consultation on marine mammals. Food and Agric. Organ. of U.N., Bergen, Norway, 13 August–9 September 1976. ACMRR/MM/SC/4.

Freeman, H. C., and Horne, D. A. 1973. Mercury in Canadian seals. Bull. Environ. Contam. Toxicol. 10:172–180.

Freeman, H. C.; Sangalang, G.; and Uthe, J. F. 1975. A study of the effects of contaminants on steroidogenesis in Canadian grey and harp seals. Int. Counc. Explor. Sea, Counc. Meet., Marine Mammals Comm., N:7. 8pp.

Gallivan, G. J. 1977. Temperature regulation and respiration in the freely diving harp seal (*Phoca groenlandica*). M.S. Thesis. Univ. Guelph. 60pp.

Gentry, R. L., and Withrow, D. E. 1978. Steller sea lion. Pages 166–171 *In* D. Haley, ed. Marine mammals of Eastern North Pacific and arctic waters. Pacific Search Press, Seattle. 256pp.

George, J. C., and Ronald, K. 1973. The harp seal, Pagophilus groenlandicus (Erxleben, 1777). Part 25: Ultrastructure and metabolic adaptation of skeletal muscle. Can. J. Zool. 51:833–839.

———. 1975. The harp seal, *Pagophilus groenlandicus* (Erxleben, 1777). Part 27: Structure and metabolic adaptation of the caval sphincter muscle with some observations on the diaphragm. Acta Anat. 93:88–99.

George, R. E. 1979. The senior manager's self-reported use of information sources for selected decision types. Ph.D. Thesis. Univ. Waterloo. 112pp.

Geraci, J. R., and Smith, T. G. 1976. Direct and indirect effects of oil on ringed seals (*Phoca hispida*) of the Beaufort Sea. Fish. Res. Board Can. J. 33:1976–1984.

Gilbert, J. R. 1977. Past and present status of grey seals in New England. Int. Counc. Explor. Sea., Counc. Meet., Marine Mammals Comm., N:14.

Gilmartin, W. G.; Sweeney, J. C.; and Gunnels, R. D. 1975. Effects of certain environmental pollutants on the California sea lion. Marine Mammals Comm., Contract MM5AC014, 1 January 1975–30 October 1975, Final Rep. Manuscript. 10pp.

Grav, H. J.; Blix, A. S.; and Pasche, A. 1974. How do seal pups survive birth in arctic winter? Acta Physiol. Scand. 92:427–429.

Gray, D. F., and Beck, B. 1979. Eastern Canadian grey seal: 1978 research report and stock assessment. Can. Atlantic Fish. Sci. Adv. Comm., Res. Doc. 79/1. 8pp.

Green, R. F. 1972. Observations on the anatomy of some cetaceans and pinnipeds. Pages 247–297 *in* S. H. Ridgway ed. Mammals of the sea: biology and medicine. C. C. Thomas, Springfield, Ill.

Green-Hammond, K. A. 1980. Fisheries management under the Fishery Conservation and Management Act, the Marine Mammal Protection Act, and the Endangered Species Act. Manuscript. 52pp.

Gulland, J. A. 1976. A note on the strategy of the management of marine mammals. *In* Symposium: Scientific consultation on marine mammals. Food and Agric. Organ. of U.N., Bergen, Norway, 13 August–9 September 1976. Advis. Comm. Mar. Resour. Res., Mar. Mammal, Sci. Consult. 82.

Harrison, R. J., and King, J. E. 1965. Marine mammals. Hutchinson & Co., London. 192 pp.

Hecht, S.; Hendley, C. D.; Ross, S.; and Richmond, P. N. 1948. The effect of exposure to sunlight on night vision. Am. J. Ophthal. 31:1573–1580.

Hempleman, H. V., and Lockwood, A. P. M. 1978. The physiology of diving in man and other animals. Inst. Bio., Stud. Bio. 99. Edward Arnold Publ., London. 58pp.

Hewer, H. R. 1964. The determination of age, sexual maturity, longevity, and a life table in the grey seal

(*Halichoerus grypus*). Zoo. Soc. London Proc. 142:593–624.

Hobson, E. S. 1966. Visual orientation and feeding in seals and sea lions. Nature (London) 210:326–327.

Hol, R.; Bliz, A. S.; and Myhre, H. O. 1975. Selective redistribution of the blood volume in the diving seal (*Pagophilus groenlandicus*). Pages 423–431 *in* K. Ronald and A. W. Mansfield, eds. Biology of the seal. Int. Counc. Explor. Sea, Rapp. & P. V. Reun., 169. 557pp.

Holden, A. V. 1978. Pollutants and seals. Mammal. Rev. 8:53–66.

Holt, S. J., and Talbot, L. M. 1978. New principles for the conservation of wild living resources. Wildl. Monogr. 59:1–33.

Howell, A. B. 1970. Aquatic mammals, their adaptions to life in the water. Reprint ed. Dover Publ., New York. 388pp.

Huey, L. M. 1930. Capture of an elephant seal off San Diego, California, with notes on stomach contents. J. Mammal. 11:229–231.

Hyland, J. L.; Schneider, E. D.; and Narranyunsett, E. R. L. 1977. Petroleum hydrocarbons and their effects on marine organisms, populations, communities and ecosystems. Int. Counc. Explor. Sea, Counc. Meet., Marine Mammals Comm., E:64. 42pp.

Innes, S.; Stewart, R. E. A.; and Lavigne, D. M. 1980. Growth in Northwest Atlantic harp seals, *Pagophilus groenlandicus*. Manuscript 36pp.

International Union for the Conservation of Nature and Natural Resources. 1980. World conservation strategy: living resource conservation for sustainable development. Int. Union for Conserv. of Nature and Nat. Resour., Gland, Switzerland.

Irving, L. 1938. Vascular adjustments of diving animals during apneoa. Science N.S. 88:502.

———. 1972. Arctic life of birds and mammals including man. Springer-Verlag, New York. 192pp.

Jamieson, G. S., and Fisher, H. D. 1972. The pinniped eye: a review. Pages 245–261 *in* R. J. Harrison, ed. Functional anatomy of marine mammals. Vol. 1. Academic Press, New York. 451pp.

Johnson, M. L., and Jeffries, S. J. 1977. Population evaluation of the harbour seal (*Phoca vitulina richardii*) in the waters of the state of Washington. U.S. Marine Mammal Comm., Rep. no. MMC-75/05. 27pp.

Kapel, F. O. 1973. Some second-hand reports on the food of harp seals in west Greeland waters. Int. Counc. Explor. Sea, Counc. Meet., Marine Mammal Comm., N:8. 7pp.

Kenyon, K. W. 1973*a*. The Guadalupe fur seal (*Arctocephalus townsendi*). Pages 82–87 *in* Proc. Workshop Meet. Seal Spec. Threatened Depleted Seals World, Survival Serv. Comm. I.U.C.N. Publ. New Ser., Suppl. Pap. 39. 176pp.

———. 1973*b*. The Hawaiian monk seal (*Monachus schauinslandi*). Pages 88–97 *in* Proc. Workshop Meet. Seal Spec. Threatened Depleted Seals World, Survival Serv. Comm. I.U.C.N. Publ. New Ser., Suppl. Pap. 39. 176pp.

———. 1976. Critical habitat of the Hawaiian monk seal, including a review of the status of the Caribbean and Mediterranean monk seals. U.S. Marine Mammals Comm. Manuscript. 41pp.

———. 1978*a*. Walrus. Pages 178–183 *in* D. Haley, ed. Marine mammals of Eastern North Pacific and arctic waters. Pacific Search Press, Seattle. 256pp.

———. 1978*b*. Hawaiian monk seal. Pages 212–216 *in* D. Haley, ed. Marine mammals of Eastern North Pacific and arctic waters. Pacific Search Press, Seattle. 256pp.

Kenyon, K. W., and Fiscus, C. H. 1963. Age determination in the Hawaiian monk seal. J. Mammal. 44:280–282.

Kenyon, K. W., and Rice, D. W. 1959. Life history of the Hawaiian monk seal. Pacific Sci. 13:215–252.

King, J. E. 1964. Seals of the world. Br. Mus. (Nat. Hist.), Trustees, London. 154pp.

———. 1972. Observations on phocid skulls. Pages 81–115 in R. J. Harrison, ed. Functional anatomy of marine mammals. Vol. 1. Academic Press, London. 451pp.

Koeman, J. H.; Peters, W. H. M.; Smit, C. J.; Tjioe, P. S.; and de Goeij, J. J. M. 1972. Presistent chemicals in marine mammals. TNO-Nieuws 27:570–578.

Koeman, J. H.; Van de Ven, W. S. M.; de Goeij, J. J. M.; Tjoie, P. S.; and Van Haaften, J. L. 1975. Mercury and selenium in marine mammals and birds. Sci. Total Environ. 3:279–287.

Kooyman, G. L.; Hammond, D. D.; and Schroeder, J. P. 1970. Bronchograms and tracheograms of seals under pressure. Science 168:82–84.

Kulu, D. D. 1972. Evolution and cytogenetics. Pages 503–527 in S. H. Ridgway, ed. Mammals of the sea: biology and medicine. C. C. Thomas, Springfield, Ill.

Lavigne, D. M. 1975. Harp seal, Pagophilus groenlandicus, production in the Western Atlantic during March 1975. Int. Comm. Northwest Atlantic Fish., Res. Doc. 75/XII/150, Ser. 3728. 2pp.

———. 1978. The harp seal controversy reconsidered. Queen's Q. 85:377–388.

———. 1979. Management of seals in the Northwest Atlantic Ocean. Trans. North Am. Wildl. Nat. Resourc. Conf. 44:488–497.

Lavigne, D. M.; Bernholtz, C. S.; and Ronald, K. 1977. Functional aspects of pinniped vision. Pages 135–173 in R. J. Harrison, ed. Functional anatomy of marine mammals. Vol. 3. Academic Press, New York. 428pp.

Lavigne, D. M., and Ronald, K. 1972. The harp seal, Pagophilus groenlandicus (Erxleben, 1777). Part 23: Spectral sensitivity. Can. J. Zool. 50:1197–1206.

Laws, R. M. 1956. Growth and sexual maturity of aquatic mammals. Nature (London) 178:193–194.

———. 1959. Accelerated growth in seals, with special reference to the phocidae. Norsk Hvalf.-Tid. 48:425–452.

———. 1962. Age determination of pinnipeds with special reference to growth layers in the teeth. Z. Saeugetierk. 27:129–146.

LeBoeuf, B. J., and Panken, K. J. 1977. Elephant seals breeding on the mainland in California. California Acad. Sci. Proc. 41:267–280.

Lett, P. F., and Benjaminsen, T. 1977. A stochastic model for the management of the northwestern Atlantic harp seal (Pagophilus groenlandicus) population. Fish. Res. Board Can. J. 34:1155–1187.

Lett, P. F.; Gray, D. F.; and Mohn, R. K. 1977. New estimates of harp seal production on the Front and in the Gulf of St. Lawrence and their impact on herd management. Int. Comm. Northwest Atlantic Fish., Res. Doc. 77/xi/68. 18pp.

Lett, P. F.; Mohn, R. K.; and Gray, D. F. 1978. Density-dependent processes and management strategy for the Northwestern Atlantic harp seal. Int. Comm. Northwest Atlantic Fish., Spec. Meet. Standing Comm. on Res. and Stat., Res. Doc. 78/xi/84. 49pp.

Ling, J. K. 1970. Pelage and molting in wild animals with special reference to aquatic forms. Q. Rev. Bio. 45:16–54.

———. 1974. The integument of marine mammals. Pages 1–44 in R. J. Harrison, ed. Functional anatomy of marine mammals. Vol. 2. Academic Press, New York. 366pp.

Ling, J. K., and Button, C. E. 1975. The skin and pelage of grey seal pups (Halichoerus grypus Fabricus), with a comparative study of foetal and neonatal moulting in the pinnipedia. Pages 112–132 in K. Ronald and A. W. Mansfield, eds. Biology of the seal. Int. Counc. Explor. Sea, Rapp. & P-V. Reun., 1969. 557pp.

Lipps, J. H., and Mitchell, E. 1976. Trophic model for the adaptive radiations and extinctions of pelagic marine mammals. Paleobiology 2:147–155.

McClung, R. M. 1978. Hunted mammals of the sea. Morrow & Co., New York. 191pp.

McLaren, I. A. 1958. The biology of the ringed seal (Phoca hispida Schreber) in the eastern Canadian arctic. Fish. Res. Board Can. Bull. 118. 97pp.

Mansfield, A. W. 1967a. Distribution of the harbour seal, Phoca vitulina, Linnaeus, in Canadian waters. J. Mammal. 48:249–257.

———. 1967b. Seals of arctic and eastern Canada. Fish. Res. Board Can. Bull. 137. 35pp.

———. 1978. Reproduction of the grey seal, Halichoerus grypus, in eastern Canada. Int. Counc. Explor. Sea, Counc. Meet., Marine Mammals Comm., N:13.

Mansfield, A. W., and Beck, B. 1977. The grey seal in eastern Canada. Environ. Can., Fish. Marine Serv., Tech. Rep. no. 704. 81pp.

Mansfield, A. W., and Sergeant, D. E. 1973. Seals as indicators of pollution. Int. Counc. Explor. Sea, Counc. Meet., Marine Mammals Comm., E:25. 7pp.

Marine mammal news. 1977. Nautilus Press, Washington, D.C. 8pp.

Mate, B. R. 1976. Zalophus californianus californianus. In Symposium: scientific consultation on marine mammals. Food and Agric. Organ. of U.N., Bergen, Norway, 31 August–9 September 1976. ACMRR/Ad Hoc III/19.

———. 1978. California sea lion. Pages 172–177 in D. Haley, ed. Marine mammals of Eastern North Pacific and arctic waters. Pacific Search Press, Seattle. 256pp.

Maxwell, G. 1967. Seals of the world. Constable & Co., London. 151pp.

Miller, E. H. 1976. Walrus ethology. Part 2: Herd structure and activity budgets of summering males. Can. J. Zool. 54:704–715.

Møhl, B. 1964. Preliminary studies on hearing in seals. Dan. Naturhist. Foren., Vidensk., Medd., 127:283–294.

———. 1968a. Hearing in seals. Pages 172–195 in R. J. Harrison, R. C. Hubbard, R. S. Peterson, C. E. Rice, and R. S. Schusterman, eds. The behaviour and physiology of pinnipeds. Appleton-Century-Crofts, New York. 411pp.

———. 1968b. Auditory sensitivity of the common seal in air and water. J. Aud. Res. 8:27–38.

Møhl, B., and Ronald, K. 1975. The peripheral auditory system of the harp seal, Pagophilus groenlandicus (Erxleben, 1777). Pages 514–523 in K. Ronald and A. W. Mansfield, eds. Biology of the seal. Int. Counc. Explor. Sea, Rapp.-V. Reun., 169. 557pp.

Møhl, B.; Ronald, K.; and Terhune, J. M. 1975. Underwater calls of the harp seal, Pagophilus groenlandicus. Pages 533–543 in K. Ronald and A. W. Mansfield, eds. Biology of the seal. Int. Counc. Explor. Sea, Rapp. & P.-V. Reun., 169. 557pp.

Mohr, E. 1950. Behaarung und haarwechsel der robben. Neue Ergeb. Probl. Zool. (Klatt.-Festchr.), pp. 602–614.

Morejohn, G. V., and Baltz, D. M. 1970. Contents of the stomach of an elephant seal. J. Mammal. 51:173–174.

Munz, F. W. 1971. Vision: visual pigments. Pages 1–32 in W. S. Hoar and D. J. Randall, eds. Fish physiology.

Part 5: Sensory systems and electric organs. Academic Press, New York. 600pp.

Myers, C. A. 1967. The impact of computers on management. MIT Press, Cambridge, Mass.

Nagy, A. R., and Ronald, K. 1975. A light and electromicroscopic study of the structure of the retina of the harp seal, *Pagophilus groenlandicus* (Erxleben, 1777). Pages 92–96 *in* K. Ronald and A. W. Mansfield, eds. Biology of the seal. Int. Counc. Explor. Sea, Rapp. & P.-V. Reun., 169. 557pp.

National Energy Board. 1977. Reasons for decision on northern pipeline. Supply and Services, Ottawa. Vol. 1. 372pp.

Newby, T. C. 1978. Pacific harbor seal. Pages 184–191 *in* D. Haley, ed. Marine mammals of eastern North Pacific and arctic waters. Pacific Search Press, Seattle. 256pp.

Norris, K. S. 1969. The echolocation of marine mammals. Pages 391–423 *in* H. T. Anderson, ed. The biology of marine mammals. Academic Press, New York. 511pp.

Øritsland, N. 1978. Some applications of thermal values of fur samples in expressions for *in vivo* heat balance. Univ. Oslo, Inst. Zoophysiol., Oslo, Norway.

Øritsland, N. A.; Lavigne, D. M.; and Ronald, K. 1978. Radiative surface temperatures of harp seals. Comp. Biochem. Physiol. 61A:9–12.

Øritsland, N. A., and Ronald, K. 1977. A simulation programme for heat balance in marine mammals. Int. Counc. Explor. Sea, Counc. Meet., Marine Mammals Comm., N:8. 15pp.

————. 1978*a*. Aspects of temperature regulation in harp seal pups evaluated by *in vivo* experiments and computer simulations. Acta Physiol. Scand. 103:263–269.

————. 1978*b*. Solar heating of mammals: observations of hair transmittance. Int. J. Biometeor. 22:197–201.

Øritsland, T. 1978. The status of Norwegian studies of harp seals at Newfoundland. Int. Comm. Northwest Atlantic Fish. Redbk. 1971 (3):185–209.

Payne, M. R. 1979. Growth in the Antarctic fur seal, *Arctocephalus gazella*. J. Zool. 187:1–20.

Percy, J. A. 1977. Effects of oil on arctic marine organisms: a review of studies conducted by the Arctic Biological Station. Int. Counc. Explor. Sea, Counc. Meet., Fish. Improv. Comm., E:26.

Pianka, E. R. 1978. Evolutionary ecology. Harper & Row, New York. 397pp.

Platt, N. E.; Prime, J. H.; and Witthames, T. R. 1974. The age of the grey seal at the Farne Islands. Int. Counc. Explor. Sea, Counc. Meet., Marine Mammals Comm., N:3. 7pp.

Popov, L. A. 1966. (On an ice floe with the harp seals: ice drift of biologists in the White Sea.) Priroda 9:93–101. (In Russian.) Fish. Res. Board Can. Transl. Ser. 814, 1907.

————. 1976. Status of main ice forms of seals inhabiting waters of the U.S.S.R. and adjacent to the country marine areas. *In* Symposium: Scientific consultation on marine mammals. Food and Agric. Organ. of U.N., Bergen, Norway, 13 August–9 September 1976. ACMRR/MM/SC/51.

Purkinje, J. E. 1825. Neue Beitrage zur Kenntnis des Sehen. Berlin.

Ramprashad, F. 1975. Aquatic adaptations in the ear of the harp seal (*Pagophilus groenlandicus*) (Erxleben, 1777). Pages 102–111 *in* K. Ronald and A. W. Mansfield, eds. Biology of the seal. Int. Counc. Explor. Sea, Rapp. & P.-V. Reun., 169. 557pp.

Ramprashad, F.; Corey, S.; and Ronald, K. 1973. Anatomy of the seal's ear (*Pagophilus groenlandicus* Erxleben, 1777). Pages 264–305 *in* R. J. Harrison, ed. Functional anatomy of marine mammals. Vol. 1. Academic Press, New York.

Ramprashad, F.; Money, K. E.; and Ronald, K. 1972. The harp seal, *Pagophilus groenlandicus* (Erxleben, 1777). Part 21: The structure of the vestibular apparatus. Can. J. Zool. 50:1357–1361.

Ray, G. C., and Watkins, W. A. 1975. Social function of underwater sounds in the walrus *Odobenus rosmaris*. Pages 524–526 *in* K. Ronald and A. W. Mansfield eds. Biology of the seal. Int. Counc. Explor. Sea, Rapp. & P.-V. Reun., 169. 557pp.

Ray, G. C.; Watkins, W. A.; and Burns, J. J. 1969. The underwater song of *Erignathus* (bearded seal). Zooogica (New York) 54:79–83.

Reeves, R. R. 1978. Atlantic walrus (*Odobenus rosmarus rosmarus*): a literature survey and status report. Wildl. Res. Rep. 10. U.S. Dept. Interior, Fish Wildl. Serv., Washington, D.C. 41pp.

Repenning, C. A.; Peterson, R. S.; and Hubbs, C. L. 1971. Contributions to the systematics of the southern fur seals, with particular reference to the Juan Fernandez and Guadalupe species. Pages 1–34 *in* W. H. Burt, ed. Antarctic Pinnipedia. Antarct. Res. Ser. 18, Am. Geophys. Union. 226pp.

Ricker, W. E. 1975. Mortality and production of harp seals with reference to a paper of Benjaminsen and Øritsland (1975). Int. Comm. Northwest Atlantic Fish, Ser. Res. 3716. 41pp.

Ronald, K.; Hanly, L. M.; Healey, P. J.; and Selley, L. J. 1976. An annotated bibliography on the Pinnipedia. Int. Council Explor. Sea, Charlottenlund, Denmark. 785pp.

Scheffer, V. B. 1958. Seals, sea lions, and walruses: a review of the Pinnipedia. Stanford University Press, Stanford, Calif. 179pp.

————. 1977. A lesson in survival. Anim. Kingdom 80:6–12.

Scheffer, V. B., and Johnson, A. M. 1963. Molt in the northern fur seal. U.S. Fish Wildl. Serv., Spec. Sci. Rep., Fish. Ser. 450. 34pp.

Schevill, W. E.; Watkins, W. A.; and Ray, G. C. 1966. Analysis of underwater *Odobenus* calls with remarks on the development and function of the pharangeal pouches. Zoologica (New York) 51:103–106.

Scholander, P. F.; Irving, L.; and Grinnell, S. W. 1942. On the temperature and metabolism of the seal during diving. Cell Comp. Physiol. J. 19:67–78.

Schusterman, R. J. 1967. Underwater sound production by captive California sea lions, *Zalophus californianus*. Zoologica (New York) 52:21–24.

————. 1972. Visual acuity in pinnipeds. Pages 469–492. *in* H. E. Winn and B. L. Olla, eds. Behaviour of marine animals: current perspectives in research. Vol. 2. Plenum Press, New York. 503pp.

————. 1975. Pinniped sensory perception. Pages 165–169 *in* K. Ronald and A. W. Mansfield, eds. Biology of the seal. Int. Counc. Explor. Sea, Rapp. & P.-V. Reun., 169. 557pp.

Sergeant, D. E. 1966. Reproductive rates of harp seals, *Pagophilus groenlandicus* (Erxleben). Fish. Res. Board Can. J. 23:757–766.

————. 1969. On the population dynamics and size of stocks of harp seals in the northwestern Atlantic. Int. Comm. Northwest Atlantic Fish., Res. Doc. 69/31, Ser. 2171.

————. 1971. Calculation of production of harp seals in the western North Atlantic. Int. Comm. Northwest Atlantic Fish. Redbk. 1971(3):157–184.

————. 1973. Feeding, growth, and productivity of Northwest Atlantic harp seals (*Pagophilus groenlandicus*). Fish. Res. Board Can. J. 30:17–29.

_____. 1975. Estimating numbers of harp seals. Pages 274–280 *in* K. Ronald and A. W. Mansfield, Eds. Biology of the seal. Int. Counc. Explor. Sea, Rapp. & P.-V. Reun., 169. 557pp.

_____. 1976. History and present status of populations of harp and hooded seals. *In* Symposium: Scientific consultation marine mammals. Food and Agric. Organ. of U.N., Bergen, Norway, 13 August–9 September, 1976. Advis. Comm. Mar. Resour. Res., Mar. Mammals, Sci. Consult. 1.

Sergeant, D. E., and Armstrong, F. A. J. 1973. Mercury in seals from Eastern Canada. Fish. Res. Board Can. J. 30:843–846.

Sivertsen, E. 1941. On the biology of the harp seal, *Phoca groenlandica* Erxl., investigations carried out in the White Sea. Hvalrad, Skr. 26. 166pp.

Smirnov, N. A. 1924. On the eastern harp seal *Phoca (Pagophoca) groenlandica*. Tromso Mus., Arsb, Nat. Adv. 47:3–11. 11pp.

Smith, T. G., and Stirling, I. 1975. The breeding habits of the ringed seal (*Phoca hispida*): the birth, air, and associated structures. Can. J. Zool. 53:1297–1305.

Stewart, R. E. A., and Lavigne, D. M. 1980. Neonatal growth in Northwest Atlantic harp seals, *Pagophilus groenlandicus*. Manuscript. 36pp.

Stirling, I. 1975. Factors affecting the evolution of social behaviour in the pinnipedia. Pages 205–212 *in* K. Ronald and A. W. Mansfield, eds. Biology of the seal. Int. Counc. Explor. Sea, Rapp. & P.-V. Reun., 169. 557pp.

Sumner-Smith, G.; Pennock, P. W.; and Ronald, K. 1972. The harp seal, *Pagophilus groenlandicus* (Erxleben, 1777). Part 16: Epiphyseal fusion. J. Wildl. Dis. 8:29–32.

Sweeney, J. C., and Geraci, J. R. 1979. Medical care and strandings. Pages 275–296 *in* J. R. Geraci and D. J. St. Aubin, eds. Biology of marine mammals: insights through strandings. U.S. Dept. Commerce Natl. Tech. Info. Serv. PB-293 890; U.S. Marine Mammals Comm., Rep. no. MMC/771 13. 343pp.

Tandler, J. 1899. Ueber ein corpus cavernosum tympanicum beim seehund. Monatsschr. Ohrenheilkd. Kehldopfheilkd., Nasen-, Rachenkr., Organ Oesterr. Otol. Ges. 33:437–440.

Terhune, J. M. 1973. Aspects of hearing and acoustical communication of seal. L. S. Thesis. Aarhus Univ., Aarhus, Denmark.

Terhune, J. M., and Ronald, K. 1970. The audiogram and calls of the harp seal (*Pagophilus groenlandicus*) in air. Pages 133–143 *in* Symposium: biological sonar and diving mammals. Proc. 7th Annu. Conf. Bio. Sonar Diving Mammals, 23–24 October 1970, Stanford Res. Inst., Menlo Park, Calif.

_____. 1971. The harp seal, *Pagophilus groenlandicus* (Erxleben, 1777). Part 10: The air audiogram. Can. J. Zool. 49:385–390.

_____. 1972. The harp seal, *Pagophilus groenlandicus*. Part 3: The underwater audiogram. Can. J. Zool. 50:565–569.

_____. 1973. Some hooded seal (*Cystophora cristata*) sounds in March. Can. J. Zool. 51:319–321.

_____. 1974. Underwater hearing of phocid seals. Int. Counc. Explor. Sea, Counc. Meet., Marine Mammal Comm., N:5. (Restricted.) 9pp.

Terhune, J. M.; Terhune, M. E.; and Ronald, K. 1979. Location and recognition of pups by adult female harp seals. Appl. Anim. Ethol. 5:375–380.

Tikhomirov, E. A. 1968. (Body growth and reproductive organs development of the North Pacific phocid.) Pages 216–243 *in* V. A. Arseniev and K. I. Panin, eds. Lastonogie severnoy chasti tikhogo okeana [Pinnipeds of the North Pacific]. Moscow Vses. Nauchno-Issled. Inst. Morsk. Rybn, Khoz. Okeanogr. (Vniro), Tr. 68; Vladivostok. Tikhookean. Nauchno-Issled. Inst. Rybn. Khoz. Okeanogr. (Tinro), Izv. 62. 274pp.

Toffler, A. 1970. Future shock. Random House, New York. 505pp.

United Nations Environment Programme. 1977. Register of international conventions and protocols in the field of the environment. UNEP/GC/Information, 7 February 1977. Governing Counc., 5th sess., Nairobi, 9–25 May 1977. No. 76-3464. 96pp.

_____. 1978. Register of international conventions and protocols in the field of environment. UNEP/GC/Information, Supplement, 15 December 1977. Governing Counc., 6th sess., Nairobi, 9–25 May 1978. No. 77-5230. 37pp.

U.S. Department of Commerce. 1977. The story of the Pribilof fur seals. Gov. Printing Office, Washington, D.C. 13pp.

U.S. Marine Mammal Commission. 1979. Annual report of the marine mammal commission, calendar year 1978. Rep. to Congr., 31 January 1979, Washington, D.C. 108pp.

U.S. Office of Federal Register. 1979. Government manual, 1979–80. Natl. Arch. and Rec. Serv., Gen. Serv. Admin., Washington, D.C. 737pp.

Walls, G. L. 1942. The vertebrate eye. Cranbrook Press, Bloomfield Hills, Mich.

KEITH RONALD, College of Biological Science, University of Guelph, Guelph, Ontario.

JANE SELLEY, College of Biological Science, University of Guelph, Guelph, Ontario.

PAMELA HEALEY, College of Biological Science, University of Guelph, Guelph, Ontario.

41

West Indian Manatee

Trichechus manatus

Daniel K. Odell

NOMENCLATURE

COMMON NAMES. West Indian manatee, manatee, sea cow, Caribbean manatee, Florida manatee
SCIENTIFIC NAME. Family: Trichechidae *Trichechus manatus*
SUBSPECIES. *T. m. manatus, T. m. latirostris*. Most biologists feel that there is insufficient data at the present time to warrant subspecific designations of *T. manatus* (Moore 1951*a*; Gunter 1954). The original division (Harlan 1824) may simply reflect individual variation. Other living members of the Trichechidae are the Amazonian manatee (*T. inunguis*) and the West African manatee (*T. senegalensis*); the fourth living sirenian is the dugong (Dugongidae: *Dugong dugon*) (see Husar 1977*a*, 1977*b*, 1978*a*, 1978*b*, 1978*c*; Ronald et al. 1978).

DISTRIBUTION

The West Indian manatee is currently distributed in the southern United States (latitude 30° N), throughout the Caribbean region, along the east coast of Central America, and on the northeast coast of South America as far south as Marque Seca, Brazil (Latitude 12° S) (Husar 1977*b*, 1978*c*; U.S. Fish and Wildlife Service 1978; Odell et al. 1978) (figure 41.1). This species is coastal in distribution and can be found in fresh, brackish, and salt-water habitats. It moves freely between the salinity extremes. Florida is essentially the northern end of the range, although individuals are often reported as far north as New Jersey on the Atlantic coast (latitude 40° N) and as far west as Texas on the Gulf coast (U.S. Fish and Wildlife Service 1978). Similarly, at the southern end of the range, manatees have been reported as far south as Espirito Santo, Brazil (latitude 20° S) (U.S. Fish and Wildlife Service 1978). The northern and southern extremes of distribution are apparently limited by air and water temperatures. The exact distribution of the manatee is not known because many areas within the overall range have not been surveyed.

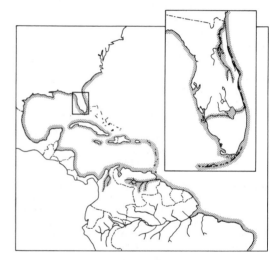

FIGURE 41.1. Distribution of the West Indian manatee (*Trichechus manatus*). Inset: primary distribution of the Florida population. After U.S. Fish and Wildlife Service 1978.

DESCRIPTION

Published records indicate that adult manatees may reach a length of 450 cm (Gunter 1941). The longest manatee collected in recent years was a 387-cm female (Beck, Bonde, and Odell in press) and the heaviest was a 1000-kg, 350-cm female (E. Asper personal communication 1978). (The 387-cm animal was not weighed.) The overall body shape is fusiform and flattened dorsoventrally (figure 41.2). A large, rounded, spatulate tail is the major organ of propulsion. In accordance with streamlining, external protuberances that would lead to turbulent water flow are reduced or absent. Manatees lack auditory pinnae and pelvic appendages. The long bones of the pectoral appendages are shortened but the flippers remain flexible. Nails are present. Hairs are sparsely scattered over the body surface, but the facial vibrissae are prominent. The color of the adult skin is best described as "grayish," while

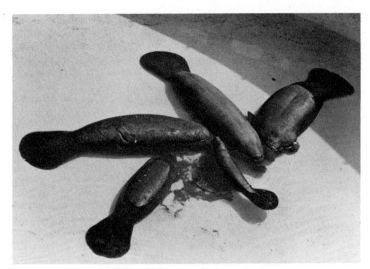

FIGURE 41.2. The West Indian manatee (*Trichechus manatus*) colony at the Miami Seaquarium. The animal on the left has a healed propeller scar. The photograph was taken when the water level was lowered to facilitate tank cleaning.

newborn calves are nearly black. The epidermis is continually sloughing off, perhaps to prevent "excessive" accumulation of algae or epizoites. The only externally apparent, sexually dimorphic characteristic is the placement of the genital openings. They are slightly posterior to the umbilicus in males and slightly anterior to the anus in females. Females appear to attain larger weights and lengths than males (Odell 1977; Odell et al. in press).

Skeleton. The skeleton is massive, composed of dense pachyostotic bone (Fawcett 1942a). There are no marrow cavities in the ribs or the long bones of the pectoral appendage. The vertebrae formula is unusual: there are 6 cervical, 17–19 thoracic, and 27–29 lumbocaudal vertebrae (Hatt 1934; Jones and Johnson 1967; Kaiser 1974). The bones of the skull are as dense as those of the postcranial skeleton (figure 41.3). Functional incisors and canines are absent, although an unerupted pair of vestigial upper incisors occurs in young animals. Adults usually have 6 or 7 erupted (functional) molariform teeth in each quadrant. The crowns are bilophodont with two major and two minor cusps. Maxillary teeth have tripartite roots and mandibular teeth bipartite roots. Tooth replacement is in the horizontal plane, posterior to anterior (Domning and Magor 1977). It is not clear if tooth movement is a discontinuous or a continuous function. Bone mineral is apparently resorbed at the anterior edge of alveoli. Miller et al. (1980) suggest that the forces of occlusion are the impetus behind tooth movement. As a particular tooth moves forward the crown is worn down and the root is resorbed. In adults, the anterior tooth in each row is often a thinly worn crown with no root, loosely held in place by the gums. This extreme degree of wear seems to be found only in the Florida population and may be due to the nature of the sub-

strate on which the food plants are growing (D. Domning personal communication 1979). The tooth pattern in the ramus of a young manatee, in which there has been no tooth replacement, may be contrasted with the dental pattern in an adult (figure 41.4).

As in cetaceans, the otic bullae are not fused to the skull. They are massive and dense, and lie in intimate contact with the squamosal. It is noteworthy that the zygomatic process of the squamosal is the only part of the skull containing oil. The auditory sensitivity of the Amazonian manatee is greatest in those areas of the head in close proximity to the zygomatic process (Bullock et al. 1980; D. Domning personal communication 1978).

PHYSIOLOGY AND ANATOMY

The physiology of *T. manatus* has not been examined in any detail. Speculation is extensive and based on limited data. New studies are highly restricted because of the manatee's endangered status. Scholander and Irving (1941) documented the response of restrained individuals to forced submersion. The animals exhibited the typical "diving response" (bradycardia and apparent peripheral vasoconstriction) found in marine mammals subjected to similar conditions (see Elsner 1969). There are few data on manatee body temperature. A newborn calf had a deep rectal temperature of about 35.5° C (Odell unpublished data 1974). Irvine (1979) reported a mean deep body temperature of 36.4° C. Basal metabolic rate (300 cc O_2/kg/min) apparently is considerably lower than in other marine mammals (Scholander and Irving 1941; Irving 1969). The low metabolic rate may be linked with hypothyroidism (Fawcett 1942a; Harrison and King 1965), but has not been verified. Harrison (1969) reported thyroid glands ranging in weight from 0.11 to 0.13 g/kg body mass.

Odell et al. (in press) reported values ranging from 0.06 to 0.32 ($\bar{x} = 0.22 \pm 0.10$) g/kg. There is also limited evidence that manatees in Florida may undergo seasonal weight changes. They are heaviest in late sum-

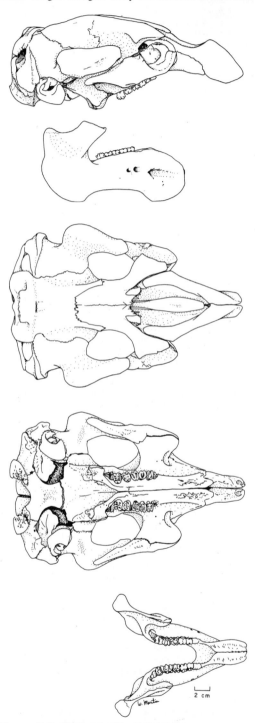

mer, fall, and early winter (Odell et al. in press). If this is the case, organ weights should be compared with fat-free body weight. Low metabolic rate has also been linked with winter mortality in Florida (Campbell and Irvine in press). It may prove most interesting to examine manatee thyroid structure and function on a seasonal basis and in comparison with detailed metabolic studies.

Odell et al. (in press) derived the following weight-length relationship: $W_{kg} = 4.28 \times 10^{-6} L_{cm}^{3.26}$, $r^2 = 0.98$. They also gave the following organ weight-body weight relationships ($\bar{x} \pm$ S.D.): heart 3.25 $\pm$ 0.87 g/kg, liver 15.4 $\pm$ 5.68, kidneys 4.08 $\pm$ 1.52, adrenal glands 0.06 $\pm$ 0.04, spleen 0.06 $\pm$ 0.08, pancreas 0.31 $\pm$ 0.11. The spleen is extremely small and the location of hematopoetic tissue is not known, although it is probably the vertebrae. The lungs are long (exceeding 1 m in adults), unlobed, and dorsoventrally flattened. They extend the length of the body cavity (Wislocki 1935*a*). The major bronchi extend nearly the full length of the lungs, as in other marine mammals. Similarly, the diaphragm extends the length of the body cavity. The diaphragm is actually composed of two hemidiaphragms that are parallel to the body axis, in contrast to the oblique situation in other mammals (Reynolds 1980). This unique condition unquestionably relates to buoyancy regulation and the horizontal attitude of manatees in the water.

The digestive tract occupies a major portion of the body cavity, as one would expect in a herbivore. On first inspection the stomach appears to be multicompartmental (see plates in Murie 1872). In fact, the stomach proper is a single, highly muscular compartment with a single-lobed "digestive gland" located along the greater curvature near the esophageal opening. The second "stomach compartment" is the duodenal ampulla with two small diverticulae at the proximal end. The small and large intestines are of nearly equal length, joined by a cecum with two hornlike diverticulae. The combined length of the intestines can exceed 40 m (Reynolds 1979*b*) and the weight of the hindgut, including contents, may represent 14 percent of the total body weight (Reynolds 1979*a*, 1980). Most, if not all, of the cellulose digestion probably occurs in the hind gut, most likely with the aid of a bacterial and/or protozoan flora/fauna. This has been suggested for the West African manatee (Lemire 1968) and has been documented for the dugong (Murray et al. 1977).

The structure of the pancreas has not been fully described, but it appears similar to that of other mammals, and at least three morphologically distinct types of endocrine cell have been found in the Islets of Langerhans (Hinkley et al. 1979).

Lomolino (1977) studied the digestive efficiency of manatees (*T. manatus*) fed water hyacinth (*Eichornia* sp.) and lettuce. Mean digestive efficiencies for hyacinth in terms of dry weight, energy, and nitrogen were 82.6, 80.0, and 78.1 percent, respectively. Similar values for lettuce were 91.4, 88.8, and 93.8 percent. However, Best (1979) reported digestibility (percentage dry matter) of *Panicum, Cabomba,* and *Pistia*

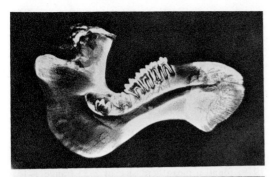

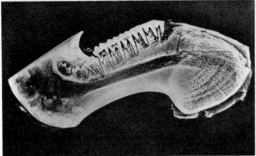

FIGURE 41.4. Xeroradiographs of the mandibles of West Indian manatee (*Trichechus manatus*) young (181-cm male, M-76-30, collected on 28 December 1978 in Dade County, Florida), top, and adult (316-cm female, M-76-7, collected on 28 March 1976 in Brevard County, Florida), bottom, showing the pattern of tooth development, crown wear, and root resorption. The young animal had not lost any teeth.

to be 44, 55, and 68 percent, respectively, in *T. inunguis*.

The manatee kidney is a multilobulated, flattened organ sandwiched between the peritoneum and the diaphragm. It is not reniculate, as in cetaceans and pinnipeds. The manatee's ability to move between the salinity extremes and apparently function perfectly well in fresh or salt water suggests that the kidneys may play a role. Manatee renal physiology has not been studied. However, they will drink fresh water and Hartman (1974, 1979) suggested that their distribution was restricted by the availability of fresh water.

The myology of the manatee has not been detailed (Murie 1872; Ronald et al. 1978), but is probably grossly similar to that of *T. inunguis* (Domning 1978).

Cardiovascular anatomy and physiology are also poorly documented. The heart is unique in that the ventricular apices are separate (Quiring and Harlan 1953). Murie (1872) and Fawcett (1942*b*) described the general vascular network and the vascular bundles, rather than a major artery and vein, that extend in to the flippers. Similar bundles (one on each side) extend posteriorly from the thoracic region, in the ventrolateral body wall, to coalesce in the pelvic region and enter the hemal canal of the posterior vertebrae. White et al. (1976) described the basic hematological parameters and Farmer et al. (1979) compared some of these parameters with their study on *T. inunguis*.

The manatee nervous system has partially been

described (Murie 1872; Quiring and Harlan 1953; Harrison and King 1965; Verhaart 1972). The brain has relatively unconvoluted cerebral hemispheres (see plates in Murie 1872). Schneyer et al. (1979) have examined the anterior pituitary at the light microscopic and ultrastructural levels. It is similar to that of other mammals, and at least six distinct cell types have been found.

Manatees have extremely small eyeballs (about 15 mm diameter) for such a large animal. Cohen, et al. (1979) found that the manatee retina contains both rods and cones. Their visual capabilities have not been studied.

The diploid chromosome number is 48 (White et al. 1976).

REPRODUCTION

The female has a bicornuate uterus, and the ovaries are located inguinally within a bursa (see Hill 1945). The adult ovaries are sheetlike (Mossman and Duke 1973: 381), often with numerous, large follicles up to 5 mm in diameter (Odell unpublished data 1976). The corpus luteum is not grossly apparent in the pregnant animal. The smallest female known to have given birth was 260 cm long. This may correspond with an age of about eight years (Odell et al. in press), although Hartman (1979) feels that sexual maturity may occur at four years. The estrous cycle has not been examined. Gestation is thought to be on the order of one year (Dekker 1977). The placenta is hemichorial and deciduate (Wiskocki 1935*b*).

The testes are large and located abdominally, posteriolateral to the kidneys. In mature males they may weigh up to 1000 g combined (Odell et al. in press). Testis weight increases rapidly at a body length of about 275 cm long (probably 8–10 years of age). Sperm have been found in animals as small as 250 cm long. There is some evidence that testicular activity may be seasonal (Odell et al. in press).

Neonates occur year round but, at least in Florida, more calves seem to be born in spring and summer (Irvine et al. in press). The newborn are black in color and rely on their flippers for locomotion more than on their tails. Average birth length is probably in the 120–140 cm range and weight about 30 kg. Calves suckle from the axillary teats of the mother. While the newborn may ingest small amounts of vegetation within a few days of birth, the age at nutritional independence is not known (Odell in press). They appear to remain with the mother for up to two years. Twinning has been reported on the basis of circumstantial evidence (Hartman 1971). The calving interval is unknown, but seems to be approximatley three to five years.

ECOLOGY

The West Indian manatee generally occurs in habitats occupied by people. This is particularly true in Florida, where inevitable conflicts arise. Manatees are restricted to near-shore tropical and subtropical areas where seagrasses or fresh-water vegetation occur. Environmental temperatures delineate the northern and southern ends of the range. Most studies concerning manatee ecology have been done in Florida and may not be an accurate representation of the situation in other parts of the range. The size of the population of *T. manatus* is not known. The Florida population is conservatively estimated to be about 1000 (Irvine and Campbell 1978) and its stability is uncertain.

Movements. In Florida some manatees move northward and westward in the summer. They have been found in Georgia, the Carolinas, Virginia, Mississippi, Louisiana, and Texas. The extent of this seasonal movement is not known, nor are seasonal movements in other parts of the range.

Manatees do move into natural and artificial warm water areas in Florida during winter (Moore 1951*b*). These movements are probably cued by decreasing air and/or water temperatures. The best-known natural warm spring is Crystal River on the gulf coast (Hartman 1971, 1979). Power plants are the major artificial source of warm water.

FOOD HABITS

Manatees eat a variety of submerged, emergent, floating, and overhanging vegetation. Marine vegetation consumed includes *Thalassia* (turtle grass), *Syringodium* (manatee grass), *Diplanthera, Halophila,* and *Halodule* (shoalgrass), and *Ruppia* (widgeongrass). Freshwater vegetation includes *Hydrilla, Eichornia* (water hyacinth), *Cabomba* (fanwort), *Alternathera, Vallisneria* (wild celery), *Elodea, Ceratophyllum* (coontail), *Myriophyllum* (watermilfoil), *Potamogeton* (pond weed), and *Najas* (naiad) (Hartman 1974, 1979). Manatees have also been known to eat the leaves and seeds of mangroves (*Rhizophora*) (Maynard 1872; Odell unpublished data 1977), as well as a great diversity of bank vegetation that is within their reach. In captivity, they will eat a variety of lettuce, cabbage, apples, bananas, carrots, and other fruits and vegetables. Manatees essentially are herbivores, but they undoubtedly ingest a variety of organisms that live on the vegetation they consume. However, there is evidence that they occasionally will eat fish (Powell 1978).

Bachman and Irvine (1979) gave the composition of manatee milk (mean values) as: 80 percent water, 20 percent solids, 13 percent lipid, 7 percent protein, 0.3 percent carbohydrate, and 1 percent ash.

BEHAVIOR

Few field studies have been done on the behavior of manatees because they are difficult to observe (see Husar 1978*c*). Hartman (1971, 1979) has made the most extensive set of field observations on manatees that congregate during the winter at the clear, warm springs in Crystal River, Florida. He described many of the routine behavior patterns exhibited by manatees. These studies are being continued by the U.S. Fish and Wildlife Service. Hartman's study was limited because the manatees only congregated during the winter and

were apparently widely dispersed during the rest of the year. Reynolds (1977, 1979*c*, in press) conducted a similar study on a semiisolated colony of manatees in a fresh-water lake system. These animals could be observed year round. The latter situation is probably close to what one might expect in other parts of the range. However, these studies were done in Florida, at the northern end of the range, and may not be representative of the species.

Manatees are perhaps best described as being "weakly social" and do not occur in large herds, with the exception of winter congregations in Florida. Odell (1979) found a mean herd size of 2.55 (range 1–15) based on 302 herds observed from the air in the Everglades National Park over a 2.5-year period. Reynolds (1977) found a mean herd size of 2.54. Both Hartman (1971, 1979) and Reynolds (1977) agree that the most stable social group is the female-calf pair. In captivity, a calf may suckle for more than three years (Odell unpublished data 1979). From the air, I have often seen groups of three animals: a small calf, a large adult (presumably the calf's mother), and an animal of intermediate size that might be an older calf of the same female. The period of social dependence could be quite long.

A commonly observed grouping is the "estrous herd" (Hartman 1971, 1979). This consists of one female in estrus and a variable number of males trying to mate with her. Presumably only the most persistent male mates.

Manatees produce a simple, squeaklike vocalization that is audible to the human ear. Most of the sound energy is below 5 kHz and there are apparently no ultrasonic components (Schevill and Watkins 1965). There is circumstantial evidence that manatees have excellent hearing capabilities (Hartman 1979). Since they often inhabit turbid waters, hearing probably plays a large role in herd cohesion, particularly of the female-calf group. Reynolds (1977) recorded a lengthy vocal interchange between a female and her calf.

Chemoreception probably plays an important role in manatee behavior (Hartman 1971, 1979). Animals often "nibble" each other's skin and anal and genital regions. Feces ingestion is not uncommon. Chemoreception may be important for males in order to locate estrous females. There is confusion in the literature as to whether or not manatees have a vomeronasal organ (S. Barrett personal communication 1979). The turbinate bones are reduced. The histological structure of the olfactory epithelium has not been examined. Taste buds have been found on the tongue (Barrett 1979).

West Indian manatees feed on bottom, midwater, and floating vegetation, and on overhanging and bank vegetation. They have split, "prehensile" upper lips covered with stiff vibrissae. The highly flexible forelimbs are often used to hold vegetation (see Hartman 1971, 1979). Manatees spend a large portion of their time feeding (6–8 hrs/day) and may consume 10–15 percent of their body weight per day. Food selection and the nutritional content of the various food items have not been examined in detail.

MORTALITY

Manatee mortality is of great concern because of the endangered status of the species. Aside from hunting throughout the range, mortality factors have been studied only in Florida (Irvine et al. in press). Large predators such as crocodiles, alligators, and sharks may take an occasional small manatee, but it is not documented. Human activities, particularly boating, are the major identifiable cause of manatee mortality in Florida (Hartman 1971) (figure 41.5). Although many manatees have scars caused by boat propellers, the incidence of mortality due to boats was not documented until recently. In 1974 a carcass salvage program was started in Florida. From April 1974 through December 1979, 337 manatee carcasses were recovered. Thirty-six percent of the deaths were attributed to human activities, as follows: boats/barges 22 percent; human structures (including flood control dams (figure 41.6) (Odell and Reynolds 1979) 8 percent; and other human causes, including drowning in nets, ingestion of fish hooks (Forrester et al. 1975), hunting/shooting, and entanglement in ropes 6.0 percent. The causes of the remainder of the deaths were undetermined, but included natural mortality.

Some manatee mortality is related to cold weather. This was suggested by earlier researchers (e.g., Cahn 1940; Krumholz 1943; Layne 1965), but documentation was not available until a comprehensive carcass salvage program was undertaken. In 1977–78, the weather was colder than in previous years and the mortality in January and February was considerably higher (Irvine et al. in press; Campbell and Irvine in press). The nature of the relationship between manatee mortality and cold weather is unclear. Speculation includes increased susceptibility to disease (e.g., pneumonia), and inability to produce enough heat metabolically to compensate for increased heat loss, resulting in death from hypothermia. In any event, the trend of increased mortality during cold weather is clear.

Parasites. Three species of endoparasite have been described: a stomach nematode (*Plicatolabia hagenbecki*) (Radhakrishnan and Bradley 1970) and two trematodes (*Chiorchis fabaceus*) from the large intestine (Fischoeder 1901) and *Opisthotrema cochleotrema* from the nasal passages (Price 1932). Forrester et al. (1979) reported the manatee as a new host for several species of microphallid trematode that occur in the small intestine. An adult cestode has been found in the duodenum of one animal from south Florida (D. Forrester personal communication 1979). Epizoites include a copepod (*Harpacticus pulex*) (Humes 1964), the diatoms *Zygnema* and *Navicula*, the remora (*Remora*) (J. Cardona personal communication 1979), and numerous other commensals living in the cracks and crevices of the skin (Husar 1978c). At least two species of algal epiphyte (*Lyngbya* and *Compsapogon*) (Hartman 1971) and a barnacle (*Chelonibia manati*) (Ross and Newman 1967; W. Newman personal communication 1979) have been found associated with the manatee. Darwin (1854) reported the barnacle *Platylepas bissexlobata* (de Blainville 1824) (= *P.*

FIGURE 41.5. A West Indian manatee (*Trichechus manatus*) killed by a boat propeller (305-cm male, M-75-1, collected on 19 January 1975 in Dade County, Florida).

FIGURE 41.6. A West Indian manatee (*Trichechus manatus*) (275-cm male, M-76-35, collected on 15 September 1978 in Dade County, Florida) killed in an automatic flood control dam in southern Florida. Impressions left by the dam gate are clearly visible. (See Odell and Reynolds 1979 for details.)

hexastylos [Fabricius 1798]) on the manatee from "Honduras" (H. R. Spivey personal communication 1979). Stubbings (1965) reported *P. hexastylos, C. manati*, and *Balanus trigonus* from *T. senegalensis* from Senegal.

Diseases. Naturally occurring diseases in the manatee have not been studied, primarily because fresh carcasses are difficult to obtain. Osteomyelitis, resulting from a harpoon wound, was documented in *T. inunguis* (Frye and Herald 1969). Manatees (*T. manatus*) have been found in a bloated condition, unable to submerge (Hartman 1971). Death in one case was attributed to an intestinal infection (J. R. White personal communication 1978). Pneumonia has been implicated in several deaths, particularly in the winter (Husar 1978*c*). A skin condition of unknown etiology, possibly a fungus, has been found on three manatees in south Florida (J. R. White personal communication 1978; E. Asper personal communication 1980). Most certainly a number of diseases will be identified as fresh carcasses become available.

AGE DETERMINATION

A quantitative technique for determining the age of manatees has not been found. Manatees lack tusks used for age determination of dugongs (Mitchell 1973, 1976, 1978), and the molar teeth are not permanent. Growth layers have been found in the ribs, but the early (oldest) layers are probably obliterated by bone mineral recycling (Odell 1977). Growth layers have been found in the ramus (jaw) and may be the preferred aging technique (A. Myrick personal communication 1979). Epiphyseal closure in the flipper bones, eye lens weight, and aspartic acid racemization studies have been proposed. A crude alternative is the correlation of body length and age using known age animals. However, only one known-age animal is in captivity in the United States (see Odell in press), and it may not grow at the same rate as free-ranging animals. Also, individual variation limits the precision of this method. Manatees have reached over 25 years of age in captivity and probably reach 40–50 years in the wild.

ECONOMIC STATUS AND MANAGEMENT

The manatee cannot be considered a game or pest animal at the present time because of its endangered status, but hunting for meat, bone, hides, and fat has caused severe population reduction (Bertram and Bertram 1973; Peterson 1974). Although the manatee is protected by law throughout most of its range (Husar 1978*c*), it is impossible to prevent all poaching. Even in Florida, where the manatee has been protected since 1893, poaching still occurs (Irvine et al. in press). Florida legislation enacted in 1978 created regulations for boating activities in areas of manatee congregations during the winter months. In the United States the manatee is also protected by the Marine Mammal Protection Act of 1972 and the Endangered Species Act of 1973 (U.S. Fish and Wildlife Service 1978).

Perhaps the greatest potential economic use of manatees is as a biological agent for the clearance of aquatic weeds (Vietmeyer 1976). The effectiveness of their voracious appetites has been demonstrated in Guyana (Allsopp 1960) and in Florida (Sguros 1966). The Florida studies were terminated because some of the manatees were killed by vandals and others apparently died from pneumonia during cold weather. Before manatees can be seriously considered for extensive aquatic weed clearance, their biology must be understood and the species must be removed from the endangered list. In time, it may be possible to match the food intake of a group of manatees with the vegetation growth rate in a particular lake, river, or canal. If the group reproduces, the offspring could be removed for use in other areas. Assuming that manatee physiology has been taken into consideration, human activities (e.g., vandalism) would be the most serious threat to the weed clearance proposals. It has also been suggested that manatees could be produced commercially as a source of meat (Guyana National Research Science Council 1974). Their endangered status attests to the fact that they are good to eat (Bertram and Bertram 1973). Presently, this plan is purely speculative.

CURRENT RESEARCH AND MANAGEMENT NEEDS

Before effective management, conservation, and recovery plans can be implemented, more must be known about the biology of the manatee. Great progress has been made in identifying major mortality factors. Power plants and other industrial activities have created artificial warm springs where manatees congregate in the winter (Moore 1951*b*; Hartman 1974). These artificial springs may keep manatees north of their historical winter range, thus exposing them to weather conditions more severe than they would normally encounter and resulting in an increased mortality rate (Campbell and Irvine in press). What would happen to manatees that were using a heated effluent if the discharge suddenly stopped during a period of severe cold weather? This question cannot be answered until the manatee's metabolic physiology is better understood. Certainly manatees are attracted to warm water, both natural and artificial, and one might expect to find increased mortality in these areas because marginal or sick animals, as well as normal animals, would seek the warm water. These congregations would also seem to increase the chances of manatees being killed by boats, although this has not been documented. Studies are under way to eliminate mortality caused by flood control dams (Odell and Reynolds 1979) and boat speeds have been reduced by Florida law in areas where manatees congregate in winter. The questions of age at sexual maturity, gestation period length, calving interval, and longevity remain unanswered. Refuges cannot properly be established until migratory patterns and food habits are known and thermoregulatory and osmoregulatory physiology have been studied. Finding survey techniques for monitoring changes in the population is complicated by the fact that manatees can remain submerged for extended periods over 20 minutes (Reynolds 1977), and often inhabit turbid waters. That manatees inhabit waters under the jurisdiction of many countries surely complicates the matter.

However, management plans for certain critical areas are required now, before all the biological data are in. These plans should be concerned with the possible effects of rapid coastal development (e.g., Crystal River) and with the resulting increases in dredging, boats, and people.

U.S. Fish and Wildlife Service contracts DI FSW-14-16-0008-930, DI FSW-14-16-0009-78-030, and DI FSW-14-16-0009-79-935 have provided funding for this chapter.

LITERATURE CITED

Allsopp, W. H. L. 1960. The manatee: ecology and use for weed control. Nature 188:762.

Bachman, K. C., and Irvine, A. B. 1979. Composition of milk from the Florida manatee, *Trichechus manatus latirostris*. Comp. Biochem. Physiol. 62A:873–878.

Barrett, S. 1979. Taste receptors in the West Indian manatee, *Trichechus manatus*. Abstr. 3rd Biennial Conf. Bio. Marine Mammals. Seattle, Wash., 7–11 October 1979.

Beck, C. A.; Bonde, R. K.; and Odell, D. K. In press. Manatee mortality in Florida during 1978. *In* R. L. Brownell, Jr., and K. Ralls, eds. Proc. West Indian Manatee Workshop.

Bertram, G. C. L., and Bertram, C. K. R. 1973. The modern Sirenia: their distribution and status. Bio. J. Linn. Soc. 5:297–338.

Best, R. 1979. Food and feeding habits of wild and captive Sirenia. Projeto Peixe-Boi, I.N.P.A., Manaus, Brazil. 78pp. Mimeogr.

Bullock, T. H.; Domning, D. P.; and Best, R. C. 1977. Hearing in a manatee (Sirenia: *Trichechus inunguis*). Page 72 *in* Proc. (abstr.) 2nd Conf. Bio. Marine Mammals.

———. 1980. Evoked potentials demonstrate hearing in a manatee (*Trichechus inunguis*). J. Mammal. 61:130–133.

Cahn, A. R. 1940. Manatees and the Florida freeze. J. Mammal. 21:222–223.

Campbell, H. W., and Irvine, A. B. In press. Manatee mortality during the unusually cold winter of 1977–1978. *In* R. L. Brownell, Jr., and K. Ralls, eds. Proc West Indian Manatee Workshop.

Cohen, J. L.; Tucker, G. S.; and Odell, D. K. 1979. Structure and function of the retina of the West Indian manatee, *Trichechus manatus*. Abstr. 3rd Biennial Conf. Bio. Marine Mammals. Seattle, Wash., 7–11 October 1979.

Darwin, C. 1854. Monograph on the subclass Cirripedia. Vol. 2. R. Soc. London Publ. 25. 684pp.

Dekker, D. 1977. Zeekoegeborte. Artis 23(4):111–119.

Domning, D. P. 1978. The myology of the Amazonian manatee, *Trichechus inunguis* (Natterer) (Mammalia: Sirenia). Acta Amazonica 8, suppl. 1, 81pp.

Domning, D. P., and Magor, D. 1977. Taxa de substituição horizontal de dentes no peixe-boi. Acta Amazonica 7:435–438.

Elsner, R. 1969. Cardiovascular adjustments to diving. Pages 117–145 *in* H. T. Andersen, ed. The biology of marine mammals. Academic Press, New York. 511pp.

Farmer, M.; Weber, R. E.; Bonaventura, J.; Best, R. C.; and Domning, D. P. 1979. Functional properties of hemoglobin and whole blood in an aquatic mammal, the Amazonian manatee (*Trichechus inunguis*). Comp. Biochem. Physiol. 62A:231–238.

Fawcett, D. W. 1942a. The amedullary bones of the Florida manatee (*Trichechus latirostris*). Am. J. Anat. 71:271–309.

———. 1942b. A comparative study of blood-vascular bundles in the Florida manatee (*Trichechus latirostris*) and in certain cetaceans and edentates. J. Morph. 71:105–133.

Fischoeder, F. 1901. Die Paramphistomiden der Saugethiere. Zool. Anz. 24:367–375.

Forrester, D. J.; Black, D. J.; Odell, D. K.; Reynolds, J. E.; Beck, C. A.; and Bonde, R. K. 1979. Parasites of manatees (*Trichechus manatus*) in Florida. Abstr., Proc. 10th Annu. Conf. and Workshop, Int. Assoc. Aquatic Anim. Med., St. Augustine, Fla., April.

Forrester, D. J.; White, F. H.; Woodard, J. C.; and Thompson, N. P. 1975. Intussception in a Florida manatee. J. Wildl. Dis. 11:566–568.

Frye, F., and Herald, E. S. 1969. Osteomyelitis in a manatee. J. Am. Vet. Med. Assoc. 155:1073–1076.

Gunter, G. 1941. Occurrence of the manatee in the United States, with records from Texas. J. Mammal. 22:60–64.

———. 1954. Mammals of the Gulf of Mexico. Pages 543–551 *in* P. Galtsoff, ed. Gulf of Mexico, its origin, waters and marine life. Fish Bull. 55:1–604.

Guyana National Science Research Council. 1974. An international centre for manatee research. Report of a workshop held 7–13 February 1974, Georgetown, Guyana. Nat. Sci. Res. Counc. Guyana (Georgetown). 34pp.

Harlan, R. 1824. On a species of lamantin (*Manatus latirostris* n.s.) resembling the *Manatus senegalensis* (Cuvier) inhabiting the east coast of Florida. J. Philadelphia Acad. Nat. Sci. 3:390–394.

Harrison, R. J. 1969. Endocrine organs: hypophysis, thyroid, and adrenal. Pages 349–390 *in* H. T. Andersen, (ed.) The biology of marine mammals. Academic Press, New York. 511pp.

Harrison, R. J., and King, J. E. 1965. Marine mammals. Hutchinson & Co., London. 192pp.

Hartman, D. S. 1971. Behavior and ecology of the Florida manatee, *Trichechus manatus latirostris* (Harlan), at Crystal River, Citrus County. Ph.D. Dissertation. Cornell Univ., Ithaca, N. Y. 285pp.

———. 1974. Distribution, status and conservation of the manatee in the United States. U.S. Fish Wildl. Serv., Natl. Fish Wildl. Lab. Rep., Contract 14-16-0008-748. 246pp.

———. 1979. Ecology and behavior of the manatee (*Trichechus manatus*) in Florida. Am. Soc. Mammal. Spec. Publ. 5. 153pp.

Hatt, R. T. 1934. A manatee collected by the American museum Congo expedition, with observations on the recent manatees. Bull. Mus. Nat. Hist. 66:533–566.

Hill, W. C. O. 1945. Notes on the dissection of two dugongs. J. Mammal. 26(2):153–175.

Hinkley, R. E.; Schneyer, A.; Reynolds, J.; and Odell, D. K.; 1979. Structural organization of the pancreas of the West Indian manatee. Abstr., 3rd Biennial Conf. Bio. Marine Mammals. Seattle, Washington, 7–11 October 1979.

Humes, A. G. 1964. *Harpacticus pulex*, a new species of copepod from the skin of a porpoise and a manatee in Florida. Bull. Marine Sci. Gulf Caribbean 14:517–528.

Husar, S. L. 1977a. *Trichechus inunguis*. Mammal. Species 72: 1–4.

———. 1977b. The West Indian manatee (*Trichechus manatus*). U.S. Fish Wildl. Serv., Wildl. Res. Rep. 7. 22pp. Washington, D.C.

———. 1978a. *Dugong dugon*. Mammal. Species 88: 1–7, 4 figs.

———. 1978b. *Trichechus senegalensis*. Mammal. Species 89: 1–3, 3 figs.

———. 1978c. *Trichechus manatus*. Mammal. Species 93: 1–5, 3 figs.

Irvine, A. B. 1979. The possible influence of metabolic rate on the winter distribution of the West Indian manatee, *Trichechus manatus*, in Florida. Abstr. 3rd Biennial

Conf. Bio. Marine Mammals. Seattle, Wash., 7-11 October 1979.

Irvine, A. B., and Campbell, H. W. 1978. Aerial census of the West Indian manatee, *Trichechus manatus,* in the southeastern United States. J. Mammal. 59:613-617.

Irvine, A. B.; Odell, D. K.; and Campbell, H. W. In press. Manatee mortality in the southeastern United States from 1974 to 1977. *In* R. L. Brownell, Jr., and K. Ralls, eds. Proc. West Indian Manatee Workshop.

Irving, L. 1969. Temperature regulation in marine mammals. Pages 147-174 *in* H. T. Andersen, ed. The biology of marine mammals. Academic Press, New York. 511pp.

Jones, J. K., Jr., and Johnson, R. R. 1967. Sirenians. Pages 366-373 *in* S. Anderson and J. K. Jones, eds. Recent mammals of the world: a synopsis of families. Ronald Press Co., New York. 453pp.

Kaiser, H. E. 1974. Morphology of the Sirenia: A macroscopic and x-ray atlas of the osteology of recent species. S. Karger, Basel. 76pp.

Krumholz, L. A. 1943. Notes on manatees in Florida waters. J. Mammal. 24:272-273.

Layne, J. N. 1965. Observations on marine mammals in Florida waters. Bull. Florida State Mus. Bio. Sci. 9:131-181.

Lemire, M. 1968. Particularités de l'estomac du lamantin *Trichechus manatus* Link (Sireniens, Trichechides). Mammalia 32:475-524.

Lomolino, M. V. 1977. The ecological role of the Florida manatee (*Trichechus manatus latirostris*) in water hyacinth-dominated ecosystems. M.S. Thesis. Univ. of Florida, Gainesville. 169pp.

March, H.; Spain, A. V.; and Heinsohn, G. E. 1978. Physiology of the dugong. Comp. Biochem. Physiol. 61A:159-168.

Maynard, C. J. 1872. Catalogue of the mammals of Florida, with notes on their habits, distribution, etc. Bull. Essex Inst. 4:135-150.

Miller, W. A.; Sanson, G. D.; and Odell, D. K. 1980. Molar progression in the manatee (*Trichechus manatus*). Anat. Rec. 196:128A.

Mitchell, J. 1973. Determination of relative age in the dugong *Dugong dugon* (Müller) from a study of skulls and teeth. Zool. J. Linn. Soc. 53:1-23.

———. 1976. Age determination in the dugong, *Dugong dugon* (Müller). Bio. Conserv. 9:25-28.

———. 1978. Age growth layers in the dentine of dugong incisors (*Dugong dugon* [Müller]) and their application to age determination. Zool. J. Linn. Soc. 62:317-348.

Moore, J. C. 1951*a*. The status of the manatee in the Everglades National Park, with notes on its natural history. J. Mammal. 32:22-36.

———. 1951*b*. The range of the Florida manatee. Q. J. Florida Acad. Sci. 14(1):1-19.

Mossman, H. W., and Duke, K. L. 1973. Comparative morphology of the mammalian ovary. Univ. Wisconsin Press, Madison. 461pp.

Murie, J. 1872. On the form and structure of the manatee (*Manatus americanus*). Trans. Zool. Soc. London 8:127-202, pls. 17-26.

Murray, R. M.; March, H.; Heinsohn, G. E.; and Spain, A. V. 1977. The role of the mid-gut caecum and the large intestine in the digestion of sea grasses by the dugong (Mammalia: Sirenia). Comp. Biochem. Physiol. 56A: 7-10.

Odell, D. K. 1977. Age determination and biology of the manatee. U.S. Fish Wildl. Serv., Natl. Fish Wildl. Lab. Rep., Contract 14-16-0008-930. 124pp.

——— 1979. Distibution and abundance of marine mammals in the waters of the Everglades National Park. Pages 673-678 *in* R. M. Linn, ed. Proc. 1st Conf. Sci. Res. Natl. Parks, U.S. Dept. Inter., Natl. Park Serv., Trans. Proc. Ser. 5. 1325pp.

———. In press. Growth of a West Indian manatee (*Trichechus manatus*) born in captivity. *In* R. L. Brownell, Jr., and K. Ralls, eds. Proc. West Indian Manatee Workshop.

Odell, D. K.; Forrester, D.; and Asper, E. D. In press. Preliminary analysis of organ weights and sexual maturity in the West Indian manatee, *Trichechus manatus. In* R. L. Brownell, Jr., and K. Ralls, eds. Proc. West Indian Manatee Workshop.

Odell, D. K., and Reynolds, J. E. 1979. Observations on manatee mortality in south Florida. J. Wildl. Manage. 43:572-577.

Odell, D. K.; Reynolds, J. E.; and Waugh, G. 1978. New records of the West Indian manatee (*Trichechus manatus*) from the Bahama Islands. Bio. Conserv. 14:289-293.

Peterson, S. L. 1974. Man's relationship with the Florida manatee, *Trichechus manatus latirostris* (Harlan): an historical perspective. M. A. Thesis. Univ. Michigan, Ann Arbor. 78pp.

Powell, J. A., Jr. 1978. Evidence of carnivory in manatees (*Trichechus manatus*). J. Mammal. 59:442.

Price, E. W. 1932. The trematode parasites of marine mammals. Proc. U.S. Natl. Mus. 81. 68pp.

Quiring, D. P., and Harlan, C. F. 1953. On the anatomy of the manatee. J. Mammal. 34:193-203.

Radhakrishnan, C. V., and Bradley, R. E. 1970. Some helminths from animals at Busch Gardens Zoological Park. ASB Bull. 17:58. Abstr.

Reynolds, J. E. 1977. Aspects of the social behavior and herd structure of a semiisolated colony of Florida manatees (*Trichechus manatus*). M.S. Thesis. Univ. Miami, Coral Gables, Fla. 113pp.

———. 1979*a*. Functional morphology of the gastrointestinal tract of the West Indian manatee, *Trichechus manatus.* Poster presented at 3rd Conf. Bio. Marine Mammals. Seattle, Wash., 7-11 October 1979.

———. 1979*b*. Internal and external morphology of the manatee (sea cow). Anat. Rec. 193:663.

———. 1979*c*. The semisocial manatee. Nat. Hist. 88:44-52.

———. 1980. Aspects of the structural and functional anatomy of the gastrointestinal tract of the West Indian manatee, *Trichechus manatus.* Ph.D. Dissertation. Univ. Miami, Coral Gables, Fla. 107pp.

———. In press. Manatees of Blue Lagoon Lake, Miami, Florida: biology and effects of man's activities. *In* R. L. Brownell, Jr., and K. Ralls, eds. Proc. West Indian Manatee Workshop.

Ronald, K.; Selley, L. J.; and Amoroso, E. C. 1978. Biological synopsis of the manatee. IDRC-TS13e. Int. Development Res. Centre, Ottawa. 112pp.

Ross, A., and Newman, W. A. 1967. Eocene Balanidae of Florida, including a new genus and species with a unique plan of ''turtle-barnacle'' organization. Am. Mus. Novit. 2288. 21pp.

Schevill, W. E., and Watkins, W. A. 1965. Underwater calls of *Trichechus* (Manatee). Nature 205:373-374.

Schneyer, A.; Hinkley, R.; and Odell, D. 1979. Structural organization of the anterior pituitary gland of the West Indian manatees. Abstr. 3rd Biennial Conf. Bio. Marine Mammals, Seattle, Wash., 7-11 October 1979.

Scholander, P. F., and Irving, L. 1941. Experimental investigations on the respiration and diving of the Florida manatee. J. Cell. Comp. Physiol. 17:169-191.

Sguros, P. L. 1966. Research report and extension proposal

submitted to the Central and Southern Florida Flood Control Board on the use of the Florida manatee as an agent for suppression of aquatic and bankweed growth in essential inland waterways. Dept. Bio. Sci., Florida Atlantic Univ., Boca Raton, Fla. 57pp.

Stubbings, H. G. 1965. West African Cirripedia in the collections of the Institute Français d'Afrique Noire, Dakar, Senegal. Bull. I.F.A.N. 27:876–907.

U.S. Fish and Wildlife Service. 1978. Administration of the Marine Mammal Protection Act of 1972, June 22, 1977 to March 31, 1978. U.S. Fish Wildl. Serv., Washington, D.C. 80pp.

Verhaart, W. J. C. 1972. The brain of the seacow *Trichechus*. Psychiat. Neurol. Neurochir. (Amst.) 75:271–292.

Vietmeyer, N., ed. 1976. Making aquatic weeds useful: some perspectives for developing countries. Natl. Acad. of Sci., Washington, D.C. 175pp.

White, J. R.; Harkness, D. R.; Isaaks, R. E.; and Duffield, D. A. 1976. Some studies on blood of the Florida manatee, *Trichechus manatus latirostris*. Comp. Biochem. Physiol. 55A:413–417.

Wislocki, G. B. 1935a. The lungs of the manatee (*Trichechus manatus*) compared with those of other aquatic mammals. Bio. Bull. 68:385–396.

———. 1935b. The placentation of the manatee (*Trichechus latirostris*). Mem. Mus. Comp. Zool. Harvard 54:159–178, 7 pls.

DANIEL K. ODELL, Division of Biology and Living Resources, Rosenstiel School of Marine and Atmospheric Science, University of Miami, 4600 Rickenbacker Causeway, Miami, Florida 33149.

VII

Artiodactyla

42

Collared Peccary

Dicotyles tajacu

John A. Bissonette

NOMENCLATURE

COMMON NAMES. Collared peccary, javelina

SCIENTIFIC NAME. *Dicotyles tajacu*

SUBSPECIES. *D. t. angulatus, D. t. bangsi, D. t. crassus, D. t. crusnigrum, D. t. humeralis, D. t. nanus, D. t. nelsoni, D. t. nigrescens, D. t. sonoriensis,* and *D. t. yucatanensis.*

Subspecies designations are from Hall and Kelson (1959) and apply to North and Central American populations only. The subspecies *D. t. angulatus* occurs in Texas and possibly southeastern New Mexico, and *D. t. sonoriensis* in Arizona and southwestern New Mexico (Mearns 1907; Miller and Kellogg 1955). The collared, the white-lipped (*Tayassu pecari*), and the newly rediscovered Chacoan peccary, provisionally considered to be conspecific with *Catagonus wagneri* by Wetzel et al. (1975) and also called the tagua or pagua peccary, comprise the only living species in the family Tayassuidae (Wetzel 1977*a*, 1977*b*). Only *Dicotyles* is found in North America; all three species coexist sympatrically in the Chaco region of Paraguay (Wetzel and Lovett 1974; Wetzel et al. 1975).

DISTRIBUTION

Collared peccaries are found in a wide range of habitats from Argentina northward to Texas, New Mexico, and Arizona (Sowls 1966, 1978). They inhabit the southern part of Texas from the Gulf Coast county of Refugio, northwest to southern Nolan County, southwest to northern Ward County, northwest to the New Mexico border in central Loving County, west along the border to western Culberson County, and south to the Mexican border in southeastern Hudspeth County (figure 42.1). Peccary distributions are disjunct around the larger metropolitan areas. Total area occupied in Texas equals roughly 227,369 km².

In New Mexico, peccaries occur mainly in the southwestern counties. Their current range encompasses Hildago County from the Mexican border north through western Grant County on a line from the town of Continental, and east of Lordsburg to northeast of Silver City, then directly west to Cliff and northwest

FIGURE 42.1. Distribution of the collared peccary (*Dicotyles tajacu*) in the continental United States.

along Route 180 to south of Luna in Catron County (Donaldson personal communication).

There are isolated populations of peccaries in the Cedar, Tres Hermanas, and Florida mountains in south-central New Mexico, and they probably occur in southeastern New Mexico near the Texas border. Attempts in the mid-1960s to reestablish peccaries on former range in the Guadalupe Mountains west of Carlsbad and in the San Andres Mountains northeast of Las Cruces failed (Donaldson personal communication). Presently, about 13,768 km² in the state are occupied by peccaries.

Arizona populations are found primarily in the southeastern quarter of the state, and are generally contiguous with those in New Mexico. Peccaries occur from Alpine, Arizona, in southern Apache County northwest along the Apache Stigreaves National Forest boundary on the Mogollon Rim in Navajo County to Route 87, then west-southwest in Coconino County to just north of Strawberry, northwest to Page Springs in

Yavapai County, and north across the Coconino-Yavapai County line to the southern tip of the Kaibab National Forest. Peccaries are distributed south to Maricopa County and Phoenix and then east around the metropolitan area, where their populations are disjunct. From Phoenix, peccaries are found west to Gila Bend, and then south and west through the Cabeza Prieta Game Range in Yuma County to the Mexican border. They occur throughout most of Pima, Santa Cruz, Cochise, Greenlee, Graham, Pinal, and Gila counties in southeastern Arizona. Five isolated pockets of javelina occur in western Arizona: four in Yuma County near Peach Springs, Wikieup, and the Harcuvar and Tank mountains, and the fifth in Yavapai County between Wagner, Constellation, and Crown King. Total range occupied equals about 110,427 km².

Peccaries in the northernmost parts of their range in all three states, but especially in the mountains of west Texas, New Mexico, and Arizona, are subject to occasional periods of snow and very cold weather and probably undergo repeated local extinction and repopulation. No work has been conducted to document the dynamics of this phenomenon.

DESCRIPTION

Peccaries are small ungulates with short legs and a typical piglike snout. The tail is about 30 mm long (Sowls 1966) and barely noticeable except upon close inspection. Total length of adult animals is approximately 86 to 96 cm. Height at the shoulders ranges from 50 to 60 cm. The front feet have four digits, the rear have three. Weights of adult animals range from 13 to 27 kg. and depend largely upon nutrition. From the side view, the head and forequarters appear overly large for the delicate legs and small hindquarters. No sexual dimorphism in size or coloration has been reported.

Peccaries have enlarged upper and lower canines approximately 30 to 35 mm in length. Only the tips of the upper tusks protrude beyond the lips. Unlike wild suids, the tusks are not sharply curved and flared but are straight or very slightly curved. Four pairs of mammae are apparent; only the two posterior pairs appear to be functional (Neal 1959; Sowls 1966). However, Sowls (1965) found that the second anterior pair of mammae secreted some milk through galactophores (milk ducts). Peccaries have a subcutaneous scent gland, described by Epling (1956), located on the dorsal midline 20 to 25 cm anterior to the tail.

Most adults have knee calluses, perhaps because of bedding postures and kneeling while eating. Torn noses and lacerated ears are common.

In adults, pelage is light grey or brown to almost black with a salt and pepper effect caused by distinctive banding of individual bristles. When the dark-colored dorsal bristles are erect, a white stripe formed by the coordination of the white bands of individual bristles is apparent (Schweinsburg 1969). A light-colored collar, for which the animals are named, is variable in shade and pattern. Occasionally, reddish individuals occur.

Individual bristles are bicolored with white or beige, and black or dark brown bands (Wetzel 1977*a*, 1977*b*). Four to six distinct bands are usual for each hair shaft. Band width is variable and appears to be related to the location of the bristle either on the collar, dorsal midline, or body. Bristles surrounding the dorsal scent gland are often stained; the white bands appear yellow. Peccaries less than two to three months old are reddish brown with a dark dorsal midline stripe and distinctive collar.

Skull and Dentition. Adult peccary skulls are roughly triangular shaped in side view (figure 42.2). The rostrum slopes upward to a prominent lambdoidal

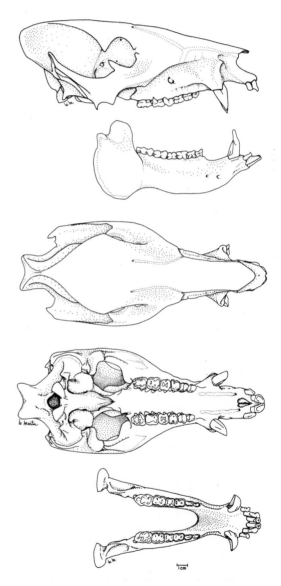

FIGURE 42.2. Skull of the collared peccary (*Dicotyles tajacu*). From top to bottom: lateral view of cranium, lateral view of mandible, dorsal view of cranium, ventral view of cranium, dorsal view of mandible.

crest. The zygomatic arches are thick and flare posteriorly. The mandible is sturdy. Herring (1974) discovered that the sequence of suture closure in peccaries differs from that in most other mammals in the early fusion of the palatal and facial sutures. She postulated that the early fusion strengthens the snout and is related to rooting and feeding activities.

A normal peccary has the following dental formula:

$$I\frac{2-2}{3-3} C\frac{1-1}{1-1} P\frac{3-3}{3-3} M\frac{3-3}{3-3} = \frac{18}{20} = 38$$

The form I 2 - 2/3 - 3 means there are 2 incisors on each side of the upper jaw and 3 on each side of the lower jaw. When specific teeth are referred to, uppercase letters will refer to teeth of the upper jaw and lower-case letters to teeth of the lower jaw. The upper third pair of incisors and the first pair of premolars are missing. The upper and lower canines are well developed in adults of both sexes. The posterior portion of the lower canines occludes with the anterior portion of the uppers, resulting in flat contact surfaces with sharp edges. In aged animals, canines are well worn, often less than 15 mm long. Molars are bunodont and have changed little from ancestral times, in contrast to suid molars, which have changed substantially from their original pattern (Herring 1972). Diastemas are present between the second upper incisors and the canines, and between the upper and lower canines and anterior premolars. Herring (1972) compared the functional anatomy of suid and tayassuid skulls and dentition and discussed the role of canine morphology in suid-tayassuid evolution. She postulated that peccary canines evolved for use mainly as weapons. However, Woodburne (1968) found slight canine dimorphism, possibly indicating an additional function in competitive interactions between males.

PHYSIOLOGY

Peccaries are well adapted to the arid environments of the desert and can survive for periods of up to six days without any water. Turnover rates for peccaries with access to water averaged 1.58 l of water per day, while dehydrated peccaries averaged 0.55 l per day, indicating adaptive physiological mechanisms for conserving fluids under stress. Free-ranging peccaries lost 1.35 l of water per day during the summer and 1.17 l in winter (Zervanos and Day 1977). Zervanos and Hadley (1973) demonstrated that respiratory evaporation was the main avenue of water loss. During dehydration, peccaries reduced evaporative water loss by 68 percent and urinary water loss by 93 percent. They reported that peccaries were unable to produce dry feces. I have noticed that dry feces were produced during the fall and winter. Consumption of different food resources may account for observed differences in feces; whether a water-saving mechanism is involved is uncertain, but this would be relatively easy to test in the laboratory.

Peccaries seem capable of maintaining full hydration in desert environments without free water. Minnamon (1962) reported that absence of water did not alter home range movements or activities. Prickly pear (*Opuntia* sp.) cladophylls appear to supply most water requirements. Zervanos and Day (1977) reported that an 18.2-kg peccary required approximately 1.5 kg wet weight (0.3 kg dry weight) of opuntia (78 percent water by weight) per day to maintain water balance. In addition, Zervanos (1972) and Zervanos and Hadley (1973) reported skin temperatures of free-ranging peccaries were labile and ranged from 37.5 to 40.9° C during all seasons. Skin temperatures were always above ambient temperatures (Zervanos and Day 1977).

REPRODUCTION

Courtship. The dominant, or alpha, male copulates with most of the estrous females in the herd. Bissonette (1976, in press) reported that the alpha male was involved in five of six successful intromissions observed in free-ranging peccaries. On only one occasion was a subordinate male successful. Subordinates may be successful in copulating only when two or more females are in estrus synchronously and perhaps only when ovulation is synchronized. However, data are incomplete and more research is needed on this aspect of peccary biology. During the breeding season, an adult male peccary forms a short tending bond with the estrous female that lasts from a few hours to several days (Low 1970; Bissonette 1976). During this time the male remains within a few meters and does not allow other males near the female; however, other females may approach. Subordinate males do not leave the group and do not form separate male groups. Sowls (1966) and McCulloch (1955) reported that conflict among males was not noticeably greater when receptive females were present than at other times. Although no clear yearly pattern of interaction rates is apparent, there is a tendency for more interactions to occur just before and during parturition in early summer, and during the breeding season in late fall and winter. Highly aggressive interactions such as fights were typically observed in west Texas peccaries only when a female was in estrus (Bissonette 1976, in press).

Parturition. Collared peccaries evolved in South America (Woodburne 1968) in what were probably mesic and relatively predictable environments. Breeding and parturition may have occurred without significant peaks throughout the year in these climates. In the unpredictable and arid climate of the American southwest, although peccaries may breed throughout the year, strong seasonal peaks in farrowing have been reported (Neal 1959; Sowls 1965, 1974; Schweinsburg 1969; Low 1970; Bissonette 1976). Low (1970) reported that breeding and parturition are strongly related to forage conditions; during years of good forage a greater proportion of young were born early in April or May. However, females are capable of repeated births during any one year (Sowls 1965, 1978). Number of pregnancies and parturitions is related to forage conditions within a year as well as to the success of previous parturitions that year. Females whose young are lost soon after parturition may breed again. Females are capable of continually repeating the estrous cycle if they fail to become pregnant (Sowls 1966).

In the southwest it is possible for dry and wet periods to alternate within the same year. Successful parturition can occur throughout the year, although young born during the dry periods are less likely to survive. Usually, peccaries in Texas were born from May to July (Low 1970; Bissonette 1976, in press). Most peccaries in Arizona were born in July and August, when vegetation was most abundant following the rains (Knipe 1957; Neal 1959; Sowls 1966, 1978). The gestation period is about 145 days (Low 1970; Sowls 1961; 1965). Usually two young are produced (Jennings and Harris 1953; Knipe 1957; Sowls 1966, 1978; Low 1970; Smith and Sowls 1975; Bissonette 1976), but litters may include three or four young. Halloran (1945) reported a female with five fetuses. Seldom do more than two young in the same litter survive. Smith and Sowls (1975) described the pattern of fetal development in peccaries.

Sowls (1965) reported that captive female peccaries in Arizona bred as early as 33 weeks of age, and the earliest parturition observed was in a female 54 weeks of age. Low (1970) stated the earliest observed female copulation for Texas animals occurred at 44 weeks of age; however, the earliest conception occurred at 48–49 weeks. In Texas, ovarian analysis backdated the first corpora lutea of nonpregnancy scars to 32–33 weeks. Captive males were capable of breeding at 46–47 weeks of age.

Brown et al. (1963), Sowls et al. (1961), and Sowls (1963) studied the chemical composition and physical properties of mature and colostrum milk and milk fat of collared peccaries. Comparisons with domestic sows indicated that peccary milk was lower in fat, with a mean of 4.7 percent versus a mean of 6.9 percent. Mean total protein levels were higher for peccary (5.6 percent) than for domestic sows (4.5 percent).

ECOLOGY

Activity. Activity patterns of peccary are strongly influenced by temperature (Bigler 1964, 1974; Bissonette 1978). In west Texas, the phenological periods of November–February (winter), March–April and September–October (spring and fall), and May–August (summer) were characterized by increasingly warmer temperatures; mean maximum temperatures equaled 18.8° C, 26.2° C, and 32.0° C, respectively (Bissonette 1976). During the summer months peccaries were primarily crepuscular and nocturnal in their habits (Elder 1956; Eddy 1959; Schweinsburg 1969). With cooler fall and winter temperatures the herds were active for longer periods during midday (figure 42.3). In midwinter when night temperatures approached freezing, they often huddled for warmth during at least part of the night (Zervanos 1972; Zervanos and Hadley 1973). In Texas, most activity during winter was concentrated on the open bajada, or flat desert country. Significantly less activity occurred on the bajada dur-

ing the daylight hours in summer. Spring and fall appeared to be transitional, with periods of activity and resting during both day and night.

Social Organization and Group Formation. Peccaries are territorial (Ellisor and Harwell 1969; Schweinsburg 1969; Bissonette 1976, in press) and are organized into herds that appear highly stable. Unlike in other animals, where the dominant male defends the area, more than one peccary herd member may contribute to territorial defense (Bissonette 1976). Young animals remain with the herd (Bissonette 1976). High genetic relatedness within groups may occur, but this remains to be studied. Herds are comprised of roughly equal numbers of males and females (Bissonette 1976), although Sowls (1974, 1978) reported some variation in sex ratio. At birth, females outnumber males. Sowls (1966) and Smith and Sowls (1975) in Arizona reported a male:female ratio of roughly 40:60 at parturition. Low (1970) reported a 42:58 *in utero* sex ratio for Texas populations. As peccaries mature, the sex ratio appears to approximate 50:50 (Sowls 1978).

Peccaries exhibit linear dominancy heirarchies that include males and females (figure 42.4). Separate male or female hierarchies do not appear to exist. Sowls (1974) suggested that among penned peccaries, females were usually dominant over males in most situations. Dominance hierarchies in free-ranging peccaries were stable and closely related to the size of the animal (Bissonette 1976). Exceptions occurred during squabbles over food items. Possession of a food item seemed to confer an advantage unless there was a large disparity in size between interacting animals. Subordinates won approximately half of the encounters with more dominant animals when they were in possession of the disputed food item.

Although territorial groups were generally stable, feeding subgroups formed and remained apart for periods lasting more than two weeks (Bissonette 1976, in press; Day 1977). Subgroup membership was usually consistent but some interchange occurred. Interchange between territorial groups was infrequent, although Schweinsburg (1969, 1971) documented its occurrence.

Peccary territories were found to be roughly equivalent to home range. Little overlap between

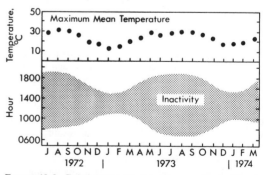

FIGURE 42.3. Relationship of ambient temperature to diurnal and crepuscular peccary activity patterns. From Bissonette 1976.

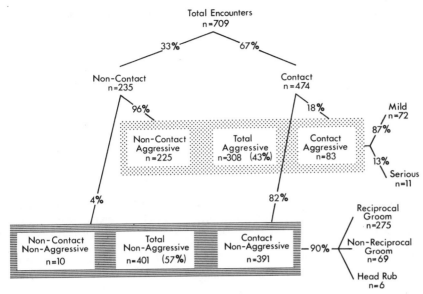

FIGURE 42.4. A typical linear dominance hierarchy for collared peccaries (*Dicotyles tajacu*). Adapted from Bissonette 1976.

groups occurred, and the areas were defended. Herds were not seen more than 185 m inside the boundary of another herd (Schweinsburg 1971). There is evidence of scent marking by peccaries (Schweinsburg 1971; Bissonette 1976).

Formation of territorial groups in peccaries is influenced by the existence of subgroups. Bissonette (1976) hypothesized that subgroups form the nucleus for new territorial groups. As subgroups spend more time apart, each group restricts its movements to a portion of the territory, and range extensions into adjacent territories become evident. Interactions between subgroups are characterized by increasing aggression. The decisive steps for territorial group formation occur when subgroups do not rejoin during the breeding season, in effect creating an additional breeding alpha male. Bissonette (1976, in press) documented the components of territorial group formation.

Herd Dynamics. It is important to distinguish between territorial groups and subgroups when discussing peccary herd dynamics. Territorial groups often split into two or more subgroups. Day (personal communication) has termed this phenomenon *fragmentation*. Although territorial group size changes with mortality and recruitment, little successful interchange appears to occur between territorial groups (but see Schweinsburg 1971). However, subgroups change in size and composition throughout the year. These changes may be influenced by female behavior, habitat, or season of the year and its related temperature.

West Texas peccaries breed primarily in late November to mid-January. Males no not herd females, yet subgroups tend to coalesce into the larger territorial groups during this time. Small subgroups are seen less frequently. It may be that females are influencing herd size by selecting the alpha male with which they will breed.

Habitat may also influence group size. Mean peccary subgroup size on the bajada in Texas was significantly larger (13) than that observed in dense vegetation (10). Since I followed peccaries after they left heavy cover, accurate counts were obtained.

Season of the year and its related temperature regime appeared to influence peccary group dynamics. Peccaries used more open areas during the cooler periods and subgroups were larger ($\bar{X} = 14.5$), while more dense, shaded areas were used during the summer when mean group sizes were smaller ($\bar{X} = 9$). In contrast, mean territorial herd size for west Texas peccaries over a two-year period was 17.5 animals.

Herd sizes in Texas averaged 15.8 animals, while Schweinsburg (1971) reported 8 to 9.5 animals per herd in southern Arizona. Day (personal communication) estimated a mean of 7.8 (range 3–14) animals in 10 unhunted herds versus a mean of 14.0 (range 6–25) animals from 14 hunted herds at 3-Bar Wildlife Area, Arizona. The combined average for the 24 herds was 11.4 peccaries. Better habitat existed on the hunted range.

Home Range. Ellisor and Harwell (1969) reported that home range sizes in Texas ranged from 72.9 to 255.4 ha ($\bar{X} = 153.4$ ha) but gave no population density estimates. Bissonette's (1976) estimates for west Texas ranged from 201 to 245 ha ($\bar{X} = 216.2$ ha). Minnamon (1962) reported home ranges of peccary herds in the Tucson Mountains did not exceed 390 ha. Home ranges in the nearby Tortolita Mountains measured 267–808 ha (Bigler 1964).

Measurements of home range sizes for individual peccaries are seldom given in the literature; instead, home ranges for groups are reported. These animals

TABLE 42.1. Summary of representative herd densities for peccary (*Dicotyles tajacu*) populations in the southwestern United States

Density (animals/km²)	Location	Source
2.6–3.9	Welder Wildlife Refuge, Texas	Low (1970)
3.9–10.9	King Ranch, Texas	Low (1970)
1.2	Watson Ranch, Texas	Low (1970)
2.5	Black Gap Management Area, Texas	Low (1970)
3.4–10.9	Big Bend National Park, Texas	Bissonette (1976)
1.3–19	Central Arizona	Day (1977)
5.2–19	Southern Arizona	Schweinsburg (1970)

form close-knit social groups, with both sexes occupying and using the habitat in a similar manner. Therefore, home range sizes are equivalent to the area occupied by the group. As mentioned, these areas are defended and hence group home ranges are roughly equivalent to territories.

Density. The chief paradigm of density dependency, namely, that animal density is positively correlated with food resources or other life requisites, logically leads to predictive statements about the factors controlling animal numbers. If close relationships exist, assessments of resource abundance should give a close estimation of the density of the population. Certain species, because of their life history characteristics, are better suited to investigations of this kind. Territorial herbivores with clearly defined food preferences are easily studied.

Peccary densities vary by habitat (table 42.1). Densities in west Texas and Arizona were correlated ($r^2 = 0.95$) with percentage abundance of available forage (Texas: prickly pear, *Agave lechuguilla,* and forbs; Arizona: prickly pear and forbs) on their territories and in Texas negatively correlated ($r^2 = 0.91$) with percentage woody cover (Bissonette 1976, in press). These correlations indicate how closely peccary numbers at carrying capacity are adjusted to the environment. Low (1970) suggested that predation and parasitism are probably of minor importance in regulating populations. These effects are density dependent, having proportionately greater influence at high densities. According to Low, drought, with its concomitant effects on plant quality and availability, appeared to be the ultimate factor controlling javelina populations in Texas. Because abundance of prickly pear appears little affected by varying rainfall from year to year, peccaries use it heavily when other more nutritious foods are not available. Preference for opuntia appears to be a matter of its availability, not high nutrient value. Availability of forbs and highly nutritious seeds, fruits, and nuts probably are most important in maintaining a positive energy balance for successful reproduction.

Harvest. During the 1978–79 season, 27,448 javelina were harvested legally in the United States (table 42.2). Texas accounted for the largest number of animals, with 22,647 peccaries killed from 1 October 1978 to 1 January 1979. Ninety-seven javelina were harvested in New Mexico during the December 1978 season, while Arizona accounted for the balance of 4,744 javelina during the January 1979 archery season and the March 1979 gun season. In total, an excess of 76,000 hunters went afield hunting for collared peccaries. Approximately 36 percent were successful.

In Arizona, peccary harvest, not including archery kills, totaled 2,236 animals (table 42.3) in 1957, and increased to a high of 6,602 animals killed in 1970. Hunting by permit was initiated in 1972, and harvest appears to have stabilized below 5,000 animals annually (Arizona Game and Fish Department 1979). Records for bow harvest of javelina are available from 1962. Kills increased steadily from 72 animals in 1962 to a high of 1,085 in 1976. Bow harvest in 1979 was 738 peccaries (see table 42.3). From 1957 to 1979, a total of 112,428 peccaries have been legally harvested by bow and gun hunters in Arizona.

FOOD HABITS

Collared peccaries are generalist herbivores. Contrary to popular belief, little if any animal matter is taken

TABLE 42.2. Summary of peccary (*Dicotyles tajacu*) harvest statistics for the continental United States 1978–79

State	Range (km²)		Number Harvested		Number of Permits or Hunters Afield		Percentage Hunter Success
Texas	227,369	(64.7)[a]	22,647	(82.4)	54,021	(71.0)	41.9
New Mexico	13,768	(3.9)	97	(0.4)	273	(0.4)	35.5
Arizona	110,427	(31.4)	4,744	(17.2)	21,824	(28.6)	21.7
Total	351,564 (100)		27,488 (100)		76,118 (100)		$\bar{X} = 33–36$

[a] Values in parentheses are percentages of column total.

TABLE 42.3. Archery and gun harvest of peccaries (*Dicotyles tajacu*) in Arizona, 1957–79

Year	Gun Harvest	Cumulative Gun Harvest	Archery Harvest	Cumulative Archery Harvest	Total Archery and Gun Harvest	Total Cumulative Harvest
1957	2236	2236			2236	2236
1958	2511	4747			4511	4747
1959	3010	7757			3010	7757
1960	3098	10855			3098	10855
1961	4191	15046			4191	15046
1962	4343	19389	72	72	4415	19461
1963	4867	24256	145	217	5012	24473
1964	5898	30154	191	408	6089	30562
1965	5231	35385	176	584	5407	35969
1966	5267	40652	201	785	5466	41437
1967	5310	45962	163	948	5473	46910
1968	5082	51044	213	1161	5295	52205
1969	5903	56947	218	1379	6121	58326
1970	6602	63549	229	1608	6831	65157
1971	5959	69508	373	1981	6332	71489
1972	3966	73474	332	2313	4298	75787
1973	4746	78220	486	2799	5232	81019
1974	5195	83415	513	3312	5708	86727
1975	3792	88208	671	3983	4463	92191
1976	4522	92730	1085	5068	5607	97798
1977	4576	97306	840	5908	5416	103214
1978	3594	100900	876	6783	4470	107684
1979	4006	104906	738	7522	4744	112428

Total gun and archery harvest, 1957–79 = 112,428

regularly. Prickly pear, *Agave lechuguilla,* century plant, forbs, grass, and various seeds and nuts are eaten by peccaries (Jennings and Harris 1953; Eddy 1961; Low 1970; Bissonette 1976).

Peccaries fed *ad libitum* only on opuntia consumed a daily portion amounting to one-third of their body weight and were able to exist on this starvation diet for about five months (Sowls 1966). Prickly pear is at best a maintenance food resource, available to peccaries when other, more nutritious, forage is absent. High densities, as influenced by rapid growth, healthy maintenance, and successful reproduction, are probably related to the seasonal availability of protein-rich food resources, but these relationships need further study.

BEHAVIOR

Peccaries are social animals. Partially as a result of year-round group living and perhaps high genetic relatedness, they have evolved complex vocal and behavioral repertoires. Schweinsburg (1969) classed peccary behavior into broad categories including intraspecific aggressive behavior, sexual behavior, and maternal behavior, among others, and further subdivided aggressive behavior into distinct categories. Sowls (1974) described behavioral patterns for Arizona peccaries and mentioned that tooth chattering was associated with threat interactions. Bissonette (1976, in press) described 3 broad classes of vocalization, including aggressive, submissive, and alarm calls, that included 15 distinct, graded, or linked vocal patterns. In addition, 6 behavioral categories were described that

included 31 distinct, graded, or linked behavioral patterns.

The most common behavioral pattern exhibited by peccaries is reciprocal grooming (figure 42.5). The animals approach and stand alongside one another, head to tail. Each peccary rubs its head along the hind legs, rump, and scent gland of the other. In west Texas, 43 percent of all interactions involved reciprocal grooming. This interaction probably is most important in maintenance of the social bond, including intragroup recognition of territory members and their social rank. Byers (1978) noted that the preorbital glands were involved in reciprocal grooming and suggested the need for additional studies on scent marking in peccaries. Presence of preorbital glands has not been reported for Texas or Arizona peccaries.

Sowls (1974) described peccaries as contact animals. Few large mammals, especially ungulates, show

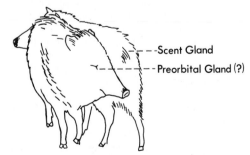

FIGURE 42.5. Reciprocated grooming by two collared peccaries (*Dicotyles tajacu*). From Bissonette 1976.

the high percentage of contact interaction exhibited by peccaries. Living in closely knit social groups the year round probably has a direct relationship on this phenomena, as does the hypothesized high genetic relationship between herd members. Bissonette (1976, in press) reported that of 31 different behavior patterns, 55 percent were characterized by contact. Of 709 recorded interactions involving these patterns, 67 percent involved contact, and of these, 82 percent were not aggressive in nature. Grooming, including reciprocal and nonreciprocal rubbing, and head rubbing were involved in 49 percent (350 of 709) of all encounters and comprised 90 percent of all contact nonaggressive interactions.

AGE DETERMINATION

Kirkpatrick (1957) and Kirkpatrick and Sowls (1962) described the pattern of tooth replacement in peccaries. Peccaries are born with all four deciduous canine teeth and the posterior pair of lower incisors. In sequence, eruption of the third pairs of upper (P3) and lower (p3) premolars, the first pair of lower incisors (i1), the second pairs of upper (P2) and lower (p2) premolars, the first pair of upper incisors (I1), the fourth upper (P4) and lower (p4) pairs of premolars, and the second upper (I2) and lower (i2) pairs of incisors complete the temporary dentition between two and three months of age. The first upper (M1) and lower (m1) permanent molars erupt at about five months of age, followed by replacement of the canines. The third pair of lower incisors (i3) then appear, and the second upper (M2) and lower (m2) pairs of molars are replaced. The remaining incisors are replaced and the permanent premolars erupt. Permanent dentition is complete at about 84 weeks of age with eruption of the third upper and lower pairs of molars (M3m3) (Kirkpatrick and Sowls 1962).

ECONOMIC STATUS AND MANAGEMENT

Javelina in Texas were not an important game animal until recently. In numerous areas across the state they were considered a menace to sheep and goats and were killed at every opportunity (Clark 1974). The Texas Parks and Wildlife Department issued an interim Fish and Wildlife Action Plan in 1972–73 suggesting javelina be promoted more intensively as a game species. As a result, studies concerning life history, ecology, and potential as a game animal were initiated on the state wildlife management areas. The 1974–75 Fish and Wildlife Plan reiterated the need to promote javelina as a game animal. Investigations of herd dynamics and javelina biology are continuing on the wildlife management areas in Texas.

No research on peccaries was in progress or planned by the New Mexico Game and Fish Department in 1979. Several releases of peccary onto former ranges in the Guadalupe Mountains west of Carlsbad and in the San Andres Mountains northeast of Las Cruces were attempted in the mid-1960s. None was successful (Donaldson personal communication). It is unclear if studies were conducted to determine why the releases were not successful. Peccary range in New Mexico is limited largely to the southwestern part of the state, and total population is probably less than 3,000 animals. Research monies presently are being used for more pressing resource management problems.

The Arizona Game and Fish Department has an active research and management program, and has been studying peccary for over two decades. Peccary in Arizona are considered an important game animal. Hunters afield since the early 1960s have exceeded 20,000 annually.

A significant management problem in Arizona is the apparently localized depletion of peccary herds due to hunting. Herds near access roads receive heavy hunting pressure. As a consequence, some herds have been reduced severely, while other, more distant, herds have received little hunting pressure. As a result, new management procedures are being implemented. Experimental pistol hunts have been instituted and are being evaluated. The reduced effectiveness of short-barreled weapons offers a possible solution for lowering the kill in some areas (Day 1977). Limiting hunters to the use of primitive weapons such as muzzle loaders and bows is another potential solution. Day (1977) suggested the use of a rest-rotation system of hunting, where certain heavily harvested hunt units are closed for a number of years to allow repopulation. He cautions that this method would be effective only if javelina herds increase rapidly in the absence of hunting. Concurrent work by the Arizona Game and Fish Department has demonstrated that peccaries do not increase rapidly in the absence of hunting.

A related problem involves herd dynamics relative to herd reduction. Do reduced herds restrict their movements to a portion of their former range? Do adjacent herds begin to occupy the vacated range, and if so, how rapidly? Day (1977) suggested that it is unlikely that severely depleted herds can regain their former numbers through recruitment, herd interchange, or invasion of unoccupied range by a neighbor herd in one year's time. If this is correct, closing heavily hunted areas may be the preferred management strategy for sustained harvest and maintenance of adequate population numbers.

Ellisor and Harwell (1979) suggest that habitat type is very important in determining the rate of recovery in hunted herds. They report that two herds, occupying habitats of much different quality, were hunted with the same pressure and that the herd on the better habitat area recovered more rapidly, with a survival rate of 1.08 young per adult female compared to 0.36 young per female in poorer habitat. In Texas, good habitat consists of heavy brush with an abundance of prickly pear (Ellisor and Harwell 1979).

Forbs are also a critical food item. Livestock grazing can have serious deleterious effects on peccary populations because competition for limited resources is most likely to occur with ephemeral, protein-rich forbs. Although forbs comprise only a small proportion of total plant biomass even in the best of years in west Texas and Arizona, they may be of inordinate value to ungulate populations, including peccaries. Not surprisingly, parturition of most ungulates generally coincides with the spring flush of green vegetation, especially forbs.

Management decisions involving habitat manipulation for competing species can have a profound effect upon javelina populations. In south Texas, land is increasingly being cleared of brush and converted to pasture for livestock. As mesquite and other brushy species are removed, habitat for javelina is likewise reduced. It is unlikely that displaced herds can maintain their numbers when their territories are reduced or destroyed.

CURRENT RESEARCH AND MANAGEMENT NEEDS

At the present time the Texas Parks and Wildlife Department is involved in studies at several wildlife management areas in the state. Restocking studies are being conducted statewide. A population dynamics study has recently been completed. The effect of hunting on peccary populations is being investigated at Chaparral Wildlife Management Area. At this writing, a mobility, home range, and habitat preference project is planned for Black Gap Wildlife Management Area, while a javelina food habit study is planned for the western Edwards Plateau.

Management needs for the state include information regarding habitat requirements, population dynamics, and aspects of life history. Winkler (personal communications) suggests that javelina may be one of the least studied game animals indigenous to Texas.

The Arizona Game and Fish Department is cooperating with the Wildlife Research Unit at the University of Arizona in Tucson on a project involving the dynamics of peccary dispersion. A study involving peccary nutrition has recently been completed at the unit. Projects conducted by the state Game Department involving trapping methods and the effects of hunting on javelina populations are continuing. Current studies involve the behavioral ontogeny of javelina, and the relationships between habitat and vegetation quality and javelina population density in Arizona. As indicated earlier in this chapter, additional research is needed on scent marking, genetic relatedness within and between groups, and other aspects of peccary behavior and ecology.

Peccaries have received relatively little attention in the literature, perhaps due to their limited distribution in the United States. The completed research, however, has provided a solid biological basis for peccary management, explaining much of the qualitative biology of these animals. Much remains to be done. Ecological relationships must be quantified. As with most research, as many new questions arise as are solved. This is as it should be.

LITERATURE CITED

Arizona Game Commission. 1979. Game survey and harvest data summary. Arizona Game and Fish Dept., Phoenix. Pp. 76–82.

Bigler, W. J. 1964. The seasonal movements and herd activities of the collared peccary (*Pecari tajacu*) in the Tortolita Mountains. M.S. Thesis. Univ. Arizona, Tucson. 52pp.

———. 1974. Seasonal movements and activity patterns of the collared peccary. J. Mammal. 55(4):851–855.

Bissonette, J. A. 1976. The relationship of resource quality and availability to social behavior and organization in the collared peccary. Ph.D. Thesis. Univ. Mich., Ann Arbor. 134pp.

———. 1978. The influence of extremes of temperature on activity patterns of peccaries. Southwest Nat. 23(3):339–346.

———. In press. Social behavior and ecology of the collared peccary in Big Bend National Park. Natl. Park Serv. Sci. Monogr. 16, Washington, D.C.

Brown, W. H.; Stull, J. W.; and Sowls, L. K. 1963. Chemical composition of the milk fat of the collared peccary. J. Mammal. 44(1):112–113.

Burt, W. H., and Grossenheider, R. P. 1964. A field guide to the mammals. Houghton Mifflin Co., Boston. 284pp.

Byers, J. A. 1978. Probable involvement of the preorbital glands in two social behavioral patterns of the collared peccary, *Dicotyles tajacu*. J. Mammal. 59(4):855–856.

Clark, T. L. 1974. Javelina status report. Texas Parks Wildl. Dept., Austin, 2pp. Mimeogr.

Day, G. I. 1977. Javelina activity patterns. P-R Rep. W78R29. Arizona Game and Fish Dept., Phoenix. 13pp.

Eddy, T. A. 1959. Foods of the collared peccary, *pecari tajacu sonoriensis* (Mearns) in southern Arizona. M.S. Thesis. Univ. Ariz., Tucson. 102pp.

———. 1961. Foods and feeding patterns of the collared peccary in southern Arizona. J. Wildl. Manage. 25(3):248–257.

Elder, J. B. 1956. Watering patterns of some desert game animals. J. Wildl. Manage. 20(4):368–378.

Ellisor, J. E., and Harwell, W. F. 1969. Mobility and home range of collared peccary in southern Texas. J. Wildl. Manage. 33:425–527.

———. 1979. Ecology and management of javelina in south Texas. Texas Parks Wildl. Dept. F.A. Rep. Ser. 16:1–25.

Epling, G. P. 1956. Morphology of the scent gland of the javelina. J. Mammal. 37(2):246–248.

Hall, E. R., and Kelson, K. R. 1959. The mammals of North America. Vol. 2. Ronald Press Co., New York. Pp. 994–999.

Halloran, A. R. 1945. Five fetuses reported for *Pecari angulatus* from Arizona. J. Mammal. 26(4):434.

Herring, S. W. 1972. The role of canine morphology in the evolutionary divergence of pigs and peccaries. J. Mammal. 53(3):500–512.

———. 1974. A biometric study of suture fusion and skull growth in peccaries. Anat. Embryol. 146:167–180.

Jennings, W. S., and Harris, J. T. 1953. The collared peccary in Texas. Texas Game and Fish Comm. F. A. Rep. Ser. 12:1–31.

Kirkpatrick, R. D. 1957. A method of age determination for the collared peccary, *Pecari tajacu sonoriensis* (Mearns). M.S. Thesis. Univ. Ariz., Tucson. 40pp.

Kirkpatrick, R. D., and Sowls, L. K. 1962. Age determination of the collared peccary by tooth-replacement pattern. J. Wildl. Manage. 26(2):214–217.

Knipe, T. 1957. The javelina in Arizona. Wildl. Bull. 2. Ariz. Game and Fish Dept., Phoenix. 96pp.

Low, W. A. 1970. The influence of aridity of reproduction of the collared peccary (*Dicotyles tajacu* [Linn.]) in Texas. Ph.D. Thesis. Univ. British Columbia, Vancouver. 170pp.

McCulloch, C. Y. 1955. Breeding record of javelina, *Tayassu angulatus* in southern Arizona. J. Mammal. 36(1):146.

Mearns, E. A. 1907. Mammals of the Mexican boundary of the United States. Pt. 1. U.S. Natl. Mus. Bull. 56:159–169.

Miller, G. S., Jr., and Kellogg, R. 1955. Last of North American mammals. U.S. Natl. Mus. Bull. 205. 954pp.

Minnamon, P. S. 1962. The home range of the collared peccary *Pecari tajacu* (Mearns) in the Tucson Mountains. M.S. Thesis. Univ. Arizona, Tucson. 42pp.

Neal, B. J. 1959. A contribution on the life history of the collared peccary in Arizona. Am. Midl. Nat. 61(1):177–190.

Schweinsburg, R. E. 1969. Social behavior of the collared peccary (*Pecari tajacu*) in the Tucson Mountains. Ph.D. Thesis. Univ. Arizona, Tucson. 115pp.

——. 1971. Home range, movements, and herd integrity of the collared peccary. J. Wildl. Manage. 35(3):455–460.

Schweinsburg, R. E., and Sowls, L. K. 1972. Aggressive behavior and related phenomena in the collared peccary. Z. Tierpsychol. 30:132–145.

Smith, N. S., and Sowls, L. K. 1975. Fetal development of the collared peccary. J. Mammal. 56(3):619–625.

Sowls, L. K. 1961. Gestation period of the collared peccary. J. Mammal. 42(3):425–426.

——. 1963. Chemical composition of the milk fat of the collared peccary. J. Mammal. 44(1):112–113.

——. 1965. Reproduction in the collared peccary, *Tayassu tajacu*. J. Reprod. Fert. 9:371–372.

——. 1966. Reproduction in the collared peccary (*Tayassu tajacu*). Pages 155–172 *in* I. W. Rowlands, ed. Comparative biology of reproduction in mammals. Symp. Zool. Soc. London. 15. Academic Press, London.

——. 1974. Social behavior of the collared peccary *Dicotyles tajacu*. Pages 114–165 *in* V. Giest and F. Walther, eds. The behaviour of ungulates and its relation to management. Vols. 1 and 2. IUCN Publ. 24. Morges, Switzerland.

——. 1978. Collared peccary. Pages 191–205 *in* J. L. Schmidt and D. L. Gilbert, eds. Big game of North America. Stackpole Books, Harrisburg, Pa.

Sowls, L. K.; Smith, V. R.; Jenness, R.; Sloan, R. E.; and Regehr, E. 1961. Chemical composition and physical properties of the milk of the collared peccary. J. Mammal. 42(2):245–251.

Wetzel, R. M. 1977*a*. The extinction of peccaries and a new case of survival. Ann. New York Acad. Sci. 288:538–544.

——. 1977*b*. The chacoan peccary. Bull. Carnegie Mus. Nat. Hist. 3. 36pp.

Wetzel, R. M., and Lovett, J. W. 1974. A collection of mammals from the chaco of Paraguay. Univ. Connecticut Occas. Pap. 2(13):203–216.

Wetzel, R. M.; Dubos, R. E.; Martin, R. L.; and Myers, P. 1975. Catagenus, an ''extinct'' peccary, alive in Paraguay. Science 189:379–381.

Woodburne, M. O. 1968. The cranial myology and osteology of *Dicoytles tajacu,* and its bearing on classification. Mem. California Acad. Sci. 7:1–48.

Zervanos, S. M. 1972. Thermoregulation and water relations of the collared peccary (*Tayassu tajacu*). Ph.D. Thesis. Arizona State Univ. 160pp.

Zervanos, S. M., and Day, G. I. 1977. Water and energy requirements of captive and free-living collared peccaries. J. Wildl. Manage. 41(3):527–532.

Zervanos, S. M., and Hadley, M. F. 1973. Adaptational biology and energy relationships of the collared peccary (*Tayassu tajacu*). Ecology 54(4):759–774.

JOHN A. BISSONETTE, Oklahoma Cooperative Wildlife Research Unit, 404 Life Sciences West, Oklahoma State University, Stillwater, Oklahoma 74074.

43

Elk

Cervus elaphus

James M. Peek

NOMENCLATURE

COMMON NAMES. Elk, wapiti

SCIENTIFIC NAME. *Cervus elaphus*

SUBSPECIES. *C. e. nelsoni:* Rocky mountain elk; *C. e. canadensis:* Eastern elk (extinct); *C. e. manitobensis:* Manitoba elk; *C. e. merriami:* Merriam elk (extinct); *C. e. roosevelti:* Roosevelt elk; *C. e. nannodes:* Tule elk.

The elk is a group of subspecies of the red deer complex that is distributed across North America. The entire species is distributed across Eurasia as well as North America. Walker (1975), McCullough (1966), and Jones et al. (1975), and most recently Bryant and Maser (in press) have treated the taxonomy of this group. Major changes from earlier classifications (Murie 1951) include the Tule elk as a part of the species, and revert to the original Linnaean classification, which treats the whole red deer complex, including the elk, as one species.

DISTRIBUTION

Geist (1971) noted that Siberian and North American elk are very similar, agreeing with Guthrie's (1966) suggestion that dispersal across the Bering Sea land bridge to North America occurred during Illinoian glaciation times. Subsequent reverse migration to Siberia occurred during interglacial periods. Elk are known to have occupied the nonglaciated center of Alaska during the Wisconsin glaciations (Guthrie 1966). Subsequently, elk colonized virtually all of temperate North America (Murie 1951), excluding the Great Basin and the southeastern United States. Merriam's elk and the eastern elk are now extinct and the prairie and deciduous hardwood portions of the original range are no longer occupied, except for small, isolated, introduced populations. Populations and ranges were severely reduced by the early 1900s. However, through restocking programs elk have gradually been reintroduced to virtually all suitable range (figure 43.1). Peek (in press) reported that in British Columbia, Colorado, Oregon, and Washington, more range is currently occupied by elk than in 1800. Virtually all states and provinces that

FIGURE 43.1. Distribution of the elk (*Cervus elaphus*).

have elk populations sustaining hunting activity report more occupied range now than in 1930. The only exception is in Manitoba, where agricultural activities have reduced elk habitat. Elk are currently expanding into previously unoccupied range in northern British Columbia, and have recently been observed on shrub-steppe in central Washington on historically unoccupied range (Rickard et al. 1977).

Increasing current population levels of Tule elk on presently occupied range appears possible, as do reintroductions to unoccupied range (McCullough 1966).

A common misconception is that the elk was primarily a "plains" animal, being forced into mountainous terrain by the onslaughts of advancing human occupation and disruption of their native habitat. Elk were present in the area north of Yellowstone National Park prior to any significant intrusion by people (Lovaas 1970). The Roosevelt elk obviously occupied forested, mountainous terrain prior to the transgressions of civilization. Although undoubtedly there were shifts in range use in forested areas in response to presence or absence of suitable habitat (such as seral brushfields created by fire), the wapiti must be considered both a plains dweller and a mountain dweller. It was extirpated from plains habitats but is not a recent colonizer of mountainous, forested terrain.

DESCRIPTION

Major differences among *C. e. nannodes*, *C. e. nelsoni*, and *C. e. roosevelti* are represented in table 43.1 (McCullough 1966).

Size. The elk include the largest subspecies of the red deer complex, but individual weights are affected by environmental factors. These in turn affect nutritional status and growth. Newborn calves weigh about 15 kg (Johnson 1951), and virtually double their weight within two weeks after birth (table 43.2). Yearlings in their second fall weigh upward of two-thirds of what adults weigh. Adult cows weigh about 80 percent of what adult bulls weigh. Authenticated records indicate that bulls may weigh over 450 kg; an Afognak Island bull weighed by Troyer (1960) was estimated to be 590 kg.

There is substantial overlap in range of sizes between subspecies; however, the Tule elk is significantly smaller. Average live weight of seven adult Tule elk bulls was 194 kg, but some large individuals may exceed 318 kg. Weights of adult bulls of other subspecies average over 300 kg. Body weights of adult red deer bulls from Scotland are heaviest in late September just prior to or early in the rut, and are least in March (Mitchell et al. 1976). Cows without calves continue to increase in weight through November, and reach lows in late April. Scottish red deer adult bulls weigh approximately 110 kg, one-third the average of adult bull elk from Montana (Quimby and Johnson 1951), although neonates of both subspecies weigh about the same.

Pelage. Calves are born with cream-colored spots on a russet coat, with spots becoming progressively less apparent through the summer, and finally disappearing in August. The rump patch is deep yellow to almost orange, fading to a cream color in winter. Adults are dark brown to brownish red in summer; however, winter pelage, acquired in late summer, is characteristically dimorphic. The bull is easily recognized by the light cream colored coat contrasted with a darker mane. Even yearling bulls exhibit this characteristic, but it is not as pronounced. Cows remain darker

colored, although coat color fades during the winter. Roosevelt elk tend to be darker than Rocky Mountain elk.

Antlers. The elk is a "six-point" deer (Geist 1971), reflecting the most common number of tines on a mature bull. Male calves commonly grow "buttons" that are 2–3 cm long and recognizable only upon close examination. The yearling bull may grow a "spike" or exhibit up to five tines, but only rarely has a brow tine. Colorado yearling bulls examined by Boyd (1970) were 72 percent spikes, while 28 percent had two or more tines per antler. Again, nutritional status affects antler growth, with yearling males that experience difficult winter or other conditions as calves growing fewest tines and smallest spikes. Spike antlers may vary widely in the range of 15 cm to 90 cm.

Two-year-old bulls, often called "raghorns," produced antlers with up to five tines, including a brow tine. Blood and Lovaas (1966) reported three Manitoba elk of this age that had antlers with three to four tines. They weighed 1.8–2.8 kg, and measured 48–85 cm long.

Adult bull antlers weigh 13.2 kg or more per pair (Blood and Lovaas 1966). There is evidence that Roosevelt elk antlers tend to be shorter than those of Rocky Mountain or Manitoba elk but are thicker and may be heavier (Troyer 1960), though again variation between individuals is greater than between subspecies. Bulls with up to eight tines per antler are not uncommon.

Antlers are shed primarily in late March and early April. Red deer bulls shed antlers earlier after mild winters than after severe winters. Bulls in good condition shed earlier than those in poor condition, and younger bulls shed earlier than older bulls. Antler regrowth becomes apparent in late May; they are fully formed by early August, at which time rubbing of velvet begins (Watson 1971).

Records of the Boone and Crockett Club indicate that most trophy class bull elk are of the subspecies *nelsoni* (Nesbitt and Parker 1977). Antler lengths in the 150 cm range are exhibited by the largest bulls.

Dentition. The development and wear patterns of teeth in elk have traditionally served as indices to age. The dental formula for an elk with a complete set of adult teeth is: $i\frac{000}{123}$ $C\frac{1}{1}$ $pm\frac{234}{234}$ $m\frac{123}{123}$, for a total of 34 teeth (figure 43.2). Neonates have deciduous or "milk" incisors, milk premolars erupted, and the first molar

TABLE 43.1. Some morphological comparisons of Tule elk (*C. e. nannodes*), Rocky Mountain elk (*C. e. nelsoni*), and Roosevelt elk (*C. e. roosevelti*)

Character	Tule Elk	Rocky Moutain Elk	Roosevelt Elk
Antler shape	Light and spreading Branches curved	Light and spreading Branches straight	Heavy and crowning Branches short
Pelage color	Light	Dark	Dark
Skull form	Short and broad	Intermediate	Long and slender
Length of toothrow	Longest	Shorter	Shortest

SOURCE: After McCullough 1966.

TABLE 43.2. Weights and measurements of North American elk

Subspecies	Location	Number	Age	Sex	Whole Weight	Length (cm)				Reference
						Total	Tail	Hind Foot	Ear	
nelsoni	Montana	23	1 day	Both	14 (9–20)	97 (76–112)	5 (4–7)	39 (35–43)	11 (10–13)	Johnson (1951)
nelsoni	Montana	47	5–7 days	Both	20 (15–27)	108 (97–117)	6 (5–7)	41 (39–44)	12 (11–13)	Johnson (1951)
nelsoni	Wyoming	1	2 months	♂	59					Murie (1951)
nelsoni	Colorado	2	8–9 months	♂	114 (107–123)					Boyd (1970)
nelsoni	Colorado	4	8–9 months	♀	111 (91–120)	176 (169–184)	8 (6–10)	43 (29–57)	13 (9–17)	Boyd (1970)
manitobensis	Manitoba	1	5 months	♂	134					Blood & Lovaas (1966)
manitobensis	Manitoba	1	8 months	♀	133					Blood & Lovaas (1966)
nelsoni	Missouri	3	5–6 months	♂	123 (118–132)					Murphy (1963)
nelsoni	Missouri	2	5–6 months	♀	104 (91–118)					Murphy (1963)
roosevelti	Alaska	2	15–16 months	♂	272 (250–293)[a]					Troyer (1960)
nelsoni	Missouri	4	16–17 months	♂	157 (143–195)					Murphy (1963)
nelsoni	Missouri	3	16–17 months	♀	197 (182–222)					Murphy (1963)
nelsoni	Colorado	2	18–20 months	♀	146 (137–163)	210 (205–215)	12 (11–12)	64 (63–65)	19 (19–20)	Boyd (1970)
nelsoni	Colorado	3	18–20 months	♂	178 (143–220)					Boyd (1970)
nelsoni	Montana	2	19 months	♂		201 (190–211)	13	62 (61–62)	21 (20–21)	Quimby & Johnson (1951)
nelsoni	Montana	4	19 months	♀		200 (196–207)	12 (9–13)	60 (57–62)	19 (18–20)	Quimby & Johnson (1951)
manitobensis	Manitoba	1	18 months	♀	220					Blood & Lovaas (1966)
nelsoni	Colorado	16	3 years + winter	♂	237 (193–285)	221 (207–239)	11 (9–19)	64 (62–67)	20 (18–22)	Boyd (1970)
nelsoni	Montana	11	3 years + winter	♀	255 (245–292)	227 (208–248)	14 (10–18)	63 (60–67)	20 (18–22)	Quimby & Johnson (1951)
nelsoni	Montana	10	3 years + winter	♂	331 (298–373)	242 (231–251)	14 (13–16)	67 (64–69)	21 (19–23)	Quimby & Johnson (1951)
roosevelti	California	9	3 years + winter	♀	215 (171–292)	221 (206–234)	13 (10–17)	66 (64–69)	22 (20–22)	Harper et al. (1967)
roosevelti	California	9	3 years + winter	♂	254 (178–326)[a]	234 (206–246)	12 (11–13)	69 (66–74)	21 (19–22)	Harper et al. (1967)
roosevelti	Alaska	10	4 years + winter	♂	381 (336–497)					Troyer (1960)
manitobensis	Manitoba	8	3 years + winter	♂	353 (288–478)	241 (234–262)	12 (10–17)	69 (68–72)	22 (21–22)	Blood & Lovaas (1966)
manitobensis	Manitoba	4	3 years + winter	♀	275 (258–289)	224 (198–239)	11 (8–14)	67 (62–69)	20 (20–21)	Blood & Lovaas (1966)

[a] Field dressed weights.

either covered by membrane or just protruding (Johnson 1951). From 8 days to 4 months, the incisors, canines, premolars, and first molar are evident. At age 4 months, the calf dentition is $di_{1,2,3}$ $dc\frac{1}{1}$ $dm\frac{234}{234}$ $m\frac{1}{1}$. By 16 months, the yearling dentition is i, d $l_{2,3}$ $dc\frac{1}{1}$ $dm\frac{234}{234}$ $m\frac{12}{12}$ to $i_{1,2}$ di_3 $dc\frac{1}{1}$ $dm\frac{234}{234}$ $m\frac{12}{12}$, with variation in replacement occurring in the first two incisors and in whether or not $m\frac{2}{2}$ is erupted completely (Quimby and Gaab 1957). At age 28 months, dental formulas range from i_{123} $c\frac{1}{1}$, $pm\frac{234}{234}$ $m\frac{123}{123}$ to $i_{1,2,3}$ $dc\frac{1}{1}$ $pm\frac{234}{234}$ $m\frac{123}{123}$, with

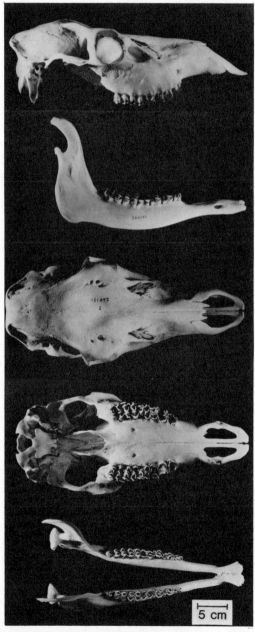

FIGURE 43.2. Skull of the elk (*Cervus elaphus*). From top to bottom: lateral view of cranium, lateral view of mandible, dorsal view of cranium, ventral view of cranium, dorsal view of mandible.

$m\frac{3}{3}$ partially erupted. Complete permanent dentition occurs at three years of age. Although wear patterns of premolars and molars may be used to estimate the age of adults, examination of annulations in dental cementum is more reliable (Low and Cowan 1963).

REPRODUCTION

The female reproductive organs of members of the genus *Cervus* are generally similar to those of the Bovidae. The uterus is bipartite, similar to that of a cow (*Bos taurus*) (Eckstein and Zuckerman 1956:75). The average conception period for Montana elk was the first week in October (Morrison et al. 1959). McCullough (1966) reported variations in the rut of Tule elk from August to September. Flook (1970) found mean conception dates of 11, 19, and 28 September in Banff and Jasper National Parks. One five-year-old cow apparently conceived in November.

The number of estrous periods that may occur in elk during the rut has not been studied, but may be four, similar to that of deer (Cheatum and Morton 1946). Yearling elk may be bred more commonly during later estrous periods. The first estrus may be "silent," without the usual manifestations of rutting behavior or resulting pregnancy. Parkes (1952) postulates that silent estrus is attributable to an alteration in the balance between luteinizing hormone (LH) and follicle-stimulating hormone (FSH). If adenopituitary activity is low, FSH may fade at a greater rate than LH, resulting in an alternation between estrogen and progesterone secretion, which would cause no visible manifestation of estrus. A silent estrus may occur in early September (Morrison 1960).

Elk exhibit a postconception ovulation (Halazon and Buechner 1956), as does the conspecific red deer (Douglas 1966). Ovaries of anestrous elk are characterized by the absence of corpora lutea or follicles with a diameter greater than 2 millimeters (Halazon and Buechner 1956). Proestrous ovaries, collected in August and September, were characterized by the presence of several follicles 7–11 mm in diameter. Estrous ovaries had one freshly ruptured follicle, but McDonald (1969) reported that ovulation occurred in metestrus. After ovulation, and during pregnancy, a second period of follicular development occurs, and one follicle ovulates to form a secondary corpus luteum, which is smaller than the primary corpus luteum formed by the initial ovulation. Primary corpora lutea range from 10 to 18 mm in diameter, while secondary corpora lutea average 5–10 mm. About 55 percent of the secondary corpora lutea appear in the same ovary as the primary one. About 60 percent of 288 female elk examined contained secondary corpora lutea. The following mechanism for post conception ovulation has been postulated: the second ovulation apparently occurs before the embryo is firmly implanted into the maternal blood supply, so it is possible that placental hormones are required to suppress the secretion of gonadotropins. If the attachment does not occur before the secretion of gonadotropin, presumably luteinizing hormone, the ovaries may be stimu-

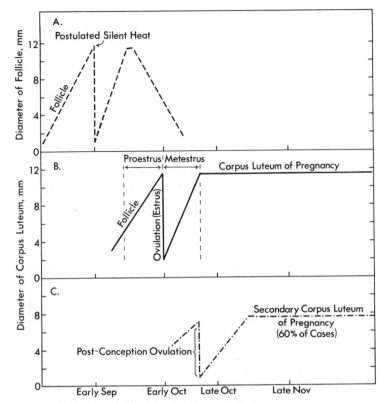

FIGURE 43.3. Ovarian cycle of the elk (*Cervus elaphus*).

lated, resulting in a postconception ovulation and subsequent formation of a secondary corpus luteum. The postconception ovulation described for elk is different from that described for the mare (*Equus caballus*) (McDonald 1969:355), because the primary corpus luteum does not regress before the next ovulation takes place. Instead, both corpora lutea are present at the same time in elk.

The ovarian cycle may be considered as a circadian rhythm in cervids, which is initiated in response to the declining light in fall. The proestrous period, the period of follicular growth and estrogen production in the ovary, is regulated by an increase in FSH production from the adenopituitary. Proestrus is also characterized by growth in the rest of the genital tract, especially the uterus, and lasts two to three days in the cow. The estrous period, when copulation occurs, may be less than 24 hours long in cervids (Markgren 1969), and is characterized by increased estrogen and LH production, along with decreased FSH production. The metestrous period is when ovulation occurs, and is characterized by formation of the corpus luteum function and postconception ovulation in the elk. Production of progesterone occurs during metestrus and it is the period of pregnancy. The ovarian cycle, as far as it is known in elk, is illustrated in figure 43.3.

One calf is produced annually, usually in late May or early June; twinning is rare (Greer 1966). Yearling cows may be fertile (Buechner and Swanson 1955), but this is highly dependent upon their weight, which in turn is related to nutritional status (Mitchell and Brown 1973). Pregnancy rates of adult cows are high, usually above 90 percent. Older cows, over eight years of age, may be less fertile, especially during less favorable nutritional conditions (Greer 1966). Thorne et al. (1976) reported that cows that lost more than 3 percent of their body weight between January and late May produced small calves that were less likely to survive. Mitchell and Brown (1973) report that lactating cows in Scotland are less apt to breed the following year, again depending upon nutritional status of the individual.

Yearling males may be fertile (Conaway 1952) and contribute substantially to breeding, especially when adult bulls are absent or greatly reduced in a population, as through hunting. Bulls over three years old were major participants in rutting activities in the Canadian National Parks, where adult bulls were common (Flook 1970; Struhsaker 1967). However, in the Blue Mountains of Oregon, yearling bulls predominated and contributed substantially to breeding. There is no evidence that a bull's fertility declines in old age (Flook 1970).

FOOD HABITS

The diet of elk is highly variable and depends upon local availability of forage. Food habits tend to overlap those of all other ungulates, and the elk is able to vary its diet on the same area according to season and availability. Kufeld (1973) found that 159 forbs, 59 grasses, and 95 shrubs have been reported as elk forage.

Grasses or shrubs constitute the major winter diet. The diet of Roosevelt elk in Oregon consisted of 56 percent browse with trailing blackberry (*Rubus* spp.), grasses, sedges, and salal (*Gaultheria* spp.) preferred species (Harper 1971). The White River elk in Colorado also used 56 percent browse in winter, including oak brush (*Quercus gambelli*), aspen (*Populus tremuloides*), serviceberry (*Amelanchier alnifolia*), big sage (*Artemesia tridentata*), and snowberry (*Symphoricarpos albus*) (Boyd 1970). Hash (1974) reported that browse constituted 92 percent of the elk diet in the Lochsa River area of Idaho. Redstem ceanothus (*Ceanothus sanguineus*), mountain maple (*Acer glabrum*), scouler willow (*Salix scouleriana*), shinyleaf ceanothus (*Ceanothus velutinus*), and serviceberry were preferred items. Montana studies summarized by Rognrud and Janson (1971) are representative of elk wintering on grasslands. Climax grasses such as bluebunch wheatgrass (*Agropyron spicatum*), Idaho fescue (*Festuca idahoensis*), western wheatgrass (*Agropyron smithii*), and rough fescue (*Festuca scabrella*) were preferred forage species. Grass constituted from 65 to 100 percent of the winter diet on these grassland ranges.

The spring diet reflects a transition from winter to summer foods, with grasses often being most important. As summer nears, forbs become important, although leaves of browse species may be readily taken. Pale agoseris (*Agoseris glauca*), wild onion (*Allium* spp.), arnicas (*Arnica cordifolia*), fireweed (*Epilobium angustifolium*), geranium (*Geranium viscossissimum*), balsamroot (*Balsamorhiza sagitata*), dandelions (*Taraxacum* spp.), and Lupines (*Lupinus* spp.) are among the important species. Fall diets often revert to predominately grass or browse.

POPULATION CHARACTERISTICS

Elk are known to live beyond 20 years of age (Quimby and Gaab 1957), but mean life expectancies are much lower, and are significantly different between the sexes (Flook 1970; Peek et al. 1967; Kimball and Wolfe 1974). Overall sex ratios approach 1:1 at birth, but progressively favor females as population cohorts are followed through time. Sex ratios of adults were 36:100 on Banff National Park (Flook 1970) and 31:100 on the National Elk Refuge, Wyoming (Cole 1969). The White River, Colorado, bull population was so extensively harvested that sex ratios prior to the hunting season were 10.3:100 (Boyd 1970). Peek et al. (1967) reported progressive decreases in life expectancies of bulls as compared to cows as forage conditions on winter range progressively deteriorated in the Gallatin River area, Montana. Thus, as forage conditions deteriorate, or as exploitation of bulls increases, life expectancies of bulls decrease more rapidly than those of cows.

Flook (1970) reported that populations where exploitation was low and that were characterized by a preponderance of older animals could be expected to show greater disparity in sex ratios because bulls did not live as long as cows. The reason for the greater mortality rate of bulls was related to rutting activities. These activities cause great energy expenditures and predispose bulls to malnutrition and death, especially when winter energy intake is inadequate. Trends in fat reserves, as evidenced by weight, show that bulls do fluctuate more than cows through the annual cycle (Mitchell et al. 1976). Life expectancies of bulls are highly variable between populations, ranging from 2.67 years (Peek et al. 1967) to 3 years (Picton 1961) for two Montana populations in terms of "half-life" (the time required for one-half of a cohort to die). Mean length of life for bulls in a Colorado population was 1.21 years (Boyd 1970). Cow life expectancies range from 2.1 to 4.9 (half-life) in the Gallatin River, Montana, herd. Boyd (1970) reported that 86 percent of the female population in the White River area was killed in 4 years, as compared to 2 years for bulls.

Calf survival is equally variable between years and populations. Observations to determine percentages of cows and calves in populations in winter periods provide data on cow:calf ratios. These reflect both production and survival of calves to six months. These cow:calf ratios range from 100:71 (Boyd 1970) to 100:18 (Cole 1969) between populations, and from 100:45 to 100:18 within one population in Jackson Hole, Wyoming, over a 10-year period. These variations may be attributable more to fluctuating survival patterns of neonates than to prepartum mortality. Pregnancy rates of the Northern Yellowstone population fluctuated little over a 6-year period (Greer 1966) in comparison to cow:calf ratios (Houston 1974). Schlegel (1976) reported that predation, especially by black bears, on calves less than one month old was responsible for the rapid declines in calves in the Lochsa River area of Idaho. However, Thorne et al. (1976) demonstrated that calves weighing less than 11.4 kg at birth had only a 50 percent chance of survival, because of malnutrition-related mortality. Thus, even if postpartum losses are more critical than prepartum mortality, the nutritional status of the pregnant cow is again involved.

Predation on young calves can be high in some instances regardless of their physical condition. Calves were taken in greater proportions than their occurrence by mountain lions (*Felis concolor*) in winter in Idaho (Hornocker 1970). Cowan (1947) reported that calves were as prevalent as adults in the summer diet of wolves (*Canis lupus*), indicating high calf predation.

Elk may constitute a primary food source for wolves in some areas, and individuals that are preyed upon are not necessarily "culls" (diseased, young, or old), according to Cowan (1947). Thus, it is possible that in some instances wolves and other large predators may affect population dynamics of elk, especially through their impact on calf survival. It is emphasized that elk have evolved patterns of behavior and survival that allow them to coexist with predators. The influence of predation is an integral and essential component of the species ecology.

The major mortality factor on most elk populations is hunter harvest and associated extralegal losses such as crippling and poaching. Losses due to malnutrition can be significant during severe winters in some populations, while predation may be important in some instances. Some mortality due to combat between bulls occurs during the rut (Flook 1970), and accidents, such as drowning (Martinka 1969*a*, 1969*b*), have been reported. Diseases and parasites are also proximate causes of mortality, often in relation to malnutrition. Ultimately elk populations appear to be naturally regulated on a basis of forage quality and quantity, with other regulating mechanisms being proximate causes.

MOVEMENT

Elk are known to undertake long seasonal migrations between summer and winter ranges, but some populations are essentially nonmigratory. The populations in and around Yellowstone National Park provide examples of both. Movements between the upper limits of summer range and lower limits of winter range for the Gallatin population extend over about 80 km (Brazda 1953). The Jackson Hole population wintering on the National Elk Refuge may migrate as far as 88 km to summer range (Cole 1969). The Northern Yellowstone population wintering near the north-central boundary will also migrate similar distances (Craighead et al. 1972). Other migratory populations include the Sun River, Montana, elk (Picton 1960) and the White River, Colorado, elk (Boyd 1970).

Conversely, a population in the Madison River drainage inside Yellowstone National Park is essentially nonmigratory and exhibits local shifts in habitat use rather than pronounced seasonal movements (Craighead et al 1973). The Roosevelt elk occupying Boyes Prairie, California, are nonmigratory, but other populations in the Siskiyou Mountains were migratory before they were extirpated (Harper et al. 1967). The Manitoba elk occupying Riding Mountain National Park are apparently nonmigratory, but do exhibit movements of at least 8 km into agricultural areas during more severe winters (Blood 1966). The Wind Cave National Park population is confined by fence and survives well without any migratory behavior (Varland et al. 1978). Elk in the Missouri River Breaks of northeastern Montana are nonmigratory, but utilize different portions of their range at different times of the year (Mackie 1970). Some populations, such as those in the Selway River area, Idaho, have segments that exhibit only small seasonal shifts in habitat use, while other segments exhibit long movements between summer and winter ranges (Dalke et al. 1965a). Almost all populations that are migratory contain segments that are associated with winter ranges all year long, for example, Jackson Hole elk (Martinka 1969*b*) and the White River, Colorado, population (Boyd 1970). Thus, elk may be migratory or nonmigratory, depending upon the seasonal availability of suitable habitats.

Calving areas in the Gallatin area (Johnson 1951), the Selway River area, Idaho (Young and Robinette 1939), and the Sun River area, Montana (Picton 1960), are on the upper limits of winter ranges. Johnson (1951) reported that newborn calves on the Gallatin River area of Montana were located predominantly in interspersed sagebrush-timber areas, but within 74 yards of timber if in sagebrush and within 10 yards of sagebrush if in timber. The importance of the ecotone between open and dense cover for calving was emphasized.

Mackie (1970) reported cows and calves using Douglas fir (*Pseudotsuga menziesii*)-juniper (*Juniperus horizontalis*) stands in late May and early June in the Missouri River Breaks, Montana. Elk calves were hidden in sagebrush-grass areas in central Idaho (Davis 1970). Harper et al. (1967) found that 10 calves were born in meadows and grasslands, 6 in riverside hardwoods or brush, and 2 along a spruce-salmonberry/meadow edge in the Boyes Prairie area of northern California.

The interspersion of cover to open areas appears to be critical. Sagebrush or other shrubs or taller herbaceous vegetation is used by newborn calves to hide under. The actual calving site may be less important than subsequent sites used by calves to hide in during the "hiding" period.

Movement to the upper limits of summer range in mountainous terrain appears to be related to the presence of tabanid flies and to vegetation development (Brazda 1953). Cows with calves appear to lag behind other members of a population in their timing of spring movements. Natural and artificial salt licks on winter range do not appear to affect the timing of movements to summer range in the Selway River area of Idaho (Dalke et al. 1965*b*). Movements onto winter range are often initiated with the first snowstorms. Mature bulls frequently appear first, but most of a population will migrate as snow progressively eliminates higher elevations as foraging areas. Snow accumulations elsewhere may force elk onto south-facing slopes and associated river bottoms at the lowest elevations of the annual range. However, some population segments in the Yellowstone and in the East Fork of the Salmon River, Idaho (Wittinger 1978), remain at very high elevations on grassy, windblown slopes in winter.

BEHAVIOR

The elk is very gregarious, but great variation in degree of sociality exists between seasons, sexes, and populations. Altmann (1956) and Murie (1951) observed that cows sought seclusion from other elk prior to and during parturition. The calf hides for the first 18–20 days of its life, associating with the cow primarily for periods to nurse (Lent 1974). Calves join "nursery herds" a few weeks after parturition, and then remain in cow-calf groups of varying sizes through the summer (Altmann 1956). Martinka (1969) indicated that by mid-August a decrease in group size and an attachment of adult bulls to cow-calf-yearling bull groups occurred in the Jackson Hole, Wyoming, area.

Aggregations appear to be related to vegetation, with the largest groups occurring in the most open habitats (Knight 1970). In summer, cow-calf bands of up to 400 individuals with young bulls have been observed in the White River, Colorado, area (Boyd 1970) and in Yellowstone (Murie 1951). Conversely, mean group sizes of the Sun River, Montana, population, which occupies more densely forested terrain, were 7.2 in open areas and 2.5 in timbered areas (Picton 1960). Varland et al. (1978) reported that aggregations were lowest in summer and highest in winter in Wind Cave National Park—a typical observation.

A possible reason for variation in aggregation size relative to vegetative cover is that a large group in open terrain may provide security for individuals included that substitutes for more dense cover (Crook 1970). Protection against predators, availability of forage, population density, breeding activities, and weather-snow conditions are other influences on aggregation patterns.

Struhsaker (1967) reported that rutting groups in fall consisted of 2–26 animals, with 1 adult bull associated with a harem of cows, calves, and occasionally a yearling bull. Large groups of yearling and young adult bulls, apparently nonbreeding, were observed during the rut. Solitary adult bulls also occur, often those searching for harems or displaced from rutting activities by other adult bulls. Combat between adult bulls is intensive and can result in death (Flook 1970).

Sexes tend to occupy separate areas (Peek and Lovaas 1968). Adult bulls on winter range in the Gallatin tended to concentrate on the fringes of the range, while cows, calves, and younger bulls most frequently occupied the central portions. This segregation is also evident at other times of the year, except during the breeding season. Geist and Petocz (1977) argued that separation of sexes in bighorn sheep (*Ovis canadensis*) serves to minimize competition for forage, which may increase survival, and this may be true for elk as well.

Elk are polygamous, the bull collecting a harem of cows and calves. Altmann (1952) reported that bulls actively searched for and joined existing cow-calf groups. Toleration by adult bulls of yearling bulls in harems occurs but not of other adult bulls. Struhsaker (1967) reported no indication of territorial behavior for the wapiti in Banff National Park, but Knight (1970) found that bulls with and without harems present defended areas during the rut in the Sun River area. Scent may be used to identify territories or the presence of the bull (Graf 1956).

Rutting behavior of bulls consists of bugling, thrashing, digging, and rubbing of antlers, wallowing, sparring, and a series of other aggressive displays reflecting various levels of emotional intensity. Struhsaker (1967) reported the following displays: head extension (a means of herding the harem), sexual approach (approach to a cow with head held upright and muzzle tilted downward), muzzling (the perineum of a cow or calf), flehmen (lip curling or holding the mouth open), jawing (by cows, opening and closing the mouth rapidly), head lowering (head lowered to ground, ears pressed back against the head), throat over back (bull over a cow), and mounting for pur-

poses of copulation. Agonistic behavior not necessarily associated with one sex or with breeding activity included one individual displacing another from a site, kicking forward with stiff foreleg (primarily by cows and calves), elevated muzzle, and raising of hind legs. Alarm calls are made by a bark (Struhsaker 1967). A squeal, emitted by all sex and age classes, serves to retain group cohesion.

Although highly gregarious, aggregations of elk are not consistently composed of the same individuals (Knight 1970). Considerable interchange between groups occurs, and associations appear to be short termed. A cow with a calf up to three months old may be the most consistent, stable social group in this species.

MANAGEMENT

Management of elk has undergone dramatic changes over the past four decades. First efforts to reestablish populations on suitable ranges and to protect existing populations from poaching helped to restore the species to virtually all suitable habitats. Present populations total at least 440,000 individuals (Swanson et al. 1969). They may be increased in many areas if encroachment on habitats is prevented, exploitation is more intensively regulated, and coordination with other resource management activities is improved.

In the decades of 1940–60, concern over the deteriorated condition of winter range forage, especially around the national parks and wilderness areas was an overriding issue. Efforts to reduce population levels to allow recovery of vigor and condition of critical winter foraging areas and minimize often spectacular malnutrition-related losses were common. Public opposition to large reduction of populations, coupled with the vagaries of hunter success as related to availability of the migratory populations were frequent occurrences influencing management of elk.

In the late 1950s and early 1960s, concern over competition for forage between elk and livestock also increased. Investigations revealed that the potential for competition for forage between elk, cattle, and domestic sheep (*Ovis aries*) was indeed great (Stevens 1966; Skovlin et al. 1968; Blood 1966). Efforts to coordinate grazing programs and elk populations and to improve range conditions have since become important.

Competition is known to be especially critical when rangeland conditions are poor. Often, the reason for range deterioration is historic, a product of past heavy livestock grazing that is no longer as intensive. Very often, elk populations increased in areas where livestock grazing was ongoing. This caused concern among livestock interests when efforts to improve range condition called for modification of grazing practices or reduction in livestock numbers. Elk are usually not the cause for such programs, but since they do compete for forage, they have been blamed. Cooperative efforts to accomplish management goals for livestock, elk, and the ultimately critical rangeland resource both depend upon are now bearing fruit, as in east-central Oregon (Anderson and Scherzinger 1975). However, such efforts must be intensified in the future for the benefit of all concerned.

Elk populations in northern Idaho and adjacent areas originally proliferated following the holocaustic fires of the 1910–30 period, reaching peak populations in the 1950s or earlier (Leege 1968). Subsequently, heavy browsing of shrubs on winter range and spectacular die-offs were common. Efforts have been made to rehabilitate winter browse ranges through burning, which stimulates resprouting and seedling establishment of palatable shrubs. These programs have not been extensive enough to do more than slow the decline of the elk populations. Shrubs are being replaced by conifers that shade out the intolerant browse species at a more rapid rate than burning programs can create high-quality forage. The programs need to be intensified if elk populations are to be stabilized or increased.

Other efforts in elk management, initiated in the 1960s, were directed at more intensive control and manipulation of harvests, primarily to provide maximum opportunities for sport hunting while still maintaining populations. As numbers of hunters increased, it became apparent that new types of hunting regulations that restricted the taking of cows were needed. Colorado initiated a system of validating a limited number of hunting licenses for the taking of cows, which allowed all other license holders to take only bulls (Hunter 1959).

Other states have also restricted the legal cow harvest through permit systems or bull-only regulations. Currently, efforts to manage populations intensively through manipulation of hunter harvest and more intensive data-gathering activities to obtain population data (age structure, sex ratios, numbers harvested, production/survival criteria, population trends) are expanding. The use of population models (Gross 1969) to predict consequences of various management strategies is in an advanced stage of development, and is being applied to the management of some populations. Hunter harvest, while being the major mortality factor for most elk populations, is intensively regulated and is not jeopardizing any population. However, as demand for hunting opportunities increases, the need to monitor population trends more intensively, to regulate the harvest, and to predict population performance increases as well.

Elk populations in forests subject to logging have also been affected by human activities. Although timber harvest in the 1950s and early 1960s was not extensive enough to cause concern, and often served to create needed foraging areas, as logging intensified in the late 1960s it became apparent that coordination was necessary. Extensive logging on important elk summer–fall range in the Bitterroot River drainage of Montana was perhaps the first situation to illustrate the conflict (Popovich 1975). Thiessen (1976) demonstrated progressive shifts in elk distribution and decreases in harvest as more summer range was logged in

south-central Idaho. Important investigations were initiated in Montana in the early 1970s to define logging-associated relationships and provide recommendations to coordinate forestry and logging activities (Lyon 1975). Subsequently, Thomas et al. (1976) developed a system of guidelines for predicting elk habitat use in timber management areas of eastern Oregon.

A major problem associated with logging is access by hunters to areas previously not accessible by vehicle prior to the establishment of road systems. Elk populations have commonly shifted to areas where activity on roads is minimal and hunter access is poorest. Currently road closures and associated access policies are being implemented that will minimize dislocations of elk from prime habitat. As the U.S. and Canadian hunters have become progressively better equipped with all-terrain vehicles and have increased in number, the need to provide elk areas where vehicle access is restricted has become widely recognized.

Elk have traditionally been associated with wilderness, and indeed the more famous populations of the Yellowstone, Selway, Sun River, White River, and elsewhere live in wilderness of one category or another. Efforts to restore a natural fire regime to these areas in order to ensure that natural dynamic processes are allowed to exist will no doubt benefit elk. However, in some cases, prescribed burning may be more useful, since it can be regulated and does not depend upon chance occurrence. It is hoped that fire management policies in the wilderness will be flexible enough to accommodate the use of prescribed burning. A number of restrictions on human use of the wilderness are both beneficial and detrimental to elk. Restriction of human access can serve to prevent heavy hunter harvest and maintain elk populations closer to their wild state than in other areas. However, guides and outfitters also come under restrictions that often affect traditional ways of operating. Also, if elk populations do increase to levels that cause excessive forage deterioration, they are more difficult to control. Introduction and/or proper management of predators such as the wolf, cougar, and grizzly bear (*Ursus arctos*) may well serve to benefit elk populations and ultimately reestablish more nearly the true wilderness condition. Obviously, fire management and predators are highly controversial issues, but both are to be supported by those interested in elk and the wilderness.

The elk is one of the most popular and sought-after wildlife species in western North America. It is highly adaptable but sensitive to human activities in its habitat. The tasks ahead are: first, to ensure that habitats are retained and made more productive for elk; second, to coordinate elk habitat management with other uses of the land on which they exist; and third, to ensure that direct human exploitation of populations is intensively regulated. As more demands upon forests and rangeland occur, these tasks will become more complicated and demand increased skills and initiative of those entrusted with the conservation and management of this species.

LITERATURE CITED

Altmann, M. 1952. Social behavior of elk, *Cervus canadensis nelsoni,* in the Jackson Hole area of Wyoming. Behaviour 4(2):116–143.

———. 1956. Patterns of social behavior in big game. Trans. North Am. Wildl. Conf. 21:538–545.

Anderson, E. W., and Scherzinger, R. J. 1975. Improving quality of winter forage for elk by cattle grazing. J. Range Manage. 28(1):120–125.

Blood, D. A. 1966. Range relationships of elk and cattle in Riding Mountain National Park, Manitoba. Can. Wildl. Serv. Wildl. Manage. Bull. Ser. 1, no. 19. 62pp.

Blood, D. A., and Lovaas, A. L. 1966. Measurements and weight relationships in Manitoba elk. J. Wildl. Manage. 30(1):135–140.

Boyd, R. J. 1970. Elk of the White River plateau, Colorado. Colorado Div. Game Fish Parks 121pp.

Brazda, A. R. 1953. Elk migration patterns and some of the factors affecting movements in the Gallatin River drainage, Montana. J. Wildl. Manage. 17(1):9–23.

Bryant, L. D., and Maser, C. In press. Taxonomic description, distribution and status. *In* J. W. Thomas, ed. Elk of North America. Wildl. Manage. Inst., Washington, D.C.

Buechner, H. K., and Swanson, C. V. 1955. Increased natality resulting from lowered population densities among elk in southeastern Washington. Trans. North Am. Wildl. Conf. 20:561–567.

Cheatum, E. L., and Morton, G. H. 1946. Breeding season for white-tailed deer in New York. J. Wildl. Manage. 10:249–263.

Cole, G. F. 1969. The elk of Grand Teton and southern Yellowstone National Parks. U.S.D.I. Natl. Park Serv. Res. Rep. GRTE-N-1. 192pp.

Conaway, C. P. 1952. The age at sexual maturity of male elk. J. Wildl. Manage. 16(3):313–315.

Cowan, I. McT. 1947. The timber wolf in the Rocky Mountain National Parks of Canada. Can. J. Res. 25 (sec. D):139–174.

Craighead, J. J.; Atwell, G.; and O'Gara, B. W. 1972. Elk migration in and near Yellowstone National Park. Wildl. Monogr. 29. 48pp.

———. 1973. Home ranges and activity patterns of nonmigratory elk of the Madison drainage herd as determined by biotelemetry. Wildl. Monogr. 33. 50pp.

Crook, J. H. 1970. The socio-ecology of primates. Pages 103–166 *in* J. H. Crook, ed. Social behavior in birds and mammals. Academic Press, Inc., New York. 492pp.

Dalke, P. D.; Beeman, R. D.; Kindel, F. J.; Robel, R. J.; and Williams, T. R. 1965*a.* Seasonal movements of elk in the Selway River drainage, Idaho. J. Wildl. Manage. 29(2):333–338.

———. 1965*b.* Use of salt by elk in Idaho. J. Wildl. Manage. 29(2):319–332.

Davis, J. L. 1970. Elk use of spring and calving range during and after controlled logging. M.S. Thesis. Univ. Idaho. 51pp.

Douglas, M. J. W. 1966. Occurrence of accessory corpora lutea in red deer, *Cervus elaphus.* J. Mammal. 47(1):152–153.

Eckstein, P., and Zuckerman, S. 1956. Morphology of the reproductive tract. Pages 43–155 *in* Parkes, A. S., ed. Marshall's physiology of reproduction. Vol. 1 (1). Longman's, Green & Co., London.

Flook, D. R. 1970. A study of sex differential in the survival of wapiti. Can. Wildl. Serv. Rep. Ser. 11. 71pp.

Geist, V. 1971. The relation of social evolution and dispersal in ungulates during the Pleistocene, with emphasis on the old world deer and the genus *Bison.* Q. Res. 1(3):285–315.

Geist, V., and Petocz, R. G. 1977. Bighorn sheep in winter: do rams maximize reproductive fitness by spatial and habitat segregation from ewes? Can. J. Zool. 55(11):1802–1810.

Graf, W. 1956. Territorialism in deer. J. Mammal. 37(2):165–170.

Greer, K. R. 1966. Fertility rates of the northern Yellowstone elk populations. 46th Annu. Conf. Western Assoc. State Fish Game Comm. Pp. 123–128.

Gross, J. E. 1969. Optimum yield in deer and elk populations. Trans. 34th North Am. Wildl. Nat. Res. Conf. Pp. 373–386.

Guthrie, R. D. 1966. The extinct wapiti of Alaska and Yukon territory. Can. J. Zool. 44(1):47–57.

Halazon, G. C., and Buechner, H. K. 1956. Postconception ovulation in elk. Trans. North Am. Wildl. Conf. 21:545–554.

Harper, J. A. 1971. Ecology of Roosevelt elk. Oregon Game Comm., Portland. 44pp.

Harper, J. A.; Harn, J. H.; Bentley, W. W.; and Yocom, C. F. 1967. The status and ecology of the Roosevelt elk in California. Wildl. Monogr. 16. 49pp.

Hash, H. S. 1974. Movements and food habits of the Lochsa elk. M.S. Thesis. Univ. Idaho, Moscow. 76pp.

Hornocker, M. G. 1970. An analysis of mountain lion predation upon mule deer and elk in the Idaho Primitive Area. Wildl. Monogr. 21. 39pp.

Houston, D. G. 1974. The northern Yellowstone elk. Vols. 1 and 2: History and demography. U.S. Natl. Park Serv. Mammoth, Wyoming. 185pp. Manuscript.

Hunter, G. N. 1959. Management values of Colorado's elk validation system. Proc. Western Assoc. State Game and Fish Comm. 39:209–212.

Johnson, D. E. 1951. Biology of the elk calf, *Cervus canadensis nelsoni.* J. Wildl. Manage. 15(4):396–410.

Jones, J. K.; Carter, D. C. and Genoways, H. H. 1975. Revised checklist of North American mammals north of Mexico. Texas Tech. Univ. Mus. Occas. Pap. 28:14pp.

Kimball, J. F., and Wolfe, M. L. 1974. Population analysis of a northern Utah elk herd. J. Wildl. Manage. 38(8):161–174.

Knight, R. R. 1970. The Sun River elk herd. Wildl. Monogr. 23. 66pp.

Kufeld, R. C. 1973. Foods eaten by the Rocky Mountain elk. J. Range Manage. 26(2):106–113.

Leege, T. A. 1968. Prescribed burning for elk in northern Idaho. Tall Timbers Fire Ecol. Conf. 8:235–253.

Lent, P. C. 1974. Mother-infant relationships in ungulates. Pages 14–55 *in* The behavior of ungulates and its relation to management. IUCN Publ. New Ser. 24.

Lovaas, A. L. 1970. People and the Gallatin elk herd. Montana Fish Game Dept. 44pp.

Low, W. A., and Cowan, I. McT. 1963. Age determination of deer by annular structure of dental cementum. J. Wildl. Manage. 27(3):466–471.

Lyon, L. J. 1975. Coordinating forestry and elk management in Montana: initial recommendations. Trans. North Am. Wildl. Nat. Res. Conf. 40:193–201.

McCullough, D. R. 1966. The tule elk: its history, behavior, and ecology. Ph.D. Dissertation. Univ. California, Berkeley. 389pp.

McDonald, L. E. 1969. Veterinary endocrinology and reproduction. Lea & Febinger. 406pp.

Mackie, R. J. 1970. Range ecology and relations of mule

deer, elk and cattle in the Missouri River Breaks, Montana. Wildl. Monogr. 20. 79pp.

Markgren, G. 1969. Reproduction of moose in Sweden. Viltrevy 6(3):127–199.

Martinka, C. J. 1969*a*. An incident of mass elk drowning. J. Mammal. 50(3):640–641.

_____. 1969*b*. Population ecology of summer resident elk in Jackson Hole, Wyoming. J. Wildl. Manage. 33(3):465–481.

Mitchell, B., and Brown, D. 1973. The effects of age and body size on fertility of female red deer (*Cervus elaphus* L.) Int. Congr. Game Bio. 11:89–98.

Mitchell, B.; McCowan, D.; and Nicholson, I. A. 1976. Annual cycles of body weight and condition in Scottish red deer, *Cervus elaphus*. J. Zool. London 180:107–127.

Morrison, J. A. 1960. Ovarian characteristics in elk of known breeding history. J. Wildl. Manage. 24(3):197–207.

Morrison, J. A.; Trainer, C. E.; and Wright, P. L. 1959. Breeding season in elk as determined from known-age embryos. J. Wildl. Manage. 23(1):27–34.

Murie, O. J. 1951. The elk of North America. Stackpole Books, Harrisburg, Pa. 376pp.

Murphy, D. A. 1963. A captive elk herd in Missouri. J. Wildl. Manage. 27(3):411–414.

Nesbitt, W. H., and Parker, J. S. 1977. North American big game. 7th ed. Boone and Crockett Club and Natl. Rifle Assoc. of Am., Washington, D.C. 367pp.

Parkes, A. S., ed. 1952. Marshall's physiology of reproduction. 3rd ed. Vol. 2. Longman Publishing Co., London. 880pp.

Peek, J. M. In press. The future of elk and elk hunting. *In* J. W. Thomas, ed. Elk of North America. Wildl. Manage. Inst., Washington, D.C.

Peek, J. M., and Lovaas, A. L. 1968. Differential distribution of elk by sex and age on the Gallatin winter range, Montana. J. Wildl. Manage. 32(3):553–557.

Peek, J. M.; Lovaas, A. L.; and Rouse, R. A. 1967. Population changes within the Gallatin elk herd, 1932–1965. J. Wildl. Manage. 32(2):304–316.

Picton, H. D. 1960. Migration patterns of the Sun River elk herd. J. Wildl. Manage. 24:279–290.

_____. 1961. Differential hunter harvest of elk in two Montana herds. J. Wildl. Manage. 25(4):415–421.

Popovich, L. 1975. The bitterroot: remembrances of things past. J. For. 73:791–793.

Quimby, D. C., and Gaab, J. E. 1957. Mandibular dentition as an age indicator in Rocky Mountain elk. J. Wildl. Manage. 21(4):435–451.

Quimby, D. C., and Johnson, D. E. 1951. Weights and measurements of Rocky Mountain elk. J. Wildl. Manage. 15(1):57–62.

Rickard, W. H.; Hedlund, J. D.; and Fitzner, R. E. 1977. Elk in the shrub-steppe region of Washington: an authentic record. Science 196:1009–1010.

Rognrud, M., and Janson, R. 1971. Elk. Pages 39–51 *in* T.

W. Mussehl and F. W. Howell, eds. Game management in Montana. Montana Fish Game Dept., Helena. 238pp.

Schlegel, M. 1976. Factors affecting calf elk survival in northcentral Idaho. Proc. 56th Annu. Conf. Western Assoc. State Game Fish Comm. pp. 342–355.

Skovlin, J. M.; Edgerton, P. J.; and Harris, R. W. 1968. The influence of cattle management on deer and elk. Trans. North Am. Wildl. Nat. Res. Conf. 33:169–181.

Stevens, D. R. 1966. Range relationships of elk and livestock, Crow Creek drainage, Montana. J. Wildl. Manage. 32(2):349–363.

Struhsaker, T. T. 1967. Behavior of elk (*Cervus canadensis*) during the rut. Z. Tierpsychol. 24:80–114.

Swanson, G. A.; Shields, J. T.; Olson, W. H.; Decker, E.; Held, R. B.; Strayer, J. A.; Hill, R. R.; Hunter, G. N.; and Purdue, G. 1969. Fish and wildlife resources on public lands: a study for the public land law review commission. Clearinghouse Fed. Sci. and Tech. Inf. U.S. Dept. Commerce, Washington, D.C.

Thiessen, J. 1976. Some elk-logging relationships in southern Idaho. Proc. Elk-Logging-Roads Symp., Univ. Idaho, Moscow. Pp. 3–5.

Thomas, J. W.; Miller, R. J.; Black, H.; Rodiek, J. E.; and Maser, C. 1976. Guidelines for maintaining and enhancing wildlife habitat in forest management in the Blue Mountains of Oregon and Washington. Trans. North Am. Wildl. Nat. Res. Conf. 41:452–476.

Thorne, E. T.; Dean, R. E.; and Hepworth, W. G. 1976. Nutrition during gestation in relation to successful reproduction in elk. J. Wildl. Manage. 40(2):330–335.

Troyer, W. A. 1960. The Roosevelt elk on Afognak Island, Alaska. J. Wildl. Manage. 24(1):15–21.

Varland, K. L.; Lovaas, A. L.; and Dahlgren, R. B. 1978. Herd organization and movements of elk in Wind Cave National Park, South Dakota. U.S.D.I. Natl. Park Serv. Nat. Res. Rep. 13. 28pp.

Walker, E. P. 1975. Mammals of the world. 3rd ed. Johns Hopkins Univ. Press, Baltimore. 1500pp.

Watson, A. 1971. Climate and the antler-shedding performance of red deer in northeast Scotland. J. Appl. Ecol. 8:53–67.

Wittinger, W. T. 1978. Habitats, food habits, and range use of mule deer, elk, and cattle on the Herd Creek restrotation grazing system, East Fork Salmon River, Idaho. M.S. Thesis. Univ. Idaho, Moscow. 125pp.

Young, V. A., and Robinette, W. L. 1939. A study of the range habits of elk in the Selway River game preserve. Univ. Idaho Bull. 34(16):1–48.

JAMES M. PEEK, College of Forestry, Wildlife, and Range Sciences, University of Idaho, Moscow, Idaho 83843.

44

Mule Deer

Odocoileus hemionus

Richard J. Mackie
Kenneth L. Hamlin
David F. Pac

NOMENCLATURE

COMMON NAMES. Mule deer, black-tailed deer
SCIENTIFIC NAME. *Odocoileus hemionus*
SUBSPECIES. *O. h. hemionus*, Rocky Mountain mule deer; *O. h. californicus*, California mule deer; *O. h. fuliginatus*, southern mule deer; *O. h. peninsulae*, Peninsula mule deer; *O. h. crooki*, desert mule deer; *O. h. columbianus*, Columbian black-tailed deer; and *O. h. sitkensis*, Sitka black-tailed deer.

The mule deer and black-tailed deer comprise two groups of subspecies or races of mule deer, characterized broadly by the extremes in external appearance and behavior within the species. Thus, Wallmo (1978) noted, "mule deer and black-tailed deer are sufficiently different to justify distinct common names, yet similar enough to be included in one species." Among the two groups combined, Cowan (1956) recognized 11 subspecies. Recent taxonomical treatments (Wallmo 1978, 1981), however, indicate general agreement on only the 7 listed above.

Mule deer are members of the family Cervidae, which in North America includes the elk, or wapiti (*Cervus elaphus*), moose (*Alces alces*), caribou (*Rangifer tarandus*), and white-tailed deer (*Odocoileus virginianus*). This family dates from the Miocene in the Old World and probably reached North America during the latter part of that epoch. However, the genus *Odocoileus* apparently has its phylogenetic foundation based strictly in the New World. Geist (1981) has observed that the biology of *Odocoileus* is so different from that of Old World deer that any similarity is likely to be analogous rather than homologous. In North America, two successful experiments in evolution produced the white-tailed deer and the mule deer–blacktail group (Cowan 1956). The former evolved in the relatively mesic, deciduous forest areas of the eastern portion of the continent, the latter in the dry, rugged badlands and mountains of the west. Today, though the two species overlap broadly in geographic range and to some extent in local distribution, where occasional hybridization may occur, they remain remarkably distinct in their biological, ecological, and behavioral attributes. Mule deer may be distinguished readily

FIGURE 44.1. Distribution of the black-tailed deer and mule deer (*Odocoileus hemionus*).

from white-tails by a number of characteristics. Externally, these include the form and color of the tail, the shape and position of metatarsal glands, the form of the antlers among males, and overall appearance. Cranial and tooth characteristics, especially the size and form of the lower incisors, also differ. When alarmed *O. hemionus* moves with either a stilted, stiff-legged gait or a four-footed bound with its tail held either below the horizontal or not wagging. The white-tail, on the other hand, moves more leisurely in a slow, quiet walk or in graceful leaps and bounds with the tail held erect and wagging.

DISTRIBUTION

The geographic range of mule and black-tailed deer (figure 44.1) encompasses most of temperate North America between the Pacific coast and the 100th meridian. In the eastern part it extends into central North Dakota, east-central South Dakota and Nebraska, west-central Kansas, and extreme northwestern Oklahoma and Texas. Scattered populations or individuals, however, have been reported as far east as Minnesota and Iowa. Mule deer occur as far south as central Mexico along interior highlands and throughout the Baja Peninsula. The distributional limits in the north are the extreme southwest corner of Manitoba, the southwest half of Saskatchewan, all but the most northerly extremes of Alberta and British Columbia, and coastal southwestern Alaska. Local populations also occur further north in Alaska, on the Afognak and Kodiak Islands, and in Prince William Sound, as a result of introductions (Wallmo 1978).

O. h. hemionus is the most widely distributed of the subspecies, with a range larger than that of all other races combined. The range includes essentially all of the Rocky Mountain, intermountain, Great Basin, and Great Plains regions within the geographic range of the species. The California mule deer occurs only in southern California, where it overlaps with Columbian black-tails, Rocky Mountain mule deer, and southern mule deer, the last occurring only in extreme southern California and approximately the northern one-third of the Baja Peninsula. Peninsula mule deer (*O. h. peninsulae*) are restricted in distribution to the southern two-thirds of the Baja Peninsula. The desert mule deer is distributed from approximately central Arizona and New Mexico south to the distributional limits of the species in Mexico. The black-tailed subspecies are narrowly distributed in a strip along the Pacific Coast; the Columbian race from approximately Monterey County, California, north to central British Columbia; and the Sitka race northward to Alaska. Except in southeastern Alaska, black-tails overlap somewhat throughout the length of their range with Rocky Mountain mule deer.

DESCRIPTION

Body Size and Growth. Mule deer are quite variable in size. The largest individuals occur in the Rocky Mountains among *O. h. hemionus*. Anderson et al. (1974) listed average field carcass weights of 51 males one and a half years of age and older as 74.04 kilograms. For 91 females the average was 58.99 kg. Hunter (1924) reported maximum eviscerated carcass weights of wild bucks to be 172 kg, while Mackie (1964) reported similar weights for does as 59 kg. Body lengths of 51 males and 90 females, age one and a half years and older, averaged 152.3 cm and 142.4 cm, respectively, while shoulder heights averaged 96.6 cm for males and 90.8 cm for females (Anderson et al. 1974). The smallest individuals occur among the black-tailed deer. Field carcass weights of adult males and females have been recorded as low as 50 and 32 kg, respectively (Cowan and Guiguet 1965). Body size and weight may vary greatly among individual mule and black-tailed deer depending upon age, time of year, reproductive status, and environmental factors that affect the nutritional status and growth of animals.

Newborn mule deer fawns range in weight from about 2.3 to 5.0 kg. The mean for 285 Rocky Mountain mule deer fawns, weighed during nine different studies, ranged from 2.7 kg to about 4.0 kg.

Both prenatal growth, reflected in birth weights, and postnatal growth are influenced by environmental, ecological, and physiological factors that influence the nourishment received by fetuses and growing fawns. Although Verme (1963) found that deficiencies in the diet of pregnant does substantially reduced fetal and birth weights in white-tailed deer, Robinette et al. (1973) indicated that the quality of the mother's diet had no measurable effect on birth weights of mule deer fawns. Mule deer fawns grow rapidly and by five to six months of age average about 30 kg, with males tending to be slightly heavier than females. Yearlings generally weigh 50 to 60 kg, with females averaging about 10 percent less than males. Anderson et al. (1974) indicated that males continued to gain weight and presumably grow throughout life, while females achieved their maximum weight at about eight years of age. Mackie (1964) concluded that males gained weight at least to the age of seven and a half years, while weights of females changed very little after two and a half years.

Body weights fluctuate seasonally among adult mule deer of all races and both sexes, increasing during summer and early autumn and decreasing during late fall and winter. Carcass weights of mature males peaked during October and reached a low in March (Anderson et al. 1974). Female weights also peaked in October but were lowest in April. Mean seasonal weight losses were 19 percent and 22 percent of peak weights for males and females, respectively. Wood et al. (1962) reported overwinter weight losses of 20–22 percent for a young adult male mule deer on a high-quality diet.

Dentition. The teeth of mule deer are heterodont. The dental formula for adults with complete permanent dentition is: 0/3, 0/1, 3/3, 3/3 = 32 (figure 44.2). Upper incisors, upper canines in most deer, and upper and lower first premolars are missing. In mule deer, as in other cervids, the lower canines are incisiform.

In fawns, deciduous or milk incisors and canines are fully erupted 10 days after birth, while mineralization and eruption of premolars requires 2 1/2–3 months (Rees et al. 1966). Points of M_1 first become evident at approximately 2 1/2–3 months and eruption is complete by 12–14 months of age (Robinette et al. 1957). Points of M_2 first appear at about 8 months, with eruption complete at 20–23 months, while M_3 emerges at 15 months and is complete at 28 months when the full permanent adult dentition is achieved. These tooth eruption and replacement rates are somewhat slower than those reported for white-tailed deer by Severinghaus (1949).

Tooth replacement patterns provide a reliable means of determining age in mule deer through about two and a half years of age. For older deer, age may be estimated approximately by evaluating wear on mandibular cheek teeth (Robinette et al. 1957). Age may also be determined by sectioning the root of an incisor in the laboratory and enumerating annual rings in the cementum (Low and Cowan 1963). Age of deer cannot be determined by the number of antler points.

Antlers and Antler Growth. Male mule deer, like most other cervids, have deciduous antlers. When mature and completely ossified, antlers consist of a thick sheath and base of hard compact bone around a core of spongy bone. Antlers apparently evolved as a sociological adaption associated with the reproductive success of individuals (Geist 1968) and may provide the major stimulus for social interaction (Bubenik 1968).

In mule deer, the antlers of adults branch equally, or dichotomously, above the base typically to form four major tines, or points, per side. A relatively short brow tine may also occur. In contrast, the antlers of adult white-tailed deer consist of a series of tines arising from a continuous, forecurving main beam.

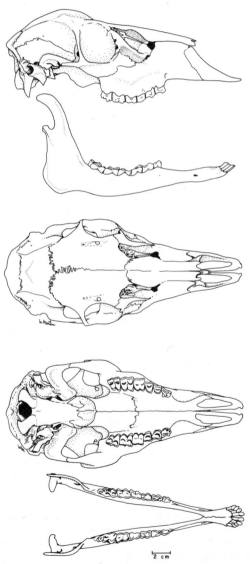

FIGURE 44.2. Skull of the mule deer (*Odocoileus hemionus*). From top to bottom: lateral view of cranium, lateral view of mandible, dorsal view of cranium, ventral view of cranium, dorsal view of mandible.

Antler development in Rocky Mountain mule deer begins with the formation of antler primordia, or pedicels, in the fetus 73–83 days after conception (Hudson and Browman 1959). Male fawns typically have "buttons" that first appear at about three months of age (Davis 1962) but are recognizable only by close examination. In subsequent years, particularly among yearlings, growth and development appear to be sensitive to the nutritional quality of the diet, with body growth taking precedence over antler growth (French et al. 1955). Yearling males on adequate diets will normally have "forked" antlers with 2 points on a side, while deficient diets result in thin "spike" antlers. Fawns born as singletons may also develop larger antlers as yearlings than those born as one of a set of twins (Robinette et al. 1977). Yearling males having 3 or more points on a side are uncommon. The "typical" 4-point antler of adults frequently is first obtained at two years, though 3 points may be more common and larger 2-point antlers may yet occur at that age. Subsequent growth is most prominent in the diameter and length of the antler, while brow tines and additional, nontypical tines may also appear. Nontypical antlers having 20 or more points on a side have been recorded.

The annual antler cycle begins in April or May with the appearance of velvety antler "bulbs." Initially, growth is slow, but by summer it becomes extremely rapid and may exceed 1 cm per day (Goss 1963). Growth is completed by late summer. The velvet dries and is usually shed in late August and September. Antler shedding dates are broad and variable. Northern Rocky Mountain mule deer usually drop their antlers in late January and early February, though individual bucks may shed as early as mid-December or as late as the end of March. Mature bucks may shed antlers earlier than yearlings, and bucks that are undernourished or in poor physical condition often shed earlier than those on good diets and in good condition.

Pelage. Adult mule deer molt twice annually, in late spring and in early fall. The characteristic dark gray winter coat with a woolly undercoat is acquired in fall, growing from September through January. It is shed during May to early June and is replaced by the reddish coat of summer. The autumnal molt begins on the outer ear pinna and moves caudad to the head, neck, shoulders, and flanks before proceeding to the abdomen, legs, and rump patch, which remains white throughout the year. The spring molt occurs as approximately the reverse of the autumnal. The phenology of molting may vary locally and between individual deer. Barren does molt earlier than pregnant individuals, and deer in poor condition may experience an extended molt (Cowan and Raddi 1972).

Fawns are born spotted, with the pelage replaced between July and October. Mule deer fawns typically have a medium-dark brown coat in contrast to the lighter, orange-colored coat of white-tailed deer fawns. Robinette et al. (1977) noted that Rocky Mountain mule deer fawns lost their spots at an average age of 86 days.

Vocalizations. Mule deer emit a variety of sounds involving communication between deer. Studies suggest that communication between adults of either sex include low-pitched vocalizations of undisturbed animals, whistling snorts of alarm, and coughing grunts of bucks challenging rivals. When deer are distressed, they may emit very loud sounds ranging from bleats to sharp barks to deep guttural roars depending upon age, sex, and degree of disturbance (Cowan 1956).

Gait. The mule deer has a characteristic type of gait, referred to as stotting. The stot is a four-footed bound that may carry the animal 60 cm or more above the ground and span a distance of 3–5 m. Speeds up to 40 km per hour can be attained for short durations. More importantly, perhaps, this gait enables mule deer to move sure-footedly with extremely high maneuverability in the steep rocky terrain in which they typically occur. If necessary, they can turn in any direction or completely reverse the direction of movement in the course of a single bound.

Another characteristic gait of mule deer is a deliberate stilted, stiff-legged walk, which seems to be used mostly when an animal is alarmed but not aware of the cause or curiously investigating an unknown element in its environment. Undisturbed mule deer walk normally, trot, lope, and even gallop occasionally in the course of normal movements associated with feeding, travel, play, and courtship.

PHYSIOLOGY

Nutrition and Food Habits. Mule deer are herbivores, possessing a four-chambered ruminating stomach in which vegetation is reduced to usable form by microbial fermentation. This process, though efficient in extracting energy and nutrients from plant tissue, is slow. Both the amount of vegetation that can be ingested, which is determined by the size of the rumen, and the rate at which it can be passed through the gut, which is determined by the digestibility of the plant tissues consumed, are extremely important in the nutritional physiology of ruminants. Deer, due to their small body size, possess a rumen of small capacity. This dictates a high turnover rate of ingested vegetation, utilization of easily digestible forage, and, consequently, very selective feeding behavior as compared with larger herbivores.

In the digestive process, the carbohydrates in plant tissues ingested by deer are converted to volatile fatty acids (VFA). These are absorbed from the stomach directly into the blood to provide a major energy source for the animal. Crude fats, mostly unsaturated fatty acids in plants, are hydrogenated by microbes in the rumen and to a lesser extent in the lower gastrointestinal tract. Additional fatty acids are synthesized in the rumen from plant lipids and carbohydrates. Fats have more than twice the caloric value of carbohydrates and, when highly digestible, provide much readily available energy. Body fat deposits of deer consist mainly of saturated fatty acids. Plant proteins are digested by rumen microbes that synthesize or use the available nitrogen to build amino acids to synthesize microbial proteins, which are later digested in the abomasum and small intestines.

In addition to energy and protein, ingested plant tissues provide dietary vitamins and minerals as well as water. Deer apparently can store or synthesize all vitamins needed for metabolism such that nutritional deficiencies can seldom, if ever, be traced to this source. Mineral elements important in growth, development, and metabolism include calcium, phosphorus, sodium, potassium, chlorine, magnesium, sulfur, iron, iodine, copper, selenium, zinc, and manganese. Of these, phosphorus, iodine, and selenium could be limiting for mule deer on some western rangelands. Thus, mule deer commonly visit salt and other mineral "licks," especially during spring, where some minerals are obtained by direct ingestion.

The specific foods eaten by mule deer are extremely varied. Kufeld et al. (1973) listed 788 species of plant, including 202 shrubs and trees, 484 forbs, 84 grasses, sedges and rushes eaten by Rocky Mountain mule deer alone. Even this list is incomplete because of studies not cited. Also, some plants used in only small amounts often are not listed in reports. Not all plants are good deer food, but many different plants are used at some time, in some places by mule deer. Some plants may be eaten in one area and not in another, or only in certain seasons or stages of growth. Some species may be eaten only in association with, or in the absence of, other forage species, or only when deer may periodically use the sites on which certain plants occur. Deer seem to have the ability to select plant parts and plants from certain soil types or sites that are highest in nutritional content (Bissell 1959; Swift 1948). The kinds and amounts of different plants eaten may also vary between individual mule deer (Willms and McLean 1978). Because of this, it usually is not feasible to generalize about the kinds and quantities of forage utilized, even by broad forage classes or habitat types.

The various nutritional requirements of mule deer differ seasonally and according to the sex, age, activity, condition, and reproductive status of the animal as well as with environmental conditions. The ability of deer to meet these needs and to maintain a positive energy balance over an extended period of time similarly differs. The natural environments in which mule deer occur are almost infinitely variable; and they are never static. Foods eaten vary drastically in kind, quantity, and nutritional quality as well as in digestibility from one season to another, from one year to the next, and from place to place. Thus, the costs of obtaining energy and ultimately the net energy gained also vary. Sometimes only small and seemingly insignificant differences in the kinds, amounts, growth, and chemical composition of plant tissues available may make the difference between a negative and a positive energy balance for deer feeding on them.

Generally, in spring and early summer, when herbaceous vegetation is immature and succulent, plant tissues are easily fermented, the total concentration of VFA is greatest, and usable energy is high. Plant protein contents are also high. In late summer and autumn, herbaceous plants become dry and woody vegetation ceases to grow, resulting in diets progressively lower in protein, higher in carbohydrates, and much higher in lignin and cellulose content. Thus, digestibility also decreases, rumen turnover time and energy costs for fermentation increase, and the quantity of usable energy and nutrients derived from forage is reduced. However, deer have adapted physiologically and behaviorally to this change. At least in more severe northern environments, the metabolic rate is reduced, food intake is voluntarily limited, and activity becomes restricted in winter. During this period when mule deer may predominantly utilize woody browse and other low-quality forages, they typically experience an energy deficit that must be met by drawing upon body fat deposits accumulated in late summer and fall. Their survival is governed not by the potential of winter forage supplies to meet energy needs, but by temperature and snow conditions. These environmental factors determine forage availability, energy expenditures, and the costs for energy gained in feeding. The amount of fat that an animal carries and the severity and duration of the winter determine whether finite fat deposits will carry it through winter (Wallmo et al. 1977). When fat reserves are expended before a positive energy balance is again achieved in spring, the animal will begin to assimilate structural body proteins. If this source of maintenance is required for more than a short time, then death will result.

Although winter is usually the most critical period in the annual nutritional cycle of mule deer, negative energy deficits leading to malnutrition and starvation losses may occur at almost any time of the year. On some very arid and desert ranges, drought conditions may be more critical during late summer and fall than during winter, when rain and snow typically lead to more favorable forage conditions.

Thermal Requirements and Climatic Relations. Across their range, mule deer endure climates with average temperatures ranging from $-15°$ C ($5°$ F) or less in January in the northern Rocky Mountains to over $30°$ C ($86°$ F) in July on hot deserts. Extreme temperatures may be below $-60°$ C ($-76°$ F) and above $50°$ C ($122°$ F) (Wallmo 1981).

There are no published data on body temperatures in wild, free-ranging mule deer. Results of studies on captive deer indicate an average body core temperature ranging from $37.1°$ to $40.6°$ C ($98.8°-105.1°$ F) (Leopold et al. 1951; Cowan and Wood 1955; Thorne 1975).

Mule deer are homoiothermal and lack an extensive system of sweat glands that essentially prevents heat loss due to evaporation (Loveless 1967). However, thermoregulation is achieved by shivering, changes in posture, erecting or fluffing hairs, altering food habits, and the use of vegetative cover and physiographic sites that afford shade in summer and protection from wind and radiant heat loss in winter. Daily and seasonal activity patterns and movements may be strongly influenced or induced by air tempera-

ture regimes (Loveless 1967); however, wind, relative humidity, and snow depths, individually or in combination with temperature and each other, may also be important. In Colorado, mule deer tended to move about in their principal winter habitat, seeking out the most "comfortable" temperature zones, which appeared to be from $-9°$ C ($15°$ F) to $7°$ C ($45°$ F).

The reactions of mule deer to climate or weather factors (except for snow depth) seem to vary, especially in relation to ambient temperatures. Wind, for example, has little apparent influence on movements and activities except during periods of cold weather when high winds force deer to seek shelter and reduce their activities. Humidity may influence the comfort of deer and thus the extent to which they are active (Linsdale and Tomich 1953), but again it probably does not act independently of other weather factors. Loveless (1967) observed that deer appeared slightly more "nervous" than usual during periods of low atmospheric pressure. This is related to temperature gradients and, to some extent, to precipitation and wind currents, which also influence movements and activity. Precipitation, either rain or snow, apparently has little direct effect on mule deer except when severe storms force animals to seek shelter.

Snow depth probably has more influence on the behavior of mule deer than any other single weather element. In autumn, depths of 15–30 cm may be sufficient to initiate major migratory movements or shifts in habitat use. Depths of 25–30 cm may impede movements, especially among young animals, while more than 50 cm will essentially preclude the use of an area by deer (Loveless 1967). The time, rate, and amount of snow accumulation thus become major factors in determining both the time and the extent of concentration of mule deer on winter ranges throughout the mountainous area. In the Bridger Mountains of Montana, "normal" snow depths and patterns restrict mule deer to less than 20 percent of their total year-long range in winter, and under more severe snow conditions only 20–50 percent of the winter range may be usable (Mackie et al. 1978). Generally in these situations, mule deer move down from higher elevations and up from areas of low relief onto steep slopes of south and west exposure. They also may move to windswept ridges or into timber cover, where snow accumulations are less and/or clearing occurs more rapidly after storms. In addition to these influences, snow depth greatly affects the kinds and amounts of food plants available to mule deer.

Water Requirements and Relations. The water requirement of mule deer apparently varies across their range. Because they evolved in a generally arid region, it would follow that mule deer are well adapted to cope with a scarcity of free water in their environments. In the semiarid Missouri Breaks region of Montana, seasonal changes in the availability of free water did not influence the distribution of mule deer (Mackie 1970). Mule deer do make use of free water when available, especially during dry conditions. In the very arid, desert southwest, however, the availability of free water can greatly influence local distributions. In most regions, water contained in succulent forage or occurring as dew on forage plants is sufficient to meet metabolic needs during spring, summer, and early fall, while ingestion of snow fulfills this requirement during late fall and winter.

REPRODUCTION

Breeding and Gestation. Most mule deer attain sexual maturity and become capable of breeding at approximately one and a half years of age. Rare instances of pregnancy among fawns have been reported. The minimum breeding age of females is influenced by their physical condition, and in some severe or nutritionally impoverished environments successful conception may not occur until the third fall or even later. Also, while yearling males are physically capable of breeding, the presence of many older, more mature males in a population may limit their participation in the rut.

Breeding occurs in autumn and early winter, the exact dates varying by subspecies and location. Rocky Mountain mule deer breed primarily during November and December, with parturition occurring mostly in June. Southern races tend to breed somewhat later, in December and January, with fawns being born in July and August. The gestation period for mule deer varies from 183 to 218 days, with a mean of 203 days (Robinette et al. 1977). Breeding among black-tailed deer begins in September and October, a month or more earlier than in most mule deer, and extends into mid-December. It has been hypothesized that the timing of reproductive activity in deer is controlled at least broadly by day length. However, variations occur across regions too small to be affected by photoperiodism. This may reflect local adaptations to environmental conditions such that fawning occurs at the time most favorable to early growth and survival.

The estrous cycle in female mule deer is 22-28 days. The period of estrus, when the doe is receptive to the buck, lasts 24-36 hours. More than one ovum may be shed during ovulation. If conception does not occur, the cycle may be repeated several times. The annual reproductive cycle in males begins with the regrowth of antlers in spring and summer. Antler growth and other reproductive activity is initiated and controlled by levels of testosterone in the blood. With the onset of the rut in autumn, the neck swells and aggressive behavior becomes apparent. Also at this time, the males become hyperactive, movements increase greatly, and food intake declines. In Rocky Mountain mule deer the most intensive rutting behavior and breeding occur from mid-November through mid-December, though potency spans the entire period of female fertility and viable, mature sperm may be present in the seminiferous tubules of some males each month of the year (Anderson and Medin 1964).

Litter Sizes. Adult female mule and black-tailed deer commonly conceive twins, while yearlings usually carry a single fetus. Robinette et al. (1955) reported litter sizes for 492 females as: no fetuses, 2 percent; one fetus, 37 percent; twins, 60 percent; and triplets, 1 percent. Reproductive potential, however, may vary quite widely between populations on different ranges as well as from year to year on the same range according to local environmental conditions and the nutritional status of does. Pregnancy rates range from 75 to 100 percent and fetal rates range from about 1.4 to 1.8 fetuses per doe. Pregnancy and fetal rates are generally lower for yearling females than for older adults. Fetal rates for black-tails may be slightly lower than those reported for mule deer.

The full reproductive potential of mule deer is only rarely reflected in the recruitment of fawns and, especially, yearlings into a population. The reproductive performance of does may be strongly influenced by their physical condition and nutritional status at the time of breeding and during the gestation period, especially the last two to three months. Total fawn production in a population will also be determined by the age distribution of females, including the proportions of fawns, yearlings, and older adults. There is, however, no substantial evidence to support the occasionally stated assumption that populations comprised of a high proportion of very old does are less productive than populations with a predominantly young age structure, and indeed the reverse may sometimes be true. Postpartum survival of fawns may be influenced by many factors. The most important of these may be their physical condition at birth as determined by the quality of the diet of pregnant does on winter and spring ranges, during lactation by the nutritional plane of the mothers on summer range, and after weaning by the quality of late summer, fall, and winter diets. Other factors regulating fawn survival include predation, disease, weather extremes, hunting, and accidents. Overall, it is not uncommon that 25-30 percent of the fawns produced by a mule deer population are lost by autumn, 50 percent or more by early winter, and up to 75 percent or more by spring.

Longevity. Records for tame and captive mule deer indicate that does may live as long as 22 years and bucks may live to 16 years (Cowan 1956). Data on the maximum life expectancy of wild deer in natural populations are scarce, but both does and bucks may live only about half as long as their captive counterparts. Records for individually tagged mule deer in the Bridger Mountains of Montana show that females seldom live beyond 10-12 years of age, while males seldom live beyond 8 years (Mackie et al. 1978). Only 1 percent of 1,790 adult mule deer bucks harvested on a Utah range were placed in age classes 8 years and older (Robinette et al. 1977).

Tooth wear, which influences feeding efficiency, apparently is one of the more important factors determining the maximum potential life span of deer, and may be responsible for differences between captive and wild mule deer, and between wild mule deer on different ranges. Differences in soil texture and food types may affect the rate of tooth wear. Reasons for the apparently greater maximum longevity of mule deer females as compared with males have not been studied. However, males in all age classes have a higher natural mortality rate than females of their respective cohorts, even in the absence of hunting.

In wild populations, extensive mortality accrues annually to both sexes such that most individuals die from one cause or another before reaching "old" age.

BEHAVIOR

Socializations. The degree of sociability in mule deer varies according to season, sex, population, and subspecies. In general, these animals are neither highly gregarious nor strictly solitary. During much of the year, as environmental conditions permit, they tend to occur rather widely dispersed across suitable habitat, either individually or in small groups. When circumstances dictate, as in the use of a common feeding area or cover type, or when forced to concentrate on winter or other range areas, mule deer will readily aggregate in large groups, sometimes numbering several hundred animals. Such large groups rarely endure, however.

Mule deer are most dispersed during the summer, especially during the fawning period, when does seek isolation and yearlings are driven off. During this period, groups consist primarily of adult males sharing common or broadly overlapping home ranges, yearlings wandering together, and nonparous does and other deer sharing common feeding areas within overlapping home ranges. Does with young fawns are generally intolerant of other deer, though their tolerance of others gradually increases as summer progresses. After six to eight weeks, yearlings may be accepted back into a "family group" and does and fawns from adjacent home ranges may at least temporarily band together while foraging. Average observed group size thus increases slowly through late summer and autumn into winter. The largest groups typically occur in the north during midwinter to late winter, when snow depths greatly restrict available range and forage supplies. Large groups also occur during early spring, when the animals frequently concentrate their feeding on certain open range sites where new green forage first appears and is most available.

The nature of the social bond, if any, between individual mule deer is not well known. Miller (1974) reported that black-tailed deer showed a strong tendency to form cohesive groups within which individuals were socially bonded and each held distinct social ranking. Groups led by females were distinct social units from winter into the prefawning period, while male groups were distinct from antler casting in winter to the late prerut period. It is possible that similar social bonding occurs generally in mule deer, though available evidence is sketchy and sometimes contradictory. Mackie et al. (1976) found that the occurrence of individually marked deer in groups on winter range changed frequently. Indexes of association between individual does were generally very low. High degrees of association were characteristic only of doe and fawn, and of certain mature bucks that were distributed

loosely in "groups" associated with certain areas of the winter range. Steerey (1979) indicated that at least some does may be associated with one another quite closely during the winter, and generally during other seasons.

It has been hypothesized for white-tailed deer that female groups represent associations of maternally related individuals (Hawkins and Klimstra 1970). Evidence for this in mule deer is limited, though observations over several years of individually marked animals in the Missouri River Breaks, Montana, indicate that some groups forming during fall and winter consist entirely of "family" members, including does, fawns, and yearlings spanning several generations. This apparently is not the case among mature males, groupings of which may be related more to distribution and habitat usage.

Adults of both sexes establish and traditionally utilize relatively small seasonal home ranges within suitable habitat. Males older than yearlings, however, tend to occur somewhat grouped, with closely overlapping if not common home ranges, within certain "buck habitats" (Mackie 1970). In some areas at least, buck habitats may be located peripheral to areas used by does. Does tend to be more widely and uniformly distributed across available habitat. Because of this, individual mature bucks may be seen in the company of other males more frequently or consistently from winter through early autumn than is the case with does. The apparent socialization among males breaks down during the rut, when individuals become highly aggressive toward one another and widely dispersed across all available habitat.

Miller (1974) believed that black-tailed deer exhibited both individual and group territoriality in their distribution and use of range resources. This phenomenon has not been evident or documented in studies of the behavior and habitat usage of mule deer.

Reproductive Behavior. Mule and black-tailed deer are polygamous, the males wandering about extensively, seeking and pursuing individual does in estrus. Mature bucks in particular are highly aggressive during the rut and are antagonistic toward others. Miller (1974) suggested that superior, dominant black-tailed males established rutting territories within which other bucks were not tolerated.

Typical rutting behavior of bucks includes snorting, urine marking, thrashing and rubbing antlers in shrubs and trees, sparring and antler fighting with other males, and, occasionally, herding individual does. Does may also be more active during the rut. Females may utilize various behaviors such as urine marking to signal their location and the onset of estrus to the male. Actual courtship apparently involves a number of different behavioral patterns on the part of the buck and the doe, culminating in breeding. A single doe may be bred several times during the estrous period, and a single buck may breed many, if not all, receptive does over a large area during the period of rut. Mature black-tailed bucks may court and breed primarily with mature does, while yearling does are courted and bred mainly by yearling and younger adult males (Miller 1974).

Other Behavior. Mule deer tend to be creatures of habit in many aspects of their activities. As indicated above, adults of both sexes establish and traditionally utilize relatively small seasonal or, where environmental conditions permit, year-long home ranges, though females may be more faithful to these areas than males. The timing, extent, and other features of seasonal and year-long movements, including the size, shape, and location of these home ranges, all appear to reflect individual behavior patterns. Daily activity patterns also tend to be habitual, as are the kinds and amounts of forage selected and the particular habitat or cover type used.

Mule deer apparently are quite capable of learning and adjusting their activity patterns and habits to accommodate at least some changes in their environment. Thus, in the Missouri River Breaks, mule deer became extremely alert and relatively few could be observed feeding following a period of heavy hunting pressure (Mackie 1970). Similarly the alertness of mule deer and their response to disturbance may vary between populations or habitats. Variance may be in relation to the occurrence of natural predators, previous disturbing experiences, or the security of an area.

Unlike white-tails, mule deer do not necessarily attempt to hide from predators. Rather, they attempt to detect danger at long range with their large ears, excellent vision, and use of generally open habitats to outmaneuver their enemies. In doing so, they effectively utilize characteristics of the terrain, including steep slopes, boulders, ledges, trees, brush, and deadfalls, together with their stotting gait, to place obstacles between themselves and the predator. This strategy requires precise timing and very calm, unexcitable individual behavior, and probably explains the greater use of open habitats by mule deer as compared with white-tailed deer (Geist 1981).

Habitat. Mule deer are broadly adapted and may be found in all major climatic and vegetational zones of western North America except the arctic, tropics, and most extreme desert. Thus, their habitats are many and extremely diverse, and complete description is a complex if not elusive task.

Generally, mule deer frequent semiarid, open forest, brush, and shrub lands associated with steep, broken, or otherwise rough terrain. Their stronghold may be the mountain-foothill habitats that extend from northern New Mexico and Arizona north into British Columbia and Alberta along the Rocky Mountains and other mountain ranges. However, extensive populations also occur in prairie habitats, especially in the Great Plains along the eastern and northeastern limits of their distribution, and in semidesert shrub habitats of the southwest.

The mountain-foothill habitats occupied by mule deer span a broad range of latitudes and elevations. Climates and topography are diverse and, correspondingly, vegetational components vary considerably. Mule deer associated with these habitats are primarily migratory. In summer, they usually occur widely distributed over all suitable or available habitat within a mountain-foothill complex. In winter, however, extreme snow depths usually preclude use at higher elevations and mule deer are forced to concentrate, often at very high densities, on lower south-facing slopes where snow depths are less and stands of shrubby vegetation provide winter forage. Generally, montane and subalpine forest communities dominate summer ranges, while open, shrub-dominated slopes and ridges characterize the primary wintering areas.

In the prairie province, level and rolling plains, dominated by grasslands, provide little habitat for mule deer. Rather, it is rough, timbered or nontimbered "breaks" along river drainages, heavily dissected "badlands," and brushy streamcourses and draws that provide deer habitat (Hamlin 1978a, 1978b; Severson and Carter 1978). Mule deer do not occur at high densities over broad areas, though locally high densities may occur in rough breaks or badlands and along streamcourses. In these environments, deer tend to be nonmigratory. Use of more open prairie is usually patchy and may involve extensive movements. Ponderosa pine (*Pinus ponderosa*) and Rocky Mountain juniper (*Juniperus scopulorum*) provide cover in timbered breaks and badlands, while cottonwoods (*Populus* sp.), green ash (*Fraxinus pennsylvanica*), and box elder (*Acer negundo*) are important along streamcourses and in draws. Common and important shrubs for mule deer food and cover include sagebrush (*Artemisia* sp.), rabbitbrush (*Chrysothamnus* sp.), skunkbush sumac (*Rhus trilobata*), snowberry (*Symphoricarpos* sp.), rose (*Rosa* sp.), chokecherry (*Prunus virginiana*), buffaloberry (*Shepherdia* sp.), and willow (*Salix* sp.).

In the Southwest, mule deer are associated with two types of semidesert range; both are arid, sparsely vegetated environments dominated by shrubs. One, occurring in southern Arizona and New Mexico, western Texas, and parts of Mexico, is characterized by creosote bush (*Larrea* sp.), mesquite (*Prosopis* sp.), greasewood (*Sarcobatus* sp.), and several species of cactus. In some areas, various species of oak (*Quercus* sp.) and chaparral occur. The other, the northern or Great Basin type, occurs in parts of Nevada, western Utah, and southeastern Oregon. Common plants are sagebrush, saltbush (*Atriplex* sp.), cliffrose (*Cowania stansburiana*), and winterfat (*Eurotia* sp.). Juniper/pinyon woodlands and pine forests may occur at higher elevations.

Black-tailed deer inhabit temperate, coniferous forests occurring along the northern Pacific Coast from northern California north to southeastern Alaska, though important populations also occur in woodland-chaparral habitats of central coast ranges in California.

Coastal rain forest habitats are characterized by dense coniferous forests and a marine climate with cool temperatures, many cloudy days, and high precipitation. In California and southern Oregon, redwoods (*Sequoia* sp.) and Douglas fir (*Pseudotsuga menziesii*) typify these habitats. To the north, Sitka spruce (*Picea sitchensis*), western red cedar (*Thuja plicata*), western hemlock (*Tsuga heterophylla*), and Douglas fir predominate. Forest succession is very rapid, and the kinds and amounts of understory present vary widely depending upon the state of succession. Habitat values and use of black-tailed deer also vary with successional stage. Different successional stages may be important in different areas. For example, in western Washington, deer use was highest generally between 10 and 30 years after the forest was opened (Brown 1961). On Vancouver Island, British Columbia, and in coastal forests of southeastern Alaska, however, deer use of mature forests has been shown to be generally much higher than use of logged habitats (Schoen and Wallmo 1978).

California woodland-chaparral habitats consist of two major vegetation types, oak woodlands and chaparral. The woodlands, characterized by oak and pine, include a variety of shrubs and contain numerous grassy openings. The chaparral type is dominated by numerous shrubs, including chamise (*Adenostoma* sp.) and manzanita (*Arctostaphylos* sp.). In the absence of fire, chaparral becomes extremely dense and is little used by deer. Use of such areas can only be restored through fire or mechanical disruptions of the vegetation.

Within any broad habitat or area, mule deer usually occupy a variety of different "habitat types" or local vegetation-topographic complexes. Within these, they will utilize a variety of different plant communities or vegetation types. They will live and reproduce more successfully or occur at higher densities in some habitats than in others. Also, occupancy and use of different types is not necessarily constant, but may vary seasonally, with sex and age of the animals, population density, and other variables, including environmental conditions. Exactly how, as well as how successfully, mule deer use any habitat is determined by both the resource needs of the animals (food, cover, water, space, etc.) and where or how these resources are distributed in the environment. Mule deer require a diversity of plant species as food, at any one time and during the course of the year, such that several individual vegetation types must be available and used. A diversity of vegetation may also be required to meet their various needs for hiding, escape, and thermal cover. Because of this, the juxtaposition and/or the degree of interspersion of different food and cover or "habitat" types may be more important than the occurrence of individual types.

MOVEMENTS AND HOME RANGE

Mule deer generally confine themselves, at least seasonally, to small, individual home ranges within which only very short daily movements are necessary. Extreme movements typically occur only during migration or among males during the rut. More extensive movement also may result from unusually severe environmental conditions, such as very deep snow or extreme drought, and during the wanderings or dispersal movements of yearlings or other young adults. The size of the areas mule deer use and the amount of

movement they make vary widely between individuals, sexes, and populations and especially with the habitat occupied.

Migratory movements are characteristic of mule deer associated with mountain-foothill habitats. Distances covered in migration may vary from a few km to over 160 km. The timing of migratory movements may also vary. Fall migrations usually are related to the occurrence and depths of snow on summer and intermediate ranges, while the timing of spring migration typically has been associated with snow melt and the availability of succulent forage on summer areas. Studies in the Bridger Mountains, Montana, however, have also indicated that both the timing of migration and the areas used by adults may be traditional and vary widely between individuals (Mackie and Knowles 1977). In migratory situations, distinct summer and winter home ranges are established and used. In the Bridger Mountain study, winter home ranges varied from 40 to 819 ha for bucks and from 34 to 307 ha for does (Steerey 1979; Youmans 1979). Average winter home range size varied inversely with winter severity. Summer home ranges of does averaged 92 ha (37–207 ha) (Pac 1976; Steerey 1979), while adult males ranged over areas from 52 to 66 ha (Steerey 1979). Other studies on mountainous summer ranges have also indicated home ranges 40–100 ha in size (White 1960; Leopold et al. 1951).

Mule deer are not considered migratory in most prairie habitats, but they may migrate locally or use some parts of their year-long range more intensively than others. Also, movement out of normal home ranges may occur during periods of unusually severe environmental conditions. Year-long home ranges of 10.6 km² and 12.4 km² have been reported for females and males, respectively, on semidesert range (Rogers et al. 1978). In timbered, prairie "breaks" habitat, Hamlin (1978a) found the average year-long home range of does to be 7.0 km² (3.6–10.1 km²), while a mature buck had an annual range of 21.7 km². Severson and Carter (1978) reported that movements and home ranges of mule deer in open, flat prairie habitats were much larger than those in prairie "badlands." Home ranges in "badlands" apparently were greater than those reported for mule deer in mountain-foothill habitats.

In general, movements and home range size vary among individual deer using the same general habitat; bucks use larger areas and move more widely than does, particularly when rutting movements are included. Movements and home range size increase as distances between food, cover, and water sources increase, as well as with decreasing complexity or diversity of habitats.

Yearlings and occasionally other younger adult mule deer may move very widely, or leave a population entirely, in the process of establishing permanent home ranges. Such movements occur among young of both sexes, but appear to be especially prevalent in males. This leads to continuous genetic interchange between populations and among segments of a population.

Home range and movement characteristics of

black-tailed deer apparently are very similar to those described above for mule deer (Dasmann and Taber 1956; Taber and Dasmann 1958; Brown 1961).

POPULATION ECOLOGY

Historic Trends. Seton (1929) estimated that as many as 10 million mule deer and 3 million black-tailed deer may have existed in North America in presettlement times. Most authorities today view these estimates as too high. Deer thrive on disturbed ranges in intermediate stages of plant succession and may have become more abundant and widely distributed in North America during the past half-century than ever before. Early settlers found them scarce in many parts of the west and archeological evidence (Jennings 1957) has suggested that mule deer may not have occurred historically in much of the Great Basin area of Nevada and Utah where they have occurred abundantly in recent years. Because of this, Wagner (1978) estimated that no more than 5 million, and possibly even fewer, mule and black-tailed deer occurred in the western United States during pre-Columbian times.

Preexisting populations of both forms, but especially mule deer, declined drastically and became quite localized over most of the west during or immediately following settlement. This was the result of unrestricted hunting and use of wild animals for food, and because of disturbance and preemption of deer habitats by agriculture. By the turn of the century mule deer were generally scarce, though ''lows'' in some areas in the north may not have occurred until the 1920s and 1930s. The disturbances of settlement, especially widespread and often abusive livestock grazing, logging, and burning, ultimately proved beneficial, however, as range and forest vegetation was opened up and became more diversified. Many plants more palatable to mule deer than those that dominated the original vegetation (Longhurst et al. 1968, 1976) either invaded or increased greatly in abundance. Mule deer responded rapidly to these changes and, under restrictive hunting regulations, increasingly effective law enforcement, widespread predator control, and perhaps generally favorable weather conditions, increased in numbers and distribution throughout the west. By the early 1920s, mule deer were extremely abundant in some parts of the southwest, for example, the Kaibab Plateau, Arizona (Rasmussen 1941). From the 1920s through the 1950s increases spread north and west to the point where mule deer were of unprecedented abundance (Wagner 1978). This abundance could not be sustained, however, and within a few years after peak populations occurred in most areas, numbers began to decline. Overpopulation in many areas often reduced the abundant and nutritious forage plants on which the deer had originally thrived. Fawn production and/or survival declined and malnutrition and starvation losses during winter or other critical periods became common (Leopold et al. 1947). Hunting regulations were generally liberalized in most areas in management efforts to ''balance'' deer numbers with food supplies, prevent depredations to agricultural crops and

products, and/or harvest the large surplus of deer that prevailed. Beginning during the mid-1960s, and especially during the early 1970s, mule deer appeared to decline very sharply in numbers and distribution over most of the west. This decline, like the increases, occurred somewhat later in the north than in the south. Efforts have been made to tie this decline to many factors, including hunting, range deterioration or changes due to overbrowsing, succession and livestock grazing, predation, competition with livestock and other animals, destruction or loss of habitat due to human development, and climate/weather changes. None of these, individually or combined, satisfactorily explains population declines in all areas where they apparently occurred.

Population Characteristics and Dynamics. The abundance of mule deer, like that of other animals, is determined both by the number of deer that can occur or be supported per unit area of habitat and by the total amount of habitat available. Thus, any factor or factors that influence either local densities or the amount of area available to or usable by mule deer are important in determining population size and trends.

The habitats occupied by mule and black-tailed deer vary greatly in their ability to support deer, at any given time as well as through time, depending upon how well each habitat satisfies the environmental requirements, adaptations, and tolerances of the animals (Mackie 1978a). Mule deer densities in open, prairie-plains habitat are typically very low: usually less than 2/km² and frequently less than 0.5/km². These habitats are characterized by low relief and relatively few different vegetational types. They also tend to be quite variable with respect to meeting seasonal resource needs of deer. Because of this, mule deer must occur either widely dispersed on large individual home ranges or concentrated in the few areas where their diverse needs can be met more locally. Broken prairie ''breaks'' and ''badlands'' habitats, typified by greater topographic and vegetational diversity, meet the diverse resource needs of mule deer more consistently; higher densities, 1.5–4.5/km² or more, usually occur. Mountain-foothill habitats generally seem to support relatively high densities. Recent winter population estimates for the Bridger Mountains, Montana, indicate 4–7 deer/km² of year-long habitat at a time when mule deer numbers are relatively low. Robinette et al. (1977) estimated that the average posthunting season density of mule deer on a mountain-foothill–type range in Utah was about 16/km² during the years 1947–56, a period of extreme mule deer abundance. Mountain-foothill–type habitats tend to be highly complex and diverse environments in which the seasonal resource needs of mule deer can be consistently met within very small areas and many such areas are available to deer (Mackie 1978a). Swank (1958) estimated densities of about 4 mule deer/km² for Arizona chaparral (semidesert shrub) habitats just prior to the hunting season.

Although mule deer typically do not occur in extremely high densities based on the broad areas they inhabit year-long, winter concentrations consisting of 30–50 deer/km², and occasionally up to 200/km² or more, are not uncommon, especially in mountain-foothill habitats.

Black-tails may occur in higher-density populations than mule deer. Dasmann (1956) reported summer densities in California chaparral ranging from a high of about 55 deer/km² immediately following a fire to about 30/km², which he considered more normal for this habitat, several years later. Winter densities of two populations inhabiting unmanaged and managed (burned and seeded) chaparral were about 10 deer/km² and 23 deer/km², respectively (Taber 1956). Black-tailed deer densities in coastal forest habitats are dependent upon the time and extent of logging or burning (Brown 1961). For cne area, Brown (1961) estimated summer-fall densities to range from 13 to 22 deer/km², while on another area they ranged from about 6 to 10 deer/km², the higher values representing more favorable successional stages.

Most mule deer populations are almost constantly fluctuating. At least four different patterns of change can be recognized: seasonal, annual, periodic, and long term. Seasonal changes result from the annual dynamics of populations. Populations are highest in late spring or early summer as fawns are added. Mortality accrues through summer, fall, winter, and spring to reduce them to their lowest level immediately prior to fawning the following year. The magnitude of seasonal change depends upon the numbers of fawns born and seasonal mortality patterns and rates. Fawn mortality is often high at or immediately following parturition. Generally lower mortality prevails among both fawns and adults through summer and early fall. Exceptions occur on some southern, semidesert mule deer ranges and black-tail ranges in California woodland-chaparral, where drought conditions during fall may sharply reduce numbers of fawns and adults during that period. Most mule and black-tailed deer populations are hunted during fall; harvests tend to reduce numbers through that period. In northern environments, winter is often a critical factor in the survival of mule deer and some mortality, especially among fawns and old adults, is typical. Occasionally, it may be extensive, resulting in the loss of as many as one-third or one-half of all animals present in early winter. Early spring may also be an important mortality period in some areas or during some years.

Annual, or year-to-year, fluctuations in population size occur as a result of differences in fawn production and/or seasonal mortality rates. Because of this, density during a particular season may be higher or lower than during the previous or succeeding year.

Periodic fluctuations are general population increases and/or decreases spanning several years. They generally reflect short-term, accrued effects of high or low fawn production and recruitment and/or mortality rates. That is, several years of relatively high recruitment and/or below-average adult mortality tend to result in general population increases, while low recruitment and/or above-average mortality result in a general population decline. This type of fluctuation is illustrated in figure 44.3, which shows the general trend in one mule deer population, in the Missouri River Breaks, Montana, from 1960 through 1980. Data for this population (Mackie 1976a) show that the average annual mortality rate, from winter to winter, has been about 30–33 percent. Above-average mortality was associated with downward population trends, especially those occurring during years or periods of severe environmental conditions, such as extreme drought during 1961 and 1962 and severe winters in 1964–65 and 1971–72. Fawn production and survival, together with this "catastrophic" mortality, were the major factors influencing trends. Low fawn recruitment, such that less than 30–33 percent of the early winter population was comprised of this age class, was invariably associated with a decreasing trend (see figure 44.3). Higher recruitment led to increases, and where it occurred over several years, as from 1967 to 1970, the population grew rapidly. Because of this, any factor or combination of factors that affects fawn production and/or survival, or adult mortality rates, will also be involved in the determination of the population trend. It should also be noted, however, that periodic trends can be influenced directly by changes or fluctuations in the amount of habitat available to deer. Habitat availability is affected by the presence and activities of other animals and people or by environmental factors and conditions.

Long-term trends in mule deer populations are those that accrue as broad changes in numbers and distribution over time. Thus, a population or several populations in an area or region may fluctuate annually or periodically within either an expanding trend or a shrinking trend. Such trends reflect long-term, often very slow and subtle changes in the kinds, quality, and amounts of habitat available to deer.

The sex and age structure of a deer population provides an ongoing record of the dynamics of that population in the particular environment it occupies. Any variation in environmental conditions that affects natality and/or mortality patterns or rates can also affect population structure. For this reason, there is no "typical" sex or age structure that will generally characterize all populations or any one population over time.

An example of the manner and extent to which

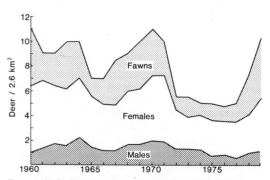

FIGURE 44.3. Population trends of mule deer (*Odocoileus hemionus*) in the Missouri River Breaks, Montana, 1960–78.

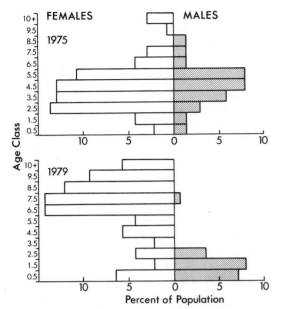

FIGURE 44.4. Change in sex-age structure of a mule deer (*Odocoileus hemionus*) population during late winter in the Bridger Mountains, Montana, 1975–79.

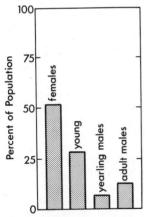

FIGURE 44.5. Average sex and age composition of mule deer (*Odocoileus hemionus*) herds in Glacier National Park during early winter, 1967/68 through 1976/77. After Martinka 1978.

population structure may change through time is shown in figure 44.4, which depicts structural changes in a Rocky Mountain mule deer population in the Bridger Mountains, Montana, from 1975 to 1979. These data also illustrate the manner in which population structure and structural changes reflect past events in the dynamics of these animals and may forecast future population trends. The predominance of middle or "prime" age animals in the spring of 1975 reflected extensive overwinter mortality of young and old animals and a 30 percent decrease in total deer numbers. Subsequent low recruitment in 1975–78 resulted in a population structure dominated by progressively older

females, a gradual elimination of older males as the year classes that remained after the 1974–75 winter were lost, and a shift to young males from recruitment in 1977–79. If recruitment remains low, the prevalence of "older" females in the 1979 population structure could foreshadow a future general decline in the number and proportion of females as these animals reach their maximum life span and leave the population.

In natural populations, adult females generally outnumber adult males by more than 2:1, even where no hunting occurs (figure 44.5). Heavily hunted populations often show winter sex ratios of only 5–25 adult males:100 females, while light to moderately hunted populations may have ratios of 20–40 males:100 females. However, natural changes in age structure (see figure 44.4) can also result in ratios of only 10–20 males:100 females in winter populations.

Extrinsic Factors Influencing Populations. Many factors play a role in the population ecology of mule and black-tailed deer, and may limit populations in one place or another and at one time or another. While habitat and nutritional limitations probably are of foremost importance, generally, weather, diseases, parasites, predation, competition with other wild and domestic ungulates, and hunting and other activities of people may also play important roles.

Weather exerts an important yet highly variable influence on deer populations. It can influence deer populations both as a stress factor during severe winter weather and as a nutritional-forage factor through precipitation. Because of its variability, weather will often influence population changes on a seasonal and annual basis, and less frequently over a period of several years. Unlike many other factors, weather can temporarily affect population levels over extensive areas.

The role of disease and parasites in limiting deer numbers is not well understood. Hibler (1981) discusses most of the diseases and parasites known to occur in mule deer and black-tailed deer. While all may cause, directly or indirectly, the death of individual animals, there has been no documentation of massive population impacts of these factors in wild populations. Mule deer and black-tailed deer have coevolved with most of the diseases and parasites infecting them, thereby decreasing the likelihood that important population consequences of these factors can occur. Since wild animal populations cannot be "hospitalized" and treated, populations present today are likely to contain a high proportion of resistant animals that are the result of thousands of years of natural selection. Diseases or parasites introduced from some other area or species with which the deer did not coevolve could possibly have dramatic population consequences. Some diseases and parasites are always present within deer populations, but their importance seems to vary in relation to other predisposing factors. For example, an animal on a poor nutritional regime is more likely to die as a result of infection by disease or parasites than an animal living on a good nutritional regime. Mule deer live in dry environments that are not as conducive to chronic disease and parasitic problems as are humid environments.

While the deaths of individual deer due to predation have been well documented, the population effects of those deaths is less certain. There is much current research on this subject, but most has not been published at this time. Some recent work has indicated that in at least some situations and at some time, predation may have population impacts. Coyotes (*Canis latrans*) primarily kill young deer, but may kill adults in some situations. Mountain lions (*Felis concolor*) are capable of killing any deer, given the right situation. Bobcats (*Felis rufus*) and golden eagles (*Aquila chrysaëtos*) occasionally kill young deer. Whether these losses to predation are merely compensatory or affect the ultimate population levels has yet to be determined in most cases.

The influences of sport hunting usually depend upon how well the harvest is regulated in relation to population size and dynamics, and on the vulnerability of deer to hunting. Flexible and opportunistic regulations allow for the harvest of surpluses when they are available. When segments of the population with high natural mortality rates are not available, liberal regulations and heavy hunting pressure may shift the harvest to population segments with low natural mortality rates. Persistent heavy hunting pressure in low-security habitats may eliminate deer from some areas and also directly limit the total population size. Security of deer has traditionally been related only to habitat characteristics. Vulnerability of deer to hunters may be increased through changes in leisure time, mobility, financial status, and equipment used by hunters.

Mule deer may compete for food, cover, and space with all other wild and domestic ungulates on western rangelands (Mackie 1976*b*, 1978*b*). The extent to which competition actually occurs or is important in the population ecology of mule deer depends upon many factors, so generalization is difficult, if not impossible.

Humans and their related activities have been, and continue to be, influential in mule deer population changes. As described earlier, widespread population declines can be traced to human activities during the middle to late 1800s. Conversely, extensive population increases in more recent times can be partially traced to human management activities combined with the beneficial consequences of natural systems working on habitats formerly altered by people. The significance of mankind's past influence on mule deer lies in the fact that the alterations were all within the limits of tolerence of the animal. In modern times, people have the ability to alter environments beyond these limits by completely usurping space needed by deer, which can effectively eliminate their basis for existence.

MANAGEMENT

Management actions and practices applied to mule and black-tailed deer have varied greatly over the past century as well as from one part of their range to another. In general, the earliest efforts, in the late 1800s and early 1900s, were directed to the protection of dwindling or remnant populations. These included close regulation or prohibition of hunting, establishment of refuges or protective game ranges, and predator control programs. Later, in some places, efforts to restore populations in suitable habitat included trapping and transplanting deer. The role and importance of any one of these approaches in influencing historical population trends in any one area may be debated. Collectively, they are credited with at least helping to speed the recovery and spread of mule deer throughout the west.

As populations increased, often to the point where damage to range forage plants and agricultural crops was apparent, management emphasis shifted from rigid protection to population control. Almost everywhere, efforts were directed toward liberalizing hunting regulations to "balance" deer populations with their forage supplies on important or "key" habitats, to prevent depredations to agricultural crops and products, and to provide recreational hunting and the use of the large annual surplus of deer that might otherwise be lost to starvation and other natural mortality. As a basis for management, emphasis was also placed on research studies and surveys to learn more about the biology and ecology of mule deer and their habitats. This new "science" gradually provided a framework of methods, concepts, and criteria for deer management based on facts. Thus, management efforts came to focus on the systematic collection of data on deer population characteristics and trends, utilization and condition trends of important forage plants, especially on winter ranges, and deer harvests. These data, interpreted in light of existing knowledge of the habitat relationships of deer, provided the factual basis for ascertaining management needs and developing and "selling" harvest recommendations. At the same time, information and education programs to inform sportsmen and the general public about management needs and efforts and to develop public understanding and acceptance of these efforts became an integral part of mule deer management.

Generally, these same approaches are followed in management of mule and black-tailed deer today. Individual programs have become more refined with accumulating knowledge of the biology, behavior, and ecology of deer and the application of new technology (for example, radio-telemetry and computer-based simulation modeling) in management.

Population control and regulation of hunting continue to dominate the management efforts of state wildlife agencies. In part, this reflects their basic legal responsibility; however, it is equally important that hunting is the only factor influencing deer populations that state wildlife agencies can directly control. Although most of these agencies include habitat acquisition in their deer management programs, few control more than a very small fraction of the total habitat occupied by mule and black-tailed deer. Most winter range in mountainous areas is privately owned. The remainder is federal land administered by the U.S. Forest Service, Bureau of Land Management, and other federal agencies for many different uses. The manner and the extent to which these lands are used for other purposes determine the amount and quality of habitat available and, ultimately, the distribution and abundance of deer.

The general decline of mule deer populations throughout the west during the late 1960s and early 1970s, together with a growing public concern for management of all natural resources on public lands in particular, paved the way for increased emphasis on habitat management. With the decline, concern was not with too many deer on most ranges but with too few, and with the reasons for this. As a result, much effort has been directed to assessing or reassessing the relationships between deer populations and habitat factors and to protecting or maintaining important habitats threatened by conflicting land uses. Public concerns fostering increased involvement of federal agencies in wildlife management have led to greater consideration of the habitat needs of mule and black-tailed deer and to expanded opportunities for deer habitat management on public lands. Recently federal agencies have hired more field-level biologists to evaluate habitat conditions and problems and develop land management policies. As a result, cooperative efforts to actively manage, maintain, and enhance deer habitats have been established and expanded in most areas.

Effective deer management includes many different programs and activities. Different objectives may be developed, depending upon the species, time, place, and agency involved. The essential components are: (1) a sound research program to provide and expand knowledge about the biology, behavior, and ecology of deer and the application for management; (2) management surveys to determine and assess habitat and population characteristics and trends; (3) information and education to further public understanding and support of management programs; (4) law enforcement to assure that laws and regulations developed to manage deer populations are followed; and (5) habitat management, including acquisition, maintenance, and physical alteration.

Within this broad framework, there is no generally accepted philosophy or widely applicable formula for management of mule and black-tailed deer. Just as specific management needs and opportunities can vary widely—between subspecies, between areas and broad habitat types, and even between local populations and habitats—specific methods or techniques that may be employed in management will also vary. Discussions of the use and merits of the many approaches and specific techniques that have been or may be applied fill a voluminous literature and are similar in many respects to those discussed for white-tailed deer (chapter 45). It should be noted, however, that what is useful or may provide benefits in one place or under one set of circumstances may be worthless or even harmful to others. Thus, each potential approach or technique must be carefully selected and evaluated with respect to the particular management situation or need to which it is to be applied.

This chapter was a contribution of the Montana Department of Fish, Wildlife and Parks, Federal Aid in Wildlife Restoration Project W-120-R.

LITERATURE CITED

Anderson, A. E., and Medin, D. E. 1964. Reproductive studies. Job Completion Rep., Fed. Aid Proj. W-105-R-3. Pages 239-269 *in* Colorado Game and Fish Dept., Game Res. Rep., January 1964.

Anderson, A. E.; Medin, D. E.; and Bowden, D. C. 1974. Growth and morphometry of the carcass, selected bones, organs, and glands of mule deer. Wildl. Monogr. 39. 122pp.

Bissell, H. 1959. Interpreting chemical analyses of browse. California Fish Game 45:57-58.

Brown, E. R. 1961. The black-tailed deer of western Washington. Washington Dept. Game Bio. Bull. 13. 124pp.

Bubenik, A. G. 1968. The significance of the antlers in the social life of the Cervidae. Deer 1:208-214.

Cowan, I. McT. 1956. Life and times of the coast black-tailed deer. Pages 523-617 *in* W. P. Taylor, ed. The deer of North America. Stackpole Co., Harrisburg, Pa., and Wildl. Manage. Inst., Washington, D.C. 668pp.

Cowan, I. McT., and Guiguet, C. J. 1965. The mammals of British Columbia. British Columbia Prov. Mus. Handb. 11. 414pp.

Cowan, I. McT., and Raddi, A. G. 1972. Pelage and molt in the black-tailed deer (*Odocoileus hemionus*, Rafinesque). Can. J. Zool. 50:639-647.

Cowan, I. McT., and Wood, A. J. 1955. The growth rate of the black-tailed deer (*Odocoileus hemionus columbianus*). J. Wildl. Manage. 19:331-336.

Dasmann, R. F. 1956. Fluctuations in a deer population in California chaparral. Trans. North Am. Wildl. Conf. 21:487-499.

Dasmann, R. F., and Taber, R. D. 1956. Behavior of Columbian black-tailed deer with reference to population ecology. J. Mammal. 37:143-164.

Davis, R. W. 1962. Studies on antler growth in mule deer (*Odocoileus hemionus hemionus*, Rafinesque). Pages 61-64 *in* Proc. 1st Natl. White-tailed Deer Dis. Symp., Univ. Georgia, Athens.

French, C. E.; McEwen, L. C.; Magruder, N. D.; Ingram, R. H.; and Swift, R. W. 1955. Nutritional requirements of white-tailed deer for growth and antler development. Pennsylvania Agric. Exp. Stn. Bull. 600. 50pp.

Geist, V. 1968. Horn-like structures as rank symbols, guards and weapons. Nature 220:813-814.

———. 1981. Behavior: adaptive strategies in mule deer. Pages 157-223 *in* O. C. Wallmo, ed. Mule and black-tailed deer of North America. Univ. Nebraska Press, Lincoln.

Goss, R. J. 1963. The deciduous nature of deer antlers. Pages 339-369 *in* P. Sognnaes, ed. Mechanisms of hard tissue destruction. Am. Assoc. Advancement Sci. Publ. 75. 764pp.

Hamlin, K. L. 1978*a*. Mule deer population ecology, habitat relationships, and relations to livestock grazing management and elk in the Missouri River Breaks, Montana. Pages 141-183 *in* Montana Deer Studies, Prog. Rep., Fed. Aid in Wildl. Restor., Proj. W-120-R-9, Montana Dept. Fish Game, Helena. 217pp.

———. 1978*b*. Population ecology and habitat relationships of mule deer and white-tailed deer in the prairie-agricultural habitats of eastern Montana. Pages 185-197 *in* Montana Deer Studies, Prog. Rep., Fed. Aid in Wildl. Restor., Proj. W-120-R-9, Montana Dept. Fish Game, Helena. 217pp.

Hawkins, R. E., and Klimstra, W. D. 1970. A preliminary study of the social organization of white-tailed deer. J. Wildl. Manage. 34:407-419.

Hibler, C. P. 1981. Diseases. Pages 129-155 *in* O. C. Wallmo, ed. Mule and black-tailed deer of North America. Univ. Nebraska Press, Lincoln.

Hudson, P., and Browman, L. G. 1959. Embryonic and fetal development of the mule deer. J. Wildl. Manage. 23:295–304.

Hunter, J. W. 1924. Deer hunting in California. California Fish Game 10:18–24.

Jennings, J. D. 1957. Danger cave. Univ. Utah Anthropol. Pap. 27:328pp.

Kufeld, R. C.; Wallmo, O. C.; and Feddema, C. 1973. Foods of the Rocky Mountain mule deer, U.S.D.A. For. Serv. Res. Pap. RM-111. 31pp.

Leopold, A.; Riney, T.; McCain, R.; and Tevis, L., Jr. 1951. The Jawbone deer herd. California Dept. Fish Game Bull. 4. 139pp.

Leopold, A.; Sowls, L. K.; and Spencer, D. L. 1947. A survey of overpopulated deer ranges in the United States. J. Wildl. Manage. 11:162–177.

Linsdale, J. M., and Tomich, P. Q. 1953. A herd of mule deer: a record of observations made on the Hastings Natural History Reservation. Univ. California Press, Berkeley. 567pp.

Longhurst, W. M.; Garton, E. O.; Heady, H. F.; and Connoly, G. E. 1976. The California deer decline and possibilities for restoration. Annu. Meet., Western Sect. Wildl. Soc., Fresno, Calif. 41pp. Multilith.

Longhurst, W. M.; Oh, H. K.; Jones, M. B.; and Kepner, R. E. 1968. A basis for the palatability of deer forage plants. Trans. North Am. Wildl. Conf. 33:181–189.

Loveless, C. M. 1967. Ecological characteristics of a mule deer winter range. Tech. Bull. 20. Colorado Game, Fish and Parks Dept., Denver. 124pp.

Low, W. G., and Cowan, I. McT. 1963. Age determination of deer by annular structure of dental cementum. J. Wildl. Manage. 27:466–471.

Mackie, R. J. 1964. Montana deer weights. Montana Wildl. (Winter 1964), 9–14.

———. 1970. Range ecology and relations of mule deer, elk and cattle in the Missouri River Breaks, Montana. Wildl. Monogr. 20. 79pp.

———. 1976a. Mule deer population ecology, habitat relationships, and relations to livestock grazing management and elk in the Missouri River Breaks, Montana. Pages 68–94 in Montana Deer Studies, Prog. Rep., Fed. Aid in Wildl. Restor., Proj. W-120-R-7, Montana Dept. Fish Game, Helena. 170pp.

———. 1976b. Interspecific competition between mule deer, other game animals and livestock. Pages 49–54 in Mule deer decline in the West: a symposium. Utah State Univ. Coll. Nat. Resour., Agric. Exp. Stn., Logan.

———. 1978a. Natural regulation of mule deer populations. 1978 N. W. Sect. Symp. Nat. Regulation of Wildl. Populations. Vancouver, B.C.

———. 1978b. Impacts of livestock grazing on wild ungulates. Trans. North Am. Wildl. and Nat. Res. Conf. 43:462–476.

Mackie, R. J.; Hamlin, K. L.; and Mundinger, J. G. 1976. Habitat relationships of mule deer in the Bridger Mountains, Montana. Prog. Rep., Fed. Aid in Wildl. Restor., Proj. W-120-R-6, Montana Dept. Fish Game. 46pp. Multilith.

Mackie, R. J.; and Knowles, C. J. 1977. Population ecology and habitat relationships of mule deer in the Bridger Mountains, Montana. Pages 47–74 in Montana Deer Studies. Prog. Rep., Fed. Aid in Wildl. Restor., Proj. W-120-R-8. Montana Dept. Fish Game, Helena. 168pp.

Mackie, R. J.; Pac, D. F.; and Jorgensen, H. E. 1978. Population ecology and habitat relationships of mule deer in the Bridger Mountains, Montana. Pages 83–128 in

Montana Deer Studies. Prog. Rep., Fed. Aid in Wildl. Restor., Proj. W-120-R-9. Montana Dept. Fish Game, Helena. 217pp.

Martinka, C. J. 1978. Ungulate populations in relation to wilderness in Glacier National Park, Montana. Trans. North Am. Wildl. Nat. Resour. Conf. 43:351–357.

Miller, F. L. 1974. Four types of territoriality observed in a herd of black-tailed deer. Pages 644–660 in V. Geist and F. R. Walther, eds. The behavior of ungulates and its relation to management. I.U.C.N. Publ., New Ser. 24, Morges, Switzerland.

Pac, D. F. 1976. Distribution, movements and habitat use during spring, summer and fall by mule deer associated with the Armstrong winter range, Bridger Mountains, Montana. M.S. Thesis. Montana State Univ., Bozeman. 120pp.

Rasmussen, D. I. 1941. Biotic communities of Kaibab Plateau, Arizona. Ecol. Monogr. 3:229–275.

Rees, J. W.; Kainer, R. A.; and Davis, R. W. 1966. Chronology of mineralization and eruption of mandibular teeth in mule deer. J. Wildl. Manage. 30:629–631.

Robinette, W. L.; Baer, C. H.; Pillmore, R. E.; and Knittle, C. E. 1973. Effects of nutritional change on captive mule deer. J. Wildl. Manage. 37:312–326.

Robinette, W. L.; Gashwiler, J. S.; Jones, D. A.; and Crane, H. S. 1955. Fertility of mule deer in Utah. J. Wildl. Manage. 19:115–136.

Robinette, W. L.; Hancock, N. V.; and Jones, D. A. 1977. The Oak Creek mule deer herd in Utah. Utah State Div. Wildl. Res., Salt Lake City. 194pp.

Robinette, W. L.; Jones, D. A.; Rogers, G.; and Gashwiler, J. S. 1957. Notes on tooth development and wear for Rocky Mountain mule deer. J. Wildl. Manage. 21:134–153.

Rogers, K. J.; Ffolliott, P. F.; and Patton, D. R. 1978. Home range and movement of five mule deer in a semidesert grass-shrub community. U.S.D.A. For. Serv. Res. Pap. RM-355. 6pp.

Schoen, J. W., and Wallmo, O. C. 1978. Timber management and deer in southeast Alaska: current problems and research direction. Pages 69–85 in O. C. Wallmo and J. W. Schoen, eds. Sitka black-tailed deer: proceedings of a conference in Juneau, Alaska. U.S.D.A. For. Serv., Alaska Region, and Alaska Dept. Fish Game, Juneau. Series no. R10-48. 231pp.

Seton, E. T. 1929. Lives of game animals. Vol. 3, pt. 1. Doubleday, Doran, & Co., Inc., Garden City, N.Y. 412pp.

Severinghaus, C. W. 1949. Tooth development and wear as criteria of age in white-tailed deer. J. Wildl. Manage. 13:195–216.

Severson, K. E., and Carter, A. V. 1978. Movements and habitat use by mule deer in the northern Great Plains, South Dakota. Proc. 1st Int. Rangelands Congr., Denver, Colo.

Short, H. L. 1981. Nutrition and metabolism. Pages 99–127 in O. C. Wallmo, ed. Mule and black-tailed deer of North America. Univ. Nebraska Press, Lincoln.

Steerey, W. F. 1979. Distribution, range use, and population characteristics of mule deer associated with the Schafer Creek winter range, Bridger Mountains, Montana. M.S. Thesis. Montana State Univ., Bozeman. 119pp.

Swank, W. G. 1958. The mule deer in Arizona chaparral. Arizona Game and Fish Dept. Wildl. Bull. 3. 109pp.

Swift, R. W. 1948. Deer select most nutritious forages. J. Wildl. Manage. 12:109–110.

Taber, R. D. 1956. Deer nutrition and population dynamics in the North Coast Range of California. Trans. North Am. Wildl. Conf. 21:159–172.

Taber, R. D., and Dasmann, R. F. 1958. The black-tailed deer of the chaparral; its life history and management in the North Coast Range of California. Game Bull. 8, California Fish Game Dept., Sacramento. 163pp.

Thorne, E. T. 1975. Normal body temperature of pronghorn antelope and mule deer. J. Mammal. 56:697–698.

Verme, L. J. 1963. Reproduction studies on penned white-tailed deer. J. Wildl. Manage. 29:74–79.

Wagner, F. H. 1978. Effects of livestock grazing and the livestock industry on wildlife. Pages 121–145 *in* H. Brokaw, ed. Wildlife and America. Council on Environ. Quality, Washington, D.C. 532pp.

Wallmo, O. C. 1978. Mule and black-tailed deer (*Odocoileus hemionus*). Pages 31–42 *in* D. L. Gilbert and J. L. Schmidt, eds. Big game of North America: ecology and management. Wildl. Manage. Inst., Stackpole Books, Inc., Harrisburg, Pa. 494pp.

———. 1981. Mule and black-tailed deer distribution and habitats. Pages 1–25 *in* O. C. Wallmo, ed. Mule and black-tailed deer of North America. Univ. Nebraska Press, Lincoln.

Wallmo, O. C.; Carpenter, L. H.; Regelin, W. L.; Gill, R. B.; and Baker, D. L. 1977. Evaluation of deer habitat on a nutritional basis. J. Range Manage. 30:122–127.

White, K. L. 1960. Differential range use by mule deer in the spruce-fir zone. Northwest Sci. 34(4):118–126.

Willms, W., and McLean, A. 1978. Spring forage selection by tame mule deer on big sagebrush range, British Columbia. J. Range Manage. 31:192–199.

Wood, A. J.; Cowan, I. McT.; and Nordan, H. C. 1962. Periodicity of growth in ungulates as shown by deer of the genus *Odocoileus*. Can. J. Zool. 40:593–603.

Youmans, H. B. 1979. Habitat utilization by mule deer of the Armstrong winter range, Bridger Mountains, Montana. M.S. Thesis. Montana State Univ., Bozeman. 66pp.

RICHARD J. MACKIE, Department of Biology, Montana State University, Bozeman, Montana 59717.

KENNETH L. HAMLIN, Montana Department of Fish and Game, Lewistown, Montana 59457.

DAVID F. PAC, Montana Department of Fish and Game, Bozeman, Montana 59717.

45

White-tailed Deer

Odocoileus virginianus

William T. Hesselton
RuthAnn Monson Hesselton

NOMENCLATURE

COMMON NAMES. Virginia deer, white-tail(ed) deer, white-tail
SCIENTIFIC NAME. *Odocoileus virginianus*
SUBSPECIES NORTH OF MEXICO. *O. v. virginianus, O. v. borealis, O. v. dacotensis, O. v. ochrourus, O. v. leucurus, O. v. couesi, O. v. texanus, O. v. macrourus, O. v. mcilhennyi, O. v. taurinsulae, O. v. venatorius, O. v. hiltonensis, O. v. nigribarbis, O. v. osceola, O. v. seminolus,* and *O. v. clavium.*

There are only two species of *Odocoileus* in the Americas, indeed in the entire world: *Odocoileus virginianus* and *O. hemionus,* the mule deer or blacktailed deer. These two species belong to the order Artiodactyla, family Cervidae.

It is generally accepted that North American cervids, including elk (*Cervus elaphus*), moose (*Alces alces*), and caribou (*Rangifer tarandus*), arrived on this continent at various times from the middle Miocene to the late Pleistocene, between 1 million and 18 million years ago (Goodwin 1961; Kellogg 1956; Severinghaus and Cheatum 1956). Larger members of the deer family in North America (elk, moose, caribou) have similar counterparts in Europe and Asia. There are no similar European or Asian counterparts to *Odocoileus* spp.; they are peculiar to the Western hemisphere.

DISTRIBUTION

The white-tailed deer is one of the most adaptable animals in the world. Its range extends from the southern tip of Hudson Bay well into South America. It is present in 45 of the contiguous 48 states, probably totally absent in Utah and rare at best in Nevada and California. It is found in all Canadian provinces except Newfoundland-Labrador, Prince Edward Island, Yukon, and Northwest Territories (figure 45.1). To the south, the range encompasses all of Mexico and Central America, including several islands off the Panamanian Coast (Kellogg 1956; Goodwin 1961).

It had long been believed that this was the southern range limit of *O. virginianus* and that the deer of

FIGURE 45.1. Distribution of the white-tailed deer (*Odocoileus virginianus*).

South America were *O. cariacou.* More recent data indicate that the *Odocoileus* of the northwestern half of South America are in fact *O. virginianus,* with eight new subspecies being described (Halls 1978).

Needless to say, the remarkably adaptive white-tailed deer lives in areas where climatic factors vary from humid, tropical jungle to dry, hot desert to northern subarctic conditions. The northern limits of distribution are probably a combination of the duration of severe cold and snow depths that effectively immobilize deer for extended periods of time.

The white-tailed deer probably has extended its range northward as a result of forestry and agricultural practices. These range extensions are also temporary in the sense that if these practices ceased or were altered so that the habitat returned to primeval conditions, deer distribution would decline.

In addition to adapting to a variety of range conditions, white-tails are very tolerant of people and their practices. There are white-tails living almost within the shadows of New York's skyscrapers. Conflicts between deer and agricultural plantings and backyard gardens are common.

DESCRIPTION

Body Size and Growth. The range in size of white-tailed deer is extreme. A mature *O. v. borealis* on good range can attain a weight of 192 kg live weight or 160 kg hog-dressed weight (Seton 1929). Every year bucks with dressed weights in the vicinity of 115–135 kg are taken in many northern states and provinces. Female deer seldom achieve a whole weight greater than 90 kg. In contrast to these large deer are Florida Keys white-tails (*O. v. clavium*). Large specimens of this subspecies may weigh no more than 22–25 kg (Madson 1961).

White-tail bucks achieve their maximum size at four to five years of age. Females probably achieve their maximum size about one year earlier. Total length of a large male specimen of *O. v. borealis* may be 2.4 m and shoulder height may be about 1 m (Kellogg 1956). *O. v. clavium* may have a total length of approximately half that size.

The weight and skeletal size of fawns at birth vary considerably depending on several factors. Nutritional status of the doe and number of embryos in utero are the two most important factors. In general, male fawns of the northern subspecies will weigh between 3 and 4 kg at birth; female fawns will weigh approximately 2.5–3.5 kg. Haugen and Davenport (1950) reported a male fawn with a weight of close to 7 kg at birth. Severinghaus reported a female fawn of 4.5 kg at birth (Severinghaus and Cheatum 1956).

General Description. The white-tail is distinguished by its long bushy tail, long legs with narrow pointed hooves, conspicuous ears, naked nose pad, and antlers (in males). The tail is white on the underside and varies in color from gray to brown to a reddish brown on the dorsal surface. The tail is carried erect like a flag when the animal is disturbed.

Like other cervids, the white-tail walks on the extreme tips of its toes, an adaptation for cursorial locomotion (Goodwin 1961). The toes form split hooves of keratinaceous material.

Pelage. The coat of the white-tail exhibits seasonal variation; animals undergo two complete molts per year. The summer coat is a reddish color; the hairs are short, thin, and somewhat wiry. In northern latitudes it does not provide much protection from insects. The summer coat is usually replaced in late summer or early fall with the winter or blue gray coat. Winter hair is much different from summer hair. The hairs are longer, thicker, and more brittle. Their insulating qualities are excellent. The hairs continue to grow in length and diameter, probably achieving their greatest size in midwinter. Toward spring, the hairs seem to become more brittle and the ends break off easily. Coupled with the normal molting that occurs in early spring, this accounts for the ragged appearance of deer during this period.

Fawns are reddish brown with white dorsal spots.

They begin to lose their spots at about three to four months of age. Fawns molt in late summer or early fall, with their new coat being the same as the adult winter coat.

Adult deer during all seasons have a distinctive white band across the nose and a less noticeable white eye ring. There is also a large white patch or bib in the throat region. All underparts, including the insides of the legs, are white.

Recorded incidents of melanistic white-tails are very rare. Seton (1929) reported, "Melanism has not yet been recorded for the species." Three cases of partial melanism were noted in the Adirondack Mountains of New York by Townsend and Smith (1933).

Albinism or partial albinism, however, appears to be much more common. Hesselton (1969b) worked for several years with a confined deer herd that included several hundred partial albinos. Of several hundred white white-tails examined, we can recall seeing only one true albino. This particular deer had pink eyes and was noticeably whiter than the other deer—all of the deer were completely white, with no "piebalds" or "paints." The appearance of these deer at Seneca Army Depot is similar to that of those studied and reported on Grand Island, Michigan, during the early years of this century (Shiras 1935). White deer occasionally appear in many places. There is a persistent population on the New York–Vermont border and another in central New Hampshire.

Glands. White-tails have four prominent sets of cutaneous glands: the preorbitals, at the corners of the eyes; the metatarsals, on the outsides of the hind legs between the ankles and the hooves; the tarsals, on the insides of the hind legs at the hock, or heel, joint; and the interdigitals, between the toes on each foot. The preorbitals are tear glands that lubricate and cleanse the eyes. The tarsal gland consists of an aggregation of enlarged sebaceous (oil) glands and some enlarged sudoriferous (sweat) glands. It secretes an oily material with a strong smell of ammonia. This ammonia smell results in part from the habit of adult animals of both sexes of deliberately urinating on the gland tufts. When excited, the deer raises the hairs of the tufts, and the glands then emit the musky odor. The metatarsal glands are composed essentially of coiled, tubular, sudoriferous glands. These glands also secrete an oily substance with a pungent, musky odor. The interdigital glands consist of a shallow pouch opening on the dorsal side of the interungual ligament. The surface of the skin lining the gland bears numerous papillae. From the summit of each papilla arises a single long hair that projects from the opening of the gland and conducts the secretion between the hoofs. Both sebaceous and sudoriferous glands are present, with the latter in the majority (Severinghaus and Cheatum 1956). All of these glands except the preorbitals probably are used as sensors or communicators of varying degrees.

Antlers. The antlers of deer are deciduous bony structures that grow out from the permanent pedicles of the frontal bones (Waldo and Wislocki 1951). Antlers are covered with skin and hair only as they grow, and are continuous with the frontal bones. The skin and hair appears as a soft, velvety membrane that is highly vascularized. The vessels transport the required nutrients for antler growth. Antler growth on white-tailed deer occurs on an annual cycle. Antlers are shed or cast each year following the rut or breeding season. New antler growth usually starts in early April; increasing daylight presumably causes the pituitary to secrete a hormone that stimulates antler growth. As spring and summer elapse, testosterone is secreted, which stimulates further growth of antlers and growth of testes during late summer. As testes "mature" they begin secreting increasing levels of testosterone into the bloodstream, with a peak occurring about early November. Prior to the peak of testosterone secretion, antler growth ceases and the antlers harden. The flow of blood is stopped and the velvet membrane dries up, cracks, splits, and falls off. This is usually associated with prerut activity involving the rubbing of antlers on bushes and small trees. All of this acts to polish the antlers, and they soon become the sparring weapons needed for rutting activity.

Antlers are retained in this form for a period of only a few months. Antler shedding associated with rapid decreases in testosterone levels occurs as early as December in some instances and continues until about March. Most antlers are shed by early January.

Antler growth is directly related to genetics and nutrition, with nutrition playing the greatest role in the size of antlers. Numerous studies have documented the association between antler size and nutrition, and work is continuing (French et al. 1955; Long et al. 1959; Hesselton and Sauer 1973).

Many provinces and states routinely measure and record antler data at deer checking stations during fall hunting seasons. The relationship between poor antler growth and combinations of poor range and/or overpopulated deer ranges is commonly recognized. The variation in antler size between deer of the same age can be pronounced (figure 45.2). Young deer are more sensitive to habitat deficiencies than are older deer, and therefore reflect such deficiencies in their own physical condition. Yearling male white-tails on excellent range are capable of growing large multibranched antlers. In many instances yearling males on good range exhibit better antler growth than older males on relatively poor range. It is not uncommon on poor or overstocked range never to see a yearling deer with multibranched antlers. Severinghaus and Cheatum (1956) documented that in such areas it is not unusual for 50 percent of the yearling males, about 15 percent of 2½-year-old males, and perhaps 5 percent of the 3½-year-old males to have spike antlers less than 8 cm in length.

Occasionally, an antlered female occurs. Usually these antlers are unbranched and covered with velvet. Perhaps 1 in every 1,000 to 1,100 does produce antlers (Hesselton unpublished data). These antlers are not annual growth, but generally remain for the life of the animal.

On deer with multibranched antlers, the main beam of the antler originates from the permanent pedi-

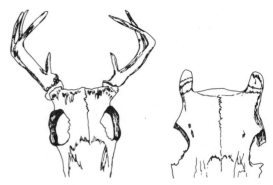

FIGURE 45.2. Contrast in antler development of white-tailed deer (*Odocoileus virginianus*) yearling bucks on good range (left) versus poor range (right).

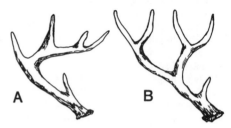

FIGURE 45.3. Contrast between the antler structure of white-tailed deer (*Odocoileus virginianus*) (A) and that of mule deer (*Odocoileus hemionus*) (B).

cles of the frontal bone, rises from back of the head behind the eyes, then curves slightly upward, forward, outward, and then inward over the face. A series of unbranched tines or points project upward from the main beam. Usually there is a small, short spike, or brow tine, projecting upward from just above the base of the beam.

A buck produces its first set of antlers when it is one year old. Actually, the deer is 16–18 months old when the antlers are polished or hardened. Though their condition is directly related to nutrition, the form and configuration of antlers in an individual animal are usually the same year after year, with the size increasing as the deer attains maximum size (approximately four to six years old).

Antlers on white-tails differ from those of mule deer or black-tails in that all points arise directly from the main beam, as opposed to the dichotomous branching on mule deer and black-tails (figure 45.3).

Skull and Dentition. "The skull of the white-tailed deer may be distinguished from skulls of other cervids by the vomer dividing the nares into two separate chambers posteriorly, and by the width of the slightly inflated auditory bullae, which is equal to the length of the bony tube leading to the meatus. The lacrimal vacuity is very large, and the lacrimal fossae are small and shallow" (Godin 1977:262) (figure 45.4).

The dental formula for adult deer with permanent dentition is: 0/3, 0/1, 3/3, 3/3 = 32 teeth. Upper incisors are missing and the lower canines are incisiform. Full, permanent dentition is attained at approximately 16–18 months, with full eruption of the upper third molars occurring at 22–24 months. In fawns, eruption of the incisors through the gum takes place prior to birth. By 4 weeks of age, four incisiform teeth and two premolars are partially or fully erupted. By 10 weeks the third premolar has erupted. All of these teeth are deciduous. By 7 months, the first permanent molar has erupted. By 13 months, the second molar has erupted. The third molar appears shortly after, but is not fully erupted in the mandible until 19 months of age.

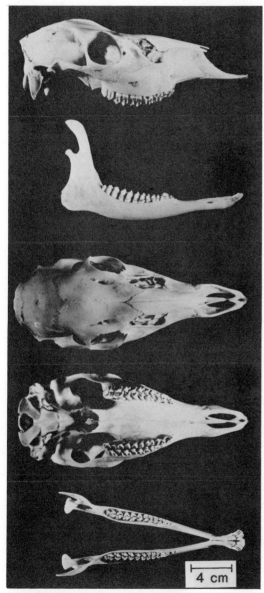

FIGURE 45.4. Skull of the white-tailed deer (*Odocoileus virginianus*). From top to bottom: lateral view of cranium, lateral view of mandible, dorsal view of cranium, ventral view of cranium, dorsal view of mandible.

REPRODUCTION

Reproductive Rates. Female white-tails on good range are capable of breeding during their first year, i.e., ovulating and successfully breeding at approximately six to seven months of age. As many as 60–70 percent of female deer in this age class do breed (Haugen and Trauger 1962; Hesselton and Jackson 1974). Unquestionably, successful reproduction among females of this age class is directly related to nutritional condition. Occasionally these fawn females have multiple "litters." Hesselton and Jackson (1974) found a total of 169 embryos in utero from a total of 130 pregnant females examined, a mean of 1.3 embryos per female. In contrast, does of the same age class from poor or overstocked range never breed.

Reproductive rates of older females also vary considerably. Healthy, mature females on good range almost always bear twins every year; 10–15 percent of them will bear triplets. In contrast, W. T. Hesselton has examined reproductive tracts of does that were 3½ –4½ years of age from very poor range (Vermont and Antacosti Island, Province of Quebec) that exhibited no evidence of ovulation. Their condition was so poor that estrus never occurred. Hesselton and Jackson (1974) indicated the tremendous difference in reproductive rates that can occur between does from poor range and does of the same ages from good range. Yearlings averaged 1.01 embryos on poor range versus 1.8 on good range; does over 2½ years old averaged 1.4 embryos on poor range versus 1.9 on good range.

Verme (1965), working in Michigan with pen-raised deer subjected to diets of varying nutritional quality, substantiated this contrast. He also indicated that does on poor quality diets were unable to sustain their fawns even after they were born.

Breeding and Gestation. White-tails are autumn breeders, with the peak of breeding occurring during the first two weeks of November in northern provinces and states. There appears to be a gradation in breeding dates related to latitude. Breeding occurs in January and February among white-tails in Arizona and Mexico (Nichol 1938; McCabe and Leopold 1951). It could be anticipated that white-tails living even closer to the equator might breed almost on a year-round basis; however, there is no evidence to support this.

Estrus usually lasts about 24 hours, and, if the doe is not bred, will recur in about 28 days. White-tails have about a 201-day gestation period. There are many records of does being bred one to two months beyond the peak of the breeding season. Almost invariably these are seven- to nine-month-old does.

The breeding season of white-tails is a frenzied period. Almost everyone has seen pictures or mounted heads of adult bucks with their antlers intertwined from fighting. A large white-tail is a dangerous animal during the rutting season. There have been numerous incidents of deer "keepers" or caretakers in zoos and parks being mauled, and occasionally killed, by a rutting buck.

The question of the ability of buck fawns (six to nine months old) to breed in the wild is still unan-

TABLE 45.1. Fetal sex ratios of white-tailed deer (*Odocoileus virginianus*) in the northeastern region of the United States and Canada, 1958–61

Age of Dam (Years)	Number of Male Embryos	Number of Female Embryos	Males per 100 Females
1	148	102	145.1
2	409	342	119.6
3	326	305	106.9
4	205	202	101.5
5	150	143	104.9
6	106	91	116.5
7	64	55	116.4
8	24	26	92.3
9	9	19	47.4
10	7	7	100.0
11+	11	8	137.5
Total	1,459	1,300	112.2

SOURCE: McDowell 1962.

swered. The only incidents of this occurring that we are aware of involved captive deer (Lambiase et al. 1972; Silver 1965). At least one doe was bred in a pen that contained only does and fawns (Lambiase et al. 1972).

It is interesting to speculate as to the possibility of buck fawns actually doing any breeding in the wild. There are areas in New York state, for example, where circumstances exist that might allow this to happen. In these areas, where deer range conditions are excellent, seven- to nine-month-old male fawns with weights of 50–55 kg (whole weight) and 2 cm of hardened antler are observed.

These areas are very accessible to hunters, and deer populations are in balance with the food supply—in fact, there is an abundance of food, both "natural" and agricultural crops. There is a tremendous turnover of the deer population—a 3½-year-old buck or a female more than 5½ years old is rare. The adult male age composition during hunting seasons indicates that 1½-year-old males comprise 85–90 percent of the herd. Thus, if 85–90 percent of the adult male segment is harvested during the deer season (late November and early December), and those remaining probably are shedding their antlers in early to mid-December, possibly many seven- to eight-month-old females being bred in January and February may be bred by male fawns. Although this is speculative, we think the possibility exists.

Sex Ratios. In utero, or primary, sex ratio data are difficult to acquire. These data are usually acquired from deer killed in motor vehicle collisions. Severinghaus and Cheatum (1956) reported a sex ratio of 117.2 males per 100 females based on a sample of 2,096 embryos from 11 states. White (1968) recorded a primary sex ratio of 114 males per 100 females based on a sample of 62 fetuses.

McDowell (1962) compiled primary sex ratio data from the northeastern states and Canada, and from other areas. Based on a sample of 2,749 embryos, the primary sex ratio was 112.2 males per 100 females for the northeastern states and Canada. The sample for other areas was much smaller (361), but the sex ratio was similar (111.1:100).

Data from the northeast and Canada were further examined by the age of the dam. Fetal sex ratios, by age of dam, are shown in table 45.1. The sex ratio from age class 1 (145.1 males per 100 females) is significantly different from that of the combined older age classes. Further, there is a significant linear correlation between fetal sex ratio and the age of the dam through the 11 age classes.

A more recent study conducted in New York also examined fetal sex ratios (Rasmussen et al. 1979) (tables 45.2–45.5). The data in table 45.5 are probably comparable to those of McDowell (1962). It is interesting to note the similarity between singleton births of age class 1 for each study.

TABLE 45.2. Fetal sex ratios of white-tailed deer (*Odocoileus virginianus*) by age of dam, singleton births, New York State

Age Class	Number of Females Examined	Number of Male Fetuses	Number of Female Fetuses	Males per 100 Females
1	90	53	37	143.2
2	131	62	69	89.9
3+	122	73	49	148.9
Total	343	188	155	121.3

SOURCE: Compiled from Rasmussen et al. 1979.

TABLE 45.3. Fetal sex ratios of white-tailed deer (*Odocoileus virginianus*) by age of dam, twin births, New York State

Age Class	Number of Females Examined	Number with Two Males	Number with Two Females	Number with One of Each Sex	Total Number of Males	Total Number of Females	Males per 100 Females
1	33	5	6	22	32	34	94.1
2	156	42	36	78	162	150	108.0
3+	347	73	80	194	340	354	96.1
Total	536	120	122	294	534	538	99.3

SOURCE: Compiled from Rasmussen et al. 1979.

In the Rasmussen study, the majority of females in age class 1 came from good deer range, whereas the older females almost all came from poor or over-stocked range. Virtually all age class 1 females in tables 45.3 and 45.4 came from good range, while the older deer in table 45.3 came from ranges varying from poor to good. Virtually all of the females in table 45.4 came from good range.

The management implications based on these data are quite interesting. It appears as though an intensively managed deer herd with a rapid population turnover will result in greater numbers of male deer being produced than will an older population with fewer younger-breeding females.

FOOD HABITS

As indicated earlier, the white-tailed deer is one of the most adaptable animals in the world. The species lives from the near arctic to the tropics. This tremendous adaptability is also reflected in the diversity of foods that the white-tail eats. It has been called a "browser and a grazer." In fact, it is both, and more.

We might call the white-tailed deer the "red fox" of the herbivores. In spite of the fact that it can, and does, eat a tremendous variety of foods, it has an uncanny ability to select the most nutritious foods available when it has the opportunity to be selective. Every researcher, including these authors, probably has at one time or another placed a variety of foods (in our case, some 10–12 browse species) at a deer's disposal, and watched it quite methodically reject species of lesser nutritive value to select the most nutritious ones. Agriculturists tell of observing deer select green

bean seedlings that were fertilized over unfertilized ones.

The white-tail, of course, is an ungulate with a four-part stomach, dependent as all ungulates are on very specific rumen bacteria and protozoans for digestion.

One of the first studies conducted using the scientific approach to deal with white-tail food habits was conducted in New York (Maynard et al. 1935). The objective of this study was to determine some nutritional requirements of deer and the value of several wild browse plants and artificial foods. The study was prompted by many annual deer die-offs that finally were attributed to starvation. This early study determined that "various species of browse, notably white cedar (*Thuja occidentalis*), yellow birch (*Betula lutea*) and soft maple (*Acer rubrum*) will maintain animals in satisfactory weight and vigor during the winter months." The authors also found that alfalfa hay could be used successfully, but that balsam fir (*Abies balsamea*) was useless. They tested many species of browse plant, and compiled a table of "Relative Preference of Deer for Various Browse Species."

A great deal of work has been done on the food habits of white-tails, yet the list by Maynard et al. (1935) is still quite valid. The food habits of deer are inextricably associated with what is available. Calhoun and Loomis (1975) noted that deer in the corn belt states have drastically modified their feeding habits since 1900. Today they prefer cultivated crops to browse, the traditional white-tail food of the past. The authors also pointed out that corn, soybeans, and other legumes exceed the deer's use of wild foods and that acorns and dogwood fruits are utilized in abundance

TABLE 45.4. Fetal sex ratios of white-tailed deer (*Odocoileus virginianus*) by age of dam, triplet births, New York State

Age Class	Number of Females Examined	Number with All Males	Number with All Females	Number Two and One Female	Number One and Two Females	Total Number of Males	Total Number of Females	Males per 100 Females
1	3	0	1	0	2	2	7	28.6
2	7	1	0	5	1	14	7	200.0
3+	39	3	5	20	11	60	57	105.2
Total	49	4	6	25	14	76	71	107.0

SOURCE: Compiled from Rasmussen et al. 1979.

TABLE 45.5. Fetal sex ratios of white-tailed deer (*Odocoileus virginianus*) by age of dam (all deer combined from tables 45.2, 45.3, and 45.4), New York State

Age Class	Number of Females Examined	Total Number of Males	Total Number of Females	Males per 100 Females
1	126	87	78	111.5
2	294	238	226	105.3
3+	508	473	460	102.8
Total	928	798	764	104.5

SOURCE: Compiled from Rasmussen et al. 1979.

when available. The latter is certainly used throughout the range of white-tails, especially acorn mast.

The role that browse plays in the diet of deer varies in relation to its availability, relative quality of species, and relation to other available foods. Miller (1961) noted that the white-tailed deer is essentially a browsing ruminant, and identified 60 plant species occurring from Virginia to Texas, their seasonal use by deer, and the effects on the plants of various browsing levels. The diet is chiefly tender shoots, twigs, and leaves, a wide assortment of herbaceous food stuffs, mast, and certain fruits. Browse by far makes up the bulk of its diet. In east Texas, Lay (1969) found that the principal foods of white-tailed deer were browse, fruits, succulent herbage, mushrooms, and agricultural crops. Browse accounted for less than half of the forage types utilized.

Working on several national forests in the northeastern states, Stiteler and Shaw (no date) pointed out that 100 of 106 woody plants growing there had been browsed. Eighty-five percent of the browsing occurred on noncommercial plant species. They determined a palatability factor as an expression of browse index (browse index being the ratio of the twigs browsed by deer to the total twigs available to deer). An index of 1.0 indicates that the twigs are browsed in direct proportion to their availability. Eleven species had browse indexes greater than 1.0. Regionally, greenbrier spp. (*Smilax*), birch spp. (*Betula*), hobble bush (*Viburnum alnifolium*), blackberry-raspberry (*Rubus* spp.), mountain maple (*Acer spicatum*), and flowering dogwood (*Cornus florida*) showed the heaviest browsing. Red maple was used more in northern forests than in southern. This study also resulted in an estimate of browse utilized in the total diet. The estimates varied from a low of 2.7 percent to a high of 17.1 percent. They conclude that deer were consuming large amounts of something besides browse. Other studies have reached the same conclusion. Watts (1964), working with captive deer in an oak forest, observed a significantly higher preference for dry leaves during winter than for woody browse. Dunkeson (1955) reported that herbaceous plants formed a large part of the deer diet throughout the year in Missouri. Crawford and Leonard (1965) found that deer in Arkansas made more use of wood twigs, when oak mast yield was low. Stiteler and Shaw concluded that "over most of the deer range in the Northeast, woody twigs probably make up less than 10 percent of the total food consumed." They suggested, and we heartily concur, that

habitat managers should give more attention to grasses and other herbaceous plants as winter food for white-tailed deer in the northeast. The serious problem with that, of course, is snow depth. The problem of providing wild foods to deer on northern ranges when the ground is snow covered is awesome.

The tremendous variety of forbs and herbaceous material consumed by deer in the summer was illustrated by Sauer et al. (1969). Some 72 species were eaten by deer in an enclosure. No change in relationship between numbers of deer present and the use of herbaceous growth was detected. However, there was a relative preference for plant species based on availability and consumption. Examples of preferred species were early goldenrod (*Solidago juncea*), woodland goldenrod (*Solidago nemoralis*), sensitive fern (*Onoclea sensibilis*), bracken fern (*Pteridium aquilinum*), whorled loose strife (*Lysimachia quadrifolia*), timothy (*Phleum pratense*), and wrinkled goldenrod (*Solidago rugosa*).

Cushwa et al. (1970), examined rumen samples from deer collected on a year-round basis. These deer were collected in conjunction with a study undertaken by the Southeastern Cooperative Wildlife Disease Center. Deer were collected in Alabama, Florida, Georgia, North Carolina, South Carolina, and Virginia. Seasonally, they found the following:

In Spring, green succulent leaves and stems of both woody and herbaceous species were the dominant food items. Fruit of prickly pear (*Opuntia humifusia*) and hawthorn (*Crataegus* spp.) were also important in the coastal plain; yellow poplar flowers (*Liriodendron tulipifera*), mushrooms, and acorns were other important items.

In summer, materials from succulent green plants continued to dominate foods taken. Mushrooms were the next dominant item, followed again by acorns.

In fall, acorns were the dominant food item. Japanese honeysuckle (*Lonicera japonica*) was the most common item among deer collected from the piedmont region, however. Other important food items were mushrooms, grapes (*Vitis* spp.), apples (*Malus* spp.), prickly pear fruits (*Opuntia* spp.), sumac (*Rhus* spp.), blueberry (*Vaccinium* spp.), and honey locust fruits (*Gleditsia triacanthos*). Leaves of woody species occurred frequently, but no woody twigs were detected.

In winter, acorns, grasses, and Japanese honeysuckle appeared to be the most common food items. Mushrooms were also important, as were grapes and sumac fruits in early winter. Rhododendron (spp.) leaves were the dominant food item in the more southerly Appalachian Mountains. Succulent woody twigs were noted in the coastal plain, but hardened woody twigs were not detected anywhere. Cushwa et al. (1970) concluded that woody twigs are browsed during spring and early summer when this part of the plant is actually more of a succulent. Browsing of hardened woody twigs in this part of the white-tail's range therefore appears to be a very minor part of the diet.

Deer have been known to eat rather bizarre food items, including marine kelp on the rocky shores of Antacosti Island, Quebec. We observed a deer eating highly toxic mayapple fruits (*Podophyllum peltatum*).

Most food habits studies that have been conducted on white-tailed deer have been conducted during the winter. This was probably necessitated by the fact that winter food problems (overpopulations of deer with resulting starvation) dominated the thinking of biologists concerned with deer nutrition matters. Winter food habits studies conducted across northern North America all have indicated similarities in species browsed by white-tails, i.e., white cedar (*Thuja occidentalis*), yew (*Taxus canadensis*), red maple (*Acer rubrum*), sumac (*Rhus* spp.), *Viburnum* spp., mountain maple (*Acer spicatum*), mountain ash (*Sorbus americana*), sugar maple (*Acer saccharum*), hemlock (*Tsuga canadensis*), *Lonicera* spp., and *Cornus* spp.

These species appear again and again on lists of foods used from the Canadian Maritimes to the Lake states. Food habits studies further west (Black Hills and Rocky Mountains) indicate a completely different diet:bearberry (*Arctostaphylos* spp.), *Populus* sp., snowberry (*Symphoricarpos* spp.), *Salix* spp., serviceberry (*Amelanchier* spp.), ponderosa pine (*Pinus ponderosa*), and Douglas fir (*Pseudotsuga menziesii*) (Martin et al. 1951). Martin et al. also listed hundreds of plants, both wild and cultivated, with notations as to whether or not deer utilize them throughout North America.

As indicated previously, the white-tail is truly a remarkably adaptable animal in its food habits. This fact creates conflicts between white-tails and people. Deer depredate numerous agricultural crops, including apples, soybeans, growing apple trees, vineyards, nursery stock of all kinds, cauliflower, beans, buckwheat, and forest species. The magnitude of this problem was reflected by Harlen (1977), who cited several cases involving deer damage to corn and alfalfa. He noted instances in which deer actually destroyed (ate) half of the growing hay crop. Corn yield may be reduced from a potential of 33 tons/hectare to 11 tons/ha because of deer damage to growing crops.

Aimonetti (1977) noted the lack of regeneration of desirable tree species, including oak (*Quercus* spp.), ash (*Fraxinus* spp.), basswood (*Tilia* spp.), cucumber tree (*Magnolia acuminata* L.), tulip poplar, maples (*Acer* spp.), white pine (*Pinus strobus*), and hemlock (*Tsuga canadensis*), because these species are preferred browse of deer. He also noted the lack of shrubby species such as hobble bush (*Viburnium al-nifolium*), american yew (*Taxus canadensis*), wild raisin (*Viburnum cassinoides*), and elderberry (*Sambucus canadensis*), caused by the browsing by excessive numbers of deer. Future timber crops may be jeopardized because the deer prevent seedlings from ever growing beyond the seedling stage. White-tails thrive in good agricultural areas, and their food habit preferences thus conflict with the interests of people.

French et al. (1955) studied the daily energy requirements for white-tails. They found a daily requirement of about 0.9 kg. (3,600 calories) of good quality, air-dry feed for a deer weighing 23-27 kg; 1.5-1.7 kg (6,300 calories) for a 46-kg deer; and 2.5-2.7 kg (9,900 calories) for a 69-kg deer. This latter energy requirement is the equivalent of at least 3.0-3.4 kg of good deer browse of usual moisture content. The protein requirement for optimal growth was found to be 13 to 16 percent of the ration.

BEHAVIOR

Basically, the white-tail is a shy, secretive, and usually elusive animal. From its very first moments in life a deer learns to hide and to be elusive. Its activity depends on a number of factors, including the season of year, relationship and proximity to humans, weather conditions, and numbers of deer in the area.

Reproductive. As mentioned previously, the behavior of adult males changes drastically as the rutting season approaches. As antlers become polished and testosterone levels approach their peak, bucks become very aggressive. During the rut, bucks move about seeking estrous females by smell. The buck will stay with the doe until she has been serviced and then he loses interest and seeks out another one. Often several bucks will be trailing a doe in heat, with the dominant buck in front. As they trail along behind, their noses are close to the ground following her scent. They utter a low moaning sound as they move along, associated with an occasional "click." The moaning sound is definitely auditory, but the clicking appears to be caused by the hocks rubbing together as the deer walks along. Bucks are definitely receptive to the "antler rattling" ploy of hunters at this time. Bucks also create scrapes during prerutting activity. Scrapes are created by pawing the ground until it is torn up—usually an area of less than a square meter. The buck probably urinates in it, and on the tarsal glands, while standing over the scrape. Usually there are rubbed trees in the immediate vicinity from antler polishing and sparring. These are breeding season territories, which are defended when approached by other bucks.

Females at this time of year have in effect chased away their fawn(s) of the previous year and have become solitary, coy creatures. It is interesting to note that all of this activity generally occurs in conjunction with the annual hunting season. Yet the presence of millions of hunters sharing the wilds with deer has little effect on breeding. Hunters who recognize scrapes and ruts, and have a little patience, usually succeed in killing a deer. In spite of all the human activity there is no interference with breeding—all the does that come into estrus are bred.

As the breeding season ends and bucks shed their antlers, deer tend to become gregarious. Family groups reform and in turn join with other family groups. Occasionally, large herds are seen during the winter.

Yarding. In northern ranges deer concentrate, or "yard up" during severe weather. This might begin as early as December and last until early May if the winter is severe. Deer concentrate either in streambottoms or along the edges of ponds, bogs, and lakes where there is conifer cover, or on the southerly exposure of hills with or without conifer cover. Yarding activity is basically an energy-saving mechanism. Conifers intercept falling snow on the branches. Much of this snow on the branches either evaporates shortly after a snowfall or falls to the ground as water, and this results in lesser amounts of snow under conifers than in open hardwoods within the same area. Ideally, the deer yard would have conifers with branches growing close to the ground because this reduces wind and cold air flow, and results in a more comfortable environment for deer. Deer yards usually are located with a southerly exposure allowing the deer full advantage of the winter sun. Deer return to the same areas within a deer yard from year to year.

White-tails in northern areas are migratory. The results of deer trapping efforts illustrate this. During the course of trapping and ear tagging deer in New York, many deer were retrapped during subsequent years in the same box trap (Hesselton 1969a). In the wintertime on northern ranges and probably on overstocked ranges in areas where winter severity is less of a problem, a definite "pecking order" becomes established. As deer become more and more restricted to trails because the snow continues to get deeper, the bigger, older females become dominant. They are usually the first deer in a group as they move around seeking food, and therefore they dominate the available food. As winter conditions worsen, this behavior develops to the extent that adult does will drive their own fawns away from food.

Attempts have been made to drive deer from their chosen wintering site, where the food supply may be depleted, to areas with abundant food. Such attempts usually fail. Deer will, however, eat artificial food such as alfalfa hay if it is brought to areas of their choosing.

As the fawning season approaches, the winter-spring groups disperse. Pregnant does will seek out a quiet place to have their fawns. Sites are variable and may be in the woods or in the middle of a hay field. Newborn fawns move about very little during their first few days of life. The doe does not stay with the fawn continuously. She returns every few hours to nurse and groom the fawn. This procedure does not vary much for the first few weeks of a fawn's life. By two to three weeks of age, the fawn is experimenting with food items such as grass and new browse. Fawns can be easily caught by humans during the first three to four days, but they become extremely difficult to catch after

that. By the time they are four months old they are usually weaned, and their behavior parallels that of their mother. They will stay together as a group until the onset of the rut, when once again the doe will separate from the fawn.

Herd Behavior. The basic grouping of white-tails appears to be a familial one, often encompassing the offspring of two breeding seasons. However, in winter concentration areas these groups and groups of a few adult bucks utilize the same general area. Deer will concentrate in large groups in apple orchards, and in hay fields during late summer. Even though it appears to be one large group of deer to the casual observer, especially when they are startled and bound off in different directions, they are actually "organized" and families will regroup shortly.

Playing. Deer seem to spend time simply playing. This behavior is not peculiar to fawns; adults also participate. It usually entails chasing or racing or jumping. Sometimes this sort of behavior appears to be caused by attempts at avoiding insects or some other unpleasant thing. But often it appears as though the deer are racing around simply for the sheer enjoyment of it.

Avoiding Unpleasantness. Deer behavior is affected by cold and snow, heat, insects, and, of course, humans and other predators. Deer try to remain comfortable and conserve energy as weather conditions vary. During the hot time of year they tend to stay in the woods. In hilly or mountainous country they will take advantage of hilltops where there is more apt to be air flow; this is of advantage in avoiding insects also.

The senses of white-tails are acute, especially the auditory and olfactory. Wild predators, including dogs, probably are not too successful at killing healthy, adult white-tails under normal conditions. However, in northern deer yards when deer are concentrated and limited to deep, packed trails, predation is fairly common. Snow conditions can vary tremendously, giving predators even more of an advantage—crusts that will support running dogs but not deer are devastating. Deer will usually panic and run from a predator; occasionally one will turn and fight using the front legs.

Behavior toward Aircraft. Deer behavior toward aircraft helps explain why deer hunters utilizing elevated blinds (tree stands) are successful. Deer very rarely look up. When an aircraft approaches at low elevations, deer become alert and agitated. They do not always run, however. Sometimes they continue lying down even when a helicopter is directly overhead. Sometimes they start running when the helicopter is still several hundred meters away—and not always in a direction away from the approaching noise.

MORTALITY

Legal Harvest. Legal "in-season" hunting accounted for about two million white-tailed deer in North America in 1978. Legal harvest is fully discussed in the section "Management." An additional unknown number of deer die as a result of wounding during the legal season. These unretrieved deer are believed to equal 40 percent of the legal harvest in Mississippi, and 0.3 percent (Indiana) to 12 percent (Iowa) in other states.

In some states and provinces, additional deer are legally taken by native peoples under treaty provisions. The largest harvest appears to be in Manitoba, where it totals about 4 percent of the legal harvest.

Additional legal deer are taken in many states as a result of crop depredation by farmers, orchardists, nurserymen, and others.

Illegal Harvest. State and provincial biologists give widely varying estimates of illegal kill within their areas of responsibility. In a recent survey by Hesselton and Hesselton (1979), state and provincial deer biologists indicated that illegal harvest could be insignificant or could be equal to or greater than the legal harvest. In a number of states, the illegal kill is estimated to be 2–4 percent of the legal kill (e.g., Oklahoma, Ohio, West Virginia, Massachusetts). In other states and provinces (Michigan, South Dakota, New Brunswick) estimates are much higher. It is believed that in most states actual illegal harvest exceeds the estimates.

Trauma. By far the largest number of traumatic losses of deer occur as the result of car-deer collisions. In high population (high traffic) states, such car-deer accidents may account for numbers of deer equal to 15–100 percent of the legal kill. In at least one state, Connecticut, car-kills exceed the legal harvest. Pennsylvania has, for several years, recorded known car-kills in excess of 25,000 deer. In more remote, less populated states, significantly fewer deer die on the highways. Interstate high-speed highways through areas of high deer populations kill extremely high numbers of white-tails. Train-deer collisions are probably, like car-deer collisions, a function of the number of kilometers of track (road) and the density of white-tail populations.

Other traumatic causes of death include accidents with farm equipment such as mowers and combines, fence accidents, falls, and drownings. The first is a problem where large farm operations and high deer populations coincide. Accidents involving fences (broken legs, entanglement) are largely a problem in the western United States where large tracts of prairie are fenced for grazing by cattle or sheep. Falls and drownings may often be involved with attempted predation or deer being chased or pushed by dogs, humans, and recreational vehicles.

Nettles et al. (1976) found that traumatic injuries result in very little chronic debilitation in deer that survive. However, most of these injuries are fatal.

Predation. The most important predator of white-tailed deer in North America is undoubtedly the domestic dog. Most cases of deer killed by dogs occur in the winter and early spring in northern states when deer are weakened by malnutrition and hampered by deep or crusted snow. Numbers of deer killed may equal 10 percent of the legal harvest. Significant native predators of the white-tail in western and southern North America are the coyote (*Canis latrans*) and the mountain lion (*Felis concolor*). The wolf (*Canis lupus*) in eastern Canada undoubtedly utilizes white-tailed deer to some extent.

Nutritional Deficiencies. In some areas of the western United States (notably Texas) deer are periodically lost due to drought. Far more deer are lost in northern states and provinces every year to malnutrition, which results from extreme overpopulations of deer on limited winter range. Biologists in most northern states conduct dead-deer surveys each year to assess such losses. However, there is really no good indication of actual numbers of deer that die each year of starvation. In a severe winter, up to 50 percent of the fawns born the previous summer may die of malnutrition. Starvation is generally determined by examination of the bone marrow in the femur of deer found dead. Normal, healthy deer have bone marrow that is 90 percent fat and appears whitish and solid. As body fat stores are depleted, fat levels in the marrow decrease. The marrow becomes darker in color and less "greasy." When it reaches 25 percent fat or less, the texture becomes gelatinous (Cheatum 1949). At this red gelatinous stage, malnutrition is severe and can be assumed to have been the primary or secondary cause of death.

Toxicity. The term *toxicity* encompasses a multitude of problems. It can be defined as the influence of deleterious effects on an organism or tissue. This poisoning can be of several types based on the source and type of poison or toxin. The major sources of toxicity of deer (and indeed all wildlife species) are soil, plants (natural toxins), industrial effluents, and pesticides.

The most familiar soil-associated toxicity factor is selenium poisoning, encountered in parts of the western United States where soil selenium levels are considerably higher than normal. Some plants concentrate the selenium; the plants are then eaten. The resulting disease in deer can be either acute (blindness, internal congestion, rumen impaction) or chronic (emaciation, abnormal hoof growth, cirrhosis). The chronic form of selenium poisoning is better known as alkali disease. Reed and Shave (1976) reported a number of cases in South Dakota.

Plant toxicity generally occurs in deer in a sporadic fashion. Although a number of common plants are potentially toxic to deer, they are not normally taken at levels that cause death. Plants that are toxic and are occasionally taken by deer are lupin (*Lupinus* spp.), larkspur (*Delphinium* spp.), chokecherry (*Prunus virginiana*), mountain mahogany (*Cercocarpus* spp.), Russian thistle (*Salsola kali*), foxglove (*Digitalis* spp.), and milkweed (*Asclepias* spp.). The toxic manifestations of these plants differ based on the type of poison found in the plant. Difficulty in breathing is one symptom most cause. Wyand et al. (1971) reported a case of kidney failure due to oxalosis in a deer in Connecticut. They speculated that the oxalosis may have been the result either of ingestion of plants containing oxalate or of ethylene glycol (antifreeze) poisoning. This deer also had aspergillosis.

Ethylene glycol poisoning is an example of environmental toxicity due to industrial effluents. Discarded materials such as used antifreeze, or old paint buckets that contain remains of lead-base paint, are often found some distance from human activity and constitute a hazard to deer and other curious wildlife. Most industrial effluent situations are localized problems where toxic discharges from an industrial plant or other commercial operation are released in an area where they are accessible to deer. Karstad (1967) reported a situation where an industrial discharge contained very high levels of fluoride. Some deer in this area had pitting and black discoloration of the teeth. Other dental abnormalities were associated with the high fluoride intake.

Often more widespread than discharge situations are insecticide/pesticide toxicity situations. The pesticides (especially chlorinated hydrocarbons) are generally slow to degrade naturally and therefore build up in water, soil, and plant material. The buildup of these toxins in the environment is a problem for all wildlife, including deer. A similar pattern of environmental contamination is seen with some industrial effluents—polychlorinated biphenyls (PCBs) are a good example. Because white-tails are at an intermediate trophic level, however, the problem is less acute for them than it is for carnivores.

Parasites. White-tailed deer serve as hosts for a variety of endoparasites and ectoparasites. The most commonly found ectoparasite on white-tailed deer is the deer ked or louse, *Tricholipeurus* sp. Two species of this chewing louse are commonly found on deer throughout the United States. An individual deer may be infested with thousands of lice. Since the lice are not transmittors of any disease, they are only a problem in otherwise debilitated animals. Deer suffering from malnutrition may often have higher body burdens of lice than deer in good condition. Turner (1971) indicated that high populations of lice in the spring are probably due to weakening of deer by the harsh conditions of winter. Species of the genus *Solenoptes* (bloodsucking lice) are found only rarely on white-tailed deer.

Several genera of tick occur on white-tails. The most common are *Dermacentor albipictus,* the winter tick of deer and elk, and *D. variabilis,* the American dog tick. *D. albipictus* is common on deer in the northern, western, and southwestern United States. The larval ticks molt on the deer during the winter and drop off in the spring after mating on the host. *D. variabilis,* which is common in the eastern two-thirds of the continent, attaches to the deer in early spring, mates, and drops by midsummer. This species of tick is a potential vector for several diseases, none of which are important in deer, however. *Boophilus annulatus* (the Texas cattle fever tick) and *B. microplus* (the cattle fever tick) once occurred on white-tails in Texas and Florida, respectively. Because the ticks carry diseases that are important to the livestock industry,

large-scale deer reductions in Florida were undertaken in the process of tick eradication (completed in 1945). Kistner and Hayes (1970) found white-tailed deer infested with *B. microplus* on St. Croix, in the U.S. Virgin Islands. The Texas cattle fever tick was eradicated from Texas but is still found in Mexico and South America.

Nose bots (*Cephenemyia* spp.) are commonly found in deer. The bots are found in the retropharyngeal pouches and nasal passages of the deer. The adult fly deposits its eggs in the nasal passages of the deer. The first larval stage moves to the retropharyngeal pouches. The third instar stage is generally expelled by the sneezing of the host. Two dozen or more larvae may be found in one deer. Deer generally withstand the nasal bot with no deleterious effects. In Michigan, Whitlock (1939) reported an incidence of infestation in deer of approximately 25 percent. During the same period of time, the incidence in New York was somewhat less, 11.6 percent of fawns and 14.5 percent of adults (Severinghaus and Cheatum 1956). In the southeastern United States, the incidence was significantly higher during winter and summer (41 percent and 31 percent, respectively) than during the spring and fall (9 percent and 6 percent, respectively) (Nettles and Doster 1975). About one-third of the deer collected in fall and winter at Welder Foundation in Texas were infested with nose bots (Glazener and Knowlton 1967).

The filarial worm (*Setaria yehi*) is found in the lowland deer of the southeast. This worm, as an adult, is found in the abdominal cavity. It is a large roundworm that is easily seen.

Footworms (*Wehrdikmansia cervipedis*) are found subcutaneously in deer. The location of the adult worms may include the feet and hocks, bases of the ears, brisket, or rump. Microfilaria radiate out where they are picked up by biting insects. The tongue worm and gullet worm (*Gongylonema* sp.) are found at fairly high incidences in the mucosa of the tongue and esophagus, respectively, in deer of the eastern United States. *Setaria,* the footworms, and *Gongylonema* sp. are prevalent in deer but are of no importance to the health of the animals.

White-tailed deer are host to a number of species of lungworm. *Protostrongylus coburni* is commonly found in the bronchi and bronchioles of deer in eastern North America. *Leptostrongylus alpenae* is found in the bronchi, bronchioles, and parenchyma of the lung. *Dictyocaulus* sp., the cattle lungworm, is found in the upper respiratory tract of deer. When parasite numbers are low, the deer is unaffected. Large numbers of any species of lungworm or of a combination of species can cause bronchitis or, in the case of *Protostrongylus,* interstitial pneumonia. A fourth "lungworm" in deer, *Pneumostrongylus tenuis* (*Parelaphostrongylus tenuis*), is not found as an adult in the lung at all, but is more commonly known as the moose brainworm or meningeal worm. This parasite has been extensively studied in recent years. White-tailed deer are the normal host of the parasite, and in deer it rarely causes any neurologic disorder. The adult worm, in deer as well as moose, is found on the meninges of the brain. Occasionally, the worms penetrate into the brain itself. It is in these cases that disease may occur. F. Gilbert (1973) found incidences of 47 percent in fawns and 81 percent in adult deer in Maine. The average number of adult worms was two in fawns and slightly less than four in adults. Earlier estimates in Maine (Behrend and Witter 1968) indicated an incidence of 66 percent for fawns to 100 percent for adults in areas of dense deer populations. In all cases, incidences were higher for female than for male deer. Behrend (1970) found an incidence of 77 percent infestation in deer of the Adirondack region of New York. A second species, *Parelaphostrongylus andersoni,* occurs as an adult in the thigh and loin muscles (Nettles and Prestwood 1976). Another filarial worm, the arterial worm *Elaeophora schneideri,* is found primarily in the large arteries of the neck and head and in other arteries of the body. Deer are not affected by this parasite, but it has a devastating effect on domestic sheep and on elk.

Two flukes occur in the liver of white-tailed deer. *Fasciola hepatica* is less common (and less noticeable) than *Fascioloides magna.* The latter is the "giant" liver fluke commonly seen by hunters when removing the liver of the deer. In deer, the flukes become encased in fibrous cysts and there is little resultant liver damage. Severinghaus and Cheatum (1956) speculated that higher incidences of this fluke in the Adirondacks of New York are the result of more feeding by deer on aquatic and low-growing plants.

Numerous species of parasite are found in the gastrointestinal tract of the white-tail. None, however, is of major significance to the adult deer. One potential exception is *Strongyloides* sp. in fawns in Florida (Forrester et al. 1974). This intestinal parasite causes diarrhea and bloody stools. Another strongyloid (*Eucyathostomum webbi*) was described from deer in Georgia (Pursglove 1976). A survey of intestinal nematodes in deer in the southeastern United States found six species at incidences of greater than 5 percent of 961 deer examined (Pursglove et al. 1976): *Capillaria bovis, E. webbi, Monodontus louisianensis, Nematodirus odocoilei, Oesophagostomum venulosum,* and *Trichuris* sp. *Capillaria* sp., *Nematodirus* sp., and *O. venulosum* are commonly found in deer over most of North America (Beaudoin et al. 1970). Abomasal parasites commonly reported are *Haemonchus contortus* and several species of *Ostertagia.* Eve and Kellogg (1977) found that abomasal parasite counts were directly related to overpopulation of deer. Higher levels of parasites occurred where deer populations were in excess of the optimum for their range.

Several tapeworms are also found in the intestines of deer. Extremely high incidences of *Thysanosoma actinoides* occur in deer in the Rocky Mountains region. This worm may also be found in the bile duct. The tapeworm *Moniezia* sp. also occurs at high incidences in some areas of North America.

Protozoan parasites include *Sarcocystis* spp., harmless in deer, and *Babesia cervi.* The latter species is closely related to that which causes cattle fever, a disease that devastated cattle herds in the south in the late 19th and early 20th centuries. *Theileria cervi* also occurs in deer in Texas, although *Babesia* is most prevalent. In addition to the latter two intraerythrocytic parasites, *Anaplasma marginale* is occasionally found in the red blood cells of white-tailed deer.

Bacterial Diseases. The bacterial disease of deer most often seen by wildlife managers is actinomycosis, or "lumpy jaw." Generally, the results of the disease are seen, perhaps while examining mandibles for age determination. *Actinomyces bovis* causes abscesses in the bony tissue of the mandible, maxilla, or turbinates, and in the surrounding soft tissue. The most common location is the mandible behind the diastama. The necrotic process and subsequent osteogenic activity result in the familiar bony mass. In live deer, a characteristic swelling (generally unilateral) of the mandible occurs. A purulent exudate and difficulty in breathing may be noticed if the nasal turbinates are involved. Morbidity is probably low. Difficulty in eating and subsequent emaciation are the most severe result when mandibles are affected, and difficulties in breathing are the most serious result of turbinate infection.

Anthrax is a dreaded disease both because of its economic impact and because it is transmissible to man. It is caused by *Bacillus anthracis,* a distinctive rod-shaped, encapsulated bacteria. Upon exposure to air, the active bacilli sporulate and become extremely resistant. They may remain in the soil, potentially infective for years under the proper conditions. For this reason, suspect cases should not be opened and necropsied in the field without extreme care. An outbreak of the disease occurred in June 1963 on Beulah Island, Arkansas, in white-tailed deer, cattle, pigs, and horses (Kellogg et al. 1970). As often occurs, the outbreak followed a period of comparative drought. Lesions in deer were similar to those in domestic animals—edema and swelling, dark unclotted blood, enlarged spleen, tissue congestion, and frothy exudate from the nostrils. Deer died of anthrax only in overpopulated areas. Deer in adjacent areas where cattle were dying of anthrax, but where deer populations were more optimal, did not die from the disease.

Brucellosis (*Brucella* sp.) is an important disease of cattle. Livestock growers in the western United States have long been concerned that deer may act as reservoirs for the disease, thus making eradication impossible. While white-tails may carry the infection, the incidence seems to be low enough to vindicate them as the transmittor of the disease to cattle. Infection is generally established by reaction to the cattle agglutination test. Witter and O'Meara (1970) reviewed a number of studies and concluded that the disease is of little importance to deer.

Another bacterial disease that affects humans, domestic animals, and deer, along with other wildlife species, is tuberculosis. The causative agent of this disease is the bacillus *Mycobacterium tuberculosis.* The most characteristic symptom is the development of small nodules, or tubercles, in various parts of the body. The tubercles may be distributed along the interior of the rib cage, along the diaphragm, or in the lungs. Externally, deer exhibit emaciation, and European deer with the disease often show abnormal antler growth (Francis 1958). The disease in deer is a chronic one. Little information on incidences in deer is available, however. Friend et al. (1963) described a case of the disease in a deer in New York State.

Three diseases that occur in deer are caused by anaerobic bacteria entering the body through openings in the mucous membranes or through wounds. Blackleg and malignant edema are caused by different species of *Clostridium;* still another species of *Clostridium* is the causative agent of botulism. Species that cause the two former diseases are soil contaminants that infect the deer as opportunists—through wounds, etc. Blackleg is generally found in younger animals. Fawns and older deer that are affected show extreme swelling and edema, drainage of dark, frothy blood, and gas bubbles in the musculature. Only one incidence of extensive deer losses to this disease has been reported (Armstrong and MacNamee 1950). Malignant edema is a very acute disease. The infected portion of the body swells and becomes hot. As this condition subsides, toxemia develops (Howe 1970). This disease occurs only sporadically. The third anaerobic bacteria is *Spherophorus necrophorus,* the causative agent of necrobacillosis. This disease is a periodic problem in mule deer in California, but is less important in whitetails. The major lesions, purulent in nature, may occur in the feet, mouth and throat, rumen, liver, or lungs. If death results, the lungs are generally found to be involved (Rosen 1970).

Another purulent infection involving bacteria commonly encountered is *Corynebacterium pyogenes.* This bacterium causes abscesses at sites of old trauma. Often, at the end of a long hunting season, hunters will dress their deer and find abscesses that have formed at the site of a superficial grazing by a bullet earlier in the season. Arrows are occasionally found embedded in an abscessed area. Occasionally multiple abscesses form with no apparent cause. These abscesses produce a thick, creamy green-tinged pus. Rosen and Holden (1961) reviewed incidences of *C. pyogenes* in deer from California and described the disease. Another species, *Corynebacterium pseudotuberculosis,* was reported by Stauber et al. (1973) in Idaho. The disease produced is called "caseous lymphadenitis" (or sometimes "pseudotuberculosis"). Several diseases caused by other bacterial species are also called "pseudotuberculosis." The causative agents of these diseases (*Yersinia enterocolitica* and *Pasteurella pseudotuberculosis*) have been isolated from whitetailed deer. Both morbidity and mortality are very low, so these diseases are of little importance in deer.

Other bacterial diseases that have been reported in white-tailed deer include pasteurellosis (Minnesota, South Dakota), listeriosis (New York, Michigan, Illinois), psittacosis (New York, Quebec), and leptospirosis (New York, Pennsylvania, Louisiana, South Dakota, Ontario).

Fungal Diseases. One instance has been mentioned of the occurrence of aspergillosis. This disease is gener-

ally the result of ingesting moldy hay. It can also occur naturally, however.

A subcutaneous infection of the fungus *Alternaria* was reported from a deer in New York (Salkin and Stone 1974). Attempts to transmit the infection were equivocal. The infection probably seldom occurs naturally.

Dermatophilosis—also known as streptothricosis and caused by fungi of the genus *Dermatophilus*—is more commonly encountered in deer. It was encountered in New York in 1960 (Dean et al. 1961) and has been recorded in that state several times since then. It was also reported from a South Carolina deer (Kistner et al. 1970). In that case, the cutaneous *Streptothricosis* was determined to be the cause of death. The disease causes scaly lesions with loss of hair, exudate, and extensive dermatitis.

Viral Diseases. Bluetongue is a disease that is an increasing problem in North America. It has become a commercially important disease in livestock (especially sheep) since its introduction from Africa in the mid-1940s. It has been reported in white-tails in Texas on a number of occasions, apparently enzootic in deer on the Welder Wildlife Refuge (Hoff et al. 1974). These authors did extensive serological screening of the deer on the refuge. Only a few animals examined had any signs of the patent disease. The disease in deer is characterized by anorexia, difficulty in breathing, salivation, and swelling and cyanosis of the tongue—hence the name bluetongue. Bloody diarrhea and lameness may occur. In fatal cases in deer, the spleen and lymph nodes are congested (Trainer 1970). Bluetongue may be confused with epizootic hemorrhagic disease (EHD), as the signs are similar.

EHD is a viral disease of major importance in white-tails in a number of areas of North America. Outbreaks have occurred in the eastern, central, and western parts of the United States and in Canada. The first documented outbreak was in New Jersey in 1955 (Shope et al. 1955). In that incident approximately 700 deer died. Many earlier deer die-offs in various states may well have been from EHD. A more recent outbreak in New Jersey was reported by McConnell et al. (1976). In that die-off an estimated 1,000 deer were lost. The disease appears to be endemic in South Dakota. Outbreaks have occurred almost yearly since 1952 that have been attributed to EHD (Reed and Shave 1976). Outbreaks in all parts of the continent occur during the time of year when insect vectors are available to transmit the virus. The disease is not transmissible from deer to deer. The clinical signs of the disease, beginning about a week after exposure, are sudden in onset. Deer lose their wariness, stop eating, become weak, and show mucosal congestion and excessive salivation. There may be blood in the urine, feces, and saliva. Internal lesions are characteristic—extensive hemorrhaging of any and all organs and thoracic edema.

Mucosal disease, which may be the same disease as malignant catarrhal fever (MCF), has been described in white-tails in North Dakota (Richards et al. 1956). Mucosal disease and MCF in cattle and some other species of deer have clinical signs and pathology very similar to those seen in EHD. However, the incubation period of the virus is longer than that of EHD.

Serologic surveys have been done on deer in a large number of geographic areas for evidence of exposure to a wide variety of viruses. Viruses that have been identified in deer as a result of positive results on serologic screens include infectious bovine rhinotracheitis, California encephalitis, St. Louis encephalitis, and the Bunyamivera group of viruses. No diseases have been reported in conjunction with these findings.

Neoplasia. A variety of neoplastic lesions can occur in deer, as in all species. Only one is of major importance in white-tailed deer—the fibromas found quite regularly on the skin. The incidence of fibroma in deer in New York was 1.4 percent (Friend 1967). It was much higher in males of all age groups than in females. Most affected deer have more than one tumor on the body. Small tumors exist as raised, dark areas, generally on the head. Some deer have dozens of tumors of various sizes. Large fibromas are pendulous and pedunculated. Externally, they are dark and tough. Internally, the tumors are glossy white and are strictly innocuous. When many large tumors are involved, problems can occur. Severinghaus and Cheatum (1956) recorded one case of a fatal abscess resulting from laceration of a tumor and one case of a fawn with tumors of the mouth and eyes that mechanically rendered the animal unable to feed.

Congenital Abnormalities. A variety of congenital abnormalities have been reported in white-tailed deer. Miller and Cawley (1970) reported a case of polydactylism in a white-tailed fawn in Ontario. The right forefoot of this animal had five full digits instead of the normal four digits and hoof material on each digit.

Microopthalmia has been reported on several occasions. This condition occurs with or without the presence of other ocular abnormalities.

R. M. Hesselton examined a fawn from Vermont that had severe scoleosis. The deformity was severe enough to cause pressure on the spinal cord. The fawn had extensive traumatic injury that appeared to be self-inflicted in attempts to walk. The problem apparently became more intense as the fawn grew and gained weight. It was several months old when found dead.

Miscellaneous Pathology. Reed and Shave (1976) reported a number of cases of deer with polioencephalomalacia; they did not speculate on the cause of the lesion. A similar lesion is encountered on occasion in horses given moldy grain. A number of other causes may be possible. Deer, like all mammals, are subject to pathology of all body systems. They may die of heart attacks, strokes, endocrine dysfunction, or any of a variety of conditions familiar in humans and other species.

AGE DETERMINATION

It is necessary to know the age of a deer for a number of reasons. It is possible to determine the rate of replenishment and also the rate of decline in deer populations by monitoring herd age compositions. The rate of harvest is reflected in changes in age compositions: a young, healthy, heavily harvested deer herd will have very few old animals in it (figure 45.5A), in contrast to an older, lightly harvested population (Figure 45.5B). The ability to compare deer herds from different areas, using physiological parameters such as reproductive rates, carcass size (usually a ''dressed'' weight), and antler size, is dependent upon accurate age determinations (Hesselton and Sauer 1973).

The standard technique used for age determination in white-tailed deer has been tooth eruption and wear (Severinghaus 1949). This technique has been modified and refined somewhat to make it even more useful (principally by John E. Tanck, New York Fish and Game Department). This technique utilized the known schedule of tooth eruption, which is basically complete at about 21 months. Deer that are older than one and a half years, therefore, are aged by the relative amount of tooth wear. This was initially determined by examining deer of known age and determining a number of recognizable tooth wear or erosion characteristics. The technique is a subjective one, which causes some problems. In order for this technique to be employed to its fullest potential, known age jaws from deer representative of all habitat types should be utilized for comparison purposes. Wear patterns are dissimilar between deer on different habitats. Deer eating herbaceous vegetation on sandy soils exhibit different wear patterns from deer browsing exclusively on hardwood browse. Detractors of the technique usually

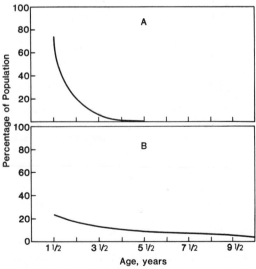

FIGURE 45.5. A, percentage of various classes in a heavily harvested white-tailed deer (*Odocoileus virginianus*) herd; B, percentage of age classes in a lightly harvested herd.

fail to take these facts into consideration when using the technique.

In an attempt to overcome the problems inherent in the eruption/wear technique, the use of cementum annuli is a more precise method. The phenomenon of variable seasonal growth in some animals results in tooth cementum "layering" similar to growth rings in trees. The technique is basically the same as that developed by Scheffer (1950) and Laws (1952) in their studies of northern fur seals, (*Callorhinus ursinus*) and elephant seals (*Mirounga angustirostris*), respectively. Since then, the technique has been applied to many species of mammal.

While it is possible to use virtually any permanent tooth for sectioning, most biologists use the first incisor. Incisors are easy to pull and they have a large root, making them easy to handle.

Several studies have been conducted, and others are currently being conducted, to determine the relative amount of variability between tooth eruption/wear and tooth sectioning. Gilbert and Stolt (1970) found that "more deer were assigned to the younger age-classes (<5½ years) and less to the older age-classes (>5½ years) by tooth wear characteristics than by annuli counts. The percent difference between deer assigned to each age-class by the two methods increased from 6.3 percent for yearlings to 100 percent for deer 6½ years of age or older." Personal observations of tooth age assignments by tooth replacement/wear techniques in several states confirm these findings. That is, the technician tends to overestimate the age of younger deer and underestimate the age of older deer. The overestimating of age in younger deer is more critical than age discrepancies among older deer. Younger deer, especially yearlings, are much more susceptible to range condition differences, and therefore are used for comparison purposes between different ranges.

Studies have been conducted to determine the reliability of the tooth section technique using known-age samples. Gilbert (1966) worked with a small sample (10 deer), but the deer ranged in age from 1½ to 11 years. He concluded that the technique was valid. Sauer (1971) worked with a larger sample of 55 deer ranging in age from 1 to 6½ years. She found that the technique was 84 percent accurate for assigning age.

Sauer (1971) also made some comparisons between the two techniques using samples from three areas in New York State and Antacosti Island, Quebec, Canada. She found that "only 40 percent of the animals were assigned to the same age by both techniques; by tooth sections 12 percent were aged younger and 48 percent were aged older than tooth wear ages." She also noted that there was less correlation between the two techniques in animals from relatively poorer range than in those from better range.

Those provinces and states with deer seasons in December or later should use the tooth section technique on all yearling deer that have completed their premolar eruptions. We do not believe it is necessary to use the relatively expensive and time-consuming tooth section technique to age all deer every year. A complete study of all age groups should be done periodically to "smooth out" the errors inherent in the tooth wear/replacement technique. A correction factor should be determined and then annually applied to age compositions determined by the relatively quicker and less expensive tooth wear/replacement technique.

ECONOMIC STATUS

Any discussion concerning economics and white-tailed deer is comparable to a two-edged sword: it works both ways. As mentioned previously, white-tails are capable of doing incredible amounts of economic damage—to motor vehicles, to agricultural crops, to commercial nurseries, and to future timber crops. In contrast, however, there are positive economic benefits that are easily worth billions of dollars annually. The 1975 National Survey of Hunting, Fishing and Wildlife-Associated Recreation (USFWS 1977) indicated that expenditures by big game hunters totaled $2.56 billion in 1975, exclusive of license purchases. Approximately 80–82 percent of the days expended on big game hunting was devoted to deer hunting. The survey does not allow a distinction between mule deer/black-tailed deer hunting and white-tailed deer hunting. However, there undoubtedly are far more participants in white-tailed deer hunting than in hunting for other species of deer.

The average expenditure per deer harvested in New Jersey in 1975 was $2,170, and deer hunters spent an average of $36 per recreation day for deer hunting. Projecting further, 134,429 licensed deer hunters spent between $27,423,516 and $30,649,812 in New Jersey in 1975 (Burke et al. 1978). Extrapolating from the previously cited national survey of 1975 indicates that the average deer hunter expended approximately $225 on deer hunting in 1975; it seems reasonable that inflation would raise this figure to at least $250 by 1978. The 1975 national survey also indicated that there were 12,403,000 deer hunters in the United States. License sales have tended to plateau in recent years, so it is doubtful if there were many more deer hunters in 1978 than there had been in 1975. Our previous data indicated a legal kill of approximately 125,000 white-tails in Canada and 1,875,000 in the United States in 1978. If we assume that there were approximately 13 million deer hunters in the United States in 1978, and that perhaps 75 percent of them hunted white-tailed deer (9,750,000), then the success rate was approximately 20 percent. Applying these figures to the $250 per hunter expenditure (above), then we have a value of approximately $1,250 per deer legally killed! Note that this figure is in contrast to the previously cited figure of $2,170 per deer for New Jersey.

Similar data from Vermont (Gilbert 1973*a*, 1973*b*) indicated that deer hunters spent $26,760,327 in 1970 in that state. A total of 9,665 legal deer were reported taken in 1970 by approximately 135,000 deer hunters. Thus, Vermont hunters had a success rate of approximately 7 percent. The average expenditure per hunter approximated $195, and the value of each deer taken was $2,770.

The examples cited certainly do not reflect what conditions are in every state or province. The two states cited are quite different in many respects—sociologically, economically, numbers of inhabitants, land use practices, deer herd management practices, etc. Actually they represent two extremes in many respects, yet they both indicate a value per deer taken substantially higher than that reflected in the national survey.

We previously noted that data collected in 1978 indicated that as many as 1,000,000 white-tails succumb to causes other than legal hunting. If we can assume that the financial data cited above are reasonable, then it is obvious that deer carcasses worth between $1,250,000,000 and $2,770,000,000 are wasted annually.

In attempting to illustrate further the economic impacts that white-tails have, let us consider some studies relative to deer and motor vehicle collisions. Connecticut, with a legal deer kill of 948 deer, had 1,161 deer reported killed by motor vehicle collision. Insurance claim statistics indicate that the average amount of damage to a motor vehicle approximates $400 (Herig personal communication). A recent study conducted in Wisconsin (Pils and Martin 1979) indicated that mean insurance claims for vehicle damage increased from $407 in 1976 to $503 in 1978, while average personal repair estimates jumped from $189 in 1976 to $318 in 1978. Total state-wide damage estimates were $22.1 million during this time period. Pils and Martin estimated that as many as 18,000 deer were being killed on the highways by 1978. As mentioned in the "Mortality" section of this chapter, the numbers of deer killed in motor vehicle collisions in some states and provinces is astronomical—for example, Pennsylvania reported 25,000± annually. Insurance companies pay out over $15 million annually in Pennsylvania to repair automobiles struck by deer ($600/deer). Nation-wide figures for either Canada or the United States are unavailable. Only a few wildlife agencies indicated that they attempt to keep track of these figures. A few that do, though, in addition to those mentioned here, are: Delaware (600–700), Florida (500–1,200), Illinois (2,175), Indiana (2,114), Iowa (2,872), Kansas (1,266), Massachusetts (384), Michigan (17,155), New Hampshire (485), New Jersey (3,105), Ohio (8,400), Oklahoma (500), South Carolina (885), Vermont (2,546), West Virginia (2,369), Nova Scotia (871), and New York (15,000 estimated). It is obvious that there are widespread discrepancies in reporting these kinds of data, but we believe that even the limited data above give the reader some idea of the magnitude of this problem. Deer in only those states reported here (approximately 106,000 deer) are responsible for between $40 million and $60 million worth of damage. Undoubtedly, collective damage to motor vehicles amounts to more than $100 million in damage claims annually in Canada and the United States.

While economic losses to agricultural crops are common and recognized to be substantial, similar losses to the timber industry are less commonly acknowledged. Aimonetti (1977) stated that in Pennsylvania "an average stand of Allegany hardwood managed on a sound basis with four intermediate thinnings over a 100 year period will yield about 4,600 board feet in thinnings and 9,200 board feet per acre at the time of the regeneration cut. With only a minimal deer problem the market value of the harvested trees, discounted back to seedling size or age one is about $55 per acre. If the deer delay the establishment of regeneration for 30 years, because of this longer time to grow the stand the value decreases to a *minus* $5 per acre—a $60 loss which converts an asset into a liability." He also noted that there are areas in northwestern Pennsylvania where cuts made in the 1930s and 1940s are still not producing a timber crop.

Timber growers throughout North America are very much concerned with the problem of deer overpopulation. Not only do deer retard growing trees, but their selection of certain browse species affects the species composition of trees that grow to maturity. Unquestionably, then, white-tails can and do have tremendous negative economic impacts that probably collectively amount to hundreds of millions, if not billions of dollars annually.

Considering the positive aspect, the value of deer meat amounts to about $300 to $400 million annually, assuming that the average carcass has 34 kg of edible meat and that its value is the current value of beef ($1.20/kg).

Deer hides can be sold for between $5 and $10 depending on size; thus 2 million legally taken hides are worth $10–20 million annually.

Satellite industries that are involved with meat, hides, and heads of deer are, of course, the leather garment industry, taxidermists, butchers, and meat lockers. It costs $10–25 to have a deer butchered and the meat prepared for the freezer. There are no data on how many deer are treated in this manner, certainly the deer harvest must generate several millions of dollars for meat processors. A good deerskin jacket of medium length might utilize four to six deer hides and sell for well over $100.

Other positive economic aspects are such things as hunter utilization of hotels, motels, restaurants, and grocery stores. Farmers lease their deer-hunting rights for a fee in many areas. In Texas, deer hunters pay up to $300/gun/season or $20/day for deer hunting privileges (Moody 1969). In Georgia, Almand (1968) cited a rural six-county area that receives more than $500,000 annually during the deer season; three-quarters of this from nonresidents of the area. This, of course, reflected economic values of the mid-60s!

In many states and provinces, secondary schools are closed for the opening day of the deer season, and sometimes for several days, to allow students and faculty to go deer hunting. While this certainly has little economic value, it does illustrate the importance of deer to local economies in many areas of North America.

MANAGEMENT

Probably no other wild animal in the world has been the subject of as much study as the white-tailed deer.

Its tremendous range in distribution and its ability to survive under all manner of conditions strongly suggest a need for continual studies in many areas.

Management of white-tailed deer has three phases: (1) manipulation of populations via hunting season length or bag limits; (2) manipulation of habitat to increase, decrease, or maintain deer abundance; and (3) people management! Managing white-tailed deer undoubtedly is the most controversial topic facing wildlife managers today, as it has been for the past fifty years. Every interested individual considers himself an authority on the subject. To varying degrees that certainly is true. The problem facing various wildlife agency deer managers is satisfying all of these "experts." This "people management" has been addressed repeatedly with varying degrees of success or failure. We believe that the success stories outnumber the failures: many states and provinces now regularly harvest female deer and fawns. All states and provinces harvest some antlerless deer, either by archery, special muzzle loader or "primitive weapons" seasons, or more traditional methods—that is, antlerless quotas superimposed on the regular buck season (Vermont, New Jersey, New York) or regular "any deer" seasons (Maine, Illinois).

Several studies have addressed the problem of people's attitudes toward hunting, taking antlerless deer, and other aspects of "people management" (Applegate 1973, 1975; W. Shaw 1974; Shaw and Gilbert 1974; Brown and Thompson 1976; Kellert 1976). A study sponsored by the Northeast Deer Study Group and conducted by a New York City–based consulting/marketing firm resulted in the preparation of a communications plan for deer management. This plan identifies all of the various "publics" that must be recognized and convinced of the need for a deer management program. The communications plan recommends the approach or technique(s) to be employed regarding each respective "public." Identified "publics" are state and local politicians, children and adolescents, deer hunters, "young" hunters, "other" hunters and fishermen, antihunting organizations, large landowners, hunting community residents, voters, other wildlife users, conservation-environmental groups, farmers, and wildlife management professionals. Ultimately, of course, the decision to implement a deer communications plan (or not) is a decision that each respective provincial state resource agency director must make.

We believe that far too little effort is invested in this phase of most programs. The necessary expertise is lacking in most agencies, and universities do not produce the kinds of individuals that are needed for this type of work—that is, promoting wildlife management, and in particular white-tailed deer management.

Manipulating white-tail numbers is relatively easy compared to the problem of "people management." Most resource agencies employ professional wildlife biologists with the capability of managing white-tailed deer resources. A number of factors govern the degree or relative precision with which respective agencies manage their deer herds. Generally speaking, prov-

inces or states with low human populations in relation to the number of deer available manage their herds with less precision than heavily populated provinces or states. The ultimate in precision is probably achieved by New Jersey. The New Jersey system involves dividing the state into 36 deer management zones ranging in size from 230 to 880 km². Each zone represents an area with similar deer herd characteristics, including weight, antler growth, reproductive rate, land ownership, land use and trends, and soils and vegetation (Burke et al. 1979). Many other agencies are managing their herds in a similar manner, but usually with less precision.

Basically, the New Jersey system works in the following manner. All of the factors mentioned, as well as trends in deer population, by unit, as measured by legal harvest, and "other" kills such as automobile and starvation, and success rates from previous seasons, are built into a system that determines the number of antlerless deer kills needed to accomplish the deer unit objective. In some units the population has become too large and therefore an increased number of antlerless permits is issued.

Virtually all provinces and states are harvesting some antlerless deer. There are many types of antlerless seasons, such as early archery, postseason archery, muzzle loader, antlerless quotas superimposed on the "regular" buck season, and others. But though there remains a reluctance toward killing antlerless deer in some provinces and states, the need for antlerless seasons is becoming more and more accepted.

Numbers of Deer. A survey of all provinces and states indicates that there was a minimum of about 15,000,000 white-tails in Canada and the United States in 1978 (table 45.6). Some states and provinces are still experiencing population increases, but most have probably reached, or even exceeded, their maximum level. Bartlett (1949) estimated that white-tails inhabited approximately 3,840,000 km² of range in 1948, and Seton (1929) estimated an inhabited range of 5,112,000 km². Today's population must be closer to the latter figure, considering their range expansion, especially in the Canadian prairie country. So, roughly, in 1978 there were at least 3 deer per km² of inhabited range in Canada and the United States. On a state-wide basis, Alabama has the most dense population of deer (approximately 9 deer/km² of entire state). Other states with populations equaling or exceeding 6

TABLE 45.6. Numbers of white-tailed deer (*Odocoileus virginianus*) in Canada and the United States in precolonial times, 1908, 1948, and 1978.

Year	Number of White-Tailed Deer
Precolonial	40,000,000
1908	500,000
1948	6,000,000
1978	15,000,000

SOURCE: Seton 1929; Bartlett 1949.

deer per km² on a state-wide basis are Michigan, Mississippi, Pennsylvania, Vermont, and Wisconsin. Populations exceeding 10 deer/km² are common in many states on a local basis. For example, biologists in New Jersey estimate populations of 30 deer/km² in some counties (George Howard personal communication).

Nine provinces/states reported that they had deer populations greater than 500,000. Texas reported the greatest population—3,075,000 deer. In addition to the 9 provinces/states above were 7 others with populations between 250,000 and 500,000 deer, and still 10 others with populations between 100,000 and 250,000 deer.

Deer Kills. Collectively, approximately 2,000,000 deer were killed by legal hunting in 1978. Six states had kills in excess of 100,000 deer, with Texas leading (272,076 deer). Methods used to measure "legal" deer kills are quite varied. Some states/provinces require that successful deer hunters mail in a report tag, usually within 24 hours of killing the deer. Some require that all deer be brought to an agency-operated deer check station. Some require that all deer be brought to designated registration areas, usually rural grocery stores or service stations. Some states/provinces are quite casual about this entire procedure, while others use fairly sophisticated systems for determining the numbers of successful hunters who do not bother to report their deer. New York has used a sophisticated reporting system for many years, and the reported kill is increased by a factor that compensates for nonreporters. Studies done in New York have shown some interesting behavioral traits of deer hunters in this regard (Jackson 1973). Apparently the transient, urban hunter reports killing a deer more frequently than does a hunter from a smaller community who is less transient. This latter hunter, in turn, is more apt to report killing a deer than is the rural hunter who is not transient at all. The so-called reporting percentage for hunters of antlered bucks varies from 65 to 80 percent for different areas in New York, and the reported kills are increased accordingly.

In 1947, approximately 640,000 white-tailed deer were harvested in Canada and the United States. This represented about 11 percent of the estimated available numbers of deer (Bartlett 1949). Bartlett pointed out that "game men say 1,100,000 deer should have been taken." This would represent 18 percent of the population, certainly a very conservative figure. In 1978 about 13 percent of the available estimated numbers were killed by legal hunting, not too much of an improvement!

We attempted to collect information concerning "other than legal" losses. Only 22 provinces and/or states attempt to make "guesstimates" of "other" losses. Of those that responded, 7 indicated that "other" losses are equal to or greater than the legal kill. Two indicated that they are equal to one-half the legal kill, and 1 indicated that they are equal to one-half the total population. Of the remaining 12 respondents, all but 1 indicated that these "other" losses varied from 15 percent of the legal kill to 30 percent of the population.

"Other" losses were identified as due to collision with automobiles, dog predation, "wild" predation, malnutrition, illegal (including poaching), crippling, drowning, fence accidents, diseases, fawn mortality, mowers, and falls. These losses are discussed under "Mortality."

Those provinces/states that responded were a good cross-section representing deer populations of all sizes, dense human populations, and sparse human populations. Because of this we feel that it is appropriate to suggest that "other losses" collectively are at least one-half as great as legal kills, or approximately one million deer. This represents approximately 6½ percent of the total population. Thus, collectively, legal kills and "other losses" total about three million deer, or 20 percent of the total population.

One of the biggest challenges facing deer biologists is developing programs to "convert" as many "other" losses as possible to legal kills. Techniques need to be developed that allow agencies to be more precise in identifying the degree and types of "other" losses. This will aid the agencies in deciding whether or not to attempt corrective measures.

Some agencies are using the trap-tag-recovery technique in an attempt to determine relative numbers of deer dying from causes other than legal harvesting. Deer are usually caught in a "Michigan"-type box trap (figure 45.6) or some modification of it. After capture, deer are usually ear tagged and in some instances radio-telemetry collars are attached.

In order to achieve the maximum amount of information attainable from tagging studies such as this, additional surveys and/or censuses should be conducted simultaneously. Intensive efforts should be made to locate and identify *all* causes of death in the area. At the conclusion of such studies, the ratio of tagged to untagged deer can reliably be shown for all causes of death. This kind of information should aid administrators in making better management decisions.

Population modeling, utilizing such data as known kills from a variety of sources but especially legal hunting season kills as a minimum, and reproductive rates are being used to determine populations and population projections. This technique is less expensive than using a combination of other kinds of surveys, such as spring dead deer searches or motor vehicle collision data collections. The cost of population modeling varies directly with the amount of precision desired. There are two problems with population modeling. The first is simply a communications problem, resulting from the difficulty many people have in understanding the dynamics of wild animal population fluctuation and how collected data reflect those populations. The second problem is that a population model will aid in identifying "other" losses collectively, but not as to the specific cause of death. These kinds of data can only be determined with complementary surveys and censuses.

Management Techniques. Many techniques have been developed over the years in attempts to census white-tails. However, most of these techniques are ap-

FIGURE 45.6. "Michigan," or Stephenson, box trap.

plicable only to the census area and cannot be applied to an entire state or province, or even to some lesser subdivision. It is beyond the scope of this chapter to describe the many census techniques and variations available. We will describe some of the more commonly used techniques.

Direct Count Census Types. AERIAL COUNTS. The use of aerial censusing has been revolutionized by the helicopter. However, most biologists feel that aerial counts, even under the best of conditions, leave something to be desired. It is extremely difficult to see deer under conifer cover, even with snow on the ground. Even in open hardwood habitats it is difficult to get accurate counts if the terrain is uneven. When groups of deer are encountered, shadows confound counts. However, certain situations do lend themselves to aerial censusing. For a number of years W. T. Hesselton was involved in a study of the population dynamics of a controlled deer herd at Seneca Army Depot in western New York (Hesselton et al. 1965; Hesselton 1969b). The Seneca Army Depot herd is a mixture of normal-colored white-tails and a mutant white-coated variety. The white deer were protected during early years, and their numbers increased to the point where they constituted 25–35 percent of the total herd, or as many as 250 deer. The white deer were censused annually by two counts, usually conducted on the same day, and of course prior to snow cover. The entire depot was surveyed twice, once flying a north-south pattern and once flying an east-

west pattern. Subsequent removal of deer over the years indicated that between 85 and 90 percent of the white deer were seen during the helicopter counts.

Similar surveys were conducted when snow cover was present, to census the brown segments of the population. These usually accounted for 60–70 percent of the population. There was no conifer cover present at all, and approximately 20 percent of the depot habitat consists of open fields and gray-stemmed dogwood (*Cornus racemosa*) patches. The remaining area is covered with stands of large, mature hardwoods and hedgerows.

Other than in situations similar to Seneca Army Depot, actual counts are impractical. However, the use of aerial surveys by helicopter may be practical to acquire trend data from year to year. The state of Connecticut has been using this technique for several years. The same grid lines are flown each year, under approximately the same weather conditions, that is, winter with snow on the ground (Paul Herig personal communication). Aerial surveys are very practical for locating winter concentration areas, and many states and provinces utilize this technique for that purpose.

FLUSHING COUNTS. This technique has several varieties. The most familiar involves walking preselected lines, either permanently marked or following a compass, and recording the flushing distance from the observer for each deer flushed. Then, one assumes that twice the average flushing distance for all deer flushed is the average width of the surveyed strip. The total population is calculated from these data (Erickson 1940).

A modification of this technique involves marking animals and applying Lincoln index techniques. This system was used at Seneca Army Depot. Ratios of brown/white deer were determined, and figures derived from previously described aerial surveys were applied to the sample from the flushing survey.

DRIVES OR COMPLETE SURVEYS. These surveys require a great deal of labor and need good control of personnel. Basically, all deer are forced out of randomly selected areas—areas that are far enough apart to prevent duplicating counts—and these data are applied to a larger area encompassing sampled areas. Again at Seneca we used a modification of this technique and used brown/white ratios in conjunction with surveyed areas.

AUTOMOBILE COUNTS. These are used as population trend indicators by surveying the same route annually and counting all deer seen.

Indirect Counts. PELLET COUNTS. This is a technique that is applicable only in winter-spring situations. It is based on the assumption that deer defecate at an average rate of 12.7 times per day. It is also based on the assumption that the pellet groups being counted are deposited within a certain time span. The average number of pellet groups per acre is computed for each cover type known to be inhabited by deer. This is usually done by counting all pellet groups within a series of circles along predetermined lines, in a statistically designed manner. The number of pellet groups per hectare for each cover type is multiplied by the number of hectares in each type, resulting in the total number of pellet groups per habitat type. One other factor that must be at least closely approximated is the number of days that deer have been on the range. Knowing this, the number of pellets per hectare for each habitat type, and the daily defecation rate allows an estimation of the number of deer in the area.

This technique probably only has application in northern or mountainous areas where deer are known to concentrate for a period of time during the winter.

TRACK COUNTS. This technique was originally developed to census migratory mule deer on western ranges (Rasmussen and Doman 1943). Basically, deer migrating to wintering range at lower elevations were censused as they crossed dirt roads or fire lanes or left tracks in new snow. A modification of this technique has been used in association with drives: deer are driven out of an area and their tracks are counted in new snow or in the dirt of freshly dragged roads surrounding the driven area.

KILL FIGURES AND PHYSICAL CONDITION PARAMETERS. Changes in relative age compositions reflect changes in population sizes as well, of course, as changes in hunting pressure. Kill figures are usually derived from examination of hunter-killed deer carcasses at deer check stations during the fall hunting seasons. Data collected at such stations vary from simply recording kill location and sex of deer to very sophisticated operations, including having the entire deer brought in to be field dressed by agency personnel. All manner of data are usually collected at these more sophisticated operations. Food habits studies (rumen collection) can be conducted; pathologic studies involving blood, urine, and various tissue samples can be conducted; and the more routine data, such as whole weight/dressed weight, various measurements, antler measurements, reproductive data (both uterine/ovarian analyses from females and testes from males), and age data, can be collected.

We believe that recording and analyzing the following data are absolutely necessary for properly monitoring the physical condition of a deer herd.

1. Kill location. This is necessary for managing populations with some degree of precision beyond mere state-wide regulatory seasons.
2. Kill date. This is necessary in order to be able to compare the physical condition of any one herd to another. That is, there is little value in comparing five-month-old deer killed in early October to seven-month-old deer killed in December.
3. Sex of deer (self-explanatory).
4. Age of deer. This is necessary for physical condition parameter comparisons.
5. Antler beam diameter. This measurement is an excellent condition indicator, and is easy to do.
6. Teat size on yearling females. This is a relatively new technique reported by Sauer and Severinghaus (1977). Measurement of teat size from yearling

females examined at fall–early winter deer check stations separates females that had suckled a fawn (pregnant) from those that had not, with approximately 80 percent accuracy.
7. Carcass size on fawns and yearlings. Actually, carcass size (either whole weight or field dressed weight) on fawns is probably sufficient, if taken in conjunction with other physical parameters cited above, as a range condition indicator.

Habitat Management. Contrary to popular belief, the white-tail is not an animal of the mature woods. It does best in subclimax or ''temporary'' habitat. Ideal white-tail habitat encompasses the following: a combination of active dairy farms with hay fields and woodlots, intermixed with abandoned farms with hay fields reverting to brush lots, and a good commercial timber-harvesting operation working in the woodlots. If this habitat was located in northern states or provinces, conifer cover in woodlots would be necessary for winter shelter. The white-tail is a creature that thrives in an interspersed habitat.

It is not possible within the framework of this chapter to describe the many kinds and combinations of habitat management practices that relate to white-tailed deer management. We will briefly mention some that do have wide application.

The degree or intensity of habitat manipulation for white-tails varies, of course, with the overall objective of the responsible agency. Wildlife management practices designed to improve an area for one species probably are no longer acceptable on public lands, at least in the United States. Wildlife managers must bear this fact in mind.

Techniques that are employed in many areas include the following:

UNEVEN-AGED FOREST MANAGEMENT. The objective of this type of management is to maintain a forest with all age classes mixed. This will provide a constant supply of browse and mast, and, if the forest is in a northern state or province, special consideration must be directed toward maintaining conifer covers used by deer in the winter. Practices such as maintaining logging roads by seeding with legumes and grasses, and even creating small grassy openings scattered throughout the forest, will help.

CLEAR-CUTTING. This technique has applicability in many situations. It probably produces the greatest amount of food per unit cut of any technique. It has widespread use in the lake states, especially on the upper peninsula of Michigan. Ideally, if this technique is to be used, the clear cuts should be made in such a manner that the food-producing plots are interspersed with the uncut plots that provide shelter and escape cover.

There are, of course, many cutting practices that fall somewhere between uneven-aged forest management and clear-cutting. These include partial cuttings, thinning, weeding, and timber stand improvement. One other technique is known as release cutting. This is used to open up the overhead canopy in situations such as old apple orchards that are deteriorating because of being overtopped by other species of tree.

BRUSH MANAGEMENT. The key to deer abundance and deer health in southwestern states, and probably in Mexico, is to maintain a habitat interspersed with "brush" and grasses. This is accomplished by manipulating invading brush with bulldozers, burning, or applying herbicides (Knipe 1977; Hailey 1979).

PLANTING. This technique does not have a great deal of merit simply because of the great costs involved actually to affect an area large enough to be meaningful. In fact, many of the problems associated with deer and agriculture can be related to plantings—plantings *not* intended for deer. These include orchard and ornamental planting damage to actual suppression of future timber or pulp, and especially Christmas tree plantations.

New York is experimenting with a novel approach involving planting. White cedar (*Thuja occidentalis*) seedlings are being planted in areas known to be used by deer in the winter for shelter (usually hillsides with a southerly exposure), but with an inadequate food supply. The seedlings are being protected from overutilization by deer with a wire cylinder completely surrounding the seedling. The objective is to allow annual growth to come through the wire mesh and be available to deer.

Herbaceous food plantings are made in many areas in association with existing forestry practices, usually seeding logging roads, or log landings, and clear-cut areas that are to be maintained by mowing.

FERTILIZING. Compared to other techniques, this one has been neglected as a method for directly manipulating deer habitat. However, early work by Mitchell and Hosley (1936) involved nutritive values of flowering dogwood (*Cornus florida*) and the white-tail's ability to seek out fertilized plants as opposed to controls. Wood and Lindzey (1967) experimented with several fertilizer components on five species of browse. They concluded that "the findings suggest that forest fertilization may have a significant impact on deer management techniques." They also speculated on the potential for shifting herds from one parcel of land to another (at least locally) by fertilizing.

We believe that further research is needed in these areas.

Any habitat manipulation efforts that are designed to improve an area's carrying capacity, prior to achieving deer herd control, are wasted. White-tails are capable of widespread habitat damage, both in wild, forested situations and in semirural or urban situations. Unless excluded by fencing, which is usually prohibitive, herd control is a prerequisite for habitat improvement work.

RESEARCH NEEDS

A long list of research needs each of which is designed to overcome some relatively local problem or a problem peculiar to one or more agencies can easily be developed. Such a list is given below. However, it is our opinion that there are just two very basic overriding problems that desperately need research, and they are closely related. The first is to design systems for converting to legally harvested deer those one million deer that are being lost to causes other than legal hunting. The second and most important is to develop a communications plan designed to convince the various publics of the need for a sound deer management program. This, of course, is a prerequisite for gaining public confidence and support for the deer management program of each respective agency. Details for implementing such a program were given in the Northeast Deer Study Group's "Communications Plan for Deer Management" (Harkins 1979).

The wildlife management profession is terribly weak in the area of communications. Universities and wildlife agencies have done a superb job in training wildlife biologists—biologists who have been highly successful in resolving many of the problems associated with the biological aspects of wildlife management. The science of wildlife management has been developed to a very high level regarding animals and habitat. However, we are terribly weak when it comes to communicating our thoughts and findings to the public and to politicians in such a way that we can properly manage our wildlife resources, in particular, the white-tailed deer.

Aside from this one overriding problem, there are a number of areas where additional research or refinement of existing techniques is needed. One of the first is the design of adequate, inexpensive fencing systems to keep deer off highways, especially high-speed, limited-access highways. Another important need is to develop a range-wide system for monitoring the physical health of deer and their susceptibility to various diseases and parasites. As indicated previously, deer are susceptible to many diseases and parasites, some of which have livestock and human health implications. For example, there is no range-wide plan to control deer populations should there be an extensive outbreak of anthrax.

Another research area that has implications far beyond white-tailed deer is the effects on habitat, both short term and long term, of the tremendous increase in woodlot cutting for firewood.

Research is needed to identify the causes and magnitude of all significant factors that fall in the category of "other than legal" deaths in deer. Identifying the losses should be a prerequisite to any attempts to overcome such factors.

Research in the area of repellents and attractants is needed, especially with regard to controlling damage to agricultural crops and nursery stock. Studies of the competition between deer and other wild big game animals and domestic stock are needed.

A list of research needs can encompass perhaps hundreds of specific items, but we do not think that it is necessary to provide such details here. Those research needs mentioned are the ones with broadest applications and implications. We cannot overemphasize the need for developing good, sound deer communications plans. Accomplishing this necessitates producing wildlife biologists within traditional fish and wildlife agencies with expertise in the fields of communications, sociology, and psychology.

The authors wish to dedicate this chapter to the many individuals belonging to the Northeast Deer Study Group (now called the Northeast Deer Technical Committee) who have contributed so much to our collective body of knowledge concerning the magnificent white-tail. We also wish to thank the many unnamed members of each provincial and state wildlife agency who responded to our questionnaire.

LITERATURE CITED

Aimonetti, M. 1977. How a forester views our deer problem. Pennsylvania Farmer (December). P. 10.

Almand, J. D. 1968. Wildlife: how valuable? Georgia Game and Fish 1:10–12.

Applegate, J. E. 1973. Some factors associated with attitude toward deer hunting in New Jersey residents. Trans. North Am. Wildl. Nat. Resour. Conf. 38:267–273.

_____. 1975. Attitudes toward deer hunting in New Jersey: a second look. Wildl. Soc. Bull. 3:3–6.

Armstrong, H. L., and MacNamee, J. K. 1950. Blackleg in deer. J. Am. Vet. Med. Assoc. 117:212.

Bartlett, I. H. 1949. White-tailed deer resources: United States and Canada. Trans. 14th North Am. Wildl. Conf. Pp. 543–552.

Beaudoin, R. L.; Samuels, W. M.; and Strome, C. P. A. 1970. A comparative study of the parasites of two populations of white-tailed deer. J. Wildl. Dis. 6:56–63.

Behrend, D. F. 1970. The nematode *Pneumostrongylus tenuis* in white-tailed deer in the Adirondacks. New York Fish Game J. 17:45–49.

Behrend, D. F., and Witter, J. F. 1968. *Pneumostrongylus tenuis* in white-tailed deer in Maine. Pittman-Robertson Proj. W-37-R. 8pp.

Brown, T. L., and Thompson, D. Q. 1976. Changes in posting and landowner attitudes in New York state, 1963–1973. New York Fish Game J. 23:101–137.

Burke, D.; Deatly, A.; Eriksen, R. E.; Lund, R. C.; McConnell, P. A.; and Winkel, R. P. 1978. An assessment of deer hunting in New Jersey. New Jersey Div. Fish, Game and Shell Fisheries and Pittman-Robertson Proj. W-45-R-13. 28pp.

Burke, D.; McConnell, P. A.; Hawkinson, B. J.; Erikson, R. E.; and Lund, R. C. 1979. New Jersey's white-tailed deer. Deer Rep. 6. Pittman-Robertson Proj. W-45-R-15. 15pp.

Calhoun, J., and Loomis, F. 1975. Prairie white-tails. Illinois Dept. Conserv., Springfield, Ill. 49pp.

Cheatum, E. L. 1949. Bone marrow as an index of malnutrition in deer. New York State Conserv. 3:19–22.

Crawford, H. S., and Leonard, R. G. 1965. The Sylamore deer study. Mimeogr. 5pp.

Cushwa, C. T.; Downing, R. L.; Harlow, R. F.; and Urbston, D. F. 1970. The importance of woody twig ends to deer in the southeast. U.S.D.A. For. Serv. Res. Pap. SE67. S.E. For. Exp. Stn. Asheville, N.C. 11pp.

Dean, D. J.; Gordon, M. A.; Severinghaus, C. W.; Kroll, E. T.; and Reilly, J. R. 1961. Streptothricosis: a new zoonotic disease. New York State J. Med. 61:1283–1287.

Dunkeson, R. R. 1955. Deer range appraisal for the Missouri Ozarks. J. Wildl. Manage. 19:358–364.

Erickson, A. B. 1940. Notes on a method for censusing white-tailed deer in the spring and summer. J. Wildl. Manage. 4:15–18.

Eve, J. H., and Kellogg, F. E. 1977. Management implications of abomasal parasites in southeastern white-tailed deer. J. Wildl. Manage. 41:169–177.

Forrester, D. J.; Taylor, W. J.; and Humphrey, P. P. 1974. Strongyloidiasis in white-tailed deer fawns in Florida. J. Wildl. Dis. 10:146–148.

Francis, J. 1958. Tuberculosis in animals and man. Cassell and Co., Ltd., London. 244pp.

French, C. E.; McEwen, L. C.; Magruder, N. D.; Ingram, R. H.; and Swift, R. W. 1955. Nutritional requirements of white-tailed deer for growth and antler development. Bull. 600. Agric. Exp. Stn. Pennsylvania State Univ. 50pp.

Friend, M. 1967. Skin tumors in New York deer. Bull. Wildl. Dis. Assoc. 3:102–104.

Friend, M.; Kroll, E. T.; and Gruft, H. 1963. Tuberculosis in a wild white-tailed deer. New York Fish Game J. 10:118–123.

Gilbert, A. H. 1973a. Expenditure patterns of non-resident sportsmen in Vermont, 1970. Univ. Vermont Agric. Exp. Stn. Rep. 55pp.

_____. 1973b. Expenditure patterns of resident sportsmen in Vermont, 1970. Univ. Vermont Agric. Exp. Stn. Rep. 52pp.

Gilbert, F. F. 1966. Aging deer by incisor cementum. J. Wildl. Manage. 30:200–202.

_____. 1973. *Parelaphostrongylus tenuis* (Dougherty) in Maine. Part 1: The parasite in white-tailed deer (*O. virginianus*) Zimmerman. J. Wildl. Dis. 9:136–143.

Gilbert, F. F., and Stolt, S. L. 1970. Variability in deer age determination. J. Wildl. Manage. 34:532–535.

Glazener, W. C., and Knowlton, F. F. 1967. Some endoparasites found in Welder Refuge deer. J. Wildl. Manage. 31:595–597.

Godin, A. J. 1977. Wild mammals of New England. Johns Hopkins Univ. Press, Balitmore. 304pp.

Goodwin, G. G. 1961. The big game animals of North America. Outdoor Life and E. P. Dutton and Co., Inc., New York. 104pp.

Hailey, T. L. 1979. Basics of brush management for white-tailed deer production. Texas Parks and Wildl. Dept., Austin. 8pp.

Halls, L. K. 1978. White-tailed deer. Pages 43–65 *in* J. L. Schmidt and D. Gilbert, eds. Big game of North America: ecology and management. Stackpole Co., Harrisburg, Pa., and Wildl. Manage. Inst., Washington, D.C. 494pp.

Harkins, J. P. 1979. Communications plan for deer management. Pub. Northeast Deer Study Group, Educ. Comm. 130pp.

Harlen, C. 1977. Farmers talk about deer damage. Pennsylvania Farmer (Dec). P. 6.

Haugen, A. O., and Davenport, L. A. 1950. Breeding records of white-tailed deer in the Upper Peninsula of Michigan. J. Wildl. Manage. 14:290–295.

Haugen, A. O., and Trauger, D. L. 1962. Ovarian analysis for data on corpora lutea changes in white-tailed deer. Iowa Acad. Sci. 69:231–238.

Hesselton, W. T. 1969a. Deer trapping and tagging. New York State Conserv. 23:14–17.

_____ 1969b. The incredible white deer herd. New York State Conserv. 24:18–19.

Hesselton, W. T.; and Hesselton, R. A. M. 1979. Questionnaire of all provinces/states concerning deer kill statistics. 2pp.

Hesselton, W. T., and Jackson, L. W. 1974. Reproductive rates of white-tailed deer in New York State. New York Fish Game J. 21:135-152.

Hesselton, W. T., and Sauer, P. R. 1973. Comparative physical condition of four deer herds in New York according to several indices. New York Fish Game J. 20:77-107.

Hesselton, W. T.; Severinghaus, C. W.; and Tanck, J. E. 1965. Deer facts from Seneca Depot. New York State Conserv. 20:28.

Hoff, G. L.; Trainer, D. O.; and Jochim, M. M. 1974. Bluetongue virus and white-tailed deer in an enzootic area of Texas. J. Wildl. Dis. 10:158-163.

Howe, D. L. 1970. Miscellaneous bacterial disease. Pages 376-381 in J. W. Davis, L. H. Karstad, and D. O. Trainer, eds. Infectious diseases of wild mammals. Iowa State Univ. Press, Ames. 421pp.

Jackson, L. 1973. The Malone deer study. Pittman-Robertson Proj. W-89-R. 20pp.

Karstad, L. 1967. Fluorosis in deer (Odocoileus virginiana). Bull. Wildl. Dis. Assoc. 3:42-46.

Kellert, S. R. 1976. Attitudes and characteristics of hunters and anti-hunters and related policy suggestions. Working Pap. presented to Fish and Wildl. Serv., U.S. Dept. Interior. 64pp.

Kellogg, F. E.; Prestwood, A. K.; and Noble, R. E. 1970. Anthrax epizootic in white-tailed deer. J. Wildl. Dis. 6:226-228.

Kellogg, R. 1956. Where and what are the white-tails? Pages 31-55 in W. P. Taylor, ed. Deer of North America. Stackpole Co., Harrisburg, Pa., and Wildl. Manage. Inst., Washington, D.C. 668pp.

Kistner, T. P., and Hayes, F. A. 1970. White-tailed deer as hosts of cattle fever-ticks. J. Wildl. Dis. 6:437-440.

Kistner, T. P.; Shotts, E. B.; and Greene, E. W. 1970. Naturally occurring cutaneous streptothricosis in a white-tailed deer (Odocoileus virginianus). J. Am. Vet. Med. Assoc. 157:633-635.

Knipe, T. 1977. The Arizona white-tail deer. Special Rep. no. 6. Arizona Game and Fish Dept. 108pp.

Lambiase, J. T., Jr.; Amann, R. P.; and Lindzey, J. S. 1972. Aspects of reproductive physiology of male white-tailed deer. J. Wildl. Manage. 36:868-875.

Laws, R. M. 1952. A new method of age determination for mammals. Nature 169:972-973.

Lay, D. W. 1969. Foods and feeding habits of white-tailed deer. White-tailed Deer in Southern Forest Habitat, 25-26 March 1969, Proc. Symp. Nacogdoches, Texas.

Long, T. A.; Cowan, R. L.; Wolfe, C. W.; Rader, T.; and Swift, R. W. 1959. Effect of seasonal feed restriction on antler development of white-tailed deer. Prog. Rep. 209. Agric. Exp. Stn. Pennsylvania State Univ. 11pp.

McCabe, R. A., and Leopold, A. S. 1951. Breeding season of the Sonora white-tailed deer. J. Wildl. Manage. 15:433-434.

McConnell, P. A.; Lund, R. C.; and Boss, N. R. 1976. The 1975 outbreak of hemorrhagic disease among white-tailed deer in northwestern New Jersey. Trans. 33rd Northeast Fish Wildl. Conf. Pp. 35-44.

McDowell, R. D. 1962. Relationship of maternal age to prenatal sex ratios in white-tailed deer. Rep. 4. Proc. Northeast Sect. Wildl. Soc. Monticello, N.Y. Mimeogr.

Madson, J. 1961. The white-tailed deer. Olin Mathieson Chemical Corp., East Alton, Ill. 100pp.

Martin, A. C.; Zim, H. I.; and Nelson, A. L. 1951. American wildlife and plants. Dover Publications, Inc., New York 484pp.

Maynard, L. A.; Bump, G.; Darrow, R.; and Woodward, J. C. 1935. Food preferences and requirements of the white-tailed deer. Bull. 1, New York State Conserv. Dept. and New York State Coll. Agric. 35pp.

Miller, F. L., and Cawley, A. J. 1970. Polydactylism in a white-tailed deer from eastern Ontario. J. Wildl. Dis. 6:101-103.

Miller, H. A. 1961. Page 1 in K. Halls and R. H. Ripley, eds. Deer browse plants of southern forests. Southern and Southeastern For. Exp. Stns, For. Serv., U.S.D.A. For. Game Res. Com. Southeastern Sect. Wildl. Soc.

Mitchell, H. L., and Hosley, N. W. 1936. Differential browsing by deer on plots variously fertilized. Black Rock For. Pap. Harvard Univ. 1:24-77.

Moody, R. D. 1969. The goals of private forest holding in deer management. Pages 90-92 in Proc. Symp. White-tailed deer in the southern forest habitat. Southern For. Exp. Stn. For. Serv. U.S.D.A. in cooperation with For. Game Com. Southeastern Sect. Wildl. Soc. and School For., Stephen F. Austin State Univ.

Nettles, V. F., and Doster, G. L. 1975. Nasal bots of white-tailed deer in the southeastern United States. Proc. 29th Annu. Conf. Southeastern Assoc. Game and Fish Commissioners Pp. 651-655.

Nettles, V. F.; Hayes, F. A.; and Martin, W. M. 1976. Observation of injuries in white-tailed deer. Proc. 30th Annu. Conf. Southeastern Assoc. Game and Fish Commissioners Pp. 474-480.

Nettles, V. F., and Prestwood, A. K. 1976. Experimental Parelaphostrongylus andersoni infections in white-tailed deer. Vet. Pathol. 13:381-393.

Nichol, A. A. 1938. Experimental feeding of deer. Univ. Ariz., Coll. Agric., Agric. Exp. Stn. Tech. Bull. 75:1-39.

Pils, C., and Martin, M. A. 1979. The cost and chronology of Wisconsin deer: vehicle collisions. D.N.R. Res. Rep. 103. Wisconsin Dept. Nat. Resour. 5pp.

Pursglove, S. R., Jr. 1976. Eucyathostomum webbi sp. N (Strongyloidea: Cloacinidae) from white-tailed deer (Odocoileus virginianus). J. Parasitol. 62:574-578.

Pursglove, S. R.; Prestwood, A. K.; Nettles, V. F.; and Hayes, F. A. 1976. Intestinal nematodes of white-tailed deer in southeastern United States. J. Am. Vet. Med. Assoc. 169:896-900.

Rasmussen, D. I., and Doman, E. R. 1943. Census methods and their application in the management of mule deer. Trans. 8th North Am. Wildl. Conf., Pp. 369-379.

Rasmussen, G. P.; Hesselton, W. T.; Jackson, L. W.; Severinghaus, C. W.; and Free, S. L. 1979. Fetal sex ratio of white-tailed deer in New York State. New York Fish Game J. In press.

Reed, D., and Shave, H. 1976. Deer diseases in South Dakota, 1972-76. Pittman-Robertson Projs. W-75-R-15, 16, 17, 18. 32pp.

Richards S. H.; Schipper, I. A.; Eveleth, P. F.; and Shumard, R. F. 1956. Mucosal disease of deer. Vet. Med. 51:358-62.

Rosen, M. N. 1970. Necrobacillosis. Pages 286-292 in J. W. Davis, L. H. Karstad, and D. O. Trainer, eds. Infectious diseases of wild mammals. Iowa State Univ. Press, Ames. 421pp.

Rosen, M. N., and Holden, F. F. 1961. Multiple purulent abscess (Corynebacterium pyogenes) of deer. California Fish Game 47:293-300.

Salkin, I. F., and Stone, W. B. 1974. Subcutaneous mycotic infection of a white-tailed deer. J. Wildl. Dis. 10:34-38.

Sauer, P. R. 1971. Tooth sectioning vs. tooth wear for assigning age to white-tailed deer. Trans. Northeast Fish Wildl. Conf. 28:9-20.

Sauer, P. R., and Severinghaus, C. W. 1977. Determination and application of fawn reproductive rates from yearling teat length. Joint Northeast-Southeast Deer Study Group Meet., Fort Pickett, Va. Pp. 81–85.

Sauer, P. R.; Tanck, J. E.; and Severinghaus, C. W. 1969. Herbaceous food preferences of white-tailed deer. New York Fish Game J. 16:145–157.

Scheffer, V. B. 1950. Growth layers on the teeth of Pinnipedia as an indication of age. Science. 112:309–311.

Seton, E. T. 1929. Lives of game animals. Volume 3, part i. Doubleday, Doran, New York. 412pp.

Severinghaus, C. W. 1949. Tooth development and wear as criteria of age in white-tailed deer. J. Wildl. Manage. 13:195–216.

Severinghaus, C. W., and Cheatum, E. L. 1956. Life and times of the white-tailed deer. Pages 57–186 *in* W. P. Taylor, ed. The deer of North America. Stackpole Company, Harrisburg, Pa., and Wildl. Manage. Inst., Washington, D.C. 668pp.

Shaw, D. L., and Gilbert, D. L. 1974. Attitudes of college students toward hunting. Trans. North Am. Wildl. and Nat. Resour. Conf. 39:157–162.

Shaw, W. 1974. A survey of hunting opponents. Wildl. Soc. Bull. 5:19–24.

Shiras, G., III. 1935. Hunting wildlife with camera and flashlight. Volume 1. Lake Superior region. Natl. Geogr. Soc., Washington, D.C. 450pp.

Shope, R. E.; MacNamara, L. G.; and Mangold, R. 1955. Report on the deer mortality: epizootic hemorrhagic disease of deer. New Jersey Outdoors 6(5):16–21.

Silver, H. 1965. An instance of fertility in a white-tailed buck fawn. J. Wildl. Manage. 29:634–636.

Stauber, E.; Armstrong, P.; Chamberlain, K.; and Gorgen, B. 1973. Caseous lymphadenitis in a white-tailed deer. J. Wildl. Dis. 9:56–57.

Stiteler, W. M., and Shaw, S. P. n.d. Use of woody browse by white-tail deer in heavily forested areas of northeastern United States. U.S. For. Serv. Rep. 17pp. Mimeogr.

Townsend, M. T., and Smith, M. W. 1933. The white-tailed deer of the Adirondacks. Bull. New York State Coll. For. Syracuse Univ. 6:153–385.

Trainer, D. O. 1970. Bluetongue. Pages 55–59 *in* J. W. Davis, L. H. Karstad, and D. O. Trainer, eds. Infectious diseases of wild mammals. Iowa State Univ. Press, Ames. 421pp.

Turner, E. C. 1971. Fleas and lice. Pages 65–77 *in* J. W. Davis and R. C. Anderson, eds. Parasitic diseases of wild mammals. Iowa State Univ. Press, Ames. 364pp.

U.S. Fish and Wildlife Service. 1977. 1975 National survey of hunting fishing and wildlife: associated recreation. U.S. Dept. Interior, Washington, D.C. 91pp.

Verme, L. J. 1965. Reproductive studies on penned white-tailed deer reproduction. J. Wildl. Manage 29:74–79.

Waldo, C. M., and Wislocki, G. B. 1951. Observations on the shedding of the antlers of Virginia deer (*Odocoileus virginianus borealis*). Am. J. Anat. 88:351–396.

Watts, R. C. 1964. Forage preferences of captive deer while free ranging in a mixed oak forest. Pennsylvania Coop. Wildl. Res. Unit Pap. 112. 16pp.

White, C. M. 1968. Productivity and dynamics of the white-tailed on the Crane Naval Depot. Ph.D. Thesis. Purdue Univ., Lafayette, Ind. 171pp.

Whitlock, S. C. 1939. The prevalence of disease and parasites in white-tail deer. Trans. 4th North Am. Wildl. Conf. pp. 244–249.

Witter, J. F., and O'Meara, D. C. 1970. Brucellosis. Pages 249–255 *in* J. W. Davis, L. H. Karstad, and D. O. Trainer, eds. Infectious diseases of wild mammals. Iowa State Univ. Press, Ames. 421pp.

Wood, G. W., and Lindzey, J. S. 1967. The effects of forest fertilization on the crude protein, calcium and phosphorous content of deer browse in a mixed oak forest. Trans. Northeast Sect. Wildl. Soc. Quebec City, P.Q. 24:14.

Wyand, D. S.; Langheinrich, K.; and Helmboldt, C. F. 1971. Aspergillosis and renal oxalosis in a white-tailed deer. J. Wildl. Dis. 7:52–56.

WILLIAM T. HESSELTON, 43 Hall Road, Oxford, Mass. 01540.

RUTHANN MONSON HESSELTON, 43 Hall Road, Oxford, Mass. 01540.

46

Moose

Alces alces

John W. Coady

NOMENCLATURE

COMMON NAME. Moose, elk (in Europe)

SCIENTIFIC NAME. *Alces alces*

SUBSPECIES. Moose belong to the family *Cervidae*, which also includes caribou (*Rangifer tarandus*) and several species of deer (*Odocoileus* spp.) in North America. Living representatives of the genus *Alces* Gray belong to one species (Peterson 1952). Four subspecies of moose are distinguished in North America, based primarily on skull characteristics: *A. a. americana* occurs from Ontario eastward; *A. a. andersoni* occurs from Ontario to British Columbia, *A. a. shirasi* occurs in the mountains of Wyoming, Idaho, Montana, and southeast British Columbia; and *A. a. gigas* occurs in Alaska, western Yukon, and northwestern British Columbia (figure 46.1). Three subspecies are recognized in Eurasia, where they were formerly known as elk.

DISTRIBUTION

Peterson (1955) and Kelsall and Telfer (1974) reviewed the postglacial dispersal of moose in North America. Moose are Holarctic in distribution, having emigrated from Eurasia across the Bering land bridge to unglaciated refugia in Alaska during the Illinoian glaciation (Péwé and Hopkins 1967). The Bering land bridge and continental glaciations appeared and vanished repeatedly during the Pleistocene Epoch, accompanied perhaps by successive emigrations of moose from Eurasia to North America. During interglacial intervals, moose dispersed from Alaska refugia to Canada and the continental United States. During the maximum stage of the Wisconsin glaciation three distinct refugia for moose south of the ice field and one refugium in Alaska north of it existed. These areas may have provided the isolation necessary for the development of the four North American forms recognized today. Following the retreat of glaciers and growth of forests, moose dispersed from their southern and northern refugia to colonize extensive areas in Canada and the United States.

The systematic status of moose in North America and Eurasia is not entirely clear. *A. a. shirasi* and *A.*

FIGURE 46.1. Distribution of the moose (*Alces alces*).

a. americana (Hsu and Benirschke 1969) and *A. a. gigas* (Rausch 1977) have fundamental chromosome numbers $2n = 70$, while *A. a. alces* from Europe has $2n = 68$ (Aula and Kaariainen 1964). Therefore, polymorphism occurs with respect to the karyotype of some subspecies from Eurasia and North America. Egorov (1972) concluded on morphologic grounds that *A. a. pfizenmayeri* from eastern Siberia should be designated *A. a. gigas*. If subsequent studies reveal a chromosome number $2n = 70$ and confirm Egorov's (1972) conclusion, this would suggest a late Pleistocene reverse migration west from the Alaska refugium (Rausch, personal communication).

During the past 100 years moose have expanded their breeding range in several areas (Peterson 1955). In particular, both *A. a. americana* and *A. a. andersoni* have colonized the area between Lake Superior and the lowlands south of Hudson Bay (Krefting 1974) and *A. a. americana* has dispersed throughout most of Quebec south of Ungava Bay (Brassard et al. 1974). *A. a. andersoni* has probably extended its range north to tundra areas of the Northwest Territories and west to the coastal mountains of British Columbia (Kelsall and Telfer 1974). *A. a. shirasi* has extended north along the Rocky Mountains into Alberta and as far south as northern Utah (Peek 1974). *A. a. gigas* has extended its breeding range toward coastal areas in most of Alaska and northwest Canada (LeResche et al. 1974*a*). The recent increase in observations of moose in some remote areas may result from greater opportunity for sightings because of the increased presence of people. Further, moose may be only occasional immigrants to some areas, and sightings do not necessarily imply the presence of established populations. Therefore, apparent range extensions must be interpreted with caution.

Recent reduction in range of moose has occurred in two areas. In New England *A. a. americana* apparently extended south into Pennsylvania during the 18th century, but retreated to northern New York in the 19th century, and is now limited to Maine and northern New Hampshire and Vermont (Dodds 1974; Godin 1977). In the Midwest, *A. a. andersoni* formerly occurred in northern Wisconsin and the peninsula of Michigan but is now virtually absent from this area (Krefting 1974). Habitat loss and excessive hunting may have caused the extirpation of moose from both regions.

Moose have been introduced into areas of their historical range and into areas where they have not previously occurred. Most introductions have probably not succeeded in meeting intended goals and have neither made significant contribution to the welfare of moose nor benefited mankind. Exceptions to this poor record occurred in Newfoundland, where transplants totaling six moose between 1878 and 1904 resulted in the colonization of the entire island by 1945 (Pimlott 1953), and near Cordova, Alaska, where the introduction of approximately 25 calves between 1949 and 1958 resulted in a particularly viable moose population (Burris and McKnight 1973). Notable disappointments were attempts to transplant moose to Anticosti Island near Newfoundland (Pimlott 1961) and Kodiak Island, Alaska (Burris and McKnight 1973). Serious consideration was recently given to reintroducing moose into the Tug Hill region of upstate New York, although other attempts to reestablish moose in New York during this century have been unsuccessful (Dodds 1974).

Although moose populations have fluctuated primarily in response to changes in habitat, only excessive exploitation or habitat alteration by humans has posed a serious threat. Population growth or decline in local areas in response to various environmental influences may not be synchronous, but regional trends in population size are sometimes evident. Densities of *A. a. americana* are increasing slightly in many areas, with the exception of Newfoundland, where populations are stable or declining slightly (Dodds 1974; Brassard et al. 1974). Both *A. a. andersoni* and *A. a. shirasi* appear to be generally stable or slowly increasing throughout most of their range. Exceptions to this trend are evident in Manitoba and some areas of Saskatchewan, where population declines may be due to habitat loss from timber harvests. *A. a. gigas* populations appear to be high and stable in western and northern portions of their range, but low and stable or still declining from interior and southern Alaska eastward (LeResche et al. 1974*a*). Moose population levels are relatively high in some areas along the western periphery of their range, such as in Utah and Colorado, the coastal mountains of British Columbia, western and northern Alaska, and northwestern Canada. In most central and eastern regions population levels have decreased from the high densities promoted by habitat improvement following extensive fire and logging earlier in the century.

DESCRIPTION

Written accounts of moose in North America date to the time of early explorers and settlers. Fantastic descriptions and superstitions persisted into the last century, when authors such as Merrill (1920) and Seton (1927) provided more accurate yet casual and subjective accounts of the species. The first scientific study of moose was conducted by Murie (1934) on Isle Royale, Ontario, and for the next 20 years factual information on moose was collected from several areas. This considerable body of data was enhanced and summarized by Peterson (1955). His monograph still serves as a basic reference on moose. Since then, studies of moose have increased at an exponential rate, and information was again summarized by Bédard et al. (1974) in a comprehensive volume and by Franzmann (1978) in a review chapter. The most current information on moose is available in scientific journals and in annual proceedings of the North American Moose Conference and Workshop, state-prepared Federal Aid in Wildlife Restoration Project Reports, and provincial research reports.

The physical appearance of moose has been well described by Peterson (1955, 1974) and others. Being the largest representative of the family Cervidae and second only to bison (*Bison bison*) in weight among terrestrial species in North America, the moose is indeed an imposing animal. A massive body supported by exceptionally long and slender legs, a long muzzle with a large and pendulous nose, a conspicuous dewlap extending from the throat, and an inconspicuously short tail are notable features of moose.

Weight. Peterson (1974) summarized reports of weights and measurements of moose. Recorded weights of newborn calves are scarce, although they probably range between 11 and 16 kg. Verme (1970) reported a weight of 11.2 kg for a neonate in Michigan, and Coady (1973) found that five fetuses collected during the peak of calving in interior Alaska averaged 15.3 kg, and ranged from 13.6 to 17.2 kg. Birth weights of 6 to 10 kg have been reported for moose in the Soviet Union (Knorre 1961), although light neonates probably have a decreased potential for survival. Growth rates for *A. a. gigas* calves were presented by Franzmann et al. (1978). Seventy-six measurements of different moose showed that the growth rate of calves was approximately linear during their first five months. They reached an average weight of 180 kg by October. A 15-kg neonate must therefore gain over 1 kg per day to weigh 180 kg by October. Weights remain stable during winter before growth begins again in spring.

Whole weights of adult moose are also scarce. Most values are derived from dressed weights and may not be accurate approximations of whole weight (Peterson 1974). Blood et al. (1967) reported mean whole weights for female and male *A. a. andersoni* from Alberta of 418 kg and 414 kg, respectively. Franzmann et al. (1978) reported mean whole weights for free-ranging female and male *A. a. gigas* from Alaska of 400 kg and 455 kg, respectively. The heaviest moose from Alaska reported by Franzmann et al. (1978) and Gasaway and Coady (1974) were a 549-kg male in September and a 546-kg male in October, respectively. Moose undergo pronounced seasonal fluctuations in body weight, with maximum weight occurring in fall and minimum weight occurring in spring. Franzmann et al. (1978) found that the average weight of penned females and males increased 55 percent and 47 percent, respectively, between May and their maximum weight in fall. Gasaway and Coady (1974) found a 48 percent increase in the weight of unrestrained females from interior Alaska between May and October. Therefore, with some variation due to gestation and lactation among females, the seasonal pattern of body weight is one of gain during summer and loss during winter. This striking seasonal variation in weight must be considered when comparing values for moose from different areas and seasons.

Measurements. Skeletal measurements of moose are more readily available in the literature. They probably provide a better measure of growth rates than do body weights because they are not subject to seasonal variation. Peterson (1974) illustrated the standard external skeletal measurements for moose, including total length, tail length, hind foot length, ear length, and shoulder height. In addition, chest girth, measured as the circumference of the body immediately posterior to the front legs, may be used as an index to body size, and chest height above the ground may be measured as the distance from the sternum to the tip of the hoof. Tall chest height is a decided advantage to moose in moving through deep snow, and limited data suggest that a selective advantage for long legs may be evident in moose occupying deep snow areas (Coady 1974*a*). Franzmann et al. (1978) presented measurements of moose from Alaska and concluded that total length was the best indicator of growth rate. He found that total length, chest height, and leg length of moose from several areas in Alaska reflected growth in body size for the first seven or more years of life. Average morphometric measurements for five seven-year-old male *A. a. gigas* reported by Franzmann et al. (1978) are: total length 318 cm, shoulder height 190 cm, chest girth 205 cm, and hind foot length 81 cm.

Skull and Dentition. Moose are true heterodonts, having a dental formula of 0/3, 0/1, 3/3, 3/3, for a total of 32 teeth. As with many cervids, upper incisors are absent. The lower incisors and incisiform canines bite

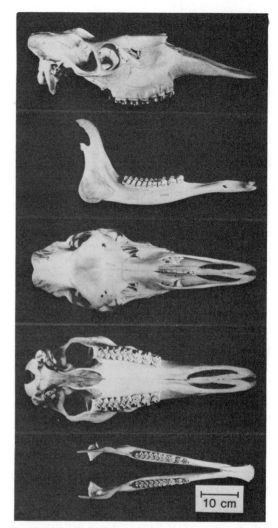

FIGURE 46.2. Skull of the moose (*Alces alces*). From top to bottom: lateral view of cranium, lateral view of mandible, dorsal view of cranium, ventral view of cranium, dorsal view of mandible.

against a callous pad in the upper jaw. The postorbital bar is complete, antorbital pits are present, and the nasal bones are short while the maxilla and premaxilla are elongate (figure 46.2). Both enumeration of cementum layers and tooth replacement and wear have been used for age appraisal in moose. During the past two decades analysis of cementum layers has become the preferred technique for age determination (Sergeant and Pimlott 1959), although errors of minus one to plus three years occur using this method in Alaskan moose (Gasaway et al. 1978).

Antlers. Antlers are strongly palmate and present only in males. They are a bonelike calcification that forms from an outgrowth of the frontal bone called the pedicle. The antler, which begins to grow as early as March or April, is covered by skin, or "velvet," with a profuse vascular system. Antler growth stops by late August or September and the velvet is rubbed off against branches of trees and shrubs. Newly exposed white but blood-stained antlers later become dark tan in color. The bone is reduced between the pedicle and the base of the antler until each antler is shed, usually by November on large bulls and as late as April on young bulls. Although calves may develop small "buttons," antlers are usually first obvious in yearlings, and may vary from "spikes" several centimeters long to multiple-spike or slightly palmate forms. Maximum antler development usually occurs in 8- to 13-year-old bulls in Alaska, and large antlers may weigh between 25 and 35 kg (Gasaway 1975). One of the largest sets of moose antlers scored by Boone and Crockett came from interior Alaska and had a spread of 196 cm and scored 253 points. Size and form of antlers vary greatly between individuals, but the largest antlers are grown by *A. a. gigas*. Regardless of maximum size, antlers grown by an adult moose represent a prodigious achievement in calcium metabolism.

Pelage. The pelage of moose is variable in color. Neonates may be light reddish, changing to rust color, then to shades of brown after one to three months of life. Adults range from light tan to almost black, depending on age and season. Adults begin molting in spring and appear progressively more ragged because of initial hair loss on the flanks and shoulders that exposes patches of darkly pigmented skin. Growth of short hair and dark-tipped guard hair is generally complete by midsummer, although guard hair probably continues to lengthen, reaching 20 cm or more on some areas of the body by winter. Bleaching and hair breakage during winter result in gradual lightening of the pelage. I am not aware of melanism in moose, although individuals with white pelage are occasionally observed in Alaska. Females usually have a white patch of hair near the vulva that may be useful for sex identification (Mitchell 1970). This is absent from the inguinal region of males.

REPRODUCTION

Breeding occurs during fall, with the peak of rutting activity between late September and early October. Remarkable synchrony in the timing of rut exists among moose in different areas and years in North America (Lent 1974). This is reflected in the consistency in calving dates observed throughout the range of moose. The gestation period is approximately 243 days (Peterson 1955), and most calving occurs between late May and early June. Moose do not form harems (Altmann 1959; Geist 1963) in which a male controls movements of several females, as is seen in some other cervids. Rather, they form rutting groups ranging in size from male and female pairs to 30 or more adults. Movement of both bulls and cows to and from groups may occur. The neck and shoulders of bulls swell during the rut, and individuals eat little for several days during the excitement of courtship. Agonistic encounters between bulls involve protracted displays, charges, violent fights, and occasional death. Agonistic behavior among females may also occur during the rut.

Recurrence of estrus, or delayed estrus, may be evident in some females. If breeding does not occur during an initial estrus, successive estrous cycles may occur at 20- to 30-day intervals (Edwards and Ritcey 1958; Markgren 1969). Conception during late estrous periods results in summer or fall parturitions, which are disadvantageous to calves because of their shortened growth period during summer and small size at the beginning of winter. Evidence for late pregnancies is limited in North America (Coady 1974b), although they commonly occur in Sweden (Markgren 1969). Factors contributing to late conceptions may be low bull/cow ratios (so that a bull is not available during the initial estrous period), breeding by young females, poor nutritional condition of females, and the tendency of cows with calves to avoid large rutting groups early in the rut.

The reproductive performance of a population is determined by several factors, including ovulation, pregnancy, and natality rates. The ovulation rate, expressed as the number of ovulations per 100 females of breeding age per breeding season, is determined by counting corpora lutea in pairs of ovaries collected after the breeding season. Pregnancy rate is determined from the number of embryos or fetuses per 100 females of breeding age per year, and natality rate is the number of live births per 100 females of breeding age per year. Since not all ovulations result in conception, and not all pregnancies result in the birth of viable calves, in utero counts overestimate natality rates.

Pimlott (1959), Schladweiler and Stevens (1973), and Simkin (1974) reviewed reproductive parameters of moose populations in North America. They found that ovulation rates for yearlings and for adults 2.5 years and older ranged from 0 to 0.54 and 0.71 to 1.27, respectively, and that pregnancy rates for yearlings and adults ranged from 0 to 0.47 and 1.00 to 1.20, respectively. Eleven to 29 percent of pregnant adults bore twins. Twin fetuses in yearlings (Pimlott 1959; Blood 1974), triplet fetuses in adults, and triplet calves with adults (Pimlott 1959; Hosley and Glaser 1952) are unusual. However, I have seen two sets of triplet short yearlings during April in northern Alaska. Since 80–90 percent of adult females in most moose populations in North America become pregnant annually, variation in the reproductive performance of populations results primarily from the number of yearlings, the incidence of pregnancy in yearlings and twinning in adults.

Natality rates for moose populations are difficult to determine because of the secretive behavior of parturient females and neonatal calves. Loss of embryos and fetuses does occur, although it is unusual (Pimlott 1959; Markgren 1969). Most loss of neonates is probably due to accidents, predation, or occasional abandonment (LeResche 1968). Age-specific natality rates are greatest among middle and old age class females (Markgren 1969). First births, regardless of the age of the female, are generally singles, and a decrease in the incidence of twinning and overall reproductive function may occur in very old females.

Considerable data on wild ungulates have accumulated since the work by Cheatum and Severinghaus (1950), which indicated that reproductive performance is closely influenced by nutritional plane. McCullough's (1979) work is especially constructive; he demonstrated that reproductive performance and survival of white-tailed deer (*Odocoileus virginianus*) on the George Reserve, Michigan, are strongly influenced by population density, particularly by the total number of females in the population. Moose populations on high-quality ranges where food is not limiting consist of a high proportion of females that first breed as yearlings and conceive twins as adults. Conversely, dense moose populations existing where food is limiting display lowered reproductive performance because of the delayed age of first breeding and low incidence of twinning. This relationship is illustrated by the reproductive performance of moose on Elk Island National Park, Canada (Blood 1974) (figure 46.3). Between 1960 and 1964, when there was low but increasing population density, the percentages of yearlings pregnant and of adults with twin fetuses increased. However, during a five-year period of population decline because of herd reduction between 1967 and 1972, the percentages of yearlings pregnant and adults with twin fetuses remained low. Implicit in these observations is that food is limiting at high population density but not at low density. Similar relationships between nutritional status and reproductive performance have been documented for other moose populations (Pimlott 1959, 1961; Markgren 1969).

Factors other than food resources may influence the reproductive performance of moose populations. Low pregnancy rates are possible resulting from low bull/cow ratios in which all estrous females are not bred. Sex ratios in unhunted moose populations usually approximate 50:50 (Bubenik 1972; Peterson 1977). However, in the Matanuska Valley, Alaska, intensive hunting of bulls skewed sex ratios to as low as 4 bulls:100 cows during the late 1950s and early 1960s (Bishop and Rausch 1974). Nevertheless, pregnancy rates remained near 90 percent during this period. In other areas of North America where large rutting aggregations do not occur, extremely low bull/cow ratios may result in reduced pregnancy rates or in con-

ception during a second or later estrus. Although skewed sex ratios apparently do not reduce the reproductive capability of populations in some areas, the minimum bull/cow ratio in different habitats that will insure the breeding of all receptive females is not known. Genotypic differences in the reproductive potential of populations are also possible (Geist 1974), although no empirical data to support this hypothesis are available.

ECOLOGY

Movements. Movements of moose may consist of local travel within seasonal ranges, migrations between seasonal ranges, and dispersal to new ranges. Precise patterns are often difficult to define because of variable movements by individuals comprising most populations. Individuals occupying local areas during a particular season may seasonally migrate during different times and to different locations, and some animals may remain resident in areas throughout the year. Nevertheless, movement patterns can be identified. Successful management of moose requires that identities and movements of population segments be determined to better understand the environmental factors that influence animals seasonally occupying different areas.

A seasonal home range is the area in which an individual conducts its normal activities during a period of time. LeResche (1974) reviewed the home range concept as it applies to moose and found that seasonal home ranges throughout North America are generally small, seldom exceeding 5–10 km². Home ranges are usually smaller in winter than in summer and tend to constrict with increasing snow depth (Coady 1974a). In the Soviet Union, Knorre (1959) found that the average winter home range size of moose decreased from 225 ha during mild winters to 97 ha during moderate winters and 5 ha during severe winters. A female in Minnesota occupied a 2.4-ha balsam fir (*Abies balsamea*) stand during 25 days of rapid snow accumulation (Van Ballenberghe and Peek 1971). I found that some radio-collared adult moose during winters of low snow depth in interior Alaska did not establish small home ranges, contrary to most observations, but traveled throughout an area of 300 km² during the winter. However, most studies indicate that the seasonal home range of individuals consists of a small area used intensively for a period of time, after

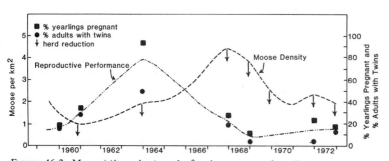

FIGURE 46.3. Moose (*Alces alces*) per km² and percentage of yearlings and adults.

which the moose move to and intensively use another area.

Migration is considered to be regular annual movements that involve return to at least one common area each year (LeResche 1974). Migrations may consist of relatively short movements with little change in elevation. Such movements have been documented in northern Minnesota (Berg 1971; Van Ballenberghe and Peek 1971), on the north slope of Alaska (Mould 1979), and in other regions with little relief. Distances moved are usually 2 to 10 km and consist primarily of a shift to adjacent, more seasonally favorable habitat. Migrations in mountainous areas usually involve a significant change in elevation. Horizontal distances between ranges may be as little as 2 km (Knowlton 1960) or as great as 170 km (Barry 1961) or more. Change in elevation usually occurs and results in selection for a different habitat type. Some moose migrate between high-elevation summer range and low-elevation winter range, as in Wells Gray Park, British Columbia (Edwards and Ritcey 1956). Other individuals may spend the summer at low elevation, move to high elevation during fall and early winter, and perhaps return to low elevation during midwinter to late winter, as in interior Alaska (Bishop 1969).

Population segments with different migration patterns commonly occupy the same range during certain periods of the year. For example, on a lowland summer range in interior Alaska one group of moose migrates south and another group migrates north to high elevations in late summer and fall to breed and remain for part or all of the winter. A third group remains resident in the lowland area throughout the year, moving a relatively short distance between adjacent summer and winter ranges. The population segments migrating north and south return to the lowland the following summer to share a common range again with the resident population segment. Diversity of movement patterns is probably the rule rather than the exception, particularly on western ranges, where migrations may be extensive. Interactions between resident and nonresident migratory population segments produce seasonally changing aggregations and segregations, and therefore profoundly affect population dynamics and management strategies.

Migration in moose is an adaptation that optimizes survival through utilization of the most favorable environments available. Annual horizontal movements between adjacent areas of little relief result in changes in habitat type. These seasonal shifts may provide more favorable forage supply and cover where microclimatic conditions are moderated. More lengthy movements involving change in elevation also result in habitat selection for different forage and for different climatic conditions. Weather, particularly snow condition, is the most frequently reported factor mediating moose migrations (LeResche 1974). Coady (1974*a*) reviewed numerous studies from North America that correlated increasing snow depths on summer range with migration to winter range, and snow melt and appearance of bare ground with movement to summer range. Winter severity may also influence the distance

moved by some individuals as well as the proportion of animals in a population segment that migrate to different areas. For example, VanBallenberghe (1977) found that during a winter of low snow depth in south-central Alaska some moose remained on the summer range and others migrated to an adjacent area during winter. However, during winters of deep snow most moose migrated longer distances to another winter range. The timing of other migrations may not be directly related to weather conditions, because movement from lowland summer ranges may begin as early as midsummer; movement from riparian winter ranges may begin in late winter before a decrease in snow depth or amelioration in temperature occurs. These movements may be related to changes in quantity or phenology of forage or to other environmental stimuli, or perhaps to an internal timing mechanism (LeResche 1974).

Numerous studies from Minnesota, Wyoming, Montana, British Columbia, and Alaska reviewed by LeResche (1974) confirm the traditional use of seasonal ranges and migration routes by some moose. Traditional movement patterns may even persist over several generations, presumably learned by calves when they accompany their dams. The concept of fidelity to seasonal ranges is supported by studies that indicate that moose may not readily reoccupy adjacent areas that are depleted by hunting (Goddard 1970; LeResche 1974). Intensive hunting can depress population segments that are annually resident in an area during the hunting season, while subsequent migrations may cause a seasonal increase in moose numbers without actually contributing to the hunted population segment.

When deviations from traditional movement patterns occur they are most common among subadult moose. Although data are limited, it appears that yearlings are less traditional in their movements, undertake more lengthy migrations, and have larger home ranges than adults (Houston 1968, 1974; Roussel et al. 1975; Lynch 1976). Pimlott (1959) and Simkin (1965) suggested that the high vulnerability of yearlings to hunting in Newfoundland and Ontario perhaps resulted from their more extensive movements compared with those of adults. Similarly, Cumming (1974) believed that yearlings immigrated to heavily hunted areas where the harvest consisted of over 40 percent yearlings. Therefore, the rate and degree to which available habitat is occupied may be influenced by the proportion of yearlings in adjacent areas.

Dispersal by moose is an obvious mechanism that has facilitated finding and exploiting new habitat (Geist 1971). Some moose habitat is ephemeral, consisting of seral shrub communities. Therefore, consistently traditional movement patterns would not favor the exploitation of new habitat created by fire or other natural phenomena. Nevertheless, dispersal by moose has been observed in some regions and not in others. Mercer and Kitchen (1968) reported that the dispersal rate of moose introduced to Labrador in 1953 was approximately 30 km per year, while dispersal from an indigenous population was 18 to 33 km per year.

Moose population density frequently increases rapidly following fire, as it did after the 1947 burn on the Kenai Peninsula, Alaska (Spencer and Chatelain 1953; Spencer and Hakala 1964). Population growth in this new habitat could have been initiated by immigrants or by moose previously resident in the area. However, little or no population increase occurred on many interior Alaska ranges created by fire in the late 1960s and early 1970s. Moose population levels were already moderately high preceding the 1947 burn on the Kenai (Spencer and Hakala 1964), while population density was low and declining and calf survival was low during the 1970s in interior Alaska. The low density and, in particular, the scarcity of young animals to disperse in interior Alaska may have prevented the occupation of these new ranges. Therefore, production of new range does not always insure that colonization and population growth will follow. Even when population increase follows range improvement, it may be due in part or entirely to improved reproduction by moose resident in the area rather than to immigration.

Social Patterns and Activity. Moose are the least gregarious North American cervid (Altmann 1956; Denniston 1956). The most durable bonds are formed between females and their calves, and normally persist from birth to shortly before parturition the following year. Occasionally a calf will remain with its dam during a second year, presumably if a new calf is not born or does not survive. Aggregations containing bulls during the fall are the largest. They apparently serve to bring males together for social interaction and males and females together for breeding (Peek et al. 1974; Lent 1974). Group sizes in Alaska may reach 30 or more moose during fall, particularly in alpine habitat. Aggregations may also be large during winter, although individuals may be attracted to specific areas by favorable browse supply rather than by social interaction. Winter aggregations appear to be loosely knit and transitory, with individuals seemingly oblivious to each other. During periods of deep snow, moose in eastern North America aggregate in upland "yards," while in western North America they aggregate in lowland riparian areas.

Moose are not vocal during most of the year. However, during the rut a variety of vocalizations are emitted by both males and females, including "croaks" and "barks" by bulls and quavering "moans" by cows (Lent 1974). Cows may "grunt" at calves, and young calves may "cry" when frightened or distressed. The most common nonvocal signals are produced by antlers, either during rubbing or thrashing on shrubs or during fights. Both sounds attract other males during the rut (Lent 1974).

The response of individuals when disturbed is variable, although the species is generally mild tempered and not excitable. Animals frequently tolerate the close approach of vehicles or humans on foot without retreating. Exceptions to this temperance occur during the rut, when males become aggressive toward each other as well as occasionally toward humans, and following calving, when cows may vigorously protect their calves from any apparent danger. At other times moose are seldom belligerent, although harassment or stress may provoke conflict (Denniston 1956).

Moose may be active at any time, although several activity peaks seem to occur throughout a 24-hour period. Geist (1963) believed that changing light at dawn and at dusk during summer synchronized activity peaks at those times, while two other activity peaks during the day were due to an endogenous feeding rhythm. Sigman (1977) found three to five activity peaks during daylight hours in *A. a. gigas,* although individual activity patterns were not well synchronized. She also noted considerable variation in activities among periods of observation that may have been due to changing environmental conditions. Geist (1963) found that the duration of both active and rest periods was greater in winter than in summer, and that fewer active and rest periods occurred per day in winter than in summer.

Habitat. Moose are a species of the boreal forest. Their distribution is more closely related to the range of northern trees and shrubs than to any other factor. Moose eat a variety of plants, ranging from mosses to trees. Forage diversity is greatly reduced during winter, and moose feed primarily on woody browse that extends above the snow. Therefore, shrubs and trees are the most important winter forage for moose and are a principal limitation to populations throughout North America.

FIGURE 46.4 Seral growth winter moose (*Alces alces*) habitat dominated by balsam fir and white birch trees in eastern Minnesota. Note that the moose is instrumented with a radio transmitter collar.

FIGURE 46.5. Riparian shrub winter moose (*Alces alces*) habitat dominated by willows in interior Alaska. Note the antler pedicle visible in front of the ear and above the eye of the moose.

Minimum requirements for winter food and cover are satisfied by a great diversity of vegetation types across North America, although regional similarities do exist. In the east, balsam fir (*Abies balsamea*) in climax forests and white birch (*Betula papyrifera*) and quaking aspen (*Populus tremuloides*) in seral stands (figure 46.4) created by fire and logging are trees of universal importance to moose (Pimlott 1953; Des Meules 1962; Dodds 1974; Brassard et al. 1974; Krefting 1974; Crête and Bédard 1975). The best moose habitats in mountainous areas are mixed stands of balsam fir and white birch, particularly where they are interspersed with patches of muskeg. In nonmountainous areas mixed forests including willows (*Salix* spp.) and quaking aspen are important (Joyal 1976). The use of balsam fir forests increases during winter when snow depths exceed 75 cm in deciduous stands (Des Meules 1964; Telfer 1970). This occurs because the retention of snow by trees produces bowl-shaped depressions, or "quamaniq," beneath the crowns and considerably reduces snow depth on the ground in closed-canopy coniferous forests. Moose utilize aquatic areas extensively when available during early and midsummer, although these areas are not essential summer habitat (Dodds 1974).

A transition zone between prairies and forests occurs in central North America from northwestern Minnesota to northern Alberta. This broad, flat mosaic of willow, quaking aspen, and marsh is important moose habitat throughout the year (Berg and Phillips 1974). The use of willow stands during early winter gradually decreases with increasing snow depth as moose shift to tall deciduous tree habitats. Coincident with snow melt in spring, habitat use shifts to sparse or low-growth willow stands, where moose tend to remain until winter. Aquatic areas are important throughout the summer.

In western North America, shrub communities are the most important winter habitat for moose (LeResche

et al. 1974*a*; Peek 1974). From Wyoming to the north slope of Alaska, riparian willow stands provide winter range for increasing numbers of moose throughout the winter (figure 46.5). Maximum use of these areas occurs from midwinter to late winter and during those winters of greatest snow depth. Coniferous trees adjacent to shrub stands enhance the value of riparian habitats in many areas by providing cover. The importance of these areas to moose cannot be overstated, because although they are seral communities they are self-renewing through alluvial action. Therefore, they provide permanent seral habitat. Important seral shrub habitat is also created by wildfire, clear-cutting, and other perturbations that remove climax vegetation (LeResche et al. 1974*a*; Davis and Franzmann 1979). Revegetation of disturbed areas by shrubs produces the most ephemeral winter moose habitat because the growth of shrubs and changes in vegetation composition gradually decrease the value of these areas for moose. Climax communities are also important winter habitat for moose in some western regions (LeResche et al. 1974*a*). These diverse communities occur both in upland areas near timberline and in lowland areas. They are dominated by willow or shrub birch (*B. glandulosa*) in upland areas and stands of frequently decadent willow interspersed with deciduous tree stands and muskeg in lowland areas.

Habitat use in western North America is most extensive during summer and fall and is gradually restricted during winter (LeResche et al. 1974*a*). Both lowland and upland climax shrub habitats are heavily used during summer and fall, although they may be occupied throughout the year by some moose. By early winter a shift to upland and lowland seral habitats is evident. This trend continues throughout the winter as snow depth increases, although during winters of particularly deep snow upland seral ranges are abandoned in favor of lowland riparian areas. As snows recede in spring, moose disperse to other habitats, particularly lowland and upland climax stands, and make intensive use of aquatic areas during early summer when they are available.

The boreal forests of North America are historically fire-dependent ecosystems, fire being the predominant factor determining their species composition and age structure (Rowe and Scotter 1973; Viereck 1973; Viereck and Schandelmeier 1980). Lightning and aboriginal mankind were important causes of fire. With the arrival of contemporary humans in the boreal forest, the incidence of fire resulting from accidental causes and from land clearing increased. Disturbance to forests does not necessarily insure the production of moose browse, because the pattern of revegetation depends on numerous factors, such as seed source and soil characteristics (Lutz 1956). Leopold and Darling (1953) noted that not all wildfire in Alaska favors moose through the regrowth of browse species. Nevertheless, fire and timber harvest have produced vast amounts of seral moose range across North America during this century. Buckley (1958) concluded that forest disturbance in Alaska during the first half of the century resulted in more moose in Alaska than ever before. This condition likely prevailed in other regions also. More recently, an intensified policy of fire suppression and a decrease in forest perturbation from other causes have greatly diminished the amount of seral moose range in some regions, with the inevitable consequence of reducing the size of moose populations.

The seral ranges most valuable to moose are those in which regrowth of palatable browse species following disturbance is interspersed with mature or unaltered stands. Browse regrowth is usually superior in quantity and quality to that existing before disturbance (Cowan et al. 1950). Discontinuity of fire or timber harvest that enhances "edge effect" is beneficial because it provides cover near feeding areas, increases the variety of forage species, and staggers the maturation rate (Dodds 1955; LeResche et al. 1974a). However, logging from particularly small units may not benefit moose because of the restricted size and rapid maturation of seral stage vegetation (Peek et al. 1976). The optimum age of moose browse regrowth following fire in Alaska is less than 50 years, and moose density usually peaks 20–25 years after burning (LeResche et al. 1974a). The growth of shrubs and invasion by less desirable browse species are inevitable. Maturation eventually decreases the value of seral range to moose unless ranges are renewed by repeated disturbance. Therefore, most seral habitats and moose populations that utilize those habitats are ephemeral.

Interspecific Interactions. Competition for forage between moose and other cervids may be locally intense under some conditions. Although the distribution of white-tailed deer and mule deer (*O. hemionus*) overlaps that of moose along the southern portion of boreal forests, direct competition for food probably occurs only during periods of deep snow (Prescott 1974). White-tailed deer and, to a lesser extent, mule deer tend to congregate, or "yard," during winter at low elevations in areas providing the most favorable climatic and shelter conditions. Moose are not as restricted by the snow depths that cause deer to yard, and tend to remain more dispersed and at higher elevations during winter. However, with persistent deep snow, moose may be forced into areas used by deer, and, because overlap in food preference exists, direct competition for food may occur. Competition between elk (*Cervus elaphus*) and moose where the two species are sympatric in the Rocky Mountains is usually slight (Stevens 1974). Moose prefer browse in riparian shrub "habitats" during winter, and elk graze in more open areas, feeding largely on grass. Competition is probably intensive only when food is limited by depletion or severe winter weather. Davis and Franzmann (1979) believed that competition between caribou (*Rangifer tarandus*) and moose on sympatric ranges was insignificant. In western North America moose may seasonally use the same range as domestic livestock, and conflicting observations are reported from different areas (Wolfe 1974). Competition may be locally intense for particular food items. However, in most areas there appears to be negligible competition. Interactions between moose and small herbivores may occur throughout the range of moose in North America. Beavers (*Castor canadensis*) probably benefit moose where the species are sympatric (Wolfe 1974). Tree cutting and dam construction by beavers stimulate the growth of new shoots from mature *Populus* trees, and thereby increase available moose browse. Competition for the same food does occur, although most studies indicate that it operates to the detriment of beavers rather than to moose. The extent of competition between moose and hares (*Lepus* spp.) in North America is unclear. Hares utilize many of the same browse species as moose. However, the level on the plant browsed by each species is usually different, with hares utilizing the lower portions of plants depending on snow depth, and moose browsing upper portions (Telfer 1974). Dodds (1960) believed that competition in Newfoundland occurred only in cutover areas for balsam fir and paper birch, and that primarily snowshoe hares (*Lepus americanus*) were adversely affected. Nevertheless, Wolff (1980) concluded that during a peak of snowshoe hare abundance in interior Alaska, moose browse was considerably reduced, and that this may have contributed to a moose population decline.

FOOD HABITS AND ENERGY REQUIREMENTS

Moose are primarily browsers, feeding on trees and shrubs during winter. During summer they may graze on emergent and herbaceous plants, although leaves and succulent leaders on shrubs and trees are also used. Peek (1974) reviewed food habits of moose in North America and concluded that willows are the most important winter food of moose in western areas, and balsam fir, quaking aspen, and paper birch are most important in central and eastern areas. However, he cautioned that numerous exceptions exist, and that preferred species vary locally.

Food habits of moose are influenced by forage availability. Species composition and relative abundance of browse species change because of natural succession and previous utilization, and this may result in temporal changes in forage preference of moose. For example, regrowth following wildfire in 1947 on the Kenai Peninsula, Alaska, produced a multispecies range of willow, birch, and aspen. However, subsequent succession and intensive browsing by moose in most areas has resulted in a winter range now dominated by white birch (Oldemeyer et al. 1977). Similarly, in many areas of southern Newfoundland during the past 50 years, moose have nearly prevented regeneration of balsam fir and white birch by intensive browsing, resulting in spruce-dominated forests (Mercer and Manuel 1974). Snow may also influence the availability of forage species. In some areas of eastern North America where deep snow restricts moose to coniferous forests, diets shift from predominately deciduous to coniferous browse. Some forage species may be buried by snow. In Minnesota, for example, red osier (*Cornus stolonifera*) is important to moose during fall until snow covers most plants (Peek et al. 1976). On the Kenai Peninsula, lowbush cranberry (*Vaccinium vitis-idaea*) is an important winter forage until it becomes unavailable under the snow (LeResche and Davis 1973). In many areas of Alaska, however, moose may paw through snow up to 40 cm deep to graze herbaceous plants.

Peek (1974) concluded that all methods for determining food habits of moose have serious limitations, and recommended the use of several techniques to provide the most useful information. The two most common techniques are examination of shrubs in the spring to determine browse removed and examination of rumen contents. However, browse surveys do not indicate changes in forage preference that may occur during winter, and rumen content analysis is biased by differential digestion of various items and particle sizes. Direct observations of feeding moose may be useful in some situations, particularly in determining summer food habits. Peek (1974) suggested that food habits data are most useful for determining the adequacy of a diet when combined with habitat condition and population performance information.

Moose are ruminants, with a stomach developed for pregastric microbial fermentation of cellulose. Plant material is fermented and digested to form chiefly carbon dioxide, methane, and volatile fatty acids. The gases are expelled and the acids are absorbed and oxidized to provide energy for maintenance, growth, and reproduction. Energy requirements for moose vary greatly during the year (Gasaway and Coady 1974). Although the maintenance energy necessary for minimum physical and physiological functions remains relatively constant, the energy needed for the production of new tissue changes. For example, the energy required by pregnant cows increases significantly in late winter due to rapid fetal growth and subsequent lactation. Superimposed upon this is the energy required for weight gain during summer and fall by both cows and bulls.

Moose eat more food of higher quality in summer, when energy requirements are high, than in winter, when requirements are low. Gasaway and Coady (1974) estimated that the daily food intake by adult *A. a. gigas* is approximately 4.5 to 5.5 kg dry weight during winter and 10.0 to 12.0 kg dry weight during summer. Lower food intake during winter is not the result of decreased availability but of increased retention time for the fermentation of low-quality food in the rumen and perhaps of a voluntary reduction in food intake. The energy produced by the fermentation and digestion of food ranges from approximately 30 percent less to over 200 percent more than that required for maintenance during winter and summer, respectively. Consequently, the energy derived from food during winter is less than that required for survival, and stored fat and protein are catabolized, resulting in weight loss. During summer more energy is produced from food than is required, and the excess energy is stored as body tissue and results in weight gain. The considerable seasonal fluctuations in body weight of moose reflect their positive energy balance during summer and negative energy balance during winter.

Malnutrition occurs when the ingestion of low-quality browse is inadequate to prevent catabolism of all energy reserves. It invariably occurs during winter on depleted range, or even on high-quality range when moose are congregated and snow depths restrict movements to obtain adequate forage. Substantial winter mortality across North America has been attributed to malnutrition (Hatter 1949; Dodds 1974; Bishop and Rausch 1974), and latent effects of malnutrition during winter may also predispose moose to subsequent mortality factors at other seasons (Peterson 1977).

MORTALITY

Causes of mortality to moose vary in magnitude among different areas. Some combination of hunting, predation, and malnutrition is probably the major cause of mortality to moose in most areas of North America. Accidents and disease may also be significant in some areas.

Predators. The gray wolf (*Canis lupus*) is the most effective natural predator of moose in North America. Although wolves may not exert a long-term limit on moose population levels, they can temporarily regulate moose numbers (Frenzel 1974; Wolfe 1974; Peterson 1977). The degree of control exerted by predation is determined by the ratio of predators to prey and by the mediating influence of factors such as weather, food supply, and hunting. Wolf predation is selective for calves and adults, and therefore may be a significant source of mortality to ungulate populations (Mech 1966; Pimlott 1967). Selective predation may prevent growth of moose population by reducing the recruitment rate of yearlings below the mortality rate of adults. Moose in their prime are less likely to be killed by wolves, except under stressful circumstances. For example, deep snow may reduce the selective advantage enjoyed by prime age moose.

Studies reviewed by Pimlott (1967) and Frenzel (1974) consistently support the position that prime age class moose are relatively immune to predation under most conditions. On Isle Royale, moose calves were the age class most often killed by wolves (Peterson and Allen 1974; Peterson 1977). Calves comprised 34 percent of all kills located during several winters, although the average proportion of calves in the herd was only 14 percent (Peterson 1977). In some areas of Alaska during some winters moose calves comprise the majority of wolf kills (Burkholder 1959; Stephenson and Johnson 1973). They comprise almost all of the identifiable material in wolf scats at some dens during spring and summer (Rausch and Bishop 1968). On Isle Royale, moose between one and seven years of age were killed in lower proportion than they occurred in the population, while adults older than seven were taken in greater proportion (Peterson 1977). This trend of increasing vulnerability of older adults to wolf predation is illustrated by the age distribution of wolf-killed adult moose on Isle Royale compared to the hypothetical age distribution of a population (figure 46.6). Mortality of one- to five-year-old moose is relatively low, but it increases rapidly after eight years of age to exceed the hypothetical percentage occurrence of each age class in the population. Females predominated in samples of adult moose killed by wolves during winter on Isle Royale (Wolfe 1977). On Isle Royale, the sex ratio among adults in the population was approximately equal, suggesting that if selection for females by wolves occurred, it was compensated by a disproportionate mortality of males from other factors. Differential mortality of adult males over adult females is common in ungulate populations (Robinette et al. 1957; Flook 1970).

Both black bears (*Ursus americanus*) and brown bears (*U. arctos*) are predators of moose in North America (Hosley 1949; Peterson 1955). Ballard et al. (1979) determined that black bears or brown bears were responsible for 77 percent (N = 84) of observed instances of moose calf mortality in the Nelchina Basin and the Kenai Peninsula, Alaska. In the Nelchina Basin brown bears were most numerous and accounted

for most of the mortalities, while on the Kenai Peninsula black bears predominated and were the major predator (Franzmann et al. 1980). Wolves were present in both regions, but accounted for only a small portion of the observed mortality. Bears occasionally kill adult moose (LeResche 1968), but they prey primarily on calves. Although bears may be an important cause of mortality to moose in some areas during summer, their impact as predators on moose in North America is not as extensive as that of wolves.

Other wild carnivores, such as the wolverine (*Gulo gulo*), coyote (*Canis latrans*), and lynx (*Felis canadensis*), probably have little influence as predators on moose in North America, and their predation is probably limited to occasional opportunistic encounters. However, domestic dogs (*Canis familiaris*) may adversely impact moose in some areas near human settlements. In Alaska they commonly kill young calves, and during winters with deep snow they may harass adults to the point of exhaustion.

Parasites and Diseases. Anderson and Lankester (1974) reviewed diseases and parasites that infect moose in North America. "Moose disease," caused by the meningeal worm *Parelaphostrongylus tenuis*, has the greatest detrimental impact on the species. *P. tenuis* is a clinically silent nematode infesting white-tailed deer over most of their range in North America. Snails are the intermediate host. When infective larvae are accidentally ingested by moose, they inflict a neurological disorder usually resulting in paraplegia and eventual death. Considerable evidence shows that this problem occurs when moose and white-tailed deer occupy the same range. However, ecological separation that restrict range overlap limits the incidence of infection (Telfer 1967; Kelsall and Prescott 1971). "Moose disease" has been implicated as the cause of significant declines in moose populations in Nova Scotia and New Brunswick, and perhaps in Maine and Minnesota (Anderson and Lankester 1974). Disease outbreaks were coincident with significant increases in deer populations in these areas, and the disease appears to be moving west with the extension of the range of white-tailed deer. *P. tenuis* may be the principal factor limiting southeastern expansion of moose populations (Kelsall and Telfer 1974; Gilbert 1974).

Numerous other infections and parasitic diseases are found in moose, and some may be locally detrimental to populations. In addition to *P. tenuis,* the liver fluke (*Fascioloides magna*) and winter tick (*Dermacentar albipictus*) are probably derived from deer of the genus *Odocoileus* (Anderson and Lankester 1974). The importance of these parasites as pathogens of moose is unclear, although dead individuals with massive infections have been reported. Other diseases of moose, such as brucellosis and anthrax, are acquired from livestock. Both may be highly pathogenic to individuals, although their effect on populations is unknown. Several helminths reviewed by Anderson and Lankester (1974), Samuel et al. (1976), and Addison et al. (1979) infect moose when suitable canid final hosts are present. Most are probably not significant mortality

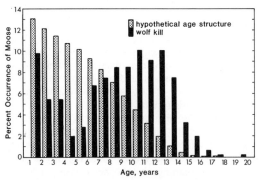

FIGURE 46.6. Percentage occurrence of moose (*Alces alces*) in wolf kills.

factors, although the presence of cysticerci in flesh makes meat less attractive for human consumption. Because several diseases and parasites of moose in North America are acquired from native cervids and livestock, environmental changes that alter the distribution of deer and the importation of livestock may expose moose populations to additional pathogens.

Hunting. Hunting of moose populations in North America where bulls-only hunting occurs alters sex ratios in favor of females and locally reduces population levels. Data from hunter harvests and population surveys indicate that sport hunting is usually selective for males (Karns 1972; Bishop and Rausch 1974; MacLennan 1975). Even where either may be shot, harvests are usually skewed toward males, and harvest skewed toward females may indicate overhunting (Cumming 1974). Although some selection for large-antlered, prime age class bulls and avoidance of calves may occur, hunters generally kill moose in proportion to their age distribution in the population (Addison and Timmermann 1974; Karns et al. 1974). Local harvest levels are usually related to the ease of access. Heavy hunting in the most accessible areas adjacent to roads and watercourses and the generally traditional nature of seasonal moose movements have contributed to local population declines (LeResche 1974; Ritcey 1974; Mercer and Manuel 1974). Presently the greater mobility provided by off-road vehicles, snow machines, and aircraft has helped to distribute hunters more widely.

Malnutrition. Malnutrition in moose results from inadequate digestion of browse or from a deficiency of particular nutrients. It is usually precipitated by environmental stress during winter. Because moose usually exist during winter in a negative energy balance, factors that accentuate weight loss are potentially serious. Deep snow is probably the most common environmental factor contributing to malnutrition in moose. Snow depths greater than 70 cm impede movement, and depths greater than 90 cm, or approximately equal to the chest height of adult *A. a. gigas,* restrict movement to the extent that adequate food intake may be impossible (Coady 1974*a*).

Deep snow not only increases energy requirements for movement but also restricts the area of movement, and results in moose utilizing browse of decreasing quality as confinement continues. Malnutrition is most likely to result in population decline on poor-quality winter range, such as on the Kenai Peninsula, Alaska (Oldemeyer et al. 1977). However, it may also occur on good-quality range with deep snow conditions. Shallow snow that does not impede movement but that conceals important, low-growing food items in some areas may cause malnutrition (LeResche and Davis 1973).

Differential mortality from malnutrition is best explained on the basis of energy requirements and reserves. Calves have less fat and protein available for catabolism than do adults and are therefore less able to maintain themselves during periods of low-quality forage intake. Because of their relatively short legs, they require more effort for movement in deep snow. Calves are consistently the first age class to succumb to malnutrition during stressful winters (Houston 1968; Coady 1973; LeResche and Davis 1973; Peterson 1977). Malnutrition may be a significant cause of death to neonates if they are unable to nurse adequately. Insufficient milk production, or abandonment and subsequent starvation, may contribute to the mortality of neonates, particularly among twins. Adults dying from malnutrition are generally older aged animals, and most studies indicate that males die in greater proportion than their occurrence in the population (Peterson 1977). Old animals are probably less able to tolerate nutritional stress than are prime age class adults. Males lose substantial weight during the rut and are unable to regain that weight before winter (Lent 1974). Therefore, males may begin winter with lower body reserves than females.

Accidents. In some areas the most significant source of accidents is motor vehicles. During the winter of 1977, 38 moose were reported killed on highways near Fairbanks, Alaska. Many others were probably injured and eventually died, but were not reported. This mortality occurred from a population of approximately 1,000 moose that spent the winter in the area. During some winters several hundred moose are killed by trains between Anchorage and Fairbanks, Alaska. The extent of collisions with moose on railroads and highways is influenced by snow depths. During winters with deep snow, plowed railroad and highway rights-of-way provide an attractive corridor for movement. This mortality is probably not selective for particular sex or age individuals.

Other accidents appear to be a relatively minor source of mortality to moose. Drowning is most common in calves and probably occurs when young calves attempt to follow their dam across swift streams or choppy lakes or are unable to exit the water because of steep banks. Falls are probably nonselective for all sex and age class moose. Fatal combat injuries occur to bulls during the rut. I have observed three sets of moose skulls locked together by the antlers (figure 46.7) and other bulls killed by puncture wounds in the abdomen in interior Alaska. Puncture wounds to the body, lacerations to the head, shredded ears, and eye injuries were observed during the rut on numerous adult males on the Kenai Peninsula, Alaska (J. Davis, ADF&G, personal communication). The extent of many injuries suggests that contestants may be killed during the rut with regularity in some areas. Similar observations were provided by Peterson (1955) and Geist (1971).

POPULATION REGULATION

Components of the environment or intrinsic mechanisms act to prevent population increase and stop declines, otherwise moose populations would expand indefinitely or become extinct. Most theories of natural regulation of ungulate populations consider either predation (Keith 1974) or environmental carry-

FIGURE 46.7. Moose (*Alces alces*) antlers and skulls locked together during the rut in interior Alaska. The spread of the small antlers was about 120 cm, and that of the large antlers was about 150 cm.

ing capacity (Caughley 1970) to be the principal regulating factor. It is difficult to identify which factors have been most important in regulating moose populations because: (1) population growth in response to increased food supply resulting from fire and logging has frequently been coincident with a decline of predator populations; (2) hunting, which is not equivalent to natural predation, has been the major mortality factor contributing to population decline in many areas; (3) reliable quantitative information about moose and predator populations and range conditions that would accurately document population changes is generally unavailable; (4) few areas contain naturally regulated populations of both predators and moose; and (5) these factors often interact.

Although quantitative population data may be limited, certain well-known empirical relationships form the basis of the following evolutionary theory hypothesized by Geist (1971). Moose inhabit both short-lived seral and permanent climax communities. Therefore, the amount of moose habitat has repeatedly fluctuated between low, consisting primarily of permanent habitat, and high, consisting primarily of seral shrub growth in disturbed areas. Moose populations have frequently responded to changes in habitat by population growth and decline. Populations persist at low densities in reservoirs of permanent habitat. When new habitat becomes available some moose are able to occupy those areas and achieve high reproductive potential, which results in population growth. Either rapid or gradual population decline usually follows, ultimately terminating in the virtual disappearance of moose from an area as the habitat becomes unsuitable. The repeated population growth in response to fluctuations in habitat implies either that some animals are resident in or migrate through an area and are able to establish viable populations, or that moose in adjacent areas are capable of immigrating to and exploiting new habitat.

Moose population fluctuations may be characterized by three phases: irruption, decline, and equilibrium. Caughley (1970) defined an irruptive fluctuation as an increase in numbers over at least two generations followed by a marked decline. Irruptions have occurred in areas where food supply exceeded that needed by resident ungulates (Caughley 1976). Dramatic moose population growth on new range occurred in Newfoundland (Pimlott 1953), Isle Royale (Murie 1934; Aldous and Krefting 1946), British Columbia (Hatter 1949), Kenai Peninsula, Alaska (Spencer and Chatelain 1953; Spencer and Hakala 1964), interior Alaska and the Copper River Delta, Alaska (LeResche et al. 1974a), and western and northern Alaska (Coady 1980). Moose population growth results primarily from an increased natality rate attributable to increased twinning by adults and breeding by yearlings in response to abundant forage. Also, the most dramatic population increases have usually occurred when predators were absent or at very low density (Keith 1974). The moose population growth on new range cited above occurred when predation by wolves was limited or absent.

Population decline frequently follows rapid population growth and may occur regardless of the presence of predators. A population decline in Newfoundland and repeated irruptive fluctuation on Isle Royale and the Kenai Peninsula occurred in the absence of wolves, while those in central British Columbia and interior Alaska occurred in the presence of wolves. Overbrowsing on winter range and malnutrition of moose were usually evident during population decline. Although spectacular population crashes are associated with overbrowsing and malnutrition, most such declines are less pronounced and are accompanied by a variety of mortality factors, including parasites, severe weather, predation, and hunting. Decreased natality rates may accelerate the decline of moose populations when overbrowsing and malnutrition occur.

The third phase of population fluctuation that may occur is equilibrium. Equilibrium is most closely approximated when moose density and range conditions stabilize. It may occur following population growth with no population decline, as in northern Alaska (Coady 1980), or more frequently after a population decline. While the concept of equilibrium between moose and their food supply is useful, even short-term stability of moose populations is unusual. This is particularly true for populations exploiting early seral range in which the quantity and quality of forage change rapidly. Further, pronounced differences in food availability occur annually because of varying snow conditions, while changes in moose density can be induced by high mortality during severe winters or by changing rates of harvest or predation. Even on Isle Royale, where moose are supported largely by mature forests and no hunting occurs, the population has fluctuated from about 600-800 in 1960 to about 1,200-1,400 in 1974 (Peterson 1977). Normal fluctuations in population size are thought to result from both gradual changes in habitat quality and annual changes in winter conditions that affect food availability. These nutritional changes alter both natality and mortality rates.

Two factors that influence the regulation of moose populations are the extent of perturbation to and the size of ecosystems. Virtually all moose populations have been impacted by mankind, although some have been altered considerably more than others. Absence of wolves from the Kenai Peninsula, Alaska (Bishop and Rausch 1974), may have contributed to the spectacular irruptive fluctuations in that area. The irruptions on Isle Royale in the absence of wolves (Peterson 1977) and on the isolated range at Yakutat, Alaska, in the presence of wolves (Klein 1965) may have been amplified by the limited extent of the system and the inability of moose to disperse. In other systems excessively high levels of harvest have contributed greatly to abrupt population declines. On the Tanana Flats in interior Alaska the moose population declined from over 10,000 to approximately 3,000 animals between 1965 and 1975 (Bishop and Rausch 1974; Coady 1976). Although a number of factors contributed to this decline, annual hunting mortality of 10-20 percent between 1972 and 1974 clearly was a significant source of loss (W. Gasaway, ADF&G, personal communication).

In systems less influenced by people, irruptive fluctuations have usually not been spectacular. Since wolves became established on Isle Royale in the late 1940s (Mech 1966), the irruptions evident earlier in the century have not recurred. Peterson (1977) hypothesized that in the absence of wolves, moose populations on Isle Royale would fluctuate more dramatically, increasing during periods of mild winters and sharply declining during severe winters. In Mt. McKinley (Denali) National Park, Alaska, moose are not hunted, and population fluctuations have been less dramatic than in adjacent regions where hunting has been a significant mortality factor (Haber 1977). On the north slope of Alaska moose recently became established and increased in numbers between the late 1940s and early 1960s (Coady 1980). Wolf predation in this region is selective primarily for caribou, and hunting mortality on moose has probably ranged between 3 and 5 percent of the population per year. Nevertheless, the population has not declined and appears to have stabilized at approximately 2,000 animals since the late 1960s.

Most data support the hypothesis that food is the ultimate factor limiting the size of moose populations (Pimlott 1961; Peterson 1977). Food available to moose is dependent upon the quantity and quality of range, although its accessibility is modified by snow, and nutrients available per individual are determined by population density. The mechanism of control acts both to increase natality and decrease mortality rates when adequate quality food is abundant, and to decrease natality and increase mortality rates as food becomes limiting. Mortality rates are mediated through such causes as predation, disease, and hunting, although the relative impact of these mortality factors depends on the population status. Fluctuations can be minimized only by careful assessment of population parameters and responsive management that regulates all components in the system to strive for equilibrium between food supply and animal density. However, it is highly unlikely that even timely and intensive management will achieve population stability in most areas, particularly where food resources vary because of forest succession and changing snow conditions.

MANAGEMENT

There is a prevailing attitude that ecosystems, including those of which moose are a part, do not need to be managed. This belief ignores the fact that the pervasive influence of mankind has extended to virtually every area on earth. Further, this influence has been generally negative through altering species composition and community structure and changing relative numbers of individuals and even the occasional elimination of species from areas. Moose have been greatly affected by humans throughout North America, and their distribution and population status in many areas clearly reflect the impact of modern human activities. As a result, positive management actions rather than benign neglect are now necessary to perpetuate healthy moose habitats and populations over the broadest possible area if moose are to prosper and humans are to benefit from their success. Management should not be more intensive than that required to achieve established goals, and, whenever possible, the realization of goals should be through natural mechanisms. Nevertheless, many of the valued and desirable features of moose populations can be preserved only by deliberate and scientific manipulation by people.

The legal authority to manage moose as well as other resident wildlife species in North America is vested largely in state and provincial wildlife management agencies. Their conservation responsibilities include ensuring perpetuation of the species, establishing long-range management plans and goals for the species, collecting and analyzing biological information, and developing and enforcing management regulations for different management districts. Direction to most management agencies in the United States is provided by a complex interaction of the legislatures, the governors, appointed regulatory boards or commissions, local advisory committees, the public, and federal policies in national parks, refuges, and other conservation units. Superimposed on this matrix are pressures resulting from political expediencies that may or may not be based on facts, and federal legislation that has diminished and continues to diminish the traditional management authority of the states.

Planning. The need for systematic planning to manage wildlife has increased greatly in recent years. In the past when relatively few demands were placed upon moose as a resource and when landownership and land use patterns were less complex, planning efforts were equally simple. During the past few decades, however, loss of habitat in some areas, accompanied by moose population declines, enlarging and more mobile human populations, and increasing demands for multiple use of the resource, require a scientific and systematic approach to management that only planning and long-range goals can provide. In Alaska, for example, the state has proposed numerous moose management plans for different populations or areas. One or more of nine different management objectives have been identified for each plan, in addition to the basic goal of maintaining healthy moose populations. Objectives such as "to provide an opportunity to view, photograph, and enjoy wildlife," "to provide for subsistence use of wildlife," "to provide the greatest opportunity to participate in hunting," and "to provide for certain commercial uses of wildlife" illustrate the broad spectrum of demands placed on moose in Alaska. Each plan is accompanied by a description of the management actions required to achieve and maintain the objective. Public support of management plans is essential to their success in the United States, where, by law, wildlife is common property held in trust by the states or the federal government.

Data Collection. The increasingly intensive and diverse demands placed upon moose as a renewable resource dictate the need for accurate and precise information so that management actions will produce a predictable response. Intuition and supposition too frequently formed the basis for past decisions resulting in mismanagement of populations. For example, some of the most agonizing moose management errors in Alaska resulted from continued liberal and frequently either-sex hunting seasons in the absence of timely data to justify large harvests. With the benefit of hindsight it is apparent that substantial hunting kill did not achieve the expected results in many cases, but contributed to unwanted population declines. Subsequently, conservative males-only hunting seasons often did not facilitate the expected population growth. The lessons to be learned are that expedient collection and evaluation of data are needed to manage moose. Population dynamics are exceedingly complex, and our knowledge of moose populations and the influence of variables in the system is frequently limited. Population management cannot be very precise in the absence of adequate data, and the manipulation of variables requires caution and restraint to avoid serious mistakes.

While information needs are diverse, one of the greatest obstacles to precise management of moose is the inability to census populations accurately. Timmermann (1974) divided moose inventory methods into ground and aerial surveys. Ground surveys may consist of direct enumeration of moose or indirect indexes, such as pellet group counts. Difficulties with these techniques are associated with the inability to see all animals or pellet groups present and the very limited size of the area that may be surveyed. At best, these techniques may reflect annual changes in relative abundance of moose in small areas.

Aerial surveys are the most widely used method of estimating the abundance or density of moose in North America. Basically, two aerial procedures have been used. The transect method, described by Banfield et al. (1955), involves following parallel flight lines at low altitude and counting all moose observed within a given distance of the aircraft. Although used extensively, this procedure has a number of limitations. The intensity of coverage and time spent viewing per unit of area are low, and therefore a substantial but unknown proportion of the moose present are not seen. This is particularly true in the more densely vegetated habitats and along the margins of the transect, where the angle of view is most oblique. Determining the width of coverage and the position of flight lines is difficult, particularly where few topographic or other identifying features are available for reference. Therefore, incomplete coverage because of widely spaced flight lines is likely. Moose censuses using this procedure almost always underestimate the actual population size because many animals are not seen.

Intensive search of quadrats is the second aerial moose census procedure that has been used in North America. The technique, described by Evans et al. (1966), consists of dividing the census area into strata of high, medium, and low abundance of moose, gridding each stratum into sampling quadrats, and randomly selecting some quadrats for intensive searching. Quadrats are searched intensively until no additional moose can be located. Results of quadrat searches from each stratum are then analyzed statistically and an estimate of the total population size is made. The increased counting effort per unit of area results in a higher proportion of moose being located in quadrats than in areas searched using transects. However, some moose are still undetected during quadrat searches. For example, LeResche and Rausch (1974) found that during quadrat searches of 2.6 km² pens containing known numbers of moose, experienced observers saw an average of only 68 percent of the moose present. They determined that the accuracy of counts reflected observer experience, snow conditions, habitat type, and time of day.

Difficulty in determining the number of moose present but not observed during aerial searches has been a major obstacle to making accurate estimates of population size. Recently, Gasaway (1977, 1978, 1980) and Gasaway et al. (1979) have estimated the sightability of moose under different environmental and habitat conditions found in many areas of Alaska. Gasaway's procedure involved determining the approximate location of a radio-collared moose from high altitude. A quadrat approximately 2.6 km² in size was then established in which the instrumented moose was thought to occur. The quadrat was intensively searched for all moose present. If the radio-collared individual was not observed, it subsequently was located electronically and the probable reason for not sighting it during the intensive search was noted. Repetitive sightability trials under different environmental conditions have resulted in calculating estimates of the sightability of different sex and age classes of moose during aerial surveys in relation to habitat selection, snow conditions, and the activity of individuals. Results from this work have provided more accurate estimates of population size from aerial surveys using the intensive search of quadrats method.

The second primary need for management information is to determine the number of moose a range can support. Prudent population manipulation requires an understanding of the relationship between moose and their food supply. This problem has been approached by studying both the range and the animals.

The relationship between animal density and range conditions is exceedingly complex and dynamic. Measurement of browse consumption by moose in relation to annual production by plants is important to determine the proportion of key browse species utilized. Crête and Bédard (1975) found that less than 15 percent by weight of all accessible browse was removed during winter in areas of relatively high moose density in Quebec. In Alaska, however, Milke (1969) and Wolff (1976) found that approximately 38 percent and 50 percent by weight, respectively, of available browse was utilized by moose during winter. Wolff (personal communication) believed that removal of 50 percent of available browse was near the maximum utilization possible in his study area. This relationship between browse utilization and moose density is complicated further by such factors as alternate sources of forage, which are not easily measured, and changes in availability of forage species under different snow conditions. Further, the quality of browse is influenced by several factors, and little work has been done to determine the nutrient content of many foods eaten by moose. A better understanding of the numerous factors influencing the relationship between moose and their food supply is required before range surveys or measurements of browse consumption can be used to reflect accurately the carrying capacity of a range. At this time range studies are probably most useful to reflect trends in annual use by moose. Knowledge of these trends helps interpret other indexes to moose population changes.

Range evaluation techniques are usually very time consuming and result in data that are difficult to interpret. However, the physical and physiological condition of an individual moose provides a direct indication of its nutritional condition, which, in turn, is determined by the relationship between the moose and the food supply. A subjective assessment of such factors as growth rate, maximum body or antler size, and fat deposition were once adequate to determine the nutritional status of individuals. Nevertheless, the need for more precise management of moose populations in many areas dictates that more quantitative data now be collected.

Long-term studies in Alaska have been directed toward assessing the condition of individuals and the range relative to the nutritional requirements of moose. Body weights and morphometric measurements of moose were related to nutritional condition by LeResche and Davis (1971). They found that moose from a high-density population had low growth rates and large seasonal fluctuations in body weight, and believed this reflected poor nutritional conditions. A similar relationship between growth rate and food quality is apparent in deer in North America (Klein 1970). Subsequent studies to relate physiological parameters to the nutritional condition of moose included the assessment of hematology and blood chemistry (LeResche et al. 1974b; Franzmann and LeResche 1978), mineral composition of moose hair (Franzmann et al. 1975), and volatile fatty acid production during microbial fermentation in the rumen (Gasaway and Coady 1974). The objectives of these studies were to establish baseline physiological values for moose under different nutritional regimes that could be used to determine the intensity of interaction between populations and their food supply. Franzmann (1977) applied several condition-assessment criteria to moose on different ranges in Alaska and compared them to the high- and the low-quality populations known to occur there. The quantitative condition assessment he developed will help assess the quality of ranges used by other populations in Alaska.

Management of moose habitat in North America requires decisions to maintain, reduce, or expand existing range. Most seral range has been the fortuitous result of fire, timber harvest, or other disturbance. Humans can readily diminish moose habitat by disregarding the role of fire and timber harvest in ecosystems. Intentional enhancement of moose range has been limited to a few small, prescribed fires, decisions to allow occasional wildfires to burn, and mechanical land clearing (Oldemeyer et al. 1977). However, the total acreage of seral range produced in this manner has been small, and benefits to moose localized. In eastern and central areas of North America where timber harvest is a principal source of habitat diversity and seral range (Krefting 1974), moose habitat and populations may be maintained. However, in fire-dependent western forests the current policy of fire suppression has resulted in the unnatural development of extensive homogeneous stands of mature timber. The consequence has been a maturation of forests and a reduction in the natural mosaic of vegetation diversity created and maintained by fire. Unless the public and the land management agencies are willing to recognize that fire is a natural and necessary ecological process, and federal land managers adapt a more flexible program of fire control, moose populations will continue to decline as habitat is lost.

Implementation. Implementation of plans to achieve goals is the final stage of the management process. Management of moose involves both minimizing adverse impacts caused by the use and development of other resources, and providing for uses of moose that benefit people. Some problems may require the protection or alteration of habitat, while others may require manipulation of a population itself. Frequently, treatment of both the habitat and the population is the most expedient means of achieving a management goal. Habitat perpetuation is probably the most urgent need for moose in most areas of North America. This is illustrated by the fact that many accidental extirpations of wildlife populations throughout the world resulted from habitat alteration and not from direct manipulation of populations (Caughley 1977). Most wildlife populations, including those of moose, are as vulnerable to a change in habitat as to a direct change in numbers caused by people. Habitat perpetuation is achieved through designation of land classes that are afforded varying degrees of protection under federal and state laws. National parks, preserves, and wildlife refuges, and state parks, refuges, and wildlife habitat lands are examples of these classes. Management plans that include provisions for the perpetuation of fish and wildlife habitat are required for most state and federal lands, even when those lands are classified for other uses. Some private land owned by individuals or organizations such as the Nature Conservancy also ensure preservation of wildlife habitat. Laws such as the Sikes Act also help insure perpetuation of wildlife habitat on public land. Unfortunately, except for the loss of critical moose habitat such as some winter range, and calving or rutting areas, the adverse influence of habitat loss or alteration is probably cumulative. Single events such as agricultural development and road or pipeline construction may not have measurable and immediately apparent effects on moose populations. However, the cumulative influence of events that result in detrimental habitat alteration may become apparent after a period of time. Moose have clearly demonstrated their ability to tolerate, adapt to, and even benefit from some change. However, continuous erosion of the quality and quantity of habitat or disruption of movement patterns between ranges has been detrimental to populations. Protection of critical habitat when conflicts with other land use occur, or recovery of lost habitat, is probably the greatest challenge confronting the moose manager. It therefore behooves moose managers and land managers alike to consider carefully the probable short- and long-term consequences of events that will alter habitat.

Population management requires one of three actions: (1) manipulation of a small or declining population to increase density; (2) manipulation of a large or increasing population to reduce or control density; and (3) exploitation to remove a sustained yield (Caughley 1977). The obvious first step in resolving a moose management problem is to identify the problem. Small or declining populations are limited by low natality rates and/or high mortality rates. A low natality rate is usually a manifestation of limited food, and the solution probably lies with improving the quality or quantity of food to increase the reproductive capability. An excessive mortality rate may act on adults and/or calves. High adult mortality is often a result of overhunting or predation. The solution is to limit moose harvests by imposing more restrictive hunting regulations or manipulating predator populations. High juvenile mortality is usually the result of excessive predation or malnutrition. Removal of some predators may reduce predator-induced mortality, although predator control is not always technically, economically, or politically feasible. Malnutrition may be reduced by improving the food resource or by reducing the population size. However, malnutrition resulting from severe winter weather is largely a density-independent factor, and is subject to the vagaries of nature.

The intentional reduction or control of a moose population is uncommon except in insular areas or where other natural regulating factors have been removed. In accessible areas sport and subsistence hunting may be effective in limiting a population. In other areas, prevention of irruptions by population control would probably not be technically possible or economically feasible. Because moose irruptions are usually a result of the exploitation of abundant food resources, intentional habitat improvement through burning or logging should be undertaken only when the expected population increase can be accommodated by hunting, natural mortality, or dispersal without prematurely reducing the carrying capacity of the range. Because population stability is usually a management goal, continuous habitat improvement of seral range is required to maintain a stable food base.

Management to provide a sustained yield (SY) entails the designation of an annual harvest level that can be maintained without causing a decline in the population. Many values of SY exist for a population, depending on the uses of the population identified in management goals. The SY from a population that optimizes each management goal is termed the optimum sustained yield (OSY). This is the basis for moose management in North America today. Optimum sustained yield may range from zero in areas such as national parks, where the management goal is to provide for viewing without harvesting moose, to the maximum sustained yield (MSY) in areas such as subsistence use areas where the management goal is to provide the largest possible harvest of moose. OSY values between these two extremes may be designated for areas in which management goals provide for hunting under uncrowded conditions or an opportunity to take large or trophy-sized moose.

Harvest. Regulations establishing hunting seasons and bag limits, restricting the number of hunters, or requiring selective harvesting of moose are means of controlling harvests. Seasons may be shortened or closed to restrict or halt harvest, or they may be changed to direct harvests to population segments that are accessible at different times. The number of hunters can be controlled by requiring hunters to obtain one of a limited number of permits to hunt in an area. Harvest levels can be controlled through registration hunts, where hunters check out of an area by reporting their take at the end of a hunt. When a prescribed number of moose are killed the season is closed. Harvest may be directed to population segments by requiring selective harvests in which all or portions of seasons are limited to one or both sexes. Less frequently, selection for certain age groups is regulated by limiting harvests to or protecting calves, yearlings, or bulls with a certain antler size.

Because moose populations are dynamic and fluctuate in size in response to changes in their environment, population sizes and trends must be considered annually when establishing harvest levels. When annual population fluctuations are small, variability may be ignored, and harvest levels may be based on average population size over several years. Frequently, perceived changes in population size may be largely due to a severe winter, an exceptionally large harvest during a previous year, or a population trend that was undetected for several years. Under such circumstances, a "tracking strategy" (Caughley 1977) is most suitable, in which the rate of harvesting is increased when the population increases, and is decreased or suspended when the population declines.

Management of moose in North America must be responsive to public desires and should maintain the integrity of ecosystems. Many management problems are compounded by mismanagement or unresponsive management and by sociopolitical pressures. Frequently, population trends are available from existing data, but an unwillingness to believe survey data, inadequate evaluation of other sources of information, or incomplete analyses of data prevent a corrective response by managers to a problem. On more than one occasion in Alaska, sport hunters near urban areas correctly detected moose population declines and demanded more restrictive harvests well before managers accepted that fact and initiated appropriate management actions. Conversely, subsistence hunters in rural areas of the state have been reluctant to accept the assessment of low or declining moose populations and have continued large harvests in spite of restrictive regulations or even closed seasons.

Moose managers also need to consider all factors affecting populations. Too often regulation of harvests through restricting seasons and hunters is the only response by managers to a problem. This occurs because harvests are usually the only factor readily controlled by managers. However, other environmental influences that affect both the natality and mortality rates of moose populations and the role of moose in the ecosystem should be assessed and changed if possible. Habitat should be managed and preserved to maintain an adequate food resource. Restrictive harvests will never result in the growth of a population that is declining because of range deterioration. Range improvement or the acceptance of reduced population density are the only two alternatives. Excessive mortality is frequently due to a combination of factors, such as hunting, predation, and malnutrition, and manipulation of all factors may be desirable. Restricted harvests and range improvement are corrective management actions, but some form of predator control may also be useful to expedite population recovery. However, before initiating predator control, the probability of reducing predator density and promoting the desired response in the moose population, the cost-benefit ratio, the impact to other components of the ecosystem, and the sociopolitical considerations must be carefully evaluated.

Management of moose populations is both a biological and a social science. Biological limits of systems must be understood so that management ac-

tions produce the desired response. However, management goals must reflect not only biological limits of populations but also human needs and desires. A firm commitment to conserve and manage moose and the ecosystems of which they are a part, and recognition of the needs of society, will insure that moose persist as an integral part of our resident fauna and contribute to our culture, esthetic enjoyment, and recreation in the future.

R. Bishop, W. Gasaway, D. McKnight, and V. VanBallenberghe kindly reviewed the manuscript. Work was funded in part by Federal Aid in Wildlife Restoration.

LITERATURE CITED

Addison, E. M.; Fyvie, A.; and Johnson, F. J. 1979. Metacestodes of moose, *Alces alces*, of the Chapleau Crown Game Preserve, Ontario. Can. J. Zool. 57:1619–1623.

Addison, R. B., and Timmermann, H. R. 1974. Some practical problems in the analysis of the population dynamics of a moose herd. North Am. Moose Conf. and Workshop 10:76–106.

Aldous, S. E., and Krefting, L. W. 1946. The present status of the Isle Royale moose. Trans. North Am. Wildl. Conf. 11:296–308.

Altmann, M. 1956. Patterns of social behavior in big game. Trans. North Am. Wildl. Conf. 21:538–545.

———. 1959. Group dynamics in Wyoming moose during the rutting season. J. Mammal. 40:420–424.

Anderson, R. C., and Lankester, M. W. 1974. Infectious and parasitic diseases and arthropod pests of moose in North America. Nat. can. 101:23–50.

Aula, P., and Kaariainen, L. 1964. The karyotype of the elk (*Alces alces*). Hereditas 51:274–278.

Ballard, W. B.; Franzmann, A. W.; Taylor, K. P.; Spraker, T.; Schwartz, C. C.; and Peterson, R. O. 1979. Comparison of techniques utilized to determine moose calf mortality in Alaska. Proc. North Am. Moose Conf. and Workshop 15:362–387.

Banfield, A. W. F.; Flook, D. R.; Kelsall, J. P.; and Loughrey, A. G. 1955. An aerial survey technique for northern big game. Trans. North Am. Wildl. Conf. 20:519–530.

Barry, T. W. 1961. Some observations of moose of Wood Bay and Bathurst Peninsula N.W.T. Can. Field Nat. 75:164–165.

Bédard, J.; Telfer, E. S.; Peek, J.; Lent, P. C.; Wolfe, M. L.; Simkin, D. W.; and Ritcey, R. W., eds. 1974. Alces: moose ecology, écologie de l'original. Les Presses de l'Université Laval, Quebec. 741pp.

Berg, W. E. 1971. Habitat, movements, and activity patterns of moose in northwestern Minnesota. M.S. Thesis. Univ. Minnesota. 98pp.

Berg, W. E., and Phillips, R. L. 1974. Habitat use by moose in northwestern Minnesota with reference to other heavily willowed areas. Nat. can. 101:101–116.

Bishop, R. H. 1969. Preliminary review of changes in sex and age ratios of moose and their relation to snow conditions on the Tanana Flats, Alaska. Pap. presented at 6th Annu. North Am. Moose Com. Meet., 3–5 Feb., Kamloops, B.C. 14pp. Mimeogr.

Bishop, R. H., and Rausch, R. A. 1974. Moose population fluctuations in Alaska, 1950–1972. Nat. can. 101:559–593.

Blood, D. A. 1974. Variation in reproduction and productivity of an enclosed herd of moose (*Alces alces*). Proc. Int. Congr. Game Bio. 11:59–66.

Blood, D. A.; McGillis, J. R.; and Lovaas, A. L. 1967. Weights and measurements of moose in Elk Island National Park, Alberta. Can. Field Nat. 81:263–269.

Brassard, J. M.; Audy, E.; Crête, M.; and Greiner, P. 1974. Distribution and winter habitat of moose in Quebec. Nat. can. 101:67–80.

Bubenik, A. B. 1972. North American moose management in light of European experiences. Proc. North Am. Moose Conf. and Workshop 8:276–295.

Buckley, J. L. 1958. Wildlife in arctic and subarctic Alaska. Pages 185–205 *in* H. P. Hansen, ed. Arctic biology. Oregon State Univ. Press, Corvallis. 318pp.

Burkholder, R. L. 1959. Movements and behavior of a wolf pack in Alaska. J. Wildl. Manage. 23:1–11.

Burris, O. E., and McKnight, D. E. 1973. Game transplants in Alaska. Wildl. Tech. Bull. No. 4. Alaska Dept. Fish and Game, Juneau. 57pp.

Caughley, G. 1970. Eruption of ungulate populations, with emphasis on Himalayan thar in New Zealand. Ecology 51:53–72.

———. 1976. Wildlife management and the dynamics of ungulate populations. Pages 183–246 *in* T. H. Coaker, ed. Applied biology Vol. 1. Academic Press, New York. 358pp.

———. 1977. Analysis of vertebrate populations. John Wiley & Sons, New York. 234pp.

Cheatum, E. L., and Severinghaus, C. W. 1950. Variations in fertility of white-tailed deer related to range conditions. Trans. North Am. Wildl. Conf. 15:170–190.

Coady, J. W. 1973. Interior moose studies. Fed. Aid. Wildl. Rest. Proj. Prog. Rep. W-17-4 and W-17-5. Alaska Dept. Fish and Game, Juneau. 53pp.

———. 1974*a*. Influence of snow on behavior of moose. Nat. can. 101:417–436.

———. 1974*b*. Late pregnancy of a moose in Alaska. J. Wildl. Manage. 38:571–572.

———. 1976. Status of moose populations in interior Alaska. Wildl. Infor. Leafl. No. 2. Alaska Dept. Fish and Game, Fairbanks. 4pp.

———. 1980. History of moose in northern Alaska and adjacent regions. Can. Field Nat. 94:61–68.

Cowan, I. McT.; Hoar, W. S.; and Hatter, J. 1950. The effect of forest succession upon the quantity and upon the nutritive values of woody plants used as food by moose. Can. J. Res. 28, Sect. D:249–271.

Crête, M., and Bédard, J. 1975. Daily browse consumption by moose in the Gaspé Peninsula, Quebec. J. Wildl. Manage. 39(2):368–373.

Cumming, H. G. 1974. Annual yield, sex and age of moose in Ontario as indices to the effects of hunting. Nat. can. 101:539–558.

Davis, J. L., and Franzmann, A. W. 1979. Fire-moose-caribou interrelationships: a review and assessment. Proc. North Am. Moose Conf. and Workshop 15:80–118.

Denniston, R. H. 1956. Ecology, behavior and population dynamics of the Wyoming or Rocky Mountain moose, *Alces alces shirasi*. Zoologica 41:105–118.

Des Meules, P. 1962. Intensive study of an early spring habitat of moose (*Alces alces americana* Cl.) in Laurentides Park, Quebec. Pap. presented at Northeastern Sect. Wildl. Conf., Monticello, N.Y. 12pp. Mimeogr.

———. 1964. The influence of snow on the behaviour of moose. Trans. Northeastern Sect. Wildl. Conf. 21. 17pp.

Dodds, D. G. 1955. A contribution to the ecology of the moose in Newfoundland. M.S. Thesis. Cornell Univ., Ithaca, N.Y. 106pp.

———. 1960. Food competition and range relationships of moose and snowshoe hare in Newfoundland. J. Wildl. Manage. 24:52–60.

———. 1974. Distribution, habitat and status of moose in the Atlantic provinces of Canada and northeastern United States. Nat. can. 101:51–65.

Edwards, R. Y., and Ritcey, R. W. 1956. The migrations of a moose herd. J. Mammal. 37:486–494.

———. 1958. Reproduction in a moose population. J. Wildl. Manage. 22:261–268.

Egorov, O. V. 1972. Sistematicheskoe polozhenie losia (*Alces alces*) iz basseinov rek Kolymi i Indigirki. Teriologiia 1:38–44.

Evans, C. D.; Troyer, W. A.; and Lensink, C. J. 1966. Aerial census of moose by quadrat sampling units. J. Wildl. Manage. 30:767–776.

Flook, D. R. 1970. A study of sex differential in the survival of wapiti. Can. Wildl. Serv. Rep. Series II. 71pp.

Franzmann, A. W. 1977. Condition assessment of Alaskan moose. Proc. North Am. Moose Conf. and Workshop 13:119–127.

———. 1978. Moose. Pages 67–81 *in* J. L. Schmidt and D. L. Gilbert, eds. Big game of North America. Stackpole Books, Harrisburg, Pa. 494pp.

Franzmann, A. W.; Flynn, A.; and Arneson, P. D. 1975. Levels of some mineral elements in Alaska moose hair. J. Wildl. Manage. 39:374–378.

Franzmann, A. W., and LeResche, R. E. 1978. Alaskan moose blood studies with emphasis on condition evaluation. J. Wildl. Manage. 42:344–351.

Franzmann, A. W.; LeResche, R. E.; Rausch, R. A.; and Oldemeyer, J. L. 1978. Alaskan moose measurements and weights and measurement-weight relationships. Can. J. Zool. 56:298–306.

Franzmann, A. W.; Schwartz, C. C.; and Peterson, R. O. 1980. Moose calf mortality in summer on the Kenai Peninsula. J. Wildl. Manage. 44:764–768.

Frenzel, L. D. 1974. Occurrence of moose in food of wolves as revealed by scat analysis: a review of North American studies. Nat. can. 101:467–479.

Gasaway, W. C. 1975. Moose antlers: how fast do they grow? Misc. Publ., Alaska Dept. Fish and Game, Fairbanks. 2pp.

———. 1977. Moose survey procedures development. Fed. Aid Wildl. Rest. Proj. Rep. W-17-9. Alaska Dept. Fish and Game, Juneau. 69pp.

———. 1978. Moose survey procedures development. Fed. Aid Wildl. Rest. Proj. Prog. Rep. W-17-10. Alaska Dept. Fish and Game, Juneau. 47pp.

———. 1980. Interior moose studies. Fed. Aid Wildl. Rest. Proj. Final Rep. W-21-1. Alaska Dept. Fish and Game, Juneau. 45pp.

Gasaway, W. C., and Coady, J. W. 1974. Review of energy requirements and rumen fermentation in moose and other ruminants. Nat. can. 101:227–262.

Gasaway, W. C.; Harbo, S. J.; and DuBois, S. D. 1979. Moose survey procedures development. Fed. Aid Wildl. Rest. Proj. Prog. Rep. W-17-11. Alaska Dept. Fish and Game, Juneau. 47pp.

Gasaway, W. C.; Harkness, D. B.; and Rausch, R. A. 1978. Accuracy of moose age determinations from incisor cementum layers. J. Wildl. Manage. 42:558–563.

Geist, V. 1963. On the behavior of North American moose (*Alces alces andersoni* Peterson 1950) in British Columbia. Behavior 20:377–416.

———. 1971. Mountain sheep. Univ. Chicago Press, Chicago. 383pp.

———. 1974. On the reproductive potential in moose. Nat. can. 101:527–537.

Gilbert, F. F. 1974. *Parelaphostrongylus tenuis* in Maine. Part 2: Prevalence in moose. J. Wildl. Manage. 38:42–46.

Goddard, J. 1970. Movements of moose in a heavily hunted area of Ontario. J. Wildl. Manage. 34:439–445.

Godin, A. J. 1977. Wild mammals of New England. Johns Hopkins Univ. Press, Baltimore. 304pp.

Haber, G. C. 1977. Socio-ecological dynamics of wolves and prey in a subarctic ecosystem. Ph.D. Dissertation. Univ. British Columbia. 785pp.

Hatter, J. 1949. The status of moose in North America. Proc. North Am. Wildl. Conf. 14:492–501.

Hosley, N. W. 1949. The moose and its ecology. U.S. Fish Wildl. Serv., Leafl. 317. 51pp.

Hosley, N. W., and Glaser, F. S. 1952. Triplet Alaskan moose calves. J. Mammal. 33:247.

Houston, D. B. 1968. The Shiras moose in Jackson Hole, Wyoming. Grand Teton Nat. Hist. Assoc. Tech. Bull. No. 1. 110pp.

———. 1974. Aspects of the social organization of moose. Pages 690–696 *in* V. Geist and F. Walther, eds. The behaviour of ungulates and its relation to management. Vol. 2. IUCN Publication No. 24. 941pp.

Hsu, T. C., and Benirschke, K. 1969. An atlas of mammalian chromosomes. Springer-Verlag, New York.

Joyal, R. 1976. Winter foods of moose in La Vérendrye Park, Québec: an evaluation of two browse survey methods. Can. J. Zool. 54:1765–1770.

Karns, P. D. 1972. Minnesota's 1971 moose hunt: a preliminary report on the biological collections. Proc. North Am. Moose Conf. and Workshop 8:115–123.

Karns, P. D.; Haswell, H.; Gilbert, F. F.; and Patton, A. E. 1974. Moose management in the coniferous-deciduous ecotone of North America. Nat. can. 101:643–656.

Keith, L. B. 1974. Population dynamics of mammals. Int. Congr. Game Bio. 11:2–58.

Kelsall, J. P., and Prescott, W. 1971. Moose and deer behavior in snow in Fundy National Park, New Brunswick. Can. Wildl. Serv. Rep. Ser. No. 15, Ottawa. 27pp.

Kelsall, J. P., and Telfer, E. S. 1974. Biogeography of moose with particular reference to western North America. Nat. can. 101:117–130.

Klein, D. R. 1965. Post-glacial distribution patterns of mammals in the southern coastal regions of Alaska. Arctic 18:7–20.

———. 1970. Food selection by North American deer and their response to overutilization of preferred plant species. Pages 25–46 *in* A. Watson, ed. Animal populations in relation to their food resources. Br. Ecol. Soc. Symp. No. 10. Blackwell Sci. Publ., Oxford, 477pp.

Knorre, E. P. 1959. Ecology of the moose. Trudy Pechora-Ilych gos. Zapov. 7:5–167.

———. 1961. The results and perspectives of domestication of moose. Trudy Pechora-Ilych gos. Zapov. 9:1–263.

Knowlton, F. F. 1960. Food habits, movements and populations of moose in the Gravelly Mountains, Montana. J. Wildl. Manage. 24:162–170.

Krefting, L. 1974. The ecology of the Isle Royale moose. Tech. Bull. No. 297, For. Ser. No. 15. Agric. Exp. Stn., Univ. Minnesota. 75pp.

Lent, P. C. 1974. A review of rutting behavior in moose. Nat. can. 101:307–323.

Leopold, A. S., and Darling, F. F. 1953. Effects of land use on moose and caribou in Alaska. Trans. North Am. Wildl. Conf. 18:553–562.

LeResche, R. E. 1968. Spring-fall calf mortality in an Alaska moose population. J. Wildl. Manage. 32:953–956.

———. 1974. Moose migrations in North America. Nat. can. 101:393–415.

LeResche, R. E.; Bishop, R. H.; and Coady, J. W. 1974*a*.

Distribution and habitats of moose in Alaska. Nat. can. 101:143–178.

LeResche, R. E., and Davis, J. L. 1971. Moose research report. Fed. Aid Wildl. Rest. Proj. Prog. Rep. W-17-3. Alaska Dept. Fish and Game, Juneau. 156pp.

———. 1973. Importance of nonbrowse foods to moose on the Kenai Peninsula, Alaska. J. Wildl. Manage. 37:279–287.

LeResche, R. E., and Rausch, R. A. 1974. Accuracy and precision of aerial moose censusing. J. Wildl. Manage. 39:175–182.

LeResche, R. E.; Seal, U. S.; Karns, P. D.; and Franzmann, A. W. 1974*b*. A review of blood chemistry of moose and other cervidae, with emphasis on nutritional assessment. Nat. can. 101:263–290.

Lutz, H. J. 1956. Ecological effects of forest fires in interior of Alaska. U.S.D.A. Tech. Bull. 1133. 121pp.

Lynch, G. M. 1976. Some long-range movements of radio-tagged moose in Alberta. Proc. North Am. Moose Conf. and Workshop 12:220–235.

McCullough, D. R. 1979. The George Reserve deer herd. Univ. Michigan Press, Ann Arbor, 271pp.

MacLennan, R. R. 1975. An analysis of fluctuating moose populations in Saskatchewan. Tech. Bull. No. 2, Saskatchewan Dept. Tourism and Renewable Resour., Regina. 16pp.

Markgren, G. 1969. Reproduction of moose in Sweden. Viltrevy 6:127–299.

Mech, L. D. 1966. The wolves of Isle Royale. Fauna of Natl. Parks of U.S., Fauna Ser. 7. Washington, D.C. 210pp.

Mercer, W. E., and Kitchen, D. A. 1968. A preliminary report on the extension of moose range in the Labrador Peninsula. Proc. North Am. Moose Conf. and Workshop 5:62–81.

Mercer, W. E., and Manuel, F. 1974. Some aspects of moose management in Newfoundland. Nat. can. 101:657–671.

Merrill, S. 1920. The moose book. E. P. Dutton & Co., New York. 378pp.

Milke, G. C. 1969. Some moose-willow relationships in the interior of Alaska. M.S. Thesis. Univ. Alaska, Fairbanks. 79pp.

Mitchell, H. 1970. Rapid aerial sexing of antlerless moose in British Columbia. J. Wildl. Manage. 34:645–646.

Mould, E. 1979. Seasonal movements related to habitat of moose along the Colville River, Alaska. Murrelet 60:6–11.

Murie, A. 1934. The moose of Isle Royale. Univ. Michigan Mus. Zool., Misc. Publ. No. 25. Univ. Michigan Press, Ann Arbor. 44pp.

Oldemeyer, J. L.; Franzmann, A. W.; Brundage, A. L.; Arneson, P. D.; and Flynn, A. 1977. Browse quality and the Kenai moose population. J. Wildl. Manage. 41:533–542.

Peek, J. M. 1974. On the nature of winter habitats of Shiras moose. Nat. can. 101:131–141.

Peek, J. M.; LeResche, R. E.; and Stevens, D. R. 1974. Dynamics of moose aggregations in Alaska, Minnesota, and Montana. J. Mammal. 55:126–137.

Peek, J. M.; Urich, D. L.; and Mackie, R. J. 1976. Moose habitat selection and relationships to forest management in northeastern Minnesota. Wildl. Monogr. No. 48. 65pp.

Peterson, R. L. 1952. A review of the living representatives of the genus *Alces*. Contrib. R. Ontario Mus. Zool. and Paleontol. No. 34. 30pp.

———. 1955. North American moose. Univ. Toronto Press, Toronto. 280pp.

———. 1974. A review of the general life history of moose. Nat. can. 101:9–21.

Peterson, R. O. 1977. Wolf ecology and prey relationships on Isle Royale. Natl. Park Serv. Sci. Monogr. Ser. No. 11. 210pp.

Peterson, R. O., and Allen, D. L. 1974. Snow conditions as a parameter in moose-wolf relationships. Nat. can. 101:481–492.

Péwé, T. L., and Hopkins, D. M. 1967. Mammal remains of pre-Wisconsin age in Alaska. Pages 266–270 in D. M. Hopkins, ed. The Bering land bridge. Stanford Univ. Press, Stanford. 495pp.

Pimlott, D. H. 1953. Newfoundland moose. Trans. North Am. Wild. Conf. 18:563–581.

———. 1959. Reproduction and productivity of Newfoundland moose. J. Wildl. Manage. 23:381–401.

———. 1961. The ecology and management of moose in North America. Terre Vie 2:246–265.

———. 1967. Wolf predation and ungulate populations. Am. Zool. 7:267–278.

Prescott, W. H. 1974. Interrelationships of moose and deer of the genus Odocoileus. Nat. can. 101:493–504.

Rausch, R. A., and Bishop, R. H. 1968. Report on 1966–67 moose studies. Fed. Aid Wildl. Rest. Proj. Prog. Rep. W-15-2. Alaska Dept. Fish and Game, Juneau. 263pp.

Rausch, R. L. 1977. O zoogeografii nekotorykh Beringiiskikh mlekopitaiushchikh. Pages 162–177 in V. E. Sokolov, ed. Uspekhi sovremennoi teriologii. Nauka, Moscow. 227pp.

Ritcey, R. W. 1974. Moose harvesting programs in Canada. Nat. can. 101:631–642.

Robinette, W. L.; Gashwiler, J. S.; Low, J. B.; and Jones, D. A. 1957. Differential mortality by sex and age among mule deer. J. Wildl. Manage. 21:1–16.

Roussel, Y. E.; Audy, E.; and Potvin, F. 1975. Preliminary study of seasonal moose movements in Laurentides Provincial Park, Quebec. Can. Field Nat. 89:47–52.

Rowe, J. S., and Scotter, G. W. 1973. Fire in the boreal forest. Quartern. Res. 3:444–464.

Samuel, W. M.; Barrett, M. W., and Lynch, G. M. 1976. Helminths in moose of Alberta. Can. J. Zool. 54:307–312.

Schladweiler, P., and Stevens, D. R. 1973. Reproduction of Shiras moose in Montana. J. Wildl. Manage. 37:535–544.

Sergeant, D. E., and Pimlott, D. H. 1959. Age determination in moose from sectioned incisor teeth. J. Wildl. Manage. 23:315–321.

Seton, E. T. 1927. Lives of game animals. 4 vol. Doubleday Doran, New York. 780pp.

Sigman, M. 1977. The importance of the cow-calf bond to overwinter moose calf survival. M.S. Thesis. Univ. Alaska, Fairbanks. 185pp.

Simkin, D. W. 1965. Reproduction and productivity of moose in northwestern Ontario. J. Wildl. Manage. 29:740–750.

———. 1974. Reproduction and productivity of moose. Nat. can. 101:517–525.

Spencer, D. L., and Chatelain, E. F. 1953. Progress in the management of the moose of southcentral Alaska. Trans. North Am. Wildl. Conf. 18:539–552.

Spencer, D. L., and Hakala, J. B. 1964. Moose and fire on the Kenai. Proc. Tall Timbers Fire Ecol. Conf. 3:11–32.

Stephenson, R. O., and Johnson, L. 1973. Wolf report. Fed. Aid. Wildl. Rest. Proj. Prog. Rep. W-17-4. Alaska Dept. Fish and Game, Juneau. 52pp.

Stevens, D. R. 1974. Rocky Mountain elk-Shiras moose range relationships. Nat. can. 101:505–516.

Telfer, E. S. 1967. Comparison of moose and deer winter range in Nova Scotia. J. Wildl. Manage. 31:418–425.

———. 1970. Winter habitat selection by moose and white-tailed deer. J. Wildl. Manage. 34:553–559.

———. 1974. Vertical distribution of cervid and snowshoe hare browsing. J. Wildl. Manage. 38:944–946.

Timmermann, H. R. 1974. Moose inventory methods: a review. Nat. can. 101:615–629.

VanBallenberghe, V. 1977. Migratory behavior of moose in southcentral Alaska. Int. Congr. Game Bio. 13:103–109.

VanBallenberghe, V., and Peek, J. M. 1971. Radiotelemetry studies of moose in northeastern Minnesota. J. Wildl. Manage. 35:63–71.

Verme, L. J. 1970. Some characteristics of captive Michigan moose. J. Mammal. 51:403–405.

Viereck, L. A. 1973. Wildfire in the taiga of Alaska. Quartern. Res. 3:465–495.

Viereck, L. A., and Schandelmeier, L. A. 1980. Effects of fire in Alaska and adjacent Canada: a literature review. Tech. Rep. 6. U.S. Dept. Interior, Bureau of Land Manage., Anchorage, Alaska. 124pp.

Wolfe, M. L. 1974. An overview of moose coactions with other animals. Nat. can. 101:437–456.

———. 1977. Mortality patterns in the Isle Royale moose population. Am. Midl. Nat. 97:267–279.

Wolff, J. O. 1976. Utilization of hardwood browse by moose on the Tanana flood plain of interior Alaska. U.S.D.A. For. Serv. Res. Note. Portland, Oreg. 7pp.

———. 1980. Moose-snowshoe hare competition during peak hare densities. Proc. North Am. Moose Conf. and Workshop 16:238–254.

John W. Coady, Alaska Department of Fish and Game, 1300 College Road, Fairbanks, Alaska 99701.

47

Caribou

Rangifer tarandus

Frank L. Miller

NOMENCLATURE

COMMON NAMES. Caribou, reindeer, ''deer''
SCIENTIFIC NAME. *Rangifer tarandus*
SUBSPECIES. *R. t. groenlandicus*, barren-ground
caribou or American tundra reindeer; *R. t. granti*,
Alaskan barren-ground caribou or Grant's caribou; *R.
t. caribou*, American woodland caribou or woodland
caribou; *R. t. pearyi*, Peary caribou or Peary reindeer
(Banfield 1961).

The Queen Charlotte Island's caribou (*R. t. dawsoni*) probably became extinct shortly after 1910 (Banfield 1963). The causes for its extinction are unknown,
but it is suggested that habitat deterioration through
amelioration of the climate and loss of genetic plasticity through isolation were more important than hunting
or other human interference (Banfield 1963).

The caribou is the only North American cervid
that has established year-round populations north of the
treeline, into some of the harshest lands in North
America. The caribou first appeared in Europe in the
mid-Pleistocene about 440,000 years ago and is a
primitive member of the deer family Cervidae (Banfield 1961, 1974). In North America, caribou were
probably in Alaska and the Yukon before the Wisconsin glaciation, but their earlier occurrences and origin
are, as yet, undescribed (Banfield 1961). The features
that suggest the primitive nature of *Rangifer* include
the possession of antlers by both sexes, long metapo-
dial bones, well-marked tarsal and interdigital glands,
and relatively simple crests on the cheek teeth (Banfield 1974). Banfield's (1961) revision of *Rangifer* is
an excellent source for the Paleontological record, vernacular names, previous revisions, and the current
taxonomy. Supplementary information is in Kelsall
(1968).

DISTRIBUTION

The Peary caribou has ranged as far south as the mainland coast of the Arctic Ocean, and on occasion has
intergraded with the Canadian mainland form of
barren-ground caribou (Banfield 1954*b*, 1961; Manning 1960). In turn, some mainland barren-ground

FIGURE 47.1. Distribution of caribou (*Rangifer tarandus*
spp.): 1, *R. t. caribou*; 2, *R. t. dawsoni*, E = extinct; 3, *R. t.
granti*; 4, *R. t. groenlandicus*; 5, *R. t. pearyi*. The actual
boundaries between the subspecies are not well defined; seasonal overlap did and probably still does occur in many areas.
Arrows indicate areas of probable subspecific overlap. After
Banfield 1974.

caribou have migrated north to the more southerly islands of the arctic archipelago and shared portions of
summer ranges of Peary caribou. Most of the interchange had apparently ceased by the end of the 1920s,
a period during which native hunters took excessive
harvests from migrating herds (Hoare 1927; Manning
1960). Little is known of the persistence of movements
between southern islands and the coastal mainland and
of the present genetic interchange between Peary and
barren-ground caribou. On the Canadian arctic archipelago, the Peary caribou shares its range with only
one other ungulate, the muskox (*Ovibos moschatus*)
(Tener 1963).

The total number of Peary caribou in the Canadian
High Arctic probably does not exceed 10,000–15,000
(Gunn et al. 1981). Currently, Melville and Prince Patrick islands—the Queen Elizabeth group—are the
heartland of Peary caribou north of 70°00″ N latitude

(Miller et al. 1977a). Peary caribou on the western Queen Elizabeth Islands declined from about 24,000 in 1961 (Tener 1963) by about 89 percent in 1974 (Miller et al. 1977a). Peary caribou supposedly occur in higher numbers on Banks, Victoria, Prince of Wales, and Somerset islands (Miller 1978) (figure 47.1).

The barren-ground caribou of Alaska and Canada ranges over thousands of square kilometers of arctic tundra. Some herds remain on the tundra year-round, while others migrate south into the boreal forest (taiga) for the winter. Those barren-ground caribou that are in the boreal forests in autumn may intergrade with northern populations of woodland caribou. Barren-ground caribou share portions of their range with moose (*Alces alces*) and muskoxen.

In Canada the barren-ground caribou occurs on the arctic islands of Baffin, Bylot, Southampton, and Coats, and possibly still on Victoria and King William (Banfield 1961). On the mainland it occurs from west of Hudson Bay in the District of Keewatin, Northwest

Territories, and northern Manitoba westward across the District of Mackenzie, Northwest Territories, northern Saskatchewan, and occasionally northeastern Alberta to the Mackenzie River, Northwest Territories (fig. 47.1).

The barren-ground caribou of the Northwest Territories occur in nine herds or populations (Calef 1978). Those include the barren-ground caribou that seasonally range into the provinces to the south. Recent estimates of densities based on aerial surveys (Calef 1978) suggest that there are currently about 550,000 barren-ground caribou west of Hudson Bay and east of the Mackenzie River (table 47.1).

At the Mackenzie River the Canadian form of barren-ground caribou (*groenlandicus*) gives way to and sometimes intergrades with the Alaskan form of barren-ground caribou (*granti*). The Alaskan barren-ground caribou ranges west and north across the Yukon Territory, onto the north slope of Alaska, and throughout arctic Alaska. Its survival as a pure stock on the

TABLE 47.1. Recent estimates of sizes of major herds of caribou (*Rangifer tarandus*) in North America

Herd	Estimate	Year	Trend
Alaska[a]			
Adak	250	1977	stable
Alaska Peninsula	18,000[b]	1977	increasing
Andreafsky	1,500–5,000	1977	unknown
Beaver	2,000	1970	stable
Central arctic	5,000	1977	unknown
Chisana	1,000–1,500	1977	stable
Delta	2,500	1977	decreasing
Fortymile	4,000	1977	decreasing
Granite Mountain	100	1977	unknown
Kenai	430–445	1977	increasing
Kilbuck Mountain	1,000	1977	unknown
Macomb	800–1,000	1977	unknown
McKinley	1,000	1977	decreasing
Mentasta	2,500	1977	stable
Mulchatna	10,000	1977	increasing
Nelchina	14,000	1977	decreasing
Rainy Pass/Farewell	3,000	1977	unknown
Ray Mountains	200	1977	unknown
Sunshine-Cloudy Mountains	500–1,000	1977	unknown
Teshekpuk	500	1977	unknown
Porcupine[c]	100,000	1977	stable
Western arctic	75,000	1977	decreasing
Canada[d]			
Porcupine[c]	93,000–100,000	1972	stable
Bluenose	90,000	1975	increasing
Bathurst	150,000	1977	decreasing
Beverly	94,000	1980	decreasing
Kaminuriak	38,000	1980	decreasing
Wagner Bay	29,000	1976	increasing
Melville Peninsula	52,000	1976	increasing
Lorrilard	17,000	1976	unknown
Baffin Island	20,000	1974	decreasing
George River	185,000	1978	increasing

[a] Data taken from Davis (1978).

[b] Extrapolated from 1975 survey data (Davis 1978).

[c] Porcupine estimates from Alaska and from Canada are for the same herd.

[d] Data taken from Calef (1978); except Beverly and Kaminuriak estimates from A. Gunn (personal communication) and George River estimate from S. Luttich (personal communication).

Alaskan Peninsula is questionable, however, because of extensive interbreeding with introduced domestic reindeer (Banfield 1961). The Alaskan form of barren-ground caribou (and some likely intergrades with woodland and/or free-ranging reindeer) has been divided into 22 herds (Davis 1978). Earlier estimates placed the number of Alaskan barren-ground caribou at about 600,000 (Skoog 1968; Hemming 1971). Subsequent work (table 47.1) indicated a marked decline to about 250,000 (Davis 1978). Overhunting and predation by wolves (*Canis lupus*) have been suggested as predominent causes for that decline.

The woodland caribou occupies the boreal forest and alpine tundra extensions of suitable mountainous habitats. Woodland caribou once occurred in Maine, New Hampshire, Vermont, Michigan, Minnesota, New Brunswick, Nova Scotia, and Prince Edward Island. The range of the woodland caribou has decreased considerably since the 1800s, probably due to destruction of the climax forests and overhunting. The northward extension of white-tailed deer (*Odocoileus virginianus*) following the loss of the climax forests likely eliminated woodland caribou from the southern portion of their range. An additional factor may have been the deer's role as a carrier of "moose sickness," the parasitic meningeal worm (*Parelaphostrongylus tenuis*) (Smith et al. 1964; Anderson 1971) that also affects woodland caribou.

Now their range has been pushed northward to Newfoundland, Quebec, north of the St. Lawrence River, except in the Shickshock Mountains of the Gaspe Peninsula, Ontario, Manitoba, Saskatchewan, Alberta, remnant populations in the Salmo River–Selkirk area (Freddy and Erickson 1975; Johnson 1976) of Idaho, and possibly Montana, Washington State, and British Columbia (fig. 47.1). The woodland caribou range extends north in the east to the Ungava-Labrador Peninsula and Newfoundland, in the west to the District of Mackenzie, Northwest Territories, the Yukon Territory, and the Copper River area of Alaska. The woodland caribou no longer exists on the Kenai Peninsula (Banfield 1961). The ranges of woodland caribou are shared with several other members of the deer family: white-tailed deer, black-tailed deer (*O. hemionus columbianus*), mule deer (*O. h. hemionus*), moose, and elk (*Cervus elaphus*).

DESCRIPTION

Caribou are medium-sized deer with relatively long legs, large hooves, and broad muzzles. Their heads are elongated, with forehead-nose profiles that vary from almost straight to the "roman noses" of some mature males. The muzzle is blunt, except in Peary caribou, and well haired except for the small oval rhinarium (Pocock 1923). The physical appearances of caribou differ among subspecies; many of the differences are given in Banfield (1961). Detailed descriptions of caribou are given by Dugmore (1913) for woodland caribou of Newfoundland, by Murie (1935) for Alaskan-Yukon caribou, and by Harper (1955) and

Kelsall (1968) for barren-ground caribou of the Northwest Territories.

Size. Males weigh from about 110 kilograms (kg) for Peary caribou (season unknown, Banfield 1961) to 299 kg for Alaskan barren-ground caribou (prerut, Skoog 1968). Seasonal differences in body weights, especially for adult males, are so great that much of the range of differences between males and females is lost. Generally, however, mature female caribou are about 10–15 percent smaller and weigh 10–50 percent less than adult males.

Pelage. The forest-dwelling woodland caribou have the darkest pelage and the Peary caribou the lightest, with more gray than brown in their coats. Barrenground caribou have many variations in intermediate shades of color relative to the other two subspecies. The underparts of caribou are lighter, leg coloring and socks are pronounced, and flank stripes vary from prominent to lacking. Descriptions of pelages using standard color guides are given in Manning (1960) and Banfield (1961).

Each spring, caribou lose their winter pelage, which falls away in patches, revealing darker hair beneath and leaving the caribou with a ragged appearance. Prime bulls are usually the first to develop sleek summer coats, followed by juveniles and yearlings, by calfless cows, and, last, by maternal cows. By August, all but cows in poor condition are in new coats, which are the darkest. In autumn, the hair lengthens and longer, white-tipped guard hairs grow to form the winter coat. By prerut (September–October), the bulls have developed handsome white manes and the pelage shows the maximum contrast of an individual's coloration. As winter progresses, the guard hairs are bleached and many of the hair tips are broken off, resulting in a general lightening and loss of contrast in the pelage. Finally, the winter coat falls away and the cycle begins again.

Calves are born with a light brown or red brown body and a dark brown to black dorsal stripe from the neck to the tail. Their undersides are white or light gray (Kelsall 1968).

Antlers. Both males and females have antlers, although the male's are larger and can be of impressive proportions. The main beams are long and curved toward the animal's posterior with a cylindrical to flattened cross-section. Usually one and sometimes two brow tines are widely palmated (the "shovels") and the dominant one extends vertically over the face. The brow and second tines extend anteriorly and the terminal tines extend posteriorly, and can be either distinct or palmate.

The antlers of females and young caribou are smaller and simpler. Females do, however, sometimes grow miniature replicas of the male's large antlers. Calves grow "spike" antlers that remain in velvet for their first winter and spring.

The shapes of caribou antlers vary greatly, and no two sets are the same. One antler of a pair is not the mirror image of the other (Banfield 1954a). The variation among populations usually negates the use of antlers in detailed taxonomic work (Banfield 1961), although Bubenik (1975*b*) argued for their potential use in taxonomy. The annual cycle of antler development (e.g., Banfield 1954*a;* Harper 1955; Moisan 1959; Bubenik 1975*b*) varies with the sex, reproductive status, and age of the caribou.

Males shed their antlers during or after the rut. The role of testosterone in the antler cycle of caribou is minor, and under deep physical stress some older bulls can drop their antlers before the end of the rut (Murie 1935; Bubenik 1975*b*). Those bulls without antlers immediately lose their sex drive and dominance and do not further participate in rutting activities (Espmark 1964*a*). The antlerless bulls turn to foraging and recover their physical condition more rapidly than other prime bulls that participated in the entire rut; such a process could have survival value.

Female caribou retain their antlers through the winter and often shed them about the time of calving (e.g., Lent 1965*a;* Skoog 1968; Bergerud 1976). The presence of osteolytic resorption in antlers in December suggests that the retained antlers of females and juveniles could serve as a calcium bank during the winter (Belanger et al. 1967).

The annual cycle of antler velvet and antler shedding varies with sex and age. Bergerud (1976) illustrated the antler cycle for woodland caribou in Newfoundland. Skoog (1968) and Miller (unpublished data) observed slight subspecific variations from woodland caribou in the annual antler cycle of barren-ground caribou: (1) mature barren-ground bulls begin to grow antlers in early March, although a few carry hard antlers until late February; (2) yearling barren-ground caribou begin to grow new antlers in late April and/or early May, although some carry old antlers through to April or May; (3) many nonpregnant barren-ground females begin growing antlers as early as April, but some carry last year's antlers until May or June; (4) many pregnant barren-ground females start antler growth about one week after calving, and, while many carry old antlers to calving, some cast their antlers as early as March or April.

Hooves. The hooves of caribou are well adapted to life on spongy muskegs and frozen snow-covered ground. The hoof is as wide as or wider than it is long; dew claws are large and add greatly to the bearing surface of the foot. The hoof is curved inward and abrupt at the tip. In winter, the soft pads of the hoof deteriorate and the pad area becomes concave. The hoof becomes sharp edged and hair between the hooves elongates to cover the pad. Those changes result in a spreading, "nonskid" support that is effective on snow and ice (Kelsall 1968). A polydactylous hoof was found on a Canadian barren-ground caribou (Miller and Broughton 1971).

Skull and Dentition. Banfield (1961:26–27) diagnosed the skull characteristics (figure 47.2) of *Rangifer* as follows: "Cranium moderately expanded, extreme postorbital position of the pedicels which encroach on the parietal bone, nasals prominent and

expanded proximally, preorbital pit in the lachrymal bone moderate in size, lachrymal vacuity generally moderate in size. Premaxillae prevented from reaching nasals by a small lobe on the maxillae."

Caribou are heterodonts (Loomis 1925; Frick 1937); the permanent dentition is subhypsodont, selenodont, and deerlike in general, but seemingly more adapted for grazing than for browsing (Kelsall 1968; Skoog 1968). This adaptation of the teeth has, however, been questioned by Miller (1974*c*) in the light of the known varied food habits of caribou. The dental formula is 0/3, 1/1, 3/3, 3/3 = 34.

Neonatal caribou have a set of functional deciduous milk teeth that are smaller versions of their perma

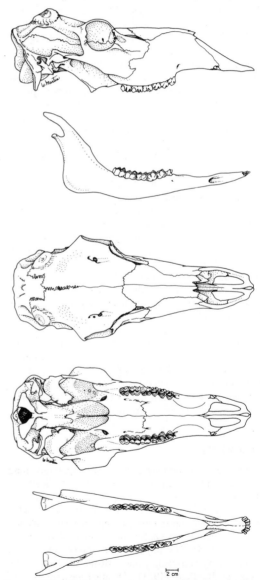

FIGURE 47.2. Skull of the caribou (*Rangifer tarandus*). From top to bottom: lateral view of cranium, lateral view of mandible, dorsal view of cranium, ventral view of cranium, dorsal view of mandible.

nent counterparts. The first permanent tooth to erupt is the first molar at 3–5 months, followed by the incisors and the second molar at 10–15 months, then the premolars and third molar at 22–29 months (Miller 1972, 1974c). Wear on the teeth is slight until after the third year of life. Further attrition is not marked until the animal is 5–7 years old; by 10 years the wear is appreciable on most teeth and the molars appear to be losing their usefulness.

Dental anomalies include morphological variations, supernumeraries, and agenesis (Banfield 1954a, 1961; Miller and Tessier 1971). Many dental anomalies are congenital, while others may be caused by trauma and subsequent repair.

Skin Glands. Caribou have preorbital, tarsal, interdigital, and caudal (tail) glands that all secrete odorous material (Quay 1955; Lewin and Stelfox 1967; Kelsall 1968; Anderson et al. 1975; Muller-Schwarze et al. 1978). The preorbital glands are located medial to the eye sockets, over the lachrymal pit in the lachrymal bone. They are almost hidden by a dense covering of hair. A tarsal gland is located on each hind leg, on the inside of the hock at the tarsal joint; the glands are fringed by tufts of stiff, whitish hair, usually stained yellowish by the secretion. Deep-socketed interdigital glands occur on each hind foot, and a smaller one on each front foot, medial and posterior to the primary digits. The caudal gland is located underneath the distal portion of the tail.

These scent glands most likely play an important role in social organization. The fact that caribou can track each other by scent makes it possible for separated members to relocate the group if they lose sight of each other. Future work will undoubtedly show that scent glands function in many ways to provide olfactory signals to caribou under various conditions and in different situations. If individual caribou can track each other, why should they not recognize each other as individuals, rather than by sex and age class only?

Longevity. Caribou are moderately long-lived ungulates. Some extremes of longevity for North American caribou are 18–20 years (McEwan 1963), 15+ (Skoog 1968), 17 (Bergerud 1971c), and 17+ (Miller 1974c).

PHYSIOLOGY

Nutrition. Physiological studies of nutrition of caribou are beginning to unravel the mechanisms of the adaptation of caribou to their arctic and subarctic environments. Lichens (*Ascomycetes*), usually the main winter forage, are low in protein and high in carbohydrates (Scotter 1965, 1972; McEwan and Whitehead 1970). The latter is fermented by specifically adapted rumen microorganisms (Dehority 1975a, 1975b) and during its fermentation heat is produced. A low concentration of rumen ammonia also results, which assists the recycling of nitrogen by increasing the rate of transfer of urea to the rumen from the plasma (Wales et al. 1975). The increase of urea recycling in winter (Wales et al. 1975) compensates for the low protein content of lichens and also has a

marked effect on the water flux (White 1975). The total amount of body water increases principally because water replaces decreasing body fat (Cameron and Luick 1972; Cameron et al. 1975). The increased volume of body water could function as a thermal buffer, especially to the temperature changes caused by food and snow intake (Cameron et al. 1975). The decrease in nitrogen intake and the increase in urea recycling are also accompanied by increases in glucose resynthesis (White 1975): the ability to have a high rate of glucose resynthesis is particularly important in pregnant or lactating females (Luick and White 1975).

The reduced forage intake during winter (McEwan 1968), probably in response to lower energy requirements (White 1975), in itself reduces energy expenditure. Thing (1977) calculated that the energy costs of foraging in snow increase markedly in late winter. By eating less, the caribou saves the energy required both to find food and to warm it once eaten—an additional cost estimated to be 20–25 percent of the fasting metabolic rate (White and Yousef 1974; White 1975; Young and McEwan 1975).

Caribou conserve energy in winter through the insulating qualities of their pelage and by heating the blood returning from the distal joints of their legs with outgoing blood in the proximal arteries (Jacobi 1931; Scholander et al. 1950a, 1950b, 1950c; Irving and Krog 1955; Moote 1955; Lentz and Hart 1960; Hart et al. 1961; Øritsland 1974). The feet of caribou remain flexible in subzero temperatures because the fatty tissues formed there remain soft. Marrow fats further up the leg are solid even at room temperature.

In summer, caribou lose heat by panting, from patches of bare skin on their bodies, and from the sparsely haired extremities (McEwan et al. 1965; Irving 1966; Krog and Wika 1975; Yousef and Luick 1975). Stonehouse (1968) suggested that the growing antlers in velvet served a primary function in the dissipation of body heat. However, Krog and Wika (1975) believed that heat loss from the antlers was unavoidable because of the blood flow necessary for rapid antler growth.

Growth. Caribou calves grow rapidly in the first five months after birth. Their body weights increase by an average of 800 percent (Dauphine 1976), from about 6 kg to 48 kg. Variations in body weight may reflect the condition of the cow in late gestation: nutritionally stressed cows gave birth to small calves that had low survival rates (Skoog 1968).

Caribou normally grow between April–June and October–December of each year (Dauphine 1976). The rate of growth depends on their nutritional plane in relation to other demands such as hair and antler growth, reproduction, weather, and movements, including responses to insect harassment (Skoog 1968).

The age at which skeletal growth stops depends on which particular bone is measured (McEwan and Wood 1966; Dauphine 1976). Mandibles continue to lengthen until 5½ years of age in cows and 6 years in bulls (Miller and McClure 1973), but body weights reach maximums at 4½ and 6 years for cows and bulls respectively (Dauphine 1976). There are marked seasonal fluctuations in body weight because of changes in fat deposits (Dauphine 1976). Female caribou gain weight annually during late summer through early winter. Then their weights stabilize or decline in winter and are lowest during early summer due to pregnancy and lactation. Adult males gain the most weight from summer to the beginning of the rut. Mature prime males typically lose up to 25 percent of their weight during the rut; in winter their weight stabilizes or declines (Dauphine 1976). Young animals reach maximum weights between early summer and early winter. Their weights decline during winter, as they usually lack sufficient fat reserves to maintain body weights until spring.

REPRODUCTION

Caribou rely on high-quality forage on their summer range for their reproduction, growth, and winter survival. Female caribou with low fat reserves in the autumn do not breed but build up their fat reserves and breed the following autumn. Body size and fat reserves of females are also related to the age of first conception; although most cows conceive at 3½ years of age, a few will conceive at 1½ years of age in good condition (Dauphine 1976). This varies among populations: more Alaskan than Canadian caribou conceive at early ages (Skoog 1968). Caribou are relatively slow to mature and do not bear twins, with rare exceptions (McEwan 1971; Shoesmith 1976). The adult females, however, are very fertile, with pregnancy rates of about 80 percent or more. The chief cause of reproductive failure is a failure to conceive, as *in utero* mortality is rare (Dauphine 1976).

Insufficient build-up of fat reserves in summer and subsequent failure to conceive may explain the fluctuations in pregnancy rates of some Peary caribou (Thomas and Broughton 1978). Although females can be above the threshold of physical condition necessary to conceive, subsequent deterioration of the cow's condition can contribute to calf mortality at birth or shortly thereafter. Death of the calf soon after birth relieves the cow of the metabolic cost of lactation. This relief allows it to build up fat reserves sufficiently to breed again the following autumn, but sometimes not sufficiently to produce and rear a calf successfully. As a result, Dauphine (1976) suggests a direct relationship between high calf mortality and high pregnancy rates.

Breeding and Gestation. One of the most striking features of the reproduction of caribou is the synchronization of the rut and thus of calving (Dauphine and McClure 1974). Caribou have two or three full estrous cycles, and one or more of those result in ovulation without overt estrous behavior. Such "silent heats" could generate and synchronize the endocrine system and thus mating (McEwan and Whitehead 1972; Bergerud 1975). Although there is variation in the dates of rutting between different caribou populations, the rut is usually in October and November, and local synchrony of breeding is apparent.

The average gestation period is 225–235 days according to Skoog (1968), or 227–229 days according to Bergerud (1978), and the synchronized breeding results in a strongly peaked distribution of births. Usually between 80 and 90 percent of caribou calves are born in a 10-day period in late May or early June (Dauphine and McClure 1974; Bergerud 1975).

Parturition. Parturition has been observed by several workers (Kelsall 1957; de Vos 1960; Lent 1966*a;* Bergerud 1974*d*). The newborn calf usually stands and walks within the first hour of life and nurses shortly thereafter. The mother vigorously licks and cleans the neonate. She nibbles and pulls the afterbirth from around the calf and often eats it (Pruitt 1960*a;* Miller and Parker 1968).

Miller and Parker (1968) suggested that maternal caribou frequently consume the afterbirth. There are two possible reasons: (1) removal of the tissue reduces odors associated with the birth site and, thus, the chances of predation; and (2) the maternal females derive nutritional benefit from eating the afterbirth. The latter supposition seems most tenable for caribou. Neonates follow their mothers and the birthplace is usually abandoned within hours. Also, the attending cow would be sighted by a predator long before the predator was close enough to detect the odors of the birth site, especially on tundra calving grounds.

ECOLOGY

Habitat. The geographical distribution of caribou encompasses two of the largest biomes in North America: tundra and taiga (and their southern extensions in mountainous areas). Variations in climate, geology, and topography from Alaska to Newfoundland result in differences in the plant communities of these biomes, but there are also similarities in habitat types used by caribou. The typical range of eastern woodland caribou is climax stands of northern boreal forest with mixed-age stands of black and white spruces (*Picea mariana, P. glauca*), balsam fir (*Abies balsamea*), and white birch (*Betula papyrifera*) with tree and ground lichens (Cringan 1957). The habitat of many western woodland caribou often includes mountain summits above the timber line or elevated table lands with alpine meadows and open subalpine forest (Edwards 1958).

In the taiga toward the treeline the trees become more widely spaced and the ground cover of lichens, ericaceous shrubs, willows (*Salix* spp.), and dwarf birch (*B. glandulosa*) increases. North of the treeline the black spruce–moss muskeg and sedge (*Carex* spp.) bogs of poorly drained areas are replaced by willow and alder (*Alnus* spp.) thickets along watercourses, and by sedge in grassland communities on wet areas. Dwarf shrub–heath communities occupy large areas of the tundra except on drier sites, where fructose lichens and mosses dominate with dwarf willow, birch, and rhododendron (*Rhododendron lapponicum*) (Rowe 1959; Miller 1976). In the High Arctic, Peary caribou show a preference for polar "desert" and similar dry to mesic range types with sparse vegetation of willow, sedges, grasses, and forbes (Parker and Ross 1976; Russell et al. 1978).

The climate of caribou ranges is characterized by long, cold winters; short, cool summers; and low precipitation. Much of the precipitation falls as snow, which can cover the ground for seven to nine months. Snow cover data are of particular importance to understanding caribou biology but show great variation according to locality. The caribou contend with snow depths usually exceeding 50 cm except in the High Arctic and on the open tundra. Autumn and winter thaws and rain storms can occur in almost any northern area. These sometimes cause serious restrictions on forage availability (Miller et al. 1977a). Spring melting and refreezing of snow leading to crust formation often restrict caribou feeding and movements (Pruitt 1959; Henshaw 1968b; Bergerud 1974b; Stardom 1975; Miller 1976; Miller and Gunn 1978).

The short growing season on caribou ranges is partially offset by the increased solar radiation and increased day length characteristic of high altitudes and latitudes (Klein 1964). In addition, plants growing on the ranges have higher nitrogen, phosphorus, and carbohydrate levels than plants at lower latitudes (Chapin et al. 1975). Some arctic plants are adapted to the short growing season by having wintergreen and evergreen leaves (Bell and Bliss 1977), which are preferred winter forage (Kelsall 1968). Caribou sometimes expose, trample, and eat feeding push-ups constructed on ice in winter by muskrats (*Ondatra zibethicus*) (Skoog 1968; Kelsall 1970). Caribou will use mineral licks (Calef and Lortie 1975).

Summer and Winter Ranges. Most caribou populations have distinct summer and winter ranges; the latter are characterized by tree and/or shrub cover and lichens. Caribou are not solely dependent upon lichens in winter and may take other forage if lichens are unavailable (Murie 1935; Skoog 1968; Bergerud 1972). Winter range is usually regarded as the limiting factor for caribou populations; therefore, any factors modifying or reducing the availability of winter range are of particular interest to wildlife managers.

All Peary caribou remain on tundra ranges throughout the year. Some barren-ground caribou herds remain all year on the tundra, while other herds move to forested ranges in winter. Woodland caribou, even the migratory herds of the Ungava Peninsula, usually move to forested winter ranges. However, even the forest-dwelling herds will move in winter to tundra or tundralike areas if excessively deep snow cover restricts forest forage supplies.

Movements between the tundra and forested ranges are learned traditions of particular herds that have resulted in favorable long-term survival. The yearly differences in movements reflect the annual variations in snow cover and icing conditions, which influence availability of forage supplies.

The effect of forest fires on winter ranges of caribou has long been a controversial subject. There are two theories concerning the importance of forest fires in limiting or reducing numbers and changing distributions of caribou. One is that forest fires are a paramount consideration because of the extensive detrimental impact on forage supply and the subsequent restorations of such stocks (Leopold and Darling 1953; Edwards 1954; Cringan 1957; Banfield and Tener 1958; Scotter 1964, 1967a; Kelsall 1968). Other theories suggest that only certain fires under certain conditions are important to caribou. Finally, some researchers feel that fires are not only beneficial to caribou but also are necessary to promote the heterogeneity of vegetation needed to perpetuate mixed forage supplies (Skoog 1968; Bergerud 1971a, 1971b, 1974a; Rowe and Scotter 1973; Bunnell et al. 1975; Johnson and Rowe 1975; Miller 1976). Probably, the moderate views better reflect the realities of range caribou relations, although production and availability of forage are the ultimate factors in governing numbers and distributions of caribou. Therefore, managers must remain aware that wildfires can be important in the ecology of caribou. If rates of burning approach the time required for rotation of successional ranges of vegetation to provide sufficient forage supplies for existing and future numbers of caribou, protective actions should be taken to minimize destruction of ranges.

Movements and Migrations. If one aspect characterizes the ecology of caribou, it is survival through adaptive movements and migrations. Bergerud (1974d) believed that interaction with wolves led to gregarious behavior. Movements and migrations followed as a result, so that caribou could maintain themselves in relation to their varying forage supplies. The caribou's movements and migrations are further governed to varying degrees in time and space by weather, especially snow cover and icing conditions, bloodsucking and biting insects, and various physiological and psychological drives. Snow cover most often is the dominant influence on movements by caribou during most of the year. Evaluations of the snow (nival) environment of caribou have been made by many workers (Banfield 1949; Pruitt 1959; Edwards and Ritcey 1960; Henshaw 1968b; Bergerud 1971a; Stardom 1975; LaPerriere and Lent 1977; Thing 1977).

The annual cycle of movements of migratory barren-ground caribou begins in spring. As the amount of daylight increases and snow begins to recede, wintering bands of caribou begin to coalesce, and these large aggregations move northward. In the early stages of the spring migration, the caribou move sporadically until the apparent urge to return to the calving ground grips the parturient cows and they move steadily north. The number of juveniles, and especially yearlings, accompanying the females to the calving ground varies from year to year depending on the difficulties of migration. In some years, deep snow and slush prevent most young animals from migrating with the cows. They drop behind and move northward with the bulls at a more leisurely pace as traveling conditions improve. Once the traditional calving ground is reached, the parturient cows disperse. They may calve in relative isolation or close to groups, depending on the timing of their arrival on the calving grounds and the peak of calving.

The maternal cows and their newborn calves form nursery bands after the peak of calving (Pruitt 1960a).

The nursery bands then merge into large postcalving aggregations that move off the calving ground to summer ranges and mix with the bulls and the rest of the herd. During the summer months they move extensively, often 500 km or more. Some herds return south into the boreal forest, while others move northward to tundra and coastal areas. In autumn, there are prerut movements and the prime bulls establish their dominance hierarchies. With the beginning of the rut, bulls join cow-juvenile groups and remain in their company until the cows become receptive. Breeding often takes place while the aggregations are on the move, sometimes even during the autumn migration to the wintering grounds (Henshaw 1970). After the rut, the caribou move to their winter ranges. Adult bulls often separate from the cow-juvenile groups, and some groups of bulls move farther into the boreal forest.

Seasonal movements of the more sedentary forest-dwelling woodland caribou and Peary caribou are small scaled by comparison with those of migratory barren-ground and woodland caribou. Local alterations in movement patterns of some woodland and Peary caribou reflect seasonal variations in distribution and availabilities of forages, and specific physical differences among the habitats.

On western ranges, where the elevation can vary by over 2,000 m, the habitat influences seasonal movements of woodland caribou (Edwards 1954, 1958; Edwards and Ritcey 1959, 1960). The 25 to 30 remaining woodland caribou in the Selkirk Mountains of northern Idaho, northeastern Washington, and southern British Columbia have stopped making distinct seasonal shifts in elevation (Freddy 1979). Woodland caribou on mountainous western ranges are found on high meadows and in adjacent, open subalpine forests above 2,000 m during summer and fall. The first deep snow of winter forces them down to mature forests in valleys, where they prefer poorly drained sites interspersed with open bogs, meadows, and ponds. As winter progresses and the snow settles, hardens, and becomes crusted, the caribou return to their summer haunts on the high, wind-swept ridges and remain there until spring. Softening snow drives them into lowland forest again, where they remain until they can return to the snow-free uplands in May and June.

Most, if not all, Peary caribou make migrationlike trips at about calving time. They may travel between islands, or around or across one island. Movements are not necessarily well in advance (days or weeks) of calving, as they are with woodland and barren-ground caribou (Miller et al. 1977*a*, 1977*b;* Miller and Gunn 1978). Peary caribou seem to prefer certain areas of different islands for calving, but to date no specific calving grounds are known.

The Peary caribou constantly move and feed on their summer ranges until early fall. There appears to be a prerut movement, or "fall shuffle," in late August to September. Nothing is known concerning the rut in Peary caribou. Cast male antlers found along coastal areas suggest that there is some preference for the coast during the rut or shortly thereafter. This would facilitate male-female contact.

Herds that use more than one island annually probably return to winter range shortly after freeze-up in late autumn following the rut. The rut most likely occurs on the island where the caribou summered. Peary caribou that remain on the same islands yearlong shift from summer range to winter ranges on different parts of the islands (Freeman 1975; Miller et al. 1977*a*). Even the sedentary individuals must sometimes move to adjacent islands to survive when deep snow reduces forage availability on large areas of their home ranges.

Biologists have described caribou movements and migrations as nomadic (Skoog 1968; Bergerud 1974*d*). That description should not connote herds of caribou wandering aimlessly about tundra and taiga without purpose. Their wanderings are structured and carried out in an orderly sequence (Heape 1931). Caribou herds, or segments of herds, prefer specific sections of ranges: not only calving grounds but also wintering areas; spring, summer, and fall staging areas; rutting areas; and migrational paths.

To maintain such traditional range use requires a refined state of *Ortstreue*, or fidelity of the offspring to the land of the parents. The apparent wanderings of individuals and groups of caribou are strongly orientated and directed.

Wise management of caribou herds demands that the movements and migrations of caribou be studied in detail to determine if individual caribou show strong seasonal affinities for specific sections of their ranges. If different caribou from one segment of a herd return to the same areas every year, continued heavy annual harvest of them would not be detrimental (assuming that the population could sustain it), because different individuals would be filling the voids each year. If, however, individual caribou did show strong affinities for specific sections of the population's ranges, continued heavy annual harvest of them would soon markedly reduce or destroy that segment of the population.

Herd Composition. Caribou are highly gregarious. The zenith of their gregarious behavior is reached in the postcalving herds composed of tens of thousands of barren-ground caribou. Those herds are truly "living tides flowing over the arctic prairies." No one can ever forget the sights and sounds of such a mass of moving animals. Thousands of grunting cows are answered by their bleating calves. There is a continuous clicking of hooves, and a myriad of coughs, sneezes, and belches coming from the moving mass. Such aggregations sometimes bunch so tightly as they mill about that from the air they appear like a swarm of bees.

Large herds of migrating caribou are temporary gatherings of many social units (groups or bands). Some social order may be maintained within aggregations by an interacting hierarchy of dominant animals from the bands that form the aggregations.

Social Structure. Caribou occur in different groupings, but I consider the band the primary distinguishable social unit. I characterize bands of barren-ground caribou by their representation, by sex and age composition: cow, bull, subadult, juvenile, cow-juvenile, cow-juvenile-bull, and bull-cow-juvenile bands (Miller 1974c). Lent (1966a) and Bergerud (1974d) considered caribou groups as open social units with no long-term stability. I suggest, however, that at least the cores of barren-ground caribou bands represent closed or semiclosed social units. There is evidence of group cohesion and leadership (Miller et al. 1972) and of long-term social bonds in barren-ground caribou on the mainland of northern Canada (Parker 1972b; Miller et al. 1975a). If long-term social bonds do exist among members of a social unit, one must consider how group unity is perpetuated from one year to the next.

The basic social unit of barren-ground caribou is the winter band, of which there are four types: bull, cow-juvenile, juvenile, and subadult. The juvenile and subadult bands are usually mixed or in close association with the cow-juvenile bands. Juvenile and subadult bands may be the result of the loss of maternal cows. Their locations with respect to adults may be governed by the antagonism of the remaining mature cows or dominant subadults.

The core of the winter cow-juvenile bands is formed by caribou related through a matriarchal bloodline, supplemented occasionally by neighboring cows and their young or juveniles. The winter bull band recruits most three- and four-year-old males from subadult groups in the fall, during the prerut period. Most band members have similar body and antler size. Although antlers may vary, they are not usually as large as the antlers of prime bulls. The bull band maintains a nucleus of breeding bulls from year to year with common learned behavioral habits. Under normal conditions, this assures a supply of breeders in the traditional rutting areas.

In spring, the bonds between members of the cow-juvenile bands weaken as the caribou bands come together and the caribou move north to the treeline. On the tundra, the parturient cows intensify their antagonism, and many of the juveniles and subadults drop behind. This response is reinforced by deep, wet snow along the migration route, which slows the young animals.

After their calves are born, the cows move about the calving ground and join other cows, calves, and those juveniles and subadults that have arrived. Small groups are attracted to larger groups until, finally, they form postcalving herds of several thousand caribou.

A postcalving herd has a distinct structure: at the core are the maternal cows with their calves, equally spaced throughout and occupying an elliptical or round area. Along the periphery are groups of yearlings and juveniles in constant motion. Some caribou rejoin fellow members of previous winter bands during formation of postcalving aggregations. Large numbers of juveniles and subadults left behind may rejoin the cow component of their winter bands when the postcalving groups move into the area occupied by those young animals and mature bulls that have moved northward after the cows. The cows and calves merge with the bulls, remaining juveniles, and subadults, and move off on their midsummer migration. The animals continuously shuffle and reshuffle. By the onset of rut the winter bands have reformed. They remain generally in larger groupings until arrival on the wintering ground. The bull bands separate from the cow-juvenile bands and move off to their particular wintering areas.

I believe that socialization during the postcalving aggregations leads to the regrouping of previous winter bands. This link is necessary for perpetuating beneficial behavioral traits (Miller 1974c), which is of particular concern to managers of caribou populations.

Barren-ground caribou usually form larger groups and aggregations than woodland or Peary caribou form, although some herds of migratory woodland caribou—for example, the George River herd, in Quebec-Labrador—might be as large during population highs. The sizes of groups reflect the overall number of caribou in the herd, the size of their ranges, the dispersion of available forage, and possibly subspecific differences in behavior. If the groups are more or less "closed" social units, their sizes would also depend on the survival of group members. Group sizes could vary considerably (usually 5–30), but could still include a core of related animals. The smallest groups of caribou usually occur in midwinter, except for cow-calf pairs during calving. The small winter groupings probably reflect the distribution and availability of forage supplies and possibly a high degree of social intolerance for nongroup members.

Forest-dwelling woodland caribou form relatively small social units, probably because of their seclusive nature and the mosaic pattern of their habitat. During late winter, those small units merge into small aggregations, probably because foraging sites and resting areas are restricted.

The average sizes of Peary caribou groups are small at calving time and then, as with the other subspecies, their groups increase markedly in size during postcalving movements. Such aggregations can exceed 100 caribou but average much less. Winter groups of Peary caribou are small—almost always fewer than 10 individuals, and usually averaging only 3 to 4 animals (Miller et al. 1977a)—because forage is relatively limited on most of their wintering areas.

Interspecific Relations. Most interspecific interactions between caribou and the other animals on their ranges are either predator-prey or competitive. Animals that do not interact directly with caribou benefit by scavenging upon them.

The wolf (*Canis lupus*) is the principal predator of caribou, but on some ranges, especially where wolves have been reduced, the lynx (*Felis lynx*) and the grizzly bear (*Ursus arctos*) are more important. The black bear (*Ursus americanus*), the wolverine (*Gulo gulo*), the coyote (*Canis latrans*), the red fox (*Vulpes vulpes*), the bobcat (*Felis rufus*), the golden eagle (*Aquila chrysaetos*), and the raven (*Corvus corax*) sometimes prey on caribou, especially newborn calves.

Most potential competitors include muskoxen, lemmings, (*Dicrostonyx* spp. and *Lemmus* spp.), arctic hares (*Lepus arcticus*), and snowshoe hares (*L. americanus*). Minor or theoretical competitors could include other ungulates, other rodents, and grazing waterfowl (Kelsall 1968; Skoog 1968).

Scavengers other than the previously named predators would include the polar bear (*Ursus maritimus*), the arctic fox (*Alopex lagopus*), gulls (*Larus* spp.), and jaegers (*Stercorarius* spp.). There are a host of other potential scavengers, such as mustelids and carrion-eating birds (Kelsall 1968; Skoog 1968).

Fly Season. Perhaps no other aspect of the caribou's relationship to its environment is so vivid as its stressful encounters with bloodsucking and biting insects during summer. A year when such insects are numerous is indeed a time of madness for the caribou. There are no areas free of harassing insects on the inland tundra. Caribou must constantly dash about wildly to escape the ever-present hordes of flies during the diurnal peaks of insect activity. Fortunately, in all but the worst fly years, there is some temporary relief for the caribou during the cooler early and late hours of the day. Caribou occupying coastal summering areas often seek relief from flies by moving out onto the mud flats and simply standing motionless with their heads down, muzzles nearly to the ground.

Windswept ridges, glaciers, and lingering snow drifts on tundra ranges of some barren-ground caribou and most western woodland caribou and land areas above the timber line are free of flies. On those sites, caribou usually remain huddled closely together throughout much of the day, dispersing to feed in the cooler late and early hours. In Newfoundland, Bergerud (1974*d*) observed that woodland caribou dispersed into the forest, because flies were less active in the shade. This option appears to be the only one available to most of the woodland caribou of eastern North America, unless they are on coastal areas or the shores of large lakes.

Peary caribou on the more southern islands of the Canadian arctic archipelago experience fly seasons of short duration that would seldom, if ever, have much impact on their physical condition. Flies in certain years could stress them, but this would be of minimal impact compared to mainland situations.

Only the Peary caribou of the more northerly High Arctic islands are continuously free of harrassment by insects. These animals might not survive the rigors of the High Arctic if they had to withstand the additional stress of harassment by hordes of insects.

FOOD HABITS

Caribou, like most North American cervids, feed on a broad range of plants, including lichens, fungi, sedges, grasses, forbs, and twigs and leaves of woody plants. Their preference for lichens is unique among North American ungulates, and is the key to caribou survival in many areas. Descriptions of feeding habits are based on direct observations of free-ranging and captive animals, and also upon examination of fecal and rumen samples, fistula experiments, and examinations of craters and other feeding areas. Feeding habits and preferences vary according to range type. Much in-

formation has been compiled for woodland caribou in Newfoundland (Bergerud and Noland 1970; Bergerud 1971*a*), in Ontario (Cringan 1957; Simkin 1965), and in British Columbia (Edwards and Ritcey 1960; Edwards et al. 1960); for barren-ground caribou in Canada (Scotter 1967*b*; Kelsall 1968; Miller 1974, 1976) and in Alaska (Skoog 1968; Klein 1970*a*, 1970*b*; White et al. 1975; Kuropat 1978); and for Peary caribou (Parker and Ross 1976; Wilkinson et al. 1976; Parker 1978; Thomas et al. 1976, 1977; Shank et al. 1978).

Most caribou ranges are fragile and slow to recover from misuse. Many of those ranges have short growing seasons with low productivity. Also, lichens are especially susceptible to trampling. It is unlikely, however, that caribou would destroy their own forage supplies (Skoog 1968): (1) their cursory feeding behavior prevents excessive use of individual plants; (2) they select mostly the newer growth parts of plants, which are readily replaced; (3) they have diverse feeding habits and use a wide range of plants; (4) they use primitive plant (lichen) communities; and (5) snow often covers large areas of ranges causing caribou to move to other areas.

The wide range of habitats and snow conditions that determines caribou feeding habits limits the value of generalizations. Nevertheless, some comments are applicable to many caribou. Foliose lichens dominate their diet in the fall and winter and, in deep snow areas, the terrestrial lichens of early winter are replaced by arboreal lichens as snow depth increases. Many woodland caribou in the western states and in western Canada rely on arboreal lichens for winter survival (Edwards and Ritcey 1960; Edwards et al. 1960). Stardom (1975) reported that woodland caribou in Manitoba fed mainly on arboreal lichens in open tamarack (*Larix laricina*) bogs during early winter until snow cover appeared to hinder travel. They then moved to mature pine (*Pinus divaricata*) lichen ridges and fed mainly on the ground lichens (*Cladonia* spp.). During spring the woodland caribou fed along shorelines on *Carex* spp. and on ground lichens on southeast-facing slopes of rocky lake shores. Barren-ground caribou select fungi in fall (Kelsall 1968; Miller 1976). The winter diet includes woody twigs of shrubs such as *Vaccinium* spp. and some trees (*Salix* spp.), evergreen leaves, and graminoids that have retained some green leaves (Kelsall 1968).

As the snow melts, caribou seek exposed sites to feed on leaves and graminoids, which tend to dominate the diet to the exclusion of lichens as summer progresses. In particular, at that time of year, caribou select plant species according to phenology of greening leaf buds and flower buds. The selectivity is closely tied to the nutritive status and chemical defense posture of the part of the plant (Kuropat 1978). The caribou feed while on the move, and tend to concentrate on one or two species at a particular time (Kuropat 1978). They use sight and smell to select preferred plant parts (Wright, *in* Klein and White 1978).

Little is known about how selectively caribou feed during fall and winter, but limited availability may reduce selectivity. Captive caribou and reindeer prefer different lichen species (Des Meules and Heyland 1969*a*, 1969*b*; Holleman and Luick 1977). Caribou probably detect the lichens under snow by their keen sense of smell. In deeper snow they may smell lichens through air vents caused by shrubs reaching the snow's surface (Bergerud 1974*b*). Caribou will move soft shallow snow with their noses, but as snow depths increase they dig craters of varying sizes with their front feet to expose vegetation. If the snow is crusted, caribou may break the crust with their front feet. The thresholds of depth and crust hardness—caused by wind, surface melting and refreezing, or freezing rain—that prevent cratering vary according to habitat. In the taiga, where snow tends to be deep and soft, caribou stop cratering when snow hardness exceeds 50 g/cm² (Pruitt 1959). On the tundra, however, caribou continued to crater until snow hardness reached 6,500–9,000 g/cm² in Alaska (Henshaw 1968*b*; Thing 1977) and 10,000 g/cm² in the High Arctic (Thomas et al. 1977). Critical snow depths for cratering by caribou vary from 60 cm (Pruitt 1959) to 75 cm (Thing 1977).

BEHAVIOR

Behavioral messages can be passed by visual, auditory, and olfactory modes either singly or in combination. They are manifested as acts of aggression, displays of dominance, vocalizations, skeletal sounds, and secretions of pheromones. Supposedly, all these messages have the meaning of either "come closer" (affin type) or "move away" (diffug type) (Lent 1974).

The antlers of caribou deserve special consideration in descriptions of their behavior, as they have high social significance. Behavioral posturing or gestures in which antlers have a role have been described by Dugmore (1913), Pruitt (1960*a*), Espmark (1964*b*), Lent (1965*a*, 1965*b*), Henshaw (1968*a*), Bergerud (1973, 1974*c*) and Bubenik (1975*b*).

Pruitt (1966) believed that the well-developed brow tine (shovel) protects the eyes during bouts of "bush-thrashing" (stereotyped movements of the head, swinging antlers back and forth through bush-type vegetation). Bubenik (1975*b*) believed that the brow tine is used as an offensive weapon and that the second (bez) tine protects the eyes and facial region.

Henshaw (1968*a*) believed that retention of antlers by pregnant females enhances their social ranking and, therefore, their ability to compete for restricted forage supplies, thus contributing to their survival and that of the fetus. Although the occurrence of antlers on females is acceptably explained by Henshaw, one might ask (1) why are one- to three-year-old males that often retain their antlers throughout most or all of the winter permitted socially to remain in close association with wintering groups of pregnant females and their young? and (2) why do many females in populations of caribou, especially in Newfoundland, not possess antlers (Bergerud 1971*a*)? Henshaw (1968*a*) also believed that possession of antlers by calves allowed them to exhibit a marked degree of

precocity and independence from maternal care during winter. Although such behavior by calves is likely necessary, as pregnant maternal cows offer little or no care, their small spike antlers are probably of little significance in encounters. Cows are often even antagonistic toward their previous year's offspring during gestation, especially in winter if forage becomes restricted. Female caribou without antlers often attack by kicking with their forelegs.

Vocalization and Other Sounds. Caribou of both sexes and all ages produce many sounds, especially when in large postcalving aggregations and during the rut (Murie 1935; Banfield 1954*a*; de Vos 1960; Pruitt 1960*a*; Espmark 1964*a*; Lent 1965*b*, 1966*b*, 1975; Kelsall 1968; Bergerud 1973, 1974*c*; and Erickson 1975). Cows give coughlike grunts and calves have bleating cries; rutting males grunt, snort, slurp, cough, sneeze, and pant. When caribou move, their hooves make a unique clicking sound that is caused by the sesamoid bones slipping over each other. This noise is particularly noticeable when a large number of caribou are on the move and, in calm weather, at a distance over 100 m, even when mixed with assorted vocalizations.

Locomotion. Caribou have several gaits, including a walk, fast walk, trot, pace, and gallop. They walk while feeding and moving between feeding areas. Although walking during maintenance activities seems leisurely, it has been timed at 7 km per hour over long distances and on rough terrain (Pruitt 1960*b*). Fast walks often occur during migration and periods of initial alertness. Trotting often follows an alarm or stresses such as predators or insect harassment. If the alarm or stress situation continues and/or increases, caribou may pace or, more commonly, gallop.

Caribou are adept climbers and will ascend cliffs and traverse glacial snowfields. They climb sheer walls of hard-packed snow by digging steps with their front hooves as they make their way up and over the barriers. Surprisingly, they often make those steep climbs when there are pathways available that offer little or no resistance.

Caribou are strong swimmers and readily cross swift rivers and swim lakes during their travels, again even when not apparently necessary. Their broad hooves and dew claws act as efficient paddles. They swim with their heads held high out of the water at speeds ranging from 3 to 11 km/h (Seton 1927; Banfield 1954*a*; Kelsall 1968).

Reproductive Behavior. PRERUT. In September and early October, the bulls strip the velvet from their antlers, become irritable, and engage in sparring bouts with other similar-sized males to establish social rank. Bulls that appear dominant seldom actually fight subordinates, but effectively use their antlers and posturing to reinforce their status (Bubenik 1975*a*). Other common breeding displays seen during this period of sorting out socially are: bush gazing (stand motionless, with a fixed distant gaze) (Lent 1965*b*), bush thrashing, hock rubbing or tramping (take up a hunched position and in most cases urinate on the hocks) (Espmark 1964a; Lent 1965b), mock battles, rearing, and flailing.

RUT. The intensities and kinds of behavior exhibited during the rut vary according to subspecies, habitat, and possibly the nutritional state of the breeders. Open habitat appears to be preferred, if available. Barren-ground bulls do not gather harems (Pruitt 1960a; Lent 1965b; Skoog 1968). Dugmore (1913) suggested that male woodland caribou gather harems; however, Bergerud (1973, 1974c) stated that such groupings are not true harems because members of such groups are often transient.

The preliminary phase of the rut involves testing the estrous state of females; males drive females, who flee before them. A modified form of threat (figures 47.3 and 47.4) is often exhibited by the male as it drives a female (Pruitt 1960a; Lent 1965b). Pursuits of females by males are often interrupted by fights with other males. In contrast to sparring during prerut, fights are often vigorous encounters (Lent 1965b). Banfield (1954a) reported mortality from rutting fights and Bergerud (1971c) attributed high mortality among prime bulls to fighting during the rut. Males also stop pursuing the females to bush-thrash and bush-gaze.

As the rut peaks, prime bulls concentrate on tending estrous females and markedly reduce their foraging. The bull tends only one cow at a time and follows it wherever it goes (Lent 1965b). Tending bulls usually show agonistic behavior only when another animal approaches the tended cow. Tending bulls tramp and bush-gaze.

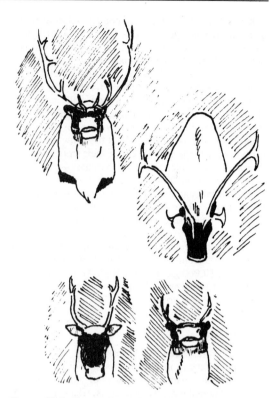

FIGURE 47.4. Top, outline and tonal pattern of normal (left) and threat (right) poses as presented by a cow to another caribou (*Rangifer tarandus*); bottom, outline and tonal pattern of threat (left) and attack (right) poses as presented by a bull to another caribou.

Copulation is rapid and has rarely been observed (Kelsall 1968). Bergerud (1974c) suggested that the much greater weight of the male requires that copulation be brief. Espmark (1964a) believed that for Swedish reindeer copulation occurred only once with each cow and mostly at dawn or dusk.

CALVING. On the calving grounds, the pregnant cows space themselves in smaller groups (maternity bands) (Pruitt 1960a). In Alaska, calving groups were open and transient, but showed both intragroup and intergroup social facilitation (Lent 1966a). Females that drop their calves early or late tend to separate from other caribou, probably because the groups remain on the move. Lent (1966a), Kelsall (1968), Skoog (1968), and Bergerud (1975) all suggested that, during peak calving time, cows about to give birth do not actively seek isolation. Pruitt (1960a) stated, however, that females giving birth during the peak of calving move off a short distance to calve and then rejoin their groups.

The caribou calf moves off with the mother within hours after birth. The maternal cow uses head bobbing (moves its head up and down on a vertical plane from the ground to about the height of the calf) (figure 47.5) and vocalizations to strengthen the calf's following response (Pruitt 1960a; Lent 1966a). After the calves are mobile, "nursery bands" composed almost en-

FIGURE 47.3. Top, threat pose, exhibited by a female caribou (*Rangifer tarandus*); bottom, modification of threat pose by male caribou in courtship display.

tirely of the calves and their maternal cows are formed (Pruitt 1960*a*).

MOTHER-YOUNG BOND. The mother licks and feeds its calf within the first minutes of the calf's life, initiating the mother-young bond. As the calf grows it solicits care from the mother and further develops that bond. Bonding and individual recognition involve visual, auditory, and olfactory stimuli. A strong mother-young bond is necessary for the survival of offspring during the first six months of life. In the absence of a strong mother-young bond, most newborn calves would die during the rapid and extensive post-calving movements of the large herds. The mother-young bond assures to a great extent that the maternal caribou will provide its young with (1) passive immunity through colostrum and milk, (2) nutrition, (3) thermoregulation by direct contact and indirectly by licking and drying, (4) assistance in traversing difficult terrain, (5) defense against predators, (6) optimum environment for rapid development and learning, and (7) behavior patterns that shield the infant from extremes of social and nonsocial stimuli (de Vos 1960; Pruitt 1960*a;* Lent 1966*a;* Kelsall 1968; Skoog 1968).

After the death of a mother or its young, the surviving member will show a retention of the bond, sometimes for several days (Banfield 1954*a;* Lent 1966*a;* Miller and Broughton 1973). Cows with calves do not readily accept other calves during the postpartum period (de Vos 1960; Pruitt 1960*a;* Lent 1966*a*). Pruitt (1960*a*) observed, however, that cows deprived of their calves will accept strange calves.

NURSING. During the first two days after a calf is born, it nurses about every 18 minutes, usually for periods of less than 1 minute. Calves over one week of age nurse only about one-third as frequently as younger calves (Lent 1966*a*).

The most common position for nursing is reverse parallel, which allows the mother to lick and nibble the calf's anogenital and back regions. Such care-giving behavior from the mother prolongs the nursing period (Lent 1966*a*). Calves also nurse from the rear, sometimes while the mother is walking.

Little is known of the actual weaning process in caribou. Kelsall (1968) stated that suckling was greatly reduced by early July. Biting insects would greatly disrupt nursing after July. He concluded, therefore, that weaning must occur at about that time. Although his conclusion is based on considerable observation, it is difficult to accept because of the high energy demands of the rapid growth of the calf during its first six months of life. Skoog (1968) suggested that calves are weaned between September and December, most likely before November. The calf probably associates with its mother during most of the first year of life, not only because the mother is a source of high nutritional energy but also because the mother-young bond fosters psychological well-being. Formation of "peer groups" by short-yearlings (calves of that year) contributes to the deterioration of the mother-young bond and facilitation of the calf's assimilation into its mother's social group.

Agonistic Behavior. THREAT AND ATTACK. Pruitt (1960*a*) described and illustrated threat and attack poses by caribou (figure 47.6). He suggested that threat posturing (muzzle extended, antlers back) is used by females to ward off strange calves or adults, by calfless females showing antagonism toward males during the calving period, and by males to challenge other males. Threat posturing also is part of courtship displays during the rut, and is used by dominant animals to defend or scare subordinates from feeding craters in winter. Attacks (head lowered, antlers presented) occur when the threat pose fails to intimidate or when the individuals are more highly motivated. Attacks commonly occur when two females compete over a single calf, when juvenile animals are testing their strength or playing, and when two or more males are sparring before or during the rut.

DOMINANCE-SUBORDINATION. A system of rank order probably exists involving all individuals within a social unit. Age, sex, size, and possession of antlers are all important criteria in determining rank order. Prime males are usually dominant over all other animals during the rut: older, larger bulls with bigger antlers are

FIGURE 47.5. Head-bobbing pose, exhibited by a maternal cow to its newborn calf caribou (*Rangifer tarandus*).

FIGURE 47.6. Attack pose, exhibited by a male caribou (*Rangifer tarandus*).

dominant over younger, smaller bulls with small antlers.

Subordinate animals are not allowed into the feeding craters of dominant animals in winter. Also, dominant caribou will displace subordinate animals from feeding craters. Maternal cows are usually dominant over other caribou in winter. Adult animals usually dominate all juveniles (18–22 months old) and yearlings (6–10 months old). As winter progresses, the social status of short-yearlings changes as the mother-young bond weakens.

Lent (1966a) argued that caribou recognize sex and age classes and that the dominance hierarchies are based on classes rather than on individuals. Parker (1972b) and Miller et al. (1975a), however, obtained evidence for long-term associations between caribou, suggesting that caribou do indeed recognize individuals. Bergerud (1973) suggested that the establishment of a hierarchy depends on frequent reinforcement through interactions, but it may be necessary to have intensive interactions only for relatively brief periods of the year.

ALARM. Pruitt (1960a) described and illustrated the unique alarm stance (one hind leg spread, head erect) and excitation jump (rearing, pivoting, interdigital glands discharge). Both acts are highly noticeable to other members of the group and, thus, warn animals that do not actually see the danger. The alarm stance occurs when a caribou is alerted to some unidentified stimulus. The excitation jump usually occurs after the caribou has investigated the foreign stimulus by sight or smell and has begun to flee. Low-intensity alarms include head low alerts (brief poses during feeding activities) and head high alerts (more intensive and prolonged poses made by moving, bedded, or feeding caribou).

Predator-Prey Behavior. The caribou has evolved with the wolf, and undoubtedly much of the caribou's behavior has developed as a result of that relationship. The gregariousness of caribou, especially after calving, probably resulted from wolf-caribou encounters (Bergerud 1971b, 1974d; Miller 1974c). The caribou's use of openings, frozen bogs, and lakes for resting areas is also a learned behavior that seemingly resulted from wolf-caribou interactions. The caribou's apparent reluctance to enter riparian willows and other heavy brush, and its state of alertness when passing through, suggest that it associates such cover with attack by wolves and bears. Caribou have learned to distinguish between hunting and nonhunting wolves. The development of specific flight releasers (Pruitt 1965) and threshold distances allows the caribou to conserve energy by not taking unnecessary escape during wolf-caribou encounters.

POPULATION DYNAMICS

Reproduction, mortality, and movements (emigrations) regulate the number of caribou within the population. Mass emigrations are not predictable and it is not known why they occur. Dispersal may be density related, that is, a threshold density triggers social intolerance and leads to dispersal of some segment of the population. It is difficult, however, to perceive how a density concept works, because the clumping behavior of caribou on winter feeding sites and postcalving aggregations suggests that caribou are tolerant of high densities.

Basically, a population increases when recruitment exceeds mortality during a time period; it decreases when the opposite occurs.

Natality. The fertility and birth rates in caribou are usually about the same from year to year (Skoog 1968; Parker 1972a; Bergerud 1974a; Dauphine 1976). However, in some years disease or poor nutrition can cause high intrauterine mortality in some herds (Skoog 1968; McGowen 1966; Neiland et al. 1968) and the birth rate drops relative to the fertility rate.

The number of females reaching sexual maturity governs the potential rates of natality within a population. Free-ranging caribou calves are seldom bred. It is unlikely that females bred as calves could produce and rear young successfully, because they are physiologically and, probably, psychologically unsuited for the task. Even the yearling females in many populations are not ready to be mothers. They usually lack the necessary fat reserves and they are still growing during the second winter of life. Also, young, primiparous mothers may lack the psychological adjustment for calving, and the subsequent mother-young bond either does not develop or is weak (Skoog 1968; Miller and Broughton 1974). Most females have to live at least three years before they successfully produce and rear their young.

Although the reproductive rate fluctuates only slightly, annual yearling increments to the population often fluctuate markedly. Most of the information on mortality in *Rangifer* has been obtained from studies of reindeer because of commercial interest and the relative ease of investigating semidomestic herds (Skoog 1968).

Forage. The ultimate limiting factor on caribou populations is theoretically their maximum seasonal forage supplies. Forage production and availability give a true approximation of the range's ability to support caribou.

Bergerud (1974a) stated that a winter shortage of forage for free-ranging caribou has not changed birth or death rates in any populations of caribou. However, data on those rates are difficult to obtain and the lack of evidence is not conclusive that such changes have not occurred. Forage shortages occur in winter due to adverse snow cover and icing conditions. Forage production as a limiting factor would be density dependent. It is unlikely that other mortality factors allow caribou numbers to reach such critically high population levels on most ranges that forage becomes the controlling factor.

Weather. Weather, alone or combined with mortality factors, could hold a caribou population at a low level where density-dependent factors are not effective. Ultimately, caribou are governed more by precipitation in the form of snow cover, icing, freezing rains, and cold rains than by all other environmental factors. To some degree, snow and ice determine where, how, and if caribou can live for most of the year. Precipitation that greatly restricts forage availability combined with low temperatures and high winds is most likely to result in high mortality. Prolonged and severe restrictions of forage supplies lead to malnutrition and subsequent death by starvation or predation. Although there is little evidence for starvation caused by adverse weather conditions in free-ranging caribou on the mainland, there is evidence for it among Peary caribou on the High Arctic islands (Miller et al. 1977*a*). I believe, as did Skoog (1968), that even if mortality is attributed to starvation caused indirectly by weather conditions, the regulating factor is the existing weather, not the absolute food supply.

MORTALITY

Calf Mortality. High calf mortality on calving grounds is common in caribou and is caused mostly by adverse weather and predation (Pruitt 1961; Kelsall 1968; Skoog 1968; Bergerud 1971*a*; Parker 1972*a*; Miller and Broughton 1974). Precipitation, cold, and wind are a deadly combination for newborn calves, often resulting in hypothermia (Banfield 1954*a*) or respiratory problems such as pneumonia (Miller and Broughton 1974), and subsequent death. Soft, deep snow traps calves and they subsequently die (Kelsall 1968). Weather also severely limits the availability of forage to parturient cows during gestation and indirectly causes calf mortality (Miller et al. 1977*a*). This condition is critically important to Peary caribou in the Canadian High Arctic (Thomas et al. 1976, 1977; Thomas and Broughton 1978).

Predation on some calving grounds can be severe and often exceeds the wolves' needs (Miller and Broughton 1974). Wolves, bears, lynxes, and eagles seek calves as easy prey and often have marked impacts on calf crops. Predation on newborn calves varies from population to population and within populations from year to year. It has been suggested that wolves kill mostly those young that would not survive anyway. Although wolves would readily take sick and lame calves, they must take healthy calves when there are few sick or lame animals available. An interesting supposition proposed by K. A. Neiland (Klein and White 1978) is that wolves perpetuate the occurrence of pathological agents in caribou populations by their interactions with caribou and, thus, assure themselves of an easily obtainable food source.

Most accidental deaths and some predation of calves result from the calf's investigative behavior, its lack of maternal care, or the agonistic or escape behaviors of other animals (intraspecific factors, Skoog 1968). Other deaths are due to drownings, trampling, injuries caused by hostile adults (foreleg kicks and/or antler punctures), falls from cliffs, and injuries such as sprained or broken limbs that prevent the calves from keeping up with their group's movements. Maternal cows usually remain with their injured calves, but such

infants usually fall easy prey to wolves, bears, and sometimes eagles.

Poor nutrition in mothers can result in the death of the mother before parturition, malpresentation, stillbirth, physiological disorders of various organ systems, physical abnormalities of the calf, or neonatal death through malnutrition of the fetus. Mothers can transmit fatal diseases to their young. The frequency of disease in calves is not known, but epidemic brucellosis has occurred in barren-ground caribou of the arctic herds in Alaska (Neiland et al. 1968).

Adult Mortality. Factors that predispose caribou to death by accident or predation are their investigative behavior, their nutritional state, adverse weather—precipitation, cold, and wind—predation by wolves and bears, and limited forage. Forage is limited possibly by production, but usually by availability truly the result of weather. Much of the adult mortality due to accidents and predation is the direct or indirect result of intraspecific behavioral and physiological characteristics. Investigative behavior can lead to drowning by falling through weak ice, falls from cliffs, and snow slides. Sprained or broken limbs from travel on dangerous terrain leads to subsequent death by predation. Agonistic and reproduction behaviors may result in encounters that lead to death by kicking or antler punctures, or cause injuries that lead to subsequent death by predation. Examples of mortality proximally caused by wolf predation but ultimately the result of intraspecific physiological characteristics include:

1. as bulls build up fat layers before the rut, they can run only with great difficulty and tire quickly, making them more vulnerable to predation
2. prime bulls are often very lean after the rut and they do not build up sufficient reserves to carry them through to spring; thus, they are more vulnerable to predation
3. near-term cows suffer from the large size of the fetus and high energy requirements in late winter; thus, they are more vulnerable to predation (Skoog 1968).

Environmental (interspecific, Skoog 1968) mortality factors should take the greatest toll on caribou numbers over the long term. There is a great deal of debate over what are the principal mortality factors operating in caribou populations. Sometimes the same biologists advanced different reasons for the various herds. Many early workers (Leopold and Darling 1953; Edwards 1954; Scotter 1964, 1967*a*) believed that range deterioration, caused mostly by fires and land clearing and logging, on southern ranges was the principal cause for the decline of caribou numbers throughout much of North America. Cringan (1957) and Banfield and Tener (1958) blamed much of the decline on overhunting or range deterioration. Banfield (1954*a*), Banfield and Tener (1958), Sonnenfield (1960), and Bergerud (1974*a*) all believed that predation by wolves or humans was the principal cause. Only one thing is certain: at some time or another all of these mortality factors, along with weather and disease, have had marked impacts on caribou numbers.

PREDATION. Caribou evolved with the wolf, and only in relatively recent times has the wolf been extirpated or seriously reduced in numbers on the southern and eastern ranges of caribou. Wolf densities remain high among most barren-ground caribou populations. They are possibly high relative to the low numbers of Peary caribou, and muskoxen are a buffer prey species in High Arctic regions. Moose (*Alces alces*), sheep (*Ovis*), and goats (*Oreamnos americanus*) are buffer prey species for wolves on most Alaskan and Yukon ranges. The lynx has replaced the wolf as a major predator of caribou calves in Newfoundland and is possibly important on other southern ranges. The grizzly bear is an important predator on calves, and occasionally adults, on some ranges in Alaska, the Yukon Territory, and the western District of Mackenzie in the Northwest Territories.

This chapter is concerned with the management and conservation of caribou as a renewable resource, not merely with caribou as an entity in arctic and subarctic natural systems. Biologists are interested in the sustained use of the resource by humans as well as the continued perpetuation of the species (which includes wolf predation on caribou). Therefore, the concept of dynamic equilibrium is not a functional basis on which to accept the wolf-caribou relationship.

I suggest that wolves should be a part of the caribou's environment. But wolves are large carnivores that require considerable amounts of meat and protein for their reproductive success and survival. A wolf requires 11–14 caribou annually (Clarke 1940; Kelsall 1968; Parker 1972b), if caribou are its major prey. Food habits of wolves on barren-ground caribou range in Canada have been reported on by Kuyt (1972).

Large numbers of wolves have a potentially large impact on a caribou population, based on estimates of a wolf's required annual kill of caribou and the fact that wolves sometimes kill more than they need, especially of caribou calves. Less than 1 wolf per 100 caribou could control the growth of a barren-ground caribou population (Parker 1972a). Skoog (1968) stated that predation was the greatest single mortality factor in most wild populations of *Rangifer*.

Although Crisler (1956) and Kelsall (1968) stated that caribou, except the incapacitated and calves, could normally outrun single wolves, Murie (1944), Banfield (1954a), Skoog (1968), and other field workers observed that wolves are capable of overtaking healthy caribou. Such feats are not always necessary, as wolves often take prey by surprise or ambush. When in pursuit, wolves also take advantage of any tactical mistakes the caribou may make, for example, leaving hard snow and attempting to cross soft snow areas. The most interesting aspect of wolf hunting tactics is their ability to run relays and to involve several members of the pack in herding and ambushing prey. Skoog (1968) suggested that wolves can obtain their prey as needed from whatever caribou are available, and that the final selection is made more by chance than by design. Sick and lame animals may be selected, if available, but if the caribou population is healthy and there are few

incapacitated individuals, the wolves will take healthy animals. Wolves can and sometimes do control and even depress caribou populations.

PATHOLOGY. The importance of disease and parasitism upon caribou and as a control of caribou numbers is not yet well known (Kelsall 1968; Skoog 1968). A detailed bibliography of parasites, diseases, and disorders of *Rangifer* and other ungulates was compiled by Neiland and Dukeminier (1972). Disease and parasite burdens in caribou are endemic and of normal incidence in most populations. Sporadic or possibly periodic outbreaks of disease of epidemic proportions probably occur in some caribou populations. The only epidemic disease in a free-ranging population of caribou occurred in Alaska (Neiland et al. 1968; Skoog 1968). Brucella organisms were associated with orchitis-epididymitis, bursitis-synovitis, and metritis, singly or in combination in Alaskan caribou (Neiland et al. 1968). Examination of barren-ground caribou in Canada led to the conclusion that brucellosis was not a serious threat in caribou (Broughton et al. 1970).

Another important disease in Alaskan caribou is necrobacillosis (Skoog 1968). The causative agent (*Spherophorus necrophorus*) usually enters through lesions in the feet or mouth, and results in hoof rot or necrotic stomatitis. Banfield (1954a) considered that actinomycosis (lumpy jaw) was widespread in Canadian caribou. More recent work by Miller et al. (1975b) and by Doerr and Dieterich (1979) has suggested, however, that most, if not all, mandibular deformities are not of actinomycotic origin. Fibropapillomas have been found in Alaskan (Skoog 1968) and Canadian caribou (Broughton et al. 1972).

The caribou is host to relatively few species of external parasite, but the species that do parasitize them are common and cause considerable harassment during summer. Harassment by bloodsucking and parasitic flies can lead to injuries when the caribou react by flight, and subsequently to death by predation. Severe attacks by mosquitoes (*Culicidae*) and black flies (*Simuliidae*) have reportedly resulted in the deaths of caribou (Skoog 1968).

The warble fly (*Oedamagena tarandi*) and the nose bot (*Cephenomyia trompe*) parasitize caribou (Banfield 1954a; Skoog 1968; Kelsall 1975), but they are more important in influencing the ecology of caribou than as pathologic agents (see "Ecology" section). The effects of the warble flies on caribou are unknown, but they likely cause general debilitation of the host. Heavy infestations of nose bot larvae severely distress caribou and may weaken infected animals.

Internal parasites include various cestodes, some nematodes, and one trematode (Banfield 1954a; Skoog 1968). The common forms are *Taenia hydatigena*, found mostly in the liver; *T. krabbei*, mostly in the muscle; and *Echinococcus granulosus*, mostly in lungs and occasionally in the liver. Again, the importance of these internal parasites is not known but they have general debilitating effects on the host and could predispose the host to predation. The thread lungworm (*Dictyocaulus hadweni*) is associated with debilitated caribou and supposedly causes their death. Skoog (1968) suggested that this parasite might cause heavy mortality to nutritionally stressed caribou, but there is no evidence for it. Besnoitiosis has been found in barren-ground caribou in Canada (Choquette et al. 1967) and Q fever in Alaskan caribou (Hopla 1975).

ACCIDENTS. Although accidents usually contribute little to the overall mortality within a caribou population, on some occasions accidents can result in a substantial number of deaths. Animals may drown while crossing treacherous stretches of river, by walking onto weak ice, or by becoming trapped in the water by high ice shelves (Kelsall 1968 and others). Many accidents occur when caribou are in large migrating aggregations and traveling over unfamiliar terrain. They often result in injuries that increase vulnerability to predation.

PHYSIOLOGICAL. Physiological disorders cause very few deaths among adult caribou. Most such deaths occur among females experiencing complications in fetal delivery. Additional mortality can result from organ failures brought on by stress in caribou of either sex or any age.

SEX AND AGE DETERMINATION

To evaluate fully the status of any group of caribou, the sex and age of individuals must be determined by aerial and ground surveys (Davis et al. 1978*b*). Segregations based on physical appearances of free-ranging caribou by sex and age at different seasons of the year involve a high degree of subjectivity.

The observer must note: (1) the presence or absence of antlers; (2) the relative size and development of antlers; (3) the age of antlers—new, when in velvet, and old, when hard and polished; (4) the relative body size of individuals; (5) the state of pelage condition and molt, presence of white mane on adult males in autumn; (6) the shape of the face, forehead to nose line, roman nosed or straight faced; (7) the presence and development of the udder; (8) the presence or absence of the vulva; (9) the presence of a penis sheath; and (10) indirectly, the probable sex and age classes of companion animals (other group members). Surveys are made during spring migrations, precalving, calving, postcalving, midsummer migration, prerut, rut, postrut, fall migration, and winter. Aerial surveys of and ground counts on small land units might give critically misleading ratios of sex and age classes in terms of the entire herd. The problem of nonrandom distribution of caribou by sex and age can be overcome through (1) repeated aerial surveys of large land units and (2) ground counts of high proportions of the herds during the rut, when the sex and age classes are the most mixed.

Currently, the most accurate method for determining the age of a caribou involves microscopic examination of histologically prepared sections of mandibular teeth (McEwan 1963; Miller 1974*a*, 1974*c*). Maxillary teeth could be used but management techniques are usually restricted to the collection of only the dentary bone.

Cementum annuli are counted but results can be validated only by examining teeth of known-aged caribou; this has not been done as yet. More data must be obtained on the possibility of individual variation in apposition of dental cementum and controlling internal and external factors of cemental growth. Many tooth sections cannot be read without some degree of subjectivity and annulus counts may not be exact. Most age assignments should be thought of as the age, plus or minus one year. The cost, time, and elaborate equipment required to determine the age of caribou by histological techniques make routine use of those procedures on large samples, such as annual hunter kills, questionable.

A second technique—the visual, gross examination of the state of eruption and attrition of the mandibular tooth row (Chatelain 1954; Watson and Keough 1954; Bergerud 1970; Miller 1972, 1974*c*)—is all that is required for most management procedures. Visual examination can be supplemented with linear tooth measurements to help improve the accuracy of the age determination (Miller and McClure 1973; Miller 1974*c*).

The permanent mandibular teeth erupt during the first 3 years of the caribou's life. Most caribou have a full set of mandibular teeth by the 29th month of life (see Miller 1974*c*). Differentiation between 2 and 3 year olds can be made in the field on the basis of lack of wear on the posterior cusp of the third molar. Also, open apical areas occur on the roots of the fourth premolar and third molar at 2 years (versus wear and closure at 3 years) and can be checked with ease in the field on dried mandibles and in the laboratory on fresh mandibles.

Because the ability to evaluate wear patterns varies among investigators, it would be more realistic to assign caribou to age classes rather than specific years. First the caribou's age should be estimated to the closest year (1–10+ years), then the individuals placed in an age class: calf, 1, 2, 3, 4–5, 6–9, and 10+ years. It is most difficult to differentiate between 3 and 4 year olds; however, I suggest that it should be done to minimize the chance of lumping some 2 year olds in a 3 to 5 year olds class. In Alaska, Chatelain (1954) developed age class groupings of 1, 2, 3, 4–6, 7–9, and 10+ years; Skoog (1968) also produced the same age classes but included a calf category.

Misleading estimates of age based on visual examination of mandibular teeth could result from (1) malocclusion from misalignment of the maxillary and mandibular tooth rows, (2) variations in the curvatures of the rami, (3) and variations in primary and secondary axes of the mandibular blades. Some caribou would show extreme variations in attrition because of missing teeth and/or abnormal alignment of the tooth rows, caused by orientation of the teeth in their sockets (Miller 1974*c*).

Sex can be determined between older (4–5+ years) bulls and similar aged cows and all younger animals of either sex by comparing mandible lengths

(Bergerud 1964b; Miller and McClure 1973; Miller 1974c). The mandibular blade is measured from the posterior rim of the ramus to the anteriormost position of the alveolar bone below the first incisor. A series of such measurements must be established for each herd studied. Ideally, those measurements would be obtained from jaws collected from animals of known sex and dates of death and known age as determined by annuli counts in the dental cementum.

ECONOMIC STATUS

In the past, caribou were all-important to arctic and subarctic native cultures: caribou meat was the staple diet for both humans and dogs; caribou hide and sinew provided the materials for clothing, tents, sleeping bags, etc.; and hides, sinews, antlers, and bone were used for many tools and weapons. Caribou, especially the migratory herds, were the key to the very existence of Inuit and Indian cultures on the central tundra and adjacent taiga (Kelsall 1968:206–216, 278–281). Indians in the northern boreal forests might have survived without the caribou, but to do so they would have had to alter their cultures markedly. An abundance of caribou meant good times and a shortage meant hardships and sometimes slow death from malnutrition and starvation.

Caribou remain to this day a staple in the diets of many northern people and are still valued as an integral part of their chosen (traditional) ways of life. Until those native people choose an alternative life style that does not have a great cultural dependency on caribou, it will be difficult to measure the replacement value of caribou. Current evaluations of the worth of caribou must include moralistic and esthetical considerations. The caribou's potential to reproduce and in turn the potential of the offspring to reproduce and expand in numbers make today's value of caribou only a mere fraction of the projected value 100 or more years from now, and at best a poorly based guess.

Another nebulous costing exercise is to estimate the esthetical value of caribou. The price would be appreciable, even at today's rates. But is this market-price approach to the value of caribou a legitimate consideration beyond the point of sustained annual yields and/or recreational pursuits? The caribou is a natural entity in arctic and subarctic environs. Would those regions not lose their very character with the passing of caribou from their landscapes?

MANAGEMENT AND RESEARCH

(Much of the following text on management and research is liberally awash with my personal opinions and beliefs; the material does not reflect the opinions, beliefs, or policies of the Canadian Wildlife Service, the Department of the Environment, or the government of Canada.)

Most attempts at managing big game species in North America on a biological basis have been plagued by overriding sociopolitical considerations. No other ungulate species has suffered as much from sociopolit-

ical pressures in the face of opposing biological concerns as the migratory caribou of North America.

It is impossible truly to manage caribou until there is control of the harvests. If we cannot regulate the harvests of caribou on a biological basis, we cannot hope to manage the resource. The management of caribou in North America is unique in that, in addition to being seasonally sought after by meat, trophy, and white subsistence hunters, caribou remain a staple in the diets of many northern natives who have built their cultures around the species. The native harvest of caribou in Canada is almost unrestricted, but in Alaska there is some regulation (Davis et al. 1978a). The freedom to harvest an unrestricted number of caribou has proven detrimental to caribou populations since the introduction of the rifle (e.g., Banfield 1957; Banfield and Tener 1958; Kelsall 1968). Most native populations are growing and are now concentrated in settlements from which organized hunts by snow machines or aircraft are increasing the kill. The shift in most native communities to a partially wage-based economy has increased the native ownership of rifles, snow machines, and even aircraft to be used in hunting. In particular, aircraft hunting leads to excessive harvests of caribou because of the range that can be covered and the ease and speed with which carcasses can be returned to the settlement.

Native people often state their desire to continue their traditional ways of life, and few people would deny them that wish. Unfortunately, however, the natives' tradition now includes the use of power boats to replace kayaks, snow machines and aircraft instead of dog teams, and high-powered rifles with telescopic scopes instead of bows and spears. Caribou populations cannot withstand unregulated harvests employing such modern equipment and logistics.

Native hunters are not convinced that they are misusing the resource, and instead have blamed industrial exploration activities and the activities of caribou biologists for declining numbers and changes in the distribution of the herds. Current sociopolitical stands by native groups over aboriginal rights and land claims require public maintenance of this conviction. In Canada, the problems of overharvest are unlikely to be resolved in the near future. At present, the only acceptable approach is to attempt to educate native peoples on the ability of caribou populations to maintain their numbers at various levels of survival and mortality. Natives must understand that caribou are a future resource that can be maintained only if replacement equals or exceeds losses. The dilemma is whether the migratory herds of caribou can survive until the natives accept the truth!

A second, somewhat unique consideration in the management of migratory caribou is that many migrations take caribou across national and international boundaries. For example, (1) barren-ground caribou of the Porcupine herd (100,000) move among Alaska, Yukon, and Northwest Territories; (2) barren-ground caribou of the Beverly (94,000) and Kaminuriak (38,000) herds move among the Northwest Territories, Manitoba, and Saskatchewan and in some years northeastern Alberta; and (3) woodland caribou on the George River herd (185,000) move between Quebec and Labrador. Remnants of the Selkirk Mountains herd (20–30) still move among southern British Columbia, northern Idaho, and northwestern Washington. Those transboundary movements necessitate agreement and cooperation at all levels of government and between the various wildlife and other resource agencies and the people concerned with the welfare of caribou, their habitat, and the people that use those resources. Well-balanced programs are difficult, if not impossible, to achieve when confronted with a multitude of social, economic, and political concerns among the different jurisdictions.

Management Regions. One of the first problems faced by managers of caribou is to determine what unit of caribou of what range used by caribou represents a manageable entity. Fortunately, because of their gregarious nature and general affinities for migration paths and calving grounds, most groupings of woodland caribou and barren-ground caribou are discrete enough to be recognized by managers as herds or populations. Unfortunately, caribou biologists are not always in agreement as to what constitutes a herd or a population. Although definitions do exist, they are not explicit enough, without considerable qualification, to justify a strong stand on the matter.

In Alaska, biologists have believed that movements among their caribou units are frequent and great enough to classify them as herds, and they think of all caribou in Alaska as a population. In Canada, biologists working on barren-ground caribou first thought of them as herds; then after learning more, they thought of them as populations (infrequent and negligible movements among units). Now some biologists in Canada have reverted to the herd category, based on no apparent biological reasoning, but just a matter of preference in definitions or literary style.

The distributions of caribou can be considered on a geographical rather than a subspecific basis, although distributions of subspecies do, for the most part, parallel geographical distributions. I suggest three management regions for caribou.

First are the arctic islands of the Canadian archipelago, where a rigorous polar desert environment with unique climatic conditions periodically reduces survival of Peary caribou. Those stringent climatic factors maintain numbers of Peary caribou below theoretical carrying capacities by causing considerable variation in forage availability from year to year. Restrictive snow and ice conditions make estimates of carrying capacities based on standing crops and annual productions of little or no value. It is likely that the key to springtime survival for Peary caribou is the amount of bare, windblown beach ridges and slopes that is available to them, though the vegetation on such sites is relatively sparse.

Peary caribou have been thought of as island populations. But recently, Miller et al. (1977a, 1977b) and Miller and Gunn (1978) have given evidence for extensive interisland movements (migrations) and have suggested that Peary caribou have actually established interisland populations.

Second, the mainland tundra and taiga sections of the open boreal forests of Alaska and Canada form a region for barren-ground caribou and migratory populations of woodland caribou. Environmental and social pressures are common throughout that vast area of North America, and managers seek answers to basically the same questions whether the caribou are in Alaska or Canada. Migratory caribou form large aggregations that create such special range problems as trampling or possible local overgrazing or continued access to ranges.

Third is the management region for the more sedentary woodland caribou of the boreal forests. The geographical distributions of woodland caribou, their differences in foraging habits, their socialization, and their year-round habitation of forests set them apart from populations of barren-ground and woodland caribou which inhabit the tundra during all or part of the year. The general lack of large social aggregations and a greater dependency on arboreal lichens for winter forage are characteristic of forest woodland caribou.

I suggest that this subdivision into three management regions is a desirable preliminary step for managers of caribou in their thinking and planning of proper management procedures and goals. Most management problems within each of those areas would be similar enough to warrant exchange of information and ideas by all North American managers of caribou populations. Biologists must not lose sight, however, of the important differences between each area, such as: (1) seasonal quality of forage; (2) seasonal availability of forage; (3) levels of harassing insects; (4) importance of climate; (5) variation in factors causing or potentially causing range deterioration; (6) kinds and numbers of animal predators; (7) levels of hunting pressures; and (8) amount of industrial and resource development.

Land areas within states, provinces, and territories are most often divided into wildlife management regions, zones, areas, or units with little or no ecological consideration given to the divisions. Usually such jurisdictions exist mainly for the supposed expedition of administrative and managerial concerns. Therefore, while the bookkeeping of renewable resources may be based on such units of land, about the only management policies that may reflect biological considerations are the seasons and bag limits. In reality, the biological reason(s) for established seasons and bags may extend well beyond the boundaries of a jurisdictional unit but is usually confined to the unit for ease of application and enforcement. Other biological considerations do not necessarily relate to the jurisdictional boundaries but are actually tied to different ranges and management regions as I have previously defined them.

Management Practices. Two important conditions are necessary for the proper management of caribou: (1) the authority (ability) to regulate the harvests of all users, and (2) a biological basis upon which to set the regulations. For the most part, neither has been obtained; until the harvest of caribou by native peoples is regulated, there can be no true management of caribou as a renewable resource on a sustained yield basis. This belief was strongly put forth for barren-ground caribou in Canada by Banfield (1957) and Kelsall (1968)—and apparently fell on deaf ears, as no concerted actions have been taken during the last 20 years. However, we must continue to work at obtaining a better biological basis from which managers can draw, when and if they are given the opportunity to manage.

In Canada, education of natives to the realities of current caribou utilization will be a long-term process. It is unlikely, if the educational process is successful, that it will be accomplished without some caribou populations first being almost extinguished. Even a catastrophic loss of caribou would not necessarily convince the natives of the need for regulation of harvests, for because of their beliefs, they most likely would not recognize or accept the cause of the loss. Therefore, educational programs, out of necessity, may become the major concern of caribou managers in the 1980s.

Management of caribou, like that of other game species, has generally followed a five-step sequence (Kelsall 1968): (1) regulation of hunting; (2) instituting predator control; (3) giving special status to land areas for wildlife; (4) transplanting and reintroductions; and (5) placing controls on the environment.

REGULATION OF HUNTING. This step has been only partially implemented, as native peoples are not subject to such regulations, at least in Canada. The situation appears better in Alaska, but it is still far from being totally satisfactory. Kelsall (1968) noted that most caribou hunters traditionally, and in disregard for the law, have habitually killed caribou in excess and often wasted their kills. Enforcement of existing regulations on caribou hunting has been, with few exceptions, inadequate or nonexistent.

In Alaska prior to 1925 little control was exerted on the killing of caribou. Passage of the Alaska Game Law in 1925 created the Alaska Game Commission. Subsequent regulations were basically restrictive in permitting human harvest of caribou until 1959. Statehood in 1959 brought in largely defensible relaxation of regulations in view of an apparent increase in the caribou population from 1947 to 1960. The liberal seasons and bags remained in effect until it was too late for the Fortymile caribou herd in 1973, the Nelchina herd in 1972, and the western arctic herd in 1976–77. That is, human harvests exceeded the annual increments in those herds, either alone or in conjunction with wolf predation (J. L. Davis personal communication). Harvests were monitored in some areas south of the Yukon River as early as 1963 by mandatory harvest report cards, hunter check stations, and village harvest logs. The same systems of monitoring harvests are still in effect, with the addition of some permit hunts. After 1975 harvest permits or harvest reports were also required north of the Yukon River, except by residents of management units numbers 25 and 26.

In Canada some control of hunting on woodland caribou has been in effect throughout ranges during the mid-1900s. Most provinces and territories closed or drastically shortened their seasons and reduced their bag limits in or after the 1950s. Most of those restrictions were only partially effective, at best, as they actually restricted only the nonresident hunters. Local residents, Metis and natives, continued their traditional hunting practices in disregard of the regulations.

Peary caribou have been relatively free of hunting pressures until recently, solely because of their remoteness from human population centers, except for two Inuk settlements. Residents other than Inuk in the Canadian High Arctic are subject to seasons and bag limits on Peary caribou. But in reality there is virtually no enforcement. The exception would probably be in the two settlements, mainly because everyone's awareness of each other's activities would dampen such illegal taking of caribou.

Barren-ground caribou in Canada were hunted without constraint until the 1950s. Restrictive measures in the form of seasons and bag limits have affected mainly nonresident hunters and resident nonnative hunters, the latter to a lesser degree. Enforcement of the restrictions in a manner that would control residents, especially Metis, actually is lacking, with few exceptions. Treaty Indians in Canada remain above all laws and regulations pertaining to seasons and bags for caribou! Although the Inuk do not have a treaty, they are accorded the same rights where hunting is concerned.

In the 1960s, as a result of the caribou crisis, the barren-ground caribou was designated an endangered species under the Northwest Territories Act. Caribou hunting was restricted by law to only those people that depended on them for subsistence. Regulations were passed against the wasting and abandoning of caribou and against the feeding of caribou to dogs. Wildlife officers tried to persuade natives not to kill cows and calves. The Canadian Department of Indian and Northern Affairs during the 1960s encouraged natives to reduce their take of caribou by providing them with fish nets and with commercial meats.

In the 1970s much of the good was undone by the Department of Indian and Northern Affairs and the government of the Northwest Territories as they shifted their emphases to participation in utilization of caribou through organized hunts. Although the barren-ground caribou remained on the endangered species list, its meat was legally offered for sale in the Northwest Territories. Caribou management programs in Canada through the 1960s and the 1970s, for the most part, lacked direction and common goals. The attitude that wildlife had to pay for itself became prevalent in many agencies, and the caribou populations have borne the brunt of this approach supposedly to justify their guardianship. I believe that this poor-man farming attitude is detrimental to the well-being of the resource and is essentially a violation of the caribou manager's responsibility to his charge. The manager's role is twofold: essentially he should manage for (1) annual sustained yields of the resource for those that need it and for those that desire to use it, including maintenance of large herds for nonconsumptive users; and (2) conservation and, when necessary, preservation of the resource. The common goal should be the maximum use and enjoyment of the resource, not rural enterprise!

One very important factor in the setting of harvest constraints is control of the land over which the caribou range. Again, Alaska has also had the advantage in this matter. In the past, U.S. Park Service lands were the only areas over which the Alaska Department of Fish and Game did not have wildlife management jurisdiction. Now, however, with the D2, BLM Organic Act, Endangered Species Act, etc., the matter of who has jurisdiction over wildlife management is in question in Alaska. In Canada most ranges of barren-ground caribou are controlled not by wildlife agencies but by the Department of Indian and Northern Affairs. That the best interest of the caribou is being served is questionable, because the department is also responsible for the welfare of the native peoples of Canada and for control of resource development of the North. As a consequence, neither the N.W.T. government nor the Department of the Environment can act effectively to curtail native hunting without full support from the Department of Indian and Northern Affairs—which support has not been offered.

Basic to all caribou management plans is control of the resource. Without such control, any management program will fall short of its ultimate desired goal: rational, sustained use of the resource and its conservation. Also of prime importance in the management of caribou is the protection of the large herds and their migratory habits. Unfortunately, procedures for obtaining these goals extend far beyond biological considerations. Social and political pressures at all levels are acting against a successful outcome of the problem of control of the resource.

Somehow, the natives must be convinced that game management agencies are acting on their behalf, that suggested courses of action to manage caribou are necessary, so that their great grandchildren will at least have caribou around them and be able to hunt them, even if they probably will not be able to subsist on them on a year-round basis. At the same time, caribou must be conserved as a natural entity in their environment. This will be by no means an easy task in view of the current resource development on some of their ranges.

PREDATOR CONTROL. Predator control in the past has essentially meant control of wolves (the currently favored expression is "maintenance of the predator population"). Wolf control on caribou ranges has followed the usual sequence of bounty payments, systematic government control through employment of predator control officers, the dropping of bounty payments (Pimlott 1961; Kelsall 1968), and then no control. Bounty payments are at best a form of rural welfare; they have never proven effective as a means by which predators can be controlled. It is also likely that no good professional bounty hunter ever put himself out of business!

In Alaska, the U.S. Predator and Rodent Control Program began an active campaign of wolf control on Alaskan caribou ranges in 1947. The predator control work (i.e., wolf control) included poisoning, shooting from airplanes, and the payment of bounties. It was continued throughout most caribou ranges until 1960. When Alaska gained statehood in 1959 there was a dramatic shift in management emphasis. In general, wolf control ceased, but some taking of wolves was still allowed by issue of "aerial wolf permits." Aerial hunting of wolves was curtailed north of the Brooks Range in 1970 and statewide in 1972. Localized, limited, and closely regulated aerial wolf control has been pursued since 1976 to help certain depressed moose and caribou populations to increase.

In Canada, government control of wolves began in the early 1950s, when the decline of barren-ground caribou appeared to be catastrophic. In western Canada a serious and widespread rabies epidemic gave it a big boost in 1957 (see Pimlott 1961 and Kelsall 1968 for more details). Poisoned baits (alkaloidal strychnine and sometimes the poison 1080) were used almost exclusively in the provinces and the Northwest Territories from 1951 to the mid-1960s. The Canadian Wildlife Service ran the control program in the Northwest Territories from 1956 to 1959, then it was taken over by the Northwest Territories Northern Administration Branch. The main thrust of the control program took place in Manitoba and the Northwest Territories from 1955 to 1960. The wolf control program was considered successful in giving the barren-ground caribou a reprieve (Pimlott 1961 and Kelsall 1968). The degree of success of the program, however, could not be measured quantitatively: there was a known take of 6,890 wolves in the Northwest Territories alone between 1952 and 1961.

Wolf control on caribou ranges in Canada during the late 1960s and throughout the 1970s has been unsystematic, local, and sporadic. Wolf bounties have been reinstated, dropped, and reinstated solely as a result of sociopolitical pressures, with no real concern for the biological implications of such actions. The sustained annual kill of wolves by Inuk hunters in the Northwest Territories gives evidence that the bounty system is not an effective way of reducing populations of wolves. As stated by Pimlott (1961), the bounty system does not actually control wolves, it does not concentrate effort where it is really needed, and animals are killed in vast areas where no killing is justified.

I believe that a well thought out and properly executed wolf maintenance program is a justifiable management tool, when annual net losses of a caribou population constantly exceed annual recruitment. Such a program would have long-term benefits for both the caribou and the wolves. Pimlott (1961:150) summed up the matter nicely when he said, "as long as control of tundra wolves is judicious it will have little or no effect on their status decades hence. Their numbers are so closely related to those of the barren-ground caribou that their ultimate fate hinges more on the caribou than on control. If the caribou survive and continue their migrations, the tundra wolves will survive. For this reason, intensive, short-term control of tundra wolves can conceivably benefit the wolves themselves.'' This same view was developed and held by the members of the Alaska Conservation Society (Weeden 1976) when they called for a carefully executed wolf reduction program on the range of the western arctic caribou herd by the Alaska Department of Fish and Game in 1975.

From a wildlife management standpoint it is unfortunate that wolf control as a management tool has become such a misunderstood and distasteful exercise to usually well-meaning but often misinformed and misguided individuals. Not only the lay public but also many wildlife biologists have apparently lost their objectivity in the matter. Emotions run high, from moralistic stands to rhetoric taught by biologists about how wolves kill only sick, old, and some young prey. There is no quicker way to cause a look of shocked disbelief than to tell a defender of the wilderness that wolves quite often kill healthy adult caribou. Disbelief will likely turn to horror if you add that the killing by those wolves often appears unnecessary, as the wolves sometimes feed little or not at all on the kill.

It will be a long road back for both layman and biologist. But I think that the sooner we view the matters of predation and controls in a realistic manner, the sooner we will realize that both wolves and caribou benefit from the maintenance of wolf numbers. This consideration must, of course, be qualified to point out that wolf maintenance is only desirable when and where caribou populations are in steady decline caused or accelerated by wolf predation or hunting by humans. Of course, constraints on the harvests by humans in such situations are also a necessary and valid management procedure. In reality, there are only two factors in the ecology of caribou that we can actually control when given the proper authority—wolves and humans (hunters, nonconsumptive recreationalists, developers, and exploiters).

LAND AREAS WITH SPECIAL STATUS FOR WILDLIFE. Caribou have benefited, at least indirectly, from the establishment of special wildlife management areas, sanctuaries, and preserves. One of the best examples is the Arctic National Wildlife Range: 3.6 million hectares in northeastern Alaska, set aside in 1960. Calving grounds and summer ranges of the Porcupine caribou herd occur within the Arctic National Wildlife Range. But Prudhoe Bay lies only about 100 km to the west of the refuge, and oilmen believe they can smell oil underneath it. So the future of the Arctic National Wildlife Range as a wildlife refuge is doubtful in the face of strong pressures by the petroleum industry to explore and develop any reserves within the refuge.

In 1926 the Arctic Islands Game Reserve was created. It covered all of the Peary caribou ranges on the arctic islands of Canada and some barren-ground caribou ranges on a large section of central mainland tundra in the Northwest Territories. Natives could hunt within it but no others. However, it was rescinded in 1966.

The Thelon Game Sanctuary was established in 1927 in the central mainland barren-grounds of the Northwest Territories. Its primary purpose was to protect muskoxen, but barren-ground caribou of the Beverly population summer and sometimes calve within the sanctuary (Hoare 1930; Clarke 1940). Currently, the mineral industry is lobbying to rescind the Thelon Game Sanctuary so that they can explore it and develop any worthwhile finds.

Unfortunately, such areas actually offer little year-round benefit to caribou populations: (1) the caribou use them only seasonally, then range over unprotected areas; (2) natives are not restricted from hunting caribou on those areas; and (3) most of the established refuges and sanctuaries are in areas that were not heavily hunted by natives. Migratory caribou cannot be protected effectively by reserves, except seasonally and locally, because of the huge areas over which they range (Kelsall 1968). However, protection of calving grounds, postcalving areas, and migration routes by special reserve land status or by special land use regulations could be beneficial in giving maternal cows and newborn calves an added degree of protection during those time periods (Miller 1974b). Such areas were traditionally afforded protection by their remoteness and ruggedness, but the airplane and snowmobile now make them readily available.

Encroachment by resource exploiters and the added pressures of a looming North American energy crisis make the special status of wildlife areas tenuous at best. If we lose the large migratory herds of barren-ground caribou in Canada and Alaska, the arguments for keeping out or restricting exploitation and development activities on tundra ranges will be greatly weakened. It is indeed a vicious circle: the natives are jeopardizing the herds by overuse, but their dependence on caribou gives the species a high political profile that they would not otherwise have.

TRANSPLANTS AND REINTRODUCTIONS. As Kelsall (1968) mentioned, to help assure success when transplanting barren-ground caribou, the restocking should be done in an area where (1) the caribou cannot stray; (2) there is little or no harvesting; and (3) there is a nonconflicting need for caribou. Few such places exist for barren-ground caribou in Canada. Bergerud (1974d) noted that, in the reintroductions of 226 woodland caribou to 18 sites in Newfoundland between 1961 and 1965, the greatest problem was preventing adult caribou from straying. He observed that hand-reared calves did not attempt to migrate or stray from the transplant sites. Bergerud (1978) reported that success had varied among the transplants and to date caribou had been harvested from four transplant sites; he suggested that releases should succeed on islands free of disease and predators (Bergerud 1978).

Caribou calves were introduced to Adak Island in the Aleutians during 1958 and 1959. Bergerud (1978) reported that harvests have been as high as 30 percent on the Adak population in 1973 and that the 1973 harvest was a record for sustained yield. This assumes that the Adak population maintained its size after 1973.

Kelsall (1968) pointed out that Southampton Island in northwestern Hudson's Bay was an excellent site for the reintroduction of caribou. Caribou were abundant on the island until 1924 but were almost totally gone by 1930. The Canadian Wildlife Service at the request of the government of the Northwest Territories captured 52 caribou on Coats Island (about 80 km south of Southampton Island) and transplanted them to Southampton Island in summer 1967. No adequate survey has been done of the reintroduced caribou to date, but sporadic observations suggest that they are apparently doing well on Southampton.

Restocking of islands and possibly small areas of the mainland with caribou may have some limited potential as a management practice. The main aim should be to prevent local loss of caribou so that such practices are not necessary. If restocking or creating new herds is justifiable, we should remember that domestic reindeer are not a legitimate substitute for caribou. I believe that the introduction of reindeer on former caribou range by any wildlife agency is indeed a dereliction of their charge and an irrational act.

CONTROLS ON THE ENVIRONMENT. Controls of environmental factors that adversely affect forage supplies or promote disease or parasites may at times be important and even necessary for sustained harvests of caribou. Of even more importance is the control of activities by all users, consumptive and nonconsumptive, and especially of industrial uses of caribou ranges. Most attempts at any physical control of the environment should be made only during the most intensive management programs, with the probable exception of forest fire control. Forest fire suppression on winter ranges of caribou will be necessary if the rates of wildfires increase over past rates and the areas burned are excessive in size. Kelsall (1968) believed that the need to suppress fires on caribou winter ranges was well recognized, but opposition to that belief has been raised by Bergerud (1971*a*, 1972, 1974*a*, 1978) and others. Kelsall's (1968) plan called for mapping unburned winter ranges and ranking them in relation to (1) quantity and quality of lichens present; (2) their location relative to caribou use; (3) potential effects of fires on each mature forest area; and (4) relative potential of fire-damaged immature forest. As yet, the topic of the impact of forest fires on caribou winter range remains open to debate.

Forested winter ranges could be fertilized, but the value of such action and the possible undesirable side effects are as yet unknown. Such extreme management practices demand study on small areas before extensive programs are considered. Fencing of hazardous water crossings has been suggested by many interested groups. Such ventures would be very costly and probably would create as many hazards as they removed. Kelsall (1968) believed that the long-term average of caribou lost annually by drowning did not exceed the number taken by only a few native hunters. He further pointed out that caribou also drown on quiet water crossings when struck by sudden gales or by breaking through unsafe ice (Kelsall 1968). Diversion of caribou

from one area may simply cause their deaths elsewhere.

At the current and foreseeable future levels of caribou management it is hard to conceive of extensive use of environmental controls in caribou management. In fact, it is difficult to see any, unless rapid industrial development brings networks of roads and pipelines onto caribou ranges. Alterations will be made then only if the funds come from industry. In such an event, it may be necessary to build crossing devices to allow the free flow of caribou past roads and pipelines. In that event, we will find out if caribou can live with modern man and his activities.

I think that proper management of caribou should include (1) control (regulations) of all persons utilizing or in some way affecting the resource; (2) periodic, accurate estimates of the number of births and deaths in the population; (3) periodic, accurate estimates of the size, sex, and age structures of the population; (4) samples of physical condition, when and where possible, to gain insight into caribou performances under various environmental conditions and to detect stress situations, when they occur; and (5) ongoing studies to detect herd splintering or dispersals that lead to egress or ingress. To date, such a well-rounded program on a periodic basis has not been maintained for all the major caribou herds or populations.

Problems inherent in most field work are lack of funds, lack of continuity in personnel and programs, shortage of experienced personnel, and insufficient preparatory training in procedures for field techniques. Many field techniques demand powers of observation. Most observers can improve their skills with practice and proper training, but some people are incapable of becoming accurate observers regardless of their formal training, specialized training, or time in the field. Practical training courses and field supervision often lead to the detection of people who are potentially good observers, and those who are "bad" can be assigned other tasks.

Many, if not most, previous estimates of the different parameters for measuring and evaluating caribou population dynamics have suffered from two basic shortcomings: (1) the sample sizes relative to the size of the population were too small; and (2) samples were often localized and represented only a single or, at best, a few segments out of all the segments that made up the population. Caribou are usually strongly segregated and, I think, probably remain segregated to some degree at virtually all times. Therefore, samples that are taken from only one or a few restricted areas are not going to reflect accurately the condition being measured. This has been well documented in the central arctic caribou herd of Alaska (Cameron and Whitten 1979; Cameron et al. 1979). Future work must stress the need for larger samples and better representation of the population in sampling procedures. I suggest that any sample that is less than 10 percent of the population size and not taken from, at least, most segments of the population be used with extreme caution; larger, seemingly more representative samples should still be suspect when setting management policies pertaining to harvest regimes. The caribou manager should never forget that he is working only with estimates that are subject to wide variances, usually of unknown magnitudes.

Bergerud (1978), among others, has stated that census results should not be accepted without two independent methods showing agreement. I basically agree with this belief but at the same time realize that it is often difficult to obtain one good estimate under field conditions. There is little wisdom in reducing the effort in one sample just to take a second sample. When funds and manpower prohibit two sampling procedures at high levels of effort, I think it would be better to go for one sample at a high level of accuracy than two at relatively low levels. While the use of two independent methods is always desirable, many management decisions will often have to be made on the basis of only one such estimate. Therefore, field workers must strive for continued improvements in methodology.

DETERMINING SEX AND AGE STRUCTURE OF THE POPULATION. The best time to perform ground classification of caribou for sex and age composition is in the autumn, usually late October through to mid-November. The actual degree of segregation of the caribou at that time is unknown. Many workers believe that the caribou are not segregated near the beginning of the rut. It is unlikely that this assumption is true, as I have found that groups of caribou killed during prerut were still highly segregated in late September (Miller 1974c). It is probably true, however, that there is less segregation at that time than at other times of the year. Sex and age classification can also be done along with calf production counts (from the ground) just after calving. The results can be compared to autumn compositions for further insight into evaluations of estimates of sex and age composition of the population. Caribou are classified as bulls, cows, yearlings, and calves. An unknown category is usually necessary if you are going to account for all animals examined. If time and experience permit, you should separate bulls into mature, prime bulls (full-sized bodies and antlers; often roman-nosed head profiles) and young bulls (smaller bodies, with small antlers; lacking face profiles of older bulls). Some attempt should be made at identifying juvenile males (one to three years old), which may be confused with adult cows. You can only be sure of adult cows that have been sexed by external genitalia.

Identification of sex and age classes should be based on external body characteristics, especially external genitalia, as outlined by Bergerud (1961, 1964a), Skoog (1968), and others. Experience, if not essential, is very valuable, and the same people should do these counts each year.

Bergerud (1974c) suggested that the sex ratio of one mature bull per two mature cows is a species characteristic for caribou. Thus, he suggested that sex ratio should be sought if the quality of the stock is a primary consideration in the management program.

Populations of caribou that are heavily sport hunted for trophy animals are not likely to maintain this sex ratio of one mature male per two females unless conservative seasons and bags are imposed. Sex and age compositions of native kills are most likely as much by chance as by design. Availability to the natives usually dictates what is taken each year. Natives did and probably would still follow the fat cycle in the caribou, if possible; that is, they would select for mature males in summer until the rut, then they would select for mature cows throughout the winter until just before parturition. It is obvious that there was no room for any concept of conservation as we understand it in the native's use of caribou on an annual basis. The killing of pregnant cows throughout the winter was reasonable when the natives had to live by the old rule of "a bite of fat for each bite of lean." This rule was based on sound judgment, acquired over many years of coping with the extreme cold of the arctic and subarctic winters—intake of fats were mandatory to survival. Unfortunately, this practice of killing cows throughout the winter persists in the absence of a true need—old habits die slowly.

Information on sex and age of hunter kills could be used to gain some insight into the composition of the population. However, the accuracy of such information must always be questioned and treated with caution. Data on sex and age composition can be obtained by setting up hunter check stations on roads that give limited access to areas where large sport or nonnative subsistence kills of caribou occur.

In the past and currently, collecting of sex and age statistics from native kills of caribou was usually not even attempted. This was the case for good reason, as most often the obtainment of figures for the annual harvest of caribou by natives was done only once a year. Such exercises relied on the natives' having good memories and a willingness to be cooperative and accurate in their accounts. Such reliance was seldom fruitful. More recently, some attempts have been made at collecting kill statistics several times a year or even monthly. Future efforts at obtaining statistics on the native harvest of caribou should include sex and age information, when possible. However, we must first get accurate figures of the annual harvest of caribou by natives before we give much concern to the sex and age of their kills. This seemingly can best be done by employing some respected native resident of each settlement to make weekly door-to-door checks of the hunters' success. It is unreasonable to expect most native hunters to provide accurate information on sex and age compositions of their annual kill of caribou if they are asked only once each year.

When and if such information is obtained, the biologist must remember that the sex and age compositions are measures of the kill and not necessarily of the population as a whole. The constant segregation of caribou makes it highly unlikely that the kills will represent the entire population. An exception to this condition may occur when many entire social groups are killed seasonally. Some biologists subscribe to the use of relatively small shot samples of caribou to evaluate population dynamics. As mentioned previously, I question such a practice, even though I have done it out of necessity. We really never know when we are right or wrong until we have sampled most or all of the population under consideration. However, I accept the unfortunate fact that we often have to work with small samples because they are the best that we can obtain. But that is no reason for losing sight of the dangers inherent in such use, especially when applied at the population level.

Banfield (1955) and Miller (1974c) have both built life tables for barren-ground caribou in Canada. Banfield constructed a time-specific life table from 292 mandibles found in 1948 and 1949 on summer and winter ranges of caribou. I used 943 caribou killed between March 1966 and July 1968 (Miller 1974c). All of my samples were taken from the Kaminuriak population on both winter and summer ranges. Banfield determined the ages of his specimens based essentially on the pattern of wear and eruption known for white-tailed deer at that time. I sexed all of my specimens in the field, did histological sections of mandibular teeth to determine their ages by counting annulations in the dental cementum, then determined eruption and wear patterns for the mandibular teeth on a seasonal basis. Banfield (1955) presented only one curve for both sexes, whereas my sample allowed the construction of a curve for each sex (Miller 1974c). Total mortality was 92 percent for the first 10 years in Banfield's (1955) life table and 92 percent for females and 99 percent for males in my life table (Miller 1974c). When I think about the apparent quality of Banfield's (1955) sample of caribou lower jaws used in construction of his life table and the supposed quality of the shot sample of caribou that I used in my life table, I ponder whether I should conclude that the two samples were equally good or equally bad. Although the resultant information, for the most part, looks reasonable, there is no way of knowing whether or not it represents the living population. Therefore, even though such exercises are interesting to do and talk about, they must be treated as suspect with regard to whether or not they truly reflect what they are supposed to represent.

EVALUATING PHYSICAL CONDITION

Whenever the opportunity presents itself, biologists should necropsy caribou and record information on physical condition, cause(s) of death, and occurrence of disease and parasitism. Bergerud (1978) suggested that animals should be collected in May and June to evaluate the annual nutritional impact of overwintering and again in September for statistics on conditions developed in summer. I agree that such small samples provide the biologist with additional feeling for his work and may reflect conditions being experienced at the population level. I do not share Bergerud's opinion, however, that such samples reflect conditions applicable to the entire population. My lack of faith in such small samples is especially strong when the information is used for evaluating population dynamics. The recent histories of all animals in a caribou popula-

tion are not the same: (1) all caribou in a population do not winter or summer on the same areas, so they do not necessarily obtain the same quality and quantity of forage; (2) they are not necessarily exposed to the same levels of disease and parasitism; (3) all caribou in a population are not exposed to the same level of predator and insect harassment; and (4) caribou are usually segregated by sex and age. Therefore, information from a small sample, especially from a single segment of the population, may be quite misleading. Granted, if unfavorable conditions are detected in a small sample, then greater efforts can be made to determine occurrence throughout the population.

For these reasons I suggest that biologists may justifiably use small samples for gaining insight into subject areas of concern or possibly in the hope of detecting unfavorable conditions that are suspected because of the prevailing environmental conditions. But if managers use the information from such samples as the sole means of making management decisions, they run the risk of being seriously wrong. Caribou biologists and managers must remember that they are working with a living entity and thus some, possibly many, detrimental impacts brought on by unsound judgments in management decisions may be irreversible.

Sponsoring organized hunts in Canada for the sole purpose of obtaining caribou for natives that had no caribou readily available to them tends to be a blemish on our records as professionals. I think this because those hunts were carried out with no knowledge of and probably no real thought for the ability of the caribou being harvested to sustain the additional kill; no biological information was obtained from the kills; and on some occasions, the carcasses or portions of the meat were left behind and never subsequently retrieved. Organized hunts are by no means traditional or esthetic and are offered to the natives as privileged services, not as a fulfillment of their native rights. Therefore, such hunts should be allowed only when the wildlife agency concerned knows that the caribou population receiving the additional harvest can sustain it. The hunts should be controlled by wildlife personnel in a manner that allows the collection of as much biological information as possible. When a wildlife agency (or any other government agency) provides such an expensive service, maximum benefit to caribou biologists as well as the natives should be the outcome of the venture.

LIVE-CAPTURE AND MARKING OF CARIBOU

Many projects require live-capture and marking of or telemetric attachments on individual caribou. Such activities are often the most difficult and sometimes the most expensive phase of a project, and they are always integral to the success or failure of the study. Long- and short-term studies of the discreteness of herds, subpopulations, and populations for specific ranges and calving grounds; dispersals of individuals or groups from their major social units; seasonal movements (rates of travel) and distributions; group cohesion; companion animals; causes of natural mortality; hunting pressures; reproductive success of individual females—all these studies require individually identifiable caribou (or group identification) and often telemetrically equipped caribou for detailed, useful results. Capture and marking techniques are in a constant state of development, as is telemetry. Capture techniques for caribou can be considered under two major headings: (1) water crossing, and (2) on land.

Live-Capture at Water Crossings. Capture is greatly facilitated if the caribou under study have known, traditional places of water crossings on rivers or lakes that are wide enough (usually 400 m or more) to allow pursuit and overtaking of the caribou while they are swimming.

The basic technique that has proven successful for capture of caribou at water crossings is as follows. First, the capture team must know approximately when, where, and in which direction the caribou will be crossing the water body.

Stable boats, 7-m freighter canoes with 20-hp outboard motors or the equivalent, are positioned several hundred meters (or closer if cover is afforded) from one or both sides of the landing site on the far shore of the crossing. The capture crew remains out of sight, or at least inactive and on a vantage point, if one exists. When the caribou have entered the water, the crew waits until the caribou have swum well out into the crossing. Once the caribou are far enough from shore that they can be overtaken before they can retreat from the water or make it across and out of the water, the crew(s) take to their boat(s). The ideal team size is three four-person crews each consisting of one motorman, one recorder, one crooker and tagger (the person who initially catches and subsequently ear tags or affixes the collar, etc., to the caribou), and one holder (the person who takes the caribou from the crook and holds it in place while it is being worked on) in three boats.

One boat races to the far side of the crossing from where the caribou entered the water to prevent a retreat; the other boats flank the swimming caribou. If necessary, one of the flanking boats races ahead to turn the caribou back if they are getting too close to their landing. Each motorman singles out an individual caribou, pulls along side of it, and throttles back to the swimming speed of the caribou. The crooker reaches out with his shepherd's staff and crooks the caribou around the neck and pulls it to the gunwale at the bow of the boat. (A lasso can be used instead of a shepherd's crook.) The holder reaches over the gunwale and grabs the caribou by the head, lifts the caribou slightly, and holds it securely against the top of the gunwale. The tagger first reaches over the side and grabs the caribou's tail with one hand and the outward hind leg with the other hand. He then rolls the caribou up and toward the boat to expose its anal region and genitalia, and then calls out the sex to the recorder. Both the tagger and the holder determine the age class of the animal and call it out to the recorder. The tagger then releases the rear of the animal and moves forward to attach the eartags, collar, etc. The caribou is then released and the motorman pulls the boat away and

pursues another animal, or maneuvers to prevent animals from leaving the water.

There can be many variations on this theme. If manpower is in short supply, a three-man crew without a separate recorder can be used. Even a two-man crew could do the job on wide crossings: one motorman (with a special device for locking the boat on course after the caribou is caught), who moves forward to do the tagging, and a crooker who also serves as holder and possibly recorder.

If there is a need for capturing and holding animals for subsequent work or transport, corrals can be built out of sight on the far shore and boom-type fences extended out on the water. The caribou could then be herded inside the booms and run into the corral. Entire groups could be captured if there was need for such a large-scale operation.

Fixed-wing airplanes and helicopters on floats may be used when the capture of only a few caribou is desired and mobility is a primary consideration. When using an aircraft, the pilot lands and taxies up behind the caribou so that the animal comes between the floats. The tagger works from one of the floats or on some fixed-wing airplanes from the cross-arm attachment between floats (staying out of the prop is mandatory for the success of the operation). Often the animal can be worked on without being restrained, but, if necessary, a lasso rope can be dropped over its head and the animal secured in place by it.

There are many advantages to capturing caribou at water crossings: (1) a large number of caribou can be captured within a relatively short period of time; (2) the technique is apparently less stressful to caribou than other capture techniques (minimal restraining time, no drugs are used, and individuals are allowed to remain with or catch up with other group members immediately after release); (3) manpower can be as low as two men or as large as resources permit; (4) the cost per animal caught is usually much lower than by other techniques; and (5) the technique can be made fairly mobile, especially if the capture of a large number of caribou at each location is not a goal. The main disadvantage of reliance of capture of caribou at water crossings is that the technique is restricted to use during the open water period, which is usually only from July into September of most years. Also, the caribou to be studied must use suitable water crossings with reasonable annual regularity.

There is one sociopolitical disadvantage to all methods of live-capture and marking or radio-equipping of caribou: most northern natives object strongly to such management procedures. Although most, if not all, northern natives believe that the killing of caribou is a noble activity, they also, unfortunately, believe that the marking and release of caribou for whatever purpose is degrading to those caribou. Many northern natives hold a mystical belief that marked caribou will tell others where they were debased and thus the herds will not return to those locations. With a strand of logic, they conclude that the killing of caribou near traditional water crossings is alright—because dead caribou tell no tales.

The free intermixing of marked or radio-equipped

caribou shortly after release with uncaptured caribou should argue against any degradation of the handled individuals. Also, recapture of marked caribou at their original capture location during the same seasonal movements indicates that the experience does not have any more lasting effect on them than would an unsuccessful wolf attack. Convincing the natives of this is a necessary task if we want to continue such work (at least in Canada).

Caribou normally exhibit caution and nervousness at water crossings, which is likely predator orientated, as both wolves and bears often frequent the shores of water crossing areas.

A technique well worth investigation is to decoy caribou into the water. I have thought about taking some object such as a 10-gallon fuel drum and attaching two together end to end. I would paint them dark brown with a nonglare duck boat type of paint and attach pairs of caribou antlers of various sizes to the front end of each pair of drums. Painted empty plastic detergent jugs could be used as heads. The decoys would then be anchored in the water off the landing site in patterns simulating groups of swimming caribou. The idea could be applied to the use of boom fences for corralling caribou by attaching individual decoys in a manner that creates a physical barrier to the swimming caribou. If the decoys were effective, they would likely also serve as a "psychological barrier;" as swimming caribou would be less likely to fight the boom fences if they thought that the decoys were real caribou.

Live-Capture on Land. Tested and potential methods of live-capture of caribou on land are many and varied. The three main categories that come to mind are (1) nets, (2) corrals, and (3) immobilization by chemical restraint. Discussions of all of the variations of all of the possible techniques under each of the above categories are beyond the scope of this chapter. I will, therefore, consider only those that I know have been proven successful or that I think are most likely to have the greatest promise.

NETS AND CORRALS. Nets and corrals will be considered together because many of their potential uses are overlapping. Corrals are better suited for long-term studies that require the live-capture of many caribou on a seasonal or annual basis. Nets have the primary advantage of being portable and the associated costs of use are usually much lower.

The use of nets includes set tangle nets, propelled tangle nets, nets fired from cannons mounted obliquely on the ground or mounted obliquely or horizontally on poles or trees at varying heights, and tangle nets fired from cannons mounted on pursuing helicopters. Corrals can be built out of secured nets, wire fencing, or logs, using metal or wooden posts or trees for uprights. Holding pens in corrals could be covered with burlap, canvas, or even plywood. Corrals can be built with set lines or automatic eyes that when triggered would cause a drop gate or swinging gate to close and lock.

Set tangle nets, tangle nets propelled from stationary mounts, and corrals can be used along trails, at water crossings that are too narrow for capture of caribou while they are still in the water, or in "blind

spots" on terrain where caribou can be herded. Tangle nets fired from cannons mounted on helicopters can be used anywhere and at any time that a helicopter can operate. Such captures would, however, most likely be less traumatic to the captured animals in deep snow, where the impact of netting would be buffered to some extent by the snow cover.

Fixed-wing airplanes, helicopters, snowmobiles, boats, and people on horseback or afoot could be used to herd and harass the caribou into the nets or corrals. Most netting operations using set tangle nets would probably be more successful during periods of deep snow cover, but they can be used at all times of the year. Cannon netting and corrals can be used year-round with good success.

Corrals, extensive drift fences, and mazes built of chopped and uprooted trees and boulders were constructed by aboriginal natives to direct caribou to people lying in wait with spears and bows and arrows. Snares were also set along the courses of the fence lines and in mazes to catch and hold caribou for killing. The effectiveness of such structures along migrational paths was most likely proportional to the length of the drift fences and the numbers of physical mazes and hunters. The existence of such traps argues well for the long-term fidelity of caribou to traditional migrational paths. It also suggests that automatic tagging or collaring devices (snares) could be used with some success along forest trails or along trails in deep snow.

Set tangle nets have been used successfully to live-capture caribou (Des Meules 1965, 1968; Miller et al. 1971). The approach is best used during winter or during the early period of spring migrations when trails in deep snow are regularly used by caribou. The best situations for sets are on forested river or lake shores where caribou are coming onto the frozen water bodies along well-defined trails in the snow.

Likely sites for setting nets are located by aerial search (or can be done by snowmobile treks). Cessna 185s are well suited for the work, but other light, fixed-wing airplanes can be used. Use of a helicopter would give an additional degree of flexibility to the operation. Helicopters would be mandatory on terrain that lacked large bodies of frozen water for landing fixed-wing aircraft. Fixed-wing airplanes on big wheels can, however, be used on High Arctic islands and on some mainland tundra habitat types that are suitable for big wheel landings. Snowmobiles can be used to transport netting crews and to chase caribou into the nets instead of using aircraft, especially if the crew were to work one general area for an extended time period.

When a likely netting site is located, the network of trails is then examined and a place is chosen where a major trail leaves the forest and comes onto the frozen water surface. The plane is then landed near that point, making sure not to taxi over the major trail, and the crew deplanes with the net(s). The net (a good size is 3–4 m by 30–40 m with about 25-cm^2 mesh) is set across the trail just inside the trees from the shoreline. The upper edge of the net is hung on cut-off limbs and tied every 3–6 m at a height of about 2 m to secure it. The loose bottom of the net is then pulled (pursed) toward the direction from which the caribou will enter, so that their feet will be entangled within the mesh when their body strikes the net and causes it to move backward. Antlered individuals are more apt to be caught than anterless ones. The use of a second net set 3–4 m behind the first will increase the catch of antlerless (and antlered) animals. The effectiveness of set design varies directly with the configuration of the net (figure 47.7). Use of double nets with a narrow entrance that could be blocked off with one end of a net (fig. 47.7) would increase the catch per set and thus reduce the cost per animal caught considerably, especially if large catches were desired.

When the net is in place, two trees should be cut and placed in the snow about 25–40 m out on the ice to guide the pilot(s) in driving the caribou into the net(s). The pilot then take the plane aloft, relocates the caribou as quickly as possible, and herds them toward the net. If the caribou have left the ice during the time required for setting the net(s), the pilot dives the plane in front of them to turn them back to the ice. Once the caribou are back to the ice, or if they have remained there, the pilot lands the plane on the far side of the

caribou relative to the netting site and begins driving the caribou toward the net by taxiing after them. The pilot should hold down the taxiing speeds to allow the caribou to get onto their main back trail and follow it into the net.

If the caribou attempt to leave the main trail and break for the ice on side trails, the pilot should cut them off and herd them back to the main trail. When within 30–60 m of the net, or when the shoreline is coming up, the pilot should race the engine to induce the caribou to make a final rush into the net. Caribou appear to see the net just before they hit it and if they are not encouraged to jump into it, they sometimes stop and walk or run around it.

During herding, the netting crew remains hidden as close as possible to the entrance of the set without being detected (often 20 m or less, depending on cover afforded). Crew members on snowshoes cover any side trails close by the set that caribou might attempt to use if they break from the main trail as they approach the shoreline. If caribou do break onto those side trails, the snowshoers rush out toward the caribou and frighten them back onto the main trail.

Once the caribou are entangled in the net, the crew members rush in and hog-tie each animal by the two hind legs and one foreleg. The restrained animals are then processed as quickly as possible, held, and released as nearly as possible at the same time to reduce the chances of group break-up due to the capture and restraint. A special effort should be made to see that all calves are released with their mothers, if known, or at least with an adult cow.

The capture of caribou when using set tangle nets should be about proportional to (1) the size, number, and configuration of nets used; (2) the amount of manpower employed; and (3) the number of aircraft or snowmobiles used for herding the caribou into the net(s). Recent and prevailing weather conditions, especially the character of the snow, are important to netting success, as are pilot abilities. Once temperature fluctuations cause snow to thaw by day and refreeze at night, frozen river and lake surfaces become too hard packed for airplanes to land and pursue the caribou. Also, once the bearing surface of the snow will support running caribou, they can freely leave their trails and the operation becomes more difficult. At those times helicopters or snowmobiles can be used for herding the caribou.

A Cessna 185 or similar airplane can carry up to a three-man netting crew, one net (3 × 30 m size class), associated equipment and supplies, plus one pilot. As little as one man and a helpful pilot can do the job, but with much more handling time and much less efficiency. The ideal team size when using fixed-wing airplanes is three planes with three three-man netting crews: two planes to serve as "flankers" and one on drag to help keep the caribou from breaking from or turning back on their direction of travel. A Bell-206B helicopter or its equivalent can carry a two-man crew and necessary materials, plus pilot. When using a helicopter, manpower is necessarily reduced, but a single helicopter with a skilled pilot can perform nearly

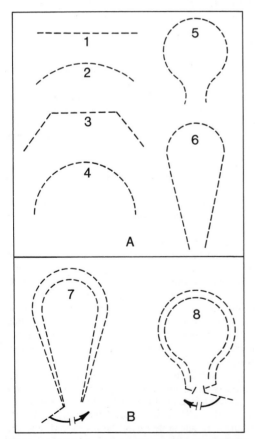

FIGURE 47.7. Tangle net set design. A, design used for live capture of caribou (*Rangifer tarandus*) (Miller et al. 1971): 1, line; 2, crescent; 3, angle; 4, horseshoe; 5, obovate; 6, oblanceolate. B, designs thought to be more effective but untested: 1, double oblanceolate; 2, double obovate.

as well as three fixed-wing airplanes when herding the caribou.

The use of drop tangle nets and cannon-fired tangle nets has not yet been developed specifically for the capture of caribou. I think that both netting approaches hold promise for the live-capture of caribou, as does the use of automatic tagging devices. The main advantage of these techniques would be that small field parties, two-man crews or larger, could be left for long periods to work caribou intensively within specific areas such as mineral licks, winter trails, and migrational paths. In restricted areas, crews can move about on snowshoes and hand pull their equipment and supplies on sledges. When working in larger areas where disturbance of the caribou is not as much of a consideration, the crews could travel by snowmobiles and haul their camps, equipment, and supplies by sledges attached to their snowmobiles.

The use of baits and decoys for attracting caribou should be investigated. Kelsall (1968:215–216) summarizes findings of several authors on the use of baits and decoys by primitive northern natives. Primitive natives used sprinklings of dog urine, vegetation, and saltwater ice blocks to attract caribou to pitfall traps. They also decoyed caribou in summer by imitating the grunts of cow-calf pairs and during rut by bellowing like bulls. Caribou can even be attracted to a person(s) moving about in the open: waving white cloth often causes caribou to come in close to investigate. Bubenik (1975a) used dummies of caribou heads made of styrofoam or PVC-foam partly covered with nylon fur, replaceable antlers of various sizes, and movable ears, eyes, and neck to approach caribou in his studies. Therefore, it should be possible for biologists with small cannon nets and a lot of patience to develop sets and procedures for attracting caribou to the sets.

CHEMICAL IMMOBILIZATION OF CARIBOU. No other technique for live-capture of free-ranging or captive animals has been as widely used by as many biologists under differing conditions and on different species as immobilization by chemical restraint. Books (e.g., Harthoorn 1970, 1976), a Wildlife Society monograph (Harthoorn 1965), and hundreds of articles—e.g., on M99 (Houston 1970; Presnell et al. 1973; Coggins 1975), succinylcholine chloride (Miller 1968; Allen 1970; Jacobsen et al. 1976), rompun (Bauditz 1972; Harrington 1974), phencyclidine hydrochloride (Stelfox and Robertson 1976), and nicotine (Behrend 1965)—have been written on the subject.

It is obvious that the subject cannot be covered in detail here. I recommend that interested biologists obtain a copy of *Chemical Immobilization of North American Game Mammals,* by Hebert and McFetridge (1978) as a basic guide to the subject. Hebert and McFetridge (1978) provide information on (1) drugs used for immobilization; (2) modes of delivery and safe handling of drugs; (3) reactions and sensitivity to immobilization; (4) care of animals during capture, transport, and release procedures; and (5) legal considerations. They give 86 tables from the literature containing information on dosage rates, induction time,

immobilization time, recovery time, etc.: 21 tables for deer, 11 for moose, 19 for elk, 6 for sheep, 1 for goats, 1 for pronghorn antelope, 19 for bears, 2 for wolves, and 6 for native felids. Their work contains lists of 29 different drugs that are sold under 45 different trade names and addresses of 28 suppliers. They also cite over 150 references that relate to chemical restraint of mammals and provide a glossary of terms, plus appendixes of other valuable information on the subject.

Many biologists have captured caribou with different drugs in projectile syringes (Cap-Chur or Pneudart trajectories) fired from CO_2- or powder-operated rifles and pistols. Some have used syringes (darts) attached to the tips of arrows propelled by longbows or attached to bolts shot by crossbows. However, little has been published in readily available journals on the results of immobilization by chemical restraint of caribou. I assume that this lack of literature relating to the immobilization of caribou is due for the most part to the fact that other workers found, as I did, that they could capture caribou using the same approximate dosages of succinylcholine chloride as reported by Bergerud et al. (1964). Why more recent efforts using rompun and M99 have not reached the literature, I do not know.

Woodland caribou were immobilized with succinylcholine chloride delivered by CO_2 Cap-Chur equipment in Newfoundland (Bergerud et al. 1964). Free-ranging caribou required relatively heavier dosages of 0.099–0.143 mg of succinylcholine chloride per kilogram of body weight than captive caribou, which required only 0.055–0.088 mg/kg. The caribou were located on foot and by helicopter searches during the calving season and the rut, then stalked on foot by the shooter. Female caribou appeared to have higher tolerance to the drug than males: captive females, 0.077 mg/kg; captive males, 0.055–0.077 mg/kg; free-ranging females, 0.110–0.154 mg/kg; and freeranging males 0.088 mg/kg.

Alaskan biologists have shot caribou from helicopters using a Cap-Chur 28-gauge powder dart gun and 3-cc syringes loaded with 12-20 mg of succinylcholine chloride (R. D. Cameron personal communication). They varied their dosages of the drug by season of the year based on estimated body weights but continued to find, as others have, that caribou exhibit considerable individual variation in their tolerance to the drug. Biologists in Alaska have also used 2.5-4.0 mg of M99 with 20-30 mg of rompun to immobilize caribou (R. D. Cameron and J. L. Davis personal communication). Rompun was used in combination with M99 to reduce the excitant effects of the M99 on the drugged caribou. It is likely that 5[+] mg in combination with rompun would be better dosages for adult male caribou (J. L. Davis personal communication).

Haigh (1976) used 0.282 mg/kg of rompun to immobilize captive caribou. Many of the other immobilizing drugs (Hebert and McFetridge 1978) that have not been tested on caribou or for which results have not been reported will likely be usable.

The live-capture of caribou by immobilization with projectile syringes has the advantage that the equipment is highly portable and thus the method is applicable as an opportunistic technique. Concerted efforts can be made to capture caribou by shooting with darts or the dart equipment can be regularly transported in the field with the biologist and used whenever an opportunity presents itself. This technique is well suited for the recapture of caribou for remeasurements or refitting of telemetric packages. Biologists should remember that there is almost always some mortality associated with the use of this technique and plan accordingly.

Marking Caribou. Materials and methods for marking large ungulates are constantly changing and considerations for caribou are indeed numerous. There is no practical way of covering the subject here. Biologists must become familiar with the possible materials and methods for marking caribou through the literature and discussions with other biologists. Possible materials range from natural materials through space-age synthetics. Elasticized synthetics can be used where expansion of collars is necessary or desirable: e.g., on rutting bulls and young, fast-growing caribou. However, if collars for adult males are attached at a single upper point by material such as polyethylene braided rope, elasticization of the collar is not necessary. But I think all collars put on young individuals should be elasticized.

Materials used in marking must be able to withstand extreme fluctuations in temperature and varying degrees of wetness and dryness. Many of the space-age synthetics appear best suited for prolonged (two to four years) use under such environmental conditions. The basic methods of marking caribou are: (1) ear tags, (2) ear streamers, (3) collars, (4) coloring with dyes or paints, (5) tattooing, (6) freeze branding, and (7) branding.

EAR TAGS. Many different sizes and shapes of commercial cattle and sheep ear tags are available for use on caribou. Custom stamping of lettering or numbering is done on the tags by the manufacturer or retailer. Such ear tags are usually metal or plastic, puncture types, and are attached with special pliers or by interlocking parts. Most ear tags serve mainly as semipermanent markers, as they are usually not readily visible. Some larger, colored plastic ear tags would be visible to ground observation, at least over short distances. The biologist must remember that the large metal tags could lead to the freezing of ear tissue at the extremely low temperatures commonly experienced by caribou in winter and thus subsequent loss of ear tags.

EAR STREAMERS. Most ear streamers are made of plasticized materials but other space-age synthetics will probably do the job as well, as most are resistant to deterioration at low temperatures. They can be color coded for identification of individuals. Ear streamers can be attached by passing the streamer through a slit in the ear, then back through a slit in the upper end of the material (see Giles 1969:310, figs. 18, 20) or they can be attached with rivets or interlocking or self-locking fixtures.

COLLARS. Most of the collaring materials and collar designs used on other large North American ungulates can be adapted for use on caribou. The material used and the design and size of the collar would be governed mainly by the use that was going to be made of it. The two basic considerations are its visual qualities and its physical properties. If the primary use of the collar is going to be for visual identification of individuals, the body of the collar should be wide, lettered or numbered with a penetrating paint, or color coded. If the primary use is to be as a carrier for some kind of telemetric package, strength and position (flexibility) mainly should be sought. In such cases a narrow belt type of collar may be more satisfactory. Of course, a combination of high visibility and strength for retention of a telemetric package might be necessary or desirable. I have found vinyl-coated, nylon-webbed materials most suitable for collars, but any of the space-age synthetics would probably do as well or better. We have successfully used leather belting, industrial canvas belting, vinyl-coated nylon fabrics, neoprene-coated fabrics, and hypalon-coated fabrics. Other fabrics that can be used are neoprene-hypalon-coated nylon, cool-top marine fabric, vinyl/nylon laminate, riviera vinyl-coated dacron, fluorescent vinyl-coated cotton, fluorescent/luminescent vinyl-coated cotton, and fluorescent vinyl-coated nylon. Polyethylene ropes make excellent collar attachments; they are simply tied and the ends heated and melted back into a section of the rope above the knot. Collars can also be affixed around the caribou's neck with snaps through grommets in the ends of the collars, or by sewing, riveting, nuts and bolts, interlocking or self-locking fixtures, staples, and adhesives.

COLORING. Caribou can be marked by coloring them with various dyes or paints. The dyes or paints can be applied by (1) hand throwing, (2) being propelled from CO_2 and powder guns, (3) using automatic spraying devices, and (4) dropping from aircraft. How, when, where, and how many caribou you want to mark will largely determine the kinds of dyes or paints and methods used for applying them. There are commercial paint balls available for marking animals that are fired by Cap-Chur rifles and pistols, bursting on impact and leaving colored blotches on the animals. Automatic spraying devices could be used for marking caribou at mineral licks or along established trails. Hand throwing or shooting color markers at caribou would be practical only when a small sample of marked caribou was needed.

A relatively cheap and effective method of color-marking large, gregarious ungulates has been used by Simmons (1971) on Dall sheep and by Miller et al. (1977b) on Peary caribou. The method employs the use of a dye solution holding tank carried in a small aircraft with a quick-release mechanism controlled by the pilot—much in the fashion of a crop-dusting operation.

This method of dye spraying caribou has great potential for use in studies requiring the marking of large numbers of animals. Studies of social or range affinities or fidelities by caribou can benefit greatly from this method of marking groups of caribou by aerial dye spraying.

Many—probably most—caribou biologists think of caribou as being in open social units with no true long-term permanency (e.g., Lent 1966*a;* Skoog 1968; Bergerud 1974*d*). Their subsequent conclusion appears to be that as there is no real need for recognition of individuals if social units lack stability—the caribou merely recognize each other at the sex and age class level and not as specific individuals. I disagree with those biologists on this matter, both as to the degree of openness of caribou social units and as to the animals' abilities to recognize each other as individuals (Miller 1974*c*). I think that this method of dye spraying from aircraft would give an excellent opportunity of testing the discreteness of wintering groups of caribou during the spring migrations. Entire groups of wintering caribou on the same range could be dye sprayed various colors (each group a different color). Then they could be observed as they move into large migratory aggregations and their subsequent basic wintering group cohesiveness determined. Therefore, I will summarize the aerial dye spraying method as I have used it on Peary caribou (Miller et al. 1977*b*).

The dye spray tank was designed by and built to order for Simmons (1971). The tank that we used contained about 200 liters (l) of solution. It was mounted on the floor of the airplane behind the pilot's seat with lug bolts. Two Helio Courier aircraft were used for the spraying. Only the pilot rode in the airplane during the dye spraying flights (Canadian MOT regulations governing the use of the tank in light aircraft). The pilot controlled the load of dye solution with an "all or nothing" type of quick-release line. That is, whenever he pulled the line, 200 liters of solution was dumped; no partial dumping was possible.

A biodegradable, nontoxic dye was mixed in fuel drums by melting snow and heating the water to about 40–60° C, then adding the dye and stirring it into solution, allowing the solution to cool, and then adding about 20 liters of isopropyl alcohol. The isopropyl alcohol aided in the rapid absorption of the dye into the caribou pelage.

The dye solution was then pumped into the holding tank in the airplane. The pilot took off, located a group of caribou as quickly as possible, and made a 10–20-m altitude pass over them. The plane overshot the group of caribou and as the pilot judged his lead and passed by, he pulled the quick-release line and dumped all 200 liters of dye solution. The timing of the release of the dye solution requires experience, so some trial and error exercises are necessary, and the flying requires a skilled pilot. Hundreds or even thousands of caribou can be marked by using this technique whenever caribou are in large groups, especially while they are on migration. The technique could also be used on small groups during periods when the caribou are relatively sedentary.

The cost per 200-liter load of dye solution should be between $100 and $150, not accounting for labor and aircraft time. Each load could mark tens of caribou

under favorable conditions. Therefore, the cost per animal marked should be much lower than that by any other such highly visible marking technique currently known.

TATTOOING, FREEZE BRANDING, AND BRANDING. It seems that tattooing, freeze branding, or branding would be methods of limited use in marking caribou. These possible methods of marking caribou should not, however, be totally forgotten, as there might be limited occasions when they would be applicable to specific situations. For example, lip tattooing gives a permanent hidden identification, if one is ever needed.

Ecological Biotelemetry. Perhaps no single tool has as many applications to techniques for obtaining biological data as telemetry. Ecotelemetry has made possible many otherwise not practical studies and has allowed us to obtain a wealth of hitherto unavailable biological information. New horizons await us with future development of the art. Future development in miniaturization of telemetric parts, computer analyses of telemetric data, satellite packages, declassification of military technology, etc., will greatly expand the field of ecotelemetry.

One version has it that the art of ecotelemetry as an applied science began in 1957 with the telemetering of the temperature of an incubating penguin egg in the Antarctic (Adams 1965). In less than two decades great strides have been made in the field of telemetry. Examples of biological data that can be obtained from telemetry are many, the potential probabilities countless: e.g., body temperature, respiration rates, respiratory tidal volumes, blood pressure, heart sounds, intestinal pressures, pH, insulative quality of fur and feathers, ectoparasite counts, information on behavior and behavioral responses to natural and foreign stimuli, navigation (homing), migration, daily and seasonal movements (rates, distances, locations, etc.), and macroclimatic and microclimatic data.

I recommend that any biologist not well versed in electronics find an expert in electronics to work with him at the start of the study. My own experiences with radiotelemetry suggest that equipment failures are not uncommon and that correction usually requires the expertise of an experienced electronics trouble shooter.

Interested biologists can familiarize themselves with the field of ecotelemetry by reading such materials as "A Special Report on Bio-Telemetry," *Bio-Science* (February 1965), for background and more recent proceedings, such as the first (Long 1977) and second (Long 1979) international conferences on wildlife telemetry at Laramie, Wyoming, for information on the current state of the art.

CURRENT RESEARCH AND MANAGEMENT NEEDS

Caribou biologists, managers, and systems modellers from North America and Scandinavia took part in a symposium-workshop at the University of Alaska, 17–19 November 1977, to review their current state of knowledge, viewpoints, and research priorities. I believe that those proceedings (Klein and White 1978) serve as the most complete and up-to-date thinking on research needs for caribou. It was agreed that knowledge of the following parameters was necessary for an understanding of population dynamics of caribou: (1) population size, (2) population structure, (3) age-specific birth and death rates, (4) mortality to predators, (5) other natural mortality, (6) dispersal, (7) human harvest, (8) distribution (seasonal movements), and (9) status of other interacting herbivores (competitors and alternate prey of predators).

Bergerud (1978) stated that, in general, there is reason to be hopeful for the future of caribou, as they are highly adapted and adaptable. I truly hope he is right. Currently I can see little cause for optimism, especially with regard to the future of migratory herds of caribou. Subsistence users want to utilize the caribou beyond the maintenance capabilities of the herds. Exploitation of petroleum, gas, minerals, and water has the potential for disrupting migrations and reducing availability of caribou ranges. Politicians and pressure groups through misguided beliefs or lack of understanding or indifference are willing either to not act at all or to take the wrong actions on behalf of the caribou. Even some wildlife people suggest turning the management of caribou over to the native peoples— that indeed would not be a panacea for the caribou.

If we are to manage and conserve caribou properly, a hitherto unknown, concerted, all-out effort must be made—now. Federal, provincial, state, and territorial agencies charged with the responsibility for management and conservation of caribou and their ranges must vigorously press for the constraints necessary to protect the caribou. I hope that such an effort is in the offing—and is offered in time.

To paraphrase Harper (1955), no other large North American land mammal is of such primary importance as the caribou as a source of food to natives and nonnatives; no other carries out such extensive and spectacular migrations; no other can be seen in such vast herds; and no other exhibits so close an approach to a "Garden of Eden" trustfulness in the presence of man. Thus, perhaps, no other is more worthy of being cherished and safeguarded in its natural haunts for use and enjoyment of future generations than the caribou of North America.

Working with caribou has given me the opportunity to witness some of the most marvelous sights in nature, and has reinforced in me the belief that the arctic would indeed be an empty land without the migratory caribou.

I am particularly grateful to other caribou biologists who have taken the time over the years to debate matters pertaining to ecological and behavioral relations of caribou to their environment. I offer special thanks to those colleagues who argued the "gray areas" of those considerations, as much of what I think I know about caribou has been developed out of those conversations.

I am especially thankful to J. L. Davis, Alaska State Department of Fish and Game, for providing me with a great deal of information on caribou and their management in Alaska.

I thank Dr. A. Gunn, Northwest Territories Wildlife Service; M. C. S. Kingsley, R. H. Russell, and Dr. W. E. Stevens, Canadian Wildlife Service, for critically reading earlier versions of the manuscript for this chapter.

I also offer a special thanks to G. D. Hobson, director, Polar Continental Shelf Project, Energy, Mines and Resources Canada, for over a decade of financial and logistical support of my work on caribou. Without that support, some of the work would never have been done, and other studies would have been carried out at much lower levels of effort.

LITERATURE CITED

Adams, L. 1965. Progress in ecological biotelemetry. Bio-Science (February), pp. 83–156.

Allen, T. J. 1970. Immobilization of white-tailed deer with succinylocholine chloride and hyaluronidase. J. Wildl. Manage. 34:207–209.

Anderson, G.; Andersson, K., Brundin, A.; and Rappe, C. 1975. Volatile compounds from the tarsal scent gland of reindeer (*Rangifer tarandus*). J. Chem. Ecol. 1:275–281.

Anderson, R. C. 1971. Neurologic disease in reindeer (*Rangifer tarandus*) introduced into Ontario. Can. J. Zool. 49:159–166.

Banfield, A. W. F. 1949. The present status of North American caribou. Trans. North Am. Wildl. Conf. 14:477–491.

———. 1954a. Preliminary investigation of the barren-ground caribou. Can. Wildl. Serv. Wildl. Manage. Bull., Ser. 1, no. 10B. 112pp.

———. 1954b. The role of ice in the distribution of mammals. J. Mamml. 35:104–107.

———. 1955. A provisional life table for the barren-ground caribou. Can. J. Zool. 33:143–147.

———. 1957. The plight of the barren-ground caribou. Oryx 4:5–20.

———. 1961. A revision of the reindeer and caribou genus *Rangifer*. Nat. Mus. Can. Bull. 177, Bio. Ser. no. 66. 137pp.

———. 1963. The disappearance of the Queen Charlotte Island's caribou. Nat. Mus. Can. Bull. 185:40–49.

———. 1974. The mammals of Canada. Univ. Toronto Press, Toronto. 438pp.

Banfield, A. W. F., and Tener, J. S. 1958. A preliminary study of the Ungava caribou. J. Mammal. 39:560–573.

Bauditz, R. 1972. Sedation, immobilization, and anesthesia with Rompun in captive and free-living wild animals. Vet. Med. Rev. 3:204–221.

Behrend, D. R. 1965. Notes on field immobilization of white-tailed deer with nicotine. J. Wildl. Manage. 29:889–890.

Belanger, L. F.; Choquette, L. P. E.; and Cousineau, J. G. 1967. Osteolysis in reindeer antlers: sexual and seasonal variations. J. Calc. Tiss. Res. 1:37–43.

Bell, K. L., and Bliss, L. C. 1977. Overwinter phenology of plants in a polar semidesert. Arctic 30:118–121.

Bergerud, A. T. 1961. Sex determination of caribou calves. J. Wildl. Manage. 25:205.

———. 1964a. A field method to determine annual parturition rates for Newfoundland caribou. J. Wildl. Manage. 28:477–480.

———. 1964b. Relationship of mandible length to sex in Newfoundland caribou. J. Wildl. Manage. 28:54–56.

———. 1970. Eruption of permanent premolars and molars for Newfoundland caribou. J. Wildl. Manage. 34:962–963.

———. 1971*a*. Abundance of forage on the winter range of Newfoundland caribou. Can. Field Nat. 85:39-52.

———. 1971*b*. Hunting of stag caribou in Newfoundland. J. Wildl. Manage. 35:71-75.

———. 1971*c*. The population dynamics of Newfoundland caribou. Wildl. Monogr. no. 25. 55pp.

———. 1972. Food habits of Newfoundland caribou. J. Wildl. Manage. 36:913-923.

———. 1973. Movement and rutting behavior of caribou (*Rangifer tarandus*) at Mount Albert, Quebec. Can. Field Nat. 87:357-369.

———. 1974*a*. Decline of caribou in North America following settlement. J. Wildl. Manage. 38:757-770.

———. 1974*b*. Relative abundance of food in winter for Newfoundland caribou. Oikos 25:379-387.

———. 1974*c*. Rutting behaviour of Newfoundland caribou. Pages 394-435 *in* V. Geist and F. Walters, eds. The behaviour of ungulates and its relation to management. Vol. 1. IUCN New Ser. Publ. 24, Morges, Switzerland. 940pp.

———. 1974*d*. The role of the environment in the aggregation, movement, and disturbance behaviour of caribou. Pages 552-584 *in* V. Geist and F. Walters, eds. The behaviour of ungulates and its relation to management. Vol. 2. IUCN New Ser. Publ. 24, Morges, Switzerland. 940pp.

———. 1975. The reproductive season in Newfoundland caribou. Can. J. Zool. 53:1213-1221.

———. 1976. The annual antler cycle in Newfoundland caribou. Can. Field Nat. 90:449-463.

———. 1978. Caribou. Pages 83-101 *in* J. L. Schmidt and D. L. Gilbert, eds. Big game of North America: ecology and management. Stackpole Books, Harrisburg, Pa. 494pp.

Bergerud, A. T.; Butt, A.; Russell, H. L.; and Whalen, H. 1964. Immobilization of Newfoundland caribou and moose with succinylcholine chloride and Cap-Chur equipment. J. Wildl. Manage. 28:49-53.

Bergerud, A. T., and Nolan, M. J. 1970. Food habits of hand reared caribou *Rangifer tarandus* L. in Newfoundland. Oikos 21:348-350.

Broughton, E.; Choquette, L. P. E.; Cousineau, J. G.; and Miller, F. L. 1970. Brucellosis in reindeer, *Rangifer tarandus* L., and the migratory barren-ground caribou, *Rangifer tarandus groenlandicus* (L.), in Canada. Can. J. Zool. 48:1023-1027.

Broughton, E.; Miller, F. L.; and Choquette, L. P. E. 1972. Cutaneous fibropapillomas in migratory barren-ground caribou. J. Wildl. Dis. 8:138-140.

Bubenik, A. B. 1975*a*. Significance of antlers in the social life of barren-ground caribou. Pages 436-461 *in* J. R. Luick, P. C. Lent, D. R. Klein, and R. G. White, eds. Proc. 1st Int. Reindeer and Caribou Symp. Bio. Pap. Univ. Alaska, Spec. Rep. No. 1. 551 pp.

———. 1975*b*. Taxonomic value of antlers in genus *Rangifer*, H. Smith. Pages 41-63 *in* J. R. Luick, P. C. Lent, D. R. Klein, and R. G. White, eds. Proc. 1st Int. Reindeer and Caribou Symp. Bio. Pap. Univ. Alaska, Spec. Rep. no. 1. 551pp.

Bunnell, F.; Dauphine, T. C.; Hilborn, R.; Miller, D. R.; Miller, F. L.; McEwan, E. H.; Parker, G. R.; Peterman, R.; Scotter, G. W.; and Walters, J. C. 1975. Preliminary report on computer simulation of barren-ground caribou management. Pages 189-193 *in* J. R. Luick, P. C. Lent, D. R. Klein, and R. G. White, eds. Proc. 1st Int. Reindeer and Caribou Symp. Bio. Pap. Univ. Alaska, Spec. Rep. no. 1. 551pp.

Calef, G. W. 1978. Population status of caribou in the Northwest Territories. Pages 9-16 *in* D. R. Klein and R. G. White, eds. Parameters of caribou population ecology in Alaska. Proc. Symp. and Workshop. Bio. Pap. Univ. Alaska, Spec. Rep. no. 3. 49pp.

Calef, G. W., and Lortie, G. M. 1975. A mineral lick of the barren-ground caribou. J. Mammal. 56:240-242.

Cameron, R. D., and Luick, J. R. 1972. Seasonal changes in total body water, extra-cellular fluid, and blood volume in grazing reindeer. Can. J. Zool. 50:107-116.

Cameron, R. D.; White, R. G.; and Luick, J. R. 1975. The accumulation of water in reindeer during winter. Pages 374-378 *in* J. R. Luick, P. C. Lent, D. R. Klein, and R. G. White, eds. Proc. 1st Int. Reindeer and Caribou Symp. Bio. Pap. Univ. Alaska, Spec. Rep. no. 1. 551pp.

Cameron, R. D., and Whitten, K. R. 1979. Seasonal movements and sexual segregation of caribou determined by aerial survey. J. Wildl. Manage. 43:626-633.

Cameron, R. D.; Whitten, K. R.; Smith, W. T.; and Roby, D. D. 1979. Caribou distribution and group composition associated with construction of the Trans-Alaskan pipeline. Can. Field Nat. 93:155-162.

Chapin, F. S. III; Van Cleve, K.; and Tieszen, L. L. 1975. Seasonal nutrient dynamics of tundra vegetation at Barrow, Alaska. Arctic Alpine Res. 8:209-226.

Chatelain, E. F. 1954. Antler-jaw study. Fed. Aid Wildl. Rest., Progr. Rep. Proj. W-3-R. 8 vols. U.S. Fish and Wildl. Serv., Juneau, Alaska. 8:4-14.

Choquette, L. P. E.; Broughton, E.; Miller, F. L.; Gibbs, H. C.; and Cousineau, J. G. 1967. Besnoitiosis in barren-ground caribou in northern Canada. Can. Vet. J. 8:282-287.

Clarke, C. H. D. 1940. A biological investigation of the Thelon Game Sanctuary. Nat. Mus. Can. Bull. 96, Bio. Ser. no. 25. 135pp.

Coggins, V. L. 1975. Immobilization of Rocky Mountain elk with M99. J. Wildl. Manage. 39:814-816.

Cringan, A. T. 1957. History, food habits, and range requirements of the woodland caribou of continental North America. Trans. North Am. Wildl. Conf. 22:485-501.

Crisler, L. 1956. Observations of wolves hunting caribou. J. Mammal. 37:337-346.

Dauphine, T. C., Jr. 1976. Biology of the Kaminuriak population of barren-ground caribou. Part 4: Growth, reproduction, and energy reserves. Can. Wildl. Serv. Rep. Ser. no. 38. 71pp.

Dauphine, T. C., Jr., and McClure, R. L. 1974. Synchronous mating in Canadian barren-ground caribou. J. Wildl. Manage. 38:54-66.

Davis, J. L. 1978. History and current status of Alaska caribou herds. Pages 1-8 *in* D. R. Klein and R. G. White, eds. Parameters of caribou population ecology in Alaska. Proc. Symp. and Workshop. Bio. Pap. Univ. Alaska, Spec. Rep. no. 3. 49pp.

Davis, J. L.; Grauvogal, C.; Reynolds, H.; and Valkenburg, P. 1978*a*. Human utilization of the Western Arctic caribou herd. Fed. Aid Wildl. Rest. Proj. W-17-8 and W-17-9, Job 3.20R. Alaska Dept. Fish and Game, Juneau. 43pp.

Davis, J. L.; Reynolds, H. V.; Valkenburg, P.; and Shideler, R. T. 1978*b*. Sex and age composition of the Porcupine caribou herd. Fed. Aid Wild. Rest. Proj. W-17-9 and W-17-10, Job 3.23R. Alaska Dept. Fish and Game, Juneau. 23pp.

Dehority, B. A. 1975*a*. Characterization studies of rumen bacteria isolated from Alaskan reindeer (*Rangifer tarandus*). Pages 228-240 *in* J. R. Luick, P. C. Lent, D. R. Klein, and R. G. White, eds. Proc. 1st Int. Reindeer and

Caribou Symp. Bio. Pap. Univ. Alaska, Spec. Rep. no. 1. 551pp.

———. 1975b. Rumen ciliate protozoa of Alaskan reindeer and caribou (*Rangifer tarandus* L.). Pages 241–250 *in* J. R. Luick, P. C. Lent, D. R. Klein, and R. G. White, eds. Proc. 1st Int. Reindeer and Caribou Symp. Bio. Pap. Univ. Alaska, Spec. Rep. no. 1. 551pp.

Des Meules, P. 1965. Operation caribou. Les Carnets de Zoologie 25(2):20–23.

———. 1968. Bringing back the caribou. Animals 10(12):560–563.

Des Meules, P., and Heyland, J. 1969a. Contributions to the study of the food habits of caribou. Part 2: Daily consumption of lichens. Nat. Can. 96:333–336.

———. 1969b. Contributions to the study of the food habits of caribou. Part 1: Lichen preferences. Nat. Can. 96:317–331.

de Vos, A. 1960. Behavior of barren-ground caribou on their calving grounds. J. Wildl. Manage. 24:250–258.

Doerr, J. G., and Dieterich, R. A. 1979. Mandibular lesions in the western arctic caribou herd of Alaska. J. Wildl. Dis. 15:309–318.

Dugmore, A. A. R. 1913. The romance of the Newfoundland caribou. J. P. Lippincott Co., Philadelphia. 186pp.

Edwards, R. Y. 1954. Fire and the decline of a mountain caribou herd. J. Wildl. Manage. 18:521–526.

———. 1958. Land form and caribou distribution in British Columbia. J. Mammal. 39:408–412.

Edwards, R. Y., and Ritcey, R. W. 1959. Migrations of caribou in a mountainous area in Wells Gray Park, British Columbia. Can. Field Nat. 73:21–25.

———. 1960. Foods of caribou in Wells Gray Park, British Columbia. Can. Field Nat. 74:307.

Edwards, R. Y.; Soos, J.; and Ritcey, R. W. 1960. Quantitative observations on epidendric lichens used as food by caribou. Ecology 41:425–431.

Erickson, C. A. 1975. Some preliminary observations on interspecific acoustic communication of semi-domestic reindeer, with emphasis on the mother-calf relationship. Pages 387–397 *in* J. R. Luick, P. C. Lent, D. R. Klein, and R. G. White, eds. Proc. 1st Int. Reindeer and Caribou Symp. Bio. Pap. Univ. Alaska, Spec. Rep. no. 1. 551pp.

Espmark, Y. 1964a. Rutting behavior in reindeer (*Rangifer tarandus* L.). Anim. Behav. 12:159–163.

———. 1964b. Studies in dominance-subordination relationship in a group of semi-domestic reindeer (*Rangifer tarandus* L.). Anim. Behav. 12:420–426.

Freddy, D. J. 1979. Distribution and movements of Selkirk caribou, 1972–1974. Can. Field Nat. 93:71–74.

Freddy, D. J., and Erickson, A. W. 1975. Status of the Selkirk Mountain caribou. Pages 221–227 *in* J. R. Luick, P. C. Lent, D. R. Klein, and R. G. White, eds. Proc. 1st Int. Reindeer and Caribou Symp. Bio. Pap. Univ. Alaska, Spec. Rep. no. 1. 551pp.

Freeman, M. M. R. 1975. Assessing movement in an arctic caribou population. J. Environ. Manage. 3:251–257.

Frick, C. 1937. Horned ruminants of North America. Am. Mus. Nat. Hist. Bull. 69. 669pp.

Giles, R. H., Jr., ed. 1969. Wildlife management techniques. 3rd ed. rev. Wildl. Soc., Washington, D.C. 623pp.

Gunn, A.; Miller, F. L.; and Thomas, D. C. 1981. The current status and future of Peary caribou (*Rangifer tarandus pearyi*) on the arctic islands of Canada. Bio. Conserv. 19:283–296.

Haigh, J. C. 1976. Fentanyl-based mixtures in exotic animal neuroloptanalgesia. Pages 164–180 *in* M. E. Fowler, ed. Proc. Am. Assoc. Zoo. Vet. Congr., St. Louis. 265pp.

Harper, F. 1955. The barren-ground caribou of Keewatin. Univ. Kansas Mus. Nat. Hist. Misc. Publ. no. 6. 163pp.

Harrington, R. 1974. Immobilon-Rompun in deer. Vet Rec. 94:362–363.

Hart, J. S.; Heroux, O.; Cottle, W. H.; and Mills, C. A. 1961. The influence of climate on metabolic and thermal responses of infant caribou. Can. J. Zool. 39:845–856.

Harthoorn, A. M. 1965. Application of pharmacological and physiological principles of restraint of wild animals. Wildl. Mongr. no. 14. 78pp.

———. 1970. The flying syringe. Geoffrey Bles, London. 287pp.

———. 1976. The chemical capture of animals. Balliere Tindall, London. 416pp.

Heape, W. 1931. Migration, emigration, and nomadism. W. Heffer & Sons, Cambridge, England. 369pp.

Hebert, D. M., and McFetridge, R. J. 1978. Chemical immobilization of North American game mammals. Alberta Recreation, Parks, and Wildl., Fish and Wildl. Div., Edmonton. 84pp.

Hemming, J. E. 1971. The distribution and movement patterns of caribou in Alaska. Alaska Dept. Fish and Game, Game Tech. Bull. no. 1. 60pp.

Henshaw, J. 1968a. A theory for the occurrence of antlers in females of the genus *Rangifer*. J. Brit. Deer Soc. 1:222–226.

———. 1968b. The activities of wintering caribou in northwestern Alaska in relation to weather and snow conditions. Int. J. Biometeor. 12:21–27.

———. 1970. Consequences of travel in the rutting of reindeer and caribou (*Rangifer tarandus*). Anim. Behav. 18:256–258.

Hoare, W. H. B. 1927. Report on investigations affecting Eskimo and wild life, District of Mackenzie, 1925–1926, together with general recommendations. Dept. of Interior, Northwest Territories and Yukon Br., Ottawa. 44pp. Mimeographed.

———. 1930. Conserving Canada's musk-oxen (being an account of an investigation of Thelon Game Sanctuary, 1928–29, with a brief history of the area and an outline of known facts regarding the musk-ox). Dept. of Interior, Northwest Territories and Yukon Br., Ottawa. 53pp.

Holleman, D. R., and Luick, J. R. 1977. Lichen species preference by reindeer. Can. J. Zool. 55:1368–1369.

Hopla, C. E. 1975. Q Fever and Alaskan caribou. Pages 498–506 *in* J. R. Luick, P. C. Lent, D. R. Klein, and R. G. White, eds. Proc. 1st Int. Reindeer and Caribou Symp. Bio. Pap. Univ. Alaska, Spec. Rep. no. 1. 551pp.

Houston, D. B. 1970. Immobilization of moose with M99 etorphine. J. Mammal. 51:396–399.

Irving, L. 1966. Adaptation to cold. Sci. Am. 214:94–101.

Irving, L., and Krog, J. 1955. Temperature of skin in the arctic as a regulator of heat. J. Appl. Physiol. 7:355–364.

Jacobi, A. 1931. Das Rentier: eine zoologische Monographie der Gattung *Rangifer*. Akad. Verlag. Leipzig. 264pp.

Jacobsen, N.; Armstrong, W.; and Moen, A. 1976. Seasonal variation in succinylcholine immobilization of captive white-tailed deer. J. Wildl. Manage. 40:447–453.

Johnson, D. R. 1976. Mountain caribou: threats to survival in the Kootenay Pass region, British Columbia, Northwest Sci. 50:97–101.

Johnson, E. A., and Rowe, J. S. 1975. Fire in the subarctic wintering ground of the Beverly caribou herd. Am. Midl. Nat. 94:1–14.

Kelsall, J. P. 1957. Continued barren-ground caribou studies. Can. Wildl. Serv. Manage. Bull. Ser. 1, no. 12. 148pp.

_____. 1968. The migratory barren-ground caribou of Canada. Can. Wildl. Serv. Monogr. no. 3, Ottawa. 340pp.

_____. 1970. Interaction between barren-ground caribou and muskrats. Can. J. Zool. 48:605.

_____. 1975. Warble fly distribution among some Canadian caribou. Pages 509–517 *in* J. R. Luick, P. C. Lent, D. R. Klein, and R. G. White, eds. Proc. 1st Int. Reindeer and Caribou Symp. Bio. Pap. Univ. Alaska, Spec. Rep. no. 1. 551pp.

Klein, D. R. 1964. Range-related differences in growth of deer reflected in skeletal ratios. J. Mammal. 45(2):226–235.

_____. 1970*a*. Food selection by North American deer and their response to overutilization of preferred plant species. Pages 25–46 *in* A. Watson, ed. Animal populations in relation to their food resources. Br. Ecol. Soc. Symp. no. 10. Blackwell Sci. Pub., Oxford and Edinburgh. 447pp.

_____. 1970*b*. Interactions of *Rangifer tarandus* (reindeer and caribou) with its habitat in Alaska. 8th Int. Congr. Game Bio., Helsinki, Finland. Pp. 289–293.

Klein, D. R., and White, R. G. 1978. Parameters of caribou population ecology in Alaska. Proc. Symp. and Workshop, Bio. Pap. Univ. Alaska, Spec. Rep. no. 3. 49pp.

Krog, J. O., and Wika, M. 1975. The circulation in the growing reindeer antlers. Pages 368–373 *in* J. R. Luick, P. C. Lent, D. R. Klein, and R. G. White, eds. Proc. 1st Int. Reindeer and Caribou Symp. Bio. Pap. Univ. Alaska, Spec. Rep. no. 1. 551pp.

Kuropat, P. 1978. Range interrelationships of the western arctic herd. Alaska Coop. Wildl. Res. Unit, Univ. Alaska, Semiannual Prog. Rep. 30:52–55.

Kuyt, E. 1972. Food habits of wolves on barren-ground caribou range. Can. Wildl. Serv. Rep. Ser. no. 21. 36pp.

LaPerriere, A. J., and Lent, P. C. 1977. Caribou feeding sites in relation to snow characteristics in northeastern Alaska. Arctic 30:101–108.

Lent, P. C. 1965*a*. Observations on antler shedding by female barren-ground caribou. Can. J. Zool. 43:553–558.

_____. 1965*b*. Rutting behavior in a barren-ground caribou population. Anim. Behav. 13:259–264.

_____. 1966*a*. Calving and related social behavior in the barren-ground caribou. Z. Tierpsychol. 23:701–756.

_____. 1966*b*. The caribou of northwestern Alaska. Pages 481–517 *in* N. J. Wilimovsky and J. N. Wolfe, eds. Environment of the Cape Thompson region, Alaska. U.S. Atomic Energy Comm., Washington, D.C. 1,250pp.

_____. 1974. Mother-infant relationships in ungulates. Pages 14–55 *in* V. Geist and F. Walthers, eds. The behaviour of ungulates and its relation to management. Vol. 1. IUCN New Ser. Publ. no. 24, Morges, Switzerland. 940pp.

_____. 1975. A review of acoustic communication in *Rangifer tarandus*. Pages 398–408 *in* J. R. Luick, P. C. Lent, D. R. Klein, and R. G. White, eds. Proc. 1st Int. Reindeer and Caribou Symp. Bio. Pap. Univ. Alaska, Spec. Rep. 1. 551pp.

Lentz, C. P., and Hart, J. S. 1960. The effect of wind and moisture on heat loss through the fur of newborn caribou. Can. J. Zool. 38:679–688.

Leopold, A. S., and Darling, F. F. 1953. Wildlife in Alaska. Ronald Press Co., New York. 129pp.

Lewin, V., and Stelfox, J. G. 1967. Functional anatomy of the tail and associated behavior in woodland caribou. Can. Field Nat. 1:63–66.

Long, F. M., ed. 1977. First International Conference on Wildlife Biotelemetry. Pages 1–159 *in* Proc. Int. Conf. Wildl. Biotelemetry, Laramie, Wyoming, 27, 28, 29 July 1977.

_____. 1979. Second International Conference on Wildlife Biotelemetry. Pages 1–259 *in* Proc. Int. Conf. Wildl. Biotelemetry, Laramie, Wyoming, 30, 31 July and 1 August 1979.

Loomis, F. G. 1925. Dentition of artiodactyls. Bull. Geol. Soc. Am. 36:583–604.

Luick, J. R., and White, R. G. 1975. Glucose metabolism in female reindeer. Pages 379–386 *in* J. R. Luick, P. C. Lent, D. R. Klein, and R. G. White, eds. Proc. 1st Int. Reindeer and Caribou Symposium. Bio. Pap. Univ. of Alaska, Spec. Rep. 1. 551pp.

McEwan, E. H. 1963. Seasonal annuli in the cementum of the teeth of barren-ground caribou. Can. J. Zool. 41:111–113.

_____. 1968. Hematological studies of barren-ground caribou. Can. J. Zool. 46:1031–1036.

_____. 1971. Twinning in caribou. J. Mammal. 52:479.

McEwan, E. H., and Whitehead, P. E. 1970. Seasonal changes in the energy and nitrogen intake in reindeer and caribou. Can. J. Zool. 48:905–913.

_____. 1972. Reproduction in female reindeer and caribou. Can. J. Zool. 50:43–46.

McEwan, E. H., and Wood, A. J. 1966. Growth and development of the barren-ground caribou. Part 1: Heart girth, hind foot length, and body weight relationships. Can. J. Zool. 44:401–411.

McEwan, E. H.; Wood, A. J.; and Nordan, H. C. 1965. Body temperature of barren-ground caribou. Can. J. Zool. 43:683–687.

McGowen, T. A. 1966. Caribou studies in northwestern Alaska. Proc. Western Assoc. State Game and Fish Commissioners 46:57–66.

Manning, T. H. 1960. The relationship of the Peary and barren-ground caribou. Arctic Inst., North Am. Tech. Pap. 4. 52pp.

Miller, D. R. 1974. Seasonal changes in the feeding behavior of barren-ground caribou on the Taiga winter range. Pages 744–755 *in* V. Geist and F. Walters, eds. The behaviour of ungulates and its relation to management. Vol. 2. IUCN New Ser. Publ. no. 24. Morges, Switzerland. 940pp.

_____. 1976. Biology of the Kaminuriak population of barren-ground caribou. Part 3: Taiga winter range relationships and diet. Can. Wild. Serv. Rep. Ser. no. 36. 42pp.

Miller, F. L. 1968. Immobilization of free-ranging black-tailed deer with succinycholine chloride. J. Wildl. Manage. 32:195–197.

_____. 1972. Eruption and attrition of mandibular teeth in barren-ground caribou. J. Wildl. Manage. 36:606–612.

_____. 1974*a*. Age determination of caribou by annulations in dental cementum. J. Wildl. Manage. 38:47–53.

_____. 1974*b*. A new era: are migratory barren-ground caribou and petroleum exploitation compatible? Trans. Northeastern Sec. Wildl. Soc. 31:45–55.

_____. 1974*c*. Biology of the Kaminuriak population of barren-ground caribou. Part 2: Dentition as an indicator of sex and age; composition and socialization of the Population. Can. Wildl. Serv. Rep. Ser. no. 31. 88pp.

_____. 1978. Numbers and distribution of Peary caribou on the arctic islands of Canada. Pages 16–19 *in* D. R. Klein and R. G. White, eds. Parameters of caribou population ecology in Alaska. Proc. Symp. and Workshop. Bio. Pap. Univ. Alaska, Spec. Rep. no. 3. 49pp.

Miller, F. L.; Anderka, F. W.; Vithayasai, C.; and McClure, R. L. 1975a. Distribution, movements, and socialization of barren-ground caribou radio-tracked on their calving and post-calving areas. Pages 423–435 in J. R. Luick, P. C. Lent, D. R. Klein and R. G. White, eds. Proc. 1st Int. Reindeer and Caribou Symp. Bio. Pap. Univ. Alaska, Spec. Rep. no. 1. 551pp.

Miller, F. L.; Behrend, D. R.; and Tessier, G. D. 1971. Live capture of barren-ground caribou with tangle nets. Trans. Northeastern Sec. Wildl. Soc. 28:83–90.

Miller, F. L., and Broughton, E. 1971. Polydactylism in a barren-ground caribou from northwestern Manitoba. J. Wildl. Dis. 7:307–309.

———. 1973. Behaviour associated with mortality and stress in maternal-filial pairs of barren-ground caribou. Can. Field Nat. 87:21–25.

———. 1974. Calf mortality on the calving ground of Kaminuriak caribou. Can. Wildl. Serv. Rep. Ser. no. 26. 26pp.

Miller, F. L.; Cawley, A. J.; Choquette, L. P. E.; and Broughton, E. 1975b. Radiographic examination of mandibular lesions in barren-ground caribou. J. Wildl. Dis. 11:465–470.

Miller, F. L., and Gunn, A. 1978. Interisland movements of Peary caribou south of Viscount Melville Sound, Northwest Territories. Can. Field Nat. 92:327–333.

Miller, F. L.; Jonkel, C. J.; and Tessier, G. D. 1972. Group cohesion and leadership response by barren-ground caribou to man-made barriers. Arctic 25:193–202.

Miller, F. L., and McClure, R. L. 1973. Determining age and sex of barren-ground caribou from dental variables. Trans. Northeastern Sec. Wildl. Soc. 30:79–100.

Miller, F. L., and Parker, G. R. 1968. Placental remnants in the rumens of maternal caribou. J. Mammal. 49:778.

Miller, F. L.; Russell, R. H.; and Gunn, A. 1977a. Distributions, movements, and numbers of Peary caribou and muskoxen on western Queen Elizabeth Islands, Northwest Territories, 1972–74. Can. Wildl. Serv. Rep. Ser. no. 40. 55pp.

———. 1977b. Interisland movements of Peary caribou (Rangifer tarandus pearyi) on western Queen Elizabeth Islands, arctic Canada. Can. J. Zool. 55:1029–1037.

Miller, F. L., and Tessier, G. D. 1971. Dental anomalies in barren-ground caribou. J. Mammal. 52:164–174.

Moisan, G. 1959. The caribou of Gaspe. Northeastern Sec. Wildl. Soc. Conf., (Univ. Montreal) 10:201–207.

Moote, I. 1955. The thermal insulation of caribou pelts. Textile Res. J. 25:837.

Müller-Schwarze, D.; Kallquist, L.; Mossing, T.; Brundin, A.; and Andersson, G. 1978. Responses of reindeer to interdigital secretions of conspecifics. J. Chem. Ecol. 4:325–336.

Murie, A. 1944. The wolves of Mount McKinley. U.S. Natl. Park Serv., Fauna Natl. Parks U.S., Fauna Ser. no. 5. 238pp.

Murie, O. J. 1935. Alaska-Yukon caribou. U.S. Dept. Agric. Bur. Bio. Surv. North Am. Fauna no. 54. 93pp.

Neiland, K. A., and Dukeminier, C. 1972. A bibliography of the parasites, diseases, and disorders of several important wild ruminants of the northern hemisphere. Alaska Dept. Fish and Game, Game Tech. Bull. no. 3. 151pp.

Neiland, K. A.; King, J. A.; Huntley, B. E.; and Skoog, R. O. 1968. The diseases and parasites of Alaskan wildlife populations. Part 1: Some observations on brucellosis in caribou. Bull. Wildl. Dis. Assoc. 4:27–36.

Øritsland, N. A. 1974. A windchill and solar radiation index for homeotherms. J. Theor. Bio. 47:413–420.

Parker, G. R. 1972a. Biology of the Kaminuriak population of barren-ground caribou. Part 1: Total numbers, mortality, recruitment, and seasonal distribution. Can. Wildl. Serv. Rep. Ser. no. 20, 95pp.

———. 1972b. Distribution of barren-ground caribou harvest in northcentral Canada. Can. Wildl. Serv. Occas. Pap. no. 15. 20pp.

———. 1978. The diets of muskoxen and Peary caribou on some islands in the Canadian High Arctic. Can. Wildl. Serv. Occas. Pap. no. 35. 21pp.

Parker, G. R., and Ross, R. K. 1976. Summer habitat use by muskoxen (Ovibos moschatus) and Peary caribou (Rangifer tarandus pearyi) in the Canadian High Arctic. Polarforshung 46:12–25.

Pimlott, D. H. 1961. Wolf control in Canada. Can. Audubon Mag. (November–December), pp. 145–152.

Pocock, R. I. 1923. On the external characters of Elaphurus, Hydropotes, Pudu, and other Cervidae. Proc. Zool. Soc. London, pp. 181–207.

Presnell, K. R.; Presidente, P. J. A.; and Rapley, W. A. 1973. Combination of etorphine and xylazine in captive white-tailed deer. Part 1: Sedative and immobilization properties. J. Wildl. Dis. 9:336–341.

Pruitt, W. O., Jr. 1959. Snow as a factor in the winter ecology of barren-ground caribou (Rangifer arcticus). Arctic 12:159–179.

———. 1960a. Behavior of the barren-ground caribou. Univ. Alaska Bio. Pap. no. 3. 44pp.

———. 1960b. Locomotor speeds of some large northern mammals. J. Mammal. 41:112.

———. 1961. On postnatal mortality in barren-ground caribou. J. Mammal. 42:550–551.

———. 1965. A flight releaser in wolf-caribou relations. J. Mammal. 46:350–351.

———. 1966. The function of the brow tine in caribou antlers. Arctic 19:111–113.

Quay, W. B. 1955. Histology and cytochemistry of skin gland areas in the caribou, Rangifer, J. Mammal. 36:187–201.

Rowe, J. S. 1959. Forest regions of Canada. Dept. Northern Affairs Nat. Resour., Ottawa. For. Br. Bull. no. 123. 71pp.

Rowe, J. S., and Scotter, G. W. 1973. Fire in the boreal forest. Quaternary Res. 3:444–464.

Russell, R. H.; Edmonds, E. J.; and Roland, J. 1978. Caribou and muskoxen habitat studies. Environmental-Social Program, Northern Pipelines, ESCOM no. A1-26. Minister of Indian and Northern Affairs and Minister of State, Ottawa. 140pp.

Scholander, P. F.; Hock, R.; Walters, V.; and Irving, L. 1950a. Adaptations to cold in arctic and tropical mammals and birds in relation to body temperature, insulation and basal metabolic rate. Bio. Bull. 99:259–271.

Scholander, P. F.; Hock, R.; Walters, V.; Johnson, F.; and Irving, L. 1950b. Heat regulation in some arctic and tropical mammals and birds. Bio. Bull. 99:237–258.

Scholander, P. F.; Walters, V.; Hock, R.; and Irving, L. 1950c. Body insulation of some arctic and tropical mammals and birds. Bio. Bull. 99:225–236.

Scotter, G. W. 1964. Effects of forest fires on the winter range of barren-ground caribou in northern Saskatchewan. Can. Wildl. Serv. Wildl. Manage. Bull. Ser. 1, no. 18. 111pp.

———. 1965. Chemical composition of forage lichens from northern Saskatchewan as related to use by barren-ground caribou. Can. J. Plant Sci. 45:246–250.

———. 1967a. Effects of fire on barren-ground caribou and their forest habitat in northern Canada. Trans. North Am. Wildl. Nat. Resour. Conf. 32:246–259.

_____. 1967*b*. The winter diet of barren-ground caribou in northern Canada. Can. Field Nat. 81:33–39.

_____. 1972. Chemical composition of forage plants from the Reindeer Reserve, Northwest Territories. Arctic 25:21–27.

Seton, E. T. 1927. Hoofed animals. Vol. 3, pages 53–150 *in* Lives of game animals. Doubleday Page, New York. 4 vols. 3,115pp.

Shank, C. C.; Wilkinson, P. F.; and Penner, D. F. 1978. Diet of Peary caribou, Banks Island, NWT. Arctic 31:125–132.

Shoesmith, M. W. 1976. Twin fetuses in woodland caribou. Can. Field Nat. 90:498–499.

Simkin, D. W. 1965. A preliminary report of woodland caribou study in Ontario. Ontario Dept. Lands and Forests Sec. Rep. (Wildl.) no. 59. 75pp.

Simmons, N. M. 1971. An inexpensive method of marking large numbers of Dall sheep for movement studies. Trans. 1st North Am. Wild Sheep Conf., Fort Collins, Colo. 1:116–126.

Skoog, R. O. 1968. Ecology of the caribou (*Rangifer tarandus granti*) in Alaska. Ph.D. Thesis. Univ. California, Berkeley. 699pp.

Smith, H. J.; Archibald, R. M.; and Corner, A. H. 1964. Elaphostrongylosis in Maritime moose and deer. Can. Vet. J. 5:287–296.

Sonnenfeld, J. 1960. Changes in an Eskimo hunting technology: an introduction to implement geography. Ann. Ass. Am. Geog. 50:172–186.

Stardom, R. R. P. 1975. Woodland caribou and snow conditions in southeast Manitoba. Pages 324–334 *in* J. R. Luick, P. C. Lent, D. R. Klein, and R. G. White, eds. Proc. 1st Int. Reindeer and Caribou Symp. Bio. Pap. Univ. Alaska, Spec. Rep. no. 1. 551pp.

Stelfox, J., and Robertson, J. 1976. Immobilizing bighorn sheep with succinylocholine chloride and phencyclidine hydrochloride. J. Wildl. Manage. 40:174–176.

Stonehouse, B. 1968. Thermoregulatory function of growing antlers. Nature 218:870–872.

Tener, J. S. 1963. Queen Elizabeth Islands game survey, 1961. Can. Wildl. Serv. Occas. Pap. no. 4 50pp.

Thing, H. 1977. Behaviour, mechanics, and energetics associated with winter cratering by caribou in northwestern Alaska. Bio. Pap. Univ. Alaska no. 18. 41pp.

Thomas, D. C., and Broughton, E. 1978. Status of three Canadian caribou populations north of 70 in winter 1977. Can. Wildl. Serv. Prog. Note no. 85. 12pp.

Thomas, D. C.; Russell, R. H.; Broughton, E.; Edmonds, E. J.; and Gunn, A. 1977. Further studies of two populations of Peary caribou in the Canadian arctic. Can. Wildl. Serv. Prog. Note no. 80. 13pp.

Thomas, D. C.; Russell, R. H.; Broughton, E.; and Madore, P. L. 1976. Investigations of Peary caribou populations on Canadian arctic islands. Can. Wildl. Serv. Prog. Note no. 64. 13pp.

Wales, R. A.; Milligan, L. P.; and McEwan, E. H. 1975. Urea recycling in caribou, cattle and sheep. Pages 297–307 *in* J. R. Luick, P. C. Lent, D. R. Klein, and R. G. White, eds. Proc. 1st Int. Reindeer and Caribou Symp. Bio. Pap. Univ. Alaska, Spec. Rep. no. 1. 551pp.

Watson, G. W., and Keough, E. P. 1954. Caribou jaw study. U.S. Fish Wildl. Serv., Fed. Aid Wildl. Restor. Proj. W-3-R-9, Q. Prog. Rep. 9:50–60.

Weeden, R. 1976. ACS and western arctic caribou-wolf problem. Alaska Conserv. Rev. 17:6–7.

White, R. G. 1975. Some aspects of nutritional adaptations of arctic herbivorous mammals. Pages 239–268 *in* F. J. Vernberg, ed. Adaptations to the environment. Educational Publishers, New York. 576pp.

White, R. G.; Thomson, B. R.; Skogland, R.; Person, S. J.; Russell, D. E.; Holleman, D. F.; and Luick, J. R. 1975. Ecology of caribou at Prudhoe Bay, Alaska. Pages 150–201 *in* J. Brown, ed. Ecological investigations of the tundra biome in the Prudhoe Bay region, Alaska. Bio. Pap. Univ. Alaska, Spec. Rep. no. 2. 215pp.

White, R. G., and Yousef, M. K. 1974. Energy cost of locomotion in reindeer. Pages 38–42 *in* Studies on the nutrition and metabolism of reindeer-caribou in Alaska with special interest in nutritional and environmental adaptation. Prog. Rep. July 1973–December 1974, Inst. Arctic Bio., Univ. Alaska, Fairbanks. 189pp.

Wilkinson, P. F.; Shank, C. C.; and Penner, D. F. 1976. Muskox-caribou summer range relations on Banks Island, N.W.T. J. Wildl. Manage. 40:151–162.

Young, B. A., and McEwan, E. H. 1975. A method for measurement of energy expenditure in unrestrained reindeer and caribou. Pages 355–359 *in* J. R. Luick, P. C. Lent, D. R. Klein, and R. G. White, eds. Proc. 1st Int. Reindeer and Caribou Symp. Bio. Pap. Univ. Alaska, Spec. Rep. no. 1. 551pp.

Yousef, M. K., and Luick, J. R. 1975. Responses of reindeer, *Rangifer tarandus*, to heat stress. Pages 360–367 *in* J. R. Luick, P. C. Lent, D. R. Klein, and R. G. White, eds. Proc. 1st Int. Reindeer and Caribou Symp. Bio. Pap. Univ. Alaska, Spec. Rep. no. 1. 551pp.

FRANK L. MILLER, Canadian Wildlife Service, Western and Northern Region, Room 1000, 9942, 108 Street, Edmonton, Alberta, Canada T5K 2J5.

48

Pronghorn

Antilocapra americana

David W. Kitchen
Bart W. O'Gara

NOMENCLATURE

COMMON NAMES. Pronghorn, antelope, prongbuck, pronghorn antelope, berrendos (Mexican). Cree and Sioux names for the pronghorn meant "small caribou" and "little pale deer."
SCIENTIFIC NAME. *Antilocapra americana*
SUBSPECIES. *A. a. americana, A. a. mexicana, A. a. peninsularis, A. a. oregona,* and *A. a. sonoriensis.*

The pronghorn is the only living species of the subfamily Antilocaprinae, family Bovidae (O'Gara and Matson 1975). The subfamily dates from the Miocene in North America (Frick 1937), the genus from the middle Pliocene (Webb 1973), and the species from early Pleistocene.

A vast majority of the present-day pronghorns belong to the subspecies *A. a. americana.* Goldman (1945) noted that "three somewhat isolated, finger-like southern extensions carry the general range of the pronghorn antelope as a species into Mexico. These peripheral extensions represent geographic races differing from the typical form and from one another only in comparatively slight details of size, color, and structure." The same can be said for the western race, *A. a. oregona.* Bailey (1932), in naming that subspecies, stated that "the animals show only slight and gradual variation over their entire range, and no sharp lines of difference between described forms can be found." Whether all five subspecies are valid is a moot question further complicated by transplants of *A. a. americana* into the ranges of other subspecies.

DISTRIBUTION

Pronghorns formerly roamed north to a little beyond the South Saskatchewan River in Saskatchewan, the Red Deer River in Alberta, and southwestern Manitoba, Canada; southward through Chihuahua, and Coahuila to northeastern Durango, Mexico; the desert plains of central and western Sonora, to about 29° 30' N latitude on the gulf side of Baja and 27° N on the west coast; eastward to western Minnesota, western Iowa, northwestern Missouri, Kansas, Oklahoma, and western Texas; and westward to western Montana,

FIGURE 48.1. Distribution of the pronghorn (*Antilocapra americana*) in North America.

southern Idaho, eastern Oregon, Nevada, and California (Hall 1946; Miller and Kellogg 1955). Pronghorn still occupy shrub and grasslands through much of their former range (Russell 1964). Limited use is also made of deserts and agricultural lands, especially where grain or hay fields are interspersed with native vegetation (figure 48.1). Antelope have been introduced outside of their historic range to eastern Washington and the Hawaiian Islands, but with little success.

DESCRIPTION

Pronghorns have robust, somewhat chunky bodies that are predominantly white and rusty brown to tan with black and dark brown markings on the head and neck (figure 48.2). The large head is carried upright. The ears have curved tips and are covered inside and out with hair; they are usually erect, giving the animal an alert appearance. The eyes are large for the size of the animal and black, with long, black eyelashes that serve as sun visors. The mucous membranes of the nose and mouth also are coal black.

The legs and feet are long and slim, and lack the dewclaws common to most ruminants. The pointed hoofs are cloven and black with cartilaginous padding to cushion the shock of running over hard ground and rocks. The front hoofs are slightly larger than the rear ones and seem to carry most of the animal's weight when it is running.

Skin and Pelage. Pronghorn skin is thin and almost useless as leather. Bucks have nine skin glands and does have six. Both sexes have four interdigital and two rump, or ischiadic, glands. The former apparently produce sebum to condition the hoofs and the latter discharge airborne scent when the animals are excited (Moy 1970, 1971; O'Gara and Moy 1972). In addition, each buck has a subauricular gland below each ear and a median gland in the middle of its back about 30 cm anterior to the tail. All three function in courtship and the subauriculars are used for marking vegetation on territories (Bromley and Kitchen 1974; Kitchen 1974).

Coloration differs according to season, sex, and age. Color patterns usually conform to the following description in northern races but southern animals are slightly paler.

White hair covers the undersurface of the body and extends down the inner limbs and up the sides of the body, forming a rectangular area between the shoulder and the hip. The throat is marked with a white crescent above a white shield; the point of the shield joins the white underparts at the base of the neck. The lips, chin, areas around the bases of the horns and on the cheeks, and inner surfaces of the ears are whitish. White hairs of the two rump patches are about 8 cm long and, when erected, fan beyond the normal contours of the body. A narrow, tan to rusty strip connects or nearly connects the upper surface of the tail with the colored back. Rusty brown to tan hairs, 3 to 4 cm long, cover the back, most of the neck, and the outer sides of the upper parts of the limbs. Lower parts of the limbs are covered with short, tan hair. The color becomes more rusty on the neck; erectile 8- to 10-cm hairs of the mane are russet, usually tipped with varying amounts of black. The head is creamy white on the sides to wood brown on top, and is marked in the male with brownish black patches starting just below the ears and extending downward 8 to 10 cm. Sometimes the entire face below a line connecting the horns is blackish. Often, this mark is T-shaped, with a horizontal line between the horns and a vertical line down the nose, yet other animals have only black near the end of the nose.

Newborn fawns are paler and grayer than their parents; white areas are stained with buff and blackish parts are faintly indicated. Whorls of dusky hair mark the location of coming horns.

Pronghorns begin shedding in late winter or early spring. By August, the old hair of healthy animals is completely shed and the new coat is sleek, displaying the darkest colors of the year. The new hair is short and flexible, but it lengthens as winter approaches. During winter, the hair bleaches and the once rusty brown areas become light tan. Fawns molt their drab-colored coats at about three weeks of age, acquiring adult colors (Bromley 1977).

The stalks of the long winter hairs are thicker than the tips. When once bent, a hair does not straighten again, and hairs are easily rubbed off. Coarse and lifeless as the long winter hair may look, it provides excellent protection from the cold. The hairs have tough surfaces and large central air cells (Murie 1870). These cells provide dead-air spaces like those in a down-filled jacket. The ability to erect the hair on any part of its body probably adds to the efficiency of thermoregulation in the pronghorn.

Horns. Spots where horns will develop are obvious on fetuses about four months before birth. In Montana, all adult bucks and about 70 percent of the adult does had horns, although the horns of does averaged only 42 mm long. The horns of females began growing during the second year of life, whereas those of the males started in the second month (O'Gara 1968).

Blackish horns with anterior prongs and superior hooks, usually having whitish tips, project from ridges just above the eyes of adult bucks. The prongs are triangular. Below the prongs, the sheaths are compressed laterally; above the prongs, the hooks are cylindrical (figure 48.3). The visible horns rest on bony cores like sheaths on knife blades; the two are separated by a living layer of skin. Lengths of horns vary somewhat from area to area. During autumn in Montana, buck fawns have 4- to 5-cm cone-shaped horns, yearlings have 13- to 23-cm horns that may or may not be pronged, and males two years old or older have pronged horns 30 to 40 cm long. Maximum horn development is generally reached at four to five years of age. Montana pronghorns generally shed their horns during November (O'Gara 1968); Buechner (1950a) observed horn shedding in Texas from mid-October to early November.

Doe horns usually do not have prongs or the prongs are rudimentary. Horn sizes, as well as the time of shedding, are much more variable among females than males (O'Gara 1969a).

The horns of buck antelope grow and harden dur-

FIGURE 48.2. Pronghorn (*Antilocapra americana*), showing black jaw patch and unique horn structure.

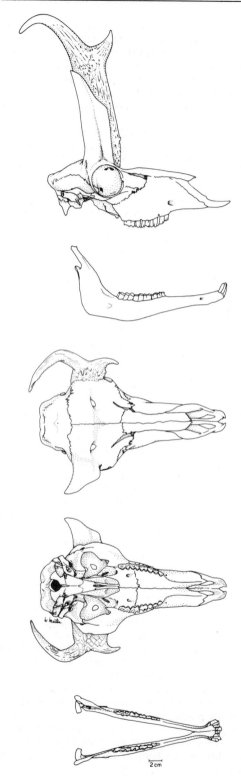

FIGURE 48.3. Skull of the pronghorn (*Antilocapra americana*). From top to bottom: lateral view of cranium, lateral view of mandible, dorsal view of cranium, ventral view of cranium, dorsal view of mandible. The right horn sheath has been removed.

ing 9 to 10 months of the year. Early investigators reported that the horns were composed of agglutinated hair (Bailey 1920; Seton 1953). Hair is incorporated into the rapidly growing horns near the bases, but pronghorn horns are similar to those of other bovids. Pronghorns are the only bovids that shed horns annually, but, contrary to some American textbooks, all subfamilies of Bovidae have members that shed horns at least once or twice during their lives (O'Gara and Matson 1975).

Skull and Dentition. The frontal sinuses of pronghorn skulls open to the outside by two large, longitudinal fossae in the dorsal surface of the frontal bones. The lacrimal bones do not articulate with the nasal bones (Hall and Kelson 1959). Suborbital depressions are not present and the supraorbital foramina are large. The nasal bones are furcate and widest posteriorly. The masseteric ridge is low and the auditory bullae are moderate, compressed, and angular. The supraoccipital is perpendicular and concave, and the basioccipital tubercles are abortive. The mandibular angles are widely rounded (Murie 1870). Horn cores in males, and when present in females, are directly superior to the orbits, which are very large and slightly elevated above the face. The brain is located in the posterior one-third of the skull and the nose is elongated (see figure 48.3).

Antelope teeth are hypsodont and selenodont; they have long crowns and either no roots or roots that develop months or years after the teeth erupt. Such teeth continue to grow and are adapted for grinding rough vegetation and compensating for wear from grit and dirt. The six incisors and two canines in the lower jaw (together called incisiform teeth) are simple, spadelike teeth with rounded tips that become flatter with wear. Measurements of tooth heights above the gum lines do not aid in determining age because the extent of exposure often remains nearly constant throughout life. The dental formula is 0/3, 0/1, 3/3, 3/3 = 32.

PHYSIOLOGY

Body Weight, Size, and Temperature. Members of the southern subspecies are generally considered smaller than the northern ones, and the Oregon antelope is supposedly slightly larger than the American (Bailey 1932). However, measurements available for the subspecies indicate slight differences in size. Southern animals are lighter in weight than northern ones. Average body measurements (mm) of adult females/males collected in Alberta (Mitchell 1971) were: total length 1405.6/1415.7, height at shoulder 860.3/874.5, length of tail 96.9/105.2, and length of ear 141.6/143.1. Weights vary seasonally; adult *A. a. americana* does collected throughout the year in Alberta averaged 50.5 kg (range 46.9 to 56.2), while males in the same collection averaged 56.4 kg (range 46.5 to 70.3) (Mitchell 1971). Adult *A. a. mexicana* does and bucks in Texas averaged 40 and 41 kg (Buechner 1950a). Northern fawns weigh from 3 to 5 kg at birth. In the northern parts of the pronghorn's range, bucks are heaviest during late summer (Hepworth and Blunt 1966; O'Gara 1968; Mitchell 1971). In Colorado, bucks stored little fat until after the breeding season (Bear 1971). The stresses of parturition and lactation keep does thin until autumn; maximum weights were reached in November and December in Montana and Colorado (O'Gara 1968; Bear 1971). Both sexes were generally thinnest during May in the north. In the southwest, antelope weights are undoubtedly depressed by the same stresses of the rut and lactation, but for desert animals, times of weight gains probably correlate with rainy seasons.

An average body temperature of 38.5° C was recorded via telemetry units implanted in the flanks of two pregnant pronghorn does (Lonsdale et al. 1971). Thorne (1975) recorded slightly lower temperatures of 36.9° C and 38.3° C for a mature and a yearling doe. He found that antelope temperatures were erratic and changes of 1.5° to 2° C were common.

Visceral Organs. Antelope stomachs are only about one-half as large as those of domestic sheep. The heart, lungs, liver, and kidneys are all larger than those of sheep (Sundstrom et al. 1973). The small stomach necessitates the use of nutritious, high-protein forage, but is an advantage to the fast-running pronghorn, as are the large heart and lungs. The large liver may aid the pronghorn in using plants higher in selenium, alkaloids, and essential oils than livestock can. The large liver also must provide a ready source of energy in stored glycogen. Large kidneys may be an adaptation to conserve water in desert areas, but also may allow greater excretion of toxic substances from plants. Pronghorns open their mouths and gulp air as soon as they start to run. They have large trachea, allowing rapid exchange of oxygen.

Water and Salt Needs. Hoover et al. (1959) reported that some pronghorn herds in Colorado were never observed drinking water. Later studies indicated that antelope need water during dry weather in some areas. Beale and Smith (1970) found that, when forbs were abundant and their moisture content was 75 percent or more, Utah pronghorns did not drink although water was readily available. However, during extremely dry periods, water consumption reached three liters per day per animal. Sundstrom (1968) observed a close relationship between antelope distribution and water during a census that began during the last week of July in the Red Desert of Wyoming. Most pronghorns were within 5-6 km of water. Occasionally, adult bucks were found 10 km from any apparent source of water. However, Carr (ND) observed that in some areas of Sonora, Mexico, pronghorns survived where free water was not available. He postulated that during the dry months these antelope must obtain moisture from the chain fruit cholla (*Opuntia fulgida*) and other desert succulents.

Water flux is similar to that in domestic sheep and mule deer (*Odocoileus hemionus*), but noticeable differences exist between the water kinetics of males and females. Under test conditions, pronghorns had slightly higher content of body water than was reported for other ruminants, possibly because of lower fat content than most domestic or laboratory animals (Wesley et al. 1970).

Audubon and Bachman (1851) and Skinner (1922) noted that pronghorns used natural salt and salt put out for other ungulates. Because they choose a succulent diet, pronghorns should use more salt than most other ungulates.

Blood. Barrett and Chalmers (1976) analyzed blood from trapped adults and neonate pronghorn in Alberta. Significant differences were not noted between singles and twins at birth, but considerable differences existed between sexes. Leucocyte count, sodium, potassium, and phosphorus had higher values in the blood of pronghorn neonates when the ambient temperature was <10° C than when it was >20° C, whereas blood urea nitrogen was higher at >20° C than at <10° C.

Trout (1976) and Barrett and Chalmers (1976, 1977) provided data on blood from trapped animals in Idaho and Alberta, but those hematological values should probably be regarded as preliminary information until additional studies are conducted on pronghorns from different areas.

Metabolic Rates. Fasting metabolic rates of four pronghorns ranging from 108 to 182 days of age ranged from 61 to 110 and averaged 92 kcal/kg/day times $_w0.75$, where W is total weight in kilograms. Heat production was comparatively high, possibly related to the high metabolism of young animals (Wesley et al. 1970). The concentration and percentage distribution of volatile fatty acids in pronghorn rumens correspond to those of domestic ruminants on somewhat comparable diets, but some statistically significant differences ($P < 0.05$) were obtained on the percentage distribution of the acids of adults and fawns (Nagy and Williams 1969). The average vitamin A potency of fresh pronghorn liver during winter in Oregon was 1024 International Units per gram; during late spring and early summer, potency dropped 20 percent; in September, the average of 25 livers was more than 2200 I.U. per gram (Weswig 1956).

REPRODUCTION

Pronghorns are polygamous; females usually become sexually mature at 16 months of age but occasionally conceive at about 5 months (Wright and Dow 1962; Mitchell 1967; O'Gara 1968). The breeding period lasts from mid-September to early October in the north and from late July to early October in the south (Lehman and Davis 1942; Buechner 1950a). The gestation period in captivity averages 252 days (Hepworth and Blunt 1966). Twins are much more common than single births, ovulation and breeding are almost simultaneous, and does on good range usually ovulate four to seven ova. Upon reaching the uterus, fertilized ova rapidly expand into blastocysts approximately 3 mm in diameter. During the third stage walls of each blastocyst begin to elongate and form a tube about 125 mm long and 0.5 mm in diameter. At that time, the uterus is active and the fragile thread stage blastocysts are kneaded together and often tangle, forming overhand knots, or two threads may form granny and/or square knots. One-fourth to one-third of the ova generally die of malnutrition during the thread stage because their fetal membranes are so reduced by knotting and breaking off that they cannot absorb enough nutrition. Next, tubal walls thicken and diameters increase to nearly those of the uterine lumens. As the inner cell masses become embryos, they sink through the walls of the tubes, acquire amnions, and then float free within the chorions. Cells on the outside of the chorions acquire brush borders to assist in absorption of food, and the embryos develop rapidly as new fetal membranes bud. The yolk sac, from the midgut, is a Y-shaped, ribbon-like tube several times longer than the embryo before implantation. The forks of the Y lie tightly against the chorion, apparently absorbing nutrients and transferring them to the gut. Each horn of the uterus has two expansions of the lumen, one near the *corpus uteri* (proximal chamber) and one near the oviduct (distal chamber). A rich supply of blood serves the uterine wall at the proximal chamber; fewer blood vessels supply the wall near the distal chamber (O'Gara 1969b).

Quadruplets often survive the thread stage, and as many as seven embryos have been reported (Mitchell 1965). If only two embryos survive, they both locate in the proximal chambers. Implantation begins about a month after fertilization, when embryos are approximately 5 mm long. Fetal membranes of embryos in proximal chambers grow extremely fast. When a proximal embryo's necrotic tip reaches the membranes of a distal embryo, the tip usually pierces the distal chorion, folds in most of the length of the distal allantois, and carries the distal membranes and embryo to the oviduct, where the embryo perishes from lack of nutrition. Pronghorns are the only animals known to reduce the number of embryos during pregnancy by the two methods described above (O'Gara 1969b).

As implantation proceeds, chorionic villi in cotyledons give off short, leaflike secondary villi at right angles to the primary villi and the structure becomes dendritic, but the placenta remains epitheliochorial (Wislocki and Fawcett 1949). Because maternal tissue is not destroyed during placentation, there is no bleeding at birth unless the birth canal is torn. Fetal membranes of pronghorn twins join and fuse in the *corpus uteri* when the fetuses are approximately 50 mm long and 75 days old. After fusion, the ends of chorions, in apposition to one another, degenerate and the amnions fuse. Allantoes persist through midgestation as vesicles containing nitrogenous wastes from the fetuses. The amnion extrudes from the vulva for some time before the first fawn is born. The first part of a fawn to appear is usually the white-tipped front hoofs (O'Gara 1968).

The *corpora lutea* grow to approximately 5 mm in diameter during the first week after ovulation; if the doe becomes pregnant, slow but steady growth continues until the birth of the fawns. At birth, *corpora lutea* measure more than 8 mm; one or two days later, they are reduced to spheres about 5 mm in diameter;

six weeks later, the *corpora albicantia* appear as orange bodies less than 2 mm in diameter or as thin lines (O'Gara 1968).

Yearling males are capable of breeding (Wright and Dow 1962), but fawns generally are considered incapable. However, testes of fawns are in the late prepubertal stage by November (O'Gara et al. 1971). Yearling males seldom get a chance to breed because older bucks hold territories where most of the breeding takes place (Kitchen 1974).

Testes of mature males vary in weight annually, averaging more than 40 g in July and August and less than 20 g in January and February. In Montana, spermatogenesis by yearlings and adults continues throughout the year but is limited from December through February. By late March, tubules of the epididymides contain scattered to numerous spermatozoa. From April through November, epididymides are packed with spermatozoa; during December, they again become scattered and leucocytes appear, apparently "cleaning up" degenerating spermatozoa. Male pronghorns apparently are capable of ejaculating spermatozoa during much of the year, and the short estrous period of females determines the breeding season (O'Gara et al. 1971).

ECOLOGY

Habitat. Pronghorns use 26 different prairie and shrubland habitats within their range (Sundstrom et al. 1973). Although specific plant species vary with the region, certain characteristics are common to habitats with high pronghorn populations. Key factors are: a low, rolling topography; a mixture of forage types (i.e., forbs, grasses, and shrubs); and an annual precipitation of 25-35 cm (Yoakum 1974; Autenrieth 1978). Specific forage requirements for the presence and reproduction of pronghorns are: a strong forb component (25-35 percent by composition); high-quality winter browse that is above snow level (10-20 percent by composition); a mixture of native grasses; and at least 50 percent ground cover of plants (Fichter and Nelson 1962; Bayless 1969; Sundstrom et al. 1973; Autenrieth 1978).

The present distribution of pronghorns by habitat shows that 67 percent (N = 390,000 animals) occur in various shortgrass to midgrass habitats, 32 percent occur in mixed grass-shrub habitats, and 1 percent are in deserts (Yoakum 1972; Sundstrom et al. 1973; Yoakum 1974). The grassland habitats contain the best mixed vegetative conditions in proper amounts and consequently receive the highest use by pronghorns. All management guidelines stress the mixed vegetative conditions required by pronghorns for good health and reproduction (Autenrieth 1978).

Group Size, Population Density, and Sex Ratios. Pronghorns are social and form herds. The reproductive state of an individual, its sex, and the season of the year influence herd size. Winter herds contain all sex and age classes and range in size from 2 to thousands of animals (Einarsen 1948; Buechner

1950*a*, 1950*b;* see Yoakum 1967 for a summary). As winter abates, large winter herds break up into smaller herds according to sex and age. Young males (one to four years old) form bachelor herds of 2–40 animals; females form herds of 5–20 animals and may associate with older males (more than five years old). As does approach parturition, they leave the herd and scatter over large areas to give birth (McLean 1944; Bromley 1968, 1977; Kitchen 1974; Autenrieth and Fichter 1975). When fawns are three to six weeks old and capable of sustained, rapid flight, does form nursery herds of 2–20 does accompanied by their fawns and yearling does. Older males are solitary during summer, but associate with nearby nursery herds (Buechner 1950*a;* 1950*b;* Cole and Wilkins 1958; Bromley 1968, 1977; Kitchen 1974). As the rut begins, bachelor herds break up and young males wander alone or in small groups that move from area to area harassing large males with does (Bromley 1969, 1977; Kitchen 1974; Kitchen and Bromley 1974). During the rut, females are aggregated or alone and frequently move from one large male to another (Kitchen 1974). Large males are usually with female groups during the rut, but may be solitary if no females are foraging on their territories.

The apparent population density is a function of the total population, herd size and the amount of suitable habitat in the area. For pronghorns, group size and habitat suitability vary with the season and the reproductive state of individuals. Population densities, therefore, should be computed by season and habitat type with regard to reproductive state. No comprehensive data that consider all of these factors are available, but reports of densities range from 0.6 to 3.27 animals/100 ha.

Pronghorn sex ratios approximate 1:1 at birth (Bromley 1968, 1977; Kitchen 1974; Autenrieth and Fichter 1975). Adult sex ratios vary from 1 male:1.10 females to 1 male:5 females (Einarsen 1948; Foree 1972). How much of this variability is due to natural differential mortality and how much is due to male-only hunting seasons is unknown.

Activity Patterns. Pronghorns are active day and night, but activity is greatest just after sunrise and before sunset. These peaks are most evident during the hottest days of the year (Kitchen 1974). Pronghorns follow a cycle of feeding, bedding, ruminating, and, in the driest areas, movements to and from free water. Actual schedules of these activities vary from day to day and show regional and seasonal variations as well (Skinner 1922; Kautz 1942; McLean 1944; Einarsen 1948; Buechner 1950*a*, 1950*b;* see Yoakum 1967 for summary). Nocturnal bouts of feeding are shorter than diurnal bouts and bedding periods are longer at night (Einarsen 1948; Buechner 1950*a*, 1950*b;* Kitchen 1974).

Home Range, Territory, and Movements. Home ranges vary from 440 to over 1200 ha for nursery and bachelor herds in Palouse Prairie habitat (Kitchen 1974). These figures are consistent with home range sizes in many grassland and grass-shrubland habitats. The variability of home range size depends on the

habitat quality, past history of domestic animal grazing, overall population size, herd size, and season of the year. Therefore, no single home range size is typical of pronghorns when in herds. Individual home ranges may be the same as or smaller than, but rarely larger than, the herd's home range. In general, winter home ranges for individuals are smaller than summer ranges because of restrictions on movement caused by snow. Individual winter home ranges in central Montana were from 165 to 2300 ha (Bayless 1969), which appears to be typical for pronghorns.

Large males are usually territorial (Cole and Wilkins 1958; Bromley 1969, 1977; Gilbert 1973; Kitchen 1974; Kitchen and Bromley 1974). Territory size at the National Bison Range in Montana ranged from 23 to 434 ha (Bromley 1969; Kitchen 1974). This size is typical of areas with good range conditions and high populations. Territories are larger in areas with poor range conditions and low populations.

Daily movements vary with the season: they are shortest (1 km/day) in spring, when forage is succulent and abundant, and longest (3–10 km/day) in fall, when forage is dry and not abundant. Such movements are in response to the phenology of certain plants and to changes in weather conditions. In winter, shifts in wind direction or new snowfall caused Montana pronghorns to move 1–2 km to areas protected from wind and having taller vegetation (Kitchen 1974). Lactating does shift summer foraging areas in response to plant succulence and plant phenology. Bromley (1977) showed how plant distributions influenced buck and doe movements in Wind Cave National Park, South Dakota.

Pronghorns may migrate between summer and winter ranges. Such migrations involve distances of 18 to over 160 km, but can be less than 3–4 km (Beer 1944; McLean 1944; Einarsen 1948; Baker 1953; Martinka 1967; Kitchen 1974; Hoskinson 1977). In Idaho, pronghorns migrated to and from winter ranges in response to forage succulence and not to weather conditions (i.e., snow depth, cold, wind, etc.). In spring, pronghorns moved to summer range as forage became more succulent; and in winter, they followed a reverse gradient of forage succulence back to their winter range (Hoskinson 1977).

FOOD HABITS

The diversity of pronghorn foods is a reflection of their broad geographic range and use of many habitat types. Food habits vary locally, regionally, and range wide and can be summarized only on a forage-type basis of forbs, grasses, shrubs, and others (i.e., domestic crops, cacti, etc.). Data on food habits were summarized from studies throughout the pronghorns' range (table 48.1). Browse is important year-round and it is critical in winter (Bayless 1969). Forbs are used in all seasons except fall; this low use probably is due to a lack of succulence and key nutrients. Grasses are used least, but are important spring forage. Domestic crops (i.e., wheat, alfalfa, etc.) are important locally, particularly in winter and spring (Cole and Wilkins 1958;

TABLE 48.1. Food habits of the pronghorn (*Antilocapra americana*) by season

Season	Browse		Forbs		Grass		Other[a]	
	$\overline{X}\%$	Range%	$\overline{X}\%$	Range%	$\overline{X}\%$	Range%	$\overline{X}\%$	Range%
Spring	61.8	5.9–100	26.5	4–74	12.2	0–49	24.8	0–88
Summer	54.4	1.0–100	37.9	0–85	6.6	0–30	5.9	0–42
Fall	72.1	0.0–98	23.2	1.7–82	3.8	0–20	32.0	0–100
Winter	71.3	4.5–100	9.1	0–47	3.4	0–30	40.4	0–100
Yearlong[b]	62.4	5.5–74	44.0	0–93.7	7.0	0–26	9.8	0–19

SOURCE: Summarized from studies done in all portions of pronghorn range.
NOTE: $\overline{X}\%$ = the mean percentage of the food type in the pronghorn's diet.
[a] Includes wheat, other grain crops, alfalfa, cacti, and miscellaneous.
[b] From studies carried out on a year-long basis only, not a mean derived from the above data.

Hoover et al. 1959). Succulence of vegetation is considered important in pronghorn food selection and may be related to good reproduction (Fichter and Nelson 1962; Autenrieth 1978), migrations (Hoskinson 1977), and the placement and quality of territories (Kitchen 1974).

BEHAVIOR

The Breeding System. The older literature describes the pronghorn as a harem-breeding species (McLean 1944; Einarsen 1948; Buechner 1950a, 1950b; and many others) like elk (*Cervus elaphus*) (McCullough 1969; Bowyer 1974) and horses (*Equus caballus*) (Berger 1977; Feist 1971; Feist and McCullough 1975, 1976; Green and Green 1977; Kitchen et al. 1977). Recent studies, however, have demonstrated that most pronghorns are territorial (Cole and Wilkins 1958; Bromley 1969, 1977; Gilbert 1973; Kitchen 1974; Kitchen and Bromley 1974). One early worker, Buechner, pointed out at the 1971 Conference on Ungulate Behavior and Its Relation to Management, in Calgary, Alberta (Geist and Walther 1974), that pronghorns he observed during the late 1940s and early 1950s were in fact territorial, and not harem breeders as reported (Buechner 1950a, 1950b). Recent observations (Kitchen unpublished field notes) of pronghorns in Oregon showed a territorial breeding system and not a harem system as reported by Einarsen (1948).

At present, Deblinger and Ellis (1977) have published the only account of a pure harem system where males collect and defend a herd of does whose members do not change with time. In another observation (McNay personal communication) in Nevada, males apparently deserted their territories and collected harems. This occurred during rut when receding water caused plants to flush in a normally dry lake bed, and does responded to the growth of succulent plants by moving into the area to forage. The territorial males also moved onto the lake bed, but could not establish territories. They collected and defended does, and breeding took place amidst a great deal of chasing and displaying.

Kitchen and Griep (1978) reported that a population of pronghorns near Rawlins, Wyoming, had a system of dominions, as defined by Brown (1963). In this instance, dominance and breeding privileges were restricted to specific areas, but the areas were not defended from other large males. Females were herded within these areas, but harems were not formed and does frequently moved between dominions during rut. Large males even moved between areas with each other, but dominance between males shifted as they moved from dominion to dominion.

The variability of the pronghorn's breeding appears to have an ecological basis consistent with hypotheses generated by Ralls (1977) and by Owen-Smith (1977) concerning the evolution of social systems. When forage quality (especially succulence) varies between areas and the best resources are clumped, then pronghorns are territorial. In addition, males on the best territories do most of the breeding (Kitchen 1974; Kitchen and Bromley 1974). Bromley (1977) has shown similar variability in territories based on the distribution of plant species. As resources become more uniform in distribution and of poorer quality, the system shifts toward dominions. The breeding system shifts to harem formation when unusual resource conditions occur, when population levels are low, or when sex ratios are unusually skewed (very few males to many females, ratios of 1 male to 10 or more females).

Thus, the expression of the pronghorn's breeding system varies along a continuum from a classical territorial system through a system of dominions and ends with harem formation. The basis for these differences is ecological and includes resource quality and distribution, population size, and sex ratios. Deblinger and Ellis (1977) suggested that the fragmented nature of pronghorn populations could cause local differences in the use of particular behavior patterns that through the multiplier effect (Wilson 1975; Crook et al. 1976) are the proximal causes of these differences in expression of the pronghorn's breeding system. Evidence for this hypothesis for the proximal determinants of pronghorn social behavior is not available from the current pronghorn literature.

Territorial Defense. Males defend their territories from late March or early April until the end of the rut (Kitchen 1974; Kitchen and Bromley 1974). Generally, defense occurs in a graded fashion and the intensity of defense is dictated by the age of the intruder and how far he has penetrated into the territory. Thus, large males near the center of a territory elicit an intense reaction, while young males (yearlings and two year olds) or animals near a boundary cause only mild reactions. Territorial males react first with a snort-wheeze vocalization that often causes other males to leave the territory. The usual sequence of events in a territorial defense is for the territorial male to: (1) stare at the intruder; (2) make a snort-wheeze vocalization; (3) approach the intruder; (4) interact with the intruder; and (5) chase off the intruder (or the intruder withdraws from the territory) (Kitchen and Bromley 1974). Steps may be skipped, or the defense may terminate at any step if the intruder withdraws from the territory.

Descriptive Behavior. Thirty-one separate behavior patterns and several vocalizations have been described for the pronghorn (Bromley 1969, 1977; Bromley and Kitchen 1974; Kitchen 1974; Kitchen and Bromley 1974). Given the diversity of the pronghorn's behavioral repertoire, it is impossible to cover all aspects of pronghorn behavior in this review. However, courtship behavior has received much attention because it is ritualized from dominance postures (figure 1 in Bromley and Kitchen 1974). The critical differences between the pronghorn male's dominance posture and courtship posture are: the ears are in a neutral position in courtship (figure 48.4), but are laid back in dominance; courting males whine and lip smack when approaching a doe, but a dominant male grinds its teeth as it approaches another male; and a courting male prances rapidly in a high-stepping gait when approaching a doe, whereas a dominant male walks slowly and stiffly toward its opponent (Bromley and Kitchen 1974; Kitchen 1974).

Antipredator Behavior. Pronghorn herds apparently function as escape cover for individuals (Hamilton 1971; Kitchen 1974), and frightened individuals seek a herd when escaping from predators. Pronghorn herds flee in an eliptically shaped formation and may run at speeds of 64–72 kmph (Bullock 1974; Kitchen 1974). Individuals, however, run faster than a herd and may reach speeds of 78.5–86.5 kmph. Pronghorns use a strategy of speed and the cover of the herd to escape from predators. The pronghorn's body markings may act, in part, as visual distraction to predators and keep

the predator from singling out an individual to pursue (Kitchen 1974).

Mother-Young Behavior and Fawn Socialization. Mother-young behavior is limited to a few grooming, orientation, and nursing behaviors (Bromley 1968, 1977; Kitchen 1974; Autenrieth and Fichter 1975). Although the behavioral repertoire is limited, the change in the mother-young relationship is complex and is critical to fawn socialization (Bromley 1968, 1977; Autenrieth and Fichter 1975). Does nurse and groom their fawns on roughly a two- to three-hour schedule for the first three weeks of life. After the establishment of the doe-fawn bond and when the fawns are mature enough to flee without rapid exhaustion, does join nursery herds. The frequency and duration of doe-fawn contacts are reduced in the nursery herd, and by late July does rebuff up to 40 percent of their fawns' nursing attempts. Male fawns are often weaned by the rut in September, and female fawns are usually weaned by November. The parent-offspring bond lasts only four to five months for does and their male fawns, but may last a lifetime for does and their female fawns. The longer relationship between the females may be due to doe fawns joining their mother's herd and remaining on the parent's home range (Pyrah 1970; Kitchen 1974).

Fawns interact almost exclusively with each other in the nursery herds and begin the establishment of lifelong dominance relationships (Kitchen 1974). Of 36 individually known male fawns from the 1969 cohort, none changed its relative dominance with others over a nine-year period. The three top males in the 1969 cohort have done 42 percent of all observed breeding for this age class (Kitchen in preparation).

MORTALITY

Few data are available but most pronghorns do not live beyond 9 years (Hepworth 1965). Hepworth and Blunt (1966) stated that the oldest jaw in the Wyoming Fish and Game Department Collection was from a 10.5-year-old doe and O'Gara (1968) collected a doe in Montana that had been tagged as an adult and had to be over 10 years old. Kerwin and Mitchell (1971) found, by the cementum annuli method, that Alberta antelope ranged up to 15.5 years of age.

In studied populations that were not expanding, poor reproduction, rather than diseases or other problems with adults, has generally been the depressing agent. High fawn losses, rather than depressed fecundity, were found in all instances, and mortality was usually highest within the first month of life.

Human predation, both legal hunting and poaching, is the greatest cause of pronghorn mortality. Hunters' efforts and money brought antelope back from the brink of extinction and a regulated harvest is now needed to keep the herds in balance with available habitat. Thus, legal hunting is an asset, not a liability, to pronghorn populations. Poaching, like many other causes of mortality, may be locally detrimental but is not generally important to the species.

FIGURE 48.4. Courtship approach of a pronghorn buck (*Antilocapra americana*). Note the elevated head; ears are not laid back. In a dominance display a male stands in the same position but the ears are laid back.

Fences and other man-made barriers to pronghorn movements are probably the second greatest decimating factor to antelope populations. Such barriers are especially damaging when combined with overgrazing, other types of habitat destruction, unusual weather conditions, and high densities of predators.

Many studies have been conducted on the causes of mortality among young fawns. Most of the earlier studies were inconclusive, but the investigators invariably indicated that predation was not important. Two exceptions were studies in Arizona (Arrington and Edwards 1951) and Utah (Udy 1953) that indicated that coyote (*Canis latrans*) control greatly increased pronghorn fawn survival. During each of those studies, coyote control was practiced in one study area and not in the other. Since the advent of light-weight radio equipment, studies in Montana, Idaho, and Utah have indicated fawn mortalities of 12 to 90 percent from coyotes, bobcats (*Felis rufus*), or golden eagles (*Aquila chrysaetos*), or a combination of the three (Beale 1970; Reichel 1976; Bodie 1978; Von Gunten 1978). Although recent studies indicate heavy predation, they were conducted because of poor reproduction and should not be interpreted as representing conditions in most antelope herds. Generally, herds that are small, are restricted in their movements, or have been recently translocated appear to be most susceptible to predation. Fences to control livestock apparently facilitate coyote and bobcat predation on pronghorns.

Severe winters affecting antelope populations have been reported in a number of areas (Rand 1947 in Canada; Udy 1953 in Utah; Martinka 1967 in Montana; Riddle and Oakley 1973 in Wyoming; Hailey 1977 in Texas). Deaths were attributed to freezing storms that caused chilling, deep snows that impeded travel, and man-made barriers that restricted movements. The winter of 1977/78 was especially severe on the northern plains and thousands of antelope perished in Alberta and Montana.

At the end of a drought in 1957, the pronghorn population of the Trans-Pecos region of Texas was about 7,300 animals (Hailey 1977). The herd increased to over 12,000 between 1957 and 1961, during years of normal precipitation. Another drought from 1961 to 1964 dropped the herds to 5,000. The animals were forced to subsist on inadequate and sometimes toxic vegetation because they could not cross woven-wire fences that were constructed to control domestic livestock. Buechner (1950a) noted that, historically, pronghorns of the Trans-Pecos region had moved to ranges with better forage conditions during years of low precipitation.

Antelope mortalities have been reported from complications at parturition, prenatal abnormalities, congenital deformities, old age, lightning, hailstorms, miring in mud, drowning, locking horns, and fighting. None of these sources of mortality is considered important to the population as a whole.

The extent to which traffic on roads, railroads, and highways contributes to antelope mortality is not documented. Although occasional spectacular kills are made by a speeding vehicle, the results are probably of little significance to the populations involved. During February 1976 in southern Idaho, a train killed 132 antelope as the wintering animals crowded onto the tracks; 43 were adult does carrying 85 fetuses. Thus, the actual loss to the antelope population of that area was 217 animals.

Epizootic hemorrhagic disease, bluetongue, necrobacillus, vibriosis, actinomycosis, and keratitus have all been diagnosed in pronghorns. At least 30 species of nematode, 4 cestodes, 1 trematode, and 2 coccidia, 6 species of tick, and a hippoboscid fly have also been found in or on antelope. Occasional extensive die-offs have been reported, but the impact on the species has been small. Usually when diseases or parasites have killed many pronghorns, other factors, such as crowding, concentrations at waterholes, malnutrition, or heavy range use by domestic sheep, were involved. In such situations, the disease or parasite may actually be beneficial in thinning out a population before conditions worsen.

AGE DETERMINATION

The sectioning of teeth provides the most accurate method of aging pronghorns (McCutchen 1969; Kerwin and Mitchell 1971). Dow and Wright (1962) developed an easy going scheme that is commonly used by biologists and game managers. By this method, antelope shot in the fall can be aged by simply looking at their teeth.

During hunting seasons, pronghorns are generally four to five months older than the whole-year ages used here. Fawns, which should be recognizable by their size anyway, have eight uniformly small front teeth. About 85 percent of the yearlings have two large permanent teeth (first incisors) in the center flanked by three small teeth on each side. If the first incisors are not erupted, the milk teeth are somewhat spread by jaw growth. Approximately 67 percent of the 2 year olds have four large center teeth flanked by two small ones on each side. If the second incisors are not erupted, the milk teeth are usually pushed out or loose. About 67 percent of the 3 year olds have eight large teeth, and 33 percent have six flanked by two peglike canines. If all eight front teeth are permanent, the cheek teeth must be checked. Deciduous premolars are shed at 2.3 years, when 12 infundibula are visible in the molariform teeth on each side of the jaw. Disappearance of these cavities provides the clue to distinguishing age for the older age classes. Over half of the 3.5 year olds still have 12 infundibula, which are then lost by wear at the rate of about 2 per year. Animals with no remaining infundibula are estimated to be 9 years old or older.

ECONOMIC STATUS AND MANAGEMENT

Pronghorns reached a low of only 13,000 animals in the 1920s (Hoover et al. 1959), but with careful conservation, restricted hunting, and reintroductions, they recovered to 406,000 by 1976. They are a major big game resource in the western United States. Their true economic and esthetic value to the west is hard to estimate. They generate a great deal of local income, and hunts where guides get $500 to $600 fees for seven days were common in Montana in the early 1970s (Atcheson 1972). One can only guess how much total income is generated by hunters in the local economies, but figures in the millions of dollars certainly are not unrealistic.

Pronghorns do not compete for forage with cattle, but do with sheep (Buechner 1950*a*). Competition with sheep is principally for forbs; in areas overgrazed by sheep, pronghorn populations have declined. Much of the decline in pronghorns is caused by sheep-proof fences that restrict pronghorn movements, resulting in direct pronghorn losses to starvation and accidents. Fences have been designed to allow pronghorns to move over areas but to restrict domestic livestock (Mapston 1972).

The Pronghorn Antelope Workshop has developed a series of "Guidelines for the Management of Pronghorn Antelope," edited by Autenrieth (1978). This document contains specific recommendations and provides management guidelines for the animal, the habitat, and specific problems (i.e., industrial development, fences, etc.). The key factors with regard to habitat are discussed in the "Habitat" section of this chapter, and habitat is considered the key pronghorn management problem today.

CURRENT RESEARCH AND MANAGEMENT NEEDS

Information on optimal foraging behavior is needed to assess food habits data in a more useful manner. Many areas of prime pronghorn habitat may be strip mined. Unless better data are available on food habits and food distribution relevant to pronghorns, reestablishing pronghorn use of reclaimed lands may fail. The role of forage quality in reproductive success has been little explored and research is needed. Studies of predator-prey relations in large pronghorn populations are only just getting under way and much information is needed. Although many studies of social behavior are available, more are needed in diverse habitats to determine the full flexibility of the pronghorn's social system. The genetics of pronghorns have received scant attention and could be an important area of study.

LITERATURE CITED

Arrington, O. N., and Edwards, A. E. 1951. Predator control as a factor in antelope management. Trans. North Am. Wildl. Conf. 16:179-190.

Atcheson, J. 1972. Antelope from a consumer's point of view. 5th Biennial Antelope States Workshop. Pp. 2-9.

Audubon, J., and Bachman, J. 1851. The viviparious quadrupeds of North America. V. G. Audubon, New York. Vol. 11. 334pp.

Autenrieth, R., ed. 1978. Guidelines for the management of pronghorn antelope. Pronghorn Antelope Workshop. 53pp.

Autenrieth, R., and Fichter, E. 1975. On the behavior and socialization of pronghorn fawns. Wildl. Monogr. 42. 111pp.

Bailey, V. 1920. Old and new horns of the prong-horned antelope. J. Mammal. 1(3):128-130.

_____. 1932. The Oregon antelope. Proc. Bio. Soc. Washington, D.C. 45:45-46.

Baker, T. 1953. Antelope movement and migration studies. Wyoming Wildl. 17(10):31-36.

Barrett, M. W., and Chalmers, C. A. 1976. Baseline hematologic and clinical chemistry values for pronghorns. Proc. Antelope States Workshop 7:104-117.

_____. 1977. Hematological values for adult free-ranging pronghorns. Can. J. Zool. 55:448-455.

Bayless, S. 1969. Winter food habits, range use, and home range of antelope in Montana. J. Wildl. Manage. 33(3):538-551.

Beale, D. M. 1970. Some possible predator-prey relationships between bobcats and pronghorn antelope on desert ranges. Proc. Antelope States Workshop 4:75-77.

Beale, D. M., and Smith, A. D. 1970. Forage use, water consumption, and productivity of pronghorn antelope in western Utah. J. Wildl. Manage. 34(3):570-582.

Bear, G. D. 1971. Seasonal trends in fat levels of pronghorns, *Antilocapra americana*, in Colorado. J. Mammal. 52(3):583-589.

Beer, J. 1944. Distribution and status of pronghorn antelope in Montana. J. Mammal. 25(1):43-46.

Berger, J. 1977. Organizational systems and dominance in feral horses in the Grand Canyon. Behav. Ecol. Sociobio. 2:131-146.

Bodie, W. L. 1978. Pronghorn fawn mortality in the Upper Pahsimeroi River drainage of Central Idaho. Proc. Antelope States Workshop 8:417-428.

Bowyer, R. T. 1974. Social behavior of Roosevelt elk during rut. M.S. Thesis. Humboldt State Univ., Arcata, Calif. 122pp.

Bromley, P. T. 1968. Pregnancy, birth, behavioral development of the fawn, and territoriality in the pronghorn (*Antilocapra americana* Ord) on the National Bison Range, Moiese, Montana. M.S. Thesis. Univ. Montana, Missoula. 137pp.

_____. 1969. Territoriality in pronghorn bucks on the National Bison Range, Moiese, Montana. J. Mammal. 50(1):81-89.

_____. 1977. Aspects of the behavioral ecology and sociobiology of the pronghorn (*Antilocapra americana*). Ph.D. Dissertation. Univ. Calgary, Alberta. 370pp.

Bromley, P. T., and Kitchen, D. W. 1974. Courtship in the pronghorn (*Antilocapra americana*). Pages 356-364 *in* V. Geist and F. Walther, eds. The behaviour of ungulates and its relation to management. IUCN Publ. 24. Vol. 1. Morges, Switzerland.

Brown, J. 1963. Aggressiveness, dominance and social organization in the Steller's jay. Condor 65:460-484.

Buechner, H. K. 1950*a*. Life history, ecology, and range use of the pronghorn antelope in Trans-Pecos, Texas. Am. Midl. Nat. 43(2):257-354.

_____. 1950*b*. Range ecology of pronghorn on the Wichita Mountains Wildlife Refuge. Trans. North Am. Wildl. Conf. 15:627-644.

Bullock, R. E. 1974. Functional analysis of locomotion in pronghorn antelope. Pages 274-305 *in* V. Geist and F. Walther, eds. The behavior of ungulates and its relation to management. IUCN Publ. 24. Vol. 1. Morges, Switzerland.

Carr, J. N. No date. Arizona Game and Fish Dept. Endangered Species Investigations Sonoran Pronghorn Proj. W-53-R-23, Work Plan 7, Job 1, Performance Rep. 1 July 1972 to 30 June 1973. 11pp.

Cole, G. F., and Wilkins, B. T. 1958. The pronghorn antelope: its range use and food habits in Central Montana with reference to wheat. Montana Fish Game Dept. Tech. Bull. 2:1–39.

Crook, J. H.; Ellis, J. E.; and Goss-Custard, J. D. 1976. Mammalian social systems: structure and function. Anim. Behav. 24(2):261–274.

Deblinger, R. D., and Ellis, J. E. 1977. Aspects of intraspecific social variation in pronghorns. 7th Biennial Antelope States Workshop.

Dow, S. A., Jr., and Wright, P. L. 1962. Changes in mandibular dentition associated with age in pronghorn antelope. J. Wildl. Manage. 26(1):1–18.

Einarsen, A. S. 1948. The pronghorn antelope and its management. Wildl. Manage. Inst., Washington, D.C. 235pp.

Feist, J. D. 1971. Behavior of feral horses in the Pryor Mountain Wild Horse Range. M.S. Thesis. Univ. Michigan, Ann Arbor. 96pp.

Feist, J. D., and McCullough, D. R. 1975. Reproduction in feral horses. J. Reprod. Fert. Suppl. 23:13–18.

———. 1976. Behavior and communication patterns in feral horses. Z. Tierpsychol. 41:337–371.

Fichter, E., and Nelson, A. E. 1962. Study of a pronghorn population. P-R Proj. W-85-R-13. Idaho Dept. Fish Game, Boise. 17pp.

Foree, B. 1972. Northwestern Nevada antelope studies. Pages 25–27 in 5th Biennial Antelope States Workshop.

Frick, C. 1937. Horned ruminants of North America. Bull. Am. Mus. Nat. Hist. 69:1–699.

Geist, V., and Walther, F., eds. 1974. The behaviour of ungulates and its relation to management. IUCN Publ. 24. Vols. 1 and 2. Morges, Switzerland.

Gilbert, B. K. 1973. Scent marking and territoriality in pronghorn (Antilocapra americana) in Yellowstone National Park. Mammalia 37(1):25–33.

Goldman, E. A. 1945. A new pronghorn antelope from Sonora. Proc. Bio. Soc. Washington, D.C. 58:3–4.

Green, N. F., and Green, H. D. 1977. The wild horse population of Stone Cabin Valley, Nevada: a preliminary report. Pages 59–65 in Proc. Natl. Wild Horse Forum, Reno, Nevada.

Hailey, T. L. 1977. Handbook on pronghorn antelope management. Texas Parks Wildl. Dept., Austin. 47pp.

Hall, E. R. 1946. Mammals of Nevada. Univ. California Press, Berkeley. 710pp.

Hall, E. R., and K. R. Kelson. 1959. The mammals of North America. Ronald Press Co., New York. 1083 + 79pp.

Hamilton, W. D. 1971. Geometry for the selfish herd. J. Theor. Bio. 31(2):295–311.

Hepworth, B. 1965. Investigations of pronghorn antelope in Wyoming. Proc. Antelope States Workshop. 1:1–12.

Hepworth, W., and Blunt, F. 1966. Research findings on Wyoming antelope. Pages 24–29 in Wyoming Wildl., Spec. Antelope Issue.

Hoover, R. L.; Till, E. E.; and Ogilivie, S. 1959. The antelope of Colorado. Colorado Game Fish Tech. Bull. 4:110.

Hoskinson, R. L. 1977. Migration behavior of pronghorn antelope and summer movements and fall migration of pronghorn fawns in southeastern Idaho. Ph.D. Dissertation. Univ. Minnesota. 133pp.

Kautz, L. G. 1942. Antelope survey. Colorado Wildl. Res. Q. Prog. Rep. 8:1–26.

Kerwin, M. L., and Mitchell, G. J. 1971. The validity of the wear-age technique for Alberta pronghorns. J. Wildl. Manage. 35(4):743–747.

Kitchen, D. W. 1974. Social behavior and ecology of the pronghorn. Wildl. Monogr. 38. 96pp.

———. In preparation. Early dominance and breeding success in pronghorn males.

Kitchen, D. W., and Bromley, P. T. 1974. Agonistic behavior of territorial pronghorn bucks. Pages 365–381. In V. Geist and F. Walther, eds. The behaviour of ungulates and its relation to management.

Kitchen, D. W.; Green, N.; and Green, H. 1977. Research needed on wild horse ecology and behavior to develop adequate management plans for public lands. Proc. Natl. Wild Horse Forum 54–58.

Kitchen, D. W., and Griep, P. 1978. Variation in breeding behavior of pronghorn bucks. Annu. Meet. Anim. Behav. Soc., Seattle, Washington.

Lehman, V. W., and Davis, J. B. 1942. Experimental wildlife management in the south Texas chaparral. P-R Q. Rep. 1-R C1, Texas Game, Fish Oyster Comm., Austin. 11pp.

Lonsdale, E. M.; Bradach, B.; and Thorne, E. T. 1971. A telemetry system to determine body temperatures in pronghorn antelope. J. Wildl. Manage. 35(4):747–751.

McCullough, D. R. 1969. The tule elk: its history, behavior and ecology. Univ. California Publ. Zool. 88:1–209.

McCutchen, H. E. 1969. Age determination of pronghorns by the incisor cementum. J. Wildl. Manage. 33(1):172–175.

McLean, D. D. 1944. The prong-horned antelope in California. California Fish Game 30(4):221–241.

Mapston, R. D. 1972. Guidelines for fencing on antelope ranges. Proc. 5th Biennial Antelope States Workshop 5:167.

Martinka, C. 1967. Mortality of northern Montana pronghorn in a severe winter. J. Wildl. Manage. 31(1):159–164.

Miller, G. S., and Kellogg, R. 1955. List of North American recent mammals. U.S. Natl. Mus. Bull. 205:1–954.

Mitchell, G. J. 1965. Natality, mortality, and related phenomena in two populations of pronghorn antelope in Alberta, Canada. Ph.D. Thesis. Washington State Univ., Pullman. 221pp.

———. 1967. Minimum breeding age of female pronghorn antelope. J. Mammal. 48(3):489–490.

———. 1971. Measurements, weights, and carcass yields of pronghorns in Alberta. J. Wildl. Manage. 35(1):76–85.

Moy, R. F. 1970. Histology of the subauricular and rump glands of the pronghorn (Antilocapra americana Ord). Am. J. Anat. 129:65–88.

———. 1971. Histology of the forefoot and hindfoot interdigital and median glands of the pronghorn. J. Mammal. 52(2):441–446.

Murie, J. 1870. Notes on the anatomy of the prongbuck, Antilocapra americana. Proc. Zool. Soc. London 334–368.

Nagy, J. G., and Williams, G. W. 1969. Rumino reticular VFA content of pronghorn antelope. J. Wildl. Manage. 33(2):437–439.

O'Gara, B. W. 1968. A study of the reproductive cycle of the female pronghorn (Antilocapra americana Ord). Ph.D. Thesis. Univ. Montana, Missoula. 161pp.

———. 1969a. Horn casting by female pronghorns. J. Mammal. 50(2):373–375.

———. 1969b. Unique aspects of reproduction in the female pronghorn (Antilocapra americana Ord). Am. J. Anat. 125:217–231.

O'Gara, B. W., and Matson, G. 1975. Growth and casting of horns by pronghorns and exfoliation of horns by bovids. J. Mammal. 56(4):829–846.

O'Gara, B. W., and Moy, R. F. 1972. Histology and morphology of scent glands and their possible roles in pronghorn behavior. Proc. Antelope States Workshop 5:192–208.

O'Gara, B. W.; Moy, R. F.; and Bear, G. D. 1971. The annual testicular cycle and horn casting in the pronghorn (*Antilocapra americana*). J. Mammal. 52:537–544.

Owen-Smith, N. 1977. On territoriality in ungulates and an evolutionary model. Q. Rev. Bio. 52:1–38.

Pyrah, D. 1970. Antelope herd ranges in Central Montana. Proc. 4th Antelope States Workshop, Scottsbluff. 16–20.

Ralls, K. 1977. Sexual dimorphism in mammals: avian models and unanswered questions. Am. Nat. 3(981):917–938.

Rand, A. L. 1947. The 1945 status of the pronghorn antelope, *Antilocapra americana* (Ord) in Canada. Nat. Mus. Can., Ottawa Bull. 106, Bio. Ser. 34. 34pp.

Reichel, J. D. 1976. Coyote-prey relationships on the National Bison Range. M.S. Thesis. Univ. Montana, Missoula. 94pp.

Riddle, P., and Oakley, C. 1973. The impact of a severe winter and fences on antelope mortality on Southcentral Wyoming. Western Assoc. State Game and Fish Commissioners 53:174–188.

Russell, T. P. 1964. Antelope of New Mexico. Bull. New Mexico Dept. Game and Fish 12:1–103.

Seton, E. T. 1953. Lives of game animals. C. T. Bradford Co., Boston. Vol. 3, pt. 11, pp. 413–467.

Skinner, M. P. 1922. The prong-horn. J. Mammal. 3(2):82–105.

Sundstrom, C. 1968. Water consumption by pronghorn antelope and distribution related to water in Wyoming's Red Desert. Proc. Antelope States Workshop 3:39–46.

Sundstrom, C.; Hepworth, W. G.; and Diem, K. L. 1973. Abundance, distribution and food habits of the pronghorn. Wyoming Game Fish Comm. Bull. 12. 61pp.

Thorne, E. T. 1975. Normal body temperature of pronghorn antelope and mule deer. J. Mammal. 56(3):697–698.

Trout, L. E. 1976. Blood analysis of Idaho pronghorns. Proc. Antelope States Workshop 7:122–126.

Udy, J. R. 1953. Effects of predator control on antelope populations. Fed. Aid Div. Utah Dept. Fish Game Publ. 5. 43pp.

Von Gunten, B. L. 1978. Pronghorn fawn mortality on the National Bison Range. Proc. Antelope States Workshop 8:413–416.

Webb, S. D. 1973. Plieocene pronghorns in Florida. J. Mammal. 54(1):203–221.

Wesley, D. E.; Knox, K. L.; and Nagy, J. G. 1970. Energy flux and water kinetics in young pronghorn antelope. J. Wildl. Manage. 34(4):908–912.

Weswig, P. H. 1956. Vitamin A storage. Oregon State Game Bull. 11(12):7.

Wilson, E. O. 1975. Sociobiology: the new synthesis. Belknap Press, Cambridge, Mass. 697pp.

Wislocki, G. B., and Fawcett, D. W. 1949. The placentation of the pronghorned antelope (*Antilocapra americana*). Bull. Mus. Comp. Zool. 101:545–558.

Wright, P. O., and Dow, S. A., Jr. 1962. Minimum breeding age in pronghorned antelope. J. Wildl. Manage. 26(1):100–101.

Yoakum, J. D. 1967. Literature of the American pronghorn antelope. U.S. Dept. Int., Bur. Land Manage., Reno. 82pp.

———. 1972. Antelope-vegetative relationships. Proc. 5th Antelope States Workshop 5:171–177.

———. 1974. Pronghorn habitat requirements for sagebrush-grasslands. Proc. 6th Antelope States Workshop 6:16–25.

DAVID W. KITCHEN, Department of Wildlife Management, School of Natural Resources, Humboldt State University, Arcata, California 95521.

BART W. O'GARA, Cooperative Wildlife Research Unit, University of Montana, Missoula, Montana 59312.

49

Bison

Bison bison

H. W. Reynolds
R. D. Glaholt
A. W. L. Hawley

NOMENCLATURE

COMMON NAMES. American bison, bison, buffalo, plains bison, prairie bison, wood bison, woodland or mountain bison
SCIENTIFIC NAME. *Bison bison*
SUBSPECIES. *B. b. bison, B. b. athabascae*

The bison is a member of the family Bovidae, to which domestic cattle, sheep, and goats belong. Both sexes possess true horns that are never shed. These horns are composed of a bony core and a hard, outer sheath of horny epidermis. Fossil bovids date to the Lower Miocene (Burt and Grossenheider 1952; Walker et al. 1964). Bovids are grazers primarily and browsers secondarily, and they possess a four-chambered, ruminating stomach. They inhabit major grassland, shrubland, forest, and tundra ecosystems. They feed by biting off forage material with forward-projecting incisor teeth that are present only in the lower jaw.

The genus *Bison* is characterized by a short, broad forehead with a narrowed muzzle and pointed nasal bones (Hall and Kelson 1959). Bison are particularly noted for a massive head, a short neck, a high hump at the shoulders, and short, curved, rounded horns that exhibit annual growth patterns (Soper 1964). The short tail is haired and tufted at the tip. The hair on the head, neck, and shoulders is brownish black. There is a distinct beard formed by long, woolly hair on the chin. Body pelage is generally brown, varying moderately with season to light brown. True buffalo, native only to Africa and Asia, belong to genera distinctly different from bison and do not possess the characteristic shoulder hump of bison. Although it is a misnomer, the popular name *buffalo* has been used interchangeably for bison since early explorers first discovered the species in North America.

The earliest known fossil records of the genus *Bison* appeared in the Villafranchian deposits of India and China (Geist 1971). These late Pliocene forms eventually gave rise to the present-day bison. It is believed that bison first immigrated to North America over the Bering land bridge (Beringia) during the early and middle Pleistocene near the time of the Illinoian glacial and Sangamon interglacial (mid-Pleistocene)

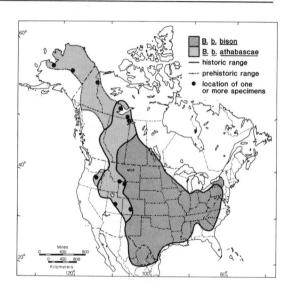

FIGURE 49.1. Recent and prehistoric distribution of plains bison (*Bison bison bison*) and wood bison (*Bison bison athabascae*). Recent distribution is based on historical accounts; prehistoric distribution is based on fossil evidence of horn core occurrence. After Skinner and Kaisen 1947.

periods and then again with the recession of the Wisconsin glaciation during the late Pleistocene. This latter glaciation is believed to have mediated much of bison evolution and zoogeography. Fossil bison have been found throughout northeastern Asia, Alaska, and most of western and central North America, and show evidence of considerable variation in body size, conformation, and horn growth. Theories have been proposed to elucidate the evolutionary history of bison, and some have stimulated controversial discussions. Definitive works on the evolution and zoogeography of the genus have been done by Flerov and Zablotski (1961), Guthrie (1970, 1980), Geist (1971), Geist and Karsten (1977), Harington (1980), Hillerud (1980), and Wilson (1980). A few others have reported their findings regarding fossil bison, theories on evolution of bison, and relationships of bison and humans in North America (Fuller and Bayrock 1965; Schultz and Hillerud 1977; Frison 1980).

TAXONOMY

Bison taxonomy has been a controversial issue for many years, and classification to the subspecies level remains a matter for debate. At the present time, two subspecies of North American bison are recognized. Conflicting evidence about the size and color of wood bison can be found in the literature (Peden and Kraay 1979). Allen (1876), Seton (1886), and Ogilvie (1893) were among the first to describe wood bison as distinct from plains bison by their larger size and darker color. Rhoads (1897) was the first to scientifically describe wood bison as a separate subspecies. Many other authors (Raup 1933; Soper 1941; Skinner and Kaisen 1947; Banfield and Novakowski 1960; Flerov 1965; Karsten 1975; Geist and Karsten 1977) have acknowledged subspecific status for wood bison in agreement with Rhoads. However, Graham (1923), Seibert (1925), and Garretson (1927) recognized the larger size and darker color of wood bison but attributed this to environmental differences and not to genetics. Hornaday (1889) did not recognize wood bison as a separate subspecies. More recently, work with cranial characteristics of plains bison suggested that a revision of the genus may be necessary (Shackleton et al. 1975). In addition, Peden and Kraay (1979) questioned the present taxonomy of bison on the basis of their research on blood characteristics.

Osteology. Skinner and Kaisen (1947) made a preliminary revision of the genus *Bison*, based on skull characteristics, measurements, and identifiable patterns of horn core growth. They classified mountain bison as a southern extension of the woodland race into mountainous habitat along the Rockies. However, Skinner and Kaisen (1947:165) apparently recognized an important shortcoming in their work when they wrote, "the population sample of true *athabascae* skulls is too small to present a comprehensive understanding of the amount of variation possible within this mountain or woodland race." Bayrock and Hillerud (1964) used the standard skull measurements (Skinner and Kaisen 1947) to describe three *Bison bison athabascae* (Rhoads) skulls that had been collected in Wood Buffalo Park in 1925. Two of these skulls exhibited measurements exceeding both the maximum and the minimum for the subspecies, thereby appreciably extending the range of variation for cranial measurements of wood bison. Skinner and Kaisen (1947) concluded that horn cores of fossil bison appeared to provide the best distinguishing criterion after allowing for growth, age, and individual variation. In contrast, horn core dimensions were the most variable of the craniometric measurements taken from 157 known-age skulls of plains bison (Shackleton et al. 1975). However, before any judgment can be made as to the value of horn core data for use in taxonomic classification, further studies are required to determine intraspecific and interspecific variability of these and other skull characteristics (Shackleton et al. 1975). Shackleton et al. concluded that a revision of the genus will likely be necessary.

The National Museum of Natural Sciences in Ottawa is currently involved in an osteological study to clarify: (1) systematics of extant forms of *Bison*, (2) evolutionary trends in late Pleistocene and postglacial fossil *Bison*, and (3) possible origins of extant forms. Preliminary results indicate that wood bison differ demonstrably from plains bison in cranial and morphological characters (C. van Zyll de Jong personal communication). These initial analyses support subspecific status for wood bison.

Genetics and Blood Parameters. Several researchers have used serological characteristics of bison to elucidate phylogenetic relationships of bovids (Braend 1963; Ying and Peden 1977; Buckland and Evans 1978*a*, 1978*b*; Fulton et al. 1978; Peden and Kraay 1979). However, the usefulness of blood characteristics in taxonomic studies has not yet been thoroughly assessed and is presently being developed.

In a study of karyotypes of wood bison and plains bison, Ying and Peden (1977) reported that both subspecies were characterized by the same number of chromosomes (2n = 60). When G-banding patterns were compared for wood bison and plains bison, 20 pairs plus the sex chromosomes were noted to be homologous, but patterns for the 9 remaining pairs of chromosomes were not distinguishable (Ying and Peden 1977). The question of homology or nonhomology between wood bison and plains bison will remain unanswered until research can provide a high-quality G-band with subbanding patterns for all chromosomes.

Peden and Kraay (1979) studied 10 blood characteristics from each of five herds of North American bison. They subjected bison blood samples to hemolytic tests using 13 cattle blood-typing reagents, and used discriminate analyses to interpret the data. They concluded that current taxonomic classification for North American bison is questionable. It is also clear that development of further tests for red cell antigens and blood enzymes, and additional measurements of morphological and skeletal characteristics, are needed to clarify the taxonomic status of bison.

The Status of Wood Bison. The Canadian Wildlife Service has taken the position that wood bison are a separate subspecies; however, it is generally recognized that there are still unresolved aspects of bison phylogeny and taxonomy. It is hoped that ongoing research will shed new light on some of these questions. If the weight of scientific opinion, backed by adequate data, indicates that subspecific status is not justified, the various gene pools of North American bison could be managed separately or as one continental gene pool, depending upon goals and objectives of management. Presently, however, we have the option to manage the purest of wood bison available as a subspecies and/or as distinct subpopulations of a common species. Once a decision has been made to withdraw the subspecific status and genetic mixing has occurred, that decision becomes irreversible. Considerable thought, research, and support are prerequisite to a decision of that significance.

DISTRIBUTION

Plains Bison. The American bison ranged throughout much of North America (figure 49.1). Ernest Thompson Seton estimated that 75 million were in North America before white settlers arrived (Dary 1974). However, McHugh (1972) estimated that 30 million buffalo was the maximum number that available range could support. Although millions of bison once roamed this region, few free-ranging herds of North American bison remain. Descriptions of historical distribution patterns for plains bison are given by Allen (1876), Hornaday (1889), and Skinner and Kaisen (1947).

The northeastern boundary for the historic range of plains bison is roughly outlined by a line extending from northcentral Saskatchewan in a southeastward direction to the southern shore of the Great Lakes. The northern boundary for bison in central Canada is also approximated by this line (figure 49.1).

The eastern boundary was that of the Allegheny Mountains in the United States, extending south through the states of Maryland, Virginia, North Carolina, and South Carolina. In the south, plains bison range extended from Alabama across southern Mississippi and Louisiana westward along the southeastern coast of Texas and into Mexico. It was in southeastern Texas, near present-day Houston, that the discovery of the American bison was made by a European, Cabeza de Vaca, in 1530 (Hornaday 1889).

The western boundary of North American plains bison distribution generally extended northward from north-central Mexico and merged with the historic range for wood bison along the eastern foothills of the Rocky Mountains (figure 49.1). In Alberta, Canada, the western and northern boundaries of plains bison range approximated the boundary of the ecotone between grassland and forest habitat. To the north of this interface lies boreal forest and the historic range of wood bison.

Wood Bison. In Canada, historical distribution of wood bison included most of the boreal regions of British Columbia, Alberta, Saskatchewan, Northwest Territories, and Yukon Territory. Wood bison also ranged along the eastern slopes of the Rocky Mountains from northern Canada to Colorado in the United States (figure 49.1). The prehistoric range extended northward to include most of the Yukon Territory and northern Alaska (Skinner and Kaisen 1947). During the late 1700s wood bison were distributed throughout northern Alberta to points east of the Slave River and north to eastern Great Slave Lake, Yellowknife Bay, Lac la Martre, Fort Liard, and Fort Simpson in the Northwest Territories (Soper 1964). Areas in British Columbia where wood bison specimens have been recorded are at the mouth of the Kitimat River, Atlin, Cecil Lake, and Fort Saint John (Smith 1977). The western boundary of wood bison range is not well defined but appears to have been close to the western edge of the Rocky Mountain chain (figure 49.1). The eastern boundary approximated a line from the southeastern end of Great Slave Lake along the east side of the Slave River, south to the west end of Lake Athabasca, and then irregularly south and southeast to the Clearwater River and Peter Pond Lake in northern Saskatchewan (Soper 1941). The southeastern boundary continued from Saskatchewan across central Alberta, at about the same line as the interface of the boreal forest region and the fescue grasslands, to the east slopes of the Rockies, and from there south, encompassing the mountain chain to New Mexico. The coniferous forests and aspen parkland with interspersed meadows and prairies, typical of this area, formed the main habitat for the wood bison.

DESCRIPTION

Plains Bison. The unmistakable appearance of bison is characterized by a massive, heavy head with a short, broad nasal area. A short neck and a high shoulder hump leave the impression that the forequarters are out of proportion to the smaller-appearing hindquarters. There is a tufted tail of moderate length (Banfield 1974). The short, round, black horns rise laterally from the side of the head (figure 49.2, lateral view) and curve inward over the head. The horns of the female are more slender and tend to curve inward more than those of the male. Bison have rather short legs and large, rounded hooves. Tracks are similar to those of domestic cattle (*Bos taurus*). Sexual dimorphism is apparent among adults, but, in general, females resemble males in color, body configuration, and presence of permanent horns.

SIZE. Bison are the largest native terrestrial mammal in North America. Plains bison are considered to be slightly smaller than the woodland race. Sex- and age-specific body weights and measurements for bison from different localities are given in table 49.1. Banfield (1974) stated that male plains bison reached adult size at six years of age, while females attained maximum size at about four years. In studying the relationship of weight to chest girth, Kelsall et al. (1978) discovered that males were 9.1 percent heavier than females of equal chest girth. The relationship of weight to girth approximated a linear relationship. These authors believed that their regression data would prove useful to estimate bison weights from chest girth measurements if caution was exercised in extrapolating results.

PELAGE. The pelage of bison is composed of long, coarse guard hairs with a thick, woolly undercoat (Banfield 1974). The hair on the head, shoulders, and forelegs is long, shaggy, manelike, and dark brown to black in color. The hindquarters are covered with short, straight, light-colored hair. Chin hair usually resembles the shape of a goatee-type beard. The head is very dark, almost black, with little or no color contrast. There are usually two seasonal molts: one in spring and one in summer (Banfield 1974). Albino and gray hair colors are rarely observed in bison. Historically, the former type was held in great reverence by the Plains Indians. Newborn calves are reddish to orange brown, but this changes to the typical dark

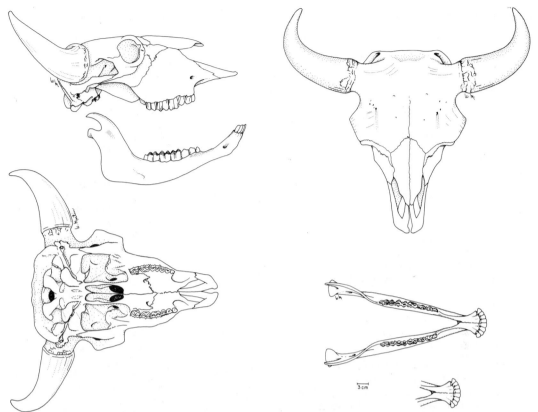

FIGURE 49.2. Skull of the bison (*Bison bison*). From top to bottom: lateral view of cranium and mandible (left), and dorsal view of cranium (right); ventral view of cranium (left), and dorsal view of mandible and incisor bar (right).

brown at about two to three months of age (Meagher 1978).

SKULL AND DENTITION. Skulls of male bison are larger and more massive, and have longer, more pronounced horn cores and burrs than those of females (Skinner and Kaisen 1947). The muzzle is narrow with long, pointed nasal bones that do not reach the premaxillae (fig. 49.2, dorsal view). The nasal opening is composed of premaxillae, maxillae, and nasals; the orbits are tubular, composed of frontals, lacrimals, and jugals. Unlike the Cervidae and the Antilocapridae, bison do not have preorbital vacuities in the skull. The length of 12 male bison skulls ranged from 500 to 600 mm and width varied from 240 to 280 mm (Allen 1876). Overall length of 29 plains bison skulls varied from 491 to 570 mm, and the greatest postorbital width ranged from 271 to 343 mm (Skinner and Kaisen 1947). Hall and Kelson (1959) reported that the greatest skull length of male plains bison ranged between 491 and 595 mm. Length for 81 specimens of adult male plains bison from Elk Island National Park, Alberta, ranged from 476 to 570 mm, and the greatest postorbital width ranged between 282 and 352 mm (Shackleton et al. 1975). The overall length for 33 adult female skulls from the park population varied from 425 to 500 mm, and the greatest

postorbital width for 35 specimens ranged from 237 to 275 mm (Shackleton et al. 1975).

Bovids have hypsodont cheek teeth and no upper incisors. The upper canines are reduced or absent (fig. 49.2, lateral view). The selenodont molars of bison have a median style or enamel fold between the anterior and posterior lobes (fig. 49.2, dorsal view of mandible) which tends to disappear with age and wear (Skinner and Kaisen 1947). These molars are used to section and grind vegetation finely (Vaughan 1972). The dentition and limbs of bovids are phylogenetically advanced and probably developed in association with grazing habits (Vaughan 1972). The dental formula for bison and all other bovids is: 0/3, 0/1, 3/3, 3/3 = 32.

Wood Bison. Wood bison possess the same general characteristics as plains bison except for minor differences in morphology (general conformation), pelage, and skeletal measurements. For example, wood bison have larger horn cores and exhibit differences in other cranial elements. Karsten (1975) reported that wood bison possess denser fur than plains bison; he also described wood bison as larger, more elongated in front, heavier, darker in color, and having a squarish hump with a more gently sloping back contour than the plains bison. Geist and Karsten (1977) have documented in detail how the wood bison bull and cow

TABLE 49.1. Body measurements and weights of plains bison from several locations in North America

Sex	Age Class	Body Measurements (cm)				Weight (km)	Reference
		Total Length	Tail Length	Hind Foot	Height at Shoulder		
M	12.5 years	318			186	814[a]	Halloran 1961
F	6.5 years				157[b]	488[a]	Halloran 1961
M	adult	340	43	61	178	816–998	Soper 1964
F	adult		25–30% smaller			363–544	Soper 1964
	adult	210–350	50–60		260–280	450–1350	Walker et al. 1964
M	adult					907[c]	Meagher 1973
F	adult					363–499	Meagher 1973
M and F	yearling					227–318	Meagher 1973
M and F	calves					136–181	Meagher 1973
M	adult	304–380	43–48	56–68	167–182	460–720	Banfield 1974
F	adult	213	45	53	152	360–460	Banfield 1974
Mixed[d]	mixed[d]					450[d]	Telfer and Scotter 1975

[a] Heaviest in sample of 510.

[b] 5.5 year old female.

[c] Maximum weight was estimated.

[d] Majority of animals in sample were adult females.

differ significantly in descriptive characteristics from their prairie counterparts. A summary of the description by Geist and Karsten (1977) follows.

1. The hair on top of the head, around the horns, in the beard, and in the midventral neck area is significantly shorter and less dense in wood bison bulls than in their plains counterparts of the same age. Thus, the head of the wood bison appears smaller, the horns longer, and the ears more noticeable (figure 49.3). The beard of the wood bison is smaller and more pointed, and the long mane extending from the beard to the brisket on the plains bison is short or absent in the wood bison. The head and neck of the wood bison is generally darker in color than that of the plains bison.

2. The long hair in the area of the "chaps" on the front legs forms a skirt on the plains bison but is virtually absent on the wood bison (fig. 49.3). This most striking difference in pelage between wood and plains bison partly accounts for the more massive appearance of the plains bison in the front quarters.

3. The "robe" or cape of the shoulders, hump, and neck of the plains bison is more distinct and lighter (golden) colored than that of the wood bison. The cape of the plains bison is composed of longer hair that forms an obvious boundary with the rest of the body fur just posterior to the shoulders. It is not as well developed in the wood bison.

4. The tail of the wood bison appears to be longer and more heavily haired than that of its plains cousin.

5. The penis sheath tuft of the wood bison appears shorter and thinner than that of the plains bison.

6. Wood bison tend to be taller at the hump, which is squarer than the hump of the plains bison. The back contour of the wood bison has a more abrupt change at the hump, but is more gently sloping or flat overall (fig. 49.3). There is less sexual dimorphism in wood bison cows with respect to body size, horn structure, pelage characteristics, and body proportions.

Roe (1970) cited observations of two early frontiersmen that contradicted numerous claims that wood bison were larger than plains bison. Regardless, most

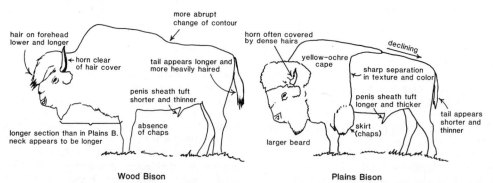

FIGURE 49.3. Basic pelage and morphological differences between a wood bison bull (*Bison bison athabascae*) (left) and a plains bison bull (*Bison bison bison*) (right), shown diagrammatically.

biologists who have had the opportunity to view both wood bison and plains bison at Elk Island agree with the descriptions presented by Karsten (1975) and by Geist and Karsten (1977). Historical accounts of wood bison, most of which related to size and color, were usually compiled by explorers. Therefore, one can expect some degree of controversy in reports of these untrained observers, especially when describing characteristics that are highly variable. Regardless, the number of contradictory reports are in the minority, while the majority of the historical accounts agree with more recent descriptions of living wood bison.

Specific research information on wood bison is lacking. Therefore, we must rely on the data available for plains bison for basic biological principles and assume that little or no difference exists between the subspecies. Unless otherwise stated, the information reported in this chapter has resulted from studies with plains bison.

PHYSIOLOGY

Metabolism. Bison and cattle differ in their metabolic response to cold. Christopherson et al. (1979) observed that in winter at −30° C, the metabolic rate of bison (718 kJ per kg metabolic body weight per day) was less than that of cattle (830 kJ per kg metabolic body weight per day). At 10° C the metabolic rate of bison (934 kJ per kg metabolic body weight per day) was greater than that of cattle (659 kJ per kg of metabolic body weight per day). In winter temperatures of 10° C apparently exceeded the upper critical temperature of bison. An increase in metabolic rate is considered the normal response to cold exposure, although this response may be attenuated by cold acclimation (Slee 1972). However, one-year-old bison calves decreased their metabolic rate in spring from 748 kJ per kg metabolic body weight per day at 10° C to 584 kJ per kg metabolic body weight per day at −30° C (Christopherson et al. 1979). The reduced metabolic rate was attributed largely to a reduction in activity. The bison calves were as cold-tolerant at 6 months of age as Hereford yearlings were between 13 and 17 months of age. One-year-old Hereford calves that were acclimatized to ambient spring temperatures increased their metabolic rate from 760 kJ per kg metabolic body weight per day at 10° C to 938 kJ per kg metabolic body weight at −30° C.

The greater cold tolerance of bison can be attributed primarily to their greater pelage insulation. When comparing winter hair coats, Peters and Slen (1964) observed that bison had a greater weight of hair per unit area, a greater density of hair, and a greater fineness of hair than Hereford, Angus, or Shorthorn cattle. Cattle × bison hybrids were intermediate in these pelage characteristics. The greater cold tolerance thereby imparted to hybrids resulted in a greater capacity for foraging on open range in the winter by hybrids when compared with cattle (Smoliak and Peters 1955).

Blood Chemistry and Hematology. Marler (1975) analyzed jugular and tail vein blood samples from 77 bison in two herds in Kansas. No differences in chemistry or hematology were noted between sexes,

but there were differences between bison less than two years old and those two or more years of age. Most blood components measured were at levels comparable to domestic cattle but hematologic values were generally greater in bison. The mean packed cell volume (PCV) for all bison was 50 percent and the mean hemoglobin (Hgb) concentration was 17 g per 100 ml, compared with values of approximately 35 percent and 11 g per 100 ml, respectively, for domestic cattle (Schalm et al. 1975). Mehrer (1976) observed a mean hematocrit of 47 percent and a Hgb concentration of 17g percent (equivalent to 17 g/100 ml) for 163 bison ranging in age from one to five years and sampled from five different herds in five states. The density of red blood cells (RBC), PCV, and Hgb can increase with excitement (Searcy 1969; Swenson 1970). Because wild animals are likely to be more excited than domestic animals during sampling, the higher erythrocytic values of bison might be attributable in part to excitation. Haines et al. (1977) recorded an oxygen-carrying capacity for adult bison of 22.2 ml per 100 ml blood and a Hgb concentration of 17.1 g per 100 ml blood. These values were equivalent to values reported for some cervids and exceeded values reported for several domestic ungulates.

Blood urea nitrogen (BUN) was higher in bison than in cattle under a variety of ration and season conditions (Hawley 1978). Nitrogen recycling may be related to the level of BUN over a wide range of concentrations (Houpt 1970). A greater level of BUN could thereby contribute to greater recycling of nitrogen, and this has been suggested as one reason for the greater digestive capacity of bison with low protein rations (Peden et al. 1974). However, BUN is not the only parameter influencing nitrogen recycling. Keith (1977) reported that rumen ammonia levels and salivary urea concentrations were more strongly influenced by factors other than BUN.

Means and standard deviations of serum glutamate oxaloacetate transaminase (GOT) levels reported by Marler (1975) for bison were 99 ± 18 mU per ml in adults and 128 ± 31 mU per ml in animals under two years of age. Keith et al. (1978) observed levels of 57 ± 25 IU per l to 121 ± 63 IU per l in adults. These values are somewhat higher than the normal range for cattle (Kaneko and Cornelius 1971). Serum GOT levels can increase markedly with rough handling of untractable animals, which could be one explanation for higher GOT levels in bison.

Alkaline phosphatase (ALP) was higher in young bison than in adults (Marler 1975). Higher levels in young animals have also been observed in pronghorn antelope (*Antilocapra americana*) by Barrett and Chalmers (1977), and in cattle and sheep by Kaneko and Cornelius (1970).

Digestion. Most studies of the digestion of forage by bison have involved comparisons with cattle. The poorest-quality ration compared was comprised of native winter forages containing less than 6 percent crude protein (CP) content (Peden et al. 1974). The highest-quality ration was an alfalfa hay of 18.7 percent CP content (Richmond et al. 1977). In general, digestibility coefficients for all nutrients were greater in bison than in cattle for poor-quality, low-protein feeds. With high-quality, high-protein feeds, differences in digestibility between bison and cattle were not as great (table 49.2).

Peden et al. (1974) used a nylon bag technique to measure the digestion of native forages grazed by bison and cattle on shortgrass prairie. Digestibilities were greater in bison than in cattle for fall and winter forages in which the crude fiber (CF) content was high and the CP content was less than 7 percent. For spring and summer forages, in which the CP content exceeded 7 percent and the CF content was low, differences in digestibilities between bison and cattle were not evident.

Richmond et al. (1977) used acid-insoluble ash as an indicator to compare the digestion of alfalfa, grass, and sedge hays in bison, Hereford, and yak (*Bos grunniens*) yearlings. The CP and acid detergent fiber (ADF) contents of the alfalfa, sedge, and grass hays were, respectively, 18.7 percent and 30.5 percent, 8.3 percent and 39.2 percent, and 6.6 percent and 40.3 percent. Grass and sedge hays were digested more by bison than by cattle, but there was little difference in the digestion of alfalfa.

Young et al. (1977) used chromic oxide as an indicator to compare the digestibility of a pelleted alfalfa–bromegrass mixture (15.1 percent CP, 25.1 percent ADF) in pairs of bison, yak, and three breeds of cattle. The greatest digestibility was observed in Holstein cattle, while Hereford cattle, Highland cattle, and bison displayed digestibilities similar to one another.

Hawley et al. (1981a) compared the digestion of five native forages in bison and cattle using a nylon bag technique. The forage samples ranged from 36.4 percent CF to 70.1 percent CF. The mean dry matter (DM) digestibilities of forages for bison and cattle were, respectively: willow (*Salix* spp.), 58 percent and 52 percent; slough sedge (*Carex atherodes*), 59 percent and 37 percent; baltic rush (*Juncus balticus*), 55 percent and 32 percent; aleppo avens (*Geum aleppicum*), 45 percent and 43 percent; and northern reedgrass (*Calamagrostis inexpansa*), 46 percent and 24 percent. The weighted average DM digestibility values were 52 percent and 39 percent in bison and cattle, respectively.

The digestion of slough sedge has also been compared in bison and cattle through total fecal collection (Hawley et al. 1981b). The CP and ADF contents of the sedge were about 8 percent and 46 percent, respectively. All nutrients tested were digested to a greater extent by bison than by cattle; organic matter digestibilities averaged 55 percent and 49 percent for bison and cattle, respectively.

The reason for greater digestibilities in bison than in cattle when the experimental rations are low in protein and high in fiber is unknown. Because major differences are apparent only with low-quality forages, differences in digestion may be related to the low nitrogen or high fiber contents characteristic of poor-quality forages. This suggests that animal species differences in nitrogen recycling or passage rate could be contributing to species differences in forage digestion.

When nitrogen in the rumen is limiting, greater recycling of nitrogen to the rumen may enhance micro-

TABLE 49.2. Digestibility comparisons between bison and cattle

Feed	Component measured[a]	Digestibility (%)		Reference
		Bison	Cattle	
Five native forages	NBDMD	52	39	Hawley et al. 1981a
Sedge	DM	51	46	Hawley et al. 1981b
	OM	55	49	
	DE	51	45	
	CP	38	28	
Winter forage[b]	NBDMD	63	42	Peden et al. 1974
Summer forage[b]	NBDMD	80	80	Peden et al. 1974
Sedge	DM	64	58	Richmond et al. 1977
	N	54	47	
Grass	DM	74	62	Richmond et al. 1977
	N	70	58	
Alfalfa	DM	78	76	Richmond et al. 1977
	N	84	83	
Alfalfa-bromegrass	DM	50	52[d]	Young et al. 1977
	DE	49	50	
	CP	67	65	

[a] NBDMD = nylon bag dry matter disappearance, DM = dry matter, OM = organic matter, DE = digestible energy, CP = crude protein, N = nitrogen.

[b] Extrapolated from graph at approximate maximum and minimum overall digestibilities.

[c] The data in table 2 of Richmond et al. (1977) are incorrectly tabulated. The data are correctly presented in table 2 of Richmond et al. (1976).

[d] Average of three breeds.

bial fermentation and hence forage digestion (Peden et al. 1974). Crude protein levels below about 6 percent appear to limit rumen metabolism in domestic ruminants (Gilchrist and Clark 1957; Glover and Dougall 1960). Peden et al. (1974) observed greater forage digestibilities in bison than in cattle only when CP levels were less than about 7 percent and suggested that these differences might be attributable to animal species differences in nitrogen recycling. This hypothesis has not been tested. The research of el-Shazly et al. (1961) suggested that the availability of nitrogen markedly affects the digestion of fibrous substances within the rumen. The extent of nitrogen recycling and the impact of this recycled nitrogen on digestion in ruminants is under discussion (Engelhardt 1978).

Differences in nitrogen recycling would presumably produce different rumen environments and, therefore, different rumen microbial populations. Pearson (1967) observed that the rumen bacteria and ciliate protozoa from range-fed bison killed in fall were similar in kind and number to those found in domestic livestock. Jones and Hawley (unpublished data) found only small differences in the number of rumen protozoa in bison and cattle receiving a high-quality finishing ration under feedlot conditions. Comparative studies using low-quality rations and involving population studies of individual microbial species are needed to determine if animal species differences in forage digestion are related to rumen microbial differences.

A lower rumen passage rate generally increases the digestion of fibrous feedstuffs (Church 1975). With a lower passage rate, the feed is maintained in the presence of gastrointestinal microflora for a longer period of time. Young et al. (1977) observed forestomach retention times of 38.7 h in bison and an average of 30.6 h in three breeds of cattle. The longer retention time in bison was not manifested by greater forage digestion. However, the ration tested was of relatively good quality and the effect of increased retention time on digestion would therefore be somewhat muted. In cattle and sheep, an increase in circulating thyroid hormones increases gut motility and passage rate (Miller et al. 1974; Kennedy et al. 1977). In the experiments of Young et al. (1977), the level of plasma thyroxine was lower in bison than in cattle. However, the level of plasma triiodothyronine (a more potent equivalent) was higher in bison than in cattle. Data are insufficient to relate bison and cattle differences in circulating thyroid hormone to species differences in passage rates and forage digestion. Dziuk (1965) studied reticulorumen motility of bison using sampling periods of 2 to 3 h and concluded that the motility in bison was similar to that in domestic cattle. Data on the amount of time per day which bison spend ruminating would be useful in relating long-term reticulorumen motility, rumen retention time, and feed passage rate. The contribution of these factors to species differences in digestibility is unknown. During *in situ* nylon bag experiments (Peden et al. 1974; Hawley et al. 1981*a*), samples remained within the rumen for equal lengths of time, thereby obviating the direct effect of the rate at which material leaves the rumen on the digestion of that material. In these experiments, all forages were digested more by bison than by cattle.

Intake. In general, reducing DM intake increases the digestibility of fibrous foods (Schneider and Flatt 1975). In some instances, intake has been observed to be less in bison than in cattle (Peters 1958; Christopherson et al. 1976, 1978; Hawley et al. 1981*b*). Thus, greater forage digestion in bison might be associated with lower rates of intake. However, greater digestibility coefficients were observed in bison than in cattle when intake rates did not differ significantly (Richmond et al. 1977; Hawley et al. 1981*b*).

Rice et al. (1974) estimated greater intake rates for bison than for cattle or sheep on native range. Forage grazed by bison was less digestible than that grazed by cattle, which, in turn, was less digestible than forage grazed by sheep. They suggested that intake was greater in order to compensate for a poorer-quality ration. Because bison were less selective grazers (Peden et al. 1974; Rice et al. 1974), the availability of acceptable forage may have been greater for bison than for the other, more selective grazers.

Nutrient Requirements. Data are insufficient to permit identification of the specific requirements for nutrients in the feed of bison. Based on existing data it seems reasonable to assume that for most purposes the nutrient requirements of bison are similar to those of cattle. Differences in digestive capacities between bison and cattle could be considered if the feed provided to bison is based on published feed requirements for cattle. These requirements can be adjusted for bison depending on the feedstuff, the season, and the feeding regime (see "Economic Status and Management" section). There are insufficient data to develop exact conversion factors for individual feeds. Differences between bison and cattle in the availability of feed nutrients are greatest when feed quality is lowest.

Growth and Performance. Bison weigh 14–18 kg at birth (McHugh 1972). In studies of calf production by Hereford and cattalo cows, Peters and Slen (1966) reported birth weights of calves surviving to weaning as 32.7 kg for Herefords, 31.5 kg for cattalo from dams less than ¼ bison, 30.5 kg for cattalo from ¼ bison dams, and 26.4 kg for cattalo from F_1 hybrid dams. Thus, birth weight decreased with an increased proportion of bison in the dam. The rate of gain of calves was inversely correlated with the birth weight. Average daily gain of calves from Hereford dams was 0.69 kg, from less than ¼ bison dams was 0.74 kg, from ¼ bison dams was 0.77 kg, and for calves from F_1 hybrid dams was 0.79 kg. Keller (1980), working with data from subsequent generations, calculated that the effect of bison percentage in the dam on calf average daily gain was negligible despite the fact that the percentage of bison in the dam had a negative effect on total milk yield. Total milk yield decreased by over 10 kg for each percentage increase of bison in the dam.

McHugh (1958) reported that plains bison bulls approach maximum size by five to six years of age, with small yearly increments for a few years thereafter.

Observations by various bison producers indicate that there are considerable differences among herds and that bulls sometimes continue growing for as long as nine years and cows for seven years. Approximate adult body sizes of plains bison are listed in table 49.1.

Under conditions that are favorable for cattle growth, bison have a lower rate of gain than do cattle. When fed a complete finishing ration under feedlot conditions, male and female bison calves gained 0.64 kg per day and 0.50 kg per day, respectively, while Hereford calves gained 0.91 kg per day and 0.82 kg per day, respectively (Peters 1958). Cattalo calves were intermediate in performance, gaining 0.86 kg per day and 0.68 kg per day for males and females, respectively. Feed conversion efficiencies were higher for cattle. Young et al. (1977) recorded an average gain from November to October of 215 kg in female Hereford calves, as compared to 160 kg in a female bison calf and 163 kg in a male bison calf. The ration was an alfalfa-bromegrass mixture. Christopherson et al. (1979) fed one male and one female bison calf and two male and two female Hereford calves a ration comprised of 40 percent alfalfa-brome hay and 60 percent concentrate at a rate of 100 g of feed per unit metabolic weight per day. The average daily gain over one year of the bison calves as approximated from graphically presented data was 0.5 kg per day, while that of the cattle calves was 0.9 kg per day.

Hawley (unpublished data) observed that Hereford steers exceeded bison steers in rate of gain from approximately 4 months to 31 months of age while being held under feedlot conditions. The average growth rate of six bison steer calves from September 1975 to November 1977 was 0.4 kg per day. The greatest rate of gain in the bison was 0.5 kg per day from March to June 1976, inclusive. Feed intake and feed conversion were 5.5 kg per day and 9.8 percent, respectively. The ration, fed *ad libitum*, contained 12.4 percent CP, 68 percent total digestible nutrients (TDN), 0.7 percent calcium, and 0.4 percent phosphorus. The rate of gain, feed intake, and feed conversion of eight Hereford steer calves during this period and receiving the same ration were 1.1 kg per day, 9.9 kg per day, and 10.9 percent, respectively. When receiving a finishing ration of 13.4 percent CP and 72 percent TDN at approximately 20 months of age, cattle gained 1.1 kg per day and bison 0.4 kg per day. These data may not be representative of maximum animal performance because of the intensive handling of animals and the interspersion of experiments involving different rations. Hawley et al. (1981b) observed that bison steers gained 0.4 kg per day on sedge hay in the summer, while Hereford steers did not show any appreciable growth. When ration composition and intake only marginally meet the nutrient requirements for growth of cattle, as was the case in this experiment, bison rate of gain may exceed cattle rate of gain. However, rate of gain in this experiment was measured over only 16 days. Dry matter intake rates on a body weight basis were similar. The intake of digestible energy on a metabolic body weight basis was slightly greater for bison because of their greater ability to digest the hay.

Hawley et al. (1981b) observed a significantly (P < 0.05) lower DM intake of sedge hay and a numerically lower gain in bison steers than in cattle steers during winter. Inappetence, reduced metabolism, and reduced growth in the winter are common in wild ungulates (Wood et al. 1962; Silver et al. 1969; Ozoga and Verme 1970; Kirkpatrick et al. 1975; Westra and Hudson 1979), and are viewed as adaptive strategies developed to reduce nutritional requirements.

Seasonal inappetence and reduced growth rate may be dependent on the age of the animal and on ration quality. Bison calves that were fed high-quality rations were observed to grow considerably in their first winter (Christopherson et al. 1979). Richmond et al. (1977) suggested that seasonal changes in growth and the stress of handling and confinement could have contributed to the relatively poor performance they observed in bison fed sedge and grass hays and the good performance of bison relative to cattle when fed alfalfa.

Carcass Characteristics. The massive foreshoulder of bison gives the impression that a large proportion of the body weight is in the forequarter. Yearling bull bison had approximately 54 percent of the carcass weight in the forequarters, compared with 52.5 percent in Hereford bulls that were fed in the same manner (Peters 1958). However, Berg and Butterfield (1976) observed that the proportion of body muscle in the proximal pelvic limb area of a bison bull was greater than that of domestic cattle bulls. Meat from carcasses of 2.5-year-old bison was compared to meat from A2, C1, and D1 grade cattle carcasses for quality and consumer acceptance (Cox 1978). In general, taste panelists could not tell the difference between bison and cattle meat.

REPRODUCTION

Breeding Season. In most regions, the breeding season for bison generally occurs between July and October (Garretson 1927; Soper 1941; Fuller 1966; Lott 1972; Meagher 1973; Banfield 1974; Haugen 1974). The temporal variability in onset and duration of the rut may be related to variation in climate, photoperiod, habitat, population density, and genetic expression. Female bison tend to be seasonally polyestrous, with a cycle of approximately three weeks' duration (Fuller 1966; Banfield 1974); however, unseasonal matings sometimes occur (Soper 1941; McHugh 1958; Banfield 1974).

The breeding season, or rut, varies in length depending on herd location and has been observed to last from 15 June to 30 September at Hayden Valley, Yellowstone National Park, Wyoming (McHugh 1958); from mid-July to mid-August in Yellowstone National Park (Meagher 1978); from 23 June to 14 September at Wind Cave National Park, South Dakota (McHugh 1958); from 21 July to 15 August for peak breeding activity and from 25 August to 3 September for a second smaller peak of activity in Wind Cave National Park and Custer State Park, South Dakota (Haugen

1974); from 17 July to 24 August at Fort Niobrara National Wildlife Refuge, Nebraska (Mahan 1978); from 1 June to 30 July at the Wichita Mountains Wildlife Refuge, Oklahoma (Halloran and Glass 1959); and from 1 July to 30 September in Wood Buffalo National Park, Alberta and the Northwest Territories (N.W.T.) (Fuller 1960; Banfield 1974).

Age at Sexual Maturity and Conception. The age at which bison cows first conceive varies considerably among locations and often within herds from the same region (Meagher 1973). In Nebraska and South Dakota, bison cows ordinarily conceive for the first time at two years of age (Haugen 1974). A few bison cows conceive as yearlings, giving birth to a calf at two years of age (Fuller 1966). In Wood Buffalo National Park, 5 percent of yearling cows were breeding (Fuller 1961); in the Wichita Mountains Wildlife Refuge, 13 percent conceived as yearlings (Halloran 1968); and in a sample of plains bison cows from Nebraska and South Dakota, 6 percent of yearlings showed a corpus luteum but no sign of embryo development (Haugen 1974). In Wood Buffalo National Park in the Hay Camp herd, 38 percent of two year olds and 65 percent of three year olds were pregnant. In the Lake Claire herd to the south, pregnancy rates in 1952 were higher than in the Hay Camp herd: 59 percent of two year olds and 81 percent of three year olds were pregnant (Fuller 1966). In both herds, 52 percent of the cows conceived for the first time as three year olds. Sexual maturity was attained earlier in bison cows at the Wichita Mountains Wildlife Refuge and in herds from Nebraska and South Dakota than in herds from Wood Buffalo National Park. From a sample of 35 cows in the Wichita Mountains Wildlife Refuge, 73 percent conceived as two year olds, producing first calves at age three (Halloran 1968). In the Nebraska and South Dakota bison herds, 87 percent of the two-year-old cows were carrying embryos (Haugen 1974).

Bison in Yellowstone National Park first conceived at a greater age when compared with bison of other herds. In a semidomesticated herd studied during 1940–41 at that park, 50 percent of 3.5-year-old females were pregnant and 92 percent of the 4.5 year olds were pregnant. During the 1964–66 study, pregnancy rates had declined to 27 percent of 3.5 year olds and 71 percent of the 4.5 year olds (Meagher 1973). In Nebraska and South Dakota, 89 percent of plains bison cows that were 5 years of age and older had developed corpora lutea, an indication of having undergone estrus in the current breeding season (Haugen 1974).

Attainment of sexual maturity in male bison was similar to that for female bison in Wood Buffalo National Park. A small proportion of the yearlings, approximately one-third of the two year olds, and virtually all bison three years of age and older were sexually mature (Fuller 1961). In the Wichita Mountains Wildlife Refuge, two experimental bulls were not effective herd sires as yearlings but were effective at two years of age (Halloran 1968). In small samples of bison from Nebraska and South Dakota, breeding condition, as indicated by presence of sperm in the epididymis, was evident in three of six yearling males and in more than 75 percent of each of the subsequent age classes to five year olds and older (Haugen 1974). In the Hayden Valley bison herd in Yellowstone Park, males eight years old and older were the most active sexually, as indicated by the proportion of males establishing tending bonds (McHugh 1958). Only mature males tended cows in the bison herds in Wood Buffalo National Park (Fuller 1960). Male bison attain sexual maturity well in advance of becoming part of the active breeding population (Meagher 1973).

Longevity. The longevity of bison is not well documented. However, reports do exist of bison living beyond 20 years of age and even up to age 41 (Dary 1974). In wild populations, by the time a bison has reached age 15 it can be considered to have entered old age; in captivity, life span increases.

Gestation. The gestation period for bison is usually 9 to 9.5 months and is similar to that for domestic cattle (Garretson 1927; Soper 1941; Walker et al. 1964; Halloran 1968). Haugen (1974) reported that the gestation period was about 285 days for bison herds in South Dakota and Nebraska, and Banfield (1974) indicated that, in general, the gestation period is between 270 and 300 days.

Pregnancy Rate and Reproductive Rate. Pregnancy rates for bison vary according to age, with reproductive vigor highest in animals between age 3 and the onset of old age (12–15 years) (Fuller 1961). Wild bison normally produce two calves every 3 years (Fuller 1966; Banfield 1974).

In Yellowstone National Park, pregnancy rates in wild bison varied from 94 percent during 1931–32 to 52 percent during 1964–66 (Meagher 1973). During the latter study, there was a lower rate in all age classes over 2.5 years old than in the semidomesticated herd of 1940–41. Fuller (1966) reported that the average pregnancy rate for the Hay Camp bison herd at Wood Buffalo National Park was 67 percent during the period 1952–56.

The reproductive rate (calf crop expressed as a percentage of all herd females less calves and yearlings) for the bison herd in the Wichita Mountains Wildlife Refuge varied from 47 to 60 percent between 1960 and 1966, with an average of 52 percent (Halloran 1968). In that study, the rate for 198 experimental animals from six different age classes was 67 percent. The higher value for the experimental animals compared with the entire herd was likely attributable to the higher percentages of younger animals (prime breeding age) in the former group (Halloran 1968).

Disease such as brucellosis may influence pregnancy rate. Brucellosis causes abortion of calves and temporary sterility in cattle (Choquette et al. 1978). At Wood Buffalo National Park, brucellosis infection rates were 27 percent and 36 percent in 1956 and 1958, respectively. Assuming that the effects of this disease are similar in bison, then both conditions of the disease could have contributed to the low incidence of pregnancy at midterm in the Hay Camp herd (Fuller 1966).

The physiological stress of pregnancy may, at the same time, predispose an animal to disease such as tuberculosis (Fuller 1961). At Hay Camp, 51 percent of the pregnant cows had tuberculosis. However, pregnancy rates for tubercular and healthy bison (55 percent and 51 percent, respectively) were not significantly different (Fuller 1966).

At Yellowstone National Park, brucellosis appeared to have little effect on pregnancy rates, as lower rates of infection did not result in higher pregnancy rates (Meagher 1973). Tunnicliff and Marsh (1935:751) concluded that "while the infection is evidently active in the herd, any reproductive inefficiency which may exist is not entirely due to Brucella infection in the cows." The low pregnancy rate in the Yellowstone herd may be influenced more by environmental factors, especially those related to the severity of winter (Meagher 1973).

At the Wichita Mountains Wildlife Refuge, Halloran (1968) attributed the low reproductive rate (52 percent) to high calf survival. In this herd, calves were observed nursing into their second year of life. This prolonged the physiological stress on the cows, which could adversely affect both pregnancy and reproductive rates. Lactating cows at Wood Buffalo National Park carried smaller midwinter fetuses than did nonlactating cows, which suggested that bison cows with calves bred later in the year than did dry cows (Fuller 1961). Survival of a calf apparently reduces the female's chance of conceiving and the vigor of the calf in utero should she conceive that season. Loss of a calf prior to the end of the breeding season probably increases the chances of breeding for the cow (Fuller 1966). Loss of a calf at any time prior to weaning may enhance the vigor of a calf in utero by improving the energy balance of the cow during the critical winter period. Calf survival did not appear to be an important influence on reproductive rate in Yellowstone bison (Meagher 1973). Population age structure differences may also account for some of the variation in pregnancy rates among herds.

Number of Young. One calf is usual, and twins are rare (Garretson 1927; Fuller 1961; Banfield 1974). One instance of twin bison calves was reported by McHugh (1958) in the Lamar herd in Yellowstone and one set of twins was recorded in the plains bison herd at Wichita Mountains Wildlife Refuge in 1965 (Halloran 1968). In Wood Buffalo National Park from 1952 to 1956, no twins were observed in 481 gravid uteri examined (Fuller 1966), nor were twins observed in the 1964–68 productivity study at Yellowstone National Park (Meagher 1973). In Elk Island National Park during the period 1945 to 1960, three sets of plains bison twins were observed (R. Jones personal communication).

Calving Season. Garretson (1927) described calving in bison as lasting from April to June. However, conception and, therefore, parturition can occur at any time of year. From 1937 to 1950, calving at Yellowstone had commenced by mid-April, while more recent studies indicated a later calving season, with most calves being born in the first half of May (Meagher 1973). McHugh (1958) noted that a few calves were born from June through October in herds at Yellowstone Park, the Crow Reservation, Wind Cave National Park, and the National Bison Range, Montana. The calving season at Yellowstone has become more prolonged as the size of the population has increased (Meagher 1973).

In the Wichita Mountains Wildlife Refuge herd, the first bison calf was always recorded within the period from 10 March to 7 April (Halloran 1968). In Wood Buffalo National Park, calving was observed from mid-April until the beginning of June (Egerton 1962; Banfield 1974). Soper (1941) noted that most calving in this park occurred around mid-May. Egerton (1962) observed that the calving period in the confined bison herd in Waterton Lakes National Park lasted from April through July.

In Elk Island National Park, the calving period for wood bison appears to commence during the latter part of April and lasts through to the latter part of August. That for plains bison usually begins about the first week in May and lasts until the latter part of August (R. Jones personal communication). The majority of wood bison and plains bison calves are born between 1 May and 15 June; occasionally late calves are born in September. However, more births of plains bison are occurring later in the season, even into November, and the calving season has become noticeably prolonged as herd size has increased.

In assessing McHugh's data from the United States, Egerton (1962) suggested that the calving season is about two weeks later in northern herds, a phenomenon likely related to the variations in climate and photoperiod between these regions.

Primary Sex Ratio. At Yellowstone National Park during the 1930s, Rush (1932) reported that primary sex ratios for bison ranged from 108 to 163 males per 100 females. The average in a sample of 294 fetuses was 56 percent males. At the Wichita Mountains Wildlife Refuge between 1908 and 1966, 51 percent of the 5,633 bison born there were males (Halloran 1968). In bison herds in Nebraska and South Dakota, the sex ratio for 101 embryos examined was 54.5 percent males (Haugen 1974). Between 1952 and 1956 in Wood Buffalo National Park, the primary (in utero) sex ratio for 472 fetuses examined was 112 males per 100 females, or 53 percent males (Fuller 1966). Palmer (1916) reported 119 males per 100 females (54 percent males) in a sample of 460 plains bison fetuses.

The reported sex ratios from the above five herds of bison are similar. With the exception of one Yellowstone sample, they vary only within a 5 percent range. A slight excess of males in utero is common among mammals (Fuller 1961).

Calf Percentages. In Yellowstone National Park, spring calf percentages (expressed as percentages of mixed herds) are normally between 18 and 20 percent (Meagher 1973). In the spring of 1965, after herd reductions had removed large numbers of cows, the lowest calf percentages of the study (7–14 percent) were recorded for three herds. Calf percentages for these populations increased after 1965. Pooled percentages for the three herds in 1967 (20 percent) and 1968 (19 percent) suggested that calf production for the population was leveling and if so, then the proportion of newborn calves in mixed bison herds in Yellowstone approximated 20 percent (Meagher 1973). Nonselective herd reductions may alter calf percentages by taking an imbalanced harvest from the more easily slaughtered mixed herd groups rather than over the entire population.

In Wood Buffalo National Park, a potential calf crop of 20–25 percent of the herd is expected during the latter part of June and early July. The calf crop declines by approximately 2 percent per month until December, when calves make up less than 10 percent of the herd (Fuller 1966). Survival to the yearling category in the park ranged between 5 and 8 percent of the herd per year.

ECOLOGY

Interspecific Relationships. ELK. In studying species interaction at the Jackson Hole Wildlife Park, Wyoming, McHugh (1958) placed bison at the top of the interspecific dominance hierarchy, followed by elk (*Cervus elaphus*), mule deer (*Odocoileus hemionus*), pronghorn antelope, moose (*Alces alces*), and white-tailed deer (*Odocoileus virginianus*). Bison are usually dominant over elk, and McHugh (1958) noted that bison calves could displace six-point bull elk. However, at the edge of bison herds five- and six-point bull elk could displace bison cows and yearling bulls. Aggression by bison reversed any elk dominance. Bison occasionally forced elk into deep snow in winter and chased them from feed in summer. The usually wary elk were sometimes caught and butted. Bison have been known to harass and kill elk calves at Jackson Hole Wildlife Park, Fort Niobrara National Wildlife Refuge, Wind Cave National Park, and Yellowstone National Park (Rush 1942; McHugh 1958; Mahan 1977). At the National Bison Range, a bottle-raised bull elk calf formed an attachment to two cow bison and when mature was observed to round up a herd of cow bison and bugle. On another occasion a harem bull elk killed a yearling bison (McHugh 1958). At Yellowstone National Park, elk and bison will associate within 10 m of each other despite their seeming intolerance. At Elk Island National Park and Yellowstone National Park, bison interfere with elk live-trapping programs, in part, because elk will not enter traps when bison are near. Like other North American cervids, elk are frequent browsers and as such are not in direct competition with bison for forage.

MULE DEER AND WHITE-TAILED DEER. Bison have been observed to charge and strike mule deer at Jackson Hole Wildlife Park and at Yellowstone National Park (McHugh 1958). Bison and white-tailed deer at Colorado National Monument did not appear to compete for food (Capp 1964).

PRONGHORN ANTELOPE. In Wyoming, Bryant (1885) noted that pronghorns and deer would seek protection from wolves in bison herds. At Wind Cave National Park, pronghorn, at times, pass near or through bison herds unhindered; however, at the Jackson Hole Wildlife Park bison disturbed a group of resting pronghorns when passing within 50 m (McHugh 1958). Bison will occasionally charge pronghorns. McHugh (1958) observed a bison kill an eight-month-old pronghorn buck.

Pronghorn are highly selective feeders (Schwartz et al. 1977), while bison are more flexible in choice of diet. Where pronghorn diets are 50–80 percent forbs, bison diets are usually less than 20 percent forbs (Peden 1972). Bison compete less intensively for food with pronghorn than with cattle or sheep (Peden 1972). The theory that large and small ruminants will not compete with each other for food resources (Bell 1971) is further affirmed by similarity in sheep and pronghorn diets and their dissimilarity to bison diets (Peden 1972).

MOOSE. At Jackson Hole Wildlife Park, bison killed a seven-month-old moose calf that had recently been placed in the park (McHugh 1958). Moose are primarily browsers and probably do not compete with bison for food.

The intense aggression of bison toward other wildlife species observed at the Jackson Hole Wildlife Park may be more frequent than that occurring in the wild as a result of the semiconfined nature of the animals at the park.

BIGHORN SHEEP. Three instances of an older bighorn ram being associated with a bison bull were observed at the National Bison Range (McHugh 1958).

DOMESTIC LIVESTOCK. Draft horses (*Equus caballus*) at the Jackson Hole Wildlife Park were dominant over bison cows and yearlings at a salt lick during October; however, by March they had lost some dominance (McHugh 1958). All bison exhibited dominance over a saddle horse. In Yellowstone National Park, the head animal keeper observed only one horse killed by bison in 23 years. Numerous instances of horses being killed or charged when used to pursue bison have been reported from other areas (McHugh 1958, 1972; F. Dixon personal communication). Cattle ranged with bison during summer on a ranch in South Dakota and appeared to be compatible; the two species essentially ignored each other (Colman 1978).

Bison diets more closely resemble those of cattle than those of sheep under light grazing conditions; however, the differences become less distinct under heavy grazing (Peden 1972). Diets of feral horses near Sundre, Alberta, had a higher sedge component than did diets of free-ranging cattle (Salter 1978). In this respect, feral horse diets show greater similarity to diets of northern bison than do cattle diets. Herbivores, however, appear to select for higher proportions of plant groups that are best digested (Peden et al. 1974). This may provide a mechanism whereby bison and other herbivores can avoid competition, particularly at low to moderate stocking rates. In the past there seems to have been a reasonably good balance between bison and their forage supply, and competition with other species for food has not been a major limiting factor to population growth (Longhurst 1961).

WOLF. In some regions bison may form the prey base on which wolf populations depend. Bison are, for the most part, indifferent to the presence of wolves until attacked (Fuller 1960; McHugh 1972). When bison populations drop below a certain critical level, wolf predation may in the absence of alternate prey take more than the annual increment and effectively reduce the bison population (L. Carbyn personal communication).

GRIZZLY BEAR. The importance of scavenged or killed bison in the diets of grizzly bears is considerably reduced from what it likely was when the plains grizzly and large bison herds coexisted on the Great Plains. Winter-killed bison and other ungulates may be important food sources to bears in early spring after they emerge from dens (Meagher 1978). In Yellowstone National Park, three grizzly bears were observed capturing an elk calf within 213 m of a bison herd without causing alarm to the bison (McHugh 1958). On occasion, bison prove too large an adversary for grizzly bears and may kill their would-be predators (McHugh 1972).

COYOTE. At Yellowstone National Park, coyotes stayed closer to bison in winter than in summer (McHugh 1958). These coyotes may have caught small mammals that were trapped in snow craters made by feeding bison or they may have been waiting for scavenging opportunities. On two occasions, coyotes wandering into a herd of bison at the National Bison Range were horned and trampled to death (McHugh 1958).

In consideration of the previous historic abundance of bison, canid and ursid predation appears to have been insignificant relative to other mortality factors. In the past, bison may have been a prominent prey species and as such supported a large and diverse fauna of scavengers and predators.

RODENTS. Bison at Wind Cave National Park frequented prairie dog (*Cynomys ludovicianus*) towns more than surrounding grasslands. Such sites are used for wallowing and often provide a variety of forbs (McHugh 1958).

HUMANS. People have hunted North American bison for more than 12,000 years (McHugh 1972). The ecological relationship between the two species has remained that of a classic predator and its prey for approximately 11,900 of those years. In this regard, an early pioneer stated that for the Indian, bison were "meat, drink, shoes, houses, fire, vessels, and their Master's whole substance" (McHugh 1972:4). Bison influenced the settlement of North America more than any other endemic species (Roe 1970). While aboriginal cultures spiritually embraced the bison, Europeans often viewed them as "an insufferable nuisance"

(McHugh 1972:xxii). It was this dichotomy that, in part, led to the disruption of what had been an ecological balance between humans and bison. In less than 100 years an estimated 40-60 million North American bison had been reduced to just over 1,000 animals (Hornaday 1889). It was only the dedication of a few conservationists in Canada and the United States and the physical isolation of bison in the Mackenzie basin (Raup 1933) that saved the bison from extinction.

Vegetation. Evidence suggests that bison helped maintain the short-grass prairie and its associated fauna (Weaver and Clements 1938; Larson 1940). In Montana, the Lewis and Clark expedition observed vast numbers of bison in areas floristically dominated by short-grass species and found the soil dark, rich, and fertile. The presence and activity of the vast herds observed by that expedition would scarify and enrich the soil (particularly in arid regions), thus allowing recovery of the nutritious short-grass species (buffalo grass [*Buchloe dactyloides*] and blue grama [*Bouteloua gracilis*]). Larson (1940) observed that short-grass vegetation thrived under moderately heavy grazing, unlike the taller bunch grasses (e.g., sand dropseed [*Sporobolus cryptandrus*] and needle and thread [*Stipa comata*]). Following the decimation of bison in Nebraska, the distinguished botanist C. E. Bessey theorized that these short grasses were being replaced by ranker species as a result of reduction in grazing by bison, seed vectoring, and incidence of prairie fire.

The thick fur on the head and forequarters of a bison is ideally suited for dispersal of awned, barbed, or sticky seed-bearing structures. The seed of buffalo grass, cockle burs (*Xanthium italicum*), and St. Johnswort (*Hypericum perforatum*) readily adhere to bison fur. The dissemination of the last throughout the National Bison Range is thought to have been caused by bison (McHugh 1958).

Localized stands of timber, particularly those not tightly clumped together, may be considerably affected by the horning and thrashing of bison during the rut and at other times. McHugh (1958) estimated that 51 percent of the lodgepole pine in some areas of Yellowstone National Park had been horned by bison. Such activity by bison and/or elk may inhibit succession of prairie to forest and thus serve to maintain grasslands (Moss 1932; Patten 1963; Meagher 1973).

Soil. Where bison trails or wallows are cut into steep hillsides, considerable water and wind erosion can occur (McHugh 1958; Capp 1964; Meagher 1973). Hillside trails can serve as drainage channels, effectively lowering the water table in upland areas and subsequently causing a change in the vegetation. Where trails cut near the top of steep sandy hills, erosion and slippage may produce barren areas. At the Colorado National Monument, Capp (1964) believed that poor soil conditions would occur in this area regardless of the presence of bison. On flat terrain, bison wallows can serve as water catchments. Such small ponds become available to both invertebrates and vertebrates for a number of uses and may enhance growth of specific vegetation. By creating trails through different habitats, bison help provide access corridors for many species of mammals, including humans.

FOOD HABITS

In the majority of situations, North American bison are grazers. Because bison are located in widely varying habitats throughout their range, it is most useful to identify their diets by association with geographic area. Seasonal diets from four different populations of bison have been compared by forage class (table 49.3). Grasses and sedges were the most important foods of free-roaming bison in three of these herds. Peden (1976) confirmed that bison on the short-grass plains in northeastern Colorado consume mainly grasses at all seasons. Similarly, on semidesert ranges in southwestern Colorado, grass was the dominant forage utilized by bison during summer and exceeded 27 percent during all seasons. Sedges (*Carex* spp.) have also been noted as common dietary items of several bison populations (Fuller 1966; Allison 1973; Meagher 1973; Reynolds et al. 1978). In Yellowstone National Park and in northern Canada, sedges comprised the highest proportion of bison diets in all seasons, while grasses were second in importance. In northeastern Colorado, sedges were important to bison only during spring. In other bison herds located at Wood Buffalo National Park (Soper 1941) and Elk Island National Park (Holsworth 1960), bison were observed feeding on grasses in summer and sedges in winter. In some areas, forbs are seasonally important to foraging bison (Soper 1941; Nelson 1965; Banfield 1974; Wasser 1977). In semidesert range in southwestern Colorado, forbs were common food items during all seasons but never exceeded 17 percent in any one season (Wasser 1977). In Yellowstone National Park and in northern Canada, forbs appeared to be important to bison only during summer, whereas in northeastern Colorado forbs were important dietary items during fall and winter. Bison in the Colorado National Monument primarily forage on grass in summer, browse in winter,

and eat both in spring and fall (Wasser 1977). Unlike other populations, these bison utilized browse as the major source of food (67 percent) during winter. Browse was not eaten by bison on the short-grass plains in northeastern Colorado and was of only minor importance to bison in Yellowstone National Park and in northern Canada.

Yellowstone National Park. In Yellowstone National Park, much of the habitat is forested by lodgepole pine (*Pinus contorta*). Interspersed throughout the area are meadows with considerable sedge and grass. In the park, sedges were the main bison forage in all seasons and grasses were the next most common forage (Meagher 1973). Sedge content in the diet varied from 37 percent in fall to 56 percent in winter, while the grass content varied from 30 percent to 46 percent in fall and spring, respectively. Minor quantities of forbs (6 percent) and browse (2 percent) were consumed, mainly in summer.

Northeastern Colorado. On the short-grass plains in Colorado, on which blue grama is the dominant species, Peden (1976) observed 36 plant species in the diets of bison; only 11 contributed significantly to the total. Blue grama and buffalo grass were the most abundant plants in the habitat and also in the diet. A preference for western wheatgrass (*Agropyron smithii*) over blue grama was noted. Other commonly consumed species were red three-awn (*Aristida longiseta*), sun sedge (*Carex heliophila*), scarlet mallow (*Sphaeralcea coccinea*), sand dropseed, and needle and thread.

Southwestern Colorado. On semidesert range in the Colorado National Monument, Wasser (1977) reported strong selectivity by bison for preferred forages. The major plant communities in the study area were: sagebrush (*Artemisia* spp.), Utah juniper (*Juniperus osteosperma*), mixed sagebrush and juniper, and semidesert grass-saltbush (Graminoid-*Atriplex* sp.). The most common plant species in the bison diet during most seasons was four-wing saltbush (*A. canescens*), fol-

TABLE 49.3. Percentage composition in the diet of bison by forage class and season as summarized from 4 studies in different areas of the species range

	Season															
	Winter				Spring				Summer				Fall			
Forage Class	A	B	C	D	A	B	C	D	A	B	C	D	A	B	C	D
Grass	34	87	36	27	46	79	16	57	32	88	24	72	30	89	21	57
Sedge and rush	65	*a*	63	*a*	50	16	81	1	59	1	59	4	69	—	71	2
Forbs	*a*	9	—	3	3	*a*	1	17	6	3	8	7	*a*	9	4	11
Browse	1	—	1	67	*a*	—	2	25	2	—	8	17	*a*	—	2	29

A: Yellowstone National Park—Forested with interspersed grass-sedge meadows (after Meagher 1973).

B: Northeastern Colorado—Shortgrass plains (after Peden 1976).

C: Northern Canada—Boreal forest interspersed with large grass-sedge meadows (after Reynolds 1976).

D: Colorado National Monument—Semi-desert range, southwestern Colorado. Average of samples collected from three canyons (after Wasser 1977).

a Less than 1 percent.

lowed by needle and thread, which was important during cooler months. Sand dropseed and Galleta grass (*Hilaria jamesii*) were prominent in the diet in warmer seasons. Prickly pear cacti (*Opuntia* spp.) were among the 10 top forages during all seasons except summer (Wasser 1977). The only forbs significantly utilized during all seasons except winter were mallows. Some of the most common plants in the habitat, cheatgrass (*Bromus tectorum*), Utah juniper, and big sagebrush (*Artemisia tridentata*), were the least preferred forages.

Northern Canada. In the Northwest Territories, the bison habitat along the Slave River lowlands is within the Boreal Forest region of Canada, where white spruce (*Picea glauca*) forests separate vast open meadows supporting sedge and grass communities. Bison diets from this area contained 29 different plants, of which 12 were present in quantities exceeding 1 percent in any one season. Slough sedge (*Carex atherodes*) was by far the most abundant plant in the diet, varying from 42 percent in winter to 77 percent in spring. The second most common food was reedgrass (*Calamagrostis* spp.), which varied from 15 percent of the diet in spring to 35 percent in winter. These two forages contributed more than 70 percent of the bison diet at all seasons (Reynolds et al. 1978). Although slough sedge was the most abundant plant in the bison diet, it was second to reedgrass in abundance in the habitat. Forbs and browse appeared to be of minor importance and were consumed only in summer and fall.

Forage Selection. Bison generally are less selective in what they eat when compared with other ungulates under similar environmental conditions. In the Henry Mountains of Utah, bison selected similar forage to cattle but tended to move more, did not overgraze preferred feeding areas, and made greater use of steep slopes (Nelson 1965). Thus, bison more uniformly utilized their range with less damage to forage species. The diets of bison on the short-grass plains of Colorado resembled cattle diets more than sheep diets in areas of light grazing (Peden 1972). Bison were less selective than cattle and are, therefore, better adapted to utilize the herbage resource of the short-grass prairie more fully (Peden et al. 1974). In Yellowstone National Park, forage availability appeared to regulate bison feeding patterns within the preferred sedge and upland feeding sites (Meagher 1973). Bison were usually the least selective and cattle were the most selective during feeding trial experiments involving bison, yak, and cattle when provided sedge, grass, and alfalfa hays (Richmond et al. 1977). In northern Canada, forage consumption by bison was not directly proportional to plant availability, indicating light to moderate feeding selection (Reynolds et al. 1978).

It is apparent that where grasses and sedges are available in the habitat, they are selectively grazed by bison, and where they are sparse, browse may be substituted. Dietary shifts from grasses to sedges and back again within a habitat type are usually associated with plant phenology.

BEHAVIOR

Herding Behavior. Bison are gregarious animals. Group size does vary, however; over the course of a year three types of groups can be observed: matriarchal groups (cows, calves, yearlings, and sometimes a few older bulls), bull groups (including solitary bulls), and breeding groups (a combination of the first two groups).

Individual matriarchal groups vary little in size for most of the year. During the rut they are joined by breeding bulls and other matriarchal groups to form breeding groups (table 49.4). In Wood Buffalo National Park, Fuller (1960) suggested that the average group size for matriarchal groups ranged from 11 to 20 individuals and that there was considerable flexibility in the size of groups. Smaller subgroups may be observed within matriarchal and breeding groups. These may include nursery groups composed of cow-calf pairs, calf groups, bull groups, barren cows, and yearlings, accompanied by some two year olds (McHugh 1958). There is some controversy over whether matriarchal groups are consanguineous; Seton (1929) and Soper (1941) suggested that they are, while Garretson (1938), McHugh (1958), and Fuller (1960) believed otherwise.

Mature bulls seldom form groups of more than a few animals and seem to become less gregarious with increasing age. Solitary bulls are common even during the rut. Cows are occasionally found in bull groups as well.

McHugh (1958) reported that breeding herds in Yellowstone ranged in size from 19 to 480 and averaged 175 animals (table 49.4). Egerton (1962) suggested that the average size of herds in Wood Buffalo National Park was between 11 and 30 animals. At the National Bison Range, Lott (1974) observed an average group size during the rut of 57 with a maximum of 174, while on Catalina Island, California, he found an average breeding group size of 17. Shackleton (1968) observed smaller breeding groups at Elk Island National Park than on the National Bison Range.

Bison herds often show a remarkable degree of herd fidelity even after temporarily mixing with other herds (Fuller 1960; L. Carbyn personal communication). Fences separating groups of animals will not discourage their attempts to form a single herd. McHugh (1958) and Meagher (1973) observed that disturbance by aircraft and certain physical phenomena such as shade and rain would increase herd cohesiveness.

Dominance Relationships. The expression of intraspecific dominance in bison appears after the first few weeks of life (McHugh 1958; Mahan 1978). McHugh (1958) did not find any correlation between the position of dominance of the calf and its seniority in the calf group, morphological differences, or the mother's position in the dominance hierarchy (derived dominance). However, he did observe derived dominance in a few cases. Male calves tend to dominate female calves, as is the norm for older animals. As calves

TABLE 49.4. Size and composition of matriarchal and breeding groups at Lamar and Hayden valleys in Yellowstone National Park, Wyoming, and at Wind Cave National Park, South Dakota

	Group Location and Season			
Item	Lamar (January–March)	Lamar (May)(Calving)	Wind Cave (Calving)	Hayden (Rut)
Group Size				
Number of groups	18	15	17	36
Mean	23	23.6	21.9	175.3
Standard deviation	11.4	18.4	21.2	108
Range	10–50	4–63	3–76	19–480
Group Composition				
Number of groups censused	12	10	14	3
Mean size	20.3	17.3	16.8	115.2
Mean grouping tendency				
Cows two or more years old	12.1[a]	8.6[b]	7.4	39.3
Yearlings	4.4	4.8	4.0	20.0
Calves	0	3.9	3.9	25.3
Bulls two to three years old	3.8	3.7	1.1	16.3
Bulls four or more years old	0	0.2	0.4	14.3
Percentage of bulls four years old	0	1.0	2.1	9.2[c]

SOURCE: After McHugh 1958.
[a] 19.3 percent two year olds.
[b] 19.5 percent two year olds.
[c] Computed from census of 8 groups.

mature, differences in size, weight, seniority, and sex (within similar age groups) become increasingly important in the establishment of dominance relationships.

It would appear that dominance among calves is primarily a function of inherent disposition. Early disposition may predispose an individual bison to develop specific physical attributes significant in maintaining or advancing dominance position later in life.

McHugh (1958) observed that dominance expressions by bison at Jackson Hole Wildlife Park were either "passive" or "aggressive." Passive dominance did not involve the use of force, whereas aggressiveness involved the use of force or threat. Of the 1,027 dominance interactions observed, 73 percent were passive and 27 percent were aggressive.

After three years of studying a herd of 14 bison, McHugh (1958) did not observe any permanent reversals in hierarchial position between the herd's two mature bulls and among the seven most dominant cows. At the National Bison Range, Lott (1974) observed that 12 percent of aggressive interactions resulted in hierarchial reversals in a breeding herd containing 35 mature bulls. A number of dominance triangles were also observed. Instability in dominance relationships may be a function of habitat attributes, size of age class, and fatigue (Lott 1974). The expression of dominance tends to be more intense between animals whose positions are close together in dominance hierarchy.

Expressions of dominance can be extreme. McHugh (1958) observed a bull bison hook a cow on

his horns and throw her over his back; subsequent autopsy revealed two broken ribs and a collapsed, punctured lung. Bulls, and cows with calves, are the most prone to be involved in intraspecific interaction (Egerton 1962). Almost any desired resource can elicit expression of dominance. The expression of dominance may occur as a result of competition for some obvious resource or may be precipitated by sudden disturbance. Bull groups at Elk Island National Park have been observed to initiate fighting, mounting, and horning when vehicles stopped near them. Similar occurrences have been observed in Wyoming (McHugh 1958) and in Wood Buffalo National Park (Fuller 1960).

Migrations. Bison often undertake annual migrations that may be altitudinal or directional. In the American northwest, Garretson (1927) noted that there was a definite movement of herds from the plains region in the east to the foothill areas of the Rockies in the winter. The reverse occurred in spring. Regarding these migrations, Hornaday (1889:423, 424) stated that "the buffalo had settled migratory habits. . . . At the approach of winter the whole great system of herds which ranged from the Peace River to the Indian Territory moved south a few hundred miles, and wintered under more favorable circumstances than each band would have experienced at its farthest north." This could explain why bison were seen on the range all year (Garretson 1927). Bison, particularly cows, show strong affinity to winter range (Meagher 1973). In mountainous areas, altitudinal movements to lowland winter range in fall and to higher summer range in spring are quite common. Snipe flies (*Symphoromyia* sp.) may be responsible for some altitudinal movements in Yellowstone bison herds during the summer (Meagher 1973). Large, wind-swept prairies may also be chosen in summer for similar relief. Directional movements occur annually at Wood Buffalo National Park (Reynolds 1976) and at Yellowstone National Park (Meagher 1978).

Temporal and spatial variations in range use are likely related to such factors as tradition, forage availability and nutritional quality, macroclimatic and microclimatic variations, open water, shelter, and insect harassment.

Traveling Behavior. Bison usually travel via the most practical direct route and will rapidly establish trails to do so (Garretson 1927; McHugh 1958). Forest and shrub areas are often used as daily and seasonal travel corridors in northern Canada (Reynolds 1976). When crossing rivers, bison usually take shallow fords with gradual approaches; however, they will not hesitate to cross large, swift-flowing rivers in northern Canada, such as the Peace, Liard, and Nahanni.

During inclement weather, bison often head into the wind, unlike domestic cattle, a behavior perhaps related to the greater amount of hair insulating a bison's head and forequarters. If unable to avoid traveling in deep snow, bison will form a line, with the lead animals plunging to create deep trenches (Meagher 1973).

Foraging Behavior. McHugh (1958) observed that feeding activity was mostly a diurnal occurrence, with the night spent loafing or occasionally feeding and traveling. Similar behavior has been reported in cattle (Hancock 1953). On the Slave River lowlands, Reynolds (1976) observed that foraging was done in open meadows, while loafing and ruminating occurred in forest habitat. During winter, bison in that area foraged in small, sheltered meadows and along river and creek beds. Such areas offered less severe snow conditions and tended to support preferred forage. Unlike all other North American bovids and cervids, bison prefer to use their massive heads to sweep snow away from forage, rather than to paw.

When lakeshores are free of ice and emergent vegetation is available, bison often forage in chest-deep water, similar to feeding behavior in moose. Bison usually prefer to seek open waters or break through ice to get water rather than to eat snow (McHugh 1958).

Stampedes. Stampedes occur as a relatively disorderly movement involving a number of animals. Disturbance of a wary group within or near the herd may precipitate such sudden movements. The required stimulus can be seemingly insignificant, making the stampede appear spontaneous. Stimulus to stampede may be caused by the sudden running of one animal toward the herd after being alarmed or by any other sudden external stimulus. Bison are more wary of humans on foot than of vehicles (Fuller 1960). Older cows tend to be the most wary and often take the lead in sudden herd movements (McHugh 1958; Fuller 1960). Observations at Elk Island National Park and Wood Buffalo National Park substantiate the role of cows as leaders during herd movements (F. Dixon personal communication; L. Carbyn personal communication), while other observers have ascribed this role to bulls (Seibert 1925; Soper 1941). When fleeing down a winding road or trail in a forested area, a mature bull may lag behind, possibly to see if the source of disturbance continues to follow. Bison are capable of fleeing at speeds up to 60 km per hour (McHugh 1958; Fuller 1960).

The Rut. During the rut there is a marked increase in herd size and activity. Bull bison, like the males of many other ungulate species, are particularly active at this time. The activities of rutting bulls may include sexual investigation, Flehmen, tending cows, incomplete and fertile mountings, threat posturing, fighting, horning, wallowing, and loud vocalizations.

SEXUAL INVESTIGATION. Mature bulls tend to form their own groups apart from the cow-calf herd. However, they often enter the cow-calf herd to investigate cows sexually and temporarily stay within the herd to tend a cow approaching or in estrus during the rut (McHugh 1958; Fuller 1960; Egerton 1962; Shackleton 1968; Shult 1972; Meagher 1973).

Bulls methodically check herd cows by sniffing their vulvas and often prod resting animals to stand for a more thorough examination (McHugh 1958). Petropavlovskii and Rykova (1958) suggested that stimulation by bulls may induce estrus in cows.

FLEHMEN. Flehmen refers to a reflexive facial expression manifest by bison and many other ungulates (Egerton 1962; Geist 1963; Alexander et al. 1974; Mahan et al. 1978). It often occurs during the rut when a bull sniffs and/or licks the vulva or urine of a cow, however, bulls, females, and immature animals will initiate Flehmen over other odors. During Flehmen the upper lip is curled upward and the neck is extended, an expression that may last several seconds. It is believed that Flehmen makes the vomeronasal organ more effective (Estes 1972). A cow's urine on the ground, a bloody wound, amniotic fluid, rotted skeletons, bison hair, new calves, and human urine may also stimulate Flehmen (Egerton 1962; Herrig and Haugen 1969; Shult 1972; Lott 1974).

TENDING. Tending is defined as a temporary bond between a cow and a bull (McHugh 1958) that can last from a few seconds to several days. While tending, the bull usually tries to keep the tended cow peripheral to the main herd by keeping himself between the herd and the cow. A bull will occasionally use considerable force in order to keep the cow sexually isolated (Lott 1974). Close proximity of other males is not tolerated. Not all tending leads to copulation, and a cow may aggressively resist and thus effectively select the sire of its calf.

In consideration of the short-term nature of the tending bond, Seton (1929) and Soper (1941) described the mating system in bison as polygamous, while McHugh (1958:24) referred to it as "temporary monogamous mateship." Atypical tending bonds, where bulls tended cows for very short periods or tended calves or young bulls, have also been observed (McHugh 1958). In the latter case, he observed a bull tend a yearling for approximately four hours and attempt mounting with penis unsheathed.

COPULATION. Immediately prior to copulation both animals may engage in amatory behavior such as mutual licking and butting. The cow may also attempt to mount the bull. A cow mounting another cow or mounting the tending bull is often a sign of estrus in domestic cattle (Dukes 1937; Schein and Fohrman 1955).

The bull indicates his intention to mount by swinging his head up onto the rump of the cow. He next rears up to embrace the lower ribcage of the cow with his forelimbs and follows with penetration. The cow and bull may start to run while copulating. Insemination is usually achieved after the first few thrusts, after which, the cow often displays "servicing symptoms" (Jaczewski 1958) that may last up to several hours (McHugh 1958; Shult 1972). Typically, the cow arches her back and holds her tail at some angle to the body. Urine and/or semen are often voided. Most breeding is done by "prime bulls," generally those animals between six and nine years of age (Egerton 1962; Lott 1974). Copulation is usually a crepuscular or nocturnal activity, a behavioral phenomenon that may enhance physiological performance, such as increased semen viability, or coincide with a time when the animals are less conspicuous to predators.

THREAT POSTURING AND FIGHTING. Both sexes and all age groups of bison may engage in threat displays and fighting. The most frequent and dramatic participants are usually bulls over four years of age (McHugh 1958; Fuller 1960; Lott 1974). Although agonistic behavior may occur at any time of year, it is much more common during the rut when herds are larger.

Threat postures may be the prelude to fighting, although they are usually sufficient to terminate an encounter prior to serious physical contact. Such postures include elevation of the tail, broadside threats, pawing and wallowing, aggressive advances and lunges, and the nod threat (McHugh 1958; Lott 1974). During the broadside threat one or both animals stand broadside to each other, presumably to display their size, disposition, and intent. Aggressive advances involve one or both animals approaching the other using a slow foot-by-foot walk that may, if not resulting in displacement, lead to physical exchange. Following such advances and between fights, the pair of combatants may bob their heads up and down in what Lott (1974) described as the nod threat.

Fighting may involve butting, horn locking, shoving, and hooking. The thick cushion of hair on the head helps to reduce the impact from butting (McHugh 1958). Hooking can result in serious injury. In this respect, bison differ somewhat from other North American bovids and cervids. The frequency with which serious injury occurs suggests that bison have better perfected offensive strategies than defensive. Bison occasionally take advantage of exposure of the flanks of an opponent and, rather than resume head-to-head combat, follow through to gore the animal in the side and belly. Such goring may result in broken bones, lacerations, punctured organs, general trauma, and often death (McHugh 1958; Lott 1974). Approximately 50 percent of the bison carcasses examined at a herd-reduction slaughter at Elk Island National Park during the winter of 1971 showed evidence of rehealed broken ribs. The majority of these were assumed to be a direct result of fighting.

Fighting between sexes or involving more than two animals may also occur (McHugh 1958; Lott 1974). Bulls may temporarily stop tending a cow to cross the herd and fight. Disturbance by an external source such as an approaching vehicle or placing animals in confinement may also provide the stimulus for fighting.

Submissive display can include turning and/or running away, backing up with head swinging side to side, and sudden resumption of grazing; the victor may also attempt to mount the loser (Lott 1974; Meagher 1978).

HORNING. During the rut bulls frequently horn trees, often pine (*Pinus* spp.) or spruce (*Picea* spp.) (McHugh 1958). Rutting bulls at Elk Island National Park thrash and horn shrubs and saplings of several species. Following horning, the trees may be used for rubbing and/or be simply uprooted. Rubbing posts are frequently sought by both sexes throughout the year. Horning has also been observed in cows just prior to parturition (McHugh 1958). Although the activity is not restricted to calving and rutting seasons, the increased frequency at those times suggests that it could have importance in physically conditioning the animal for particularly stressful periods.

WALLOWING. Wallowing is practiced by both sexes and all age classes of bison. Wallows are usually in dry sites, although wet, muddy wallows may be used. Wallowing may have a role in grooming, sensory stimulation, alleviating skin irritations, and reproductive behavior. As with horning, wallowing may help precondition the animal for periods of physical stress. Dust, which packs into the hair as a result of wallowing, appears to minimize the effect of insects (Lott 1974). Bulls wallow more frequently during the rut and will occasionally urinate in the wallow prior to engaging in the activity (McHugh 1958). Such deliberate application of urine to the individual has been observed in the European wisent (*Bison bonasus*) (Hediger 1968), caribou (Espmark 1964; Lent 1965), and both sexes of black-tailed or mule deer (Müller-Schwarze 1971). The marked item may be an inanimate object such as a tree or an animate object, for instance, where the bull or cow are moving "territories." Urine odor may advertise the physical condition of the rutting bull. Lott (1974) suggested that the odor of urine from a bull may permit it to use preestablished dominance relations more effectively in the dark, a tenable hypothesis, considering the crepuscular nature of rutting activity and the highly developed olfactory sense of bison.

VOCALIZATION. The repertoire of sounds made by bison includes soft to loud grunts, bleats, roars, snorts, sneezes, foot stamping, and tooth grinding (McHugh 1958; Fuller 1960; Lott 1974; Gunderson and Mahan 1980). Calves bleat and issue piglike grunts in response to grunting by or separation from the dam, when playing, and in response to other stimuli. When searching for a calf, cows will often snort or give a loud grunt similar to their "threat grunt." Bulls are prone to giving loud, lionlike roars or bellows, particularly during the rut (Shult 1972; Lott 1974; Meagher 1978; Gunderson and Mahan 1980). Such roars may be audible from 5 km (McHugh 1958) to 16 km (Audubon and Bachman 1849). The roar produced by bison is the result of a single forceful exhalation over the vocal cords, in contrast to the two-way system in domestic cattle (Gunderson and Mahan 1980). Bulls often roar while tending cows, prior to fighting, while moving through or approaching bull subgroups or mixed herds, in answer to a roar from another bull, and less commonly at other times (e.g., when loafing, when disturbed by vehicles, in response to imitated roars or distant thunder). Occasionally cows with newborn calves will roar when approached. Bulls also use snorts and foot stamping as part of their agonistic behavior. McHugh (1958) observed bison producing a squeaking noise by grinding their teeth. Tooth grinding has been observed in elk at Elk Island National Park when individuals were angered by handling procedures. Bison are usually more vocal during herd movements, as are domestic cattle.

Parturition and Early Life. Immediately prior to parturition, the behavior of the cow and nearby members of the herd often changes. McHugh (1958) observed that cows close to parturition became restless and excitable as well as exhibited marked physical changes such as viscous, mucous discharge from the vagina, swelling of the vulva into a heart-shaped flaccid mass, and filling of the udder. Prior to calving, cows often wander away from the herd for one or more days, although calving may occur within the herd (Audubon and Bachman 1849; McHugh 1958; Egerton 1962).

Bison usually give birth while lying down. The amniotic membranes, portions of the umbilical cord, and the placenta are often eaten by the cow (McHugh 1958; Egerton 1962; Mahan 1978). Egerton (1962) noted that cows licked their calves frequently for several hours postpartum and that such licking seemed to stimulate activity in the calf. Licking may also serve to dry and warm the calf and thus lessen the stress imposed by harsh climatic conditions. Calves have been observed standing within a few minutes (Egerton 1962) to 85 minutes (Mahan 1978) after birth.

Suckling behavior appears to be the first directed action of the newborn calf (Egerton 1962; Mahan 1978), and may motivate the calf to stand and gain mobility. First suckling may take anywhere from 12 to 95 minutes (Mahan 1978).

Calves usually nurse by standing parallel to the cow and facing the cow's posterior. Suckling periods last an average of 6.3 minutes (Mahan 1978) to 10 minutes (McHugh 1958). Suckling periods are usually short and erratic in newborn calves and similar for those older than three months (McHugh 1958; Egerton 1962; Mahan 1978). Disturbance near a calf often induces suckling behavior (Egerton 1962). Yearlings may suckle occasionally (Hornaday 1889; McHugh 1958; Egerton 1962; Mahan 1978), although most bison are weaned prior to this stage. McHugh (1958) reported weaning within 7–8 months, while Mahan (1978) believed it occurred within 9–12 months. Bison cows do not appear to force wean their calves.

Cow-calf pairs remain in close contact for the first few weeks, although, occasionally, calves younger than two or three weeks may rest in calf subgroups. McHugh (1958) believed that, during the first year of life, cohesion between the calf and cow was sufficiently evident to identify them during most periods of the day. Recognition between the cow and calf may include the use of scent, sight, and/or sound, although calves may occasionally follow the wrong cow. Other herd members may focus considerable interest on new cow-calf pairs shortly following parturition and will occasionally come to sniff and lick the calf (Egerton 1962; Engelhard 1970). However, cows will not hesitate to defend their calves against intruding bison or other animals. Egerton (1962) and Mahan (1978) observed that cows usually try to keep themselves between their calf and other herd members. The ability of a cow to defend its calf from investigation by other bison may depend on the position of the cow in the dominance hierarchy (Egerton 1962). McHugh (1958) observed cows defending their calves by quick charges or slow advances when confronted by other species. Similar defensive behavior was reported by Hornaday (1889) and Garretson (1938).

Protection of calves may be shared by other herd members. At Wood Buffalo National Park a small, mixed herd of bison defended a calf against a wolf attack for 36 hours (L. Carbyn personal communication). In the same park, Fuller (1960) observed a small bison herd, with no calves, generally ignore a pack of wolves that were harassing a female member of this herd. It may be that the presence of a calf induces special behavior in bison herds. Grinnell (1904) believed that when bison are attacked or threatened by wolves while they are close together, all will stand by and defend each other. He did not believe that bulls make it their business to defend calves. Although he never observed a bull obviously defend a calf, McHugh (1958) did observe one incident where a mixed herd, including two older bulls, had clustered around a corral housing a lone calf and could not be chased away.

A cow will occasionally abandon its calf when the calf drops behind after a long chase or when it is roped from a herd. When the precipitating disturbance has stopped, the cow usually returns in search of its calf (Seton 1929; Soper 1941; McHugh 1958).

Several instances have been reported where calves as young as two days old tried to defend themselves

(Hornaday 1889; Inman 1899; McHugh 1958). Bison calves will occasionally hide in foliage as a defensive behavior (Allen 1876; Grinnell 1904; McHugh 1958).

Play. Play in bison, as in other mammals, appears to occur with a frequency inversely proportional to age. It is manifest by seemingly purposeless frolicking, including chasing, battling, butting, mounting, kicking, and racing. To the casual observer the motivation to engage in play appears to be "for the sake of the activity itself"; however, its inverse relationship with age suggests that this temporary predisposition may have another function. Such activity would no doubt hasten muscle development and coordination essential in later life and any "enjoyment" experienced in the process would serve as a psychological incentive.

Disposition. The disposition and approachability of individual bison and of herds are a function of the many environmental, genetic, and sociological conditions impinging on the individual. Bison have been described as having personalities (Shinn 1978). Although bison have been trained to do tricks and pull carts, they become more prone to aggression and intolerance with maturity and should be treated with considerable respect. "Pet" bison, once mature, have been known to kill their owners (Garretson 1927). At Elk Island National Park, a park warden, who was experienced in working with bison, was nearly gored to death by a bull bison after releasing it from a squeeze chute. The attack came without any warning and minimal threat posturing (F. Dixon personal communication). Many similar attacks have been reported (McHugh 1958, 1972). During handling, bison can become enraged and can inflict serious damage on other animals, on themselves, and on property. At other times, these animals can be docile and shy. As a general rule, bison should be given the same respect due any wild animal.

MORTALITY

Diseases. Anthrax is an infectious disease of humans and animals caused by the bacterium *Bacillus anthracis*. Choquette (1970) reviewed the occurrence of anthrax in wildlife and described its etiology, transmission, signs, pathology, diagnosis, immunity, treatment, and control. The disease is nearly universal in distribution and has been reported in a variety of mammals, the majority of which are herbivores (Choquette et al. 1972). Anthrax was introduced to North America in Louisiana during the early 1700s at the time of settlement by the French (Cousineau and McClenaghan 1965). It may have infected bison of the Great Plains, but few reports are available (Novakowski et al. 1963). McNary (1948) reported anthrax in captive bison in Pennsylvania in 1947. It was first reported in Canadian wildlife in 1962, when an outbreak was discovered in bison in the Northwest Territories (Novakowski et al. 1963). Further outbreaks have continued in this region throughout the last two decades (E. Broughton personal communication). Between 1962 and 1971 at least 1,003 bison from

wild herds in the Northwest Territories and Wood Buffalo National Park were killed by anthrax (Choquette et al. 1972). During July and August 1978, another outbreak caused the death of at least 78 bison from these herds (B. Stephenson personal communication). Anthrax has contributed significantly to mortality in these populations (table 49.5). Local environmental conditions may favor survival of the disease. Anthrax spores have continued to contaminate and persist in the soil and water of northern Canada, resulting in unpredictable but recurring explosive outbreaks (Choquette et al. 1972). Carrion eaters and scavengers have probably helped disseminate the disease. Migratory birds can also contribute to the spread of the disease by carrying spores in their intestinal tract (E. Broughton personal communication). Since 1962, control measures have included depopulation in areas of former outbreaks (Novakowski et al. 1963); continual summer surveillance of bison herds, especially in areas of former outbreaks; disposal of infected bison carcasses by incinerating; liming and burying the carcasses; and mass vaccination during June. Anthrax vaccination programs in the Northwest Territories and in Wood Buffalo National Park were terminated in 1975 and 1977, respectively, because the effect of mass vaccination under field conditions was difficult to evaluate, these programs were expensive, logistics of round-ups were great, and round-ups resulted in mortality of unknown numbers of bison. In view of the 1978 outbreak the value of these programs should be reassessed.

Tuberculosis is a chronic infectious disease of humans and animals caused by species of *Mycobacterium*, such as *M. tuberculosis* (humans), *M. bovis* (other mammals), and *M. avium* (poultry). *M. bovis* has been identified as the causative agent of tuberculosis in bison of northern Canada (Choquette and Stewart 1959). In the bovine, the organism may enter the body through any of five routes: (1) respiratory, (2) alimentary, (3) congenital, (4) genital, and (5) cutaneous (Stamp 1959). The respiratory and alimentary tracts are the usual routes of infection.

Tuberculosis was first reported in bison at Wainwright Buffalo Park, Alberta, in 1923 (Cameron 1924). The source of infection of the Wainwright herd was probably local domestic cattle rather than bison herds in Montana from which the Wainwright bison were obtained (Hadwen 1942). In Wainwright Buffalo Park during 1923–39, 12,005 bison were slaughtered and an average of 54 percent of the bison killed had tubercular lesions (determined by meat inspection) (Hadwen 1942). This postmortem lesion rate varied between 30 and 77 percent. Evidence indicates that tuberculosis was brought to Wood Buffalo National Park in 1925 with the introduction of plains bison from Wainwright (Fuller 1966). It is not known if the disease was endemic among northern wood bison. In Wood Buffalo National Park, the percentage of bison reactors to tuberculin testing was 14.5 percent in 1957, 19 percent in 1958, and 13.5 percent in 1959 (Choquette et al. 1961). In this same study, the postmortem lesion rate, based on a total of 436 bison examined, was 50.2 percent. The prevalence of tuberculosis

TABLE 49.5. Number of bison deaths attributed to anthrax in the Northwest Territories, Canada, 1962–1979

Year	Location			Reference
	Hook Lake[a]	Grand Detour[b]	Wood Buffalo National Park	
1962	281			Novakowski et al. 1963
1963	12	269		Cousineau and McClenaghan 1965
1964	44	202	53	Choquette et al. 1972
1965[c]				Choquette et al. 1972
1966[c]				Choquette et al. 1972
1967			120	Choquette et al. 1972
1968			1	Choquette et al. 1972
1969[c]				Choquette et al. 1972
1970[c]				Choquette et al. 1972
1971	37[d]			Choquette et al. 1972
subtotal	374	471	174	
1972–77[c]				E. Broughton personal communication
1978	12	27	39	B. Stephenson personal communication
1979[c]				E. Broughton personal communication
Total	386	498	202	

[a] East side of Slave River, NWT.
[b] West side of Slave River, NWT.
[c] No deaths attributed to anthrax.
[d] Choquette et al. (1972:130) lists 31 deaths; however, 37 are now believed attributable to anthrax (E. Broughton personal communication).

in the bison of Wood Buffalo National Park was also reported by Fuller (1966). During 1952–56, 40 percent of the 1,508 bison slaughtered at Hay Camp showed tubercular lesions (determined by postmortem examination). During 1938 and 1939, 800 bison slaughtered in Elk Island National Park were examined for tuberculosis (Hadwen 1942), and during 1959 and 1960, 500 postmortem examinations were made (Choquette et al. 1961). No tubercular lesions were observed in either study. Tuberculosis was later discovered in both plains bison and wood bison at Elk Island National Park during the 1960s but stringent control measures eradicated the disease by 1971. Tuberculosis has never been detected in bison at Yellowstone National Park (Meagher 1973).

During 1965, in conjunction with a slaughter program to supply meat, a minor testing program attempted to reduce the incidence of tuberculosis in bison at Wood Buffalo National Park. However, a control program per se never developed in the park or in the Northwest Territories. Since 1966, testing was discontinued because the negative animals were mingling with the remainder of the park population, and the recurring high incidence (> 20 percent) of tuberculosis in the herds indicated that the program was not controlling the disease. Round-ups of wild bison in remote areas to test for and control tuberculosis are not practical.

The gregarious nature of bison tends to perpetuate the maintenance and dissemination of tuberculosis. Tuberculosis in the Wainwright herd for more than 26 years did not appear to interfere with herd productivity (Hadwen 1942). However, the importance of tuberculosis as a mortality factor is difficult to determine for wild, free-roaming bison herds.

Brucellosis (Bang's disease, infectious abortion, undulant fever in humans) is an infectious disease caused by species of bacteria of the genus *Brucella*. The six species are *B. melitensis*, usually pathogenic for goats and sheep but can affect other species, including cattle and humans; *B. abortus*, usually pathogenic for cattle, causing abortion, but can also affect other species, including humans; *B. suis*, usually pathogenic in pigs but can also affect hares (*Lepus americanus*), reindeer, humans, and other species; *B. neotomae*, occurring in the desert wood rat (*Neotoma lepida*); *B. ovis*, pathogenic in sheep; and *B. canis*, pathogenic in dogs (*Canis familiaris*) (Brinley-Morgan and McCullough 1974). *B. abortus* is the usual causative agent of brucellosis in North American bison. It is not known whether brucellosis was endemic or introduced among North American bovids (Meagher 1973). It is suspected that brucellosis was present in wildlife populations prior to settlement of North America by white people because of its 1962 discovery in barren-ground caribou in Alaska (Neiland et al. 1968) and also in northern Canada in 1968 (E. Broughton personal communication). These northern caribou had not been in contact with domestic livestock and, therefore, were not infected by that means. In addition, the brucella strain isolated in both Alaska and northern Canada was *Brucella suis* type 4.

Brucellosis is of particular economic importance, as it is a serious disease in cattle (Rush 1932; Meagher 1973, 1974). It was first tested for and reported in bison of Yellowstone National Park in 1917 (Tunnicliff and Marsh 1935) and has since been recognized as a problem among five other bison herds in North America. Brucellosis causes abortion, temporary sterility, frequent returns to service, metritis (inflammation of the uterus), and lowered milk production in cattle (Meagher 1973; Choquette et al. 1978). Abortion caused by brucellosis has been reported in bison (E. Broughton personal communication). It is presumed that infected bison shed brucella organisms, thereby contaminating feed and water. Dissemination of the disease is enhanced due to the gregarious nature of bison. In wild herds, other animals such as wolves (*Canis lupus*), coyotes, and foxes (*Vulpes vulpes*) contribute to the spread of the disease by acting as vectors (Choquette et al. 1978).

The rate of brucellosis infection in wild bison herds varies considerably. In 1931, Rush (1932) reported an infection rate of 53 percent in Yellowstone National Park. During 1932–34, Tunnicliff and Marsh (1935) reported an infection rate that varied from 54 to 74 percent, and in 1964–65, Meagher (1973) reported that the rate for three subpopulations of bison in Yellowstone varied from 28 to 59 percent. On the National Bison Range, in 1932 and 1933, brucella infection rates were 68 and 56 percent, respectively (Tunnicliff and Marsh 1935). Bison infection rates in Elk Island National Park in 1946–47 and in 1956–57 were 16 to 32 percent, respectively, and in 1956–57, 10 percent of the 17 bison tested in Riding Mountain National Park were infected with brucellosis (Corner and Connell 1958). In Wood Buffalo National Park, infection rates varied from 27 to 36 percent between 1956 and 1958 (Fuller 1966). Between 1958 and 1974, rates varied from 6 to 62 percent (Choquette et al. 1978). In the bison at Hook Lake, N.W.T., in 1970 and 1974 the rates of brucella infection were 26 and 39 percent, respectively (Choquette et al. 1978).

The role of the disease and its effect on reproductive activity in bison is difficult to ascertain due to the lack of data on the incidence of abortion in bison. In 1930, variation in size of fetuses in the Yellowstone herd suggested a long breeding season that may have been an effect of brucellosis (Rush 1932). Fuller (1966) believed that the low pregnancy rates in bison at Wood Buffalo National Park appeared to be related to the incidence of brucellosis. Pregnancy rates for bison in Yellowstone National Park did not appear to be influenced by the occurrence of brucellosis (Meagher 1973). Brucellosis has probably existed in the Yellowstone bison for a long time (Meagher 1973). But in spite of the prevalence of the disease in wild herds, brucellosis does not appear to be a threat to survival of bison (Choquette et al. 1978).

Brucellosis has been nearly eradicated in confined herds of plains bison in western Canadian national parks (Choquette et al. 1978). However, the presence of brucellosis in wildlife jeopardizes any attempt to raise cattle in contaminated areas and also poses a threat to public health in areas where sport hunting is permitted. It has been assumed, although not proven, that bison can transmit active brucella to cattle (DeYoung 1973).

During the early 1970s, a controversial issue involving esthetics and economics developed concerning the possible transmission of brucellosis from bison in Yellowstone National Park to cattle occupying lands adjacent to the park (DeYoung 1973; Meagher 1973, 1974). The high rates of infection in Yellowstone bison (59 percent in 1964–65) caused concern among cattle ranchers in the region. The cattle industry in that area argued that bison should not be exempt from the brucellosis eradication program of the U.S. Department of Agriculture. The park administration countered that such a program would not be compatible with the policy of maintaining wild bison populations under natural conditions. A brucellosis eradication program for wild bison is not practical as long as numerous other alternate hosts remain in close association with cattle. In an effort to appease the cattle industry and to remain compatible with national park policy, the administration of Yellowstone National Park decided to operate boundary patrols to minimize bison-cattle interactions to control transmission of the disease to cattle directly from park bison. Boundary patrols have been effective at minimizing bison-cattle encounters (Meagher 1974).

Parasites. Many species of ectoparasites and endoparasites have been found on or in North American bison (table 49.6). The greatest number of those species belong to the class Nematoda. In several instances, the death of bison in captive herds has been attributed to parasitism. For example, nine cases of ostertagiosis resulting in death were diagnosed in bison from three farms in New York State (Wade et al. 1979). At necropsy, stomach worms identified as *Ostertagia ostertagi* were recovered in large numbers from eight of those animals. In Kansas, extensive infections of lungworms (*Dictyocaulus viviparus*) and nodular worms (*Oesophagostomum radiatum*), and evidence of lesions from stomach worms (*Haemonchus contortus*) were considered the cause of death of one bison from a herd where seven similar deaths had occurred. *D. viviparus* has a well-documented pathogenicity, and may be a mortality factor in bison at Elk Island National Park. At Elk Island, *D. viviparus* was found in 2 of 24 and 5 of 21 bison examined in 1959 and 1974–77 (J. Holmes and W. Samuel personal communications). Thus, bison are hosts to many species of parasites and, on occasion, can be very "wormy." High stocking densities and the conditions of captivity may contribute to more extensive parasitic infection and reinfection.

The resistance of bovids to the hydatid worm (*Echinococcus granulosus*) may account for its absence in bison at Wood Buffalo National Park despite its occurrence in dogs, wolves, caribou, and moose in the same area (Fuller 1966). Hydatids have been observed regularly in the cervids of Elk Island National Park but have not been seen in the resident bison (E. Broughton personal communication).

Cattle, elk, humans, moose, mule deer, sheep, white-tailed deer, and other animals can be host to many of the same parasites as bison. Depending on the species of parasite, the effects of such parasitism can range from minimal irritation to acute or chronic dis-

TABLE 49.6. Parasites reported from *Bison bison*

Parasite	Site of Infection[a]	Herd Location[b]	Reference[c]
Protozoa			
Eimeria bovis	I	EINP	20
E. bukidnonensis	I		21
E. auburnensis	I	EINP, W	20, 25
E. brasiliensis	I	W	25
E. canadensis	I	W	25
Eimeria sp.	I	LZ	22
Sarcocystis sp.	HM	WBP, NY	5, 6, 27
Trematoda			
Fasciola hepatica	BD/LV	NBR, W	12, 18, 2
Fascioloides magna	LV	WBP	5, 6, 10
Cestoda			
Echinococcus granulosus	LV	WBP	5 (suspect)
Moniezia benedeni	SI	SD, M, YNP, NBR, WMWR, EINP	3, 9, 10, 12, 15, 18, 24, 26
M. planissima (probably)	SI	WBP	5
Nematoda			
Chabertia ovina	C/CO	SD, ZGP	3, 16
Cooperia bisonis	SI	WBP, EINP	5[d], 10, 15
C. oncophora	AB, SI	SD, EINP, NY	3, 4, 12, 26, 27
C. surnabada = (*C. mcmasteri*)	AB, SI	SD, EINP	3, 4, 12, 26
Cooperia sp.	SI	WBP	6
Dictyocaulus filaria	LG	WBP	5
D. viviparus (*hadweni* syn)	LG	SD, WBP, K, NBR, EINP, NY, WMWR	3, 7, 8, 10, 12, 13, 15, 18, 24, 26, 27
Dictyocaulus sp.	LG	WBNP, YNP	14, 19
Haemonchus contortus	AB	SD, K, NY, WMWR	3, 10, 12, 13, 23, 24, 27
Nematodirella longispiculata	SI	EINP	15
Nematodirus helvetianus	SI	EINP, NY	15, 27
Oesophagostomum radiatum	CO	K, EINP, WMWR	10, 12, 13, 24, 26
Oesophagostomum sp.	CO	WBP	5
Oesophagostomum sp.	SI,LI	Y	27
Ostertagia bisonis	AB	SD, WBP, K, EINP	3, 7, 10, 12, 13, 26
O. lyrata = (*Grosspiculagia lyrata*)	AB, SI	EINP	12, 26
O. ostertagi	AB	EINP, NY	10, 12, 26, 27

continued

ease resulting in death directly or indirectly (e.g., abortion, increased vulnerability to predation, thermoregulatory imbalance).

Other Pathological Conditions. In the majority of cases, other pathological conditions are of incidental occurrence rather than serious factors affecting mortality rates in bison populations. At Wood Buffalo National Park, conditions that have been reported are arteriosclerosis, lymphosarcoma, pneumonia, renal calculi (Fuller 1961), peritonitis, pleuritis, and mucoid degeneration (Choquette et al. 1961). Multiple abscesses and hepatic lesions were also observed in animals slaughtered at Wood Buffalo and Elk Island national parks (Choquette et al. 1961). Actinobacillosis was detected in three animals at Elk Island National Park. Ophthalmia and enlarged granulated livers may also be found in wild buffalo (Garretson 1927). Arthritis often occurs in association with tuberculosis or brucellosis (Fuller 1961) and tends to afflict the stifle joint. Evidence of arthritis was found in about 2 percent of carcasses examined in Wood Buffalo Na-

tional Park (Fuller 1966) and in several animals from Elk Island National Park (Choquette et al. 1961). Orchitis (inflammation of the testicles) has been observed in bison from Wainwright, Alberta (Hadwen 1942), Elk Island National Park (Corner and Connell 1958), and Wood Buffalo National Park (Choquette et al. 1961; Fuller 1966). Metritis was noted in nine animals from Wood Buffalo National Park, three of which were positive to the tuberculin test (Choquette et al. 1961). When the uterus of one of the animals and two of the dead fetuses were examined bacteriologically, no brucella organisms were recovered but a *Mycobacterium* sp. was isolated.

Predation and Hunting. Predation contributes to bison mortality in Yellowstone National Park, in Wood Buffalo National Park, and in the Slave River lowland, N.W.T., herds. Authorized hunting occurs only in a few locations in North America.

In Yellowstone National Park, predation on bison by wolves has never been a problem, as evidenced by the long survival time of injured and solitary animals

TABLE 49.6—*Continued*

Parasite	Site of Infection[a]	Herd Location[b]	Reference[c]
O. trifurcata	AB	SD	3
Setaria labiatopapillosa	CE	ND, M, NZP, WBP, WBNP, EINP, NY	1, 5, 12, 14, 15, 26, 27
S. yehi	CE	ND	1
Setaria sp.	CE	WBP	6, 10
Trichostrongylus axei	CE, SI	EINP, NY	10, 12, 26, 27
T. lerouxi	SI	NY	27
Trichuris discolor	C	NY	27
T. ovis[e]	C	NBR	12, 18
Trichuris sp. (eggs)	F	EINP	26
Arthropoda			
Damalinia (Bovicola) sedecimdecembrii	S	WBNP	14
Dermacentor albipictus	S	EINP	26
D. andersoni	S	M	17
D. nigrolineatus = (D. albipictus)	S	WMWR	24
Hypoderma lineatum	S, E	WMWR, NY	24, 27
Hypoderma sp.[f]	S, E, D	WBP, NBR	5, 6, 18
Speleognathus australis	NS	WMWR	11

NOTE: We are following the tabular format of Wade et al. 1979.

[a] AB = abomasum, BD = bile duct, C = cecum, CE = coelom, CO = colon, D = diaphragm, E = esophagus, F = feces, HM = heart muscle, I = intestine, LI = large intestine, LV = liver, LG = lung, NS = nasal sinus, S = skin, SI = small intestine.

[b] EINP = Elk Island National Park, K = Kansas, LZ = Leningrad Zoo, M = Montana, NBR = National Bison Range, ND = North Dakota, NY = New York State, NZP = National Zoological Park (District of Columbia), SD = South Dakota, WBP = Wainwright Buffalo Park, WMWR = Wichita Mountains Wildlife Refuge, WBNP = Wood Buffalo National Park, W = Wyoming, YNP = Yellowstone National Park, ZGP = Zoological Garden of Prague.

[c] 1 = Becklund and Walker 1969, 2 = Bergstrom 1967, 3 = Boddicker and Hugghins 1969, 4 = Burtner and Becklund 1971, 5 = Cameron 1923, 6 = Cameron 1924, 7 = Chapin 1925, 8 = Corner and Connell 1958, 9 = Dikmans 1934, 10 = Dikmans 1939, 11 = Drummond and Medley 1964, 12 = Dunn 1968 (literature review only), 13 = Frick 1951, 14 = Fuller 1966, 15 = J. Holmes personal communication, 16 = Jaros et al. 1966, 17 = Kohls and Kramis 1952, 18 = Locker 1953, 19 = Meagher 1973, 20 = L. Morgantini unpublished data, 21 = Pellerdy 1963, 22 = Pellerdy 1974, 23 = Ransom 1911, 24 = Roudabush 1936, 25 = Ryff and Bergstrom 1975, 26 = W. Samuel personal communication, 27 = Wade et al. 1979 (this paper includes a good literature review).

[d] *Cooperia bisonis* (Cram 1925) was reported as *Haemonchus osteragi* by Cameron (1923).

[e] Identification suspect, since it was based on examination of female worms only (Locker 1953).

[f] Other species of parasitic Diptera (e.g., those within the Culicidae, Muscidae [Burger and Anderson 1970], Rhagionidae, and Tabanidae), have been observed in association with bison (Meagher 1973).

(Meagher 1973). During the Yellowstone study, wolves were rare and those present were never observed in packs. However, a few circumstances have suggested occasional predation by the grizzly bear (*Ursus arctos*) on calves and adults (McHugh 1958; Meagher 1973).

Wolf predation on bison has been recognized as an important mortality factor in Wood Buffalo National Park (Fuller 1961, 1966). Examination of scat samples in 1932 and 1933 indicated that wolves seldom preyed on bison during summer and fall (Soper 1941). Soper did not estimate total wolf predation on bison but concluded that it was not destructive to the bison population. Fuller (1961) concluded that bison form the staple diet of the park wolves in summer and winter. He found that 80 percent of the summer wolf scats examined contained bison hair. In an early winter sample of 59 wolf stomachs, 65 percent of the food of those wolves was bison (Fuller 1966). Wolves selectively preyed on calves or aged animals and predation was not detrimental to the bison population (Fuller 1961, 1966).

The three most frequently occurring food items in a sample of 433 wolf scats collected from May to October 1978 at Wood Buffalo National Park were hare (37.3 percent), bison (18.1 percent), and muskrat (*Ondatra zibethicus*) (15.5 percent). Hare occurred more frequently than bison or muskrat in scats collected in June (41 percent), July (42 percent), and September (61.4 percent), while bison was more frequent than hare or muskrat in scats collected in May (44.4 percent) and October (32 percent). Muskrat was more frequent than bison and hare only in August (31 percent) (L. Carbyn personal communication). A wolf pack composed of 10 individuals killed an average of 1 bison every 7.8 days from 12 February to 31 March at Wood Buffalo National Park. Similar consumption rates have been observed on the Slave River Lowlands (J. Van Camp personal communication). Prey species, such as bison, are likely more vulnerable in winter and represent a greater amount of resource per unit effort at that time than do smaller prey species. At Wood Buffalo National Park, a high proportion of the bison killed by wolves were older males (L. Carbyn personal communication). On the Slave River lowlands it appeared that cows and calves were more often killed (J. Van Camp personal communication).

On the Slave River lowlands, N.W.T., free-roaming bison herds numbered approximately 2,050 animals in 1971. By March 1974, these herds had declined slightly to about 1,900. During the severe winter of 1974/75, the population declined by approximately 35 percent to 1,200. Numbers continued to dwindle to an estimated 750 by March 1977. Wolf numbers in the region concurrently increased from approximately 30–40 animals in 1974 to an estimated population of between 65 and 75 in 1976 (J. Van Camp personal communication). During the winter of 1976/77, six packs of wolves were observed on the lowlands. The unusual severity of the winter of 1974/75 could have increased vulnerability of bison to wolf predation and subsequently led to greater wolf productivity. During the period 1975 to 1977, bison was the principle food of wolves on the Slave River lowlands (J. Van Camp personal communication). Evidence from radio-collared wolves during the winter of 1976/77 indicated that wolf predation was exerting a major role in the continued decline of the bison population (Van Camp 1978b). A wolf control program, which selectively removed 44 wolves from the bison ranges on the Slave River lowlands, was initiated during the winter of 1977/78. The decline in the bison population appeared to have subsided by March 1978, as the estimate remained unchanged at 750. The wolf control program was continued during the winter of 1978/79 and resulted in the removal of 28 wolves (Jalkotzy 1979). The bison population declined to approximately 550 in March 1979 even though the wolf population had been reduced to extremely low numbers (Jalkotzy 1979). After two winters of reduction programs so few wolves remained on the bison ranges that the control program was abandoned in 1979. Despite the greatly reduced wolf population, bison numbers monitored on the east side of the Slave River declined by approximately 17 percent from 440 to 370 (V. Hawley personal communication). This continued decline in the bison population suggested that wolf predation was not the entire problem.

During regulated seasons from 1968 to 1977, sport hunting of bison was permitted on the Slave River lowlands north of Fort Smith, N.W.T., and outside of Wood Buffalo National Park. For the 10-year period, at least 1,230 animals were killed as a result of hunting. The reported number of bison kills (from license returns) averaged 179 per year between 1969 and 1974 (Van Camp 1978a). The hunter harvest rate for the period 1973 to 1976 was between 9 and 12 percent per year, while the average annual recruitment to the bison population was estimated at only 3 percent. Overhunting appeared to have caused the bison population to enter into a period of negative annual increments (Van Camp 1978a). During the fall of 1977, the recreational (sport) hunting season was closed because of a declining population. The general hunting license (GHL, a license issued to certain residents of the Northwest Territories) holders were permitted to retain their hunting rights but voluntarily agreed to reduce their harvest of bison to 25 annually. However, during 1977/78 after the close of the sport hunting season, the GHL holders were believed to have harvested at least 41 bison and during 1978/79 at least 31 bison were harvested (J. Van Camp personal communication). According to kill return data, the GHL holders have removed approximately 5 percent of the total bison herd since the 1970s. It would appear that harvest rates from hunting, at least during the period of major decline in the bison population (1973–76), exceeded the average annual recruitment rate. In combination with the pressure exerted by wolf predation, these were the two major mortality factors responsible for the continuous decline in the Slave River lowland bison population. Today, most wild bison herds are located in parks and wildlife sanctuaries and are protected from hunting.

Accidents. Accidental drowning often occurs as a result of animals falling through thin ice in spring and fall. Whole herds of bison have succumbed to such fatalities (Raup 1933; Meagher 1973). Drowning was considered an important mortality factor of the plains bison (Roe 1970). Mortality as a result of animals voluntarily crossing flooded rivers or by being trapped during abnormal spring flooding does occur. In Wood Buffalo National Park, losses of 40–50 animals per year, amounting to less than 1 percent, were reported by Soper (1941) and Fuller (1961, 1966). However, spring flooding in the Peace-Athabasca River delta in 1958 caused the death of about 500 bison, and autumn flooding in 1959 resulted in the death of an estimated 3,000 (Fuller 1966). Spring floods in 1961 resulted in the death of more than 1,100 bison, with estimates as high as 3,000. In 1971 only 48 known drownings occurred on the delta (Allison 1973). Several thousand bison were also drowned in the delta region during a nontypical spring flood in 1974 (S. Cooper personal communication). Drowning as a result of miring in bogs and falling into sinkholes and ponds was of incidental occurrence in Wood Buffalo National Park (Soper 1941). In Yellowstone National Park, a few bison from all age classes have drowned in bogs or by falling into hot pools (Meagher 1973). In Elk Island National Park, occasionally a few bison fall through the ice and drown as a result of traveling too close to beaver (*Castor canadensis*) houses when crossing frozen lakes (W. Walburger personal communication). In certain situations, accidental drownings can result in considerable mortality.

Forest fires commonly occur in northern bison ranges without causing appreciable mortality (Soper 1941; Fuller 1966). The main effect on bison is loss of cover; however, in many situations feeding habitat is improved and sometimes created by fire. Wildfires can be catastrophic to herds of buffalo, although such events are rare in occurrence. Cole (1954:454) discovered a report in which it was stated that "a slope of the prairie burned and it had killed hundreds of buffaloes. We saw as many as three hundred lying together with the hair all burned off them while many were roaming around deprived of their eyesight by the fire.'' In addition, two other historical reports of buffalo being destroyed by fire were discovered by Roe (1970) and in both cases, many animals were killed, blinded, or badly burned.

One case of a bison fatality from a motor vehicle was reported by Fuller (1966). A few similar instances have since occurred along access roads within Wood Buffalo National Park and also with wood bison in the Mackenzie Bison Sanctuary and other reintroduced herds in the Northwest Territories.

Climatic Conditions. Above-average snowfall, long periods of low temperatures, and midwinter thaws followed by severe subzero temperatures create conditions conducive to winter-caused mortality in bison. Snowfall in excess of 4 m on bison ranges in northern Canada during the 1800s resulted in the death of thousands of animals (Soper 1941). An early spring thaw followed by freezing temperatures in 1928 in Wood Buffalo Park resulted in crusting conditions that forced bison to remain in forests and sheltered areas to feed (Raup 1933). In Yellowstone National Park, Meagher (1973) defined winterkill as the combined effects of climatic stress, reduced forage availability, and physiological condition of individual animals and considered it the main cause of bison mortality at Yellowstone. Weather, independent of other factors, is not normally an important cause of mortality, but is an additional physiological stress that, in combination with predation and disease, can increase the rate of mortality (Fuller 1961). Periodic extremes of winter climates are unpredictable factors that can elevate mortality rates in free-roaming bison populations.

AGE DETERMINATION

Dentition. Changes as a result of tooth wear (particularly changes in size and shape of the style of molars) and time of molar eruption are important criteria in assigning an age class to individual bison. Skinner and Kaisen (1947) developed a system of wear classification for bison teeth by recognizing six general age categories based on tooth wear and the sequence of molar eruption: immaturity, early adolescence, late adolescence, early maturity, full maturity, and old age.

In attempting to develop a system for age determination in bison, Fuller (1954) discovered that the first three incisiform teeth were replaced rapidly by permanent incisors. The fourth incisiform tooth, the canine, was replaced after a distinct time lapse, indicating a possible separate origin. Fuller (1959) identified five yearly age classes based on sequence of tooth eruption and replacement and three general age classes based on tooth wear (young adult, adult, and aged). The precise age of animals older than four years could not be determined using Fuller's tooth replacement technique. A more precise technique for determining the age of bison older than four years was developed by Novakowski (1965). This method was based on counting annual cementum layers in the roots of the fourth premolar. Cementum deposition on PM4 in bison begins at age four, making yearly age estimates possible for animals older than this.

Frison and Reher (1970) compared patterns of mandibular tooth eruption and wear with known-age samples and by this method were able to establish seven broad age categories for bison specimens collected from the Glenrock Jump in Wyoming. Those categories were: 0.5 years, 1.5 years, 2.5 years, 3.5 years, 4.5 years, mature (5.5 to 9.5 years), and old age (10.5 to 13.5 years). On the basis of tooth eruption, the first five age groups from the Glenrock specimens were distinctly classified with no overlap and, on the basis of tooth wear, recognition of the mature and old age categories was possible. Because of the subjective nature of wear classifications, specific ages could not be determined within mature and old age categories.

Horns. On the basis of horn growth, Fuller (1959) recognized four age classes for female bison and five age classes for male bison: (1) calf—less than 1 year, (2) yearling—1–2 years, (3) spike-horn—2, 3–4 year olds, (4) young adult–4 years to 7 or 8, and (5) adult and aged—more than 7 or 8 years. The spike-horn stage was not recognized in classification of females. Counting of annual growth rings on horns was not a useful aging criterion.

Osteology. Age determination from the postcranial elements of bison has always been difficult. Because of the need to age and sex such elements from archaeological sites, Duffield (1973) developed an age determination technique using a table showing the age of the epiphyseal closure. With this table, it was possible to assign skeletal remains into yearly age categories up to 11 years. However, precision was dependent on the degree of articulation of the elements examined (Duffield 1973).

Morphology. Fuller (1959) was able to distinguish calf and yearling categories based on differences in body size and conformation.

Gravimetric Parameters. Eye lens weight is not a useful indicator of age for bison due to the large variation within age classes and the overlap between year classes (Novakowski 1965). For similar reasons, body weights were not useful as age criteria in bison older than one year.

Conclusions. Further experimentation using teeth and other postcranial material may lead to new and better methods for age determination of bison. Miller (1974), working with barren-ground caribou, concluded that histological examination of dental cementum for annuli is the best technique for age determination. Thomas (1977:209) reported that "the most rapid, effective, and efficient technique for mammalian age determination is decalcification, neutralization, cryostat sectioning, metachromatic staining and interpretation in an aqueous state." Unfortunately, Novakowski (1965) and Bourque et al. (1978) did not prepare histological sections or use metachromatic staining techniques to determine ages of bison. Both of these methods have improved accuracy in quantifying cementum annuli

(Miller 1974; Thomas 1977). Only gross age classification is possible for free-roaming bison as determined by body conformation and horn development.

ECONOMIC STATUS AND MANAGEMENT

Bison management practices in North America vary considerably depending on the objectives of the individuals and agencies controlling the animals. Bison have been managed in order to preserve the species and subspecies, for commercial meat and hide production, as a game species, as tourist attractions, and for their historical significance.

Jennings (1978) estimated that there are more than 65,000 bison in North America. Only a very small proportion of these animals are free-ranging and wild. The rest are confined in fenced areas of various sizes in parks, in nature preserves, and on private lands.

Rehabilitation Program for Wood Bison. Most early observations of northern bison were recorded by explorers during expeditions throughout northern Canada. Samuel Hearne was probably the first European to view northern wood bison in native habitat during his journey through the Lower Slave River region in the Northwest Territories in 1772 (Raup 1933). The northern wood bison declined in numbers at about the same time (1840–1900) as the great demise of the southern plains bison (Raup 1933). By 1891 the population south of Great Slave Lake and west of the Slave River was probably reduced to about 300 animals (Ogilvie 1893). Subsequently, a federal law providing protection from hunting for the few remaining bison was passed as the first of several legislative steps to control and regulate the harvest of wood bison. Enforcement was probably nonexistent until about 1897, and then only a handful of wardens were appointed and assigned the task of patroling hundreds of square kilometers of range. Wood bison began to increase and numbered approximately 500 by 1914 (Banfield and Novakowski 1960). However, it is unknown whether the increase resulted from their protection, several years of good reproduction, or a combination of both factors.

In 1922, Wood Buffalo Park was established by an Order in Council under the Forest Reserves and Parks Act in an attempt to save the northern population of wood bison (Raup 1933). The number of wood bison at that time was estimated at between 1,500 and 2,000 (Seibert 1925). The subspecies suffered a setback in 1925 when the federal government decided to move excess plains bison from Wainwright Buffalo Park to Wood Buffalo Park in Alberta and the Northwest Territories. Between 1925 and 1928, 6,673 plains bison were shipped north and released into the park. They were released along the west side of the Slave River between the Peace River and Fort Fitzgerald into range occupied by wood bison. Subsequent herd mixing resulted in rapid hybridization of the resident wood bison, as they were outnumbered by the introduced plains bison, approximately 4:1. Tuberculosis and brucellosis, bovine diseases introduced with the plains

bison, have played an important role in management of the hybrid population. The number of bison increased to an estimated 12,000 by 1934 (Soper 1941). It was generally believed that pure wood bison had become extinct by 1940, despite speculation on the existence of isolated small herds in the northwestern portion of Wood Buffalo Park.

THE REHABILITATION PROGRAM. During an aerial survey in 1957, N. S. Novakowski, of the Canadian Wildlife Service, discovered an isolated population of bison in the Nyarling River and Buffalo Lake area (figure 49.4) in the northwest corner of Wood Buffalo National Park. Subsequent investigations and collection of five specimens, of which skins and skulls of three adults were sent to the National Museum, resulted in the conclusion that these samples were of wood bison stock (Banfield and Novakowski 1960). This Nyarling herd had probably remained isolated from the hybrid herds to the east and south at Hay Camp and Lake Claire (160–325 km) and from the Grand Detour herd on the lower Slave River (120 km) (figure 49.5). In 1963, 18 animals were trapped from this herd and transferred to an area northeast of Fort Providence, N.W.T., to prevent them from hybridizing with the introduced plains bison in the south part of the park, and to isolate them from an anthrax outbreak in the bison along the Slave River. Continued concern regarding anthrax led to a second transfer of 23 wood bison in 1965 to an isolation area in Elk Island National Park to save the subspecies from extinction and to establish a source breeding herd for future transplants.

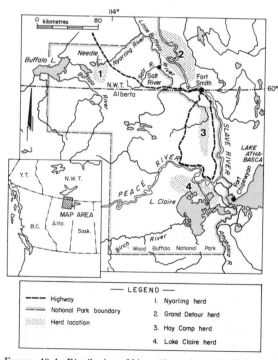

— LEGEND —	
‐‐‐‐ Highway	1. Nyarling herd
—— National Park boundary	2. Grand Detour herd
░░░ Herd location	3. Hay Camp herd
	4. Lake Claire herd

FIGURE 49.4. Distribution of bison (*Bison bison*) herds in Wood Buffalo National Park and vicinity in 1957 and location of Nyarling wood bison (*Bison bison athabascae*) herd.

During the first few years at Elk Island, disease was a problem. Since 1971, however, this population has been classified as disease free. The Fort Providence and Elk Island herds now number approximately 750 and 120, respectively.

Since 1973, representatives of the Canadian Wildlife Service, Parks Canada, and territorial and provincial wildlife agencies from western Canada have met annually to develop criteria for management of wood bison. In this context, the wood bison rehabilitation project commenced in 1975. The primary objective of the project is to establish a minimum of three free-ranging, self-perpetuating populations of wood bison in areas of historic range. The second objective is to protect and preserve the gene pool by dispersing small breeding herds to zoological gardens and parks. Animals are transferred according to the terms of a lease agreement whereby the federal government retains ownership of all wood bison and their progeny. Breeding stock has not been made available to private or commercial enterprises because of problems associated with record keeping and the lack of capability to meet the expected demand. The ultimate goal of the program is the reestablishment of free-ranging and captive populations of wood bison in sufficient numbers to warrant the animal's removal from the lists of endangered species. When this occurs, the project will terminate and commercial trade in wood bison will be possible.

ENDANGERED SPECIES RECOGNITION. The wood bison is currently recognized as an endangered subspecies and is classified as an appendix I animal in the Convention on International Trade in Endangered Species of Wild Flora and Fauna (CITES), which means that all commercial trade is prohibited except for research designed to increase population size. Export and import permits are required to transport appendix I animals across international boundaries. In addition, wood bison are listed as endangered by the Committee on the Status of Endangered Wildlife in Canada, a federal-provincial group established in 1977 by a resolution of the Federal-Provincial Wildlife Conference. This national committee, composed of representatives from government and nongovernment agencies, is responsible for establishing an official status (with supporting information) for wildlife species considered in jeopardy in Canada. The International Union for the Conservation of Nature also lists the wood bison as endangered in the Red Data Book.

STATUS OF THE PROGRAM. Successful transfers of wood bison to eight institutions (figure 49.5) were designed to enhance preservation of the gene pool. The institutions with captive wood bison and the numbers given to them are: Calgary Zoo, Alberta, 4 calves in November 1976; the Wildlife Reserve of Western Canada, Cochrane, Alberta, 2 calves in May 1977 for behavioral research; Metro Toronto Zoo, Toronto, Ontario, 10 adults in November 1977; Moose Jaw Wild Animal Park, Moose Jaw, Saskatchewan, 10 adults in March 1978; San Diego Zoo, San Diego, California, 3 calves in January 1979; Valley Zoo, Edmonton, Al-

FIGURE 49.5. Distribution of wood bison (*Bison bison athabascae*) herds in North America in February 1981. 1, Fort Providence-Mackenzie Bison Sanctuary; 2, Wood Buffalo National Park; 3, Elk Island National Park; 4, Cochrane Wildlife Reserve; 5, Calgary Zoo; 6, Moose Jaw Wild Animal Park; 7, Metro Toronto Zoo; 8, San Diego Zoo; 9, Valley Zoo-Edmonton; 10, Alberta Wildlife Park; 11, transferred herd, Nahanni Butte; 12, Banff National Park. Source wood bison originated in Wood Buffalo National Park (2); however, this herd is now composed of free-roaming hybrid animals. At present, the Fort Providence herd (1) is the largest free-ranging wood bison herd in the world. Other locations indicate captive herds.

berta, 1 yearling male in January 1980 and 1 yearling female in May 1980; The Alberta Wildlife Park, Bon Accord, Alberta, 5 adults and 9 calves in May 1980; and Banff National Park, Alberta, a mixed group of 5 wood bison in January 1981. Wood bison at Calgary Zoo, Toronto Zoo, Moose Jaw Wild Animal Park, The Alberta Wildlife Park, and Banff National Park are producing offspring each year.

The recommendation to maintain a permanent herd of approximately 200–250 wood bison at Elk Island National Park has been approved by Parks Canada. This decision has resulted in a significant contribution to preservation of the subspecies and to the rehabilitation program.

The first attempt to reestablish a free-roaming population of wood bison using the disease-free source stock from Elk Island National Park was not successful. In summer 1978, wood bison from the source herd at Elk Island were released to the wild in Jasper National Park, Alberta. This experiment failed when the herd moved out of the park onto provincial land in an area of agricultural development. By contrast, the wood bison transplanted to Fort Providence, N.W.T., in 1963 moved only short distances from the release site. Remoteness of the area, the natural barrier (Great Slave Lake) to the east, and good-quality range probably contributed to successful establishment of that herd. Generally speaking, free-ranging bison are not compatible with agriculture; that is why few such herds exist in the world today.

The third transfer of wood bison to the wild occurred during June 1980 when 28 animals were released near Nahanni Butte, N.W.T. Ten bison were equipped with radio collars. This herd is currently being monitored to determine when and where it will establish a new home territory.

PROBLEMS WITH THE REINTRODUCTION OF WOOD BISON. In attempting to establish new herds of wood bison, it is feasible to relocate animals only into isolated areas where conflict with agriculture and urban activities will be minimized. Adult wood bison should not be transplanted into areas of potential conflict due to their tendency to wander long distances when released into new range. Availability of appropriate sites is the major limiting factor to establishment of wild populations. Such sites are restricted mainly to the more northern regions of Canada. Experience gained from the transplant attempt to Jasper Park has led to consideration of alternate methods of reintroduction. Three methods that would appear to offer some promise for success are: (1) to release wood bison into remote areas; (2) to hold young animals in a corral at the chosen release site for a period of time in anticipation that they will locate nearby when released; and (3) to tame, train, and precondition newborn calves to a selected home range.

There are other problems associated with the reestablishment of wood bison. If movements of a population of transplanted wildlife are transboundary—national or international—the question of management authority and responsibility must be cooperatively addressed and resolved. It is doubtful that all provinces have sufficient legislation to protect and manage endangered species such as wood bison, even within their own boundaries. The possibility of a status change for wood bison is being pursued in anticipation of relieving legislative restrictions that may impede progress.

Transfers to the wild are negotiated under the authority of the Canada Wildlife Act as cooperative agreements. In the past, ownership status and responsibility of transferred wood bison have created some legal problems. Presently, surrender of ownership of transferred wood bison to the recipient agency is carried out by legal agreement. Total responsibility for management and control of transferred animals is then under the recipient's jurisdictional authority. Surrendering ownership of wood bison to recipient provinces and territories should add impetus to establish populations of wild wood bison. Also, an amendment to the current lease agreement whereby ownership of all wood bison in captivity may be transferred from the crown to the respective lessees has been prepared. This amendment will assist in the exchange of surplus wood bison and further disseminate the subspecies through the zoological garden and park community.

FUTURE PLANS. Future priorities of the program are to continue with transplants both to the wild and to captivity, with emphasis on the establishment of two more populations in the wild. Monitoring of the wood bison

herd released near Nahanni Butte, N.W.T., during June 1980 will be continued to determine where this herd will establish its home range. Negotiations are under way for possible transfers to sites in the Yukon Territory and in the province of Alberta. Field evaluation of potential range in the Yukon Territory occurred during July 1980, and if an agreement can be reached, a release to the Yukon will be possible. The possibility for experimental work to reintroduce wood bison in Alberta is being considered. Project completion is expected by 1986.

Free-Roaming Herds. The only large, free-roaming herds of bison in North America are in Yellowstone and Wood Buffalo national parks, the Mackenzie Bison Sanctuary, and the Slave River lowlands. Smaller free-roaming herds occur in Alaska, in northeastern British Columbia, near Nahanni Butte, N.W.T., and in northwestern Saskatchewan.

At Yellowstone National Park, management policy has changed from intensive bison ranching and predator control between 1902 and 1930 to one of minimal interference that began in the mid-1960s whereby populations were allowed to fluctuate under natural conditions (Meagher 1973). However, the animals are still subject to regulation by the U.S. Department of Agriculture and the national parks administration. A boundary control program to prevent dissemination of brucellosis outside of the park is still maintained and bison research and monitoring continues.

At Wood Buffalo National Park, a Technical Advisory Bison Committee was formed in 1975 to coordinate bison research and management. The park management objective is to maintain in a free-roaming state as large a disease-controlled herd as possible within the ecological limits of park resources. Reports on the status of park bison are compiled annually. In recent years, disease has been of minor importance and disease control programs have been relaxed.

Bison on the Slave River lowlands have been managed as a game species. A yearly census of the bison population is taken and disease testing and inoculation have occurred in the past. Management efforts to date have been unsuccessful in preventing the continued decline in herd numbers, first evidenced in 1974 (see "Predation and Hunting" section).

Management of wood bison in the Mackenzie Bison Sanctuary is directed toward preservation of the subspecies. Hunting is not yet allowed and the animals are protected by law under the Northwest Territories Wildlife Ordinance. International transport of endangered species such as wood bison is regulated by CITES (see "Wood Bison Program"). Two aerial surveys are conducted per year: one in March for total count and another in summer to determine the calf crop. No disease testing has occurred on the wood bison in the Mackenzie Bison Sanctuary.

Economic and Management Considerations. Although difficult to evaluate, the direct and indirect human benefits derived from free-roaming bison herds should not be underestimated. In captivity, the operational cost for raising bison requires a greater initial expenditure than for cattle. When designing proper enclosures, handling facilities, and types of handling practices, one must consider the wild and powerful nature of this species. Handling should be minimized during the calving and rut seasons, when the animals are more sensitive and irritable. Disturbance of corraled animals can stimulate aggressive behavior, resulting in serious injury to animals and damage to property. Although not essential, special game fencing is recommended and could cost in excess of $8,400/km. Increasing interest in bison ranching has resulted in development of specialized handling facilities and the publication of speciality magazines by The National Buffalo Association, Custer, South Dakota, and The American Buffalo Association, Coffeyville, Kansas. The National Buffalo Association has also published a book on bison history and husbandry which has been designed and intended for use as a buffalo rancher's handbook (Jennings 1978).

Despite the higher initial cost of proper fencing and handling facilities, the returns from bison ranching may be greater than those from cattle ranching. Prices for live animals in Canada and the United States during the 1978 to 1980 period ranged from $500+ for calves to $1,000+ for mature bulls. Meat prices for bison in the United States are roughly equivalent to those for beef, while more of a specialty market occurs in Canada. Heads, robes, hides, wool, hair, skulls, and horns can provide considerable additional revenue to the bison rancher. A market also exists for hooves, teeth, bones, and bladders. Harvest of bison by hunting combines meat production with recreation and, in the United States, can generate considerable money through trophy fees. The money generated from the trophy head, cape, and meat of a large bison bull is potentially much greater than that for those parts of a large domestic bull. The esthetic value of bison also makes them a popular tourist attraction. Grazing bison on public lands would make these areas more productive and potentially more attractive to public users. It is in this respect that the merits of game ranching become evident. In a study on the potential for game ranching in the boreal aspen forest of western Canada, Telfer and Scotter (1975) reported that land utilized by wild herbivores (moose, elk, deer, and bison) at Elk Island National Park supported 18 percent more grazing (measured in animal unit months) than did adjacent land grazed mainly by cattle along with some wild herbivores (moose, elk, and deer). Provided that the problems of systematic harvest are overcome, game ranching, using complementary species coupled with appropriate silvicultural practices, may provide a viable alternative on those lands that are only marginally suited to domestic livestock and crop production. In the Wichita Mountains Wildlife Refuge, the carrying capacity for bison has been increased by the artificial creation of lakes and ponds and the control of pest plants (Fuller 1961). An attempt has also been made to breed selectively for larger bison as part of the management scheme for the refuge.

Cross-breeding bison to cattle in order to take advantage of the hardiness of bison under adverse climatic and range conditions and also the meat characteristics of domestic cattle was first initiated about 1750 in Carolina and Virginia (McHugh 1972). Since that time, bison have been crossed with Herefords, Charolais, Angus, Highland, and Brahman cattle, as well as with yak. Commencing in 1924, the government of Canada conducted forty years of research into cattle-bison crosses at Wainwright and Manyberries, Alberta. Peters (1978:5) provided the following conclusion regarding the Canadian cattalo experiment,

> The experiment had demonstrated that outstanding winter hardiness, as measured by hair coat density and performance of cattle-bison hybrids and ¼-bison cattalo on winter range, could be obtained through combining the two species. The overall performance of the cattalo was still somewhat inferior to that of the control Herefords, though by 1962 both the lowest and the highest average daily feedlot gains in tests of cattalo and Hereford calves were made by cattalo. To achieve the full potential of the cross for breed development it would have been most desirable to make further introductions of bison and domestic stock, carefully selecting foundation animals on the basis of high growth rate in terms of lean meat content, and expanding the female breeding herd from approximately 100 to several hundred cows. However, there were other problems of beef cattle breeding that needed attention and would require substantial resources. The project was therefore discontinued with the slaughter of the herd in 1964.

The characteristics of bison that make them desirable as a source of meat primarily involve their ability to be productive under range conditions that are not optimal for cattle. As outlined in various sections of this chapter, bison readily consume rations of poorer quality than do cattle, they have a greater capacity to digest these rations, and they have the ability to forage on range under severe winter conditions. Although bison on range consume poorer-quality forage than do cattle under similar conditions, they will readily consume grain and high-quality forages (McHugh 1958; Peters 1958). When fed hay that was coarsely chopped, bison readily consumed the poorer-quality portions (Hawley et al. 1981*b*).

There are too few data on which to base precise feeding recommendations for bison. The following are, therefore, general guidelines subject to revision when more data are available. The nutritional requirements for bison can be considered similar to those of cattle. However, the greater assimilation of crude protein and energy by bison can be taken into account when feeding poor-quality rations. The digestibility of crude protein and gross energy in poor-quality feeds might be considered to be about 5 percentage units greater for bison than for cattle, and feed requirements could be adjusted accordingly. Possible winter inappetence and growth depression in bison living in cold climates suggests that there would be little advantage, in terms of weight gain, to feeding yearling or older animals at above maintenance levels during winter.

There may be some productive advantage to feeding bison calves above maintenance over winter. Productive responses to high-quality rations are greater in cattle than in bison. Increased management problems and decreased feed conversion make it inefficient to feed high-quality rations to bison year-round. Maximum growth should occur during summer, and high-energy, high-protein supplements should be most effective for augmenting growth at that time. Care must be taken to avoid ruminitis when feeding high-energy rations to bison.

Bison can be raised under feedlot conditions, but this obviates some of the productive advantages of bison over cattle. Because of their intractable nature, cold hardiness, and ability to digest poor-quality forages, bison appear to be most efficiently used as commercial meat animals under range conditions.

Transport and Relocation. In most cases bison will take feed and bed down during transport, similar to cattle. Covered transport should be used in order to prevent bison from jumping out. Due to the large amount of hair on the head and forequarters of bison, the potential for heat stress during transport is greater for bison than for cattle. As a means of minimizing this, bison should be transported when temperatures are relatively cool and water should always be available. If overnight road travel is anticipated, it may be useful to illuminate compartments in order to avoid any disturbance caused by on-coming vehicle headlights and other lighting.

During the transport of wood bison from Elk Island National Park to Nahanni Butte (approximately 1500 km), in June 1980, one animal appeared to suffer from heat exhaustion and would not stand until soaked with cold water. All animals frequently drank water provided to them. Fighting occurred between two bulls that occupied the same compartment and one animal suffered a lacerated shoulder and a belly wound. Two cows harassed a yearling bull as a result of disturbance during unloading. Disturbance of any kind stimulated bison to kick the walls of the compartment.

Despite apparent suitability of habitat and historical record of use, relocated bison may leave the area where they are released. This can result in animal movements of several hundred kilometers through unfamiliar and less suitable habitat. Corraling bison for several months prior to transport and release does not always ensure herd cohesion after release.

CURRENT RESEARCH NEEDS

Research into the question of bison taxonomy is required. Further osteological studies will provide additional data on morphological and skeletal measurements to help clarify the taxonomic status of bison. There is a need to develop further tests for red cell antigens and blood enzymes to assess the usefulness of blood characteristics in taxonomic studies. Additional knowledge on the evolution of bison would certainly assist this process. A revision of the genus *Bison* may prove necessary.

Continued studies of the basic physiology of bison are required. Sample size has been a limitation in most physiological experiments with bison, partly because of the difficulties of conducting intensive research with this species. The seasonal nutritional requirements of bison need to be identified and studies of digestive physiology should be continued in an attempt to elucidate the digestibility differences between bison and cattle.

A considerable amount of ethological observation has been made of bison in parks but little objective research has been conducted into the behavior of bison under conditions of intensive handling. Animal behavior in response to handling is an important aspect of bison production and requires research. Precise data on behavior and production under range conditions are also required. Peters (1978) concluded that animal breeds with outstanding winter hardiness and high productivity could be developed from bison × cattle crosses. Under feedlot conditions, the hybrids were inferior in performance compared with the control Herefords. However, there has been little selection for desirable traits in the hybrids, nor has there been much experimentation to investigate the production potential of the hybrids under adverse range conditions. It is under adverse range conditions that the contribution of certain bison traits to a hybrid cross would be desirable for animal production. Further hybridization and selection studies to evaluate the productive potential of hybrids under varying environmental conditions would be important and fruitful areas of research.

A large experimental trial incorporating the principles of game ranching, using bison as one of the native ungulates to be collectively managed (Telfer and Scotter 1975), is now required to determine the economic feasibility of such an enterprise. More research data are required to determine if the carrying capacity of rangeland is greater when native ungulates are managed under the multispecies, resource use concept, compared with the conventional agricultural practice of grazing only domestic herbivores. Comparative studies of animal production strategies in northern regions are essential to obtain information that will lead to optimum use of the rangeland resource. As the world demand for meat production continues to increase and more pressure to develop marginal agricultural land confronts government agencies, bison should be closely scrutinized as an alternative animal species to raise for meat production.

The absence of certain intestinal parasites in bison at Wood Buffalo National Park (Fuller 1966) and in bison at Yellowstone National Park (Meagher 1973) may be attributed to inadequate postmortem examinations. More detailed parasitological research would undoubtedly increase our knowledge of this aspect of bison biology. Reported low incidence of cancer in bison has become a recent research interest; however, more thorough investigation is essential if this finding is to be verified.

Continuation of bison-wolf population dynamics research on the Slave River lowlands would improve our understanding of the predator-prey relationship and the effect that a large predator such as the wolf can have on a bison population.

Specific research data concerning the general biology of wood bison are needed, especially information regarding methods of reintroduction that include preconditioning experiments to aid in the establishment of transplanted herds.

We gratefully acknowledge the Canadian Wildlife Service and Alberta Environmental Centre for supporting and encouraging us during preparation of this chapter. E. Broughton and W. Samuel provided valuable assistance through their constructive criticism of the disease and parasite sections, respectively. Special thanks are also due R. Russell for reviewing drafts of the manuscript, A. Kennedy for assistance with the age determination section, D. Robinson for library assistance, H. Breen and C. Ridewood for typing, P. Karsten for artwork, and the warden's staff at Elk Island National Park for their cooperation. We further express our great appreciation of those many individuals and agencies who have assisted with the preservation of both subspecies of bison.

LITERATURE CITED

Alexander, G.; Signoret, J. P.; and Hafez, E.S.E. 1974. Sexual and maternal behavior. Pages 222–254 in E.S.E. Hafez, ed. 1975. Reproduction in farm animals. 3rd ed. Lea & Febiger, Philadelphia. 480pp.

Allen, J. A. 1876. The American bisons, living and extinct. Mem. Mus. Comp. Zool. Harvard College. 1974. Arno Press, New York. 245pp.

Allison, L. 1973. The status of bison on the Peace-Athabasca delta. Pages M1–M27 in Peace-Athabasca Delta Proj. Ecol. Invest. and Tech. Appendixes. Vol. 2. Information Canada, Ottawa. 551pp.

Audubon, J. J., and Bachman, J. 1849. Bos americanus. Pages 292–295 in The quadrupeds of North America. V. G. Audubon, New York. 334pp.

Banfield, A.W.F. 1974. The mammals of Canada. Univ. Toronto Press, Toronto. 438pp.

Banfield, A.W.F., and Novakowski, N. S. 1960. The survival of the wood bison (*Bison bison athabascae* Rhoads) in the Northwest Territories. Natl. Mus. Can. Nat. Hist Pap., no. 8. 6pp.

Barrett, M. W., and Chalmers, G. A. 1977. Clinicochemical values for adult free-ranging pronghorns. Can. J. Zool. 55:1,252–1,260.

Bayrock, L. A., and Hillerud, J. M. 1964. New data on *Bison bison athabascae* Rhoads. J. Mammal. 45:630–632.

Becklund, W. W., and Walker, M. L. 1969. Taxonomy, hosts, and geographic distributions of the *Setaria* (nematoda:Filarioidea) in the United States and Canada. J. Parasitol. 55:359–368.

Bell, R.H.V. 1971. A grazing ecosystem in the Serengeti. Sci. Am. 225:86–93.

Berg, R. T., and Butterfield, R. M. 1976. New concepts of cattle growth. Sydney Univ. Press, Sydney. 240pp.

Bergstrom, R. C. 1967. Sheep liver fluke, *Fasciola hepatica* L., 1758, from buffalo, *Bison bison* (L. 1758) in western Wyoming. J. Parasitol. 53:724.

Boddicker, M. L., and Hugghins, E. J. 1969. Helminths of big game mammals in South Dakota. J. Parasitol. 55:1,067–1,074.

Bourque, B. J.; Morris, K.; and Spiess, A. 1978. Determining the season of death of mammal teeth from archeological sites: a new sectioning technique. Science 199:530–531.

Braend, M. 1963. Haemoglobin and transferrin types in the American buffalo. Nature 197:910–911.

Brinley-Morgan, W. J., and McCullough, N. B. 1974. Genus *Brucella*. Pages 278–282 in N. E. Gibbons et al., eds Manual of determinative bacteriology. 8th ed. Williams & Wilkens, Baltimore. 1,246pp.

Bryant, E. 1885. Rocky Mountain adventures. Worthington Co., New York. 452pp.

Buckland, R. A., and Evans, H. J. 1978a. Cytogenetic aspects of phylogeny in the Bovidae. Part 1: G-banding. Cytogenet. Cell Genet. 21:42–63.

———. 1978b. Cytogenetic aspects of phylogeny in the Bovidae. Part 2: C-banding. Cytogenet. Cell Genet. 21:64–71.

Burger, J. F., and Anderson, J. R. 1970. Association of the face fly, *Musca autumnalis*, with bison in western North America. Ann. Entomol. Soc. Am. 63:635–639.

Burt, W. H., and Grossenheider, R. P. 1952. A field guide to the mammals. Riverside Press, Cambridge, 284pp.

Burtner, R. H., and Becklund, W. W. 1971. Prevalence, geographic distribution, and hosts of *Cooperia surnabada* Antipin, 1931, and *C. oncophora* (Raillet, 1898) Ransom, 1907, in the United States. J. Parasitol. 57:191–192.

Cameron, A. E. 1923. Notes on buffalo: anatomy, pathological conditions, and parasites. Br. Vet. J. 79:331–336.

———. 1924. Some further notes on buffalo. Br. Vet. J. 80:413–417.

Capp, J. C. 1964. Ecology of the bison of Colorado National Monument. Natl. Park Serv., U.S. Dept. Interior. Government Printing Off., Washington, D.C. 30pp.

Chapin, E. A. 1925. New nematodes from North American mammals. J. Agric. Res. 30:677–681.

Choquette, L.P.E. 1970. Anthrax. Pages 256–266 in J. W. Davis, L. H. Karstad, and D. O. Trainer, eds. Infectious diseases of wild mammals. 1st ed. Iowa State Univ. Press, Ames. 421pp.

Choquette, L.P.E.; Broughton, E.; Cousineau, J. G.; and Novakowski, N. S. 1978. Parasites and diseases of bison in Canada. Part 4: Serologic survey for brucellosis in bison in Canada. J. Wildl. Dis. 14:329–332.

Choquette, L.P.E.; Broughton, E.; Currier, A. A.; Cousineau, J. G.; and Novakowski, N. S. 1972. Parasites and diseases of bison in Canada. Part 3: Anthrax outbreaks in the last decade in northern Canada and control measures. Can. Field Nat. 86:127–132.

Choquette, L.P.E.; Gallivan, J. F.; Byrne, J. L.; and Pilipavicius, J. 1961. Parasites and diseases of bison in Canada. Part 1: Tuberculosis and some other pathological conditions in bison at Wood Buffalo and Elk Island National Parks in the fall and winter of 1959–60. Can. Vet. J. 2:168–174.

Choquette, L.P.E., and Stewart, R. C. 1959. Report on studies on bison in Canada, 1959. Rep. Can. Wildl. Serv. C.W.S.C. 852. Ottawa, Ontario. 28pp. Manuscript.

Christopherson, R. J.; Hudson, R. J.; and Christophersen, M. K. 1979. Seasonal energy expenditures and thermoregulatory responses of bison and cattle. Can. J. Anim. Sci. 59:611–617.

Christopherson, R. J.; Hudson, R. J.; and Richmond, R. J. 1976. Feed intake, metabolism, and thermal insulation of bison, yak, Scottish Highland, and Hereford calves during winter. Pages 51–52 in 55th Annu. Feeder's Day Rep. Univ. Alberta, Edmonton. 75pp.

———. 1978. Comparative winter bioenergetics of American bison, yak, Scottish Highland, and Hereford calves. Acta Theriol. 23:49–54.

Church, D. C. 1975. Digestive physiology and nutrition of ruminants. Vol. 1: Digestive physiology. 2nd ed. D. C. Church, Corvallis, Ore. 350pp.

Cole, J. E. 1954. Buffalo (*Bison bison*) killed by fire. J. Mammal. 35:453–454.

Colman, D. 1978. Roy Phillips has a "home" for buffalo. Buffalo! 6(4):14.

Corner, A. H., and Connell, R. 1958. Brucellosis in bison, elk, and moose in Elk Island National Park, Alberta, Canada. Can. J. Comp. Med. Vet. Sci. 22:9–21.

Cousineau, J. G., and McClenaghan, R. J. 1965. Anthrax in bison in the Northwest Territories. Can. Vet. J. 6:22–24.

Cox, B. L. 1978. Comparison of meat quality from bison and beef cattle. Undergraduate Thesis. Dept. Home Economics, Univ. Saskatchewan, Saskatoon. 35pp.

Cram, E. B. 1925. *Cooperia bisonis*, a new nematode from the buffalo. J. Agric. Res. 30:571–573.

Dary, D. A. 1974. The buffalo book: the saga of an American symbol. Avon Books/Swallow Press, Chicago. 374pp.

DeYoung, H. G. 1973. The bison is beleaguered again. Nat. Hist. 82:48–55.

Dikmans, G. 1934. New records of helminth parasites. Proc. Helminth. Soc. Washington. 1:63–64.

———. 1939. Helminth parasites of North American semidomesticated and wild ruminants. Proc. Helminth Soc. Washington 6:97–101.

Drummond, R. O., and Medley, J. G. 1964. Occurrence of *Speleognathus australis* Womersley (Acarina:Speleognathidae) in the nasal passages of bison. J. Parasitol. 50:655.

Duffield, L. F. 1973. Aging and sexing the post-cranial skeleton of bison. Plains Anthropol. 18:132–139.

Dukes, H. H. 1937. The physiology of domestic animals. Comstock Publ. Co., Ithaca, N.Y. 695pp.

Dunn, A. M. 1968. The wild ruminant as reservoir host of helminth infection. Symp. Zool. Soc. London 24:221–248.

Dziuk, H. E. 1965. Eructation, regurgitation, and reticuloruminal contraction in the American bison. Am. J. Physiol. 208:343–346.

Egerton, P.J.M. 1962. The cow-calf relationship and rutting behavior in the American bison. M.S. Thesis. Univ. Alberta, Edmonton. 155pp.

el-Shazly, K.; Dehority, B. A.; and Johnson, R. R. 1961. Effect of starch on the digestion of cellulose *in vitro* and *in vivo* by rumen micro-organisms. J. Anim. Sci. 20:268–273.

Engelhard, J. G. 1970. Behavior patterns of American bison calves of the National Bison Range, Moiese, Montana. M.S. Thesis. Central Michigan Univ., Mt. Pleasant. 151pp.

Engelhardt, W. V. 1978. Adaption to low protein diets in some mammals. Pages 110–115 *in* Proc. Zodiac Symp. on Adaptation, 24–26 May, Wageningen, The Netherlands. 158pp.

Espmark, Y. 1964. Rutting behavior in reindeer (*Rangifer tarandus* L.). Anim. Behav. 12:159–163.

Estes, R. D. 1972. The role of the vomeronasal organ in mammalian reproduction. Mammalia 36:315–341.

Flerov, C. C. 1965. Comparative craniology of recent representatives of the genus *Bison*. Byull. Mosk. O Va Ispyt Prir. Old Bio. 70:1–17.

Flerov, C. C., and Zablotski, M. A. 1961. On the causative factors responsible for the change in the bison range. Byull. Mosk. O Va Ispyt Prir. Old Bio. 66, no. 6:99–109. (Transl. W. A. Fuller.)

Frick, E. J. 1951. Parasitism in bison. J. Am. Vet. Med. Assoc. 119:386–387.

Frison, G. C. 1980. Man and bison relationships in North America. Can. J. Anthropol. 1:75–76.

Frison, G. C., and Reher, C. A. 1970. Age determination of buffalo by teeth eruption and wear. Pages 46–50 *in* G. Frison. The Glenrock buffalo jump: Late Prehistoric period buffalo procurement and butchering. Plains Anthropol. Memoir 7. 66pp.

Fuller, W. A. 1954. The first premolar and the canine tooth in bison. J. Mammal. 35:454–456.

———. 1959. The horns and teeth as indicators of age in bison. J. Wildl. Manage. 23:342–344.

———. 1960. Behavior and social organization of the wild bison of Wood Buffalo National Park, Canada. Arctic 13:3–19.

———. 1961. The ecology and management of the American bison. Terre et la Vie 2:286–304.

———. 1966. The biology and management of the bison of Wood Buffalo National Park. Can. Wildl. Serv. Wildl. Manage. Bull. Ser. 1(16). 52pp.

Fuller, W. A., and Bayrock, L. A. 1965. Late Pleistocene mammals from central Alberta, Canada. Pages 53–63 *in* Vertebrate paleontology in Alberta. Rep. conf., Univ. Alberta, Edmonton, 29 August–3 September 1963. 76pp.

Fulton, R. D.; Caldwell, J.; and Weseli, D. F. 1978. Methemoglobin reductase in three species of Bovidae. Biochem. Genet. 16:635–640.

Garretson, M. S. 1927. A short history of the American bison. Am. Bison Soc., New York. 42pp.

———. 1938. The American bison: the story of its extermination as a wild species and its restoration under federal protection. New York Zool. Soc., New York. 254pp.

Geist, V. 1963. On the behaviour of the North American moose (*Alces alces andersoni* Peterson 1950) in British Columbia. Behaviour 20:377–416.

———. 1971. The relation of social evolution and dispersal in ungulates during the Pleistocene, with emphasis on the Old World deer and the genus *Bison*. Quaternary Res. 1:285–315.

Geist, V., and Karsten, P. 1977. The wood bison (*Bison bison athabascae* Rhoads) in relation to hypotheses on the origin of the American bison (*Bison bison* Linnaeus). Z. Saugetierk. 42:119–127.

Gilchrist, F.M.C., and Clark, R. 1957. Refresher courses in physiology. Part 3: The microbiology of the rumen. J. South Afr. Vet. Med. Assoc. 28:295–309.

Glover, J., and Dougall, H. W. 1960. The apparent digestibility of the non-nitrogenous components of ruminant feeds. J. Agric. Sci. 55:391–394.

Graham, M. 1923. Canada's wild buffalo. Canada Dept. Interior, Ottawa. 12pp.

Grinnell, G. B. 1904. The bison. Pages 111–166 *in* Caspar Whitney, George Bird Grinnell, and Owen Wister. Musk-ox, bison, sheep, and goat. MacMillan Co., New York. 284pp.

Gunderson, H. L., and Mahan, B. R. 1980. Analysis of sonograms of American bison (*Bison bison*). J. Mammal. 61:379–381.

Guthrie, R. D. 1970. Bison evolution and zoogeography in North America during the Pleistocene. Q. Rev. Bio. 45:1–15.

———. 1980. Bison and man in North America. Can. J. Anthropol. 1:55–73.

Hadwen, S. 1942. Tuberculosis in the buffalo. J. Am. Vet. Med. Assoc. 100:19–22.

Haines, H.; Chichester, H. G.; and Landreth, H. I., Jr. 1977. Blood respiratory properties of *Bison bison*. Respir. Physiol. 30:305–310.

Hall, E. R., and Kelson, K. R. 1959. The mammals of North America. Vol. 2. Ronald Press Company, New York.

Halloran, A. F. 1961. American bison weights and measurements from the Wichita Mountains Wildlife Refuge. Proc. Oklahoma Acad. Sci. 41:212–218.

———. 1968. Bison (Bovidae) productivity on the Wichita Mountains Wildlife Refuge, Oklahoma. Southwestern Nat. 13:23–26.

Halloran, A. F., and Glass, B. P. 1959. The carnivores and ungulates of the Wichita Mountains Wildlife Refuge, Oklahoma. J. Mammal. 40:360–370.

Hancock, J. 1953. Grazing behavior of cattle. Commonwealth Bur. Anim. Breeding and Genetics 21:1–13.

Harington, C. R. 1980. Faunal exchanges between Siberia and North America: evidence from Quaternary land mammal remains in Siberia, Alaska, and the Yukon Territory. Can. J. Anthropol. 1:45–49.

Haugen, A. O. 1974. Reproduction in the plains bison. Iowa State J. Res. 49:1–8.

Hawley, A.W.L. 1978. Comparison of forage utilization and blood composition of bison and Hereford cattle. Ph.D. Thesis. Univ. Saskatchewan, Saskatoon. 246pp.

Hawley, A.W.L.; Peden, D. G.; Reynolds, H. W.; and Stricklin, W. R. 1981a. Bison and cattle digestion of forages from the Slave River Lowlands, Northwest Territories, Canada. J. Range Manage. 34:126–130.

Hawley, A.W.L.; Peden, D. G.; and Stricklin, W. R. 1981b. Bison and Hereford steer digestion of sedge hay. Can. J. Anim. Sci. 61:165–174.

Hediger, H. 1968. The psychology and behavior of animals in zoos and circuses. Dover Publ., New York. 314pp.

Herrig, D. M., and Haugen, A. O. 1969. Bull bison behavior traits. Iowa Acad. Sci. 76:245–262.

Hillerud, J. M. 1980. Bison as indicators of geologic age. Can. J. Anthropol. 1:77–80.

Holsworth, W. N. 1960. Interactions between moose, elk, and buffalo in Elk Island National Park, Alberta. M.S. Thesis. Univ. British Columbia, Vancouver. 92pp.

Hornaday, W. T. 1889. The extermination of the American bison, with a sketch of its discovery and life history. Pages 367–548 in Part 2 [1889]. Rep. U.S. Natl. Mus., 1886–87.

Houpt, T. R. 1970. Transfer of urea and ammonia to the rumen. Pages 119–131. in A. T. Phillipson, ed. Physiology of digestion and metabolism in the ruminant. Proc. 3rd Int. Symp. Oriel Press, Cambridge. 636pp.

Inman, H. 1899. Buffalo Jones' forty years of adventure. Crane & Co., Topeka. 469pp.

Jaczewski, Z. 1958. Reproduction of the European bison, *Bison bonasus* (L.), in reserves. Acta Theriol. 1:333–376.

Jalkotzy, M. 1979. Wolf-bison project Slave River lowlands, N.W.T. Rep. Northwest Territories Wildl. Serv. Yellowknife, N.W.T. 17pp. Manuscript.

Jaros, Z.; Valenta, Z.; and Zajicek, D. 1966. A list of helminths from the section material of the Zoological Garden of Prague in the years 1954–1964. Helminthologia 7:281–290.

Jennings, D. C. 1978. Buffalo history and husbandry. Pine Hill Press, Freeman, N.D. 392pp.

Kaneko, J. J., and Cornelius, C. E. 1970. Clinical biochemistry of domestic animals. 2nd ed. Vol. 1. Academic Press, New York. 439pp.

———. 1971. Clinical biochemistry of domestic animals. 2nd ed. Vol. 2. Academic Press, New York. 352pp.

Karsten, P. 1975. Don't be buffaloed ... by a bison: history of Alberta herds. Dinny's Digest, Calgary Zool. Soc. 2:3–13.

Keith, E. O. 1977. Urea metabolism of North American bison. M.S. Thesis. Colorado State Univ. Fort Collins. 73pp.

Keith, E. O.; Ellis, J. E.; Phillips, R. W.; and Benjamin, M. M. 1978. Serologic and hematologic values of bison in Colorado. J. Wildl. Dis. 14:493–500.

Keller, D. G. 1980. Milk production in cattalo cows and its influence on calf gains. Can. J. Anim. Sci. 60:1–9.

Kelsall, J. P.; Telfer, E. S.; and Kingsley, M.C.S. 1978. Relationship of bison weight to chest girth. J. Wildl. Manage. 42:659–661.

Kennedy, P. M.; Young, B. A.; and Christopherson, R. J. 1977. Studies on the relationship between thyroid function, cold acclimation, and retention time of digestion in sheep. J. Anim. Sci. 45:1,084–1,090.

Kirkpatrick, R. L.; Buckland, D. E.; Abler, W. A.; Scanlon, P. E.; Whelan, J. B.; and Burkhart, H. E. 1975. Energy and protein influences on blood urea nitrogen of white-tailed deer fawns. J. Wildl. Manage. 39:692–698.

Kohls, G. M., and Kramis, N. J. 1952. Tick paralysis in the American buffalo *Bison bison* (Linn.). Northwest Sci. 26:61–64.

Larson, F. 1940. The role of the bison in maintaining the shortgrass plains. Ecology 21:113–121.

Lent, P. C. 1965. Rutting behavior in a barren-ground caribou population. Anim. Behav. 13:259–264.

Locker, B. 1953. Parasites of bison in northwestern U.S.A. J. Parasitol. 39:58–59.

Longhurst, W. M. 1961. Big game and rodent relationships for forests and grasslands in North America. Terre et la Vie 2:305–326.

Lott, D. F. 1972. Bison would rather breed than fight. Nat. Hist. 81:40–45.

———. 1974. Sexual and aggressive behavior of adult male American bison (*Bison bison*). Pages 382–393 in V. Geist and F. Walther, eds. IUCN New Ser. 24, vol. 1. Morges, Switzerland. 940pp.

McHugh, T. 1958. Social behavior of the American buffalo (*Bison bison bison*). Zoologica 43, part 1. 40pp.

———. 1972. The time of the buffalo. Alfred A. Knopf, New York. 339pp.

McNary, D. C. 1948. Anthrax in American bison "Bos bison L." J. Am. Vet. Med. Assoc. 112:378.

Mahan, B. R. 1977. Harassment of an elk calf by bison. Can. Field Nat. 91:418–419.

———. 1978. Aspects of American bison (*Bison bison*) social behavior at Fort Niobrara National Wildlife Refuge, Valentine, Nebraska, with special reference to calves. M.S. Thesis. Univ. Nebraska, Lincoln. 154pp.

Mahan, B. R.; Munger, M. P.; and Gunderson, H. L. 1978. Analysis of the Flehmen display in American bison (*Bison bison*). Prairie Nat. 10:33–42.

Marler, R. J. 1975. Some hematologic and blood chemistry values in two herds of American bison in Kansas. J. Wildl. Dis. 11:97–100.

Meagher, M. M. 1973. The bison of Yellowstone National Park. Natl. Park Serv. Sci. Monogr. Ser. 1. 161pp.

———. 1974. Yellowstone's bison a unique wild heritage. Natl. Parks and Conserv. Mag. (May), pp. 9–14.

———. 1978. Bison. Pages 123–133 in J. L. Schmidt and D. L. Gilbert, eds. Big game of North America: ecology and management. Stackpole Books, Harrisburg, Pa. 494pp.

Mehrer, C. F. 1976. Some hematologic values of bison from five areas of the United States. J. Wildl. Dis. 12:7–13.

Miller, F. L. 1974. Age determination of caribou by annulations in dental cementum. J. Wildl. Manage. 38:47–53.

Miller, J. K.; Swanson, E. W.; Lyke, W. A.; Moss, B. R.; and Byrne, W. F. 1974. Effect of thyroid status on digestive tract fill and flow rate of undigested residues in cattle. J. Dairy Sci. 57:193–197.

Moss, E. H. 1932. The vegetation of Alberta. J. Ecol. 20:380–415.

Müller-Schwarze, D. 1971. Pheromones in black-tailed deer (*Odocoileus hemionus columbianus*). Anim. Behav. 19:141–152.

Neiland, K. A.; King, J. A.; Huntley, B. E.; and Skoog, R. O. 1968. The diseases and parasites of Alaskan wildlife populations. Part 1: Some observations on brucellosis in caribou. Bull. Wildl. Dis. Assoc. 4:27–36.

Nelson, K. L. 1965. Status and habits of the American buffalo (*Bison bison*) in the Henry Mountain area of Utah. Utah State Dept. Fish and Game Publ. 65-2. 142pp.

Novakowski, N. S. 1965. Cemental deposition as an age criterion in bison, and the relation of incisor wear, eye-lens weight, and dressed bison carcass weight to age. Can. J. Zool. 43:173–178.

Novakowski, N. S.; Cousineau, J. G.; Kolenosky, G. B.; Wilton, G. S.; and Choquette, L. P. E. 1963. Parasites and diseases of bison in Canada. Part 2: Anthrax epizootic in the Northwest Territories. Pages 233–239 *in* Trans. 28th North Am. Wildl. and Nat. Resour. Conf. 551pp.

Ogilvie, W. 1893. Report on the Peace River and tributaries in 1891. Annu. Rep. Dept. Interior Canada for 1892, pt. 7. 44pp.

Ozoga, J. J., and Verme, L. J. 1970. Winter feeding patterns of penned white-tailed deer. J. Wildl. Manage. 34:431–439.

Palmer, T. S. 1916. Our national herds of buffalo. Annu. Rep. Am. Bison Soc. 10:40–62. Page 17 *in* W. A. Fuller. 1966. The biology and management of the bison of Wood Buffalo National Park. Can. Wildl. Serv. Wildl. Manage. Bull. Ser. 1(16). 52pp.

Patten, D. T. 1963. Vegetational pattern in relation to environments in the Madison Range, Montana. Ecol. Monogr. 33:375–406.

Pearson, H. A. 1967. Rumen micro-organisms in buffalo from southern Utah. Appl. Microbiol. 15:1450–1451.

Peden, D. G. 1972. The trophic relations of *Bison bison* to the shortgrass plains. Ph.D. Thesis. Colorado State Univ., Fort Collins. 134pp.

———. 1976. Botanical composition of bison diets on shortgrass plains. Am. Midl. Nat. 96:225–229.

Peden, D. G., and Kraay, G. J. 1979. Comparison of blood characteristics in plains bison, wood bison, and their hybrids. Can. J. Zool. 57:1,778–1,784.

Peden, D. G.; Van Dyne, G. M.; Rice, R. W.; and Hansen, R. M. 1974. The trophic ecology of *Bison bison* L. on shortgrass plains. J. Appl. Ecol. 11:489–498.

Pellerdy, L. P. 1963. Catalogue of Eimeriidea (Protozoa; Sporozoa). Akad. Kiado, Budapest. 160pp.

———. 1974. Coccidia and coccidiosis. 2nd ed. Verlag Paul Parey, Berlin. 959pp.

Peters, H. F. 1958. A feedlot study of bison, cattalo, and Hereford calves. Can. J. Anim. Sci. 38:87–90.

———. 1978. Utilization of bison as a genetic resource for meat production. 13th Int. Symp. Zootechny. Accademia Nazionale Di Agricoltura, Milan. 8pp.

Peters, H. F., and Slen, S. B. 1964. Hair coat characteristics of bison, domestic X bison hybrids, cattalo, and certain domestic breeds of beef cattle. Can. J. Anim. Sci. 44:48–57.

———. 1966. Range calf production of cattle, bison, cattalo, and Hereford cows. Can. J. Anim. Sci. 46:157–164.

Petropavlovskii, V. V., and Rykova, A. I. 1958. The stimulation of sexual functions in cows. Tr. Vljjarov. Sel.-Hoz.

Inst. 5:193–199. (In Russian.) Anim. Breed. Abstr. 29(1961):168.

Ransom, B. H. 1911. The nematodes parasitic in the alimentary tract of cattle, sheep, and other ruminants. U.S.D.A. Bur. Anim. Indust. Bull. 127. 132pp.

Raup, H. M. 1933. Range conditions in the Wood Buffalo Park of western Canada with notes on the history of the wood bison. Spec. Publ. Am. Comm. Int. Wildl. Prot. 1, no. 2. 52pp.

Reynolds, H. W. 1976. Bison diets of Slave River lowlands, Canada. M.S. Thesis. Colorado State Univ., Fort Collins. 66pp.

Reynolds, H. W.; Hansen, R. M.; and Peden, D. G. 1978. Diets of the Slave River lowland bison herd, Northwest Territories, Canada. J. Wildl. Manage. 42:581–590.

Rhoads, S. N. 1897. Notes on living and extinct species of North American Bovidae. Proc. Acad. Nat. Sci. Philadelphia 49:483–502.

Rice, R. W.; Dean, R. E.; and Ellis, J. E. 1974. Bison, cattle, and sheep dietary quality and food intake. J. Anim. Sci. 38:1332. Abstr.

Richmond, R. J.; Hudson, R. J.; and Christopherson, R. J. 1976. Comparison of forage intake and digestibility by bison, yak, and cattle. Pages 49–50 *in* 55th Annual Feeders' Day Rep. Univ. Alberta, Edmonton. 75pp.

———. 1977. Comparison of forage intake and digestibility by American bison, yak, and cattle. Acta Theriol. 22:225–230.

Roe, F. G. 1970. The North American buffalo: a critical study of the species in its wild state. 2nd ed. Univ. Toronto Press, Toronto. 991pp.

Roudabush, R. L. 1936. Arthropod and helminth parasites of the American bison (*Bison bison*). J. Parasitol. 22:517–518.

Rush, W. M. 1932. Bang's disease in the Yellowstone National Park buffalo and elk herds. J. Mammal. 13:371–372.

———. 1942. Wild animals of the Rockies. Harper & Brothers, New York. 296pp.

Ryff, K. L., and Bergstrom, R. C. 1975. Bovine coccidia in American Bison. J. Wildl. Dis. 11:412–414.

Salter, R. E. 1978. Ecology of feral horses in western Alberta. M.S. Thesis. Univ. Alberta, Edmonton. 239pp.

Schalm, O. W.; Jain, N. C.; and Carroll, E. J. 1975. Veterinary hematology. 3rd ed. Lea & Febiger, Philadelphia. 807pp.

Schein, M. W., and Fohrman, M. H. 1955. Social dominance relationships in a herd of dairy cattle. Br. J. Anim. Behav. 3:45–55.

Schneider, B. H., and Flatt, W. P. 1975. The evaluation of feeds through digestibility experiments. Univ. Georgia Press, Athens. 423pp.

Schultz, C. B., and Hillerud, J. M. 1977. The antiquity of *Bison latifrons* (Harlan) in the Great Plains of North America. Trans. Nebraska Acad. Sci. 4:103–116.

Schwartz, C. C.; Nagy, J. G.; and Rice, R. W. 1977. Pronghorn dietary quality relative to forage availability and other ruminants in Colorado. J. Wildl. Manage. 41:161–168.

Searcy, R. L. 1969. Diagnostic biochemistry. McGraw-Hill Book Co., New York. 660pp.

Seibert, F. V. 1925. Some notes on Canada's so-called wood buffalo. Can. Field Nat. 39:204–206.

Seton, E. T. 1886. The wood buffalo. Proc. R. Can. Inst. Ser. 3:114–117.

———. 1929. The buffalo. Pages 639–703 *in* Lives of game animals. Vol. 3, pt. 2. Doubleday, Doran & Co., Garden City, N.Y.

Shackleton, D. M. 1968. Comparative aspects of social organization of American bison. M.S. Thesis. Univ. Western Ontario, London. 67pp.

Shackleton, D. M.; Hills, L. V.; and Hutton, D. A. 1975. Aspects of variation in cranial characters of plains bison (*Bison bison bison* Linnaeus) from Elk Island National Park, Alberta. J. Mammal. 56:871–887.

Shinn, R. 1978. Buffalo in the news. Buffalo! 6(4):26.

Shult, M. J. 1972. American bison behavior patterns at Wind Cave National Park. Ph.D. Thesis. Iowa State Univ., Ames. 178pp.

Silver, H.; Colovos, N. F.; Holter, J. B.; and Hayes, H. H. 1969. Fasting metabolism of white-tailed deer. J. Wildl. Manage. 33:490–498.

Skinner, M. F., and Kaisen, O. C. 1947. The fossil Bison of Alaska and preliminary revision of the genus. Bull. Am. Mus. Nat. Hist. 89:123–256.

Slee, J. 1972. Habituation and acclimatization of sheep to cold following exposures of varying length and severity. J. Physiol. 227:51–70.

Smith, H. C. 1977. A fossil bison skull from western British Columbia. Syesis 10:167–168.

Smoliak, S., and Peters, H. F. 1955. Climatic effects on foraging performance of beef cows on winter range. Can. J. Agric. Sci. 35:213–216.

Soper, J. D. 1941. History, range, and home life of the northern bison. Ecol. Monogr. 11:349–412.

———. 1964. The mammals of Alberta. Hamly Press, Edmonton, Alberta. 402pp.

Stamp, J. T. 1959. Tuberculosis: epidemiology and pathology. Pages 687–713 in A. W. Stableforth and I. A. Galloway, eds. Diseases due to bacteria. 2 vols. Butterworths Sci. Publ., London. 810pp.

Swenson, M. J. 1970. Physiologic properties, cellular and chemical constituents of blood. Pages 21–61 in M. J. Swenson, ed. Dukes' physiology of domestic animals. 8th ed. Comstock Publ. Assoc., Cornell Univ. Press, Ithaca, N.Y. 1,463pp.

Telfer, E. S., and Scotter, G. W. 1975. Potential for game ranching in boreal aspen forests of western Canada. J. Range Manage. 28:172–180.

Thomas, D. C. 1977. Metachromatic staining of dental cementum for mammalian age determination. J. Wildl. Manage. 41:207–210.

Tunnicliff, E. A., and Marsh, H. 1935. Bang's disease in bison and elk in the Yellowstone National Park and on the National Bison Range. J. Am. Vet. Med. Assoc. 39:745–752.

Van Camp, J. 1978a. Summary of progress, wolf-bison project, Hook Lake area, N.W.T. Rep. Northwest Territories Wildl. Serv., Yellowknife, N.W.T. 29pp. Manuscript.

———. 1978b. Summary of progress, wolf-bison project, Slave River lowlands, N.W.T. Rep. Northwest Territories Wildl. Serv., Yellowknife, N.W.T. 22pp. Manuscript.

Vaughan, T. A. 1972. Mammalogy. W. B. Saunders Co., Philadelphia. 463pp.

Wade, S. E.; Haschek, W. M.; and Georgi, J. R. 1979. Ostertagiosis in captive bison in New York State: report of nine cases. Cornell Vet. 69:198–205.

Walker, E. P.; Warnick, F.; Hamlet, S. E.; Lange, K. I.; Davis, M. A.; Uible, H. E.; and Wright, P. F. 1964. Genus: Bison. Page 1433 in Vol. 2. Mammals of the world. 3rd ed. (1975). 2 vols. Johns Hopkins Press, Baltimore. 1500pp.

Wasser, C. H. 1977. Bison induced stresses in Colorado National Monument. Pages 28–36 in Final report. Natl. Park Serv. Contract PX 120060617. 120pp.

Weaver, J. W., and Clements, F. E. 1938. Plant ecology. McGraw-Hill Book Co., New York. 413pp.

Westra, R., and Hudson, R. J. 1979. Urea recycling in wapiti. Pages 236–239 in M. S. Boyce and L. D. Hayden-Wing, eds. North American elk: ecology, behavior, and management. Univ. Wyoming, Laramie. 294pp.

Wilson, M. 1980. Morphological dating of late Quaternary bison on the northern plains. Can. J. Anthropol. 1:81–85.

Wood, A. J.; Cowan, I. McT.; and Nordan, H. C. 1962. Periodicity of growth in ungulates as shown by deer of the genus *Odocoileus*. Can. J. Zool. 40:593–603.

Ying, K. L., and Peden, D. G. 1977. Chromosomal homology of wood bison and plains bison. Can. J. Zool. 55:1,759–1,762.

Young, B. A.; Schaefer, A.; and Chimwano, A. 1977. Digestive capacities of cattle, bison, yak. Pages 31–34 in 56th Annu. Feeders' Day Rep. Univ. Alberta, Edmonton. 76pp.

H. W. REYNOLDS, Canadian Wildlife Service, #1000, 9942—108 Street, Edmonton, Alberta, Canada T5K 2J5.

R. D. GLAHOLT, Canadian Wildlife Service, #1000, 9942—108 Street, Edmonton, Alberta, Canada T5K 2J5.

A.W.L. HAWLEY, Alberta Environmental Center, Box 4000, Vegreville, Alberta, Canada T0B 4L0.

50

Mountain Goat

Oreamnos americanus

Ronald A. Wigal
Victor L. Coggins

NOMENCLATURE

COMMON NAMES. Mountain goat, Rocky Mountain goat, white goat, white buffalo
SCIENTIFIC NAME. *Oreamnos americanus*
SUBSPECIES. *O. a. americanus*, *O. a. columbiae*, *O. a. kennedyi*, and *O. a. missoulae*.

The mountain goat, *Oreamnos americanus*, is in the family *Bovidae*. The monotypic genus is endemic to North America. The mountain goat is not a true goat, but rather a type of antelope. Its nearest relatives are the Old World chamois (*Rupicapra rupicapra*) of Europe and the goral (*Naemorhedus* sp.), takin (*Budorcus taxicolor*), and serow (*Capricornus* sp.) of Asia.

DISTRIBUTION

No fossil evidence of the mountain goat has been found in Asia or Europe, yet it is probable that this goat is of Asiatic origin and migrated to North America. Goats may have crossed to the New World via the Bering land bridge during the mid-Pleistocene and were able to survive the glacial age (Trefethen and Miracle 1972).

Vaughan (1972) felt that the Bering Strait was under the influence of several boreal climatic conditions during the mid-Pleistocene. The land bridge functioned as a "filter" such that only those animals adapted to those conditions traversed into America. The mountain goats may have sought refuge on the upper mountain ridges and peaks while the glaciers spread out below and covered a large portion of the Northern Hemisphere (O'Connor and Goodwin 1961).

Mountain goats historically were found in rugged coastal and mountainous areas of western North America from southeastern Alaska to south-central Washington. They ranged as far south as central Idaho and east to western Alberta and Montana (Johnson 1977*a*). In contrast to that of most other North American ungulates, the distribution of mountain goats has been extended from the historic range. Transplant programs initiated as early as 1920 have expanded the range both in states where goats historically occurred

FIGURE 50.1. Distribution of the mountain goat (*Oreamnos americanus*).

(Alaska, Idaho, Washington, and Montana) and in states outside the natural range (Dalrymple 1970).

Goats have been introduced successfully to Oregon, Nevada, Utah, Colorado, Wyoming, and South Dakota. Ranges have also been extended to unoccupied habitat in most of the states and provinces where the species was native as well (figure 50.1). Table 50.1 summarizes the current estimated abundance of mountain goats.

DESCRIPTION

The mountain goat has a stocky build and a slight hump on the withers (figure 50.2). It is about 1 m tall and about 1.2 m in total length. The legs are short and terminate with evenly paired hoofs, separated by a deep interdigital cleft, and well-developed dewclaws. Horns are present on both sexes and are of moderate size, conical, and unbranched. They are rigid at the base, diverge laterally, and incline posteriorly. Immediately posterior to the base of the horns is a pair of

TABLE 50.1. Summary of mountain goat (*Oreamnos americanus*) abundance in North America by state or province

State or Province	Estimate of Numbers Based on Past Records	Current Estimate of Numbers
British Columbia	100,000 (1964)	20,000–60,000
Alaska	no estimate	15,000–25,000
Washington	10,000 (1961)	no estimate
Montana	no estimate	no estimate
Idaho	2,785 (1955)	2,200–2,500
Yukon Territory	no estimate	1,400
Alberta	no estimate	1,200
Colorado	(introduced 1948)	575
Northwest Territories	no estimate	400+
South Dakota	(introduced 1924)	300–400
Wyoming	no estimate	70
Nevada	(introduced 1964)	30
Oregon	(introduced 1950)	28–30
Utah	(introduced 1967)	no estimate
Canadian parks		1,670–1,770

SOURCE: Johnson 1977*a*.

dark, crescent-shaped horn glands. Goats are predominantly white or yellowish white all over except for the horns, hoofs, nose, and eyelids, which are black. The pelage is long, with a coarse mane, beard, and chaps, and the woolly underfur is dense (Bailey 1936).

Size. At birth, kids are approximately 56 cm long and 34 cm high at the shoulder, and weigh about 3 kg. After two weeks, the weight increases to 4.0–4.5 kg. In Idaho, Brandborg (1955) found little difference in size between yearling males and females. They were about 96 cm in total length and 62 cm high, and weighed 18–20 kg. Thirteen yearling goats examined in South Dakota were considerably larger than those from Idaho. The average weight was 34.5 kg, total length 104–124 cm, and height at the shoulder 71–86 cm (Richardson 1971). Body weights in Montana were comparable to those in South Dakota, but Lentfer (1955) noted that yearling males averaged 35 kg, while yearling females averaged 30 kg. The mean weight of adults four years old or older in Montana was found to be 71 kg for females and 82 kg for males. Adult goats in Idaho were considerably smaller, but showed similar weight differentiation between sexes. Males averaged 70 kg and females 53 kg.

Horns. Mountain goats have true horns that are never shed. Horns grow continously and reach the peak of development sometime after the goat reaches maturity but before old age sets in.

The horn consists of two parts: the core, which is a bony outgrowth of the frontal bone of the skull, and the sheath, which is of epidermal origin and grows from the base. The average horn length for mountain goats is about 23 cm (Anderson 1940), and a greater circumference is readily apparent for males. Horns of males also have a slightly greater curvature than those of females. Horn characteristics are poor criteria for

delineating sex in the field unless the observer is experienced and very close to the goat.

Hooves. The hoof of the mountain goat is unlike that of other ungulates in that it has a slightly convexed and pliable pad that extends beyond the outer cornified shell. This adaptation provides for greater traction and affords the goat a higher degree of dexterity on rough terrain than that noted for other ungulates.

Integument and Glands. There are few integumental glands on mountain goats. Dark, crescent-shaped supraoccipital glands are present on both sexes, but are larger on males and are "particularly active" during the rut (Anderson 1940). The smaller horn glands present on females are thought to be rudimentary. During the rut, males excrete an oily exudate from these glands that is deposited by brushing the horns slowly from side to side over bunches of grass or along twigs. This act serves to mark territories and is believed to attract females (Anderson 1940) and intimidate rival males (Geist 1964).

Geist (1967) examined a young male mountain goat and found a "dermal shield" over the rump area where the majority of strikes from the horns of a conspecific would be directed. He felt that this thickened dermis would possibly provide some protection to older males during minor confrontations. However, severe punctures and lacerations would still result from major rivalries. The dermis was thickest just below the anus (22 mm), and tapered for 15 mm midway between the anus and the hock. The shield continued for about 25 mm ventrally from the anal area and decreased to 6 mm at the brisket. A thickness of 3 mm was recorded just above the tarsi on the median side of the hind leg and also for the neck just below the lower jaw.

Skull and Dentition. The horn cores are nearly straight, and are round in cross-section. Ethmoid vacuities are absent, the premaxilla does not contact the nasals, and the postorbital bar is complete. The pos-

FIGURE 50.2. Mountain goat (*Oreamnos americanus*), showing rocky habitat and steep terrain typical of range.

terior edge of the palate terminates opposite the last upper molar (figure 50.3).

The cheek teeth are hypsodont and selenodont; lower canines are incisiform. The dental formula is 0/3, 0/1, 3/3, 3/3 = 32.

Pelage. Mountain goats are the only big game animal with white pelage in the lower 48 states. They are very conspicuous on rocky slopes in summer, but become nearly invisible against snow cover in winter.

Coarse guard hairs often reach lengths of more than 18 cm on the back and legs, which give the goat its characteristic high-shouldered profile and exaggerated size. The dense underfur is fine textured and is comparable in quality to cashmere (Brandborg 1955). The pelage protects well against dry cold, but protec-

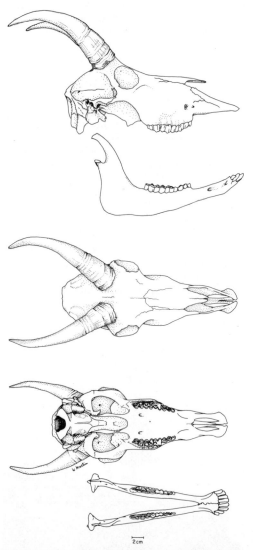

FIGURE 50.3. Skull of the mountain goat (*Oreamnos americanus americanus*). From top to bottom: lateral view of cranium, lateral view of mandible, dorsal view of cranium, ventral view of cranium, dorsal view of mandible.

tion against cold rain is less complete. Goats frequently seek shelter under ledges or trees during heavy rains (Anderson 1940).

The annual molt begins from late April to early May, depending on the locality. The first evidence of shedding is denoted by loose strands of hair hanging from the animal. The old hair, usually densely matted and discolored, appears to annoy the goats. They have been observed scratching with their horns and rubbing against rocks and trees. The molt begins at the top of the shoulders and continues along the back and up the neck. From the dorsal region, the molt line gradually moves down over the sides, buttocks, and shoulders (Anderson 1940).

Individuals appear ragged until July, after which they become increasingly smooth. The molting process is complete by November (Swift 1940).

Chadwick (1974) found that barren adult females and adult males completed the molting process three to four weeks prior to yearlings and pregnant nannies. He found that pregnant nannies initially begin to shed at the same time as other females and yearlings. However, the process appeared to be delayed during the month of June and as a result these females were most often the last to complete the molting process. It is suggested that parturition and the maintenance of newborn young may significantly divert the nanny's attention to retard the completion of the molt. Since some adult females without kids were observed to be late in molting, possibly due to the loss of their young, Chadwick (1974) speculated that this phenomenon may be useful as an estimator of early postpartum mortality.

Brandborg (1955) reported a frequent scattering of coarse brown hairs along the back, rump, and tail of goats. At one time goats may have been darker in color and inhabited wooded and rocky areas at lower elevations (Trefethen and Miracle 1972).

Senses. Olfaction in mountain goats is well developed and heavily relied upon for the detection of danger (Anderson 1940). Brandborg (1955), however, felt that the olfactory process was not nearly as well developed as that of other big game animals based upon the reaction of goats to human scent. These observations may have been a result of individual behavior rather than an indication of the acuteness of their sensory perception. Olfactory senses probably are important in the identification of kids by nannies and in the location of estrous females by males during the rut.

Vision is keen for detecting movement, but the recognition of stationary objects is thought to be poor. A person traveling below an animal situated on a vantage point rarely is unobserved. However, mountain goats cannot identify a person in the open, even at close range, provided he remains motionless (Anderson 1940).

The reaction of mountain goats to loud noises is variable, but the auditory alertness demonstrated suggests a dependence upon this sense for detecting other animals. An awareness of unfamiliar noises is frequently exhibited, whereas more common sounds, such as the rolling of rocks, cause relatively little concern (Brandborg 1955).

PHYSIOLOGY

The oxygen consumption of mountain goats, a mean value of 0.26 ml O_2/gm/hr, was fairly constant in ambient temperatures from 20° C to −20° C. At −30° C, the oxygen consumption increased about 23 percent and at −50° C an increase of approximately 130 percent occurred (Krog and Monson 1954). Krog and Monson suggested that the goat's dense pelage does not provide as much insulative protection against heat dissipation as does that of true arctic mammals. Even though temperatures within goat range often reach −30° C, the northernmost areas occupied by mountain goats are not considered to be arctic latitudinally or climatically (Krog and Monson 1954).

REPRODUCTION

The breeding season of mountain goats throughout their range is from the middle of November through early December. Brandborg (1955) reported the last week of October as the earliest observed mating and 8 December as the latest. In September, testes may have mature sperm in the epididymis and ovaries may have a corpus luteum, suggesting that earlier breeding is possible (Lentfer 1955). Estrus appears to last from 48 to 72 hours (Chadwick 1974).

Follicular development begins in September and October prior to the mating season. Reproductive data collected by Lentfer (1955) and Peck (1972) in Montana suggested that females first breed at about 2.5 years of age. Henderson and O'Gara (1978) found two yearling males that produced mature spermatozoa; whether they bred was not determined. Richardson (1971) found no yearling females with kids and assumed that they do not breed their first year. Yearling males were observed participating in the rut.

Parturition generally occurs from 15 May through 15 June, following a gestation period of approximately 180 days. Generally only one kid is born, but twins are not uncommon. Foss (1962) and Lentfer (1955) in Montana reported 8 and 30 percent, respectively, of the total number of kids observed as twins. Anderson (1940) felt that twinning probably occurred more frequently than was recorded. He suggested that twins may have a higher mortality rate, since nannies could not care for two kids as easily as one. Reports of twins sometimes may result from more than one kid following a single female after a group is frightened or from kids at play becoming separated from their mothers and following another kid and nanny (Brandborg 1955).

It is generally accepted that mountain goats are polygamous, and males constantly search for receptive females during the rut.

Kids are usually born in rough terrain and often negotiate steep slopes when only a few hours old. Females are especially watchful over their young during the first few days and often lick and prod them with the nose. Nursing regularly occurs at intervals shorter than one hour and lasts from a few seconds to more than 10 minutes, depending upon the nanny's cooperation. The feeding periods are usually terminated by the mother's walking away, often knocking the kid over if continued nursing is attempted. Brandborg (1955) observed kids eating forage and ruminating within a few days after birth. At about six weeks of age, the kids forage regularly near their mothers. Nursing periods become less frequent and shorter in duration in August or September. Weaning occurs in August or September and kids are ignored by females when nursing is attempted.

Productivity was expressed by Foss (1962) and Peck (1972) as kid/adult ratios. Foss reported an overall ratio of 24 percent, while Peck found ratios ranging from an average of 23 percent to 56 percent depending on the area. Both authors suggested that the variability in these ratios between areas was probably attributed to differences in hunting pressure. Peck (1972) also suggested that higher nutrient levels on some winter ranges may be responsible for the higher kid/adult ratios in those areas.

Mountain goats have low rates of increase on most ranges and have stable populations. As a result, Hjelford (1971) reported that mountain goats never exceed the carrying capacity of their range. However, Olmsted (1977) reported that mountain goats introduced to Olympic National Park in the late 1920s reached a population size that affected the vegetation. Various influences from feeding, trampling, bedding, and dust bathing were noted.

ECOLOGY

Mountain goats inhabit areas within the subalpine and arctic alpine zone of the northern Rockies and coast ranges. The rugged terrain is comprised of cliffs, ledges, projecting pinnacles, and talus slopes. Goat range is associated with areas having well-defined glaciation at elevations between 1,524 m and 3,353 m. While Rocky Mountain habitats are characterized by rough terrain above the timberline, areas in South Dakota occupied by introduced goats are of a considerably lower ecological level. This transitional zone, however, is well suited for goat utilization (Swift 1940).

Saunders (1955) divided mountain goat habitat in Montana into four major groups: grassy slide-rock slopes, ridge tops, alpine meadows, and timber. Grassy slide-rock slopes were major use areas in spring, summer, and fall. These slopes were often blown free of snow during winter. Alpine meadows were frequented most from July through August when vegetation on cliffs was sparse. Timber areas were utilized during summer and fall by scattered single goats and by nannies in spring prior to parturition.

Mountain goats usually remain at or above the timberline and within reach of rocky outcrops utilized for retreat from danger. They prefer to remain high on mountains and were often observed seeking windblown slopes where they could forage. Deep snows frequently force them to lower elevations

(MacGregor 1977). In general, goats tend to make use of higher elevations during summer and lower ones in winter. Some herds do winter on high, windblown alpine slopes, while others descend to lower elevations in the manner of most other western ungulates.

Watering areas may be necessary for suitable goat range and summer distribution is sometimes limited by its absence, although Brandborg (1955) felt that water availability was not a limiting factor in Idaho. Sufficient water from springs and snowbanks was available at the higher elevations; however, as summer progressed, sources at lower elevations became more scarce. Lentfer (1955) concurred that moisture requirements are satisfied by snow and free water from melting snow. Goats are frequently close to snow and have been observed to spend a good amount of time in it. They were occasionally observed eating snow and drinking from the streams of run-off. Richardson (1971) also found that water was not a limiting factor on mountain goat ranges in South Dakota, except during extended periods of drought. The actual need for water was not determined, but goats using salt invariably sought water.

Goats were observed licking salt most often during early morning and late evening. Those situated near salt grounds frequent them regularly. Salt artificially placed in old mine shafts and between ledges, where the effects of adverse weather are reduced, are most frequently visited during late summer and winter (Richardson 1971). Mountain goats often travel as far as 24 km to visit both natural and artificial licks during spring and summer. Analysis of dry soil taken from a large mineral lick near Glacier National Park, Montana, indicated the presence of 600 parts per million of soluble salts and high concentrations of phosphorus. Similar analysis of other natural licks showed somewhat lesser concentrations of soluble salts, high availability of phosphorus, and varying amounts of magnesium, sodium, chlorides, sulfates, calcium, and potassium. The utilization of natural mineral licks may indicate a mineral deficiency in the diet that is supplemented by these soils (Brandborg 1955).

FOOD HABITS

Saunders (1955) described mountain goats as snip feeders that rarely graze intently at any particular spot. A variety of plant species are utilized, particularly during summer and fall. These included foliage and seed heads of grasses, sedges, and rushes; foliage, stems, and flowers of forbs; leaves and twigs of shrubs and trees; leaves of ferns; and the entire aerial portion of mosses and lichens.

In summer feeding areas in Montana, 56 percent of the plants taken were grasses, sedges, and rushes. The most important representatives included tufted hairgrass (*Deschampsia caespitosa*), sheep fescue (*Festuca ovina*), alpine bluegrass (*Poa* sp.), sedges (*Carex* sp.), and rushes (*Juncus* sp.). Forbs comprised 24 percent of the grazed plants and included lupine (*Lupine* sp.), mountain bluebells (*Mertensia* sp.), and polemonium (*Polemonium* sp.). Dwarf huckleberry

(*Vaccinium* sp.), willow (*Salix* sp.), swamp current (*Ribes* sp.), and shrubby cinquefoil (*Vaccinium scoparium*) represented the shrubs utilized and comprised about 16 percent of the summer diet. Tree species comprised less than 4 percent of the observed food species. Mosses, lichens, and ferns occurred in trace amounts. Rocky Mountain woodsia (*Woodsia scopulina*) was the most important fern and was readily utilized by kids (Saunders 1955).

The number of direct feeding observations made in fall, winter, and spring periods were too few for consideration, but stomachs from animals collected during these periods were analyzed. There was about a 3 percent increase in the utilization of forbs and a slight decrease in the occurrence of shrubs. Other vegetation types showed essentially no change. During winter, the consumption of grasses, sedges, and rushes decreased by about 5 percent from the rate in summer and fall. Shrub species were considered unimportant at this time of year, whereas conifer utilization increased from a trace amount in summer to about 25 percent in winter. Douglas fir (*Pseudotsuga menziesii*) and alpine fir (*Abies* sp.) represented the bulk of conifer consumption. Mosses, lichens, and ferns remained insignificant. The utilization of grasses, sedges, and rushes was greater in spring than in winter, but the number of species utilized remained low. Shrub utilization increased slightly from winter to spring, but remained far below the rate observed for summer and fall. The relative volume of conifers declined, but they persisted as an important food item. Mosses, lichens, and ferns maintained trace status (Saunders 1955).

Studies in Alaska suggest that forage utilized in fall may be transitive between summer and winter. When considering rumen physiology, this gradual shift from the summer diet to the winter diet would seem feasible (Hjeljord 1971).

Seasonal variation in the utilization of certain plants by mountain goats also occurred in Washington (Anderson 1940). Goats browse more than they graze, but under certain circumstances grasses may comprise up to 95 percent of the diet. During winter months when bunch grasses (*Agropyron* sp.) were not available because of deep snow, the diet consisted almost entirely of browse. In less severe winters, cured bunch grasses comprised about 90 percent of the winter diet. From spring through summer, green grasses were heavily utilized. Thirty-four species of plant were food items for goats during this period; 16 species were food items during winter months. Major summer food species included mountain laurel (*Ceanothus veluntinus*), low-bush huckleberry (*Vaccinium* sp.), quaking aspen (*Populus tremuloides*), bunch grass, cheat grass (*Bromus tectorum*), and bluegrass (*Poa* sp.). Bunchgrass (*Erigonium heracleoides*), tall Oregon grape (*Berberis aquifolium*), and pentstemon (*Pentstemon* sp.) were important winter forage species (Anderson 1940).

In Colorado, grasslike species made up approximately 96 percent of the summer diet, while less than 4 percent of the vegetation consumed consisted of forbs. The utilization of grasslike species in winter decreased slightly, to about 88 percent, and about 12 percent of the diet consisted of shrub species. Forbs were not an important component in the winter diet (Hibbs 1967).

In South Dakota, the composition of vegetation in the Black Hills is considerably different from that found in Rocky Mountain ranges of Colorado. Lichens were abundant throughout the area and were highly utilized in all seasons of the year. Leaves and terminal branches of choke cherry (*Prunus virginiana*) also were utilized throughout the year, to the extent that this abundant species showed serious signs of overbrowsing. Various grasses and sedges were consumed in spring and summer and were often pawed free of snow and grazed upon in winter. Bearberry (*Arctostaphylos uva-ursi*) was also an important food source in fall and winter (Richardson 1971). Harmon (1944) noted that species utilization by mountain goats in the Black Hills was variable. During winter months, the diet consisted of about 60 percent mosses and lichens, 20 percent bearberry, 10 percent pine needles and twigs (*Pinus ponderosa*), and 10 percent miscellaneous ferns, grasses, and woody species.

A summary of the forage types utilized by mountain goats on summer and winter ranges is presented in tables 50.2 and 50.3, respectively.

BEHAVIOR

Rutting. During September and October, females are usually in groups, called nursery bands, consisting of an adult female, a kid or a yearling, and occasionally a two year old. Adult males sometimes join these groups, but never remain long. From late October, females are almost always with males. While females remain within their respective ranges, males actively move from range to range. Adult males are discernible from females and nonbreeding males at this time because their pelage becomes soiled while pawing "rutting pits." Before pawing, the male goat assumes a sitting position similar to that of a dog. Then with the powerful front legs soil and often snow are thrown repeatedly at the belly, flanks, and hind legs. The marking of twigs and clumps of grass with the horn glands is sometimes associated with pawing activities (Geist 1964).

By the end of October, females begin to permit males to join the nursery bands; however, courtship approaches are not tolerated. After the first week of November, the antagonistic behavior of the females toward the courting males diminishes and by the end of the month, females become receptive. Chadwick (1974) noted that female receptivity was often indicated by the reversal of sexual roles, such as females mounting the courting males and their kids.

Courting males usually approach females from the rear and assume a low, crouching position with the head and neck extended level with the ground. When nearing the female, the male often has his tail elevated, ears in a forward position, and nostrils flared. During this period of excitement, the male's tongue may flick in and out of his mouth and his head will jerk from side to side. At this point, the male often licks the female's flanks and sometimes taps her flank or haunch with his front leg. Even though the females are very tolerant, some still discourage copulation efforts by crowding against rock walls.

During the peak of the rut, courting males rapidly approach females from the rear and kick them in the haunches, often with enough force to knock them forward. Occasionally this behavior is immediately followed by the male mounting the female, clasping her flanks with his front legs, and placing his chin upon her back. Pelvic thrusts are rapid and the male usually dismounts without indication of ejaculation. Courtship is repeated prior to additional mounting attempts (Geist 1964).

Fighting. Geist (1964) suggested that fighting would be highly disadvantageous, as animals could be severely injured. Instead, a threat display by adult males usually suppresses aggressive behavior. When fighting does occur, the antagonists spin about side to side, thrusting their horns at each other's rump and belly, unlike the head-to-head posture assumed by other bovids (Geist 1967).

TABLE 50.2. Forage utilization by mountain goats (*Oreamnos americanus*) on various summer ranges in North America

		Forage Utilized					
Source	Location	Sedges, Grasses, and Rushes	Herbs (Forbs)	Coniferous Trees	Deciduous Trees (and Shrubs)	Evergreen Shrubs	Mosses, Lichens, and Ferns
Anderson 1940	Washington	12%	17%	0%	30%	41%	0%
Casebeer 1948 (from Hjeljord 1971)	Montana	3	2	0	1	94	0
Cowan 1944 (from Brandborg 1955)	Jasper and Banff national parks	63	14	0	23	0	0
Hibbs 1967	Colorado	82	14	trace	3	0	0
Hjeljord 1971	Alaska	36	64	0	0	0	0
Kerr 1965	Alberta	82	5	2	7	4	0
Saunders 1955	Montana	72	17	trace	3	trace	7
Smith 1976	Montana	72	26	trace	0	2	0

TABLE 50.3. Forage utilization by mountain goats (*Oreamnos americanus*) on various winter ranges in North America

		Forage Utilized						
Source	Location	Sedges, Grasses, and Rushes	Herbs (Forbs)	Coniferous Trees	Deciduous Trees (and Shrubs)	Evergreen Shrubs	Mosses, Lichens, and Ferns	Unidentified
Anderson 1940	Washington	75%	10%	1%	9%	5%	0%	0%
Brandborg 1955	Idaho & Montana	54	trace	trace	42	2	trace	2
Casebeer 1948	Red Butte, Montana	63	2	0	35	0	0	0
(from Brandborg 1955)	Rattlesnake Range	68	3	0	29	0	0	0
Chadwick 1974	Montana	61	5	9	11	0	14	0
Hibbs 1967	Colorado	88	0	0	11	1	0	0
Hjeljord 1971	Alaska	0	0	0	10	0	90	0
Kerr 1965	Alberta	8	trace	73	17	2	0	0
Saunders 1955	Montana	59	16	25	trace	trace	trace	0
Smith 1976	Montana	74	trace	12		trace	not accounted for	14

Play. Play behavior exhibited by kids resembles that of adults fighting. Kids often circle one another, attempting to butt the other's rump. Other play activities noted for kids included running toward each other, rearing on the hind legs, or simply running and jumping, often in snow banks. Kids playing together are often twins (Lentfer 1955).

Wariness. Mountain goats are usually well composed and not as wary as other big game species (Lentfer 1955). Humans approaching goats are usually detected, but no distress is noted. Also, human scent does not appear to alarm goats. Rideout (1974) reported that human urine was used to attract goats to a central location during live-trapping operations in Montana.

Bedding. Mountain goats spend a great deal of time loitering. Lentfer (1955) found that resting goats could be observed during most times of the day. Resting sites include rocks, snowbanks, and vegetated areas. Goats often utilize high points, especially during good weather, but seek the protection of overhanging rocks and caves during storms or on unusually hot days. In the Wallowa Mountains, up to five goats have been seen using a cave for shelter during the winter. The cave is located above the timberline and is about 30 feet deep and 15 feet high at the entrance. The cave floor is covered with a thick layer of goat droppings, indicating heavy use. Bedding sites often consist of a pawed area in loose soil excavated prior to lying down. These sites are often utilized by more than one goat, and an individual may use up to four beds a day. Thus, the number of beds in an area is not a reliable indication of the number of goats in an area (Anderson 1940).

Dusting. Goats dig dry wallows with their forefeet where they often spend hours, particularly in June and July, rolling about and pawing dirt over their bodies. They also may dust in the dry, loose soil of floors of caves (Swift 1940). Dusting probably relieves irritations caused by ectoparasites and shedding and keeps the pelage from becoming oily. Similar behavior was noted in patches of snow.

Daily Movements. The daily activities of mountain goats chiefly consist of bedding and feeding. Feeding usually begins at dawn and goats tend to work their way from lower to higher elevations throughout the day. The afternoons of warm, sunny days are usually spent lying down until feeding resumes in late afternoon and evening. Mountain goats generally remain in one locality, but sometimes travel to a different area several kilometers away, returning to the initial area a few days later (Anderson 1940). Goats may remain in restricted areas for days, while at other times they may move several kilometers in only a few hours. Typical summer daily movement varies from a few hundred meters to 0.5 km, although longer distances are not uncommon. The route taken is often determined as animals alternately move ahead of one another while feeding (Brandborg 1955).

In Alaska, goats remain at or near the timberline during summer, move up into the higher snow fields on warm days to rest on snow patches, and descend again in the evening to feed in the lush alpine meadows. Movements on winter ranges are usually much more restricted than those noted in summer, and goats sometimes stay in small areas for up to several weeks. Brandborg (1955) reported that a group of about 10 goats remained on an area of about 81 ha from February through May. Lentfer (1955) found that several marked animals remained within the area where they were captured. Some showed no patterns of movement, while others moved about seasonally. One small group traveled 4.0 km one day and 4.8 km the following day. Home ranges are generally less than 25 km² (O'Conner and Goodwin 1961).

Social Order. Adult nannies, particularly those with kids at their side, obtained the dominant position within goat social order (Kuck 1977). The dominant position obtained by nannies serves to protect kids from other aggressive goats and allows them to select winter habitat with lower snow accumulation to reduce energy expenditures. Food supplies are maintained in balance for the survival of dominant nannies on preferred winter cliffs by the dispersion of subdominant individuals through aggressive behavior. The aggressive behavior tends to disperse goat populations in relation to both physical habitat and available food supplies. With goats dispersing themselves in relation to both food and space, preventing self-destruction of their habitat, Kuck reported artificial control by hunting unnecessary to protect habitat.

Seasonal Movements. Mountain goats migrate each fall and spring between summer and winter habitats throughout most of their range. Brandborg (1955) found that goats in Montana and Idaho occupied "the lowest available winter ranges that provide preferred combinations of broken terrain and vegetative cover." These wintering areas were usually located on south-facing slopes where snow accumulations were less severe, but animals were sometimes observed on high mountain ridges foraging on small areas blown free of snow by strong winds.

Fall migrations appear to be influenced by snow-fall on the upper ranges and usually begin around the time of the first snowfall of the year. Fall movements generally involve only 5 to 6.5 km but migrations of up to 16 km are not uncommon (Anderson 1940). Brandborg (1955) felt that distances as far as 24 km are covered in Idaho and Montana.

Upward movements in spring are much more gradual than the fall migrations. Lone males are generally the first to arrive at the higher elevations, and females with kids and yearling goats usually remain at lower levels until the melting snow allows for more suitable feeding conditions.

MORTALITY

The greatest limiting factor and cause of mortality in mountain goats is the lack of suitable forage during the winter months when weather increases the susceptibility to predation, parasites, disease, and accidents.

Accidental Death. Snowslides during late winter and early spring probably account for more deaths than any other natural cause (Brandborg 1955). The skeletal remains of mountain goats have been found at the bottom of slides, and small groups of goats are sometimes killed by avalanches.

Goats often walk out on the overhangs of snow cornices that sometimes break away under the animal's weight. While not as important as snowslides, landslides during periods of heavy precipitation occasionally result in injury or death to goats. Slides are sometimes started by goats as they move across steep slopes. Goats usually pay little attention to rolling rocks, but they often flee when nearby slides become large enough to be heard plainly. Falls among mountain goats have not been readily observed, but

they do occasionally slip, especially in winter when rocks are icy and snow covered.

In the Oregon Department of Fish and Wildlife files, there are records of five goat carcasses being found in the Wallowa Mountains. There is sufficient evidence that at least four of these died in falls or avalanches. Vaughan (1975) reported the following observations while working on a goat study in the Wallowa Mountains in 1973. "On 12 April several avalanches occurred in a steep area that four goats were attempting to cross. At one point a goat slipped and was covered by snow, starting an avalanche which nearly took two other goats. All, however, scrambled to safety."

Predation. Coyotes (*Canis latrans*) are common throughout much of the range inhabited by mountain goats, but their importance as predators is generally considered insignificant. Coyote scats found in goat range in Idaho by Brandborg (1955) and in Washington by Anderson (1940) often contained goat hairs. Many of the animals fed upon by coyotes may have died from other causes and were consumed as carrion. Anderson (1940) twice observed coyotes chasing goats, and on both occasions the goats escaped easily and showed little concern. He did suggest, however, that kids and even adults could be killed by coyotes in deep snow. Brandborg (1955) felt that "occasional kills would most likely occur on open ranges where several coyotes working together could bring a goat down." Kids separated from their mothers also are likely prey. Cougars (*Felis concolor*) and bobcats (*Felis rufus*) may be important predators in remote ranges.

Predation on goats by Golden Eagles (*Aquila chrysaetus*) may be the most important form of predation, but again the total impact is thought to be insignificant.

Several observations have been made of Golden Eagles knocking goats from cliffs and carrying kids away (Anderson 1940; Brandborg 1955). Presumably, most eagle predation occurs in spring, following parturition, when kids are very small. It is not known whether those preyed upon are healthy, diseased, or deserted.

Parasites. Kerr and Holmes (1966) found only two ectoparasites on goats collected from typical habitat in Alberta in April and June. Both of these were ticks (*Dermacentor albipictus* and *D. andersoni*) and were thought to cause goats to rub off hair. Brandborg (1955) also found heavy infestations of *D. andersoni* and felt that most animals become infected when woodticks (*Dermacentor* sp.) become abundant during warm spring weather. He also noted three nymphs of the spinose ear tick (*Otobius megnini*) and a heavy infestation of the foot louse (*Linognathus pedalis*), previously unreported for mountain goats.

D. albipictus may be additive to mortality in wildlife, particularly following severe winters or during periods of low food availability (Kerr and Holmes 1966). Infestations of this tick may result in weakness, emaciation, and sometimes serious nervous disorders and severe anemia. Because they leave winter ranges in April, earlier than pregnant females, males may have a lesser degree of infestation of *D. andersoni,* which is more concentrated at lower elevations.

Richardson (1971) noted that flies constantly bother goats during warm weather and that they readily deposit eggs in the open cuts of goats held in an enclosure.

Endoparasites. Ten animals were examined by Kerr and Holmes (1966), and 11 species of endoparasite were noted, 9 of which were nematodes. These included stomach worms (*Ostertagia circumcincta, O. occidentalis, O. trifurcata, Marshallagia marshalli* and *Teladorsagia davtiani*), a thread-necked worm (*Nematodirus maculosus*), a pin worm (*Skrjabinema ovis*), a whipworm (*Trichuris ovis*), and a lungworm (*Protostrongylus stilesi*). In addition, they found adult fringed tapeworms (*Thysanosoma actinioides*) and the larval stage of the cestode *Taenia hydatigena.*

Brandborg (1955) reported endoparasites in Idaho and Montana not found by Kerr and Holmes (1966). These included lungworms (*Muellerius minutissimus*), eggs of stomach worms (*Tricholstrongylus* sp.), threadworms (*Strongyloides* sp.), lancet and liver flukes (*Dicrocoelium dendriticum* and *Fasciola hepatica*), and tapeworms (*Monezia expansa* and *Thysaniezia giardi*).

Most of the endoparasites found in mountain goats in Alberta probably are not harmful to the host except when heavy infestations are coupled with adverse environmental conditions (Kerr and Holmes 1966). However, the lungworm (*P. stilesi*), an important parasite of both wild and domestic animals, often causes physical damage to lung tissues. In addition, it is often accompanied by bacterial infections that result in pneumonic zones about the parasites. Verminous pneumonia may ensue, resulting in the death of the animal. No obvious difference was noted in parasite loads between males and females, but the intensity of infection did vary with age. Kids usually had much lower parasite numbers, and one two-week-old kid was found to be free of infestations.

Large numbers of lungworms (*P. stilesi* and *P. rushi*) were found in the 28 mountain goats examined in South Dakota (Richardson 1971). Cooley (1977) found that 60.6 and 78.4 percent of the mountain goats he examined were infected with lungworms (*Protostronglyus sp.*) in Colorado and Alberta, respectively. The frequent occurrence of parasitism is an indication that goats are injesting highly infested herbaceous vegetation and that there is repeated overutilization of vegetative species by goats. Terrestrial snails (*Vallonia gracilicosta* and *Succinea stretchiana*) are the intermediate host for both varieties of lungworm found in goats from the Black Hills. The former is the most common of 13 species of snail in the area.

Disease. Contagious ecthyma (CE), first found in mountain goats in the Kootenay National Park in British Columbia, Canada, is a viral disease most often associated with domestic sheep and goats. Common names for CE include sore mouth, contagious pustular dermatitis, scabby mouth, infectious labial dermatitis,

and orf. Samuel et al. (1975) stated that CE "is characterized by proliferative, crusted and sometimes pustular lesions of the lips, muzzle and occasionally the udder, feet and vulva." CE infections are most commonly found in kids and are generally associated with poor body condition and feeding difficulties.

Artificial salting areas may be important in the transmission of this disease, as all of the infected herds frequently visited these areas. The number of infections declined when salt blocks were removed (Samuel et al. 1975).

Hebert and Cowan (1971) found white muscle disease, a potential cause of mortality in mountain goats, in the southeastern portion of British Columbia, Canada. This disease commonly occurs in domestic calves and lambs whose mothers, during gestation, ingested foods grown in areas where sulfur and selenium compounds compete during absorption across plant cell membranes. As a result, the selenium may become less active biologically and the animal may appear to be selenium deficient. The symptoms of white muscle disease in mountain goats are varying degrees of edema and hemorrhage of muscle tissue, high levels of serum transaminase, and lesions in the cardiac and skeletal muscles (Herbert and Cowan 1971). While no evidence of muscular disorders was noted, Hebert and Cowan suggested that the combination of stress and physical exhaustion, such as from harrassment by predators or hunters, could accelerate the symptoms and lead to death.

Paratuberculosis is a specific infectious enteritis important to the livestock industry of northern North America and portions of Europe. While this disease has previously been reported in a variety of captive wild animals, it has only recently been documented in free-ranging bighorn sheep (*Ovis canadensis*) and a mountain goat from the Mt. Evans area in Colorado. Because of the possibility of transmission to other wild free-ranging ruminants and domestic livestock, paratuberculosis is a potentially important disease (Williams et al. 1979).

Brandborg (1955) noted three animals infected with the fungus actinomycosis (*Actinomyces israeli*). Two of these cases were associated with tooth abscesses and the third was actinomycotic foot rot. He also noted one mountain goat with pasteurellosis and coccidiosis infections and another with an abscess on a hind leg, but found little other evidence of disease in goats transplanted in Montana.

Other diseases found in goats included pseudotuberculosis and a form of jaw and hoof abscess sometimes called lump or lumpy jaw (MacGregor 1977).

Under present population conditions, however, disease probably is not an important factor in mountain goats.

AGE DETERMINATION

At birth, kids have three pairs of milk incisors and three upper and lower premolars visible under a thin layer of protective tissue from which they erupt within a few days. A few weeks later the milk canines erupt, and the kid has a full complement of deciduous teeth. The first permanent molars begin to erupt near the end of the summer. These molars are nearly fully erupted by midwinter, when the second molars have begun to emerge (Brandborg 1955).

At 15–16 months of age, the first medial pair of deciduous incisors are being replaced by permanent incisors. The second and third premolars show slight wear, and the fourth premolars show little wear. At about 16 to 17 months, the deciduous premolars are replaced by permanent ones, and the second lower molars are nearly fully erupted. The third molar also begins to erupt at this time. Goats examined between the ages of 26 and 29 months still retained the outer two pairs of deciduous incisors and the pair of deciduous canines; however, the second pair of permanent incisors were erupting beneath the milk teeth. At this time, the permanent second premolars are erupting and the third and fourth premolars are nearly fully erupted. The first cusp of the third molar is just beginning to erupt.

Three-year-old mountain goats have two pairs of permanent incisors and one pair each of deciduous incisors and canines in addition to a full complement of fully erupted permanent premolars and molars. The lateral pairs of permanent incisors and permanent canines erupt during the fourth summer. Full permanent dentition is attained at this time.

Extreme wear, loss, or malocclusion of the teeth in older goats may adversely affect health. In goats over eight years old, teeth are occasionally missing and severe wear is evident. The crowns of the incisors in some older animals are often worn off completely. The narrow teeth of the mandible are more susceptible to erosion than the wider, maxillary toothrow. This unequal wear on the occluding surfaces of the upper and lower toothrows often results in a series of alternating depressions and protrusions that interfere with chewing (Brandborg 1955).

Annual growth rings on the horns are useful in estimating the age of mountain goats and for indicating conditions under which the animal lived. Trefethen and Miracle (1972) attributed the formation of these rings to the softening of the horns by the oily exudate excreted from the horn glands during the rut. Brandborg

(1955) suggested, however, that the narrow rings resulted from a recession of growth due to marginal food conditions in winter in conjunction with increased sexual activity during the rut. The first annular ring does not appear until the second winter. This ring is not as distinct as subsequent rings and most often appears as a slight wrinkling on the horn's surface.

Lentfer (1955) suggested age groupings based on horn length. These groups were defined as kids, yearlings, two year olds, and three year olds or older. Although there was some overlap, age determination by horn length was quite accurate through two years.

At birth, a kid's horns are not apparent, but the areas from which they grow appear as dark protuberances slightly more than 1 cm in diameter. A few days after birth, a small knob appears in the center of these areas (Anderson 1940), and at about two weeks of age, small, round knobs about 0.5 cm in length are present (Lentfer 1955). By late fall, the horns of juveniles are from 2.5 to 6.5 cm in length. Beyond this period, horn growth is much slower and by the end of the first year, the horns are almost 10 cm long. At the end of the second and third years, horn lengths are from 14 to 16.5 cm and 19 to 23 cm, respectively. After the third year, horn growth slows, and there is usually less than 6 cm additional growth during the remainder of the animal's life (Anderson 1940; Brandborg 1955).

ECONOMIC STATUS

Mountain goats were not greatly exploited by Indians and early settlers because of the inaccessibility and remoteness of their habitat. Indians utilized mountain goats as a source of food and made blankets from the fine underwool; Indians of the Alaska coast utilized hides as breast armor (Trefethen and Miracle 1972; Geist 1964).

Currently, goats are mainly considered a trophy species and numbers of animals harvested have been low. A summary of the mountain goat harvest in North America from 1972 to 1976 is included in table 50.4. Even though the number of hunters in any one area is not great, demand for hunting supplies, lodges, and hunting guides does contribute to the economics of the

TABLE 50.4. Summary of the mountain goat (*Oreamnos americanus*) harvest in North America, 1972–76

State or Province	1972	1973	1974	1975	1976
British Columbia	1,184	1,412	859	1,057	884
Alaska	630	822	619	569	392[a]
Washington	253	266	272	238	291
Montana	234	280	306	237	[a]
Idaho	152	128	121	102	90[a]
Colorado	season closed	12	18	35	34
Alberta	14	13	28	29	29
Yukon Territory	54	42	30	25	17
Northwest Territories	(5–10 annually by residents, 46 total from 1965 to 1975 by nonresidents)				
South Dakota	season closed	12	season closed	season closed	14

SOURCE: Johnson 1977a.
[a] Incomplete return.

community. Monies generated by nature photographers and tourists who visit Colorado's mountain ranges are significant (Denny 1977). Competition between mountain goats and other game species or livestock is uncommon and no detrimental economic effects are noted.

MANAGEMENT

The major management tool for mountain goats is population control through hunting, protection of habitat, and preventing overharvest.

Hunting and Harvest. Hunting restrictions began about the turn of the century in most states and provinces. In 1897, Washington limited hunters to a three-month season with a bag limit of two goats (Johnson 1977b). Montana established a season in 1905 with a one-goat bag limit (Foss and Rognrud 1971). In the same year, British Columbia began charging a $100 fee for nonresident licenses and established a bag limit of five (Foster 1977). Generally low demand for mountain goat meat and trophies, and the inaccessibility of most ranges and hunting restrictions, as liberal as they were, apparently prevented the overexploitation of most herds for a number of years.

In recent years, increased road access, mining activity, timber harvesting, and demand for the species has led to the overexploitation of many herds. Hebert and Turnbull (1977) reported that harvest systems for goats, instituted under traditional biological reasoning appropriate to other big game species, may be adversely affecting goat populations in northern British Columbia. Declines related to overharvest have been reported in the East Kootenay, British Columbia (Phelps et al. 1976), in Idaho's Pahsimeroi herd (Kuck 1977), in southern Yukon Territory (Hoefs et al. 1977), and in the Wallowa Mountains of Oregon (Coggins unpublished data). Hebert and Turnbull (1977) and Kuck (1977) indicated that hunter harvests were probably additive rather than compensatory, as expected. When herds were exploited, they progressively selected steeper winter ranges. Traditional concepts of stimulating the production of young by increasing forage supplies did not prove to be valid.

Kuck (1977) found that, on the Pahsimeroi (Idaho), when dominant animals were removed from preferred cliffs by hunters, these voids were filled by subordinate animals from adjacent but shallower cliffs. This recessional behavior by the Pahsimeroi mountain goat herd therefore precludes the resting of available range resources and the corresponding stimulation of young normally expected following exploitation.

Most states and provinces have adopted more stringent regulations for goat hunting. The most satisfactory approach seems to be the setting of harvest quotas by individual herds. Goats of either sex are legally taken in all states and provinces except in the Yukon Territory, where nannies with kids are protected (Johnson 1977a). Hebert and Turnbull (1977) suggested that hunter harvests on coastal British Columbia herds should not exceed 4 percent of the total goat population and that regulations and harvest regimes should be based on the ecotype. Productivity varies considerably between goat herds, and hunting regulations need to be based on current local population trends. Also, because goats taken by hunting are additive to other mortality factors in some herds, conservative harvest quotas should be set.

Special management considerations may be necessary in recreational areas experiencing increased visitation by backpackers, photographers, and other nature enthusiasts. Bansner (1976) felt that the recent influx of humans into mountain goat range has created an artificial atmosphere where goats have become desensitized to people and often associate them as a source of salt. In an effort to restore a more natural human–mountain goat relationship, she recommended the implementation of an educational program dealing with goat ecology and behavior and a reduction of the amount of salt on or near areas where human-goat interactions occur, such as near campsites and along trails.

Census. Most states rely on fixed-wing aircraft or helicopters for goat surveys and composition counts. Because of the difficulty in determining the sex of animals, ratios are usually expressed as kids or yearlings per 100 adults. In Oregon, summer production count flights are generally made during late July or August and a winter count flight in January and February, using a Piper Supercub (150-hp engine). Winter counts are usually conducted at least two days following the last snowfall, so tracks and areas of feeding activity can be located. This greatly aids in the location of goats from the air. Known wintering areas are overflown at regular elevational contours to provide as complete coverage as possible.

Kuck (1977) reported that the helicopter has been Idaho's principle inventory tool and that most aerial censuses have taken place in winter during periods of peak concentrations. Restricted by time and funding, most mountain goat inventories have been irregular and incomplete. Hebert (1978) recommended that detailed surveys are best accomplished in specific mountain ranges or blocks that are representative of larger land units. He suggested preliminary surveys using fixed-wing transect surveys if the land base is large ($2,500+$ km^2), the population distribution uneven, or the density varied. Fixed-wing transect surveys should be undertaken at 150–500-meter intervals in subalpine and alpine areas depending on the area and number of mountain ranges or blocks to be surveyed and the terrain type. Surveys should provide information on total numbers, distribution, elevational and seasonal density, age ratios (kids, yearlings, possibly subadults, and adults), sex ratios (adult male to adult females), habitat types, and physical features (elevation, aspect, slope, terrain type, or topographic-moisture regime, etc.). Ground surveys should accompany detailed aerial surveys (on a representative mountain block basis) in order to complete data collection groups, sex and age ratios, and especially more accurate identification of yearlings and subadults.

Chadwick (1974) recommended that censuses be flown two hours before sunset following hot, clear summer days. Mornings were not as good, as goats were often active during bright moonlit nights. Goats in the Wallowa mountains of Oregon are very shy of aircraft and frequently attempt to hide from the survey plane.

In the area of Lake Chelan, Washington, a goat count is made by boat during January when goats are visible just below snowline (Johnson 1977*b*). In Olympic National Park, counts are conducted on the ground by hiking along preplanned census routes.

Most states and provinces also use questionnaires or harvest reports to determine the annual take of goats. Questions on the number of goats seen by hunters are frequently included on the questionnaire.

Transplant programs are also an important aspect of mountain goat management. These programs have increased goat distribution both within states where populations previously existed and in states outside the historic range. Introductions should be preceded by intensive feasibility studies to determine what physiological and economic impact will result. The feasibility study is perhaps the most important aspect of any major wildlife management activity. These studies should be completed prior to the implementation of a project and are generally necessary for the acquisition of authorization and funding.

The introduction of a feasibility study should clearly state the reasoning behind and the objectives of the proposed project. A thorough literature review of the life history of the species to be introduced should follow. Emphasis should be placed upon reputable studies conducted within habitat types similar to where the introduction is to be made and especially those within the same vicinity. Should particular aspects of the life history review be vague or incomplete, it may be desirable to initiate a research program before proceeding. The purpose of including the literature review within the feasibility study is to insure that biologists have current information and to provide administrators and legislators with a good background to decide on the project.

A detailed description of the area where the introduction is to take place should follow the literature review. This section should also give information regarding food resources, shelter, escape cover, water supply, and other such factors for both summer and winter ranges. This should include both the positive and negative effects that apply to the introduced species.

Other important areas to consider when determining the feasibility of an introduction are competition for food and shelter with other wildlife species and domestic stock; predation; disease and parasites; positive and negative economic impacts, including the opinions of local landowners and possible conflicts of interest; accessibility to the area (for ease of releasing animals and their protection once released); a complete summary of cost estimates projected to the date of implementation; and an initial 5–10-year management prospectus.

While the compilation of a complete and thorough feasibility study is a very time-consuming and difficult task, it may eliminate extracostly and poorly planned projects. For a more complete outline of topics to consider when developing a feasibility study, see Ripley (1971).

CURRENT RESEARCH AND MANAGEMENT NEEDS

In a survey of 28 northwestern North American biologists, Eastman (1977) reported that developing inventory techniques, the impact of hunting on population dynamics, and methods for predicting the carrying capacity of ranges were the three top-priority areas for mountain goat research. Home range and the role of tradition, migration and movements, the impact of forestry/mining/settlement/hydro, descriptive habitat utilization, and the impact of access are also important research and management areas.

Johnson (1977*a*) reported that most management agencies expressed concern with accessibility, harassment, and habitat destruction as a result of a spreading network of roads associated with logging and mining. Most states and provinces indicated that they lacked good information on mountain goat population dynamics as well as movement and migration patterns. It is hoped that the recent First International Mountain Goat Symposium will stimulate the interest and monetary considerations necessary to answer some of these questions.

LITERATURE CITED

Anderson, N. A. 1940. Mountain goat study/progress report. Bio. Bull. 2. 21pp.

Bailey, V. 1936. The mammals and life zones of Oregon. North Am. Fauna 55. 416pp.

Banfield, A.W.F. 1977. The mammals of Canada. Univ. Toronto Press, Toronto. 438pp.

Bansner, U. 1976. Mountain goat-human interactions in the Sperry-Gunsight Pass area, Glacier National Park. Univ. Montana, Missoula. 46pp. Mimeogr.

Brandborg, S. M. 1955. Life history and management of the mountain goat in Idaho. Idaho Dept. Fish Game. Wildl. Bull. 2. 142pp.

Chadwick, D. H. 1974. Mountain goat ecology-logging relationships in the Bunker Creek drainage of western Montana. M.S. thesis. Univ. Montana, Missoula. 262pp.

Cooley, T. M. 1977. Lungworms in mountain goats. M.S. Thesis. Colorado State Univ., Fort Collins. 175pp.

Dalrymple, B. 1970. Complete guide to hunting across North America. Harper & Rowe Publishers, Inc., New York. 848pp.

Denny, R. N. 1977. The status and management of mountain goats in Colorado. 1st Annu. Symp. Mountain Goats. 15pp.

Eastman, D. S. 1977. Research needs for mountain goat management. 1st Annu. Symp. Mountain Goats. 9pp.

Foss, J. A. 1962. A study of the Rocky Mountain goat in Montana. M.S. Thesis. Montana State College, Bozeman. 26pp.

Foss, J. A., and Rognrud, M. 1971. Rocky Mountain goat. Pages 106–113 *in* T. W. Mussehl and F. W. Howell, eds. Game management in Montana. Montana Fish Game Dept., Missoula. 238pp.

Foster, B. R. 1977. Historical patterns of mountain goat harvest in British Columbia. 1st Annu. Symp. Mountain Goats. 14pp.

Geist, V. 1964. On the rutting behavior of the mountain goat. J. Mammal. 45(4):551–568.

———. 1967. On fighting injuries and dermal shields of mountain goats. J. Wildl. Manage. 31(1):192–194.

Harmon, W. H. 1944. Notes on mountain goats in the Black Hills. J. Mammal. 25(2):149–151.

Hebert, D. M. 1978. A systems approach to mountain goat management. Proc. 1978 Northern Wild Sheep and Goat Conf. 17pp.

Hebert, D. M., and Cowan, I. M. 1971. White muscle disease in the mountain goat. J. Wildl. Manage. 35(4):752–756.

Hebert, D. M., and Turnbull, W. G. 1977. A description of southern interior and coastal mountain ecotypes in British Columbia. 1st Annu. Symp. Mountain Goats. 21pp.

Henderson, R. E., and O'Gara, B. W. 1978. Testicular development of the mountain goat. J. Wildl. Manage. 42(4):921–922.

Hibbs, L. D. 1967. Food habits of the mountain goat in Colorado. J. Mammal. 48(2):242–248.

Hjeljord, O. G. 1971. Feeding ecology and habitat preference of the mountain goat in Alaska. M.S. Thesis. Univ. Alaska, Fairbanks. 126pp.

Hoefs, M.; Lortie, G.; and Russell, D. 1977. Distribution, abundance and management of mountain goats in the Yukon. 1st Annu. Symp. Mountain Goats. 7pp.

Johnson, R. L. 1977a. Distribution, abundance and management status of mountain goats in North America. 1st Annu. Symp. Mountain Goats. 7pp.

———. 1977b. Status and management of the mountain goat in Washington. 1st Annu. Symp. Mountain Goats. 6pp.

Kerr, R. G. 1965. The ecology of mountain goats in West Central Alberta. M.S. Thesis. Univ. Alberta, Edmonton. 96pp.

Kerr, R. G., and Holmes, J. C. 1966. Parasites of mountain goats in West Central Alberta. J. Wildl. Manage. 30(4):786–790.

Krog, H., and Monson, M. 1954. Notes on the metabolism of a mountain goat. Am. J. Physiol. 178(3):515–516.

Kuck, L. 1977. The impacts of hunting on Idaho's Pahsimeroi mountain goat herd. 1st Annu. Symp. Mountain Goats. 12pp.

Lentfer, J. W. 1955. A two year study of the Rocky Mountain goat in the Crazy Mountains, Montana. J. Wildl. Manage. 19(4):417–429.

MacGregor, W. G. 1977. Status of mountain goats in British Columbia. 1st Annu. Symp. Mountain Goats. 5pp.

O'Connor, J., and Goodwin, G. G. 1961. The big game animals of North America. E. P. Dutton & Co., Inc., New York. 264pp.

Olmsted, I. 1977. Interrelationships of introduced mountain goats and subalpine habitat in the Olympic National Park. 1st Annu. Symp. Mountain Goats. 1p.

Peck, S. V. 1972. The ecology of the Rocky Mountain goat in the Spanish Peaks area of Southwestern Montana. M.S. Thesis. Montana State Univ., Bozeman. 54pp.

Phelps, D. E.; Jamieson, B.; and Demarchi, R. A. 1976. Mountain goat management in the Kootenays. Part 1, The history of goat management. Part 2, A goat management plan, 1975–1985. British Columbia Fish Wildl. Branch Rep.

Richardson, A. H. 1971. The Rocky Mountain goat in the Black Hills. S. Dakota Dept. Game, Fish and Parks Bull. 2. 25pp.

Rideout, C. B. 1974. Comparison of techniques for capturing mountain goats. J. Wildl. Manage. 38(3):573–575.

Ripley, T. H. 1971. Planning wildlife management investigations and projects. Pages 5–12 in R. H. Giles, ed. Wildlife Management Techniques. Wildlife Society, Washington, D.C.

Samuel, W. M.; Chalmers, G. A.; Stelfox, F. G.; Foewen, A.; and Thomsen, J. J. 1975. Contagious Ecthyma in bighorn sheep and mountain goats in western Canada. J. Wildl. Dis. 11(1):26–31.

Saunders, J. K., Jr. 1955. Food habits and range use of the Rocky Mountain goat in the Crazy Mountains, Montana. J. Wildl. Manage. 19(4):429–437.

Smith, B. L. 1976. Ecology of Rocky Mountain goats in the Bitterroot Mountains, Montana. M.S. Thesis. Univ. Montana, Missoula. 203pp.

Swift, L. W. 1940. Rocky Mountain goats in the Black Hills of South Dakota. Trans. North Am. Wildl. Conf. 5:441–443.

Trefethen, J. B., and Miracle, L. 1972. The new hunters encyclopedia, 3rd ed. Galahad Books, New York. 1054pp.

Vaughan, M. R. 1975. Aspects of Mountain goat ecology, Wallowa Mountains, Oregon. M.S. Thesis. Oregon State Univ., Corvallis. 113pp.

Vaughan, T. A. 1972. Mammalogy. W. B. Saunders Co., Philadelphia. 463pp.

Williams, E. S.; Spraker, T. R.; and Schoonveld, G. G. 1979. Paratuberculosis (Johne's Disease) in bighorn sheep and a Rocky Mountain goat in Colorado. J. Wildl. Dis. 15(2):221–227.

RONALD A. WIGAL, Appalachian Environmental Laboratory, Center for Environmental and Estuarine Studies, University of Maryland, Frostburg State College Campus, Frostburg, Maryland 21532.

VICTOR L. COGGINS, Oregon Department of Fish and Wildlife, Route 1, Box 228E, Enterprise, Oregon 97828.

51

Muskox

Ovibos moschatus   Anne Gunn

NOMENCLATURE

COMMON NAME. Muskox
SCIENTIFIC NAME. *Ovibos moschatus*
SUBSPECIES. *O. m. wardi*, high arctic muskoxen; and
O. m. moschatus, mainland Canadian muskoxen.

Taxonomy. The earliest names given by northern explorers to the muskox (musk or polar cattle, musk or arctic bison) emphasize its superficial resemblance to members of the genus *Bos*. Although the generic name also refers to sheeplike as well as cattlelike traits, the muskox is now classified in the tribe Ovibosini, of the subfamily Caprinae. The closest living relative is the golden takin (*Burdorcas taxicolor*), a goatlike mammal of the Himalayas.

Allen (1913) originally described three subspecies, but a more recent taxonomic study (Tener 1965) recognized only two subspecies (*O. m. wardi* of the High Arctic and *O. m. moschatus* of continental North America). Tener (1965) described the tendency of *O. m. wardi* to have whiter faces, saddles, stockings, and horns, as well as longer toothrow and styles on the upper molars, than *O. m. moschatus*. The differences are relatively slight, however. The separation of the two subspecies took place during the Wisconsin glaciation: *O. m. wardi* evolved in the refugia of northern Ellesmere and Greenland and *O. m. moschatus* evolved south of the Wisconsin icesheet (Harington 1961).

Muskoxen probably originated during the Late Miocene on the cool grasslands of central Asia. As the various muskox genera twice successively radiated out and retreated from central Asia into Asia, Europe, and Alaska, they were associated with steppe or tundra habitats. *Ovibos* evolved on the tundra of central Siberia in the Late Pliocene or Early Pleistocene and, although there was a primary invasion across to Alaska in the Late Nebraskan, the successful invasion of *Ovibos moschatus* into Alaska occurred during the Late Illinoian.

DISTRIBUTION

Muskoxen are presently found on the arctic tundra of the North American mainland, most of the arctic is-

FIGURE 51.1. Distribution of the muskox (*Ovibos moschatus*).

lands, and Greenland (figure 51.1). In the past, muskoxen were more widely distributed. During the ice ages they were found as far south as Kansas. However, as the ice and the peripheral tundra withdrew northward, so did the muskoxen (Hone 1934; Harington 1961). Muskoxen were present in northern Alaska, although not apparently numerous, prior to their extirpation in 1850-60 (Hone 1934). A herd introduced from Greenland to Nunivak Island in 1935 thrived (Spencer and Lensink 1970) and subsequent introductions were made from Nunivak Island to mainland Alaska; the current Alaskan population is estimated at 1,000 (Lent 1978). The Nunivak Island population reached 750 in 1968, but removal for transplants and sport hunting maintain the current population at about 500.

On the Canadian mainland, prior to the mid-19th century, muskoxen were found north of the treeline from the Mackenzie River east to the coast of Hudson Bay. As in Alaska, the introduction of firearms and the demands of fur traders and overwintering whalers combined to lead to the inexorable decline of muskoxen (Hone 1934; Tener 1965; Burch 1977). Concern

1021

over the future of muskoxen led to the establishment of the Thelon Game Sanctuary in 1927, specifically to protect muskoxen. By 1930, however, Anderson (1934) estimated that the numbers had dropped to 500 on the mainland. The protective measures together with the exodus of natives from the land into settlements were effective, and once more muskoxen inhabit many areas where they were plentiful in the past.

A lack of adequate surveys precludes reliable estimates of numbers; Tener (1965) estimated 1,500 muskoxen on the Canadian mainland, and the population has continued to increase since then. V. Hawley (personal communication) estimated 2,000 north of Great Bear Lake in 1975. In 1978, Northwest Territories Wildlife Service personnel believed that there were no more than 600 muskoxen in the Thelon Game Sanctuary and that the population outside of the sanctuary was an expanding population of about 500 muskoxen (B. Stephenson personal communication). On the mainland coast, in the Queen Maud Migratory Bird Sanctuary, 3,520 adult muskoxen and 448 calves were actually counted in July and August 1979 (R. Decker personal communication) and the sanctuary population was estimated to be about 5,000.

The Queen Elizabeth Islands (the high arctic islands north of M'Clure Strait, Viscount Melville Sound, and Barrow Strait) were surveyed in 1961 (Tener 1963), and the western islands of the group were surveyed in 1972-74 (Miller et al. 1977). The population of muskoxen on the western Queen Elizabeth Islands was estimated to be 2,161 in 1961 (Tener 1963) and 2,700 in 1974 (Miller et al. 1977).

Differences in the methods of estimating numbers on the western Queen Elizabeth Islands prevent a direct comparison of Tener's (1963) results with those of Miller et al. (1977). The latter suggests that muskoxen have declined on Bathurst Island but are now recolonizing Prince Patrick and Mackenzie King islands. Southwestern Melville Island, in particular the Bailey Point area, is the heartland for muskoxen on the western Queen Elizabeth Islands.

Most of the muskox counts on islands south of Viscount Melville Sound have been on Banks Islands, where 15,000-18,000 were estimated in March 1980 (B. Stephenson personal communication) from surveys of the southern half of the island in 1979 and the northern half in 1980. Muskoxen likely were once abundant on Banks Island but were reduced to a remnant population by hunting practices of the Copper Eskimos from Victoria Island. The population subsequently has grown at an estimated rate of 10 percent a year (Urquhart 1973) for a few years at least, which may reflect the virtual absence of wolves. Little is known of muskoxen on Victoria Island, which is the largest arctic island and was not surveyed until 1980. Numbers of muskoxen are probably relatively high: during a survey of only 6 percent coverage in July and August 1980, about 12,000 muskoxen were estimated (R. Jakimchuk personal communication). On Prince of Wales Island, Fischer and Duncan (1976) estimated 900 muskoxen from 185 observed. The island was surveyed again in July 1980 by the Northwest Territories Wildlife Ser-

vice, and 1,300 were estimated from 417 muskoxen counted. Hunting associated with the presence of whalers likely exterminated muskoxen on Somerset in the late 1800s. However, in 1976 about 15 muskoxen were seen by Canadian Wildlife Service field personnel and 30 muskoxen were counted by the Northwest Territories Wildlife Service in 1980.

Although muskoxen are found on Melville Peninsula, they are not currently known from Southampton Island (Harington 1961). On Baffin Island, the apparent absence of muskoxen has been related to snow depths commonly in excess of 76 cm (Harington 1961). On the mainland of Canada, the muskox distribution is largely in areas with mean annual maximum snow depths of 50 cm or less.

In Greenland the numbers of muskoxen are known to have fluctuated considerably (Vibe 1967). The current estimate is 1,500, situated mainly on the east coast (H. Thing in Lent 1978). The current population of muskoxen in Scandinavia is a remnant of several previous attempts at introductions that did not fare well, and probably now numbers 100 or fewer (Alendal in Lent 1978).

The distribution of muskoxen within the arctic regions is determined by climate, especially by snow depths. Pedersen, in Tener (1965), called the muskox a "dry climate species" to characterize its need for areas with low precipitation. Vibe (1967) correlated the declines in muskox numbers on Greenland with increased snowfall following changes in the flow patterns of the major offshore currents. On the Canadian mainland, extensions and retreats of the treeline are known. It is possible that the climatic phenomena responsible could also have contributed to fluctuations in muskox numbers. Although the arctic islands have an almost desert level of precipitation, years of heavy and/or late snowfall occur and their detrimental impact on muskox populations has been documented (Parker et al. 1975; Miller et al. 1977).

DESCRIPTION

Size. The thick, long coat of the muskox creates a misleading impression of massiveness. Relatively few body measurements and weights have been recorded for wild muskoxen. The average of 11 bulls was about 137 centimeters at the shoulder and 245 cm in length. Cows are smaller: the shoulder height of one cow was 123 cm and the average length of three cows was 199 cm (Tener 1965). The weights of four bulls from the Canadian mainland averaged 340 kilograms. Three bulls from Nunivak Island (Alaska) averaged 269 kg, with cows about 50 percent of the weight of bulls (Lent 1978).

Appearance. The compact appearance of the muskox is characterized by its shoulder hump, long brown hair falling almost to the ground, and sharp curved horns. The head is relatively short and broad and is generally held low on the short neck. From the shoulder hump the back slopes slightly to the hindquarters. The legs are short, stocky, and end in large, rounded hooves.

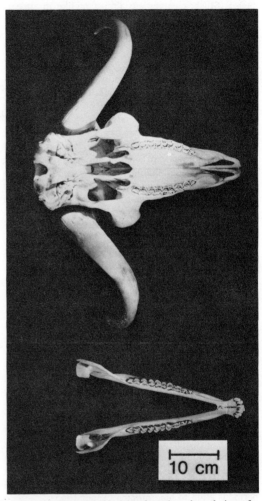

FIGURE 51.2. Skull of the muskox (*Ovibos moschatus*). Left, from top to bottom: lateral view of cranium, lateral view of mandible, dorsal view of cranium. Right: ventral view of cranium (above), dorsal view of mandible (below).

The outer rims of the hooves are hard and sharp, while the two lateral heels are softer; the combination results in the muskox's being a surprisingly swift and agile climber on snow or rock.

Pelage. The coat of the muskox is generally dark brown but some mainland animals tend to be almost black. The lower legs have distinct "stockings" of light brown to white and the face and "saddle" are white. White saddles are particularly distinctive on muskoxen from the High Arctic and Greenland. Albino muskoxen rarely occur: a white muskox cow with a normal-colored calf was observed on Melville Island in 1853 (McDougall 1857); a pale-colored calf was seen near Atkinson Point River on the Canadian mainland (Kuyt et al. 1972) and two pale-colored muskoxen were observed in 1979 during a survey of the Queen Maud Gulf Migratory Bird Sanctuary (R. Decker personal communication). The coarse guard hairs form a dense coat in which the individual hairs can exceed 60 cm in length. Under the chin the hairs are developed as a beard, and over the shoulder hump they form a distinct mane in the adult bulls. The short (10-14 cm) inconspicuous tail, the scrotum, the udder, the ears, and the head are all densely haired. Only a small patch between the nostrils and the lips is hairless.

The coarse guard hairs overlie an exceptionally effective insulating layer of fine brown wool that has remarkably long fibers, the finest wool grown by a mammal. In spring and early summer, the wool is shed in clumps that cling to the guard hairs, imparting a distinctive "moth-eaten" appearance before the mats of wool are blown away. New wool starts to grow in August and at the same time the guard hairs are renewed. Muskox calves are almost piglike in appearance, partly because the guard hairs are short and there is no mane, skirt, or beard. Yearlings start to acquire longer, coarse guard hairs but the pelage is not completely developed until three years of age.

Horns. The muskox horn is a fibrous, keratinous sheath laid down over a bone core, as in other bovids. The horns start to grow when a muskox calf is about four to five weeks old and they continue to grow until the muskox is about six years of age (Pedersen in Tener 1965). The horns of an adult bull are more robust and sweeping than those of a cow, but the most striking distinction is the enlargement and merging of the horn bases to form a massive, heavily ridged and furrowed boss in bulls. In cows, the boss is smaller and divided in the middle by a tuft of white hair. The horns darken with age (Tener 1965), and, in addition, may be stained brown from the habit of bulls horning banks and peat mounds in summer. The skull is massive; shallow antorbital pits are present (figure 51.2).

REPRODUCTION AND GROWTH

Muskox cows are probably seasonally polyestrous but there has been no research on reproductive physiology. Within a herd there appears to be no precise synchrony of estrus; breeding can occur in August and the beginning of September. The herd bull continually tests and tends the cows as they approach estrus (Smith 1977). It probably mates with most of, if not all, the receptive cows in a herd, except in large herds. In larger herds, the herd bull may become exhausted before the end of the rut from continually fending bulls away from the cows and tending the cows. The herd bull's exhaustion would allow another bull to assume the role and breed the cows that came into estrus later. The gestation period is eight to nine months (Tener 1965).

Most births probably occur in April and May, although newborn calves have been observed between late March and early June. Calves born in April and May have to contend with temperatures as low as −40° to −20° C, but they then also have the advantage of a relatively long growth period before their first winter. There are no published accounts of calving in the wild, but it is unlikely that the parturient cow leaves the protection of the herd to give birth. At birth, a muskox calf weights 10–14 kg (Lent 1978) and is relatively precocious. Lent (1974) observed calves standing 16–60 minutes after birth. He described birth and subsequent cow-calf relations in a comparative treatment of mother-infant relations in ungulates. The muskox calf is a "follower," not a "hider," and stays close to the cow for the first month. It gradually becomes more venturesome, but at the first hint of disturbance the calf will promptly return to its mother. Although calves begin to forage at two to three weeks of age, they are suckled by the cow possibly until after their first winter. Male muskoxen tend to grow at a slightly faster rate and for longer than females. Females reach adult weight in their fifth or sixth summer, compared to the sixth or seventh summer for males (Hubert 1977). Using captive muskoxen, Hubert (1977) described seasonal weight losses and gains, and in particular noted a weight loss during late winter and spring, despite the availability of fodder. A similar reduced metabolic rate in winter has been described for reindeer (Segal 1962). Hubert (1977) found that bulls and cows were in a negative energy balance for six and seven months a year, respectively. This emphasizes the importance of the relatively short summer to regain condition and build up energy reserves.

The extreme cold and severe winds, coupled with snow-covered forage, impose a severe test of the muskox's winter strategy of energy conservation. The compact conformation, with ears and tail buried in the thick, insulating layer of wool are obvious adaptations to reduce heat loss. Some of the physiological adaptations described for caribou (*Rangifer tarandus*), such as water conservation and urea recycling, are likely common to muskoxen. The reduced forage intake conserves energy, as foraging has high energetic costs. The costs of foraging include the heat required to warm forage to body temperature, the cost of cratering in snow, and the heat loss associated with the increase in body surface area of a standing (or moving) muskox compared to a bedded one. Below a critical temperature of about −40° C, muskoxen have to increase their metabolic rate to maintain their body temperatures. As the cost of foraging prohibits the strategy of increasing forage intake to compensate increased maintenance costs, the muskoxen have to use behavioral adaptations to minimize the effects of cold. Muskoxen will lie with their backs or sides to the wind and choose sheltered valleys or slopes during storms. Most observers have commented on the slow and deliberate movements of muskoxen when undisturbed: slow movements conserve energy in winter and reduce the likelihood of overheating in summer.

The same adaptations that minimize heat loss in winter can cause a tendency to overheat in summer, even with the early summer molt of the insulating wool. Allen (1913) described captive muskoxen panting during hot weather. Wild muskoxen will bed on lingering snow patches in preference to bare ground.

ECOLOGY

Habitat. Muskoxen in the High Arctic are usually found on well-vegetated sedge (*Carex* spp.) slopes on low-elevation coastal sites and valleys of watercourses (Miller et al. 1977; Russell et al. 1978). Little of the arctic landmass is well vegetated: bare ground, gravel, and rock predominate and the hydric meadow habitat preferred by muskoxen is restricted to poorly drained areas and areas of late snow cover. There are a few areas where fohn-type winds effect an ameliorative influence (Courtin and Labine 1977) and vegetation growth is more lush and the growing season possibly longer. Three such areas are known: (1) Truelove lowlands, Devon Island, (2) Fosheim Peninsula, Ellesmere Island, and (3) Bailey Point, Melville Island. These areas may be recognized as muskox heartlands.

On the mainland, the vegetation of muskox habitat is most often willow (*Salix* spp.) and birch (*Betula* spp.) thickets associated with sedges, grasses, and forbs. Muskoxen are rarely found south of the treeline. On Nunivak Island the vegetation is tundra type, but more lush and abundant (Spencer and Lensink 1970). The winter range of beach rye grass (*Elymus arenarius*)–covered sand dunes is a unique muskox habitat.

Over most of the range the climate is characterized by extremely cold, long winters; short, cool summers; and low precipitation. By contrast, Nunivak Island has considerably higher precipitation and warmer winters (Spencer and Lensink 1970). In the High Arctic, snow covers the ground for about 10 months, compared to about 8 months on the Canadian mainland. Snowfall is usually less than 50 cm in the High Arctic and about 100–130 cm on the mainland (Tener 1965).

Movements. Muskoxen are relatively sedentary animals with limited movements between winter and summer ranges. Movements of muskoxen are limited, usually no more than 1–10 km per day (Hone 1934; Parker and Ross 1976; Miller and Gunn 1979). Seasonal movements from coastal areas along drainages to the interior in summer have been observed on some of the western Queen Elizabeth Islands (Miller et al. 1977), but those movements probably do not often exceed 50 km.

Interisland movements across the sea ice occur (Miller et al. 1977). Freeman (1971) described muskox tracks crossing Jones Sound, and between Graham Island and the west coast of Ellesmere Island. The cause of such movements is unknown but it is unlikely that they are frequent and regular.

The lack of studies on marked animals limits the understanding of herd affinity to their ranges. Observations of three muskox groups on Prince of Wales Island in 1976–79 suggest that the groups did demonstrate affinities to particular drainage systems (Miller and Gunn 1979).

POPULATION DYNAMICS

The present data are inadequate to describe muskox population dynamics quantitatively. In fact, almost all the information available to date has been collected during aerial surveys. However, some data on productivity and sex and age composition have been collected during ground surveys of the Nunivak Island population and from captive muskoxen.

Productivity. Lent (1978) corrected the tendency to regard muskoxen as having one of the lowest productivities of any ungulate species. That belief arose from the suppositions that cows could not breed until four years of age and then could only calve in alternate years. Only the generalization regarding single births is valid, as twinning is unrecorded. In Norway, two-year-old cows have bred (Alendal 1976) and on Nelson Island, Alaska (Lent 1978), three-year-old cows breed occasionally, and even two year olds on Nunivak Island (Lent 1978) and in Greenland (Pederson in Tener 1965). Miller and Gunn (1979) have observed some cows breeding in three successive years in the High Arctic. The age of breeding and whether breeding occurs in successive or alternate years appear to depend on the condition of the animals and, possibly, on social factors.

Muskoxen released on Nunivak Island in 1935 increased at a relatively steady annual rate of 16 percent from 1947 to 1968, with an average annual calf increment of 19 percent, as percentages of total numbers of animals segregated (Spencer and Lensink 1970). By contrast, in the High Arctic, calf production and rearing show marked fluctuations, and the survival rate to one year of age is sometimes zero. There is a correlation between snow and ice conditions, and hence forage availability, and productivity and mortality. For example, on eastern Melville Island, muskox calf crops in July–August were 10.8 percent in 1972, 16.6 percent in 1973, and 5.6 percent in 1974 (Miller et al. 1977). These percentages reflect the variation in the severity of the three winters: 1971/72 was moderately severe; 1972/73 was favorable; and 1973/74 was severe to catastrophic. On Bathurst Island, there was apparently no calf production 1968–70 (Gray 1973) and probably no calf survival in 1974 (Miller et al. 1977). There is little evidence as to how a low nutritional plane reduces or suppresses productivity. Gray (1973) noted little or no rutting activity on Bathurst Island during 1968–70, years with severe winter weather. After a severe winter, cows that do not build up sufficient fat reserves may fail to come into estrus, which may suppress the rutting behavior of bulls.

There are few published data of calf survival except from Nunivak Island. Lent (1978) considered that an average survivorship of 85 percent was possibly an underestimate because of the difficulty of seeing small calves during summer aerial surveys on Nunivak Island. Tener (1965) believed that calf survival was more than 50 percent on the Canadian mainland but less than 50 percent on the arctic islands. On Banks Island, surveys indicated calf survival of 72 percent from June 1970 to March 1971 and of 44 percent from September 1971 to July 1972 (Urquhart 1973).

Hubert (1977) tabulated the numbers of calves and yearlings counted on northeastern Devon. A comparison of calves and yearlings to other muskoxen seen in August 1970 to May 1971, 1972, and 1973 suggests that the survival of calves over their first winter was 33 percent in 1970/71, 78 percent in 1971/72, and 75 percent in 1972/73. Northeastern Devon is particularly productive because of its microclimate. The survival of yearlings may be lower than that of calves because the cow-calf bond protects and sustains the calf. The yearling is not protected and ranks lowest in the social hierarchy. During winter, when competition is greater at crater sites in the snow, yearlings would be at a disadvantage.

Population Structure. There is variation among different areas in the sex ratio of muskox herds, but the only substantive data on the age structure of different populations are from Nunivak Island. An equal ratio of bulls to cows was reported following a period of stress from severe winters (Gray 1973).

Tener (1965) found more bulls than cows in the Hazen Lake area (Ellesmere Island), an area that is favorable to muskoxen, as did Spencer and Lensink (1970) in the expanding Nunivak Island population. In other areas, Tener (1965) found that the ratio favored cows, as did Hubert (1977) and Miller and Gunn (1979). The comparison of sex ratios between areas suggests that under more severe conditions, especially winter weather, bulls may be more vulnerable than cows. The activities of bulls during the rut likely deplete their fat reserves before the onset of winter.

FOOD HABITS

The food habits of muskoxen have been among the most studied aspects of their biology. On the arctic islands Tener (1965), Wilkinson et al. (1976), Parker and Ross (1976), and Russell et al. (1978) have studied food habits by documenting the use of range types and by direct observations. Parker (1978) found from rumen and fecal sample analyses that muskoxen prefer a sedge-willow diet in summer and sedges in winter. If snow conditions are unfavorable, the muskoxen move to ridges and slopes to feed on willow, grasses, and forbs. Parker (1978) discussed the relationship between malnutrition and winter intake of willow.

Differences in the diets of muskoxen on different stands reflect the differences in snow cover and plant communities, especially between the Canadian mainland and islands. In the Thelon Game Sanctuary, the diet of muskoxen in late winter included a greater proportion of woody plants, probably because of the tendency of woody plants to grow on more exposed hummocks. Also, the snow depths are greater than on the arctic islands. On Nunivak Island, the most accessible vegetation is on the coastal sand dunes, and beach rye grass is the major species on the dunes and in the muskox winter diet (Lent 1978).

The muskoxen scrape deep or crusted snow to expose the vegetation with their front hooves and push shallow soft snow aside with their muzzles. They tend to feed more intensively on one area where the snow is absent. On Nunivak Island, winter range receives more intensive pressure than summer range (Spencer and Lensink 1970).

ANNUAL CYCLE AND SOCIAL BEHAVIOR

Muskoxen are social ungulates and spend most of their life in the mixed-sex herd, usually led by a bull, but sometimes by a cow. Bulls also occur in small (usually 2–5), single-sex groups and as solitary animals. The size and composition of the herds vary by season, range conditions, and the number of bulls in the population. On Melville Island, Miller et al. (1977) observed a decrease in average herd size from 17.2 in March–April to 10.0 in July–August 1973. After the rut (July–September), antagonism between bulls decreases, and the herd sizes increase as bulls and/or mixed groups join. In severe winters such as 1973–74, the largest winter herds may fragment in response to competition for limited forage.

Fragmentation of herds may also result from harassment, such as by vehicles and aircraft. The presence of wolves, however, can stimulate the temporary formation of large herds. In 1976, a herd of 110 muskoxen accompanied by nine wolves was seen on Melville Island, but five days later the muskoxen were in three herds (Miller et al. 1977) and the wolves were not observed.

Observations of three muskox herds on Prince of Wales Island during 1976, 1977, and 1978 suggest that, with the exception of bulls leaving and joining, herd composition can be relatively stable (Miller and Gunn 1979). In contrast, Gray (1973) described the apparent merging and splitting of herds on Bathurst Island, although he also noted the relative stability of herds in summer.

During late spring, bulls likely leave the herds and either form bull-only herds or spend the summer as solitary animals. The lone bulls have often been regarded as "senile" or surplus bulls (Hone 1934) but the accuracy of the classification remains speculative. Observations of marked and unmarked bulls (Jonkel et al. 1975; Smith 1977; Miller and Gunn 1979) revealed that some interchange took place among solitary bulls, bulls in herds, and herd leaders. Those interchanges often led to aggressive clashes during the early rut as herd bulls strived to maintain their position of dominance in herds, and to prevent other bulls from entering the herd or cows from leaving the herd.

Most published photographs of muskoxen portray them in one of their most unique and impressive behavioral patterns, the circular or crescent-shaped defense formation taken up in response to predators or other harassment. Originally, the defense formation probably evolved as protection against predation by wolves to allow offensive use of the horns while protecting the rear of the animal and the subadults in a herd. Approaching aircraft, ground vehicles, and humans on foot all can induce muskoxen to take up a defense formation, so the formation can be used as an indicator of stress from harassment (Gray 1974; Miller and Gunn 1979).

Vocalizations. Bulls will roar at each other and occasionally at cows, particularly during the rut. Lent (1974) noted that neonatal muskoxen vocalize. I have observed calves about two months of age bleating. Lent (1974) observed individual recognition between cow-calf pairs from vocalizations. No analyses of these vocalizations have been made, however.

Other Behaviors. The gathering of muskoxen into cohesive social groupings directs much of their behavior. The unique synchronized defense behavior has already been described, but synchrony is also apparent in the foraging and resting patterns of muskoxen, especially in smaller herds. Within a herd, leadership is provided by the dominant bull in times of stress. The lead bull is often the last to respond to approaching danger, and although often stationed to the side of a defense formation, it plays a key role in maintaining herd cohesiveness. The lead bull will charge both the predator and other muskoxen attempting to leave the formation. When a muskox herd is confronted by an obstacle such as a river or steep slope the lead bull usually chooses the route and is followed by the rest of the herd, often in single file. The cows in a herd probably also have a dominance hierarchy, and a cow will adopt the role of the lead animal if bulls are absent.

One of the most evident behavioral patterns associated with dominance and defensive displays is the rubbing of the postorbital gland on the inside of the animal's extended foreleg. The function of gland rubbing has been a source of speculation, but it may be the ritualized release of a pheromone associated with stress. Other muskoxen may perceive the behavior as a threat or warning.

A slow, deliberate walk is the usual pace of undisturbed muskoxen, but when disturbed they can break into a powerful gallop. Their agility and swiftness in turning are equally impressive. The sure-footedness and strength of their stocky legs is apparent by the ease with which muskoxen will scale steep rock or snow slopes. They are powerful swimmers and will enter water readily. Solitary muskox will stand in water when confronted by predators.

Comfort behavior includes shaking to rid the coat of adhering snow, ice, water, or biting and blood-sucking insects. On rising from their beds, muskoxen may stretch their back and neck, though the thick coat often obscures the movements. Grooming behavior is dominated by rubbing against rocks, banks, or even other individuals and is especially prevalent during the molt of the wool undercoat. Scratching is rare (Gray 1973) and licking has been observed only during cow-calf interactions.

Muskox calves are often playful. Gamboling, aggressive play including butting, and sexual play with mounting of other calves or even attempts to mount the maternal cow are all common sequences. Play is sometimes more structured, with several individuals participating in contagious play patterns such as "tag" and "king of the castle." Peat or earth mounds and banks and small running streams are sometimes attractive focal points to playful muskox calves.

MORTALITY

Weather is the dominating influence on muskox populations. The severity of winters and early springs affects not only productivity but also mortality. Forage availability can be restricted by deep snow, crusted snow, ice lenses, and ground-fast ice caused by freezing rain storms and brief thaws. In addition, snow in the High Arctic can cause the formation of ground-fast

ice in the spring (Miller and Gunn 1978). Forage unavailability results in malnutrition, eventually leading to death. The winter of 1973–74 was severe on the western Queen Elizabeth Islands and the muskox population declined by 35 percent, mainly due to mortality associated with malnutrition (Miller et al. 1977). Parker et al. (1975) examined winter-killed muskox carcasses on Bathurst Island in 1974; 19 of 22 femurs had a lipid content of 7 percent or less. The high mortality imposed by severe winters is likely responsible for the periodic disappearance of muskoxen from some arctic islands, such as Prince Patrick (Miller et al. 1977).

Great mortality inflicted by weather has occurred on Greenland. Vibe (1967) described a climatic shift to mild winters with deep snow and rain freezing as a crust on the snow. Muskoxen succumbed to malnutrition and some were also trapped in their beds by ice formation on their coats (Spencer and Lensink 1970). In the winter of 1976–77, deep snow and freezing rain probably caused the death of 100 animals on Nunivak Island. In severe winters on Nunivak Island, mortality was greater for females than for males, and it may peak in the spring. Miller et al. (1977) also found that on the arctic islands most winter mortality occurred in late winter to early spring. The springtime mortality may reflect the gradual depletion of fat reserves at the end of winter, or springtime ground-fast ice formation restricting food availability.

Predation by wolves (*Canis lupus*) is a significant cause of mortality. Again as with starvation, subadults and animals older than 10 years may be the most vulnerable age classes (Parker et al. 1975; Miller et al. 1977). Wolves are opportunistic predators, and a calf or yearling even briefly exposed outside the herd's protection is susceptible to a bold dash by a wolf. Solitary muskoxen, despite their formidable horns and rapidity of movement, may often be the prey of wolves. Wolves will cut off a muskox even briefly exposed by a stampeding herd (Gray 1973). Hone (1934) and Tener (1965) presented anecdotal accounts of wolf predation. Gray (1970) described a successful attack on a solitary bull. Miller and Gunn (1977) described one wolf apparently testing a muskox herd. Based on examination of 16 carcasses in 1966–67, Freeman (1971) estimated that wolf predation was involved in about 50 percent of muskox mortality on Devon Island in the Canadian arctic. Wolf predation on muskoxen has not been quantified but its importance can be deduced from the evolution of the unique defense group formation against wolf attack (Hone 1934; Tener 1965; Gray 1974; Miller and Gunn 1979). In the Thelon Game Sanctuary, Kuyt (1972) found that 19 of 595 wolf scats contained remains of muskoxen. Included were the remains of a 15–18-year-old bull and a 1-week-old calf. Male muskoxen may die from injuries incurred during the rut. Although muskox bulls have elaborate rituals leading to headlong charges, mistiming and misplacings can occur. Injuries can also be incurred during defeats or surprise attacks. Wilkinson and Shank (1976) described six muskox bulls injured during aggressive encounters. On Prince of Wales Island, I observed bulls chasing and butting cows, and bulls slipping or being butted unexpectedly, but did not observe any apparent injuries. S. D. MacDonald (personal communication) observed traumatic injuries in a bull carcass suggestive of an impact from another bull. Wilkinson and Shank (1976) suggested that on Banks Island the mortality from bulls fighting may reach 5 to 10 percent.

Winter movements out onto the pack ice and subsequent starvation or drowning as the ice breaks up may cause mortality of muskoxen on Nunivak Island (Spencer and Lensink 1970). In addition, falls from cliffs, especially by cows attracted to the enriched vegetation above bird colonies, is a cause of mortality on Devon and Nunival islands (Freeman 1971; Lent 1978).

Parasite-induced mortality has not been well documented in the wild, but has been a concern with captive muskoxen, especially their vulnerability to parasites of domestic mammals (Tener 1965; Samuel and Gray 1974). Almost nothing is known of diseases of muskoxen. Exostotic lesions of the type caused by *Actinomyces bovis* have been found in muskox skulls (Tener 1965) from the High Arctic and from Nunivak Island.

The effects of biting flies are unlikely to be a severe problem because only the ears and eyes appear to be vulnerable (Tener 1965). In the High Arctic, only mosquitoes (*Culex* spp.) are present and they do not appear to be a problem. Tener (1965) did not find any evidence of warble fly infestation, though it has been recorded from hides of muskoxen collected near Great Bear Lake (Hone 1934).

The oldest reported muskox in the wild is 19 years of age (Parker et al. 1975).

AGE DETERMINATION

At close range on the ground, muskoxen can be aged up to about four years by horn development, boss development, and size and degree of curvature (Tener 1965; Smith 1977). From aircraft, calves and mature bulls can be distinguished. Once the muskoxen have taken up a group defense formation, however, it is difficult to count exact numbers or to observe calves that may be standing under their maternal cows.

When dental material such as mandibles are available, the pattern of tooth emergence can be used to age muskoxen up to six years old (Tener 1965). Sectioning of teeth to count annuli in the cementum of incisors or molars has been used (Parker et al. 1975) but may not be completely reliable, as the cementum lines are laid down close together. Use of a fluorescent scope may be a more appropriate technique than hematoxylin stain for counting annuli (Hinman 1979).

ECONOMIC STATUS

Muskoxen have been associated with the hunting cultures of early mankind since the end of the Middle Paleolithic. In North America, the Paleo-Indian fluted blade hunters sought game south of the Wisconsin ice

sheet, and Harington (1961) has hypothesized that those hunting activities influenced the distribution of *Ovibos*. Indian hunting could explain the apparent failure of muskoxen to reach Southampton or Baffin islands. The muskoxen were hunted by indigenous peoples for meat, sleeping robes, and horn implements.

The arrival of explorers, traders, and whalers dramatically changed the status of muskoxen because of the almost insatiable demand for meat and hides. The numbers of muskoxen slaughtered by whalers and explorers or purchased from natives were often not recorded. The Hudson Bay Company recorded 1,681 muskox hides traded in 1861 on the Canadian mainland. During the period that the company traded hides (from 1862 to 1916), more than 15,000 hides were handled.

Many explorers of Greenland and arctic Canada depended on muskoxen for meat supplies until 1917, when the muskox became a protected species. In Canada, controlled quotas provide a limited source of income to some settlements for the tourist trade. The value of a pair of bull muskox horns has recently increased, and even horns from muskoxen found dead are collected for subsequent sale.

The greatest potential economic value of muskoxen may lie in sport hunting, which can cost $4,000–6,000 per muskoxen in Canada. The collection, spinning, and sale of their wool and the tourist attraction of living muskoxen in such areas as national wildlife areas or parks will likely remain of limited importance. The wool, however, being of exceptionally fine quality and relatively available from captive herds such as in Alaska, could support cottage industries specializing in luxury-type woven or knitted articles.

MANAGEMENT

The demands for meat and hides in the nineteenth century, together with the introduction of firearms, led to declines and even extirpation of some muskox populations (Hone 1934). Regulations were imposed in 1917 in Canada to prohibit trading in hides. Full protection was obtained in 1926 and a sanctuary was established in 1927 on mainland Canada in the Thelon River drainage of the Northwest Territories.

It is difficult to separate the effects of the legislative protection from the effects of concentrating native peoples in settlements. The move to settlements probably increased the effectiveness of the protection because of greater ease of enforcement. In addition, the focus of a relatively scattered native population into settlements left large tracts of land virtually uninhabited, and has probably allowed muskox numbers to increase.

The current status of muskox management is the result of the losses and declines of muskox populations and their subsequent recoveries. The initial stand of protection has been cautiously modified to allow limited harvests based on a quota system. In Alaska, transplants from an initial, carefully fostered population on Nunivak Island have been the dominant management activity. The extent of the apparent recoveries of muskox populations in Canada has only been revealed by surveys in the last few years. Unfortunately, the gathering of relevant population data to manage the expanding populations has seriously lagged. This lag more or less nullifies the advantages of managing expanding populations that are only lightly harvested.

Management Units. The first question that management of muskoxen raises is What should be the unit of management? Muskoxen living on islands, except those on larger islands, could be managed as an island population. On larger islands and the mainland, it is difficult from our current levels of knowledge to base units on anything except geographical considerations. How geographical units such as those established in Canada (or Alaska) relate to populations and, in fact, what constitutes a discrete muskox population remain unknown.

In Canada, muskoxen can be hunted only within the system of management units that has developed since 1967. The units are educated guesswork centered on known areas of muskox habitat and in response to demands by native peoples. Within each management unit, the assignment of quotas is arbitrary and conservative. Absence of population estimates for the management units prevents efforts to relate quotas to population levels. As the quotas are extremely low, the lack of correlation with population size is not detrimental at this time. However, it is unsatisfactory as a desirable management situation. As demands for quotas increase, so will the impact of the harvest on the populations. Without knowledge of population size and recruitment, the impact of the quotas and their desirable levels will remain unknown.

Quotas. Quotas potentially are an efficient and sensitive method of managing a population for a sustained yield. They are particularly applicable, as muskoxen likely have affinity to home ranges. The herd is the basic social unit and, therefore, can be used as the basic management unit. The affinity of herds for particular ranges can result in a concentration of the harvest on a few herds, leading to local extinctions. Removal of only a certain proportion of animals from each herd would spread the harvest effort and lessen the chance of local extinctions. In the absence of appropriate knowledge, the reductions of individual herds should be conservative, to avoid fragmentation of the social structure of herds.

The harvest at the herd level should not be restricted to any one sex and/or age class. The objective of the harvest with respect to increasing, decreasing, or stabilizing population size should be a consideration in the sex-age structure of the harvest. For example, in a stable or declining herd, the greatest proportion of the harvest should be males. For each bull killed, however, one should be left. The herd bull also should not be killed—the behavior of the herd bull, especially when the herd is in a defense formation, usually identifies it. The extent to which the herd bull contributes to the stability of a herd and traditional range use is unknown, but is likely important.

Quotas for certain native settlements in the Canadian arctic were established in 1967. In 1979, 10 settlements were permitted to take a total of 252 muskoxen (139 males, 113 females). The quotas have to be used from the end of the rut to the beginning of calving (October to March). The quotas are given by management unit for native people. The meat is either locally consumed or used in other settlements. Tags of male muskoxen can, however, be used for sport hunting by decision of the Settlement's Hunters' and Trappers' Association. Holman (Victoria Island) and Sachs Harbour (Banks Island) are the only settlements to date to use 6 and 20 tags, respectively, for sport hunting. The license costs $500 (1979 Canadian dollars). From Holman, the hunt of about four days with two guides cost $4,500. The hunter takes the head, horns, and hide and the meat goes to the settlement.

In 1979, Canadian quotas of each management zone were less than 10 muskoxen, except for 18 in southern Ellesmere Island and 150 from Banks Island (table 51.1). The return of the specimens and information from the quota is not compulsory and depends on the settlement. The 1978/79 quota returns with incisor teeth and age data from Banks Island were from only 6 of 28 hunters. They represented 36 of the 96 muskoxen killed. Adult males are preferred and herds nearest the settlement are hunted the most, with up to 25 percent of the herd being killed.

Information about the harvest and enforcement of a quota, especially taking only a certain number of animals by age and sex from each herd, requires considerable cooperation from the hunters in each community. Sustained effort is required to explain the need and reasons for the quotas and return of specimens. Similarly, hunter education is necessary to explain the consequences and causes of harassment from current hunting techniques that often include the circling and driving of herds by snow machines.

In Alaska, sport hunting has been permitted since 1976 on Nunivak Island, though not without controversy (Lent 1971*a*). The size of the quota varies and is based on annual snowmobile counts in late winter. In 1978/79, there was a quota of 5 bulls to be hunted between 1 and 30 September. There was a further season between 15 February and 31 March for 20 bulls and 5 cows, but no cows were killed. To encourage the use of female tags, the cost of the hunting license was reduced for females, which are not preferred by hunters wanting the large horns of the bulls.

The muskoxen in Greenland are protected in the national park, but there is a quota hunt from Scoresbysund Inlet. The kill of 50–75 muskoxen is not thought to threaten the population, although local herds may suffer disproportionately (F. Kapel unpublished data).

The basis for decisions on the overall size of quotas depends on accurate estimates of numbers, recruitment levels, and natural mortality. Unfortunately, in Canada the Northwest Territories Wildlife Service is only beginning to investigate the first two variables. The slow start of management is partly the result of the size of the NWT and the relatively small population

TABLE 51.1. Population sizes, management units, and quotas for muskoxen in Canada, 1979

Area	Number of Management Zones	Total Size of Quota	Recent Population Estimate	Date	Source
Arctic Islands					
Northwest Victoria	3	20	12,000	1980	R. Jakimchuk unpublished data
Southeast Victoria	2	8		1979	Unpublished data
Prince of Wales	2	9	1,300	1980	A. Gunn unpublished data
Northeastern Devon	1	4	970[a]	1970	Freeman 1971
South Ellesmere	2	20			
Melville Island		0	2,390	1974	Miller et al. 1975
Banks	2	150	15,000–18,000	1980	B. Stephenson personal communication
Mainland Canada					
Bathurst Inlet and Queen Maud Gulf	4	20	5,000	1979	R. Decker unpublished data
Baker Lake	2	6	200–300[a]	1978	R. Decker unpublished data
North of Great Bear Lake	3	14	2,000–3,000	1976	V. Hawley unpublished data

[a] Based on reconnaissance-type surveys, otherwise linear transect surveys.

located in scattered settlements. As muskoxen almost solely occur in the Northwest Territories, with only a few recent records in the Yukon, the NWT Wildlife Service is the only agency with the mandate to manage muskoxen in Canada. In Alaska, the situation is different, with a consistently monitored key population on Nunivak Island, and hence with data on recruitment mortality.

Census. The estimation of numbers of muskoxen is hampered by the remoteness of their habitats, which results in high costs of aerial surveys. For example, an aerial survey of Prince of Wales Island using linear transects at 25 percent coverage planned for 1980 will cost $25,000 (1980 Canadian dollars). The island has a quota of only nine muskoxen. The design of surveys to count effectively those muskoxen whose distribution is usually restricted to drainages and coastal plains has not been perfected. Basically, reconnaissance-type surveys have predominated over linear transect surveys. Only for the arctic islands, except Ellesmere, Axel Heiberg, and Devon, are counts based on muskoxen observed on linear transects, and extrapolated to estimates depending on the level of coverage. The clumped distribution of muskoxen in discrete herds also can confound the reliability of surveys, although the clumped distribution can be described using 1.6-km strips divided into an inner and an outer strip for each side of the plane. On the other hand, the dark coats of muskoxen render them highly visible against a snow background and aerial surveys can be flown at 300 to 600 m above ground level. Surveys should be flown in March–April as day length increases but there is usually still snow-covered ground. The high altitude minimizes disturbance of the animals and increases the off-transect area that can be scanned during surveys. Relatively small areas of management interest, such as Bailey Point (Melville Island), can be covered totally

during such relatively high-altitude survey flights. Large groups should be photographed with a 35-mm camera to check visual counts.

The greatest difficulties of survey design occur on the Canadian mainland. The vast areas and the problems of delineating survey areas in relation to discrete muskox populations are not insurmountable. As we acquire knowledge of distributions and population characteristics, we will be able to define appropriate survey areas and stratify them for linear or block surveys. Aerial photographs can be used to identify likely habitat and to aid in the choice of strata. Distances from coasts and drainages would also be useful in delimiting strata (F. L. Miller personal communication).

On Nunivak Island, snow machines are used in annual counts, which include sex and age composition. In Canada, distances generally preclude such an approach; helicopters could be used. Ground observers could be positioned 500 m from herds and approach to within 200–300 m of a herd. The use of spotting scopes would allow counts by sex and age classes with minimal disturbance to herds. Disturbance, resulting in herds closing together in a group defense formation, creates considerable difficulties in obtaining accurate counts—let alone counts of calves and yearlings. The knowledge of numbers from counts conducted in June–July of cows with calves (calf production), yearlings (yearling recruitment), and juveniles accumulated over several years will allow the adjustment of quotas according to the status of the population.

To understand the effects of localized hunting pressures and to predict the effects of industrial exploration and development, knowledge of dispersal of muskoxen is needed. Unfortunately, we know little about the dispersal or movements of muskoxen, although techniques to capture and mark muskoxen are available and have been tested. Jonkel et al. (1975) used a helicopter to isolate an animal from a herd and dart the muskox with a powder-charge dart gun. Darts with 5- and 6-cm barbed needles were used so that the dart could be seen in the long hair. Succinylcholine chloride at dosages of about 60 mg for cows and 100 mg for bulls was used together with 50–110 mg of sparine. Difficulties of weighing animals in the field prevented accurate determinations of the dosage per kilogram. Jonkel et al. (1975) drugged muskoxen in April and May. However, muskoxen drugged with succinylcholine chloride in July and August on Melville Island required higher dosages and usually recovered in less than 10 minutes (F. L. Miller personal communication). Muskoxen drugged in April and May were immobile for 30–45 minutes. The differences may reflect the poorer physical condition of muskoxen in late winter.

Other drugs used include Sernylan (phencyclidine hydrochloride) in late winter on Nunivak Island with success. Tranvet (propiopromazine hydrochloride) was injected after the Sernylan to minimize convulsions, at dosages of about 20 mg/45 kg. The dosages of Sernylan varied from 0.7 to 1.2 mg/kg, and females required at least twice the dosage given to males (Jennings and Burris 1971).

Several techniques have been tested to mark muskoxen (Jonkel et al. 1975; F. L. Miller personal communication). Ear tags were not easily visible because the ears are nearly buried in the mane. Nylon streamers attached to the ear tags would be visible but their permanency is untested. Plastic streamers attached to the horns with metal hose clamps and enamel paint of the horn bosses were both successful methods of marking for up to three years. Jonkel et al. (1975) used paint-pellet pistols to fire plastic-encased balls of paint to mark horn bosses. The pistols are effective up to 20 m (F. L. Miller personal communication). Marking was successful from the ground when muskoxen were among rocks and boulders and would allow a relatively close approach.

Jonkel et al. (1975) and F. L. Miller (personal communication) placed radio-collars on muskoxen. The former had technical problems with radio-transmitters, and the latter was unable to locate any radio-collared muskoxen the following winter after marking. Miller believed that high mortality and movement during the winter of 1973/74 on Melville Island, an exceptionally severe winter, were responsible for the failure to relocate the animals. Also, the clashing and butting of muskoxen during the rut possibly damaged the collars and broke the whip antennae.

Radio-collars allow the collection of information on dispersal, movements, and interchange of animals among groups. The great expense of subsequent monitoring, especially from aircraft, is one of the main reasons limiting the use of radio-collars. Satellite radio-packages have been successfully fitted to polar bears (*Ursus maritimus*) (R. E. Schweinsburg personal communication). The initial cost is high but surveillance is regular and considerably less expensive than monitoring conventional radio-collars. Satellite monitoring may be practical for monitoring muskox movements. The telemetric package likely requires little modification to reduce its weight (5 kg) from that carried by the polar bear.

The stability of social subgroups and the need for them within larger groups are important considerations in harvesting individual herds. The interchange of solitary bulls, or bulls from bull-only groups, for lead animals in groups may reveal a mechanism by which muskox groups return to or colonize habitats. There is a possibility that solitary bulls and bull groups travel further afield. Such bulls may become lead bulls and lead their group to ranges found during their travels. The potential of the harvest to cause local extirpations near settlements underlies the importance of understanding movements and dispersal of muskoxen.

Knowledge of movements and dispersal patterns is important in predicting the results of transplanting muskoxen. In Canada, there have been no transplants to initiate recolonization of historic ranges. Transplants have been suggested to speed up recoveries of depleted populations or to extend ranges, such as to Baffin Island. To date, such suggestions have been negated by the costs involved, rather than by decisions based on whether muskoxen would thrive. In Alaska, the muskox management program is aimed at reintroducing muskoxen to their historic ranges to provide harvestable populations.

The original purpose of introducing and increasing a muskox population on Nunivak Island has proved successful in providing initial stock to reintroduce elsewhere in Alaska. Since 1969, muskoxen have been introduced to locations within the historic range. Considerable experience and expertise has been amassed in developing techniques for capturing, holding, transporting, and releasing muskoxen. Observations of muskoxen subsequent to being transplanted indicate desirable sex and age class combinations for success.

The perfected techniques resulted in the capture of 54 muskoxen in seven days in late winter 1975, with only one fatality, a pregnant two year old that probably died of ketosis (Hout 1975). The capture crews operated from snow machines. Two-year-old muskoxen either were driven into nets or had the net dropped over them. Yearlings were "bulldogged." One important precaution was to protect the horns of yearlings with burlap bags stuffed with hay, as the horns are fragile and separate from the bony core. The captured animals were hobbled, then tied onto sleds for transport to base camps for release into temporary corrals (snow pits). The captives were fed timothy hay until transported to their final destination. In this case, 40 of the muskoxen were flown to the Soviet Union; 13 were released in Alaska as animals of known sex and age identified by ear tags and color-coded nylon streamers.

Lent (1971*b*), in his discussion of movements, dispersal, and group composition of transplanted muskoxen, described the relatively long distances traveled by yearlings, and the presence of solitary yearlings and yearling groups, as resulting from the transplant's being dominated by the yearling age class. He also noted shifts in ranges and movements likely caused by harassment from snow machines. The fate of the 36 muskoxen released in 1970 at the Feather River (Seward Peninsula) illustrates some of the problems of transplants. The release was mainly of yearlings. The lack of natural barriers, together with disturbance by humans on snow machines, likely caused the wide dispersal. Muskoxen were observed 480 km away from the release site. On Nelson Island, 22 yearlings and 1 two-year-old male were released in 1967 and 1968 (Lent 1978) but did not widely disperse. There was little or no human disturbance on the island. In 1978, the Seward Peninsula muskoxen were believed to number 45 in two or more herds. The growth of the populations has been slower than expected, primarily because of wandering by adult bulls and also grizzly bear (*Ursus arctos*) predation on calves.

The Nelson Island herd numbered 107 muskoxen in 1978. The anticipated carrying capacity is 100–150 animals, so a harvest and/or transplants will be needed in the near future. The Nelson Island transplant was the first from the Nunivak Island stock and appears to have been the most successful. The range is excellent and in 1968 two yearlings were bred and calved as two year olds.

Wide dispersal was also noted (Lent 1978) as a result of transplants of 52 muskoxen released on Barter Island in 1969 and 14 released at the mouth of the Kavik River (128 km west of Barter Island) in 1970. By July, the population was thought to number 94 (Hinman 1979). The release of 36 muskoxen in March 1970 at Cape Thompson resulted in wide dispersal in small groups. This was possibly due to snowmobile activity associated with caribou hunting. Only one group of 8–10 animals became established, and it numbered 30 in 1976. An additional 34 muskoxen were released at Cape Thompson in April 1977. The Alaska Fish and Game Department adopted a policy of attempting releases in two consecutive years (as was used for Nelson Island) for future transplants.

The management of muskoxen will probably continue to use regulated harvests and removal of animals for transplants. Many of the considerations usually implicit in wildlife management (such as habitat manipulation) are impractical and currently unnecessary because of the isolated ranges of muskox. However, habitat studies may still be necessary to determine whether muskoxen are overusing their range, particularly winter range, especially in high-density situations, such as on Nunivak and Banks Island, or before transplant to new areas are undertaken.

Competition between muskoxen and reindeer or caribou is usually considered nonexistent (Wilkinson et al. 1976). However, under certain snow conditions caribou and muskoxen are forced onto the same windblown bare slopes during winter (Miller et al. 1977). Alendal (1976) thought that the decline of muskoxen on Spitzbergen Island may have been partly due to competition with caribou on the winter ranges. Inuit on Banks Island have suggested declines in caribou populations from competition with the increased muskoxen population. Observations of winter distributions and feeding habits of muskox and caribou at various densities could resolve the question of competition.

In Alaska, densities of muskoxen were too high for the winter range on Nunivak Island and damage to the habitat occurred. The muskoxen were feeding on grasses growing on sand dunes and the dunes were vulnerable to trampling damage.

Certain muskox ranges in Canada are of particular interest and importance. Those areas are heartlands for muskoxen and probably act as reservoirs when surrounding populations decline because of weather and other factors. Such areas should be considered as special management zones to ensure the continuity of their populations. Enactment as special wildlife reserves or some equivalent status would aid in their preservation. Those areas include Fosheim Peninsula (Ellesmere Island), Bailey Point (Melville Island), and Thompson River (Banks Island). For example, 26.3 percent of the muskox population of Melville Island in March 1974 was on Bailey Point, although it represents only 2 percent of the total area of the island (Miller and Russell in Russell 1976). Industrial and other human activities should be restricted in such areas.

Concerns about the effects of human activities, including industrial effects on muskox populations, have been raised. Biologists have a special responsibility to ensure that their activities in studying and managing muskoxen are designed to minimize stress. In particular, activities such as capturing animals for marking or transplanting should not be carried out shortly before or after calving (Jonkel et al. 1975).

The effects of harassment include disruption of daily activity patterns, social stresses such as increased aggressive behavior by bulls, calf desertion, pathological effects from overexertion during flight, and range desertion. Any harassment causing an increase of energy expenditure has the potential to be deleterious to the long-term well-being of muskoxen (Miller and Gunn 1979). The rapid increase in the aircraft activity associated with exploration and exploitation of mineral and hydrocarbon reserves in the arctic must be paralleled by effective regulations to minimize harassment. The defense formation of muskoxen that so increased their vulnerability to overhunting also increases their vulnerability to current human harassment. Too often the assumption is made that because the muskoxen stand their ground and allow close approach they are not stressed (Miller and Gunn 1979).

Initially, most of the concern about the impact of industrial exploration on muskoxen was centered on seismic activity, especially on Banks Island. Studies (Urquhart 1973; Beak Consultants 1975; Riewe 1973; Russell 1976) described observations of muskox herds during seismic activities and distributions before and after the activities. The results are difficult, if not impossible, to interpret in the context of how harmful such disturbances may or may not be to populations. The difficulties of adequate experimental design with suitable controls under field conditions should not be underestimated. Without controls and careful environmental monitoring, to suggest that distributional changes are only the results of disturbance can be misleading.

We can describe responses such as moving into defense formations, galloping, or alertness as responses to seismic activities, aircraft, and human disturbance. To extrapolate what the responses mean to the well-being of the population or even the individual is outside the range of our knowledge. Therefore, we can only suggest some of the possible effects (Miller and Gunn 1979) and take a conservative approach in establishing guidelines for industry. This conservative approach should be adhered to despite scattered observations on the adaptability of muskoxen, because of our ignorance about the process of adaptation and habituation under field conditions. We observed the habituation of muskoxen to a turbohelicopter on Prince of Wales Island (Miller and Gunn 1980). The presence of muskoxen on air strips at Eureka (Ellesmere Island) and Mould Bay (Prince Patrick Island) also indicates habituation to aircraft.

Some industrial developments may affect muskox ranges. In particular, oil and gas pipelines, especially elevated pipelines, have the potential to act as barriers preventing free movements on or between seasonal ranges. The proposed various pipelines that will cross muskox ranges will be buried gas lines. The activities associated with the construction and maintenance of those lines, however, have the potential to prevent free movements across the pipeline corridor.

Muskox management in Canada is in its infancy, and the use of quotas, especially at the herd level, as a regulatory tool is untested. The apparent tendency for quotas in particular and wildlife resources in general to be political bargaining points cannot be underestimated. Acceptance of native peoples' relationships with wildlife will be necessary for all concerned about the management of muskoxen. The management of muskoxen by regulated harvests and the minimization of disturbance by mineral and hydrocarbon exploratory and exploitation activities are complicated by the presently unsettled native land claims and lack of planning of integrated resource management in Canada.

It is vital to the future well-being of muskox populations that a flexible but restricted harvest system continue. The effects of uncontrolled hunting on muskoxen have already been seen once, and although under certain conditions populations can recover rapidly, a recovery is not guaranteed. The muskox is one of the most unique and interesting of North America's mammals. Its survival in one of the world's harshest environments will require sensitive management to develop the compatability of muskoxen and human activities in the arctic.

CURRENT RESEARCH AND MANAGEMENT NEEDS

Alaska and the Northwest Territories have widely different management needs for their muskox populations. Alaska is concentrating on reintroductions and establishing new populations. The NWT Wildlife Service is now faced with an apparent general increase in muskox populations and demands for larger and new quotas but has only scanty data on which to base management decisions. The Wildlife Service is currently drafting a species management plan for muskoxen to move toward integrated management in the territories.

Initial requirements for the management of muskoxen in the Northwest Territories are to complete the inventory of numbers and distributions of the various populations and to quantify the trends of population growth. On the basis of the distributions of the various populations the muskox management unit areas should be evaluated and, if necessary, redrawn to correspond with natural boundaries. At an administrative level, goals and objectives need to be established for the different management unit areas. Levels of harvest, present and future, have to be evaluated and balanced among subsistence, sport hunting, and nonconsumptive uses such as tourism. The level of harvest within each unit has to be related to population growth and whether the management objective in that unit is for a stable, increasing, or decreasing population. As it will take several years to acquire recruitment data and population growth rates, conservative quotas should be continued in the interim.

Information from the muskoxen currently harvested is necessary but not usually available. Voluntary returns are preferable, and a sustained effort will be necessary to convince hunters of the need for and use of harvest data. The sex and age class of the shot animal, a subjective estimate of the body condition, the location, and the herd size are the minimal information that should be collected, stored, and analyzed. As population dynamics of muskoxen become documented, the harvest data will be available for use in simulation models to evaluate different harvest strategies. Models are a potentially useful tool but are only as reliable as the input data. Currently, the necessary data for even a crude model are lacking.

One of the most severe restrictions on the management of muskoxen is the extremely high logistical cost of obtaining even basic data. The gathering of preliminary management data provides opportunities for the research necessary to answer basic questions about management techniques, the effects of management policies (such as the effects of harvest strategies and levels), and population dynamics. Surveys of numbers and distributions should include testing of survey designs and techniques to develop a standard methodology. Studies of movements and distributions using radio-collared muskoxen could also be used to study the effects of harvest techniques and industrial disturbance, natural mortality, social dynamics and stability of herds, and basic behavioral patterns. These are all topics immediately pertinent to management needs. Age-specific conception and pregnancy rates, diet, seasonal changes in body condition, and the incidence of disease are all topics on which data are lacking. Specimens for these studies could be collected from muskoxen killed on quotas if close cooperation with native hunters was pursued.

Muskox heartlands have been described and they need protection by the enactment of regulations to prevent or minimize disturbance and ensure their future integrity. Studies are necessary to monitor the ranges in those sanctuaries, particularly if their muskox populations increase and other herbivores are present. In particular, the apparent increase in the moose population in the Thelon Game Sanctuary requires study, as both moose and muskoxen feed on willow.

In Alaska, muskoxen were extirpated; therefore, reintroductions were a necessary first step. In the NWT, muskox populations are apparently widespread and vigorous; therefore, the need is to document their population dynamics at a level for effective management before spending time and resources on transplants.

The muskox is a largely untapped source of research topics on the behavioral and physiological adaptations of a large mammal to a uniquely cold and dry environment. A thorough understanding of the relationship of the muskox with its environment, its strategies to maintain and reproduce itself, and its adaptive abilities to cope with environmental changes (including human-induced changes) can only foster more effective management.

Most of the chapter was written while I worked for the Canadian Wildlife Service (Western and Northern Region), and the editorial help of R. H. Russell, I. G. Stirling, and D. C. Thomas is acknowledged. Special thanks are owed to F. L. Miller for patient help and suggestions.

The chapter was finished while I was working for the Northwest Territories Wildlife Service, and thanks are due to D. R. Urquhart for helpful suggestions.

The Canadian studies of muskoxen in the 1970s have most all been supported by Polar Continental Shelf Project, Department of Energy, Mines and Resources. The unstinting and understanding help of G. Hobson, Director of PCSP, is warmly acknowledged.

LITERATURE CITED

Alendal, E. 1976. The muskox population (*Ovibos moschatus*) in Svalbard. Norsk Polarinst. Arbok 1974: 159–174.

Allen, J. A. 1913. Ontogenetic and other variations in muskoxen, with a systematic review of the muskox group, recent and extinct. Mem. Am. Mus. Nat. Hist. No. 1, Part 4:102–336.

Anderson, R. M. 1934. The distribution, abundance, and economic importance of the game and fur-bearing mammals of Western North America. Proc. 5th Pacific Sci. Congr. 1938:4055–5075.

Beak Consultants, Ltd. 1975. Seismic activities and muskoxen and caribou on Banks Islands, N.W.T. Rep. prepared for Panarctic Oils, Ltd. Calgary, Alta. 15pp. Mimeogr.

Burch, E. S. 1977. Muskox and man in central Canadian subarctic, 1689–1874. Arctic 39:135–154.

Courtin, G. M., and Labine, C. L. 1977. Microclimatological studies on Truelove Lowland. Pages 73–106 *in* L. C. Bliss, ed. Truelove Lowland, Devon Island, Canada: a High Arctic ecosystem. Univ. Alberta Press, Edmonton. 714pp.

Fischer, C. A., and Duncan, E. A. 1976. Ecological studies of caribou and muskoxen in the Arctic Archipelago and northern Keewatin. Rep. to Polar Gas Environmental Program by Renewable Resources Consulting Services, Ltd. Edmonton, Alta. 194pp. Mimeogr.

Freeman, M.M.R. 1971. Population characteristics of muskoxen in the Jones Sound region of the Northwest Territories. J. Wildl. Manage. 35:103–108.

Gray, D. R. 1970. The killing of a bull muskox by a single wolf. Arctic 23:197–198.

————. 1973. Social organization and behaviour of muskoxen (*Ovibos moschatus*) on Bathurst Island, N.W.T. Ph.D. Thesis. Univ. Alberta, Edmonton. 212pp.

————. 1974. The defense formation of the muskox. Muskox 14:25–29.

Harington, C. R. 1961. History, distribution and ecology of the muskoxen. M. Sc. Thesis. McGill Univ., Montreal, Que. 489pp.

Hinman, R. A., ed. 1979. Annual report of survey-inventory activities. Part 4: Sheep, mountain goat, bison, muskoxen, marine mammals. Alaska Dept. Fish and Game, Juneau. Vol. 9. 123pp.

Hone, E. 1934. The present status of the muskox. Spec. Publ. Am. Comm. Int. Wildl. Protection No. 5. 87pp.

Hout, J. L. 1975. US-USSR muskox transplant. Nunivak Natl. Wildl. Refuge, Alaska Dept. Fish and Game, Juneau. Manuscript.

Hubert, B. A. 1977. Estimated productivity of muskox on Truelove Lowland. Pages 467–492 *in* L. C. Bliss, ed. Truelove Lowland, Devon Island, Canada: a High Arctic ecosystem. Univ. Alberta Press, Edmonton. 714pp.

Jennings, L. B., and Burris, O. E. 1971. Muskox report. Alaska Dept. Fish and Game, Job Progr. Rep. 11: 12pp.

Jonkel, C. J.; Gray, D. R.; and Hubert, B. 1975. Immobilizing and marking wild muskoxen in Arctic Canada. J. Wildl. Manage. 39:112–117.

Kuyt, E. 1972. Food habits of wolves on barren-ground caribou range. Can. Wildl. Serv. Rep. Ser. 21: 36pp.

Kuyt, E.; Schroeder, C. H.; and Brazda, A. R. 1972. An albino muskox near the Atkinson Point River, Northwest Territories. Arctic 25:239–240.

Lent, P. C. 1971a. Muskox management controversies in North America. Bio. Conserv. 3:255–263.

————. 1971b. A study of behaviour, and dispersal, in introduced muskox populations. Pages 34–50 *in* Final Rep. to Arctic Inst. of North Am., Dept. Fish and Game, Juneau, Alaska.

————. 1974. Mother-infant relationships in ungulates. Pages 14–55 *in* V. Geist and F. Walther, eds. The behaviour of ungulates and its relationship to management. Int. Union Conserv. Nature and Nat. Resour., Morges, Switzerland. Publ. new ser. no. 24. 940pp.

————. 1978. Muskox. Pages 135–147 *in* J. L. Schmidt and D. L. Gilbert, eds. Big game of North America: ecology and management. Wildl. Manage. Inst. and Stackpole Books, Ltd., Harrisburg, Pa. 490pp.

McDougall, G. F. 1857. The eventful voyage of H. M. discovery ship "Resolute" to the Arctic regions, in search of Sir John Franklin and the missing crew of H. M. discovery ships "Erebus" and "Terror," 1852, 1853, 1854. Longman, Brown, Green, Longmans and Roberts, London. 530pp.

Miller, F. L., and Gunn, A. 1977. Group of muskoxen attacked by a solitary arctic wolf, Prince of Wales Island, Northwest Territories. Muskox 21:87–88.

————. 1978. Inter-island movements of Peary caribou south of Viscount Melville Sound, Northwest Territories. Can. Field Nat. 92:327–333.

————. 1979. Responses of Peary caribou and muskoxen to helicopter harassment. Can. Wildl. Serv. Occas. Pap. 40. 90pp.

————. 1980. Responses of three herds of muskoxen to simulated slinging by helicopter, Prince of Wales Island, Northwest Territories. Can. Field Nat. 94:52–60.

Miller, F. L.; Russell, R. H.; and Gunn, A. 1977. Distributions, movements and numbers of Peary caribou and muskoxen on western Queen Elizabeth Islands, Northwest Territories, 1972–74. Can. Wildl. Serv. Rep. Ser. No. 49. 55pp.

Parker, G. R. 1978. The diets of muskoxen and Peary caribou on some islands in the Canadian High Arctic. Can. Wildl. Serv. Occas. Pap. 35. 21pp.

Parker, G. R., and Ross, R. K. 1976. Summer habitat use by muskoxen (*Ovibos moschatus*) and Peary caribou (*Rangifer tarandus pearyi*) in the Canadian High Arctic. Polarforschung 46:12–25.

Parker, G. R.; Thomas, D. C.; Broughton, E.; and Gray, D. R. 1975. Crashes of muskox and Peary caribou populations in 1973–74 on the Parry Islands, Arctic Canada. Gen. Wildl. Serv. Prog. Note 56:1–10.

Riewe, R. R. 1973. Final report on a survey of ungulate populations on the Bjorne Peninsula, Ellesmere Island: determination of numbers and distribution and assessment of the effects of seismic activities on the behaviour of these populations. Rep. to Dept. Indian and Northern Affairs. 63pp.

Russell, J. 1976. Aerial surveys of Bailey Point, Melville Island, N.W.T. Northwest Territories Fish and Wildl. Serv., Yellowknife. 11pp.

Russell, R. H.; Edmonds, J.; and Roland, J. 1978. Caribou (*Rangifer tarandus*) and muskoxen (*Ovibos moschatus*) habitat studies on Prince of Wales and Somerset Islands and Boothia Peninsula and Northern District of Keewatin, Northwest Territories, 1975 and 1976. Report prepared for Arctic Islands Pipeline Program by Can. Wildl. Serv., Edmonton, Alta. 141pp.

Samuel, W. R., and Gray, D. R. 1974. Parasitic infection in muskoxen. J. Wildl. Manage. 38:775–782.

Segal, A. N. 1962. The periodicity of pasture and physiological functions of reindeer. Pages 130–150 *in* Reindeer in the Karelian A.S.S.R., Adkad, Nauk, Petrozavodska, Can. Wildl. Serv., Edmonton, Alta. Unedited trans.

Smith, T. E. 1977. Reproductive behaviour and related social organization of the muskox on Nunivak Island. M.S. Thesis. Univ. Alaska, Fairbanks. 138pp.

Spencer, D. L., and Lensink, C. J. 1970. The muskox of Nunivak Island. J. Wildl. Manage. 34:1–15.

Tener, J. S. 1963. Queen Elizabeth Islands game survey, 1961. Can. Wildl. Serv. Occas. Pap. No. 4:1–50.

———. 1965. Muskoxen in Canada: a biological and taxonomic review. Can. Wildl. Serv. Monogr. 2. 166pp.

Urquhart, D. 1973. Oil exploration and Banks Island Wildlife. Rep. Wildl. Serv., Govt. Northwest Territories, Yellowknife. Mimeogr. 105pp.

Vibe, C. 1967. Arctic animals in relation to climatic fluctuations. Meddel. on Gronland. 170:1–227.

Wilkinson, P. F., and Shank, C. C. 1976. Rutting-fight mortality among muskoxen on Banks Island, Northwest Territories, Canada. Anim. Behav. 24:756–758.

Wilkinson, P. F.; Shank, C. C.; and Penner, D. E. 1976. Muskox-caribou summer range relations on Banks Island, N.W.T. J. Wildl. Manage. 40:151–162.

ANNE GUNN, Canadian Wildlife Service, 1000 9942–108th Street, Edmonton, Alberta TSK 3J5. Current address: Wildlife Service, Department of Renewable Resources, Government of Northwest Territories, Yellowknife, Northwest Territories XIA 2L9.

52

Mountain Sheep

Ovis canadensis and *O. dalli*

Bruce Lawson
Rolf Johnson

NOMENCLATURE

COMMON NAME. Thinhorn sheep
SCIENTIFIC NAME. *Ovis dalli*
SUBSPECIES. *O. d. dalli*, Dall's or Alaskan white sheep; *O. d. stonei*, Stone's or black thinhorn sheep.

COMMON NAME. Bighorn sheep
SCIENTIFIC NAME. *Ovis canadensis*
SUBSPECIES. *O. c. auduboni*, Audubon's bighorn, Black Hills bighorn or badland bighorn; *O. c. californiana*, California bighorn, lava bed bighorn or rimrock bighorn; *O. c. canadensis*, Rocky Mountain bighorn; *O. c. nelsoni*, *O. c. mexicana*, *O. c. texiana*, *O. c. cremnobates*, *O. c. weemsi*, desert bighorn.

DISTRIBUTION

Ovis dalli is found in extensive regions of Alaska's Brook's and Alaskan ranges, the Kenai Peninsula, and the MacKenzie and Rocky mountains of western Canada (figure 52.1). Their southern distribution extends only to the Peace River in British Columbia.

Ovis canadensis is found in relatively isolated pockets in the Coastal, Cascade, and Sierra Nevada ranges and the Rockies south of the Peace River to Mexico (figure 52.1).

DESCRIPTION

Mountain sheep are members of the family bovidae and genus Ovis. The most distinguishing feature of mountain sheep is the massive horns attained by older rams. Wild sheep are medium-sized, stout ungulates. The hooves are modified for gripping rocky terrain: the posterior half of each toe is formed into a round, rubbery pad and the toes are independently movable. Lateral hooves and interdigital and suborbital glands are present.

Large Rocky Mountain bighorn rams weigh between 300 and 325 pounds (Clark 1970); Dall's rams average about 20-40 lbs less, and desert bighorns average about 10 lbs less than Dall's sheep (Bunnell and Olsen 1976). However, normal adult weight fluctuates according to the season of the year and reproductive

FIGURE 52.1. Distribution of mountain sheep (*Ovis canadensis* and *O. dalli*).

parameters. Woolf (1971) found that captive ewes lost as much as 22 percent of their spring weight while lambing.

Although there is sexual dimorphism—for example, Dall's rams average 33 percent heavier than ewes (Bunnell and Olsen 1976)—the characteristic that exhibits the least disparity between sexes is the total length (table 52.1).

Pelage. Desert races of the bighorns are lightest in color but the pelage of all races fades as each spring approaches (Buechner 1960).

The basic color pattern of wild sheep is brown with a white muzzle, underparts, rump patch, and edgings down the rear of the legs (Clark 1970). Dall's sheep are all white except for a few black-tailed types in the Yukon-Tanana uplands (Guthrie 1972). Neonatal coloration of Dall's sheep is highly variable and includes some dark forms (Nichols 1978). The body color of Stone's sheep varies from brown black to stone gray (Guthrie 1972; O'Connor 1974), but generally whitens with age (Clark 1970). The rare Fannin

TABLE 52.1. Mean measurements of wild sheep

	Ovis dalli		*Ovis canadensis*	
	Males	Females	Males	Females
Total length	1540 mm	1350 mm	1640 mm	1527 mm
Tail length	95	88	110	100
Hind foot length	423	0	420	348
Height at shoulder		991		991
Basilar length of skull	255	227	266	244

SOURCE: Compiled from Clark (1970) and Hall and Kelson (1959).

sheep, an intergrade between *O. d. dalli* and *O. d. stonei,* have a gray saddle patch over the back and tail but otherwise are white (Guthrie 1972).

Wild sheep have unbanded awn-type guard hairs. The hairs are round to oval and reach a length of 322 mm and a diameter of 4.8 mm (Moore et al. 1974). The underhair is angora and provides excellent insulation.

The annual spring molt lasts approximately one to two months. Mature healthy rams generally molt first, followed by ewes and sheep in poor condition. (Cowan 1940; Schmidt and Gilbert 1978).

Skull and Dentition. The average weight of a mature ram skull, excluding the lower jaw, is 40 lbs (Clark 1970). The lambdoidal suture forms an approximately straight line. The upper ends of the premaxillae do not meet the nasals and maxillae. The infraorbital foramen is small with a well-defined rim. The brain case is pneumatic and the occipital condyles are enlarged. Figures 52.2 and 52.3 detail *O. dalli* and *O. canadensis,* respectively.

Wild sheep have true horns that, in proportion to their body size, are the largest of all ruminants, comprising 8–12 percent of a ram's body weight (Geist 1966*b*). The horns are rugose and unforked. Ram horns are massive and grow in a tight spiral; those of ewes are much smaller and straighter. They are a light, dirty ochre color. Dall's and Stone's horns are, on the average, somewhat thinner than the bighorn's but there is much intergradation (O'Connor 1974). The horns of the desert varieties typically display the most divergent growth, i.e., the distance between tips is the greatest (Clark 1970). Horns first begin to appear at approximately 2 months of age (Hansen 1965). By 5½ months they are 5–7 cm in length (Jones 1959). After 1 year the basal circumference of male horns is greater than that of female horns and increases at approximately the same growth rate as the length. During the first 2 years the horn is basically triangular in cross-section; thereafter, the base swells and the horn loses its flat-sided shape. Growth is greatest in the summer. This period of accelerated growth produces annual rings; however, such influences as drought, disease, and other interruptions in feeding habits are indicated by fissures between the rings. During the first 3–4 years horn rings are several inches apart; thereafter, they occur progressively closer and become harder to differentiate from the normal corrugations (see figure 52.2).

Ninety-five percent of brooming, breaking, or wearing away of the tips is probably the result of fighting, while the remainder is due to accidents and digging (Welles and Welles 1961; Hemming 1969; Clark 1970). The most pronounced brooming is usually associated with the most massive horns. Bunnell (1978) found that the extent of horn growth during the initial 5 years of age was related to the health of the mother and not to neonatal body size. By 5 years of age horns have reached 2/3 of their maximum length and 90 percent of their greatest basal circumference. The remaining growth takes up to 10 more years. The maximum basal circumference is usually about 39 cm for thinhorn sheep and about 46 cm for bighorns (Hall and Kelson 1959).

All forms of North American wild sheep have similar, if not identical, tooth development. The incisors are spatulate. The molars are long and broad—"subhypsodont" (Geist 1971), as they do not contain an open pulp cavity throughout their life. The deciduous dentition is complete at birth or by the first week of age. Unlike that of most other wild ungulates, the permanent dentition of bighorn and thinhorn sheep is not fully erupted until four years of age. The tooth replacement patterns of Dall's, Rocky Mountain, and desert bighorns are similar (table 52.2). The adult formula is 0/3, 0/1, 3/3, 3/3 = 32. Dental anomalies include malocclusions, deformations, lost teeth, incomplete sets, where alveoli are not even present, and vestigial upper canines (Murie 1944; Deming 1952; Welles and Welles 1961).

PHYSIOLOGY

Temperature and Blood Chemistry. The normal range of rectal temperature is 38.6°–38.9° C, but it varies according to the ambient temperature, stress, and the season of the year. It has been found to follow a seasonal cycle (Franzmann and Hebert 1971).

Chappel and Hudson (1978) found that the metabolic rate of Rocky Mountain bighorns is lowest at an ambient temperature of −10° C and the thermoneutral zone is −20° to +10°C. They also observed that moderate wind affected metabolic rate only at ambient temperatures below −20° C.

The blood transferrins and hemoglobins were found to be unique in four subspecies of North American wild sheep: *O. c. canadensis, O. c. mexicana, O.*

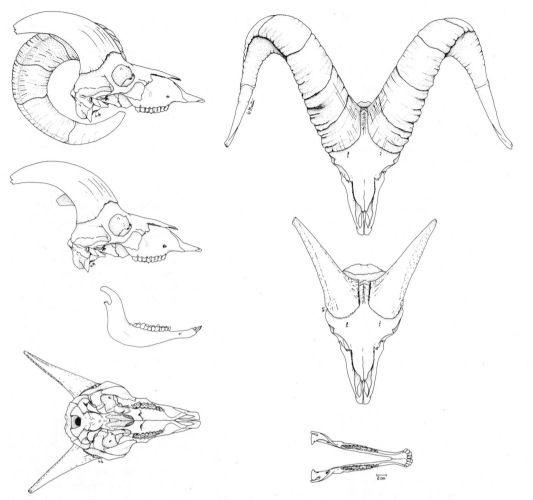

FIGURE 52.2. Skull of the thinhorn sheep (*Ovis dalli*). From top to bottom: lateral view of cranium, with one horn removed (left), and dorsal view of cranium with horns (right); lateral view of cranium, with both horns removed to show permanent bony core, and lateral view of mandible (left), and dorsal view of cranium, with horns removed (right); ventral view of cranium, with horns removed (left), and dorsal view of mandible (right).

d. dalli, and *O. d. stonei* (Nadler et al. 1971). Other physiologic values, including blood proteins and minerals, white blood cell count, red blood cell count, and packed cell volume, were reported by Franzmann and Thorne (1970), Woolf and Kradel (1970), and Franzmann (1971, 1972).

Neural. Mountain sheep depend more on their visual capabilities than on their auditory or olfactory senses. Hunters, in particular, have found mountain sheep very difficult to approach without being seen.

REPRODUCTION

Anatomy. There is a paucity of information concerning the reproductive anatomy and physiology of wild sheep. The following account for domestic sheep, derived from Blom (1968) and Hafez (1968), is probably applicable to wild sheep.

The ovary, oviduct, and uterus are supported by a broad dorsolateral ligament in the region of the ilium. The bipartite uterus resembles a ram's horns with the convexity dorsal. The almond-shaped ovaries are located laterally and in close apposition to the fusion of the uteri in open ovarian bursae. The bursae are pouches derived from the same tissue as the ligament and attach the suspended oviducts to the uteri. The right ovary is most active. Mature corpora lutea are spheroid or oval and the oviduct is pigmented. The endometrium of the uterus is characterized by numerous pigmented caruncles, each generating many cotyledons during pregnancy. Annular folds form the lumen of the cervix, and the hymen is well developed.

There are two inguinal, functional mammaries; supranumerary teats, if present, are located anteriorly to the normal ones. Fine hair covers the teats, and the connective tissue closing the orifice is elastic.

The prostate gland of the ram is disseminate. The scrotum is pendulous and the testes inguinal. Rams have a fibroelastic penis with a filiform appendage, the

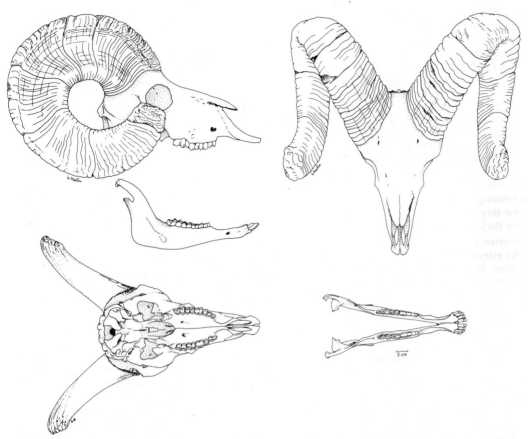

FIGURE 52.3. Skull of the bighorn sheep (*Ovis canadensis*). From top to bottom: lateral view of cranium and mandible (left) and dorsal view of cranium (right); ventral view of cranium, with horns removed (left), and dorsal view of mandible (right). Note the horn annuli, the normal corrugations, and the increment lost to brooming.

precesses uretherae, located at the tip, which rotates rapidly during ejaculation to spray the semen in the vagina.

Physiology. Ewes are monestrous; in northern regions they rut in November and December (Geist 1971; Nichols 1978). However, for desert bighorns in more southerly regions, the rut may persist for up to nine months, although it reaches a peak in August and September (Welles and Welles 1961). Prior to estrus, one to four Graafian follicles develop simultaneously. Estrus lasts two days and receptive females are recognizable by their swollen vulva and lumen. At the end of estrus, ovulation is spontaneous. Corpora lutea grow until they obscure all other ovarian bodies and persist up to five months after parturition (Nichols 1978).

With our present state of knowledge, it is almost impossible to determine reproductive history during the fall, for degenerating corpora lutea of pregnancy cannot be differentiated from degenerating corpora lutea of ovulation (Johnson 1974). Also, there is no apparent correlation between the numbers of ovarian scars and age (Nichols 1978).

Breeding. An anestrous ewe may be courted throughout the year, particularly by young rams, but it tends to avoid them and withdraw from any mounting attempts (Welles and Welles 1961; Geist 1968, 1971). During estrus, the ewe becomes more aggressive. It searches for the largest horned rams, accepts their mounting attempts, and may even court another ram if its partner

TABLE 52.2. Wild sheep permanent tooth eruption patterns

Tooth	Dall's	Rocky Mountain Bighorn	Desert Bighorn
M_1	1–4 months	1–4 months	6 months
M_2	8–13	8–13	16
I_1	13–16	13–16	12
I_2	25–28	25–28	24
P_2	27–32	25	24
P_3	25–30	25	24
P_4	25–30	25–30	24
M_3	22–40	22–40	30
I_3	33–36	33–36	36
C	45–48	45–48	48

SOURCE: Compiled from Deming (1952) and Hemming (1969).

is exhausted. A ewe in estrus may or may not leave its band (Jones 1959; Geist 1971).

The position in the dominance hierarchy of rams in rut essentially determines whether or not they will breed. Fights or jousts for dominance become most serious during the rut, but even then serious injuries are rare (Smith 1954). Rams that establish their dominance (see "Social Organization") do most of the courting and breeding (Geist 1966b, 1971).

Breeding is polygamous. Rams move freely among bands of ewes seeking those in estrus (Smith 1954; McCann 1956; Leslie and Douglas 1979). Ram bands of up to eight mature animals may roam together (except desert bighorns, which usually travel singly). Although Murie (1944) suggested that Dall's rams exhibited harem-gathering behavior, it is generally believed that they do not (Hansen 1967; Geist 1971).

After finding a receptive ewe, dominant rams attempt to drive away other rams. Typically, the ram checks for estrus by sniffing the ewe's vulva and tasting her urine. If she is in heat he often delivers a stiff foreleg kick to stimulate a chase, or the ewe may initiate the chase. It may or may not be strenuous. If one of the pair tires, the partner usually waits until the other is ready to resume. When the ewe is sufficiently stimulated she assumes a position of lordosis. If the ram becomes exhausted, nearby subordinates may usurp his position at any time (Blood 1963; Geist 1971).

The age at which ewes attain puberty is quite variable and is dependent mainly on their physical condition. Nichols (1978) found 75 percent of yearling Dall's ewes collected after the rut to be pregnant. Although Woodgerd (1964) reported that yearling Rocky Mountain bighorn ewes in good condition or in captivity become mature and breed at 18 months, most wild bighorns become mature at 2½ years of age (Cowan and Geist 1971). Likewise, large-bodied rams may reach sexual maturity within 18 months but smaller rams may take as long as 36 months (Nichols 1978; Woodgerd 1964; and others).

Very old ewes generally do not breed (Jones 1959); however, Nichols (1978) found two 13-year-old Dall's ewes that were pregnant. A 15-year-old ewe died immediately after parturition. Nichols felt that males were reproductively active throughout their lives, but McCann (1956) reported that a few very old rams remained on the summer range and never participated in the fall rut.

Homosexuality is normal in North American wild sheep, for the physical and behavioral characteristics of estrous ewes and young rams are quite similar, including submissiveness to mounting. It has been suggested this is an adaptation that aids in the survival of immature sheep by permitting their existence with the older rams (Geist 1968, 1971).

All sheep, including domestic varieties, are interfertile. Bighorn rams occasionally enter farmyards and breed with domestic ewes; the offspring have a mixture of traits (Senger and Forrester 1960; O'Connor 1974). In Nevada, a desert bighorn ram was experimentally bred to a domestic ewe. The ewe gave birth to two female lambs but, contrary to expectation, the offspring proved fertile (Pulling 1945). Pulling also reported that the ram was bred to these hybrid daughters and that they both had lambs that were three-quarters wildstock. While successful interbreeding has occurred between domestic and Rocky Mountain as well as California subspecies, the occurrence is rare and not well documented.

Parturition. Gestation lasts from 5½ to 6 months for mountain sheep. The average duration is probably 175 days (Geist 1971). The rate of pregnancy is variable. Nichols (1978) found that 100 percent of adult Dall's ewes studied were pregnant, compared to 75 percent of yearlings. The probability of pregnancy, particularly for yearlings, is lowered when the number of rams declines. However, too many rams may also cause a decline of pregnancy because of interference (Smith 1954).

Lambing, for northern sheep, occurs between late April and late June, with the majority of lambs being born before the end of May (Pitzman 1970; Nichols 1974b, 1978). Desert bighorn lambs occasionally are dropped throughout the year, with the exception of during the rut; the peak, however, is from January to April (Leslie and Douglas 1979). The majority of ewes give birth to only one lamb a year (Welles and Welles 1961; Geist 1971; Nichols 1978; and others). Few instances of twinning have been recorded; however, Spalding (1966) found that 4 of 12 ewes killed by autos in British Columbia were carrying healthy twin fetuses. Recent studies with mountain sheep in captivity (Eccles and Shackleton 1979) reveal that the incidence of twinning is far greater than was previously realized.

The birth of a bighorn lamb takes relatively little time. Approximately 45 minutes prior to parturition, the placental membranes begin to appear. Ten minutes before birth the ewe commences with heavy panting. The cephalic birth takes approximately 10–15 minutes, at the end of which the ewe stands up to facilitate the final expulsion.

Newborn. Immediately following parturition the ewe licks off the placental fluids from the neonate, beginning with the head and face. Most ewes begin eating the placental membranes soon afterward.

Lambs are precocial and within a day or so climb almost as well as their mother (Murie 1944; Pitzman 1970). By their second spring bighorns are totally independent of their mother (Geist 1971). Within two weeks lambs are able to nibble grass (Murie 1944; Welles and Welles 1961), and they are weaned between one and seven months of age, most likely at four to five months (Blood 1963; and others) (figure 52.4). Neonatal *O. dalli* weigh between 3 and 4 kg (Geist 1971; Bunnell and Olsen 1976). In October they may weigh around 32 kg (Blood et al. 1970). Ewes reach their adult weight by four to five years of age, while rams do not achieve maximum weight until they are six or seven years old (Blood et al. 1970; Bunnel and Olsen 1976).

FIGURE 52.4. Immature bighorn sheep (*Ovis canadensis*), showing rudimentary horn development.

ECOLOGY

Mountain sheep inhabit remote mountain and desert regions, where human contacts are limited. They have become adapted to a wide range of climates from the frigid arctic mountains of Alaska and the Yukon to the scorching deserts of the southwestern United States and Mexico. Although the temperatures often vary greatly, sheep prefer drier regions. Northern races typically are more fit and enjoy enhanced reproduction and lowered mortality when winter temperatures are cold, snowfall moderate, and wind persistent (Nichols 1973, 1974c, 1976; Heimer and Smith 1975). Of course, water must be available for the desert bighorns to thrive (Jones et al. 1957). Bighorns for the most part are more heavily impacted by humans than are the more remote thinhorns. Most thinhorn populations inhabit relatively undisturbed ecosystems, and diseases cause few problems.

All North American wild sheep are restricted to semiopen, precipitous terrain with rocky slopes, ridges, and cliffs or rugged canyons (Todd 1972). It has been suggested that human activities and wolf predation have restricted sheep to rougher terrain, but it is generally considered unlikely that they ever roamed out onto plains (Buechner 1960). McCann (1956) felt that sheep remain on rough, craggy terrain because they find a great deal of forage and little competition from other ungulates. Escape terrain is also a very important habitat requirement. While sheep are not always found in precipitous mountain areas, ewes and lambs rely on these areas for escape cover, especially during the lambing period.

The vegetation of areas inhabited by bighorns and thinhorns generally consists of few trees, some low-growing shrubs, and either natural or burned grasslands (Buechner 1960; Geist 1971). Sheep avoid thick forests, although they temporarily seek refuge from adverse weather in tree stands or when frightened (Spalding and Mitchell 1970; Bunnell 1978). Heimer (1973) suggested that mineral licks be considered a habitat requirement of Dall's sheep; their presence certainly affects the local distribution of sheep. Except for introduced populations it is probably rare when habitat use ever reaches 100 percent of the available area, because of the traditional gregariousness and home range instincts of sheep or environmental pressures (Shannon et al. 1975; Hudson et al. 1976). For example, Stelfox and Taber (in Etter 1973) found that a majority of Canadian bighorns winter on less than 10 percent of the available winter range, and Leslie and Douglas (1979) found that desert bighorn ewes used no more than 55 percent of their total home range during any one season.

Lambing success varies considerably. Generally, 50 lambs:100 ewes is considered adequate for a population increase. While Murie (1944) found only 16 lambs:100 ewes in Mt. McKinley National Park, Alaska, Woodard et al. (1974) reported a ratio of 83 lambs:100 ewes in Colorado.

The sex ratio of sheep is assumed to be even at birth, but normal adult ratios vary widely. Constan (1972) reported a ratio of 31 rams:100 ewes in Wyoming, while Buechner (1960) reported 137 rams:100 ewes in an Arizona population. Various local factors are apparently responsible for the observed different rate of survival between the sexes.

The lowest sheep population densities occur on the poorest ranges of the desert, where densities may be only 3 sheep per km². The highest densities typically occur on northern winter ranges where there may be 19–23 sheep per km² (Blood 1963).

Activity. Sheep are diurnal and usually exhibit little nocturnal activity (Welles and Welles 1961; Blood 1963). Woolf et al. (1970) reported an interesting exception: the sheep of Yellowstone National Park, probably in response to heavy tourist pressure, fed long after sunset. Although the sheep fed at intervals throughout the night, they rarely wandered far from their bedgrounds. The daily activity pattern consists of alternate feeding and resting periods. The periods are not synchronous either within or between groups, as some sheep will be resting while others are feeding (Jones 1959; Todd 1972). Feeding periods are shortest when the quantity and quality of forage are high (Welles and Welles 1961). Activity follows a seasonal pattern that is probably related more to photoperiod than to temperature. In the summer, approximately half of the day is spent feeding during three to four periods (Blood 1963; and others). In the winter, or during severe weather, less time is spent feeding: usually just one midday period (Smith 1954; Geist 1971). During the resting periods sheep bed down, chew their cud

(Jones 1959), and occasionally rise to stretch and defecate. Bighorns can rest almost equally well standing in a rather stiff posture with both hind legs somewhat outstretched (Welles and Welles 1961).

Beds. Beds are made in loose surface material that is cleared away by several swipes of the forefeet. An oval depression is formed measuring approximately 0.6 × 0.4 m and varying from a few cm to over 0.3 m in depth (Jones 1959; Clark 1970). A new bed is made each time a sheep lies down; however, old beds may be reused after they have been pawed out deeper (Jones 1959; Welles and Welles 1961). Favorite spots are along ridges but may also be in erosion sites formed by grizzly bears (*Ursus arctos*) digging for ground squirrels (Geist 1971). During the day, beds are made at random locations in the open near the feeding site except during inclement weather, when they are located in more rugged terrain. Toward evening, beds are more carefully chosen. They are always near rugged terrain with cliffs or large rocks in back (Smith 1954; Jones 1959). Caves, where available, also provide excellent shelter for bedding (Geist 1968; Nichols and Erickson 1969).

Home Range and Movements. Sheep are not territorial; rather, they follow season-specific home ranges. While some sheep, especially introduced populations, spend the entire year in one area, most wild sheep follow established migration routes. Young sheep learn movement patterns by following the older ewes or larger-horned rams. However, the associations between young and old individuals are labile; thus, a young sheep may acquire parts of the patterns of many older sheep. By four years of age, individuals have established home ranges that they utilize throughout life (Geist 1967, 1971).

The size of the home range is generally smallest in the winter when forage is scarce. However, the summer home range of desert bighorns is smaller than the winter range due to their dependence on water sources (Jones et al. 1957). Woolf et al. (1970) found that bands of sheep in Yellowstone National Park usually roamed 3–5 km daily on their summer range, less often up to 16 km. Welles and Welles (1961) found that desert bighorns remained within a radius of 32 km throughout life.

The local distribution of sheep depends on the habitat conditions, sex, and age of the animals and the season of the year. Shannon et al. (1975) found that slope, distance to escape terrain, salt availability, elevation, aspect, forest cover, shrub productivity, biomass and nitrogen content of palatable grasses, and snow all affected local seasonal distribution. Except during the rut the sexes remain separate, with ewes preferring more precipitous terrain (Blood 1963; Woolf et al. 1970; Geist 1971; Geist and Petocz 1977).

Generally, sheep have two distinct, separate ranges in summer and winter, with corresponding spring and fall migrations. In the fall ewes and lambs normally precede rams down from the summer range. In the spring ewes that are going to lamb on their summer range leave earliest; otherwise rams usually are the first to leave the winter range (Smith 1954; Blood 1963).

Geist (1971) felt that rams have six distinct ranges (prerut, rut, winter, late winter–spring, salt lick, and summer), while ewes have four (winter, spring, lambing, and summer). On the prerut, or fall, range, rams assemble around partly exposed level spaces in preparation for the rut. There is often considerable social interaction and jousting.

The "rut range" overlaps the ewe's winter range where the sexes mingle for approximately two weeks. The oldest rams are the first to leave for their winter range and eventually the others follow (Geist 1971; Petocz 1973).

Most of the year is spent on the winter range where the elevation is typically below 3,300 m (Smith 1954; Geist 1971). The aspect is usually south or southwest (Shannon et al. 1975; Hudson et al. 1976; Stelfox 1976). Rugged terrain is always nearby, although rams often venture out onto the more open slopes (Geist and Petocz 1977). Jones et al. (1957) found the elevation of the desert bighorns' winter range to be indeterminate. They noted that the animals never strayed far from the base of the mountains, usually the eastern aspect, where they used dry gullies. During severe winter weather if snow becomes unusually deep or crusted, both sexes of northern sheep move to slightly higher elevations where wind and sunshine have cleared the more exposed slopes and ridges (Nichols and Erickson 1969; Geist 1971; Geist and Petocz 1977). This occurs most frequently during midwinter.

The spring range is primarily characterized by the same environmental parameters as the winter range. However, sheep begin to respond to local green-ups along streambanks and valleys (Hudson 1976; Stelfox 1976); hence, their distribution is somewhat broadened.

Salt licks are select areas with the greatest activity around them in the spring (McCann 1956; Samuel et al. 1975).

The lambing range has been classified as part of the winter range or as intermediate between winter and summer range. Preferred lambing range is in the most precipitous, inaccessible cliffs near forage, and generally has a dry, southern exposure (Pitzman 1970; Geist 1971).

In the summer most sheep are found grazing on the grassland meadows and plateaus above the timber (Shannon et al. 1975; and others). The aspect of the summer grazing range changes depending upon forage succulence (Smith 1954; Stelfox 1976). In the early summer the southern and southwestern exposures are most frequently utilized; however, in the case of desert bighorns (Jones 1959) the eastern aspect is preferred. By late summer the more northerly exposures become preferred.

Minor, but frequent, sheep movements are caused by temporary environmental conditions. Unlike some other ungulates, migratory sheep will leave a part of their range before it becomes overgrazed or dessicated (Welles and Welles 1961). They follow the best foraging conditions, especially the green-up of succulent grasses and forbs (Blood 1963). Weather often causes changes in the local distribution of sheep. During periods of drought, desert bighorns wander from waterhole to waterhole much more frequently (McCann 1956). Stelfox (1976) found a high correlation between movements and barometric changes. Sheep moved away from exposed slopes before storms arrived.

Seasonal migrations may be affected by the weather or the availability of food and water, but ultimately are timed by evolved, internal mechanisms, such as the reproductive urge or photoperiod response (Geist 1971). The distance of seasonal migration may vary from 0.8 to 60 km, depending on weather conditions and forage availability (Spalding and Mitchell 1970).

During the spring migration sheep leisurely drift along, basically foraging their way up the slope (Stelfox 1976). In the fall, however, during the rut, their movements are much more determinate and hasty (Geist 1971). They avoid timbered areas but move quickly when forced to cross them. Mountain sheep may spend hours, or even days, watching low country routes before crossing (Murie 1944). When they do cross they adhere to streambeds wherever possible, following in single file (Geist 1971).

Competition. The effect of domestic livestock grazing on mountain sheep range is very controversial and depends on the proximity and population size of competing species. The forage resource is usually considered in competition studies but social intolerance of domestic livestock by sheep is also a form of competition. Domestic livestock have been reported to have little deleterious effect if they do not graze on critical winter ranges of sheep (McCann 1956; Jones 1959; Hudson et al. 1976). Nevertheless, extensive competition by livestock, especially on public lands, persists and is one of the reasons for the decline in density of wild sheep populations (Buechner 1960; Woodard et al. 1974).

Domestic sheep eat the same forage as wild sheep and are direct competitors when on the same range. However, there is comparatively little spatial overlap of the species, except where bighorns migrate through domestic sheep range, or where domestic sheep have usurped bighorn summer alpine ranges, as in portions of Colorado and Wyoming (Buechner 1960).

Competition between bighorn sheep and cattle has been reported to be a serious problem in British Columbia (Demarchi 1962) and in Idaho (Morgan 1973). Drewek (1970) and Hudson et al. (1976), however, found that mountain sheep distributions were only weakly influenced by grazing cattle. Halloran (1949) found that cattle in Arizona often deprived desert bighorn of range near critical water supplies in narrow mountain canyons. McCann (1956) suggested that cattle may indirectly affect mountain sheep by competing with elk (*Cervus elaphus*) and forcing them onto sheep ranges.

Competition from burros is a serious problem where feral burros are found on desert bighorn range. Desert sheep biologists (in Trefethen 1975) described the serious bighorn:burro situation and established guidelines for management of desert bighorn sheep. In fact, these biologists recommended: "Wild, free roaming burros, horses and livestock should be removed from desert bighorn habitats."

Elk can be serious competitors with bighorn sheep where winter ranges overlap. Competition studies in Washington State (Estes 1979) revealed that California bighorn and Rocky Mountain elk food habits were similar in forage class composition throughout the year. Estes (1979) concluded, however, that although forage class composition of both elk and sheep showed remarkable similarity, many differences in species lessened their competition. When the winter weather is mild and population levels are normal, elk remain in the valleys. Severe weather, often coupled with overpopulation, forces them up to the exposed forage at higher elevations where the sheep are found (McCann 1956; Oldemeyer et al. 1971). Elk may become so numerous that huge areas of potential forage for sheep become unavailable due to the icy crust that forms as a result of elk trampling the snow (Cowan 1947). Likewise, Stelfox (1976) found that sheep winter ranges were grazed heavily by elk during the summer. Because there is extreme similarity of the preferred forage of sheep and elk (Cowan 1947; McCann 1956; Oldemeyer et al. 1971; Constan 1972; Stelfox 1976), where their ranges are sympatric, the two species are serious competitors.

Deer (*Odocoileus virginianus, O. hemionus*) are other potentially important ungulate competitors. Deer use of sheep ranges varies according to season, weather, range condition, and population density. Blood (1967) noted that few deer used winter ranges of sheep in British Columbia. However, Stelfox (1976) found that the percentage utilization of winter ranges by deer surpassed that of sheep in the fall, winter, and spring. Several investigators found that desert bighorns and deer used the same ranges (Halloran and Kennedy 1949; Smith 1954; Jones et al. 1957) without detrimental effects.

FOOD HABITS

Wild sheep have relatively large rumens because of the abrasiveness of the forage in their diet. Their rumen digests 90 percent of the cellulose and 69 percent of the organic matter that passes through it (Geist 1971).

The various races of sheep range over a variety of habitats, and forage preferences reflect changes in species availability. While mountain sheep are basically grazers of grass and forbs, they will eat other vegetation depending on availability (McCann 1956; Clark 1970; Shannon et al. 1975). One study conducted in Washington State (Estes 1979) compared forage availability with consumption. This study revealed that of the abundant grass species, *Poa sandbergii* and *Bromus* spp. were preferred and *Agropyron spicatum* and *Festuca idahoensis* were avoided. Forbs are probably the most palatable form of vegetation but are only seasonally available (Todd 1972). On the average, forbs are higher in these important nutrients than grasses, but Demarchi (1968) found that bluebunch wheatgrass (*Agropyron spicatum*) is the most nutritious forage item. Phosphorus and calcium levels of range plants decline in the winter (Nichols 1974*c*), but Hebert (1972) found that they remain at acceptable levels in sheep all year.

Seasonal forage consumption depends upon plant succulence, nutrient content, and availability (Todd 1975). During warm months sheep use areas of forage green-up, which maximizes their nutrient intake and probably is more important than any specific plant species. In northern populations browse is used most extensively in the fall and winter, particularly during heavy snow accumulations. Desert bighorns, unlike the other races, rely on browse most of the year (Todd 1972).

The staple foods of Dall's sheep are fescues (*Festuca* spp.) and sedges (*Carex* spp.), particularly the seed heads. Willow (*Salix* spp.), horseweed (*Erigeron* spp.), and alpine fireweed (*Epilobium latifolium*) are important throughout the spring and summer. Horsetails (*Equisetum* spp.) and Richardson's saxifrage (*Therefon richardsonii*) are highly relished (Murie 1944; Viercek 1963). Nichols (1974*a*) noted that lichens and mosses were ingested frequently, although possibly incidental to grazing for grasses.

Bighorns eat a variety of grasses, including sedges, bluegrasses (*Poa* spp.), wheatgrasses (*Agropyron* spp.), bromes, and fescues. Browse species include sagebrush (*Artemisia* spp.), willow, rabbitbrush (*Chrysothamnus* spp.), curlleaf mountain mahogany (*Cercocarpus ledifolius*), winterfat (*Eurotia lanata*), bitterbrush (*Purshia* spp.), and green ephedra (*Ephedra* spp.). Forbs include phlox (*Phlox* spp.), cinquefoils (*Potentilla* spp.), twinflower (*Linnaea americana*), and clover (*Trifolium* spp.) (Constan 1972; Todd 1972, 1975; Stelfox 1976; Thorne 1976).

Because of the dry climate, browse is the dominant food of the desert bighorn, and includes desert holly (*Atriplex hymenelytra*), honeysweet (*Tidestromia oblongifolia*), brittlebush or encelia (*Encelia* spp.), hairy mountain mahogany (*Cercocarpus breviflorus*), silk tassel (*Garrya wrightii*), desert mallow (*Sphaeralcea ambigua*), Russian thistle (*Salsola perifera*), false mesquite (*Calliandra eriophylla*), goatnut (*Simmondsia chinensis*), white ratany (*Krameria canescens*), bursage (*Hyptis emoryi*), mesquite (*Prosopis juliflora*), catclaw (*Acacia greggii*), ironwood (*Olneya tesota*), paloverde (*Cercidium* spp.), and coffeeberry (*Rhamnus purshiana*). During the dry season cacti pulp and fruit are eaten, including opuntia (*Opuntia* spp.), barrel cacti (*Cylindropuntia* spp.), pincushion (*Mammillaria* spp.), and giant saguaro (*Cereus giganteus*). Graminae dry leafage is eaten throughout the year and is an important food reserve, especially near waterholes (Halloran and Kennedy 1949; Halloran and Crandell 1953; Jones et al. 1957; Welles and Welles 1961; Todd 1972; Leslie and Douglas 1979).

A few unusual plants have been recorded as eaten by mountain sheep. Conifers may be eaten in the winter (Todd 1972). Blood (1967) found that death camas (*Zygadenus venenosus*) and lupine (*Lupinus* spp.), both poisonous to domestic livestock, were eaten with no observed ill effects. Nichols (1973) reported that Dall's sheep were attracted to the false Hellebore plant (*Veratrum* spp.), which is poisonous to domestic sheep. Dall's sheep licked the roots and behaved much as they do at salt licks.

The mineral requirements of mountain sheep are still unclear. They are attracted to salt licks even years after all the salt is gone, but it is unknown whether they seek them for a condiment or out of necessity (Smith 1954). Palatability tests show that they prefer sodium chloride to other compounds, but Murie (1944) reported a salt lick in McKinley National Park, Alaska, containing only calcium and iron phosphates in soluble form. Salt licking may be a means of replenishing bone mineral reserves after winter's depletion (Geist 1971).

Snow and succulents provide an almost year-round water source for northern sheep (Geist 1971; Todd 1972). However, for the desert bighorn, drinking water is critical, especially if competition is keen (Buechner 1960). Desert bighorns primarily utilize ephemeral water sources. They may drink every day if water is nearby, particularly the lambs and ewes, but may go without water for up to 14 days in the dry season (Welles and Welles 1961). During drought ewes are restricted to within 1–3 km of water, but rams roam farther (Blong and Pollard 1968; Leslie and Douglas 1979). Waterholes with repeated human disturbances, as well as those that are foul or alkaline, are avoided (Jones et al. 1957; Blong and Pollard 1968). Visits to waterholes are most frequent in the early morning or late evening (Jones 1959). Since water is one of the major limiting factors of desert bighorns, management agencies have installed cisterns and other water developments in critical areas.

BEHAVIOR

Foraging and Feeding. While foraging, sheep are continually on the move. Their average daily movement is between 0.4 and 0.8 km, but may reach 3.2 km (Woolf et al. 1970). They appear to wander aimlessly but steadily. They avoid foraging in deep snow, but Nichols (1974*a*) found that forage covered with snow often was more succulent than that which was exposed and sheep preferred digging for it if the snow was not deep or crusted.

Sheep may just nip the heads and tips of the tenderest plants, such as forbs, or paw up entire plants through dirt or snow (Jones 1959; Todd 1972; Estes 1979). At times they will use their horns to pry up vegetation wedged under stones or roots (Norris 1955). Feeding efficiency appears to be poorest for sheep in the smallest bands. Bunnell and Olsen (1976) found that single sheep constantly interrupt their feeding to survey the landscape, presumably for predators.

Social Organization. The basic unit of sheep social organization is the band. Adult bands are segregated by sex (Geist 1971). One to several bands constitute a

FIGURE 52.5. Typical herd composition of ewes and lambs.

herd (Murphy and Whitten 1976). The size of the band varies depending upon environmental disturbances, habitat space, distribution of available resources, predation, season, and sex. In winter, when available habitat is limited and bands are close together, large fluctuations in band size occur (Woolf et al. 1970). Berger (1978) felt that the most efficient band size with respect to predator avoidance was five; little advantage is secured with any more individuals. Ram bands are largest in the spring and smallest during the rut. Ewe bands are largest after the spring lambing season and smallest during the fall rut and winter (Blood 1963). Leslie and Douglas (1979) and others found that desert bighorn bands dispersed during drought.

The largest bands are composed of ewes and lambs (figure 52.5). Rams are more solitary. Generally, a mature ewe leads the band but this may vary, particularly for short movements. Ewe band size varies from 5 to 15 sheep, on the average, and often consists of a family group (Jones 1959; Welles and Welles 1961). Young rams one to two years old disassociate themselves from ewe bands, and after a brief period of wandering, associate themselves with a ram band.

Ram bands may be as large as 12 individuals but 2 to 5 is more common. The coexistence of rams in a band depends upon a dominance hierarchy, established by jousts or fights (Geist 1966b). The more severe challenging fights occur when two similar-sized rams meet for the first time, as when one attempts to enter another's band (Blood 1963; Geist 1971). The rams stand from 5 to 14 m apart, then run at each other,

rising on their hind legs just before butting heads. Hansen (1967) estimated that animals each attain speeds up to 54 kph. They normally butt each other's horns but occasionally contact the opponent's sides (Geist 1971). Jousts are not always limited to large rams: subordinate rams may become involved too. Eventually one of the rams runs away or begins appeasement behavior characteristic of subordinates, and the joust ends.

Communication. Wild sheep vocalize in a manner similar to domestic sheep, but with more vibrato and deeper tone. Bleating occurs most frequently in large maternal ewe bands, and serves to maintain lamb and mother contacts (Welles and Welles 1961).

Play. There is some play between adults but usually lambs play with each other or their mothers (Murie 1944; Jones 1959). There are two types of play: contact play, during which similar-aged lambs butt each other, and locomotor play, during which all ages participate in chasing each other (Berger 1978). The frequency of play is highest in the early morning and evening (Welles and Welles 1961).

Snow. Snow produces little noticeable effect on the intensity of social interaction during the rut, but depresses all behavior afterward. Under severe snow conditions, rams, regardless of dominance order, become uncommonly aggressive toward each other (Petocz 1973).

Horning. Sheep "horn" shrubs and small trees by pushing and rubbing them with their horns. The reason for this behavior is unknown, although it may remove irritants or provide a stimulant. In either case, it occurs most frequently during intensive social interactions or immediately prior to long movements (Geist 1971).

Wariness. Sheep respond to disturbances in one of three ways: they assume an attention posture, they assume an alarm posture, or, if startled at close range, they run. In the attention posture, they stare in the direction of the disturbance. When alarmed they may snort, paw the ground, bow their head, or, in the presence of wolves, huddle in a tight circle facing out (Murie 1944; Geist 1971). Ewes with young lambs are the most wary (Murie 1944). If not hunted, some populations are tolerant of humans and easily become habituated (Geist 1971).

MORTALITY

Survivorship curves of sheep are commonly cited as classic examples of K-selected populations. Mortality is high for sheep 1-2 years of age, drops to a relatively low rate for 2-8 year olds, then increases to a maximum for those older than 8 or 9 years (Geist 1966*b*, 1971; Hansen 1967; and others). Leslie and Douglas (1979), however, reported that desert bighorns appear to have nearly equal mortality among all age groups and speculated that this is characteristic of sheep populations limited by factors other than predation or disease. Fetal absorption is common for populations on poor range associated with severe winter weather, drought, or high densities (Heimer 1976; and others). Sheep that survive may live to be 15-17 years of age (Hansen 1967; Clark 1970), but 10-12 years is more realistic (Bunnell 1978). Geist (1971), and many other researchers believed that the stress placed on rams during the rut may be responsible for differential mortality between rams and ewes. Personal observations in nonhunted populations also reflect a much greater natural mortality in the ram segment of the herd.

Under certain conditions, mountain sheep may become vulnerable to the rigors of harsh weather. Long periods of deep or crusted snow deplete their energy reserves. Drought reduces forage availability and consequently milk production in lactating ewes, which may affect lamb mortality. Drought also may increase predation by concentrating sheep near the remaining waterholes (Halloran and Deming 1958).

Accidents occur periodically. Broken limbs and death from falls are not uncommon and avalanches in early spring cause some mortality (Nichols and Erickson 1969; Pitzman 1970). Natural rain catchments, whose sides are too steep for escape, cause many desert bighorn drownings (Mensch 1969).

Predators. Golden Eagles (*Aquila chrysaetos*) occasionally threaten lambs but are rarely successful in taking one. Mountain sheep are only an incidental food item in the diet of grizzly or black bears (*Ursus arctos, U. americanus*) and wolverines (*Gulo gulo*), and are eaten only as carrion (Murie 1944; Pitzman 1970). Coyotes (*Canis latrans*) are frequently seen near

mountain sheep in local areas and could be an efficient predator under certain conditions. Thorne (1976), for example, found that 7 of 11 coyote scats contained bighorn remains. In the far north, Murie (1944) felt that coyotes were outcompeted by the wolf (*Canis lupus*). In their range, wolves are the major predator of sheep (Scott et al. 1950; Geist 1971). In McKinley National Park, Alaska, Murie (1944) indicated that populations of wolves and sheep were in equilibrium and the wolves were a limiting factor. This study was conducted from 1939 to 1941 on land where wolf hunting was forbidden. In a later study, however, Nichols and Erickson (1969) stated that in Alaska, wolf predation was a relatively minor factor of sheep mortality. Wolves chase many bands of sheep until by their persistence they isolate the weaker individuals (Murie 1944; Geist 1971). Thus, sick or infirm animals are culled from the herd. However, Heimer (1976) found that the normal incidence of disease in Dall's sheep in the McKinley National Park was equivalent to the incidence in those sheep preyed upon by wolves. Therefore, he felt that wolves do not selectively cull sheep, at least not with respect to diseased animals.

Parasites and Diseases. Four viral agents have been associated with mountain sheep. One of these, contagious ecthyma, or soremouth, is easily recognized. Lesions appear in the spring and summer, commonly on the lips but also on the face, ears, feet, around the eyes, and even on the udders of ewes nursing infected lambs. Soremouth usually runs a benign course and uncomplicated cases heal in about a month. Secondary infections, however, often result in death (Davis 1956). Contagious ecthyma is quite debilitating and may result in stunted growth. Since mountain sheep are concentrated around salting sites in the spring, contagious ecthyma is most likely found in these areas. In western Canada where artificial salting has been a common practice, the incidence of soremouth is high around salting sites (Blood 1971; Samuel et al. 1975). In other bighorn ranges, however, where artificial salting has been carried out for years, no soremouth has been reported.

Bluetongue, or malignant catarrhal fever, is relatively new to the United States. Bluetongue is transmitted by a *Culicoides* gnat, hence, it is most troublesome in the wet season or around watering areas in the dry season. It is noncontagious. Yearling sheep appear to be most susceptible. Symptoms include fever, depression, lack of appetite, and inflammation and laceration of mucous membranes of the nasal cavities, mouth, and tongue. The tongue may become swollen and cyanotic (bluish). The acute phase is short but may be prolonged if the animal is debilitated (McKercher et al. 1956). Although the susceptibility of the desert bighorn is unknown, this disease has become a problem in Texas and may have become a potential limiting factor since its introduction (Robinson et al. 1967, 1974).

Encephalitis, an arbovirus, also occurs in sheep. Little is known of its relationship to sheep, but neutralizing antibodies of the western and California varieties were detected in 8 of 65 bighorns (12 percent) from Wyoming, Montana, and New Mexico (Trainer and Hansen 1969).

Parks et al. (1972) isolated PI-3 virus from a bighorn with an acute respiratory illness. Of 30 bighorns from Wyoming tested, almost all were found to have been exposed to PI-3 (Thorne 1976).

Diseases of bacterial origin may destroy bone and occasionally soft tissue. They originate in lesions, either those associated with changes in dentition or those that occur when sheep are forced to browse on woody plants for extended periods of time. Actinomycosis—"lumpy jaw" or "big jaw"—is caused by the mycelial bacterium *Actinomyces bovis*. It primarily attacks bones of the skull (Monlux and Davis 1956). It is prevalent among Dall's sheep (Ericson and Neiland 1972), but desert bighorns are essentially unaffected (Welles and Welles 1961). Necrotic stomatitis is caused by *A. necrophorus*, which primarily attacks the soft mucous oral tissues (Murie 1944). Actinobacillosis, or "wooden tongue," is caused by the bacterium *Actinobacillus lignieresi*. It attacks the soft tissues, primarily the lymph nodes of the neck or the tongue (Monlux and Davis 1956). Bacterial infections are frequently found in lesions developed from other causes. One of the more virulent bacterial organisms, *Corynebacterium pryogenes*, has been found in bighorns from Montana (Woodgerd 1964) and Washington State.

Sporozoan diseases include infection by *Sarcocystis tenella*, a muscle parasite (Becklund and Senger 1967), and coccidiosis. Eleven species of *Eimeria* are responsible for coccidiosis in sheep, which causes diarrhea or "scours" (Lotze 1956). It is assumed that all 11 species are equally contagious but *E. ninaekohlyakimovae* is probably the most virulent and common.

Cestodes that parasitize mountain sheep include the thin-necked bladderworm (*Taenia hydatigena*), two species of double-pored ruminant tapeworm (*Moniezia* spp.), the fringed tapeworm (*Thysanosoma actinoides*), and the skirted tapeworm (*Wyominia tetoni*) (Becklund and Senger 1967).

Abdominal nematodes include one species of spiruroid (*Gongylonema* sp.) (Jones et al. 1957) and two species of filarium (*Setaria* spp.) (Buechner 1960; Becklund and Senger 1967; Becklund and Walker 1967, 1969).

Because of the intricate ruminant digestive system, numerous species of gastrointestinal nematode have been found in mountain sheep (table 52.3). Where known, the life cycle is direct (Allen 1956; Kates et al. 1956). Symptoms of gastrointestinal nematode infestation include diarrhea, anemia, edema, emaciation, appetite loss, and weakness (Kates et al. 1956).

The lungworm-pneumonia complex, or hemorrhagic septicemia, can be caused by at least seven species of nematode, including hair lungworms (*Muellerius capillaris*, *M. minutissimus*), thread lungworms (*Dictyocaulus vivaparus*), and the protostrongylids (*Protostrongylus frosti*, *P. rufescens*, *P. rushi*, *P. stilesi*) (Smith 1954; Becklund and Senger 1967; Demartini and Davies 1977). The two most important are *P. rushi*, which attacks the bronchioles, and *P. stilesi*, which attacks lung parenchyma (Forrester and Senger 1964). Clinical symptoms include fever, cough, hyperpnea (rare), nasal discharge, diarrhea, loss of weight, and an unthrifty appearance (Woolf et al. 1970). Hudson et al. (1971) described a procedure for detecting lungworms, in vitro, by the presence of homocytotropic antibodies. Most commonly, however, lungworm is detected by examination of lung tissue or fecal matter.

The incidence of lungworm infestation approaches 100 percent in some herds, although the level

TABLE 52.3. Gastrointestinal nematodes of North American wild sheep

Genus	Number of Species Identified	Common Name	Site(s) of Infection
Capillaria	1 unknown	Capillarids	small intestine
Cooperia	*oncophora, surnabada*	Cooperias	middle of small intestine
Haemonchus	*contortus, placei*	stomach worms	abomasum
Nematodirus	*abnormalis, archari, davtiani, filicollis, helvetianus, lanceolatus, maculosus, odocoilei, oratianus, spathiger*	thread-necked stongyles	first third of small intestine
Oesophagostomum	1 (*venulosum*)	nodular worms	large intestine
Ostertagia	*circumcincta, lyrata, marshalli, accidentalis, ostertagi trifurcata*	medium stomach worms	stomach lining
Pseudostertagia	1 (*bullosa*)	medium stomach worm	stomach lining
Skrjabinema	1 (*ovis*)	pinworm	large intestine
Telodorsagia	1 (*davtiana*)		
Trichostrongylus	*axei, colubriformis, rugatus*	stomach and intestinal hair-worms	stomach glands or near pyloric valve
Trichuris	*discolor, ovis schumakovitschi*	ruminant whipworms	large intestine

SOURCE: Compiled from Becklund and Senger 1967; Becklund and Walker 1967; Knight and Uhazy 1973; Smith 1954; Uhazy and Holmes 1971; Kates et al. 1956; Allen 1956.

of individual infection varies depending upon sheep and domestic livestock densities, range conditions, climate, season, and age. During the summer, lambs and yearlings may not have infestations (Murie 1944; Uhazy et al. 1973; Woodard et al. 1974), but by their first winter most sheep have contracted lungworms (Forrester and Senger 1964; Uhazy and Holmes 1971). Desert bighorns appear to have lighter infestations, possibly due to climate or low density. Uhazy et al. (1973) found a significantly greater proportion of heavy infestations of lungworms in sheep inside the borders of national parks. High parasite load probably reflected the greater density of sheep within the park. Forrester and Senger (1964) believed that diet may influence mortality caused by lungworms, but Buechner (1960) stated that it is probably independent of the range vegetative condition. A significant correlation exists between the intensity of the lungworm infestation and the amount of precipitation in the spring of the previous year (Forrester and Senger 1964; Wilson and Honess 1965; Forrester and Littell 1976). Uhazy et al. (1973) attributed this seasonal variation of lungworm infestation to migration patterns and the concomitant changes in density. Forrester and Senger (1964) concluded that seasonal variations were due to changes in diet; stress associated with winter hardships, breeding, pregnancy, and lambing; or some biological characteristic of lungworms themselves, such as life span or infectiousness. Conclusive evidence of prenatal lungworm infestation has been described by Gates and Samuel (1977).

Pneumonia is often associated with lungworms, but the exact relationship is unknown (Demartini and Davies 1977). It is assumed that the incidence and virulence of pneumonia are enhanced by heavy lungworm levels. Numerous species of bacterium have been isolated from infected lungs of mountain sheep. Some of the more predominant include: *Acinetobacter calcoaceticus, Corynebacterium hemolyticum, C. pyogenes, Haemophilus heamolyticus, Klebsiella pneumoniae, Moraxella nonliquifaciens, Mycoplasma arginini*, a *Neisseria* sp., *Pastuerella hemolytica, P. multocida, Staphylococcus aureaus*, and a *Streptococcus* sp. (Woolf et al. 1970; Al-Aubaidi et al. 1972; Woodard et al. 1974; Thorne 1975a, 1975b, 1976). It is possible that these organisms are merely opportunistic and not the cause of pneumonia.

Mites cause mange and scabies. The sheep scab mite (*Psoroptes equivar ovis*) causes common scabies (Becklund and Senger 1967). This highly contagious disease usually starts on the back and sides and becomes extensive within three months of the initial infestation. Bedding grounds harbor the highest densities of these mites (Kemper and Peterson 1956). Existing evidence indicates that a *Psoroptes* mite was responsible for the epidemic in the latter half of the 19th century (Smith 1954; Buechner 1960). The sarcoptic mange mite (*Sarcoptes scabiei* var. *ovis*) attacks the barer areas of the body and face, burrowing into the skin (Kemper and Peterson 1956). Dipping of domestic sheep is probably the chief reason for today's low incidence of mange and scabies (Jones 1959).

Ticks that infest wild sheep are the winter tick (*Dermacentor albipictus D. hunteri*), the wood tick (*D. venustus*), and the spinose ear tick (*Otobius megnini*). Lice include *Bovicola jellisoni* and *B. ovis* (Becklund and Senger 1967; KeChung 1977).

Two species of dipterid infest wild sheep. The female botfly (*Oestrus ovis*) lays its eggs in the nostrils of sheep. The larvae burrow into the nasal membranes and subsist on the mucus secreted. Besides the physical damage caused by the larvae, sheep become very nervous and disturbed by the adult fly, which distracts them from their normal activities (Gobbett 1956). The sheep tick (*Melophagus ovinus*), not a true tick but a wingless fly, is widely distributed on domestic sheep but little is known of its natural levels on wild sheep (Imes and Babcock 1942).

Amyloidosis is a metabolic disorder most prevalent in stressed animals (Hadlow and Jellison 1962). Sheep appear to be unusually susceptible. DeMartini and Davies (1977) found that the adrenals of sheep that died from pneumonia were greatly enlarged. It is not surprising that amyloidosis is often found in association with pneumonia.

White muscle disease occurs in mountain sheep just as in domestic and some other wild animals. In domestic stock, the disease is also known as stiff lamb or nutritional myopathy. The heavy weight-bearing muscles of the thigh are most often affected in sheep. The origin of the disease is unknown and muscle degeneration remains latent until the animal is subjected to considerable exertion and stress (Smith et al. 1972). White muscle disease occurs in selenium-deficient areas and is treated by selenium shots. Clinical signs of the disease are a stiff, arched back or inability to use affected muscles. Since much of the mountain sheep range lies in a selenium-deficient zone, selenium shots are frequently given for preventative maintenance to all captured animals.

AGE DETERMINATION

The age of bighorns and thinhorns up to 1 year old may be fairly accurately determined by size and external criteria (Hansen 1965). The most common method of determining the age of all wild sheep is to count horn annuli. Each year of growth is terminated by a visible ring around the horns (see "Skull and Dentition" section) (figure 52.6). The horn ring method is quite reliable for rams up to 7 years old but horn annuli on ewes and older rams are quite indistinct and may be covered with hair (Geist 1966a; Hemming 1969). Bunnell (1978) found that due to brooming 94 percent of the ram and 45 percent of the ewe lamb increments are lost by the age of 12. He also found that in 50 percent of the rams 7 years old or older, the second year's horn ring was lost in brooming.

Tooth replacement patterns provide an accurate guide of age up to four years (see "Skull and Dentition" section). After four years of age molar length:width ratios and wear may be used to estimate age. Young sheep have molars that reach the base of the jawbone. As the animal matures, the molars are

FIGURE 52.6. Mature bighorn (*Ovis canadensis*) ram, showing horn annuli and brooming at distal end of horn.

pushed out and the total tooth length becomes shorter because of tooth wear. However, because of considerable individual variation in diet, and hence tooth wear, this method is not very reliable (Murie 1944; Hemming 1969). The most accurate method of age determination is to judge the annuli deposited in tooth cementum. Each annuli represents one year's growth. Thus, age can be estimated to the nearest month if the date of death is known and a median birth date assumed (Hemming 1969).

Sex of sheep older than lambs is easily distinguished because the horns of males are noticeably larger. The ram's horn becomes much larger at the base and develops a curl with age. Horns on ewes retain the lamb shape, with only slight growth in length each year.

POPULATION STATUS AND MANAGEMENT

Estimates of sheep numbers on the North American continent in the recent past have been as high as 1½–2 million (Buechner 1960). Presently, there is less than a tenth of that number (table 52.4). Diseases in conjunction with severe winter weather caused large fluctuations in populations (Stelfox 1971, 1976).

During the latter half of the 19th century diseases such as scabies, livestock competition and winter range restriction, and indiscriminate hunting by settlers, Indians, railmen, and traders caused a drastic reduction in numbers. During the early 1900s, hunting seasons were gradually eliminated, allowing a partial recovery.

However, from 1920 to 1950 another series of die-offs occurred. These were attributed to the lungworm-pneumonia complex, heavy livestock competition and the concomitant range deterioration, and unusually severe weather (Stelfox 1971, 1976).

In an effort to alleviate the violent population fluctuations, "no hunting" laws were gradually repealed. Hunting, however, has not been a panacea because of public opposition and evidence that herds of hunted sheep increase as fast as unhunted ones (Buechner 1960; Trefethen 1975; Nichols 1976).

In Alaska over 1,000 thinhorn rams were taken annually from 1967 to 1974 (Nichols 1975). In the

TABLE 52.4. North American wild sheep population estimates

Population	Status
Bighorns	
Rocky Mountain	20,410– 23,430
California	3,220– 3,420
desert	13,080– 14,705
Subtotal	36,710– 41,555
Thinhorns	
Dall's sheep	72,200– 92,500
Stone sheep	9,000– 15,000
Subtotal	81,200–107,500
Total	117,910–149,055

SOURCE: Compiled from Trefethen 1975.

lower states the total yearly harvest was fewer than 300 bighorns during this period. Buechner (1960) estimated that approximately 1,000 bighorn rams could be harvested on a sustained-yield basis in the United States. If the natural rate of increase of sheep is conservatively estimated at 0.2 (Buechner 1960) and the minimum numbers of thinhorns and bighorns are assumed to be 81,200 and 36,710, respectively, then the annual replacement rate of rams, assuming an even sex ratio, is 8,120 thinhorns and 3,671 bighorns.

Mountain sheep hunting has traditionally been for rams only and was further restricted by a 3/4 or full horn curl policy. In the last few years most states and provinces have adopted more stringent horn curl regulations. One state, Nevada, adopted a rather unique method for restricting the harvest to old rams: rams must be at least seven years old or have a Boone and Crockett score of 144 points using the horn with the most points doubled. While the overall trend has been for more restrictive hunting seasons, in some cases local situations have dictated either sex or ½ curl ram seasons.

Management. One of the most accurate census techniques for sheep is aerial counts, using either fixed-wing aircraft or helicopters (Smith 1954; Tsukamoto 1975). The percentage of sheep sighted from aerial surveys varies depending on topography and vegetation cover. Heimer (1976) stated that experienced observers can spot 93–98 percent of Dall's sheep actually present. Pitzman (1970) found that 86 percent of Dall's lambs and 96–99 percent of adults could be identified from the air. For each census, 39–51 percent of desert bighorns were counted from helicopters. It is best to conduct two surveys, one before lambing in the spring and one afterward in the summer, in order to determine most accurately survival, annual production, and sex ratios (Nichols and Erickson 1969). According to Tsukamoto (1975), time-lapse photo census is as accurate as aerial surveys.

Waterhole counts are primarily used for surveying desert bighorns. However, they are influenced by the weather and the need for water. Therefore, they are most accurate when taken during the dry season (Hansen 1967).

Other census methods are: mineral lick counts, which Heimer (1973) found to be the most economical for relative indexes of production and survival; foot and horseback transects; and boat surveys, which are useful for canyon-inhabiting sheep.

While mountain sheep, as well as most other wild ungulates, tend to crave salt in the spring, no definitive need for salt has been established. Skipworth (1974) discovered that heavy amounts of grit were being ingested by bighorns, presumably at natural salt licks. He concluded that this might impair metabolic performance and suggested that artificial licks be provided. Other studies have revealed that the salt consumption at natural licks may be not for salt but for some trace mineral. In several states, trace mineral salt blocks are provided for sheep as a management tool. The salt

concentrates sheep for survey purposes and trace minerals are available to fulfill a possible mineral deficiency in their diet. Contagious ecthyma is generally transmitted at salting sites, however, and where it is found, no artificial salting should be initiated.

Water development for the desert bighorn leads to more uniform distribution of sheep over their range but may promote disease (Halloran and Deming 1958). Development of springs should protect existing water from evaporation, debris, and cloudbursts. Covered troughs are usually all that is necessary. If economical, semiactive mines may be pumped. Rain catchments should be designed for improved efficiency. This may include constructing earthen or reinforced concrete dams and/or sealing the basins. Arroyo sides may be blasted out to form underground water tanks. Any steep sides should be blasted away to improve the structure's safety and prevent drownings. It has been shown that sheep show little reluctance to using artificial structures except for systems where the water is piped a long distance (Halloran and Deming 1958).

Wild sheep are very susceptible to diseases, and in recent years a determined effort has been made to treat some diseases such as the lungworm-pneumonia complex. In Washington State both wild and captive bighorns have been successfully treated with the experimental drug albendazole. The practical application of a treatment program, of course, depends on getting the medication into wild sheep. A semiwild population of Rocky Mountain bighorns in Washington State has been treated with medicated alfalfa pellets, while lungworm medication has been mixed with apple mash in Colorado. These experiments have been successful, but not all sheep populations may be treatable because of their wariness or remoteness. Further research is needed to determine the feasibility of treating remote populations.

Other research experiments have attempted to treat bluetongue in sheep. Robinson et al. (1974) found that *Culicoides* gnats, the natural vectors of bluetongue virus, could be used to inoculate sheep with bluetongue vaccine but the technique was too costly. Likewise, they felt that *Culicoides* control was not yet practical. One objective that is both feasible and even necessary is the control of disease in domestic livestock, as 70 percent of all the parasites of wild sheep also occur in domestic sheep and 30 percent in cattle.

Prior to transplanting wild sheep, the conditions of the prospective range must be assessed for suitability and future developments. Wilson (1975) presented a comprehensive outline of techniques for capturing and transplanting desert bighorns. Because the stress of capture may impair wild sheep immunity to diseases (Hudson 1973), precautions must be taken to monitor and control the health of captives. A number of authors have reported various chemical and physiological values of wild sheep and their applications. Serum protein electrophoresis can be used to detect subclinical disease (Woolf and Kradel 1970). Blood urea nitrogen values, glucose, rectal temperature, and packed cell volume values have been shown to reflect protein intake, excitability, and general condition, respectively (Franzmann 1972). Physiologic values at capture, after handling, and during captivity were presented by Franzmann and Thorne (1970) and are useful for determining stress and nutritional status.

The illegal taking of wild sheep, especially desert bighorns, has resulted in strict regulations and new law enforcement techniques. A permanent-type seal fastened to the horn of harvested sheep is used by many management agencies to identify legal rams permanently. Remains of mountain sheep can also be accurately identified for law enforcement purposes in a number of ways. Hemoglobin patterns provide accurate identity of fresh blood (Bunch et al. 1976). Keiss and Morrison (1956) described the precipitin reaction for identification of blood, bloodstains, and meat, and Belden (1976) outlined the hemagglutination of uncooked blood and meat.

Research studies have been conducted with a number of marking devices to identify free-roaming wild sheep individually. The techniques are nearly as numerous as the researchers and often are designed to accomplish specific management objectives, i.e., to determine migration routes. In recent years radiotelemetry has been used effectively to monitor movements and facilitate rather complicated research. Telemetry studies are expensive, however, and require considerable time to monitor collared animals. Other less sophisticated marking techniques use colored neck collars, ear tags, or dye dropped at a low altitude from an airplane.

The future of wild sheep depends on the preservation and improvement of critical native ranges. Wild sheep are poor competitors with other wild and domestic ungulates, and native ranges are diminishing. Other than curtailment of development, the most important step is to reduce grazing pressure by all ungulates, domestic and wild. Stelfox (1976) recommended that for an improvement of Canadian bighorn ranges, grazing must be reduced such that only 40 percent of the range is utilized.

Fire is a natural phenomenon that has had a substantial impact on bighorn ranges. Many traditional sheep ranges frequently burned from wild fires but in the last 50 years these fires have been suppressed. Fire is often an important factor in range regeneration, a detriment to forest encroachment on grasslands, and possibly an aid in parasite control. Recent experiments in British Columbia (Elliot 1978) on Stone sheep document the value of fire in improving ranges. In these experiments, prescribed burning was used as a range enhancement technique for the production of trophy Stone sheep. Sheep on the burned range grew faster and produced more trophy sheep than the unburned control area. One of the most practical ways to revegetate strip mines is by burning followed by hydroseeding (Etter 1973; Trefethen 1975). While fertilization on sheep ranges may have some merit, results of fertilizing in conjunction with the application of herbicides are dubious and appear to have little utility (Bear 1975).

CURRENT RESEARCH AND MANAGEMENT NEEDS

Management problems of wild sheep are quite diverse and differ between races and local situations. An excellent treatment of the status and needs of wild sheep is described in *The Wild Sheep in Modern North America,* edited by James B. Trefethen (1975). This publication is a compendium of a workshop on the management biology of North American wild sheep by management and research biologists. The most comprehensive description of research needs for each race of sheep is presented in this book.

Since management agencies are responsible for ensuring the well-being of wild sheep, research projects should be guided by management needs. Many of the current research projects are aimed at solving management problems. Bighorn sheep appear to be more vulnerable to parasite- and disease-induced die-offs than are other ungulates and a great deal of research has been conducted in this area. Drugs are currently being tested for control of lungworms, and recent experiments are encouraging.

Protection and enhancement of sheep habitat are imperative for the survival of wild sheep. Many ranges have been abused by too many ungulates of various domestic and wild species for too long. Research studies have been done and others need to be done on ways to enhance wild sheep ranges. Fire has long been ignored as a management tool in enhancing range conditions. Recent studies by Elliot (1978) on Stone sheep document the value of fire on thinhorn ranges. Now land managers should consider fire as a management tool in range enhancement.

In the last 20 or 30 years many resource agencies have inaugurated an active transplant program to reintroduce wild sheep to ranges where native species have been extirpated. In many cases initial releases were very successful but were followed by declines. One possible explanation for decreased productivity in isolated populations with a small gene pool is inbreeding.

The various thinhorn and bighorn races of wild sheep have specific research needs, and few generalizations can be made with regard to research priorities for all species. The resource manager must establish research priorities and assign research projects accordingly.

The future of wild sheep depends most on the preservation and improvement of critical native ranges. Wild sheep are poor competitors with other wild and domestic ungulates and native ranges are diminishing. Other than curtailment of development, the most important step is to reduce grazing pressure by all ungulates, domestic and wild. Stelfox (1976) recommended that, for an improvement of Canadian bighorn ranges, grazing be reduced such that only 40 percent of the range is utilized.

LITERATURE CITED

Al-Aubaidi, J. M.; Taylor, W. D.; Bubash, G. R.; and Dardiri, A. H. 1972. Identification and characterization of

Mycoplasma arginini trom bighorn sheep (*Ovis canadensis*) and goats. J. Am. Vet. Med. Assoc. 33:87–90.

Allen, R. W. 1956. Nodular worms of sheep and goats. Pages 399–401 *in* A. Stefferud, ed. Animal diseases, U.S.D.A. 1956 Yearb. Agric. Gov. Printing Off., Washington, D.C. 591pp.

Bear, G. D. 1975. Range improvement studies. Pages 22–23 *in* O. B. Cope, ed. Colorado game research review, 1972–1974. Colorado Div. Wildl., Denver.

Becklund, W. W., and Senger, C. M. 1967. Parasites of *Ovis canadensis canadensis* in Montana with a checklist of the internal and external parasites of the Rocky Mountain bighorn sheep in North America. J. Parsitol. 53:157–165.

Becklund, W. W., and Walker, M. L. 1967. *Nematodirus odocoilei* sp. n. (Nematoda: Trichostrongylidae) from the black-tailed deer, *Odocoileus hemionus,* in North America. J. Parasitol. 53:392–394.

———. 1969. Taxonomy, hosts, and geographic distribution of the *Setaria* (Nematoda: filarioidea) in the United States and Canada. J. Parasitol. 55:659–660.

Belden, E. L. 1976. Improved methods for serological identification of native Wyoming ungulates. Fed. Aid in Wildl. Restor. Proj. F-W-3-R-22, work plan 2, job 9W. Wyoming Game and Fish Dept.

Berger, J. 1978. Group size, foraging, and antipredator ploys: an analysis of bighorn sheep decisions. Behav. Ecol. Sociobio. 4:91–99.

Blom, E. 1968. Male reproductive organs. Pages 27–37 *in* E.S.E. Hafez, ed. Reproduction in farm animals. 2nd ed. Lea & Febiger, Philadelphia. 402pp.

Blong, B., and Pollard, W. 1968. Summer water requirements of desert bighorn in the Santa Rosa mountains, California, in 1965. California Fish Game. 54:289–296.

Blood, D. A. 1963. Some aspects of behavior of a bighorn herd. Can. Field Nat. 77:77–94.

———. 1967. Food habits of the Ashnola bighorn sheep herd. Can. Field. Nat. 81:23–29.

———. 1971. Contagious ecthyma in Rocky Mountain bighorn sheep. J. Wildl. Manage. 35:270–275.

Blood, D. A.; Flook, D. R.; and Wishart, W. D. 1970. Weights and growth of Rocky Mountain bighorn sheep in western Alberta. J. Wildl. Manage. 34:451–455.

Buechner, H. K. 1960. The bighorn sheep in the United States, its past, present and future. Wildl. Monogr. 4. 174pp.

Bunch, T. D.; Meadows, R. W.; Foote, W. C.; Egbert, L. N.; and Spillett, J. J. 1976. Identification of ungulate hemoglobins for law enforcement. J. Wildl. Manage. 40:517–522.

Bunnell, F. L. 1978. Horn growth and population quality in Dall sheep. J. Wildl. Manage. 42:764–775.

Bunnell, F. L., and Olsen, N. A. 1976. Weights and growth of Dall sheep in Kluane Park Reserve, Yukon Territory. Can. Field Nat. 90:157–162.

Chappel, R. W., and Hudson, R. J. 1978. Winter bioenergetics of Rocky Mountain bighorn sheep. Can. J. Zool. 56:2388–2393.

Clark, J. L. 1970. The great arc of the wild sheep. Univ. Oklahoma Press, Norman. 274pp.

Constan, K. J. 1972. Winter foods and range use of three species of ungulates. J. Wildl. Manage. 36:1068–1076.

Cowan, I. McT. 1940. Distribution and variation in the native sheep of North America. Am. Midl. Nat. 24:505–580.

———. 1947. Range competition between mule deer, bighorn sheep, and elk in Jasper Park, Alberta. Trans. North Am. Wildl. Conf. 12:223–227.

Cowan, I. McT., and Geist, V. 1971. The North American wild sheep. Pages 58–83 *in* Water, R., ed. North American big game. 1971 ed. Boone & Crockett Club, Pittsburgh, 403pp.

Davis, C. L. 1956. Sore mouth in sheep and goats. Pages 414–416 *in* A. Stefferud, ed. Animal diseases. U.S.D.A. 1956 Yearb. Agric. Gov. Printing Off., Washington, D.C.

Demarchi, R. A. 1962. An ecological study of the Ashnola bighorn winter ranges. B.S.A. Thesis. Univ. British Columbia, Vancouver.

———. 1968. Chemical composition of bighorn winter forages. J. Range Manage. 21:385–388.

Demartini, J. C., and Davies, R. B. 1977. An epizootic of pneumonia in captive bighorn sheep infected with *Muellerius* sp. J. Wildl. Dis. 13:117–124.

Deming, O. V. 1952. Tooth development of the Nelson bighorn sheep. California Fish Game J. 38:523–529.

Drewek, J. 1970. Population characteristics and behavior of introduced bighorn sheep in Owyhee County, Idaho. M.S. Thesis. Univ. Idaho, Moscow. 46pp.

Eccles, T. R., and Shackleton, D. M. 1979. Recent records of twinning in North American mountain sheep. J. Wildl. Manage. 43:974–976.

Elliot, John P. 1978. Range enhancement and trophy production in stone sheep. Pages 113–118 *in* Proc. 1978 Northern Wild Sheep and Goat Conf. 2–4 April 1978, Penticton, British Columbia.

Ericson, C., and Neiland, K. A. 1972. Dall sheep diseases and parasites. Fed. Aid in Wildl. Restor. Proj. W-17-4 and 5, job 6.6R. Alaska Dept. Fish Game.

Estes, R. D. 1979. Ecological aspects of bighorn sheep populations in southeastern Washington. M.S. Thesis. Washington State Univ., Pullman. 124pp.

Etter, H. M. 1973. Mined-land reclamation studies on bighorn sheep range in Alberta, Canada. Bio. Conserv. 5:191–195.

Forrester, D. J., and Littell, R. C. 1976. Influence of rainfall on lungworm infections in bighorn sheep. J. Wildl. Dis. 12:48–51.

Forrester, D. J., and Senger, C. M. 1964. A survey of lungworm infection in bighorn sheep of Montana. J. Wildl. Manage. 28:481–491.

Franzmann, A. W. 1971. Physiologic values of stone sheep. J. Wildl. Dis. 7:139–141.

———. 1972. Environmental sources of variation of bighorn sheep physiologic values. J. Wildl. Manage. 36:924–932.

Franzmann, A. W., and Hebert, D. M. 1971. Variation of rectal temperature in bighorn sheep. J. Wildl. Manage. 35:488–494.

Franzmann, A. W., and Thorne, E. T. 1970. Physiologic values in wild bighorn sheep (*Ovis canadensis canadensis*) at capture, after handling and after captivity. J. Am. Vet. Med. Assoc. 157:647–650.

Gates, C. C., and Samuel, W. M. 1977. Prenatal infection of the Rocky Mountain bighorn sheep (*Ovis c. canadensis*) of Alberta with the lungworm *Protostrongylus* spp. J. Wildl. Dis. 13:248–250.

Geist, F. 1966*a*. Validity of horn segment counts in aging bighorn sheep. J. Wildl. Manage. 30:634–635.

———. 1966*b*. The evolutionary significance of mountain sheep horns. Evolution 20:558–566.

———. 1967. A consequence of togetherness. Nat. His (October), pp. 24–30.

———. 1968. On delayed social and physical maturation in mountain sheep. Can. J. Zool. 46:899–904.

_____. 1971. Mountain sheep. A study in behavior and evolution. Univ. Chicago Press, Chicago. 383pp.

Geist, V., and Petocz, R. G. 1977. Bighorn sheep in winter: do rams maximize reproductive fitness by spatial and habitat segregation from ewes? Can. J. Zool. 55:1802–1810.

Gobbett, N. G. 1956. Head grubs of sheep. Pages 407–411 *in* A. Stefferud, ed. Animal diseases. U.S.D.A. 1956 Yearb. Agric. Gov. Printing Off., Washington, D.C.

Guthrie, R. D. 1972. Fannin's color variation of the Dall sheep, *Ovis dalli,* in the Mentasta mountains of eastern Alaska. Can. Field Nat. 86:288–289.

Hadlow, W. J., and Jellison, W. L. 1962. Amyloidosis in Rocky Mountain bighorn sheep. J. Am. Vet. Med. Assoc. 141:243–247.

Hafez, E.S.E. 1968. Female reproductive organs. Pages 61–80 *in* E.S.E. Hafez, ed. Reproduction in farm animals. 2nd ed. Lea & Febiger, Philadelphia. 402pp.

Hall, E. R., and Kelson, K. R. 1959. The mammals of North America. 2 vols. Ronald Press, New York. 1083pp.

Halloran, A. F. 1949. Desert bighorn management. Trans. 14th North Am. Wildl. Conf. 14:527–536.

Halloran, A. F., and Crandell, H. B. 1953. Notes on bighorn food in the Sonoran zone. J. Wildl. Manage. 17:318–320.

Halloran, A. F., and Deming, O. V. 1958. Water development for desert bighorn sheep. J. Wildl. Manage. 22:1–9.

Halloran, A. F., and Kennedy, C. A. 1949. Bighorn-deer food relationships in southern New Mexico. J. Wildl. Manage. 13:417–419.

Hansen, C. G. 1965. Growth and development of desert bighorn sheep. J. Wildl. Manage. 29:387–391.

_____. 1967. Bighorn sheep populations of the desert game range. J. Wildl. Manage. 31:693–706.

Hebert, D. M. 1972. Forage and serum phosphorus values for bighorn sheep. J. Range Manage. 25:292–296.

Heimer, W. E. 1973. Dall sheep movements and mineral lick use. Final report. Fed. Aid in Wildl. Restor. Proj. W-17-2, 3, 4, 5; job 6.1R. Alaska Dept. Fish Game.

_____. 1976. Interior sheep studies. Vol. 2. Fed. Aid in Wildl. Restor. Proj. W-17-8, job 6.9R, 6.10R, 6.11R and 6.12R. Alaska Dept. Fish Game.

Heimer, W. E., and Smith, A. C., III. 1975. Ram horn growth and population quality: their significance to Dall sheep management in Alaska. Wildlife Tech. Bull. 5. Alaska Dept. Fish Game.

Hemming, J. E. 1969. Cemental deposition, tooth succession, and horn development as criteria of age in Dall sheep. J. Wildl. Manage. 33:552–558.

Hudson, R. J. 1973. Stress and in vitro lymphocyte stimulation by phytohemagglutinin in Rocky Mountain bighorn sheep. Can. J. Zool. 51:479–482.

_____. 1976. Resource division within a community of large herbivores. Can. Nat. 103:153–167.

Hudson, R. J.; Bandy, P. J.; and Kitts, W. D. 1971. In vitro detection of homocytotropic antibody in lungworm-infected Rocky Mountain sheep. Clin. Exp. Immunol. 8:345–354.

Hudson, R. J.; Hebert, D. M.; and Brink, V. C. 1976. Occupational patterns of wildlife on a major East Kootenay winter-spring range. J. Range Manage. 29:38–42.

Imes, M., and Babcock, O. G. 1942. Sheep ticks. Pages 912–916 *in* G. Hambridge, ed. Keeping livestock healthy. U.S.D.A. 1942 Yearb. Agric. Gov. Printing Off., Washington, D.C.

Johnson, R. L. 1974. Bighorn sheep, 1973: a biological evaluation of the Tucannon bighorn with notes on other Washington sheep. Washington Dept. Game, Olympia. 69pp.

Jones, F. L. 1959. A survey of the Sierra Nevada bighorn. Sierra Club Bull. 35:29–76.

Jones, F. L.; Flittner, G.; and Gard, R. 1957. Report on a survey of bighorn sheep in the Santa Rosa mountains, Riverside County. California Fish Game J. 43:179–191.

Kates, K. C.; Allen, R. W.; and Turner, J. H. 1956. Roundworms of the digestive tract. Pages 389–399 *in* A. Stefferud, ed. Animal diseases. U.S.D.A. 1956 Yearb. Agric. Gov. Printing Off., Washington, D.C.

KeChung, K. 1977. Notes on populations of *Bovicola jellisoni* on Dall's sheep (*Ovis dalli*). J. Wildl. Dis. 13:427–428.

Keiss, R. W., and Morrison, S. W. 1956. Identification of Colorado big game animals by the precipitin reaction. J. Wildl. Manage. 20:169–172.

Kemper, H. E., and Peterson, H. O. 1956. Scabies in sheep and goats. Pages 403–407 *in* A. Stefferud, ed. Animal diseases. U.S.D.A. 1956 Yearb. Agric. Gov. Printing Off., Washington, D.C.

Knight, R. A., and Uhazy, L. S. 1973. Redescription of *Trichuris* (=Trichocephalus) *Schumakovitschi* (Savinkova, 1967) from Canadian Rocky Mountain bighorn sheep (*Ovis canadensis canadensis*). J. Parasitol. 59:136–140.

Leslie, D. M., Jr., and Douglas, C. L. 1979. Desert bighorn sheep of the River Mountains, Nevada. J. Wildl. Manage. Wildl. Monogr. 66.

Lotze, J. C. 1956. Coccidiosis of sheep and goats. Pages 387–389 *in* A. Stefferud, ed. Animal diseases. U.S.D.A. 1956 Yearb. Agric. Gov. Printing Off., Washington, D.C.

McCann, L. J. 1956. Ecology of the mountain sheep. Am. Midl. Nat. 56:297–324.

McKercher, D. G.; McCrory, B. R.; and Kouvigan, J. L. 1956. Bluetongue of sheep. Pages 418–422 *in* A. Stefferud, ed. Animal diseases. U.S.D.A. 1956 Yearb. Agric. Gov. Printing Off., Washington, D.C.

Mensch, J. L. 1969. Desert bighorn (*Ovis canadensis nelsoni*) losses in a natural trap tank. California Fish Game J. 55:237–238.

Monlux, A. W., and Davis, C. L. 1956. Actinomycosis and Actinobacillosis. Pages 265–268 *in* A. Stefferud, ed. Animal diseases. U.S.D.A. 1956 Yearb. Agric. Gov. Printing Off., Washington, D.C.

Moore, T. D.; Spence, L. E.; and Dugnolle, C. E. 1974. Identification of the dorsal guard hairs of some mammals of Wyoming, ed. W. G. Hepworth. Wyoming Game and Fish Dept. Bull. 4.

Morgan, J. K. 1973. Last stand for the bighorn. Nat. Geogr. 144:383–399.

Murie, A. 1944. The wolves of Mt. McKinley. Fauna of Natl. Parks of the U.S. Fauna Series 5.

Murphy, E. C., and Whitten, K. R. 1976. Dall sheep demography in McKinley Park and a reevaluation of Murie's data. J. Wildl. Manage. 40(4):597–609.

Nadler, C. F.; Woolf, A.; and Harris, K. E. 1971. The transferrins and hemoglobins of bighorn sheep (*Ovis canadensis*), Dall sheep (*Ovis dalli*) and mouflon (*Ovis musimon*). Comp. Biochem. Physiol. 40B:567–570.

Nichols, L. 1973. Dall sheep winter range and climate. Fed. Aid in Wildl. Restor. Proj. W-17-4,5; job 6.7R. Alaska Dept. Fish Game.

_____. 1974a. Dall sheep food habits and body condition during winter. Fed. Aid in Wildl. Restor. Proj. W-17-4,5; job 6.3R. Alaska Dept. Fish Game.

_____. 1974b. Productivity in unhunted and heavily exploited Dall sheep populations. Fed. Aid in Wildl. Restor. Proj. W-17-4,5; job 6.4R. Alaska Dept. Fish Game.

———. 1974c. Dall sheep winter range and climate. Fed. Aid in Wildlife Restor. Proj. W-17-4,5; job 6.7R. Alaska Dept. Fish Game.

———. 1975. Report from Alaska. Pages 8–13 in J. B. Trefethen, ed. The wild sheep in modern North America. Boone & Crockett Club, New York. 302pp.

———. 1976. An experiment in Dall sheep management: progress report. Pages 16–34 in Trans. 2nd North Am. Wild Sheep Conf.

———. 1978. Dall sheep reproduction. J. Wildl. Manage. 42:570–580.

Nichols, L., and Erickson, J. A. 1969. Dall sheep. Fed. Aid in Wildl. Restor. Proj. W-15-R-3 and W-17-1; Work Plan N, Job nos. 3, 4, 5, 6, 7. Alaska Dept. Fish Game.

Norris, C. B. 1955. The bighorns of Joshua Tree. Am. For. (September), pp. 54–55.

O'Connor, J. 1974. Sheep and sheep hunting. Winchester Press, New York. 308pp.

Oldemeyer, J. L.; Marmore, W. L.; and Gilbert, D. L. 1971. Winter ecology of bighorn sheep in Yellowstone National Park. J. Wildl. Manage. 35:257–269.

Parks, J. B.; Post, G.; Thorne, T.; and Nash, P. 1972. Parainfluenza: 3 virus infections in Rocky Mountain bighorn sheep. J. Am. Vet. Med. Assoc. 161:669–672.

Petocz, R. G. 1973. The effect of snow cover on the social behavior of bighorn rams and mountain goats. Can. J. Zool. 51:987–993.

Pitzman, M. S. 1970. Birth behavior and lamb survival in mountain sheep in Alaska. M.S. Thesis. Univ. Alaska, College.

Pulling, A. V. 1945. Hybridization of bighorn and domestic sheep. J. Wildl. Manage. 9:82–83.

Robinson, R. M.; Hailey, T. L.; Livingston, C. W.; and Thomas, J. W. 1967. Bluetongue in the desert bighorn sheep. J. Wildl. Manage. 31:165–168.

Robinson, R. M.; Hailey, T. L.; Marburger, R. G.; and Weishuhn, L. 1974. Vaccination trials in desert bighorn sheep against bluetongue virus. J. Wildl. Dis. 10:228–231.

Samuel, W. M.; Chalmers, G. A.; Stelfox, J. G.; Loewen, A.; and Thomsen, J. J. 1975. Contagious ecthyma in bighorn sheep and mountain goat in western Canada. J. Wildl. Dis. 11:26–31.

Schmidt, J. L., and Gilbert, D. L., eds. 1978. Big game of North America: ecology and management. Wildl. Manage. Inst. and Stackpole Books, Harrisburg, Pa. 494pp.

Scott, R. F.; Chatelain, E. F.; and Elkins, W. H. 1950. The status of the Dall sheep and caribou in Alaska. Trans. 15th North Am. Wildl. Conf. 15:612–625.

Senger, C. M., and Forrester, D. J. 1960. Experimental infestation of a Rocky Mountain bighorn lamb with *Melophagus ovinis* (Diptera: Hippoboscidae). J. Parasitol. 46:598.

Shannon, N. H.; Hudson, R. J.; Brink, V. C.; and Kitts, W. D. 1975. Determinants of spatial distribution of Rocky Mountain bighorn sheep. J. Wildl. Manage. 39:387–401.

Skipworth, J. P. 1974. Ingestion of grit by bighorn sheep. J. Wildl. Manage. 38:880–883.

Smith, D. 1954. The bighorn sheep in Idaho. Idaho Dept. Fish Game, Wildl. Bull. 1.

Smith, H. A.; Jones, T. C.; and Hunt, R. D. 1972. Veterinary pathology. 4th ed. Lea & Febiger, Philadelphia. 1521pp.

Spalding, D. J. 1966. Twinning in bighorn sheep. J. Wildl. Manage. 30:207.

Spalding, D. J., and Mitchell, H. B. 1970. Abundance and distribution of California bighorn sheep in North America. J. Wildl. Manage. 34:473–475.

Stelfox, J. G. 1971. Bighorn sheep in the Canadian Rockies: a history, 1800–1970. Can. Field Nat. 85:101–122.

———. 1976. Range ecology of Rocky Mountain bighorn sheep in Canadian National Parks. Can. Wildl. Rep. Ser. N. 39.

Thorne, E. T. 1975a. Diagnosis of diseases in wildlife. Fed. Aid in Wildl. Restor. Proj. FW-3-R-21, work plan 1, job 1-W. Wyoming Game and Fish Dept.

———. 1975b. The status, mortality and response to management of the Whiskey Basin bighorn sheep herd. Fed. Aid in Wildl. Restor. Proj. FW-3-R-21, work plan 3, job 15W. Wyoming Game and Fish Dept.

———. 1976. The status, mortality and response to management of the Whiskey Basin bighorn sheep herd. Fed. Aid in Wildl. Restor. Proj. FW-3-R-22, work plan 3, job 15W. Wyoming Game and Fish Dept.

Todd, J. W. 1972. A literature review on bighorn sheep food habits. Colorado Div. Game, Fish, Parks. Spec. Rep. 27.

———. 1975. Foods of Rocky Mountain bighorn sheep in southern Colorado. J. Wildl. Manage. 39:108–111.

Trainer, D. O., and Hansen, R. P. 1969. Serologic evidence of arbovirus infections in wild ruminants. Am. J. Epidem. 90:354–358.

Trefethen, J. B., ed. 1975. The wild sheep in modern North America. Boone & Crockett Club, New York. 302pp.

Tsukamoto, G. K. 1975. A brief history and status of bighorn sheep and its management in Nevada. Pages 48–58 in J. B. Trefethen, ed. The wild sheep in modern North America. Boone & Crockett Club, New York. 302pp.

Uhazy, L. S., and Holmes, J. C. 1971. Helminths of the Rocky Mountain bighorn sheep in western Canada. Can. J. Zool. 49:507–512.

Uhazy, L. S.; Holmes, J. C.; and Stelfox, J. G. 1973. Lungworms in the Rocky Mountain bighorn sheep of western Canada. Can. J. Zool. 51:817–824.

Viercek, L. H. 1963. Sheep and goat investigations. Range survey. Fed. Aid in Wildl. Restor. Proj. W-6-R-3, work plan E, Job 2-A. Alaska Dept. Fish Game.

Welles, R. E., and Welles, F. B. 1961. The bighorn of Death Valley. U.S. Natl. Park Serv. Fauna Ser. 6. 242pp.

Wilson, L. O., ed. 1975. Guidelines for reestablishing and capturing desert bighorn. Pages 269–296 in J. B. Trefethen, ed. The wild sheep in modern North America. Boone & Crockett Club, New York. 302pp.

Wilson, L. O., and Honess, R. F. 1965. Some internal parasites from fecal examinations of bighorn sheep in southeastern Utah. Proc. Utah Acad. Sci. 42:284–286.

Woodard, T. N.; Gutierrez, R. J.; and Rutherford, W. H. 1974. Bighorn lamb production, survival and mortality in south-central Colorado. J. Wildl. Manage. 38:771–774.

Woodgerd, W. 1964. Population dynamics of bighorn sheep on Wildhorse Island. J. Wildl. Manage. 28:381–391.

Woolf, A. 1971. Influences of lambing and morbidity on weights of captive Rocky Mountain bighorns. J. Mammal. 52:242–243.

Woolf, A., and Kradel, D. C. 1970. Hematological values of captive Rocky Mountain bighorn sheep. J. Wildl. Dis. 6:67–68.

Woolf, A.; Kradel, D. C.; and Bubash, G. R. 1970. Mycoplasma isolates from pneumonia in captive Rocky Mountain bighorn sheep. J. Wildl. Dis. 6:169–170.

BRUCE LAWSON, Appalachian Environmental Laboratory, Center for Environmental and Estuarine Studies, University of Maryland, Frostburg State College Campus, Frostburg, Maryland 21532.

ROLF JOHNSON, Washington Department of Game, 600 N. Capitol Way, Olympia, Washington 98504.

VIII

Exotic Species

53

Nutria

Myocastor coypus

Gale R. Willner

NOMENCLATURE

COMMON NAMES. Coypu, nutria, swamp beaver
SCIENTIFIC NAME. *Myocastor coypus*
SUBSPECIES. *M. c. bonariensis*, Argentina; *M. c. coypus*, Chile; *M. c. melanops*, southern Chile; *M. c. popelairi*, Bolivia; and *M. c. santacruzae*, Patagonia.

Most authors believe that *M. c. bonariensis* was brought from Argentina to the United States, and was the primary source of nutria introduced into North America (Atwood 1950; Evans 1970). Nutria of all subspecies have been introduced at one time or another (Evans 1970). Subspecific differentiation is based on color variation and geographic location (Murúa et al. 1981). Early published accounts refer to the nutria as *Myopotamus bonariensis* or *Myocastor bonariensis*.

The nutria is sometimes treated as a member of the family Capromyidae (Walker et al. 1975), although it is more accurate to place it in the monotypic family Myocastoridae (Woods and Howland 1979). The suborder, Hystricognatha, includes the rodents with an enlarged infraorbital foramen in the cranial portion of the skull and a flared lower mandible (Woods and Howland 1979) (figure 53.1). The genus *Myocastor* is monotypic. The name *coypus* is Indian in origin and translates as "water sweeper" (Murúa et al. 1981). Nutria, in Spanish, means otter (Evans 1970).

DISTRIBUTION

Nutria were first introduced into the United States for fur farming in 1899 at Elizabeth Lake, California (Evans 1970). However, it was not until the late 1930s, when unscrupulous promoters sold nutria as "weed cutters," that large numbers of nutria were imported to North America. Many fur farms were started but soon collapsed due to poor reproduction performance by their stock, low fur prices, and competition with other fur bearer pelts such as beaver (Evans 1970). The high costs of food and fence enclosures also contributed to the failure of nutria fur farms (Coreil personal communication). As a result, animals were released into the wild or escaped from inadequate pen facilities. Additional information on nutria introductions into

North America and Europe may be found in Bourdelle (1939), Laurie (1946), Ashbrook (1948), Atwood (1950), Rawley (1956), Presnall (1958), Aliev (1966*a*), Gosling (1974, 1981), Willner et al. (1979), and Safonov (1981). Nutria fur farming was discussed by Maurice (1931), Aliev (1956), and Kinsel (1958).

Feral populations of nutria are now established in at least 15 states and are considered an important fur resource or pest mammal in 9 of them. Substantial populations occupy fresh and brackish wetlands from Texas to Alabama, North Carolina to Maryland, and Oregon to Washington (figure 53.2). Taking each state in turn, nutria have been reported in California (Howard 1953; Schitoskey et al. 1972), Florida (Griffo 1957), Kansas (Hoffmeister and Kennedy 1947), Illinois (Hoffmeister 1958), Louisiana (O'Neil 1949; Harris 1956; Evans 1970), Maryland (Harman and Thoerig 1968; Willner et al. 1979), Minnesota (Gunderson 1955), Montana (Jellison 1945), North Carolina (Milne and Quay 1967), Ohio (Petrides and Leedy 1948), Oregon (Kebbe 1959), Texas (Davis 1958), Utah (Low 1946), New Mexico (Ashbrook 1948), Washington (Larrison 1943), and Wisconsin (Hale 1950). Adams (1956) listed an additional 15 states.

DESCRIPTION

The nutria has a robust, high-arched body with short legs. The forefeet have five toes, four prominent with long claws and one reduced; the hind feet are webbed between four of the five toes. The tail is long, round, and scantily haired. The pelage is composed of two kinds of hair: long, coarse guard hair and soft, dense underfur. The underfur functions as insulation and, as the hair dies, it takes on a gray color. The density of down hair is greatest on the abdomen (111–129 hairs/mm^2) and is thickest during the winter months. According to Ehrlich (1958) glands located at the bases of the sensory bristles by the lips and near the anus lubricate the hair as the nutria grooms itself. However, the main function of the anal gland is to mark points within the home range. The mean weight of the anal gland is heavier in males than in females, 12.2 g and 4.1 g, respectively (Gosling 1977). Sexes are colored alike, with the dorsal pelage varying from dark brown to

1059

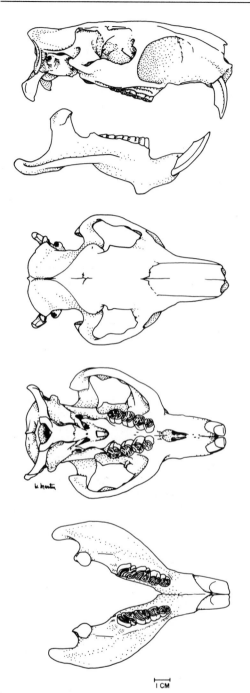

FIGURE 53.1. Skull of the nutria (*Myocastor coypus*). From top to bottom: lateral view of cranium, lateral view of mandible, dorsal view of cranium, ventral view of cranium, dorsal view of mandible.

yellow brown (Chabreck and Dupuie 1970). The belly contains a lighter gray underfur and is the commercially valuable portion of the pelt. The head is large, rectangular in shape, with a tapering muzzle that is covered by white hairs.

Like many other aquatic mammals, the nutria has a valvular mouth: the lips close behind the incisors to allow the animal to gnaw while submerged. The dental formula is 1/1, 0/0, 1/1, 3/3 = 20. The cheek teeth are typical of herbivorous mammals: rooted but hypsodont, with numerous folds in the enamel on the occlusal surface. The incisors are arc shaped, red orange in color, and rootless, like those of a beaver (*Castor canadensis*). The female has four or five pairs of mammae located dorsolaterally, which may be a unique adaptation for suckling young while swimming (Newson 1966) or, more likely, for maintaining an "alert" posture while nursing young on the nest (Weir 1974).

In Oregon, males 10 months or older average 5.94 kg (range 3–10.9) and females (including pregnant animals) 5.26 kg (range 4.76–11.34). The largest male captured had a body length of 660 mm, tail length 406 mm, and hind foot length 157.5 mm; the largest female body length was 635 mm, tail length 406 mm, and hind foot length 154.9 mm (Peloquin 1969). In Louisiana, Atwood (1950) recorded similar body lengths for both sexes; however, the tail lengths were slightly longer, 431.8 mm in males and 419.1 mm in females. The weight for the largest male was 8.16 kg, for the largest female 7.93 kg. In England, fully grown males (≥ 100 weeks old) average 6.7 ± 0.09 kg and females 6.36 ± 0.12 (Gosling 1977). This degree of sexual dimorphism also occurs in body and foot lengths.

The feces of nutria are particularly distinctive; individual scats are oblong in shape with fine longitudinal grooves. Nutria scats are approximately 5 cm long; those of the native muskrat (*Ondatra zibethicus*) are much smaller, generally 1.8 cm long, and lack grooves (Evans 1970). Nutria are often confused with muskrat but are easily distinguished by the round tail and, except when very young, by their larger size. The muskrat's tail is smaller and is compressed laterally.

Woods and Howland (1979) described the masticatory apparatus of nutria and other capromyid rodents and Wagner (1963) described the digestive system in nutria. Dixon (1960) described the skull and the vertebral and appendicular skeleton of the nutria, noting its similarity with the guinea pig and beaver. Detailed descriptions of the anatomy of nutria may be found in Koch (1953). Rossolimo (1958) measured over 700 skulls of nutria from different age classes and developed a growth index. Langenfeld (1977*a*, 1977*b*, 1977*c*, 1977*d*) described the anatomy and histology of the sympathetic trunk. The adrenals are large in relation to body weight, when compared with other mammals, and the left adrenal is significantly larger than the right (Wilson and Dewees 1962; Wilson et al. 1964; Katomski and Ferrante 1974). Two recent papers reported anomalies. Schitoskey (1971) found numerous anomalies and pathological conditions in skulls collected from Louisiana, including dental diseases, dental anomalies, and skull injuries. Willner and Chapman (1977) found two Maryland nutria with six toes; such animals have also been trapped in England (Gosling unpublished data).

FIGURE 53.2. Introduced range of the nutria (*Myocastor coypus*) in North America. Inset, approximate native range in South America. Inset adapted from Packard 1967.

PHYSIOLOGY

Growth. In Maryland, Dixon et al. (1979) examined the effects of trapping and weather on body weights of nutria. They found: (1) that growth rates were lowest during the winter and highest during the summer; (2) no significant difference in growth rates between leg-trapped and live-trapped nutria, although the growth rates were consistently lower in leg-trapped animals; (3) that the lower growth rates of nutria were attributable to the severity of the winter weather; (4) that males had a faster initial rate of increase in weight and reached a greater weight than females; and (5) that the daily growth rates for males and females were 0.0120g.g.$^{-1}$ day^{-1} and 0.0116g.g.$^{-1}$ day^{-1}, respectively.

In Louisiana, Robicheaux (1978) observed a decline in growth rates in summer for all age and sex classes. Adults of both sexes showed a greater decrease in total length growth and a greater weight loss than immatures.

Willner et al. (1980) found that it was possible to estimate the weight gain pattern of feral nutria in Maryland by obtaining body weights at successive times and derived and used the equation:

$$w = (w_{max} - w_0)(1 - e^{-bt}) + w_0,$$

where w_{max} = maximum body weight (grams)
  b = a constant (0.00244 ± 0.000060 day^{-1} for
      males and 0.00247 ± 0.00012 day^{-1} for
      females)
  w_0 = initial weight at birth (227.36 grams)
and t = time (days).

The weight of males could be predicted reliably up to age six but the weight of females could be predicted accurately only up to two years (figure 53.3).

Variation in Condition. To determine the physiological condition of nutria collected from marshes on the Eastern Shore of Maryland, Willner et al. (1979) used an adrenal index (adrenal weight/body weight), a condition index (body weight/body length), and a spleen index. These indexes were assessed to factors such as season, sex, age, weather variables, and reproductive status. The authors found that adrenal responses of males, pregnant females, and nonpregnant females were similar seasonally, with the highest values recorded in January and February. Age was directly correlated with the adrenal index, suggesting that nutria may have been under more stress as they grew older. Variation in the condition index showed that the nutria were in best condition from May to September and at their poorest in January and February. Fat reserves vary in a similar way in England, with the highest values in August–September and lowest in late winter (Gosling 1974). The spleen index of pregnant females was directly correlated with freeze-free days, temperature, and total precipitation. These results may have been influenced by the fact that Maryland is the northernmost portion of the range of nutria in eastern North America.

Willner et al. (1979) also observed severe physical reactions to the cold (bob tails, frostbitten ears and feet) (figure 53.4), and an increase in resorptions, particularly during the severe freeze of 1976-77. In Louisiana, Coreil (personal communication) captured 30 nutria during the winter of 1976/77 and held them in captivity for several months. At least three animals lost

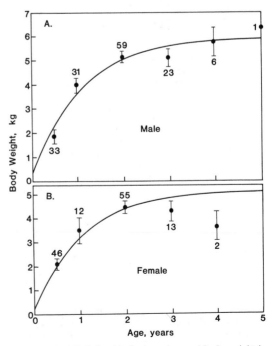

FIGURE 53.3. Relationship between age and body weight in nutria (*Myocastor coypus*). Adapted from Willner et al. 1980.

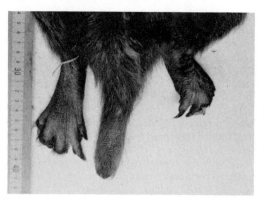

FIGURE 53.4. Severe frostbite injury to the feet and tail of a nutria (*Myocastor coypus*). The animal was collected on Maryland's Eastern Shore during the severe winter of 1976/77.

a 76.2–101.6-mm section at the tip of the tail due to frostbite injury. Ehrlich (1962) noted that mortality of feral nutria increased in lakes and swamps that froze during the winter months. To maintain feral populations established, he suggested that ponds should be drained and a dense growth of vegetation encouraged, to provide shelter.

In Russia, Aliev (1965a) observed that periods of freezing conditions often resulted in physical debilitation (lost toes, frostbitten tail) and mass mortality. Similar responses to cold weather have been recorded in England (Axell 1963; Norris 1967a).

GENETICS AND BLOOD ANALYSIS

Very little information is available on the genetics of nutria. The diploid number of chromosomes is 42 (Tsigalidou et al. 1966; George and Weir 1974). Kasumov et al. (1976) found that the karotype is represented by dibrachial chromosomes; the Y chromosome is a small acrocentric. One pair of chromosomes carried satellites. Tsigalidou et al. (1967) did not observe any chromosomes with satellites; they record 7 pairs of metacentric, 4 pairs of telocentric, and 10 pairs of submetacentric chromosomes.

Szynkiewicz (1968) studied the blood groups of nutria and found two types of antibody (CO_1 and CO_2). Szynkiewicz (1971) continued a series of studies on the differentiation of beta-globulin subfractions in nutria sera. The nutria examined in this study came from six breeding centers in Poland. Each breeding center appeared to have different gene frequencies of beta-globulin (presumably transferins).

Brown (1966) and Kimura and Johnson (1970) investigated the serum proteins of nutria. Adults contained an additional serum protein, the concentration of globulins and albumin were greater in adults, and a difference in lipoprotein between fetal and adult nutria existed (Brown 1966). More recently, Morgan et al. (1981) examined the genetic variation in Maryland nutria by analyzing serum and eye lens proteins, and serum and liver enzymes. They concluded that this population is homogeneous in its genetic composition, as evidenced by the lack of variation in soluble proteins and enzymes. Based on 100 samples, they found that all serum protein systems were monomorphic and were characterized by the typical mammalian albumin and transferrin.

REPRODUCTION

Male Anatomy. The external genitalia of the male nutria consists of a penis, a prepus, and a glans penis that contains an os baculum. The os baculum is cylindrical in shape and is composed of three parts: ossified bone, a transition zone, and a cartilaginous tip. The total length of the baculum varies between 15 and 23 mm and the diameter varies between 1 and 4 mm. The penis is projected posteriorly (Weir 1974), and the surface of the glans is covered by minute scales (Hillemann et al. 1958). As males grow, the testes descended from the abdominal cavity to the inguinal canal (Peloquin 1969). Mann and Wilson (1962) found that the secretions of the vesicular gland contained large amounts of fructose; the prostate contained more citric acid. The secretions of the bulbourethral contained more sialic acid and phosphorus than those of other glands. The secretions from these glands cause the seminal fluid to gel and form a copulatory plug, which could possibly be used to indicate that mating has occurred. Plugs were found in a small proportion of the females killed during the control operation in England (Gosling unpublished data). Stanley and Hillemann (1960) described in detail the histology of the male and female reproductive organs. Pietrzyk-Walknowska (1956) noted that, as the male becomes sexually active at the age of four to five months, the ketosteroid hormones and cholesterol content increased in the testes as well as in the adrenal cortex. Spermatogenesis declines in animals four to five years old, but large quantities of cholesterol and ketosteroid are present (Pietrzyk-Walknowska 1956). The mean testis weight for adult males was 4.8 ± 0.76 g (range 4.0–6.5 g). Sperm is present throughout the year (Brown 1975).

Female Anatomy. The female genitals consist of a vaginal orifice and a prominent papilla. The vaginal orifice is below but contiguous with the urinary papilla. A clitoris is present and contains a cartilaginous os clitoris. The uterus of the nutria is duplex, having two distinct horns and cervixes. Ovaries are not encapsulated and may attain an average length, diameter, and weight of 12 mm, 7.5 mm, and 186 mg, respectively (Hilleman et al. 1958). Weir (1974) asserts that no vaginal closure membrane is formed in the nutria. Peloquin (1969) observed that the vaginal orifice opened in females between the ages of four and nine months. Atwood (1950) reported that the aperature remained open through the life of the female, never closing between estrus and parturition. However, some constriction was noted during the last month of pregnancy (Newson 1966; Peloquin 1969), and in contrast to the previous observations, total or near-total closure of the orifice is normal for feral nutria in England (Gosling unpublished data).

The placenta of coypus has a well-developed decidua basalis and is connected to the uterine wall by a number of large blood vessels running through it (Newson 1966). Hillemann and Gaynor (1961) measured the allantoic placenta of 12 near-term nutria and found that the mean weight was 14.5 g (range 7.5–45.6 g), the mean length was 34.4 mm (range 25–46), the mean width was 27.8 mm (range 22–34), and the mean thickness was 18.9 mm (range 10–28). Hillemann and Gaynor (1961) and Hillemann and Ritschard (1967) gave detailed descriptions of the structure of the placenta. Histological studies of the ovaries obtained from pregnant female nutria were made by Gluchowski and Maciejowski (1958) and Rowlands and Heap (1966).

The mammary glands are approximately circular in dorsal aspect. The left anterior gland averages 7.04 ± 0.16 cm in diameter. When inactive the gland is white and less than a millimeter deep, but as parturition approaches it assumes a cream color and a swollen, lobular appearance; during lactation the mean depth of the left anterior gland is 6.38 ± 0.12 mm (Gosling 1980*a*).

Breeding Season. Nutria are nonseasonal breeders. In Maryland, Willner et al. (1979) found that feral nutria were in breeding condition throughout the year, although from a sample of 277 females collected in one year, 180 (64.9 percent) were pregnant. In Oregon, Peloquin (1969) found nutria breeding throughout the year, with peak birth periods occurring in January, March, and May and a smaller peak birth period occurring in October. Feral nutria also were found to breed continuously throughout the year in Louisiana (Atwood 1950) and in England (Newson 1966; Gosling 1974). Evans (1970) found that 85 percent of sexually mature females from the southwestern United States were pregnant even though they suspected that less than half of the population produced young that survived. Factors that affect nutria reproductive potential are food type and availability, weather, predators, and disease (Evans 1970).

Gestation Period and Fetal Development. In the natural range of the nutria, the gestation period varies from 127 to 132 days (Cabrera and Yepes 1940). In the United States and in Europe, the gestation period was found to be similar (Atwood 1950; Weir 1974). Newson (1966) obtained 19 litters of known age and correlated fetal weight with stage of gestation. Figure 53.5 can be used to estimate the age of embryos (Newson 1966; Chapman et al. 1980), and table 53.1 gives the approximate ages of embryos less than 50 days gestation using crown-to-rump length as the criterion (Newson 1966; Chapman et al. 1980). Embryonic development is slow for the first month of gestation, while the embryo is imbedded in a tissue mass formed at each implantation site (Newson 1965, 1966). After day 30 the placental disc starts to develop and embryonic growth accelerates. At a length of 7 mm the embryo's external features are well developed. The de-

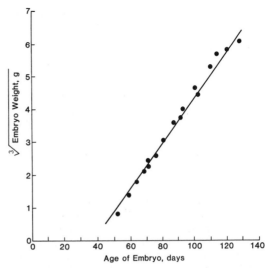

FIGURE 53.5. Relationship between embryo weight and age in nutria (*Myocastor coypus*). Adapted from Chapman et al. 1980.

velopment of the digestive, circulatory, nervous, urinary, and respiratory systems of the nutria embryo was discussed in detail in Chapman et al. (1980).

Breeding Age. Females begin breeding during the first year of life. In Maryland, Willner et al. (1979) found embryos in young that were 6 months old. In England, some female nutria reach sexual maturity at three months of age during the summer (Newson 1966). However, Konieczna (1956) found that female nutria in Poland were not sexually mature until 10–12 months of age. Cholesterol and ketosteroid levels increased as ovaries of young females developed. Large amounts of these substances were found in the ovary and cortex of the adrenal in animals during estrus; in contrast, only small amounts were found in pregnant females (Konieczna 1956). In Oregon, Peloquin (1969) noted that seven females gave birth to their first litters when 12 to 15 months of age. Four females gave birth to their second litter when 18 months old and one gave birth to its second litter when about 16 months old.

TABLE 53.1. Relationship between age and length in nutria (*Myocastor coypus*) embryos

Length (mm)	Approximate Age in Days
3–5	15 or less
5+ to 7	15+ to 20
7+ to 9	20+ to 25
9+ to 10	25+ to 30
10+ to 12	30+ to 35
12+ to 14	35+ to 40
14+ to 20	40+ to 45
20+ to 25	45+ to 55
25+ to 30	55+ to 70

SOURCE: Newson 1966; Chapman et al. 1980.

TABLE 53.2. Mean litter size of nutria (*Myocastor coypus*)

Mean Litter Size	Range	Location	Source
3.96	1–7	Maryland	Willner et al. 1979
4.23	1–9	southwestern Louisiana	Atwood 1950
4.4		southwestern Louisiana	Harris 1956
4.50	1–9	southern Louisiana	Evans 1970
4.56–4.81		southwestern Louisiana	Chabreck et al. 1981
5.0	3–8	Oregon	Peloquin 1969
5.03	1–11	southwestern Louisiana	Adams 1956
5.63	2–11	southwestern Louisiana	Atwood 1950
5–6		western Louisiana, eastern Texas	Evans 1970

Estrous Cycle. The nutria is a polyestrous mammal (Skowron-Cendrzak 1956). Reports of the length of the estrous cycle are extremely variable: Wilson and Dewees (1962) recorded intervals of 5–28 days, with one female maintaining estrus for up to 14 days. Newson (1966) recorded intervals of up to 60 days, with healthy females occasionally showing no cycles over several months. As noted by Asdell (1964), this variation suggests that ovulation is coitus induced. Estrus persists for a day in young females and seldom exceeds two days in multiparous animals (Skowron-Candrzak 1956). A postpartum estrus occurs within two days of parturition (Matthias 1941; Gosling 1980*a*), as, indeed, might be expected from the very high proportion of females that are found to be pregnant in most postmortem samples (Atwood 1950; Newson 1966). The mean postpartum interval is estimated to be 2.1 ± 0.8 weeks (Gosling 1980*a*).

Litter Size and Productivity. The mean litter size of nutria varies from 3.96 to 6.0 in areas where they were introduced (table 53.2). The mean litter size reported for the nutria in South America is about 5 (Cabrera and Yepes 1940). As many as 11 young were reported by Adams (1956) in Louisiana; Newson (1966) reported 13 conceptuses per litter for multiparous nutria in England. Many investigators found that litter size increases with age. In Maryland three-year-old nutria had a maximum of 7 in a litter, while younger animals had about 4 in a litter (Willner et al. 1979). In England, Newson (1966) reported that first litters averaged 5.06, while second litters averaged 5.63. In Louisiana, Atwood (1950) and Evans (1970) also reported that litter size increased with age.

Willner et al. (1979) found that of 211 viable nutria fetuses for which sexes were determined in Maryland, 88 were implanted in the left uterine horn and 123 in the right horn. The difference was statistically significant ($x^2 = 5.80$, $P < 0.025$). Rowlands and Heap (1966) similarly found that the mean number of embryos was 3.4 in the right horn and 2.7 in left (N = 43). In contrast, the results of post-mortem examinations of 223 pregnant females in Louisiana (Adams 1956) yield average values of 2.48 and 2.56 in the right and left horn, respectively.

In Oregon, Peloquin (1969) examined 55 pairs of ovaries and found a mean of 7.6 (3–17) corpora lutea. The same value was obtained by Gluchowski and

Maciejowski (1958). Rowlands and Heap (1966) found 8.8 corpora lutea per female in England. The majority of the ovaries examined by Peloquin (1969) contained an excess of corpora lutea. Rowlands and Heap (1966) found that these additional corpora lutea persisted throughout pregnancy although they declined progressively. The volume of luteal tissue grew rapidly after week 5, peaked at week 15, and then degenerated rapidly. Nutria might be polyovular, but the evidence is slight (Gluchowski and Maciejowski 1958).

Willner et al. (1979) estimated the annual productivity for a population of nutria in Maryland. They found that the mean number of young produced annually per female was 8.1, based on procedures developed by Lechleitner (1959) and Chapman et al. (1977). Severe winter weather had a profound impact on nutria reproduction. The smallest litter sizes were reported during the winter of 1976 (Willner et al. 1979).

Resorption and Abortion. If an embryo dies in an early stage of development, the junction between the placental disc and the decidua basalis breaks, leaving the embryo and placental disc to resorb. The decidua atrophies and becomes necrotic, persisting until parturition (Newson 1966). The embryos are expelled but others become compressed and mummified (Newson 1966). In Maryland, 9.8 percent of the embryos were resorbed. The resorption rate for litters was highest in December and January in Maryland during the most inclement weather (Willner et al. 1979). In Louisiana, deterioration of habitat appeared to be a major factor in reducing litter size (Atwood 1950). In Oregon, Peloquin (1969) found a resorption rate of 24.6 percent. In England, Newson (1966) found 6 percent resorption in first litters, which averaged 5.36 embryos, and 9 percent resorption in subsequent litters, which averaged 6.22 embryos.

Nutria appear to abort quite readily. Newson (1966) suggested that 26–28 percent of litters were lost in this way. Partial abortion has also been detected (Gluchowski and Maciejowski 1958).

Neonates and Lactation. The young are precocial, born fully furred and active. The mean body weight at birth is about 225 g (range 175–332 g, N = 45) (Newson 1966). Nutria gained weight rapidly during the first five months of life (Peloquin 1969). The young are covered by soft and downy hair; the tail hair appears silky until the end of the first month, when it is replaced by coarse hair (Peloquin 1969). Ehrlich (1966) described the changes in color of the incisor as nutria approached sexual maturity. The duration of lactation in feral English nutria is estimated to be 7.7 ± 1.0 weeks but young can survive if weaned at five days (Newson 1966; Gosling 1980*a*). Nutria milk was found to contain 41.5 percent dry matter, 27.9 percent fat, 13.7 percent protein, 0.50 percent sugar, and 1.3 percent ash (Ehrlich 1958).

ECOLOGY

Movement. Nutria generally remain in one area throughout their life. Kays (1956) found that 50 percent of the tagged males were recaptured within 91 m of the release site in a study in Louisiana. Under ideal conditions, Adams (1956) noted that the daily cruising range of nutria is less than 45 m. In a brackish marsh in Louisiana, Robicheaux (1978) found that 80 percent of the recaptured marked nutria moved less than 0.4 km; 20 percent moved between 0.4 and 1.35 km. Most movement occurs along water routes (Robicheaux 1978). Linscombe et al. (1981) reported that the farthest distance traveled by a recaptured nutria in a Louisiana brackish marsh was 3.2 km.

It is extremely difficult to determine movements of nutria by standard radio-collar telemetry techniques. In most cases, nutria develop a severe physiological reaction to collars: dermatitis, abrasions, or lesions appear around the neck (Coreil and Perry 1977). The only collar material that was found suitable was a nylon-covered rubber tubing collar with a 60-g transmitter attached at 14 percent larger than the circumference of the neck (Coreil and Perry 1977). In Louisiana, Coreil (unpublished data) radio-tracked seven adult females between 1977 and 1978. The average home range size analyzed by the minimum area method was 60 ha, in comparison to the 12.5 ha analyzed by the modified minimum area method (table 53.3). Miura (1976) discussed the dispersal of nutria in Japan.

Nests and Burrows. De Soriano (1960) described a complex burrow system in Argentina that extended 6 m

from entrances next to a watercourse. There were several entrances and a number of underground nest chambers. In North America, Peloquin (1969) found that the mean length of 20 burrows was 2.44 m, while Atwood (1950) recorded a range of 1.2–1.5 m. Nutria preferred to burrow into banks with a 45–90° slope (Peloquin 1969). The temperature inside one burrow in Argentina remained between 8° and 10° C while ambient temperature ranged from −4° to 24° C (de Soriano 1963).

Nutria build a variety of nests, from small flattened areas of plant material to large platforms up to a meter high in shallow water. Gosling (1979) saw captive animals collecting nest material (straw) then arranging it with a flattened central area and a raised, loosely packed rim; nest-building behavior was most common on cold nights. In the wild, nests are constructed of locally available vegetation. When this becomes seasonally scarce, competition for existing platforms increases (Coreil personal communication). Ehrlich (1966) described the preparturition nest-building behavior of nutria in captivity.

FOOD HABITS

Feeding Behavior. Hailman (1961) and Milne and Quay (1967) described the foraging behavior of nutria in the wild. After the food is cut, it is brought to feeding platforms or stations that are supported by floating logs, to shallow water areas, or to matted vegetation (Atwood 1950). Nutria chew their food in a modified propalinal manner (Woods 1976). They consume approximately 25 percent of their weight daily (Gosling 1974). Agricultural crops like corn as well as matured vegetation are in the diet of nutria (figure 53.6) (see "Economic Status" section).

Food Preferences. Like the muskrat, the nutria is an opportunistic feeder, consuming a variety of plants throughout its natural and introduced ranges. In Chile, Murua et al. (1981) studied the food habits of nutria by analyzing fecal material collected from a marsh and a

TABLE 53.3. Telemetric derived home range estimates for nutria (*Myocastor coypus*) using minimum and modified minimum area methods, Rockefeller State Wildlife Refuge, Cameron Parish, Louisiana

			Home Range (ha)	
Adult Sex	Tracking Period	Number of Triangulation Fixes	Minimum Area	Modified Minimum Area
Female	6/29/77–7/2/78	600	21.87	12.22
Female	7/1/77–8/28/77	127	9.66	7.26
Female	7/30/77–9/18/77	97	4.79	3.16
Female	8/6/77–11/27/77	187	15.13	6.15
Female	10/10/77–1/15/78	130	176.07	16.07
Female	12/21/77–3/18/78	142	99.99	25.98
Female	4/22/78–7/2/78	112	93.08	16.89
Average			60.09	12.53

SOURCE: Coreil unpublished data.

FIGURE 53.6. Nutria (*Myocastor coypus*) often feed on corn in fields adjacent to the marshes of Blackwater National Wildlife Refuge, Dorchester County, Maryland.

man-made lagoon. Throughout the year, particularly in winter, *Limnobium stoloniferum* and grasses were important in the diet of animals from the lagoon. Animals from the marsh favored grasses and *Scirpus californicus* for most of the year, but during summer *Typha angustifolia* was the dominant plant found in the feces. During winter, nutria from the marsh fed heavily on roots, while nutria from the lagoon consumed primarily stems and leaf portions of available plants.

In the United States, the major plants consumed by nutria in brackish-water marshes in southwestern Louisiana were *Spartina cynosuroides* and *Scirpus olneyi* (Chabreck et al. 1981; Harris and Webert 1962). *Spartina patens,* an abundant plant in the marsh, was used very little, if at all. Willner et al. (1979) found that nutria stomachs collected from a brackish marsh in Dorchester County, Maryland, contained mostly three-square rush, *S. olneyi,* the dominant species in the marsh. (See Willner et al. [1975] to compare muskrat food habits in Maryland.) In fresh-water marshes in southeastern Louisiana, *Eleocharis palustris* was found in nearly all the stomachs, but *Alternanthera philoxeroides* was considered the most preferred plant food, particularly in July (Shirley et al. 1981). Atwood (1950) examined food platforms in a fresh-water marsh in Louisiana and found that, of 16 plants, pickerel weed (*Pontederia lanceolata*) was used the most. Swank and Petrides (1954) found that nutria in ponds in eastern Texas preferred cattail (*Typha latifolia*), giant cutgrass (*Zizaniopsis milacea*), arrowhead (*Sagittaria* sp.), panic grass (*Panicum* sp.), white waterlily (*Nymphaea elegans*), pickerel weed (*Pontederia cordata*), and rice cutgrass (*Leersia oryzoides*). Warkentin (1968) listed food plants of nutria in a Louisiana pond. In Oregon, Wentz´(1971) listed 40 species of plant in the diet of nutria, with

Salix spp. being used the most, especially in winter. Based on forage ratios, *Sagittaria latifolia* and *Polygonum* sp. were the most favored but least available in the environment. In North Carolina, Milne and Quay (1967) summarized the summer and winter food habits of nutria from five acres along Hatteras Island by availability index, dietary importance, and forage ratios. Of 22 plants fed to captive nutria, Gainey (1949) found that the most preferred was bur reed (*Sparganium americanum*); the least preferred was water shield (*Brasenia schreberi*). For food habits studies in Russia, see Vinogradov (1965); for studies in England, see Ellis (1963) and Gosling (1974).

The nutria is entirely herbivorous in the United States, consuming animal matter (insects) only incidentally (Shirley et al. 1981). However, along the coast of South America and England, nutria occasionally consume crustaceans and fresh-water mussels (see Atwood 1950; Gosling 1974).

BEHAVIOR

Some aspects of nutria behavior have been described in studies of captive animals. Lomnicki (1957) found that animals in a fur farm were primarily nocturnal although they became conditioned to diurnal activity when fed during the day. In England, Gosling (1979) found that nutria became active just before sunset and returned to their nests a few hours before dawn. Most of the active period was spent in feeding, grooming, and swimming. Nutria groom by scratching and by "nibbling" movements of the mouth. They swim using alternate propulsive thrusts of the webbed hind feet and sometimes float immobile on the water surface for short periods. Eighty to 86 percent of eliminated feces is produced while swimming. Like rabbits, nutria reingest their feces, usually by eating directly from the anus. Reingestion, or coprophagy, occurs after the animal has returned to the nest at the end of the active period (Gosling 1979).

Chabreck (1962) used recorders placed along the trails of feral nutria and confirmed that they were mainly nocturnal. Although he found no relationship between activity and temperature, Gosling (1979) found that nutria returned to the nests earlier on cold nights. They spent the predawn hours in close contact with other animals, perhaps to maintain body temperature. A consequence of this attenuation of the active period is that feral nutria in England are forced to become diurnal during very cold periods in order to make up lost feeding time (Gosling et al. 1980*a*). Warkentin (1968) found that daytime activity in Louisiana was also influenced by temperature. Sunning and sleeping were the main activities below 28° C; above this temperature most animals fed, groomed, or slept. Above 34° C no animals were found sunning.

A definitive study of the social behavior of wild nutria is lacking, and there is some disagreement in the literature. Ryszkowski (1966) concluded that adult coypus were solitary. However, his study was of overcrowded animals in captivity. The concensus is that nutria are highly gregarious (Ehrlich 1966; Warkentin 1968; Gosling 1977). Warkentin (1968) found that most lived in groups of 2–13 animals with occasional solitary males. She also identified an alpha male and an alpha female in each group that were dominant over the other animals. Gosling (1977) observed a system that was similar in some respects: groups were composed essentially of related adult females and their offspring and a large male. Other males were expelled as young adults. Females are dominant over males except at mating (Warkentin 1968) and they vigorously defend the nest area for a period after a new litter is produced (Carill-Worsley 1932; Ehrlich 1966; Ryszkowski 1966).

MORTALITY

Chapman et al. (1978) compared the differential survival rates among leg-trapped and live-trapped nutria in a study area located on the Blackwater National Wildlife Refuge in Maryland. They found that a significantly greater mortality occurred in leg-trapped animals than in live-trapped animals—74 and 53 percent, respectively. In England Newson (1969) found a consistent year-to-year survival rate of 26 percent for males and 33 percent for females (74 and 67 percent mortality, respectively). Both studies included natural mortality as well as mortality from trapping.

Diseases. Equine encephalomyelitis is endemic in nutria in South America (Page et al. 1957). Three diseases—leptospirosis, hemorrhagic septicemia, and paratyphoid—were found in feral nutria from Louisiana. These diseases may cause embryo mortality, although no studies have been conducted to confirm this (Evans 1970). In Maryland, a yellow white vaginal pus was observed in many feral female nutria (Willner unpublished). It was identified as salmonelloses.

Skin Irritations. Chabreck et al. (1977) found that the skin of nutria became infected by achenes of the smooth beggar tick (*Bidens laevis*). Kinler and Chabreck (1978) noted that as a result of this condition, chronic dermatitis, the value of the pelt was significantly reduced. The most commonly infected area was the thorax region; the least was the neck. Animals showed loss of appetite and appeared depressed. Smooth beggar ticks can be controlled by spraying with 2, 4-D amine (Kinler and Linscombe 1981).

Diseases in Captive Nutria. Pridham and Thackeray (1959) diagnosed *Streptococcus durans* in ranch nutria. Nutria kept in pens commonly suffered from bacterial pneumonia, hepatitis-nephritis, *Strongyloides* spp. infection, and neoplasms. The major cause of these diseases was related to poor sanitary conditions of the pens (Pridham et al. 1966). *Salmonella typhimurium* was also reported in captive nutria (Steffen 1955). External symptoms were listlessness, anorexia, emaciation, and death. The principal source of infection was water infected with excrements of diseased individuals, particularly during warm weather. Internal symptoms include enlargement of the spleen, swollen lymph nodes, and hemorrages in the mucosa; the kidneys and liver are in a degenerated condition, and the intestinal mucosa is edamatous and shows changes to catarrhal, hemorrhagic, or diphtherial inflammation (Steffen 1955). Better food and cleaner pen facilities would reduce the spread of infection.

Karkalitskii et al. (1970) studied pyridoxine deficiency in nutria in the laboratory and found that the blood serum showed a decline in alpha- and beta-globulins, an increase in albumins, decreased activity of aspartate-aminotransferase, and an increase in beta lipoproteins; the liver and kidneys degenerated. External symptoms included poor appetite, weight reduction, and lesions on the nose. These symptoms disappeared when animals were fed vitamin B. In a zoo in Arkansas, Cross and Thomas (1966) noted that five nutria succumbed to hydatid disease. Cysts were found in the lungs and livers.

Parasites. In South America, Babero et al. (1979) examined 11 nutria in Chile and found the following internal parasites: three nematodes (*Graphidioides myocastoris*, *Trichuris myoscastoris*, and *Dipetalonema travassosi*), one trematode (*Hippocrepis myocastoris*), and one cestode (*Rodontolepis* sp.).

The most common endoparasites found in nutria were summarized by Babero and Lee (1961). They listed 11 trematodes, 21 cestodes, and 31 nematodes. Prasad (1960) reviewed the literature on the coccidia of the nutria. Little (1965) observed that *Strongyloides myopotami*, a parasite of nutria, caused "creeping eruption" in the skin of humans. This infection is commonly referred to as "nutria itch" by local trappers in Louisiana (Evans 1970). *Strongyloides* sp. may reduce fecundity and may also be fatal to nutria. The cestode *Taenia taeniformes* was found in livers of animals that were collected from the marshes of the eastern shore of Maryland (Willner unpublished data). In Japan, Nagahana et al. (1977) noted that the nutria was not a reservoir host for the liver fluke, *Clonorchis sinensis*. In Louisiana, biting lice (*Pitrufquenia coypus*) were found on the snout of a nutria (Miller 1956). In England, Newson and Holmes (1968) noted a seasonal change in the rate of infection, particularly by the biting louse *Pitrufquenia coypus*. The change may be related to the molt sequence, which occurs primarily between March and April and again in early autumn, as the highest incidence of infection occurred in the late winter. George (1964) identified the flea *Ceratophyllus gallinae* from a nutria in England. Tick infestation (*Ixodes ricinus, I. arvicolae, I. hexagonus, I. trianguliceps*) occurred primarily on the lips, eyelids, and, most often, ears (Newson and Holmes 1968). In Maryland, ticks (*Dermacentor variabilis*) were found in the ears of feral nutria, sometimes as many as 27 in one ear (Willner unpublished data).

Predators. In South America, the major predator on nutria are the caymans (*Caiman latirostris, C. sclerops,* and *C. niger*), although the jaguar (*Felis onza*), puma (*Felis concolor*), ocelot (*Felis pardalis*), and little spotted cat (*Felis tigrina*) are also listed as important predators (Aliev 1966b).

In Louisiana, the alligator (*Alligator mississippiensis*) is a major predator, particularly when densities of nutria are high. Valentine et al. (1972) examined the stomachs of 413 alligators from the Sabine National Wildlife Refuge, Louisiana, and noted that nutria occurred in 56 percent of the stomachs examined in 1961. During other years (1962, 1964), nutria constituted less than 7 percent of the diet of the alligator. Estimated densities of nutria in 1961 peaked at 74,000. Densities dropped to less than 10,000 between the years 1962 and 1965. McNease and Joanen (1977) found that nutria were in the fall diet of alligators from fresh, intermediate, and brackish marshes of Louisiana in 1972 and 1973. Warkentin (1968) found that young nutria were taken as prey by garfish (*Lepisosteus* sp.), cottonmouths (*Agkistrodon piscivorous*), and Redshouldered Hawks (*Buteo lineatus*) in a Louisiana pond. In Dorchester County, Maryland, 2 of 69 Bald Eagle (*Haliaeetus leucocephalus*) nests visited in 1978 contained remains of nutria; in 1979, 2 of 77 nests contained nutria remains. Nutria bones were not found in any nests in 1977 or 1980 (Pramstaller personal communication). Bald Eagle nests in Louisiana also contained remains of nutria (Dugoni 1980). Other potential predators include turtles, large snakes, and some birds of prey that may prey on immature or sick and injured animals (Evans 1970). Jemison and Chabreck (1962) examined barn owl pellets collected from a Louisiana marsh area where nutria were abundant, but found no evidence of owls feeding on nutria. In England young coypus are taken by a variety of predators, including the fox (*Vulpes vulpes*), stoat (*Mustela erminea*), heron (*Ardea cinerea*), marsh harrier (*Circus aeruginosus*), and owls (Ellis 1965). In Russia, Aliev (1966b) noted that nutria are particularly

vulnerable during droughts or freezing conditions and as a result are often prey items for jackals (*Canis aureus*), feral dogs, wolves (*Canis lupus*), jungle cats (*Felis chaus*), harriers (*Circus aeruginosus*), and Tawny Owls (*Strix aluco*). Avian predators such as the Magpie (*Pica pica*) and the Hooded Crow (*Corvus cornix*) will take young nutria.

AGE DETERMINATION

Tooth Wear and Pelage Characteristics. Several criteria have been used to determine the age of nutria. Aliev (1965b) described in detail the growth and development of ranch nutria from birth to six years of age. Included in the discussion is a description of molar wear and eruption that can be used to age nutria (table 53.4). Brown (1975) determined the age of nutria by body weight and pelage characteristics. Juvenile pelage was woolly in appearance and juveniles weighed less than 1.25 kg. Subadults weighed between 1 and 4.25 kg and the pelage was in the process of molting. Adults weighed 4.0 to 8.0 kg, with females weighing a maximum of 7.5 kg.

Hind Foot Measurements. Adams (1956) classified animals with hind foot lengths of 10.9 cm to be less than 3 months old (immatures); 11.2–12.4 cm, 3 to 5 months old (subadult); and 12.7 cm, more than 5 months (adults).

Eye Lens. A relationship between dry eye lens weight and age has been established from a sample of 199 known-age nutria in England (Gosling et al. 1980b). A multiple regression analysis showed that sex and nutritional status, among other variables, had negligible effects on lens weight. The calculated equation (r = 0.99) was:

$$\log_{10}(\text{age} + K) = 0.511 + 0.013 \text{ (lens weight)}$$

where K = 4.34 months (the gestation period).

Baculum Length. Schitoskey (1972) noted variation in the size of bacula of nutria of different age classes.

ECONOMIC STATUS

Three main categories of damage—to crops, to drainage systems, and to natural plant communities—have been recorded throughout the nutria's introduced range. Nutria feed on a very broad range of crops (Gosling 1974, 1980b). Schitoskey et al. (1972) conducted a survey to determine the extent of damage in California, particularly along edges of sugar cane fields, and found that 11 percent of the total area was damaged. The animals nipped the stalks of crops such as corn and sugar cane, injuring or killing the plants. In crops such as alfalfa, rice, and ryegrass, entire plants were consumed. Kuhn and Peloquin (1974) observed damage to fruit trees, nut trees, conifers, and deciduous forest trees caused by girdling. Blair and Langlinais (1960) reported damage to bald cypress plantations by pulling up seedlings and eating the roots. In England, nutria feed on almost every arable

TABLE 53.4. Tooth wear as a method to determine age of nutria (*Myocastor coypus*)

Age	Comment	Number of Molars
full-term fetus	Nutria are born with teeth	0–1
less than 6 months		2–3
6 months	Chewing surfaces of teeth notably enlarged. Three molars fully emerged, fourth is breaking through.	3–4
1 year	All four molar teeth equally developed; crowns are the same level. Teeth are of hyposodontal type; grinding surfaces of molar teeth are inclined inside. There is a distinct black ring around all the molars.	4
2 years	Simultaneous and gradual grinding off of all the molars takes place. At this age, the first molar has moderate wear.	4
3 years	The animal at this age is a fully formed adult. Heavy wear noted on first molar, moderate wear around other molars.	4
4 years	Heavy wear noted on the first and second molars. Moderate wear around other molars. Black ring around molars has faded.	4
5 years.	Heavy wear noted on first, second, and third molars. Moderate wear around fourth molar.	4
6 years	At this age, the first three molars are so worn down that they frequently fall out; the fourth molar, although very worn, is still functional.	4

SOURCE: Aliev 1965*b*.

crop that is available. They graze on sprouting cereals in the winter and spring and eat mature seed heads in the autumn; brassicas (kale, cabbage, brussels sprouts, etc.) and root crops (sugar beet, carrots, etc.) are eaten mainly in the winter, which reflects the pattern of feeding on wild plants (Ellis 1963; Gosling 1974, 1980*b*).

Nutria also cause damage to levees in commerical crawfish ponds in Louisiana. Levees must be checked almost daily to ensure no leakage due to nutria burrowing (Coreil personal communication). The most economically important damage in England is also through burrowing. Nutria dig burrows up to 6 m deep in the banks of waterways (de Soriano 1960), and these burrows sometimes penetrate or weaken river banks that protect low-lying areas of drained agricultural land from flooding (Cotton 1963).

At high densities nutria can make a tremendous impact on natural plant communities and, most conspicuously, emergent plants. In eastern England in the 1950s a high population caused a dramatic reduction in the area of reed swamp (mainly *Phragmites australis* and *Typha* spp.) (Ellis 1963, 1965). Similar reduction with a corresponding increase in open water has been seen in other parts of Europe (Hillbricht and Ryszkowski 1961; Ehrlich and Jedynak 1962) and in North America (Wentz 1971; Linscombe et al. 1981). Nutria also reduce the abundance of individual species through their selective feeding: cowbane (*Circuta virosa*) and great water dock (*Rumex hydrolapathum*) were all but eliminated in some areas during the peak nutria population of the 1950s in England (Ellis 1963, 1965).

In areas where nutria densities are kept low by trapping and/or climatic factors, damage to marshland, crops, and drainage systems is minimal. In Maryland, trapping and long periods of cold weather keep population density below a level at which significant damage could occur. In Louisiana (Harris and Webert 1962), Maryland (Chapman et al. in preparation), and Oregon (Wentz 1971), it was concluded that the overall impact of nutria on marshes was acceptably low.

CONTROL AND MANAGEMENT

Trapping Methods. In Louisiana, Evans et al. (1972) evaluated different trapping, handling, and marking methods used in nutria studies. The most effective methods were:

1. Trapping animals with a single- or double-door, noncollapsible, treadle-operated box trap with bait. Set traps on land along trails, near den entrances, or on rafts.
2. Handling nutria with a modified sling device or the tail-hold method. A modified choker was effective if needed to mark animals. Use of thick gloves was suggested.
3. Sodium pentobarbital, considered the safest and most effective anesthetic for nutria. Diazepam was suggested as a suitable tranquilizer. Ketamine hydrochloride, a tranquilizer, injected in the muscles, also works well.
4. No. 3 monel ear tags (National Band and Tag Co., Newport, Ky.) inserted in the ears and webbed portion of the hind foot, for identifying animals. Some nutria may develop a physiological reaction to these tags, however.

Palmisano and Dupuie (1975) and Linscombe (1976) evaluated the efficiency of leg-hold traps to capture fur-bearing animals in Louisiana. More nutria were caught in victor no. 2 leg-hold traps than in the Conibear 220, although no difference was observed between the victor no. 2 and the victor no 1½. The Conibear was effective in killing the nutria. Palmisano and Dupuie (1975) noted that 30 percent of nutria caught in leg-hold traps were immatures and not suitable for market.

Density Estimates. Linscombe et al. (1981) used harvest rates to show the decline of a nutria population in a *S. olneyi* marsh in Louisiana. Other methods suggested for determining population levels involve counting "sign" in the field such as active runways, fecal material, or food platforms.

Population densities at Blackwater National Wildlife Refuge were estimated by using the mark-recapture method and applying the Peterson index. Densities varied from a low of 0.5 nutria per ha to a maximum of 21.4 per ha. Maximum densities occurred on a brackish marsh dominated by *Scirpus olneyi*, *S. robustus*, and *Typha angustifolia* and bounded by stands of loblolly pine, *Pinus taeda* (Willner et al. 1979). In Louisiana, Robicheaux (1978) used the Schnabel index to estimate the seasonal population levels and found that densities varied from 1.34 nutria per ha in the summer to 6.5 nutria per ha in the winter. In Oregon, Wentz (1971) observed that densities were particularly low: 1.05 nutria per surface hectare of water during the winter when water levels were high; in summer densities increased to 22.66 nutria per surface hectare of water. Brown (1975) found 24.7 nutria per surface hectare of water in a polluted Forida pond and 5.9 nutria per surface hectare of water in an unpolluted pond. Simpson and Swank (1979) estimated the population level of nutria on a Texas marsh as 129 based on the Schnable index and 89 based on the Lincoln index. The difference in the estimates was attributed to the method of recapturing marked animals. Adult and subadult nutria became trap shy when live-traps were used to recapture them.

Mursaloglu (1981) reported that nutria density along the Karasu River in Turkey was dependent upon variations in the depth of the river, which fluctuated widely from year to year.

Control. Several methods have been developed to control nutria populations in North America, particularly in areas where they live adjacent to grain fields and cause damage. Evans (1970) described several "direct control" techniques including chemical control such as zinc phosphide baiting, shooting, live-trap, and leg-hold trap. Indirect control methods involved modifying farming practices to minimize nutria damage. Proper drainage, land grading, and vegetation control were three methods suggested. Schitoskey et al. (1972) noted that trapping was an effective control method when used on low-density populations, but that zinc phosphide baiting was more effective. Talbert (1962) noted that trapping with no. 3 leg-hold traps and shooting were effective methods for control, although no. 2 leg-hold traps were just as successful (Evans 1970). Kuhn and Peloquin (1974) found that the rodenticide prolin was effective in controlling nutria. Necropsies on animals given the poison showed that the lips of nutria turned yellow green, massive hematoma was present in the thoracic region, and bleeding occurred from the penis, nostrils, and lips. Evans and Ward (1967) found that a rodenticide anticoagulent, used to control nutria, was causing "secondary poisoning" in ranch mink and dogs that fed on nutria killed by this method. The anticoagulant accumulated in the liver of nutria.

Leg-hold traps are illegal in England for humanitarian and conservation reasons. A variety of alternative techniques have been tried, including zinc phosphide baiting, but none has proved more efficient than cage trapping, which is the sole method of control in use (Norris 1967*b*; Gosling 1974, 1981). All traps are visited every day and the numerous nontarget captives are released unharmed; any nutria trapped are shot. Cage trapping is thus acceptable to landowners with conservation and game-shooting interests. Its effectiveness has recently been demonstrated by an experimental five-year trapping program in a 30-km² area of wetland in east Norfolk that has succeeded in removing more than 99 percent of a large resident population (Gosling unpublished data).

The dynamics of the English population are intensively monitored using techniques described by Gosling (1974, 1981). Data are obtained from dissections of animals killed during control operations by 20 full-time trappers. The population is retrospectively censused from the numbers killed and their estimated ages. The results of this reconstruction together with fecundity estimates obtained from postmortem material are used for computerized simulations of the population with various control intensities. The simulations are used as a basis for recommendations about the trapper force required to achieve various control objectives (Gosling 1981).

Commercial Harvest. Nutria are rarely harvested on a systematic basis because in most of the world their populations are too small for economic exploitation. However, the very large population of the southern United States allows a substantial cropping program. The harvest generally coincides with that of the muskrat and extends from December to February or March (Chabreck and Dupuie 1970). The price of a pelt averaged between $2.50 in 1970–71 and $5.25 in 1975–76 (Deems and Pursley 1978). Louisiana has consistently reported harvests of over 1 million animals since the 1970–71 trapping season. The annual income for the 1975–76 trapping season was over $8 million for 1,525,506 pelts for Louisiana (Deems and Pursley 1978). In the 1979–80 trapping season, Louisiana trappers harvested 1,300,822 pelts, for a total value of about $9 million (Linscombe personal communication). The nutria brings in more revenues in Louisiana than the muskrat, mink (*Mustela vison*), raccoon (*Procyon lotor*), and river otter (*Lutra canadensis*) (Chabreck et al. 1981). Lowery (1974) compared the harvests of muskrats and nutria in Louisiana. Fur buyers grade pelts on the primeness of the pelt. Prime pelts were considered at least 66 cm long when dried, consisting of dark prime fur with no holes and heavy skin. Only the abdominal portion of the pelt is utilized by the fur industry (Lowery 1974). Chabreck and Dupuie (1970) found that the best pelts were those taken in the winter months. The Louisiana Department of Education (1955) discussed the proper method of skinning and preparing pelts. The fur industry markets this pelt under the name *nutria* but some of it is dyed and sold as "seal" or "beaver" (Larrison 1943).

The meat of nutria is sold to ranches for mink food and the government buys the meat for food to raise screwworm larvae for the screwworm eradication program (O'Neil and Linscombe 1977). It is illegal to sell nutria meat for human consumption, although it is eaten by trappers and other outdoorsmen (Coreil personal communication).

Habitat Management. Suggestions for brackish marsh management for nutria include control of water levels and salinity to encourage the growth of favored food items, *Spartina cynosuroides* and *Scirpus olneyi* (Chabreck et al. 1981). Dense stands of these plants grow in marshes where the salinity concentration is less than 10 ppt. *Spartina cynosuroides* thrives along elevated banks of bayous, ponds, and well-drained marshes; *S. olneyi* can withstand shallow flooding (Chabreck et al. 1981).

Management plans applied for the muskrat, such as marsh burning and ditching, benefit nutria populations in brackish-water wetlands. See Giles (1978) and Perry (1982) for a review of muskrat management practices. One study in Louisiana found that marsh areas separated by ditches spaced at 0.4-km intervals provided a more even distribution of trapping pressure. Also, this design would provide for a better dispersion of water during periods of low water as well as increase the number of elevated spoil sites and cover during floods (Robicheaux 1978).

The intensity of marsh management for fur production on the 4 million acres of Louisiana coastal marsh varies from very low to high. The productivity of a particular marsh area determines what level of management is economically feasible. In coastal Louisiana, only muskrat and nutria normally occur in numbers justifying active management. The majority of coastal Louisiana marshlands are owned by large corporations. Land managers lease the trapping rights to individuals on an annual basis. The most common lease terms require 25 percent of the income from the fur to be returned to the landowner. This percentage may vary from 10–45 percent. Generally, the higher the percentage, the more the landowner is involved with assisting the trapper in maintaining the marshland.

Brackish marsh is primarily a producer of muskrat. However, since 1970, extremely high nutria densities and production (numbers trapped) have been observed on several areas of brackish marsh with three-cornered grass (*Scirpus olneyi*) abundant. Such nutria production is periodic and may or may not occur with high densities of muskrat. The interrelationship between muskrat and nutria is unknown.

Brackish marsh fur management for nutria or muskrats is very similar. Such management is primarily vegetation management. Wakefield wiers (fixed level) are constructed and maintained in tidal bayous and set at an elevation of six inches below the floor of the marsh. These wiers moderate the change in water levels. They help moderate hurricane storm tides and

yet hold enough water to provide access during the winter months when north winds produce very low tidal stages. Some intensively managed marshes have flap gate structures. This structure allows much greater flexibility in water level control. Such control is used to encourage the growth of three-cornered grass and provide access for trapping during winter months.

Ditching in brackish marsh provides drainage of low spots (old eatouts) to encourage revegetation. O'Neil (1949) discusses this at length in his muskrat book. Ditching also provides a constant supply of water, even during a drought. Primarily, ditching provides a greater trapping pressure per trapper than would be possible without it. This pressure is very important when you are attempting to manage a population of nutria with a density of 8–10 per acre.

The practice of burning the marsh varies in timing depending on whether the management is for muskrat or nutria. Marsh burning for muskrats generally occurs in late September or October prior to trapping. Nutria trappers are reluctant to burn prior to trapping. Most trappers are convinced that nutria move out of a burned marsh and thus may be caught on an adjacent trapping lease. If muskrats are not very abundant on a brackish marsh tract, nutria are generally trapped throughout the three-month season (December–February). Then burning is done in late February or March. If muskrats are mixed with nutria and both will be trapped, then nutria are trapped first. Then the marsh is burned, and muskrat trapping commences in January or February.

Unpublished telemetry data indicate that nutria do move out of a burned area. The animals appear to move to the available cover closest to their original home range. As the burned area begins to green up, the nutria feed in this area at night and return to cover during the day. Eventually, cover is restored, and the animals remain in the burned area full time once again. Management goals in brackish marsh should be maintaining a good food supply, stabilizing water levels, and providing access for trapping.

Habitat management for nutria in fresh (or intermediate) marsh is less exact. In general, this marsh type is more diverse. The number of plant species present is much greater than in brackish marsh. Because of this diversity, food supply does not appear to be a serious limiting factor. However, water level fluctuations certainly have a great influence on nutria populations. Fresh marsh water levels in coastal Louisiana are determined by local rainfall and Mississippi River runoff. Fresh marsh management for nutria is basically water management. Using levee construction, gates, pumps, etc., the objective of stable water levels is sought. Summer droughts or winter floods definitely have detrimental effects on nutria. Having observed significant weight loss during periods of high water level in winter, it is believed that significant resorption occurs as well as low survival of young. Summer drought also concentrates animals in limited available water rather than limited higher ground. Food supplies are necessarily changed and limited. Predation by alligators certainly increase during drought.

Fur harvest records from different land companies

indicate that fresh marsh areas with floatant (floating mats of vegetation) appear to be the most productive and most stable for nutria (see Palmisano 1972: figure 4). This information offers more support for our thoughts concerning water level. Floating marsh does not experience floods or droughts; the vegetation mat simply rises and lowers with the water level, and thus the nutria's habitat remains relatively unchanged. This advantage of floating fresh marsh was observed during April 1980 in comparing a study area (floating marsh) on Lacassine National Wildlife Refuge with another fresh marsh area without floating marsh. The nutria on the Lacassine study area were relatively undisturbed by a 2-foot increase in water level following 15–20 inches of local rain in 2 days. The other area luckily had a few elevated sites created by trappers or land managers when cutting trapping ditches. In this area, nutria were crowded shoulder to shoulder on these sites. Undoubtedly, many nutria did not survive, since this water level remained for over 10 days. Surviving animals were certainly stressed. The long-term effect of this stress is unknown. Management in fresh marsh where floating marsh is not present but flooding or drought is likely should stress ditching, with the spoil material piled high, creating elevated sites. When possible, levees created by oil activity should be planned to improve water level control. In certain situations, levees should provide escape cover and elevations during floods. A location canal can provide a constant water source during a drought if not adversely affecting the hydrology of a marsh (Linscombe personal communication).

This chapter could not have been completed without the assistance of a number of people. Special thanks are due to Greg Linscombe of the Louisiana Department of Wildlife and Fisheries, who provided assistance in reviewing the manuscript and contributed a significant portion of the section dealing with habitat management of nutria. Paul Coreil of the Louisiana Cooperative Extension Service made available home range data that were unavailable in the literature. Michael Pramstaller of the Raptor Information Center of the National Wildlife Federation provided information about eagle predation on nutria in Maryland. Dr. L. Morris Gosling of the Coypu Research Laboratory in England was particularly helpful in reviewing and commenting on the drafts of this chapter. Dr. Robert Chabreck of Louisiana State University and Dr. Charles Woods of Florida State University reviewed the manuscript and made many helpful suggestions. Dr. R. P. Morgan of the Appalachian Environmental Laboratory assisted with the genetics section. Also, I would like to thank Evelyn Kirk and Kathy Twigg of the Appalachian Environmental Laboratory for typing the chapter and Frances Younger for drawing the graphs.

LITERATURE CITED

Adams, W. H., Jr. 1956. The nutria in coastal Louisiana. Proc. Louisiana Acad. Sci. 19:28–41.

Aliev, F. F. 1956. Theoretical and practical foundations of coypu (*Myocastor coypus* Molina) raising in Azerbaijan. Tranl. Inst. Zool. Acad. Sci., Azerbaijan, S. S. R. 19:5–96.

———. 1965a. Extent and causes of nutria mortality in the water bodies of the southern U.S.S.R. Mammalia 29:435–437.

———. 1965b. Growth and development of nutrias' functional features. Fur Trade J. Can. 42(11):2–3; 42(12):2–3; 43(2):2–3.

———. 1966a. Numerical changes and the population structure of the coypu, *Myocastor coypus* (Molina, 1782) in different countries. Saugetierk. Mitt. 15:238–242.

———. 1966b. Enemies and competitors of the nutria in U.S.S.R. J. Mammal. 47:353–355.

Asdell, S. A. 1964. Patterns of mammalian reproduction. 2nd ed. Constable, London. 437pp.

Ashbrook, F. G. 1948. Nutrias grow in the United States. J. Wildl. Manage. 12:87–95.

Atwood, E. L. 1950. Life history studies of nutria, or coypu, in coastal Louisiana. J. Wildl. Manage. 14:249–265.

Axell, H. E. 1963. Coypu (*Myocastor coypus*) at Minsmere during the frost of January and February, 1963. Trans. Suffolk Nat. Soc. 12:257–259.

Babero, B. B.; Cabello, C.; and Kinoed, J. 1979. Helmintofauna de Chile. Part 5: Nuevos Parasitos del Coipo *Myocastor coypus* (Molina, 1782) Bol. Chile Parasitol. 34:26–31. (In Spanish, English summary.)

Babero, B. B., and Lee, J. W. 1961. Studies on the helminths of nutria, *Myocastor coypus* (Molina), in Louisiana with check-list of other worm parasites from this host. J. Parasitol. 47:378–390.

Blair, R. M., and Langlinais, M. J. 1960. Nutria and swamp rabbits damage baldcypress plantings. J. For. 58:388–389.

Bourdelle, E. 1939. American mammals introduced into France in the contemporary period, especially *Myocastor* and *Ondatra*. J. Mammal. 20:287–291.

Brown, L. E. 1966. An electrophoretic comparison of the serum proteins of fetal and adult nutria (*Myocastor coypus*). Comp. Biochem. Physiol. 19:479–481.

Brown, L. N. 1975. Ecological relationships and breeding biology of the nutria (*Myocastor coypus*) in the Tampa, Florida, area. J. Mammal. 56:928–930.

Cabrera, A., and Yepes, J. 1940. Mamiferos SudAmericanos (oida, costumbres y description). Compania Argentina de Editores, Buenos Aires, Argentina. 344pp.

Carill-Worsley, P.E.T. 1932. A fur farm in Norfolk. Trans. Norfolk Norwich Nat. Soc. 13:105–115.

Chabreck, R. H. 1962. Daily activity of nutria in Louisiana. J. Mammal. 43:337–344.

Chabreck, R. H., and Dupuie, H. H. 1970. Monthly variation in nutria pelt quality. Proc. Annu. Conf. Southeastern Assoc. Game Fish Comm. 24:169–175.

Chabreck, R. H.; Love, J. R.; and Linscombe, G. 1981. Foods and feeding habits of nutria in brackish marsh in Louisiana. Pages 531–543 in J. A. Chapman and D. Pursley, eds. Worldwide Furbearer Conference proceedings. Vol. 1. Worldwide Furbearer Conf., Inc., Frostburg, Md. 652pp.

Chabreck, R. H.; Thompson, R. B.; and Ensminger, A. B. 1977. Chronic dermatitis in nutria in Louisiana. J. Wildl. Dis. 13:333–334.

Chapman, J. A.; Harman, A. L.; and Samuel, D. E. 1977. Reproductive and physiological cycles in the cottontail complex in western Maryland and nearby West Virginia. Wildl. Monogr. 56:1–73.

Chapman, J. A.; Lanning, J. C.; Willner, G. R.; and Pursley, D. 1980. Embryonic development and resorption in feral nutria (*Myocastor coypus*) from Maryland. Mammalia 44:371–379.

Chapman, J. A.; Willner, G. R.; Dixon, K. R.; and Pursley, D. 1978. Differential survival rates among leg-trapped and live-trapped nutria. J. Wildl. Manage. 42:926–928.

Coreil, P. D., and Perry, H. R., Jr. 1977. A collar for attaching radio transmitters to nutria. Proc. Annu. Conf. Southeastern Assoc. Game Fish Comm. 31:254-258.

Cotton, K. E. 1963. The coypu. River Board's Assoc. Year book 11:31-39.

Cross, J. H., and Thomas, R. M. 1966. Hydatid disease in the nutria. J. Parasitol. 52:1215-1216.

Davis, W. B. 1958. Distribution of nutria in Texas. Texas Game Fish 16:22.

Deems, E. F., Jr., and Pursley, D. 1978. North American furbearers: their management, research, and harvest status in 1976. Int. Assoc. Fish and Wildl. Agencies. 155pp.

de Soriano, B. S. 1960. Elementos constitutivos de una habitacion de *Myocastor coypus bonariensis* (Geoffroy) (''nutria''). Revista de la Facultad de Humanidades y Ciencias 18:257-276.

_____. 1963. La temperatura de la habitacion hipogea de *Myocastor coypus bonariensis* Geoffroy, ''Nutria,'' en relacion con la temperatura ambiental. Actas y trabajos del primer congreso sudamericano de zoologia 1:153-158.

Dixon, J. R. 1960. The skeletal system of the nutria (*Myocastor coypus*). J. Mammal. 41:89-97.

Dixon, K. R.; Willner, G. R.; Chapman, J. A.; Lane, W. C.; and Pursley, D. 1979. Effects of trapping and weather on body weights of feral nutria in Maryland. J. Appl. Ecol. 16:69-76.

Dugoni, J. A. 1980. Habitat utilization, food habits, and productivity of nesting southern Bald Eagles in Louisiana. M.S. Thesis. Louisiana State Univ. 150pp.

Ehrlich, S. 1958. The biology of the nutria. Bamidgeh Bull. Fish Cult. Israel 10:36-43, 60-70.

_____. 1962. Experiment on the adaptation of nutria to winter conditions. J. Mammal. 43:418.

_____. 1966. Ecological aspects of reproduction in nutria *Myocastor coypus* Mol. Mammalia 30:142-152.

Ehrlich, S., and Jedynak, K. 1962. Nutria influence on a bog lake in northern Pomorze, Poland. Hydrobiologia 19:273-297.

Ellis, E. A. 1963. Some effects of selective feeding by the coypu (*Myocastor coypus*) on the vegetation of Broadland. Trans. Norfolk Norwich Nat. Soc. 20:32-35.

_____. 1965. The broads. Collins, London. 401pp.

Evans, J. 1970. About nutria and their control. U.S. Fish Wildl. Ser. Res. Publ. 86:1-65.

Evans, J.; Ells, J. O.; Nass, R. D.; and Ward, A. L. 1972. Techniques for capturing, handling, and marking nutria. Trans. Annu. Conf. Southeastern Assoc. Game Fish Comm. 25:295-315.

Evans, J., and Ward, A. L. 1967. Secondary poisoning associated with anticoagulant-killed nutria. J. Am. Vet. Med. Assoc. 151:856-861.

Gainey, L. F. 1949. Comparative winter food habits of the muskrat and nutria in captivity. Louisiana State Univ., Baton Rouge. 41pp. Manuscript.

George, R. S. 1964. *Ceratophyllus g. gallinae* (Schrank) (Siphonaptera: Cetatophyllidae) from a British coypu (*Myocastor coypus* Molina) (Rodentia: Capromyidae). Entomol. Gazette 15:40-41.

George, W., and Weir, B. J. 1974. Hystricomorph chromosomes. Symp. Zool. Soc. London 34:79-108.

Giles, R. H., Jr. 1978. Wildlife management. Freeman & Co., San Francisco. 416pp.

Gluchowski, W., and Maciejowski, J. 1958. Investigations on factors controlling fertility in the coypu. Part 2: Attempts at determining the potential fertility, based on histological studies of the ovary. Ann. Univ. Mariae Curie-Sklodowska, E., 13:345-361.

Gosling, L. M. 1974. The coypu in East Anglia. Trans. Norfolk Norwich Nat. Soc. 23:49-59.

_____. 1977. Coypu, *Myocastor coypus*. Pages 256-265 *in* G. B. Corbet and H. N. Southern, eds. The handbook of British mammals. Blackwell Sci. Publ., Oxford. 520pp.

_____. 1979. The twenty-four hour activity cycle of captive coypus (*Myocastor coypus*). J. Zool. London 187:341-367.

_____. 1980*a*. The duration of lactation in feral coypus (*Myocastor coypus*). J. Zool., London 191:461-474.

_____. 1980*b*. The role of wild plants in the ecology of mammalian crop pests. *In* M. Thresh, ed. The role of wild plants in the ecology of crop pests. Pitman, London.

_____. 1981. The dynamics and control of a feral coypu population. Pages 1806-1825 *in* J. A. Chapman and D. Pursley, eds. Worldwide Furbearer Conference proceedings. Vol. 3. Worldwide Furbearer Conf., Inc., Frostburg, Md. 652pp.

Gosling, L. M.; Guyon, G. E.; and Wright, K.M.H. 1980*a*. Diurnal activity of feral coypu (*Myocastor coypus*) during the cold winter of 1978/79. J. Zool. 192:143-146.

Gosling, L. M.; Huson, L. W.; and Addison, G. C. 1980*b*. Age estimation of coypus (*Myocastor coypus*) from eye lens weight. J. Appl. Ecol. 17:641-647.

Griffo, J. V., Jr. 1957. The status of the nutria in Florida. Q. J. Florida Acad. Sci. 20:209-215.

Gunderson, H. L. 1955. Nutria, *Myocastor coypus*, in Minnesota. J. Mammal. 36:465.

Hailman, J. P. 1961. Stereotyped feeding behavior of a North Carolina nutria. J. Mammal. 42:269.

Hale, J. B. 1950. Nutria versus muskrat. Wisconsin Conserv. Bull. 15:15-16.

Harman, D. M., and Thoerig, T. 1968. Occurrence of the porcupine, *Erethizon dorsatum*, and the nutria, *Myocastor coypus bonariensis*, in western Maryland. Chesapeake Sci. 9:138-139.

Harris, V. T. 1956. The nutria as a wild fur mammal in Louisiana. Trans. North Am. Wildl. Conf. 21:474-486.

Harris, V. T., and Webert, F. 1962. Nutria feeding activity and its effect on marsh vegetation in southwestern Louisiana. Spec. Sci. Rep. Wildl. 64:1-53.

Hillbricht, A., and Ryszkowski, L. 1961. Investigations of the utilization and destruction of its habitat by a population of coypu, *Myocastor coypus* Molina, bred in semi-captivity. Ekologia Polska. Ser. A, 9:505-524.

Hillemann, H. H., and Gaynor, A. I. 1961. The definitive architecture of the placentae of nutria, *Myocastor coypus* (Molina). Am. J. Anat. 109:299-318.

Hillemann, H. H.; Gaynor, A. I.; and Stanley, H. P. 1958. The genital systems of nutria (*Myocastor coypus*). Anat. Rec. 130:515-531.

Hillemann, H. H., and Ritschard, R. L. 1967. Comparative fibroarchitecture in the mammalian placenta and adnexa. Trans. Am. Microsc. Soc. 86:184-194.

Hoffmeister, D. F. 1958. The future status of the nutria, fur-bearing rodent, in Illinois. Illinois Acad. Sci. Trans. 51:48-50.

Hoffmeister, D. F., and Kennedy, C. D. 1947. The nutria, a South American rodent, in Kansas. Trans. Kansas Acad. Sci. 49:445-446.

Howard, W. E. 1953. Nutria (*Myocastor coypus*) in California. J. Mammal. 34:512-513.

Jellison, W. L. 1945. Spotted skunk and feral nutria in Montana. J. Mammal. 26:432.

Jemison, E. S., and Chabreck, R. H. 1962. Winter barn owl (*Tyto alba*) foods in a Louisiana coastal marsh. Wilson Bull. 74:95-96.

Karkalitskii, I. M.; Karkalitskaya, G. F.; Ashikhmina, E.

M.; Kourizhnykh, N. D.; Tuzova, G. P.; Plotnikova, G. F.; and Berdnikov, M. P. 1970. Characteristics of biochemical shifts during experimental B6 hypovitaminosis. Vop Pitan 29:23–28.

Kasumov, N. I.; Radzhabli, S. I.; and Kuliev, G. K. 1976. Cytogenetic study of nutria. Part 1: Somatic and meiotic cells of standard and white nutria. Genetika 12:174–176. (In Russian, English summary.)

Katomski, P. A.; and Ferrante, F. L. 1974. Catecholamine content and histology of the adrenal glands of the nutria (*Myocastor coypus*). Comp. Biochem. Physiol. A Comp. Physiol. 48:539–546.

Kays, C. E. 1956. An ecological study with emphasis on nutria (*Myocastor coypus*) in the vicinity of Price Lake, Rockefeller Refuge, Cameron Parish, Louisiana. M.S. Thesis. Louisiana State Univ. 145pp.

Kebbe, C. E. 1959. The nutria in Oregon. Oregon State Game Comm. Bull. 14:8.

Kimura, P. T., and Johnson, M. L. 1970. Serum proteins and haptoglobins of Caribbean rodents. Comp. Biochem. Physiol. 37:277–280.

Kinler, N., and Chabreck, R. H. 1978. Nutria pelt damage from *Bidens laevis*. Trans. Southeastern Fish Wildl. Conf. 32:369–377.

Kinler, N. W., and Linscombe, G. 1981. Smooth beggartick, its distribution, control and impact on nutria in coastal Louisiana. Pages 142–154 in J. A. Chapman and D. Pursley, eds. Worldwide Furbearer Conference proceedings. Vol. 1. Worldwide Furbearer Conf., Inc., Frostburg, Md. 652pp.

Kinsel, G. V. 1958. The theory and practice of nutria raising. Fur Trade J. Can. (Toronto). 231pp.

Koch, T. 1953. Beitrage zur Anatomie des Sumpfbibers. Leipzig Verl. Hirzel. 168pp.

Konieczna, B. 1956. Sexual maturation and reproduction in the nutria (*Myocastor coypus*). Part 2: The ovary. Folia Biol. 4:139–150.

Kuhn, L. W., and Peloquin, E. P. 1974. Oregon's nutria problem. Proc. Vertebr. Pest Conf. 6:101–105.

Langenfeld, M. 1977a. Anatomical and histological structure of the sympathetic trunk in the coypu (*Myocastor coypus* Molina). Part 1: Anatomy and topography of the cervical part. Polskie Archiwum Wet. 20:143–157.

⸻. 1977b. Anatomical and histological structure of the sympathetic trunk in the coypu (*Myocastor coypus* Molina). Part 2: Macrostructure and topography of the thoracic part. Polskie Archiwum Wet. 20:145–153.

⸻. 1977c. Anatomical and histological structure of the sympathetic trunk in the coypu (*Myocastor coypus* Molina). Part 3: Macrostructure and topography of the lumbar section. Polski Archiwum Wet. 20:143–150.

⸻. 1977d. Anatomic and histologic structure of the sympathetic trunk in the coypu (*Myocastor coypus* Molina). Macrostructure and topography of the sacrocaudal segment. Folia Morphol. 36:245–255.

Larrison, E. J. 1943. Feral coypus in the pacific northwest. Murrelet 24:3–9.

Laurie, E. M. 1946. The coypu (*Myocastor coypus*) in Great Britain. J. Anim. Ecol. 15:22–34.

Lechleitner, R. R. 1959. Sex ratio, age classes and reproduction of the black-tailed jackrabbit. J. Mammal. 40:63–81.

Linscombe, G. 1976. An evaluation of the No. 2 victor and 220 conibear traps in coastal Louisiana. Trans. Annu. Southeastern Assoc. Game Fish Comm. Jackson, Miss. 30:560–568.

Linscombe, G.; Kinler, N.; and Wright, V. 1981. Nutria population density and vegetative changes in brackish marsh in coastal Louisiana. Pages 129–141 in J. A. Chap-

man and D. Pursley, eds. Worldwide Furbearer Conference proceedings. Vol. 1. Worldwide Furbearer Conf., Inc., Frostburg, Md. 652pp.

Little, M. D. 1965. Dermatitis in a human volunteer infected with *Strongyloides* of nutria and raccoon. Am. J. Trop. Med. Hyg. 14:1007–1009.

Lomnicki, A. 1957. The daily rhythm of activity in the nutria (*Myocastor coypus* Molina). Folia Biol. 5:293–306. (In Polish, English summary.)

Louisiana Department of Education. 1955. How to skin and prepare nutria pelts. Louisiana Trade School, Natchitoches, La. 35pp.

Low, J. B. 1946. Nutria introduced in Utah marshlands near Salt Lake. Utah Fish Game Bull. 3:4–5.

Lowery, G. H., Jr. 1974. The mammals of Louisiana and its adjacent waters. Louisiana State Univ. Press, Baton Rouge. 565pp.

McNease, L., and Joanen, T. 1977. Alligator diets in relation to marsh salinity. Proc. Annu. Conf. Southeastern Assoc. Fish and Wildl. 31:36–40.

Mann, T., and Wilson, E. D. 1962. Biochemical observations on the male accessory organs of nutria, *Myocastor coypus* (Molina). J. Endocrinol. 25:407–408.

Matthias, K. E. 1941. Nutria, profitable fur discovery. Am. Fur Breeder 14:18–20.

Maurice, J. A. 1931. Le ragondin. Arch. Hist. Nat. Soc. Acclim., Paris, France. 234pp.

Miller, A. 1956. The biting louse, *Pitrufquenia coypus* Marelli, on nutria in Louisiana. J. Parasitol. 42:583.

Milne, R. C., and Quay, T. L. 1967. The foods and feeding habits of nutria on Hatteras Islands, North Carolina. Proc. Annu. Conf. Southeastern Assoc. Game Fish Comm. 20:112–123.

Miura, S. 1976. Dispersal of nutria in Oklahoma Prefecture. J. Mammal. Soc. Jap. 6:231–237.

Morgan, R. P. II; Willner, G. R.; and Chapman, J. A. 1981. Genetic variation in Maryland nutria, *Myocastor coypus*. Pages 30–37 in J. A. Chapman and D. Pursley, eds. Worldwide Furbearer Conference proceedings. Vol. 1. Worldwide Furbearer Conf., Inc., Frostburg, Md. 652pp.

Mursaloglu, B. 1981. The recent status and distribution of Turkish furbearers. Pages 86–94 in J. A. Chapman and D. Pursley, eds. Worldwide Furbearer Conference proceedings. Vol. 1. Worldwide Furbearer Conf., Inc., Frostburg, Md. 652pp.

Murúa, R.; Neuman, O.; and Dropelmann, J. 1981. Food habits of *Myocastor coypus* (Molina) in Chile. Pages 544–558 in J. A. Chapman and D. Pursley, eds. Worldwide Furbearer Conference proceedings. Vol. 1. Worldwide Furbearer Conf., Inc., Frostburg, Md. 652pp.

Nagahana, M.; Hatsushika, R.; Shimizu, M.; and Kawakami, S. 1977. Does the nutria, *Myocastor coypus*, serve as the reservoir of liver fluke, *Clonorchis sinensis?* Jap. J. Parasitol. 26:41–45. (In Japanese. English summary.)

Newson, R. M. 1965. Reproduction in the feral coypu, *Myocastor coypus*. J. Reprod. Fertility 9:380–381.

⸻. 1966. Reproduction in the feral coypu (*Myocastor coypus*). Symp. Zool. Soc. 15:323–334.

⸻. 1969. Population dynamics of the coypu, *Myocastor coypus* (Molina), in eastern England. Warsaw Proc. IBP Secondary Productivity in Small Mammal Populations, Oxford, England. Pp. 203–204.

Newson, R. M., and Holmes, R. G. 1968. Some ectoparasites of the coypu (*Myocastor coypus*) in eastern England. J. Anim. Ecol. 37:471–481.

Norris, J. D. 1967a. A campaign against feral coypus (*Myocastor coypus* Molina) in Great Britain. J. Appl. Ecol. 4:191–199.

———. 1967*b*. The control of coypus (*Myocastor coypus* Molina) by cage trapping. J. Appl. Ecol. 4:167–189.

O'Neil, T. 1949. The muskrat in the Louisiana coastal marshes. Louisiana Dept. Wildl. and Fish, New Orleans. 152pp.

O'Neil, T., and Linscombe, G. 1977. The fur animals, the alligator, and the fur industry in Louisiana. Louisiana Dept. Wildl. and Fish Ed. Bull. 106. 66pp.

Packard, R. L. 1967. Octodontodid, Bathyergoid, and Ctenodactlyoid rodents. Pages 273–290 *in* S. Anderson and J. K. Jones, Jr., eds. Recent mammals of the world. Ronald Press, Co., New York. 453pp.

Page, C. A.; Harris, V. T.; and Durand, J. 1957. A survey of virus in nutria. Southwestern Louisiana J. 1:207–210.

Palmisano, A. W. 1972. Habitat preference of waterfowl and fur animals in the northern gulf coast marshes. Proc. Coastal Marsh and Estuary Manage. Symp. 2:163–190.

Palmisano, A. W., and Dupuie, H. H. 1975. An evaluation of steel traps for taking fur animals in coastal Louisiana. Proc. Annu. Conf. Southeastern Assoc. Game and Fish Comm. 29:342–347.

Peloquin, E. P. 1969. Growth and reproduction of the feral nutria *Myocastor coypus* (Molina) near Corvallis, Oregon. M.S. Thesis. Oregon State Univ. 55pp.

Perry, H. R., Jr. 1982. Muskrats (*Ondatra zibethicus* and *Neofiber alleni*). Pages 282–325 *in* J. A. Chapman and G. A. Feldhamer, eds. Wild mammals of North America: biology, management, and economics. Johns Hopkins Univ. Press, Baltimore. 1147pp.

Petrides, G. A., and Leedy, D. L. 1948. The nutria in Ohio. J. Mammal. 29:182–183.

Pietrzyk-Walknowska, J. 1956. Sexual maturation and reproduction in *Myocastor coypus*. Part 3: The testicle. Folia Biol. 4:151–160.

Prasad, H. 1960. Two new species of coccidia of the coypu. J. Protozool. 7:207–210.

Presnall, C. C. 1958. The present status of exotic mammals in the United States. J. Wildl. Manage. 22:45–50.

Pridham, T. J.; Budd, J.; and Karstad, L. H. 1966. Common diseases of fur bearing animals. Part 2: Diseases of chinchillas, nutria and rabbits. Can. Vet. J. 7:84–87.

Pridham, T. J., and Thackeray, E. L. 1959. The isolation of a streptococcus, Lancefield's group D, from nutria. Can. J. Comp. Med. and Vet. Sci. 23:81–83.

Rawley, E. V. 1956. $$Nutria?? Utah Fish Game Bull. 12:6–7.

Robicheaux, B. L. 1978. Ecological implications of variably spaced ditches on nutria in a brackish marsh, Rockefeller Refuge, Louisiana. M.S. Thesis. Louisiana Tech. Univ. 49pp.

Rossolimo, O. L. 1958. The periodic growth of the skull of nutria (*Myopotamus coypus* Molina). Nouchn Dokl Vysshei Shkoly Biol. Nauki 2:55–57.

Rowlands, I. W., and Heap, R. B. 1966. Histological observations on the ovary and progesterone levels in the coypu (*Myocastor coypus*). Symp. Zool. Soc. London 15:335–352.

Ryszkowski, L. 1966. The space organization of nutria (*Myocastor coypus*) populations. Symp. Zool. Soc. London 18:259–265.

Safonov, V. G. 1981. The status and reestablishment of fur resources in the U.S.S.R. Pages 95–110 *in* J. A. Chapman and D. Pursley, eds. Worldwide Furbearer Conference proceedings. Vol. 1. Worldwide Furbearer Conf., Inc., Frostburg, Md. 652pp.

Schitoskey, F., Jr. 1971. Anomalies and pathological conditions in the skulls of nutria from southern Louisiana. Mammalia 35:311–314.

———. 1972. Bacular variation in nutria from southern Louisiana. Southwestern Nat. 16:454–457.

Schitoskey, F., Jr.; Evans, J.; and Lavoie, G. K. 1972. Status and control of nutria in California. Proc. Vertebr. Pest Conf. 5:15–17.

Shirley, M. G.; Chabreck, R. H.; and Linscombe, G. 1981. Foods of nutria in fresh marshes of southeastern Louisiana. Pages 517–530 *in* J. A. Chapman and D. Pursley, eds. Worldwide Furbearer Conference proceedings. Vol. 1. Worldwide Furbearer Conf., Inc., Frostburg, Md. 652pp.

Simpson, T. R., and Swank, W. G. 1979. Trap avoidance by marked nutria: a problem in population estimation. Proc. Annu. Conf. Southeastern Assoc. Game Fish Comm. 33:11–14.

Skowron-Cendrzak, A. 1956. Sexual maturation and reproduction in *Myocastor coypus*. Part 1: The oestrus cycle. Folia Biol. 4:119–138.

Stanley, H. P., and Hillemann, H. H. 1960. Histology of the reproductive organs of nutria, *Myocastor coypus* (Molina). J. Morphol. 106:277–299.

Steffen, J. 1955. Observations on diseases occurring in mass breeding of coypu. Medycyna Weterynaryjna 11:270–275.

Swank, W. G., and Petrides, G. A. 1954. Establishment and food habits of the nutria in Texas. Ecology 35:172–176.

Szynkiewicz, E. 1968. Studies on antigenic differentiation of blood in the coypu (*Myocastor coypus* Molina 1792). European Conf. Anim. Blood Groups Biochem. Polymorphism 11:567–570.

———. 1971. Investigations on differentiation of betaglobulin subfractions in the blood serum of nutria (*Myocastor coypus* Molina 1792). Genetica Polonica 12:465.

Talbert, R. E. 1962. Control of nutria. California Dept. Agric. Bull. 51:156–157.

Tsigalidou, V.; Simotas, A. G.; and Fasoulas, A. 1966. Chromosomes of the coypus (*Myocastor coypus* Molina). Nature 211:994–995.

———. 1967. The chromosomes of the nutria. Riv. Zootec. (Spec. Issue, Int. Symp. Zootech.), pp. 421–422.

Valentine, J. M., Jr.; Walther, J. R.; McCartney, K. M.; and Ivy, L. M. 1972. Alligator diets on the Sabine National Wildlife Refuge, Louisiana. J. Wildl. Manage. 36:809–815.

Vinogradov, V. V. 1965. Nutria requirements in natural fodder and the determination of the extent of its range. Zool. ZH 44:1712–1721. (In Russian, English summary.)

Wagner, J. A. 1963. Gross and microscopic anatomy of the digestive system of the nutria, *Myocastor coypus bonariensis* (Geoffroy). J. Morphol. 112:219–334.

Walker, E. P. et al. 1975. Mammals of the world. 3rd ed. 2 vols. Johns Hopkins Univ. Press, Baltimore.

Warkentin, M. J. 1968. Observations on the behavior and ecology of the nutria in Louisiana. Tulane Stud. Zool. Bot. 15:10–17.

Weir, B. J. 1974. Reproductive characteristics of hystricomorph rodents. Symp. Zool. London 34:265–301.

Wentz, W. A. 1971. The impact of nutria (*Myocastor coypus*) on marsh vegetation in the Willamette Valley, Oregon. M.S. Thesis. Oregon State Univ. 41pp.

Willner, G. R., and Chapman, J. A. 1977. Polydactyly in *Myocastor coypus*. Virginia J. Sci. 22:143.

Willner, G. R.; Chapman, J. A.; and Goldsberry, J. R. 1975. A study and review of muskrat food habits with special reference to Maryland. Maryland Wildl. Adm., Publ. Ecol. 1:1–25.

Willner, G. R.; Chapman, J. A.; and Pursley, D. 1979. Reproduction, physiological responses, food habits, and abundance of nutria on Maryland marshes. Wildl. Monogr. 65:1–43.

Willner, G. R.; Dixon, K. R.; Chapman, J. A.; and Stauffer, J. R. 1980. A model for predicting age-specific body weights of nutria without age determination. J. Appl. Ecol. 7:343–347.

Wilson, E. D., and Dewees, A. A. 1962. Body weights, adrenal weights and oestrous cycles of nutria. J. Mammal. 43:362–364.

Wilson, E. D.; Zarrow, M. X.; and Lipscomb, H. S. 1964. Bilateral dimorphism of the adrenal glands in the coypu (*Myocastor coypus,* Molina). Endocrinology 74:515–517.

Woods, C. A. 1976. How hystricognath rodents chew. Am. Zool. 16:215. Abstr.

Woods, C. A., and Howland, E. B. 1979. Adaptive radiation of capromyid rodents: anatomy of the masticatory apparatus. J. Mammal. 60:95–116.

GALE R. WILLNER, Appalachian Environmental Laboratory, Center for Environmental and Estuarine Studies, University of Maryland, Frostburg State College Campus, Frostburg, Maryland 21532.

54

Norway Rat and Allies

Rattus norvegicus and Allies

William B. Jackson

NOMENCLATURE

COMMON NAMES. Norway rat, common rat, wharf rat, house rat, sewer rat, brown rat
SCIENTIFIC NAME. *Rattus norvegicus*

COMMON NAMES. Black rat, roof rat, ship rat, gray-bellied rat, fruit rat, Alexandrine rat
SCIENTIFIC NAME. *Rattus rattus*

COMMON NAMES. House mouse, field mouse
SCIENTIFIC NAME. *Mus musculus*

The term *commensal* is commonly used to refer to the three common murid rodents found in the continental United States and frequently living in close association with humans. It means "sharing the table," which seems appropriate, although some might consider the relationship more of a parasitic one. The Polynesian rat (*Rattus exulans*), common in Hawaii and across the Pacific basin, is not considered here.

The roof rat, *R. rattus* has usually been referenced by the trinomial *R. r. rattus*, *R. r. frugivorus*, and *R. r. alexandrinus*, especially in the earlier literature. In most places of the world subject to modern commerce, mixing of these supposed subspecies has been so great that I consider such taxonomy a useless exercise. Litter mates may exhibit the several color phases, and the trinomial should be reserved for major geographic populations (Johnson 1962). Certainly the efforts of Schwarz and Schwarz (1965) to deal with this genus should be ignored.

House mouse (*Mus musculus*) populations similarly have been thoroughly mixed and remixed in North America. Morphological variants can be found in field studies. Schwarz and Schwarz (1943) designated *M. m. domesticus* and *M. m. brevirostris* in the United States. I find little practical value in pursuing such taxonomic efforts at this time.

The roof rat and house mouse were early arrivals in the New World, coming with the explorers and colonists. Considerable expansion along avenues of commerce had occurred before the Norway rat reached these shores about 1775. Gradual displacement of the roof rat by the larger Norway rat took place, especially from northern and inland areas. Ecke (1954) documented the population shift in Georgia.

DISTRIBUTION

The roof rat currently is restricted to the southern, southeastern, and western United States. In the west it has expanded its range considerably, utilizing the blackberry (*Ribes* sp.) associated with old mining campsites and the lush vegetation of freeways and urban housing areas. In Arizona it may live in the sewers but range out into the vegetation at night to feed.

The house mouse is ubiquitous. Wherever man erects structures, this species is likely to be found. Even in the arid plains where grain storage structures are widely scattered, the house mouse is a common infestant (Anderson 1964). In old field successions and even in wooded environments the house mouse often can be found in association with several native rodent species.

The Norway rat is found throughout North America, almost always related to man directly or indirectly. In urban areas it thrives on garbage and stored foods; in rural environments, on plant crops, animals, and stored grain. Along ditches, rivers, and marshes sewage and flotsam as well as nesting wildlife contribute to its support. Now-isolated populations may have originated when a homestead, dump, or other structure existed and then persisted after the human support vanished. In desert or arctic environments, this species may be limited in its distribution by food and water sources. Anchorage and Nome, Alaska, have infestations related to their garbage dumps (Schiller 1956). On the Aleutian island Amchitka, the rats range widely, feeding on beach debris and nesting seabirds, although the garbage dump of the military base was the focus of the infestation (Brechbill 1977). Through intensive control efforts, the province of Alberta, Canada, is currently attempting to prevent the establishment of the Norway rat. In northern Mexico, this species exists where villages harbor it, although native rodent species may be commensal as well (Rex Lord personal communication).

DESCRIPTION

One general characteristic of rodents is their continuously growing incisor teeth. These grow 10-13 cm a year but are kept at chisel sharpness by grinding against each other. Gnawing on hard objects daily is not required by murids (despite statements in many books), for rats and mice can be maintained successfully on soft foods throughout their life. The dental formula of *R. norvegicus, R. rattus,* and *Mus musculus* is 1/1, 0/0, 0/0, 3/3 = 16. Characteristics of the three species are detailed in table 54.1, and skulls are shown in figures 54.1, 54.2, and 54.3, respectively.

PHYSIOLOGY

Rodents are color blind and have relatively poor distance vision. They respond quickly to motion, even

TABLE 54.1. Characteristics of commensal rodents

Characteristic	Norway Rat	Roof Rat	House Mouse
General appearance	large, robust	sleek, graceful	small, slender
Adult size			
weight (gm)	200–500	150–250	12–30
length			
head + body (mm)	180–255	165–205	65–90
tail (mm)	150–215	190–255	75–100
hind foot (mm)	>40	<40	<20
ear (mm)	<20	>20	13
Snout	blunt	pointed	pointed
Ears	small, covered with short hairs	large, nearly naked	large, some hair
Eyes	small	large, prominent	small
Tail	dark above, pale beneath	uniformly dark	uniformly dark
Fur	brown with scattered black (agouti); venter gray to yellow/white; shaggy	agouti to gray to black; venter white, gray, or black; smooth	light brown, light gray; smooth
Number of mammae (female)	12	8–12; usually 10	10
Dental formula	1/1, 0/0, 0/0, 3/3 = 16	1/1, 0/0, 0/0, 3/3 = 16	1/1, 0/0, 0/0, 3/3 = 16
Teeth	1st upper molar shorter than combined lengths of 2nd and 3rd molars; molar tubercles in 3 longitudinal rows	1st upper molar shorter than combined lengths of 2nd and 3rd molars; molar tubercles in 3 longitudinal rows	1st upper molar longer than combined lengths of 2nd and 3rd molars; molar tubercles in 3 longitudinal rows
Skull	parietal ridges parallel; length of parietal ridges > distance between ridges	parietal ridges bowed; length of parietal ridges < distance between ridges	supraorbital ridges faint or absent
Droppings	capsule shaped, 20 mm	spindle shaped, 12 mm	rod shaped, 3–6 mm
Senses			
sight	poor, color blind	poor, color blind	poor, color blind
smell, taste, touch, hearing	excellent	excellent	excellent
Food	omnivorous; often preference for meats (22–30 g/d)	omnivorous, especially fruits, nuts, grains, vegetables (15–30 g/d)	omnivorous, prefers cereal grains (3 g/day)
Water	15–30 ml/day (plus H_2O in food)	15–30 ml/day (plus H_2O in food)	3–9 ml/day; can subsist without free water
Feeding habits	shy (new object reaction); steady eater	shy (new object reaction); steady eater	inquisitive; nibbler
Climbing	readily climbs; limited agility	agile, active climber	good climber
Nests	usually burrows	walls, attics, vines, trees, sometimes burrows	within structures, stored food; burrows
Swimming	excellent swimmer	can swim	can swim
Home range radius (m)	30–50	30–50	3–10
Age at mating (months)	2–3	2–3	1.5–2
Breeding season	spring and fall peaks	spring and fall peaks	year long
Gestation period (days)	22	22	19
Young per litter	8–12	4–8	4–7
Litters per year	4–7	4–6	8
Young weaned/female/year	20	20	30–35
Length of life	1 year	1 year	1 year

SOURCE: Compiled from Brooks 1973; Howard and Marsh 1976; Pratt and Brown 1976; Marsh and Howard 1977; Pratt et al. 1977.

NOTE: Data are averages and not representative of extremes.

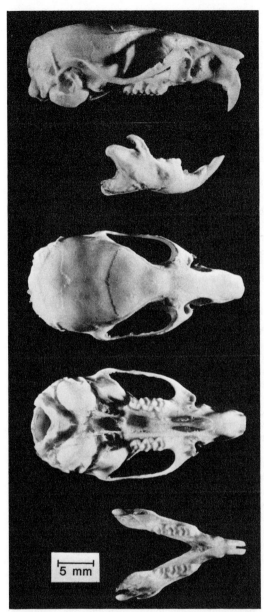

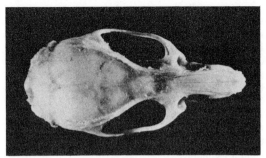

FIGURE 54.2. Skull of the black rat (*Rattus rattus*): dorsal view of cranium.

not be thoroughly cleaned; this removes familiar odors (Temme 1980).

Secreted chemicals that act as stimuli are referred to as pheromones, and Bronson (1979) refers to priming pheromones in house mouse studies. The urinary

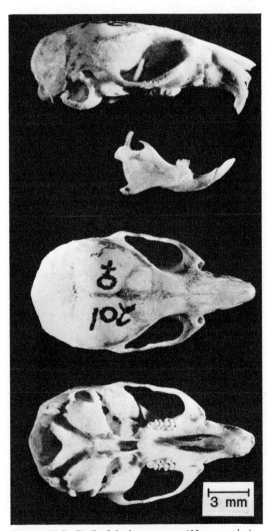

FIGURE 54.1. Skull of the Norway rat (*Rattus norvegicus*). From top to bottom: lateral view of cranium, lateral view of mandible, dorsal view of cranium, ventral view of cranium, dorsal view of mandible.

under very dim light; however, they are the most insensitive to the red end of the spectrum (Fall 1974). Their chemical senses are acute, and they will respond to concentrations of salts, sugars, or other chemicals at a few parts per million.

Odors play a part in their recognition of other animals. Trails are established with urine and secretions of body glands, and vaginal secretions attract and excite males when the female is in estrus. New objects are "marked" in the process of investigation. In fact, traps that have been "seasoned" by use should

FIGURE 54.3. Skull of the house mouse (*Mus musculus*). From top to bottom: lateral view of cranium, lateral view of mandible, dorsal view of cranium, ventral view of cranium.

cues of a socially dominant male can accelerate ovulation in females, and a female's urinary cues can elevate the pheromonal potency of an adult male, thus providing a feedback loop. This system promotes the ability of dispersing young to colonize, but the application of such concepts to control of commensal rodents is largely unexplored (Marsh and Howard 1979).

Hearing ability extends far up on the ultrasonic scale. Vocalizations of the young notably are in this range. Communications among adults also occur in these frequencies and form the basis for one repellency control technique.

REPRODUCTION

Breeding or Sexual Maturity. Sexual maturity is defined in various ways, and confusion between physiological and chronological ages makes discussion and comparison difficult. Maturity criteria, such as testes descent and vaginal orifice perforation, are convenient, because they can be determined quickly by external examination. The weight or body length at which 50 percent of the population achieves a maturational criterion provides a direct method of group comparisons (Jackson 1962).

For Norway and roof rats, vaginal orifice perforation and testes descent occur at 2–3 months after birth. For the house mouse, the time interval is 1.5–2 months. Some geographic (climatic) variations exist. Also, individuals born in the fall may not achieve sexual maturity until early spring, thus there is confusion between chronological and physiological ages. See table 54.1 for representative data.

Other criteria of sexual maturity, especially if necropsy is possible, are valuable. Even better than testis position (or size) is the presence or absence of visible, coiled tubules in the cauda epididymis; their presence is highly correlated with motile sperm in the testis (Jackson 1962). In nonbreeding seasons when gonad size is regressing, this character is especially useful. Similarly, the seminal vessicles regress quickly when androgen stimulation is withdrawn.

The presence of readily visible uterine scars can be used to distinguish between primiparous and multiparous females. Even scars from successive pregnancies can be distinguished if the uteri are bleached (A. Bowerman and J. Brooks unpublished data). Counting corpora lutea provides an estimate of ova released and can be used to estimate intrauterine (including preimplantation) losses (Hall and Davis 1950).

Development. At birth, murids are altricial: sightless, hairless, and helpless. By the end of the first week the external ears have opened, and their bodies are covered with short hair. By two weeks of age the eyes have opened, and coordinated movements can occur. Short excursions from the nest provide the first exposure to the outside environment. Soon thereafter they are following their mother to nearby food sources.

The next few weeks are critical in the behavioral development of the pups. If the mother is pregnant again from the postpartum estrus, even though the lactation will delay implantation, the new litter will be born within four weeks (table 54.1), and the now-weaned pups will be displaced. They will have followed and watched their mother and imitated its food preferences. They will explore their environment further and watch the activities of other rats. The degree to which they exploit their environment will be a function of the activities they see and imitate plus any additional extensions these young animals add on their own. By the time they reach sexual maturity, they will have passed through their critical period of learning, and their behavioral repertoire will be rather well established (Scott 1962).

Sexual relationships among rats are promiscuous. Many males will mate with an estrous female, and hundreds of mountings (most without ejaculation) per hour in artificially dense populations (on dumps, in lab colonies) have been witnessed. Sexually mature animals live in separate burrows, although a mother may allow some association with its female offspring. During the nonbreeding winter season, a female and its nearly grown offspring may share a nest site.

ECOLOGY

Movements. According to popular myth, rats move great distances and often in mass migrations. The German name *Wanderatte* for the Norway rat suggests this. R. H. Creel (1915), a Public Health Service scientist, did little to help clarify the misconception when he collected rats from all over New Orleans and then liberated them in the French Quarter. When recaptures were made several kilometers from the release point, he concluded that rats normally move great distances.

Not until decades later, when live-trapping and marking and placement of dyed food (to produce highly visible colored feces) were employed, was it determined that commensal rodents typically were sedentary. Home ranges were a hundred meters in diameter for rats, much less for mice (table 54.1). As long as the environment was stable and provided food, water, and shelter, the rodents moved little (Davis 1953a,b).

If necessary, a rat would travel ½ km to a food source. Its home range thus would be 500 m by 5 cm. If the environment were disturbed—closing a dump, demolishing a warehouse—then the rats would be forced out, and emigrants might be observed (or experienced) in the search for new infestation sites. Similarly, the seasonal influx of rodents from fields into barns and houses reflects a decrease in available food and a general decreased suitability of the outside environment.

Young males, especially as they mature, are forced from the parental burrow. If nearby harborage is already occupied, more extended exploration is required. As young rodents move into unfamiliar areas, they are highly vulnerable to attack by established rats and predators, including cars, people, and the baits of pest control operators. Mortality often is high.

Populations. Logarithmic population growth, based on minimum maturation times, maximum litter sizes, and monthly intervals between litters, cannot be maintained and rarely occurs. High mortality rates, seasonal rather than continuous breeding, and various behavioral feedback mechanisms intervene to retard growth.

Considerable energy is required for maintenance of behavioral systems, especially at high densities. Brown (1963) found that 10 times as much energy was expended by dominant male house mice in high-density populations as in low-density ones. Active huddling, a response to stress, required one-fourth of the energy budget for the high-density males yet was virtually absent in their low-density counterparts. In the high-density population cannibalism was common, and no young were weaned.

Christian (1978) and others have attempted to provide a comprehensive population control hypothesis with sufficient flexibility to accommodate the many vagaries of natural populations. Very generally, the many aspects of increasing competition and strife stimulate the neuroendocrinological system via the pituitary. In turn, the release of ACTH stimulates the adrenal cortex, and the resultant production of corticosteroids (directly or indirectly) inhibits reproduction, growth, and antibody production. The pituitary-adrenal axis is reversed through mortality and reduced reproduction. This appears to be part of the explanation for population cycles and constitutes a feedback mechanism to limit population growth.

FOOD HABITS

The Norway rat is omnivorous, although it spread from the grassland and grain-producing areas of Central Asia. Its food preferences are very nearly like those of humans; it readily feeds on meats and fruits as well as on various grains and grain products (Jackson 1965). Roof rats, though also omnivorous, readily seek fruits and nuts, even when still on the trees. The damages to tree crops, such as citrus (*Citrus* sp.), coconuts (*Cocos* sp.), macadamia nuts (*Macadamia ternifolia*), and cacao (*Theobroma* sp.), are well known. The house mouse also readily feeds on grains, and most other foods as well.

Given the opportunity, rats will utilize foods with higher protein and fat components (Hauseman 1932; Maller 1967), but responses to available foods may be quite variable. Schein and Orgain (1953) found that urban, garbage-fed rats preferred grains and grain products, meats, cooked eggs, and potatoes. Less preferred were many vegetables; spiced foods were avoided. Shuyler (1954) showed preferences for meats, sweet corn, melons, fruits, and eggs. Laboratory rat studies add many selection patterns relative to specific food components (Barnett 1975). Frequently sugar is added to baits to enhance their palatability.

Early experience from feeding on food brought to the nest seems to have little effect on subsequent food preferences (Krishnakumari 1973). However, the feeding patterns of adults, imitated by the young who follow them in excursions from the nest, may influence later preference patterns of these maturing individuals. This may result in local food "dialects."

Under lab conditions, novel foods tend to be selected over familiar ones (Munn 1950). However, in many field environments strange foods often are ignored initially by rats (neophobia), even though they may eventually be preferred. Thus, prebaiting—use of nontoxic bait for several days—often is recommended to increase the success of a toxic bait program. (The common anticoagulant baits act as their own prebait.) If a sublethal dose of a rodenticide is received and an adverse physiological response is associated with either the taste of the toxicant or the carriers, the animal will reject the substance on reexposure, although gradual extinction over several months may occur (Tongtavee 1978).

Mice tend to be nibblers; rats are more gluttonous (Barnett 1956). Rats frequently feed early in the evening and again heavily before dawn. Mice are likely to be active any time disturbance is minimal, even during the day. If the environment is disrupted or if population density is high, hoarding often occurs, with food being taken from exposed sites and secreted in nest boxes or burrow systems. Young and low-dominance animals may find food (and toxic rodent baits) this way. Rats require about 30 g of dry food daily, mice about 3 g.

Rats need 15–20 cc of water daily. Their range is restricted in arid regions, where they are forced to live in sewers, along irrigation canals, in orchards, or in structures where water can be obtained. House mice can obtain their daily water requirements (1–2 cc) from the water in their food and from production of metabolic water (Chew and Hinegardner 1957; Fertig and Edmonds 1969). Even when raised on very dry food, they can exist, although their growth and reproduction are slowed. Even so, they drink water readily, and water baits are excellent delivery systems for toxicants in dry environments.

BEHAVIOR

Behavioral events (e.g., eye opening, movements from nest, weaning) have been adequately described by many authors. Efforts to modify subsequent behavior through neonatal chemotherapy have provided useful insights into the basis of behavior (Bronson and Desjardins 1968). However, little has been done to examine critical periods as a normal component of behavioral development in the rat. For example, daily handling of pups born to newly captured feral Norway rats will result in fully socialized adult rats (Bastian 1972).

Various observations indicate that young rats are socialized to their environment, in large part by following the mother, imitating its movement patterns and food selections. Females tend to assume the social status of the mother and are tolerated more readily than nonfamilial individuals (McCartney 1971). By the time individuals achieve sexual maturity, their innovative behavior seems to decrease and their behavioral repertoire becomes relatively fixed. However, we are just beginning to understand the behavior of rats apart from caged and experimental environments.

Each species has its mode of recognition. For roof

rats it involves a nuzzling, in which the dominant animal actually engulfs the muzzle of the subordinate (McCartney and Marks 1973). The Polynesian rat, in contrast, utilizes a boxing technique. When the two species are put into forced sympatry for the first time, the larger roof rat tries to nuzzle the Polynesian rat, only to be bitten by the smaller rat. After a couple of futile attempts at recognition, the roof rat may overpower and kill the alien rat.

Norway rats and roof rats may live in the same building, but maintain spatial separation vertically. In the absence of the roof rat, however, the Norway rat may be found in attics and second floors if food is present. Given the opportunity, the Norway rat will be a predator on house mice, and "killer" rats can be identified (Pion 1969). Although both species frequently use the same runways, a time separation of activity periods may keep interspecies contacts to a minimum.

John B. Calhoun pioneered the observational study of the feral rat, and summarized considerable data in his "Ecology and Sociology of the Norway Rat" (1963). He worked with 0.1-ha pens and high-density populations. When food supplies were restricted, dominant rats controlled access routes, maintained defended territories, reproduced successfully, and harrassed the subordinate animals. In these experiments escape was not possible (obviously an artificial constraint).

Laboratory observations by Barnett (1975) of individual feral rats or small groups in large cages or pens also have facilitated our understanding of behavior sequences. Blanchard and Blanchard (1977) compared the behavior of wild and laboratory strains of rat. Their extensive and extended efforts have provided strong support for the view that aggressive and defensive behaviors of this laboratory rat strain are, in fact, typical of the feral rat. They further pointed out that defensive behavior tends to limit access of the attacker to specific bite targets, while attack behavior forces such access.

Even so, genotypic differences exist between feral and laboratory populations. Barnett et al. (1979) demonstrated less adrenal response to crowding with lab animals. Price (1978) likewise related aggression to genotype.

Social Organization. The neighborhood rat (or mouse) population has a social organization. The larger males or near-term females or those with young are likely to be dominant. These animals have initial access to food and favorable burrow sites. Sometimes—but not always—they are first caught or killed, but they may be bait or trap shy from prior experiences and thus difficult to remove. Dominant males may "patrol" the colony area, marking the perimeter with urine (Brown 1953; Desjardins et al. 1973). If subordinate animals and recently weaned individuals (especially males) cannot find food or shelter among the established population, they disperse to find suitable habitat conditions.

MORTALITY

Though a laboratory rat may live several years in a benign captive environment, a feral counterpart under natural conditions is not likely to survive its first year. Davis (1948) determined that Norway rats had an annual probability for survival of about 5 percent. In urban areas, the chances were even less.

Most pathogens and parasites, having evolved "prudent" relationships with their hosts, are rarely the direct cause of death. That they can weaken an animal, so that it is more likely to be caught by a predator or adversely affected by lack of food, is more likely. However, predators like cats are not likely to have much impact on urban rat populations (Jackson 1951). While cats may prevent invasion of rats into previously uninfested buildings, many cats are too well fed to be active predators.

Especially at high population densities or when access to food is limited, intraspecific strife increases. Alpha males or pregnant females harass or even kill subdominants. Wounding, with subsequent secondary infection, occurs. Cannibalism of dominance-contest victims and newborn young increases. Animals become behaviorally debilitated. Reproduction and resistance to acute stresses are inhibited by a complex series of adrenal cortex and pituitary changes (Christian 1978).

Parasites. Ectoparasites of rodents have been of concern and consequently studied because of their role as disease vectors. However, with the decline in incidence of murine typhus and plague, interest in fleas has declined also; and national surveys are no longer carried out by the Public Health Service. In potential plague areas, flea indices (especially for *Xenopsylla cheopis*) may be watched relative to institution of such control measures as burrow and runway dusting with insecticides or closing sections of national parks.

One parasite study of significance is that by Farhang-Azad (1976), who completed a comprehensive analysis of Norway rats at the Baltimore Zoo. Nearly 50 percent of rats had ectoparasites (80 percent mites, 16 percent lice, 4 percent fleas), and infestation rates were highest in warmer months.

Internal parasites were of greater concern to zoo personnel because of possible transmission to display animals. Sixteen percent of the rats were infested with liver tapeworm (*Taenia taeniaeformis*), and the incidence was higher in areas with feral house cats, the definitive host. The roundworm, *Capillaria hepatica*, found in nearly all adult rats, was maintained through cannibalism within the burrows (Farhang-Azad 1977). While the rats were able to exploit the zoo environment, their parasites did not appear to constitute a direct threat to the zoo population.

Murine diseases are discussed in the section "Economic Status and Management."

ECONOMIC STATUS AND MANAGEMENT

Because of their commensal nature, these rodents are particularly evident in our environment. Most people regard them, especially rats, with revulsion, although inner city inhabitants perhaps best understand the full meaning of the word "commensal" (Jackson 1980).

Rat bites are recorded at the rate of 1 per 100,000 inhabitants, but probably the actual rate is much higher (Clinton 1969). Young children and the infirm are most frequently inflicted, sometimes with fatal results. Rat bites involve the hazard of puncture wounds and can be disfiguring both physically and emotionally. Rat-bite fever may be transmitted, but its incidence is poorly documented. The bacteria causing leptospirosis are shed in the urine; those for food poisoning (salmonellosis), in the feces. Thus, contamination by rodents of our food and water with their hair, feces, urine, and pathogens is a matter of real concern.

Vectors of Disease. Fleas, either directly through their bites or by fecal contamination, transmit plague and murine typhus. This form of typhus, rather widespread in the southern United States during the first half of this century, is now largely restricted to Gulf Coast states and California (Adams et al. 1970). The reduction from 5,000 reported cases in the 1940s to less than 100 annually is related to widespread use of DDT after World War II for ectoparasite control and the more effective rat control efforts made possible when the anticoagulant rodenticides became available. Yet potentially serious incidents do occur. Werkheiser (1973) described a situation on the New Orleans docks that could have resulted in a local epidemic had not effective rat and ectoparasite control measures been utilized.

Plague does not now exist in commensal rodents in the United States. The last such outbreak was recorded in 1924 in Los Angeles. However, plague does exist in numerous species of native rodent in the western United States. Usually it is enzootic, but epizootics do occur. People coming in contact with dying rodents are at risk, and several human cases are documented annually. Concern does exist, especially in California, that plague-infected fleas from native rodents will jump land-use barriers by infecting commensal species as suburbia continues to expand into the chapparel and rangeland.

Rickettsial pox, transmitted by mites from infected mice, is known especially from New York and New England cities, where large numbers of people and mice are brought together in apartment buildlings. First recognized in 1946, a few cases continue to be identified each year.

Commensal rodents are not known to carry and transmit rabies in the United States. The United States Public Health Service (anonymous 1980) has formally recommended against specific antirabies treatments in the event of rat and mouse bites. The ignorance by physicians of this advice has forced many persons into painful and often hazardous treatments following rat bites.

Additional, perhaps less well known diseases may be associated with commensal rodents. Lymphocytic choriomeningitis (LCM) may be spread by secretions of pet rodents, including mice. Trichina worms, the cause of "measley pork," may be harbored and shed by rats. Despite requirements that garbage fed to hogs be cooked to kill such parasites, the fall butchering season on family farms and the associated sausage making produces scattered trichinosis outbreaks in the consumers of this harvest bounty. The extent that rats are actually involved is not well documented. While rats from other parts of the world, arriving as stowaways at ports or airports, may be infected with more exotic parasites or pathogens, our rather meager efforts in recent years (including the Vietnam War period) appear to have prevented incursions.

Destruction. Destruction or contamination of food, fiber, and property by these rodents is well known, but specific evaluation of economic losses seldom is accomplished. "Official guestimates" usually suffice (Jackson 1977). The common technique is to assume $10/rat/year and 1 rat/person. Both ratios are without valid statistical foundation, although they are frequently found in the technical literature. Regardless, the damage and losses, though not well quantified, are great. Loss of quality may be greater than appreciated when feces, hair, or urine-contaminated grain supplies are incorporated into food products.

Major food suppliers, processors, and distributors in the United States make strenuous efforts, in part stimulated by U.S. Department of Agriculture (USDA), Food and Drug Administration (FDA), and Environmental Protection Agency (EPA) inspectors, to maintain pest-free environments and produce clean food products. One apparently insurmountable problem is the poor condition of railroad cars. Thus, clean foods loaded into rodent and insect-infested vehicles become contaminated by the end of the journey. Because of this, in-car fumigation of shipments frequently is utilized to prevent further contamination.

Property destruction likewise is significant but poorly documented. Gnawing the insulation on wires can result in fires, power shortages, or telephone interruptions. Perhaps a quarter of the fires of undetermined origin are rodent related. When solid waste accumulations (and, correspondingly, rats) are eliminated, the number of structural fires has declined by 50 percent (Walcott and Vincent 1975).

Control and Management. These rodents need food, water, and harborage to survive. If these essential elements are removed, the commensal rodents cannot maintain themselves. Several decades ago, long before formal reference to Integrated Pest Management (IPM), this concept was demonstrated in Baltimore (Davis 1953*b*). Poisoning and trapping had only temporarily reduced rat populations in city blocks. However, when sanitary police enforced garbage storage regulations and when outbuildings were removed, board fences torn down, broken concrete taken up, and junk trucked away, rat populations decreased almost to zero (figure 54.4).

A clean environment is difficult to maintain, however. This is why rats are "people problems."

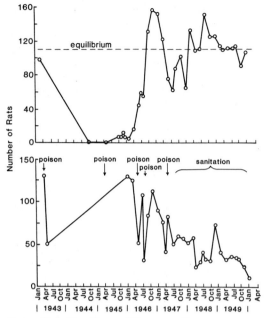

FIGURE 54.4. Top, history of Norway rat (*Rattus norvegicus*) populations in a city block in Baltimore, Maryland, subjected to poison and trapping only. Bottom, Norway rat population subjected to poison and then finally to sanitation.

Because it is easier to distribute rat poison, many urban programs emphasized this approach rather than dealing with causative conditions. With the ready availability of anticoagulant rodenticides and their general effectiveness and safety, it was hard to garner political support for an environmental approach and to enforce it.

For more than a decade, the Public Health Service has provided $12–14 million annually for urban rat control. Nearly 100 communities have been assisted. The stated program emphasis has been to reduce the food supply and harborage for rats through education of the residents. Killing of rats with fumigants and acute and chronic toxicants (such as calcium cyanide, red squill, zinc phosphide, and anticoagulants) was encouraged where rat bites had occurred and during clean-up and renovation campaigns.

But many found it easier to push a bait packet down a rat hole than to put the lid back on the garbage can. Feral dogs ripped open the easy-to-handle plastic garbage bags. Apartment residents, fearful for personal safety, simply threw garbage out of windows. Regardless, more than 13,000 of 26,000 blocks in 68 federal project cities have been designated for maintenance status, indicating that rat infestations of premises are 2 percent or less and that improved garbage storage has been achieved (anonymous 1979).

The commercial pest control operator is faced with a similar dilemma. The householder, food processor, or food service facility wants a pest-free environment without undertaking the necessary structural and management changes. The ready availability of many new and highly effective pesticides from the late 1940s often provided tools so that the environmental

deficiencies could be ignored. Then pests evolved resistance to many compounds, and government regulations became increasingly restrictive about pesticide usage.

In 1970 genetic resistance in Norway rats (and later in roof rats and house mice) was identified in the United States (Jackson et al. 1971). Since then, the phenomenon has been found in 40 communities (Jackson and Ashton 1979). Except in a few sites (such as Chicago), public safety is not threatened. New rodenticides as well as some of the older acute toxicants (table 54.2) now provide effective tools for killing even these "resistant" animals (Marsh et al. 1980).

While environmental controls should be sufficient to manage rodents in urban areas, in practice they must be supplemented by toxicants, traps, and other devices. Old buildings cannot be made rodent proof, especially to mice. Dealing with infestations on an area, rather than individual premises, basis often is difficult because residents and landowners fail to cooperate. Sewers provide underground access routes for rat movements, and water seals do not prevent their penetration of toilets and basement drains. Storage within the house of foods often makes them available to rodents.

For agricultural crops environmental controls are much more difficult to manage. When rodent damages (from both murid and native rodents) to rice or sugar cane are significant, use of rodenticides, perhaps aerially applied, may be required. This is already a routine procedure for sugar cane in Hawaii. It is also done for management of native rangeland rodents in the United States. We are just now being concerned about manipulating of planting/harvesting schedules, changing field sizes and rotations, and the impact of intercropping. The no-till corn culture has returned rodents to the grain field, a phenomenon largely absent since field shocking was abandoned several decades ago.

The roof rat increasingly has become a pest of the affluent suburbs in California (Brooks 1966). Utility lines provide travel routes and ready access to attics and other structures. The lush vegetation and fruit and nut trees provide harborage and food. Because of its arboreal habits, this "bare-tailed squirrel" is difficult to control.

Sometimes the matter of biological control is pressed. Attempts at introducing pathogens have failed to alter the population levels of rats (Davis 1951), and the World Health Organization (1967) has advised against further attempts in the use of *Salmonella* bacteria because of its ineffectiveness and the potential hazard to the human hosts.

Effective larger predators are lacking in urban environments; most dogs and cats are too well fed, and many scatter garbage by ripping open plastic bags or tipping cans and thus assist the rats to food. The introduction of the mongoose (*Herpestes* sp.) to island communities has been a disaster to native fauna with little impact on the rats.

Use of traditional chemosterilants (sex hormone derivatives) has not been successful (Marsh and Howard 1970), but the advent of a combination toxicant-sterilant offers new opportunities (table 54.2). The introduction of genetically sterile males to depress reproductive rates similarly has not been successful, in part because of the promiscuous mating behavior of rodents (Glass 1974).

Use of biological control in the sense of environmental management is the fundamental recommendation. Removal of garbage and rubbish and proper construction of residences and food storage structures prevent rodents from establishing themselves and increase the effectiveness of existing predators, both natural and artificial (traps, rodenticides). For such field crops as no-till corn the stubble can be broken down in the fall to expose the rodents to the local raptors during the winter. However, in large-scale monocultures chemical tools usually are required as well.

Laboratory Animals. The laboratory mouse and rat are albino forms of *Mus musculus* and *Rattus norvegicus,* respectively. Laboratory use of rodents was recorded in the 19th century, and albino rats trapped from wild colonies were kept for show and breeding (Altman and Katz 1979). However, it was not until the early part of this century that albino rats were imported from

Europe and laboratory colonies established. During the first quarter of the 20th century various inbred lines were established and many mutant forms recognized. Similarly, investigators began inbreeding mice to obtain genetic homogeneity for studying factors affecting tumor transplantation. Morphologic parameters were summarized by Altman and Katz (1979). Comparable roof rat stocks have not been developed, however.

CURRENT RESEARCH AND MANAGEMENT NEEDS

The only sustained research program on commensal rodents in the United States was mounted at The Johns Hopkins University School of Hygiene and Public Health in Baltimore in the 1940s and 1950s with Rockefeller Foundation and U.S. Public Health Service support. These biological studies provided much of the current knowledge on commensal rodents and the basis for planning management programs.

Certainly it is not necessary to replicate such an effort prior to launching control efforts in other geographic sectors. Environmental manipulation clearly

TABLE 54.2. Characteristics of common rodenticides

Rodenticide	LD_{50} (mg/kg)	Percentage in Bait	Bait	Acceptance	Cause of Death	Hazard
Acute (single dosage)[a]						
Zinc phosphide[b]	40	1.0–2.0	grain, fresh foods	good; bait shyness may occur	heart paralysis; GI involvement	medium
Strychnine	6–8	0.25–1.0	grain	good; mice only	CNS involvement	medium
Red Squill[b,c]	500	10	grain, fresh foods	fair; bait shyness may occur	heart paralysis	low; own emetic
1080 and 1081[d]	2–50	0.2–2.0	grain, fresh foods, water	excellent	heart paralysis, CNS involvement	high
Bromethalin (Dispatch)[e]	2–5	0.005	grain	excellent	metabolic inhibition	low
Anticoagulants (multiple dosage)						
Hydroxycoumarins[b] (warfarin, Fumarin, Pival, PMP)	1[f]	0.025	grain, water	excellent	internal hemorrhage	low
Indandiones[b] (Diphacinone, Chlorophacine)	0.5[f]	0.005	grain	excellent	internal hemorrhage	low
Second generation						
Brodifacoum (Talon)	0.2–1.0[g]	0.005	grain	excellent	internal hemorrhage	low
Bromadiolone (Maki)	1–2[g]	0.005	grain	excellent	internal hemorrhage	low

SOURCE: Compiled from Brooks 1973; Howard and Marsh 1976; Pratt and Brown 1976; Marsh and Howard 1977; Pratt et al. 1977.

[a] ANTU, phosphorus, arsenic, and barium carbonate are rarely used; thallium, Vacor (DLP-787), DDT, and Norbormide are not labeled for use in the United States.

[b] Also used in form of tracking powder.

[c] Synthetic form with greater efficacy being developed.

[d] All uses restricted to certified applicators.

[e] Experimental Use Registration (1981); delayed death (3+ days).

[f] Must be ingested in successive doses for 4–10 days.

[g] Single feeding lethal but feeding continues to death.

was discerned then (though not often implemented), and despite the fact that we now refer to Integrated Pest Management (see 1980 U.S. NRC report), the central concept has not changed. While we do have an adequate base for management programs, this is not to say that additional research is not needed.

The sustained use of anticoagulants has selected for anticoagulant resistance. Even though new, "second generation" anticoagulants are now available, the first signs of resistance to these compounds are just appearing. We know all too little about the geographic variations in the genetics and biochemistry of the initial resistance, and we are not prepared to handle this second phase of resistance evolution. We are missing the opportunity to study the dynamics of selection pressure in a mammalian population.

The development and testing of candidate rodenticides has been delegated entirely to the private sector. A generation ago the Communicable Disease Center (renamed the Center for Disease Control) of the U.S. Public Health Service was active in the evaluation of new rodenticides and formulations and provided field trials to determine efficacy and safety. The Fish and Wildlife Service was actively involved in screening potential compounds and had a significant role in developing rodenticide programs in nonurban environments. Only remnants of such efforts continue. These agencies could provide significant public service by greater direct involvement in the research and development mission, especially with compounds that lack patent protection and thus are not likely to find private sector interest.

The matter of "humane" toxicants, traps, or other devices will be increasingly vexing as interest groups propose criteria laced with emotion and promote arguments of legal animal "rights." Only when experimentally well-documented definitions are available can a reasonable resolution of this social issue be attempted.

Continuing investigations into potential toxicants, behavior-modifying chemicals, and devices are needed as part of the continuing effort in pest management. As important as new technology is, the adaptation of existing tools—for example, nonanticoagulant toxicants and ultrasonics—for efficacious use is needed. Furthermore, the importance of public education and determining effective and efficient modes of information transfer cannot be overestimated, especially relative to habitat modification and environmental manipulation as a basis for rodent control (Marsh and Jackson 1978).

Since the decline of murine typhus after World War II, concern about rodent-borne diseases has all but disappeared. Homage still is paid in textbooks, and ill-advised physicians prescribe antirabies injections following rat bites; except for these, we hear little. The role of rats as disease "elevators" from sewers is rarely investigated, yet isolated investigations suggest that leptospirosis is still a disease to be concerned about. Recent studies of parasites are distinguished by their rarity.

The roof rat, because of its restricted geographic range, has been less investigated. In California, it might be considered a "bare-tailed squirrel" ecologically; it is a frequent inhabitant of well-vegetated residential neighborhoods. The public health attitude toward this situation, including the potential for plague introduction, needs attention. In the arid Southwest, the invasion of sewer systems by roof rats needs to be documented and evaluated.

Government documents and textbooks glibly speak of economic losses due to rodent activity. Yet most of these are based on unverified "guestimates." Validating population and loss estimates should be a matter of concern, even if only for the matter of accurate cost/benefit ratio determinations.

There continue to be fascinating areas of basic research that may well have significant implications for population management. Such areas as gene flow between population demes, further elucidation of the adrenal cortex-pituitary axis, critical learning periods for these species in natural populations, persistence of bait shyness, sensory responses, and critical points in population growth need further study.

In the heyday of U.S. Public Health Service vector control efforts, a great deal of research and field implementation of rodent control programs was carried out. By the last decade this had been transformed into a grant program to urban centers but without back-up research support. The evaluation of these massive efforts over the past 40 years has not been undertaken. While most of the key figures in these programs are still able to provide oral histories, a synthesis ought to be attempted.

LITERATURE CITED

Adams, W. H.; Emmons, R. W.; and Brooks, J. E. 1970. The changing ecology of murine (endemic) typhus in southern California. Am. J. Trop. Med. Hyg. 19:311–318.

Altman, P. L., and Katz, D. D.. eds. 1979. Inbred and genetically defined strains of laboratory animals. Federation Am. Soc. Experimental Bio., Bethesda. 418pp.

Anderson, P. K. 1964. Lethal alleles in *Mus musculus:* local distribution and evidence for isolation of demes. Science 145:177–178.

Anonymous. 1979. Urban rat control, United States, April–June 1979. Morbidity and Mortality Weekly Rep. 28:505–507.

Anonymous. 1980. Rabies prevention. Morbidity and Mortality Weekly Rep. 29:265–280.

Barnett, S. A. 1956. Behaviour components in the feeding of wild and laboratory rats. Behaviour 9:24–43.

———. 1975. The rat: a study in behavior. Univ. Chicago Press, Chicago. 318pp.

Barnett, S. A.; Dickson, R. G.; and Hocking, W. E. 1979. Genotype and environment in the social interactions of wild and domestic Norway rats. Aggressive Behav. 5:105–119.

Bastian, R. K. 1972. Early development and the effects of differential early handling in the wild Norway rat. M.S. Thesis. Bowling Green State Univ., Bowling Green, Ohio. 135pp.

Blanchard, R. J., and Blanchard, D. C. 1977. Aggressive behavior in the rat. Behav. Bio. 21:197–224.

Brechbill, R. A. 1977. Status of the Norway rat. Pages 261–267 *in* M. L. Merritt and R. G. Fuller, eds. The environment of Amchitka Island, Alaska. Energy Research and Development Admin. (TID-26712). 682pp.

Bronson, F. H. 1979. The reproductive ecology of the house mouse. Q. Rev. Bio. 54:265–299.

Bronson, F. H., and Desjardins, C. 1968. Aggression in adult mice: modification of neonatal injections of gonadal hormones. Science 161:705–706.

Brooks, J. E. 1966. Roof rats in residential areas: the ecology of invasion. California Vector Views 13:69–74.

———. 1973. A review of commensal rodents and their control. Critical Rev. in Environ. Control 3:405–453.

Brown, R. Z. 1953. Social behaviour, reproduction, and population changes in the house mouse (*Mus musculus* L.). Ecol. Monogr. 23:217–240.

———. 1963. Patterns of energy flow in populations of the house mouse (*Mus musculus*). Bull. Ecol. Soc. Am. 44:129.

Calhoun, J. B. 1963. The ecology and sociology of the Norway rat. U.S. Public Health Serv. Pub. 1008. 288pp.

Chew, R. M., and Hinegardner, R. T. 1957. Effects of chronic insufficiency of drinking water in white mice. J. Mammal. 38:361–374.

Christian, J. J. 1978. Neurobehavioral endocrine regulation of small mammal populations. Pages 143–158 in D. D. Snyder, ed. Populations of small mammals under natural conditions. Spec. Publ. 5, Pymatuning Lab. Ecol., Univ. Pittsburgh, Pittsburgh. 237pp.

Clinton, J. M. 1969. Rats in urban America. Public Health Rep. 84:1–7.

Creel, R. H. 1915. The migratory habits of rats with special reference to the spread of plague. Public Health Rep. 30:1679–1685.

Davis, D. E. 1948. The survival of wild brown rats on a Maryland farm. Ecology 29:437–448.

———. 1951. The relation between the level of population and the prevalence of Leptospira, Salmonella, and Capillaria in Norway rats. Ecology 32:465–468.

———. 1953*a*. Analysis of home range from recapture data. J. Mammal. 34:352–358.

———. 1953*b*. The characteristics of rat populations. Q. Rev. Bio. 28:373–401.

Desjardins, C.; Maruniak, J. A.; and Wessells, F. H. 1973. Social rank in house mice: differentiation revealed by ultraviolet visualization of urinary marking patterns. Science 182:939–941.

Ecke, D. H. 1954. An invasion of Norway rats in southwestern Georgia. J. Mammal. 35:521–525.

Fall, M. W. 1974. The use of red light for handling wild rats. Lab. Anim. Sci. 24:686–687.

Farhang-Azad, A. 1976. Ecology of Norway rat populations and *Capillaria hepatica*. Ph.D. Diss. The Johns Hopkins Univ., Baltimore. 174pp.

———. 1977. Ecology of *Capillaria hepatica* (Bancroft 1893) (Nematoda). Part 2: Egg-releasing mechanisms and transmission. J. Parasitol. 63:701–706.

Fertig, D. S., and Edmonds, V. W. 1969. The physiology of the house mouse. Sci. Am. 221:103–108.

Glass, B. P. 1974. The potential value of genetically sterile Norway rats in regulating wild populations. Proc. 6th Vertebr. Pest Conf., Anaheim, Calif., pp. 49–54.

Hall, O., and Davis, D. E. 1950. Corpora lutea counts and their relation to the numbers of embryos in the wild Norway rat. Texas Rep. Bio. Med. 8:564–582.

Hausmann, M. F. 1932. The behavior of albino rats in choosing food and stimulants. J. Comp. Psychol. 13:279–309.

Howard, W. E., and Marsh, R. E. 1976. The rat: its biology and control. Univ. California, Div. Agric. Sci. leaflet 2896, pp. 1–22.

Jackson, W. B. 1951. Food habits of Baltimore, Maryland,

cats in relation to rat populations. J. Mammal. 32:458–461.

———. 1962. Population studies: reproduction. Pages 92–107 in T. I. Storer, ed. Pacific island rat ecology. Bishop Mus. Bull. 225. 274pp.

———. 1965. Feeding patterns in domestic rodents. Pest Control 33:12, 50.

———. 1977. Evaluation of rodent depredations to crops and stored products. European and Mediterranean Plant Protection Organization Bull. 7:503–508.

———. 1980. Rats: friend or foe? Pest Control 48:14–16, 20.

Jackson, W. B., and Ashton, A. D. 1979. Present distribution of anticoagulant resistance in the United States. Pages 392–397 in J. W. Suttie, ed. Vitamin K metabolism and Vitamin K–dependent proteins. Univ. Park Press, Baltimore. 592pp.

Jackson, W. B.; Spear, P. J.; and Wright, C. G. 1971. Resistance of Norway rats to anticoagulant rodenticides confirmed in the United States. Pest Control 39:13–14.

Johnson, D. H. 1962. Rodents and other micronesian mammals collected. Pages 21–38 in T. I. Storer, ed. Pacific island rat ecology. Bishop Mus. Bull. 225. 274pp.

Krishnakumari, M. K. 1973. Effects of early food experiences on later food preferences in adult rats. Pest Control 41:36, 38, 43.

McCartney, W. C. 1971. A comparative study of the social behavior, organization, and development of two species of the genus *Rattus* (*R. exulans* and *R. rattus*). Ph.D. Diss. Bowling Green State Univ., Bowling Green, Ohio. 142pp.

McCartney, W. C., and Marks, J. 1973. Inter- and intraspecific aggression in two species of the genus *Rattus:* evolutionary and competitive implications. Proc. Pennsylvania Acad. Sci. 47:145–148.

Maller, O. 1967. Specific appetite. Pages 201–212 in M. R. Kare and O. Maller, eds. The chemical senses and nutrition. Johns Hopkins Univ. Press, Baltimore. 488pp.

Marsh, B. T., and Jackson, W. B. 1978. Environmental control of rats. Pest Control 46(8):12–14, 16, 37–38, 43, 54; 46(9):26–29, 40–40d, 42.

Marsh, R. E., and Howard, W. E. 1970. Chemosterilants as an approach to rodent control. Proc. 4th Vertebr. Pest Conf., West Sacramento, Calif., pp. 55–63.

———. 1977. The house mouse: its biology and control. Univ. California, Div. Agric. Sci. leaflet 2945, pp. 1–28.

———. 1979. Pheromones (odors) for rodent control? Pest Control Tech. 7:22–23.

Marsh, R. E.; Howard, W. E.; and Jackson, W. B. 1980. Bromadiolone: a new toxicant for rodent control. Pest Control 48:22, 24, 26.

Munn, N. L. 1950. Handbook of psychological research on the rat. Houghton Mifflin Co., Boston. 598pp.

Pion, L. W. 1969. Early experience, social contact, and the incidence of mouse killing behavior in Norway rats. Bull. Ecol. Soc. Am. 50:88.

Pratt, H. D.; Bjornson, B. F.; and Littig, K. S. 1977. Control of domestic rats and mice. 77-8141. Dept. Health, Education, and Welfare; Public Health Service; Center for Disease Control, Atlanta. 47pp.

Pratt, H. D., and Brown, R. Z. 1976. Biological factors in domestic rodent control. 76-8144. Dept. Health, Education, and Welfare; Public Health Service; Center for Disease Control, Atlanta. 30pp.

Price, O. 1978. Genotype versus experience effects on aggression in wild and domestic Norway rats. Behaviour 64:340–353.

Schein, M. W., and Orgain, H. 1953. A preliminary analysis

of garbage as food for the Norway rat. Am. J. Trop. Med. Hyg. 2:1117–1130.

Schiller, E. L. 1956. Ecology and health of *Rattus* at Nome, Alaska. J. Mammal. 37:181–188.

Schwarz, E., and Schwarz, H. K. 1943. The wild and commensal stocks of the house mouse, *Mus musculus* Linnaeus. J. Mammal. 24:59–72.

———. 1965. A monograph of the *Rattus rattus* group. An. Esc. Nac. Cienc. Bio. Mex. 14:79–178.

Scott, J. P. 1962. Critical periods in behavioral development. Science 138:949–958.

Shuyler, H. R. 1954. The development of baits for *Rattus norvegicus,* with special reference to initial acceptability. Ph.D. Diss. Purdue Univ., Lafayette, Ind. 560pp.

Temme, M. 1980. House mouse behavior in multiple capture traps. Pest Control 48:16, 18–19.

Tongtavee, K. 1978. Zinc phosphide development for rodent control. Ph.D. Diss. Bowling Green State Univ., Bowling Green, Ohio. 142pp.

U.S. National Research Council. 1980. Urban pest management. National Academy Press, Washington, D.C. 273pp.

Walcott, R. M., and Vincent, B. W. 1975. The relationship of solid waste storage practices in the inner city to the incidence of rat infestation and fires. 530/SW/150. Environmental Protection Agency, Cincinnati. 14pp.

Werkheiser, A. C. 1973. New Orleans rids docks of rats murine typhus deferred. J. Environ. Health 36:234–236.

World Health Organization. 1967. Joint FAO/WHO Expert Committee on Zoonoses. Third report. WHO Tech. Rep. Ser. 378. 127pp.

WILLIAM B. JACKSON, Bowling Green State University, Center for Environmental Research and Services, Bowling Green, Ohio 43403.

55

Wild Horses

Equus caballus and Allies

Larry M. Slade

E. Bruce Godfrey

NOMENCLATURE

COMMON NAMES. Feral horses, wild and free-roaming horses, wild horses

SCIENTIFIC NAME. *Equus* sp.

HISTORY AND ORIGIN

The family *Equidae* includes horses, asses, zebras, and onagers in the genus *Equus,* with nine species. These species and corresponding chromosome numbers are: *E. przewalski* (Mongolian wild horse), 66; *E. caballus* (domestic horse), 64; *E. assinus* (donkey), 62; *E. hemionus* (Mongolian wild ass), 56; *E. onager* (Persian wild ass), 56; *E. kiang* (Tibetan wild ass), 56; *E. grevyi* (Somililand zebra), 46; *E. burchelli* (African zebra), 44; and *E. zebra* (Cape Colony zebra), 32 (Jones and Bogart 1971). Fossil records suggest that present-day horses originated in a small, four-toed, rodentlike creature named *Hyracotherium.* Remains of *Hyracotherium* have been found in England and North America from the eocene. Those found in North America were erroneously named *Eohippus,* and that term is commonly used in North America today although the skeletal structures of *Hyracotherium* found in England and *Eohippus* found in America were similar (Simpson 1951). There were several species, which varied greatly in size. The smallest was about 25 centimeters in height at the shoulder, and the largest about 50 cm. There were four toes on the front feet and three functional toes on the hind. During the period when Laurasia was separating, migration between Europe and North America was possible, and common ancestors for horses, which developed later in North America and Asia, could have used this route. In Europe *Hyracotherium* became extinct. In North America and Asia, development continued through 30 different species and 15×10^6 generations, to form 9 species of modern equid (Simpson 1951; Haines 1971). Major changes that marked the evolutionary development were in the limbs, teeth, and body size (figure 55.1). The numbers of toes decreased to one, and the shape and characteristics of the feet developed to facilitate speed and load bearing. Three major changes occurred in the teeth: crown height increased gradually to give longer life, a cementum layer developed to fill the pits and valleys of the teeth and prevent decay, and tooth patterns changed to permit grinding of food (Simpson 1951). The teeth of present-day horses are highly adapted for eating grass. Due to its high silica content, grass is a very harsh food and even the high-crowned, enameled teeth of modern horses wear down very rapidly as a result of grazing. The browse eater, *Hyracotherium,* could not survive today.

It is assumed that modern species of equid developed from three primitive strains: the Mongolian horse, which is thought to be the same as *Equus przewalski;* the Celtic horse, which is similar to present-day Tarpans (*E. caballus*); and the forest horse, which is represented by primitive types of Shetland pony. The Celtic horse was the first to be domesticated, and was most likely the progenitor of modern light-legged horses. The forest horse was the progenitor of modern draft horses. Of these three primitive strains, only true wild descendents of Mongolian horses remain, represented today by *Equus przewalski* (Simpson 1951; Haines 1971).

DISTRIBUTION

Wild horse herds are known to exist in the U.S.S.R., Africa, and the western United States. In every case little is known about their demography or current population. The most is known, however, about wild horses in western America. As a result, emphasis will be placed on these animals in this chapter.

Few issues associated with wild horses at the present time are more widely disputed by various interest groups than estimates of population size. The Wild Horse and Burro Act of 1971 (PL-92-195) was passed basically to save a species of animal that was near extirpation. Yet less than five years later, other interest groups were complaining that populations were expanding at alarming rates. While these divergent opinions were expected, they were commonly not based on scientifically defensible estimates of population. As a result, considerable controversy still exists today. This controversy represents one of the major reasons for

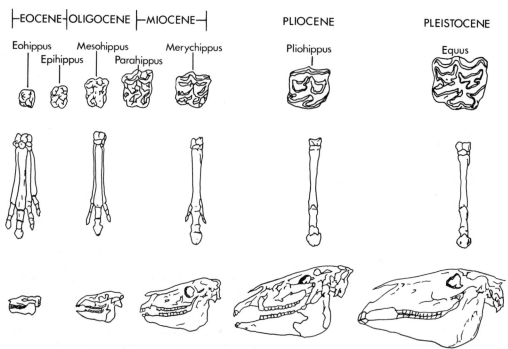

┠EOCENE┨OLIGOCENE ┠MIOCENE┨   PLIOCENE   PLEISTOCENE

Eohippus   Mesohippus   Merychippus   Pliohippus   Equus
   Epihippus   Parahippus

FIGURE 55.1. Evolution of the horse (*Equus caballus*): teeth, feet, and skull.

establishment of the research program suggested in the Rangeland Improvement Act of 1978 (PL 95-514). Historic estimates of wild horse numbers are limited in scope. However, the small amount of data available today suggests that wild horses are not dispersed equally. Essentially all wild horses and burros are found within the eleven Western states (figure 55.2). The largest number of wild horses is found in Nevada (table 55.1).

DESCRIPTION

The body color of Przewalski horses is a neutral gray with black points, a dorsal stripe, and zebra markings on the front legs. This coloration is representative of nonferal wild horses. The dunn coloration with the dorsal stripe and zebra markings on the front legs is also an authentic color pattern. Domestic coat color varieties such as bay or brown are common among feral horses, but they lack the camouflaging values of the wild body colorations (Castle 1954).

There are no records of wild horse sizes or weights. Przewalski horses are relatively small, ranging in shoulder height from 13 to 14 hands and in weight from 300 to 360 kg. Feral horses are 14 to 15 hands (132–142 cm) high. Females weigh from 270 to 340 kg., while males weigh from 360 to 390 kg. It is likely that the relatively small size of wild horses is due to their restricted diet (Beebe and Johnson 1964). We have noted that young feral horses, when taken into captivity and fed diets containing grain, achieve body weights much higher than those normally reported. Restrictive diet in the wild also results in a relatively slow

maturation rate for wild horses. Two- and three-year-old animals often appear to have the same morphology and development as yearlings in domestic herds (Ryden 1970; Hall 1972). In spite of differences in diet and lifestyle, the physiology of the gastrointestinal tract of wild horses is similar to that of domesticated horses (Ensminger 1951).

There appears to be a difference between domesticated and wild horses in the structure of the feet. Many of the problems associated with the feet of domesticated horses, such as cracks and abnormal structure, are likely due to confined stabling and restricted movement. The feet of wild horses are small in shape, tough in substance, and uniformly worn down by constant traveling over abrasive terrains (Dobie 1952). The skull is illustrated in figure 55.3.

REPRODUCTION

The breeding season is influenced by the length of daylight and by the availability of nutrients. It generally begins during late spring and ends in early autumn. Among feral horse bands, foaling by individual mares in alternate years is common. Abortions may occur during late autumn, winter, and early spring. Alternation of foaling years is most common where feed is sparse. The average gestation period is about 340 days. Postpartum estrus usually occurs 7 to 11 days after foaling. The length of the estrus period varies from 5 to 8 days but may be longer in early spring and shorter as summer approaches (Hafez et al. 1962; Berliner 1969).

Reproduction rates range from 40 foals/100 mares in relatively good conditions to 20/100 in poor condi-

FIGURE 55.2. Distribution of primary wild and free-roaming horse (*Equus caballus*) herds in the United States.

tions. Fillies comprise about 60 percent of the foal crop (Feist 1971; Hall 1972).

ECOLOGY

Few studies have dealt with the dietary habits of feral horses. Under ordinary range conditions, 80 to 95 percent of their diet consists of grasses and grass-like plants. Horses consume more browse than forbs. During the winter, feral horses may utilize brushy

TABLE 55.1. Bureau of Land Management wild horse and burro inventory data estimates of population

| | Number of Animals | | | |
| | 1974 | | 1976 | |
States	Horses	Burros	Horses	Burros
Arizona	115	10,000[a]	107	2,668
California	3,000	3,200	4,230	3,072
Colorado	500		1,035	0
Idaho	500	8	874	9
Montana	325		257	0
Nevada[b]	20,000	1,000	22,258	842
New Mexico	7,550	80	6,420	104
Oregon	5,265	16	7,493	25
Utah	1,000	50	1,803	70
Wyoming	4,411	20	8,833	0
Total	42,666	14,374	53,310	6,790

SOURCE: Second report to Congress.

NOTE: Included in total numbers each year are horses and burros claimed under section 5 of the Wild Horse and Burro Act. The total number may include some branded horses grazing in trespass that were not claimed.

[a] Estimate before aerial census of 1975.

[b] Does not include those animals in Nevada that are the responsibility of the Susanville, California, District.

species, primarily saltbrush (*Atriplex* sp.), rabbit-brush (*Chrysothamnus* spp.), big sagebrush (*Artemisia tridentata*), greasewood (*Sarcobatus vermiculatus*), black sagebrush (*Artemisia nova*), Utah juniper (*Juniperus osteosperma*), mountain mohagany (*Cercocarpus* sp.), winter fat (*Eurotia lanata*), and milk vetch (*Astragralus kentrophyta*). When available, feral horses will feed on the bark from quaking aspen (*Populus tremuloides*). Feral horses have not been observed to feed on scrubby cinquefoil (*Potentilla fruticose*) or prickly pear cactus (*Opuntia polysantha*), even though these plants may be abundant in their grazing areas (Zarn et al. 1977).

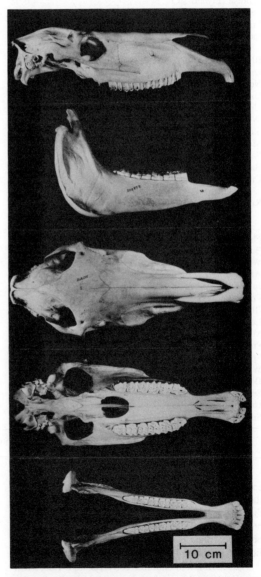

FIGURE 55.3. Skull of the horse (*Equus caballus*). From top to bottom: lateral view of cranium, lateral view of mandible, dorsal view of cranium, ventral view of cranium, dorsal view of mandible.

TABLE 55.2. Percentage relative density of discerned fragments from horse fecal samples from Pryor Mountain Wildhorse Range, Montana (based on 400 fields per sample, taken at various elevations)

Food Source	Summer '74			October '74		January '75		
	Lower	Middle	Upper	Lower	Middle	Lower	Middle	Upper
Bluebunch wheatgrass (Agropyron spicatum)	40.80	83.18	46.20	50.43	38.21	36.35	76.55	20.04
Wheatgrass (Agropyron sp.)	0.22			0.29				
Threeawn (Aristida)				0.68		1.81		
Blue grama (Boutelova gracilis)	0.22	0.88		1.65		0.33		
Brome (Bromus sp.)	3.87		0.67	0.10	0.52			
Reedgrass (Calamagrostis sp.)							0.09	
Sedge (Carex sp.)	1.77	3.15	5.25	2.27	4.02	1.70	6.19	17.09
Danthonia (Danthonia sp.)			0.22		0.52			0.12
Wildrye (Elymus sp.)		0.08				0.11		
Fescue (Festuca sp.)		0.68			1.58	3.11	0.09	1.06
Prairie junegrass (Koeleria cristata)		3.08	5.01	0.29	3.45		0.17	4.90
Indian ricegrass (Oryzopsis hymenoides)	3.17	2.02	2.14	2.68	0.52	0.90	0.52	0.47
Bluegrass (Poa sp.)	46.42	3.66	14.16	4.14	0.52	4.82	1.39	0.94
Dropseed (Sporobolus)	0.22	0.08	0.33		0.10	5.31	0.35	
Needlegrass (Stipa sp.)	0.11	0.23			23.41		13.87	0.52
Unknown grass					0.31			
Unknown sedge					0.10	0.11		
Western yarrow (Achillea millefolium)				0.19	0.21		0.12	
Sagebrush (Artemisia sp.)	0.11	0.92	0.22	6.00	0.94	3.35	4.39	8.53
Milk vetch (Astragalus sp.)	0.11		7.03				0.09	0.23
Saltbush (Atriplex sp.)		0.08		1.86	0.94	18.34	7.55	0.82
Barberry (Barberis sp.)					2.56			
Curlleaf cercocarpus (Cercocarpus ledifolius)					23.64	1.24		0.58
Rubber rabbitbrush (Chrysothamnus nauseosus						0.35		
Composite	0.11		0.78		0.31		0.09	0.12
Fleabane (Erigeron sp.)		0.30	0.11			0.11	0.09	
Buckwheat (Eriogonum sp.)	0.22			0.29	0.10	2.40		
Winterfat (Eurotia lanata)	1.77	1.70		5.22	11.23	2.64	0.26	0.12
Alumroot (Houchera sp.)							0.09	
Juniper (Juniperus sp.)				0.10	. 0.31	0.33		
Prickly phlox (Leptodactylon pungens)						0.11	0.09	0.12
Bladderpod (Lasquerella sp.)					0.21	0.22		0.12
Lupine (Lupinus sp.)		0.23		0.10				
Moss				0.10			0.09	
Phlox (Phlox)			0.67	0.10		2.40	0.78	43.80
Spruce (Picea)					4.13	0.11		
Woolly indianwheat (Plantago purshii)	0.11	0.53	16.99					
Douglas fir (Pseudotsuga menziesii)				0.10	3.12	0.11	0.26	0.35
Seed	0.11							
Unknown forb	0.55		0.22		2.45	0.22		0.47
Small soapweed (Yucca glauca)	0.11							

NOTE: There was very little seed or glume material in horse diets.

Although horses are primarily grazers, they can eat and apparently survive on a wide variety of foods (tables 55.2 and 55.3). The digestive tract is similar to that of a ruminant in that microfauna facilitate the breakdown of roughage. This allows the horse to feed on fibrous plant material and to compete relatively efficiently with ruminants (Ensminger 1951).

Bacterial fermentation converts high-fiber, low-protein plant material into organic acids, B vitamins, and bacterial protein. The B vitamins and organic acids are efficiently absorbed from the large intestines, primarily in the area of the large colon. Although mechanisms for the digestion of bacterial proteins have not been determined, experiments by the author (Slade et al. 1971; Godbee and Slade 1979) have demonstrated a relatively high capacity for domestic horses to absorb amino acids from the large intestine. Because digestion occurs both prececaly and postcecaly, the horse more efficiently utilizes highly soluble plant and animal protein and carbohydrates such as starch or sugars than do ruminants. Horses maintain relatively high blood levels of glucose and avoid problems of hypoglycemia common to ruminants. Cecal fermentation enables the horse to obtain a significant portion of its energy requirements from the products of cellulose breakdown, through bacterial fermentation similar to that in ruminants.

There are no data available on the amount of water needed or consumed by wild horses under various habitat conditions. Domestic horses consume about 46 liters per day in a normal environment (Hanauer 1973). Generally, bands of feral horses visit a water hole once a day, usually in the late afternoon. During very hot weather they may water twice a day. Over most of their range, horses are always within 6–8 km of a water hole. The time spent at water holes ranges from 30 minutes to several hours depending on the sense of security of the band. Actual drinking time normally does not exceed 3 to 5 minutes per individual (Fiest 1971). Horses have a keen sense of smell for water and it is not uncommon for them to paw and dig up to 1 m in old river beds to find subsurface water (Odberg 1972). During droughts, feral horses sometimes become so franctic upon finding water that they pile on top of each other, drowning and trampling to death weaker animals.

BEHAVIOR

Social Structure. Most free-ranging equid populations are not homogeneous social units. Two basic patterns of social behavior exist: harem groups, composed

TABLE 55.3. Percentages of fragment in diets of wild horses determined by microhistological analysis of feces (based on 400 fields, taken at two elevations), Pryor Mountain, Wild Horse Range, Montana, spring 1975

Food Source	Elevation	
	Lower	Middle
Wheatgrass	18.08	62.12
(*Agropyron* sp.)		
Grama	0.10	
(*Bouteloua* sp.)		
Sedge	1.55	7.34
(*Carex* sp.)		
Indian ricegrass	4.12	2.30
(*Oryzopsis hymenoides*)		
Needlegrass	67.02	23.34
(*Stipa* sp.)		
Sagebrush	1.16	
(*Artemisia* sp.)		
Saltbush	5.64	1.56
(*Atriplex* sp.)		
Curlleaf mountain mahogany	0.10	
(*Cercocarpus ledifolius*)		
Rubber rabbitbrush	0.48	
(*Chrysothamnus nauseosus*)		
Winterfat	1.46	2.52
(*Eurotia lanata*)		
Bladderpod		0.10
(*Lesquerella* sp.)		
Phlox	0.29	0.82
(*Phlox* sp.)		

of several mares led by a dominant male, and territorial groups. Feral horses generally form harem groups that do not establish territories but move freely over home ranges shared with other species (Hall 1972; Pelligrini 1971). In the territorial groups, found among asses and zebras, there are no bonds other than sexual relationships. Animals occur singly, in stallion groups, mare-foal groups, or in mixed herds. All of the territorial groups vary in their composition, which may change at any time (Klingel 1972).

Harems usually consist of five to six females and one or two stallions. The mares of a harem remain together even if the stallion dies or is replaced by another. Indeed, the mares of a harem group will not readily accept a new stallion until weeks after his "ownership" of the unit is recognized by other mature males. Harem social units retain their social and spatial identities in larger aggregations. Only rarely do mature females change their group memberships (Dobie 1952; Ryden 1970).

Adolescent and mature males are the primary source of exchange between harem groups. Because the harem stallion is ultimately threatened with the possibility of being displaced by a younger, stronger male, male colts are forcibly expelled from their maternal groups by their sires. The age of expulsion varies from one to three years, depending on the environmental conditions, aggressiveness of the dominant male, and rate of development of the young. The young males join bachelor groups after they leave their maternal harem. These bachelor groups are commonly controlled by a dominant stallion and are loosely organized. When stallions approach the age and size at which they can capture a harem of their own, they may leave the bachelor group and remain alone (Dobie 1952; Ryden 1970; Tyler 1972).

Adolescent females are captured when they are two to three years of age by mature males seeking to establish or add to their harems. In some populations of feral horses, mature mares forcibly expel adolescent females from the harem. Stallions often do not protect adolescent females from raids by other stallions.

Mature mares restrict their social activities to members of their own group. They avoid close contact with individuals that are not part of the harem. Feral mares seldom stray more than 275 m from other members of the harem. The approach of a strange individual or group causes them to band more closely than otherwise (Dobie 1952; Ryden 1970; Tyler 1972).

Nursing young are well integrated into the social structure. They are protected not only by their mothers but also by other members of the harem unit. The patterns of association of nursing foals closely resemble those of their dams. As they grow older they become more and more independent, seeking with increasing frequency the company of peers (Tyler 1972; Waring 1970a, 1970b).

Any wild horse seen by itself is usually either an adolescent male, a male that has reached maturity and is trying to capture its own harem, an animal unable to keep up with the group, or an old stallion that has lost his harem (Dobie 1952; Ryden 1970).

Feral mares maintain rigid dominance hierarchies within the harem. Their progeny are included in the dominance structure as long as they remain within their maternal group. There is usually an alpha (most dominant) individual in each social unit, whether a complete dominance hierarchy is maintained or not. The alpha individual is generally a mature stallion, although in many harems it may be a mature mare. The dominant individual leads the group to forage and water (Dobie 1952; Ryden 1970; Hall 1972). Stallions control their groups by biting, by kicking, and with a threatening posture that includes elongating and arching the neck and weaving the head back and forth. This threatening posture is generally all that is necessary to make a member of the group obey. Whenever a group runs to escape danger, a dominant mare takes the lead and the stallion brings up the rear. Many stallions are very domineering and keep a close watch over members of their group. Only during foaling will the stallion permit a mare to leave the harem to find a secluded spot in which to give birth (Dobie 1952; Ryden 1970).

When a strange stallion approaches a harem group it is met by the harem stallion. Heads are extended in a threatening posture. The front feet are used as weapons as the stallions rear on hind legs. During combat, the teeth may inflict considerable damage and the combatants often wheel and kick in attempts to injure each other. They may shriek, snort, and scream. When one falls or runs away, it is seldom pursued by the winning stallion. Intolerance of other males is greatest during the breeding season (Dobie 1952; Ryden 1970).

The size of the home range used by bands of feral horses is influenced by the season of the year and the availability of forage and water. Ranges seldom exceed 40 km², and are often used by several bands. Four requirements normally designate a home range: grazing area, shelter, water, and shade. Although feral horses are not territorial, they do maintain a sphere of intolerance, or an area around a harem or family group that the dominant male will defend against other males. This sphere expands during the breeding season and contracts when it is over. The spacing between bands is related to defense of the band and not defense of a geographic area (Pelligrini 1971).

Vocalizations and Postures. Feral horses use a variety of vocalizations including snorts, neighs, nickers, squeals, and screams (Feist 1971; Waring 1971). The snort is a danger signal used mostly by stallions and seldom by mares. The neigh is a distress call used primarily by mares and younger horses. However, stallions may use the neigh to call straying mares and foals, or in preparation for harem defense. The nicker is used in close communication and courtship. The squeal is commonly uttered by a mare when a stallion approaches and sniffs her genitalia. Mares also squeal when fighting or displaying aggressiveness. The scream is emitted exclusively by stallions during aggressive interactions.

Six facial expressions commonly occur in equids (Tyler 1972). They include yawning, the flehmen posture, greeting, threat, and snapping expressions.

Yawning occurs before or after resting, when mares in estrus are sniffed by stallions, by stallions after mating, or by foals after suckling.

The greeting expression occurs when two individuals such as a stallion and a mare meet. They extend their heads and touch each other's muzzles and lips. These may develop into threat postures with the ears laid back. The greeting expression is also used as a preliminary to mutual grooming. Young horses greet older horses by retracting the lips and snapping the teeth lightly.

In the flehmen posture, the animal extends its neck and curls its upper lip so that the teeth are exposed. This expression commonly takes place by the stallion, while sniffing a mare during precopulatory activity (Hafez et al. 1962). Stallions, mares, or foals may also display this posture after sniffing urine. Mares display flehmen after sniffing a fresh placenta.

The threat expression is characterized by backward-directed ears. Mild threats are exhibited by slightly laying back the ears, but intense threats are expressed with the ears flat against the head and the mouth opened. This often occurs before a dominant animal attempts to bite a subordinate. When driving mares, stallions also use a threat gesture of stretching the neck toward the ground, laying the ears flat against the head, and swaying the head back and forth (Hafez et al. 1962).

The snapping expression is a threat characterized by stretching the neck with the ears laid slightly back and down and corners of the mouth drawn back, partly exposing the teeth. The lower jaw is moved up and down. This expression is a common greeting in foals or yearlings when threatened or approached by adult mares or stallions (Hafez et al. 1962).

Feral mares become sexually active at about 3 years of age. They generally cease sexual activity at about 10 years of age. As estrus approaches, the mare allows the stallion to smell and bite her. When ready to copulate, the tail is lifted and held to the side, the pelvis lowered and the hind legs spread. The intensity of estrous behavior varies between individuals, but normally peaks just before ovulation. Adult mares in estrus normally actively seek a stallion, but become passive after a stallion shows interest. They stand quietly with the hind legs straddled and the tail raised, and often turn the head to touch the stallion's muzzle. On occasion they squeal, urinate in small amounts, or paw the air with forefeet. During diestrus the mare is nonreceptive to the stallion and will display defensive reactions varying from aggressiveness to disinterest (Hafez et al. 1962).

During courtship, the stallion exhibits the flehmen posture after smelling the mare. He sometimes snorts or whinnies and nibbles or licks the mare before mounting. True precopulatory behavior of stallions is very brief. Usually, it is longer with young mares than with older, more experienced mares. Adult stallions chase colts away from adult mares in estrus but may allow them to copulate with young females up to four years of age. Courtship is important for successful mating because the stallion depends upon erotic stimuli to achieve vascular engorgement of the penis. This is elicited by visual, auditory, tactile, and olfactory stimuli (Ryden 1970).

Mares generally seek seclusion as parturition approaches. Dominant mares tend to travel greater distances and spend longer times away from the group than subordinate mares. Some parturient mares permit another mare to accompany and remain near them at the birth site. This associate is nearly always a mare without a foal of her own.

Grooming. Grooming among feral horses includes the following activities: shaking, rubbing one body part against another, rubbing against the ground, rolling, scratching or rubbing against a tree or bush, and mutual grooming (Feist 1971; Pelligrini 1971; Tyler 1972).

Mutual grooming is a common part of the daily activities of feral bands. It is accomplished by using the incisor teeth to groom the neck, withers, and base of mane. Horses often groom one side and then switch to the other. Mutual grooming occurs between all combinations of ages, except between stallions and immature males. Mutual grooming is normally associated with insect infestations or the shedding of hair.

Dusting sights are normally scattered throughout the range and are used by individuals from many different bands. Although dusting is the most common form of scratching, individuals may also roll in mud and water. Rolling most often occurs at the end of a resting period or when patches of wet ground are encountered after a rain.

Feral horses often rub against fixed objects such as stumps or trees. They often have favorite objects to which they return daily.

Other. The practice of depositing fecal matter in one place to create piles of manure, commonly referred to as stud piles, may be interpreted in several different ways. Stallions may do this in order to establish territories. However, the observation that individuals from several bands may use the same stud pile suggests that this is not a purpose. When a feral band passes a stud pile, it is common for the stallion to defecate, with accompanying posturing, on the pile. The stud piles may simply represent a ''community bulletin board,'' because a stallion's dung certainly constitutes a visual and olfactory notice to other stallions that he has been there. Mares are much less predisposed than stallions to deposit their fecal material in a dung pile. It is also possible that the dung pile may be used as a source of nutrients when forage is scarce during periods of drought or winter. Horses are generally coprophagous (Feist 1971).

MORTALITY

Reports of diseases and parasites among feral bands of horses are extremely limited. In 1930, the Bureau of Animal Industry reported that 17 percent of the wild horses on the San Carlos Apache Indian land were infected with dourine, a chronic venereal disease of equids commonly termed equine syphillis (Roby and Mott 1956). The Federal Bureau of Animal Husbandry removed about 500 horses from Nevada wild herds in 1935 because of the presence of this disease. The Bureau of Land Management in 1974 reported a suspected outbreak of the same disease in wild horses near China Lake, California. Other diseases common to domestic horses may also affect wild horses, although data on their incidence are not available. It is also likely that the species of parasites that infest domestic horses may infest wild or feral horses, but there are no data on the degree of infestation. Because of the nature of their lifestyles, it is quite unlikely that wild horses would be as susceptible to parasitic infestations as domestic horses. They cover a wider range of territory, and the harsh environment in which they live does not predispose itself to the spread of parasites.

Historically, wild horse densities have been restricted primarily by people. Wild horse roundups in the United States were widely carried out until 1971. Other predators may take some horses but few studies to date have been conducted on predation of wild horses. However, it is generally accepted that no wild animal has a significant impact on most wild horse herds. The only exception may be the cougar (*Felis concolor*), which may take significant numbers in some horse herds. However, the low density of cougar populations near wild horse herds suggests that their total impact on horse populations is minor. Thus, people represent the primary factor affecting wild horse herds.

Very little research has been conducted concerning competition between wild horses and other species. The grazing preferences of horses and their year-round use of most ranges suggest that forage competition with other species can become severe when forage supplies are scarce (Artz 1977). Perhaps the prime area of agreement concerns the ability of wild burros to dominate scarce water supplies in the U.S. Southwest. This factor is generally accepted as being detrimental to populations of desert bighorn sheep (*Ovis canadensis nelsoni*).

AGE DETERMINATION

The age of horses is generally determined by dentition, the morphology of the teeth, and the wear on the surfaces of the teeth (Bone 1964). It is assumed that those changes that are used to determine the age of domesticated horses would be similar in wild horses.

The eruption of the teeth in the lower and upper jaws is used to determine the age of a horse from birth until five years of age. From that time on, age is determined by the wearing away of the surface of the teeth, until the horse is approximately 12 years of age. From 13 years on, the age of the individual is determined by the shape and slant of the teeth as well as the wearing away of the surface (figure 55.4).

Most foals are born with central upper and lower incisors, or these incisors erupt shortly after birth. Intermediate incisors appear at approximately 7 to 8

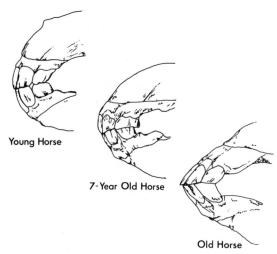

Young Horse

7-Year Old Horse

Old Horse

FIGURE 55.4. Comparing the slant of teeth for age determination in horses (*Equus caballus*).

months of age. The yearling horse has three pairs of upper and lower incisors. The deciduous central incisors are replaced at about 2½ years of age, the intermediate incisors at about 3½, and the corner incisors at about 4½. By 5 years of age, all of the permanent incisors are in use. The tushes incisiform, or canine teeth, appear at about 4½ or 5 years of age.

By the end of the 6th year the cups are worn from the surface of the central incisors, by the end of the 7th year from the intermediate incisors, and by the end of the 8th from the corner incisors. By the end of the 9th year the cups disappear from the upper central incisors, and by the end of the 10th and 11th years, respectively, from the intermediate and upper corner incisors.

At about 10 years of age, a groove appears on the tooth near the gums on the upper corner incisors and begins to extend downward. It is usually about halfway down by the end of the 15th year, and all the way down by the end of the 20th year. Thus, a 20-year-old horse would have a groove all the way down the upper corner incisors. This groove is referred to as "Galvayne's groove." Galvayne's groove begins to fill in from the top beginning with the 21st year of age. By the time a horse is 25 years old it is one-half filled. If the horse lives to its 30th year, the groove is completely filled (figure 55.5).

The surfaces of the teeth of young horses are elliptical in shape, but as the horse grows older they wear down. The surface, or "table," of the tooth takes on the characteristics of that particular region of the shaft of the tooth. By middle age the teeth are rounded in shape, while in old age the surface of the teeth is triangular in shape (figure 55.6). Teeth may wear more rapidly than normal in horses grazing on forage high in silica or more slowly in animals that eat less harsh and abrasive forage.

ECONOMIC STATUS AND MANAGEMENT

Wild horses management problems are important only in specific areas because most herds are located in remote areas in the western United States. Most of these lands are managed by the Bureau of Land Management (BLM). The BLM oversees nearly 18 times as many horses and nearly 22 times as many burros as the Forest Service. While horses exist in other places, the problems are particularly acute for BLM districts located within the Great Basin and in western Wyoming (Godfrey 1979).

Until passage of the wild horse and burro acts (PL 86-234 and 92-195) most wild horse herds were managed and controlled by local ranchers or wranglers. The Wild Horse and Burro Act of 1971 (PL 92-195) changed the status of wild horse and burro populations. Some of the major provisions of this act include:

1. All "wild and free-roaming horses and burros" (WFHB) that use public lands during any portion of the year are to be administered by the Forest Service (FS) and the Bureau of Land Management.
2. Excess numbers can be destroyed in a "humane" manner or "captured and removed for private maintenance under humane conditions and care."
3. No WFHB herd nor any part thereof can be sold for any consideration.
4. WFHB using privately owned lands can be removed by personnel of the responsible agency if requested by the landowner.
5. Private citizens may not do any of the following:
   a. remove or attempt to remove any WFHB from public lands,
   b. convert any WFHB to private use without authority,
   c. maliciously cause the death or harrassment of any WFHB,
   d. process or permit any WFHB to be processed into any commercial product, or
   e. sell any WFHB held under private maintenance.
6. No WFHB herd is to be relocated to areas where they did not exist when PL 92-195 was passed in 1971.

Recently, some of the restrictions found in PL 92-195 and PL 86-234 have been modified. For example, PL 94-579, which was passed on 21 October 1976, provided that helicopters and other motor vehicles could be used to capture and transport WFHB provided: (*a*) a public hearing is held, and (*b*) their use is under the direct supervision of BLM or FS employees. This change allowed agency personnel to accelerate roundups in areas where excess numbers of WFHB were judged to exist. Further changes were passed in PL 95-514 (1979), which provided that transfer of ownership of a WFHB to a private citizen may occur if the WFHB has been held under private maintenance for one year, and that no more than four WFHB may be adopted by any one individual in any year. While no information is available concerning the impact possible ownership has had on the demand by potential adopters for additional animals, it is not likely that their desires have changed significantly. As one would expect, most adopters desire young females (less than three years of age), while most animals captured are commonly older animals. In addition, at least 50 percent of the animals captured in most roundups

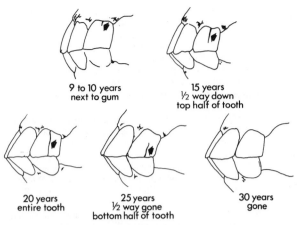

9 to 10 years
next to gum

15 years
½ way down
top half of tooth

20 years
entire tooth

25 years
½ way gone
bottom half of tooth

30 years
gone

FIGURE 55.5. Galvayne's groove occurs only on the upper incisors and is of importance in the age determination of horses.

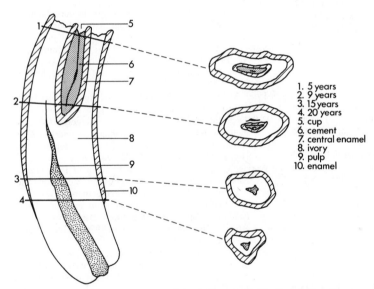

1. 5 years
2. 9 years
3. 15 years
4. 20 years
5. cup
6. cement
7. central enamel
8. ivory
9. pulp
10. enamel

FIGURE 55.6. Change in the shape of the surface, or "table," of teeth with age in horses.

are male. As a result, many of the older stallions are not adoptable and are destroyed in a humane manner (Godfrey 1979).

The management of wild horse herds by the BLM is relatively expensive. Data from the BLM indicate that in 1978 it cost at least $400 to place an animal in an acceptable adopter's care (Godfrey 1979). Undoubtedly these costs have risen since that time.

Sales by agencies that are not subject to the wild horse laws (for example, the Park Service, Defense Department, or Bureau of Sport Fisheries and Wildlife) suggest that the sales value is often not sufficient to cover capture costs, implying that the value of wild horses is not high. Yet, essentially no other species has received as much political support.

Though some interest groups do not agree, it is generally accepted that wild horse herds are not in danger of being eliminated (Zarn et al. 1977; U.S.D.I. and U.S.D.A. 1976). Rather, the major management problem centers around reducing and maintaining numbers to levels that are supportable by the habitats used. Agency personnel are the only people who can authorize a roundup of horses that use BLM or FS lands in order to reduce overgrazing or competition with other species. As a result, state fish and game departments and other interest groups have had a difficult time getting populations reduced when conflicts arose.

Historically, wild horses have been used as beasts of burden, pet food, rodeo stock, pleasure animals, and breeding stock. The wild horse acts specifically eliminate some of these uses until ownership can be obtained. As a result, most wild horses that are adopted can legally be used only for riding or as pets. This status differs significantly from those of other domestic or wild animals because use in any "commercial product" is specifically excluded. After ownership is obtained, these commercial uses are not excluded, but legal restrictions may effectively eliminate such uses in the short run.

CURRENT RESEARCH AND MANAGEMENT NEEDS

One of the primary reasons why management of wild horse herds has been so controversial stems from a lack of scientific information. As a result, Public Law 95-514 required that research on wild horses be undertaken by groups not directly associated with the agencies charged with their management. This research is currently being formulated and funded. Most of it emphasizes forage requirements and nutrition, population dynamics, competition with other species, and alternative methods of controlling populations. This research will help solve some of the problems faced by management personnel. However, much remains unknown and it is likely that wild horse management will be governed by emotion rather than fact for some time in the future.

LITERATURE CITED

Artz, J. L., ed. 1977. Proceedings of the national wild horse forum. Nevada Agric. Exp. Stn. Bull. R-127. 110pp.

Beebe, B. F., and Johnson, J. R. 1964. American wild horses. David McKay, New York. 180pp.

Berliner, V. R. 1969. The estrous cycle of the mare. Page 267 in H. H. Cole and P. T. Cupps, eds. Reproduction in domestic animals. Academic Press, New York. 657pp.

Bone, J. R. 1964. The age of the horse. Southwest Vet. 17:269-272.

Castle, W. E. 1954. Coat color inheritance in horses and in other mammals. Genetics 39(1):35-44.

Dobie, J. F. 1952. The mustangs. Little, Brown & Company, Boston. 376pp.

Ensminger, M. E. 1951. Horse husbandry. Interstate Printers & Publishers, Danville, Ill. 336pp.

Feist, J. D. 1971. Behavior of feral horses in the Pryor Mountain wild horse range. M.S. Thesis. Univ. Michigan. 129pp.

Godbee, R. G., and Slade, L. M. 1979. Nitrogen absorption from the cecum of a mature horse. 5th Equine Nutr. and Physiol. Proc.

Godfrey, E. B. 1979. The economic role of wild and free roaming horses and burros on rangelands in the western United States. Final Rep. Submitted to Intermountain For. and Range Exp. Stn. 39pp.

Hafez, E.S.E.; Williams, M.; and Wierzbowski, S. 1962. The behaviour of horses. Pages 370-396 in E.S.E. Hafez, ed. The behaviour of domestic animals. Williams & Wilkins, Baltimore.

Haines, Francis. 1971. Horses in America. Thomas Y. Crowell, New York. 213pp.

Hall, R. 1972. Wild horse: biology and alternatives for management, Pryor Mountain wild horse range. Bur. Land Manage., Billings Dist. 67pp.

Hanauer, E. 1973. The science of equine feeding. A. S. Barnes, New York. 78pp.

Jones, W. E., and Bogart, R. 1971. Genetics of the horse. Caballus Publishers, Ft. Collins, Colo. 356pp.

Klingel, J. 1972. Social behaviour of African *Equidae*. Zool. Africa 7:175-185.

Odberg, F. O. 1972. An interpretation of pawing by the horse (*Equus caballus L.*): displacement activity and original functions. Rev. part of B.S. Thesis. State Univ. Ghent, Belgium. 12pp.

Pelligrini, S. W. 1971. Home range, territoriality and movement patterns of wild horses in the Wassuk Range of western Nevada. M.S. Thesis. Univ. Nevada, Reno. 39pp.

Roby, T. O., and Mott, L. W. 1956. Diseases and parasites affecting horses and mules. Pages 531-533 in Yearbook of agriculture animal diseases. U.S. Dept. Agric., Washington, D.C.

Ryden, H. 1970. America's last wild horses. Dutton, New York. 311pp.

Simpson, G. G. 1951. Horses: the story of the horse family in the modern world and through sixty million years of history. Oxford Univ. Press, New York. 247pp.

Slade, L. M.; Bishop, R.; Morris, J. G.; and Robinson, D. W. 1971. Digestion and absorption of ISN-labelled microbial protein in the large intestine of the horse. Br. Vet. J. 127:X1.

Tyler, S. 1972. The behavior and social organization of New Forest ponies. Anim. Behav. Monogr. 5:84-196.

U.S. Dept. Inter. and Dept. Agric. 1976. A report to Congress by the secretary of the Interior and the secretary of Agriculture on administration of the Wild Free-Roaming Horse and Burro Act, P.L. 92-195. 45pp.

Waring, G. H. 1970a. Perinatal behavior of foals (*Equus caballus*). Pap. presnted at 50th Annu. Meet. Am. Soc. Mammal., 18 June 1970. College Station, Texas.

———. 1970b. Primary socialization of the foal (*Equus caballus*). Pap. presented at Anim. Behav. Soc. at 21st Annu. AIBS Meet., 29 August 1970. Indiana Univ., Bloomington, Ind.

———. 1971. Sounds of the horse (*Equus caballus*). Pap. presented to Ecol. Soc. Am. 22nd Annu. Am. Inst. Bio. Sci. Meet., 2 September 1971. Colorado State Univ., Fort Collins.

Zarn, M.; Heller, T.; and Collins, K. 1977. Wild and free-roaming horses: status of present knowledge. U.S. Dept. Inter., Bur. Land Manage., and U.S. Dept. Agric., For. Serv., tech. note 294. Denver Serv. Center, Denver, Colo. 72pp.

LARRY M. SLADE, Department of Animal, Dairy, and Veterinary Sciences, Utah State University, Logan, Utah 84322.

E. BRUCE GODFREY, Department of Economics, Utah State University, Logan, Utah 84322.

56

Feral Hog
Sus scrofa

James M. Sweeney
John R. Sweeney

NOMENCLATURE

COMMON NAMES. Feral hog, feral swine, feral pig, razorback, wild boar, wild hog
SCIENTIFIC NAME. *Sus scrofa*
SUBSPECIES. *S. s. andamensis, S. s. assuricus, S. s. attila, S. s. barbarus, S. s. castillanus, S. s. chirodontus, S. s. cristatus, S. s. floresianus, S. s. koreanus, S. s. leucomystax, S. s. libycus, S. s. majori, S. s. meridionalis, S. s. moupinens, S. s. nicobaricus, S. s. nigripes, S. s. papuensis, S. s. reiseri, S. s. scrofa, S. s. timorensis,* and *S. s. vittatus.*

The number of subspecies attributed by taxonomists to *Sus scrofa* has varied over time. The above list includes those presently recognized (Bratton 1977). The feral hog in North America probably has its origin in the subspecies *S. s. scrofa.*

The family Suidae first appeared in early Oligocene in or around India (Pilgrim 1941). During the following epochs Suidae spread throughout Europe, Central Asia, and Africa. The genus *Sus* first appeared in fossil records of the middle Miocene, and occurred in Europe and Africa by the upper Pliocene (Osborn 1910). In addition to this natural migration, humans, in domesticating swine, introduced pigs throughout the remainder of the Old World and most of the New World, which resulted in the almost universal distribution of Suidae today.

This chapter will be primarily limited to the "feral hog" of domestic origin but living in a wild state, as opposed to the "European wild hog" (original wild stock). Although morphologically distinct, both the feral hog and European wild hog are recognized as *Sus scrofa.* Some authors have referred to the feral hog as *S. s. domesticus;* however, this is not a commonly accepted practice.

DISTRIBUTION

Sus scrofa is native to a wide geographic area in the Old World, but is now extirpated in Great Britain and

FIGURE 56.1. The pelage of feral hogs (*Sus scrofa*) is usually coarser and denser than that of domestic hogs, providing greater resistance to the elements. The general body shape of the feral hog lies somewhere between that of the round, "fattened" domestic hog and the streamlined, sloping shape of the European wild hog.

FIGURE 56.2. The prominent pelage color of feral hogs (*Sus scrofa*) is black, with mottled black and brown or black and white being the next most common.

the Nile Valley (Bratton 1977). Feral hogs in North America are commonly believed to have originated from domestic hogs introduced by early settlers from the European countries, with the first introductions probably being made by Columbus in 1493 in the West Indies, and DeSoto in 1593 in Florida (Towne and Wentworth 1950). However, there is now some evidence to indicate that the genus *Sus* was present in the United States long before the introduction of domestic hogs by Columbus. An incisor, two canines, three cheek teeth, and a fragment of the occiput were recently recovered from the 10 Mile Rock archeological dig in northwestern Arkansas (Quinn 1970). Carbon-14 dating of these remains indicates that *Sus* (possibly *Sus indicus*) was present prior to the 1490s. Domestic hogs were also reportedly released in California by Spanish explorers in 1769 (Hutchinson 1946) and in the Hawaiian Islands by Captain Cook in 1778 (Kramer 1971).

Feral hog populations are now found throughout the southeastern United States from Texas east to Florida and north to Virginia, and in California, Hawaii, Puerto Rico, and the Virgin Islands. A small population also exists in New Zealand (Wodzicki 1950).

The often-practiced open range policy of hog management has resulted in the continued "release" of domestic hogs into many feral hog populations. The European wild hog has also been released in several locations within the continental United States, particularly in the Appalachian Mountains (Lewis et al. 1965; Howe and Bratton 1976). Trapping and relocation of feral hogs by state wildlife agencies has also aided in the dispersion of feral hogs. Therefore, feral hog populations today are a mixture of some European wild hogs, recent domestic hogs, and feral hogs.

DESCRIPTION

Size. In general, the physical appearance of the feral hog is intermediate between those of the domestic hog

and the European wild hog (figure 56.1). The morphological characteristics of feral hogs are dependent upon the degree of continued crossbreeding with domestic stock or introduced European wild hogs, and on how long the feral population has existed in the wild. Brisbin et al. (1977a) reported significant differences in two populations of feral swine that differed in the amount of time they had been in a feral state. Hogs in a recently feral population (20 years since domestication) were heavier and had greater total body length than feral hogs that had been essentially free from the influences of domestication for several hundred years. As a group, however, feral hogs are thinner, appearing more streamlined in body shape, with longer and sharper tusks and coarser coats than domestic hogs (Hanson and Karstad 1959; Golley 1962).

Body size and weight of feral hogs are highly variable depending upon the genetic history of the population and the local environment in which the population exists. Sweeney (1970) reported that feral hogs in the older age classes ($\geq$18 months) ranged from 96 to 108 kg in mean weight and 177 to 192 cm in mean total length for boars, and from 66 to 92 kg in mean weight and 134 to 185 cm in mean total length for sows. These animals were from a recently feral population inhabiting a river swamp near Aiken, South Carolina. Feral hogs inhabiting the coast near Georgetown, South Carolina, were smaller. Boars in the older age classes averaged 49–60 kg and sows averaged 42–47 kg (Wood and Brenneman 1977). A population of extremely small hogs was reported by Brisbin et al. (1977a). Adults in this population averaged only 23 kg in weight and 96 cm in total length. These sizes are the smallest reported for the genus *Sus*, including strains of miniature swine. The population studied by Brisbin et al. (1977a) inhabited a Georgia coastal island that was extremely overpopulated. Samples of adult sows taken after this same population was significantly reduced in density yielded mean weights of 30 kg and mean total lengths of 101 cm.

In all of the above studies there was a tendency for the males to be larger than the females. However, individual variation in each age class prevented statistical verification of this trend, except for those feral hogs in the study by Wood and Brenneman (1977) that were more than three years old. Feral hogs of both sexes appeared to continue to increase in weight at least through the second (Sweeney 1970) or third (Wood and Brenneman 1977) year.

Pelage. The prominent pelage color of feral hogs is black (figure 56.2). Spotted black and brown, or black and white is the next most common coloration. Brown or roan is not uncommon, and occasionally feral hogs will be all white. Only rarely has the Hampshire color phenotype (generally gray to black with a white shoulder band) been reported (Golley 1962; Sweeney 1970; Brisbin et al. 1977a). The pelage of feral hogs is usually coarser and denser than that of domestic hogs. It provides a greater resistance to chilling, probably a selective advantage in the feral state (Foley et al. 1971; Hansen et al. 1972).

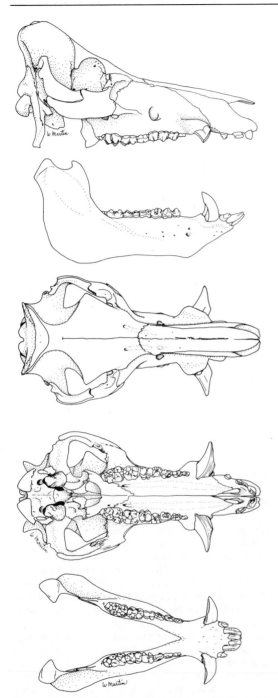

FIGURE 56.3. Skull of the feral hog (*Sus scrofa*). From top to bottom: lateral view of cranium, lateral view of mandible, dorsal view of cranium, ventral view of cranium, dorsal view of mandible.

Body. Other anatomical characteristics of feral hogs are similar to those of domestic hogs. The face is elongated and laterally compressed; the nostrils are situated in a smooth terminal disc of the mobile snout. The shoulder region is noticeably thickened, with very wide scapula. The ulna and radius of the foreleg are connected, whereas the tibia and fibula of the hind leg are not. The second and fifth digits of the feet, known as dewclaws, are accessory and do not reach the ground (Sisson and Grossman 1938; Mellen 1952).

Dentition. The dental formula of the pig is: 3/3, 1/1, 4/4, 3/3 = 44 (figure 56.3). Replacement of the deciduous teeth is usually complete by 20–22 months of age (tables 56.1 and 56.2). Cheek teeth are bunodont. The upper and lower canines of the boar are curved backward and outward, with the lower canines being aligned in front of the upper canines. Friction between the upper and lower canines produces a sharp edge on the lower tooth. Tusks (lower canine from gum line to tip along outer curve) of adult boars may grow to 8–9 cm in length (Pullar 1953; Barrett 1978). Canines of adult sows are usually somewhat less developed. The canines are rootless and capable of continued growth (Sisson and Grossman 1938).

PHYSIOLOGY

Little definitive research has been completed on the physiology of feral hogs. Inferences may be drawn from the vast amount of information available on the physiological processes of domestic swine (Frandson 1965; Mount 1968; Pond and Maner 1974; Sorenson 1979) to provide insight concerning processes of the feral hog. However, caution should be exercised. Although similarities may exist, it is probable that notable differences also occur.

Irving (1966) reported that the domestic hog was capable of maintaining skin temperatures of 10° to 12° C on the dorsal surface in a −12° C environment, while body temperature at a depth of 45.7 mm remained normal at 38.4° C. The primary mechanism involved for conservation of body heat in this study of bare-skinned swine was the restriction of peripheral blood vessels. Domestic hogs can increase their

TABLE 56.1. Average periods of tooth eruption and replacement in domestic swine

Tooth	Eruption	Change
I_1	2–4 weeks	12 months
I_2	{ upper 2–3 months	
	{ lower 1½–2 months	16–20 months
I_3	before birth	8–10 months
C	before birth	9–10 months
P_1	5 months	
P_2	5–7 weeks	
P_3	{ upper 4–8 days	
	{ lower 2–4 weeks	
P_4	{ upper 4–8 days	12–15 months
	{ lower 2–4 weeks	
M_1	4–6 months	
M_2	8–12 months	
M_3	18–20 months	

SOURCE: Sisson and Grossman 1938:488.

TABLE 56.2. Average periods of tooth eruption and replacement in European wild hogs

Age	Temporary Teeth	Permanent Teeth
0–6 days	$i\frac{3}{3}$, $c\frac{1}{1}$	
7–22 days	$i\frac{3}{3}$, $c\frac{1}{1}$, $p\frac{3}{4}$	
23–40 days	$i\frac{1\ 3}{1\ 3}$, $c\frac{1}{1}$, $p\frac{3}{3\ 4}$	
6–7 weeks	$i\frac{1\ 3}{1\ 3}$, $c\frac{1}{1}$, $p\frac{3\ 4}{3\ 4}$	
7–19 weeks	$i\frac{1\ 2\ 3}{1\ 2\ 3}$, $c\frac{1}{1}$, $p\frac{2\ 3\ 4}{2\ 3\ 4}$	
20–33 weeks	$i\frac{1\ 2\ 3}{1\ 2\ 3}$, $c\frac{1}{1}$, $p\frac{2\ 3\ 4}{2\ 3\ 4}$	$P\frac{1}{1}$ $M\frac{1}{1}$
30–51 weeks	$i\frac{1\ 2}{1\ 2}$, $p\frac{2\ 3\ 4}{2\ 3\ 4}$	$I\frac{3}{3}$, $C\frac{1}{1}$, $P\frac{1}{1}$, $M\frac{1}{1}$
12–15 months	$i\frac{2}{2}$, $p\frac{2\ 3\ 4}{2\ 3\ 4}$	$I\frac{1\ 3}{1\ 3}$, $C\frac{1}{1}$, $P\frac{1}{1}$, $M\frac{1\ 2}{1\ 2}$
14–18 months	$i\frac{2}{2}$	$I\frac{1\ 3}{1\ 3}$, $C\frac{1}{1}$, $P\frac{1\ 2\ 3\ 4}{1\ 2\ 3\ 4}$, $M\frac{1\ 2}{1\ 2}$
18–22 months	$i\frac{2}{}$	$I\frac{1\ \ \ 3}{1\ 2\ 3}$, $C\frac{1}{1}$, $P\frac{1\ 2\ 3\ 4}{1\ 2\ 3\ 4}$, $M\frac{1\ 2}{1\ 2}$
21–26 months		$I\frac{1\ 2\ 3}{1\ 2\ 3}$, $C\frac{1}{1}$, $P\frac{1\ 2\ 3\ 4}{1\ 2\ 3\ 4}$, $M\frac{1\ 2\ \ }{1\ 2\ 3}$
+26 months		$I\frac{1\ 2\ 3}{1\ 2\ 3}$, $C\frac{1}{1}$, $P\frac{1\ 2\ 3\ 4}{1\ 2\ 3\ 4}$, $M\frac{1\ 2\ 3}{1\ 2\ 3}$

SOURCE: Matschke 1967:112.

metabolism, and therefore heat production, only very slightly (Irving 1966).

In a study of a feral hog population that has been extant in a wild state for several hundred years, Graves and Graves (1977) found that feral hogs may be much more efficient than domestic hogs in storing and mobilizing energy. This would be an advantage for a wild population in coping with cold periods, and especially in meeting the selective pressures of extreme seasonal variation in food availability (Martin and Herbein 1976). Also, as noted earlier, the pelage of feral hogs is usually denser than that of domestic hogs, which aids in resistance to chilling.

Domestic hogs are generally considered to have low tolerance of food with high salt content. However, in the coastal population studied by Graves and Graves (1977), feral hogs apparently had evolved a more efficient mechanism(s) for handling salt in their diets. During the summer, much of their food was obtained in salt marshes. Food habits studies of a feral swine population near Georgetown, South Carolina, also indicated use of salt marsh food sources (Roark 1977).

Some research has been done on the use of feral hogs as a possible indicator species for monitoring environmental contaminants (Stribling 1978). Due to the omnivorous feeding habits and nomadic movements of feral hogs, they "sample" their environment more completely than animals with restricted movements and diets. Levels of radioactive contamination (Cesium-137) were monitored for a population of swine inhabiting a river swamp system that was receiving coolant waters from nuclear reactors. Radiocesium concentration in major body compartments of individual feral hogs indicated that the skeletal muscle and kidney had the highest Cesium-137 levels of all body compartments. Those of the liver and bone were among the lowest. Whole body burdens (full body radio-cesium levels) of feral hogs taken from this site were higher than those taken from a Georgia coastal island population. However, except for a few individuals taken in summer, radio-cesium levels in feral hogs on the reactor site were low and well below the minimum safety levels established for human consumption. Seasonal studies indicated that feral hogs had higher whole body burdens in the summer, when they were utilizing foods found in habitats where radioactive contamination would be more available.

GENETICS

Little definitive work has been completed on the genetics of feral hogs. However, comparisons of the karyotypes of domestic hogs, European wild hogs, and domestic × European hybrids may be instructive. McFee et al. (1966) reported that the normal complement of chromosomes for European wild hogs was 36.

Similar results were found in a study on the cytogenetics of hogs on the Tellico Wildlife Management Area in Tennessee (Rary et al. 1968). Pure strain domestic hogs consistently had 38 chromosomes, pure strain European wild hogs had 36, and their hybrid cross produced fertile offspring with a diploid chromosome number of 37.

After examination of the chromosome groups 27-36, 27-37, and 27-38, Rary et al. (1968) hypothesized that the difference in the normal diploid number of chromosomes could be accounted for by the centric fusion of two pairs of telocentric chromosomes, resulting in an evolutionary reduction of chromosomes from 38 for domestic hogs to 36 for European wild hogs, or by the centric splitting of chromosomes 35 and 36 in the European wild hog, thereby increasing the total number of chromsomes in the domestic hog. Hsu and Mead (1969) support the latter explanation, saying that the 38-chromosome condition is most likely the result of a centric split of a pair of metacentric chromosomes producing two pairs of telocentrics.

Because feral hogs are descendants of domestic hogs, one would expect their diploid chromosome count to be $2n = 38$. However, many populations of feral hogs have a history of crossbreeding to varying extents with introduced European wild hogs. The frequency of domestic and European wild hogs in a given feral hog population could be determined through cytogenetic studies and the utilization of the Hardy-Weinberg Law (Rary et al. 1968).

A detailed study of genetic material has been conducted in two populations of feral hogs in South Carolina (Brisbin et al. 1977*b*). Using horizontal starch gel electrophoresis to estimate the levels of genetic variability, banding patterns in 20 proteins were examined in feral hogs collected from the Ossabaw Island and Savannah River populations (Selander et al. 1971; Manlove et al. 1976). Allele frequencies of the feral hogs examined differed between the two populations at two variable genetic loci. These loci included Phosphoglucose isomerase and 6-Phosphogluconate dehydrogenase. Five percent of the loci were polymorphic in the Ossabaw Island population, while 20 percent were polymorphic in the Savannah River population. The average heterozygosity across all loci was $\bar{H} = 0.005$ for the Ossabaw Island population and $\bar{H} = 0.035$ for the Savannah River population.

REPRODUCTION

Puberty. Detailed studies to determine the estrous cycle and length of gestation in feral swine have not been conducted. However, feral hogs are polyestrous, and are likely similar to domestic swine, which have an estrous cycle of approximately 21 days and a gestation period of about 112–114 days. In general, feral hogs reach puberty between the ages of 5 and 10 months. In a study of feral hogs in South Carolina, Sweeney et al. (1979) reported that male feral hogs attained puberty at 5–7 months of age, while female feral hogs attained puberty at 10 months. Barrett (1978) found that feral hogs in California reached puberty around 6 months of

age. He noted that feral boars seldom played an important role in breeding until 12–18 months of age, due to the dominance of older males. Feral sows varied greatly in their age at puberty, from as early as 3 months to as late as 13 months of age.

The age at puberty for feral hogs is similar to the ages reported for domestic and European wild hogs. Spermatozoa are present in the testes of the male domestic hog as early as 4.6 months of age and as late as 5.2 months due to variations in nutrition and breed (Phillips and Zeller 1943). In the case of inbred male domestic hogs, spermatozoa are first present at 6 months of age (Hauser et al. 1952). Male European wild hogs, however, reach maturity at the age of 7 months (Henry 1966).

Litter Size. Litter sizes of feral hogs vary considerably. Sweeney et al. (1979) reported a mean fetal count of 7.4 with a range of 5 to 12 per sow for a population inhabiting the Savannah River swamp in South Carolina. A later study of this same population showed a mean fetal count of 8.4 with a range of 3 to 11. A sample of 34 sows near Georgetown, South Carolina, yielded fetal counts of 3 to 11 with a mean of 5.1. Barrett (1978) observed similar variation in California: he recorded as few as 1 to as many as 10 viable fetuses per sow, with an average of 5.6. The ratio of the fetuses to corpora lutea for populations studied by Sweeney et al. (1979) and Barrett (1978) were 69 and 66 percent, respectively. However, a later study (Sweeney 1979) involving the Savannah River population and two other populations inhabiting coastal South Carolina and Georgia yielded a ratio near 83 percent.

Average fetal counts should be adjusted to account for postnatal loss from stillborn hogs. Asdell (1964) stated that stillborn loss in domestic hogs averaged 6 percent. It may be that a similar loss occurs in feral hogs.

The average litter size for feral hogs is generally larger than the average reported for European wild hogs, but within the lower limits of the range of litter sizes recorded for domestic hogs. Asdell (1964) set the average for European wild hogs at 4 to 5 young per sow. Henry (1966) found that Tennessee European wild hogs had an average litter size of 4.8, and Pine and Gerdes (1973) reported that European wild hogs in California had an average litter size of 4.2. Nalbandov (1976) noted that the average litter size of domestic hogs ranged from 6.6 to 21.2 depending on breed. Barrett (1978), in his review of the literature, found that modern breeds of domestic hog can produce as many as 24 young per litter.

Farrowing. The feral hog is physiologically capable of breeding year-round, and farrowing has been observed throughout the year (Sweeney et al. 1979). However, two peaks in farrowing activity often occur in feral hog populations. A population of feral hogs in South Carolina (Sweeney et al. 1979) was observed to have a prominent peak in farrowing activity in mid-winter (February) and a second, less evident rise in farrowing in summer (July). Barrett (1978) also noted two peaks in farrowing in populations of feral hogs in

California. One population had two small rises, one in July and another in November. Another population showed peaks again in July and November, but the November peak was much more prominent. This is similar to the farrowing activity of European wild hogs in Tennessee, where peaks in early summer (May–June) and midwinter (January–February) occur (Henry 1966).

These peaks loosely correlate with the times that most managers cause their domestic swine to farrow, in spring or early summer and in later autumn (Smith 1952; Ensminger 1961). In domestic swine production, because the postpartum estrus in domestic swine is usually not fertile, peaks in farrowing are usually scheduled six months apart. This enables individual sows to participate in both farrowing periods. Unless feral sows exhibit a fertile postpartum estrus, the farrowing peaks observed in the wild may likely be the result of environmental factors such as weather or food supply, with different sows contributing to each of the farrowing peaks.

ECOLOGY

Range and Habitat. The range of feral hogs in North America seems to be limited by two major factors: land use, particularly as it affects cover, and climate (Hanson and Karstad 1959). Cover is essential for feral hogs, and habitat use is directly proportional to the density of cover (Barrett 1978). Besides providing food and protection from disturbance, cover can play an important role in the reduction of heat loss by decreasing air flow and by reradiating thermal energy (Tregear 1965; Moen 1968). Climate can also play an important role with respect to the feeding behavior of feral hogs. With increased frost penetration feral hogs are faced with increased difficulty in rooting (Hanson and Karstad 1959). However, during this time of year, hard mast—acorns of oaks (*Quercus* spp.) and nuts of hickories (*Carya* spp.)—becomes available and rooting activity often is greatly reduced in preference for these nutritious and palatable foods.

Within their ranges, feral hogs have adapted to a variety of habitat types. By far the most commonly used habitat is bottomland with the associated mixed hardwood forest cover and a source of permanent water. Feral hogs in South Carolina used the river swamp and associated bottomlands almost exclusively, except in late winter and early summer, when a shift in movement occurred into the associated upland pine plantations (Sweeney 1970; Kurz and Marchinton 1972). Likewise, Golley (1962) noted that feral hogs in Georgia resided primarily in the swamps and marshes of the coastal plain, ranging out to the uplands and pinewoods less frequently. The largest concentration of feral hogs in California was found in the larger creek canyons, particularly on the north slopes that supported heavy thickets of live oaks (Barrett 1978).

Seasonal changes in habitat use are apparently linked with changes in food availability (Sweeney 1970; Kurz and Marchinton 1972; Graves and Graves 1977; Barrett 1978). Feral hogs on Ossabaw Island, off the coast of Georgia, moved out of their preferred oak/hickory habitat into surrounding salt marshes only in late winter when food became scarce in the forested areas (Graves and Graves 1977). In another area, abundant acorn mast in the fall was reflected by concentrated feral hog activity in the bottomlands along the Savannah River; a distinct shift in movement into the upland pine plantations in June corresponded with the ripening of plums (Kurz and Marchinton 1972). Irrigated pastures and boulder washes were used as alternate habitats by California feral hogs for nocturnal feeding during dry periods (Barrett 1971). Pullar (1950) and Wodzicki (1950) also concluded that food supply strongly influenced movements of feral hogs in Australia and New Zealand.

Herd Composition. Feral hogs characteristically travel in groups of eight or less, with rarely more than three adults per group. Boars are usually solitary, except when associated with breeding groups (Kurz and Marchinton 1972; Graves and Graves 1977; Barrett 1978).

Groups of piglets commonly range in size from one to five with one feral sow (Kurz and Marchinton 1972), but may increase to an average of seven with two sows. The second sow in a group often had experienced high piglet mortality (Graves and Graves 1977). Feral piglets in South Carolina remained together as a group until they reached 25 to 35 kgs in weight (Kurz and Marchinton 1972). Recently, two feral sows from this same population were observed to travel with their new litters of four and five piglets as one herd for a period of eight months (Lewis Crouch personal communication). Barrett (1978) reported that feral sows in California traveled in the company of one to three generations of offspring. The social pattern described here does not preclude the congregation of large numbers of feral hogs on a particularly choice and localized food source. Barrett (1978) noted herds of as many as 97 feral hogs feeding on an irrigated pasture.

The sex ratio of domestic swine—52.8 percent male—is skewed slightly from the theoretical 50:50 ratio (Nalbandov 1976). Observations of feral hogs on Ossabaw Island indicated that sex ratios within size classes did not differ from 1:1 except for animals over 64 kg, where males predominated 3:1 (Graves and Graves 1977). A sample of feral hogs collected along the Savannah River included 53.8 percent males, but only the 7- to 12-month age group (2:1) differed significantly from a 1:1 ratio (Sweeney 1970). A more recent study of these two populations and an additional population near Georgetown, South Carolina, found that fetal sex ratios did not differ significantly from a 1:1 ratio. The sex ratio of young feral hogs (less than 13 months old) on the Dye Creek Preserve in California was 1:1, but the adult population (13 months or older) was comprised of 49.6 percent males. The uneven adult sex ratio was likely due to earlier hunting pressure on the male segment (Barrett 1978).

Movements. Movements of feral hogs are generally drifting, almost nomadic, but within a given home range area. Average home range sizes vary from 200 to 300 ha. Movement from one area to another within the home range is apparently linked with seasonal changes in food availability. Daily activity patterns, however, seem to be influenced by ambient temperature, with diurnal or crepuscular activity common throughout most of the year and nocturnal activity often predominant in summer (Kurz and Marchinton 1972). Because they lack sweat glands, hogs must rely heavily on such behavioral thermoregulation (Mount 1968).

Crouch (personal communication) observed diurnal activity in the summer as well, in feral hogs inhabiting bottomland hardwoods and river swamp edges. Such habitats probably provided relatively cool ambient temperatures and abundant water sources for wallows. Conversely, nocturnal activity may be evident, even in winter, in populations of feral hogs subjected to regular hunting pressures (Hanson and Karstad 1959).

Five feral hogs fitted with radio-transmitters exhibited elongated home ranges of 123–799 ha along the Savannah River in South Carolina, with an average of 396 ha. Two sows had much smaller ranges—17 and 30 ha—during the farrowing period. No significant differences occurred between daily movements of boars and sows, except late in pregnancy when the sow's movements were greatly reduced. The hogs were primarily diurnal from October through May; however, nocturnal movement increased significantly during summer months. There was also considerably more activity on moonlit nights than on dark nights (Kurz and Marchinton 1972). Another radio-telemetry study of six feral hogs in coastal South Carolina reported home range sizes of 146–292 ha, with an average of 203 ha (Wood and Brenneman 1980).

Repeated observations of marked feral hogs over periods of up to two years indicated that male feral hogs in California had home ranges of at least 5,000 ha. Feral sows had home ranges of 1,000 to rarely 2,500 ha. Movements of sows were much more restricted than those of boars, especially when sows were with litters. These feral hogs were believed to be crepuscular in spring and autumn, diurnal in winter, and nocturnal in summer (Barrett 1978).

FOOD HABITS

Feral hogs are omnivorous and their feeding habits can generally be described as opportunistic. They exhibit no particular preferences for a given plant or animal species. However, plant material is preferred over animal matter, with the latter making up only a minor part of the diet. Mast, both soft and hard, is probably one of the most preferred plant food items. Roots, tubers, and shoots of herbaceous vegetation vary in importance with seasonal availability. There is no major variation in food habits between sexes or age classes.

In the following discussion of food habits, the amounts of food items eaten are reported in percentage of the total volume or percentage of the aggregate dry weight. It should be noted that these two measurements are not necessarily directly comparable for all food items. Comparison of results between studies using different measures of abundance should therefore be based on the relative importance of food items within each study.

The food habits of wild hogs (feral hogs × European wild hogs on the coastal plain of Texas were detailed by Springer (1977). Green stems and leaves of herbaceous plants accounted for 45 percent of the diet in the spring and 12–18 percent in other seasons. Grasses provided over 20 percent of the total spring diet but varied from 5 percent to less than 1 percent at other times of the year. Roots were ingested throughout the year, accounting for 4–17 percent of the volume in spring to 55 percent in the winter. Acorns provided 50 and 48 percent of the fall and winter volumes, respectively. Mushrooms were the main type of fungus eaten, varying from less than 2 percent by volume in the winter to almost 15 percent in the spring. Invertebrates were taken in amounts varying from 11 percent of the winter volume to more than 21 percent in the fall. Earthworms (Lumbricidae) and March fly larvae (Bibionidae) were the predominant invertebrates consumed. Vertebrates were taken in small quantities at all times of the year, but snakes were the only type taken every season.

Similar food habits were reported for feral hogs in the coastal plain of South Carolina (Roark 1977). Herbaceous material, primarily grasses, represented 51 percent of the aggregate dry weight of all ingesta in the spring diet and 9.5–35.8 percent in other seasons. Roots were consumed throughout the year, as in Texas, but were of importance on the coastal plain of South Carolina only in the spring (29 percent) and summer (38 percent). Fruits represented from 9 percent aggregate dry weight in the summer to 84 percent in the winter. Acorns again were the most common type of fruit taken, providing 44 percent and 80 percent of the fall and winter diets, respectively. Mushrooms were ingested in all seasons, but exceeded 3 percent aggregate dry weight only in the summer (12 percent). Invertebrates were consumed in amounts approaching 3 percent in all seasons except fall, when they represented less than 1 percent of the aggregate dry weight. The most commonly taken invertebrates were Scarabid larvae (Scarabaeidae), centipedes (Geophilidae, Scolopendridae), and earthworms. Vertebrates were consumed in about the same quantities as the invertebrates—less than 3 percent in all seasons.

Similar seasonal variability in food habits of feral hogs, as the result of food availability, was documented for feral hogs at Dye Creek Ranch, California (Barrett 1978). In early spring, feral hogs grazed green grasses and forbs. As herbaceous vegetation hardened in late spring and early summer, wild oats (*Avena barbata*) and bulbs (*Bradiaea spp.*) became important constituents of the feral hog's diet. Bulbs continued to be an important food item through the summer until manzanita berries (*Arctostaphylos manzanita*) became available. In September, feral hogs fed more and more heavily on acorns, and from October to December acorns were a major portion of the diet. Animal matter made up only 1.6 percent of the total annual volume.

Feral hogs also feed on carrion (Hanson and Karstad 1959; Roark 1977; Barrett 1978). Several times hunters on St. Vincent Island National Wildlife Refuge, Florida, have tracked wounded deer and found feral hogs feeding on the dead animal (Thompson 1977). In Australia, the consumption of carrion by feral hogs indirectly assists in blowfly control (Pullar 1950).

The omnivorous food habits of feral hogs presents the possibility of competition with many other wildlife species. This competition and the ground disturbance from rooting have created problems for many resource managers. Feeding by feral hogs on the seeds and seedlings of longleaf pine (*Pinus palustris*) was considered a serious problem in the regeneration of longleaf pine (Wahlenberg 1946; Wakely 1954). However, it is not generally thought to be important in pine regeneration in the South today (Wood and Brenneman 1977). Reproduction in hardwoods can likewise be restricted through consumption of acorns (Lucas 1977). Depredation on agricultural crops is also a problem (Pine and Gerdes 1973; Springer 1977), particularly where these crops have been planted on wildlife refuges for waterfowl (Thompson 1977).

Nest predation was considered important on iguana on Mona Island, Puerto Rico (Weinwandt 1977), and in rookeries on Auk Island, New Zealand (Challies 1975), and could be a serious threat to loggerhead turtle (*Caretta caretta*) reproduction in coastal areas (Thompson 1977). However, nest predation on upland game birds in the United States is generally considered to be of only minor significance (Thompson 1977). Successful wild turkey (*Meleagris gallopavo*) reproduction has occurred in the presence of large populations of feral hogs in areas of Florida and South Carolina (Wood and Barrett 1979).

The rooting behavior of feral hogs presents a potential problem in the destabilization of surface soils along stream channels, road banks (Lucas 1977), and sand dunes on coastal areas (Wood and Barrett 1979). The disturbance of the forest floor may also be considered objectionable by some outdoor recreationists.

The most important form of competition with other wildlife is probably that centered on the use of mast. Many highly preferred species of wildlife, including deer (*Odocoileus spp.*), wild turkey, and squirrels (*Sciurus spp.*), depend on soft and/or hard mast as an important food item. Most wildlife refuges with significant feral hog populations consider mast competition a serious to very serious factor (Thompson 1977; Wood and Lynn 1977). However, one must remember that the degree of competition will depend on the relative abundance of the mast supply and whether or not the wildlife species utilizing the mast can utilize alternate sources of nutritionally adequate foods during periods of poor mast production.

BEHAVIOR

Farrowing. Movement of feral sows decreases as the pregnant sow approaches parturition. This reduction in

activity is because the sow centers its activities around a farrowing nest (Kurz and Marchinton 1972). Nests are usually relatively shallow depressions in the ground with or without bedding material, and are usually located in shaded areas on high, dry ground (Hanson and Karstad 1959). Of two nests found by Kurz and Marchinton (1972), one contained pine straw that appeared to be bedding material, while the other had no bedding material. Barrett (1978) noted that nests might be simply a large bed rooted into the soil, but were usually haphazardly lined with grasses, leaves, and other vegetation. The collection of vegetation for construction of a nest is a common behavior in feral sows; sows kept in concrete farrowing pens may often be quieted by providing them straw or other nesting materials. Sweeney (1970) described an elaborate nest made up of pine straw and broom sedge, and Wood and Brenneman (1980) reported that the majority of nests they observed were large piles of pine straw.

Detailed observations of one feral sow in South Carolina revealed that it would leave the nest, during the first two weeks after parturition, from about 9:00 A.M. to 4:00 P.M. During the sow's absence, newborn piglets remained closely huddled in the nest unless disturbed. At three weeks of age, the piglets began following the sow in its daily movements. The feral sow moved out of the farrowing range at that time (Kurz and Marchinton 1972). Observations of feral sows in California also indicated that sows moved fewer than 0.5 km from nest sites until the piglets were three weeks of age (Barrett 1978). A slightly different behavior was noted by Crouch (personal communication), who observed two feral sows and their piglets. Although the piglets usually stayed near the nest, they did explore nearby areas. At one to three weeks of age the piglets followed the sows for short distances from the nest. In fact, one sow and its litter moved about 1.6 km when the piglets were approximately one week old. This movement may have been in response to the observer's presence.

On one occasion, a feral sow was observed to nurse another sow's litter. Observations of the pseudoparent one week earlier indicated that it was very close to parturition. Although there was no evidence of its litter, its physical condition while nursing indicated that parturition had occurred. The parent feral sow did not display aggression toward this sow (Kurz and Marchinton 1972).

Breeding Groups. Barrett (1978) observed 4 different feral sows come into estrus. As many as 10 boars gathered around each sow for about two days. Once the sows were receptive, copulation took place as often as every 10 minutes. Unless the weather was very hot or very cold, breeding continued through day and night. Kurz and Marchinton (1972) observed copulation in two groups of feral hogs, each consisting of 4 boars and 4 sows. Although the boars fought among themselves, the sows appeared to be rooting for food. This feeding response was believed to be displacement behavior, as the sows moved much more rapidly than usual through the area.

Fighting between feral boars in breeding groups

can be separated into two types: the establishment of dominance, and the display or maintenance of dominance (Kurz and Marchinton 1972). In the establishment of dominance, feral boars charge head-on, pushing with great force. Two competing boars place their heads along the neck and shoulder region of the opponent. Constant circling, upward slashing with the tusks, and variable amounts of vocalization are all part of the pattern. Fighting continues until one animal squeals and retreats. Displays for the maintenance of dominance, however, last only a matter of seconds. Boars display by facing each other with erect manes and open mouths. There is often much loud roaring and coughing. These displays are usually terminated with one or two lunges by the dominant boar (Kurz and Marchinton 1972; Barrett 1978).

Breeding activity is generally a combination of courtship, male fighting, displacement behavior, and copulation. After copulation, dominant boars usually lie down nearby, permitting subordinate boars to breed the receptive sow. Homosexual activity often occurs, especially in young males (Barrett 1978).

Senses. Feral hogs apparently depend on olfaction and hearing more than vision to keep in contact with their environment (Wesley and Klopfer 1962). Populations of feral hogs in California (Barrett 1978) and South Carolina were easily approached from downwind, even in open country. A common vocalization by feral hogs is an alarm grunt given by the feral hog that first scents an intruder. Such an alarm causes almost immediate flight response in the rest of the herd (Crouch personal communication). Other vocalizations include those common to domestic hogs. Scent posts were commonly used, often in conjunction with body scratching. Travel normally occurred along well-defined paths, located primarily by smell (Barrett 1978).

MORTALITY

The single most important decimating factor responsible for mortality in feral hogs is hunting by people in one form or another. Natural predation usually plays only a minor role in the mortality of feral hogs. Intermediate in impact of these two factors is the loss due to diseases and parasites, particularly through an interaction with age.

Hunting. Hunting of feral hogs may take place for direct control of nuisance animals, for sport, or through poaching. The number of animals removed under control practices varies with the degree of the problem. Wood and Brenneman (1977) reported an annual removal of 100 feral hogs for control of population levels on the Hobcaw Barony, South Carolina. Sweeney (1970) indicated that sufficient numbers of feral hogs, an average of 50/yr, were removed each year from a population along the Savannah River to curtail damage to planted pine seedlings. The goal on most wildlife refuges with feral hog populations is to eliminate populations completely or at least to reduce them to a minimum (Thompson 1977).

Florida, California, and Hawaii have the largest harvests of feral hogs as game animals. From 1971 to 1976 sport hunting in Florida accounted for the taking of an average of 56,200 hogs per year. The highest recorded harvest was 84,100 taken in 1974 (Bob Butler personal communication). In California the annual harvest increased from 1960 to 1976, and now has leveled off at 32,000 (Vic Simpson personal communication). The estimated feral hog harvest in Hawaii is 10,000 animals annually (Wood and Barrett 1979).

Uncontrolled hunting in states where the feral hog is not protected as a game animal and illegal poaching result in the removal of additional animals from many feral hog populations. Hanson and Karstad (1959) indicated that it was a common practice for residents along riverbottoms and swamps to shoot or trap feral hogs whenever they needed pork. This practice still exists in many areas of the Southeast. Wood and Brenneman (1977) estimated that as many as 50 animals were poached annually from the Hobcaw Barony feral hog population in South Carolina. On the Dye Creek Preserve in California, poaching removed approximately the same number of feral hogs as the annual sport harvest, which represented 10 percent of the fall population (Barrett 1978).

Predation. In general, predation is of minor importance as a decimating factor, particularly in the removal of adult animals. Alligators (*Alligator mississippiensis*), black bears (*Ursus americanus*), and mountain lions (*Felis concolor*) have all been reported as predators of an occasional adult feral hog (Hamilton 1941; Hanson and Karstad 1959; Wood and Brenneman 1977). Bobcats (*Felis rufus*), however, may be a significant predator on feral pigs in some areas. They are important predators of young pigs in the Southeast (Hanson and Karstad 1959). Wood and Brenneman (1977) reported that bobcat predation occurred only on feral hogs less than three to four months old. However, Davis (1955) studied the food habits of bobcats in an area sympatric with feral hogs in Alabama and found no evidence of predation on feral hogs.

Coyotes (*Canis latrans*) were the only potential predators whose population numbers were high enough to cause significant mortality to feral hog populations in California (Barrett 1978). Of 1,042 coyote scats examined, 12.7 percent contained remains of feral hogs. Ninety-five percent of the feral hog remains were from piglets. Although fecal analysis suggested that the coyote could be an important predator of feral pigs, Barrett (1978) believed that the majority of the feral hog remains were consumed as carrion.

Parasites and Diseases. The combined mortality due to parasites and diseases, including trauma (other than hunting or predation) and starvation, impinges more heavily on feral hog populations than predation, but remains less significant than human hunting activities. Disease seems to interact with age, increasing the impact of this decimating factor in animals older than 2 years or younger than 6 months. Wood and Brenneman (1977) examined age structures for a feral hog population on Hobcaw Barony in South Carolina, and concluded that disease played an increasingly important role in the mortality of older animals. Only 12 percent of the feral hog population studied was over 3 years of age, and there were 61 percent fewer 3-year-old feral hogs than 2 year olds. They also felt that disease was an important decimating factor of feral pigs less than 6 months of age: they estimated piglet mortality from disease to be between 50 and 70 percent. Sweeney (1970) found a similar age interaction with disease in feral hogs along the Savannah River. Fifteen percent of the feral hogs necropsied showed some clinical signs of disease, with the frequency increasing in older animals. Fifty-seven percent of the animals over 18 months were diseased, whereas only 6, 15, and 25 percent of the feral hogs less than 7, 7–12, and 13–18 months of age, respectively, were considered diseased. Only 3 percent of the total population was over 24 months of age.

In several areas of the Southeast, Hanson and Karstad (1956) found that 79 percent of feral hogs tested carried antibodies to vesicular stomatitis virus. They also found antibodies to eastern equine encephalomyelitis in 25 percent of blood samples of the hogs tested. Pseudorabies, hog cholera, and a disease resembling swine pox were also suspected to occur, but not documented, in these feral hog populations.

Hanson and Karstad (1959) believed that the levels and types of parasites and diseases differed between the feral hog and domestic swine due to differences in habits, nutrition levels, and genetic background. Free-ranging feral hogs do not have as much opportunity to transmit parasites or diseases as do domestic hogs confined in pastures. Parasites that have no intermediate host, such as Ascarid worms, but require the accidental ingestion of an embryonic ovum passed to the ground by a previous host are not common in feral hogs. However, parasites whose life cycle includes passage through an intermediate host commonly consumed as food by feral hogs, such as *Metastrongylus* lungworms via earthworms, are much more prevalent.

Hanson and Karstad (1959) noted that all feral hogs over 15 kg that they examined had lungworms. Some animals had severe infestations resulting in chronic bronchitis. Also, kidney worms (*Stephanura dentatus*) were often present in these animals. Although Trichina were never found, *Sarcosporidia* were present in the muscles of every animal examined. Heavy infestations of the common sucking louse (*Haematopinus suis*) were found on pigs during the winter. Other ectoparasites were not evident, except for a few species of tick in July.

Wild hogs on the Aransas Refuge also have had a history of heavy parasitism (Coombs and Springer 1974; Ruddle 1975). One hundred percent of the wild hogs checked contained large numbers of swine kidney worms, resulting in necrosis of the lungs and liver and the associated bacterial infection of the lung, liver, and perirenal areas. Other parasites found in significant quantities included lung worms, round worms (*Ascaris summ*), hookworms (*Globocephalus urosubulatus*), miscellaneous stomach worms, and Ixodid ticks of three species, up to 800 per animal.

Barrett (1978) did not conduct a detailed study of parasites and diseases on feral hogs in California, but

believed that parasites were not a major mortality factor. The most common internal parasite found was lungworms. All hogs examined had the common sucking louse, but only young piglets had significantly heavy infestations. Diseases, however, were considered important, particularly in young and old animals.

Mortality due to trauma was considered a major factor in loss of piglets in California, while starvation was considered the most important cause of death in older animals (Barrett 1978). Accidents resulting in the loss of piglets included: crushing of piglets by the sow while in the nest, tusking or trampling by boars, blindness caused by dry grass florets impacting in the eyes, and separation from the sow and littermates resulting in exposure and starvation. Starvation can also be important in piglet mortality when sows are on a low plain of nutrition, resulting in poorer quality and/or quantity of milk. In older feral hogs, tooth deterioration was a significant mortality factor. All hogs four years of age or older had one or more periodontal abscesses, leading to bone decay and generalized septicemia.

AGE DETERMINATION

Age determination is fundamental in the analysis of wildlife populations. Although some criteria of age are more accurate than others, all are at best only estimators of the actual age of an individual. It is advantageous, therefore, to have several different techniques of age determination for any given species in order to improve the accuracy of estimation. Two techniques have been developed for determining the age of feral hogs: the use of tooth eruption pattern and wear, and the use of the eye lens weight.

Tooth Eruption and Wear. To date, a detailed schedule has not been published on the tooth eruption pattern in feral hogs. Schedules have been published for domestic hogs (Sisson and Grossman 1938) and European wild hogs (Matschke 1967). For each respective tooth, eruption often occurs at an earlier age in domestic hogs (see table 56.1) than in European wild hogs (see table 56.2), particularly $I\frac{2}{2}$ and $M\frac{3}{3}$. However, with the exception of $p\frac{4}{4}$ the sequence of eruptions is the same. In the domestic hog, p^4 erupts before $p_{\overline{4}}$, which is the reverse of the pattern seen in European wild hogs.

In feral hogs, $I^{\underline{3}}$ and $P_{\overline{1}}$ frequently do not erupt. Barrett (1978) reported that 45 percent of 100 skulls and lower jaws examined were missing one to three of these four teeth. Normally if one $I^{\underline{3}}$ incisor failed to erupt so did the other, and likewise for the $P_{\overline{1}}$ premolars. In feral hogs along the Savannah River in South Carolina, the $P_{\overline{1}}$ premolars fail to erupt more than 50 percent of the time.

Pine and Gerdes (1973) found that the European wild hog tooth eruption pattern gave satisfactory estimates when used to determine ages for feral hogs in Monterey County, California. Likewise, the true ages of tagged feral piglets on the Dye Creek Preserve in California corresponded best with the tooth eruption pattern of European wild hogs. Use of this schedule in this case slightly overaged the feral piglets (Barrett

TABLE 56.3. Estimated ages of feral hogs over 26 months of age by examination of the third molars

Third Molars	Age (Months)
Present, but <75% irrupted	26–30
75–90% irrupted	30–36
100% irrupted	36+
Cusps 25–50% worn	48+
Cusps 60–90% worn	60+
Cusps completely worn (first molars often lost)	72+

SOURCE: Barrett 1978:289.

1978). However, the tooth eruption pattern for domestic hogs provided the most accurate schedule for estimating the age of feral hogs along the Savannah River, South Carolina. Limited data suggested that in this instance the domestic schedule slightly underaged the feral hogs (Sweeney 1970). The actual time of tooth eruption and replacement in feral hogs is likely variable. In a study of known-age European wild hogs, Matschke (1967) noted that tooth eruption and replacement time varied considerably among animals, even among litter mates raised under the same conditions. This variation increased with age.

Tooth eruption patterns are primarily designed for estimating the age of animals up to two years old. Sisson and Grossman's (1938) schedule provides estimated ages for domestic hogs up to 20 months, and Matschke's (1967) schedule provides age estimates for European wild hogs up to 26 months of age. Barrett (1978) extended the technique through the use of the degree of eruption of M_3^3 as noted by Matschke (1967) and the wear on M_3^3 as described by Cabon (1959). By incorporating detailed examination of the third molars, he was able to group animals roughly into yearly age classes from three to six years. (See table 56.3.)

Eye Lens Weight. Eye lens weight offers a second criterion for age determination of feral hogs. It has been used with varying degrees of success for many different species (Friend 1968). In a controlled experiment with pen-raised European wild hogs fed three different diets, Matschke (1963) determined that the eye lens weight was a reflection of the body weight and as such was not a reliable indicator of age. However, in a study conducted on feral hogs in South Carolina a significant correlation ($r = 0.95$, $P < 0.001$) was found between eye lens weight and the age of feral hogs, even when body weight failed as an indicator of age (Sweeney et al. 1970).

Estimating the age of feral hogs by dentition tends to clump animals into established and often overlapping age classes. In many cases, the tooth eruption pattern places an animal in a zone of overlap, forcing a subjective decision in assigning the individual to a specific age class. However, the establishment of an eye lens weight-to-age curve based on known-age animals in a given area would provide a continuum of ages along the curve. Measurements of subsequent eye lens

weights would then yield age estimates without the need for subjective decisions. Eye lens weight data, therefore, may provide the best age structure estimates for feral hog populations. However, the use of eye lens weight requires detailed laboratory measurements. As a result, tooth eruption pattern remains the best field technique.

ECONOMIC STATUS

Reflecting its variable status as a game animal or pest, the feral hog could be viewed economically as either a credit or a debit. Its greatest economic impact, however, is probably the cost related to the detrimental effects of its omnivorous diet and feeding behavior. Feral hogs can seriously restrict timber reproduction and destabilize surface soils (Lucas 1977). They can also severely damage agricultural crops by knocking down or rooting up large areas (Pine and Gerdes 1973). Some national wildlife refuges have had to resort to extensive fencing to protect crops planted for waterfowl management (Thompson 1977). Feral hogs represent increased maintenance costs on management areas where rooting has resulted in the destruction of dikes, roads, trails, and recreation areas (Pine and Gerdes 1973; Belden and Frankenberger 1977; Thompson 1977).

These problems, along with the fact that feral hogs compete with other wildlife for food and in some local cases may cause significant nest depredation, have resulted in the development of continued control programs on state and federal lands. These programs represent a constant economic drain in both labor and support budgets.

Another area of economic concern is the potential of feral hog populations to serve as a reservoir of certain diseases and parasites transmissable to humans and domestic livestock. Because of their domestic origin, feral hogs may represent a greater hazard in this respect than other species of wildlife (Hanson and Karstad 1959).

Although in most situations feral hogs are not considered a financial asset, they have the potential of providing increased commercial gain because of their productivity and adaptability to different habitats. Springer (1977) noted that greater numbers of game ranchers in Texas are recognizing the feral hog as a marketable product. Some landowners charge $10–75 per head on guaranteed feral hog hunts. The Dye Creek Preserve in California uses commercial guided hunts as an economical approach to the control of feral hogs, in areas where feral hogs are not the featured species (Barrett 1978).

"Hog claims," which came into use in the 1800s (Lucas 1977) as a method of recognizing local ownerships, were much more prevalent in the past than they are today. A hog claim denoted an individual's ownership of all unfenced hogs on a particular piece of land. These claims were primarily used to raise hogs under an open range condition, for family use or for sale. Today, in the few areas that still recognize hog claims, claim owners may also charge for hunting privileges on their claims, or live-trap and sell adult boars to hunting preserves (Belden and Frankenberger 1977).

Merchants and state game agencies where feral hogs are managed as a game species may realize some economic benefits associated with an additional recreational resource. The extent of this economic gain has not been documented and is likely quite variable depending upon the recreational demand and the cost of management. The cost to the Florida Game and Fresh Water Fish Commission for restocking feral hogs on public hunting areas has varied from $10 to $55 per head (Belden and Frankenberger 1977).

MANAGEMENT

Management of feral hogs varies substantially throughout their range. In most locations, management is nonexistent until local populations reach densities high enough to cause noticeable environmental damage by rooting or significant competition for mast with other more favored wildlife species. Then management usually takes the form of control or attempted eradication. With the absence of large predators, feral hogs have the capability of quickly reaching high population densities in favorable habitat. Therefore, herd increase is rarely a management problem.

Federal. Management of feral hogs is inconsistent, even among different groups within the federal government. The U.S. Forest Service recognizes all feral hogs as domestic livestock, and as such considers them as trespassers on national forest lands (Lucas 1977). Concurrently, the U.S. Fish and Wildlife Service recognizes feral hogs as feral animals (Thompson 1977). Both agencies, however, consider the species as a nuisance and pursue programs of control to reduce populations to a minimum. The main concern of the U.S. Fish and Wildlife Service is the competition for food between feral hogs and other wildlife species such as deer, turkeys, squirrels, and waterfowl (Thompson 1977). The list of problems with feral hogs presented by the U.S. Forest Service is considerably longer, and includes food competition, destabilization of surface soils, restriction of timber reproduction, and conflicts with outdoor recreationists (Lucas 1977).

Control on national wildlife refuges has most often taken the form of live-trapping and relocation to other areas, especially to state wildlife management areas. Hunting by primitive weapons has been used on some areas with little effect and, in a few situations, hunting with dogs under special permit has been allowed. This latter method proved very effective but also could potentially cause considerable controversy.

The U.S. Forest Service's removal program was complicated at first by the presence of large numbers of domestic hogs. Although there were existing regulations prohibiting unauthorized livestock on national forest lands, the foresters were reluctant to use their authority prior to the 1960s. By 1964 the population of unauthorized hogs on national forests was estimated to be around 33,000 animals (Lucas 1977). The Forest Service finally realized that it had to begin a hog removal program. Since most of these animals were domestic livestock claimed by local landowners, the removal program had to comply with a complex series of legal procedures including notification of the owners, allowing them the opportunity to remove their hogs, notification of impoundment, impoundment of the hogs, notification of public auction, public auctions, and issuance of bills of sale. Hog populations on national forests were reduced to an estimated 3,000 animals by 1972 (Lucas 1977). Most domestic livestock have been removed. Efforts are now centered on controlling the feral populations through trapping and by allowing hunters to harvest them under either no limit or a very liberal limit.

State. Management of feral hogs is inconsistent at the state level of government as well. Although most states do not consider the feral hog as a game species, a few do. California, Florida, Hawaii, North Carolina, West Virginia, and Tennessee recognize the feral hog as a game animal, at least in some areas. Management may be passive in nature with the setting of liberal seasons, or may involve active trapping and relocation programs. California has an open season on feral hogs with a statewide bag limit of one per day (one in possession), in all except Monterey County, which is closed from April through September. Florida recognizes the feral hog as a game animal only in those areas where the public has shown an interest, the landowners are agreeable, and the county commission has passed a resolution recommending such action. Management of feral hogs on these areas (35 wildlife management areas, 3 state parks, part or all of 6 counties, 1 state wildlife refuge, and 1 private management area) includes both the setting of specific seasons and an active restocking program. Wild hogs may be taken only during the two-month open deer season with a bag limit of one per day and two per season (Belden and Frankenberger 1977).

A major problem encountered on wildlife management areas in Florida is the difficulty of maintaining hog populations under heavy hunting pressures. Several hundred hogs are relocated annually, primarily from state parks where they have become a nuisance on high-use public hunting areas. However, this has become a "put-and-take" system in some cases. Experience has shown that release areas are sometimes determined more by distance from capture site, available time, and labor than by any real need for more hogs. In addition, unless released in suitable habitat, feral hogs will disperse within a few days and not be available to hunters (Belden and Frankenberger 1977). Other management strategies employed in Florida to prevent depletion of hog populations have included shortening the season, limiting or excluding the use of dogs, and setting a limit on the total number of hogs to be harvested, after which the season is closed. These regulations generally have been successful only in those areas that contained an abundance of escape cover.

Private. Management of feral hogs is practiced to varying degrees by private landowners. These activities may be no more than periodic releases on land leased by private hunt clubs, use of "hog claims," or detailed management of existing populations on commercial hunting areas.

Releases of wild hogs by members of private hunt clubs have been for the purpose of initiating new populations or improving the local population's genetic stock. The animals released are usually purported to be European wild hogs, but often are of domestic origin. Such releases ordinarily involve only a few animals, but over time have been a major factor in range expansion.

As noted in the previous section, "hog claims" are still recognized in some areas within the range of feral hogs. These claims provide income for rural families through the sale of feral boars for relocation onto hunt clubs or through the leasing of hunting privileges on the area covered by the claim (Belden and Frankenberger 1977).

Some feral hog populations are managed much more intensively to provide commercial hunting. A prime example of this is the Dye Creek Preserve in California. All feral hog hunts are guided and closely monitored. The goal of management in this case is to create a population sex and age structure that will ensure the continued production of adult trophy boars while maintaining the population at or below the estimated carrying capacity of 6 feral hogs/km² (Barrett 1978).

Management Guidelines. There are as many management options for the feral hog as there are managers confronted with the problem. Options can vary from efforts to increase population densities to intense eradication programs. Due to its omnivorous feeding habits and rooting behavior, the feral hog, whether considered a prized game animal or a detested pest, has a very real impact on its environment and the other animal species with which it interacts. Therefore, one must carefully evaluate all positive and negative impacts of each management option before taking action.

When developing management guidelines, whether on federal, state, or private lands, one must first compare the feral hog's potential detrimental impact to its potential benefits. This should include careful consideration of such factors as: (1) local flora and fauna, especially endangered species, that might be adversely affected by the habits of feral hogs; (2) contribution to soil erosion and stream siltation; (3) interaction with and the potential spread of disease to and among other animal species, especially domestic livestock; and (4) possible confrontation with other outdoor recreational activities. These factors should be weighed against a critical review of: (1) the potential of the feral hog as another game species; (2) hunter interest; and (3) economic gain to the surrounding area. Also, local traditions must not be overlooked and costs of implementing each management option must be detailed.

CURRENT RESEARCH AND MANAGEMENT NEEDS

Further study is needed on many aspects of feral hogs. Basic research is needed to expand the data base on behavior and movements, reproduction, physiology, and mortality—particularly the differential mortality of young and old. Better census and age determination techniques are needed. New control methods need to be evaluated, including repellents, species-specific poisons, chemosterilants, and environmental manipulations. An effort should also be made to document more fully the impact of feral hogs on native flora and fauna.

LITERATURE CITED

Asdell, S. A. 1964. Patterns of mammalian reproduction. 2nd ed. Cornell Univ. Press, Ithaca, N.Y. 670pp. (Artiodactyla: Suidae, pages 537–553.)

Barrett, R. H. 1971. Ecology of the feral hog in Tehama County, California. Ph.D. Thesis. Univ. Calif., Berkley. 368pp.

———. 1978. The feral hog on the Dye Creek Ranch, California. Hilgardia 46:283–355.

Belden, R. C., and Frankenberger, W. B. 1977. Management of feral hogs in Florida: past, present and future. Pages 5–10 *in* G. W. Wood, ed. Research and management of wild hog populations: proceedings of a symposium. Belle W. Baruch For. Sci. Inst., Clemson Univ., S.C.

Bratton, S. P. 1977. Wild hogs in the United States: origin and nomenclature. Pages 1–4 *in* G. W. Wood, ed. Research and management of wild hog populations: proceedings of a symposium. Belle W. Baruch For. Sci. Inst., Clemson Univ., S.C.

Brisbin I. L., Jr.; Geiger, R. A.; Graves, H. B.; Pinder, J. E., III; Sweeney, J. M.; and Sweeney, J. R. 1977*a*. Morphological characterizations of two populations of feral swine. Acta Theriol. 22:75–85.

Brisbin, I. L., Jr.; Smith, M. W.; and Smith, M. H. 1977*b*. Feral swine studies at the Savannah River Ecology Laboratory: An overview of program goals and design. Pages 71–90 *in* G. W. Wood, ed. Research and management of wild hog populations: proceedings of a symposium. Belle W. Baruch For. Sci. Inst., Clemson Univ., S.C.

Cabon, K. 1959. Problem der Altersbestimmung beim Wildschwein *Sus scrofa* nach der Methode von Dub. Acta Theriol. 388:188–120.

Challies, C. N. 1975. Feral pigs (*Sus scrofa*) on Aukland Island: status and effects on vegetation and nesting sea birds. New Zealand J. Zool. 2:479–490.

Coombs, D. W., and Springer, M. D. 1974. Parasites of feral pig × European wild boar hybrids in southern Texas. J. Wildl. Dis. 10:436–441.

Davis, J. R. 1955. Food habits of the bobcat in Alabama. M.S. Thesis. Ala. Polytechnic Inst., Auburn. 74pp.

Ensminger, M. E. 1961. Swine science. 3rd ed. Interstate Printers and Publishers, Danville, Ill. 692pp.

Foley, C. W.; Seerley, R. W.; Hansen, W. J.; and Curtis, S. E. 1971. Thermoregulatory responses to cold environment by neonatal wild and domestic piglets. J. Anim. Sci. 32:926–929.

Frandson, R. D. 1965. Anatomy and physiology of farm animals. Lea & Febiger, Philadelphia. 501pp.

Friend, M. 1968. The lens technique. Trans. North Am. Wildl. Conf. 33:279–298.

Golley, F. B. 1962. Mammals of Georgia. Univ. Georgia Press, Athens. Pages 198–199.

Graves, H. B., and Graves, K. L. 1977. Some observations of biobehavioral adaptations of swine. Pages 103–110 *in* G. W. Wood, ed. Research and management of wild hog populations: proceedings of a symposium. Belle W. Baruch For. Sci. Inst., Clemson Univ., S.C.

Hamilton, W. J., Jr. 1941. Notes on some mammals of Lee County, Florida. Am. Midl. Nat. 25:686–691.

Hansen, W. J.; Foley, C. W.; Seerley, R. W.; and Curtis, S. E. 1972. Pelage traits in neonatal wild, domestic and crossbred piglets. J. Anim. Sci. 34:100–103.

Hanson, R. P., and Karstad, L. 1956. Enzootic vesicular stomatitis. Proc. Annu. Meet. U.S. Livest. Sanit. Assoc. 60:288–292.

———. 1959. Feral swine in the southeastern United States. J. Wildl. Manage. 23:64–74.

Hauser, E. R.; Dickerson, G. E.; and Mayer, D. T. 1952. Reproductive development and performance of inbred and crossbred boars. Missouri Agric. Exp. Stn. Res. Bull. 503. 56pp.

Henry, V. G. 1966. European wild hog hunting season recommendations based on reproductive data. Proc. Southeast Assoc. Fish Game Commissioners 20:139–145.

Howe, T. D., and Bratton, S. P. 1976. Winter rooting activity of the European wild boar in the Great Smoky Mountains National Park. Castanea 41:256–264.

Hsu, T. C., and Mead, R. A. 1969. Mechanisms of chromosomal changes in mammalian speciation. Pages 8–17 in K. Benirschke, ed. Comparative mammalian cytogenetics. Springer-Verlag, Berlin.

Hutchinson, C. B., ed. 1946. California agriculture. Univ. California Press, Berkeley. 444pp.

Irving, L. 1956. Physiological insulation in bare-skinned swine. Appl. Physiol. 9:414–420.

———. 1966. Adaptations to cold. Sci. Am. 214:94–101.

Kramer R. J. 1971. Hawaiian land mammals. Charles E. Tuttle Co., Inc., Rutland, Va. 347pp.

Kurz, J. C., and Marchinton, R. L. 1972. Radiotelemetry studies of feral hogs in South Carolina. J. Wildl. Manage. 36:1240–1248.

Lewis, J. C.; Matschke, G.; and Murry, R. 1965. Hog Subcommittee report to the chairman of the Forest Game Committee, Southeastern Section, TWS. Nashville, Tenn. 18pp. Mimeogr.

Lucas, E. G. 1977. Feral hogs: problems and control on national forest lands. Pages 17–22 in G. W. Wood, ed. Research and management of wild hog populations: proceedings of a symposium. Belle W. Baruch For. Sci. Inst., Clemson Univ., S.C.

McFee, A. F.; Banner, M. W.; and Rary, J. M. 1966. Variation in chromosome number among European wild pigs. Cytogenetics 5:75–81.

Manlove, M. N.; Avise, J. C.; Hillestad, H. O.; Ramsey, P. R.; Smith, M. H.; and Straney, D. O. 1976. Starch gel electrophoresis for the study of population genetics in white-tailed deer. Proc. Southeast Assoc. Fish Game Commissioners 29:392–402.

Martin, R. J., and Herbein, J. G. 1976. A comparison of the enzyme levels and in vitro utilization of various substrates for lipogenesis in pair-fed lean and obese pigs. Proc. Soc. Exp. Bio. Med. 151:231–235.

Matschke, G. H. 1963. An eye lens-nutrition study of penned European wild hogs. Proc. Southeast Assoc. Fish Game Commissioners 17:20–27.

———. 1967. Aging European wild hogs by dentition. J. Wildl. Manage. 31:109–113.

Mellen, I. M. 1952. The natural history of the pig. Exposition Press Inc., New York. 157pp.

Moen, A. N. 1968. Surface temperatures and radiant heat loss from white-tailed deer. J. Wildl. Manage. 32:338–344.

Mount, L. E. 1968. Adaptation of swine. Pages 277–291 in E.S.E. Hafez, ed. Adaptation of domestic animals. Lea & Febiger, Philadelphia.

Nalbandov, V. 1976. Reproductive physiology of mammals and birds. W. H. Freeman & Co., San Francisco. 334pp.

Osborn, H. E. 1910. The age of mammals in Europe, Asia, and North America. MacMillan Co., New York. 635pp.

Phillips, R. W., and Zeller, J. H. 1943. Sexual development in small and large types of swine. Anat. Rec. 85:387–400.

Pilgrim, G. E. 1941. The dispersal of the Artiodactyla. Bio. Rev. 16:134–163.

Pine, D. S., and Gerdes, G. L. 1973. Wild pigs in Monterey County, California. California Fish Game 59:126–37.

Pond, W. G., and Maner, J. H. 1974. Swine production in temperate and tropical environments. W. H. Freeman & Co., San Francisco. 646pp.

Pullar, E. M. 1950. The wild (feral) pigs of Australia and their role in the spread of infectious diseases. Aust. Vet. J. 26(5):99–110.

———. 1953. The wild (feral) pigs of Australia: their origin, distribution and economic importance. Mem. Natl. Mus. Victoria 18:7–23.

Quinn, J. H. 1970. Special note in Soc. Vert. Paleontol. News Bull. 8:33.

Rary, J. M.; Henry, V. G.; Matschke, G. M.; and Murphee, R. L. 1968. The cytogenetics of swine in the Tellico Wildlife Management Areas, Tennessee. J. Hered. 59:201–204.

Roark, D. N. 1977. Stomach analyses of feral hogs at Hobcaw Barony, Georgetown, South Carolina. M.S. Thesis. Clemson Univ., S.C. 46pp.

Ruddle, W. D. 1975. Helminth parasites of feral swine from the Arkansas National Wildlife Refuge. M.S. Thesis. Texas A&M Univ., College Station. 43pp.

Selander, R. K.; Smith, M. H.; Yang, Y.; Johnson, W. E.; and Gentry, J. B. 1971. Biochemical polymorphism and systematics of the genus Peromyscus. I, Variation in the old field mouse (Peromyscus polionotus). Studies in Genetics 6. Univ. Texas Publ. 7103:49–90.

Sisson, S., and Grossman, J. D. 1938. The anatomy of the domestic animals. 3rd ed. W. B. Saunders Co., Philadelphia. 972pp.

Smith, W. W. 1952. Pork production. MacMillan Co., New York. 616pp.

Sorenson, A. M., Jr. 1979. Animal reproduction principles and practices. McGraw Hill, New York. 496pp.

Springer, M. D. 1977. Ecologic and economic aspects of wild hogs in Texas. Pages 37–46 in G. W. Wood, ed. Research and management of wild hog populations: proceedings of a symposium. Belle W. Baruch For. Sci. Inst., Clemson Univ., S.C.

Stribling, H. L. 1978. Radiocesium concentrations in two populations of naturally contaminated feral hogs (Sus scrofa domesticus). M.S. Thesis. Clemson Univ., Clemson, S.C. 57pp.

Sweeney, J. M. 1970. Preliminary investigations of a feral hog (Sus scrofa) population on the Savannah River Plant, South Carolina. M.S. Thesis. Univ. Georgia, Athens. 58pp.

Sweeney, J. M.; Provost, E. E.; and Sweeney, J. R. 1970. A comparison of eye lens weight and tooth irruption pattern in age determination of feral hogs (Sus scrofa). Proc. Southeast Assoc. Fish Game Commissioners 24:285–291.

Sweeney, J. M.; Sweeney, J. R.; and Provost, E. E. 1979. Reproductive biology of a feral hog population. J. Wildl. Manage. 43:555–559.

Sweeney, J. R. 1979. Ovarian activity in feral swine. Bull. South Carolina Acad. Sci. 41:74.

Thompson, R. L. 1977. Feral hogs on national wildlife refuges. Page 11–16 in G. W. Wood, ed. Research and management of wild hog populations: proceedings of a symposium. Belle W. Baruch For. Sci. Inst., Clemson Univ., S.C.

Towne, C. W., and Wentworth, E. N. 1950. Pigs from cave to cornbelt. Univ. Oklahoma Press, Norman. 305pp.

Tregear, R. T. 1965. Hair density, wind speed, and heat loss in mammals. J. Appl. Physiol. 20:796–801.

Wahlenberg, W. G. 1946. Longleaf pine. Charles Lathrop Pack For. Found. Washington, D.C. 429pp.

Wakely, P. C. 1954. Planting southern pines. U.S.D.A. For. Serv. Agric. Monogr. 18. 233pp.

Weinwandt, T. A. 1977. Pigs. Pages 182–212 *in* Unit plan for management of Mono Island Forest Reserve. For. Task Force, Puerto Rico Dept. Nat. Res., San Juan.

Wesley, F., and Klopfer, F. D. 1962. Visual discrimination learning in swine. Z. Tierpsychol. 19:93–104.

Wodzicki, K. A. 1950. Wild pig (*Sus scrofa* Linn.). Pages 227–240 *in* Introduced mammals in New Zealand: an ecological and economic survey. New Zealand Dept. Sci. Indust. Res. Bull. 98.

Wood, G. W., and Barrett, R. H. 1979. Status of wild pigs in the United States. Wildl. Soc. Bull. 7:237–246.

Wood, G. W., and Brenneman, R. E. 1977. Research and management of feral hogs on Hobcaw Barony. Pages 23–35 *in* G. W. Wood, ed. Research and management of wild hog populations: proceedings of a symposium. Belle W. Baruch For. Sci. Inst., Clemson Univ., S.C.

———. 1980. Feral hog movements and habitat utilization in coastal South Carolina. J. Wildl. Manage. 42:420–427.

Wood, G. W., and Lynn, T. E., Jr. 1977. Wild hogs in southern forests. Southern J. Appl. For. 1(2):12–17.

JAMES M. SWEENEY, Department of Forestry, University of Arkansas at Monticello, Monticello, Arkansas 71655.

JOHN R. SWEENEY, Department of Entomology and Economic Zoology, Clemson University, Clemson, South Carolina 29631.

57

Sika Deer

Cervus nippon

George A. Feldhamer

NOMENCLATURE

COMMON NAMES. Axis deer (not to be confused with *Cervus axis*), Japanese deer, sika deer, speckled deer, spotted deer

SCIENTIFIC NAME. *Cervus nippon*

SUBSPECIES. *C. n. aplodontus, C. n. hortulorum, C. n. nippon, C. n. pulchellus, C. n. sichuanicus, C. n. taiouanus,* and *C. n. yesoensis.*

The taxonomic relationships of sika deer are uncertain and the number of subspecies described by recent authors ranges from 7 to 13 (cf. Fisher et al. 1969; Whitehead 1972). Much of the early systematic work on this species was extremely confusing. See Feldhamer (1980) for a review of synonymies.

DISTRIBUTION

The original range of sika deer extended on the Asian mainland from Siberia and Manchuria south to parts of Vietnam, and throughout Japan and Formosa (figure 57.1). However, five of the seven subspecies are presently regarded as either endangered or extirpated within their native range. These subspecies now may exist only in countries where they have been introduced. Numerous introductions of sika deer have been made and the species has adapted to a variety of environmental conditions. They were introduced into parks in Great Britain as early as 1860, and today several feral populations exist there as well as in Scotland and Ireland (Page 1964). Sika deer were introduced into Denmark in 1900 (Bennetsen 1976), and free-ranging populations exist there, in many other European countries, in wildlife preserves throughout the Soviet Union, and in Morocco. Free-ranging populations that were introduced around the turn of the century are also found in Australia and New Zealand (Bentley 1967).

In the United States, sika deer have been introduced in several states. In Maryland, they were released on James Island in 1916 and today are established on the eastern shore, as well as on Chincoteague Island, Virginia. Small numbers of sika deer remain from introductions in Kansas, Oklahoma, and Wiscon-

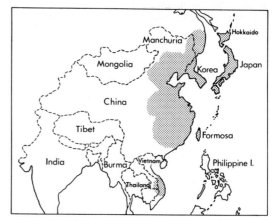

FIGURE 57.1. Original range of the sika deer (*Cervus nippon*).

sin, while fairly large herds are currently distributed throughout Texas (figure 57.2). The species occurs on game farms in several states, notably Florida and Texas.

Although sika deer are beginning to receive the attention of wildlife biologists in the United States, the majority of work on this species has been done in other countries. This information should prove valuable to the future management of sika deer in states with expanding populations of this species and to illustrate general problems often encountered with introduced species.

DESCRIPTION

Size. These are generally diminutive deer and they exhibit a large degree of variation among subspecies. Subspecific variation in size and weight is substantial: adults range in size from 86 to 188 cm at the shoulder. Sexual dimorphism is apparent. Males weigh from 50 to 140 kg, while females are generally between 40 and 60 kg.

Pelage. Coat color is chestnut brown to reddish olive, grading to a yellow brown, tan, or gray, depending on

FIGURE 57.2. Distribution of free-ranging sika deer (*Cervus nippon*) in the United States.

the subspecies (Whitehead 1972). The middorsal area is somewhat darker than the rest of the coat and forms an indistinct line from the head to the rump. Numerous white or yellowish spots are present on the upper sides. Calves are tan colored and the conspicuous white spots occur in seven or eight rows. Spots are generally more noticeable in the summer than in winter. The chin, throat, and belly are off white or gray in color. However, both sexes possess a dark throat and neck mane during the winter. The winter coat is very dense, with fine, curly wool near the hide and longer guard hair through it. Hair in the winter coat is 5 to 7 cm long and thicker than in summer pelage (Flerov 1952). The hairs of the summer coat are fine, straight, and widely spaced. In the summer, the length of hair is no more than 3 cm. In Soviet Armenia, molt occurs during March and April (Airumyan 1962), although exact timing of molt depends on the climate of the specific region. In New Zealand, males enter the rut in winter pelage, while females occasionally are in their summer coat (Kiddie 1962).

At all times of the year, a large white rump patch is evident. This erectile caudal disc is ringed with a black stripe. However, Ryukyu sika are prone to melanism, and all-black individuals sometimes occur. These melanistic animals lack the large white rump patch.

Glands. As in all cervids, metatarsal glands are present on the canon bones of sika deer. These glands are surrounded by tufts of grayish brown hairs. The tufts are about 2.5 cm in diameter. Paired suborbital or facial glands also are apparent in sika deer. There are no glands on the hind pasterns.

Hooves. The hooves of adult males are about 6 cm in length and 4 cm in width, while those of females are slightly smaller. Hooves are sharp pointed with a hard outer sheath and sharp edges. The pads are highly striated. Anomalies of the hoof may occur, and a case

of polydactylism was described by Davidson (1971) for a sika deer in New Zealand.

Antlers. The antlers are narrow, rough, and deeply grooved and stand erect over the head. The main beam has a reinforcing edge from the brow tine to the tray tine. The number of points per antler normally varies from two to five. There is a upswept brow tine that branches from the main beam about 2.5 cm above the coronet. The second brow tine is absent, or it forms only a rudimentary tubercle. A forked tine, or rarely a palmated tine, surmounts the tray tine (Flerov 1952). Antlers are relatively short, ranging from about 35 to 66 cm in height, depending upon the subspecies and local conditions. They are about 2.5 cm in diameter at the base; a spread of 40 to 50 cm is considered exceptional (Page 1964). Asymmetry, additional processes, and other abnormalities are not uncommon (Prisjazhnuk 1971).

Males are in "velvet" beginning in May and begin fraying the velvet from their antlers in late August; most will be in "hard horn" by early September. In Maryland, England, and Japan, as in most temperate areas, the antlers are shed in late April or May, and the new set begins growing immediately. The color of the velvet has been used to differentiate among subspecies of sika deer (Davidson 1973*a*).

Skull and Dentition. The maximum length of the skull is about 32 cm in adult males and about 28 cm in adult females. The head of a sika deer appears generally short, with the frontal and parietal area more rounded in appearance than that of white-tailed deer (*Odocoileus virginianus*).

The auditory bullae are smaller than the occipital condyles and the nasal bones protrude only about 0.6 cm past the maxilla. A canine tooth protrudes on each side of the maxilla (figure 57.3); the lower canines are incisiform. The molariform teeth are high crowned with cresentric ridges of enamel. The dental formula for sika deer is: 0/3, 1/1, 3/3, 3/3 = 34.

PHYSIOLOGY

Physiology of sika deer has received little attention. They apparently are able to acclimate to the cold temperatures encountered in northern Europe and the Soviet Union. Whitehead (1972) stated that sika deer in Michigan were capable of withstanding a winter temperature as low as $-40°$ C. This would depend upon the nutritional condition of individuals, availability of cover, and other factors.

Ohtaishi and Too (1974) investigated the thermoregulatory function of the "velvety" antlers of sika deer. Antler temperature paralleled often extreme changes in ambient temperatures. The temperature response of other portions of the body that were monitored did not follow this pattern. They suggested that this mechanism was important to individuals in regions where daily or seasonal fluctuations in ambient temperature were severe.

According to Flerov (1952), skeletal growth is essentially completed by 2 years of age, although male

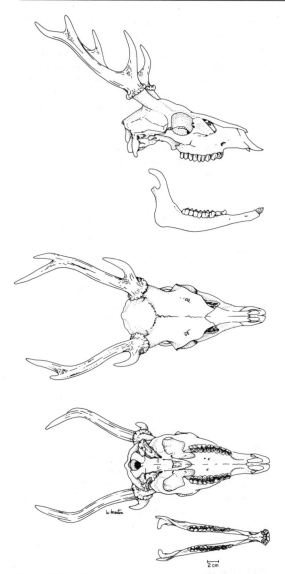

FIGURE 57.3. Skull of the sika deer (*Cervus nippon*). From top to bottom: lateral view of cranium, lateral view of mandible, dorsal view of cranium, ventral view of cranium, dorsal view of mandible.

GENETICS

Polymorphism. The karyotypes of 11 Manchurian sika deer (*C. n. hortulorum*) from a deer park in England were studied by Gustavsson and Sundt (1969). Variations between individuals in the diploid number of chromosomes ranged from 64 to 68. The normal complement of chromosomes, which was presumed to be 68, consisted almost entirely of one armed or t chromosomes. Two autosomes plus the y chromosome of males were two armed. The observed variations in number were the result of translocations of centric fusion type. Because the origins of the deer used in this study were unknown, the reason for the polymorphic systems noted in the population could not be ascertained. It was believed, however, that the park deer may have originated with individuals from different geographic areas of the native range (Gustavsson and Sundt 1969).

McDougall and Lowe (1968) conducted electrophoretic studies on the blood serum proteins of several species of deer in England. They found no variations of the transferrin patterns from the 10 sika deer examined.

Hybridization. Howard (1965) felt that competition and behavioral differences restricted hybridization between sika and red deer in New Zealand, and of 800 to 1,000 sika deer examined by Kiddie (1962), only 4 exhibited hybrid characteristics. Nonetheless, as noted by Flerov (1952), hybrids of these species apparently occur quite frequently in northern Asia and were well known to early Chinese observers. Lowe and Gardiner (1975) used multivariate analysis on measurement data from the skulls of sika deer, red deer, and their hybrids, and attempted to find distinguishing characteristics for each. They reported an extreme degree of hybridization. These contradictory viewpoints may reflect geographic differences in reproductive traits actually exhibited by the deer or may be due to the fact that "the degree of hybridization exhibited by the skulls was not reflected in the coat color or other body features, and in fact, appeared to be completely independent of them" (Lowe and Gardiner 1975:562). These investigators concluded that hybrid offspring were common between female sika deer of mainland subspecies and male red deer. The sika females of the island subspecies apparently did not hybridize with red deer, however.

REPRODUCTION

Sika deer probably attain sexual maturity between 16 and 18 months of age, and are able to breed for the first time during their second year. The pregnant sika calf described by Chapman and Horwood (1968) probably was unusual, although in England, pregnant sika calves apparently are "by no means a rare occurrence" (Horwood and Masters 1970). As in all cervids, females have a bicornate uterus and two pairs of mammae.

In Maryland, the rut generally begins in late September and lasts for one and a half or two months, with

sika deer continue to increase in weight until they are 7–10 years old, while females increase in weight until they are 4–6 years old. Calves grow rapidly and by 8 months of age may be only 5 cm shorter at the shoulder and about 9 kg lighter than their mother (Horwood and Masters 1970). Male calves produce pedicles by 6 or 7 months of age, and single "spike" antlers as yearlings. Of 136 yearling sika males examined over a 4-year period in Maryland, 127 (93 percent) had spike antlers (Feldhamer unpublished data).

Feldhamer and Chapman (1978) examined the pH of sika deer fecal droppings as a potential method for distribution or census analyses. The observed range of 3.7 in pH values was much greater than that reported for other species of deer.

the greatest activity in late October. Females remain in estrus for only a few days. If unbred, they enter estrus again following a short diestrous period. The proportion of pregnant adult females is high in most feral populations (Kaznevskii 1972; Davidson 1976), although this may not be the case in semiconfined herds.

The gestation period is about 30 weeks and calving occurs (in the Northern Hemisphere) from May through June and occasionally as late as August (Evtushevsky 1974; Maruyama et al. 1975). A single calf is the norm, although twins have been reported in the Soviet Union. At birth, the calves weigh between 4.5 and 7.0 kg and are about 50 cm in height. They do not venture from cover during the initial 3 weeks following birth. Kiddie (1962) stated that calves were weaned prior to the upcoming rut. However, Flerov (1952) claimed that nursing continued for at least 8–10 months, or almost until the next parturition. A limited number of adult females collected in the winter in Maryland were not lactating. Likewise, few adult females taken during the hunting season are lactating (Feldhamer unpublished data).

ECOLOGY

Habitat. Sika deer are found at elevations ranging from sea level to 1800 m. They prefer forested areas with dense understory, both within their native range and where introduced. However, they are quite adaptable and do well in a variety of habitat types, from deciduous and coniferous woodlands to estuarine reed beds and similar wet areas. In Maryland, they are generally found in association with freshwater marshes.

Herd Composition. The species is not gregarious and single animals may be seen almost as often as small herds. Dzieciolowski (1979) noted only one or two animals in half of the observations of aggregations of sika deer observed. The largest aggregation was 12 animals. Throughout most of the year, adult males are solitary. During the calving season, females and their young form groups of 2 to 3, possibly with the previous year's offspring, and remain separated from other members of the population. Ito (1968) felt that the temporary female groups, as opposed to the rather unstable male groups, displayed the only "clear social organization." Group size is thus very variable. During the summer in Japan, Shibata (1969) observed a total of 68 animals in 9 groups of from 3 to 15. For the 300–350 sika deer on Askold Island, Soviet Union, groups of 2 to 10 animals were observed more often than groups of more than 10 (Prisjazhnuk and Prisjazhnuk 1974). In a population studied by Ito (1968), the mean number of animals per group varied in relation to season, topography, and vegetative type. The average number per herd was 1.7 in the forest, while it was 5.8 in open fields, with occasional congregations of 40 to 50 animals. These concentrations are not because of gregariousness of the deer, but are a result of a common source of forage. The groups tend to disperse as individuals bed down for the day. Ito estimated that the density on his study area, Kinkazan Island, Japan, ranged from about 0.07 per hectare to 1.6 per ha.

The sex ratio of calves is probably 1:1 (Bennetsen 1976; Davidson 1976). The ratio of males:females:calves reported on Askold Island was 1:3.6:1.3. Horwood and Masters (1970) reported the following male to female sex ratios for three age classes of sika deer: fetal 90.0:100, calves 82.1:100, and adults 86.2:100. However, the difficulties involved in census sometimes precluded recognition of males in the older age categories. An unbalanced sex ratio observed on Kinkazan Island also was considered by Ito (1976) to be the result of misidentification by the observers.

Movements. Feldhamer et al. (in preparation) found that the home range of a radio-collared female sika deer in Maryland was 73.5 ha and was very stable during a two-year period. A male collared at the same time had a home range of 151.0 ha. The male was killed 4.2 km from the original point of capture. Both animals were about 10 months old when collared. Davidson (1979) found that the majority of 54 tagged sika deer in New Zealand remained in, or returned to, the area in which they were tagged. Mean distances moved were: yearling males, 2.5 km; yearling females, 1.9 km; adult males, 5.9 km; and adult females, 1.7 km. However, Maruyama et al. (1978) found that the home ranges of two radio-collared sika bucks were only 6.5 and 19.0 ha, respectively.

Altitudinal movements of sika deer have been investigated in mountainous areas of Japan. Summer ranges were generally larger than winter ranges and seasonal factors affecting movements throughout the year included snowfall and subsequent melt, differential development of forage, reproductive periods, and plant defoliation (Maruyama et al. 1976; Miura 1974). These movements involved an elevational range of about 700 m.

Around plantation areas in Japan, Maruyama and Sekiyama (1976) reported that more deer moved through a "passage" forest, 50 m in width, and through older aged stands, than through forested areas in earlier stages of development. With the exception of transient males, however, no migrational movements of sika deer have been noted in New Zealand (Wodzicki 1950).

Davidson (1973*a*) calculated initial dispersal rates of 0.6–0.9 km/year for sika deer in New Zealand up to 1950. Dispersal rates increased to 1.4/1.5 km/year after that, possibly due simply to increased familiarity with the species, which resulted in increased sighting records. Feldhamer et al. (1978) noted a dispersal rate of 0.8 km/year for sika deer in Maryland over a 60-year period.

FOOD HABITS

Sika deer, like all cervids, have a four-chambered, ruminating stomach. They forage primarily from dusk to dawn, although they also may be active at times during the day. They are highly adaptable in their feeding habits, and consume a wide variety of vegetative species.

The diversity and seasonal fluctuations in feeding

habits of sika deer in the Soviet Union, Japan, and New Zealand were reviewed by Feldhamer (1980).

Little work has been done on the food habits and preferences of sika deer in the United States. In Maryland, Flyger and Davis (1964) felt that poison ivy (*Rhus radicans*), Japanese honeysuckle (*Lonicera japonica*), and greenbrier (*Smilax* sp.) may be selected food species. They also reported that pokeweed (*Phytolacea americana*), wax myrtle (*Myrica* sp.), American holly (*Ilex opaca*), bark of loblolly pine (*Pinus taeda*) and large-toothed aspen (*Populus grandidentata*), and cordgrass (*Spartina patens*) were browsed, although some of these species may have been low-preference "starvation" foods. During the spring and early summer, sika deer are very partial to emerging agricultural crops, especially soybeans and corn.

The feeding strategy of sika deer was summarized by Kiddie (1962:25) as follows: "A close feeding pattern, associated with smaller body size, contributes to its superior ability to obtain sustenance. In this, it is assisted by its utilization, during late spring, summer and autumn, of grassland, particularly of weed species and shrubs. . . . This seasonal change in the use of resources results in the maximum growth of those browse-intolerant species and seedlings in the forest which are later utilized for winter feeding."

Many of these points were also noted by Davidson (1973b). Thus, although it is not an overtly aggressive colonizer, the feeding strategy and manner of exploiting the habitat demonstrated by this species have caused it to displace red deer (*Cervus elaphus*) from formerly established ranges in areas of New Zealand (Kean 1959). In a comparable manner, sika deer also may be capable of displacing native white-tailed deer where they are sympatric in Maryland, Texas, and other states (Cook 1977; Feldhamer et al. 1978).

BEHAVIOR

Reproductive Behavior. During the summer, adult males begin to establish territories for the upcoming rut. Males dig holes about a meter wide and 0.3 m deep with their antlers and forefeet, into which they urinate frequently. These holes delimit the boundaries of individual territories, as does the thrashing of surrounding ground cover with the antlers. Depending upon such factors as population density or topography, Kiddie (1962) felt that 2 ha represented the size of the largest manageable territory. Horwood and Masters (1970), however, felt that sika bucks on their study area may have defended territories from 8 to 12 ha in size. Subordinate bucks are often found around the periphery of territories. During the rut, behavioral interactions between rival males are not merely ritualistic; Kiddie (1962) reported that fights were often "spectacular" for their "speed and fierceness." Fighting with antlers and hooves sometimes resulted in fatal wounds, and blinding, torn nostrils, and severe puncture wounds were not uncommon results of territorial challenges. Following the rut, some adult bucks in Maryland have severe head wounds.

Sika deer are polygamous and the "harem" of a successful territorial male may number as many as 12 females. Bucks follow available females and attempt to drive them to their territorial areas, where mating takes place. The male may mount a female several times unsuccessfully, each attempt followed by intervals of 10 to 15 minutes of relative inactivity. A successful copulation lasts only about 4 seconds and may occur at any time of the day or night. Bucks remain on their territories and often do not feed until the latter stages of the rutting season. Females appear to feed normally throughout this period.

Pregnant or lactating females apparently engage in behavior that encourages active dispersal and minimum intrusion of their calving area by other females. During this period, adult does have been observed to face each other and strongly stomp the ground with the forefeet (Maruyama et al. 1975). Dispersal also may be facilitated by the "snorting voice" vocalizations of females.

Vocalizations. This species probably is one of the most vocal members of the Cervidae. Kawamura (1957) recorded 10 different vocalizations from a herd of sika deer, although only 2 were considered to be common. Kiddie (1962) noted up to 5 distinct calls from individual deer. Vocalizations range from soft whistles between females to "goatlike bleats" from doe to fawn and "soft horselike neighs" from fawn to doe. During the rut, males may emit "blood-curdling screams" (Flyger and Davis 1964). The alarm call, which may be uttered by either sex and is audible up to 0.8 km, has been described by various investigators as a "sharp scream," a "high-pitched whistle followed by a gutteral bark," a "chirplike sound," and a "snorting voice."

MORTALITY

Ohtaishi (1978) compared the age structures of sika deer from three different herds in Japan. The maximum (ecological) longevity of wild, unconfined deer was 12 years, with no difference between the sexes. Male sika from a protected, unhunted herd reached a maximum age of 13 years, females 17 years. In a herd that was protected, unhunted, and fed, males attained a maximum (physiological) longevity of 21 years, females 25 years.

Ryabov (1974) reported that 80 percent of the diet of wolves (*Canis lupus*) on the Khopersk Reserve consisted of sika deer. The loss of deer was not excessive, however, and the predation may have provided an important damping effect on otherwise excessive increases in density of deer populations (Ryabov 1973). Deaths of males, females, and calves on the reserve occurred in the ratio 1:1:1.5 (Kaznevskii 1974), although not all were attributed to predation.

Calves also may fall prey to smaller carnivores, including foxes (*Vulpes* sp.), lynx (*Felix lynx*), or large raptors. Isolated instances of calves being killed by flocks of crows (*Corvus levaillantii*) have been reported (Shibata 1969). Kaznevskii (1972) estimated that under normal conditions, 50 percent of each calf crop failed to survive the first year, in part because of predation. However, populations of sika deer generally do not appear to be adversely affected by normal predation pressure.

Parasites and Diseases. Ovcharenko (1963) distinguished two groups of nematodes afflicting sika deer—those such as *Setaria cervi* and *Onchocerca* sp., present at all ages, and those that are not specific for cervids and generally do not occur in deer more than three years of age. This latter group includes species such as *Nematodirus helvetianus* and *Capillaria* sp.

A new species of parasitic arthropod, *Haemaphysalis mageshimaensis,* was described by Saito and Hoogstraal (1973). They reported that this tick was specific to sika deer in Mage Island, Japan. Ohbayashi (1966) found two previously unreported species of nematode, *Rinadia japonica* and *Spiculopteragia yamashita,* in the small intestine of a sika deer from Hokkaido, Japan. Mogi (1977) reported on a new species of dipteran from a sika deer from central Honshu, Japan.

Of six sika deer examined in Czechoslovakia, one simple infection of footworm (*Wehrdickmansia cervipedis*) and one mixed infection of *W. cervipedis* and *Onchocerca flexuosa* were found (Dykova and Blazek 1972).

Sika deer also may be infected with several species of spirochete. Vysotskii and Ryashchenko (1961) examined blood samples from 450 sika deer and found that 38 reacted to *Leptospirosis pomona* antigen, 23 to *L. tarrasowi* antigens, and 4 to *L. saxkoebing* antigens. In New Zealand, however, sika deer are of little significance as reservoirs of *Leptospirosis* sp., or of *Brucella abortus* and *Salmonella* sp. (Daniel 1967), rickettsial diseases commonly associated with domestic livestock. Piroplasmosis was reported by Flerov (1952) to be extremely widespread among sika deer populations.

Another disease of domestic ruminants is malignant catarrhal fever (MCF). This viral infection has been described previously as occurring in axis deer (*Cervus axis*) in Texas, black-tailed deer (*Odocoileus hemionus*) in Colorado, and white-tailed deer in New Jersey. Sanford et al. (1977) described lesions of the head, eye, and internal organs caused by MCF in 3 captive sika deer that had been in contact with domestic species. Antibodies to *Myxovirus parainfluenza* 3, a virus associated with humans and domestic livestock, were detected in only 1 of 12 sika deer tested from the Maryland-Virginia area (Shah et al. 1965).

In Maryland, a mass mortality of sika deer on James Island was initially attributed to pine oil poisoning and subsequent malnutrition (Hayes and Shotts 1959). However, Christian et al. (1960) refuted this conclusion. Although they documented two diseases that occurred in members of the population (glomerulonephritis and inclusion hepatitis), they felt that these were effects rather than causes of population change. They attributed the mass mortality to "adrenocortical, sympatho-adrenal, and other metabolic responses" (p. 94) that resulted from extremely high population densities and associated "social strife." This has also been cited as a causative factor in the population changes of several other mammalian species.

AGE DETERMINATION

While calves up to 8 months of age may be distinguished from adults on the basis of size, a more precise estimation of the age of sika deer may be made by the degree of development and wear of dentition. Wear and development patterns for sika deer are probably the same as those determined for red deer by Lowe (1967). These patterns have not yet been determined for sika deer of known age. Calves are born with incisors and lower canines already erupted, and within a matter of days the premolars appear. By 5 months of age, the first molar is in place. The second molar appears at 14 to 15 months of age, while the central pair of deciduous incisors is replaced by permanent, markedly larger teeth. By about 2 years of age, all the "milk" dentition is replaced and the third molar erupts (Horwood and Masters 1970). The length of the jaw may also serve as a basis to distinguish calves from adults. Jaws are 167 mm or less in length in calves and 185 mm or more in adults (Chapman and Horwood 1968).

Ohtaishi (1975) estimated the age of sika deer maintained under park conditions by the degree of wear on the first incisor. Three age classes were determined for the deciduous incisor of calves, and 16 age classes for the permanent incisor of adults.

Prisjazhnuk (1968) sectioned the first incisor from 109 sika deer of known age. At birth, the cementum of the roots of these teeth has one light and one dark band. An additional light band is deposited during the first summer and a dark band during the first winter of life. Subsequent light and dark bands are deposited yearly, and serve as an indication of age for this species. In a like manner, Ueckermann and Scholz (1971) recommended a count of the layers in the cementum of the first molar to determine the age of sika deer.

Feldhamer and Chapman (1980) found a significant curvilinear correlation between estimated age and eye lens weight. However, the mean lens weight for each age class was distinct only for calves and yearlings.

ECONOMIC STATUS AND MANAGEMENT

The various management strategies employed relative to sika deer have varied according to whether the goal was to preserve an endangered subspecies, harvest a popular game animal, or reduce populations. In many areas of their native range, no active management was practiced. Populations of the mainland subspecies have been drastically reduced or extirpated, with the exception of *C. n. hortulorum.* Considering the island subspecies, since January 1955, Japan has denoted *C. n.* [*keramae*] *nippon* as a "national monument" and afforded it full protection (Kuroda 1965). Other measures taken to promote the continued survival of this population on the three small islands it inhabits included construction of water catchment basins, planting forage crops, and elimination of competing species.

One of the reasons native populations declined so rapidly was relentless harvesting of deer for their antlers. Antlers in velvet were used as aphrodisiacs; in 1929 a pair would sell for approximately $100 (Fisher et al. 1969). Currently in the Soviet Union, 10,000–20,000 deer, mainly red deer but including sika deer, are maintained in captivity. The velvet is removed from the antlers and ground up in solution. This solution is given to patients in a series of 10 injections and is believed to act as a general stimulant (V. L. Lavrov personal communication).

It is ironic that while sika deer are extirpated throughout much of their original range, the species has expanded to pest proportions where introduced. Economic disadvantages include losses of mature trees caused by girdling of bark, destruction of seedlings and threats to forest regeneration, and losses incurred through crop depredation. This is exemplified in New Zealand, with no native species of deer. Deer were first introduced there in 1851; sika deer in 1900. Presently, eight species of cervids abound. Population densities are so numerous in many areas that extreme damage to the habitat has resulted. Management efforts to minimize damage to watershed protection forests and pasturelands have included liberal hunting regulations, large-scale poisoning campaigns, and aerial culling operations (Daniel 1962). Although the situation in Denmark has not reached such critical proportions, management there also involves manipulation of bag limits and culling operations to limit population densities and reduce crop depredation. On the Khopersk Reserve, USSR, Ryabov (1973) suggested that populations of wolves be allowed to regain their former densities in order to help control populations of sika deer. In Ireland, sika deer damage mature trees by removing or scoring the bark. Larner (n.d.) reported that periods of major damage were associated with the rut and just prior to antler drop. About 30 percent of all trees showed some degree of damage.

Sika deer in Maryland illustrate many of the classical management problems associated with an introduced species. To a large extent, how sika deer are viewed from a management and economic standpoint depends on the time of year. In the spring and summer, they are often serious crop depredators. Crop depredation permits are issued and sika (as well as white-tailed) deer are shot in an effort to minimize damage. In the fall, however, sika deer are held in high esteem by many hunters and the revenues gained through the various aspects of sport hunting are a considerable boost to local economies.

Stocking. An introduction may offer a new game animal to an area, provide a "last refuge" for a threatened species, or provide a more diverse fauna for nonconsumptive users. However, the negative aspects of an introduction may outweigh any positive benefits. Several of the reasons for not introducing exotic species, summarized by Bump (1968) and Giles (1978) and noted above, include: (1) introductions usually fail, (2) population density of the exotic may increase to pest proportions and/or displace native species, (3) the exotic may alter or degrade the habitat to such an extent that it cannot be returned to the original condition, (4) there is always the potential for associated exotic diseases and parasites, (5) the costs of selection, transportation, quarantine, and subsequent monitoring of exotics is high, and (6) if the stocking is successful, a native environmental complex is significantly different—and will probably never return to its historical condition.

In Texas, sika deer are classified as nongame animals and legally have the status of domestic livestock. As such, it is feared that their densities will increase, and because of the diversity of their diets, they may displace native white-tailed deer on overgrazed ranges (Cook 1977). Generally, Box (1968:19) felt that the following habitat criteria should be met prior to the introduction of exotics: (1) there is vegetation in excess of that needed to control erosion following the grazing season, (2) excess vegetation is not being utilized by native species, (3) excess vegetation is palatable to exotics, (4) the abundance of unpalatable plants is increasing, (5) a change in animal species will increase grazing pressure on plants that are increasing in abundance, (6) exotics will not eliminate key plant species for native wildlife, (7) exotic animals can be controlled to obtain balanced grazing pressure, (8) exotics and native animals provide a balanced use of vegetation, and (9) exotics can be confined and numbers legally controlled.

Although the habitat is not overgrazed, sika deer in Maryland still may be displacing white-tails in areas of sympatry (Feldhamer et al. 1978) (figure 57.4).

Census Techniques. On Maryland's Eastern Shore, the extremely thick, tangled understory and marshy habitat occupied by sika deer preclude use of many of the standard census techniques. Direct counts are of

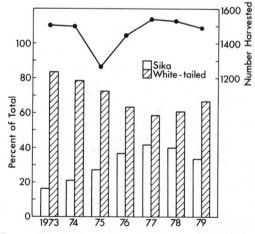

FIGURE 57.4. Relative harvests of sika deer (*Cervus nippon*) and white-tailed deer (*Odocoileus virginianus*) in Dorchester County, Maryland, 1973–79.

little value; pellet or track counts are most practical. These counts provide only a relative index of abundance through time. The habitat also limits the use of Stephenson (Michigan) traps or tranquilizer darts for the capture of deer. Food is available throughout the year and the deer are in excellent physical condition all year. Thus, it is difficult to entice them into traps regardless of the quality or palatability of the bait. Darting is also hampered by the habitat characteristics. Although sika deer graze in open agricultural fields at night, they are extremely wary and usually remain within several meters of adjoining woodland. Because of much shooting, either by poachers or as a method to control crop depredations, the deer often flee quickly at the approach of research personnel. They are somewhat less wary in heavy cover and may be approached more closely. However, darting through heavy underbrush is very difficult. Use of a transmitter dart could help alleviate some of the problems. A corral trap also was of limited value in taking sika deer.

Commercial Hunting Programs. Sika deer are hunted on a commercial basis in Texas, where they were first introduced for this purpose in 1939. They have also provided commercial revenue in Pennsylvania, Ohio, Tennessee, and other eastern states. Schreiner (1968) outlined the economics of commercial hunting and noted that the capital investment is quite high for the rancher in fencing and purchase of animals. In 1967, a sika deer could be purchased for $100–250. Trophy bucks are generally three to four years of age; the range in trophy fees charged for a mature sika at this time was $200–400.

CURRENT RESEARCH AND MANAGEMENT NEEDS

Populations of sika deer offer excellent opportunities to investigate competitive relationships between exotic and native fauna, and the impact of introductions on native habitat. As previously indicated, a taxonomic revision of the species may clarify many uncertainties that currently exist. Little or no information is available on many basic life history parameters of sika deer, either within their native range, in the United States, or in other countries in which introductions were made.

A study of the food habits, preferences, and nutritive requirements of sika deer, especially in overbrowsed habitats, could explain their ability to adapt to diverse areas and apparently to displace other sympatric species of deer.

James Peek and P. Quentin Tomich critically reviewed an earlier draft of the manuscript. Much of the research done in Maryland was supported by Federal Aid in Wildlife Restoration project W-49-R. This is contribution no. 1000, Appalachian Environmental Laboratory, Center for Environmental and Estuarine Studies, University of Maryland.

LITERATURE CITED

Airumyan, V. A. 1962. The question of acclimatization of spotted deer in Armenia. *In* Vysshaya Shkola [Problems of ecology] (Moscow) 7–8. (From Bio. Abstr. 44, no. 1 [1963], from Referat. Zhur, Bio., Trans. no. 20I322. Translation of Russian abstract.)

Bennetsen, E. 1976. Sikavildtet (*Cervus nippon*) i Danmark. Danske Vildtundersogelser no. 25. 32pp. (In Danish; English summary.)

Bentley, A. 1967. An introduction to the deer of Australia. Hawthorn Press, Melbourne. 224pp.

Box, T. W. 1968. Introduced animals and their implications in range vegetation management. Pages 17–20 *in* Introductions of exotic animals: ecological and socioeconomic considerations. 18th Meet. Am. Inst. Bio. Sci., Texas A & M Univ. 25pp.

Bump, G. 1968. Exotics and the role of the state-federal foreign game investigation program. Pages 5–8 *in* Introduction of exotic animals: ecological and socioeconomic considerations. 18th Meet. Am. Inst. Bio. Sci., Texas A & M Univ. 25pp.

Chapman, D. I., and Horwood, M. T. 1968. Pregnancy in a sika deer calf, *Cervus nippon.* J. Zool. 155:227–228.

Christian, J. J.; Flyger, V.; and Davis, D. E. 1960. Factors in the mass mortality of a herd of sika deer, *Cervus nippon.* Chesapeake Sci. 1:79–95.

Cook, R. L. 1977. Exotic big game animals in Texas. Texas Parks Wildl. Dept. Staff Rep. 18pp.

Daniel, M. J. 1962. Control of introduced deer in New Zealand. Nature 194:527–528.

———. 1967. A survey of diseases in fallow, Virginia and Japanese deer, chamois, tahr and feral goats and pigs in New Zealand. New Zealand J. Sci. 10:949–963.

Davidson, M. M. 1971. A case of polydactylism in sika deer in New Zealand. J. Wildl. Dis. 7:109–110.

———. 1973*a.* Characteristics, liberation and dispersal of sika deer (*Cervus nippon*) in New Zealand. New Zealand J. For. Sci. 3:153–180.

———. 1973*b.* Use of habitat by sika deer. Pages 55–67 *in* Assessment and management of introduced animals in New Zealand forests. New Zealand For. Serv., For. Res. Inst., Symp. 14.

———. 1976. Season of parturition and fawning percentages of sika deer (*Cervus nippon*) in New Zealand. New Zealand J. For. Sci. 5:355–357.

———. 1979. Movement of marked sika (*Cervus nippon*) and red deer (*Cervus elaphus*) in Central North Island, New Zealand. New Zealand J. For. Sci. 9:77–88.

Dykova, I., and Blazek, K. 1972. Subcutaneous filariasis in red deer. Acta Vet. 41:117–124.

Dzieciolowski, R. 1979. Structure and spatial organization of deer populations. Acta Theriol. 24:3–21.

Evtushevsky, N. N. 1974. Reproduction of *Cervus nippon hortulorum* SW. under conditions of the middle Dnieper area. Vestn. Zool. 7:23–28. (In Russian; English summary.)

Feldhamer, G. A. 1980. *Cervus nippon.* Mammal. Species no. 128:1–7.

Feldhamer, G. A., and Chapman, J. A. 1978. Fecal pH of sika and white-tailed deer. Proc. Pennsylvania Acad. Sci. 52:197–198.

———. 1980. Evaluation of the eye lens method for age determination in sika deer. Acta Theriol. 25:239–244.

Feldhamer, G. A.; Chapman, J. A.; and Miller, R. L. 1978. Sika deer and white-tailed deer on Maryland's eastern shore. Wildl. Soc. Bull. 6:155–157.

Fisher, J.; Simon, N.; and Vincent, J. 1969. Wildlife in danger. Viking Press, Inc., New York. 368pp.

Flerov, K. K. 1952. Musk deer and deer. *In* Fauna of USSR: mammals. Acad. Sci., Moscow. Vol. 1, no. 2. 257pp.

Flyger, V., and Davis, N. W. 1964. Distribution of sika deer (*Cervus nippon*) in Maryland and Virginia in 1962. Chesapeake Sci. 5:212–213.

Giles, R. H., Jr. 1978. Wildlife management. W. H. Freeman & Co., San Francisco. 416pp.

Gustavsson, I., and Sundt, C. O. 1969. Three polymorphic chromosome systems of centric fusion type in a population of Manchurian sika deer (*Cervus nippon hortulorum* Swinhoe). Chromosoma 28:245-254.

Hayes, F. A., and Shotts, E. B. 1959. Pine oil poisoning in sika deer. Southeastern Vet. 10:34-39.

Horwood, M. T., and Masters, E. H. 1970. Sika deer. British Deer Soc. 29pp.

Howard, W. E. 1965. Control of introduced mammals in New Zealand. New Zealand Dept. Sci. Industries Res. Info. Serv. no. 45. 96pp.

Ito, T. 1968. Ecological studies on the Japanese deer, *Cervus nippon centralis* Kishida, on Kinkazan Island. Part 2: Census and herd size. Bull. Marine Bio. Stn. Asamuchi, Tohoku Univ. 13:139-149.

———. 1976. Contour line transect census for sika deer on Kinkazan Island. Saito Ho-on Kai Mus. Res. Bull. no. 44:31-38.

Kawamura, S. 1957. Deer in the Nara Park. *In* K. Imanishi, ed. Nippon Dobutsuki IV. Kobunsha, Tokyo. 164pp. (In Japanese.) (Cited in Maruyama et al. 1975, original not seen.)

Kaznevskii, P. F. 1972. Reproduction of the herd of sika deer (*Cervus nippon* Temm.) in the Khopersk Reserve. Byull. Mosk. O-Va. Ispyt. Prir. Otd. Bio. 77:58-60. (In Russian; English summary.)

———. 1974. Sika deer mortality in the Khopersk Reserve. Byull. Mosk. O-Va. Ispyt. Prir. Otd. Bio. 79:132-134. (In Russian only.)

Kean, R. I. 1959. Ecology of the larger wildlife mammals of New Zealand. New Zealand Sci. Rev. 17:35-37.

Kiddie, D. G. 1962. The sika deer (*Cervus nippon*) in New Zealand. New Zealand For. Ser. Info. Ser. no. 44. 35pp.

Kuroda, N. 1965. On the protection of *Cervus nippon keramae*. J. Mammal. Soc. Jap. 2:109-110. (In Japanese; English summary.)

Larner, J. B. n.d. Sika deer damage to mature woodlands of southwestern Ireland. Cong. Game Bio. 13:192-202.

Lowe, V.P.W. 1967. Teeth as indicators of age with special reference to Red deer (*Cervus elaphus*) of known age from Rhum. J. Zool. (London) 152:137-153.

Lowe, V.P.W., and Gardiner, A. S. 1975. Hybridization between red deer (*Cervus elaphus*) and sika deer (*Cervus nippon*) with particular reference to stocks in N. W. England. J. Zool. (London) 177:553-566.

McDougall, E. I., and Lowe, V.P.W. 1968. Transferrin polymorphism and serum proteins of some British deer. J. Zool. (London) 155:131-140.

Maruyama, N.; Ito, T.; Tamura, M.; Miyaki, M.; Abe, S.; Takatsuki, S.; and Naito, T. 1978. Application of radiotelemetry to sika deer on Kinkazan Island. J. Mammal. Soc. Jap. 7:189-198.

Maruyama, N., and Sekiyama, K. 1976. Effect of passage forest on sika deer. J. Mammal. Soc. Jap. 7:9-15. (In Japanese; English summary.)

Maruyama, N.; Sugimori, F.; Totake, Y.; and Miura, S. 1975. The snorting voice of the sika deer in relation to its spacing distribution. J. Mammal. Soc. Jap. 6:155-162.

Maruyama, N.; Totake, Y.; and Okabayaski, R. 1976. Seasonal movements of sika deer in Omate-Nikko, Tochigi Prefecture. J. Mammal. Soc. Jap. 6:187-198.

Miura, S. 1974. On the seasonal movements of sika deer populations in Mt. Hinokiboramara. J. Mammal. Soc. Jap. 6:51-66. (In Japanese; English summary.)

Mogi, M. 1977. On a species of *Lipotena* (Diptera: Hippoboseidae) newly discovered from the Japanese deer. Jap. J. Sanit. Zool. 28:449-450.

Ohbayashi, M. 1966. On *Spiculopteragia yamashitai* n. sp. and *Rinadia japonica* n. sp. (Nematoda: Trichostrongylidae) from the Yeso Island deer *Cervus nippon yesoensis* (Heude). Jap. J. Vet. Res. 14:117-129.

Ohtaishi, N. 1975. Wear on incisiform teeth as an index to the age of Japanese deer at Nara Park. Annu. Rep. Nara Deer Res. Assoc. Pp. 71-82.

———. 1978. Ecological and physiological longevity in mammals from the age structures of Japanese deer. J. Mammal. Soc. Jap. 7:130-134.

Ohtaishi, N., and Too, K. 1974. The possible thermoregulatory function and its character of the velvety antlers in the Japanese deer (*Cervus nippon*). J. Mammal. Soc. Jap. 6:1-11. (In Japanese; English summary.)

Ovcharenko, D. A. 1963. Age changes of helmintho-fauna of the speckled deer (*Cervus nippon hortulorum*) under park maintenance in the Far East. Vest. Leningradsk. Univ. Ser. Bio. 18:5-11. (In Russian; English summary.)

Page, F. J. T. 1964. Order Artiodactyla. Pages 400-422 *in* H. N. Southern, ed. The handbook of British mammals. Blackwell Sci. Publ., Oxford. 465pp.

Prisjazhnuk, V. E. 1968. Determination of age of sika deer from the lamination on the incisors cement. Mosk. Obs. Ispyt. Prir. 73:51-63. (In Russian; English summary.)

———. 1971. Cases of asymmetry, abnormal structure, and horn injury in the wild axis deer (*Cervus nippon* T.). Zool. Zh. 50:1380-1387. (In Russian, English summary.)

Prisjazhnuk, V. E., and Prisjazhnuk, N. P. 1974. Sika deer (*Cervus nippon* Temm.) on Askold Island. Byull. Mosk. O-Va. Ispyt. Prir. Otd. Bio. 79:16-27. (In Russian; English summary.)

Ryabov, L. S. 1973. Wolves in the forests of the Khoper Basin. Byull. Mosk. O-Va. Ispyt. Prir. Otd. Bio. 78:12-16. (In Russian; English summary.)

———. 1974. The attitude of wolves towards domestic animals and wild ungulates in the region of the Khopersk Nature Reserve. Byull. Mosk. O-Va. Ispyt. Prir. Otd. Bio. 79:6-16. (In Russian; English summary.)

Saito, Y., and Hoogstraal, H. 1973. *Haemaphysalis* (*Kaiseriana*) *mageshimaensis* n. sp. (Ixodoidea: Ixodidae), a Japanese deer parasite with bisexual and parthenogenetic reproduction. J. Parasitol. 59:569-578.

Sanford, S. E.; Little, P. B.; and Rapley, W. A. 1977. The gross and histopathologic lesions of malignant catarrhal fever in three captive sika deer (*Cervus nippon*) in southern Ontario. J. Wildl. Dis. 13:29-32.

Schreiner, C. 1968. Uses of exotic animals in a commercial hunting program. Pages 13-16 *in* Introductions of exotic animals: ecological and socioeconomic considerations. 18th Meet. Am. Inst. Bio. Sci., Texas A & M Univ.

Shah, K. V.; Schaller, G. B.; Flyger, V.; and Sherman, C. M. 1965. Antibodies to *Myxovirus parainfluenza* 3 in sera of wild deer. Bull. Wildl. Dis. Assoc. 1:31-32.

Shibata, T. 1969. Observations of the diurnal activity of the sika deer, *Cervus nippon centralis* Kishida, on the island of Kinkazan in northeastern Japan. Sci. Rep. Yokosuka City Mus. 15:97-111. (In Japanese; English summary.)

Ueckermann, E., and Scholz, H. 1971. A note on the tooth development and aging of sika deer (*Sika nippon*). Z. Jagdwiss 17:49-52. (In German; English and French summaries.)

Vysotskii, B. V., and Ryashchenko, L. P. 1961. Leptospirosis in spotted deer, *Cervus nippon*. J. Microbio. (Moscow) 5:67-68. (In Russian; English summary.)

Whitehead, G. K. 1972. Deer of the world. Viking Press, New York. 194pp.

Wodzicki, K. A. 1950. Introduced mammals of New Zealand. New Zealand Dept. Sci. Industry Res. Bull. no. 98. 255pp.

George A. Feldhamer, Appalachian Environmental Laboratory, Center for Environmental and Estuarine Studies, University of Maryland, Frostburg State College Campus, Frostburg, Maryland 21532.

Appendixes, Glossary, Index

Appendix 1

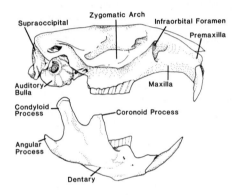

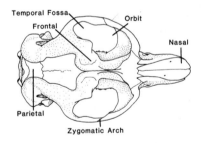

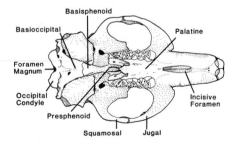

Cranium and mandible of a muskrat (*Ondatra zibethicus*),
showing the location of bones commonly cited in the text.
Top: lateral view of cranium and mandible. Middle: dorsal
view of cranium. Bottom: ventral view of cranium.

Appendix 2

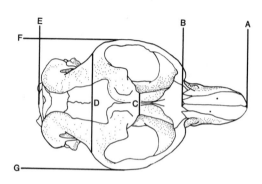

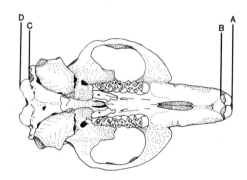

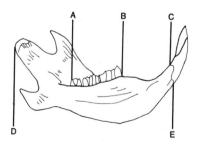

Cranium and mandible of a muskrat (*Ondatra zibethicus*), showing the common measurements taken from a mammalian skull. Top: dorsal view of cranium. Middle: ventral view of cranium. Bottom: lateral view of mandible.

top
greatest length of skull = A → E
nasal length = A → B
least interorbital breadth = C
breadth of braincase = D
zygomatic breadth = F → G

middle
condylobasal length = A → D
basal length = A → C
basilar length = B → C

bottom
mandibular diastema = B → C
mandible length = D → E
mandibular tooth row = A → B

Glossary

ABOMASUM. The fourth, or digestive, stomach of a ruminant.

ACETABULUM. The socket in the pelvic girdle at the point where the ilium, ischium, and pubis meet, and into which the head of the femur articulates.

ADULT PELAGE. The hair covering characteristic of an adult.

AGENESIS. Lack of, faulty, or incomplete development.

ALPHA FEMALE. Dominant female.

ALPHA MALE. Dominant male.

ALTRICIAL. Pertaining to newborn mammals that initially are completely dependent upon parental care for survival; usually born blind and naked.

ANESTRUS. The period of sexual quiescence between two periods of sexual activity in cyclically breeding mammals.

ANKYLOSED. United or jointed together, as bones or hind parts joined to form a single bone or part.

ANOREXIA. Loss of appetite.

ANTHROPOGENIC. Relating to natural processes of animal behavior interpreted in human terms.

ASPHYXIA. Deficiency of oxygen, excess of carbon dioxide in the blood.

ATAXIA. Lack of normal coordination of voluntary muscular movements.

BACULUM. A bone in the penis of certain mammals, principally carnivores, rodents, pinnipeds, and primates.

BALEEN. Whalebone; the cornified epithelial plates suspended from the upper jaws of mysticete whales, used to strain food from the ocean waters.

BEDDING SITES. Sites at which an animal sleeps (usually refers to ungulates).

BIFURCATED. Divided into two branches, forked.

BIODEGRADATION. The degrading of the biosphere.

BLASTOCYST. An early stage of an embryo.

BOLE. The trunk or stem of a tree.

BOSS. A knoblike process on the horns of a mammal.

BRADYCARDIA. The condition characterized by an abnormally slow heart beat.

BREACHING. The leaping of a whale out of the water.

BUREAU OF BIOLOGICAL SURVEY. The precursor of the U.S. Fish and Wildlife Service.

BURSA. A sac or saclike cavity.

CALCAR. A spurlike projection on the ankle of chiropterans.

CALLOSITIES. Barnacles and other attachments on a whale, usually in the facial region.

CARNASSIAL TEETH. The last upper premolar and first lower molar of many carnivorous mammals with secondont dentition; the largest pair of bladelike teeth that occlude with a scissorslike action.

CATABOLISM. Destructive or aerobic metabolism.

CEMENTUM. The layer of bonelike material covering the root of a tooth, sometimes termed *cement*. It is deposited throughout life and can be used to determine age in many mammals.

CERVICAL CANAL. The passage to the uterus.

CERVIX. The necklike, constricted portion of the uterus.

CHEEK TEETH. Collectively, the premolars and molars, or any teeth posterior to the canines (generally used when premolars and molars cannot be readily differentiated).

CINERADIOGRAPH. A photograph made by a process similar to an X-ray process.

CIRCADIAN. Pertaining to behavioral or physiological rhythms associated with a 24-hour cycle.

CIRQUE. Geologically, a steep, hollow excavation on a mountainside due to glacial erosion.

CLAVICLES. Bones that make up part of the pectoral girdle.

COIL-SPRING TRAP. A type of steel leg trap that utilizes coil springs to maintain pressure on the trap jaws.

COLLOID. A solution with small, insoluble particles that remain in suspension.

COLOSTRUM. A specialized secretion of the mammary glands that is produced during the first few days after parturition. It contains a high concentration of protein and antibodies.

COMMERCIAL. Pertaining to certain mammals harvested for profit; those wild mammals of monetary value.

CONDYLOBASAL. Pertaining to the length of the skull from the anterior edge of premaxilla to the posterior edge of the occipital condyles (see appendix 1).

CONIBEAR TRAP. A steel kill-trap used primarily to

collect squirrels, muskrats, and the smaller carnivores.

CONSPECIFICS. Individuals of the same species.

COPROPHAGY. Reingestion of fecal material (also referred to as refection).

COPULATION. Coitus; the union of male and female reproductive organs.

CORACOID. One of the elements of the pectoral girdle in lower vertebrates and monotremes, rudimentary and fused to the scapula in most marsupial and placental mammals.

CORPUS ALBICANS. The degenerated corpus luteum, formed after birth of the fetus or after the egg fails to implant in the uterus.

CORPUS LUTEUM. A mass of yellowish, glandular tissue formed from the Graafian follicle after ovulation.

CORPUS UTERI. The main part or body of the uterus.

CORTEX. The outer or superficial part of an organ or structure.

CREPUSCULAR. Pertaining to the twilight periods of dusk and dawn.

CURSORIAL. Pertaining to running locomotion.

CUTICLE. The thin outer layer of a hair.

DBH. Diameter at breast height (refers to measurement of trees).

DEME. An isolated population.

DEW CLAWS. Vestigial digits on the foot of certain mammals; the second and fifth toes.

DEWLAP. The pendulous fold of skin under the neck of some mammals, notably the moose.

DICHROMISM. The ability of a structure to present two different colors when viewed from different directions.

DIGITIGRADE. Pertaining to walking on the digits, with the wrist and heel bones held off the ground.

DILAMBDODONT TOOTH. A tooth characterized by a W-shaped ectoloph on the occlusal surface.

DIURNAL. Active during daylight hours.

DOMAIN. The area in which a mammal lives.

DRESSED WEIGHT. The weight of a mammal after removal of all internal organs.

ECOTONE. A zone of transition between habitat types.

EMBRYO. An unborn animal.

ENDOGENOUS. Growing from or on the inside, as catabolic products excreted by a normal organism.

ENVIRONMENTAL IMPACT STATEMENT (EIS). A written report required by governmental agencies stating the impact of various types of proposed alteration, such as a power plant or highway, on the environment.

EPIPUBIC BONES. Paired bones that project anteriorly from the pelvic girdle into the abdominal body wall of most marsupials and monotremes.

ESTIVATE. To pass the summer in a dormant state.

ESTRUS. The receptive period of female mammals; the rut; the whole estrous cycle.

EXCRESCENCE. Abnormal growth or increase.

EXOSTOSIS. A spur or growth from a bone or tooth root.

EXTINCTION. Complete destruction of a mammalian species.

EXTIRPATION. Destruction of a mammalian population in a specifically defined area.

FARROWING. The birth of a swine.

FATHOM. A measure of length containing six feet.

FEEDING CRATERS. Formed by ungulates as they dig through snow for food.

FETUS. Unborn mammal (embryo).

FLUKES. Tailfin of a whale.

FODDER. Coarse food for cattle, horses, and sheep.

FORAGE. The plant food of certain mammals, such as deer.

FOSSORIAL. Pertaining to or the adaptation for existence under ground.

FUSIFORM. Cigar or torpedo shaped.

GAMBOL. To bound or spring, to frisk or play.

GENETIC RELATEDNESS. The degree to which different populations of the same species are related to one another.

GEOLOGIC TIME TABLE. A time table based on geology which dates from the Recent Epoch (less than 600,000 years ago) to the Precambrian Period (4.5 billion years ago).

GESTATION PERIOD. The period of embryonic development during which the developing zygote is in the uterus; the period between fertilization and parturition.

GROUNDFAST ICE. Ice that forms on the bottom of an arctic ocean, lake, or stream.

GUARD HAIR. Coarse hair that extends beyond the underfur of mammals in fur-bearing species. It is often clipped when the pelt is prepared as a garment.

HABITUATION. The tendency to utilize the same area repeatedly.

HAUL-OUT SITE. An area at which pinnipeds leave the water.

HEMATOCRIT. The relative amounts of plasma and corpuscles in the blood.

"HERDING." The forced grouping of a large number of ungulates.

HETEROTHERMIC. Acting as either a homeotherm or a poikilotherm.

HIBERNACULA. A place where a bat may hibernate or spend time in a torpid state.

HIBERNATE. To pass the winter in a dormant state.

HOG-DRESSED WEIGHT. The weight of an animal with the heart and lungs remaining (often used synonymously with field-dressed).

HOLARCTIC REGION. Collectively, the Nearctic and Palearctic faunal regions.

HOMEOTHERM. A warm-blooded animal.

HOMEOTHERMIC. Able to maintain constant body temperature within the range of tolerance of ambient temperatures.

HYDROPHYTIC. Descriptive of a plant that grows in water or saturated soil.

HYPERCAPNIA. Tolerance of excess carbon dioxide in the tissues of the body.

HYPOPHYSIS. The body of the pituitary gland.

HYPOXIA. Deficiency of oxygen in the tissues of the body.

HYPSODONT. Pertaining to a high-crowned tooth (opposite of brachydont).

HYSTRICOMORPHS. Old and New World rodents in which the infraorbital foramen is greatly enlarged. Typically, it includes species such as the porcupine and nutria.

ILIUM. The most dorsal of the three bones in each half of the pelvic girdle; the pelvic bone that articulates with the sacral vertebrae.

INCISIFORM. Referring to chisel-shaped teeth, generally canines with the same structure as incisors.

INDUCED OVULATION. Ovulation that requires the act of copulation to occur, as in the Felidae.

INFUNDIBULA. The hollow, conical process of gray matter to which the pituitary body is attached (also refers to inner crests of molariform dentition).

INGUINAL. Of or near the groin.

INGUINAL CANAL. In male mammals, a small opening in the musculature of the abdominal wall on either side at the base of the scrotum through which the testes moves out of the abdominal cavity into the scrotum.

INSULAR. Refers to a population separated from other populations of the same species, usually by a barrier.

INTROMISSION. The act of passing sperm into the female during copulation.

INTUMESCENT. Swollen or inflamed; a tumor.

ISCHIUM. The most posterior and ventral of the three bones in each half of the pelvic girdle.

JUVENILE. A generalized age category between immature and adult; may or may not be sexually mature.

KERATINOUS. Impregnated with keratin, a tough, fibrous protein especially abundant in the epidermis and epidermal derivatives.

KITCHEN MIDDENS. A mound of shells, animal bones, and other refuse such as often marks the location of a prehistoric settlement.

KNOT. A unit of speed equal to one nautical mile per hour.

LANUGO. Fine, soft hair on the embryonic fetus; a type of villus.

LAPAROTOMY. A cut through the abdominal wall, as in some types of surgery.

LINEAL. Directly descended.

LONG-SPRING TRAP. A type of steel leg-trap in which the spring is a long metal strap projecting away from the jaws; often used for fox and coyote.

LOPHODONT. Pertaining to a tooth that has an occlusal surface pattern consisting of a ridge formed by the elongation and fusion of cusps.

MACULA UTRICULI. A portion of the inner ear, important in orientation.

MAMMAE. A gland for secreting milk, present in female mammals but rudimentary in males.

MANUBRIUM. The uppermost portion of the sternum which articulates with the clavicles.

MARSUPIUM. The external pouch formed by a fold of skin and supported by epipubic bones in the abdominal wall, found in most marsupials and in some monotremes (echidnas). It encloses the mammary glands and serves as an incubation chamber.

MEDULLA. The central portion of a structure composed of distinct concentric layers or regions, e.g., the medulla of a hair, ovary, kidney, or adrenal gland.

MELON. A fatty deposit on the facial area responsible for the prominently bulging forehead of many delphinids.

MESIC. Pertaining to environmental conditions with medium moisture supply, as opposed to hydric (wet) or xeric (dry).

METACONID. In the lower cheek teeth, a cusp on the posterior, lingual side of the trigonid area of the crown.

METALOPH. A cusp-formed elongate ridge on upper molariform tooth.

MYSTACIAL. Pertaining to a stripe or fringe of hairs suggestive of a mustache, as in certain mustelids.

NASOPHARYNX. The upper portion of the pharynx, continuous with the nasal passages.

NAUTICAL MILE (NM). A unit of distance for sea and air navigation, equal to 1,852 meters.

NEGRI BODY. An inclusion found in the nerve cells in rabies.

NESTLING. Neonate or young animal.

NULLIPAROUS. Never having given birth.

NURSERY GROUNDS. Traditional areas used for parturition and initial rearing of young (often used in reference to pinnipeds).

OLFACTORY. Relating to the sense of smell.

OMASUM. The division between the reticulum and the abomasum in the stomach of a ruminant; the third stomach.

OS CLITORIDIS. A small, sesamoid bone present in the clitoris of the females of some mammal species, homologous to the baculum in males.

OVIDUCT. The Fallopian tube; the duct that carries the egg from the ovary to the uterus.

OVULATION. The process by which an egg is released from the ovary into the oviduct.

PACK. A group of wild carnivores living and hunting together.

PAINTS. Patches or spots of three colors.

PAIRED SPERM. Two sperm that are joined.

PANMICTIC. Exhibiting random or nonselective mating within a breeding population.

PAROUS. Having produced young or given birth.

PARTURITION. The process by which the embryo of therian mammals separates from the mother's uterine wall and is born.

PELAGE. Collectively, all the hairs on a mammal.

PERINEUM. The area between the anus and the vulva in a female, or the anus and the scrotum in a male.

PHARYNGEAL POUCHES. Outgrowths of ectoderm on both sides of the pharynx which meet the corresponding visceral furrows and give rise to the visceral clefts in vertebrate embryos.

PIEBALD. Having patches or spots of two colors (usually used in relation to cervids).

PINNA. The external ear.

PISCIVOROUS. Fish eating.

PLANTAR. Relating to the bottom (sole) of the foot.

PODS. Groups of marine mammals (seals, whales, etc.).

POLYMORPHONUCLEAR LEUKOCYTE. A neutrophil or other leukocyte with a distinctly lobed nucleus.

POSTPARTUM. Following birth.

PREBAITING. Placing bait at a trap or station prior to trapping.

PREPUTIAL. Pertaining to modified sebaceous glands.

PRIMARY FOLLICLES. Structures in the ovary that contain eggs awaiting development.

PRIME FUR. A pelt of the highest quality.

PRIMORDIAL. Earliest formed, most primitive.

PURKINJE SHIFT. A shift of the region of apparent maximal spectral luminosity from yellow with the light-adapted eye toward violet with the dark-adapted eye, associated with predominance of cone vision in lighter and rod vision in darker illumination.

PUSH-UP. A mass of frozen vegetation over a hole in the ice used by muskrats for access to the water.

RAPHE. A seamlike joining of two lateral halves of an organ.

"RENDEZVOUS SITE." Aggregation areas of wolf pups or other immature canids.

RETE. A network (generally of blood vessels or nerve fibers).

RETE MIRABILE. ("Wonderful net") a dense network of blood vessels important in oxygen and heat exchange.

RETICULUM. The second division in the stomach of a ruminant, or "cud-chewing" mammal.

RIPARIAN. Adjacent to a body of water.

ROLLING SITES. Areas about two square meters in size matted down by frolicking otter.

RUGOSE. Ridged; full of wrinkles.

RUMEN. The first division in the stomach of a ruminant, or "cud-chewing" mammal.

SAGITTAL. Pertaining to the medial dorsal ridge of the cranium.

SCALLOPED. Having segments or projections forming an edge.

SCATS. Feces or droppings.

SCRAPES. Areas scraped bare (usually by otters), distinguished from haul-outs by the absence of food remains or scats.

SECTORIAL. Cutting or shearing (same as carnassial).

SELENODONT. Having a crown pattern of molariform teeth characterized by longitudinally oriented, crescent-shaped ridges.

SEROTINOUS. Late or delayed in development.

SERRATED. Notched along the edge.

SOMATIC TISSUE. Tissue of the body, as opposed to germ tissue.

SPERMATOGENESIS. Production of sperm.

SUBADULT. A general age category between immature and adult; juvenile.

SUBCUTANEOUS. Beneath the skin.

SUBNIVEAN. Beneath the snow.

SUCCESSION. The orderly process of replacement of one community with another.

SUPRAOCCIPITAL. Pertaining to the dorsal portion of the occipital bone.

SUSTAINED YIELD. Yield from a renewable resource that is produced by a rate of harvest that is less than or equal to the net rate of productivity of the resource.

TEMPORAL RIDGES. A pair of ridges on the top of the cranium of many mammal species.

THERMAL NEUTRAL ZONE. The range of temperatures in which an endotherm expends little or no energy in regulating body temperature.

THERMOGENIC. Related to the production of heat.

THERMOREGULATION. The maintenance of a fairly constant body temperature through heat production, heat transfer, and other physiological processes.

TORPID. Dormant or inactive, usually accompanied by decreased body temperature.

TUBOUTERINE JUNCTION. The junction of the uterus and Fallopian tubes in female mammals.

UNGULATES. Hoofed mammals of the orders Perissodactyla and Artiodactyla.

VAGINA. The portion of the female reproductive tract that receives the male's penis during copulation; the canal between the vulva and uterus.

VASOCONSTRICTION. Reduction in the diameter of blood vessels; reduced flow of blood.

VESTIBULE. A cavity or space serving as an entrance to another cavity or space.

VIXEN. A female fox.

VULVA. The external genital organs of the female.

WHELP. To give birth to; the young of any carnivore.

WISENT. The European bison, *Bison bonasus*.

XEROPHYTIC. Adapted for environmental conditions of limited water supply.

YEARLING. A general age class between immature and adult (usually used in reference to a cervid that is 1½ years of age).

ZONA PELLUCIDA. The layer surrounding the blastula during early cleavage stages.

Index

1135

The Johns Hopkins University Press

Wild Mammals of North America: Biology, Management, and Economics

This book was composed in Times Roman text and display type by the
Composing Room from a design by Alan Carter. It was printed on
50-lb. Decision Smooth paper by Universal Lithographers. The manuscript was edited by Wendy A. Harris.